2016 IEEE Applied Power Electronics Conference and Exposition (APEC 2016)

Long Beach, California, USA
20-24 March 2016

Pages 2223-2972

IEEE Catalog Number: CFP16APE-POD
ISBN: 978-1-4673-9551-9

Copyright © 2016 by the Institute of Electrical and Electronic Engineers, Inc
All Rights Reserved

Copyright and Reprint Permissions: Abstracting is permitted with credit to the source. Libraries are permitted to photocopy beyond the limit of U.S. copyright law for private use of patrons those articles in this volume that carry a code at the bottom of the first page, provided the per-copy fee indicated in the code is paid through Copyright Clearance Center, 222 Rosewood Drive, Danvers, MA 01923.

For other copying, reprint or republication permission, write to IEEE Copyrights Manager, IEEE Service Center, 445 Hoes Lane, Piscataway, NJ 08854. All rights reserved.

***This publication is a representation of what appears in the IEEE Digital Libraries. Some format issues inherent in the e-media version may also appear in this print version.

IEEE Catalog Number: CFP16APE-POD
ISBN (Print-On-Demand):   978-1-4673-9551-9
ISBN (Online):   978-1-4673-9550-2
ISSN: 1048-2334

Additional Copies of This Publication Are Available From:

Curran Associates, Inc
57 Morehouse Lane
Red Hook, NY 12571 USA
Phone: (845) 758-0400
Fax: (845) 758-2633
E-mail: curran@proceedings.com
Web: www.proceedings.com

2016 IEEE Applied Power Electronics Conference and Exposition (APEC 2016)

Long Beach, California, USA
20-24 March 2016

Pages 2223-2972

IEEE Catalog Number: CFP16APE-POD
ISBN: 978-1-4673-9551-9

TECHNICAL PAPERS

Session T01: Three-Phase AC-DC Converters
Location: 101A
March 22, 2016 8:30 - 12:00
Session Chairs: Gerry Moschopoulos, *Western University, Canada*
Patrick Wheeler, *University of Nottingham*

Hardware Implementation and Characterization of SiC-Based Hybrid Three-Phase Rectifier Employing Third Harmonic Injection 1
M. Makoschitz, *Technische Universität Wien, Austria*
M. Hartmann, *Schneider Electric SE, Austria*
H. Ertl, *Technische Universität Wien, Austria*

Voltage Oriented Control of the Three-Level Vienna Rectifier using Vector Control Method 9
Jeevan Adhikari, *National University of Singapore, Singapore*
Prasanna IV, *National University of Singapore, Singapore*
S.K. Panda, *National University of Singapore, Singapore*

Compensation of Neutral Point Deviation in 3-Level NPC Converter Under Unbalanced Grid Conditions 17
Kyungsub Jung, *Chungbuk National University, Korea, South*
Yongsug Suh, *Chungbuk National University, Korea, South*

High Power Factor Modular Polyphase AC/DC Converters with Galvanic Isolation based on Resistor Emulators 25
Javier Sebastián, *Universidad de Oviedo, Spain*
Ignacio Castro, *Universidad de Oviedo, Spain*
Diego G. Lamar, *Universidad de Oviedo, Spain*
Aitor Vázquez, *Universidad de Oviedo, Spain*
Kevin Martín, *Universidad de Oviedo, Spain*

Reduced Duty-Cycle Loss and Output Inductor Current Ripple in a ZVS Switched Three-Phase Isolated PWM Rectifier 33
Jahangir Afsharian, *Ryerson University, Canada*
Dewei David Xu, *Ryerson University, Canada*
Tao Zhao, *Ryerson University, Canada*
Bing Gong, *Murata Power Solution, Canada*
Zhihua Yang, *Murata Power Solution, Canada*

Analysis, Design, and Evaluation of Three-Phase Three-Wire Isolated AC-DC Converter Implemented with Three Single-Phase Converter Modules 38
Laszlo Huber, *Delta Products Corporation, United States*
Misha Kumar, *Delta Products Corporation, United States*
Milan M. Jovanović, *Delta Products Corporation, United States*
Dinggang Ping, *Delta Electronics Shanghai Co., Ltd., China*
Gang Liu, *Delta Electronics Shanghai Co., Ltd., China*

Startup Procedure for Three-Phase Three-Wire Isolated AC-DC Converter Implemented with Three Single-Phase Converter Modules 46
Misha Kumar, *Delta Products Corporation, United States*
Laszlo Huber, *Delta Products Corporation, United States*
Milan M. Jovanović, *Delta Products Corporation, United States*
Dinggang Ping, *Delta Electronics Shanghai Co., Ltd., China*
Gang Liu, *Delta Electronics Shanghai Co., Ltd., China*

Control of a Single-Stage Three-Phase Boost Power Factor Correction Rectifier 54
Ayan Mallik, *University of Maryland, United States*
Bryan Faulkner, *Virginia Polytechnic Institute and State University, United States*
Alireza Khaligh, *University of Maryland, United States*

A Bidirectional Single-Stage Three-Phase Rectifier with High-Frequency Isolation and Power Factor Correction 60
Bruno Ricardo de Almeida, *Universidade Federal do Ceará, Brazil*
Demercil de Souza Oliveira Jr., *Universidade Federal do Ceará, Brazil*
Paulo P. Praça, *Universidade Federal do Ceará, Brazil*

Session T02: High Frequency and Fast-Response DC-DC Converters
Location: 104A
March 22, 2016 8:30 - 12:00
Session Chairs: Olivier Trescases, *University of Toronto*
Jeff Nilles, *Texas Instruments*

A 5 MHz, 12 V, 10 A, Monolithically Integrated Two-Phase Series Capacitor Buck Converter 66
Pradeep S. Shenoy, *Texas Instruments Inc., United States*
Orlando Lazaro, *Texas Instruments Inc., United States*
Ramanathan Ramani, *Texas Instruments Inc., United States*
Mike Amaro, *Texas Instruments Inc., United States*
Wlodek Wiktor, *Texas Instruments Inc., United States*
Joseph Khayat, *Texas Instruments Inc., United States*
Brian Lynch, *Texas Instruments Inc., United States*

A 10-MHz Isolated Class-Φ_2 Synchronous Resonant DC-DC Converter 73
Yuan Zhou, *Nanjing University of Aeronautics and Astronautics, China*
Zhiliang Zhang, *Nanjing University of Aeronautics and Astronautics, China*
Xue-Wen Zou, *Nanjing University of Aeronautics and Astronautics, China*
Zhou Dong, *Nanjing University of Aeronautics and Astronautics, China*
Xiaoyong Ren, *Nanjing University of Aeronautics and Astronautics, China*

865 MHz Switching-Speed Step-Down DC-DC Power Converter for Envelope Tracking 79
Vivek Mehrotra, *Teledyne Scientific Company, United States*
Andrea Arias, *Teledyne Scientific Company, United States*
Joshua Bergman, *Teledyne Scientific Company, United States*
Charles Neft, *Teledyne Scientific Company, United States*
Miguel Urteaga, *Teledyne Scientific Company, United States*
Berinder Brar, *Teledyne Scientific Company, United States*

Current Parking Regulator for Zero Droop/Overshoot Load Transient Response 86
Sudhir S. Kudva, *Nvidia Corporation, United States*
William J. Dally, *Nvidia Corporation, United States*
Thomas H. Greer III, *Nvidia Corporation, United States*
C. Thomas Gray, *Nvidia Corporation, United States*

A 5MHz, 24V-to-1.2V, AO^{2}T Current Mode Buck Converter with One-Cycle Transient Response and Sensorless Current Detection for Medical Meters 94

Xugang Ke, *University of Texas at Dallas, United States*
Joseph Sankman, *Texas Instruments Inc., United States*
Dongsheng Ma, *University of Texas at Dallas, United States*

Capacitively-Aided Switching Technique for High-Frequency Isolated Bus Converters 98

Seungbum Lim, *Massachusetts Institute of Technology, United States*
Alex J. Hanson, *Massachusetts Institute of Technology, United States*
Juan A. Santiago-González, *Massachusetts Institute of Technology, United States*
David J. Perreault, *Massachusetts Institute of Technology, United States*

A 10 MHz, 48-to-5V Synchronous Converter with Dead Time Enabled 125 ps Resolution Zero-Voltage Switching 106

Alexander Barner, *Robert Bosch GmbH, Germany*
Jürgen Wittmann, *Hochschule Reutlingen, Germany*
Thoralf Rosahl, *Robert Bosch GmbH, Germany*
Bernhard Wicht, *Hochschule Reutlingen, Germany*

Plug-and-Play Electronic Capacitor for VRM Applications 111

Or Kirshenboim, *Ben-Gurion University of the Negev, Israel*
Alon Cervera, *Ben-Gurion University of the Negev, Israel*
Bar Halivni, *Ben-Gurion University of the Negev, Israel*
Eli Abramov, *Ben-Gurion University of the Negev, Israel*
Mor Mordechai Peretz, *Ben-Gurion University of the Negev, Israel*

Adaptive Voltage Positioning (AVP) Design of Multi-Phase Constant On-Time I^2 Control for Voltage Regulators with Ramp Compensations 118

Kuang-Yao Cheng, *Texas Instruments Inc., United States*
Yipeng Su, *Texas Instruments Inc., United States*

Session T03: Microgrids and Hybrid Systems
Location: 104B
March 22, 2016 8:30 - 12:00
Session Chairs: Yunwei Li, *University of Alberta*
Joesep Guerrero, *Aalborg University*

Reactive Power Support Capabilities of Nonsynchronous Interconnection Systems in Microgrid Applications 125

Yong-Duk Lee, *University of Connecticut, United States*
Sung-Yeul Park, *University of Connecticut, United States*

Zero Standby Power High Efficiency Hot Plugging Outlet for 380VDC Power Delivery System 132

Kai Tan, *North Carolina State University, United States*
Chang Peng, *North Carolina State University, United States*
Pengkun Liu, *North Carolina State University, United States*
Xiaoqing Song, *North Carolina State University, United States*
Alex Q. Huang, *North Carolina State University, United States*

Design of Control System for Smooth Mode Transfer in Smart Microgrid Application 138
Mingzhi Gao, *Zhejiang University, China*
Canhui Zhang, *Zhejiang University, China*
Maohang Qiu, *Zhejiang University, China*
Min Chen, *Zhejiang University, China*
Aron Levy, *Technology Dynamics Inc., United States*

Resonance Propagation Modeling and Analysis of AC Filters in a Large-Scale Microgrid ... 143
Yusi Liu, *University of Arkansas, United States*
Chris Farnell, *University of Arkansas, United States*
H. Alan Mantooth, *University of Arkansas, United States*
Juan Carlos Balda, *University of Arkansas, United States*
Roy A. McCann, *University of Arkansas, United States*
Cheng Deng, *University of Arkansas, United States*

A New Bidirectional DC-DC Converter for Fuel Cell, Solar Cell and Battery Systems 150
Ankur Patel, *Vicor Corporation, United States*

A Multiport Isolated DC-DC Converter 156
Yan-Kim Tran, *École Polytechnique Fédérale de Lausanne, Switzerland*
Drazen Dujic, *École Polytechnique Fédérale de Lausanne, Switzerland*

A Seamless Transfer Control Method with High Load Sharing Performance for Modular ESS ... 163
Jung-Hoon Ahn, *Sungkyunkwan University, Korea, South*
Won-Yong Sung, *Sungkyunkwan University, Korea, South*
Chang-Yeol Oh, *Sungkyunkwan University, Korea, South*
Byoung-Kuk Lee, *Sungkyunkwan University, Korea, South*
Yun-Sung Kim, *Dongahelecomm Corporation, Korea, South*

A Plug-and-Play Ripple Mitigation Approach for DC-Links in Hybrid Systems 169
Sinan Li, *University of Hong Kong, Hong Kong*
Albert T.L. Lee, *University of Hong Kong, Hong Kong*
Siew-Chong Tan, *University of Hong Kong, Hong Kong*
S.Y. Ron Hui, *University of Hong Kong, Hong Kong*

Active Control of Low Frequency Common Mode Voltage to Connect AC Utility and 380 V DC Grid 177
Fang Chen, *Virginia Polytechnic Institute and State University, United States*
Rolando Burgos, *Virginia Polytechnic Institute and State University, United States*
Dushan Boroyevich, *Virginia Polytechnic Institute and State University, United States*
Xuning Zhang, *Virginia Polytechnic Institute and State University, United States*

Session T04: Control Strategies for Inverters and Motor Drives
Location: 103C
March 22, 2016 8:30 - 12:00
Session Chairs: Bilal Akin, *Univeristy of Texas, Dallas*
Babak Nahid-Mobarakeh, *University of Lorraine*

A Three-Level Space Vector Modulation Scheme for Paralleled Two Converters to Reduce Zero-Sequence Circulating Current and Common Mode Voltage 185
Zhongyi Quan, *University of Alberta, Canada*
Yunwei Li, *University of Alberta, Canada*

Nonlinearity Analysis and Linear Modulation Method for Two Level Voltage Source Inverter with Low Switching to Operating Frequency Ratio 193
Yongjae Lee, *Seoul National University, Korea, South*
Jung-Ik Ha, *Seoul National University, Korea, South*

Synchronization Strategies in Cascaded H-Bridge Multi Level Inverters for Carrier based Sinusoidal PWM Techniques 199
Saroj Kumar Sahoo, *Indian Institute of Technology Kharagpur, India*
Tanmoy Bhattacharya, *Indian Institute of Technology Kharagpur, India*

Design and Implementation of a Sinusoidal Flux Controller for Core Loss Measurements 207
Burak Tekgun, *University of Akron, United States*
Ali R. Boynuegri, *University of Akron, United States*
Md Asif Mahmood Chowdhury, *University of Akron, United States*
Yilmaz Sozer, *University of Akron, United States*

Implementation of Deadbeat-Direct Torque and Flux Control for Synchronous Reluctance Machines to Minimize Loss Each Switching Period 215
Michael Saur, *Universität der Bundeswehr München, Germany*
Francisco Ramos, *Universität der Bundeswehr München, Germany*
Aday Perez, *Universität der Bundeswehr München, Germany*
Dieter Gerling, *Universität der Bundeswehr München, Germany*
Robert D. Lorenz, *University of Wisconsin at Madison, United States*

Addressing the Unbalance Loading Issue in Multi-Drive Systems with a DC-Link Modulation Scheme for Harmonic Reduction 221
Yongheng Yang, *Aalborg University, Denmark*
Pooya Davari, *Aalborg University, Denmark*
Firuz Zare, *Danfoss Power Electronics A/S, Denmark*
Frede Blaabjerg, *Aalborg University, Denmark*

Input Current Interharmonics in Adjustable Speed Drives Caused by Fixed-Frequency Modulation Techniques 229
Hamid Soltani, *Aalborg University, Denmark*
Pooya Davari, *Aalborg University, Denmark*
Poh Chiang Loh, *Aalborg University, Denmark*
Frede Blaabjerg, *Aalborg University, Denmark*
Firuz Zare, *Danfoss Power Electronics A/S, Denmark*

Low-Frequency Voltage Ripples in the Flying Capacitors of the Nested Neutral-Point-Clamped Converter 236
Amer M.Y.M. Ghias, *University of Sharjah, U.A.E.*
Josep Pou, *University of New South Wales, Australia*
Salvador Ceballos, *TECNALIA, Spain*
Vassilios G. Agelidis, *University of New South Wales, Australia*

DC Bus Capacitor Discharge of Permanent Magnet Synchronous Machine Drive Systems for Hybrid Electric Vehicles 241
Ziwei Ke, *Oregon State University, United States*
Julia Zhang, *Oregon State University, United States*
Michael W. Degner, *Ford Motor Company, United States*

Session T05: Si Devices and Power Module Packaging
Location: 101B
March 22, 2016 8:30 - 12:00
Session Chairs: Iulian Nistor, *Corporate Research, ABB Inc.*
Brian Rowden,

C_{OSS} Hysteresis in Advanced Superjunction MOSFETs 247
J.B. Fedison, *Enphase Energy, Inc., United States*
M.J. Harrison, *Enphase Energy, Inc., United States*

Compact Electrothermal Models for Unbalanced Parallel Conducting Si-IGBTs 253
Roozbeh Bonyadi, *University of Warwick, United Kingdom*
Olayiwola Alatise, *University of Warwick, United Kingdom*
Ji Hu, *University of Warwick, United Kingdom*
Zarina Davletzhanova, *University of Warwick, United Kingdom*
Yeganeh Bonyadi, *University of Warwick, United Kingdom*
Jose Ortiz-Gonzalez, *University of Warwick, United Kingdom*
Li Ran, *University of Warwick, United Kingdom*
Philip Mawby, *University of Warwick, United Kingdom*

**General 3D Lumped Thermal Model with Various Boundary Conditions for High Power
IGBT Modules** 261
Amir Sajjad Bahman, *Aalborg University, Denmark*
Ke Ma, *Aalborg University, Denmark*
Frede Blaabjerg, *Aalborg University, Denmark*

Improved 6.5kV FREEMD-Pair based on SiC JFET and Si IGBT 269
Xiaoqing Song, *North Carolina State University, United States*
Alex Q. Huang, *North Carolina State University, United States*
Chang Peng, *North Carolina State University, United States*
Liqi Zhang, *North Carolina State University, United States*

**On the Comparative Assessment of 1.7 kV, 300 a Full SiC-MOSFET and Si-IGBT
Power Modules** 276
Muhammad Nawaz, *ABB Corporate Research, Sweden*
Kalle Ilves, *ABB Corporate Research, Sweden*

**Suppression of Reverse Recovery Ringing 3.3kV/450A Si/SiC Hybrid in Low Internal
Inductance Package Next High Power Density Dual; nHPD^2** 283
Katsuaki Saito, *Hitachi Europe Ltd., United Kingdom*
Daisuke Kawase, *Hitachi Power Semiconductor, Ltd., Japan*
Masamitsu Inaba, *Hitachi Power Semiconductor, Ltd., Japan*
Keiichi Yamamoto, *Hitachi Power Semiconductor, Ltd., Japan*
Katsunori Azuma, *Hitachi Power Semiconductor, Ltd., Japan*
Seiichi Hayakawa, *Hitachi Power Semiconductor, Ltd., Japan*

**New Layout Concepts in MW-Scale IGBT Modules for Higher Robustness during Normal
and Abnormal Operations** 288
Paula Diaz Reigosa, *Aalborg University, Denmark*
Francesco Iannuzzo, *Aalborg University, Denmark*
Stig Munk-Nielsen, *Aalborg University, Denmark*
Frede Blaabjerg, *Aalborg University, Denmark*

Design, Package, and Hardware Verification of a High Voltage Current Switch 295
Ankan De, *North Carolina State University, United States*
Adam Morgan, *North Carolina State University, United States*
Vishnu Mahadeva Iyer, *North Carolina State University, United States*
Haotao Ke, *North Carolina State University, United States*
Xin Zhao, *North Carolina State University, United States*
Kasunaidu Vechalapu, *North Carolina State University, United States*
Subhashish Bhattacharya, *North Carolina State University, United States*
Douglas C. Hopkins, *North Carolina State University, United States*

Investigation of Short Circuit in a IGBT Power Module with Three-Level Neutral Point Clamped Type 2 (NPC2, T-NPC, Mixed Voltage) Topology 303
Kevin Lenz, *Danfoss Silicon Power, Germany*
Vladan Jerinic, *Danfoss Silicon Power, Germany*
Reiner Hinken, *Danfoss Silicon Power, Germany*

Session T06: DC-DC Converter Control
Location: 102AB
March 22, 2016 8:30 - 12:00
Session Chairs: Sombuddha Chakraborty, *Texas Instruments*
Rafael Pena Alzola, *University of British Columbia*

Closed-Loop Design and Time-Optimal Control for a Series-Capacitor Buck Converter 308
Timur Vekslender, *Ben-Gurion University of the Negev, Israel*
Ofer Ezra, *Ben-Gurion University of the Negev, Israel*
Yevgeny Bezdenezhnykh, *Ben-Gurion University of the Negev, Israel*
Mor Mordechai Peretz, *Ben-Gurion University of the Negev, Israel*

Unified Constant On/Off-Time Hybrid Compensation for Fast Recovery in Digitally Current-Mode Controlled Point-of-Load Converters 315
K. Hariharan, *Indian Institute of Technology Kharagpur, India*
Santanu Kapat, *Indian Institute of Technology Kharagpur, India*
Siddhartha Mukhopadhyay, *Indian Institute of Technology Kharagpur, India*

Digital Implementation of Adaptive Synchronous Rectifier (SR) Driving Scheme for LLC Resonant Converters 322
Chao Fei, *Virginia Polytechnic Institute and State University, United States*
Fred C. Lee, *Virginia Polytechnic Institute and State University, United States*
Qiang Li, *Virginia Polytechnic Institute and State University, United States*

Digital Synchronous Rectification Controller for LLC Resonant Converters 329
Maryam S. Amouzandeh, *University of Toronto, Canada*
Behzad Mahdavikhah, *University of Toronto, Canada*
Aleksandar Prodić, *University of Toronto, Canada*
Brent McDonald, *Texas Instruments Inc., United States*

A Novel Adaptive Synchronous Rectification Method for Digitally Controlled LLC Converters ... 334
Fan Wang, *Texas Instruments Inc., United States*
Brent A. McDonald, *Texas Instruments Inc., United States*
Jeff Langham, *Texas Instruments Inc., United States*
Bo Fan, *Texas Instruments Inc., China*

Influence of the ADC Zero Bin on the Performance of an Integrated DC-DC Converter 339
S. Vesti, *Infineon Technologies Austria AG, Austria*
M. Agostinelli, *Infineon Technologies Austria AG, Austria*
H. Koltsov, *Infineon Technologies Austria AG, Austria*
S. Marsili, *Infineon Technologies Austria AG, Austria*

Improved Current-Mode Control with Single-Cycle Load Transient 343
Virginia Li, *Virginia Polytechnic Institute and State University, United States*
Pei-Hsin Liu, *Virginia Polytechnic Institute and State University, United States*
Qiang Li, *Virginia Polytechnic Institute and State University, United States*
Fred C. Lee, *Virginia Polytechnic Institute and State University, United States*

A Mixed-Signal Ripple-Based Controller for a 16 V, 10 MHz Integrated Buck Converter 350
Sergii Tkachov, *Infineon Technologies Austria AG, Austria*
Matteo Agostinelli, *Infineon Technologies Austria AG, Austria*

**New Control Concept for Soft-Switching Flyback Converters with Very High
Switching Frequency** 355
A.M. Connaughton, *Technische Universität Graz, Austria*
K. Krischan, *Technische Universität Graz, Austria*
K.K. Leong, *Infineon Technologies AG, Austria*
A. Muetze, *Technische Universität Graz, Austria*

Session T07: Solar Energy Systems
Location: 104C
March 22, 2016 8:30 - 12:00
Session Chairs: Babak Fahimi, *UT- Dallas*
Morgan Kiani, *Texas Christian University*

**Analysis, Modeling and Control of an Interleaved Isolated Boost Series Resonant
Converter for Microinverter Applications** 362
Luciano A. Garcia-Rodriguez, *University of Arkansas, United States*
Cheng Deng, *University of Arkansas, United States*
Juan Carlos Balda, *University of Arkansas, United States*
Andrés Escobar-Mejía, *Universidad Tecnologica de Pereira, Colombia*

**Benchmarking of Constant Power Generation Strategies for Single-Phase Grid-
Connected Photovoltaic Systems** 370
Ariya Sangwongwanich, *Aalborg University, Denmark*
Yongheng Yang, *Aalborg University, Denmark*
Frede Blaabjerg, *Aalborg University, Denmark*
Huai Wang, *Aalborg University, Denmark*

**Advanced Slip Mode Frequency Shift Islanding Detection Method for Single Phase Grid
Connected PV Inverters** 378
Bahador Mohammadpour, *Queen's University, Canada*
Majid Pahlevani, *Queen's University, Canada*
Sajjad Makhdoomi Kaviri, *Queen's University, Canada*
Praveen Jain, *Queen's University, Canada*

Direct MPPT Control of PWM Converters for Extreme Transient PV Applications 386
Ignacio Galiano Zurbriggen, *University of British Columbia, Canada*
Francisco Paz, *University of British Columbia, Canada*
Martin Ordonez, *University of British Columbia, Canada*

Feeding Partial Power into Line Capacitors for Low Cost and Efficient MPPT of Photovoltaic Strings 392
Ali Elrayyah, *Qatar Environment and Energy Research Institute, Qatar*
Mohammed Badawey, *University of Akron, United States*
Yilmaz Sozer, *University of Akron, United States*

Single Phase Cascaded H5 Inverter with Leakage Current Elimination for Transformerless Photovoltaic System 398
Xiaoqiang Guo, *Yanshan University, China*
Xiaoyu Jia, *Yanshan University, China*
Zhigang Lu, *Yanshan University, China*
Josep M. Guerrero, *Aalborg University, Denmark*

Optimal Low Switching Frequency Pulse Width Modulation of Current-Fed Three-Level Inverter for Solar Integration 402
Gnana Sambandam Kulothungan, *National University of Singapore, Singapore*
Akshay K. Rathore, *National University of Singapore, Singapore*
Amarendra Edpuganti, *National University of Singapore, Singapore*
Dipti Srinivasan, *National University of Singapore, Singapore*

Low Leakage Current Single-Phase PV Inverters with Universal Neutral-Point-Clamping Method 410
Liwei Zhou, *Shandong University, China*
Feng Gao, *Shandong University, China*

Modular Subpanel Photovoltaic Converter System: Analysis and Control 417
Yuan Li, *Sichuan University / Northeastern University, China*
Yue Zheng, *Northeastern University, United States*
Su Sheng, *Northeastern University, United States*
Brad Scandrett, *PowerFilm, Inc., United States*
Brad Lehman, *Northeastern University, United States*

Session T08: Advanced Converter for Power Systems used in Transportation
Location: 103AB
March 22, 2016 8:30 - 12:00
Session Chairs: Omer Onar, *Oak Ridge National Laboratory*
Khurram Afridi, *University of Colorado, Boulder*

Integrated DC-DC Converter Design for Electric Vehicle Powertrains 424
Saeed Anwar, *University of Tennessee, United States*
Weimin Zhang, *University of Tennessee, United States*
Fred Wang, *University of Tennessee, United States*
Daniel J. Costinett, *University of Tennessee, United States*

A 1 MHz Bi-Directional Soft-Switching DC-DC Converter with Planar Coupled Inductor for Dual Voltage Automotive Systems 432
Chenhao Nan, *Arizona State University, United States*
Raja Ayyanar, *Arizona State University, United States*

A Bridgeless Totem-Pole Interleaved PFC Converter for Plug-In Electric Vehicles 440
Yichao Tang, *University of Maryland, United States*
Weisheng Ding, *University of Maryland, United States*
Alireza Khaligh, *University of Maryland, United States*

Stability Analysis of Hybrid AC/DC Power Systems for More Electric Aircraft 446
Mehdi Karbalaye Zadeh, *Norwegian University of Science and Technology, Norway*
Roghayeh Gavagsaz-Ghoachani, *Université de Lorraine, France*
Babak Nahid-Mobarakeh, *Université de Lorraine, France*
Serge Pierfederici, *Université de Lorraine, France*
Marta Molinas, *Norwegian University of Science and Technology, Norway*

On the Concept of the Multi-Source Inverter 453
Lea Dorn-Gomba, *McMaster University, Canada*
Pierre Magne, *McMaster University, Canada*
Clement Barthelmebs, *McMaster University, Canada*
Ali Emadi, *McMaster University, Canada*

Time-Domain Analysis of a Wide-DC-Range Series Resonant Dual-Active-Bridge Bidirectional Converter with a New Passive Auxilliary Circuit 460
Alireza Safaee, *Queen's University, Canada*
Praveen Jain, *Queen's University, Canada*
Alireza Bakhshai, *Queen's University, Canada*

A New High Capacity Compact Power Modules for High Power EV/HEV Inverters 468
Seiichiro Inokuchi, *Mitsubishi Electric Corporation, Japan*
Shoji Saito, *Mitsubishi Electric Corporation, Japan*
Arata Izuka, *Mitsubishi Electric Corporation, Japan*
Yuki Hata, *Mitsubishi Electric Corporation, Japan*
Shinji Hatae, *Mitsubishi Electric Corporation, Japan*
Toshiya Nakano, *Powerex, Inc., United States*
Eric R. Motto, *Powerex, Inc., United States*

Modular Pet, Two-Phase Air-Cooled Converter Cell Design and Performance Evaluation with 1.7kV IGBTs for MV Applications 472
Frederick Kieferndorf, *ABB Switzerland Ltd, Switzerland*
Uwe Drofenik, *ABB Switzerland Ltd, Switzerland*
Francesco Agostini, *ABB Switzerland Ltd, Switzerland*
Francisco Canales, *ABB Switzerland Ltd, Switzerland*

A Phase Shift Full Bridge based Reconfigurable PEV Onboard Charger with Extended ZVS Range and Zero Duty Cycle Loss 480
Haoyu Wang, *ShanghaiTech University, China*

Session T09: Gate Drives, Failure Analysis, and Protection
Location: 102C
March 22, 2016 8:30 - 12:00
Session Chairs: Zhiliang Zhang, *Nanjing University of Aeronautics and Astronautics*
Indumini Ranmuthu, *Texas Instruments*

Series Arc Fault Detection Method based on Statistical Analysis for DC Microgrids 487
Gab-Su Seo, *Seoul National University, Korea, South*
Jung-Ik Ha, *Seoul National University, Korea, South*
Bo-Hyung Cho, *Seoul National University, Korea, South*
Kyu-Chan Lee, *Smart Power Supply Co., Ltd., Korea, South*

Arc Welding Inverter with Embedded Digital Active EMI Controller 493
Junpeng Ji, *Xi'an Jiaotong University, China*
Wenjie Chen, *Xi'an Jiaotong University, China*
Xu Yang, *Xi'an Jiaotong University, China*

A Thermo-Sensitive Electrical Parameter with Maximum dI_C/dt during Turn-Off for High Power Trench/Field-Stop IGBT Modules 499
Yuxiang Chen, *Zhejiang University, China*
Haoze Luo, *Zhejiang University, China*
Wuhua Li, *Zhejiang University, China*
Xiangning He, *Zhejiang University, China*
Jun Ma, *Shanghai Electric, China*
Guodong Chen, *Shanghai Electric, China*
Ye Tian, *Shanghai Electric, China*
Enxing Yang, *Shanghai Electric, China*

A Software Frequency Response Analysis Method to Monitor Degradation of Power MOSFETs in Basic Single-Switch Converters 505
Serkan Dusmez, *University of Texas at Dallas, United States*
Manish Bhardwaj, *Texas Instruments Inc., United States*
Lei Sun, *University of Texas at Dallas, United States*
Bilal Akin, *University of Texas at Dallas, United States*

A New Capacitance Estimation Method of Supercapacitor Bank using a Bank Impedance and Current Injection 511
Junwon Lee, *Chungnam National University, Korea, South*
Hyunsik Jo, *Chungnam National University, Korea, South*
Hanju Cha, *Chungnam National University, Korea, South*

Gate Driver Design for 1.7kV SiC MOSFET Module with Rogowski Current Sensor for Shortcircuit Protection 516
Jun Wang, *Virginia Polytechnic Institute and State University, United States*
Zhiyu Shen, *Virginia Polytechnic Institute and State University, United States*
Christina Dimarino, *Virginia Polytechnic Institute and State University, United States*
Rolando Burgos, *Virginia Polytechnic Institute and State University, United States*
Dushan Boroyevich, *Virginia Polytechnic Institute and State University, United States*

2 MHz High-Density Integrated Power Supply for Gate Driver in High-Temperature Applications 524

Remi Perrin, *Université Claude Bernard Lyon 1, France*
Bruno Allard, *Université Claude Bernard Lyon 1, France*
Cyril Buttay, *Université Claude Bernard Lyon 1, France*
Nicolas Quentin, *Université Claude Bernard Lyon 1, France*
Wenli Zhang, *Virginia Polytechnic Institute and State University, United States*
Rolando Burgos, *Virginia Polytechnic Institute and State University, United States*
Dushan Boroyevich, *Virginia Polytechnic Institute and State University, United States*
Philippe Preciat, *Labinal Power Systems, France*
Donatien Martineau, *Labinal Power Systems, France*

Design Consideration of Gate Driver Circuits and PCB Parasitic Parameters of Paralleled E-Mode GaN HEMTs in Zero-Voltage-Switching Applications 529

Juncheng Lu, *Kettering University, United States*
Hua Bai, *Kettering University, United States*
Alan Brown, *Hella Corporate Center USA Inc., United States*
Matt McAmmond, *Hella Corporate Center USA Inc., United States*
Di Chen, *GaN Systems Inc., Canada*
Julian Styles, *GaN Systems Inc., Canada*

A Gate Driver of SiC MOSFET for Suppressing the Negative Voltage Spikes in a Bridge Circuit 536

Qi Zhou, *Shandong University, China*
Feng Gao, *Shandong University, China*

Session T10: Control of AC-DC Converters
Location: 102AB
March 23, 2016 8:30 - 10:10
Session Chairs: Tsorng-Juu Liang, *National Cheng-Kung University (Taiwan)*
Laszlo Balogh, *Fairchild Semiconductor*

Interleaved Boost based AC/DC Bidirectional Converter with Four Quadrant Power Control based on One-Cycle Controller (OCC) 544

Snehal Bagawade, *Queen's University, Canada*
Praveen Jain, *Queen's University, Canada*

A New Control Scheme to Improve Load Transient Response of Single Phase PWM Rectifier with Auxiliary Current Injection Circuit 552

Naga Brahmendra Yadav Gorla, *National University of Singapore, Singapore*
Sandeep Kolluri, *National University of Singapore, Singapore*
Pritam Das, *National University of Singapore, Singapore*
Sanjib Kumar Panda, *National University of Singapore, Singapore*

Active Capacitor with Ripple-Based Duty Cycle Modulation for AC-DC Applications 558

Ching-Chieh Yang, *National Taiwan University, Taiwan*
Yang-Lin Chen, *National Taiwan University, Taiwan*
Yaow-Ming Chen, *National Taiwan University, Taiwan*

Novel Approach to Current-Mode Control in DCM/CCM Boundary Boost PFC 564

Giovanni Gritti, *STMicroelectronics, Italy*
Claudio Adragna, *STMicroelectronics, Italy*

Reducing the Switching Frequency Variation Range for CRM Buck PFC Converter by Variable On-Time Control 572
Xiaoping Wang, *Nanjing University of Science and Technology, China*
Kai Yao, *Nanjing University of Science and Technology, China*
Junfang Zhang, *Nanjing University of Science and Technology, China*

Session T11: GaN-Based DC-DC Converters
Location: 104A
March 23, 2016 8:30 - 10:10
Session Chairs: Alexis Kwasinski, *University of Pittsburgh*
Regan Zane, *Utah State*

High Efficiency 20-400 MHz PWM Converters using Air-Core Inductors and Monolithic Power Stages in a Normally-Off GaN Process 580
Alihossein Sepahvand, *University of Colorado at Boulder, United States*
Yuanzhe Zhang, *University of Colorado at Boulder, United States*
Dragan Maksimović, *University of Colorado at Boulder, United States*

Thermal Evaluation of Chip-Scale Packaged Gallium Nitride Transistors 587
David Reusch, *Efficient Power Conversion Corporation, United States*
Johan Strydom, *Efficient Power Conversion Corporation, United States*
Alex Lidow, *Efficient Power Conversion Corporation, United States*

Over 300kHz GaN Device based Resonant Bidirectional DCDC Converter with Integrated Magnetics 595
Gang Liu, *Fudan University, China*
Dan Li, *Fudan University, China*
Yungtaek Jang, *Delta Products Corporation, United States*
Jianqiu Zhang, *Fudan University, China*

Effective Control & Software Techniques for High Efficiency GaN FET based Flexible Electrical Power System for Cube-Satellites 601
Ashish Shrivastav, *North Carolina State University, United States*
Shikhar Singh, *IBM, United States*
Anirudh Mahajan, *North Carolina State University, United States*
Subhashish Bhattacharya, *North Carolina State University, United States*

A 98.8% Efficient Bidirectional Full-Bridge Isolated DC-DC GaN Converter 609
Rakesh Ramachandran, *University of Southern Denmark, Denmark*
Morten Nymand, *University of Southern Denmark, Denmark*

Session T12: Electric Machines
Location: 101A
March 23, 2016 8:30 - 10:10
Session Chairs: Bilal Akin, *Univeristy of Texas, Dallas*
Bulent Sarlioglu, *University of Wisconsin - Madison*

Comparison of Lateral- and Cylindrical-Stator Electrical Machines for High-Speed Direct-Drive Applications in Confined Spaces 615
Arda Tüysüz, *ETH Zürich, Switzerland*
Johann W. Kolar, *ETH Zürich, Switzerland*

Novel Contactless Axial-Flux Permanent-Magnet Electromechanical Energy Harvester 623
Michael Flankl, *ETH Zürich, Switzerland*
Arda Tüysüz, *ETH Zürich, Switzerland*
Ivan Subotic, *Liverpool John Moores University, United Kingdom*
Johann W. Kolar, *ETH Zürich, Switzerland*

Design of Rare-Earth Free Five-Phase Outer-Rotor IPM Motor Drive for Electric Bicycle 631
Md. Zakirul Islam, *University of Akron, United States*
Seungdeog Choi, *University of Akron, United States*

Transverse Flux Machines with Rotary Transformer Concept for Wide Speed Operations without using Permanent Magnet Material 638
Iftekhar Hasan, *University of Akron, United States*
Md Wasi Uddin, *University of Akron, United States*
Yilmaz Sozer, *University of Akron, United States*

Field Oriented Modeling and Control of Six Phase, Open-Delta Winding, Interior Permanent Magnet Synchronous Machines Considering Current Unbalance and Zero Sequence Currents 643
Murat Senol, *RWTH Aachen University, Germany*
Michael Schubert, *RWTH Aachen University, Germany*
Georges Engelmann, *RWTH Aachen University, Germany*
Rik W. De Doncker, *RWTH Aachen University, Germany*
Thorben Grosse, *RWTH Aachen University, Germany*
Kay Hameyer, *RWTH Aachen University, Germany*

Session T13: Advances in Magnetics
Location: 101B
March 23, 2016 8:30 - 10:10
Session Chairs: Matthew Wilkowski, *Enpirion*
Charles Sullivan, *Dartmouth*

Passive Integration using FMLF Technique for Integrated Boost Resonant Converters 651
Cheng Deng, *University of Arkansas, United States*
Luciano Andres Garcia Rodriguez, *University of Arkansas, United States*
Juan Zou, *Xiangtan University, China*
Juan Carlos Balda, *University of Arkansas, United States*

Magnetic Characterization Technique and Materials Comparison for Very High Frequency IVR 657
Dongbin Hou, *Virginia Polytechnic Institute and State University, United States*
Fred C. Lee, *Virginia Polytechnic Institute and State University, United States*
Qiang Li, *Virginia Polytechnic Institute and State University, United States*

Large-Signal Power Circuit Characterization of On-Silicon Coupled Inductors for High Frequency Integrated Voltage Regulation 663
S. Kulkarni, *Tyndall National Institute, Ireland*
Z. Pavlovic, *Tyndall National Institute, Ireland*
S. Kubendran, *Tyndall National Institute, Ireland*
C. Carretero, *Universidad de Zaragoza, Spain*
N. Wang, *Tyndall National Institute, Ireland*
C. O'Mathuna, *Tyndall National Institute / University College Cork, Ireland*

Point-of-Load Inductor with High Swinging and Low Loss at Light Load 668
Ting Ge, *Virginia Polytechnic Institute and State University, United States*
Khai Ngo, *Virginia Polytechnic Institute and State University, United States*
Jim Moss, *Texas Instruments Inc., United States*

Iron Loss Evaluation of Three-Phase Inductor for Three-Phase PWM Inverter 676
Hiroaki Matsumori, *Tokyo Metropolitan University, Japan*
Toshihisa Shimizu, *Tokyo Metropolitan University, Japan*
Koushi Takano, *Iwatsu Test Instrument Corporation, Japan*
Ishii Hitoshi, *Iwatsu Test Instrument Corporation, Japan*

Session T14: System Design and Layout for Improved Performance
Location: 102C
March 23, 2016 8:30 - 10:10
Session Chairs: Jeff Nilles, *Texas Instruments*
Ernie Parker, *Crane Aerospace & Electronics*

CMOS Gate Drive IC with Embedded Cross Talk Suppression Circuitry for SiC Devices 684
Jeffery Dix, *University of Tennessee, United States*
Zheyu Zhang, *University of Tennessee, United States*
Benjamin J. Blalock, *University of Tennessee, United States*

Optimal Design of a Voltage Regulator based Resonant Switched-Capacitor Converter IC 692
Eli Abramov, *Ben-Gurion University of the Negev, Israel*
Alon Cervera, *Ben-Gurion University of the Negev, Israel*
Mor Mordechai Peretz, *Ben-Gurion University of the Negev, Israel*

Novel Highly Integrated Current Measurement Method for Drive Inverters 700
N. Langmaack, *Technische Universität Braunschweig, Germany*
G. Tareilus, *Technische Universität Braunschweig, Germany*
M. Henke, *Technische Universität Braunschweig, Germany*

**A Novel DBC Layout for Current Imbalance Mitigation in SiC MOSFET Multichip
Power Modules** 704
Helong Li, *Aalborg University, Denmark*
Stig Munk-Nielsen, *Aalborg University, Denmark*
Szymon Bęczkowski, *Aalborg University, Denmark*
Xiongfei Wang, *Aalborg University, Denmark*

**A Double-End Sourced Multi-Chip Improved Wire-Bonded SiC MOSFET Power
Module Design** 709
Miao Wang, *Ohio State University, United States*
Fang Luo, *Ohio State University, United States*
Longya Xu, *Ohio State University, United States*

Session T15: Modeling of AC Energy Converters and Systems
Location: 104B
March 23, 2016 8:30 - 10:10
Session Chairs: Jaber Abu Qahouq, *The University of Alabama*
Xiongfei Wang, *Aalborg University*

Comparing Extended Kalman Filter and Particle Filter for Estimating Field and Damper Bar Currents in Brushless Wound Field Synchronous Generator for Stator Winding Fault Detection and Diagnosis 715
Sivakumar Nadarajan, *National University of Singapore, Singapore*
S.K. Panda, *National University of Singapore, Singapore*
Bicky Bhangu, *Rolls-Royce Singapore Pte. Ltd., Singapore*
Amit Kumar Gupta, *Rolls-Royce Singapore Pte. Ltd., Singapore*

Analytical Determination of Conduction Power Losses for Active Neutral-Point-Clamped Multilevel Converter 720
Vahid Dargahi, *Clemson University, United States*
Arash Khoshkbar Sadigh, *Extron Electronics, United States*
Keith Corzine, *Clemson University, United States*

Multifrequency Small-Signal Model of Voltage Source Converters Connected to a Weak Grid for Stability Analysis 728
Xing Li, *Huazhong University of Science and Technology, China*
Hua Lin, *Huazhong University of Science and Technology, China*

A New Approach to Control the Modified LinVerter for High Frequency Applications 733
Peyman Farhang, *University of Southern Denmark, Denmark*
Stefan Mátéfi-Tempfli, *University of Southern Denmark, Denmark*

Small-Signal Terminal Characteristics Modeling of Three-Phase Boost Rectifier with Variable Fundamental Frequency 739
Zeng Liu, *Xi'an Jiaotong University, China*
Jinjun Liu, *Xi'an Jiaotong University, China*
Dushan Boroyevich, *Virginia Polytechnic Institute and State University, United States*

Session T16: Manufacturing, Test, and Reliability
Location: 103C
March 23, 2016 8:30 - 10:10
Session Chairs: Jim Marinos, *Payton Group*
Brian Narveson, *Narveson Innovative Consulting*

Reliability Analysis of a High-Efficiency SiC Three-Phase Inverter for Motor Drive Applications 746
Juan Colmenares, *KTH Royal Institute of Technology, Sweden*
Diane-Perle Sadik, *KTH Royal Institute of Technology, Sweden*
Patrik Hilber, *KTH Royal Institute of Technology, Sweden*
Hans-Peter Nee, *KTH Royal Institute of Technology, Sweden*

RCP Evaluation of Electrolytic Capacitor Degradation for SMPS Failure Prediction 754
Hiroshi Nakao, *Fujitsu Laboratories Ltd., Japan*
Yu Yonezawa, *Fujitsu Laboratories Ltd., Japan*
Yoshiyasu Nakashima, *Fujitsu Laboratories Ltd., Japan*
Fujio Kurokawa, *Nagasaki University, Japan*

Modular Test System Architecture for Device, Circuit and System Level Reliability Testing 759
Roland Sleik, *Kompetenzzentrum Automobil- und Industrieelektronik GmbH, Austria*
Michael Glavanovics, *Kompetenzzentrum Automobil- und Industrieelektronik GmbH, Austria*
Sascha Einspieler, *Kompetenzzentrum Automobil- und Industrieelektronik GmbH, Austria*
Annette Muetze, *Technische Universität Graz, Austria*
Klaus Krischan, *Technische Universität Graz, Austria*

**EMI Noise Cancelation by Optimizing Transformer Design without Need for the
Traditional Y-Capacitor** 766
Yongjiang Bai, *Xi'an Jiaotong University, China*
Wenjie Chen, *Xi'an Jiaotong University, China*
Ruirui He, *Xi'an Jiaotong University, China*
Dan Zhang, *Silergy Corp., China*
Xu Yang, *Xi'an Jiaotong University, China*

**Manufacturing, Assembly and Production Qualifications of High Density, High
Reliability POL DC-DC Converters** 772
Fariborz Musavi, *CUI Inc., United States*

Session T17: Soft-Switching Converters in Renewable Energy Systems
Location: 104C
March 23, 2016 8:30 - 10:10
Session Chairs: Khurram Afridi, *University of Colorado at Boulder*
Katherine Kim, *Ulsan NIST*

**Power Flow Control and ZVS Analysis of Three Limb High Frequency Transformer
based Three-Port DAB** 778
Ritwik Chattopadhyay, *North Carolina State University, United States*
Subhashish Bhattacharya, *North Carolina State University, United States*

**A Novel Multi-Input Converter using Soft-Switched Single-Switch Input Modules with
Integrated Power Factor Correction Capability for Hybrid Renewable Energy Systems** 786
Sanjida Moury, *York University, Canada*
John Lam, *York University, Canada*
Vineet Srivastava, *Cistel Technology Inc., Canada*
Ron Church, *Cistel Technology Inc., Canada*

**Analysis and Design of Impulse Commutated ZCS Three-Phase Current-Fed Push-Pull
DC/DC Converter** 794
Radha Sree Krishna Moorthy, *National University of Singapore, Singapore*
Akshay Kumar Rathore, *National University of Singapore, Singapore*

ZCS Resonant Converter based Parallel Balancing of Serially Connected Batteries String 802
Ilya Zeltser, *Rafael Advanced Defense Systems Ltd., Israel*
Or Kirshenboim, *Ben-Gurion University of the Negev, Israel*
Nadav Dahan, *Ben-Gurion University of the Negev, Israel*
Mor Mordechai Peretz, *Ben-Gurion University of the Negev, Israel*

A Novel Topology of High Voltage and High Power Bidirectional ZCS DC-DC Converter based on Serial Capacitors 810
Lejia Sun, *Xi'an Jiaotong University, China*
Fang Zhuo, *Xi'an Jiaotong University, China*
Feng Wang, *Xi'an Jiaotong University, China*
Tianhua Zhu, *Xi'an Jiaotong University, China*

Session T18: Solid State Lighting
Location: 103AB
March 23, 2016 8:30 - 10:10
Session Chairs: Jim Spangler, *Spangler Prototype Inc*
Nan Chen, *ABB*

Control Scheme for TRIAC Dimming High PF Single-Stage LED Driver with Adaptive Bleeder Circuit and Non-Linear Current Reference 816
Weizhong Ma, *Hangzhou Dianzi University, China*
Xiaogao Xie, *Hangzhou Dianzi University, China*
Yang Han, *Hangzhou Dianzi University, China*
Hao Deng, *Hangzhou Dianzi University, China*

Three Phase Converter with Galvanic Isolation based on Loss-Free Resistors for HB-LED Lighting Applications 822
Ignacio Castro, *Universidad de Oviedo, Spain*
Diego G. Lamar, *Universidad de Oviedo, Spain*
Manuel Arias, *Universidad de Oviedo, Spain*
Javier Sebastián, *Universidad de Oviedo, Spain*
Marta M. Hernando, *Universidad de Oviedo, Spain*

A ZV-ZCS Electrolytic Capacitor-Less AC/DC Isolated LED Driver with Continous Energy Regulation 830
John Lam, *York University, Canada*
Nader A. El-Taweel, *York University, Canada*

High Efficiency and Power Density GaN-Based LED Driver 838
Eric Faraci, *Texas Instruments Inc., United States*
Michael Seeman, *Texas Instruments Inc., United States*
Bin Gu, *Texas Instruments Inc., United States*
Yogesh Ramadass, *Texas Instruments Inc., United States*
Paul Brohlin, *Texas Instruments Inc., United States*

A Novel LED Drive System based on Matrix Rectifier 843
Baoping Shi, *Nanjing University of Aeronautics and Astronautics, China*
Bo Zhou, *Nanjing University of Aeronautics and Astronautics, China*
Jiadan Wei, *Nanjing University of Aeronautics and Astronautics, China*
Xianhui Qin, *Nanjing University of Aeronautics and Astronautics, China*
Yuanyu Yang, *Nanjing University of Aeronautics and Astronautics, China*
Bing Liu, *Nanjing University of Aeronautics and Astronautics, China*

Session T19: Resonant and Soft Switching DC-DC Converters
Location: 101A
March 23, 2016 14:00 - 17:30
Session Chairs: Mahshid Amirabadi, *Northeastern University*
Ray Orr, *Solantro*

LLC Synchronous Rectification using Coordinate Modulation 848
Mehdi Mohammadi, *University of British Columbia, Canada*
Navid Shafiei, *University of British Columbia, Canada*
Martin Ordonez, *University of British Columbia, Canada*

Low Parasitics Planar Transformer for LLC Resonant Battery Chargers 854
Mohammad Ali Saket, *University of British Columbia, Canada*
Navid Shafiei, *University of British Columbia, Canada*
Martin Ordonez, *University of British Columbia, Canada*
Marian Craciun, *Delta-Q Technologies Corporation, Canada*
Chris Botting, *Delta-Q Technologies Corporation, Canada*

New Symmetrical Bidirectional L3C Resonant DC-DC Converter with Wide Voltage Range 859
Minjae Kim, *Seoul National University of Science and Technology, Korea, South*
Shinyoung Noh, *Seoul National University of Science and Technology, Korea, South*
Sewan Choi, *Seoul National University of Science and Technology, Korea, South*

**Influence of the Junction Capacitance of the Secondary Rectifier Diodes on Output
Characteristics in Multi-Resonant Converters** 864
Stefan Ditze, *Fraunhofer Institute for Integrated Systems and Device Technology, Germany*
Thomas Heckel, *Fraunhofer Institute for Integrated Systems and Device Technology, Germany*
Martin März, *Fraunhofer Institute for Integrated Systems and Device Technology, Germany*

A Triple Active Bridge DC-DC Converter Capable of Achieving Full-Range ZVS 872
Ling Jiang, *University of Tennessee, United States*
Daniel Costinett, *University of Tennessee, United States*

A Novel High Gain Step-Up Resonant DC-DC Converter for Automotive Application 880
Fei Shang, *Illinois Institute of Technology, United States*
Mahesh Krishnamurthy, *Illinois Institute of Technology, United States*
Alexander Isurin, *Vanner Inc., United States*

**Series Injection Enabled Full ZVS Light Load Operation of a 15kV SiC IGBT based Dual
Active Half Bridge Converter** 886
Awneesh Tripathi, *North Carolina State University, United States*
Sachin Madhusoodhanan, *North Carolina State University, United States*
Krishna Mainali, *North Carolina State University, United States*
Kasunaidu Vechalapu, *North Carolina State University, United States*
Subhashish Bhattacharya, *North Carolina State University, United States*

**Soft Switching for Half Bridge Current Doubler for High Voltage Point of Load Converter
in Data Center Power Supplies** 893
Yutian Cui, *University of Tennessee, United States*
Weimin Zhang, *University of Tennessee, United States*
Leon M. Tolbert, *University of Tennessee, United States*
Daniel J. Costinett, *University of Tennessee, United States*
Fred Wang, *University of Tennessee, United States*
Benjamin J. Blalock, *University of Tennessee, United States*

An Algorithm to Analyze Circulating Current for Multi-Phase Resonant Converter 899
Hongliang Wang, *Queen's University, Canada*
Yang Chen, *Queen's University, Canada*
Zhiyuan Hu, *Queen's University, Canada*
Laili Wang, *Queen's University, Canada*
Tianshu Liu, *Queen's University, Canada*
Wenbo Liu, *Queen's University, Canada*
Yan-Fei Liu, *Queen's University, Canada*
Jahangir Afsharian, *Murata Power Solutions, Canada*
Zhihua Yang, *Murata Power Solutions, Canada*

Session T20: Control Applications and Modulation Schemes
Location: 102C
March 23, 2016 14:00 - 17:30
Session Chairs: Masoud Karimi Ghartemani, *Mississippi state University*
Paul Bauer, *University of Lorraine*

A Simple Active Damping Method for Active Power Filters 907
Huawei Yuan, *Tsinghua University, China*
Xinjian Jiang, *Tsinghua University, China*

**Simultaneous Voltage and Current Compensation of the 3-Phase Electric Spring with
Decomposed Voltage Control** 913
Shuo Yan, *University of Hong Kong, Hong Kong*
Tianbo Yang, *University of Hong Kong, Hong Kong*
C.K. Lee, *University of Hong Kong, Hong Kong*
Siew-Chong Tan, *University of Hong Kong, Hong Kong*
S.Y. Ron Hui, *University of Hong Kong / Imperial College London, Hong Kong*

**Self-Synchronization Operation of Global Synchronous Pulsewidth Modulation with
Communication Fault Tolerant and Simplified Calculation Capabilities** 921
Tao Xu, *Shandong University, China*
Feng Gao, *Shandong University, China*

**Design Considerations and Predictive Direct Current Control of Active Regenerative
Rectifiers for Harmonic and Current Ripple Reduction** 928
Alberto Berzoy, *Florida International University, United States*
A.A.S. Mohamed, *Florida International University, United States*
Osama Mohammed, *Florida International University, United States*

**A Robust Controller for Medium Voltage AC Collection Grid for Large Scale
Photovoltaic Plants based on Medium Frequency Transformers** 936
Bahaa Hafez, *Texas A&M University, United States*
Prasad Enjeti, *Texas A&M University, United States*
Shehab Ahmed, *Texas A&M University at Qatar, Qatar*

**Optimal Low Switching Frequency Pulse Width Modulation of Current-Fed Five-Level
Inverter for Solar Integration** 943
Gnana Sambandam Kulothungan, *National University of Singapore, Singapore*
Akshay K. Rathore, *National University of Singapore, Singapore*
Amarendra Edpuganti, *National University of Singapore, Singapore*
Dipti Srinivasan, *National University of Singapore, Singapore*

Design and Implementation of D-Σ Digital Controlled Multi-function Inverter to Achieve APF, Active Power Injection and Rectification 951
T.-F. Wu, *National Tsing Hua University, Taiwan*
H.-C. Hsieh, *National Chung Cheng University, Taiwan*
L.-C. Lin, *National Tsing Hua University, Taiwan*
C.-H. Chang, *National Tsing Hua University, Taiwan*

Operation and Analysis of an Improved Transformerless Unified Power Flow Controller 959
Yang Liu, *Michigan State University, United States*
Shuitao Yang, *Michigan State University / Ford Motor Company, United States*
Fang Zheng Peng, *Michigan State University, United States*

Design Consideration of Converter based Transmission Line Emulation 966
Bo Liu, *University of Tennessee, United States*
Shuoting Zhang, *University of Tennessee, United States*
Sheng Zheng, *University of Tennessee, United States*
Yiwei Ma, *University of Tennessee, United States*
Fred Wang, *University of Tennessee, United States*
Leon M. Tolbert, *University of Tennessee, United States*

Session T21: Advances in Wide BandGap Devices
Location: 104A
March 23, 2016 14:00 - 17:30
Session Chairs: Doug Hopkins, *North Carolina State University*
Alex Huang, *North Carolina State University*

Short-Circuit Characterization of 10 kV 10A 4H-SiC MOSFET 974
Emanuel-Petre Eni, *Aalborg University, Denmark*
Szymon Bęczkowski, *Aalborg University, Denmark*
Stig Munk-Nielsen, *Aalborg University, Denmark*
Tamas Kerekes, *Aalborg University, Denmark*
Remus Teodorescu, *Aalborg University, Denmark*

Record-Low 10mΩ SiC MOSFETs in TO-247, Rated at 900V 979
Vipindas Pala, *Wolfspeed, A Cree Company, United States*
Gangyao Wang, *Wolfspeed, A Cree Company, United States*
Brett Hull, *Wolfspeed, A Cree Company, United States*
Scott Allen, *Wolfspeed, A Cree Company, United States*
Jeffrey Casady, *Wolfspeed, A Cree Company, United States*
John Palmour, *Wolfspeed, A Cree Company, United States*

Performance Evaluation of Multiple Si and SiC Solid State Devices for Circuit Breaker Application in 380VDC Delivery System 983
Kai Tan, *North Carolina State University, United States*
Pengkun Liu, *North Carolina State University, United States*
Xijun Ni, *North Carolina State University, United States*
Chang Peng, *North Carolina State University, United States*
Xiaoqing Song, *North Carolina State University, United States*
Alex Q. Huang, *North Carolina State University, United States*

Evaluation of High Voltage Cascode GaN HEMTs in Parallel Operation 990
He Li, *Ohio State University, United States*
Xuan Zhang, *Ohio State University, United States*
Lucheng Wen, *Ohio State University, United States*
John Alex Brothers, *Ohio State University, United States*
Chengcheng Yao, *Ohio State University, United States*
Ke Zhu, *Ohio State University, United States*
Jin Wang, *Ohio State University, United States*
Liming Liu, *ABB Inc., United States*
Jing Xu, *ABB Inc., United States*
Joonas Puukko, *ABB Inc., United States*

A New Driving Concept for Normally-On GaN Switches in Cascode Configuration 996
Bernhard Zojer, *Infineon Technologies Austria AG, Austria*

**Avoiding Divergent Oscillation of Cascode GaN Device Under High Current Turn-Off
Condition** 1002
Weijing Du, *Virginia Polytechnic Institute and State University, United States*
Xiucheng Huang, *Virginia Polytechnic Institute and State University, United States*
Fred C. Lee, *Virginia Polytechnic Institute and State University, United States*
Qiang Li, *Virginia Polytechnic Institute and State University, United States*
Wenli Zhang, *Virginia Polytechnic Institute and State University, United States*

Temperature-Dependent Turn-On Loss Analysis for GaN HFETs 1010
Edward A. Jones, *University of Tennessee, United States*
Fred Wang, *University of Tennessee, United States*
Daniel Costinett, *University of Tennessee, United States*
Zheyu Zhang, *University of Tennessee, United States*
Ben Guo, *United Technologies Research Center, United States*

**Analysis of Parasitic Elements of SiC Power Modules with Special Emphasis on
Reliability Issues** 1018
Diane-Perle Sadik, *KTH Royal Institute of Technology, Sweden*
Juan Colmenares, *KTH Royal Institute of Technology, Sweden*
Hans-Peter Nee, *KTH Royal Institute of Technology, Sweden*
Konstantin Kostov, *Acreo Swedish ICT AB, Sweden*
Florian Giezendanner, *Alstom Power Sweden AB, Sweden*
Per Ranstad, *Alstom Power Sweden AB, Sweden*

**Static and Dynamic Characterization of GaN HEMT with Low Inductance Vertical Phase
Leg Design for High Frequency High Power Applications** 1024
Nidhi Haryani, *Virginia Polytechnic Institute and State University, United States*
Xuning Zhang, *Virginia Polytechnic Institute and State University, United States*
Rolando Burgos, *Virginia Polytechnic Institute and State University, United States*
Dushan Boroyevich, *Virginia Polytechnic Institute and State University, United States*

Session T22: Motor Drive Design and Inverter Topologies
Location: 101B
March 23, 2016 14:00 - 17:30
Session Chairs: Yingying Kuai, *Caterpillar Inc.*
Jin Wang, *The Ohio State University*

A Family of Single-Phase Current Source Converters with Double Outputs 1032
Louelson A. Costa, *Universidade Federal de Campina Grande, Brazil*
Maurício B.R. Corrêa, *Universidade Federal de Campina Grande, Brazil*
Montiê A. Vitorino, *Universidade Federal de Campina Grande, Brazil*
Gutemberg G. Dos Santos, *Universidade Federal de Campina Grande, Brazil*
Darlan A. Fernandes, *Universidade Federal da Paraíba, Brazil*

Multiple-Output Boost Resonant Inverter for High Efficiency and Cost-Effective
Induction Heating Applications 1040
Hector Sarnago, *Universidad de Zaragoza, Spain*
Oscar Lucia, *Universidad de Zaragoza, Spain*
José M. Burdío, *Universidad de Zaragoza, Spain*

Development of 2-kW Interleaved DC-Capacitor-Less Single-Phase Inverter System 1045
Runruo Chen, *Michigan State University, United States*
Hulong Zeng, *Michigan State University, United States*
Deepak Gunasekaran, *Michigan State University, United States*
Yunting Liu, *Michigan State University, United States*
Fang Z. Peng, *Michigan State University, United States*

Single Stage Transformer Isolated High Frequency AC Link based Open End Drive 1051
Srikant Gandikota, *University of Minnesota, United States*
Ned Mohan, *University of Minnesota, United States*

A Quasi-Z-Source Integrated Multi-Port Power Converter with Reduced Capacitance for
Switched Reluctance Motor Drives 1057
Fan Yi, *University of Texas at Dallas, United States*
Wen Cai, *University of Texas at Dallas, United States*

A Fault-Tolerant Topology of T-Type NPC Inverter with Increased Thermal
Overload Capability 1065
Jiangbiao He, *Marquette University, United States*
Nathan Weise, *Marquette University, United States*
Lixiang Wei, *Rockwell Automation, United States*
Nabeel A.O. Demerdash, *Marquette University, United States*

A Novel Analysis and Design Method of Phase Lead Filters in Repetitive Controllers for
Pulse-Width Modulated Inverters 1071
Shunfeng Yang, *Nanyang Technological University, Singapore*
Peng Wang, *Nanyang Technological University, Singapore*
Yi Tang, *Nanyang Technological University, Singapore*
Michael Zagrodnik, *Rolls-Royce Singapore Pte. Ltd., Singapore*
Xiaolei Hu, *Nanyang Technological University, Singapore*
King Jet Tseng, *Nanyang Technological University, Singapore*

Research on the Filter of Load Side Converter in BDFG based Ship Shaft Power Generation System 1078

Meilin Wang, *Huazhong University of Science and Technology, China*
Hua Lin, *Huazhong University of Science and Technology, China*
Hongbin Yang, *Huazhong University of Science and Technology, China*
Xingwei Wang, *Huazhong University of Science and Technology, China*

Investigation of Common Mode Current Related DC-Bus Overvoltage in Multiple Converter Systems 1084

Jiangbiao He, *Rockwell Automation, United States*
Zoran Vrankovic, *Rockwell Automation, United States*
Patrick E. Ozimek, *Rockwell Automation, United States*
Craig Winterhalter, *Rockwell Automation, United States*

Session T23: Modeling of Magnetic Circuits and Systems
Location: 102AB
March 23, 2016 14:00 - 17:30
Session Chairs: Ed Herbert,
Jin Ye, *San Francisco State University*

High Frequency AC Inductor Analysis and Design for Dual Active Bridge (DAB) Converters .. 1090

Zhe Zhang, *Technical University of Denmark, Denmark*
Michael A.E. Andersen, *Technical University of Denmark, Denmark*

A Comprehensive Assessment of PM Motor Topology Impact on Magnet Defect Fault Signatures 1096

Mohsen Zafarani, *University of Texas at Dallas, United States*
Taner Goktas, *University of Texas at Dallas, United States*
Bilal Akin, *University of Texas at Dallas, United States*

High Frequency Modeling for Transformer Common Mode Noise Coupling Path based on Multiconductor Transmission Line Theory 1102

Peipei Meng, *Wuhan University of Technology, China*
Xiangming Zhang, *Naval University of Engineering, China*

Leakage Flux Modelling of Multi-Winding Transformer using Permeance Magnetic Circuit ... 1108

Min Luo, *École Polytechnique Fédérale de Lausanne, Switzerland*
Drazen Dujic, *École Polytechnique Fédérale de Lausanne, Switzerland*
Jost Allmeling, *Plexim GmbH, Switzerland*

Modeling Magnetic Devices using SPICE: Application to Variable Inductors 1115

J. Marcos Alonso, *Universidad de Oviedo, Spain*
Gilberto Martínez, *Continental Automotive R&D, Mexico*
Marina Perdigão, *Universidade de Coimbra, Portugal*
Marcelo Cosetin, *Universidade Federal de Santa Maria, Brazil*
Ricardo N. do Prado, *Universidade Federal de Santa Maria, Brazil*

Investigation of a Thermal Model for a Permanent Magnet Assisted Synchronous Reluctance Motor 1123

Joseph Herbert, *University of Akron, United States*
A.K.M. Arafat, *University of Akron, United States*
Guo-Xiang Wang, *University of Akron, United States*
Seungdeog Choi, *University of Akron, United States*

Design Procedure for Multi-Phase External Rotor Permanent Magnet Assisted Synchronous Reluctance Machines 1131
Sai Sudheer Reddy Bonthu, *University of Akron, United States*
Seungdeog Choi, *University of Akron, United States*

Applicability and Limitations of an M2Spice-Assisted "Planar-Magnetics-in-the-Circuit" Simulation Approach 1138
Samantha J. Gunter, *Massachusetts Institute of Technology, United States*
Minjie Chen, *Massachusetts Institute of Technology, United States*
Stephanie A. Pavlick, *Massachusetts Institute of Technology, United States*
Rose A. Abramson, *Massachusetts Institute of Technology, United States*
Khurram K. Afridi, *University of Colorado at Boulder, United States*
David J. Perreault, *Massachusetts Institute of Technology, United States*

Session T24: Inverter/Converter Control
Location: 103C
March 23, 2016 14:00 - 17:30
Session Chairs: Siavash Pakdelian, *UMass Lowell*
Behrooz Mirafzal, *Kansas State University*

Solution of Input Double-Line Frequency Ripple Rejection for High-Efficiency High-Power Density String Inverter in Photovoltaic Application 1148
Xiaonan Zhao, *Virginia Polytechnic Institute and State University, United States*
Lanhua Zhang, *Virginia Polytechnic Institute and State University, United States*
Rachael Born, *Virginia Polytechnic Institute and State University, United States*
Jih-Sheng Lai, *Virginia Polytechnic Institute and State University, United States*

Fractional-Order Phase Lead Compensation for Multi-Rate Repetitive Control on Three-Phase PWM DC/AC Inverter 1155
Zhichao Liu, *University of South Carolina, United States*
Bin Zhang, *University of South Carolina, United States*
Keliang Zhou, *University of Glasgow, United Kingdom*

A Robust Modified Model Predictive Control (MMPC) based on Lyapunov Function for Three-Phase Active-Front-End (AFE) Rectifier 1163
M. Parvez, *University of Malaya, Malaysia*
S. Mekhilef, *University of Malaya, Malaysia*
Nadia M.L. Tan, *Universiti Tenega Nasional, Malaysia*
Hirofumi Akagi, *Tokyo Institute of Technology, Japan*

Adaptive Reference Model Predictive Control for Power Electronics 1169
Yun Yang, *University of Hong Kong, Hong Kong*
Siew-Chong Tan, *University of Hong Kong, Hong Kong*
Shu-Yuen Ron Hui, *Imperial College London, United Kingdom*

Power Switch Lifetime Extension Strategies for Three-Phase Converters 1176
Serkan Dusmez, *University of Texas at Dallas, United States*
Enes Ugur, *University of Texas at Dallas, United States*
Bilal Akin, *University of Texas at Dallas, United States*

Current Controller Modeling for an Interleaved Boost with Voltage Multiplier Cells for PV Applications 1183
Alessandro Pevere, *Katholieke Universiteit Leuven, Belgium*
Urmimala Chatterjee, *Katholieke Universiteit Leuven, Belgium*
Johan Driesen, *Katholieke Universiteit Leuven, Belgium*

New Active Capacitor Voltage Balancing Method for Five-Level Stacked Multicell Converter 1191
Arash Khoshkbar Sadigh, *Extron Electronics, United States*
Vahid Dargahi, *Clemson University, United States*
Keith Corzine, *Clemson University, United States*

Gate Signal Jitter Elimination and Noise Shaping Modulation for High-SNR Class-D Power Amplifiers 1198
M. Mauerer, *ETH Zürich, Switzerland*
A. Tüysüz, *ETH Zürich, Switzerland*
J.W. Kolar, *ETH Zürich, Switzerland*

Analysis and Compensation of Inverter Nonlinearity for Three-Level T-Type Inverters 1206
Hyeon-Sik Kim, *Seoul National University, Korea, South*
Yong-Cheol Kwon, *Seoul National University, Korea, South*
Seung-Jun Chee, *Seoul National University, Korea, South*
Seung-Ki Sul, *Seoul National University, Korea, South*

Session T25: Topics in Renewable Energy Systems I
Location: 104B
March 23, 2016 14:00 - 17:30
Session Chairs: Fei Gao, *University of Technology of Belfort-Montbéliard*
Kent Wanner, *John Deere*

Front-End Isolated Quasi-Z-Source DC-DC Converter Modules in Series for Photovoltaic High-Voltage DC Applications 1214
Yushan Liu, *Texas A&M University at Qatar, Qatar*
Haitham Abu-Rub, *Texas A&M University at Qatar, Qatar*
Baoming Ge, *Texas A&M University, United States*

Analysis of Non Detection Zone for Multiple Distributed PCS based on Equivalent Single PCS using Reactive Power Approach 1220
Byeong-Heon Kim, *Seoul National University, Korea, South*
Seung-Ki Sul, *Seoul National University, Korea, South*

Optimal Power Scheduling for a Grid-Connected Hybrid PV-Wind-Battery Microgrid System .. 1227
Adriana Luna, *Aalborg University, Denmark*
Nelson Diaz, *Aalborg University, Denmark*
Mehdi Savaghebi, *Aalborg University, Denmark*
Juan C. Vásquez, *Aalborg University, Denmark*
Josep M. Guerrero, *Aalborg University, Denmark*
Kai Sun, *Tsinghua University, China*
Guoliang Chen, *Shanghai Solar Energy & Technology Co., Ltd., China*
Libing Sun, *Shanghai Solar Energy & Technology Co., Ltd., China*

High Efficiency Power Converter for a Doubly-Fed SOEC/SOFC System 1235
Kevin Tomas-Manez, *Technical University of Denmark, Denmark*
Alexander Anthon, *Technical University of Denmark, Denmark*
Zhe Zhang, *Technical University of Denmark, Denmark*

A Hierarchical Active Balancing Architecture for Li-Ion Batteries 1243
Han-Dong Gui, *Nanjing University of Aeronautics and Astronautics, China*
Zhiliang Zhang, *Nanjing University of Aeronautics and Astronautics, China*
Dong-Jie Gu, *Nanjing University of Aeronautics and Astronautics, China*
Yang Yang, *Nanjing University of Aeronautics and Astronautics, China*
Zhouyu Lu, *Nanjing University of Aeronautics and Astronautics, China*
Yan-Fei Liu, *Queen's University, Canada*

A Series-DG based Autonomous Islanding Microgrid 1249
Beihua Liang, *Tianjin University, China*
Yun Wei Li, *University of Alberta, Canada*
Jinwei He, *Tianjin University, China*
Chengshan Wang, *Tianjin University, China*

An Enhanced Droop Control Scheme for Resilient Active Power Sharing in Paralleled Two-Stage PV Inverter Systems 1253
Hongpeng Liu, *Harbin Institute of Technology, China*
Yongheng Yang, *Aalborg University, Denmark*
Xiongfei Wang, *Aalborg University, Denmark*
Poh Chiang Loh, *Aalborg University, Denmark*
Frede Blaabjerg, *Aalborg University, Denmark*
Wei Wang, *Harbin Institute of Technology, China*
Dianguo Xu, *Harbin Institute of Technology, China*

Voltage Closed-Loop Virtual Synchronous Generator Control of Full Converter Wind Turbine for Grid-Connected and Stand-Alone Operation 1261
Yiwei Ma, *University of Tennessee, United States*
Liu Yang, *University of Tennessee, United States*
Fred Wang, *University of Tennessee, United States*
Leon M. Tolbert, *University of Tennessee, United States*

DC Voltage Ripple Quantification for a Flywheel-Battery based Hybrid Energy Storage System 1267
Christopher R. Lashway, *Florida International University, United States*
Ahmed T. Elsayed, *Florida International University, United States*
Osama A. Mohammed, *Florida International University, United States*

Session T26: Electric Vehicle Charging Systems
Location: 104C
March 23, 2016 14:00 - 17:30
Session Chairs: Jim Spangler, *Spangler Prototype Inc*
Hadi Malek, *Ford*

Adaptive Loss Reduction Charging Strategy Considering Variation of Internal Impedance of Lithium-Ion Polymer Batteries in Electric Vehicle Charging Systems 1273
Nari Kim, *Sungkyunkwan University, Korea, South*
Jung-Hoon Ahn, *Sungkyunkwan University, Korea, South*
Dong-Hee Kim, *Sungkyunkwan University, Korea, South*
Byoung-Kuk Lee, *Sungkyunkwan University, Korea, South*

A Pulse Width Modulated LLC Type Resonant Topology Adpated to Wide Output Voltage Range 1280
Haoyu Wang, *ShanghaiTech University, China*

A Series Resonant Circuit for Voltage Equalization of Series Connected Energy Storage Devices 1286
Yanqi Yu, *University of British Columbia, Canada*
Raed Saasaa, *University of British Columbia, Canada*
Wilson Eberle, *University of British Columbia, Canada*

Implementation of 3.3-kW GaN-Based DC-DC Converter for EV On-Board Charger with Series-Resonant Converter that Employs Combination of Variable-Frequency and Delay-Time Control 1292
Yungtaek Jang, *Delta Products Corporation, United States*
Milan M. Jovanović, *Delta Products Corporation, United States*
Juan M. Ruiz, *Delta Products Corporation, United States*
Misha Kumar, *Delta Products Corporation, United States*
Gang Liu, *Delta Electronics Shanghai Co., Ltd., China*

Dual Active Bridge-Based Full-Integrated Active Filter Auxiliary Power Module for Electrified Vehicle Applications with Single-Phase Onboard Chargers 1300
Ruoyu Hou, *McMaster University, Canada*
Ali Emadi, *McMaster University, Canada*

All-SiC Inductively Coupled Charger with Integrated Plug-In and Boost Functionalities for PEV Applications 1307
M. Chinthavali, *Oak Ridge National Laboratory, United States*
O.C. Onar, *Oak Ridge National Laboratory, United States*
S.L. Campbell, *Oak Ridge National Laboratory, United States*
L.M. Tolbert, *Oak Ridge National Laboratory, United States*

Switching Condition and Loss Modeling of GaN-Based Dual Active Bridge Converter for PHEV Charger 1315
Lingxiao Xue, *Virginia Polytechnic Institute and State University, United States*
Dushan Boroyevich, *Virginia Polytechnic Institute and State University, United States*
Paolo Mattavelli, *Università degli Studi di Padova, Italy*

Analysis of Cascaded Multi-Output-Port Converter for Wireless Plug-In Hybrid/On-Board EV Chargers 1323
Erdem Asa, *Hevo Power Inc. / New York University, United States*
Kerim Colak, *Istanbul Ulasim A.S., Turkey*
Dariusz Czarkowski, *New York University, United States*

Comparative Analysis of High Step-Down Ratio Isolated DC/DC Topologies in PEV Applications 1329
Zhiqing Li, *ShanghaiTech University, China*
Haoyu Wang, *ShanghaiTech University, China*

Session T27: Utility Interface and Inverter Applications
Location: 103AB
March 23, 2016 14:00 - 17:30
Session Chairs: Akshay Kumar Rathore, *Concordia University*
Yichao Tang, *Texas Instruments*

DC to Single-Phase AC Voltage Source Inverter with Power Decoupling Circuit based on Flying Capacitor Topology for PV System 1336
Hiroki Watanabe, *Nagaoka University of Technology, Japan*
Keisuke Kusaka, *Nagaoka University of Technology, Japan*
Keita Furukawa, *Nagaoka University of Technology, Japan*
Koji Orikawa, *Nagaoka University of Technology, Japan*
Jun-Ichi Itoh, *Nagaoka University of Technology, Japan*

GaN FET and Hybrid Modulation based Differential-Mode Inverter 1344
Sudip K. Mazumder, *NextWatt LLC, United States*
Ankit Gupta, *University of Illinois at Chicago, United States*
Shirish Raizada, *University of Illinois at Chicago, United States*
Harshit Soni, *University of Illinois at Chicago, United States*
Nikhil Kumar, *University of Illinois at Chicago, United States*
Paromita Mazumder, *NextWatt LLC, United States*
Parijat Bhattachaarjee, *NextWatt LLC, United States*

Thermal and Electrical Co-Design of a Modular High-Density Single-Phase Inverter using Wide-Bandgap Devices 1350
Steven Chung, *University of Toronto, Canada*
Miad Nasr, *University of Toronto, Canada*
David Guirguis, *University of Toronto, Canada*
Masafumi Otsuka, *University of Toronto, Canada*
Shahab Poshtkouhi, *University of Toronto, Canada*
David K.W. Li, *University of Toronto, Canada*
Vishal Palaniappan, *University of Toronto, Canada*
David Romero, *University of Toronto, Canada*
Cristina Amon, *University of Toronto, Canada*
Ray Orr, *Solantro Semiconductor, Canada*
Olivier Trescases, *University of Toronto, Canada*

Reactive Power Compensation with Improvement of Current Waveform Quality for Single-Phase Buck-Type Dynamic Capacitor 1358
Xinwen Chen, *Huazhong University of Science and Technology, China*
Ke Dai, *Huazhong University of Science and Technology, China*
Chen Xu, *Huazhong University of Science and Technology, China*
Ziwei Dai, *Huazhong University of Science and Technology, China*
Li Peng, *Huazhong University of Science and Technology, China*

Circulating Current Reduction for a D-Σ Digital Controlled Transformerless UPS 1364
T.-F. Wu, *National Tsing Hua University, Taiwan*
T.-H. Shiu, *National Tsing Hua University, Taiwan*
P.-H. Lin, *National Tsing Hua University, Taiwan*
L.-C. Lin, *National Tsing Hua University, Taiwan*
J.-W. Huang, *Industrial Technology Research Institute, Taiwan*

A Multi-Function Three-Level Dynamic Voltage Corrector with Wide Correction Range and Short Circuit Fault Isolation 1371

Jiankun Cao, *Nanjing University of Aeronautics and Astronautics, China*
Pengling Ding, *Nanjing University of Aeronautics and Astronautics, China*
Haichun Liu, *Nanjing University of Aeronautics and Astronautics, China*
Shaojun Xie, *Nanjing University of Aeronautics and Astronautics, China*

Effects and Analysis of Minimum Pulse Width Limitation on Adaptive DC Voltage Control of Grid Converters 1376

Bo Sun, *Aalborg University, Denmark*
Ionut Trintis, *Aalborg University, Denmark*
Stig Munk-Nielsen, *Aalborg University, Denmark*
Josep M. Guerrero, *Aalborg University, Denmark*

Improved Three-Phase Micro-Inverter using Dynamic Dead Time Optimization and Phase-Skipping Control Techniques 1381

S. Milad Tayebi, *University of Central Florida, United States*
Xianmin Mu, *University of Central Florida, United States*
Issa Batarseh, *University of Central Florida, United States*

Correcting Current Imbalances in Three-Phase Four-Wire Distribution Systems 1387

Vinson Jones, *University of Arkansas, United States*
Juan Carlos Balda, *University of Arkansas, United States*

Session T28: Isolated DC-DC Converters
Location: 104A
March 24, 2016 8:30 - 11:20
Session Chairs: Dragan Maksimovic, *UC Boulder*
Zhong Ye, *Texas Instruments*

New Design Methdology for Megahertz-Frequency Resonant DC-DC Converters using Impedance Control Network Architecture 1392

Yushi Liu, *University of Colorado at Boulder, United States*
Ashish Kumar, *University of Colorado at Boulder, United States*
Jie Lu, *University of Colorado at Boulder, United States*
Dragan Maksimovic, *University of Colorado at Boulder, United States*
Khurram K. Afridi, *University of Colorado at Boulder, United States*

Dual Voltage Regulations of Single Switch Flyback Converter using Variable Switching Frequency 1398

Jin-Woong Kim, *Seoul National University, Korea, South*
Jung-Ik Ha, *Seoul National University, Korea, South*

On-Chip PLL-Based Methods for Synchronizing Active Switches Across the Isolation Boundary in DC-DC Converters 1403

Shahab Poshtkouhi, *University of Toronto, Canada*
Miad Fard, *University of Toronto, Canada*
Olivier Trescases, *University of Toronto, Canada*

An Isolated Soft-Switching Buck-Boost Converter Utilizing Two Transformers and Embedded Bidirectional Switches on Secondary-Side for Wide Voltage Applications 1410

Tingting Liu, *Nanjing University of Aeronautics and Astronautics, China*
Hongfei Wu, *Nanjing University of Aeronautics and Astronautics, China*
Yan Xing, *Nanjing University of Aeronautics and Astronautics, China*
Kai Sun, *Tsinghua University, China*

Effect of Transformer Design on Operation of Fundamental Duty Modulation for Dual-Active-Bridge Converter 1416

Wooin Choi, *Seoul National University, Korea, South*
Moonhyun Lee, *Seoul National University, Korea, South*
Bo-Hyung Cho, *Seoul National University, Korea, South*

A High Step-Up Bidirectional Isolated Dual-Active-Bridge Converter with Three-Level Voltage-Doubler Rectifier for Energy Storage Applications 1424

Xiaohai Zhan, *Nanjing University of Aeronautics and Astronautics, China*
Hongfei Wu, *Nanjing University of Aeronautics and Astronautics, China*
Yan Xing, *Nanjing University of Aeronautics and Astronautics, China*
Hongjuan Ge, *Nanjing University of Aeronautics and Astronautics, China*
Xi Xiao, *Tsinghua University, China*

Digitized Self-Oscillating Loop for Piezoelectric Transformer-Based Power Converters 1430

Marzieh Ekhtiari, *Technical University of Denmark, Denmark*
Thomas Andersen, *Technical University of Denmark, Denmark*
Zhe Zhang, *Technical University of Denmark, Denmark*
Michael A.E. Andersen, *Technical University of Denmark, Denmark*

Session T29: Multilevel Converters
Location: 101A
March 24, 2016 8:30 - 11:20
Session Chairs: Maryam Saeedifard, *Georgia Tech*
                      Julia Zhang, *Oregon State University*

An Isolated Topology for Reactive Power Compensation with a Modularized Dynamic-Current Building-Block 1437

Hao Chen, *Georgia Institute of Technology, United States*
Anish Prasai, *Varentec, Inc., United States*
Deepak Divan, *Georgia Institute of Technology, United States*

Design and Control of a Compact MMC Submodule Structure with Reduced Capacitor Size using the Stacked Switched Capacitor Architecture 1443

Yuan Tang, *University of Warwick, United Kingdom*
Minjie Chen, *Massachusetts Institute of Technology, United States*
Li Ran, *University of Warwick, United Kingdom*

Fundamental Frequency Sorting Strategy for Capacitor Voltage Balance of Modular Multilevel Converters with Phase Disposition PWM 1450

Kun Wang, *Zhejiang University, China*
Yan Deng, *Zhejiang University, China*
Wenyu Li, *Zhejiang University, China*
Hao Peng, *Zhejiang University, China*
Guipeng Chen, *Zhejiang University, China*
Xiangning He, *Zhejiang University, China*

Active Voltage Balancing Control for 10kV Three-Level Converter using Series-Connected HV-IGBTs 1456

Shiqi Ji, *Tsinghua University, China*
Ting Lu, *Tsinghua University, China*
Zhengming Zhao, *Tsinghua University, China*
Hualong Yu, *Tsinghua University, China*
Fred Wang, *University of Tennessee, United States*

Average-Value Model of Modular Multilevel Converters Considering Capacitor Voltage 1462

Heya Yang, *Zhejiang University, China*
Yuxiang Chen, *Zhejiang University, China*
Wuhua Li, *Zhejiang University, China*
Xiangning He, *Zhejiang University, China*
Wei Sun, *China Electric Power Research Institute, China*
Yongning Chi, *China Electric Power Research Institute, China*
Yan Li, *China Electric Power Research Institute, China*

New Submodule Circuits for Modular Multilevel Current Source Converters with DC Fault Ride through Capability 1468

Xinyu Yu, *Tsinghua University, China*
Yingdong Wei, *Tsinghua University, China*
Qirong Jiang, *Tsinghua University, China*

Voltage and Power Balance Control Strategy for Three-Phase Modular Cascaded Solid Stated Transformer 1475

Zhiyu Zhang, *Zhejiang University, China*
Hengyang Zhao, *Zhejiang University, China*
Shihang Fu, *Zhejiang University, China*
Jianjiang Shi, *Zhejiang University, China*
Xiangning He, *Zhejiang University, China*

Session T30: Multilevel and Matrix Converters for Motor Drives
Location: 102C
March 24, 2016 8:30 - 11:20
Session Chairs: SeonHwan Hwang, *Kyungnam University, Korea*
Xiaohu Liu, *GE*

New Flying-Capacitor-Based Multilevel Converter with Optimized Number of Switches and Capacitors Controlled with a New Logic-Form-Equation based Active Voltage Balancing Technique 1481

Vahid Dargahi, *Clemson University, United States*
Arash Khoshkbar Sadigh, *Extron Electronics, United States*
Keith Corzine, *Clemson University, United States*

New Low-Cost Five-Level Active Neutral-Point Clamped Converter 1489

Hongliang Wang, *Queen's University, Canada*
Lei Kou, *Queen's University, Canada*
Yan-Fei Liu, *Queen's University, Canada*
Paresh C. Sen, *Queen's University, Canada*
Sucheng Liu, *Anhui University of Technology, China*

Medium Voltage (≥ 2.3 kV) High Frequency Three-Phase Two-Level Converter Design and Demonstration using 10 kV SiC MOSFETs for High Speed Motor Drive Applications .. 1497
Sachin Madhusoodhanan, *North Carolina State University, United States*
Krishna Mainali, *North Carolina State University, United States*
Awneesh Tripathi, *North Carolina State University, United States*
Kasunaidu Vechalapu, *North Carolina State University, United States*
Subhashish Bhattacharya, *North Carolina State University, United States*

Novel Three Phase Multi-Level Inverter Topology with Symmetrical DC-Voltage Sources . 1505
Ahmed Salem, *Aswan University, Egypt*
Emad M. Ahmed, *Aswan University, Egypt*
Mahrous Ahmed, *Aswan University, Egypt*
Mohamed Orabi, *Aswan University, Egypt*

A 2 kW, Single-Phase, 7-Level, GaN Inverter with an Active Energy Buffer Achieving 216 W/in^3 Power Density and 97.6% Peak Efficiency 1512
Yutian Lei, *University of Illinois at Urbana-Champaign, United States*
Christopher Barth, *University of Illinois at Urbana-Champaign, United States*
Shibin Qin, *University of Illinois at Urbana-Champaign, United States*
Wen-Chuen Liu, *University of Illinois at Urbana-Champaign, United States*
Intae Moon, *University of Illinois at Urbana-Champaign, United States*
Andrew Stillwell, *University of Illinois at Urbana-Champaign, United States*
Derek Chou, *University of Illinois at Urbana-Champaign, United States*
Thomas Foulkes, *University of Illinois at Urbana-Champaign, United States*
Zichao Ye, *University of Illinois at Urbana-Champaign, United States*
Zitao Liao, *University of Illinois at Urbana-Champaign, United States*
Robert C.N. Pilawa-Podgurski, *University of Illinois at Urbana-Champaign, United States*

Indirect Matrix Converter based Open-End Winding AC Drives with Zero Common-Mode Voltage 1520
Saurabh Tewari, *MTS Systems Corporation, United States*
Ranjan K. Gupta, *First Solar, Inc., United States*
Apurva Somani, *Dynapower Company LLC, United States*
Ned Mohan, *University of Minnesota, United States*

Precharging Strategy for Soft Startup Process of Modular Multilevel Converters based on Various SM Circuits 1528
Jiangchao Qin, *Arizona State University, United States*
Suman Debnath, *Oak Ridge National Laboratory, United States*
Maryam Saeedifard, *Georgia Institute of Technology, United States*

Session T31: System Design Techniques for Reduced EMI
Location: 101B
March 24, 2016 8:30 - 11:20
Session Chairs: John Vigars, *Allegro Microsystems*
Doug Hopkins, *North Carolina State University*

Conducted EMI Analysis and Filter Design for MHz Active Clamp Flyback Front-End Converter 1534
Xiucheng Huang, *Virginia Polytechnic Institute and State University, United States*
Junjie Feng, *Virginia Polytechnic Institute and State University, United States*
Fred C. Lee, *Virginia Polytechnic Institute and State University, United States*
Qiang Li, *Virginia Polytechnic Institute and State University, United States*
Yuchen Yang, *Virginia Polytechnic Institute and State University, United States*

EMC Investigation of a Very High Frequency Self-Oscillating Resonant Power Converter 1541
Jeppe A. Pedersen, *Technical University of Denmark, Denmark*
Arnold Knott, *Technical University of Denmark, Denmark*
Michael A.E. Andersen, *Technical University of Denmark, Denmark*

Numerical Optimization of Passive Line Filter Components for Suppression of Electromagnetic Interference (EMI) 1547
Carsten Henkenius, *Universität Paderborn, Germany*
Norbert Fröhleke, *Universität Paderborn, Germany*
Joachim Böcker, *Universität Paderborn, Germany*
Heiko Figge, *Delta Energy Systems GmbH, Germany*

Electromagnetic Noise Coupling and Mitigation for Fast Response On-Die Temperature Sensing in High Power Modules 1554
Chengcheng Yao, *Ohio State University, United States*
Pengzhi Yang, *Ohio State University, United States*
Mingzhi Leng, *Ohio State University, United States*
He Li, *Ohio State University, United States*
Lixing Fu, *Ohio State University, United States*
Jin Wang, *Ohio State University, United States*
Ke Zou, *Ford Motor Company, United States*
Chingchi Chen, *Ford Motor Company, United States*

Ultra-Low Inductance Vertical Phase Leg Design with EMI Noise Propagation Control for Enhancement Mode GaN Transistors 1561
Xuning Zhang, *Virginia Polytechnic Institute and State University, United States*
Zhiyu Shen, *Virginia Polytechnic Institute and State University, United States*
Nidhi Haryani, *Virginia Polytechnic Institute and State University, United States*
Dushan Boroyevich, *Virginia Polytechnic Institute and State University, United States*
Rolando Burgos, *Virginia Polytechnic Institute and State University, United States*

Decoupling of Interaction between WBG Converter and Motor Load for Switching Performance Improvement 1569
Zheyu Zhang, *University of Tennessee, United States*
Fred Wang, *University of Tennessee, United States*
Leon M. Tolbert, *University of Tennessee, United States*
Benjamin J. Blalock, *University of Tennessee, United States*
Daniel J. Costinett, *University of Tennessee, United States*

Control and Characterization of Electromagnetic Emissions in Wide Band Gap based Converter Modules for Ungrounded Grid-Forming Applications 1577
Robert Cuzner, *University of Wisconsin at Milwaukee, United States*
Rasoul Hosseini, *University of Wisconsin at Milwaukee, United States*
Andrew Lemmon, *University of Alabama, United States*
James Gafford, *Mississippi State University, United States*
Michael Mazzola, *Mississippi State University, United States*

Session T32: Modeling of DC Energy Converters and Systems
Location: 102AB
March 24, 2016 8:30 - 11:20
Session Chairs: Santanu Kapat, *IIT Kharagpur*
Sombuddha Chakraborty, *Texas Instruments*

A Practical Switching Time Model for Synchronous Buck Converters 1585
Yuan Rao, *Texas Instruments Inc., United States*
Surinder P. Singh, *Texas Instruments Inc., United States*
Taisuke Kazama, *Texas Instruments Inc., United States*

Off-Line Identification of Digitally Controlled Power Converters using an Analog Frequency Response Analyzer 1591
Marco Meola, *Zentrum Mikroelektronik Dresden AG, Germany*
Anthony Kelly, *Altera Corporation, Ireland*

Extended Wide-Load Range Model for Multi-Level DC-DC Converters and a Practical Dual-Mode Digital Controller 1597
Nenad Vukadinović, *University of Toronto, Canada*
Aleksandar Prodić, *University of Toronto, Canada*
Brett A. Miwa, *Maxim Integrated, United States*
Cory B. Arnold, *Maxim Integrated, United States*
Michael W. Baker, *Maxim Integrated, United States*

Burst Mode Control and Switched-Capacitor Converters Losses 1603
Michael Evzelman, *Utah State University, United States*
Regan Zane, *Utah State University, United States*

Equivalent Circuit Modeling of LLC Resonant Converter 1608
Shuilin Tian, *Virginia Polytechnic Institute and State University, United States*
Fred C. Lee, *Virginia Polytechnic Institute and State University, United States*
Qiang Li, *Virginia Polytechnic Institute and State University, United States*

Small Signal Modeling of the Hysteretic Modulator with a Current Ripple Synthesizer 1616
Yi Huang, *Intersil Corporation, United States*
Chun Cheung, *Intersil Corporation, United States*

A Black-Box Modeling Approach for DC Nanogrids 1624
A. Francés, *Universidad Politécnica de Madrid, Spain*
R. Asensi, *Universidad Politécnica de Madrid, Spain*
O. García, *Universidad Politécnica de Madrid, Spain*
R. Prieto, *Universidad Politécnica de Madrid, Spain*
J. Uceda, *Universidad Politécnica de Madrid, Spain*

Session T33: Gate Drive Techniques
Location: 103C
March 24, 2016 8:30 - 11:20
Session Chairs: Christopher Bridge, *SIMPLIS Technologies*
Martin Ordonez, *University of British Columbia*

Design and Evaluation of Isolated Gate Driver Power Supply for Medium Voltage Converter Applications 1632
Krishna Mainali, *North Carolina State University, United States*
Sachin Madhusoodhanan, *North Carolina State University, United States*
Awneesh Tripathi, *North Carolina State University, United States*
Kasunaidu Vechalapu, *North Carolina State University, United States*
Ankan De, *North Carolina State University, United States*
Subhashish Bhattacharya, *North Carolina State University, United States*

General-Purpose Clocked Gate Driver (CGD) IC with Programmable 63-Level Drivability to Reduce IC Overshoot and Switching Loss of Various Power Transistors 1640
Koutarou Miyazaki, *University of Tokyo, Japan*
Seiya Abe, *Kyushu Institute of Technology, Japan*
Masanori Tsukuda, *Kyushu Institute of Technology, Japan*
Ichiro Omura, *Kyushu Institute of Technology, Japan*
Keiji Wada, *Tokyo Metropolitan University, Japan*
Makoto Takamiya, *University of Tokyo, Japan*
Takayasu Sakurai, *University of Tokyo, Japan*

An Integrated SiC CMOS Gate Driver 1646
Matthew Barlow, *University of Arkansas, United States*
Shamim Ahmed, *University of Arkansas, United States*
H. Alan Mantooth, *University of Arkansas, United States*
A. Matt Francis, *Ozark Integrated Circuits, Inc., United States*

Digital Active Gate Drives using Sequential Optimization 1650
Daniel J. Rogers, *University of Oxford, United Kingdom*
Boris Murmann, *Stanford University, United States*

One Adaptive Turn-Off Method for PFC Converter with Voltage Spike Limitation 1657
Qunfang Wu, *Nanjing University of Aeronautics and Astronautics, China*
Qin Wang, *Nanjing University of Aeronautics and Astronautics, China*
Lan Xiao, *Nanjing University of Aeronautics and Astronautics, China*
Jialin Xu, *Nanjing University of Aeronautics and Astronautics, China*
Hongxu Li, *Nanjing University of Aeronautics and Astronautics, China*

A Digital Implementation for PWM Phase-Frequency Synchronization in SMPS Systems 1663
Luca Bizjak, *Infineon Technologies Austria AG, Austria*
Emanuele Bodano, *Infineon Technologies Austria AG, Austria*
Ante Gotovac, *Infineon Technologies Austria AG, Austria*
Sergii Tkachov, *Infineon Technologies Austria AG, Austria*

A High Accuracy and High Bandwidth Current Sense Circuit for Digitally Controlled DC-DC Buck Converters 1670
David Stack, *Altera Corporation, Ireland*
Anthony Kelly, *Altera Corporation, Ireland*
Thomas Conway, *University of Limerick, Ireland*

Session T34: Energy Storage Systems
Location: 104B
March 24, 2016 8:30 - 11:20
Session Chairs: Wei Qiao, *University of Nebraska Lincoln*
Yilmaz Sozer, *University of Akron*

Modular Multilevel Dual Active Bridge DC-DC Converter with ZVS and Fast DC Fault Recovery for Battery Energy Storage Systems 1675
Yuxiang Shi, *Florida State University, United States*
Rui Li, *Florida State University, United States*
Hui Li, *Florida State University, United States*

An Analytical Framework to Design a Dynamic Frequency Control Scheme for Microgrids using Energy Storage 1682
Ajit A. Renjit, *Ohio State University, United States*
Feng Guo, *NEC Laboratories America, Inc., United States*
Ratnesh Sharma, *NEC Laboratories America, Inc., United States*

Comparative Evaluation of LiFePO$_4$ Cell SOC Estimation Performance with ECM Structure and Noise Model/Data Rejection in the EKF for Transportation Application 1690
Hyun-jun Lee, *Soongsil University, Korea, South*
Joung-hu Park, *Soongsil University, Korea, South*
Jonghoon Kim, *Chosun University, Korea, South*

A Power Sharing Scheme for Series Connected Offshore Wind Turbines in a Medium Voltage DC Collection Grid 1695
Michael T. Daniel, *Texas A&M University, United States*
Prasad N. Enjeti, *Texas A&M University, United States*

Fault Ride-Through Performance Evaluation of an Interleaved Grid-Connected Converter Employing Low Switching Frequency 1702
Lorand Bede, *Aalborg University, Denmark*
Ghanshyamsinh Gohil, *Aalborg University, Denmark*
Mihai Ciobotaru, *University of New South Wales, Australia*
Tamas Kerekes, *Aalborg University, Denmark*
Remus Teodorescu, *Aalborg University, Denmark*
Vassilios G. Agelidis, *University of New South Wales, Australia*

Analysis of Two Charging Modes of Battery Energy Storage System for a Stand-Alone Microgrid 1708
Jongmin Jo, *Chungnam National University, Korea, South*
Hanju Cha, *Chungnam National University, Korea, South*

Proposition and Experimental Verification of a Bi-Directional Isolated DC/DC Converter for Battery Charger-Discharger of Electric Vehicle 1713
Ryota Kondo, *Mitsubishi Electric Corporation, Japan*
Yusuke Higaki, *Mitsubishi Electric Corporation, Japan*
Masaki Yamada, *Mitsubishi Electric Corporation, Japan*

Session T35: Topics on Inductive and Capacitive Wireless Power Transfer
Location: 104C
March 24, 2016 8:30 - 11:20
Session Chairs: Chris Mi, *San Diego State University*
Omer Onar, *Oak Ridge National Laboratory*

A CLLC-Compensated High Power and Large Air-Gap Capacitive Power Transfer System for Electric Vehicle Charging Applications 1721
Fei Lu, *University of Michigan at Ann Arbor, United States*
Hua Zhang, *Northeastern Polytechnical University, China*
Heath Hofmann, *University of Michigan at Ann Arbor, United States*
Chris Mi, *San Diego State University, United States*

A Large Air-Gap Capacitive Power Transfer System with a 4-Plate Capacitive Coupler Structure for Electric Vehicle Charging Applications 1726
Hua Zhang, *Northwestern Polytechnical University, China*
Fei Lu, *University of Michigan at Ann Arbor, United States*
Heath Hofmann, *University of Michigan at Ann Arbor, United States*
Weiguo Liu, *Northwestern Polytechnical University, China*
Chris Mi, *San Diego State University, United States*

Dynamic Wireless Power Transfer System for Electric Vehicles to Simplify Ground Facilities – Power Control and Efficiency Maximization on the Secondary Side – 1731
Katsuhiro Hata, *University of Tokyo, Japan*
Takehiro Imura, *University of Tokyo, Japan*
Yoichi Hori, *University of Tokyo, Japan*

Uniform-Gain Frequency Tracking of Wireless EV Charging for Improving Alignment Flexibility 1737
Yabiao Gao, *University of Georgia, United States*
Antonio Ginart, *University of Georgia / Sonnenbatterie GmbH, United States*
Kathleen Blair Farley, *Southern Company Services, Inc., United States*
Zion Tsz Ho Tse, *University of Georgia, United States*

Design and Optimization of a Multi-Coil System for Inductive Charging with Small Air Gap .. 1741
Christopher Joffe, *Fraunhofer Institute for Integrated Systems and Device Technology, Germany*
Andreas Roßkopf, *Fraunhofer Institute for Integrated Systems and Device Technology, Germany*
Stefan Ehrlich, *Fraunhofer Institute for Integrated Systems and Device Technology, Germany*
Christian Dobmeier, *Fraunhofer Institute for Integrated Systems and Device Technology, Germany*
Martin März, *Fraunhofer Institute for Integrated Systems and Device Technology, Germany*

Core Design for Better Misalignment Tolerance and Higher Range of Wireless Charging for HEV 1748
Mostak Mohammad, *University of Akron, United States*
Sangshin Kwak, *Chung-ang University, Korea, South*
Seungdeog Choi, *University of Akron, United States*

A 25 kW Industrial Prototype Wireless Electric Vehicle Charger 1756
Mariusz Bojarski, *Hevo Power Inc., United States*
Erdem Asa, *Hevo Power Inc. / New York University, United States*
Kerim Colak, *Istanbul Ulasim A.S., Turkey*
Dariusz Czarkowski, *New York University, United States*

Session T36: Wireless Power Transfer
Location: 103AB
March 24, 2016 8:30 - 11:20
Session Chairs: Sriram Jala Reddy, *Ford Motors*
Michael Masquelier, *WAVE*

Full-Bridge Series Resonant Multi-Inverter Featuring New 900-V SiC Devices for Improved Induction Heating Appliances 1762
Mario Pérez-Tarragona, *Universidad de Zaragoza, Spain*
Héctor Sarnago, *Universidad de Zaragoza, Spain*
Óscar Lucía, *Universidad de Zaragoza, Spain*
José M. Burdío, *Universidad de Zaragoza, Spain*

A Novel Phase Control of Single Switch Active Rectifier for Inductive Power Transfer Applications 1767
Kerim Colak, *Istanbul Ulasim A.S., Turkey*
Erdem Asa, *Hevo Power Inc. / New York University, United States*
Dariusz Czarkowski, *New York University, United States*

Optimal Shaped Dipole-Coil Design and Experimental Verification of Inductive Power Transfer System for Home Applications 1773
Duy T. Nguyen, *Korea Advanced Institute of Science and Technology, Korea, South*
Eun S. Lee, *Korea Advanced Institute of Science and Technology, Korea, South*
Byeung G. Choi, *Korea Advanced Institute of Science and Technology, Korea, South*
Chun T. Rim, *Korea Advanced Institute of Science and Technology, Korea, South*

A Novel Time-Sharing Current-Fed ZCS High Frequency Inverter-Applied Resonant DC-DC Converter for Inductive Power Transfer 1780
Kyohei Konishi, *Kobe University, Japan*
Tomokazu Mishima, *Kobe University, Japan*
Mutsuo Nakaoka, *University of Malaya, Malaysia*

Optimization of Coils for Magnetically Coupled Resonant Wireless Power Transfer System based on Maximum Output Power 1788
Dan Jiang, *Nanjing University of Aeronautics and Astronautics, China*
Yong Yang, *Nanjing University of Aeronautics and Astronautics, China*
Fuxin Liu, *Nanjing University of Aeronautics and Astronautics, China*
Xinbo Ruan, *Nanjing University of Aeronautics and Astronautics, China*
Xuling Chen, *Nanjing University of Aeronautics and Astronautics, China*

Online Regulation of Receiver-Side Power and Estimation of Mutual Inductance in Wireless Inductive Link based on Transmitter-Side Electrical Information 1795
Jeff Po Wa Chow, *City University of Hong Kong, Hong Kong*
Henry Shu-Hung Chung, *City University of Hong Kong, Hong Kong*
Chun Sing Cheng, *City University of Hong Kong, Hong Kong*

Dynamic Period Switching of PRS-PWM with Run-Length Limiting Technique for Spurious and Ripple Reduction in Fast Response Wireless Power Transmission 1802
Takahiro Moroto, *Keio University, Japan*
Toru Kawajiri, *Keio University, Japan*
Hiroki Ishikuro, *Keio University, Japan*

Session T37: Single-Phase AC-DC Converters
Location: 102AB
March 24, 2016 14:00 - 17:30
Session Chairs: Dusty Becker, *Emerson Network Power*
Pritam Das, *National University of Singapore*

A Flyback AC/DC Converter using Power Semiconductor Filter for Input Power
Factor Correction 1807
Chung-Pui Tung, *City University of Hong Kong, Hong Kong*
Henry Shu-Hung Chung, *City University of Hong Kong, Hong Kong*

Reducing the Variation Range of the Switching Frequency for CRM Boost PFC
Converter by Injecting 3rd Harmonic into the Input Current 1815
Yi Wang, *Nanjing University of Science and Technology, China*
Kai Yao, *Nanjing University of Science and Technology, China*

A Sustained Increase of Input Current Distortion in Active Input Current Shapers to
Eliminate Electrolytic Capacitor for Designing AC to DC HB-LED Drivers for Retrofit
Lamps Applications 1823
D.G. Lamar, *Universidad de Oviedo, Spain*
M. Arias, *Universidad de Oviedo, Spain*
A. Rodriguez, *Universidad de Oviedo, Spain*
J. Sebastian, *Universidad de Oviedo, Spain*
A. Fernandez, *European Space Agency, Netherlands*
J.A. Villarejo, *Universidad de Cartagena, Spain*

Reduced Current Stress Bridgeless Cuk PFC Converter with New Voltage Multiplier Circuit ... 1831
Yi-Hung Liao, *National Penghu University of Science and Technology, Taiwan*

Implementation of Multi-Level Bridgeless PFC Rectifiers for Mid-Power Single
Phase Applications 1835
Trong Tue Vu, *Eisergy Ltd., Ireland*
George Young, *Eisergy Ltd., Ireland*

US Mains Stacked Very High Frequency Self-Oscillating Resonant Power Converter with
Unified Rectifier 1842
Jeppe A. Pedersen, *Technical University of Denmark, Denmark*
Mickey P. Madsen, *Technical University of Denmark, Denmark*
Jakob D. Mønster, *Technical University of Denmark, Denmark*
Thomas Andersen, *Technical University of Denmark, Denmark*
Arnold Knott, *Technical University of Denmark, Denmark*
Michael A.E. Andersen, *Technical University of Denmark, Denmark*

Digital-Based Interleaving Control for GaN-Based MHz CRM Totem-Pole PFC 1847
Zhengyang Liu, *Virginia Polytechnic Institute and State University, United States*
Zhengrong Huang, *Virginia Polytechnic Institute and State University, United States*
Fred C. Lee, *Virginia Polytechnic Institute and State University, United States*
Qiang Li, *Virginia Polytechnic Institute and State University, United States*

A Novel AC-to-DC Adaptor with Ultra-High Power Density and Efficiency 1853
Yan-Cun Li, *Virginia Polytechnic Institute and State University, United States*
Fred C. Lee, *Virginia Polytechnic Institute and State University, United States*
Qiang Li, *Virginia Polytechnic Institute and State University, United States*
Xiucheng Huang, *Virginia Polytechnic Institute and State University, United States*
Zhengyang Liu, *Virginia Polytechnic Institute and State University, United States*

**A Single-Stage Single-Phase Isolated AC-DC Converter based on LLC Resonant Unit
and T-Type Three-Level Unit for Battery Charging Applications** 1861
Yikai Gao, *University of Texas at Dallas, United States*
Wen Cai, *University of Texas at Dallas, United States*
Fan Yi, *University of Texas at Dallas, United States*

Session T38: Non-Isolated DC-DC Converters
Location: 101A
March 24, 2016 14:00 - 17:30
Session Chairs: Pradeep Shenoy, *Texas Instruments*
Juan Rivas-Davila, *Stanford*

DC-DC Power Converter Controller for SOC Balancing of Paralleled Battery System 1868
Jaber A. Abu Qahouq, *University of Alabama, United States*
Lin Zhang, *University of Alabama, United States*
Yuan Cao, *University of Alabama, United States*
Bharat Balasubramanian, *University of Alabama, United States*

Ultra-Step-Up DC-DC Converter with Integrated Autotransformer and Coupled Inductor ... 1872
Yam P. Siwakoti, *Aalborg University, Denmark*
Frede Blaabjerg, *Aalborg University, Denmark*
Poh Chiang Loh, *Aalborg University, Denmark*

Optimal Dynamic Phase Add/Drop Mechanism in Multiphase DC-DC Buck Converters 1878
Anandha Ruban T T, *Texas Instruments India Pvt. Ltd., India*
Preetam Tadeparthy, *Texas Instruments India Pvt. Ltd., India*
Sankaran Aniruddhan, *Indian Institute of Technology Madras, India*
Vikram Gakhar, *Texas Instruments India Pvt. Ltd., India*
Muthusubramanian Venkateswaran, *Texas Instruments India Pvt. Ltd., India*

**A Universal Self-Calibrating Dynamic Voltage and Frequency Scaling (DVFS) Scheme
with Thermal Compensation for Energy Savings in FPGAs** 1882
Shuze Zhao, *University of Toronto, Canada*
Ibrahim Ahmed, *University of Toronto, Canada*
Carl Lamoureux, *University of Toronto, Canada*
Ashraf Lotfi, *Altera Corporation, United States*
Vaughn Betz, *University of Toronto, Canada*
Olivier Trescases, *University of Toronto, Canada*

**Morphing Switched-Capacitor Step-Down DC-DC Converters with Variable
Conversion Ratio** 1888
Song Xiong, *University of Hong Kong, Hong Kong*
Ying Huang, *University of Hong Kong, Hong Kong*
Siew-Chong Tan, *University of Hong Kong, Hong Kong*
Shu-Yuen Ron Hui, *University of Hong Kong, Hong Kong*

Compact Modular Switched-Capacitor DC/DC Converters with Exponential Voltage Gain 1894
Ying Huang, *University of Hong Kong, Hong Kong*
Song Xiong, *University of Hong Kong, Hong Kong*
Siew-Chong Tan, *University of Hong Kong, Hong Kong*
Shu-Yuen Ron Hui, *University of Hong Kong, Hong Kong*

Study and Implementation of a High Step-Up Voltage DC-DC Converter using Coupled-Inductor and Cascode Techniques 1900
Tsorng-Juu Liang, *National Cheng Kung University, Taiwan*
Yung-Ting Huang, *National Cheng Kung University, Taiwan*
Jian-Hsing Lee, *National Cheng Kung University, Taiwan*
Lo Pang-Yen Ting, *National Cheng Kung University, Taiwan*

20 mV Input, 4.2 V Output Boost Converter with Methodology of Maximum Output Power for Thermoelectric Energy Harvesting 1907
Taichi Ogawa, *Toshiba Corporation, Japan*
Takeshi Ueno, *Toshiba Corporation, Japan*
Takayuki Miyazaki, *Toshiba Corporation, Japan*
Tetsuro Itakura, *Toshiba Corporation, Japan*

Clarification of Relationship between Current Ripple and Power Density in Bidirectional DC-DC Converter 1911
Hoai Nam Le, *Nagaoka University of Technology, Japan*
Koji Orikawa, *Nagaoka University of Technology, Japan*
Jun-Ichi Itoh, *Nagaoka University of Technology, Japan*

Session T39: Inverter Applications and Technologies
Location: 101B
March 24, 2016 14:00 - 17:30
Session Chairs: Ali Khajehoddin, *University of Alberta*
Wen Cai, *University of Texas, Dallas*

Grid-Voltage Feedforward based Control for Grid-Connected LCL-Filtered Inverter with High Robustness and Low Grid Current Distortion in Weak Grid 1919
Jinming Xu, *Nanjing University of Aeronautics and Astronautics, China*
Qiang Qian, *Nanjing University of Aeronautics and Astronautics, China*
Shaojun Xie, *Nanjing University of Aeronautics and Astronautics, China*
Binfeng Zhang, *Nanjing University of Aeronautics and Astronautics, China*

Evaluation of PV Frequency-Watt Function for Fast Frequency Reserves 1926
J. Neely, *Sandia National Laboratories, United States*
J. Johnson, *Sandia National Laboratories, United States*
J. Delhotal, *Sandia National Laboratories, United States*
S. Gonzalez, *Sandia National Laboratories, United States*
M. Lave, *Sandia National Laboratories, United States*

A Systematic Design Method and Verification for a Zero-Ripple Interface for PV/Battery-to-Grid Applications 1934
Suvankar Biswas, *University of Minnesota, United States*
Ned Mohan, *University of Minnesota, United States*
William Robbins, *University of Minnesota, United States*

Grid-Voltage-Feedforward Active Damping for Grid-Connected Inverter with LCL Filter 1941
Minghui Lu, *Aalborg University, Denmark*
Xiongfei Wang, *Aalborg University, Denmark*
Frede Blaabjerg, *Aalborg University, Denmark*
S.M. Muyeen, *Petroleum Institute, U.A.E.*
Ahmed Al-Durra, *Petroleum Institute, U.A.E.*
Siyu Leng, *Petroleum Institute, U.A.E.*

A High Power Density Single-Phase Inverter using Stacked Switched Capacitor Energy Buffer 1947
Colin McHugh, *University of Colorado at Boulder, United States*
Sreyam Sinha, *University of Colorado at Boulder, United States*
Jeffrey Meyer, *University of Colorado at Boulder, United States*
Saad Pervaiz, *University of Colorado at Boulder, United States*
Jie Lu, *University of Colorado at Boulder, United States*
Fan Zhang, *University of Colorado at Boulder, United States*
Hua Chen, *University of Colorado at Boulder, United States*
Hyeokjin Kim, *University of Colorado at Boulder, United States*
Usama Anwar, *University of Colorado at Boulder, United States*
Ashish Kumar, *University of Colorado at Boulder, United States*
Alihossein Sepahvand, *University of Colorado at Boulder, United States*
Scott Jensen, *University of Colorado at Boulder, United States*
Beomseok Choi, *University of Colorado at Boulder, United States*
Daniel Seltzer, *University of Colorado at Boulder, United States*
Robert Erickson, *University of Colorado at Boulder, United States*
Dragan Maksimovic, *University of Colorado at Boulder, United States*
Khurram K. Afridi, *University of Colorado at Boulder, United States*

A Novel Single-Stage Dual-Active Bridge based Isolated DC-AC Converter 1954
Shiladri Chakraborty, *Indian Institute of Technology Kharagpur, India*
Souvik Chattopadhyay, *Indian Institute of Technology Kharagpur, India*

Ultra-Low Ripple Inverters for Distributed Generation Applications 1962
Ang Shen, *Missouri University of Science and Technology, United States*
Pourya Shamsi, *Missouri University of Science and Technology, United States*
Mehdi Ferdowsi, *Missouri University of Science and Technology, United States*

A 15 kV SiC MOSFET Gate Drive with Power Over Fiber based Isolated Power Supply and Comprehensive Protection Functions 1967
Xuan Zhang, *Ohio State University, United States*
He Li, *Ohio State University, United States*
John A. Brothers, *Ohio State University, United States*
Jin Wang, *Ohio State University, United States*
Lixing Fu, *Texas Instruments Inc., United States*
Mico Perales, *MH GoPower Co., Ltd., Taiwan*
John Wu, *MH GoPower Co., Ltd., Taiwan*

A 15-kV Class Intelligent Universal Transformer for Utility Applications 1974
Jih-Sheng Lai, *Virginia Polytechnic Institute and State University, United States*
Wei-Han Lai, *Enertronics, Inc., United States*
Seung-Ryul Moon, *Virginia Polytechnic Institute and State University, United States*
Lanhua Zhang, *Virginia Polytechnic Institute and State University, United States*
Arindam Maitra, *Electric Power Research Institute, United States*

Session T40: Modeling, Modulation and Control of Motor Drive
Location: 102C
March 24, 2016 14:00 - 17:30
Session Chairs: Jin Wang, *The Ohio State University*
River-TinHo Li, *ABB*

Modulation Technique for Common Mode Voltage Reduction in a Matrix Converter Drive Operating with High Voltage Transfer Ratio 1982
Varsha Padhee, *Rockwell Automation, United States*
Ashish Kumar Sahoo, *University of Minnesota, United States*
Ned Mohan, *University of Minnesota, United States*

Soft-Switched Discontinuous Pulse-Width Pulse-Density Modulation Scheme 1989
Arash Rahnamaee, *University of Illinois at Chicago, United States*
Alireza Mojab, *University of Illinois at Chicago, United States*
Hossein Riazmontazer, *University of Illinois at Chicago, United States*
Sudip K. Mazumder, *University of Illinois at Chicago, United States*
Milos Zefran, *University of Illinois at Chicago, United States*

A Novel Flux Estimator based on SOGI with FLL for Induction Machine Drives 1995
Rende Zhao, *China University of Petroleum, China*
Zhen Xin, *Aalborg University, Denmark*
Poh Chiang Loh, *Aalborg University, Denmark*
Frede Blaabjerg, *Aalborg University, Denmark*

Performance Characterization of Random Pulse Width Modulation Algorithms in Industrial and Commercial Adjustable Speed Drives 2003
Kevin Lee, *Eaton Corporation, United States*
Guangtong Shen, *Purdue University, United States*
Wenxi Yao, *Zhejiang University, China*
Zhengyu Lu, *Zhejiang University, China*

Stability Analysis and Controller Synthesis for Digital Single-Loop Voltage-Controlled Inverters 2011
Xiongfei Wang, *Aalborg University, Denmark*
Poh Chiang Loh, *Aalborg University, Denmark*
Frede Blaabjerg, *Aalborg University, Denmark*

High Efficiency, Hybrid Selective Harmonic Elimination Phase-Shift PWM Technique for Cascaded H-Bridge Inverters to Improve Dynamic Response and Operate in Complete Normal Modulation Indices 2019
Amirhossein Moeini, *University of Florida, United States*
Zhao Hui, *University of Florida, United States*
Shuo Wang, *University of Florida, United States*

Implementation and Experimental Validation of Efficiency Improvement in PMSM Drives through Switching Frequency Reduction 2027
Parag Kshirsagar, *United Technologies Research Center, United States*
Krishnan Ramu, *Virginia Polytechnic Institute and State University, United States*

Sensorless Speed Control of Symmetrical Triple-Star Nine-Phase Interior Permanent Magnet Machines 2035
Olorunfemi Ojo, *Tennessee Technological University, United States*
Medhi Ramezani, *Tennessee Technological University, United States*

Mitigation of Common-Mode Noise in Wide Band Gap Device based Motor Drives 2043

Sneha Narasimhan, *Rockwell Automation, United States*
Saurabh Tewari, *MTS Systems Corporation, United States*
Eric Severson, *University of Minnesota, United States*
Rohit Baranwal, *University of Minnesota, United States*
Ned Mohan, *University of Minnesota, United States*

Session T41: Gate Drivers and Integrated Packaging
Location: 103C
March 24, 2016 14:00 - 17:30
Session Chairs: Qiang Li, *Virginia Tech*
Jean-Luc Schanen, *Ecole Nationale Supérieure de l'Energie*

A High-Efficient Driving Isolated Drive-by-Microwave Half-Bridge Gate Driver for a GaN Inverter 2051

Shuichi Nagai, *Panasonic Corporation, Japan*
Yasufumi Kawai, *Panasonic Corporation, Japan*
Osamu Tabata, *Panasonic Corporation, Japan*
Songbaek Choe, *Panasonic Corporation, Japan*
Noboru Negoro, *Panasonic Corporation, Japan*
Tesuzo Ueda, *Panasonic Corporation, Japan*

Sensing Gallium Nitride HEMT Junction Temperature using Gate Drive Output Transient Properties 2055

He Niu, *University of Wisconsin at Madison, United States*
Robert D. Lorenz, *University of Wisconsin at Madison, United States*

Design and Application of a 1200V Ultra-Fast Integrated Silicon Carbide MOSFET Module ... 2063

Suxuan Guo, *North Carolina State University, United States*
Liqi Zhang, *North Carolina State University, United States*
Yang Lei, *North Carolina State University, United States*
Xuan Li, *North Carolina State University, United States*
Wensong Yu, *North Carolina State University, United States*
Alex Q. Huang, *North Carolina State University, United States*

Active Gate Charge Control Strategy for Series-Connected IGBTs 2071

Fan Zhang, *Xi'an Jiaotong University, China*
Xu Yang, *Xi'an Jiaotong University, China*
Yu Ren, *Xi'an Jiaotong University, China*
Ying Chen, *Xi'an Jiaotong University, China*
Ruifeng Gou, *Xi'an XD Power Systems Co., LTD, China*

A MV Intelligent Gate Driver for 15kV SiC IGBT and 10kV SiC MOSFET 2076

Awneesh Tripathi, *North Carolina State University, United States*
Krishna Mainali, *North Carolina State University, United States*
Sachin Madhusoodhanan, *North Carolina State University, United States*
Akshat Yadav, *North Carolina State University, United States*
Kasunaidu Vechalapu, *North Carolina State University, United States*
Subhashish Bhattacharya, *North Carolina State University, United States*

Linear Temperature Sensors in High-Voltage GaN-HEMT Power Devices 2083
Richard Reiner, *Fraunhofer Institute for Applied Solid State Physics, Germany*
Patrick Waltereit, *Fraunhofer Institute for Applied Solid State Physics, Germany*
Beatrix Weiss, *Fraunhofer Institute for Applied Solid State Physics, Germany*
Matthias Wespel, *Fraunhofer Institute for Applied Solid State Physics, Germany*
Dirk Meder, *Fraunhofer Institute for Applied Solid State Physics, Germany*
Michael Mikulla, *Fraunhofer Institute for Applied Solid State Physics, Germany*
Rüdiger Quay, *Fraunhofer Institute for Applied Solid State Physics, Germany*
Oliver Ambacher, *Fraunhofer Institute for Applied Solid State Physics, Germany*

An Innovative Power Module with Power-System-in-Inductor Structure 2087
Laili Wang, *Sumida Corporation, Canada*
Doug Malcolm, *Sumida Corporation, Canada*
Yan-Fei Liu, *Queen's University, Canada*

Thermal Analysis of a Magnetic Packaged Power Module 2095
Laili Wang, *Sumida Corporation, Canada*
Doug Malcolm, *Sumida Corporation, Canada*
Wenbo Liu, *Queen's University, Canada*
Yan-Fei Liu, *Queen's University, Canada*

**Analysis of a Low-Inductance Packaging Layout for Full-SiC Power Module Embedding
Split Damping** 2102
Yu Ren, *Xi'an Jiaotong University, China*
Xu Yang, *Xi'an Jiaotong University, China*
Fan Zhang, *Xi'an Jiaotong University, China*
Linlin Tan, *Xi'an Jiaotong University, China*
Xiangjun Zeng, *Xi'an Jiaotong University, China*

Session T42: Component Modeling
Location: 103AB
March 24, 2016 14:00 - 17:30
Session Chairs: Sheldon Williamson, *University of Ontario Institute of Technology*
Abhijit Pathak, *Infineon/IR*

**Comprehensive Parametric Analyses of Thermally Aged Power MOSFETs for Failure
Precursor Identification and Lifetime Estimation based on Gate Threshold Voltage** 2108
Serkan Dusmez, *University of Texas at Dallas, United States*
Bilal Akin, *University of Texas at Dallas, United States*

Modeling and Design Guidelines of High Density Power Inductor for Battery Power Unit 2114
Zhigang Dang, *University of Alabama, United States*
Jaber A. Abu Qahouq, *University of Alabama, United States*

Degradation of Low Voltage Metal Oxide Varistors in Power Supplies 2122
Dawood Talebi Khanmiri, *Northeastern University, United States*
Roy Ball, *Mersen USA, United States*
Jerry Mosesian, *Mersen USA, United States*
Brad Lehman, *Northeastern University, United States*

Characterization and Modeling of SiC MOSFET Body Diode 2127
Kang Peng, *University of South Carolina, United States*
Soheila Eskandari, *University of South Carolina, United States*
Enrico Santi, *University of South Carolina, United States*

A Simple Behavioral Electro-Thermal Model of GaN FETs for SPICE Circuit Simulation 2136

Liyao Wu, *Georgia Institute of Technology, United States*
Maryam Saeedifard, *Georgia Institute of Technology, United States*

Decomposition and Electro-Physical Model Creation of the CREE 1200V, 50A 3-Ph SiC Module 2141

Adam J. Morgan, *North Carolina State University, United States*
Yang Xu, *North Carolina State University, United States*
Douglas C. Hopkins, *North Carolina State University, United States*
Iqbal Husain, *North Carolina State University, United States*
Wensong Yu, *North Carolina State University, United States*

A Three-Legged MATLAB/Simulink Transformer Model using a Fictitious Delta Winding 2147

Thomas A. Nondahl, *Rockwell Automation, United States*
Jingbo Liu, *Rockwell Automation, United States*
Peter B. Schmidt, *Rockwell Automation, United States*

A Lifetime Prediction Method for LEDs Considering Mission Profiles 2154

Xiaohui Qu, *Southeast University, China*
Huai Wang, *Aalborg University, Denmark*
Xiaoqing Zhan, *City University of Hong Kong, Hong Kong*
Frede Blaabjerg, *Aalborg University, Denmark*
Henry Shu-Hung Chung, *City University of Hong Kong, Hong Kong*

Enhanced Li-Ion Battery Modeling using Recursive Parameters Correction 2161

Jae-Gu Kim, *Sungkyunkwan University, Korea, South*
Jung-Hoon Ahn, *Sungkyunkwan University, Korea, South*
Byoung-Kuk Lee, *Sungkyunkwan University, Korea, South*

Session T43: Grid and Utility Interface
Location: 104A
March 24, 2016 14:00 - 17:30
Session Chairs: Manish Bhardwaj, *Texas Instruments*
Nan Chen, *ABB*

Robust Sensorless Control of Grid Connected Converters with LCL Line Filters using Frequency Adaptive Observers as AC Voltage Estimators 2167

Vlatko Miskovic, *Danfoss Drives, United States*
Vladimir Blasko, *United Technologies Research Center, United States*
Thomas Jahns, *University of Wisconsin at Madison, United States*
Robert Lorenz, *University of Wisconsin at Madison, United States*
Haojiong Zhang, *Danfoss Drives, United States*

Active Stabilization of Direct Matrix Converter Input Side Filter through Grid Current Control 2175

Martin Leubner, *Technische Universität Dresden, Germany*
Nico Remus, *Technische Universität Dresden, Germany*
Marc Stübig, *Technische Universität Dresden, Germany*
Wilfried Hofmann, *Technische Universität Dresden, Germany*

Impedance-Based Stability Analysis of Single-Phase Inverter Connected to Weak Grid with Voltage Feed-Forward Control 2182
Jiangfeng Wang, *Nanjing University of Aeronautics and Astronautics, China*
Jianhui Yao, *Nanjing University of Aeronautics and Astronautics, China*
Haibing Hu, *Nanjing University of Aeronautics and Astronautics, China*
Yan Xing, *Nanjing University of Aeronautics and Astronautics, China*
Xiaobin He, *Shanghai Institute of Space Power-Sources, China*
Kai Sun, *Tsinghua University, China*

New Configuration of Dynamic Voltage Restorer for Medium Voltage Application 2187
Arash Khoshkbar Sadigh, *Extron Electronics, United States*
Vahid Dargahi, *Clemson University, United States*
Keith Corzine, *Clemson University, United States*

Studies on the Clustered Voltage Balancing Mechanism for Cascaded H-Bridge STATCOM ... 2194
Daorong Lu, *Nanjing University of Aeronautics and Astronautics, China*
Haibing Hu, *Nanjing University of Aeronautics and Astronautics, China*
Yan Xing, *Nanjing University of Aeronautics and Astronautics, China*
Xiaobin He, *Shanghai Institute of Space Power-Sources, China*
Kai Sun, *Tsinghua University, China*
Jianhui Yao, *Nanjing University of Aeronautics and Astronautics, China*

Design of a Fast Response Time Single-Phase PLL with DC Offset Rejection Capability ... 2200
Abhijit Kulkarni, *Indian Institute of Science, India*
Vinod John, *Indian Institute of Science, India*

Four New Applications of Second-Order Generalized Integrator Quadrature Signal Generator 2207
Zhen Xin, *Aalborg University, Denmark*
Rende Zhao, *China University of Petroleum, China*
Xiongfei Wang, *Aalborg University, Denmark*
Poh Chiang Loh, *Aalborg University, Denmark*
Frede Blaabjerg, *Aalborg University, Denmark*

Three-Phase Multiple Harmonic Sequence Detection based on Generalized Delayed Signal Superposition 2215
Yong Lu, *Xi'an Jiaotong University, China*
Guochun Xiao, *Xi'an Jiaotong University, China*
Xiongfei Wang, *Aalborg University, Denmark*
Frede Blaabjerg, *Aalborg University, Denmark*

Hybrid Modelling and Control of Single-Phase Grid-Connected NPC Inverters 2223
Xingda Yan, *University of Southampton, United Kingdom*
Zhan Shu, *University of Southampton, United Kingdom*
Suleiman M. Sharkh, *University of Southampton, United Kingdom*

Session T44: Topics in Renewable Energy Systems II
Location: 104B
March 24, 2016 14:00 - 17:30
Session Chairs: Akshay Kumar Rathore, *Concordia University*
Yichao Tang, *Texas Instruments*

Stability Criterion and Controller Parameter Design of Radial-Line Renewable Systems with Multiple Inverters 2229
Wenchao Cao, *University of Tennessee, United States*
Xuan Zhang, *University of Tennessee, United States*
Yiwei Ma, *University of Tennessee, United States*
Fred Wang, *University of Tennessee, United States*

Stability Analysis and Improvement of Solid State Transformer (SST)-Paralleled Inverters System using Negative Impedance Feedback Control 2237
Qing Ye, *Florida State University, United States*
Hui Li, *Florida State University, United States*

Compensator-Less Structures for Droop Control of Single Phase Inverters in a Flexible Microgrid 2245
Onkar Vitthal Kulkarni, *Indian Institute of Technology Bombay, India*
Suryanarayana Doolla, *Indian Institute of Technology Bombay, India*
B.G. Fernandes, *Indian Institute of Technology Bombay, India*

Comparative Evaluation of the Loss and Thermal Performance of Advanced Three Level Inverter Topologies 2252
Alexander Anthon, *Technical University of Denmark, Denmark*
Zhe Zhang, *Technical University of Denmark, Denmark*
Michael A.E. Andersen, *Technical University of Denmark, Denmark*
Grahame Holmes, *RMIT University, Australia*
Brendan McGrath, *RMIT University, Australia*
Carlos Teixeira, *RMIT University, Australia*

Dual Buck Inverter with Series Connected Diodes and Single Inductor 2259
Liwei Zhou, *Shandong University, China*
Feng Gao, *Shandong University, China*

Magnetic Integration of the Harmonic Filter Inductor for Dual-Converter Fed Open-End Transformer Topology 2264
Ghanshyamsinh Gohil, *Aalborg University, Denmark*
Lorand Bede, *Aalborg University, Denmark*
Remus Teodorescu, *Aalborg University, Denmark*
Tamas Kerekes, *Aalborg University, Denmark*
Frede Blaabjerg, *Aalborg University, Denmark*

Mechanism Analysis and Mitigation of Instability in Grid-Connected Voltage Source Inverter with LCL Filters based on Terminal Impedance 2272
Teng Liu, *Xi'an Jiaotong University, China*
Zeng Liu, *Xi'an Jiaotong University, China*
Jinjun Liu, *Xi'an Jiaotong University, China*
Qingyun Dou, *Xi'an Jiaotong University, China*

Seven-Switch Five-Level Active Neutral-Point Clamped Converter and Optimal Modulation Strategy 2278

Hongliang Wang, *Queen's University, Canada*
Lei Kou, *Queen's University, Canada*
Yan-Fei Liu, *Queen's University, Canada*
Paresh C. Sen, *Queen's University, Canada*
Sucheng Liu, *Anhui University of Technology, China*

A Simple Variable Step Size Method for Maximum Power Point Tracking using Commercial Current Mode Control DC-DC Regulators 2286

Su Sheng, *Northeastern University, United States*
Brad Lehman, *Northeastern University, United States*

Session T45: Envelope Tracking and Resonant Conversion
Location: 104C
March 24, 2016 14:00 - 17:30
Session Chairs: Brian Zahnstecher, *PowerRox*
Davide Giacomini, *Infineon*

Envelope Tracking GaN Power Supply for 4G Cell Phone Base Stations 2292

Yuanzhe Zhang, *University of Colorado at Boulder, United States*
Johan Strydom, *Efficient Power Conversion Corporation, United States*
Michael de Rooij, *Efficient Power Conversion Corporation, United States*
Dragan Maksimović, *University of Colorado at Boulder, United States*

Envelope Tracking Power Supply for Volume-Sensitive Low-Power Applications based on a Resonant Switched-Capacitor Converter 2298

Alon Cervera, *Ben-Gurion University of the Negev, Israel*
Mor Mordechai Peretz, *Ben-Gurion University of the Negev, Israel*

A Passive-Impedance-Matching Concept for Multi-Phase Resonant Converter 2304

Hongliang Wang, *Queen's University, Canada*
Yang Chen, *Queen's University, Canada*
Yan-Fei Liu, *Queen's University, Canada*

LLC Converter with Auxiliary Switch for Hold Up Mode Operation 2312

Yang Chen, *Queen's University, Canada*
Hongliang Wang, *Queen's University, Canada*
Yan-Fei Liu, *Queen's University, Canada*
Jahangir Afsharian, *Murata Power Solutions, Canada*
Zhihua Yang, *Queen's University, Canada*

A Common Capacitor Multi-Phase LLC Resonant Converter 2320

Hongliang Wang, *Queen's University, Canada*
Yang Chen, *Queen's University, Canada*
Zhiyuan Hu, *Queen's University, Canada*
Laili Wang, *Queen's University, Canada*
Yajie Qiu, *Queen's University, Canada*
Wenbo Liu, *Queen's University, Canada*
Yan-Fei Liu, *Queen's University, Canada*
Jahangir Afsharian, *Murata Power Solutions, Canada*
Zhihua Yang, *Murata Power Solutions, Canada*

LLC Resonant Converter Design for Bendable Power Converter 2328
Kwun Yuan Godwin Ho, *University of Hong Kong, Hong Kong*
M.H. Bryan Pong, *University of Hong Kong, Hong Kong*
Shu-Yuen Ron Hui, *University of Hong Kong, Hong Kong*

**Design Consideration of MHz Active Clamp Flyback Converter with GaN Devices for
Low Power Adapter Application** 2334
Xiucheng Huang, *Virginia Polytechnic Institute and State University, United States*
Junjie Feng, *Virginia Polytechnic Institute and State University, United States*
Weijing Du, *Virginia Polytechnic Institute and State University, United States*
Fred C. Lee, *Virginia Polytechnic Institute and State University, United States*
Qiang Li, *Virginia Polytechnic Institute and State University, United States*

**A New Capacitor Voltage Balancing Control for Hybrid Modular Multilevel Converter
with Cascaded Full Bridge** 2342
Mahendra B. Ghat, *Indian Institute of Technology Bombay, India*
Anshuman Shukla, *Indian Institute of Technology Bombay, India*
Richa Mishra, *Indian Institute of Technology Bombay, India*

**Sensorless Scheduling of the Modular Multilevel Series-Parallel Converter: Enabling a
Flexible, Efficient, Modular Battery** 2349
Stefan M. Goetz, *Duke University, United States*
Zhongxi Li, *Duke University, United States*
Angel V. Peterchev, *Duke University, United States*
Xinyu Liang, *North Carolina State University, United States*
Chengduo Zhang, *North Carolina State University, United States*
Srdjan M. Lukic, *North Carolina State University, United States*

Session D01: AC-DC Converters
Location: Poster Area
March 24, 2016 11:30 - 14:00
Session Chairs: Nathan Weise, *Marquette*
Daniel Costinett, *University of Tennessee-Knoxville*

**An Input Current Calculation Switching Driver for High Power-Factor and Phase-Cut
Dimmer Compatibility** 2355
Hyunchul Eum, *Fairchild Semiconductor International, Inc., Korea, South*
Youngjong Kim, *Fairchild Semiconductor International, Inc., Korea, South*
Kuohsien Huang, *Fairchild Semiconductor International, Inc., Taiwan*

High Frequency Range Conducted Common-Mode Noise Suppression in SMPS 2360
Jinping Zhou, *Delta Electronics Shanghai Co., Ltd., China*
Yicong Xie, *Delta Electronics Shanghai Co., Ltd., China*
Min Zhou, *Delta Electronics Shanghai Co., Ltd., China*

**Improved Medium Voltage AC-DC Rectifier based on 10kV SiC MOSFET for Solid State
Transformer (SST) Application** 2365
Qianlai Zhu, *North Carolina State University, United States*
Li Wang, *North Carolina State University, United States*
Liqi Zhang, *North Carolina State University, United States*
Wensong Yu, *North Carolina State University, United States*
Alex Q. Huang, *North Carolina State University, United States*

Suppression of Circulating Current in Parallel Operation of Three-Level Converters 2370
Young-Kwang Son, *Seoul National University, Korea, South*
Seung-Jun Chee, *Seoul National University, Korea, South*
Younggi Lee, *Seoul National University, Korea, South*
Seung-Ki Sul, *Seoul National University, Korea, South*
Changjin Lim, *LG Electronics, Korea, South*
Sungjae Huh, *LG Electronics, Korea, South*
Jaeyoon Oh, *LG Electronics, Korea, South*

Hybrid Bridgeless DCM SEPIC Rectifier Integrated with a Modified Switched Capacitor Cell ... 2376
Paulo Junior Silva Costa, *Universidade Federal de Santa Catarina, Brazil*
Telles Brunelli Lazzarin, *Universidade Federal de Santa Catarina, Brazil*
Carlos Henrique Illa Font, *Universidade Tecnológica Federal do Paraná, Brazil*

**LCL Filter Design for Three-Phase Two-Level Power Factor Correction using Line
Impedance Stabilization Network** 2382
Alireza Kouchaki, *University of Southern Denmark, Denmark*
Morten Nymand, *University of Southern Denmark, Denmark*

Sensorless Current Rebuilding Strategy in a Single Phase Bridgeless PFC 2389
Felipe López, *Universidad de Cantabria, Spain*
Paula Lamo, *Universidad de Cantabria, Spain*
Alberto Pigazo, *Universidad de Cantabria, Spain*
F.J. Azcondo, *Universidad de Cantabria, Spain*

A Compact Electrolytic-Free Two-Stage Universal Input Offline LED Driver 2395
Saad Pervaiz, *University of Colorado at Boulder, United States*
Ashish Kumar, *University of Colorado at Boulder, United States*
Khurram K. Afridi, *University of Colorado at Boulder, United States*

Session D02: DC-DC Converters I
Location: Poster Area
March 24, 2016 11:30 - 14:00
Session Chairs: Charles Sullivan, *Dartmouth*
Mahshid Amirabadi, *Northeastern University*

**Design Methodology for a High Insulation Voltage Power Transmission Function for
IGBT Gate Driver** 2401
Sokchea Am, *Grenoble Institute of Technology, France*
Pierre Lefranc, *Grenoble Institute of Technology, France*
David Frey, *Grenoble Institute of Technology, France*
Mahmoud Ibrahim, *Grenoble Institute of Technology, France*

**Optimized Design of GaN Switching Capacitor based Envelope Tracking Power Supply
for Satellite Applications** 2409
Qian Jin, *Nanjing University of Aeronautics and Astronautics, China*
M. Vasić, *Universidad Politécnica de Madrid, Spain*
O. Garcia, *Universidad Politécnica de Madrid, Spain*
P. Alou, *Universidad Politécnica de Madrid, Spain*
J.A. Oliver, *Universidad Politécnica de Madrid, Spain*
J.A. Cobos, *Universidad Politécnica de Madrid, Spain*

An Isolated High Step-Up Converter with Continuous Input Current and LC Snubber 2415
K.I. Hwu, *National Taipei University of Technology, Taiwan*
W.Z. Jiang, *National Taipei University of Technology, Taiwan*
Y.T. Yau, *National Taipei University of Technology, Taiwan*

Output-Inductor-Less Full-Bridge Converter with SiC-MOSFETs for Low Noise and ZVS Operation 2422
Kazuhide Domoto, *Nagasaki University, Japan*
Yoichi Ishizuka, *Nagasaki University, Japan*
Seiya Abe, *Kyushu Institute of Technology, Japan*
Tamotsu Ninomiya, *Green Electronics Research Institute, Kitakyushu, Japan*

Reduction Technique of Leakage Flux Effects on GaN-HEMTs in 5 MHz / 100 W Isolated DC-DC Converters 2430
Akinori Hariya, *Nagasaki University, Japan*
Tomoya Koga, *Nagasaki University, Japan*
Ken Matsuura, *TDK Corporation, Japan*
Hiroshige Yanagi, *TDK-Lambda Corporation, Japan*
Satoshi Tomioka, *TDK-Lambda Corporation, Japan*
Yoichi Ishizuka, *Nagasaki University, Japan*
Tamotsu Ninomiya, *City of Kitakyushu, Japan*

A High-Voltage Level Shifter with Sub-Nano-Second Propagation Delay for Switching Power Converters 2437
Ahmed Abdelmoaty, *Ohio State University, United States*
Mohammad Al-Shyoukh, *TSMC Inc., United States*
Ayman Fayed, *Ohio State University, United States*

Dual-Output, Three-Level GaN-Based DC-DC Converter for Battery Charger Applications 2441
Ren Ren, *Nanjing University of Aeronautics and Astronautics, China*
Bo Liu, *University of Tennessee, United States*
Edward A. Jones, *University of Tennessee, United States*
Fred Wang, *University of Tennessee, United States*
Zheyu Zhang, *University of Tennessee, United States*
Daniel Costinett, *University of Tennessee, United States*

Quadruple Active Bridge DC-DC Converter as the Basic Cell of a Modular Smart Transformer 2449
Levy F. Costa, *Christian-Albrechts-Universität zu Kiel, Germany*
Giampaolo Buticchi, *Christian-Albrechts-Universität zu Kiel, Germany*
Marco Liserre, *Christian-Albrechts-Universität zu Kiel, Germany*

Analytical Model of a Phase-Shift Controlled Three-Level Zero-Voltage Switching Converter 2457
Cas Bakker, *Prodrive Technologies, Netherlands*
Bas Vermulst, *Technische Universiteit Eindhoven, Netherlands*
Anton Driessen, *Prodrive Technologies, Netherlands*

High Efficiency Design for ISOP Converter System with Dual Active Bridge DC-DC Converter 2465

Masaki Sato, *Nagasaki University, Japan*
Kazuhide Domoto, *Nagasaki University, Japan*
Yoichi Ishizuka, *Nagasaki University, Japan*
Masahiro Yamaguchi, *Tohoku University, Japan*
Shinya Manabe, *RICOH Electronic Devices Co., Ltd., Japan*
Hiizu Okubo, *RICOH Electronic Devices Co., Ltd., Japan*
Atsushi Itagaki, *Ryowa Electronics Co., Ltd., Japan*

Wide Input Range Power Converters using a Variable Turns Ratio Transformer 2473
Ziwei Ouyang, *Technical University of Denmark, Denmark*
Michael A.E. Andersen, *Technical University of Denmark, Denmark*

Design Approaches for Fast Supercapacitor Chargers for Applications like SCATMA, SRUPS 2479
Nicoloy Gurusinghe, *University of Waikato, New Zealand*
Nihal Kularatna, *University of Waikato, New Zealand*
W. Howell Round, *University of Waikato, New Zealand*
D. Alistair Steyn-Ross, *University of Waikato, New Zealand*

Stack Multiphase Asymmetrical Half-Bridge Topology Offering Advance Performance and Efficiency 2485
Trong Tue Vu, *Eisergy Ltd., Ireland*
George Young, *Eisergy Ltd., Ireland*

Session D03: DC-DC Converters II
Location: Poster Area
March 24, 2016 11:30 - 14:00
Session Chairs: Jason Stauth, *Dartmouth*
Yan-Fei Liu, *Queens*

Design of a Novel APWM Half-Bridge DC-DC Resonant Converter with Load-Independent Soft-Switching and Reduced Circulating Current 2491
Kawsar Ali, *National University of Singapore, Singapore*
Sandeep Kolluri, *National University of Singapore, Singapore*
Naga Brahmendra Yadav Gorla, *National University of Singapore, Singapore*
Pritam Das, *National University of Singapore, Singapore*
Sanjib Kumar Panda, *National University of Singapore, Singapore*

A Low-Volume Hybrid Step-Down DC-DC Converter based on the Dual use of Flying Capacitor 2497
S.M. Ahsanuzzaman, *University of Toronto, Canada*
Yingxian Ma, *University of Toronto, Canada*
Abrar Ahmed Pathan, *University of Toronto, Canada*
Aleksandar Prodić, *University of Toronto, Canada*

Fractional Pulse Skipping in Digitally Controlled DC-DC Converters for Improved Light-Load Efficiency and Power Spectrum 2504
Bipin Chandra Mandi, *Indian Institute of Technology Kharagpur, India*
Santanu Kapat, *Indian Institute of Technology Kharagpur, India*
Amit Patra, *Indian Institute of Technology Kharagpur, India*

A New Compact and High Efficiency Resonant Converter 2511
Sheng-Yang Yu, *Texas Instruments Inc., United States*

A 10-MHz eGaN FETs based Isolated Class-Φ_2 DCX 2518
Xuewen Zou, *Nanjing University of Aeronautics and Astronautics, China*
Zhiliang Zhang, *Nanjing University of Aeronautics and Astronautics, China*
Zhou Dong, *Nanjing University of Aeronautics and Astronautics, China*
Yuan Zhou, *Nanjing University of Aeronautics and Astronautics, China*
Xiaoyong Ren, *Nanjing University of Aeronautics and Astronautics, China*
Qianhong Chen, *Nanjing University of Aeronautics and Astronautics, China*

Multi-Level Capacitor Clamped DC-DC Multiplier/Divider with Variable and Fractional Voltage Gain – An (n/m)X DC-DC Converter 2525
Deepak Gunasekaran, *Michigan State University, United States*
Liang Qin, *Wuhan University, China*
Ujjwal Karki, *Michigan State University, United States*
Yuan Li, *Sichuan University, China*
Fang Z. Peng, *Michigan State University, United States*

Multi-Mode Quasi-Z-Source Series Resonant DC/DC Converter for Wide Input Voltage Range Applications 2533
Dmitri Vinnikov, *Ubik Solutions LLC, Estonia*
Andrii Chub, *Tallinn University of Technology, Estonia*
Indrek Roasto, *Ubik Solutions LLC, Estonia*
Liisa Liivik, *Tallinn University of Technology, Estonia*

Hybrid Serial-Output Converter for Integrated LED Lighting Applications 2540
T. McRae, *University of Toronto, Canada*
A. Prodić, *University of Toronto, Canada*
G. Lisi, *Texas Instruments Inc., United States*
W. McIntrye, *Texas Instruments Inc., United States*
A. Aguilar, *Texas Instruments Inc., United States*

Analysis and Modeling of a Modular ISOP Full Bridge based Converter with Input Filter ... 2545
P. Zumel, *Universidad Carlos III de Madrid, Spain*
E. Oña, *Universidad Carlos III de Madrid, Spain*
C. Fernandez, *Universidad Carlos III de Madrid, Spain*
M. Sanz, *Universidad Carlos III de Madrid, Spain*
A. Lazaro, *Universidad Carlos III de Madrid, Spain*
A. Barrado, *Universidad Carlos III de Madrid, Spain*
A. Vazquez, *Universidad de Oviedo, Spain*
D.G. Lamar, *Universidad de Oviedo, Spain*

Wide-Input High Power Density Flexible Converter Topology for DC-DC Applications 2553
Parth Jain, *University of Toronto, Canada*
Aleksandar Prodić, *University of Toronto, Canada*
Alexander Gerfer, *Würth Elektronik eiSos GmbH & Co. KG, Germany*

High Efficiency LLC Converter Design for Universal Battery Chargers 2561
Navid Shafiei, *University of British Columbia, Canada*
Ali Arefifar, *University of British Columbia, Canada*
Mohammad Ali Saket, *University of British Columbia, Canada*
Martin Ordonez, *University of British Columbia, Canada*

A New High Power Density Modular Multilevel DC-DC Converter with Localized Voltage Balancing Control for Arbitrary Number of Levels 2567
Ahmed Morsy, *Texas A&M University, United States*
Yong Zhou, *Texas A&M University, United States*
Prasad Enjeti, *Texas A&M University, United States*

Design and Control of a Fault Tolerant Soft Switching DC-DC Converter for High Power High Voltage Applications 2573
Tao Li, *Rensselaer Polytechnic Institute, United States*
Leila Parsa, *Rensselaer Polytechnic Institute, United States*

Accurate Parametric Steady State Analysis and Design Tool for DC-DC Power Converters 2579
Mohammad Daryaei, *University of Alberta, Canada*
Mohammad Ebrahimi, *University of Alberta, Canada*
S. Ali Khajehoddin, *University of Alberta, Canada*

Analysis of Multi-Output Half-Wave Semi-Synchronous Rectifier with a Uniform Magnetic Field Transmitter 2587
Erdem Asa, *Hevo Power Inc. / New York University, United States*
Kerim Colak, *Istanbul Ulasim A.S., Turkey*
Dariusz Czarkowski, *New York University, United States*

High Gain QZS DC/DC Converter with Coupled Inductor 2592
Rafael V. Silva, *Universidade Federal do Ceará, Brazil*
Antônio A.A. Freitas, *Universidade Federal do Ceará, Brazil*
Marcus R. Castro, *Universidade Federal do Ceará, Brazil*
Fernando L.M. Antunes, *Universidade Federal Rural do Semi-Árido, Brazil*
Edilson M. Sá Jr., *Universidade Federal do Ceará, Brazil*

Session D04: Utility Interface
Location: Poster Area
March 24, 2016 11:30 - 14:00
Session Chairs: Ali Khajehoddin, *University of Alberta*
                 Julia Zhang, *Oregon State University*

A Power Decoupling Method with Small Capacitance Requirement based on Single-Phase Quasi-Z-Source Inverter for DC Microgrid Applications 2599
Dingyi He, *University of Texas at Dallas, United States*
Wen Cai, *University of Texas at Dallas, United States*
Fan Yi, *University of Texas at Dallas, United States*

Operation Analysis of High Efficiency Grid Connected Bi-Directional Power Conversion System for Various Storage Battery Systems with Bi-Directional Switch Circuit Topology 2607
Go Yamada, *Panasonic Corporation, Japan*
Takaaki Norisada, *Panasonic Corporation, Japan*
Fumito Kusama, *Panasonic Corporation, Japan*
Keiji Akamatsu, *Panasonic Corporation, Japan*
Masakazu Michihira, *Kobe City College of Technology, Japan*

Fault Tolerant Control of MMC with Redundant Sub-Modules based on Carrier Phase Shift Modulation 2613
Kai Li, *Tsinghua University, China*
Zhengming Zhao, *Tsinghua University, China*
Liqiang Yuan, *Tsinghua University, China*
Sizhao Lu, *Tsinghua University, China*
Bing Pan, *State Grid Smart Grid Research Institute, China*
Zhengang Lu, *State Grid Smart Grid Research Institute, China*

A New Topology of Multilevel VSC Converter for Hybrid HVDC Transmission System 2620
Jae-Jung Jung, *Seoul National University, Korea, South*
Shenghui Cui, *RWTH Aachen University, Germany*
Seung-Ki Sul, *Seoul National University, Korea, South*

Performance of Solid State Transformers Under Imbalanced Loads in Distribution Systems 2629
Tao Yang, *University College Dublin, Ireland*
Ronan Meere, *University College Dublin, Ireland*
Cathal O'Loughlin, *University College Dublin, Ireland*
Terence O'Donnell, *University College Dublin, Ireland*

Steady-State Analysis of Modular Multilevel Converter (MMC) Under Unbalanced Grid Conditions 2637
Xiaojie Shi, *University of Tennessee, United States*
Yalong Li, *University of Tennessee, United States*
Zhiqiang Wang, *University of Tennessee, United States*
Bo Liu, *University of Tennessee, United States*
Leon M. Tolbert, *University of Tennessee, United States*
Fred Wang, *University of Tennessee, United States*

Design and Control of a Compensated Submodule Testing Scheme for Modular Multilevel Converter 2645
Yuan Tang, *University of Warwick, United Kingdom*
Li Ran, *University of Warwick, United Kingdom*
Olayiwola Alatise, *University of Warwick, United Kingdom*
Philip Mawby, *University of Warwick, United Kingdom*

A Voltage Independent Islanding Detection Method and Low Voltage Ride through of a Two-Stage PV Inverter 2652
Partha Pratim Das, *Indian Institute of Technology Kharagpur, India*
Souvik Chattopadhyay, *Indian Institute of Technology Kharagpur, India*
Shiladri Chakraborty, *Indian Institute of Technology Kharagpur, India*

Low Cost and High Efficiency Topology for Flexible Integration of Multi-PV and Batteries in Resonant-Based Converters 2660
Ali Elrayyah, *Qatar Environment and Energy Research Institute, Qatar*

Real-Time Integrated Model of a Micro-Grid with Distributed Clean Energy Generators and their Power Electronics 2666
Weiqiang Chen, *University of Connecticut, United States*
Ali M. Bazzi, *University of Connecticut, United States*
James Hare, *University of Connecticut, United States*
Shalabh Gupta, *University of Connecticut, United States*

Minimization of Inter-Module Leakage Current in Cascaded H-Bridge Multilevel Inverters for Grid Connected Solar PV Applications 2673

V.V.S. Pradeep Kumar, *Indian Institute of Technology Bombay, India*
B.G. Fernandes, *Indian Institute of Technology Bombay, India*

Effect of Grid Inductance on Grid Current Quality of Parallel Grid-Connected Inverter System with Output LCL Filter and Closed-Loop Control 2679

Wooyoung Choi, *University of Wisconsin at Madison, United States*
Woongkul Lee, *University of Wisconsin at Madison, United States*
Bulent Sarlioglu, *University of Wisconsin at Madison, United States*

Small Signal Modeling and Control of a Grid Tied Converter without a Syncronization Unit 2687

Subhajyoti Mukherjee, *Missouri University of Science and Technology, United States*
Pourya Shamsi, *Missouri University of Science and Technology, United States*
Mehdi Ferdowsi, *Missouri University of Science and Technology, United States*

Bridgeless SEPIC PFC Converter for Low Total Harmonic Distortion and High Power Factor ... 2693

Yasemin Onal, *Bilecik Seyh Edebali University, Turkey*
Yilmaz Sozer, *University of Akron, United States*

Effectiveness of Pareto-Front Analysis Applied to the Design of a Single-Phase PFC Rectifier 2700

Mahmoud Ibrahim, *Eaton Corporation, France*
Luc Gonnet, *Eaton Corporation, France*
Pierre Lefranc, *Grenoble Institute of Technology, France*
David Frey, *Grenoble Institute of Technology, France*
Jean-Paul Ferrieux, *Grenoble Institute of Technology, France*
Sokchea Am, *Grenoble Institute of Technology, France*

State Space Analysis and Duty Cycle Control of a Switched Reactance based Center-Point-Clamped Reactive Power Compensator 2706

Pankaj Kumar Bhowmik, *University of North Carolina at Charlotte, United States*
Somasundaram Essakiappan, *University of North Carolina at Charlotte, United States*
Madhav Manjrekar, *University of North Carolina at Charlotte, United States*

A SiC-Based Power Converter Module for Medium-Voltage Fast Charger for Plug-In Electric Vehicles 2714

Srdjan Srdic, *North Carolina State University, United States*
Chi Zhang, *North Carolina State University, United States*
Xinyu Liang, *North Carolina State University, United States*
Wensong Yu, *North Carolina State University, United States*
Srdjan Lukic, *North Carolina State University, United States*

Shunt Active Power Filter based on Cascaded Transformers Coupled with Three-Phase Bridge Converters 2720

Gregory A. de Almeida Carlos, *Universidade Federal de Campina Grande, Brazil*
Cursino B. Jacobina, *Universidade Federal de Campina Grande, Brazil*
João Paulo R. Méllo, *Universidade Federal de Campina Grande, Brazil*
Euzeli C. dos Santos Jr., *Indiana University - Purdue University, United States*

Independent DC Link Voltage Control of Cascaded Multilevel PV Inverter 2727

Qingyun Huang, *North Carolina State University, United States*
Wensong Yu, *North Carolina State University, United States*
Alex Q. Huang, *North Carolina State University, United States*

New Active Damping Method for LCL Filter Resonance based on Two Feedback System 2735
Mahmoud A. Gaafar, *Kyushu University, Japan*
Gamal M. Dousoky, *Minia University, Egypt*
Masahito Shoyama, *Kyushu University, Japan*

Static Synchronous Generator Model for Investigating Dynamic Behaviors and Stability Issues of Grid-Tied Inverters 2742
Liansong Xiong, *Xi'an Jiaotong University, China*
Xiaokang Liu, *Xi'an Jiaotong University, China*
Feng Wang, *Xi'an Jiaotong University, China*
Fang Zhuo, *Xi'an Jiaotong University, China*

Session D05: Motor Drives and Inverters: Modeling and Control I
Location: Poster Area
March 24, 2016 11:30 - 14:00
Session Chairs: Liming Liu, *ABB Inc.*
Thomas Gietzold, *United Technologies Aerospace Systems*

Initial Orientation and Sensorless Starting Strategy of Wound-Rotor Synchronous Starter/Generator 2748
Jichang Peng, *Northwestern Polytechnical University, China*
Weiguo Liu, *Northwestern Polytechnical University, China*
Jinhao Meng, *Northwestern Polytechnical University, China*
Tao Meng, *Northwestern Polytechnical University, China*
Guangzhao Luo, *Northwestern Polytechnical University, China*

A Novel Method for Polarity Detection of Non-Salient PMSMs in Initial Position Estimation 2754
Bing Liu, *Nanjing University of Aeronautics and Astronautics, China*
Bo Zhou, *Nanjing University of Aeronautics and Astronautics, China*
Jiadan Wei, *Nanjing University of Aeronautics and Astronautics, China*
Long Wang, *Nanjing University of Aeronautics and Astronautics, China*
Tianheng Ni, *Nanjing University of Aeronautics and Astronautics, China*

A Speed Adaptive Sensorless Flux Observer for the Induction Motor Drive using Sylvester Criterion Design 2759
Mihai Comanescu, *Penn State Altoona, United States*

Discontinuous PWM for Low Switching Losses in Indirect Matrix Converter Drives 2764
Yeongsu Bak, *Ajou University, Korea, South*
Kyo-Beum Lee, *Ajou University, Korea, South*

Model Predictive Control for Extended Kalman Filter based Speed Sensorless Induction Motor Drives 2770
Jie Li, *Xi'an University of Technology, China*
Li-Heng Zhang, *Xi'an University of Technology, China*
Ying Niu, *Xi'an University of Technology, China*
Hai-Peng Ren, *Xi'an University of Technology, China*

Research on Excitation Control Methods for the Two-Phase Brushless Exciter of Wound-Rotor Synchronous Starter/Generators in the Starting Mode 2776
Ningfei Jiao, *Northwestern Polytechnical University, China*
Weiguo Liu, *Northwestern Polytechnical University, China*
Tao Meng, *Northwestern Polytechnical University, China*
Jichang Peng, *Northwestern Polytechnical University, China*
Shuai Mao, *Northwestern Polytechnical University, China*

A High Performance Speed Regulator Design for AC Machines 2782
Adil Khurram, *American University of Sharjah, U.A.E.*
Habibur Rehman, *American University of Sharjah, U.A.E.*
Shayok Mukhopadhyay, *American University of Sharjah, U.A.E.*

Zero-Sequence Current Suppression for Open-End Winding Induction Motor Drive with Resonant Controller 2788
Hajime Kubo, *Meidensha Corporation, Japan*
Yasuhiro Yamamoto, *Meidensha Corporation, Japan*
Takeshi Kondo, *Meidensha Corporation, Japan*
Kaushik Rajashekara, *University of Texas at Dallas, United States*
Bohang Zhu, *University of Texas at Dallas, United States*

Optimized Control of High-Performance Servo-Motor Drives in the Field-Weakening Region 2794
Jack Bermingham, *Moog Ireland Ltd, Ireland*
Gerard O'Donovan, *Moog Ireland Ltd, Ireland*
Ray Walsh, *Moog Ireland Ltd, Ireland*
Michael Egan, *University College Cork, Ireland*
Gordon Lightbody, *University College Cork, Ireland*
John G. Hayes, *University College Cork, Ireland*

Motor Current Reference Generation for Reducing Motor Currents in Drive Systems with Single-Phase Diode Rectifier and Small DC-Link Capacitor 2801
Young-Ho Chae, *Seoul National University, Korea, South*
Jung-Ik Ha, *Seoul National University, Korea, South*

A Simple Double Mapping based SVPWM Method for Balancing DC-Link Capacitor Voltages of Five-Level Diode-Clamped Converters 2806
Aparna Saha, *University of Akron, United States*
Ali Elrayyah, *Qatar Environment and Energy Research Institute, Qatar*
Yilmaz Sozer, *University of Akron, United States*

Session D06: Motor Drives and Inverters: Modeling and Control II
Location: Poster Area
March 24, 2016 11:30 - 14:00
Session Chairs: Bulent Sarlioglu, *University of Wisconsin - Madison*
Yichao Tang, *Texas Instruments*

Capacitor-Clamped Inverter based Transient Suppression Method for Azimuth Thruster Drives 2813
Shantha Gamini Jayasinghe, *Australian Maritime College, University of Tasmania, Australia*
Viknash Shagar, *Australian Maritime College, University of Tasmania, Australia*
Hossein Enshaei, *Australian Maritime College, University of Tasmania, Australia*
Danyal Mohammadi, *Boise State University, United States*
Mahinda Vilathgamuwa, *Queensland University of Technology, Australia*

Active Common-Mode Voltage Reduction in a Fault-Tolerant Three-Phase Inverter 2821
Danyal Mohammadi, *Boise State University, United States*
Said Ahmed-Zaid, *Boise State University, United States*

Power Cycling Lifetime Improvement of Three-Level NPC Inverters with an Improved DPWM Method 2826
Jiangbiao He, *Marquette University, United States*
Lixiang Wei, *Rockwell Automation, United States*
Nabeel A.O. Demerdash, *Marquette University, United States*

Synchronous Optimal Pulsewidth Modulation Digital Implementation Concept for Multilevel Converters 2833
Jackson Lago, *Universidade Federal de Santa Catarina, Brazil*
Marcelo Lobo Heldwein, *Universidade Federal de Santa Catarina, Brazil*

Analytical Determination of Conduction Losses for Modified Flying Capacitor Multicell Converters 2840
Vahid Dargahi, *Clemson University, United States*
Arash Khoshkbar Sadigh, *Extron Electronics, United States*
Keith Corzine, *Clemson University, United States*

Comparison of Electrical Losses in an Inverter-Fed Five-Phase and Three-Phase Permanent Magnet Assisted Synchronous Reluctance Motor 2847
Akm Arafat, *University of Akron, United States*
Seungdeog Choi, *University of Akron, United States*

A Hybrid Adaptive Observer for the Speed and Flux Estimation of Induction Motors 2855
Mihai Comanescu, *Penn State Altoona, United States*

Determination of CM Choke Parameters for SiC MOSFET Motor Drive based on Simple Measurements and Frequency Domain Modeling 2861
Di Han, *University of Wisconsin at Madison, United States*
Casey Morris, *University of Wisconsin at Madison, United States*
Woongkul Lee, *University of Wisconsin at Madison, United States*
Bulent Sarlioglu, *University of Wisconsin at Madison, United States*

An Improved Model Predictive Current Control of Permanent Magnet Synchronous Motor Drives 2868
Yongchang Zhang, *North China University of Technology, China*
Sugu Gao, *North China University of Technology, China*
Wei Xu, *Huazhong University of Science and Technology, China*

Analysis of Magnet Defect Faults in Permanent Magnet Synchronous Motors through Fluxgate Sensors 2875
Taner Goktas, *University of Texas at Dallas, United States*
Kun Wang Lee, *University of Texas at Dallas, United States*
Mohsen Zafarani, *University of Texas at Dallas, United States*
Bilal Akin, *University of Texas at Dallas, United States*

Session D07: Motor Drives and Inverters: Topologies

Location: Poster Area
March 24, 2016 11:30 - 14:00
Session Chairs: Amirnaser Yazdani, *Ryerson University*
Babak Nahid-Mobarakeh, *University of Lorraine*

Performance Comparison of Transfer Switch Topologies in Switched-Doubly-Fed Machine Drives 2881
Arijit Banerjee, *Massachusetts Institute of Technology, United States*
Steven B. Leeb, *Massachusetts Institute of Technology, United States*
James L. Kirtley, *Massachusetts Institute of Technology, United States*

Multilevel Converter Topologies for High-Power High-Speed Switched Reluctance Motor: Performance Comparison 2889
Devendra Patil, *University of Texas at Dallas, United States*
Shiliang Wang, *University of Texas at Dallas, United States*
Lei Gu, *University of Texas at Dallas, United States*

Bidirectional Magnetically Coupled T-Source Inverter for Extra Low Voltage Application 2897
Thomas Baier, *Friedrich-Alexander-Universität Erlangen-Nürnberg, Germany*
Bernhard Piepenbreier, *Friedrich-Alexander-Universität Erlangen-Nürnberg, Germany*

Active Virtual Ground: Single Phase Grid-Connected Voltage Source Inverter Topology ... 2905
River Tin-Ho Li, *ABB China Ltd., China*
Carl Ngai-Man Ho, *University of Manitoba, Canada*

Design and Evaluation of 30kVA Inverter using SiC MOSFET for 180°C Ambient Temperature Operation 2912
Feng Qi, *Ohio State University, United States*
Miao Wang, *Ohio State University, United States*
Longya Xu, *Ohio State University, United States*
Bo Zhao, *State Grid Corporation of China, China*
Zhe Zhou, *State Grid Corporation of China, China*
Xizhou Ren, *State Grid Corporation of China, China*

A DC to Three-Phase Boost-Buck Inverter with Stored Energy Modulation and a Tiny DC Link Capacitor 2919
Mahima Gupta, *University of Wisconsin at Madison, United States*
Giri Venkataramanan, *University of Wisconsin at Madison, United States*

Drive Circuits for Ultra-Fast and Reliable Actuation of Thomson Coil Actuators used in Hybrid AC and DC Circuit Breakers 2927
Chang Peng, *North Carolina State University, United States*
Alex Huang, *North Carolina State University, United States*
Iqbal Husain, *North Carolina State University, United States*
Bruno Lequesne, *E-Motors Consulting, LLC, United States*
Roger Briggs, *Energy Efficiency Research, LLC, United States*

Improved Transformerless Dual Buck Inverters with Buffer Inductors 2935
Liwei Zhou, *Shandong University, China*
Feng Gao, *Shandong University, China*

A 99% Efficiency SiC Three-Phase Inverter using Synchronous Rectification 2942
Shan Yin, *Nanyang Technological University, Singapore*
K.J. Tseng, *Nanyang Technological University, Singapore*
C.F. Tong, *Nanyang Technological University, Singapore*
Rejeki Simanjorang, *Rolls-Royce Singapore Pte. Ltd., Singapore*
C.J. Gajanayake, *Rolls-Royce Singapore Pte. Ltd., Singapore*
Amit K. Gupta, *Rolls-Royce Singapore Pte. Ltd., Singapore*

Comparison and Evaluation of Common Mode EMI Filter Topologies for GaN-Based Motor Drive Systems 2950
Casey T. Morris, *University of Wisconsin at Madison, United States*
Di Han, *University of Wisconsin at Madison, United States*
Bulent Sarlioglu, *University of Wisconsin at Madison, United States*

Analysis of Thermal Cycling Stress on Semiconductor Devices of the Modular Multilevel Converter for Drive Applications 2957
Xiangyu Han, *Georgia Institute of Technology, United States*
Qichen Yang, *Georgia Institute of Technology, United States*
Liyao Wu, *Georgia Institute of Technology, United States*
Maryam Saeedifard, *Georgia Institute of Technology, United States*

Fault Tolerant Topologies of Five-Level Active Neutral-Point-Clamped Converters 2963
Jun Li, *ABB Inc., United States*

Session D08: Advanced Components and Devices
Location: Poster Area
March 24, 2016 11:30 - 14:00
Session Chairs: Abhijit Pathak, *Infineon/IR*
Doug Hopkins, *North Carolina State University*

Dynamic Characterization of the Input and Reverse Transfer Capacitances in Power MOSFETs under High Current Conduction 2969
Cristino Salcines, *Universität Stuttgart, Germany*
Ingmar Kallfass, *Universität Stuttgart, Germany*
Hisao Kakitani, *Keysight Technologies International, Japan*
Atsushi Mikata, *Keysight Technologies International, Japan*

Medium Voltage Power Switch based on SiC JFETs 2973
Xueqing Li, *United Silicon Carbide, Inc., United States*
Hao Zhang, *United Silicon Carbide, Inc., United States*
Peter Alexandrov, *United Silicon Carbide, Inc., United States*
Anup Bhalla, *United Silicon Carbide, Inc., United States*

Numerical Model and Experimental Study on Comparison of Semiconductor Pulsed Power Devices 2981
Lin Liang, *Huazhong University of Science and Technology, China*
Changdong Chen, *Huazhong University of Science and Technology, China*
Fang Luo, *Ohio State University, United States*

A Normalization Procedure of DC-Side Stray Inductance for High-Speed Switching Circuit 2986
Masato Ando, *Tokyo Metropolitan University, Japan*
Keiji Wada, *Tokyo Metropolitan University, Japan*

Thermal Network Parameter Identification of IGBT Module based on the Cooling Curve of Junction Temperature 2992

Xiong Du, *Chongqing University, China*
Tengfei Li, *Chongqing University, China*
Jun Zhang, *Chongqing University, China*
Heng-Ming Tai, *University of Tulsa, United States*
Pengju Sun, *Chongqing University, China*
Luowei Zhou, *Chongqing University, China*

Design and Evaluation of High Current PCB Embedded Inductor for High Frequency Inverters 2998

Mehrdad Biglarbegian, *University of North Carolina at Charlotte, United States*
Neel Shah, *University of North Carolina at Charlotte, United States*
Iman Mazhari, *University of North Carolina at Charlotte, United States*
Johan Enslin, *University of North Carolina at Charlotte, United States*
Babak Parkhideh, *University of North Carolina at Charlotte, United States*

Prognosis of Wire Bond Lift-Off Fault of an IGBT based on Multisensory Approach 3004

Moinul Shahidul Haque, *University of Akron, United States*
Jeihoon Baek, *Korean Rail Research Institute, Korea, South*
Joseph Herbert, *University of Akron, United States*
Seungdeog Choi, *University of Akron, United States*

Electrical Parasitics and Thermal Modeling for Optimized Layout Design of High Power SiC Modules 3012

Amir Sajjad Bahman, *Aalborg University, Denmark*
Frede Blaabjerg, *Aalborg University, Denmark*
Atanu Dutta, *University of Arkansas, United States*
Alan Mantooth, *University of Arkansas, United States*

Calculation of Losses in PCB Windings for Multi-Coil Contactless Charging Systems 3020

J. Serrano, *Universidad de Zaragoza, Spain*
J. Acero, *Universidad de Zaragoza, Spain*
I. Lope, *BSH Home Appliances Group, Spain*
C. Carretero, *Universidad de Zaragoza, Spain*
J.M. Burdío, *Universidad de Zaragoza, Spain*
R. Alonso, *Universidad de Zaragoza, Spain*

Design of Efficient Loads for Domestic Induction Heating Applications by Means of Non-Magnetic Thin Metallic Layers 3026

Jesús Acero, *Universidad de Zaragoza, Spain*
Claudio Carretero, *Universidad de Zaragoza, Spain*
Rafael Alonso, *Universidad de Zaragoza, Spain*
José Miguel Burdío, *Universidad de Zaragoza, Spain*

A New Evaluation Circuit with a Low-Voltage Inverter Intended for Capacitors used in a High-Power Three-Phase Inverter 3032

Kazunori Hasegawa, *Kyushu Institute of Technology, Japan*
Ichiro Omura, *Kyushu Institute of Technology, Japan*
Shin-Ichi Nishizawa, *Kyushu Institute of Technology / National Institute of Advanced Industrial Science and Technology, Japan*

Energy Absorption Capability of Low Voltage Metal Oxide Varistors in AC and Impulse Currents 3038

Dawood Talebi Khanmiri, *Northeastern University, United States*
Roy Ball, *Mersen USA, United States*
Craig McKenzie, *Mersen USA, United States*
Brad Lehman, *Northeastern University, United States*

Optimization and Experimental Validation of Medium-Frequency High Power Transformers in Solid-State Transformer Applications 3043

M.A. Bahmani, *Chalmers University of Technology, Sweden*
T. Thiringer, *Chalmers University of Technology, Sweden*
M. Kharezy, *SP Technical Research Institute of Sweden, Sweden*

Evaluation of Core Loss in Magnetic Materials Employed in Utility Grid AC Filters 3051

Remus Beres, *Aalborg University, Denmark*
Xiongfei Wang, *Aalborg University, Denmark*
Frede Blaabjerg, *Aalborg University, Denmark*
Claus Leth Bak, *Aalborg University, Denmark*
Hiroaki Matsumori, *Tokyo Metropolitan University, Japan*
Toshihisa Shimizu, *Tokyo Metropolitan University, Japan*

A Novel Gate Assisted Circuit to Reduce Switching Loss and Eliminate Shoot-Through in SiC Half Bridge Configuration 3058

Shan Yin, *Nanyang Technological University, Singapore*
K.J. Tseng, *Nanyang Technological University, Singapore*
C.F. Tong, *Nanyang Technological University, Singapore*
Rejeki Simanjorang, *Rolls-Royce Singapore Pte. Ltd., Singapore*
C.J. Gajanayake, *Rolls-Royce Singapore Pte. Ltd., Singapore*
Amit K. Gupta, *Rolls-Royce Singapore Pte. Ltd., Singapore*

Session D09: System Design Considerations for Power Electronics
Location: Poster Area
March 24, 2016 11:30 - 14:00
Session Chairs: John Vigars, *Allegro Microsystems*
Ernie Parker, *Crane Aerospace & Electronics*

Methods to Enhance the Thermal Performance of a 3D Power Package 3065

Jonathan Noquil, *Texas Instruments Inc., United States*
Ozzie Lopez, *Texas Instruments Inc., United States*
Tianyi Luo, *Lehigh University, United States*

Highly Reliable and Cost Effective Thick Film Substrates for Power LEDs 3069

Paul Gundel, *Heraeus Deutschland GmbH & Co. KG, Germany*
Ryan Persons, *Heraeus Deutschland GmbH & Co. KG, Germany*
Melanie Bawohl, *Heraeus Deutschland GmbH & Co. KG, Germany*
Mark Challingsworth, *Heraeus Deutschland GmbH & Co. KG, Germany*
Christoph Czwickla, *Heraeus Deutschland GmbH & Co. KG, Germany*
Virginia Garcia, *Heraeus Deutschland GmbH & Co. KG, Germany*
Christina Modes, *Heraeus Deutschland GmbH & Co. KG, Germany*
Ilias Nikolaidis, *Heraeus Deutschland GmbH & Co. KG, Germany*
Jessica Reitz, *Heraeus Deutschland GmbH & Co. KG, Germany*
Caitlin Shahbazi, *Heraeus Deutschland GmbH & Co. KG, Germany*
Torsten Nowak, *Fraunhofer-Institut für Zuverlässigkeit und Mikrointegration, Germany*

Design and Evaluation of SiC-Based High Power Density Inverter, 70kW/Liter, 50kW/kg ... 3075
Koji Yamaguchi, *IHI Corporation, Japan*

An Improved Automatic Layout Method for Planar Power Module 3080
Puqi Ning, *Chinese Academy of Sciences, China*
Xuhui Wen, *Chinese Academy of Sciences, China*
Yaohua Li, *Chinese Academy of Sciences, China*
Xiongxuan Ge, *Chinese Academy of Sciences, China*

Practical Implementation Schemes of Motor Speed Measurement by Magnetic Encoder on Electric Power Steering Applications 3086
Jae-Hyun Lee, *Hyundai Mobis, Korea, South*

Low-Cost Input Impedance Estimator of DC-to-DC Converters for Designing the Control Loop in Cascaded Converters 3090
M. Sanz, *Universidad Carlos III de Madrid, Spain*
A. Lázaro, *Universidad Carlos III de Madrid, Spain*
M. Bermejo, *Universidad Carlos III de Madrid, Spain*
D. López del Moral, *Universidad Carlos III de Madrid, Spain*
P. Zumel, *Universidad Carlos III de Madrid, Spain*
C. Fernández, *Universidad Carlos III de Madrid, Spain*
A. Barrado, *Universidad Carlos III de Madrid, Spain*

On-Chip High Performance Magnetics for Point-of-Load High-Frequency DC-DC Converters 3097
Dragan Dinulovic, *Würth Elektronik eiSos GmbH & Co. KG, Germany*
Mahmoud Shousha, *Würth Elektronik eiSos GmbH & Co. KG, Germany*
Martin Haug, *Würth Elektronik eiSos GmbH & Co. KG, Germany*
Alexander Gerfer, *Würth Elektronik eiSos GmbH & Co. KG, Germany*
Mike Wens, *MinDCet NV, Belgium*
Jef Thone, *MinDCet NV, Belgium*

Effects of Auxiliary Source Connections in Multichip Power Module 3101
Helong Li, *Aalborg University, Denmark*
Stig Munk-Nielsen, *Aalborg University, Denmark*
Szymon Bęczkowski, *Aalborg University, Denmark*
Xiongfei Wang, *Aalborg University, Denmark*
Emanuel-Petre Eni, *Aalborg University, Denmark*

Session D10: Modeling and Simulation
Location: Poster Area
March 24, 2016 11:30 - 14:00
Session Chairs: Marco Meola, *ZMD AG*
Mehdi Ferdowsi, *Missouri University of Science & Technology*

Modelling Technique Utilizing Modified Sigmoid Functions for Describing Power Transistor Device Capacitances Applied on GaN HEMT and Silicon MOSFET 3107
H.L. Yeo, *Nanyang Technological University, Singapore*
K.J. Tseng, *Nanyang Technological University, Singapore*

Design and Precise Modeling of a Novel Digital Active EMI Filter 3115
Junpeng Ji, *Xi'an Jiaotong University, China*
Wenjie Chen, *Xi'an Jiaotong University, China*
Xu Yang, *Xi'an Jiaotong University, China*

Development of a Hybrid Emulation Platform based on RTDS and Reconfigurable Power Converter-Based Testbed 3121
Shuoting Zhang, *University of Tennessee, United States*
Yiwei Ma, *University of Tennessee, United States*
Liu Yang, *University of Tennessee, United States*
Fred Wang, *University of Tennessee, United States*
Leon M. Tolbert, *University of Tennessee, United States*

Online Temperature Estimation for Phase Change Composite – 18650 Lithium Ion Cells based Battery Pack 3128
Mohamad Salameh, *Illinois Institute of Technology, United States*
Ben Schweitzer, *AllCell Technologies, United States*
Peter Sveum, *AllCell Technologies, United States*
Said Al-Hallaj, *AllCell Technologies, United States*
Mahesh Krishnamurthy, *Illinois Institute of Technology, United States*

Modeling and Fault Diagnosis of Inter-Turn Short Circuit for Five-Phase PMSM based on Particle Swarm Optimization 3134
Jianwei Yang, *Northwestern Polytechnical University, China*
Manfeng Dou, *Northwestern Polytechnical University, China*
Zhiyong Dai, *Northwestern Polytechnical University, China*
Dongdong Zhao, *Northwestern Polytechnical University, China*
Zhen Zhang, *Northwestern Polytechnical University, China*

Comprehensive Modeling, Testing, and Experimental Validation of Ultracapacitor Open Circuit Voltage Characteristics 3140
Amandeep Singh, *University of Ontario Institute of Technology, Canada*
Najath Abdul Azeez, *University of Ontario Institute of Technology, Canada*
Sheldon S. Williamson, *University of Ontario Institute of Technology, Canada*

Novel SPICE Model for Common Mode Choke Including Complex Permeability 3146
Katsuya Nomura, *Toyota Central R&D Labs., Inc., Japan*
Naoto Kikuchi, *Toyota Central R&D Labs., Inc., Japan*
Yoshitoshi Watanabe, *Toyota Central R&D Labs., Inc., Japan*
Shuntaro Inoue, *Toyota Central R&D Labs., Inc., Japan*
Yoshiyuki Hattori, *Toyota Central R&D Labs., Inc., Japan*

Session D11: Control I
Location: Poster Area
March 24, 2016 11:30 - 14:00
Session Chairs: Bilal Akin, *Univeristy of Texas, Dallas*
Brian Zahnstecher, *PowerRox LLC*

Analysis and Design of Capacitive Power Transmission System Employing Out-of-Band Wireless Feedback Link 3153
Sung-Jin Choi, *University of Ulsan, Korea, South*
Hee-Su Choi, *University of Ulsan, Korea, South*

Introducing Fourier-Based Modeling and Control of Active-Bridge Converters 3158
B.J.D. Vermulst, *Technische Universiteit Eindhoven, Netherlands*
J.L. Duarte, *Technische Universiteit Eindhoven, Netherlands*
C.G.E. Wijnands, *Technische Universiteit Eindhoven, Netherlands*
E.A. Lomonova, *Technische Universiteit Eindhoven, Netherlands*

A Stability Analysis and Efficiency Improvement of Synchronverter 3165
Prasanna Piya, *Mississippi State University, United States*
Masoud Karimi-Ghartemani, *Mississippi State University, United States*

Compensation of Switching Dead-Time Effects in Voltage-Fed PWM Inverters using FPGA-Based Current Oversampling 3172
Bastian Weber, *Leibniz Universität Hannover, Germany*
Tobias Brandt, *Leibniz Universität Hannover, Germany*
Axel Mertens, *Leibniz Universität Hannover, Germany*

Control Strategy of High Power Converters with Synchronous Generator Characteristics for PMSG-Based Wind Power Application 3180
Yuzhi Zhang, *University of Arkansas, United States*
Haoyan Liu, *University of Arkansas, United States*
H. Alan Mantooth, *University of Arkansas, United States*

Phase Compensation, ZVS Operation of Wireless Power Transfer System based on SOGI-PLL 3185
Pingan Tan, *Xiangtan University, China*
Haibing He, *Xiangtan University, China*
Xieping Gao, *Xiangtan University, China*

A Novel Low-Cost Online State of Charge Estimation Method for Reconfigurable Battery Pack 3189
Ni Lin, *University of Nebraska at Lincoln, United States*
Song Ci, *University of Nebraska at Lincoln, United States*
Dalei Wu, *University of Tennessee at Chattanooga, United States*

Effect of Decoupling Terms on the Performance of PR Current Controllers Implemented in Stationary Reference Frame 3193
Sizhan Zhou, *Xi'an Jiaotong University, China*
Jinjun Liu, *Xi'an Jiaotong University, China*

Fuzzy Predictive DTC of Induction Machines with Reduced Torque Ripple and High Performance Operation 3200
Alberto Berzoy, *Florida International University, United States*
Osama Mohammed, *Florida International University, United States*
Johnny Rengifo, *Universidad Simon Bolivar, Venezuela*

Session D12: Control II
Location: Poster Area
March 24, 2016 11:30 - 14:00
Session Chairs: Martin Ordonez, *University of British Columbia*
Jiangbiao He, *GE Global Research*

Fixed-Frequency Generalized Peak Current Control (GPCC) for Inverters 3207
Mohammad Ebrahimi, *University of Alberta, Canada*
S. Ali Khajehoddin, *University of Alberta, Canada*

Improved Control Strategy of 1 MHz LLC Converter for High Frequency Resolution 3213
Hwa-Pyeong Park, *Ulsan National Institute of Science and Technology, Korea, South*
Jee-Hoon Jung, *Ulsan National Institute of Science and Technology, Korea, South*

Bumpless Control for Reduced THD in Power Factor Correction Circuits 3219
Joel Steenis, *Microchip Technology, United States*
Alex Dumais, *Microchip Technology, United States*

Mixed-Signal Hysteretic Internal Model Control of Buck Converters for Ultra-Fast Envelope Tracking 3224
V. Inder Kumar, *Indian Institute of Technology Kharagpur, India*
Santanu Kapat, *Indian Institute of Technology Kharagpur, India*

A Continuous Actor-Critic Maximum Power Point Tracker Applied to Low Power Wind Turbine Systems 3231
J.L. Wattes, *Universidade Federal do Ceará, Brazil*
A.J.S. Dias Jr., *Universidade Federal do Ceará, Brazil*
A.P.S. Braga, *Universidade Federal do Ceará, Brazil*
P.P. Praça, *Universidade Federal do Ceará, Brazil*
A.U. Barbosa, *Universidade Federal do Ceará, Brazil*
D.S. de Souza Oliveira Jr., *Universidade Federal do Ceará, Brazil*

Multi-Band Mixed-Signal Hysteresis Current Control for EMI Reduction in Switch-Mode Power Supplies 3237
Arindam Mandal, *Indian Institute of Technology Kharagpur, India*
V. Inder Kumar, *Indian Institute of Technology Kharagpur, India*
Santanu Kapat, *Indian Institute of Technology Kharagpur, India*

A Parabolic Current Control based Digital Current Control Strategy for High Switching Frequency Voltage Source Inverters 3243
Lanhua Zhang, *Virginia Polytechnic Institute and State University, United States*
Rachael Born, *Virginia Polytechnic Institute and State University, United States*
Xiaonan Zhao, *Virginia Polytechnic Institute and State University, United States*
Jih-Sheng Jason Lai, *Virginia Polytechnic Institute and State University, United States*
Hongbo Ma, *Southwest Jiaotong University, China*

Finite Control Set Model Predictive Control of Dual-Output Four-Leg Indirect Matrix Converter Under Unbalanced Load and Supply Conditions 3248
Ozan Gulbudak, *University of South Carolina, United States*
Enrico Santi, *University of South Carolina, United States*

A Silicon Carbide Integrated Circuit Implementing Nonlinear-Carrier Control for Boost Converter Applications 3255
Richard Kyle Harris, *University of Tennessee, United States*
Benjamin M. McCue, *University of Tennessee, United States*
Benjamin D. Roehrs, *University of Tennessee, United States*
Charles Roberts II, *University of Tennessee, United States*
Benjamin J. Blalock, *University of Tennessee, United States*
Daniel J. Costinett, *University of Tennessee, United States*
Kouros Sariri, *Frequency Management International, United States*
George Megyei, *Frequency Management International, United States*
Cheng-Po Chen, *GE Global Research, United States*
Avinash Kashyap, *GE Global Research, United States*
Reza Ghandi, *GE Global Research, United States*

A New Current Mode Constant on Time Control with Ultrafast Load Transient Response 3259
Syed Bari, *Virginia Polytechnic Institute and State University, United States*
Qiang Li, *Virginia Polytechnic Institute and State University, United States*
Fred C. Lee, *Virginia Polytechnic Institute and State University, United States*

A Web-Based Tool for Compensation Design of Power Converters using Hybrid Optimization 3266
Srikanth Pam, *Texas Instruments Inc., India*
Yudhister Satija, *Texas Instruments Inc., India*
Pradeep Chawda, *Texas Instruments Inc., United States*
Makram Mansour, *Texas Instruments Inc., United States*
Robert Hanrahan, *Texas Instruments Inc., United States*
Jeff Perry, *Texas Instruments Inc., United States*

Second Order Sliding Mode Controlled Point of Load Power Supply 3273
Prasanta K. Achanta, *University of Colorado at Boulder, United States*
David C. Jones, *University of Colorado at Boulder, United States*
Dragan Maksimovic, *University of Colorado at Boulder, United States*
Serhii M. Zhak, *Linear Technology Corporation, United States*
Brett Miwa, *Maxim Integrated, United States*
Cory Arnold, *Maxim Integrated, United States*

Vibration and Torque Ripple Reduction of Switched Reluctance Motors through Current Profile Optimization 3279
Cong Ma, *University of Nebraska at Lincoln, United States*
Liyan Qu, *University of Nebraska at Lincoln, United States*
Rakesh Mitra, *Nexteer Automotive, United States*
Prerit Pramod, *Nexteer Automotive, United States*
Rakib Islam, *Nexteer Automotive, United States*

Modified Predictive Current Control of Neutral-Point Clamped Converter with Reduced Switching Frequency 3286
Dinto Mathew, *Indian Institute of Technology Bombay, India*
Anshuman Shukla, *Indian Institute of Technology Bombay, India*
Santanu Bandyopadhyay, *Indian Institute of Technology Bombay, India*

Implicit Finite Control Set Model Predictive Current Control for Modular Multilevel Converter based on IPA-SQP Algorithm 3291
Hamed Nademi, *ABB AS, Norway*
Lars Einar Norum, *Norwegian University of Science and Technology, Norway*

Resolution Requirements to Avoid Limit Cycling in LLC Resonant Converter 3297
Shadi Dashmiz, *University of Toronto, Canada*
Behzad Mahdavikhah, *University of Toronto, Canada*
Aleksandar Prodić, *University of Toronto, Canada*
Brent McDonald, *Texas Instruments Inc., United States*

Session D13: Renewable Energy Systems I
Location: Poster Area
March 24, 2016 11:30 - 14:00
Session Chairs: Akshay Kumar Rathore, *Concordia University*
Xiaoqiang Guo, *Yanshan University, China*

Reduction of Storage Capacity in DC Microgrids using PV-Embedded Series DC Electric Springs 3302
Ming-Hao Wang, *University of Hong Kong, Hong Kong*
Siew-Chong Tan, *University of Hong Kong, Hong Kong*
Shu-Yuen Ron Hui, *University of Hong Kong, Hong Kong*

A Vector Control Strategy of Grid-Connected Brushless Doubly Fed Induction Generator based on the Vector Control of Doubly Fed Induction Generator 3310
Sheng Hu, *Wuhan University of Technology, China*
Guorong Zhu, *Wuhan University of Technology, China*

An Energy Router based on Multi-Winding High-Frequency Transformer 3317
Xianzhuo Liu, *Tsinghua University, China*
Zedong Zheng, *Tsinghua University, China*
Kui Wang, *Tsinghua University, China*
Yongdong Li, *Tsinghua University, China*

Noise Suppression of the DWT-Based MRA on Mother Wavelet and Decomposition Level Optimization for a Robust Adaptive SOC Estimator in Multi-Cell Battery String 3322
Jonghoon Kim, *Chosun University, Korea, South*
Chang Yoon Chun, *Seoul National University, Korea, South*
Woonki Na, *California State University, Fresno, United States*

A Feedforward Control based Power Decoupling Scheme for Voltage-Controlled Grid-Tied Inverters 3328
Baojin Liu, *Xi'an Jiaotong University, China*
Zeng Liu, *Xi'an Jiaotong University, China*
Jinjun Liu, *Xi'an Jiaotong University, China*
Teng Wu, *Xi'an Jiaotong University, China*
Shike Wang, *Xi'an Jiaotong University, China*

Light Load Efficiency Improvement of Solar Farms Three-Phase Two-Stage Module Integrated Converter 3333
Ahmadreza Amirahmadi, *University of Central Florida, United States*
Utsav Somani, *University of Central Florida, United States*
Mahmood Alharbi, *University of Central Florida, United States*
Charlie Jourdan, *University of Central Florida, United States*
Issa Batarseh, *University of Central Florida, United States*

Switching System Stability Analysis of DC Microgrids with DBS Control 3338
Na Zhi, *Xi'an University of Technology, China*
Hui Zhang, *Xi'an University of Technology, China*
Xi Xiao, *Tsinghua University, China*

A Grid-Connected WECS with Power Limiting Control 3346
Jéssica Santos Guimarães, *Universidade Federal do Ceará, Brazil*
Demercil de Souza Oliveira Jr., *Universidade Federal do Ceará, Brazil*
Juliano de Oliveira Pacheco, *Universidade Federal do Ceará, Brazil*
Paulo P. Peixoto, *Universidade Federal do Ceará, Brazil*

Overshoot Control of the Electromagnetic Torque during Fault Recovery for an SCIG with a STATCOM 3353
Zahra Mahmoodzadeh, *Washington State University, United States*
Mehrdad Yazdanian, *Washington State University, United States*
Hooman Ghaffarzadeh, *Washington State University, United States*
Ali Mehrizi-Sani, *Washington State University, United States*

A Self-Adaptive Power Balance Control Strategy for PV Inverters in Islanded Microgrids 3358
Zhenxiong Wang, *Xi'an Jiaotong University, China*
Hao Yi, *Xi'an Jiaotong University, China*
Fang Zhuo, *Xi'an Jiaotong University, China*
Zhigang Zhang, *Xi'an Jiaotong University, China*

High Performance ZVT with Bus Clamping Modulation Technique for Single Phase Full Bridge Inverters 3364
Yinglai Xia, *Arizona State University, United States*
Raja Ayyanar, *Arizona State University, United States*

Small AC Signal Droop based Secondary Control for Microgrids 3370
Teng Wu, *Xi'an Jiaotong University, China*
Zeng Liu, *Xi'an Jiaotong University, China*
Jinjun Liu, *Xi'an Jiaotong University, China*
Baojin Liu, *Xi'an Jiaotong University, China*
Shike Wang, *Xi'an Jiaotong University, China*

Mode Transition Control Strategy for Multiple Inverter based Distributed Generators Operating in Grid-Connected and Stand-Alone Mode 3376
Onkar Vitthal Kulkarni, *Indian Institute of Technology Bombay, India*
Suryanarayana Doolla, *Indian Institute of Technology Bombay, India*
B.G. Fernandes, *Indian Institute of Technology Bombay, India*

An Autonomous Power Management Strategy based on DC Bus Signaling for Solid-State Transformer Interfaced PMSG Wind Energy Conversion System 3383
Rui Gao, *North Carolina State University, United States*
Iqbal Husain, *North Carolina State University, United States*
Alex Q. Huang, *North Carolina State University, United States*

An Isolated Buck-Boost Type High-Frequency Link Photovoltaic Microinverter 3389
Shiladri Chakraborty, *Indian Institute of Technology Kharagpur, India*
Souvik Chattopadhyay, *Indian Institute of Technology Kharagpur, India*

Energy Management and Stabilization of a Hybrid DC Microgrid for Transportation Applications 3397
Mehdi Karbalaye Zadeh, *Norwegian University of Science and Technology, Norway*
Louis-Marie Saublet, *Université de Lorraine, France*
Roghayeh Gavagsaz-Ghoachani, *Université de Lorraine, France*
Babak Nahid-Mobarakeh, *Université de Lorraine, France*
Serge Pierfederici, *Université de Lorraine, France*
Marta Molinas, *Norwegian University of Science and Technology, Norway*

A Low-Cost Solar Micro-Inverter with Soft-Switching Capability Utilizing Circulating Current 3403
Xiaohu Liu, *GE Global Research, United States*
Mohammed Agamy, *GE Global Research, United States*
Dong Dong, *GE Global Research, United States*
Maja Harfman-Todorovic, *GE Global Research, United States*
Luis Garces, *GE Global Research, United States*

Session D14: Renewable Energy Systems II
Location: Poster Area
March 24, 2016 11:30 - 14:00
Session Chairs: Haoyu Wang, *Shanghai Tech University*
Robert Pilawa-Podgurski, *University of Illinois at Urbana-Champaign*

Design and Stability Analysis for an Autonomous DC Microgrid with Constant Power Load ... 3409
Qianwen Xu, *Nanyang Technological University, Singapore*
Xiaolei Hu, *Nanyang Technological University, Singapore*
Peng Wang, *Nanyang Technological University, Singapore*
Jianfang Xiao, *Nanyang Technological University, Singapore*
Leonardy Setyawan, *Nanyang Technological University, Singapore*
Changyun Wen, *Nanyang Technological University, Singapore*
Lee Meng Yeong, *Rolls-Royce Singapore Pte. Ltd., Singapore*

MPC-SVM Method for Vienna Rectifier with PMSG used in Wind Turbine Systems 3416
June-Seok Lee, *Korea Railroad Research Institute, Korea, South*
Yeongsu Bak, *Ajou University, Korea, South*
Kyo-Beum Lee, *Ajou University, Korea, South*
Frede Blaabjerg, *Aalborg University, Denmark*

An Equivalent Circuit Model for State of Energy Estimation of Lithium-Ion Battery 3422
Kaiyuan Li, *Nanyang Technological University, Singapore*
King Jet Tseng, *Nanyang Technological University, Singapore*

Distributed Optimal Control of Reactive Power and Voltage in Islanded Microgrids 3431
Yanbo Wang, *Aalborg University, Denmark*
Xiongfei Wang, *Aalborg University, Denmark*
Zhe Chen, *Aalborg University, Denmark*
Frede Blaabjerg, *Aalborg University, Denmark*

New Start-Up Scheme for HF Transformer Link Photovoltaic Inverter 3439
Abhijit Kulkarni, *Indian Institute of Science, India*
Vinod John, *Indian Institute of Science, India*

Analysis and Improvement of Harmonic Quasi Resonant Control for LCL-Filtered Grid-Connected Inverters in Weak Grid 3446
Qiang Qian, *Nanjing University of Aeronautics and Astronautics, China*
Jinming Xu, *Nanjing University of Aeronautics and Astronautics, China*
Shaojun Xie, *Nanjing University of Aeronautics and Astronautics, China*
Lin Ji, *Nanjing University of Aeronautics and Astronautics, China*

Model Predictive Control Method to Reduce Common-Mode Voltage and Balance the Neutral-Point Voltage in Three-Level T-Type Inverter 3453
Xiangyang Xing, *Shandong University, China*
Alian Chen, *Shandong University, China*
Zicheng Zhang, *Shandong University, China*
Jie Chen, *Shandong University, China*
Chenghui Zhang, *Shandong University, China*

Convergence Analysis of Distributed Control for Operation Cost Minimization of Droop Controlled DC Microgrid based on Multiagent 3459
Chendan Li, *Aalborg University, Denmark*
Juan C. Vásquez, *Aalborg University, Denmark*
Josep M. Guerrero, *Aalborg University, Denmark*

A Novel Model Predictive Control Algorithm to Suppress the Zero-Sequence Circulating Currents for Parallel Three-Phase Voltage Source Inverters 3465
Zicheng Zhang, *Shandong University, China*
Alian Chen, *Shandong University, China*
Xiangyang Xing, *Shandong University, China*
Chenghui Zhang, *Shandong University, China*

Design of Dynamic Voltage Restorer and Active Power Filter for Wind Power Systems Subject to Unbalanced and Harmonic Distorted Grid 3471
Woei-Luen Chen, *Chang Gung University, Taiwan*
Meng-Jie Wang, *Chang Gung University, Taiwan*

Dynamic Variable Coupling Analysis and Modeling of Proton Exchange Membrane Fuel Cells for Water and Thermal Management 3476
Daming Zhou, *Université de Technologie de Belfort-Montbéliard, France*
Elena Breaz, *Université de Technologie de Belfort-Montbéliard, France*
Alexandre Ravey, *Université de Technologie de Belfort-Montbéliard, France*
Fei Gao, *Université de Technologie de Belfort-Montbéliard, France*
Abdellatif Miraoui, *Université de Technologie de Belfort-Montbéliard, France*
Ke Zhang, *Northwestern Polytechnical University, China*

Voltage and Frequency Control of Electric Spring based Smart Loads 3481
Yun Yang, *University of Hong Kong, Hong Kong*
Siew-Chong Tan, *University of Hong Kong, Hong Kong*
Shu-Yuen Ron Hui, *University of Hong Kong, Hong Kong*

Second Harmonic Current Compensator with Improved One-Cycle-Control 3488
Li Zhang, *Nanjing University of Aeronautics and Astronautics, China*
Xinbo Ruan, *Nanjing University of Aeronautics and Astronautics, China*
Xiaoyong Ren, *Nanjing University of Aeronautics and Astronautics, China*

Frequency Adaptive Control of a Smart Transformer-Fed Distribution Grid 3493
Zhi-Xiang Zou, *Christian-Albrechts-Universität zu Kiel, Germany*
Giovanni De Carne, *Christian-Albrechts-Universität zu Kiel, Germany*
Giampaolo Buticchi, *Christian-Albrechts-Universität zu Kiel, Germany*
Marco Liserre, *Christian-Albrechts-Universität zu Kiel, Germany*

A Synchronization Scheme for Single-Phase Grid-Tied Inverters under Harmonic Distortion and Grid Disturbances 3500
Lenos Hadjidemetriou, *University of Cyprus, Cyprus*
Elias Kyriakides, *University of Cyprus, Cyprus*
Yongheng Yang, *Aalborg University, Denmark*
Frede Blaabjerg, *Aalborg University, Denmark*

Series-Parallel Connection of Low-Voltage Sources for Integration of Galvanically Isolated Energy Storage Systems 3508
Ramy Georgious, *Universidad de Oviedo, Spain*
Jorge Garcia, *Universidad de Oviedo, Spain*
Angel Navarro, *Universidad de Oviedo, Spain*
Sarah Saeed, *Universidad de Oviedo, Spain*
Pablo Garcia, *Universidad de Oviedo, Spain*

Saturation Controller-Based Direct Power Control for Doubly-Fed Induction Generator 3514
Chun Wei, *University of Nebraska at Lincoln, United States*
Zhe Zhang, *Nexteer Automotive, United States*
Wei Qiao, *University of Nebraska at Lincoln, United States*
Liyan Qu, *University of Nebraska at Lincoln, United States*

Inductance-Simulating Control for DFIG-Based Wind Turbine to Ride-Through Grid Faults 3521
Donghai Zhu, *Huazhong University of Science and Technology, China*
Xudong Zou, *Huazhong University of Science and Technology, China*
Yong Kang, *Huazhong University of Science and Technology, China*
Lu Deng, *Wuhan NARI Limited Company of State Grid Electric Power Research Institute, China*
Qingjun Huang, *State Key Laboratory of Disaster Prevention & Reduction for Power Grid Transmission and Distribution Equipment, China*

Session D15: Transportation Power Electronics
Location: Poster Area
March 24, 2016 11:30 - 14:00
Session Chairs: Ted Bohn, *Argonne National Labs*
Khurram Afridi, *University of Colorado, Boulder*

Misalignment Effect on Efficiency of Wireless Power Transfer for Electric Vehicles 3526
Yabiao Gao, *University of Georgia, United States*
Antonio Ginart, *University of Georgia / Sonnenbatterie GmbH, United States*
Kathleen Blair Farley, *Southern Company Services, Inc., United States*
Zion Tsz Ho Tse, *University of Georgia, United States*

Genetic Algorithm Design of a 3D Printed Heat Sink 3529
Tong Wu, *University of Tennessee, United States*
Burak Ozpineci, *Oak Ridge National Laboratory, United States*
Curtis Ayers, *Oak Ridge National Laboratory, United States*

Evaluation of Power Flow Control for an All-Electric Warship Power System with Pulsed Load Applications 3537
J. Neely, *Sandia National Laboratories, United States*
L. Rashkin, *Sandia National Laboratories, United States*
M. Cook, *Sandia National Laboratories, United States*
D. Wilson, *Sandia National Laboratories, United States*
S. Glover, *Sandia National Laboratories, United States*

Reduced Active Switch AC to DC Rectifier with High Frequency Isolation for Electric Vehicle Chargers 3545
José Juan Sandoval, *Texas A&M University, United States*
Taeyong Kang, *Texas A&M University, United States*
Prasad Enjeti, *Texas A&M University, United States*

A Wide Bandgap Device based Multilevel Switched-Capacitor Converter 3553
Diogo Cesar Santos de Moura, *North Dakota State University, United States*
Boris Curuvija, *North Dakota State University, United States*
Dong Cao, *North Dakota State University, United States*

Session D16: Power Topologies, Distribution, and Control
Location: Poster Area
March 24, 2016 11:30 - 14:00
Session Chairs: Tiefu Zhao, *Eaton*
Xiaonan Lu, *Argonne National Laboratory*

Novel Circulating Current Suppression Strategy for MMC based on Quasi-PR Controller 3560
Shengbao Geng, *Shanghai Jiao Tong University, China*
Yiliang Gan, *Shanghai Jiao Tong University, China*
Yungui Li, *Shanghai Jiao Tong University, China*
Lijun Hang, *Shanghai Jiao Tong University, China*
Guojie Li, *Shanghai Jiao Tong University, China*

Assymetric Duty-Cycle Phase-Shift Modulation for Power Management in Double Half-Bridge Inverter with Partly Coupled Inductive Loads 3566
C. Carretero, *Universidad de Zaragoza, Spain*
H. Sarnago, *Universidad de Zaragoza, Spain*
O. Lucia, *Universidad de Zaragoza, Spain*
J. Acero, *Universidad de Zaragoza, Spain*
J.M. Burdío, *Universidad de Zaragoza, Spain*

Control Implementation for a Wide Voltage Range High Efficiency Power Supply Utilizing Low Voltage MOSFETs 3570
Werner Konrad, *Technische Universität Graz, Austria*
Gerald Deboy, *Infineon Technologies AG, Austria*
Annette Muetze, *Technische Universität Graz, Austria*

A Single-Phase Dual Frequency Inverter based on Multi-Frequency Selective Harmonic Elimination 3577
Chongwen Zhao, *University of Tennessee, United States*
Daniel Costinett, *University of Tennessee, United States*
Brad Trento, *University of Tennessee, United States*
Daniel Friedrichs, *Medtronic, United States*

Grid Connected DC Distribution Network Deploying High Power Density Rectifier for DC Voltage Stabilization 3585
Danillo B. Rodrigues, *Universidade Federal do Triângulo Mineiro, Brazil*
Paulo R. Silva, *Universidade Federal de Uberlândia, Brazil*
Gustavo B. Lima, *Universidade Federal do Triângulo Mineiro, Brazil*
Ernane A.A. Coelho, *Universidade Federal de Uberlândia, Brazil*
Luiz C.G. Freitas, *Universidade Federal de Uberlândia, Brazil*

Even-Harmonic Repetitive Control for Circulating Current Suppression in Modular Multilevel Converters 3591
Shunfeng Yang, *Nanyang Technological University, Singapore*
Peng Wang, *Nanyang Technological University, Singapore*
Yi Tang, *Nanyang Technological University, Singapore*
Michael Zagrodnik, *Rolls-Royce Singapore Pte. Ltd., Singapore*
Xiaolei Hu, *Nanyang Technological University, Singapore*
King Jet Tseng, *Nanyang Technological University, Singapore*

A New DSC-PLL using Recursive Discrete Fourier Transform for Robustness to Frequency Variation 3598
Jaedo Lee, *Korea Institute of Nuclear Safety, Korea, South*
Hanju Cha, *Chungnam National University, Korea, South*

A Four-Quadrant Modulation Technique for Cascaded Multilevel Inverters to Extend Solution Range for Selective Harmonic Elimination/Compensation 3603
Hui Zhao, *University of Florida, United States*
Shuo Wang, *University of Florida, United States*

Online Battery Impedance Spectrum Measurement Method 3611
Jaber A. Abu Qahouq, *University of Alabama, United States*

Analysis and Control of a Reduced Switch Converter for Active Magnetic Bearings 3616
Dong Jiang, *Huazhong University of Science and Technology, China*
Parag Kshirsagar, *United Technologies Research Center, United States*

A Novel Balanced Winding Topology to Mitigate EMI without the Need for a Y-Capacitor 3623
Yongjiang Bai, *Xi'an Jiaotong University, China*
Xu Yang, *Xi'an Jiaotong University, China*
Xinlei Li, *Silergy Corp., China*
Dan Zhang, *Silergy Corp., China*
Wenjie Chen, *Xi'an Jiaotong University, China*

Topology and Control Strategy for Accelerated Lifetime Test Setup of DC-Link Capacitor of Wind Turbine Converter 3629
Youngjong Ko, *Christian-Albrechts-Universität zu Kiel, Germany*
Holger Jedtberg, *Christian-Albrechts-Universität zu Kiel, Germany*
Giampaolo Buticchi, *Christian-Albrechts-Universität zu Kiel, Germany*
Marco Liserre, *Christian-Albrechts-Universität zu Kiel, Germany*

Voltage Droop Compensation based on Resonant Circuit for Generalized High Voltage Solid-State Marx Modulator 3637
Hiren Canacsinh, *Instituto Superior de Engenharia de Lisboa, Portugal*
Luís M. Redondo, *Instituto Superior de Engenharia de Lisboa, Portugal*
J. Fernando Silva, *Instituto Superior Técnico, Portugal*
Beatriz Borges, *Instituto Superior Técnico, Portugal*

Four H-Bridge based Shunt Active Power Filter for Three-Phase Four Wire System 3641
Edgard L.L. Fabricio, *Universidade Federal da Paraíba, Brazil*
Cursino B. Jacobina, *Universidade Federal de Campina Grande, Brazil*
Gregory A.A. Carlos, *Universidade Federal de Campina Grande, Brazil*
Maurício B.R. Correa, *Universidade Federal de Campina Grande, Brazil*

High-Frequency AC Distributed Power Delivery System 3648
Mengqi Wang, *University of Michigan at Dearborn, United States*
Qingyun Huang, *North Carolina State University, United States*
Wensong Yu, *North Carolina State University, United States*
Alex Q. Huang, *North Carolina State University, United States*

Effect of the Capacitance Distribution on the Output Impedance of the Half-Wave Cockcroft-Walton Voltage Multiplier 3655
Liran Katzir, *Tel Aviv University, Israel*
Doron Shmilovitz, *Tel Aviv University, Israel*

Session D17: Emerging and Renewable Power
Location: Poster Area
March 24, 2016 11:30 - 14:00
Session Chairs: Katherine Kim, *Ulsan NIST*
Dimitri Torregrossa, *EPFL*

A Cost Effective High Performance LED Driver Powered by Electronic Ballasts 3659
Jianwen Shao, *STMicroelectronics, United States*
Thomas Stamm, *STMicroelectronics, United States*

Model Predictive Control of Z-Source Four-Leg Inverter for Standalone Photovoltaic System with Unbalanced Load 3663
Sertac Bayhan, *Gazi University, Turkey*
Mohamed Trabelsi, *Texas A&M University at Qatar, Qatar*
Haitham Abu-Rub, *Texas A&M University at Qatar, Qatar*

Efficiency Optimization of an Integrated Wireless Power Transfer System by a Genetic Algorithm 3669
Rosario Pagano, *Integrated Device Technology Inc., United States*
Siamak Abedinpour, *Integrated Device Technology Inc., United States*
Angelo Raciti, *Università degli Studi di Catania, Italy*
Salvatore Musumeci, *Università degli Studi di Catania, Italy*

Loss Analysis of a High Efficiency GaN and Si Device Mixed Isolated Bidirectional DC-DC Converter 3677
Fei Xue, *North Carolina State University, United States*
Ruiyang Yu, *North Carolina State University, United States*
Alex Q. Huang, *North Carolina State University, United States*

Dynamic Efficiency Tracking Controller for Reconfigurable Four-Coil Wireless Power Transfer System 3684
Yuan Cao, *University of Alabama, United States*
Zhigang Dang, *University of Alabama, United States*
Jaber A. Abu Qahouq, *University of Alabama, United States*
Evan Phillips, *University of Alabama, United States*

Wireless Power and Data Transfer System for Smart Bridge Sensors 3690
Yujin Jang, *Korea Advanced Institute of Science and Technology, Korea, South*
Jung Kyu Han, *Korea Advanced Institute of Science and Technology, Korea, South*
Shin Young Cho, *Korea Advanced Institute of Science and Technology, Korea, South*
Gun-Woo Moon, *Korea Advanced Institute of Science and Technology, Korea, South*
Ji-Min Kim, *Korea Advanced Institute of Science and Technology, Korea, South*
Hoon Sohn, *Korea Advanced Institute of Science and Technology, Korea, South*

Inrush Transient Current Analysis and Suppression of Photovoltaic Grid-Connected Inverters during Voltage Sag 3697
Zhongyu Li, *China University of Petroleum, China*
Rende Zhao, *China University of Petroleum, China*
Zhen Xin, *Aalborg University, Denmark*
Josep M. Guerrero, *Aalborg University, Denmark*
Mehdi Savaghebi, *Aalborg University, Denmark*
Peide Li, *Shandong Jinan Power Equipment Factory Co., LTD, China*

A Highly Reliable Single-Stage Converter for Electric Vehicle Applications 3704
S.A.Kh. Mozaffari Niapour, *Northeastern University, United States*
Mahshid Amirabadi, *Northeastern University, United States*

Simple and Efficient Low Power Photovoltaic Emulator for Evaluation of Power Conditioning Systems 3712
Jesus Gonzalez-Llorente, *Universidad Sergio Arboleda, Colombia*
Andres Rambal-Vecino, *Universidad Sergio Arboleda, Colombia*
Luciano A. Garcia-Rodriguez, *University of Arkansas, United States*
Juan C. Balda, *University of Arkansas, United States*
Eduardo I. Ortiz-Rivera, *University of Puerto Rico at Mayaguez, Puerto Rico*

Data Transmission Method without Additional Circuits in Bidirectional Wireless Power Transfer System 3717
Yeongrack Son, *Seoul National University, Korea, South*
Jung-Ik Ha, *Seoul National University, Korea, South*

Improved Impedance Source Inverter for Hybrid/Electric Vehicle Application with Continuous Conduction Operation 3722
Thilak Senanayake, *University of Tsukuba, Japan*
Ryuji Iijima, *University of Tsukuba, Japan*
Takanori Isobe, *University of Tsukuba, Japan*
Hiroshi Tadano, *University of Tsukuba, Japan*

Hybrid Modelling and Control of Single-Phase Grid-Connected NPC Inverters

Xingda Yan
Eletro-mechanical Research Group
University of Southampton
SO17 1BJ, United Kingdom
Email: xy1g12@soton.ac.uk

Zhan Shu
Eletro-mechanical Research Group
University of Southampton
SO17 1BJ, United Kingdom

Suleiman M. Sharkh
Eletro-mechanical Research Group
University of Southampton
SO17 1BJ, United Kingdom

Abstract—**In this paper, a grid-connected single-phase neutral point clamped inverter is represented by a hybrid model with both continuous and discrete dynamics. A novel state-feedback switching control law is proposed in terms of linear matrix inequalities (LMIs) which can drive the output current to track a sinusoidal reference. The main advantage of this method is that the switching states of the inverter are directly determined by the controller. Another advantage is that the switching law can be obtained off-line, which significantly reduces the online computation burden. Then the neutral point voltage ripple is minimized by a capacitors voltage balancing controller. Simulation results confirm the feasibility and performance of the proposed control scheme.**

I. INTRODUCTION

In recent years, distributed generation (DG) systems, including renewables, are increasingly used to reduce the reliance on fossil fuel and cut emission [1][2]. A DC voltage is usually supplied by the distributed energy resources, such as photovoltaic (PV), fuel cells, and wind power. Hence, DC/AC inverters are used to interface these sources to the grid supplying high quality AC current in accordance with national standard [3]. Among many kinds of inverter topologies, the Neutral Point Clamped (NPC) inverter with lower dV/dt and switch stress has attracted great interest [4][5] and it is the most widely used multilevel inverter in industry.

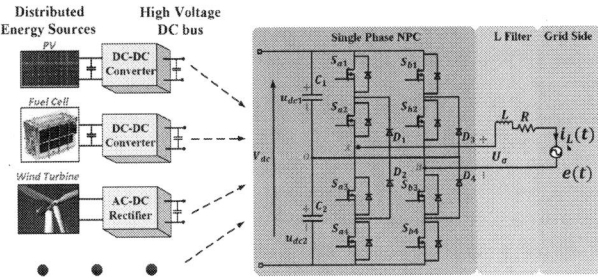

Fig. 1: Single-phase grid-connected NPC inverter for distributed energy source application

In order to deliver high quality electric power into the grid, current controllers are commonly used in grid-connected inverters, such as linear controllers (i.e. PI controller) in conjunction with pulse width modulation (PWM) [6][7]. Due to the number of the power switches, complicated modulation techniques are adopted, such as, carrier-based sinusoidal PWM [8], and space vector PWM [9]. Instead of considering the inverter as a linear system, another interesting alternative is model predictive control, which includes the nonlinearities and constraints into the control law and does not need any modulator [5][10][11]. But in order to predict the behavior of the system in real time, a powerful microprocessor is necessary to handle the computational burden, and the accuracy of the prediction are highly sensitive to the variation of system parameters. Hence, more research effort is needed to develop simple, robust and high performance current control strategies for grid-connected inverters.

In this paper, the grid-connected NPC inverter shown in Fig. 1 is represented by a hybrid model including both continuous and discrete dynamics. The main contribution of this paper is that a novel state-feedback switching control law is designed to force the output current to track a desired sinusoidal current reference. Furthermore, the neutral point voltage ripple is minimized by a capacitors voltage balancing controller. Similar control techniques were proposed for DC-DC converters [12][13][14]. However, the problem in this paper differs considerably from [12][13][14] due to the non-linearity of the reference signal and the grid, and also due to the number of power switches. The proposed switching control law is designed by solving linear matrix inequalities (LMIs), which guarantees the global asymptotic stability in the sense of Lyapunov theory [15]. Another advantage of the proposed approach is that the switching law coefficients can be calculated off-line, which results in a very light online computation burden. Simulation results validate the performance of the proposed control scheme in different operation conditions including reference tracking, reference variation, DC-link voltage variation, and parameters tolerances.

II. SWITCHING STATES OF SINGLE-PHASE GRID-CONNECTED NPC INVERTER

A two-leg single-phase NPC inverter is shown in Fig. 1. Each leg (A or B), which consists of four IGBTs ($S_{x1}, S_{x2}, S_{x3}, S_{x4}$, $x = A$ or B) connected in series and two clamped diodes, has three possible states as shown in Table I(a), where P, O and N stand for positive, neutral, negative

978-1-4673-9551-9/16 $31.00 © 2016 IEEE

TABLE I: Switching states for each leg and the whole inverter

(a) State definition for each leg

Leg state	P	O	N
S_{x1}	ON	OFF	OFF
S_{x2}	ON	ON	OFF
S_{x3}	OFF	ON	ON
S_{x4}	OFF	OFF	ON

(b) Voltage output and redundant switching states

Voltage Output U_σ	Index σ	Switching states
V_{dc}	1	PN
$V_{dc}/2$	2	PO,ON
0	3	PP,OO,NN
$-V_{dc}/2$	4	OP,NO
$-V_{dc}$	5	NP

state. The operation of each IGBT for different states is also defined in the table. For the specific inverter described in this paper, there are nine switching states due to the combination of the two legs, which are PP (Leg A in P state, and leg B in P state), PO, PN, OP, OO, ON, NP, NO, NN. However, if we neglect the unbalance issue of the neutral point, only five voltage levels can be output to the filter part, which are V_{dc}, $V_{dc}/2$, 0, $-V_{dc}/2$ and $-V_{dc}$. Table I(b) shows the possible switching states for given voltage output U_σ.

III. Hybrid Modelling and Switching Controller Design

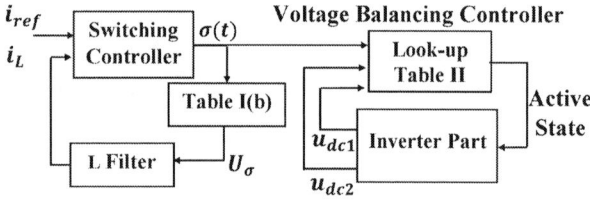

Fig. 2: A Proposed Control Structure

A novel control scheme is proposed as shown in Fig. 2. The whole system is separated as the inverter part and the filter part. A switching controller will only choose the proper voltage output U_σ, which can force the output current i_L to track the time-variant current reference i_{ref}; then an auxiliary voltage balancing controller will take advantage of the redundant switching states to balance the neutral-point voltage. For example, if $U_2 = V_{dc}/2$ is chose by the switching controller, two different switching states can be activated: PO or ON. The voltage balancing controller will decide which is more suitable based on the measurement of u_{dc1} and u_{dc2}. The details of the design are as follow.

A. Hybrid Model of the filter part

In this subsection, it assumes the dc-link capacitors are balanced ($u_{dc1} = u_{dc2} = \frac{V_{dc}}{2}$).

Define $\sigma : [0, \infty) \to \Xi = \{1, 2, ..., 5\}$ as a piecewise constant function, which is a mapping from the state-space to the index of the active voltage output U_σ. The relation between the voltage output U_σ and the index is shown in Table I(b).

Then a hybrid model with both continuous and discrete dynamics can be defined as:

$$\dot{i}_L(t) = A i_L(t) + B U_\sigma + D(e(t)), \qquad (1)$$

where $A = -\frac{R}{L}$, $B = \frac{1}{L}$, $D(e(t)) = -\frac{e(t)}{L}$, and U_σ is defined in Table I(b).

B. Switching controller design

In order to describe the dynamic of the sinusoidal reference $i_{ref}(t)$, a reference model is proposed in which $z(t)$ is the state of the model, $i_{ref}(t)$ is the output, ω is the angular frequency, and I is the magnitude.

$$\dot{z}(t) = H z(t), \quad z^T(t) N z(t) = I^2, \qquad (2)$$
$$i_{ref}(t) = Z z(t), \qquad (3)$$

where

$$z(t) = \begin{bmatrix} I \sin(\omega t) \\ I \cos(\omega t) \end{bmatrix} \qquad H = \begin{bmatrix} 0 & \omega \\ -\omega & 0 \end{bmatrix}$$
$$N = \begin{bmatrix} 1 & 0 \\ 0 & 1 \end{bmatrix} \qquad Z = \begin{bmatrix} 1 & 0 \end{bmatrix}.$$

The relation $z^T(t) N z(t) = I^2$ specifies the magnitude since $(I \sin(\omega t))^2 + (I \cos(\omega t))^2 = I^2$.

Furthermore, if the tracking error is defined as $\varepsilon(t) = i_L(t) - i_{ref}(t)$, the tracking error system is

$$\dot{\varepsilon}(t) = A\varepsilon(t) + B U_\sigma + D(e(t)) + (AZ - ZH)z(t). \qquad (4)$$

Hence, the switching law $\sigma(t)$ should be able to drive the error state $\varepsilon(t)$ to the origin. In this paper, the following switching law $\sigma(\varepsilon(t))$ and a mode-dependent Lyapunov function $V(\varepsilon(t), \sigma)$ are considered.

$$\sigma(\varepsilon(t)) \triangleq \arg\max_{i \in \Xi}\{v_i(\varepsilon(t))\}, \qquad (5)$$
$$V(\varepsilon(t)) \triangleq \max_{i \in \Xi}\{v_i(\varepsilon(t))\}, \qquad (6)$$
$$\text{where} \quad v_i(\varepsilon(t)) \triangleq \varepsilon(t)^T P_i \varepsilon(t) + 2\varepsilon(t)^T Q_i(z(t)),$$
$$Q_i(z(t)) \triangleq Q_{i0} + Q_{i1} z(t).$$

Note: *The* $\arg\max\{v_1, ...v_5\}$ *is used to denote the index of the maximum element* v_i. *For example,* $v_1 = 1$, $v_2 = 3$, $v_3 = 4$, $v_4 = 5$, $v_5 = 2$. *Hence,* $V(\varepsilon(t)) = v_4$. $\sigma(\varepsilon(t))$ *will equal to the index of the maximum* $v_i(\varepsilon(t))$, *which is* $\sigma = 4$. *Then the voltage output* $-V_{dc}/2$ *will be activated according to Table I(b).*

The performance requirement is proposed as:

$$\dot{V}(\varepsilon(t)) + \alpha_\sigma V(\varepsilon(t)) < 0, \qquad (7)$$
$$V(\varepsilon(t)) > 0, \qquad (8)$$

where α_σ are given non-negative constants.

Then the LMI conditions, which can guarantee the closed-loop globally asymptotically stability of the tracking error system (4) is presented below. Before it, two technical lemmas are given first.

Lemma 1. *(Bilinear Matrix Decompositions)[16] Consider two same dimension matrices $X, Y \in \mathbb{R}^{n*m}$,*

1) *The mappings $f(X) = X^T X$ and $g(X) = XX^T$ are positive semidefinite and convex on $\mathbb{R}^{n*m}$.*

2) *The bilinear matrix form $b(X,Y) = X^T Y + Y^T X$ can be decomposed as the difference of two positive semidefinite and convex parts*

$$b(X,Y) = \frac{1}{2}\left[(X+Y)^T(X+Y) - (X-Y)^T(X-Y)\right] \tag{9}$$

Lemma 2. *(Schur Complement) [17] Suppose that matrix A is symmetric. Then the following two matrices inequality is equivalent.*

1)
$$BB^T - A < (\leq)0 \tag{10}$$

2)
$$\begin{bmatrix} A & B \\ B^T & I \end{bmatrix} > (\geq)0 \tag{11}$$

Theorem 1. *Assume $z(t)$ and $e(t)$ are bounded by a given polytope Π. Consider the tracking error system (4) and given non-negative constants α_i, $i \in \Xi$, if the following parameter-dependent LMIs with variables P_i, $Q_i(z(t))$, $\forall i \in \Xi$ are satisfied at the vertices of Π.*

$$P_i > 0, \quad \sum_{i=1}^{\Xi} Q_{i0} = 0, \quad \sum_{i=1}^{\Xi} Q_{i1} = 0, \tag{12}$$

$$\begin{bmatrix} \Delta_{i1}^L + \alpha_i \Delta_{i2} + \Delta_{i3} & \frac{1}{\sqrt{2}}(X+Y)^T \\ * & I \end{bmatrix} > 0, \tag{13}$$

where

$$\Delta_{i1}^L = \begin{bmatrix} A^T P_i + P_i A & P_i(AZ - ZH) & \theta_{13} \\ * & 0 & (AZ - ZH)^T Q_i(z(t)) \\ * & * & \theta_{33} \end{bmatrix},$$

$$\theta_{13} = P_i B_i + P_i D(e(t)) + A^T Q_i(z(t)) + Q_{i1} Hz(t),$$

$$\theta_{33}^L = Q_i^T(z(t))B_i + B_i^T Q_i(z(t)),$$

$$X = \begin{bmatrix} 0 & 0 & 0 \\ * & 0 & 0 \\ * & * & Q_i(z(t)) \end{bmatrix},$$

$$Y = \begin{bmatrix} 0 & 0 & 0 \\ * & 0 & 0 \\ * & * & D(e(t)) \end{bmatrix},$$

$$\Delta_{i2} = \begin{bmatrix} P_i & 0 & Q_i(z(t)) \\ * & 0 & 0 \\ * & * & 0 \end{bmatrix},$$

$$\Delta_{i3} = \begin{bmatrix} 0 & 0 & 0 \\ * & N & 0 \\ * & * & -I^2 \end{bmatrix}.$$

Then the tracking error system (4) is globally asymptotically stable with the switching law (5) and the conditions (7)(8) are satisfied.

Remark 1. *The controller variables P_i, Q_{i0}, Q_{i1}, $\forall i \in \Xi$ can be obtained off-line by solving the LMI conditions (12)(13) through Matlab combined with YALMIP interface [18]. Hence, the online computation of the switching rule (5) is simple and straightforward.*

C. Auxiliary voltage balancing controller

In order to maintain the balance of the neutral-point O and relieve the stress on each power switch. An auxiliary voltage balancing controller is proposed in the form of a look-up table (Table II).

TABLE II: Auxiliary voltage balancing controller

Voltage Output Index	Criteria		Active State
$\sigma = 1$	NA		PN
$\sigma = 2$	$u_{dc1} > u_{dc2}$		ON
	$u_{dc1} < u_{dc2}$		PO
$\sigma = 3$	Previous Switching State	PP,PO	PP
		ON,OP	OO
		NO,NP	NN
$\sigma = 4$	$u_{dc1} > u_{dc2}$		OP
	$u_{dc1} < u_{dc2}$		NO
$\sigma = 5$	NA		NP

TABLE III: NPC inverter parameters

NPC inverter Configuration	
	in S.I
V_{dc}	400V
L	50mH
C	750μF
R	0.1Ω
I	10A
Grid Voltage	230V
Grid Frequency	50Hz
Switching/Sampling Frequency	20kHz

IV. SIMULATION RESULTS

The proposed control scheme is verified through simulation. Both the grid-connected NPC inverter and the controllers are implemented in Matlab Simulink, and the parameters of the nominal operation in shown in Table III.

A. Nominal Operation

First of all, the controller is tested when the inverter is in nominal mode. The responses are shown in Fig. 3. As we can seen, the inductor current i_L is able to track the sinusoidal reference i_{ref} with reasonable ripples. The capacitor voltages u_{dc1} and u_{dc2} are generally balanced with the proposed auxiliary voltage balancing controller.

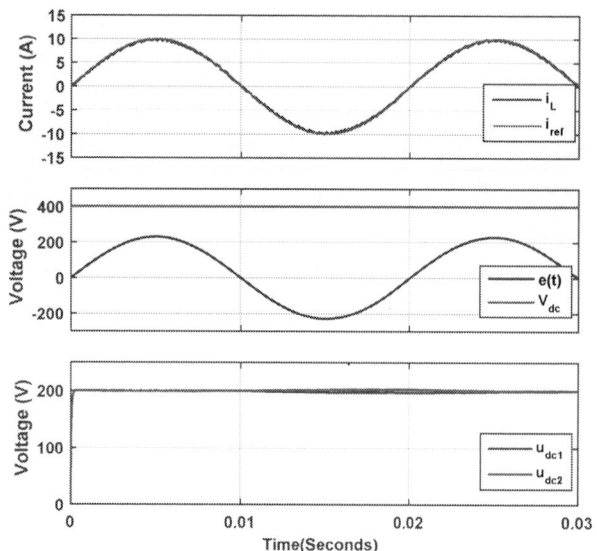

Fig. 3: **Nominal operation**

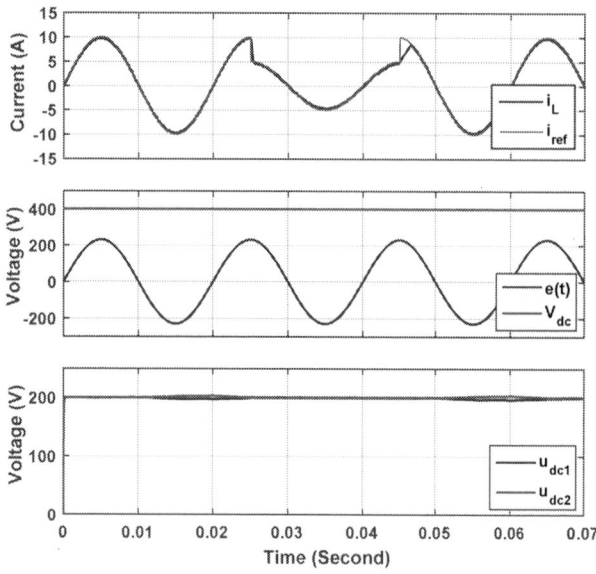

Fig. 4: **Reference magnitude changed to** 5V **during** 0.025s **to** 0.045s

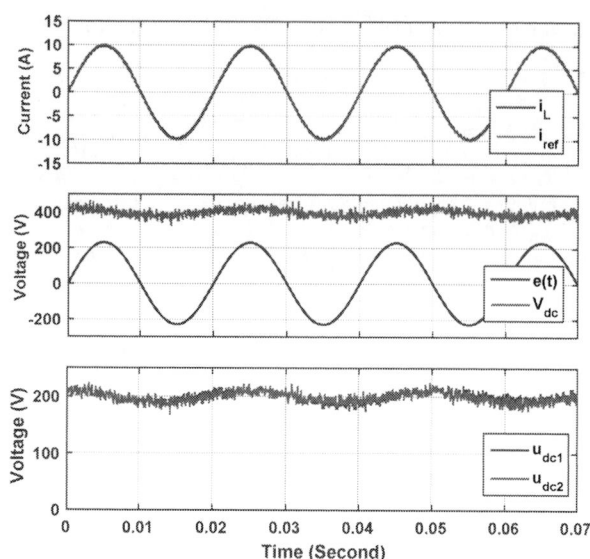

Fig. 5: **DC-link voltage with large ripple and variation**

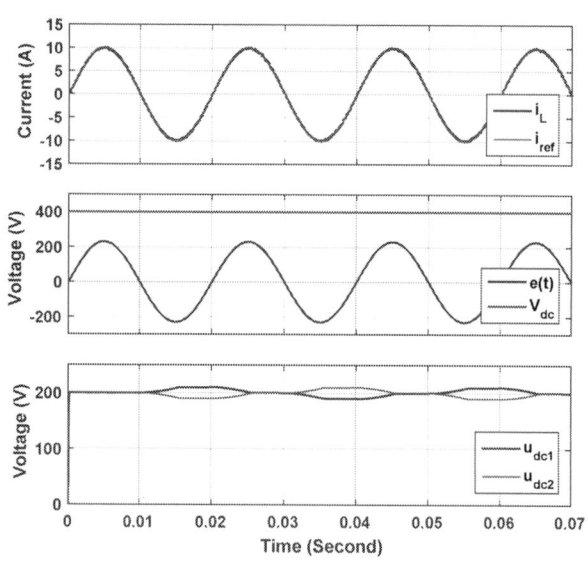

Fig. 6: **Inductor value changed to** 40mH

B. Reference variation

In order to test the robustness of the proposed method, the current reference is changed as shown in Fig. 4. At 0.025s, the reference signal i_{ref} is changed to 5V peak to peak from 10V peak to peak, then at 0.045s i_{ref} is changed back to 10V peak to peak. Under the proposed control scheme, the inductor current i_L followed the change of the reference signal with fast dynamic responses. At the same time, the balance of the neutral point is kept under the proposed auxiliary voltage balancing controller.

C. DC-link Voltage variation

In practice, the DC-link voltage may suffer from ac noise and variation. A robust controller should be able to handle this realistic situation. In this part, some ac noise is added to the DC-link voltage V_{dc} as shown in Fig. 5. However, the inductor current i_L is still tracking the reference current with good performance. The capacitor voltages u_{dc1} and u_{dc2} are balanced but with similar ac noise which is acceptable.

D. model parameters tolerances

In this part, the inductor value is changed to $40mHz$ from $50mH$ to test if the proposed controllers can handle the uncertainty of the model parameter. As shown in Fig.6, the inductor current i_L is generally not affected. The difference of the capacitor voltages u_{dc1} and u_{dc2} is more obvious but is still varied around the balanced point.

V. CONCLUSION

In this paper, a new modelling and control scheme is proposed for a single-phase NPC inverters that can interface distributed generation systems to the grid. Compared with conventional PWM control approaches, the controller determines the switching states of the inverter directly. Furthermore, compared with predictive control approaches, the proposed method can be implemented on basic microprocessors with low sampling rates since the coefficients of the switching law are computed off-line and the sampling frequency is equal to switching frequency. The robustness of the presented control scheme is also verified with respect to the current reference variation, the DC-link voltage variation and model parameters tolerances. In the future, the proposed method may be extended to three phase case and experiment validation will be carried out.

APPENDIX

Proof of Theorem 1

Proof. Due to the convexity of the LMIs, if the conditions are satisfied at the vertices of the Π, it can also be fulfilled by $\forall z(t)$ and $\forall e(t)$ in the polytope Π.

The condition (7) holds if

$$\dot{v}_i(\varepsilon(t)) + \alpha_i v_i(\varepsilon(t)) < 0 \qquad (14)$$

holds.

Due to $z^T(t)Nz(t) = I^2$, (14) holds if

$$\dot{v}_i(\varepsilon(t)) + \alpha_i v_i(\varepsilon(t)) + (z^T(t)Nz(t) - I^2) < 0 \qquad (15)$$

Choose an augmented variable as $\xi = [\varepsilon(t)\ z(t)\ 1]$, after calculating the derivative of $v_i(\varepsilon(t))$, the above equation (15) yields:

$$\Delta_{i1} + \alpha_i \Delta_{i2} + \Delta_{i3} < 0 \qquad (16)$$

where

$$\Delta_{i1} = \begin{bmatrix} A^T P_i + P_i A & P_i(AZ - ZH) & \theta_{13} \\ * & 0 & (AZ - ZH)^T Q_i(z(t)) \\ * & * & \theta_{33} \end{bmatrix}$$

$$\theta_{13} = P_i B_i + P_i D(e(t)) + A^T Q_i(z(t)) + Q_{i1} Hz(t)$$
$$\theta_{33} = Q_i^T(z(t))B_i + B_i^T Q_i(z(t))$$
$$\quad + Q_i^T(z(t))D(e(t)) + D^T(e(t))Q_i(z(t))$$

$$\Delta_{i2} = \begin{bmatrix} P_i & 0 & Q_i(z(t)) \\ * & 0 & 0 \\ * & * & 0 \end{bmatrix}$$

$$\Delta_{i3} = \begin{bmatrix} 0 & 0 & 0 \\ * & N & 0 \\ * & * & -I^2 \end{bmatrix}$$

One can notice the bilinear term $Q_i^T(z(t))D(e(t)) + D^T(e(t))Q_i(z(t))$ in Δ_{i1}. So Δ_{i1} is separated as two parts: linear part and bilinear part:

$$\Delta_{i1}^L = \begin{bmatrix} A^T P_i + P_i A & P_i(AZ - ZH) & \theta_{13} \\ * & 0 & (AZ - ZH)^T Q_i(z(t)) \\ * & * & \theta_{33} \end{bmatrix},$$
$$\qquad (17)$$

$$\theta_{13} = P_i B_i + P_i D(e(t)) + A^T Q_i(z(t)) + Q_{i1} Hz(t),$$
$$\theta_{33}^L = Q_i^T(z(t))B_i + B_i^T Q_i(z(t));$$
$$\Delta_{i1}^B = X^T Y + Y^T X, \qquad (18)$$

$$X = \begin{bmatrix} 0 & 0 & 0 \\ * & 0 & 0 \\ * & * & Q_i(z(t)) \end{bmatrix},$$

$$Y = \begin{bmatrix} 0 & 0 & 0 \\ * & 0 & 0 \\ * & * & D(e(t)) \end{bmatrix}.$$

Applying Lemma 1 on (18) , (16) holds if

$$\Delta_{i1}^L + \alpha_i \Delta_{i2} + \Delta_{i3}$$
$$+ \frac{1}{2}\left[(X+Y)^T(X+Y) - (X-Y)^T(X-Y) \right] < 0 \qquad (19)$$

holds. Then (19) holds if

$$\Delta_{i1}^L + \alpha_i \Delta_{i2} + \Delta_{i3} + \frac{1}{2}(X+Y)^T(X+Y) < 0 \qquad (20)$$

holds. Now applying Lemma 2, (20) yields

$$\begin{bmatrix} \Delta_{i1}^L + \alpha_i \Delta_{i2} + \Delta_{i3} & \frac{1}{\sqrt{2}}(X+Y)^T \\ * & I \end{bmatrix} > 0 \qquad (21)$$

In order to prove condition (8), we have:

$$\sum_{i=1}^{\Xi} v_i(\varepsilon(t)) = \varepsilon^T(t)\left(\sum_{i=1}^{\Xi} P_i(z(t)) \right) \varepsilon(t)$$
$$+ 2\varepsilon^T(t)\left(\sum_{i=1}^{\Xi} Q_{i0} \right) + 2\varepsilon^T(t)\left(\sum_{i=1}^{\Xi} Q_{i1} \right) z(t)$$
$$\qquad (22)$$

Substituting (12),

$$\sum_{i=1}^{\Xi} v_i(\varepsilon(t)) > 0 \qquad (23)$$

hold. One can conclude that at least one of the $v_i(\varepsilon(t)$ is positive definite. Hence,

$$V_\sigma(\varepsilon(t)) = \max_{i \in \Xi} v_i(\varepsilon(t)) > 0 \qquad (24)$$

hold. For the global stability, an additional condition is required, which $V_\sigma(\varepsilon(t))$ is radially unbounded. Since $\sum_{i=1}^{\Xi} v_i(\varepsilon(t))$ is positive definite, it is radially unbounded.

Because

$$\sum_{i=1}^{\Xi} v_i(\varepsilon(t)) < 5V_\sigma(\varepsilon(t)) \tag{25}$$

hold, $V_\sigma(\varepsilon(t))$ is radially unbounded.

Thus, the tracking error system (4) is globally asymptotically stable with the switching law (5). $\square$

REFERENCES

[1] G. Pepermans, J. Driesen, D. Haeseldonckx, R. Belmans, and W. Dhaeseleer, "Distributed generation: definition, benefits and issues," *Energy policy*, vol. 33, no. 6, pp. 787–798, 2005.

[2] S. Abu-Sharkh, R. Arnold, J. Kohler, R. Li, T. Markvart, J. Ross, K. Steemers, P. Wilson, and R. Yao, "Can microgrids make a major contribution to uk energy supply?" *Renewable and Sustainable Energy Reviews*, vol. 10, no. 2, pp. 78–127, 2006.

[3] S. M. Sharkh, M. A. Abu-Sara, G. I. Orfanoudakis, and B. Hussain, *Power electronic converters for microgrids*. John Wiley & Sons, 2014.

[4] U.-M. Choi, H.-G. Jeong, K.-B. Lee, and F. Blaabjerg, "Method for detecting an open-switch fault in a grid-connected npc inverter system," *Power Electronics, IEEE Transactions on*, vol. 27, no. 6, pp. 2726–2739, 2012.

[5] A. Calle-Prado, S. Alepuz, J. Bordonau, J. Nicolas-Apruzzese, P. Cortes, and J. Rodriguez, "Model predictive current control of grid-connected neutral-point-clamped converters to meet low-voltage ride-through requirements," *Industrial Electronics, IEEE Transactions on*, vol. 62, no. 3, pp. 1503–1514, 2015.

[6] J. Rodriguez, J.-S. Lai, and F. Z. Peng, "Multilevel inverters: a survey of topologies, controls, and applications," *Industrial Electronics, IEEE Transactions on*, vol. 49, no. 4, pp. 724–738, 2002.

[7] D. G. Holmes and T. A. Lipo, *Pulse width modulation for power converters: principles and practice*. John Wiley & Sons, 2003, vol. 18.

[8] L. M. Tolbert and T. G. Habetler, "Novel multilevel inverter carrier-based pwm method," *Industry Applications, IEEE Transactions on*, vol. 35, no. 5, pp. 1098–1107, 1999.

[9] G. I. Orfanoudakis, M. A. Yuratich, and S. M. Sharkh, "Nearest-vector modulation strategies with minimum amplitude of low-frequency neutral-point voltage oscillations for the neutral-point-clamped converter," *Power Electronics, IEEE Transactions on*, vol. 28, no. 10, pp. 4485–4499, 2013.

[10] T. Geyer, G. Papafotiou, and M. Morari, "Model predictive control in power electronics: A hybrid systems approach," in *Decision and Control, 2005 and 2005 European Control Conference. CDC-ECC'05. 44th IEEE Conference on*. IEEE, 2005, pp. 5606–5611.

[11] S. Kouro, P. Cortés, R. Vargas, U. Ammann, and J. Rodríguez, "Model predictive control: simple and powerful method to control power converters," *Industrial Electronics, IEEE Transactions on*, vol. 56, no. 6, pp. 1826–1838, 2009.

[12] M. Senesky, G. Eirea, and T. J. Koo, "Hybrid modelling and control of power electronics," in *Hybrid Systems: Computation and Control*. Springer, 2003, pp. 450–465.

[13] G. S. Deaecto, J. C. Geromel, F. Garcia, and J. Pomilio, "Switched affine systems control design with application to dc–dc converters," *IET control theory & applications*, vol. 4, no. 7, pp. 1201–1210, 2010.

[14] X. Yan, Z. Shu, and S. M. Sharkh, "Prediction-based sampled-data control for dc-dc buck converters," in *Smart Grid and Renewable Energy (SGRE), 2015 First Workshop on*. IEEE, 2015, pp. 1–6.

[15] G. E. Dullerud and F. Paganini, *A course in robust control theory: a convex approach*. Springer Science & Business Media, 2013, vol. 36.

[16] Q. Tran Dinh, S. Gumussoy, W. Michiels, and M. Diehl, "Combining convex-concave decompositions and linearization approaches for solving BMIs, with application to static output feedback," *IEEE Transactions on Automatic Control*, vol. 57, no. 6, pp. 1377–1390, 2012.

[17] S. P. Boyd, L. El Ghaoui, E. Feron, and V. Balakrishnan, *Linear matrix inequalities in system and control theory*. SIAM, 1994, vol. 15.

[18] J. Löfberg, "Yalmip: A toolbox for modeling and optimization in matlab," in *Computer Aided Control Systems Design, 2004 IEEE International Symposium on*. IEEE, 2004, pp. 284–289.

Stability Criterion and Controller Parameter Design of Radial-Line Renewable Systems with Multiple Inverters

Wenchao Cao, Xuan Zhang, Yiwei Ma and Fred Wang

Center for Ultra-Wide-Area Resilient Electric Energy Transmission Networks (CURENT)
The University of Tennessee
Knoxville, TN 37996, USA
wcao2@vols.utk.edu

Abstract—To address the instability issue in renewable systems of a radial-line structure with multiple current-controlled interface inverters, this paper proposes a practical stability criterion to easily analyze the system stability and a controller parameter design method to guarantee stable system operation with good oscillation damping performance. The proposed impedance-based sufficient stability criterion does not need the pole calculation of the return ratio matrices, while the phase margin of the system can still be obtained for system dynamic performance evaluation. Based on the phase margin information, design rules of inverter controller parameters are further proposed for system stability. The output admittance model of current-controlled inverters in an arbitrary *d-q* frame is also derived to facilitate the stability analysis. Simulation and experimental results verify the effectiveness of the proposed stability criterion and controller parameter design method.

Keywords—stability; renewable energy; impedance; inverters

I. INTRODUCTION

For the power grids with multiple renewable energy sources, one commonly seen system structure is the radial-line structure, as shown in Fig. 1, such as feeders of photovoltaic (PV) inverters in distribution systems [1] and the series daisy-chain collector system within a renewable power plant [2]. Adoption of interface inverters may cause instability issues to the renewable energy systems, due to the interactions among multiple inverters and the grid [3]-[8].

Regarding the small-signal stability analysis of systems with multiple inverters, compared with the approaches using the closed-loop transfer function matrix model [4]-[6] or the state-space model [7] of the whole system, the impedance-based approach has the advantage to clearly interpret the impact of individual components or subsystems on the system stability [9] by the impedance-based stability criteria, such as the Nyquist stability criterion [9]-[13], and the generalized Nyquist stability criterion (GNC) [14]-[17].

For radial-line renewable systems with multiple inverters, the impedance-based approach can predict the system stability by the impedance ratio between the output impedance of each inverter and the equivalent impedance of the rest system [12]-[13], or by the impedance ratio between the grid impedance

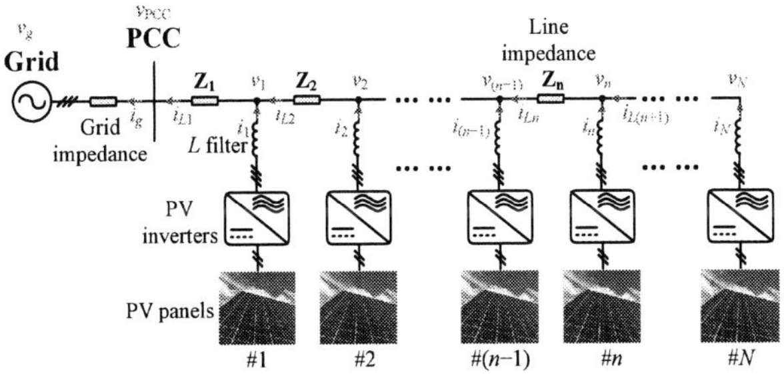

Fig. 1. Simplified one-line diagram of a grid-connected radial-line PV system with multiple PV inverters.

and the equivalent impedance on the other side at the point of common coupling (PCC) [17]. When using the Nyquist stability criterion or GNC, not only the encirclement of the Nyquist plot around $(-1, j0)$ should be examined, but also the right-half-plane (RHP) poles of the impedance ratio or return ratio matrix should be checked [14], [18]. If the number of inverters is large and the order of the impedance ratio is high, the pole calculation may suffer from large computation effort or poor accuracy. To avoid the pole calculation of the impedance ratio, other stability criterion can be adopted, such as the impedance-sum-type criterion [19] or the Nyquist criterion for multi-loop system [20], but the stability margin required for system dynamic performance evaluation cannot be easily obtained. The stability of a system with two converters connected to a non-ideal power grid at different coupling points is investigated in [17], by examining the roots of the determinant of the return difference matrix of the whole system. The focus of [17] is the effect of system parameters on the stability regions in the parameter planes, instead of the traditional gain margin or phase margin. Moreover, the analysis has not been generalized to a general radial-line system with more inverters. Reference [21] proposed to analyze the stability of a radial distribution network in a sequential procedure by adding the inverters to the passive component network one by one, while an improved method is proposed in [22] to analyze the system stability step by step from the simplest entity to the entire network. However, the system stability margin and controller parameter design have not been discussed in [21] and [22]. In addition, the stability is

---

This work was supported primarily by the Engineering Research Center Program of the National Science Foundation and DOE under NSF Award Number EEC-1041877 and the CURENT Industry Partnership Program.

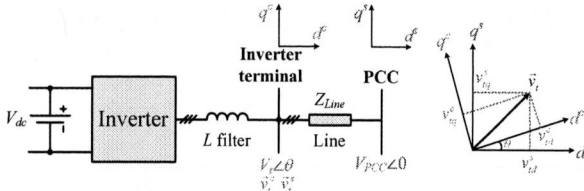

Fig. 2. Relation between different *d-q* frames.

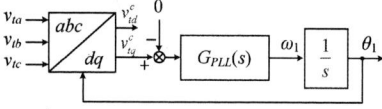

Fig. 3. Block diagram of the SRF PLL loop.

analyzed using phase impedances instead of impedances in the *d-q* domain.

For the small-signal stability analysis in the *d-q* domain, the system model needs to be built in a common *d-q* frame [23]. But the output admittance model of current-controlled inverters is usually expressed in the *d-q* frame aligned to the inverter terminal voltage [24], while the model in an arbitrary *d-q* frame has not been discussed in references yet.

In this paper, an impedance-based sufficient stability criterion of general radial-line renewable systems with multiple current-controlled inverters in the *d-q* domain is proposed. The system stability can be examined by checking the encirclements of the point $(-1, j0)$ by the characteristic loci of the return ratio matrix at each bus successively from the farthest bus to the PCC, without the need for pole calculation of the return ratio matrices. The phase margin of the system can also be obtained while applying the proposed stability criterion, based on which some design rules of the controller parameters are proposed. The output admittance model of current-controlled inverters in an arbitrary *d-q* frame is also derived to facilitate the system stability analysis.

II. OUTPUT ADMITTANCE MODEL OF CURRENT-CONTROLLED INVERTERS IN AN ARBITRARY D-Q FRAME

A. Calculation of the Steady-State Point

Assume the arbitrary *d-q* frame is chosen to be aligned with the voltage at PCC v_{PCC}, and it is selected as the common system *d-q* frame with the superscript *s*. The steady-state point regarding bus voltage v_n, inverter current i_n and current i_{Ln} through the line impedance $\mathbf{Z_n}$ (with inductance L_n and resistance R_n) in this common *d-q* frame can be obtained by solving the steady-state equations of the radial-line system as shown in Fig. 1, including KCL equation (1), feeder line equation (2), inverter output current and voltage equations (3) and (4) (assuming unity power factor), PCC voltage equation (5) as well as the grid voltage equation (6). Then, the bus angle θ_n can be calculated in (7).

$$\begin{bmatrix} I_{Lnd} \\ I_{Lnq} \end{bmatrix} = \begin{bmatrix} I_{L(n+1)d} \\ I_{L(n+1)q} \end{bmatrix} + \begin{bmatrix} I_{nd} \\ I_{nq} \end{bmatrix} \tag{1}$$

$$\begin{bmatrix} R_n & -\omega L_n \\ \omega L_n & R_n \end{bmatrix} \begin{bmatrix} I_{Lnd} \\ I_{Lnq} \end{bmatrix} = \begin{bmatrix} V_{nd} \\ V_{nq} \end{bmatrix} - \begin{bmatrix} V_{(n-1)d} \\ V_{(n-1)q} \end{bmatrix} \tag{2}$$

$$\sqrt{I_{nd}^2 + I_{nq}^2} = I_n^* \tag{3}$$

$$\frac{I_{nd}}{I_{nq}} = \frac{V_{nd}}{V_{nq}} \tag{4}$$

$$V_{PCCq} = 0 \tag{5}$$

$$\sqrt{V_{gd}^2 + V_{gq}^2} = V_g \tag{6}$$

$$\theta_n = \arctan \frac{V_{nq}}{V_{nd}} \tag{7}$$

B. Output Admittance Model of Current-Controlled Inverters

In the following analysis, the dc-link voltage V_{dc} is assumed constant for simplicity. The inverter *d-q* frame with the superscript *c* is aligned with the inverter terminal voltage v_t. As shown in Fig. 2, let the angle between the inverter terminal voltage and v_{PCC} be θ, then the relation between $\vec{v}_t^c$ and $\vec{v}_t^s$ can be expressed as (8). By adding small-signal perturbation and considering $\cos\tilde\theta \approx 1$ and $\sin\tilde\theta \approx \tilde\theta$, the small-signal model is derived in (9), where the symbol ~ denotes small-signal variables. Considering the open-loop relation (10) and the closed-loop relation (11) of the conventional phase-locked loop (PLL) in the synchronous reference (*d-q*) frame (SRF) as shown in Fig. 3, the small-signal model can be further derived as (12) and (13). Similarly, the relations for currents ($\vec{\tilde{i}}^c$ and $\vec{\tilde{i}}^s$) and inverter controller outputs ($\vec{\tilde{v}}_c^c$ and $\vec{\tilde{v}}_c^s$) can be obtained in (14).

$$\begin{bmatrix} v_{td}^c \\ v_{tq}^c \end{bmatrix} = \mathbf{T_\theta} \begin{bmatrix} v_{td}^s \\ v_{tq}^s \end{bmatrix} \quad \text{with} \quad \mathbf{T_\theta} = \begin{bmatrix} \cos\theta & \sin\theta \\ -\sin\theta & \cos\theta \end{bmatrix} \tag{8}$$

$$\begin{bmatrix} \tilde{v}_{td}^c \\ \tilde{v}_{tq}^c \end{bmatrix} = \mathbf{T_\theta} \begin{bmatrix} \tilde{v}_{td}^s \\ \tilde{v}_{tq}^s \end{bmatrix} + \begin{bmatrix} V_{tq}^c \\ -V_{td}^c \end{bmatrix} \tilde\theta \tag{9}$$

$$\begin{cases} \tilde\theta = \tilde{v}_{tq}^c G_{PLL} \dfrac{1}{s} \\[2mm] G_{PLL} = K_{PLLp} + \dfrac{K_{PLLi}}{s} \end{cases} \tag{10}$$

$$\tilde\theta = \frac{G_{PLL}}{s + V_{td}^c G_{PLL}} \mathbf{T_\theta} \begin{bmatrix} \tilde{v}_{td}^s \\ \tilde{v}_{tq}^s \end{bmatrix} = T_{PLL} \mathbf{T_\theta} \begin{bmatrix} \tilde{v}_{td}^s \\ \tilde{v}_{tq}^s \end{bmatrix} \tag{11}$$

$$\begin{bmatrix} \tilde{v}_{td}^c \\ \tilde{v}_{tq}^c \end{bmatrix} = \mathbf{T_\theta} \begin{bmatrix} \tilde{v}_{td}^s \\ \tilde{v}_{tq}^s \end{bmatrix} + T_{PLL} \begin{bmatrix} 0 & V_{tq}^c \\ 0 & -V_{td}^c \end{bmatrix} \mathbf{T_\theta} \begin{bmatrix} \tilde{v}_{td}^s \\ \tilde{v}_{tq}^s \end{bmatrix} = \mathbf{G_{vt}} \mathbf{T_\theta} \begin{bmatrix} \tilde{v}_{td}^s \\ \tilde{v}_{tq}^s \end{bmatrix} \tag{12}$$

$$\vec{\tilde{v}}_t^c = \mathbf{G_{vt}} \mathbf{T_\theta} \vec{\tilde{v}}_t^s \quad \text{with} \quad \mathbf{G_{vt}} = \mathbf{I} + T_{PLL} \begin{bmatrix} 0 & V_{tq}^c \\ 0 & -V_{td}^c \end{bmatrix} \tag{13}$$

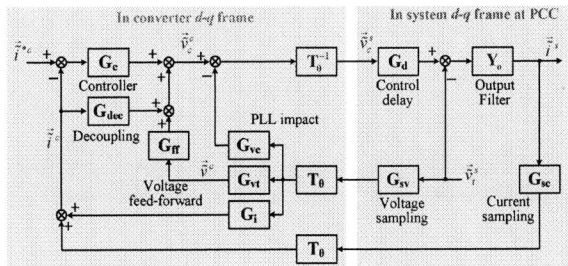

Fig. 4. Control block diagram of the current-controlled inverter.

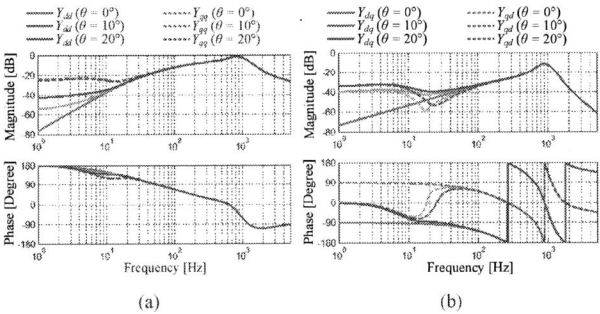

Fig. 5. Output admittances of the inverter: (a) Y_{dd}, Y_{qq}; (b) Y_{dq}, Y_{qd}.

Electrical Parameters		Values
L filter	L_f	0.575 mH
	R_{Lf}	0.2 Ω
DC-link voltage	V_{dc}	400 V
AC grid voltage	V_g	170 V (phase peak)
Inverter rated current	I_{rated}	10 A
Fundamental frequency	ω_1	60 Hz
Grid impedance	L_g	0.575 mH
Each line impedance	L_{Line}	0.7 mH

TABLE I. SYSTEM ELECTRICAL PARAMETERS

TABLE II. CONTROLLER PARAMETERS OF THE INVERTER

Controller Parameters		Values
Switching frequency	f_s	10 kHz
Switching period	T_s	100 μs
Current controller	K_{cp}	2.6
	K_{ci}	2275
PLL	K_{PLLp}	0.312
	K_{PLLi}	5.294
Delay time	T_d	150 μs

$$\mathbf{G_{ff}} = \frac{1}{1+s/\omega_{ff}}\mathbf{I} \qquad (18)$$

$$\mathbf{G_d} = e^{-T_d s}\mathbf{I} \qquad (19)$$

$$\mathbf{T_c} = \mathbf{Y_o}\mathbf{G_d}\mathbf{T_\theta}^{-1}\left(\mathbf{G_c}-\mathbf{G_{dec}}\right)\mathbf{T_\theta}\mathbf{G_{sc}} \qquad (20)$$

$$\mathbf{G_{clc}} = \left(\mathbf{I}+\mathbf{T_c}\right)^{-1}\mathbf{Y_o}\mathbf{G_d}\mathbf{T_\theta}^{-1}\mathbf{G_c} \qquad (21)$$

$$\begin{aligned}\mathbf{Y_{oc}} &= \left(\mathbf{I}+\mathbf{T_c}\right)^{-1}\mathbf{Y_o}\\ &\cdot\left\{\mathbf{I}-\mathbf{G_d}\mathbf{T_\theta}^{-1}\left[\mathbf{G_{ff}}\mathbf{G_{vt}}-\left(\mathbf{G_c}-\mathbf{G_{dec}}\right)\mathbf{G_i}-\mathbf{G_{vc}}\right]\mathbf{T_\theta}\mathbf{G_{sv}}\right\}\end{aligned} \qquad (22)$$

$$\begin{cases}\vec{\tilde{i}}^c = \mathbf{T_\theta}\vec{\tilde{i}}^s + \mathbf{G_i}\mathbf{T_\theta}\vec{\tilde{v}}_t^s \quad \text{with} \quad \mathbf{G_i} = T_{PLL}\begin{bmatrix} 0 & I_q^c \\ 0 & -I_d^c \end{bmatrix}\\[2mm] \vec{\tilde{v}}_c^c = \mathbf{T_\theta}\vec{\tilde{v}}_c^s + \mathbf{G_{vc}}\mathbf{T_\theta}\vec{\tilde{v}}_t^s \quad \text{with} \quad \mathbf{G_{vc}} = T_{PLL}\begin{bmatrix} 0 & V_{cq}^c \\ 0 & -V_{cd}^c \end{bmatrix}\end{cases} \qquad (14)$$

Then the control block diagram of the current-controlled inverter can be depicted in Fig. 4. $\mathbf{Y_o}$ is the admittance matrix of the inverter output L filter. $\mathbf{G_c}$ represents the proportional plus integral (PI) current controller matrix and $\mathbf{G_{dec}}$ is the decoupling term. A first-order low-pass filter with the cut-off frequency ω_{ff} is adopted in the voltage feed-forward transfer function matrix $\mathbf{G_{ff}}$. The PWM modulation is modelled as a delay component $\mathbf{G_d}$. $\mathbf{G_{sc}}$ and $\mathbf{G_{sv}}$ are the transfer function matrices for current and voltage measurement. The current open-loop gain $\mathbf{T_c}$, closed-loop gain $\mathbf{G_{clc}}$ and output admittance $\mathbf{Y_{oc}}$ in the arbitrary d-q frame can be derived. Fig. 5 illustrates the bode plots of $\mathbf{Y_{oc}}$ for the inverter with parameters listed in Table I and Table II, $\omega_{ff}=20\times2\pi$ rad/s and the output current (I_d =10 A, I_q =0 A) in its own inverter d-q frame, considering three different values of θ (namely, 0°, 10° and 20°). It can be seen that the inverter terminal voltage angle θ mainly affects the inverter admittances in the low frequency range.

$$\mathbf{Y_o} = \begin{bmatrix} L_f s + R_{Lf} & -\omega_1 L_f \\ \omega_1 L_f & L_f s + R_{Lf} \end{bmatrix}^{-1} \qquad (15)$$

$$\mathbf{G_c} = \left(K_{cp}+\frac{K_{ci}}{s}\right)\mathbf{I} \qquad (16)$$

$$\mathbf{G_{dec}} = \begin{bmatrix} 0 & -\omega_1 L_f \\ \omega_1 L_f & 0 \end{bmatrix} \qquad (17)$$

III. SMALL-SIGNAL STABILITY CRITERION

The parameters of the radial-line system with three current-controlled inverters under study in this paper are listed in Table I and Table II. Three inverters have the same electrical and controller parameters. The impedance-based system equivalent circuit can be obtained in the system d-q domain, as shown in Fig. 6. The grid current $\mathbf{I_g}$ can be expressed as (23), and the system stability can be examined by analyzing the poles of the transfer function matrices $\mathbf{A}$, $\mathbf{B_1}$, $\mathbf{B_2}$ and $\mathbf{B_3}$. These matrices can be obtained by deriving the equivalent Norton circuit at each bus and simplifying the system structure successively from the farthest point (Inverter 3) to the PCC (Inverter 1), as shown in Fig. 7. The result is expressed in (24), where the equivalent admittances are defined in (25) and the return-ratio matrix $\mathbf{T_{m\_x}}$ and its closed-loop gain $\mathbf{T_{clm\_x}}$ at each bus are defined in (26). The grid or line impedance $\mathbf{Z_n}$ is expressed in (27). According to GNC, $\mathbf{T_{m\_B3}}$ has no RHP poles. If the characteristic loci of $\mathbf{T_{m\_B3}}$ have zero encirclement around $(-1, j0)$, $\mathbf{T_{clm\_B3}}$, $\mathbf{Y_{eq3}}$ and $\mathbf{Y_{B2R}}$ are stable, and as a result $\mathbf{T_{m\_B2}}$ has no RHP poles. The analysis can be extended to other return-ratio matrices. Therefore, if the characteristic loci of all the return-ratio matrices $\mathbf{T_{m\_x}}$ at all the buses have zero encirclement around

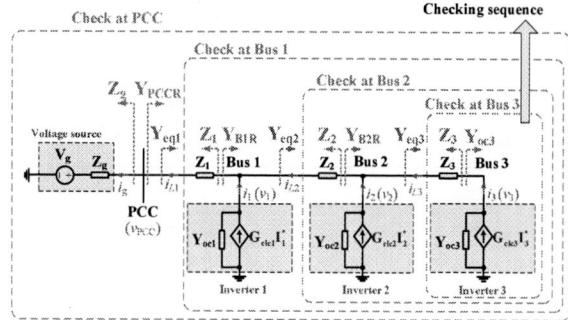

Fig. 6. Impedance-based system equivalent circuit with illustration of stability check at each bus.

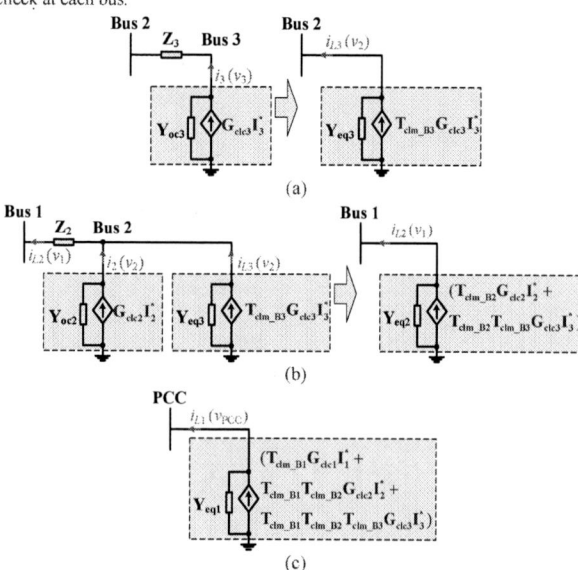

Fig. 7. Derivation of the equivalent Norton circuit at each bus: (a) Bus 2, (b) Bus 1, (c) Bus PCC.

$(-1, j0)$, all the transfer function matrices $\mathbf{A}$, $\mathbf{B_1}$, $\mathbf{B_2}$ and $\mathbf{B_3}$ are stable, and thus the grid current $\mathbf{I_g}$ and the total system are stable. And there is no need to calculate the poles of the return ratio matrices. It should be noted that all the inverters are assumed to be stable stand alone, and thus the current closed-loop gains $\mathbf{G_{clcn}}$ and output admittances $\mathbf{Y_{ocn}}$ are stable.

$$\mathbf{I_g} = \mathbf{A}\mathbf{V_g} + \mathbf{B_1}\mathbf{I_1^*} + \mathbf{B_2}\mathbf{I_2^*} + \mathbf{B_3}\mathbf{I_3^*} \qquad (23)$$

$$\mathbf{I_g} = \mathbf{T_{clm\_PCC}}\,\mathbf{Y_{PCCR}}\,\mathbf{V_g} - \mathbf{T_{clm\_PCC}}\,\mathbf{T_{clm\_B1}}\,\mathbf{G_{clc1}}\mathbf{I_1^*}$$
$$- \mathbf{T_{clm\_PCC}}\,\mathbf{T_{clm\_B1}}\,\mathbf{T_{clm\_B2}}\,\mathbf{G_{clc2}}\mathbf{I_2^*} \qquad (24)$$
$$- \mathbf{T_{clm\_PCC}}\,\mathbf{T_{clm\_B1}}\,\mathbf{T_{clm\_B2}}\,\mathbf{T_{clm\_B3}}\,\mathbf{G_{clc3}}\mathbf{I_3^*}$$

$$\begin{cases} \mathbf{Y_{eq3}} = \left(\mathbf{Z_3} + \mathbf{Y_{oc3}^{-1}}\right)^{-1} = \mathbf{T_{clm\_B3}}\,\mathbf{Y_{oc3}} \\[4pt] \mathbf{Y_{eq2}} = \left(\mathbf{Z_2} + \mathbf{Y_{B2R}^{-1}}\right)^{-1} = \mathbf{T_{clm\_B2}}\,\mathbf{Y_{B2R}} \\[4pt] \mathbf{Y_{PCCR}} = \mathbf{Y_{eq1}} = \left(\mathbf{Z_1} + \mathbf{Y_{B1R}^{-1}}\right)^{-1} = \mathbf{T_{clm\_B1}}\,\mathbf{Y_{B1R}} \\[4pt] \mathbf{Y_{B2R}} = \mathbf{Y_{oc2}} + \mathbf{Y_{eq3}} \\[4pt] \mathbf{Y_{B1R}} = \mathbf{Y_{oc1}} + \mathbf{Y_{eq2}} \end{cases} \qquad (25)$$

$$\begin{cases} \mathbf{T_{clm\_B3}} = \left(\mathbf{I} + \mathbf{T_{m\_B3}}\right)^{-1} = \left(\mathbf{I} + \mathbf{Y_{oc3}}\mathbf{Z_3}\right)^{-1} \\[4pt] \mathbf{T_{clm\_B2}} = \left(\mathbf{I} + \mathbf{T_{m\_B2}}\right)^{-1} = \left(\mathbf{I} + \mathbf{Y_{B2R}}\mathbf{Z_2}\right)^{-1} \\[4pt] \mathbf{T_{clm\_B1}} = \left(\mathbf{I} + \mathbf{T_{m\_B1}}\right)^{-1} = \left(\mathbf{I} + \mathbf{Y_{B1R}}\mathbf{Z_1}\right)^{-1} \\[4pt] \mathbf{T_{clm\_PCC}} = \left(\mathbf{I} + \mathbf{T_{m\_PCC}}\right)^{-1} = \left(\mathbf{I} + \mathbf{Y_{PCCR}}\mathbf{Z_g}\right)^{-1} \end{cases} \qquad (26)$$

$$\mathbf{Z_n} = \begin{bmatrix} L_n s + R_n & -\omega_1 L_n \\ \omega_1 L_n & L_n s + R_n \end{bmatrix} \qquad (27)$$

The result can be generalized to the proposed impedance-based sufficient stability criterion of radial-line systems with N current-controlled inverters as follows:

(1) Assume PCC is Bus B_0, the ideal grid is Bus B_{-1}, and the grid impedance is $\mathbf{Z_0}$. Define the return ratio matrix $\mathbf{T_{m\_Bn}}$ at each bus B_n ($n=0 \sim N$) as $\mathbf{T_{m\_Bn}} = \mathbf{Y_{BnR}}\mathbf{Z_n}$, where $\mathbf{Z_n}$ is the line impedance between Bus B_{n-1} and Bus B_n on the left-side and $\mathbf{Y_{BnR}}$ is the total admittance on the right side of Bus B_n.

(2) The checking sequence of the return ratio matrices at all buses is from the farthest bus B_N to the PCC bus B_0.

(3) If the characteristic loci of each return ratio matrix have zero encirclement around $(-1, j0)$, the total system is stable. If the characteristic loci of the return ratio matrix at Bus B_n have non-zero encirclement around $(-1, j0)$, the total system is probably unstable, and there is no need to check the remaining buses.

For the radial-line system with three current-controlled inverters under study in this paper, assume the grid voltage magnitude is V_g=170 V, and the output currents of all inverters are the same: I_d=10 A, I_q=0 A in their own inverter d-q frames. According to the steady-state point calculation in Section II, the magnitude and phase angle of each bus in the common d-q frame aligned with the PCC voltage v_{PCC} are as follows: V_1=175.91 V, θ_1=2.83°; V_2=176.46 V, θ_2=4.63°; V_3=176.77 V, θ_3=5.51°. The cut-off frequency parameter ω_{ff} of the first-order low-pass filter in the voltage feed-forward gain $\mathbf{G_{ff}}$ is selected as an example to investigate its effect on the system stability [25]. Three cases are investigated by the proposed stability criterion. Case 1: ω_{ff}=20×2π rad/s; Case 2: ω_{ff}=50×2π rad/s; Case 3: ω_{ff}=300×2π rad/s. The Nyquist plots of the characteristic loci (λ_1 and λ_2) of the return ratio matrix at each bus are shown in Fig. 8. For Case 1 and Case 2, all the characteristic loci have zero encirclement around $(-1, j0)$, so the system is stable. For Case 3, λ_1 of the return ratio matrix $\mathbf{T_{m\_B2}}$ at Bus 2 has two encirclements around $(-1, j0)$, so the system is probably unstable.

IV. CONTROLLER PARAMETER DESIGN FOR STABILITY

A. Controller Parameter Design for Normal Operation

In addition to the requirement of stable operation, stability margin is also an important concern when designing the controller parameters of the inverters for good system dynamic performance. While using the proposed stability criterion for stability check, the phase margin ϕ_m of the return ratio matrix at each bus can also be obtained, which is the angle difference

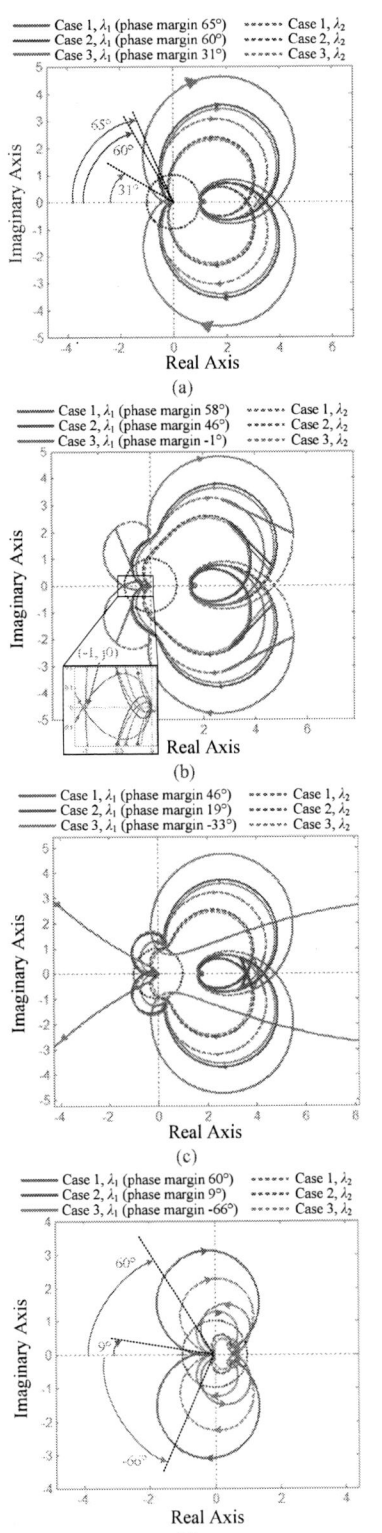

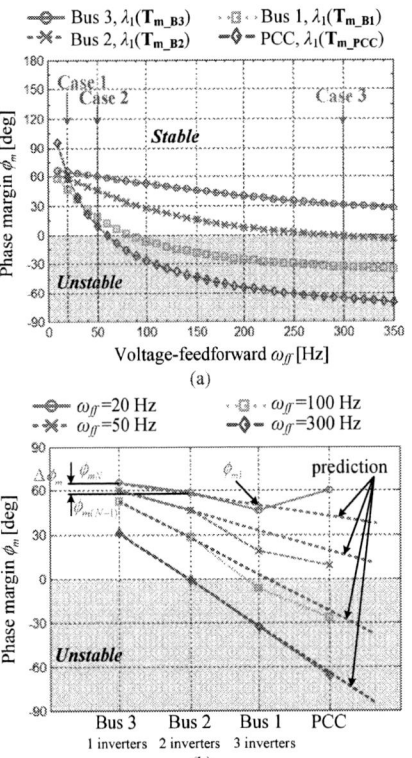

Fig. 9. Impact of voltage-feedforward ω_{ff} on stability and phase margin ϕ_m at each bus: (a) ϕ_m versus ω_{ff}; (b) ϕ_m versus Bus number (or inverter numbers).

Fig. 8. The characteristic loci of the return ratio matrices. (a) T_{m_B3} at Bus 3; (b) T_{m_B2} at Bus 2; (c) T_{m_B1} at Bus 1; (d) T_{m_PCC} at PCC.

between the unit-circle intersection point and the negative real axis, as shown in Fig. 8. Considering that there are two characteristic loci (λ_1 and λ_2), the smaller angle of these two intersection points is chosen as ϕ_m. It can be seen from Fig. 8(d) that Case 1 is a stable case with enough phase margin and good oscillation damping performance, and Case 2 is a stable case with limited phase margin and poor oscillation damping performance, while Case 3 is an unstable case with negative phase margin and resonance.

The impact of the increase of the voltage-feedforward cut-off frequency ω_{ff} on the stability and phase margin ϕ_m at each bus is further investigated, as shown in Fig. 9. Several characteristics can be observed and used as controller parameter design rules:

(1) With the increase of ω_{ff}, the phase margin of each return ratio matrix deceases. So, ω_{ff} should be selected sufficiently small to achieve stability and good oscillation damping performance.

(2) The phase margin gradually decreases from the farthest bus (Bus 3) to the nearest bus (Bus 1). When assuming all the line impedances ($Z_1 \sim Z_N$) are the same and the size and parameters of all the inverters are the same (such as the system under study in this paper), the decrease of phase margin is generally linear with the bus number (or equivalently the increase of the inverter numbers), as shown in Fig. 9(b).

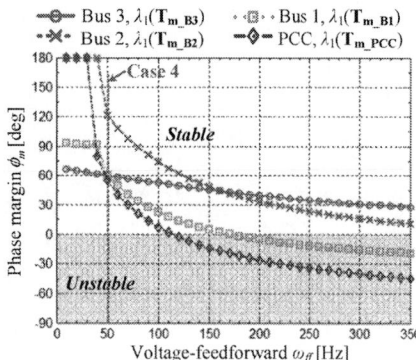

Fig. 10. Impact of ω_{ff} on stability and phase margin ϕ_m at each bus, when Inverter 2 is disconnected.

Therefore, the phase margins (ϕ_{mN} and $\phi_{m(N-1)}$) obtained from the first and second stability checks of return ratio matrices at Bus B_N and Bus B_{N-1} can be utilized to predict the approximate phase margin ϕ_{m1} at Bus B_1 as (28). Conversely, a sufficient phase margin ϕ_{m1} at Bus B_1 (e.g. 30°) can be required to keep stability at PCC under the grid impedance variation, and then the phase margin ϕ_{mN} at Bus B_N and the controller parameter (ω_{ff}) can be designed. For example, as shown in Fig. 9(b), when $\omega_{ff}=300\times2\pi$ rad/s, the phase margin ϕ_{m2} at Bus 2 is already 0°, so the system is unstable even when only two inverters are connected to PCC, and ω_{ff} should be re-designed.

$$\phi_{m1} \approx \phi_{mN} - (N-1)\Delta\phi_m = \phi_{mN} - (N-1)\left(\phi_{mN} - \phi_{m(N-1)}\right) \quad (28)$$

B. Design Considering Inverter Disconnection

During the system operation, there is a possibility that one or several inverters are not in service and disconnected from the system, and the system stability and margin might change when still using the designed parameters. Assuming Inverter 2 is disconnected, the stability and phase margin of the return ratio matrix at each bus are examined again by applying the stability criterion proposed in Section III, as shown in Fig. 10. It can be noticed that the phase margins at Bus 2, Bus 3 and PCC are improved under the same controller parameter (ω_{ff}) with Inverter 2 disconnected, compared with Fig. 9(a). It can be understood in the following way. If Inverter #n is disconnected from Bus B_n, the magnitude of the right-side admittance $\mathbf{Y_{BnR}}$ of Bus B_n is reduced. Then the magnitude interaction between the left-side line admittance $\mathbf{Y_n}$ ($\mathbf{Y_n} = \mathbf{Z_n}^{-1}$) and $\mathbf{Y_{BnR}}$ is weakened, or the magnitude intersection point moves to a higher frequency where the angle of $\mathbf{Y_{BnR}}$ is closer to the passive region ($-90°\sim90°$) [26] as shown in Fig. 5(a) and the phase difference between $\mathbf{Y_{BnR}}$ and $\mathbf{Y_n}$ at the intersection point is reduced. Therefore, the proposed controller parameter design method can guarantee the stable operation of the system even with the disconnection of several inverters.

C. Impact of Operating Point Changes

The system is not always at the rated operating point due to the output power or current variations of the PV inverters as well as the possible disconnection of several inverters. Strictly speaking, for controller parameter design, the system stability should be examined using the proposed criterion multiple times for all the possible operating points. Nevertheless, the operating point changes (such as inverter output current changes and bus voltage angle changes) mainly affect the inverter output admittances in the low frequency range within the PLL bandwidth or outer power control loop bandwidth as shown in Fig. 5. Therefore, the impact of operating point changes on grid synchronization stability and low frequency oscillation should be considered, while the impact on inner control loop parameter design is small when only the harmonic stability is concerned [9]. The impact and design of other controller parameters (such as current controller bandwidth ω_c, and active damping parameters if LCL filters are used) can be analyzed in a similar way.

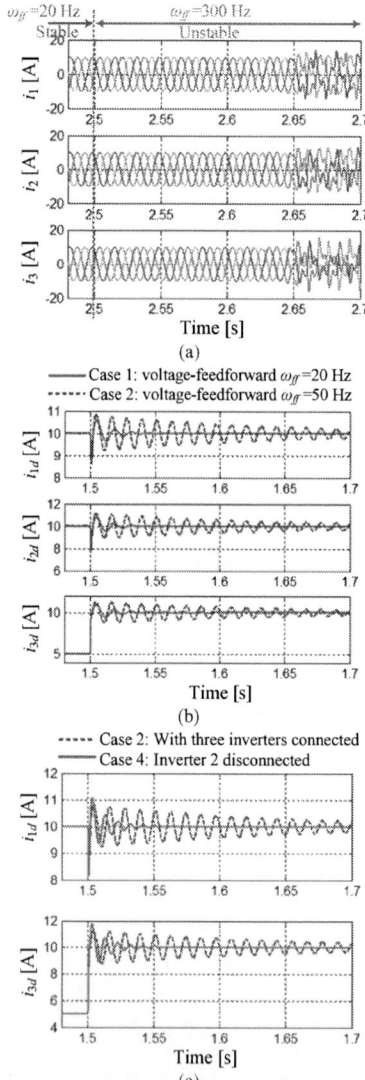

Fig. 11. Simulation results. (a) Change from Case 1 to Case 3. (b) Comparison between Case 1 and Case 2 under current reference change of Inverter 3. (c) Impact of disconnection of Inverter 2: Case 2 versus Case 4.

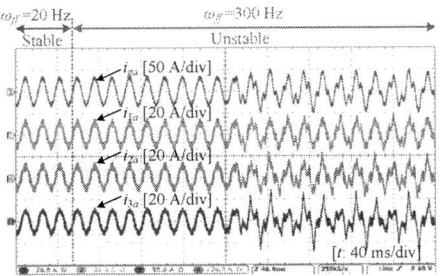

Fig. 12. Experimental results of the radial-line system with three inverters when ω_{ff} changes from $20\times2\pi$ rad/s (Case 1) to $300\times2\pi$ rad/s (Case 3).

V. SIMULATION AND EXPERIMENTAL RESULTS

The radial-line system in the above analysis is simulated using MATLAB/Simulink. The aforementioned cases are investigated. Fig. 11(a) shows that with the change of ω_{ff} from Case 1 to Case 3, the inverter currents change from stable to unstable. Fig. 11(b) shows that Case 1 has a better oscillation damping performance than Case 2 under the d-axis current reference change of Inverter 3 from 5 A to 10 A. Fig. 11(c) shows that the oscillation damping performance in Case 4 with Inverter 2 disconnected is improved as compared to Case 2.

The same radial-line system has been set up and investigated in experiments. Fig. 12 shows that the phase-a currents of three inverters go from stable to unstable when ω_{ff} changes from $20\times2\pi$ rad/s (Case 1) to $300\times2\pi$ rad/s (Case 3). Fig. 13 displays the comparison of the phase current and d-axis current responses of Case 1, Case 2 and Case 4 under the step change of the Inverter 3 d-axis current reference from 5 A to 10 A. With a higher voltage feed-forward cut-off frequency, the current response in Case 2 is worse with a longer oscillation period than that in Case 1. In addition, the disconnection of Inverter 2 results in better oscillation damping performance in Case 4 in contrast to Case 2. These simulation and experimental results have verified the above analysis.

VI. CONCLUSIONS

In this paper, an impedance-based sufficient stability criterion is proposed to analyze the small-signal stability of radial-line systems with multiple current-controlled inverters in d-q domain. The system stability can be examined by checking the encirclements of the point $(-1, j0)$ by the characteristic loci of the return ratio matrix at each bus successively from the farthest bus to the PCC. The pole calculation of return ratio matrices is avoided, compared to GNC, while the phase margin of the system can still be obtained. Design rules of inverter controller parameters are also proposed for stable system operation with the consideration of inverter disconnection. The proposed stability criterion and controller parameter design method are verified by simulation and experiments.

ACKNOWLEDGMENT

This work was supported primarily by the Engineering Research Center Program of the National Science Foundation and the Department of Energy under NSF Award Number EEC-1041877 and the CURENT Industry Partnership Program.

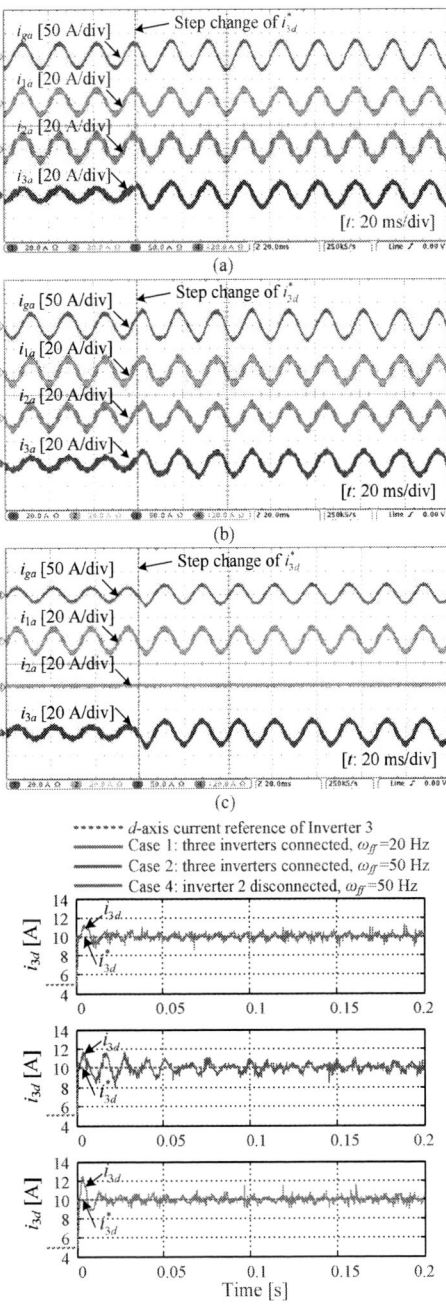

Fig. 13. Experimental results of the radial-line system during the step change of the d-axis current reference i_{3d}^* of Inverter 3 from 5 A to 10 A. (a) Case 1, $\omega_{ff} = 20\times2\pi$ rad/s; (b) Case 2, $\omega_{ff} = 50\times2\pi$ rad/s; (3) Case 4, $\omega_{ff} = 50\times2\pi$ rad/s; (4) comparison of the Inverter 3 d-axis current i_{3d} responses in these three cases.

REFERENCES

[1] Y. Zhou, H. Li, and L. Liu, "Integrated autonomous voltage regulation and islanding detection for high penetration PV applications," *IEEE Trans. Power Electron.*, vol. 28, no. 6, pp. 2826–2841, June 2013.

[2] E. Muljadi, C. P. Butterfield, A. Ellis, J. Mechenbier, J. Hochheimer, R. Young, N. Miller, R. Delmerico, R. Zavadil, and J. C. Smith, "Equivalencing the collector system of a large wind power plant," in *Proc. IEEE Power Engineering Society General Meeting*, 2006, pp. 1–9.

[3] J. H. R. Enslin, and P. J. M. Heskes, "Harmonic interaction between a large number of distributed power inverters and the distribution network," *IEEE Trans. Power Electron.*, vol. 19, no. 6, pp. 1586–1593, Nov. 2004.

[4] J. L. Agorreta, M. Borrega, J. Lopez, and L. Marroyo, "Modeling and control of N-paralleled grid-connected inverters with LCL filter coupled due to grid impedance in PV plants," *IEEE Trans. Power Electron.*, vol. 26, no. 3, pp. 770–785, March 2011.

[5] P. Brogan, "The stability of multiple, high power, active front end voltage sourced converters when connected to wind farm collector system," in *Proc. EPE Wind Energy Chapter Seminar*, 2010, pp. 1–6.

[6] J. He, Y. W. Li, D. Bosnjak, and B. Harris, "Investigation and active damping of multiple resonances in a parallel-inverter-based microgrid," *IEEE Trans. Power Electron.*, vol. 28, no. 1, pp. 234–246, Jan. 2013.

[7] N. Pogaku, M. Prodanovic, and T. C. Green, "Modeling, analysis and testing of autonomous operation of an inverter-based microgrid," *IEEE Trans. Power Electron.*, vol. 22, no. 2, pp. 613–625, March 2007.

[8] F. Wang, J. L. Duarte, M. A. M. Hendrix, and P. F. Ribeiro, "Modeling and analysis of grid harmonic distortion impact of aggregated DG inverters," *IEEE Trans. Power Electron.*, vol. 26, no. 3, pp. 786–797, March 2011.

[9] X. Wang, F. Blaabjerg, and W. Wu, "Modeling and analysis of harmonic stability in an ac power-electronics-based power system," *IEEE Trans. Power Electron.*, vol. 29, no. 12, pp. 6421–6432, Dec. 2014.

[10] J. Sun, "Impedance-based stability criterion for grid-connected inverters," *IEEE Trans. Power Electron.*, vol. 26, no. 11, pp. 3075–3078, 2011.

[11] M. Cespedes, and J. Sun, "Impedance modeling and analysis of grid-connected voltage-source converters," *IEEE Trans. Power Electron.*, vol. 29, no. 3, pp. 1254–1261, March 2014.

[12] X. Wang, F. Blaabjerg, M. Liserre, Z. Chen, J. He, and Y. Li, "An active damper for stabilizing power-electronics-based ac systems," *IEEE Trans. Power Electron.*, vol. 29, no. 7, pp. 3318–3329, July 2014.

[13] X. Wang, F. Blaabjerg, and P. C. Loh, "Proportional derivative based stabilizing control of paralleled grid converters with cables in renewable power plants. " in *Proc. IEEE ECCE* 2014, pp. 4917–4924.

[14] M. Belkhayat. Stability Criteria for AC Power Systems with Regulated Loads. Ph.D. thesis, Purdue University, Dec. 1997.

[15] B. Wen, D. Boroyevich, R. Burgos, P. Mattavelli, and Z. Shen, "Small-signal stability analysis of three-phase ac systems in the presence of constant power loads based on measured d-q frame impedances," *IEEE Trans. Power Electron.*, vol. 30, no. 10, pp. 5952–5963, Oct. 2015.

[16] A. Radwan and Y. Mohamed, "Analysis and active-impedance-based stabilization of voltage-source-rectifier loads in grid-connected and isolated microgrid applications," *IEEE Trans. Sustain. Energy*, vol. 4, no. 3, pp. 563–576, Jul. 2013.

[17] C. Wan, M. Huang, C. K. Tse, and X. Ruan, "Effects of interaction of power converters coupled via power grid: a design-oriented study," *IEEE Trans. Power Electron.*, vol. 30, no. 7, pp. 3589–3600, July 2015.

[18] B. Wen, D. Boroyevich, P. Mattavelli, R. Burgos, and Z. Shen, "Impedance-based analysis of grid-synchronization stability for three-phase paralleled converters," in *Proc. IEEE APEC* 2014, pp. 1233–1239.

[19] F. Liu, J. Liu, H. Zhang, and D. Xue, "Stability issues of $Z + Z$ type cascade system in hybrid energy storage system (HESS)," *IEEE Trans. Power Electron.*, vol. 29, no. 11, pp. 5846–5859, Nov. 2014.

[20] X. Wang, F. Blaabjerg, and P. C. Loh, "An impedance-based stability analysis method for paralleled voltage source converters," in *Proc. IEEE IPEC* 2014, pp. 1529–1535.

[21] C. Yoon, X. Wang, C. L. Bak, and F. Blaabjerg, "Stabilization of multiple unstable modes for small-scale inverter-based power systems with impedance-based stability analysis," in *Proc. IEEE APEC* 2015, pp. 1202–1208.

[22] C. Yoon, H. Bai, X. Wang, C. L. Bak, and F. Blaabjerg, "Regional modeling approach for analyzing harmonic stability in radial power electronics based power system," in *Proc. IEEE PEDG* 2015, pp. 1–5.

[23] A. Radwan, and Y. Mohamed, "Stabilization of medium-frequency modes in isolated microgrids supplying direct online induction motor loads," ," *IEEE Trans. Smart Grid*, vol. 5, no. 1, pp. 358–370, Jan. 2014.

[24] B. Wen, D. Boroyevich, R. Burgos, P. Mattavelli, and Z. Shen, "Analysis of d-q small-signal impedance of grid-tied inverters," *IEEE Trans. Power Electron.*, vol. 31, no. 1, pp. 675–687, 2016.

[25] X. Zhang, F. Wang, W. Cao, and Y. Ma, "Influence of voltage feed-forward control on small-signal stability of grid-tied inverters," in *Proc. IEEE APEC* 2015, pp. 1216–1221.

[26] L. Harnefors, M. Bongiorno, and S. Lundberg, "Input-admittance calculation and shaping for controlled voltage-source converters," *IEEE Trans. Ind. Electron.*, vol. 54, no. 6, pp. 3323–3334, Dec. 2007.

Stability Analysis and Improvement of Solid State Transformer (SST)-Paralleled Inverters System using Negative Impedance Feedback Control

Qing Ye and Hui Li
Center for Advanced Power Systems
Florida State University
Tallahassee, FL - 32310, USA
hli@caps.fsu.edu

Abstract— **In this paper, the stability of solid state transformer (SST)–paralleled inverters system is investigated by using the unified impedance-based stability criterion (UIBSC). Theoretical analysis reveals that the output impedance characteristics of SST are frequency-dependent and multiple resonances can happen in the SST-paralleled inverters system. A lead-lag controller and a negative impedance feedback controller are developed to mitigate the resonances within and beyond the current control loop respectively. Without additional sensors, the proposed negative impedance feedback control is able to achieve better damping function in a much wider frequency range compared to other methods. In addition, the proposed control method is less sensitive to the time delay. Simulation results are provided to validate the functionality of proposed methods.**

Keywords—solid state transformer; paralleled inverters; harmonics instability; multiple resonances; negative impedance feedback control

I. INTRODUCTION

Solid state transformer (SST) plays a key role in the future renewable electric energy distribution and management (FREEDM) systems to integrate multiple distributed renewable energy resources (DRER), distributed energy storage devices (DESD), and load [1]. The stability of SST-paralleled inverters system faces different challenges compared to grid-tied paralleled inverters. First, unlike the inductive grid impedance in the low voltage distribution system, the output impedance characteristic of SST is frequency-dependent. Second, in order to achieve good power quality and transient performance, harmonics compensation control and some feedforward controls have been applied to SST [2], which further shape the SST output impedance to be distinct from the grid impedance. Moreover, since the SST output impedance characteristic is frequency-dependent, the interactions among the output impedances of inverters and SST could trigger multiple resonances with frequencies below and beyond the bandwidth of current control loop. Thus, dedicated damping techniques are required to mitigate these resonances in SST-paralleled inverters system. The control loop redesign method can be used to damp out the resonance within the current control loop bandwidth [3-4]. However, it cannot mitigate the resonance with frequency beyond the bandwidth in SST-paralleled inverters system. The feedback control method is a good candidate for high frequency resonance mitigation. Nevertheless, the performance of

feedback control degrades when the dominant harmonics frequency is much higher than current control loop bandwidth [5] Moreover, its damping effect deteriorates due to the impact of control delay at the high frequency range [6]. A high-pass-filtered active damping technique is introduced in [7] to overcome the limitation caused by control delay in grid-tied inverter system. However, the output impedance characteristics of SST vary at the frequency domain, which can degrades the functionality of the high-pass-filtered active damping technique.

The impedance based stability criterion (IBSC) has been proposed in [8] to assess the grid-tied inverter system stability. In this paper, an improved IBSC method [9] has been applied to evaluate the stability issue of SST-paralleled inverters system, which will not only save computation efforts but also to provide a physical insight on how to dampen the potential multiple resonance. In addition, a negative impedance feedback control is proposed to mitigate the high frequency resonance. Compared to the existing feedback control, the proposed method is effective in a much wider frequency range and less sensitive to the control delay. In addition, a lead-lag controller is utilized to suppress the low frequency resonance. Therefore, the improvement of system stability and integration capability is achieved.

II. STABILITY ANALYSIS OF SST-MULTIPLE PARALLELED INVERTERS SSYTEM

A. Impedance Derivation

The system configuration and control diagrams of SST-paralleled inverters are depicted in Fig. 1. The SST is a three stages AC-AC converter, which utilizes an LC filter to attenuate the switching harmonics at the low voltage AC terminal. The output voltage of SST v_{sst} is regulated to interface the paralleled inverters. The harmonics compensation controller (HCC) $G_{i,1}(s)$ and voltage reference feedforward control are applied to achieve good power quality and transient performance, as shown in the control diagram of SST inverter stage. Since this research is focused on the harmonics instability owing to the fast dynamics of control loop, the dc-link voltage V_{dc_low} is considered to be constant [10-11]. Therefore, the output impedance of SST inverter stage represents its terminal characteristics at the low voltage side, which is derived in (1).

978-1-4673-9551-9/16 $31.00 © 2016 IEEE

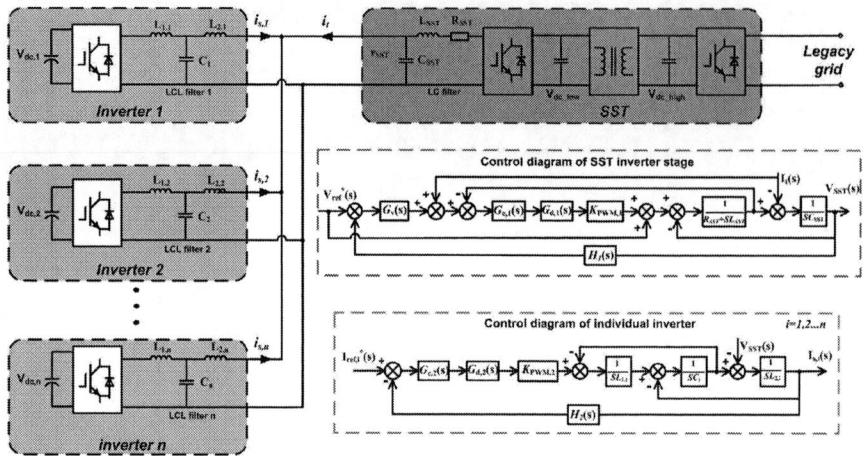

Fig. 1. System configuration and control diagrams of SST-paralleled inverters system

All the parameters in (1) can be found in Fig.1 and the corresponding values are shown in TABLE I in Appendix. For the inverters, the same assumption has been made that the dc-link voltage is constant. Also, the grid synchronization loop for the paralleled inverters usually has the bandwidth lower than the resonance frequency [11], so it is neglected when the harmonics instability is addressed in this paper. Then the output admittance of single inverter is derived as:

$$Y_{inv}(s) = \frac{(s^2 L_{1,i} C_i + 1)}{s^3 L_{1,i} L_{2,i} C + s(L_{2,i} + L_{1,i}) + G_{c,2}(s) G_{d,2}(s) K_{PWM,2}} \quad (2)$$

All the parameters in (2) can be found in Fig.1 and the corresponding values are shown in TABLE II in Appendix.

The bode plots of SST output impedance Z_{sst} and single inverter output admittance Y_{inv} are depicted in Fig. 2. Z_{sst} exhibits inductive feature at the middle frequency range, and capacitive behavior when frequency is higher. In addition, the HCC implemented in SST induces the phase boost at the middle frequency range. One the other hand, Y_{inv} reveals the capacitive characteristics at the high frequency range similar to Z_{sst}; but at middle frequency range, the phase of Y_{inv} decreases gradually from nearly 90° to a negative value.

B. Stability Analysis of SST-multiple Paralleled Inverters

The unified impedance-based stability criterion (UIBSC) has been proposed in [9], which is applied to assess the SST-paralleled inverters system stability. Compare to the traditional impedance-based stability criterion (IBSC), this method provides reduced computation efforts. Additionally, it is able to provide a physical insight if multiple resonances has been formed. In this paper, each inverter has the same parameters, and the global minor loop gain (GMLG) is derived in (3):

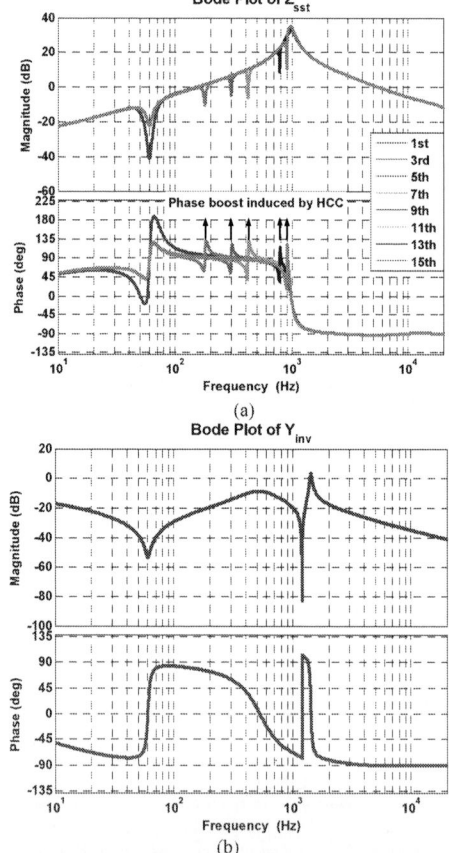

Fig. 2. Bode plot of (a) SST output impedance, and (b) Single inverter output admittance.

$$Z_{sst} = \frac{sL_{sst} + R_{sst}}{sC \cdot (sL_{sst} + R_{sst} + G_{c,1}(s) K_{PWM,1} G_{d,1}(s)) + G_v(s) G_{c,1}(s) K_{PWM,1} G_{d,1}(s) + 1} \quad (1)$$

$$GMLG = n \cdot Z_{sst} \cdot Y_{inv} \quad , \tag{3}$$

where n is the number of inverters connected to the SST, and Z_{sst} and Y_{inv} are derived in (1) and (2) respectively.

The phase of GMLG can be 180° at both middle and high frequency range according to Fig. 2, thus multiple possible resonances can be triggered as n increases. It is worth mentioning that although the resonant controls beyond 5*th* order in HCC also introduce the phase boost, the phase descent of Y_{inv} at the corresponding frequency guarantees that the phase of GMLG is smaller than 180°. In other word, these high order resonant controls will not cause instability issues. Therefore, they are neglected when analyze the effect of HCC on system stability.

The Nyquist plots of 1*st* ,3*rd* and 5*th* order GMLGs are depicted in Fig. 3. The 5*th* order GMLG encircles the point of (-1, j0), which means that the HCC has a negative impact on system stability, and the resonance with frequency around 300Hz occurs. Under this case, the maximum number of integrated inverter is 3. The simulation results in Fig. 4(a) and (b) validate the theoretical analysis. There are two other possible resonances at different frequency ranges as denoted in Fig. 3. The 200 Hz resonance is induced by in the 3*rd* order resonant control of HCC, while the 3.36 kHz one is resulted from the capacitive output impedance of both SST and inverters at the high frequency range. The theoretical analysis is consistent with the hypothesis that multiple resonances could occur.

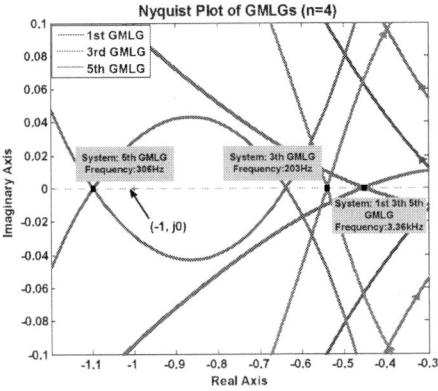

Fig. 3. Theoretical analysis of SST-paralleled inverters, system is unstable when $n = 4$.

III. PROPOSED HARMONICS INSTABILITY MITIGATION TECHNIQUES

In section II, The stability analysis of SST-paralleled inverters system demonstrates the possibility of multiple resonances. Thus, dedicated resonance mitigation techniques are required to improve system stability. However, it is difficult to dampen out all the possible resonances with one particular mitigation technique due to the wide frequency range. Besides, the selection of appropriate damping methods depends on the resonance frequency. Therefore, the combination of different

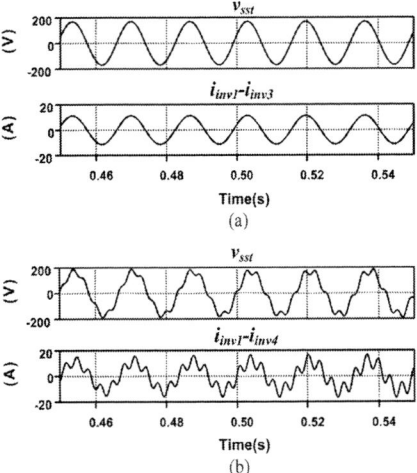

Fig. 4. Simulation results when (a) $n = 3$, system is stable;
(b) $n = 4$, system is unstable.

damping methods is necessary to achieve the effective mitigation function. Since the control systems of inverters are much simpler than that of SST, it's easier and more accurate to apply the damping controllers in inverters without deteriorating the stability and performance of the original system.

A. Resonance Mitigation within Current Control Bandwidth

The middle frequency resonances depicted in Fig. 3 is within the bandwidth of inverters' current control loop, and the lead-lag controller is integrated with the pre-designed current controller to realize the damping function. In order to compensate the phase boost function introduced by the 3rd and 5th order resonant control in HCC, the lead-lag controller is designed in (4):

$$G_{lead\_lag}(s) = \frac{f_1}{f_2} \cdot \frac{1 + s/2\pi f_1}{1 + s/2\pi f_2} \quad , \tag{4}$$

where f_1=100Hz, f_2=350Hz.

Fig. 5 depicts the comparison of single inverter output admittance with and without lead-lag controller. The phase drop achieved by the lead-lag controller fully compensates the negative effect of HCC, and the phase of GMLG is always smaller than 180° at the middle frequency range. Thereby the possible resonance within the control bandwidth is well-damped. Besides, the parameter of inverter resonant control $K_{r,v}$ is increased to 100 so that the magnitude of Y_{inv} has trivial change at the line frequency, so good reference tracking ability is maintained. Fig. 6 demonstrates the Nyquist plot of GMLGs after the lead-lag controller implemented. Compared to the Fig. 3, each GMLG now only has one intersection with the negative real axis at the frequency around 3.33 kHz. Therefore, the resonances induced by HCC have been effectively suppressed. Furthermore, the 3.33 kHz resonance has been successfully decoupled from the middle frequency resonances, which will be addressed in the next section. Simulation results are presented in Fig. 7 to validate the theoretical analysis.

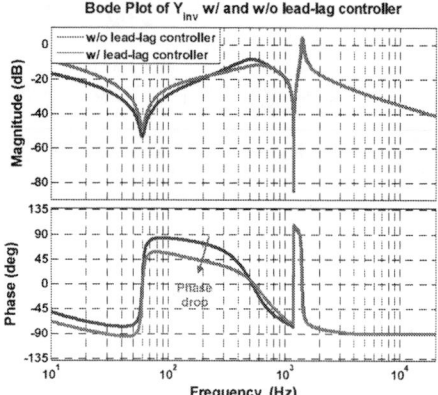

Fig. 5. Comparison of single inverter output admittance w/ and w/o lead-lag controller.

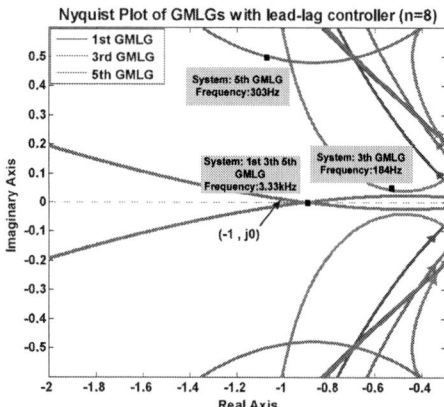

Fig. 6. Nyquist plot of GMLGs with lead-lag controller, system is stable when $n = 8$.

B. Resonance Mitigation beyond Current Control Bandwidth

The controller redesign method only shapes the inverter output admittance effectively within the current control bandwidth, as shown in Fig. 4. In this section, a feedback control is proposed to mitigate the resonance beyond the current control bandwidth. Since this resonance has been successfully decoupled from those within the bandwidth, only the impedances beyond the bandwidth are of interest here. Therefore, the single inverter output admittance can be simplified in (5).

Fig. 8 depicts the comparison of single inverter output admittance before and after simplification. The admittances match well beyond 3 kHz, where the possible resonance could happen. Thus, the simplified admittance is accurate enough to address the resonance issue beyond the bandwidth. Moreover, it also validates that the controller redesign method cannot

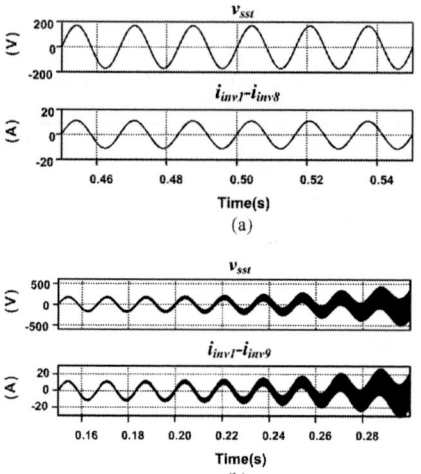

Fig. 7. Simulation results with lead-lag controller implemented, when (a) $n = 8$, system is stable; (b) $n = 9$, system is unstable.

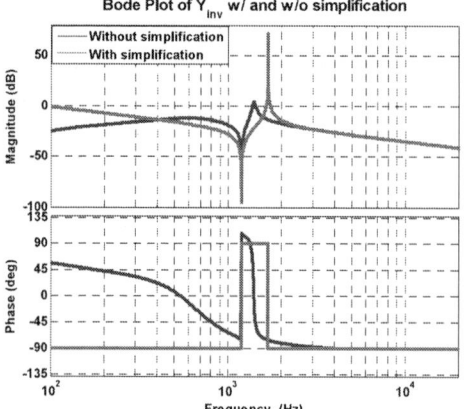

Fig. 8. Bode plot of single inverter output admittance before and after simplification.

shape the inverter output admittance at the concerned frequency ranges. By neglecting the low order terms in both numerator and denominator of (5), the inverter output admittance can be further simplified as $1/sL_{2,i}$. Therefore, there are two methods to enhance system stability by applying virtual impedance control. The first method is to decrease the magnitude of inverter output admittance. The greater the value of $L_{2,i}$, the smaller the magnitude of Y_{inv}, while the phase remains the same. Therefore, by adding a virtual inductor in series with $L_{2,i}$, the gain margin of GMLG can be increased, and system stability can thereby be improved. The second method is to boost the phase at the resonance frequency. This can be

$$Y_{inv}(s) = \frac{(s^2 L_{1,i} C_i + 1)}{s^3 L_{1,i} L_{2,i} C + s(L_{2,i} + L_{1,i}) + G_{c,2}(s) G_{lead\_lag}(s) G_{d,2}(s) K_{PWM,2}} \approx \frac{(s^2 L_{1,i} C_i + 1)}{s^3 L_{1,i} L_{2,i} C + s(L_{2,i} + L_{1,i})}. \tag{5}$$

achieved by implementing a virtual resistors in series with $L_{2,i}$. As a consequence, the resonance frequency will be pushed even higher. Since the magnitude of single inverter output admittance is decreasing with -20dB/decades, the system stability can be enhanced.

The conceptual diagram of inverter control block with virtual inductor L_v is depicted in Fig. 9(a). The feedback node is moved to the output of controller $G_{c,2}(s)$ for implementation, as shown in Fig. 9(b). The feedback term $G_x(s)$ is derived in (6):

$$G_x(s) = \frac{sL_v(s^2 L_{1,i} C_i + 1)}{G_{d,2}(s) K_{PWM,2}} \quad , \qquad (6)$$

where the definitions of all the parameters can be found in Fig. 9.

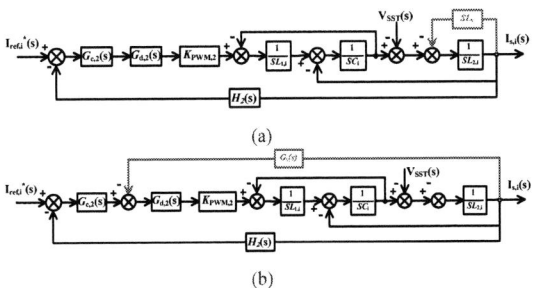

Fig. 9. (a) Conceptual diagram and (b) Realization diagram of negative inductor feedback control.

There is a prediction component in (6), given by

$$G_p(s) = \frac{1}{G_{d,2}(s)} = e^{1.5sT_s} \quad , \qquad (7)$$

which cannot be realized physically, and is usually neglected [6]. However, neglecting $G_p(s)$ here can cause significant phase shaping error at the concerned frequency range. Considering that the sampling period for each inverter is 100 μs, the frequency of the prediction component is around twice of the resonance frequency, so it is approximated to be -1 instead of being neglected so that more accurate impedance shaping function can be achieved. With this approximation, the feedback term $G_x(s)$ is further expressed as (8):

$$G_x(s) = \frac{-sL_v(s^2 L_{1,i} C_i + 1)}{K_{PWM,2}} \quad , \qquad (8)$$

which can be treated as a negative inductor feedback control (NIFC) term without the prediction component.

Although a large value of L_v is desired for the sake of system stability, the stability of single inverter needs to be guaranteed

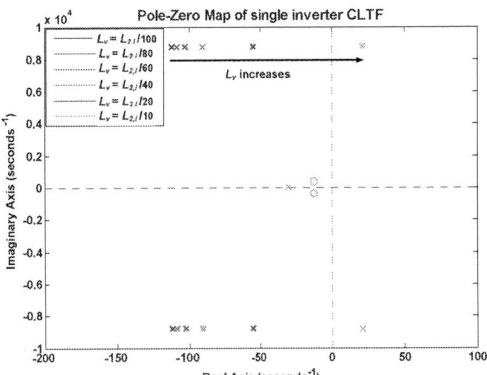

Fig. 10. Single inverter stability analysis with different value of virtual inductor.

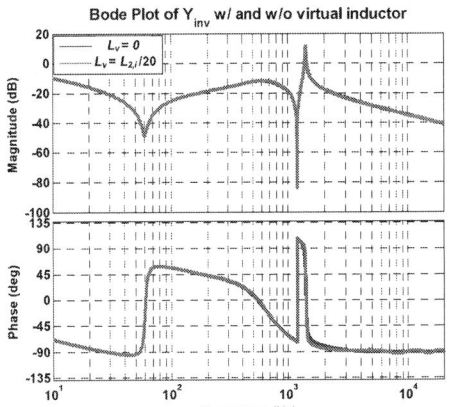

Fig. 11. Single inverter output admittance w/ and w/o negative virtual inductor.

so that the UIBSC can be applied. The pole-zero map of single inverter closed-loop transfer function (CLTF) is depicted in Fig. 10, and stability of single inverter is evaluated with different value of L_v. It is found that when L_v increases to $L_{2,i}/10$, the single inverter is unstable. In other word, the assurance of single inverter stability poses limitations on the choice of L_v.

With the virtual inductor implementation, the output admittance of single inverter is modified as in (9), where L_v denotes the virtual inductor, and all the other parameters are defined in Fig. 9(a).

The comparison of single inverter output admittance with and without NIFC is shown in Fig. 11. When $L_v = L_{2,i}/20$, the magnitude difference of single inverter output impedance is negligible. Under this case, the Nyquist plot of GMLG with NIFC is depicted in Fig. 12, and n_{max} is still 8. Therefore, the desired resonance mitigation capability is not achieved with the negative virtual inductor implemented.

$$Y_{inv\_NIFC}(s) = \frac{(s^2 L_{1,i} C_i + 1)}{s^3 L_{1,i} L_{2,i} C + s(L_{2,i} + L_{1,i}) + G_{c,2}(s) G_{lead\_lag}(s) G_{d,2}(s) K_{PWM,2} - sL_v(s^2 L_{1,i} C_i + 1) \cdot G_{d,2}(s)} \qquad (9)$$

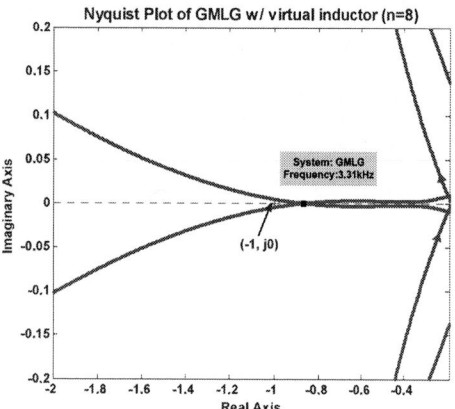

Fig. 12. Nyquist plot of GMLGs with NIFC, $n_{max} = 8$.

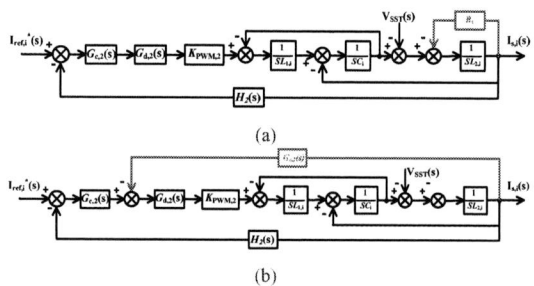

(a)

(b)

Fig. 13. (a) Conceptual diagram and (b) Realization diagram of negative resistor feedback control.

Fig. 13 depicts the conceptual diagram and the realization of negative resistor feedback control (NRFC). The feedback term is derived in (10):

$$G_{x,1}(s) = \frac{-R_v(s^2 L_{1,i} C_i + 1)}{K_{PWM,2}}. \tag{10}$$

Where R_v is the virtual resistor and the other parameters has been defined in Fig. 13(a). The value of R_v is derived to be 1.2 with consideration of single inverter stability. The design method is the same as the one used to find the value of the virtual inductor.

With the virtual resistor implemented, the output admittance of single inverter is derived in (11), where all the parameters are defined in Fig.13(a). The resonance mitigation methods in [5] and [6] are used for comparison. The control diagrams can be found in the reference, and here only the final derivation of output admittances are shown in (12) and (13) respectively. Z_v

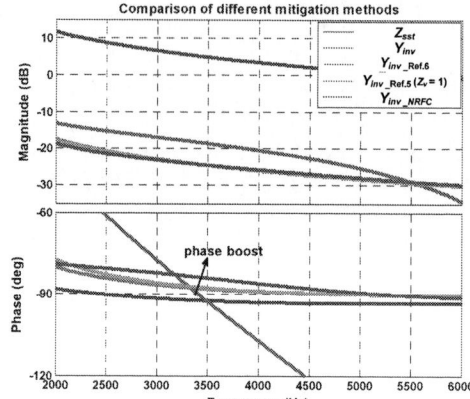

Fig. 14. Comparison of single inverter output admittance with different mitigation methods.

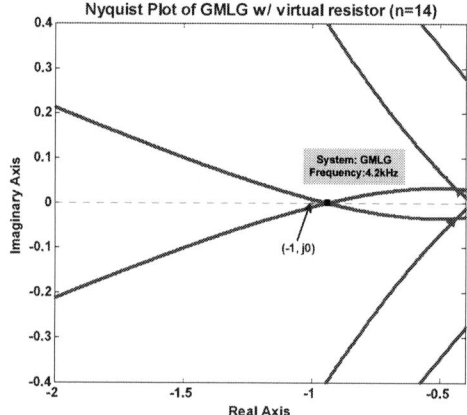

Fig. 15. Nyquist plot of GMLG with proposed method, system is stable when $n = 14$.

in (12) is the virtual output impedance. In this case, it is selected to be 1 with consideration of single inverter stability.

The comparison of single inverter output admittance with different mitigation methods is shown in Fig. 14. By applying the proposed method, Y_{inv_NRFC} has a phase boost at the resonance frequency compared to original Y_{inv}. So the resonance frequency is pushed to a higher value. Meanwhile, the magnitude of Y_{inv_NRFC} is descending with -20 dB/decades. Therefore, with the same value of n, the magnitude of GMLG is decreased at the new resonance frequency, and system stability is improved. On the other hand, $Y_{inv\_Ref.5}$ is consistent

$$Y_{inv\_NRFC}(s) = \frac{(s^2 L_{1,i} C_i + 1)}{s^3 L_{1,i} L_{2,i} C + s(L_{2,i} + L_{1,i}) + G_{c,2}(s) G_{lead\_lag}(s) G_{d,2}(s) K_{PWM,2} - R_v(s^2 L_{1,i} C_i + 1) \cdot G_{d,2}(s)}. \tag{11}$$

$$Y_{inv\_Ref.5}(s) = \frac{(s^2 L_{1,i} C_i + 1)}{s^3 L_{1,i} L_{2,i} C + s(L_{2,i} + L_{1,i}) + G_{c,2}(s) G_{lead\_lag}(s) G_{d,2}(s) K_{PWM,2} + Z_v \cdot G_{d,2}(s)}. \tag{12}$$

$$Y_{inv\_Ref.6}(s) = \frac{(1 - G_{d,2}(s))(s^2 L_{1,i} C_i + 1)}{s^3 L_{1,i} L_{2,i} C + s(L_{2,i} + L_{1,i}) + G_{c,2}(s) G_{lead\_lag}(s) G_{d,2}(s) K_{PWM,2}}. \tag{13}$$

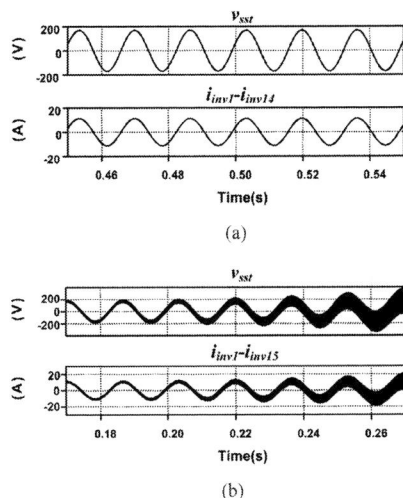

Fig. 16. Simulation results with proposed control, (a) $n = 14$,
system is table; (b) $n = 15$, system is unstable.

with the Y_{inv}; Y_{inv_Ref}.6 only experiences a phase boost before the resonance frequency, and it has a greater magnitude compared to Y_{inv}. Therefore, both methods cannot provide resonance mitigation function. The Nyquist plot of GMLG ($n = 14$) with proposed method is depicted in Fig. 15, and the system is stable. Simulation results are shown in Fig. 16 to validate the theoretical analysis.

IV. CONDLUSION

This paper investigates the stability of SST-paralleled inverters system and develops the dedicated active damping techniques to enhance the system stability and integration capability. The UIBSC is applied to assess SST-paralleled inverters system stability, and it is found that multiple possible resonances can happen. The harmonics compensation controller utilized in SST induces a phase boost, which causes the resonance within the current control loop bandwidth; a lead-lag controller is designed to provide a phase drop so that the resonance is well-damped. Meanwhile, a negative impedance feedback control (NIFC) is proposed to mitigate the resonance beyond the current control loop. This damping technique can be represented by a negative impedance in series with grid-side inductor of LCL filter. On one hand, the greater value of negative impedance, the better damping function can be achieved; on the other hand, a large value of the negative impedance could cause single inverter instability, and the presumption of UIBSC is not solid any more. Therefore, the design of negative impedance is a trade-off problem. In addition, the phase lag introduced by the control delay changes with frequency, which deteriorates the active damping function. With consideration of the aforementioned problems, the negative resistor feedback control is developed with approximation of control delay. The comparison of inverter output impedance with different damping method shows that the proposed method can achieve better damping function in a much wider frequency range; In addition, the proposed controller is less sensitive to the control delay.

APPENDIX A

TABLE I
SST INVERTER STAGE PARAMETERS

System parameters		Values
Power rating	P	20kW
Switching frequency	f_{sw}	10kHz
Sampling period	T_s	50μs
DC-link voltage	Vdc	200V
Voltage loop PR controller	$K_{p,v}$	0.2
	$K_{r,v}$	50
Current loop HCC	$K_{p,i}$	3.5
	$K_{pr,i}$,(i=1,3,5...15)	750,600,500,900, 10,10,250,250
LC filter	L_{sst}	1.686mH
	R_{sst}	0.07Ω
	C_{sst}	30μF

TABLE II
SINGLE INVERTER PARAMETERS

System parameters		Values
Power rating	P	1kW
Switching frequency	f_{sw}	10kHz
Sampling period	T_s	100μs
DC-link voltage	Vdc	200V
Current loop PR controller	$K_{p,v}$	0.22
	$K_{r,v}$	60
LCL filter	L_1	900μH
	C	20μF
	L_2	900μH

REFERENCES

[1] X. She, A. Huang, and R. Burgos, "Review of solid-state transformer technologies and their application in power distribution systems," *IEEE J. Emerging Sel. Topics Power Electron.*, vol. 1, no. 3, pp. 186–198, Sep. 2013.

[2] J. Ge, Z. Zhao, L. Yuan, and T. Lu, "Energy feed-forward and direct feed-forward control for solid-state transformer," *IEEE Trans. Power Electron.*, vol. 30, no. 8, pp. 4042–4047, Aug. 2015.

[3] M. Cespedes and J. Sun, "Impedance shaping of three-phase grid-parallel voltage-source converters," in *Proc. IEEE APEC*, Feb. 2012, pp. 754–760

[4] S. Zhang, S. Jiang, X. Lu, B. Ge, and F. Peng, "Resonance issues and damping techniques for grid-connected inverters with long transmission cable," *IEEE Trans. Power Electron.*, vol. 29, no. 1, pp. 110–120, Jan. 2014.

[5] J. He, Y. W. Li, D. Bosnjak, and B. Harris, "Investigation and active damping of multiple resonances in a parallel-inverter-based microgrid," *IEEE Trans. Power Electron.*, vol. 28, no. 1, pp. 234–246, Jan. 2013.

[6] D. Yang, X. Ruan, and H. Wu, "Impedance shaping of the grid-connected inverter with LCL filter to improve its adaptability to the weak grid condition," *IEEE Trans. Power Electron.*, vol. 29, no. 11, pp. 5795–5805, Nov. 2014.

[7] X. Wang, F. Blaabjerg, and P. Loh, "Grid-current-feedback active damping for LCL resonance in grid-connected voltage source

converters," *IEEE Trans. Power Electron.*, vol. pp, no. 99, pp. 1, Mar. 2015.

[8] J. Sun, "Impedance-based stability criterion for grid-connected inverters," *IEEE Trans. Power Electron.* , vol. 26, no. 11, pp. 3075–3078, Nov.2011.

[9] Q. Ye, R. Mo, Y. Shi, and H. Li., "A unified impedance-based stability criterion (UIBSC) for paralleled grid-tied inverters using global minor loop gain (GMLG)," in *Proc. IEEE ECCE*, Sep. 20–24, 2015, pp. 5816-5821.

[10] M. Cespedes and J. Sun, "Renewable energy systems instability involving grid-parallel inverters," in *Proc. IEEE APEC*, Feb. 2009, pp. 1971–1977.

[11] X. Wang, F. Blaabjerg, and W. Wu, "Modeling and analysis of harmonic stability in an AC power-electronics-based power system," *IEEE Trans. Power Electro*n., vol. 29, no. 12, pp. 6421–6432, Dec. 2014

Compensator-less Structures for Droop Control of Single Phase Inverters in a Flexible Microgrid

Onkar Vitthal Kulkarni
Department of
Energy Science and Engineering,
Indian Institute of Technology, Bombay
Email: onkarvkulkarni@gmail.com

Suryanarayana Doolla
Department of
Energy Science and Engineering,
Indian Institute of Technology, Bombay
Email: suryad@iitb.ac.in

B. G. Fernandes
Department of
Electrical Engineering,
Indian Institute of Technology, Bombay
Email: bgf@ee.iitb.ac.in

Abstract—**Droop control techniques reported in the literature mainly use compensator based multi-loop feedback structure. To tune these compensators either trial method or inverter average model based complex compensator design method is used. It is cumbersome to tune these compensators especially when large number of inverters having different power ratings are present in the system. In this paper, two novel and simplified compensator-less control structures are proposed for droop control. These controllers utilize existing filter structure of grid-connected inverters. The proposed controllers are validated using MATLAB/SIMULINK simulation results. Finally, their performances are compared.**

Index Terms—**Multiple inverters, Automatic mode transition, PLL, State machine**

I. INTRODUCTION

It is desirable to have the inverter interfaced distributed generators (DGs) able to operate in both the grid-connected as well as stand-alone modes [1]. In stand-alone scenario, if multiple such inverters are present in the system, it is essential to operate them in proportional power sharing mode without using any communication. To achieve this objective the droop control technique is generally preferred. Many techniques are proposed in the literature for improving transient and steady state performance of conventional droop control. These techniques use compensator based complex multi-loop feedback structures having outer voltage and inner current control loops. The limitations of compensator-less virtual flux droop method reported in the literature [8] are also discussed in this paper. In most cases related to stand-alone operation, inverters with LC filters are considered for control design. However, in grid-connected inverters, LCL filter with damping resistors are preferred over L and LC filters. Hence, in a flexible microgrid, the power sharing control must work satisfactorily for inverters having LCL filter. In this paper, based on analysis of structures of various droop control methods available in the literature, two compensator-less controllers are proposed for parallel single-phase inverters operating with conventional droop control method. Droop control technique for proportional power sharing among multiple three phase inverters is originally proposed in [2]. This strategy is same as that used for power sharing among multiple synchronous generators wherein a droop is introduced in the frequency of

each generator with the real power delivered by the generator. The control implementation of this method is achieved using four main stages viz. (1) power calculation, (2) generation of reference voltage magnitude and frequency using appropriate droop characteristics, (3) voltage and frequency regulation using PI compensators and (4) hysteresis based direct flux control (DFC). Further improvements to this structure are done by introducing inner current control loop and constant frequency PWM technique [3]. The other variant of conventional droop controller for better performance in low voltage microgrid is inverse droop. The disadvantages of both the conventional and inverse droop controllers such as less accurate reactive power sharing, sluggish transient response, voltage drop due to load effect etc. are overcome by techniques presented in [4], [5] and [6], respectively. Constant frequency droop control is proposed in [7] by introducing droop in phase angle with active power output. Unavailability of initial phase angle information of all the sources is the main drawback of this method which can be overcome using GPS signal. These methods use complex multifeedback loops to achieve output voltage and inner current control loops. Considering three-phase systems these loops may be essential for balanced operation of the complete system. Hence, if large number of inverters are present in the system it is complex to achieve proper tuning. To overcome this issue, virtual flux droop control is proposed in the literature for three-phase system wherein all the regulators are eliminated and droop terms are introduced in inverter flux angle and magnitude with respect to active and reactive power, respectively [8].

For parallel operation of single phase inverters, such compensator-based control structures are proposed in the literature using PI and PR compensators. Some improvements include software based output current feedback for minimizing number of sensors required [9] etc. These techniques are tested for satisfactory performance on both the linear and non-linear loads. They use compensator based complex multifeedback structure for achieving outer voltage and inner current regulation. The tuning of these compensators becomes complex especially when large number of inverters are present in the system.

This paper is organized as follows. In Section II, the details

978-1-4673-9551-9/16 $31.00 © 2016 IEEE

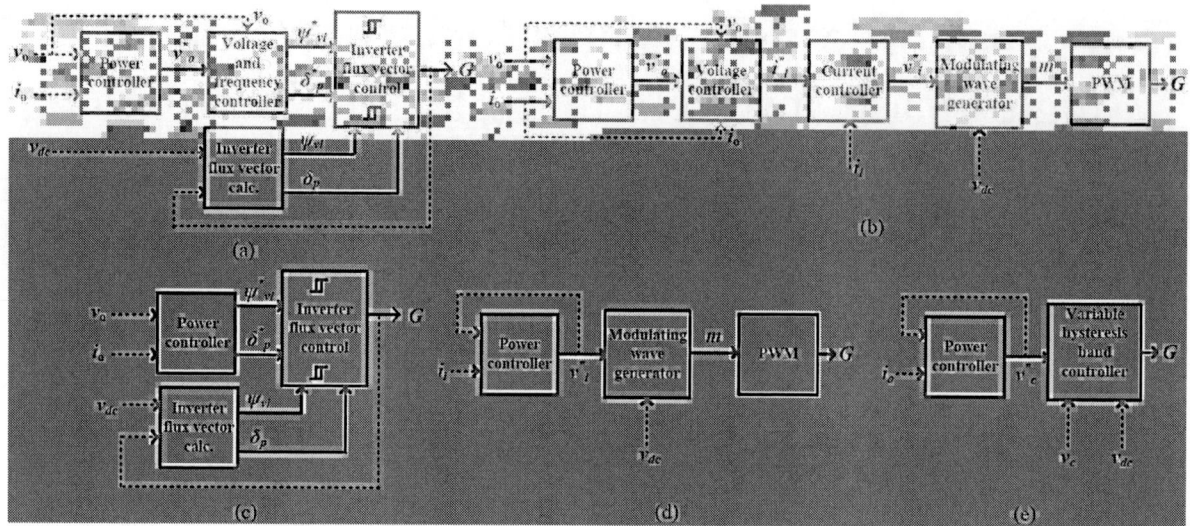

Fig. 1: Droop control structures (a) virtual flux based conventional droop control (b) outer voltage and inner current control (c) virtual flux droop control (d) proposed direct modulating wave controller and (d) proposed adaptive hysteresis band controller

of the compensator-less droop control structures are provided. In Section III, simulation results confirming the validity of the proposed controllers considering a multiple inverter system are presented and from these results the conclusion is drawn in Section IV.

II. COMPENSATOR-LESS DROOP CONTROL STRUCTURES

In this paper, two control structures are proposed for eliminating compensators and thus minimize tuning efforts. Both the methods are compared for their performance with linear as well as non-linear loads.

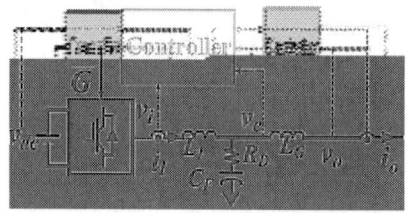

Fig. 2: Measurements associated with inverter in a flexible microgrid

In Fig. 1(a), (b) and (c), various control structures reported in the literature are shown considering measurement notations of Fig. 2. The power controller stage in these block diagrams consists of power calculation as shown in Fig. 3 (for single phase inverters) and droop characteristics shown in Fig. 4. In the structure shown in Fig. 1(a) proposed in [2], inverter output power (associated with v_o and i_o) is measured and used for generating output voltage (across capacitor) reference using characteristics shown in Fig. 4. This reference is then tracked using PI compensators, flux vector estimation and hysteresis flux control as shown in Fig. 1(a). In the structure shown in Fig. 1(b) [3], outer voltage and inner current control loops

are used to track the capacitor voltage reference generated by power controller. This structure has four compensators in total. Additionally, in [3], a coupling inductor is placed after LC filter to achieve better power sharing dynamics. It also uses constant switching frequency based PWM technique. In Fig. 1(c), block diagram of virtual flux droop controller is shown. In this controller, power sharing among parallel operating inverters is achieved by drooping magnitude and phase angle of inverter virtual flux [8]. In this method, virtual flux (ψ_{vi}) associated with inverter voltage (v_i) is controlled. Hence, in this method need of compensators is eliminated. This method implements direct flux control (DFC), which requires inverter flux vector estimation from inverter switching state (v_i) and hysteresis based control of inverter internal flux (ψ_{vi}) which in turn contains sector detection algorithm. This method is computation extensive. The presence of hysteresis control introduces variable switching frequency thus may not achieve satisfactory power quality performance on existing LCL filter of grid-connected inverter. Additionally, phase angle based droop control also requires GPS signal to get information regarding initial phase angles of other inverters [7], [10]. To overcome these problems, two simple structures are proposed for conventional droop control.

A. Direct modulating wave control (DMWC)

In DMWC, magnitude and frequency of v_i is directly altered unlike virtual flux controller (depicted in Fig. 1(c)) which alters the magnitude and phase angle of virtual flux (ψ_{vi}). The overview of DMWC is shown in Fig. 1(d). In this way, the voltage magnitude and frequency from power controller can be directly utilized for generating modulating wave. This modulating wave can be directly fed to existing PWM controller of the grid-connected inverter. This reduces mathematical complexity and by using $P-f$ and $Q-V$ droop

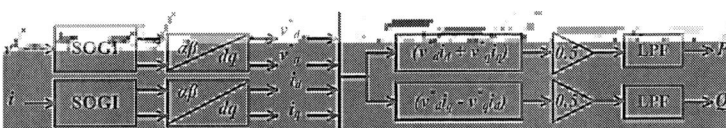
Fig. 3: Power calculation

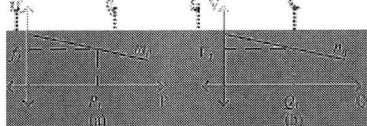
Fig. 4: Droop characteristics (a) P-f (b) Q-V

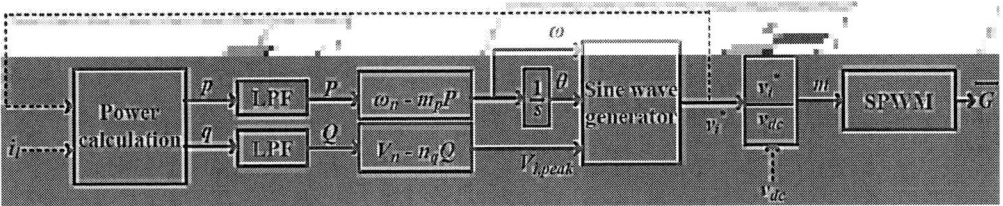
Fig. 5: Direct modulating wave controller

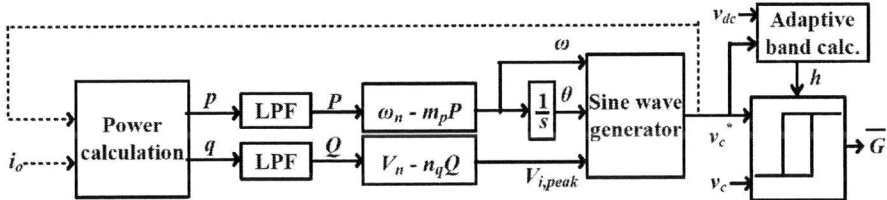
Fig. 6: Adaptive hysteresis band voltage controller

need of GPS signal is eliminated. Other advantages include constant switching frequency operation, low current ripple and hence low THD.

The details of the modified controller are shown in Fig. 5. As the inverter itself is acting as a directly controlled voltage source, its power output is calculated using appropriate voltage (v_i) and current (i_l) quantities as shown in Fig. 5. The output voltage magnitude and frequency from power controller are used for calculation of instantaneous value of inverter fundamental voltage and is directly fed to the power controller. The power calculation block contains second order generalized integrator (SOGI), $\alpha\beta - dq$ transformation, power calculation in $dq-$ reference frame and appropriate low pass filters (LPFs) as shown in Fig. 3. The existing unipolar PWM algorithm in the grid-connected controller is used for generating gate pulses which results in doubling of equivalent switching frequency.

B. Adaptive hysteresis band voltage control (AHBVC)

In [11], a constant switching frequency band controller for dynamic voltage restorer (DVR) is presented alongwith a modified filter structure. This structure is similar to that of the filter structure for grid-connected inverters shown in Fig. 2 [1] i.e. LC filter alongwith damping resistor R_D. This controller can be used to regulate inverter output voltage to the reference value without using compensators and still achieve constant inverter switching frequency. Thus, the reference voltage from droop controller can be tracked using adaptive hysteresis band based capacitor voltage control. Thus, voltage v_c can be regulated to its reference value using constant switching

frequency band controller and the grid side inductance (L_G) can be used to implement actual inductive output impedance. The expression for switching frequency is given by (1). The minimum and maximum values of switching frequency are given by (2) and (3), respectively. Thus, the value of hysteresis band for desired constant switching frequqency f_{sw} is given by eq. (4) [11].

$$f_{sw} = \frac{R_D}{4V_{dc}hL_I}\left(V_{dc}^2 - V_{ref}^2 sin^2(\omega t)\right) \quad (1)$$

$$f_{swmin} = \frac{R_D}{L_I}\frac{V_{dc}^2 - V_{ref}^2}{4hV_{dc}} \quad (2)$$

$$f_{swmax} = \frac{R_D}{L_I}\frac{V_{dc}}{4h} \quad (3)$$

$$h = \frac{R_D}{4V_{dc}f_{sw}L_I}\left(V_{dc}^2 - V_{ref}^2 sin^2(\omega t)\right) \quad (4)$$

where, L_I is inverter-side inductance, V_{dc} is dc link voltage. The inverter and LC filter combination acts as a controlled voltage source. Thus, signals v_c and i_o are required for hysteresis control and power calculation. The instantaneous voltage for power calculation can be directly generated from power controller outputs V_{ref} and ω in a similar manner as that of DMWC. The same can also be used in eq. (4). The overview of the controller is shown in Fig. 1(e). The details of the control structure are given in Fig. 6. Extra sensors (which are generally not present in grid-connected inverters) need to be added to measure v_c and i_o. Voltage THD might be higher due to the existing filter designed for $2f_{sw}$.

978-1-4673-9551-9/16 $31.00 © 2016 IEEE

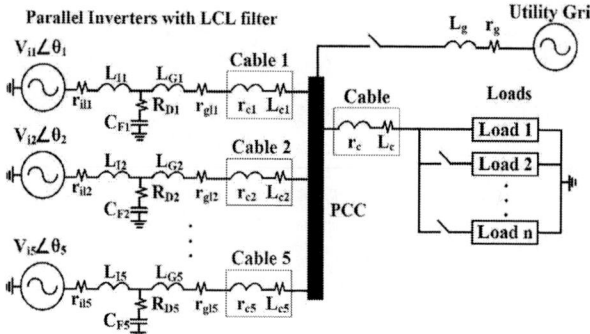

Fig. 7: System under consideration for simulation study

III. SIMULATION RESULTS AND DISCUSSIONS

The system configuration for a 230 V, 50 Hz system considered for the simulation study is shown in Fig. 7. It consists of five single phase inverters. Switching frequency is 15 kHz for all the inverters. The droop coefficients m_p and n_q for all the inverters are selected in inverse proportion to their ratings and large enough to ensure accurate power sharing as indicated in Table I. Appropriate values of LCL filter alongwith daming resistor are considered.

Table I: Droop coefficients for various inverters

Inverter no.	m_p	n_q	LCL values
1	1.8e-5	0.004	
2	2.25e-5	0.005	L_I = 1.25 mH
3	3e-5	0.0067	L_G = 1.5 mH
4	4.5e-5	0.01	R_d = 4.1 Ω
5	9e-5	0.02	C_F = 8 μF

A. Simulation results for non-linear load

In this scenario, the load in the system is a 2.2 kVA non-linear load as shown in Fig. 8. The values of the components are $L_{in} = 1\,mH$, $C_{dc} = 500\,\mu F$ and $R_{load} = 40\,\Omega$.

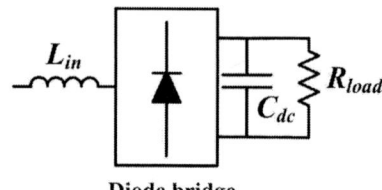

Diode bridge

Fig. 8: Non-linear load

The simulation results for DMWC are shown in Fig. 9, 11, 13 and 15, respectively. Similarly, results for AHBVC are shown in Fig. 10, 12, 14 and 16, respectively. These results can be used for performance comparison of two controllers.

B. Simulation results for linear load

In this scenario, parallel inverters are connected to a linear load of $9000 + j4000$ Var. The power sharing dynamics for step load changes for DMWC and AHBVC are shown in Fig. 17 and 18, respectively.

C. Performance comparison

The merits and demerits of these methods are as follows

1) Load voltage THD: From Figs. 15 and 16, it can be observed that DMWC has better THD (3.53%) than AHBVC (5.32%) and thus it is compliant with the standard IEC 62040-3 for class-I UPS [9]. It can also be observed that in case of DMWC load voltage has double switching frequency (30 kHz) component whereas AHBVC has a prominent component at the switching frequency ($f_{sw} = 15\,kHz$). This is due to the fact that AHBVC utilizes two-level hysteresis controller which does not use two extra possible switching states as that of unipolar PWM.

2) Number of sensors: Grid-connected inverters, generally contain inverter-side inductor current control loop to inject desired amount of active and reactive power into the grid. Hence, this loop requires one current sensor for i_l. In addition to that, it requires one voltage sensor for v_o to achieve synchronization using PLL and other to measure dc link voltage for implementing desired PWM technique. Voltage v_o is also beneficial in giving voltage feed-forward to the current control loop. In this way, grid-connected inverter generally consists of two voltage and a current sensor.

As can be seen from Fig. 5, DMWC requires current i_l for power calculation and v_{dc} for PWM implementation. Hence, there is no need of introducing any extra sensor for implementing DMWC.

On the other hand, AHBVC requires output current i_o for power measurement and capacitor voltage v_c for hysteresis controller as can be seen from Fig. 6. Hence, two extra sensors need to be installed in an existing grid-connected inverter for implementing AHBVC.

3) Power sharing dynamics: It can be seen by comparing Figs. 9 and 10 as well as Figs. 17 and 18 that AHBVC has better dynamic power sharing performance than DMWC due to the presence of voltage regulator. In case of non-linear load, as the load injects reactive power, the inverters absorb it in an approximate proportion to their ratings as can be seen from Figs. 9(b) and 10(b). In case of DMWC, filter capacitors also become part of the load with respect to inverters. Thus, reactive power absorbed by inverters is more for DMWC than AHBVC. Hence, magnitude of the reference voltage is also more in case of DMWC than AHBVC due to droop characteristics shown in Fig. 4.

4) Load voltage drop: In Fig. 19, RMS values of load voltage for the system feeding non-linear load are shown. From this figure, it can be concluded that DMWC offers the highest load voltage magnitude as compared to AHBVC. This is mainly due to the reactive power injection by filter capacitor as explained in comparison of power sharing dynamics.

978-1-4673-9551-9/16 $31.00 © 2016 IEEE   2248

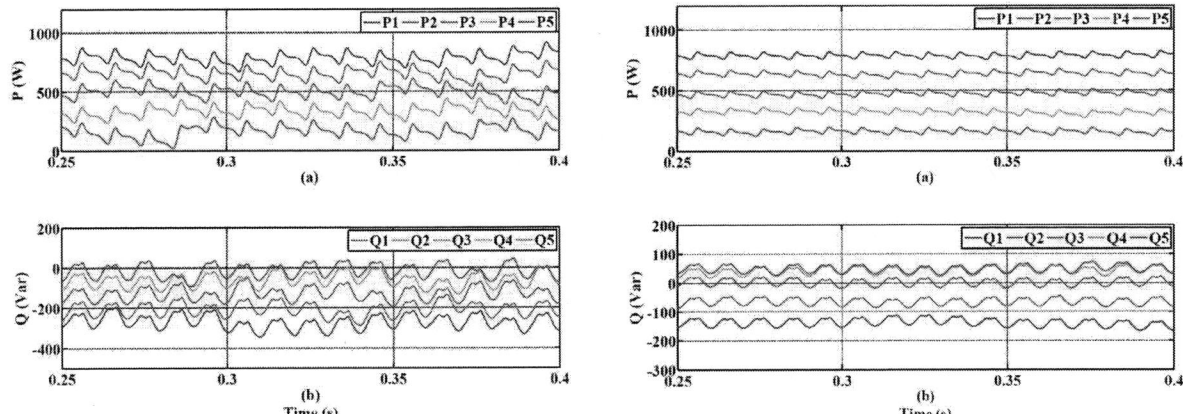

Fig. 9: Power outputs of inverters under non-linear load conditions for DMWC (a) active power and (b) reactive power

Fig. 10: Power outputs of inverters under non-linear load conditions for AHBVC (a) active power and (b) reactive power

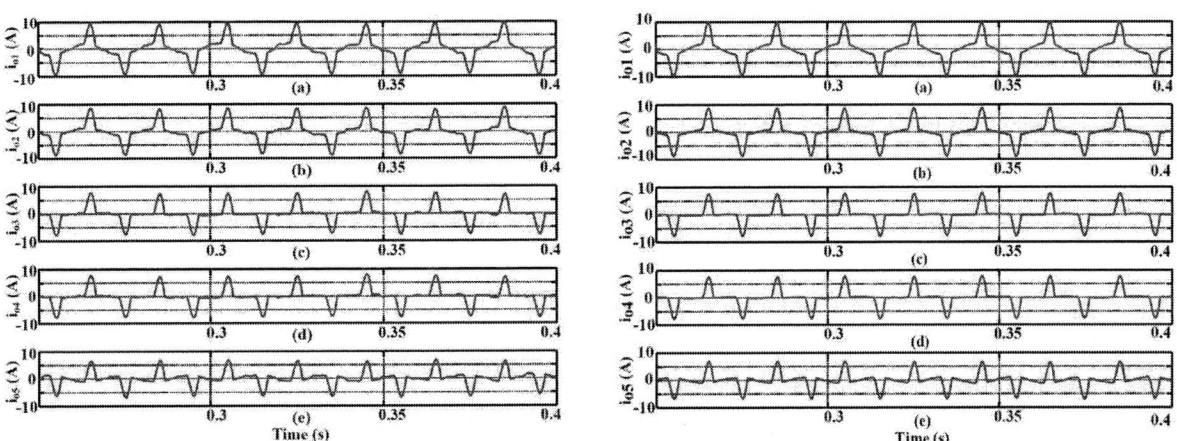

Fig. 11: Inverter output currents under non-linear load conditions for DMWC

Fig. 12: Inverter output currents under non-linear load conditions for AHBVC

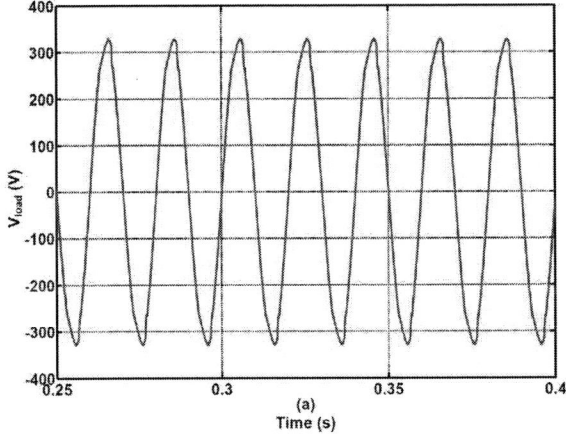

Fig. 13: Load voltage using DMWC

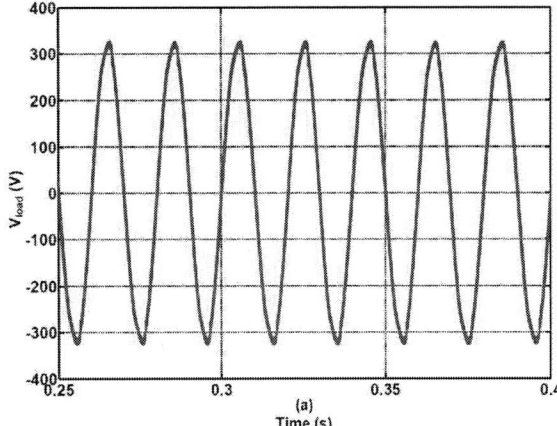

Fig. 14: Load voltage using AHBVC

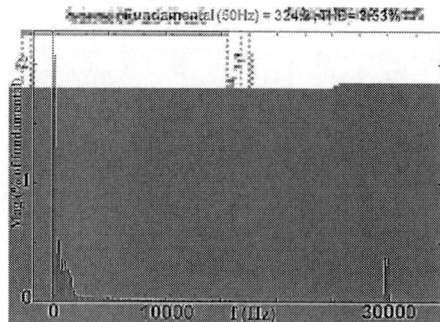

Fig. 15: Load voltage FFT for DMWC

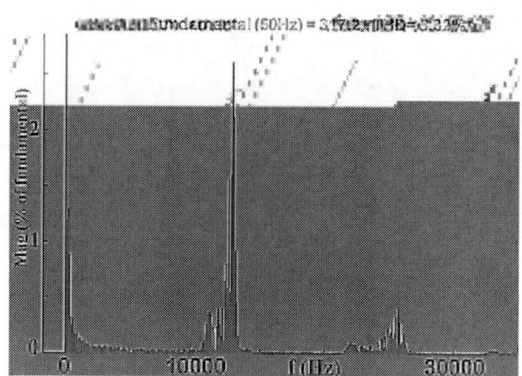

Fig. 16: Load voltage FFT for AHBVC

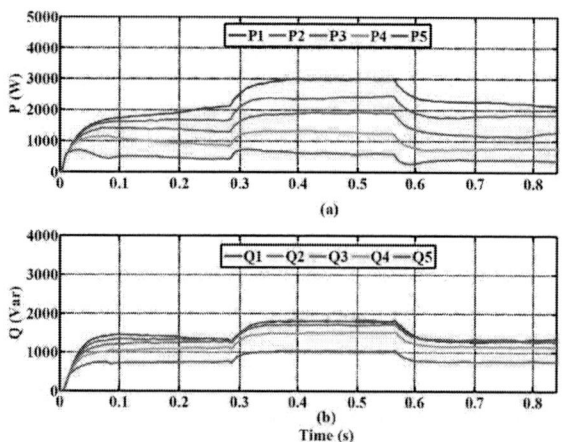

Fig. 17: Power sharing dynamics for DMWC under linear load condition (a) active power (b) reactive power

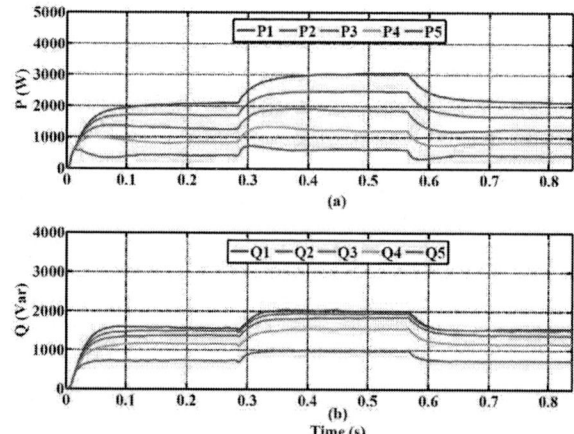

Fig. 18: Power sharing dynamics for AHBVC under linear load condition (a) active power (b) reactive power

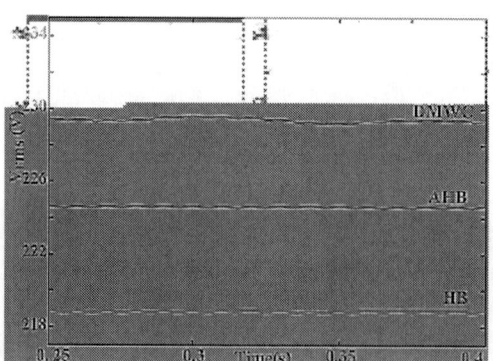

Fig. 19: RMS value of load voltages for DMWC and AHBVC

978-1-4673-9551-9/16 $31.00 © 2016 IEEE

IV. CONCLUSION

In this paper, two simple compensator-less controllers are proposed for operating existing grid-tie inverters in stand-alone power sharing mode. The proposed structures eliminate drawbacks of virtual flux droop control such as mathematical complexity, need of GPS signal for power sharing accuracy and variable switching frequency. The effectiveness of controllers is validated and compared using simulation results for linear as well as non-linear loads. DMWC technique is beneficial because of lower THD, lower load voltage drop, sufficiency of existing sensors. AHBVC has better power sharing dynamics due to the presence of output voltage controller.

V. ACKNOWLEDGMENTS

The authors would like to thank the support of NCPRE, IIT Bombay a project funded by MNRE, India.

REFERENCES

[1] R. Teodorescu and F. Blaabjerg, "Flexible control of small wind turbines with grid failure detection operating in stand-alone and grid-connected mode," *IEEE Trans. Power Electron.*, vol. 19, no. 5, pp. 1323–1332, Sept. 2004.

[2] M. C. Chandorkar, D. M. Divan, and R. Adapa, "Control of parallel connected inverters in standalone ac supply systems," *IEEE Trans. Ind. Appl.*, vol. 29, no. 1, pp. 136–143, Jan. 1993.

[3] N. Pogaku, M. Prodanovic, and T. C. Green, "Modeling, analysis and testing of autonomous operation of an inverter-based microgrid," *IEEE Trans. Power Electron.*, vol. 22, no. 2, pp. 613–625, Mar. 2007.

[4] J. M. Guerrero, J. Matas, L. de Vicua, M. Castilla, and J. Miret, "Wireless-control strategy for parallel operation of distributed-generation inverters," *IEEE Transactions on Industrial Electronics*, vol. 53, no. 5, pp. 1461–1470, Oct. 2006.

[5] J. M. Guerrero, J. Matas, L. G. de Vicuna, M. Castilla, and J. Miret, "Wireless-control strategy for parallel operation of distributed-generation inverters," *IEEE Trans. Ind. Electron.*, vol. 53, no. 5, pp. 1461–1470, Oct. 2006.

[6] Q.-C. Zhong, "Robust droop controller for accurate proportional load sharing among inverters operated in parallel," *IEEE Trans. Ind. Electron.*, vol. 60, no. 4, pp. 1281–1290, Apr. 2013.

[7] R. Majumder, B. Chaudhuri, A. Ghosh, G. Ledwich, and F. Zare, "Improvement of stability and load sharing in an autonomous microgrid using supplementary droop control loop," *IEEE Trans. Power Syst.*, vol. 25, no. 2, pp. 796–808, May 2010.

[8] J. Hu, J. Zhu, D. G. Dorrell, and J. M. Guerrero, "Virtual flux droop method - a new control strategy of inverters in microgrids," *IEEE Trans. Power Electron.*, vol. 29, no. 9, pp. 4704–4711, Sep. 2014.

[9] A. Hasanzadeh, O. Onar, H. Mokhtari, and A. Khaligh, "A proportional-resonant controller-based wireless control strategy with a reduced number of sensors for parallel-operated upss," *IEEE Trans. Power Del.*, vol. 25, no. 1, pp. 468–478, Jan. 2010.

[10] M. S. Golsorkhi and D. D. C. Lu, "A control method for inverter-based islanded microgrids based on v-i droop characteristics," *IEEE Trans. Power Del.*, vol. 30, no. 3, pp. 1196–1204, Jun. 2015.

[11] S. Sasitharan and M. K. Mishra, "Constant switching frequency band controller for dynamic voltage restorer," *IET, Power Electron.*, vol. 3, no. 5, pp. 657–667, Sep. 2010.

Comparative Evaluation of the Loss and Thermal Performance of Advanced Three Level Inverter Topologies

Alexander Anthon, Zhe Zhang, Michael A. E. Andersen
Dept. of Electrical Engineering
Technical University of Denmark
Kgs. Lyngby, Denmark
Email: jant@elektro.dtu.dk

Grahame Holmes, Brendan McGrath, Carlos Teixeira
School of Electrical and Computer Engineering
RMIT University
Melbourne, Australia

Abstract—This paper presents a comparative evaluation of the loss and thermal performance of two advanced three-level inverter topologies, namely the SiC based T-Type and the Hybrid-NPC, both of which are aimed at reducing the high switching losses associated with a conventional Si based T-Type inverter. The first solution directly replaces the 1200 V primary Si IGBT switches with lower loss 1200 V SiC MOSFETs. The second solution strategically adds 600 V CoolMos FET devices to the conventional Si T-Type inverter to reduce the primary commutation losses. Semiconductor loss models, experimentally verified on calibrated heat sinks, are used to show that both variations can significantly reduce the semiconductor losses compared to the Si based T-Type inverter. The results show that both alternatives are attractive if high efficiencies and reduced thermal stress are major requirements for the converter design.

Index Terms—T-Type, Hybrid-NPC, SiC MOSFET, Si IGBT, CoolMos

I. INTRODUCTION

Transformerless photovoltaic (PV) systems are becoming favored in the residential sector due to their reduced size, cost and higher efficiencies compared to transformer based alternatives [1]. To further improve low cost PV systems, previous research has intensively investigated the trade-offs between two- and three-level inverters and has found that three-level inverters have lower total semiconductor losses as the switching frequency increases, and also allow a significant size reduction in the AC filter [2], [3]. Within the three-level inverter alternatives, the Neutral-Point-Clamped (NPC) [4] and the T-Type [5] topologies are widely used, each with particular advantages and drawbacks. For example, since the NPC inverter can use semiconductor devices that need to block only half the DC link voltage, its switching losses are always lower at any given switching frequency compared to the T-Type inverter, whose outer switches must block the whole DC link voltage and hence incur higher switching losses. Nevertheless, the T-Type converter can still achieve lower total semiconductor losses compared to the NPC alternative due to its reduced conduction losses. Hence switching frequency is clearly a crucial parameter in this comparison [3]. Due

to recent advances in new semiconductor devices such as silicon carbide (SiC), switching losses in a power converter can be significantly reduced compared to standard Si IGBT alternatives using these devices [6], [7]. However, while the benefits and potential of these devices have been well reported [8]–[12], they are not yet in commonplace usage within commercial converter systems.

A further way to reduce the high switching losses in the T-Type inverter is to strategically add lower voltage switching devices in addition to the conventional T-Type circuit in order to manage the primary commutation events. This approach, called a Hybrid-NPC inverter, has been found to achieve higher efficiencies compared to a conventional T-Type structure with higher voltage (1200 V) Si IGBTs [13]. But to date, only few references are available on this topology alternative [14], [15]. In particular a topological comparative evaluation of the loss and thermal performance between the Hybrid-NPC and the T-Type inverter using next generation switching devices such as SiC under exactly the same operating conditions is not known to the authors. This work therefore presents such a detailed loss comparison for these two advanced inverter alternatives, using semiconductor loss models based on datasheet information (to calculate conduction losses), switching transition measurements (to calculate switching losses) and verification of the loss models thermally on calibrated heat sinks.

II. T-TYPE AND HYBRID-NPC INVERTER

The three inverter alternatives considered in this paper are shown in Fig. 1, with the conventional Si based T-Type structure shown in Fig. 1a as a reference. Its operational principle is illustrated in Fig. 1d-Fig. 1e. Initially, as shown in Fig. 1d, when a zero output voltage is required with a positive output current, diode D_2 and switch S_2 conduct this load current and the blocking voltage across both S_1 and S_4 is $V_{DC}/2$.

Then, to achieve a positive output voltage, switch S_1 turns on with a commutation voltage of $V_{DC}/2$ and the

978-1-4673-9551-9/16 $31.00 © 2016 IEEE

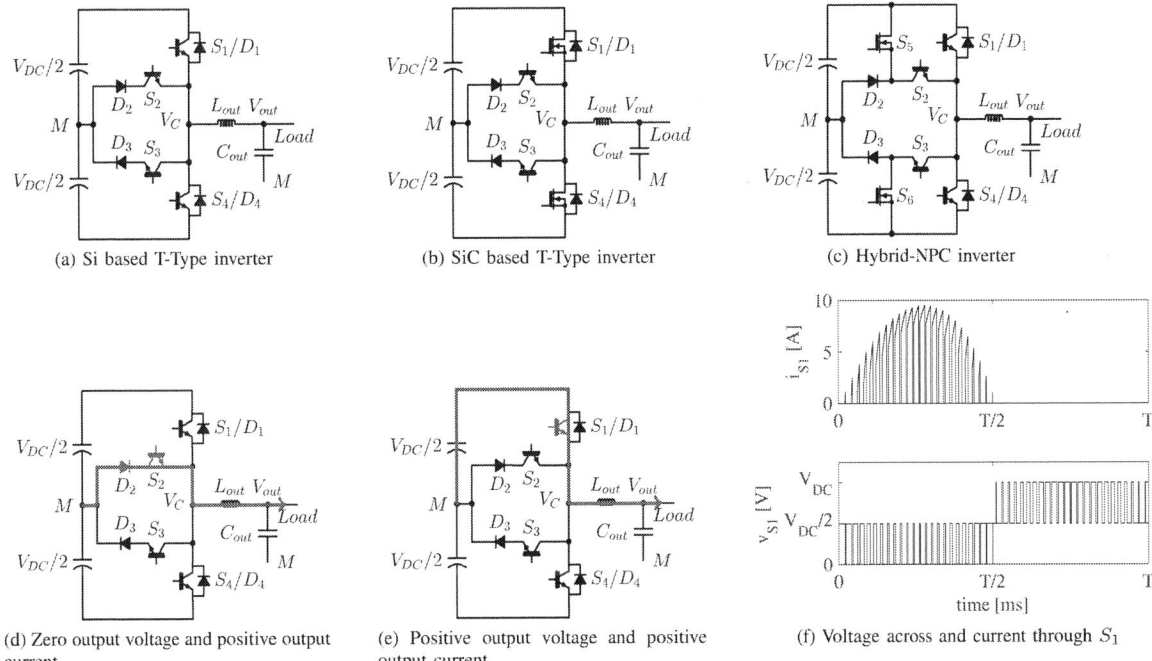

(a) Si based T-Type inverter   (b) SiC based T-Type inverter   (c) Hybrid-NPC inverter

(d) Zero output voltage and positive output current

(e) Positive output voltage and positive output current

(f) Voltage across and current through S_1

Fig. 1. Inverter alternatives used in this study in (a) - (c), commutation from zero output voltage to positive output voltage in (d)-(e) and voltage and current through device S_1 in (f)

switching losses associated with this transition. Finally, a zero output voltage is re-established by turning switch S_1 off, with associated turn off losses for this transition. This process repeats throughout the positive fundamental half cycle as shown in Fig. 1f. Note that when the converter output voltage is switched to the positive DC rail, switch S_4 must block the whole DC link voltage, i.e. V_{DC}, which therefore requires S_4 to be rated to accommodate the full DC link voltage.

A similar process occurs for the negative fundamental half cycle, with diode D_3 and switch S_3 conducting current to achieve a zero output stage and switch S_4 turning on to achieve a negative converter output stage. Note that when the converter is switching during the negative half cycle, switch S_1 must now block the whole DC link voltage, as shown on

the right half side of Fig. 1f. Since S_1 and S_4 need a higher voltage rating to block the whole DC link voltage, in contrast to the inner bi-directional devices D_2/D_3 and S_2/S_3, which need to block only half the DC link voltage, their switching losses are a major contributor to the overall semiconductor losses. Hence they can be directly replaced with SiC switching devices as shown in Fig. 1b to reduce these switching losses, with the inverter's topological structure and thus its modulation principles unchanged.

Alternatively, additional low voltage rated switching devices S_5 and S_6 can be added into the circuit, as shown in Fig. 1c, to make a Hybrid-NPC structure. The switching principle of this inverter is a little different as shown in Fig. 2, in that one of either S_5 or S_6 turn on first to create the positive or negative output voltage as required. Since these devices need only be

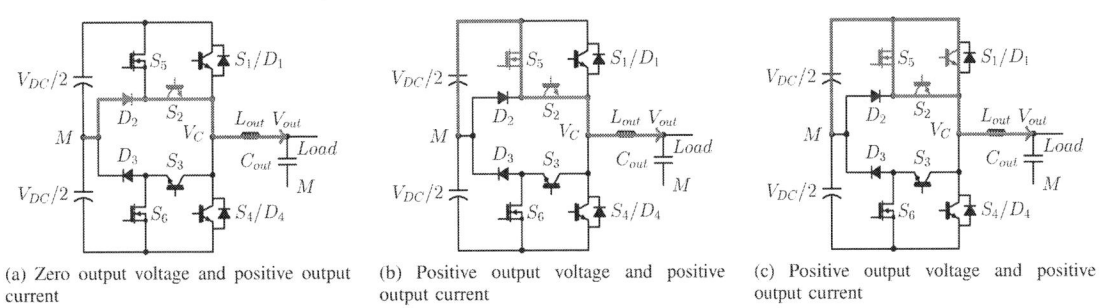

(a) Zero output voltage and positive output current

(b) Positive output voltage and positive output current

(c) Positive output voltage and positive output current

Fig. 2. Operation principle of Hybrid-NPC converter

rated to half the DC link voltage, their switching losses will be less than for a conventional T-Type inverter (600 V CoolMos FET devices are used in this work to minimize these switching losses). Once the switching transition is complete, current flows through the two devices S_5 and S_2 as shown in Fig. 2b (for a positive output voltage and current), which increases their conduction losses to a level similar to a conventional NPC inverter. Switch S_1 is then turned on (with almost zero switching losses), and the current flow changes to share between the two conduction paths as shown in Fig. 2c to achieve a similar conduction loss as for a standard T-Type inverter (since the forward voltage drop across S_1 is much the same as before).

The turn-off sequence for the Hybrid-NPC is in the reverse order, i.e. S_1 first turns off with essentially zero switching losses, and then S_5 turns off with appropriate losses against a commutation voltage of $V_{DC}/2$.

III. SEMICONDUCTOR DEVICE SELECTION

With the operation principles of the three inverter topologies identified, the selection of appropriate semiconductor devices for the topology comparison can now proceed. Since the targeted application for this topology is a grid-connected PV inverter system, the DC link voltage can go up to over 800 V. Thus a 1200 V rated device for S_1/S_4 is required. For this voltage range, the usual semiconductor device choice is Si IGBTs, which are known to have higher switching losses than either SiC or CoolMos devices, particularly because of their relatively large turn off energies caused by their long delay tail currents. Fig. 3b and Fig. 3c illustrate this difference, showing the turn on and turn off switching energies for a 1200 V Si IGBT (S_1/S_4 in Fig. 1a), a 1200 V SiC MOSFET (S_1/S_4 in Fig. 1b) and a 600 V CoolMos (S_5/S_6 in Fig. 1c) that were directly measured at appropriate voltages and currents for their T-Type inverter context, using the laboratory prototype shown in Fig. 3a. It can be seen from these results that while the 1200 V Si IGBT turn on energies are not so much larger than the CoolMos device, both the CoolMos FET and the SiC MOSFET show a superior turn off switching loss behavior. This is a particularly interesting observation since the turn off energies have been found to be the limiting factor for high efficient high switching frequency operation of the T-Type inverter [16]. Note also that since PV inverters operate mainly at unity power factor [17], the inner bi-directional device (S_2/S_3 in all topologies) switching losses will be essentially negligible and are therefore not included in this switching energy comparison.

To complete the switching device loss comparison, their forward conduction voltages can be taken from the manufacturer's datasheets. The results are presented in Fig. 4, and show that the SiC MOSFET as a direct replacement to the 1200 V Si IGBT can also greatly reduce conduction losses over the current range of interest. Particularly at low currents, the SiC MOSFET shows a large voltage drop reduction due to its low on-state resistance, while the Si IGBT has a bipolar output characteristic and therefore a more constant and larger voltage

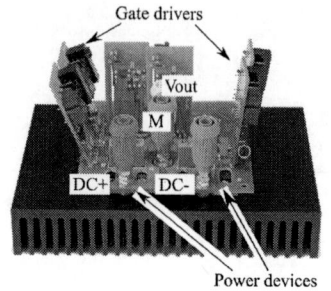

(a) Laboratory prototype

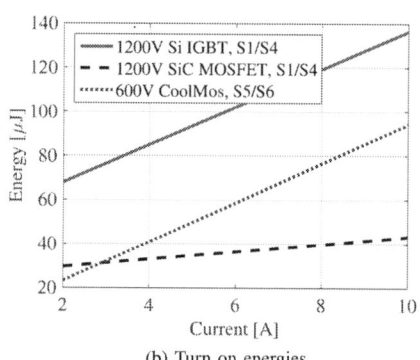

(b) Turn on energies

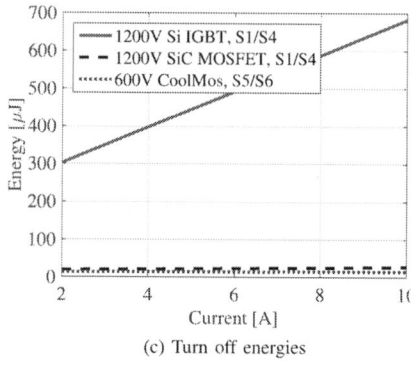

(c) Turn off energies

Fig. 3. Laboratory prototype and measured switching energies

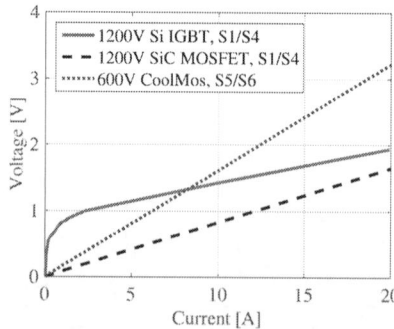

Fig. 4. Forward voltages of the primary devices

drop. Fig. 4 also shows that the 600 V CoolMos device has a relatively large forward voltage compared to the SiC MOSFET due to its Si based semiconductor substrate. Table I lists all semiconductor devices used in this comparison evaluation.

IV. LOSS BREAKDOWN ANALYSIS

Once the device forward conduction and switching losses have been characterized, a loss breakdown analysis for their operation in the T-Type and Hybrid-NPC converter structures can be conducted. The IGBT conduction loss model is obtained using its dynamic on-resistance r_{on} and zero on-state voltage V_0, i.e.

$$P_{con,IGBT} = V_0 I_{AV} + r_{on} I_{rms}^2 \quad , \quad (1)$$

where I_{AV} and I_{rms} are the average and root-mean-square currents through the device. For the SiC MOSFET and the CoolMos FET, only their on resistance $R_{DS(on)}$ is needed to determine conduction losses, i.e.

$$P_{con,FET} = R_{DS(on)} I_{rms}^2 \quad . \quad (2)$$

The conduction losses for the diodes are based on their threshold voltage V_T and dynamic on-resistance r_{on}, i.e.

$$P_{con,Diode} = V_T I_{AV} + r_{on} I_{rms}^2 \quad . \quad (3)$$

For the switching energies, Fig. 3b and Fig. 3c show that the switching losses for each device have a linear relationship to the switched current. Therefore, all switching energies can be modeled as a linear equation according to

$$E_{on,S1,4,5,6} = a_{on} i_{out}(t) mod(t) + b_{on} \quad (4)$$
$$E_{off,S1,4,5,6} = a_{off} i_{out}(t) mod(t) + b_{off} \quad (5)$$

where a_{on}, a_{off}, b_{on} and b_{off} are curve fitting constants for each device derived from the plots shown in Fig. 3. $i_{out}(t)$ is the AC load current and $mod(t)$ is the output voltage modulation function which is defined in the usual way as

$$mod(t) = M sin(\omega t) \quad (6)$$

where M is the modulation index. The overall averaged

TABLE I
SEMICONDUCTOR DEVICES USED

	Si T-Type	SiC T-Type	Hybrid-NPC
S_1/S_4	IKW15N120T2	C2M0080120D	IKW15N120T2
S_2/S_3	IKP15N60T	IKP15N60T	IKP15N60T
D_2/D_3	C3D10060A	C3D10060A	C3D10060A
S_5/S_6			SPP20N60S5

switching losses can then be calculated as

$$P_{sw,S1,4,5,6} = f_{sw} \frac{1}{T} \int_{0+\varphi}^{T/2} (E_{on,S1,4,5,6} + E_{off,S1,4,5,6}) \, dt \quad (7)$$

Once these equations are established and the average and rms currents are determined either analytically or via simulations, the total semiconductor losses can be calculated for any given operating point, with an associated device loss breakdown. Fig. 5 shows this loss breakdown for the Si based T-Type, the SiC MOSFET based T-Type and the Hybrid-NPC inverters with the specifications given in Table II, and operating at an output power of 1.5 kW.

From this result, it can immediately be seen that even though the outer switch commutation voltage is only $V_{DC}/2$, switching losses in the 1200 V Si IGBT are the largest loss contributor to the overall semiconductor losses. Obviously, this effect becomes more severe as the switching frequency increases. Both the SiC based T-Type and the Hybrid-NPC substantially reduce these switching losses as shown in Fig. 5b and Fig. 5c. In fact, for this particular example, at a switching frequency of 16 kHz, the switching losses in the 1200 V Si IGBT are 7.4 W while the switching losses in the 1200 V SiC MOSFET are only 0.8 W and the switching losses using the 600 V CoolMos FET device are 1.1 W. Note also that semiconductor losses are more evenly distributed among the devices for these two more advanced arrangements. Thus, both inverter variations are attractive alternatives compared to a conventional T-Type inverter structure when reduced semiconductor losses are an important factor.

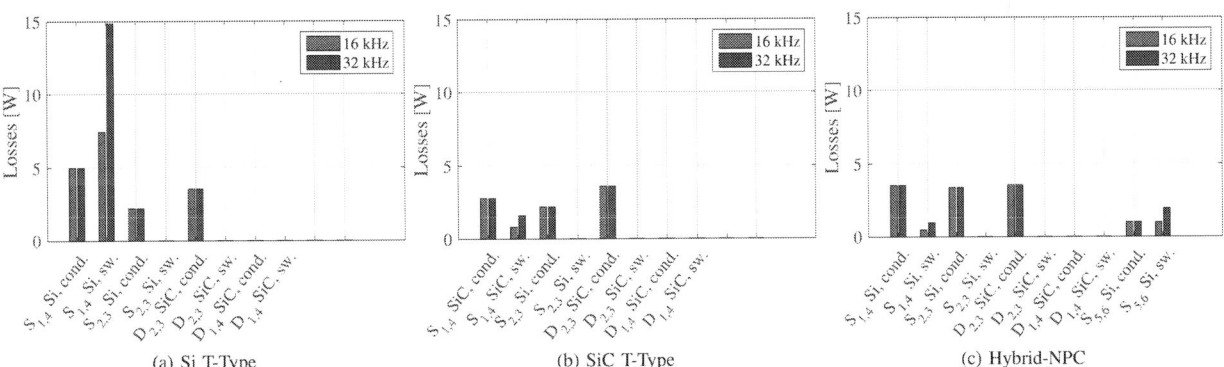

Fig. 5. Loss breakdown analysis for different inverter alternatives. DC link voltage $V_{DC} = 800$ V, filtered output voltage $V_{out,RMS} = 230$ V, output power $P_{out} = 1500$ W

TABLE II
INVERTER SPECIFICATIONS

Symbol	Meaning	Value
V_{DC}	DC link voltage	800 V
V_{out}	Filtered output voltage, rms	230 V
f_{out}	Fundamental frequency	50 Hz
L_{out}	Filter inductor	3 mH
C_{out}	Filter capacitor	4.4 µF
M	Modulation index	0.85

V. LOSS MODEL VALIDATION BY THERMAL MEASUREMENTS

Since the losses and the loss reduction discussed in this paper relate only to the semiconductor devices, they can be readily validated experimentally. This was done using thermal measurements on the device heat sink since semiconductor device losses lead directly to an increased heat sink temperature. To accurately match these temperature measurements to the semiconductor losses, the converter power stage was located inside an open ended chimney as shown in Fig. 6a. To minimize any thermal influence from the surrounding of the power stage (for instance gate driver circuitry), the heat sink was thermally decoupled from the rest of the power stage circuitry using a wooden panel as shown in Fig. 6b. Then, two temperatures are measured, one at the top of the heat sink T_{HS} and one below the heat sink giving T_{amb}, as shown in Fig. 6b. The difference between these readings gives the relative heat sink rise according to

$$\Delta T = T_{HS} - T_{amb} \qquad . \qquad (8)$$

The measurement was used to carefully calibrate the heat sink using known DC loads. This was achieved by supplying the inverter with a known DC voltage and current (and hence power) with inverter switch states selected such that the semiconductor devices absorb all of the power supplied from the controlled DC source. This is illustrated in Fig. 7 for the switches of the Hybrid-NPC converter, i.e. S_5, S_1 and S_4. Similar results were taken for as many different switch pair combinations as possible (e.g. S_1, S_3 and D_3 as a combination and D_2, S_2 and S_4 as another combination), to achieve a well-defined temperature profile of the heat sink. The injected power corresponds to the thermal energy forced into the heat

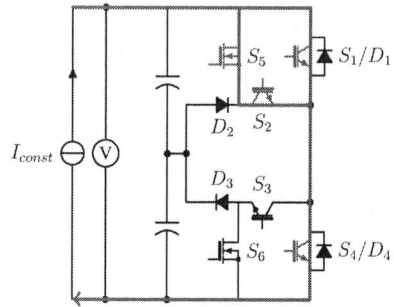

Fig. 7. Switch pair S_5, S_2, S_1 and S_4 are conducting

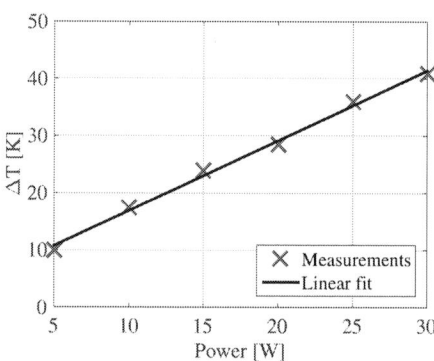

Fig. 8. Device losses versus heat sink temperature rise

sink, and is thus responsible for the heat sink temperature rise. Note that several calibration runs are necessary for different power levels to achieve a relation between the injected power and the heat sink temperature rise over a wide range of power loss points, as shown in Fig. 8. The resultant loss profile is linear, as could be expected for a constant heat sink thermal impedance.

VI. EXPERIMENTAL RESULTS

Once the calibration procedure was completed, the converter was then operated at a number of operating conditions to determine the aggregate semiconductor device losses. Operating the converter using phase disposition (PD) PWM [18], [19] with the parameter specifications provided in Table II, the resulting experimental output waveforms for a 230 V, 50 Hz system at 1.5 kW are shown in Fig. 9. The loss results for different operating conditions such as varying output power and switching frequency are shown in Fig. 10, where the predicted semiconductor losses are compared against the measured semiconductor losses. The results are clearly well within the measurement bounds of the experimental thermal measurement technique, and confirm that both the SiC based T-Type inverter and the Hybrid-NPC inverter achieve a major loss reduction compared to the conventional Si based T-Type inverter. More specifically, at 1.5 kW and 16 kHz, the Si T-Type inverter has total semiconductor losses of 22 W while the SiC based alternative

 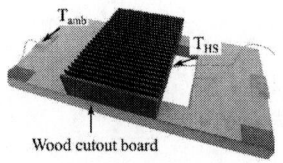

(a) Converter placed in chimney

(b) Thermal measurements location

Fig. 6. Thermal measurement setup

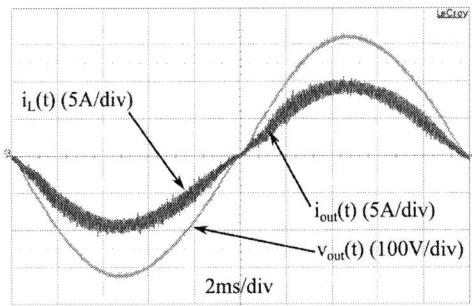

Fig. 9. Experimental output waveforms

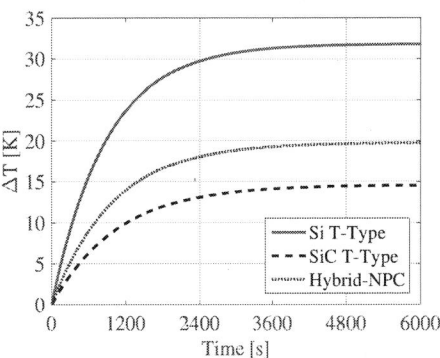

Fig. 11. Heat sink temperature rise of different inverter alternatives

has only 9 W losses and the Hybrid-NPC converter shows semiconductor losses of about 13 W. This results in a loss reduction of around 60 % for the SiC based converter and 42 % for the Hybrid-NPC. Hence the Si based T-Type inverter has the highest heat sink temperature rise above ambient at that operating point, shown in Fig. 11, where the heat sink temperature rises for each alternative are presented. In particular, for the conventional T-Type inverter, the temperature rise of the heat sink above ambient is 31.8 °C compared to only 14.6 °C for the SiC alternative and 19.8 °C

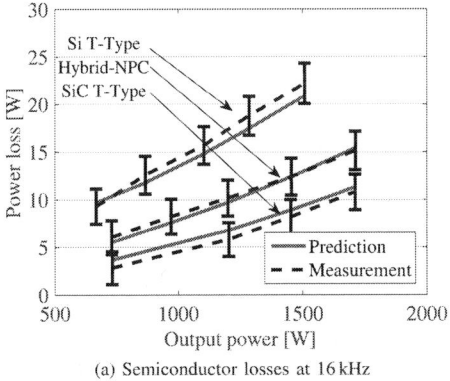

(a) Semiconductor losses at 16 kHz

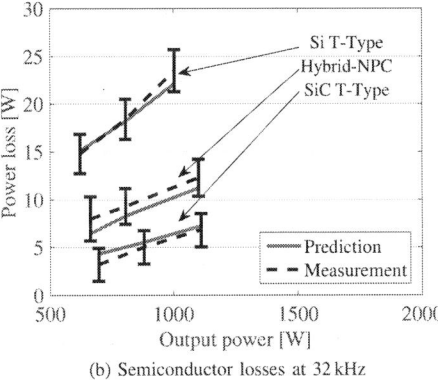

(b) Semiconductor losses at 32 kHz

Fig. 10. Semiconductor losses experimentally obtained via thermal measurements

for the Hybrid-NPC. Thus the loss reduction can not only be interpreted in terms of higher efficiency, but there is potential for further cost reduction by using a smaller heat sink.

VII. DISCUSSION

Two observations from Fig. 10 are worthy of further comment regarding the two converter alternatives. Firstly, while the Hybrid-NPC can substantially reduce its total semiconductor losses compared to the conventional T-Type inverter, its loss reduction is not as good as the SiC based T-Type structure. This can be explained by recognizing that although the switching losses are greatly reduced for the Hybrid-NPC converter, its total semiconductor conduction losses are larger compared to the SiC based T-Type inverter because of the very low on-state resistance of the SiC MOSFETs as shown in Fig. 4. Furthermore, from Fig. 2c, the conduction losses in the inner bi-directional switches S_2/S_3 are increased because they conduct current during both the zero converter output period and positive/negative converter output period.

The second observation relates to switching losses. As the switching frequency is increased, the power loss increase is larger for the Hybrid-NPC alternative compared to the SiC based T-Type converter. This can be explained from Fig. 3b, which identifies larger turn on energies for the CoolMos FET relative to the SiC MOSFET. Therefore, at any particular switching frequency, switching losses in the Hybrid-NPC will be higher than the SiC MOSFET based T-Type structure.

VIII. CONCLUSION

This paper has compared two promising three-level inverter topologies that aim to reduce switching losses compared to a conventional T-Type inverter structure. The first alternative is to simply replace the lossy 1200 V Si IGBTs with low loss 1200 V SiC MOSFETs. The second alternative strategically adds 600 V CoolMos FET devices to better support the switching transitions. A loss breakdown analysis using a loss model obtained from datasheet information and in-circuit measurement of switching events quantifies the loss reduction for both alternatives. In order to verify these semiconductor loss models, a simple thermal measurement technique was used based on calibrated heat sinks. The experimentally confirmed results show that a total semiconductor loss reduction of up to 60 % can be achieved using SiC MOSFETs and 42 % for the Hybrid-NPC inverter. Furthermore, this loss reduction for both alternatives has the additional benefit of operating at a significantly lower temperature, which offers further potential for reduced heat sink costs and/or increased inverter life expectancy.

REFERENCES

[1] T. Kerekes, R. Teodorescu, P. Rodriguez, G. Vazquez, and E. Aldabas, "A new high-efficiency single-phase transformerless PV inverter topology," *IEEE Trans. Ind. Electron.*, vol. 58, no. 1, pp. 184–191, 2011.

[2] R. Teichmann and S. Bernet, "A comparison of three-level converters versus two-level converters for low-voltage drives, traction, and utility applications," *IEEE Trans. Ind. Appl.*, vol. 41, no. 3, pp. 855–865, May 2005.

[3] M. Schweizer, T. Friedli, and J. Kolar, "Comparative evaluation of advanced three-phase three-level inverter/converter topologies against two-level systems," *IEEE Trans. Ind. Electron.*, vol. 60, no. 12, pp. 5515–5527, Dec 2013.

[4] A. Nabae, I. Takahashi, and H. Akagi, "A new neutral-point-clamped pwm inverter," *IEEE Trans. Ind. Appl.*, vol. IA-17, no. 5, pp. 518–523, Sept 1981.

[5] M. Schweizer and J. Kolar, "Design and implementation of a highly efficient three-level t-type converter for low-voltage applications," *IEEE Trans. Power Electron.*, vol. 28, no. 2, pp. 899–907, Feb 2013.

[6] A. Lemmon, M. Mazzola, J. Gafford, and K. Speer, "Comparative analysis of commercially available silicon carbide transistors," in *Applied Power Electronics Conference and Exposition (APEC), 2012 Twenty-Seventh Annual IEEE*, 2012, pp. 2509–2515.

[7] A. Rodriguez, M. Fernandez, A. Vazquez, D. Lamar, M. Arias, and J. Sebastian, "Optimizing the efficiency of a dc-dc boost converter over 98% by using commercial SiC transistors with switching frequencies from 100 kHz to 1MHz," in *Applied Power Electronics Conference and Exposition (APEC), 2013 Twenty-Eighth Annual IEEE*, 2013, pp. 641–648.

[8] F. Xu, B. Guo, L. Tolbert, F. Wang, and B. Blalock, "An all-sic three-phase buck rectifier for high-efficiency data center power supplies," *IEEE Trans. Ind. Appl.*, vol. 49, no. 6, pp. 2662–2673, Nov 2013.

[9] C.-M. Ho, H. Breuninger, S. Pettersson, G. Escobar, and F. Canales, "A comparative performance study of an interleaved boost converter using commercial si and SiC diodes for PV applications," *IEEE Trans. Power Electron.*, vol. 28, no. 1, pp. 289–299, 2013.

[10] T. Friedli, S. Round, D. Hassler, and J. Kolar, "Design and performance of a 200-khz all-sic jfet current dc-link back-to-back converter," *IEEE Trans. Ind. Appl.*, vol. 45, no. 5, pp. 1868–1878, Sept 2009.

[11] H. Zhang, L. Tolbert, and B. Ozpineci, "Impact of SiC devices on hybrid electric and plug-in hybrid electric vehicles," *IEEE Trans. Ind. Appl.*, vol. 47, no. 2, pp. 912–921, 2011.

[12] X. Huang, G. Wang, Y. Li, A. Q. Huang, and B. Baliga, "Short-circuit capability of 1200v sic mosfet and jfet for fault protection," in *Applied Power Electronics Conference and Exposition (APEC), 2013 Twenty-Eighth Annual IEEE*, March 2013, pp. 197–200.

[13] T. B. Soeiro and J. W. Kolar, "The new high-efficiency hybrid neutral-point-clamped converter," *IEEE Trans. Ind. Electron.*, vol. 60, no. 5, pp. 1919–1935, May 2013.

[14] W. Wu, F. Wang, and Y. Wang, "A novel efficient t type three level neutral-point-clamped inverter for renewable energy system," in *Power Electronics Conference (IPEC-Hiroshima 2014 - ECCE-ASIA), 2014 International*, May 2014, pp. 470–474.

[15] A. Anthon, Z. Zhang, M. Andersen, and T. Franke, "Efficiency evaluation on a coolmos switching and igbt conducting multilevel inverter," in *Applied Power Electronics Conference and Exposition (APEC), 2015 IEEE*, March 2015, pp. 2251–2255.

[16] H. Uemura, F. Krismer, and J. Kolar, "Comparative evaluation of t-type topologies comprising standard and reverse-blocking IGBTs," in *Energy Conversion Congress and Exposition (ECCE), 2013 IEEE*, 2013, pp. 1288–1295.

[17] *IEEE Application Guide for IEEE Std 1547(TM), IEEE Standard for Interconnecting Distributed Resources with Electric Power Systems*, IEEE Std. 1547.2-2008, 2009.

[18] G. Carrara, S. Gardella, M. Marchesoni, R. Salutari, and G. Sciutto, "A new multilevel pwm method: a theoretical analysis," *IEEE Trans. Power Electron.*, vol. 7, no. 3, pp. 497–505, Jul 1992.

[19] B. McGrath and D. Holmes, "A comparison of multicarrier pwm strategies for cascaded and neutral point clamped multilevel inverters," in *Power Electronics Specialists Conference, 2000. PESC 00. 2000 IEEE 31st Annual*, vol. 2, 2000, pp. 674–679 vol.2.

Dual Buck Inverter with Series Connected Diodes and Single Inductor

Liwei Zhou, Feng Gao
School of Electrical Engineering
Shandong University
Jinan, China
18769785783@163.com

Abstract—**In a DC-AC system, some problems may threaten the reliability of the whole system, such as the shoot through issue and the failure of reverse recovery. Some methods are proposed to improve the reliability of the converters. The dual buck inverters can solve the above problems without adding dead time but the dual buck topology has a main drawback of low magnetic utilization which increases the volume and weight of the system. This paper firstly summarizes the traditional dual buck topologies including a kind of single inductor dual buck inverter which can make full use of the inductance. Then a method to improve the reliability of the MOSFET inverter is proposed. A kind of novel dual buck inverter with series connected diodes and single inductor is introduced. The novel inverter retains the dual buck topologies' advantage of high reliability and can make full use of the inductance. Also, compared to the traditional single inductor dual buck topology, the controlling strategy of the proposed inverter is simpler. Finally, the simulation and experimental results verified the theoretical analysis.**

Keywords—dual buck inverter; shoot through problem; photovoltaic inverter; leakage current

I. INTRODUCTION

The fast development of the clean energy power generation requires the inversion system, especially the inverters, to be more and more reliable. Yet shoot through problem of the power devices is a major threaten to the reliability. As is known, a traditional method to solve the shoot through issue is by setting dead time. However, the dead time will cause a distortion of the output current. Also, during the dead time, the current may flow through the body diode of the switch which can cause the failure of the reverse recovery [1].

For the purpose of solving the above problems, the dual buck topologies are proposed in a lot of research. By combining two unidirectional buck circuits, the dual buck inverters will not suffer the threaten of shoot through problem and the freewheeling current will flow through the independent diodes which can solve the reverse recovery problem of the MOSFET's body diodes. However, the major drawback of the dual buck topologies is the magnetic utilization. Only half of the inductance is used in every working mode. And it will obviously increase the weight and volume of the system [2]-[4].

In order to improve the magnetic utilization of the dual buck inverter, a kind of single inductor dual buck topology was proposed in [5]. Compared with the traditional full bridge inverter, two extra switches are applied in the proposed topology. The single inductor topology can make full use of the

inductance, but the conducting loss is largely increased because four switches are flown through during the power delivering modes.

This paper proposed a kind of novel phase leg topology with series connected diodes and single inductor to highly improve the reliability of the inverter, especially for the MOSFET inverter [6]. Applying the phase leg to the single phase inverter, an improved single inductor dual buck inverters are proposed in this paper. The novel topology has the following advantages: firstly, retains the advantages of the traditional dual buck inverters, secondly, makes full use of the inductance, thirdly, the proposed inverter saves two switches compared to the traditional single inductor topology, which makes a lower conducting loss and a simpler controlling strategy. The simulation and experimental results have verified

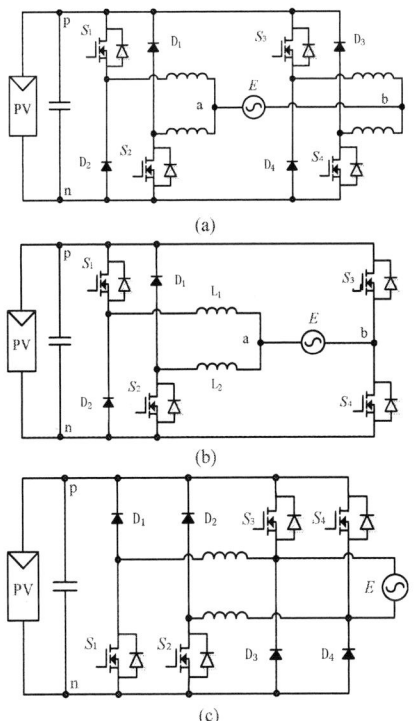

Fig. 1. Traditional Dual buck and dual boost full bridge inverters.

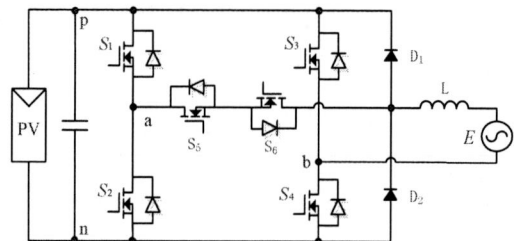

Fig. 2. Traditional Dual buck full bridge inverter with single inductor.

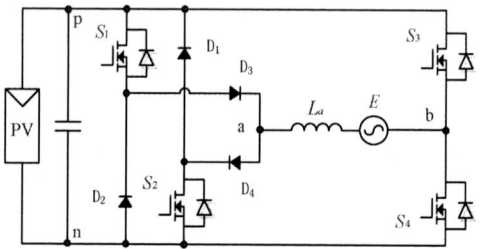

Fig. 4. Proposed dual buck full bridge inverters with single inductor.

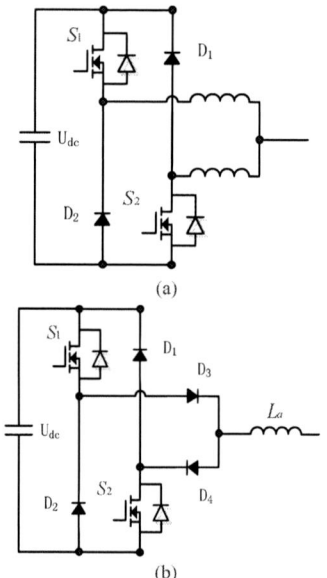

(a)

(b)

Fig. 3. (a)Traditional dual buck phase leg (b) proposed dual buck phase legs with series connected diodes and single inductor.

the analysis.

II. TRADITIONAL DUAL BUCK TOPOLOGIES

Fig. 1 shows the traditional dual buck and dual boost inverters [7]-[8]. The most attractive advantage of the dual buck topologies is the high reliability. Firstly, without adding the extra dead time, the dual buck topologies can solve the shoot through problem. Secondly, compared to the traditional H-bridge inverter, the current will not flow through the body diodes of the switches in the dual buck topologies which means no reverse recovery problem exists in the MOSFET phase legs. Considering the above two aspects, the dual buck topologies can achieve high reliability without the shoot through and reverse recovery issues.

However, the main drawback of the dual buck topologies is the low magnetic utilization. In each power delivering and freewheeling modes, the current only flow through half of the inductance, which means the other half of the inductance is wasted in each working condition. The low utilization of the inductance makes the increasing of the weight and volume for the whole system. To solve this problem, a concept of single inductor dual buck full bridge inverter [5] is proposed. Fig. 2

shows the single inductor topology. The novel topology includes six switches and two diodes. Comparing to the traditional dual buck full bridge inverter, the single inductor topology can save half of the inductance. And the novel topology retains the original advantages of high reliability. Also, there is no need to add the dead time in the high frequency unipolar switching strategy. The inductance can be fully utilized in the single inductor inverter. However, a high level of conduction loss is the main drawback of the novel topology. During the power delivering mode, the current flows through four switches which is a lot more than the traditional full bridge inverters. Besides, compared to the traditional H-bridge inverters, the extra two switches make controlling strategy more complex. And in the dual buck single inductor inverter, the current will flow through the body diodes of the series MOSFET switches which can cause the problem of reverse recovery.

To solve the problem of traditional H-bridge inverter, including the shoot through issue and the reverse recovery of the MOSFET, a kind of dual buck inverter with series connected diodes and single inductor is proposed in this paper. The newly proposed topology retains the advantage of traditional dual buck inverter and solve the problem of low magnetic utilization. Also, the proposed topologies will not invite extra switches which means a simpler controlling strategy compared to the traditional dual buck single inductor full bridge inverter in [5].

III. HIGHLY RELIABLE MOSFET INVERTER WITH SINGLE INDUCTOR

This section proposes a kind of novel MOSFET phase leg which maintains the high reliability of the dual buck topology and also makes full use of the dual buck's inductance. Fig. 3 shows the traditional dual buck phase leg and the proposed novel MOSFET phase leg. The two inductors in Fig. 3(a) are replaced by two diodes and one inductor just as shown in Fig. 3(b). Applying the proposed phase leg to the full bridge inverter, a novel dual buck MOSFET inverter with series connected diodes and single inductor is proposed then. The novel dual buck inverter is shown in Fig. 4. Compared to the traditional single inductor dual buck inverter in Fig. 2, the proposed topologies save two switches which means a simpler control strategy. Meanwhile, in the power delivering mode, the current of the novel topology only flows through one switch and two diodes which is less than the traditional one in Fig. 2 [5].

978-1-4673-9551-9/16 $31.00 © 2016 IEEE

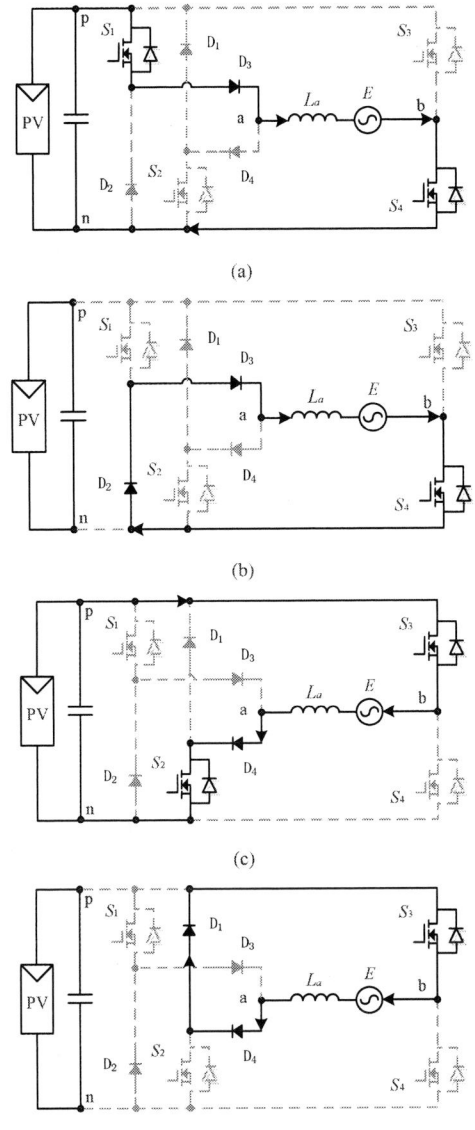

(a)

(b)

(c)

(d)

Fig. 5. Four working modes of the proposed dual buck full bridge inverter with single inductor in Fig. 4.

So, the proposed single inductor dual buck topologies have the advantages in the aspect of efficiency, control complexity and system cost and size. The operational principle of proposed single inductor dual buck inverter can be illustrated with four operation modes. Fig. 5 shows the specific current flow paths during the energy transferring modes and the freewheeling modes. A unipolar SPWM strategy is applied to control the four switches of the novel inverter.

A. *Operational Principle of the Proposed Inverter*

Mode 1: During positive half period, S_1 is modulated in high frequency, while S_4 is always ON. When S_1 and S_4 are on, the current flows through S_1, D_3, grid and S_4 successively.

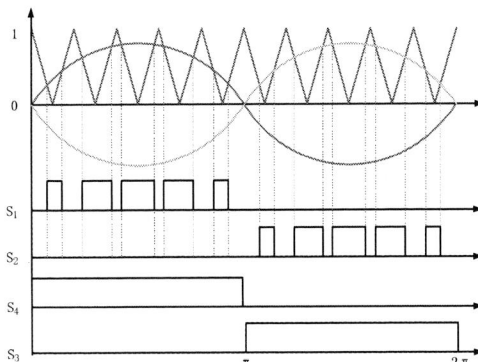

Fig. 6. The switching signals of the proposed inverters.

Mode 2: When S_1 is off, the current flows through D_2, D_3, grid and S_4 successively. As shown in Fig. 5(b), in this freewheeling mode, the diode D_4 prevents the current from flowing through the body diode of S_2, which avoid the failure of the MOSFET's reverse recovery.

Mode 3: During negative half period, S_2 is modulated in high frequency, while S_3 is always ON. When S_2 and S_3 are on, the current flows through S_3, grid, D_4 and S_2 successively.

Mode 4: When S_2 is off, the current flows through S_3, grid, D_4 and D_1 successively. As shown in Fig. 5(d), in this freewheeling mode, the diode D_3 prevents the current from flowing through the body diode of S_1, which can also avoid the failure of the MOSFET's reverse recovery.

The switching signals of the proposed inverter are shown in Fig. 6. Without the extra two switches of the traditional dual buck single inductor inverter [5] in Fig. 2, the proposed dual buck topology with series connected diodes can achieve the high reliability. No dead time is needed in the high frequency of the switches. Thus, the distortion rate of the output current can be decreased.

B. *Analysis of the Common-mode Characteristic*

The transformerless photovoltaic (PV) grid-connected system is an important application for the single phase inverter. However, in a transformerless PV system, the fluctuation of the common mode voltage will excite leakage current in the common mode path which may cause the safety problems and

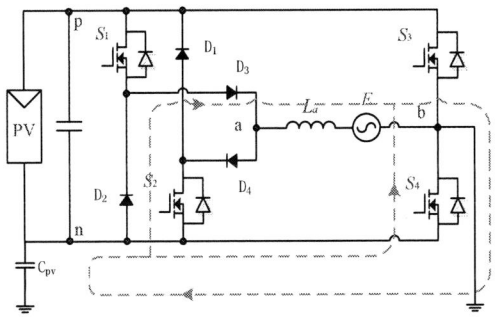

Fig. 7. The equivalent common-mode circuit of the proposed single inductor dual buck inverter.

978-1-4673-9551-9/16 $31.00 © 2016 IEEE   2261

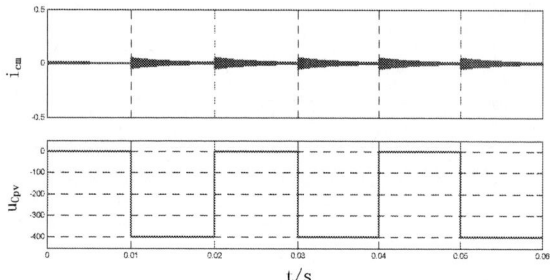

Fig. 8. The simulated common mode waveforms of the proposed dual buck inverters.

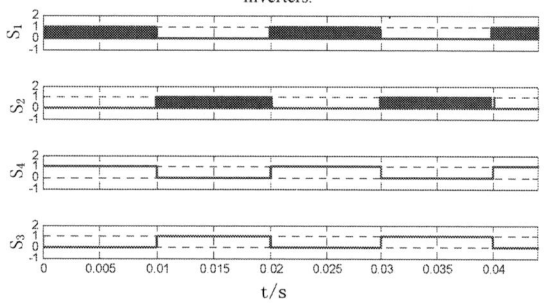

Fig. 9. The simulated switching signals of the proposed dual buck inverters.

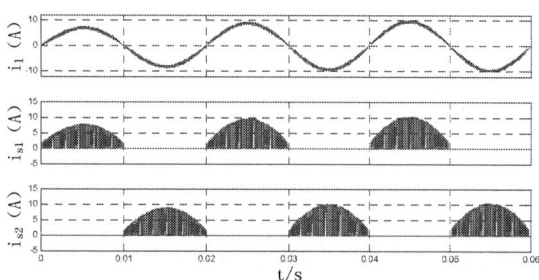

Fig. 10. The filtering current and the switching current of the proposed inverter in Fig. 4.

distort the output current. The equivalent common mode circuit of the proposed inverter is shown in Fig. 7. The red lines represent the flowing path of the leakage current. The value of the leakage current depends on the fluctuating frequency of the common-mode voltage, $u_{C_{pv}}$ where the C_{pv} represents the equivalent stray capacitance of the PV panel.

As is shown in Fig. 7, in the positive grid period, the upper potential of C_{pv} is equal to the ground. So, the $u_{C_{pv}}$ is zero in this situation. On the other hand, in the negative grid period, the potential of point p is equal to the ground. Thus, the upper potential of C_{pv} is lower than the ground by u_{dc}. In this situation, the $u_{C_{pv}}$ is $-u_{dc}$. So, whether in the positive grid period or in the negative grid period, the voltage of the PV stray capacitance, $u_{C_{pv}}$, is kept constant. The common-mode current i_{cm} is mainly induced by the fluctuation of u_{cm} as expressed in (1)

$$i_{cm} = C_{pv}\frac{du_{cm}}{dt} \qquad (1)$$

So, the common mode leakage current can be limited to a low level in the proposed topology. The relevant common-

mode waveforms are shown in Fig. 8 where the leakage current, i_{cm}, is lower than 0.05A. Thus, the common-mode characteristic of the proposed topology satisfies the standards, e.g. DIN VDE 0126-1-1.

IV. SIMULATION AND EXPERIMENTAL RESULTS

The simulation and experimental results are shown in this section. The proposed inverters in Fig. 4 were simulated in Matlab/Simulink. The DC voltage is 400V, and the grid voltage is 220V/50Hz. The switching frequency is 10kHz. The output inductor is 2mH. The grid current is controlled by a conventional PR controller. Fig. 9 shows the simulated switching signals of the proposed inverter. Fig. 10 shows the filtering current and switching current of the proposed inverter. The current waveforms of the switches are all unidirectional which indicate that no freewheeling current is flowing through the body diodes of the MOSFET. So the proposed inverter will not be threatened by the reverse recovery issue, thus the reliability of the inversion system is largely improved.

In order to further validate the proposed topology and the modulation strategies, the 1kW experimental prototype of the proposed inverter in Fig. 4 was built with the same parameters as the simulation model. The experimental results are shown in Fig. 11 and Fig. 12. Fig. 11 represents the experimental switching signals of the proposed inverter. Fig. 12 shows the grid current, the common mode leakage current and the output voltage of the proposed inverter from the top to the bottom of the picture respectively.

V. CONCLUSION

This paper reviews the already published dual buck topologies. The advantages and disadvantages of the dual buck inverters are specifically analyzed. In order to solve the main drawback of low magnetic utilization, a kind of phase leg topology is proposed. By applying the novel phase leg to the full bridge inverter, the new topology maintain the high reliability of the traditional dual buck inverter and the magnetic utilization is largely improved. Also, compared to the traditional single inductor dual buck inverter, the novel topology has the advantages in conducting loss and controlling complexity. The simulation and experimental results verified the performance of proposed inverter.

REFERENCES

[1] T. Kerekes, R. Teodorescu, P. Rodriguez, G. Vazquez, E. Aldabas, "A new high-efficiency single-phase transformerless PV inverter topology," IEEE Trans. Ind. Electron., vol. 58, no. 1, pp. 184-191, Jan. 2011.

[2] Zhu, Chenghua, Fanghua Zhang, and Yangguang Yan, "A novel split phase dual buck half bridge inverter", in Proc. 20th IEEE Applied Power Electronics Conference and Exposition, 2005, vol.2, pp.845-849.

[3] Hong Feng, Ying Pei-pei, Wang Cheng-hua, "Decoupling Control of Input Voltage Balance for Diode-Clamped Dual Buck Three-Level Inverter", in Proc. 28th Annual IEEE Applied Power Electronics Conference and Exposition, Long Beach, California, USA, March 17-21, 2013, pp.482-488.

[4] Liu Miao, Hong Feng, Wang Cheng-hua. A Novel Flying-Capacitor Dual Buck Three-Level Inverter [C], in Proc. 28th Annual IEEE Applied Power Electronics Conference and Exposition, Long Beach, California, USA, March 17-21, 2013, pp.502-506.

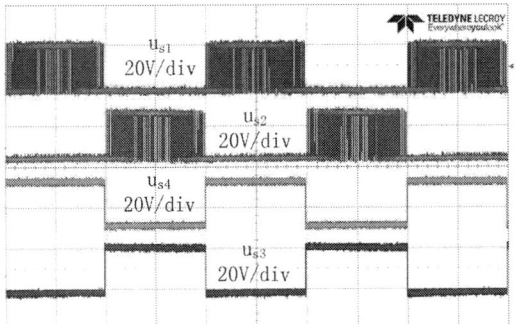

Fig. 11. The experimental driving voltage of the proposed inverter.

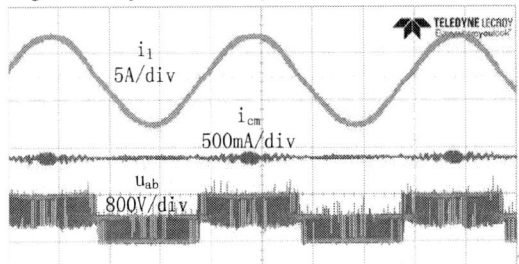

Fig. 12. The experimental output waveforms of the proposed inverter.

[5] Hong Feng, Liu Jun, Ji Baojian, Zhou Yufei, and Wang Jianhua, "Single Inductor Dual Buck Full-Bridge Inverter", IEEE Trans. Ind. Electron., vol. 62, no. 8, pp. 4869–4877, Aug 2015.

[6] B. F. Chen, B. Gu, L. H. Zhang, Z. U. Zahid, Z. L. Liao, J.-S. Lai, Z. L. Liao and R. X. Hao, "A High Efficiency MOSFET Transformerless Inverter for No-isolated Micro-inverter Application," *IEEE Trans. Power Electron.*, vol. 30, no. 7, pp. 3610–3622, July. 2015.

[7] B. F. Chen, P. W. Sun, C. Liu, C-L. Chen, J.-S. Lai, and W. Yu, "High efficiency transformerless photovoltaic inverter with wide-range power factor capability", in Proc. of IEEE 27th Applied Power Electronics Conference and Exposition, Orlando, FL, Feb. 2012.

[8] B. Gu, J. Dominic, J.-S. Lai, C-L. Chen, T, LaBella, and B. F. Chen, "High Reliability and Efficiency Single-Phase Transformerless Inverter for Grid-Connected Photovoltaic Systems," *IEEE Trans. Power Electron*, vol.28, no. 5, pp.2235-2245. May, 2013.

[9] T. Kerekes, R. Teodorescu, P. Rodriguez, G. Vazquez, and E. Aldabas, "A new high-efficiency single-phase transformerless PV inverter topology," IEEE Trans. Ind. Electron., vol. 58, no. 1, pp. 184–191, Jan. 2011.

Magnetic Integration of the Harmonic Filter Inductor for Dual-Converter Fed Open-End Transformer Topology

Ghanshyamsinh Gohil, Lorand Bede, Remus Teodorescu, Tamas Kerekes, Frede Blaabjerg

Department of Energy Technology, Aalborg University, Denmark

gvg@et.aau.dk

Abstract—Many high power converter systems are often connected to the medium voltage network using a step-up transformer. In such systems, the converter-side windings of the transformer can be configured as an open-end and multi-level voltage waveforms can be achieved by feeding these open-end windings from both ends using the dual-converter. An LCL filter with separate converter-side inductors for each of the converter is commonly used to attenuate the undesirable harmonic frequency components in the grid current. The magnetic integration of the converter-side inductors is presented in this paper, where the flux in the common part of the magnetic core is completely canceled out. As a result, the size of the magnetic component can be significantly reduced. A multi-objective design optimization is presented, where the energy loss and the volume are optimized. The optimization process takes into account the yearly load profile and the energy loss is minimized, rather than minimizing the losses at a specific operating point. The size reduction achieved by the proposed inductor is demonstrated through a comparative evaluation. Finally, the analysis is supported through simulations and experimental results.

I. INTRODUCTION

Many high power converter systems are often connected to the medium voltage network and a step-up transformer is used to match the voltage levels of the converter with the medium voltage grid. In some applications, the transformer is also required for providing galvanic isolation. In such systems, the converter-side windings of the transformer can be configured as an open-end. This open-end transformer winding can be fed from both the ends using the two-level Voltage Source Converters (VSCs) [1], as shown in Fig. 1. The number of levels in the output voltages is the same as that of the three level Neutral point clamped (NPC) converter and each of the two-level VSC operates with the half of the dc-link voltage than the dc-link voltage required for the three level NPC. However, Common Mode (CM) circulating current flows through the closed path if both the VSCs are connected to a common dc-link. The CM circulating current can be suppressed either by using the CM choke [2], [3] or by employing a proper Pulse Width Modulation (PWM) scheme to ensure complete elimination of the CM voltage [4], [5]. However, in many applications, isolated dc-links can be derived from the source itself and such extra measures for CM circulating current suppression may not be required. For example, the isolated dc-links can be obtained in

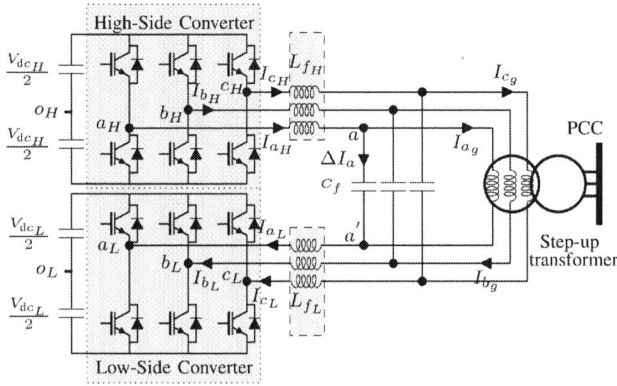

Fig. 1. The system configuration of the dual converter fed open-end winding transformer topology with two separate dc-links. The open-end primary windings are fed from two-level voltage source converters.

1) PhotoVoltaic (PV) systems by dividing the total number of arrays into two groups to form separate dc-links [1].
2) Wind Energy Conversion System (WECS): isolated dc-links can be obtained by using a dual stator-winding generator [6].

Therefore the analysis presented in this paper is mainly focused on the dual converter fed open-end transformer topology with two separate dc-links.

An LCL filter is commonly used in high power grid-connected applications [7] and one of the possible arrangement of the LCL filter for the dual-converter fed open-end transformer topology is shown in Fig. 1. The leakage inductance of the transformer is used as a grid-side inductor of the LCL filter. Two VSCs (denoted as High-Side Converter (HSC) and Low-Side Converter (LSC) in Fig. 1) are connected to a common shunt capacitive branch of the LCL filter through the converter-side inductors L_{f_H} and L_{f_L}, respectively. The magnetic integration of the L_{f_H} and L_{f_L} is presented in this paper. As a result of this magnetic integration, the flux in the common part of the magnetic core is completely canceled out. This leads to substantial reduction in the size of the converter-side inductor.

This paper is organized as follows: The operation principle of the dual-converter fed open-end transformer topology is

978-1-4673-9551-9/16 $31.00 © 2016 IEEE

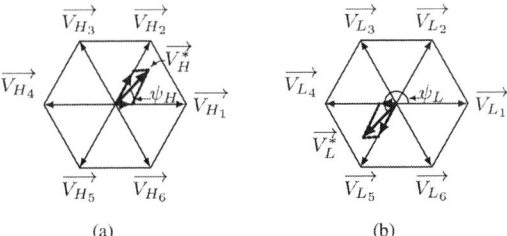

(a)   (b)

Fig. 2. Reference voltage space vector and its formation by the geometrical summation. (a) Reference voltage space vector for the HSC, (b) Reference voltage space vector for the LSC.

briefly discussed in Section II. The magnetic structure of the proposed integrated inductor is described in Section III. Section IV discusses the design procedure in general and the multi-objective optimization of the integrated inductor. The size reduction achieved by the magnetic integration is also demonstrated by comparing the volume of the integrated inductor with the separate inductor case for the 6.6 MVA, 3.3 kV WECS and it is presented in Section V. The simulation and the experimental results are finally presented in Section VI.

II. DUAL-CONVERTER FED OPEN-END TRANSFORMER TOPOLOGY

The operation of the dual-converter fed open-end transformer converter is briefly described in this section. The dual-converter system consists of the HSC and the LSC is shown in Fig. 1. A Two-level VSC is used for both the HSC and the LSC. However, the discussion presented in this paper is also applicable to other converters as well.

The reference voltage space vector $\overrightarrow{V^*}$ is synthesized by modulating the HSC and the LSC. The magnitude of the reference voltage space vectors of the HSC and the LSC is half than that of the desired reference voltage space vector $\overrightarrow{V^*}$ ($|\overrightarrow{V_H^*}| = |\overrightarrow{V_L^*}| = |\overrightarrow{V^*}|/2$). The reference voltage space vector angle of the HSC is kept the same as that of the desired voltage space vector ($\psi_H = \psi$), whereas the reference voltage space vector angle of the LSC is shifted by an angle $180°$ ($\psi_L = \psi_H + 180°$), as shown in Fig. 2.

From Fig. 1, the voltage across the shunt capacitive branch C_f of the LCL filter is given as

$$V_{xx'} = (V_{x_H o_H} - V_{x_L o_L}) - L_{f_H} \frac{dI_{x_H}}{dt} - L_{f_L} \frac{dI_{x_L}}{dt} + V_{o_H o_L} \quad (1)$$

where the subscript x represents the phases $x = \{a, b, c\}$. As the dc-links are separated, the common-mode components of the voltages in (1) do not drive any common-mode circulating current. As a result, only the differential mode current would flow through the inductors. For the dual converter system, these currents are equal.

$$I_{x_H} = I_{x_L} = I_x \quad (2)$$

where I_{x_H} and I_{x_L} are the currents in the phase x of the HSC and LSC, respectively. Assuming $L_{f_H} = L_{f_L} = L_f/2$

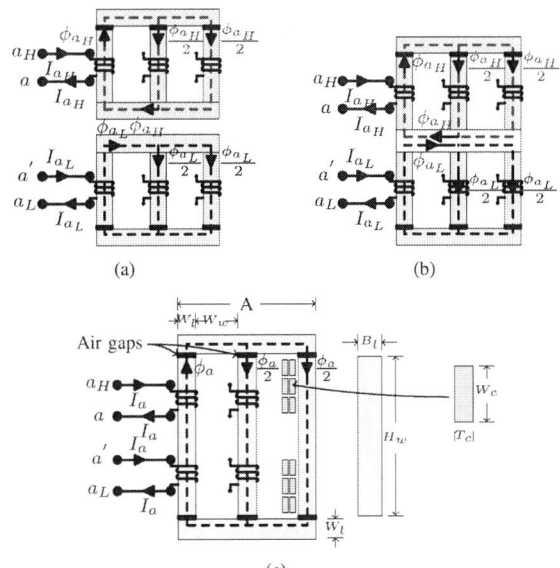

(a)   (b)

(c)

Fig. 3. Magnetic structure. (a) Separate inductors for the HSC and the LSC, (b) Flux cancellation through magnetic integration, (c) Proposed integrated inductor.

and using (1) and (2), the voltage across the converter-side inductor is given as

$$L_f \frac{dI_x}{dt} = (V_{x_H o_H} - V_{x_L o_L}) - V_{xx'} + V_{o_H o_L} \quad (3)$$

where L_f is the equivalent converter-side inductance of the LCL filter. A single magnetic component with the inductance L_f is realized by the magnetic integration of the L_{f_H} and L_{f_L} and the structure is discussed in the following section.

III. INTEGRATED INDUCTOR

In this paper, a three-phase inductor is chosen for the illustration due its wide-spread use in the high power applications. However, it is important to point out that the same analysis can be used for the single-phase inductor as well. The three-phase three-limb converter-side inductor for both the HSC and the LSC are shown in Fig. 3(a). These two inductors can be magnetically integrated as shown in Fig. 3(b), where both the inductors share a common magnetic path. The magnetic structure has six limbs, on which the coils are wound. The upper three limbs belong to the L_{f_H}, whereas the lower three limbs receive the coils corresponding to the L_{f_L}. The upper limbs are magnetically coupled using the top bridge yoke, whereas the lower three limbs are magnetically coupled using the bottom bridge yoke. The upper and the lower limbs share a common yoke, as shown in Fig. 3(b).

Considering three-phase three-wire system

$$I_a + I_b + I_c = 0 \quad (4)$$

and at a particular instance

$$I_a = -(I_b + I_c) \quad (5)$$

978-1-4673-9551-9/16 $31.00 © 2016 IEEE   2265

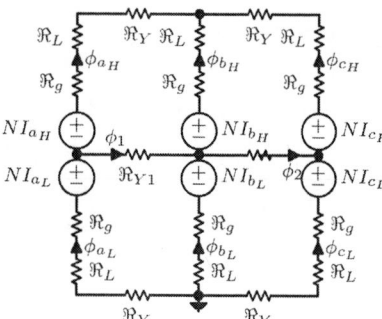

Fig. 4. Simplified reluctance model of the magnetic structure shown in Fig. 3(b).

The flux distribution in the magnetic structure for this case (the positive value of the I_a and the negative values of the I_b and I_c) is shown in Fig. 3(b). The simplified reluctance model of this magnetic structure is shown in Fig. 4, where $\Re_L$, $\Re_g$, and $\Re_Y$ are the reluctances of the limb, the air gap, the top and the bottom bridge yokes, respectively. The reluctance of the common yoke is represented as $\Re_{Y1}$. ϕ_{a_H}, ϕ_{b_H}, and ϕ_{c_H} are the fluxes in the upper three limbs whereas ϕ_{a_L}, ϕ_{b_L}, and ϕ_{c_L} are the fluxes in the lower three limbs. ϕ_1, and ϕ_2 represent the fluxes in the common yokes, as shown in Fig. 4. By solving the reluctance network, the fluxes in various parts of the integrated inductor are obtained and they are given as

$$\phi_{a_H} = \frac{NI_{a_H}}{\Re_g + \Re_L + \Re_Y}, \quad \phi_{a_L} = \frac{NI_{a_L}}{\Re_g + \Re_L + \Re_Y} \quad (6)$$

$$\phi_{c_H} = \frac{NI_{c_H}}{\Re_g + \Re_L + \Re_Y}, \quad \phi_{c_L} = \frac{NI_{c_L}}{\Re_g + \Re_L + \Re_Y} \quad (7)$$

and

$$\phi_{b_H} = \frac{NI_{b_H}}{\Re_g + \Re_L}, \quad \phi_{b_L} = \frac{NI_{b_L}}{\Re_g + \Re_L}, \quad \phi_1 = 0, \quad \phi_2 = 0 \quad (8)$$

The flux components in the common yoke (ϕ_1 and ϕ_2) are zero and therefore the common yoke can be completely removed, as shown in Fig. 3(c). The integrated inductor has only two yokes, compared to four in the case of the separate inductors. As a result, substantial reduction in the volume of the inductor can be achieved through magnetic integration of L_{f_H} and L_{f_L}.

By analyzing the magnetic structure shown in Fig. 3(c) and neglecting the asymmetry introduced by the three-limb structure (by neglecting $\Re_Y$), the value of the L_f is obtained as

$$L_f = \frac{4N^2}{2\Re_g + 2\Re_L} \quad (9)$$

The reluctance of the air gap is generally large compared to the reluctance of the magnetic material. Therefore, simplified expression for the converter side filter inductor is given as

$$L_f \approx \frac{2N^2}{\Re_g} \approx \frac{2\mu_0 N^2 A_g'}{l_g} \quad (10)$$

where μ_0 is the permeability of the free space, l_g is the length of the air gap (of upper/bottom limb) and A_g' is the effective

TABLE I
SYSTEM SPECIFICATIONS AND PARAMETERS FOR SIMULATION AND HARDWARE STUDY

Parameters	Simulations	Experiment
Power S	6.6 MVA (6 MW)	11 kVA (10 kW)
Switching frequency f_{sw}	900 Hz	900 Hz
AC voltage (line-to-line) V_{ll}	3300 V	400 V
Rated current	1154 A	15.8 A
DC-link voltage ($V_{dc_H} = V_{dc_L}$)	2800 V	330 V
Modulation index range	$0.95 \leq M \leq 1.15$	$0.95 \leq M \leq 1.15$
Transformer leakage L_g	525 μH (0.1 pu)	4.2 mH (0.1 pu)

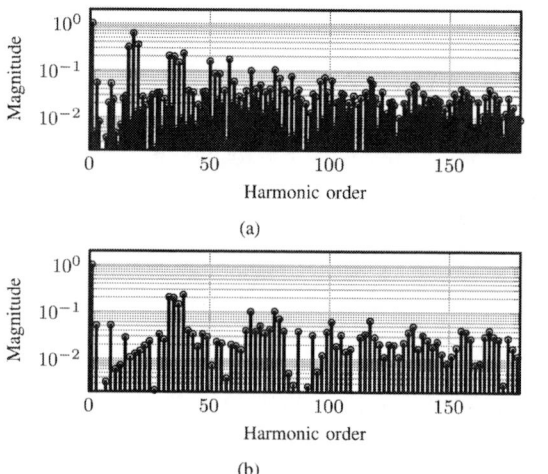

Fig. 5. Simulated harmonic spectrum of the switched voltages with the modulation index $M = 1$ and the switching frequency of 900 Hz. (a) Switched output voltage of the high-side converter $V_{a_H o_H}$ (normalized with respect to the $V_{dc}/2$), (b) Resultant voltage ($V_{a_H o_H} - V_{a_L o_L}$) (normalized with respect to the V_{dc}).

cross-sectional area of the air gap after considering the effects of the fringing flux. The effective cross-sectional area of the air gap $A_{g'}$ is obtained by evaluating the cross-section area of the air gap after adding l_g to each dimension in the cross-section.

IV. DESIGN OF THE INTEGRATED INDUCTOR

A design methodology is demonstrated in this section by carrying out the design of the integrated inductor for the high power WECS. The system specifications of the WECS is given in Table I. The WECS operates with the power factor close to unity and in this case, the 60° Discontinuous Pulse-Width Modulation (commonly referred to as a DPWM1 [8]) scheme could result up to 50% switching loss reduction compared to the continuous modulation scheme. Therefore DPWM1 is used to modulate the HSC and the LSC. Using this specifications, the design of the integrated inductor is carried out and the design steps are illustrated hereafter.

A. Obtain the value of the converter-side inductor L_f

The harmonic spectra of the switched output voltage of the HSC is shown in Fig. 5(a). The major harmonic components in the switched output voltage of the individual converter appears at the carrier frequency, whereas these components are substantially reduced in the resultant voltage, as shown in Fig. 5(b). The spectrum comprises the maximum values of the individual voltage harmonic components of the resultant voltage, over the entire operating range is obtained and it is defined as a Virtual Voltage Harmonic Spectrum (VVHS) [9]. Using VVHS, the required admittance for the hth harmonic component is obtained as

$$Y_h^* = \frac{I_{h,BDEW}^*}{V_{h,VVHS}} \quad (11)$$

where $I_{h,BDEW}^*$ is the specified BDEW current injection limit of the hth harmonic component (refer to [10]) and $V_{h,VVHS}$ is the maximum values of the hth harmonic components over the entire operating range. The value of the filter parameters are then chosen such that the designed filter has a lower admittance than the required value of the filter admittance for all the harmonic frequency components of interest (upto 180th harmonic frequency component in case of the BDEW standard) [11]. Once the value of L_f is obtained, an optimized design can be carried out.

B. Core Loss Modeling

The Improved Generalized Steinmetz Equation (IGSE) [12], [13] is used to calculate the core losses. The core losses per unit volume is given as

$$P_{fe,v} = \frac{1}{T} \int_0^T k_i \left| \frac{dB(t)}{dt} \right|^\alpha (\Delta B)^{\beta - \alpha} dt \quad (12)$$

where α, β and k_i are the constants determined by the material characteristics. ΔB is the peak-to-peak value of the flux density and T is the switching interval. The flux waveform has major and minor loops and these loops are evaluated separately.

1) Major Loop: Assuming the inductance value to be constant, the flux density in the limb corresponding to the phase x is given as

$$B_x(t) = \frac{L_f I_{x,f}(t)}{2N A_c} \quad (13)$$

where A_c is the cross-sectional area of the limb, $I_{x,f}$ is the fundamental component of the current.

2) Minor Loop: The reference space vector $\vec{V}_H^*$ and $\vec{V}_L^*$ are synthesized using active and zero voltage vectors and the volt-second balance is maintained by choosing appropriate dwell time of these vectors. The application of the discrete vectors results in an error between the applied voltage vector and the reference voltage vector, as shown in Fig. 6 for the HSC. Similarly, the error voltage vectors for the LSC also exists due to the finite sampling. These error voltage vectors lead to the minor loop in the flux density waveform and it is evaluated by performing time integral of the error voltage vector.

The time integral of the error voltage vector is known as the harmonic flux vector [14], [15] and the difference of the harmonic flux vectors of the HSC and the LSC are

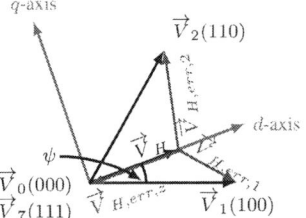

Fig. 6. The active and zero vectors to synthesize given reference vector and corresponding error voltage vectors.

directly proportional to the flux in the integrated inductor. In the reference frame, rotating synchronously at the fundamental frequency, the instantaneous error voltage vectors can be decomposed into d-axis and the q-axis components and they are for the HSC given as (see Fig. 6)

$$\vec{V}_{H,err,1} = \frac{2}{3} V_{dc} \{ (\cos \psi_H - \frac{3}{4} M) - j \sin \psi_H \}$$
$$\vec{V}_{H,err,2} = \frac{2}{3} V_{dc} \{ [\cos(60° - \psi_H) - \frac{3}{4} M] \quad (14)$$
$$+ j \sin(60° - \psi_H) \}$$
$$\vec{V}_{H,err,z} = -\frac{1}{2} V_{dc} M$$

Similarly, the d-axis and the q-axis components of instantaneous error voltage vectors of the LSC are also obtained. Then the difference of the d-axis components of the harmonic flux vectors of the HSC and LSC and the difference of the q-axis components of the harmonic flux vectors of the HSC and LSC are evaluated separately as

$$B_{ac,d}(t) = \frac{1}{2N A_c} \int (\vec{V}_{H,err,d} - \vec{V}_{L,err,d}) dt$$
$$B_{ac,q}(t) = \frac{1}{2N A_c} \int (\vec{V}_{H,err,q} - \vec{V}_{L,err,q}) dt \quad (15)$$

Using the d-axis and the q-axis components, the ripple component of the flux density in the limb corresponding to phase a is obtained as

$$B_a = B_{ac,d} \cos \psi - B_{ac,q} \sin \psi \quad (16)$$

The VSCs are assumed to be modulated using the asymmetrical regularly sampled PulseWidth Modulation (PWM), where the reference voltage space voltage vector is sampled twice in a carrier cycle.

Using this information, the core loss calculations have been carried out for the major loop and each of the minor loops.

C. Copper Loss Modeling

The copper loss is evaluated by considering the ac resistance of the winding, which takes into account the skin and proximity effects [16]. The total winding losses due to the harmonic frequency components of the coil current is [17]

$$P_{cu} = R_{dc} \sum_{h=1}^{\infty} k_{p_h} I_{x_h}^2 \quad (17)$$

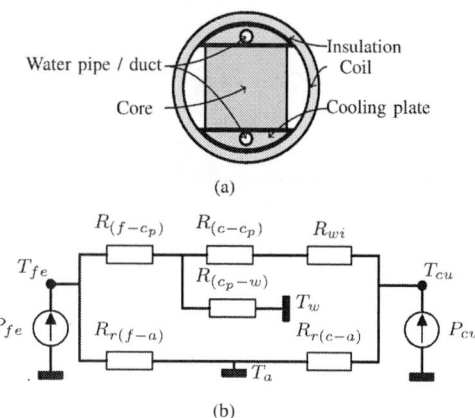

(a)

(b)

Fig. 7. Simplified thermal model of the integrated inductor. P_{fe} and P_{cu} are the core and copper losses, respectively.

where

$$
\begin{aligned}
k_{p_h} = \sqrt{h}\Delta \Big[& \frac{\sinh(2\sqrt{h}\Delta) + \sin(2\sqrt{h}\Delta)}{\cosh(2\sqrt{h}\Delta) - \cos(2\sqrt{h}\Delta)} \\
& + \frac{2}{3}(m^2 - 1)\frac{\sinh(\sqrt{h}\Delta) - \sin(\sqrt{h}\Delta)}{\cosh(\sqrt{h}\Delta) + \cos(\sqrt{h}\Delta)} \Big]
\end{aligned} \tag{18}
$$

and $\Delta = T_c/\delta$ and R_{dc} and R_{ac} are the dc and the ac resistance of the coil, respectively. m is the number of layers in the coil, T_c is the thickness of the conductor, and δ is the skin depth. I_{x_h} is the hth harmonic frequency component of the line current I_x. The harmonic spectrum of the resultant voltage is obtained analytically [18] and the hth harmonic frequency component of the line current I_x is obtained as

$$
I_{x_h} = Y_{H,LCL}V_h \tag{19}
$$

where V_h is the hth harmonic frequency component of the resultant voltage and $Y_{H,LCL}$ is the admittance offered by the LCL filter to the hth harmonic frequency component.

D. Thermal Modeling

The liquid cooling of the inductor is considered. The cooling arrangement is shown in Fig. 7(a). The semi-circular aluminum cooling plates with the duct to carry the coolant is considered. This cooling plate is electrically insulated using the epoxy resin. As the heat transfer is anisotropic for the laminated steel, two cooling plates along the edges that are perpendicular to the lamination direction are considered. The hot spot temperature in both the core and the coil (T_{fe} and T_{cu}, respectively) is estimated using the equivalent thermal resistance network [19], shown in Fig. 7(b). For the simplicity of the analysis, the temperature in the core and the coil is assumed to be homogeneous.

The heat transfer mechanism due to the convection and the radiation is considered, where $R_{cv(f-w)}$ and $R_{cv(c-w)}$ are the convection thermal resistance between the core and coolant (water) and between the coil and coolant, respectively. Similarly, $R_{r(f-a)}$ and $R_{r(c-a)}$ represent the radiation thermal

resistance between the core and the ambient and between the coil and the ambient, respectively. The radiation thermal resistance value is obtained using the formulas presented in [19].

The thermal resistance between the cooling plate and the coolant is given as

$$
R_{(c_p-w)} = \frac{1}{h_{c_p-w}A_{c_p-w}} \tag{20}
$$

where h_{c_p-w} is the heat transfer coefficient and A_{c_p-w} is the coolant contact surface. The heat transfer coefficient is

$$
h_{c_p-w} = 3130 \Big(\frac{q}{785.4D_d^2}\Big)^{0.87}(100D_d)^{-0.13} \tag{21}
$$

where q is the coolant flow rate in [l/s] and D_d is the diameter of the duct in [m]. The thermal resistance between the core and duct surface is given as

$$
R_{(f-c_p)} = \frac{2L_{eq}}{\lambda_{cp}(A_{c_pf} + \pi D_dL_d)} + \frac{T_i}{\lambda_iA_{c_pf}} \tag{22}
$$

where L_{eq} is the equivalent distance from the cooling surface to the duct, A_{c_pf} is the contact area of the cooling plate with the core, L_d is the length of the duct, and λ_{cp} is the thermal conductivity of the aluminum. T_i is the thickness of the insulation and λ_i is the thermal conductivity of the insulation. Using (20) and (22), the thermal resistance between the core and the coolant is obtained as

$$
R_{cv(f-w)} = R_{(f-c_p)} + R_{(c_p-w)} \tag{23}
$$

In a similar manner, the thermal resistance between the coil and the coolant $R_{cv(c-w)}$ can be also obtained. However, in the heat flow path of the copper losses, there is an additional layer of the insulation material, which is represented as R_{wi} in Fig.7(b).

E. Loading Profile and Energy Yield

The typical wind profile and the power output of a wind turbine over an one year span is shown in Fig. 8. As it is evident from Fig. 8, the power processes by the converter varies in large range and optimizing the inductor for a specific loading condition may result in the suboptimal overall performance. Therefore, instead of optimizing the inductor efficiency at specific loading condition, the energy loss is minimized. In addition to the energy loss minimization, the volume minimization is also considered and multi-objective optimization has been carried out. The energy loss (kWh) per year is calculated using the loading profile and loss modeling and it is used into the optimization algorithm.

F. Optimization Process

The multi-objective optimization has been performed, which minimize a vector of objectives $F(X)$ and returns the optimal parameters values of X.

$$
\min F(X) \tag{24}
$$

where

$$
F(X) = [F_1(X), F_2(X)] \tag{25}
$$

978-1-4673-9551-9/16 $31.00 © 2016 IEEE

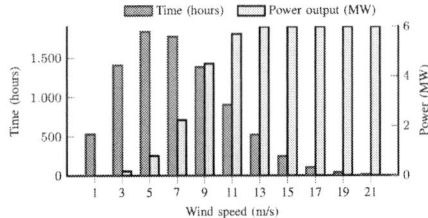

Fig. 8. Wind profile and associated output power of a typical 6 MW wind turbine.

TABLE II
VALUES OF L_f AND C_f OF THE LCL FILTER

Parameters	Values
L_f	1200 μH (0.22 pu)
Shunt capacitance $C_f + C_d$	289 μF (0.15 pu)

where $F_1(X)$ returns the energy loss (kWh) and $F_2(X)$ returns the volume of the active parts of the inductor (ltr.). The parameters that are optimized are

$$X = [N\ B_m\ J\ m\ W_c]^T \qquad (26)$$

where B_m is the maximum flux density and J is the current density. W_c is the width of the coil (refer Fig. 3(c)). Once the system specifications and the constraints are defined, the optimization has been carried out. As the number of turns N and the number of layers m only take the integer values, mixed-integer optimization problem has been formulated. The steps followed for the optimization process is shown in Fig. 9 and explained briefly hereafter.

1) Step 1: Value of the converter-side inductor L_f: The leakage inductance of the transformer is considered as a part of the grid-side inductor L_g and the use of any additional inductor is avoided. Therefore, the value of the L_g is fixed and the values of the L_f and c_f are obtained while observing the following constraints:

1) $I_h < I_h^*$ where h is the harmonic order ($2 \le h \le 180$)
2) Reactive power consumption in shunt branch $\le 15\%$ of rated power.

The calculated values of the L_f and C_f are listed in Table II.

2) Step 2: Derive dependent design variables: The dependent design variable are derived from the free design variables and the system specifications. The cross-sectional area of the core is obtained as

$$A_c = \frac{L_f I_{max}}{2NB_m} \qquad (27)$$

where I_{max} is the rated current. The dimensions of the core is then obtained as $W_l D_l = A_c/k_s$, where k_s is the stacking factor. For simplicity, $W_l = B_l$ is assumed in this study. The cross-section area of the conductor is obtained as

$$A_{cu} = W_c T_c = \frac{I_{max}}{J} \qquad (28)$$

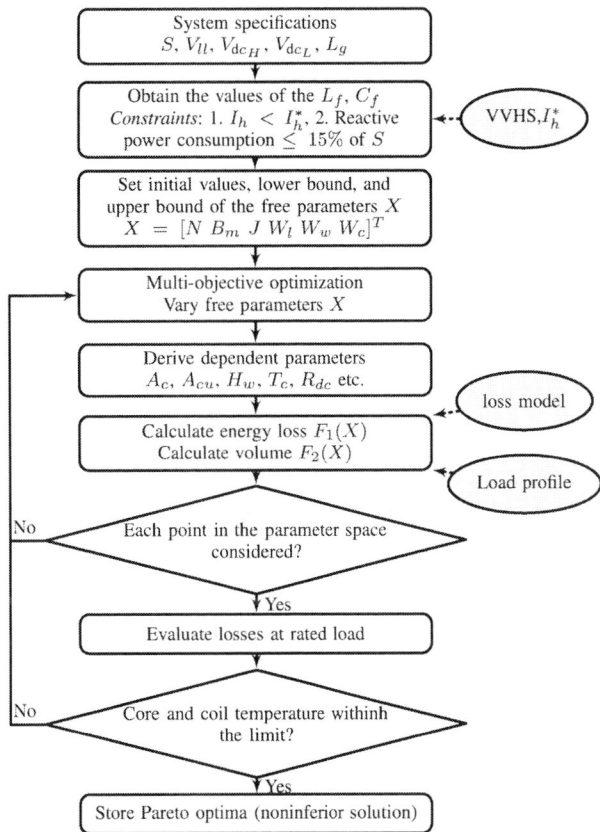

Fig. 9. Block diagram, which illustrates the steps of the optimization procedure.

3) Step 3: Objective function evaluation: The objective functions $F_1(x)$ and $F_2(x)$ are evaluated for the given set of parameters and specific mission profile. The core losses and the copper loss for each of the specific loading conditions, shown in Fig. 8, are evaluated. Using this information, the energy loss over one year period is evaluated. Similarly, the volume of both the core and the copper is also calculated.

4) Air gap length: The liquid cooling effectively removes the heat generated due to the copper losses and allows designers to reduce the constant losses (mostly core losses) by increasing the number of turns N. This leads to an improvement in the energy efficiency. However, a larger number of turns also results in larger air gap, which leads to higher fringing flux. The solution is to use several small air gaps, which is achieved by using the discrete core blocks. The length of each of these air gaps and core blocks is limited to 2.5 mm and 30 mm, respectively. If any of these quantities is violated, the solution is discarded.

5) Temperature estimation: The core and the copper losses are evaluated at the rated load conditions and the results are fed to the thermal network shown in Fig. 7(b). By solving the thermal network, temperature of the core (T_{fe}) and the coil (T_{cu}) is obtained. This gives the worst case temperature rise. If

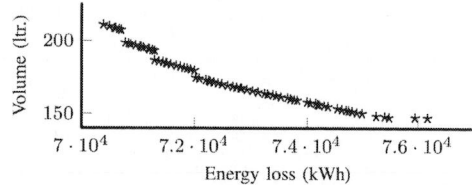

Fig. 10. Calculated volume and energy loss of the integrated inductor for different Pareto optimal solutions.

TABLE III
PARAMETER VALUES OF THE SELECTED DESIGN.

Item	Value	Item	Value	Item	Value
N	25	A_c	30200 mm^2	B_m	1.36 T
W_w	140 mm	J	4.08 A/mm^2	W_c	33 mm
A_{cu}	396 mm^2	H_w	980 mm	k_s	0.92
k_w	0.6	T_c	12 mm	m	2
k_i	0.96	α	1.55	β	1.87

the temperature rise is above the prescribed value, the solution is discarded and the optimization steps are again executed for a new set of free variables.

V. DESIGNED PARAMETERS AND VOLUMETRIC COMPARISON

The energy loss and the volume of the inductor are closely coupled and competes with each other. For example, In a given system, the reduction in the volume often leads to the rise in the losses. As a result, there is no unique solution to the optimization problem. A non-inferior (Pareto optimal) solution is obtained as shown in Fig. 10, where the reduction in the energy loss requires increase in the volume. Out of these several possible design solutions, one that suits the application the most, has been selected. The parameter values of the selected design are given in Table III.

The volume of the inductor is 187.1 ltr. and the energy loss over one year span is 72899 kWh. The coils are designed to carry the rated current (1154 A) and can be wound using copper bars. The major harmonic component in the coil current is at 1.8 kHz and at this frequency, the increase in the ohmic losses in the ac resistance of the coil due to skin effect is insignificant. Therefore, the use of the copper bars for the coils is considered.

The core losses and the copper losses at the full load conditions are 7.09 kW and 9.67 kW, respectively. The coolant flow in each of the duct is taken to be 0.06 l/s and the duct diameter D_d is 0.01 m. The inlet temperature of the coolant is assumed to be 20°C. The core temperature at the rated load is calculated to be 74 °C, whereas the temperature of the coil is found to be 86 °C.

A. Volumetric comparison

The magnetic integration leads to a reduction in the size of the inductor. This has been demonstrated by comparing the volume of the integrated inductor with the volume of the inductors in a separate inductor case. The values of the current

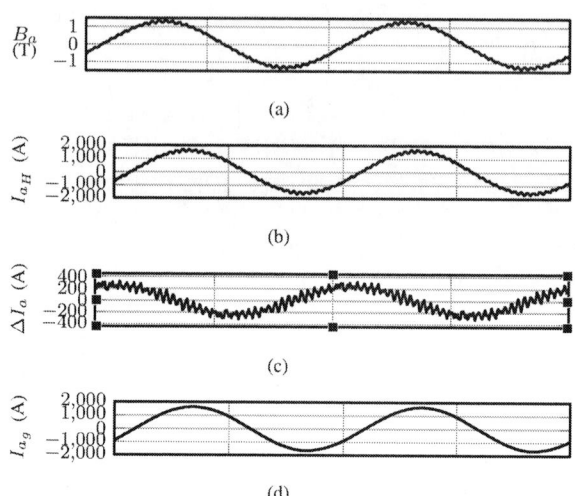

Fig. 11. Simulation results. (a) Flux density waveform in the limb of phase a, (b) Output current of the high-side converter I_h, (c) Current through the shunt branch of the LCL filter, (d) Current through the open-end transformer windings.

Fig. 12. Photo of the implemented inductor.

density of the copper, the maximum flux density in the core and the number of turns N are taken to be the same in both the cases. The volume of the magnetic material of the integrated inductor is calculated to be 132.2 ltr, compared to the 177.3 ltr. for the separate inductors. This demonstrates around 25.4% reduction in the magnetic material, which translates to 314 kg reduction in the weight of the magnetic material (assuming the use of the grain oriented steel).

VI. SIMULATION AND HARDWARE RESULTS

The time domain simulations have been carried out using PLECS. The parameters used in the simulations are specified in Table I. The integrated inductor is modeled using the magnetic toolbox, which uses the permanence model.

The converter is operated at the rated power and the simulated flux density waveform in one of the limb is shown in Fig. 11(a). The output current of the HSC is shown in Fig. 11(b), which has a major harmonic component at 2nd carrier frequency harmonic. The shunt branch of the LCL filter offers low impedance path to the harmonic components, as shown in Fig. 11(c). As a result, the injected grid currents have the desired waveform quality, as shown in Fig. 11(d).

A small scale prototype has been built to verify the effectiveness of the proposed inductor. The specifications of

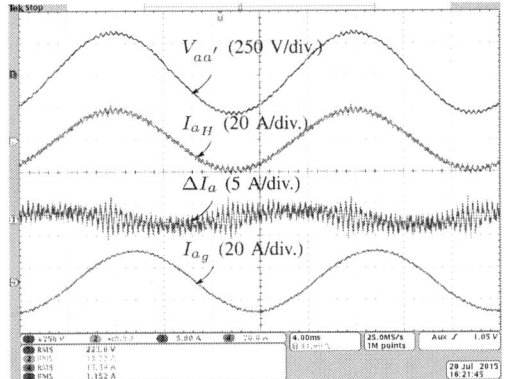

Fig. 13. Experimental results with the system operated at the rated load. Ch1: Voltage across the shunt capacitive branch of phase a of an LCL filter, Ch2: Phase a current of the high side converter, Ch3: Current through the shunt branch of the LCL filter, Ch4: Phase a grid current.

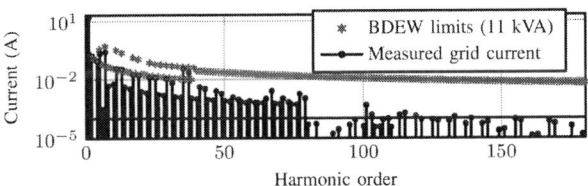

Fig. 14. Harmonic spectra of the measured grid current and associated BDEW harmonic injection limits.

the small scale system is given in Table II. The integrated inductor is realized using the standard laminated steel (0.35 mm) and the coils are wound using the AWG 11. Each coils have 102 turns and the length of the air gap is 2.014 mm. The photo of the inductor prototype is shown in Fig. 12. The converter-side inductance is 9 mH (0.195 pu) and the capacitance of the shunt branch is 24 μF (0.11 pu). The dc-link voltage was derived from the dc power supply and the ac power source from the California Instruments (MX-35) was used to emulate the grid. The converter was operated at the rated load conditions and the experimental results are shown in Fig. 13. The harmonic spectra of the injected current is shown in Fig. 14, where it is evident that the harmonic injection limits are obeyed.

VII. CONCLUSION

An integrated inductor for the the dual-converter fed open-end transformer topology is proposed. The dual-converter system often comprises of two identical VSCs. These two VSCs use two separate converter-side inductors for the LCL filter implementation. For the dual-converter system, the output currents of the given phase of both VSCs are equal. This property of the dual-converter system is exploited to cancel out the flux in one of the yokes of both the inductors through the magnetic integration. Moreover, a multi-objective optimization has been performed to identify best possible solutions which leads to the minimization of the energy loss and minimization

of the volume of the inductor. The size reduction achieved through magnetic integration is demonstrated by comparing the volume of the proposed solution with the separate inductor case. The integrated inductor leads to 25.4% reduction in the volume of the magnetic material. This translates to 314 kg reduction in the weight of the magnetic component for the 6.6 MVA, 3.3 kV WECS system. The performance of the filter has been verified by simulation and experimental studies.

REFERENCES

[1] G. Grandi, C. Rossi, D. Ostojic, and D. Casadei, "A new multilevel conversion structure for grid-connected pv applications," *IEEE Trans. Ind. Electron.*, vol. 56, no. 11, pp. 4416–4426, Nov 2009.

[2] H. Stemmler and P. Guggenbach, "Configurations of high-power voltage source inverter drives," in *Proc. of Fifth European Conference on Power Electronics and Applications*, Sep 1993, pp. 7–14 vol.5.

[3] T. Boller, J. Holtz, and A. Rathore, "Optimal pulsewidth modulation of a dual three-level inverter system operated from a single DC link," *IEEE Trans. Ind. Appl.*, vol. 48, no. 5, pp. 1610–1615, Sept 2012.

[4] N. Bodo, M. Jones, and E. Levi, "A space vector pwm with common-mode voltage elimination for open-end winding five-phase drives with a single DC supply," *IEEE Trans. Ind. Electron.*, vol. 61, no. 5, pp. 2197–2207, May 2014.

[5] M. R. Baiju, K. Mohapatra, R. S. Kanchan, and K. Gopakumar, "A dual two-level inverter scheme with common mode voltage elimination for an induction motor drive," *IEEE Trans. Power Electron.*, vol. 19, no. 3, pp. 794–805, May 2004.

[6] E. Levi, "Multiphase electric machines for variable-speed applications," *IEEE Trans. Ind. Appl.*, vol. 55, no. 5, pp. 1893–1909, May 2008.

[7] F. Blaabjerg, M. Liserre, and K. Ma, "Power electronics converters for wind turbine systems," *IEEE Trans. Ind. Appl.*, vol. 48, no. 2, pp. 708–719, March 2012.

[8] D. G. Holmes and T. A. Lipo, *Pulse Width Modulation for Power Converters: Principles and Practice*. Hoboken, NJ: Wiley-IEEE Press, 2003.

[9] A. Rockhill, M. Liserre, R. Teodorescu, and P. Rodriguez, "Grid-filter design for a multimegawatt medium-voltage voltage-source inverter," *IEEE Trans. Ind. Electron.*, vol. 58, no. 4, pp. 1205–1211, 2011.

[10] "Technical guidline: Generating plants connected to the medium-voltage network." BDEW Bundesverband der Energie- und Wasserwirtschaft e.V., [Online]. Available: http://www.bdew.de, 2008.

[11] G. Gohil, L.Bede, R.Teodorescu, T. Kerekes, and F. Blaabjerg, "Line filter design of parallel interleaved VSCs for high power wind energy conversion system," *IEEE Trans. Power Electron.*, [Online early access], DOI: 10.1109/TPEL.2015.2394460, 2015.

[12] K. Venkatachalam, C. Sullivan, T. Abdallah, and H. Tacca, "Accurate prediction of ferrite core loss with nonsinusoidal waveforms using only steinmetz parameters," in *Computers in Power Electronics, 2002. Proceedings. 2002 IEEE Workshop on*, 2002, pp. 36–41.

[13] J. Li, T. Abdallah, and C. Sullivan, "Improved calculation of core loss with nonsinusoidal waveforms," in *Industry Applications Conference, 2001. Thirty-Sixth IAS Annual Meeting. Conference Record of the 2001 IEEE*, vol. 4, 2001, pp. 2203–2210 vol.4.

[14] A. Hava, R. Kerkman, and T. Lipo, "Simple analytical and graphical methods for carrier-based PWM-vsi drives," *IEEE Trans. Power Electron.*, vol. 14, no. 1, pp. 49–61, Jan 1999.

[15] G. Narayanan and V. T. Ranganathan, "Analytical evaluation of harmonic distortion in PWM AC drives using the notion of stator flux ripple," *IEEE Trans. Power Electron.*, vol. 20, no. 2, pp. 466–474, 2005.

[16] P. Dowell, "Effects of eddy currents in transformer windings," *Proceedings of the Institution of Electrical Engineers,*, vol. 113, no. 8, pp. 1387–1394, August 1966.

[17] W. Hurley, E. Gath, and J. Breslin, "Optimizing the ac resistance of multilayer transformer windings with arbitrary current waveforms," *IEEE Trans. Power Electron.*, vol. 15, no. 2, pp. 369–376, Mar 2000.

[18] G. Gohil, L. Bede, R. Teodorescu, T. Kerekes, and F. Blaabjerg, "An integrated inductor for parallel interleaved three-phase voltage source converters," *IEEE Trans. Power Electron.*, [Online early access], DOI: 10.1109/TPEL.2015.2459134, 2015.

[19] A. V. d. Bossche and V. C. Valchev, *Inductors and Transformers for Power Electronics*. Boca Raton, FL: CRC Press, 2004.

Mechanism Analysis and Mitigation of Instability in Grid-Connected Voltage Source Inverter with *LCL* Filters Based on Terminal Impedance

Teng Liu, Zeng Liu, Jinjun Liu, Qingyun Dou

State Key Laboratory of Electrical Insulation and Power Equipment, School of Electrical Engineering
Xi'an Jiaotong University
Xi'an China
teng.liu@stu.xjtu.edu.cn

Abstract—Grid-connected Voltage Source Inverter (VSI) with *LCL* filters, controlled by proportional-resonant compensator, is very popular in utility application for its better power quality feature, while it is prone to instability especially under weak grid condition. This paper deals with this issue utilizing the terminal impedances of the inverter and the grid. Firstly, the instability mechanism is revealed by modeling the inverter output impedance, and it is observed that the item, contributing to the instability, is a sudden peak in the magnitude of the inverter output admittance. Moreover, it is found that this peak is inevitable, and is introduced by a stable designed inverter with high current control loop bandwidth. Secondly, to dampen out this peak for mitigating the instability, a novel control method based on adding lag compensator directly into the current control loop is proposed without extra needs on the current and voltage sensors. Lastly, the effectiveness of the proposed instability mitigation scheme is verified both by the simulation and experimental results.

Keywords—grid-connected voltage sourece inverter; LCL filters; instability mechanism; impedance-based stability criterion; lag compensator.

I. INTRODUCTION

The grid-connected voltage source inverters (VSIs) have been widely used for integrating renewable energy into power grid [1]. To meet the grid codes, the power filters are always needed. *LCL* filters used as interface between VSI and grid can provide higher attenuation of the switching ripples with smaller volumes and size compared with the conventional *L* filters [2]. However, the *LCL* resonance characteristic imposes constraints on the current controller design and limits the control loop bandwidth for a stable designed VSI [3].

To dampen out *LCL* resonant peak, the active damping methods mainly based on the feedback of filter state variables are preferred compared with passive dampers which cause the extra power losses [4-8]. In [5], the resonance damping was achieved by the proportional feedback of the capacitor current, whose function is equivalent to a virtual resistor connected in parallel with the filter capacitor. Similarly, the capacitor voltage can also be used as the feedback variable to realize the active damping [6-7]. However, these schemes often need extra sensors to detect state variables, and the formed dual loop control is somewhat complicated. To tackle with these issues, the inherent damping effect of system transport delay was revealed and a stable VSI with grid current feedback can be achieved without any damping only requiring that the *LCL* resonance frequency is above one-sixth of the system sampling frequency [9]. It makes the control structure simple and the extra sensors are also omitted, but the cost is to sacrifice the current control loop bandwidth for enough stability margin. Moreover, such stable designed inverter, without any damping, is still prone to instability when it connects to the power grid, especially in weak grid condition, due to the interaction between the inverter and the grid.

To deal with the instability caused by the interaction between the grid-connected voltage source inverter with *LCL* filters and the weak grid, this paper adopts the stability criterion based on terminal impedance of the inverter and grid [10], and the contribution of this paper can be highlighted as below. Firstly, the instability mechanism is revealed by modeling the inverter output impedance. A sudden peak in the magnitude of the inverter output admittance, which causes the system to be unstable is observed. Then, the cause for this peak is investigated. It is found that this peak, rather than the *LCL* resonant peak, results from the particular characteristic of the current control loop gain. To make the analysis more complete, the universality for the sudden peak is also identified. It reveals that this peak is inevitable under the condition that a good system performance is pursued on the premise of stability. Further, for mitigating the instability, this paper proposes a novel control method to dampen this peak, which is based on adding a lag compensator directly into the current control loop aiming at reshaping the loop gain characteristic. Finally, the simulation and experimental verifications are presented to validate the theoretical analyses and proposed method.

This work was supported by the National Natural Science Foundation of China under Grant 51437007, and the Power Electronics Science and Education Development Program of Delta Environmental & Educational Foundation under Grant DREM2014002.

978-1-4673-9551-9/16 $31.00 © 2016 IEEE

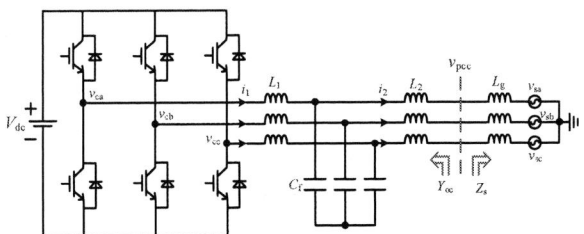

Fig. 1. Three-phase grid-connected VSI with *LCL* filters.

TABLE I.   MAIN CIRCUIT PARAMETERS

Item	Symbol	Value
DC-link voltage	V_{dc}	200V
Grid voltage (*l-g*, rms)	v_s	50V
Grid fundamental frequency	f_0	50Hz
Switching frequency	f_{sw}	10kHz
Sampling frequency	f_s	10kHz
Converter-side filter inductor	L_1	3.5mH
Grid-side filter inductor	L_2	1.75mH
Filter capacitor	C_f	3μF
Grid inductor	L_g	3.5mH

II.   INSTABILITY MECHANISM ANALYSIS

A. System Description

A three-phase grid-connected VSI with *LCL* filters are shown in Fig. 1. L_g represents the grid impedance. The dc-link voltage V_{dc} is assumed to be constant for simplicity, which means the outer voltage control loop can be neglected. This assumption is reasonable because the outer voltage control loop bandwidth is always designed much smaller than the inner current control loop bandwidth, so that the outer voltage control loop won't have much influence on the high-frequency instability [11]. Besides, the grid voltage v_s is regarded as three-phase balanced, so the per-phase control block diagram can be applied for the following analysis. The main parameters of the system studied in this paper are listed in Table I.

Fig. 2 shows the per-phase block diagram of the grid current control loop. $G_c(s)$ is the current compensator, which is implemented with a proportional-resonant (PR) controller shown in (1).

$$G_c(s) = k_p + \frac{k_r s}{s^2 + \omega_0^2} \tag{1}$$

where ω_0 is the grid fundamental angular frequency. $G_d(s)$ accounts for the digital computation and PWM delay, which can be approximated as follow [12]:

$$G_d(s) = e^{-1.5 T_s s} \tag{2}$$

where T_s is the system sampling period.

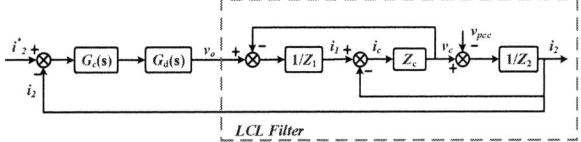

Fig. 2. Per-phase control block diagram of the grid current control loop.

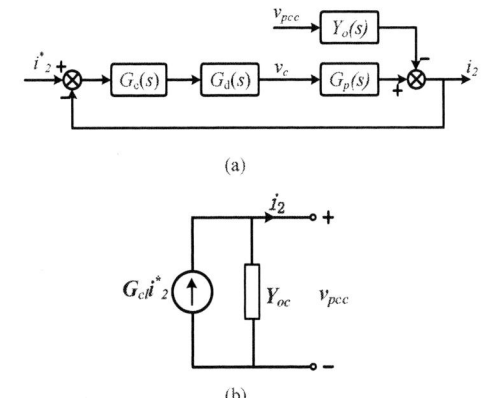

Fig. 3. Equivalent control block diagram and the equivalent circuit of the VSI with *LCL* filter. (a) Control block diagram. (b) Norton equivalent circuit.

B. Impedance-Based Stability Analysis

According to Fig. 2, the transfer function from the output voltage of the inverter v_o to the grid current i_2 and the transfer function from the point of common coupling (PCC) voltage v_{pcc} to i_2 are derived as shown in (3) and (4).

$$G_p(s) = \frac{i_2(s)}{v_o(s)} = \frac{Z_c}{Z_1 Z_c + Z_2 Z_c + Z_1 Z_2} \tag{3}$$

$$Y_o(s) = \frac{-i_2(s)}{v_{pcc}(s)} = \frac{Z_1 + Z_c}{Z_1 Z_c + Z_2 Z_c + Z_1 Z_2} \tag{4}$$

where Z_1, Z_2 and Z_c are the impedances for the converter-side inductor L_1, the filter capacitor C_f, and the grid-side inductor L_2, respectively.

Based on (3) and (4), the equivalent per-phase grid current control block diagram can be obtained as shown in Fig. 3 (a). Then, the response of the grid current can be given by

$$i_2 = G_{cl} i_2^* - Y_{oc} \cdot v_{pcc} \tag{5}$$

$$G_{cl} = \frac{T}{1 + T} \tag{6}$$

$$Y_{oc} = \frac{Y_o}{1 + T} \tag{7}$$

$$T = G_c(s) \cdot G_d(s) \cdot G_p(s) \tag{8}$$

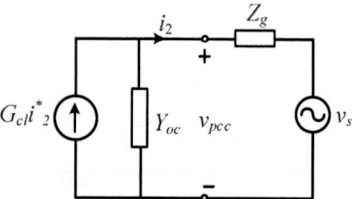

Fig. 4. The Norton equivalent circuit of the whole system.

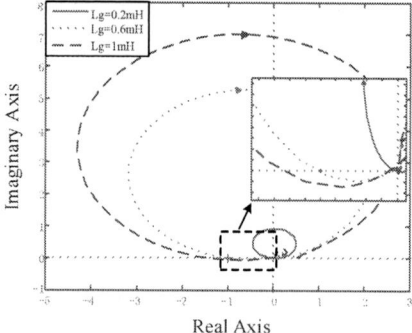

Fig. 5. Nyquist plots of T_m with different grid impedances.

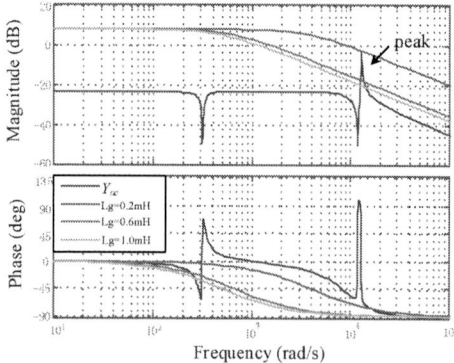

Fig. 6. Bode diagrams of Y_{oc} and different Y_g.

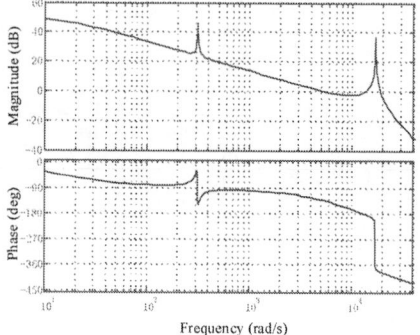

Fig. 7. Bode diagrams of the current control loop gain T.

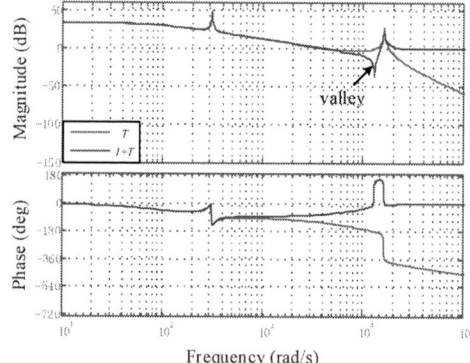

Fig. 8. Bode diagrams of T and $1+T$.

where i_2^* is the grid current reference, G_{cl} is the closed-loop gain of the current control loop, Y_{oc} is the closed-loop output admittance of the VSI with *LCL* filters seen from PCC, and T is the control loop gain. Thus, the terminal characteristic of the VSI with *LCL* filters can be represented by the Norton equivalent circuit, as shown in Fig. 3 (b).

Further, the equivalent circuit of the whole system representing the terminal behavior can be obtained, as shown in Fig. 4. Z_g is the grid impedance. For applying the impedance-based stability analysis, the whole system can be divided into two subsystems from the PCC. The VSI with *LCL* filters makes up the inverter subsystem, and grid subsystem contains the power sources and grid impedance. Consequently, the minor loop gain can be derived as (9) [13].

$$T_m = Y_{oc} \cdot Z_g = \frac{Y_{oc}}{Y_g} \qquad (9)$$

where Y_g is the grid admittance.

Because the grid subsystem can be treated as a passive network, the stability of the whole system is decided by T_m only when the inverter subsystem is standalone stable [14]. As mentioned before, the inverter subsystem is purposely designed stable by setting the *LCL* resonance frequency larger than one-sixth of the system sampling frequency, rather than using any damping. Thus, the Nyquist criterion can be applied to T_m.

Fig. 5 depicts the Nyquist plots of T_m with different grid impedances. It is seen that the system becomes unstable with the increase of the grid impedance. To figure out the essential instability mechanism, the bode diagrams of Y_{oc} and Y_g are also shown in Fig. 6. It can be found that not only the grid impedance but also a sudden peak in the magnitude of Y_{oc} is responsible for the unstable phenomenon.

To further figure out the cause for this sudden peak, the characteristic of Y_{oc} is analyzed. Fig. 7 shows the bode diagrams of the current control loop gain T. It is obvious that the magnitude of T is much larger than one around the *LCL* resonance frequency range. Consequently, the equation of Y_{oc} in this particular frequency range can be approximated as

$$Y_{oc} = \frac{Y_o}{1+T} \approx \frac{Y_o}{T} = \frac{Z_1 + Z_c}{G_c(s)G_d(s)Z_c} \qquad (10)$$

978-1-4673-9551-9/16 $31.00 © 2016 IEEE   2274

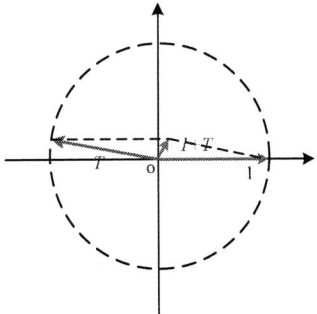

Fig. 9. The relationship between T and $1+T$.

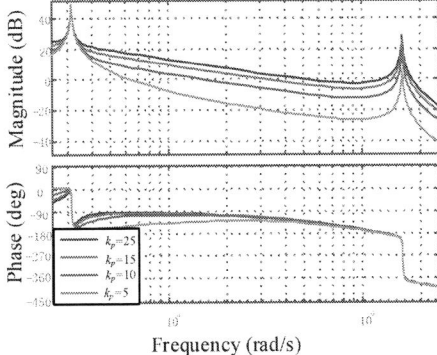

Fig. 10. Bode diagrams of T under different proportional gain.

Based on (10), the LCL resonant peak is eliminated, which means the peak existed in Y_{oc} is not caused by the LCL filters. Then, it is observed that there is a valley in the magnitude of $1+T$ which exactly changes to that sudden peak in Y_{oc} as shown in Fig. 8. This valley is totally dependent on the characteristic of the control loop gain. Notice that the magnitude of T reaches close to 0dB when the phase angle is around -180°, this feature leads to a sum vector $1+T$ with a very small magnitude as shown in Fig .9, which causes the sudden peak in Y_{oc} and leads to the unstable system.

C. Universality of The Sudden Peak

After finding out the instability mechanism, the universality of this sudden peak should be identified. That's to say, whether such loop gain characteristic causing the peak in the magnitude of Y_{oc} is common or not should be judged.

If the design requirements for the inverter subsystem are recalled, it can be found that there are two compromises existed in the design process. Firstly, a stable inverter subsystem is obtained by making the LCL resonance frequency larger than one-sixth of the system sampling frequency, and a larger resonance frequency means a larger stability margin. However, increasing the resonance frequency will decrease the ability of switching ripple attenuation. Thus, there is a tradeoff between stability margin and switching ripple attenuation in the aspect of resonance frequency. Secondly, it is known that a small proportional gain of the current compensator means a good stability margin and at the same time brings a poor dynamic performance. Therefore, there is another tradeoff between

stability and dynamic response in the aspect of the proportional gain. It should also be noticed that a small proportional gain makes the magnitude of T around the LCL resonance frequency range not much larger than one, which means the LCL resonant peak would still appear in Y_{oc} and threaten system stability as shown in Fig. 10.

Based on the above analysis, the design criterion for the inverter subsystem is decided that a better system performance is pursued on the premise of stability. Thus, the LCL resonance frequency is chosen not only larger than but close to one-sixth of the system sampling frequency. For the proportional gain, the value is setting as large as possible while the system stability is guaranteed. Consequently, such designed current control loop gain characteristic makes the sudden peak in the magnitude of Y_{oc} inevitable.

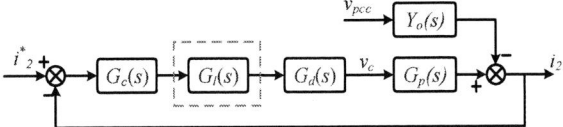

Fig. 11. Modified per-phase current control block diagram.

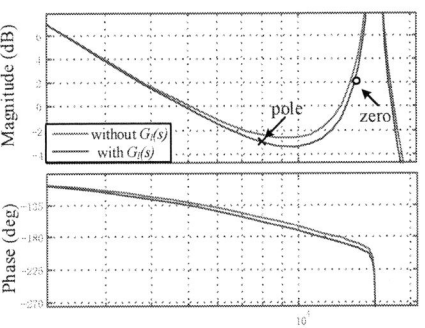

(a)

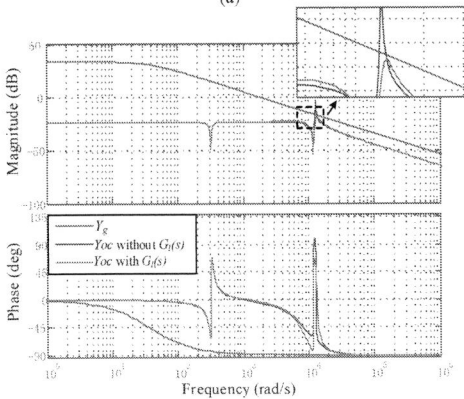

(b)

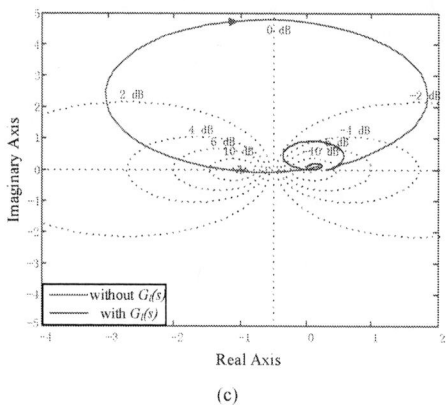

(c)

Fig. 12. Frequency characteristics of (a) T, (b) Y_g and Y_{oc} and (c) T_m with and without lag compensator $G_l(s)$ respectively.

III. CONTROL SCHEME FOR INSTABILITY MITIGATION

A. Proposed Control Scheme

Based on the instability mechanism analyzed above, the key point of the stability improvement is to modify the control loop gain characteristic for dampening the sudden peak in Y_{oc}. To preserve the merit of the simple control structure, a lag compensator $G_l(s)$ is directly added into the original current control loop as shown in Fig. 11. $G_l(s)$ is implemented with a simple first-order transfer function which is expressed as

$$G_l(s) = \frac{1+\tau_1 s}{1+\tau_2 s} \quad \tau_2 > \tau_1 > 0 \tag{11}$$

The purpose of the lag compensator is to make the phase angle of the current control loop gain have the extra phase delay within the particular frequency range where the original phase angle is around -180°. After adding the lag compensator, the loop gain characteristic will be shaped. Consequently, the sudden peak in the magnitude of Y_{oc} can be dampened, and the sharply changed phase angle also be softened.

It is obvious that there are two parameters to be determined for the lag compensator. One is the location of the zero which is related to τ_1, another is the location of the pole related to τ_2. It is expected that the phase angle of the control loop gain has the enough extra phase delay within that particular frequency range for dampening the sudden peak effectively. Moreover, the lag compensator is required to have as little influence as possible on the control loop gain characteristics in other frequency ranges. To satisfy the above requirements, the pole of the lag compensator should be located between first two crossover frequencies of the control loop gain, and the location of the zero is set between the second crossover frequency and the LCL resonance frequency as shown in Fig. 12 (a).

B. Performance Verification in Frequency-Domain

To confirm the effectiveness of the proposed instability mitigation method, a certain set of lag compensator parameters are chosen to be simulated in the frequency-domain. Fig. 12 (b)

Fig. 13. Photographs of the experimental setup.

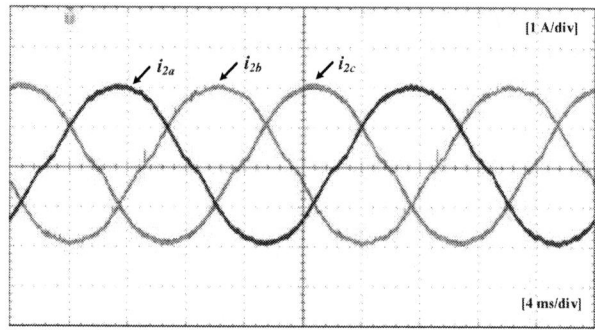

Fig. 14. Measured three-phase grid currents when the inverter subsystem is connected to the ideal grid (L_g=0).

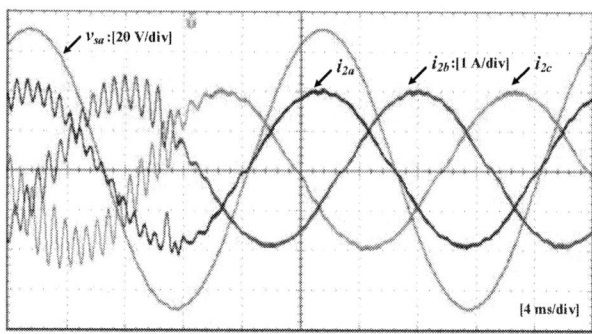

Fig. 15. Measured grid voltage and three-phase grid currents when the inverter subsystem is connected to the unideal grid (L_g=3.5mH) with and without the lag compensator.

shows bode diagrams of Y_g and Y_{oc} with and without the lag compensator. It is clear to see that the magnitude of Y_g has intersection points with the sudden peak of Y_{oc} without lag compensator, which causes the unstable potential. With the phase angle information of Y_{oc}, it can be judged that the whole system is unstable. After adding the lag compensator, the sudden peak in Y_{oc} is largely dampened and the magnitude of Y_g is larger than that of Y_{oc} in all frequency ranges, which indicates a stable system. The stabilizing performances can also be seen in Fig. 12 (c), which depicts the Nyquist plots of T_m

with and without the lag compensator. It is obvious that the whole system becomes stable by using the lag compensator.

IV. EXPERIMENTAL RESULTS

In order to further verify the effectiveness of the proposed stability improvement method, an experimental prototype of three-phase grid-connected VSI with *LCL* filters is built up as shown in Fig. 13. The main circuit parameters for the experimental setup are listed in Table I. The ideal power grid is emulated by a *Chroma Programmable AC Source*. Besides, the grid impedance is represented by an inductor whose inductance is 3.5mH. The control algorithms for the inverter subsystem are implemented in TMS320F28335 system.

Firstly, it should be guaranteed that the inverter subsystem is standalone stable, which is verified by connecting the inverter subsystem to the ideal grid. In this condition, Fig. 14 shows the measured three-phase grid currents. It is obvious that the inverter subsystem works in the stable state.

Further, the inverter subsystem is connected to the unideal grid by inserting the inductors emulating the grid impedance. The whole system turns out to be unstable as shown in Fig. 15. After enabling the lag compensator, it is clear to find that the whole system becomes stable, which validates the proposed stability improvement method.

V. CONCLUSION

This paper has analyzed the instability mechanism of the grid-connected VSI with *LCL* filters based on the terminal impedance characteristic. The inverter subsystem without any damping is designed stable by setting the *LCL* resonance frequency larger than one-sixth of the system sampling frequency. The output admittance of the inverter subsystem seen from PCC has been derived, and a sudden peak in the magnitude of the inverter output admittance which is responsible for the instability has been revealed. This peak caused by the particular characteristic of the current control loop gain is inevitable if a better system performance which means faster dynamic response and better switching ripple attenuation capacity is pursued on the premise of stability. Further, for mitigating the system instability, adding the lag compensator into the grid current control loop has been proposed to dampen this sudden peak. The proposed control method is easily to be implemented and no extra sensors are needed. Meanwhile, the stability is improved while keeping a high current control loop bandwidth. Experimental results have validated the effectiveness of the proposed instability mitigation method. It should also be noticed that there is a limitation in this method that the parameters of the system need to be well known.

REFERENCES

[1] J. Agorreta, M. Borrega, J.Lopez, and L. Marroyo, "Modeling and control of *N*-paralleled grid-conneted inverters with LCL filter coupled due to the grid impedance in PV plants," *IEEE Transactions on Power Electronics*, vol. 26, no. 3, pp. 770-785, Mar. 2011.

[2] M. Liserre, F. Blaabjerg, and S. Hansen, "Design and control of an *LCL*-filtered-based three-phase active rectifiers," *IEEE Transactions on Industry Applications*, vol. 41, no. 5, pp. 1281-1291, Sept./Oct. 2005.

[3] K. Jalili and S. Bernet, "Design of *LCL* filters of active-front-end two-level voltage-source converters," *IEEE Transactions on Industry Electronics*, vol. 56, no. 5, pp. 1674-1689, May 2009.

[4] R. N. Beres, X. Wang, F. Blaabjerg, C. L. Bak, and M. Liserre, "Comparative evaluation of passive damping topologies for parallel grid-connected converters with LCL fitlers," in *IEEE 2014 International Power Electronics Conference*, 2014, pp. 3320-337.

[5] J. He and Y. W. Li, "Generalized closed-loop control schemes with embedded virtual impedances for voltage source converters with LC or LCL filters," *IEEE Transactions on Power Electronics*, vol. 27, no. 4, pp. 1850-1861, Apr. 2012.

[6] M. Prodanovic and T. C. Green, "Control and filter design of three-phase inverters for high power quality grid connection," *IEEE Transactions on Power Electronics*, vol. 18, no. 1, pp. 373–380, Jan. 2003.

[7] M. Malinowski and S. Bernet, " A simple voltage sensorless active damping scheme for three-phase PWM converters with an LCL filter," *IEEE Transactions on Industry Electronics*, vol. 55, no. 4, pp. 1876-1880, Apr. 2008.

[8] D. Pan, X. Ruan, C. Bao, W. Li, and X. Wang, "Capaciotr-current-feedback active damping with reduced computation delay for improving robustness of LCL-type grid-connected inverter," *IEEE Transactions on Power Electronics*, vol. 29, no. 7, pp. 3414-3427, Jul. 2014.

[9] S. G. Parker, B. P. McGrath, and D. G. Holmes, "Regions of active damping control for LCL filters," *IEEE Transactions on on Industry Applications*, vol. 50, no. 1, pp. 424-432, Jan./Feb. 2014.

[10] J. Sun, "Impedance-based stabiliy criterion for grid-connected converters," *IEEE Transactions on Power Electronics*, vol. 26, no. 11, pp. 3075-3078, Nov. 2011.

[11] X. Wang, F. Blaabjerg, M. Liserre, Z. Chen, J. He, and Y. W. Li, "An active damper for stabilizing power-electronics-based AC systems," *IEEE Transactions on Power Electronics*, vol. 29, no. 7, pp. 3318-3329, Jul. 2014.

[12] S. Buso and P. Mattavelli, *Digital Control in Power Electronics*, San Francisco, CA: Morgan & Claypool Publ., 2006.

[13] A. Riccobono, E. Santi, "Stability analysis and design of stable DC distribution systems through positive feed-forward control using a novel passivity-based stability criterion," in *IEEE 2013 Applied Power Electronics Conference and Exposition*, 2013, pp. 1693-1701.

[14] R. D. Middlebrook, "Input filter considerations in design and application of switcing regulators," *IEEE IAS Aunual Meeting*, 1976.

Seven-Switch Five-Level Active Neutral-Point Clamped Converter and Optimal Modulation Strategy

Hongliang Wang, *Senior Member, IEEE*, Lei Kou,
Yan-Fei Liu, *Fellow, IEEE* and Paresh C. Sen, *Life Fellow,IEEE*
Department of Electrical and Computer Engineering
Queen's University
Kingston, Canada
hongliang.wang@queensu.ca, kou.lei@queensu.ca,
yanfei.liu@queensu.ca, senp@queensu.ca

Sucheng Liu
School of Electrical and Information Engineering
Anhui University of Technology
Ma' anshan, China
liusucheng@gmail.com

Abstract—The five-level active-neutral-point-clamped (ANPC) converters are newly hybrid topologies which have the advantages of traditional neutral-point-clamped (NPC) and flying-capacitor (FC) multilevel converters. A novel seven-switch five-level ANPC (7S-5L-ANPC) inverter topology is proposed, which reduces the number of active semiconductor switches over traditional 5L-ANPC inverter topologies. It is capable of operating under any power factor conditions. The special modulation strategy for 7S-5L-ANPC is proposed and simulation shows only reactive current is flowing through the seventh switch. Thus, low current rating switch can be selected for the seventh switch. The proposed 7S-5L-ANPC inverter is very suitable for unity and high power factor applications such as photovoltaic (PV) application. A 1KW single-phase inverter prototype is built to verify the validity and flexibility of the proposed topology and modulation.

I. INTRODUCTION

Multilevel converters (or inverters) have been used for power conversion in high-power applications such as medium voltage grid (2.3KV, 3.3KV, or 6.9KV) to reduce the switch voltage stress, and photovoltaic (PV) application to reduce the filter size [1, 2]. Compared to two-level voltage source inverters, the advantages of multilevel inverters are lower voltage stress, higher efficiency, smaller filter size and lower common-mode voltage [3].

There are three traditional multilevel topologies [4-8]: the neutral-point-clamped (NPC) type [4, 5], flying-capacitor (FC) type [6], and cascaded H-bridge (CHB) type [7-8]. Many five-level NPC topology has been derived in [9]. NPC

type generates the voltage levels from the neutral point voltage by adopting diodes. The drawback is the increased number of switching devices when voltage level increases. FC type outputs the voltage level by summing the flying-capacitor voltage. However, higher voltage level leads to more flying-capacitors and the complexity of control strategy to balance the voltages of each flying-capacitor is then increased. The CHB multilevel inverters use series-connected H-bridge cells with an isolated dc voltage sources connected to each cell [10]. Similarly, to have more output levels, more cells are needed. This will lead to impracticality of this type of topology since more DC sources are required.

Active-neutral-point-clamped (ANPC) which is one of the multilevel topology has been proposed in Refs [11-15]. Three 5L-ANPC topologies are shown in Fig.1. The ANPC type converter combines the features of NPC and FC topology. The ANPC topologies is receiving more and more attentions nowadays because of high efficiency and multi-level output. In this paper, a novel seven-switch five-level ANPC (7S-5L-ANPC) inverter topology is proposed. Compared to traditional ANPC topologies, the proposed topology reduces the number of active semiconductor switches. In addition, a low current rating switch can be chosen for the seventh switch. This paper is organized as follows: Section II describes working principles of proposed topology; Section III discusses the modulation strategy of 7S-5L-ANPC topologies; Section IV and V show the simulation and experimental results and Section VI gives the conclusion.

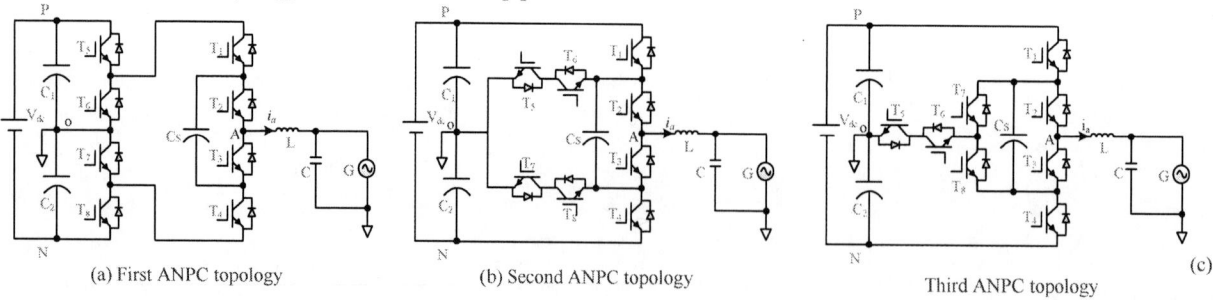

(a) First ANPC topology   (b) Second ANPC topology   Third ANPC topology   (c)

Fig. 1. Five-level ANPC topology

II. THE PROPOSED 7S-5L-ANPC INVERTER

A. Introduction of 7S-5L-ANPC Inverter

For PV grid-connection application, the output current and grid voltage are in phase. For this reason, some reactive power current paths can be ignore. Then, some active switches can be replaced by diodes in order to increase the efficiency. Because of this, a novel 5-level ANPC inverter is proposed. It consists of 7 switches (T_1 to T_7), 2 independent-diodes (D_{F1}, D_{F2}) and 1 flying-capacitor (C_S). The configuration of 7-switches-5-level ANPC (7S-5L-ANPC) inverter is shown in Fig.2.

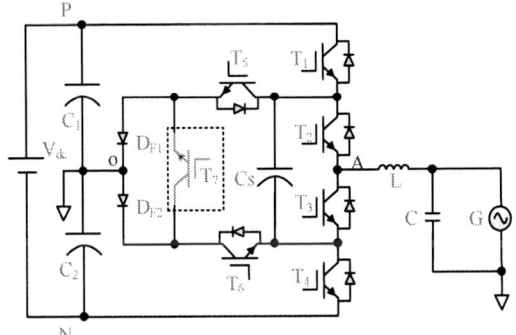

Fig. 2. Proposed 7S-5L-ANPC inverter

Based on half bridge inverter, this topology is very suitable for both single-phase and three-phase transformer-less PV system application because of no leakage current generation. In addition, only two series-connected bulk capacitors, C_1 and C_1, are connected to DC-link in which NPC type and FC type need four in series. So the difficulty of balancing DC capacitor voltages is greatly reduced. Compared to traditional 5L-ANPC which needs 8 switches and other types of 5-level inverter which need more, the proposed 5L-ANPC inverter only has 7 active switches and 2 discrete diodes. And from the analysis in the following section, it can be obtained that under appropriate modulation method, only reactive current will flow through the seventh switch T_7, which means under unity power factor and high power factor applications, the current through T_7 is very small. Thus switch with low current rating can be selected for T_7. It can be concluded that the main advantage of 7S-5L-ANPC over traditional ANPC and other types of 5-level inverter topology is the reduced number of active semiconductors.

The seventh switch T_7 is employed under reactive power operation. The current stress of T_7 is decided by power factor $cos\theta$. A smaller IGBT without anti-diode or MOSFET can be used for T_7. The specific modulation strategy will be discussed in the following section.

B. Operation of 7S-5L-ANPC Topology

Define the input DC voltage as V_{dc}. The proposed 7S-5L-ANPC outputs five voltage levels which are $+V_{dc}/2$, $+V_{dc}/4$, 0, $-V_{dc}/4$ and $-V_{dc}/2$ (+2 level, +1 level, 0 level, -1 level and -2 level). The five voltage levels are achieved by summing the flying-capacitor voltage and DC capacitor voltage. Two DC-link capacitors in series are to provide $\pm V_{dc}/2$ voltage levels. The voltage of flying-capacitor is controlled to be kept at $V_{dc}/4$ to provide $\pm V_{dc}/4$ voltage levels.

As a result, there are eight switching states for 7S-5L-ANPC inverter topology: A, B, C, D, E, F, G and H, outputting $+V_{dc}/2$, $+V_{dc}/4$, $+V_{dc}/4$, $+0$, -0, $-V_{dc}/4$, $-V_{dc}/4$ and $-V_{dc}/2$ respectively. The specific switching states and current path are shown in Fig.3. Red line is the active power branch, and Green line is the reactive power branch. The switching pattern, voltage of flying-capacitor and output voltage are presented in Table I.

From Table I, it is observed that for $+V_{dc}/4$, 0 and $-V_{dc}/4$ output voltage levels, each level owns a pair of redundant switching states. When the switching pattern is $\pm V_{dc}/4$, the flying capacitor is in charge mode or discharge mode according to the direction of output current.

Without switch T_7, switching states C, D, E and F can only provide single current flowing path due to the presence of diode, which sacrifice the reactive power current path. Therefore, under reactive power operation (e.g. $cos\theta < 0.95$): in switching states C, D, E and F, T_7 is switched on to provide bi-directional current path.

TABLE I.   SWITCHING STATES OF 7S-5L-ANPC INVERTER

Switching state	Switch number							Output voltage V_{Ao}	Flying capacitor C_S	
	T_1	T_2	T_3	T_4	T_5	T_6	T_7		$i_a>0$	$i_a<0$
A	1	1	0	0	0	1	0	$+V_{dc}/2$	--	--
B	1	0	1	0	0	1	0	$+V_{dc}/4$	Charge	Discharge
C	0	1	0	0	0	1	1	$+V_{dc}/4$	Discharge	Charge
D	0	0	1	0	0	1	1	$+0$	--	--
E	0	1	0	0	1	0	1	-0	--	--
F	0	0	1	0	1	0	1	$-V_{dc}/4$	Discharge	Charge
G	0	1	0	1	1	0	0	$-V_{dc}/4$	Charge	Discharge
H	0	0	1	1	1	0	0	$-V_{dc}/2$	--	--

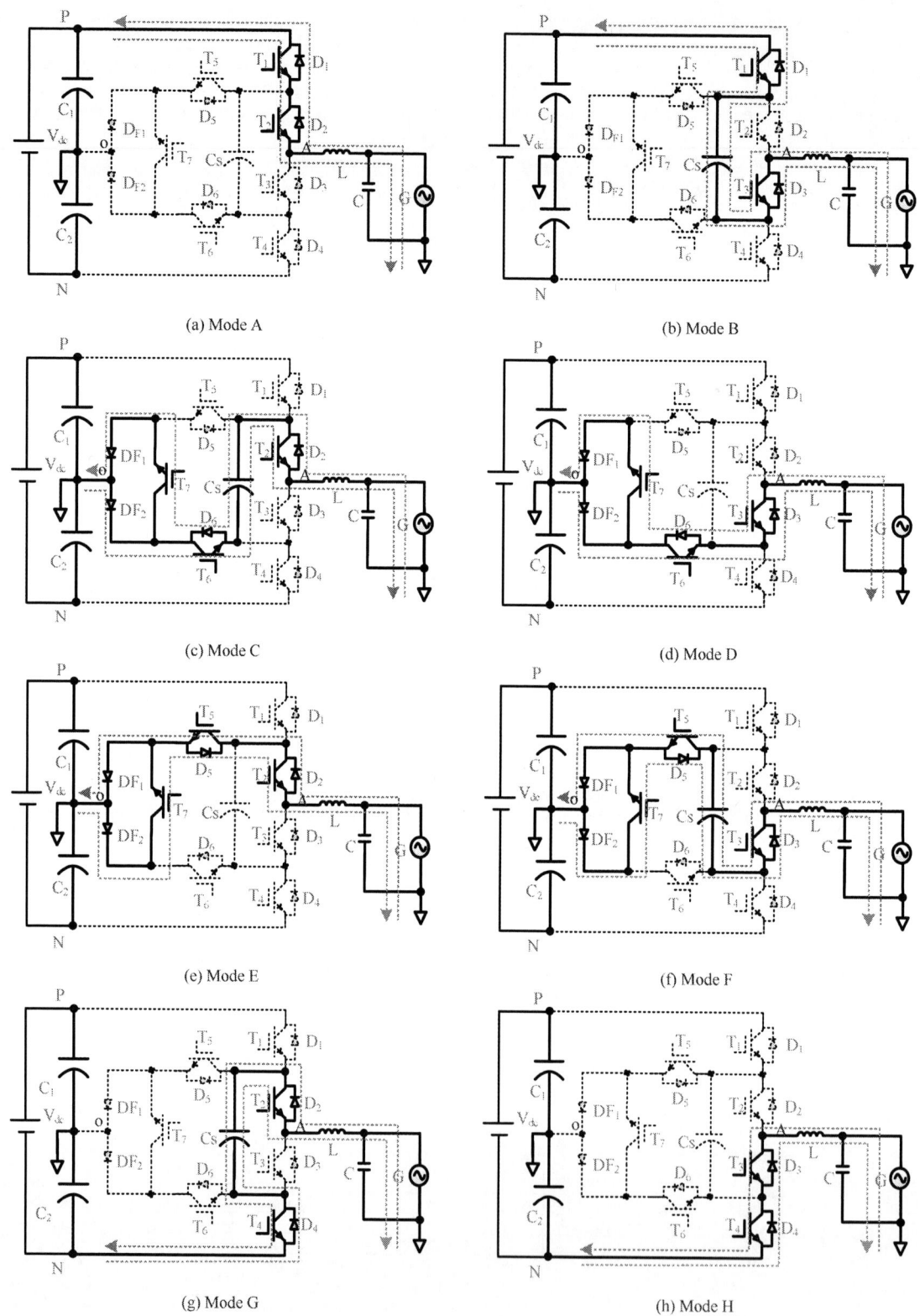

(a) Mode A   (b) Mode B

(c) Mode C   (d) Mode D

(e) Mode E   (f) Mode F

(g) Mode G   (h) Mode H

Fig. 3. Eight switching states and current-flow path for 7S-5L-ANPC

III. MODULATION STRATEGY

This section investigates the modulation method for proposed 7S-5L-ANPC inverter. Same as traditional ANPC topologies, the new topology has eight switching states. However, the existence of three pairs of redundant switching modes will result in diverse modulation strategies. Therefore, appropriate switching states selection is very important because it leads to reduced switching and conduction losses.

A. Optimized zero output voltage states selection

For two redundant switching states D and E which provide zero output voltage, combination of two discrete diodes D_{F1}, D_{F2} and switch T_7 allows bi-directional current flowing path. Consequently, both modes can be used during positive, negative or whole grid period. Thus, four possible combinations of zero output modes are available for modulation. Four conditions are: (1) mode D for positive reference cycle, and mode E for negative cycle; (2) mode E for positive reference cycle, and mode D for negative cycle; (3) mode D for all zero output switching states; (4) mode E for whole reference period. Among four different situations, the currents through T_7 are different. In optimal situation, only reactive current is passing through T_7 and one discrete diode while all active current is flowing through complementary diode. In situation 1, it is only reactive current is going through switch T_7. In situation 2, all active current is passing through T_7. In situations 3 and 4, half active and reactive current is flowing through T_7. Consequently, mode D for positive grid period and mode E for negative grid cycle is the best combination for proposed 7S-5L-ANPC inverter. Simulation verification is provided in the following section.

Additionally, under situation 1, switch T_5 and T_6 will be turned on and off at line frequency. This results in the reduced switching loss. In conclusion, mode D should be used in positive grid cycle, which is called positive freewheeling state. Similarly, mode E is called negative freewheeling state and used during negative grid cycle.

B. Modulation method for 7S-5L-ANPC inverter

Fig.4 shows the diagram of modulation method for 7S-5L-ANPC inverter. There are four zones according to whether the output current and voltage are in same direction: zone 1 and 3 are reactive power zone; zone 2 and 4 belong to active power zone. In PV application, the power factor is usually larger than 0.9. An example of power factor $cos\theta > \sqrt{3}/2$ is used in our case.

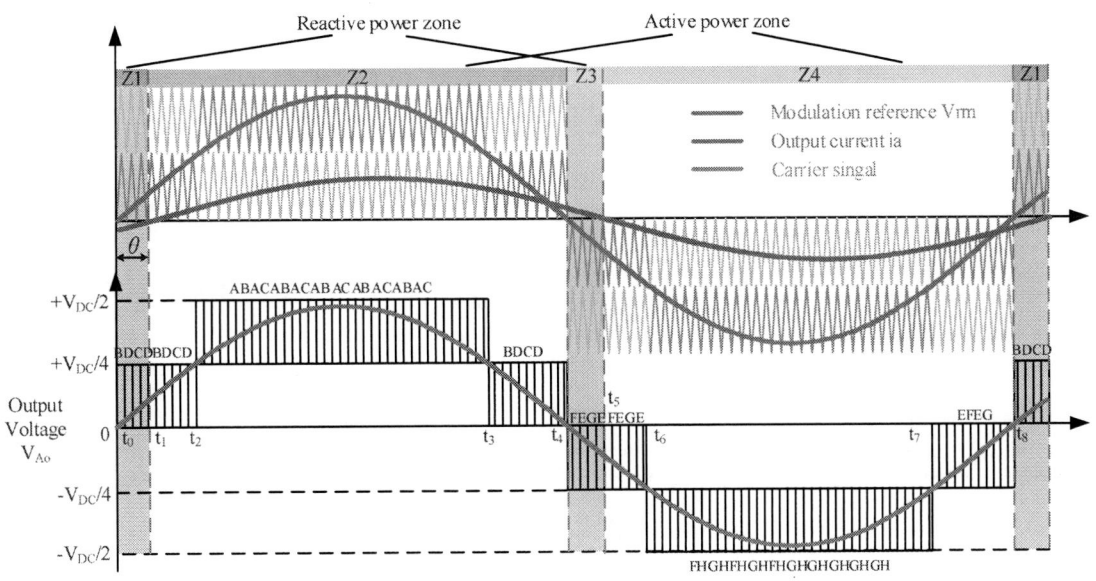

Fig. 4. PWM modulation of 7S-5L-ANPC inverter ($cos\theta > \sqrt{3}/2$)

As shown in Fig.4, the reference is compared to four carrier signals, generating five level outputs. Phase Disposition (PD) PWM modulation method (all carrier signals are in phase) [16-17] is adopted because of its lower THD (total harmonic distortion, THD).

From t_0 to t_2 and t_3 to t_4, reference voltage signal is between $+V_{dc}/4$ and 0. When reference is larger than carrier signal, the inverter outputs $+V_{dc}/4$ (+1 level), which can be achieved by states B and C. Although the redundant switching states (B, C) generate the same output voltage level, their effect on the flying-capacitor is opposite to each other. This gives an opportunity to regulate the voltage across flying-capacitor. One way is to adopt the redundant switching states alternately. Consequently, over a grid period or longer, the charging time and discharging time of flying-capacitor is the same. The flying capacitor can achieve voltage self-balancing. Additionally, this alternation will decrease the flying-capacitor voltage ripple to lowest. Therefore, smaller capacitor can be used for flying-capacitor. When reference is less than carrier signal, the inverter gives +0 output, which is achieve by positive freewheeling state D. As a result, the switching state sequence of (B, D, C, D) is generated.

From t_2 to t_3, the output reference voltage is between $+V_{dc}/4$ and $+V_{dc}/2$. Mode A is required to generate $+V_{dc}/2$ output level. Modes B and C can be selected alternately to provide $+V_{dc}/4$ output. In this way, sequence of (B, A, C, A) guarantees output voltage and flying-capacitor voltage balancing.

Similarly, when reference signal is negative, it also compares to two carrier signals. When outputting $-V_{dc}/4$ voltage levels, redundant switching states (F, G) are employed alternately to keep flying-capacitor balanced. Consequently, during t_4 to t_6 and t_7 to t_8, switching state sequence (F, E, G, E) is acquired; from t_6 to t_7, switching state sequence (F, H, G, H) is adopted.

As discussed before, in reactive power zones: Z1 (t_0 to t_1) and Z3 (t_4 to t_5), the output voltage and current are in opposite directions. For example, in reactive power zone Z1, the current is negative, which is the green line in Fig.3 (b), (c) and (d). So in modes C and D, the current is no longer flowing through discrete diode D_{F2}; instead, it passes through switch T_7 and another discrete diode D_{F1}. Similarly, in Z3, the output current is flowing through T_7 and D_{F2}. In other cases, no current will go through T_7. Therefore, current rating of switch T_7 can be selected according to the system requirement. If the system is working under unity power factor, then the current through T_7 is closed to zero; under high power factor ($\cos\theta > \sqrt{3}/2$), small current will pass through T_7 so that a low current rating semiconductor can be selected; in low power factor case, T_7 can be selected to be the same as T_5 and T_6 due to the same voltage and current ratings. According to this, it can be concluded that the proposed 7S-5L-ANPC inverter is very suitable for PV application.

IV. SIMULATION VERIFICATION

To verify the selected zero output states combination is the optimal selection, simulation tests are carried out in two cases: one is under active power operation, the other is reactive power operation. Simulation parameters are shown in Table II.

TABLE II. SYSTEM PARAMETERS

Input voltage	400V	Grid voltage	110V_RMS
DC capacitor	2000μF	Grid frequency	60Hz
Flying capacitor	310μF	Output current	4.6A_RMS
Output filter inductor	1.6mH	Switching frequency	15kHz

A. Active Power operation ($\cos\theta = 1$)

Under unity power factor, the simulation waveforms of inverter output voltage, flying-capacitor voltage, grid voltage, output current and T_7 current under four switching conditions are shown in Fig.5.

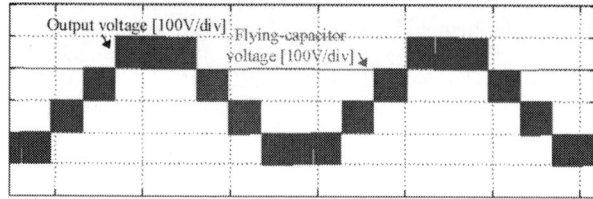

(a)Inverter output voltage and flying-capacitor voltage

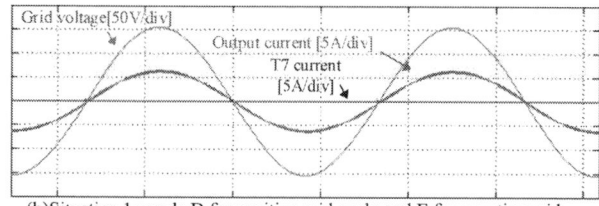

(b)Situation 1: mode D for positive grid cycle and E for negative grid cycle

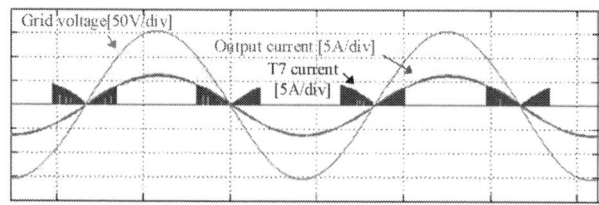

(c)Situation 2: mode E for positive grid cycle and D for negative grid cycle

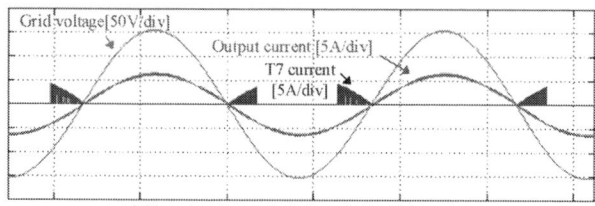

(d)Situation 3: mode D for whole grid cycle

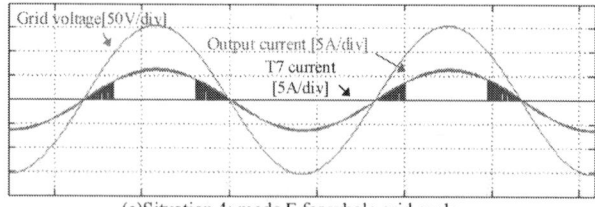

(e)Situation 4: mode E for whole grid cycle

Fig. 5. Active power operation: simulation waveforms of inverter output voltage, flying-capacitor voltage, grid voltage, output current and T_7 current under four switching conditions

Fig.5 (a) shows the five-level bridge voltage and flying-capacitor voltage. As can be seen, the flying-capacitor is capable of obtaining self-balancing at 100V, and the measured ripple voltage is 2V (2%).

The waveforms of currents through T_7 in four different zero switching states combinations are shown in Fig.5 (b) to (e) respectively. In Fig.5 (b), the system is working under situation 1 (mode D for positive grid cycle and E for negative grid cycle): the T_7 current is almost zero, which means no active current is flowing through T_7. Therefore, zero switching states combination in situation 1 is the optimal one among four combinations.

B. Reactive Power operation ($cos\theta = 0.9$)

Under reactive power operation: $cos\theta = 0.9$, simulation waveforms of inverter output voltage, flying-capacitor voltage, grid voltage, output current and T_7 current under four switching conditions are shown in Fig.6.

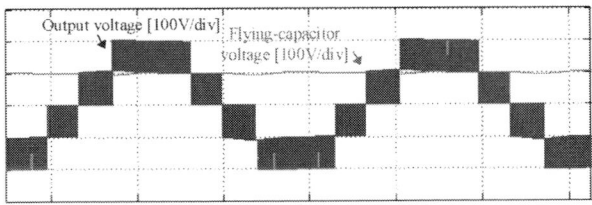

(a)Inverter output voltage and flying-capacitor voltage

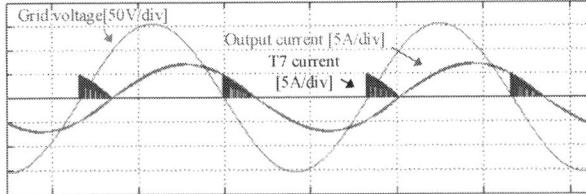

(b)Situation 1: mode D for positive grid cycle and E for negative grid cycle

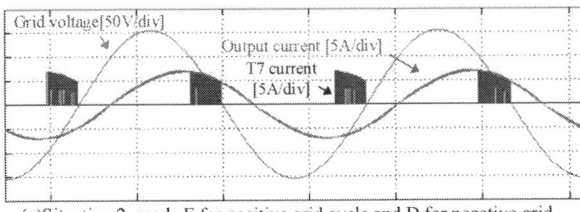

(c)Situation 2: mode E for positive grid cycle and D for negative grid cycle

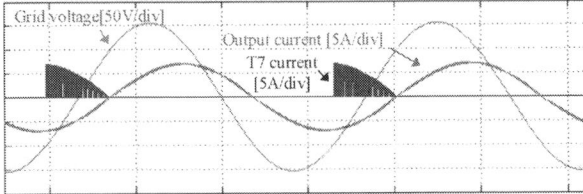

(d)Situation 3: mode D for whole grid cycle

(e)Situation 4: mode E for whole grid cycle

Fig. 6. Reactive power operation $cos\theta = 0.9$: simulation waveforms of inverter output voltage, flying-capacitor voltage, grid voltage, output current and T_7 current under four switching conditions

Fig.6 (a) shows the five-level inverter output voltage and FC voltage, which are closed the waveforms in active power operation. The ripple voltages of flying-capacitor in this case is 5V (5%).

Fig.6 (b) to (e) shows the output current, grid voltage and T_7 currents in four situations. According to the analysis in section III, we know that in optimal situation, only reactive current is passing through T_7. Under reactive power condition, there is phase shift between grid voltage and output current. When current and voltage are in opposite direction, the system is entering reactive power zone. In situation 1, which is shown in Fig.6 (b), it is observed that only reactive current is going through switch T_7 (current is flowing through T_7 only in reactive power zones Z1 and Z3). In situation 2, all active current is passing through T_7, which is not desirable. In situations 3 and 4, half active and reactive current is flowing through T_7, which is also not what we need. With simulation results under active and reactive power conditions, it is concluded that mode D for positive grid period and mode E for negative grid cycle is the best combination for proposed 7S-5L-ANPC inverter.

V. EXPERIMENTAL RESULTS

To demonstrate the ability of proposed topology and its modulation strategy. A 500W single-phase 7S-5L-ANPC inverter grid-connection experimental prototype is built, as shown in Fig.7. The system includes main circuit, DSP and FPGA control board, DC source, output filter and measuring instruments. The control board employs a combination of the Texas Instruments TMS320F28335 control card and the Altera Cyclone IV EP4CGX22 FPGA card to provide powerful real-time mathematical calculations and control functions. The experimental parameters are the same as parameters used in simulation, shown in Table II.

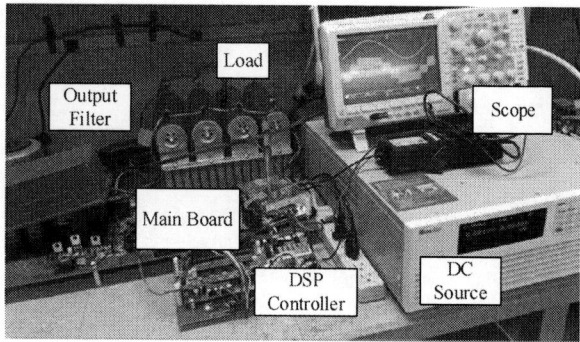

Fig. 7. Experimental prototype

The experimental tests are carried out in two cases: active power operation and reactive power operation.

A. Active Power operation

Fig.8 and Fig.9 show the experimental results under unity power factor condition. Fig.8 shows inverter output voltage, grid voltage, flying-capacitor voltage and output current: channel 1 is the output bridge voltage; channel 2 is $110V_{RMS}$ grid voltage; channel 3 is the flying-capacitor voltage, which is balanced at 100 volts; channel 4 is the output current, which is sinusoidal without distortion and in phase with grid voltage.

Fig.9 shows two DC-link capacitors voltages, flying-capacitor voltage and output current: channel 1 is lower DC-link capacitor voltage and channel 2 is upper DC-link capacitor voltage, which both have a line-frequency fluctuation. The measured flying-capacitor ripple voltage is 3V (3%) and DC-link capacitor ripple voltage is 8V (4%).

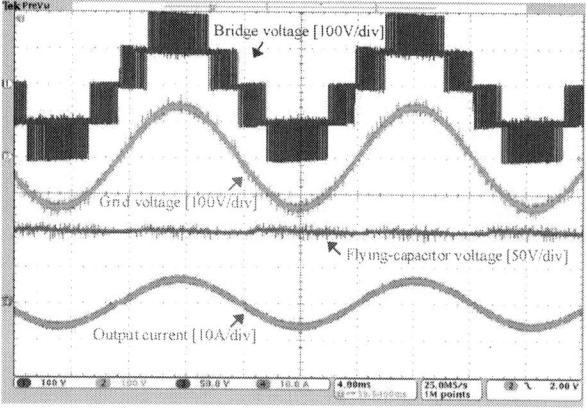

Fig. 8. Experimental results under unity power factor condition: waveforms of inverter bridge voltage, grid voltage, flying-capacitor voltage and output current

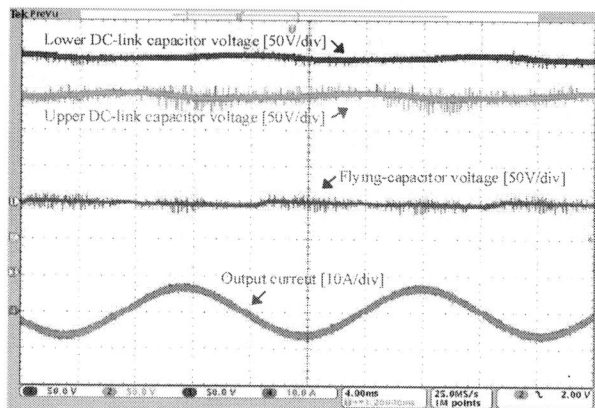

Fig. 9. Experimental results under unity power factor condition: waveforms of lower DC-link capacitor votlage, upper DC-link capacitor voltage, grid voltage and output current

B. Reactive Power operation ($\cos\theta = 0.9$)

To testify the proper system operation under reactive power condition, experimental works are carried out. The power factor is 0.9. Experimental results are shown in Fig.10 and Fig.11.

Fig.10 shows inverter output voltage, grid voltage, flying-capacitor voltage and output current. It is observed that the waveform of inverter output voltage V_{Ao} is closed to the waveform in active power condition. The flying-capacitor voltage is also balanced at 100 volts, and measured ripple voltage is 6V (6%). Under 0.9 power factor, the output current and grid voltage has a 25 degree phase shift. In this situation, the inverter still produces good quality current waveform without distortion.

Fig.11 shows two DC-link capacitors voltages, flying-capacitor voltage and output current. The DC-link capacitors voltages waveforms in this situation are almost the same as ones under active power condition. The measured DC-link capacitor ripple voltage is 8V (4%).

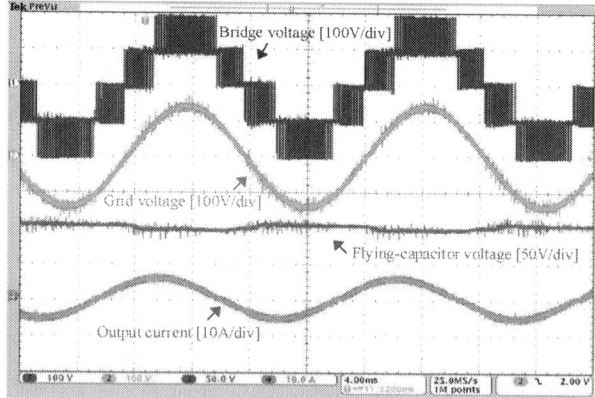

Fig. 10. Experimental results under reacitve power operation $cos\theta = 0.9$: waveforms of inverter bridge voltage, grid voltage, flying-capacitor voltage and output current

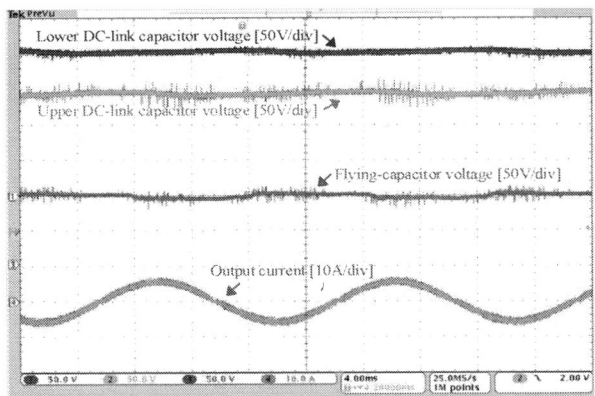

Fig. 11. Experimental results under unity power factor $cos\theta = 0.9$: waveforms of lower DC-link capacitor votlage, upper DC-link capacitor voltage, grid voltage and output current

From the experimental results achieved in two situations, it can be concluded that the with proposed modulation strategy applied on proposed 7S-5L-ANPC inverter, the system is capable of outputting good quality AC current. The flying-capacitor voltage is self-balanced. The ripple voltages of FC voltage and DC-link capacitor voltage are kept within a reasonable range. The effectiveness of proposed topology and modulation method is verified.

VI. CONCLUSION

A novel 7S-5L-ANPC inverter topology with fewer active semiconductor switches is proposed. Compared to traditional five-level ANPC topologies. Only seven active switches are employed. Switch with low current rating can be selected for the seventh switch T_7 because only reactive current is flowing through it. The basic working principles and modulation strategy are explained in this paper. The simulation and experimental results demonstrated the effectiveness of proposed topology and modulation method

REFERENCE

[1] P Hammond. A New Approach to Enhance Power Quality for Medium Voltage AC Drives. IEEE Trans. on Industry Applications, 1997, 33(1): 202-208.

[2] G Beinhold, R Jakob and M Nahrstaedt. A New Range of Medium Voltage Multilevel Inverter Drives with Floating Capacitor Technology[C]. In: Conference Proceedings of the 9th European Conference on Power Electronics, EPE 2001, Austria, CD-ROM.

[3] J. Rodriguez, J. S. Lai, and F. Z. Peng, "Multilevel inverters: A survey of topologies, controls, and applications," IEEE Trans. Ind. Electron. Vol. 49, no. 4, pp. 724–738, Aug. 2002.

[4] A. Nabae, I. Takahashi, and H. Akagi, "A new neutral-point clamped PWM inverter," IEEE Trans. Ind. Applicant., vol. IA-17, pp. 518–523,Sept./Oct. 1981. J. Rodriguez, S. Bernet, B. Wu, J. O. Pontt, and S. Kouro, "Multilevel voltage-source-converter

topologies for industrial medium-voltage drives," IEEE Trans. Ind. Electron., vol. 54, no. 6, pp. 2930–2945,Dec. 2007.

[5] J. Rodriguez, S. Bernet, B. Wu, J. O. Pontt, and S. Kouro, "Multilevel voltage-source-converter topologies for industrial medium-voltage drives," IEEE Trans. Ind. Electron., vol. 54, no. 6, pp. 2930–2945,Dec. 2007.

[6] F. Richardeau, P. Baudesson, and T. A. Meynard, "Failure-tolerance and remedial strategies of a PWM multi-cell inverter," IEEE Trans. Power Electron., vol. 17, no. 6, pp. 905–912, Nov. 2002.

[7] J. I. Leon, S. Kouro, S. Vazquez, R. Portillo, L. G. Franquelo, J. M.Carrasco, and J. Rodriguez, "Multidimensional modulation technique for cascaded multilevel converters," IEEE Trans. Ind. Electron., vol. 58, no. 2,pp. 412–420, Feb. 2011.

[8] Kagarlu,M.F and Babaei.E" A Generalized Cascaded Multilevel Inverter Using Series Connection of Sub-multilevel Inverters," IEEE Trans. Power Electron. Vol. 28, no. 2, pp. 625–636, Aug. 2013.

[9] H. Wang, Y.-F. Liu, and P. C. Sen, "A neutral point clamped multilevel topology flow graph and space NPC multilevel topology," in Energy Conversion Congress and Exposition (ECCE), 2015 IEEE, 2015, pp. 3615-3621.

[10] Kouro. S, Malinowski. M, Gopakumar. K, Pou. J, Franquelo. L, "Recent Advances and Industrial Applications of Multilevel Converters," IEEE Trans. Ind. Electron., vol. 57, no. 8,pp. 2553-2580, Feb. 2010.

[11] P.Barbosa, P. Steimer, J. Steinke, J. meysenc, "Active Neutral-point-clamped Multilevel Converter" Power Electronics Specialists Conference, 2005. PESC' 05. IEEE 36th 16-16 June 2005. (2001), 2001, pp. 2296-2301.

[12] T. Brückner, S. Bernet, and H. Güldner, "The active NPC converter and its loss-balancing control," IEEE Trans. Ind. Electron., vol. 52, no. 3, pp. 855–868, 2005.

[13] L.A. Spepa, P.M. Brbosa, P.K. Steimer, and J. W. Kolar, "Five-Level virtual –flux direct power control for the active neutral-point-clamped multilevel inverter," in Proc. IEEE PESC, Jun. 2008, PP. 1668-1674.

[14] K.Wang, Y.Li and Z. Zheng, "A neutral-point potential balancing algorithm for five-level ANPC converters," in Proc. ICEMS, Aug. 2011, pp. 1-5.

[15] Soeiro.T.B, Carballo. R, Moia. J, Garcia. G. O, Heldwein. M. L, "Three-phase five-level active-neutral-point-clamped converters for medium voltage applications," in Power Electronics Conference (COBEP), Brazilian, 2013, pp. 85

[16] V. G. Agelidis, "Non - Deterministic AM-PWM Strategy for Three - Phase VSI," Ind. Electron. Control Instrumentation, 1994. IECON '94., 20th Int. Conf., 1987.

[17] A. Alesina and M. Venturini, "Solid-state power conversion: A Fourier analysis approach to generalized transformer synthesis," IEEE Trans. Circuits Syst., vol. 28, no. 4, 1981.

A Simple Variable Step Size Method for Maximum Power Point Tracking Using Commercial Current Mode Control DC-DC Regulators

Su Sheng, Brad Lehman*
Electrical and Computer Engineering
Northeastern University
Boston, Massachusetts, USA
sheng.s@husky.neu.edu

Abstract—This paper proposes a method to achieve maximum power point tracking (MPPT) function by "tricking" inexpensive commercial-off-the-shelf (COTS) current mode control regulators. In the commercial market, there are many such inexpensive regulators, but they are designed for the output voltage control. The proposed method utilizes a voltage signal from an additional external microcontroller to regulate the compensation pin of the DC-DC regulator to achieve the MPPT function by using perturb and observe (P&O) with variable step size. A prototype is designed, built, and tested. Because the COTS hardware is simple to implement, MPPT can be achieved without complete system design, making MPPT accessible to hobbyists and non-experts in power supplies.

Keywords—MPPT; perturb and observe (P&O), current mode control; boost converter

I. INTRODUCTION

Extensive research, development and demonstration efforts have focused on DC-DC converters with MPPT [1]-[8]. However, almost all the DC-DC converters with MPPT are designed by using discrete components, such as low side and high side gate drives. They also need additional circuits for controlling and maintaining maximum output power all the time. As a result, the design of the DC-DC converters with MPPT is complicated and the products are often not compact especially in the low power. On the other hand, in the commercial market there are many good solutions for the DC-DC converters with output voltage control. The commercial DC-DC regulators are delicately designed for different voltage and current ranges, and high efficiency can be achieved for high current applications. Because of the compact package utilized, small size and high reliability of the DC-DC converters can be realized at the same time. Unfortunately, the regulators are designed for the output voltage control instead of the input voltage control. For a DC-DC converter using a PV panel as the input source, these regulators cannot be implemented directly to perform MPPT function.

In this paper, a method is proposed to achieve MPPT by injecting a control voltage into the compensation pin of a commercial current mode control DC-DC regulator. The method demonstrates that it is economically feasible to use

standard current mode control DC-DC regulator to regulate the input voltage and current of a DC-DC converter, whose input is a PV panel. The advantages of this method are: 1) Because the commercial DC-DC regulators are typically in small sizes, the volume of the DC-DC converter with MPPT can sometimes be reduced. 2) In the commercial market there are so many current mode control DC-DC regulators for output voltage control, that the potential candidates that can perform MPPT can be greatly extended. 3) Typically the commercial DC-DC regulators have complete solutions for the output voltage control design, the features can also be included in the design of DC-DC converters with MPPT and the design time can be shortened.

Specifically, the contributions of this paper include:

- Section II presents the new system level architecture by using COTS current mode control DC-DC regulators to achieve MPPT function, in a simple and low cost manner.

- Section III introduces, for the first time in the literature, methods to use commercial current mode control DC-DC converters (VRMs, Bricks, etc.) to perform input voltage control suitable for MPPT. A control voltage from an additional microcontroller is injected to the compensation pin of the commercial regulator to tune the inductor current, so as to regulate the input voltage. The small signal transfer functions for the MPPT system are analyzed and the variable step MPPT algorithm is introduced

- Experimental results are provided in Section IV to show the effectiveness of the proposed low cost method.

II. SYSTEM ARCHITECTURE

Conventionally, a digital controlled DC-DC converter with MPPT has the architecture shown in Fig. 1(a). The microcontroller will generate the PWM signals to the gate driver so as to control the duty cycle of the switches and to achieve MPPT. In this paper we propose a system architecture for MPPT shown in Fig. 1(b). We use the commercial current mode control DC-DC regulator for the DC-DC converter that extracts energy from a PV panel to the DC load. For most of

---

*This work was partially funded through grants by PowerFilm and DoD contract *W56KGU-14-C0014*

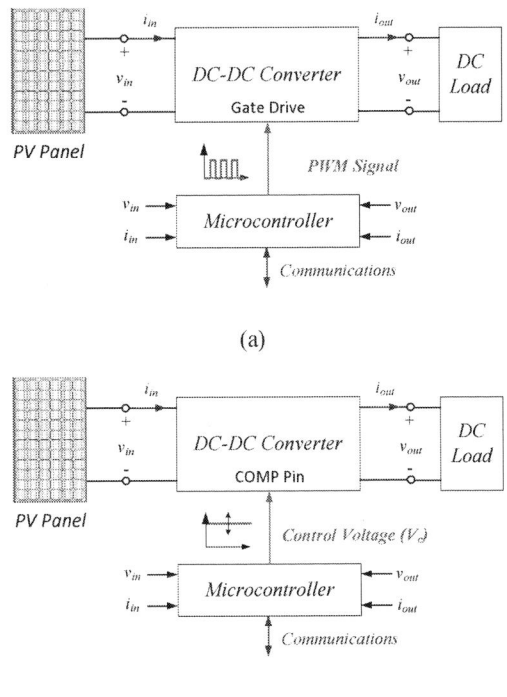

(a)

(b)

Fig. 1 (a) System diagram of a conventional DC-DC converter with the function of MPPT; (b) Proposed system diagram by using a commercial DC-DC regulator.

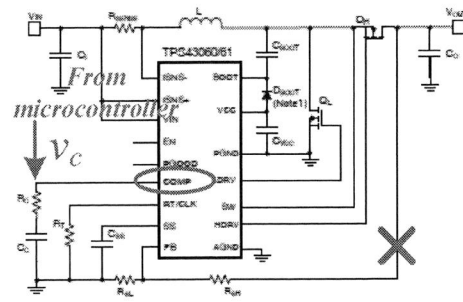

Fig. 2 Diagram of a DC-DC boost converter using a commercial DC-DC regulator.

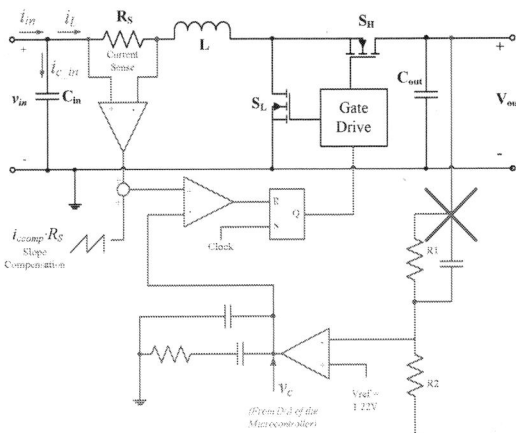

Fig. 3 Proposed method to inject a control voltage to the DC-DC regulator.

the commercial current mode control DC-DC regulators, there is a compensation pin and the voltage of this pin (v_c) will be compared with the sensed current signal to generate the gate drive signals of the switches. For most of the COTS DC-DC converters, v_c would be controlled by the output voltage feedback loop to achieve a constant output voltage level. However, in the proposed architecture the output feedback loop is cut off and a voltage control signal from a microcontroller adjusts its voltage of v_c, so as to realize MPPT.

As an example, Fig. 2 shows the schematic of a DC-DC boost converter with a commercial current mode control DC-DC regulator [9]. With the switch gate drivers integrated in the regulator, only a few external components are added to build to converter. The regulator is designed for output voltage control so that the output voltage is constant over a certain output current range. As shown in Fig. 3, there are two control loops in the regulator. The inner control loop is current control that regulates the inductor current; the outer loop is voltage control that compares the output voltage with a voltage reference to generate the reference at the compensation pin for the current loop. However, if we want to design a DC-DC converter that can perform input voltage or current control, the converter cannot be used directly because it does not have an interface to accept a changing PWM signal for the gate drive signals of the switches, so as to regulate the input voltage of the DC-DC converter at the desired voltage or current level. Further, the input PV power varies according to the irradiance, temperature,

and operating conditions, an algorithm should be developed in the microcontroller to perform MPPT.

III. PROPOSED METHOD TO PERFORM MPPT WITH A COMMERCIAL CURRENT MODE CONTROL DC-DC REGULATOR

Fig. 4 illustrates the overall control diagram of the proposed method. The PV voltage and current are sensed and the corresponding information is transferred to the microcontroller through analog to digital conversion (ADC). The MPPT algorithm designed will generate the control signal based on the current and previous voltage and current measurements. A salient difference between the MPPT algorithm designed and the conventional ones is that the controlled variable is a voltage (v_c) instead of the duty cycle. v_c will be injected into the compensation pin of the DC-DC regulator. The increased or decreased v_c will force the inductor current and consequently force the input voltage of the DC-DC converter to decrease or increase, so as to tune the PV power point. Each main block will be explained below.

It should be pointed out that the developed method can be applied not only for the boost converter but also for other topologies. However, for simplicity the following analysis is for the boost converter.

978-1-4673-9551-9/16 $31.00 ©2016 IEEE   2287

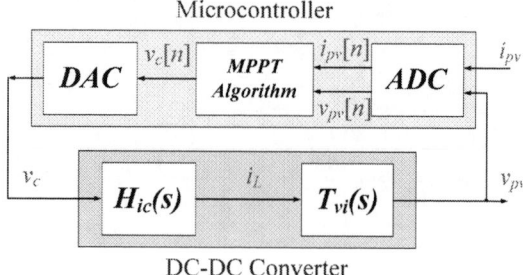

Fig. 4   Control diagram of the proposed method.

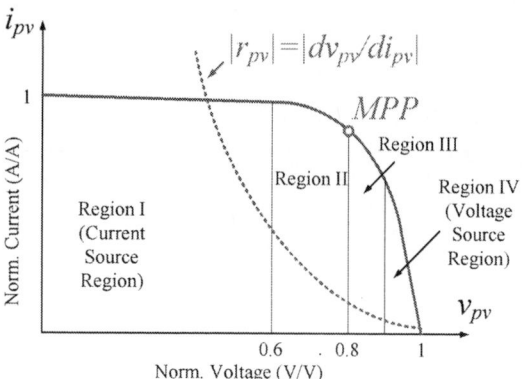

Fig. 7   Dynamic resistance of the PV panel and the classification of different Regions on the PV I-V curve.

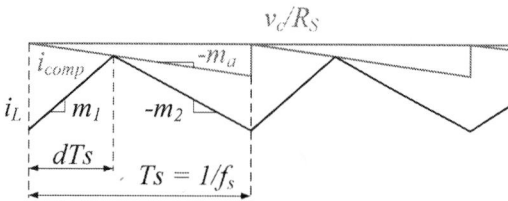

Fig. 5   Peak current mode control modulation waveforms

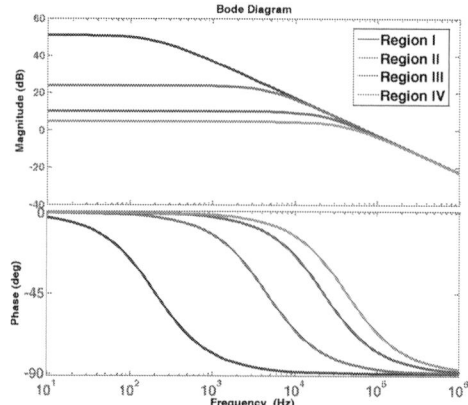

Fig. 8   Bode plot of $|T_{vi}(s)|$ in different Regions.

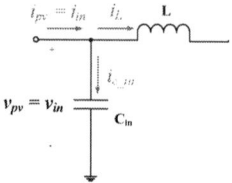

Fig. 6   Equivalent circuit to derive $T_{vi}(s)$.

A. Control Voltage to Inductor Current Transfer Function

As shown in Fig. 4, in the proposed approach, the control voltage (v_c) is from the digital to analog conversion (DAC) of a digital signal processor. The inductor current can be directly controlled by the control voltage and the closed loop control voltage to inductor current transfer function in the s-domain [10] – [11] is

$$H_{ic}(s) = \frac{\hat{i}_L(s)}{\hat{v}_C(s)} = \frac{1}{R_S} \frac{\omega_n^2}{s^2 + \omega_{sh}s + \omega_n^2} \qquad (1)$$

where

$$\omega_n = \sqrt{12}f_s, \ \omega_{sh} = 6\frac{1-\alpha}{1+\alpha}f_s, \ \alpha = \frac{m_2 - m_a}{m_1 + m_a} \qquad (2)$$

and $\hat{x}$ represents the small signal component of variable x. R_S is the current sense resistance, f_s is the switching frequency of

the converter. For the calculation of α, m_1 is the on-slope of the inductor current, m_2 is the off-slope of the inductor current, and m_a is the compensation slope, as shown in Fig. 5. It should be noted that (1) is valid for the frequency range $0 < f < f_s/2$, and (1) is topology independent, meaning it can be applied not only to the boost converter but also for the other topologies. From PV voltage control perspective, the frequencies of the two poles of $H_{ic}(s)$ are typically close the half of the switching frequency, which is much higher than the frequency of MPPT. So we can focus on the low frequency gain of $H_{ic}(s)$, which is $1/R_s$, meaning the inductor current is proportional to v_c at the low frequency range.

B. Inductor Current to Input Voltage Transfer Function

As shown in Fig. 6, because a PV panel is connected to the input of the DC-DC boost converter, we have $v_{in} = v_{pv}$ and $i_{in} = i_{pv}$, the inductor current to input voltage transfer function is given by

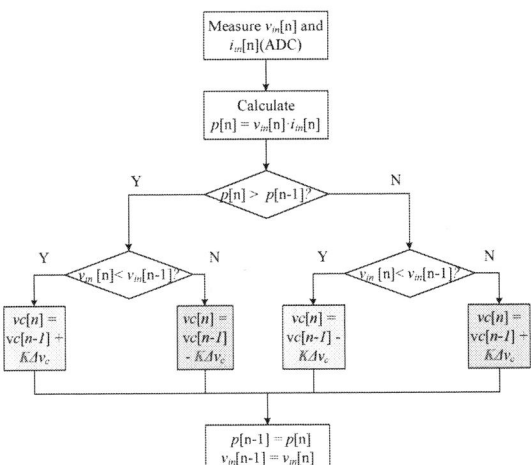

Fig. 9   Proposed MPPT Algorithm.

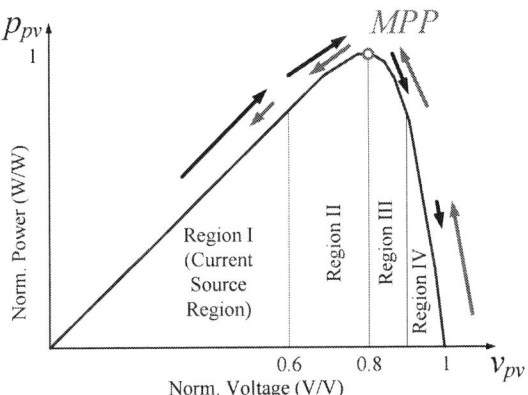

Fig. 10 P&O with variable steps.

$$T_{vi}(s) = \frac{\widehat{v}_{pv}(s)}{\widehat{i}_L(s)} = -\frac{1}{C_{in}s + (1/r_{pv})} = r_{pv}\cdot\frac{1}{\left|r_{pv}\right|C_{in}s+1} \qquad (3)$$

where C_{in} is the input capacitance of the DC-DC converter, r_{pv} is dynamic resistance of the PV panel [12]–[16], which is equal to dv_{pv}/di_{pv} and always has a negative value, which means when an increasing PV current will cause a decreasing PV voltage. Furthermore, r_{pv} is a nonlinear function of the PV voltage (v_{pv}), and it can be classified to the four regions on the I-V curve of the PV panel for easy analysis [13], as shown in Fig. 7. $|r_{pv}|$ is large in the Region I (current source region) and small in the Region IV (voltage source region). Region II and III can be considered as MPP region with the boundary of V_{MPP}, where $|r_{pv}|$ has values between Region I and IV

Fig. 8 shows the Bode plot of $-T_{vi}(s)$. It is a first order system with a pole at $1/(2\pi\,C_{in}\cdot|r_{pv}|)$. As a result the controller can be relatively easy to design. However, because $|r_{pv}|$ has different values through the PV's I-V curve, the gain of each transfer function is different with the respect of the PV voltage and the effect of this large variation cannot be ignored when design the MPPT algorithm.

C. Variable Step MPPT Algorithm Design

Using the above analysis, we can derive the transfer function of $\widehat{v}_{pv}/\widehat{v}_C$ as:

$$\frac{\widehat{v}_{pv}(s)}{\widehat{v}_C(s)} = -\frac{\left|r_{pv}\right|}{R_S}\frac{\omega_n^2}{s^2+\omega_{sh}s+\omega_n^2}\cdot\frac{1}{\left|r_{pv}\right|C_{in}s+1} \qquad (4)$$

Notice that there is a negative sign in (4) because positive increments of v_c cause negative variations of the PV voltage. The DC gain of (4) is: $H_{ci}(0)T_{vi}(0)=|r_{pv}|/R_s$, in which R_s is a constant while r_{pv} has different values in different Regions. A control voltage change step Δv_c will cause different PV voltage change in different Regions. The same Δv_c will cause a much higher voltage change in Region I than in Region IV because $|r_{pv}|$ is much larger in Region I than in Region IV. To compensate this nonlinear relationship between v_{pv} and v_c, in this paper, we adopt a variable steps perturb and observe (P&O) MPPT algorithm, as shown in Fig. 9. The MPPT algorithm designed has the ability to deal with different gains over the I-V curve by adding different control voltage step size and it can be described as

$$v_C[n] = v_C[n-1] \pm K(n)\cdot\Delta v_C \qquad (5)$$

where $K(n)$ is variable magnitude and adjusted base on the P&O direction and the Region of I-V curve, and stored in a lookup table The qualitative idea of the variable step size MPPT [17]-[19] is shown in Fig. 10. For example, suppose the observed P&O direction is counter clockwise ($v_{pv}[n] < v_{pv}[n-1]$), and the power point is entering Region I. This is not wanted because the converter may shut down due to the low input voltage. So a very low $K(n)$ should be adopted to resist the power point to enter this region. Another example is at the

TABLE I. $K(N)$ SELECTION

	Region I	**Region II**	**Region III**	**Region IV**
range	$<0.6Voc_c$	$0.6\text{-}0.8Voc$	$0.8\text{-}0.9Voc$	$>0.9Voc$
$\|r_{pv}\|$	Extra Large (EL)	Large (L)	Normal (N)	Small (S)
direction	0 1	0 1	0 1	0 1
$K(n)$	ES L	S N	N N	EL N

Note: direction = 0 means counter clock wise in Fig. 10 (red arrows), direction = 1 means clock wise in Fig. 10 (black arrows).

Fig. 11 Testing circuit, using MSP430F5338 DSP control board to perform MPPT with a commercial DC-DC boost converter (TPS43061 evaluation board).

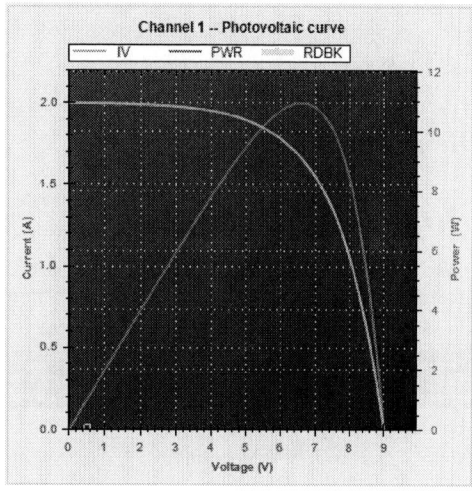

Fig. 12 PV simulator setup in the software.

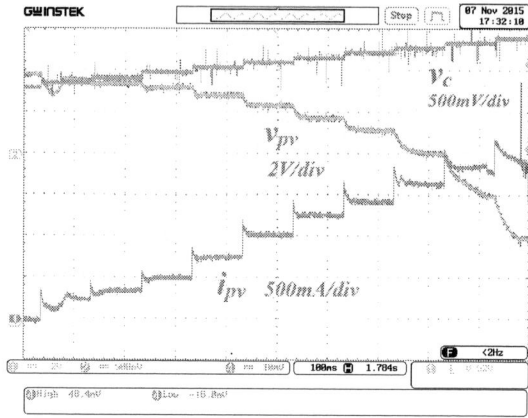

Fig. 13 PV voltage and current variations with same Δvc.

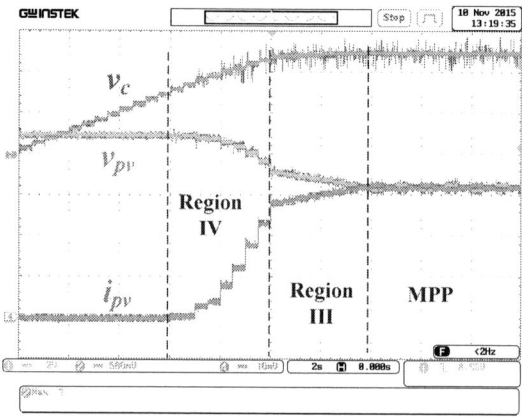

Fig. 14 MPPT procedure with variable Δvc.

startup of the converter the input voltage is close to the PV open circuit voltage, the power point is in Region IV which has a very low DC gain, as a result a very high $K(n)$ should be utilized to accelerate the climbing speed to find the MPP. The complete $K(n)$ selection is listed in Table I.

IV. EXPERIMENTAL VERIFICATION

The experimental setup is built to verify the effectiveness of the proposed method. As shown in Fig. 11, TPS43061 evaluation board is utilized as the DC-DC boost converter to extract energy from the PV simulator to the electronic load. The DSP (MSP430F5338) in a DSP evaluation board measures input voltage, input current and output voltage of the DC-DC converter through the voltage and current sensing circuit, perform the MPPT algorithm, and then the digital to analog conversion (DAC) output of the DSP can inject the control

voltage (v_c) to the compensation pin of a commercial DC-DC regulator (TPS43061). . The DSP evaluation board in Fig. 11 could be replaced by a small microcontroller for reduced size.

In the experiments the P-V and I-V curves of the PV simulator output are shown in Fig. 12. The MPP is when V_{in} is ~6.6V, I_{in} is 1.66A and the MPP is ~10.95W. Electronic load together with a DC power supply are utilized to create a constant output voltage to be 15V.

Fig. 13 shows the PV voltage and current variations with the same Δv_c. Although the control voltage step is the same, for different PV operating voltage, the voltage drop becomes different. The lower the PV voltage, the higher the PV voltage drop is generated. Further the PV voltage curve is nonlinear with a linear control voltage rise. The results coincide with the analysis in Section IV. It should be pointed out that in Fig. 13 the Δv_c is setup to be large to generate significant PV voltage and current variations. In the real MPPT algorithm Δv_c is much smaller.

With the proposed variable MPPT algorithm implemented, the PV simulator can work near the vicinity of maximum power point. Fig. 14 illustrates the proposed MPPT procedure from the startup of the DC-DC converter. At the beginning v_{pv}

is the open circuit voltage (9V), and the power point is in Region IV (voltage source region), so the designed gain is high. When the power point enters Region III, there is a significant gain reduction to acquire stable power point approaching to MPP.

V. CONCLUSIONS AND FUTURE WORK

In this paper a method has been derived to perform the MPPT function incorporating a commercial DC-DC regulator, which is originally designed for output voltage control. An external control voltage generated from the microcontroller is injected into the compensation pin of the current mode control DC-DC regulator to tune the inductor current, so as to regulate the input PV voltage and current at the desired power point. Experimental results have been reported to validate the proposed approach.

At present the Region boundaries are pre stored in the microcontroller. In the future the boundaries can be on line recognized and can be modified automatically. Also the value of Δv_c and MPPT period can be optimized to achieve better performance.

ACKNOWLEDGMENT

The authors would like to thank Prof. Yuan Li for several helpful discussions and Yue Zheng for the setup of the experimental bench.

REFERENCES

[1] Yihua Hu; Weidong Xiao; Wenping Cao; Bing Ji; Morrow, D.J., "Three-Port DC–DC Converter for Stand-Alone Photovoltaic Systems," *IEEE Trans. Power Electron.* , vol.30, no.6, pp.3068-3076, June 2015.

[2] Olalla, C.; Clement, D.; Rodriguez, M.; Maksimovic, D., "Architectures and Control of Submodule Integrated DC–DC Converters for Photovoltaic Applications," *IEEE Trans. Power Electron.*, vol.28, no.6, pp.2980-2997, June 2013.

[3] Gonzalez Montoya, D.; Ramos-Paja, C.A.; Giral, R., "Improved Design of Sliding-Mode Controllers Based on the Requirements of MPPT Techniques," *IEEE, Trans. Power Electron.*, vol.31, no.1, pp.235-247, Jan. 2016.

[4] Khanna, R.; Qinhao Zhang; Stanchina, W.E.; Reed, G.F.; Zhi-Hong Mao, "Maximum Power Point Tracking Using Model Reference Adaptive Control," *IEEE, Trans. Power Electron.*, vol.29, no.3, pp.1490-1499, March 2014.

[5] W. Feng, W. Xinke, F. C. Lee, W. Zijian, K. Pengju, and Z. Fang, "Analysis of Unified Output MPPT Control in Subpanel PV Converter System," *IEEE Trans. Power Electron.*, vol. 29, pp. 1275-1284, 2014.

[6] M. S. Agamy, M. H. Todorovic, A. Elasser, S. Chi, R. L. Steigerwald, J. A. Sabale, A. J. McCann, Li Zang, and F. Mueller, "An efficient partial power processing DC/DC converter for distributed PV architectures," *IEEE Trans. Power Electron.*, vol. 29, no. 12, pp. 674–686, Feb. 2014.

[7] G. R. Walker and P. C. Sernia, "Cascaded dc-dc converter connection of photovoltaic modules," *IEEE Trans. Power Electron.*, vol. 19, no. 4, pp. 1130–1139, Jul. 2004.

[8] C. Hua, J. Lin, and C. Shen, "Implementation of a DSP-controlled photovoltaic system with peak power tracking," *IEEE Trans. Ind. Electron.*, vol. 45, no. 1, pp. 99–107, Feb. 1998.

[9] TPS43061 Datasheet, http://www.ti.com/product/tps43061.

[10] Ridley, R.B., "A new, continuous-time model for current-mode control [power convertors]," *IEEE Trans. Power Electron.*, vol.6, no.2, pp.271-280, Apr 1991.

[11] Kondrath, N.; Kazimierczuk, M.K., "Control-to-output transfer function of peak current-mode controlled PWM DC-DC boost converter in CCM," *Electronics Letters*, vol.47, no.17, pp.991-993, August 2011.

[12] T. Suntio et. al., "Issues on solar-generator interfacing with current-fed MPP-tracking converters", *IEEE, Trans. Power Electron.*, vol. 25, no. 9, pp. 2409-2419, September 2010.

[13] Weidong Xiao; Dunford, W.G.; Palmer, P.R.; Capel, A., "Regulation of Photovoltaic Voltage," *IEEE Trans. Ind. Electron.*, vol.54, no.3, pp.1365-1374, June 2007.

[14] J. Thongpron, K. Kirtikara, and C. Jivicate, "A method for determination of dynamic resistance of photovoltaic modules under illumination," *Solar Energy Mater. Solar Cells*, vol. 90, no. 18/19, pp. 3078–3084, Nov. 2006.

[15] M. G. Villalva and E. R. Filho, "Dynamic analysis of the input-controlled buck converter fed by a photovoltaic array," *Sba Controle Automac, ͂ao*, vol. 19, no. 4, 2008.

[16] A. Urtasun, P. Sanchis, and L. Marroyo, "Adaptive voltage control of the dc/dc boost stage in PV converters with small input capacitor," *IEEE Trans. Power Electron.*, vol. 28, no. 11, pp. 5038–5048, Nov. 2013.

[17] Yuncong Jiang; Qahouq, J.A.A.; Haskew, T.A., "Adaptive Step Size With Adaptive-Perturbation-Frequency Digital MPPT Controller for a Single-Sensor Photovoltaic Solar System," *IEEE Trans. Power Electron.*, vol.28, no.7, pp.3195-3205, July 2013.

[18] Fangrui Liu; Shanxu Duan; Fei Liu; Bangyin Liu; Yong Kang, "A Variable Step Size INC MPPT Method for PV Systems," *IEEE Trans. Ind. Electron.*, vol.55, no.7, pp.2622-2628, July 2008.

[19] Qiang Mei; Mingwei Shan; Liying Liu; Guerrero, J.M., "A Novel Improved Variable Step-Size Incremental-Resistance MPPT Method for PV Systems," *IEEE Trans. Ind. Electron.*, vol.58, no.6, pp.2427-2434, June 2011.

Envelope Tracking GaN Power Supply for 4G Cell Phone Base Stations

Yuanzhe Zhang*, Johan Strydom†, Michael de Rooij†, Dragan Maksimović*

*Colorado Power Electronics Center
Department of Electrical, Computer and Energy Engineering
University of Colorado, Boulder, CO, USA
Email: yuanzhe.zhang@colorado.edu

†Efficient Power Conversion Corporation, El Segundo, CA, USA
Email: michael.derooij@epc-co.com

Abstract—This paper introduces an envelope tracking (ET) power supply for 4G cell phone base stations using EPC eGaN® FETs. An analytical loss model is developed for design optimization and verified by a single phase synchronous buck converter using the zero-voltage switching (ZVS) technique. The model is then extended to four phases and is used to design a 60 W ET power supply with 20 MHz large signal bandwidth (BW). At 25 MHz per-phase switching frequency, measured static power stage efficiency peaks at 96.5% with 68 W output power delivered from 30 V. Experimental results demonstrate accurate tracking of 20 MHz 7 dB Peak-to-Average Power Ratio (PAPR) LTE envelope with 92.3% total efficiency, delivering 67 W average power from 30 V.

I. INTRODUCTION

RF power amplifiers (RFPAs) have low efficiency when transmitting signals with high peak-to-average power ratio (PAPR), as in the case of 4G LTE signals. To improve system efficiency, the drain supply for the RFPA needs to be modulated based on the input signal, where an envelope tracking (ET) power supply, or a drain supply modulator, is required to generate time-varying envelope waveforms. Two basic categories of ET power supplies have been studied: linear power amplifiers [1], [2] and switching converters [3]–[8]. Switching converters are preferred for high efficiency operation, but their bandwidth is often limited as a result of limited switching frequency, especially with silicon technology. Therefore, hybrid ET power supplies, or switching converter assisted linear amplifiers, were introduced to combine the advantages of the linear amplifier and the switching converter [9]–[13].

Wide band-gap device technologies have led to breakthrough in switching converters in terms of tracking bandwidth and efficiency [13]–[22]. For instance, single phase GaN ET power supplies are capable of 20 MHz bandwidth and up to 83% average efficiency for tracking LTE envelope [5]–[7]. Circuit design techniques such as multi-phase topology have also achieved record performance in ET supplies [3], [23]. Even with silicon Trench MOS technology, an 8-phase ET switching power supply has demonstrated 10 MHz large signal bandwidth and 93% average efficiency when tracking

6.8 dB PAR WCDMA envelope [3]. A two-phase GaN ET power supply has shown 89% average tracking efficiency [23].

This paper proposes an eGaN® FET based switching ET power supply for 4G LTE cell phone base stations, offering 20 MHz bandwidth, operating at 30 V, and delivering 60 W average output power. The design includes a four-phase synchronous buck converter power stage, with synchronous FET bootstrap supply [24] and a zero-voltage switching (ZVS) compatible fourth order output filter [23]. The design guidelines for the power stage, output filter and gate driver are given in Section II, where an analytical loss model is developed and used to optimize the 60 W ET power supply with 20 MHz large signal bandwidth. Experimental results are presented in Section III for the four phase prototype operating at 25 MHz per-phase switching frequency, having peak steady-state power stage efficiency of 96.5% with 68 W output power delivered from 30 V. The results also demonstrate accurate tracking of 20 MHz 7 dB PAPR LTE envelope with 92.3% efficiency, delivering 67 W average power from 30 V. Conclusions are presented in Section IV.

II. DESIGN GUIDELINES

Figure 1 shows a simplified diagram of an N-phase ET power supply with a fourth order output filter. The half-bridge power stage in each phase is driven by its own pulse-width modulated (PWM) control signals. The triangular dual edge PWM carriers are phase shifted by $2\pi/N$ (in radians). In this configuration, the ripple cancellation effect provides extra attenuation at the switching frequency and its harmonics [3], [25], allowing lower switching frequency for the same bandwidth compared to single phase converter. A resistive load is used to emulate the behavior of a PA in saturation (or gain compression) mode. The design objectives are: 30 V input voltage, 60 W average output power, 300 W peak output power and 20 MHz bandwidth.

ZVS operation is a popular soft-switching technique where a filter inductor resonates with a switching node capacitance during switching transitions [26]. In multi-phase designs, ZVS

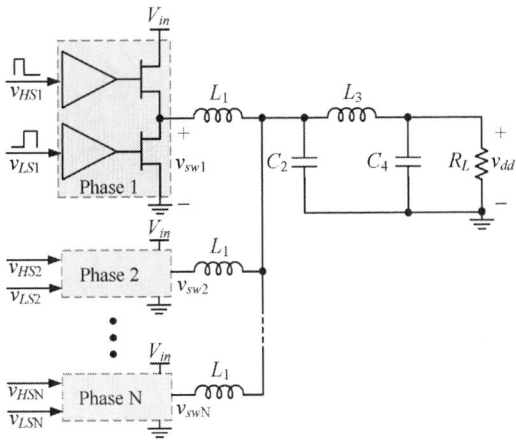

Figure 1. N phase synchronous buck converter with fourth-order filter.

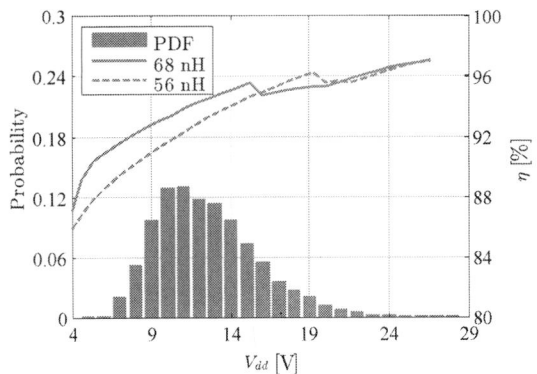

Figure 2. Model predicted steady-state efficiency for the output voltage range of interest, with the probability density function (PDF) of a 20 MHz 7 dB PAPR LTE envelope superimposed.

is preferable as it provides inherent phase current balancing [27].

A. Power Stage

To achieve ZVS, the inductor current ripple (defined as the difference between the peak and the average current) must be large enough, usually at least the average current, and depends on various operating conditions such input voltage, duty cycle, load, etc. As a result, there is the trade-off between conduction loss and switching loss. In [4], an optimization method based on the analytical loss model was introduced, where the device size was selected to balance the trade-off and achieve the highest efficiency.

Three different device sizes are available from the EPC8000 series each rated at 40 V: EPC8004 (110 mΩ), EPC8007 (160 mΩ) and EPC8008 (325 mΩ) [28]–[30]. The choice of four phase is based on the current rating of the devices. The per-phase switching frequency of 25 MHz is selected. An analytical loss model is developed for selecting optimum power stage device size, or EPC part. The power stage model employed is similar to the model presented in [4]. The loss mechanisms considered in the model are: FET conduction loss, FET switching loss and ZVS inductor loss. The inductor is selected so that the ZVS condition is satisfied for duty cycles up to approximately 0.75. The model suggests that the EPC8004 is the best candidate for the required power level, with the ZVS inductor L_1 in each phase of 56 nH.

The device size optimization only considers a single steady-state operation point. In the ET applications, however, the optimization goal is the averaged efficiency when tracking a certain envelope signal. Therefore, the steady-state operating point where the output power is equal to the averaged output power in tracking condition is used for device size optimization. Furthermore, the selected ZVS condition for duty cycles up to 0.75 may not result in the best averaged efficiency under tracking conditions. The inductor, or consequently, the ZVS range can be further adjusted for a given target envelope signal.

Figure 2 shows the model predicted steady-state efficiency of the four-phase converter for a range of output voltages, for two ZVS inductors L_1: 56 nH and 68 nH. The amplitude probability density function (PDF) of a 20 MHz 7 dB PAPR LTE envelope is superimposed. The slight drop in each efficiency traces indicates the point where partial ZVS occurs. With the 56 nH inductor, the ZVS range is wider and the peak efficiency is slightly higher. However, at lower output voltages the 68 nH inductor yields higher efficiency. Under quasi-static approximation where the losses under tracking conditions are assumed to be the same as in steady-state, the averaged tracking efficiency can be estimated. For this particular envelope signal, the predicted averaged tracking efficiency is 93.7% for 56 nH, and 94.5% for 68 nH. Therefore, 68 nH is used in the final prototype.

B. ZVS Output Filter

Output filter of a switching converter has to meet the switching ripple attenuation requirement. In ET applications, the bandwidth of the output filter needs to be carefully considered as well. With a higher-order filter, the trade-off between filter bandwidth and switching ripple attenuation is much easier to address. Therefore, fourth order filters are considered in this work. While classical filter designs (e.g. Butterworth, Bessel, Legendre, etc) already have standard normalized transfer functions and the corresponding circuit realizations, they do not support ZVS operation [23]. The incompatibility comes from the constraint on the first filter inductor L_1, which also acts as the ZVS inductor. Some other filters based on a π approach that have a capacitor as the first element are also not suitable implementations, due to the increased switching loss associated with the capacitor. Following the design procedure presented in [23], a fourth order ZVS-compatible output filter is designed for the four-phase converter.

As an example, the component values for a fourth-order ZVS filter design with 20 MHz bandwidth are listed in Table I.

Table I
ZVS 4TH-ORDER FILTER DESIGNS FOR A 4 PHASE CONVERTER.

BW	L_1	C_2	L_3	C_4	R_L
20 MHz	68 nH	5 nF	22 nH	1.6 nF	2.6 Ω

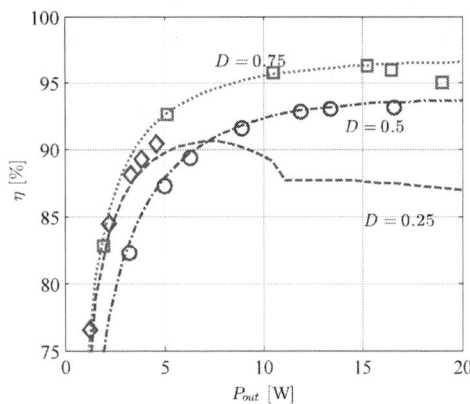

Figure 3. Power stage efficiency of the single phase ET prototype: $V_{in} =$ 30 V, $f_s =$ 25 MHz, model predicted (dashed lines) and measured (markers).

Note that in Fig. 1, each phase has its own inductor L_1 and shares the rest of the filter components.

C. Gate Driver

The design of half-bridge gate driver that supports high-efficiency operation at tens of megahertz range is particularly challenging. Conventional bootstrap diode based gate drivers suffer from the reverse recovery loss of the bootstrap diode. An isolated driver design is introduced in [4], capable of driving depletion mode GaN HEMTs at switching frequencies up to 40 MHz. However, the isolated auxiliary power supply and the transformer need to be carefully designed and are large in size.

The gate driver implemented in this work features a synchronous FET bootstrap supply, where a GaN FET is used to replace the function of the bootstrap diode and eliminate the associated reverse recovery loss [24]. The synchronous bootstrap FET only processes the power for the gate driver, and therefore FETs with small gate periphery (size) and low gate charge work best. EPC2038 [31] is selected as the synchronous FET, as it has the lowest gate charge (44 pC) lowest parasitic capacitances and smallest footprint (0.9x0.9 mm) for a very compact design and minimal impact on the ZVS operation.

III. EXPERIMENTAL VERIFICATION

A single phase synchronous buck converter operating at 25 MHz switching frequency was built and tested first. Power stage efficiency was measured at three different duty cycles (0.25, 0.5 and 0.75) with 30 V input voltage, for a range of output power. The efficiency result is represented by the markers in Fig. 3, with the model predicted efficiency indicated

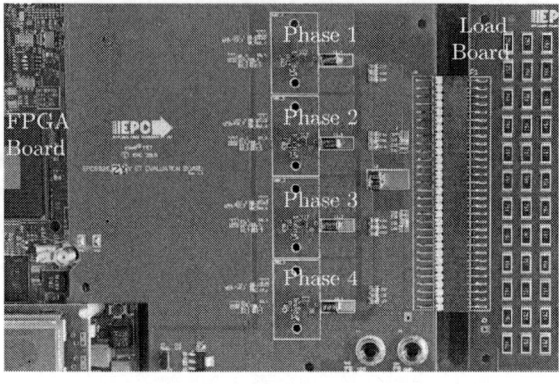

(a)

(b)

Figure 4. (a) Four phase ET prototype; and (b) zoomed details of one phase with EPC8004 as power stage devices and EPC2038 in the synchronous FET bootstrap supply.

by dashed lines. The experimental results indicate a good correlation with the prediction from the analytical loss model. The efficiency peaks at 96% and remains above 90% for output power in the 5-20 W range.

A four phase synchronous buck converter was then built. Figure 4 shows a photo of the four phase prototype with EPC8004 as power stage devices and EPC2038 in the synchronous FET bootstrap supply. The four phases are separated well from each other for improved heat dissipation distribution. High Q air core inductors from Coilcraft and high frequency capacitors from Johanson Technology are used in the fourth order ZVS-compatible output filter, as listed in Table I.

High frequency pulse-width modulated control signals are generated using an Altera Stratix IV FPGA with a duty cycle and dead-time resolution of 200 ps. The eight control signals (high-side and low-side control signals are separated in each phase) are configured in bonded mode for low channel to channel skew [32].

The four phase converter was tested in a similar manner, at three different duty cycles (0.25, 0.5 and 0.75) with 30 V input voltage, for a range of output power (with variable load resistor). The efficiency results are shown by the markers in Fig. 5, with the model predicted efficiency indicated by dashed

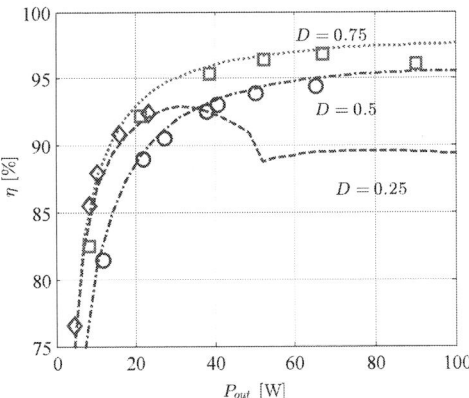

Figure 5. Power stage efficiency of the four phase ET prototype: $V_{in} = 30$ V, $f_s = 25$ MHz, model predicted (dashed lines) and measured (markers).

lines. The shape of the efficiency curves resemble those in the single phase, as expected, and the efficiency peaks at 96.7%.

In order to estimate the averaged tracking efficiency, the four phase converter was further measured at static operating points with a fixed load resistor of 2.6 Ω for a range of duty cycles from 0.1 to 0.6. The measured power stage efficiency and total efficiency (including gate drive loss) are shown in Fig. 6a. The measured gate driver loss is 1.1 W. Since the designed averaged output power is only 60 W, the maximum measured steady-state output power is 140 W, corresponding to the duty cycle of 0.56. Figure 6b shows the same measured power stage efficiency, compared to the loss model prediction, with the PDF of a 20 MHz LTE envelope superimposed. It further verifies the loss model particularly in the ZVS range. At higher output voltages where ZVS is lost, the model predicts higher efficiency because of the approximations in the loss model and the ignored thermal effect. The discrepancy is acceptable in this case because the probability distribution at higher output voltages is very low.

The ET control signals with the required pre-filtering and pre-distortion [23] are processed and generated using MAT-LAB, and are stored in the memories of the Stratix IV FPGA. The signals are then transmitted to the gate drivers, as in [4], [7].

Figure 7a shows a sample of the output voltage $v_{dd}(t)$ generated by the four phase ET prototype, as well as the switching node voltages v_{sw1} through v_{sw4}, where ZVS transitions can be observed for a range of duty cycles. Occasional partial ZVS or hard switching is caused by the signal dynamic and limited dead time resolution. Figure 7b shows a comparison of the experimental versus target envelope waveforms demonstrating accurate tracking and low output voltage ripple. The measured normalized RMS error (NRMSE) is 1.2%. The power stage and total average efficiencies measured under dynamic tracking condition are 93.7% and 92.3%, respectively, with 67 W output power.

Since the converter is operated in open-loop, phase current

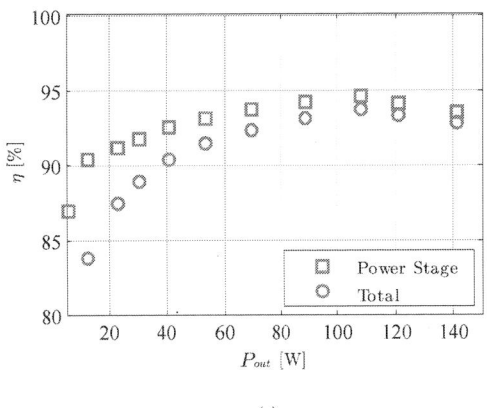

(a)

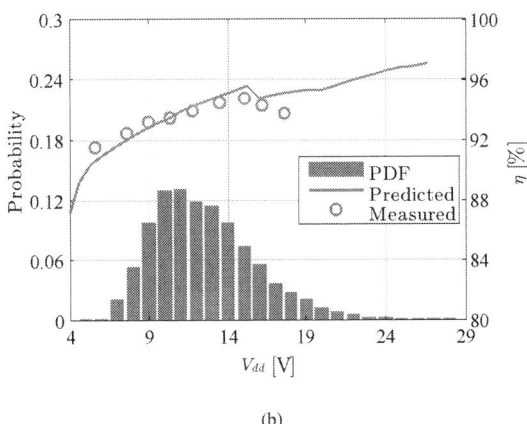

(b)

Figure 6. (a) Measured static power stage and total efficiencies for the four phase ET prototype, $V_{in} = 30$ V, per-phase $f_s = 25$ MHz; and (b) measured and model predicted power stage efficiencies and the probability distribution function (PDF) of the target 20 MHz LTE envelope signal.

balancing relies only on ZVS. According to Fig. 6b, approximately 85% of the time the converter is operating with ZVS, and therefore phase current should be balanced. The infrared image shown in Fig. 8a for the four phase converter shows similar temperature among all phases, indicating each phase has approximately the same loss and therefore the current is balanced.

IV. CONCLUSIONS

This paper introduced a 20 MHz bandwidth four phase envelope tracking (ET) power supply built using eGaN® FETs that is suitable for 4G cell phone base stations. An analytical technique was introduced that was used to design an optimal four-phase ET power supply operating at 25 MHz per-phase switching and which included a fourth order ZVS filter. Experimental results demonstrated accurate tracking of a 20 MHz bandwidth LTE envelope signals with 92.3% total efficiency at 67 W output power. The design is fully scalable by using smaller devices for lower power base stations.

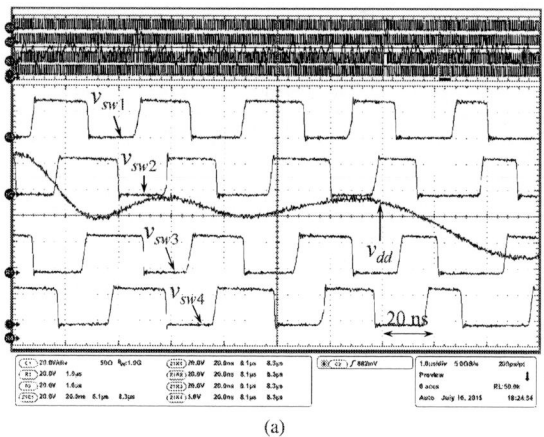

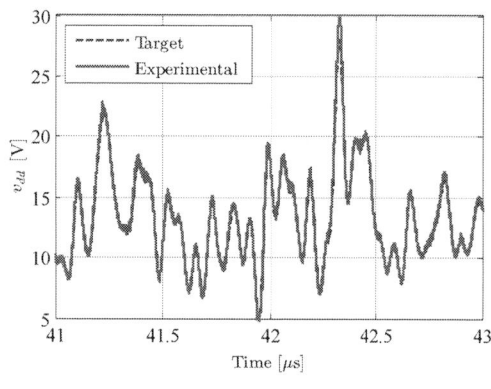

(a)

(b)

Figure 7. (a) experimental LTE envelope waveforms $v_{dd}(t)$ and switching node voltages v_{sw1} through v_{sw4}; and (b) Experimental and target 20 MHz LTE envelope waveforms v_{dd}.

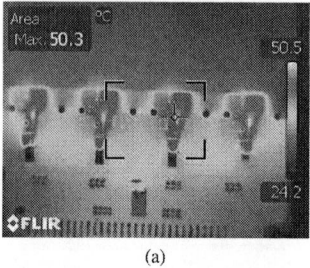

(a)

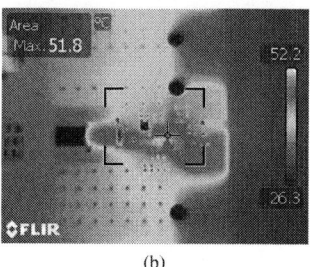

(b)

Figure 8. Infrared thermal image of the four phase prototype in operation: (a) heat is evenly distributed acorss all phases; and (b) zoomed details of one phase.

REFERENCES

[1] D. Li, M. Rodriguez, A. Zai, D. Sardin, D. Maksimovic, and Z. Popovic, "RFPA supply modulator using wide-bandwidth linear amplifier with a GaN HEMT output stage," in *Proc. IEEE Workshop Contr. Modl.*, 2013.

[2] P. Theilmann, J. Yan, C. Vu, J. sun Moon, H. Moyer, and D. Kimball, "A 60mhz bandwidth high efficiency X-band envelope tracking power amplifier," in *Proc. IEEE Compound Semicond. Integr. Circuit Symp.*, Oct. 2013, pp. 1–4.

[3] M. Norris and D. Maksimovic, "10 MHz large signal bandwidth, 95% efficient power supply for 3G-4G cell phone base stations," in *Proc. IEEE 27th Annu. Appl. Power Electron. Conf. Expo.*, Feb. 2012, pp. 7 –13.

[4] M. Rodriguez, Y. Zhang, and D. Maksimovic, "High-frequency PWM buck converters using GaN-on-SiC HEMTs," *IEEE Trans. Power Electron.*, vol. 29, no. 5, pp. 2462–2473, 2014.

[5] Y.-P. Hong, K. Mukai, H. Gheidi, S. Shinjo, and P. Asbeck, "High efficiency GaN switching converter IC with bootstrap driver for envelope tracking applications," in *Proc. IEEE Radio Freq. Integr. Circuits Symp.*, 2013, pp. 353–356.

[6] S. Shinjo, Y.-P. Hong, H. Gheidi, D. Kimball, and P. Asbeck, "High speed, high analog bandwidth buck converter using GaN HEMTs for envelope tracking power amplifier applications," in *Proc. IEEE Wireless Sens. and Sens. Netw.*, 2013, pp. 13–15.

[7] Y. Zhang, M. Rodriguez, and D. Maksimovic, "100 MHz, 20 V, 90%

efficient synchronous Buck converter with integrated gate driver," in *Proc. IEEE Energy Convers. Congr. Expo.*, 2014.

[8] P. Cheng, M. Vasic, O. Garcia, J. Oliver, P. Alou, and J. Cobos, "Minimum time control for multiphase buck converter: Analysis and application," *IEEE Trans. Power Electron.*, vol. 29, no. 2, pp. 958–967, Feb 2014.

[9] D. Li, Y. Zhang, M. Rodriguez, and D. Maksimovic, "Band separation in linear-assisted switching power amplifiers for accurate wide-bandwidth envelope tracking," in *Proc. IEEE Energy Convers. Congr. Expo.*, Sep. 2014, pp. 1113–1118.

[10] V. Yousefzadeh, E. Alarcon, and D. Maksimovic, "Band separation and efficiency optimization in linear-assisted switching power amplifiers," in *Proc. IEEE 37th Power Electron. Spec. Conf.*, Jun. 2006, pp. 1–7.

[11] P. Miaja, M. Rodriguez, A. Rodriguez, and J. Sebastian, "A linear assisted DC/DC converter for envelope tracking and envelope elimination and restoration applications," *IEEE Trans. Power Electron.*, vol. 27, no. 7, pp. 3302–3309, Jul. 2012.

[12] M. Vasic, O. Garcia, J. Oliver, P. Alou, and J. Cobos, "Theoretical efficiency limits of a serial and parallel linear-assisted switching converter as an envelope amplifier," *IEEE Trans. Power Electron.*, vol. 29, no. 2, pp. 719–728, Feb 2014.

[13] P. Miaja, A. Rodriguez, and J. Sebastian, "Buck-derived converters based on gallium nitride devices for envelope tracking applications," *IEEE Trans. Power Electron.*, vol. 30, no. 6, pp. 2084–2095, Jun. 2015.

[14] J. Rivas, D. Jackson, O. Leitermann, A. Sagneri, Y. Han, and D. Perreault, "Design considerations for very high frequency dc-dc converters," in *Proc. IEEE 37th Power Electron. Spec. Conf.*, Jun. 2006, pp. 1 –11.

[15] D. Perreault, J. Hu, J. Rivas, Y. Han, O. Leitermann, R. Pilawa-Podgurski, A. Sagneri, and C. Sullivan, "Opportunities and challenges in very high frequency power conversion," in *Proc. IEEE 24th Annu. Appl. Power Electron. Conf. Expo.*, Feb. 2009, pp. 1–14.

[16] R. Pilawa-Podgurski, A. Sagneri, J. Rivas, D. Anderson, and D. Perreault, "Very-high-frequency resonant boost converters," *IEEE Trans. Power Electron.*, vol. 24, no. 6, pp. 1654–1665, Jun. 2009.

[17] T. Andersen, S. Christensen, A. Knott, and M. A. E. Andersen, "A VHF class E DC-DC converter with self-oscillating gate driver," in *Proc. IEEE 26th Annu. Appl. Power Electron. Conf. Expo.*, Mar. 2011, pp. 885–891.

[18] J. Hu, A. Sagneri, J. Rivas, Y. Han, S. Davis, and D. Perreault, "High-frequency resonant SEPIC converter with wide input and output voltage ranges," *IEEE Trans. Power Electron.*, vol. 27, no. 1, pp. 189–200, Jan. 2012.

[19] A. Sagneri, D. Anderson, and D. Perreault, "Optimization of integrated transistors for very high frequency dc-dc converters," *IEEE Trans. Power Electron.*, vol. 28, no. 7, pp. 3614–3626, Jul. 2013.

978-1-4673-9551-9/16 $31.00 © 2016 IEEE

[20] J. Garcia, R. Marante, and M. de las Nieves Ruiz Lavin, "GaN HEMT Class E2 resonant topologies for UHF DC/DC power conversion," *IEEE Trans. Microw. Theory Tech.*, vol. 60, no. 12, pp. 4220–4229, Dec. 2012.

[21] J. Garcia, R. Marante, M. Ruiz, and G. Hernandez, "A 1 GHz frequency-controlled class E2 DC/DC converter for efficiently handling wideband signal envelopes," in *Proc. IEEE MTT-S Int. Microw. Symp. Dig.*, Jun. 2013, pp. 1–4.

[22] Y. Zhang, M. Rodriguez, and D. Maksimovic, "High frequency synchronous buck converter using GaN-on-SiC HEMTs," in *Proc. IEEE Energy Convers. Congr. Expo.*, 2013, pp. 488–494.

[23] Y. Zhang, M. Rodriguez, and D. Maksimovic, "Output filter design in high-efficiency wide-bandwidth multi-phase buck envelope amplifiers," in *Proc. IEEE 30th Annu. Appl. Power Electron. Conf. Expo.*, March 2015, pp. 2026–2032.

[24] M. A. de Rooij, *Wireless Power Handbook*, 2nd ed. El Segundo: Power Conversion Publications, 2015.

[25] J. Sebastian, P. Fernandez-Miaja, F. Ortega-Gonzalez, M. Patino, and M. Rodriguez, "Design of a two-phase buck converter with fourth-order output filter for envelope amplifiers of limited bandwidth," *IEEE Trans. Power Electron.*, vol. 29, no. 11, pp. 5933–5948, Nov 2014.

[26] D. Maksimovic, "Design of the zero-voltage-switching quasi-square-wave resonant switch," in *Proc. IEEE 24th Power Electron. Spec. Conf.*, Jun. 1993, pp. 323 –329.

[27] D. Costinett, D. Seltzer, R. Zane, and D. Maksimovic, "Analysis of inherent volt-second balancing of magnetic devices in zero-voltage switched power converters," in *Proc. IEEE 28th Annu. Appl. Power Electron. Conf. Expo.*, Feb. 2013, pp. 9–15.

[28] Efficient Power Conversion Corporation. (2015) Enhancement mode power transistor EPC8004 - Datasheet. [Online]. Available: http://epc-co.com/epc/documents/datasheets/EPC8004_datasheet.pdf

[29] Efficient Power Conversion Corporation. (2013) Enhancement mode power transistor EPC8007 - Datasheet. [Online]. Available: http://epc-co.com/epc/documents/datasheets/EPC8007_datasheet.pdf

[30] Efficient Power Conversion Corporation. (2013) Enhancement mode power transistor EPC8008 - Datasheet. [Online]. Available: http://epc-co.com/epc/documents/datasheets/EPC8008_datasheet.pdf

[31] Efficient Power Conversion Corporation. (2015) Enhancement mode power transistor EPC2038 - Datasheet. [Online]. Available: http://epc-co.com/epc/documents/datasheets/EPC2038_preliminary.pdf

[32] Altera. (2014) Stratix IV device handbook volume 2: Transceivers. [Online]. Available: https://www.altera.com/products/fpga/stratix-series/stratix-iv/support.html

978-1-4673-9551-9/16 $31.00 © 2016 IEEE

Envelope Tracking Power Supply for Volume-Sensitive Low-Power Applications Based on a Resonant Switched-Capacitor Converter

Alon Cervera, *Student Member, IEEE* and Mor Mordechai Peretz, *Member, IEEE*

The Center for Power Electronics and Mixed-Signal IC, Department of Electrical and Computer Engineering
Ben-Gurion University of the Negev, P.O. Box 653, Beer-Sheva, 8410501 Israel
cervera@bgu.ac.il , and morp@ee.bgu.ac.il
http://www.ee.bgu.ac.il/~pemic

Abstract– **This paper introduces a new envelope tracking power supply to enhance the performance and reduce the overall volume of transmitter modules. The realization is based on a gyrator resonant switched-capacitor converter (GRSCC) that acts as a controlled bi-directional current source with a virtually instantaneous response to create the desired envelope. A quasi-non-casual control method is further applied to closely track high-rate signals without further increasing the switching frequency. The resultant dynamic performance for a given envelope reference signal is significantly improved, providing envelope supply voltage without clipping distortion. Design example and simulations are detailed, and the operation of envelope tracker is verified on a 150mW prototype switched at 2.3MHz, able to track a 1MHz envelope with less than half the losses of a linear regulator.**

I. INTRODUCTION

With the proliferation of portable electronics, the issue of power management has become one of the primary factors to achieve small and light equipment, and in particular prolonging the device operability (battery life) and maintaining satisfactory performance. The two main power consumers of a portable system are the processor and the communication (wireless) transmitter. The main objective of the processor's supply is to sustain a well-regulated constant output voltage under wide range of load changes. On the other hand, the main challenge in the design of the transmitter supply is for its voltage to be modulated at base-band signal rate (10MHz range), requiring extremely high control bandwidth [1-4]. Therefore, design efforts are not as much in efficiency but in size of passive components [1].

Envelope Elimination and Restoration (EER) or Envelope Tracking (ET) methods [2-16] realize envelope modulation of the power supply by either linear regulators, switch-mode regulators, or both. To achieve high efficiency of the switched-mode converter, switched-inductor converter topologies are prominent. Ideally there, the efficiency characteristics can be maintained high for wide range of conversion ratios. However, to minimize the necessity of additional linear regulation to achieve the required bandwidth, ultra-fast switching converters are used [16-18] where their switching losses may affect the efficiency. In addition, the total size of these solutions is not necessarily reduced since now inductors are required, which may also prohibit on-chip integration.

Switched-capacitor (SC) technology has a unique benefit of being a 'perfect' candidate for miniaturization and on-chip integration [19-21]. In the context of ET, SC is a poor solution since it can produce high efficiency in singular conversion ratios [22, 23]. Fortunately, a new family of gyrator resonant switched-capacitor converters (GRSCC) with continuous conversion ratio was recently introduced in [24]. In addition to a wider efficiency-curve, the converter has current sourcing capability and has been shown to respond without delay to line and load variations [19]. These attributes make the GRSCC an attractive candidate for ET applications.

The objective of this study is to introduce a *new rapid adaptive voltage scaling envelope tracking system that is realized by a GRSCC for volume-sensitive low-power applications*. The solution, as detailed in Fig. 1, provides a high-efficiency, *volume-saving* alternative to the conventional switched-inductor approach. Combined with a newly developed non-linear controller, the ET power supply minimizes the

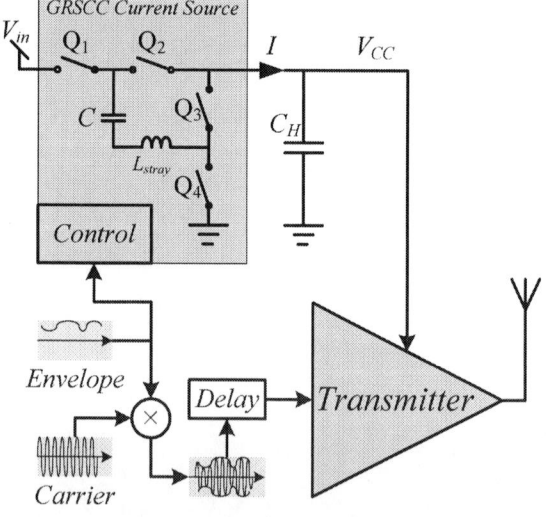

Fig. 1   GRSCC-based envelope tracker.

978-1-4673-9551-9/16 $31.00 © 2016 IEEE

tracking mismatch and significantly reduces losses related to the linear regulator shape adjuster.

The rest of the paper is organized as follows: Section II describes the new ET principle of operation, portrays the effect of design constraints on the tracking capability, and presents a control scheme to minimize tracking mismatches. Section III delineates the realization of a GRSCC as an envelope tracker. Next, Section IV provides a design example to track a typical orthogonal frequency-division multiplexing (OFDM) RF signal. Experimental results and the conclusion are then provided in Sections V and VI, respectively.

II. ENVELOPE TRACKING BY CURRENT SOURCING

A generic behavior of the circuit in Fig. 1 can be conceptually described by a controlled current-source that produces $I(t)$ and mimics the operation of the GRSCC, capacitance C_H and a load resistance R_{PA}, as illustrated in Fig. 2. As described in [3,6], a constant resistor is used to emulate the effective loading by a transmitter. For $V_{CC}(t)$ to perfectly match a desired envelope signal $V_E(t)$, the current source is to output a signal $I(t)$ with the linear dependency of the form:

$$I(t) = \frac{V_E(t)}{R_{PA}} + C_H \frac{dV_E(t)}{dt} . \quad (1)$$

A physical converter is limited by current boundaries of I_{min} and I_{max}. As a result, in some cases $V_{CC}(t)$ would be lower than the desired $V_E(t)$, a prohibited scenario in ET applications. Derived from (1) and the circuit in Fig. 2, the response of $V_{CC}(t)$ has a first-order nature due to the current limitation of I_{max}. In most ET applications, the information for the required voltage is available by pseudo non-casual data (e.g. broadcast signal). Therefore, skewing the current injection sequence can be applied by a factor ΔT_{max} which can be expressed as:

$$\Delta T_{max} = - R_{PA} C_H \ln\left(1 - \frac{V_{peak}}{I_{max} R_{PA}}\right), \quad (2)$$

where V_{peak} is the maximum value for $V_{CC}(t)$. A profile vector with length ΔT_{max} of the desired target envelope is used as the reference signal to the tracking system.

To account for the current output limitations and to satisfy $V_{CC}(t) > V_E(t)$ at all times, the reference vector is reconstructed based on the voltage ramp up or down capabilities of the current source. The resulting $V_E^*(t)$ deviates from $V_E(t)$ only when the desired envelope's slews are higher than can be obtained by the current source applied on C_H.

The procedure to generate $V_E^*(t)$ is carried out in three steps and is conceptually illustrated in Fig. 3 which describes an arbitrary portion of a given envelope signal $V_E(t)$. A preliminary step based on the system parameters (i.e., the current ratings, output capacitance and the load), calculates the ramping up/down capabilities in the system as follows:

$$V_{ramp\text{-}up}(t) = I_{max} R_{PA}\left(1 - e^{-\frac{t}{R_{PA}C_H}}\right) \quad (3a)$$

$$V_{ramp\text{-}down}(t) = -I_{max} R_{PA}\left(1 - 2e^{-\frac{t}{R_{PA}C_H}}\right) . \quad (3b)$$

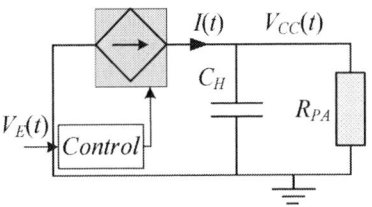

Fig. 2   An average model to illustrate the tracking concept. The GRSCC is modelled by a controlled current source with saturation.

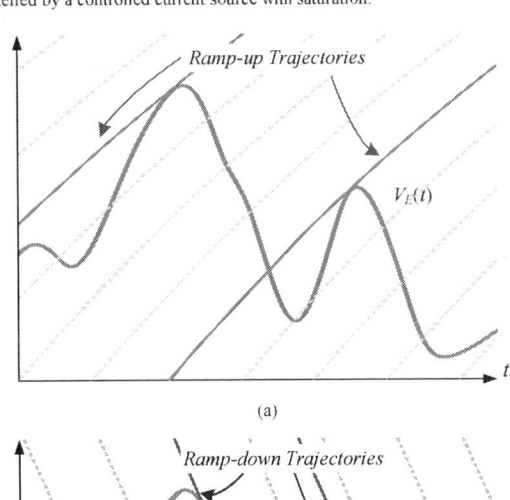

(a)

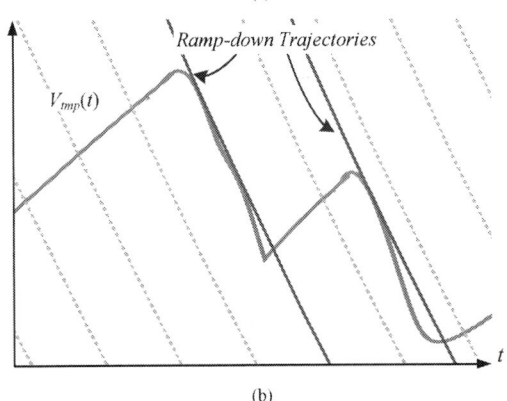

(b)

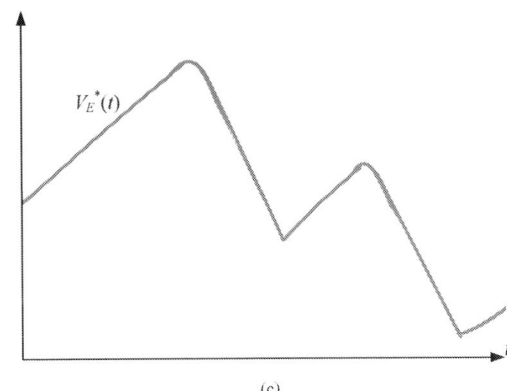

(c)

Fig. 3   Step-by-step tracking refernce vector generation procedure. (a) accommodating ramp-up limitations; (b) ramping down; (c) the resultant reference signal compared to the desired one.

In the second step (Fig. 3a), for a given vector length of $\Delta T_{\max}$, the possible ramp-up trajectories are compared with the slews of $V_E(t)$ and the reference vector is reconstructed by tangential trajectories to $V_E(t)$ (i.e., with equal slope) to assure that $V_{CC}(t) = V_E^*(t) > V_E(t)$. In the final step (Fig. 3b), the procedure is repeated with the ramp-down trajectories. The resulting reference vector $V_E^*(t)$ is depicted in Fig. 3c. For evaluation purposes, the current-limited injection sequence can be derived using (1) with $V_E^*(t)$. Fig. 4 shows a wider view of the tracking operation, demonstrating generation of a continuous reference vector with severe current output limitations, as well as comparison with constant supply setting. It can be observed that in spite of the limited ramping capabilities, significant portion of the losses is reduced thanks to the tracking method.

III. REALIZATION OF A CURRENT SOURCE BY A GYRATOR RESONANT SWITCHED CAPACITOR CONVERTER

The GRSCC topology (Fig. 1) recently presented in [19] has evolved from the conventional soft-switched resonant switched-capacitor converter (SCC) configuration. As can be seen in Fig. 5, a third switching state is added to balance the charge difference between the input and output rather than introducing losses. By doing so, the converter allows higher efficiency over a wide and continuous step-up/down conversion ratio. Thanks to its soft-switching resonant nature it is applicable at high frequencies and features near-ideal bi-directional pulsed current source behavior. The pulsed behavior can be expressed by average pulse current amplitude, I_{pulse}, of the converter, given by:

$$I_{\text{pulse}} = \frac{2CV_{\text{in}}}{T_{\text{pulse}}}, \qquad (4)$$

where C is the flying capacitor value, V_{in} is the supply voltage, and $T_{\text{pulse}} = 3\pi\sqrt{LC}$ is the pulse duration of the converter, set by the resonant network of L and C. Bi-directional operation is realized by reversing the three-state switching scheme, and is expressed in the context of (4) by the expression having positive values for sourcing current and negative values for sinking current operation. As described in detail in [19], immediate response can be achieved with the GRSCC by pulse-density modulation (PDM).

Voltage regulation using ripple-based PDM [19,25] applies valley comparison to the output voltage, i.e. a comparator that triggers the pulsed source each time a reference value is met. Ideally, this method can be applied to envelope tracking by comparing $V_{CC}(t)$ to the synthesized $V_E^*(t)$. Practically, in envelope tracking applications the high-rate envelope variation and, in this study, the need to track a synthesized $V_E^*(t)$ prohibits to use a physical comparator due to the intrinsic delay between the valley detection to the pulse output. To remedy this, pre-calculation and comparison is conducted in advance as follows:

Assuming that the resultant output voltage deviation for a given pulse magnitude of I_{pulse} is relatively small, $V_{CC}(t)$ can be approximated by first order to (Fig. 6):

$$V_{CC}(t) = \begin{cases} \left(I_{\text{pulse}} - \dfrac{V_{CC}(t_0)}{R_{PA}}\right)\dfrac{(t-t_0)}{C_H} & , \ t_0 < t < t_1 \\[2mm] -\dfrac{V_{CC}(t_1)}{R_{PA}}\dfrac{(t-t_1)}{C_H} & , \ t_1 < t < t_2 \end{cases}, \qquad (5)$$

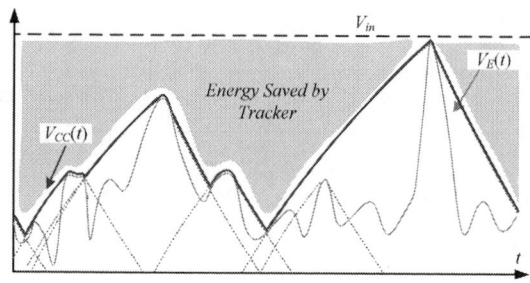

Fig. 4   Illustration describing the tracking operation and saved energy for an information segment in $V_E(t)$ under severe current limitation conditions.

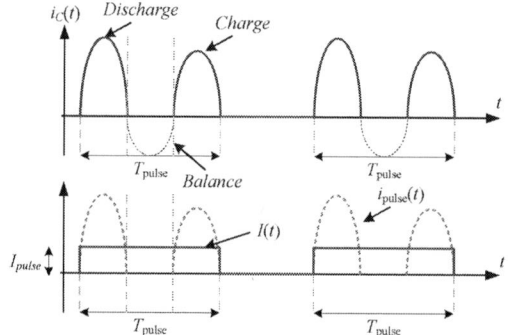

Fig. 5   Current waveforms on the GRSCC depicted in Fig.1 for two pulses. The top graph displays the current on the resonant tank, $i_c(t)$, bottom graph displays the resonant pulse as seen by thhe output, namely $i_{\text{pulse}}(t)$ and its' average representation of $I(t)$. switch activity is as follows: discharge - Q_2,Q_4; balance - Q_2,Q_3; charge - Q_1,Q_3.

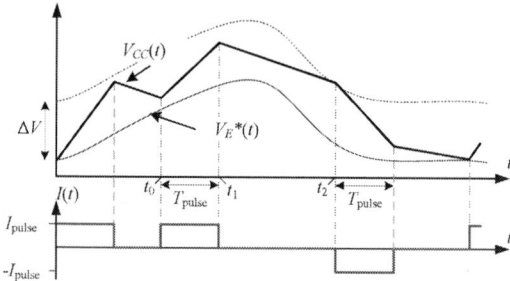

Fig. 6   The methodology for using a constant-pulse source for tracking $V_E^*(t)$

where t_0 is the time of an initiated pulse, $t_1 = t_0 + T_{\text{pulse}}$ is the pulse end and t_2 is the trigger point for the next pulse, when $V_{CC}(t) = V_E^*(t)$ (see Fig. 6). $V_{CC}(t)$ is actively reduced when needed by sinking excess charge back to the source using indication from an upper threshold. This threshold is set to $V_{CC}(t) = V_E^*(t) + \Delta V_{CC}$, where ΔV_{CC} is the impact of a single pulse, expressed as:

$$\Delta V_{CC} = \frac{T_{\text{pulse}} I_{\text{pulse}}}{C_H}. \qquad (6)$$

It should be noted that one potential drawback of this method is lack of feedback and as a result, sensitivity to parameter variations. This is resolved by either parameter estimation, or worst-case based design with minor effect on the tracking capability.

IV. SIMULATION CASE STUDY

General design guidelines for the construction of the GRSCC have been previously delineated in [19]. Therefore only specific considerations for the ET application are described through a simulation example.

The target parameters for the envelope tracker are listed in Table I. Given the output power and peak to average power ratio (PAPR), the peak current is $I_{peak} \approx 0.45$ A at $V_{CC} \approx 2.24$ V. Based

TABLE I – EXAMPLE DESIGN SPECIFICATIONS

Component	Value
Input voltage V_{in}	3.3 V
Minimum input voltage $V_{in,min}$	3 V
Average PA output power P_{out}	250 mW
Reflected load impedance R_{PA}	4.7 Ω
Peak average power ratio PAPR	4
Estimated envelope frequency	2.5 MHz

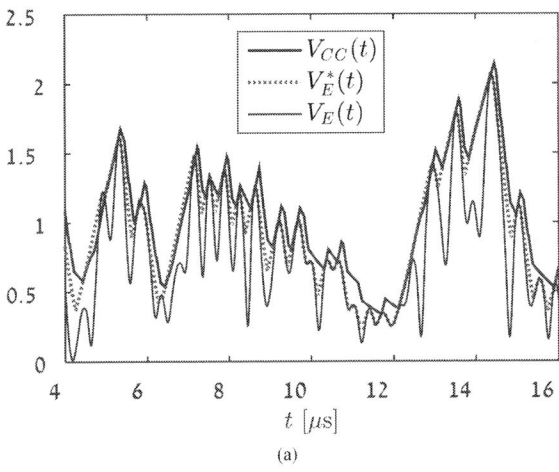

(a)

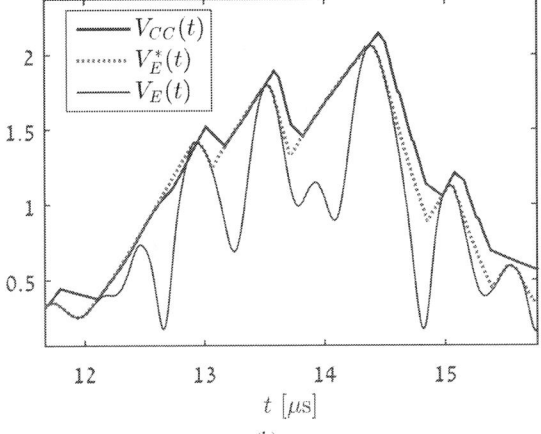

(b)

Fig. 7   Simulation results of the converter tracking a reconstructed envelope above $V_E(t)$. (a) wide view, (b) zoom.

on (4), the converter is designed to output current of $I_{pulse} = 1$ A to satisfy ramping capabilities of 2 V/μs. The maximum effective frequency is set at $F_{max} = 1/T_{pulse} = 10$ MHz. The resulting resonant network components' values are $C_f \approx 15$ nF, $L \approx 8$ nH. Choosing $C_H = 0.4$ μF results in $\Delta V_{CC} \approx 0.25$ V.

The tracking results have been produced using MATLAB and are shown in Fig. 6, which also shows the reference vector generated as described earlier. The design example results in 250% tracking efficiency improvement compared to a constant supply voltage ($\eta_{ET} = V_E / V_{CC} = 70\%$ versus 24%). It should be noted that the efficiency estimation included conservative design of the GRSCC with series loop resistances of $R_S = 50$mΩ, and power-stage efficiency of $\eta = 85\%$.

V. EXPERIMENTAL RESULTS

To verify the operation of the envelope tracker and demonstrate the control scheme solution, a 250mW prototype was built and tested, and is depicted in Fig. 7. Table I lists the power-stage specifications and components' values. The converter was realized by a GRSCC as described in Section III with the resonant tank constructed using capacitors alone. To complete the resonant tank, air-core inductance of 10nH was added to reach the total of 20nH. The envelope reference signal was synthesized using MATLAB, based on generic OFDM symbols, and then translated to pulse timing and direction vectors. GRSCC control was implemented on an Altera Cyclone IV FPGA [26] to create timed source/sink gate-signals for the transistor drivers. Soft-switching was achieved with pre-calibration of switching timings.

Fig. 8 demonstrates the converter's output, recorded from oscilloscope (Teledyne LeCroy HD4016) and compared to the original envelope and the predicted trace through MATLAB. The tracking results are in very good agreement with the theoretical

TABLE II – EXPERIMENTAL SPECIFICATIONS

Component	Value
Input voltage V_{in}	3 V
Average PA output power P_{out}	250 mW
Load impedance R_{PA}	4.7 Ω
Tank capacitance C	60 nF
Estimated tank inductance L	20 nH
Output capacitance C_H	0.6 nF
Estimated loop resistance R_S	0.05 Ω
Maximum switching frequency Fs	3 MHz
Estimated envelope frequency	1 MHz
MOSFET type	SiA436DJ

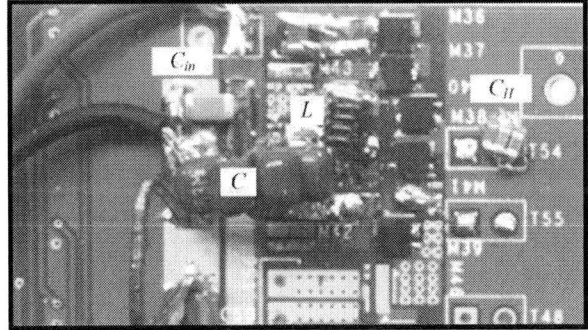

Fig. 8   250 mW 3MHz Experimental prototype. Zoomed in is the resonant tank.

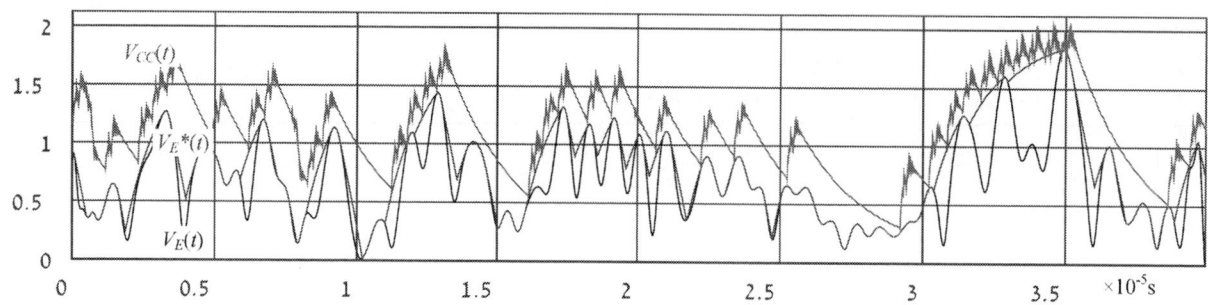

Fig. 9   Experimental results comparing a recorded signal to the original reconstructed envelope and $V_E(t)$.

predictions, verifying the capabilities of the converter as well as the design procedure. The tracking efficiency is measured to be $\eta_{ET} \approx 65\%$ versus 20% in the linear mode approach, with $1/T_{pulse}$ just marginally above the envelope frequency, an improvement of more than 200%. Furthermore, the presented solution is with similar performance to a switched-inductor approach. It should be noted that in some cases the current pulse is triggered before $V_{CC}(t)$ reaches $V_E^*(t)$, and is due to a slight deviation of the resonant parameters and efficiency estimation in relation to the virtual boundaries discussed in Section III.

VI. CONCLUSION

A new, rapid adaptive voltage scaling envelope tracking system based on a high-efficiency gyrator resonant-switched capacitor converter has been presented. Detailed analysis of the system detailed a new method for envelope tracking by current sourcing which facilitates tight tracking under limited design considerations (e.g. power and switching frequency). A design example has been provided to emphasize considerations oriented toward the target values and performance. The analysis has been methodically verified by simulations and experiments and the results are in excellent agreement with the theoretical predictions providing tracking efficiency improvement of more than 200% using switching frequency of less than 2.5 times higher than the envelope signal.

Combined with the topology benefits, the volume-saving simple GRSCC voltage regulation scheme presents an attractive alternative to the switch-inductor converters, in particular in area sensitive applications, and establishes the foundations for better power delivery concepts for envelope tracking applications.

REFERENCES

[1] R.Y.L. Zhu, D. Prikhodko, Y. Tkachenko, "LTE power amplifier module design: Challenges and trends," *IEEE International Conference on Solid-State and Integrated Circuit Technology (ICSICT)*, pp.192-195, Nov. 2010.

[2] V. Yousefzadeh, E. Alarcon, D. Maksimovic, "Efficiency optimization in linear-assisted switching power converters for envelope tracking in RF power amplifiers," *IEEE International Symposium on Circuits and Systems, ISCAS 2005*, pp.1302-1305 Vol. 2, May 2005.

[3] V. Yousefzadeh, E. Alarcon, D. Maksimovic, "Band Separation and Efficiency Optimization in Linear-Assisted Switching Power Amplifiers," *IEEE Power Electronics Specialists Conference, PESC '06*, pp.1,7, June 2006

[4] Yusoff, "The auxiliary envelope tracking RF power amplifier system," *PhD diss.*, Cardiff University, 2012.

[5] F. Wang, D. Kimball, J. Popp, A. Yang, D.Y.C. Lie, P. Asbeck, and L. Larson, "Wideband envelope elimination and restoration power amplifier

with high efficiency wideband envelope amplifier for WLAN 802.11 g applications," *Microwave Symposium Digest, 2005 IEEE MTT-S International*, pp.345-348, 2005.

[6] M. Vasic, O. Garcia, J.A. Oliver, P. Alou, D. Diaz, J.A. Cobos, "Multilevel Power Supply for High-Efficiency RF Amplifiers," *IEEE Trans. on Power Electronics*, vol.25, no.4, pp.1078-1089, April 2010.

[7] M. Vasic, O. Garcia, J.A. Oliver, P. Alou, J.A. Cobos, "Serial or parallel linear-assisted switching converter as envelope amplifier: Optimization and comparison," *Energy Conversion Congress and Exposition (ECCE), 2011 IEEE*, pp.2488-2494, Sept. 2011.

[8] O. Garcia, M. Vasic, P. Alou, J.A. Oliver, J.A. Cobos, "An overview of fast DC-DC converters for envelope amplifier in RF transmitters," *IEEE Applied Power Electronics Conference and Exposition (APEC)*, pp.1313-1318, Feb. 2012.

[9] A. Soto, J.A. Oliver, J.A. Cobos, J. Cezon, F. Arevalo, "Power supply for a radio transmitter with modulated supply voltage," *IEEE Applied Power Electronics Conference and Exposition,. APEC '04*, pp.392-398, 2004.

[10] J. T. Stauth, "Energy Efficient Wireless Transmitters: Polar and Direct-Digital Modulation Architectures," ProQuest, 2008.

[11] P. Markowski, J. Ronnie, A. Stiedl, "New Achievements in Envelope Tracking Technology," Emerson Network Power, September 2011.

[12] V. Yousefzadeh, E. Alarcon, D. Maksimovic, "Three-level buck converter for envelope tracking applications," *IEEE Trans. on Power Electronics*, vol.21, no.2, pp.549,552, March 2006.

[13] N. Wang, V. Yousefzadeh, D. Maksimovic, S. Pajic, Z.B. Popovic, "60% efficient 10-GHz power amplifier with dynamic drain bias control," *IEEE Trans. on Microwave Theory and Techniques*, vol.52, no.3, pp.1077-1081, March 2004.

[14] M. Norris, D. Maksimovic, "10 MHz large signal bandwidth, 95% efficient power supply for 3G-4G cell phone base stations," *IEEE Applied Power Electronics Conference and Exposition (APEC)*, pp.7-13, Feb. 2012.

[15] F. Wang, D.F. Kimball, J.D. Popp, A.H. Yang, D.Y. Lie, P.M. Asbeck, L.E. Larson, "An Improved Power-Added Efficiency 19-dBm Hybrid Envelope Elimination and Restoration Power Amplifier for 802.11g WLAN Applications," *IEEE Trans. on Microwave Theory and Techniques*, vol.54, no.12, pp.4086-4099, Dec. 2006.

[16] D. Diaz, O. Garcia, J.A. Oliver, P. Alou, Z. Pavlovic, J.A. Cobos, "The Ripple Cancellation Technique Applied to a Synchronous Buck Converter to Achieve a Very High Bandwidth and Very High Efficiency Envelope Amplifier," *IEEE Trans. on Power Electronics*, vol.29, no.6, pp.2892-2902, June 2014.

[17] L. Marco, A. Poveda, E. Alarcon, D. Maksimovic, "Bandwidth limits in PWM switching amplifiers," *IEEE International Symposium on Circuits and Systems (ISCAS 2006) Proceedings*, pp. 5326, May 2006.

[18] K. S. Leung and H. S. H. Chung, "A comparative study of the boundary control of buck converters using first- and second-order switching surfaces—Part I: Continuous conduction mode," *in Proc. IEEE Power Electron. Spec. Conf.*, pp. 2133 -2139, Jun. 2005.

[19] A. Cervera and M.M. Peretz, "Resonant switched-capacitor voltage regulator with ideal transient response," *IEEE Trans. on Power Electronics*, vol.30, no.9, pp.4943-4951, Sep. 2015.

[20] O. Kirshenboim, A. Cervera, and M. M. Peretz, "Improving Loading and Unloading Transient Response of a Voltage Regulator Module Using a

Load-Side Auxiliary Gyrator Circuit," *IEEE Applied Power Electronics Conf., APEC-2015*, pp.913-920, March 2015.

[21] R.C.N. Pilawa-Podgurski and D.J. Perreault, "Merged two-stage power converter with soft charging switched-capacitor stage in 180 nm CMOS," *IEEE Journal of Solid-State Circuits*, vol. 47, no. 7, pp. 1557-1567 2012.

[22] S. Ben-Yaakov, "Behavioral average modeling and equivalent circuit simulation of switched capacitors converters," *IEEE Trans. on Power Electronics*, vol.27, no.2, 632-636, 2012.

[23] S. Ben-Yaakov, "On the influence of switch resistances on switched-capacitor converter losses," *IEEE Trans. on Industrial Electronics, Letters*, vol.59, no.1, 638-640, 2012.

[24] A. Cervera, M. Evzelman, M.M. Peretz, and S. Ben-Yaakov, "A high efficiency resonant switched capacitor converter with continuous conversion ratio," *IEEE Trans. on Power Electronics*, vol.30, no.3, pp.1373-1382, March 2015.

[25] R. Redl and Jian Sun, "Ripple-based control of switching Regulators—An overview," *IEEE Transactions on Power Electronics*, vol. 24, no. 12, pp. 2669-2680 2009.

[26] DE2 Development and Education Board user manual, Altera Corporation, 2006.

A Passive-Impedance-Matching Concept for Multi-phase Resonant Converter

Hongliang Wang, *Senior Member, IEEE*, Yang Chen, Yan-Fei Liu, *Fellow,IEEE*

Department of Electrical and Computer Engineering

Queen's University

Kingston, Canada

hongliang.wang@queensu.ca, yang.chen@queensu.ca, yanfei.liu@queensu.ca

Abstract— **A passive impedance matching (PIM) concept is proposed for multi-phase resonant converters application. The new concept can achieve load current sharing between every unit automatically. A passive element such as an inductor or a capacitor is connected in the common branch of the two units (or multi-units), such that several sets of equivalent positive resistors and negative resistors are distributed into these units accordingly. The equivalent virtual resistors are of variable value, which automatically adjust the DC gain in each unit to achieve load sharing. Three equivalents LLC resonant converters with virtual variable resistor are analyzed. The theoretical analysis result shows that the first and third types have good load current sharing performance. A 600W, 12V two-phase LLC converter prototype is built based on the first type. The prototype verified the feasibility and demonstrated advantages of PIM multiphase resonant converter.**

I. INTRODUCTION

LLC resonant converter has been widely used due to its high efficiency achieved by zero voltage switching (ZVS) on the primary-side MOSFETs and zero current switching (ZCS) on secondary-side diodes [1][2]. For high power applications, current stress of power devices increase with the power rating, so multiphase parallel technique is a good choice to solve this problem [3][4][5]. But, components tolerances may cause each LLC unit to have different resonant frequency. This will lead to the deviation of current stress in each LLC unit [6][7][8]. Small component tolerances will cause large current imbalance. Thus, the key problem is load sharing.

Three technologies have been used to achieve current sharing in multiphase LLC converter. One is the active method in which passive components tolerance can be compensated by adjusting the variable capacitor [6][9] or inductor [10] in an additional circuit. This method has prefect load sharing performance, but it has large cost, complex control and non-excellent dynamic performance because of

sensing the circulating current and controlling the additional switches. The second method is self-balanced DC voltage based on series bus capacitors[11][12]. Take two-phase LLC converter as an example, the mid-point voltage is changed according to two unit's power. Thus, the system has low cost and good load current sharing performance. However, it has poor reliability because the DC gain is halved when one unit is broken. Besides, it is hard for modularization design since the DC voltage stress is reduced with module number increased. The third method is built in three-phase three-wire structure for three-phase LLCs, which has good load current sharing near resonant frequency as all of three-phase resonant current is zero[13][14].It is only suitable for three-phase - the load current does not share when the number of module exceeds three. In a nutshell, existing studies have limitation on cost, complex control, modularization and dynamic performance.

In this paper, a passive impedance matching concept is proposed for multi-phase resonant converter. A passive element, such as an inductor or a capacitor, is put in the common branch of the multiple units. A set of virtual resistors (positive and negative) are yielded through the common branch inductor or capacitor, which changes the gain curve in different converter. Three resonant converters with variable resistor have been analyzed. A 600W, 12V two-phase LLC converters prototype based on sharing resonant inductor is built to verify the feasibility and demonstrate the advantages. The paper is organized as follows. Section II introduces virtual impedance concept; Section III analyzes the three resonant converters with variable resistor. Section IV shows the experimental results. Conclusion is given in Section V.

II. VIRTUAL IMPEDANCE CONCEPT

Fig.1 shows the basic circuit about the common branch impedance. Fig.1 (a) shows that K current source units $I_1(s)$, $I_2(s)$... $I_k(s)$ in parallel are connected to the common branch. The voltage of common impedance Z_s is V_s. The total current is defined as $I_s(s)$

$$I_s(s) = \sum_1^k I_j(s) \tag{1}$$

The impedance Z_s can be separated into Z_{s1}, Z_{s2}... Z_{sk} for each unit based on current $I_1(s)$, $I_2(s)$... $I_k(s)$ to achieve 'virtual open' multiple units, which is shown in Fig.1(b).Three kinds of passive elements (inductors, capacitors and resistors) can be selected as impedance Z_s. Resistors will introduce power loss and lower efficiency of converter. Thus, inductors and capacitors are better candidates in this application. Fig.1(c) and Fig.1 (d) illustrate two examples of sharing an inductor and a capacitor in the common branch.

Fig.1 (e) is the vector plot of sharing an inductor in Fig.1 (c). Vector $I_s(s)$ is composed of vector $I_1(s)$ and $I_2(s)$. The current $I_1(s)$ is leading current $I_2(s)$, so the impedance angle of $Z_1(s)$ is smaller than 90^0 and that of $Z_2(s)$ is larger than 90^0. In consequence, impedance $Z_1(s)$ is equivalent to an inductor k_1L_s

and a positive resistor R_{s1} connected in series. Similarly, impedance $Z_2(s)$ is equivalent to one inductor k_2L_s and a negative resistor R_{s2} connected in series. Where k_1, k_2 is the coefficient parameters. They can be found from (2).

$$\frac{(sk_1L_s + R_{s1})(sk_2L_s + R_{s2})}{(k_1 + k_2)sL_s + (R_{s1} + R_{s2})} = sL_s \tag{2}$$

The circuit in Fig.1(c) can be replaced by the circuit in Fig.1 (g). In the same analysis, two separated impedances can be given by virtual open. The circuit in Fig.1 (d) can be replaced by the circuit in Fig.1 (h). Where k_1, k_2 is the coefficient parameters. They can be found from (3).

$$\frac{(\frac{1}{sk_1C_s} + R_{s1})(\frac{1}{sk_2C_s} + R_{s2})}{(\frac{1}{k_1} + \frac{1}{k_2})\frac{1}{sC_s} + (R_{s1} + R_{s2})} = \frac{1}{sC_s} \tag{3}$$

No matter for the sharing inductor or capacitor, an equivalent circuit with virtual positive resistor and negative resistor can always be derived. Three different LLC converters with variable resistors will be analyzed.

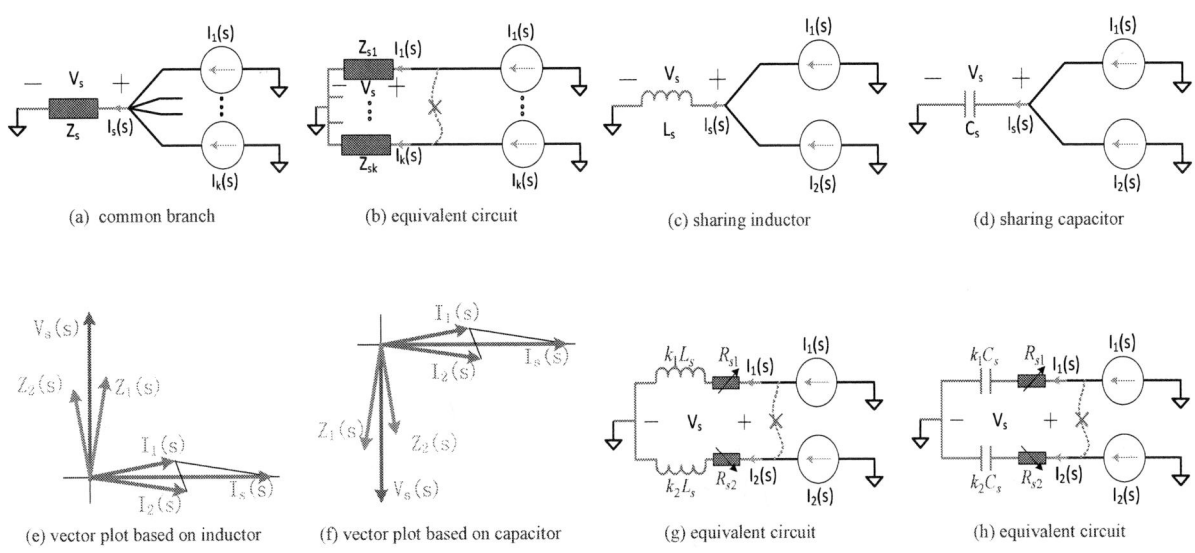

(a) common branch (b) equivalent circuit (c) sharing inductor (d) sharing capacitor

(e) vector plot based on inductor (f) vector plot based on capacitor (g) equivalent circuit (h) equivalent circuit

Fig.1 basic circuit about the common branch

III. ANALYSIS OF THREE LLCS WITH VARIABLE RESISTOR

From the analysis in the above section, the common branch impedance is equivalent to the new impedance

network with virtual positive resistor or negative resistor. For easy understanding, LLC resonant converter is made as an example. Three LLC converters with virtual resistors based on different common branch will be analyzed next. The first common branch is series branch; the second one is parallel branch and the last one is the primary-side branch of transformer.

A. FIRST LLC CONVERTER WITH VARIABLE RESISTOR

Fig.1 shows the first LLC with a virtual resistor, R_s. If the value of resistor R_s is zero, the converter in Fig.2 (a) is traditional LLC converter. Fig.2 (b) shows the equivalent circuit based on fundamental harmonic analysis (FHA).

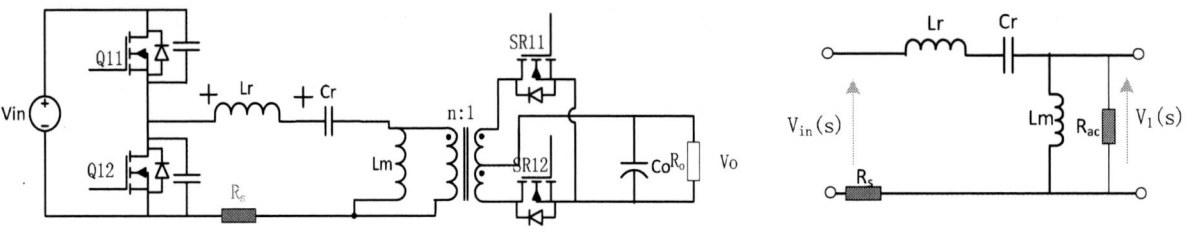

(a) first LLC Converter with virtual resistor

(b) FHA equivalent circuit

Fig.2 first LLC converter with virtual resistor

The gain expression of the first LLC converter with resistor is shown in (4):

$$G_{DC} = \frac{nV_o}{V_{in}/2} = \left| \frac{V_1(s)}{V_{in}(s)} \right|$$

$$= \frac{1}{\sqrt{[\frac{1}{m}(\frac{f_r}{f_s})^2 - \frac{(1+m+mk)}{m}]^2 + [(\frac{k}{mQ}+Q)(\frac{f_r}{f_s}) - Q(\frac{f_s}{f_r})]^2}} \quad (4)$$

Where

$$Q = \sqrt{\frac{L_r}{C_r}}\frac{1}{R_0}, f_r = \frac{1}{2\pi\sqrt{L_r C_r}}, m = \frac{L_m}{L_r}, R_{ac} = \frac{8n^2}{\pi^2}R_o, k = \frac{R_s}{R_{ac}},$$

$$f = \frac{f_s}{f_r}$$

Fig.3 shows gain curve with different switching frequency. $k=0$ represents the traditional LLC converter gain curve; $k=0.05$ represents an LLC converter with a positive resistor; $k=-0.05$ represents an LLC converter with a negative resistor. The gain value reduces significantly when k increases from zero to positive at switching frequency f_{s1}. Likewise, the gain value increases significantly when k decreases from zero to negative at switching frequency f_{s1}. Thus, adding virtue resistors into the series branch can improve the performance of load current sharing significantly.

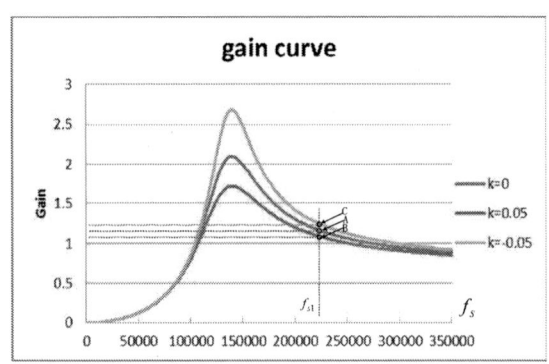

Fig.3 gain curve of first LLC converter with resistor

B. SECOND LLC CONVERTER WITH VARIABLE RESISTOR

Fig.4 shows the second LLC with a virtual resistor. A resistor R_s is added into parallel branch. If the value of resistor R_s is zero, the converter in Fig.4 (a) is a traditional LLC converter. Fig.4 (b) shows the FHA equivalent circuit.

The gain expression of the first LLC converter with resistor is shown in (5):

$$G_{DC} = \frac{nV_o}{V_{in}/2} = \left|\frac{V_1(s)}{V_{in}(s)}\right|$$

$$= \frac{\sqrt{m^2(\frac{f_s}{f_r})^4 + \frac{k^2}{Q^2}(\frac{f_s}{f_r})^2}}{\sqrt{[1+k-(1+k+m)(\frac{f_s}{f_r})^2]^2 + \{(\frac{f_s}{f_r})[\frac{k}{Q}+mQ-mQ(\frac{f_s}{f_r})^2]\}^2}} \quad (5)$$

Where

$$Q = \sqrt{\frac{L_r}{C_r}}\frac{1}{R_0}, f_r = \frac{1}{2\pi\sqrt{L_rC_r}}, m = \frac{L_m}{L_r}, R_{ac} = \frac{8n^2}{\pi^2}R_o, k = \frac{R_s}{R_{ac}},$$

$$f = \frac{f_s}{f_r}$$

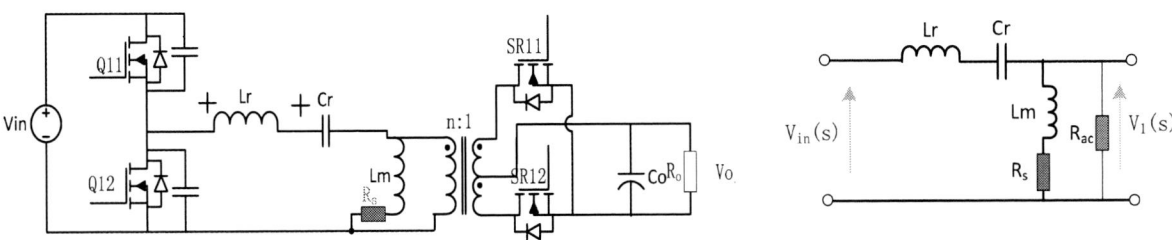

(a) second LLC Converter with virtual resistor   (b)FHA equivalent circuit

Fig.4 second LLC converter with virtual resistor

Fig.5 shows gain curve with switch frequency. *k*=0 represents the traditional LLC converter gain curve; *k*=0.05 represents an LLC converter with positive resistor; *k*=-0.05 represents an LLC converter with a negative resistor. The gain value almost doesn't change no matter increasing or decreasing k at switching frequency f_{s1}. Thus, adding virtue resistors into the parallel branch cannot improve the performance on current load sharing.

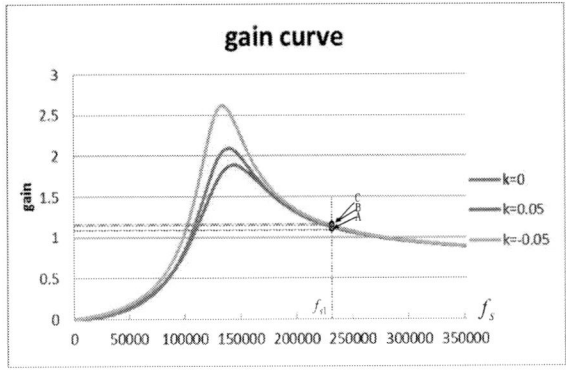

Fig.5 gain curve of second LLC converter with resistor

C. *THIRDRESONANT CONVERTER WITH VARIABLE RESISTOR*

A CLL resonant converter has been reported [15][16] and it has similar performance with traditional LLC converter. A CLL resonant converter has a passive inductor in the primary-side of transformer. Thus, this topology can be used to build virtual resistor for multi-phase CLL application. Fig.6 shows the second LLC with virtual resistor. A resistor R_s is added into parallel branch. If the value of resistor R_s value is zero, the converter in Fig.6 (a) is the traditional LLC converter. Fig.6 (b) shows the equivalent circuit based on fundamental harmonic analysis (FHA).

The gain expression of the first LLC converter with resistor is shown in (6):

$$G_{DC} = \frac{nV_o}{V_{in}/2} = \left|\frac{V_1(s)}{V_{in}(s)}\right|$$

$$= \frac{(\frac{f_s}{f_r})^2}{\sqrt{[\frac{(1+k)}{m} - (1+k)(\frac{f_s}{f_r})^2]^2 + \{\frac{m-1}{m}Q(\frac{f_s}{f_r})[1-(\frac{f_s}{f_r})^2]\}^2}} \quad (6)$$

Where

$$L_{eq} = \frac{L_1 L_2}{L_1 + L_2}, f_r = \frac{1}{2\pi\sqrt{L_{eq}C_1}}, Q = \sqrt{\frac{L_{eq}}{C_1}}\frac{1}{R_{ac}}, m = L_1 / L_{eq},$$

$$R_{ac} = \frac{8n^2}{\pi^2}R_o, k = \frac{R_s}{R_{ac}}$$

Fig.7 shows gain curve with switch frequency. $k=0$ represents the traditional LLC converter gain curve $k=0.05$

represents the LLC converter with a positive resistor; $k=-0.05$ represents the LLC converter with a negative resistor.

The gain value reduces significantly when k increases from zero. Likewise, the gain value increases significantly when k decreases from zero to negative. Thus, in this situation adding virtue resistors can improve the performance on current load sharing.

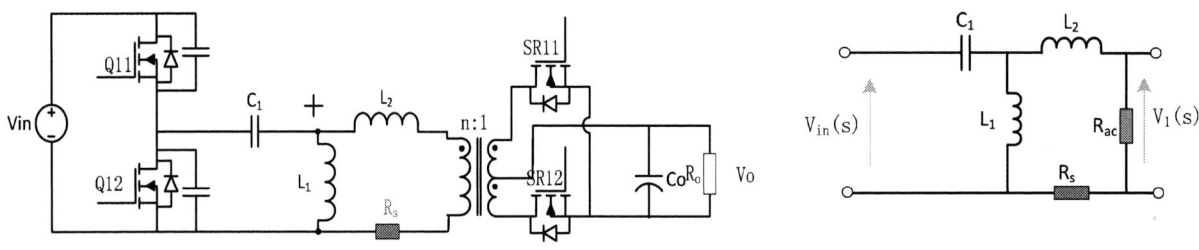

(a) third LLC Converter with virtual resistor   (b) FHA equivalent circuit

Fig.6 third LLC converter with virtual resistor

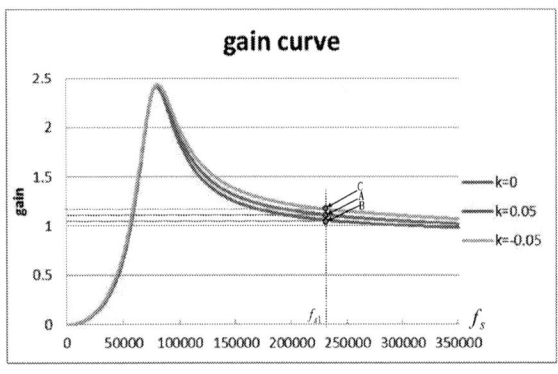

Fig.7 gain curve of third CLL converter with resistor

IV. EXPERIMENTAL RESULTS

A 600W two-phase LLC converter in section II (A) is built to verify the feasibility and demonstrate the advantages of the proposed method. The experiment prototype is shown in Fig.8[17]. Fig.8 (a) shows two-phase LLC resonant converter sharing series inductor. Fig.8 (b) shows FHA equivalent circuit based on Fig.8 (a). According to passive impedance matching concept, the equivalent circuit based on separated impedance is shown in Fig.8 (c). Assuming the virtual opening when the separated impedance is distributed by the equivalent current $i_{Lr1}(s)$ and $i_{Lr2}(s)$. Thus, another equivalent

circuit can be achieved as shown in Fig.8 (d). A virtual positive and negative resistor R_{s1} or R_{s2} are added into the series branch. From the FHA circuit at Fig. 8 (d), Two-phase LLC converter shown in Fig. 8 (a) has be transferred into Fig. 8 (e). The virtual resistors, R_{s1} and R_{s2}, have a different sign. One is positive resistor, other is a negative resistor.

The resonant current sharing error is defined as follows

$$\sigma_{resonant} = \frac{\left|rms(i_{Lr1}) - rms(i_{Lr2})\right|}{\left|rms(i_{Lr1}) - rms(i_{Lr2})\right|} \qquad (7)$$

The prototype parameter is shown in Table.1. The rated power of each phase is 300W. Output voltage is 12V. The leakage inductances value of each phase are 6uH, 6.5uH, respectively. Series inductances are 22.5uH and 24.5uH, respectively. Series capacitances are 12nF and 13nF, respectively.

Fig.9 shows the resonant current waveform at 180W, 300W operation without load sharing method. The phase two i_{Lr2} is almost a triangular waveform. Phase one converter resonant current, delivers nearly all of power to load. As only phase one provides total load power, it is not possible to provide the experiment results at total load 600W because phase one will be over-current and damage.

Fig.10 shows the resonant current waveform at 180W, 300W operation with propose method. The resonant current iLr1, iLr2 are almost same. There is only a small angle difference. Thus, the load current can shared. The experiment results at total load 600W load current is shown in Fig. 10 (c).

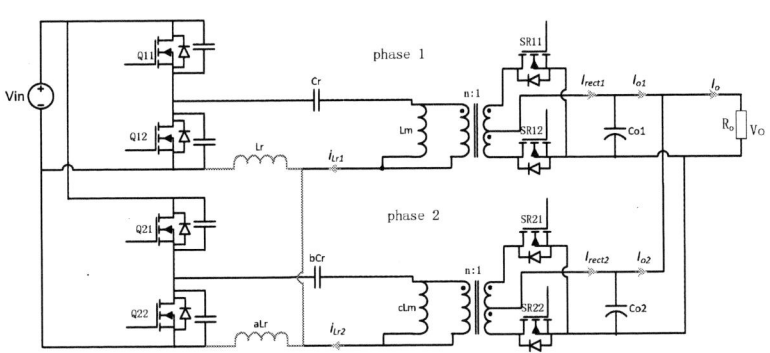

(a) two-phase LLC with first common branch using inductor

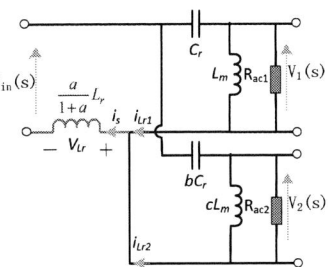

(b) FHA equivalent circuit

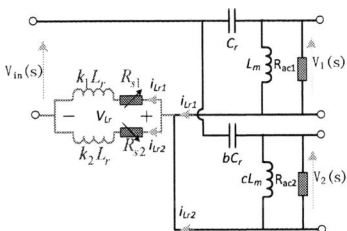

(c) equivalent circuit based on separated impedance

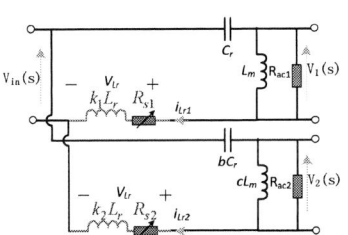

(d) equivalent circuit with independent impedance

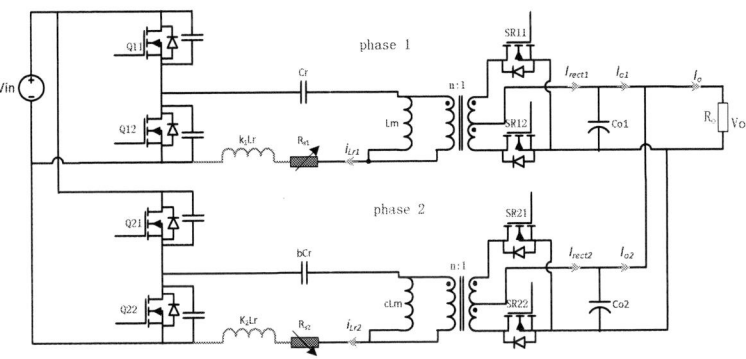

(e) equivalent two-phase LLC converter with virtual positive and negative resistor

Fig.8 two-phase first LLC converter by sharing resonant inductor

Tab.1 Prototype parameters

Switching frequency	180kHz-300kHZ	Output Capacitance	1790μF
Input Voltage	340V-400V	Series Capacitance(Cr)	12nF(phase1) 13nF (phase 2)
Output Voltage	12V	Resonant Inductance(Lr)	22.5μH(Phase1) 24.5μH(Phase2)
Output Power	300W×2	Magnetizing Inductance(Lm)	95μH(Phase1) 92μH(Phase2)
Transformer Ratio n	20:1	Leakage Inductance(Le)	6μH(Phase1) 6.5μH(Phase2)

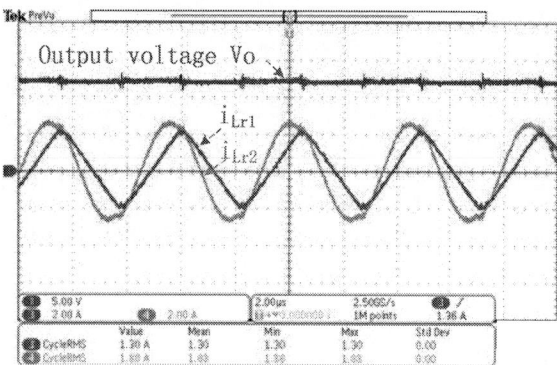

(a) Steady state at 180W load

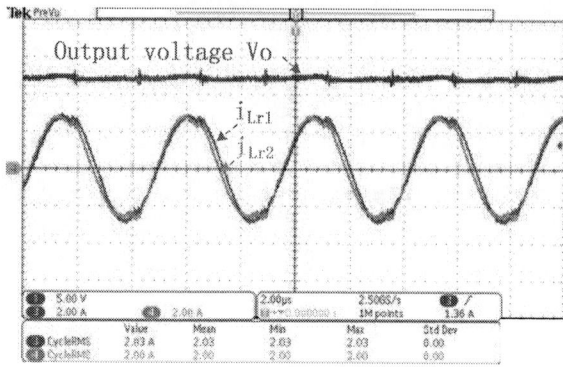

(b) Steady state at 300W load

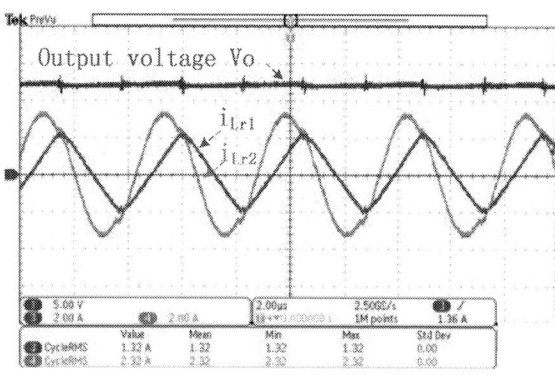

(b) Steady state at 300W load

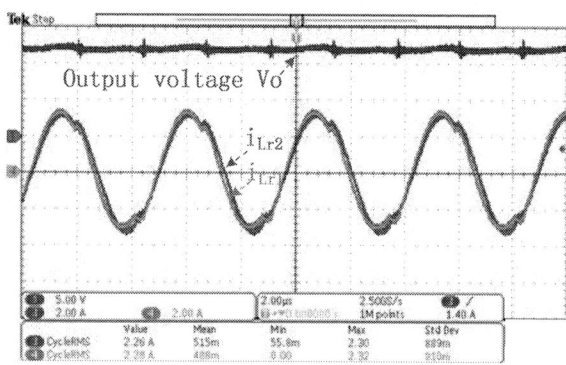

(c) Steady state at 600W load

Ch1: output voltage; Ch3: resonant current of phase one; Ch4: resonant current of phase two.

Fig.9 experiment waveform of two-phase conventional LLC converter

Ch1: output voltage; Ch3: resonant current of phase one; Ch4: resonant current of phase two.

Fig.10 experiment waveform of two-phase conventional LLC converter

To express the resonant current sharing error between the two phases according to (7), the resonant current and resonant current sharing error are shown in Fig. 11, Fig. 12. The resonant current sharing error increases from 10% to 28% with load power from 60W to 300W according to Fig.11. The resonant current sharing error reduced from 5.5% to 0.44% when load power changes from 60W to 600W in Fig.12. The current sharing performance is better with load increasing, which is useful parallel operation. The resonant current sharing error reduces 63 times and is only 0.44% at 600W total load power. Circulating current is significantly reduced using PIM multiphase resonant converter.

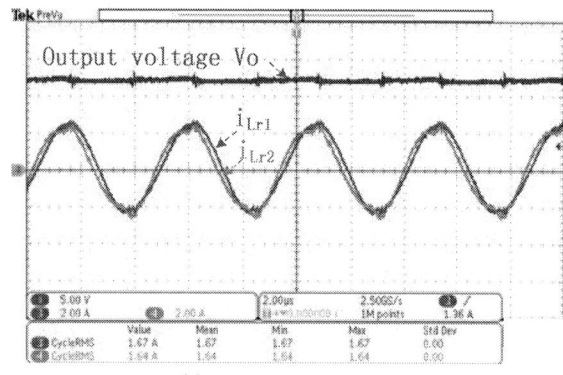

(a) Steady state at 180W load

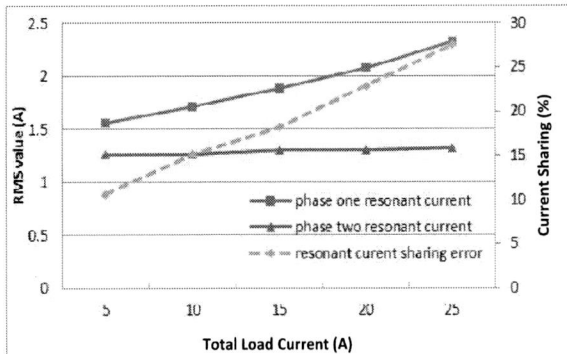

Fig.11 resonant current of two-phase conventional LLC converter

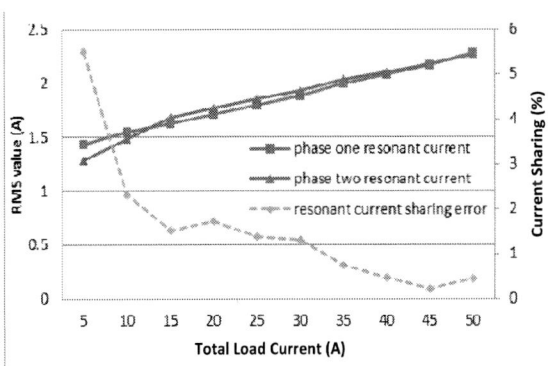

Fig.12 resonant current of two-phase proposed LLC converter

V. CONCLUSION

A passive impedance matching concept is proposed in this paper to achieve load sharing automatically for multi-phase resonant converter. The new concept views the impedance network as passive components in addition with virtual positive resistors and negative resistors. The first and third type of converters can achieve nearly perfect load sharing. Besides, the new concept does not require active control. A 600W two-phase LLC converter sharing the common series inductor is built to verify the feasibility and demonstrate the advantages of the proposed method. The experiment results show that the resonant current sharing error reduces 63 times and is only 0.44% at 600W total load power. This passive impedance matching concept for parallel technique can be extended any phases that is more than two, and any resonant converters, such as series resonant converter, parallel resonant converter, LCC and so on.

References

[1] Y. Bo, "Topology Investigation for Front End DC/DC Power Conversion for Distributed Power System," Virginia Polytechnic Institute and Stage University, 2003.

[2] Y. Z. Y. Zhang, D. X. D. Xu, M. C. M. Chen, Y. H. Y. Han, and Z. D. Z. Du, "LLC resonant converter for 48 V to 0.9 V VRM," *2004 IEEE 35th Annu. Power Electron. Spec. Conf. (IEEE Cat. No.04CH37551)*, vol. 3, 2004.

[3] M. T. Zhang, M. M. Jovanović, and F. C. Y. Lee, "Analysis and evaluation of interleaving techniques in forward converters," *IEEE Transactions on Power Electronics*, vol. 13, no. 4. pp. 690–698, 1998.

[4] R. Hermann, S. Bernet, Y. Suh, and P. K. Steimer, "Parallel connection of integrated gate commutated thyristors (IGCTs) and diodes," *IEEE Trans. Power Electron.*, vol. 24, no. 9, pp. 2159–2170, 2009.

[5] J. Rabkowski, D. Peftitsis, and H. P. Nee, "Parallel-operation of discrete SiC BJTs in a 6-kW/250-kHz DC/DC boost converter," *IEEE Trans. Power Electron.*, vol. 29, no. 5, pp. 2482–2491, 2014.

[6] Z. Hu, Y. Qiu, L. Wang, and Y. F. Liu, "An interleaved LLC resonant converter operating at constant switching frequency," *IEEE Trans. Power Electron.*, vol. 29, no. 6, pp. 2931–2943, 2014.

[7] H. Figge, T. Grote, N. Froehleke, J. Boecker, and P. Ide, "Paralleling of LLC resonant converters using frequency controlled current balancing," *PESC Rec. - IEEE Annu. Power Electron. Spec. Conf.*, pp. 1080–1085, 2008.

[8] B. C. Kim, K. B. Park, and G. W. Moon, "Analysis and design of two-phase interleaved LLC resonant converter considering load sharing," in *2009 IEEE Energy Conversion Congress and Exposition, ECCE 2009*, 2009, pp. 1141–1144.

[9] Z. Hu, Y. Qiu, Y. F. Liu, and P. C. Sen, "A control strategy and design method for interleaved LLC converters operating at variable switching frequency," *IEEE Trans. Power Electron.*, vol. 29, no. 8, pp. 4426–4437, 2014.

[10] E. Orietti, P. Mattavelli, G. Spiazzi, C. Adragna, and G. Gattavari, "Two-phase interleaved LLC resonant converter with current-controlled inductor," *2009 Brazilian Power Electron. Conf. COBEP2009*, pp. 298–304, 2009.

[11] B. C. Kim, K. B. Park, C. E. Kim, and G. W. Moon, "Load sharing characteristic of two-phase interleaved LLC resonant converter with parallel and series input structure," *2009 IEEE Energy Convers. Congr. Expo. ECCE 2009*, pp. 750–753, 2009.

[12] F. Jin, F. Liu, X. Ruan, S. Member, and X. Meng, "Multi-Phase Multi-Level LLC Resonant Converter with Low Voltage Stress on the Primary-Side Switches," pp. 4704–4710, 2014.

[13] E. Orietti, P. Mattavelli, G. Spiazzi, C. Adragna, and G. Gattavari, "Analysis of multi-phase LLC resonant converters," *2009 Brazilian Power Electron. Conf. COBEP2009*, pp. 464–471, 2009.

[14] E. Orietti, P. Mattavelli, G. Spiazzi, C. Adragna, and G. Gattavari, "Current sharing in three-phase LLC interleaved resonant converter," *2009 IEEE Energy Convers. Congr. Expo. ECCE 2009*, pp. 1145–1152, 2009.

[15] V. Tech, W. Hall, and F. C. Lee, "High Switching Frequency , High Efficiency CLL Resonant Converter with Synchronous Rectifier Daocheng Huang Dianbo Fu," pp. 804–809, 2009.

[16] D. Huang, D. Fu, F. C. Lee, and P. Kong, "High-frequency high-efficiency CLL resonant converters with synchronous rectifiers," *IEEE Trans. Ind. Electron.*, vol. 58, no. 8, pp. 3461–3470, 2011.

[17] H. Wang, Y. Chen, Y.-F. Liu, J. Afsharian, and Z. A. Yang, "A common inductor multi-phase LLC resonant converter," in Energy Conversion Congress and Exposition (ECCE), 2015 IEEE, 2015, pp. 548-555.

LLC Converter with Auxiliary Switch for Hold Up Mode Operation

Yang Chen, Hongliang Wang, *Senior member,*
Yan-Fei Liu, *Fellow,IEEE*

Department of Electrical and Computer Engineering
Queen's University, Kingston, Canada
yang.chen@queensu.ca, hongliang.wang@queensu.ca,
yanfei.liu@queensu.ca

Jahangir Afsharian and Zhihua (Alex) Yang
Murata Power Solutions
Toronto, Canada
jafsharian@murata.com, ZYang@murata.com

Abstract— **This paper proposes a new half bridge (HB) LLC resonant converter with pulse width modulated (PWM) auxiliary switch (sLLC converter) for hold-up mode operation. The proposed sLLC converter works with synchronous rectifier (SR), which is suitable for low output voltage applications. The magnetizing inductor can be designed with large value to reduce the circulation loss. In the proposed sLLC converter, the auxiliary switch and diode branch will provide charging path for the resonant inductor. For nominal 400V input, sLLC achieves same performance as conventional LLC converter, and all the good features such as soft switching are naturally retained. For slight input voltage fluctuation, frequency modulation is used to regulate the output voltage. When the input voltage drops further, HB switches will operate at constant minimum frequency, and the auxiliary switch will operate in PWM mode to energize the resonant inductor with 400V bus directly during hold up period, thus the output voltage can be maintained at desired level. To verify the effectiveness of the proposed sLLC converter, operational principle and equivalent circuit will be carefully explained and analyzed in this paper. A 300W prototype is built for 250V-400V input, 12V output application.**

Keywords— LLC; hold up; high voltage gain; auxiliary switch; PWM; telecom power supply

I. INTRODUCTION

Recently, considerable research has been conducted on LLC-based telecommunication power supplies. PFC stage converts the AC line into 400V DC, which is further converted by the DC/DC stage to 12V DC. The typical structure is shown in Fig. 1. LLC resonant converter is widely used as the DC/DC stage because it achieves high efficiency as well as low EMI performance due to the inherent zero voltage switching (ZVS) on the primary MOSFETs and zero current switching (ZCS) on the secondary rectifiers.

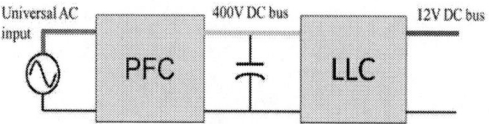

Fig. 1. Structure of the front end converter

A critical issue for telecom power supply is the hold up problem illustrated in Fig. 2. When the AC line fails, the 400V bus voltage drops continuously. It is desired the 12V DC be maintained for several tens of milliseconds, until the UPS takes over [1]. Conventional LLC is not suitable for hold up operation. If an LLC converter is designed to achieve high voltage gain, the efficiency at nominal 400V input will be severely penalized [2], [3]. Therefore, to solve the hold-up problem, the primary objective is:

1) to increase the operational input voltage range of the LLC converter;

2) to maintain high efficiency for the nominal 400V input operation.

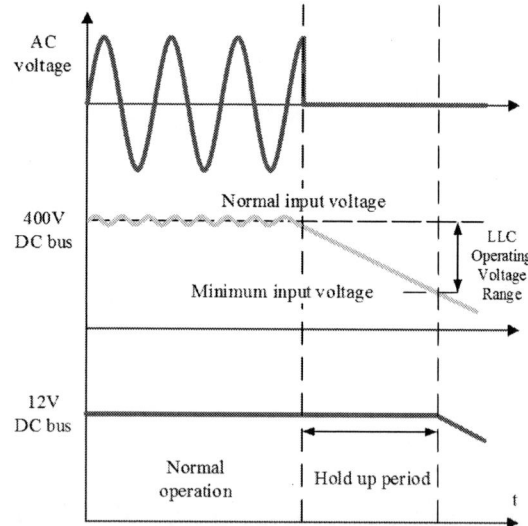

Fig. 2. Hold-up problem process

To achieve high efficiency at 400V operation, the transformer turns ratio needs to be properly designed to ensure that the converter operates at resonant frequency at 400V input. Moreover, the magnetizing inductance should be optimized to reduce the circulating current in the resonant tank as much as possible while maintain ZVS for the HB FETs. At last, for 12V output applications, on the secondary side, synchronous rectifiers (SR) should be used instead of diode rectifiers, as the forward voltage drop of diodes would be a deal breaker of the whole efficiency.

To improve the LLC converter's operational voltage range to meet hold-up time requirement, quite a few methods have been proposed.

Among them, the most straightforward way is to employ a cascaded structure with a baby boost converter [4]. However, the additional power diode will reduce the efficiency at nominal 400V operation. Besides, the two stage configuration is complicated, and consequently costly.

A category of approaches solving the hold-up problem by utilizing auxiliary windings on the secondary side of the main transformer [5], [6]. Generally speaking, as long as the LLC converter reaches the peak gain, the switch-controlled auxiliary windings will take over the secondary side power transfer. Increased transformer turns-ratio helps to achieve higher voltage gain of the LLC converter during the hold-up period. Also, the discrete design between nominal 400V operation and hold-up state operation can maintain high efficiency for 400V input. However, usually the main transformer is the most bulky and lossy part of the converter, thus adding extra windings makes it even more difficult to optimize the transformer from both efficiency and power density improvement point of view.

By driving the half-bridge MOSFETs with asymmetric pulse-width modulation (APWM) rather than conventional frequency modulation (FM), LLC converter can improve output-to-input voltage gain without any additional components [7]. This method, however, suffers from limited peak gain enhancement. Besides, once the resonant tank is designed, the maximum gain that APWM control could achieve is also determined, which is not practical in terms of design flexibility.

A critical insight was revealed in [8] that if the resonant tank can be charged with more energy during one switching cycle, LLC converter achieves higher gain. To charge the resonant tank more, the secondary windings are short circuit for a certain period of time in every switching cycle. The downside of this method is that quite a few components need to be added in the power train on the secondary side, which causes size increasing and efficiency reducing.

Based on [8], a few improving methods propose to adopt either Boost PWM discontinuous current mode (DCM) control [9] or phase shift control on LLC topology[10], [11]. The common principle of these methods is that, in each switching cycle, the resonant tank will be short circuit on either primary side or secondary side by auxiliary switches for a period of time, so that the resonant inductor can be energized more quickly, hence store and transfer more power. The nominal 400V efficiency remains uninfluenced as compared to a conventional LLC optimized for 400V input voltage. However, for [9], [10] and [11], none of the method is compatible with SR, thus cannot be used in low output voltage applications.

In this paper, sLLC converter is proposed to solve the hold-up problem while avoiding the aforementioned issues. The proposed sLLC converter works with SR, thus it is particularly suitable for low output voltage applications in telecommunication field. The design for hold up operation is independent of nominal 400V operation design, in which the magnetizing inductor can be designed with large value to reduce the circulation loss, such that the converter achieves optimal efficiency at nominal 400V. During hold up period, the auxiliary switch operates in PWM mode. When the auxiliary switch is turned ON, the input DC bus voltage, rather than the capacitor voltage in conventional LLC converter, is constantly applied to the resonant inductor, thus the resonant inductor accumulates more energy in shorter time, and high voltage gain is achieved. This paper is organized as follows: Section II presents the operating principle and mode analysis; Section III gives the detailed circuit analysis; Section IV demonstrates the experimental results; and Section V concludes the paper.

II. OPERATION PRINCIPLE OF THE PROPOSED CONVERTER

Fig. 3 shows the proposed LLC converter with auxiliary switch (sLLC) for hold-up operation. The resonant tank is connected to 400V bus and middle point of the half bridge respectively. The magnetizing inductor could be integrated into the main transformer, while the resonant inductor L_r and resonant capacitor C_r are external. The charging branch consisted of one diode D_a and one MOSFET Q_a connects the resonant inductor L_r and the primary ground. C_a is the capacitor paralleled to Q_a. It should be noted that the Q_a is ground referenced, which enables simple driving.

When the input voltage is around nominal 400V, the auxiliary switch is kept OFF. The sLLC converter operates just

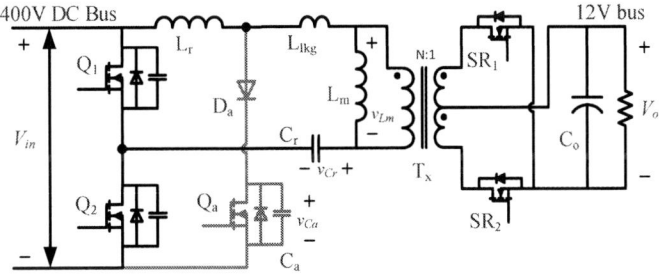

Fig. 3. Proposed sLLC topology with SR

like the conventional half bridge LLC converter. Thus, all the desirable features of LLC converters are automatically retained. Besides, the parameter design of the resonant components (L_r, C_r, L_m) only considers nominal 400V efficiency, which means that magnetizing inductor could be relatively large value to reduce the circulating magnetizing current, and to improve the efficiency. Frequency modulation (FM) is still used to regulate the output voltage when the input voltage drops slightly (i.e. switching frequency reduces with input voltage reducing). This is good to address the low frequency fluctuation on 400V bus from the PFC stage.

If the input voltage reduces to a level that the LLC converter cannot maintain the required output voltage level, the auxiliary switch will start to operate. During the positive half cycle when Q_2 is ON (source injects energy to the resonant tank), the auxiliary switch turns ON for certain period of time of a switching period, allowing the resonant inductor to be charged directly by the bus. It is more effective to boost the stored energy in the resonant inductor with the 400V bus than the resonant capacitor (the case in conventional LLC converter), thus sLLC transfers more energy in one switching cycle to achieve higher DC gain. The longer the auxiliary switch is ON, the more energy will be stored, and the higher gain can be achieved. During the hold-up period, Q_a will control the charging time of L_r to regulate the output voltage. C_a is used to absorb the voltage spike from the leakage inductor L_{lkg} when turning off Q_a. D_a provides unidirectional charging path, preventing C_a and the parasitic capacitor of Q_a from resonating during the normal operation. The key waveforms of the proposed sLLC converter during hold up period is shown in Fig. 4, in which G_{Q1}, G_{Q2} and G_{Qa} are the gate signals; i_{Lr} and i_{Lm} are the resonant inductor current and the magnetizing current in the primary side; v_{Cr} is the voltage stress on resonant capacitor; v_{Lm} is the voltage across the magnetizing inductor; i_{Da}, i_{Qa} and i_{Ca} are the current stresses in the auxiliary diode, auxiliary switch and paralleled capacitor; v_{Ca} is the voltage stress across Q_a and C_a; i_{SR1} and i_{SR2} are the secondary current; and I_o is average load current.

During hold up period, the operation in each switching cycle can be divided into 8 modes (M1~M8). The equivalent circuits are shown in Fig.5.

Mode 1 (M1): Q_2 and Q_a are turned ON at t_0. L_r will be charged by the bus voltage, so that i_{Lr} increases linearly and sharply. L_{lkg}, L_m and C_r will short circuit and resonate. The magnetizing current is negative during M1. No current will go through the transformer, and the output capacitor C_o discharges.

Mode 2 (M2): Q_a turns OFF at t_1 while Q_2 remains ON. L_r current will be released through the transformer. SR2 will conduct to charge C_o. Sudden change of current path will cause voltage spike on L_{lkg}, which will be absorbed by C_a. The capacitor voltage reaches its maximum at t_2. Large C_a and small L_{lkg} will reduce the voltage spike.

Mode 3 (M3): L_r, L_{lkg}, and C_r will resonate in M3 with Q_2 ON. SR2 conducts to charge C_o. The transformer will be clamped by the output voltage V_o, thus magnetizing inductor current i_{Lm} increase linearly. At t_3, i_{Lr} meets i_{Lm}, and the SR current drops to zero.

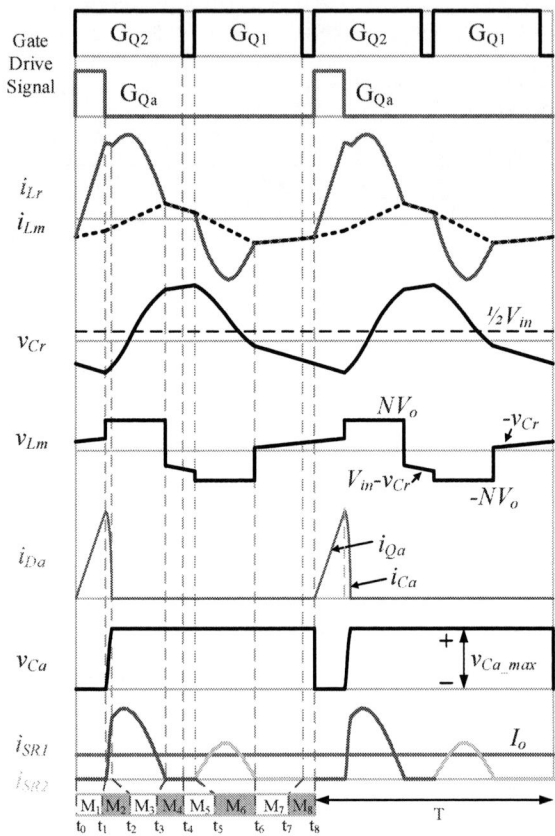

Fig. 4. Key waveforms of sLLC during holdup period

Mode 4 (M4): In M4, L_m will resonate together with L_r, L_{lkg}, and C_r. Transformer is in idle mode while the load current is provided by C_o. At t_4, Q_2 is turned OFF.

Mode 5 (M5): Both Q_1 and Q_2 are OFF during M5, the dead time. The resonant current will charge the parasitic capacitor of Q_2 while discharge that of Q_1. When the voltage across Q_1 drops to zero or so, the body diode of Q_1 conducts, clamping the voltage of Q_1 at around zero volts, thus, ZVS is achieved for Q_1 when it is turned on at t_5.

Mode 6 (M6): L_r, L_{lkg}, and C_r will resonate in M6 with Q_1 ON. SR1 conducts to charge C_o and powering the load. The transformer will be reversely clamped by the output voltage V_o, thus magnetizing inductor current i_{Lm} decrease linearly. At t_6, i_{Lr} meets i_{Lm}, and i_{SR1} drops to zero.

Mode 7 (M7): L_m will resonate jointly with L_r, L_{lkg}, and C_r in M7. Transformer power transfer is cutoff while the load current is provided by C_o. At t_7, Q_1 is turned OFF.

Mode 8 (M8): M8 is the dead time that the HB switches are OFF. The parasitic capacitor of Q_1 will be charged and that of Q_2 discharged by the resonant current. ZVS is achieved for Q_2 when it is turned on at t_8.

After M8, Q_2 and Q_a will be turned ON, and the converter operates in M1 again.

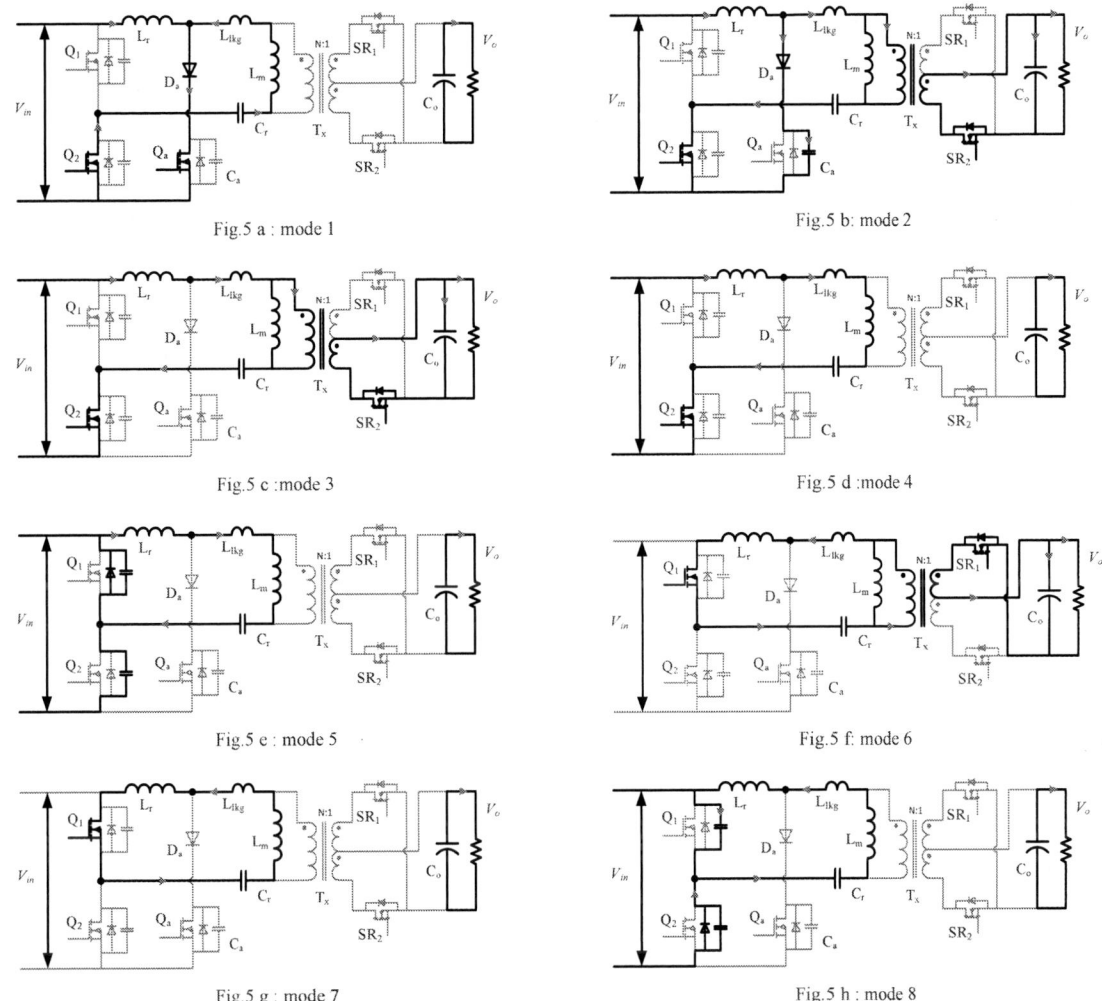

Fig.5 a : mode 1

Fig.5 b: mode 2

Fig.5 c :mode 3

Fig.5 d :mode 4

Fig.5 e : mode 5

Fig.5 f: mode 6

Fig.5 g : mode 7

Fig.5 h : mode 8

Fig. 5.   Equivalent circuit of the proposed sLLC converter in hold-up state

III.   CIRCUIT ANALYSIS OF THE PROPOSED CONVERTER

In this section, key characteristics and design considerations of the proposed sLLC converter during hold up period will be presented. Behaviors of the equivalent circuit in each mode will be analyzed and expressed in fundamental physics and mathematics. LLC converter naturally has no close-form solution of currents and voltages due to the nonlinear characteristics. To relieve the complexity of the design process, the equivalent circuit of sLLC converter in each mode is simplified to obtain the close-form solution in this analysis. The approximations are made based on the assumptions below:

1)   Assumption of extreme case behavior

As stated in Section II, the resonant tank absorbs power only in the positive half cycle. It can be inferred that, for given load power, when the duty cycle of the auxiliary switch increases, the resonant inductor current i_{Lr} increases in the positive half cycle, while decreases in the negative half cycle. It is observed that when the duty cycle is beyond certain value, the power transfer in the negative half cycle will reduce to zero. The resonant inductor current i_{Lr} will always be equal to the magnetizing current i_{Lm} during the negative half cycle. The analysis in Section III will be based on the assumption that all load power is transferred during the positive half cycle, as it indicates the extreme operation condition in terms of the voltage gain capacity, and the worst case in terms of component stresses.

Fig. 6 shows the waveforms of simplified sLLC converter in extreme case. In addition to the nomenclature used in Fig. 4, i_{Cr} is the current in the resonant capacitor; I_{Lm_bias} is the DC bias in the magnetizing inductor; D is the duty cycle of the auxiliary switch; T_s is the period of one switching cycle; T_r is the period of a resonant cycle that $T_r = 2\pi\sqrt{L_r C_r}$. To distinguish between the simplified circuit and the actual circuit, the time instant will be τ_0, τ_1... instead of t_0, t_1... used in Fig. 4.

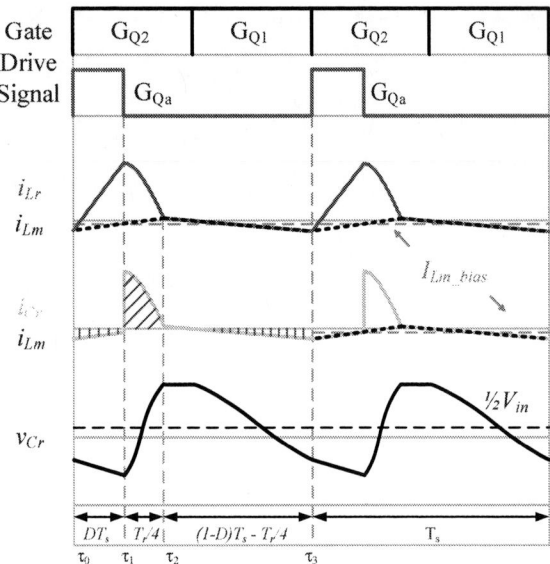

Fig. 6. Key waveforms of simplified sLLC converter in extreme case

2) Assumption of timing

In one switching cycle, timing can be defined by the resonant inductor current. During τ_0 to τ_1, the time interval DT_s, the auxiliary switch Q_a is ON. The resonant inductor current i_{Lr} increases linearly. Instantaneously Q_a turns OFF, i_{Lr} resonates in a sinusoidal shape until it meets i_{Lm}. This period of time (τ_1 to τ_2) is roughly $T_r/4$. It could be understood in the means that i_{Lr} drops from the peak value to 0 in a quarter of one resonant cycle of L_r and C_r. The actual time length may be influenced by the duty cycle D. Generally speaking, when D is very large, the resonant period is slightly below $T_r/4$, and vice versa. In this section, the resonant period is fixed at $T_r/4$. In the rest of the switching cycle (τ_2 to τ_3), no power will be transferred to the load, and i_{Lr} is always equal to i_{Lm}.

3) Assumption of resonant tank currents

In real case, the magnetizing current i_{Lm} is linear only when the power stage transfers power to the load; otherwise i_{Lm} should be sinusoidal. In this approximation, the magnetizing current i_{Lm} will have a triangular shape. More details will be included in the following part.

When the duty cycle of the auxiliary switch D is high, resonant current i_{Lr} is of much higher value than the magnetizing current i_{Lm}, so that i_{Lm} can be neglected in i_{Lr} dominated calculation such as the input charge calculation. Currents in the parasitic components will also be neglected.

4) Assumption of resonant tank parameters

As the parameter design for hold up mode is decoupled with 400V case, it is also assumed that a set of well-designed resonant tank parameters are already obtained to achieve high efficiency for 400V operation. In this section, parameters and specifications in Table I will be used for the analysis.

A. Current stress of switches

During the hold up operation period, the HB bottom switch, the auxiliary switch and diode will have the same peak current as that in the resonant inductor.

The magnetizing current i_{Lm} is very low compared with i_{Lr}. Considering $i_{Lm} = 0$, the initial resonant current i_{Lr} will also be zero. The peak value I_{Lr_pk} can be calculated by (1).

$$I_{Lr\_pk} = \frac{V_{in}}{L_r} \cdot DT_s \tag{1}$$

For the HB top switch, the current is low during hold up process, so that once 400V design if completed, the selected bottom switch will survive for the hold up period.

B. Output-to-input DC Voltage Gain (conversion ratio)

For sLLC converter (also for HB LLC converter), energy exchange between source and resonant tank occurs only in the positive half cycle [12]. In each switching cycle, during Q_2 is ON, the time integral of the resonant inductor current i_{Lr} is equal to the total input charge. The input energy can then be obtained by multiplying the input charge by the input bus voltage. The output energy in one switching cycle is determined by the average output power and the switching frequency. Assuming 100% efficiency, the input energy should be equal to the output energy according to energy conservation law. Considering constant load current, the relation can be expressed in (2):

$$V_{in} \cdot \int_0^{\frac{T_s}{2}} i_{Lr}(t)dt = \frac{V_o \cdot I_o}{f_s} \tag{2}$$

Equation (3) shows the total input charge during the positive half cycle.

$$\int_0^{\frac{T_s}{2}} i_{Lr}(t)dt = \int_0^{DT_s} \frac{I_{Lr\_pk}}{DT_s} \cdot t dt + \int_{\frac{T_r}{4}}^{\frac{T_r}{2}} I_{Lr\_pk} \sin(\frac{2\pi}{T_r}t)dt \tag{3}$$

Combining (1), (2), and (3), the analytical expression of the output voltage V_o can be found in (4):

$$V_o = \frac{V_{in}^2 D}{L_r I_o}\left(\frac{D}{2f_s} + \sqrt{L_r C_r}\right) \tag{4}$$

It could be seen that the output voltage will increase monotonically with duty cycle D increases.

C. Magnetizing current with DC bias

The asymmetrical current waveform between positive half cycle and negative half cycle will introduce a DC bias I_{Lm_bias} on the magnetizing inductor current. In steady state, ampere-second balance must be achieved on the resonant capacitor. If the average value is i_{Lm} is 0 during τ_1 to τ_2, I_{Lm_bias} can be found in (5):

$$I_{Lm\_bias} = \frac{1}{T_s} \int_{\frac{T_r}{4}}^{\frac{T_r}{2}} I_{Lr\_pk} \sin(\frac{2\pi}{T_r}t)dt$$
$$= I_{Lr\_pk} f_s \sqrt{L_r C_r} \tag{5}$$

In real case, the magnetizing current reach the maximum value I_{Lm_max} at time instant τ_2, and the minimum value I_{Lm_min} occurs near the time instant τ_1. The magnetizing inductor current is sinusoidal and near symmetrical during time interval τ_1 to τ_2 and τ_2 to τ_3. Thus it is reasonable to assume the slope of the linearized magnetizing current be of the same absolute value. Assuming the slope is k, there is:

$$\Delta I_{Lm} = I_{Lm\_max} - I_{Lm\_min}$$
$$= kDT_s + \frac{nVo}{L_m}\frac{T_r}{4} = k\left[(1-D)T_s - \frac{T_r}{4}\right] \tag{6}$$

Then the peak value of the magnetizing current can be obtained by (7):

$$I_{Lm\_pk} = I_{Lm\_min} = I_{Lm\_bias} - \frac{\Delta I_{Lm}}{2}$$
$$= I_{Lr\_pk}\sqrt{L_r C_r} f_s - \frac{nV_o}{L_m}\left(\frac{\pi\sqrt{L_r C_r}D}{2(1-2D) - \pi\sqrt{L_r C_r} f_s} + \frac{\pi\sqrt{L_r C_r}}{2}\right) \tag{7}$$

D. Resonant capacitor voltage stress

During the positive half cycle, the positive portion of the resonant inductor current i_{Cr} (i.e. i_{Lr}) will charge the resonant capacitor C_r from the minimum value V_{Cr_min} to the maximum value V_{Cr_max}. Neglecting i_{Lm}, V_{Cr_min} occurs near the time instant τ_1, then i_{Lr} will charge C_r to V_{Cr_max} at around time instant τ_2. This can be expressed by (8):

$$\Delta V_{Cr} = V_{Cr\_max} - V_{Cr\_min}$$
$$= \frac{1}{C_r} \int_{\frac{T_r}{4}}^{\frac{T_r}{2}} I_{Lr\_pk} \sin(\frac{2\pi}{T_r}t)dt = I_{Lr\_pk}\sqrt{\frac{L_r}{C_r}} \tag{8}$$

Considering the $V_{in}/2$ bias on C_r, the peak voltage stress on the resonant capacitor is given in (9):

$$V_{Cr\_pk} = V_{Cr\_max} = \frac{V_{in}}{2} + \Delta V_{Cr} = \frac{V_{in}}{2} + I_{Lr\_pk}\sqrt{\frac{L_r}{C_r}} \tag{9}$$

IV. EXPERIMENT RESULTS

A 250V-400V input, 12V/300W output prototype was built to verify effectiveness the proposed sLLC converter and its hold up ability. The detailed design requirement and power train parameters are given in Table 1.

TABLE I. SPECIFICATIONS OF SLLC CONVERTER

Input voltage	250V-400V
Output voltage/power	12V/300W
Transformer turns ratio	17:1
Resonant inductor	24μH
Resonant Capacitor	12nF
Magnetizing inductor	250μH
Leakage inductor	6μH
Output capacitor	2mF
HB MOSFETs	IPW60R190C6
SR MOSFETs	SiRA00DP
Auxiliary MOSFET	C2M0160120D
Auxiliary Diode	APT60D60BG
Auxiliary capacitor	2nF

Fig. 7 shows the steady state waveforms of sLLC converter under 300W full load operating at 400V input voltage. The switching frequency is 240 kHz. It is lower than the designed resonant frequency due to the impact of the leakage inductor of the transformer. The resonant current is close to sinusoidal. The peak value is 3A.

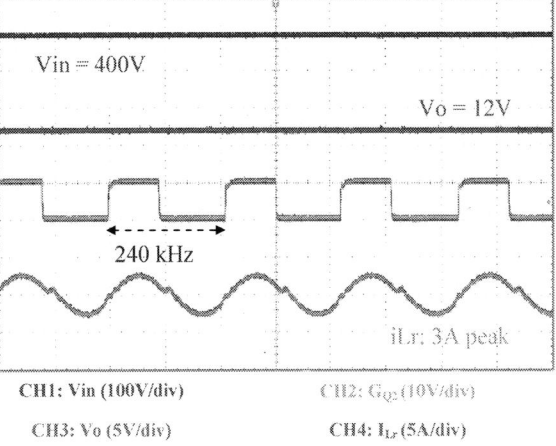

CH1: Vin (100V/div)   CH2: G$_{Q2}$ (10V/div)

CH3: Vo (5V/div)   CH4: I$_{Lr}$ (5A/div)

Fig. 7. 400V input, 12V/25A full load steady state waveform

Fig. 8 shows the steady state waveforms of sLLC converter under 300W full load operating at 250V input voltage. HB switches are operating at the minimum frequency at 140kHz. The duty cycle of the auxiliary switch is around 0.12. The peak value of the inductor current is 8A.

Fig. 10 shows the steady state waveforms of sLLC converter under 12V/15A 60% load operating at 400V input voltage. The switching frequency is 250 kHz. The resonant current is close to sinusoidal. The peak value is 1.8A.

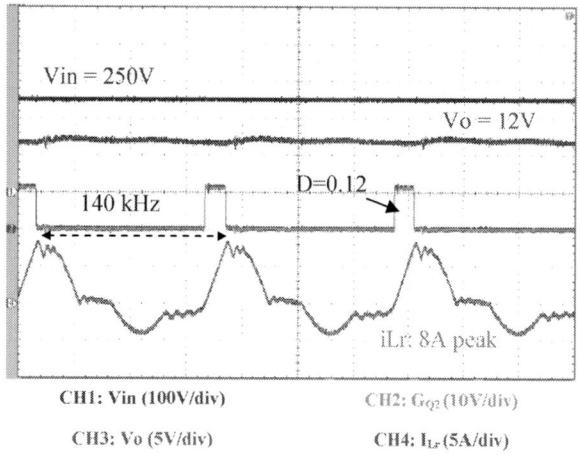

CH1: Vin (100V/div)   CH2: G$_{Q2}$ (10V/div)

CH3: Vo (5V/div)   **CH4: I$_{Lr}$ (5A/div)**

Fig. 8. 250V input, 12V/25A full load steady state waveform

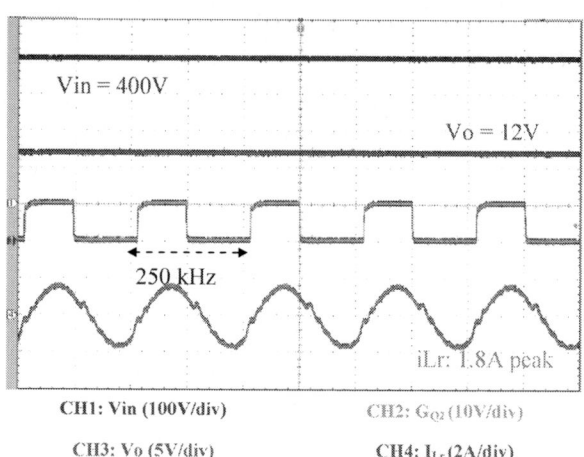

CH1: Vin (100V/div)   CH2: G$_{Q2}$ (10V/div)

CH3: Vo (5V/div)   **CH4: I$_{Lr}$ (2A/div)**

Fig. 10. 400V input, 12V/15A 60%load steady state waveform

Fig. 9 shows the hold-up process under full load. When the input voltage is between 400V and 320V, the auxiliary switch is OFF, and frequency modulation of HB switches regulates the output voltage. When the input voltage drops below 320V, Q_a starts to take over the output voltage regulation. It could be observed that the output voltage does not lose regulation until the input voltage drops below 250V.

Fig. 11 shows the steady state waveforms of sLLC converter under 12V/15A 60% load operating at 250V input voltage. HB switches are operating at the minimum frequency at 140kHz. The duty cycle of the auxiliary switch is around 0.09. The peak value of the inductor current is 5A.

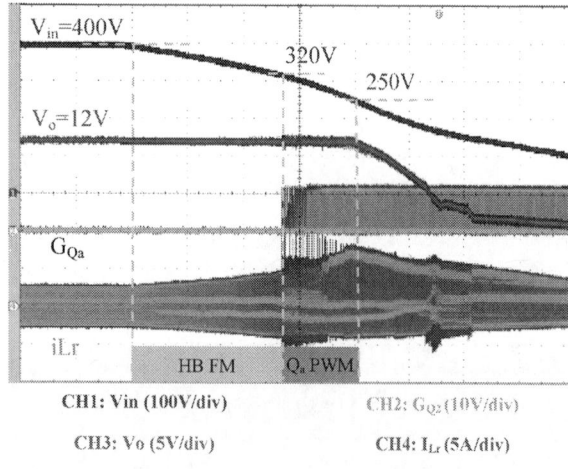

CH1: Vin (100V/div)   CH2: G$_{Q2}$ (10V/div)

CH3: Vo (5V/div)   **CH4: I$_{Lr}$ (5A/div)**

Fig. 9. 12V/25A full load dynamic waveform

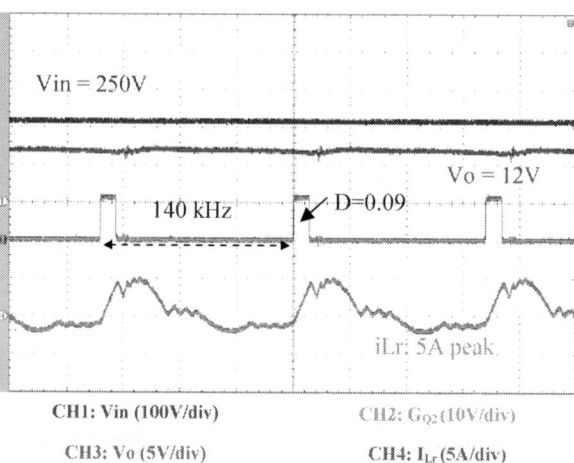

CH1: Vin (100V/div)   CH2: G$_{Q2}$ (10V/div)

CH3: Vo (5V/div)   **CH4: I$_{Lr}$ (5A/div)**

Fig. 11. 250V input, 12V/15A 60%load steady state waveform

Fig. 12 shows the hold-up process under 15A (60% load). Output voltage can be regulated by frequency modulation of HB switches till 300V input. Between 300V and 210V, Q_a operates in PWM mode to hold the output voltage at 12V.

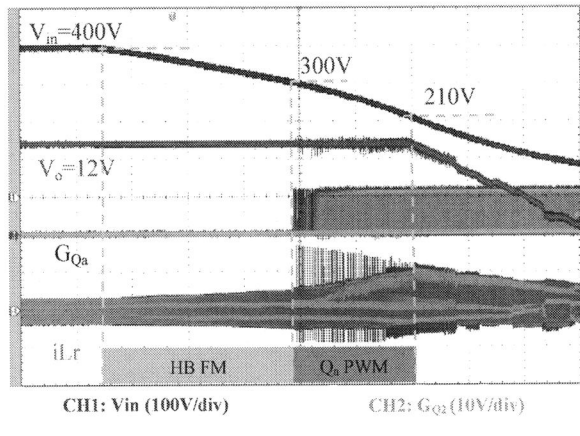

CH1: Vin (100V/div)   **CH2: G_{Q2} (10V/div)**

CH3: Vo (5V/div)   **CH4: I_{Lr} (5A/div)**

Fig. 12. 12V/15A 60% load dynamic waveform

V. CONCLUSION AND FUTURE WORK

In this paper, a new HB LLC converter with auxiliary switch (sLLC) is proposed to solve the hold-up problems in server and telecommunication power applications. The proposed sLLC converter is suitable for low voltage application where SR is needed. The magnetizing inductor L_m can be integrated into the transformer core to reduce the converter size. Also, L_m can be designed with large value to reduce the circulation loss. During 400V input, the auxiliary switch will not conduct and the circuit operation is same as the conventional LLC converter. When input voltage is low, the auxiliary switch is turned on to increase the resonant inductor energy in one switching cycle, thus hold up operation is achieved that the output voltage can be maintained at 12V. A 250V-400V input, 300W prototype has been built to verify the feasibility and effectiveness of the proposed sLLC converter.

REFERENCES

[1] B. Yang and F. C. Lee, "LLC Resonant Converter for Front End DC / DC Conversion," in Applied Power Electronics Conference and Exposition, 2002. APEC 2002. Seventeenth Annual IEEE, 2002, vol. 2, pp. 1108–1112.

[2] Y. Chen, H. Wang, Z. Hu, Y.-F. Liu, J. Afsharian, and Z. A. Yang, "LCLC resonant converter for hold up mode operation," in Energy Conversion Congress and Exposition (ECCE), 2015 IEEE , vol., no., pp.556-562, 20-24 Sept. 2015

[3] H. Wang, Y. Chen, Y.-F. Liu, J. Afsharian, and Z. A. Yang, "A new LLC converter family with synchronous rectifier to increase voltage gain for hold-up application," in Energy Conversion Congress and Exposition (ECCE), 2015 IEEE , vol., no., pp.5447-5453, 20-24 Sept. 2015

[4] Y. Xing, L. Huang, X. Cai, and S. Sun, "A Combined Front End DC / DC Converter," in Applied Power Electronics Conference and

Exposition, 2003. APEC '03. Eighteenth Annual IEEE, 2003, vol. 00, no. C, pp. 1095–1099.

[5] B. Yang, P. Xu, and F. C. Lee, "Range winding for wide input range front end DC/DC converter," APEC 2001. Sixth. Annual. IEEE Appl. Power Electron. Conf. Expo. (Cat. No.01CH37181), vol. 1, pp. 476–479, 2001.

[6] M.-Y. Kim, B.-C. Kim, K.-B. Park, and G.-W. Moon, "LLC series resonant converter with auxiliary hold-up time compensation circuit," 8th Int. Conf. Power Electron. - ECCE Asia, pp. 628–633, May 2011.

[7] B.-C. Kim, K.-B. Park, and G.-W. Moon, "LLC resonant converter with asymmetric PWM for hold-up time," in Power Electronics and ECCE Asia (ICPE & ECCE), 2011 IEEE 8th International Conference on, 2011, pp. 38 – 43.

[8] B.-C. Kim, K.-B Park, S.-W. Choi, and G.-W. Moon, "LLC series resonant converter with auxiliary circuit for hold-up time," INTELEC 2009 - 31st International Telecommunications Energy Conference, 1–4. doi:10.1109/INTLEC.2009.5351749

[9] I.-H. Cho, Y.-D. Kim, and G.-W. Moon, "A Half-Bridge LLC Resonant Converter Adopting Boost PWM Control Scheme for Hold-Up State Operation," IEEE Trans. Power Electron., vol. 29, no. 2, pp. 841–850, Feb. 2014.

[10] J. Kim and G. Moon, "A New LLC Series Resonant Converter with a Narrow Switching Frequency Variation and Reduced Conduction Losses," IEEE Trans. Power Electron., vol. 29, no. 8, pp. 4278–4287, 2014.

[11] H. Wu, T. Mu, X. Gao, and Y. Xing, "A Secondary-Side Phase-Shift-Controlled LLC Resonant Converter With Reduced Conduction Loss at Normal Operation for Hold-Up Time Compensation," Application. Power Electronics, IEEE Transactions on, 30(10), 5352 − 5357. doi:10.1109/TPEL.2015.2418786

[12] Z. Hu; Y.-F. Liu; P.C. Sen, "Cycle-by-cycle average input current sensing method for LLC resonant topologies," Energy Conversion Congress and Exposition (ECCE), 2013 IEEE , vol., no., pp.167,174, 15-19 Sept. 2013

A Common Capacitor Multi-Phase LLC Resonant Converter

Hongliang Wang, *Senior Member, IEEE*, Yang Chen, Zhiyuan Hu, Laili Wang, Yajie Qiu, Wenbo Liu, Yan-Fei Liu, *Fellow,IEEE*
Department of Electrical and Computer Engineering
Queen's University, Kingston, Canada
hongliang.wang@queensu.ca, yang.chen@queensu.ca, zhiyuan.hu@queensu.ca,
l.l.wang@queensu.ca,yajie.qiu@queensu.ca,wenbo.liu@queensu.ca,
yanfei.liu@queensu.ca

Jahangir Afsharian and Zhihua (Alex) Yang
Murata Power Solutions
Toronto, Canada
jafsharian@murata.com, ZYang@murata.com

Abstract— In this paper, a new common capacitor current sharing method is proposed for multi-phase LLC resonant converter. Automatic current sharing is achieved by using a common resonant capacitor for all the LLC resonant stages, by connecting the resonant capacitors in each phase in parallel. The proposed method can automatically share the load current without any additional circuits and control strategy. The current sharing performance of the proposed common capacitor current sharing method is analyzed under Fundamental Harmonic Analysis (FHA) assumption. A 600W two-phase LLC converter prototype based on the proposed method is built to verify the feasibility. Excellent current sharing performance (6.5% current sharing error at a wide load range) has been achieved.

I. INTRODUCTION

Resonant converter is attractive for isolated DC/DC application, such as flat-panel TVs, laptop adapters, server power supplies and so on, due to its high efficiency and high power density. LLC resonant converter can naturally achieve zero voltage switching (ZVS) for the primary-side MOSFET and zero current switching (ZCS) for the secondary-side diodes [1], [2]. For high power applications, the current stress of power devices increases with the power rating increasing. Multiphase parallel technique is a good choice to solve this problem [3][4][5]. However, due to the tolerance of resonant components, the resonant frequency of each individual LLC stage will be different, thus the output currents will be different [6][7][8]. It is observed that small component tolerance (such as 5%) can cause significant current imbalance. Therefore, current sharing is essential in order to achieve multiphase operation for LLC converter.

Some technologies have been developed to achieve current sharing for multiphase LLC converters. A category of active methods adjust the equivalent resonant capacitor [9], [10] or inductor [11] to compensate the resonant tank components' tolerances using additional MOSFETs. The circuit diagram for switched capacitor is shown in Fig. 1. The circuit diagram for variable inductor is shown in Fig. 2. Excellent load sharing performance can be achieved using these active methods. However, these methods suffer from high cost, complex control and non-excellent dynamic performance because sensing and control circuits have delays. DC voltage self-balanced method based on series input capacitors is one of the passive methods used to achieve current sharing [12][13]. Fig. 3 shows the circuit diagram of a two-phase LLCs as an example. The mid-point voltage is changed according to two phase's power. The system is of low cost and good load current sharing performance. However, it is hard to achieve modularization design and hot swap, because once the parameter design is finished, the module counts cannot be changed. There is another current sharing method based on three-phase three-wire structure. The three-phase LLCs have a 120° phase-shift between each phase. The load current sharing performance is good near resonant frequency, as all of three-phase resonant current is zero [14][15]. However, it can only be applied to three LLC modules in parallel. The load current will not share when the numbers of parallel modules is more than three.

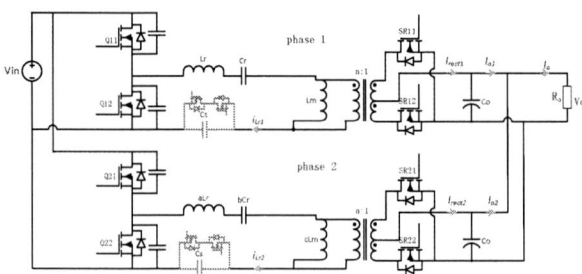

Fig. 1. Switched capacitor multi-phase LLC converter

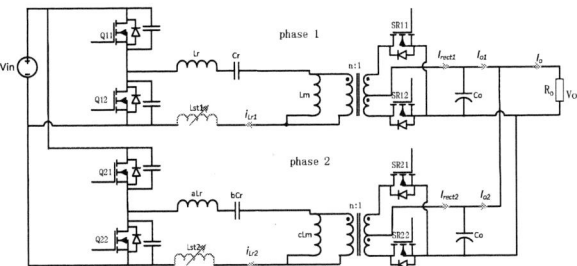

Fig. 2. Variable inductor multi-phase LLC converter

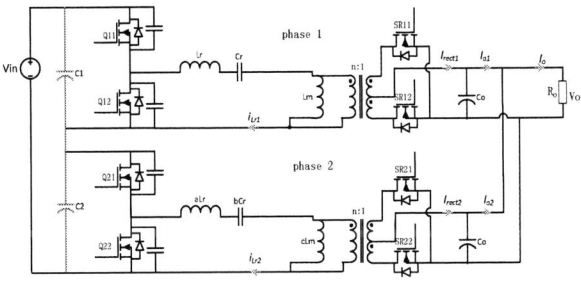

Fig. 3. Series DC-capacitor multi-phase LLC converter

Therefore, the existing technologies cannot provide cost effective and flexible current sharing technologies for multi-phase LLC resonant converters.

In this paper, a new common capacitor multi-phase LLC resonant converter is proposed to achieve load sharing without any addition components or control. In this method, the resonant capacitor in each LLC phase is connected in parallel. As a result, the load current is automatically shared. This technology is simple to implement with no additional cost. It can be expanded to arbitrary module counts without redesign the resonant parameters. This paper is organized as following: Section II describes the load sharing characteristic with/ without the proposed common capacitor method; Section III provides simulation and experimental results of a two-phase 600W prototype with common capacitor; and Section IV concludes the whole paper.

II. LOAD SHARING CHARACTERISTIC OF COMMON INDCTOR MULTI-PHASE LLC RESONANT CONVERTER

Mathematic model of LLC converter is needed for analyzing the current sharing characteristics. For simple understanding, two-phase LLC converter will be used as example in this paper. Fig. 4 shows the circuit diagram and its FHA equivalent circuit for two-phase LLC converter in conventional structure without current sharing [16]. L_r, C_r, L_m are the series resonant inductor, series resonant capacitor, magnetizing inductor of phase #1 respectively. aL_r, bC_r, cL_m are the resonant inductor, resonant capacitor, magnetizing inductor of phase #2. The values, a, b, c indicate the

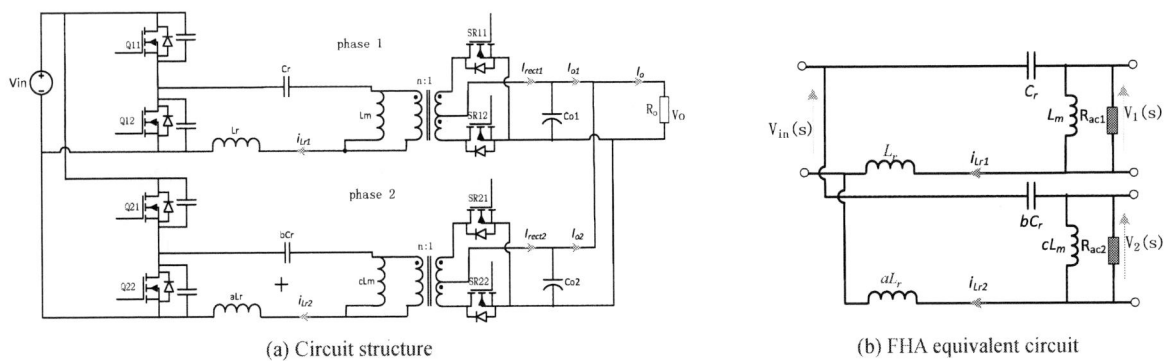

| (a) Circuit structure | (b) FHA equivalent circuit |

Fig. 4. Conventional two-phase LLC resonant converter

components' tolerances between the two LLC phases. n is transformer turn ratio. i_{Lr1}, i_{Lr2}, I_{rect1}, I_{rect2}, I_{o1}, and I_{o2} are the resonant currents, rectifier currents and load currents of two phases. Fig. 4 (b) is the equivalent circuit based on Fundamental Harmonic Analysis (FHA). In steady-state, the load resistor R_o is virtually divided into R_{o1} and R_{o2} according to the actual load current of each phase. The primary-side equivalent ac resistors R_{ac1} and R_{ac2} are determined accordingly and shown in (1).

$$\begin{cases} R_{o1} = \dfrac{1}{k} R_o, R_{o2} = \dfrac{1}{(1-k)} R_o, k \in [0,1] \\[2mm] R_{ac} = \dfrac{8n^2}{\pi^2} R_o, R_{ac1} = \dfrac{8n^2}{\pi^2} R_{o1}, R_{ac2} = \dfrac{8n^2}{\pi^2} R_{o2} \end{cases} \quad (1)$$

In (1), k is the impedance sharing error. The value of k should be between 0 and 1. k=0.5 means the load power can be equally shared by two phases. k=0 or 1 means the load power can only be provided by one phase.

The output side of the two phases are connected together, thus the ac voltage magnitude should always be the same, while the angles should be different because of the parameter tolerance. The relationship is described in (2).

$$|V_1(\text{s})| = |V_2(\text{s})| \tag{2}$$

The transfer function of $V_1(s)$ and $V_2(s)$ can be derived from the components impedance in (3):

$$
\begin{cases}
V_1(\text{s}) = \dfrac{R_{ac1}//sL_m}{R_{ac1}//sL_m + sL_r + 1/sC_r}V_{in}(s) \\[3mm]
V_2(\text{s}) = \dfrac{R_{ac2}//scL_m}{R_{ac2}//scL_m + saL_r + 1/sbC_r}V_{in}(s)
\end{cases} \tag{3}
$$

Fig. 5 (a) shows two-phase LLC resonant converter with proposed current sharing technology. The only difference from the conventional structure is that the series resonant capacitors of the two LLC converters are connected together. Fig.5 (b) shows the FHA equivalent circuit.

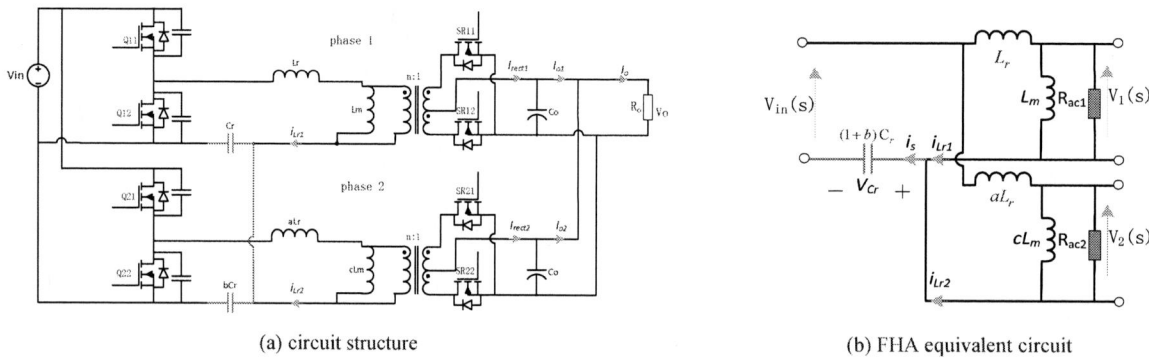

(a) circuit structure   (b) FHA equivalent circuit

Fig.5   Proposed two-phase LLC resonant converter

Similarly, the transfer function of $V_1(s)$ and $V_2(s)$ can be derived in (4):

$$
\begin{cases}
V_1(\text{s}) = \dfrac{R_{ac1}//sL_m}{R_{ac1}//sL_m + sL_r}\left(V_{in}(s) + V_{Cr}(s)\right) \\[3mm]
V_2(\text{s}) = \dfrac{R_{ac2}//s(cL_m)}{R_{ac2}//s(cL_m) + s(aL_r)}\left(V_{in}(s) + V_{Cr}(s)\right)
\end{cases} \tag{4}
$$

According to (1), (2) and (3) or (4), a quadratic equation (5) can be found for the impedance sharing error k for both the conventional and the proposed parallel LLC converters:

$$Ak^2 + Bk + C = 0 \tag{5}$$

For two-phase proposed LLC converter, the parameter A, B, C can be expressed in (6):

$$
\begin{cases}
A = \omega^4(a^2-1)c^2 L_r^2 L_m^2 \\[2mm]
B = -2\omega^4 a^2 c^2 L_r^2 L_m^2 \\[2mm]
C = \omega^4 a^2 c^2 L_r^2 L_m^2 + \omega^2[(a^2-c^2)L_r^2 + 2(ac-c^2)L_r L_m]R_{ac}^2
\end{cases} \tag{6}
$$

For two-phase conventional LLC converter, the parameter A, B, C can be expressed in (7):

$$
\begin{cases}
A = \omega^2(1-b^2)c^2 L_m^2 - \omega^4(2ab-2b^2)c^2 L_r L_m^2 C_r \\[2mm]
\quad + \omega^6(a^2-1)b^2 c^2 L_r^2 L_m^2 C_r^2 \\[2mm]
B = -2\omega^2 c^2 L_m^2 + 4\omega^4 abc^2 L_r L_m^2 C_r - 2\omega^6 a^2 b^2 c^2 L_r^2 L_m^2 C_r^2 \\[2mm]
C = \omega^2 c^2 L_m^2 - 2\omega^4 abc^2 L_r L_m^2 C_r + \omega^6 a^2 b^2 c^2 L_r^2 L_m^2 C_r^2 \\[2mm]
\quad + (1-b^2 c^2)R_{ac}^2 - \omega^2[(2ab-2b^2 c^2)L_r + (2bc-2b^2 c^2)L_m]C_r R_{ac}^2 \\[2mm]
\quad + \omega^4(ab-bc)[(ab+bc)L_r^2 + 2bcL_r L_m]C_r^2 R_{ac}^2
\end{cases} \tag{7}
$$

For quadratic equations, the roots, which are the current sharing error in this case, can be found in (8),

$$
k = \begin{cases}
-\dfrac{C}{B} & A=0, B\neq0 \\[4mm]
\dfrac{-B\pm\sqrt{B^2-4AC}}{2A} & A\neq0, \sqrt{B^2-4AC}\geq0
\end{cases} \quad and\ k\in[0,1] \tag{8}
$$

The current sharing error k is valid when k is between 0 and 1. Conditions $k=0$ and $k=1$ mean one phase provides all the power and the other phase does not provide power. Conditions $k<0$ and $k>1$ does not exist because this means one phase absorbs the power. Accordingly, the load current sharing error σ_{load} is defined in (9); and the resonant current sharing error $\sigma_{Resonant}$ is defined in (10).

978-1-4673-9551-9/16 $31.00 © 2016 IEEE   2322

$$\sigma_{load} = \frac{|I_{01} - I_{02}|}{|I_{01} + I_{02}|} = abs(1-2k), k \in [0,1] \qquad (9)$$

$$\sigma_{Resonant} = \frac{|rms(i_{Lr1}) - rms(i_{Lr2})|}{|rms(i_{Lr1}) + rms(i_{Lr2})|} \qquad (10)$$

Table.1 shows the resonant parameters of the phase #1, serving as the reference, to which the component tolerances of phase #2 will be compared.

TABLE I.   NOMINAL PARAMETER

L_r	C_r	L_m	n	f_r	V_o	$P_o(total)$
29µH	12nF	95µH	20	270kHz	12V	600W

If $a>1$ or $b>1$, it means phase #2 has lower resonant frequency compared with phase #1. Assuming same load, at given switching frequency, phase #2 will have lower voltage gain. If the two phases achieve same output voltage, phase #1 will output more power. If $c>1$, phase #2 will have higher inductor ratio which results in lower voltage gain. Phase #1

will output more power to keep the output voltage same. And vice versa, if $a<1$, $b<1$, or $c<1$, phase #2 will output more power. Thus, the worst situation is that parameters a, b, c deviates in the same direction.

Fig. 6 shows load current sharing error with different parameter tolerances in conventional two-phase LLC converter. Fig. 6 (a), (b), (c) shows the current sharing error at +5% L_r, +5% C_r, +5% L_m tolerance respectively. The current sharing error reduces with total load current increasing and input voltage decreasing. The worst case is shown in Fig.6 (d), in which three resonant parameters have +5% deviation simultaneously. The current sharing error is 60% at 50A load current for nominal 400V input voltage.

Fig. 7 shows the load current sharing error with different parameter tolerances in proposed two-phase common capacitor LLC converter. Fig.7 (a), (b), (c) shows the current sharing error at +5% L_r, +5% C_r, +5% L_m, respectively. Specifically, for +5% C_r case, L_r and L_m have no tolerance, thus there is no current difference as the two capacitors are paralleled. For simultaneous +5% tolerance on L_r, C_r, L_m case shown in Fig.7 (d), the current sharing error is about 2% at 50A load current, 400V input voltage.

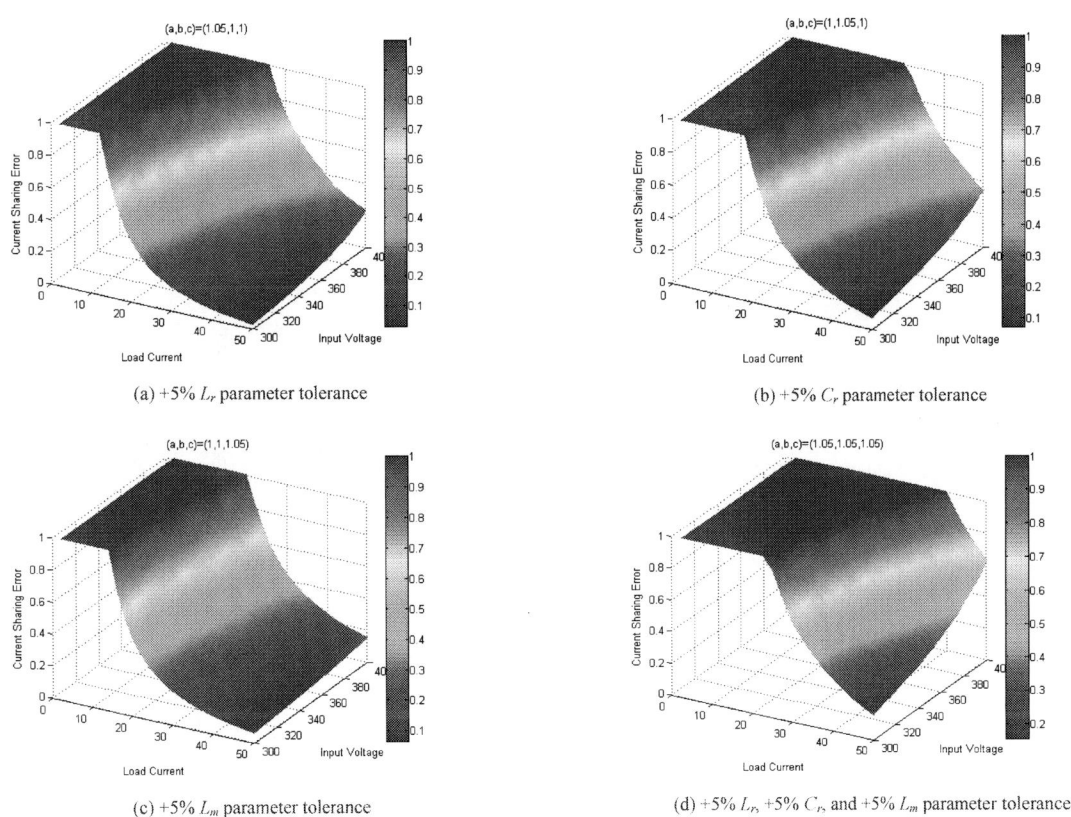

(a) +5% L_r parameter tolerance

(b) +5% C_r parameter tolerance

(c) +5% L_m parameter tolerance

(d) +5% L_r, +5% C_r, and +5% L_m parameter tolerance

Fig. 6. Current sharing error under two-phase independent LLC converter.

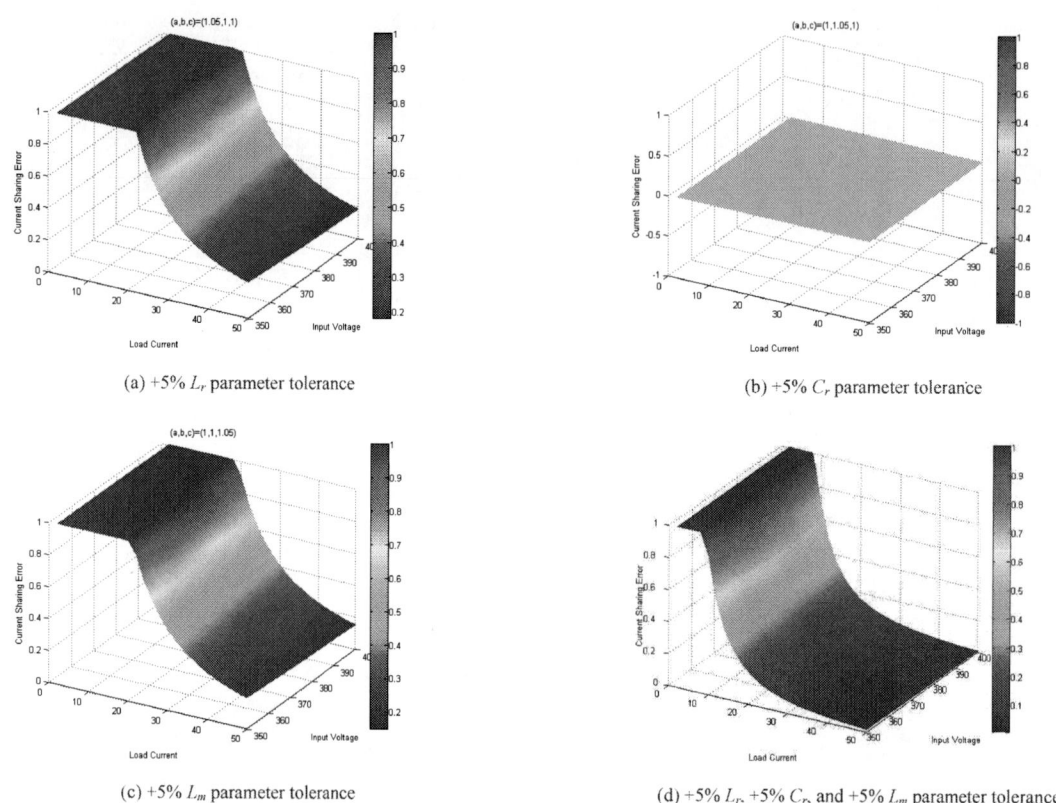

(a) +5% L_r parameter tolerance

(b) +5% C_r parameter tolerance

(c) +5% L_m parameter tolerance

(d) +5% L_r, +5% C_r, and +5% L_m parameter tolerance

Fig. 7. Current sharing error under two-phase common capacitor LLC converter

III. SIMULATION AND EXPERIMENT RESULTS

A 600W two-phase LLC converter prototype using common capacitor current sharing technology is built to verify the feasibility and to demonstrate the advantages of the proposed method. The circuit diagram is shown in Fig.5 (a). The parameters are shown in Table 2.

TABLE II. PROTOTYPE PARAMETERS

Input Voltage range	350V-400V
Output Voltage	12V
Output Power	300W × 2
Transformer Ratio n	20:1
Output Capacitance	1790μF
Series Capacitance(C_r)	13nF(Phase1) 12nF(Phase2)
Resonant Inductance(L_r)	24.5μH(Phase1) 22.5μH(Phase2)
Leakage Inductance(L_e)	6.5μH(Phase1) 6μH(Phase2)
Magnetizing Inductance(L_m)	95μH(Phase1) 92μH(Phase2)

Fig.8 show simulation waveforms of conventional two-phase LLC converter without current sharing at 15A, 25A

total load under 400V input voltage. The designed rated load current is 25A for each phase, thus, in conventional parallel structure, only 25A total load current experiment is provide to avoid the overcurrent of each phase.

In Fig. 8, the load current I_{o1} and I_{o2} are measured after the output capacitor. I_{o2} may have negative current at switching frequency level. This is because the output voltage has switching frequency ripple, the load current I_{o2} also has a switching frequency ripple to charge or discharge the output capacitor C_{o2}. On the other side, the average value of I_{o2} is zero. Thus, only phase #1 provides the load power.

Fig.9 shows simulation waveforms of two-phase LLC converter using the proposed common capacitor current sharing method at 15A, 25A, 50A total load under 400V input voltage. The load current difference is reduced from 15A to 0.6A between Fig. 8 (a) and Fig. 9 (a). The load current difference is reduced from 25A to 1A between Fig.8 (b) and Fig. 9 (b). Fig.9 (c) shows that at 50A total load, the load difference between the two phases is around 1A.

The resonant current, rectifier current are almost same for two phases. Thus, the load current is shared by two phases. It is believed that good resonant inductor current sharing guarantees good load current sharing as indicated according to Fig. 8 and Fig. 9.

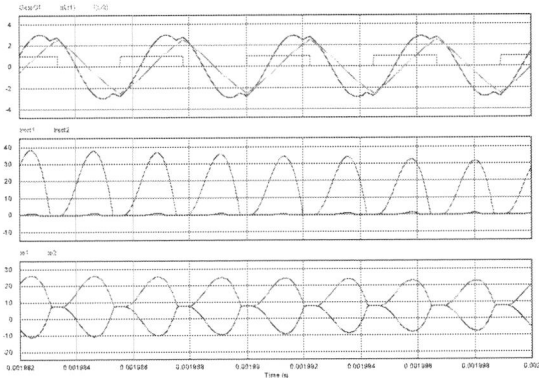

(a) Steady state at 15A load

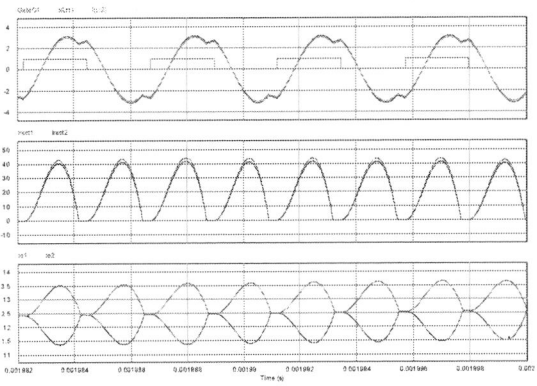

(b) Steady state at 25A load

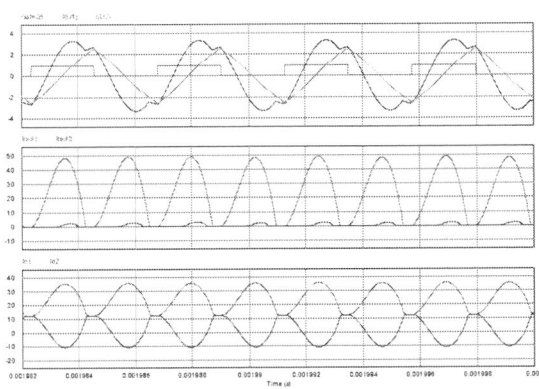

(b) Steady state at 25A load

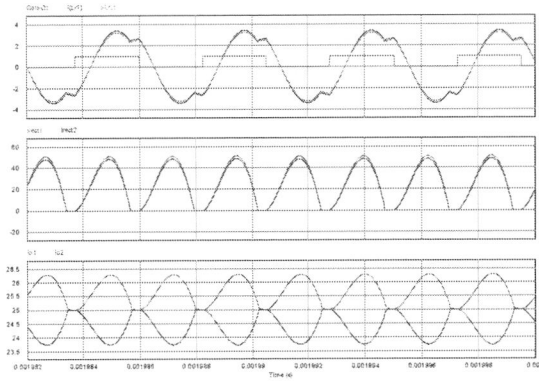

(c) Steady state at 50A load

Fig. 8. Simulation waveform of conventional two-phase LLC converter

Fig. 9. Simulation waveform of proposed two-phase LLC converter

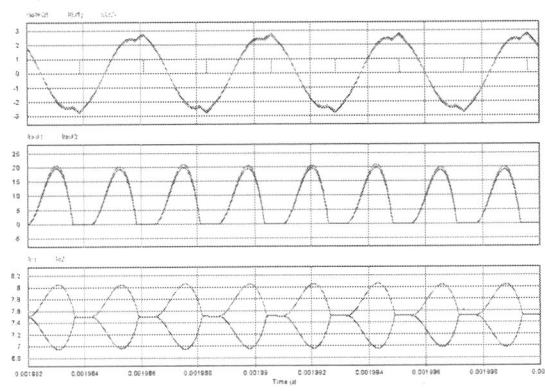

(a) Steady state at 15A load

Fig.10 shows the experiment waveform of conventional two-phase LLC converter for 15A and 25A load under 400V input voltage. Channel 1 is the output voltage. Channel 3, channel 4 are the resonant current of two phases. The resonant current i_{Lr_ch3} is almost triangular waveform, which means that phase almost does not provide the power for output load. Fig. 11 shows the experiment waveform of proposed two-phase LLC converter under 400V input voltage. The resonant current i_{Lr_ch3} and i_{Lr_ch4} is almost same. This indicates the two phases have shared loads.

978-1-4673-9551-9/16 $31.00 © 2016 IEEE

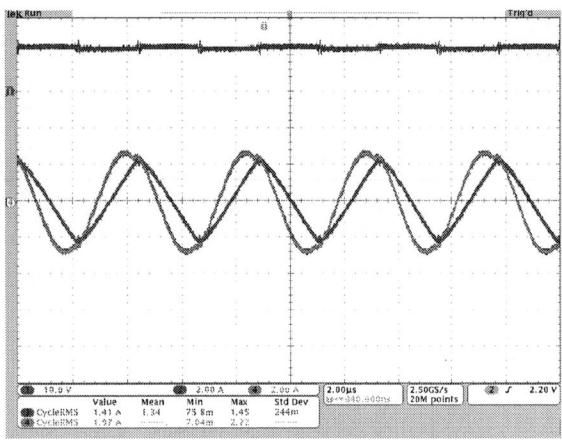

(a) Steady state at 15A load

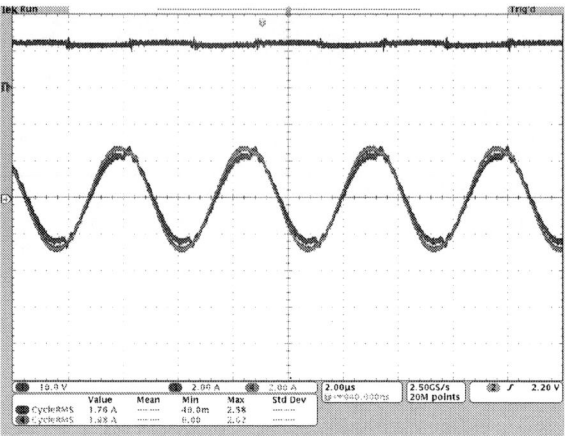

(b) Steady state at 25A load

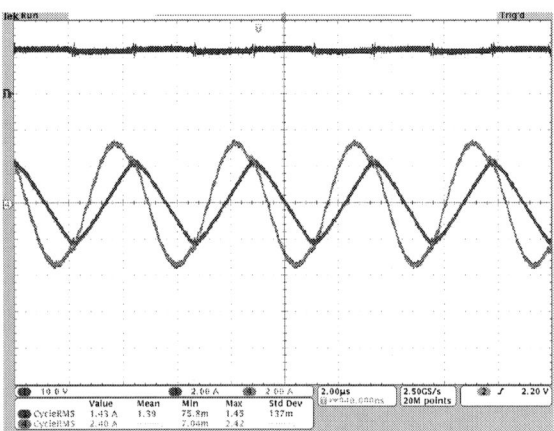

(b) Steady state at 25A load

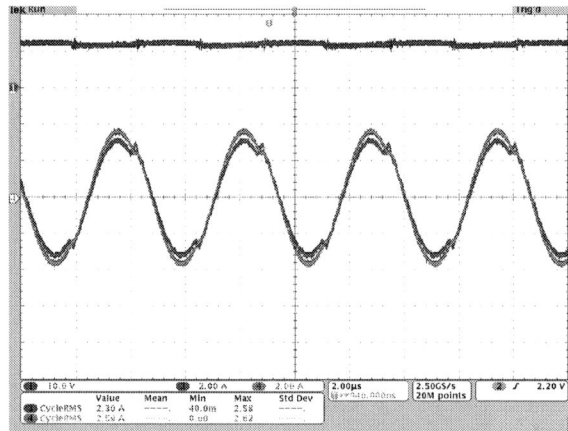

(c) Steady state at 50A load

Fig. 10. Experiment waveform of conventional two-phase LLC converter

Fig. 11. Experiment waveform of proposed two-phase LLC converter

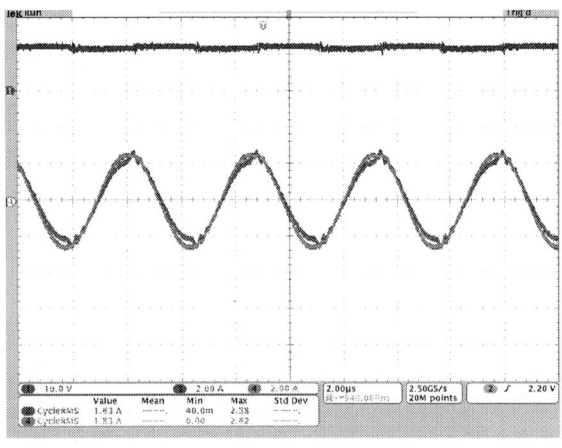

(a) Steady state at 15A load

To express circulating resonant current according to (5), the resonant current and relative resonant current error are shown in Fig. 12 and Fig.13.

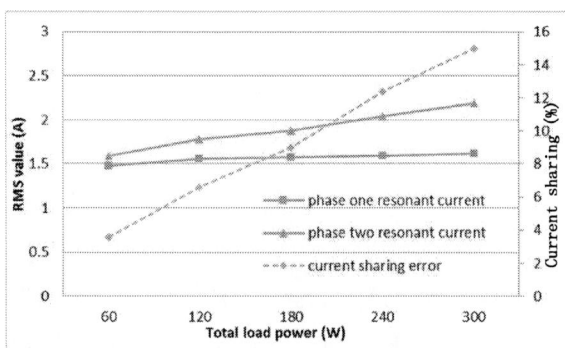

Fig. 12. Resonant current and relative error of conventional two-phase LLC converter

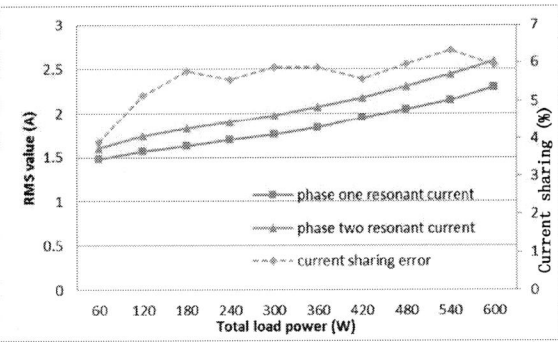

Fig. 13. Resonant current and relative error of common capacitor two-phase LLC converter

For conventional two-phase LLC converter, the resonant current error increases from 10% to 28% with load current increasing from 5A to 25A according to Fig. 12.

For the two-phase LLC converter with common capacitor technology, the resonant current error is at around 6.5% for the whole load range from Fig.13. Thus, the circulating current between phases is significantly reduced using the proposed method.

IV. CONCLUSION AND FUTURE WORK

A new, common capacitor current sharing strategy for multi-phase LLC resonant converter is proposed. The series resonant capacitors in each LLC converter are connected in parallel. No additional components are needed to achieve current sharing. Mathematical model is built based on FHA to analyze the current sharing characteristics of a two-phase LLC converter. The analysis results shows that the circulating current is significantly reduce using the proposed method. A two-phase LLC converter prototype with 300W per phase is built using the common capacitor current sharing method. The simulation and experiment results show that the relative circulating resonant current can be maintained at 6.5% for all load conditions with proposed common capacitor method.

REFERENCES

[1] Y. Bo, "Topology Investigation for Front End DC/DC Power Conversion for Distributed Power System," Virginia Polytechnic Institute and Stage University, 2003.

[2] Y. Z. Y. Zhang, D. X. D. Xu, M. C. M. Chen, Y. H. Y. Han, and Z. D. Z. Du, "LLC resonant converter for 48 V to 0.9 V VRM," 2004 IEEE 35th Annu. Power Electron. Spec. Conf. (IEEE Cat. No.04CH37551), vol. 3, 2004.

[3] M. T. Zhang, M. M. Jovanović, and F. C. Y. Lee, "Analysis and evaluation of interleaving techniques in forward converters," IEEE Transactions on Power Electronics, vol. 13, no. 4. pp. 690–698, 1998.

[4] R. Hermann, S. Bernet, Y. Suh, and P. K. Steimer, "Parallel connection of integrated gate commutated thyristors (IGCTs) and diodes," IEEE Trans. Power Electron., vol. 24, no. 9, pp. 2159–2170, 2009.

[5] J. Rabkowski, D. Peftitsis, and H. P. Nee, "Parallel-operation of discrete SiC BJTs in a 6-kW/250-kHz DC/DC boost converter," IEEE Trans. Power Electron., vol. 29, no. 5, pp. 2482–2491, 2014.

[6] Z. Hu, Y. Qiu, Y. F. Liu, and P. C. Sen, "An interleaving and load sharing method for multiphase LLC converters," Conf. Proc. - IEEE Appl. Power Electron. Conf. Expo. - APEC, no. 1, pp. 1421–1428, 2013.

[7] H. Figge, T. Grote, N. Froehleke, J. Boecker, and P. Ide, "Paralleling of LLC resonant converters using frequency controlled current balancing," PESC Rec. - IEEE Annu. Power Electron. Spec. Conf., pp. 1080–1085, 2008.

[8] B. C. Kim, K. B. Park, and G. W. Moon, "Analysis and design of two-phase interleaved LLC resonant converter considering load sharing," in 2009 IEEE Energy Conversion Congress and Exposition, ECCE 2009, 2009, pp. 1141–1144.

[9] Z. Hu, Y. Qiu, L. Wang, and Y. F. Liu, "An interleaved LLC resonant converter operating at constant switching frequency," IEEE Trans. Power Electron., vol. 29, no. 6, pp. 2931–2943, 2014.

[10] Z. Hu, Y. Qiu, Y. F. Liu, and P. C. Sen, "A control strategy and design method for interleaved LLC converters operating at variable switching frequency," IEEE Trans. Power Electron., vol. 29, no. 8, pp. 4426–4437, 2014.

[11] E. Orietti, P. Mattavelli, G. Spiazzi, C. Adragna, and G. Gattavari, "Two-phase interleaved LLC resonant converter with current-controlled inductor," 2009 Brazilian Power Electron. Conf. COBEP2009, pp. 298–304, 2009.

[12] B. C. Kim, K. B. Park, C. E. Kim, and G. W. Moon, "Load sharing characteristic of two-phase interleaved LLC resonant converter with parallel and series input structure," 2009 IEEE Energy Convers. Congr. Expo. ECCE 2009, pp. 750–753, 2009.

[13] F. Jin, F. Liu, X. Ruan, S. Member, and X. Meng, "Multi-Phase Multi-Level LLC Resonant Converter with Low Voltage Stress on the Primary-Side Switches," pp. 4704–4710, 2014.

[14] E. Orietti, P. Mattavelli, G. Spiazzi, C. Adragna, and G. Gattavari, "Analysis of multi-phase LLC resonant converters," 2009 Brazilian Power Electron. Conf. COBEP2009, pp. 464–471, 2009.

[15] E. Orietti, P. Mattavelli, G. Spiazzi, C. Adragna, and G. Gattavari, "Current sharing in three-phase LLC interleaved resonant converter," 2009 IEEE Energy Convers. Congr. Expo. ECCE 2009, pp. 1145–1152, 2009.

[16] H. Wang, Y. Chen, Y.-F. Liu, J. Afsharian, Z. Yang, "A common inductor multi-phase LLC resonant converter," 2015, IEEE Energy Convers. Congr. Expo. ECCE 2015, pp. 548-555, 2015

LLC Resonant Converter Design for Bendable Power Converter

Godwin Kwun Yuan Ho, Bryan Man Hay Pong, and Ron Shu Yuen Hui
Electrical and Electronic Engineering Department
The University of Hong Kong

Abstract—**In this paper, a design approach of LLC resonant converter with bendable transformer is presented. First, the electrical parameters of the LLC resonant converter are calculated based on the first-harmonic-approximation (FHA) equivalent circuit. Second, the bendable transformer design approach based on bendable transformer model gives the physical structure of transformer from the electrical parameters calculated. This design approach is adopted in an LLC resonant converter with an isolated output with ratings of 5 V and 2 A for a USB power supply. The design approach is described in details and experimental measurements of a hardware prototype are included to confirm the design approached.**

Keywords—LLC resonant converter; bendable transformer

I. INTRODUCTION

Wearable electronic products are becoming popular Different form the traditional portable electronics, wearable electronics require not only small in size and light in weight, but also comfortable to wear. This extra requirement introduces a new challenge: physically flexible. Although there a lot of work on flexible electronics [1] and many flexible electronic components are available today [2]-[7], there is a lack of flexible power converter. In many applications such as medical equipment, the converter output is directly in contact with the human body and isolation is mandatory for safety. In this paper a bendable isolated converter is presented.

Transformer is an essential passive component in switched mode power converter but traditional ferrite core based transformer prevents flexibility of the power converter. Coreless printed-circuit-board (PCB) transformer [8]-[14] offers an attractive solution and applications of power transfer has been demonstrated [8]-[12]. Coreless PCB transformer fabricated on a flexible substrate provides the flexibility for bendable power converter. Also it is easy to manufacture [15]. However, PCB transformer has relatively high leakage inductance and low self-inductance compare with the traditional core-based transformer. The LLC resonant converter topology makes use of the inherent high leakage inductance as the resonant inductor. Also planer transformer has been incorporated into resonant converter since its inception [10], [11]. LLC converter is a good foundation to proceed to bendable converter.

The coreless PCB is an attractive solution to bendable converter but the design approach is not well documented [15]-[19]. Unlike core-base transformer, coupling factor of the coreless PCB transformer can be very low. The magnetizing inductance is determined by the coil pattern itself. The equivalent turns ratio is not dominated by the physical turns

ratio but the mutual inductance and the secondary inductance of the transformer. Therefore, the core-base transformer design procedure is not suitable for the coreless PCB transformer. Hence, a more appropriated coreless PCB transformer design approach is needed to design the LLC converter with bendable transformer.

II. LLC CONVERTER AND ITS FHA

A schematic diagram of an LLC resonant converter is shown in fig. 1. An ideal transformer mode is used in the circuit. A resonant tank with magnetizing inductor L_{mag}, resonant inductor L_r and resonant capacitor C_r (LLC) is included in the circuit. The first-harmonic-approximation (FHA) approach [24] is based on the assumption that the power transfer from the source to the load through the resonant tank is almost completely associated with the fundamental component of the Fourier expansion of the currents and voltage involved [20]-[27].

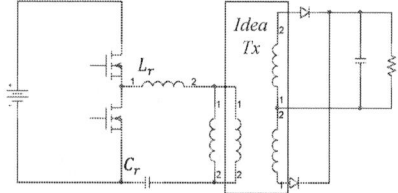

Fig. 1. LLC resonant converter equivalent circuit

III. BENDABLE TRANSFORMER MODEL

Inductance equations for standard regular shapes coil have been well documented [28]-[34] but they are very limited in flexible winding structures. Parameters change as the structure is bent. A bendable transformer model based on the PEEC theory has been developed and used in this paper [15].

Each coil is represented by a finite set W with N segments. Each segment is represented by a matrix with segment-current vector $\vec{C_s}$ and position vector $\vec{P_s}$. Segment-current vector describes the length and the direction of the conductor. Position vector describes the position of the conductor segment in the 3D space. Figure 2 shows a coil in 3D by segment current vector and position vector.

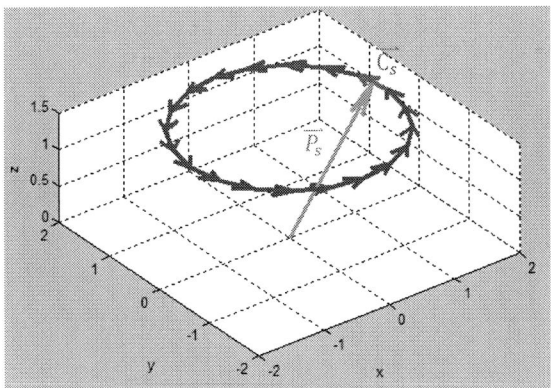

Fig. 2. A Coil by current-segment vector and position vector

Define set $W_1 = \left\{ \left[\overrightarrow{C_{1s1}}\; \overrightarrow{P_{1s1}} \right], \left[\overrightarrow{C_{1s2}}\; \overrightarrow{P_{1s2}} \right], \ldots, \left[\overrightarrow{C_{1sN1}}\; \overrightarrow{P_{1sN2}} \right] \right\}$ as the primary coil

and $W_2 = \left\{ \left[\overrightarrow{C_{2s1}}\; \overrightarrow{P_{2s1}} \right], \left[\overrightarrow{C_{2s2}}\; \overrightarrow{P_{2s2}} \right], \ldots, \left[\overrightarrow{C_{2sN1}}\; \overrightarrow{P_{2sN2}} \right] \right\}$ as the secondary coil. The mutual inductance is

$$L_{mutual} = \sum_{i=1}^{N1} \sum_{j=1}^{N2} \frac{\overrightarrow{C_{1si} \cdot C_{2sj}}}{\left| \overrightarrow{P_{1si} - P_{2sj}} \right|} \tag{1}$$

and the self-inductance is

$$L_{self} = \sum_{i=1}^{N1} \sum_{j=1}^{N1} f(i,j) \tag{2}$$

$$f(i,j) = \begin{cases} \dfrac{\overrightarrow{C_{1si} \cdot C_{1sj}}}{\left| \overrightarrow{P_{1si} - P_{1sj}} \right|} & for\ i \neq j \\[2ex] 2l \left[\begin{array}{c} log\left(\frac{2l}{\alpha+\beta} \right) + 0.5 \\ + \frac{0.2235(\alpha+\beta)}{l} \end{array} \right] (10^{-6}) & for\ i = j \end{cases} \tag{3}$$

IV. LLC RESONANT CONVERTER WITH BENDABLE TRANSFORMER

Three steps are proposed to design the bendable transformer. First, calculate the ideal transformer turns ratio, primary and secondary self-inductance and mutual inductance. Second, calculate the bendable transformer structure. Third, verify the design when the transformer is bent. A design flow chart is shown in fig. 3.

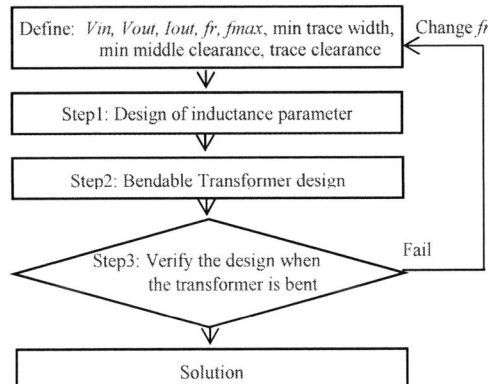

Fig. 3. LLC resonant converter with bendable transformer design flow chat.

A. Step 1: Design of Inductance Parameter

Most of the parameters of the LLC converter is calculated by the FHA approach after the electrical specifications of the converter are defined [27]. The primary self-inductance L_p, secondary self- inductance L_s and mutual inductance L_m of the bendable transformer are calculated by the following equation with respect to fig. 4.

$$L_p = \frac{L_r}{(1 - K^2)} \tag{4}$$

$$L_s = L_{mag} \frac{N_s}{N_p} \tag{5}$$

$$L_m = K \sqrt{L_p L_s} \tag{6}$$

$$K = \left(\sqrt{\frac{L_r}{L_{mag}}} + 1 \right)^{-1} \tag{7}$$

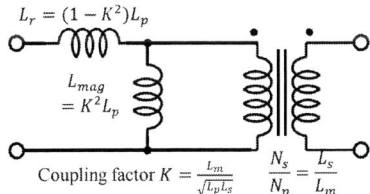

Fig. 4 Ideal transformer model.

B. Step 2: Bendable Transformer Design

The structure of the bendable transformer is calculated based on the: primary self-inductance L_p, secondary self-inductance L_s, mutual inductance L_m and maximum DC resistance. Numerical method and bendable transformer model is applied. Rectangular coil with the golden ratio (long: width = 1.618:1) is chosen in here. A MatLab program is developed for the proposed design approach. The program has three parts : primary coil design, secondary coil design and transformer structure design.

Here the target inductance, minimum trace width, trace clearance and the minimum middle clearance shown in fig. 5 are defined to start with. There are three design loops as shown in fig. 6: calculate the number of turns, fine tune the dimensions and fulfill the maximum DC resistance. The turns start with a small center turn. The turns then increases outwards whereby the inductance and the size grow until the target is researched. The primary coil is then fine-tuned by increasing the middle clearance. The third loop repeats comparing the coil DC resistance and the maximum resistance by increasing the trace width by 1 mm each step until the coil DC resistance is reduced. The primary coil structure design approach ends with a primary coil structure.

978-1-4673-9551-9/16 $31.00 © 2016 IEEE

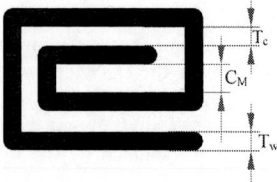

Fig. 5 Primary Coil structure.

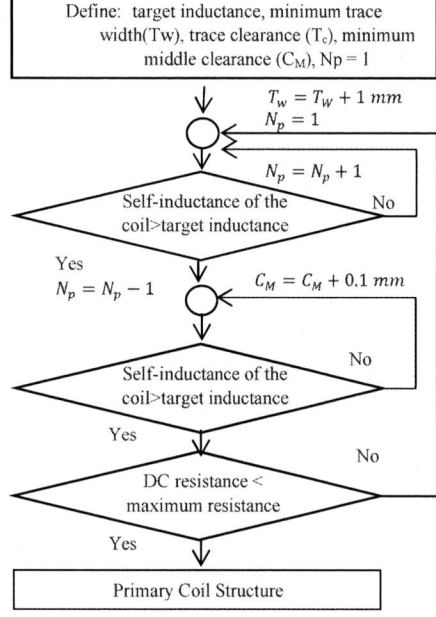

Fig. 6 Primary coil design approach.

The secondary coil design flow is shown in fig. 7. In order to produce maximum coupling between the coils, the secondary coil is made to have similar size on top of the primary coil as much as possible. A first secondary winding starts from a position according to the outermost primary coil. The turns are then increased inwards until the secondary inductance is reached. The secondary coil is then fine-tuned by shrinking the coil towards the center.

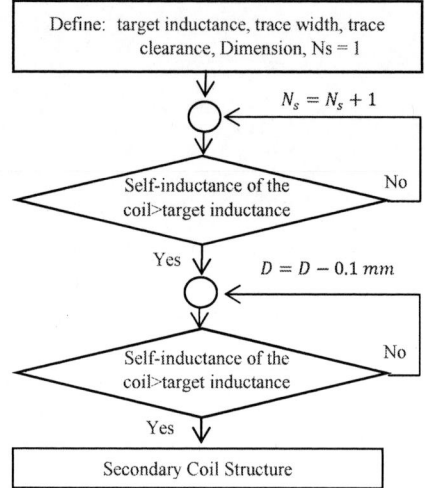

Fig. 7 Secondary coil design approach.

The transformer structure design completes with the determination of the coils alignment which is shown in fig. 8. Predefined gap G is the thickness of the substrate between the primary and secondary coil such that the positions of primary coil is [0 0 0] and secondary coil is [0 0 G]. The y position of the secondary coil is then shifted until the mutual inductance falls towards the target value.

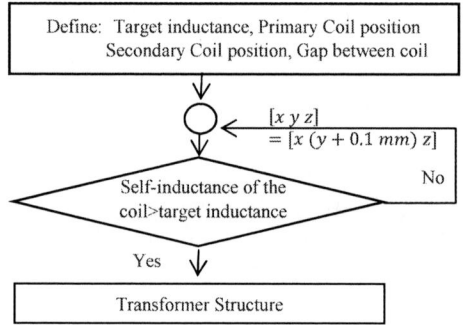

Fig. 8 Transformer structure design approach.

C. Step 3: Verify the design when the transformer is bent

The program repeats to verify the LLC converter with bendable transformer by increasing the resonant frequency until the output of the converter fulfil the design specifications. This completes the transformer design with all parameters collected.

V. LLC RESONANT CONVERTER WITH BENDABLE TRANSFORMER DESIGN EXAMPLE AND EXPERIMENTAL RESULT

A practical 12 V to 5 V, 2 A converter is built to verify the LLC converter with bendable transformer design approach and demonstrate the bendable transformer in fig. 9. The maximum switching frequency is 1.2MHz. GaN (Gallium Nitride) MOSFET are used in the prototype. The input capacitance and gate capacitance of GaN MOSFET is relatively low compared with silicon based MOSFET. This allows higher switching frequency and lower switching loss.

The bendable transformer design detail is shown in table I. The target inductances and measured inductances are shown in table II.

TABLE I
Transformer structure

Primary turns	5
Secondary turns	4
Primary outer rectangle dimension:	129 mm x 79 mm
Secondary outer rectangle dimension:	122 mm x 73 mm
Trace Width	7 mm
Trace thickness	0.5 oz
Trace Clearance	0.5 mm
Gap distance between conductor:	0.3 mm
Y offset	20 mm

TABLE II
Transformer Inductances

	Target	Measured	Err
Primary self-inductance	1.63 uH	1.68 uH	-3.1 %
Secondary-self inductance	0.96 uH	0.96 uH	0 %
Mutual Inductance	0.927 uH	0.93 uH	-0.03 %

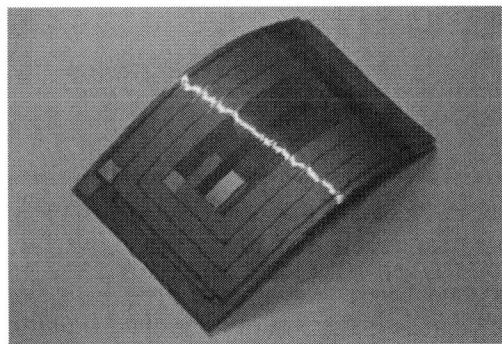

Fig. 9a. The bendable transformer prototype.

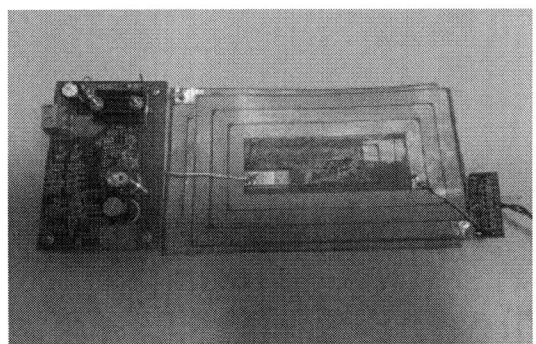

Fig. 9b. The bendable transformer with evaluation board.

Figure 10 shows the measured waveforms of the prototype at 1 A output when the transformer is flat. The resonant current shows that the switching frequency is a little bit higher than the resonant frequency. The low side gate drive is turn on after the switching node voltage drop to zero. Such that the LLC converter is operating with zero voltage switching.

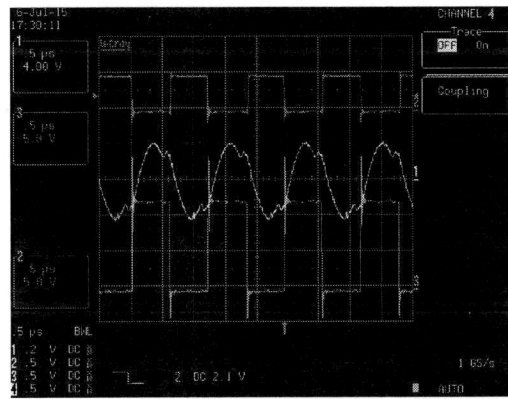

Fig. 10. Waveform of the prototype (Ch1:Resonat tank current (1A/div); Ch2: Low side gate voltage; Ch3: Switching node voltage).

The converter is operated with load current from 0.1 A to 2 A. The frequency is adjusted so that the output voltage lies around 5 V. Figure 11 shows the output voltage is regulated at 5 V within 1% error at different loading.

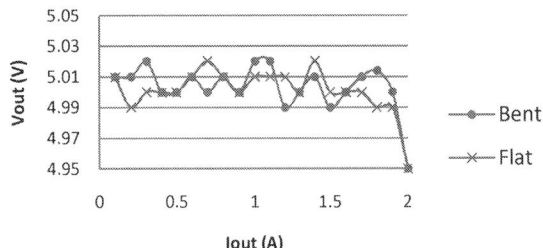

Fig. 11. Output voltage against Output Current.

Figure 12 shows the switching frequency of the transformer is flat with a bending angle equal to π. The operation frequency changes in a relatively small range between 700 kHz to 1.2 MHz. The parameters of the transformer are changed when that transformer is in different bending angles [15]. The transformer coil pattern has been chosen carefully in this paper in order to limit these parameters change. The switching frequency and the resonant frequencies change within 5% and the bent transformer request a higher switching frequency.

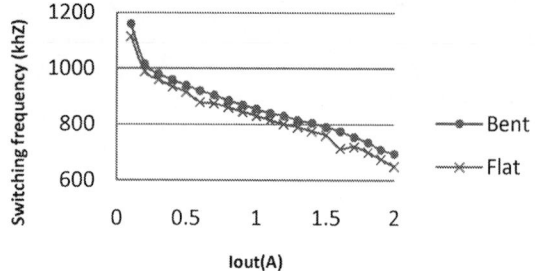

Fig. 12. Switching frequency against Output Current.

Figure 13 shows the efficiency of the converter. The maximum efficiency of the converter is 70%. This is relatively low. The transformer has the highest loss in the whole converter. The thickness of the transformer copper trace is 0.5 oz. The power efficiency of the converter in flat is very close to the when bent. Although the resonant frequency of the converter is changed after bent, the change lies within 5%. Also the resistance of the transformer is not changed.

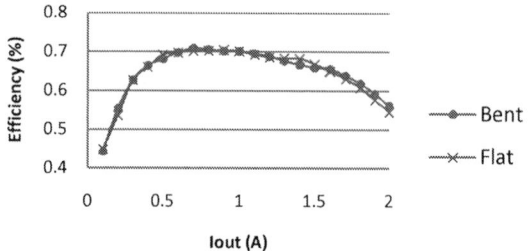

Fig. 13. Efficiency against output current

VI. CONCLUSION

In this paper, a new design approach of bendable converter based is presented. The proposed design approach is based on the FHA equivalent circuit and the bendable transformer model. The structure of the bendable transformer is given by the proposed design approach with numerical method.

A practical 12 V to 5 V, 2 A converter is built to verify the LLC converter with bendable transformer design approach. The maximum switching frequency is 1.2 MHz. GaN (Gallium Nitride) MOSFET are used in the prototype. The experimental results show that output voltage is regulated at 5 V within 1% error at different loading. The operation frequency changes in a relatively small range between 700 kHz to 1.2 MHz. The parameters of the transformer are changed when that transformer is bent to different bending angles. The switching frequency and the resonant frequencies change within 5%. The bent transformer requests a higher switching frequency. The maximum efficiency of the converter is 70%.

REFERENCES

[1] Hu, J., "Overview of flexible electronics from ITRI's viewpoint,"*VLSI Test Symposium (VTS), 2010 28th*", pp.84,84, 19-22 April 2010

[2] Smith, J.T.; O'Brien, B.; Yong-Kyun Lee; Bawolek, E.J.; Christen, J.B., "Application of Flexible OLED Display Technology for Electro-Optical Stimulation and/or Silencing of Neural Activity", *Display Technology, Journal of*, vol.10, no.6, pp.514-520, June 2014

[3] Jungsuek Oh; Kyusang Lee; Hughes, T.; Forrest, S.; Sarabandi, K., "Flexible Antenna Integrated With an Epitaxial Lift-Off Solar Cell Array for Flapping-Wing Robots ", *Antennas and Propagation, IEEE Transactions on*, vol.62, no.8, pp.4356-4361, Aug. 2014

[4] Chien-Yi Peng; Dhakal, T.P.; Rajbhandari, P.; Garner, S.; Cimo, P.; Lu, S.; Westgate, C.R., "Flexible CZTS solar cells on flexible Corning®Willow®Glass substrates",*Photovoltaic Specialist Conference (PVSC), 2014 IEEE 40th*, pp.0409-0412, 8-13 June 2014

[5] Kayes, B.M.; Ling Zhang; Twist, R.; I-Kang Ding; Higashi, G.S., "Flexible Thin-Film Tandem Solar Cells With >30% Efficiency,"*Photovoltaics, IEEE Journal of*, vol.4, no.2, pp.729-733, March 2014

[6] Sungryul Yun; Suntak Park; Bongjae Park; Seung Koo Park; Prahlad, H.; von Guggenberg, P.; Ki-Uk Kyung, "Polymer-Based Flexible Visuo-Haptic Display," *Mechatronics, IEEE/ASME Transactions on* , vol.19, no.4, pp.1463-1469, Aug. 2014

[7] Cheol Jang; Kukjoo Kim; Kyung Cheol Choi, "Toward Flexible Transparent Plasma Display: Optical Characteristics of Low-Temperature Fabricated Organic-Based Display Structure," *Electron Device Letters, IEEE* , vol.33, no.1, pp.74-76, Jan. 2012

[8] Zhang, J.; Hurley, W.G.; Wolfle, W.H., "Design of the planar transformer in LLC resonant converters for micro-grid applications," *Power Electronics for Distributed Generation Systems (PEDG), 2014 IEEE 5th International Symposium on* , pp.1-7, 24-27 June 2014

[9] Hui, S.Y.; Chung Henry Shu-Hung; Tang, S.C., 'Coreless printed circuit board (PCB) transformers for power MOSFET/IGBT gate drive circuits', *IEEE Transactions on Power Electronics*, vol. 14, pp.422-430, May 1999

[10] Tang, S.C.; Hui, S.Y.; Chung H.S.H, 'Coreless planar printed-circuit-board (PCB) transformers-a fundamental concept for signal and energy transfer', *IEEE Transactions on Power Electronics*, vol.15, pp.931-941, Sept. 2000

[11] S.Y.R. Hui and S.C. Tang, "Planar printed-circuit-board transformers with effective electromagnetic interference (EMI) shielding", US patent 6,501,364, Dec. 2002.

[12] S.Y.R. Hui and S.C. Tang, "Method of operating a coreless printed-circuit-board (PCB) transformer", European Patent EP 0935263B, May 2004

[13] S.C. Tang, S.Y.R. Hui and H. Chung, 'A low-profile low-power converter using coreless PCB transformer with ferrite polymer composite', *IEEE Transactions on Power Electronics*, vol.16, No.4 , pp.493-498, July 2001

[14] Eberhard Waffenschmidt, Bernd Ackermann, "Size advantage of coreless transformers in the MHz range", *EPE 2001*, paper DS2-9.

[15] Ho, Godwin K.Y.; Zhang, Cheung, Pong, Bryan M.H.; Hui, S.Y., " Bendable Transformer for Wearable Electronics", in *Energy conversion congress &Expo 2015, pp.* pp.5865-5871, 20-24 Sept 2015

[16] Fernández, C.; Prieto, R.; Garcia, O.; Cobos, J.A., "Coreless Magnetic Transformer Design Procedure,"*Power Electronics Specialists Conference, 2005. PESC '05. IEEE 36th*, vol., no., pp.1548,1554, 16-16 June 2005

[17] Bouabana, A.; Sourkounis, C., "Design and analysis of a coreless flyback converter with a planar printed-circuit-board transformer,"*Optimization of Electrical and Electronic Equipment (OPTIM), 2010 12th International Conference on*, vol., no., pp.557,563, 20-22 May 2010

[18] Bouabana, A.; Sourkounis, C.; Mallach, M., "Design and analysis of different structure of a coreless planar transformer for a flyback converter," *Power Electronics, Electrical Drives, Automation and*

Motion (SPEEDAM), 2012 International Symposium on, vol., no., pp.827,831, 20-22 June 2012

[19] Zhang, J.; Hurley, W.G.; Wolfle, W.H., "Design of the planar transformer in llc resonant converters for micro-grid applications,"*Power Electronics for Distributed Generation Systems (PEDG), 2014 IEEE 5th International Symposium on* , vol., no., pp.1,7, 24-27 June 2014

[20] T. Duerbaum: *First harmonic approximation including design constraints*, Telecommunications Energy Conference, 1998. INTELEC. Pages: 321 – 328

[21] M.B. Borage; S.R. Tiwari; S. Kotaiah, "*Design Optimization for an LCLType Series Resonant Converter*", www.powerpulse.net

[22] G. Ivensky; S. Bronstein; S. Ben-Yaakov, "*Approximate analysis of the Resonant LCL Converter*", 23rd IEEE Israel Convention, Tel-Aviv 2004, Pages 44 – 48

[23] B. Lu; W. Liu; Y. Liang; Lee, F.C.; van Wyk, J.D., "*Optimal design methodology for LLC resonant converter*", Applied Power Electronics Conference and Exposition, 2006. APEC '06. Twenty-First Annual IEEE, 19-23 March 2006. Pages: 533 – 538

[24] S. De Simone; C. Adragna; C. Spini; G. Gattavari, "*Design-oriented steady-state analysis of LLC resonant converters based on FHA*", Power Electronics, Electrical Drives, Automation and Motion, 2006. SPEEDAM 2006. International Symposium, May, 23rd - 26th, 2006, Page(s):200 – 207

[25] H. Choi, "*Analysis and Design of LLC Resonant Converter with Integrated Transformer*", Applied Power Electronics Conference, APEC, 2007 - Twenty Second Annual IEEE Feb. 2007 Page(s):1630 – 1635

[26] Y. Fang; D. Xu; Y. Zhang; F. Gao; L. Zhu, "*Design of High Power Density LLC Resonant Converter with Extra Wide Input Range*", Applied Power Electronics Conference, APEC 2007 - Twenty Second Annual, IEEE Feb. 2007 Page(s):976 - 981

[27] *Silvio De Simone. (2014), LLC reosnant half-bridge converter design guideline*, [Online]. Available : http://www.st.com/web/en/resource/technical/document/application_not e/CD00143244.pdf

[28] F.W. Grover, "Inductance calculations:Working formulas and Tables", New York: Dover Publications Inc., 2004 (reprint).

[29] W.G. Hurley and M.C. Duffy, "Calculationof self and mutual impedances in planar sandwich inductors", *IEEE Transactions on Magnetics*, vol.33, no.3, pp. 2282-2290, May 1997

[30] W.G. Hurley, M.C. Duffy, S. O'Reilly and S.C. O'Mathuna, "Impedance formulas for planar magnetic structures with spiral windings", *IEEE Industrial Electronics*, vol. 46, no.2, pp.271-278, April 1999

[31] S.I. Babic and C. Akyel, "New analytic-numerical solutions for the mutual inductance of two coaxial circular coils with rectangular cross section in air," *IEEE Transactions on Magnetics*, vol. 42, no. 6, pp. 1661-1669, June 2006.

[32] S. Babic and C. Akyel, "Improvement in calculation of the self- and mutual inductance of thin-wall solenoids and disk coils", *IEEE Transactions on Magnetics*, vol.36, no.4, Part: 2, pp.1970- 1975, July 2000

[33] G. Hurley, M. Duffy, J. Zhang, I. Lope, B. Kunz and W. Wolfle, "A Unified Approach to the Calculation of Self and Mutual Inductance for Coaxial Coils in Air", *IEEE Transactions on Power Electronics*, (early access)

[34] Babic, S.; Sirois, F.; Akyel, C.; Girardi, C., "Mutual Inductance Calculation Between Circular Filaments Arbitrarily Positioned in Space: Alternative to Grover's Formula," *Magnetics, IEEE Transactions on* , vol.46, no.9, pp.3591-3600, Sept. 2010

[35] E. B. Rosa and L. Cohen, "Formulae and Tables for the Calculation of Mutual and Self-inductance," *Bulletin of the Bureau of Standards*, vol. 5, pp. 35-50, 1907

Design Consideration of MHz Active Clamp Flyback Converter with GaN Devices for Low Power Adapter Application

Xiucheng Huang, Junjie Feng, Weijing Du, Fred C. Lee, and Qiang Li

Center for Power Electronics Systems, the Bradley Department of Electrical and Computer Engineering
Virginia Polytechnic Institute and State University, Blacksburg, VA, 24061, USA

Abstract— With ever increasing demands of smaller size, lighter weight for all forms of consumer electronics, efficient power conversion with higher operating frequency has always being pursued rigorously. This paper demonstrates high frequency, high efficiency and high power density design of active clamp flyback converter for adapter application. Both the primary and secondary switches are gallium nitride (GaN) devices which can significantly reduce the device related conduction and switching loss. The design procedures, including the selection of active clamping capacitor, optimization of flyback transformer, and EMI filter design, are presented in detail. A 65W (19.5V/3.3A) prototype of active clamp flyback front end converter is developed to verify the feasibility of the system design. The prototype efficiency is 1~2% higher than the state of art product and the power density (exclude case) is more than 40W/in^3.

Keywords—Gallium nitride device, active clamp flyback, soft-switching, high frequency, EMI filter

I. INTRODUCTION

A growing demand for the size reduction of the power supplies systems has stimulated substantial development and research effort in high efficiency and high power density power conversion. As the silicon-based semiconductor devices approach their theoretical performance limit, the ability to improve the performance of the next generation of power supplies by optimization of topology, magnetic and power management technique is diminished. The emerging wideband-gap device, such as gallium nitride (GaN) devices will certainly bring out significant incremental efficiency improvement. GaN devices have a much lower gate charge and lower output capacitance, therefore, are capable of operating at a considerably higher switching frequency than the silicon MOSFET while maintaining high efficiency [1-8].

One of the biggest market of power supplies, in both volume and revenue, is the ac-dc adapter/charger for consumer electronics, including laptop, tablet, smart phone, mobile devices, game console, printer, and stereo sound-bars, etc. The market is projected to surpass \$8 billion in 2015 and reach \$9 billion by 2018; with much of this growth being driven by smart phones, tablets and a number of emerging applications [9]. Adapters are cost driven and are customized for each mobile device for each generation. Disposal of these out-of-date adapters becomes a grief environmental concern. Most of the adapters are operating at relative low frequency (<100 kHz) and with an efficiency below 92% and power density below 10W/in^3.

Flyback converters are dominant topology for low power offline application due to its simplicity and low cost. Several literatures [6, 7, 8, 10] have demonstrated flyback converter with GaN devices operating over 1MHz in order to reduce passive components volume. Usually, an RCD clamp circuit is necessary to dissipate the leakage energy and suppress the voltage spike when the switch is off. However, the leakage energy loss is proportional to the switching frequency and it is quite considerable at MHz frequency range. Moreover, the voltage ringing causes high voltage slew rate which has significant impact on EMI noise, especially at 10~30MHz range. The active clamp flyback converter can clamp the voltage without any ringing and recycle the transformer leakage energy [11-13]. Zero-voltage-switching (ZVS) for both main switch and clamping switch can be realized by proper design as well. Different from low frequency operation, leakage inductance energy is not sufficient to realize ZVS for the main switch at MHz. The magnetizing current need to reverse direction to help achieve ZVS. It is worthwhile to point out that the clamping capacitor plays an important role on reducing converter conduction loss and switching loss. The design trade-off is analyzed in section II.

On the other hand, the traditional flyback transformer is in hand-made fashion which is intensive labor involved manufacturing process. The cost is a concern and the parameter variation is another circuit design issue. PCB winding based transformer is only feasible when the switching frequency is over several hundreds of kHz due to less turns number and smaller core size. The leakage inductance and parasitic capacitance of transformer can be well controlled by PCB manufacture. Moreover, shielding can be easily integrated in the PCB winding to reduce the CM noise [14]. The optimization of flyback transformer with integrated shielding is presented in section III.

High frequency operation has significant impact on EMI filter design as well. The corner frequency of CM and DM noise shifts to higher frequency which requires smaller CM choke and DM choke. Moreover, one-stage filter can be used to achieve the required attenuation over whole conducted EMI noise testing frequency range (150kHz ~ 30MHz) due to much higher corner frequency. The analysis of EMI performance in active clamp flyback front end converter and the filter design are discussed in section IV. A 65W (19.5V/3.3A) prototype of active clamp flyback front end converter is developed to verify the feasibility of the system design. The switching frequency is 900 kHz to 1.2 MHz over wide input voltage range. ZVS is achieved for both main switch and clamp switch by controlling the negative magnetizing current. The prototype efficiency is

1~2% higher than the state of art product and the power density (exclude case) is more than 40W/in^3.

II. OPTIMIZATION OF ACTIVE CLAMP CIRCUIT

A. Benefits of flyback converter with active clamp circuit

Fig. 1 shows the circuit configuration of the active clamp flyback converter. The basic operation principle is well-known to power electronics researchers/engineers, which will not be discussed in this paper. However, it is worthwhile to point out that the leakage inductance energy is not sufficient to realize ZVS at high frequency (e.g. MHz). The magnetizing current need to reverse direction to help achieve ZVS for the primary switch. The main benefits of flyback converter with active clamp circuit are summarized as follows.

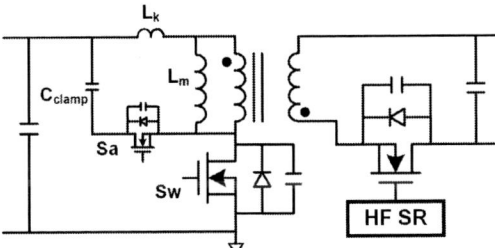

Fig. 1 Active clamp flyback converter with all parasitic

- Drain-source voltage of main switch is clamped when it is turned off. Low voltage rating device could be used as the primary switches.

- Leakage inductance energy is absorbed by clamping capacitor, and part of it is utilized to achieve ZVS for primary switch and part of it is transferred to the output.

- ZVS can be achieved for both main switch and clamp switch by magnetizing current. The dv/dt is significantly reduced and therefore the common-mode noise is also reduced.

- The secondary side current is free of high frequency oscillation and SR driving is much easier [6].

In addition to the common knowledge on the active clamp flyback converter, it is interesting to find out that the active clamp circuit modifies the transformer behavior. Fig. 2 shows the ideal transformer winding current comparison of the traditional flyback and active clamp flyback converter. It is well-known that the traditional flyback transformer is actually an inductor with two windings conducting current in different dime period. There is no flux cancellation and therefore the winding loss is typically large. However, there is flux cancellation effect in active clamp flyback transformer marked as the shaded zone in Fig. 2. When the primary current goes negative, the magnetic field strength is reduced and the current can be more evenly distributed in the winding. The total winding loss reduction is about 20% based on FEA simulation result.

All of these features mentioned above results in higher efficiency and lower noise which makes active clamp flyback topology very suitable for high density adapter application.

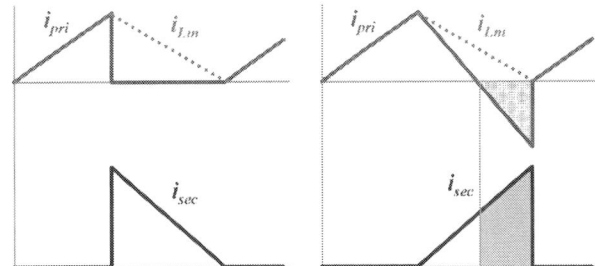

(a) Traditional flyback   (b) Active clamp flyback
Fig. 2 Comparison of ideal transformer winding current

B. Selection of clamping capacitance

The clamping capacitance impacts the resonant frequency as well as the initial condition of the resonant tank. As a result, it determines the current waveforms of the resonant tank and the secondary side.

Fig. 3 shows the simulated current waveforms considering all the parasitics at 90V$_{ac}$ input and full load output condition. With smaller C$_{clamp}$, half of the resonant period formed by L$_k$ and C$_{clamp}$ is close to the main switch S$_w$ off period. i$_{Lk}$ is close to i$_{Lm}$ when clamping switch is turn off, which means the turn off current of clamping switch is small. On the other hand, the current ripple of i$_{Lk}$ is large due to the complete resonant. The secondary side current i$_{SR}$ is the difference between i$_{Lm}$ and i$_{Lk}$, and therefore, i$_{SR}$ also has large ripple. On the contrast, the turn off current of clamping switch increases with larger C$_{clamp}$, while the current ripple of both primary and secondary side reduces significantly. Fig. 4 summarize the impact of C$_{clamp}$ on the related converter power loss. It shows that increasing C$_{clamp}$ in the range of below 100nF significantly reduces the conduction loss and slightly increases the turn off switching loss. Therefore, the total C$_{clamp}$ related loss reduces with larger capacitance. Further increasing C$_{clamp}$ has diminish return in terms of power loss reduction, but the penalty is larger size and relatively slow dynamic performance.

In this design, 100nF is chosen in the converter design from lower loss, smaller size and faster dynamic perspective.

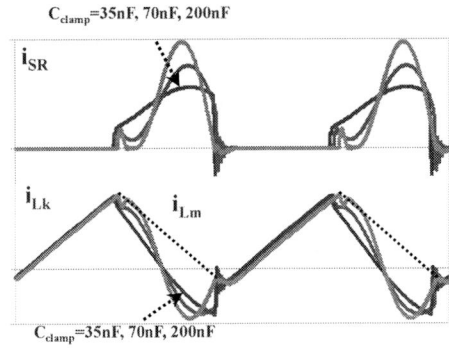

Fig. 3 Impact of clamping capacitor on primary and secondary current waveforms

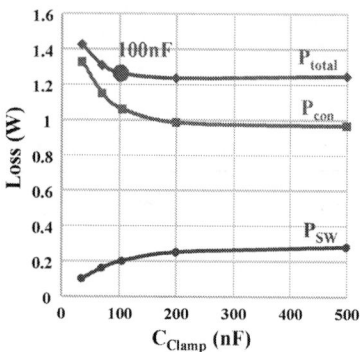

Fig. 4 C_{clamp} impact on converter loss

III. OPTIMIZATION OF ACTIVE CLAMP FLYBACK TRANSFORMER

The traditional low frequency flyback transformer is typically in a hand-made fashion with solid wire. The parasitics are hard to control and the parameters varies from piece by piece. It is also difficult to implement shielding layer in order to reduce common-mode noise. PCB winding based transformer is practical when the switching frequency is close to MHz. It is easier to control the parasitics and also more standardized for automation manufacture. Shielding layer can be easily added in the PCB layer [14]. On the other hand, transformer loss is the major part of flyback converter total loss. The design target is to minimize the loss within 2W, which is 3% of total power of a 65W adapter. Some key design points are discussed in this section. The optimization process shown below is based on $90V_{ac}$ input and full load output condition.

A. Core material

It is critical to select a proper core material for MHz operation. The core loss density is the key criteria. Table I lists the core materials that may be suitable for MHz application. All the core materials are Mn-Zn ferrite. The core loss data are derived from datasheet which are measured under sinusoid excitation without DC bias. However, the voltage waveform across the active clamp flyback transformer is rectangular shape which will impact the core loss data significantly. Fig. 5 shows the measured core loss under 1MHz rectangular voltage excitation with Mu's method [15, 16]. It clearly shows that the core loss data is different from the data listed in Table I. The measured data indicates that ML90S has minimal loss at 1MHz with B_m lower than 100mT.

TABLE I. CORE MATERIAL COMPARISON

	3F45	N49	P61	ML90S
Manufactory	Ferroxcube	TDK	ACME	Hitachi
Initial Permeability	900	1500	900	900
Core Loss (kW/m^3) 1MHz, 50mT	300	400	150	200
Core Loss (kW/m^3) 1MHz, 100mT	1600	2300	3000	4000

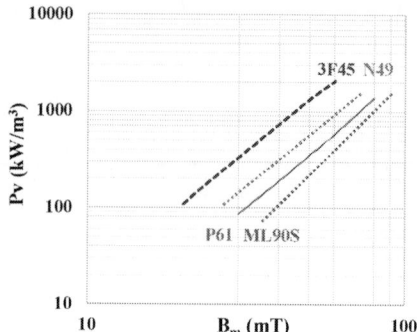

Fig. 5 Measured core loss of 4 materials under 1MHz rectangular voltage excitation.

B. Secondary side turns number: N_S

The voltage second of the flyback transformer significantly reduced when the switching frequency increased to MHz range. The flux density of core is determined by voltage second and cross section area as well as the transformer turns number. Since the output voltage is constant, it is used to calculate the voltage second of the active clamp flyback transformer. Accordingly, the relationship between core loss density and turns number is shown in Fig. 6. The cross section area sweeps from $30mm^2$ to $50mm^2$ which can be considered as a reasonable range from both loss and volume perspective. It clearly shows that the core loss density is too high for $N_S=1$ case even $Ae=50mm^2$. The calculated core loss is over 1W which is unacceptable in 65W adapter application. The core loss density of $N_S=2$ case is in the range of 200~700 kW/m^3 with reasonable Ae. This is more practical and chosen for the converter design.

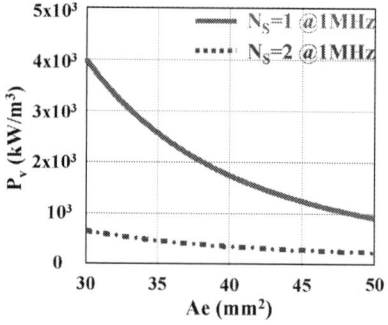

Fig. 6 Core loss density vs. N_S

C. Core cross section area: Ae

In addition to the impact on the core loss density, the cross section area also impacts the core volume and the length of winding. The relationship between transformer loss and cross section area is shown in Fig. 7 which assumes the window width is constant. The winding structure is shown in Fig. 11(a). The turns ratio is preset as 10:2 and the optimization will be shown in the next sub-section. The winding loss is based on Maxwell FEA 2D simulation. The core loss reduces when Ae increases since the impact of ΔB reduction is overwhelming than the increase of core volume. The winding loss increases with larger Ae due to increased winding length. The optimal

Ae to achieve minimal total loss is around 50mm^2. The standard core shape ER23/3/6/13 is used in this converter design.

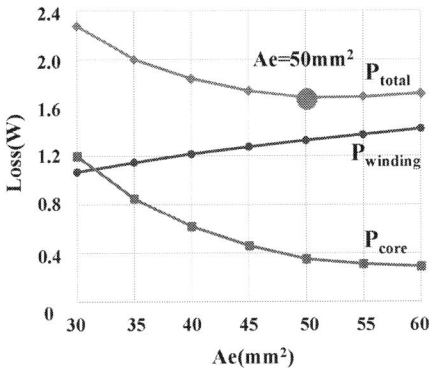

Fig. 7 Impact of Ae on transformer loss

D. Turns ratio

The transformer turns ratio has several impacts on the converter design. The voltage stress of primary and secondary switch is determined by turns ratio. The voltage rating of available high voltage GaN switch that used as the primary switch is 600V, and this sets the maximum turns ratio to be 7 with 15% margin. The secondary GaN switch can select 100~200V eGaN according to the turns ratio. The transformer turns ratio also impacts the root mean square (rms) value of primary and secondary current as well as the duty cycle of primary and secondary switch. The rms value of the primary and secondary side current are plotted in Fig. 8. It shows that the primary current decreases with larger turns ratio since the primary side main switch duty cycle increases. On the contrary, the secondary side current increases with larger turns ratio.

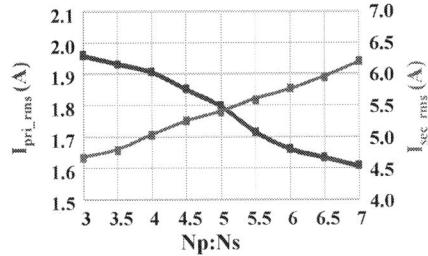

Fig. 8 Impact of turns ratio on converter current

The relationship of converter loss and turns ratio is shown in Fig. 9. The core loss reduces with larger turns ratio since the duty cycle of secondary side reduces which results in a reduction of voltage second applied to the transformer. Basically, the total device conduction loss reduces with larger turns ratio since better device could be used in secondary side. The conduction loss including devices and windings increase with larger turns ratio. Overall, the turns ratio between 4 and 5 can achieve minimal loss, and 5 is chosen for the prototype demonstration.

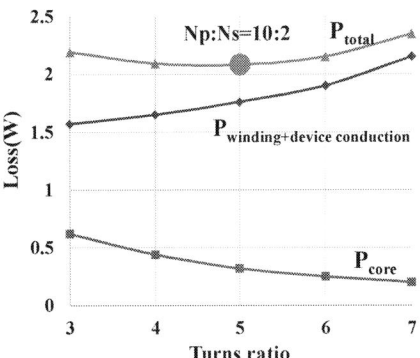

Fig. 9 Impact of turns ratio on converter loss

E. Integrate shielding layer

The complete shielding can be easily implemented in the PCB winding based transformer [14]. Fig. 10 shows the circuit diagram of flyback converter with shielding and Fig. 11 shows the transformer structure with shielding. The shielding is connected to the primary ground. Therefore the CM noise current coming from the primary noise source is circulating within the primary side. Making the shielding layer exactly the same as the secondary side winding generates no voltage difference between secondary side winding and shielding, which means no CM current flowing from either side. Rotating the shielding layer with any angle creates displacement current circulating between secondary side winding and shielding, but it does not create CM current from either side [14].

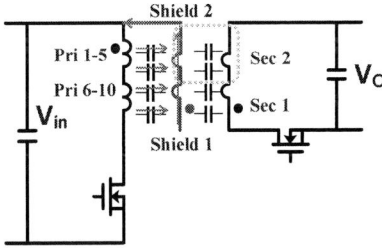

Fig. 10 Flyback transformer with shielding

One of the general concern of adding shielding layer to the flyback transformer is the winding loss. The 3D FEA simulation is carried out to evaluate the impact of the shielding layer on the total transformer winding loss as shown in Fig. 12. It shows that the current distribution in the primary and secondary winding almost remains the same and the current density in shielding layer is also very small. The total winding loss with the shielding layer is 1.35W and it is only 0.05W larger than the case without shielding layer, which is negligible loss increase.

To summarize the flyback transformer design, the ML90S from Hitachi is chosen for the core material; ER23/3.6/13 core shape is selected with optimized cross section area; the primary to secondary winding turns ratio is 10:2; the shielding layer is integrated in the PCB winding based transformer with negligible increase of total winding loss.

978-1-4673-9551-9/16 $31.00 © 2016 IEEE

Pri 1-5
Shielding 1
Sec 1
Sec 2
Shielding 2
Pri 6-10

(a) Transformer winding cross section view

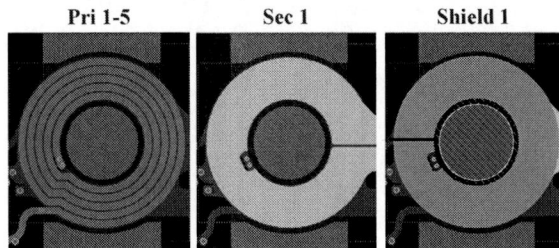

Pri 1-5   Sec 1   Shield 1

(b) detailed PCB layout view

Fig. 11 Flyback transformer winding structure with shielding

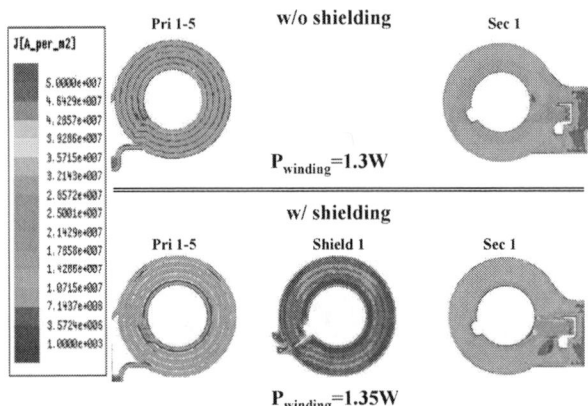

Fig. 12 3D FEA simulation of flyback transformer with and without shielding layer

IV. EMI ANALYSIS AND FILTER DESIGN OF ACTIVE CLAMP FLYBACK CONVERTER

A. Analysis of shielding impact on CM/DM noise

The electromagnetic interference (EMI) noise, especially the common-mode (CM) noise is more severe due to high dv/dt induced by GaN devices. Fig. 13 shows the CM noise propagation path in the active clamp Flyback converter. The primary switches induce high dv/dt which dominates the overall CM noise magnitude, and it is represented as a noise source V_S. Compared to the primary to the power earth parasitic capacitance C_{PE}, the interwinding capacitance of the flyback transformer C_{PS} and the secondary ground to power earth parasitic capacitance C_{SE} are much larger and it is the major coupling path of the CM noise.

With the complete shielding layer integrated in the PCB winding based transformer, the CM noise path through C_{PS} is cut off. As a result, the total CM noise current reduces signifi-

cantly due to much reduced capacitance. Fig. 14 shows the CM noise comparison between the converter prototypes with and without shielding. The switching frequency is near 1MHz. It clearly shows that CM noise has 27dB reduction with shielding. More importantly, the high frequency noise, in the range of 10~30MHz, is also significantly reduced. It is usually difficult to be attenuated by filters since the characteristic of CM choke turns to be poor due to core material permeability drop and parasitic impact at that frequency range.

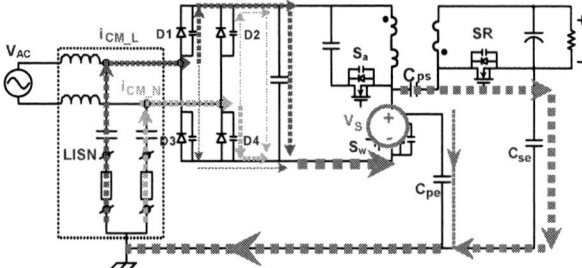

Fig. 13 CM noise path of active clamp flyback converter

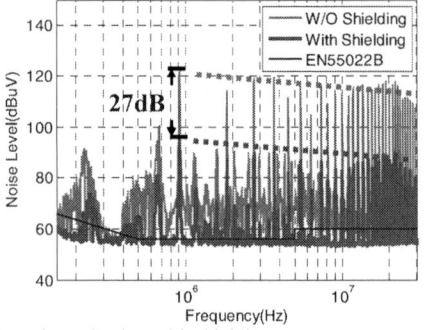

Fig. 14 CM noise reduction with shielding

It is interested to notice that the shielding also helps to reduce the differential mode (DM) noise, as shown in Fig. 15. The cause of this phenomenon is the CM/DM noise transformation when the L and N line impedance is unbalanced. The reduction of total CM noise current results in a reduction of DM noise which is transformed from CM noise. The detailed mechanism of this issue is presented in [17].

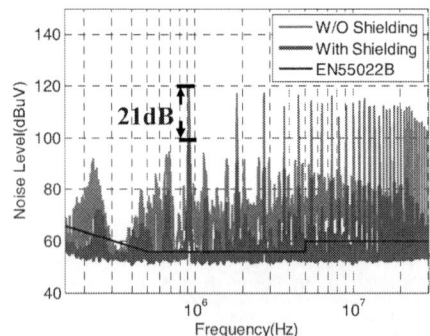

Fig. 15 DM noise reduction with shielding due to CM/DM transformation

It is essential to understand the impact of shielding on the total EMI noise in order to design compact EMI filter with required attenuation.

B. High frequency EMI filter design

The CM and DM filter corner frequency can be calculated by the desired attenuation at switching frequency according to the conducted EMI standard EN55022B. By increasing the switching frequency to MHz range, the corner frequency of CM and DM increases to around 100kHz which is much higher than current industry practice. The simple one-stage filter can be used to achieve the required attenuation. The parameters of the EMI filter is summarized in Table II.

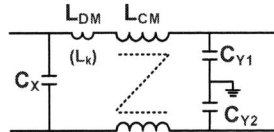

Fig. 16 One stage EMI filter

TABLE II. PARAMETERS OF THE EMI FILTER

Parameters	L_{CM}	L_{DM}	$C_{Y1}=C_{Y2}$	C_X
Value	1.4mH	0.03mH	1nF	130nF

The conventional core materials for CM/DM chokes have high permeability at few hundreds of kHz and drops sharply at frequency above 500kHz. They are good enough for the converter operateing below few hundreds of kHz. However, these kinds of materials are not suitable for MHz application. The desired features for high frequency chokes includes relative high initial permeability and relative stable permeability over the frequency range from 1MHz to 5MHz. The ferrite materials 3D3 from Ferroxcube is chosen for the prototype design.

V. EXPERIMENTAL RESULTS AND DISCUSSIONS

In order to verify the feasibility of MHz converter design, a 65W (19.5V/3.3A) prototype is developed as shown in Fig. 17. The authors intend to implement MCU based control since there are no commercial available controllers that are targeted at MHz switching frequency. The close loop control is still under development and the results shown below are all based on open loop test which presets PWM frequency and duty cycle at certain input and output condition.

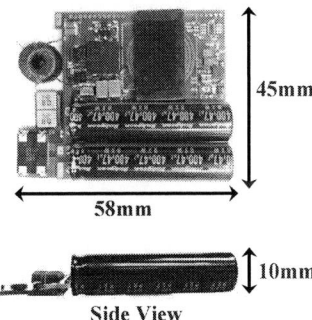

Fig. 17 1MHz 65W active clamp flyback converter prototype

GaN devices are used for both primary and secondary switches to reduce conduction and switching losses. The 600V GaN HEMT from Transphorm is used as the primary main switch and active clamp switch. The 150V eGaN from EPC is used as the synchronous rectifier. The PCB board is 6-layer which includes two layers shielding as shown in Fig. 11. The power density excludes the case is over 40W/in^3, which is at least two times higher than the state-of-the-art products.

Fig. 18 shows the key experimental waveforms at low line and high line input condition. There is no voltage spike across the main switch which indicates that the leakage energy is well absorbed by clamping circuit. The primary switches can achieve ZVS over wide input range. It eliminates the turn on switching loss and avoid high dv/dt at the turn on instant.

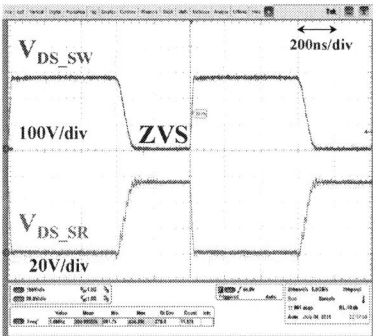

(a) Key waveforms at 110V$_{AC}$

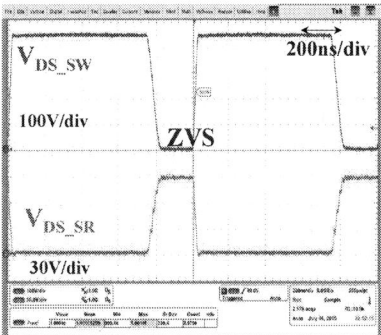

(b) Key waveforms at 230V$_{AC}$

Fig. 18 Prototype experiment waveforms

Fig. 19 shows the CM/DM noise spectrum with peak mode measurement under 110V$_{AC}$ input full load output condition which is the worst case for the prototype design. The red curves are the results with shielding but without filters. The blue curves are the final results with EMI filter. It clearly shows that the blue curves are already lower than the quasi-peak standard.

The measured full load efficiency over wide input range is shown in Fig. 20. The efficiency of the prototype is 1~2% higher than the state-of-the-art product. It is worthwhile to point out that the power density improvement is accompanied with efficiency improvement due to thermal restriction. Based on the thermal simulation carried on a converter which is enclosed by case, the converter efficiency should be above 92%

at worst case to achieve 25W/in³ power density without violating the thermal standard IEC60950. The prototype full load efficiency is above 92.8% over wide input range which makes it possible to achieve high density goal. More testing will be done to verify the prototype design can meet the general specifications of adapters.

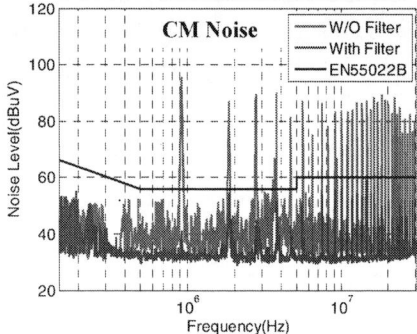

(a) CM noise measurement

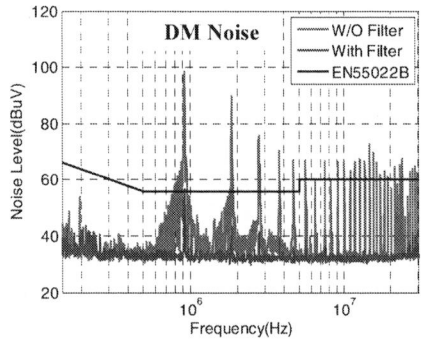

(b) DM noise measurement

Fig. 19 Prototype EMI noise measurement with and without filter

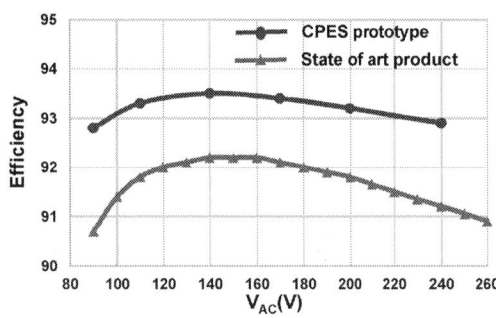

Fig. 20 Efficiency measurement of the prototype

VI. CONCLUSION

This paper presents the design consideration of MHz active clamp flyback converter for adapter application. Each design procedure targets at minimal power loss as well as smaller size. The PCB winding based transformer has the advantage of high density, controllable parasitics, and easy integration of shielding. The prototype verifies the feasibility of the system design. With high frequency operation and dedicated design,

much higher density and higher efficiency are achieved compared with state of art product. Future work will focus on light load efficiency and thermal condition evaluation.

ACKNOWLEDGMENT

The information, data, or work presented herein was funded in part by the Office of Energy Efficiency and Renewable Energy (EERE), U.S. Department of Energy, under Award Number DE-EE0006521 with North Carolina State University, PowerAmerica Institute.

This work was supported in part by the Power Management Consor-tium in CPES, Virginia Tech.

This work was conducted with the use of GaN device samples by Transphorm and EPC of the CPES Industry Consortium Program.

DISCLAIMER

The information, data, or work presented herein was funded in part by an agency of the United States Government. Neither the United States Government nor any agency thereof, nor any of their employees, makes any warranty, express or implied, or assumes any legal liability or responsibility for the accuracy, completeness, or usefulness of any information, apparatus, product, or process disclosed, or represents that its use would not infringe privately owned rights. Reference herein to any specific commercial product, process, or service by trade name, trademark, manufacturer, or otherwise does not necessarily constitute or imply its endorsement, recommendation, or favoring by the United States Government or any agency thereof. The views and opinions of authors expressed herein do not necessarily state or reflect those of the United States Government or any agency thereof.

REFERENCES

[1] B. Hughes, J. Lazar, S. Hulsey, D. Zehnder, D. Matic, and K. Boutros, "GaN HFET switching characteristics at 350V-20A and synchronous boost converter performance at 1MHz," in *proc. IEEE APEC*, 2012, pp 2506-2508.

[2] J. Delaine, P. Olivier, D. Frey, and K. Guepratte, "High frequency DC-DC converter using GaN device," in *proc. IEEE APEC*, 2012, pp 1754-1761.

[3] X. Huang, Z. Liu, Q. Li, and F C. Lee, "Evaluation and application of 600 V GaN HEMT in cascode structure," *IEEE Trans. on Power Electron.*, vol. 29, no. 5, pp.2453-2461, May. 2014.

[4] X. Huang, Z, Liu, F. C. Lee, and Q. Li, "Characterization and enhancement of high-votlage cascode GaN devices," *IEEE Trans. on Electron Devices*, vol. 62, no. 2, pp.270-277, Feb. 2015.

[5] X. Zhang, C.Yao, X. Lu, E. Davidson, M. Sievers, M. J. Scott, P. Xu, and J. Wang, "A GaN transistor based 90W AC/DC adapter with a buck-PFC stage and an isolated Quasi-switched-capacitor DC/DC stage," in *proc. IEEE APEC*, 2014, pp 109-116.

[6] X. Huang, W. Du, F. C. Lee, and Q. Li, "A novel driving scheme for synchronous rectifier in MHz CRM flyback converter with GaN devices, " in *proc. IEEE ECCE*, 2015, pp 5089-5095.

[7] Z. Zhang, K. D. T. Ngo, and J. L. Nilles, "A 30-W flyback converter operating at 5 MHz," in *proc.IEEE APEC*, 2014, pp 1415-1421.

[8] T. Labella, B. York, C. Hutchens, and J. S. Lai, "Dead time optimization through loss analysis of an active-clamp flyback converter utilizing GaN devices," in *proc.IEEE ECCE*, 2012, pp 3882-3889.

[9] External power adapters and chargers report-2015, IHS Technology.

[10] H. Jia, O. A. Rahman, K. Padmanabhan, P. Shea, I. Batarseh, and Z. J. Shen, "MHz-frequency operation of flyback converter with monolithic

self-synchronized rectifier (SSR)," in *proc.IEEE INTELEC*, 2010, pp 1-6.

[11] C. T. Choi, C. K. Li, and S. K. Kok, "Control of an active clamp discontinuous conduction mode flyback converter," in *Proc. IEEE Power Electronics and Drive Systems Conf.*, 1999, vol. 2, pp. 1120-1123.

[12] R. Watson, F. C. Lee, and G. Hua, "Utilization of an Active-Clamp Circuit to achieve Soft Switching in Flyback Converters," *IEEE Trans.on Power Electron.*, vol.11, pp162-169, January 1996.

[13] J. Zhang, X. Huang, X. Wu, and Z. Qian, "A high efficiency flyback covnerter with new active clamp technique," *IEEE Trans.on Power Electron.*, vol.25, No. 7, pp1775-1785, Jul. 2010.

[14] Y. Yang, D. Huang, F. C. Lee, and Q. Li, "Analysis and reduction of common mode EMI noise for resonant converters," in *proc. IEEE APEC*, 2014, pp 566-571.

[15] M. Mu, F. C. Lee, "A new core loss model for rectangular AC voltages," *in proc. IEEE ECCE*, 2014, pp 5214-5220.

[16] M. Mu, F. C. Lee, Q. Li, D. Gillham, and K. Ngo, "A high frequency core loss measurement method for arbitrary excitations," in *proc. IEEE APEC*, 2011, pp 157-162.

[17] X. Huang, J. Feng, F. C. Lee, and Q. Li, "Conducted EMI analysis and filter design for MHz active clamp flyback front-end converter," in *proc. IEEE APEC*, 2016, to be published.

A New Capacitor Voltage Balancing Control for Hybrid Modular Multilevel Converter with Cascaded Full Bridge

Mahendra B. Ghat
Electrical Engineering Department,
IIT Bombay,
Mumbai, India
mahendraghat@ee.iitb.ac.in

Anshuman Shukla
Electrical Engineering Department,
IIT Bombay,
Mumbai, India
ashukla@ee.iitb.ac.in

Richa Mishra
Electrical Engineering Department,
IIT Bombay,
Mumbai, India.
richam@ee.iitb.ac.in

Abstract — Recently, a hybrid modular multilevel converter with cascaded full bridges (HMMC-CFB) is proposed. This converter has mainly two parts: a director part (DP) and a wave shaping part (WSP). The DP consists of a conventional modular multilevel converter (MMC) with half bridge submodules (HBS). The WSP is a series connection of full bridge submodules (FBS) and attenuates the voltage harmonics produced by DP. The energy exchanged by WSP changes with modulation index and causes improper capacitor voltage regulation. For satisfactory operation of HMMC-CFB, the total capacitor voltage of WSP should remain constant. This paper presents an analysis of energy exchanged by WSP and proposes a new control technique to regulate the total dc voltage of WSP. In the proposed control, the energy exchanged by WSP is controlled by changing the slope of the voltage reference signal of DP. In addition, analysis of circulating current of HMMC-CFB is presented. A circulating current suppressing controller (CCSC) is used to control the second harmonic component. The performance of the proposed control technique for the standalone model of HMMC-CFB is evaluated by using time domain simulation studies in the PSCAD software environment. To test fault blocking capability of converter, a grid connected model is built in PSCAD and analyzed for dc fault tolerant capability.

Keywords—HVDC system, hybrid modular converter, modular multilevel converter, dc fault tolerant

I. INTRODUCTION

The modular multilevel converter (MMC) is one of the preferred choices for VSC-HVDC systems because of its modularity and scalability [1]-[4]. However, the inability to block or limit the dc fault current is one of its main limitations, especially for the applications like the dc-grid and the HVDC system with overhead lines. In case of dc side fault, the MMC acts as an uncontrolled diode rectifier which allows the ac system to feed the load with very high fault current through the freewheeling diodes connected across the IGBTs [11]-[12], [20]. To tackle this problem different techniques have been proposed in literatures [5]-[10]. One way is to use the conventional ac and dc circuit breaker. This process is slower which forces the converter ratings to be overrated so that it can handle high fault current until the time when the breaker operates. Recently, a hybrid dc circuit breaker has been proposed which can operate faster [5]-[7]. Another approach as presented in [8], uses the thyristors to bypass the cells in case of dc fault. This method requires the thyristors to be rated to carry the fault current. Another approach is use full bridge submodules (FBS) in MMC instead of the half bridge submodules (HBS) [9]. The FBSs can be controlled to inject the capacitor voltages of opposite polarity which can be used to limit or block the fault current. However, this method requires doubling of the number of semiconductor switches in the conduction path. Recently, a family of hybrid multilevel converter is proposed with dc fault blocking capability [10]. These hybrid converter topologies are the combination of a two level VSC and the wave-shaping part consisting of a series connection of FBSMs. These converters have the inherent fault tolerant capability and less device count, as compared to than in FBS based MMC. Some of these hybrid topologies proposed are alternate arm modular converter [11], hybrid cascaded multilevel converter [12], and hybrid multilevel converter with cascaded H-bridges on the dc side [20]. The hybrid cascaded multilevel converter presented in [12], uses lowest number of devices and capacitors. However, the major drawbacks of this converter are higher losses in director part (DP) because of hard switching, requirement of nearly identical switching characteristics of individual semiconductors with dynamic voltage sharing because of series connection of IGBTs in the DP, and requirement of low order harmonic filter to mitigate low energy spikes due to mis-synchronization between DP and wave shaping part (WSP) [13]-[14].

A new topology, hybrid MMC with cascaded full bridges (HMMC-CFB) is presented in [15]. The single phase structure of the HMMC-CFB is shown in Fig. 1(a). This converter can block or control the dc fault current and slow down the change rate of output voltage for DP. This improves the synchronization between DP and WSP [12], [15]-[16]. However, in HMMC-CFB, a change in modulation index correspondingly results in the change in energy exchanged by the WSP. This may deviate the overall capacitor voltage of WSP from its reference value [17]. The existing methods proposed to resolve this issue are asymmetrical square wave modulation (ASWM) for DP [17], [21] and modulating the converter using third harmonic subtraction method [17], [22]. In this paper, the analysis for energy exchanged by WSP of HMMC-CFB is presented and a new technique is proposed to control the energy exchanged by WSP. The energy exchanged by WSP is controlled by changing the slope of the trapezoidal voltage reference signal of DP. In addition, for HMMC-CFB,

978-1-4673-9551-9/16 $31.00 © 2016 IEEE

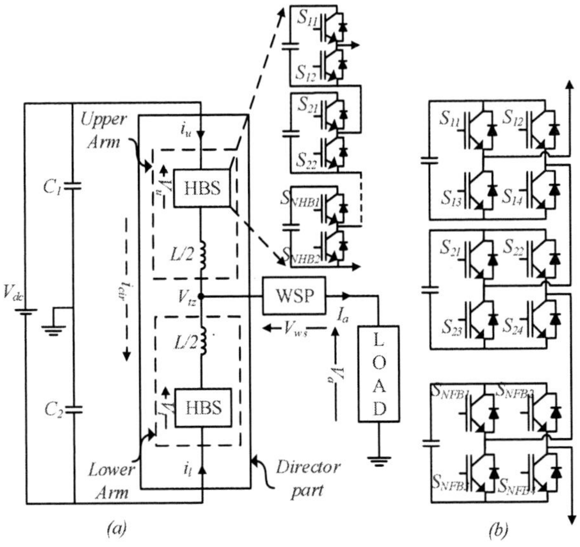

Fig. 1. (a) Single phase hybrid modular multilevel converter with cascaded FBS, (b) wave shaping part (WSP) of the converter.

the circulating current analysis and circulating current control by second order harmonic injection [18] is presented. The studies are carried out with proposed control based on simulations in PSCAD.

II. Hybrid MMC with Cascaded Full Bridge

A. Converter Topology and operating principles

In [15], the authors have explained the detailed converter topology, the average model of converter, operation principle, and dc side fault tolerant capability of converter. As shown in Fig. 1(a), the converter is mainly divided into two parts, one is the DP, which consist of two arms and two inductors per phase. In each arm N_{HB} HBS are connected in series. The DP controls the magnitude and phase of the converter output voltage to effectively control the active and reactive power exchanged by the converter. The other part is the WSP, which is a series connection of N_{FB} FBS shown in Fig. 1(b). The WSP acts as a series active filter, to attenuate harmonics produced by the DP. The HBS of each arm are configured to block the total dc link voltage, whereas the FBS of each phase are configured to block only half of the dc link voltage. Consequently the total number of FBS are the quarter number as that of HBS per phase, provided the capacitor and switching device voltage ratings are same.

The voltage reference signals used for converter are shown in Fig. 2. The trapezoidal reference signal is used for the DP. The reference for WSP is obtained by subtracting the output voltage reference from the trapezoidal voltage reference signal of DP. With the use of MMC instead of the two-level converter, the rate of change of the output voltage of DP slows down. This overcomes the problem of fast tracking of the sharp edges, as observed in [16]. In this way, it is possible to get improved synchronization between high and low power stages. Moreover, the converter output voltage waveform is free from spikes avoiding low order harmonic filter requirement.

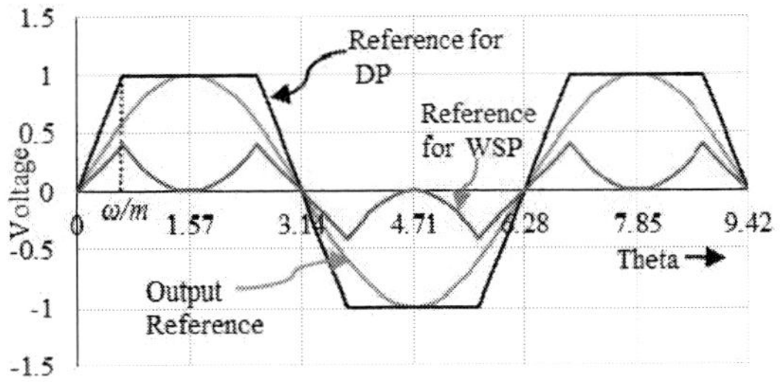

Fig. 2. Reference waveforms of output voltage, DP voltage, and WSP

For proper operation of the converter, the total and individual capacitor voltages of FBS in the WSP should be monitored and maintained around their nominal values. To control these voltages a new control technique is proposed with the block diagram shown in Fig. 3. The overall proposed control of converter consists of controlling the slope of trapezoidal reference signal used for DP and reducing the ac component in circulating current by circulating current suppression control (CCSC) [18], as shown in Fig. 3. The detailed control technique is described in following sub-sections.

B. Capacitor Voltage Balancing of WSP

For satisfactory operation of HMMC-CFB, the capacitor voltage of the WSP should remain constant. However, since the load current is flowing through the FBS, depending upon current direction and switching states of FBS, the charge of capacitor will fluctuate over time. The overall capacitor voltage depends on the energy exchanged by WSP. To analyze energy exchanged by WSP, consider the output voltage (V_a) and output current (I_a) for phase a as,

$$\begin{aligned} V_a(t) &= V_m \sin(\omega t) \\ I_a(t) &= I_m \sin(\omega t - \phi) \end{aligned} \tag{1}$$

where V_m is the peak value of output voltage, I_m is the peak value of output current, ϕ is the phase difference between voltage and current, ω is the fundamental frequency in radian and t is the time in second. From Fig. 1, the voltage across WSP (V_{ws}) can be described as,

$$V_{ws}(\mathrm{t}) = V_{tz}(\mathrm{t}) - V_a(\mathrm{t})$$

$$V_{ws}(t) = V_{tz}(t) - V_m \sin(\omega t) \tag{2}$$

In (2), V_{tz} is the output voltage of DP. The voltage across the WSP (V_{ws}) is the summation of switched capacitor voltages of FBS as shown in Fig. 1 and given as

$$V_{ws}(t) = \sum_{i=1}^{N} S_i * V_{c(t)} \tag{3}$$

where S_i is the switching state of FBS (S_i = -1, 0, or 1), V_c is the individual capacitor voltage of FBS. From Fig. 1 and reference signal of DP shown in Fig. 2, the output of DP (V_{tz}) can be described as,

$$
V_{tz}(t) = \begin{cases}
\dfrac{V_{dc}}{2} mt & o \leq t \leq \dfrac{1}{m} \\[2ex]
\dfrac{V_{dc}}{2} & \dfrac{1}{m} \leq t \leq \dfrac{\pi}{\omega} - \dfrac{1}{m} \\[2ex]
\dfrac{V_{dc}}{2}\left(-mt + \dfrac{m\pi}{\omega}\right) & \dfrac{\pi}{\omega} - \dfrac{1}{m} \leq t \leq \dfrac{\pi}{\omega} + \dfrac{1}{m} \\[2ex]
-\dfrac{V_{dc}}{2} & \dfrac{\pi}{\omega} + \dfrac{1}{m} \leq t \leq \dfrac{2\pi}{\omega} - \dfrac{1}{m} \\[2ex]
\dfrac{V_{dc}}{2}\left(mt - \dfrac{2m\pi}{\omega}\right) & \dfrac{2\pi}{\omega} - \dfrac{1}{m} \leq t \leq \dfrac{2\pi}{\omega}
\end{cases} \quad (4)
$$

where m is the slope of trapezoidal reference signal. The average instantaneous power P_{ws} flowing into the WSP can be expressed as,

$$P_{ws} = V_{ws}(t) * I_a(t)$$

$$= \frac{V_m I_m}{2}\left[\cos(2\omega t - \phi) - \cos(\phi)\right] + V_{tz}(t) I_m \sin(\omega t - \phi) \quad (5)$$

Thus, the energy exchanged of WSP (W_{ws}) with the load over one cycle is given by,

$$W_{ws} = \int_0^{2\pi/w} P_{ws}(t)\, dt \quad (6)$$

By using (4) and (5), (6) can be rewritten as,

$$
W_{ws} = \begin{vmatrix}
\dfrac{V_m I_m}{2} \displaystyle\int_0^{2\pi/w} \left[\cos(2\omega t - \phi) - \cos(\phi)\right] dt + \\[3ex]
\begin{pmatrix}
\displaystyle\int_0^{1/m} mt * \sin(\omega t - \phi)\, dt + \int_{1/m}^{\pi/\omega - 1/m} \sin(\omega t - \phi)\, dt \\[3ex]
+ \displaystyle\int_{\pi/\omega - 1/m}^{\pi/\omega + 1/m} \left(-mt + \dfrac{m\pi}{\omega}\right) * \sin(\omega t - \phi)\, dt \\[3ex]
+ \displaystyle\int_{\pi/\omega + 1/m}^{2\pi/\omega - 1/m} \sin(\omega t - \phi)\, dt \\[3ex]
+ \displaystyle\int_{2\pi/\omega - 1/m}^{2\pi/\omega} \left(mt - \dfrac{2m\pi}{\omega}\right) * \sin(\omega t - \phi)\, dt
\end{pmatrix} \dfrac{V_{dc} I_m}{2}
\end{vmatrix}
$$

$$W_{ws} = \frac{V_m I_m}{\omega}\cos(\phi)\left[\frac{4m}{\omega * mi}\sin\left(\frac{\omega}{m}\right) - \pi\right] \quad (7)$$

Eqn. (7) shows that energy exchanged by WSP depends upon the modulation index (mi) which is given by $2V_m/V_{dc}$ and slope of trapezoidal output (m) of DP. For positive value of W_{ws}, WSP voltage increases and vice versa. To keep the capacitors voltage constant, the net energy exchanged by WSP should be zero. Hence, from (7),

$$\left[\frac{4m}{\omega * mi}\sin\left(\frac{\omega}{m}\right) - \pi\right] = 0 \Rightarrow mi = \frac{4m}{\omega * \pi}\sin\left(\frac{\omega}{m}\right) \quad (8)$$

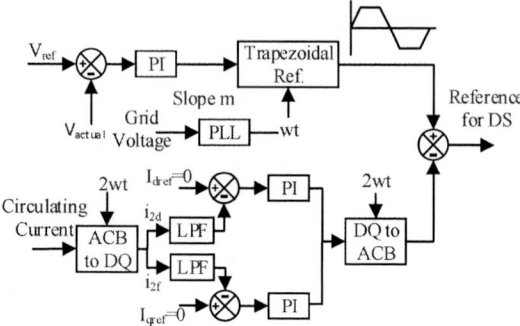

Fig. 3. Proposed control technique for voltage control of WSP and circulating current control.

Thus from (8), by controlling the slope (m) for a specific modulation index (mi), it is possible to control the energy exchanged by WSP and hence net capacitor voltage of WSP, shown in Fig. 3. In Fig. 3, the average capacitor voltage of WSP (V_{actual}) which is proportional to energy stored, is compared with reference voltage (V_{ref}). The error goes through a PI controller. The output of PI controller gives the value of slope m for trapezoidal reference signal. From (8), for $0 \leq m < \infty$, the limits of mi over which the control is possible, is expressed as,

$$0 \leq mi \leq \frac{4}{\pi} \quad (9)$$

In the range, as expressed by (9), by controlling the slope of DP reference signal (m), it is possible to control the capacitor voltage of WSP.

C. Circulating current control of DP:

The DP of HMMC-FBS consists of N_{HB} numbers of HBS in each arm. Source of these HBS are capacitors. The instantaneous values of these capacitors are not same, which leads to circulating current in the arms. If this circulating current is not properly controlled, then it has adverse effect on the rating of the devices and losses of converter. Moreover this circulating current increases ripple in the capacitor voltage. To reduce ripple, either capacitor size is increased which makes system bulky and costly, or reduce the circulating current. The second option is preferred and many circulating current control techniques for MMC are presented in literature [18], [24]-[28].

If i_u is the upper arm current and i_l is the lower arm current of the converter, then the circulating arm current is expressed as [28],

$$i_{cir} = \frac{i_u - i_l}{2} \quad (10)$$

These arm currents in terms of line current and output voltage of DP of converter are expressed as [28],

$$i_u = I_a \frac{1 + V_{tz}}{2} \quad and \quad i_l = I_a \frac{1 - V_{tz}}{2} \quad (11)$$

From (10) and (11), the circulating current can be rewritten as,

$$i_{cir} = \frac{I_a V_{tz}}{2} \quad (12)$$

978-1-4673-9551-9/16 $31.00 © 2016 IEEE

The Fourier series expansion of V_{tz} expressed in (4) is given as,

$$V_{tz} = \frac{2V_{dc}m}{\pi w}\left[\begin{array}{l}\sin\left(\dfrac{w}{m}\right)\sin\left(wt\right)+\dfrac{1}{9}\sin\left(\dfrac{3w}{m}\right)\sin\left(3wt\right)+\\[2mm]\dfrac{1}{25}\sin\left(\dfrac{5w}{m}\right)\sin\left(5wt\right)+...\end{array}\right] \quad (13)$$

Using (1) and (13), Eqn. (12) can be rewritten as,

$$i_{cir} = -\frac{I_m V_{dc} m}{2\pi\omega}\left[\begin{array}{l}\sin\left(\dfrac{\omega}{m}\right)\cos\left(\varphi\right)-\sin\left(\dfrac{\omega}{m}\right)\cos\left(2\omega t-\varphi\right)+\\[2mm]\dfrac{1}{9}\sin\left(\dfrac{3\omega}{m}\right)\cos\left(2\omega t+\varphi\right)-\\[2mm]\dfrac{1}{9}\sin\left(\dfrac{3\omega}{m}\right)\cos\left(4\omega t-\varphi\right)+..\end{array}\right] \quad (14)$$

Eqn. (14) shows that the circulating current has a predominant dc component and a second order ac component. The values of other harmonic components (fourth, sixth, eight,...) are small and can be neglected. The dc component is essential for the active power exchange. Thus by suppressing the predominant second order component, it is possible to reduce the circulating current [18]. The circulating currents in phases-b and c are respectively 120° and 240° phase shifted with respect to phase-a circulating current. *The dynamics of circulating current at DP of HMMC-FCM is given by* [15],[18]

$$L\frac{di_{cir}}{dt}+R\frac{di_{cir}}{dt}=\frac{V_{dc}}{2}-\frac{V_u+V_l}{2}=V_{diff} \quad (15)$$

where V_u and V_l are the upper and lower arm capacitor voltages of HBS, R is the equivalent arm resistance, and V_{diff} is the inner unbalance voltage of converter. Rewriting (15) in *a-c-b* sequence and after transformation of three phase circulating currents to negative sequence and twice the frequency rotating reference frame we have [18],

$$\begin{bmatrix}v_{diffd}\\v_{diffq}\end{bmatrix}=L\frac{d}{dt}\begin{bmatrix}i_{2d}\\i_{2q}\end{bmatrix}+\begin{bmatrix}0 & -2wL\\2wL & 0\end{bmatrix}\begin{bmatrix}i_{2d}\\i_{2q}\end{bmatrix}+R\begin{bmatrix}i_{2d}\\i_{2q}\end{bmatrix} \quad (16)$$

where i_{2d} and i_{2q} are the dq components of circulating current, and V_{diffd} and V_{diffq} are the dq components of the inner unbalance voltage V_{diff} in the double frequency, negative sequence rotating reference frame. These dq components of circulating currents are compared with reference values which are set to zero to minimize second order component (Fig. 3). The voltage reference obtained from CCSC are subtracted from the trapezoidal reference. Then the output of this controller is used as a reference to obtain switching signals for DP.

D. Individual Capacitor Voltage Balancing:

As described above in Sub-sec. II B, the net capacitors voltage of FBS in WSP can be held constant by controlling *m*. However, this may not guarantee the equal distribution of voltage across all the FBS, which is a necessary condition for proper and satisfactory operation of the converter. Similarly the capacitor voltages of HBS should also be equal and constant. To keep the individual capacitor voltages of HBS and FBS equal to $V_{dc}/2N_{HB}$ and $V_{dc}/2N_{FB}$, respectively, the capacitor voltages are sorted and depending upon the current direction, for charging mode the submodules with lower voltages are inserted and for discharging mode the submodules with higher voltages are inserted [14, 19]. The sorting algorithms for each of the HBS and FBS are implemented in manner described in the following sub-sections.

1) Capacitor volage control of HBS:

The director part of the converter uses HBS. They are either switched on (capacitor comes in current path) or bypassed (capacitor is bypassed). To control the capacitors voltage of HBS, for the positive direction of current *(*direction of i_u indicated in the Fig. 1(a)), the HBS with lower values of capacitor voltages are inserted first, and for negative direction of current the HBS with higher capacitor voltages are inserted first. The sorting algorithm used for DP is summarized in the flow chart shown in Fig. 4. Using this method the continuous capacitor voltage balancing of HBS is achieved.

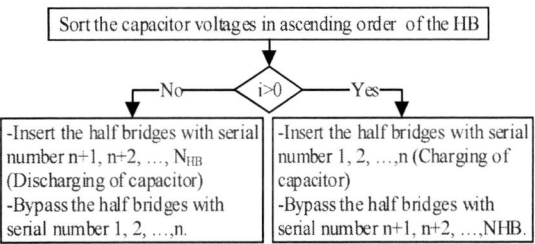

Fig. 4. Sorting technique for DP

2) Capacitor volage control of FBS:

Unlike the HBS, the FBS can be either positively inserted, negatively inserted, or bypassed. To generate the gate signals for SMs, first all the capacitor voltages of FBS are measured and sorted in ascending or descending order. Depending on the current direction, either the capacitors with maximum voltages or with minimum voltages are inserted. If one of the two current directions results in charging of the capacitors then the SMs with lowest capacitor voltages are inserted, and the SMs with highest capacitor voltages are inserted for the other current direction. A block diagram illustrating sorting

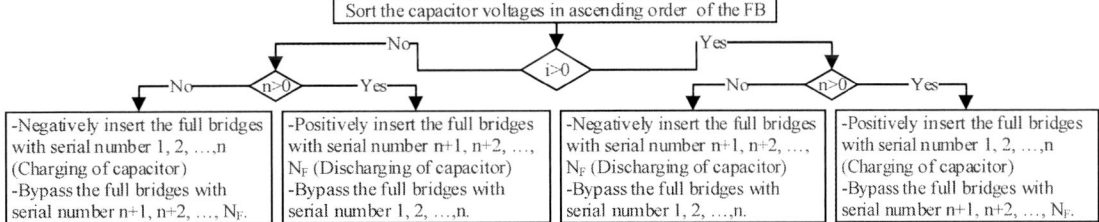

Fig. 5. Sorting technique for WSP

algorithm technique for WSP is shown in Fig. 5. This technique insures the equal charge distribution over all the SM capacitors

III. SIMULATION RESULTS

A. Standalone operation:

To validate the proposed control technique, a standalone test circuit of a three phase HMMC-CFB connected to RL load is simulated in PSCAD. The simulation parameters are V_{dc}= 40 kV, N_{FB}= 4, FBS capacitance = 5 mF, N_{HB}= 8, HBS capacitance= 5 mF, and the converter is controlled with SPWM with switching frequency of 2 kHz. Fig. 6 shows the results of standalone operation of HMMC-CFB with a modulation index mi=1.1, and supplies a passive load of 50 Ω resistance and 100 mH inductance, which is equivalent to 0.85 lagging power factor at 50 Hz. Fig. 6(a) shows the average capacitor voltage of WSP with and without proposed voltage control. The system starts without voltage control and at the instant t = 0.3 s, the proposed voltage control is activated. The voltage increases while control is deactivated and settles to reference value after activation of control. The individual capacitor voltages of WSP are shown in Fig. 6(b), which are equal and validate the sorting technique. Fig. 6(c) and 6(d) shows the output voltage and current waveform for the RL load. Fig. 6(e) and 6(f) are the frequency spectrum of output voltage without and with proposed voltage control. From the figures, it is clear that the THD of output voltage reduces with proposed controller. Fig. 7 shows the results with and without circulating current control. Circulating current control is started at t = 0.5 s. Fig. 7(a) shows the results of circulating

current for phase a. Fig. 7(b) shows the average upper arm capacitor voltage. The capacitor voltage fluctuations are reduced after the activation of CCSC. The upper and lower arm currents of phase a are shown in Fig. 7(c). Fig. 7(d) shows the second order dq component of circulating currents, which becomes nearly zero after activation of controller. Fig. 7(e) and 7(f) show the frequency spectrum for arm current without and with circulating current control. The second order harmonics and hence THD has reduced with CCSC.

B. Grid connected system:

To demonstrate usefulness of converter for HVDC system and test the dc fault blocking capability, a grid connected system is built in PSCAD as shown in Fig. 8. The parameters used for this system are listed in Table I.

To control active power or dc voltage and reactive power or ac voltage, decoupled current control scheme is used [12], [16].

Table I: Simulation parameters of HMMC-CFB

Parameters	Values
Transformer(Star/Delta)	400/100 kV
DC link Voltage	200 kV
No. of HBSM per arm	8
No. of HBSM per phase	4
HBSM voltage	25 kV
FBSM voltage	25 kV
DC link capacitance	150 uF
HBSM capacitance	4.5 mF
FBSM capacitance	2.5 mF
AC grid voltage	400 kV

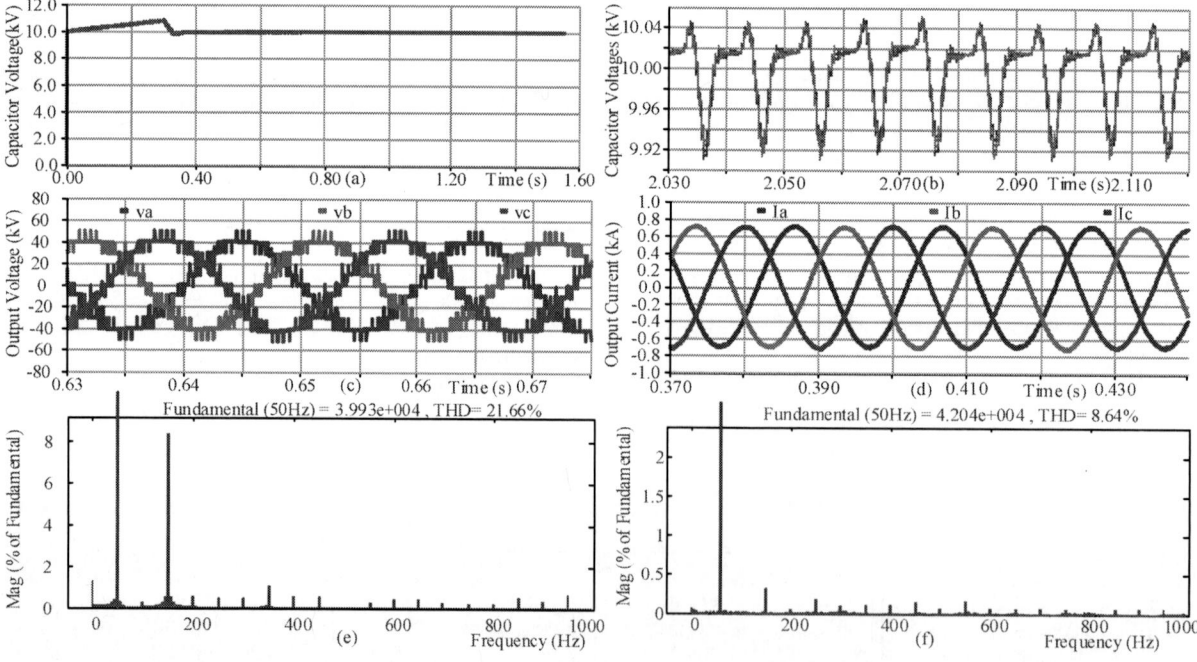

Fig. 6. Standalone converter (a) Capacitor voltage of WSS when proposed control is applied at t=0.3sec, (b) Individual capacitor voltages of WSS, (c) Output voltage of converter, (d) Output current of the converter, (e) Harmonic spectrum of output voltage without capacitor voltage control, and (f) Harmonic spectrum of output voltage without capacitor voltage control.

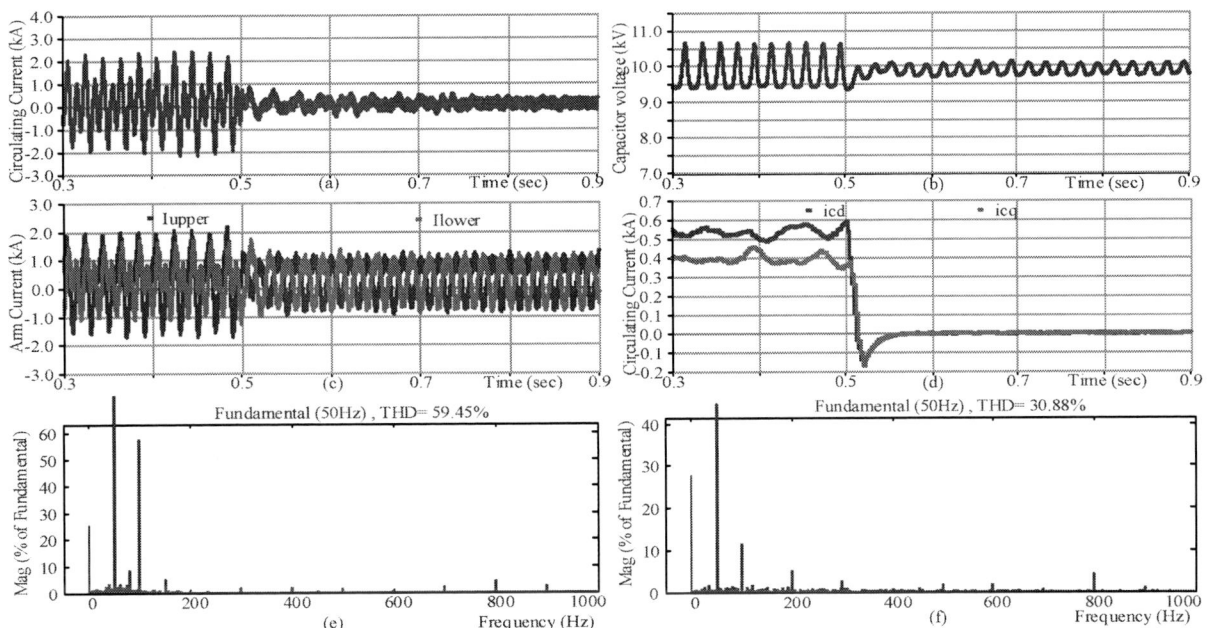

Fig. 7. Converter waveforms when circulating current control is applied at t=0.5sec, (a) Circulating current of phase a, (b) Average upper arm capacitor voltage of phase a, (c) Upper and lower arm currents, (d) dq component of arm current in rotating reference frame of twice frequency, (e) Harmonic spectrum of arm current without circulating current control, and (f) Harmonic spectrum of arm current with circulating current control

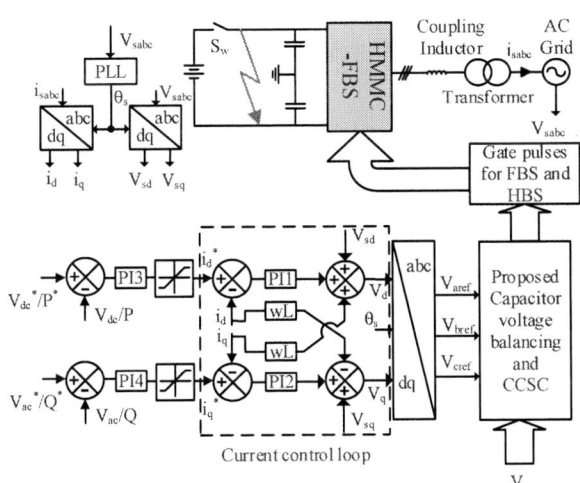

Fig. 8. Schematic diagram summarizing the control technique used for grid connected HMMC-CFB

The reference signals generated from decoupled current control is given to proposed controller. This insures voltage control and circulating current control of HMMC-CFB. To test inherent fault blocking capability of converter, switch S_w is opened and pole to pole dc fault is created. Fig. 9 shows the results of grid connected converter system. Fig. 9(a) shows the results of active and reactive power reversal from 40MW to -40MW and -40MVAr to 40MVAr respectively. The pole to pole fault occurs from t = 3.0 s to 3.2 s. To control the high dc fault current, the gate pulses of switching devices are blocked

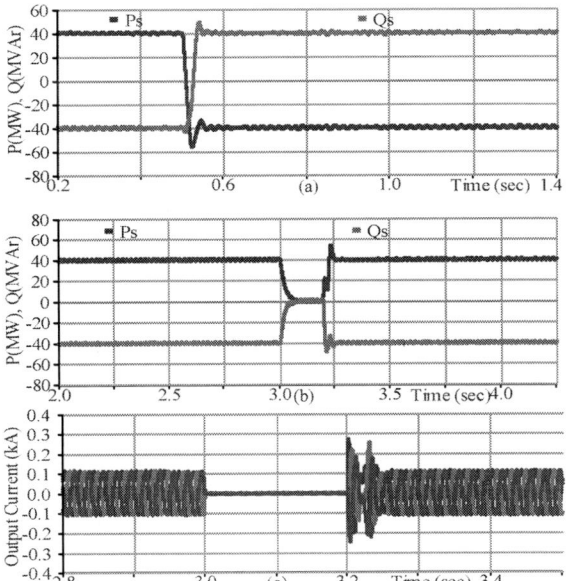

Fig. 9. Grid connected converter, (a) Active and reactive power reversal of converter, (b) Active and reactive power of converter when there is pole to pole fault from t=3.0 to 3.2sec, (c) Output current of current when there is pole to pole fault from t=3.0 to 3.2sec.

from time t = 3.0 s to t = 3.2 s. As gate pulses of FBS are blocked, fault current flows through freewheeling diodes and capacitor voltage of FBS applied across the ac voltage which controls the fault current. The active and reactive power and

978-1-4673-9551-9/16 $31.00 © 2016 IEEE   2347

output current waveforms for the dc fault are shown in Fig. 9(b) and 9(c). During fault condition currents are nearly zero, which validate the fault blocking capability of converter.

IV. CONCLUSION

In this paper, the analysis and a new control technique to regulate the energy exchanged by WSP of HMMC-CFB is proposed. The control of energy exchange is achieved by changing slope of the trapezoidal reference of the DP. Also the analysis of circulating current and its control by CCSC .is presented. With these technique the circulating current is reduced considerably. These two control techniques are tested with real time simulation model in PSCAD/EMTDC. The simulation results highlight satisfactory results for proposed controller in terms of capacitor voltage regulation and reduction of circulating current. Grid connected converter model is simulated and tested for dc fault current blocking capability.

ACKNOWLEDGMENT

The authors would like to thank IRCC-IIT Bombay and MHRD, Government of India for the financial support.

REFERENCES

[1] A. Lesnicar and R. Marquardt, "An innovative modular multilevel converter topology suitable for a wide power range," *in Proc. Power Tech Conf.*, 2003, pp.1-6.

[2] S. Allebrod, R. Hamerski, and R. Marquardt, "New transformerless, scalable modular multilevel converters for hvdc-transmission," *in Proc. IEEE Power Electron. Specialists Conf.*, 2008, pp. 174–179.

[3] J. Dorn, H. Huang, and D. Retzmann, "Novel voltage sourced converters for hvdc and facts applications," *in Proc. CIGRE, Osaka, Japan*, 2007.

[4] R. Marquardt, "Modular multilevel converter: An universal concept for hvdc-networks and extended dc-bus-applications," *in Proc. Int. Power Electron. Conf.*, 2010, pp. 502–507.

[5] A. Shukla and G. D. Demetriades, "A Survey on hybrid circuit-breaker topologies," *IEEE Trans. on Power Del.*, Vol. 30, no. 2, pp. 627-641, April 2015.

[6] M. Callavik, A. Blomberg, J. Ha□fner, and B. Jacobson, "The hybrid HVDC breaker," *ABB Grid Systems Technical Paper*, Nov. 2012.

[7] P. van Gelder and J. Ferreira, "Zero volt switching hybrid DC circuit breakers," *in Proc. IEEE Ind. App. Conf.*, 2000, pp. 2923–2927.

[8] L. Xiaoqian, S. Qiang, L. Wenhua, R. Hong, X. Shukai, and L. Licheng, "Protection of Nonpermanent Faults on DC Overhead Lines in MMC-Based HVDC Systems," *IEEE Trans. on power Del.*, vol. 28, pp. 483-490.

[9] M. Merlin, T. Green, P. Mitcheson, D. Trainer, D. Critchley, and R. Crookes, "A new hybrid multi-level Voltage-Source Converter with DC fault blocking capability," *in Proc. IET 9 th Int. Conf. on AC and DC Power Transm.*, 2010, pp. 1–5.

[10] C. Davidson and D. Trainer, "Innovative concepts for hybrid multi-level converters for hvdc power transmission", *in Proc.9th IET Int. Conf. AC DC Power Transm.*, 2010, pp.1 -5.

[11] M. Merlin, T. Green, P. Mitcheson, D. Trainer, R. Critchley, W. Crookes, and F. Hassan, "The alternate arm converter: A new hybrid multilevel converter with dc-fault blocking capability," *IEEE Trans. Power Del.*, vol. 29, no. 1, pp. 310–317, Feb. 2014.

[12] G. Adam, K. Ahmed, S. Finney, K. Bell, and B. Williams, "New Breed of Network Fault-Tolerant Voltage-Source-Converter HVDC

Transmission System," *IEEE Trans. on Power Systems*, vol. 28, no. 1, pp. 335– 346, 2013.

[13] Z. Yushu, G. P. Adam, T. C. Lim, S. J. Finney, and B. W. Williams, "Hybrid Multilevel Converter: Capacitor Voltage Balancing Limits and its Extension," *IEEE Trans. on Ind. Inform.*, vol. 9, pp. 2063-2073, Nov. 2013.

[14] X. Yinglin, X. Zheng, and T. Qingrui, "Modulation and Control for a New Hybrid Cascaded Multilevel Converter With DC Blocking Capability," *IEEE Tran. on Power Del.*, vol. 27, pp. 2227-2237, 2012.

[15] R. Li, G. Adam, D. Holliday, J. Fletcher and B. Williams, "Hybrid Cascaded Modular Multilevel Converter with DC Fault Ride-Through Capability for HVDC Transmission System," *IEEE Trans. Power Del.*, vol. PP, no.99, pp.1-1.

[16] G. P. Adam, S. J. Finney, and B. W. Williams, "Hybrid converter with ac side cascaded H-bridge cells against H-bridge alternative arm modular multilevel converter: Steady-state and dynamic performance," *IET Generation, Transmiss. Distrib.*, vol. 7, p. 318-328, Mar. 2013.

[17] G. P. Adam, I. Abdelsalam, S. J. Finney, D. Holliday, B. W. Williams, and J. Fletcher, "Comparison of two advanced modulation strategies for a hybrid cascaded converter," *in Proc. ECCE Asia Downunder*, 2013, pp. 1334-1340.

[18] Q. Tu , Z. Xu and L. Xu "Reduced switching-frequency modulation and circulating current suppression for modular multilevel converters", *IEEE Trans. Power Del.*, vol. 23, no. 3, pp.2009 -2017 2011 .

[19] G. T. Son, H.-J. Lee, T. S. Nam, Y.-H. Chung, U.-H. Lee, S.-T. Baek, K. Hur and J.-W. Park, "Design and control of a modular multilevel HVDC converter with redundant power modules for noninterruptible energy transfer", *IEEE Trans. Power Del.*, vol. 27, no. 3, pp.1611 -1619 2012.

[20] G. Adam, I. Abdelsalam, K. Ahmed, and B. Williams, "Hybrid Multilevel Converter With Cascaded H-bridge Cells for HVDC Applications: Operating Principle and Scalability," *IEEE Trans. on Power Electronics*, vol. 30, no. 1, pp. 65-77, Jan. 2015.

[21] Y. Xue, Z. Xu, and Q. Tu, "Modulation and Control for a New Hybrid Cascaded Multilevel Converter With DC Blocking Capability," *IEEE Trans. on Power Del.*, vol. 27, no. 4, pp. 2227–2237, 2012.

[22] Y. Zhang, G. Adam, S. Finney, and B. Williams, "Improved pulsewidth modulation and capacitor voltage-balancing strategy for a scalable hybrid cascaded multilevel converter," *IET Journ. on Power Electron.*, vol. 6, no. 4, pp. 783–797, 2013.

[23] J. Pou, S. Ceballos, G. Konstantinou, G. Capella and V. Agelidis, "Control strategy to balance operation of parallel connected legs of modular multilevel converters", *in Proc. IEEE Int. Symp. Ind. Electron*, 2013, pp. 1-7.

[24] S. P. Engel and R. W. De Doncker, "Control of the modular multi-level converter for minimized cell capacitance," in Proc. European Conference on Power Electronics and Applications (EPE), 30 Aug.-1 Sept. 2011, Birmingham, UK

[25]] R. Picas, J. Pou, S. Ceballos, V. G. Agelidis, and M. Saeedifard, "Minimization of the Capacitor Voltage Fluctuations of a Modular Multilevel Converter by Circulating Current Control," in Proc. IEEE Industrial Electronics Conference (IECON), 25-28 June 2012, Montreal, Canada

[26] J. Pou, J. Zaragoza, G. Capella, I. Gabiola, S. Ceballos, and E. Robles, "Current balancing strategy for interleaved voltage source inverters," EPE Journal, vol. 21, no. 1, pp. 29-34, March 2011.

[27] A. Antonopoulos, L. Angquist, and H. P. Nee, "On dynamics and voltage control of the modular multilevel converter," in Proc. Eur. Conf. Power Electron. Appl., Barcelona, Spain, 2009, pp. 1–10.

[28] P. Josep, S. Ceballos, G. Konstantinou, V. G. Agelidis, R. Picas, and J. Zaragoza. "Circulating current injection methods based on instantaneous information for the modular multilevel converter." *Industrial Electronics, IEEE Transactions on* 62, no. 2 (2015): 777-788.

978-1-4673-9551-9/16 $31.00 © 2016 IEEE

Sensorless Scheduling of the Modular Multilevel Series-Parallel Converter:

Enabling a Flexible, Efficient, Modular Battery

Stefan M. Goetz, Zhongxi Li, Angel V. Peterchev[1]
Duke University
Durham, NC, USA
[1] These authors contributed equally

Xinyu Liang, Chengduo Zhang, Srdjan M. Lukic[1]
North Carolina State University
Raleigh, NC, USA

Abstract—**We present a control approach for the modular multilevel converter (MMC) with series and parallel module connectivity (MMSPC) that provides natural module balancing, reduced conduction losses, and enhanced robustness afforded by the parallel mode. In conventional MMC control, the voltage of each module's storage element has to be measured or estimated to enable the controller to equalize the voltages across all modules. This requirement has been one of the key barriers for MMCs in low and medium power applications. In contrast, we use the parallel connectivity of the MMSPC for module voltage balancing. It also enables robust operation in the presence of battery failure by supporting the module voltage with frequent parallelizations of the residual module capacitance. The parallel connectivity further reduces conduction losses at voltage levels below the system maximum by decreasing the effective source impedance. This approach renders attractive for the first time low and medium power MMC applications as well as MMC-based battery storage. These are illustrated with an experimental MMSPC system comprising eight battery modules that generates high-quality ac output without filtering magnetics.**

Keywords—Modular multilevel converter; parallel mode; switched-capacitor converter; open-loop; sensorless; balancing; battery.

I. INTRODUCTION

Modular multilevel converters (MMC) are currently in use in many power electronics applications [1]. The converter operation is based on the principle of generating higher voltages by adding up the output of series-connected lower-voltage modules. These modules, each equipped with a module energy storage—e.g., a capacitor and/or a battery—can either contribute to the output or be bypassed and thus generate a high-quality output voltage with efficiencies above 99% [2]. A long list of advantages established the MMC in a number of applications ranging from grid facilities and renewable energy integration [1-12] to electro-mobility and medical device technology [13-17]. In addition to the series connection

of MMC, several recent articles introduced a parallel mode so that the capacitors of the modules can be dynamically arranged not only in series, but also in parallel [18-20]. The parallel connection mode of modules was found to achieve better module utilization and significantly reduce conduction loss due to a lower total internal resistance.

However, besides their advantages, modular multilevel converters present a control challenge. The high control effort complicates their application particularly in smaller systems, e.g., automotive converters, small and medium drives, as well as battery management systems. Specifically, the sophisticated control has to manage the output and balance the charge of the module storages. The closed-loop control for balancing of the module storages requires accurate and fast monitoring of each module to avoid, on the one hand, draining any module storage and subsequently halting the system and, on the other hand, an overflow, which would damage the storage itself and the semiconductors [1, 21]. The controller cost drivers are therefore the sensors and the signal isolators. The latter have to shift the signals from the variable module potential to the potential of the controller.

Although several approaches for balancing with reduced module monitoring have been described [22-25], none of them could guarantee stable long-term operation. As these methods typically rely on a time integral of a current to estimate the charge inflows and outflows of each individual module, smallest offsets of the current sensor—a common issue of most current sensors—will almost necessarily drain or overload some module storage elements. In consequence, unavoidable measurement offsets halt or damage the system after sufficient time.

In this paper, we describe and demonstrate a method for sensorless balancing the module voltages, i.e., without monitoring the individual modules. The method uses the new parallel connectivity option of modular multilevel technology to match the module voltages, as well as to reduce conduction losses. By using the modular multilevel series-parallel converter (MMSPC) topology and battery storage in the modules, we demonstrate how to create a battery system with ac output

978-1-4673-9551-9/16 $31.00 © 2016 IEEE

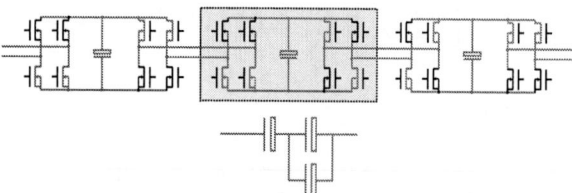

Fig. 1. Circuit topology of three neighboring MMSPC modules, illustrating a representative connection state. The color code indicates parts of the circuit that are connected together, resulting in the connection pattern of the simplified wiring diagram shown at the bottom. The storage element can be a capacitor, a battery, or a combination of both. The prototype in this paper uses a battery as the main storage element in parallel with a filtering capacitor.

and intrinsic balancing among the modules. The simplicity of the presented approach could bring the advantages of modular multilevel converters to low- and medium-power applications.

II. Topology

The MMSPC topology, which was presented and described in detail in the literature [18], is shown in Fig. 1. A MMSPC module shares two connections with each of its neighboring modules; each of the four module terminals is fed by a half-bridge connected to the module storage element (see Fig. 1). The interconnection of two modules through a pair of module terminals enables series connectivity of the module energy storages in two polarities and bypass as known from MMC [26] and parallel connectivity across several modules. The topology has several features that our control method exploits systematically:

1) As an extension that subsumes a conventional MMC, the circuit can operate identical to an MMC without a parallel mode.

2) Although the number of distinct switches is increased compared to MMC, the circuit does not require more silicon. The reason is that in every module state the MMSPC can use two switches in parallel to carry the current that the MMC would handle with a single switch. Consequently, each MMSPC switch can have half of the current rating of a MMC switch.

3) The parallel mode reduces the source impedance of the converter and therefore conduction loss. During low-amplitude sections of the sinusoidal output voltage, modules are successively reconfigured from a series to a parallel electric connection to minimize loss.

4) The parallel mode further allows energy transfer between modules. For example, it enables the MMSPC to operate as a switched-capacitor converter with respect to both the converter output as well as local energy transfer among modules.

III. Module Scheduling

A. Constraints and Objectives

Our sensorless scheduler has to fulfill four constraints during operation. For each switch update cycle, the controller has to schedule the modules and set their states such that they (a) generate the specified output voltage, (b) balance the module

Fig. 2. Picture of the prototype's eight module boards containing the power FETs, gate drives, isolation, and current sensors (used only for system characterization, and not by the scheduler). The batteries are off board (not shown).

storage elements, (c) use the parallel mode for small conduction loss, and (d) limit the switching loss. In the presented controller, the output voltage is generated by a separate multi-carrier modulator using established technology [27], which provides a quantized output to the scheduler to balance objectives (b) – (d). Except for modulation indices equal to one or the number of modules, N, each quantized output level can be generated by a number of different series/parallel combinations of modules, which can be used to optimize the given objectives. Medium modulation indices of approximately $N/2$ leave the highest number of alternatives with equal output.

1) *Balancing:* Although the system does not monitor the individual modules' charge levels, long-term operation requires that the scheduler can guarantee a balanced state with absolute certainty. This requirement is achieved by using the new parallel mode of the MMSPC for charge equilibration between modules. The scheduler has to assure that every module is turned parallel to each of its neighbors as often as possible and at least once for a specified duration of time. If this duration of time passes without a parallel state, parallelization is enforced. If each module is regularly parallelized with its immediate neighbors for equilibration, it is indirectly also equilibrated with their neighbors, and so on. Thus, the method does not need to keep track of the individual charge inflows and outflows. It ensures a balanced state by minimizing the time in which an individual module is either directly or indirectly parallelized to each other module in the setup. The loss caused by equalizing two modules with unequal voltages

$$E_{\text{loss,par}} = \tfrac{1}{4} C_{\text{m}} \Delta V_{\text{m}}^2 \leq \tfrac{1}{4} C_{\text{m}} \left(\frac{I \tau_e}{C_{\text{m}}} \right)^2 \propto \frac{1}{C_{\text{m}}} \tag{1}$$

with load current I and interval τ_e between equalization, falls quadratically with capacitance C_{m} per module, favoring bal-

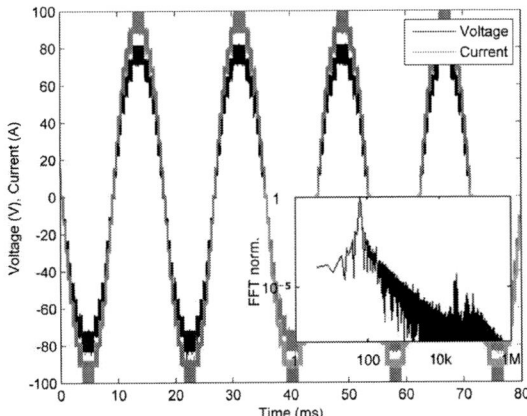

Fig. 3. Measured voltage and current of the demonstrator with 3.8 kW resistive load, a basic frequency of 60 Hz, and an overall switching rate of 30 kHz. The spectrum in the inset exhibits low harmonics at the overtones of the switching frequency.

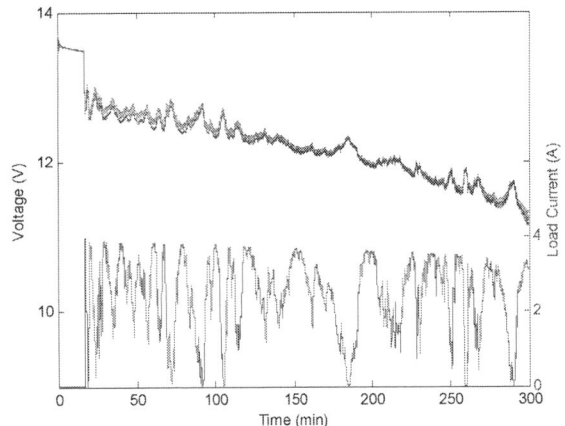

Fig. 4. Module balancing performance demonstrated with long-term measurement of the eight individual battery voltages (left y-axis) during ac operation with a highly variable load (right y-axis).

ancing of systems with large module energy storages, such as double-layer capacitors or batteries.

2) Conduction loss: For a minimum overall source impedance of the converter, the scheduler should use the parallel mode such that series paths with only a small number of parallel modules are minimized. A series connection of similarly sized units of parallel-connected modules has a lower overall resistance than a less even distribution of the available parallel modules into series-connected units. For example, seven paralleled modules connected in series to a single module would generate a much higher impedance than four paralleled modules connected in series to the other four in parallel, while generating the same voltage output. This can be translated to a relatively simple equivalent rule. For the best use of the parallel mode, at each time those alternatives are preferred whose parallel blocks are most equal in size, which is equivalent to

$$\min_{N=\sum_{i=1}^{m} n_i} \sum_{i=1}^{m} \frac{1}{n_i} \qquad (2)$$

for m parallel blocks, each with n_i modules.

3) Switching loss: The switching loss is controlled by minimizing the switches that have to be toggled for the transition from the current state of the entire system to the next one. The switching distance between two alternative states can be evaluated easily. Except for the transition from a negative series connection to a positive series connection and vice versa, which toggles all eight switches, any other transition between different states requires toggling four switches, i.e., half of the switches. The current load per toggled switch is assumed to be equal to half of the module current I so that the number of toggled switches of each interconnection can be used to calculate the loss of each switching option, which equals

$$E_{\text{loss,sw}} = N_{\text{sw}} \cdot \left(\underbrace{\tfrac{1}{2} V_m I t_c}_{\text{saturation loss}} + \underbrace{\tfrac{1}{2} V_m^2 C_o}_{\text{capacitive loss}} \right) \qquad (3)$$

with the number of toggled switches N_{sw}, the module voltage V_m, the load current I, the switch commutation time t_c, and the switch output capacitance C_o.

B. Combination of Objectives

Each individual objective still leaves a high number of alternatives. Thus, to avoid concurrent optimization of these objectives, we applied the objectives subsequently to reduce the number of alternatives in four steps. The priority of each objective can be chosen according to the switching rates. For the ac battery presented here, we have an average switching frequency per module of 800 Hz to above 2 kHz, while the module capacity divided by the rated peak module current exceeds 4 min. Therefore, we prioritize the reduction of switching loss over balancing, which is temporarily less critical. In that priority scheme, the scheduler initially generates a list of all module state alternatives that would provide the specified voltage output but require less than a certain number of toggled switches from the state of the modules in the previous switch update cycle. From this short list, those alternatives are selected that would cause no more than 5% higher conduction loss than the best option in that list.

Among the remaining options, the scheduler shuffles for improving charge balancing. Whereas it could pick the switching state that turns module interconnections to the parallel state with the longest time since the last parallelization (the equivalent of earliest-deadline-first real-time scheduling [28, 29]), we demonstrate that due to the high switching rate, shuffling by randomization is sufficient and reduces the required computational effort.

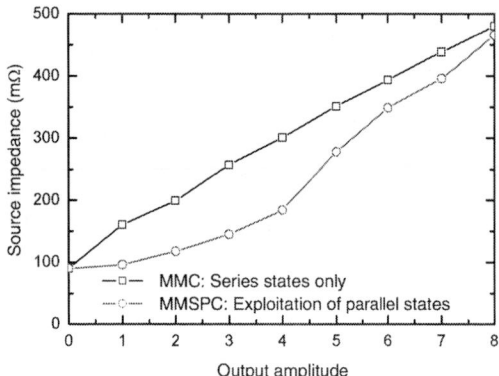

Fig. 5. Measured source impedance of the converter using either only series combinations of modules (as in the MMC) or both series and parallel states (enabled by the MMSPC). The measurements are for the possible eight different output voltage levels of either polarity as well as zero output.

IV. EXPERIMENTAL RESULTS

A. Demonstrator

We implemented a MMSPC demonstrator system operating at 12 V, with eight modules, each incorporating a lead-acid battery (12 V, 7 Ah, Enersys Genesis NP7-12), aluminum electrolytic capacitors (1 mF, ZLH, Rubycon, Inc.), and low-impedance ceramic capacitors (200 µF, X7R, Taiyo Yuden Co. Ltd.) as charge storage. Each module switch comprised a silicon FET (IRFP4310, International Rectifier) and a Schottky diode (STPS 8H100D, STMicroelectronics) anti-parallel with the FET without the need for further snubbing of the switching transients. The demonstrator has an rms current rating of 200 A and a peak voltage of 96 V. The converter control is implemented on an Opal real-time controller system (OP5607, OPAL-RT Technologies, Inc.).

The carrier frequency of the PWM is varied up to 30 kHz. The controller distributes the switching activity among all modules so that the switching rate per module amounts to up to 3.75 kHz on average. As outlined above, the scheduler carries out sequential optimization of the performance objectives. First, it selects the module configurations that fulfill the output voltage level command from the modulator. Subsequently, the scheduler eliminates all alternatives that would require more than eight toggled switches relative to current module states in order to keep switching loss low. From this short list, those states are selected that would cause no more than 5% higher conduction loss than the best option to keep conduction loss low. The scheduler randomly shuffles these options to select the final configuration, in lieu of optimizing balancing. The parallelization loss is not addressed explicitly, since it is very small as discussed previously [18].

B. Output Quality

Figure 3 presents ac output at unity power factor and peak modulation index. The 17 voltage levels of the demonstrator

allow low harmonics of the voltage of less than −86 dBc (see Fig. 3 inset).

C. Module Balancing

The control is able to keep the module storages balanced over a long duration by the use of the switched-capacitor functionality in frequent parallelization as demonstrated in Fig. 4, despite an intentionally reduced overall switching rate of only 7 kHz. The individual traces in Fig. 4 represent the voltages of the individual module storages over time while the system is facing a fluctuating grid load profile. The largest duration without parallelization of any interconnection was detected to be shorter than 3 ms and therefore by orders of magnitude lower than a discharging cycle of the implemented module storages, which is in the minute range. In consequence, the standard deviation of the module voltages stays below 77 mV (with a median of 22 mV) for the five-hour trace in Fig. 4.

D. Source Impedance

In contrast to MMC with exclusive series and bypass connectivity, the parallel connectivity of modules reduces the effective source impedance of the module string and thus the loss. With the changing module configuration, also the source impedance dynamically changes with the output voltage. We quantified the reduction of the source impedance with the use of parallel module connectivity and compared it to the converter operated without a parallel mode. Without the use of the parallel mode, the setup emulates a conventional MMC. We extracted the source impedance from four-wire measurements for various load currents between 0 A and 93 A for each output level and hence module configuration separately. The results are depicted in Fig. 5. For zero voltage and peak voltage output, both MMC and MMSPC use the same module configuration. Consequently, MMC and MMSPC also share the same impedance for those two configurations. The intermediate output levels, however, increasingly introduce parallel connectivity with reducing amplitude for the MMSPC so that the impedance is up to 40% lower than in MMC. For ac output, the output voltage is dominantly in that intermediate range, increasing the advantage of the reduced source impedance. Furthermore, the reduction for intermediate voltages increases the converter efficiency compared to MMC especially for inductive loads and low output amplitudes. Thus, the use of the parallel mode reduces the conduction loss by 20% for $\cos \varphi = 1$ and by 37% for $\cos \varphi = 0.8$ at equal apparent output power for peak ac output voltage equal to the system maximum (96 V).

E. Robustness to Module Battery Failure

In addition to sensorless balancing, the energy transfer across modules using the parallel mode enables further redundancy. As Fig. 6 demonstrates, the converter can tolerate the failure of a battery without losing the respective module and the corresponding voltage step in the output. During the recording displayed in Fig. 6, one module's battery is disconnected, simulating an opened battery contactor due to a failure of the corresponding module battery. Whereas a conventional MMC has to bypass any defective modules, which reduces the peak voltage rating, the demonstrator uses the neighbors for cyclic re-

978-1-4673-9551-9/16 $31.00 © 2016 IEEE

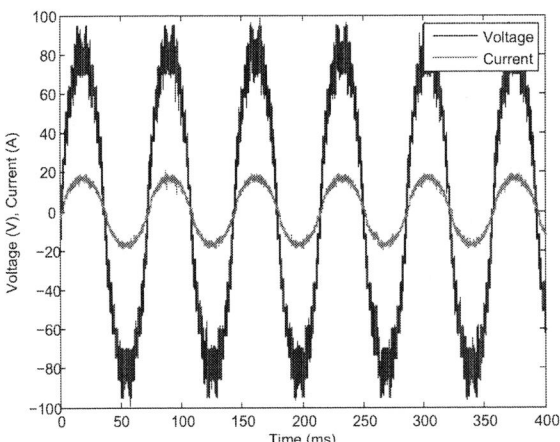

Fig. 6. Measured output voltage and current for one failed and subsequently disconnected module battery. Despite the battery failure, all distinct 16 voltage steps are present in the recording.

charging of the remaining capacitance of the affected module without battery.

The maximum current the setup can provide in case of broken module batteries, $I_{max} = f_{parallel} Q_{balance} = f_{parallel} C_{residual} V_{mod}$, simply depends on the balancing charge, $Q_{balance}$, given by the product of the respective module's remaining capacitance, $C_{residual}$, and the module voltage, V_{mod}, as well as on the rate of turning the respective module in parallel to neighbors for recharging, $f_{parallel}$. Upon detection of a failed battery, the control method can adjust its scheduling to maximize $f_{parallel}$ for the affected module. Consequently, parallelization, which has been the least important constraint for operation without battery failures for the specific demonstrator presented here, can now be prioritized over conduction loss and switching loss minimization. Furthermore, the overall switching frequency can be increased to raise the current limit linearly, at the price of also linearly increasing switching loss. Since the maximum current discharges an affected module without battery completely before it is recharged through a neighbor so that the effective average voltage of the module would be $V_{mod}/2$ only due to switching modulation, derating may be suggested. In Fig. 6, for instance, the demonstrator is operated at 37% of the maximum theoretical current for a single-battery failure of 54 A.

V. CONCLUSIONS

We presented a robust method for sensorless scheduling of the MMSPC, reducing the high monitoring effort of conventional MMC systems, which is a key barrier for in low and medium power applications. The parallel mode was used for transferring energy between modules to equilibrate their voltages, obviating the need for closed-loop monitoring and balancing. The elimination of excessive monitoring, together with the increased efficiency associated with the parallel connectivity mode of MMSPC, enables several applications that are not possible or not efficient with the MMC. As one particularly promising application, we

demonstrated a modular battery system with ac output. The prototype produces a high-quality output with low harmonic profile. Moreover, the system is robust to failure of a battery unit within a module as a result of the concurrent switched-capacitor functionality that recharges the residual module capacitance. The same system could output other waveforms including dc (not shown here), while providing battery unit balancing in the background by dynamic re-arrangement of its battery units. In summary, the demonstrated MMSPC system breaks up the conventional hardwired battery pack structure to render the interconnections of the battery units dynamically reconfigurable, providing a practical integrated converter/battery system.

ACKNOWLEDGMENT

This work was funded in part by the Duke University Energy Initiative.

REFERENCES

[1] A. Lesnicar and R. Marquardt, "An innovative modular multilevel converter topology suitable for a wide power range," *Power Tech Conference Proceedings, 2003 IEEE Bologna*, vol. 3, p. 6, 23-26 June 2003 2003.

[2] S. Allebrod, R. Hamerski, and R. Marquardt, "New transformerless, scalable Modular Multilevel Converters for HVDC-transmission," in *Power Electronics Specialists Conference, 2008. PESC 2008. IEEE*, 2008, pp. 174-179.

[3] M. Glinka and R. Marquardt, "A new AC/AC multilevel converter family," *Industrial Electronics, IEEE Transactions on*, vol. 52, pp. 662-669, 2005.

[4] Y. Wang and R. Marquardt, "Future HVDC-grids employing modular multilevel converters and hybrid DC-breakers," *Power Electronics and Applications (EPE), 2013 15th European Conference on*, pp. 1-8, 2-6 Sept. 2013 2013.

[5] D. Hong, S. Bai, and S. M. Lukic, "Closed Form Expressions for Minimizing Total Harmonic Distortion in 3-Phase Multilevel Converters," *Power Electronics, IEEE Transactions on*, vol. PP, pp. 1-1, 2013.

[6] B. Sanzhong and S. Lukic, "Modular design of cascaded H-bridge for community energy storage systems by using secondary traction batteries," *Applied Power Electronics Conference and Exposition (APEC), 2014 Twenty-Ninth Annual IEEE*, vol. 29, pp. 3297-3304, 16-20 March 2014 2014.

[7] B. Gultekin and M. Ermis, "Cascaded Multilevel Converter-Based Transmission STATCOM: System Design Methodology and Development of a 12 kV ± 12 MVAr Power Stage," *Power Electronics, IEEE Transactions on*, vol. 28, pp. 4930-4950, 2013.

[8] J. Rodriguez, L. G. Franquelo, S. Kouro, J. I. Leon, R. C. Portillo, M. A. M. Prats, and M. A. Perez, "Multilevel Converters: An Enabling Technology for High-Power Applications," *Proceedings of the IEEE*, vol. 97, pp. 1786-1817, 2009.

[9] S. Kouro, M. Malinowski, K. Gopakumar, J. Pou, L. G. Franquelo, W. Bin, J. Rodriguez, M. A. Perez, and J. I. Leon, "Recent Advances and Industrial Applications of Multilevel Converters," *Industrial Electronics, IEEE Transactions on*, vol. 57, pp. 2553-2580, 2010.

[10] R. Marquardt, "Modular Multilevel Converter: An universal concept for HVDC-Networks and extended DC-Bus-applications," *Power Electronics Conference (IPEC), 2010 International*, pp. 502-507, 21-24 June 2010 2010.

[11] G. R. Walker and P. C. Sernia, "Cascaded DC-DC converter connection of photovoltaic modules," *Power Electronics, IEEE Transactions on*, vol. 19, pp. 1130-1139, 2004.

[12] M. J. Duran, S. Kouro, W. Bin, E. Levi, F. Barrero, and S. Alepuz, "Six-phase PMSG wind energy conversion system based on medium-voltage multilevel converter," in *Power Electronics and Applications (EPE 2011), Proceedings of the 2011-14th European Conference on*, 2011, pp. 1-10.

[13] L. Leclere and C. Galmiche, "A transformerless full redundant electrical propulsion solution to enhance power density, availability and low noise signature," in *Electric Ship Technologies Symposium (ESTS), 2011 IEEE*, 2011, pp. 296-299.

[14] S. Bhattacharya, D. Mascarella, and G. Joos, "Modular multilevel inverter: A study for automotive applications," in *Electrical and Computer Engineering (CCECE), 2013 26th Annual IEEE Canadian Conference on*, 2013, pp. 1-6.

[15] A. V. Peterchev, Z.-D. Deng, and S. M. Goetz, "Advances in transcranial magnetic stimulation technology," in *Brain Stimulation: Methodologies and Interventions*, I. M. Reti, Ed., ed Hoboken, NJ: John Wiley & Sons, 2015.

[16] L. Lambertz, R. Marquardt, and A. Mayer, "Modular converter systems for vehicle applications," in *Emobility - Electrical Power Train, 2010*, 2010, pp. 1-6.

[17] S. M. Goetz, M. Pfaeffl, J. Huber, M. Singer, R. Marquardt, and T. Weyh, "Circuit topology and control principle for a first magnetic stimulator with fully controllable waveform," in *Engineering in Medicine and Biology Society (EMBC), 2012 Annual International Conference of the IEEE*, 2012, pp. 4700-4703.

[18] S. Goetz, A. V. Peterchev, and T. Weyh, "Modular Multilevel Converter with Series and Parallel Module Connectivity: Topology and Control," *IEEE Transactions on Power Electronics*, vol. 30, pp. 203-215, 2015.

[19] K. Ilves, F. Taffner, S. Norrga, A. Antonopoulos, L. Harnefors, and H.-P. Nee, "A Submodule Implementation for Parallel Connection of Capacitors in Modular Multilevel Converters," *Energy Conversion Congress and Exposition (ECCE), 2013 IEEE European*, 2013.

[20] A. Nami, L. Jiaqi, F. Dijkhuizen, and G. D. Demetriades, "Modular Multilevel Converters for HVDC Applications: Review on Converter Cells and Functionalities," *Power Electronics, IEEE Transactions on*, vol. 30, pp. 18-36, 2015.

[21] B. Gemmell, J. Dorn, D. Retzmann, and D. Soerangr, "Prospects of multilevel VSC technologies for power transmission," in *Transmission and Distribution Conference and Exposition, 2008. T&D. IEEE/PES*, 2008, pp. 1-16.

[22] D. Siemaszko, A. Antonopoulos, K. Ilves, M. Vasiladiotis, L. Ängquist, and H. P. Nee, "Evaluation of control and modulation methods for modular multilevel converters," in *Power Electronics Conference (IPEC), 2010 International*, 2010, pp. 746-753.

[23] L. Ängquist, A. Antonopoulos, D. Siemaszko, K. Ilves, M. Vasiladiotis, and H. P. Nee, "Open-Loop Control of Modular Multilevel Converters Using Estimation of Stored Energy," *Industry Applications, IEEE Transactions on*, vol. 47, pp. 2516-2524, 2011.

[24] M. A. Perez, J. Rodriguez, E. J. Fuentes, and F. Kammerer, "Predictive Control of AC-AC Modular Multilevel Converters," *Industrial Electronics, IEEE Transactions on*, vol. 59, pp. 2832-2839, 2012.

[25] L. M. Tolbert and T. G. Habetler, "Novel multilevel inverter carrier-based PWM method," *Industry Applications, IEEE Transactions on*, vol. 35, pp. 1098-1107, 1999.

[26] A. Lesnicar and R. Marquardt, "An innovative modular multilevel converter topology suitable for a wide power range," in *Power Tech Conference Proceedings, 2003 IEEE Bologna*, 2003, p. 6 pp. Vol.3.

[27] B. P. McGrath and D. G. Holmes, "Multicarrier PWM strategies for multilevel inverters," *Industrial Electronics, IEEE Transactions on*, vol. 49, pp. 858-867, 2002.

[28] R. K. Abbott and H. Garcia-Molina, "Scheduling real-time transactions: a performance evaluation," *ACM Trans. Database Syst.*, vol. 17, pp. 513-560, 1992.

[29] M. Kargahi and A. Movaghar, "A Method for Performance Analysis of Earliest-Deadline-First Scheduling Policy," *The Journal of Supercomputing*, vol. 37, pp. 197-222, 2006/08/01 2006.

An Input Current Calculation Switching Driver for High Power-Factor and Phase-cut Dimmer Compatibility

Hyunchul Eum, Youngjong Kim and Kuohsien Huang

OPS AE & RD, Fairchild Semiconductor

Gyeonggi-do, Republic of Korea / Hsinchu, Taiwan

Abstract— **A new switching mode technique based on input current calculation has been proposed for accurate control of input current in single stage AC-DC converter. The input current shape is precisely managed to a desired pattern by calculating averaged switching current without a compensation method and a fine tuning sequence. This switching mode scheme provides sinusoidal input current with excellent power factor in any operational mode such as DCM, BCM and CCM. Input current can be also controlled higher than TRIAC holding current for better phase-cut dimmer compatibility. The proposed switching mode driver has been verified in 8 W dimmable LED lighting system.**

Keywords—Power factor, Dimmer compatibility, LED driver

I. INTRODUCTION

In phase-cut dimming system, the key technical issue has been mostly TRIAC holding current maintenance with low power LED lamp. In order to continue TRIAC conduction till the line voltage zero crossing, there have been several trials to flatten the driver input current higher than the holding current. In the previous compensation methods to provide constant input current, the input current is not precisely constant with a compensation error [1].

Power factor (PF) is also required in case that dimmer is not connected. When dimmer absence is detected, several phase-cut dimming LED driver solutions change the operating mode for better PF and THD. However, PF is worse due to additional high voltage bleeder capacitor [2], [3]. Various technical approaches have been introduced to improve PF especially when buck and buck-boost topologies are designed in boundary conduction mode [4-8]. However, there has not been a general-purpose switching method to optimize PF and THD at all conduction modes such as DCM, BCM and CCM.

To overcome the limitation of constant input current control for dimmer compatibility and the degradation of PF and THD in boundary conduction mode, the proposed switching scheme excellently can control the driver input current as a flat shape when dimmer is present. PF and THD are optimized in any conduction conditions (DCM, BCM and CCM) when phase-cut dimmer is not connected. By fundamental approach for input current control in this paper, the complicated compensation methods are not necessary and input current shape is precisely adjusted.

II. INPUT CURRENT CALCULATION SWITCHING TECHNIQUE

An input current calculation (ICC) switching determines switch turn-on time based on an averaged switching current to implement accurate input current control in buck or buck-boost topology. Fig.1 shows a switching block diagram for ICC.

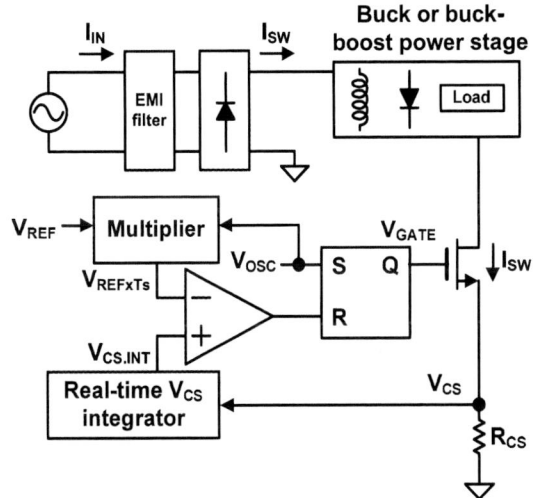

Fig. 1 Input current calculation switching

The input current (I_{IN}) is same as the averaged switching current ($I_{SW.AVG}$) filtered by EMI filter and $I_{SW\ AVG}$ is proportional to an averaged current sense voltage ($V_{CS.AVG}$), which is given as

$$I_{IN} = I_{SW.AVG} = \frac{1}{T_S}\int_0^{T_{ON}} I_{SW}(t)dt = \frac{1}{T_S}\int_0^{T_{ON}} \frac{V_{CS}(t)}{R_{CS}}dt = \frac{V_{CS.AVG}}{R_{CS}} \quad (1)$$

Switch turn-on time is determined by comparing V_{REFxTs} and $V_{CS.INT}$. V_{REFxTs} is the multiplication of V_{REF} and switching period as like

$$V_{REFxTs} = k_{REFxTs} \cdot V_{REF} \cdot T_S \qquad (2)$$

where k_{REFxTs} is the multiplication coefficient and T_S is switching period extracted from the oscillator (V_{OSC}).

$V_{CS.INT}$ is real-time integration value of V_{CS} and given as

$$V_{CS.INT} = k_{CS.INT} \cdot \int_0^t V_{CS}(t)dt \qquad (3)$$

where $k_{CS.INT}$ is the integration coefficient.

From the equations (1), (2) and (3), the proportional relation between I_{IN} and V_{REF} is obtained as

$$I_{IN} = \frac{k_{REFxTs}}{k_{CS.INT}} \cdot \frac{V_{REF}}{R_{CS}} \qquad (4)$$

Therefore, based on the calculation block diagram, input current shape follows V_{REF} which is proportional to $V_{CS.AVG}$ in (1).

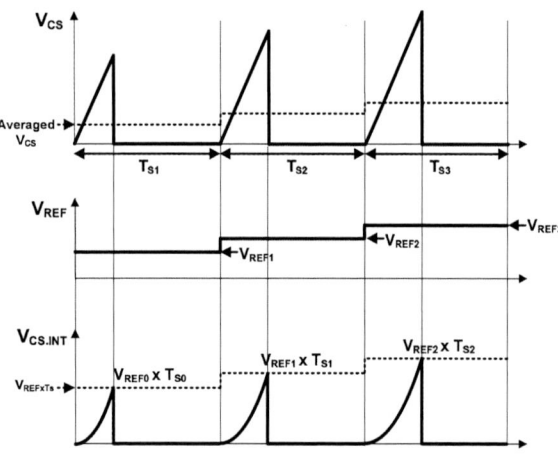

Fig. 2 Switching block waveforms

Fig. 2 shows calculation sequence in ICC block. V_{GATE} is turned off when $V_{CS.INT}$ reaches to $V_{REF.Ts}$ which is the multiplication result from previous switching cycle. Therefore, the input current quickly follows V_{REF} with only one switching delay which is negligible in 50 ~ 60 Hz line frequency. The novel switching mode technique based on switching current calculation provides excellent input current control by simply applying V_{REF} and setting appropriate R_{CS}.

III. CONSTANT TRIAC CONDUCTION CURRENT CONTROL

The major problem in the phase-cut dimming has come out as a visible flicker due to small input current of LED lamp and unstable TRIAC conduction accordingly. As shown in Fig. 3, TRIAC turn-off time is fluctuated every half line cycle when the input current reaches to TRIAC holding current (I_{HOLD})

with gentle slope before input voltage zero crossing. With the TRIAC turn-off time fluctuation, driver input voltage is slightly changed so that the amount of power delivery to LED load is also varied with visible flicker.

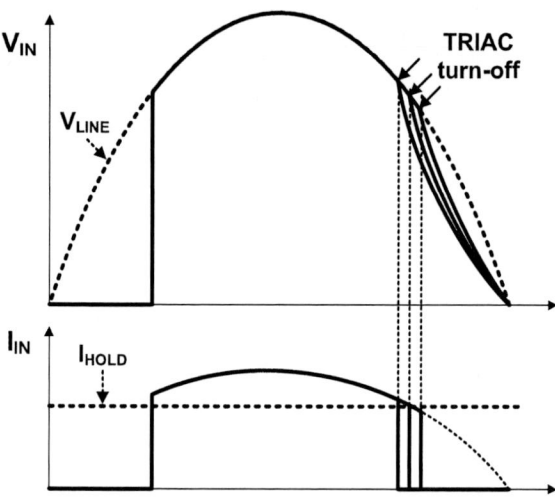

Fig. 3 Conventional phase-cut dimming system

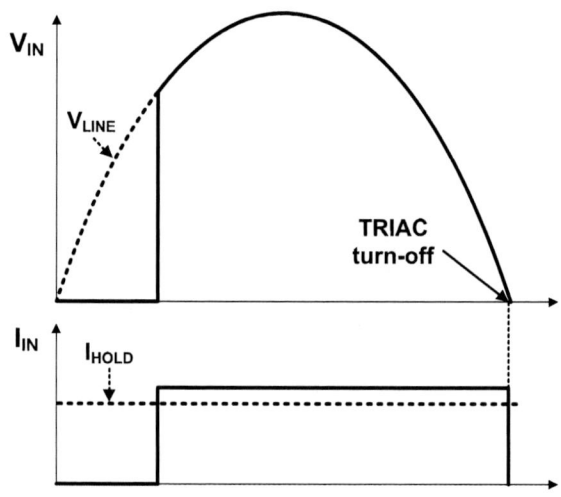

Fig. 4 Constant conduction current control

Fig. 4 shows stable TRIAC conduction by constant TRAIC conduction current control. Once the input current is constantly managed regardless of the input voltage change, TRIAC conduction can be maintained till the input voltage zero crossing and TRIAC turn-off fluctuation is eliminated with no visible flicker. In the conventional peak current mode control, it is hard to shape constant input current since peak current doesn't represent input current in buck or buck-boost topology. However, ICC switching mode can be utilized to control constant TRIAC conduction current as shown in Fig. 5. In phase-cut dimming condition, V_{REF} is the output of feedback

978-1-4673-9551-9/16 $31.00 © 2016 IEEE   2356

loop with narrow loop bandwidth similar to conventional PFC control loop. $V_{REF.DIM}$ is the reference for output current regulation and V_{ILED} is sensed LED current information. As phase angle is reduced, $V_{REF.DIM}$ decreases to regulate smaller LED current corresponding to the phase angle. In order to guarantee TRIAC conduction at low phase angle condition, minimum V_{REF} is limited by clamping circuit with $V_{REF.MIN}$ not to allow the input current lower than TRIAC holding current.

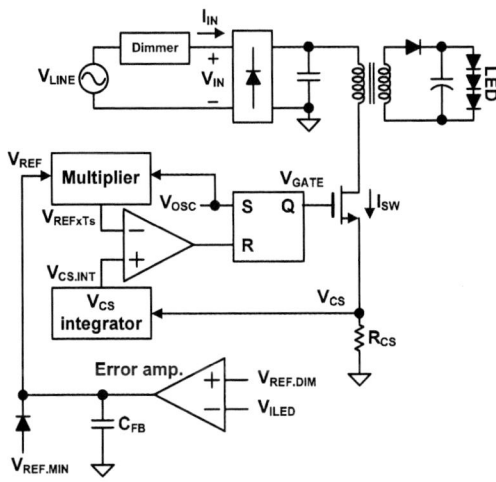

Fig. 5 Constant input current control block

IV. POWER FACTOR CORRECTION OPTIMIZED FOR ALL CONDUCTION MODES

In the single stage flyback topology, DCM with fixed turn-on time and frequency easily provides good PF and THD while BCM reduces switching loss with quasi-resonant switching for better efficiency. In order to implement both good PF and high system efficiency, ICC switching method can be utilized.

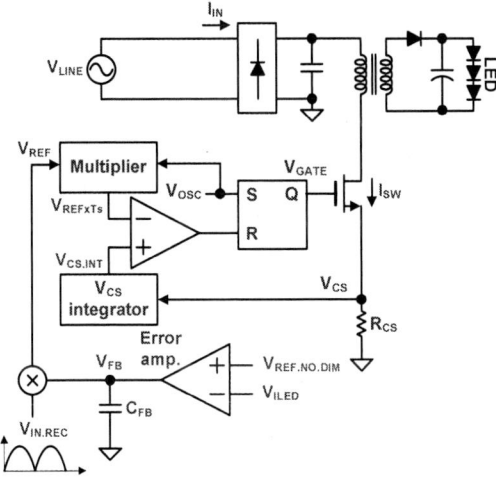

Fig. 6 PFC control block

As shown in Fig. 6, V_{REF} is multiplied result of a rectified input voltage ($V_{IN.REC}$) and feedback voltage (V_{FB}). When V_{FB} is flat in a half line cycle by narrow loop bandwidth, I_{IN} follows the sinusoidal $V_{IN.REC}$ shape with optimized PF and THD regardless of conduction modes. Fig. 7 shows the simulation result of the sinusoidal input current controlled by ICC switching mode. Whether f_{SW} is varied at BCM, the input current is not distorted in the peak line voltage region. Even in CCM, the input current is maintained same as the sinusoidal waveform by the ICC switching mode.

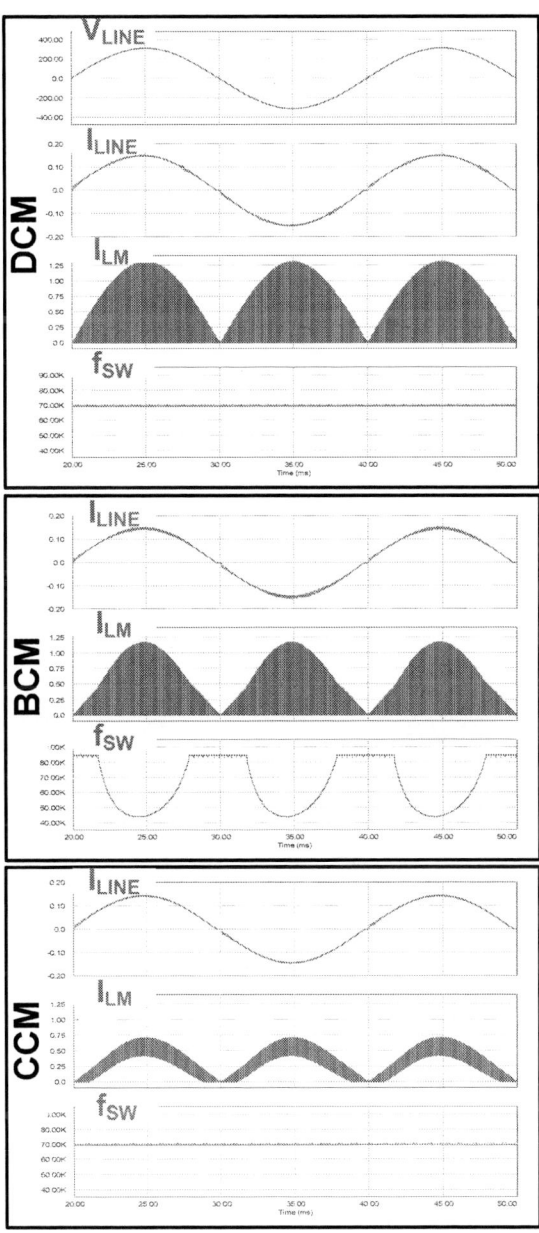

Fig. 7 Simulation result of ICC switching mode at DCM, BCM and CCM

V. EXPERIMENTAL RESULTS

The proposed ICC switching system has been verified in 8W phase-cut dimming driver. Fig. 8 shows constant TRIAC conduction current based on the control block in Fig. 5. Once TRIAC is fired and conduction begins, input current is flat to maintain TRIAC conduction current higher than the holding current till the line voltage zero crossing. Fig. 9 and 10 show the comparison of the input current shape in the conventional voltage mode and the proposed ICC mode. In order to optimize PF and THD when dimmer is not connected, the test has been implemented based on the block diagram in Fig. 6. Operational mode in the comparison test is basically BCM with max. frequency limit so that test result shows BCM at V_{LINE} peak area and DCM at V_{LINE} zero crossing area. As shown in Fig. 9, conventional voltage mode at BCM region distorts line current which degrades PF and THD. The proposed ICC switching mode shows excellent sinusoidal line current shape whether frequency drops at BCM. Fig. 11 and 12 show PF and THD performance comparison between the conventional voltage mode and ICC switching mode. Especially in THD test result, the ICC mode significantly improves THD lower than 3% in the whole input voltage.

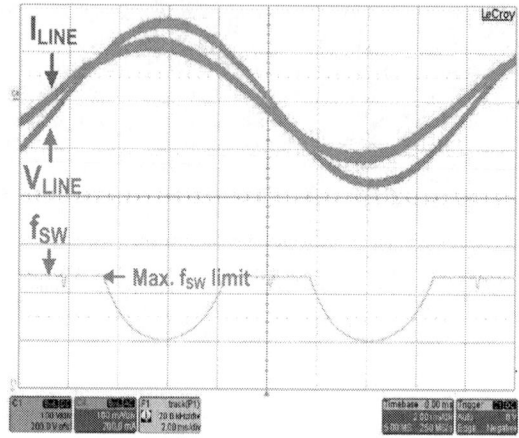

Fig. 10 Input current at ICC mode

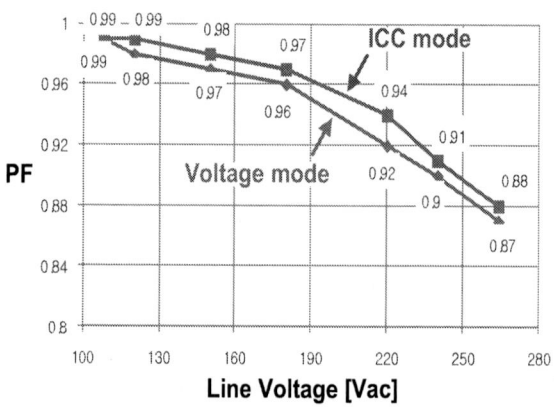

Fig. 11 PF performance comparison
(Conventional voltage mode vs ICC mode)

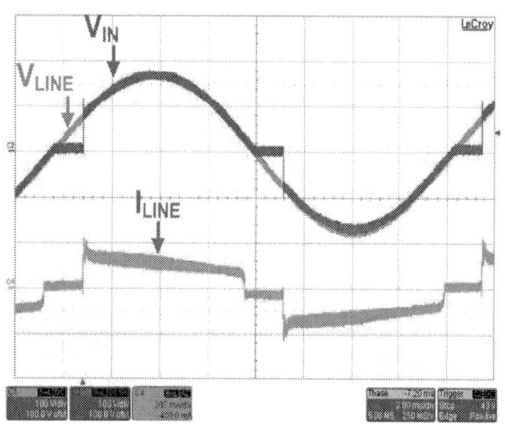

Fig. 8 Constant TRIAC conduction current

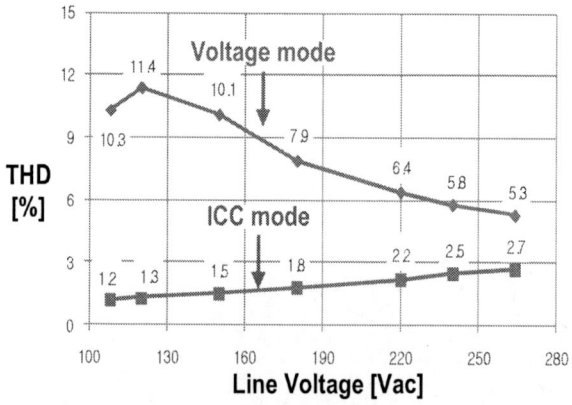

Fig. 12 THD performance comparison
(Conventional voltage mode vs ICC mode)

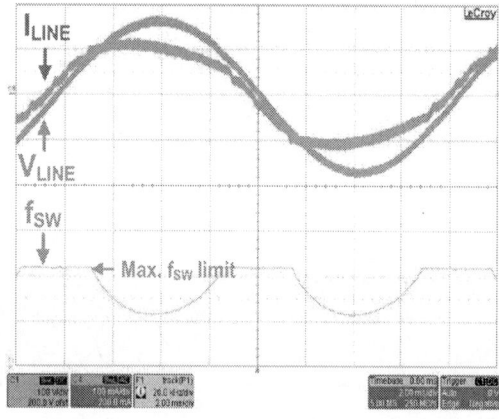

Fig. 9 Input current at conventional voltage mode

VI. CONCLUSION

A novel input current calculation switching technique for high PF and stable TRIAC dimmer compatibility has been proposed and constructed in 8W LED driver system. The input current calculation (ICC) switching is implemented by calculating averaged switching current to accurately control the input current shape without the conventional compensation methods. For high TRIAC dimmer compatibility, ICC switching is utilized to shape the constant TRIAC conduction current to stabilize TRIAC conduction. In order to optimize PF and THD, ICC switching controls the excellent sinusoidal line current in all conduction mode such as DCM, BCM and CCM. The proposed switching mode shows high TRIAC dimmer compatible performance and significant improvement of PF and THD.

REFERENCES

[1] "Using the TPS92075 Buck-Boost Converter" Texas Instrument Application Note, SLVU813A.

[2] M. Doshi and J. Patterson, "Input Filter Design for TRIAC Dimmable LED Lamps," in IEEE Energy Conversion Congress and Exposition, 2013, pp. 4631-4638.

[3] "Design Guidance for TRIAC Dimmable LED Driver Using FL7730" Fairchild Semiconductor Application Note, AN-9745.

[4] T. Yan, J. Xu, F. Zhang, J. Sha and Z. Dong, "Variable-On-Time-Controlled Critical-Conduction-Mode Flyback PFC Converter," in IEEE Transactions on Industrial Electronics, vol. 61, no. 11, pp. 6091-6099, Nov. 2014

[5] C. Adragna and G. Gritti, "High-Power-Factor Quasi-Resonant Flyback Converters Draw Sinusoidal Input Current," in IEEE Applied Power Electornics Conference and Exposition, 2015, pp. 498-505.

[6] X. Wu, J. Yang, J. Zhang and Z. Qian, "Variable On-Time (VOT)-Controlled Critical Conduction Mode Buck PFC Converter for High-Input AC/DC HB-LED Lighting Applications," in IEEE Transactions on Power Electronics, vol. 27, no. 11, pp. 4530-4539, Nov. 2012

[7] L. Huber, L. Gang and M. Jovanovic, "Design-Oriented Analysis and Performance Evaluation of Buck PFC Front End," in IEEE Transactions on Power Electronics, vol. 25, no. 1, pp. 85-94, Jan. 2010

[8] Y. Wang, Y. Zhang, Q. Mo, M. Chen and Z. Qian, "An Improved Control Strategy Based on Multiplier for CRM Flyback PFC to Reduce Line Current Peak Distortion," in IEEE Energy Conversion Congress and Exposition, 2010, pp. 901-905.

High Frequency Range Conducted Common-Mode Noise Suppression in SMPS

Jinping Zhou, Yicong Xie and Min Zhou

Delta Power Electronics Center
Delta Electronics (Shanghai) Co.,Ltd.
Shanghai, China
Email: jinping.zhou@deltaww.com.cn

Abstract —In power electronics system, high frequency range CM noise suppression is a great challenge on the issue of conducted EMI. People may think that a perfect EMI filter with little self parasitics and mutual parasitics would suppress HF noise well. However, it will be shown that the practice story is different. In this paper, besides the parasitics in EMI filter, some other factors which increase the difficultly of high frequency range CM noise suppression are discussed. HF noise model and noise transmitting path are given. Based on the noise model, noise suppression techniques are analyzed and verified with experimental results.

Keywords —Conducted EMI; Common mode Noise; SMPS;

I. INTRODUCTION

Many techniques have been developed on conducted EMI noise suppression, such as EMI filter, noise cancellation based on noise balance [1], active EMI filter [2]. However, these researches mainly focus on low frequency range, usually below several mega hertz. At present, High Frequency (HF) range (several mega hertz to 30 mega hertz) Common-Mode (CM) noise suppression is becoming the greatest challenge on the issue of conducted EMI.

In an isolated EMI filter, the effects of self parasitics and mutual parasitics on filter's performance in HF range are pointed out and analyzed in recent years [3]. However, when the EMI filter is placed in a real power electronics system, the situation will change. Parasitics in EMI filter may not be the most important factors that related to HF range noise suppression. Some other external parasitics induced by system requirement and test setup must be taken into account, including the stray capacitance from load to earth ground, impedance of metal case and parasitic parameters of input cable. This paper will firstly analyze the significant effects of these parasitic parameters on HF range CM noise suppression and then a detailed CM noise model is given. Finally, HF range noise suppression techniques are disclosed and verified with experimental results.

II. CONDUCTED CM NOISE MODEL

A conventional solution for conducted CM noise suppression of Switch Mode Power Supply (SMPS) is shown in Fig.1. The dot frame denotes grounded metal case; its EMI filter topology adopts two LC stages; all Y

capacitors at input side and output side are connected to grounded case. Based on prior research, its CM noise model is shown in Fig.2 [1]. Where, Model I is the model without considering the parasitics of component, model II is the model when considering the parasitics of filter component; C_{PS} is the stray capacitance from transformer primary side to secondary side, C_{Y0} is the Y capacitor between primary minus and secondary minus. The small-signal frequency analysis (1v AC sweep) results are shown in Fig.3. It shows that self parasitics of filter components do have significant effects on filter performance in this simple CM noise model.

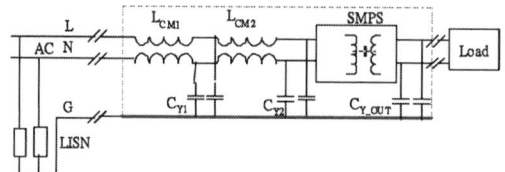

Fig.1 A conventional solution of conducted CM noise suppression

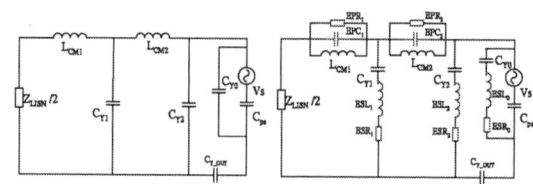

(a) Model I:ideal component (b) Model II:nonideal component

Fig.2 Simple CM noise model

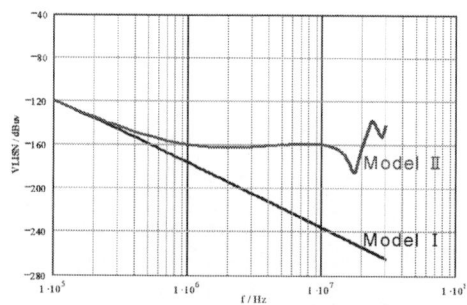

Fig.3 Influence of parasitics of filter component

978-1-4673-9551-9/16 $31.00 © 2016 IEEE

However, when the frequency reaches tens of mega hertz, some other external parasitic parameters induced by measurement setup must be taken into account.

a) Stray capacitor from load to earth ground C_{L_G}

Usually, power supply and its load are not in one case, they are connected to each other through a piece of cable. Thus, there will have a stray capacitor C_{L-G} between load and earth ground. This stray capacitance will introduce a portion of HF CM current flowing into earth ground, which leads more difficulty in noise suppression, as shown in Fig.4.

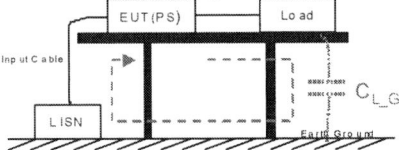

Fig.4 Setup of conducted EMI measurement

The value of C_{L-G} is decided by the type of load, its size, and the distance from load to earth ground. For example, if the load is a large load box placed on ground, C_{L_G} can have several hundreds of pico-farad. If the load is small cement resistor placed on table. C_{L_G} will only have several tens of pico-farad.

b) Case impedance

Case impedance L_{CASE} mentioned here means the impedance between grounding points of input side Y capacitor and output side Y capacitor. This impedance will worsen the performance of noise backflow, and more CM current flows into earth ground, as shown in Fig.5. Usually, this parasitic inductance will have several nano henries to tens of nano henries. Though its value is very small, its influence is significant. Here, Equivalent Series Inductor (ESL) of output Y capacitor has the same role as L_{CASE} since they are in series with each other.

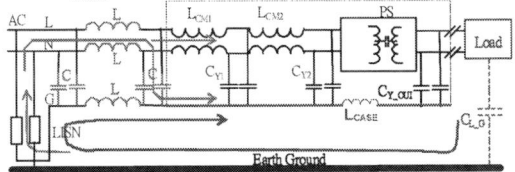

Fig.5 Significant parasitics induced by measurement setup

b) Self parasitics and mutual parasitics of input cable

As impedance induced by input cable will be even larger than $1/2 * Z_{LISN}$ at high frequencies, it also has detrimental effects on CM noise suppression. Usually, input cable used in EMI test is a standard three-pin cable including L, N and G lines [4], as shown in Fig.6. Because the highest frequency we are interested in is 30MHz, and the cable length is about 1.5 meter, less than 1/6 wavelength, one stage lumped-pi circuit is used to approximately model it. The parasitic parameters can be calculated as [5]:

$$L = 0.002 \cdot l \cdot (\ln(\frac{4l}{d} - 0.75))uH \qquad (1)$$

$$M = 0.002 \cdot l \cdot (\ln(\frac{2l}{W}) - 1 - \frac{W}{l})uH \qquad (2)$$

$$2C = \frac{0.0885 \cdot l \cdot \pi \cdot \varepsilon_r}{\ln(2\frac{W}{d})} pF \qquad (3)$$

Where 'l' is the length of the cable (cm), 'd' is the diameter of the line (cm), 'W' is distance apart between two lines (cm).

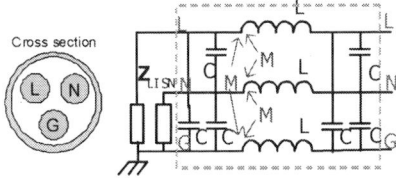

Fig.6 Structure and lumped-circuit model of input cable

Most of the CM current flowing in earth ground flows back through the ground line within input cable and can not be detected by LISN, as shown in Fig.5. However, parasitic inductance of ground line will block the CM current backflow, and a portion of CM current is shunted to LISN branch. On the other hand, the stray capacitor between the G line and L, N lines will make the noise coupled to LISN more easily.

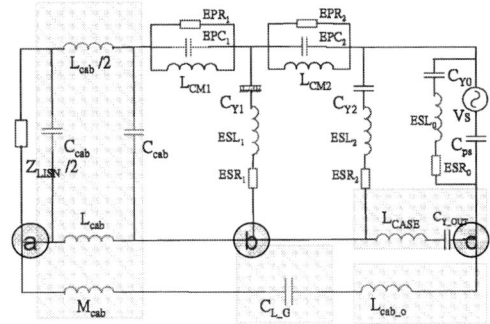

Fig.7 Model □: Detailed CM noise model when considering the

effects of load, case, cable

Considering all those external parasitic parameters together, a detailed CM noise model is shown in Fig.7. Denote it as Model □. Where,

L_{cab}, C_{cab}, M_{cab} is the parasitic parameter of input cable. They are decoupled from lumped-circuit based on circuit theory and the concept of CM noise.

$$L_{cab} = L - M, M_{cab} = M, C_{cab} = 2C \qquad (4)$$

L_{CASE} denotes the impedance between grounding points of input side Y capacitors and output side Y capacitors.

C_{L_G} is the stray capacitor from load to earth ground. L_{cab_o} is the parasitic inductance of output cable.

III. ANALYSIS OF THE EFFECTS OF EXTERNAL PARASITICS

To simplify the analysis, model □ is firstly simplified to model □-a, with □/Y transformation, as shown in Fig.8(a). Impedance Z_a, Z_b, Z_c is from the network a-b-c in Fig.7, which includes L_{CASE}, C_{Y_OUT}, L_{Cab_O}, C_{L_G}, M_{Cab} and L_{Cab}. Comparing model □-a to model □, the main differences are Z_a, Z_b, Z_c. Model □-a is further simplified to model □-b with two steps of □/Y transformation as shown in Fig.8(b). Impedance Z_A, Z_B, Z_C is determined by CM choke and Y capacitor.

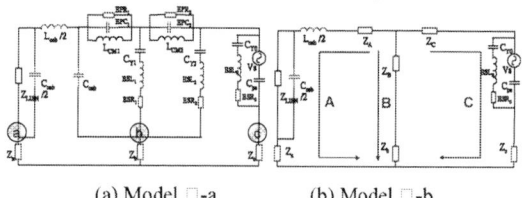

(a) Model □-a   (b) Model □-b

Fig.8 □/Y transformation of model □

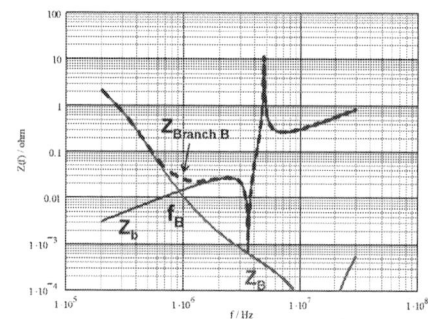

(a) Impedance of branch B

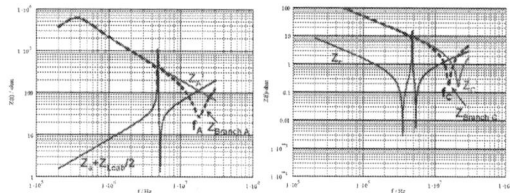

(b) Impedance of branch A   (c) Impedance of branch C Fig.9

Influence of Z_a, Z_b, Z_c in branch A, B, C

In Fig.8(b), model □-b contains three series branches. Effects of load, cable, case can be investigated by analyzing the effects of Z_a, Z_b, Z_c in branch A, B, C. Comparing to impedance Z_A, Z_B, Z_C, if impedance Z_a, Z_b, Z_c, is very small, their effects can be neglected. Otherwise, the effects of Z_a, Z_b, Z_c is significant.

Branch B is firstly analyzed as an example. B is a bypass branch that the higher branch impedance, the worse filtering performance. It is obvious that Z_b is in series with Z_B. Usually, Z_b behaves like an inductor,

while Z_B behaves like a capacitor, Fig.9(a) shows the impedance of Z_b and Z_B. At low frequencies, Z_B is much larger than Z_b, the impedance of branch B is determined by Z_B, effects of Z_b can be neglected. However, at high frequencies, after the inflexion frequency f_B, Z_b dominates the branch impedance since Z_b is larger than Z_B. That means at high frequencies, external parasitics must be taken into account. The effects of Z_a in branch A and Z_c in branch C are similar to Z_b. The only difference is that their influence range is higher (f_A&f_C), as shown in Fig.9(b) and Fig.9(c), where Z_A' is the impedance of branch A except $Z_a+Z_{Lcab}/2$ (external parasitic part), Z_C' is the impedance of branch C except Z_c (external parasitic part).

The simulated result can verify the analysis. Plot the result of small-signal frequency analysis based on Model □ and Model □ respectively, as show in Fig.(10). In high frequency ranges, the noise level calculated by model □ is totally higher than that by model □ since the consideration of load, cable, case.

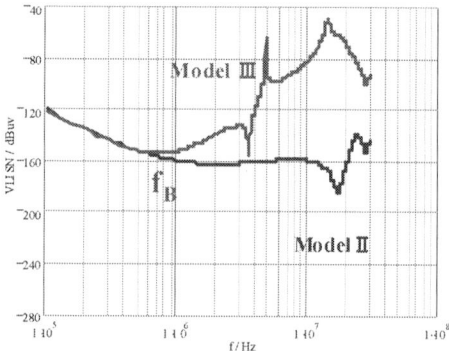

Fig.10 Effects of load, case, cable on HF range CM noise

Another evaluation method can also verify the analysis above. In model □, assume common mode choke 1, 2 and Y capacitor 1 have ideal behavior, as shown in Fig.11(a)(b). Fig.11(c) is the impedance of L_{CM1}, with or without considering EPC&EPR, two curves are of great difference in HF range. However, its noise level almost has no improvement, as shown in Fig.11(d). The results show that at high frequency range, external parasitics dominate filtering performance, but not the self parasitics in EMI filter components.

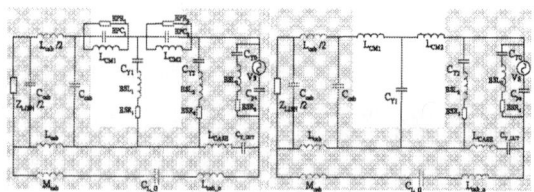

(a) Non-ideal choke & Y cap.   (b) Ideal choke & Y cap.

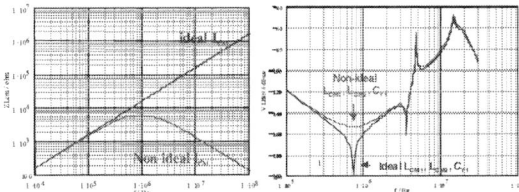

(c) Impedance of L_{CM1} (d) Effects of parasitics in Filter

Fig.11 Dominant factors influence HF CM noise transmitting

The analysis in this section can explain the following noise transmitting path. High frequency common mode noise mainly flows through external loops and has no relation with filter components L_{CM1}, C_{Y1} and L_{CM2}, as shown in Fig.12. To sum up, high frequency common mode noise is transmitted from noise source to LISN through three loops. The first is noise source loop, formed by noise source, C_{Y_OUT}, L_{CASE}, C_{Y2}. Most noise current flows within this loop. It generates a voltage drop V_{CASE} on case impedance. The second is noise transmitting loop (Ground loop). Current I_G, caused by V_{CASE}, will transmit the noise from source side to LISN side. The last one is noise receiving loop. The voltage drop on L_{cab} will be coupled to LISN trough stray capacitor between ground wire and L/N wire.

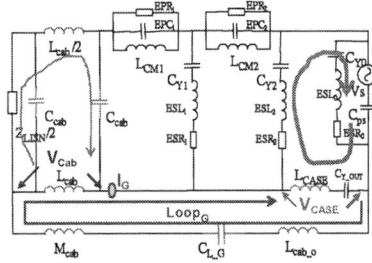

Fig.12 MF-HF CM noise transmitting path in model □

IV. SOLUTIONS ON HF CM NOISE SUPPRESSION

Though the parasitics induced by load, case, cable have great effects on high frequency range noise suppression, they can not be controlled since they are decide by system requirement. Solutions mentioned in this section are to place filter components in the best position to maximize its performance. Based on the noise transmitting path mentioned above, it's clear that lower V_{CASE} will produce smaller I_G. Thus, V_{Cab} will also be lower and less noise will be coupled to LISN. According to this principle, several noise suppression techniques are disclosed.

a) Filter topology

In order to reduce the voltage drop on case, we should ensure the noise current flowing through it as small as possible. One way is to parallel C_{Y2} with C_{Y0}, as shown in Fig.13. Thus the noise current flows through C_{Y2} will not pass L_{CASE}. The voltage drop on case

impedance V_{CASE} can be reduced, which benefits the HF CM noise reduction. Fig.14. shows the effect of this solution.

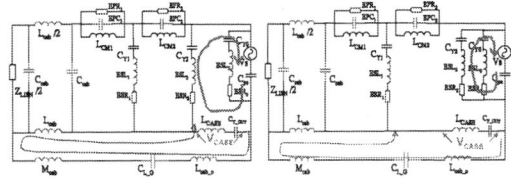

Fig.13 Parallel C_{Y2} with C_{Y0}

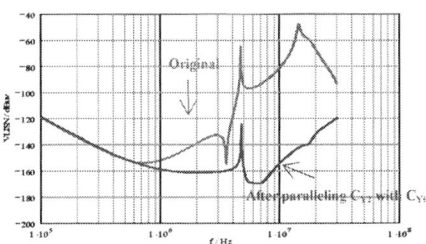

Fig.14 Effects of parallel C_{Y2} with C_{Y0}

Another way to reduce V_{CASE} is to add a small HF CM choke in noise source loop, as L_{CM2} shown in Fig.15(a). The effect is also very obvious, as Fig.15(b) shows. But if add it before C_{Y1} in series with L_{CM1}, then no effect.

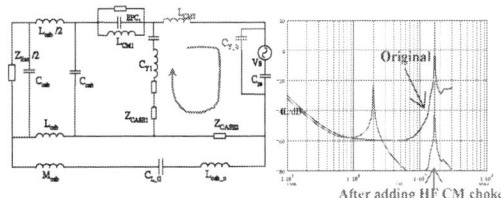

(a) Add HF CM choke in noise source loop   (b) Effect

Fig.15 Add HF CM choke in noise source loop

b) Bead core on ground wire

To get lower V_{Cab}, the noise current in noise transmitting loop I_G must be reduced. Adding choke or bead core on ground line is an effective way, as shown in Fig.16. The inductance of such choke or bead core can't be too large with the consideration on safety issue. Usually several micro henries is enough.

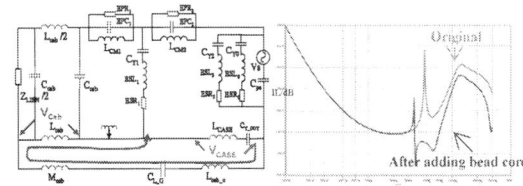

Fig.16 Add ground wire choke or bead core

V. EXPERIMENTAL VERIFICATION

978-1-4673-9551-9/16 $31.00 © 2016 IEEE

To verify the CM noise model and solutions mentioned above, several tests are done based on two different prototypes.

a) Prototype A -- Parallel C_{Y2} with C_{Y0}

Prototype A is a 2000w telecom power with two stages L-C CM filter. After paralleling C_{Y2} with C_{Y0}, the HF EMI is obviously improved, as shown in Fig.17.

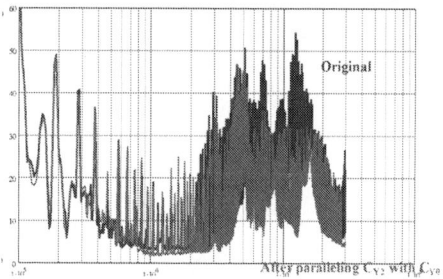

Fig.17 Parallel C_{Y2} with C_{Y0}

b) Prototype B -- Add HF CM choke and ground line bead core

Prototype B is an 180w sever power supply with one stage L-C CM fitter. Its CM noise model is shown in Fig.18(a). Firstly add a 3uH HF common mode choke L_{CM2} before C_{Y1}, in series with low frequency CM choke L_{CM1}. Fig.18(b) shows that it has no effect. Then add it in noise source loop. The result is in accord with expectation, as shown in Fig.18(c), it has good performance. Then add a bead core on ground line to reduce noise current I_G in noise transmitting loop, the HF peak is even lower, as shown in Fig.18(d).

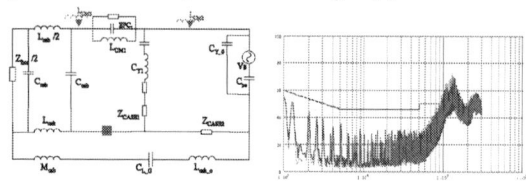

(a) Noise model of Prototype B (b) HF CM choke before C_{Y1}

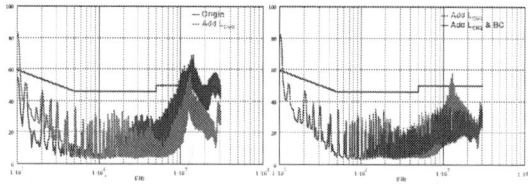

(c) HF CM choke after C_{Y1} (d) Ground wire bead core

Fig.18 HF CM choke & ground line bead core

VI. CONCLUSION

This paper discussed the significant effects of three aspects of external parasitics on high frequency range common mode noise suppression. They are stray capacitor from load to earth ground, self parasitics and mutual parasitics of input cable and case impedance. High frequency range common mode noise may be conducted to LISN via those parasitics, which makes improving the behavior of filter components useless. In this case, the following solutions will be effective:

1. Avoid Y capacitor facing to power stage directly in input EMI filter;

2. Add high frequency CM choke in noise source loop;

3. Add proper bead core on input ground line.

These solutions are verified with experimental results on CM noise reduction in high frequency range.

REFERENCE

[1] Sicong Lin, Min Zhou, Jianping Ying et al. Novel Methods to Reduce Common-mode Noise Based on Noise Balance [C]. 37th IEEE Power Electronics Specialists Conference. Jeju Korea June, 2006: 2728-2733.

[2] Thomas Farkas, Martin F. Schlecht, "Viability of Active EMI Filters for Utility Applications" IEEE Transactions on Power Electronics, Vol. 9, No. 3, pp. 328-337, May 1994.

[3] Shuo Wang, Fred C. Lee, D. Y. Chen and W. G. Odendaal, "Effects of Parasitic Parameters on EMI Filter Performance," Power Electronics, IEEE Transactions, Volume 19, Issue 3, pp. 869 – 877, May 2004.

[4] Li Ran, Sunil Gokani, Jon Clare, "Conducted Electromagnetic Emissions in Induction Motor Drive Systems Part II: Frequency Domain Models," IEEE TRANSACTIONS ON POWER ELECTRONICS, VOL. 13, NO. 4, JULY 1998.

[5] C.R. Paul, Introduction to Electromagnetic Compatibility. New York: Wiley, 1992

Improved Medium Voltage AC-DC Rectifier Based on 10kV SiC MOSFET for Solid State Transformer (SST) Application

Qianlai Zhu, Li Wang, Liqi Zhang, Wensong Yu, Alex Q. Huang
FREEDM Systems Center, North Carolina State University, Raleigh, US
qzhu4@ncsu.edu

Abstract— An improved bidirectional medium voltage AC-DC converter based on 10kV silicon carbide (SiC) MOSFETs for SST (Solid State Transformer) application is presented in this paper. Avalanche breakdown of the reverse blocking silicon diode and bridge arm shoot-through problems in traditional high voltage bridge-type AC-DC converters are solved. Shoot-through currents are limited to low di/dt events that are readily controlled, allowing zero dead-time operation. The reverse recovery dissipation of the SiC MOSFET is eliminated because no freewheeling current will flow through the body diode. This increases the efficiency as well as the reliability of the SiC MOSFET. Detailed power stage operating principles and energy transfer mechanism are described. A unique customized 10kV SiC MOSFET/JBS diode power module is developed and tested, which further reduces parasitic parameters and simplifies converter wire connection. This topology is therefore a very good choice for median voltage applications.

Keywords—SST; Avalanche; SiC MOSFET; shoot-through; reliability; medium voltage

I. INTRODUCTION

Solid-state transformer (SST) is a revolutionary smart grid device based on high frequency power electronic technology and is designed to replace traditional 60 Hz copper and iron based transformers and to provide very much needed smart grid functionalities in the areas of power management and fault management. Comparing to traditional transformer, the SST also has smaller volume, lighter weight while having the features of power factor correction, fault isolation, harmonic isolation and instantaneous voltage regulation [1-3]. It is being considered as the next key interface device in future smart grids.

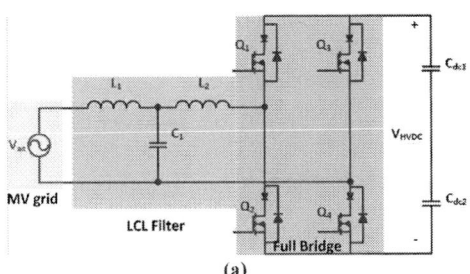

(a)

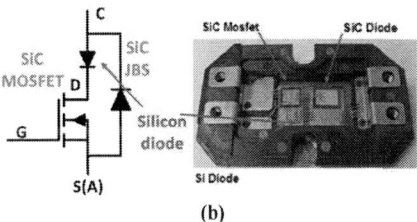

(b)

Fig. 1. GEN II SST (a) full bridge AC-DC Topology (b) device structure and photo

For connecting SST to medium voltage distribution grid, achieving high reliability is a critical factor. In traditional medium voltage bridge-type AC-DC converters such as the one used in NC state university FREEDM GEN II SST [4] shown in Fig. 1, at least two problems reduce its reliability:

The first one is the potential shoot-through problem, which might lead to fatal destroy in medium voltage applications. Longer dead time is needed to alleviate this problem, causing larger distortion and harmonics in input current and more losses as current will flow through the SiC MOSFET's body diode during the dead time.

The second problem is the avalanche of the integrated silicon diode. As shown in Fig. 1(b), a silicon Schottky diode is integrated in series with the SiC MOSFET to prevent the conducting of body diode of the SiC MOSFET and a SiC JBS diode is paralleled to conduct the reverse direction current [4-6]. The silicon diode is needed since the current through the body diode will cause degradation of the MOSFET[11-12] and this problem is currently being addressed by many device manufactures. However, this silicon diode will experience avalanche break down every switching duty cycle. As shown in Fig. 2:

$t_0 \sim t_1$: Q_2 is on, Q_1 is off, The C_{oss} of the MOSFET is charged to bus voltage and keep constant , silicon diode forward voltage V_{CD} is around zero.

$t_1 \sim t_2$: At time t_1, Q_2 turns off, Q_1 is still off because of the dead time, current start to discharge JBS diode's parasitic junction capacitor C_{JBS}. V_{C_JBS} decreases while V_{DS_Q1} remains high.

$$V_{CD} = V_{C\_JBS} - V_{coss} \qquad (1)$$

V_{CD} goes negative and reaches the silicon diode's avalanche voltage at t_2. At time t_2, the silicon diode goes into avalanche mode and starts to conduct current inversely and C_{oss} starts to discharge. The silicon diode stays in avalanche mode until V_{coss} is discharged to a level that can no longer force the silicon diode into avalanche mode.

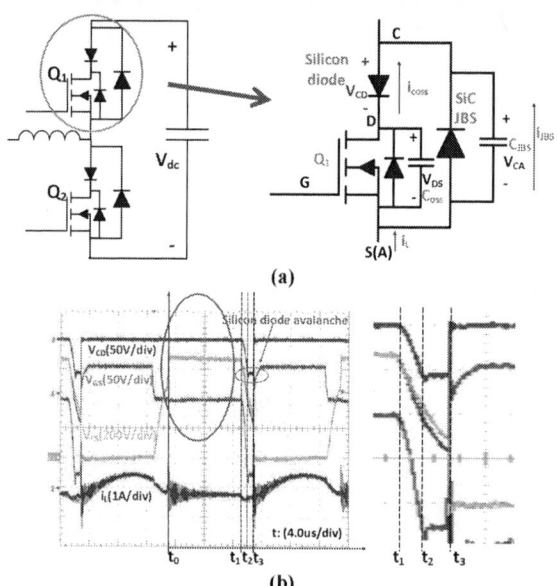

(a)

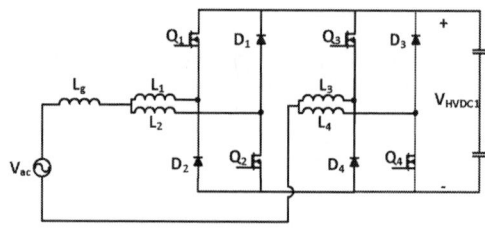

(b)

Fig. 2. GEN II SST (a) Coss discharging path before turn on (b) avalanche test waveform

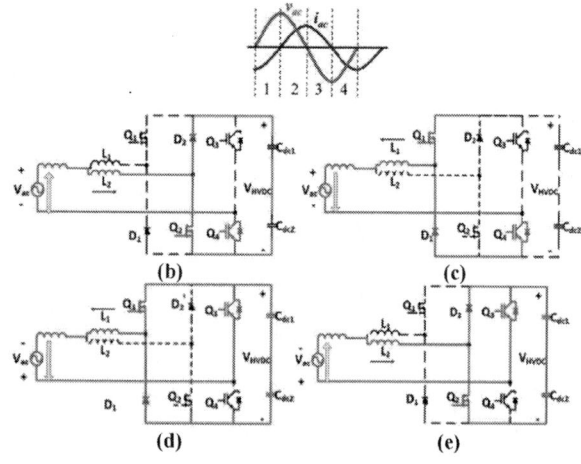

Fig. 3. Proposed AC-DC topology for next generation SST

Improved topology is proposed to solves the above two major issues. The concept of improved topology shown in Fig. 3 comes from dual buck topology, which naturally solves the shoot-through and avalanche of silicon diode problems. Shoot-through currents are limited by two seriously connected inductors L_1 and L_2 to a safe and low di/dt rate that is possible to control, allowing basically zero dead-time operation. No free-wheeling current will go through the body diode of the SiC MOSFET since the reverse current is blocked by D_1 or D_2 so no integrated silicon diode is needed anymore. It is there for expected that both efficiency and reliability will improve in the proposed topology [7-10].

II. CONVERTER OPERATION PRINCIPLE

A. Bidirectional AC-DC Topology Operation Principle

This bidirectional AC-DC topology has the ability to transfer both active and reactive power at the same time. The detail operating principle are shown in Fig. 4 and Table I gives the switch operating schemes. Base on the input voltage and input current direction, the operating can be divided into four operating region. We define the i_{ac} current positive when the current goes into the ac source.

Fig. 4. Active and reactive power transferring principle. (a) Conceptual voltage and current waveform. (b) Q2, D2 and Q4 work together. (c) Q1, D1 and Q4 work together. (d) Q1, D1 and Q3 work together. (e) D2, Q2 and Q3 work together.

1). $V_{ac} > 0\ \&\ I_{ac} < 0$; within this region, the current flows from ac to dc, converter works as rectifier, as shown in Fig. 4 (b), Q_4 is always on, Q_2 and D_2 works as PWM as boost pair.

2). $V_{ac} > 0\ \&\ I_{ac} > 0$; under this condition, the converter works like inverter, Q_4 keeps on, Q_1 and D_1 work as PWM buck pair to transfer power from dc to ac.

3). $V_{ac} < 0\ \&\ I_{ac} > 0$; the converter works as a rectifier under this mode, Q_3 is always on, Q_1 and D_1 pair work as a boost pair to transfer power from the ac to dc.

4). $V_{AC} < 0\ \&\ I_{AC} < 0$; during this time, the convert works as inverter as the current goes into the ac side. Q_3 is always on, while Q_2 and D_2 work as a PWM buck pair.

Table I. Different switch combinations.

Status		Q1&D1	Q2&D2	Q3	Q4
Rectifier	$V_{ac} > 0\ \&\ I_{ac} < 0$	OFF	PWM	OFF	ON
Inverter	$V_{ac} > 0\ \&\ I_{ac} > 0$	PWM	OFF	OFF	ON
Rectifier	$V_{ac} < 0\ \&\ I_{ac} > 0$	PWM	OFF	ON	OFF
Inverter	$V_{ac} < 0\ \&\ I_{ac} < 0$	OFF	PWM	ON	OFF

B. Inductor Design Consideration 1

Two additional inductors L_1 and L_2 are added to limit the shoot-through di/dt rate. In order to achieve small size while obtaining acceptable di/dt rate when Q_1 and Q_2 are both turned on, the value of the inductors should be carefully designed. The current raising rate is decided by voltage and inductance,

$$\frac{di}{dt} = \frac{V_{dc}}{L_1 + L_2} = \frac{V_{dc}}{2L_1} \tag{2}$$

If a short circuit happens when V_{dc} is 6kV, the over current protection execution delay Δt is 5us, and it is needed to control the peak current within 60A so that the device is protected, then,

$$L_1 = \frac{\Delta t \times V_{dc}}{2\Delta i} \tag{3}$$

Each inductance should be larger than 0.25mH. Consider a margin, the inductances are chosen to be 0.3mH each.

C. Inductor Design Consideration II

The design of inductance L_1 and L_2 will also affect the potential conduction of the body diode of MOSFET. As shown in Fig. 5, during positive half cycle of input voltage, when JBS D_1 conducts the current, the voltage on the MOSFET Q_1 should be

$$V_{DQ1} = V_{D2} + V_{L2} - V_{L1} \tag{4}$$

Since no current goes through L2 during this time interval,

$$V_{L1} = 0 \tag{5}$$

While L_2 and L_s share the voltage difference between V_{ac} and V_{dc}

$$V_{ac} - V_{dc} = V_{L1} + V_{Ls} \tag{6}$$

The voltage on the inductors are decided by their inductance,

$$\frac{V_{L1}}{V_{Ls}} = \frac{L_1}{L_s} = k \tag{7}$$

$$V_{L1} = \frac{V_{ac} - V_{dc}}{1 + k} \tag{8}$$

As ac voltage is always lower than dc voltage, V_{L1} will always lower than zero, which will help to keep the MOSFET body diode staying off.

Worst condition happens at the peak of ac voltage, while the absolute value of V_{L1} is at the lowest point. Assumes V_{L1} lower than -3V is needed to keep the body diode off under 6kV condition, with a Ls value of 80mH, the inductance ratio is

$$k > \frac{V_{L1}}{V_{dc} - V_{ac\_peak} - V_{L1}} = \frac{1}{299} \tag{9}$$

In design rule 1, the inductor L_1 is designed to be 0.3mH, the k value now is 3/800, which satisfy the k range requirement.

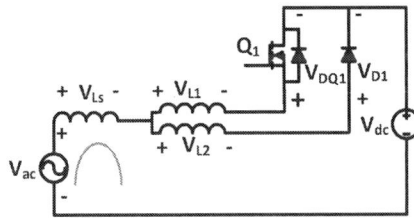

Fig. 5. Equivalent Circuit

Fig. 6 shows an example of inductors L_s and $L_1 \& L_2$ The volume of L_s is 40 times the volume of L_1, which means, although two additional inductors are needed in the proposed topology, only 5% extra magnetic volume is added to the whole system.

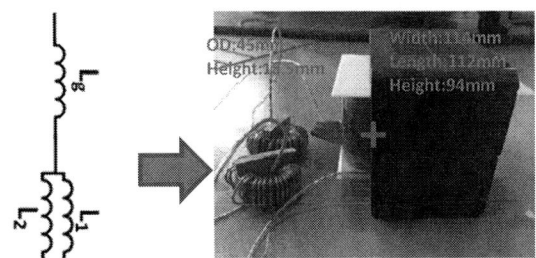

Fig. 6. Designed inductors

III. CUSTOMIZED 10KV SIC MODULE

A unique customized 10kV SiC module with high voltage isolation and without shoot-through of the high voltage DC link is designed, fabricated and tested for this application.

It has 80% volume reduction compared to previous packaged SiC module used in GEN-II SST [4] by POWEREX and CREE.

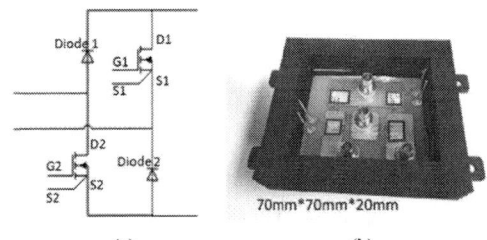

(a)   (b)

Fig. 7. Unique high voltage power module: (a) topology (b) 4 in 1 packaged module

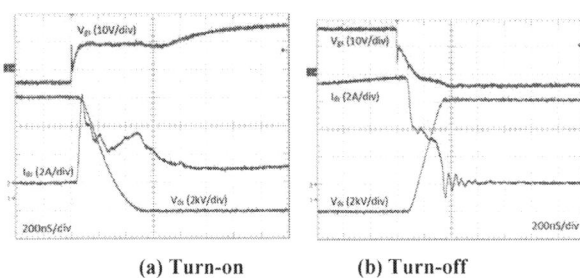

(a) Turn-on   (b) Turn-off

Fig. 8. Testing results of the MOSFET under 8kV: (a) turn-on and (b) turn-off waveforms

The module is fabricated using AlN Direct Bonded Copper as the substrate of the module, on which 10kV SiC bare dies are directly soldered for heat transfer and high voltage isolation. Kelvin connection of the MOSFETs is deployed to ensure the gate driver capability of the noise immunity. The module is encapsulated with SEMICOSIL® 915HT to provide high voltage isolation and enhance mechanical strength of the wire bonds. 10kV/10A SiC MOSFET and JBS diode [4] developed by Cree are used to construct this module.

The dynamic turn-on and turn-off waveforms for 10kV 10A SiC MOSFET under 8kV using double pulse test are displayed in Fig. 8.

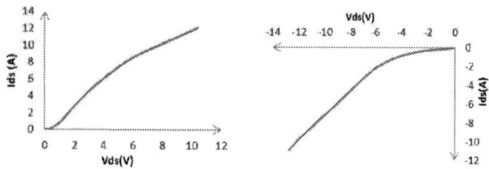

(a) MOSFET Forward characteristics   (b) IV curve of the body diode

Fig. 9. Testing results of the static characteristics: (a) Forward characteristics and (b) IV curve of body diode

Fig. 9 shows the IV curve of the body diode of the 10kV SiC MOSFET, which can be used to design the inductor.

IV. EXPERIMENTAL RESULTS

A 3.6kVac input 10kW medium voltage AC-DC prototype based on the proposed topology and novel power module are constructed to verify the performance. The specifications of prototype are listed in Table II. .

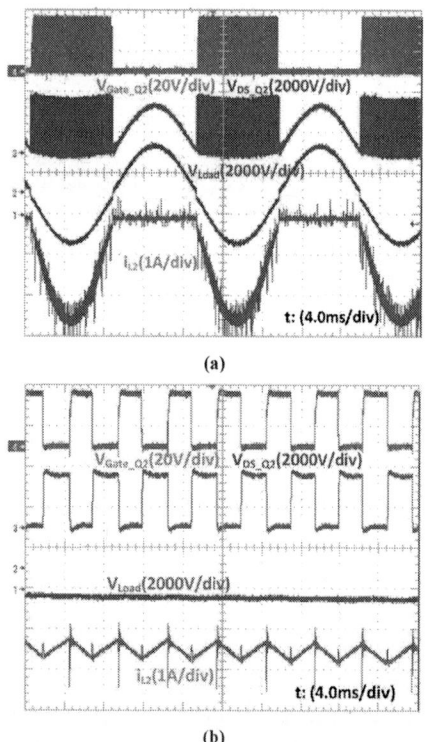

(a)

(b)

Fig. 10. Testing waveforms of inverter mode under 3kVdc, 2500W

TABLE II
AC-DC CONVERTER SPECIFICATION(TESTED VALUE)

Components	Parameters(Tested)
V_{in} (Input voltage)	3600V(1800V)
V_{out} (Output voltage)	6000V(3000V)
P_{out} (Output current)	10Kw(2.5kW)
f_s (Switching frequency)	6kHz
L_s (input inductance)	80mH
L_1&L_2	300uH
C_{in} (input capacitor)	33nF
C_o (Output capacitance)	31uF
Digital Controller IC	TMS320F28377D
Current Sensor	HASS 50-S
Local driver IC	IXDN609SI

Fig. 10 (b) and (c) shows the voltage and current waveforms tested in the inverter mode at V_{dc}= 3kV. The two buck circuits work alternately based on the input ac voltage. The controlled AC current follows the reference very well.

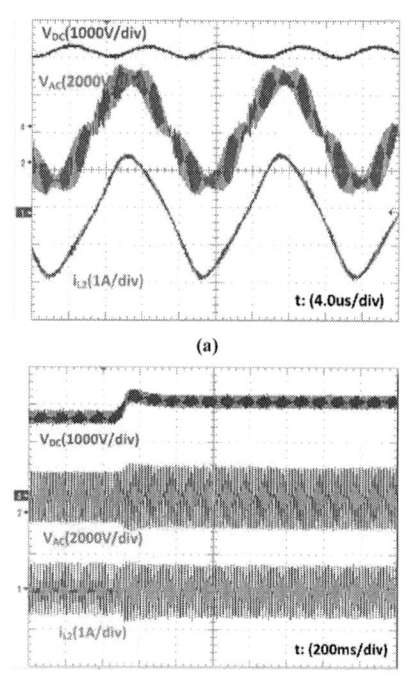

(a)

(b)

Fig. 11. Testing waveforms of the rectifier mode under 3kVdc, 2500W

The rectifier mode waveforms under 3kVdc and 2.5kW load are displayed in Fig. 11. Fig. 11(a) shows the detail voltage and current waveforms. The green waveform is the input ac voltage, it has high voltage ripple is because that this voltage is sensed on the capacitor of the input LCL filter. It is sensed to verify the PLL can work under different grid impedance condition

978-1-4673-9551-9/16 $31.00 © 2016 IEEE   2368

For rectifier mode operation, in order to achieve enough gain for the current loop under heavy load and light load condition, DQ reference frame transformation and feedforward are used in the current loop.

Fig. 11(b) shows the soft start of the DC voltage. We can find there is little overshoot of the DC bus voltage.

V. CONCLUSIONS

In this paper, an improved bidirectional AC-DC topology based on 10kV SiC MOSFET for SST applications is proposed. It solves shoot-through and integrated silicon diode avalanche problems, hence enhancing efficiency and reliability. No freewheeling current will go through the body diode of the SiC MOSFET therefore removing one of the bottleneck in using SiC MOSFET in bridge-type converters. A unique customized four-in-one customized 10kV SiC module is packaged and tested under 8kVdc with double pulse test. A 3.6kVac input 10kW prototype was built and. Experimental results verified the operation principle and the customized power module.

REFERENCE

[1] T. Zhao, G. Wang, S. Bhattacharya, and A. Q. Huang, "Voltage and power balance control for a cascaded h-bridge converter-based solid-state transformer," IEEE Trans. Power Electron., vol. 28, no. 4, pp. 1523–1532, Apr. 2013.

[2] T. Zhao, J. Zeng, S. Bhattacharya, M. E. Baran, and A. Q. Huang, "An averagemodel of solid state transformer for dynamic system simulation," in Proc. IEEE Power Energy Soc. Gen. Meeting, Jul. 2009, pp. 1–8.

[3] X. She, A. Q. Huang, and R. Burgos, "Review of solid-state transformer technologies and their application in power distribution systems," IEEE J. Emerg. Sel. Topics Power Electron., vol. 1, no. 3, pp. 186–198, Sep. 2013

[4] Fei Wang, Gangyao Wang, Alex Huang, Wensong Yu, Xijun Ni, "Design and operation of A 3.6kV high performance solid state transformer based on 13kV SiC MOSFET and JBS diode," in IEEE 2014 Energy Conversion Congress and Exposition (ECCE), 2014, pp. 4553 –4560.

[5] J. Richmond, S. Leslie, B. Hull, M. Das, A. Agarwal, and J. Palmour, "Roadmap for megawatt class power switch modules utilizing large area silicon carbide MOSFETs and JBS diodes," in Proc. IEEE ECCE, Sep. 20–24, 2009, pp. 106–111.

[6] H. Mirzaee, A. De, A. Tripathi, S. Bhattacharya, "Design comparison of high-power medium-volage converters based on a 6.5-kV Si-IGBT/SiPiN diode, a 6.5-kV Si-IGBT/SiC-JBS diode, and a 10-kV SiCMOSFET/SiC-JBS diode," IEEE Trans. Ind. Appl., vol. 50, no. 4, July/Aug. 2014.

[7] Z. Yao, L. Xiao, and Y. Yan, "Dual-buck full-bridge inverter with hysteresis current control," IEEE Trans. Ind. Electron., vol. 56, no. 8, pp. 3153– 3160, Aug. 2009

[8] H. Qian, J.-S. Lai, and W. Yu, "Novel bidirectional ac–dc MOSFET converter for energy storage system applications," in Proc. IEEE ECCE, Sep. 2011, pp. 3466–3471.

[9] P. W. Sun, C. Liu, J.-S. Lai, and C.-L. Chen, "Cascade dual buck inverter with phase-shift control," IEEE Trans. Power Electron., vol. 27, no. 4, pp. 2067–2077, Apr. 2012.

[10] Stanley G R, Bradshaw K M. "Precision DC-to-AC power conversion by optimization of the output current waveform-The half-bridge revisited ," In: Proceeding of the 28th IEEE power Electronics Specialists conference, St. Louis, Missouri, USA, 1997, 35(2): 993- 999.

[11] A. Agarwal, H. Fatima, S. Haney, and S.-H. Ryu, "A new degradation mechanism in high-voltage SiC power MOSFETs," IEEE Electron Device Lett., vol. 28, no. 7, pp. 587–589, Jul. 2007

[12] Funaki, T., "A study on performance degradation of SiC MOSFET for burn-in test ofbody diode

[13] ," Power Electronics for Distributed Generation Systems (PEDG), 2013 4th IEEE International Symposium on, pp. 1 - 5, July 2013

Suppression of Circulating Current in Parallel Operation of Three-Level Converters

Young-Kwang Son, Seung-Jun Chee, Younggi Lee,
Seung-Ki Sul
Department of Electrical and Computer Engineering
Seoul National University
Seoul, Republic of Korea

Changjin Lim, Sungjae Huh, Jaeyoon Oh
R&D Institute
LG Electronics Co.
Seoul, Republic of Korea

Abstract— The Zero Sequence Circulating Current (ZSCC) flows inevitably in parallel converters sharing common DC and AC sources. The currents flowing commonly in all converters increase losses and decrease the overall capacity of parallel converters. This paper proposes a simple and effective ZSCC suppression method based on the Space Vector PWM (SVPWM) method with the ZSCC controller. The zero sequence voltage for the proposed SVPWM is calculated based not on the phase voltage references, but on the grid voltage. The limit of linear modulation index of the converters with the proposed method is analyzed compared to other methods, and it is proven that the limit of linear modulation index can be maximized with the proposed method. The effectiveness of the proposed method has been verified through the experimental set-up consisted of the four parallel three-level converters. It has been confirmed that the ZSCC is well suppressed and larger linear modulation index is achieved at the same time with the proposed method. Moreover, the proposed control method does not require any communication between converters to suppress the ZSCC unlike the conventional methods.

Keywords—Zero Sequence Circulating Current; Parallel Converters; Circulating Current Suppression; Limit of Linear Modulation Index; Voltage Utilization Rate;

I. INTRODUCTION

With the rapid growth of renewable energy markets over recent years, the grid-connected converters have become indispensable to interface the distributed generators with the grid. Though two-level converters are still dominant as a topology for the grid-connected converters, the three-level converters are getting popular by virtue of its higher efficiency and smaller filter size compared to those of two-level converters.

Meanwhile, if converters are operated in parallel, which is a form of modular structure, the efficiency at a light load would be enhanced by turning on-off each converter according to the load condition. Furthermore, the production and maintenance cost of overall converters would be also reduced [1]-[4]. To save the cost of the system in parallel converters, the converters usually share same DC and AC sources as shown in Fig. 1. However, in this case, a considerable amount of Zero Sequence Circulating Current (ZSCC) inevitably flows between the converters [5]-[9]. The ZSCC does not just lower the system's efficiency and overall capacity of the parallel converters but

also distorts the current, which causes EMI and harmonic problems. Therefore, it is desirable to minimize the ZSCC as much as possible.

Many researches have been carried out on the ZSCC control in parallel converters [5]-[12]. The PI type ZSCC controller was utilized to eliminate the ZSCC in most researches, and some of them presented the feed-forward control method to enhance the performance of the controller [10]-[12].

In this paper, an improved ZSCC suppression control method for parallel three-level converters is proposed. With the proposed control method, the ZSCC is almost perfectly suppressed even at transients or in the situation where the parallel converters transfer different amounts of active or reactive power from each other. This characteristic cannot be achieved with the ZSCC suppression control method based on the conventional SVPWM. Moreover, since the proposed control scheme does not require any communication between converters to control the ZSCC, the proposed controller is more scalable and reliable compared to the conventional controllers proposed in [10]-[12] which set zero sequence voltage through communications between converters. In addition, the limit of the linear modulation index of parallel converters according to each PWM method is analyzed. It is also verified that the limit of linear modulation index with the proposed method is larger than those of the conventional ZSCC suppression control methods.

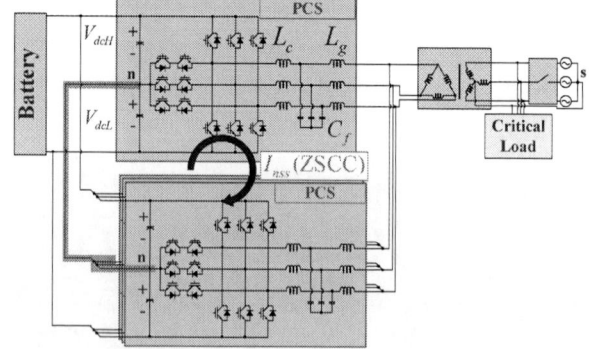

Figure 1. Four parallel three-level converters sharing common DC and AC sources.

II. ZSCC IN THREE-LEVEL PARALLEL CONVERTERS

A. Zero Sequence Voltage and DC-link Voltage Balancing in Three-Level Converters

The ZSCC flows between parallel converters owing to the difference between the corresponding pole voltages of each converter. The difference naturally comes from a slight difference of the switching characteristics of each power semiconductor switches of converters even if the same gating signals have been applied to each converter. Due to the difference of the output filter impedances of the converters, the corresponding pole voltages of each converter would be different not only in transient state, but also in steady state to share equal portion of total load current. Because of these differences, the ZSCC would flow and the efficiency and total capacity of the converters would be degraded.

To suppress the ZSCC, several kinds of ZSCC controller have been used and the output of the ZSCC controller was added to the output voltages of the current regulator of each converter. The output voltages of the current controller of each converter would be different to regulate its own current, resulting in different phase voltage references for each converter. In this case, the Zero Sequence Voltage (ZSV) of each converter, defined as instantaneous average of the pole voltages of each converter, would be different under Space Vector PWM (SVPWM) and the ZSCC flows because of the difference in the ZSVs.

In converters with three-level topology, the ZSV of each converter also influences the balancing control of its DC-link voltage [13]. To set the ZSV for balancing the DC-link voltage equally while considering the suppression of the ZSCC, the upper capacitor voltage (V_{dcH}) and the lower capacitor voltage (V_{dcL}) of all the converters should be same respectively. For that reason, it is recommended that the DC-link neutral points of parallel converters be tied together as shown in Fig. 1 [14].

The ZSV for the DC-link balancing control can be set as (1).

$$V_{sn\_balancing} = sign(I_{qs}^{e*})\frac{V_{dcH} - V_{dcL}}{2}. \qquad (1)$$

B. Relationship of the Output Power and ZSV

If the active power and reactive power references are set, the d, q-axes current references of each converter can be calculated. Then, d, q-axes voltage references of the k-th converter among n-parallel converters are determined as (2) in steady state.

$$V_{dqs\_k}^{e*} = j\omega_e L i_{dqs\_k}^{e*} + e_{dqs}^e, \ where \ k = 1, 2, \cdots, n \qquad (2)$$

, where e_{dqs}^e and ω_e stand for the grid voltage and angular frequency, respectively, in d-q reference frame, and L for the summation of L_c and L_g in Fig. 1.

The active power and reactive power of parallel converters would be different not only in transient state, but also in steady state; because the filter inductances of the converters are different from each other due to tolerance of L, C components. These differences between the converters cause the difference in current references, which lead to the mismatch of the voltage references (V_{dqs_k}*) as shown in Fig. 2. If each converter uses the conventional SVPWM method, the ZSV of the k-th converter would be calculated as (3) based on its own voltage references (V_{dqs_k}*) [15]-[18].

$$V_{sn\_ff\_k} = -\frac{\max(V_{as\_k}^*, V_{bs\_k}^*, V_{cs\_k}^*) + \min(V_{as\_k}^*, V_{bs\_k}^*, V_{cs\_k}^*)}{2}. \qquad (3)$$

Finally, the different power references cause the different ZSVs, which are a main contributing factor of the ZSCC. To avoid different ZSVs, Sine PWM (SPWM), where no ZSV is used at the cost of 15% reduction of linear modulation index range, could be a candidate in the view-point of the ZSCC. However, 15% reduction of the linear modulation index range would not be tolerable in most cases of the applications.

C. Conventional Methods for Suppressing ZSCC

In general, the ZSCC has been suppressed by the zero-sequence current controller; to overcome the limited bandwidth of the zero-sequence controller, several methods using the feed-forward voltage, $V_{sn\_ff\_k}$, have been proposed as shown in Fig. 3 [10]-[12].

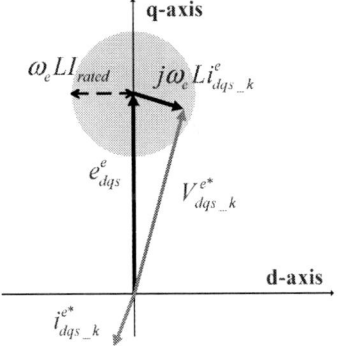

Figure 2. d, q-axes vector diagram of k-th parallel converter.

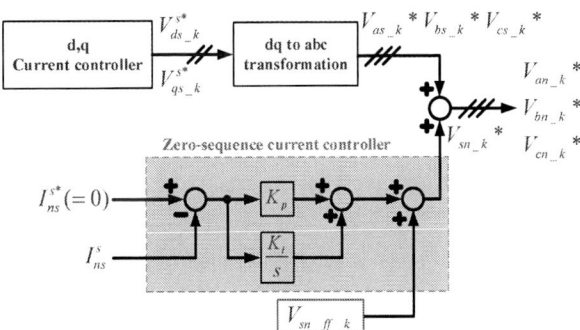

Figure 3. Zero sequence current controller of k-th parallel converter.

The ZSCC can be dramatically suppressed if all parallel converters use the identical V_{sn_ff}. V_{sn_ff} of all the converters can be set as the SVPWM ZSV of the master converter as (4) [10]. This method is termed as Method 1 (M1) in this paper. The other option for the identical V_{sn_ff} is the average value of the SVPWM ZSVs of the converters as (5), and this method is termed as Method 2 (M2) [11].

$$V_{sn\_ff\_M1} = -\frac{\max(V^*_{as\_master}, V^*_{bs\_master}, V^*_{cs\_master}) + \min(V^*_{as\_master}, V^*_{bs\_master}, V^*_{cs\_master})}{2}. \quad (4)$$

$$V_{sn\_ff\_M2} = -\frac{\sum_{k=1}^{n} \{\max(V^*_{as\_k}, V^*_{bs\_k}, V^*_{cs\_k}) + \min(V^*_{as\_k}, V^*_{bs\_k}, V^*_{cs\_k})\}/2}{n}. \quad (5)$$

III. PROPOSED ZSCC SUPRESSION METHOD

As mentioned in the previous section, the ZSV of each converter would be different from each other if the conventional SVPWM method is applied to all converters as in (3) [15]-[18]. Then, the ZSCC flows inevitably as described before. In this paper, the grid voltage, which are common to all parallel converters and the major portion of output voltages of current controller of each converter as seen from (2), are used for the calculation of the ZSV for SVPWM of each converter as (6). This method is termed as Grid based SVPWM (GSVPWM). Because all parallel converters are directly connected to the grid and measure the grid voltage for its own Phase Lock Loop (PLL), all converters would have the same ZSV with this arrangement.

$$V_{sn\_ff} = -\frac{\max(e^*_{as\_grid}, e^*_{bs\_grid}, e^*_{cs\_grid}) + \min(e^*_{as\_grid}, e^*_{bs\_grid}, e^*_{cs\_grid})}{2}. \quad (6)$$

To suppress the ZSCC, several methods such as M1, M2, and the method compensating the difference of zero vectors between converters have been introduced in [10]-[12]. But those are less scalable and reliable than GSVPWM due to the fact that the communication between converters at every sampling period is necessary for them to determine the feed-forward term of the zero sequence current controller (V_{sn_ff}).

IV. ANALYSIS OF VOLTAGE UTILIZATION RATE

In the case of SVPWM, the largest Limit of Linear Modulation Index (LLMI) of a single converter is considered as $\frac{2}{\sqrt{3}}$ [19], under the definition of the modulation index (MI) as (7).

$$MI \equiv \frac{|V_{as}|_{peak}}{V_{dc}/2}. \quad (7)$$

The ZSCC would be shrunken if all parallel converters use the identical V_{sn_ff} like the cases of M1, M2, and GSVPWM. To have the capability of the ZSCC suppression, the LLMIs of those methods of converters are inevitably degraded compared to that of the independently operating converter based on SVPWM. However, the maximum LLMI of each method would be different. The ZSVs of M1, M2, and GSVPWM are

not based on their own phase voltage references so there exists angular and magnitude mismatch of ZSV between them and the conventional SVPWM.

Fig. 4 shows the LLMI according to the mismatch of the angle and magnitude from its conventional SVPWM ZSV. The angular mismatch of the SVPWM ZSV degrades the LLMI more than the mismatch on magnitude does. If the angular mismatch is larger than 15.5°, SVPWM where the ZSV is not based on each converter's own output voltages is no more superior to SPWM in the view point of the LLMI. The LLMI is the lowest at 60° angular mismatch as expected

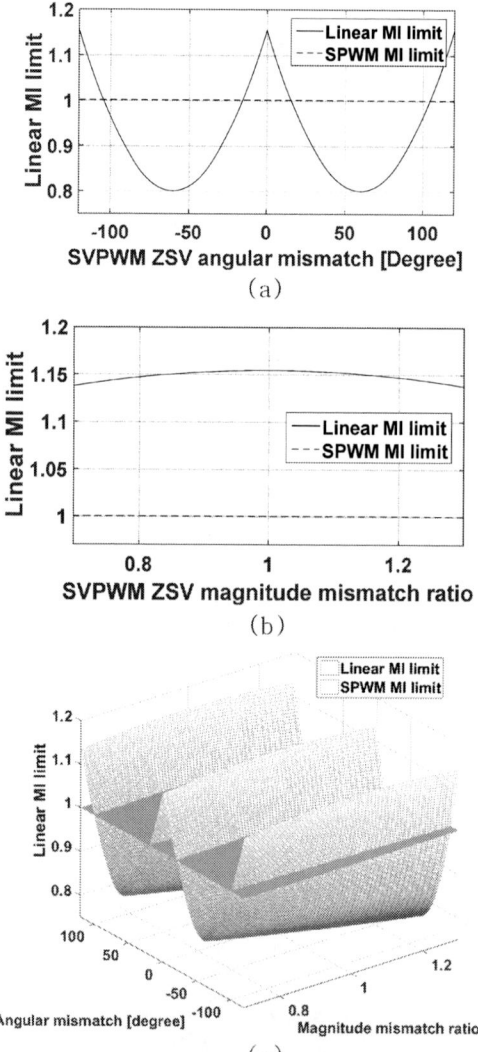

(a)

(b)

(c)

Figure 4. Linear modulation index limit (LLMI) (a) LLMI under angular mismatch without mismatch on magnitude, (b) LLMI under mismatch on magnitude without angular mismatch, and (c) three dimensional view of LLMI under angular mismatch and mismatch on magnitude.

Table I. Experimental Parameters.

Item	Value	Unit
S_{rated}	5	kVA
f_{sw}	5	kHz
$L_c + L_g$	0.075	$p.u.$
V_{ll_grid}	220	V_{rms}
I_{rated}	18.6	A_{peak}

Table II. LLMI according to V_{sn_ff}.

V_{sn_ff}	Maximum angular mismatch of ZSV	LLMI	Communication at each PWM sampling period
SPWM	0	1	No
M1	$2\tan^{-1}(\frac{\omega_e LI_{rated}}{V_{rated}})$	1.06	Yes
M2	$2\frac{n-1}{n}\tan^{-1}(\frac{\omega_e LI_{rated}}{V_{rated}})$	1.08 (when $n = 4$)	Yes
GSVPWM	$\tan^{-1}(\frac{\omega_e LI_{rated}}{V_{rated}})$	1.11	No

because the phase voltage reference and the ZSV have the maximum values at the same moment.

In the following analysis, the LLMIs of M1, M2 and GSVPWM are compared. Prior to the analysis, it should be noted that the LLMI at the worst case of each PWM method determines the DC-link voltage at the system design stage to guarantee the satisfactory regulation of the grid current, and it consequently affects the cost of the converter and the reliability of the overall system.

If M1 is applied, the angular mismatch of the ZSV reaches up to $2\tan^{-1}(\frac{\omega_e LI_{rated}}{V_{rated}})$ where power references of one converter and another converter are rated values and have the opposite signs in the worst case transient. If M2 is applied, the angular mismatch can reach up to $2\frac{n-1}{n}\tan^{-1}(\frac{\omega_e LI_{rated}}{V_{rated}})$ in the worst case. With GSVPWM, the maximum angular mismatch in the worst case is $\tan^{-1}(\frac{\omega_e LI_{rated}}{V_{rated}})$.

With the parameters given in Table I, the LLMI of each method was calculated and listed in Table II. The voltage utilization rate of GSVPWM is superior compared to that of others. It can be noted that $V_{sn\_balancing}$ in (1) is neglected in above comparison since the magnitude of $V_{sn\_balancing}$ is much smaller than the magnitude of the ZSV.

V. EXPERIMENTAL RESULTS

The experimental set-up is implemented as Fig. 5. Four three-level T-type converters are connected in parallel between a battery bank and the grid in common, and the neutral points of three-level converters are tied together. The parameters of the experimental set-up are listed in Table I.

Fig. 6 (a) and (b) shows the A-phase total current (I_{as_total}), the upper capacitor voltage (V_{dcH}), the lower capacitor voltage (V_{dcL}), and the ZSCCs of all four converters (I_{zscc_PCS1}, I_{zscc_PCS2}, I_{zscc_PCS3}, I_{zscc_PCS4}) simultaneously. These four converters are connected to the grid in sequence with the rated power reference. When connecting to the grid, the power reference of each converter increases with limited slew rate, 0.2 p.u./s. In

5 kVA PCS × 4

Figure 5. Experimental Setup: 5kVA Four Parallel three-level T-type Converters.

the case of the conventional SVPWM, which sets $V_{sn\_ff\_k}$ as (3), the difference of the power references causes the considerable ZSCC at transient as shown in Fig. 6 (a). The magnitude of the ZSCC of the converter lastly connected to the grid is the largest as shown in Fig. 6 (a), and it reaches up to 54% of the rated current of the converter. It should be definitely suppressed for the practical application of parallel converters. As shown in Fig. 6 (b), the ZSCC is well suppressed with the GSVPWM method which sets V_{sn_ff} as (6). In both PWM methods, the DC-link voltage is well balanced.

To be more specific, the waveform at the moment where PCS1 outputs the rated power (5kW), and PCS2 outputs zero power (0kW), was depicted in Fig. 7. This situation occurs when PCS2 just started switching while PCS1 was running.

Fig. 7 (a), (b), (c), and (d) show the ZSVs of PCS1 and PCS2 ($V_{sn\_ff\_PCS1}$, $V_{sn\_ff\_PCS2}$), the difference between the ZSVs of PCS1 and PCS2 ($V_{sn\_ff\_PCS1} - V_{sn\_ff\_PCS2}$), and the ZSCCs of PCS1 and PCS2 (I_{zscc_PCS1}, I_{zscc_PCS2}) simultaneously. As analyzed in II-B. and shown in Fig. 7 (a), different active power between parallel converters cause different ZSVs between converters and it inevitably generates ZSCCs. The dominant ZSV component is 3rd harmonic, and the ZSCC on PCS1 flows only into PCS2 since ZSCC cannot flow into the grid. With GSVPWM, the ZSCC is almost perfectly suppressed even in the situation where each PCS outputs totally different amounts of power as shown in Fig. 7 (b). These experimental results confirmed the previous analysis on the ZSCC which claimed that different ZSVs between parallel

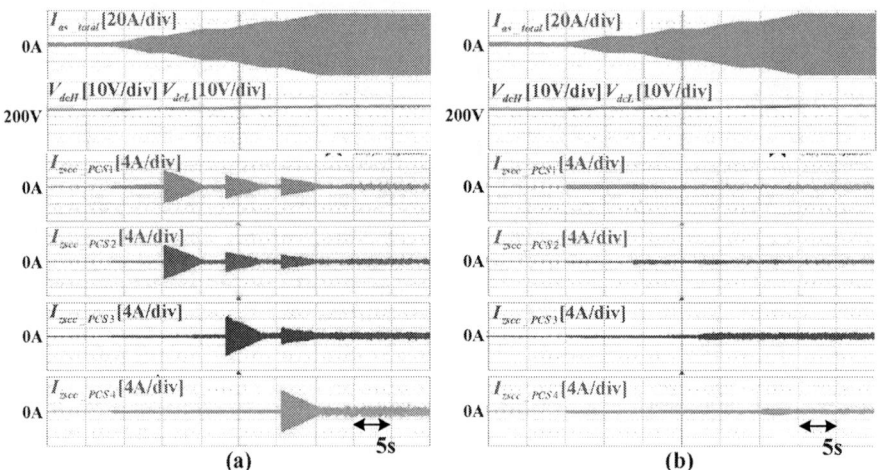

Figure 6. Experimental results – ZSCC (a) with the conventional SVPWM, and (b) with GSVPWM.

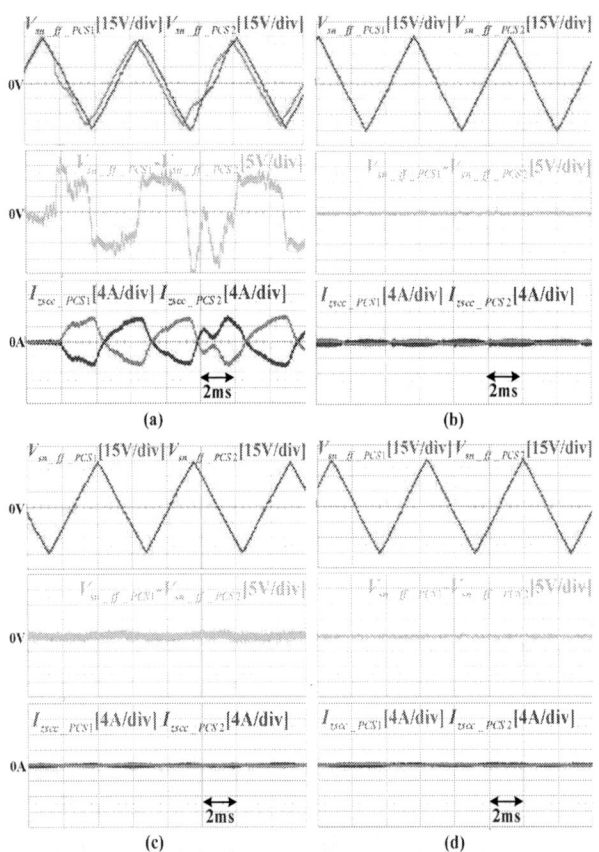

Figure 7. Experimental results – V_{sn_ff} and ZSCC (a) at transient with the conventional SVPWM, (b) at transient with the GSVPWM, (c) at steady state with the conventional SVPWM, and (d) at steady state with the GSVPWM.

converters are the source of ZSCC. As shown in Fig. 7 (c) and (d), the ZSCC at steady state is negligible in both the conventional SVPWM and GSVPWM.

It can be noted that M1 nor M2 could not be applied to the experimental parallel converter systems because both methods require communication between the converters at each PWM sampling period to transfer their common ZSV (V_{sn_ff}), but each experimental converter was made with the form of modular structure consists of a control board and a power board communicating each other through CAN (Controller Area Network). CAN, which is one of the most common communication system whose bandwidth is up to several Mbps, is insufficient to support the communication for M1 or M2. Since M1 and M2 with powerful communication capability, and GSVPWM method share the same ZSCC suppression principle, their ZSCC suppression performance would be the same. However, due to no need of powerful communication link between converters GSVPWM method is still meaningful because it is more scalable and reliable than M1 or M2.

VI. CONCLUSIONS

This paper has proposed a simple and effective ZSCC suppression method, which utilizes zero sequence controller with a feed-forward term. For enhanced linear modulation index range of the PWM, the ZSV for SVPWM in the proposed method has been calculated based on the grid voltage, which are common to all parallel converters and the major portion of the output voltages of the current regulator of each converter. Since the proposed method does not require any communication to suppress the ZSCC unlike the conventional methods, it can alleviate the burden of communication systems and enhance the scalability and reliability of overall system. To verify the validity of the proposed control method, the experiments have been conducted with four parallel three-level converters. Furthermore, the LLMIs with ZSCC controller were analyzed. According to the analysis, without

978-1-4673-9551-9/16 $31.00 © 2016 IEEE

communication between converters, the proposed method can extend the LLMI by 3~11% compared to the conventional methods still keeping the ZSCC virtually null.

REFERENCES

[1] J. L. Agorreta, M. Borrega, J. Lopez, and L.Marroyo, "Modeling and control of N-paralleled grid-connected inverters with *LCL* filter coupled due to grid impedance in PV plants," *IEEE Trans. Power Electron.*, vol. 26, no. 3, pp. 770–785, Mar. 2011.

[2] X. Wang, F. Zhuo, J. Li, L. Wang, and S. Ni, "Modeling and control of dual-stage high-power multifunctional PV system d-q-o coordinate," *IEEE Trans. Ind. Electron.*, vol. 60, no. 4, pp. 1556–1570, Apr. 2013.

[3] R. Li and D. Xu, "Parallel operation of full power converters in permanent-magnet direct-drive wind power generation system," *IEEE Trans. Ind. Electron.*, vol. 60, no. 4, pp. 1619–1629, Apr. 2013.

[4] F. Bovolini and H. Pinheiro, "Flexible arrangement of static converters for grid connected wind energy conversion systems," *IEEE Trans. Ind. Electron.*, vol. 61, no. 9, pp. 4707–4721, Sep. 2014.

[5] C. T. Pan and Y. H. Liao, "Modeling and coordinate control of circulating currents in parallel three-phase boost rectifiers," *IEEE Trans. Ind. Electron.*, vol. 54, no. 2, pp. 825–838, Apr. 2007.

[6] C. T. Pan and Y. H. Liao, "Modeling and coordinate control of circulating currents for parallel three-phase boost rectifiers with different load sharing," *IEEE Trans. Ind. Electron.*, vol. 55, no. 7, pp. 2776–2785, Jul. 2008.

[7] S. K. Mazumder, "Continuous and discrete variable-structure controls for parallel three-phase boost rectifiers," *IEEE Trans. Ind. Electron.*, vol. 52, no. 2, pp. 340–354, Apr. 2005.

[8] Z. Ye, D. Boroyevich, J. Y. Choi, and F. C. Lee, "Control of circulating current in two parallel three-phase boost rectifiers," *IEEE Trans. Power Electron.*, vol. 17, no. 5, pp. 609–615, 2002.

[9] T. P. Chen, "Circulating zero-sequence current control of parallel three-phase inverters," *IEE Proceedings-Electric Power Appl.*, vol. 153, no. 2, pp. 282–288, 2006.

[10] T. P. Chen, "Dual-modulator compensation technique for parallel inverters using space-vector modulation," *IEEE Trans. Ind. Electron.*, vol. 56, no. 8, pp. 3004–3012, 2009.

[11] J. S. S. Prasad, R. Ghosh, and G. Narayanan, "Common-Mode Injection PWM for Parallel Converters," *IEEE Trans. Ind. Electron.*, vol. 62, no. 2, pp. 789–794, 2015.

[12] Z. Shao, X. Zhang, F. Wang, and R. Cao, "Modeling and Elimination of Zero-Sequence Circulating Currents in Parallel Three-Level T-Type Grid-Connected Photovoltaic Inverters," *IEEE Trans. Power Electron.*, vol. 30, no. 2, pp. 1050 – 1063, 2015.

[13] J. Shen, S. Schroder, B. Duro, and R. Roesner, "A Neutral-Point Balancing Controller for a Three-Level Inverter With Full Power-Factor Range and Low Distortion," IEEE Trans. Ind. Appl., vol. 49, no. 1, pp. 138–148, Jan. 2013.

[14] Z. Shao, X. Zhang, F. Wang, and R. Cao, "Modeling and Elimination of Zero-Sequence Circulating Currents in Parallel Three-Level T-Type Grid-Connected Photovoltaic Inverters," *IEEE Trans. Power Electron.*, vol. 30, no. 2, pp. 1050 – 1063, 2015.

[15] D. W. Chung, J. S. Kim, and S. K. Sul, "Unified voltage modulation technique for real-time three-phase power conversion," *IEEE Trans. Ind. Appl.*, vol. 34, no. 2, pp. 374–380, 1998.

[16] H. W. van der Broeck, H.-C. Skudelny, and G. V. Stanke, "Analysis and realization of a pulsewidth based on voltage space vectors," *IEEE Trans. Ind. Appl.*, vol. 24, no. 1, pp. 142–150, Jan. 1988.

[17] J. Holtz, "Pulsewidth modulation—A survey," *IEEE Trans. Ind. Electron.*, vol. 39, no. 5, pp. 410–420, Oct. 1992.

[18] J. H. Kim and S. K. Sul, "A carrier-based PWM method for three-phase four-leg voltage source converters," *IEEE Trans. Power Electron.*, vol. 19, no. 1, pp. 66–75, 2004.

[19] S.-J. Chee, S. Ko, H.-S. Kim, and S.-K. Sul, "Common-mode Voltage Reduction of Three Level Four Leg PWM Converter," *IEEE Trans. Ind. Appl.*, vol. 51, no. 5, pp. 4006-4016, 2015.

978-1-4673-9551-9/16 $31.00 © 2016 IEEE

Hybrid Bridgeless DCM SEPIC Rectifier Integrated with a Modified Switched Capacitor Cell

Paulo Junior Silva Costa and Telles Brunelli Lazzarin,
IEEE Member

Dept. of Electrical Engineering, Campus Florianópolis,
Federal University of Santa Catarina (UFSC)
Florianópolis, Brazil
junior.paulocosta@gmail.com, telles@inep.ufsc.br

Carlos Henrique Illa Font, *IEEE Member*

Dept. of Electronics Engineering, Campus Ponta Grossa,
Federal University of Technology – Paraná (UTFPR)
Ponta Grossa, Brazil
illafont@utfpr.edu.br

Abstract — **In this paper is proposed a novel single-phase PWM bridgeless rectifier, based on SEPIC converter topology, integrated with a modified switched capacitor cell. The structure has the absence of the diode bridge at the input port reducing the number of components and conduction losses. Besides, it has the presence of a switched capacitor cell, providing double gain at the output voltage. A comparison with the conventional SEPIC shows that the proposed converter has lower voltage stress on the semiconductors when both converters are designed for the same output voltage, and the same voltage stress across the semiconductors when the output voltage of the proposed converter is twice bigger than the conventional SEPIC. Therefore, the proposed structure can be applied in DCM SEPIC rectifiers improving the converter static gain, making it suitable for higher voltage applications. The paper also proposes a modified switched capacitor cell, which does not change the storage capacitor operation mode of the SEPIC rectifier. To validate the theoretical analyses a prototype of 500 W was built considering an input and an output voltage of 220 V and 400 V, respectively, and a switching frequency of 50 kHz.**

Keywords — *Bridgeless, DCM, high power factor, SEPIC, single-phase rectifier, switched capacitor.*

I. INTRODUCTION

The single-phase bridgeless rectifiers have been broadly employed at industry due to their reduced losses and number of components. Most of the current single-phase bridgeless rectifiers are based on conventional DC-DC converters, for instance, Boost, Buck-Boost, Cúk, Zeta and SEPIC, where high power factor can be obtained without the use of current control when they operate in DCM (discontinuous conduction mode) [1].

SEPIC rectifiers have continuous input current even in discontinuous conduction mode making them attractive for this operation mode. However, a conventional SEPIC rectifier has high voltage stress on the semiconductors. Therefore, it is not attractive in electronic equipments, which demand high output voltages [2]-[9]. On the other hand, converters based on switched capacitors can double the output voltage without increasing the voltage stress across the semiconductors [10]-[13].

Based on the SEPIC rectifier approached in [9], multiplier SEPIC DC-DC converter in [10], switched capacitor cell in

[11], hybrid rectifiers in [13] and in the studies [3]-[8],[9],[12], in [14] was proposed a hybrid single-phase/-switch rectifier (SEPIC + switched capacitor), in which the switched capacitor cell is modified to work on the DCM SEPIC rectifier.

Therefore, in this paper is proposed the integration of the switched capacitor cell (derived from [14]) with a bridgeless SEPIC rectifier, resulting in a novel structure designed hybrid single-phase bridgeless SEPIC rectifier. The resulting topology has as main characteristics double static gain and all advantages of bridgeless rectifiers.

Experimental results to validate the theoretical analyses were also carried out for a prototype of 500 W and they are also presented in this paper.

II. PROPOSED BRIDGELESS SEPIC RECTIFIER WITH A MULTIPLER STAGE

The proposed topology of the single-phase bridgeless SEPIC rectifier integrated with a voltage multiplier stage, which is based on the switched capacitor cell, is shown in Fig. 1. The bridgeless rectifier structure is composed of the following components: L_i, $D1$, $D3$, $S1$, $S2$, C_{i1}, D_{o1}, L_o and C_{o1}. The absence of an input diode bridge reduces the number of components and the presence of only one diode and one switch in the flowing current path, during each switching cycle, can result in less conduction losses. The rectifier employs two switches, but both use the same gate signal.

The multiplier stage is built with C_S, C_{o2}, D_{e1}, D_{e2}, D_{e3} and D_{o2}. In the conventional DCM SEPIC rectifier, C_{i1} has to reproduce the harmonics of the rectified input voltage to provide a high quality input current [8]. Therefore, in this paper is proposed an addition of two diodes (D_{e1} and D_{o2}) on the conventional switched capacitor cell (C_S, D_{e2}, D_{e3} and C_{o2}) used in [10]-[13] in order to charge and discharged the switched capacitor, C_S, without modifying the voltage across C_{i1}. Hence, the new switched capacitor cell maintains the voltage characteristics in C_{i1} similar to the conventional SEPIC rectifier. Consequently, the proposed structure ensures an input current with lower distortion, reduced number of components, double static gain and same voltage stress across semiconductors when compared with the conventional SEPIC.

978-1-4673-9551-9/16 $31.00 © 2016 IEEE

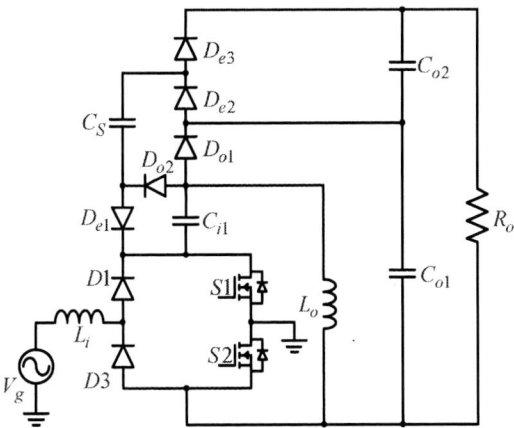

Fig. 1. Proposed hybrid bridgeless rectifier.

A. Operation Stages

The proposed rectifier has eight topological stages for the discontinuous conduction mode. The operation stages are shown in Fig. 2 considering the input voltage positive. The other four topological states have similar path, however, considering negative voltage at the input. In steady state operation, the voltages across C_{i1}, C_S, C_{o1} and C_{o2} are, respectively, V_g, $0.5V_o$, $0.5V_o$ and $0.5V_o$, where V_g is the input voltage and V_o is the average value of output voltage.

On the first topological stage in Fig. 2, when the switches $S1$ and $S2$ are turned on the diodes D_{e1} and D_{e2} are forward-biased, while the diodes D_{e3}, D_{o1} and D_{o2} are reversed-biased. Thus, the currents through the inductors L_i and L_o increase according to the relations given by V_g/L_i and V_g/L_o, respectively. The capacitor C_S in this stage is fed by C_{o1} and the current does not flow through C_{i1} due to modifications carried out in the switched capacitor cell, as proposed in this paper. The resistance load, R_o, at this stage is supplied by the capacitors C_{o1} and C_{o2}.

The second topological stage starts when the switches $S1$ and $S2$ are turned off, hence, D_{e3}, D_{o1} and D_{o2} are forward-biased and D_{e1} and D_{e2} are blocked. The previously energy stored in capacitor C_S and in the inductors L_i and L_o are, therefore, transferred to the capacitors C_{o1}, C_{o2} and to R_o. Capacitors C_S and C_{o2} are parallel connected and the currents in L_i and L_o decrease linearly according to the relations $-V_o/L_i$ and $-V_o/L_o$, respectively.

The third topological stage starts when the forward current in D_{o1} becomes null. In this stage, the capacitor C_S and the inductors L_i and L_o keep charging C_{o1}, C_{o2} and supplying energy to the load.

The fourth topological stage starts when the currents IL_{imin} and IL_{omin} become equal, however, with opposite sign. Therefore, the forward current in diode D_{o2} is null and the converter is in the discontinuous stage of the SEPIC. During this stage the capacitors C_{o1} and C_{o2} fed the load.

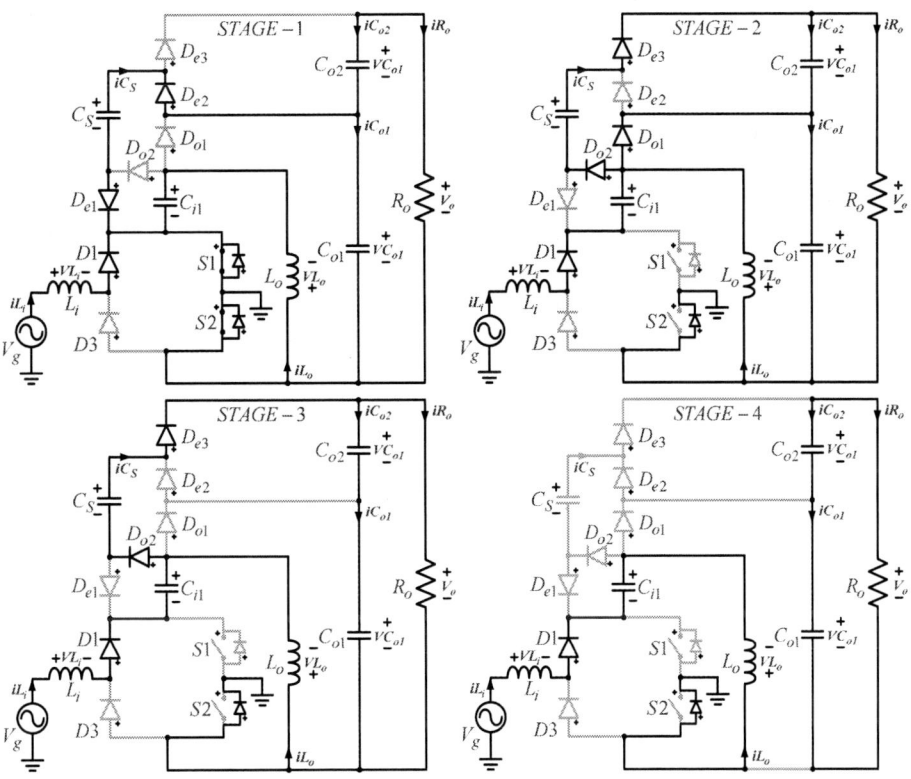

Fig. 2. Topological stages considering a positive input voltage.

B. Proposed Converter Waveforms

The main waveforms of the proposed converter are illustrated in Fig. 3 and Fig. 4 for the grid and switching periods, respectively. The voltage (V_g) and current (iL_i) at the input of the converter are shown in the upper graph of Fig. 3, while the output voltage (V_o) and voltages across C_{o1} and C_{o2} are shown in lower graph in Fig. 3. The rectifier has the input current in phase with the input voltage and it ensures the multiplication of the output voltage ($VC_{o1} = VC_{o2}$ and $VC_o = 2.VC_{o1}$). Currents and voltages in all components for one switching period are shown in Fig. 4.

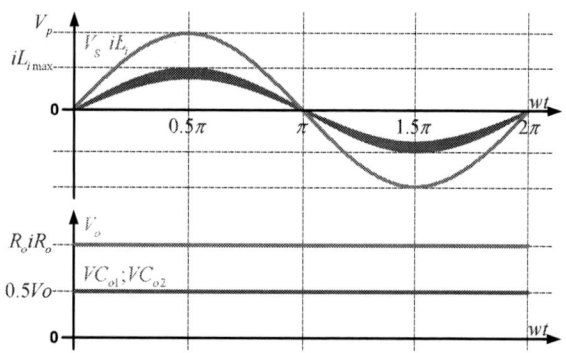

Fig. 3. The proposed rectifier waveforms for the grid period.

III. THEORETICAL ANALYSIS

The static gain and the main equations used in the passive components design for the proposed hybrid rectifier are presented in this section.

A. Static Gain

The static gain of the proposed rectifier is defined by the ratio between the output (V_o) and input peak (V_p) voltages, given by

$$G = \frac{V_o}{V_p} = D \sqrt{\frac{R_o (L_i + L_o)}{4 L_i L_o f_s}} . \tag{1}$$

B. Inductors Design

The inductor values L_i and L_o are obtained by

$$L_i = \frac{V_p D}{\Delta i_{Li} f_s} , \tag{2}$$

and

$$L_o = \frac{L_i R_o V_p^2 D^2}{2 L_i V_o^2 f_s - R_o V_p^2 D^2} , \tag{3}$$

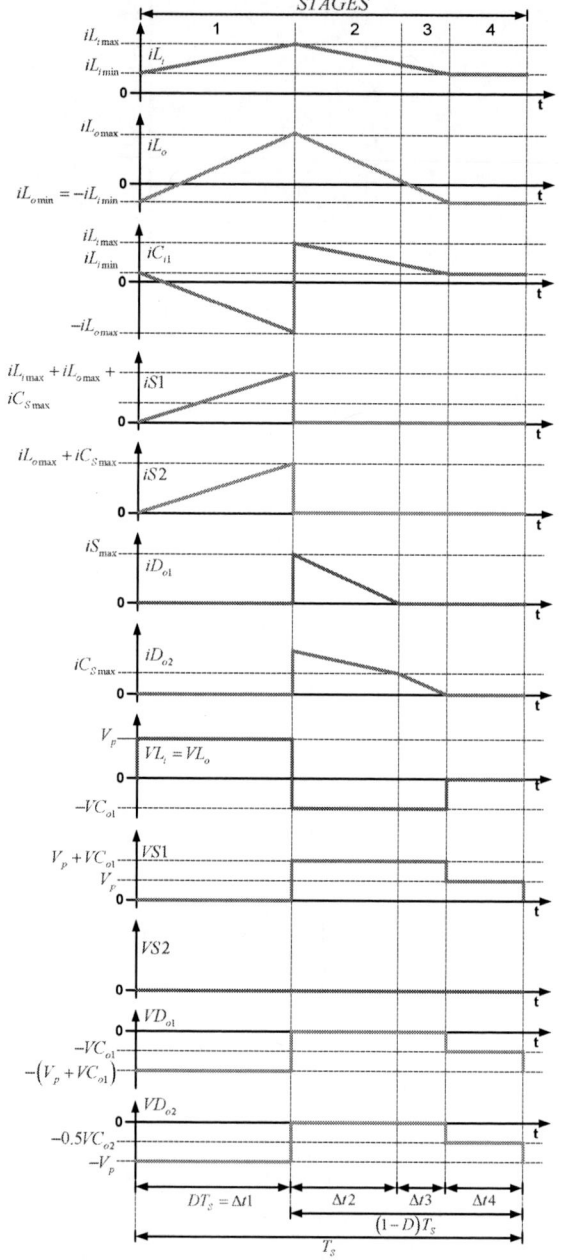

Fig. 4. The proposed rectifier waveforms for the switching period.

respectively. The inductor L_i is designed based on the ripple current specification, as shown in (2) and L_o is derived from L_i. This procedure design ensures the discontinuous conduction mode operation.

The average and the RMS current values of the input and the output inductors are given by

$$I_{Lo\_Avg} = \frac{D^2 V_p^2 (L_i + L_o)}{4 V_o L_i L_o f_s} , \tag{4}$$

$$I_{Loef} = \frac{1}{24}\sqrt{\frac{2D^3V_p^2\begin{pmatrix} V_pV_oL_i^2\left(128-192D\right) \\ +V_p^2L_oD\pi\left(-27L_{oa}+54L_i\right) \\ +V_o^2L_i^2\pi\left(48-36D\right) \end{pmatrix}}{V_o^2L_i^2L_o^2f_s^2\pi}} \quad , \quad (5)$$

and

$$I_{Lief} = \frac{\sqrt{6}}{24}\sqrt{\frac{D^3V_p^2\begin{pmatrix} V_O^2L_iD\left(12L_i+24L_o\right) \\ +L_o^2\left(16V_o^2-9V_p^2D^2\right) \end{pmatrix}}{V_o^2L_i^2L_o^2f_s^2}} \quad . \quad (6)$$

C. Capacitors Design

The capacitance values of C_{i1}, C_{o1}, C_{o2} and C_S are given by

$$C_{i1} = \frac{D^2V_p\left[D\left(V_pL_o-V_oL_i\right)+2V_oL_i\right]^2}{8V_o^2L_i^2L_o\Delta V_{Ci1}f_s^2} \quad , \quad (7)$$

and

$$C_S = C_{o1} = C_{o2} = \frac{2P_ot_{hut}}{V_o^2-\left(0,9V_o\right)^2} \quad . \quad (8)$$

The capacitor C_{i1} is designed from the ripple voltage specification and capacitors C_{o1}, C_{o2} and C_S are designed from the hold-up time specification.

IV. PROTOTYPE AND EXPERIMENTAL RESULTS

A prototype was built to verify the theoretical results of the proposed hybrid bridgeless SEPIC rectifier. The prototype was implemented according to the specifications shown in Table I and its photograph is illustrated in Fig. 5. The list of the components used in the prototype is described in Table II.

Experimental waveforms of the input voltage (V_g) and the input current (iL_i) are presented in Fig. 6 at the rated power (500 W). The results show that the input current has sinusoidal waveform and it is in phase with the input voltage.

The input current harmonic spectrum is illustrated in Fig. 7, presenting a total harmonic distortion (THD) around 3.20%, leading the converter power factor equal to 0.998.

The output voltage and current are shown in Fig. 8. Their average values were 400.3 V and 1.28 A, respectively. The output power in this test was measured approximately 512.38 W.

Experimental voltage waveforms across C_S, C_{o1} and C_{o2} are shown in Fig. 9. The results demonstrate the voltage balance on the capacitors with the voltage values around 200 V. This feature ensures a voltage multiplication, i.e, a converter with double static gain.

TABLE I. DESIGN SPECIFICATION

Specification	Values
Input peak voltage – V_p	311 V
Output voltage - V_o	400 V
Output power - P_o	500 W
Switching frequency – f_S	50 kHz
Maximum duty cycle - D	0.35
Hold-up-time - t_{hut}	8.33 ms
C_{i1} ripple voltage	20 %
L_i ripple current	10 %

TABLE II. LIST OF COMPONENTS

Components	Values
Inductor L_i	Inductance: 6.67mH Turns: 220 Wire: 18 AWG Core: APH46P60
Inductor L_o	Inductance: 120.6µH Turns: 39 Wire: 64 x 32 AWG Core: EE42/20 3c94
Transistors $S1$ and $S2$	SPW47N60C3 (650V/47A)
Diodes D_{o1}, D_{o2}, D_{e1}, D_{e2} and D_{e3}	MUR860 (600V/8A)
Diodes $D1$ and $D2$	1N5408 (1000V/3A)
Capacitor C_{i1}	2 x 470nF/400V
Capacitors C_{o1}, C_{o2} and C_S	2 x 1mF/250V
Control circuit	UC3525A

Fig. 5. Picture of the prototype.

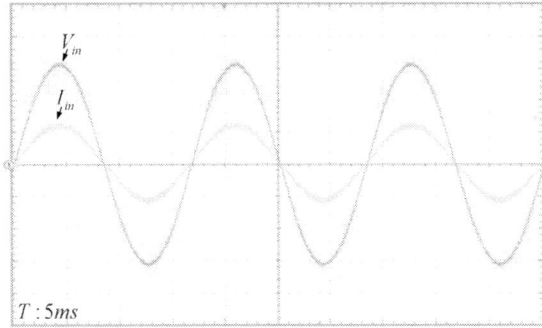

Fig. 6. Input voltage (100V/div) and current (3A/div).

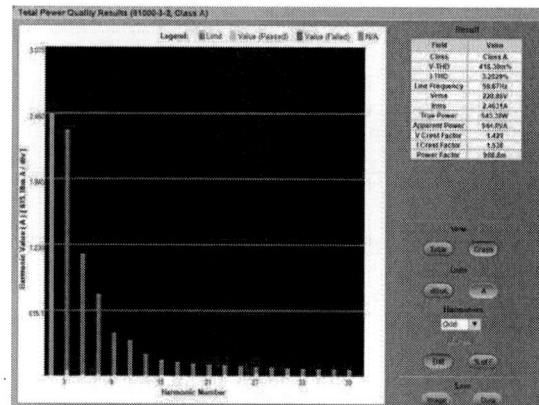

Fig. 7. Input current harmonic spectrum.

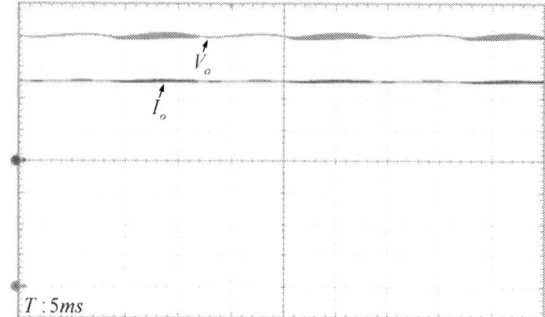

Fig. 8. Output voltage (50V/div) and output current (500mA/div).

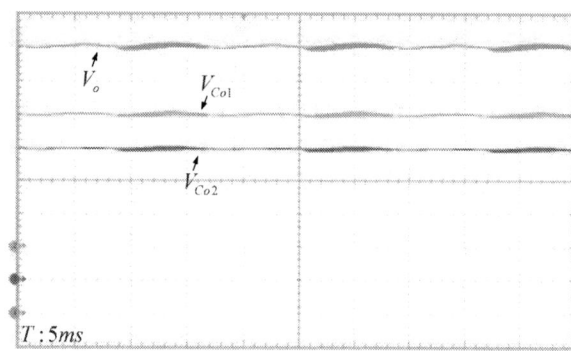

Fig. 9. Voltage in the load (50V/div) and across C_{o1} and C_{o2} (50V/div).

The current and ripple voltage waveforms on switched capacitor, C_s, are shown in Fig. 10. The peak value and RMS value of current is about 9.55 A and 3.79 A, respectively.

Fig. 11 shows the voltage across switches $S1$ and $S2$, where one can see that the maximum voltage across switch $S1$ is equal to 534 V and switch $S2$ is equal to 523 V. In the conventional SEPIC rectifier the theoretical maximum voltage across switches is equal to the sum of the input peak voltage

and output voltage. Thus, for the design specifications, the theoretical maximum voltage across switches would be equal to 711 V. In the proposed hybrid rectifier, the theoretical maximum voltage across the switches is equal to the sum of peak input voltage and half of output voltage, i.e., 511 V for the design specifications, which enables the use of lower voltage devices and lower conducting loss.

The efficiency and the total harmonic distortion of the proposed converter are shown in Fig. 12 and Fig. 13. The efficiency and the total harmonic distortion at the rated power are approximately 93.8% and 3.22%, respectively.

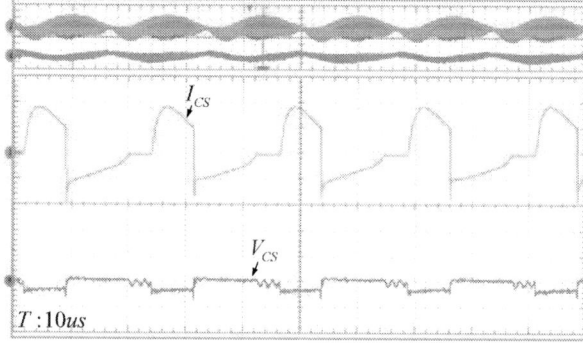

Fig. 10. Current through Cs (5A/div) and ripple voltage across Cs (3V/div).

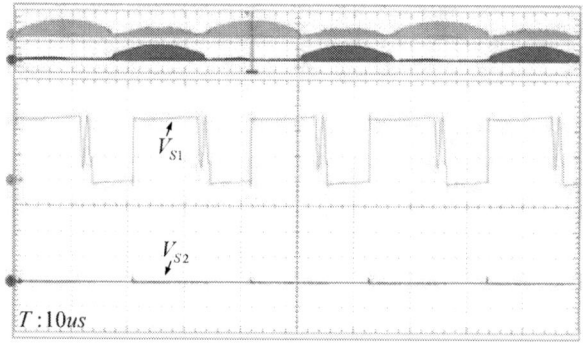

Fig. 11. Voltage across the switches (200V/div).

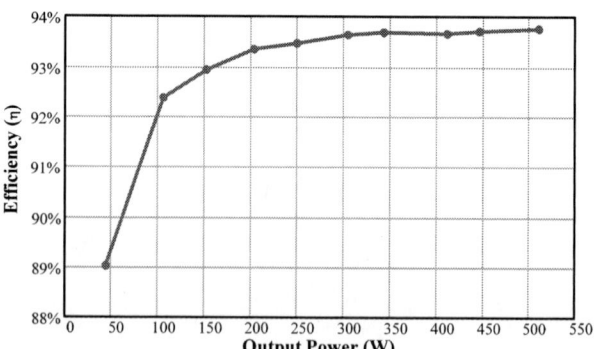

Fig. 12. Measured efficiency of the proposed converter.

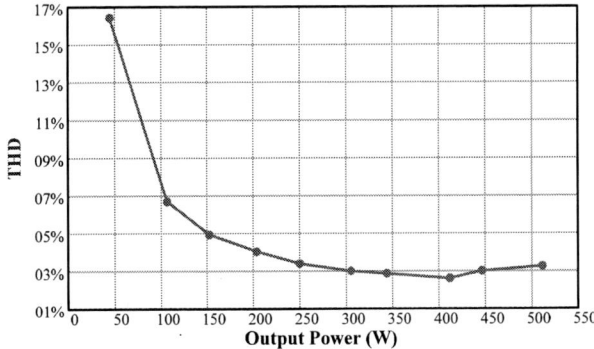

Fig. 13. Measured THD of the proposed converter.

V. CONCLUSIONS

This paper proposes a hybrid single-phase bridgeless SEPIC rectifier with a modified switched capacitor cell. The resulting topology maintains the SEPIC bridgeless rectifier characteristics adding the multiplier characteristic of switched capacitor.

Comparing the proposed converter with the conventional SEPIC rectifier, it presents double static gain, preserving the same voltage stress across the semiconductors.

When the proposed converter is compared with the SEPIC rectifier integrated with a switched capacitor cell introduced in [14], the bridgeless SEPIC reduces the number of components decreasing the conduction losses because there is only one diode and one switch in the current path during each switching cycle.

The experimental results corroborate the adequate operation of the proposed hybrid bridgeless rectifier, supplying, at the rated power, a current THD equal to 3.2%, a power factor of 0.9988, and an efficiency of 93.8%.

Hence, the theoretical analysis and the experimental results shown that the proposed converter is a potential candidate for applications which require higher output voltages and lower voltage stresses.

ACKNOWLEDGMENT

The authors would like to thank CNPq (National Council for Scientific and Technological Development) for their contribution to this work in the form of a grant provided to Paulo Junior Silva Costa.

REFERENCES

[1] B. Singh, B. N. Singh, A. Chandra, K. Al-Haddad, A. Pandey, D. P. Kothari, "A review of single-phase improved power quality AC-DC converters", IEEE Transactions on Industrial Electronics, vol. 50, no. 5, pp. 962 - 981, October 2003.

[2] M. S. Ortmann, S. A. Mussa, M. L. Heldwein, "Concepts for high efficiency single-phase three-level PWM rectifiers", in Proc. of IEEE Energy Conversion Congress and Exposition, pp. 3768 - 3775, 2009.

[3] D. S. L. Simonetti, J. Sebastian, F. S. dos Reis, J. Uceda, "Design criteria for SEPIC and Cuk converters as power factor preregulators in discontinuous conduction mode", in Proc. of International Conference on Industrial Electronics, Control, Instrumentation and Automation, vol. 1, pp. 283 - 288, 1992.

[4] D. S. L. Simonetti, J. Sebastian, J. Uceda, "The discontinuous conduction mode Sepic and Cuk power factor preregulators: analysis and design", IEEE Transactions on Industrial Electronics, vol. 44, no. 5, pp. 630 - 637, October 1997.

[5] E. H. Ismail, "Bridgeless SEPIC rectifier with unity power factor and reduced conduction losses", IEEE Transactions on Industrial Electronics, vol. 56, no. 4, pp. 1147 - 1157, April 2009.

[6] A. J. Sabzali, E. H. Ismail, M. A. Al-Saffar, A. A. Fardoun, "New bridgeless DCM Sepic and Cuk PFC rectifiers with low conduction and switching losses", IEEE Transactions on Industrial Electronics, vol. 58, no. 9, pp. 4153 - 4160, September 2011.

[7] M. Mahdavi, H. Farzanehfard, "Bridgeless SEPIC PFC rectifier with reduced components and conduction losses", IEEE Transactions on Industry Applications, vol. 47, no. 2, pp. 873 - 881, March/April 2011.

[8] G. Tibola, I. Barbi, "Isolated three-phase high power factor rectifier based on the SEPIC converter operating in discontinuous conduction mode", IEEE Transactions on Power Electronics, vol. 28, no. 11, pp. 4962 - 4969, November 2013.

[9] T. B. Lazzarin, F. A. B. Batista, P. J. S. Costa, C. H. Illa Font, "Proposal of a modular three-phase SEPIC-DCM rectifier for small wind energy conversion systems", in Proc. of 24th IEEE International Symposium on Industrial Electronics (ISIE), pp. 439 - 445, 2015.

[10] J. C. Rosas-Caro, J. C. Mayo-Maldonado, J. E. Valdez-Resendiz, "Multiplier SEPIC converter", in Proc. of 21st International Conference on Electrical Communications and Computers (CONIELECOMP), pp. 232 - 238, 2011.

[11] R. L. Andersen, T. B. Lazzarin, I. Barbi, "A 1-kW step-up/step-down switched-capacitor AC–AC converter," IEEE Transactions on Power Electronics, vol. 28, no. 7, pp. 3329 - 3340, July 2013.

[12] T. B. Lazzarin, R. Andersen, I. Barbi, "A switched-capacitor three-phase AC-AC converter," IEEE Transactions on Industrial Electronics, vol. 62, no. 2, pp. 735 -745, February 2015.

[13] D. F. Cortez, I. Barbi, "A family of high-voltage gain single-phase hybrid switched-capacitor PFC rectifiers", IEEE Transactions on Power Electronics, vol. 30, no. 8, pp. 4189 - 4198, September 2014.

[14] P. J. S. Costa, C. H. Illa Font, T. B. Lazzarin, "Single-Phase SEPIC rectifier with double voltage gain provided by a switched capacitor cell", in Proc. of 13th Brazilian Power Electronics Conference and 1st IEEE Southern Power Electronics Conference – COBEP/SPEC, 2015, in press.

LCL Filter Design for Three-phase Two-level Power Factor Correction using Line Impedance Stabilization Network

Alireza Kouchaki[1], Morten Nymand[2]

Maersk Mc-Kinney Moller Institute, University of Southern Denmark
Odense, Denmark
[1]alko@mmmi.sdu.dk, [2]mny@mmmi.sdu.dk

Abstract—This paper presents *LCL* filter design method for three-phase two-level power factor correction (PFC) using line impedance stabilization network (LISN). A straightforward *LCL* filter design along with variation in grid impedance is not simply achievable and inevitably lead to an iterative solution for filter. By introducing of fast power switches for PFC applications such as silicon-carbide, major current harmonics around the switching frequency drops in the region that LISN can actively provide well-defined impedance for measuring the harmonics (i.e. 9 kHz-30MHz). Therefore, LISN can be replaced with unknown grid impedance at high frequency, simplify the model of the filter, and provide repetitive measurements. In this paper, all the filter parameters are derived with analyzing the behavior of the converter at high frequency with presence of LISN impedance. The minimum required filter capacitor is derived using the current ripple behavior of converter-side inductor. The grid-side inductor is achieved as a function of LISN impedance to fulfill the grid regulation. To verify the analyses, an *LCL* filter is designed for a 5 kW SiC-based PFC. The simulation and experimental results support the validity of the method.

Keywords—LCL filter; power factor correction; line impedance stablization network;

I. INTRODUCTION

These days, three-phase grid-connected PWM voltage source converters (VSCs) like two-level or multilevel converters are widely used in many applications. Trying to improve the power quality and attenuating the current harmonics generated by these converters leads to different approaches such as filter design and harmonic elimination/mitigation methods [1]-[7]. To attenuate the harmonic contents at high frequencies one possible solution is relying on the inductor of three-phase boost VSC as a filter. Nevertheless, this solution leads to a bulky inductor with high power inductor losses. Besides, the large inductance value degrades the performance of the controller [5]. Employing high order filters such as *LCL*, *LLCL* filters to fulfill the grid regulations (e.g. IEEE 519) are highly attractive solution and have been studied in many researches [5]-[18].

In [5], a step by step method has been proposed for design of *LCL* filter with passive damping. The position of the sensors and their impacts on the control system and filter capacitor are analyzed in [6]. In [7], margins for the filter inductor are analytically defined with active damping solution. The final method in [7] is based on the iteration.

One of the main parameter in the *LCL* filter design is the converter-side inductor which is chosen to limit the current ripple produced by the PWM converter. Reference [8] and [9] have analyzed the maximum current ripple in two-level and three-level grid-connected converters, respectively.

Resonant of *LCL* filters degrades the performance of the converter. Exciting of the resonance in PFC applications may happen due to injection of unwanted harmonics from the grid and/or the converters harmonics. Therefore, damping of the resonance is essential by active or passive methods. For stiff grids, the solution of passive damping is more attractive due to its simplicity and low cost [11] and different passive damping methods have been analyzed in many publications [9]-[14]. In [9], different configurations for passive damping have been analyzed. In [11], the dissipated power in different passive damping configurations is investigated. In [10], the *LCL* and *LLCL* filter have been designed and the damping resistor is defined based on the performance of the filter at difference frequencies. On the other hand, [10] has used the quality factor of the filter as a parameter for finding the damping resistor. In [12], different *LCL* filter designs with different passive damping methods for high and low power applications have been investigated. The optimal design of the *LCL* filter with passive damping is investigated in [16]. On the other hand, the proposed solution is not offering optimum values for the converter side and grid side inductors. Moreover, the final algorithm is based on the iterative solution.

The designed filter is tested under different grid conditions which may lead to different results. In state-of-the-art, stabilization of the converter at high and low frequencies is analyzed by considering the grid impedance [19]-[21]. However, the impact on the filter parameters caused by the worst case condition (weak grid) has not been studied.

Among all aforementioned methods for designing the filter parameters, the parameters are mainly chosen using iterations. More importantly, the behavior of the current waveform on both the grid-side and converter-side are not analyzed. The question is that how designing of the filter can be initiated and tested without effect of the grid. In other words, how the experimental results of a grid-connected converter can be repetitive in spite of the grid impedance variation. Constant impedance for a wide range of frequencies is required to design and examine the performance of the filter. To prevent the dependency of the filter on the grid impedance and optimally calculate the filter parameters, the line impedance stabilization network (LISN) can be employed in the test setup [22]. Fig. 1 shows the complete setup of a grid connected converter with LISN. LISN is high impedance towards the high frequency harmonics of the grid and low impedance towards the high frequency harmonics of the converter.

This paper presents a method for designing an *LCL* filter for two-level PFCs using LISN. Using the equivalent circuit of the converter, the effect of LISN on measurement is studied. Filter parameters are, then, calculated by analyzing the equivalent circuit. In this paper, a passive damping method also is employed for improving the dynamic performance of the converter. Finally, a 5 kW three-phase PFC setup is used to verify the performance of the designed filter.

II. EFFECT OF LISN ON HIGH FREQUENCY HARMONICS

A. Single-phase equivalent circuit

The single-phase equivalent circuit not only simplifies designing the filter, but also helps to investigate the effect of LISN on circuit. To do that, the noise source must be defined. For an SPWM grid-connected VSC, using double Fourier analysis, the amplitude of ac link voltage at multiples of switching frequency (carrier frequency) is expressed as follows [23]:

$$V_{xN}{}^{m,n} = \frac{4V_{dc}}{\pi} \sum_{m=1}^{\infty} \sum_{n=-\infty}^{\infty} \frac{1}{m} J_n \left(m \frac{\pi}{2} M \right) \sin\left([m+n] \frac{\pi}{2} \right), (1)$$

where m is the carrier index variable, n is the baseband index variable, J is the Bessel function, and x represents the phase name as a, b, or c.

FFT of the inductor current shows the major harmonics happen at switching side-band (i.e. $m = 1$, $n = \pm2$), while AC link voltage does not have component at these specific harmonics. Another words, ac link voltage does not individually contribute in generating the largest current harmonic. Instead, $v_{xn}(t)$ which is a function of all three ac link voltages (e.g. $v_{an}(t) = 1/3(2v_{aN}(t)-(v_{bN}(t)-v_{cN}(t)))$) generates the most significant voltage harmonics at switching side-band harmonics based on (2) (at $m = 1$, $n= \pm2$):

$$V_{xn}{}^{m=1,n=\pm2} = -\frac{4V_{dc}}{\pi} J_{\pm2} \left(\frac{\pi}{2} M \right). \quad (2)$$

Ideally without using LISN, having stiff grid leads to a simplified equivalent circuit which is shown in Fig. 2. In practice, the grid impedance changes and it may affect performance of the filter [20].

LISN provides high impedance network for the grid at high frequencies (9 kHz-30 MHz) and at the same time provides low impedance network for the high frequency harmonics of the converter. All the high frequency harmonics of the converter flow through the capacitor (C_{LISN}) and the resistor (R_{LISN}) as shown in Fig. 1. On the other hand, the high frequency harmonics generated by the grid cannot flow through the converter due to presence of the inductance (L_{LISN}). The fundamental harmonic of the grid, however, sees very low impedance on the path to the converter and therefore gives the power to the converter. Typically, the impedance in series with the grid at fundamental frequency is about 15 mΩ which is negligible. Based on (2) and the operation principle of LISN, the equivalent circuit of the three-phase PFC can be depicted as Fig. 3.

B. High frequency current harmonics

As it is clear from the equivalent circuit of the converter, the current harmonics flowing in the grid-side inductance (L_g) is affected by the LISN resistance. From the circuit theory, the effect of this resistor at any frequency can be calculated.

The current harmonic using equivalent circuit without LISN can be achieved as follows:

$$I_g{}^h(s) = \frac{V_{an}{}^h}{LL_g C_f s^3 + (L+L_g)s}. \quad (3)$$

Similarly, the current harmonic with presence of LISN ($I_{g,LISN}$) can be calculated in (4) where the LISN impedance is added to the circuit.

$$I_{g,LISN}{}^h(s) = \frac{V_{an}{}^h}{LL_g C_f s^3 + R_{LISN} C_f L s^2 + (L+L_g)s + R_{LISN}}. \quad (4)$$

As it can be seen from (3) and (4), the transfer function changes using LISN, clearly. Therefore, the difference between the current with and without LISN can be obtained as follows:

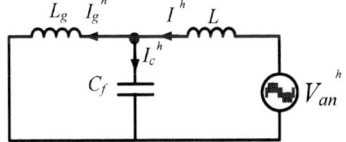

Fig. 2. Single phase equivalent circuit of the converter at hth harmonic (h >> 1).

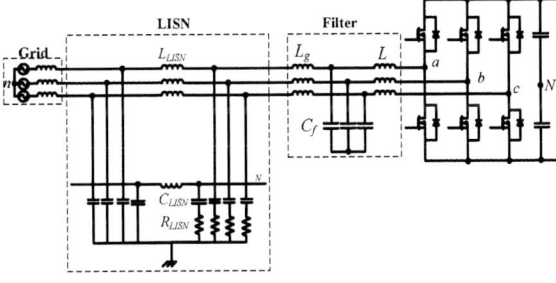

Fig. 1. The schematic of the grid-connected VSC with LISN.

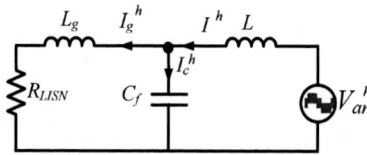

Fig. 3. Single phase equivalent circuit of the converter at h^{th} harmonic with LISN ($h \gg 1$).

$$I_g^h(s) / I_{g,LISN}^h(s) = 1 + \frac{R_{LISN}\left(C_f L s^2 + 1\right)}{L L_g C_f s^3 + \left(L + L_g\right) s}. \quad (5)$$

This equation implies that the transfer function of the *LCL* filter employing LISN has changed. Accordingly, for studying the stability of the converter at different frequencies, LISN should be removed from the test setup. However, (5) shows that at a certain frequency the difference between the current harmonic is only a constant value added to the measured current (e.g. in the case of designing filter, the interesting harmonic is f_s-$2f_{grid}$). In steady state analysis, the constant value at this specific frequency can be calculated.

$$k = \left| I_g^h / I_{g,LISN}^h \right|_{s = j\omega_h}, \ \omega_h = 2\pi\left(mf_s \pm nf_{grid}\right). \quad (6)$$

where k is:

$$k = \sqrt{1 + \frac{R_{LISN}^2\left(1 - C_f L \omega_h^2\right)^2}{\left(L L_g C_f \omega_h^3 - \left(L + L_g\right)\omega_h\right)^2}}. \quad (7)$$

According to (7), k is always larger than one which implies that LISN adds a constant attenuation to the current. On the other hand, LISN emulates well-defined and constant impedance for high frequency applications which can be treated as grid impedance at high frequency.

III. DESIGNING THE FILTHE PARAMETERS

In this section, designing filter parameters is presented without considering filter damping. After choosing the filter parameters, the passive damping method is also employed to provide good dynamic performance.

A. Converter-side inductance (L)

The converter-side inductor regardless of the filter configuration is chosen to limit the current ripple at converter side. For calculating the converter-side inductance value, [8] has been used in this paper. According to the conclusion from [8], the maximum current ripple at peak current is more interested with respect to the practical issues such as saturation of the core material. Therefore, only the equation related to the peak current is used. The minimum required inductance value as a function of the maximum current ripple at peak current is derived using (8) [8].

$$L = \frac{1}{2\Delta i_{peak@\,\text{peak current}} f_s}\left[M_{index}V_{dc} - V_m\left(1 + 0.5M_{index}\right)\right], \quad (8)$$

where M_{index} is the modulation index and $\Delta i_{peak@\text{peak current}}$ is the maximum current ripple at peak current. V_{dc} and V_m are the dc link voltage and peak of the grid voltage, respectively.

B. Filter capacitor (C_f)

In the literature, normally, the required filter capacitor is calculated using the limited reactive power absorbed by the filter capacitor. As a result, the maximum filter capacitor size is determined as (9) [5].

$$\left|Q / P\right| = \omega_g Z_b C_{f,\max} \leq 0.05, \ Z_b = V_b^2 / S, \quad (9)$$

where Q is the maximum absorbed reactive power by the capacitor, Z_b is the base impedance which is derived using base voltage (V_b) and apparent power of the converter (S).

However, the minimum required filter capacitance also needs to be determined to have a valuable margin. To find the minimum filter capacitance, voltage ripple across the capacitor needs to be calculated. It can be assumed that the whole current ripple from the converter-side inductor ideally flows through the filter capacitor.

The required filter capacitor, then, can be calculated by the stored charge in the filter capacitor caused by the maximum current ripple of the converter-side inductor. The converter-side inductance current is shown in Fig. 4. The current ripple behavior at peak current and zero crossing are also shown in Fig. 4(b) and (c). The peak current and zero crossing current are the main interest of the analysis since the maximum current ripple can happen at either one of them [8]. Therefore, the maximum charge is also dependent on one of these two points. The charge area (ΔQ) is shown as shaded area in Fig. 4(b) and (c) needs to be found. The charge at zero crossing and peak current are expressed as follows, respectively:

$$\Delta Q_{\text{zero current}} = \Delta i_{peak} T_2, \quad (10)$$

$$\Delta Q_{\text{peak current}} \simeq \Delta i_{peak}\left(T_2 + T_3\right)/4, \quad (11)$$

where Δi_{peak}, T_2, and T_3 are defined in Fig. 4 for zero and peak current.

Δi_{peak} for peak current can be extracted from (8) and for zero current it is equal to [8]:

$$\Delta i_{peak@\,\text{zero current}} = \frac{\sqrt{3}}{24} M_{index} \frac{V_{dc}}{Lf_s} \quad (12)$$

As it is shown in Fig. 4, the important time intervals such as T_2 can be achieved using the relation between the reference and carrier waveforms. It should be noticed that the modulation index is also taken into account for finding the time interval equation. For calculating the charge in zero current, the time interval T_2 at zero current is derived as follows [see Fig. 4(b)]:

$$T_{2\,@\,\text{zero current}} = M_{index}\,\frac{\sqrt{3}}{8f_s} \qquad (13)$$

Similarly, the effective time intervals at peak current (T_2 and T_3) are obtained as (14).

$$\left(T_2 + T_3\right)_{@\text{peak current}} = \left(1 + M_{index}\right)\frac{1}{4f_s} \qquad (14)$$

Therefore, from (12), (13) and (8), (14) the charge at zero current and peak current are derived, respectively.

$$\Delta Q_{\text{zero current}} = \frac{M_{index}^{\,2}}{64Lf_s^{\,2}}V_{dc} \qquad (15)$$

$$\Delta Q_{\text{peak current}} \simeq \frac{\left(1 + M_{index}\right)}{32Lf_s^{\,2}}\left[M_{index}V_{dc} - V_m\left(1 + 0.5M_{index}\right)\right] \quad (16)$$

Acceptable voltage ripple of filter capacitor (ΔV_{ripple}) can give the requirement for the filter capacitor.

$$C_{f,\text{min}} = \frac{\Delta V_{ripple}}{\Delta Q} \qquad (17)$$

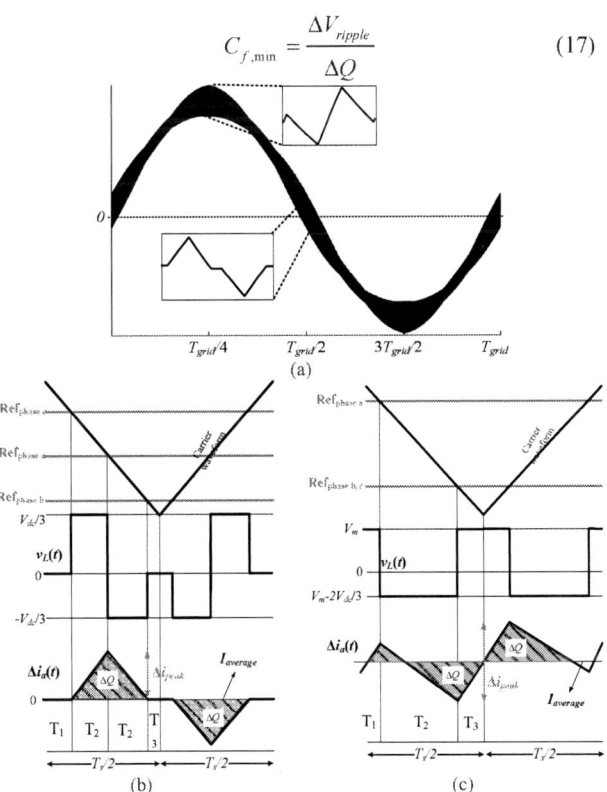

Fig. 4. (a) The converter-side inductor current, (b) current at zero crossing, and (c) at peak current.

C. Grid-side inductance (L_g)

Now, all the filter parameters except L_g are known. The purpose of using L_g is to attenuate the current harmonic at grid side and fulfill the grid regulations. Moreover, it also protects the filter capacitor from unwanted harmonics from the grid side. The current harmonic at switching frequency and its side-band must be lower than a specific value (I_g^{h*}) which has been dictated by the grid standard (i.e. IEEE-519). Modifying (4) and replacing the current ($I_{g,LISN}$) with the standard current, the required grid-side inductance value is calculated as (18).

$$L_g = \left(\sqrt{\left(\left|\frac{V_{an}}{I_g^{*h}}\right|\right)^2 - \left(R_{LISN}\,\alpha\right)^2} - L\omega_h\right)\Big/\alpha\omega_h \qquad (18)$$

where α is defined as follows:

$$\alpha = 1 - C_f L\omega_h^{\,2} \qquad (19)$$

IV. DAMPING OF THE FILTER

Fig. 5 shows the solution preferred out of many options for damping the resonance peak of the filter [9], [11]-[14]. This solution incorporates an R_d-C_d damping branch parallel to the main filter capacitor. At the switching frequency, the impedance of filter capacitor is smaller than the total impedance presented by the parallel damping branch. As it is expected, all high frequency ripple flows through the filter capacitor. R_d and C_d are responsible for damping the resonance of the filter which is smaller than the switching frequency. According to [11], C_d can be chosen equal to C_f. For choosing R_d, there should be a tradeoff between the loss in damping branch and proper damping [12].

V. COMPONENT SELECTION

The specifications of the converter are listed in TABLE I. This converter is a two-level three-phase SiC-based VSC. The switching frequency is considerably high, because of comparatively better performance of SiC against Si at high frequencies.

According to (8), the required inductance value to have 30% of the nominal current as a maximum current ripple is 580 μH. The current ripple at peak of the current and zero current can be calculated using (8) and (13), respectively. Then the charges in the filter capacitor for both with 0.93 as modulation index are:

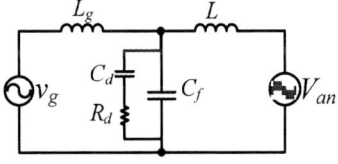

Fig. 5. Filter with passive damping.

978-1-4673-9551-9/16 $31.00 © 2016 IEEE

TABLE I. Specification of the converter

Rated power	5 kW
Grid voltage (*rms*)	230 V
Grid frequency (f_g)	50 Hz
DC link voltage	700 V
Maximum current ripple$^{\text{converter side}}$	30% I_n
Switching frequency (f_s)	50 kHz

$$\begin{cases} \Delta Q_{\text{zero current}} = 6.5 \mu C \\ \Delta Q_{\text{peak current}} = 7.3 \mu C \end{cases} \rightarrow C_f = \frac{\Delta Q_{\text{peak current}}}{1\% V_g} = 2.5 \mu F \quad (20)$$

Using (9), the maximum filter capacitor for limiting the reactive power is 5 µF. This value shows that the calculated capacitance using the stored charge in the capacitor is below the limit.

The inductance values should be 100 µH based on (18). Note that for calculating the inductance value, V_{an}^{*h} is 33% of fundamental value. I_g^{*h} is according to the standard should be lower than 0.3% of the nominal input current. However, to calculate the required inductance value, the limit should be further decreased due to the effect of LISN. So, the limit for the current harmonic is calculated to 0.14% of nominal current.

The resonant frequency of the filter is 10 kHz without employing $R_d.C_d$ parallel branch. To damp the resonant of the filter, the required damping resistor is calculated 10 Ω and C_d as it was mentioned is equal to filter capacitor which is 2.5 µF. Therefore, in this case the resonant frequency is replaced at 7.7 kHz.

VI. Simulation and Experimental Results

To demonstrate the filter design, its performance has been simulated in MATLAB/Simulink environment. The results have been shown for the 5kW PFC in Fig. 6. Fig. 6(a) shows the converter-side current with the maximum inductor current of 3 A. The performance of the filter is examined using FFT of the current. The current harmonic at m_f -2 is about 0.22% of the nominal current which shows complying of the filter with grid standard.

The frequency response of the filter has been studied as shown in Fig. 6(d). The frequency responses of the filter without damping, with $R_d.C_d$ damping are shown. The frequency response of one of the most common damping methods is also shown in this figure to compare the performance of damping used in this paper and damping when R_d is employed in series with C_f. The slope of the attenuation in $R_d.C_d$ method has not been changed while in R_d method the slope has decreased 20 dB/dec.

Fig. 7(a) shows the experimental setup for verifying the designed filter. The current at full load condition is also shown in Fig. 7(b) with FFT of the converter and grid side currents. The major harmonics is around the switching frequency. The attenuation that the filter has provided is 40 dB around the switching frequency. The power loss in the damping branch is calculated using the method mentioned in [11]. The *rms* current flowing through the damping branch is 0.18 A. The power loss

using the current harmonics of damping branch is about 0.4 W (each phase). If instead of using damping branch, the damping resistor is put in series with the filter capacitor, the attenuation after the resonant frequency decreases and the power losses also increases. In this case the power loss in the damping resistor is about 4 W (each phase). Therefore, this method of damping which is used in this paper is effective and the damping losses are very small.

If the attenuation of the switching side-band harmonics is calculated using (2), they are in match with the experimental results. TABLE II demonstrates the comparison of analyses and experimental.

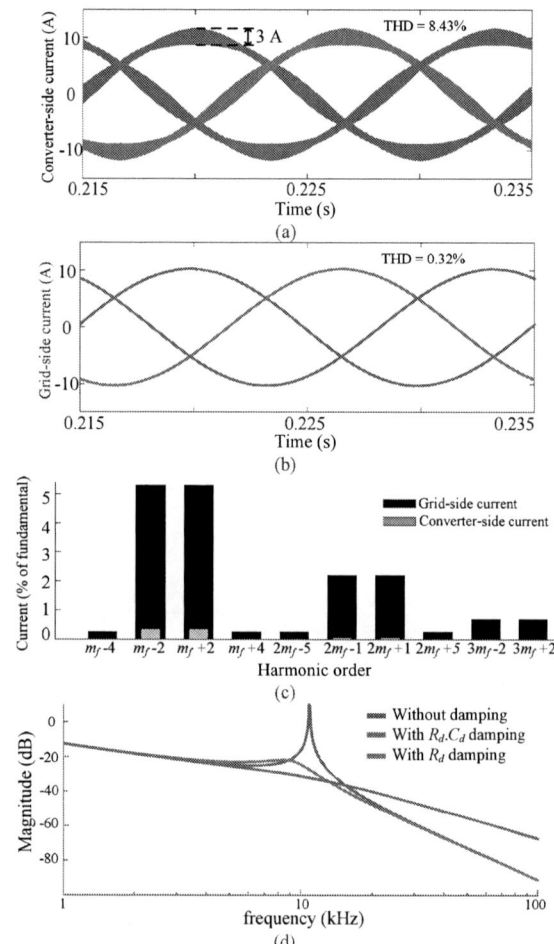

Fig. 6. (a) The converter-side current, (b) the grid-side current, (c) FFT of the current ($m_f = f_s/f_g$), and (d) frequency response of the filter with and without damping.

TABLE II. Comparison of measured and calculated harmonic at f_s-2f_g with and without filter.

	Calculated (dB)	Measured (dB)
Without filter	-8.5	-8
With filter	-54	-48

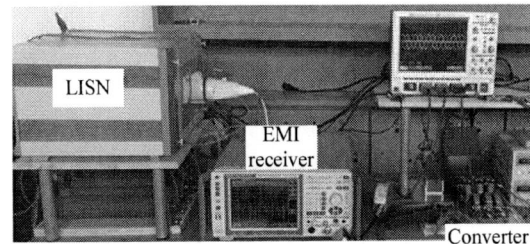

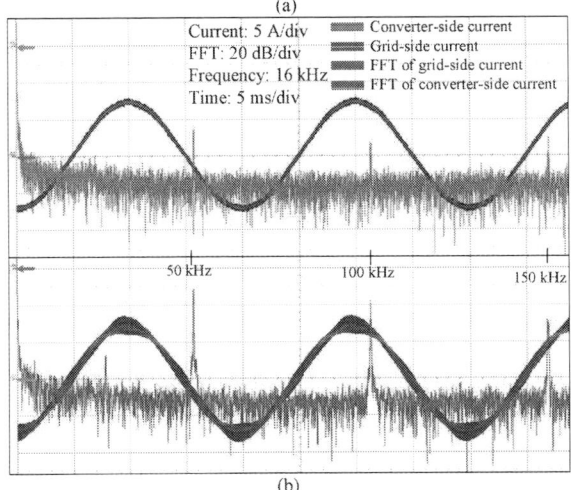

Fig. 7. Experimental results (a) the experiment setup, (b) the measured current at converter and grid side with FFT.

VII. CONCLUSION

This paper has presented *LCL* filter design method for three-phase two-level power factor correction (PFC) using line impedance stabilization network (LISN). Interacting of filter performance with the grid impedance (which is not constant) inevitably leads to an iterative solution. However, by introducing fast power switches for PFC applications such as silicon-carbide (SiC), major current harmonics around the switching frequency drops in the region that LISN can provide well-defined impedance for measuring the harmonics (i.e. 9 kHz-30MHz). Therefore, the filter design gets independent from the grid impedance and the repetitive measurements are achievable. In this paper, the simplified single-phase equivalent circuit of three-phase PFC with LISN model is provided. The converter-side inductor has been calculated to limit the current ripple. Using the current ripple analyses for the converter-side inductor, the minimum required filter capacitor has been obtained. The provided high frequency impedance by LISN, simplifies the last step of the filter design. The grid-side inductance value as a function of the LISN resistor, converter-side inductor, and filter capacitor is derived to fulfil the grid regulations. To verify the validity of the derived equations for the filter, a filter has been designed for a 5 kW SiC-based three-phase PFC with switching frequency of 50 kHz. Simulation and experimental results show a good agreement with the analyses. The designed filter is able to fulfil the grid regulation.

Although, the presented method is capable of designing filter without any iteration, the stability of the converter cannot be examined with help of LISN. On the other hand, the stability analyses should be done by removing LISN from the setup. In spite of having well-defined high frequency impedance, the method is not applicable for low frequency switching converter (below 9 kHz).

ACKNOWLEDGMENT

The project is sponsored by the Danish National Advanced Technology Foundation under Intelligent Efficient Power Electronics (IEPE), strategic research center between the industries and universities in Denmark.

REFERENCES

[1] A. Marzoughi, H. Iman-Eini, "Selective harmonic elimination for cascaded H-bridge rectifiers based on indirect control," *Power Electronics and Drive Systems Technology* (PEDSTC), pp.79-85, 15-16 Feb. 2012.

[2] M. Aleenejad, H. Iman-Eini, S. Farhangi, "A minimum loss switching method using space vector modulation for cascaded H-bridge multilevel inverter," *Electrical Engineering* (ICEE), 20th Iranian Conference on, pp.546-551, May 2012.

[3] A. Marzoughi, H. Iman-Eini, "An optimal selective harmonic mitigation technique for high power converters," *Electrical power and Energy Systems*, vol. 49, pp. 34-39, 2013.

[4] M. Aleenejad, R. Ahmadi, P. Moamaei, "Selective harmonic elimination for cascaded multicell multilevel power converters with higher number of H-Bridge modules," *Power and Energy Conference at Illinois* (PECI), 2014.

[5] M. Liserre, F. Blaabjerg, S. Hansen, "Design and control of an *LCL*-filter-based three-phase active rectifier," Industry Applications, IEEE Trans. on, vol. 41, no. 5, pp. 1281-1291, 2005.

[6] M. Liserre, F. Blaabjerg, A. Dell Aquila, "Step-by-step design procedure for a grid-connected three-phase PWM voltage source converter," International Journal of Electronics, vol. 91, no. 8, pp. 445-460, Aug 2004.

[7] K. Jalili, S. Bernet, "Design of *LCL* filters of active-front-end two-level voltage-source converters," Industrial Electronics, IEEE Trans. on, vol.56, no.5, pp.1674-1689, May 2009.

[8] A. Kouchaki, F. Javidi, F. Haase, M. Nymand, "An Analytical Inductor Design Procedure for Three-phase PWM Converters in Power Factor Correction Applications," Power Electronics and Derive Systems (PEDS), June 2015.

[9] Y. Jiao, F.C. Lee, "*LCL* filter design and inductor current ripple analysis for 3-level NPC grid interface converter," Energy Conversion Congress and Exposition (ECCE), pp.1911-1918, Sept. 2014.

[10] Weimin Wu, Yuanbin He, Tianhao Tang, F. Blaabjerg, "A New Design Method for the Passive Damped *LCL* and *LLCL* Filter-Based Single-Phase Grid-Tied Inverter," Industrial Electronics, IEEE Trans. on, vol.60, no.10, pp.4339-4350, Oct. 2013.

[11] R. Pena-Alzola, M. Liserre, F. Blaabjerg, R. Sebastián, J. Dannehl, F. W. Fuchs, "Analysis of the passive damping losses in *LCL*-filter based grid converters," Power Electron., IEEE Trans. on, vol. 28, no. 6, pp. 2642-2646, 2013.

[12] R. Beres, Xiongfei Wang, F. Blaabjerg, C.L. Bak, M. Liserre, "A review of passive filters for grid-connected voltage source converters," Applied Power Electronics Conference and Exposition (APEC), pp. 2208-2215, March 2014.

[13] R. Beres, Xiongfei Wang, F. Blaabjerg, C.L. Bak, M. Liserre, "Comparative evaluation of passive damping topologies for parallel grid-connected converters with *LCL* filters," Power Electronics Conference (ECCE-ASIA), pp.3320-3327, May 2014.

[14] R. Beres, Xiongfei Wang, F. Blaabjerg, C.L. Bak, M. Liserre, "New optimal design method for trap damping sections in grid-connected *LCL*

filters," Energy Conversion Congress and Exposition (ECCE), pp. 3620-3627, Sept. 2014.

[15] Byung-Geuk Cho, Seung-Ki Sul, "Non-iterative *LCL* filter design for three-phase two-level voltage-source PWM converters," Power Electronics Conference (ECCE-ASIA), pp.2802-2809, May 2014.

[16] Y. Patel, D. Pixler, A. Nasiri, "Analysis and design of TRAP and *LCL* filters for active switching converters," Industrial Electronics (ISIE), IEEE International Symposium on , pp.638-643, July 2010.

[17] D. Jovcic, Lu Zhang, M. Hajian, "*LCL* VSC Converter for High-Power Applications," Power Delivery, IEEE Trans. on, vol.28, no.1, pp.137-144, Jan. 2013.

[18] J. Muhlethaler, M. Schweizer, R. Blattmann, J.W. Kolar, A. Ecklebe, "Optimal Design of *LCL* Harmonic Filters for Three-Phase PFC Rectifiers," Power Electronics, IEEE Trans. on, vol.28, no.7, pp.3114-3125, July 2013.

[19] A. Rockhill, M. Liserre, R. Teodorescu and P. Rodriguez, "Grid-filter design for a multimegawatt medium-voltage voltage-source inverter," Industrial Electronics, IEEE Trans. on, vol. 58, pp. 1205-1217, 2011.

[20] M. Liserre, R. Teodorescu, F. Blaabjerg, "Stability of photovoltaic and wind turbine grid-connected inverters for a large set of grid impedance values," Power Electronics, IEEE Transactions on, vol.21, no.1, pp.263-272, Jan. 2006.

[21] M. Liserre, A. Dell'Aquila, F. Blaabjerg, "Stability improvements of an *LCL*-filter based three-phase active rectifier," Power Electronics Specialists Conference, vol.3, pp.1195-1201, 2002.

[22] I.A. Makda, M. Nymand, "Differential mode EMI filter design for isolated DC-DC boost converter," Power Electronics and Applications (EPE'14-ECCE Europe), 16th European Conference on, pp.1-8, Aug. 2014.

[23] D.G. Holmes, T.A. Lipo, Pulse width modulations for power converters: principles and practice, Wiley publication, 2003.

Sensorless Current Rebuilding Strategy in a Single Phase Bridgeless PFC

Felipe López[1], Paula Lamo[2], Alberto Pigazo[2], *Senior Member, IEEE*, and F.J. Azcondo[1], *Senior Member, IEEE*

[1] Dept. Electronics Technology, Systems and Automation Engineering
University of Cantabria
39005, Santander, Spain
[2] Computer Science and Electronics Department
University of Cantabria
39004, Santander, Spain
Email: lopezvfe@unican.es

Abstract— Bridgeless PFC is a one-stage AC/DC rectifier. Such a solution achieves higher efficiency compared to traditional PFC topologies (diode bridge + DC/DC converter). However, the input AC voltage must be measured before the input inductor. Line current sensing is expensive or produces load dependent loss. This paper presents a solution that avoids the use of an input current sensor. Instead, the current is rebuilt inside a digital device (FPGA) by indirectly measuring the voltage across the input inductor.

Index Terms— PFC, bridgeless, current sensorless, FPGA, harmonic distortion

I. INTRODUCTION

THE use of an active AC/DC front-end stage is becoming mandatory in AC line connected equipment. Most home appliances, like TV, desktop PC, battery chargers, etc. make use of it. The traditional diodes based AC/DC rectifiers cause harmonic pollution and poor power factor, and consequently increase losses, reduce the capability of the line to transport active power and result in premature aging in distribution cables and transformers [1], [2].

This fact and the need to comply with standards like IEC 61000-3-2 require using PFC stages in off-line power supplies [3].

Many topologies are suitable for this task and different configurations are available [4]. The most extended one consists of a diode Bridge rectifier plus a Boost converter. However, this solution is a two-stage configuration so the power loss penalizes its performance due to the double power processing. It can be shown that the Bridgeless topology improves the efficiency because only two devices conduct in each instant, considering the continuous conduction mode (CCM), while three devices conducts in the Bridge plus Boost topology [5], [6].

Input voltage and current sensing methods are more complicated in the Bridgeless converter than in the conventional Bridge + Boost PFC because the reference of these signals is not the common ground. Hence, input voltage measurement requires an isolated transformer or a differential amplifier.

To sense the line current, the problem is even greater because the available solutions are expensive (Hall-effect sensor), dissipate power (shunt resistor) or need a complicated circuitry if they are based on a current transformer [5], [7]–[9].

This paper presents a solution that avoids the use of a sensor for input current sensing. Instead, the input current (same as the inductor current) is rebuilt in a digital device (Spartan XC7A100T) measuring the voltage across the input inductance [10].

II. BRIDGELESS PFC CONVERTER

Among the different bridgeless topologies available and already presented [4], [11], the one in the Fig. 1 is used in this paper because of its simplicity and ease implementation, because the gate signals, *G1* and *G2*, share the common ground reference.

The duty cycle, *d*, is defined in the positive line semi period as the time in which transistor *Q1* is in the ON-state over the switching period. Similarly, *d*, is defined in the negative line semi period as the time in which transistor *Q2* is in the ON-state over the switching period.

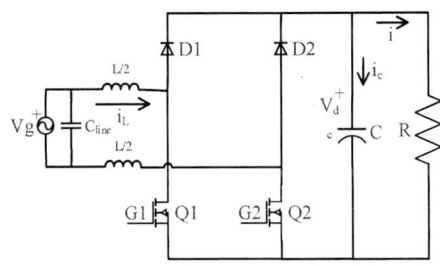

Fig. 1. Bridgeless PFC power converter

During the positive half line cycle ($v_g > 0$), two states are possible in each switching cycle (T_S):

1. When $Q1$ is ON, during dT_s, the current flows through $Q1$ and $Q2$, as represented in Fig. 2a.

The equations that represent this state are:

$$v_L = v_g - i_L(r_L + 2R_{DS}) \tag{1}$$
$$i_c = -i, \tag{2}$$

where i is the load current.

To better model the behavior of the bridgeless converter, the resistances r_L and R_{DS} are taken into account:

- r_L represents the parasitic resistance of the input inductor
- R_{DS} represents the parasitic resistance of the MOSFET

2. When $Q1$ is in the OFF-state, the current flows through $D1$ and $Q2$, as can be seen in Fig. 2b.

The equations that represent this state are:

$$v_L = v_g - i_L r_L - V_F - V_{DC} - i_L R_{DS} \tag{3}$$
$$i_c = i_L - i \tag{4}$$

- V_F represents the voltage drop across the upper diode ($D1$) in the off-state.

In the same way, during the negative half-line cycle ($v_g < 0$), also two states occur in each switching cycle:

1. During the ON-state, dT_s, the current flows through $Q2$ and $Q1$ as represented in Fig. 2c.

The equations in this state are:

$$v_L = v_g - i_L(r_L + 2R_{DS}) \tag{5}$$
$$i_c = -i \tag{6}$$

2. Similarly, during the OFF-state, $(1-d)T_s$, the current flows through $D2$ and $Q1$ as represented in Fig. 2d.

In this state, the equations are:

$$v_L = v_g - i_L r_L - i_L R_{DS} - V_F + V_{DC} \tag{7}$$
$$i_c + i_L + i = 0 \tag{8}$$

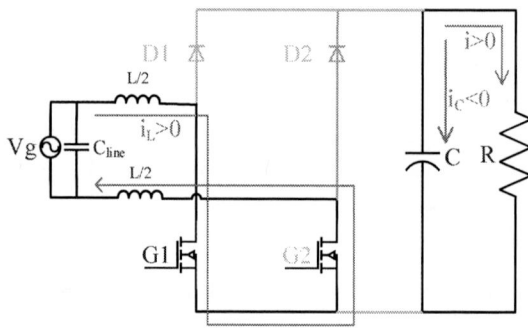

a) ON-state during the positive half-line cycle

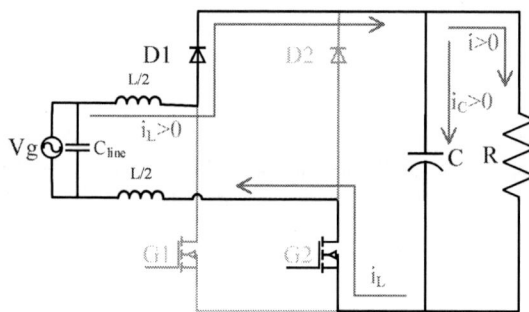

b) OFF-state during the positive half-line cycle

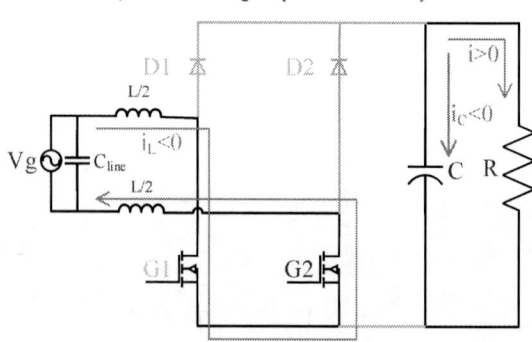

d) OFF-state during the negative half-line cycle

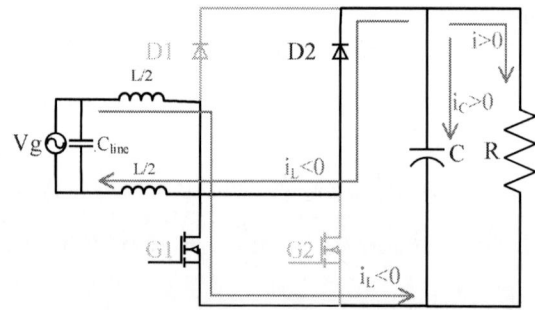

d) OFF-state during the negative half-line cycle

Fig. 2. Possible states of the selected topologies based on the input voltage and the active devices ($Q1$, $Q2$).

To sum up, during the ON-state, the equation that models the behavior of the converter is:

$$v_L = v_g - i_L(r_L + 2R_{DS}),\qquad(9)$$

while, during the OFF-state, the equation that models the converter is:

$$v_L = v_g - sign(v_g)V_{DC} - i_L(r_L + R_{DS}) - V_F\qquad(10)$$

where

$$sign(v_g) = \begin{cases} 1 & when\ v_g > 0 \\ -1 & when\ v_g < 0 \end{cases}\qquad(11)$$

Averaging through the switching period, with d obtained by comparing the estimated current with a carrier signal, using the Non-Linear Carrier Control (NLC) technique [12], as shown in Fig. 3 the following equation is obtained, where $\langle v_L \rangle_{Ts}$ is average the voltage across the inductor over the switching period:

$$\langle v_L \rangle_{T_S} = \langle v_g \rangle_{T_S} - sign(v_g)V_{DC}(1-d) - i_L(r_L + R_{DS}) - i_L R_{DS}d - V_F(1-d))\qquad(12)$$

The equation shown in (12) is valid when the converter operates in continuous conduction mode (CCM). However, close to the zero-crossing point of the grid voltage, discontinuous conduction mode (DCM) appears, especially under light load condition. In this mode, as can be seen in Fig. 4, a third state appears at the end of the switching period. This new state must be taken into account in the current rebuilding algorithm. In this third state, the voltage across the input inductor is $v_L = 0V$.

Finally, as can be seen in (12), the sign of the input voltage must be known by the estimation algorithm. To properly acquire the input voltage sign, a digital PLL is used to detect the zero-crossing instant. This module will be deeper explained in the next section.

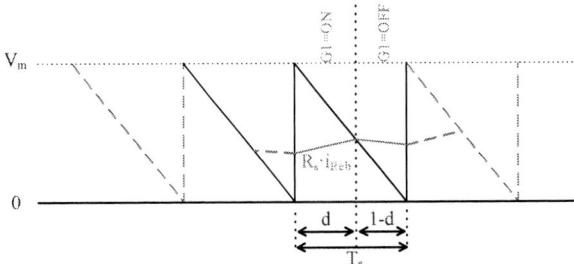

a) Positive half-line cycle

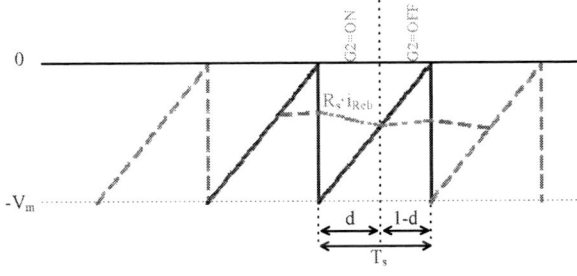

b) Negative half-line cycle

Fig. 3. Switching sequence a) positive half-line cycle b) negative half-line cycle

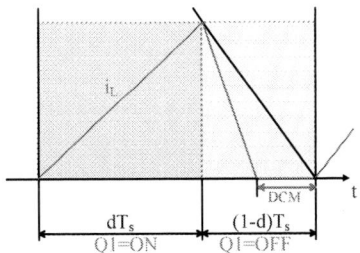

Fig. 4. DCM effect

III. CONTROL SCHEME

The block diagram of the NLC current control implemented in the FPGA, for the bridgeless converter to operate as a PFC, is depicted in Fig. 5.

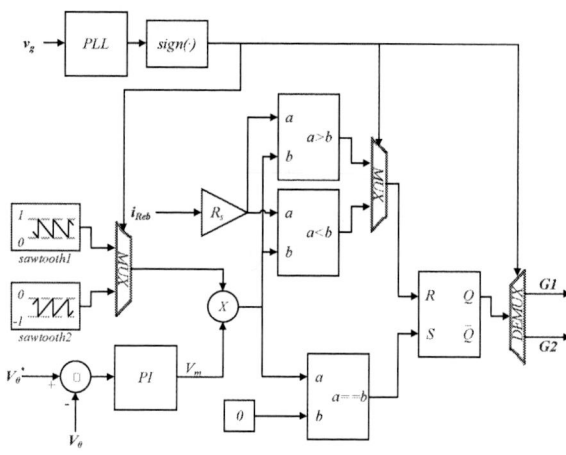

Fig. 5. Control scheme block diagram

This controller assumes positive current. Therefore, a rectification action is computed inside the FPGA once the sign of the input voltage is known, see (12).

There are several ways to achieve the synchronization with the line to acquire the voltage sign. One option is based on an analog zero-cross detector [13]. The main problem of zero-cross detectors are low immunity to harmonics and noise, as it is well summarized in [14]. Even though this circuit obtains good results for lots of applications, the growing up concern about power quality problems makes it advisable to consider other higher performance solutions [15]. Taking that into account, a digital Synchronous Rotary Frame PLL (SRF-PLL) is adopted [16] in order to accurately determine the sign of the input voltage. The input voltage is measured through a sigma-delta analog-to-digital converter ($\Sigma\Delta$ ADC) and rebuilt by means of a cascaded integrator–comb (CIC) filter, then, a virtual β component is generated by means of a T/4 delay buffer. The obtained $\alpha\beta$ components are normalized before applying the transformation to synchronously rotating frame (Park transform). This frame is rotated forcing the imaginary axis to become orthogonal to α, and hence, synchronized to the grid. This is carried out by applying a low pass filter after the Park transform [17].

The last step is the current rebuilt computation. To do so, the equations (9) and (11) are implemented inside the FPGA in order to estimate the voltage across the inductor (v_L) (taking also into account the periods in which the converter operates in DCM). Knowing the voltage across the inductor, the estimated current through it (i_{Reb}) is calculated:

$$v_L(t) = L \cdot \frac{di_L(t)}{dt} + i_L(t) \cdot r_L \qquad (13)$$

That turns into (14) in the Laplace domain:

$$Y_L(s) = \frac{I_L(s)}{V_L(s)} = \frac{1}{r_L}\frac{1}{1 + s\frac{L}{r_L}} \qquad (14)$$

To implement (14) in a digital device, it has to be discretized, using the following parameters:

- Clock period: 10ns
- Discretization method: zero-order hold ('zoh')

Which results in (15):

$$Y_L(s) = \frac{1.101 \cdot 10^{-5} \cdot z^{-1}}{1 - 0.999989899 \cdot z^{-1}} \qquad (15)$$

IV. SIMULATIONS AND EXPERIMENTAL RESULTS

To validate the proposed control, simulations have been carried out under the conditions shown in Table 1.

Table 1. Parameters of the bridgeless PFC converter

Input Voltage	50Vrms
Output Voltage	150V
Load (R)	500Ω
Input Frequency	50Hz
V_F	1.5V
R_{DS}	400mΩ
r_L	1Ω
L	1mH
Switching Frequency	48kHz
DC-link	940µF

The simulations have been carried out using MATLAB/Simulink and PLECS, and the results obtained are shown next.

Fig. 6 includes the grid voltage (v_g) and the current (i_L) impressed by converter working as a PFC.

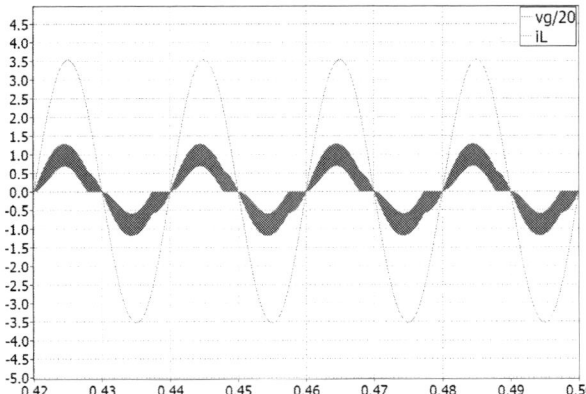

Fig. 6. Waveforms obtained from the simulations: a) grid voltage (green), b) grid current (red).

It can be seen that the rebuilt current is not perfectly sinusoidal. This is due to differences between the estimated inductor volt-seconds in the rebuilding algorithm and the actually applied in the converter model. Keep in mind that converter parasitics (inductor resistance, r_L, MOSFET-ON resistance R_{DS}, etc.) are taken into account, but not dynamically compensated yet, since the purpose of this paper is to demonstrate the viability of the current sensorless control applied to the bridgeless topology. However, these effects cannot be directly extrapolated to the laboratory results since more factors are involved like turn-on and turn-off delays, ADC accuracy and non-linearities, etc.

In the same way, and under the same conditions shown in Table 1, some experimental tests were carried out in the laboratory setup. As explained in the previous section and expressed in (12), still two variables need to be measured: v_g and v_{DC}. To do so, a simple resistor divider and an optically isolated ΣΔ ADC are used for each one. In the same way, to drive the MOSFETs from the FPGA, and optically isolated driver is used. The scheme described above is shown in Fig. 7.

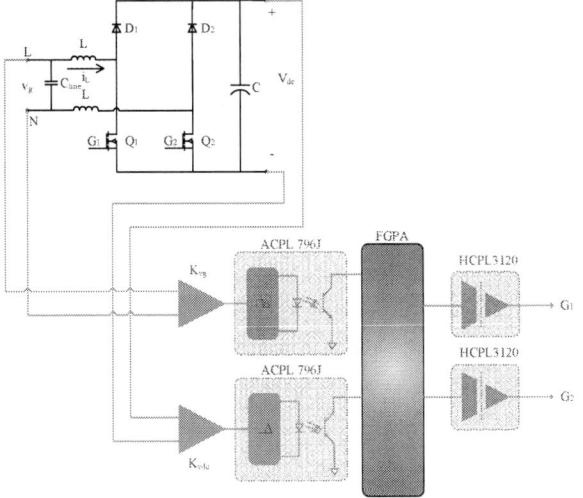

Fig. 7. Scheme used in the laboratory setup

In the AC side, an Agilent 6813B is used as a power source. In the DC side, a 500Ω resistor is used as a load. In Fig. 8, the obtained waveforms are shown.

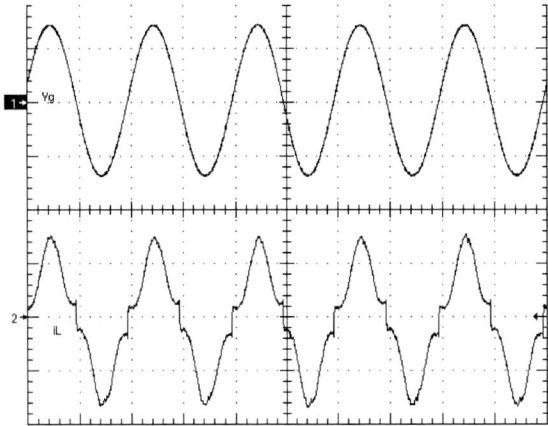

Fig. 8. Waveforms obtained from the oscilloscope: a) voltage grid (channel 1, 50V/div) b) grid current (channel 2, 1A/div)

To prove that the converter fulfill the standard IEC 61000-3-2 Class C, an application included in the same scope was ran. The results are shown in Fig. 9 in a bar graph format. Yellow bars represent the limits established by the standard IEC 61000-3-2 and green bars shown the harmonic content of the input current demanded by the PFC. It can be seen that, even some harmonic content is present (mainly third harmonic), the proposed solution matches the limits established by the standard.

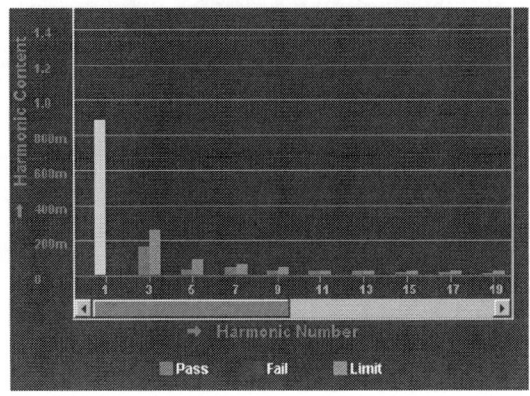

Fig. 9. Line current harmonic content

V. CONCLUSIONS

A current control method based in the NLC control that does not need to sense the input current has been proposed for the Bridgeless PFC converter. The control is implemented in an FPGA. Simulations and experimental results have been performed to assess the proposed method according to values included in Table 1. Results proof the viability of the proposal, although some distortion is observed, especially when the

978-1-4673-9551-9/16 $31.00 © 2016 IEEE

converter operates in DCM. Further algorithm modifications and experimental tests will be carried out in the lab to extend it to a standard input voltage and to higher powers rate. Additionally, dynamic compensation will be implemented in order to minimize the errors due to non-ideal behavior of the converter due to parasitic resistances and non-linear characteristic of the input inductor, on-resistance (R_{DS}) of the MOSFETs, turn-on and turn-off delays in the driving signals, etc.

ACKNOWLEDGMENT

This work is funded by the Spanish Government and the European Union through the project TEC2014-52316-R: "Estimation and Optimal Control for Energy Conversion with Digital Devices": ECOTRENDD.

REFERENCES

[1] M. Yazdani-Asrami, M. Mirzaie, and A. A. S. Akmal, "Investigation on impact of current harmonic contents on the distribution transformer losses and remaining life," in *2010 IEEE International Conference on Power and Energy*, 2010, pp. 689–694.

[2] D. M. Said, K. M. Nor, and M. S. Majid, "Analysis of distribution transformer losses and life expectancy using measured harmonic data," in *Proceedings of 14th International Conference on Harmonics and Quality of Power - ICHQP 2010*, 2010, pp. 1–6.

[3] O. Garcia, J. A. Cobos, R. Prieto, P. Alou, and J. Uceda, "Single phase power factor correction: a survey," *IEEE Trans. Power Electron.*, vol. 18, no. 3, pp. 749–755, May 2003.

[4] B. Singh, B. N. Singh, A. Chandra, K. Al-Haddad, A. Pandey, and D. P. Kothari, "A review of single-phase improved power quality ac~dc converters," *IEEE Trans. Ind. Electron.*, vol. 50, no. 5, pp. 962–981, Oct. 2003.

[5] L. Huber and M. M. Jovanovic, "Performance Evaluation of Bridgeless PFC Boost Rectifiers," *IEEE Trans. Power Electron.*, vol. 23, no. 3, pp. 1381–1390, May 2008.

[6] K. Masumoto, K. Shi, M. Shoyama, and S. Tomioka, "Comparative study on efficiency and switching noise of bridgeless PFC circuits," in *2013 IEEE 10th International Conference on Power Electronics and Drive Systems (PEDS)*, 2013, pp. 613–618.

[7] M. Reddig, W. Zhou, and M. Schlenk, "True bridgeless PFC - stages with advanced current measuring circuit," in *2011 IEEE 33rd International Telecommunications Energy Conference (INTELEC)*, 2011, pp. 1–6.

[8] T. Qi, L. Xing, and J. Sun, "Dual-Boost PFC Converter Control Without Input Current Sensing," in *2009 Twenty-Fourth Annual IEEE Applied Power Electronics Conference and Exposition*, 2009, pp. 1855–1861.

[9] C.-M. Wang, "A Novel ZCS-PWM Power-Factor Preregulator With Reduced Conduction Losses," *IEEE Trans. Ind. Electron.*, vol. 52, no. 3, pp. 689–700, Jun. 2005.

[10] F. Javier Azcondo, A. de Castro, V. M. Lopez, and O. Garcia, "Power Factor Correction Without Current Sensor Based on Digital Current Rebuilding," *IEEE Trans. Power Electron.*, vol. 25, no. 6, pp. 1527–1536, Jun. 2010.

[11] Bin Su and Zhengyu Lu, "An Interleaved Totem-Pole Boost Bridgeless Rectifier With Reduced Reverse-Recovery Problems For Power Factor Correction," *IEEE Trans. Power Electron.*, vol. 25, no. 6, pp. 1406–1415, Jun. 2010.

[12] D. Maksimovic and R. W. Erickson, "Nonlinear-carrier control for high-power-factor boost rectifiers," *IEEE Trans. Power Electron.*, vol. 11, no. 4, pp. 578–584, Jul. 1996.

[13] V. M. Lopez, F. J. Azcondo, A. de Castro, and R. Zane, "Universal Digital Controller for Boost CCM Power Factor Correction Stages Based on Current Rebuilding Concept," *IEEE Trans. Power Electron.*, vol. 29, no. 7, pp. 3818–3829, Jul. 2014.

[14] F. Blaabjerg, R. Teodorescu, M. Liserre, and A. V. Timbus, "Overview of Control and Grid Synchronization for Distributed Power Generation Systems," *IEEE Trans. Ind. Electron.*, vol. 53, no. 5, pp. 1398–1409, Oct. 2006.

[15] F. Lopez-Colino, A. Sanchez, G. Alvarez, A. de Castro, and J. Garrido, "Handling input voltage frequency variations in power factor correctors with precalculated duty cycles," in *2014 IEEE 15th Workshop on Control and Modeling for Power Electronics (COMPEL)*, 2014, pp. 1–5.

[16] V. Kaura and V. Blasko, "Operation of a phase locked loop system under distorted utility conditions," *IEEE Trans. Ind. Appl.*, vol. 33, no. 1, pp. 58–63, 1997.

[17] P. R. Remus Teodorescu, Marco Liserre, *Grid Converters for Photovoltaic and Wind Power Systems*. Wiley, 2011, p. 416.

978-1-4673-9551-9/16 $31.00 © 2016 IEEE

A Compact Electrolytic-Free Two-Stage Universal Input Offline LED Driver

Saad Pervaiz, Ashish Kumar, Khurram K. Afridi

University of Colorado Boulder, Boulder, CO, 80309, USA

Email: saad.pervaiz@colorado.edu

Abstract—This paper presents a single phase, two-stage electrolytic-free offline LED driver that utilizes a hybrid film-ceramic stacked switched capacitor (SSC) energy buffer (EB) in place of limited-life electrolytic capacitors for twice-line-frequency energy buffering. The proposed LED driver comprises a front-end power factor correction (PFC) stage, followed by an isolated dc-dc conversion stage, with an SSC energy buffer connected across the intermediate dc bus. Compared to a single capacitor, the use of SSC energy buffer reduces the passive volume of the energy buffer by a factor of two. A simple and robust control scheme is proposed to interface the SCC energy buffer with the PFC converter. A prototype 300 W LED driver for universal input voltage range (90-265 Vrms) and 12 V output voltage is designed, built and tested based on the proposed architecture. The SSC energy buffer achieves a peak efficiency of 98.5% and maintains an efficiency of above 96% across a wide output power range, thus minimally impacting the overall system efficiency. In addition, the impact of the intermediate dc bus voltage ripple on overall system efficiency is investigated. It is found that there is negligible change in overall system efficiency when the intermediate bus voltage ripple ratio is varied between 3% and 7.5%.

Keywords—ac/dc power conversion, dc/ac power conversion, energy buffer, switched capacitor, twice line frequency.

I. INTRODUCTION

Single-phase, high power factor ac-dc converters require energy storage to buffer the difference in instantaneous power between their input and output ports. The instantaneous power at the input ac port varies at twice the line frequency, whereas constant power is required at the output dc port, warranting an intermediate energy storage element. Traditionally, electrolytic capacitors are used as the twice-line-frequency energy storage elements, but their relatively short lifetime limits the life of the overall converter [1]. Electrolytic capacitors can be replaced by longer life film and ceramic capacitors. However, these alternatives have lower energy density, and a direct replacement of electrolytic capacitors by film or ceramic capacitors results in an increase in the passive volume. With the ever increasing high energy density demand on power electronics, this is not a desirable solution [2]. However, film and ceramic capacitors do have a lower ESR compared to electrolytic capacitors. This allows them to be efficiently charged/discharged across a much wider voltage range enabling a larger fraction of their total stored energy to be utilized. Hence, film and ceramic capacitors can be employed for twice-line-frequency energy buffering, provided a mechanism exists to maintain the bus voltage ripple within prescribed limits. A number of approaches have been presented in literature to increase the energy utilization of capacitors, including the use of bi-directional dc-dc converters [3], [4], energy buffers incorporated in the power stage design

[5]-[7], and use of switched capacitor energy buffers [8]-[15]. Amongst the switched capacitor approaches, the stacked switched capacitor (SSC) energy buffer has the least complexity [10]-[15]. However, all the SSC energy buffers presented in the past have been designed for single-stage ac-dc converters.

Figure 1 shows the topology of a conventional single phase, two-stage ac-dc converter. It consists of a front end PFC stage followed by a second isolated dc-dc stage. For twice line frequency energy buffering, a relatively large electrolytic capacitor is connected at the intermediate dc bus between the two converters. This paper presents a two-stage electrolytic-free off-line LED driver that utilizes a stacked switched capacitor energy buffer to store the twice line frequency energy. The proposed LED driver consists of a front-end power factor correction (PFC) stage with the SSC energy buffer at its output, followed by an isolated dc-dc conversion stage that regulates the output voltage. The SSC energy buffer presented here uses both film and ceramic capacitors to achieve a favorable tradeoff between size and manufacturing complexity. In addition, this paper introduces a novel, simple and robust control strategy to interface the SSC energy buffer with the PFC converter. The proposed scheme eliminates the need for an artificial voltage feedback, which has been needed in past designs [10], thus reducing system complexity. A prototype 300 W LED driver for universal input voltage range (90-265 Vrms) and 12 V output voltage has been designed, built and tested. The SSC energy buffer achieves a peak efficiency of 98.5% while maintaining efficiency above 96% across a wide output power range, thus minimally impacting the overall system efficiency. The effect of the intermediate dc bus voltage ripple on the overall system efficiency is also investigated. It is found that the change in the efficiency of the overall converter is negligible when the intermediate bus voltage ripple ratio is varied from 3% to 7.5%. This indicates that the system size can be significantly reduced by designing it with a higher intermediate bus ripple without compromising efficiency.

The remainder of this paper is organized as follows: Section II presents the proposed system architecture, along with

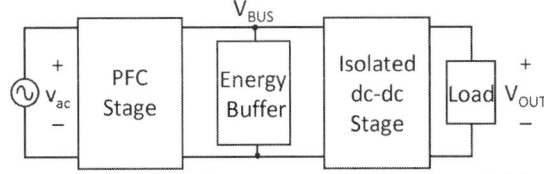

Figure 1: Architecture of a two stage, single phase offline LED driver.

a discussion on the topology selection for the SSC energy buffer. The proposed control scheme to interface the SSC energy buffer with the PFC stage is presented in section III. Details regarding the prototype design and component selection are presented in section IV. Section V presents the experimental results. Finally, section VI presents the conclusions.

II. PROPOSED ARCHITECTURE

The proposed architecture of a two-stage electrolytic-free LED driver, utilizing a stacked switched capacitor (SSC) energy buffer is shown in Fig. 2. It comprises a front end PFC stage, followed by an isolated dc-dc stage with the SSC energy buffer connected across the intermediate dc bus. The following subsection describes the working of the SSC energy buffer.

A. SSC Energy Buffer – Operational Principle

An SSC energy buffer consists of two series connected blocks (referred to as backbone and supporting blocks) of switches and capacitors. Its basic operational principle is that the voltage across the individual blocks and individual capacitors is allowed to vary across a relatively wide range. However, the voltage variation across one block compensates for the variation across the other, such that the total voltage across both blocks, which is the intermediate dc bus voltage, remains within the desired range. By allowing its individual capacitors to charge and discharge over a wide voltage range, the SSC energy buffer utilizes the total energy stored in its capacitors effectively; hence, minimizing the total passive volume of the energy buffer. Therefore, even relatively low energy density film and ceramic capacitors can be used without a substantial increase in the passive volume relative to a single high-energy density electrolytic capacitor. Several implementations of the SSC energy buffer have been proposed, including unipolar and bipolar designs. Each of these have enhanced variants, which can be further optimized by appropriately selecting their capacitance values [10]-[15]. Compared to their bipolar counterparts, enhanced unipolar SSC energy buffers offer reduced switch complexity and are the focus of the rest of this paper.

An enhanced unipolar SSC energy buffer (shown in Fig. 2) has one backbone capacitor, m supporting capacitors, $m + 1$ switches and is referred to as a 1-m enhanced unipolar SSC energy buffer. This SSC energy buffer operates as follows: before the buffer starts normal operation, all the capacitors are charged to specific voltage levels using a pre-charge sequence. Figure 3 shows the steady-state operational waveforms of a 1-m enhanced unipolar SSC energy buffer. During the

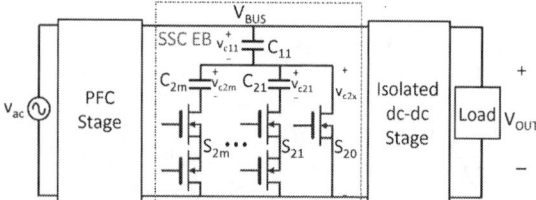

Figure 2: Proposed architecture of a two-stage LED driver utilizing a stacked switched capacitor (SSC) energy buffer.

Figure 3: Operational waveforms for a 1-m enhanced unipolar SSC energy buffer.

discharge phase, first the switch S_{20} is turned on and the backbone capacitor C_{11} is connected across the bus. When the bus voltage decreases to the minimum allowed voltage, S_{20} is turned off and S_{21} is turned on. This connects the supporting capacitor C_{21} in series with the backbone capacitor. The voltage across the backbone and the supporting capacitors now add up to the maximum allowed bus voltage and the discharge process continues. C_{21} is disconnected once the bus voltage reaches its minimum allowed value again, and C_{22} is connected in series with the backbone capacitor by turning on S_{22}. The discharge process continues until all the supporting capacitors have been discharged, and the backbone capacitor reaches its minimum voltage level. At this point, the charging cycle begins, which is simply the reverse of the discharging cycle. It can be seen from Fig. 3 that the backbone capacitor, C_{11}, which stores the major part of the energy of the buffer, experiences a large voltage swing. Hence, a large fraction of the total energy stored in the SSC energy buffer is extracted.

B. SSC Energy Buffer Topology and Voltage Ripple Ratio Selection

The SSC energy buffer utilizes more of its total stored energy than a single capacitor, allowing substantial reduction in passive volume. The passive volume reduction achieved by an SSC energy buffer relative to a single capacitor depends on its topology and the allowed intermediate dc bus voltage ripple ratio, R_v which is defined as the ratio of the peak voltage ripple amplitude to the nominal value of the intermediate dc bus voltage [15]. For an enhanced unipolar SSC energy buffer, designed for a voltage ripple ratio of 5%, the passive volume of the buffer as a function of the number of switches, relative to a single capacitor (0 switches) is shown in Fig. 4. It can be seen that even a simple 1-1 enhanced unipolar SSC energy buffer, containing 2 switches, can result in greater than 40% reduction in passive volume. As a trade-off between simplicity and passive volume reduction, a 1-2 enhanced unipolar SSC energy buffer is selected for the proposed architecture.

The volume reduction achieved by an SSC energy buffer also depends on the designed ripple ratio, R_v. The higher the allowed voltage ripple ratio, the greater the energy that can be extracted from the energy buffer and the smaller the size. The

978-1-4673-9551-9/16 $31.00 © 2016 IEEE

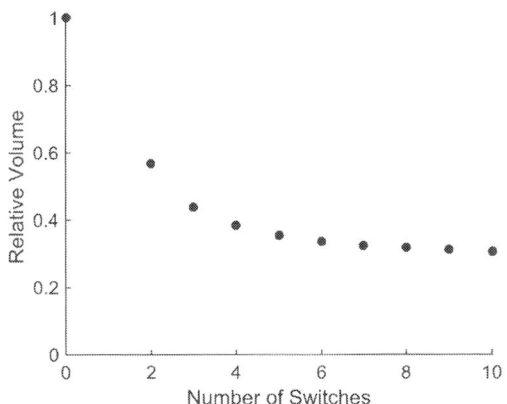

Figure 4: Passive volume of enhanced unipolar SSC energy buffer relative to a single capacitor, as a function of number of switches for a voltage ripple ratio of 5%.

volume reduction achieved by a 1-2 enhanced unipolar SSC energy buffer (having 3 switches), relative to its volume at 1% voltage ripple ratio, as a function of ripple ratio is shown in Fig. 5. It can be seen that a ripple ratio of 5% results in more than 70% reduction in passive volume, compared to a 1% ripple ratio design. However, a larger intermediate dc bus voltage ripple imposes additional stress on the second stage dc-dc converter, which now has to operate over a wider input voltage range, potentially reducing its efficiency. Hence, the voltage ripple ratio needs to be selected based on a trade-off between passive volume reduction and the efficiency of the converter, as is discussed further in Section V.

III. CONTROL DESIGN

A variable ac input voltage and/or variable output power PFC stage generally employs a voltage control loop that maintains the intermediate dc bus voltage at a nominal value. In a conventional two-stage offline LED driver, a single

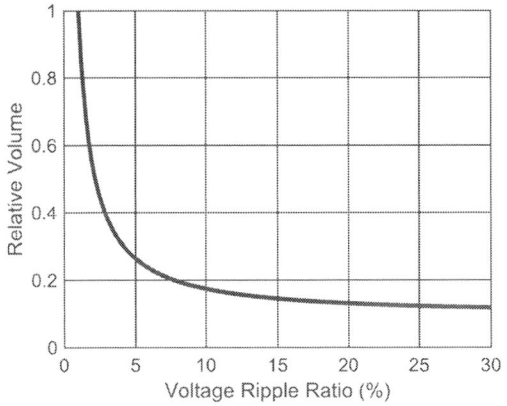

Figure 5: Passive volume reduction achieved by 1-2 enhanced unipolar SSC energy buffer as a function of R_v, relative to 1% ripple ratio case.

capacitor is connected at the output of the PFC stage to buffer twice-line-frequency energy. The PFC stage uses the intermediate bus voltage, (i.e., the voltage across the buffering capacitor) to control its average output power. Since this voltage is a good measure of the energy stored in the capacitor, this feedback mechanism ensures that the average output power from the PFC matches the average input power drawn by the isolated dc-dc stage and the intermediate bus voltage remains stable. However, when the single capacitor is replaced by the SSC energy buffer, the bus voltage is no longer a true representation of the actual energy stored in the SSC energy buffer. All previous designs of variable ac input voltage/variable output power ac-dc converters with an SSC energy buffer have utilized an artificial voltage feedback to maintain a stable intermediate bus voltage [10]. In this method, an artificial voltage is generated based on the state of the SSC energy buffer that represents the stored energy. This computed voltage value is then fed to the PFC stage in place of the bus voltage. This method typically requires an additional micro-controller, dedicated to computing the energy in the SSC energy buffer and converting it to an artificial voltage. This adds complexity to the overall system, increases cost and makes it less reliable.

The control approach proposed here incorporates the effect of the SSC energy buffer into the design of the voltage control loop of the PFC stage. The first step in the design is to compute the control-to-output voltage transfer function of the PFC stage. This transfer function depends on the topology of the PFC stage and the effective capacitance connected at its output. A generic low-frequency small signal model for a PFC stage is shown in Fig. 6 [16]. The model parameters g_2, j_2, and r_2 depend on the type of the PFC stage and the control scheme being used. From Fig. 6, the control-to-output voltage transfer function of the PFC stage can be computed and is given as:

$$\frac{\hat{v}}{\hat{v}_c} = j_2 R_o \parallel r_2 \frac{1}{1+sC_oR_o\parallel r_2}. \qquad (1)$$

It can be seen from (1) that the system dynamics depend on the output capacitance C_o, which determines the closed loop pole location. In traditional systems, C_0 is the capacitance of the twice-line-frequency energy buffering capacitor. However, in the case of the SSC energy buffer, an effective capacitance of the SSC energy buffer needs to be computed, which can then be used in the design of the feedback loop. This effective capacitance can be determined from the total energy that flows into the buffer during a complete charge cycle and the corresponding change in the bus voltage. Over a charging period, the bus voltage increases from its minimum allowed value to its maximum allowed value. Hence, the equivalent capacitance of the SSC energy buffer with a ripple ratio of R_v can be expressed as:

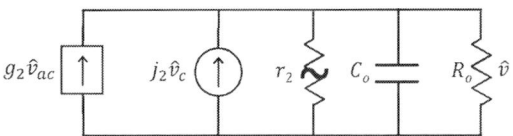

Figure 6: Low frequency small-signal model of a PFC converter.

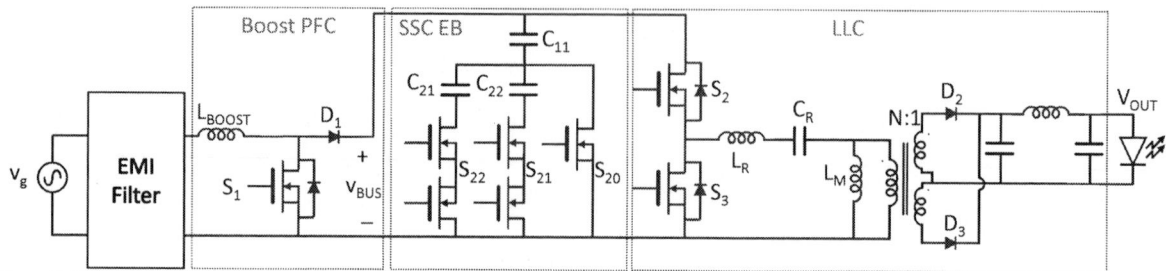

Figure 7: Proposed two-stage LED driver, consisting of an input PFC boost stage, followed by SSC energy buffer and a half-bridge LLC resonant dc-dc stage.

$$C_{eq} = \frac{P_{dc}}{2R_v V_{nom}^2 \omega_{line}}, \qquad (2)$$

where P_{dc} is the average power that flows into the isolated dc-dc stage, V_{nom} is the nominal intermediate dc bus voltage, R_v is the intermediate bus voltage ripple ratio (defined as: $R_v = \frac{V_{max}-V_{nom}}{V_{nom}}$, where V_{max} is the maximum allowed bus voltage), and ω_{line} is the ac line radial frequency. The control-to-output voltage transfer function of the PFC boost stage with an SSC energy buffer can now be obtained by replacing C_0 in (1) with the equivalent capacitance C_{eq} given in (2). Hence, a compensator can now be designed using conventional control design techniques that directly sense the intermediate bus voltage and feed it back to the controller without any additional processing. This eliminates any inaccuracies and delays that would result from the artificial voltage feedback loop, making the control loop more robust.

IV. PROTOTYPE DESIGN

A 300 W, universal input voltage (90-265 V$_{rms}$), 12 V output voltage, two-stage electrolytic-free LED driver has been designed, built and tested. The topology of the prototype system is shown in Fig. 7. It comprises a front-end PFC boost stage, a half-bridge LLC resonant converter as the second stage, and a 1-2 enhanced unipolar SSC energy buffer connected at the intermediate dc bus. A photograph of the prototyped LED driver is shown in Fig. 8.

For the proposed LED driver, a continuous conduction mode boost converter is used as the PFC stage. The boost converter is operated using fixed off-time control (FOT). In this control scheme, the inner current loop uses peak current control. Once the inductor current (which is the same as the switch current) reaches the desired peak value, the active switch S_1 is turned off and kept off for a fixed time period. The outer voltage loop regulates the dc bus voltage using a PI compensator. This control scheme has the advantage that its inner current loop is unconditionally stable and no additional ramp is required [17]. For the designed PFC stage, the boost off-time is set to be 5 μs, the maximum switching frequency is 80 kHz, while the inductance value is chosen to be 500 μH. Since the boost converter is designed for an input voltage range of 90-265 V$_{rms}$, the nominal bus voltage, V_{nom}, is regulated at 400 V.

For the energy buffer, a 1-2 enhanced unipolar SSC energy buffer, with a ripple ratio of 5% is designed. This

Table I: System Design Parameters

Component	Values
L_{BOOST}	500 μH
C_{11}	30 μF
C_{21}	150 μF
C_{22}	100 μF
L_R	55 μH
C_R	24 nF
L_M	280 μH

implementation of the SSC energy buffer has 3 switches. From Fig. 4, it can be seen that this design results in more than 50% decrease in passive volume as compared to a single capacitor, and thus presents a good compromise between switch complexity and volume reduction. The designed SSC energy buffer, shown in Fig. 8, uses both film and ceramic capacitors. The backbone capacitor (C_{11}) is realized using a single film capacitor, since the ceramic capacitors at high voltage levels (> 400 V) are available in small denominations only. As a result, a ceramic capacitor based solution would consist of a large number of capacitors, presenting difficulties in manufacturing. For the supporting capacitors, which are low voltage capacitors, class-II ceramic capacitors are used. The non-linear capacitance variation of these capacitors with applied voltage is taken into account when selecting an appropriate capacitance value for these capacitors. The capacitancs used in the prototype 1-2 enhanced unipolar SSC energy buffer are given in Table I.

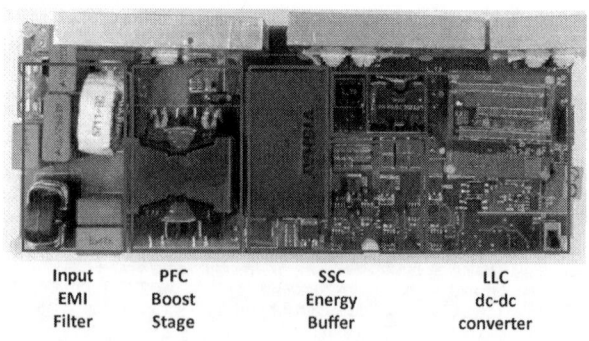

Figure 8: Photograph of the prototype 300 W electrolytic-free offline LED driver.

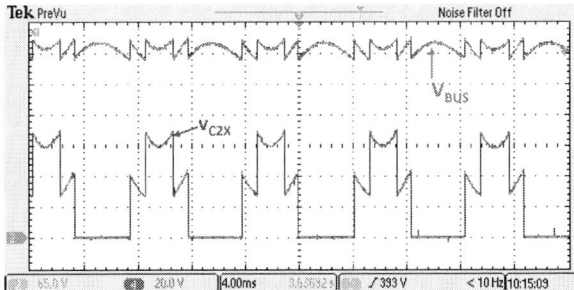

Figure 9: Intermediate bus voltage (V_{BUS}) (blue) and center node voltage (v_{c2x}) (green) of the SSC energy buffer at full load.

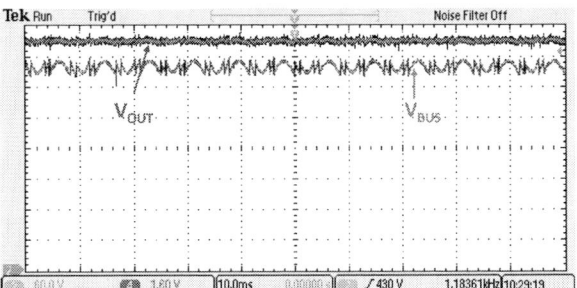

Figure 10: Intermediate bus voltage, V_{BUS} (blue) and the output voltage, V_{OUT} (green).

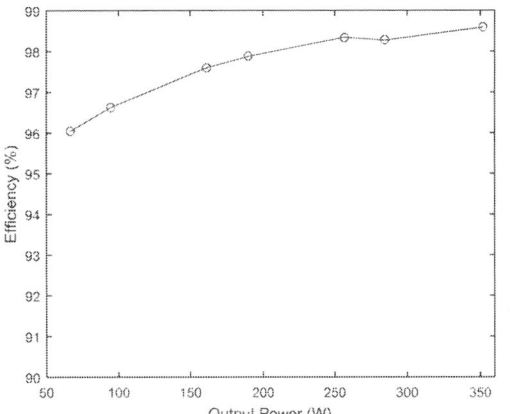

Figure 11: Round-trip efficiency of the SSC energy buffer as a function of output power.

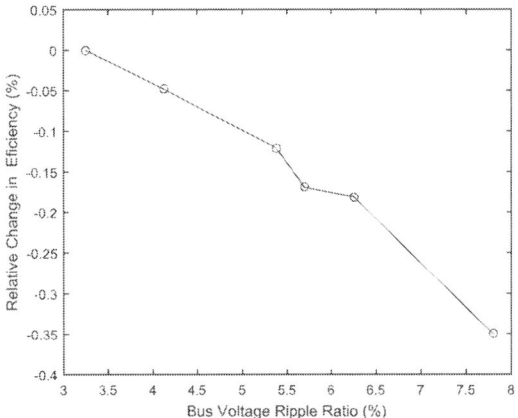

Figure 12: Relative change in overall efficiency of the converter as a function of intermediate bus voltage ripple ratio.

For the second stage dc-dc converter, an LLC resonant converter is chosen. An LLC resonant converter is known to maintain high efficiency across wide input voltage range [18], and conveniently provides isolation and large step down capability. Since this stage is required to provide a high step down ratio, a half bridge inverter and a full-bridge rectifier are used. Frequency control is utilized for output voltage regulation. Component values and other operational parameters of the prototyped LED driver are listed in Table I.

V. EXPERIMENTAL RESULTS

The prototype electrolytic-free offline LED driver is tested up to 300 W with an LED load. The intermediate bus voltage, V_{BUS} and the center node voltage of the SSC energy buffer, v_{c2x}, working at full load are also shown in Fig. 9. It can be seen that the SSC energy buffer maintains the bus voltage ripple within the specified bounds, with a nominal value of 400 V and a ripple ratio of 5%. Figure 10 shows the intermediate bus voltage, V_{BUS}, and the system output voltage, V_{OUT}, at the rated output power (300 W). It can be seen that the ripple on the output voltage is much smaller.

The efficiency of the SSC energy buffer operating at various power levels is shown in Fig. 11. It can be seen that the SSC energy buffer achieves a peak efficiency of 98.5% at full load, and maintains an efficiency of above 96% across a wide (5:1) output power range. As a result, the SSC energy buffer has minimal impact on the efficiency of the overall system.

In this work, the impact of the intermediate bus voltage ripple on the efficiency of the overall system is also investigated. Intermediate bus voltage ripple ratio, R_v, is varied

from 3% to 7.5%, corresponding to a peak-to-peak ripple variation of 23 V to 66 V. The maximum system efficiency is measured for a ripple ratio of 3%. As the ripple ratio increases, a small decrease in the efficiency is observed. The change in the efficiency of the system relative to the 3% ripple ratio case is shown in Fig. 12. It can be seen that the system efficiency decreases by a maximum of only 0.35% across the full range of ripple variation. This indicates that the ripple ratio used in the design of the SSC energy buffer may be increased significantly, with minimal impact on overall system efficiency. Designing for higher ripple ratios would result in further decrease in the size of the SSC energy buffer and a corresponding increase in system power density.

VI. CONCLUSIONS

This paper presents a single phase, two-stage electrolytic-free LED driver that utilizes a stacked switched capacitor energy buffer to buffer the twice-line-frequency energy. The proposed system consists of a front-end boost PFC stage followed by a dc-dc conversion stage, with an SSC energy buffer connected across the intermediate dc bus. The SSC energy buffer uses longer life film and ceramic capacitors along with a simple switch structure to replace the limited life electrolytic capacitors. A simple and robust control scheme is proposed to interface the SSC energy buffer with the PFC stage. A prototype 300 W LED driver for universal input voltage range (90-265 Vrms) and 12 V output voltage has been designed, built and tested. The SSC energy buffer achieves a peak efficiency of 98.5% and is able to maintain an efficiency above 96% across a 5:1 output power range, thus minimally impacting the efficiency of the overall system. This paper also investigates impact of the intermediate dc bus voltage ripple on the overall converter efficiency. It is found that there is a negligible change in efficiency when the intermediate bus voltage ripple ratio is varied between 3% and 7.5%. Hence further size reduction could be achieved without compromising overall converter efficiency if the SSC energy buffer is designed for higher ripple ratios.

ACKNOWLEDGMENT

The authors would like to thank GRE Alpha Electronics for supporting this work.

REFERENCES

[1] M.L. Gasperi, "Life Prediction Model for Aluminum Electrolytic Capacitors," *Proceedings of the IEEE Industry Applications Society (IAS) Annual Meeting*, pp. 1347-1351, San Diego, CA, Oct. 1996.

[2] J. Shao and T. Stamm, "A Low Cost High Power Factor Primary Regulated Offline LED Driver," *Proceedings of the IEEE Industrial Electronics Society Annual Conference (IECON)*, pp. 4498-4502, Montreal, Canada, Oct. 2012.

[3] A.C. Kyritsis, N. Papanikolaou and E. Tatakis, "Enhanced Current Pulsation Smoothing Parallel Active Filter for Single Stage Grid Connected AC-PV Modules," *Proceedings of the International Power Electronics and Motion Control Conference (EPE-PEMC)*, pp.1287-1292, Poznan, Poland, 2008.

[4] H. Wang and H. Chung, "A Novel Concept to Reduce the DC-Link Capacitor in PFC Front-End Power Conversion Systems," *Proceedings of the IEEE Applied Power Electronics Conference (APEC)*, pp. 1192-1197, Orlando, FL, 2012.

[5] P.T. Krein and R.S. Balog, "Cost-Effective Hundred-Year Life for Single-Phase Inverters and Rectifiers in Solar and LED Lighting Applications Based on Minimum Capacitance Requirement and a Ripple Power Port," *Proceedings of the IEEE Applied Power Electronics Conference (APEC)*, pp. 620-625, Washington, DC, Feb. 2009.

[6] B.J. Pierquet and D.J. Perreault, "Single-Phase Photovoltaic Inverter Topology with Series-Connected Power Buffer," *Proceedings of the IEEE Energy Conversion Congress and Exposition (ECCE)*, pp. 2811-2818, Atlanta, GA, Sept. 2010.

[7] M. Chen, KK. Afridi, D.J. Perreault, "A Multilevel Energy Buffer and Voltage Modulator for Grid-Interfaced Micro-Inverters", *Proceedings of the IEEE Energy Conversion Congress and Exposition (ECCE)*, Denver, CO, Sep. 2013.

[8] S. Sugimoto, S. Ogawa, H. Katsukawa, H. Mizutani and M. Okamura, "A Study of Series-Parallel Changeover Circuit of a Capacitor Bank for an Energy Storage System Utilizing Electric Double Layer Capacitors," *Electrical Engineering in Japan*, vol. 145, no. 3, pp. 33-42, Nov. 2003.

[9] X. Fang, N. Kutkut, J. Shen and I. Batarseh, "Ultracapacitor Shift Topologies with High Energy Utilization and Low Voltage Ripple," *Proceedings of the International Telecommunications Energy Conference (INTELEC)*, Orlando, FL, Jun. 2010.

[10] M. Chen, K.K. Afridi and D.J. Perreault, "Stacked Switched Capacitor Energy Buffer Architecture," *IEEE Transactions on Power Electronics*, vol. 28, no. 11, pp. 5183-5195, Nov. 2013.

[11] A.H. Chang, J.J. Colley and S.B. Leeb, "A Systems Approach to Photovoltaic Energy Extraction," *Proceedings of the IEEE Applied Power Electronics Conference (APEC)*, pp. 59-70, Orlando, FL, Feb. 2012.

[12] K.K. Afridi, M. Chen and D.J. Perreault, "Enhanced Bipolar Stacked Switched Capacitor Energy Buffers," *IEEE Transactions on Industry Applications*, vol. 50, no. 2, pp. 1141-1149, Mar./Apr. 2014.

[13] Y. Ni, S. Pervaiz, M. Chen and K.K. Afridi, "Energy Density Enhancement of Unipolar SSC Energy Buffers through Capacitance Ratio Optimization," *IEEE Workshop on Control and Modelling for Power Electronics (COMPEL)*, Santander, Spain, June, 2014.

[14] M. Chen, Y. Ni, C. Serrano, B. Montgomery, D.J. Perreault, K.K. Afridi, "An Electrolytic-Free Offline LED Driver with a Ceramic-Capacitor-Based Compact SSC Energy Buffer," *Proceedings of the IEEE Energy Conversion Congress and Exposition (ECCE)*, Pittsburgh, PA, Sep. 2014.

[15] S. Pervaiz, Y. Ni, KK. Afridi, "Improved capacitance ratio optimization methodology for stacked switched capacitor energy buffers", *Proceedings of the IEEE Applied Power Electronics Conference (APEC)*, Charlotte, SC, Mar. 2015.

[16] R.W. Erickson, D. Maksimovic, "Fundamentals of Power Electronics," 2nd Edition.

[17] C. Adragna, S. De Simone and G. Gattavari,"New Fixed Off-Time PWM Modulator Provides Constant Frequency Operation in Boost PFC Pre-regulators", *International Symposium on Power Electronics, Electrical Drives, Automation and Motion (SPEEDAM)*, Ischia, Jun. 2008.

[18] W. Inam, K.K. Afridi and D.J. Perreault, "Variable Frequency Multiplier Technique for High Efficiency Conversion Over a Wide Operating Range," *Journal of Emerging and Selected Topics in Power Electronics*, September, 2015.

978-1-4673-9551-9/16 $31.00 © 2016 IEEE

Design Methodology for a High Insulation Voltage Power Transmission Function for IGBT Gate Driver

Sokchea Am[1,2], Pierre Lefranc[1,2], David Frey[1,2], Mahmoud Ibrahim[1,2]

[1]Univ. Grenoble Alpes, G2ELab, F38000 – Grenoble, France
[2]CNRS, G2ELab, F38000 – Grenoble, France
e-mail: sokchea.am@g2elab.grenoble-inp.fr

Abstract—In this article, a design methodology for *DC-DC* converters with high insulation capabilities (up to *30kV*) is proposed. The objective is to provide a power transmission function for *IGBT* drivers. To achieve this, a *DC-DC* full-bridge series-series (*FB-SS*) resonant converter with a high air gap transformer is selected and studied. This high air gap transformer (loosely coupled transformer (*k*)) is used for a high galvanic insulation system. Some transformer geometries are analyzed and compared for this specific application field. Therefore, in term of coupling coefficient of transformer (*k*), the analysis and the proposed investigations prove that pot core based transformers are suitable choices. Simulation results are presented and analyzed and experimental works are briefly described. Finally, the comparison between the simulation and experimental results are illustrated to validate the proposed methodology.

Keywords— Resonant converters, planar transformer, power supply, IGBT gate driver, high galvanic insulation capabilities.

I. INTRODUCTION

As presented in [1]-[2] and depicted in Fig. 1a, gate drivers for *IGBT* and *MOSFET* power modules involve, in general, five functions: (*i*) switching pulse-width modulation (*PWM*) signal transmission, (*ii*) default information, (*iii*) power transmission or power supply, (*iv*) protection and (*v*) buffer circuit. For the application of *IGBTs* in medium voltage converters or multilevel converters, all the driver functions require a high insulation voltage capability for the separation of low-side and high-side switches. The advantages of such system are safety, and long product life.

The galvanic insulation system can be performed through: a magnetic transformer [1], [3], optical insulation system [4], capacitive couplings [5] and piezoelectric coupling [6]. According to the article [3], optical insulation devices such as optocoupler have limited range of operating temperature which is less than 100 ^{0}C. Moreover, capacitive and piezoelectric coupling cannot response to the high insulation voltage applications [7]. Thus up to *50kV* insulation voltage for *IGBT* gate driver applications, from industrial and financial point of view, the printed-circuit-board (*PCB*) planar transformer is a suitable choice. According to [8], planar transformer has become very popular in high-frequency power converters because of their advantages that they achieve in terms of low profile, excellent thermal characteristic, modularity and manufacturing simplicity, increased reliability and power density. Therefore, in this paper, the high insulation capability is based on a high air gap length (up to *3mm*) of a pot core planar transformer. In the final applications, the air gap will be

replaced by high insulation capability materials [1]: this aspect is not addressed in this paper

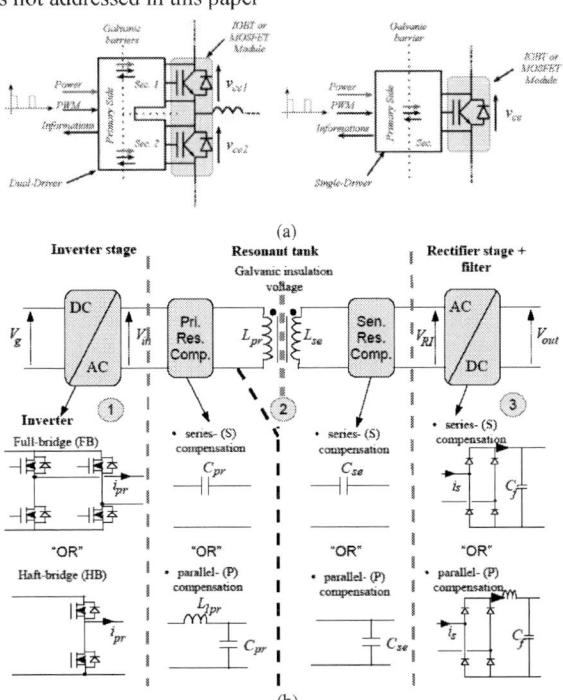

Fig. 1. (a) Synopsis of a dual and a single channel IGBT gate driver. (b) Possible DC-DC resonant converter topologies with magnetical isolation (based on a magnetic transformer)

In [1], we proposed and optimized a high galvanic insulation *IGBT* gate driver *PWM* signal transmission function. In this article, the authors address the power supply function with a very high insulation voltage capability. The aim is to reach an insulation level higher than former papers [9], [10] up to *30kV*. In order to achieve this, a *DC-DC* insulated converter topology with a high air gap transformer is proposed and studied. The insulation technique used for the windings and the core material is only suitable for few hundred of volts to few kilo-volts of insulation voltage levels [11]. However, when dealing with high power converters, the high insulation voltage is required. The main drawback of high air gap transformers is the leakage inductances. Therefore, several resonant topologies are proposed to deal with these leakage inductances.

As presented in [11], a quasi-resonant topology with one *MOSFET* and one capacitor in the primary side of the transformer is proposed for this power transmission application with just few kilo-volts of insulation level. But for a very high insulation system with a loosely coupled transformer, this topology is not a suitable choice. In [12]-[15] and Fig. 1b, the resonant converters suitable for this kind of transformer are mentioned for hundred watts to several kilowatts applications. In this paper, these resonant topologies are studied, investigated and implemented for a low power application (up to *10W*). First, these resonant converters are briefly mentioned and compared. Finally, the full-bridge series-series (*FB-SS*) resonant converter (4 *MOSFETs*, 2 capacitors, 1 planar transformer and 4 diodes) is selected. This topology is used for several applications such as: wireless charging system for mobile phones [16], wireless power supply to implantable devices (medical uses) [17]-[18], contactless battery charging of electric vehicles [12]-[13], etc.

This article is organized as follows: in Section II, the *DC-DC* converter topologies are described and analyzed. Brief comparisons between these topologies are mentioned. The design of the selected one is also presented. The electrical modeling is presented: inductances, mutual inductance coefficient, and resistance. In Section III, few high air gap transformers are shown and compared for a very high insulation voltage application objective. In Section IV, the comparison between the simulation and experimentation results are illustrated. Moreover, a high air gap pot core transformer is built and experimental results are presented and compared to simulation ones. Finally, conclusion and perspectives are proposed in Section V.

II. HIGH INSULATION VOLTAGE IGBT GATE DRIVER POWER TRANSMISSION TOPOLOGIES

There are two main categories of *DC-DC* power converters. One is non-insulated topologies (buck, boost …) and the other one is insulated topologies (flyback …). The article [19] shows several insulated topologies and the power ranges are also mentioned. In this work, a high galvanic insulation *IGBT* gate driver power transmission function based on a magnetic transformer is studied. The insulation voltage level basically depends on the transformer air gap length and its air gap material [1]. In order to increase the barrier insulation level, the transformer air gap must be increased. This means that the high insulation voltage barrier transformer design will lead to have a low coupling coefficient (*k*). For this loosely coupled transformer, the main drawbacks are the leakage inductances on the primary and secondary sides. Thus, the power supply designers proposed the *DC-DC* converter topologies which can take these leakage inductances as an advantage. Finally, resonant converter topologies are proposed to mitigate these leakage inductances.

A. DC-DC Resonant Converter Topologies

In this paper, the converter transmitter side converts the *DC* signal (V_g) to *AC* signal (V_{in}) and can be performed by a full-bridge (*FB*) or a half-bridge (*HB*) structure. As shown in Fig. 2a-b, the input voltage V_{in} of a resonant tank of these two bridges are illustrated. According to [20], the first harmonic approximation is sufficiently accurate for this system analysis. With *HB* configuration, this input voltage V_{in}, is a square waveform oscillating between *0* and $+V_g$. And with *FB*

configuration, the input voltage of resonant tank V_{in}, is a square waveform oscillating between $-V_g$ and $+V_g$. Therefore, for a brief comparison between them, if the same bridge output power is desired, the half-bridge input voltage needs to be twice as high as the full-bridge input. Thus, in this article, the *FB* structure is a suitable choice in term of low input voltage (V_g).

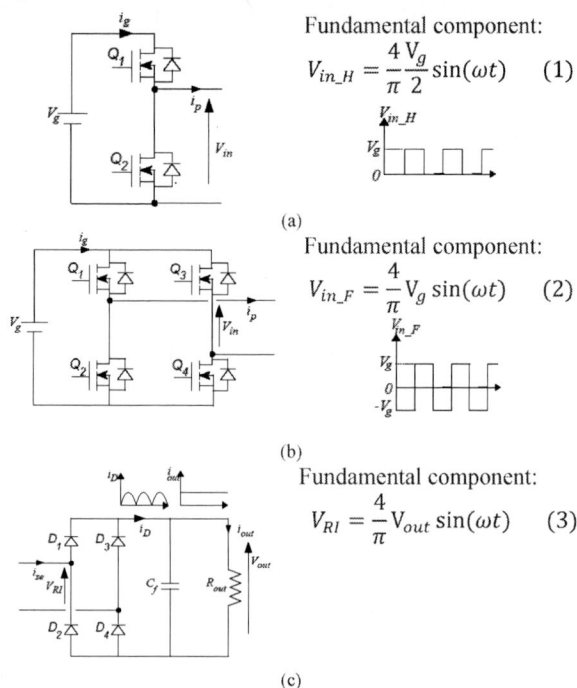

Fundamental component:

$$V_{in\_H} = \frac{4}{\pi}\frac{V_g}{2}\sin(\omega t) \quad (1)$$

(a)

Fundamental component:

$$V_{in\_F} = \frac{4}{\pi}V_g\sin(\omega t) \quad (2)$$

(b)

Fundamental component:

$$V_{RI} = \frac{4}{\pi}V_{out}\sin(\omega t) \quad (3)$$

(c)

Fig. 2. (a) a half-bridge (HB) configuration. (b) a full-bridge (FB) configuration. (c) a rectifier network

The resonant tank configurations can be series resonant (*S*) or parallel resonant (*P*) between the transformer and the compensation capacitors. To enhance the power transfer and achieve high efficiency, *DC-DC* insulated resonant converters should resonate on both sides of the transformer. In [14] (see Fig. 1b), several resonant tanks (series-series: *SS*, series-parallel: *SP*, parallel-series: *PS*, parallel-parallel: *PP*) and its equivalent impedance models are presented for a contactless electric vehicle battery charger application. Moreover, the article [21] presents the complexities and disadvantages of a parallel compensation of the transmitter and receiver coils. According to [7], [21], [22], the parallel configuration at the inverter side requires an additional inductor connected in series between the resonant tank and the power converter side. This series inductor is used to regulate the inverter current i_{pr} flowing into the parallel resonant tank. As presented in [23], the parallel resonant configuration at the transmitter coil side is useful for contactless power distribution networks in industrial sites, where the high circulating current is controlled in a track to supply multi receivers. But for only one receiver, as considered in this article (cf. Fig. 1b and Fig. 3a), this inductor will leads to increase the total converter losses. A parallel compensation in the secondary side coil requires the high circulating current at the primary side of the transformer and an additional inductor in the output filter at the receiver side is needed (cf. Fig. 1b) [21]. This normally leads to increase the winding losses in the transformer as well as the converter total losses. In addition, the voltage transfer ratio of the parallel compensations topology is always dependent to the magnetic coupling (k) of the transformer and the load. In [12], [14], the advantages of *SS* resonant tank with loosely coupled transformer are presented: the voltage ratio can be independent of the magnetic coupling of the transformer and the load; an additional inductor is not needed; that ensures a high efficiency by minimizing the circulating current through a transformer. Thus, in this paper, *SS* resonant tank is selected. Next, as shown in Fig. 2c, the rectifier network produces *DC* voltage by rectifying the *AC* current with rectifier diodes and a capacitor C_f.

Finally, the complete *DC-DC* full-bridge series-series (*FB-SS*) resonant converter is illustrated in Fig. 3a.

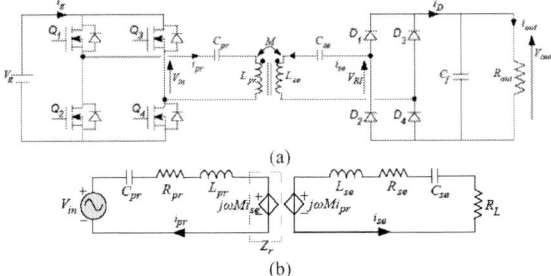

(a)

(b)

Fig. 3. (a) DC-DC *FB-SS* resonant converter. (b) Mutual coupling model

B. SS Resonant Tank Modelling

L_{pr}, L_{se} and M are the primary side inductance, the secondary side inductance and the mutual inductance of the transformer, respectively. R_{pr} and R_{se} are the winding resistances of the transformer which depend on the operating frequency due to skin effect. C_{pr} and C_{se} are the primary and secondary external compensation capacitors, to enhance the energy transfer from input voltage tank V_{in} to an equivalent output loading resistance R_L. These capacitors are calculated as expressed in (4) and (5). The mutual coupling inductance model of the *SS* resonant tank is presented in Fig. 3b.

$$C_{pr} = \frac{1}{\omega^2 L_{pr}} \qquad (4)$$

$$C_{se} = \frac{1}{\omega^2 L_{se}} \qquad (5)$$

TABLE I
PARAMETERS' CALCULATIONS FROM FIG.3B

Definition	Equations
Primary side impedance	$Z_{pr} = j\omega L_{pr} + 1/j\omega C_{pr} + R_{pr}$
Secondary side impedance	$Z_{se} = j\omega L_{se} + 1/j\omega C_{se} + R_{se}$
Equivalent impedance	$Z_r = (\omega M)^2/(Z_{se} + R_L)$
Primary side current	$i_{pr} = V_{in}/(Z_{pr} + Z_r)$
Secondary side current	$i_{se} = j\omega M i_{pr}/(Z_{se} + R_L)$
Equivalent load	$R_L = 8R_{out}/\pi^2$
Output voltage of resonant tank	$V_{RI} = R_L i_{se}$

The current dependent source $\omega M i_{se}$ in Fig. 3b can be replaced by an equivalent impedance Z_r which is calculated by

dividing $\omega M i_{se}$ with i_{pr}. And other equivalent impedances for the resonant system are described in [12], [14], and summarized in Table I.

C. Voltage Transfer Ratio

From the parameter calculations in Table I, by replacing i_{pr} into i_{se}, the expression of fundamental output voltage V_{RI} divided by input voltage V_{in} is defined. Next, the voltage transfer function (G_v) (output voltage V_{out} divided by V_g) is derived.

$$i_{se} = \frac{j\omega M V_{in}}{\left(Z_{pr} + Z_r\right)\left(Z_{se} + R_L\right)} \qquad (6)$$

$$V_{RI} = R_L \frac{j\omega M V_{in}}{\left(Z_{pr} + Z_r\right)\left(Z_{se} + R_L\right)} \qquad (7)$$

$$G_v = \frac{V_{out}}{V_g} = \frac{j\omega M}{\dfrac{Z_{pr} Z_{se} + (\omega M)^2}{R_L} + Z_{pr}} \qquad (8)$$

From (1)-(2)-(3) and (7), the expression of G_v of *FB-SS* is expressed in (8). According to [12]-[13], and (8), the G_v can operate independently of the magnetic coupling (k) and the load (R_L) at two different specific frequencies ($f_{v,l} = f_{res}/\sqrt{1+k}$ is lower than resonance frequency and $f_{v,h} = f_{res}/\sqrt{1-k}$ is higher than resonance one). As an example, the capacitors of *22nF* are selected to resonate with transformer inductances ($L_{pr}=L_{se}=4.66\mu H$) at *500kHz*. Loading conditions vary from *45Ω* to *105Ω*. In Fig. 4, their magnitudes G_v are plotted. As shown in this figure, $f_{v,l}$ and $f_{v,h}$ are the two frequencies where the G_v is independent of R_L.

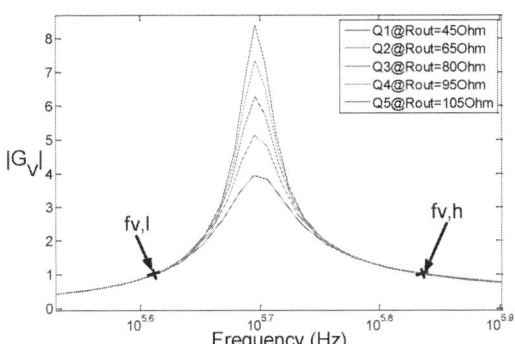

Fig. 4. Magnitude of voltage transfer ratio G_v

D. ZVS Consideration

In order to reduce the switching losses, the designer must pay attention for the choice of the converter topology which can achieve zero voltage switching (*ZVS*) and/or zero current switching (*ZCS*) conditions. In this case, the *MOSFETs* turn on when the voltage and/or current across/through it is zero. Theoretically, it will imply zero switching losses. However, zero switching losses are not strictly achievable in practice. Therefore, even with a small deviation from the ideal case, the converter with high efficiency will be achieved.

To achieve *ZVS* on the inverter switches for *FB-SS* resonant converter, the input impedance Z_{in} of the resonant circuit must exhibit an inductive behavior. Where, the phase angle (φ_{in}) of input impedance must be greater than zero ($\varphi_{in} > 0$).

The equations below present the analysis of Z_{in}:

$$Z_{in} = Z_{pr} + Z_r \qquad (9)$$

$$R_{in} = \frac{\omega^2 M^2 R_L}{\left(\omega L_{se} - \frac{1}{\omega C_{se}}\right)^2 + R_L^2} \qquad (10)$$

$$X_{in} = \left(\omega L_{pr} - \frac{1}{\omega C_{pr}}\right) - \frac{\left(\omega L_{se} - \frac{1}{\omega C_{se}}\right)}{\left(\omega L_{se} - \frac{1}{\omega C_{se}}\right)^2 + R_L^2} \qquad (11)$$

$$\varphi_{in} = \frac{180}{\pi} tan^{-}\left(\frac{X_{in}}{R_{in}}\right) \qquad (12)$$

As shown in Fig. 4, at the resonant frequency point ($f_{res.}$), the input phase angle $\varphi_{in} = 0$, thus Z_{in} is always resistive. In the frequency range $f_{v,l}$ to $f_{res.}$, the input impedance Z_{in} is always capacitive. In the frequency range f_{res} to $f_{v,h}$, the input impedance Z_{in} is always inductive. Thus, in order to achieve *ZVS*, the operating frequency f_p of the converter must be in the range f_{res} to $f_{v,h}$.

E. Converter Loss Analysis

The losses of the converter mainly come from three parts: primary inverter full-bridge stage (4 *MOSFETs*), series-series resonant tank stage (mainly from transformer), and rectifier network (4 diodes). The losses calculations are based on the simulated or measured current waveforms and the manufacturer's datasheet. The losses identifications are used in the converter efficiency (η_{con}) calculation.

Inverter full-bridge losses: Power losses of semiconductor components can be divided into two groups: conduction losses and switching losses. In (13), the switching loss is expressed as a function of the switching actions of the *MOSFET* current ($i_{SW,M(i)}$), the drain-source voltage ($V_{DS} = V_g$), the rise time (t_r), the fall time (t_f) and the operating frequency (f_p). The conduction losses can be calculated using the *MOSFET* drain-source on-state resistance (R_{DS}) and the root mean square value of *MOSFET* simulated or measured current waveforms ($i_{M,rms(i)}$) ((14)).

$$P_{SW,M(i)} = V_{DS} i_{SW,M(i)} (t_r + t_f) f_P \qquad (13)$$

$$P_{CON,M(i)} = R_{DS} \cdot i_{M,rms(i)}^2 \qquad (14)$$

Rectifier network losses: according to [24], the switching losses of the rectifier are almost eliminated in the case of Schottky diodes. The main losses are the conduction losses which can be expressed as:

$$P_{CON,D(i)} = V_D \frac{V_{out}}{R_o} + R_D \cdot i_{D,rms(i)}^2 \qquad (15)$$

Transformer losses: the transformer losses are a large part of the total losses of the converter. The transformer losses can be divided into two main parts: core losses and winding losses. The core losses can be described as in (16) and the winding losses are expressed in (17).

$$P_{C,T} = \rho f^\alpha B_m^\beta V_e \qquad (16)$$

$$P_{cu,T} = R_{Pr} i_{P,rms}^2 + R_{Se} i_{S,rms}^2 \qquad (17)$$

Finally, the converter efficiency can be computed (18):

$$\eta_{con} = \frac{P_{out}}{P_{out} + P_L} \qquad (18)$$

where, $P_{out} = \frac{V_{out}^2}{R_{out}}$ and P_L is the total losses of converter.

III. HIGH AIR GAP TRANSFORMER INVESTIGATIONS

After a brief description of the high insulation *DC-DC* converter topologies, the authors present the investigation of the transformer geometries to answer this high insulation voltage level up to *30kV*.

A. High Efficiency Transformer Considerations

The *FB-SS* converter efficiency (η_{con}) is expressed as a function of the coupling coefficient (k) and the quality factor of inductor (Q): $\eta_{con} \approx 1 - 2/(kQ)$ (details in [18] and [21]). Where, the inductive quality factor Q is proportional to the inductance values and frequency but inverse to the winding resistance (as expressed in [19]). Therefore, the higher magnetic coupling k, L_{pr} and L_{se} values provide the high converter efficiency.

$$Q_i \approx \frac{2\pi f L_i}{R_i} \quad ; i = \{pr, se\} \qquad (19)$$

Winding shape: according to [25], several coil geometries are studied and compared in term of transformer coupling factor (k) for this kind of converter. As a result, the circular coil shapes achieve higher coupling coefficient which provide higher efficiency than other coils for the same configuration. Thus, the circular windings on *PCB* are selected here.

Magnetic cores: as illustrated in [26]-[27], the comparison between a pot core transformer and other magnetic cores are illustrated. As a conclusion, the pot core transformer gives better performance indexes in terms of flux density, high magnetic coupling, minimal core and winding losses.

For these reasons, we selected pot core planar transformers with circular coils on *PCB* for this gate driver application.

In order to validate our choice, the pot core geometrical transformer is firstly simulated by finite element software (*FEMM^{TM}*) with designed transformer specification in Table II and geometrical description in Fig. 5a. The insulation barrier level of the system depends on the length of its air gap (ep_l) and the insulation material. As an example, theoretically, the breakdown voltage of printed circuit board (*PCB*) FR4 is about *10kV/mm*. Fig. 6 and Table IV present the axi-symmetrical simulation of designed transformer with *FEMM^{TM}* and simulation results (inductances, coupling efficient, etc.), respectively.

B. Experimental Verification of Transformers Parameters

To justify our choice on the transformer selection, few transformer geometries are selected and compared experimentally. As shown in Fig. 5, Pot Core (PC18/11), toroidal core and UU core will be investigated as transformer core geometries. The parameter ep_l (air gap) varies from *1mm* to *3mm* for each transformer core. The experimental results with Agilent *4294A* precision impedance analyzer are presented in Table III: As a result, ferrite pot core has the best coupling factor compared to the two other geometries. Furthermore, to validate our design, the comparison between the simulation and experimental results is summarized in Table IV.

In the next section, the simulation and experimental results of *FB-SS* with a pot core are illustrated and described.

TABLE II
SELECTED DESIGN FOR PROTOTYPE TRANSFORMER

Parameter	Value	Description
ep_1	3 mm	galvanic insulation layer
ep_2	35 μm	copper layer
ep_3	1.6 mm	PC18/11 ferrite thickness
ep_{3-1}	1.425 mm	PC18/11 ferrite thickness
ep_{3-2}	2.175 mm	PC18/11 ferrite thickness
ep_4	0.65 mm	reserve length of PCB from last coils
ep_5	0.5 mm	insulation layer
ep_6	0.5 mm	PCB for multi-layer
ep_7	0.745 mm	insulation layer between multi-layer
ep_8	1.55 mm	radius internal air of PC18/11
x_1	0.3 mm	distance between the coppers
x_2	0.63 mm	copper's width
n_1	3 turns	number of turns
n_{layer}	4 layers	number of multi-layers

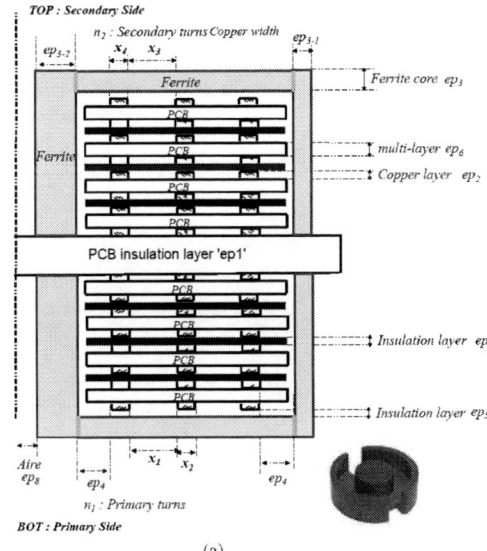

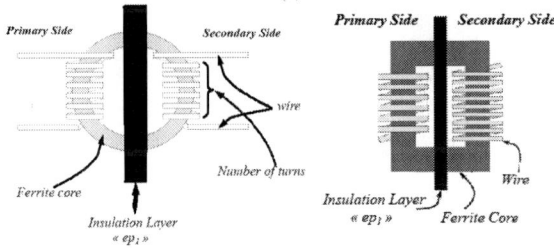

(b)   (c)

Fig. 5. Transformer cores. (a) Pot Core 2D with number of turns on PCB. (b) Toroidal core with wire turns. (c) UU core with wire turns

TABLE III
TRANSFORMER CHARACTERISTIC VALUES FROM IMPEDANCE ANALYZER "AGILENT 4294A"

Transformer Core	Insu. layer (ep₁)	Lp [μH]	Ls [μH]	M [μH]	k
"UU"		31	30	10.25	0.336
"Toroidal"	1mm	5.14	5.2	2.13	0.411
"PC18/11"		**19.11**	**18.1**	**12.34**	**0.664**
"UU"		29.8	30	9.17	0.312
"Toroidal"	1.5mm	4.98	4.9	1.94	0.392
"PC18/11"		**16.53**	**15.4**	**8.97**	**0.562**
"UU"		28.7	28	8.77	0.309
"Toroidal"	2mm	4.6	4.7	1.39	0.298
"PC18/11"		**15.2**	**13.9**	**7.18**	**0.495**
"UU"		27.8	27	6.53	0.238
"Toroidal"	2.5mm	4.47	4.7	1.39	0.276
"PC18/11"		**14.25**	**13.28**	**5.66**	**0.411**
"UU"		27.3	26.8	6.12	0.226
"Toroidal"	3mm	4.29	4.4	1.07	0.246
"PC18/11"		**13.72**	**12.88**	**4.76**	**0.358**

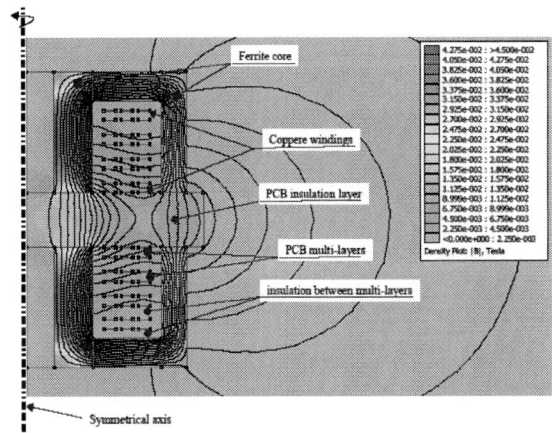

Fig. 6. *FEMM^{TM}* screenshot of the *2D* axi-symmetrical

TABLE IV
COMPARISON RESULTS OF POT CORE TRANSFORMER

Parameters	Simulated	Measured
L_{pr}	13.65 μH	13.72 μH
L_{se}	13.30 μH	12.88 μH
M	4.95 μH	4.76 μH
k	0.367	**0.358**

IV. SIMULATION AND EXPERIMENTAL RESULTS

The results of a pot core transformer with *a 3mm* air gap are presented for this high insulation voltage system design specification. The converter is designed for an output power (P_{out}) of *10W*. The input voltage (V_g) is set to *12V*. The output voltage (V_{out}) and output current (i_{out}) are *20V* and *0.5A* respectively.

A. Simulation Results

The compensation capacitors $C_{pr}=C_{se}=22nF$ are selected to resonate with $L_{pr}=13.7μH$ and $L_{se}=12.9μH$ (from Table III) at about 300kHz. The magnitude of voltage transfer ratio ($|G_v|$) of the designed specification is presented in Fig. 7

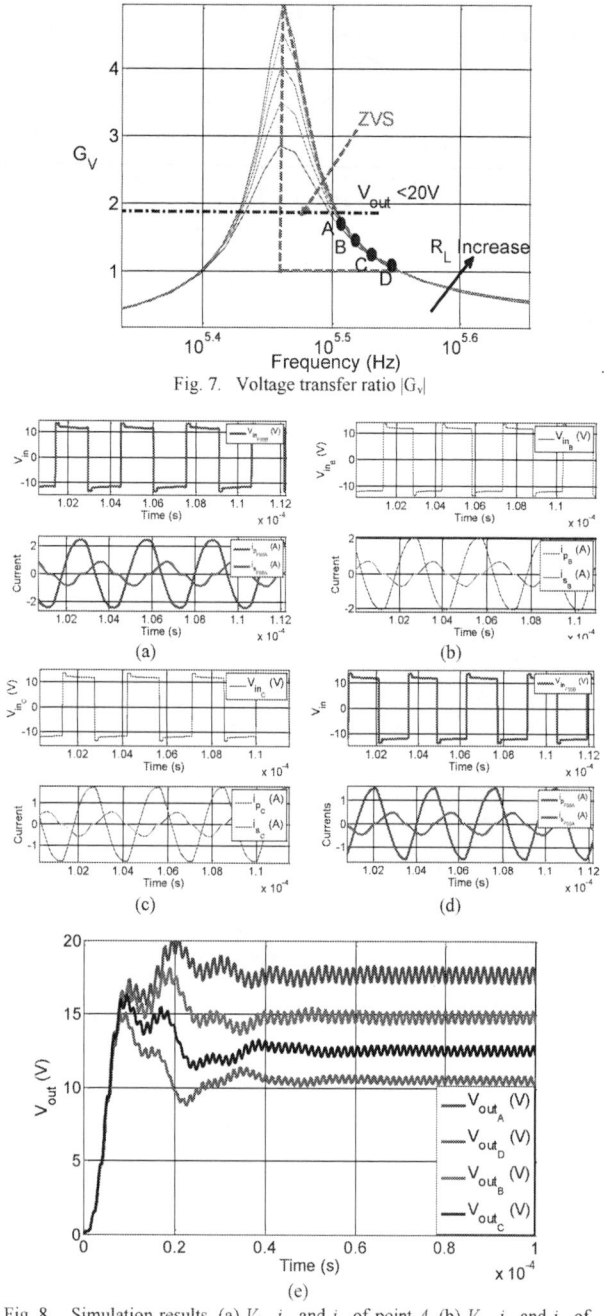

Fig. 7. Voltage transfer ratio $|G_v|$

Fig. 8. Simulation results. (a) V_{in}, i_{pr} and i_{se} of point A. (b) V_{in}, i_{pr} and i_{se} of point B. (c) V_{in}, i_{pr} and i_{se} of point C. (d) V_{in}, i_{pr} and i_{se} of point D. (e) V_{out}

As shown in Fig. 7, the *ZVS* can be achieved by choosing the operating frequency f_p higher than the resonant frequency (f_{res}). Thus, to achieve the objective ($V_{out} \leq 20V$), the operating frequency (f_p) range is from *325 kHz* to *358 kHz* (see Fig. 7). The input voltage (V_{in}) of the resonant tank, the primary side (i_{pr}) and secondary side (i_{se}) currents are the main waveforms analyzed to validate the design. The simulated waveforms at points A, B, C and D are presented in Fig. 8a, Fig. 8b, Fig. 8c

and Fig. 8d, respectively. Fig. 8e shows the output voltages, output currents and output powers of these four points. The output power of point A ($P_{out,A}$) is around twice more than point D ($P_{out,D}$) that correspond to the transformer current waveforms of each points (see Fig. 8a and 8d). The output power of point B is higher than point C and D but lower than point A. Transient simulations confirm the analytical approach.

These results clearly show that the *FB-SS* resonant converter can be used in a very high insulation voltage *IGBT* gate driver power transmission application (power transfer is lower *than 10W*) by a loosely coupled pot core transformer.

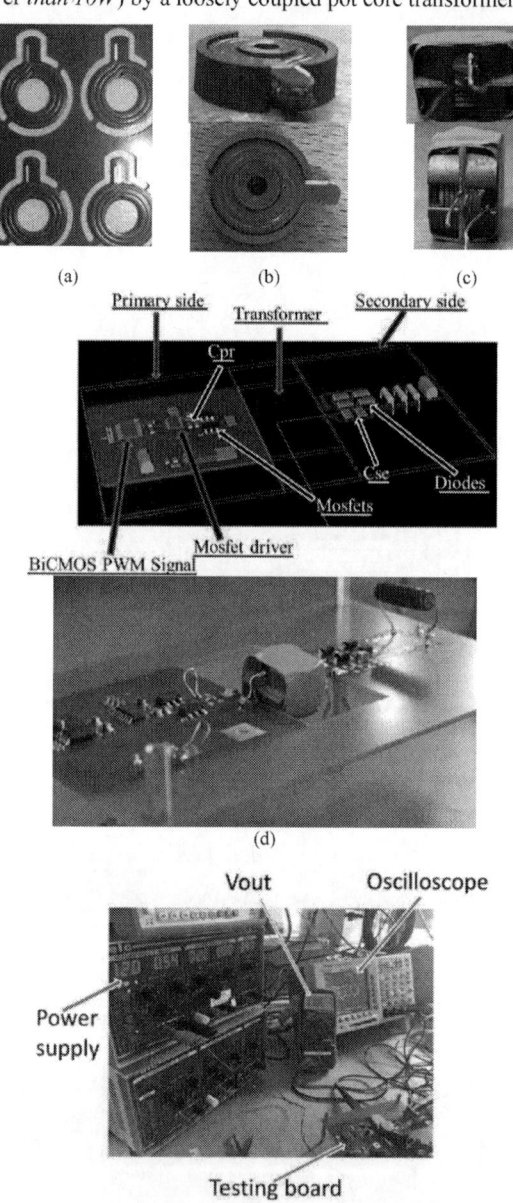

Fig. 9. *FB-SS* resonant converter prototype: (a) PCB winding. (b) Transformer one side assembly. (c) Physical transformers. (d) Experimental testing board and *3D* layout with *Kicad^{TM}* software. (e) Experimental instruments.

978-1-4673-9551-9/16 $31.00 © 2016 IEEE

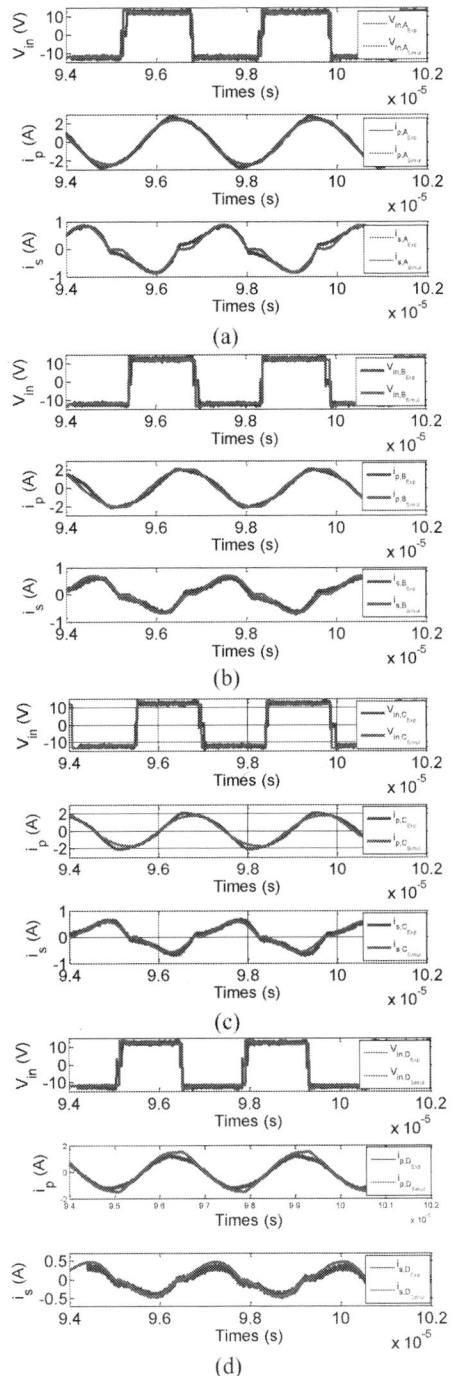

Fig. 10. Comparison between the simulation and experimental results of point (a) A. (b) B. (c) C. (d) D

B. Comparison Results

Fig. 9 shows a pot core transformer, its windings on PCBs and the experimental testing board. Fig. 9a to 9c present the circular *PCB* windings and assembled transformer prototypes with four layers per side. Fig. 9d illustrates *3D* layout design

and physical prototype of testing board with components on both sides (top and bottom) of PCB. The *N-MOSFET BSP030* is selected for the experimental setup. The high side and low side gate drivers *IR21814* from *IR* are selected as *MOSFET* gate drivers. The *BiCMOS* advanced phase-shift Pulse-Width Modulation (*PS-PWM*) controller *UCC3895DW* is used to generate the *PWM* signals with dead-time management. The diodes *ES1AL* are used in the rectifier network stage. And Fig. 9e presents all the experimental setup. Finally, an input voltage V_{in}, a primary side (i_{pr}) and a secondary side currents (i_{se}) of point *A* through point *D* are measured in order to be compared to simulation results: Fig. 10.

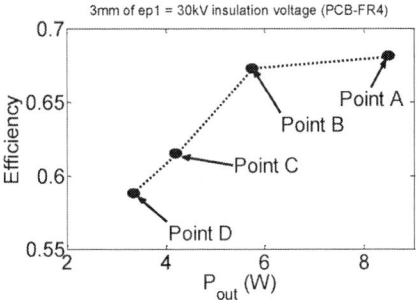

Fig. 11. Experimental efficiency data plot

Fig. 11 presents the experimental efficiency approximation (from *59%* to *69%*) of the design specifications. From the transformer experimental works, its winding resistances are around *750mΩ* in the frequency ranges. Then, for these frequency ranges, we notice that more than 50% of converter losses come from the winding losses of the pot core transformer. Therefore, in our perspectives, the transformer winding resistances must be minimized and they depend on the material technologies: *PCB* material and its thickness, copper thickness, insulation material, etc. Moreover, the optimization considerations (for our goal: $\eta_{con} > 85\%$) are briefly described in the conclusion and perspective.

Thus, these experimental results and the comparisons with simulation results clearly demonstrate that the *FB-SS* with a pot core transformer is suitable for a design of a *DC-DC* converter with a high galvanic capability up to *30kV* of an *IGBT* gate driver power transmission application.

V. CONCLUSIONS

IGBT gate driver functions are: power transmission, *PWM* signal transmission, protections, default transmission, and buffer circuit. In this article, the power transmission function for a high galvanic insulation *IGBT* gate driver system is proposed and studied. A *DC-DC* full-bridge series-series (*FB-SS*) resonant converter is selected and analyzed. A high insulation voltage system is achieved by a high air gap transformer (loosely coupled transformer).

Few transformers technologies are proposed and compared for this application. The coupling coefficient (k) is the first parameter analyzed to achieve high galvanic insulation up to *30kV* for gate drivers systems. Finally, we

978-1-4673-9551-9/16 $31.00 © 2016 IEEE   2407

demonstrate that a pot core transformer is a suitable choice for this application design.

Therefore an *IGBT* gate driver power transmission function specification is defined. A *3mm* air gap pot core transformer with *PCB* windings (about *30kV* of insulation level) is prototyped. Their simulation and experimental results are compared in order to validate this design methodology. We demonstrate that we can transmit *3.5W* to *9W* with efficiency from *59%* to *69%* with an air gap of *3mm*.

As perspectives, the *FB-SS* converter optimization design for a very high galvanic insulation *IGBT* gate driver power transmission function will be proposed and studied. The converter losses modeling, the optimization of transformer geometry based on finite element simulator, a virtual prototyping will be proposed in the next future. Maximize the converter efficiency and minimize its volume, cost, etc. can be set as optimization objectives. Furthermore, the studies of the insulation material are also considered for the future works.

ACKNOWLEDGMENT

The authors wish to acknowledge the financial support provided by Institut Carnot Energie Futur, Grenoble, France.

REFERENCES

[1] Am S., Lefranc P., Frey D., "Design methodology for optimizing a high insulation voltage insulated gate bipolar transistor gate driver signal transmission function" *IET Power Electronics*, vol. 8, no 6, p.1035-1042, June 2015. doi: 10.1049/iet-pel.2014.0434

[2] Lefranc, P., Jannot, X., Dessante, P., "Optimised design of a transformer and an electronic circuit for IGBT drivers signal impulse transmission function based on a virtual prototyping tool", *IET Power Electronics*, vol. 6, no. 4, pp. 625–633, April 2013. doi: 10.1049/iet-pel.2012.0401

[3] Muhammad K.S., Lu D.D.C, "Magnetically Isolated Gate Driver with Leakage Inductance Immunity", *IEEE Tran. on Power Electronics*, vol. 29, No. 4, pp. 1567-1572, April 2014

[4] Rouger N., Crébier J.-C., Lesaint O., "Integrated low power and high bandwidth optical isolator for monolithic power MOSFETs driver," in *23rd ISPSD'11*, 18–22 May 2011

[5] Zeltner S., "Zero voltage switching performance of 1200 v SiC MOSFET, 1200 v silicon IGBT and 900 v CoolMos MOSFET," in *CIPS'10*, pp. 1–6, 16–18 March 2010

[6] Vasic D., Costa F., Sarraute E., "Piezoelectric transformer for integrated MOSFET and IGBT gate driver," *IEEE Trans. Power Electron.*, pp. 56–65, 2006

[7] Heinemann L., Mast J., Scheible G., Heizel T., Zuelling T., "Power supply for very high insulation requirements in IGBT gate-drivers," in *33rd IAC*, vol.2,no.,pp.1562-1566, 12-15 Oct. 1998

[8] Ouyang Z., Thomsen O.C., Anderson M.A.E, "Optimal design and tradeoffs analysis of planar transformer in high power DC-DC converters," *IEEE Trans. on Ind. Electron.*, vol.59, no. 7, pp. 2800-2810, Jul. 2012

[9] Kadavelugu, A., Bhattacharya, S., "Design considerations and development of gate driver for 15kV SiC IGBT", *29th Annual IEEE Applied Power Electronics Conference and Exposition (APEC)*, vol., no., pp. 1494-1501, 16-20 March 2014

[10] Schmitt G., Kusserow W., Kennel R., "Power supply for a IGBT-driver with high insulation voltage based on a printed planar transformers," in *13th EPE-PEMC'08*, vol., no., pp.1239-1242, 1-3 Sept. 2008

[11] Lefranc, P., "Etude, conception et réalisation de circuit de commande d'IGBT de forte puissance", *PhD. Dissertation INSA Lyon, France*, 2005

[12] Zhicong Huang, Siu-Chung Wong, and Chi K. Tse, "Design methodology of a series-series inductive power transfer system for electric vehicule battery charger application," *2014 IEEE Energy Conversion Congress and Exposition (ECCE)*, vol., no., pp. 1778-1782, 14-18 Sept. 2014. Doi: 10.1109/ECCE.2014.6953633

[13] Wei Zhang, Siu-Chung Wong, Chi k. Tse, Qianhong Chen, "Design for efficiency optimization and voltage controllability of Series-Series Compensated Inductive Power Transfer Systems", *IEEE Transaction on Power Electronics*, vol. 29, no. 1, pp. 191-200, Jan. 2014. doi:10.1109/TPEL.2013.2249112

[14] Chwei-Sen Wang, Oskar H. Stielau, Grant A. Covic, "Design considerations for a contactless electric vehicle battery charger", *IEEE Transaction On Industrial Electronics*, vol. 52, no. 5, Oct. 2005. doi: 10.1109/TIE.2005.855672

[15] Hangseok Choi, "Half-bridge LLC resonant converter design using FSFR-series Fairchild power switch (FPS™)", *Application Note*, Strategic R&D, Fairchild Semiconductor, October 2014

[16] Achterberg J., Lomonova E.A., and J. de Boeif, "Coil array structures compared for contactless battery charging platform", *IEEE Transactions on Magnetics*, vol. 44, no. 5, pp. 617-622, May 2008. doi: 10.1109/TMAG.2008.917022

[17] Sato F., Nomoto T., Kano G., Matsiki H., and Sato T., "A new contactless power singnal transmission device for implanted function electrical stimulation (FES)", *IEEE Transaction on Magnetics*, vol. 40, no. 4, pp. 2964-2966, July 2004. doi: 10.1109/TMAG.2004.830416

[18] Knecht, O., Bosshard, R., and Kolar, J., "High-efficiency transcutaneous energy transfer for implantable mechanical heart support systems", *IEEE Transaction on Power Electronics*, vol.PP, no.99, pp., Jan. 2015. doi: 10.1109/TPEL.2015.2396194

[19] "Switch-Mode Power Supply: reference manual", *Energy Efficiency Innovations Group*, ON Semiconductor, 2014. www.onsemi.com

[20] Bosshard. R, Kolar J.W., Wunsch B., "Accurate finite-element modeling and experimental verification of inductive power transffer coil design" in *29th APEC*, vol., no., pp. 1648-1653, 16-20 March 2014

[21] Bosshard R., Kolar J.W., Muhlethaler J., Stevanovic I., Wunsch B., and Canales F., "Modeling and η-α-Pareto optimization of inductive power transfer coils for electric vehicles", *IEEE Journal of Emerging and Selected Topics in Power Electronics*, vol. 3, no. 1, march 2015

[22] Hong Huang, "Design an LLC Resonant Half-Bridge Power Converter", *Texas Instrument Application Note SLUP*, 2005

[23] A. Green, "10 kHZ inductively coupled power transfer-cencpt and control", *IEEE Int. Conf. Power Electronic. Variavle-Speed Drives*, pp. 694-699, 1994

[24] Fang, Z., Cai, T., Duan, S., Chen, C., "Optimal design methodology for LLC resonant in battery charging applications based on Time-Weighted Average Efficiency", *IEEE Transaction on Power Electronics*, vol. 30, no. 10, pp. 5469-5483, Oct. 2015. doi: 10.1109/TPEL.2014.2379278

[25] Bosshard R., Kolar J.W., Muhlethaler J., Stevanovic I., "Optimized magnetic design for inductive power transfer coils," in *28th APEC*, vol., no., pp. 1812-1819 17-21 march 2013

[26] Legranger J., Friedrich G., Vivier S., Mipo J.C., "Comparison of two optimal rotary transformer designs for highly constrained applications", *IEEE International Electric Machines & Drives Conference (IEMDC '07),vol.2*, pp. 1546-1551, 3-5 May 2007. doi: 10.1109/IEMDC.2007.383658

[27] Smeets J.P.C., Krop D.C.J., Jansen J.W., Hendrix M.A.M, Lomonova E.A, "Optimal design of a pot core rotating transformer", *IEEE Energy Conversion Congress and Exposition (ECCE)*, pp. 4390-4397, 12-16 Sept. 2010. doi: 10.1109/ECCE.2010.5618455

Optimized Design of GaN Switching Capacitor Based Envelope Tracking Power Supply for Satellite Applications

Qian Jin
College of Automation Engineering
Nanjing University of Aeronautics and Astronautics
Nanjing, China
qian_jin@nuaa.edu.cn

Vasić M., Garcia O. Alou P. , Oliver J.A., Cobos J.A.
Centro de Electrónica Industrial
Universidad Politécnica de Madrid
Madrid, Spain

Abstract—In modern communication system, both the bandwidth and peak-to-average power ratio of the transmitted signal are increasing rapidly. As a result, the power amplifiers based on linear power amplifiers such as class A, B or AB suffer from very low efficiency. The most promising solutions are the Envelope Tracking (ET) and Envelope Elimination and Restoration (EER) techniques, which can greatly improve the efficiency by employing a dynamic power supply that tracks the envelope of the transmitted signal. This paper presents an implementation of the ET power supply based on a multilevel converter in series with a linear regulator for a satellite application. The multilevel converter consists of a switching capacitors based converter in order to have highly efficient and light solution because the weight of power supply, together with its efficiency, is of the outmost importance for the application. A prototype capable of following a 5MHz RF envelope signal with maximum output power of 40W is fabricated. The experimental results verify the effectiveness of the proposed solution with 96.15% efficiency for the multilevel converter and 75% for the ET power supply. Furthermore, an experimental comparison between Silicon and GaN transistors is conducted in order to verify the benefits of GaN transistors.

Keywords—envelope tracking; switching capacitor; power amplifier; GaN device

I. INTRODUCTION

With the development of the modern telecommunication, it is imperative to accelerate the data transmitting rate [1]. The radio frequency signals in those efficient modulations are featured with increasing bandwidth and higher peak-to-average power ratio (PAPR). As a consequence, the linear power amplifier (PA) (class A, B or AB) suffers from low efficiency. The most promising solutions are the Envelope Elimination and Restoration (EER) [2] and Envelope Tracking (ET) [3-4] techniques, which can greatly improve the efficiency of PAs. In the case of satellite applications, the weight that has to be launched to the space is of the outmost importance, and the reduced power losses can be seen as a decrease of the system weight, due to less heatsinking. Due to this reason, these techniques are gaining on importance in this area as well. In both techniques, ET and EER, the key element is a dc-dc converter called envelope amplifier or envelope

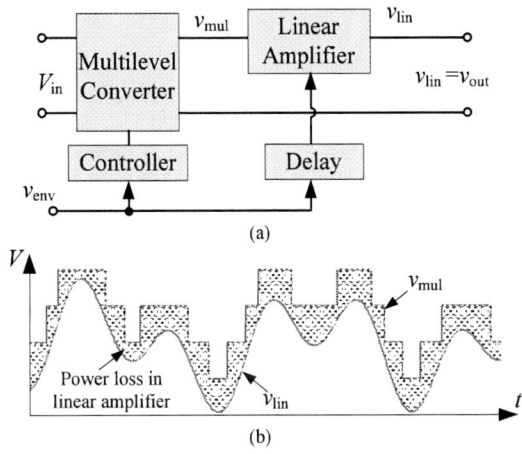

Fig. 1. Simplified schematic of the proposed envelope amplifier.

tracker. Its task is to provide a varying voltage proportional to the envelope of the transmitted signal.

In the state of the art, there are various solutions for ET power supply, such as single switching-mode converter [5], switching-mode converters in parallel [6-7], linear assisted switching-mode converter [8-9]. Due to the limited switching frequency, those solutions face the bottleneck to exceed bandwidth of several tens of kHz. Usually, to achieve 1 MHz tracking bandwidth, more than 5 MHz switching frequency is required [10]. Multilevel structure is a good candidate to increase equivalent switching frequency for a higher bandwidth. One of the solutions adopting a multilevel converter in series with a high slew rate linear amplifier [11] is shown in Fig. 1. Based on this idea, great effort is dedicated to explore a more practical multilevel converter in this paper, which helps reducing power loss and system weight.

A dc-dc converter based on switching capacitor is a good solution for high efficiency, as well as low weight and low occupied space. In [12], a switching capacitor based voltage divider is implemented with eight transistors, however, the three output voltage levels are dependent on each other. In this

978-1-4673-9551-9/16 $31.00 © 2016 IEEE

paper, a novel switching capacitor based voltage divider is proposed to provide four alterable voltage levels with only four transistors, which offers more flexibility to optimize the design. GaN transistors have been selected as the switching device due to their resistance to single event upsets and, additionally, possibility to operate at high switching frequencies, which should decrease the weight and size of the envelope tracker. Two prototypes have been made, one with Si transistors and the other with GaN transistors, in order to compare the influence of these two technologies. The experiments were performed by tracking RF signals with bandwidth of 5MHz, while the maximum power was 40W.

II. PROPOSED SOLUTION

A. Structure of the proposed multilevel converter

The proposed multilevel converter consists of a switching capacitor based voltage divider and an analog multiplexer. A simplified schematic is presented in Fig. 2. The voltage divider can provide four voltage levels, and two of these are fixed as Vin (V_4) and Vin/2 (V_2). In order to maintain the middle voltage V_2 to be half of the input voltage, a flying capacitor is placed between two switching nodes A and B. In these nodes the voltage has a PWM shape and it changes between 0 and Vin/2 (node B) or between Vin/2 and Vin (node A). If a LC low pass filter is placed in these nodes an average voltage proportional to the duty cycle, D, can be obtained, just like in the case of a buck converter. Normally, in the case of the converters based on switching capacitors, the duty cycle is 50% [12]. But, in this solution this can be changed as it will be explained later. Thus, the four output voltage levels can be derived as D *Vin/2, Vin/2, (1+D) *Vin/2 and Vin. Obviously, those four voltage levels can be optimized for a higher overall efficiency with different duty cycle, which is a significant improvement comparing with the proposed solution in [12], where all the voltage levels are unalterable, proportional to the input voltage and obtained by combining two voltage dividers which increases the complexity of the system.

In the multiplexer, SW_i (i = 2, 3, 4) are the switches that turn on and off the corresponding voltage levels. The block diodes D_2 to D_4 are in series with SW_2 to SW_4, respectively, to avoid short circuiting of different voltage levels through the body diode of multiplexer transistors. As seen in Fig. 1(b), the power loss in the linear amplifier is proportional to the voltage difference between the output voltage of the multilevel converter and linear amplifier. It can be minimized if the voltage provided by the multilevel converter is close enough to the output voltage of the linear amplifier [13]. To realize this, the multiplexer is controlled to turn on and off these voltage levels depending on the characteristics of the envelope signals. Thus, it can supply the linear amplifier with voltage levels that are as close as possible to the output voltage of the linear amplifier. Since the multiplexer operates each switch only when the signal envelope crosses certain threshold level, its switching frequency is much lower than that in the traditional PWM multilevel converter. In this way, the proposed solution has a potential to achieve a higher tracking bandwidth applying relatively low switching frequency.

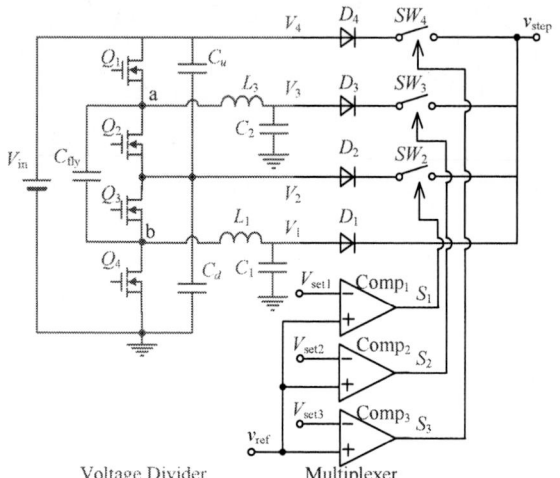

Fig. 2. Multilevel converter realized with switching capacitor based voltage divider and analog multiplexer.

In the proposed solutions, neither the voltage divider nor the analog multiplexer need precise control of its output voltage, because the linear amplifier that is put in series will perform the precise regulation, which means convenient open loop control can be adopted.

B. Power loss model

A detailed power loss model of the proposed structure is implemented to optimize all the design parameters for the highest efficiency. By combining the power loss model with the probability function of the transmitted signal ρ, the overall efficiency can be modelled precisely.

The power losses of the switching capacitor based voltage divider consist of three parts:

1. the switching losses in the employed transistors;

2. the power losses in the LC filter;

3. power losses in the switching capacitors.

The first two power loss mechanisms can be easily described as in the case of a fast-switching buck converter [15], and only the third mechanism will be analyzed in details in this paper.

The equivalent circuit of the voltage divider is illustrated in Fig. 3 (a), where I_1, I_2, I_3 are the average currents of correspond voltage levels. When Q_1 and Q_3 turn on, C_{fly} is connected in parallel with C_u, otherwise, C_{fly} is connected with C_d, as shown in Fig.3 (b) and Fig.3 (c), respectively. Therefore, when Q_1 and Q_3 are on, the currents in C_u, C_d and C_{fly} can be expressed as :

$$I_{cu\_on} = (I_1 + I_2) \cdot C_u / (C_{fly} + C_u + C_d) \qquad (1)$$

$$I_{cd\_on} = (I_1 + I_2) \cdot C_d / (C_{fly} + C_u + C_d) \qquad (2)$$

$$I_{cfly\_on} = (I_1 + I_2) \cdot C_{fly} / (C_{fly} + C_u + C_d) \qquad (3)$$

And when Q_1 and Q_3 are off, the currents in C_u, C_d and C_{fly} can be expressed as :

$$I_{cu\_off} = (I_2 + I_3) \cdot C_u / (C_{fly} + C_u + C_d) \qquad (4)$$

$$I_{cd\_off} = (I_2 + I_3) \cdot C_d / (C_{fly} + C_u + C_d) \qquad (5)$$

$$I_{cfly\_off} = (I_2 + I_3) \cdot C_{fly} / (C_{fly} + C_u + C_d) \qquad (6)$$

Therefore, the conduction loss in the switching capacitors can be derived as

$$P_{c\_con} = [D \cdot I_{c\_on}^2 + (1-D) \cdot I_{c\_off}^2] \cdot R_{esr} \qquad (7)$$

For the transient when two capacitors or voltage source and one capacitor are connected together, the charge balance always exists. Here, we take the turn-on transient for example. And the power losses due to turn-off transient can be obtained accordingly.

Before the switching transient, C_{fly} is connected in parallel with C_d, and it is switched to C_u when Q_1 and Q_3 turn on. Therefore, the charge through C_{fly} and voltage source can be expressed as:

$$Q_{cfly} = Q_{cu} + Q_{cd} \qquad (8)$$

$$Q_{vin} = Q_{cd} \qquad (9)$$

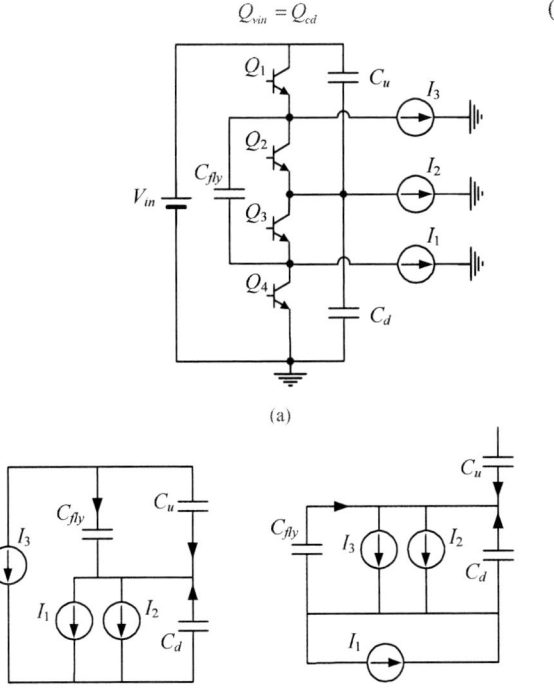

(a)

(b)   (c)

Fig.3 Equivalent circuits of the voltage divider.

As mentioned in [14], the inherent energy loss due to the voltage difference can be expressed as:

$$\Delta E_{tra} = \frac{C_{fly} \cdot (v_{cfly0+}^2 - v_{cfly0-}^2)}{2} + \frac{C_u \cdot (v_{cu0+}^2 - v_{cu0-}^2)}{2}$$
$$+ \frac{C_d \cdot (v_{cd0+}^2 - v_{cd0-}^2)}{2} + V_{in} \cdot Q_{vin} \qquad (10)$$

Where the subscripts 0+ and 0− indicate the instants right before and after the switching transient, respectively.

As for the power losses in multiplexer and linear amplifier, the detail calculation can be found in [16]. Once the power losses of the complete converter have been modeled, the optimization algorithm can be started. Fig. 4 illustrates the theoretical efficiency of the proposed envelope amplifier for a 64QAM signals. As seen, the overall efficiency decreases slightly when the switching frequency in voltage divider increases. Therefore, the switching frequency in the voltage divider can be increased for a smaller volume. Besides, since most of the power are consumed in the linear amplifier, the voltage levels that are provided to the linear amplifier are critical for the overall efficiency. In the case of a 64QAM signal, the highest efficiency always presents at duty cycle of 0.4, regardless of the switching frequency in voltage divider.

The relationships between the efficiency and the capacitance of C_u and C_{fly} are illustrated in Fig 5. As seen, large capacitance of C_u benefits the overall efficiency, but, it goes against the requirement of the small volume and weight in satellite application. Moreover, according to (5), the capacitance of C_{fly} affects the amplitude of conduction current through the transistors. The smaller capacitance of C_{fly} can decrease the conduction current, which reduces the switching losses, however, increases the conduction loss in C_u and C_d. Thus, the selection of C_{fly} should be a result of the optimization of the overall losses.

Fig.6 illustrates the efficiency depending on the ripple currents of L_1 and L_3, where Δi_{L1} and Δi_{L3} refer to the ratios of the ripple current and average current. As seen, the highest efficiency presents when ripple currents equal to two times of the average current. It is the condition for high side transistors to achieve zero current/voltage switching.

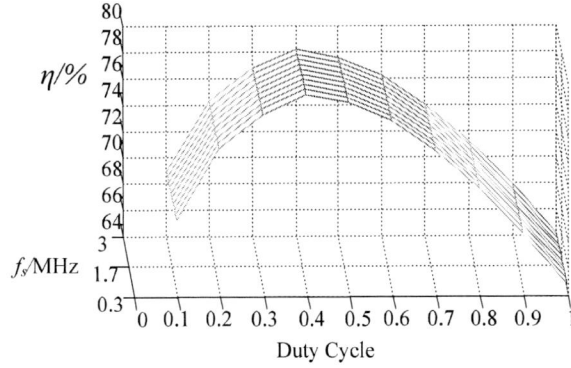

Fig. 4. Efficiency of the envelope amplifier depending on the duty cycle and switching frequency in voltage divider, in the case of a 5MHz 64QAM signal.

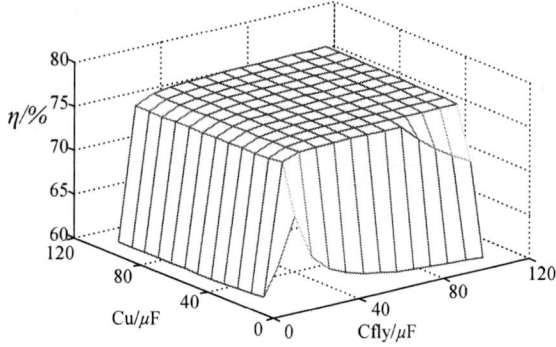

Fig. 5. Efficiency of the envelope amplifier depending on the capacitance of C_u and C_{fly}, in the case of a 5MHz 64QAM signal.

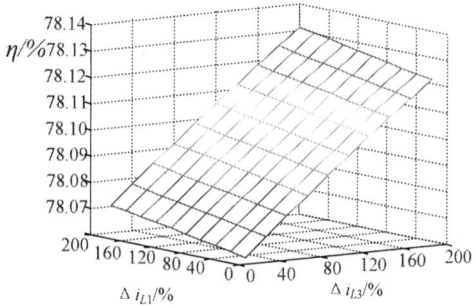

Fig. 6. Efficiency of the envelope amplifier depending on the ripple currents of L1 and L3, in the case of a 5MHz 64QAM signal.

III. DESIGNED SYSTEM

In order to verify the proposed switching capacitor based ET power supply, two prototypes have been fabricated (GaN and Si based) shown in Fig. 7. The specifications are as follows:

- Output voltage: 0 ~ 24 V;

- Peak output power : 40 W;

- Maximum tracking frequency: 5 MHz;

- Load resistance: 15 Ω.

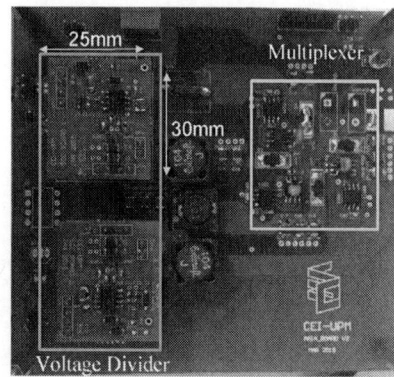

Fig. 7. Photograph of the implemented multilevel converter

As it is aforementioned, the voltage divider works with open loop control, and, in order to guarantee stable voltages at their outputs, at each output there are two ceramic capacitors in parallel (each one of 22 μF). Also, as mentioned above, the selection of C_{fly} should be a trade off between the power loss and system volume and in this design three 10 μF ceramic capacitors are placed in parallel.

To make a fair comparison, all the parameters except the transistors are kept the same in two prototypes. The detail implementations are listed in Table 1.

The triggering logic is implemented in a FPGA that is used as a source of the digital signal reference. The digitalized signal reference is sent to a D/A converter and from there to the linear amplifier. The same reference signal is sent to the triggering logic and to the linear amplifier. To synchronize these two voltages, a digital delay filter is implemented in the FPGA as well, in order to compensate the delays in the system and synchronize the multilevel output voltage with linear amplifier's reference.

IV. EXPERIMENTAL RESULTS

Fig. 8 and Fig. 9 illustrate the comparison between the measured and estimated efficiency of the first stage (voltage divider stage with buck converters) over different switching frequencies and duty cycles. The solid lines indicate the efficiency curve of calculated results and the dotted lines indicate the measured results. As seen, there is a good agreement between the theoretical efficiency and the measurement, which proves the effectiveness of the power loss model. It also can be seen that the efficiency is inversely related to the switching frequency, however, is almost independent of duty cycle. Therefore, for a higher efficiency the switching frequency of the voltage divider can be lowered. On the other hand, the selection of the duty cycle mainly depends on the efficiency of the linear amplifier. It is important to note that there is little difference between GaN and Silicon devices when switching frequency is lower than 500 kHz. Thus, the GaN devices are more suitable for designs with switching frequencies in MHz range in which the converter should work in order to minimize the overall size.

TABLE I: PARAMETERS OF THE PROTOTYPE

	Si prototype	Gan prototype
Transistors in Voltage divider	EPC2014	FDMS7620S
Transistors in analog multiplexer	EPC2014	Si4840DBY
Schottky diodes	VSSAF3L45-M3	
Drivers in voltage divider	LM2726	
Drivers in multiplexer	EL7158	
Isolator	ISO721	

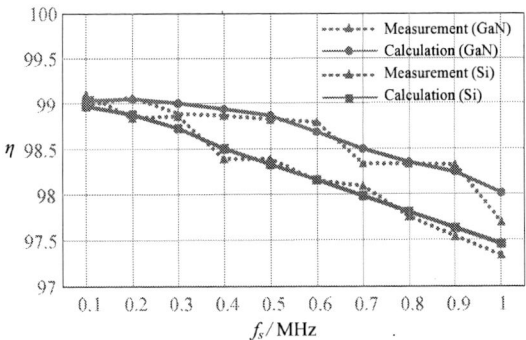

Fig. 8. Efficiency comparison over different switching frequencies when the duty cycle is selected as 0.5

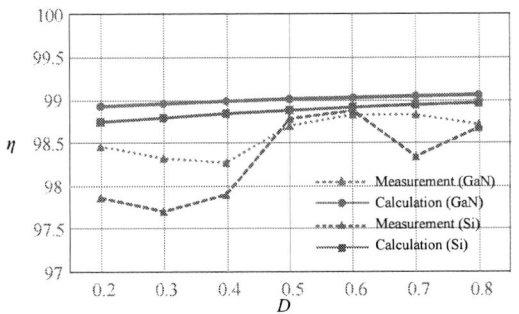

Fig. 9. Efficiency comparison over different duty cycles when the switching frequency is selected as 150 kHz.

To verify the effectiveness of the proposed solution, the multilevel converter has been tested with different sine-wave signals. Fig. 10 and Fig. 11 show the multilevel output voltage in the case when a reference is a sine-wave of 600 kHz and 3.6 MHz, respectively. The duty cycle for voltage divider is selected as 0.4 in this case. As seen, due to the stray inductors and parasitic capacitors, there is always an additional voltage spike when voltage is switched from one to another level. To reduce those parasitic parameters, special consideration has been paid to the PCB layout:

1. All switches in multiplexer are kept close to each other;

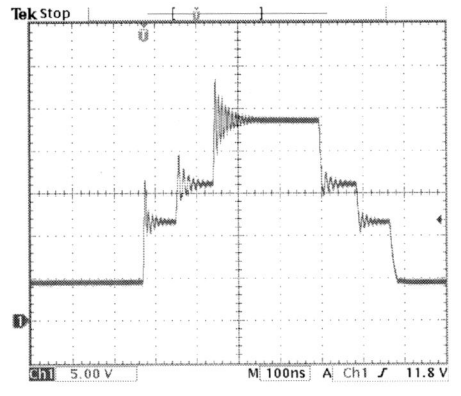

Fig. 10. Multilevel converter output voltage in the case of a 600kHz sine wave

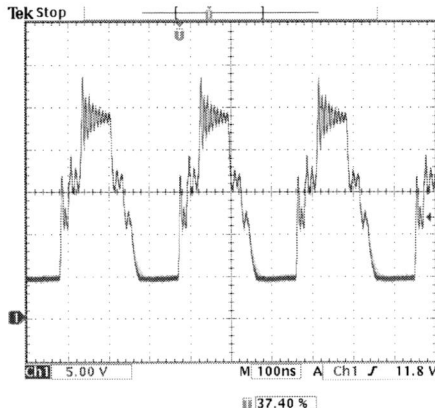

Fig. 11. Multilevel converter output voltage in the case of a 3.6 MHz sine wave.

2. The coper fill of step voltage is reduced for smaller inter-layer distributed capacitors.

Unfortunately, the oscillations are always present and may compromise the functionality of the envelope tracker.

Fig. 12 depicts the efficiency of the multilevel converter over different tracking frequencies in the case of a sine wave envelope. The efficiency is measured using a resistor as the load of multilevel converter. Since the switching loss for the multiplexer is related to the tracking frequency, the overall efficiency decreases along with the rising tracking frequency. However, the lowest efficiency, occurs at 3.6 MHz tracking frequency, is still higher than 90%. Furthermore, the implemented prototype was tested with a 5 MHz bandwidth 64QAM signal, as shown in Fig.13. The measured efficiency of the multilevel converter was 96.15% while the efficiency of the overall system was 75%, which shows a great improvement comparing with the multilevel converter based on independent voltage cells [11]. Fig.14 illustrates the measured overall efficiency over different duty cycles, when switching frequency in voltage divider is 150 kHz. As seen, the highest efficiency is obtained when the duty cycle of the first stage is 0.4, which shows a good agreement with the expected maximum obtained by theoretical analysis. The average output power was, approximately, 8W.

Fig. 12. Efficiency of multilevel converter for different frequencies of the tracked sine waves.

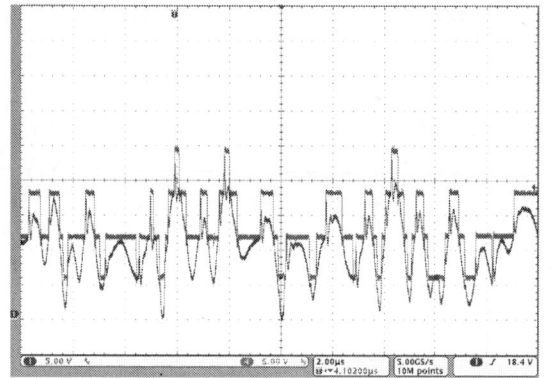

Fig. 13. Multilevel converter and envelope tracker output voltages in the case of a 5 MHz 64QAM signal.

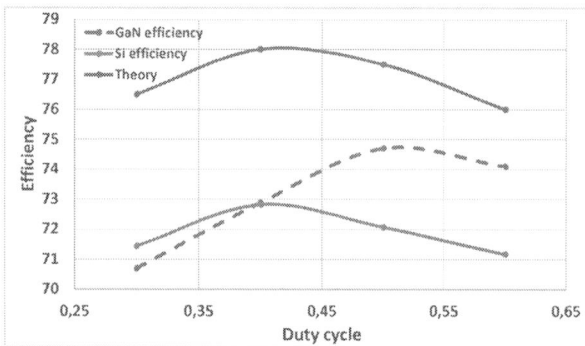

Fig. 14. Measured efficiencies of envelope amplifier over different duty cycles in the case of a 5 MHz 64QAM signal.

V. CONCLUSIONS

In this paper, a solution for envelope tracking power supply in ET technique is presented. It consists of a multilevel converter in series with a linear amplifier. A switching capacitor based voltage divider in combination with two LC filters is proposed to provide four individual voltage levels. The output voltage levels can be conveniently modulated by alterable duty cycle to maximize the efficiency of the linear amplifier, according to the tracking signals' characteristics. All the transistors are implemented with GaN device to further enhance the overall efficiency. Two prototypes (GaN and Si based) are built to verify the effectiveness of the proposed multilevel converter, which is capable of following a 5 MHz RF signal with maximum output power of 40W. The measured efficiency of the multilevel converter was 96.15% and the overall efficiency was around 75%. The experimental results show that for the switching frequencies in MHz range GaN transistors offer better efficiency than Si devices. This is of crucial importance due to the fact that the size and weight of the proposed converter is dramatically reduced using switching frequencies in this range

ACKNOWLEDGEMENT

The authors would like to thank Vladan Lazarević for his help during the final experiments that were conducted in this paper.

References

[1] Z. Wang, "Demystifying envelope tracking: use for high-efficiency power amplifiers for 4G and beyond," IEEE Microwave Magazine, vol. 16, no. 3, pp. 106-129, Apr. 2015.

[2] C. Chen, Y. Lin, T. Horng, K. Peng, C. Li, "Kahn envelope elimination and restoration technique using injection-locked oscillators," in IEEE MTT-S, 20012, pp. 1-3.

[3] O. Lazaro and G. A. Rincón-Mora, "Comparative efficiency analysis of dynamically supplied power amplifiers (PA)," in IEEE ICECS, 2010, pp. 607-610.

[4] F. Wang, D. F. Kimball, J. D. Popp, A. H. Yang, D. Y. Lie, P. M. Asbeck, and L. E. Larson, "An improved power-added efficiency 19-dBm hybrid envelope elimination and restoration power amplifier for 802.11g WLAN applications," IEEE Trans. Microw Theory Tech., vol. 54, no. 12, pp. 4086-4099, Dec. 2006.

[5] D. Diaz, O. Garcia, J. A. Oliver, P. Alou, Z. Pavlovic and J. A. Cobos, "Ripple cancellation technique applied to a synchronous buck converter to achieve a very high bandwidth and very high efficiency envelope amplifier,". IEEE Trans. Power Electron., vol. 29, no. 6, pp. 2892-2902, Jan. 2013.

[6] A. Kanbe, M. Kaneta, F. Yui, H. Kobayashi, N. Takai, T. Shimura, H. Hirata, and K. Yamagishi, "New architecture for envelope-tracking power amplifier for base station," in IEEE APCCAS, 2008, pp. 296-299.

[7] P. Y. Wu and P. K. T. Mok, "A two-phase switching hybrid supply modulator for RF power amplifiers with 9% efficiency improvement," IEEE J. Solid-State Circuits, vol. 45, no. 12, pp. 2543-2556, Dec. 2010.

[8] P. F. Miaja, M. Rodriguez, A. Rodriguez, and J. Sebastian, "A linear assisted DC/DC converter for envelope tracking and envelope elimination and restoration applications," IEEE Trans. Power Electron., vol. 27, no. 7, pp. 3302-3309, Jul. 2012.

[9] H. Xi, Q. Jin, X. Ruan, and X. Xiong, "Full feed-forward of the output voltage to improve efficiency for envelope-tracking power supply using switch-linear hybrid configuration," IEEE Trans. Power Electron., vol. 28, no. 1, pp. 451-456, Jan. 2013.

[10] L. Marco, E. Alarcon and D. Maksimovic, "Effects of switching power converter nonidealities in Envelope Elimination and Restoration technique", in IEEE International Symposium on Circuits and Systems, 2006, pp. 21-24.

[11] M.Vasić, O.Garcia, J.A.Oliver, P.Alou, D.Diaz and J.A.Cobos, "Multilevel power supply for high efficiency RF amplifier", in 24th Annual IEEE Applied Power Electronics Conference, 2009, pp. 25-30.

[12] M. Vasic, O. Garcia, J. A. Oliver, P. Alou, D. Diaz and J. A. Cobos, "Switching capacities based envelope amplifier for high efficiency RF amplifiers," in IEEE APEC, 2010, pp. 723-728.

[13] J. Hoversten and Z. Popovic. "Envelope tracking transmitter system analysis method," in IEEE RWS 2010, 2010, pp. 180-183.

[14] S. Ben-Yaakov, "Switched capacitors converters," in Professional Education Seminar at the 24th Annual IEEE Applied Power Electronics Conference, 2009.

[15] M Paolucci,"Advanced design for fast switching power mosfets," in Professional Education Seminar at the 26th Annual IEEE Applied Power Electronics Conference, 2011.

[16] M. Vasic, O. Garcia, J. A. Oliver, P. Alou and J. A. Cobos, ")Serial or parallel linear-assisted switching converter as envelope amplifierptimization and comparison," in IEEE ECCE, 2011, pp. 2488-2494.

An Isolated High Step-Up Converter with Continuous Input Current and LC Snubber

K. I. Hwu, *Member, IEEE*, W. Z. Jiang, *Student Member, IEEE*, and Y. T. Yau, *Member, IEEE*
Department of Electrical Engineering, National Taipei University of Technology, Taiwan

Abstract—In this paper, a novel galvanic isolated high step-up converter is presented, which is suitable for renewable energy applications and integrates a boost converter, a coupled inductor, a charge pump capacitor cell, and an LC snubber. Since the proposed converter possesses an input inductor, the input current is continuous. Furthermore, the proposed converter can achieve high voltage gain without extremely large duty cycle and turns ratio of the coupled inductor by using the charge pump capacitor cell. Moreover, the leakage inductance energy can be recycled to the output capacitor of the boost converter via the LC snubber, and then the leakage inductance energy stored in the capacitor of the boost converter can be transferred to the output load. As a result, the voltage spike can be suppressed to a low voltage level. Finally, the basic operating principles and some experimental results are provided to verify the effectiveness of the proposed converter.

I. INTRODUCTION

In recent years, high step-up dc-dc converters are widely used in many renewable energy systems such as fuel cells, photovoltaic (PV) panels, etc. However, the output voltages of these renewable energy systems are usually not only unstable but also not high enough to supply the output load or to be linked to an AC power grid via a DC-AC inverter [1-3]. Consequently, a high step-up converter is required to obtain a high output voltage. For non-isolated converters to be considered, the traditional boost and buck-boost converters [4] are widely used because of their simple structures. However, it is hard to achieve higher output voltages with moderate duty cycle due to the parasitic resistances of the inductors. Besides, the switches have to block high output voltages. Accordingly, the switches with high on-resistance are required, resulting in high conduction loss. Thus, in order to obtain a higher output voltage, many non-isolated step-up converters using different voltage-boosting techniques have been presented. These voltage-boosting techniques include coupled inductor [5-10], switched-capacitor [11, 12], voltage multiplier [13, 14], etc. However, in some applications, an isolated converter is preferred to meet the safety the requirements of galvanic isolation [15]. Therefore, the traditional flyback converter is quite appealing in industrial applications because of its low component number, simple structure and low cost. However, it suffers from low-voltage gain. Therefore, some isolated step-up converters have been presented [16-19]. In [16], the presented converter uses both coupled inductor and isolated

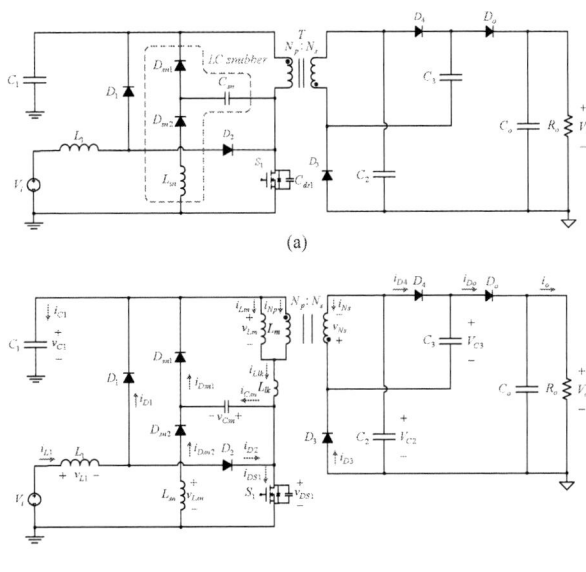

(a)

(b)

Fig.1. Proposed isolated step-up converter: (a) without current or voltage symbols; (b) with current or voltage symbols.

transformer to achieve high voltage gain and an active clamp circuit is adopted to reduce the switch voltage spike. However, this converter requires two switch drivers, which leads to the complexity of circuit. In [16], a converter integrating an active clamp flyback converter and a voltage multiplier is presented. However, the input current is pulsating, which results in high input current ripple. In [17], an isolated using a tapped-inductor and isolated-switched-capacitor is presented. However, the voltage gain is not high enough, the voltage spike across switch is pretty high, and the input current is pulsating. In [18], an isolated converter consisting a boost converter and a series resonant converter is presented. Even though the input current is non-pulsating, this converter has two stages and six switches, which increases the complexity of the overall circuit and the number of drivers. In [19], an LC sunber used in the SR flyback converter to clamp the switch voltage stress is presented because the LC snubber does not dissipate the energy theoretically and does not use any active switch.

Based on the mentioned above, in order to achieve a high output voltage, possess a non-pulsating input current, satisfy the galvanic isolation, and maintain a high efficiency. A novel isolated high step-up converter is presented, which integrates a boost converter, a coupled inductor, a charge pump capacitor cell, and an LC snubber. Therefore, in the proposed converter, the input current supplied from the source is continuous, a high voltage gain can be realized without high turns ratio and large duty cycle, the leakage inductance energy can be recycled to the output load, the voltage stresses across switches are low, and galvanic isolation exists between the input terminal and the output terminal.

II. PROPOSED CONVERTER

Fig. 1 (a) shows the proposed converter, which contains one input inductor L_1, one boost capacitor C_1, one boost diode D_1, one MOSFET switch S_1, one coupled inductor T composed of the primary winding with N_p turns and the secondary winding with N_s turns, two snubber diodes D_{sn1} and D_{sn2}, one snubber inductor L_{sn}, one snubber capacitor C_{sn}, two charge pump capacitors C_2 and C_3, two charge pump diodes D_3 and D_4, and one output capacitor C_o. In addition, the input voltage is denoted by V_i, the output voltage is signified by V_o, and the output resistor is represented by R_o.

III. BASIC ANALYSIS

For analysis convenience, there are some assumptions, and voltage symbols and current symbols to be given as follows.
(1) The coupled inductor is modeled as an ideal transformer except that one magnetizing inductor L_m is connected in parallel with the primary winding.
(2) The MOSFET switch and the diodes are viewed as ideal components.
(3) The values of all the capacitors are large enough such that the voltages across them are kept constant at some values.

The following analysis contains the operating principles, voltage gain, and boundary conditions for input inductor and magnetizing inductor. There are eight operating modes in the proposed converter. The gate driving signal v_{gs1} of the switch S_1 has the duty cycle of D, where D is the dc quiescent duty cycle created from the controller.

Furthermore, as shown in Fig. 1(b), the currents flowing through L_1, C_1, D_1, S_1, L_m, and the windings N_p and N_s are signified as i_{L1}, i_{C1}, i_{D1}, i_{DS1}, i_{Lm}, i_{Np}, and i_{Ns}, respectively. The currents flowing through D_{sn1}, D_{sn2} or L_{sn}, C_{sn}, D_2, D_3, D_4, D_o and R_o are represented as i_{Dsn1}, i_{Dsn2}, i_{Csn}, i_{D2}, i_{D3}, i_{D4}, i_{Do}, and i_o, respectively. Moreover, the voltages across L_1, C_1, L_{sn}, C_{sn}, and S_1 are denoted as v_{L1}, v_{C1}, v_{Lsn}, v_{Csn}, and v_{DS1}, respectively. The voltages across L_m or N_p, N_s, C_2, and C_3 are represented as v_{Lm}, v_{Ns}, V_{C2}, and V_{C3}, respectively.

(a) Operating principles

There are eight operating modes below. Fig. 2 shows the illustrated waveforms over one switching period.

Mode 1 [$t_0 \leq t \leq t_1$]: During this interval, as shown in Fig. 3(a), S_1 is turned on, and D_2 is forward-biased. The currents i_{D3} and i_{D4} keep charging C_2 and C_3, respectively. Since the voltage across L_m, v_{Lm}, is a negative value, which is induced from the secondary winding N_s, L_m keeps demagnetized. In the meantime, the voltage $V_{C1} + (N_p / N_s) \times V_{C3}$ is imposed on L_{lk}, which makes i_{Llk} increases rapidly. Furthermore, D_1 becomes reverse-biased, D_{sn2} becomes forward-biased, D_{sn1} remains reverse-biased. Therefore, the input voltage V_i is

imposed on L_1, thus causing L_1 to be magnetized. Moreover, C_{sn} releases energy to L_{sn}. D_o remains reverse-biased. Thus, only C_o supplies energy to the load. This mode ends when i_{Llk} is equal to i_{Lm}.

Mode 2 [$t_1 \leq t \leq t_2$]: During this interval, as shown in Fig. 3(b), S_1 keeps turned on, D_3 and D_4 become reverse-biased, and D_o becomes forward-biased. Therefore, C_1, C_2 and C_3 provide energy to the load. Meanwhile, the voltage V_{C1} is imposed on L_m and L_{lk}, making L_m and L_{lk} magnetized continuously. Moreover, the energy stored in C_{sn} is continuously released to L_{sn} until v_{Csn} reaches zero. After that, energy stored in L_{sn} is released to charge C_{sn}, causing the voltage across C_{sn}, v_{Csn}, to go up in the opposite direction. This mode ends when v_{Csn} reaches $-V_{C1}$.

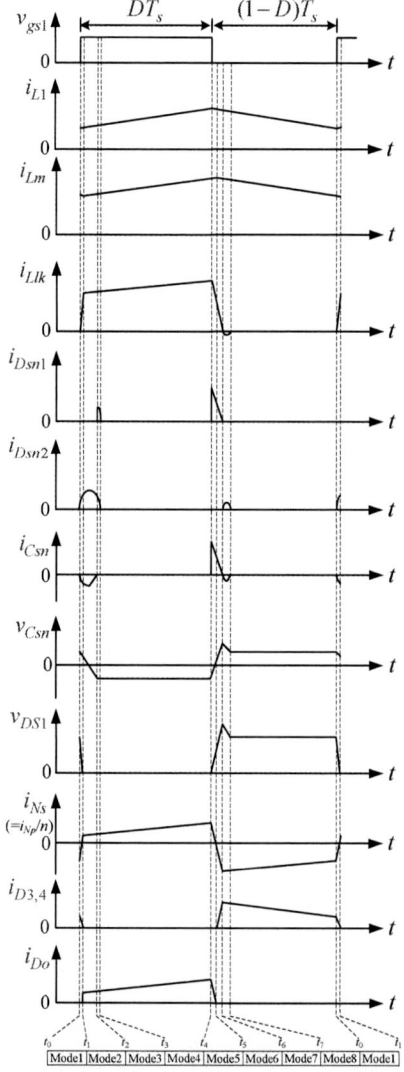

Fig. 2. Illustrated waveforms of the proposed converter.

Mode 3 [$t_2 \leq t \leq t_3$]: During this interval, as shown in Fig. 3(c), S_1 keeps turned on, D_1 and D_3 and D_4 keep reverse-biased, and D_{sn2} and D_o keep forward-biased. The voltage across the snubber capacitor, v_{Csn}, is clamped to $-V_{C1}$. Moreover, the diode D_{sn1} is forced to conduct by the current from L_{sn}. Therefore, the energy stored in L_{sn} is transferred to the output load via the coupled inductor. This mode ends when the energy from L_{sn} drops to zero, i.e., i_{Dsn2} reaches zero.

Mode 4 [$t_3 \leq t \leq t_4$]: During this interval, as shown in Fig. 3(d), S_1 keeps turned on, D_1, D_2 and D_3 keep reverse-biased, and D_o keeps forward-biased. Since i_{Dsn2} reaches zero, the diodes D_{sn1} and D_{sn2} become reverse-biased. This mode ends when S_1 becomes turned off.

Mode 5 [$t_4 \leq t \leq t_5$]: During this interval, as shown in Fig. 3(e), S_1 is turned off, but C_2, C_3 and coupled inductor keep supplying energy to the load. Moreover, the energy stored in L_1 is released to C_1 via D_1, and the leakage inductance current i_{Llk} charges the snubber capacitor C_{sn} via D_{sn1} and the parasitic capacitor C_{ds1}. Therefore, v_{Csn} increases linearly from $-V_{C1}$. This mode ends when i_{Ns} reaches zero.

Mode 6 [$t_5 \leq t \leq t_6$]: During this interval, as shown in Fig. 3(f), S_1 remains turned off. The only difference between the previous mode and this mode is that the energy stored in the magnetizing inductor L_m is transferred via D_3 and D_4 to C_2 and C_3, which are connected in parallel. Meanwhile, the output diode D_o becomes reverse-biased. Therefore, only the output capacitor C_o provides the energy to the load. Moreover, the leakage inductance current i_{Llk} continuously releases the energy to the snubber capacitor C_{sn}, which makes v_{Csn} increase gradually. The voltage across S_1 is also increasing, which is equal to $V_{C1} + v_{Csn}$. This mode ends when i_{Llk} reaches zero.

Mode 7 [$t_6 \leq t \leq t_7$]: During this interval, as shown in Fig. 3(g), S_1 remains turned off. Since the snubber capacitor voltage, v_{Csn}, reaches the maximum value, which is higher than $V_{C1} + V_{C2}/n$, the diode D_{sn2} becomes forward-biased. Therefore, the energy stored in C_{sn} is discharged to C_1 and the secondary side. During this period, the snubber capacitor C_{sn}, the snubber inductor L_{sn}, and the leakage inductor L_{lk} will resonate together. This mode ends when i_{Dsn2} reaches zero.

Mode 8 [$t_7 \leq t \leq t_0$]: During this interval, as shown in Fig. 3(h), S_1 remains turned off. Since i_{Dsn2} reaches zero, D_{sn2} becomes reverse-biased. Moreover, L_m continuously delivers energy to charge C_2 and C_3. Therefore, i_{Lm} decreases gradually. This mode ends when S_1 is turned on, and the next cycle is repeated.

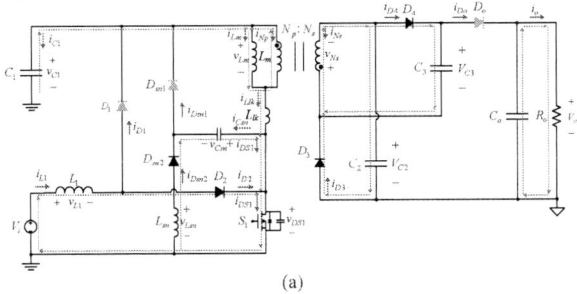

(a)

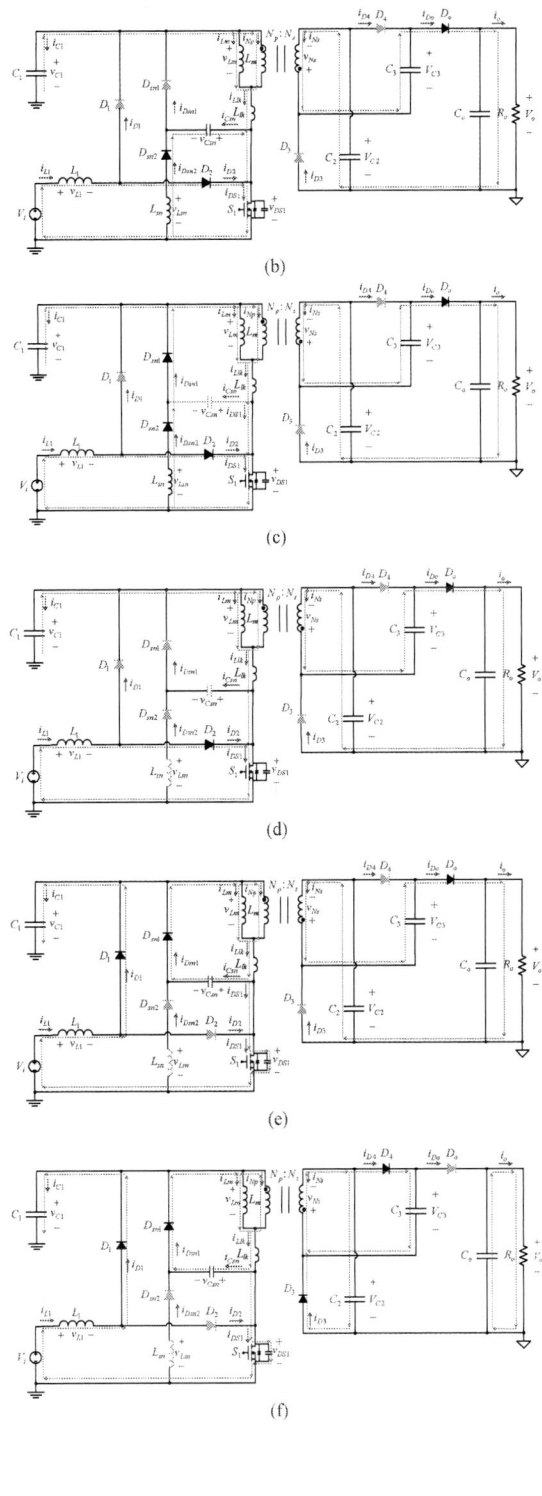

(b)

(c)

(d)

(e)

(f)

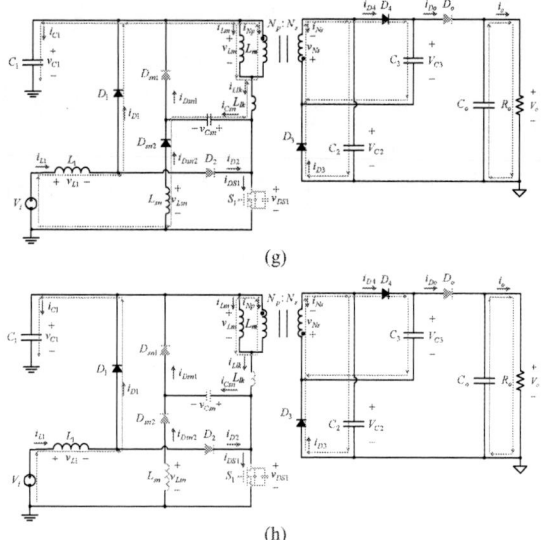

(g)

(h)

Fig. 3. Power flow paths over one switching period: (a) mode 1; (b) mode 2; (c) mode 3; (d) mode 4; (e) mode 5; (f) mode 6; (g) mode 7; (h) mode 8.

(b) Voltage gain

In order to get the voltages across C_1, C_2, and C_3, and the voltage gain, only Fig. 3(b) and Fig. 3(f) are considered here, and the leakage inductor L_{lk} and the LC snubber are ignored. In Fig. 3(b), the voltage across L_1 and L_m are as follows:

$$v_{L1} = V_i \qquad (1)$$

$$v_{Lm} = V_{C1} \qquad (2)$$

In Fig. 4(b), the voltage across L_1 and L_m are as follows:

$$v_{L1} = V_i - V_{C1} \qquad (3)$$

$$v_{Lm} = -V_{C2}/n \qquad (4)$$

First, by applying the voltage-second balance principle to L_1 over one switching period, the following equation can be obtained:

$$V_i \times D + (V_i - V_{C1}) \times (1 - D) = 0 \qquad (5)$$

$$V_{C1} = V_i \times \frac{1}{1 - D} \qquad (6)$$

Second, by applying the voltage-second balance principle to L_m over one switching period, the following equation can be obtained:

$$V_{C1} \times D + (-V_{C2} \times n) \times (1 - D) = 0 \qquad (7)$$

Also, by rearranging the above equation, the voltages across C_1 and C_2 can be obtained to be

$$V_{C2} = V_{C3} = n \times \frac{D}{1 - D} \times V_{C1} = n \times \frac{D}{(1 - D)^2} \times V_i \qquad (8)$$

From Fig. 3(b), the output voltage can be found as follows.

$$V_o = V_{C2} + V_{C3} + v_{Ns}$$

$$= \frac{2nD}{(1 - D)^2} V_i + \frac{nD}{(1 - D)^2} V_i = \frac{n(1 + D)}{(1 - D)^2} V_i \qquad (9)$$

The corresponding voltage gain can be expressed to be

$$\frac{V_o}{V_i} = \frac{n(1 + D)}{(1 - D)^2} \qquad (10)$$

Fig. 4 shows the curves of voltage gain versus duty cycle of the proposed converter considering different turns ratios.

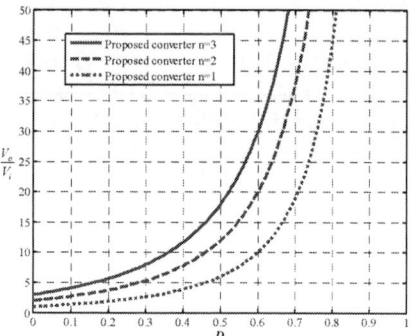

Fig. 4. Curves of voltage gain versus duty cycle for the proposed converter with different values of turns ratio n.

(c) Boundary condition for input inductor

The condition for the magnetizing inductor L_1 operating in what region will be described as follows:

$$\begin{cases} 2I_1 \geq \Delta i_{L1}, & \text{for CCM} \\ 2I_1 < \Delta i_{L1}, & \text{for DCM} \end{cases} \qquad (11)$$

where I_{L1} and Δi_{L1} are the dc and ac components of i_{L1}, respectively.

For analysis convenience, it is assumed that the input power is equal to the output power. Therefore, the input current I_{L1} can be expressed as:

$$I_{L1} = \frac{n(1 + D)}{(1 - D)^2} \times I_o \qquad (12)$$

Substituting V_o/R_o into I_o in (12) yields the following equation

$$I_{L1} = \frac{n(1 + D)}{(1 - D)^2} \times \frac{V_o}{R_o} \qquad (13)$$

Also, Δi_{L1} can be represented by

$$\Delta i_{L1} = \frac{v_{L1} \Delta t}{L_1} = \frac{V_i D T_s}{L_1} \qquad (14)$$

As $2I_{L1} \geq \Delta i_{L1}$, L_1 operates in the CCM. Moreover, the further deduction is shown as follows:

$$2I_{L1} \geq \Delta i_{L1}$$

$$\Rightarrow 2 \times [\frac{n(1 + D)}{(1 - D)^2} \times \frac{V_o}{R_o}] \geq \frac{V_i D T_s}{L_1}$$

$$\Rightarrow \frac{2L_1}{R_o T_s} \geq \frac{D(1 - D)^4}{[n(1 + D)]^2} \qquad (15)$$

$$\Rightarrow K_1 \geq K_{crit1}(D)$$

where $K_1 = \frac{2L_m}{R_o T_s}$ and $K_{crit1}(D) = \frac{D(1 - D)^4}{[n(1 + D)]^2}$

From (15), the relationship between $K_{crit1}(D)$ versus D is shown in Fig. 5 under the condition that n is set at three. It can be seen that if K_1 is larger than $K_{crit1}(D)$, L_1 operates in the CCM; otherwise, L_1 operates in the DCM.

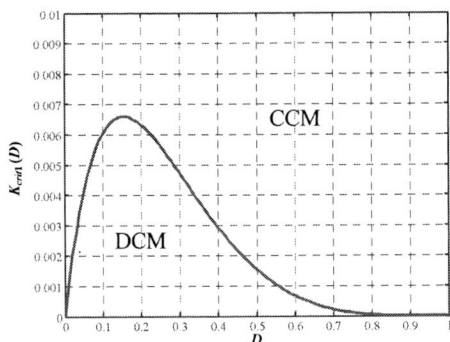

Fig. 5. Boundary condition for input inductor L_1.

(d) Boundary condition for magnetizing inductor

The condition for the magnetizing inductor L_m operating in what region will be described as follows:

$$\begin{cases} 2I_{Lm} \geq \Delta i_{Lm}, & \text{for CCM} \\ 2I_{Lm} < \Delta i_{Lm}, & \text{for DCM} \end{cases} \quad (16)$$

where I_{Lm} and Δi_{Lm} are the dc and ac components of i_{Lm}, respectively.

The expression of I_{Lm} can be obtained from (17) to (20). For analysis convenience, it is assumed that the input power is equal to the output power. According to the voltage-second balance for the inductor and the ampere-second balance for the capacitor, the dc component of the inductor voltage and the dc component of the capacitor current are both zero over one switching period in the steady state.

Therefore, as shown in Fig. 6(a), the dc component of i_{Ns}, I_{Ns}, is equal to the output current I_o; likewise, as shown in Figs. 6(a) and 6(b), the dc component of i_{Lm}, I_{Lm}, is equal to the current I_x, entering the primary side of the coupled inductor, plus the dc component of i_{Np}, I_{Np}. Therefore,

$$I_x = (1-D) \times I_{L1} \quad (17)$$

Next, based on (13) and (17), I_x is derived to be

$$I_x = (1-D) \times \frac{n(1+D)}{(1-D)^2} \times I_o \quad (18)$$

$$I_{Np} = n \times I_{Ns} = n \times I_o \quad (19)$$

$$I_{Lm} = I_x + I_{Np} = [\frac{n(1+D)}{(1-D)} + n] \times I_o = \frac{2n}{1-D} \times \frac{V_o}{R_o} \quad (20)$$

Also, Δi_{Lm} can be represented by

$$\Delta i_{Lm} = \frac{v_{Np}\Delta t}{L_m} = \frac{V_{C1}DT_s}{L_m} = \frac{V_iDT_s}{(1-D)L_m} \quad (21)$$

As $2I_{Lm} \geq \Delta i_{Lm}$, L_m operates in the CCM. Moreover, the further deduction is shown as follows:

$$2I_{Lm} \geq \Delta i_{Lm}$$

$$\Rightarrow 2 \times (\frac{2n}{1-D} \times \frac{V_o}{R_o}) \geq \frac{V_iDT_s}{(1-D)L_m}$$

$$\Rightarrow \frac{2L_m}{R_oT_s} \geq \frac{D(1-D)^2}{2n(1+D)} \quad (22)$$

$$\Rightarrow K_2 \geq K_{crit2}(D)$$

where $K_2 = \frac{2L_m}{R_oT_s}$ and $K_{crit2}(D) = \frac{D(1-D)^2}{2n(1+D)}$.

From (22), the relationship between $K_{crit2}(D)$ versus D is shown in Fig. 7 under the condition that n is set at three. It can be seen that if K_2 is larger than $K_{crit2}(D)$, L_m operates in the CCM; otherwise, L_m operates in the DCM.

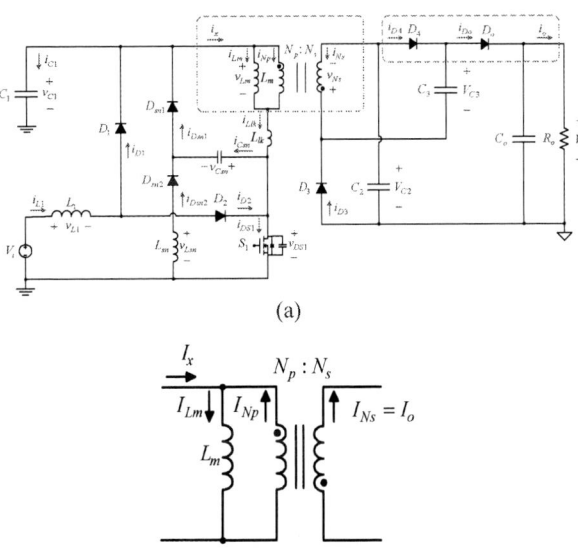

(a)

(b)

Fig. 6. DC component: (a) marked areas in the proposed converter used to analyze I_{Lm}; (b) equivalent model for the dc analysis of the coupled inductor.

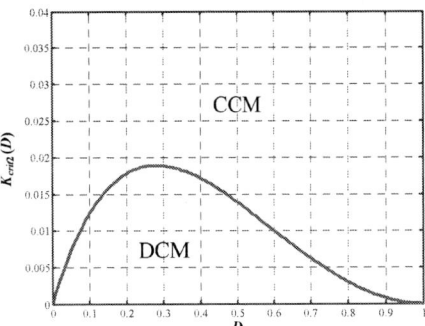

Fig. 7. Boundary condition for magnetizing inductor L_m.

IV. EXPERIMENTAL RESULTS

A prototype with 12V input voltage, 200V output voltage, 100W output power, and 100kHz switching frequency is provided to verify the feasibility. Figs. 8 to 11 show the measured waveforms at rated load. Fig. 8 shows the gate driving signal for S_1, v_{gs1}, the voltage across S_1, v_{ds1}, and the input current, i_i. Fig. 9 shows the gate driving signal for S_1, v_{gs1}, the current flowing through L_{lk}, i_{Lk}, and the secondary side current, i_{Ns}. Fig. 10 shows the gate driving signal for S_1, v_{gs1}, the voltage across C_{sn}, v_{Csn}, and the voltage on C_1, V_{C1}. From Fig. 8, one can see that the spike voltage of v_{ds1} is clamped at 56V, and the input current is continuous. Also, in Fig. 9, during the turn-off period of S_1, i_{Ns} is a negative current, which matches the waveform of i_{Llk} in Fig. 9(b). Furthermore, in Fig. 10, $v_{Csn,min}$ is about $-24V$, equal to V_{C1}, and $v_{Csn,max}$ is about 38V. Moreover, in Fig. 11, the output voltage is stabilized at 200V under rated load.

Aside from these, Fig. 12 shows the curve of efficiency versus load current. From Fig. 12, it can be seen that the efficiency all over the load range is above 82% and can be up to 89%, and the rated load efficiency is about 85%.

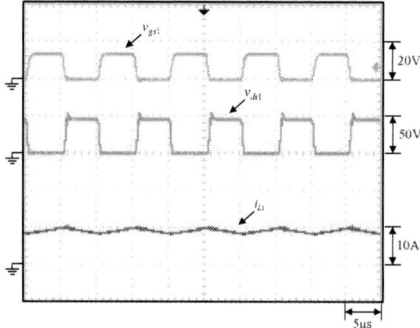

Fig. 8. Waveforms at rated load: (1) v_{gs1}; (2) v_{ds1}; (3) i_{L1}.

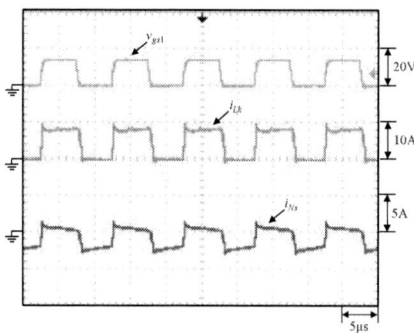

Fig. 9. Waveforms at rated load: (1) v_{gs1}; (2) i_{Lk}; (3) i_{Ns}.

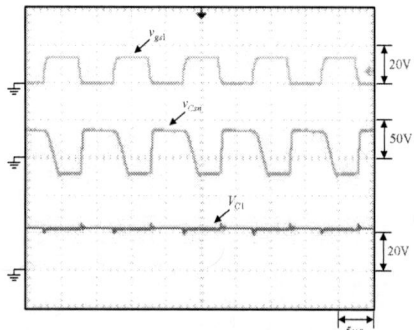

Fig. 10. Waveforms at rated load: (1) v_{gs1}; (2) v_{Csn}; (3) V_{C1}.

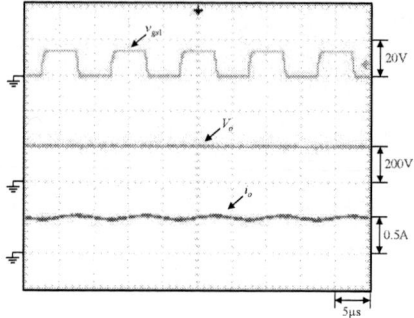

Fig. 11. Waveforms at rated load: (1) v_{gs1}; (2) V_o; (3)i_o.

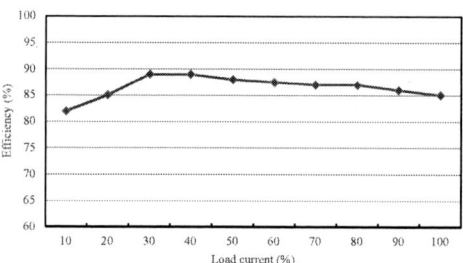

Fig. 12. Efficiency versus load current.

V. CONCLUSION

An isolated high step-up converter with continuous input current is presented. By combining a boost converter, a coupled inductor, and a charge pump capacitor cell, a high step-up voltage gain can be achieved with a relatively low duty cycle. Moreover, the leakage inductance energy can be recycled to the output capacitor of the boost converter by the LC snubber, and then the leakage inductance energy stored in the capacitor of the boost converter can be transferred to the output load. Therefore, the voltage spike of the switch can be clamped at a low value, and the switch with low turn-on resistance can be used, resulting in improving the efficiency. The operating principle analysis, designs and experimental results are provided to verify the effectiveness of the proposed converter.

REFERENCES

[1] H. Hu, S. Harb, N. Kutkut, I. Batarseh, and Z. J. Shen, "Power decoupling techniques for micro-inverters in PV systems-a review," *IEEE ECCE' 10*, pp. 3235-3240, 2010.

[2] Q. Li, and P. Wolfs, "A review of the single phase photovoltaic module integrated converter topologies with three different dc link configurations," *IEEE Trans. Power Electron.*, vol. 23, no. 3, pp. 1320-1333, 2008.

[3] M. David, N. Scholten, and W. L. Soong "Mirco-inverters in small scale PV systems: a review and future directions," *IEEE AUPEC' 13*, pp. 1-6, 2013.

[4] R. W. Erickson and D. Maksimovic, *Fundamentals of Power Electronics*, 2nd ed., Norwell: KLuwer Academic Publishers, 2001.

[5] K. I. Hwu and T. J. Peng, "High-voltage-boosting converter with charge pump capacitor and coupling inductor combined with buck-boost converter," *IET Electron.*, vol. 47, no. 1, pp. 177-188, 2014.

[6] G. M. L, Chu, D. D. C. Lu, and V. G. Agelidis, "Flyback-based high step-up converter with reduced power processing," *IET Electron.*, vol. 5, no. 3, pp. 349-357, 2012.

978-1-4673-9551-9/16 $31.00 © 2016 IEEE

[7] S. Dwari and L. Parsa, "An efficient high-step-up interleaved dc-dc converter with a common active clamp," *IEEE Trans. Power Electron.*, vol. 26, no. 1, pp. 66-78, 2011.

[8] K. I. Hwu and Y.T. Yau, "High step-up converter based on coupling inductor and booststrap capacitors with active clamping," *IEEE Trans. Power Electron. Lett.*, vol. 29, no. 6, pp. 2655-2660, 2014,

[9] Y. Berkovich, B. Axelrod, "High step-up dc-dc converter with coupled inductor and reduced switch-voltage stress," *IEEE IECON' 12*, pp. 453-458, 2012.

[10] K. I. Hwu and W. Z. Jiang, "Voltage gain enhancement for a step-up converter constructed by KY and buck-boost converters," *IEEE Trans. Ind. Electron.*, vol. 61, no. 4, pp. 1758-1768, 2014.

[11] Y. Tang, T. Wang, and Y. He, "A switched-capacitor-based active network converter with high voltage gain," *IEEE Trans. Power Electron.*, vol. 29, no. 6, pp. 2959-2968, 2014,

[12] K. I. Hwu and W. C. Tu, "Voltage-boosting converters with energy pumping," *IET Power Electron.*, vol. 5, no. 2, pp. 185-195, 2012.

[13] Y. J. A. Alcazar, D. de Souza Oliveira, F. L. Tofoli, and R. P. Torrico-Bascope, "DC-DC nonisolated boost converter based on the three-state switching cell and voltage multiplier cells," *IEEE Trans. Ind. Electron.*, vol. 60, no. 10, pp. 4438-4449, 2013.

[14] M. Prudente, L. L. Pfitscher, G. Emmendoerfer, E. F. Romaneli, and R. Gules, "Voltage multiplier cells applied to non-isolated dc-dc converters," *IEEE Trans. Power Electron.*, vol. 23, no. 2, pp. 871-887, 2008.

[15] M. Pavlovsky, S. W. H. de Hann, and J. A Ferreira, "Reaching high power density in multikilowatt dc-dc converters with galvanic isolation," *IEEE Trans. Power Electron.*, vol. 24, no. 3, pp. 603-612, 2009.

[16] G. Spiazzi, P. Mattavelli, and C. Alessandro, "High step-up ratio flyback converter with active clamp and voltage multiplier," *IEEE Trans. Power Electron.*, vol. 26, no. 11, pp. 603-612, 2011.

[17] J. H. Jang, D. H. Kim, J. W. Seo, and J. H. Park, "Series-connected isolated switched-capacitor tapped-inductor boost converter," *IEEE PEDS' 13*, pp. 277-279, 2013.

[18] M. Kasper, M. Ritz, D. Bortis, and J. W. Kolar, "PV panel-ingerated high step-up high efficiency isolated GaN dc-dc boost converter," *IEEE INTELEC' 13*, pp. 1-7, 2013.

[19] K. I. Hwu and Y. T. Yau, "Powering LED using high-efficiency SR flyback converter," *IEEE Trans. Ind. Appl.*, vol. 47, no. 1, pp. 376-386, 2011.

Output-Inductor-less Full-Bridge Converter with SiC-MOSFETs for Low Noise and ZVS Operation

Kazuhide Domoto and Yoichi Ishizuka
Nagasaki University
Graduate School of Engineering
Bunkyo-machi, Nagasaki, 852-8521, Japan

Seiya Abe
Kyushu Institute of Technology
Graduate School of Life Science and
Systems Engineering
Wakamatsu-ku, Kitakyushu, 808-0196
Japan

Tamotsu Ninomiya
Green Electronics Research Institute,
Kitakyushu
Wakamatsu-ku, Kitakyushu, 808-0135
Japan

Abstract— In this paper, a high power-density output-inductor-less full-bridge converter for the HVDC power distribution system is proposed. The new reduction techniques of switching loss and noise in the output-inductor-less full-bridge converter by using SiC-MOSFETs are applied. As the result, the power loss is reduced by 20W in a wide load range, and the noise is also reduced. Furthermore, a high power density of 15W/cm3 is achieved by the experimental breadboard with power level of 1kW.

Keywords—HVDC Power Distribution system; Full-Bridge converter; ZVS Operating; Low Noise; High Power-Density

I. INTRODUCTION

Recently, the rapid growth of internet traffic has increased the number of ICT equipment in a data center, and the electric power consumption has also been increased. Therefore, the energy-saving techniques are required in a data center. In order to satisfy the energy saving and to increase the data-center scale, the High-Voltage Direct-Current (HVDC) power distribution system is effective[1-5]. This system has an advantage of higher efficiency due to the smaller number of conversion stages when compared with the conventional AC system. Furthermore, ICT equipment and its power supply units shown in Fig. 1 should be miniaturized so as to install the much larger number of equipment in a limited space of the data center[6]. Two concepts of energy saving and space saving are combined into "high power-density" of the power converter systems. A lot of high power-density converters can be installed in one rack, and a large power capacity for the data center can be realized as shown in Fig. 2[1]. From this background, the circuit topology and implementation technique of a rectifier unit which is a part of power supply unit in the HVDC distribution system have been investigated before[1-5]. This paper proposes a new technique of reducing high-frequency noise and performing ZVS operation for all switches, and discusses the technique to achieve higher power-density. Furthermore, the installment of SiC-MOSFETs is discussed, which has a fast switching speed, a low on-resistance, a high breakdown voltage, and a body diode with much shorter reverse-recovery time. In particular, the short reverse-recovery time is effective for avoiding the short circuit of the switch arm, and then the higher reliability will be available.

II. OUTPUT-INDUCTOR-LESS FULL-BRIDGE CONVERTER

In the HVDC rectifier unit shown in Fig. 1, a Full-Bridge (FB) converter is generally used. However, the FB converter has an essential problem that surge voltage occurs at secondary-side diodes[7-10]. This surge voltage breaks down the circuit components. In order to suppress the surge voltage, the snubber circuit such as RC snubber is usually used. However, the surge voltage is becoming around 1600V in case of the HVDC system, and a snubber resistor of RC snubber generates a large power loss. As a solution for this problem, an output-inductor-less FB converter shown in Fig. 3 has been proposed[11-16], where the output capacitor serves as a surge snubber. Here, L_m is the magnetizing inductance, and L_{lk} is the leakage inductance. This converter is defined as the basic FB converter in this paper.

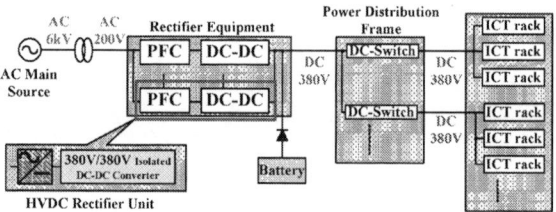

Fig. 1: HVDC power distribution system.

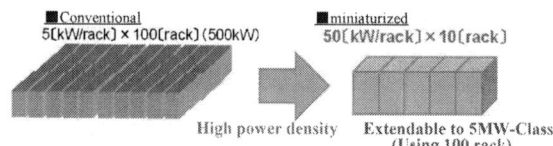

Fig. 2: The effect of high power-density.

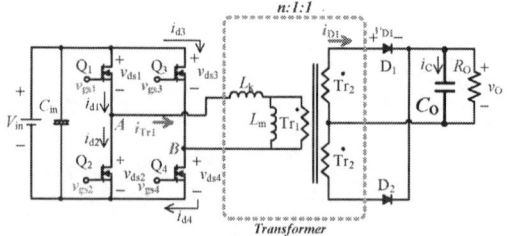

Fig. 3: Basic FB converter.

The prominent effect of surge suppression without power loss by the snubber capacitor has been confirmed so far[17]. On the other hand, this circuit topology causes problems such as increase of switching-loss of the primary switches and another high-frequency noise generation as shown in Fig. 4[17,18]. Furthermore, this noise also causes the fluctuation of load characteristics as shown in Fig. 5, which degrades the closed-loop control characteristics. The details of these problems have not been discussed in previous papers yet [11-18]. This paper reveals the causes of the problems, and proposes the solution technique.

III. PROPOSED METHOD

Figure 6 shows the proposed converter. In the secondary side, the MOSFETs Q_5 and Q_6 are installed instead of the diodes, and an external diode D is added. Furthermore, Figure 7 shows the key waveforms of the proposed control technique. The primary-side switches are controlled with the conventional phase-shift control. A proposed control method is applied to the secondary-side switching sequence, and this is the most important feature of the proposed technique. The switches Q_5 and Q_6 have a duty ratio of 50% which the phase is fixed. Furthermore, these switches turn-on and turn-off, alternately. These key waveforms are illustrated in Fig. 7. Here, the operation characteristics of proposed FB converter are analyzed. The circulating current paths corresponding to 9 states are shown in Fig. 8, and these operating states are described as follows:

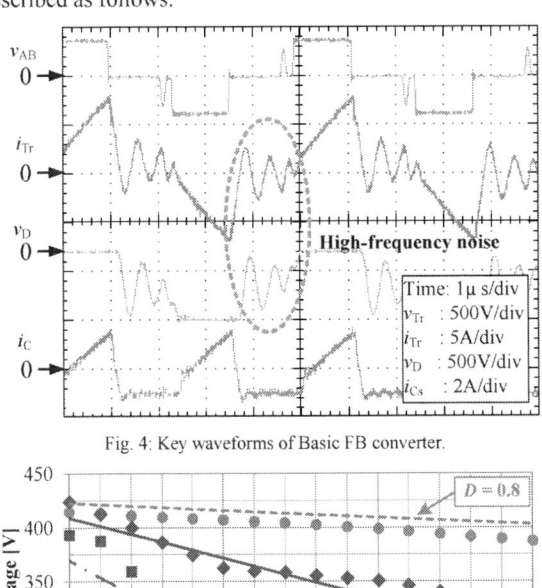

Fig. 4: Key waveforms of Basic FB converter.

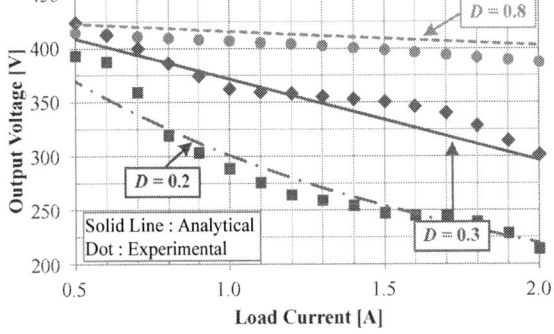

Fig. 5: Load Characteristics of open Basic FB converter.

State I : The Q_4 turns-on under ZVS condition. In secondary-side, the parasitic capacitance of Q_5 is discharged, and the parasitic capacitance C_d of diode D is charged.
State II : The parasitic capacitance C_{Q5} of Q_5 is discharged completely, and the current which flows through the transformer is reversed. Here, the C_{Q5} is charged, and the C_d is discharged.
State III : The C_{Q5} is charged, and the C_d is discharged, and is turned-on. This state will end when the Q1 turns-off.
State IV : The Q_1 turns-off. The energy of leakage inductance charges the parasitic capacitance C_{Q1} of Q1, and discharges the parasitic capacitance C_{Q2} of Q_2. When the C_{Q2} is discharged completely, the body diode is turned-on. Furthermore, the energy of L_{lk} and the electrical charge of output capacitance supply the output power.
State V : The Q_2 turns-on under ZVS condition.

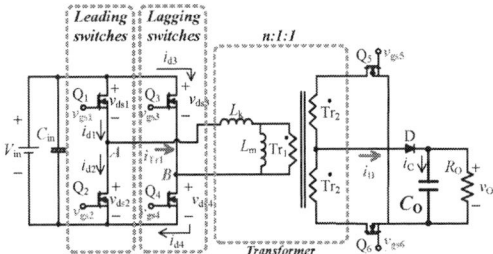

Fig. 6: Proposed FB converter.

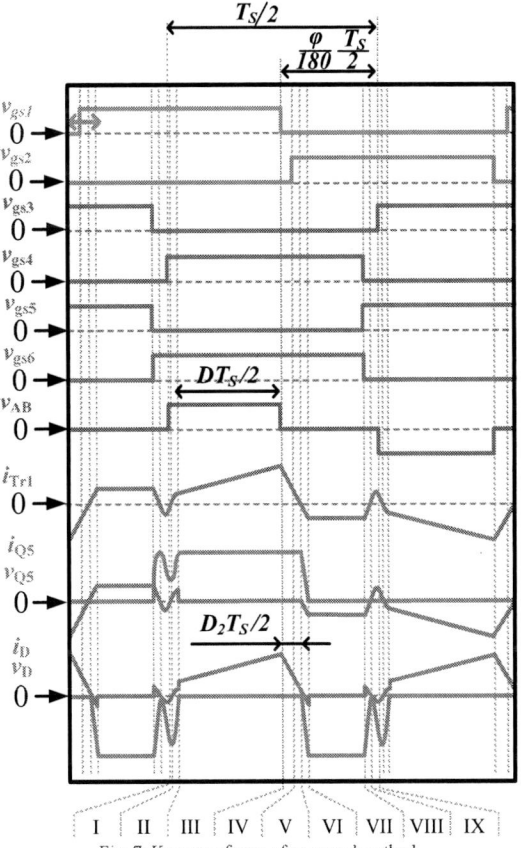

Fig. 7: Key waveforms of proposed method.

State VI : The current which flows through the D becomes zero, and then the C_d is charged. Moreover, the C_{Q5} is discharged, and the current i_{Tr1} is reversed. Here, the discharged energy charges the energy of L_{lk}. This state will end when the C_{Q5} is discharged completely.

State VII : The D is turned-off, and the body diode of Q_5 is turned-on. In this state, the energy of L_{lk} is maintained without energy loss in case of ideal which the internal power loss is zero.

State VIII : The Q_4 and Q_6 turn-off, and the Q_5 turns-on under ZVS condition. Here, the parasitic capacitance C_{Q6} of Q_6 is charged by the energy of L_{lk}, and the C_d is discharged.

State IX : The C_{Q6} is discharged, and the current i_{Tr1} is reversed, and then the body diode of Q_4 is turned-off. The parasitic capacitance C_{Q4} of Q_4 is charged, and the parasitic capacitance C_{Q3} of Q_3 is discharged. The body diode of Q_3 is turned-on under ZVS condition when the electrical charge of the C_{Q6} is larger than the electrical charge of the C_{Q3} and C_{Q4}.

A. The ZVS operating condition for lagging switches

In this section, the ZVS operating condition of the proposed FB converter is analyzed. In the proposed converter, the switch Q_3 and Q_4 are difficult to turn-on under ZVS condition. In order to turn-on under ZVS condition for these switches, the proposed method utilizes the resonance energy between the leakage inductance and the secondary-side parasitic capacitance. The resonance between parasitic components begins from state VI, and the high-frequency equivalent circuit is shows in Fig. 9. Where, the C_{st} and C_{st2} are the stray capacitance of transformer, the R_{para} is primary-side parasitic resistance.

In order to calculate the electrical charge of secondary side parasitic capacitance, the resonance current is analyzed by these equivalent circuits.

(a) From the equivalent circuit shown in Fig. 9(a), the circuit expression is

$$\begin{cases} i_{Tr} = \dfrac{C_d}{n^2}\dfrac{dnv_D}{dt} + \dfrac{4}{n^2}C_{Q5'}\dfrac{d\,nv_{Q5}/2}{dt} \\[2mm] nV_O = -L_{lk}\dfrac{di_{Tr}}{dt} - nv_D \\[2mm] 0 = L_{lk}\dfrac{di_{Tr}}{dt} + \dfrac{nv_{Q5}}{2} \end{cases} \tag{1}$$

where, $C_{Q5'} = C_{Q5} + C_{st2} + C_a$, and C_a is a ZVS auxiliary capacitor. Here, the initial value is

$$v_D(-0) = 0,\ v_{Q5}(-0) = 2V_O,\ i_{Tr}(-0) = 0 \tag{2}$$

The Eq. (1) is Laplace transformed, and the Eq. (2) is applied. Moreover, the inverse Laplace transformation is carried out, and the following equations are derived.

$$i_{Tr}(t) = -V_O\sqrt{\frac{C_d + 4C_{Q5'}}{L_{lk}}}\sin\frac{n}{\sqrt{L_{lk}\left(C_d + 4C_{Q5'}\right)}}t \tag{3}$$

$$v_D(t) = V_O\left(\cos\frac{n}{\sqrt{L_{lk}\left(C_d + 4C_{Q5'}\right)}}t - 1\right) \tag{4}$$

(b) From the equivalent circuit shown in Fig. 9(b), the circuit expression is

$$0 = L_{lk}\frac{di_{Tr}}{dt} - V_F + R_{para}i_{Tr} \tag{5}$$

The initial value is the final value of the Eq. (3). Hereafter, the following equation is derived by a similar step to (a).

$$i_{Tr}(t) = -V_O\sqrt{\frac{C_d + 4C_{Q5'}}{L_{lk}}}e^{-\frac{R_{para}}{L_{lk}}t} + \frac{V_F}{R}\left(1 - e^{-\frac{R_{para}}{L_{lk}}t}\right) \tag{6}$$

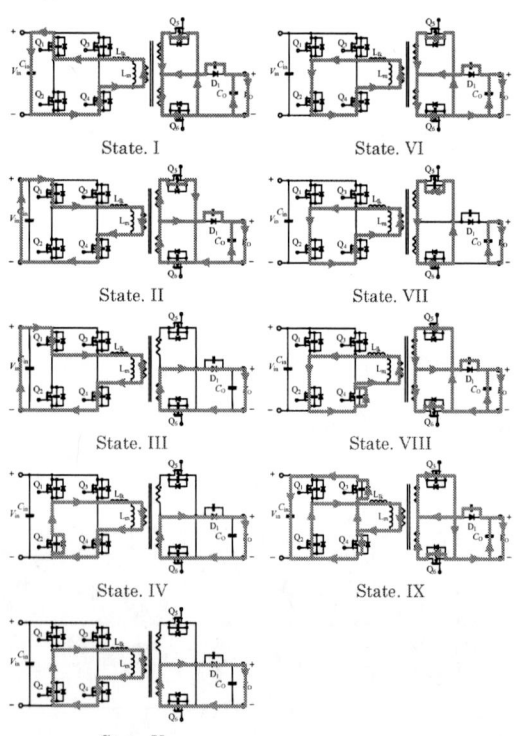

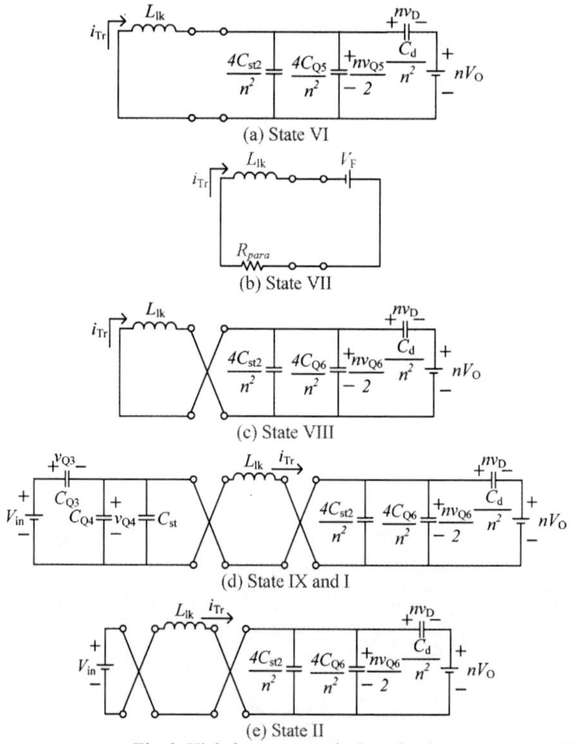

State. I

State. VI

State. II

State. VII

State. III

State. VIII

State. IV

State. IX

State. V

Fig. 8: Operating states.

(a) State VI

(b) State VII

(c) State VIII

(d) State IX and I

(e) State II

Fig. 9: High-frequency equivalent circuit

where, the time from start to end of the state VII is

$$t_{VII-end} = \frac{(1-D-D_2)T_S}{2} - \frac{\pi\sqrt{L_{lk}\left(C_d + 4C_{Q5'}\right)}}{2n} \qquad (7)$$

where, n is turn ratio, and D is defined as follows equation.

$$D = 1 - \frac{\varphi}{180} \qquad (8)$$

From Eqs. (6) and (7), when the state VII is end, the current i_{Tr} is

$$I_{tr\_VII} = -V_O \sqrt{\frac{C_d + 4C_{Q5'}}{L_{lk}}} e^{-\frac{R_{para}}{L_{lk}} t_{VII-end}} + \frac{V_F}{R}\left(1 - e^{-\frac{R_{para}}{L_{lk}} t_{VII-end}}\right) \qquad (9)$$

(c) From the equivalent circuit shown in Fig. 9(c), the circuit expression is

$$\begin{cases} i_{Tr} = -\dfrac{C_d}{n^2}\dfrac{dnv_D}{dt} - \dfrac{4}{n^2}C_{Q6'}\dfrac{d\,nv_{Q6}/2}{dt} \\[2mm] \dfrac{v_{Q6}}{2} = v_D + V_O \\[2mm] \dfrac{nv_{Q6}}{2} = L_{lk}\dfrac{di_{Tr}}{dt} \end{cases} \qquad (10)$$

where, $C_{Q6'} = C_{Q6} + C_{st2} + C_a$.
Here, the initial value is

$$v_{Q6}(-0) = 0 \,, v_D(-) = -V_O \,, i_{Tr}(-0) = I_{Tr-VII} \qquad (11)$$

From Eqs. (10) and (11), the following equations are derived.

$$i_{Tr}(t) = I_{Tr-VI}\cos\frac{n}{\sqrt{L_{lk}\left(C_d + 4C_{Q6'}\right)}}t \qquad (12)$$

$$v_D(t) = -I_{Tr-VI}\sqrt{\frac{L_{lk}}{C_d + 4C_{Q6'}}}\sin\frac{n}{\sqrt{L_{lk}\left(C_d + 4C_{Q6'}\right)}}t - V_O \qquad (13)$$

$$v_{Q6}(t) = -2I_{Tr-VI}\sqrt{\frac{L_{lk}}{C_d + 4C_{Q6'}}}\sin\frac{n}{\sqrt{L_{lk}\left(C_d + 4C_{Q6'}\right)}}t \qquad (14)$$

(d) From the equivalent circuit shown in Fig. 9(d), the circuit expression is

$$\begin{cases} V_{in} = v_{Q3} + v_{Q4} \\[2mm] -v_{Q4} = L_{lk}\dfrac{di_{Tr}}{dt} - nV_o - nv_D \\[2mm] i_{Tr} = -\dfrac{4}{n^2}C_{Q6'}\dfrac{d\,nv_{Q6}/2}{dt} - \dfrac{C_d}{n^2}\dfrac{dnv_D}{dt} \\[2mm] i_{Tr} = -C_{Q3}\dfrac{dv_{Q3}}{dt} + \left(C_{Q4} + C_{st}\right)\dfrac{dv_{Q4}}{dt} \\[2mm] \dfrac{v_{Q6}}{2} = V_o + v_D \end{cases} \qquad (15)$$

Here, the initial value is the final value of the Eqs. (12), (13), (14), and the following value.

$$i_{Tr}(-0) = 0, v_{Q3}(-0) = V_{in}, v_{Q4}(-0) = 0 \qquad (16)$$

From Eq. (15) and the initial value, the following equations are derived.

$$i_{Tr}(t) = -nI_{Tr-VII}\sqrt{\frac{2C_{Q3} + C_{st}}{n^2\left(2C_{Q3} + C_{st}\right) + 4C_{Q6'} + C_d}}\sin\omega t \qquad (17)$$

$$v_D(t) = -V_O$$
$$\quad - I_{Tr-VII}\sqrt{\frac{L_{lk}}{4C_{Q6'} + C_d}}\left[\cos\omega t + \frac{1}{\left(2C_{Q3} + C_{st}\right)L_{lk}\omega^2}\left(1 - \cos\omega t\right)\right] \qquad (18)$$

$$v_{Q6}(t) = -2I_{Tr-VII}\sqrt{\frac{L_{lk}}{4C_{Q6'} + C_d}}\left[\cos\omega t + \frac{1}{\left(2C_{Q3} + C_{st}\right)L_{lk}\omega^2}\left(1 - \cos\omega t\right)\right] \qquad (19)$$

where, $\omega = \sqrt{\dfrac{n^2\left(2C_{Q3} + C_{st}\right) + 4C_{Q6'} + C_d}{L_{lk}\left(2C_{Q3} + C_{st}\right)\left(4C_{Q6'} + C_d\right)}}$.

Here, in the Eqs. (17), (18), (19), the following relationship equation is considered, and applied.

$$C_{Q3} = C_{Q4}, C_{Q5'} = C_{Q6'} \qquad (20)$$

The Eq. (17) is the current for charging/discharging of C_{Q3} and C_{Q4}. The electrical charge which the current of Eq. (17) can charge or discharge for the C_{Q3} and C_{Q4} is derived as following equation.

$$Q_{sec} = \int i_{Tr}(t)dt = \frac{2nI_{Tr-VII}\left(2C_{Q3} + C_{st}\right)\sqrt{L_{lk}\left(4C_{Q6'} + C_d\right)}}{n^2\left(2C_{Q3} + C_{st}\right) + 4C_{Q6'} + C_d} \qquad (21)$$

The electrical charge of the primary-side parasitic capacitance is

$$Q_{pri} = V_{in}\left(2C_{Q3} + C_{st}\right) \qquad (22)$$

From Eqs. (21) and (22), the ZVS operating condition for lagging switches which is Q_3 and Q_4 is derived as following equation.

$$\left|Q_{sec}\right| > \left|Q_{pri}\right| \qquad (23)$$

The amount of Q_{sec} can be changed by the ZVS auxiliary capacitance C_a as shown in Fig. 10.

B. The output voltage at steady state condition

The power supply to the output load is examined. In order to simplify the state analysis, the primary side is assumed to be an ideal voltage source, and the state transition related to the primary side is neglected. Then converting the primary-side parameter to the secondary side, the 9 states in Fig. 8 are modified into the 3 states. The analytical model corresponding to the 3 states are shown in Fig. 11. Here, taking the output capacitor voltage as a state variable v_O and applying the extended state-space averaging method, the current i_{CO} flowing into the output capacitor during a half switching period $T_S/2$ is derived as follows.

The current I_{D0} is derived when the state III is started. From the equivalent circuit shown in Fig. 9 (e), the circuit expression is

$$\begin{cases} -V_{in} = L_{lk}\dfrac{di_{Tr}}{dt} - \dfrac{nv_{Q6}}{2} \\[2mm] \dfrac{v_{Q6}}{2} = V_o + v_D \\[2mm] i_{Tr} = -\dfrac{4}{n^2}C_{Q6'}\dfrac{d\,nv_{Q6}/2}{dt} - \dfrac{C_d}{n^2}\dfrac{dnv_D}{dt} \end{cases} \qquad (24)$$

Here, the initial value is derived by the Eqs. (17), (18) and (19).
From Eq. (24) and the initial value, the current i_{Tr} is derived as following equation.

$$i_{Tr}(t) = -\left(\begin{array}{c} \dfrac{V_{in}}{n}\sqrt{\dfrac{4C_{Q6'} + C_d}{L_{lk}}} \\[3mm] + I_{Tr-VII}\dfrac{4C_{Q6'} + C_d - n^2\left(2C_{Q3} + C_{st}\right)}{4C_{Q6'} + C_d + n^2\left(2C_{Q3} + C_{st}\right)} \end{array}\right)\sin\frac{n}{\sqrt{L_{lk}\left(4C_{Q6'} + C_d\right)}}t \qquad (25)$$

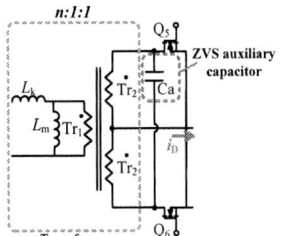

Fig. 10: The added part of the ZVS auxiliary capacitor

From the Eq. (25), the initial value I_{D0} of the state III is derived as following equation.

$$I_{D0} = -\frac{V_{in}}{n}\sqrt{\frac{4C_{Q6'}+C_d}{L_{lk}}} - I_{Tr-VII}\frac{4C_{Q6'}+C_d-n^2\left(2C_{Q3}+C_{st}\right)}{4C_{Q6'}+C_d+n^2\left(2C_{Q3}+C_{st}\right)} \quad (26)$$

Hereafter, the Eq. (26) is used to derive the current i_{CO}.

(a) During State III ($0 < t < DT_S/2$):

From the equivalent circuit shown in Fig. 11(a), the current is

$$i_{CO}(t) = \frac{n^2 t}{L_{lk}}\left(\frac{V_{in}}{n}-\hat{v}_O\right) + I_{D0} - I_O \quad (27)$$

(b) During State IV and V ($DT_S/2 < t < (D+D_2)T_S/2$):

From the equivalent circuit shown in Fig. 11(b), the current is

$$i_{CO}(t) = -\frac{n^2 \hat{v}_O}{L_{lk}}t + \frac{nDV_{in}}{2L_{lk}}T_S + I_{D0} - I_O \quad (28)$$

where, $D_2 T_S /2$ denotes the period which includes the state IV and state V, and the current $i_{CO}(t)+I_O$ of secondary diode D becomes zero at $t=(D+D_2)T_S/2$. From (28), the duty ratio D_2 is derived as follows:

$$D_2 = D\left(\frac{V_{in}}{n\hat{v}_O}-1\right) + \frac{2L_{lk}I_{D0}}{n^2\hat{v}_O T_S} \quad (29)$$

(c) During State VI, VII, VIII and IX ($(D+D_2)T_S/2 < t < T_S/2$):

From the equivalent circuit shown in Fig. 11(c), the current is

$$i_{CO}(t) = -I_O \quad (30)$$

From Eqs. (27), (28) and (30), the average current of output capacitor is expressed as follows.

$$\bar{i}_{CO}(t) = \frac{n^2 T_S}{4L_{lk}}\left\{\frac{V_{in}}{n}\left(2DD_2 + D^2\right) - (D+D_2)^2\hat{v}_O\right\} + (D+D_2)I_{D0} - I_O \quad (31)$$

As a result, the state-space averaged equation of the output voltage is derived as follows.

$$\frac{d\hat{v}_O}{dt} = \frac{\bar{i}_{CO}(t)}{C_O} \quad (32)$$

From Eqs. (26), (31) and (32), the output voltage at the steady-state condition is derived as follows.

$$V_O = \frac{\frac{DV_{in}^2}{nI_O}\sqrt{\frac{4C_{Q6}+C_d}{L_{lk}}}\left(\frac{nDT_S}{4\sqrt{L_{lk}\left(4C_{Q6}+C_d\right)}}+1\right)}{1+\frac{DV_{in}}{4L_{lk}I_O}\left(nDT_S+4L_{lk}\frac{4C_{Q6}+C_d-n^2\left(2C_{Q3}+C_{st}\right)}{4C_{Q6}+C_d+n^2\left(2C_{Q3}+C_{st}\right)}\sqrt{\frac{4C_{Q6}+C_d}{L_{lk}}}\right)} \quad (33)$$

From Eq. (33), it can be found that the voltage conversion ratio depends on the leakage inductance and the load current.

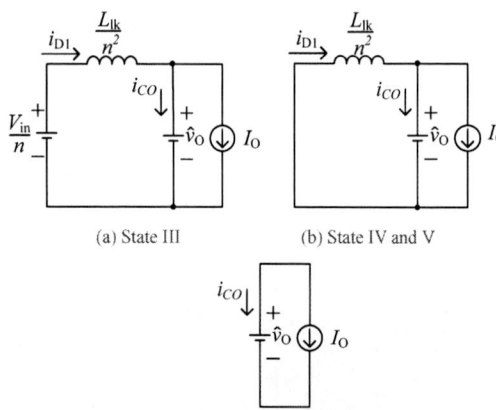

(a) State III   (b) State IV and V

(c) State VI, VII, VIII and IX

Fig. 11: Analytical model

IV. EXPERIMENTAL RESULTS

In order to examine the effectiveness of the proposed techniques, the experimental results are described as follows:

The prototype breadboard has been implemented, and the circuit parameters and specifications are shown in Table 1. In primary side, the SiC-MOSFET is applied to the lagging switches, because the body diode of SiC-MOSFET has much shorter reverse-recovery time than that of the conventional MOSFET. If the body diode of Q4 has a long reverse-recovery time, the short circuit through Q3 and Q4 occurs in the state VIII, and the state IX do not appear. Then the SiC-MOSFET is effective to avoid the short-circuit troubles, and a high-reliability converter is available. Moreover, the Si-MOSFET is applied to the leading switches. In secondary side, the both switches are applied the SiC-MOSFETs in order to reduce the on-resistance, because the Si-MOSFET which has the high-breakdown voltage potential becomes high on-resistance. Fig. 12 shows the experimental key waveforms. Comparing Fig. 7 and 12, it can be seen that the experimental key waveforms agree with the theoretical key waveforms. Furthermore, it is confirmed that the high-frequency noise, which is occurred in basic FB converter as shown in Fig. 4, is suppressed.

A. The ZVS operating condition of the prototype breadboad

The ZVS operating condition of the prototype breadboard is confirmed by Eqs. (21), (22) and (23). As a result, Figure 12 is derived. Here, the secondary-side total parasitic capacitance C_{sec_total} is

$$C_{sec\_total} = C_{Q5} + C_{st2} + Ca = C_{Q6} + C_{st2} + Ca \quad (34)$$

where, the output capacitance C_{OSS} of SCT2120AF is 149pF, and the C_{OSS} of CMF10120D is 116pF. Moreover, the C_{OSS} of C2D10120 is 88pF.

From Fig. 12, when the C_{sec_total} is larger than around 190pF, it is confirmed that the ZVS operation of lagging switches can be achieved. The C_{sec_total} of breadboard is 227pF, and the ZVS operation condition is satisfied without the auxiliary capacitor C_a. Therefore, when the input and output voltage are 400V and 380V, and the the load current is 0.25A, it is confirmed that the lagging switch has turned-on under the ZVS condition as shown in Fig. 14.

TABLE I. CIRCUIT SPECIFICATIONS

	Description	Value
	Input Voltage	380V±5%
	Output Voltage	380V
	Load Current	0.26A to 3.29A
Si- MOSFET	TK20E60W	600V/20A
SiC-MOSFET	SCT2120AF	650V/29A
	C2M10120D	1200V/24A
SiC-Diode	C2D10120	1200V/10A
Transformer	Magnetizing Inductance	500μH
	Leakage Inductance	6.9μH
	Turn ratio	25:30:30
	primary-side stray capacitance	14pF
	secondary-side stray capacitance	111pF
	Smoothing capacitor	47μF
	Switching frequency	200kHz

978-1-4673-9551-9/16 $31.00 © 2016 IEEE

B. Comparison of switching waveformes between proposed and basic FB converter

The switching waveforms of both the proposed and the basic FB converters are compared for three cases of load condition in Fig. 15. In the basic FB converter, it is confirmed that Q_1 turns-on under a hard-switching condition in all cases. On the other hand, in the proposed FB converter, Q_1 and Q_4 turn-on under the ZVS condition.

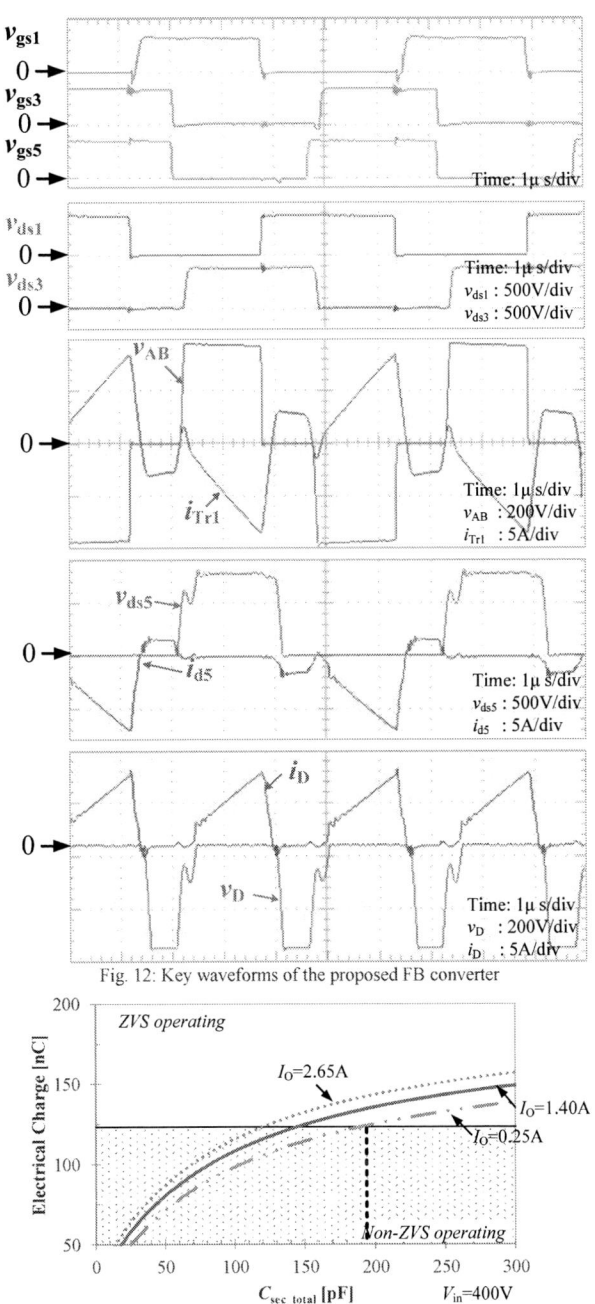

Fig. 12: Key waveforms of the proposed FB converter

Fig. 13: ZVS operating condition

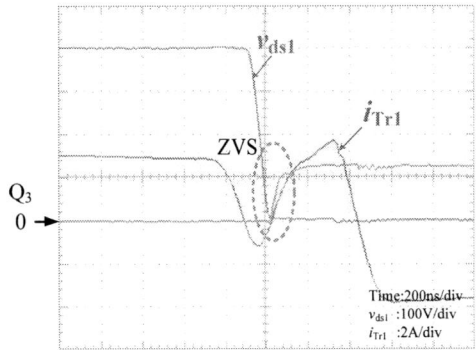

Fig. 14: the experimental waveforms of lagging switch.

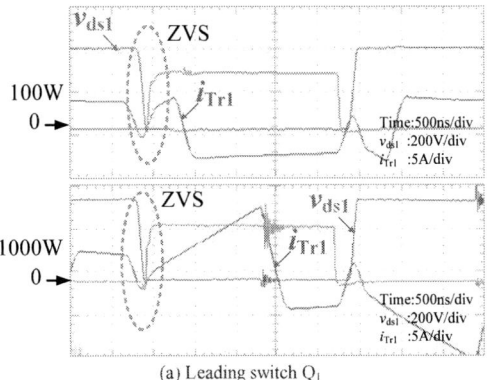

(a) Leading switch Q_1

(b) Lagging switch Q_3

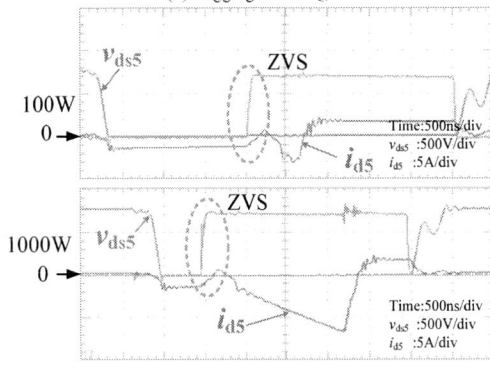

(b) Secondary-side switch Q_5

Fig. 15: Comparison of switching waveforms between proposed and basic FB converters

C. The load characteristics

By utilizing the proposed method, it is confirmed that the high-frequency noise as shown in Fig. 4 is suppressed, and then the fluctuation phenomena of the control characteristics shown in Fig.5 are removed as shown in Fig.16. Furthermore, the analytical and experimental results agree well, and the validity of Eq. (33) is confirmed. Then the dependence on the load current can be estimated.

D. The power efficiency

Figure 17 shows the power efficiency. The power efficiency in case of 100W is improved by 14%, and the efficiency in case of 1.0kW is improved by 2%. In the other word, the power loss is reduced by 20W in a wide load range by means of the proposed techniques, and then the temperature of the circuit components is decreased. As a result, a high power-density of 15W/cm3 has been achieved by the experimental breadboard shown in Fig. 18.

E. The temperture distribution patterns of the high power density FB converter with proposed method

The temperature distribution patterns around circuit components in case of the full load (V_{IN}: 380V, V_O: 380V, P_O: 1.25kW) are shown in Fig. 19. The temperature equilibrium is achieved at 25 minutes. Here, the litz wire of transformer is 104.2 °C, and this temperature is the hottest of all the other components. Subsequently, the transformer core is 64.0 °C, and the semiconductor component is 65.5 °C.

V. CONCLUSION

This paper has proposed a new operation and circuit topology for the output-inductor-less FB converter. Further, the SiC-MOSFET is installed in this converter in order to suppress the short-circuit phenomenon which is caused by a long reverse-recovery time of the body diode of MOSFET. The effectiveness of the proposed method has been examined experimentally, and the ZVS operation and the low-noise feature has been confirmed at a wide load range. As the result, the power efficiency was increased by 2 to 14 percent when compare with the basic FB converter, and then the higher power-density of 15W/cm³ was achieved.

ACKNOWLEDGMENT

The authors would like to thank Mr. S. Iwasaki, technical staff member of Nagasaki University, for his technical support.

REFERENCES

[1] N. Raxmussen: "Guidelins for Specification of Data Centers, "American Power Conversion White paper, No.120, pp.1-21, 2005

[2] A. Pratt, P. Kumar, T. V. Aldridge: "Evaluation of 400V DC Distribution in Telco and Data Centers to Improve Energy Efficiency," Proceeding of International Telecommunication Energy Conference 2007, pp.32-39, Oct. 2007.

[3] A. Matsumoto, A. Fukui, T. Takeda, M. Yamasaki: "Development of 400Vdc Power Distribution System and 400Vdc Output Rectifier," Proceeding of International Telecommunication Energy Conference 2009, Sess.PA2-1, Oct. 2009.

[4] M. Noritake, K. Hirose, M. Yamasaki, T. Osawa, H. Mikami: "Evaluation Results of Power Supply to ICT Equipment Using HVDC Distribution System," Proceeding of International Telecommunication Energy Conference 2010, pp.1-8, Jun. 2010.

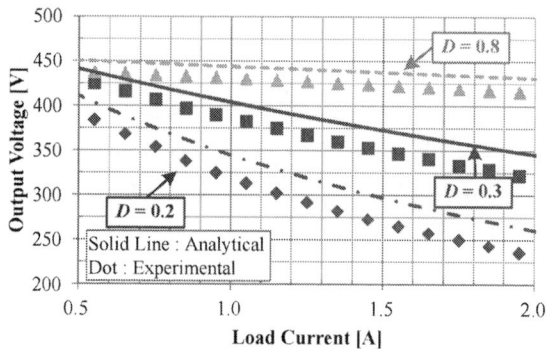

Fig. 16: Load characteristics.

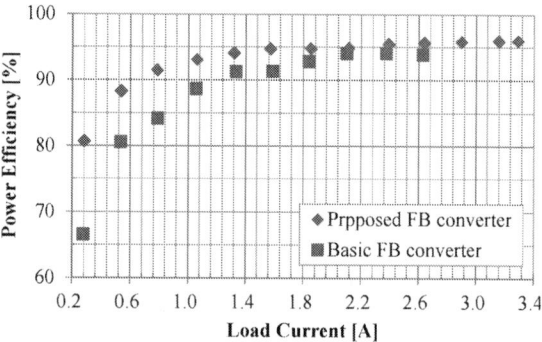

Fig. 17: Efficiency characteristics.

Input: 380V
Output: 380V
Power: 1250W
Volume: 83.9cm³
(3.6cm×4.4cm×5.3cm)
Power Density: 14.9W/cm³
without heat-sink

Fig. 18: Experimental breadboard with higher power-density of 15W/cm³

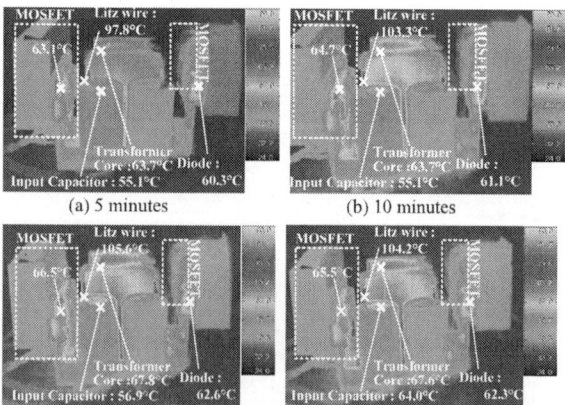

(a) 5 minutes   (b) 10 minutes

(c) 15 minutes   (d) 25 minutes

Fig. 19: Temperature distribution patterns (P_O=1.25kW).

[5] T. Yamashita, S. Muroyama, M. Yamasaki: "Technical Trends and Future Challenges of Power Supply for Information and Communication Systems," NTT Facilities Research Institute, No.24, pp.13-19, Jum. 2013.

[6] U.Badstuebner, J.Biela, J.W.Kolar:" An Optimized, 99% Efficient, 5kW, Phase-Shift PWM DC-DC Converter for Data Center and Telecom Applications," Proceeding of International Power Electronics Conference 2010, pp.626-634, June. 2010.

[7] K.Yoshida: "ZVS-PWM full-bridge converter using active current clamping with synchronous rectifiers," Proceeding of Power Electronics Specialists Conference 1999, Vol.1, pp.257-262, Aug 1999.

[8] A. Jangwanitlert, J. C. Balda.:"Phase-shifted PWM full-bridge DC-DC converters for automotive applications: reduction of ringing voltages," IEEE Transaction of Power Electronics in Transportation, pp.111-115, Oct.2004.

[9] M. Hirokawa, T. Ninomiya : "Non-Dissipative Snubber for Rectifying Diodes applied to a front-end power supply," Proceeding of Power Conversion Conference 2002, Vol.3, pp.1176-1181, Apr. 2002.

[10] K. Orikawa J. Itoh: "Analysis and Optimization Design of Snubber Circuit for Isolated DC-DC Converters in DC Power Grid," Proceeding of International Conference Renewable Energy Research and Applications 2012, Vol.133, pp.1-6, Nov. 2012.

[11] I. D. Jitaru: "A 3kW Soft Switching DC-DC Converter," Proceeding of the Applied Power Electronics Conference 2000, Vol.1, pp86-92, Feb.2005.

[12] C. Zhao, X. Wu, P. Meng, Z. Qian: "Optimum Design Consideration and Implementation of a Novel Stnchronous Rectified Soft-Switched Phase-Shift Full-Bridge Converter for Low-Output-Voltage High-Output-Current Applications," Transaction of IEEE Power Electronics, Vol.24, No.2, pp388-397, Feb.2009.

[13] R. Simanjorang, H. Yamaguchi, H. Ohashi, K. Nakao, T. Ninomiya, S. Abe, M. Kaga, A. Fukui:" High-Efficiency High-Power DC-DC Converter for Energy and Space Saving of Power-Supply System in a Data Center," Proceeding of the Applied Power Electronics Conference 2011 Record, pp.600-605, Mar. 2011.

[14] M. Pahlevaninezhad, P. Das, J. Drobnik, P. K. Jain, and A. Bakhshai: "A novel ZVZCS full-bridge DC/DC converter used for electric vehicles," IEEE Transaction of Power Electronics., vol. 27, no. 6, pp. 2752–2769, Jun. 2012.

[15] D. Gautam, F. Musavi, M. Deington, W. Eberle, W. G. Dunford: "A Zero Voltage Switching Full-Bridge DC-DC Converter with Capacitive Output Filter for Plug-in Hybrid Electronic Vehicle Battery Charging," IEEE Transaction of Power Electronics, vol. 28, No. 12, pp5728-5735, Dec.2013.

[16] P. Das, M. Pahlevaninezhad: "A Novel Full Load Range ZVS DC-DC Full-Bridge Converter with Natural Hold-Up Time Operation," Proceeding of the Applied Power Electronics Conference 2015, pp2056-2062, Mar.2015.

[17] K.Domoto, Y. Ishizuka, S. Abe, T. Ninomiya: "Control Characteristics Improvement of Full-Bridge DC-DC Converter with Snubber Capacitor," Proceeding of International Power Electronics Conference 2014, pp.3652-3658, May. 2014.

[18] K. Domoto, Y. Ishizuka, S. Abe, T. Ninomiya: "Thermal Improvement by Switching-Loss Reduction Technique in PWM-Controlled Output-Inductor-Less Full-Bridge Converter," Proceeding of International Telecommunications Energy Conference 2015, pp. 1080-1085, Oct. 2015.

Reduction Technique of Leakage Flux Effects on GaN-HEMTs in 5 MHz / 100 W Isolated DC-DC Converters

Akinori Hariya / Nagasaki University
Graduate School of Engineering
Nagasaki Univ.
Nagasaki, Japan
bb52312202@cc.nagasaki-u.ac.jp

Tomoya Koga / Nagasaki University
Graduate School of Engineering
Nagasaki Univ.
Nagasaki, Japan
bb52114215@cc.nagasaki-u.ac.jp

Ken Matsuura / TDK Corporation
Energy Device Development Section
TDK
Chiba, Japan
matsuken@jp.tdk.com

Hiroshige Yanagi / TDK-Lambda Corporation
Advanced Development Dept.
TDK-Lambda
Nigata, Japan
yanagi@jp.tdk-lambda.com

Satoshi Tomioka / TDK-Lambda Corporation
Advanced Development Dept.
TDK-Lambda
Nigata, Japan
s.tomioka@jp.tdk-lambda.com

Yoichi Ishizuka / Nagasaki University
Graduate School of Engineering
Nagasaki Univ.
Nagasaki, Japan
isy2@nagasaki-u.ac.jp

Tamotsu Ninomiya / City of Kitakyushu
Industry Promotion Dept.
City of Kitakyushu
Fukuoka, Japan
ninomiya@bell.ocn.ne.jp

Abstract— To achieve high power-density isolated dc-dc converter, gallium nitride high electron mobility transistors (GaN-HEMTs) and planar transformer have been used. Also, for the high power-density design, these components are placed close to each other. GaN-HEMTs are significantly affected by the leakage flux of the planar transformer, because of their lateral structure. Therefore, the mutual effects of the components are needed to take into account.

This paper presents the leakage flux effects of the planar transformer on GaN-HEMTs in 5 MHz isolated dc-dc converter. Moreover, to suppress the effects, the method using the magnetic shield as the part of the printed circuit board (PCB) layout has been proposed. The effects and the proposed method have been analyzed by finite element method (FEM) with Maxwell 3D. Some experiments have been done with 5 MHz unregulated LLC resonant dc-dc converter with GaN-HEMTs; the input voltage is 48 V, the output voltage is 12 V, and the size is 17 mm × 25 mm × 6.5 mm. From the experimental results, by inserting the magnetic shield, the load current range is improved from 6 to 8 A. In addition, the maximum temperature is improved 17.1 °C, and the power efficiency is increased 1.58 % at 6 A.

Keywords—high power-density dc-dc converter; MHz-level switching frequency; GaN-HEMT; FEM; electromagnetic field simulation

I. INTRODUCTION

Recently, high power-density and high power-efficiency isolated dc-dc converters have been required in information and communication technology (ICT) facilities. Therefore, MHz-level switching frequency and low power loss technique have been studied [1-6]. In particular, to achieve these requirements, LLC resonant dc-dc converter topology, Gallium Nitride high electron mobility transistors (GaN-HEMTs) and planar transformers have been used.

At kHz-level switching frequency, metal oxide semiconductor field effect transistors (MOS-FETs) which are vertical structure has been mainly used as semiconductor switches. On the other hand, GaN-HEMTs which are lateral structure have been used, because of the high switching frequency such as MHz level, in recent years.

978-1-4673-9551-9/16 $31.00 © 2016 IEEE

With the development of downsizing of converters, the components of converters are placed close to each other. Therefore, the mutual effects of the components are needed to take into account. Particularly, due to the structure of GaN-HEMT, the appearance of the novel negative effects is predicted.

The mentioned above components have been studied by several researchers: for improving the power efficiency of dc-dc converter with GaN-HEMTs, the printed circuit board (PCB) layout is optimized using finite element method (FEM) [7]; for reducing the core loss, core size and ac resistance of winding, matrix transformer has been proposed [8-11]. However, the effects of planar transformer's leakage flux on devices such as switches have not been studied so far.

In this paper, the effect of the leakage flux of the planar transformer on GaN-HEMTs in 5 MHz isolated dc-dc converter has been revealed. Moreover, to suppress the effect, the method using the magnetic shield as the part of the PCB layout has been proposed. The effect study and the proposed method have been analyzed by FEM with Maxwell 3D. The validation of the proposed method has been confirmed by the experiments with prototype PCB which have 8 layers.

In section II, the problem and analysis of lateral structure switch at high power-density dc-dc converter is described. In section III, the effect and design of the proposed magnetic shield is revealed. In section IV, the experimental results are demonstrated.

II. THE PROBLEM AND ANALYSIS OF LATERAL STRUCTURE SWITCH AT HIGH POWER-DENSITY DC-DC CONVERTER

A. The Problem of Lateral Structure Switch at High Power-Density DC-DC Converter

GaN-HEMTs are lateral-structure semiconductor switches and surface mount devices for achieving the high speed switching operation. In addition, the winding of the planar transformer is directly manufactured with a multilayered pattern in the PCB. Therefore, the direction of the magnetic flux which passes through GaN-HEMTs, is perpendicular to the direction of the current flow inside of the device. Therefore, the lateral-structure semiconductor switches are significantly affected by the magnetic flux. To confirm the effect, lateral-structure switches have been compared with vertical-structure switches by using the electromagnetic field simulation Maxwell 3D.

B. Assumptions and Equivalent Modeling

It is important to ensure the accuracy and reduce the analytical time at FEM analysis. To simplify analysis while complying with important points, the following assumptions are made:

1) Semiconductor Switches:
- The channel of the switch is conductor, which has thin thickness.
- On state is assumed because of large power loss.
- The switch device has same on-resistance. The on-resistance is set to 4 mΩ.

2) Core:
- The relative permeability is equal to the initial relative permeability. The relative permeability is set to 80.
- The geometric parameters of the core at the analysis are same as prototype core.
- The length of the air gap is same as prototype core.

3) Windings:
- The distance between primary- and secondary-side winding is equal to the thickness of PCB interlayer. The distance is 0.25 mm.
- The material of these windings is copper. The conductivity is set to 58×10^6 S/m.
- The thickness of windings is 0.1 mm.
- The turn number of primary- and secondary-side winding is 1 turn.

4) Board:
- The number of layer can be assumed 2. The reason is that the magnetomotive force of the analysis is same as that of the prototype board which has 8 layers.
- Some vias can be neglected in this analysis. The magnetic flux caused by current flow in some vias pass the cross section of channel. The cross-section area of channel is narrower than the horizontal plane area of channel. Therefore, the effect of the magnetic flux caused by via is not important.

Based on the mentioned above assumptions, the vertical- and lateral-structure model can be made as Fig. 1 (a) and (b). The detailed each switch model is shown in Fig. 2. The direction of current flow through switch model and the magnetic flux are represented. Each switch model is composed of metal and channel part. In particular, GaN-HEMTs manufactured by EPC have been used in this research. Thus, the package type of GaN-HEMT is land grid array (LGA)

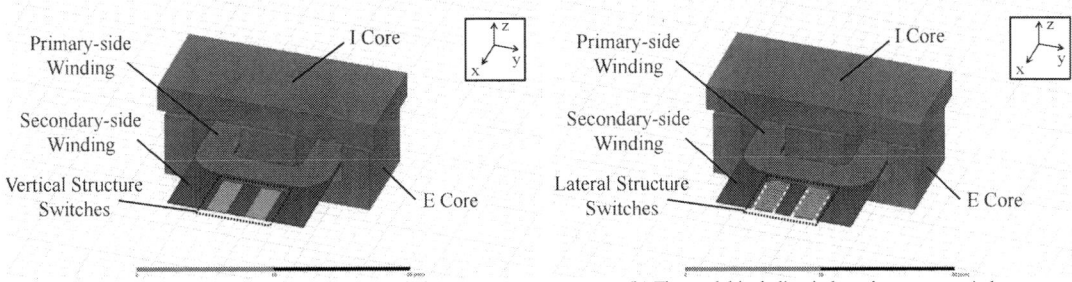

(a) The model including in vertical-structure switch.   (b) The model including in lateral-structure switch.

Fig. 1. The models for investigating the problem of lateral-structure switches.

package. Therefore, metal part of GaN-HEMTs has been created as this package type.

The difference of current direction between vertical and lateral structure is occurred at channel part. In case of vertical structure, the current direction is Z axis. On the other hand, in case of lateral structure, the current direction is X axis or Y axis. It is predicted that this difference makes the difference of the eddy current effect.

To obtain same on-resistance between vertical and lateral structure, the conductivity of channel for vertical structure is 448 S/m, and for lateral structure is 656×10^3 S/m, respectively. Also, to obtain same magnetomotive force between analysis and experiment, the amplitude of primary-side winding current is 22.1 A, and that of secondary-side winding current is 17.1 A. The phase of secondary-side winding current is delayed 27 ° as compared to that of primary-side winding current.

C. The Analytical Results

The comparison of the magnetic flux affection to metal and secondary-side winding current between vertical and lateral structure semiconductor switches is shown in Fig. 3. The current direction is represented by arrow and the current-distribution density is shown by graduations of color. From the figures, in case of both vertical and lateral structure, the direction of the channel current is X axis. Thus, the current is

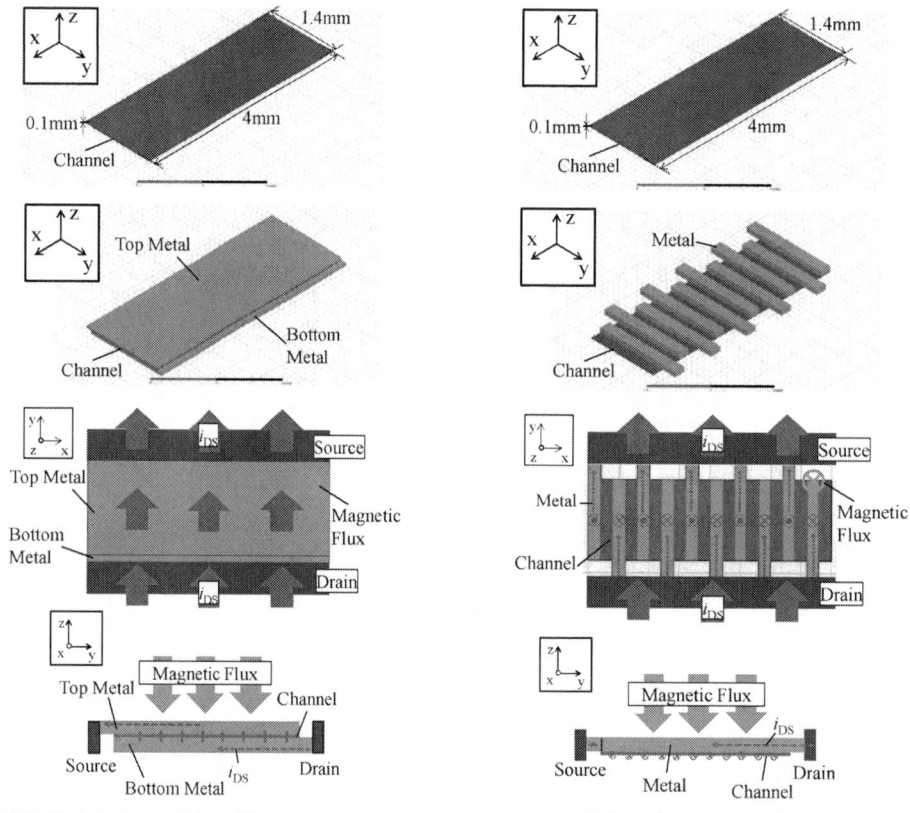

(a) Vertical-structure switch model.   (b) Lateral-structure switch model.

Fig. 2. The detailed each switch model.

(a) The model including in vertical-structure switch.   (b) The model including in lateral-structure switch.

Fig. 3. The magnetic flux affection to metal and secondary-side winding current at $f = 5$ MHz and 0° phase.

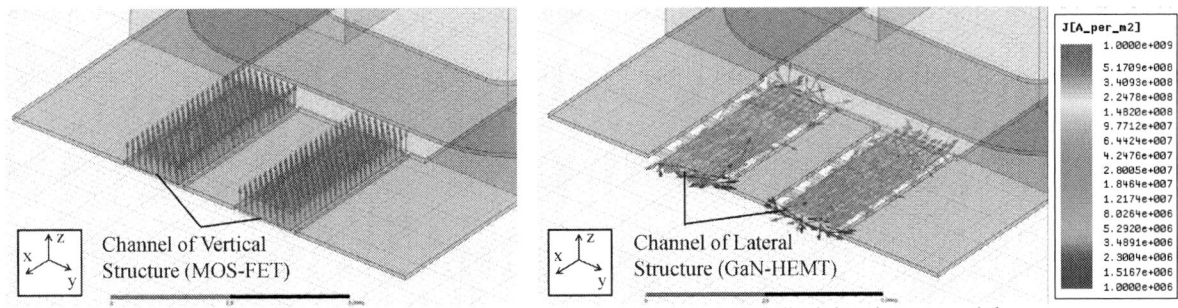

(a) The model including in vertical-structure switch.　　(b) The model including in lateral-structure switch.

Fig. 4. The magnetic flux affection to channel current at f = 5 MHz and 0° phase.

affected by the eddy current which flows through XY plane. As a result, the current flow through metal and secondary-side winding lean to edge.

The comparison of the magnetic flux affection to channel current between vertical and lateral structure semiconductor switches is shown in Fig. 4. From the figures, in case of vertical structure, the direction of the channel current is Z axis. Thus, the current is hardly affected by the eddy current. In contrast, in case of lateral structure, the direction of the channel current is X axis. Thus, the current is affected by the eddy current which flows through XY plane.

The power loss at metal and channel part of the each switch is shown in Fig. 5. From the figure, at first, the difference between vertical- and lateral-structure switch cannot be found at the power loss of metal part. In addition, the power loss of metal part in each structure is little increased by increasing operating frequency.

In contrast, the difference between vertical- and lateral-structure switch can be found at the power loss of channel part. The power loss of channel part in vertical structure is hardly increased with increasing operating frequency. On the other hand, the power loss of channel part in lateral structure is increased with increasing operating frequency. In particular, the increasing power loss of channel part in lateral structure stands out at MHz-level operating frequency region.

III. The Effect and Design of the Proposed Magnetic Shield

To reduce the power loss of channel, the magnetic shield under semiconductor switches has been proposed. By adopting the magnetic shield, the influence of the eddy current can be suppressed in semiconductor switches. The copper foil of one layer is utilized as the magnetic shield. In the proposed method, additional production process and cost are not required.

Therefore, if the inside of the switch device has copper foil pad for heat dissipation, the proposed magnetic shield is not needed. For example, in case of GaN-HEMT device with heat dissipation pad [12], the proposed method is not required. However, in case of lateral-structure silicon MOS-FET device with ball grid array (BGA) package [13-14], the proposed magnetic shield is needed for suppressing the power loss. In this research, switch device such as GaN-HEMTs of EPC

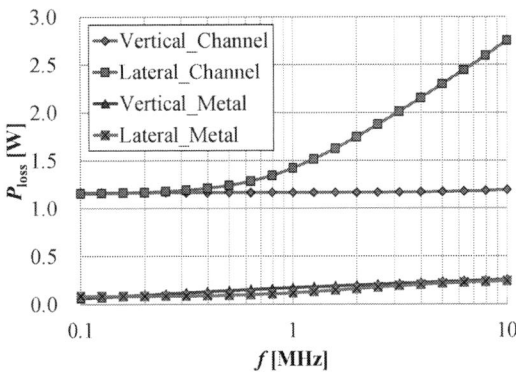

Fig. 5. The operating frequency versus the power loss at metal and channel part of each switch.

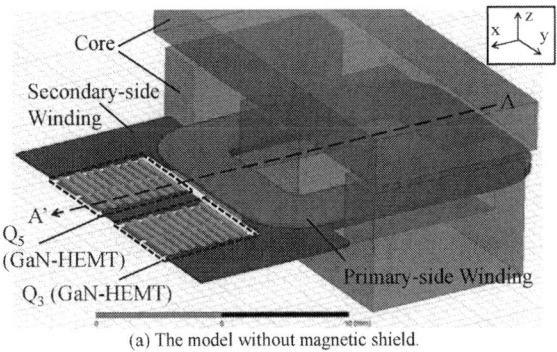

(a) The model without magnetic shield.

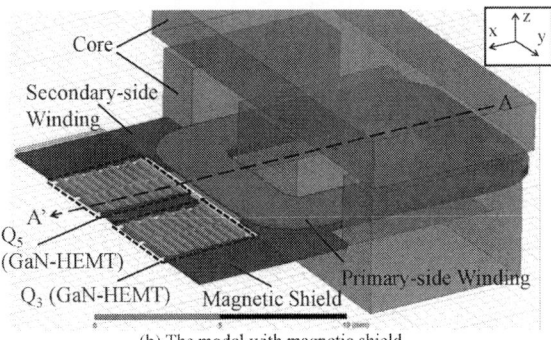

(b) The model with magnetic shield.

Fig. 6. The models for analyzing the effect of magnetic shield.

978-1-4673-9551-9/16 $31.00 © 2016 IEEE

which do not have heat dissipation pad [15-16] have been used. Thus, the proposed m ethod is effective.

In this section, the effect of the proposed magnetic shield on the power loss of channel has been revealed with Maxwell 3D. In addition, the design of the proposed method has been clarified.

A. The Model for the Analysis

The assumption, conductivity and current setting of the model are same as section 2. The model for analyzing the effect of magnetic shield is shown in Fig. 6. Figure 6 (a) is non-magnetic shield model. Figure 6 (b) is magnetic shield model. The assumptions of the magnetic shield are as follows:

- The material is copper.
- The copper foil pattern of one layer in PCB has been used as a magnetic shield.
- The thickness of magnetic shield is equal to the thickness of the copper foil pattern.

B. The Analytical Results

The current-density distribution of channel of GaN-HEMTs is shown in Fig. 7. These figures are view from the Z-axis positive direction. From these figures, by using the magnetic shield, it can be seen that the distribution is more even. As a result, the deviation of current density distribution of (b) is decreased than that of (a).

The magnetic flux density distribution is shown in Fig. 8. These figures are view from the Y-axis positive direction, and XZ-plane of arrow A-A' in Fig. 6. From these figures, by using the magnetic shield, it can be seen that the effect of magnetic flux on GaN-HEMTs can be reduced. As a result, the deviation of magnetic flux density distribution of (b) is decreased than that of (a).

C. The Design of The Magnetic Shield

To discuss the optimal size of magnetic shield, Fig. 9 shows the definition of criteria position, x_size and y_size of

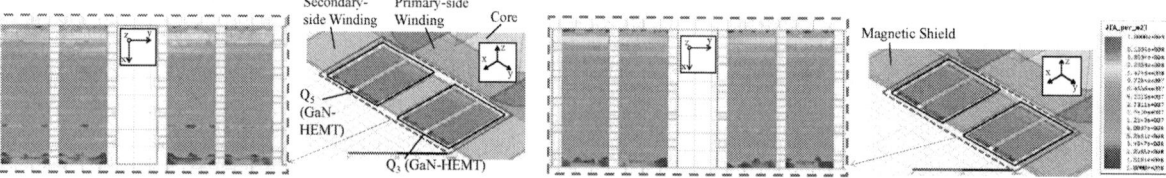

(a) The model without magnetic shield. (b) The model with magnetic shield.

Fig. 7. The current-density distribution of channel of GaN-HEMTs at f = 5 MHz and 0° phase.

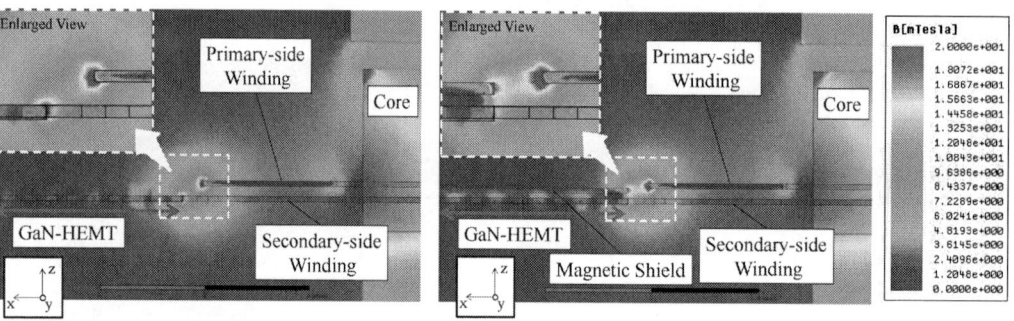

(a) The model without magnetic shield. (b) The model with magnetic shield.

Fig. 8. The magnetic flux density distribution at f = 5 MHz and 0° phase, view from the Y-axis positive direction, and XZ-plane of arrow A-A' in Fig. 6.

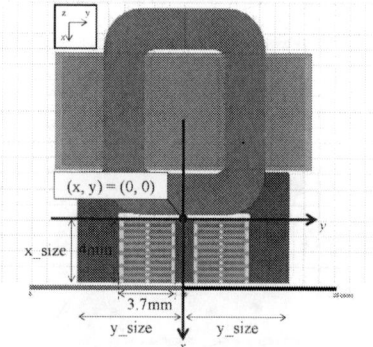

Fig. 9. The definition of criteria position, x_size and y_size of magnetic shield.

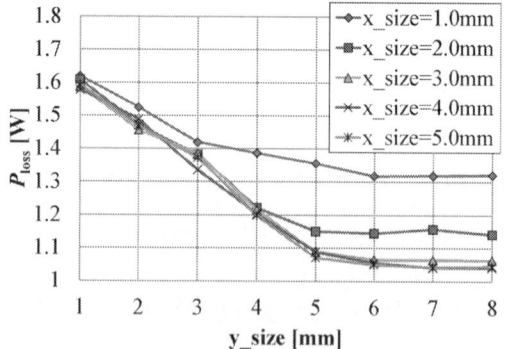

Fig. 10. The power loss at channel part of switches.

978-1-4673-9551-9/16 $31.00 © 2016 IEEE   2434

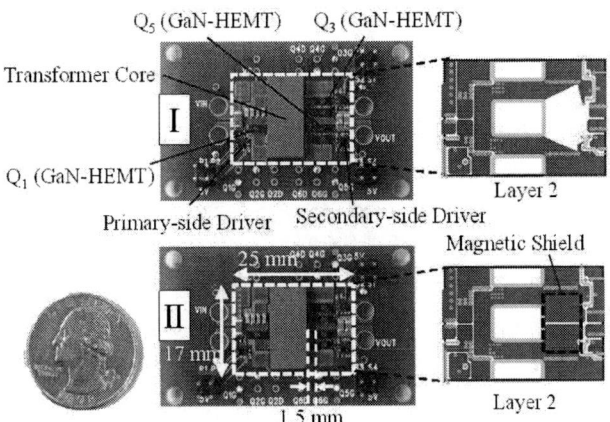

Fig. 11. The two type prototype circuits. Type I does not have magnetic shield. Type II have magnetic shield.

magnetic shield. The criteria position of magnetic shield is the origin. Also, from the figure, the area of lateral-structure switches is 4 mm × 3.7 mm.

The power loss at channel part of switches is shown in Fig. 10. The parameters are the x_size and y_size of magnetic shield. From the figure, at this model, the minimum power loss can be obtained when x_size is greater than 3 mm and 2 × y_size is greater than 12 mm. Therefore, the area of magnetic shield should be greater than 0.36 cm^2.

IV. EXPERIMENTAL RESULTS

In this section, to demonstrate the effects of the proposed method, some experiments have been done with the prototype 5 MHz and 32 W/cc unregulated LLC resonant dc-dc converter using GaN-HEMTs. The distance between GaN-HEMTs and the magnetic transformer is about 1.5 mm. The input voltage is 48 V, and the reference output voltage is 12 V. The two type prototype circuits are shown in Fig. 11, and the circuit topology is shown in Fig. 12. The prototype circuit type I does not have the magnetic shield. The prototype circuit type II have the magnetic shield which is set in 2 and 7 layers. The size of magnetic shield is about 0.36 cm^2. The components of the prototype circuit are shown in TABLE I.

The power efficiency is shown in Fig. 13. The result of type II exceeded to that of type I in the whole of load current value. At 6 A load current, the differences between the prototype I and II are 1.58 % of power efficiency and 1.29 W of power loss. In the condition of the allowable temperature of 130 °C, the prototype I can supply 6 A and the prototype II can supply 8 A load current.

The temperature distribution of the prototype dc-dc converter at I_o = 6 A is shown in Fig. 14. From the results, the large difference between (a) and (b) can be found at secondary-side components. The prototype II with magnetic shield can be suppressed the conduction loss of secondary-side switches. Furthermore, the temperature can be suppressed by heat dissipation function of magnetic shield. The maximum temperature is improved 17.1 °C at 6 A.

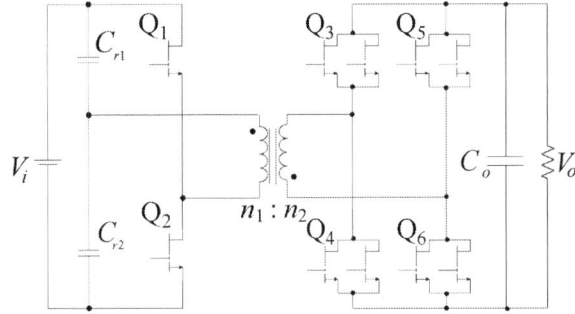

Fig. 12. The circuit topology.

TABLE I. The components of the prototype circuit.

Name	Manufacture	Part Name/ Material
Primary-side GaN-HEMT	EPC	EPC2001
Secondary-side GaN-HEMT	EPC	EPC2015
Gate Driver	TEXAS INSTRUMENTS	LM5113
Transformer Core Material	TDK	SY22 (NiZn Ferrite Core)
Resonant Capacitor	TDK	C1608C0G1H392J
Input Capacitor	TDK	C3216X7R1H105K
Output Capacitor	TDK	C2012X7R1E475M

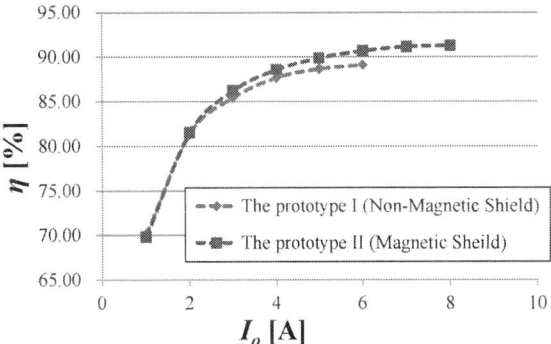

Fig. 13. The power efficiency of the converter.

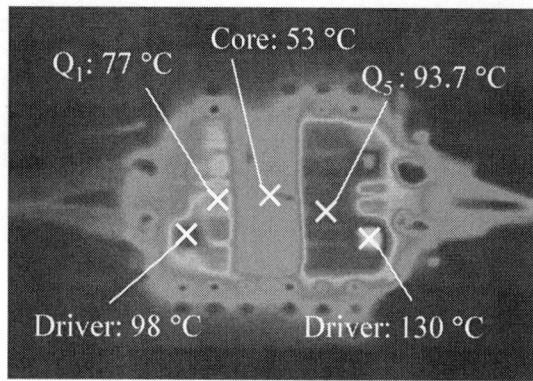 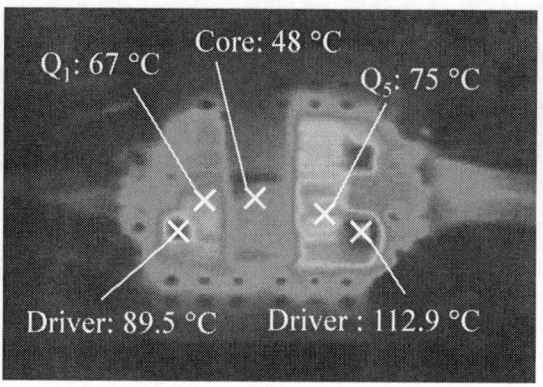 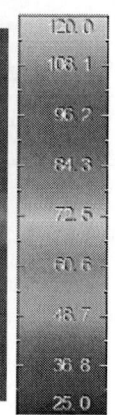

(a) The prototype I without magnetic shield. (b) The prototype II with magnetic shield.

Fig. 14. The temperature distribution of the prototype dc-dc converter at $I_o = 6$ A.

As a result, by adopting the magnetic shield, the load-current range and the power efficiency can be improved.

V. CONCLUSION

In this paper, the influence of magnetic flux on the lateral-structure-semiconductor switches has been revealed. To solve the influence, the using a magnetic shield has been proposed, and the valid of the proposed method have been confirmed by FEM with Maxwell 3D.

Some experiments have been done with the prototype 5 MHz and 32 W/cc unregulated LLC resonant dc-dc converter using GaN-HEMTs. From the experimental results, by inserting the magnetic shield, the load current range is improved from 6 to 8 A. In addition, the maximum temperature is improved 17.1 °C, and the power efficiency is increased 1.58 % at 6 A.

ACKNOWLEDGMENT

The authors would like to thank Mr. Takahiro Koga, ANSYS Japan K. K., for his advice about Maxwell 3D. Also, we would like to thank Prof. Hirotoshi Fukunaga, Nagasaki University, for his advice. Furthermore, we would like to thank Mr. Shohei Iwasaki, technical staff of Nagasaki University, for his technical support.

REFERENCES

[1] K. Matsuura, H. Yanagi, S. Tomioka, and T. Ninomiya: "Power-Density Development of a 5MHz-Switching DC-DC Converter," in Proc. 2012 IEEE Applied Power Electronics Conf., pp. 2326-2332 (Feb. 2012).

[2] D. Reusch and J. Strydom, "Evaluation of Gallium Nitride Transistors in High Frequency Resonant and Soft-Switching DC-DC Converters," IEEE Trans. Power Electron., vol. 30, no. 9, pp. 5151-5158 (Sept. 2015).

[3] D. Reusch and J. Strydom, "Evaluation of gallium nitride transistors in high frequency resonant and soft-switching DC-DC converters," in Proc. 2014 IEEE Applied Power Electronics Conf., pp. 464-470 (Mar. 2014).

[4] M. D. Seeman, S. R. Bahl, D. I. Anderson, and G. A. Shah, "Advantages of GaN in a High-Voltage Resonant LLC Converter," in Proc. 2014 IEEE Applied Power Electronics Conf., pp. 476-483 (Feb. 2014).

[5] W. Zhang, Y. Long, Y. Cui, F. Wang, L. M. Tolbert, B. J. Blalock, S. Henning, J. Moses, and R. Dean, "Impact of planar transformer winding

capacitance on Si-based and GaN-based LLC resonant converter," in Proc. 2013 IEEE Applied Power Electronics Conf., pp. 1668-1674 (Mar. 2013).

[6] S. Ji, D. Reusch, and F. C. Lee, "High-Frequency High Power Density 3-D Integrated Gallium-Nitride-Based Point of Load Module Design," IEEE Trans. Power Electron., vol. 28, no. 9, pp. 4216-4226 (Sept. 2013).

[7] W. Kangping, M. Huan, L. Hongchang, G. Yixuan, Y. Xu, Z. Xiangjun, and Y. Xiaoling, "An Optimized Layout with Low Parasitic Inductances for GaN HEMTs Based DC-DC Converter," in Proc. 2015 IEEE Applied Power Electronics Conf., pp. 948-951 (Mar. 2015).

[8] D. Reusch and F. C. Lee, "High Frequency Bus Converter with Integrated Matrix Trans-formers for CPU and Telecommunications Applications," in Proc. 2010 IEEE Energy Conversion Congress & Expo., pp. 2446-2450 (Sept. 2010).

[9] D. Reusch and F. C. Lee, "High Frequency Bus Converter with Low Loss Integrated Ma-trix Transformer," in Proc. 2012 IEEE Applied Power Electronics Conf., pp. 1392-1397 (Feb. 2012).

[10] D. Reusch and F. C. Lee, "High Frequency Isolated Bus Converter with Gallium Nitride Transistors and Integrated Transformer," in Proc. 2012 IEEE Energy Conversion Con-gress & Expo., pp. 3895-3902 (Sept. 2012).

[11] D. Huang, S. Ji, and F. C. Lee, "LLC Resonant Converter With Matrix Transformer," IEEE Trans. Power Electron., vol. 29, no. 8, pp. 4339-4347 (Aug. 2014).

[12] GaN Systems Inc. (Sep. 2015). GS61004B, 100V enhancement mode GaN transistor, Preliminary Datasheet.

[13] Great Wall Semiconductor (GWS) (Apr. 2004). Application Note: AN-504, BGA Package Design and Printed Circuit Board Guidelines for Great Wall Semiconductor's GWS12N30 lateral MOSFET.

[14] Z. J. Shen, D. Okada, F. Lin, A. Tintikakis, and S. Anderson, "Lateral Discrete Power MOSFET: Enabling Technology for Next-Generation, MHz-Frequency, High-Density DC/DC Converters," in Proc. 2004 IEEE Applied Power Electronics Conf., pp. 225-229 (Feb. 2004).

[15] Efficient Power Conversion (EPC). EPC2001 Enhancement Mode Power Transistor Datasheet.

[16] Efficient Power Conversion (EPC). EPC2015 Enhancement Mode Power Transistor Datasheet.

A High-Voltage Level Shifter with Sub-Nano-Second Propagation Delay for Switching Power Converters

Ahmed Abdelmoaty[1], Mohammad Al-Shyoukh[2], and Ayman Fayed[1]

[1]Dept. of Electrical & Computer Engineering, The Ohio State University, Columbus, Ohio, USA
[2]Austin Design Center (ADC), TSMC Inc., Austin, Texas, USA
abdelmoaty.1@osu.edu, shyoukhm@gmail.com

Abstract— A level-shifting circuit with sub-nano-second propagation delay for high input voltage switched-mode power converters is presented. The proposed circuit uses isolated low-voltage NMOS transistors and capacitive coupling to shift the control signal of the high-side power switch from a low-voltage logic domain ($V_{Logic} \sim 5V$) up to a high-voltage power domain ($V_{IN} \sim 65V$) with less than 115ps propagation delay. As a result, the non-overlap time inserted between the control signals of the high-side and low-side power switches can be minimized, leading to higher efficiency. Moreover, the control signal is shifted to the high-voltage power domain without reducing its voltage swing, which helps minimizing the on-resistance of the switch and further improves efficiency. The proposed circuit is fully integrated in a 0.18μm technology with no off-chip components. It occupies less than 0.75mm² and provides built-in protection for the low-voltage devices with no additional protection circuitry. Transistor level simulations demonstrate the circuit's functionality and performance.

Keywords—level shifting; capacitive coupling; switched-mode power converter; high-voltage gate drivers; propagation delay

I. INTRODUCTION

Many systems, such as Electrical Vehicles (EV), rely on energy sources with voltage levels up to 100V as their main source of energy. Therefore, there is an increasing demand for highly-efficient, switched-mode power converters that can operate from high voltage levels and step them down to levels where they can be used safely to power the various circuit components in these systems. A key element to maximizing efficiency in these converters is to minimize the non-overlap time between the gate control signals of the high-side and low-side power switches, while still avoiding any shoot-through current through the power switches. However, since the control signal of the high-side switch must be level-shifted from the low-voltage logic domain (V_{Logic}) used by the controller to the high-voltage power domain (V_{IN}) as shown in Fig. 1, the propagation delay (t_d) of the level-shifting circuit limits how much the non-overlap time (t_{nov}) can be reduced, especially with PVT variations and delay mismatches between the high-side and low-side gate-drive signals.

Many level-shifting circuits have been proposed in the literature, where some of them rely on pull-up and pull-down structures implemented using high-voltage or Drain-Extended MOS (DE-MOS) devices in a positive feedback loop in order to shift a logic signal from a low-voltage domain to a high-voltage domain [1]-[9]. However, these circuits suffer from a long propagation delay (10ns-200ns), and their output swing is

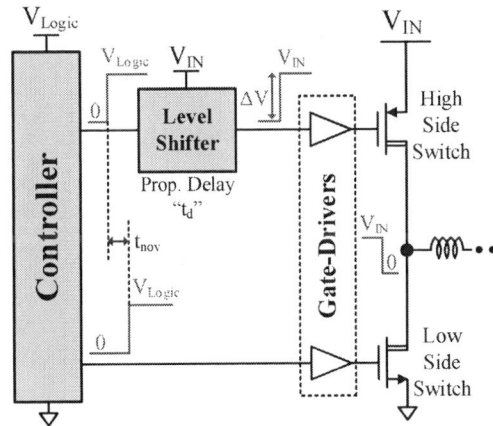

Fig. 1. Block diagram of a generic driver stage in a buck converter

typically equal to the entire input voltage, i.e. $\Delta V = V_{IN}$. This large swing is incompatible with voltage ratings of the gates of many integrated PMOS power switches and results in large dynamic power consumption and low efficiency.

Alternatively, capacitive-coupling level-shifting circuits, such as in Fig. 2(a), are more attractive as they reduce the propagation delay and limit the swing of the level-shifted signal, and hence the gate-to-source voltage of the high-side power switch, to the low-voltage logic level (i.e. $\Delta V = V_{Logic}$). This reduces the dynamic power consumption and helps meeting the voltage rating of the power switches [10], [11]. However, due to the lack of a ($V_{IN} - V_{Logic}$) supply rail, these level-shifting circuits are forced to drive the gate of the high-side power switch directly, and therefore, the charge division between the coupling capacitor (C_{C1}) and the gate capacitance (C_G) of the power switch results in reducing the swing of the level-shifted signal. Thus, a large off-chip coupling capacitor is required to minimize the reduction in the voltage swing in order to avoid degrading the on-resistance of the power switch [11]. Moreover, directly driving the large gate capacitance of the power switch limits the improvement in propagation delay.

In this paper, an improved capacitive-coupling level shifting circuit including an internal ($V_{IN} - V_{Logic}$) supply rail is presented. The proposed circuit features a sub-nano-second propagation delay and a full on-chip implementation. The design is described in details in section II, while section III is devoted for transistor level simulations. Section IV concludes the paper.

978-1-4673-9551-9/16 $31.00 © 2016 IEEE

(a)

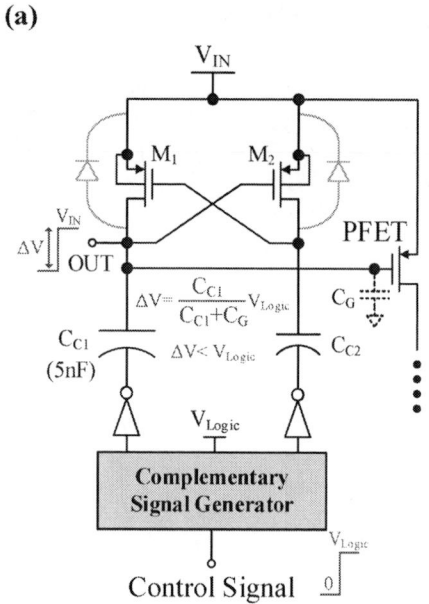

(b)

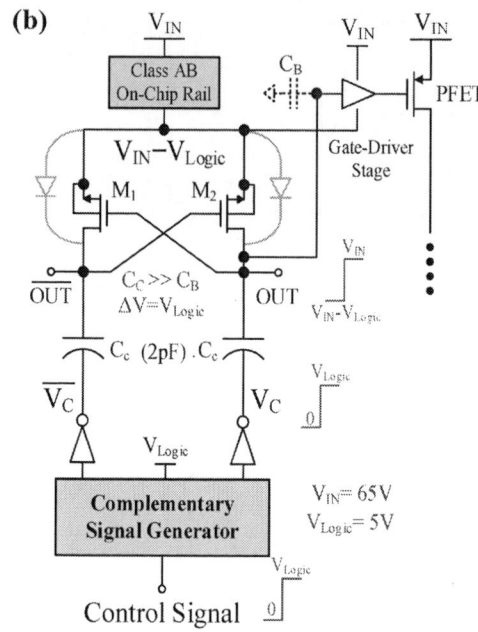

Fig. 2. Capacitive-coupling level-shifting circuits: (a) conventionanl design [11], and (b) the proposed design.

I. THE PROPOSED LEVEL-SHIFTING CIRCUIT

The proposed capacitive-coupling level-shifting circuit is shown in Fig. 2(b), where two high-voltage capacitors, two isolated low-voltage NMOS transistors, and a fully-integrated class-AB (V_{IN}–V_{Logic}) supply rail are used. The circuit shifts the control signal from the low-voltage (5V) logic domain up to the high-voltage (65V) power domain, but unlike the conventional design shown in Fig. 2(a), the proposed circuit includes an independent supply rail (V_{IN}–V_{Logic}) to enable implementing a separate gate-drive stage operating between V_{IN} and (V_{IN}–V_{Logic}). As a result, there is no need for the level-shifting circuit to drive the large gate capacitance (C_G ~200pF) of the power switch directly, and instead it only drives the small input capacitance C_B (~10fF) of the gate-drive stage. Therefore, the propagation delay is significantly reduced. Moreover, since the charge division in the proposed circuit now takes place between C_C and C_B (instead of C_{C1} and C_G in the conventional design of Fig. 2(a)), the coupling capacitor C_C no longer has to be large and off-chip to maintain the swing of the level-shifted signal at V_{Logic}. In fact, reducing the coupling capacitor (C_C) to only 2pF, which can be easily implemented on-chip, guarantees that the swing at the shifted voltage level is no less than 99.5% of V_{Logic}, and thus ensuring a small on-resistance for the power switch.

The dynamic operation of the proposed circuit relies on two matched coupling capacitors and two matched isolated low-voltage NMOS transistors (M_1 and M_2) in a cross-coupled structure as shown in Fig. 2(b). The advantage of using NMOS transistors instead of PMOS (as in conventional designs) is that the body diodes of the NMOS devices provide a built-in protection for the transistors since they prevent the nodes OUT and $\overline{OUT}$ from ever dropping more than a diode's forward-voltage below (V_{IN}–V_{Logic}). Thus, the transistors are always operating within their voltage rating limits without

employing additional protection or startup circuits. The complementary signal generator receives the low-voltage input control signal and produces two complementary versions of it (V_C and $\overline{V_C}$) that are aligned in time by matching the delays in the signals paths. The signals V_C and $\overline{V_C}$ drive the two coupling capacitors, and thus, during transitions, their complementary swing couples through the capacitors to the nodes OUT and $\overline{OUT}$, causing them to transition in a complementary fashion between V_{IN} and V_{IN}–V_{Logic}. This coupling effect activates the positive feedback loop formed by M_1 and M_2 causing one of them to fully turn ON, while the other to fully turn OFF. As a result, one of the output nodes will discharge to the supply rail (V_{IN}–V_{Logic}), while the other output node becomes floating and remains at (V_{IN}). This process is reversed once the control signal transitions in the opposite direction, and therefore two complementary signals swinging between (V_{IN}–V_{Logic}) and (V_{IN}) are generated at the output nodes (OUT and $\overline{OUT}$). Although, the proposed circuit uses an input voltage of 65V, the operation of the circuit is still valid at any input voltage (V_{IN}).

Additionally, one small current source (1µA) and two small high-voltage switches (S_1 and S_2) are added to the circuit as shown in Fig. 3. These components ensure that when one of the outputs is supposed to be at V_{IN} for an extended period of time (i.e. long duty cycles or at low frequency operation), the corresponding coupling capacitor is always charged to V_{IN}. This guarantees that there is no degradation in the output voltage swing due to charge loss or leakage across the capacitor. Moreover, this additional circuitry sets the initial conditions at the output nodes OUT and $\overline{OUT}$ such that if the control signal (V_C) is high, the node OUT is charged up to V_{IN} and the node $\overline{OUT}$ is discharged to (V_{IN}–V_{Logic}), and vice versa. This action enforces the correct complementary initial conditions at the output nodes for proper operation.

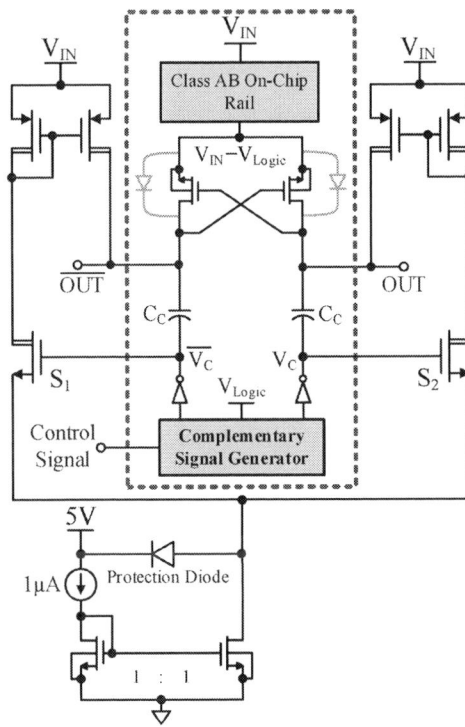

Fig. 3. Complete circuit diagram of the proposed level-shifter.

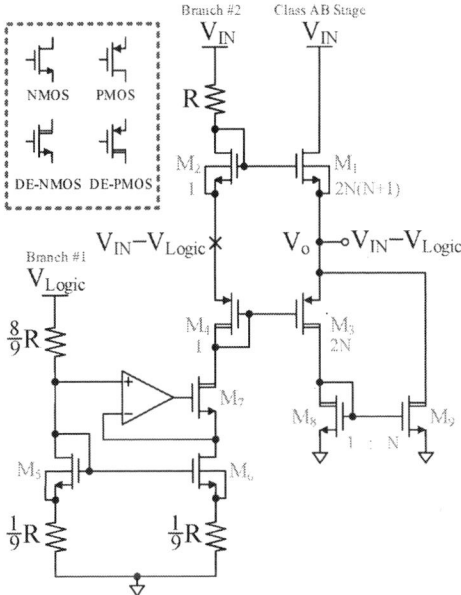

Fig. 4. Circuit diagram of the On-chip (V_{IN}-V_{Logic}) supply rail.

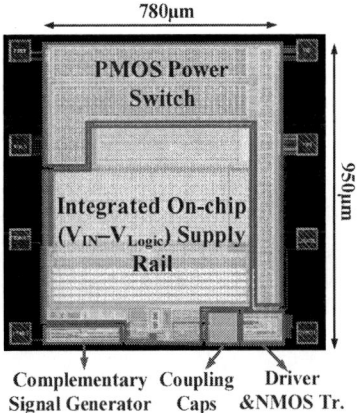

Fig. 5. Layout of the level-shifter including the full-integrated (V_{IN}-V_{Logic}) supply rail.

The fully-integrated class-AB (V_{IN}–V_{Logic}) supply rail implementation is shown in Fig. 4. The supply rail features high bandwidth in order to handle the fast transitions in the gate-driver stage. It also provides sourcing and sinking capabilities to charge the output nodes of the level shifter and discharge the gate-driver stage respectively. The (V_{IN}–V_{Logic}) supply rail is generated from the input voltage (V_{IN}) by matching the currents flowing in branches #1 and #2 shown in Fig. 4. By doing so, the voltage drop across M_2 is forced to be the same as the drop across M_5. Therefore, if branch #1 is operated from V_{Logic}, the voltage at the source of M_2 becomes (V_{IN}–V_{Logic}) provided that the resistor (R) and the transistors M_2 and M_5 are all well matched. The currents in both branches are maintained the same by using a single stage op-amp in and the transistor M_7 in a negative feedback loop to ensure identical drain-to-source voltages across M_5 and M_6. Moreover, small source-degeneration resistors are used to further improve the current matching without sacrificing voltage headroom. Branch #2 provides the necessary bias voltage for the class-AB output stage, which has an improved current-sinking capability through the action of M_8 and M_9. Additionally, the transistor M_9 is sized to be N-times larger than M_8, and the transistor M_3 is sized to be 2N-times larger than M_4 to enhance the sinking capabilities. Therefore, the transistor M_1 must be sized to be 2N(N+1) larger than M_2 in order to avoid any voltage offset at the final output node V_o. Although increasing the ratio N improves the current sinking capability of the supply rail, the quiescent current and the power consumption will increase. In this design, N is chosen to be 5 in order to keep the quiescent current below 100μA.

II. SIMULATION RESULTS

The proposed level-shifting circuit, including the high-side PMOS power switch, is implemented in TSMC 0.18μm HV technology. This process provides the option to implement NMOS transistors in separate wells from the substrate, where each transistor can have its body terminal tied to its source terminal to eliminate any body effect. Furthermore, 5V-rated devices are employed for the low-voltage portion of the proposed circuit, and therefore, the improvement reported in the propagation delay is not due to using the minimum feature-sized transistors available in the technology, but rather due to the proposed circuit topology. The layout of the proposed circuit is shown in Fig. 5 and it occupies less than 0.75 mm^2. Although the proposed circuit topology does not

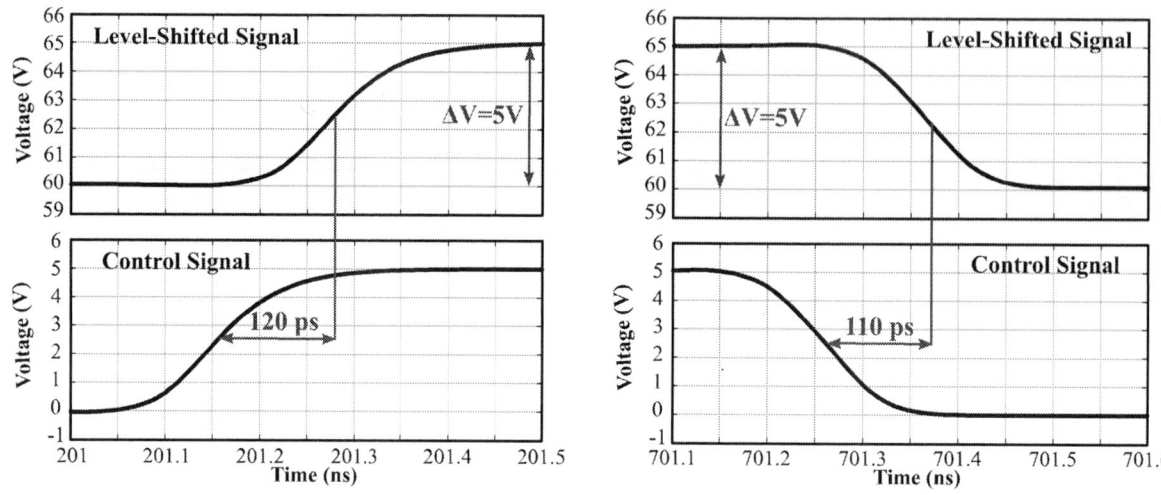

Fig. 6. Simulated propagation delay between the low-voltage control signal and the level-shifted signal at rising and falling edges.

TABLE I. PERFORMANCE COMPARISON

	[9]	[11]	This Work
Technology	CMOS 2μm	CMOS 0.5μm	CMOS 0.18μm
Prop. Delay (t_d)	80ns	0.5ns	0.115ns
Shifted-Signal Swing (ΔV)	50V	4.5V	5V
Input Voltage (V_{IN})	50V	100V	65V*
Full On-Chip Implementation	Yes	No	Yes

*Maximum input voltage is limited only by the technology, not the circuit.

impose any limitation on the input voltage level, the maximum voltage level supported by the technology used for this implementation is 65V. Therefore, the design is simulated at that input level with a 1-MHz control signal at 5V logic level. As shown in Fig. 6, the proposed circuit shifts the control signal from the 5V domain to the 65V domain with a propagation delay of 120ps and 110ps at the rising and falling edges respectively. Moreover, the swing of the level-shifted signal is maintained at the logic level of 5V. Moreover, changing the input voltage level does not have an impact on the simulated propagation delay of the proposed circuit.

III. CONCLUSION

A fully-integrated level-shifting circuit for high-voltage switching power converters has been introduced. The proposed circuit relies on capacitive coupling mechanism and a fully on-chip (V_{IN}–V_{Logic}) supply rail to achieve better propagation delay. The proposed circuit shifts a 5V logic signal to a 65V input voltage level with only 115ps propagation delay while maintaining its swing at 5V. The performance of the proposed circuit is compared with circuits reported in literature as shown in Table 1.

ACKNOWLEDGMENT

The authors would like to thank CD An, E. Ortynksa, and E. Soenen from TSMC Inc. for logistical and technical support, and NSF (ECCS-1559972) for financial support.

REFERENCES

[1] K.-H. Koo, J.-H. Seo, M.-L. Ko and J.-W. Kim, "A new level-up shifter for high speed and wide range interface in ultra deep sub-micron", *Int. Symp. Circuits and Systems (ISCAS)*, vol. 2, pp. 1063-1065, May 2005.

[2] A. Fayed, "Level shifting techniques for Mixed-Signal SoCs in low-voltage nanometer CMOS technologies," *IEEE Midwest Symposium on Circuits and Systems (MWSCAS)*, pp. 1102-1105, Aug. 2009.

[3] J. Doutreloigne, H. De Smet, J. Van den Steen and G. Van Doorselaer, "Low-power high-voltage CMOS level-shifters for liquid crystal display drivers", in *Proc. 11th Int. Conf. Microelectronics (ICM)*, pp. 213-216, Nov. 1999.

[4] J. Doutreloigne, "A fully integrated ultra-low-power high-voltage driver for bistable LCDs," in *Proc. IEEE Int. Symp. VLSI Design, Automation and Test*, pp. 1-4, April 2006.

[5] M. Rossberg, B. Vogler and R. Herzer, "600V SOI gate driver IC with advanced level shifter concepts for medium and high power applications", in *Proc. Eur. Conf. Power Electronics and Applications*, pp. 1-8, Sept. 2007.

[6] B. Serneels, M. Steyaert and W. Dehaene, "A high speed, low voltage to high voltage level shifter in standard 1.2V 0.13μm CMOS", *IEEE Int. Conf. Electronics, Circuits, and Systems*, pp. 668-671, Dec. 2006.

[7] R. Chebli, M. Sawan, Y. Savaria and K. El-Sankary, "High-voltage DMOS integrated circuits with floating gate protection technique", in *Proc. IEEE Int. Symp. Circuits and Systems (ISCAS)*, pp. 3343-3346, May 2007.

[8] Y. Moghe, T. Lehmann, and T. Piessens, "Nanosecond delay floating high voltage level shifters in a 0.35 μm HV-CMOS technology," *IEEE J. of Solid-State Circuits (JSSC)*, vol. 46, no. 2, pp. 485-497, Feb. 2011.

[9] M. J. Declerq, M. Schubert, and F. Clement, "5V-to-75 V CMOS output interface circuits," *IEEE Int. Solid-State Circuits Conference (ISSCC) Dig. Tech. Papers*, pp. 162-163, Feb. 1993.

[10] S. C. Tan and X. W. Sun, "Low power CMOS level shifters by bootstrapping technique", *Electron. Lett.*, vol. 38, no. 16, pp. 876 - 878, Aug. 2002.

[11] Z. Liu and H. Lee, "A 100V Gate Driver with Sub-Nanosecond-Delay Capacitive-Coupled Level Shifting and Dynamic Timing Control for ZVS-Based Synchronous Power Converters," in *Custom Integrated Circuits Conference (CICC)*, pp. 1-4, Sept. 2013.

Dual-output, Three-level GaN-based Dc-Dc Converter for Battery Charger Applications

Ren Ren[1,2], Bo Liu[2], Edward A. Jones[2], Fred Wang[2], Zheyu Zhang[2], Daniel Costinett[2]

Jiangsu Key Laboratory of New Energy Generation and
Power Conversion
Nanjing University of Aeronautics and Astronautics
Nanjing, China
rren3@utk.edu

Center for Ultra-wide-area Resilient Electric Energy
Transmission Networks
The University of Tennessee
Knoxville, TN 37996, USA

Abstract—Gallium Nitride (GaN) HFETS are an enabling technology for high-density converter design. This paper proposes a three-level dc-dc converter with dual outputs based on enhancement-mode GaN devices, intended for use as a battery charger in aircraft applications. The charger can output either 28 V or 270 V, selected with a jumper, which meets the two most common dc bus voltages in airplanes. It operates as an LLC converter in the 28 V mode, and as a buck converter in the 270 V mode. In both operation modes, the devices can realize zero-voltage-switching (ZVS). With the chosen modulation method, the converter can realize the frequency doubling function to act as an interleaved converter. For the LLC mode, the resonant frequency is twice the switching frequency of primary-side switches, and for buck mode, the frequency of the output inductor current is also twice the switching frequency. This helps to reduce the size of magnetics while maintaining low switching loss. Also, the converter utilizes the matrix transformer with resonant parameters designed to avoid ZVS failure. The operation principle of the converter is analyzed and verified on a 1MHz resonant frequency prototype.

Keywords—*GaN HFETS; Dual output; Battey charger; ZVS; Frequency doubling modulation*

I. INTRODUCTION

With the development of the more-electric aircraft and power electronics technology, many improvements in airplane power management have been proposed. High efficiency, high frequency and high power density are becoming a trend for converters in aerospace applications. Emerging from the latest generation of wide band-gap semiconductors, the Gallium-Nitride hetero-junction field-effect transistor (GaN HFET) has become a highly promising device for achieving high power density. In [1], the desirable features of GaN HFETS are explained from an application perspective, such as its low junction capacitance and on-resistance, high switching frequency capability, and smaller size. The GaN power device is growing in popularity, although some technical improvements are still needed before the device reaches maturity [2-5].

In this paper, a high power density converter using GaN HFETs is designed for a battery charger in an aerospace application. The output of the ac-dc power factor correction (PFC) stage is 600 V – 700 V with either 115 V or 230 V input voltage. However, the maximum voltage rating of commercial GaN HFET is currently only 600 V, so a three-level dc-dc topology is adopted to clamp the voltage stress of the switch to half of the input voltage. The difficulty in this design is realizing both the 28 V and 270 V outputs in one converter. For the low voltage output, the converter needs a transformer to step down the voltage with isolation, and an output inductor at the non-isolated high voltage output. In order to take full advantage of the GaN HFETS, a resonant topology is used in the low voltage output with fully-realized ZVS for all switches. A buck topology operated under critical conduction mode (CRM) is selected for the high voltage output. Previous papers have explored the three-level LLC converter [6-9], but they have used complex topologies with clamping diodes and voltage-dividing capacitors. Moreover, the two floating terminals of the transformer must be connected to the mid-point of the dividing capacitors and three level bridge, which makes it difficult to combine these topologies with the buck mode in the high voltage output condition.

An intuitive and simple topology has been proposed for this GaN-based battery charger, with dual output terminals to satisfy the two common voltage levels in airplanes. The magnetics component is designed for both operation modes which acting as transformer under LLC mode and output filter inductor under Buck mode. The operation and modulation method of this topology will be explained, together with design details including transformer and selection of GaN devices.

II. OPERATION PRINCIPLE OF THE CIRCUITS

A. Mode Analysis

The topology and working modes of the proposed three level converter are illustrated in Fig 1. When switch K_1 opens and K_2 closes, the converter works as an LLC resonant converter. When switch K_1 closes and K_2 opens, the converter works as a buck converter. The main waveforms of LLC mode

978-1-4673-9551-9/16 $31.00 © 2016 IEEE

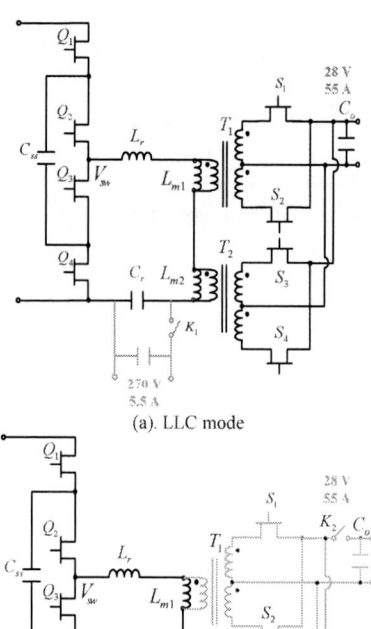

(a). LLC mode

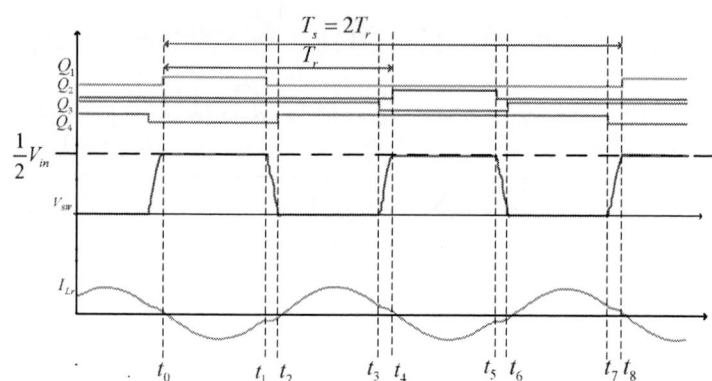

Fig. 2. The key waveform of the three level converter in LLC mode.

energy to the flying capacitor. In the secondary side, the

(b). Buck mode

Fig. 1. The dual outputs three level converter and each working mode.

are shown in Fig. 2. For the primary side switches, the voltage across each off-state switch is clamped to half of the input voltage by C_{ss}, so 600 V GaN devices can be used in the 700 V input application, and all the devices can still realize ZVS. From Fig. 2, the mid-point voltage of the three-level bridge is a square wave with a peak voltage that is also clamped to half of the input voltage. The resonant frequency is twice the switching frequency, reducing switching loss and the size of magnetics. On the secondary side, the switches must operate at the resonant frequency, but both ZVS and ZCS can be realized so that switching loss is not a dominant loss on the secondary side.

The three-level converter with flying capacitor in LLC mode has eight operation stages in each switching cycle. This simple topology can be used as LLC converter and the details of operation mode will be given below whereas the buck mode operation principle has already been analyzed in [10] and therefore will not be presented in this paper.

The equivalent circuit of LLC mode for each stage is shown in Fig. 3. The detail operating process is described detail as follows.

[t_0-t_1]: Q_1 and Q_4 are on state and Q_2 and Q_3 are off state. The excitation voltage of resonant tank equals input voltage minus voltage of flying capacitor (V_{in} - V_{css}). The input current charges the flying capacitor and input source provides the

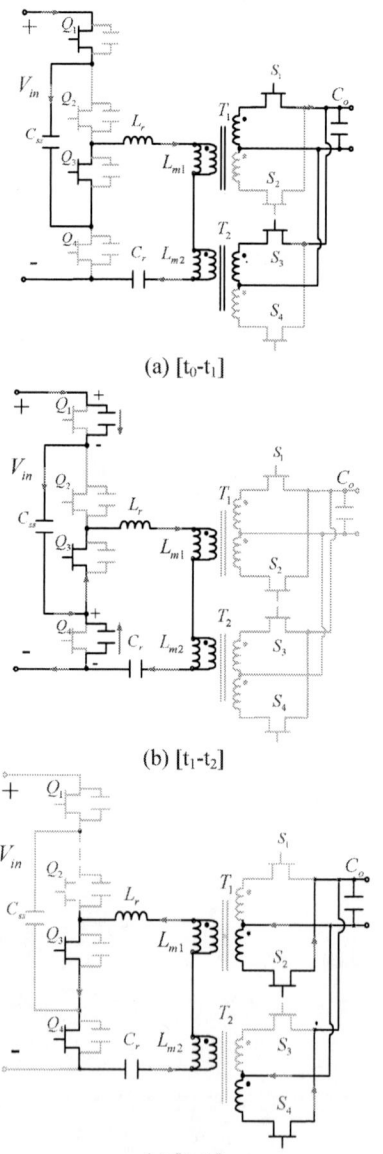

(a) [t_0-t_1]

(b) [t_1-t_2]

(c) [t_2-t_3]

978-1-4673-9551-9/16 $31.00 © 2016 IEEE   2442

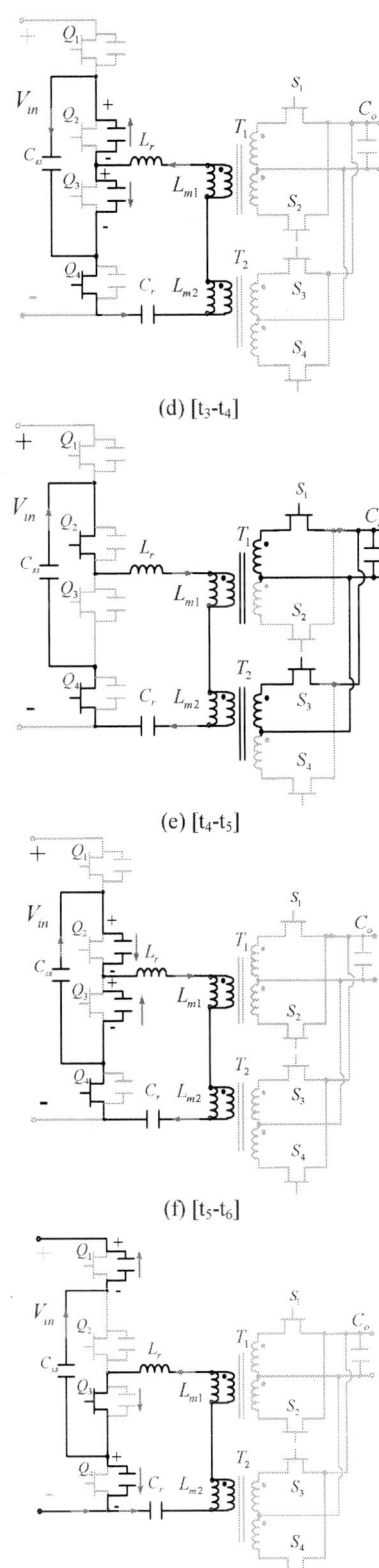

(d) [t_3-t_4]

(e) [t_4-t_5]

(f) [t_5-t_6]

(g) [t_7-t_8]

Fig. 3. The dual outputs three-level converter and its working modes.

circuits begin the first positive cycle of the resonance under one switching cycle.

[t_1-t_2]: At t_1, the resonant current equals the magnetizing current and all secondary side switches ZCS turn off. In the meantime, Q_1 turns off and the magnetizing current will charge the output capacitance of Q_1 and discharge the output capacitance of Q_4. Although the discharge current path goes through the flying capacitor, it will have no influence on the ZVS realization and total equivalent capacitance of two series capacitor due to flying capacitance is much larger than the output capacitance of GaN device.

[t_2-t_3]: When the voltage of output capacitance of Q_4 drops to zero, the equivalent body diode of Q_4 conducts. Then at t_2, Q_4 turns on with ZVS. In the secondary side, the synchronous switches S_2 and S_4 also turn on with ZVS and the circuits begin the first negative cycle of the resonance under one switching cycle.

[t_3-t_4]: Like [t_1-t_2], the negative magnetizing current will charge the output capacitance of Q_3 and discharge the output capacitance of Q_2.

[t_4-t_5]: Q_2 turns on with ZVS. The excitation voltage of resonant tank equals flying capacitor (V_{css}). The resonant current discharges the flying capacitor and flying capacitor provides the energy to the load. In the secondary side, the synchronous switches S_1 and S_3 turn on and the circuits begin the second positive cycle of the resonance under one switching cycle.

[t_5-t_6]: Like [t_1-t_2], the positive magnetizing current will discharge the output capacitance of Q_3 and charge the output capacitance of Q_2.

[t_6-t_7]: Q_3 turn on with ZVS and the circuits begin the second negative cycle of the resonance under one switching cycle. The operation and equivalent circuits are totally same with time [t_2-t_3].

[t_7-t_8]: Like [t_1-t_2], the positive magnetizing current will discharge the output capacitance of Q_1 and charge the output capacitance of Q_4. In the next time like [t_0-t_1], the Q_1 will turn on with ZVS.

Based on the above analysis, ZVS can be achieved for all primary-side switches and ZCS can be achieved for all secondary-side switches. As a result, the total switching losses of the converter are considerably reduced. High frequency operation can be achieved.

B. Automatic Voltage Balance of Flying Capacitor

From the operation of the converter circuit, one switching period has two resonant periods. The excitation voltage of two positive half cycles of the resonance are different. One is V_{in}-V_{css} and the other one is V_{css}, so the voltage of flying capacitor plays a key role for the power balance of two independent positive half cycles. If the voltage of flying capacitor is little bit far away half of the input voltage, the transferred energy to the load will be unbalanced caused by the unequal excitation voltage. In fact, this will not happen due to the automatic voltage balance function of flying capacitor of circuits. This can be demonstrated by simply qualitative analysis.

In steady state, the charge and energy of negative and positive period of resonance are the same due to the symmetrical waveforms of resonant current and voltage which can keep charge balance of flying capacitor. In the transient state, assuming the flying capacitor is higher than half of the input voltage, the excitation voltage (V_{in} - V_{css}) of first positive resonant period for charging the flying capacitor is lower than the excitation of second positive resonant period for discharging the flying capacitor in one switching period. The resonant current charge and energy are decided by the amplitude of the excitation voltage and the lower excitation voltage is, the lower resonant energy transferred to the load, so the net charge for flying capacitor is negative after above assuming unbalance condition. It will lower the voltage of flying capacitor until it reaches the half of the input voltage. This mechanism will ensure the automatic voltage balance of flying capacitor.

Based on the charge balance principle, the needed capacitance of flying capacitor can be calculated as following equation (2.1):

$$\Delta u_{css} = \int_{0}^{\frac{T_r}{2}} \frac{1}{C_{ss}} \cdot i_{Lr}(t)dt = \int_{0}^{\frac{T_r}{2}} \frac{1}{C_{ss}} \cdot I_{rms\_p}(\omega_o t + \theta)dt$$
$$= \frac{I_{rms\_p}}{\pi C_{ss} f_r} \tag{2.1}$$

where Δu_{css} is the voltage ripple of the average voltage, $i_{lr}(t)$ and I_{rms_p} are instantaneous value and RMS value of resonant current respectively and θ is phase difference between the resonant voltage and resonant current, T_r and f_r are the resonant period and frequency.

Setting Δu_{css} as the 2% of average voltage ripple under the full load condition and resonant frequency as 1MHz, the needed C_{ss} can be calculated as 1.2μF. The voltage rating of this capacitor is only half of the input voltage, so the volume and weight of this flying capacitor would be very small as a whole.

III. CONSIDERATION FOR GaN DEVICE SLECTION

For primary side, commercial GaN HFETs with 600V rating are available from several manufacturers, such as Transphorm, GaN Systems, Panasonic, IR, and HRL. For this paper, two devices were considered: the cascode GaN device TPH3006LD from Transphorm, and the enhancement-mode GS66508 from GaN Systems.

For secondary side, the considerations for choosing the suitable device are conduction loss and noise immunity. The EPC enhancement-mode device is no doubt the optimal choice with advanced LGA package and state-of-art electrical parameters through comparisons.

A. Primary Side Device Selection

Table 1 compares the parameters of these two devices. The FOM of $Q_g R_{ds\_on}$ is often used to evaluate device performance, and the GaN Systems device has a lower FOM than Tranphorm's.

Tab.1 Device comparison between GaN devices for primary devices

Parameter	TPH3006LD	GS66508
Voltage rating (V)	600	650
R_{ds_on}@25C (mΩ)	150	55
Q_g (nC)	6.2	6.5
Q_{gd} (nC)	2.2	2.8
FOM (pC·Ω)	930	357.5

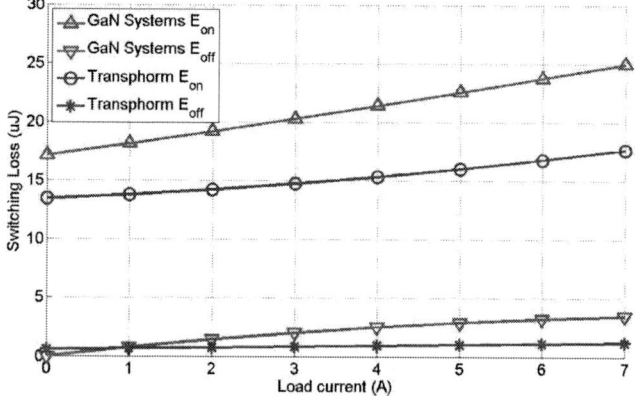

Fig. 4. Switching Loss Comparison between two devices

In order to further verify the switching loss, a double pulse test was performed with the GaN Systems device [11]. Switching loss data for the Transphorm based on double pulse testing has been previously published in [12] for comparison. Fig. 4 shows the turn-on loss E_{on} and turn-off loss E_{off} of each device for the 325V case, at an operating temperature of 125 °C. The Tranphorm device has a lower switching loss than GaN Systems in the load current range of this application, but the GaN Systems device has a much lower channel on-resistance. In fact, static testing was performed on the two devices at 125 °C, and the GaN Systems devices has much lower on-resistance than the Transphorm device at this operating temperature. Therefore, the overall device loss will be calculated in simulation. Fig. 5 compares the total loss between the two devices under the LLC mode and buck mode, respectively.

In LLC mode, due to the high RMS value of the resonant current and very low switching loss, the GaN Systems device has an obvious advantage over the Transphorm's. The total loss is 26.2W for the GaN Systems device compared with 36.2W for Transphorm's, in LLC mode at 500 kHz switching frequency. In contrast, the Transphorm device has better performance than GaN Systems device in buck mode with hard switching. Fig. 4 shows that the two devices have nearly the same total loss in the buck mode at 300 kHz. However, the Transphorm device has a much lower total buck-mode loss than GaN Systems at the higher switching frequency of 500 kHz. Overall the GaN Systems device is therefore more suitable for this application, using a control method to run the converter at different switching frequencies for each operation mode. This will be given more detailed discussion in the next section.

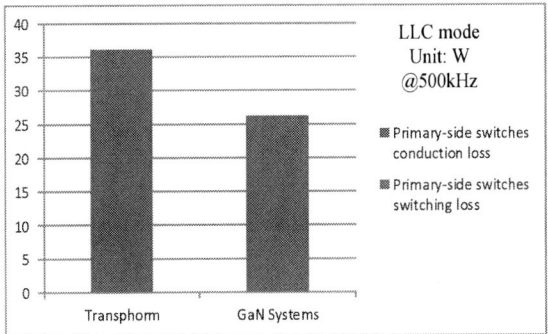
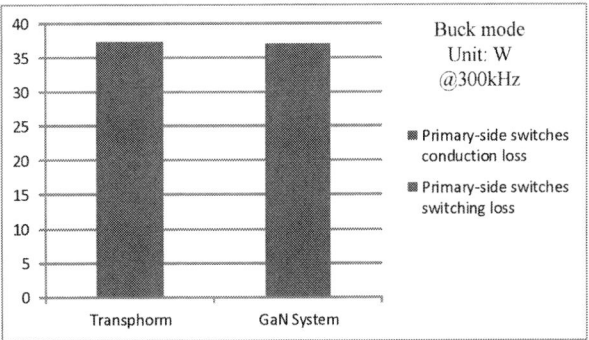

Fig. 5. Total devices loss comparison in LLC mode and Buck mode

B. Secondary Side Device Selection

The total rms value of rectified current for half resonant cycle is around 65A in full load condition. Because using of matrix transformer structure to share the output current, the rms current value to each synchronous device drops to 22A. The EPC 2022, GaN Systems GS61008p and Infineon BSB056N10NN3 are all suitable for this application with same large current rating, very similar R_{ds_on} and advanced package with 100V voltage rating. EPC2022 and GS61008p are the representative for enhancement-mode GaN device while BSB056N10NN3 is a state-of-art representative of Power MOSFET. From conduction loss view, the EPC 2022 should be better than the other two devices owing to a little bit smaller R_{ds_on}. However, from the noise immunity perspective, which device will have a less loss maybe a totally different result because the higher dV/dT noise immunity under same voltage stress means lower induced shooting through loss. This has been explained in paper [13] in detail. One intuitive way to interpret the dV/dT event induced loss is the accumulated miller charge. When Drain-Source voltage reaches to the input voltage, the miller charge should be smaller than the total charge on C_{gs} at the V_{th} level in order to avoiding the spuriously turned on. The inequality (3.1) shows relationship between the miller charge and gate charge under no induced turn-on case. The ratio between the miller charge and gate charge is called miller ratio.

$$
\begin{aligned}
& C_{gd} \cdot V_{gd} \leq C_{gs} \cdot V_{th} \\
& \rightarrow C_{gd} \cdot (V_{ds\_steady} - V_{th}) \leq C_{gs} \cdot V_{th} \\
& \rightarrow \frac{Q_{gd}}{Q_{gs1}} = \frac{C_{gd} \cdot (V_{ds\_steady} - V_{th})}{C_{gs} \cdot V_{th}} \leq 1
\end{aligned}
\tag{3.1}
$$

where C_{gd} is the miller capacitance of Gate to Drain, C_{gs} is Gate to Source capacitance, V_{th} and V_{ds_steady} are the threshold voltage of the Gate and the voltage stress under steady state respectively, Q_{gd} and Q_{gs1} are the gate to drain charge and gate charge at threshold. Here is a point should be required attention. Q_{gs1} in (3.1) is not the same definition of Q_{gs} or $Q_{gs(th)}$ in general device datasheet. Q_{gs} or $Q_{gs(th)}$ in datasheet always means the charge for C_{iss}, but for here, it only includes the charge for C_{gs}.

From above analysis, if the miller ratio is less than 1, the induced shoot-through loss can be minimized. The Fig. 6 gives this data for different voltage stress under three devices. This would help people pick up a high dV/dT noise immunity device. Also, this curve indicates the best working voltage region the device can achieve without active driver circuits. The EPC 2022 definitely a better choice both in the conduction and noise immunity view.

IV. MAGNECTICS DESIGN AND OPERATION FREQUENCY SELECTION

The matrix transformer is the combination of several individual transformers to function as one. The individual transformers have a variety of turns ratios, such as 1:1, 2:1...n:1, so the matrix transformer can realize any overall turns ratio by connecting the primary windings in series and the secondary windings in parallel. This transformer structure is typically used in high output current applications, because it can reduce the conduction loss and eddy current loss in the windings, as well as reducing conduction loss in the synchronous rectifier without paralleling devices.

Two transformers are used in this topology. The magnetizing inductance of the transformer in the LLC mode should be designed to realize the ZVS of the switch and guarantee the required voltage conversion ratio. For buck mode, the magnetizing inductance functions as the output inductor, and the switching frequency in buck mode should be chosen for low device loss and inductor loss.

A. Resonant Parameter Design In LLC Mode

Resonant parameter design is difficult with a high resonant frequency and high current output application, because a high

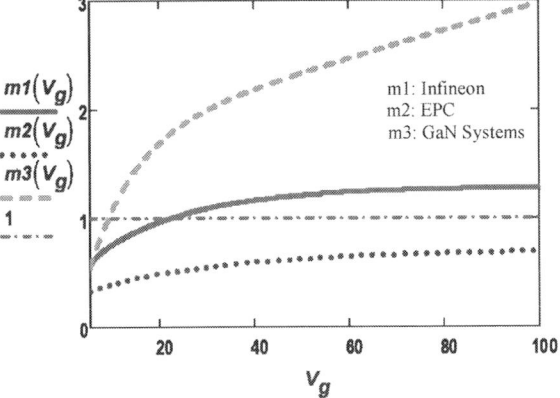

Fig. 6. Miller ratio comparison between three devices

978-1-4673-9551-9/16 $31.00 © 2016 IEEE

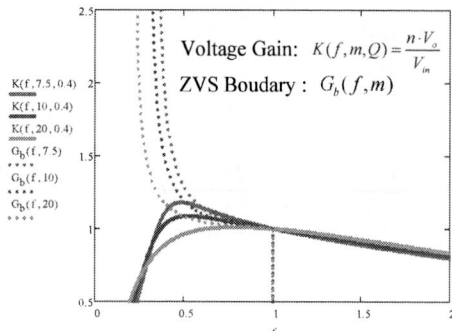

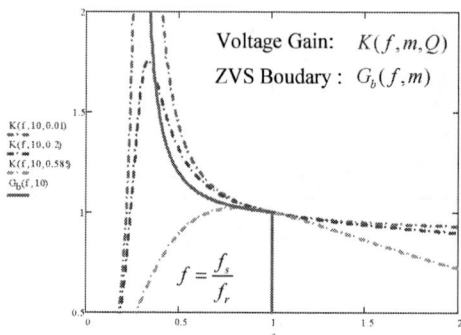

(a): ZVS realization frequency range vs. Inductance ratio m   (b): ZVS realization frequency range vs. different Q value

Fig. 7. The ZVS boundary analysis with different resonant parameters

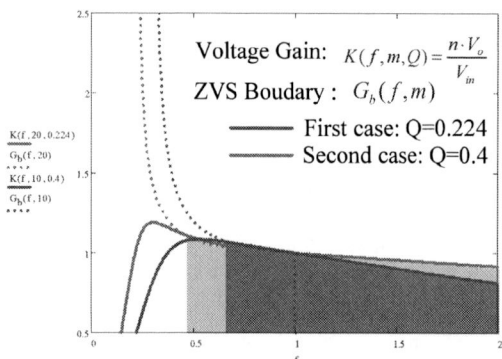

Fig. 8. Two design cases comparison

resonant tank quality factor ($Q = 2\pi f_r L_r / R_{ac}$) will narrow the frequency range of ZVS. Fig. 7(a) shows the ZVS frequency range with different ratio m between the magnetizing inductance and resonant inductance with fixed Q factor and Fig. 7(b) shows the impact of the resonant inductance with fixed ratio m on ZVS frequency range. From the Fig. 7(a), it can be seen that the smaller ratio m is, the wider ZVS frequency range becomes. Also the conclusion from Fig. 7(b) the smaller Q value is, the wider ZVS frequency range becomes, but the Q value is very hard to design to make it small under this case. The leakage inductance is the smallest value we can use for resonant inductance, so the smallest Q value under matrix transformer in this case can be calculated as:

$$Q = 2\pi f_r L_r / R_{ac} = 2\pi f_r L_r \cdot 16n^2 / (\pi^2 R_L)$$
$$= 2 \times 3.14 \times 10^6 \times 16 \times 6^2 \times 0.3 / (3.14^2 \times 0.53) \quad (4.1)$$
$$= 0.244$$

In this case, the ratio m between the magnetizing inductance and resonant inductance cannot be designed so small because if a small magnetizing inductance was chosen, the large conduction loss and high core loss will be introduced due to high magnetizing current and big air gap at the core. However, a big ratio m value is also not a good choice since voltage gain under a big ratio m is too flat which makes the frequency operation range is too wide. Even through the voltage regulation for this system is mainly depend on the front-end AC/DC converter, DC/DC converter still need to

work at the normalized voltage gain under 1 when AC/DC outputs the lowest bus voltage which still can't meet the output voltage in some low output cases. Hence, an additional 0.3 μH resonant inductance is added to obtain a proper design with a 0.4 Q value and 10 ratio m.

A comparison between two design cases is shown in the Fig. 8. A suitable ZVS and narrow voltage regulation frequency range can be both obtained in the second design case. At final, the resonant inductance is designed as 0.7uH and the magnetizing inductance is 7.6uH.

More precisely, ZVS realization has relation with the dead time and parasitic capacitance charge. Enough peak magnetizing current and dead time need to be satisfied that all the parasitic capacitance will be discharged, including the output capacitance of the primary side devices, the output capacitance of the SR devices and the transformer winding capacitance and some stray capacitance. The above conditions for ZVS realization can be expressed by inequality (4.2):

$$Q_{mag} \geq 2 \cdot C_{pri\_oss} \cdot \frac{V_g}{2} + (C_{stray} + C_W) \cdot \frac{V_g}{2} + \frac{1}{2n} 2 \cdot N \cdot C_{sec\_oss} \cdot 2V_o$$

$$(4.2)$$

where Q_{mag} is the charge provided by the magnetizing current, C_{pri_oss} and C_{sec_oss} are the primary side and secondary device's output capacitance respectively, C_{stray} and C_W are the stray capacitance of PCB layout and wingding capacitance of the transformer respectively, n and N are the transformer's turns ratio and matrix transformer amount respectively.

The magnetizing current can be treated as a constant current source in the dead time due to the LLC resonant period is much longer than the LC resonant period, so the charge provided by the magnetizing current is derived as follows equation:

$$Q_{mag} = i_{Lm\_pk}(v_g) \cdot T_d \quad (4.3)$$

where $i_{Lm\_pk}(v_g)$ is peak value of magnetizing current which is a function of input voltage v_g. T_d is the dead time. Also $i_{Lm\_pk}(v_g)$ can be expressed by a input voltage excitation:

$$i_{Lm\_pk}(V_g) = \frac{1}{2} \cdot \frac{\frac{V_g}{4} \cdot \frac{T_r}{2}}{L_m} = \frac{V_g}{16 L_m f_r} \quad (4.4)$$

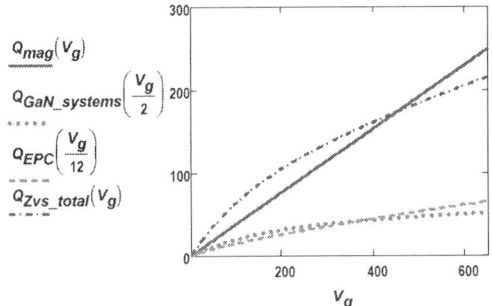

Fig. 9. ZVS realization from the charge view

where L_m is the magnetizing inductance and f_r is the resonant frequency. Substitute (4.4) to (4.3), the equation (4.4) can be rewritten as (4.5):

$$Q_{mag} = \frac{V_g T_d}{16 L_m f_r} \qquad (4.5)$$

Also because of the nonlinearity of the output capacitance of the device, the output capacitance charge can be obtained by curve fitting from the datasheet and integrated by the voltage and output capacitance function of different drain-source voltage. At last, Fig. 9 is plotted to show the provided charge and required charge under different input voltage. Here assuming the dead time is 75nS. From the Fig. 9, the ZVS can be achieved around 450V input voltage under 75nS. Therefore, different dead time can help realize ZVS under different input voltage. A dead time table can be set in the program to adjust minimum dead time to reduce the conduction loss.

B. Working Condition Selection for Buck Mode

On account of second section analysis, the GaN Systems' device is more suitable for this paper's application. However, for buck and high voltage output mode, there are still two working conditions being used for choosing. The first working condition is CCM whose advantage is low RMS inductor current, conduction loss for device and core loss. The second one is CRM whose advantage is ZVS switching and very small switching loss.

It's hard to choose one working condition without loss

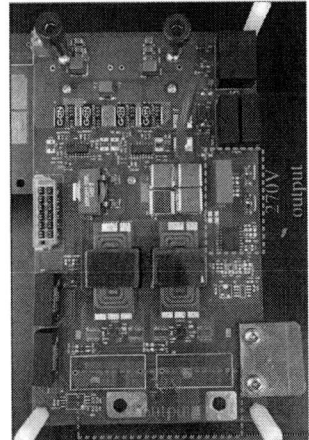

Fig. 11. Power stage without control board

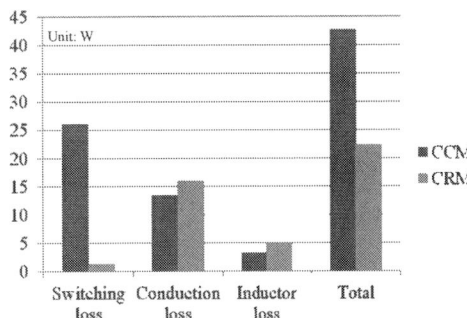

Fig. 10. CRM and CCM loss comparison under buck mode

comparison. Since the magnetizing inductance is used for output inductance for buck mode and the value can't be changed any more, the switching frequency is only thing we can do to change the way the converter working. Through the calculation, the equivalent switching frequency 280kHz and 400kHz are selected to operate the converter at CRM and CCM separately. For both cases, the maximum flux density don't exceed to 0.1T which is much lower than the saturation flux density of 3F45 0.37T at 100°C.

The Fig. 10 gives the loss comparison between the CRM and CCM mode. The conduction loss is calculated under 55°C junction temperature. In this application, CRM operation has lower loss than the CCM, so the CRM working condition is chosen for buck mode.

V. EXPERIMENTAL RESULTS

The verification prototype has been made with selected devices. The controller is the DSP TMS28377D from TI. Two magnetics core E22/6/16 made with ferrite material 3F45 are chosen. The resonant frequency is 1MHz in the LLC mode and inductor operation frequency is 280kHz in the buck mode. The Fig. 11 shows the prototype's picture.

Fig. 12 displays the key waveform of LLC operation mode and the gate signal and drain-source voltage are the switch Q_1's. Fig. 12(b) illustrates the ZVS realization for LLC mode. Fig. 13 gives the key waveforms for buck operation mode. The inductor current has negative value to help switches realize ZVS. The ZVS waveform under buck mode is shown on the Fig. 13(b). Also, as it can be seen from the Fig. 13(a), the average voltage of flying capacitor is half of input voltage which can demonstrate the automatic voltage balance function for the flying capacitor.

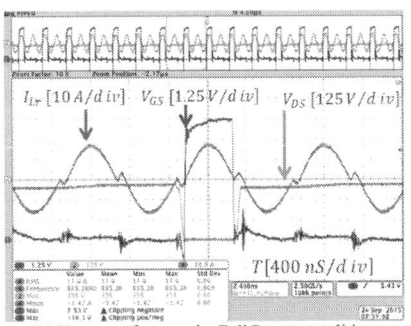

(a).Key waveform under Full Power condition

Fig. 12. Operation waveform under LLC mode

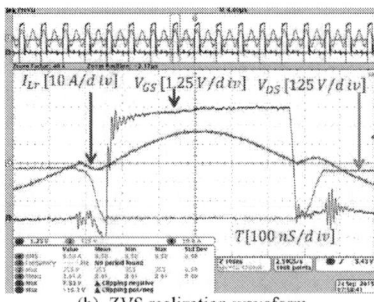

(b). ZVS realization waveform

Fig. 12. Operation waveform under LLC mode

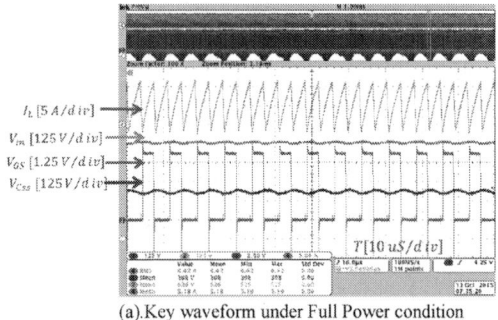

(a).Key waveform under Full Power condition

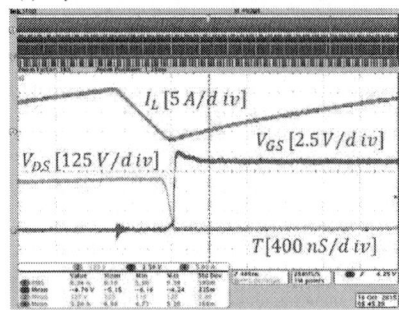

(b). ZVS realization waveform

Fig. 13. Operation waveform under Buck mode

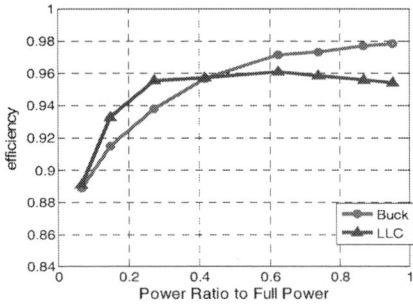

Fig. 14. The efficiency under two operation modes

The efficiency curve is given in the Fig. 14. The peak efficiencies are 96.11% and 97.83% under LLC mode and buck mode respectively.

VI. CONCLUSION

S

Overall, the proposed topology is a good candidate for the dual output battery charger application. The interleaved function takes advantage of fast switching behavior of GaN HFETS. Using the same magnetics for both stages, as either a transformer or an output inductor, helps to realize the high power density required in aerospace applications. Both operation modes can realize the ZVS for high side switches. The power stage achieved high power density including all the PCB boards and heat-sink.

ACKNOWLEDGMENT

This work was supported by Boeing Company. This work also made use of Engineering Research Center Shared Facilities supported by the Engineering Research Center Program of the National Science Foundation and the Department Of Energy under NSF Award Number EEC-1041877 and the CURENT Industry Partnership Program. The author Ren Ren gratefully acknowledges the financial support for studying abroad from China Scholarship Council.

REFERENCES

[1] Edward A. Jones, Fred Wang, Burak Ozpineci, "Application-Based Review of GaN HFETs" Wide Bandgap Power Devices and Applications (WiPDA), IEEE Workshop on, 2014,pp. 24-29

[2] Ji, S., D. Reusch and F.C. Lee, "High-Frequency High Power Density 3-D Integrated Gallium-Nitride-Based Point of Load Module Design." IEEE Transactions on Power Electronics, 2013. 28(9): p. 4216-4226.

[3] Huang, D., S. Ji and F.C. Lee, "LLC Resonant Converter With Matrix Transformer." IEEE Transactions on Power Electronics, 2014. 29(8): p. 4339-4347.

[4] Weimin Zhang, Zhuxian Xu, Zheyu Zhang, Fred Wang, Leon M.Tolbert, Benjamin J. Blalock, "Evaluation of 600 V cascode GaN HEMT in device characterization and all-GaN-based LLC resonant converter," the 19th Energy Conversion Congress and Exposition (ECCE), pp. 3571- 3578, Denver, CO, 2013.

[5] Yin Wang, Woochan Kim, Zhemin Zhang, Jesus Calata, and Khai D.T Ngo, "Experience with 1 to 3 Megahertz Power Conversion using eGaN FETs", the 28th Annual IEEE Applied Power Electronics Conference and Exposition, pp. 532- 539, Long Beach, CA, 2013.

[6] Lee, I. and G. Moon, "Analysis and Design of a Three-Level LLC Series Resonant Converter for High- and Wide-Input-Voltage Applications." IEEE Transactions on Power Electronics, 2012. 27(6): p. 2966-2979.

[7] Gu, Y., et al., "Three-Level LLC Series Resonant DC/DC Converter." IEEE Transactions on Power Electronics, 2005. 20(4): p. 781-789.

[8] Wei Chen, Yilei Gu, Zhengyu Lu, "A Novel Three Level Full Bridge Resonant Dc-Dc Converter Suitable for High Power Wide Range Input Applications," IEEE APEC Proceedings, 2007, pp. 373-379.

[9] Q. Shi, H. Hu, W. Xu, and J. Yong, "Low-order harmonic characteristics of photovoltaic inverters," Int. Trans. Electr. Energ. Syst., doi: 10.1002/etep.2085.

[10] Reusch, D.; Lee, F.C.; Ming Xu, "Three level buck converter with control and soft startup," in Energy Conversion Congress and Exposition, 2009. ECCE 2009. IEEE , vol., no., pp.31-35, 20-24 Sept. 2009

[11] Edward A. Jones, Fred Wang, "Characterization of an enhancement-mode 650-V GaN HFET," IEEE Energy Conversion Congress and Exposition (ECCE), 2015.

[12] Zhengyang Liu; Xiucheng Huang; Wenli Zhang; Lee, F.C.; Qiang Li, "Evaluation of high-voltage cascode GaN HEMT in different packages." IEEE APEC Proceedings, 2014, pp. 168-173.

[13] T. Wu, "Cdv/dt Induced Turn-On In Synchronous Buck Regulators", white paper, International Rectifier Corporatio

Quadruple Active Bridge DC-DC Converter as the Basic Cell of a Modular Smart Transformer

Levy F. Costa, Giampaolo Buticchi and Marco Liserre, *Fellow*, IEEE
Christian-Albrecht-University of Kiel (Uni-Kiel) / Power Electronics Chair (PE)
Kaiserstr. 2, 24143, Kiel, SH, Germany
Email: {lfc, gibu, ml}@tf-uni-kiel.de

Abstract—One of the main challenges of a Solid-State transformer (SST) lies in the dc-dc conversion stage. In this work, a Quadruple-Active-Bridge (QAB) dc-dc converter is investigated to be used as the basic module for the whole dc-dc stage. Besides the feature of high power density and soft-switching operation, the QAB converter provides a solution with a reduced number of high frequency transformers, since more bridges are connected to the same multi-winding transformer. To ensure soft-switching in the full operation range of the converter, two modulation strategies are investigated: the phase-shift modulation and the triangular current modulation. The theoretical analysis is developed for both modulation strategies and a comparison between them is carried out. In order to validate the theoretical analysis, a 20 kW prototype was built and tested.

I. INTRODUCTION

In recent years, the smart grid technologies have received more and more attention, as a feasible solution to manage in an efficient way the increased demand and the high penetration of distributed generation (DG). One of these technologies is the Smart Transformer, which is a Solid-State Transformer [1] - [2] with control and communication functionality [3]. This power electronics based system uses a high frequency (HF) transformer, reducing volume and weight, and it can also provide ancillary services to the grid, such as: power factor correction, active filtering, VAR compensation, electronics protection and disturbance rejection [1] - [2].

The three-stage ST is usually composed of a Medium-Voltage (MV) ac-dc stage, a HF isolated dc-dc stage and a Low-Voltage (LV) dc-ac stage. The main challenge of this architecture is the dc-dc conversion stage, since it has strict requirements, such as: high rated power, high current capability in LV side, high voltage capability in HV side, high frequency isolation and high efficiency.

To meet all these requirements, two solutions have been widely investigated: the first one is to use standard converter with high voltage rating devices [4] - [5], while the second one is based on the modular concept, in which several modules are used to share the total voltage and power among them [6] - [8]. Although the modular solution presents a high component count, it has several advantages compared to the fist solution, such as: low dv/dt (low EMI emission), possibility to use standard low voltage rating devices and also modularity, which enables fault tolerance capability.

Several converters have been investigated to be used as module of the main core of the ST, but the Dual-Active-Bridge

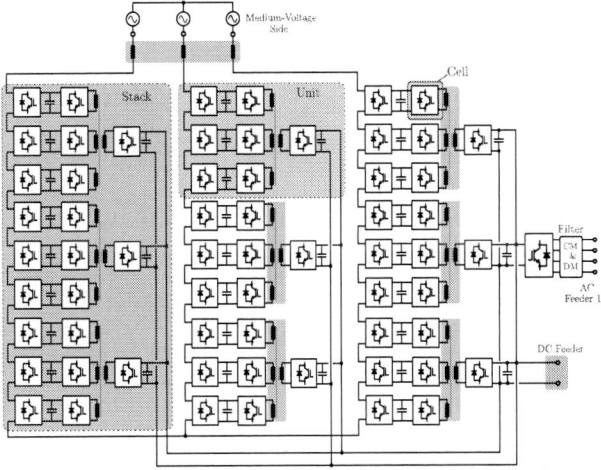

Fig. 1. Modular smart transformer architecture using the QAB converter as a basic module of the dc-dc conversion stage.

TABLE I
SPECIFICATION OF A SMART TRANSFORMER FOR DISTRIBUTION SYSTEM

Specification	Smart Transformer	Specification	QAB
Rated Power	1 MVA	Rated Power	111 kW
Input Voltage	400 V	Input voltage (LV)	700 V
Output Voltage	10 kV	Output voltage (MV)	1.13 kV
LVDC link	700 V	Switching frequency	20 kHz
MVDC link	10.2 kV	IGBT	1700 V

(DAB) and the Series-Resonant converter (SR) have received more attention, due their advantages of soft-switching, high efficiency and power density [9] - [8]. The Series-Resonant dc-dc converter presents a well regulated output voltage for wide range of load (when operating in discontinuous-conduction-mode), avoiding the requirement of control loops. For that reason, it is also called dc-transformer [6] - [8]. On the other hand, when the output voltage control or power flow control is required, the DAB converter is more advantageous, since it enables the active control of the transferred power [9] - [11].

The Multiple Active Bridge (MAB) is an alternative solution for the DAB or SRC. This kind of converter was firstly introduced in [12] and it was applied in a solid-state transformer in [13], in order to connect renewable energy

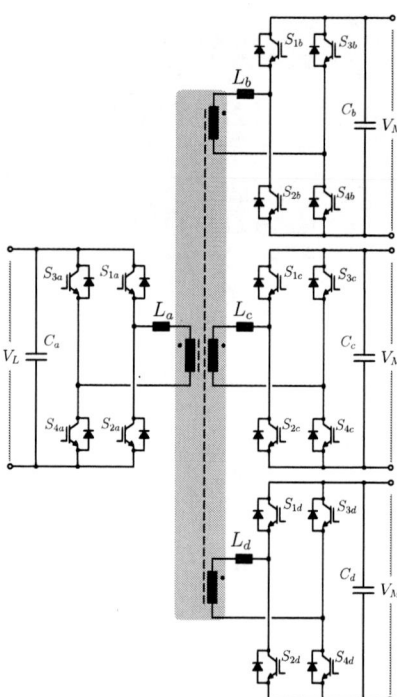

Fig. 2. Topology of the Quad-Active-Bridge dc-dc converter.

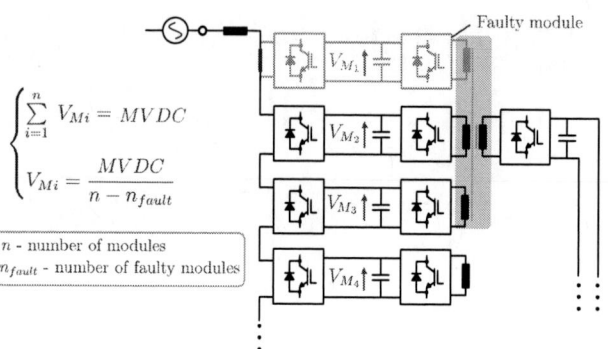

$$\begin{cases} \sum_{i=1}^{n} V_{Mi} = MVDC \\ V_{Mi} = \dfrac{MVDC}{n - n_{fault}} \end{cases}$$

n - number of modules
n_{fault} - number of faulty modules

Fig. 3. Simplified diagram of the ST, illustrating the fault operation of the system, with one faulty cell.

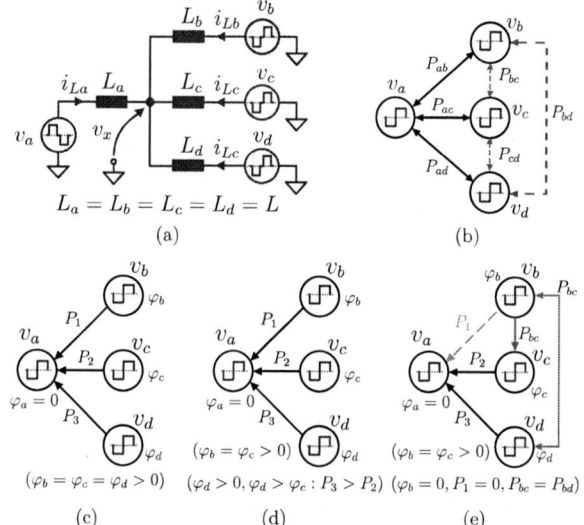

$L_a = L_b = L_c = L_d = L$

(a)

(b)

$(\varphi_b = \varphi_c = \varphi_d > 0)$

(c)

$(\varphi_d > 0, \varphi_d > \varphi_c : P_3 > P_2)$

(d)

$(\varphi_b = 0, P_1 = 0, P_{bc} = P_{bd})$

(e)

Fig. 4. Equivalent circuit of the QAB converter and possible power transfer path among the actives bridges: (a) Equivalent circuit. (b) All possible power path of the converter. (c) Normal balanced condition case, where the LV cell a receives equal power from the three MV cells (b, c and d), (load: cell a, sources: cell b, c and d). (d) Unbalanced condition case, where the LV cell a receives different power from the MV cells (b, c and d), and in this example the cell d delivers more power than the cells b and c; (load: cell a, source: cell b, c and d). (e) Unusual case, where the MV cell b operates as a source, giving energy to cells (b and c), (load: cell a, sources: cell b, c and d). In the last case, $P_1 = 0$.

sources and storage system to the grid. The MAB converter has the same features of the DAB converter, with the additional characteristic to reduce the number of the HF transformers, since the MAB converter integrates more active bridges into a single transformer. In this context, this paper investigates the application of the Quad-Active-Bridge (QAB) dc-dc converter as a basic converter to build the entire ST system.

The QAB converter is composed of four active bridges connected to the same HF transformer. Fig. 1 shows the ST architecture using the QAB converter as a basic module of the dc-dc stage, while Fig. 2 shows the topology of the QAB converter. A basic specification of the QAB converter for distribution system application is presented in Table I, where a very high Medium Voltage DC (MVDC) link is required. In order to share the MVDC link with the maximum number of modules, three bridges of the QAB converter are connected to the MV side, while only one is connected in the LV side, as illustrated in Fig. 1. Table I lists the specification of the QAB converter.

The main contribution of this paper is the investigation of the QAB converter as a basic cell for a modular ST application. To control the converter, the well-known phase-shift modulation is initially considered. Nevertheless, this modulation method has some limitations regarding the soft-switching operating range, when the output voltage varies. For this reason, an alternative scheme based on the triangular current modulation is also investigated in this paper. In section II, the operation principle of the converter is presented and both modulations schemes are described. The semiconductors

current effort and transformer current effort are presented and discussed in Section III. In section IV, the operation of the QAB converter with unbalanced load in the MV cells and the impact on the structure is discussed. Finally, experimental results obtained for both modulation strategies under investigation are presented in Section V.

II. OPERATION PRINCIPLE

The QAB is composed of four active bridges and for the analysis, each of them is denoted by the letters a, b, c and d. The elements of the bridges have sub-index $i = \{a, b, c, d\}$ to indicate the bridge the element belongs to, as depicted in Fig. 2. In ST application, the bridge a is connected to the LV side, while the bridges b, c and d are connected to the MV side. All bridges can exchange power among themselves and the possible power paths are depicted in Fig. 4 (b), where each bridge is symbolized by the voltage source v_i.

In standard operation, the power flows from the MV to the LV side and the bridges b, c and d process the same amount of power, i.e. balanced power condition, given by P_1, P_2 and P_3, respectively. This point of operation is represented in Fig. 4 (c).

The unbalanced condition, where the powers P_1, P_2 and P_3 are positive, but not equal, is also a common operation in ST application and this situation is depicted in Fig. 4 (d). This condition can happen in case of a dynamic response of the system, where the MV cells process unbalanced instantaneous power or even intentionally to perform other kinds of optimization.

An unusual operation is depicted in Fig. 4 (e), where the power flows not only from the MV to the LV side, but also from one MV cell to the others MV cells. In this situation, a MV cell operates as source and load. Although this situation is very unlikely, it is presented here to demonstrate the versatility of the QAB converter in ST application.

Similarly to the DAB, there are several possibilities to modulate the QAB converter. Originally, the classical phase-shift modulation strategy has been applied to most multiple active bridge solutions present in literature.

Nevertheless, a different modulation strategy based on the Triangular Current Modulation (TCM) was applied to the QAB converter in [14], as a possible solution to increase the soft-switching range in variable output voltage conditions. In standard operation of the ST, variation on the QAB converter output voltage is not expected. However, in fault case one or more faulty cells are disconnected from the system, as depicted in Fig. 3, and then all the remaining cells must reconfigure the system adjusting its output voltage, in order to keep the MVDC link in the correct value after the fault. For this condition, the PSM might operate with hard switching and this motivates the investigation of the TCM for QAB converter. In this paper these two modulation strategies will be compared.

To analyze the converter, an equivalent circuit based on the Y-model and depicted in Fig. 4 (a) is used, in which the bridges are replaced by rectangular voltage sources (v_a, v_b, v_c and v_d). The voltage on the central point v_x and the current slope of each inductor are given by (1) and (2), respectively, where $k = \{a, b, c, d\}$.

$$v_x = \frac{v_a + v_b + v_c + v_d}{4} \quad (1)$$

$$\frac{di_{Lk}}{dt} = \frac{(v_k - v_x)}{L} \quad (2)$$

A. Phase-Shift Modulation Strategy

For phase-shift operation, rectangular voltages v_a, v_b, v_c and v_d with phase shift φ_a, φ_b, φ_c and φ_d, respectively, and constant switching frequency f_s are applied to the transformer. The power is controlled by the phase difference among the bridges and it can be generally described as

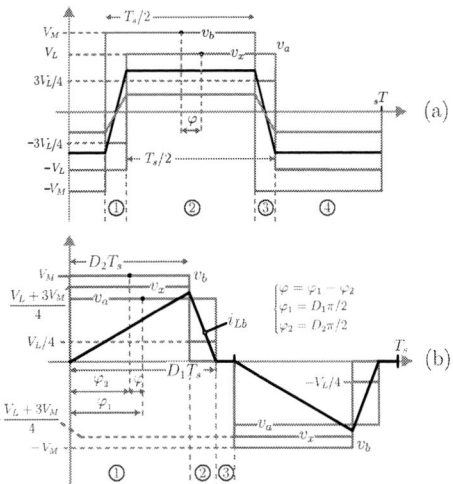

Fig. 5. Main waveforms of the QAB converter, considering positive power flow (from the MV to the LV side): (a) PSM, (b) TCM.

$$P_{ik} = \frac{V_M V_L}{2\pi f_s L n} \varphi_{ik} \left(1 - \frac{|\varphi_{ik}|}{\pi} \right), \quad \varphi_{ik} = \varphi_i - \varphi_k \quad (3)$$

where, $i = a, b, c, d$ and $k = a, b, c, d$, according to [13]. In this work, the LV bridge is used as the reference cell, thus it is defined that $\varphi_a = 0$. In the case, the power exchanged among the cells a and b is given by (4). The main waveform of the PSM is shown in Fig. 5 (a).

$$P_1 = \frac{V_M V_L}{2\pi f_s L n} \varphi_b \left(1 - \frac{|\varphi_b|}{\pi} \right) \quad (4)$$

The PSM is characterized by ZVS turn-on, but this features depends on the input and output voltages relation and also the load [15]. As the input and output voltage are considered constant, the converter can be properly designed to work with ZVS operation for its entire range of operation. As a disadvantage, a high level of reactive power that circulates in the high frequency transformer when the phase-shift operation angle is high. Therefore, to avoid high reactive circulating current, a relatively low nominal phase-shift angle must be chosen on the converters design.

B. Triangular Current Modulation Strategy

The TCM was previously applied to the DAB converter in [16] and then extended to the QAB converter in [14]. Differently from the previous modulation strategy, the TCM uses the duty-cycle to control the power transferred among the bridges. Using the TCM, the switches of the MV cells can operate with ZCS and the circulating reactive power on the converter can be reduced. However, the current on the semiconductors and transformer can have a high root-mean-square (rms) value, due to the triangular current shape with high peak, impacting on the conduction losses.

In this analysis, it is considered that the QAB converter operates in balanced condition, i.e. $P_1 = P_2 = P_3$ and the LV bridge duty-cycle is given by D_1, while the voltage of the MV cells b, c and d are given by D_2, D_3 and D_4, respectively. As the balanced condition is considered, then $D_2 = D_3 = D_4$.

The basic principle of the TCM is to impose a triangular current on the inductors, as shown in Fig. 5 (b). To achieve that, the voltages v_a and v_b should have the waveforms depicted in Fig. 5 (b) and consequently the voltage v_x is

$$
v_x = \begin{cases} \dfrac{nV_L + 3V_M}{4}, & 0 < t < d_2 T_s \\[2mm] \dfrac{nV_L}{4}, & d_2 T_s < t < d_1 T_s \\[2mm] 0, & d_1 T_s < t < T_s/2 \end{cases} \tag{5}
$$

As can seen in Fig. 5 (b), the current in the inductor starts from zero and reaches its maximum value Δi_{Lb} during the period of time between $0 < t < D_2 T_s$, where T_s is the switching period.

The current variation during this period, denoted as $\Delta i_{Lb_{(0<t<D2Ts)}}$, can be calculated by using (2), where v_x is given by (5). As the currents start from zero, the switches S_2 and S_3 of the LV side and also the switches S_5 and S_6 of the MV side turn-on at Zero-Current-Switching (ZCS).

Likewise, during the period $0 < t < D_2 T_s$, the currents decrease from Δi_{Lb} until reaches zero again. For this period, the current variation denoted as $\Delta i_{Lb_{(D2Ts<t<Ts)}}$ can also be calculated by (2). In the moment $t = D_2 T_s$, the current is zero and the switch S1 turned-off at ZCS.

To achieve ZCS operation regardless the load and input or output voltages levels, the condition $\Delta i_{Lb_{(0<t<D2Ts)}} = \Delta i_{Lb_{(D2Ts<t<Ts)}}$ must be satisfied. As a results, the relation between the duty-cycle D_1 and D_2 is found and presented in (6).

The total power transferred from the LV to the MV is given by (7) [14]. Thus, the duty-cycle of the LV bridge D_1 can be used to control the power transferred from the MV to the LV side, while the duty-cycle of the MV bridges, $D_{2,3,4}$ can be calculated to ensure the ZCS operation of the converter.

$$
D_2 = \frac{V_L \cdot n}{V_M} D_1 \tag{6}
$$

$$
P_{(tot)} = \frac{3 D_1{}^2 (V_L n)(V_M - V_L n)}{4 L f_s} \tag{7}
$$

III. Semiconductors and Transformer Effort

In order to evaluate the performance in terms of efficiency of the QAB converter with both modulation strategies, the current effort on the semiconductors and transformer, as well as the power losses in these elements are calculated and compared.

For the PSM, the current waveforms on the primary side transformer and also on the semiconductors of the LV bridge and MV bridge are depicted in Fig. 6. To calculate the rms and average value of these waveforms, the equations (8) and (9) are used. Similarly, for the TCM, the current waveforms are shown in Fig. 6. The current efforts on the semiconductors and

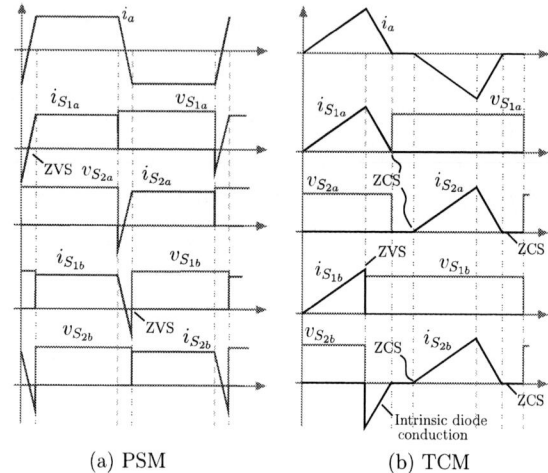

(a) PSM (b) TCM

Fig. 6. Current waveform on the semiconductors and transformer of the QAB converter, for both analyzed modulation schemes.

transformer for this modulation method were demonstrated in [14] and the equations are used in this work. The efforts are calculated taking into account the specification shown in Table I and the results are presented in Fig. 7. To evaluate the performance of the converter for the entire possible range of operation, a maximum phase shift of $\varphi = 90°$ for the PSM and a maximum duty-cycle of $D = 0.5$ for the TCM were considered. The inductor was designed in order to have the maximum power at maximum control variable (phase angle for the PSM and duty-cycle for the TCM), resulting a $L = 80\mu H$ for PSM and $L = 22.5\mu H$ for TCM.

$$
i_{S1a,rms} = \sqrt{\frac{1}{T_s} \int_0^{T_s} i_{S1a}{}^2 (t)\, dt} \tag{8}
$$

$$
i_{La,avg} = \frac{1}{T_s} \int_0^{T_s} i_{La}(t)\, dt \tag{9}
$$

As can be seen in Fig. 7 (a), the rms current in the semiconductor of the MV cell are always lower for the PSM, compared to the TCM. However, for the primary side current, there is a region where the TCM presents lower rms current. The main reason for that is the high reactive power on the converter due to the high phase shift operation angle. This effect is also observed in the rms current on the transformer primary side, depicted in Fig. 7 (c).

The average current through the semiconductors for both modulation methods are shown in Fig. 7 (b). According to this graphic, the primary side semiconductors carry the same average current for both modulation methods. However, the average current on the secondary side semiconductors are higher for the TCM method. Therefore, for this graphics, higher conduction losses for the TCM compared to the PSM are expected.

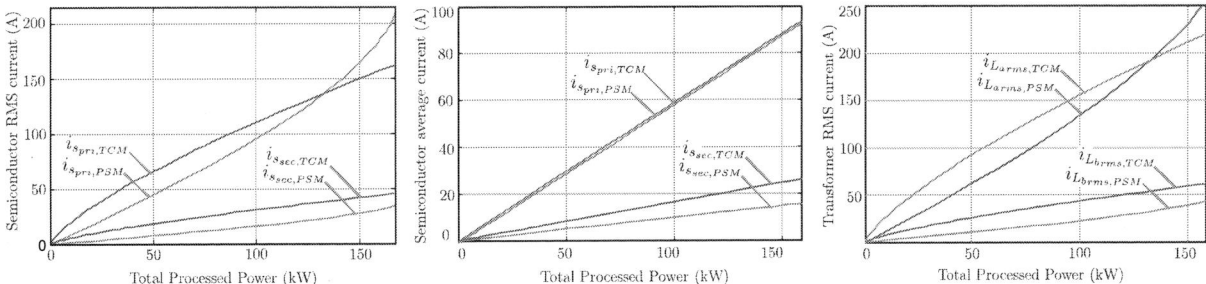

Fig. 7. Current effort on the semiconductor and transformer as function of the total transferred power for both modulation methods and considering balanced condition: (a) rms current on the semiconductors, (b) average current on the semiconductors, (d) rms current on the transformer primary side.

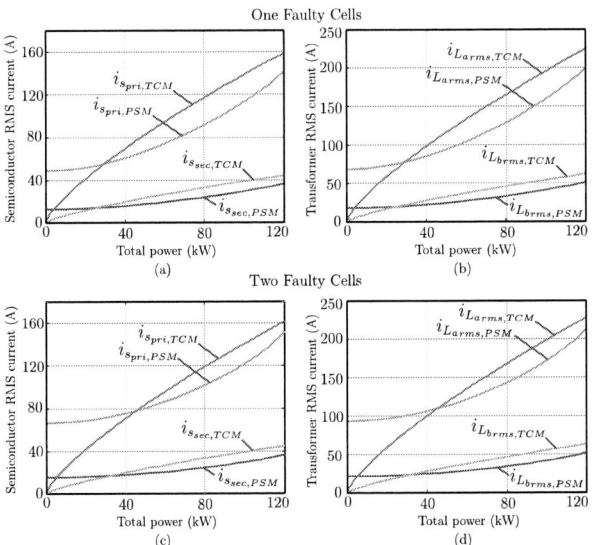

Fig. 8. Current effort on the semiconductor and transformer for fault case.

In a fault case, the output voltage of the remaining QAB converter cells must be adjusted in order to compensate the faulty cell, providing a regulated MVDC link after the fault. The variation of the QAB converter output voltage has impact on the current efforts of the semiconductors and transformer. Fig. 8 shows the current effort on the semiconductor and transformer, in case of a fault of one and two cells. As can be noticed, the current efforts increase significantly for low power (lower than 25% of the total power) when the QAB operates with the PSM and one faulty cell. In case of two faulty cells and PSM operation, the current efforts is even higher for processed power lower than 43% of the total power. Thus, in faulty operation, it is more advantageous to employ the TCM in case of light load. Moreover, in this condition the ZVS operation achieved with the PSM is not ensured anymore, and switching losses will be added in the converter losses.

IV. UNBALANCED CONDITION

In the theoretical analysis performed so far, steady-state and balanced condition is considered, i.e. $P_1 = P_2 = P_3$.

Nevertheless, situation in which different power level are processed by the MV cells will happen often in ST. For that reason, the performance of the converter during unbalanced condition is also studied in this work.

Fig. 9 shows the main voltage and current waveforms of the QAB converter for unbalanced condition. Regardless the modulation strategy, the inductor current waveform might have a high rms value in highly unbalanced operation, as those cases presented in [13] and [12]. However, for ST application, it is considered that the power flowing in each bridge does not change its direction, but only the power level is changed, as the case presented in Fig. 4 (c). Thus, the MV currents (i_{Lb}, i_{Lc} and i_{Ld}) waveforms will be modified, but the impact on the i_{La} current is minimum.

For the PSM, the main waveforms are shown in Fig. 9 (a) for a realistic situation in unbalanced condition, in which one MV cell processes around 30% more power than the other MV cells. The imbalance can be easily observed on the current waveforms i_{Lb} and i_{Lc}, but slightly noticed on the current i_{La}. This represents a great advantage of this structure, since an imbalance does not affect significantly the LV cell, which is responsible for the major losses of the converter, due to its high current. Fig. 10 (a) shows the current effort on the semiconductors of the cells b and c, when the QAB converter operates with PSM and an imbalance of 15% (i.e. one MV cell c process 15% more power then the others) and 30%. This graphic was plotted using the same specifications and parameters already presented in Section III.

For the TCM scheme, the main waveforms are shown in 9 (b) for a condition where the bridge c procesess 30% more power than the others bridges. In this case, the duty-cycle D_3 of the cell c is slightly higher than the D_2 and D_4. It is important to note in this figure that the imbalance causes the loss of the ZCS, adding switching losses to the converter, since the current starts from a small constant dc value, but different from zero.

As can be observed in Fig. 9 (b), the voltage v_x has one more level, caused by the duty-cycle difference given by $\Delta D = D_3 - D_2$, and this additional level makes the current i_b decrease and the current i_c increase. Afterward, both currents decrease with the same slope, but they reach different values, because they started from different points. As a conclusion, the

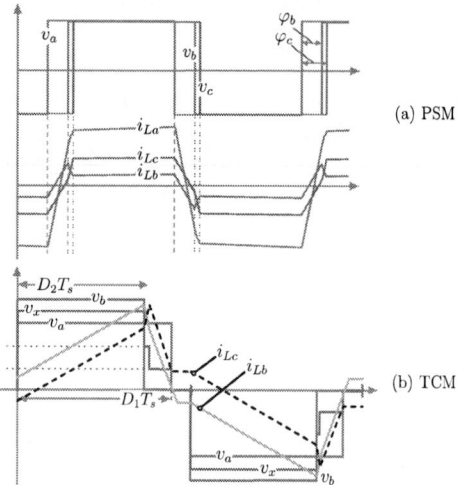

Fig. 9. Main current and voltage waveforms of the QAB converter for unbalanced condition.

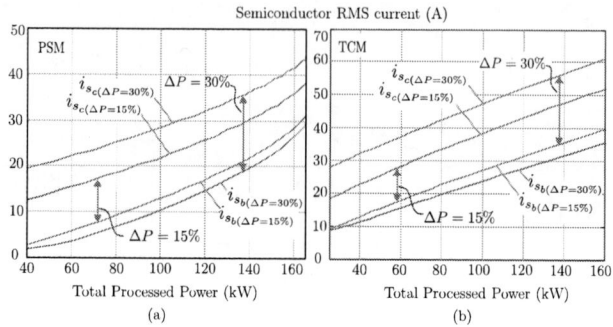

Fig. 10. Semiconductors effort of the QAB converter for an imbalance of 15% and 30%.

variable ΔD has direct impact on the dc value of the currents, and consequently the additional power delivered by the MV cell. For that reason, this new variable can be used to control the power exchange among the MV cells, as described in [14].

Fig. 10 (a) shows the current effort on the semiconductors of the cells b and c, when the QAB converter operates with TCM and an imbalance of 15% (i.e. one MV cell processing 15% more power then the others) and 30%, in order to point out the imbalance on the semiconductors effort.

V. EXPERIMENTAL RESULTS

In order to verify the operation and evaluate the performance of the QAB converter with both modulation strategies, a 20 kW prototype was designed, and the converter performance was verified experimentally.

The converter specifications, as well as the selected components used in the prototype are shown in Table II.

As can be noticed in Section III, the required inductance for the TCM is lower than the value required for the TCM. Likewise, for the PSM the transformer turn ratio should be chosen in order to have the same voltage level on the primary side and reflected secondary side. However, this situation must be avoided for the TCM. As the experimental results were obtained for both modulation methods using the same prototype, different output voltages were utilized, according to the modulation strategy, in order to have similar performance.

The experimental results consist of the relevant voltage and current waveforms for steady-state operation of the QAB converter using both presented modulation methods. The main waveforms obtained from the prototype are shown in Fig. 11. Considering the limitations of dc voltage sources, the LV port was connected to the DC supply and the ohter ports were connected to loads (reverse power flow operation).

The results of the converter operating with TCM for balanced and unbalanced loads are shown in Fig. 11 (a) and (b),

respectively. In Fig. 11 (a), the converter processes around 2.2 kW, divided equally among the MV cells. As expected, the current and voltage waveforms are very similar to the theoretical waveforms and the ZCS feature can be noticed in the figure. Fig. 11 (b) shows the results for unbalanced condition, where the bridges b and d process 640 W, while the bridge c process 480 W, resulting in a total power of approximately 1.8 kW. For safety reasons (open semiconductor modules without the insulating gel), the test was performed with voltage of 150V in the secondary side. As can be seen in Fig. 11 (b), the ZCS feature is lost.

Similarly, Fig. 11 (c) and (d) show the results for the QAB converter operating with PSM. It is important to note that the PSM modulation results were obtained with switching frequency of 40 kHz.To ensure ZVS operation, the results were obtained for the same input and output voltage of 200 V. The results for balanced load condition is depicted in Fig. 11 (c) and it is presented just to point out the proper operation of the QAB converter. In Fig. 11 (c), the imbalance can be noticed on the waveform of the current i_{Lb}. However, this effect is not very evident on the current i_{La}.

VI. CONCLUSION

In the framework of Solid-State transformer development, modular architectures will play a fundamental role. In this paper, a quadruple active bridge converter is employed as a basic cell of the dc-dc conversion stage of the ST. The theoretical analysis of the QAB converter was carried out considering the phase-shift modulation scheme and the triangular current modulation. From the comparison it is observed that in standard operation the phase-shift modulation presents lower current effort on the semiconductors and transformer, compared to the triangular current modulation, implying also reduced conduction losses. However, in a fault case, where the system must be reconfigured and the MV cells must work with higher voltage, the triangular current modulation is more advantageous.

Moreover, the theoretical analysis and experimental results showed that unbalanced load in the MV cells has an insignificant effect on the primary side current, which is responsible for the major converter losses. In other words,

TABLE II
SPECIFICATION OF THE QAB DC-DC CONVERTER PROTOTYPE

Parameter	Value
Rated Output Power	$P_o = 20kW$
LV dc-link (V_a)	$V = 200V$
MV dc-link (V_b, V_c, V_d)	$V = 250V$
Switching frequency	$f_s = 20kHz$
Leakage inductance	$L_a = L_b = L_c = L_d = 35\mu H$
Transformer turn ratio	$n = 1:1:1:1$
Output filter capacitor	$C = 500\mu F$
IGBT $S_{1,2,3,4}$	Infineon SIGC32T120R3E IGBT3 (1200V/25A)

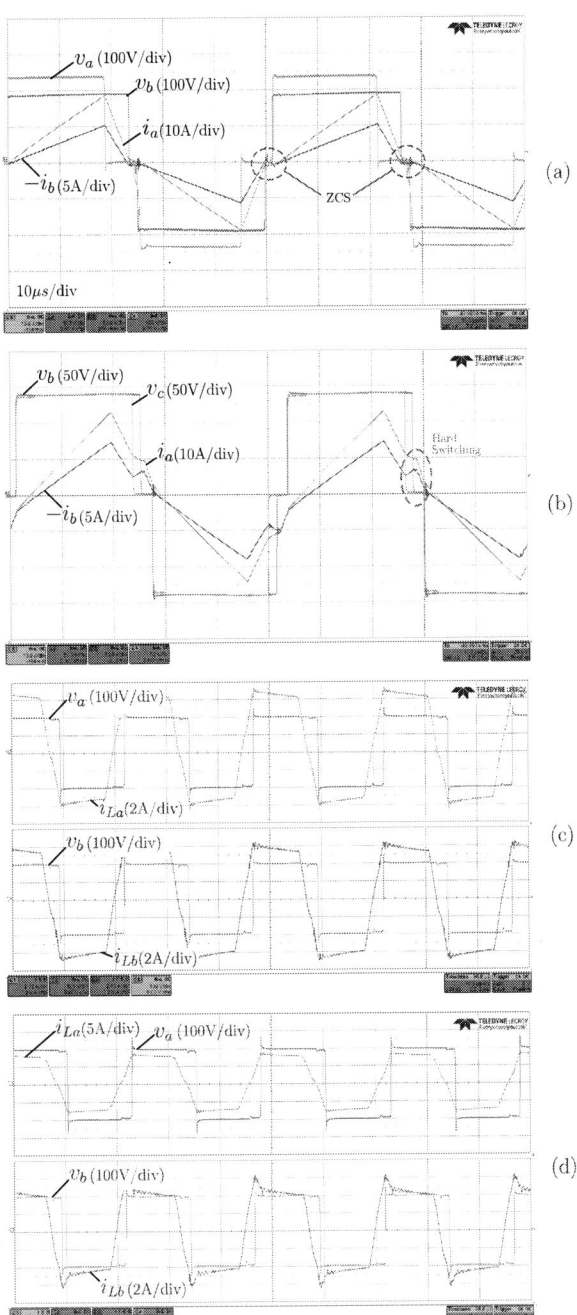

Fig. 11. Experimental results of the QAB converter using both modulation strategies and considering balanced and unbalanced conditions: (a) for the TCM and balanced load condition, (b) for the TCM and imbalance of 25%, (c) for the PSM and balanced load condition, (d) for the PSM and imbalance on the MV side.

for the configuration used in this work, the current effort on the primary side and consequently the conduction losses is unaffected by the imbalance in the MV cells.

REFERENCES

[1] X. She, R. Burgos, G. Wang, F. Wang, and A. Huang, "Review of solid state transformer in the distribution system: From components to field application," in *IEEE Energy Conversion Congress and Exposition (ECCE)*, Sept 2012, pp. 4077–4084.

[2] J. W. Kolar and G. Ortiz, "Solid-state-transformers: Key components of future traction and smart grid systems," in *Proceedings of the International Power Electronics Conference - ECCE Asia (IPEC 2014)*, May 2014.

[3] G. De Carne, M. Liserre, K. Christakou, and M. Paolone, "Integrated voltage control and line congestion management in active distribution networks by means of smart transformers," in *IEEE 23rd International Symposium on Industrial Electronics (ISIE)*, June 2014, pp. 2613–2619.

[4] R. W. D. D. N. Soltau, R. U. Lenke, "High-power dc-dc converter," Technical Report, Rheinisch-Westflische Technische Hochschule, 2013.

[5] D. Rothmund, J. Huber, and J. Kolar, "Operating behavior and design of the half-cycle discontinuous-conduction-mode series-resonant-converter with small dc link capacitors," in *Control and Modeling for Power Electronics (COMPEL), 2013 IEEE 14th Workshop on*, June 2013, pp. 1–9.

[6] D. Dujic, G. Steinke, E. Bianda, S. Lewdeni-Schmid, C. Zhao, and J. Steinke, "Characterization of a 6.5kv igbt for medium-voltage high-power resonant dc-dc converter," in *Applied Power Electronics Conference and Exposition (APEC), 2013 Twenty-Eighth Annual IEEE*, March 2013, pp. 1438–1444.

[7] D. Dujic, A. Mester, T. Chaudhuri, A. Coccia, F. Canales, and J. Steinke, "Laboratory scale prototype of a power electronic transformer for traction applications," in *Power Electronics and Applications (EPE 2011), Proceedings of the 2011-14th European Conference on*, Aug 2011, pp. 1–10.

[8] C. Zhao, D. Dujic, A. Mester, J. Steinke, M. Weiss, S. Lewdeni-Schmid, T. Chaudhuri, and P. Stefanutti, "Power electronic traction transformer:medium voltage prototype," *Industrial Electronics, IEEE Transactions on*, vol. 61, no. 7, pp. 3257–3268, July 2014.

[9] X. She, S. Lukic, A. Huang, S. Bhattacharya, and M. Baran, "Performance evaluation of solid state transformer based microgrid in freedm systems," in *Applied Power Electronics Conference and Exposition (APEC), 2011 Twenty-Sixth Annual IEEE*, March 2011, pp. 182–188.

[10] H. Fan and H. Li, "High-frequency transformer isolated bidirectional dc-dc converter modules with high efficiency over wide load range for 20 kva solid-state transformer," *Power Electronics, IEEE Transactions on*, vol. 26, no. 12, pp. 3599–3608, Dec 2011.

[11] B. Zhao, Q. Song, and W. Liu, "A practical solution of high-frequency-link bidirectional solid-state transformer based on advanced components in hybrid microgrid," *Industrial Electronics, IEEE Transactions on*, vol. 62, no. 7, pp. 4587–4597, July 2015.

[12] H. Tao, J. Duarte, and M. Hendrix, "Three-port triple-half-bridge bidirectional converter with zero-voltage switching," *IEEE Transactions on Power Electronics*, vol. 23, no. 2, pp. 782–792, March 2008.

[13] S. Falcones, R. Ayyanar, and X. Mao, "A dc-dc multiport-converter-based solid-state transformer integrating distributed generation and storage," *IEEE Transactions on Power Electronics*, vol. 28, no. 5, pp. 2192–2203, May 2013.

[14] L. F. Costa, G. Buticchi, and M. Liserre, "Quad-active-bridge as cross-link for medium voltage modular inverters," in *IEEE Energy Conversion Congress and Exposition (ECCE)*, Sept 2015, pp. 645–652.

[15] R. De Doncker, D. Divan, and M. Kheraluwala, "A three-phase soft-switched high-power-density dc/dc converter for high-power applications," *IEEE Transactions on Industry Applications*, vol. 27, no. 1, pp. 63–73, Jan 1991.

[16] N. Schibli, "Symmetrical multilevel converters with two quadrant dc-dc feeding," Ph.D. dissertation, Swiss Federal Institute of Technology, 2000.

Analytical Model of a Phase-Shift Controlled Three-Level Zero-Voltage Switching Converter

Cas Bakker
Prodrive Technologies
cas.bakker@prodrive-technologies.com

Bas Vermulst
Eindhoven University of Technology
Electromechanics and
Power Electronics group
b.j.d.vermulst@tue.nl

Anton Driessen
Prodrive Technologies
anton.driessen@prodrive-technologies.com

Abstract—Soft-switching of resonant dc-dc converters is used to enable high switching-frequencies and miniaturization of filter components. These converters are typically modeled using a first-harmonic approximation. This results in generalization of the operating modes, by only using the first harmonic of the converter. In this paper a phase-shift controlled series-resonant converter is analyzed per operating mode using an accurate analysis. For each operating mode, an analytical expression is derived resulting in a more accurate representation of the converter behavior when compared to first-harmonic approximation. Experimental results show that the model accurately predicts the converter.

I. INTRODUCTION

Switched-mode resonant power-converters are widely used in dc-dc power conversion, since these converters can be operated with higher switching frequencies by employing Zero-Voltage Switching (ZVS) and or Zero-Current Switching (ZCS). The converter proposed in [1] utilizes ZVS and maintains this over the full operating range by means of ancillary inductors. A prototype converter based on this topology, see Fig. 1, will be used to verify the proposed model. The topology can be applied in fields where a low electromagnetic emission, short response time and high efficiency are required such as automotive battery chargers and high-performance isolated medical supplies.

Resonant converters are typically modeled by applying first-harmonic approximation, as used in [2] and [3], which results in a rudimentary expression for the converter parameters such as output voltage, current and switching currents. Since approximations are used, the results are only valid within a certain operating region and for ZVS converters no certainty is acquired on whether ZVS is achieved. When using first-harmonic approximation, the information regarding different operating modes is removed by assuming that the converter response is equal over these modes. Moreover, only the first harmonic of the applied voltages is taken into account, assuming that the power transfer of the full converter is achieved through this first harmonic. The used harmonic in first-harmonic approximation is the switching frequency and not the free-oscillation frequency of the resonance of the converter. Since the actual waveforms of the converter are interrupted segments of the free-oscillation harmonic, the outcome of the analysis cannot be the same as the behavior of the converter. By only taking the first harmonic of the switching

frequency, the actual switching currents cannot be predicted with certainty. ZVS is fully based on the switching current of a converter, so based on first-harmonic approximation no certainty can be obtained on achieving ZVS.

Generalization of operating modes, can be overcome by using an extensive method as extended fundamental frequency analysis [4], [5] to find a more accurate result for resonant converters. This method introduces a separate expression per operating mode, while still assuming that the resonant current and voltages only transfer energy with a first harmonic. While this increases the validity of the model for static parameters such as output behavior (output voltage and current), prediction of the switching currents, and thus ZVS, is still not always accurate since first-harmonic approximation is used as a basis for the model. Multiple harmonics can be used to increase the model accuracy, as proposed in [6]. This increases the model complexity and computational effort, while the result always contains an error compared to the actual converter behavior. The model proposed in [7] contains an analytical description for the resonant converter operation. This model is restricted to discontinuous and boundary conduction mode, resulting in an incomplete description of the converter operation.

This paper presents an analytical model of the of the series-resonant converter depicted in Fig. 1, for all operating modes. The model incorporates no harmonic approximations and is based on piecewise linear-system analysis [8]. Per switch-stage a separate system of equations is formed, which are linked together using boundary equations at the switching moments. After which an analytical solution is found by solving the resulting set of equations per operating mode. A complete analytical model of the converter is made using the results of all operating modes. Using this method, no harmonic approximations are applied so the total behavior of the resonant converter can be accurately predicted. As a result ZVS can be verified with certainty.

In Section II the converter operation and analysis method are described. In section III the different operating modes are analyzed and modeled. In Section IV these modes will be used to find the converter parameters. Unifying the results of the separate modes results into an analytical continuous transfer function given in Section V. Here the results are compared with experimental results taken from a build 2.5kW prototype.

978-1-4673-9551-9/16 $31.00 © 2016 IEEE

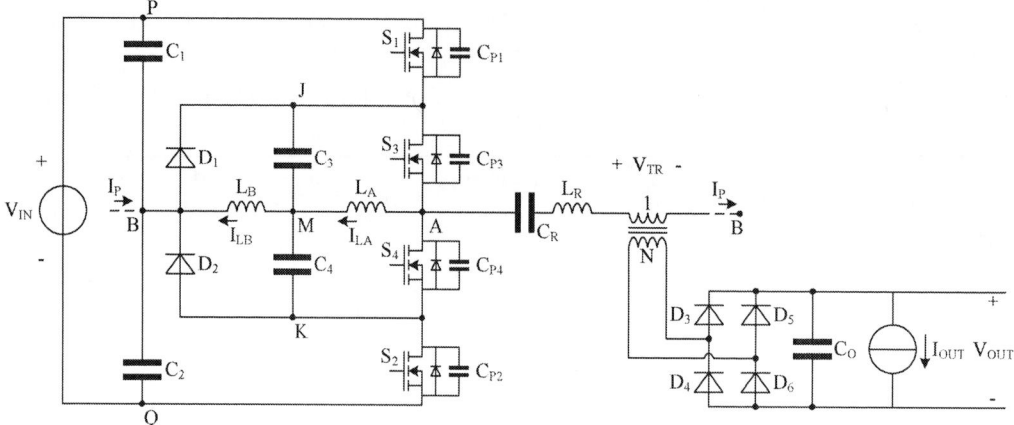

Fig. 1. The phase-shift-controlled three-level ZVS converter with full-bridge-rectifier used for analysis

II. TOPOLOGY ANALYSIS

A. Operation

The topology, as depicted in Fig. 1, contains two complimentary switching pairs (S_1, S_2 and S_3, S_4), operated at 50% duty-cycle and controlled by a given phase-shift (ϕ), as depicted in Fig. 2. As a result the voltage over C_3 and C_4, V_{JK} is alternately clamped to V_{PB} through S_1 and D_2 and to V_{BQ} through S_2 and D_1. In [1] it is concluded that the 50% duty-cycle operation combined with the ancillary inductors (L_A and L_B) result in naturally balanced mid-point voltages for the two capacitor buses, resulting in:

$$V_{PB} = V_{BQ} = \frac{1}{2}V_{IN} \qquad (1)$$

$$V_{JM} = V_{MK} = \frac{1}{2}V_{PB} = \frac{1}{4}V_{IN}, \qquad (2)$$

showing that the voltage over each individual switch is reduced to half of the input voltage. The resulting voltage waveforms are as depicted in Fig. 2, with the applied normalized phase-shift (ϕ) and the switching period (T_{SW}). The voltages V_{AM} and V_{BM} are applied over L_A and L_B respectively while the voltage V_{AB} is applied over the resonance tank (consisting of C_R and L_R) and the primary winding of the transformer. The voltage over the primary winding of the transformer (V_{TR}) is determined by the output voltage and the conduction state of the secondary diodes ($D3 - D6$). Depending on the voltages V_{AB} and V_{TR} a primary current is induced which, when rectified through the secondary diodes results into a positive average output current.

Depending on the switching current, ZVS is achieved. During ZVS the switching current charges and discharges the parasitic drain-source capacitance (C_{Px}) of the switches during the dead-time of the complimentary pairs. For example after turn-off of S_2 a positive drain-current charges C_{P2}. If this charge current is sufficient to charge C_{P2} to $\frac{1}{2}V_{IN}$ (and discharge C_{P1} to $0V$) before S_1 is turned-on, ZVS is achieved. So to achieve zero-voltage turn-on a negative I_P is required

for switching on S_1 and S_3 and a positive I_P for switching on S_2 and S_4. The requirement for ZVS operation for each switching pair then becomes:

$$i_{SW12} \geq \frac{V_{IN}(C_{P1} + C_{P2})}{2t_{DEAD}} \qquad (3)$$

$$i_{SW34} \geq \frac{V_{IN}(C_{P3} + C_{P4})}{2t_{DEAD}} \qquad (4)$$

where the suffixes $SW12$ and $SW34$ mean the switching moments of the complimentary pair switches S_1 and S_2, and S_3 and S_4 respectively and t_{DEAD} is the dead-time between the complementary switches. Since i_P is not always sufficient to satisfy requirements (3) and (4), L_A and L_B are used to ensure ZVS in any operating point. The current through these inductors are added to I_P so that:

$$i_{SW12} = i_{Lb}(t_{SW12}) + i_P(t_{SW12}) \qquad (5)$$

$$i_{SW34} = i_{La}(t_{SW34}) + i_P(t_{SW34}) \qquad (6)$$

where $i_{Lb}(t)$ and $i_{La}(t)$ are the inductor current through L_B and L_A respectively.

B. Analysis

The proposed model applies piecewise linear-system analysis. To apply the analysis it is assumed that; losses are neglected, cyclic steady-state is assumed, all capacitor voltages, except the resonant and parallel switch capacitor voltages (v_{Cr}, v_{Px}), are constant, C_R is much smaller than the remaining capacities and

$$2\pi f_{SW} > \omega_0 \qquad (7)$$

where f_{SW} is the switching frequency and ω_0 is the free-oscillation frequency of the resonance tank:

$$\omega_0 = \frac{1}{\sqrt{L_R C_R}} \qquad (8)$$

The circuit depicted in Fig. 3 is a result of applying these assumptions on the converter. This circuit is a lossless system, where the two voltage sources represent the output of the three-level stage (v_{AB}) and the voltage over the primary side of the transformer (v_{TR}). The difference of the two voltage sources is the applied voltage over the resonant tank (C_R and L_R) and results into a primary current which has to be fully determined over one half of the switching period, after which the system mirrors resulting in the same solutions. Once this current is determined, all the parameters of the converter can found.

The system consists of a second order differential equation per switch-stage. The general solution for the primary current is:

$$i_{P,k}(t_k) = A_k \sin(\omega_0 t_k + \varphi_k) \tag{9}$$

where the suffix k denotes the switch-stage, A and φ are the amplitude and phase variables that need to be calculated, and t_k the time related to the switch-stage k. Three switch-stages are present in a half of a switching period. As a result i_P is parametrized by a total of three expressions, containing two unknown variables each. The switching time of the diodes is not controlled by a parameter and is a result of the system. This is an additional unknown variable (δ). This totals the unknown variables to seven. Three different equation types can

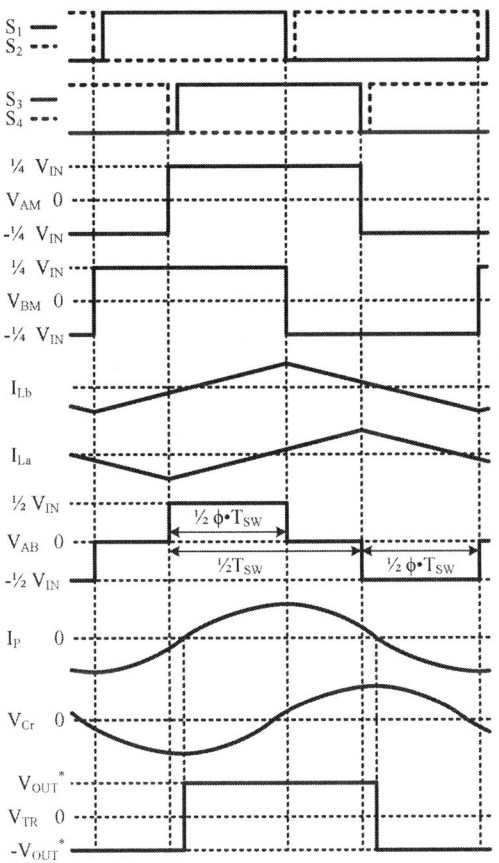

Fig. 2. Key waveforms of phase-shift-controlled three-level ZVS converter.

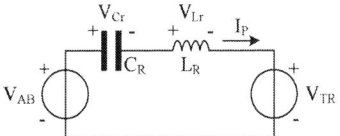

Fig. 3. Converter simplification used for model.

be used to solve these variables; inductor current continuity (boundary), capacitor voltage continuity (boundary) and power balance.

The first equation type is based on the fact that the inductor current is always continuous, so the current parametrization of two adjacent switch-stages (k and $k+1$) has to satisfy:

$$i_{P,k}(t_{k,e}) = i_{P,k+1}(t_{k+1,s}) \tag{10}$$

where suffix e denotes the end of the switch-stage and s denotes the start of the switch-stage. The second equation type ensures that the voltage over C_R is always continuous. As a result, instant voltage changes over the resonant tank are always applied over L_R. The general inductor equation:

$$v_{Lr} = L_R \frac{di_P}{dt} \tag{11}$$

is applied, where v_{Lr} contains the applied voltages (v_{AB}, v_{TR}). Using (11) the general boundary equation can be described as:

$$L_R \frac{di_{P,k}(t_{k,e})}{dt_k} - v_{AB}(t_{k,e}) + v_{TR}(t_{k,e}) =$$
$$L_R \frac{di_{P,k+1}(t_{k+1,s})}{dt_{k+1}} - v_{AB}(t_{k+1,s}) + v_{TR}(t_{k+1,s}). \tag{12}$$

The final equation type describes the power transfer between the primary side and the secondary side:

$$\frac{2}{T_{SW}} \int_0^{\frac{T_{SW}}{2}} v_{AB}(t)i_P(t)\mathrm{d}t = V_{OUT}I_{OUT} \tag{13}$$

where V_{OUT} and I_{OUT} are the average output voltage and current respectively. For each switch-stage both boundary equation types are applied. Adding the power balance equation leads to a total of seven equations to solve seven unknown variables, resulting in a solvable system.

III. OPERATING MODES

The operating mode is determined by the waveform of i_P, as depicted in Fig. 4. Here the numbers below the waveforms indicate the switch-stage. In discontinuous conduction mode (DCM), the current increases when v_{AB} is positive. After excitation the current decreases to zero and remains zero until the next transition of v_{AB}. This is due to the secondary diodes, which only conduct when forced by a positive forward voltage or current. In forced continuous conduction mode (FCCM), v_{Cr} (at the moment that i_P reaches zero) is higher than the transformed output voltage. This forces the secondary diodes in conduction, resulting in continuous conduction. In continuous conduction mode (CCM), i_P reaches zero after

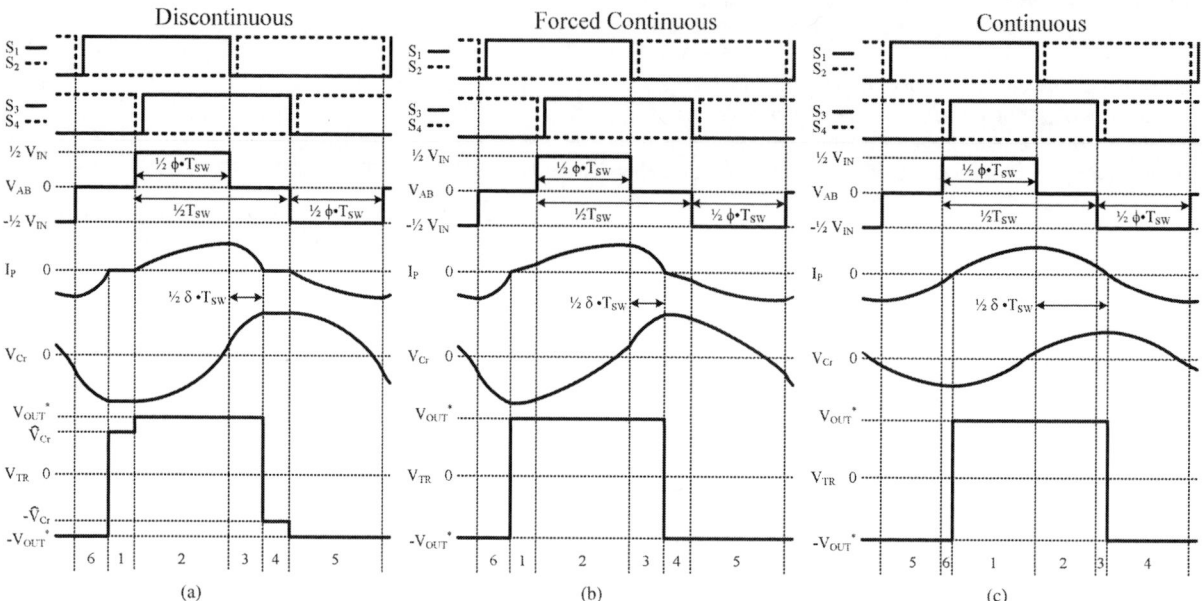

Fig. 4. Waveforms for discontinuous conduction mode (a), forced continuous conduction mode (b) and continuous conduction mode (c).

a negative v_{AB} is applied. This results into a continuous excitation of the resonance tank, and a continuous conduction.

The switching current of S_1 and S_2 is always negative during switch-on, regardless of the operating mode. If this current is sufficient, see (3), S_1 and S_2 will operate under ZVS condition. The switching current of S_3 and S_4 however, is strongly dependent on operating mode. In DCM the switch-on current of S_3 and S_4 is always zero, so ZCS is achieved. In FCCM the switch-on current of S_3 and S_4 is positive, resulting in the opposite effect as required by ZVS. In order to achieve ZVS this positive switch-on current needs to be compensated with an additional current, see (4) and (6). In CCM the switch-on current for S_3 and S_4 is always negative. If this current is sufficient, see (4), S_3 and S_4 will operate under ZVS condition. Thus the converter can achieve ZVS and or ZCS naturally in DCM and CCM, while compensation is required in FCCM to achieve ZVS.

IV. CONVERTER MODELING

The converter is modeled separately for each operating mode. The final model contains the expressions found for all operating modes. The expressions are functions of the environmental parameters (V_{IN} and V_{OUT}), the converter parameters (N, L_R, C_R and ω_0) and the control parameters (f_{SW} and ϕ).

A. Discontinuous conduction mode

In DCM I_P can be parametrized by two expressions, since i_P is zero during one switch-stage. After this switch-stage,

the current starts at zero. According to (9) the expressions for DCM then become:

$$i_{P,1}(t_1) = 0 \tag{14}$$

$$i_{P,2}(t_2) = A_2 \sin(\omega_0 t_2) \tag{15}$$

$$i_{P,3}(t_3) = A_3 \sin(\omega_0 t_3 + \varphi_3). \tag{16}$$

These expressions contain three unknown variables. Additionally the system is directly dependent on the peak resonant capacitor voltage ($\hat{v}_{Cr}$), see Fig. 4a. Since $\hat{v}_{Cr}$ is unknown, a total of five unknown variables (including δ) is found, requiring five equations to solve the system (14,15,16). The current continuity (10) from switch-stage 2 to 3 and from switch-stage 3 to 1 can be written as:

$$A_2 \sin(\omega_0 t_{2,e}) = A_3 \sin(\varphi_3) \tag{17}$$

$$A_3 \sin(\omega_0 t_{3,e} + \varphi_3) = 0 \tag{18}$$

respectively. The duration of switch-stage 2 and 3 are proportional to ϕ and δ respectively:

$$t_{2,e} = \phi \frac{T_{SW}}{2}, \qquad t_{3,e} = \delta \frac{T_{SW}}{2}. \tag{19}$$

Applying the voltage continuity (12) to all three boundaries results in:

$$-\omega_0 L_R A_2 + \frac{V_{IN}}{2} - \frac{V_{OUT}}{N} = -\hat{v}_{Cr} \tag{20}$$

$$\omega_0 L_R A_2 \cos(\omega_0 t_{2,e}) - \frac{V_{IN}}{2} = \omega_0 L_R A_3 \cos(\varphi_3) \tag{21}$$

$$-\omega_0 L_R A_3 + \frac{V_{OUT}}{N} = -\hat{v}_{Cr}. \tag{22}$$

978-1-4673-9551-9/16 $31.00 © 2016 IEEE

The power balance equation (13) is the final required equation:

$$\frac{2}{T_{SW}} \int_0^{\phi \frac{T_{SW}}{2}} \frac{V_{IN}}{2} A_2 \sin(\omega_0 t_2) \, dt_2 = V_{OUT} I_{OUT}. \quad (23)$$

Now all equations (17)-(23) are defined for DCM. The angle parameter φ_3 is expressed as a function of the parameters ϕ and δ by substitution of (19) in (18), resulting in:

$$\varphi_3 = \pi - \delta \bar{k} \quad (24)$$

with $\bar{k}$ the frequency factor given by:

$$\bar{k} = \frac{\omega_0 T_{SW}}{2} \quad (25)$$

The power balance integrals (23) can be evaluated using (24), resulting in the combined expression of:

$$\left(\frac{V_{IN}}{2} - \frac{V_{OUT}}{N} \right) A_2 \left(1 - \cos(\phi \bar{k}) \right) =$$
$$\frac{V_{OUT}}{N} A_3 \left(1 - \cos(\delta \bar{k}) \right). \quad (26)$$

Substitution of (19) and (24) in (21) results in:

$$\omega_0 L_R A_2 \cos(\phi \bar{k}) - \frac{V_{IN}}{2} = -\omega_0 L_R A_3 \cos(\delta \bar{k}). \quad (27)$$

The combination of (26) and (27) leads to elimination of δ:

$$A_2 \frac{V_{IN}}{2} \left(1 - \cos(\phi \bar{k}) \right) - \frac{V_{OUT}}{N} A_2 =$$
$$A_3 \frac{V_{OUT}}{N} - \frac{V_{OUT}}{N} \frac{V_{IN}}{2 \omega_0 L_R}. \quad (28)$$

$\hat{v}_{Cr}$ is eliminated by equating (20) to (22):

$$A_3 = A_2 + \frac{V_{IN}}{Q_{LC}} \left(\frac{2G}{N} - \frac{1}{2} \right), \quad (29)$$

here the converter gain is;

$$G = \frac{V_{OUT}}{V_{IN}}, \quad (30)$$

and the resonance quotient is:

$$Q_{LC} = \sqrt{\frac{L_R}{C_R}}. \quad (31)$$

Substitution of (29) into (28) leads to:

$$A_2 = \frac{V_{IN}}{Q_{LC}} \frac{\frac{2G}{N} - 1}{\frac{N}{2G} \left(1 - \cos(\phi \bar{k}) \right) - 2}. \quad (32)$$

δ is found by substitution of (24) in (17):

$$\delta = \frac{1}{k} \operatorname{asin} \left(\frac{A_2}{A_3} \sin(\phi \bar{k}) \right). \quad (33)$$

i_P, and so the converter behavior, is now completely described in DCM. The switching currents, capacitor voltage $V_{Cr}(t)$ and average output current;

$$I_{OUT} = \frac{V_{IN}}{2 \bar{k} Q_{LC}} \frac{\left(\frac{2}{N} - \frac{1}{G} \right) \left(1 - \cos(\phi \bar{k}) \right)}{\frac{N}{2G} \left(1 - \cos(\phi \bar{k}) \right) - 2} \quad (34)$$

are expressed in the system parameters; phase shift (ϕ), input voltage (V_{IN}), converter gain (G), transformer ratio (N), resonance quotient (Q_{LC}) and frequency factor ($\bar{k}$). These independent parameters describe the converter behavior in all modes.

B. Forced continuous conduction mode

In FCCM i_P does not remain zero for a full switch-stage. According to (9) the parametrization of the current for FCCM then becomes:

$$i_{P,1}(t_1) = A_1 \sin(\omega_0 t_1) \quad (35)$$
$$i_{P,2}(t_2) = A_2 \sin(\omega_0 t_2 + \varphi_2) \quad (36)$$
$$i_{P,3}(t_3) = A_3 \sin(\omega_0 t_3 + \varphi_3). \quad (37)$$

The above parametrization contains five unknown variables. Adding δ, the system requires six equations to solve the system (35),(36),(37). The current continuities (10) for all switch-stages are given as:

$$A_1 \sin(\omega_0 t_{1,e}) = A_2 \sin(\varphi_2) \quad (38)$$
$$A_2 \sin(\omega_0 t_{2,e} + \varphi_2) = A_3 \sin(\varphi_3) \quad (39)$$
$$A_3 \sin(\omega_0 t_{3,e} + \varphi_3) = 0. \quad (40)$$

With the switch-stage durations:

$$t_{1,e} = (1 - \phi - \delta) \frac{T_{SW}}{2}, \, t_{2,e} = \phi \frac{T_{SW}}{2},$$
$$t_{3,e} = \delta \frac{T_{SW}}{2}. \quad (41)$$

The voltage continuity equations (12) for FCCM are:

$$-\omega_0 L_R A_3 \cos(\omega_0 t_{3,e} + \varphi_3) = \omega_0 L_R A_1 + \frac{2V_{OUT}}{N} \quad (42)$$

$$\omega_0 L_R A_1 \cos(\omega_0 t_{1,e}) + \frac{V_{IN}}{2} = \omega_0 L_R A_2 \cos(\varphi_2) \quad (43)$$

$$\omega_0 L_R A_2 \cos(\omega_0 t_{2,e} + \varphi_2) =$$
$$\omega_0 L_R A_3 \cos(\varphi_3) + \frac{V_{IN}}{2}. \quad (44)$$

The power balance equation (13) for FCCM is given as:

$$\frac{2}{T_{SW}} \int_0^{\phi \frac{T_{SW}}{2}} \frac{V_{IN}}{2} A_2 \sin(\omega_0 t_2 + \varphi_2) \, dt_2$$
$$= V_{OUT} I_{OUT}. \quad (45)$$

i_P is now parametrized using six equations (38)-(45) and the six unknown variables are determined. The angle parameter φ_3 is expressed as a function of ϕ and δ by substitution of (41) in (37):

$$\varphi_3 = \pi - \delta \bar{k} \quad (46)$$

with $\bar{k}$ the frequency factor given by (25). A_1 is written as a function of A_3 by substituting (30), (31), (41) and (46) into (42):

$$A_1 = A_3 - \frac{2V_{IN}G}{NQ_{LC}}. \quad (47)$$

φ_2 can be isolated by substitution of (41) and (46) in (39). After rewriting this results into:

$$A_2 \left(\sin(\phi \bar{k}) \cos(\varphi_2) + \cos(\phi \bar{k}) \sin(\varphi_2) \right)$$
$$= A_3 \sin(\pi - \delta \bar{k}). \quad (48)$$

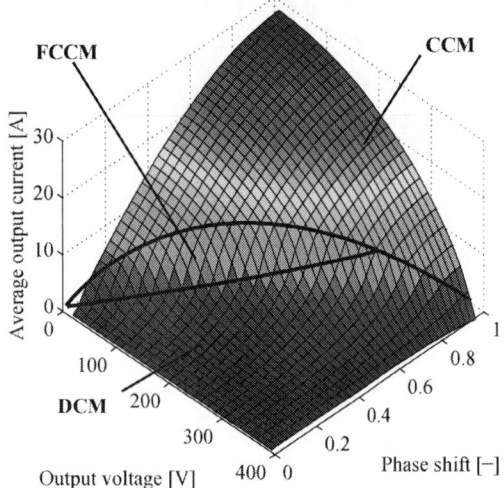

Fig. 5. Converter output current I_{OUT} vs. phase-shift ϕ and output voltage V_{OUT} over full operating range, $V_{IN} = 400V$.

Substitution of (38),(43) and (41) and rewriting leads to elimination of φ_2 and replacement of A_2 by A_1:

$$A_1 \sin\left((1-\delta)\bar{k}\right) + \frac{V_{IN}}{2\omega_0 L_R} \sin\left(\phi\bar{k}\right)$$
$$= A_3 \sin\left(\delta\bar{k}\right). \quad (49)$$

Substitution of (30), (31) and (47) and rewriting leads to the following expression for A_3:

$$A_3 = \frac{V_{IN}}{Q_{LC}} \frac{\frac{2G}{N}\sin\left((1-\delta)\bar{k}\right) - \frac{1}{2}\sin\left(\phi\bar{k}\right)}{\sin\left((1-\delta)\bar{k}\right) - \sin\left(\delta\bar{k}\right)}. \quad (50)$$

The same steps are performed, starting with (44), resulting in:

$$A_3 = \frac{V_{IN}}{Q_{LC}} \frac{\frac{2G}{N}\cos\left((1-\delta)\bar{k}\right) - \frac{1}{2}\left(\cos\left(\phi\bar{k}\right) - 1\right)}{\cos\left((1-\delta)\bar{k}\right) + \cos\left(\delta\bar{k}\right)}. \quad (51)$$

Equating these two expressions for A_3 and rewriting results in an expression for δ:

$$\delta = \frac{1}{2} - \frac{\phi}{2} + \frac{1}{k}\mathrm{acos}\left(\frac{\frac{G}{N}\sin\left(\bar{k}\right)}{\cos\left(\frac{k}{2}\right)\sin\left(\phi\frac{\bar{k}}{2}\right)}\right). \quad (52)$$

A_2 is found by squaring the results of (39) and (42):

$$A_2^2 \sin^2\left(\phi\bar{k} + \varphi_2\right) = A_3^2 \sin^2\left(\delta\bar{k}\right) \quad (53)$$

$$A_2^2 \cos^2\left(\phi\bar{k} + \varphi_2\right) = A_3^2 \cos^2\left(\delta\bar{k}\right) -$$
$$\frac{V_{IN}}{\omega_0 L_R} A_3 \cos\left(\delta\bar{k}\right) + \frac{V_{IN}^2}{4\omega_0^2 L_R^2}, \quad (54)$$

and applying the Pythagorean formula for trigonometric functions;

$$\sin^2\left(\alpha\right) + \cos^2\left(\alpha\right) = 1, \quad (55)$$

with substitution of (30) and (31) resulting in:

$$A_2 = \sqrt{A_3^2 - \frac{V_{IN}}{Q_{LC}} A_3 \cos\left(\delta\bar{k}\right) + \frac{V_{IN}^2}{4Q_{LC}^2}}. \quad (56)$$

Finally φ_2 is found by substituting (41) into (38), resulting in:

$$\varphi_2 = \mathrm{asin}\left(\frac{A_1}{A_2}\sin\left((1-\phi-\delta)\bar{k}\right)\right) \quad (57)$$

Now I_P is completely parametrized for FCCM. The output current I_{OUT} can now be calculated using the power balance (45):

$$I_{OUT} = \frac{A_2}{2\bar{k}G}\left(\cos\left(\varphi_2\right) - \cos\left(\phi\bar{k} + \varphi_2\right)\right) \quad (58)$$

C. Continuous conduction mode

In CCM i_P returns to zero after a positive v_{AB} is applied resulting in a continuous excitation of the resonance tank. This results into a continuous conduction of the secondary diodes. According to (9) the parametrization of the current for CCM becomes:

$$i_{P,1}\left(t_1\right) = A_1 \sin\left(\omega_0 t_1\right) \quad (59)$$
$$i_{P,2}\left(t_2\right) = A_2 \sin\left(\omega_0 t_2 + \varphi_2\right) \quad (60)$$
$$i_{P,3}\left(t_3\right) = A_3 \sin\left(\omega_0 t_3 + \varphi_3\right). \quad (61)$$

The above parametrization contains five unknown variables. Again by including δ, the system requires six equations to solve the system (59),(60),(61). The current continuities (10) for all switch-stages are given as:

$$A_1 \sin\left(\omega_0 t_{1,e}\right) = A_2 \sin\left(\varphi_2\right) \quad (62)$$
$$A_2 \sin\left(\omega_0 t_{2,e} + \varphi_2\right) = A_3 \sin\left(\varphi_3\right) \quad (63)$$
$$A_3 \sin\left(\omega_0 t_{3,e} + \varphi_3\right) = 0. \quad (64)$$

With the switch-stage durations:

$$t_{1,e} = (1-\delta)\frac{T_{SW}}{2}, t_{2,e} = (1-\phi)\frac{T_{SW}}{2},$$
$$t_{3,e} = (-1+\phi+\delta)\frac{T_{SW}}{2}. \quad (65)$$

The voltage continuity equations (12) for CCM are:

$$-\omega_0 L_R A_3 \cos\left(\omega_0 t_{3,e} + \varphi_3\right) = \omega_0 L_R A_1 + \frac{2V_{OUT}}{N} \quad (66)$$

$$\omega_0 L_R A_1 \cos\left(\omega_0 t_{1,e}\right) = \omega_0 L_R A_2 \cos\left(\varphi_2\right) + \frac{V_{IN}}{2} \quad (67)$$

$$\omega_0 L_R A_2 \cos\left(\omega_0 t_{2,e} + \varphi_2\right) - \frac{V_{IN}}{2} = \omega_0 L_R A_3 \cos\left(\varphi_3\right). \quad (68)$$

The power balance equation (13) for CCM becomes:

$$\frac{2}{T_{SW}}\left(\int_0^{(1-\delta)\frac{T_{SW}}{2}} \frac{V_{IN}}{2} A_1 \sin\left(\omega_0 t_1\right)\,\mathrm{d}t_1 -\right.$$
$$\left.\int_0^{(-1+\phi+\delta)\frac{T_{SW}}{2}} \frac{V_{IN}}{2} A_3 \sin\left(\omega_0 t_3 + \varphi_1\right)\,\mathrm{d}t_3\right) =$$
$$V_{OUT}I_{OUT}. \quad (69)$$

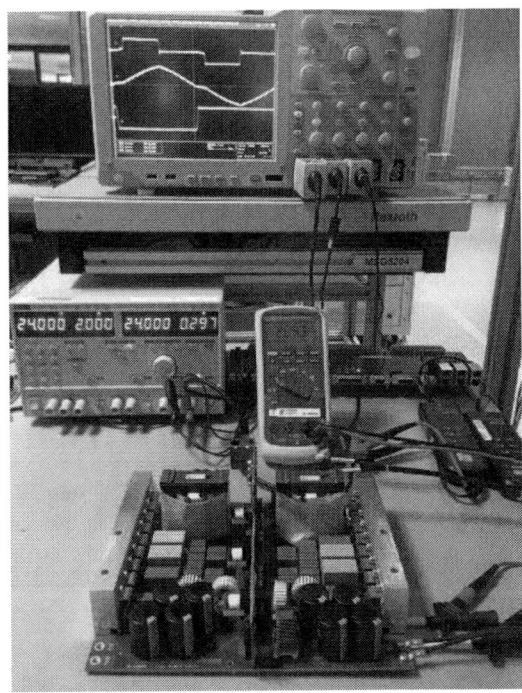

Fig. 6. Experimental setup.

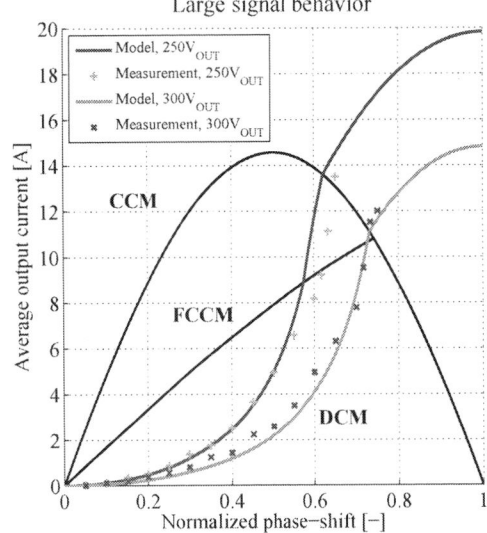

Fig. 7. Converter output current I_{OUT} vs. phase-shift ϕ and output voltage V_{OUT}, markings resemble measured data, $V_{IN} = 400V$.

i_P is now parametrized for CCM, using six equations (62)-(69) and the six variables are determined. The angle parameter φ_3 is expressed as a function of ϕ and δ by substitution of (25) and (65) in (64):

$$\varphi_3 = \pi + (1 - \phi - \delta)\,\bar{k} \qquad (70)$$

A_3 is written as a function of A_1 by substituting (30), (31), (65) and (70) into (66):

$$A_3 = A_1 + \frac{2V_{IN}G}{NQ_{LC}} \qquad (71)$$

φ_2 can be isolated by substitution of (65) and (70) in (63). Rewriting results into:

$$A_2 \sin\big((1 - \phi)\,\bar{k}\big)\cos(\varphi_2) + \\ A_2 \cos\big((1 - \phi)\,\bar{k}\big)\sin(\varphi_2) = \\ - A_3 \sin\big((1 - \phi - \delta)\,\bar{k}\big). \qquad (72)$$

Substitution of (62) and (68) and rewriting results in elimination of φ_2 and replacement of A_2 by A_1:

$$A_1 \sin\big((2 - \phi - \delta)\,\bar{k}\big) - \\ \frac{V_{IN}}{2\omega_0 L_R}\sin\big((1 - \phi)\,\bar{k}\big) = \\ - A_3 \sin\big((1 - \phi - \delta)\,\bar{k}\big). \qquad (73)$$

Substituting (71) and rewriting leads to an expression for A_1:

$$A_1 = \frac{V_{IN}}{Q_{LC}}\frac{\frac{1}{2}\sin\big((1 - \phi)\,\bar{k}\big)}{\sin\big((2 - \phi - \delta)\,\bar{k}\big) + \sin\big((1 - \phi - \delta)\,\bar{k}\big)} - \\ \frac{V_{IN}}{Q_{LC}}\frac{\frac{2G}{N}\sin\big((1 - \phi - \delta)\,\bar{k}\big)}{\sin\big((2 - \phi - \delta)\,\bar{k}\big) + \sin\big((1 - \phi - \delta)\,\bar{k}\big)}. \qquad (74)$$

The same operations are performed, starting with (68), resulting in a second expression for A_1:

$$A_1 = \frac{V_{IN}}{Q_{LC}}\frac{\frac{1}{2}\big(\cos\big((1 - \phi)\,\bar{k}\big) + 1\big)}{\cos\big((2 - \phi - \delta)\,\bar{k}\big) + \cos\big((1 - \phi - \delta)\,\bar{k}\big)} - \\ \frac{V_{IN}}{Q_{LC}}\frac{\frac{2G}{N}\cos\big((1 - \phi - \delta)\,\bar{k}\big)}{\cos\big((2 - \phi - \delta)\,\bar{k}\big) + \cos\big((1 - \phi - \delta)\,\bar{k}\big)}. \qquad (75)$$

Equating these two expressions for A_1 and rewriting results in the following expression for δ:

$$\delta = 1 - \frac{\phi}{2} - \frac{1}{k}\mathrm{asin}\left(\frac{\frac{G}{N}\sin\big(\bar{k}\big)}{\cos\big((1 - \phi)\frac{\bar{k}}{2}\big)\cos\big(\frac{\bar{k}}{2}\big)}\right). \qquad (76)$$

A_2 is found by vector addition of (62) and (67) and substituting (30) and (31):

$$A_2 = \sqrt{A_1^2 - \frac{V_{IN}}{Q_{LC}}A_1 \cos\big((1 - \delta)\,\bar{k}\big) + \frac{V_{IN}^2}{4Q_{LC}^2}}. \qquad (77)$$

Finally φ_2 is found by substitution of (65) in (62):

$$\varphi_2 = \mathrm{asin}\left(\frac{A_1}{A_2}\sin\big((1 - \delta)\,\bar{k}\big)\right) \qquad (78)$$

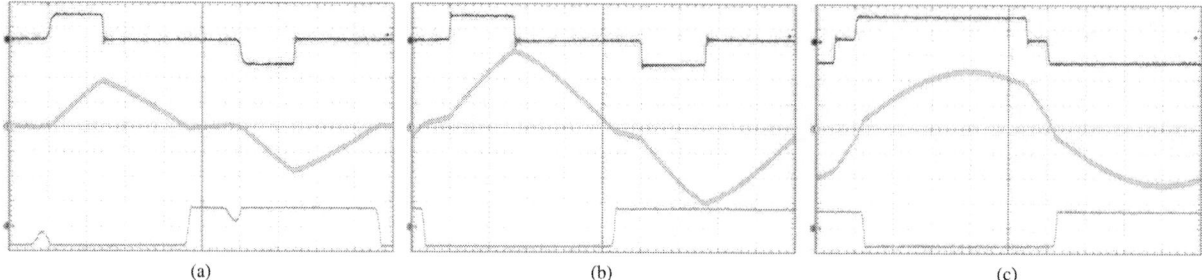

Fig. 8. Measured waveforms for discontinuous conduction mode (a), forced continuous conduction mode (b) and continuous conduction mode (c), from top to bottom: voltage v_{AB} (200V/div), primary current i_P (10A/div), and transformer voltage v_{TR} (200V/div).

Now I_P is completely parametrized for CCM. The output current I_{OUT} is calculated using the power balance equation (69):

$$I_{OUT} = \frac{1}{2\bar{k}G}\Big(A_1\left(1 - \cos\left((1-\delta)\,\bar{k}\right)\right) -$$
$$A_3\left(1 - \cos\left((1-\phi-\delta)\,\bar{k}\right)\right)\Big) \quad (79)$$

V. RESULTS

Analyzing the derived expressions, the following can be concluded for the converter; the output current (I_{OUT}) is directly proportional to the input voltage (V_{IN}), inversely proportional to the resonance quotient (Q_{LC}) and the function of the remaining parameters (N, $\bar{k}$, ϕ and G) is less straightforward. With the derived expressions a model is made to predict the converter behavior over the full operating range. The outcome of this model is depicted in Fig. 5. Here the three operating modes are indicated. A 2.5kW prototype dc-dc converter is built to verify the converter behavior in all operating modes, as well as the model description of the converter. The waveforms of the three operating modes are depicted in Fig. 8a, Fig. 8b and Fig. 8c for DCM, FCCM and CCM respectively. Here some deviation from the predicted waveforms is seen for DCM, this is due to the magnetizing inductance which is neglected at start of the analysis. The large signal transfer of the converter is measured and is depicted in Fig. 7. The calculated outcome of the model (solid lines) accurately predicts the experimentally results of the converter (markers). The error between calculation and experimental results originates from different rise and fall times for the two complimentary switch pairs, resulting in a difference in

applied phase shift and actual phase shift. The rise and fall times are directly dependent on the primary current due to ZVS, so the error is most visible in FCCM for in this mode the difference in switch current between the two complimentary pairs is the highest. The measurements were performed using the experimental setup depicted in Fig. 6, with the converter parameters listed in Table I.

VI. CONCLUSION

This paper presents an analytical model for the phase-shift controlled three-level series-resonant converter based on piecewise-linear system analysis, including an extensive analysis on both the operating modes of the converter as well as the derivation of the model itself. The model is capable to accurately predict the converter operating behavior. A separate expression is found per operating mode, ensuring that the model is accurate over the full-operating range of the converter. The converter operation in all three operating modes, and the prediction of the model are verified with a prototype converter and experimental results show that the model is an accurate prediction of the performed measurements.

REFERENCES

[1] J. L. Duarte, J. Lőkös and F. B. M. van Horck, "Phase-Shift-Controlled Three-Level Converter With Reduced Stress Featuring ZVS Over the Full Operating Range," in *IEEE Transactions on Power Electronics*, Vol. 28, no. 5, pp. 2140-2150, May 2013.

[2] C. Oeder and T. Deuerbaum, "ZVS Investigation of LLC Converters Based on FHA Assumptions," in *IEEE 2013 Applied Power Electronics Conference*, 2013, pp. 2643-2648, March 2013.

[3] M. K. Kazimierczuk and D. Czarkowski, *Resonant Power Converters*, John Wiley & Sons Inc., Hoboken, New Jersey, 2011.

[4] A. J. Forsyth, G. A. Ward and S. V. Mollov, "Extended Fundamental Frequency Analysis of the LCC Resonant Converter," in *IEEE Transactions on Power Electronics*, vol. 18, no. 6, pp. 1286-1292, Nov. 2003.

[5] J. Biela, U. Badstübner and J. W. Kolar, "Design of a 5kW, 1U, 10kW/ltr. resonant DC-DC converter for telecom applications," in *INTELEC 2007 Telecommunications Energy Conference*, 2007, pp. 824-831, Sept. 2007.

[6] X. Li, H. Li, G. Hu and Y. Xue, "A Bidirectional Dual-Bridge High-Frequency Isolated Resonant DC/DC Converter," *IEEE 2013 Industrial Electronics and Applications*, 2013, pp. 49-54, June 2013.

[7] Y. V. Singh, K. Viswanathan, R. Naik, J. A. Sabate and R. Lai, "Analysis and Control of Phase-shifted Series Resonant Converter Operating in Discontinuous Mode", in *IEEE 2013 Applied Power Electronics Conference*, 2013, pp.2092-2097, March 2013.

[8] N. B. O. L. Pettit and P. E. Wellstead, "Analyzing piecewise linear dynamical systems," in *IEEE Transactions on Control Systems*, vol. 15, no. 5, pp. 43-50, Oct. 1995.

TABLE I
PROTOTYPE CONVERTER PARAMETERS.

Parameter	Value	Unit
L_R	12.3	μH
C_R	330	nF
N	1.92	-
f_{SW}	100	kHz
V_{IN}	400	V

978-1-4673-9551-9/16 $31.00 © 2016 IEEE

High Efficiency Design for ISOP Converter System with Dual Active Bridge DC-DC Converter

Masaki Sato*, Kazuhide Domoto and Yoichi Ishizuka
Graduate School of Engineering
Nagasaki University
Nagasaki, Japan
*bb52114221@cc.nagasaki-u.ac.jp

Masahiro Yamaguchi
Graduate School of Engineering
Tohoku University
Miyagi, Japan

Shinya Manabe and Hiizu Okubo
RICOH Electronic Devices Co., Ltd.
Osaka, Japan

Atsushi Itagaki
Ryowa Electronics Co., Ltd.
Miyagi, Japan

Abstract— **The Dual Active Bridge (DAB) DC-DC converter is one of the most popular circuits for bi-directional applications because of simple structure. However, DAB DC-DC converter has few problems such as the surge voltage and hard switching in specific conditions. These problems are caused by the relationship between the output power and phase difference, which causes device breakdown and deterioration of power efficiency. For the design of Input-Series and Output-Parallel (ISOP) converter system with DAB DC-DC converter, the problem is also occurred and more severe because of its high voltage. In this paper, the design technique of ISOP with DAB for suppressing the surge voltage and realizing the high power efficiency in wide load range is proposed. As the result, this method reduces the switching surge, and the power efficiency is increased by 22.5% at light load.**

Keywords— ISOP converter system, Dual Active Bridge DC-DC converter, soft switching

I. INTRODUCTION

ISOP converter system has been attracting attention as the new power configuration [1-8]. For example, in power distribution networks, the Solid-State Transformer (SST) using ISOP converter system has been studied volume and weight reducing technique of the of converter circuit to replace the conventional huge and heavy transformer [1-4]. In general, one of the huge merits of SST reduces voltage and current stress of the switch element by series or parallel connection of a single converter module. Thus, it is possible to apply the switch elements to higher power and/or higher voltage application, and to be large buck ratio [5].

Further, the bidirectional DC-DC converter has been focused on because of the huge demand for diversification of power supply network. Above all, the DAB DC-DC converter is one of the most popular circuits for bidirectional applications because of its simple structure [9-15]. However, DAB DC-DC

converter has few problems such as the surge voltage and hard switching in specific conditions [9].

In this paper, the design technique of ISOP with DAB for suppressing the surge voltage and realizing the high power efficiency in wide load range is proposed and confirmed the design technique with experimental results.

II. DAB DC-DC CONVERTER

III. Fig. 1 shows the circuit schematic of the conventional DAB DC-DC converter. DAB DC-DC converter has the binary characteristics in bi-directions. The output power is operated by the phase-shift between the input and output bridge operation. The ideal output power equation can be represented equations as follows [10]

$$\text{IV.} \quad P_o = \frac{V_{in}V_{out}}{\omega L_r}\varphi(1-\frac{\varphi}{\pi}) \qquad (1)$$

where the input DC voltage is V_{in}, output DC voltage is V_{out}, the transformer leakage inductance is L_r, $\omega=2\pi f$, and the switching frequency is f. Also, the bridge phase difference is φ. In (1), when the phase difference $\varphi > 0$, the power transmission is forward mode. Conversely, when $\varphi < 0$, the power transmission is backward mode [14].

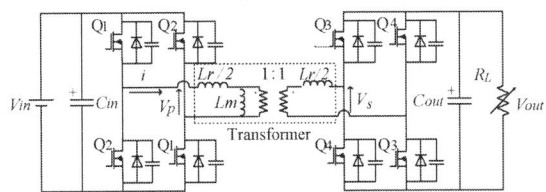

Fig. 1. The circuit schematic of the conventional DAB DC-DC converter

Fig. 2 shows operating waveform in buck mode. As shown in Fig. 2, the current waveform is varied by the load state. In this paper, a state that the current crosses the zero point at the state 1 is defined as the hard switching operation, and a state that crosses at the state 2 is defined as the soft switching operation. DAB DC-DC converter has a problem that of hard switching and current surge in hard switching operation as shown in Fig. 2 (b). The hard switching operation range is determined by the relationship between the input-output voltage conversion ratio and the phase difference as illustrated in Fig. 3 [13]. If the difference of the input and output voltage is greater, the hard switching operation range is widened and this problem is even worse. This section describes the generation mechanism of hard switching in buck mode as an example.

A. Soft switching operation

In a case of soft switching operation, the direction of the current is reversed between State 1 and State 2. Since the current path is different, the charge of the parasitic capacitance is discharged by flowing the reverse current in state 2. It is soft switching operation because Q_3 is turned ON after the charge is discharged. And the surge also does not occur since the recovery characteristics of the diode do not occur.

B. Hard switching operation

On the other hand, Fig. 4 shows an important circuit state for hard switching in State 1, State 2 and State 3. In State 1, by voltage across switch Q_3, the electrical charge is charged in the parasitic capacitance of Q_3. State 2 is the dead time, the current flows through the body diode of Q_4. Then, in State 3, Q_3 are turned ON, regardless the parasitic charge was remain. It becomes hard switching operation. In addition, by the reverse recovery characteristics of the body diode of Q_4, the leg of Q_3 and Q_4 becomes shorten. Therefore, the current surge occurs. These cause deterioration of power efficiency and device destruction.

From these analysis, the power efficiency decreases and the switching surge occurs only in hard switching operation. Several solution methodologies have been studied. For example, a method with attaching a snubber circuit [11], a method for operating as a resonant converter by mounting the capacitor [12] and a method using partial resonance by attaching LC filter [13] has reported. However, the increase of power losses, cost and weight is occurred by an additional circuit.

Therefore, in the next section, as a technique for solving these intrinsic problems of DAB in use of ISOP system is proposed.

V. ISOP CONVERTER SYSTEM

In this paper, the design technique is proposed that the ISOP converter system that connected several converters with solving the problem of DAB DC-DC converter at the hard switching operation. The ISOP converter system such as shown in Fig. 5 is the circuit method that connected in parallel at the primary side and connected in series at the secondary side with a few converters. It is possible to divide the input voltage and shunt the output current. Therefore, it is considered that the switching loss of the primary side and the conduction loss of the secondary side are reduced.

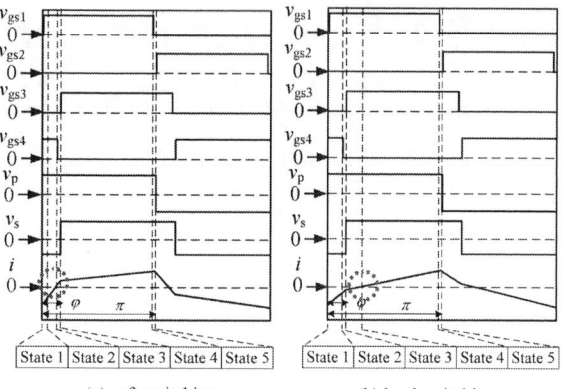

(a) soft switching   (b) hard switching

Fig. 2. Operating waveform (in buck mode)

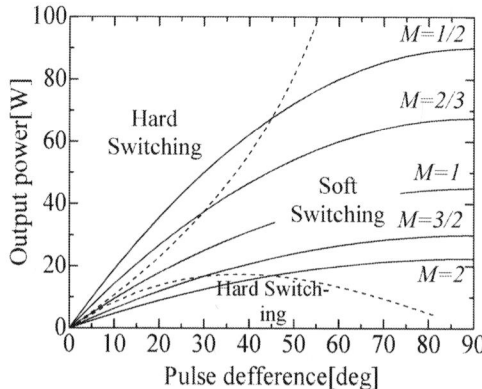

Fig. 3. The hard swithing operation boundary

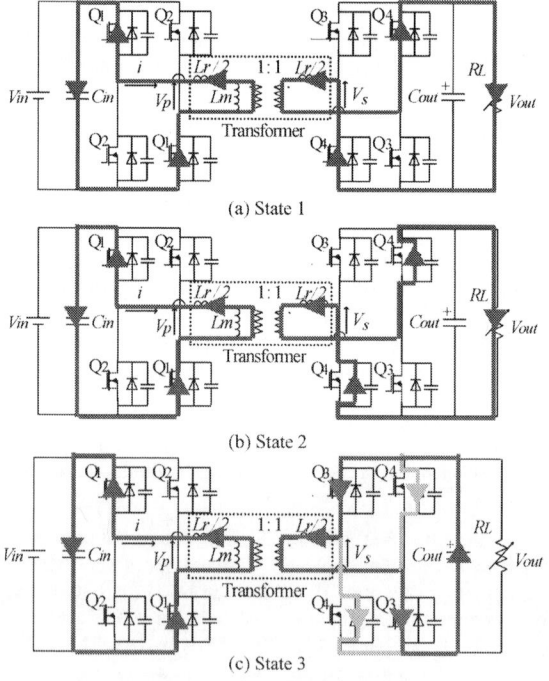

(a) State 1

(b) State 2

(c) State 3

Fig. 4. The hard switching operation ($V_{in} > V_{out}$)

978-1-4673-9551-9/16 $31.00 © 2016 IEEE

Furthermore, by dispersing the problems of withstand voltage and the maximum permissible current in the switch, it is seen that the use of high power and with the merits of large buck ratio, low volume, low weight, fault isolation, voltage regulation, and potential additional functions, such as harmonic filtering, reactive power compensation, etc. [1].

By using the ISOP converter system, and by adjusting the optimum number of modules, it is possible to equalize the input-output voltage conversion ratio. Therefore, it is assumed that to realize soft switching operation at wide load range as shown in Fig. 3.

In order to examine the power loss of the ISOP with DAB DC-DC converter, the analytical expression has been obtained in this paper. DAB DC-DC converter can be considered together with the loss of the circuit such as the conduction loss of switch and transformer copper loss in one place as shown Fig. 6 since the current flows in the leakage inductance in all state [15]. This resistance component is defined as r_{Loss}. If there is not the unevenness of the element, input voltage V_i applied to one converter module is divided V_{in}/n. In addition, magnetizing inductance: L_m >> leakage inductance: $L_{r1}+L_{r2}=L$, impedance component of the leakage inductance: ωL >> r_{Loss} >> on-resistance of switch: r_t. Therefore, it is considered that magnetizing inductance is open and $r_{Loss} = r_t = 0$.

In each state shown in Fig. 2(a), the current flowing through the leakage inductance i is the following equation [15];

for $0 \leq t \leq \dfrac{\varphi}{2\pi}T_s$ (State 1-3);

$$i(t) = \frac{V_{in}/n+\hat{v}_o}{L}t - \frac{T_s}{2\pi \times 2L}\left\{2\hat{v}_o\varphi + (\frac{V_{in}}{n}-\hat{v}_o)\pi\right\} \quad (2)$$

for $\dfrac{\varphi}{2\pi}T_s \leq t \leq \dfrac{T_s}{2}$ (State 3-5);

$$i(t) = \frac{V_{in}/n-\hat{v}_o}{L}(t-\frac{\varphi}{2\pi}T_s) + \frac{T_s}{2\pi \times 2L}\left\{2(\frac{V_{in}}{n})\varphi - (\frac{V_{in}}{n}-\hat{v}_o)\pi\right\} \quad (3)$$

for $\dfrac{\varphi}{2\pi}T_s \leq t \leq \dfrac{T_s}{2}+\dfrac{\varphi}{2\pi}T_s$;

$$i(t) = -\frac{V_{in}/n+\hat{v}_o}{L}(t-\frac{1}{2}T_s) + \frac{T_s}{2\pi \times 2L}\left\{2\hat{v}_o\varphi + (\frac{V_{in}}{n}-\hat{v}_o)\pi\right\} \quad (4)$$

for $\dfrac{T_s}{2}+\dfrac{\varphi}{2\pi}T_s \leq t \leq T_s$;

$$i(t) = \frac{-V_{in}/n+\hat{v}_o}{L}(t-\frac{\pi+\varphi}{2\pi}T_s) - \frac{T_s}{2\pi \times 2L}\left\{2(\frac{V_{in}}{n})\varphi - (\frac{V_{in}}{n}-\hat{v}_o)\pi\right\} \quad (5)$$

The effective value of the current can be obtained by using the above equations, the power loss per stage is

$$P_{loss\_per} = r_{Loss}\left(\sqrt{\frac{1}{Ts}\int_0^{Ts}i(t)^2dt}\right)^2 \quad (6)$$

$$= r_{Loss}\left\{-\frac{\varphi^3Ts^2}{6\pi^3L^2}\frac{V_{in}}{n}\hat{v}_o + \frac{\varphi^2Ts^2}{4\pi^2L^2}\frac{V_{in}}{n}\hat{v}_o + \frac{Ts^2}{48L^2}(\frac{V_{in}}{n}-\hat{v}_o)^2\right\}$$

Using $V_o = M \cdot V_i / n$, the total loss is

$$P_{loss} = P_{loss\_per} \times n$$

$$= \frac{r_{Loss}V_i^2Ts^2}{2nL^2}\left\{-\frac{M\varphi^3}{3\pi^3} + \frac{M\varphi^2}{2\pi^2} + \frac{(1-M)^2}{24}\right\} \quad (7)$$

Fig. 7 shows the power loss versus the number of modules calculated by (7). It is understood that the loss is reduced by using the ISOP converter system. However, it is desirable that the circuit configuration is the optimum number of modules since the loss is increases if the number of modules is too large.

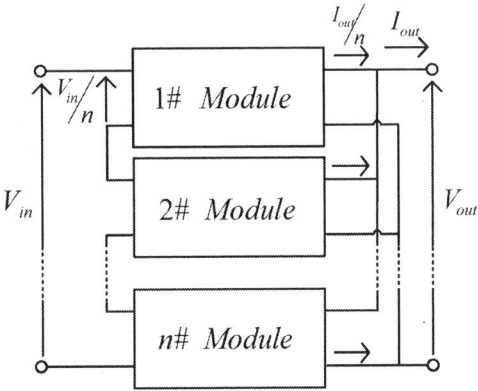

Fig. 5. ISOP converter system

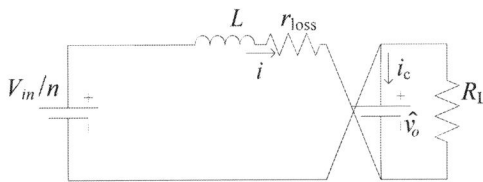

Fig. 6. Equivalent circuit of per module using ISOP converter system

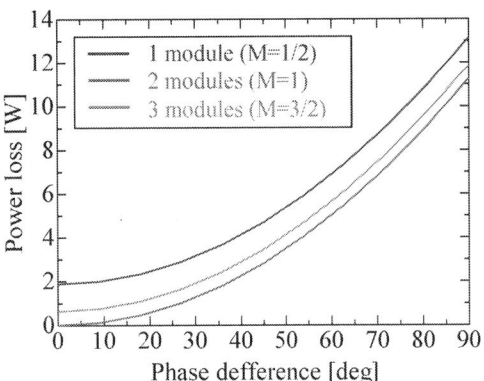

Fig. 7. Power loss characteristic

VI. Experiment Results

In order to verify the validity of the proposed ISOP converter system design, the experimental measurements are necessary. Table 1 shows experimental parameters. By V_{in}=24V and V_o=12V, 2-modules is optimum in theory. The high breakdown voltage MOSFET was used because of the surge of maximum 150V occurred when it was operating with only one-module. Three transformers are fabricated with hand winding with the target of leakage inductance of 2µH.

The main circuit is the DAB DC-DC converter without additional circuits like snubber. Experimental circuit is open loop control with UCC3895. It was verified in the case of one-module (M=1/2) with non-ISOP converter system and the case of two (M=1) or three (M=3/2) -modules with ISOP converter system. It is possible to work without problems in the buck or boost mode using the DAB DC-DC converter.

It is shown the experimental waveform of one-module in Fig. 8, two-modules in Fig. 9, and three-modules in Fig. 10. Fig. 11(a) shows the output power results and Fig. 11(b) shows the power efficiency results. First, from Fig. 8, it has become the hard switching operation since the inductor current is inverted in state 3. And surge is generated in both primary and secondary side. Usually, hard switching and serge occur in the low voltage side, and it is always usually with soft switching operation in the high voltage side with the DAB converter [10]. In the case of one-module, the surge of primary side is occurred because of the lack of the fabrication. From Fig. 9, the soft switching operation has accomplished since the inductor current is inverted in state 1. The surge of the both side has well suppressed. Finally, from Fig. 10, the hard switching operation is occurred since the inductor current is inverted in state 3 and large surge voltage is appeared. From these results, by adjusting the number of modules is such that $M = 1$, it is possible to accomplish soft-switching operation and reduce surge at DAB DC-DC converter as shows Fig. 3. In addition, converter is operating at maximum efficiency since conduction loss is small by operating voltage conversion ratio M=1.

From Fig. 11(a), by increasing the number of modules using the ISOP converter system, the output power becomes small with same phase difference. Since power per one module is dispersed by using the ISOP converter system, it is larger loss per one module than the power. Therefore, it is considered that the total power is reduced since the power transferring in one module is reduced. For example, when output power per module is 24W in one-module, the copper loss per module is 0.2Ω. When output power per module is 12W in two-modules, the copper loss per module is 0.2Ω. It is necessary to design so that the loss is reduced in order to use ISOP converter system.

By using ISOP converter system, it can be seen that the power efficiency is improved up to 22.5% than the case of non-ISOP converter system in light load as shown Fig. 11(b)

TABLE I. THE EXPERIMENTAL CIRCUIT PARAMETERS

Description		Value
Input Voltage		24V
Output Voltage		12V
Swtiching Frequency		200kHz
MOS-FET	FDP18N20F	200V / 18A 120mΩ
Transformer #1	Leakage Inductance	2.1µH
	Magnetizing Inductance	18µH
	Turn ratio	1:1
Transformer #2	Leakage Inductance	2.3µH
	Magnetizing Inductance	17.5µH
	Turn ratio	1:1
Transformer #3	Leakage Inductance	2.5µH
	Magnetizing Inductance	19µH
	Turn ratio	1:1

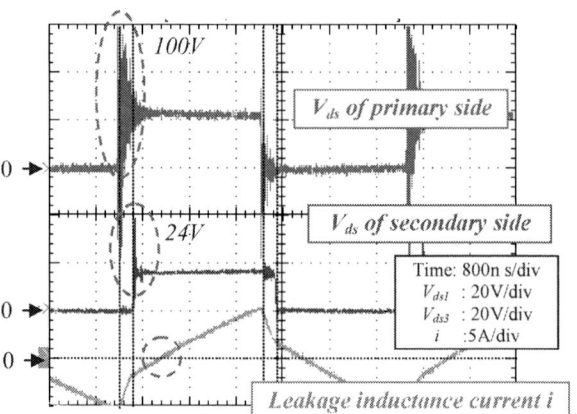

Fig. 8. The experimental waveforms in 1-module

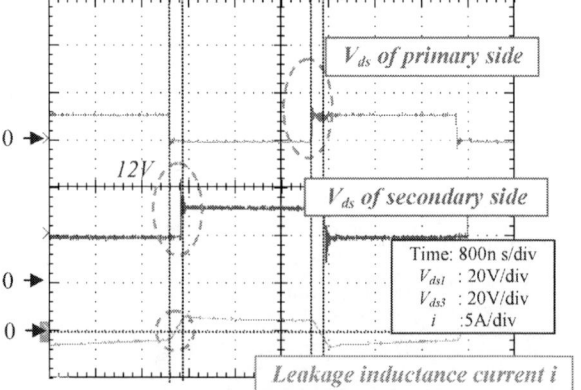

Fig. 9. The experimental waveforms in 2-modules

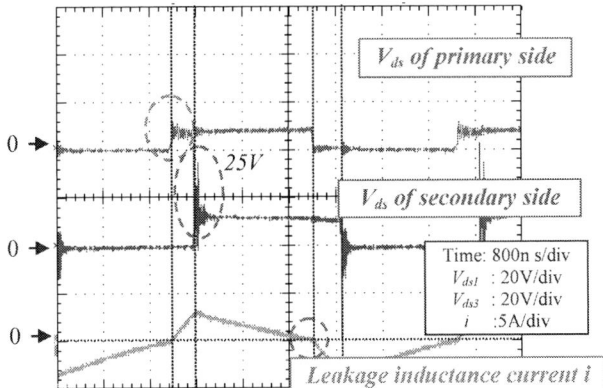

Fig. 10. The experimental waveforms in 3-modules

VII. DESIGN COMPARISON FOR INPUT CAPACITOR

When using the ISOP converter system, there are two patterns how to put the input capacitor. One case is to put the capacitance to each module as shown in Fig, 12*A*. On the other hand, it is to put only one at rail to rail of input voltage as shown in Fig. 12*B*. In this chapter, comparison of each characteristic in the case of two-module (*M*=1) as an example is discussed.

A. The case of putting input capacitor to each module

It will be described the case of putting input capacitor to each module as shown in Fig. 12(a). It is shown that the waveform of the input voltage and the current flowing through the transformer of each module in Fig. 13. From Fig. 13, it seems that the input voltage is relatively stable but uneven, and the current flowing through the transformer is also uneven. For this uneven, the first module is boost operation and the second module is buck operation at the same time. If the phase difference is small and the amount of current is small, it is possible that the hard switching operation by this uneven. Thus, it is necessary to take measures in order to give a bad influence on the circuit operation. Many papers have been solved this problem by control [1-4, 6, 7].

B. The case of putting only one input capacitor at input

It will be described the case of putting the input capacitor only one at input voltage as shown Fig. 12(b). It shows that the waveform of the input voltage and the current flowing through the transformer of each module in Fig. 14. From Fig. 14, it seems that the input voltage is oscillating but even, and the current flowing through the transformer is even. Since the current is even, it is considered that it can always be soft switching operation in $M = 1$.

The difference between these two operations will be described with reference to the state transition diagram. This paper will be described only in dead time because the voltage is vibrated at the moment of switching, namely, during dead time of state 4 of Fig. 2.

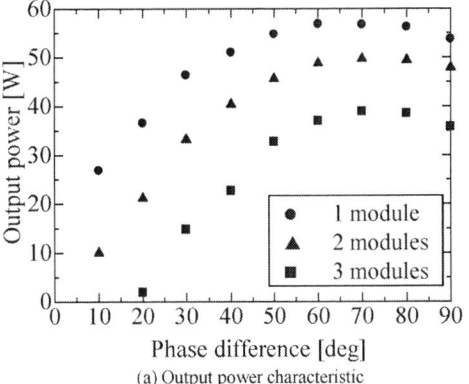

(a) Output power characteristic

(b) Efficiency characteristic

Fig. 11. The experimental resultss

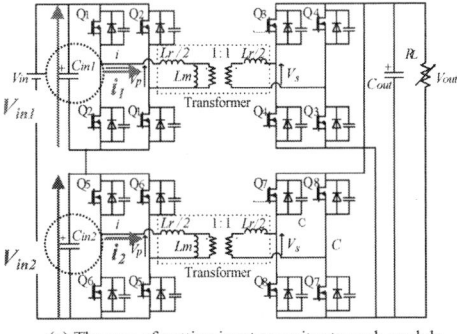

(a) The case of putting input capacitor to each module

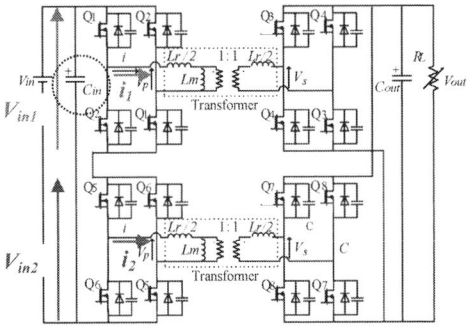

(b) The case of putting only one input capacitor at input

Fig. 12. Design comparison for input Capacitor

978-1-4673-9551-9/16 $31.00 © 2016 IEEE

Fig. 15 shows that the state transition diagram in the case of input capacitor put each module.

First, the primary side is the dead time at the switch of Q_1 is turned off. At this time, the parasitic capacitance of Q_1 and Q_5 is charged, and the parasitic capacitance of Q_2 and Q_6 is discharged. After discharging of the parasitic capacitance of Q_6 and charging of the parasitic capacitance of Q_5 is complete, the body diode of Q_6 is turned on. The current flow through the first input capacitor Cin1 since charging and discharging the parasitic capacitance of another module is not yet complete. Then, charging and discharging of the parasitic capacitance of the other module is complete, a current flows through the body diode of Q_1 and Q_2. Deviation of timing occurs since the power of each module is changed by a variation of the leakage inductance. Therefore, the current path is changed and the amperage is different for each module.

Fig. 16 shows that the state transition diagram in the case of putting only one input capacitor at input.

As with the previous case, the parasitic capacitance of Q_1 and Q_5 is charged, and the parasitic capacitance of Q_2 and Q_6 is discharged at the moment of dead time. After discharging of the parasitic capacitance of Q_6 and charging of the parasitic capacitance of Q_5 is complete, the body diode of Q_6 is turned on. In the previous case, it was discharged the energy charged in the leakage inductance through the input capacitor. However, this current path becomes open because the input capacitor is connected to a different location. Therefore, by pulling out the energy of another module, a current flows. At this time, it occurs resonance oscillation by parasitic capacitance of the switch and leakage inductance. Since the current flows through the same path in each module, there is no dispersion in voltage and current. It has been described in only primary side, but also the timing difference has occurred in the dead time of the secondary side by the operation similar to the primary side. However, the oscillation is not occurred in secondary side in all periods in order to cancel out the oscillation of each other by connecting in parallel. Therefore, the oscillation of the primary side is only observed.

Finally, the comparison is summarized. Fig. 17(a) shows the output power results and Fig. 17(b) shows the power efficiency results. Experimental and simulation conditions are both table 1. The input capacitor uses two electrolytic capacitors of 168µF. Comparing the cases A that putting the input capacitor to each module and case B that putting only one input capacitor at input. The efficiency has slightly improved 1.8% in the experiment at the maximum in the case B. It is considered that the ESR of the input capacitor is reduced because a capacitor is connected in parallel. It is necessary to use a capacitor withstand voltage and the ESR is larger in the case B in the actual equipment. Therefore, it is expected that the efficiency is deteriorated. The comparison will be continued with various points of view.

VIII. CONCLUSION & FUTURE WORK

In this paper, the design technique of ISOP with DAB for suppressing the surge voltage and realizing the high power efficiency in wide load range was proposed and confirmed the design technique with experimental results. By using the ISOP converter system, and by adjusting the optimum number of modules, it is possible to realize the soft switching operation at wide load range and high conversion ratio with DAB DC-DC converter. The optimum number of the input-output voltage conversion ratio was 1:1. In the experiment, this method reduces the switching surge and the power efficiency is increased by 22.5% at light load. Also, comparison of input capacitor design was discussed.

As the future work, more detailed analysis and the detailed transformer design will be performed.

ACKNOWLEDGMENT

The authors would like to thank Mr. S. Iwasaki, technical staff member of Nagasaki University, for his technical support.

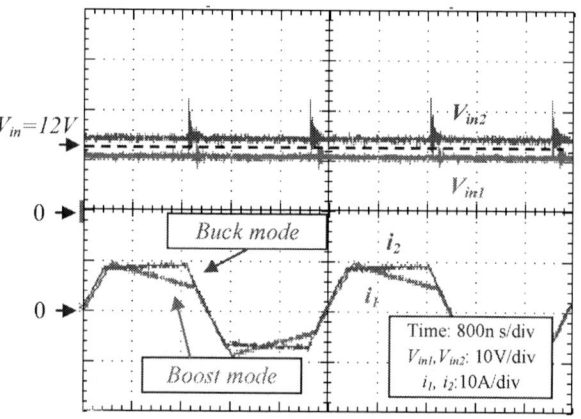

Fig. 13. The experimental waveforms in the case of A

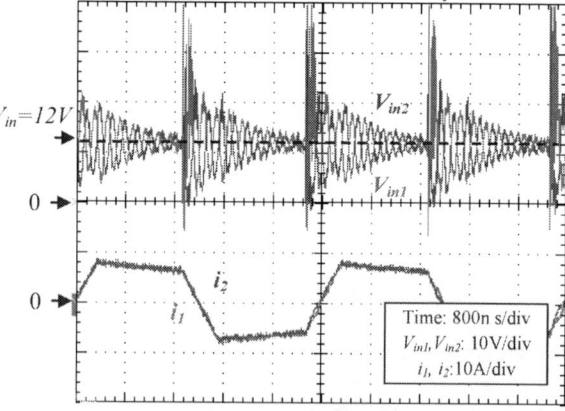

Fig. 14. The experimental waveforms in the case of B

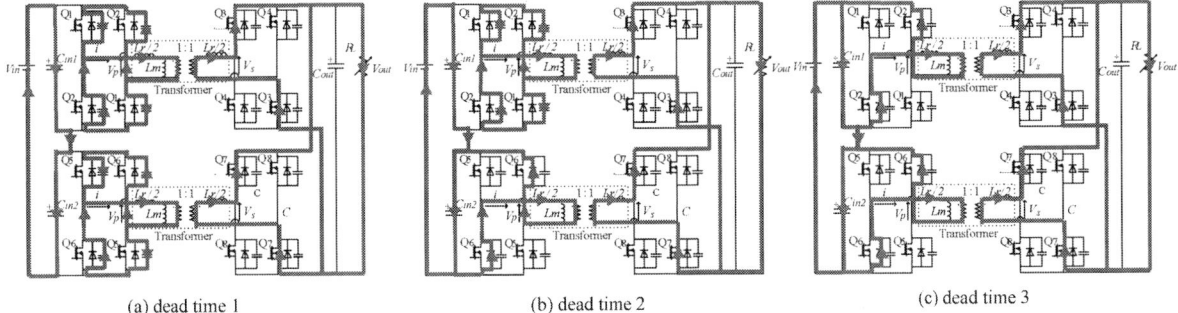

(a) dead time 1   (b) dead time 2   (c) dead time 3

Fig. 15. the state transition diagram in the case of input capacitor put each module

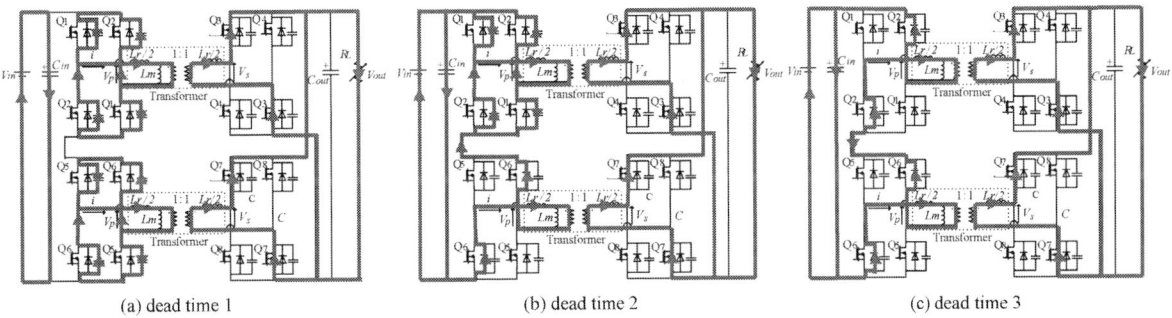

(a) dead time 1   (b) dead time 2   (c) dead time 3

Fig. 16. the state transition diagram in the case of putting only one input capacitor at input

REFERENCES

[1] Haifeng Fan and Hui Li, "Distributed Power Balance Strategy for DC/DC Converters in Solid State Transformer," Proceeding of the Applied Power Electronics Conference 2014, pp. 939 – 945, Mar. 2014.

[2] C. Fernandez, P. Zumel, M. Sanz, A. Lazaro and A. Barrado, "Combination of DCM and CCM DC/DC convert-ers for input-series output-series connection," Proceeding of the Applied Power Electronics Conference 2014, pp.2054-2060, Mar. 2014.

[3] Haifeng Fan, Hui Li, "A distributed control of input-series-output-parallel bidirectional dc-dc converter modules applied for 20 kVA solid state transformer," Proceeding of the Applied Power Electronics Conference 2011, pp. 939 – 945, Mar. 2011.

[4] Ghanshyamsinh Gohil, Huai Wang, Marco Liserre, Tamas Kerekes, Remus Teodorescu, Frede Blaabjerg, "Reduction of DC-link Capacitor in Case of Cascade Multilevel Converters by means of Reactive Power Control," Proceeding of the Applied Power Electronics Conference 2014, pp. 231 – 238, Mar. 2014.

[5] Yutian Cui, Weimin Zhang, Leon M. Tolbert, Fred Wang, Benjamin J. Blalock, "Direct 400 V to 1 V Converter for Data Center Power Supplies Using GaN FETs," Proceeding of the Applied Power Electronics Conference 2014, pp. 3460 – 3464, Mar. 2014.

[6] Xiaoyong Ren, Qiang Zhang, Zhenjin Pang, Qianhong Chen, Xinbo Ruan, " Input-Series Output-Parallel Multiple Output Converter," Proceeding of the Applied Power Electronics Conference 2014, pp. 3266 – 3272, Mar. 2014.

[7] P. Zumel, L. Ortega, A. Lazaro, C. Fernandez, A. Barrado, "Control strategy for modular Dual Active Bridge Input Series Output Parallel," Control and Modeling for Power Electronics 2013, pp. 1-7, June. 2013.

[8] Peng Zhao, Lan Xiao, Zilong Wang, "Research on a Combined Input-Series Output-Parallel DC/DC Converter," IECON 2012, pp. 725-729, Oct. 2012.

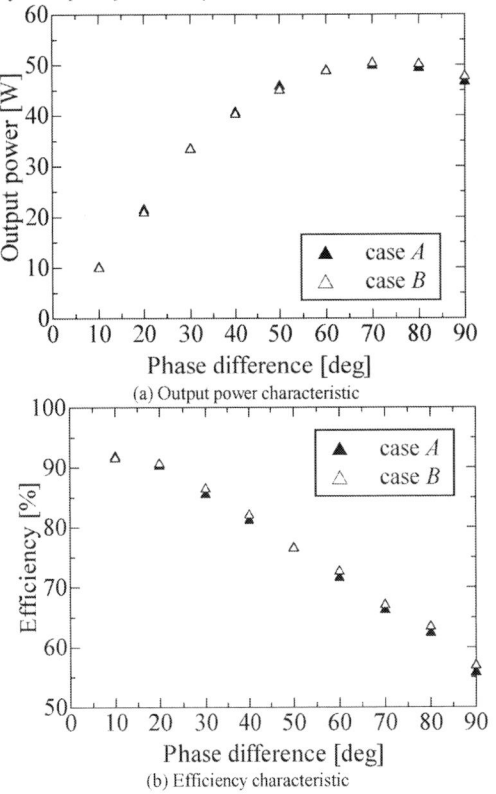

(a) Output power characteristic

(b) Efficiency characteristic

Fig. 17. The characteristic difference by how to put input capacitor

[9] Yujin Song, P. N. Enjeti, "A new soft switching technique for bidirectional power flow, full-bridge DC-DC converter," in Proc. 37th IAS Annual Meeting, pp.2314-2319, Oct. 2002.

[10] Mika Takasaki, Yoichi Ishizuka, Tamotsu Ninomiya, Yutaka Furukawa, and Toshiro Hirose "A Power Efficiency Improvement Technique for A Bi-Directional Dual Active Bridge DC-DC Converter at light load," EPE-ECCE 2013, pp.1-10, Sept. 2013.

[11] Mustansir H. Kheraluwala, Randal W. Gascoigne, Deepakraj M. Divan, and Eric D. Baumann, "Performance characterization of a high-power dual active bridge dc-to-dc converter," IEEE Trans. Industry Applications, vol.28, NO.6, pp. 1294-1301, Nov. / Dec. 1992.

[12] Shigenori Inoue and Hirofumi Akagi "A bidirectional isolated dc-dc converter as a core circuit of the next-generation medium-voltage power conversion system," IEEE Trans power Electron, vol.22, no.2, pp. 535-542, Mar. 2007.

[13] M. Pavlovsky, S. W. H. de Hann, and J. A. Ferreira, "Concept of 50kW DC/DC converter based on ZVS, quasi-ZCS topology and integrated thermal and electronic design," 2005 European Conference on Power Electronics and Applications, Sept. 2005.

[14] Rik W. A. A. De Doncker, Deepkraj M. Divan, and Mustansir H. Kheraluwala, "A Three-Phase Soft-Switched High-Power-Density dc/dc Converter for High-Power Applications," IEEE Trans. Industry Applications, vol.27, NO.1, pp.63-73, Jan. / Feb. 1991.

[15] Shun Nagata, Mika Takasaki, Yutaka Furukawa, Toshiro Hirose, Yoichi Ishizuka,"A Static Characteristic Analysis of Proposed Bi-Directional Dual Active Bridge DC-DC Converter," Power Electronics Conference 2014, pp.2252-2259, May. 2014.

Wide Input Range Power Converters Using a Variable Turns Ratio Transformer

Ziwei Ouyang, and Michael A. E. Andersen

Department of Electrical Engineering,
Technical University of Denmark, Kgs. Lyngby, Denmark
Email: zo@elektro.dtu.dk

Abstract— **A new integrated transformer with variable turns ratio is proposed to enable dc-dc converters operating over a wide input voltage range. The integrated transformer employs a new geometry of magnetic core with "four legs", two primary windings with orthogonal arrangement, and "8" shape connection of diagonal secondary windings, in order to make the transformer turns ratio adjustable by controlling the phase between the two current excitations subjected to the two primary windings. Full-bridge boost dc-dc converter is employed with the proposed transformer to demonstrate the feasibility of the variable turns ratio. 1-kW experimental prototype targeting to the PV standalone system has been built to well demonstrate a wide input voltage operation with high efficiencies.**

Keywords— wide input voltage, dc-dc, integrated transformer, variable turns ratio

I. INTRODUCTION

It is remarkable to operate a dc/dc converter with high conversion efficiency over a wide input/output voltage range for renewable applications such as solar energy, wind energy systems, and fuel cell powered systems etc. Renewable power, by nature, is varying with time, weather, and environment, and it is not stable and somehow discontinuous [1]. In other words, the output voltages of such renewable energy sources have a wide variation range. Therefore, a front-end dc-dc converter with a wide input range is often required [2]-[4]. Furthermore, there are many other applications where the power converters that can deal with a much wider range of input voltage variation are desirable. For instance, in some systems the distributed input voltages have significant transients and surges that last too long to be removed by a filter.

In a quest for converters with high efficiency and wide conversion range, many circuit topologies have been proposed. Single-transistor converter topologies, with quadratic conversion ratios, were proposed in [5] and successfully demonstrated a large step-down conversion ratio. Coupled inductors were proposed in [6] to obtain a wide conversion range. A tapped-inductor buck with soft switching and its derivations were introduced in [7]-[8]. To attain very large voltage step-up, cascaded boost converters that implement the output voltage increasing in geometric progression were introduced in [9]. These converters effectively enhance the voltage transfer ratio; however, their circuits are quite complex. In comparison, tapped-inductor boost converters proposed in [10] and [11] attain a comparable

voltage step-up preserving relative circuit simplicity. Introducing a transformer helps attaining large step-up or step-down voltage conversion ratio. Transformers' turn ratio should be chosen as to provide the desired voltage gain while keeping the duty cycle within a reasonable range for higher efficiency. However, for some resonant converters such as LLC converters and dual active bridge (DAB) converters, the converters will fail to satisfy the soft switching conditions when the ratio of input to output voltage is not close to the transformer turns ratio [12]. Variable turns ratio transformer will be a great benefit to those resonant converters to achieve the soft switching at all conditions.

This paper proposes a wider input range dc-dc converter with a new integrated transformer. The transformer winding configuration enables a variable turns ratio transformer controlled by the phase between the two paralleled power stages, and thus allowing a wider input/output range.

II. MODELING OF NEW INTEGRATED TRANSFORMER

As shown in Fig. 1(a), a new flux decoupling concept for integrated transformer design is based on 3-dimensional (3D) space orthogonal flux decoupling wherein the three flux paths extend substantially orthogonally to each other within the shared magnetically permeable core. This requires a sort of non-traditional core geometries to carry out the orthogonal flux paths, shown in Fig. 1(b) where the magnetically permeable core comprises of a base rectangular core plate with four legs situated at respective corners, and a top rectangular core plates. In order to clearly illustrate the winding arrangements, a transparent top core plate is drawn in Fig. 1(b). The first primary winding *AB* and the second primary winding *CD* are orthogonally arranged as a cross-shaped layout in-between the four legs. By using the right hand rule,

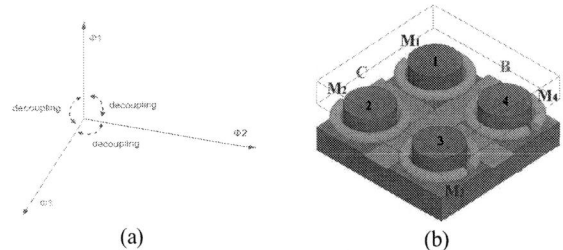

Fig. 1. (a) orthogonal flux decoupling; (b) modeling of the new integrated transformer.

the flux generated from winding AB is substantially orthogonal to the flux generated from winding CD in the base and top core plates, and thus the two primary windings are decoupled. The four legs provide a shared magnetic flux path to the two primary windings where the four secondary windings, M_1, M_2, M_3 and M_4, are wound in each respectively. Therefore, the two primary windings are individually coupled with all of the four secondary windings [13]-[14].

III. OPERATION PRINCIPLE

Magnetic flux generated from the primary windings flow through the four legs, the individual voltage will be accordingly produced on each secondary winding M_1, M_2, M_3 and M_4. The produced voltages can be varied depending on the excitation currents and the operating phase of the currents. This is due to that the magnetic flux generated from the primary windings may be cancelled or overlapped within the four legs. Therefore, variable turns ratio transformer is produced. Fig. 2 shows four stages of flux distribution in the core along the different current phases. The induced secondary voltages controlled by the phase angles between the two excitations subjected to the primary windings are shown in Fig. 3, with an assumption of the same amplitude on the two magnetizing currents.

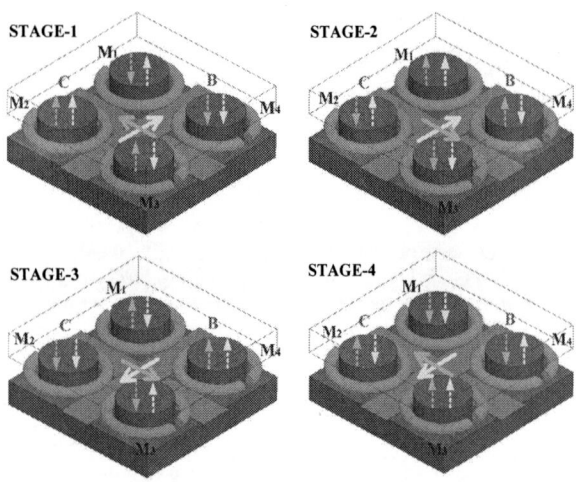

Fig. 2. Four stages of flux directions through the legs.

1) Stage 1 (t_0—t_1): The magnetizing current i_{T1} is positive and the magnetizing current i_{T2} is negative. This current state is characterized by the indicated direction shown in STAGE-1 of Fig. 2. Using the right-hand rule, the current i_{T1} leads to the magnetic fluxes in the leg 1 and leg 2 (Leg 1 is the leg corresponding with the winding M_1. The rest may be deduced by analogy) flowing along the same direction indicated by the yellow dash arrows. This is a consequence of the leg 1 and the leg 2 being arranged on the same side of the primary winding AB. Furthermore, the current i_{T1} also leads to the magnetic fluxes in the leg 3 and leg 4 flowing along the same direction. The magnetic fluxes induced by the current i_{T1} leads to oppositely directed fluxes in the leg 1 and leg 2

relative to the leg 3 and leg 4 which is a consequence of the geometry of the closed magnetic flux loop. Likewise, the current i_{T2} leads to the same flux directions in the leg 1 and leg 4 indicated by the blue dash arrows. The opposite flux direction relative to the flux direction in the leg 1 and leg 4 are induced in the leg 2 and leg 3 by the current i_{T2}. Hereby, a flux cancellation can be observed in the leg 1 and leg 3, meanwhile an overlapped flux occurs in the leg 2 and leg 4.

2) Stage 2 (t_1—t_2): The magnetizing currents i_{T1} and i_{T2}, are both positive. This currents state is characterized by the indicated direction shown in STAGE-2 of Fig. 2. The magnetic fluxes induced by the current i_{T1} through all four legs keep the same directions with those produced in the Stage 1 since the direction of the current i_{T1} is not changed. However, because the direction of the current i_{T2} is changed, the directions of the induced magnetic fluxes indicated by the blue dash arrows are accordingly changed. Hereby, a flux cancellation can be observed in the leg 2 and leg 4, meanwhile an overlapped flux occurs in the leg 1 and leg 3.

3) Stage 3 (t_2—t_3): The magnetizing current i_{T1} becomes negative and the magnetizing current, i_{T2}, still keeps positive. This current state is characterized by the indicated direction shown in STAGE-3 of Fig. 2. All magnetic fluxes induced by the currents i_{T1} and i_{T2} are reversed relative to those induced in the Stage 1.

4) Stage 4 (t_3—t_4): The magnetizing currents i_{T1} and i_{T2} are both negative. This currents state is characterized by the indicated direction shown in STAGE-4 of Fig. 2. All magnetic fluxes induced by the currents i_{T1} and i_{T2} are reversed relative to those induced in the Stage 2.

As shown in Fig. 3, not only the shapes but also the amplitudes of the voltages are changed. The maximum output voltage on the secondary windings is achieved in the cases of 0° and 180° phase shift. The minimum voltage is obtained in the case of 90° phase shift, and the case of 45° phase shift causes an output voltage in between. In fact, this characteristic of the flux cancellation and adding in each pair of diagonal legs effectively changes turns ratio of the transformer. Accordingly, the output voltage with a specific topology can be controlled by the phase shift angle. The function of voltage gain in the converter cannot simply described by the duty cycle and turns ratio since the phase shift angle has to be considered.

IV. VARIABLE TURNS RATIO TRANSFORMER

The secondary windings can be connected in different ways, for example, diagonal windings (M_1 and M_3, or M_2 and M_4) are connected in series; neighbor windings ($M1$ and $M2$, or M_3 and M_4) are connected in series; all windings are in series; and all windings are independent. Different characteristics may be created due to the different connections. The selection of the winding connection relies on the specific applications, and the full-bridge boost dc-dc converter is selected in this study. The diagonal windings are connected in series as shown in Fig. 4. The secondary windings are wound

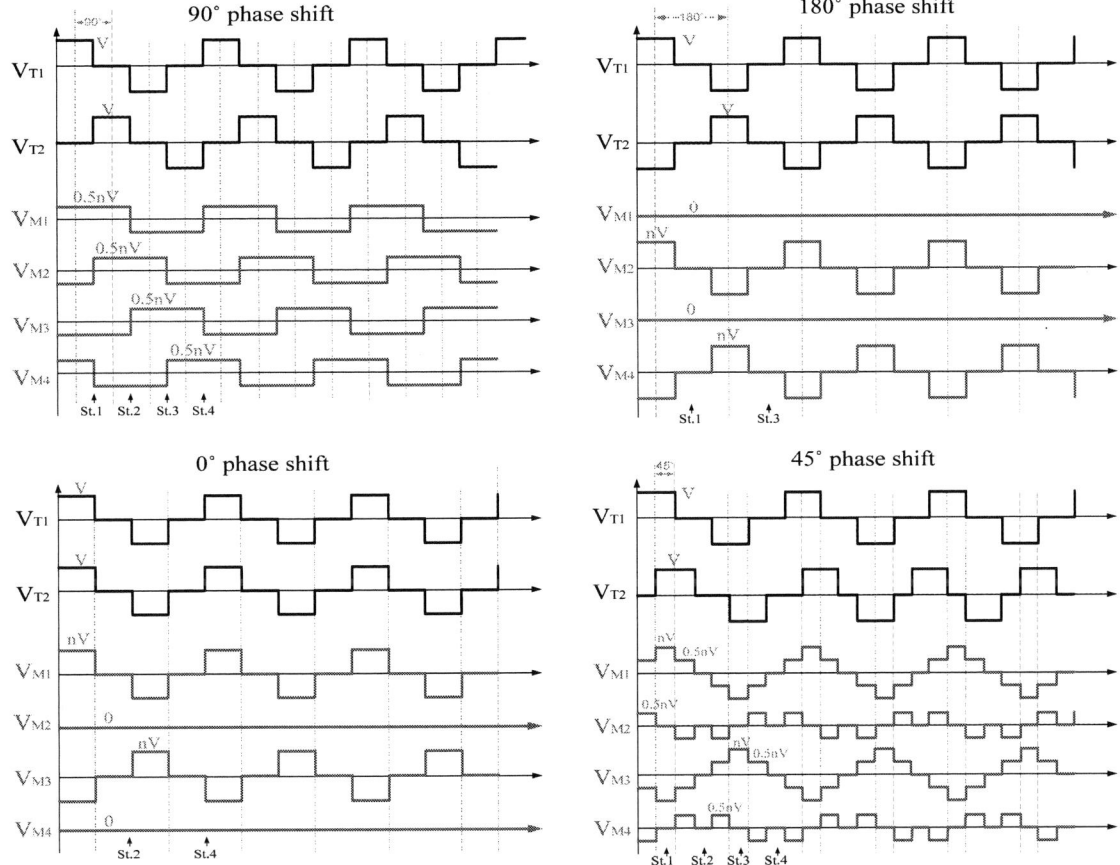

Fig. 3. Induced secondary voltages in each leg by different phase shift angles between the two primary excitations

with "8" shape in the respective diagonal legs. The equivalent circuit with the expression of traditional multiple winding transformers is shown in Fig. 5. It is noted that the secondary windings S_1, S_2, S_3 and S_4 are not identical to the individual secondary windings M_1, M_2, M_3 and M_4 shown in Fig.1. The effective turns of the windings S_1, S_2, S_3 and S_4 have been marked in Fig. 4. It can be observed that winding S_2 and S_4 are reversely connected. This is an important advantage for the

proposed dc-dc converter. Because of the reversed connection, the currents flows in the secondary windings of the transformer T_2 are cancelled when only one primary side is excited. It results in a zero current referred to the other primary side, eliminating the unnecessary circulating current when the two primary currents are shift.

Fig. 6 shows the gate signals with phase shift control where the voltages across to the transformers are shifted with

Fig. 4 Practical winding implementation on print circuit board (PCB)

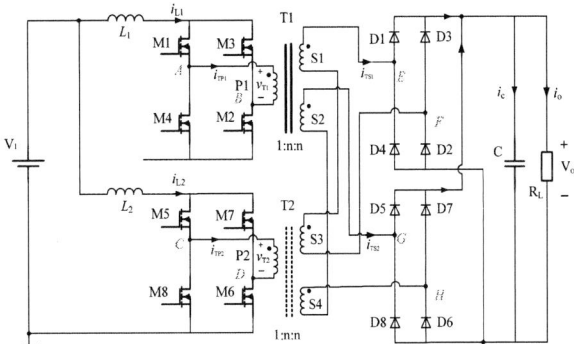

Fig. 5. Equivalent circuit of the integrated variable turns ratio transformer

978-1-4673-9551-9/16 $31.00 © 2016 IEEE

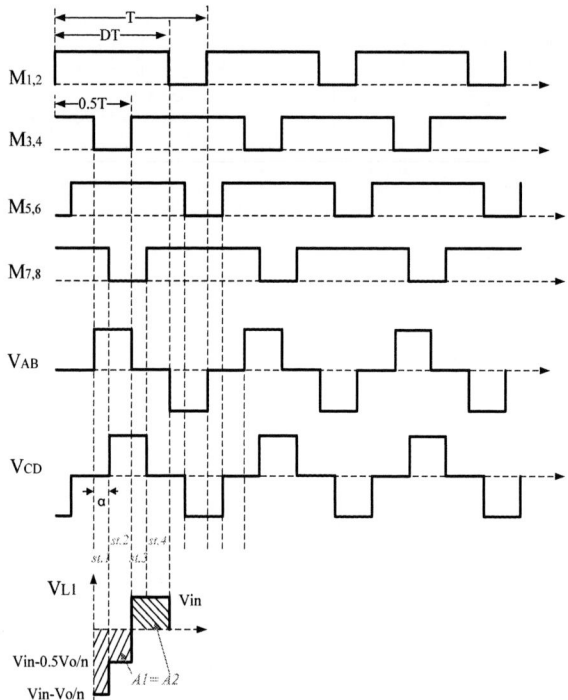

Fig. 6. Gate signals with phase shift control for the full-bridge boost dc-dc converter

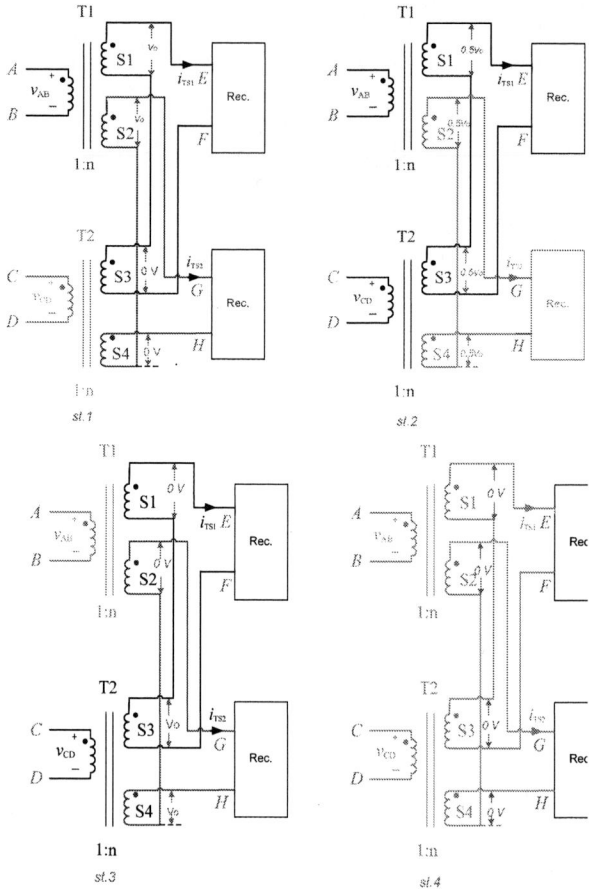

Fig. 7. Equivalent circuit of the transformer under four operation stages

the αT. The circuit run within a half period can be classified into four stages as shown in Fig. 7:

Stage 1: the inductor L_1 is discharged, and the transformer T_1 is working as normal. At the same time, the inductor L_2 is charging, and the secondary winding S_3 and S_4 are, therefore, clamped to be zero. The output voltage V_o is applied into the winding S_1 and S_2 in parallel. The winding S_3 and S_4 are still conducting in the same currents with reversed coupling. Therefore, the current referred into the primary winding CD is zero.

Stage 2: the inductors L_1 and L_2 are both discharged, and the transformers T_1 and T_2 are working as normal. The voltages applied on the winding S_2 and S_4 are cancelled due to its reversed connection and the same number of turns. Therefore, the voltage V_{GH} is zero, and the associated rectifier is reverse bias. The output voltage is equally distributed into the two series-windings S_1 and S_3. The same currents are referred to the two primary windings AB and CD.

Stage 3: the inductor L_2 is discharged, and the transformer T_2 is working as normal. At the same time, the inductor L_1 is charging, and the secondary winding S_1 and S_2 are, therefore, clamped to be zero. The output voltage V_o is applied into the winding S_3 and S_4 in parallel. The winding S_1 and S_2 are still conducting in the same currents with reversed coupling. Therefore, the current referred into the primary winding AB is zero.

Stage 4: the inductors L_1 and L_2 are both charged, and the

transformers T_1 and T_2 are not working. All the voltages applied on transformer windings are zero, and no current is running.

The voltage across on the inductor L_1 is shown in Fig. 6 where the voltage in the stage 2 is actually reduced. This is

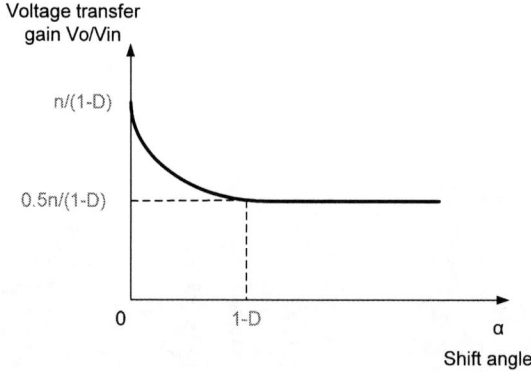

Fig. 8 Voltage transfer gain of the transformer vs. shifted angle

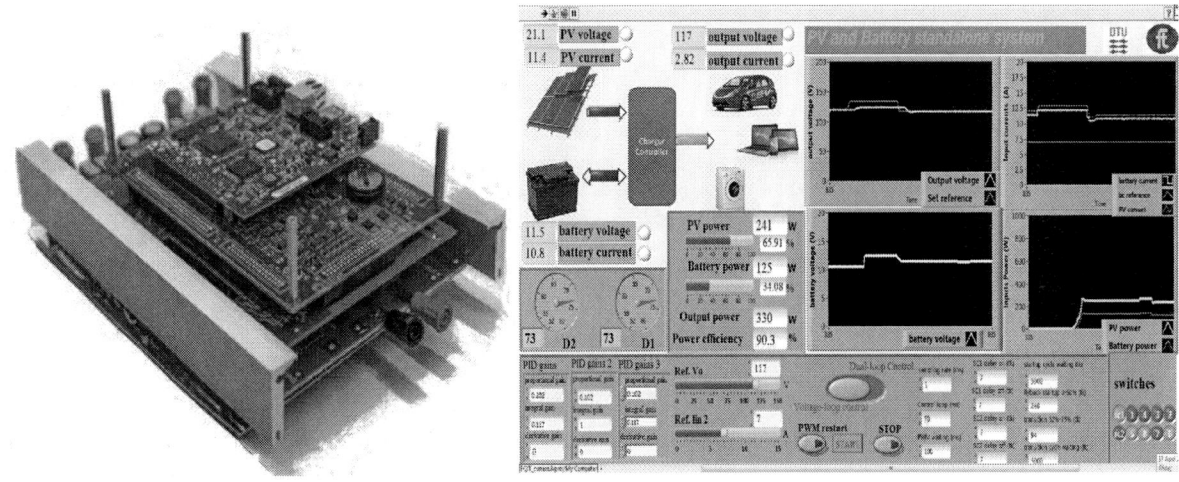

Fig. 9. Photo of experimental prototype

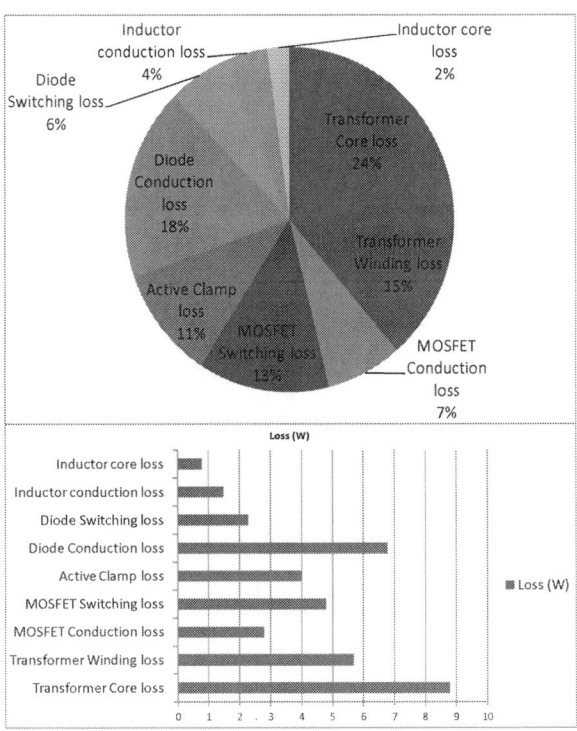

Fig. 10. Power loss breakdown when the output power is 700 W.

because only half of the output voltage is referred to the primary side. This essentially makes the transfer function different, and make it feasible to regulate the output voltage by controlling the phase shift angle. Following by the voltage-second balance of inductor, the transfer function is derived:

$$\begin{cases} \dfrac{V_o}{V_{in}} = \dfrac{n}{1-D+\alpha} & (\alpha < 1-D) \\[2mm] \dfrac{V_o}{V_{in}} = \dfrac{0.5n}{1-D} & (0.5T \geq \alpha \geq 1-D) \end{cases}$$

It is seen that the max voltage transfer gain can be achieved when the phase shift α is zero. This can be explained by the case that the two full-bridges are running in series where the windings S_2 and S_4 are not conducted. When the shift angle is increased, the gain is accordingly reduced until the angle α reaches to 1-D. When the shift angle is from 1-D to 0.5T, the gain is independent of the phase angle. This can be explained by the case that the two full-bridges are running in parallel where the minimal transfer gain is obtained.

V. EXPERIMENTAL VERIFICATION

In order to verify the correctness of the analysis, a proof-concept experimental prototype targeting to the PV standalone system has been built. Fig. 9 shows a photo of the experimental prototype. The input voltages varies from 10-40 V, the output voltage can be regulated from 60 V to 200 V (the nominal voltage is 150 V), and the nominal output power is 700 W. Mn-Zn ferrite is used as the material of magnetically permeable core. The switching frequency is operated at 100-kHz. Labview Single-RIO board is used as controller. The calculated power loss breakdown is shown in Fig. 10, in which the core loss is dominated can be observed. This is because of higher magnetic flux density caused by the low number of turns in the primary sides. The optimal design should make an increase of number of turns in the primary sides to balance the core loss and the winding loss. However, this will be at expense of PCB price due to multiple layers. The measured power efficiency is shown in Fig. 11. A peak power efficiency of 95.2% can be obtained when the two primary stages deliver power simultaneously. 45 degree phase shift gives a reduction of power efficiency at nominal power but an increase at light load. This is because 45 degree phase

978-1-4673-9551-9/16 $31.00 © 2016 IEEE

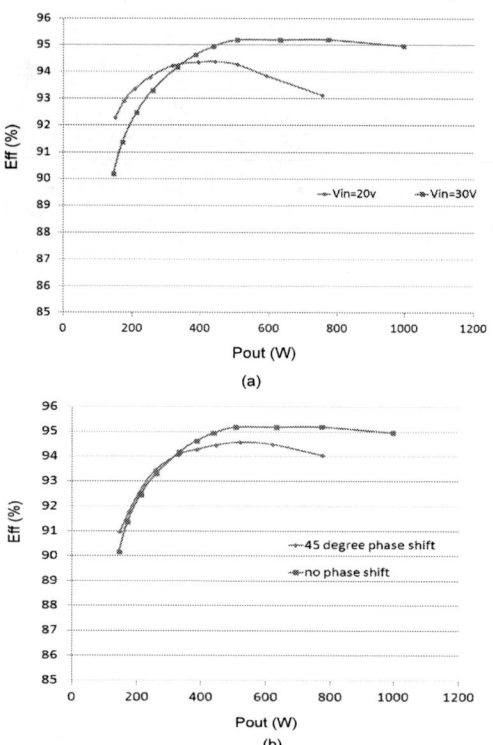

Fig. 11. Measured power efficiencies with (a) zero phase shift (b) 45 degree phase shift.

shift effectively reduces a transformer turns ratio and thus requires a higher duty cycle for voltage boosting. However, there is flux cancellation in the core and thus results in a higher efficiency in the light load.

VI. CONCLUSIONS

This paper presents a new integrated transformer with a novel magnetic core geometry of four legs, which enables variable turns ratio controlled by the phase between the two current excitations subjected to the primary windings. The modeling and operation principle have been presented. 1 kW experimental prototype has been built to demonstrate a wider input operation for photovoltaic (PV) and battery stand-alone system.

REFERENCES

[1] Guest editorial - special issue on power electronics in DC distribution systems, *IEEE Trans. on Power Electron.*, vol.28, no.4, pp.1507-1508, April 2013.

[2] L. Gao, R. A. Dougal, S. Liu, and A. P. Iotova, "Parallel-connected solar pv system to address partial and rapidly fluctuating shadow conditions," *IEEE Trans. on Industrial Electronics* , vol.56, no.5, pp.1548-1556, May 2009

[3] M. H. Todorovic, L. Palma, and P. N. Enjeti, "Design of a wide input range dc–dc converter with a robust power control scheme suitable for fuel cell power conversion," *IEEE Trans. on Industrial Electronics,*, vol.55, no.3, pp.1247-1255, March 2008

[4] H. Cheng, K. M. Smedley and A. Abramovitz, "A wide-input–wide-output (WIWO) dc–dc converter," *IEEE Trans. on Power Electron.*, vol.25, no.2, pp.280-289, Feb. 2010

[5] D. Maksimovic and S. Cuk, "Switching converter with wide dc conversion range," *IEEE Trans. Power Electron.*, vol. 6, no. 1, pp. 151–157, Jan. 1991.

[6] K. Yao,M. Ye, M. Xu, and F. C. Lee, "Tapped-inductor buck converter for high-step-down dc–dc conversion," *IEEE Trans. Power Electron.*, vol. 20, no. 4, pp. 775–780, Jul. 2005.

[7] J.-H. Park and B.-H. Cho, "Nonisolation soft-switching buck converter with tapped-inductor for wide-input extreme step-down applications," *IEEE Trans. Circuits Syst. I, Reg. Papers*, vol. 54, no. 8, pp. 1809–1818, Aug. 2007.

[8] H. Cheng and K. Smedley, "Wide input wide output (WIWO) DC-DC converter," in *Applied Power Electronics Conference and Exposition, 2008. APEC 2008. Twenty-Third Annual IEEE* , vol., no., pp.1562-1568, 24-28 Feb. 2008.

[9] F. L. Luo and H. Ye, "Positive output cascade boost converters," *Proc. Inst. Electr. Eng. Electr. Power Appl.*, vol. 151, no. 5, pp. 590–606, Sep. 2004.

[10] Q. Zhao and F. C. Lee, "High efficiency, high step-up dc–dc converters," *IEEE Trans. Power Electron.*, vol. 18, no. 1, pp. 65–73, Jan. 2003.

[11] N. Vazquez, L. Estrada, C. Hernandez, and E. Rodriguez, "The tappedinductor boost converter," in *Proc. IEEE Int. Symp. Ind. Electron.*, Jun., 4–7, 2007, pp. 538–543.

[12] Z. Wang, and H. Li, "A soft switching three-phase current-fed bidirectional dc-dc converter with high efficiency over a wide input voltage range," *IEEE Trans. on Power Electron.*, vol.27, no.2, pp.669-684, Feb. 2012

[13] Z. Ouyang, Z. Zhang, M. A. E. Andersen and O. C. Thomsen, "Four Quadrants Integrated Transformers for Dual-Input Isolated DC–DC Converters," *IEEE Trans. on Power Electron.*, vol.27, no.6, pp.2697-2702, June 2012.

[14] Z. Ouyang, M. A. E. Andersen and Z. Zhang, "An integrated magnetics component," PCT/EP2012/067422, published no.: WO 2013037696 A1, Sept., 2011.

978-1-4673-9551-9/16 $31.00 © 2016 IEEE

Design approaches for fast supercapacitor chargers for applications like SCATMA, SRUPS

Nicoloy Gurusinghe, Nihal Kularatna, W. Howell Round and D. Alistair Steyn-Ross
School of Engineering
University of Waikato
Hamilton 3240, New Zealand
Email: nihalkul@waikato.ac.nz

Abstract—**The proliferation of electric vehicles (EV) creates a need for fast battery chargers. In order for systems of supercapacitors (SC) to reach adequate energy density levels to replace battery packs, fast SC chargers will also be required. However, when a fully discharged supercapacitor pack is charged from an ideal voltage source, the five-time-constant charging bottleneck comes into play. This issue can be addressed by using the high-voltage charging topology proposed here. The new approach partitions the SC bank into two parts. One part is charged by a high voltage charging source, and the other is replenished by the energy stored in a coupled inductor within the charging path used for the first half of the bank. Early experimental results for a 30 V, 15 F SC bank are presented.**

I. INTRODUCTION

Compared to a battery with an electrochemical potential opposing the charging voltage, a completely depleted SC bank takes five time constants (approximately) to become fully charged from an ideal voltage source. The time constant is created by the capacitance and the total loop resistance. In the case of a current source, total time to charge a capacitor decreases linearly with increasing current. In either case, fast charging requires high-current capable dc sources. In comparison to a dc source powering a resistive load continuously, a SC charger can be specified by the maximum charge or the energy required by the SC bank.

Battery chargers available in literature [1]–[3], are mainly intended for fast charging auto mobile batteries with capacities in the range of several kilowatt hours. The presented charger topology is developed for special applications such as supercapacitor-assisted temperature modification apparatus (SCATMA) [4] and surge-resistant uninterrupted power supply (SRUPS) [5], [6]. In these implementations SCs are used for energy storage in an energy and power context of 20–200 Wh and 2–20 kW respectively. SC chargers found in literature [7]–[9] does not address the energy and power context of these applications.

SCATMA is an "instant" liquid-flow heating technique based on pre-stored energy to overcome the delayed delivery of hot water in domestic water taps. SRUPS uses the surge absorption capability of SCs as illustrated in [10] to come up with a surge-resistant uninterrupted power supply (UPS) which can withstand short term power outages up to several ac mains cycles. Both techniques store energy in SCs. SCATMA uses this energy for high-power energy delivery into running water

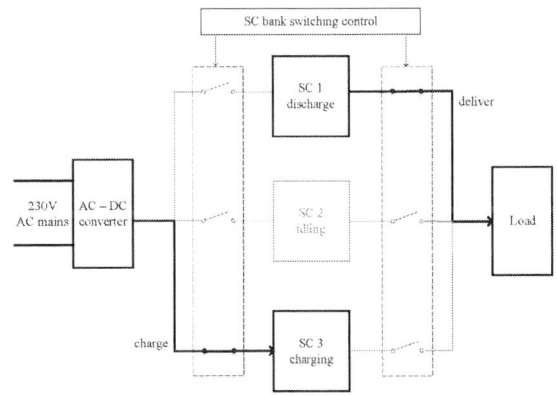

Fig. 1. Bank circulation technique: A three-bank example of the bank circulating technique. The load will be an inverter in the case of SRUPS and a resistive high power heater coil for SCATMA. Source [5].

and SRUPS uses the SCs to provide surge protection with galvanic isolation to the load.

Both SCATMA and SRUPS use a circulating SC bank to minimize the amount of pre-stored energy. In a circulating storage bank, the banks are operated independently as energy-storing elements in one of the three states of (i) discharging, (ii) idling or (iii) being replenished by the fast charger (see Fig. 1). For this approach to be feasible, the SC charging technique should charge the SCs in less than five time constants.

Section II gives an overview of the two special applications, Section III describes the bank circulation technique, Section IV analyses different capacitor charging methods, Section V compares charging a capacitor through a resistive path vs an inductive path, Section VI presents the operation of the new topology and Section VII presents the results from a prototype charger.

II. FAST CHARGING REQUIREMENTS OF SCATMA AND SRUPS

A. SCATMA

Based on the average specifications of "instant" water heating, it requires 100–150 Wh of energy for the 30 s period of operation. A SC bank up to 50 Wh is feasible based on cost

978-1-4673-9551-9/16 $31.00 © 2016 IEEE

and size. Possible energy sources for the rest of the energy requirement are,

1) to draw more than 10 A from the mains supply for a short period of time without overloading the wiring or tripping the protective circuit breaker. For example, an ABB miniature circuit breaker S201PR-K10 [11], a 240 VAC, 10 A rated supplemental protector, is designed to pass five times the nominal 10 A current for 10 s without tripping. This allows a maximum of 11.5 kW.
2) the normal mains supply can provide 2.3 kW. This can be used for the time after the mains supply has had its 10 A supply exceeded by the previous option.
3) energy stored by an alternative cheaper than SCs (such as a battery). Such a store can have a lower power density but a higher energy density than SCs.

According to option one, if a protection device designed to allow high in-rush currents during system start up is used, a maximum power of 11.5 kW for 10 s and 2.3 kW for the remaining 20 s can be drawn from the mains supply. This is a total energy of (11.5 kW×10 s) + (2.3 kW×20 s) = 32 Wh + 13 Wh = 45 Wh. Based on 150 Wh total energy, this is equivalent to 30% of the requirement. In the case of not having the option of laying a dedicated high-in-rush current subcircuit for a specific household, then the mains can supply up to 2.3 kW×30 s = 19.2 Wh, which is 12.8% of the needed total energy.

Given the availability of these options the problem lies in how this energy can be delivered on to the load with galvanic isolation and at the required 15–18 kW average power level. For both of these conditions to be satisfied, a SC bank circulation technique first proposed for SRUPS [6], [12] is used. Here SCs are used as an isolating device as well as to deliver the required high power.

B. SRUPS

The SRUPS is an UPS technique developed to withstand short power outages of several ac mains cycles up to a few hundred milliseconds. The advantage of this technique is that the inverter runs online and is continuously fed by a circulating capacitor bank as seen in Fig. 1. At a given instant a bank partitioned into three has one bank supplying the load while another is fully charged and ready to deliver and the third bank is being charged. Through this circulating principle, and by using an on-line inverter, galvanic isolation is achieved. This is advantageous for UPS surge resistance. At all times the mains supply is never directly connected to the load. The load will be fed from a separate SC bank while the mains is charging another SC bank. Any incoming transient voltage spike in the mains supply will be safely absorbed by the SC being charged.

This technique provides a new way to use SCs in the middle of an UPS where the SCs are used as both an energy storage medium and a surge absorbing element, providing an enhanced level of surge protection to the connected loads. This technique can be developed to a fully versatile topology to eliminate the requirement of a battery pack for short-term ac mains power blackouts.

Both SCATMA and SRUPS use SCs for energy storage. Due to the current cost of SCs, a SC-only solution is prohibitively expensive. SC prices are expected to drop in the future as more and more automotive applications start using SCs. But the energy density of SCs are not expected to reach that of batteries in the near future [13]. Therefore the most feasible solution is one which reduces the amount of pre-stored energy.

III. BANK CIRCULATION

The bank circulation technique can be used in any high-power application where the load power requirement is higher than that of the mains supply. Rather than purely relying on prestored energy, the amount of the energy stored can be reduced by supplementing with mains energy through a circulating bank. A circulating SC bank may consist of as many SC banks as required, each in the three states mentioned earlier. Depending on the amount of energy each bank stores, the system may be required to be circulated for several cycles during the course of operation. The optimum choice of SC bank size and number of cycles required in delivering a predetermined quantity of energy will

- minimize the leftover energy at the end of the process
- make sure the required constant power output is maintained throughout the period of operation

IV. CAPACITOR CHARGING METHODS

Most practical sources used for capacitor charging are current limited. This section compares the time taken by four differently configured voltage sources to fully charge a capacitor up to its rated voltage (V_c) starting from zero.

A. Voltage source at capacitor rated voltage with no current limit

This is the typical textbook case of charging a capacitor from a voltage source. Capacitor voltage $v_c^1(t)$ and current $i_c^1(t)$ will be

$$v_c^1(t) = V_c(1 - e^{-t/\tau}), \quad i_c^1(t) = \frac{V_c}{R}e^{-t/\tau}$$

where $\tau = RC$ is the circuit time constant. Capacitor voltage $v_c^1(t)$ reaches $0.993V_c$ after five time constants. For the capacitor to be charged within this time, the source should be capable of supplying the maximum current V_c/R. Fig. 2, case (1) illustrates the voltage and current plots.

B. Voltage source at capacitor rated voltage with current limit I

Most practical capacitor chargers will be of this nature. Since the converter is unable to deliver high currents, the operation will be in constant current mode until the current requirement of the circuit goes below I or equivalently, until the capacitor is charged up to $V_c - IR$. The constant current charging duration t_2^1 will be

$$t_1 = \frac{C}{I}(V_c - IR)$$

After t_1, the capacitor will follow an exponentially decaying current until the capacitor is fully charged. At $t = t^*$, $v_{c2}^i(t^*) = V_c - IR_L$ on a non current limited charging curve

$$t^* = -RC \ln\left[1 - \frac{V_c - IR}{V_c}\right] = RC \ln\left[\frac{V_c}{IR}\right]$$

$$v_{c,2}(t) = V_c(1 - e^{-t'/\tau}) \quad, \quad i_{c,2}(t) = \frac{V_c}{R} e^{-t'/\tau}$$

$$\text{where } t' = t - t_2^1 + t^*$$

This is shown in Fig. 2, case (2). The total time taken to fully (99.3%) charge the capacitor is

$$T_2 = t_2^1 + (5RC - t^*) = \frac{CV_c}{I} + 4RC + RC \ln\left[\frac{IR}{V_c}\right]$$

C. Voltage source at m ($m > 1$) times the capacitor rated voltage with no maximum current limit

The voltage and current curves for this charging method is shown in Fig. 2, case (5). Capacitor voltage and the current will be

$$v_c^3(t) = mV_c(1 - e^{-t/\tau}) \quad, \quad i_c^3(t) = \frac{mV_c}{R} e^{-t/\tau}$$

Time taken for $v_c^3(t)$ to reach V_c is T_3

$$T_3 = \tau \ln\left[\frac{m}{m-1}\right] \qquad (1)$$

This time can only be achieved if the source can supply mV_c/R initial current. In comparison with case A, the start-up current is m times higher. Another important point in this charging scheme is that, even when the capacitor reaches its full charge voltage the circuit current is non-zero. Therefore a voltage monitoring switch is required to extinguish the charging current once the capacitor reaches its full charge.

D. Voltage source at m times the capacitor rated voltage with current limit I

In this charging scheme, the value of source factor (m) should be higher than a critical value m^* to ensure that the capacitor will get charged from a continuous current until it reaches its rated voltage V_c

$$mV_c - IR > V_c \quad \Longrightarrow \quad m > \frac{IR}{V_c} + 1 = m^*$$

when $1 < m < m^*$
The source voltage is lower than the critical source factor m^*. The charging profile is not a continuous current. When the capacitor voltage is equal to $mV_c - IR$, at time $t_1 = \frac{C}{I}(mV_c -$

$IR)$, the charging profile will change from a constant current to a decaying current until the capacitor is fully charged. Time taken from the change-over point until the capacitor reaches its full voltage can be calculated in a similar way to case B. The time taken to charge the capacitor to V_c if the source is not current limited is

$$RC \ln\left[\frac{m}{m-1}\right]$$

The time at which the change-over occurs constant current to a decaying current on a non-current-limited charging is

$$RC \ln\left[\frac{mV_c}{IR}\right]$$

Therefore the current limited source will supply a decaying charging current for a time

$$RC \ln\left[\frac{m}{m-1}\right] - RC \ln\left[\frac{m}{m-1}\right] = RC \ln\left[\frac{IR}{V_c(m-1)}\right]$$

Hence the total time taken to charge the capacitor is

$$T_{4,1} = \frac{C}{I}mV_c + RC + RC \ln\left[\frac{IR}{mV_c}\right] \qquad (2)$$

The reduction in charging time compared to case B is,

$$\Delta T_1 = T_2 - T_{4,1} = RC \ln(m) - (m-1)\frac{CV_c}{I} \qquad (3)$$

when $1 < m < m^*$
The capacitor will continue to be charged by the continuous current I until its fully charged

$$T_{4,2} = \frac{CV_c}{I} \qquad (4)$$

The capacitor voltage will rise linearly according to

$$v_{c,4} = \frac{I}{C} t$$

Reduction in charging time compared to case B is

$$\Delta T_2 = T_2 - T_{4,2} = 4RC + RC \ln\left[\frac{IR}{V_c}\right] \qquad (5)$$

In Fig. 2, case (3 & 4) are for the critical m value with two different charging currents. It is seen that with higher charging current capable chargers the charging time is smaller. Table I compares the different charging source, time taken by each source to charge a capacitor and the maximum current required for a 100 F, 10 V SC bank with a total path resistance of $R = 250$ mΩ. By comparing the second and the last rows, a 53% charging time reduction is achieved by using a higher-voltage-rated source even with the same current rating of 15 A. The shortest charging time is achieved by using a voltage source at 20 V ($m = 2$) with no current limit. This requires the source to be capable of delivering 80 A at start up, so this charging scheme becomes less practical. In summary it is seen that by choosing a voltage source rated higher than the capacitor rated voltage, the capacitor can be charged faster even with the same continuous current rating of a source rated at the capacitor rated voltage. This is seen in Fig. 2 case (3) and (4).

TABLE I

CAPACITOR CHARGER TYPES COMPARISON FOR A 100 F, 10 V SC BANK WITH $R = 0.25\ \Omega$. CURRENT LIMIT IS KEPT BLANK FOR SOURCES WHICH CAN DELIVER MUCH HIGHER CURRENTS THAN REQUIRED.

Source factor m	Current limit I	Time to charge (s)	I (A)
1		125	40
1	15	142.1	15
2		17.3	80
$1.2 < m^* = 1.375$	15	70.7	15
$1.4 > m^* = 1.375$	15	66.6	15

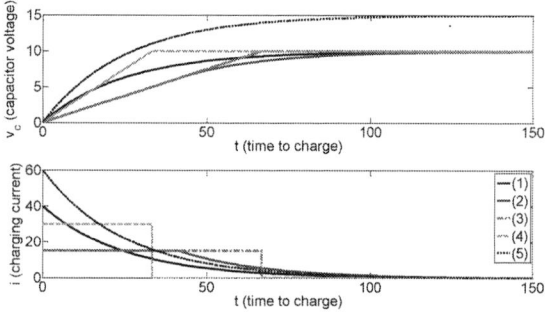

Fig. 2. Charging methods comparison for a 100 F, 10 V SC bank with 250 mΩ path resistance. (1) $m = 1$, $I = 40$ A (2) $m = 1$, $I = 15$ A (3) $m = m^*$, $I = 15$ A (4) $m = m^*$, $I = 30$ A (5) $m = 1.5$, $I = 60$ A

V. RC VS RLC CHARGING EFFICIENCY

Theoretical simulations using a higher voltage rated source to charge a capacitor show that this scheme of charging becomes more efficient with a series inductor in the charging path controlling rate of current rise. Fig. 3 compares the percentage energy loss of an RC system and an RLC system charging the same capacitor with the same path resistance. Source factor m is varied for charging a capacitor partly discharged from nV_c to V_c where $n < 1$. Percentage energy loss ($\%E_{loss}$) is calculated based on energy loss in the resistor (E_R), energy stored in the inductor (E_L) and capacitor (E_c). E_L is zero for the RC system

$$\%E_{loss} = \frac{E_R}{E_R + E_c + E_L} \times 100\%$$

These calculations assume that the higher voltage rated source is kept connected to the capacitor until the capacitor reaches its rated voltage.

In an RC system,

$$E_C = \frac{CV^2}{2}(1 - n^2), \quad E_R = \frac{CV^2}{2}(1 - n)(2m - n - 1)$$

Given the ranges for m and n

$$m > 1,\ 0 < n < 1 \quad \Leftrightarrow \quad \forall n,\ 1 - n > 0 \text{ and } 1 + n < 2$$

Therefore for efficient charging

$$E_C - E_R > 0 \implies (1 + n) - m > 0$$
$$\implies m < (1 + n) < 2 \Leftrightarrow m < 2$$

Further, for optimum performance $E_C - E_R$ should be maximised. Hence

$$\frac{d}{dn}(E_C - E_R) = 0 \implies -2n + m = 0 \Leftrightarrow m = 2n$$

Therefore the optimum will occur when $m = 2n$ and this will meet the previous condition of $m < 2$ as well, since $n < 1$.

In an RLC system, based on the size of the inductance the current response will vary. Therefore to represent common practical application, the inductor size is assumed to be small, i.e., the RLC circuit is in overdamped oscillation ($L < R^2C/4$).

Hence the energies at each component can be calculated using,

$$E_L = \frac{1}{2}L\ [I(T)]^2, \quad E_R = R\int_0^T i^2(t)\ dt$$

where T is the for the capacitor to reach its rated voltage V_c. These equations were solved numerically using MATLAB to obtain E_L and E_R. Based on these results a charger topology built to reuse the energy stored in the primary inductor to charge another part of the same capacitor bank will always perform better than a RC charging scheme. From Fig. 3 it is seen that an RLC system becomes more efficient as the source factor increases. This is mainly due to the increase in the stored energy in the inductor when using a higher source voltage. By using the proposed coupled-inductor technique, the stored energy in the inductor can be used to charge another part of the SC bank. This improves the overall charging efficiency as well as reducing the charging time.

VI. FUNDAMENTALS OF THE CHARGING TECHNIQUE

The new topology should be able utilise the following characteristics to fast-charge a SC bank by placing a series inductor in the capacitor charging path.

- ability to use the direct rectified ac mains as the high voltage input power source to the converter
- ability to extract the energy stored in the above mentioned inductor to charge a separate part of the same capacitor bank
- ability to combine the two capacitor banks together to obtain a single tapped capacitor bank (3 wires)
- lower component count for improved efficiency and cost effectiveness
- energy limited design since both the target applications of the charger are limited energy

As seen in Fig. 4, the topology utilises a coupled inductor to charge the capacitor bank partitioned into two (C_1 and C_2). The converter can be directly fed with full-wave rectified line voltage (V_g). The two parts of the capacitor bank will work as two separate capacitors in the operation of the converter as there will be no ground loops formed through any component parasitics to discharge one capacitor bank on to the other. The circuit has two states of operation based on whether the switch is on or off. This technique allows both a high-voltage source and charging a capacitor using the energy stored in an inductor. The coupled inductor is modelled as an ideal transformer with a magnetizing inductance.

The coupled inductor (L_m) is placed in the charging path of C_1 to store energy during the on period. When the switch is on, the diode is reverse-biased, thus isolating the secondary capacitor C_2 from feeding backwards into the inductor. While the input supplies energy to the inductor, the primary capacitor is also being charged.

During the off period the energy stored in the magnetic field of L_m is released from the secondary winding, through the forward-biased diode to charge C_2. But the primary capacitor will not get discharged as there is no closed loop for it to be discharged. Since V_g is rectified ac mains, an energy transfer may occur close to mains zero crossing. The effect of that on the overall operation can be neglected as the energy transferred near the zero crossing is very small.

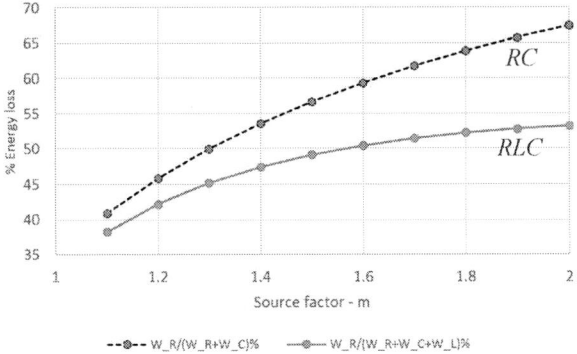

Fig. 3. Efficiency improvement of a RLC system over a RC system for m-times charging ($C = 310$ F, $L = 2$ mH, R = 50 mΩ, $V_c = 2.7$ V, $n = 0.3$)

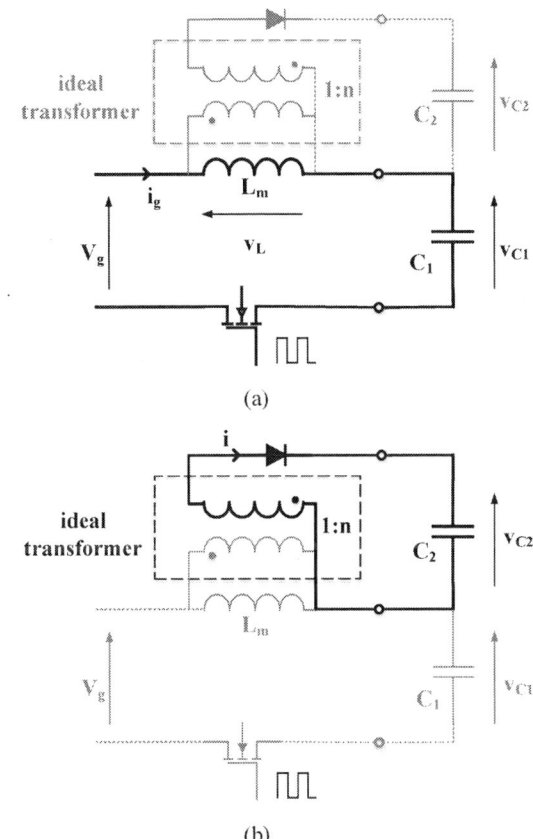

Fig. 4. Equivalent circuit for the topology. The coupled inductor is modelled by an ideal transformer in parallel with the magnetizing inductance. (a) ON state (b) OFF state. Current path in each state is indicated with thicker lines.

With the rectified ac mains used as the input voltage source, the capacitor voltages should be continuously monitored to avoid overcharging. The control signal will balance the voltage of the two capacitors and it too will provide power factor correction without using an extra choke. Since a part of the capacitor to be charged is always connected in series with the line voltage, a buffer capacitor is not required. Low parts count of the design helps to reduce total cost.

Since the converter operation is extinguished once the capacitors are fully charged, the converter operation is transient. By having the rectified mains input voltage, the resulting input current waveform peak envelope will be a fullwave rectified sine.

VII. EXPERIMENTAL RESULTS

Experiments were carried out to prove fast charging using a higher voltage rated source. Fig. 5 shows the reduction in charging time by charging a capacitor using a high-current capable power supply with the voltage rated to m times the

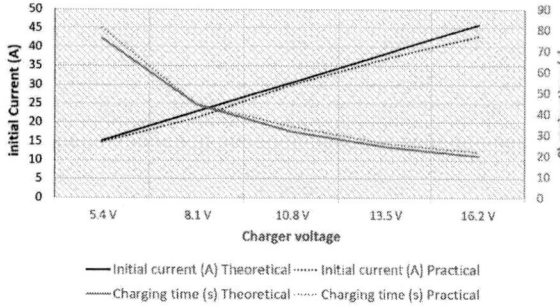

Fig. 5. Charging time improvement from high-voltage charging with terminal voltage monitoring for a 310 F, 2.7 V SC

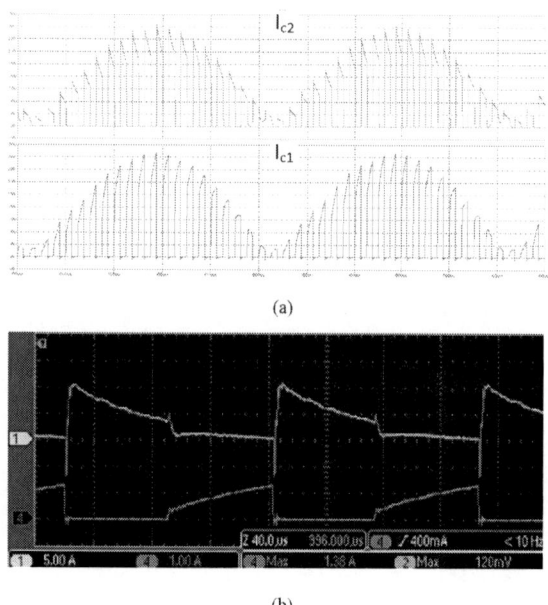

(a)

(b)

Fig. 6. Current profiles (a) primary and secondary capacitor current profiles for a fullwave rectified input voltage with a 4 A/div vertical scale and 4 ms/div horizontal scale. (b) expanded view of the switching cycle (ch1 - I_{c2} - 5 A/div and ch2 - I_{c1} - 1 A/div)

capacitor terminal voltage. As the source voltage increases the maximum current capability should also increase to achieve this time reduction.

All simulations and experiments for the new charger topology were carried out with a full-wave rectified sine input voltage source. LTSpice simulation results of primary and secondary currents enveloped inside the full-wave rectified sine wave is seen in Fig. 6(a). These are for the 600 W transformer model working with 30 V maximum input voltage. Fig. 6(b) is an expanded view of the primary and secondary currents obtained from the laboratory experiments carried out using a 30 V, 15 F SC bank with a 30 W capable transformer fed by a full-wave rectifier.

VIII. CONCLUSIONS AND FUTURE WORK

In this unique approach to charge a SC bank by dividing it into two halves, one half is charged using a higher valued DC source with capacitor bank terminal voltage monitoring. The other half is fed through a coupled-inductor based inductive energy transfer scheme. This is useful for the fast charger requirements of applications like SCATMA and SRUPS since they require limited amounts of energy at high power during each part of the charging cycle. Development of a final prototype for a 600 W system is in progress.

REFERENCES

[1] F. Musavi, M. Edington, W. Eberle, and W. Dunford, "Evaluation and efficiency comparison of front end ac-dc plug-in hybrid charger topologies," *Smart Grid, IEEE Transactions on*, vol. 3, no. 1, pp. 413–421, March 2012.

[2] A. Kuperman, I. Aharon, S. Malki, and A. Kara, "Design of a semiactive battery-ultracapacitor hybrid energy source," *IEEE Transactions on Power Electronics*, vol. 28, no. 2, pp. 806–815, Feb 2013.

[3] F. Musavi, W. Eberle, and W. Dunford, "A high-performance single-phase bridgeless interleaved PFC converter for plug-in hybrid electric vehicle battery chargers," *Industry Applications, IEEE Transactions on*, vol. 47, no. 4, pp. 1833–1843, July 2011.

[4] N. Kularatna, A. Gattuso, N. Gurusinghe, and J. Du Toit, "Pre-stored supercapacitor energy as a solution for burst energy requirements in domestic in-line fast water heating systems," in *proc of IECON 2014, USA, NOV*. IEEE, 2014.

[5] N. Kularatna, L. Tilakaratne, and P. K. Kumaran, "Design approaches to supercapacitor based surge resistant ups techniques," in *proc of IECON 2011, AUS, Nov*. IEEE, 2011.

[6] U. Madawala, D. Thrimawithana, and N. Kularatna, "An ICPT-supercapacitor hybrid system for surge-free power transfer," *Industrial Electronics, IEEE Transactions on*, vol. 54, no. 6, pp. 3287–3297, Dec 2007.

[7] R. Bodnar and W. Redman-White, "A 250W/30A fast charger for ultracapacitors with direct mains connection," in *Circuit Theory and Design (ECCTD), 2011 20th European Conference on*, Aug 2011, pp. 813–816.

[8] R. Bodnar and W. Redman-White, "An integrated ultracapacitor fast mains charger with combined power/current optimisation," in *ESSCIRC (ESSCIRC), 2013 Proceedings of the*, Sept 2013, pp. 161–164.

[9] H. Ryoo, S. Jang, Y. Jin, J. Kim, Y. Kim, S. Ahn, J. Gong, B. Lee, and D. Kim, "Design of high voltage capacitor charger with improved efficiency, power density and reliability," *Dielectrics and Electrical Insulation, IEEE Transactions on*, vol. 20, no. 4, pp. 1076–1084, August 2013.

[10] N. Kularatna, J. Fernando, A. Pandey, and S. James, "Surge capability testing of supercapacitor families using a lightning surge simulator," *Industrial Electronics, IEEE Transactions on*, vol. 58, no. 10, pp. 4942–4949, Oct 2011.

[11] "Abb miniature circuitbreakers index 1sxu000023c0202 reva15," www.abb.us/lowvoltage, accessed: 2015-11-11.

[12] N. Kularatna, L. Tilakaratne, and P. Kumaran, "Design approaches to supercapacitor based surge resistant UPS techniques," in *IECON 2011 - 37th Annual Conference on IEEE Industrial Electronics Society*, Nov 2011, pp. 4094–4099.

[13] S. Werkstetter, "Existing and future ultracapacitor-applications in the renewable energy market," in *PCIM Europe 2014; International Exhibition and Conference for Power Electronics, Intelligent Motion, Renewable Energy and Energy Management; Proceedings of*, May 2014, pp. 1–7.

Stack Multiphase Asymmetrical Half-Bridge Topology Offering Advance Performance and Efficiency

Trong Tue Vu
Icergi Ltd., Dublin, Ireland
Email: ttrongvu@icergi.com

George Young
Icergi Ltd., Dublin, Ireland
Email: georgeyoung@icergi.com

Abstract— **In quest of higher efficiency and power density, various studies have been carried out to boost the performance of conventional asymmetrical half bridge (AHB) topologies through either magnetics optimization or deployments of fully soft switching schemes. Although the reporting results from existing works are quite promising, none of them show significant breakthrough in power conversion density which is currently limited by existing technologies for high-voltage (above 500V) MOSFETs. This paper proposes a novel stacked multiphase asymmetrical half bridge (SMAHB) topology operating in a similar manner as conventional AHB converters, but enabling deployment of 250V MOSFETs for faster switching and material reduction in magnetics and EMI filters, which allows high efficient and compact implementation. The operation and performance of the proposed converter is confirmed via both simulated and experimental data.**

Keywords— AHB; multi-cell conversion; high power density; Isolated DC-DC conversion

I. INTRODUCTION

AHB converters belong to a class of soft-switching converters, but operate with a fixed switching frequency [1], [2]. The resonant tank is fundamentally formed by parasitic components of main switching devices and transformer; hence, implementation is generally simple with minimum requirement on magnetics and EMI filter design. AHB converters are mainly used after PFC rectifiers with an output voltage of around 400V for nominal operation and up to 450V for transient responses; so the main MOSFETs used for implementation must have voltage ratings above 500V if a tolerance of 20% is taken into account. Switching characteristics of high voltage MOSFETs are far from ideal and in consequence imposes a bottleneck on converter performance and power density even with optimized magnetics [3] - [5] or fully soft switching [6]. Therefore, it is desirable to develop a new AHB topology that can eliminate such constraints while retaining all desired properties of conventional ones.

Generally, boosting switching frequencies is an effective way to shrink the size of passive components, such as transformer and filters; however; switching losses are too great to be handled by existing silicon switching devices. Using GaN FETs for implementation could be a worthy solution here [7];

however it is costly and likely less reliable as compared to silicon-based ones. One possible solution is to rely on multi-cell topologies [8], [9], allowing not only magnetic volume reduction through current ripple cancelation but also the deployment of low-voltage MOSFETs for faster switching and lower losses.

Multilevel conversion techniques have been widely adopted in high voltage and high power DC-to-AC inverters and DC-to-DC choppers, where a single switching device either is unable to handle a significant voltage swing, for example over 5kV, or cannot provide adequate performance in terms of speed, efficiency and electro-magnetic interference (EMI). The key purposes of multilevel approaches are to use multiple switches for evenly sharing the voltage stress, which allows the deployment of lower voltage rating but faster switching devices.

Although many studies have been conducted to develop multi-level topologies for isolated DC-DC applications from various basic topologies including both soft-switching hard-switching converters [10] - [13], there is no research concerning multi-cell deployment for AHB topologies and making use of 250V MOSFETs as main switches for implementation of mid-power isolated DC-DC converters. Therefore, the man objective of this paper is to address all mentioned challenges.

II. PROPOSED STACKED MULTIPHASE-ASYMMETRICAL HALF-BRIDGE CONVERTER

Multilevel conversion techniques have been successfully adopted in various DC/AC and AC/DC topologies to reduce the voltage stresses on switching devices, so it is logical to borrow the same principle and apply it to conventional AHB converters. Figure 1 illustrates a result obtained by replacing a conventional voltage chopper with a three-level switching stage [9] while keeping the remaining structure of the converter unchanged. The two input capacitors C_{in1} and C_{in2} generate a bypass mid-point voltage of $V_{in}/2$ which defines the voltage stresses on the main switching devices when going through their cycle of operation. The capacitor C_b acts as an energy storage and DC blocking device while the center-tapped transformer provides galvanic isolation and voltage transformation.

978-1-4673-9551-9/16 $31.00 © 2016 IEEE

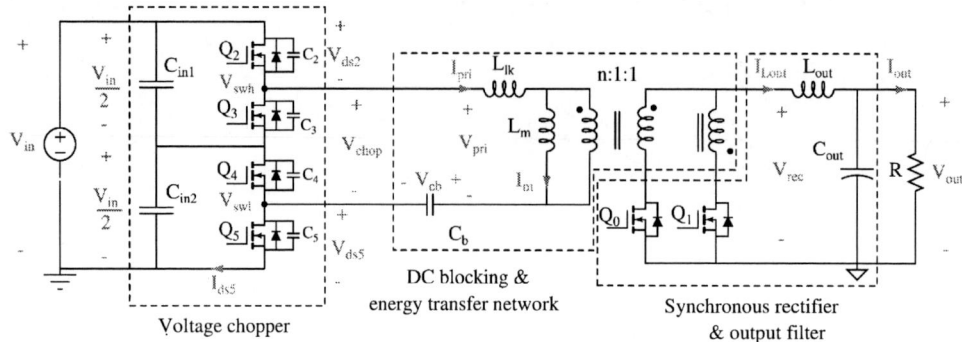

Fig. 1: Circuit diagram of the proposed stacked multiphase AHB converter

The leakage inductance L_{lk} of the transformer and internal capacitance C_2, C_3, C_4 and C_5 of the primary switching devices form a resonant tank whereby the energy stored in L_{lk} can be recycled to bring the drain-source voltages of MOSFETs to zero before they are turned on, i.e. ZVS. Two active switches at the secondary play the role of a synchronous rectifier (SR) feeding the output filter formed by an inductor and an electrolytic capacitor.

The proposed circuit as shown in Fig. 1 inherit the operating principle of the original AHB converter. Particularly, the input voltage is first processed by the voltage chopper which outputs a square wave signal having an amplitude of $V_{in}/2$ and a frequency of twice the switching frequency of the main MOSFETs. The square wave voltage then has its DC component removed when travelling further downstream and is amplified by the transformer. The pulse signals appearing at the transformer secondary terminals are then rectified by the SR, and subsequently filtered out by the output inductor L_{out} and output capacitor C_{out}.

Although the proposed converter looks very similar to that of the work presented in [11], one can easily confirm in Section III that a different modulation scheme is deployed in this paper, and, as a consequence, operating principle is not quite the same.

III. OPERATIONAL PRINCIPLES

Keeping the mid-point voltage balanced is the key point in sustaining voltage stresses evenly between primary switches; therefore, the modulation pattern should facilitate such a target. In particular, the top MOSFET Q_2 is paired with Q_3 while Q_4 is paired with the bottom MOSFET Q_5. Each of paired switches is driven in a complementary manner, i.e. when one switch of the pair is on, the other must be off and vice versa. The PWM control signals for Q_2 and Q_5 share the same duty ratio (1-D) and switching frequency f_{pwm}, but are different in phase by an angle of 180 degrees. In order to facilitate ZVS, a small dead-time is introduced between on-off transitions of each complementary pair of switches. For SR, Q_0 is set on when either Q_2 or Q_5 is on while negating the driving pulse for Q_0 gives that of Q_1. Figure 2 illustrates the driving sequence for all switches in the proposed AHB converter, and resulting voltages and currents along the power transfer path. Thanks to

the switching pattern, the output-to-input voltage ratio is simply controlled by the variable D which is defined as the duty ratio of the middle MOSFET Q_4.

The proposed converter has four main operating phases interleaved with four switch-transition phases during each switching cycle, which is highlighted by the timeline from t_0 to t_8 as highlighted in Fig. 2. Since the switching pattern is repeated within a cycle, only the first four sequential states are described in details.

- **Main phase 1 ($t_0 \sim t_1$):** power transfers from the capacitor C_{in2} to the output through Q_3, Q_5 and Q_0, and charge C_b. Voltage across the primary side of the transformer is $V_{in}/2 - V_{cb}$.

- **Transition phase 2 ($t_1 \sim t_2$):** since Q_5 is turned off at t_1, the primary current I_{pri} will start charging C_5 and discharging C_4 until $V_{C4} = V_{Cb}$. After this time instant, the secondary side of the transformer is decoupled from the primary side. The output inductor current freewheels through both SR Q_0 and Q_1 while the capacitor C_4 continues to be discharged by the energy stored in L_{lk}. only The body diode of Q_4 starts conducting when C_4 approaches zero, which allows ZVS if turning Q_4 on is triggered when such a condition is still maintained. Although both Q_4 and Q_3 now are on, there is still no energy transferring to the secondary side until $I_{pri} - I_m = -I_{out}/n$ which happens at $t = t_2$.

- **Main phase 3 ($t_2 \sim t_3$):** the blocking capacitor now connects to the primary side of the transformer in a reverse polarity fashion, which allows power stored in C_b transferring to the output. Voltage across the primary side of the transformer is $-V_{cb}$ which appears at secondary side as V_{cb}/n.

- **Transition phase 4 ($t_3 \sim t_4$):** Q_3 is switched off at $t = t_3$, which forces the primary current to go through the internal capacitor of Q_3. This action will ramp up the voltage across C_3 and simultaneously ramps down V_{C2}. When $V_{C3} > V_{Cb}$, the secondary side of the transformer is again

978-1-4673-9551-9/16 $31.00 © 2016 IEEE   2486

decoupled from the primary side until $t = t_4$. As mentioned in Transition phase 2, ZVS for Q_2 is also achieved if its gate drive signal is kicked in after the primary current change its polarity.

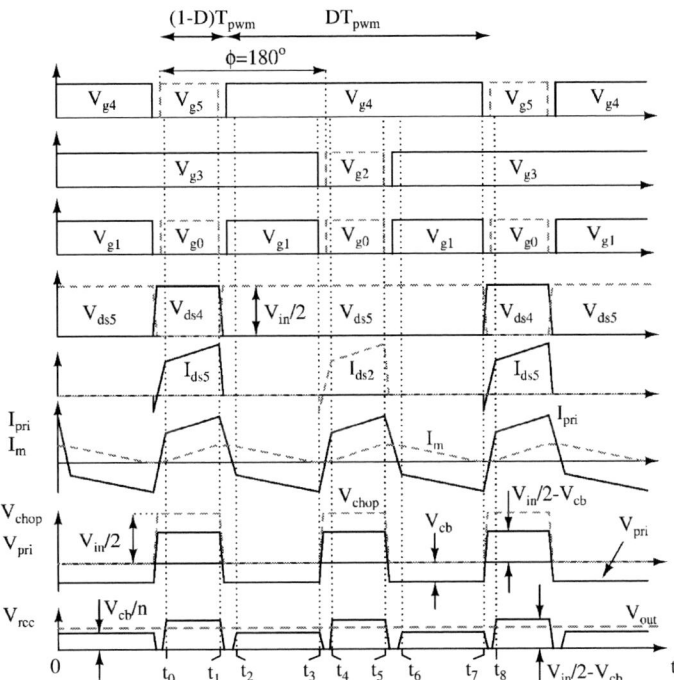

Fig. 2: Key operating waveforms of the proposed stacked multiphase AHB converters

Fig. 2 shows that the SMAHB converter has the same operational voltage and current waveforms as a conventional AHB converter with a switching frequency of $2*f_{pwm}$. and an input voltage of $V_{in}/2$ This implies that the use of a multi-level voltage chopper here does not alter the way that the converter operates, it simply doubles the operating frequency of magnetic components and reduce the converter gain by half. The effect on the conversion gain is actually trivial as one can easily compensate for such reduction by doubling the transformer ratio.

IV. CONVERTER CHARACTERISATION

Like any conventional AHB converter, the following assumptions are made to simplify the DC analysis of the proposed converter [2]:

- All components are ideal and do not incur any loss

- The leakage inductance L_{lk} is much smaller than the magnetizing inductance L_m

- The blocking capacitor voltage V_{Cb} has negligible ripples and can be considered as constant during each switching cycle.

A. Input-to-output-voltage gain

If the duty cycle loss due to transformer decoupling during transition phases is negligible, the output voltage can be calculated by averaging the rectifier voltage over time as given by

$$V_{\text{out}} = \frac{2(1-D)}{n}\left(\frac{V_{in}}{2} - V_{cb}\right) + \frac{2(D-0.5)}{n}V_{cb}. \quad (1)$$

where D denotes the duty ratio of Q_4, and is assumed to be limited to [0.5 1] while n is the turn ratio of the center-tapped transformer. V_{in}, V_{out} and V_{cb} denotes the input voltage, output voltage and blocking capacitor voltage, respectively. Applying the volt-second balance to the primary transformer winding gives

$$2(1-D)\left(\frac{V_{in}}{2} - V_{cb}\right) - 2(D-0.5)V_{cb} = 0. \quad (2)$$

Solving (2) for the blocking capacitor voltage yields

$$V_{cb} = (1-D)V_{in}. \quad (3)$$

Substituting (3) into (1) and solving for the input-to-output gain gives

$$G_{io} = \frac{V_{out}}{V_{in}} = \frac{2(1-D)(2D-1)}{n} \qquad (4)$$

Figure 3 plots the voltage gain of the proposed converter as a function of the duty ratio D and the transformer turn ratio. For any value of n, the gain curve is symmetrical around 75%, which suggests that controller design should limit the dynamic range of D to either [0.5 0.75] or [0.75 1]. Technically, operation range above 75% is preferable as this allows lower operating voltage for the blocking capacitor.

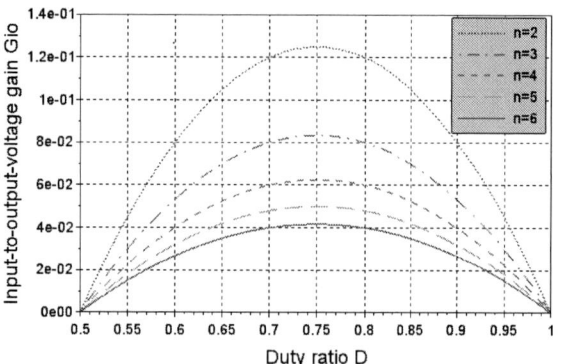

Fig. 3: DC voltage gain of the SMAHB converter

Although (4)seems to be fine for ballpark calculations, a more accurate design should include the amount of duty cycle stolen by the resonant tank for ZVS as described by (5)

$$V_{out} = \frac{2(1-D)(2D-1)}{n}V_{in} - \frac{8I_{out}L_{lk}}{n^2 T_{pwm}}, \qquad (5)$$

B. DC current and ripples

Since the sum of charge of C_b over one switching cycle equals zero at steady state, the magnetizing current is required to balance out the current withdrawn by the load, which implies

$$2(1-D)\left(I_m + \frac{I_{out}}{n}\right) + 2(D-0.5)\left(I_m - \frac{I_{out}}{n}\right) = 0. \qquad (6)$$

Solving (6) for I_m gives

$$I_m = (4D-3)\frac{I_{out}}{n}. \qquad (7)$$

Given the voltage across L_m and L_{out} as V_{pri} and $V_{rec} - V_{out}$, respectively, the magnetizing current and output current ripple can be calculated via

$$\Delta I_{m\_pp} = \frac{V_{bulk}(D-0.5)(1-D)T_{pwm}}{L_m} \qquad (8)$$

$$\Delta I_{Lout\_pp} = \frac{(1-D)T_{pwm}}{L_{out}}\left(\frac{(D-0.5)V_{in}}{n} - V_{out}\right) \qquad (9)$$

C. ZVS conditions

After Q_2 is turned off, C_3 is discharged from V_{cb} to 0V by the energy storage in L_{lk}. Therefore, the ZVS condition for Q_3 is equivalent to

$$\frac{1}{2}L_{lk}I_{pri\_peak}^2 > \frac{1}{2}(C_2+C_3)V_{Cb}^2, \qquad (10)$$

where I_{pri_peak} denotes the peak of the primary current which can be approximated by

$$I_{pri\_peak} \approx \frac{I_{out}}{n} + I_m + \frac{\Delta I_{mpp}}{2}. \qquad (11)$$

Similarly, the ZVS condition for Q2 requires that C2 is fully discharged from $V_{in}/2 - V_{cb}$ by the energy stored in the leakage inductance only Such a condition can be mathematically expressed by

$$\frac{1}{2}L_{lk}I_{pri\_valley}^2 > \frac{1}{2}(C_2+C_3)\left(\frac{V_{in}}{2}-V_{cb}\right)^2, \qquad (12)$$

where I_{pri_valley} indicates the valley of the primary current which can be approximated by

$$I_{pri\_peak} \approx -\frac{I_{out}}{n} + I_m - \frac{\Delta I_{mpp}}{2} \qquad (13)$$

V. DESIGN AND IMPLEMENTATION

The design specifications for a SMAHB converter prototype are listed in TABLE I.

TABLE I: CONVERTER DESIGN SPECIFICATIONS

Input voltage, V_{in}	360V$_{DC}$-400V$_{DC}$
Nominal output voltage, V_{out}	12V$_{DC}$
Switching frequency, f_{pwm}	200kHz
Maximum output power, P_{out_max}	200W
Maximal inductor current ripples ΔI_{Lout_max}	2.1A
Hold-up time, t_{holdup}	20ms

A. Power stage design

1) Center-tapped Transformer desgin:

As observed in (5), the presence of the duty cycle loss complicates the selection of the transformer ratio n as the leakage inductance is an also unknown variable here. Therefore, it makes sense to use (4) for initial design calculations and (5) for validation purposes.

Given the minimal input voltage of 360V and the output voltage of 12V, the transformer turn ratio should be chosen to provide a maximal voltage gain of at least 0.033. According to the set of gain curves as shown in Fig. 3, an optimal value for n should be 6.

The leakage inductance is chosen to ensure that (5) is satisfied at minimal input voltage and maximal load, i.e.

$$L_{lk} \leq \frac{n^2 T_{pwm}}{8 I_{out\_max}} \left[\frac{2(1 - D_{min})(2D_{min} - 1)}{n} V_{in\_min} - V_{out} \right]$$
$$\leq 3.8 uH$$

Considering a margin of 20% gives L_{lk} = 3uH. The magnetizing inductance value influences the peak and valley current levels which indirectly affect the ZVS condition. Therefore, L_m can be designed to ensure ZVS from full load to 20% max load. Firstly, the duty ratio for 20% load can be approximated by

$$D_{20\%} = 1 - \frac{1 - \sqrt{1 - \frac{4nV_{out}}{V_{bulk}}}}{4} = 0.88 \quad (14)$$

The maximum magnetizing inductance value ensuring ZVS is given by

$$L_{m\,max} = \frac{V_{in}(D_{20\%} - 0.5)(1 - D_{20\%}) T_{pwm}}{\sqrt{\frac{C_2 + C_3}{L_{lk}} \frac{V_{in}^2}{4}} - \frac{I_{out}}{n} + \frac{(4D_{20\%} - 3)I_{out}}{n}} \quad (15)$$
$$= 93\mu H$$

For implementation, L_m is chosen to be 65uH.

2) Output inductor design
Given the maximal ripple requirement, the output inductor can be calculated via

$$L_{out} = \frac{(1 - D)T_{pwm}}{\Delta I_{Lout\_pp}} \left(\frac{(D - 0.5)V_{in}}{n} - V_{out} \right) \quad (16)$$
$$= \frac{(1 - 0.88)5e - 6}{2.1} \left(\frac{(0.88 - 0.5)400}{6} - 12 \right)$$
$$= 3.8 uH$$

3) Switching device selection
The switching waveforms as shown in Fig. 1Fig. 2 confirms that the operating voltage of the primary FETs is only half of the input voltage, suggesting that it is possible to employ 250V MOSFETs for implementation because the nominal output voltage of an PFC stage is typically set around 400V.

B. Implementation
The design procedure as discussed in Section V.A suggests the following components: L_{out}=3.8uH, C_{out} = 1500uF – 16V

rating- electrolytic, C_{in1} = C_{in2} = 220nF - 250V rating - ceramic. The center-tapped transformer has a turn ratio of 12:2:2 with the magnetizing inductance of 65uH and the leakage inductance of 3uH. The main switches are BSC16DN25 with 250V rating and on resistance of 165mΩ, and controlled by ARM Cortex M0 (STM32F051) through proprietary isolated gate-drive circuitry

VI. EXPERIMENTAL RESULTS

Operational waveforms at 400V input voltage and load powers of 100W are illustrated in Figures 3 and 4. The experimental results show that the converter allows ZVS similar to AHB converters but has only a half of voltage stress on the main switching devices, which confirms the feasibility of the deployment of 250V MOSFETs. Although the efficiency data is not available at the time of writing this paper, it is expected to achieve a figure as high as 97%.

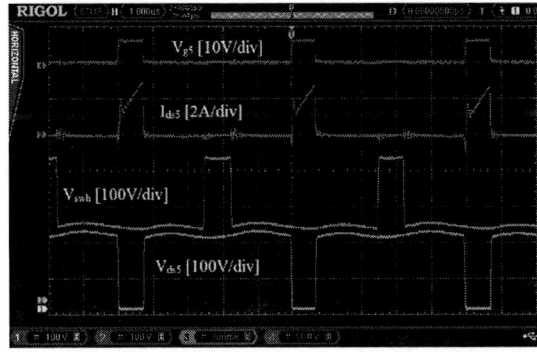

Fig. 4: Operational waveforms of the proposed SMAHB converter at V_{in} = 400V and P_{out} = 100W: (CH1) V_{ds5} – Drain source voltage Q5, (CH2) V_{swh} – High side switched node, (CH3) I_{ds5} – Q5 current, (CH4) V_{gs5} – Gate drive of Q5

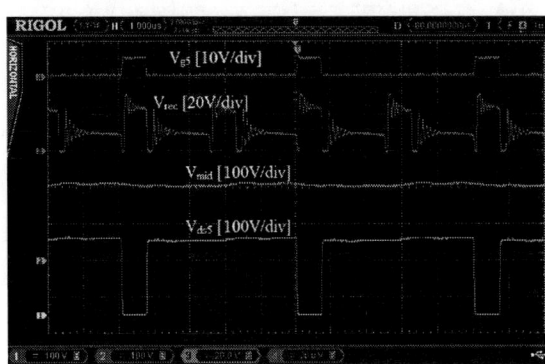

Fig. 5: Operational waveforms of the proposed SMAHB converter at V_{in} = 400V and P_{out} = 100W: (CH1) Drain source voltage Q5, (CH2) V_{mid} – Midpoint capacitor voltage, (CH3) V_{rec} – Rectified secondary voltage, (CH4) V_{gs5} – Gate drive of Q5

VII. CONCLUSIONS

This paper developed a new stacked multiphase AHB topology inheriting not only zero voltage switching from traditional AHB converters but also low voltage stresses and ripples cancelation from multilevel power conversion, allowing material reduction in magnetic sizes, EMI filters, and product volumes, and especially facilitating deployment of 250V-type MOSFETs as main switches. Although implementing the proposed converter requires more switching devices as compared with traditional solutions, the efficiency and final costs of the design are not compromised. Therefore, the new topology is well suited to economical and compact realization of isolated DC-DC converters to be operated with a maximum input voltage of 450V and output power ranging from 70W to 2kW. The common application is after the PFC stage in two-stage power supplies.

REFERENCES

[1] R. Oruganti, P. C. Heng, J. T. K. Guan, and L. A. Choy, "Soft-switched DC/DC converter with PWM control", *IEEE Trans. Power Electron.*, Vol. 13, No. 1, Jan. 1998, pp. 102-114

[2] H. Choi, "Design considerations for asymmetric half-bridge converter", Fairchild Semiconductor Power Seminar, 2008-2009

[3] I. O. Lee and G. W. Moon, "A new asymmetrical half-bridge converter with zero DC-offset current in transformer", *IEEE Trans. Power Electron.*, Vol. 28, No. 5, May 2013, pp. 2297-2306

[4] R. Miftaknudinov, A. Nemchinov, V. Meleshin, and S. Fraidlin, "Modified asymmetrical ZVS half-bridge DC-DC converter", in *Proc. IEEE Applied Power Electron. Conf. and Expo.*, Mar. 1999, pp. 567-574

[5] H. Mao, J. A. Qahouq, S. Luo, and I. Batarseh, "Zero-voltage-switching half bridge DC-DC converter with modified PWM control method", *IEEE Trans. Power Electron.*, Vol. 19, No. 4, Jul. 2004, pp. 947-958

[6] I. D. Jitaru, "99% efficiency DC-DC converter", in *Proc. 36th IEEE International Telecommunications Energy Conference*, Sept. 2014, pp. 1-6

[7] L. Zhou, and Y. Wu, "99% efficiency true-bridgeless totem-pole PFC based on GaN HEMTs", *Power Conversion Intelligent Motion* (PCIM), May 2013.

[8] J. Rodriguez, J. H. Lai, and F. Z. Peng, "Multilevel inverters: a survey of topologies, controls, and applications", *IEEE Trans. Industrial Electron.*, Vol. 49, No. 4, Aug. 2002, pp. 724-738

[9] F. Forest, T. A. Meynard, S. Faucher, F. Richardeau, J. J. Huselstein and C. Joubert, "Using the multilevel imbricated cells topologies in the design of low-power power-factor-corrector converters", *IEEE Trans. Industrial Electronics*, Vol. 52, No. 1, Feb. 2005, pp. 151-161

[10] E. S. Kim, Y. B. Byun, Y. H. Kim, and Y. G. Hong, "A three level ZVZCS phase-shifted DC/DC converter using a tapped inductor and a snubber capacitor", in *Proc. IEEE Applied Power Electron. Conf. and Expo.* (APEC), Mar. 2001, pp. 980-985

[11] I. Barbi, R. Gules, R. Redl, and N. O. Sokal, "DC-DC converter: four switches $V_{pk}=V_{in}/2$, capacitive turn-off snubbing, ZV turn-on", *IEEE Trans. Power Electron.*, Vol. 19, No. 4, Jul. 2004, pp. 918-927

[12] X. Ruan, B. Li, Q. Chen, S. C. Tan and C. K. Tse, "Fundamental Considerations of Three-Level DC-DC Converters: Topologies, Analyses , and Control", IEEE Trans. Circuits and Systems – I, Vol. 55, No. 11, Dec. 2008, pp. 3733-3743

[13] Y. Shi and X. Yang,"Zero-Voltage Switching PWM Three-Level Full-Bridge DC–DC Converter With Wide ZVS Load Range", IEEE Trans. Power Electron., Vol. 28, No. 10, Oct. 2013, pp 4511-4524

[14] Erickson, R. W. and Maksimovic, D., *Fundamental of power electronics*, 2nd Edition, Kluwer Academic Publishers, 2001

Design of a Novel APWM Half-Bridge DC-DC Resonant Converter with Load-Independent Soft-Switching and Reduced Circulating Current

Kawsar Ali, Sandeep Kolluri, Naga Brahmendra Yadav Gorla, Pritam Das and Sanjib Kumar Panda

Department of Electrical and Computer Engineering
National University of Singapore, Singapore 119077
Email:alikawsar@u.nus.edu

Abstract—**This paper presents a novel Asymmetrical-Pulse-Width-Modulated (APWM) half-bridge resonant converter with one additional active switching device at the secondary side of the transformer. The proposed converter features load-independent Zero-Voltage-Switching (ZVS) of both of the primary side switches with minimal magnetizing current. Empirical formulae are derived to design the resonant network and the high-frequency transformer systematically. The circulating current is minimized not only by operating the converter exactly at the resonant frequency, but also by reducing the magnetizing current. Moreover, the additional secondary-side switching device (which is a low-voltage and high-current rated MOSFET) achieves zero current turn-off at all load conditions. Thus the efficiency of the overall converter is significantly improved for wide load range. Simulation and experimental results are presented to validate the analysis and to prove the significance of the additional switching device at the secondary side of the transformer.**

I. INTRODUCTION

The APWM half-bridge series resonant converter is an ideal candidate for low-voltage, low-power applications because of its constant-frequency operation, limited voltage stress on the switches, soft-switching over wide load-range and the use of a simple capacitive output filter [1]–[4]. Its main drawback is the loss of ZVS at higher input voltage. To eliminate this problem, an auxiliary circuit was suggested in [1] and [3] so that the ZVS range is extended for a large input voltage range. But the auxiliary current used is quite high resulting in reduced light-load efficiency [2]. Moreover, the voltages across the auxiliary capacitors that split the DC input voltage are also not balanced because of the APWM switching pattern of the half bridge devices. An alternative solution was presented in [2] by the use of a CLL resonant circuit to maintain ZVS over all line and load conditions without excessive circulating current. However, the use of an additional inductor makes the converter bulkier.

Another important issue less addressed in APWM resonant topology is the reduction of the magnetizing current. The switching frequency is usually chosen higher than the resonant frequency to obtain a lagging current and thus minimize the requirement of magnetizing current for ZVS. However, this phase lag results in additional circulating current and especially if the maximum allowable duty ratio is less than 0.5 for maximum rated load, the required phase lag becomes more to ensure ZVS with minimum magnetizing current. This

has an adverse impact on the overall efficiency. On the other hand, if the converter is operated exactly at the resonant frequency, the magnetizing current requirement increases for ZVS at maximum load at duty ratios less than 0.5. So, there is a clear trade-off between minimization of magnetizing current and minimization of required phase lag for this converter's design.

In this paper, a novel APWM half-bridge resonant topology has been proposed which operates exactly at the resonant frequency and still requires reduced magnetizing current for achieving ZVS at maximum load. Thus, the total circulating current is reduced resulting in better overall converter efficiency for all load conditions. It is important to mention here that due to parameter variations in practical resonant circuits, it is almost impossible to operate exactly at the resonant frequency. At best, one can only achieve a very close to resonant frequency operation. However, for theoretical analysis, an exact resonant frequency operation can be safely assumed without much loss of generality. The rest of the paper is organised with section II introducing the proposed topology and its operation at critical switching instants, which leads to the design procedure of the converter as explained in section III. Section IV shows an example of design along with simulation and experimental results. Finally, section V concludes the paper.

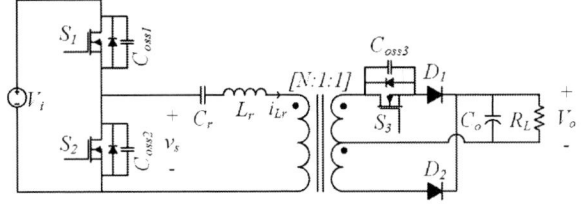

Fig. 1. Proposed Asymmetrical Pulse Width Modulated (APWM) half-bridge resonant converter.

II. PROPOSED TOPOLOGY AND ITS OPERATION

Fig. 1 shows the proposed converter which is similar to the standard half-bridge resonant topology [3], [4] except for the extra switch S_3 in series with diode D_1 at the secondary side. The half-bridge chopper circuit consisting of switches S_1 and

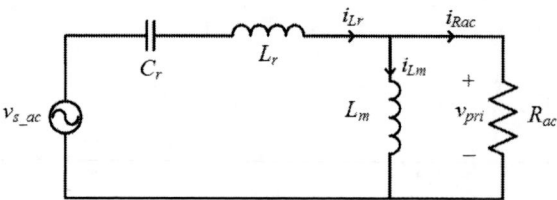

Fig. 2. Equivalent circuit based on fundamental ac component of v_s.

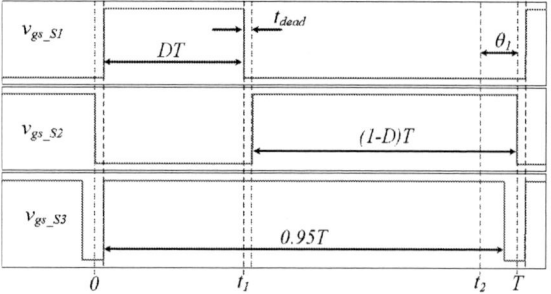

Fig. 3. Switching signals of S_1, S_2 and S_3.

S_2 produce a unipolar ac voltage v_s when switched with an asymmetric duty D as shown in Fig. 3. This voltage is incident on a resonant tank (C_r, L_r) which transfers input power to the load R_L through a high-frequency centre-tap transformer with turns ratio $N:1:1$, a diode rectifier (D_1, D_2) and an output filter C_o. It has been shown in [3], [4] that the variations of D in the range of 0 to 0.5 and in the range of 1 to 0.5 produce the same output voltage V_o. For ease of discussion here, the range of variation of D is chosen as 0 to 0.5. The pole voltage v_s is given as [4],

$$v_s = DV_i + \sum \frac{\sqrt{2}V_i\sqrt{1 - cos2n\pi D}}{n\pi}sin(n\omega_0 t + \theta_n) \quad (1)$$

where $\theta_n = tan^{-1}\left(\frac{sin2n\pi D}{1-cos2n\pi D}\right)$. D is the duty ratio for S_1 and ω_0 is the switching frequency of the converter. The capacitor C_r blocks the DC component DV_i and also forms the resonant tank with the inductor L_r. The design goal is to tune the resonant tank only for the fundamental of the ac

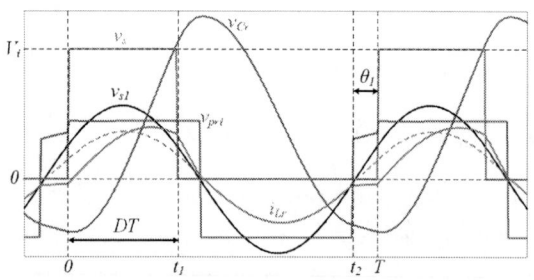

Fig. 4. Key waveforms of the converter for design of the resonant tank.

component of v_s (denoted by v_{s1} and leads v_s by an angle of θ_1). The equivalent circuit of the resonant network with the ac component of v_s (denoted as v_{s_ac}) is shown in Fig. 2. The resonant current is given as [4],

$$i_{Lr} = \frac{v_{s\_ac}}{Z_{in}} = \sum \frac{\sqrt{2}V_i\sqrt{1 - cos2n\pi D}}{n\pi |Z_{in}|}sin(n\omega_0 t + \theta_n - \phi_n)$$
$$(2)$$

where

$$|Z_{in}| = \sqrt{R_{ac}^2 + \left(n\omega_0 L_r - \frac{1}{n\omega_0 C_r}\right)^2} \quad (3)$$

$$\phi_n = tan^{-1}\left(\frac{n\omega_0 L_r - \frac{1}{n\omega_0 C_r}}{R_{ac}}\right) \quad (4)$$

$$R_{ac} = \frac{8}{\pi^2}N^2 R_L \quad (5)$$

N is the transformer turns ratio and ω_0 is the operating frequency or the switching frequency. The resonant frequency ω_r of the tank is given by,

$$\omega_r = \frac{1}{\sqrt{L_r C_r}} \quad (6)$$

Fig. 4 shows the key waveforms for the converter design and operation. It is important to mention here that although the converter is topologically similar to an LLC reonant converter, it operates in a similar way as the series resonant converter (SRC). The magnetizing inductance L_m, in parallel to the load R_{ac}, does not take part in resonance and just provide some auxiliary current to ensure Zero Voltage Switching (ZVS) of the switches, as will be seen in the coming sections. Moreover, unlike LLC resonant converter, the output voltage regulation in this case is done by variation of duty ratio and not switching frequency. The resonant tank is tuned only for the first harmonic of the APWM input voltage v_{s1} and i_{Lr} is in phase with v_{s1} since $\omega_0 = \omega_r$. The behaviour of the circuit at certain instants of the switching period which will lead to the design procedure of the system are explained below.

A. At $t=t_1$

As can be seen in Fig. 4, at $t = t_1$ when S_1 turns off and S_2 turns on, i_{Lr} has significantly large positive value. It is evident that, for APWM half bridge resonant converter as long as $\omega_0 = \omega_r$ and $D < 0.5$, for any load or input voltage condition, i_{Lr} will be positive at $t = t_1$. This means that ZVS of S_2 is always guaranteed. However, the same is not true for S_1 as seen in Fig. 4 at $t = T = \frac{2\pi}{\omega_0}$. A negative i_{Lr} is needed at $t = T$ for ZVS of S_1, which is not easy to obtain as long as i_{Lr} is in phase with v_{s1}. It might be interesting to note here that for variation of D in the range of 0.5 to 1, ZVS of S_1 will be always guaranteed and the concern will be to achieve ZVS for S_2.

978-1-4673-9551-9/16 $31.00 © 2016 IEEE

B. At $t=t_2$

As shown in Fig. 4, at $t = t_2$, i_{Lr} and v_{s1} become zero (assuming negligible magnetizing current i_{Lm}), and the transformer primary voltage is obtained as

$$v_{pri} = -v_{Cr}. \tag{7}$$

Now, if $|v_{Cr}| > NV_o$ at $t = t_2$ and the extra switch S_3 is not used, D_1 will turn-on and there will be a reflected positive load current i_{Rac} on the primary side as shown by the dotted curve in Fig. 4. Then a higher negative i_{Lm} will be required to get a resultant negative i_{Lr} in order to ensure ZVS of S_1. Thus, to ensure D_1 does not turn-on at $t = t_2$ (even in the absence of S_3), the condition maintained in this work is:

$$|v_{Cr}| \leqslant NV_o \quad at \quad t = t_2. \tag{8}$$

During the period $t = t_2$ to T, there is no power transfer from the source to the load and only the small magnetizing current i_{Lm}) flows in the resonant tank.

C. At $t=T$

At $t = T$, when the deadtime t_{dead} starts, the gate pulse of S_2 is removed and its output capacitance C_{oss2} starts charging. During this period,

$$v_{pri} = v_{Coss2} - v_{Cr} \tag{9}$$

Thus, v_{pri} will eventually exceed NV_o thereby turning on D_1 during this period. The purpose of using S_3 in series with D_1 is to prevent the turning-on of D_1 during t_{dead} so that the requirement of additional negative i_{Lm} is minimized. S_3 is turned on at $t = T + t_{dead}$, along with S_1, and it should be kept turned on as long as D_1 conducts. On the other hand, S_3 must be turned-off during the deadtime between the turn-off of S_2 and the turn-on of S_1 to prevent the turning-on of D_1. It may be noted here that since the condition $|v_{Cr}| \leqslant NV_o$ (at $t = t_2$) is already maintained in this work, S_3 may be left turned on during the period $t = t_2$ to T. A duty ratio of $0.95T$ is good enough to drive S_3. The switching signals for the three active switches are shown in Fig. 3. With such gate signal, S_3 always achieves zero current turn-off which is desired for such low-voltage high-current rated devices.

III. Design Procedure

The goal is to design the APWM resonant converter for minimum input voltage and maximum load at maximum allowable duty D_{max}. By doing so, for an increasing input voltage or a decreasing load, the output voltage can be regulated just by reducing the duty ratio D.

A. Calculation of transformer turns ratio

The rms of the fundamental ac component of v_s can be derived from (1) as [5], [6]

$$V_{s1} = \frac{V_i}{\pi}\sqrt{1 - cos2\pi D} \tag{10}$$

For $D = 0.5$ i.e., symmetrical PWM, this value is obtained as

$$V_{s1,symm} = \frac{\sqrt{2}V_{i,symm}}{\pi} \tag{11}$$

Thus, for the same V_i, V_{s1} is decreased by a factor of $\frac{\sqrt{1-cos2\pi D_{max}}}{\sqrt{2}}$ when D is changed from 0.5 to $D_{max} < 0.5$.

The transformer turns ratio N_{symm} for symmetrical PWM half-bridge resonant converter with a centre-tap transformer is given as [7],

$$N_{symm} = \frac{N_{p,symm}}{N_{s,symm}} = \frac{V_{i,symm}}{2V_o} \tag{12}$$

So, for APWM half-bridge resonant converter N should be decreased by the factor of $\frac{\sqrt{1-cos2\pi D_{max}}}{\sqrt{2}}$, and is thus given by,

$$N = \frac{N_p}{N_s} = \frac{V_i\sqrt{1 - cos2\pi D_{max}}}{2\sqrt{2}V_o} \tag{13}$$

B. Design of resonant tank

It has been already explained that the optimal design condition is: $|v_{Cr}| \leqslant NV_o$ at $t = t_2$. The voltage across the capacitor C_r is obtained as [4],

$$\begin{aligned}v_{Cr} &= DV_i + v_{Cr\_ac} \\ &= DV_i - \sum \frac{\sqrt{2}V_i\sqrt{1 - cos2n\pi D}}{\pi n^2 \omega_0 |Z_{in}| C_r}cos(n\omega_0 t + \theta_n - \phi_n)\end{aligned} \tag{14}$$

It is not easy to calculate the exact value of v_{Cr} at $t = t_2$. However, a valid approximation can be made by considering only the fundamental ac component v_{Cr_ac1}, which is at its negative peak at $t = t_2$.

$$NV_o = -v_{Cr} \approx -DV_i + |v_{Cr\_ac1}|_{t_2} \quad at \quad t = t_2 \tag{15}$$

Replacing $n = 1, \omega_0 = \omega_r = \frac{1}{\sqrt{L_r C_r}}, |Z_{in}| = R_{ac}$ and $D = D_{max}$ in (14) and then substituting into (15), the following equation is obtained.

$$\sqrt{\frac{L_r}{C_r}} = \frac{\pi R_{ac}(NV_o + DV_i)}{4NV_o} \tag{16}$$

The other obvious equation relating L_r and C_r is: $\sqrt{L_r C_r} = \frac{1}{\omega_r}$. Thus L_r and C_r can be solved as,

$$L_r = \frac{\pi R_{ac}(NV_o + DV_i)}{4NV_o\omega_r} \tag{17}$$

$$C_r = \frac{4NV_o}{\pi\omega_r R_{ac}(NV_o + DV_i)} \tag{18}$$

C. Calculation of magnetizing inductance

The choice of magnetizing inductance is a crucial issue for any resonant converter that uses the magnetizing current for achieving soft switching. This is because the magnetizing current affects the efficiency of the converter. Due importance is given in literature [8]–[11] for the optimal design of magnetizing inductance for LLC resonant converters. However, this issue is somewhat less addressed in APWM half-bridge resonant converter, primarily because so far it has been studied as a series resonant converter with the magnetizing inductance of the transformer assumed to be infinitely large. An analytical guide to choose optimal magnetizing inductance and corresponding deadtime is presented in this section.

At $t = t_2$, i_{Rac} becomes zero and the primary circuit can be approximated as an LC network (L_m, L_r and C_r in series) with initial capacitor voltage $v_{Cr\_t2} = -NV_o$ and initial inductor current $i_{Lm\_t2} = \frac{-NV_o}{L_m}\left(\frac{T}{4}\right)$. The standard solution for current in such a network is given as,

$$i_{Lm}(t) = \frac{NV_o}{\sqrt{(L_m + L_r)/C_r}}sin(\omega't)$$
$$- \frac{NV_o}{L_m}\left(\frac{T}{4}\right)cos(\omega't) \quad ; \quad t_2 < t \leqslant T \quad (19)$$

where $\omega' = \frac{1}{\sqrt{(L_m+L_r)C_r}}$. The current $i(t)$ at $t = T$ can be calculated by substituting $t = \frac{\theta_1 T}{2\pi}$ in (19).

$$i_{Lm}(T) = \frac{NV_o\sqrt{C_r}}{\sqrt{(L_m + L_r)}}sin\left(\frac{\omega'\theta_1 T}{2\pi}\right)$$
$$- \frac{NV_o}{L_m}\left(\frac{T}{4}\right)cos\left(\frac{\omega'\theta_1 T}{2\pi}\right) \quad (20)$$

At $t = T$, the gate pulse of the bottom switch S_2 is removed and the top switch S_1 will be triggered after a deadtime t_{dead}. A small negative magnetizing current $i_{Lm}(T)$ is sufficient for ZVS of S_1, as long as this current can discharge C_{oss1} (and simultaneously charge C_{oss2} to input voltage V_i) and turn on its body diode within t_{dead} [8]. Assuming $C_{oss1} = C_{oss2} = C_{oss}$, this critical condition can be expressed as follows.

$$\frac{1}{2}(L_m + L_r)i_{Lm}^2(T) = 2 \times \frac{1}{2}C_{oss}V_i^2 \quad (21)$$

Since $C_r \gg C_{oss}$, it can be assumed that only the parallel combination of C_{oss1} and C_{oss2} resonates with $(L_m + L_r)$ during t_{dead}. So, the minimum deadtime required should be one-fourth of this resonant time period to allow complete discharge of C_{oss1} before S_1 turns on [8]. Thus,

$$t_{dead\_min} = \frac{\pi\sqrt{(L_m + L_r)C_{oss}}}{\sqrt{2}} \quad (22)$$

However, t_{dead_min} calculated from the above equation will be very high and thus impractical.

A better way of choosing L_m and t_{dead} is to first fix the minimum deadtime from the turn-on/turn-off characteristics of

S_1 and S_2 and assume that the energy available in $(L_m + L_r)$ at $t = T$ is far more than that required to get C_{oss1} discharged (and C_{oss2} charged) within that fixed deadtime.

$$\frac{1}{2}(L_m + L_r)i_{Lm}^2(T) >> 2 \times \frac{1}{2}C_{oss}V_i^2 \quad (23)$$

The discharging of C_{oss1} (and charging of C_{oss2}) can now be considered as linear, instead of being resonant, with $i_{Lm}(T)$ as a constant current source. To keep a safe margin it is considered that the simultaneous charging of C_{oss1} and discharging of C_{oss2} should be completed within $0.8t_{dead}$. Thus $i_{Lm}(T)$ can be solved as,

$$i_{Lm}(T) = \frac{2C_{oss}V_i}{0.8t_{dead}} \quad (24)$$

Therefore, L_m can be calculated by equating (20) and (24).

IV. Design Example With Simulation And Experimental Results

In this section, a $300\ W$ converter with the specifications as shown in Table I is designed as per the analysis made before. The system was simulated in PSIM and also tested with a laboratory set-up.

Using the given values and the relation $\omega_r = 2\pi f_r$ in (13), (17) and (18), the values of N, L_r and C_r are calculated as $2, 13.96\ \mu H$ and $181.43\ nF$ respectively. In simulation, L_m and t_{dead} are chosen as $300\ \mu H$ and $200\ ns$ respectively that comply with (20) and (24) with sufficient margin.

In the experimental set-up, a centre-tap transformer, wound on an ETD49 ferrite core, with number of turns $10 : 5 : 5$, magnetizing inductance $292\ \mu H$ and leakage inductance $0.9\ \mu H$ was used. The values of L_r and C_r used were $12.7\ \mu H$ and $188\ nF$. A PQ26/20 ferrite core was used for the series inductor L_r. Appropriate air-gaps were provided in the inductor and the transformer to adjust the values of L_r and L_m, and also to avoid core saturation. The device used as S_1 and S_2 was IPP320N20N3G, and its maximum value of C_{oss} ($180\ pF$) for the operating condition in Table I is chosen for design purpose. The system was controlled and modulated using the TMS320F28335 Experimenter Kit from Texas Instruments. The input voltage used in the experiment ($120\ V$) had to be slightly higher than that used in the simulation ($108\ V$) for obtaining same output voltage ($24\ V$). This is attributed to the voltage drops occurring in various elements in the practical circuit.

TABLE I
CONVERTER SPECIFICATIONS FOR DESIGN

Parameter Name	Parameter Symbol	Simulation Value	Experimental Value
Minimum input voltage	V_i	$108\ V$	$120\ V$
Output voltage	V_o	$24\ V$	$24\ V$
Load resistance	R_L	$1.92\ \Omega$	$1.92\ \Omega$
Maximum duty ratio	D_{max}	0.35	0.35
Switching frequency	f_r	$100\ kHz$	$100\ kHz$
Output filter capacitance	C_o	$0.2\ mF$	$0.2\ mF$
Switch output capacitance	C_{oss}	$180\ pF$	$180\ pF$

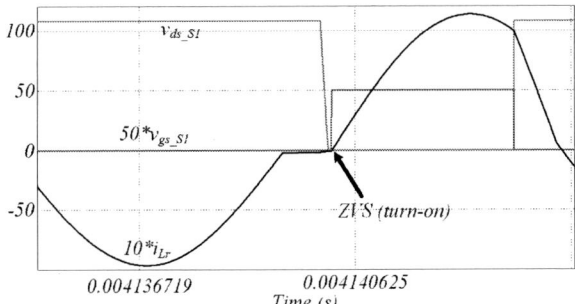

Fig. 5. ZVS turn-on of S_1 at full-load (Simulation). $V_o = 24V, D = 0.35$.

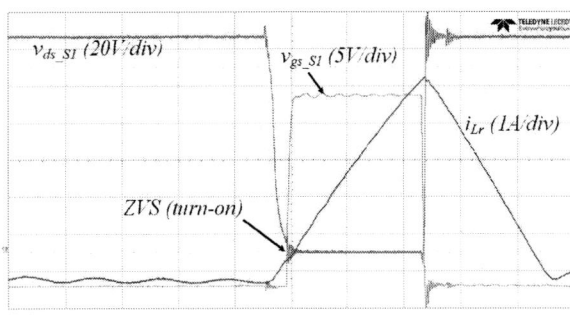

Fig. 9. ZVS turn-on of S_1 at 10% load (Experimental). $V_o = 24.2V, D = 0.16$. Time scale: $1\mu s/div$.

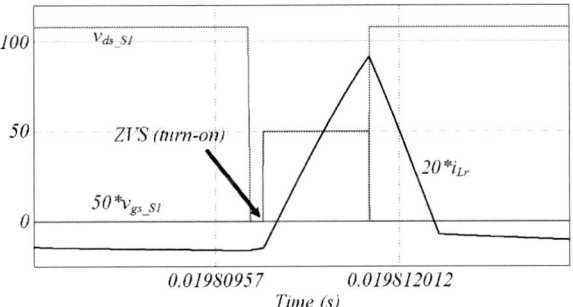

Fig. 6. ZVS turn-on of S_1 at 10% load (Simulation). $V_o = 24V, D = 0.16$.

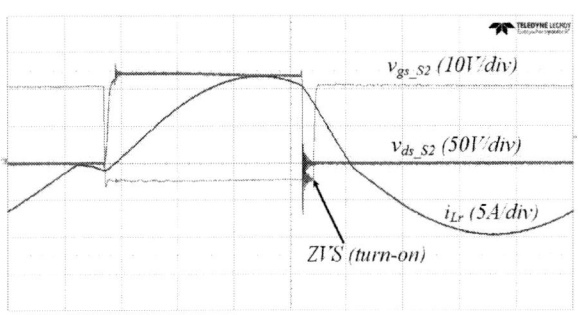

Fig. 10. ZVS turn-on of S_2 at full-load (Experimental). $V_o = 24.4V, D = 0.35$. Time scale: $1\mu s/div$.

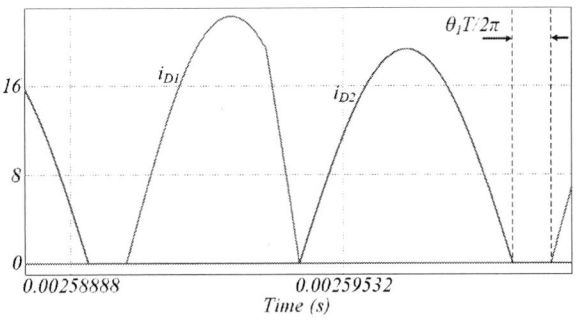

Fig. 7. Current through the output diodes at full-load (Simulation). $V_o = 24V, D = 0.16$.

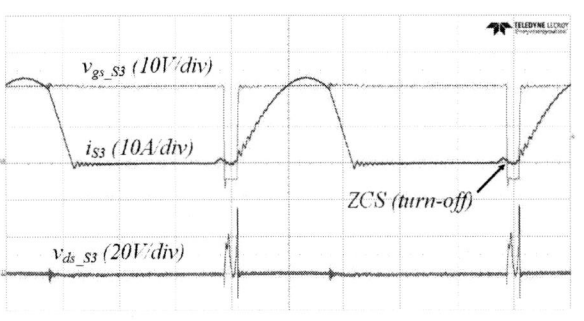

Fig. 11. Zero current turn-off of S_3 at full-load (Experimental). $V_o = 24.4V, D = 0.35$. Time scale: $2\mu s/div$.

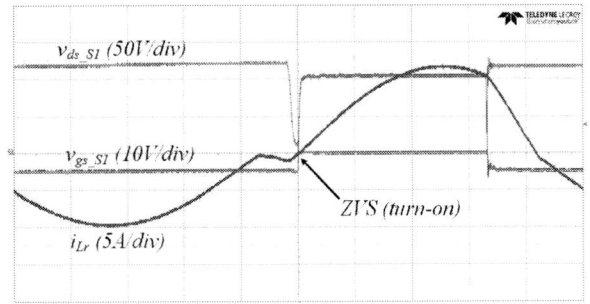

Fig. 8. ZVS turn-on of S_1 at full-load (Experimental). $V_o = 24.4V, D = 0.35$. Time scale: $1\mu s/div$.

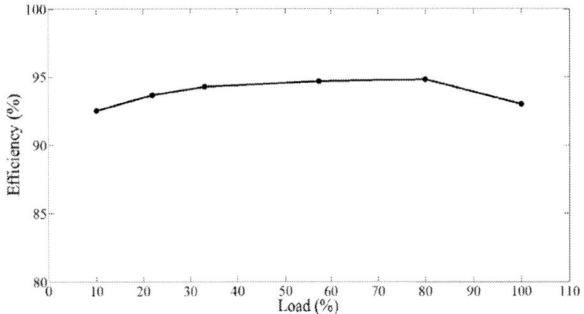

Fig. 12. Experimental efficiency plot of the proposed converter.

Figs. 5 and 6 show the key simulation waveforms at full-load and 10% load respectively, demonstrating ZVS of the top switch S_1. Experimental results for the same loading conditions are are shown in Figs. 8 and 9 and it is found that they are in close agreement with the simulation results. The peak of the magnetizing current is noted as only $0.8A$ from the experimental results.

The currents through the output diodes are shown in Fig. 7. It can be seen that, unlike conventional LLC resonant converter, the APWM converter has diode discontinuity at only one one switching transition (S_2 turn-off and S_1 turn-on). During this period, the output voltage is maintained by the output capacitor C_o and only the small magnetizing current ($0.4A$) flows in the primary side network.

The zero-voltage turn-on of the bottom switch S_2 is shown in Fig. 10. It is observed that since i_{Lr} has a high positive value before the turning-on of S_2, C_{oss2} is discharged very fast and the ZVS of S_2 is achieved very easily. Another important aspect noted from Fig. 10 is that S_2 has to break only the small magnetizing current ($0.4A$) during its turn-off, which makes the turn-off loss of S_2 smaller than that of S_1. Also, the drain-to-source voltage ringing is seen to be higher at S_1 turn-off as compared to that at S_2 turn-off because of the same reason.

The zero current turn-off of S_3 is shown in Fig.11. It can be observed that the function of S_3 is to block the transformer secondary voltage and thereby prevent the turning-on of D_1 during deadtime. This has helped in achieving the ZVS of S_1 with minimal magnetizing current.

The experimental efficiency of the converter obtained from the $300\ W$ laboratory prototype is plotted in Fig. 12 for the entire load range. It is found that the light load efficiency has improved because of minimized circulating current.

V. CONCLUSION

A mathematical design procedure is presented in this paper for choosing the values of the resonant tank parameters of an APWM half-bridge DC-DC resonant converter for a certain maximum allowable duty ratio. It is evident that, when designed for minimum input voltage and maximum load at maximum allowable duty ratio, the converter ensures zero voltage turn-on of the two primary-side switches and zero current turn-off of the secondary-side switch irrespective of load conditions. Moreover the circulating current in the primary-side is decreased significantly by operating the converter exactly at resonant frequency, as well as by minimizing the magnetizing current with the use of the additional switch at the secondary side of the transformer. Thus the designed converter achieves very high efficiency over a wide load range.

REFERENCES

[1] D. J. Tschirhart and P. K. Jain, "Design procedure for high-frequency operation of the modified series-resonant APWM converter to reduce size and circulating current," *Power Electronics, IEEE Transactions on*, vol. 27, no. 10, pp. 4181–4191, 2012.

[2] D. J. Tschirhart and P. K. Jain, "A CLL resonant asymmetrical pulsewidth-modulated converter with improved efficiency," *Industrial Electronics, IEEE Transactions on*, vol. 55, no. 1, pp. 114–122, 2008.

[3] S. Mangat, M. Qiu, and P. Jain, "A modified asymmetrical pulse-width-modulated resonant DC/DC converter topology," *Power Electronics, IEEE Transactions on*, vol. 19, no. 1, pp. 104–111, 2004.

[4] P. K. Jain, A. St-Martin, and G. Edwards, "Asymmetrical pulse-width-modulated resonant DC/DC converter topologies," *Power Electronics, IEEE Transactions on*, vol. 11, no. 3, pp. 413–422, 1996.

[5] M. Qiu, P. Jain, and H. Zhang, "An APWM resonant inverter topology for high frequency AC power distribution systems," *Power Electronics, IEEE Transactions on*, vol. 19, pp. 121–129, Jan 2004.

[6] M. Qiu, P. K. Jain, and H. Zhang, "An APWM resonant inverter topology for high frequency ac power distribution systems," in *Applied Power Electronics Conference and Exposition, 2002. APEC 2002. Seventeenth Annual IEEE*, vol. 2, pp. 1141–1147, IEEE, 2002.

[7] B. Yang, *Topology investigation for front end DC/DC power conversion for distributed power system*. PhD thesis, Virginia Polytechnic Institute and State University, 2003.

[8] J. Sabate, V. Vlatkovic, R. Ridley, F. Lee, B. Cho, *et al.*, "Design considerations for high-voltage high-power full-bridge zero-voltage-switched PWM converter," in *Proc. IEEE APEC*, vol. 90, pp. 275–284, 1990.

[9] J.-h. Jung and J.-g. Kwon, "Theoretical analysis and optimal design of LLC resonant converter," in *Power Electronics and Applications, 2007 European Conference on*, pp. 1–10, IEEE, 2007.

[10] B. Lu, W. Liu, Y. Liang, F. C. Lee, and J. D. Van Wyk, "Optimal design methodology for LLC resonant converter," in *Applied Power Electronics Conference and Exposition, 2006. APEC'06. Twenty-First Annual IEEE*, pp. 6–pp, IEEE, 2006.

[11] H.-P. Park, H.-J. Choi, and J.-H. Jung, "Design and implementation of high switching frequency LLC resonant converter for high power density," in *Power Electronics and ECCE Asia (ICPE-ECCE Asia), 2015 9th International Conference on*, pp. 502–507, IEEE, 2015.

A Low-Volume Hybrid Step-Down Dc-Dc Converter Based on the Dual Use of Flying Capacitor

S. M. Ahsanuzzaman, Yingxian Ma, Abrar Ahmed Pathan, Aleksandar Prodić

Laboratory for Power Management and Integrated SMPS, ECE Department, University of Toronto, CANADA
{ahsansm,prodic}@ece.utoronto.ca

Abstract— **This paper introduces a hybrid step-down dc-dc converter, targeted for battery powered portable applications where low-volume implementation is the key priority. The introduced architecture, combining switched-capacitor (SC) and inductor based circuits, requires low-volume for implementation by reducing the size of the filer inductor by 4 times and the output capacitor by 2 times. In addition to supporting wide input-output range for step down voltage conversions, the introduced architecture demonstrates up to 15 % power processing efficiency improvement compared to conventional buck converter and faster dynamic response. These advantages are obtained by a dual use of the flying capacitor usually existing in SC converters. The flying capacitor is used for both balancing of the front-end stage and reducing voltage swing/stress of the components. Experimental results from a 5 V, 25 W, 500 kHz prototype verify advantages of the introduced converter.**

Keywords—step-down dc-dc converter, digital controller, hybrid converter, low-volume, portable applications

I. Introduction

In modern battery powered portable applications, such as laptops, tablet computers, smart phones etc., many dc-dc switch mode power supplies (SMPS) are utilized to provide multiple voltage levels for various functional blocks [1], [2]. One of the main challenges with the implementation of the conventional SMPS is their size. In numerous portable devices SMPS are among the largest contributors to the overall size and weight of the entire device [3], [4] and large consumers of printed circuit board (PCB) space. This is primarily due to the bulky and costly reactive components of the SMPS output filters, where the inductors are the largest and heaviest components of the filters [5].

The ever increasing demand for lower volume dc-dc converters for battery powered portable electronics has primarily been met by switching at higher frequencies, up to tens of MHz [6], allowing smaller filter size. However, higher switching frequency comes with a penalty of increased switching losses [7], [8] negatively affecting the battery life.

To reduce the volume requirement of SMPS, in [9], [10] two-stage compact and power efficient solutions are presented. In these solutions, a switched-capacitor (SC) fixed-ratio front-end stage performs a large portion of voltage conversion at the peak efficiency, and an inductor based

This work of Laboratory for Power Management and Integrated SMPS is supported by Qualcomm Inc.

downstream stage then provides final regulation. This solution reduces the voltage swing at the switching node of the inductor based stage, relaxing the requirement of the filter inductor, while improving the efficiency at the same time, due to lower input-to-output voltage conversion ratio [7], [8]. However, in the previously presented series SC-inductive architectures the output voltage is limited by the conversion ratios of the front-end SC stage and hence, cannot be widely adopted for applications requiring wide input-output voltage range of operation.

This paper introduces a wide-input hybrid dc-dc converter architecture for battery-powered applications that allows volume reduction of the reactive components and efficiency improvement. In this architecture, shown in Fig.1, the first portion is identical to switched-capacitor voltage divider (SW_{1-4}, C_{in1}, C_{in2}, and C_{fly}) [9]. However, the second portion is a 2-input buck converter ($SW_{5,6}$, L, C_{out}) [11] connected across flying capacitor, C_{fly}. In this topology the flying capacitor has dual use. It provides a mean for balancing of the capacitor divider, and at the same time reduces voltage stress/swing of the components to a ½ of V_{in}.

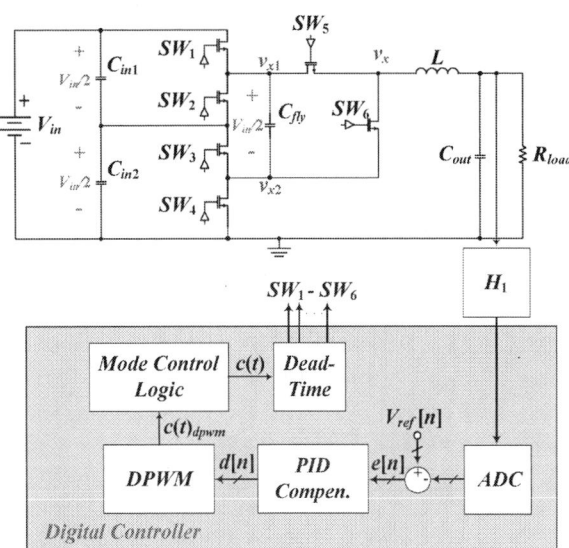

Fig. 1 Hybrid dc-dc converter with the dual use of the flying capacitor and complementary digital controller

A combined SC voltage divider and inductor based architecture was proposed in [12], where two inductors are connected to intermediate nodes of a SC voltage divider, to provide three separate output voltages. However, in that configuration the output voltages are correlated, due to their fixed input-to-output voltage ratios, and hence, cannot be controlled independently.

As shown in Section II, the introduced converter in Fig. 1 allows the switching node, v_x of the output filter to have 3 possible values of V_{in}, $V_{in}/2$ and 0 V, resulting in up to 4 times reduction of the output filter inductor, L and 2 times reduction of output capacitor, C_{out}, for approximately same semiconductor losses [13]. Furthermore, the practical implementation section shows that all the switches can be rated for a half of the voltages required for the conventional buck converter, improving power processing efficiency by up to 15 %, by reducing conduction and switching losses.

II. PRINCIPLE OF OPERATION

As shown in Fig. 1, during the circuit operation the voltages across each of the input capacitors ($C_{in1,2}$) and flying capacitor, C_{fly} are equal to $V_{in}/2$. Depending on the operating conditions, i.e. the input and output voltages, the introduced converter changes the gating sequences. Still, as described below, this voltage equilibrium across each of the input capacitors, $C_{in1,2}$, is maintained by equal charge sharing through the flying capacitor.

A. Operation for $V_{out}<1/2\ V_{in}$

In case of $V_{out} < 1/2V_{in}$, the switching node, v_x operates between $V_{in}/2$ and ground (0 V). Fig. 2 shows the gating signals for SW_{1-6}, to achieve this operation. The voltage $V_{in}/2$ at the switching node is achieved during time intervals T_1 and T_2, through two different gating sequences. During T_1, $SW_{1,3,6}$ are turned on, while during T_2, $SW_{2,4,5}$ are on. This is because in order to maintain equal voltage across input capacitors ($C_{in1,2}$), the flying capacitor, C_{fly}, needs to be placed in parallel to each of C_{in1} and C_{in2} for an equal duration of time. During the time period T_1 (Fig. 2), C_{fly} is placed across C_{in1} by turning on SW_1 and SW_3. Similarly, during the time period T_2, C_{fly} is placed across C_{in2} by turning on SW_2 and SW_4. This results in equal charge balancing across the two input capacitors while providing $V_{in}/2$ at the switching node, v_x of the output filter. Furthermore, the effective switching frequency ($1/T_{sw-eff}$) at the switching node, v_x is twice the switching frequency $f_{sw} = 1/T_{sw}$. As it will be shown later, the lower voltage swing at the switching node, combined with higher effective switching frequency results in drastic reduction of the output filter L and C_{out}.

B. Operation for $1/2V_{in}<V_{out}<V_{in}$

In case of $V_{out} > 1/2V_{in}$, the switching node, v_x operates between V_{in} and $V_{in}/2$. Fig. 3 shows the gating signals for SW_{1-6}, to achieve this operation. Similar to the previous case, $V_{in}/2$ at the switching node is achieved during time intervals T_1 and T_2, by two different gating sequences. During T_1, $SW_{1,3,6}$ are turned on, while during T_2, $SW_{2,4,5}$ are on. During the time

period T_1 (Fig. 3), C_{fly} is placed across C_{in1} by turning on SW_1 and SW_3. Similarly, during the time period T_2, C_{fly} is placed across C_{in2} by turning on SW_2 and SW_4. Again, this results in equal charge balancing across the two input capacitors while providing $V_{in}/2$ at the switching node, v_x of the output filter.

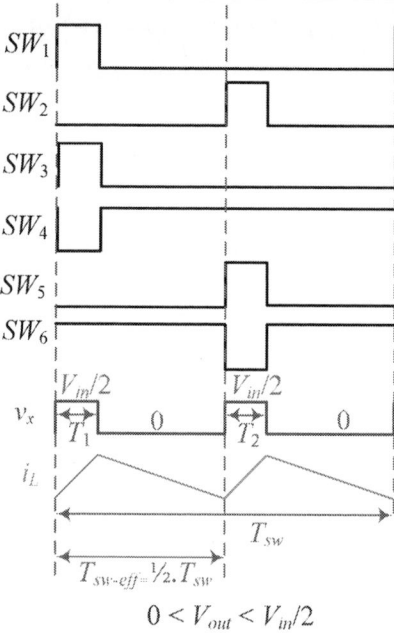

Fig. 2 Gating signals, switching node and inductor current waveforms for $V_{out}<V_{in}/2$

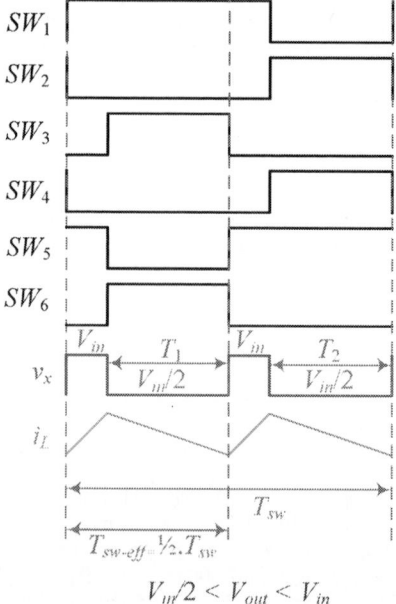

Fig. 3 Gating signals, switching node and inductor current waveforms for $V_{out}>V_{in}/2$

It can be seen that in both of these modes of operation the flying capacitor has dual use. It provides balancing of the input capacitive divider and, at the same time, reduces voltage stress/swing of the components to a ½ of V_{in}.

C. Operations for $V_{out}=1/2V_{in}$

In case of $V_{out} = 1/2V_{in}$, the switching node, v_x is at a constant voltage, $V_{in}/2$. Fig. 4 shows the gating signals for SW_{1-6}, to achieve this mode of operation. The voltage $V_{in}/2$ at the switching node is maintained during the entire switching period, T_{sw}, by two different gating sequences over subintervals T_1 and T_2. In this case, T_1 and T_2 constitute the entire switching period, T_{sw}. During T_1, $SW_{1,3,6}$ are turned on, while during T_2, $SW_{2,4,5}$ are on. During the time period T_1 (Fig. 2), C_{fly} is placed across C_{in1} by turning on SW_1 and SW_3. Similarly, during the time period T_2, C_{fly} is placed across C_{in2} by turning on SW_2 and SW_4. This results in charge balancing across the two input capacitors while providing $V_{in}/2$ at the switching node, v_x of the output filter. During this mode of operation the voltage swing at the switching node, v_x is practically zero, resulting in zero inductor current ripple.

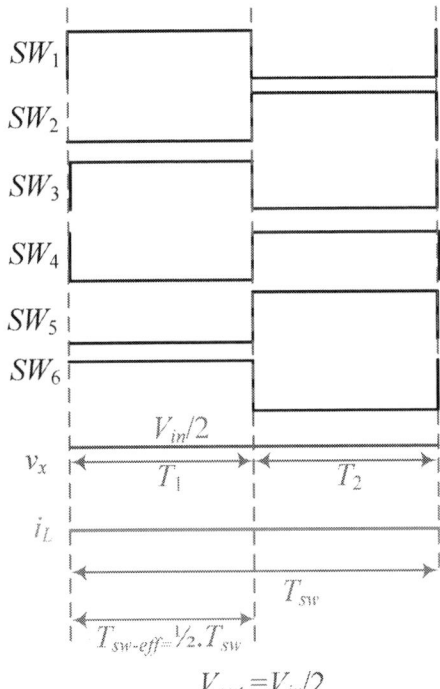

Fig. 4 Gating signals, switching node and inductor current waveforms for $V_{out}=V_{in}/2$

D. Output filter volume reduction

In a conventional buck converter the inductance is determined by the steady state ripple requirement [7], given by:

$$L_{buck} = \frac{(V_{in} - V_{out}).D}{2.\Delta i_L.f_{sw}} \quad , \qquad (1)$$

where, D in the duty ratio and Δi_L is the steady state inductor current ripple. However, as mentioned in Section II.A, due to the combined reduction of voltage swing at the switching node, v_x and 2 times higher effective switching frequency (Fig. 2, 3), the output filter inductor, L of the introduced architecture can be drastically reduced. The reduction can be explained by looking at the following equations:

$$L_{new} = \frac{V_{in}}{2.\Delta i_L.f_{sw}}.(\frac{1}{2}-D).D \quad , \quad \text{for } 0 < D < 0.5 \quad (2) \text{ and}$$

$$L_{new} = \frac{V_{in}}{2.\Delta i_L.f_{sw}}.(1-D).(D-\frac{1}{2}) \quad , \quad \text{for } 0.5 < D < 1 \quad (3)$$

Comparing Eq. (1) with (2), (3), normalized inductance values with respect to the worst case ripple condition of the conventional buck converter (D=0.5) is presented in Fig. 5. As shown in this figure, inductance reduction of the introduced converter can be at least 4 times.

This reduction is similar to multi-level (3-level) converter as shown in [14]. In addition to inductor being 4 times smaller, the introduced converter results is 2 times smaller output capacitor for the same ripple requirement. This is due to the same operating principle as 3-level converter as described in [15]. However, the introduced converter has several practical implementation advantages over the 3-level converter, which are addressed in the following section.

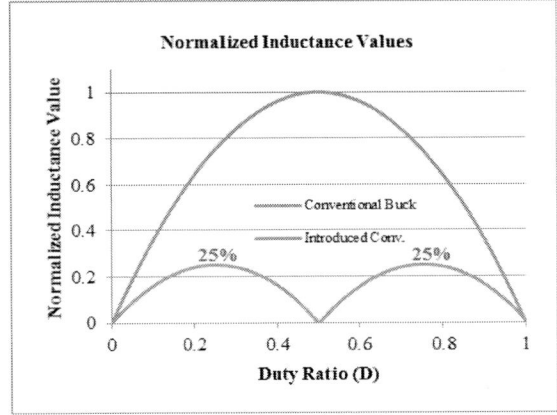

Fig. 5 Normalized output filter inductance value vs. duty ratio for conventional buck and introduced converter

III. PRACTICAL IMPLEMENTATION

As shown in Fig. 1, there are two additional switches in the introduced architecture compared to 4 switches in 3-level converters [14], [15]. The practical implementation considerations for the introduced architecture are discussed in the following subsections.

A. Semiconductor area for switches realization

In a conventional buck converter, both switches need to be rated for V_{in}, due to the blocking voltage requirement of the switches [7]. Although in the introduced converter of Fig. 1 there are 6 switches, these switches can be rated for $V_{in}/2$. Since silicon area for switch realization is approximately proportional to V_{ds}^2 [16], where V_{ds} is the blocking voltage of the switch, the semiconductor areas can be equivalent. Furthermore, in the targeted low power portable applications the volume of the output filter is significantly larger than that of the semiconductor components [13]. Since 4 times L and 2 times C_{out} reductions can be achieved as shown in Section II, the increased number of switches does not contribute to any practical implementation challenges.

For 3-level converter, the voltage swing at the switching node is also $V_{in}/2$ [14], [15] resulting in steady state blocking voltage requirement of the switches to be $V_{in}/2$. However, considering the flying capacitor could be completely discharged during startup, at least one of the switches need to be rated for full input voltage, resulting in increased silicon area for switch realization and/or additional losses [16]. Although the introduced architecture has 2 extra switches compared to 3-level converter, the overall semiconductor requirement is not significantly increased. This is due to the fact that, two series connected capacitors C_{in1} and C_{in2}, which can be considered as the input filter capacitor of the converter, result in $V_{in}/2$ voltage at the center node of the four switches (SW_{1-4}) during startup (Fig. 1).

B. Converter losses and Efficiency

Simulation test benches were developed in LTSpice for both the conventional buck and introduced hybrid step-down converters, with actual models of semiconductor switches for detailed efficiency analysis. 12 V-to-1 V operating condition was selected for the comparison, where the buck and the introduced converter operate at 1 MHz and 500 kHz, respectively. As previously described, 500 kHz operation of the introduced converter results in 1 MHz inductor current ripple. 12 V Power MOSFETs from International Rectifier (IRF7476PbF) were utilized to model the performance of the conventional buck converter under the operating condition with load currents ranging from 500 mA to 5 A. In order to compare the performance of the introduced hybrid step-down converter, two different approaches were considered. First, a conservative approach of 50 % improvement in figure-of-merit (FoM) was considerd for semiconductor switches, due to 50 % reduced blocking voltage requirement of the switches (i.e. 6 V instead of 12 V). In this case the on-resistances

(R_{ds_on}) and gate capacitances (C_g) of the switches are considerd to be 0.7 times of those of the buck switches. Second, a more realistic estimation of 70 % improvement in figure-of-merit (FoM) was considerd for semiconductor switches due to 50 % reduced blocking voltage requirement of the switches. In this case the on-resistances (R_{ds_on}) and gate capacitances (C_g) of the switches are considerd to be 0.55 times of those of the buck switches. Furthermore, in both of these approaches two times smaller inductor (220 nH) with about half parasitic resitance (6 mΩ) compared to the conventional buck converter (470 nH, 11 mΩ) were utilized. As shown in Fig. 6, more than 15 % efficiency improvement in the light load operating condition is possible, potentially making the introduced converter a very attractive solution for portable applications.

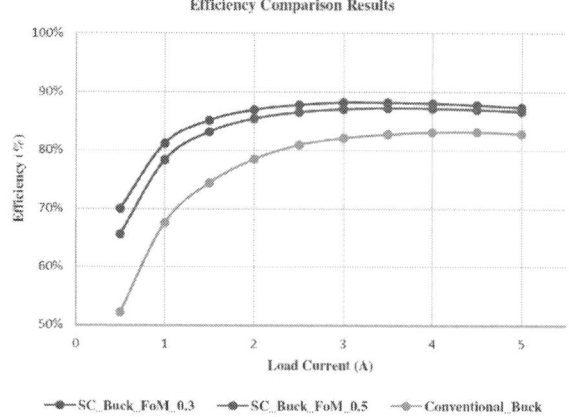

Fig. 6 Efficiency comparison with conventional 12-to-1 V buck converter

Fig. 7 shows the conduction, switching and gate drive losses breakdown for the conventional buck and introduced step-down converters for 12 V-to-1 V operating condition with the load current varying between 500 mA and 5A. As shown in this figure, most of the efficiency improvements are resulting from the significant reduction of the switching losses. This is due to improved FoM of the semiconductor switches with reduced blocking voltage requirements and operation with half the switching frequency (500 kHz instead of 1 MHz) of the introduced hybrid step-down converter. Hence, the maximum efficiency improvement is demostrated at light load operating condition, where the switching losses are the most dominant, as shown in Fig. 6.

In addition, compared to a 3-level converter, the introduced converter does not introduce the equivalent series resistance (ESR) of the flying capacitor in the conduction path of the inductor current [14]. Depending on the size and the type of the flying capacitor, the ESR value can be in the range of 5-10 mΩ, introducing aditional conduction losses in the 3-level converter. However, as explained in Section II, in the introduced hybrid step-down converter, the switched-capacitor stage operates with the principle of charge sharing between the two input capacitors C_{in1} and C_{in2} [5]. As a result, the ESR of the flying capacitor does not introduce any additional resistance in the conduction path of the inductor current.

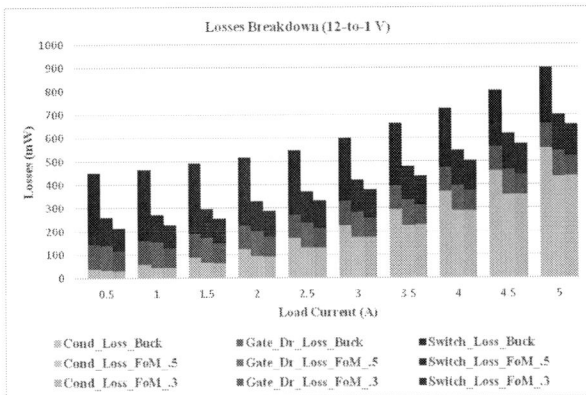

Fig. 7 Breakdown of the losses for 12-to-1 V operating condition

C. Output voltage regulation and dynamic response

The introduced converter can provide superior dynamic regulation compared to conventional buck converter. This is due to significantly smaller inductor size, which results in much higher slew rate than the conventional buck converter [7]. During light-to-heavy load transients, the voltage at the switching node, v_x can be V_{in}, by turning on SW_1 and SW_5 (Fig. 1). Similarly, during heavy-to-light load transients, 0 V appearing at v_x, by turning on SW_4 and SW_6 will result in higher slew rate than conventional buck, due to significantly smaller inductance [7].

Furthermore, the introduced converter can achieve output voltage $V_{out} = 1/2\ V_{in}$, with the gating sequence explained in Section II C. In this mode of operation the voltage swing at the switching node is practically zero, resulting in zero current ripple for the inductor current. During this mode of operation the switched-capacitor stage operates as a conventional voltage divider with 50 % duty ratio [9], where as 2-input buck operates with zero voltage swing at the switching node, v_x (Fig. 4). However, in 3-level converter operation with 50 % duty ratio can result in large voltage transient spikes and hence, special control method needs to be incorporated to eliminate them, as shown in [17]. Experimental results verifying proper operation of the introduced hybrid step-down dc-dc converter with zero current ripple is shown in the next section.

IV. EXPERIMENTAL RESULTS

A discrete prototype on printed circuit board (PCB) was developed to verify proper functionality of the introduced converter. The digital controller of Fig. 1 was developed using an FPGA system, consisting of PID Compensator, DPWM and Mode Control Logic blocks to determine the gating sequences, depending on operating conditions as described in Section II. Figs. 8 to 11 show experimental results for

different operating conditions, where the introduced converter and the buck converter operate with 500 kHz and 1 MHz switching frequencies, respectively. This results in the same effective switching frequencies for both the converters. The same inductor was used in both cases to demonstrate the reduction of the inductor current ripple, which will translate into proportional reduction of the inductance value for the same ripple requirement.

Fig. 8 shows experimental waveforms for 12 V-to-5 V ($D<0.5$) operating condition. As shown in this figure, the introduced converter, while operating at half the switching frequency results in about 65 % reduction of inductor current ripple. This is due to 6 V voltage swing reduction at the switching node, v_x (Fig. 1). Fig. 9 shows a zoomed in version of Fig. 8 to clearly demonstrate this mode of operation and gating sequences, as explained in Section II A.

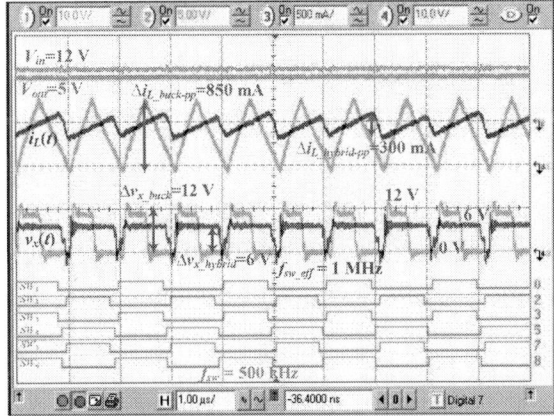

Fig. 8 Comparisons of 12 V-to-5 V operation, Ch1: Input voltage, V_{in} (10 V/div); Ch2: Output voltage, V_{out} (5 V/div); Ch3: Inductor current, i_L (500 mA /div); Ch4: Switching node, v_x (10 V/div)

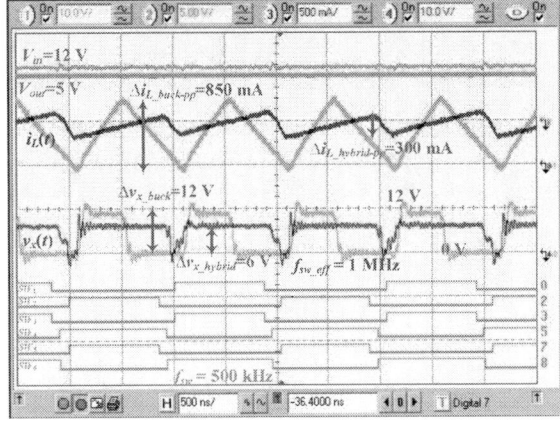

Fig. 9 Comparisons of 12 V-to-5 V operation, Ch1: Input voltage, V_{in} (10 V/div); Ch2: Output voltage, V_{out} (5 V/div); Ch3: Inductor current, i_L (500 mA /div); Ch4: Switching node, v_x (10 V/div)

Fig. 10 shows experimental waveforms for 8 V-to-5 V ($D>0.5$) operating condition. As shown in this figure, the

978-1-4673-9551-9/16 $31.00 © 2016 IEEE

introduced converter, while operating at half the switching frequency results in about 60 % reduction of inductor current ripple.

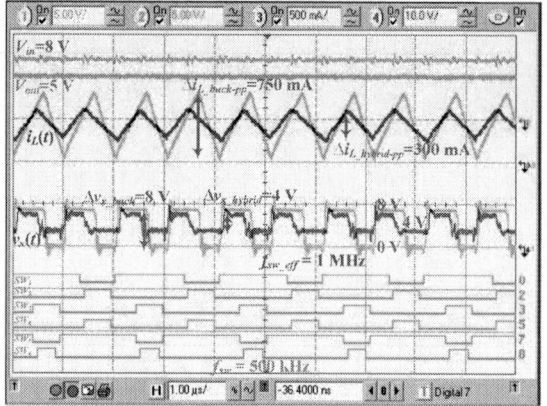

Fig. 10 Comparisons of 8 V-to-5 V operation, Ch1: Input voltage, V_{in} (10 V/div); Ch2: Output voltage, V_{out} (5 V/div); Ch3: Inductor current, i_L (500 mA /div); Ch4: Switching node, v_x (10 V/div)

Fig. 11 shows experimental waveforms for 10 V-to-5 V (D=0.5) operating condition. As shown in this figure, the introduced converter results in zero current ripple condition. This provides experimental verification of operating condition with D=0.5 from Fig. 5, which shows while the buck converter reaches its maximum current ripple, the introduced converter has zero ripple current.

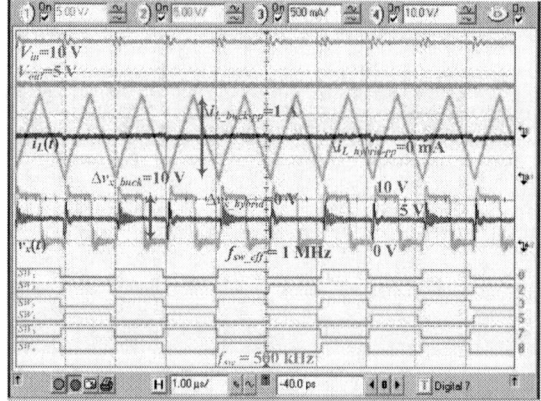

Fig. 11 Comparisons of 10 V-to-5 V operation, Ch1: Input voltage, V_{in} (10 V/div); Ch2: Output voltage, V_{out} (5 V/div); Ch3: Inductor current, i_L (500 mA /div); Ch4: Switching node, v_x (10 V/div)

V. CONCLUSIONS

This paper introduces a hybrid step-down dc-dc converter, targeted for battery powered portable applications, by combining switched-capacitor and reduced voltage based buck converter circuits. The novel converter makes dual use of the flying capacitor usually existing in SC circuits, to provide SC cell balancing and to reduce the voltage stress/swing of the components. As a result, up to four times reduction of the output filter inductor and two times reduction of the output filter capacitor, which are by far the largest components in low power applications, is achieved. In addition to supporting wide input-output range for step down voltage conversions, the introduced architecture demonstrates up to a 15 % efficiency improvement and superior dynamic regulation compared to conventional buck converter. Experimental and simulation results confirm advantages of the new topology over conventional solutions.

REFERENCES

[1] P. Henry , "New Advances in Portable Electronics," in *APEC Plenary Session*, March 2009.

[2] S.M. Ahsanuzzaman, D.A. Johns, and A. Prodić, "A configurable power management IC for low-volume dc-dc converter applications with high frequency current programmed mode control," in *Proc. IEEE Applied Power Electronics Conference and Exposition (APEC)*, March 2015.

[3] P. Kumar, and W. Proefrock, "Novel switched capacitor based Triple Output Fixed Ratio Converter (TOFRC)," in *Proc. IEEE Applied Power Electronics Conference and Exposition (APEC)*, pp.2352-2356, Feb. 2012.

[4] Y. Kaiwei, "High-frequency and high-performance VRM design for the next generations of processors," Ph.D. thesis, Virginia Polytechnic Institute and State University, 2004.

[5] M. D. Seeman, V. W. Ng, Hanh-Phuc Le; M. John, E. Alon, and S. R. Sanders, S.R., "A comparative analysis of Switched-Capacitor and inductor-based DC-DC conversion technologies," in *Proc. IEEE Workshop on Control and Modeling for Power Electronics (COMPEL)*, pp.1-7, June 2010.

[6] L. Pengfei, D. Bhatia, D. X. Lin, and R. Bashirullah, "A 90–240 MHz hysteretic controlled dc-dc buck converter with digital phase locked loop synchronization," in *IEEE Journal of Solid-State Circuits*, vol.46, no.9, pp.2108-2119, Sept. 2011.

[7] Robert W. Erickson and Dragan Maksimović , "Fundamentals of power electronics", Second Edition, New York: Springer Science+Business Media, 2001.

[8] J. Klein, "Synchronous buck MOSFET loss calculations with Excel model". Application note AN–6005, Fairchild Semiconductor, version 1.0.1, April 2006.

[9] J. Sun, M. Xu, Y. Ying, and F.C. Lee, "High power density, high efficiency system two-stage power architecture for laptop computers," in *Proc. Power Electronics Specialists Conference (PESC)*, pp.1-7, 2006.

[10] L. Seungbum, J. Ranson, D.M. Otten, and D.J. Perreault, "Two-stage power conversion architecture suitable for wide range input voltage," in *IEEE Transactions on Power Electronics*, vol.30, no.2, pp.805,816, Feb. 2015.

[11] J. Sebastian, P.J. Villegas, F. Nuno, and M.M. Hernando, "High-efficiency and wide-bandwidth performance obtainable from a two-input buck converter," in *IEEE Transaction on Power Electronics*, vol.13, no.4, pp.706-717, Jul 1998.

[12] P. Kumar, and W. Proefrock, "Novel switched capacitor based Triple Output Fixed Ratio Converter (TOFRC)," in *Proc. IEEE Applied Power Electronics Conference and Exposition (APEC)*, pp.2352-2356, Feb. 2012.

978-1-4673-9551-9/16 $31.00 © 2016 IEEE

[13] Y. Lei, W.C. Liu, and R. C. N. Pilawa, "An analytical method to evaluate flying capacitor multilevel converter and hybrid switched-capacitor converters for large voltage conversion ratios," in *Proc. IEEE Workshop on Control and Modeling for Power Electronics (COMPEL)*, Aug. 2015.

[14] T.A. Meynard, H. Foch, "Multi-level conversion: high voltage choppers and voltage-source inverters," in *Proc. IEEE PESC*, pp.397-403, 1992.

[15] V. Yousefzadeh, E. Alarcon, E., and D. Maksimovic, "Three-level buck converter for envelope tracking in RF power amplifiers," in *Proc. IEEE*

Applied Power Electronics Conference and Exposition (APEC), pp.1588-1594, Mar. 2005.

[16] B. J. Baliga, "Fundamentals of power semiconductor devices", Springer Science, 2008.

[17] N. Vukadinovic, A. Prodic, B.A. Miwa, C.B. Arnold, and M.W. Baker, "Skip-duty control method for minimizing switching stress in low-power multi-level Dc-Dc converters," in *IEEE 16th Workshop on Control and Modeling for Power Electronics (COMPEL)*, pp.1-7, July 2015.

Fractional Pulse Skipping in Digitally Controlled DC-DC Converters for Improved Light-load Efficiency and Power Spectrum

Bipin Chandra Mandi, *Student Member, IEEE*, Santanu Kapat, *Member, IEEE*
and Amit Patra, *Member, IEEE*
Embedded Power Management Laboratory
Department of Electrical Engineering
Indian Institute of Technology Kharagpur, West Bengal, India
Email: bipin0087@gmail.com, santanu.kapat@ieee.org, and amit.patra@ieee.org

Abstract—Pulse skipping modulation (PSM) helps to improve the light-load efficiency in a DC-DC converter with a stable periodic behaviour and predictable ripple parameters. A periodic steady-state operation under PSM consists of a charge pulse followed by a pre-defined count of skipped pulses, where individual pulses are referred to a fixed-frequency clock with a time period T. Thus the overall time period is simply the total count of pulses times the time period T, and the power spectrum is distributed accordingly. Although the light-load efficiency improves by increasing the number of skipped cycles, the count is limited by the specified output voltage ripple. This paper introduces a novel concept of fractional pulse skipping in a digitally controlled PSM DC-DC converter, which improves the power spectrum, without degrading the efficiency and violating the ripple constraint. In the proposed PSM, the number of charge and skipped cycles can be fully customized and the skipped cycle count needs to be periodically varied. The proposed PSM uses an existing digital-pulse-width-modulator (DPWM) architecture and adds extra features with only a little modification. Thus a seamless PSM/PWM transition can be inherently achieved. Stability analysis is carried out using discrete-time models and design guidelines are presented. The proposed scheme is implemented using an FPGA device and tested in a mixed-signal current-mode DPWM buck converter.

I. INTRODUCTION

The use of battery operated portable devices is increasing [1] with the demand of a prolonged battery life. Thus light-load efficiency in a DC-DC converter [2] needs to be improved, as the device remains idle for most of the time. It remains a challenge to develop a suitable control scheme which would also improve the power spectrum. Light-load efficiency under PWM control degrades with the reduction in load current [3], because of domination of the switching and driver losses. Although the variable switching frequency control schemes, such as pulse frequency modulation (PFM) [4]- [5], pulse train (PT) [6] and pulse regulation (PR) [7], improve the efficiency, they also pose difficulties in the design of an input filter, thus resulting in problems due to unwanted conducted EMI. A PSM scheme in [9] can selectively skip pulses, while maintaining the charge balance. The classical PSM [10] and voltage controlled PSM [11] have the limitation of spectral spreading because of difficulty in selecting the skipped cycle count. The bi-frequency PFM [12]- [13] or Dithering Skip Modulation [14] have better spectral properties than the conventional PWM

and PFM. It is therefore necessary to improve the spectral composition in voltage mode or current mode PSM. This paper introduces a fractional pulse skipping concept in a DPWM DC-DC converter, which attempts to adopt a periodic spread spectrum technique, while improving the efficiency under light-load conditions with a predictable periodic behaviour. The proposed scheme has the advantages such as: 1) current mode controlled skipped sequence of operation; 2) fast transient response with load current variation; 3) achieving predictable spectral behaviour without degrading the efficiency and ripple parameters; 4) smooth PWM/PSM transition.

This paper is organized as follows. Section II describes the proposed current-mode PSM in a buck converter and PSM logic generation. Section III presents the modeling and performance analysis of the proposed current-mode PSM scheme using discrete time modeling. The design guidelines and loss formulation for a buck converter with the current-mode PSM control scheme are discussed in Section IV. The hardware implementation and results are presented in Section V. This paper is concluded in Section VI.

II. FRACTIONAL PSM CONTROL SCHEME IN A DC-DC CONVERTER

The proposed current-mode fractional PSM scheme for a DC-DC buck converter is shown Fig. 1. The asynchronous buck converter is considered for Discontinuous Conduction Mode (DCM) operation under the light load current with the input voltage v_{in}, the output voltage v_o and the load resistor R. The design parameters are the inductor L with Equivalent Series Resistor (ESR) for Inductor denoted by r_L, the capacitor C with its ESR denoted by r_C. The semiconductor switching devices are the MOSFET and the diode to avoid the negative current in DCM operation. The gate signal u is to drive the MOSFET. In Fig. 1, v_o is sampled and digitized using an ADC at a sampling frequency $f_{c,s}$ (either the same as the switching frequency F_s or a sub-multiple of) it to produce $v_o[n]$. The error voltage $v_e[n]$ is computed by subtracting the digitized output voltage $v_o[n]$ from the reference voltage v_{ref}. The error voltage $v_e[n]$ is fed to the discrete time compensator $G_c(z)$, whose output $i_{ref}[n]$ acts as the current reference. In the mixed-signal domain, the sensed inductor current i_L and the reference current i_{ref} after DAC are fed to an analog comparator. Then

978-1-4673-9551-9/16 $31.00 © 2016 IEEE

the output from the analog comparator is used to reset an edge triggered D latch circuit to produce the control signal u_C in synchronism with F_s. The PSM control logic u_{PSM} develops one charge cycle followed by an integral number (N_1 or N_2) of skipped cycles operating in synchronism with F_s. The duty cycle u is the gate signal generated from the AND logic with the signal u_C from the current-mode controller and the output of the PSM logic block u_{PSM} as shown in Fig. 2. The circuit will be transitioning from PSM to PWM by enabling $u_T = 1$; otherwise, it will operate according to PSM logic shown in Fig. 1.

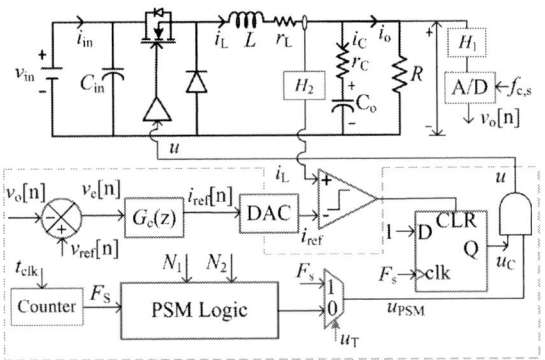

Fig. 1. Buck converter circuit diagram with the proposed fractional PSM; u is the gate signal generated from the AND logic with the signal u_C and the output of the PSM logic block u_{PSM}.

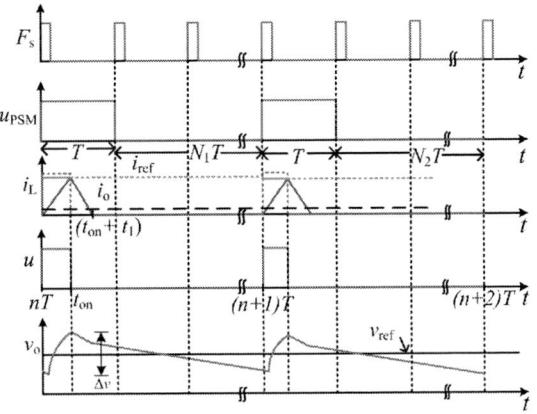

Fig. 2. Current mode fractional PSM with waveform for its operating principle, considering $(1+N_1T)$ and $(1+N_2T)$ to produce fractional skipped cycles N_sT, i_L the inductor current and v_o the output voltage.

A. Fractional PSM Logic Generation

The fractional skipped cycles denoted by N_s is realized from the PSM logic over a time period T. The effective time period is expressed as $T_{eff} = n_1(N_1+1)T + n_2(N_2+1)T = (n_1+n_2)(1+N_s)T$, where $N_s = (n_1N_1 + n_2N_2)/(n_1+n_2)$ for all $n_1, n_2 \in I$ and $N_1 \neq N_2$ with two different skipped cycles N_1, N_2, and two sequences of occurrence (n_1, n_2). For example, to achieve $N_s = 4.5$, we can choose $n_1 = n_2 = 1$, $N_1 = 4$, $N_2 = 5$. Similarly for $N_s = 4.75$ $n_1 = 3$, $n_2 = 1$, $N_1 = 5$, $N_2 = 4$. Some more

examples are given in Table I to generate the fractional skipped cycle N_s with fractional digit $1/(n_1+n_2) = 1/4$. It can also be formulated with different fractional digit $1/(n_1+n_2)$.

TABLE I. FRACTIONAL SKIPPED CYCLES LOGIC GENERATION

n_1	n_2	N_1	N_2	N_s
2	2	1	2	1.5
2	1	2	3	2.25
2	2	3	4	3.5
1	3	5	6	5.75
2	2	7	8	7.5
1	3	9	10	9.25
2	2	12	13	12.5
1	3	15	16	15.75
2	2	24	25	24.5

B. Flexibility of Fractional PSM

The control scheme has the flexibility to realize both current mode and voltage mode with extensions towards (i) PT [6], (ii) Bi-frequency PWM [12]- [13], (iii) PR [7], (iv) burst mode PFM [15] and (v) conventional PSM [9]- [11]. Considering the switching frequency $(1/T)$, the effective switching frequency $f_{eff} = 1/(N_sT)$, the charge and the skipped cycles can be varied according to the load current or the input voltage variations. The proposed strategy to find out the value of 1 charge cycles N_1 and N_2 (number of skipped cycles) is shown in the waveform of Fig. 2. The skipped cycles are calculated for a particular input voltage, output voltage, load current and switching frequency decided for optimum efficiency and the output voltage ripple within the range. For ensuring the fractional PSM, we have the following constraints:

$$T_1 = (N_1+1)T, \ T_2 = (N_2+1)T,$$
$$n_1T_1 + n_2T_2 = N_sT \ ; \ \forall \ (n_1, n_2, N_s);$$
$$n_1/N_s = \gamma \ , \ n_2/N_s = (1-\gamma) \text{ where } 0 \leq \gamma \leq 1.$$
$$\gamma T + (1-\gamma)T = N_sT = T_{eff}.$$

C. PSM/PWM Transition

Seamless transition occurs for load current changes from light load to normal load condition or vice-versa. As the current-mode fractional PSM control scheme has flexible charge and skipped cycles, it can easily transit from PSM control scheme to PWM control scheme. Again as the load current changes from normal load condition to light load condition, N_s varies and it goes to PSM mode.

III. STABILITY ANALYSIS AND MODELING OF CURRENT MODE PSM

A. Discrete-time modelling

In this section, we derive the discrete time model of a buck converter under the proposed current-mode fractional PSM. The time axis in Fig. 2 is generalised at start to be nT. The inductor current i_L increases during the ON time interval $nT < t \leq nT + t_{on}$ (named as Mode 1). Thereafter, i_L falls with a negative slope down to zero during the OFF interval $nT + t_{on} < t \leq nT + t_{on} + t_1$ (Mode 2), and i_L continues to remain at zero till the beginning of next charge cycle during the interval $nT + t_{\text{on}} + t_1 < t \le (n+1)T$ (Mode 3). The general expression of the state equation is written as in (1)

$$\dot{x} = (A_1 x + B_1 v_{in}) u + (A_2 x + B_2 v_{in})(1-u) \quad (1a)$$

$$v_{\text{o}} = C_{\text{o}} x \quad (1b)$$

where $A_1 = A_2 = \begin{bmatrix} -\frac{r_L}{L} - \frac{r_C R}{L(R+r_C)} & -\frac{R}{L(R+r_C)} \\ \frac{R}{C(R+r_C)} & -\frac{1}{C(R+r_C)} \end{bmatrix}$;

$B_1 = \begin{bmatrix} \frac{1}{L} \\ 0 \end{bmatrix}$, $B_2 = \begin{bmatrix} 0 \\ 0 \end{bmatrix}$, $C_{\text{o}} = \begin{bmatrix} \frac{R r_C}{(R+r_C)} & \frac{R}{(R+r_C)} \end{bmatrix}$.

where $u = 1$ for Mode 1 and $u = 0$ for Modes 2 and 3. The discrete time map can be derived using the solutions of state space representation.

B. Stability Analysis

Let the inductor current and the output voltage at the start of the nth clock cycle be i_{n} and v_{n}, at the $(n+1)^{\text{th}}$ clock cycle, $i_{\text{n+1}}$ and $v_{\text{n+1}}$, for $(1+N_1)T$ time period and at the $(n+2)^{\text{th}}$ clock cycle, $i_{\text{n+2}}$ and $v_{\text{n+2}}$ respectively for $(1+N_2)T$ time period. The discrete-time model of a buck under discontinuous conduction mode will be of first order consisting of the output voltage equation only [16], which can be derived using Taylor series approximation and considering terms up to the 2nd order as given below:

$$v_{\text{n+1}} = (1-\beta_1 k_1)v_{\text{n}} - (\beta_1 v_{in} t_{\text{on1}}^2/T_1{}^2)(1-v_{in}/v_{\text{n}}) \quad (2)$$

where, $T_1 = (N_1+1)T$, $\beta_1 = \left\{\rho(N_1+1)^2 T^2\right\}/2LC$,

$$\rho = R/(R+r_C), \; k_1 = 2L/R(N_1+1)T.$$

The value of on-time t_{on1} is generated using the closed-loop control at the nth clock cycle instant, which can be derived considering a discrete-time PI voltage controller as

$$t_{\text{on1}} = i_{\text{ref1}}/m_1 = [k_p(v_{\text{ref}} - v_{\text{n}}) + u_I[n]]/m_1 \quad (3)$$

where $m_1 = (v_{in} - v_{\text{n}})/L$: is the slope of the inductor current; $u_I[n] = u_I[n-1] + k_i(v_{\text{ref}} - v_{\text{n}})$, k_p and k_{int} are the proportional and integral gains. It is difficult to analytically carry out a stability study using an integral action because of the memory element; thus a proportional controller is considered and the closed-loop model becomes $v_{\text{n+1}} = (1-\beta_1 k_1)v_{\text{n}} + \alpha_1$ where,

$$\alpha_1 = \{\beta_1 k_p^2 L^2 (v_{in} - v_{\text{ref}})^2 v_{in}\}/\{v_{\text{n}} T_1{}^2 (v_{in} - v_{\text{n}})\}.$$

The perturbed linearized model can be obtained around the reference voltage $v_{\text{n}} = v_{\text{n+1}} = v_{\text{ref}}$, which is derived as

$$\frac{\partial v_{\text{n+1}}}{\partial v_{\text{n}}} = (1-\beta_1 k_1) + \left.\frac{\partial f(v_{\text{n+1}})}{\partial v_{\text{n}}}\right|_{v_{\text{n}}=v_{\text{ref}}} = (1-\beta_1 k_1)$$

Then the stability condition becomes $|(1-\beta_1 k_1)| < 1$; thus the closed-loop is stable, irrespective of the controller gain. Similarly, the discrete time map of $v_{\text{n+2}}$ can be derived with the initial point $v_{\text{n+1}}$ for time interval $(n+1)$th to $(n+2)$th clock cycle.

$$v_{\text{n+2}} = (1-\beta_2 k_2)v_{\text{n}} - (\beta t_{\text{on2}}^2 v_{in}/T_2{}^2)(1-v_{in}/v_{\text{n}}) \quad (4)$$

where, $T_2 = (N_2+1)T$, $\beta_2 = \left\{\rho(N_2+1)^2 T^2\right\}/2LC$,

$$\rho = R/(R+r_C), \; k_2 = 2L/R(N_2+1)T.$$

The value of on-time t_{on2} is generated using the closed-loop control at the $(n+1)^{\text{th}}$ clock cycle instant, which can be derived considering a discrete-time PI voltage controller as

$$t_{\text{on2}} = i_{\text{ref2}}/m_1 = [k_p(v_{\text{ref}} - v_{\text{n+1}}) + u_I[n+1]]/m_1 \quad (5)$$

where $m_1 = (v_{in} - v_{\text{n+1}})/L$: is the slope of the inductor current; $u_I[n+1] = u_I[n] + k_i(v_{\text{ref}} - v_{\text{n+1}})$. The closed-loop model at the $(n+1)^{\text{th}}$ clock cycle instant becomes $v_{\text{n+2}} = (1-\beta_2 k_2)v_{\text{n}} + \alpha_2$, where,

$$\alpha_2 = \{\beta_2 k_p^2 L^2 (v_{in} - v_{\text{ref}})^2 v_{in}\}/\{v_{\text{n+1}} T_2{}^2 (v_{in} - v_{\text{n}})\}.$$

The perturbed linearized model can be obtained around the reference voltage $v_{\text{n+1}} = v_{\text{n+2}} = v_{\text{ref}}$, which is derived as

$$\frac{\partial v_{\text{n+2}}}{\partial v_{\text{n+1}}} = (1-\beta_2 k_2) + \left.\frac{\partial f(v_{\text{n+2}})}{\partial v_{\text{n+1}}}\right|_{v_{\text{n+1}}=v_{\text{ref}}} = (1-\beta_2 k_2).$$

Then the stability condition becomes $|(1-\beta_2 k_2)| < 1$; thus the closed-loop is stable, irrespective of the controller gain. The values of N_1 and N_2 can be customized for improving the power spectrum and the efficiency.

IV. DESIGN GUIDELINES FOR THE PROPOSED CONTROLLER

A. Power loss formulation subject to skipped cycles

As the values of N_1, N_2 - are fully controllable, the losses can be minimized by changing the switching time periods. The total power loss consists of the conduction loss P_{Cond}, switching loss P_{Sw} and driver and static loss $P_{\text{Dr-st}}$ as shown in the following equations equ. (6), (7) and (8). Under the conduction losses we have P_{r_L}, the conduction loss due to the series DC resistance of the inductor during MOSFET ON and OFF times; $P_{\text{rds-on}}$, the conduction loss due to the ON resistance of the MOSFET during MOSFET ON time; P_{esr}, the conduction loss due to ESR (r_C) of the output capacitor during overall time interval; the diode conduction loss due to voltage drop of the diode (V_{drop}) during MOSFET OFF time.

$$P_{\text{Cond}} = \underbrace{r_L I_{\text{pk}}^2 \left(\frac{t_{\text{on1}} + t_1}{3(N_1+1)T} + \frac{t_{\text{on2}} + t_2}{3(N_2+1)T}\right)}_{P_{r_L}} +$$

$$\underbrace{r_{\text{ds-p}} I_{\text{pk}}^2 \left(\frac{t_{\text{on1}}}{3(N_1+1)T} + \frac{t_{\text{on2}}}{3(N_2+1)T}\right)}_{P_{rds\_on}} + \quad (6)$$

$$\underbrace{r_C \left[I_{\text{pk}}^2 \left(\frac{t_{\text{on1}} + t_1}{3(N_1+1)T} + \frac{t_{\text{on2}} + t_2}{3(N_2+1)T}\right) - i_{\text{o}}^2\right]}_{P_{\text{esr}}}$$

$$P_{\text{Sw}} = \underbrace{0.5 I_{\text{pk}} v_{in}\left[\frac{t_{\text{on1}}}{(N_1+1)T} + \frac{t_{\text{on2}}}{(N_2+1)T}\right]}_{P_{\text{tran}}} +$$

$$\underbrace{I_{\text{pk}} v_{in} \frac{t_{\text{d}}}{(N_1+N_2+2)T}}_{P_{\text{dt}}} + \underbrace{v_{in} \frac{Q_{\text{rr}}}{(N_1+N_2+2)T}}_{P_{\text{rr}}} \quad (7)$$

978-1-4673-9551-9/16 $31.00 © 2016 IEEE

$$P_{\text{Dr}-\text{st}} = \underbrace{C_{\text{gs}-\text{P}} v_{\text{in}}^2 \left[\frac{t_{\text{on}1}}{(N_1+1)T} + \frac{t_{\text{on}2}}{(N_2+1)T} \right]}_{P_{\text{driv}}} + \underbrace{I_q v_{\text{in}}}_{P_q}$$

(8)

The switching loss P_{Sw}, comprises P_{tran}: the MOSFET transition loss occurs during PMOS transition (turn ON and OFF) P_{dt}: dead time loss during the transition occurs between MOSFET and diode, and P_{rr}: diode reverse recovery loss during MOSFET OFF time interval. Under the driver losses, P_{driv} occurs due to the driver of the MOSFET switch which depends on the Gate to Source capacitance $C_{\text{gs}-\text{P}}$ of the MOSFET and the supply voltage to the gate driver V_{dd}; the static loss P_{q} depends on the quotient current I_q and the supply voltage to the gate driver V_{dd}.

B. Power loss optimization

The fractional N_{s} is calculated from the specified output voltage ripple band as it is given in equ. (9), then N_1 and N_2 can be obtained from the N_{s}. The values of $t_{\text{on}1}$ and $t_{\text{on}2}$ are given by the equ. (10). The number of skipped cycles N_1 and N_2 can also be represented in terms of on time $t_{\text{on}1}$ and $t_{\text{on}2}$ in equ. (11), where the values of $t_{\text{on}1}$ and $t_{\text{on}2}$ can limit the values of skipped cycles depending on the output voltage ripple [8] given in equ. (12).

$$N_{\text{s}} = \frac{RC\Delta v}{Tv_{\text{ref}}} + \frac{kL}{2RT} - \sqrt{\frac{2k\Delta vLC}{T^2 v_{\text{ref}}}} - 1$$

(9)

where, $k = v_{\text{in}}/(v_{\text{in}} - v_{\text{ref}})$.

$$t_{\text{on}1} = \sqrt{\left\{ 2v_{\text{ref}}^2 LT_1 \right\} / \left\{ (v_{\text{in}} - v_{\text{ref}}) v_{\text{in}} R \right\}}$$

(10a)

$$t_{\text{on}2} = \sqrt{\left\{ 2v_{\text{ref}}^2 LT_2 \right\} / \left\{ (v_{\text{in}} - v_{\text{ref}}) v_{\text{in}} R \right\}}$$

(10b)

$$N_1 = \left\lfloor \left[\left\{ (v_{\text{in}} - v_{\text{ref}}) R t_{\text{on}1}^2 v_{\text{in}} \right\} / 2v_{\text{ref}}^2 LT \right] - 1 \right\rfloor$$

(11a)

$$N_2 = \left\lfloor \left[\left\{ (v_{\text{in}} - v_{\text{ref}}) R t_{\text{on}2}^2 v_{\text{in}} \right\} / 2v_{\text{ref}}^2 LT \right] - 1 \right\rfloor$$

(11b)

$$\Delta v = \frac{(v_{\text{in}} - v_{\text{ref}})}{L} \left\{ t_{\text{k}} - \frac{Lv_{\text{ref}}}{(v_{\text{in}} - v_{\text{ref}})R} \right\} \times$$
$$\left[\frac{v_{\text{in}}}{2v_{\text{ref}}C} \left\{ t_{\text{k}} - \frac{Lv_{\text{ref}}}{(v_{\text{in}} - v_{\text{ref}})R} \right\} + r_{\text{C}} \right]$$

(12)

where, $t_{\text{k}} \in [t_{\text{on}1}, t_{\text{on}2}]$, $T_1 = (N_1+1)T$, $T_2 = (N_2+1)T$. The design parameters are given in Table II for a buck

TABLE III.   FIXED SKIPPED CYCLES, FRACTIONAL SKIPPED CYCLES AND LOAD CURRENT

(i_{o})(mA)	N_s ($\Delta v = 40$mV)
200	1.5
175	2.25
150	3.5
125	5.75
100	7.5
75	9.25
50	12.5
25	15.75
10	24.5

converter with system specifications to find the skipped cycles and the output voltage ripple trade-off. According to the input

voltage v_{in}, output voltage v_{o}, load current i_{o}, and the switching frequency F_{s}, and N_1 and N_2 can be decided for optimum efficiency satisfying the output voltage ripple specification. From this load current information the corresponding values of N_{s} will be selected. Table III shows the values of fractional skipped cycles for corresponding load current keeping the output voltage ripple within the specific band ($\Delta v = 40$mV).

V. HARDWARE IMPLEMENTATION

A buck converter prototype in a Power Circuit Board (PCB) is tested considering an FPGA device used as a digital platform. A buck converter prototype is fabricated with the specifications in Table II and the proposed control with 4 skipped cycles is implemented using an FPGA device. The converter has also the following passive parameters specifications is shown in Table II. The interfacing device is a Signal Conditioning Board (SCB) with 10 bit Analog-to-Digital Conversion (ADC), 12 bit Digital-to-Analog Conversion (DAC) and buffer ICs to interface with the sensed output voltage and inductor current of buck converter. A test condition has been taken for load current 36 mA (R = 50 ohm), the proportional gain k_{p} = 12, the integral gain k_{int} = 0.1 for the integral skipped cycles of 4 and 5 and fractional N_{s} = 4.25, 4.5 and 4.75.

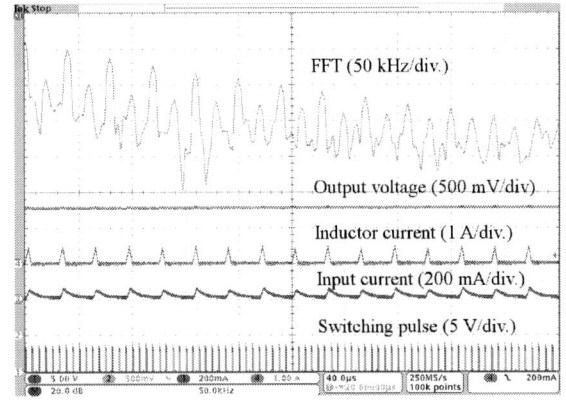

Fig. 3.   The FFT of the input current shows the spectral spreading for 4T, Ch1: Switching Pulse (5 V/div.), Ch2: the output voltage (500 mV /div.), Ch3: the input current (200 mA/div.) and Ch4: the inductor current (1 A /div.) respectively.

A. Spectral Shaping

The FFT of the input current is shown in Fig. 5. The input current spikes occur when the switching transition occurs. The hardware results depict for the proposed PSM with 1 charge cycle and 4 skipped cycles. It has periodic behavior of inductor current and output voltage. Any input voltage variation (2.7 – 4.2)V, does not give rise to non-periodic behavior. Figures 3 - 4 show the sharp FFT peaks of the input current with 5 and 4 skipped cycle counts. However, the FFT of the fractional PSM in Figures 5 - 7 with N_{s} = 4.25, 4.5 and 4.75 skipped cycles (with the effective time periods as T_{eff} = 4(1+4.25)T = 21 μs, T_{eff} = 4(1+4.5)T = 22 μs and 4(1+4.75)T = 23 μs show that the spectral peaks are considerably attenuated compared to the PSM with an integer skipped cycle count with an insignificant ripple impact.

TABLE II.   System specifications for a buck converter

$v_{in}(V)$	$v_o(V)$	$L(\mu H)$	$C(\mu F)$	$f_s(kHz)$	$i_o(A)$	$V_{dd}(V)$
[2.7-4.2]	1.8	10	470	200	0.01-1.5	5

$r_L(m\Omega)$	$r_C(m\Omega)$	$\Delta v(mV)$	$Q_{rr}(nQ)$	N	$V_d(V)$	$C_{gs}(nF)$
30	50	40	15	0-24	0.65	1

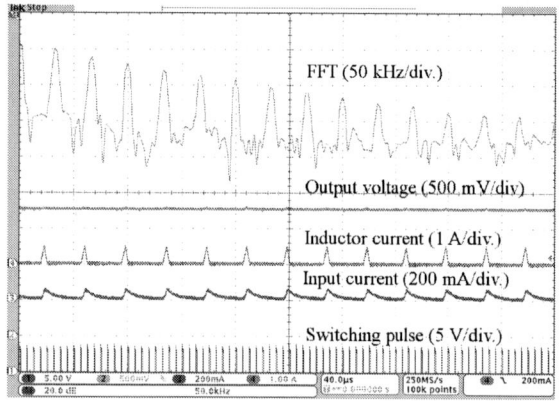

Fig. 4.   The FFT of the input current shows the spectral spreading for 5T; Ch1: Switching Pulse (5 V/div.), Ch2: the Output voltage (500 mV /div.), Ch3: the input current (200 mA/div.) and Ch4: the inductor current (1 A /div.).

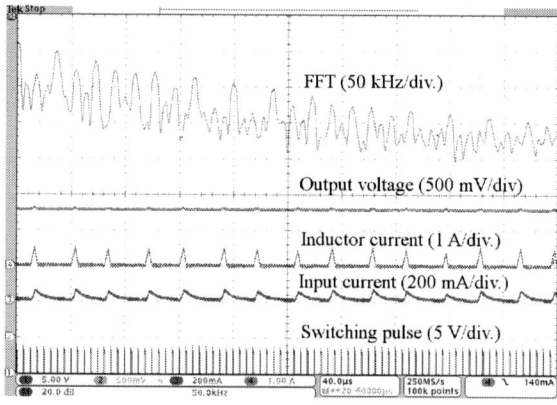

Fig. 6.   The FFT of the input current show the spectral spreading for 4.5T; Ch1: Switching Pulse (5 V/div.), Ch2: the Output voltage (500 mV /div.), Ch3: the input current (200 mA/div.) and Ch4: the inductor current (1 A /div.).

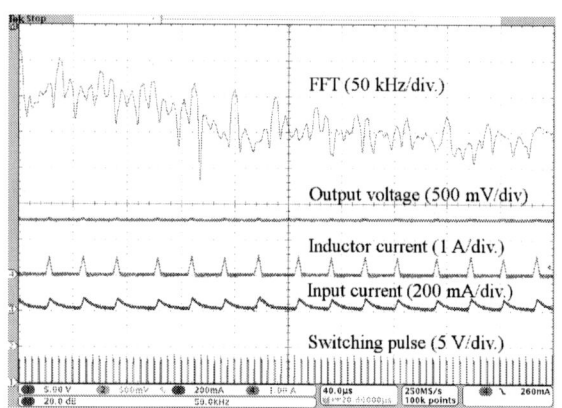

Fig. 5.   The FFT of the input current shows the spectral spreading for 4.25T; Ch1: Switching Pulse (5 V/div.), Ch2: the output voltage (500 mV /div.), Ch3: the input current (200 mA/div.) and Ch4: the inductor current (1 A /div.).

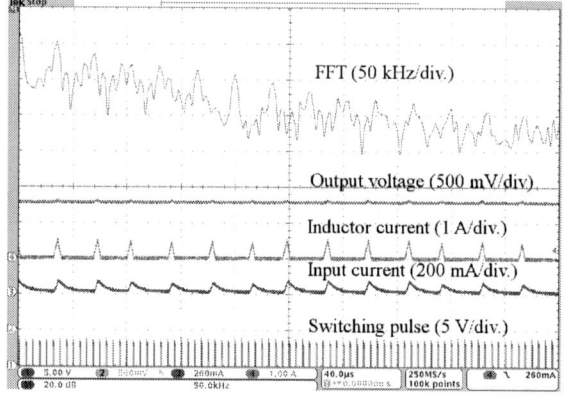

Fig. 7.   The FFT of the input current shows the spectral spreading for 4.75T; Ch1: Switching Pulse (5 V/div.), Ch2: the output voltage (500 mV /div.), Ch3: the input current (200 mA/div.) and Ch4: the inductor current (1 A /div.).

B. Smooth PWM/PSM Transition

The transient response of the system with load current variation from 1.5 A - 18 mA and 18mA - 1.5 A are shown in Fig. 8 and Fig. 9 respectively. The proposed control scheme has the PWM current control loop. So, when the power circuit operates in the normal mode of operation we can set the value of $u_T = 0$ for PWM control scheme i.e, $N_s = 0$. By setting $u_T = 1$ for fractional PSM control scheme we achieve seamless transition between the proposed PSM and PWM control schemes. The PWM control scheme has a fixed switching frequency operation with same proportional and integral controller.

Using the load current information of current-mode PSM scheme or load automation tuning [17], the number of skipped cycles can be easily changed for load transition from PSM to PWM scheme and vice versa as shown in Fig. 8 and Fig.9. The transition from PWM to PSM for 1.5A to 18 mA is considered with time period T= 5 μs and Teff = 23 μs respectively.

C. Efficiency Improvement and Stable behaviour

The proposed scheme with one charge followed by four as well as five skipped cycles is implemented, which appear alternatively to achieve fractional current-mode PSM. The fractional PSM with 4.25T, 4.5T and 4.75T at light load

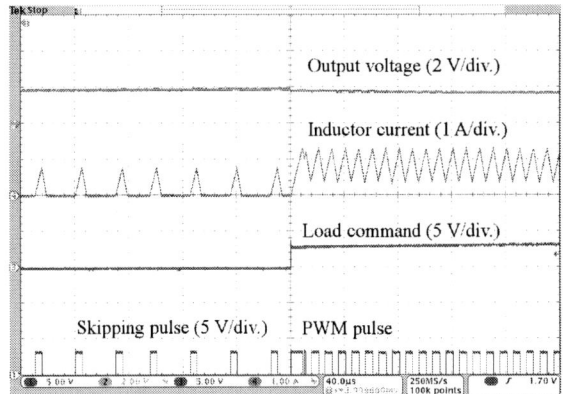

Fig. 8. Load transient Performance for load current [1.5 A - 18 mA] with PSM to PWM for 4.75 T .

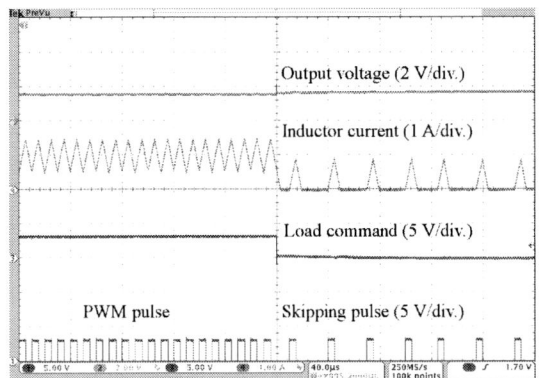

Fig. 9. Load transient Performance for load current [1.5 A - 18 mA] with PWM to PSM for 4.75 T .

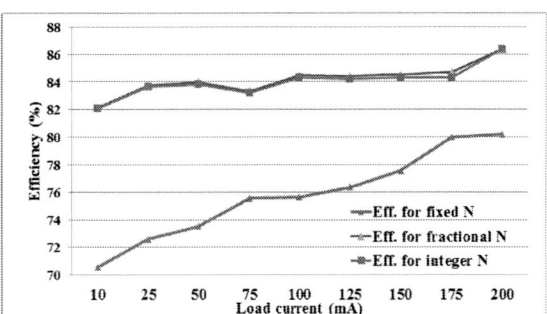

Fig. 10. Efficiency with the fixed skipped cycles, the integer skipped cycles and the fractional skipped cycles.

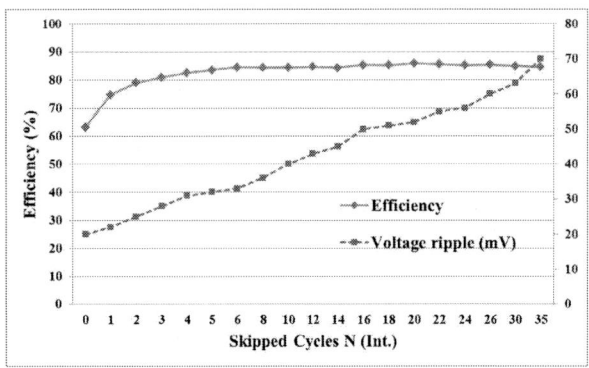

Fig. 11. Efficiency and output voltage ripple with the skipped cycles variation for a particular load current 33 mA.

exhibits stable behaviour (shown in Fig. 5 - Fig. 7). The resolution of an integer PSM scheme is limited by the time period T. Thus the effective time period may considerably differ from an optimized PFM scheme which can be better approximated using a fractional PSM scheme in the average sense. Thus the efficiency can be improved a little bit [shown in Fig. 10], while maintaining the specified voltage ripple limit [shown in Fig. 11]. The proposed control scheme can be extended to the bi-frequency PT and multi-mode DPWM control scheme [18].

VI. CONCLUSION

The proposed current-mode fractional PSM scheme uses an existing DPWM architecture with only a little modification and achieves improved spread power spectrum, and smooth PSM/PWM transition, without degrading the efficiency and the ripple parameters. In this paper, fractional PSM logic, stability analysis, compensator design and loss formulation have been discussed in detail. The proposed scheme is simple to implement and can be easily extended to single-inductor-multiple-output as well as multi-phase DC-DC converters.

REFERENCES

[1] O. Trescases and Y. Wen "A Survey of Light Load efficiency Improvement Techniques Low-Power DC-DC Converters," *in proc. IEEE ICPE-ECCE*, pp. 326–333, May/Jun. 2011.

[2] R. W. Erickson and D. Maksimovic, "Fundamentals of Power Electronics," 2nd ed., Dordrecht, Kluwer, Netherlands, 2001.

[3] K. Sheng, Y.C. Chen and C. N. Chen, "PWM control scheme under light load," US Patent 7456624 B2, Nov. 25, 2008.

[4] J. Nguyen, T. Chiang, and C Falvey, "Constant peak current minimum OFF time PFM for Switching Regulators," US Patent 7439720 B2, 2008.

[5] Y. L. Huang, K.K. Chang and M.C. Li, "Pulse Supply with PFM circuit calculating Logic State changing," US Patent 7638989 B2, Dec. 29, 2009.

[6] S. Kapat, "Configurable Multi-mode Digital Control for Light Load DC-DC Converters with Improved Spectrum and Smooth Transition," *IEEE Trans. Power Electron.*, DOI: 10.1109/TPEL.2015.2451084, 2015.

[7] M. Qin and J. Xu, "Improved pulse regulation control technique for switching DC-DC converter operation in DCM," *IEEE Trans. Ind. Electron.*, vol. 60, no. 5, May 2013.

[8] S. Kapat, B. C. Mandi, and A. Patra, "Voltage-mode Digital Pulse Skipping Control of a DC-DC Converter with Stable Periodic Behavior and Improved Light-load Efficiency," *IEEE Trans. Power Electron.*, DOI: 10.1109/TPEL.2015.2455553, 2015.

[9] S. Kapat, S. Banerjee and A. Patra, "Modeling and Analysis of DC-DC Converters Under Pulse Skipping Modulation," *IEEE Conf. TENCON*, pp. 1–6, Nov. 2008.

[10] P. Luo, B. Zhang, S. P. Wan, and Y. Feng, "Modeling and Analysis of

Pulse Skipping Modulation," *J. Electronic Science and Technology of China (JEST)*, vol. 4 no. 1. pp. 1–7, Mar. 2006.

[11] S. Kapat, S. Banerjee and A. Patra, "Discontinuous Map Analysis of a DC-DC Converter Governed by Pulse Skipping Modulation," *IEEE Trans. Circuits and Systems - I*, Jul. 2009.

[12] J. Xu and J. Wang "Bi-frequency Pulse Train Control Technique for Switching DC- DC Converters operating in DCM," *IEEE Trans. Indus. Elect.*, vol. 58, no. 8, Aug. 2011.

[13] J. Wang and J. Xu "Current Mode Bi-frequency Control Technique for Switching DC-DC Converters," *IEEE Trans. Indus. Elect.*, vol. 27, no. 4, Apr. 2012.

[14] H.W. Huang, K.H. Chen, and S. Y. Kuo,"Dithering Skip Modulation, Width and Dead Time controllers in Highly Efficient DC-DC Converters for System-On-Chip Applications," *IEEE J. Solid-State Circuits*, vol. 42, no. 11, pp. 2451-2465, Nov. 2007.

[15] J. Chen, " Buck Converter Efficiency in PFM Mode," *Power Electronics Technology*, Sep., 2007.

[16] C. K. Tse, "Complex Behavior of Switching Power Converters," New York: CRC, 2003.

[17] X. Zhang and D. Maksimovic, "Multimode Digital Controller for Synchronous Buck Converters Operating Over Wide Ranges of Input Voltages and Load Currents," *IEEE Trans. Power Electron.*, vol. 25, no. 8, Aug. 2010.

[18] S. Kapat, "Reconfigurable Periodic Bi-frequency DPWM with Custom Harmonic Reduction in DC-DC Converters," *IEEE Trans. Power Electron.*, Early access, 2015.

A New Compact and High Efficiency Resonant Converter

Sheng-Yang Yu
Power Design Services
Texas Instruments
Dallas, Texas
seanyu@ti.com

Abstract— This paper explores a new resonant converter which features compact size, high efficiency, and compatibility with self-driven synchronous rectifiers. The proposed resonant converter and the LLC series resonant converter (LLC-SRC) are in the same three-element resonant converter category and these two converters both consist of 2 inductors and one capacitor in their resonant tanks. While the LLC-SRC converter has all of its resonant elements on the input side, the proposed resonant converter has one of its resonant inductors on the output side. Utilizing the output resonant inductor allows the proposed resonant converter to implement self-driven synchronous rectifiers. When comparing the series resonant inductor in the LLC-SRC, the output inductor in the proposed resonant converter generally has lower cost and smaller size in voltage-step down applications. Operational analysis of the proposed resonant converter is made through sinusoidal approximation. A 250W prototype of the proposed resonant converter is built to verify the analysis made in this paper and evaluate its performance. With the same soft-switching characteristics as the LLC-SRC has, the proposed resonant converter can achieve 95.7% efficiency in a 430V to 27.5V/9A conversion even with diode rectifiers.

Keywords—CLL resonant converter; LLC series resonant converter; Zero voltage switching (ZVS); Zero current switching (ZCS).

I. INTRODUCTION

Among all the three-element resonant converters [1], the LLC series resonant converter (LLC-SRC) [2] (shown in Fig. 1) is the most popular topology and widely used in offline power supplies. This is because the LLC-SRC has attractive features of wide soft-switching range, high converter efficiency and easy magnetic integration.

To further improve the converter efficiency of the LLC-SRC, it is generally necessary to replace the output diodes in Fig. 1 with synchronous rectification (SR). However, SR control for the LLC-SRC is never a simple task and generally requires costly circuitry or high accuracy sensing ICs. In [3], a current transformer along with a current compensation inductor is applied to achieve primary current sensing SR control. This SR control method can provide roughly the same efficiency improvement as the commercially used MOSFET V_{DS} sensing SR technique in [4]. A phase compensation network is proposed in [5] with MOSFET V_{DS} sensing to optimize the on-time of the SR MOSFET, which gives a very good

performance in high frequency applications. When compared with the simple self-driven SR control in [6], the aforementioned SR control methods all require high circuit cost.

The reason that the LLC-SRC can't use transformer winding voltages as SR MOSFETs' on/off triggers in the self-driven SR control is because that the LLC-SRC has a current-fed, capacitor-loaded structure [5]. The voltage polarity of an output winding in a current-fed, capacitor-loaded structure is only changed when the SR is turned off, hence, SR for a current-fed, capacitor-loaded structure can't be driven by the winding voltage itself.

The CLL multi-resonant converter (CLL-MRC) in [7] (shown in Fig. 2) is another three-element resonant converter which can adopt simpler SR control than the control methods mentioned in [3-5]. The CLL-MRC in [7] consists of two inductors and one capacitor in the resonant tank and can also provide same attractive features that the LLC-SRC has— zero-

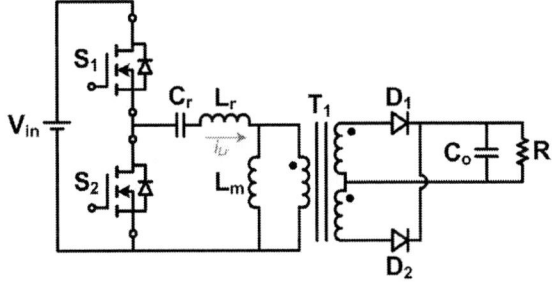

Fig. 1. LLC series resonant converter.

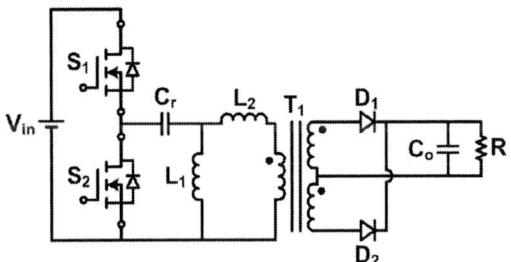

Fig. 2. CLL multi-resonant converter presented in [7].

voltage switching (ZVS) on the input switches and zero-current switching (ZCS) on the output rectifier over entire load range. It is notable that the magnetize inductance of the transformer used in the CLL-MRC in [7] is not part of the resonant element. Therefore, we are allowed to use a transformer with high magnetize inductance and low magnetizing current. Hence, when primary current sensing SR control is adopted in the CLL-MRC, there is no need to have additional circuitry to compensate transformer magnetizing current like the primary current sensing SR control scheme in [3] for the LLC-SRC.

However, the resonant inductor, L_l, in Fig. 2 of the CLL-MRC in [7] can't be integrated within a transformer. Therefore, an inductor is always needed to be placed in parallel with the transformer. In addition, a discrete inductor might be needed to connect in series with the input winding of the transformer to achieve better line regulation. Therefore, the CLL-MRC in [7] might require larger space than the LLC-SRC. Moreover, because the CLL-MRC in [7] still has a current-fed, capacitor-loaded structure, its output winding voltages still can't be utilized for SR control.

A new CLL-MRC is proposed in this paper, shown in Fig. 3. While the LLC-SRC and the CLL-MRC proposed in [7] have all of their resonant elements on the input side, the proposed CLL-MRC has one of its resonant inductors on the output side. With the inherent output inductor, the proposed CLL-MRC has a voltage-fed, inductor-loaded output structure. Hence, the output winding voltages in the proposed CLL-MRC can be used for SR control. In addition, the output inductor in the proposed CLL-MRC is generally smaller and has lower cost than the series inductor in the LLC-SRC in voltage-step down applications.

Although the two resonant inductors (L_r and L_m in Fig. 1) of the LLC-SRC can be integrated with a single transformer to further reduce component count and circuit cost, the converter efficiency of the LLC-SRC with an integrated transformer generally won't be optimized. This is because the range of L_r to L_m ratio is limited by transformer structure and the larger transformer air gap induces more wire losses. In addition, the resonant inductor current, i_{Lr}, in the LLC-SRC resonates from positive to negative; in other words, the B-H hysteresis loop of L_r covers all four quadrants. While the output resonant inductor current, i_{Lo}, in the proposed CLL-MRC is always positive, its B-H hysteresis loop remains in the first quadrant. Therefore, less AC losses in L_o of the proposed CLL-MRC than L_r of the LLC-SRC can be expected in designs with the same input/output specification. In summary, the proposed CLL-MRC is able to provide a highly efficient voltage conversion with lower cost and has smaller size than the LLC-SRC and the CLL-MRC proposed in [7].

In this paper, the voltage gain and soft-switching conditions of the proposed CLL-MRC are investigated through sinusoidal approximation [8], i.e. fundamental harmonic approximation (FHA), on the linearized CLL-MRC. Resonant inductor currents and resonant capacitor voltage equations and component stresses are explored through time domain analysis. A simple SR control method for the proposed CLL-MRC is also suggested.

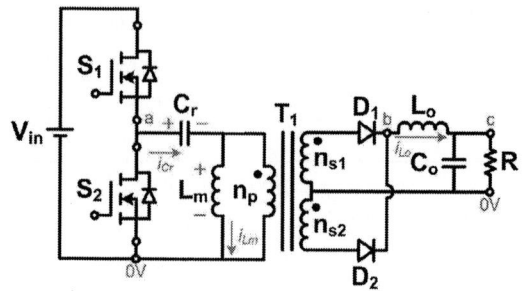

Fig. 3.   Proposed CLL multi-resonant converter.

A 250W prototype of the proposed CLL-MRC is built to verify the analysis made in this paper and evaluate its performance. ZVS on input switches and ZCS on output rectifiers of the proposed CLL-MRC are observed when the switching frequency is in between the two resonant frequencies. Efficiency of the CLL-MRC prototype is measured and compared with the efficiency of the LLC-SRC on the same prototype. When measuring the LLC-SRC efficiency, the output inductor of the CLL-MRC is shorted and a resonant inductor is inserted in series with the transformer input winding. With the same input and output conditions, the efficiency on the CLL-MRC prototype is similar to the LLC-SRC prototype. At 430V input and 27.5V/9A output, the CLL-MRC prototype can achieve 95.7% efficiency.

II.   OPERATIONAL ANALYSIS

A.   CLL Resonant Converter Linearization

To simplify the analysis, sinusoidal approximation is applied here to understand the operation of the proposed CLL-MRC. By taking the fundamental component of the square wave at node "a" in Fig. 3, the proposed CLL-MRC can be simplified to the linear circuit in Fig. 4. The AC voltage, V_s, has the following relationship with input voltage V_{in}:

$$V_s(t) = \frac{4V_{in}}{\pi} \sin(\omega_s t), \tag{1}$$

where $\omega_s = 2\pi f_s$ and f_s is the converter switching frequency. It is noticeable that the output stage is equivalent to the one shown in Fig. 5. By defining the equivalent load resistor, R_e to be

$$R_e = \frac{V_{R'}(t)}{i_R(t)} = \frac{8n^2}{\pi^2} R, \tag{2}$$

then the proposed CLL-MRC can be converted to a linear circuit shown in Fig. 6. The input to output voltage gain of the linear circuit in Fig. 6 can be derived as

$$\left| \frac{V_{R'}}{V_s} \right| = \frac{nV_o}{V_{in}/2}$$

$$= \frac{1}{\sqrt{\left(1 + \frac{\omega_2^2 - \omega_r^2}{\omega_s^2}\right)^2 + \frac{L_e}{C_r R_e^2}\left(\frac{\omega_s}{\omega_2}\sqrt{\frac{L_2}{L_e}} - \frac{\omega_r}{\omega_s}\frac{L_2}{L_e}\right)^2}}, \tag{3}$$

where $L_2 = n^2 L_o$, $L_e = L_m // L_2$, $\omega_s = 2\pi f_s$, $\omega_r = 2\pi f_r = (L_e C_r)^{-0.5}$, $\omega_2 = 2\pi f_2 = (L_2 C_r)^{-0.5}$, and V_o is the output voltage. The gain curve of the proposed CLL-MRC is shown in Fig. 7 with $V_{in} = 380$V, $n=8$, $L_m = 350\mu H$, $L_o = 1\mu H$, $C_r = 0.033~\mu F$. It can be observed that the voltage gain of the proposed CLL-MRC has similar characteristics to the voltage gain of the LLC-SRC. The peak of a gain curve decreases when the load is heavier. One difference between the gain curve of the proposed CLL-MRC and the LLC-SRC is that when f_s is equal to f_r, the voltage gain is no longer $V_o = V_{in}/2n$. Using (3) with the setting of $\omega_r = \omega_s$, the voltage gain of the CLL-MRC can be found as

$$V_o = \frac{V_{in}}{2n}\left(1 + \frac{L_2}{L_m}\right)\Big|_{f_s = f_r}. \tag{4}$$

That is, the proposed CLL-MRC can provide higher output voltage level than the LLC-SRC with the same turns ratio. In other words, the proposed CLL-MRC can be more beneficial in applications with high output voltages.

The input impedance of the linearized CLL-MRC can be expressed as

$$\begin{aligned}
Z_{in} &= \frac{1}{sC_r} + sL_m // (sL_2 + R_e) \\
&= \frac{\left(R_e - \omega_s^2 L_m C_r R_e\right) + j\left[\omega\left(L_m + L_2\right) - \omega_s^3 L_m L_2 C_r\right]}{-\omega_s^2 C_r\left(L_m + L_2\right) + j\omega_s C_r R_e}.
\end{aligned} \tag{5}$$

$\|Z_{in}\|$ can then be plotted in Fig. 8. It can be noticed that $\|Z_{in}\|$ is inductive at frequencies above $f_{\|Zin,min\|}$ and is capacitive at frequencies below $f_{\|Zin,min\|}$, where $f_{\|Zin,min\|}$ is the frequency where $\|Z_{in}\|$ is at its minimum value. This implies that zero voltage switching (ZVS) on the input switches can be achieved when the switching frequency is higher than the minimum impedance frequency.

Also, when the load is open, there are only two resonant elements, C_r and L_m, are effective in the linearized circuit. In this situation, Z_{in} can be expressed in (6):

$$Z_{in}\Big|_{R_e = \infty} = \frac{1}{sC_r} + sL_1 = \frac{1 - \dfrac{f_s^2}{f_p^2}}{j2\pi f_s C_r}, \tag{6}$$

where $f_p = 1/[2\pi~(L_m C_r)^2]$ is the parallel resonant frequency, which is the frequency that $\|Z_{in,min}\| = 0$ when the load is open.

B. Time Domain Analysis and Component Stresses

In order to have ZVS on input switches to achieve high converter efficiency, we will need to operate the proposed CLL-MRC in a frequency higher than the parallel resonant frequency. Key waveforms of the proposed CLL-MRC when $f_p < f_s < f_r$ and $f_s > f_r$ are shown in Fig. 9. With close to 50% duty cycle on the input switches, S_1 and S_2, the resonant capacitor voltage and magnetizing inductor current of the proposed CLL-MRC are similar to the LLC-SRC. The proposed CLL-MRC in

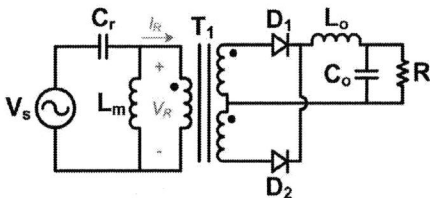

Fig. 4. CLL multi-resonant converter with sinusoidal approximation on input switches.

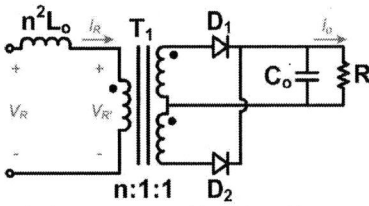

Fig. 5. Equivalent output stage of the CLL multi-resonant converter.

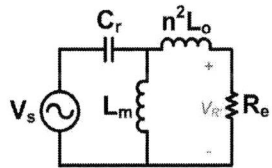

Fig. 6. Equivalent linear circuit of the CLL multi-resonant converter.

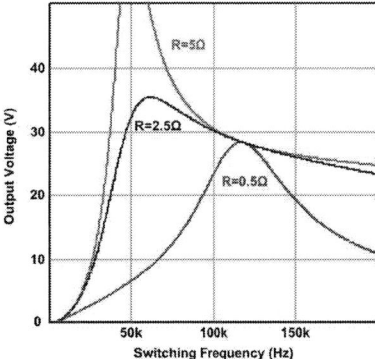

Fig. 7. Voltage gain curves of the linearized CLL multi-resonant converter.

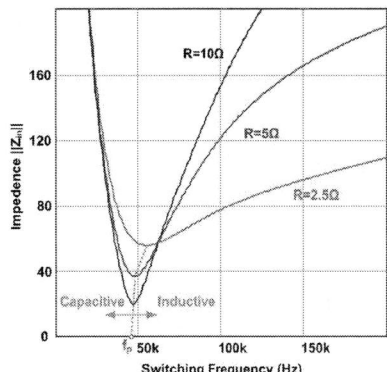

Fig. 8. Input impedance of the linearized CLL multi-resonant converter.

Fig. 3 can be simplified to Fig. 10 for time domain analysis. Notice that $v_{in}(t)=V_{in}$ during $0 < t \leq T/2$ and $v_{in}(t)=0$ during $T/2 < t \leq T$ and output voltage is assumed to be constant. The switching dynamic equations of the proposed CLL-MRC can be written as below

$$\begin{cases} C_r \dfrac{dv_{Cr}(t)}{dt} = i_{Lm}(t) + i_{L2}(t) \\[2mm] L_m \dfrac{di_{Lm}(t)}{dt} = v_{in}(t) - v_{Cr}(t) \\[2mm] L_2 \dfrac{di_{L2}(t)}{dt} = v_{in}(t) - v_{Cr}(t) - nV_o \end{cases} \quad (7)$$

Assume $v_{in}(t)$ has 50% duty cycle, inductor currents and capacitor voltages within $0<t<T/2$ will be symmetric to the waveforms within $T/2<t<T$. The average voltage of C_r will be the DC component of $v_{in}(t)$ and the average current of i_{L2} will be I_o/n. Also, L_m is the magnetizing inductor of the transformer in the proposed CLL-MRC. Therefore, the average current of L_m is zero. With above assumption and conditions, the following boundary conditions are valid and can be used to solve (7):

$$\begin{cases} i_{Lm}(0) = -i_{Lm}\left(\dfrac{T}{2}\right), i_{L2}(0) = i_{L2}\left(\dfrac{T}{2}\right) \\[2mm] \dfrac{1}{T}\int_0^T i_{Lm}(t)dt = \dfrac{2}{T}\int_0^{\frac{T}{2}} i_{Lm}(t)dt = 0 \\[2mm] \dfrac{1}{T}\int_0^T i_{L2}(t)dt = \dfrac{2}{T}\int_0^{\frac{T}{2}} i_{L2}(t)dt = \dfrac{V_o}{nR_o} = \dfrac{I_o}{n} \\[2mm] \dfrac{1}{T}\int_0^T v_{Cr}(t)dt = \dfrac{V_{in}}{2} \end{cases} \quad (8)$$

The solution to (7) can be found as

$$\begin{cases} v_{Cr}(t) = A\cos(\omega_r t - \phi) + V_{in} - nV_o \dfrac{L_e}{L_2} \\[2mm] i_{Lm}(t) = i_{Lm}(0) + \dfrac{v_{in}(t) - v_{Cr}(t)}{L_m}t \\[2mm] i_{L2} = \dfrac{i_{Lo}}{n} = -AC_r\omega_r \sin(\omega_r t - \phi) - i_{Lm}(t) \end{cases} \quad (9)$$

The values of parameters $i_{Lm}(0)$, A and ϕ can be found in the appendix. A comparison table of input/output relationship, peak value of v_{Cr} and i_{Lm}, and output rectifier voltage stresses at $f_s=f_r$ between the proposed CLL-MRC and the LLC-SRC are listed in Table I. It is notable that the proposed CLL-MRC will have higher output voltage than the LLC-SRC with the same turns ratio. This implies that if both proposed CLL-MRC and LLC-SRC have the same output voltage and L_m, the proposed CLL-MRC will have lower magnetizing current than the LLC-SRC. However, the proposed CLL-MRC will have higher voltage stresses on output rectifier diode with the same input/output spec.

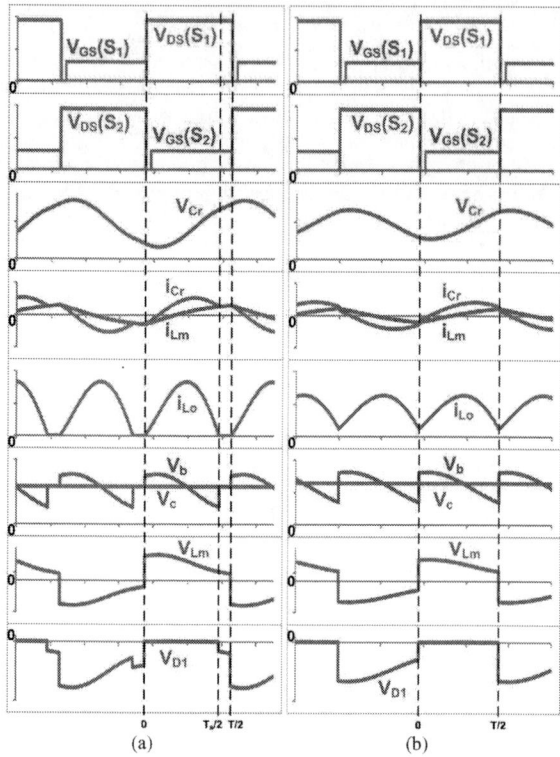

Fig. 9. Ideal CLL multi-resonant converter key waveforms: (a) $f_p<f_s<f_r$; (b) $f_s>f_r$. (V_{GS} waveforms are scaled.)

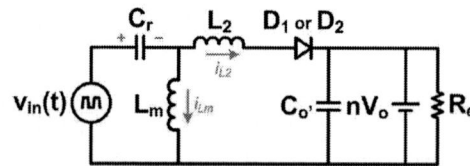

Fig. 10. Equivalent CLL multi-resonant converter circuit for time domain analysis.

Table I. Components Stresses and Input/output Relationship of the CLL-MRC and the LLC-SRC

	CLL-MRC	LLC-SRC
V_o/V_{in}	$(1+L_2/L_m)/(2n)$	$1/(2n)$
$\lvert i_{Lm,peak}\rvert$	$nV_o/(4L_m)$	$nV_o/(4L_m)$
$\lvert v_{Cr,peak}\rvert$	$\{(I_oT/(4nC_r))^2+[0.5\pi(0.5V_{in}-nV_o)]^2\}^{0.5}+0.5V_{in}$	$I_oT/(4nC_r)+0.5V_{in}$
$\lvert v_{D,peak}\rvert$	$2\lvert v_{Cr,peak}\rvert/n$	$2V_o$

C. Feasible SR Control Method for the Proposed CLL-MRC

The output inductor voltage, v_b-v_c in Fig. 3, will only be higher than zero when there is a current flow through it. In other words, either of the output rectifiers conducts current only when $v_b > v_c$. Specifically, output winding polarities and the comparison results of v_b and v_c can be used to determine the timing to turn on SR MOSFETs and which SR MOSFET should be turned on. Once the inductor current reaches its peak value, the inductor voltage will drop to zero and start to become negative. That is, by comparing v_b and v_c levels with hysteresis, we will be able to turn on and off SR MOSFETs and ensure over 50% rectifier current are conducted through MOSFET $R_{ds(on)}$. Consider the output stage in Fig. 11 is applied with aforementioned control scheme, $v_{GS(S1a)}$ will be turned on when both $v_{DS(S2a)}$ is high and $v_b > v_c$. Similar to $v_{GS(S1a)}$, $v_{GS(S2a)}$ will be turned on when both $v_{DS(S1a)}$ is high and $v_b > v_c$. The control circuit can be done with a comparator IC and low cost transistor logic gates. As shown in Fig. 12, S_{1a} or S_{2a} is turned on and off according to the transformer winding voltages and the comparison results of v_b and v_c.

This SR control method is especially beneficial for applications with low cost, high output voltage and high efficiency requirements. If wider SR conduction time is desired, a current sensing transformer can be inserted in series with the output inductor and the same control scheme used in [7] can be applied here. Unlike the CLL-MRC proposed in [7], the current sensing transformer applied for SR in the CLL-MRC proposed in this paper doesn't require safety isolation, even when safety isolation is needed between input and output. That is, the proposed CLL-MRC can have smaller footprint on SR control circuitry than the CLL-MRC proposed in [7].

III. EXPERIMENTAL RESULTS

A 250W CLL-MRC prototype, shown in Fig. 13, is built to verify the analysis made in this paper and evaluate its key features and performance. A 1μH inductor (Wurth 7443320100 with 1.85mΩ DCR) is applied as L_o. A transformer with ER28 core and a turns ratio of n=8 is applied here. The transformer's primary inductance is measured as 375μH and leakage inductance is measured as 5μH. A 0.033μF/1.25kV film capacitor is applied as C_r. For the switches and rectifiers, STMicroelectronics STP14NM50N (V_{DS}=500V, $R_{ds(on)}$=320mΩ) is used for S_1 and S_2, VF40100C-E3/4W from Vishay Semiconductor is used for D_1 and D_2, C_o=2mF, and Texas Instruments resonant mode controller UCC25600 with UCC27714 half-bridge driver is applied for frequency control and driving the MOSFETs. The power stage of the prototype is within 70mm x 42mm dimension. It is noticeable that the output inductor used in the CLL-MRC prototype has dimensions of 9.5mm x 12mm x 12mm (H x L x W), which is much smaller than resonant inductors used in the LLC-SRC at the same power level and switching frequency.

At 390Vin and 27.5V/9A output, the switching frequency f_s is at 120 kHz, which is equal to the resonant frequency f_r, as show in Fig. 14. The resonant frequency f_r can be calculated to be 118.5kHz by neglecting the transformer leakage inductance. The output inductor voltage waveform matches with the simulation results in Fig. 9. Fig. 15. provides evidence of soft-switching of the proposed CLL-MRC. When $f_s < f_r$, ZVS on input switches and ZCS on output rectifiers can be achieved. The voltage gain of the proposed CLL-MRC with the resonant parameters used in the prototype has been calculated by using (3), the results are shown in Fig. 16. Notice that the transformer leakage inductance is neglected in this calculation. The measured voltage gain most closely matches the calculation

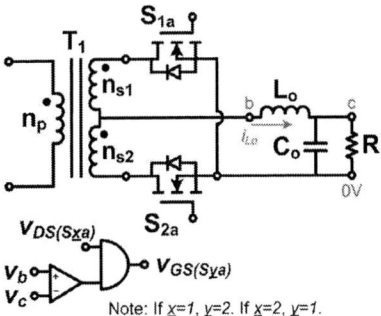

Fig. 11. CLL-MRC output circuitry with winding voltage controlled SR.

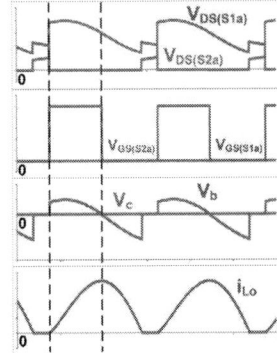

Fig. 12. CLL-MRC waveforms with winding voltage controlled SR.

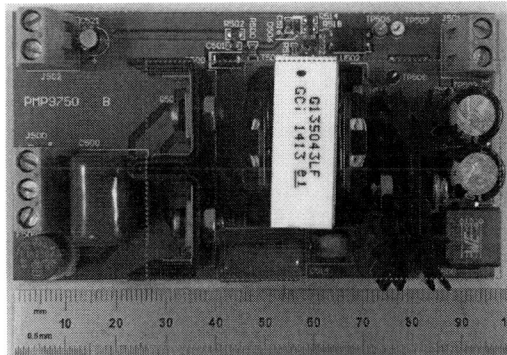

Fig. 13. 250W CLL-MRC prototype.

results when $f_{sw} \approx f_r$ because the fundamental component is most significant at $f_{sw}=f_r$.

Efficiencies of the proposed CLL-MRC and the LLC-SRC were measured on the same prototype board. When processing the efficiency measurement on the LLC-SRC, the output inductor was shorted and an EE19 inductor with 72μH inductance was placed in series with the transformer input winding as the LLC-SRC resonant inductor. It is notable that the EE19 inductor has a dimension of 17.5mm x 20mm x 20mm (H x L x W), which is 5 times larger volume than the output inductor in the CLL-MRC design. The power stages of the CLL-MRC and the LLC-SRC prototypes are shown in Fig. 17. The output voltage was fixed at 27.5V for both CLL-MRC and LLC-SRC measurements. The measurement results are shown in Fig. 18, which verifies that the proposed CLL-MRC can have similar or higher efficiency when compare to the LLC-SRC.

As mentioned in section II, the proposed CLL-MRC has higher voltage gain than the LLC-SRC with the same turns ratio. Therefore, the operational frequency of the proposed CLL-MRC is higher than the LLC-SRC under the same input voltage in this measurement. The efficiency results show that unlike the LLC-SRC, the proposed CLL-MRC efficiency at $f_{sw} > f_r$ can be higher than the efficiency at $f_{sw} = f_r$. In this prototype, a maximum 95.7% efficiency is achieved with $430V_{in}$.

IV. CONCLUSIONS

This paper proposed a compact size and high efficiency resonant converter – CLL-MRC – as an alternative choice in high performance voltage-step down applications. Its voltage gain relationship and soft-switching conditions are investigated through sinusoidal approximation. Through time-domain analysis, resonant inductor currents and resonant capacitor voltage equations and component stresses of the CLL-MRC are explored. When compared to the LLC-SRC, the proposed CLL-MRC has higher voltage gain, lower magnetizing current, and smaller B-H hysteresis loop on the resonant inductor. A simple voltage sense SR control scheme is also suggested. The voltage gain of CLL-MRC is also calculated and verified by a prototype board. Moreover, with smaller size and lower cost components, the proposed CLL-MRC can achieve similar or higher efficiency when compared to the LLC-SRC.

V. APPENDIX

The values of parameters $i_{Lm}(0)$, A and ϕ used in (9) are shown as below:

$$
\begin{cases}
\begin{aligned}
i_{Lm}(0) \\
= \frac{1}{2L_m} & \left\{ \frac{A}{\omega_r}\left[\sin\left(\frac{\omega_r T}{2} - \phi\right) + \sin\phi\right] - \frac{nV_o T\omega_2^2}{2\omega_r^2} \right\} \\
A = & -\frac{d^2 + e^2}{2\sqrt{\left(d^2 + e^2\right)\sin^2\left(\dfrac{\omega_r T}{4}\right)}} \\
\phi = & \cos^{-1}\left[\frac{e\left(1 - \cos\left(\dfrac{\omega_r T}{2}\right)\right) - d\sin\left(\dfrac{\omega_r T}{2}\right)}{2\sqrt{\left(d^2 + e^2\right)\sin^2\left(\dfrac{\omega_r T}{4}\right)}}\right] \\
d = & \left(\frac{\omega_2^2}{\omega_r^2} - 1\right)\frac{nV_o\omega_r T}{2} \\
e = & \frac{I_o T}{4nC_r}
\end{aligned}
\end{cases}
\quad \text{(A.1)}
$$

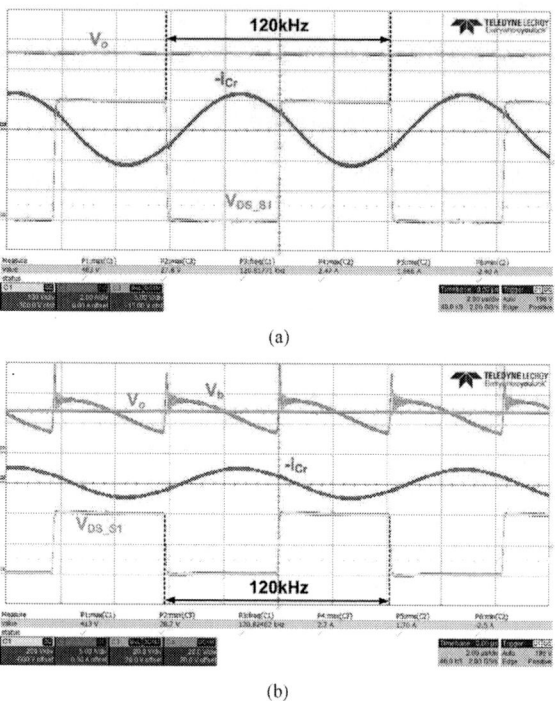

(a)

(b)

Fig. 14. Key CLL-MRC waveforms at $f_{sw}=f_r$: (a) resonant capacitor current and (b) output inductor voltage.

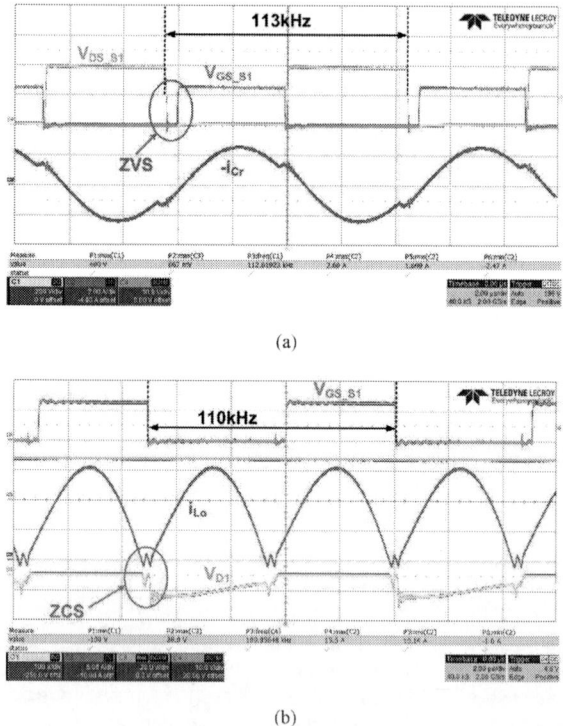

(a)

(b)

Fig. 15. Key CLL-MRC soft-switching waveforms with $V_{in} \approx 380V$, $V_o=27.5V$, and $I_o=9A$.

REFERENCES

[1] R. P. Severns, "Topologies for three-element resonant converters," *IEEE Transactions on Power Electronics,* vol. 7, pp. 89-98, Jan. 1992.

[2] B. Yang, F. C. Lee, A. J. Zhang, and G. Huang, "LLC resonant converter for front end DC/DC conversion," in *Proc. APEC*, 2002, pp. 1108-1112.

[3] X. Wu, G. Hua, J. Zhang, and Z. Qian, "A New Current-Driven Synchronous Rectifier for Series-Parallel Resonant (LLC) DC-DC Converter," *IEEE Transactions on Industrial Electronics,* vol. 58, pp. 289-297, 2011.

[4] GREEN Rectifier Controller Device [Online]. Available: http://www.ti.com/lit/ds/symlink/ucc24610.pdf

[5] D. Fu, Y. Liu, F. C. Lee, and M. Xu, "A Novel Driving Scheme for Synchronous Rectifiers in LLC Resonant Converters," *IEEE Transactions on Power Electronics,* vol. 24, pp. 1321-1329, 2009.

[6] M. M. Jovanovic, M. T. Zhang, and F. C. Lee, "Evaluation of synchronous-rectification efficiency improvement limits in forward converters," *IEEE Transactions on Industrial Electronics,* vol. 42, pp. 387-395, 1995.

[7] D. Huang, D. Fu, F. C. Lee, and P. Kong, "High-Frequency High-Efficiency CLL Resonant Converters With Synchronous Rectifiers," *IEEE Transactions on Industrial Electronics, ,* vol. 58, pp. 3461-3470, 2011.

[8] R. W. Erickson and D. Maksimovic, "Fundamentals of Power Electronics," in *Fundamentals of Power Electronics,* 2nd ed: Springer Science+Business Media Inc.

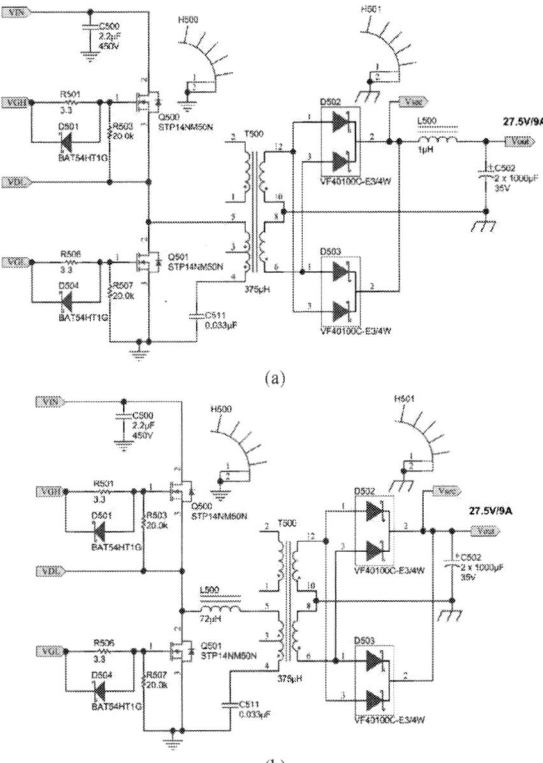

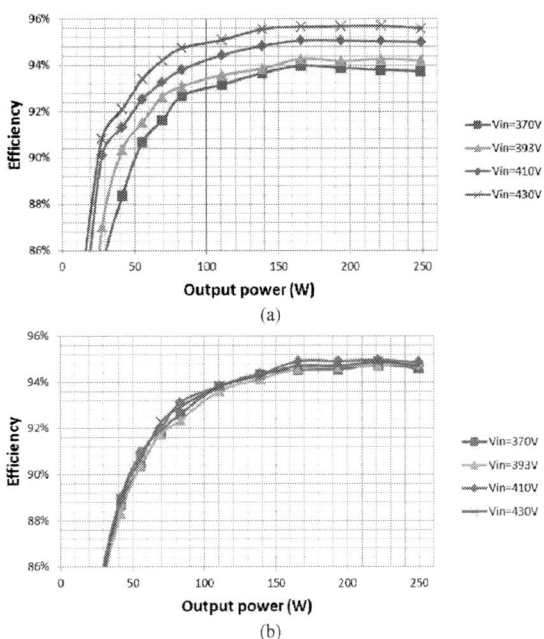

Fig. 17. Power stages of: (a) CLL-MRC prototype and (b) LLC-SRC prototype.

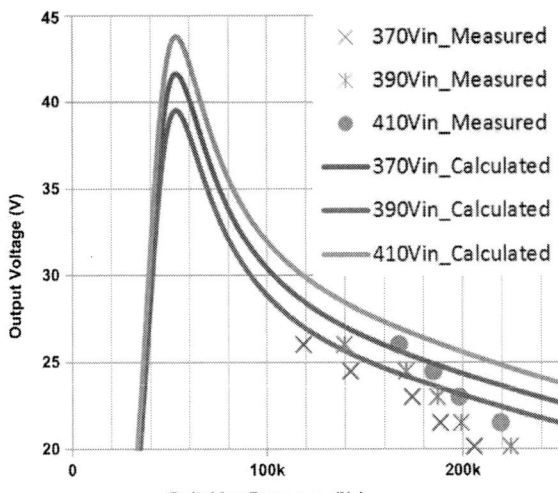

Fig. 16. Calculated and measured voltage gains of CLL-MRC.

Fig. 18. Converter efficiency with 27.5V output and different input voltages: (a) CLL-MRC, (b) LLC-SRC.

A 10-MHz eGaN FETs Based Isolated Class-Φ_2 DCX[*]

Xuewen Zou, Zhiliang Zhang, Zhou Dong, Yuan Zhou, Xiaoyong Ren, Qianhong Chen
Jiangsu Key Laboratory of New Energy Generation and Power Conversion
Nanjing University of Aeronautics andAstronautics, Nanjing, Jiangsu, P. R. China
{zouxuewen, zlzhang, bigzhou, zoezy, renxy, chenqh} @nuaa.edu.cn

Abstract- **One of the challenge of multi-MHz resonant converters is to drive the power FETs efficiently in wide input voltage range, for the reason that the phase angle between the drain-to-source voltage and gate drive voltage of the power FETs varies with the input voltage. The phase angle variation results in high reverse conduction loss and switching loss seriously at high frequency, especially for the eGaN FETs due to the reverse conduction mechanism with the switching period of hundreds of ns. In this paper, it is interesting to find that if the output voltage is controlled to follow the input voltage proportionally, the phase angle remains fixed, so that the drain-to-source voltage and gate drive voltage of the power FETs matches well in wide input voltage range. Owing to the characteristics, the reverse conduction loss of the control FET and SR FET can be minimized. Therefore, a voltage following control method was proposed for the multi-MHz resonant converters. With this voltage following control, the reference voltage of the hysteresis control follows the input voltage proportionally to have the output voltage follow the input voltage proportionally, so that a DC Transformer (DCX) can be realized efficiently. An 18-24 V input, 18 W/ 2 A output, 10-MHz prototype of the DCX with eGaN FETs was built to verify the functionality and advantages.**

Keywords—eGaN FETs; DCX; Synchronous Rectifier; voltage following control

I. INTRODUCTION

Modern power electronics applications expect power converters to achieve extremely high power density, high efficiency and faster dynamic response [1]. Normally, increasing the switching frequency of the converter is an effective way to reduce the volume of the passive devices and to increase the power density of the converter [2]. Many researches have been done based on the Class-Φ_2 resonant converter with the switching frequency of multi-megahertz, for the reason that the soft-switching of the control FET can be achieved and the voltage stress of the control FET is much reduced comparing to the Class-E resonant converter [3]. A 30 MHz isolated Class-Φ_2 resonant converter is proposed in [4], and the hysteresis control method is used to regulate output voltage precisely.

With the increase of the switching frequency, the high frequency dependent loss increases dramatically at multi-MHz. It is pointed out that the rectifier diodes in the multi-MHz resonant converter have serious impact on the efficiency of the converter. Usually, the conduction loss dissipated in the rectifier diodes accounts for 30% ~ 35% of the total loss [5]. This is because of the forward recovery phenomenon of the diodes. When the diode conducts, the current cannot flow immediately through the diode. It takes time for the injection of the carriers to establish a charge gradient in the high resistively part of the diode. Therefore, there is a large transient voltage across the diode at the turn-on instant, causing high power loss especially at multi-MHz. In order to reduce high conduction loss of the diodes, the synchronous rectifier (SR) technique is strongly desired for the multi-MHz resonant converter. Meanwhile, with the excellent Figure of Merit (FOM), the enhancement mode Gallium Nitride power transistors (eGaN FETs) are extremely suitable to the multi-MHz resonant converters [6]-[7]. Compared to the conventional silicon MOSFET, the eGaN FETs have much smaller conduction resistance and parasitic capacitance [8]-[9], which leads to further improvement of the conversion efficiency.

Normally, the hysteresis control is applied in the multi-MHz Class-Φ_2 resonant converter to regulate the output voltage for fast dynamic response. One of the major problems of the multi-MHz Class-Φ_2 resonant converter is that the reverse conduction time of the control FET and the SR FET increases with the increase of the input voltage. This is because the phase angle between the drain-to-source voltage and gate drive voltage of the power FETs varies with the input voltage. This results in high reverse conduction loss especially for the eGaN FETs as the reverse current has to follow via the channel of the eGaN FETs by the reverse conduction mechanism. Another problem of the multi-MHz Class-Φ_2 resonant converters is that with the increase of the input voltage, ZVS of the SR FET can not be achieved and the hard-switching of the SR FET happens, which will result in the high switching loss.

In order to solve the problems mentioned above, a voltage following control method is proposed. With the proposed voltage following control method, the output voltage follows the input voltage proportionally. Moreover, the phase angle between the drain-to-source voltage and gate drive voltage of the power FETs is immune to the input voltage, which leads to the nice match of the phase angle in wide input voltage range. Then, the reverse conduction loss and switching loss are minimized and a DC Transformer (DCX) with high frequency and high power density can be achieved.

Typical applications for the DCX are battery operated equipment and distributed power architectures in communication and industrial electronics, everywhere where isolated, un-regulated voltages are required. The similar architecture such as the Intermediate Bus Architecture (IBA) and the Factorized Power Architecture (FPA) are two of the typical power distribution architectures where the DCX is used. The power density of the power distribution architecture can be improved potentially if the Class-Φ_2 resonant DCX is developed.

---

[*]This work was supported by Natural Science Foundation of China (51377077) and the Fundamental Research Funds for the Central Universities (NUAA), NO. NE2014101, NO. 3082014NP2014402

978-1-4673-9551-9/16 $31.00 © 2016 IEEE

II. HARD-SWITCHING OF THE SR AND THE REVERSE CONDUCTION PROBLEMS OF THE POWER FETs IN THE CLASS-Φ_2 RESONANT CONVERTER

Normally, the frequency and the duty ratio of the drive signal of the control FET in the isolated Class-Φ_2 resonant converter are fixed. The hysteresis control is applied in the converter to regulate the output voltage to meet the specification [10].

Fig. 1 shows the isolated Class-Φ_2 resonant converter. The hysteresis control is used for the Class-Φ_2 resonant converter and the output voltage of the converter is forced to 9 V. The specifications of the isolated Class-Φ_2 resonant converter are given in Table I. Fig. 2 and Fig. 3 show the waveforms of the drain-to-source voltage of the SR FET when the input voltage is 18 V and 24 V respectively. In Fig. 2, the reverse conduction mechanism of the SR FET is not triggered, and ZVS of the SR FET is achieved at the same time. However, the resonant status of the converter varies with wide input voltage seriously. In Fig. 3, when the input voltage changes to 24 V, ZVS of the SR FET is not achieved and the hard-switching of the SR FET occurs. It leads to the high switching loss with the switching frequency of 10 MHz. The reverse conduction mechanism of the SR FET is triggered as well. The reverse conduction time of the SR FET is 9 ns, more than 15% of the total conduction time in one switching cycle which leads to high reverse conduction loss due to the high reverse conduction voltage of the eGaN FET. Fig. 4 and Fig. 5 show the waveforms of the drain-to-source voltage of the control FET when the input voltage is 18 V and 24 V respectively. Compared Fig. 6 with Fig. 7, the increasement of the reverse conduction time of the control FET can be noted. It means that with the increase of the input voltage, the reverse conduction loss of the control FET increases too.

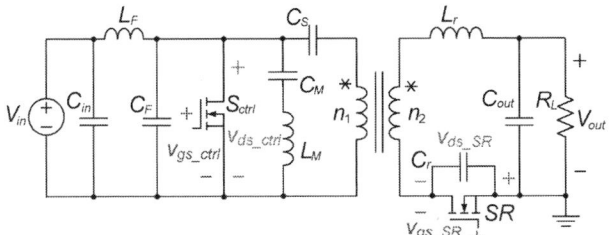

Fig. 1 The isolated Class-Φ_2 resonant converter

Based on the analysis above, when the output voltage of the isolated Class-Φ_2 resonant converter is regulated to the fixed value, ZVS of the SR FET may lose with the change of the input voltage and the hard-switching of the SR FET occurs. In addition, with the increase of the input voltage, the reverse conduction time of the control FET and SR FET increases and the reverse conduction problems happens.

Table I. SPECIFICATIONS OF THE ISOLATED CLASS-Φ_2 RESONANT CONVERTER

Input voltage V_{in}	18-24 V
Output voltage V_o	9 V
Output current I_o	2 A
Output power P_o	18 W
Switching frequency f_s	10 MHz

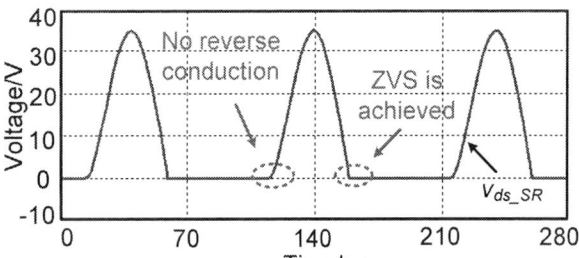

Fig. 2 The drain-to-source voltage of SR FET at 18 V input

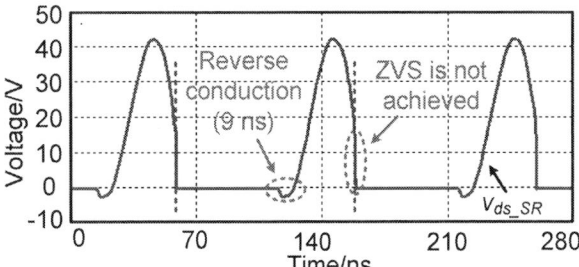

Fig. 3 The drain-to-source voltage of SR FET at 24 V input

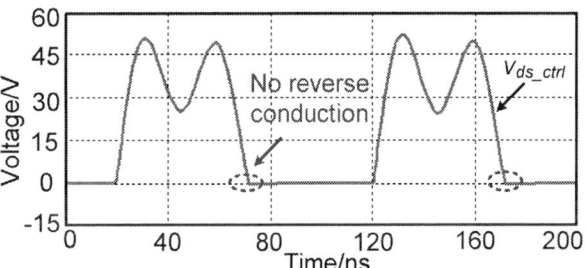

Fig. 4 The drain-to-source voltage of control FET at 18 V input

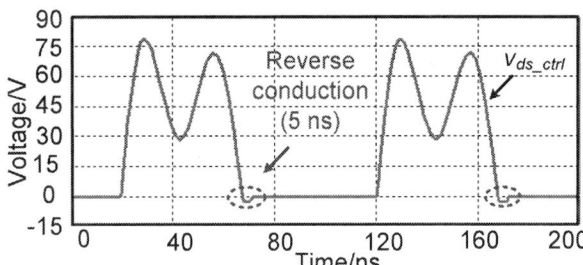

Fig. 5 The drain-to-source voltage of control FET at 24 V input

III. PHASE ANGLE RELATIONSHIP IN THE CLASS-Φ_2 RESONANT CONVERTER AND THE PROPOSED VOLTAGE FOLLOWING CONTROL METHOD

A. Phase Angle Relationship of the control FET and the SR in the Class-Φ_2 Resonant Converter

A detailed analysis is presented for the isolated Class-Φ_2 resonant converter. When the converter is in the open loop status with the rated load, the hard-switching of the SR FET and the reverse conduction problems mentioned above are minimized. The reason for this phenomenon is that in the open loop status with the rated load, the impedance characteristic of the converter is constant and the output voltage follows the input voltage proportionally. Moreover, the phase angle of the drain-to-source voltage of the control FET and the SR FET keeps unchanged with the change of the input voltage, which makes the drain-to-source voltage always match well with their drive signals. The phase angles of the drain-to-source voltage of the control FET and the SR FET are given in Fig. 6 and Fig. 7 respectively. $\varphi1$ in Fig. 6 represents the phase angle that the drain-to-source voltage v_{ds_SR} leads the drive gate voltage v_{gs_SR}. $\varphi2$ in Fig. 7 represents the phase angle that v_{ds_ctrl} leads v_{gs_ctrl}.

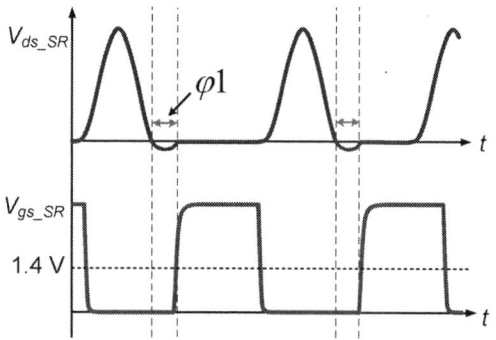

Fig. 6 The phase angle of the drain-to-source voltage of the SR FET

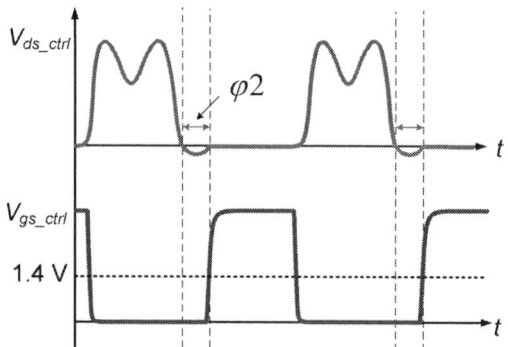

Fig. 7 The phase angle of the drain-to-source voltage of the control FET

However, in the open loop status, the change of the load will have an effect on the resonant status of the converter. That means, unlike the conventional PWM converter in the open loop status, with the change of the load, the output voltage cannot follow the input voltage proportionally and the hard-switching and reverse conduction problems cannot be avoided. So the converter with open loop status is not suitable to be used as the DCX directly.

B. The Proposed Voltage Following Control Method

A voltage following control method is proposed to make the output voltage follow the change of the input voltage proportionally. It is similar to the converter in the open loop with the rated load. The phase angle between the drain-to-source voltage and gate drive voltage of the power FETs is immune to the input voltage, which leads to the nice match of the phase angle in wide input voltage range. ZVS of the SR FET could be achieved and the reverse conduction mechanism of the control FET and SR FET is not triggered even though the input voltage of the converter is changed.

Fig. 8 shows the proposed voltage following control. The proposed control strategy is based on the voltage proportional change module and the hysteresis control. With the voltage proportional change module, the reference voltage of the hysteresis V_{ref} follows the input voltage proportionally. In addition, the isolation of the voltages V_{in} and V_{ref} is realized. The hysteresis control is used to make the output voltage follows the change of V_{ref} proportionally. So, the function that the output voltage of the converter follows the input voltage proportionally can be realized. In Fig. 8, a digital isolator is used to realize the isolation of the control signal and a time delay module is used to make sure the synch of control signals to the control FET drive and the SR FET drive. CON1 and CON2 represent the drive signals of the control FET and the SR FET respectively which have constant duty ratios.

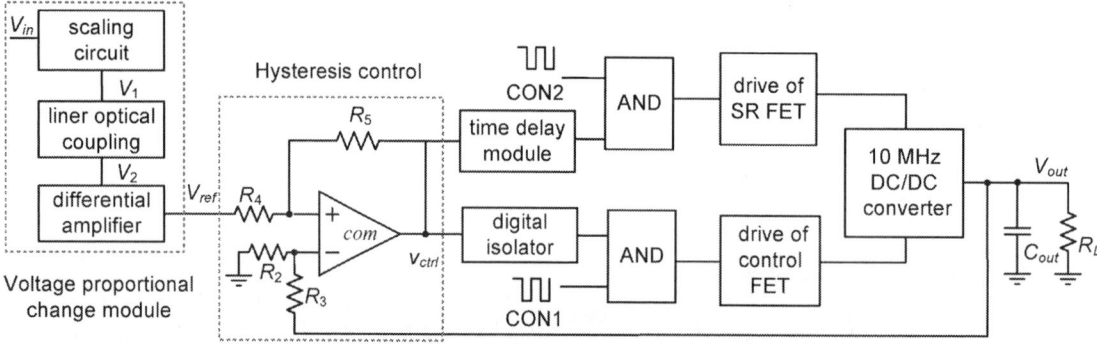

Fig. 8 The structure diagram of the voltage following control

C. The Characteristics of the Proposed Voltage Following Control

With the proposed control method, the output voltage of the converter follows the input voltage proportionally. For a certain input voltage, the value of the output voltage is decided. So the change of the load has no effect on the phase angle relationship of each node in the converter. That means the change of the load will not lead to the reverse conduction problems of the control FET and SR FET and ZVS of the SR FET could be achieved even though the input voltage is changed.

Fig. 9 and Fig. 10 give the simulation waveforms of the Class-Φ_2 resonant converter with different load when the voltage following control is applied. Compared Fig. 9 with Fig. 10, it is noted that the reverse conduction mechanism of the power FETs is not triggered and the ZVS of the SR FET can always be achieved even with the change of the load. It is noted from Fig. 9 that the modulation frequency of the converter is 24 kHz when the output power is 18 W. The modulation frequency follows the change of the load, as it can be observed from Fig. 10 that when the output power of the converter turns into 9 W the modulation frequency changes to 21 kHz. That means for the Class-Φ_2 converter with the voltage following control, the change of the load has an effect on the modulation frequency of the converter. In addition, ZVS of the SR FET can always be achieved and the reverse conduction mechanism of the power FETs is not triggered even under different load conditions.

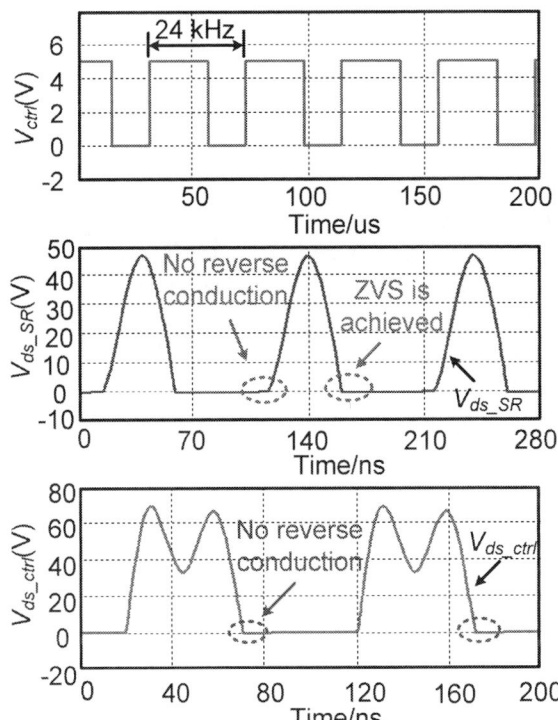

Fig. 9 The simulation waveforms of the converter with the voltage following control: V_{in}=24 V, P_o=18 W and f_s=10 MHz

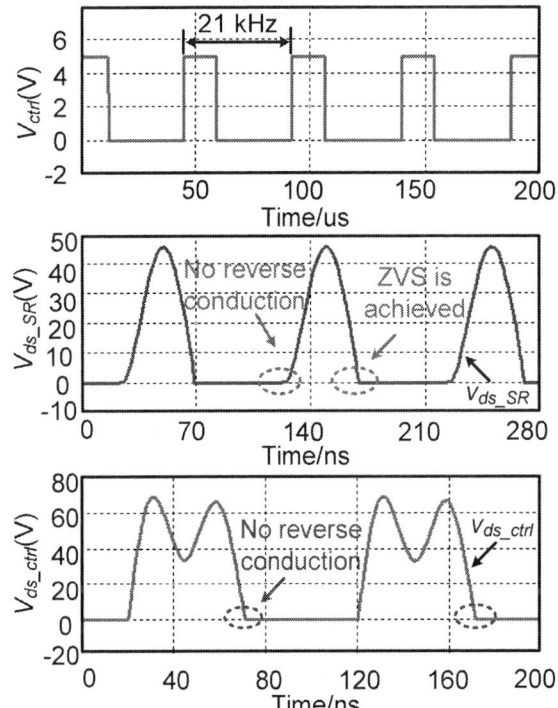

Fig. 10 The simulation waveforms of the converter with the voltage following control: V_{in}=24 V, P_o=9 W and f_s=10 MHz

D. The Realization of the Proposed Voltage Following Control

The proposed voltage following control is based on the hysteresis control. With the hysteresis control, the voltage regulation is achieved by comparing a fraction of the output voltage to the reference voltage V_{ref}. The comparator provides the control signal v_{ctrl} to modulate the converter on and off. With this control strategy, the upper limit of the output voltage V_H and the lower limit of the output voltage V_L could be obtained as follows:

$$V_H = \frac{R_2 + R_3}{R_2}\left(\frac{R_4 \cdot V_{CH}}{R_4 + R_5} + \frac{R_5 \cdot V_{ref}}{R_4 + R_5}\right) \tag{1}$$

$$V_L = \frac{R_2 + R_3}{R_2}\left(\frac{R_4 \cdot V_{CL}}{R_4 + R_5} + \frac{R_5 \cdot V_{ref}}{R_4 + R_5}\right) \tag{2}$$

where V_{CH} and V_{CL} represent the high level and low level of the comparator output voltage respectively and value of V_{CL} is 0. From equation (1) and (2), the output voltage is proportional to the reference voltage V_{ref} if the ripple of the output voltage is not taken into consideration. So, in order to make the output voltage of the converter change proportionally with the input voltage, the reference voltage V_{ref} needs to be controlled to follow the input voltage proportionally.

Fig. 11 shows the detailed circuit of the voltage proportional change module. It consists of the scaling operation circuit, the liner optical coupling and the differential amplifier. The functions of this module are as follows: 1) the reference voltage V_{ref} follows the change of input voltage proportionally; 2) realizing the isolation of the voltages V_{in} and V_{ref}. V_1 is obtained by the scaling operation circuit with the input of V_{in}. C_1 and C_2 are two filter capacitors. Through this scaling operation circuit, V_1 is proportional to the input voltage V_{in}. The functions of the liner optical coupling are to realize the isolation and the proportional amplification of V_1. It should be pointed that the value of V_1 should fall into the linear amplification area of the liner optical coupling. Then, the differential mode output voltage of the liner optical amplifier is amplified by a differential amplifier. Finally, the reference voltage V_{ref} proportional to the input voltage V_{in} is obtained.

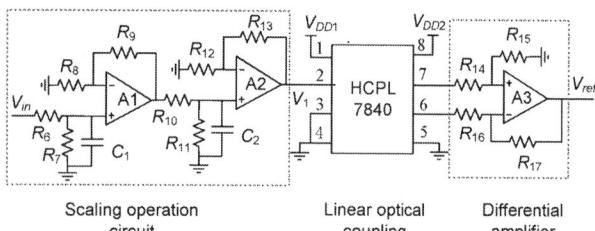

Fig. 11 Specific circuit of the voltage proportional change module

Based on the analysis above, the reference voltage of the hysteresis control follows the input voltage proportionally which leads to the output voltage changes proportionally with the input voltage. With the voltage following control, the output voltage V_{out} follows the change of input voltage V_{in} proportionally. The converter operates in a status similar to the open loop with rated load. The amplitude of the drain-to-source voltage of the control FET and the SR FET follows the input voltage proportionally. Moreover, the phase angles of the drain-to-source voltage of the control FET and the SR FET do not follow the change of the input voltage and these voltages always match well with their drive signals. The hard-switching of the SR FET and the reverse conduction problems of the control FET and SR FET can be avoided.

IV. EXPERIMENTAL VERIFICATION AND DISCUSSION

To further verify the functionality of the proposed control, an 18-24 V input, 18 W output, 10 MHz prototype of the DCX was built. The DCX is composed of the Class-Φ_2 inverter and the Class-E rectifier [11]-[12]. The specifications of the Class-Φ_2 resonant DCX are given in Table II. The parameters of the power stage and the control stage are given in Table III and Table IV respectively. The eGaN FETs of EPC2001 from EPC are used as the control FET and SR FET. The air-core inductors from Coilcraft are used in the power stage as the resonant inductance. Fig. 12 gives the photograph of the prototype.

Table II. SPECIFICATIONS OF THE CONVERTER

Input voltage V_{in}	18-24 V
Output voltage V_o	9-12 V
Output current I_o	2 A
Output power P_o	18 W
Switching frequency f_s	10 MHz

Table III. POWER STAGE COMPONENT VALUES

L_F	220 nH	L_M	220 nH
L_r	43 nH	C_M	268 pF
C_r	670 pF	C_S	4 µF
C_{in}	94 µF	C_{out}	188 µF
$n_1{:}n_2$	4:2		

Table IV. CONTROL STAGE COMPONENT VALUES

R_2	16 kΩ	R_3	56 kΩ
R_4	1 kΩ	R_5	100 kΩ
R_6, R_8	15 kΩ	R_7, R_9	4 µF
R_{10}, R_{12}	12 kΩ	R_{11}, R_{13}	1 kΩ
R_{14}, R_{16}	7.5 kΩ	R_{15}, R_{17}	3 kΩ

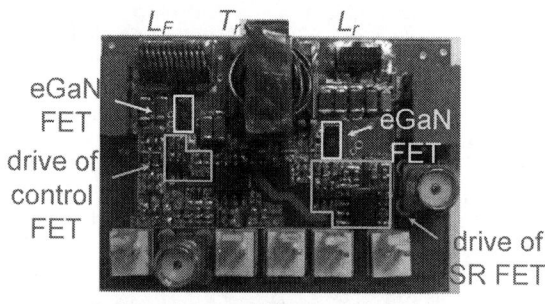

(a) Top

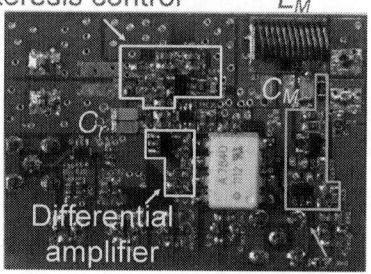

(b) Bottom

Fig. 12 The photograph of the prototype

Fig. 13 and Fig. 14 show the measured drain-to-source and gate-to-source voltage waveforms of the SR FET with the input voltage of 18 V and 24 V respectively. It is observed from Fig. 13 that the peak drain-to-source voltage of the SR FET is 34 V at 18 V input. In addition, the reverse conduction mechanism of the SR FET is not triggered and ZVS of the SR FET can be achieved in this case. In Fig. 14, when the input voltage of the converter reaches 24 V, the peak drain-to-source voltage of the SR FET increases to 45 V. It can be found that the amplitude of the drain-to-source voltage of the SR FET follows the input voltage proportionally. The reverse conduction mechanism remains to be not triggered and the ZVS still can be achieved at 24 V input. The waveforms of the SR FET agree well with the theoretical analysis.

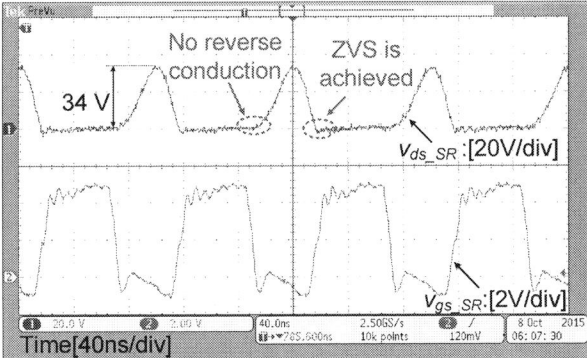

Fig. 13 Waveforms of the SR FET: V_{in}=18 V, P_o=18 W and f_s=10 MHz

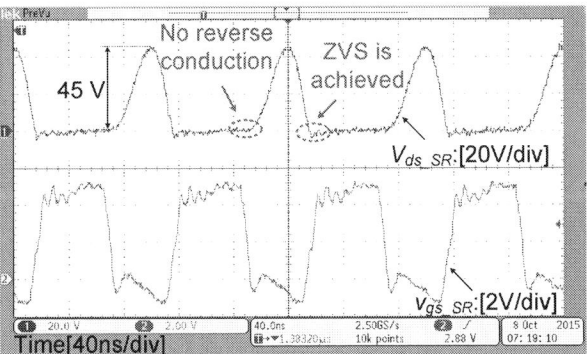

Fig. 14 Waveforms of the SR FET: V_{in}=24 V, P_o=18 W and f_s=10 MHz

The measured voltage waveforms of the control FET at 18 V input and 24 V input are given in Fig. 15 and Fig. 16 respectively. In Fig. 15, the peak drain-to-source voltage of the control FET is 51 V and the zero reverse conduction of the control FET is achieved when the input voltage is 18 V. As shown in Fig. 16, the peak drain-to-source voltage of the control FET proportionally increases to 68 V and the reverse conduction time of this control FET still remains zero, when the input voltage reaches 24 V. By comparing Fig. 15 with Fig. 16, it can be found that the amplitude of the drain-to-source voltage of the control FET follows the input voltage proportionally and the phase angle of this drain-to-source voltage does not change with the input voltage as the voltage following control is used.

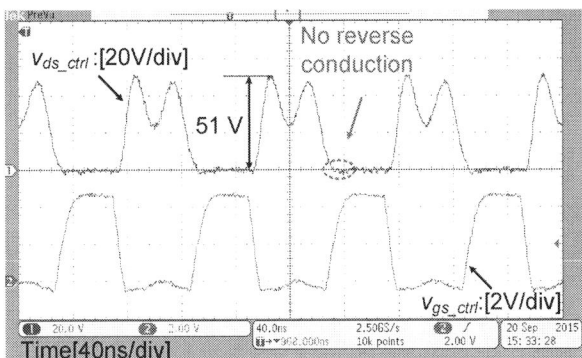

Fig. 15 Waveforms of the control FET: V_{in}=18 V, P_o=18 W and f_s=10 MHz

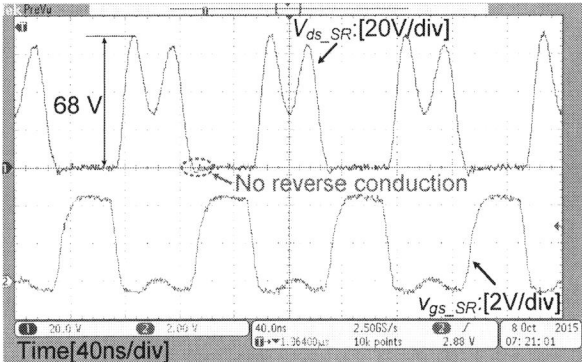

Fig. 16 Waveforms of the control FET: V_{in}=24 V, P_o=18 W and f_s=10 MHz

Fig. 17 gives the close-loop efficiency of the converter with different output power. The peak efficiency is 86.9% at full load when the input voltage is 18 V. It can be noted from Fig. 17 (a) that the efficiency decreases with the increase of the input voltage when the output power of the converter is more than 9 W. In Fig. 17 (b), when the output power of the converter is less than 9 W, the efficiency increases with the decrease of the output power. The reason for this phenomenon is that the modulation frequency f_M of the control signal v_{ctrl} is variable depending on different load and input conditions. For the Class-Φ_2 resonant DCX, the energy stored in the resonant components is dissipated during each modulation cycle, which means that higher modulation frequency, higher loss dissipation. The variable modulation frequency has a great effect on the efficiency of the converter. So, the change of the load and the input voltage leads to the change of the modulation frequency of the converter, which results in the efficiency change of the converter.

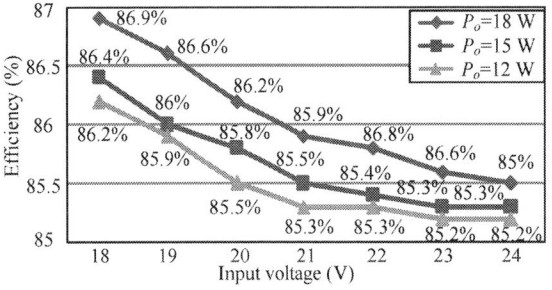

(a) P_o> 9 W

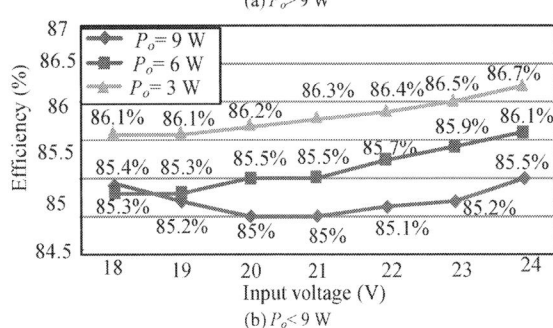

(b) P_o< 9 W

Fig. 17 Close-loop efficiency (drive loss included)

Fig 18 gives the efficiency comparison between the different control methods when the output power of the converter is 18 W. It is observed that compared to the conventional hysteresis control, the proposed voltage following control achieves significant efficiency improvement for the wide voltage input. It is found in Fig 18 that the efficiency of the converter is 79.9% at 24 V input with the conventional hysteresis control. And the efficiency increases to 85.5% as the voltage following control method is used, the efficiency improvement is 5.6% at 24 V input.

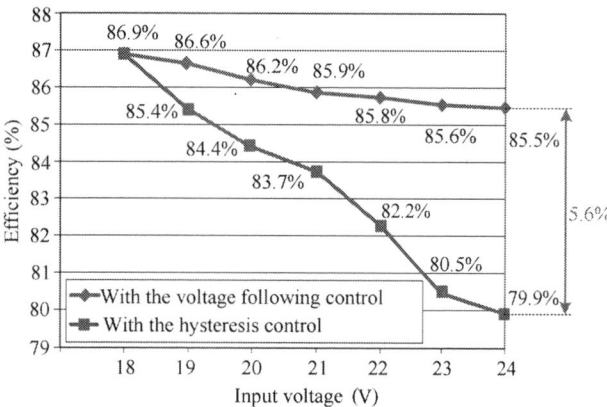

Fig. 18 Close-loop efficiency comparison (included the drive loss) : P_o=18 W

V. CONCLUSION

The voltage following control method is proposed in this paper. With this voltage following control method, the output voltage follows the input voltage proportionally. It is similar to the converter in the open loop with the rated load. Moreover, ZVS of the SR FET could be achieved and the reverse conduction loss of the control FET and SR FET is reduced. The simulation results verify the functionality of the proposed control strategy. In addition, an 18-24 V input, 18W/ 2 A output, 10 MHz prototype of DCX is built to verify the functionality and benefits of the proposed control strategy. At the rated output power 18 W, the proposed voltage following control method improves the efficiency of the converter from 79.9% to 85.5% and the efficiency is improved by 5.6% when the input voltage of the converter reaches 24 V.

REFERENCES

[1] A. Knott, T. M. Andersen, P. Kamby, J. A. Pedersen, M. P. Madsen, M. Kovacevic and M. A. E. Andersen, "Evolution of very high frequency power supplies," *IEEE Journal of Emerging and Selected Topics in Power Electronics*, Vol. 2, No. 3, pp. 386-394, Sept. 2014.

[2] D. J. Perreault, J. Hu, J.M. Rivas, Y. Han, O. Leitermann, R. C. N. Pilawa-Podgurski, A. Sagneri and C.R. Sullivan, "Opportunities and challenges in very high frequency power conversion," in Proc. IEEE APEC., pp. 1-14, 2009.

[3] J. M. Rivas, Y. Han, O. Leitermann, A. D. Sagneri and D. J. Perreault, "A high-frequency resonant inverter topology with low-voltage stress," IEEE Trans. on Power Electron., vol. 23, no. 4, pp. 1759-1771, 2008.

[4] Juan M. Rivas and Olivia Leitermann, "A very high frequency DC-DC converter based on a class Φ_2 resonant inverter", in Proc. IEEE PESC, 2008, pp. 1657 - 1666.

[5] L. C. Raymond, W. Liang and J. M. Rivas-Davila, "Performance evaluation of diodes in 27.12MHz Class-D resonant rectifiers under high

voltage and high slew rate conditions," in 2014 Workshop on Computers and Modeling in Power Electronics (COMPEL 2014), Jun. 2014.

[6] A. Lidow, "Is it the end of the road for silicon in power conversion". Application Note, EPC Co., 2010.

[7] S. L. Colino and R. A. Beach, "Fundamentals of Gallium Nitride power transistors". Application Note, EPC Co., 2010.

[8] Y. Wang, W. Kim, Z. Zhang, J. Calata and K. D. T. Ngo, "Experience with 1 to 3 megahertz power conversion using eGaN FETs," in Proc. IEEE APEC., pp. 532-539, 2013.

[9] W. Chen, R. A. Chinga, S. Yoshida, J. Lin, C. Chen and W. Lo, "A 25.6 W 13.56 MHz wireless power transfer system with a 94% efficiency GaN Class-E power amplifier," in Proc. IEEE MTT., pp. 1-3, 2012.

[10] W. Cai, Z. Zhang, X. Ren and Y. F. Liu, "A 30-MHz isolated push-pull VHF resonant converter," in Proc. IEEE APEC., pp. 1456-1460,2014.

[11] J. Rivas, "Radio frequency dc-dc power conversion," Doctor's thesis, Dept. Elect. Eng. Comput. Sci., Massachusetts Inst. Technol., Cambridge, MA, Aug. 2006.

[12] R. Pilawa-Podgurski, A. D. Sagneri, J. M. Rivas, D. I. Aderson and D. J. Perreault, "Very high frequency resonant boost converter," IEEE Trans. on Power Electron., vol. 24, no. 6, pp. 1654-1665, 2009.

Multi-level Capacitor Clamped DC-DC Multiplier/ Divider with variable and fractional voltage gain -An $\left(\frac{n}{m}\right) X$ DC –DC converter

Deepak Gunasekaran[1], Liang Qin[2], Ujjwal Karki[1], Yuan Li[3] and Fang Z. Peng[1]

[1]Dept. of Electrical and Computer Engineering, Michigan State University, East Lansing, MI, USA
[2]Department of Electrical Engineering, Wuhan University, China
[3]Department of Electrical Engineering and Information, Sichuan University, China

Abstract— Inductor-less, high gain DC-DC converter with high efficiency and high power density is a much desired circuit in Electric Vehicle (EV) powertrain and Solar photovoltaic (SPV) power converters. Modular Multi-level capacitor clamped DC-DC converter circuits (MLCCC) provide a viable solution for this case. But, their application is limited owing to their limitations in terms of fixed output voltage gains and lack of fractional output gains. The aim of this paper is to introduce an (n/m)X converter capable of both dynamically variable voltage gains and fractional voltage gains. Detailed description of the operating method along with the various stages of operation and power dissipation calculation for the proposed converter forms the main theme of the paper. Experimental results for a 1-kW prototype are presented to validate the proposed theory.

Keywords—Switched Capacitor, Multi-level DC-DC Converter, high voltage gain, High power Density, Inductor-less DC-DC converter

I. Introduction

High voltage gain, high power density bidirectional DC-DC converters have a high demand in aerospace, electric vehicle (EV) and solar photovoltaic (PV) applications. Additionally, these converters need to demonstrate high reliability levels and must be capable of withstanding high temperatures In view of this, multi-level DC-DC converters are known to provide high power density But, in order to further increase the power density and provide high temperature withstanding capability, the magnetic components in the circuit are often the bottleneck. In order to alleviate this bottleneck, a flying Capacitor based Multilevel DC-DC converter [1] with reduced inductor requirements has been proposed in [2]. This was further improved to eliminate the inductance requirement in certain cases [3]. But, the circuit is not practical for high voltage gains as the capacitor charging current passes through 'N' different switches when used to provide a gain of 'N' times the input voltage (NX) leading to low efficiency Yet another direction for research in this area has been the "switched capacitor based converters". A Modular Multi-level capacitor clamped DC-DC converter (MLCCC) circuit was first introduced in [4]. It is a direct variant of the Dickson converter [5] that was proposed in the 70's. The main advantage of the MLCCC circuit is that the capacitor charging current passes through a maximum of three

switches at any given time. Additionally, this circuit can be operated without any external input inductor. But, it has a high device count and suffers from unequal voltage stresses on different semiconductor devices used in the system. This circuit was further improved by using a double wing structure as proposed in [6]. This leads to a circuit with reduced device count and reduced capacitor voltage stress. Since, its introduction, further design improvements that enable a compact design [7] and usage of zero-current switching techniques have been introduced [8] But, the circuit can only provide fixed integer output voltage gains. In [9], an MLCCC based circuit to provide variable gains has been discussed in theory. But, it is limited to a narrow range by the load requirement. Thus, the ability to design a MLCCC based converter for a required fractional gain and the ability to provide variable output voltage gain has not been well researched and remains a coveted topic in this area. In this paper, both these important issues are addressed in order to facilitate the usage of these circuits in EV, PV and other such high power density applications. First, a new configuration of the MLCCC based double-wing structure proposed in [6] is presented. The new configuration has the ability to provide any fractional or integer voltage gain. This is termed as " $(\frac{n}{m}X)$ converter". The $\left(\frac{n}{m}\right) X$ converter is derived from a generic $\left(\frac{k}{m}\right) X$ converter. The converter is bidirectional, functioning as a "voltage multiplier" in one direction and as a "voltage divider" in the other. Additionally, a method is proposed to allow the variation of voltage gain from one available gain to another. Detailed analysis of the operation and functionality of the variable $(\frac{n}{m}X)$ converter forms the crux of this paper. A 1-kW prototype of a $(\frac{n}{m}X)$ converter has been developed to validate the design process. Simulation and experimental results are provided to validate the proposed theory.

The rest of the paper is organized as follows: Section II introduces the generic $\left(\frac{k}{m}\right) X$ converter from which the $\left(\frac{n}{m}\right) X$ converter is derived. Section III describes the various operating stages of the proposed $\left(\frac{n}{m}\right) X$ converter along with the equivalent circuit. Section IV describes the specifications for

978-1-4673-9551-9/16 $31.00 © 2016 IEEE

the developed 1-kW prototype. Section V presents the experimental results for the developed prototype. Section VI provides a theoretical framework to calculate the power dissipation and efficiency of the proposed converter.

II. A GENERIC $(\frac{k}{m}X)$ CONVERTER

Figure 1 illustrates the structure and of a generic $(\frac{k}{m}X)$ converter. It can be seen that the proposed converter has 'n' different arms and 'n' different legs leading to a total of '2n' limbs Each arm comprises of a series connection of two capacitors and two arm switches labeled with suffixes 'a' and 'b'. For e.g., arm no. 1 comprises of capacitors C_{1a} and C_{1b} connected in series with switches S_{1a} and S_{1b}. Each leg of the converter consists of a half-bridge cell consisting of two switches switched in a complimentary function. Leg 1 consists of S_{1p} and S_{1n}. In comparison to the double wing structure

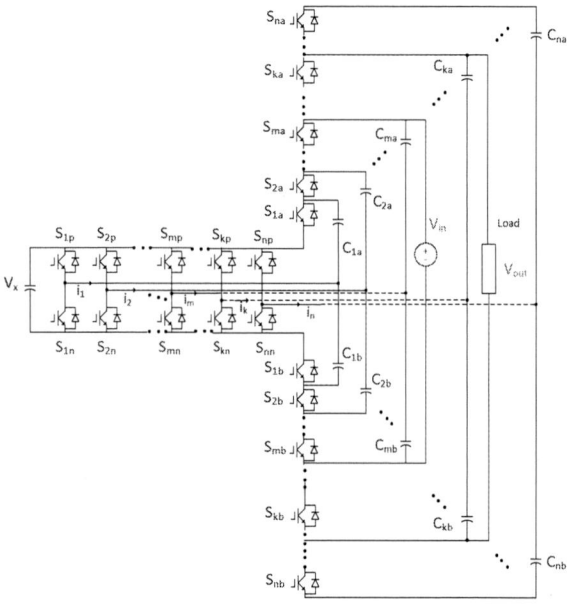

Figure 1: A generic $\left(\frac{k}{m}\right)X$ converter

proposed in [6], it can be observed that an extra degree of freedom with respect to the location of input voltage and output voltage is introduced in the $\left(\frac{k}{m}\right)X$ converter. The input voltage V_{in} can be connected across any of the 'n' arms of the converter. The output voltage can also be connected across any of the 'n' arms of the converter. For a generic location of the input voltage, V_{in} on the mth arm of the converter, the "effective input" voltage, V_X across C_X can be expressed as,

$$V_X = \frac{1}{m} * \frac{V_{in}}{2} \qquad (1)$$

For a generic location of the load across arm $'k'$, output voltage V_{out} can then be expressed as,

$$V_{out} = 2n * V_X = \frac{k}{m} * V_{in} \qquad (2)$$

Where, m is the arm across which the input voltage is connected, k is the arm across which the load is connected and $k, m \le n$

The converter is fully bidirectional. If, $k < m$, the converter functions as a buck converter. On the other hand, if $k > m$, the converter functions as a boost converter. From, Figure 1, it can also be seen that, one other degree of freedom can be obtained by placing the source or load across C_X. If the input voltage, V_{in} is placed across C_X and the load is connected across the generic arm, $'k'$ as described above, the output voltage,

$$V_{out} = 2k * V_{in} \qquad (3)$$

This presents a case of $m = \frac{1}{2}$ for the generic $\left(\frac{n}{m}\right)X$ converter. Similarly, if the load is placed across C_X and the input voltage is placed across the mth arm as described before, the output voltage V_{out} can now be described as,

$$V_{out} = \frac{1}{m} * \frac{V_{in}}{2} \qquad (4)$$

Therefore, based on the discussion in this section, the acceptable values of k and m can now be listed as below,

$$k = \frac{1}{2} \; or \; k = 1, 2 \dots n \qquad (5)$$

$$m = \frac{1}{2} \; or \; m = 1, 2, \dots. n \qquad (6)$$

The $(m, k) = \left(\frac{1}{2}, n\right) or \left(n, \frac{1}{2}\right)$ operating points lead to the nX converter proposed in [6]. For a boost converter configuration, the highest output voltage, V_{out} for a given input voltage, V_{in} can be achieved at $k = n$. Thus, $k = n$ provides the maximum possible gain. On the other hand, for a buck configuration, $m = n$ provides the minimum possible gain. Thus, the bounds for the maximum and minimum converter gains are determined by $k = n$ and $m = n$ respectively. This essentially means the $(\frac{k}{n}X)$ converter functions as an $(\frac{n}{m}X)$ boost converter or $(\frac{k}{n}X)$ buck converter within the realms of the maximum and minimum output voltage requirements. Hence, the variable gain converter in boost mode is essentially a variable $(\frac{n}{m}X)$ converter. All the analysis in future sections will be based on the $(\frac{n}{m}X)$ converter as similar analysis also holds for the buck converter.

III. THE VARIABLE $\left(\frac{n}{m}\right)X$ CONVERTER

For a generic $\left(\frac{n}{m}\right)X$ boost converter, as described above, the different gains that can be produced for $m = \frac{1}{2}$ are, $\left(\frac{n}{m}\right), \left(\frac{n-1}{m}\right), \left(\frac{n-2}{m}\right) \dots \left(\frac{1}{m}\right)$: For, $m \ge 1$, the different voltage

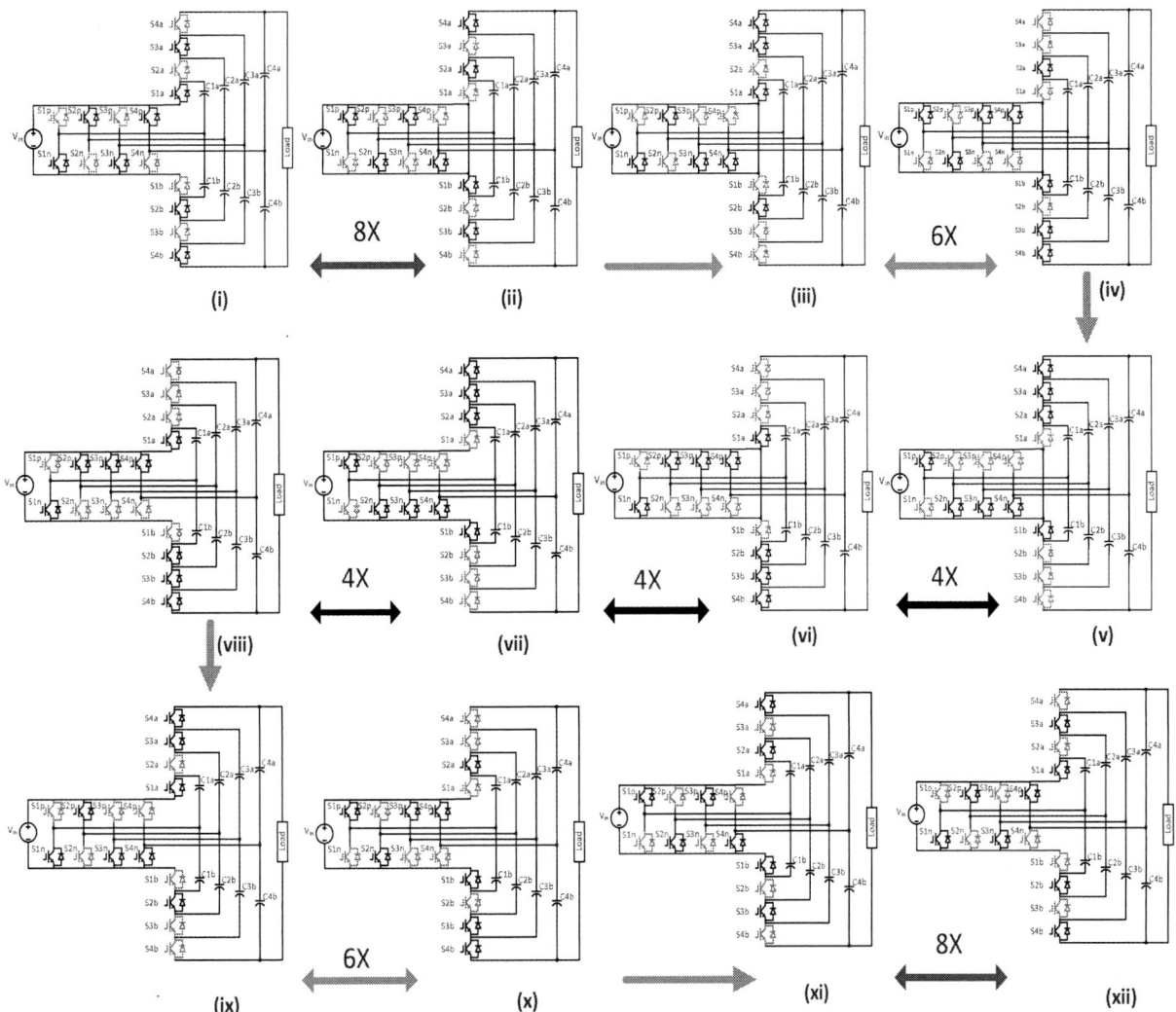

Figure 2: Operating stages for "shift-up" and "shift-down" modes

gains that can be produced are: $\left(\frac{n}{m}\right), \left(\frac{n-1}{m}\right), \left(\frac{n-2}{m}\right)....1$. The case of a variable $(\frac{4}{2})X$ converter is chosen to illustrate the various stages involved in change of gain. They are divided into two parts, viz. gain "shift-down" mode (decreasing gain) and gain "shift-up" mode (increasing gain). The gain variation is achieved by essentially forcing two arms of the converter cell to have the same corresponding capacitor voltages. The switching loss and conduction loss of the semiconductor devices during transient current determine the sizing of the switches. The efficiency and dynamic response of the overall process depends on the magnitude of energy dissipated in each cycle of the switch operation during the transient over-currents. Detailed description of these aspects is provided in later sections.

A. Gain "Shift-down" mode

The various stages involved in shifting the gain from 8 through 4 are illustrated by means of Figure 2. Stages (i) through (vi) determine the process of shifting-down from 8 to 4 through 6. The switches marked in red represent the switches that have been turned ON. The switches in black indicate the switches that have been turned OFF. As shown in the figure, the variable output gains can only be shifted one step at a time. For instance, in the $(\frac{4}{2}X)$ converter, in the "shift-down" mode, the gain can only move from 8 to 6, 6 to 4 and 4 to 2 in discrete steps.

Stages (i) and (ii)

The converter alternates between these two stages for 50% duration within a switching time period. These stages are exactly similar to the $8X$ operating states proposed in [6]. The equivalent circuits for this stage are also presented in [6].

Stages (iii) and (iv)

At the beginning of stage (iii), the switches in limb 4 (S_{4p}, S_{4n}, S_{4a} and S_{4b}) follow the corresponding switches in limb 3. Stages (iii) and (iv) alternate for 50% duration in a switching cycle. The output voltage, V_{out} reaches $6V_{in}$ after the transient over-current is damped.

The equivalent circuit to calculate the worst case transient current in the loop is shown in Figure 3. Here,

V_d is the difference in voltage before switch S_w is closed.

L_{lj} is the equivalent inductance in the jth loop

C_{lj} is the equivalent capacitance in j^{th} loop

R_{lj} is the overall resistance in j^{th} loop

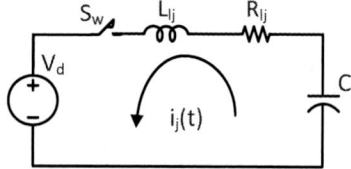

Figure 3: Approximate equivalent circuit for peak loop over-current calculation

The loop current, $i_j(t)$, for the j_{th} loop is given by the standard RLC underdamped equation. For stages (iii) and (iv), $C_l = C_{4a}$. The resistance, R_l and the inductance L_l are determined by the trace resistance, trace inductance, ESL and ESR of the capacitor, etc. The exact constituents of the loop impedances have been explained in [6].

Stages (v) and (vi)

In stages (v) and (vi), the switches in limb 3 and limb 4 follow the switches in limb 2. As a result, the output voltage, V_{out} reaches $4V_{in}$ when the transient over-current has dampened. In order to calculate the worst case transient current, the equivalent circuit in Figure 3 can be used. But, the overall loop capacitance is different in this case owing to two arms effectively being paralleled.

Here, $V_d = 30\ V$, $C_l = 2C_{4a}$. Due to double the capacitance in the equivalent circuit (compared to previous case), the magnitude of the peak transient current will also be higher.

B. Gain "Shift-up" mode

In the gain "shift-up" mode, the voltage gain undergoes a transition from $4X$ to $8X$ in sequential steps. It is achieved by travelling back from stage (vi) all the way up to stage (i). This sequence of stages is illustrated using stages (vii) through (xii). The equivalent circuit in order to calculate the peak transient current is similar to the corresponding stage in the gain "shift-down" modes. The only difference is that the sign of V_d is reversed.

IV. DESIGN OF $\left(\frac{n}{m}X\right)$ CONVERTER FOR PV APPLICATION

In case of a residential PV application, the output voltage of the nX converter is fed to the DC link of a subsequent inverter

stage. In case of a single-phase application, the inverter voltage has a peak voltage of 170 V. Thus, a DC link voltage of about 180 V should lead to minimum voltage stress on the inverter switches. A DC link voltage of 180V may not be possible using the fixed 8X converter for a given input voltage variation. In the case of fixed $8X$ converter, the peak voltage stress of the inverter will be determined as $8V_{inmax}$, where the input voltage varies between V_{inmin} and V_{inmax}. But, using the variable $\left(\frac{n}{m}\right)X$ converter, the voltage stress of the inverter switches can be significantly reduced. Table 1 illustrates a working example for this specific case. By just using three different voltage gains, the voltage stress on the inverter switches can be reduced by about 51%. Further optimization and selection of different values of n and m can provide a "finer" voltage gain variation leading to a lower voltage stress. But, the number of semiconductor devices used in the $\left(\frac{n}{m}X\right)$ converter will be increased. Thus, the selection of specific values of m and n need to be determined on a case by case basis based on application. For this paper, the working example provided in Table 1 serves as the specification for the variable 8X converter. Table 2 provides the input-output specifications for the variable $8X$ prototype. In order to keep the overall structure modular and highly efficient during normal operation, the design for the variable $8X$ converter prototype is based on the design of the compact fixed ratio nX converter in [7]. The switches and capacitors listed in [7] are also used in this prototype. But, as the voltage across the individual capacitors now varies based on the required output voltage gain, the variation in individual arm capacitances is computed based on the characteristic curves from the manufacturer. This is illustrated in Table 3.

Table 1: Gain-shift Specifications for a variable 8X converter

V_{in}	Voltage gain $\left(\frac{n}{m}\right)$	V_o
25 to 30 V	8	200 to 240 V
31 to 44 V	6	186 to 264 V
45 to 50 V	4	180 to 200 V

Table 2: Input-output specifications for a variable 8X prototype

Input DC voltage variation ,Vin	25 to 50 V (V_{inm} to V_{inmax})
Maximum load power (P_m)	1-kW
Switching frequency (f_s)	200 kHz

Table 3: Capacitance variation with change in voltage across the arm capacitors

Capacitance	Voltage Variation across capacitors	Capacitance Variation
C_{4a} and C_{4b}	90 to 132 V	6.6 to 7.92 μF
C_{3a} and C_{3b}	75 to 132 V	6.6 to 9.24 μF
C_{2a} and C_{2b}	50 to 100 V	7.92 to 10.56 μF
C_{1a} and c_{1b}	25 to 50 V	10.56 to 11.88 μF

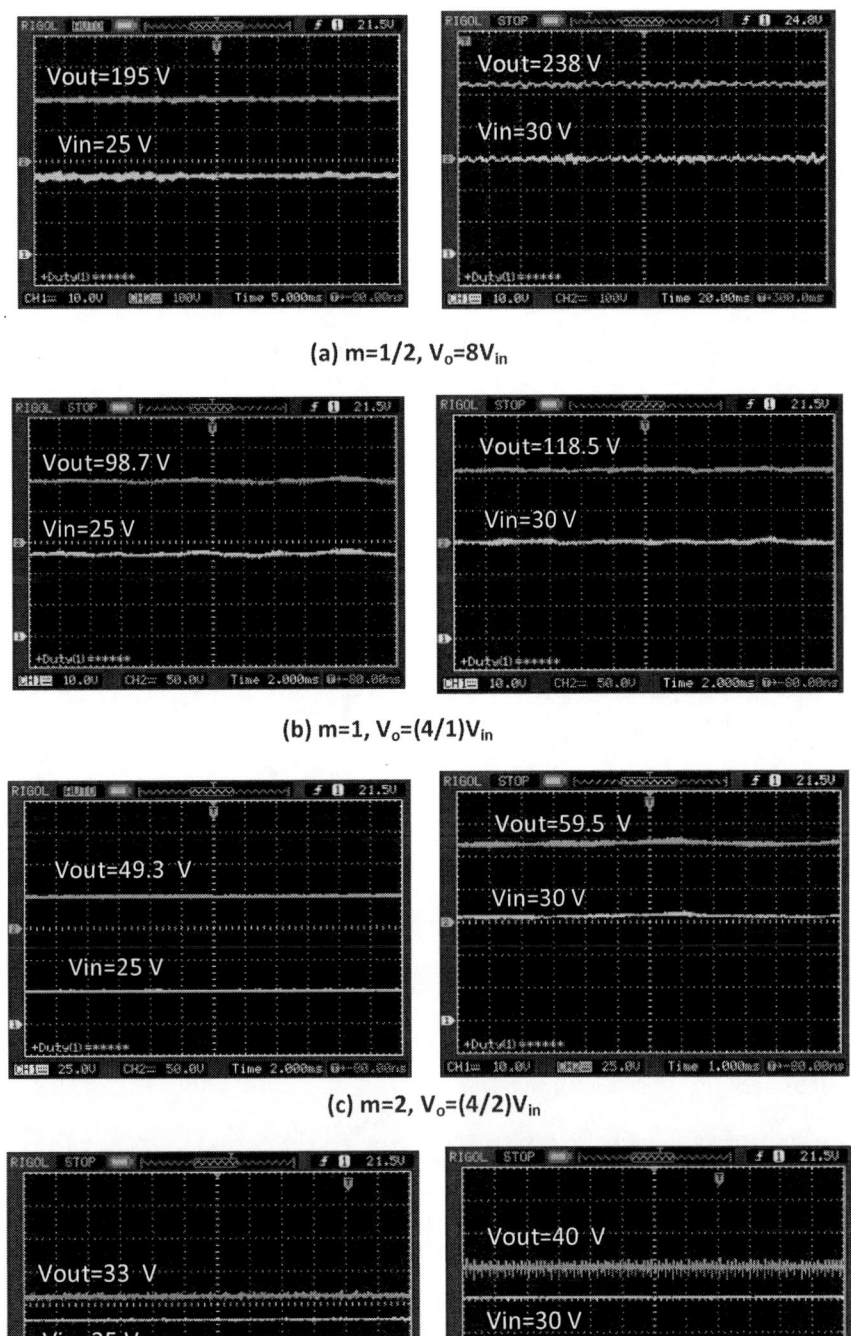

(a) m=1/2, $V_o=8V_{in}$

(b) m=1, $V_o=(4/1)V_{in}$

(c) m=2, $V_o=(4/2)V_{in}$

(d) m=3, $V_o=(4/3)V_{in}$

Figure 4: Fractional voltage gains using the $(\frac{n}{m})X$ Converter

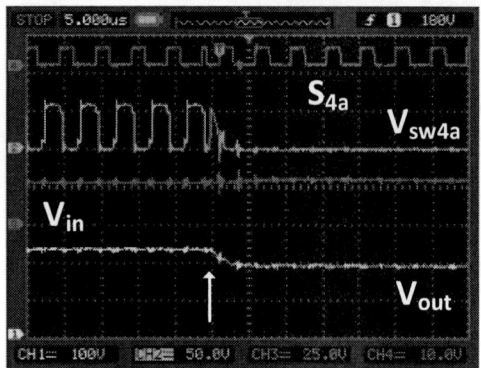

(a) Transition from 8X to 6X when V_{in} transitions from 30 V to 33 V

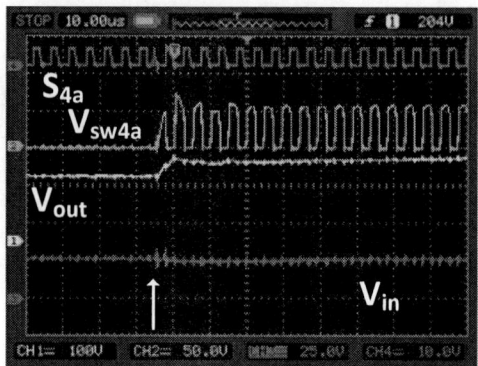

(b) Transition from 8X to 6X when V_{in} transitions from 31 V to 27 V

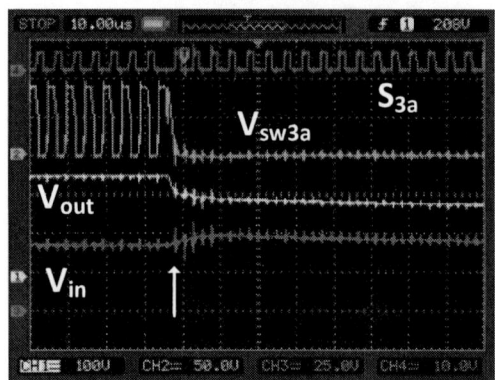

(d) Transition from 6X to 4X when V_{in} transitions from 44 V to 49 V

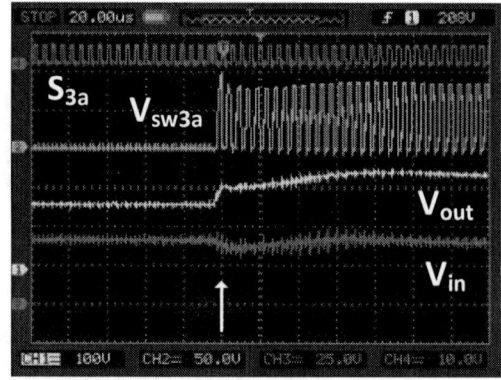

(d) Transition from 4X to 6X when V_{in} transitions from 48 V to 43 V

Figure 5: Dynamic voltage gain transitions using the $\left(\frac{n}{m}\right)X$ converter

In order to provide the variable input voltage (based on the PV output voltage range), a three-phase rectifier system is used to connect to the input of the variable $8X$ converter. In order to prevent power from being sent back to the source, a fast-recovery diode, D_1 is placed in series with the effective voltage source. The block schematic of the overall experimental setup is as shown in Figure 6.

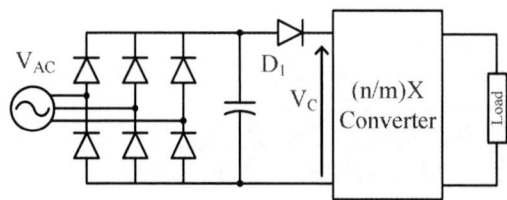

Figure 6: Block diagram of the overall experimental setup

V. EXPERIMENTAL RESULTS

The results presented are divided into two parts. At first, the ability to obtain fractional voltage gains by varying $'m'$ is demonstrated. Next, the experimental results demonstrating the variation in voltage gain in both gain "shift-up" and "shift-down" modes have been presented.

A. Fractional gains using the $\left(\frac{n}{m}X\right)$ converter

Different values of gain are obtained by varying the value of $'m'$ in the $(\frac{n}{m}X)$ converter. For a converter with $n = 4$, m can take values of $\frac{1}{2}$, 1, 2 and 3 to provide a boost ratio over 1. The results showing the input and output voltages for four different fractional gains of $(\frac{4}{\frac{1}{2}}), \frac{4}{1}, \frac{4}{2}$ and $\frac{4}{3}$ are presented in Figure 4. These results are for a load of 40 Ohms.

978-1-4673-9551-9/16 $31.00 © 2016 IEEE

B. Variable output voltage gains

The variable output gains that can be achieved for a fixed $m = \frac{1}{2}$ are presented below. These gains are dynamically varied depending on the magnitude of the input voltage as specified in Table 1. The load resistance in both these cases is fixed at $40\ ohms$. **Figure** 5 (a) illustrates the variation of variable gain from 8X to 6X and back. The white arrow indicates the time at which the transition is initiated. The voltage across switch S_{4a}, V_{Sw4a} depicts the nature of transient over-current in the loop consisting of S_{4a}. When the system reaches steady state, the voltage difference across the switch clamps the body diode. Hence, the voltage is only the diode forward voltage drop as can be seen in the figure. **Figure** 5 (b) illustrates the dynamic gain change from 6X to 4X and back when the voltage threshold is beyond that described in Table 1.

VI. POWER DISSIPATION ANALYSIS

Based on the experimental results, the method used to achieve variable output gain has been validated. But, the consequences of the over-current on the sizing of the switch and the efficiency of the overall system has to be determined.

A. Sizing of Switch to withstand transient over-current

The maximum energy dissipated by the switch is computed based on the transient over-current in the loop. Currently, the maximum current through the switch during the worst case gain variation is estimated based on the worst case equivalent circuit model described in Section 2. In this case, the worst case current would be during the variation from 6X to 4X or vice-versa. Figure 7 illustrates the model predicted peak current (Red) and the simulated transient-current (Blue) under no load condition. It can be seen that the peak current is almost accurately estimated except for the first cycle. It can also be seen that the transient over-current exists for close to $50\mu s$. But, the current is significantly damped out beyond $25\mu s$. Using the model predicted current to calculate the energy dissipation during each switching, the energy dissipated per switching can be determined as,

$$E_{sw} = \frac{V_d I_{pk}\left(t_r + t_f\right)}{6} \quad (7)$$

Where, I_{pk} is the peak current predicted by the model, t_r and is the rise time, t_f is the fall time of the MOSFET, and V_d is the peak voltage across the switch respectively.

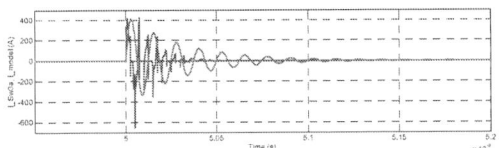

Figure 7: Worst case peak current based on equivalent circuit and simulation

For this case, $E_{sw} = 0.58\ mJ$ based on the manufacturers datasheet for the switching device. Also, using an average current for the overall transient duration, the energy dissipated due to conduction is $E_c = 1.2\ mJ$ for the entire transient

period. Thus, the overall worst case energy dissipated by the switch at the end of five switching cycles is determined to be,

$$E_{total} = 4.1\ mJ$$

Based on thermal impedance curves for the MOSFETs used, the worst case temperature rise of the MOSFET for the periodicity of such a high energy pulse lasting for $50\ \mu s$ can be calculated. For the MOSFET used in the converter arm in this case (IPB036N12N3G), the maximum temperature rise for a periodicity of $1ms$ is about 3.2 Celcius. But, this is the worst case temperature rise and in a real system, it is bound to be a few orders of magnitude lesser. Based on the current operating temperature of the switch, this temperature rise is acceptable. But, the periodicity of the high energy pulse is determined by the overall loss in efficiency it causes to the system. This is calculated in the following discussion.

B. Energy Analysis

In the case of gain variation from 6X to 4X when the input voltage is 44 V (which is the worst case in terms of energy), the overall energy lost by each capacitor in arms 3 and 4 are given using the following calculation.

$$E_c = \frac{1}{2}C(3*44)^2 - \frac{1}{2}C(2*44)^2 = 49.3\ mJ$$

Considering an average capacitance of $10.2\mu F$. There are four such capacitors that undergo this transition in voltage. Hence, the total energy that is lost from all the capacitor is 197.2 mJ. By measuring the rise in input voltage across, V_c in Figure 6, the energy stored in the capacitor during the transient can be calculated.

Based on simulation results shown in Figure 8., the energy stored by the input capacitor at no load is determined as, $E_{store} = 123.5\ mJ$. The input capacitance is $450\mu F$.

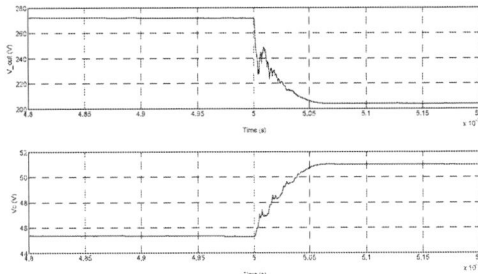

Figure 8: Energy stored at the input during gain variation

Hence, the overall energy dissipated by the converter, $E_{diss} = 73.7\ mJ$. For a $50\mu s$ transient period, this amounts to a power loss of about $1.4kW$. But, for a gain variation every one second, the $50\mu s$ energy pulse also exists only every one second. The average P_{loss} due to mode change in this case is only about 70 mW. The 1 kW (nX) converter functioning at full load at fixed gain, is about 98% [7]. But, with a mode change every 1 second, this efficiency is still predicted to be close to 98%. It must be noted that this efficiency calculation is based on theoretical model and calculations on a $(\frac{\frac{4}{1}}{2}X)$ converter. Efficient design can significantly increase the allowed periodicity of the gain variation. By, increasing the value of n and m, the energy displaced from the capacitors that causes the additional power loss can be reduced significantly. Thus, for any given application, the optimum calculation of m and n has the potential to allow frequent variation in gain without compromising efficiency. The analysis in this paper can be extended to any $\left(\frac{n}{m}\right)X$ converter.

VII. Conclusion

An inductor-less switched capacitor based (n/m)X converter has been proposed in this paper. By selecting suitable values of m and n, the converter can produce any required fractional voltage gain. The proposed (n/m)X converter has also been shown capable of providing dynamically varying voltage gains. A 1-kW prototype of the variable and fractional output gain (n/m)X converter has been built and tested. The experimental waveforms validate the proposed methods used to achieve the variable and fractional gains. A theoretical analysis on the efficiency and power dissipation of such a converter has also been presented.

References

[1] Meynard, T.A.; Foch, H., "Multi-level conversion: high voltage choppers and voltage-source inverters," Power Electronics Specialists Conference, 1992. PESC '92 Record., 23rd Annual IEEE , vol., no., pp.397,403 vol.1, 29 Jun-3 Jul 1992J.

[2] Ke Jin; Mengxiong Yang; Xinbo Ruan; Min Xu, "Three-Level Bidirectional Converter for Fuel- Cell/Battery Hybrid Power System," Industrial Electronics, IEEE Transactions on , vol.57, no.6, pp.1976,1986, June 2010 doi: 10.1109/TIE.2009.2031197.

[3] Wei Qian; Peng, F.Z.; Tolbert, L.M., "Development of a 55 kW 3X dc-dc converter for HEV systems," Vehicle Power and Propulsion Conference, 2009. VPPC '09. IEEE , vol., no., pp.433,439, 7-10 Sept. 2009.

[4] Khan, F.H.; Tolbert, L.M., "A Multilevel Modular Capacitor-Clamped DC–DC Converter," Industry Applications, IEEE Transactions on , vol.43, no.6, pp.1628,1638, Nov.-dec. 2007.

[5] Dickson, J.F., "On-chip high-voltage generation in MNOS integrated circuits using an improved voltage multiplier technique," in Solid-State Circuits, IEEE Journal of , vol.11, no.3, pp.374-378, Jun 1976.

[6] Wei Qian; Dong Cao; Cintron-Rivera, J.G.; Gebben, M.; Wey, D.; Fang Zheng Peng, "A Switched-Capacitor DC–DC Converter With High Voltage Gain and Reduced Component Rating and Count," Industry Applications, IEEE Transactions on , vol.48, no.4, pp.1397,1406, July-Aug. 2012

[7] Fang Zheng Peng; Gebben, M.L.; Baoming Ge, "A compact nX DC-DC converter for photovoltaic power systems," Energy Conversion Congress and Exposition (ECCE), 2013 IEEE , vol., no., pp.4780,4784, 15-19 Sept. 2013

[8] Dong Cao; Fang Zheng Peng, "Zero-Current-Switching Multilevel Modular Switched-Capacitor DC–DC Converter," Industry Applications, IEEE Transactions on , vol.46, no.6, pp.2536,2544, Nov.-Dec. 2010

[9] Alam, M.K.; Khan, F.H., "A high-efficiency modular switched-capacitor converter with continuously variable conversion ratio," in Control and Modeling for Power Electronics (COMPEL), 2012 IEEE 13th Workshop on , vol., no., pp.1-5, 10-13 June 2012.

Multi-Mode Quasi-Z-Source Series Resonant DC/DC Converter for Wide Input Voltage Range Applications

Dmitri Vinnikov[*], Andrii Chub[†], Indrek Roasto[*] and Liisa Liivik[†]
dmitri@ubiksolutions.eu, andrii.chub@ttu.ee
[*]Ubik Solutions LLC, Regati pst 1, 11911 Tallinn, Estonia
[†]Department of Electrical Engineering, Tallinn University of Technology, Ehitajate tee 5, 19086 Tallinn, Estonia

Abstract—This paper examines the quasi-Z-source series resonant DC/DC converter as a candidate topology for the PV module integrated converter (MIC) with an extended input voltage regulation range. The converter features multi-mode operation by combining the shoot-through pulse-width modulation and ordinary phase-shift modulation to realize the boost and buck operating modes, respectively. Our experiments confirmed that the proposed MIC is capable of ensuring the ripple-free output voltage of 400 V within the input voltage range of 10...60 V. In contrast to the traditional series resonant converter, the MIC proposed is characterized by fixed-frequency operation over the entire range of input voltage variations. It requires only generic semiconductor devices and thus is a cost-effective solution.

Keywords—resonant converter; DC/DC converter; quasi-Z-source converter; solar photovoltaic; renewable energy, module integrated converter (MIC)

I. INTRODUCTION

Growth of residential solar photovoltaic (PV) installations has been accelerated still further by the decreasing cost of PV modules as well as introduction of feed-in tariff systems for households [1]. Meanwhile, virtually all PV systems installed have connections to the grid [2]. The concept of single PV module integration became popular in residential and smart grid applications due to its high scalability and plug-and-play functionality. Microinverters were proposed for direct integration of a single PV module into the distribution grid in [3]. This approach is also known as an "AC module" concept [4].

Numerous designs of the PV microinverter have been reported recently. Commonly, the simplest approach utilizes a nonisolated, i.e. transformerless topology [5], [6]. However, this approach is feasible only for PV module integration into the low voltage (LV) distribution grids, like in the USA or Japan. Higher voltages, like 230 V in the EU, usually require isolation transformer to the step-up voltage of a single PV module to the level acceptable for the distribution grid [7]. Galvanic isolation can be used both in single stage (DC/AC) and two-stage (DC/DC/AC) microinverters [8]. In the latter approach, in the first stage, a DC/DC converter is used for maximum power point tracking (MPPT) and voltage level matching. A DC/AC converter in the second stage injects power into the distribution grid from the stabilized high voltage DC link. The two-stage microinverters enable

numerous features, such as simple energy storage integration, low frequency ripple mitigation in single-phase systems, etc. [9], [10]. Hence, it is widely adopted in industrial products.

PV module integrated converters (MICs) with possibilities of input voltage regulation in a wide range can be used as the first stage of a microinverter. Moreover, MIC suits also for the connection of a PV module to a common DC bus that supplies a grid side inverter. In this case, the inverter harvests energy from all PV modules connected to the DC bus. Hence, MIC is a versatile solution for the solar PV energy harvesting that can be adopted in residential and commercial solar PV energy systems [11]. Galvanic isolation is generally used for voltage level matching, elimination of leakage currents [12], and for safety purposes [2], [9].

As an approach, galvanically isolated current-fed DC/DC converters have proven preferable over their voltage-fed counterparts for applications with high input current and low input voltage [13]. Numerous soft-switching current-fed topologies have been proposed for PV MICs. At the same time, many of them require additional components, like active switches at the output side to control leakage inductance currents [14] - [15], or to perform the voltage step-up using leakage inductance [16]. Utilization of resonance for soft-switching of semiconductors is a modern trend in the field of PV MICs [2]. Topologies that combine current-fed input and resonant switching show superior results in PV and other renewable applications [16]-[19].

The quasi-Z-source (qZS) converters are another promising solution for the PV MICs. They also provide low input current ripple like current-fed topologies while featuring an extended input voltage range [20]. Moreover, new approaches used in the current-fed topologies can be adopted to enhance qZS converters for PV applications. In this paper, the qZS series resonant DC/DC converter is first examined as a candidate topology for a PV module integrated converter, where the wide input voltage regulation range is especially important for proper realization of a global maximum power point (MPP) tracking under PV module shaded conditions.

II. QUASI-Z-SOURCE SERIES RESONANT DC/DC CONVERTER

The quasi-Z-source series resonant DC/DC converter (qZSSRC, Fig. 1) [21]-[23] is one of the topological variations of a recently introduced galvanically isolated quasi-Z-source

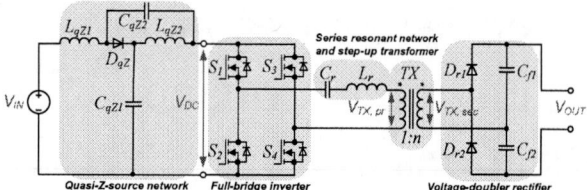

Fig. 1. Generalized topology of the series resonant quasi-Z-source DC/DC converter (qZSSRC).

DC/DC converter (qZSC) [24]-[26]. It consists of a quasi-Z-source network, a full-bridge switching stage, a series resonant tank formed by the capacitor C_r and the inductor L_r, a step-up isolation transformer, and a voltage-doubler rectifier (VDR).

Similar to the baseline qZSC topology, the qZSSRC has two main operating modes: normal (or non-shoot-through) and boost (or shoot-through) mode [24]-[27]. The resulting advantage of the qZSSRC over the qZSC is soft-switching operation. In a normal mode, it features the zero-current switching (ZCS) turn-ON and zero-voltage switching (ZVS) combined with nearly ZCS turn-OFF of the inverter switches, while VDR diodes operate with full ZCS. ZCS turn-OFF of the inverter switches is not easy to maintain in a normal mode due to variations of the resonant tank parameters caused by temperature changes and ageing. However, ZVS is possible at proper dead time to ensure charging and discharging of the parasitic output capacitances of the inverter switches with magnetizing current of the transformer. Reduced switching losses of the inverter along with the full ZCS operation of the VDR diodes can be achieved in the boost mode. The resonant capacitor C_r connected in series with the transformer primary winding eliminates a threat of transformer saturation.

Further, the buck mode is now also available in qZSSRC and provides an additional advantage gained by the implementation of the resonant network [28]. Therefore, the qZSSRC has an extended input voltage regulation range, which could be essential in such demanding applications as power conditioners for renewable energy sources.

III. MULTI-MODE OPERATION OF THE QZSSRC

To extend the input voltage regulation range without serious efficiency penalties, the proposed MIC uses a multi-mode operation. Due to the unique properties of the quasi-Z-source network, the qZSSRC is capable of combining the shoot-through pulse-width modulation (PWM) and ordinary phase-shift modulation (PSM) for the realization of the boost and buck operating modes, respectively. Moreover, in the boundary between these modes, the converter operates as a pure series resonant converter at the resonant frequency. Fig. 2 shows an example of the idealized control variables of the multi-mode qZSSRC operating within the input voltage range from $0.5V_{IN,nom}$ to $1.5V_{IN,nom}$.

During the design routine of the multi-mode qZSSRC, the nominal value of the input voltage ($V_{IN,nom}$) should be selected first. Typically, it corresponds to the MPP voltage of the PV module. In this operating point, the duty cycle of the inverter switches is close to 0.5 after the dead time deduction.

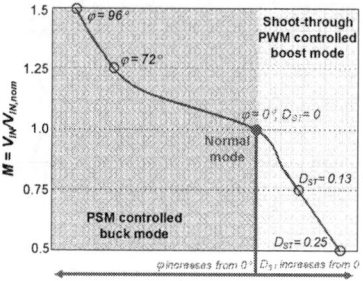

Fig. 2. An example of operating modes and idealized control variables of the multi-mode qZSSRC.

Since the switching frequency is equal to the resonant frequency ($f_{SW}=f_r$), the current of the inductor L_r and the voltage of the capacitor C_r fully resonate, which results in almost pure sinusoidal waveforms (Fig. 3a). Thus, the inverter switches and the VDR diodes can be turned ON and OFF at perfect zero current, and in this operating point, the proposed MIC will feature its maximum possible efficiency. However, ZCS turn-OFF of the inverter switches is not sustainable. The dead time can be implemented in the control of the inverter switches to ensure robust ZVS. In this case, the transformer magnetizing inductance must be dimensioned properly considering the parasitic output capacitance of the inverter switches to achieve the desired dead time. Neglecting losses in the components, the resulting DC voltage gain of the qZSSRC in the normal mode could be expressed as:

$$G_{normal} = \frac{V_{OUT}}{V_{IN,nom}} = 2 \cdot n, \qquad (1)$$

where V_{OUT} is the output voltage of the converter, and n is the transformer turns ratio.

If the input voltage drops below the nominal level, the converter starts to operate in the boost mode similarly to the traditional qZSC and the output voltage is controlled by the variation of the shoot-through duty cycle D_{ST} [27]. The switching frequency of the inverter switches f_{SW} in the boost mode remains fixed to the resonant frequency f_r. Fig. 3b shows the steady-state waveforms of the converter operating in the boost mode. Full ZCS operation of the VDR diodes is evident. Neglecting losses in the components, the resulting DC voltage gain of the qZSSRC in this mode could be expressed as:

$$G_{boost} = \frac{V_{OUT}}{V_{IN}} = \frac{2 \cdot n}{(1 - 2 \cdot D_{ST})}. \qquad (2)$$

When the input voltage is higher than the predefined nominal value, the converter starts to operate in the buck mode. In this mode, the converter could be regarded as a series resonant converter controlled by the ordinary phase-shift modulation. Fig. 3c shows the steady-state waveforms of the converter in the buck mode when the output voltage is controlled by the phase shift angle φ between the two inverter legs at the resonant frequency ($f_{SW} = f_r$) [28][29]. The idealized DC voltage gain of the qZSSRC in the buck mode could be expressed by (3) [28].

978-1-4673-9551-9/16 $31.00 © 2016 IEEE

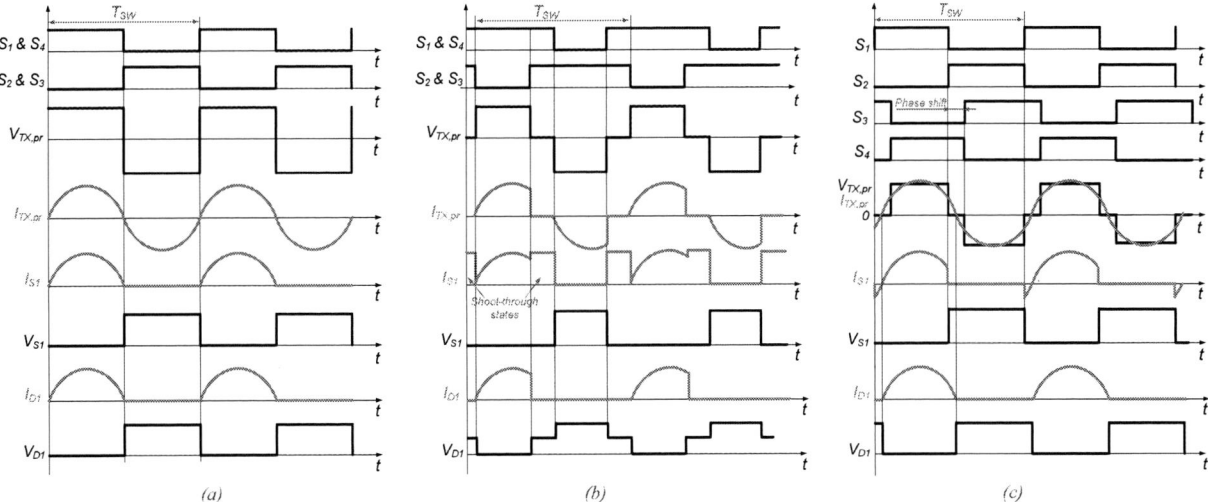

Fig. 3. Idealized steady-state waveforms of the multi-mode qZSSRC in the normal (a), boost (b), and buck (c) mode.

$$G_{buck} = \frac{V_{OUT}}{V_{IN}} = n \cdot \sqrt{2 \cdot (1 + \cos\varphi)} . \qquad (3)$$

As Fig. 3c shows, the current through the isolation transformer is assumed to be continuous over the entire operating range of the converter in the buck mode. The condition for continuous current operation results from the parameters of the resonant tank and could generally be defined by [29]:

$$\frac{\pi^3 \cdot P \cdot n^2}{8 \cdot V_{OUT}^2} \sqrt{\frac{L_r}{C_r}} > 1 , \qquad (4)$$

where P is the operating power of the converter, L_r is the resonant inductance, and C_r is the resonant capacitance.

IV. EXPERIMENTAL VERIFICATION

A. Case Study System and Converter Design

For the experimental verification of the proposed concept, the 250 W prototype of the wide input voltage range MIC based on the qZSSRC topology was developed (Fig. 4). General specifications of the prototype are listed in Table I. The proposed converter requires no specific semiconductor devices and thus it can make use of low-cost generic Si MOSFETs at the input and industrially proven SiC Schottky barrier diodes at the output. The converter was designed to operate within the input voltage range of 10...60 V and with the power profile shown in Fig. 5a. This power profile was specially synthesized to simulate the operation conditions of MIC powered by 60-cell PV modules. In such PV modules, at standard test conditions (STC), the maximum power point (250 W) typically lies in the voltage range of 30...35 V and current of roughly 8 A. This data was mixed with the typical design constraints of the MIC, like maximum input current of 10 A. Hence, the power profile can be separated into three regions. In the first region, from 10 V to 25 V, the MIC operates with a maximum input current of 10 A. Power rises linearly with the input voltage. In the second region, from

25 V to 35 V, the MIC has to handle the full power, since the maximum power point can be anywhere in this region, depending on the PV module type and its operating conditions. The third region was designed to show the start-up possibility of the converter from any open-circuit voltage within the range of the input voltages from 35 V to 60 V. Extremely high open-circuit voltage can be observed at the morning start-up of the MIC in regions with cold climate. However, the output power and duration of operation at high input voltage are small. Thus, the operating power is decreasing from 35 V to 60 V in the third region. Generally, the proposed power profile allows the operation of the MIC with 60-cell PV module to be assessed at different weather conditions, i.e. also at worst-case scenarios. Furthermore, the proposed power profile covers the operating points where the PV module is under partial shading, when the global MPP corresponds to the low voltage. Usually, 60-cell PV module includes three bypass diodes. Hence, changes of the MPP voltage could be severe when the panel is shaded by fallen leafs of trees, which is a common problem in residential PV installations.

The turns ratio of the isolation transformer was selected such that at the nominal input voltage of 35 V, the converter could provide 400 V at the output (5).

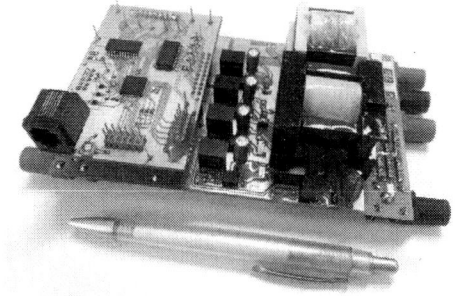

Fig. 4. Experimental prototype of the 250 W wide input voltage range MIC based on the qZSSRC topology.

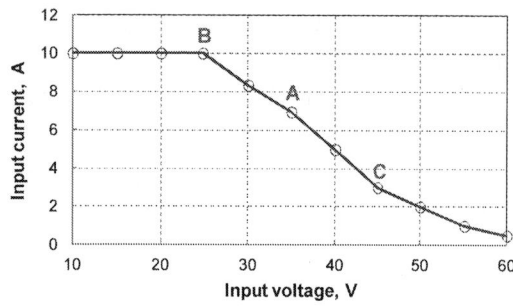

(a)

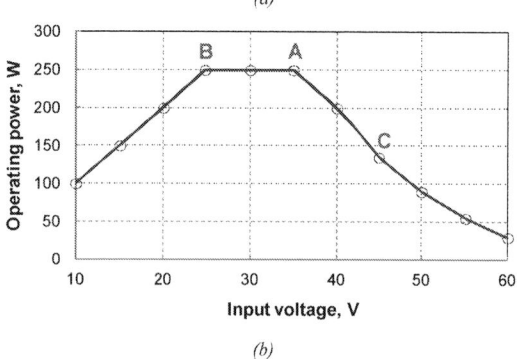

(b)

Fig. 5. Input current vs. input voltage (a) and power profile (b) of the experimental prototype.

$$n = \frac{V_{OUT}}{2 \cdot V_{IN(nom)}} = \frac{400}{2 \cdot 35} \approx 5.7 . \qquad (5)$$

The switching frequency remained fixed to the resonant frequency ($f_{SW}=f_r=100$ kHz) in all operating modes, which is one of the distinguishing features of the discussed converter. To improve the power density of the experimental prototype, the leakage inductance of the primary winding of the isolation transformer was used as a resonant inductor. According to (4), it could be found that given series resonant network (L_r=1.8 uH, C_r=1.4 uF) will feature the discontinuous current within the entire range of the buck mode of the converter.

Since the discussed converter has maximum efficiency in the normal mode, special attention should be paid to the proper selection of the dead time. During the dead time, the parasitic output capacitances of the MOSFETs are charged and discharged by the magnetizing current of the transformer, which results in the robust full ZVS operation of the inverter switches. The minimal value of the dead time could be found by:

$$T_D \geq 8 \cdot L_m \cdot f_{SW} \cdot C_{oss} , \qquad (6)$$

where L_m is the magnetizing inductance of the transformer reflected to the primary side and C_{oss} is the parasitic output capacitances of the MOSFETs. For the given case study of magnetizing inductance (L_m=33 uH), the dead time was set to 50 ns, which was sufficient to ensure robust ZVS operation of the inverter switches over a wide input voltage range.

TABLE I. GENERAL SPECIFICATIONS OF THE EXPERIMENTAL PROTOTYPE

Operating parameters	Value/type
Input voltage range, V_{IN}	10…60 V
Nominal input voltage, $V_{IN,nom}$	35 V
Maximum input current, I_{IN}	10 A
Output voltage, V_{OUT}	400 V
Switching frequency, f_{SW}	100 kHz
Components	
$S_1...S_4$	Vishay Si4190ADY
D_{qZ}	Vishay V60D100C
D_{r1}, D_{r2}	CREE C3D02060E
L_{qZ1}, L_{qZ2}	22 µH
C_{qZ1}, C_{qZ2}	26.4 µF
C_{f1}, C_{f2}	2.2 µF
L_r	1.8 µH
C_r	1.4 µF
n	5.7

B. Realization of the Control Principle

A dual modulation technique allows the converter to operate over a wide input voltage range of 10 V to 60 V. Normal mode corresponds to the input voltage of 35 V. The input voltage below the nominal value must be stepped up at the input stage using shoot-through PWM control in order to achieve nominal operating voltage of the transformer. At the same time, the input voltage above the nominal value requires PSM control to be used in order to stabilize the output voltage at the necessary level.

Shoot-through states are generated by increasing the duty cycle of active states. This causes the active states of bottom (S_2, S_4) and top (S_1, S_3) side switches to overlap with each other, as shown in Fig. 6a. During the shoot-through states, all four switches of the qZSSRC are conducting and the current through the inverter switches reaches its maximum (Fig. 3b). Therefore, the operating period of the converter in this control method consists of a combination of the shoot-through and the active states.

In PSM, the voltage is reduced by increasing the phase shift between the control signals of the inverter legs, e.g. between S_1 and S_4 (Fig. 6c). It results in a reduced duration of the active state at the isolation transformer. In addition, zero states, when only top or bottom transistors are conducting, are created.

There is a special operation mode between the boost and buck modes called the normal mode (Fig. 6b). In the normal mode, the shoot-through and phase shift are zero, i.e. the converter works as a typical voltage-fed full-bridge converter. The active state duty cycle in this state reaches its maximum of roughly 0.5, from which the dead time is subtracted. Dead time generation is essential to achieve robust ZVS operation of the inverter switches in the normal and buck modes.

The transient between the three operation modes should be smooth. The generalized block diagram of the implemented control algorithm is presented in Fig. 7. The regulation error is compensated by a PI regulator. The regulator output is connected to a special saturation block, which has two functions. It sets positive and negative saturation limits and

978-1-4673-9551-9/16 $31.00 © 2016 IEEE

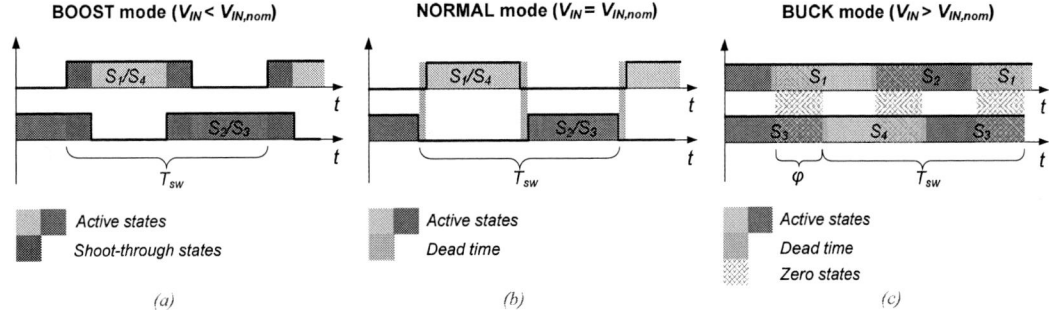

Fig. 6. Modulation methods in the boost (a), normal (b), and buck (c) mode.

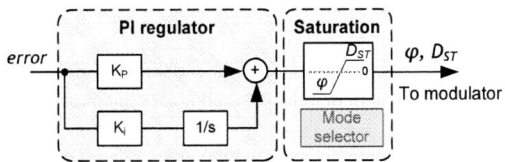

Fig. 7. Regulator and operation mode selector.

defines the operation mode. In the case of negative or positive regulator output, correspondingly the buck or the boost mode is activated. Zero output of the regulator corresponds to the normal mode. Thus, the output of this block is whether the phase shift or the shoot-through duty cycle. In this way, all three operation modes can be controlled by only one regulator.

The control system was implemented in the STM32F334 microcontroller, which is based on the cortex M4 core and has an integrated floating point unit. The internal 12 bit ADC was used for measurements. Depending on the input voltage, the qZSSRC automatically switches between the three operating modes (Fig. 8).

C. Experimental Results

Steady-state operating waveforms of the experimental multi-mode qZSSRC are presented in Figs. 8 and 9. To acquire the operating waveforms, the digital phosphor oscilloscope Tektronix DPO7254 along with the Rogowski coil current probe PEM CWTUM/015/R and high-voltage differential voltage probes Tektronix P5205A were used. The converter was supplied by the PV panel simulator Keysight E4360 and loaded by the electronic DC load Chroma 63204. Finally, the efficiency was calculated as a ratio between the input and output power of the converter, which were measured by means of the precision power analyzers Tektronix PA1000.

First, the converter was tested in the normal mode with an operating power of 250 W (point A in Fig. 5). The switching frequency was equal to the resonant frequency and the current through the isolation transformer was sinusoidal (Fig. 9a). As a result, the inverter switches and VDR diodes were all operating under ZCS, while the implemented dead time ensured the ZVS in the case of the resonance frequency variations.

Steady-state waveforms of the qZSSRC operating in the boost mode are presented in Fig. 9b. The test conditions were set according to point B in Fig. 5 (operating power 250 W

with a 25 V at the input). To step-up the input voltage, the shoot-through duty cycle was set to 0.163. It is seen that the resonance begins immediately after the shoot-through when the stepped-up input voltage is applied to the primary winding of the isolation transformer. However, with the appearance of the next shoot-through state, the resonance is ended early and the current in the resonant inductor gradually decreases to zero. It is seen from the experimental waveforms that in the boost mode, the semiconductors of the inverter bridge of the qZSSRC feature partial soft-switching – near ZCS turn-ON only, while the VDR diodes operate with ZCS. The ZCS behavior of the switches can be explained by the influence of the resonant inductor current on the switches. At the beginning of the first shoot-through state, this current is yet high and thus it is subtracted from the current of the switch S_1 and added to the current of the switch S_3. Hence, switch S_1 is turned ON at the ZCS.

At the beginning of the next shoot-through state, the resonant inductor current was added to the current of the switch S_1, while subtracted from S_3, which resulted in the ZCS turn-ON of the switch S_3. Hence, the current of the resonant inductor caused current overshoot in the switch S_1 at the beginning of the second shoot-through state.

Finally, the qZSSRC was tested in the buck mode with 45 V input and an operating power of 135 W (point C in Fig. 5). To step-down the input voltage, the phase shift angle between the two inverter legs was set to 118°. It is seen from the operating waveforms (Fig. 9c) that the current through the resonant network is discontinuous. As a result, the inverter switches and VDR diodes were all operating under the full ZCS. However, increased peak and rms currents cause higher conduction losses. Buck mode in a series resonant converter is also well known for energy circulation. Both of these factors result in an efficiency lower in the buck mode than in the boost mode.

Fig. 10a shows the experimentally obtained control variables of the wide input voltage range MIC based on the multi-mode qZSSRC topology. It was confirmed that the proposed converter is capable of ensuring the ripple free 400 V output voltage within the entire range of the input voltage variations. Fig. 10b shows the efficiency curve of the experimental MIC. As it was predicted theoretically, the peak efficiency (close to 97%) was achieved at the nominal input voltage, when the converter operates in the normal mode and features the full ZCS operation.

978-1-4673-9551-9/16 $31.00 © 2016 IEEE

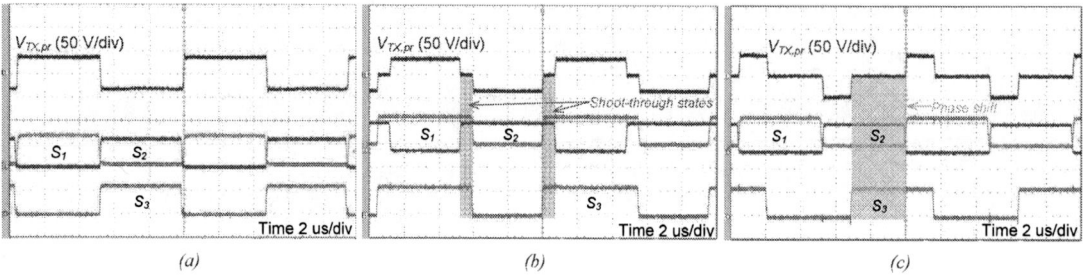

Fig. 8. Measured control signals of the inverter switches and transformer voltage of the multi-mode qZSSRC in the normal (a), boost (b), and buck (c) mode.

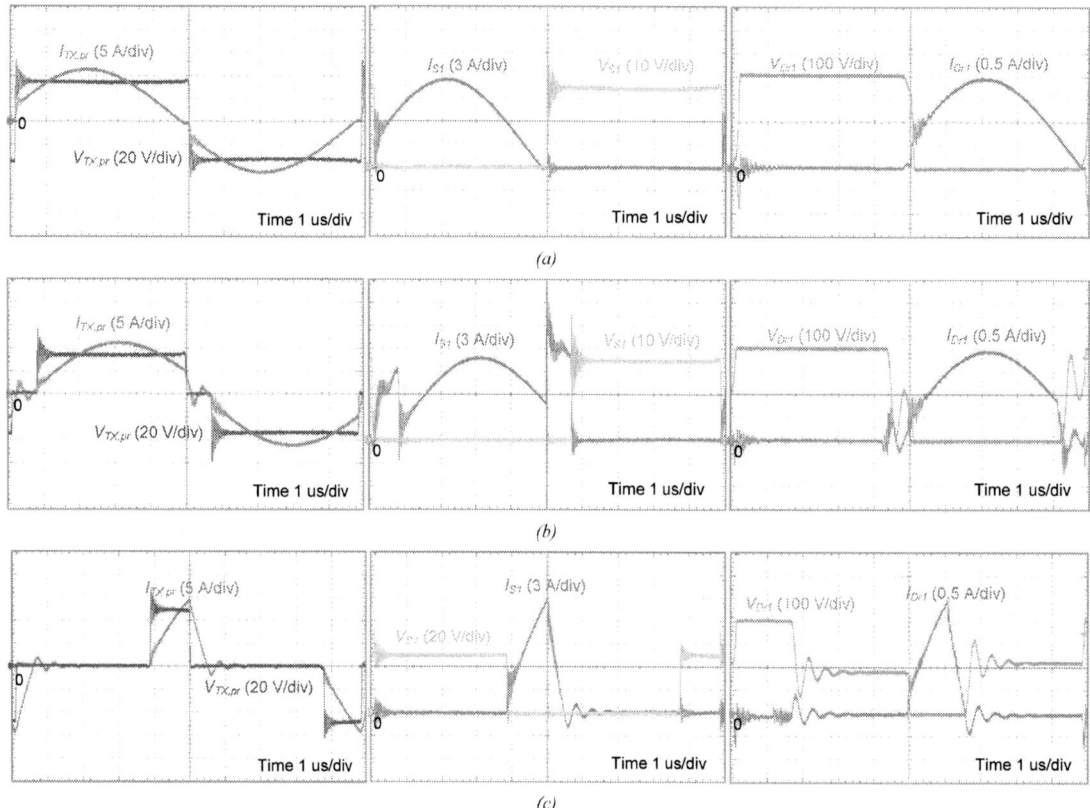

Fig. 9. Experimental voltage and current waveforms of the isolation transformer, inverter switch and VDR diode of the multi-mode qZSSRC in the normal (a), boost (b), and buck (c) mode.

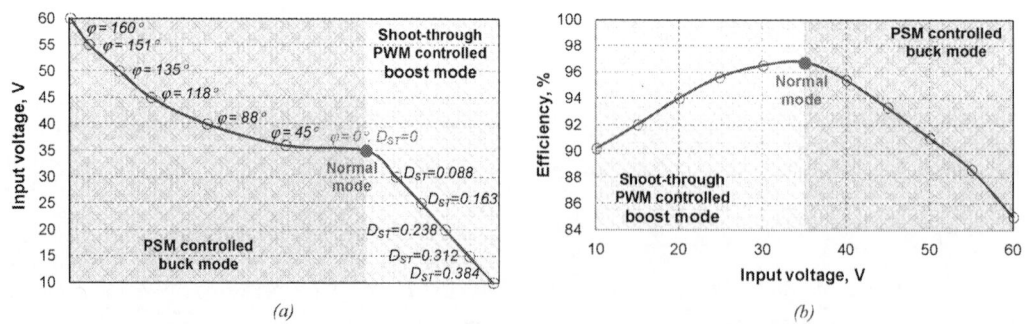

Fig. 10. Experimental control variables (a) and measured efficiency (b) of the multi-mode qZSSRC.

V. Conclusions

In this paper, a novel multi-mode quasi-Z-source series resonant DC-DC converter was examined as a candidate topology for the PV module integrated converter. To ensure a wide input voltage and load regulation, the proposed converter utilizes three different operating modes (normal, boost and buck) in a single switching stage. It was experimentally verified that the converter is capable of ensuring the 400 V output voltage within the input voltage range of 10...60 V. Moreover, the converter features the ZCS of the VDR diodes over the entire operating range, and depending on the operating mode, ZCS and/or near-ZCS of the primary side switches. The peak efficiency close to 97% was achieved at the nominal input voltage, when the converter operates in the normal mode and features the full-ZCS operation in the given case.

Acknowledgements

The authors are grateful to Ubik Solutions LLC for the support of this research. This research work was also co-financed by Estonian Ministry of Education and Research (project SF0140016s11) and Estonian Research Council (grant PUT744).

References

[1] J. Seel, G.L. Barbose, and R.H. Wiser, "An analysis of residential PV system price differences between the United States and Germany," Energy Policy, vol. 69, pp. 216-226, June 2014.

[2] S. Kouro, J.I. Leon, D. Vinnikov, and L.G. Franquelo, "Grid-connected photovoltaic systems: an overview of recent research and emerging PV converter technology," IEEE Ind. Electron. Mag., vol. 9, no. 1, pp. 47-61, March 2015.

[3] H.A. Sher, and K.E. Addoweesh, "Micro-inverters – Promising solutions in solar photovoltaics," Energy for Sustainable Development, vol. 16, no. 4, pp. 389-400, December 2012.

[4] J.J. Bzura "The ac module: An overview and update on self-contained modular PV systems," Proc. 2010 IEEE Power and Energy Soc. General Meeting, pp. 1-3, 25-29 July 2010.

[5] D. Meneses, F. Blaabjerg, O. Garcia, and J.A. Cobos, "Review and comparison of step-up transformerless topologies for photovoltaic AC-module application," IEEE Trans. Power Electron., vol. 28, no. 6, pp. 2649-2663, June 2013.

[6] T. Kerekes, R. Teodorescu, P. Rodriguez, G. Vazquez, and E. Aldabas, "A new high-efficiency single-phase transformerless PV inverter topology," IEEE Trans. Ind. Electron., vol. 58, no. 1, pp. 184-191, January 2011.

[7] E. Romero-Cadaval, G. Spagnuolo, L. Franquelo, C.A. Ramos-Paja, T. Suntio, and W.M. Xiao, "Grid-connected photovoltaic generation plants: components and operation,: IEEE Ind. Electron. Mag., vol. 7, no. 3, pp. 6-20, Sept. 2013.

[8] M.H. Todorovic, Fengfeng Tao, Rui Zhou, R. Steigerwald, M. Agamy, Yan Jiang, L. Garces, M. Schutten, and D. Marabell, "A multi-objective study for down selection of a micro-inverter topology for residential applications," Proc. 40th Photovoltaic Specialist Conf., pp. 3108-3113, 8-13 June 2014.

[9] Y. Xue, L. Chang, S.B. Kjær, J. Bordonau, and T. Shimizu, "Topologies of single-phase inverters for small distributed power generators: an overview," IEEE Trans. Power Electron., vol. 19, no. 5, pp. 1305-1314, September 2004.

[10] H. Hu, S. Harb, N. Kutkut, I. Batarseh, and Z.J. Shen, "A review of power decoupling techniques for microinverters with three different decoupling capacitor locations in PV systems," IEEE Trans. Power Electron., vol. 28, no. 6, pp. 2711-2726, June 2013.

[11] M. Kasper, D. Bortis, and J.W. Kolar, "Classification and comparative evaluation of PV panel-integrated DC–DC converter concepts," IEEE Trans. Power Electron., vol. 29, no. 5, pp. 2511-2526, May 2014.

[12] O. Lopez, R. Teodorescu, F. Freijedo, anf J. Doval-Gandoy, "Leakage current evaluation of a singlephase transformerless PV inverter connected to the grid," Proc. 22nd Annu. IEEE Appl. Power Electron. Conf. (APEC'2007), pp. 907-912, February 25 -March 1 2007.

[13] A.K. Rathore, and U. Prasanna, "Comparison of soft-switching voltage-fed and current-fed bi-directional isolated DC/DC converters for fuel cell vehicles," Proc. 2012 IEEE Int. Symp. Ind. Electron., pp. 252-257, 28-31 May 2012.

[14] Pan Xuewei, and A.K. Rathore, "Current-fed soft-switching push–pull front-end converter-based bidirectional inverter for residential photovoltaic power system," IEEE Trans. Power Electron., vol. 29, no. 11, pp. 6041-6051, November 2014.

[15] A. Blinov, D. Vinnikov, and V. Ivakhno, "Full soft-switching high step-up DC-DC converter for photovoltaic applications," Proc. 16th European Conf. Power Electron. Applicat. (EPE'14-ECCE Europe), pp. 1-7, 26-28 August 2014.

[16] T. LaBella, Wensong Yu, J.-S. Lai, M. Senesky, and D. Anderson, "A bidirectional-switch-based wide-input range high-efficiency isolated resonant converter for photovoltaic applications," IEEE Trans. Power Electron., vol. 29, no. 7, pp. 3473-3484, July 2014.

[17] X. Sun, Y. Shen, Y. Zhu, and X. Guo, "Interleaved boost-integrated LLC resonant converter with fixed-frequency PWM control for renewable energy generation applications," IEEE Trans. Power Electron., vol. 30, no. 8, pp. 4312-4326, August 2015.

[18] B. York, Wensong Yu, J.-S. Lai, "An integrated boost resonant converter for photovoltaic applications," IEEE Trans. Power Electron., vol. 28, no. 3, pp. 1199-1207, March 2013.

[19] Y.-H. Kim, S.-C. Shin, J.-H. Lee, Y.-C. Jung, and C.-Y. Won, "Soft-switching current-fed push–pull converter for 250-W AC module applications," IEEE Trans. Power Electron., vol. 29, no. 2, pp. 863-872, February 2014.

[20] Y. Zhou, L. Liu, and H. Li, "A high-performance photovoltaic module-integrated converter (MIC) based on cascaded quasi-Z-source inverters (qZSI) using eGaN FETs," IEEE Trans. on Power Electron., vol. 28, no. 6, pp. 2727-2738, June 2013.

[21] H. Cha, F.Z. Peng, and D.-W. Yoo, "Z-source resonant DC-DC converter for wide input voltage and load variation," Proc. Int. Power Electron. Conf. (IPEC'2010), pp. 995-1000, 21-24 June 2010.

[22] J. Zakis, I. Rankis, and L. Liivik, "Loss reduction method for the isolated qZS-based DC/DC converter," Electrical, Control and Communication Engineering, vol. 4, no. 1, pp. 13-18, December 2013.

[23] D. Vinnikov, J. Zakis, L. Liivik, and I. Rankis, "qZS-based soft-switching DC/DC converter with a series resonant LC circuit," Energy Saving. Power Engineering. Energy Audit, vol. 114, pp. 42-50, 2013.

[24] D. Vinnikov, and I. Roasto, "Quasi-Z-source-based isolated DC/DC converters for distributed power generation," IEEE Trans. Ind. Electron., vol. 58, no. 1, pp. 192-201, January 2011.

[25] A. Chub, D. Vinnikov, F. Blaabjerg, and F.Z. Peng, "A review of galvanically isolated impedance-source DC-DC converters," IEEE Trans. Power Electron., accepted for publication.

[26] D. Vinnikov, I. Roasto, R. Strzelecki, and M. Adamowicz, "Step-up DC/DC converters with cascaded quasi-Z-source network," IEEE Trans. Ind. Electron., vol. 59, no. 10, pp. 3727-3736, October 2012.

[27] I. Roasto, D. Vinnikov, J. Zakis, and O. Husev, "New shoot-through control methods for qZSI-based DC/DC converters," IEEE Trans. Ind. Informat., vol. 9, no. 2, pp. 640-647, May 2013.

[28] L. Liivik, A. Chub, J. Zakis, and I. Rankis, "Analysis of buck mode realization possibilities in quasi-Z-source DC-DC converters with voltage doubler rectifier," Proc. 6th Int. Conf. on Power Eng., Energy and Elect. Drives (POWERENG'2015), pp. 1-6, 11-13 May 2015.

[29] J.-P. Vandelac, P.D. Ziogas, "A DC to DC PWM series resonant converter operated at resonant frequency," IEEE Trans. Ind. Electron., vol. 35, no. 3, pp. 451-460, August 1988.

HYBRID SERIAL-OUTPUT CONVERTER FOR INTEGRATED LED LIGHTING APPLICATIONS

T. McRae, A. Prodić
Laboratory for Power Management and Integrated SMPS
ECE Department, University of Toronto, CANADA

G. Lisi, W. McIntrye, A. Aguilar
Texas Instruments, Kilby Labs, USA

Abstract-- **This paper introduces a high power density step-up converter for LED applications, based on a hybrid serial-output (HSO) architecture [1], which is suitable for on-chip implementation. In this system, the output voltage is formed by stacking the output of a switched-capacitor (SC) converter on top of a boost converter output. The high power density SC converter processes around a half of the power of the system and is left unregulated. The boost converter processes the remainder of the power and regulates the output voltage. In comparison with conventional boost-based solutions, the introduced boost-SC HSO drastically reduces the passive component volume and decreases peak voltage stress of switches.**

Experimental results obtained with a 3.7 V to 13 V, 2.6 W, 1080 kHz prototype show that the introduced SC-boost HSO converter has about three times smaller reactive component area than a conventional boost having the same power processing efficiency.

Index Terms—**Long String LEDs, Partial Power Processing, Assisting, Light Weight, Automotive, Avionic, Digital Control**

I. INTRODUCTION

Light emitting diodes (LEDs) are a popular choice for lighting applications due to their long life and high power processing efficiency [2]. Long-string LEDs are of particular interest due to the inherent current sharing between diodes, which reduces the overall number of current regulating components in the system. However, in battery powered applications, long strings require the input voltage to be stepped up to a significantly higher level. Boost converters are commonly used in practice [3] but these solutions, particularly the inductors, can occupy significant space and weight in mobile applications.

To reduce the size of the reactive components, the hybrid serial output (HSO) architecture of [1] has been proposed as an alternative to the conventional boost. In that solution, the size of reactive components of a step-up LED driver has been drastically reduced by dividing power between a switched capacitor (SC) converter and a flyback converter and connecting the outputs of these two converters in series. The flyback-SC HSO architecture requires five capacitors, and a significant increase in silicon area compared to that of the conventional boost architecture. Therefore, the flyback-SC based solution is

This work of Laboratory for Power Management and Integrated SMPS is supported by Texas Instruments, Kilby Labs, USA.

challenging to implement on chip, and does not lend itself to system on chip (SoC) and/or system in package (SinP) solutions.

The goal of this paper is to introduce the HSO-LED driver topology of Fig.1 that results in a drastic reduction of reactive components compared to the conventional boost-based solution and, at the same time, is well-suited for on chip implementation. While there are more switches compared to a boost, the peak voltage stress is lower, meaning the total silicon area can be comparable between the two solutions. Also, compared to the previously presented HSO solution, the SC-boost HSO requires one less capacitor and three fewer switches.

II. PRINCIPLE OF OPERATION

The SC-boost based HSO converter of Fig.1 operates on the partial power processing principle as the architectures presented in [1], [4], [5], [6] and [7]. The solution presented in here is inspired by the architectures presented in [8] and [9], where the full output power is divided between multiple converters by dividing the output current.

In the system of Fig.1 a regulated converter processes only a portion of the output power and the power sharing is controlled by voltage as opposed to current. In this case, the power is divided between a switched capacitor (SC) stage and a boost stage. The two converters draw current in parallel from the input and stack their outputs in series. The SC converter is left unregulated and provides roughly half of the total output power. A higher frequency boost stage processes the remaining power and provides regulation of the output voltage, as shown in Figs. 1 and 2. This hybrid architecture allows for the

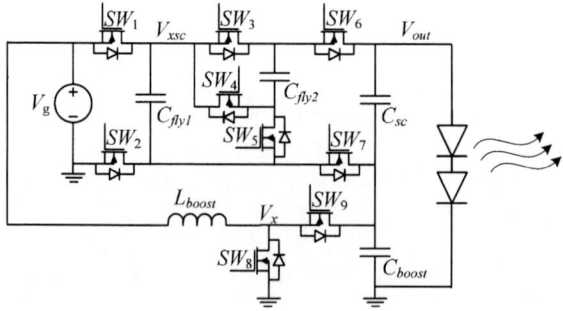

Fig.1: SC-Boost hybrid serial output (HSO) converter topology.

reduction of magnetic components, which generally take a large portion of the overall volume in conventional boost based solutions. It also improves upon the previous HSO converter presented in [1] by reducing the total number of switches and capacitors.

As discussed in [1], SC converters have much higher power density compared to their inductive counterparts [10]. The capacitors commonly used in the targeted applications can store up to a thousand times more energy per unit volume than inductors [11]. However, SC converters are highly efficient only at a few fixed voltage conversion ratios. By operating the SC converter unregulated at a fixed conversion ratio, HSO converters overcome this.

Because the outputs of two converters are being stacked on top of one another, one of them must have a floating output, to prevent the discharge of the bottom output capacitor. To achieve this, in [1] the secondary side of a flyback was placed on top of a SC converter. For the targeted applications, galvanic isolation is neither required nor desirable. For the converter of Fig. 1 the boost converter is referenced to the ground and the SC converter, which is placed on top of it, is designed to have a floating output, as described in the following section.

III. PRACTICAL IMPLEMENTATION

As derived in [1], an approximation of the power processing efficiency is:

$$\eta = \left(\frac{P_{out}}{P_{in}}\right) = \left(\frac{P_{out}}{\frac{P_{sc}}{\eta_{sc}} + \frac{P_{boost}}{\eta_{boost}}}\right) = \frac{V_{out}}{\frac{V_{sc}}{\eta_{sc}} + \frac{V_{boost}}{\eta_{boost}}} \quad , \quad (1)$$

where P_{out} is the output power of the whole system, P_{in} is the input power, P_{sc} is the output power of the SC converter, P_{boost} is the output power of the boost stage, η_{sc} is the efficiency of the SC converter, η_{boost} is the efficiency of the boost, V_{out} is the total output voltage of

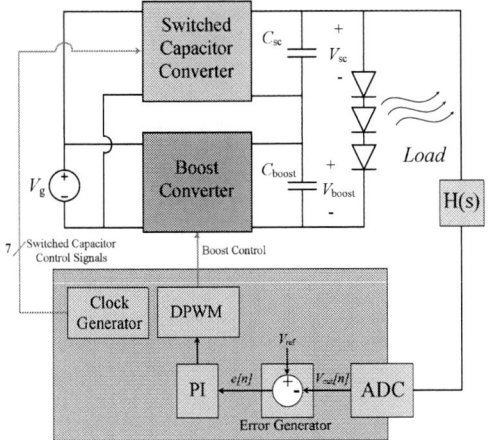

Fig. 2: Complete system with digital controller implementing voltage mode control.

the system, while V_{sc} and V_{boost} are the output voltages of the SC and of the boost, respectively.

In designing the SC and the boost converter stage this equation is taken into account and the power is divided between the two converters in such a way that high power processing efficiency, small passive volume, and low die area are achieved.

Furthermore, having an on-chip integrated solution in mind, the number of external (reactive) components needed for the system implementation is taken into account as it defines the pin count. Having a fairly low pin count in the targeted applications is often necessary to effectively utilize die area.

A. SC Converter Design

Given the common input and output voltage specifications of the system, where the LED drivers are usually required to step up the input voltage between one and two times [12], only a few choices of SC converters were considered, namely 1-1, 1-2 and 1-3. As mentioned previously, at least one of the converters in an HSO converter is required to have a floating output. Because galvanic isolation is not desired, the SC converter is designed to have a floating output. The factors taken into consideration when designing the SC converter were system efficiency, the number of switches and the passive component volume of the whole system. If the SC provides too much of the total output power, more switches and capacitors will be required. This leads to both larger pin count as the size of these capacitors necessitate the use of external components, and larger die area as more switches are required. Furthermore, the more capacitors in the SC converter generally lead to a lower required switch on-resistance for proper operation and thus a larger die area per switch. This leads to the conclusion that reducing die area means reducing the conversion ratio of the SC converter. However, if the SC does not provide a high enough output voltage; the boost will still be required to process a large portion of power and thus have a large volume. The 1-2 SC shown in Fig. 3 was found to offers a good tradeoff between these two extremes.

SC converters also have a trade-off between switching frequency, capacitor size/volume, and switch on-resistance for a given desired efficiency. For optimal performance, we want the converter to operate at the boundary between the slow switching limit (SSL), where

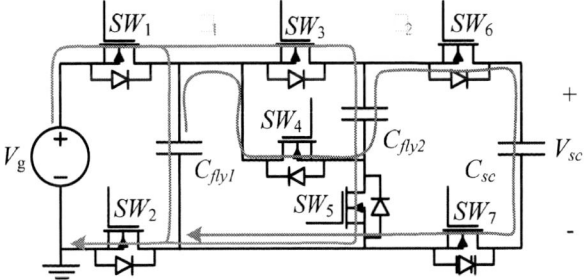

Fig. 3: 1-2 SC converter with the two alternating switching phases indicated in red and blue.

978-1-4673-9551-9/16 $31.00 © 2016 IEEE   2541

the capacitance dominates the output impedance of the converter and the fast switching limit (FSL), where the switch on-resistance dominates the output impedance as this minimizes die size for a given efficiency [11]. Optimization of the SC converter was accomplished iteratively. First a switching frequency was selected, then the output voltage ripple is reduced by increasing capacitance and the required R_{on} for the switches is determined, such that the SC converter operates at the knee point of FSL and SSL. This calculation is done until the desired efficiency is achieved based on an SC loss model including output impedance and switch output capacitance switching loss. The frequency is then incremented and the process is repeated until a plot showing the trade-off between output capacitance and switch on-resistance is generated for a given efficiency. Because, in practice, capacitors are grouped into discrete package sizes, we choose the largest capacitance that fits into a certain package to maximize SC efficiency for a given volume. From this, a particular combination of switching frequency, output capacitance, and on-resistance was chosen for a targeted efficiency.

Operation of the floating 1-2 SC converter can be explained by looking at Fig. 3. During Φ_1, switches SW_1, SW_2, SW_3, and SW_5 are turned on, connecting the flying capacitors to the input in parallel and charging them up to the V_g. During Φ_2, switches SW_4, SW_6 and SW_7 are turned on, disconnecting the flying capacitors from the input and connecting them in series to the output capacitor of the SC converter, which is stacked on top of the boost output. Because switches SW_1 and SW_2 are out of phase with switches SW_6 and SW_7, the negative terminal of the SC output capacitor is never connected to ground and thus the output capacitor of the boost is never discharged.

B. Boost Converter Design

The boost converter of this system is operated in DCM at four times higher frequency than the SC converter. Because the system generally operates at low current, operating in CCM would require large inductance to maintain ripple lower than the average inductor current. Therefore, DCM operation was chosen. Operating at higher frequency allows the inductance to be reduced even further without having a restrictively high peak current. Furthermore, DCM simplifies the dynamics of the converter, allowing a first order small signal model to be developed.

C. Controller Design

Control of this HSO converter is accomplished using a digital voltage mode controller [13], which can be seen in Fig. 2. The SC converter is left in open loop to maintain high power processing efficiency so the only control variable is the duty of the boost converter. The total output voltage is sensed, attenuated and then passed through an ADC into the digital domain. There, an error is generated and sent to a PI controller where a control signal is generated. This signal is sent to a digital pulse-width modulator (DPWM) which generates the gate control signals for the boost, thus closing the control loop

and regulating the output voltage. A clock generator is synchronized with the DPWM and generates the gating signals for the SC converter.

IV. EXPERIMENTAL RESULTS AND SIMULATIONS

Based on the diagrams of Figs. 1 and 2, an experimental prototype was created and tested. The 2.6 W prototype is designed for an input voltage range of 2.5 V to 4.5 V, a maximum output current of 200 mA at a 13 V output. A digital voltage mode controller was designed using an FPGA-based development board. To compare the passive volume to a conventional boost converter, a conventional boost prototype operating over the same range is created. To achieve similar efficiency, the inductor of the boost was selected to be 4.7 µH with an I_{sat} of 1.65 A and operated in CCM. The same switches, output capacitor and gate drivers were used for the comparison.

Following the iterative procedure described in the previous section, for the 1-2 SC converter prototype, C_{sc} is selected to be 1uF and a switching frequency of 270 kHz is chosen to optimize between capacitor size and switching loss. C_{fly} is $2C_{sc}$ but for practical reasons, the flying capacitors were chosen to be 2.2 µF. For the on-resistance of the switches, the time constant of the charging paths was selected to be 2τ, achieving around 87% charging of the capacitors. Looking at Fig. 3, we can see the two limiting charging paths in Φ_1 and Φ_2. In Φ_1, C_{fly2} has 4 switches in its path and a capacitance of 2.2 µF. Setting $\tau = T_{\Phi1}/2 = T_{sc}/4$, where T_{sc} is the switching frequency of the SC converter, yields a maximum R_{on} of 110 mΩ for SW_1, SW_2, SW_3 and SW_5. For Φ_2, there is a path resistance $3R_{on}$ and $1/2C_{out}$, yielding a maximum R_{on} for SW_4, SW_6, and SW_7 of 587 mΩ. It can be seen that this selection of a 1-2 SC converter has resulted in fewer switches compared to [1] as well as a relatively high on resistance, meaning although there are several switches in this topology, the majority of them are not subject to conduction and switching losses seen in inductive converters and, thus, can be made significantly larger without sacrificing efficiency. The boost converter is operated at 1080 kHz. DCM operation of the boost yields a peak inductor current of 0.95 A.

Using discrete converters with differing inductors and number of switches makes direct comparison challenging. All relevant differences are stated in Table 1 for clarity. It should be noted that despite being able to optimize switch on resistance for the different SC charging paths and in IC implementation design custom switches, in this discrete experimental prototype a single switch was utilized.

Table 2 shows results of comparison of the total passive component volumes of the both converters. It can be seen that the introduced HSO topology reduces the overall passive component volume approximately four times relative to the boost converter operating over the same power range.

The key experimental waveforms of the new HSO converter confirming its proper operation are shown in Figs.4 and 5. They confirm operation of the boost

Table 1: HSO and boost operating conditions and components

Input Voltage	2.5 V – 4.5 V
Output Voltage	13 V
Output Current	5 mA – 200 mA
SC Switching Frequency	270 kHz
Boost Switching Frequency	1080 kHz
HSO Converter	
MOSFET	*International Rectifier IRF2030* - Coss = 29 pF, Ciss = 110 pF, Ron = 110 mΩ, Qg = 1nC, V$_{DS}$=20V
C$_{fly1}$, C$_{fly2}$	*Murata LLL219R70J105MA01* – 2.2 μF, 0508, 8 mΩ ESR
C$_{sc}$, C$_{boost}$	*Murata LLL219R70J105MA01* - 1 uF, 0508, 8 mΩ ESR
Inductor	*Coil Craft XPL2010-222MLB* – 2.2 μH, 0.96 Arms, 156 mΩ DCR
Half Bridge Gate Driver	*Micrel MIC 4604*
High Side Gate Drive	*Linear Technology LTC 4440*
Boost Converter (Comparison)	
MOSFET	*International Rectifier IRF2030*
C$_{boost}$	*Murata LLL219R70J105MA01*
Inductor	*Wurth WE 744 042 004* – 4.7 μH, 1.65 Arms, 70 mΩ DCR
Half Bridge Gate Driver	*Micrel MIC 4604*

Table 2: Passive Component Comparison between HSO and Boost

HSO Passives	Volume	Boost Passives	Volume
Coil Craft XPL2010-222MLB	4 mm^3	**Wurth** WE 043 749 004	41.472 mm^3
C$_{fly1}$, C$_{fly2}$, C$_{sc}$, C$_{boost}$	2.125 mm^3 x 4	C$_{boost}$	2.125 mm^3
Total Passive Volume HSO	12.5 mm^3	**Total Passive Volume Boost**	43.597 mm^3

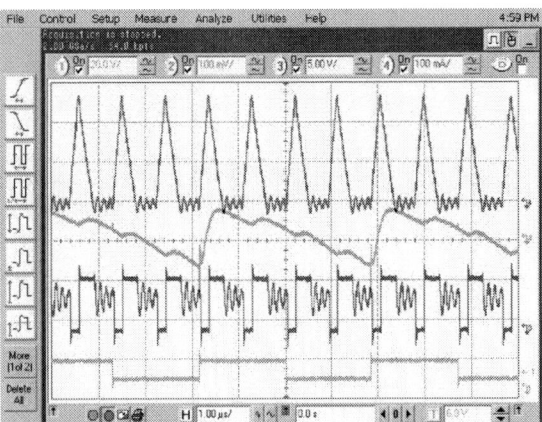

Fig. 4: Key experimental waveforms of the HSO converter: Ch.1 (yellow): SC switching node $v_{xsc}(t)$ (20 V/div); Ch.2 (green): AC component of output voltage $V_{outac}(t)$ (100 mV/div); Ch.3 (purple): Boost switching node $v_x(t)$ (5 V/div); Ch.4 (red): Boost inductor currents $i_L(t)$ (100 mA/div). Time scale is 1us/div. Operating Conditions: $V_g = 3.7$ V, $V_{out} = 13$ V, $P_{out} = 260$ mW.

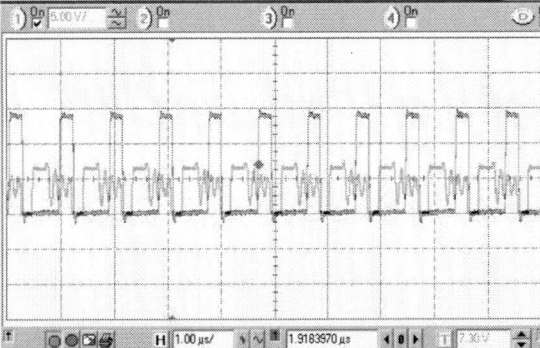

Fig. 5: Key experimental waveforms comparing the HSO to the boost converter: Ch.1 (yellow): Boost switching node $v_x(t)$ (5 V/div); Ch.2 (blue): Boost switching node fromHSO $V_{xHSO}(t)$ (5 V/div); Time scale is 1us/div. Operating Conditions: $V_g = 3.7$ V, $V_{out} = 13$ V, $P_{out} = 260$ mW.

converter in DCM at 4 times larger switching frequency than that of the SC stage, which frequency is selected through the efficiency optimization procedure.

The results of comparative efficiency measurements for the HSO and buck prototypes are shown in Figs. 6 on the left and right respectively. The results are shown for several discrete input voltage values, over entire load range. The efficiency is quite high at very low output current compared to the boost and is relatively flat over the whole range. For an easier visual comparison, the difference between the power processing efficiencies of the boost and HSO converters can be seen in Fig. 7. Averaging all the points over the entire range equally gives approximately a 3% higher power processing efficiency of the HSO.

These results confirm that the HSO topology at the same time results in drastic volume reduction (up to 4 times) and efficiency improvement, making it an attractive alternative to predominantly used boost-based solutions.

V. CONCLUSIONS

A new SC-Boost hybrid serial output (HSO) converter topology for LED drivers of mobile and other space constrained applicatons has been introduced. The SC-Boost HSO driver topology provides a drastic reduction in the size of reactive components compared to conventionally used boost-based LED driver solutions. The introduced topology is also well suited for on-chip integration of all semiconductor components and system in package (SinP) integration due to a relatively low total stress of components, lower voltage rating of transistors than that of a conventionally used boost topology, and relatively low number of reactive components, reducing required pin count. These advantages are achieved through developing a flying output SC topology that can be placed on top of

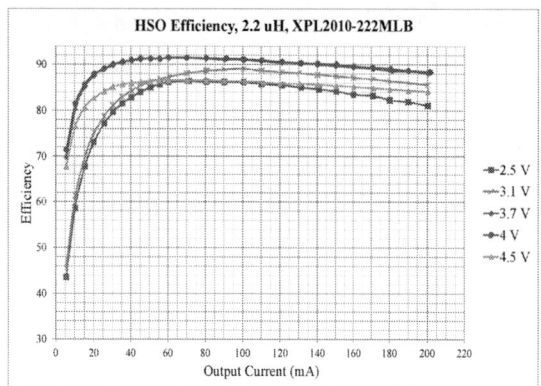 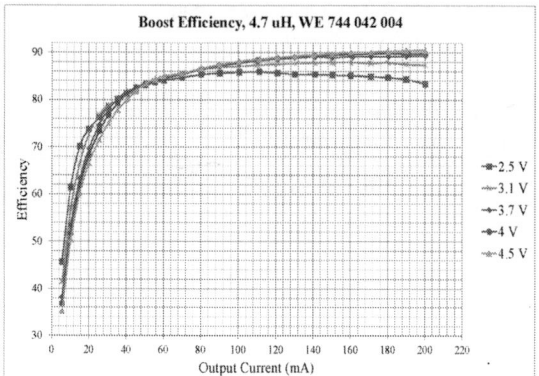

Fig. 6: HSO Efficiency using a 2.2 μH CoilCraft inductor from 5 mA to 200 mA is shown on the left. Boost efficiency with a 4.7 μH Wurth inductor shown over the same operating range.

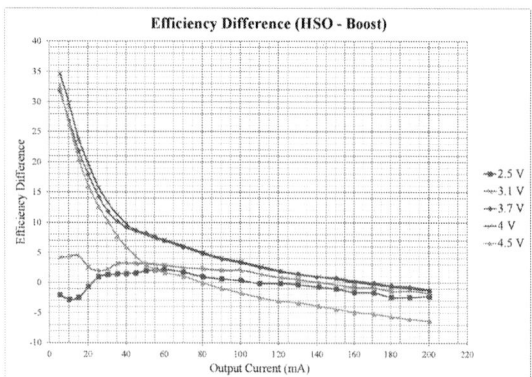

Fig. 7: Difference between HSO and boost efficiencies over the operating range.

conventional boost, which reduces the amount of power processed by magnetics in this system. The advantages of the SC-Boost LED driver topology have been experimentally verifed.

ACKNOWLEDGMENT

The authors would like to thank Bijoy Chatteerjee, Jeff Morroni, and Dave Anderson from Texas Instruments Kilby Labs and University Program for supporting this project.

REFERENCES

[1] McRae, T.; Prodic, A., "Hybrid Serial-Output Converter Topology for Volume and Weight Restricted LED Lighting Applications," In Proc. IEEE ICPE ECCE Asia 2015

[2] Steigerwald, D.A.; Bhat, J.C.; Collins, D.; Fletcher, R.M.; Holcomb, M.O.; Ludowise, M.J.; Martin, P.S.; Rudaz, Illumination with solid state lighting technology IEEE Journal of Selected Topics in Quantum Electronics Vol. 8, Issue 2, March-April 2002 Pages: 310 – 320.

[3] Van der Broeck, H.; Sauerlander, Georg; Wendt, M., "Power driver topologies and control schemes for LEDs," in Proc IEEE APEC 2007, vol., no., pp.1319,1325, Feb. 25 2007-March 1 2007

[4] Rodriguez, E.; Garcia, O.; Cobos, J.A.; Arau, J.; Uceda, J., "A single-stage rectifier with PFC and fast regulation of the output voltage," in Proc IEEE PESC, 1998. vol.2, no., pp.1642,1648 vol.2, 17-22 May 1998

[5] Shousha, M.; McRae, T.; Prodic, A.; Marten, V., "Assisting converter based integrated battery management system for low power applications," in Proc IEEE APEC 2014, vol., no., pp.1579,1583, 16-20 March 2014

[6] Junjian Zhao; Yeates, K.; Yehui Han, "Analysis of high efficiency DC/DC converter processing partial input/output power," in Proc IEEE COMPEL 2013, vol., no., pp.1,8, 23-26 June 2013

[7] "TI Tablet Solutions," Datasheet, Texas Instruments, 2013, available http://www.ti.com.

[8] Yeates, K.; Yehui Han, "Quasi-parallel switched-capacitor and regulating PWM DC-DC converter," in Proc IEEE APEC 2014, vol., no., pp.2105,2111, 16-20 March 2014

[9] Lukic, Z.; Zhenyu Zhao; Prodic, A.; Goder, D., "Digital Controller for Multi-Phase DC-DC Converters with Logarithmic Current Sharing," in Proc IEEE PESC 2007, vol., no., pp.119,123, 17-21 June 2007

[10] Seeman, M.D.; Ng, V.W.; Hanh-Phuc Le; John, M.; Alon, E.; Sanders, S.R., "A comparative analysis of Switched-Capacitor and inductor-based DC-DC conversion technologies," in Proc IEEE COMPEL 2010, vol., no., pp.1,7, 28-30 June 2010

[11] Sanders, S.R.; Alon, E.; Hanh-Phuc Le; Seeman, M.D.; John, M.; Ng, V.W., "The Road to Fully Integrated DC-DC Conversion via the Switched-Capacitor Approach," IEEE Trans. On Power Electronics, vol.28, no.9, pp.4146,4155, Sept. 2013

[12] Falin, J.; Meng, X., "Backlighting the tablet PC," Application note, Texas Instruments, 17 June 2011, available http://www.ti.com/lit/an/slyt414/slyt414.pdf

[13] Lukic, Z.; Kun Wang; Prodic, A., "High-frequency digital controller for dc-dc converters based on multi-bit Σ-Δ pulse-width modulation," in Proc IEEE APEC 2005, vol.1, no., pp.35-40 Vol. 1, 6-10 March 2005

Analysis and modeling of a modular ISOP Full Bridge based converter with input filter

P. Zumel, E. Oña, C. Fernandez, M. Sanz, A. Lazaro, A. Barrado
GSEP, Power Electronics System Group
Carlos III University of Madrid
Leganes, Spain

A. Vazquez, D.G. Lamar
Electronic Power Supply Systems Group (SEA)
University of Oviedo
33204, Gijon, Spain

Abstract— **This work presents a modular architecture based on the input series output parallel (ISOP) connection of Full Bridge Phase Shifted power converters with a common input filter. The modeling of the converter architecture taking into account the effect of the common input filter is the previous step to the controller design in order to ensure a proper regulation of external converter quantities, such as output voltage, and internal quantities, such as input voltages. In this paper a small signal model of an ISOP Full Bridge Phase Shifted converter taking into consideration the input filter is developed. Simulation and experimental results validate the theoretical predictions.**

Keywords— ISOP, full-bridge, modular converter, converter grouping

I. INTRODUCTION

The connection of power converters in modular architectures or multicell converters has been used in the last years due to the technological and economic advantages [1]-[11]. In a modular ISOP (input series output parallel) converter the input voltage and the output current are distributed among the modules. Advantages of ISOP connection are the capability to handle high voltages with low voltages devices which exhibit better figures of merit than devices rated for higher voltages, resulting in lower losses and higher switching frequencies. Modular converters have been used in Solid State Transformers, microgrids, etc.

Besides the control of the output quantities, e.g. output voltages, modular converters require the control of inner quantities, i.e., input voltages or output currents in each module in the case of ISOP connection. The control of modular converters presents several challenges [2]-[11]. If all modules are exactly equal, all electrical quantities are evenly distributed among them, and no especial care about voltages and current sharing has to be taken into account. However, in mass production, considering the component tolerances, the modules may differ one from the other, and the control has to ensure a proper distribution of voltages and current in order to avoid oversizing of the converter modules.

Previous works have dealt with full bridge ISOP connection [7][9]. A strategy based on decoupling control loops is proposed in [7] applied to ISOP full bridge converters without input filter. In this work the modeling is reviewed and the effect of the input filter is considered, which can change dramatically the behavior of the modular converter. In the ISOP connection, the input filter couples the dynamic response of the modules in such a way that the conventional decoupling techniques have to be reviewed. In this paper, the analytical model of the ISOP modular converter with common input filter is presented. Simulations are performed to illustrate theoretical predictions and control strategies. Finally, an experimental setup has been used to validate the obtained model.

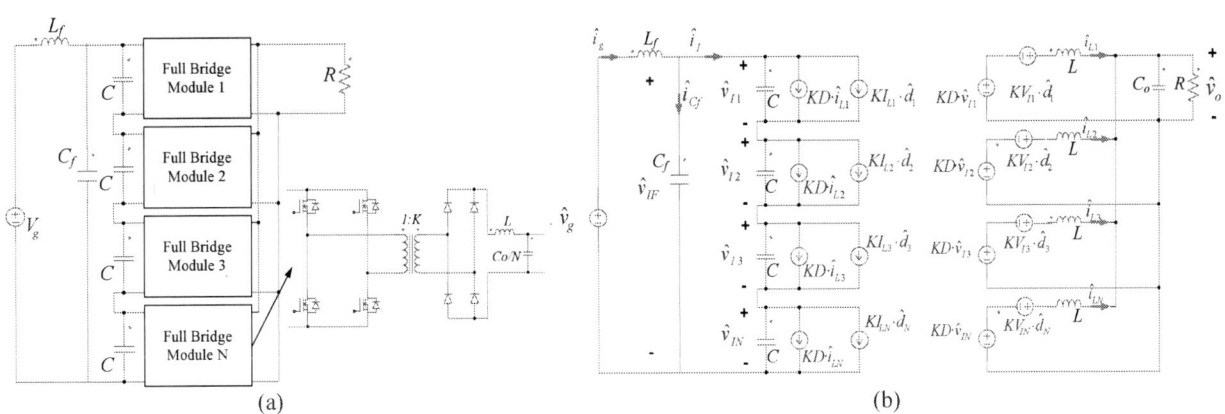

Fig. 1. a) ISOP modular converter based on Full Brdige modules with input filter, b)Small signal averaged model of ISOP with input filter

---

This work has been partially supported by the Ministry of Economy and Competitiveness and FEDER funds, through the research projects ELECTRICAR-AG- (DPI2014-53685-C2- 1-R), Consolider RUE CSD2009-00046 and DPI2013-47176-C2-2-R

II. SMALL SIGNAL MODEL OF AN ISOP MODULAR FULL BRIDGE CONVERTER

The small signal model of the modular architecture is based on the circuit presented in Fig. 1. The modular converter is based on Full Bridge Phase Shift converters. Each module has been replaced by its averaged and linearized circuit. The derivation of the small signal model is detailed in Appendix.

The control strategy for the proposed ISOP modular converter with N modules is based on the control of the total output voltage $\hat{v}_o$ and the input voltages of $N-1$ modules $\hat{v}_{I1}, \dots, \hat{v}_{I(N-1)}$. Therefore, the goal of the modeling analysis is to obtain the expressions relating $\hat{v}_{I1}, \dots, \hat{v}_{IN}$ and $\hat{v}_o$ with the control signals $\hat{d}_1, \dots, \hat{d}_N$ (normalized phase shifts or duty cycles) considering the circuits parameters of the modular converter. In order to obtain a model taking into account the effect of parasitic components, impedances have been considered for each element: Z_{Lf} for inductor filter L_f, Z_{Cf} for capacitor filter C_f, Z_C for input capacitor C, Z_L for output inductor L and Z_O for the output capacitor C_o in parallel with the load resistance R.

The assumptions in the model derivation are the following: a) all modules have the same characteristics (inductance, resistance and capacitance values); b) all modules are in the same operating point (same input voltage $\frac{V_G}{N}$, output current I_L, and duty cycle D).

A. Small signal model without input filter and decoupling control loops

From expressions derived in Appendix, when no input filter ($Z_{Lf} = 0$) is used, output voltage $\hat{v}_o$ and input voltage $\hat{v}_{Ij}$ of module j become respectively (1) and (2). It should be noted that Z_L, Z_C, Z_O, G_{vd} and A are functions of the complex variable "s", while K (transformation ratio), V_G (average input voltage) and N are constants.

$$\hat{v}_o = \frac{KV_G}{N + \frac{Z_L}{Z_O}} \frac{(\hat{d}_1 + \cdots + \hat{d}_N)}{N} = G_{vd} \frac{(\hat{d}_1 + \cdots + \hat{d}_N)}{N} \quad (1)$$

$$\hat{v}_{Ij} = -A \frac{(\hat{d}_1 + \cdots + \hat{d}_N)}{N} + A\hat{d}_j =$$

$$= \frac{A}{N} \left(\hat{d}_1 + \cdots + (N-1)\hat{d}_j + \cdots + \hat{d}_N \right) \quad (2)$$

Where A is given by (3).

$$A = -\frac{K\left(\frac{V_O}{NR} + \frac{DKV_G}{NZ_L}\right)}{\frac{1}{Z_C} + \frac{D^2K^2}{Z_L}} \quad (3)$$

The modular converter can be described by the matrix equation (4).

$$\begin{bmatrix} \hat{v}_{I1} \\ \vdots \\ \hat{v}_{IN-1} \\ \hat{v}_o \end{bmatrix} = \begin{bmatrix} \frac{N-1}{N}A & \frac{-A}{N} & \cdots & \frac{-A}{N} \\ \vdots & \ddots & & \vdots \\ \frac{-A}{N} & \cdots & \frac{N-1}{N}A & \frac{-A}{N} \\ G_{vd} & G_{vd} & G_{vd} & G_{vd} \end{bmatrix} \begin{bmatrix} \hat{d}_1 \\ \vdots \\ \hat{d}_{N-1} \\ \hat{d}_N \end{bmatrix} \quad (4)$$

Expression (4) shows that all controlled quantities ($\hat{v}_o$ and N-1 inputs voltages $\hat{v}_{Ij}$) depend on all control variables ($\hat{d}_1, \cdots, \hat{d}_N$), resulting in a MIMO (multiple input multiple output) system. A simple strategy to simplify the system and to obtain several independent SISO (single input single output) systems it based on the fact that the controlled quantities ($\hat{v}_{I1}, \dots, \hat{v}_{I(N-1)}$ and $\hat{v}_o$) depend on different linear combinations of the control variables ($\hat{d}_1, \dots, \hat{d}_N$), as shown in (1) and (2). A change of variables can be made in order to replace each linear combination of $\hat{d}_j$ by a new control quantity $\hat{x}_j$, obtaining a new description in which each controlled quantity depends on a single control signal. This decoupling transformation was first proposed in [7] and was also used in [12][13]. The new system can be described with equations (5) and (6).

$$\begin{bmatrix} \hat{v}_{I1} \\ \vdots \\ \hat{v}_{IN-1} \\ \hat{v}_o \end{bmatrix} = \begin{bmatrix} -A & 0 & \cdots & 0 \\ \vdots & \ddots & & \vdots \\ 0 & \cdots & -A & 0 \\ 0 & 0 & 0 & NG_{Vd} \end{bmatrix} \begin{bmatrix} \hat{x}_1 \\ \vdots \\ \hat{x}_{N-1} \\ \hat{x}_N \end{bmatrix} \quad (5)$$

$$\begin{bmatrix} \hat{d}_1 \\ \vdots \\ \hat{d}_{N-1} \\ \hat{d}_N \end{bmatrix} = \begin{bmatrix} -1 & 0 & 0 & \cdots & 0 & 1 \\ 0 & -1 & 0 & \cdots & 0 & 1 \\ \vdots & \vdots & \vdots & \vdots & \vdots & \vdots \\ 0 & 0 & 0 & \cdots & -1 & 1 \\ 1 & 1 & 1 & \cdots & 1 & 1 \end{bmatrix} \begin{bmatrix} \hat{x}_1 \\ \hat{x}_2 \\ \vdots \\ \hat{x}_{N-1} \\ \hat{x}_N \end{bmatrix} \quad (6)$$

Expressions (5) and (6) allow to implement N independent control loops to control the total output voltage and N-1 input voltages, as described in Fig. 2. Expression (5) defines the plant to be considered for the compensator calculation, while expression (6) defines the translation from the fictitious variables $\hat{x}_j$ calculated by compensator C1(s) and C2(s) and the real variables $\hat{d}_j$ to be applied to the actual converter. This control scheme is shown in Fig. 3.

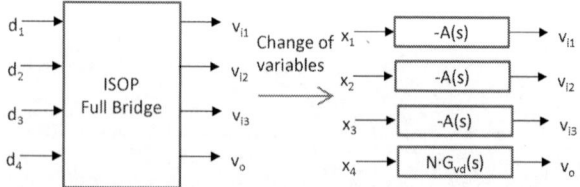

Fig. 2. Control strategy to decoupled control loop in a ISOP modular converter without input filter.

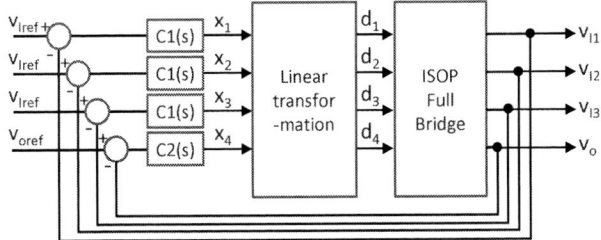

Fig. 3. Control strategy to decoupled control loop in a ISOP modular converter without input filter.

The condition to apply this strategy is that each controlled quantity depends on a linear combination of the control variables, and all these linear combinations are independent. If this condition is fulfilled, the implementation of the control system is easy since the generation of the duty cycle signals from the values of $\hat{x}_1, \dots, \hat{x}_N$ calculated in the control loops is carried out by means simple algebraic operations (6).

B. Small signal model considering input filter

From the derivations of Appendix, the expression of the output voltage and input voltage of module j are (7) and (8) respectively.

$$\hat{v}_o = G_{VdF} \frac{(\hat{d}_1 + \dots + \hat{d}_N)}{N} \tag{7}$$

$$\hat{v}_{Ij} = A_1 \frac{(\hat{d}_1 + \dots + \hat{d}_N)}{N} - A_2 \hat{d}_j \tag{8}$$

It should be noted that in this case input voltage $\hat{v}_{Ij}$ is not a linear combination of $\hat{d}_j$, since transfer functions A_1 and A_2 are not proportional. G_{VdF}, A_1 and A_2 are functions of "s".

The new system can be written in a matrix form (9).

$$\begin{bmatrix} \hat{v}_{I1} \\ \vdots \\ \hat{v}_{IN-1} \\ \hat{v}_o \end{bmatrix} = \begin{bmatrix} \frac{A_1}{N} - A_2 & \frac{A_1}{N} & \cdots & \frac{A_1}{N} \\ \vdots & \ddots & & \vdots \\ \frac{A_1}{N} & \cdots & \frac{A_1}{N} - A_2 & \frac{A_1}{N} \\ G_{VdF} & G_{VdF} & G_{VdF} & G_{VdF} \end{bmatrix} \begin{bmatrix} \hat{d}_1 \\ \vdots \\ \hat{d}_{N-1} \\ \hat{d}_N \end{bmatrix} \tag{9}$$

Due to the effect of the input filter on the input voltage transfer function (8) it is not possible to apply a simple transformation to the original MIMO system to obtain N SISO systems. The condition referred in the previous paragraphs is not met, i.e., input voltages do not depend on linear combinations of control signals $\hat{d}_j$, and therefore expression (6) is not valid in this case. In Fig. 4 frequency response of input voltage $\hat{v}_{I1}/\hat{d}_1$ and $\hat{v}_{I2}/\hat{d}_1$ are plotted. In this example the following parameters has been used: V_G=2600 V, N=4, K=1.48, L_f =39.5 mH, C_f=1mF, L=420 µH, C_o=1.32 mF, R=2.34Ω. Parasitic resistances have been considered in all elements. It should be noted that in the case without input filter, $\hat{v}_{I1}/\hat{d}_1$ and $\hat{v}_{I2}/\hat{d}_1$ are proportional, i.e., there is an offset in the

magnitude plot and a constant difference of phase equal to 180°.

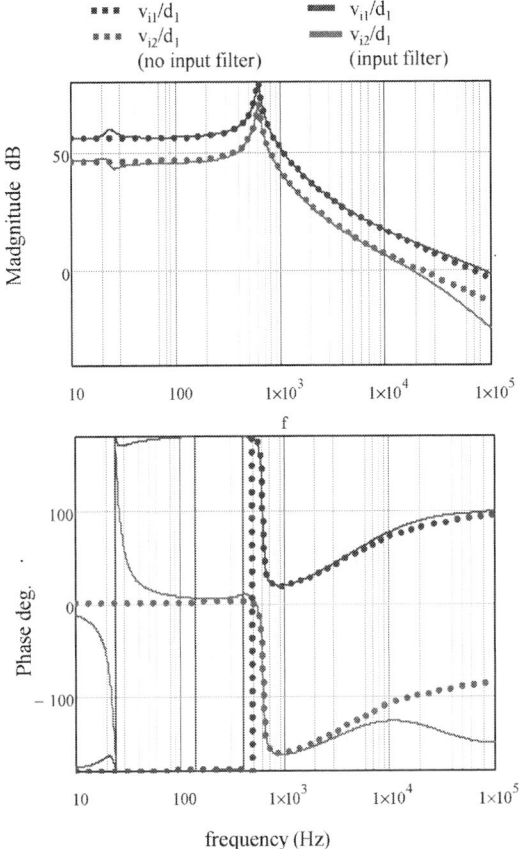

Fig. 4. Frequency response of the input voltage of module 1 $\hat{v}_{I1}/\hat{d}_1$ and input voltage of module 2 $\hat{v}_{I2}/\hat{d}_1$, with input filter (solid lines) and without input filter (dashed lines)

A similar change of variable can be applied to (9), obtaining expressions (10) and (11). In order to illustrate this calculation step, these expressions have been particularized for N=4. Expression (11) is used to generate the actual control signals $\hat{d}_1, \dots, \hat{d}_N$ to be applied to the actual converter. The main limitation of this approach is the physical implementation of this algorithm due to two reasons: 1) it is required to obtain $\frac{A_1}{A_2}$, which can be complex because of its dependence on parasitic components; 2) this transfer function is a filter with a high order in the numerator and denominator.

$$\begin{bmatrix} \hat{v}_{I1} \\ \hat{v}_{I2} \\ \hat{v}_{I3} \\ \hat{v}_o \end{bmatrix} = \begin{bmatrix} A_2 & 0 & 0 & 0 \\ 0 & A_2 & 0 & 0 \\ 0 & 0 & A_2 & 0 \\ 0 & 0 & 0 & G_{Vd} \end{bmatrix} \begin{bmatrix} \hat{x}_1 \\ \hat{x}_2 \\ \hat{x}_3 \\ \hat{x}_4 \end{bmatrix} \tag{10}$$

$$
\begin{bmatrix} \hat{d}_1 \\ \hat{d}_2 \\ \hat{d}_3 \\ \hat{d}_4 \end{bmatrix} = \begin{bmatrix} -1 & 0 & 0 & \dfrac{A_1}{4\,A_2} \\ 0 & -1 & 0 & \dfrac{A_1}{4\,A_2} \\ 0 & 0 & -1 & \dfrac{A_1}{4\,A_2} \\ 1 & 1 & 1 & 1-\dfrac{3\,A_1}{4\,A_2} \end{bmatrix} \begin{bmatrix} \hat{x}_1 \\ \hat{x}_2 \\ \hat{x}_3 \\ \hat{x}_4 \end{bmatrix} \qquad (11)
$$

The input filter has an influence on the overall modular converter similar to the case of a single module. When the output voltage is analyzed (Fig. 5), the effect of the input filter in the frequency response is similar to the classical buck converter with input filter [14]. While the magnitude plot is not very significantly affected, the phase exhibits a change of 360° at the natural frequency of the input filter. Parasitic components or additional damping circuits affect to this response.

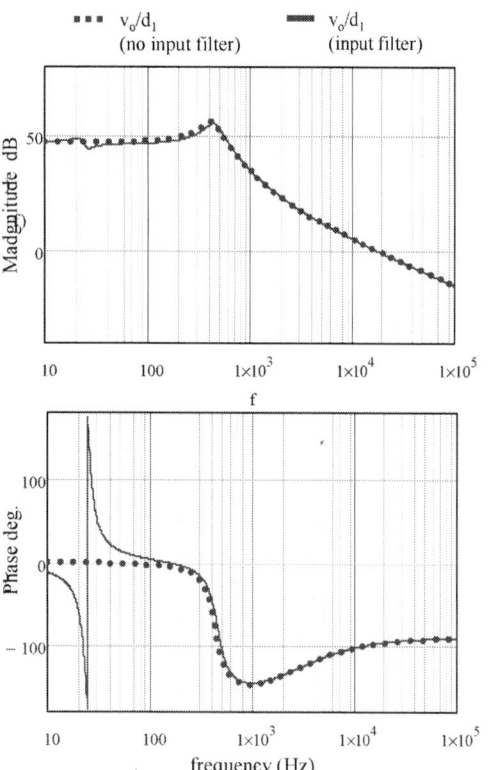

Fig. 5. Frequency response of the output $\hat{v}_o$ respect to $\hat{d}_1$ with input filter (solid line) and without input filter (dashed line).

As it has been shown, the decoupling strategy based on (6) cannot be directly used in the presence of input filter. When the input filter is used several options can be considered in order to overcome the effect on the control loops of the input filter: a) generate a decoupling strategy including transfer function $\frac{A_1}{A_2}$ in

the control variable transformation, which implies complex transformation and additional study to ensure the feasibility of the calculations; b) design the controller for cross over frequencies under the frequency range where the filter has influence; c) applying damping techniques to the filter in order to overcome the influence in the control transfer functions.

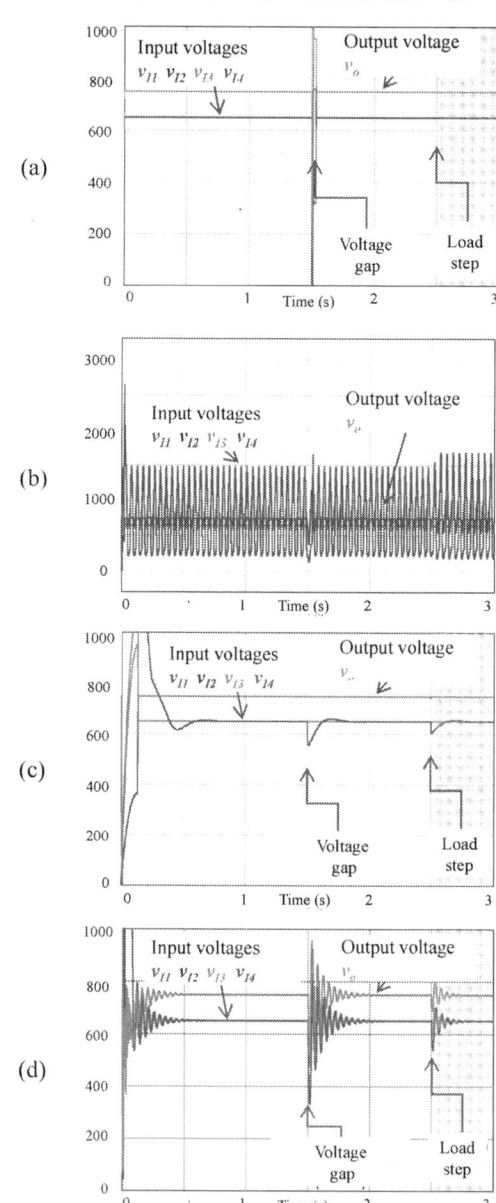

Fig. 6. Simulations of a four module ISOP modular converter

In the two last approaches, the original decoupling technique can be applied, since the controlled variable can be written as a linear combination of the control variables

These two possibilities have been simulated to illustrate this approach. Circuit parameters correspond to the example described in section II.b. In the time simulation input voltage gaps have been included in order to assess the performance of the input filter. Results are displayed in Fig. 6. A first test (Fig. 6a) is performed to the modular converter is without input filter and using the control strategy described in Fig. 3. The same regulators have been used in Fig. 6b but now there is an input filter, resulting in an unstable operation. In Fig. 6c the input filter has been damped, and the same controller than in the reference situation has been used. Finally, in Fig. 6d the regulators are designed with cross-over frequencies below the resonance frequency of the input filter.

III. EXPERIMENTAL RESULTS

In order to validate the small signal model, an experimental setup has been used to measure the transfer functions of input voltages and output voltages. In the laboratory, a set of three modules have been used in ISOP connection. Total input voltage is 50 V, output voltage 100 V, transformer ratio K=10, L_f=8 mH, C_f =440 μF, C=1720 μH, L=337 μH, C_o=3x22 μF, R=30 Ω.

The aim of the lab test is to validate the relationship among the transfer functions and the influence of the input filter. Three measurement setups have been considered: modular converter without input filter, modular converter with input filter and modular converter with the input filter and an additional parallel R-C damping branch. The modular converter operates in open loop, controlled by an FPGA. The duty cycles of two modules are constant, while in the third module the duty cycle is introduced by an analog to digital converter. The duty cycle signal is the connected to the oscillator of the Frequency Response Analyzer. The probe gain has not been corrected. Frequency response measurements are shown in Fig. 7, Fig. 8 and Fig. 9, corresponding to systems without filter, system with input filter and system with damped input filter, respectively.

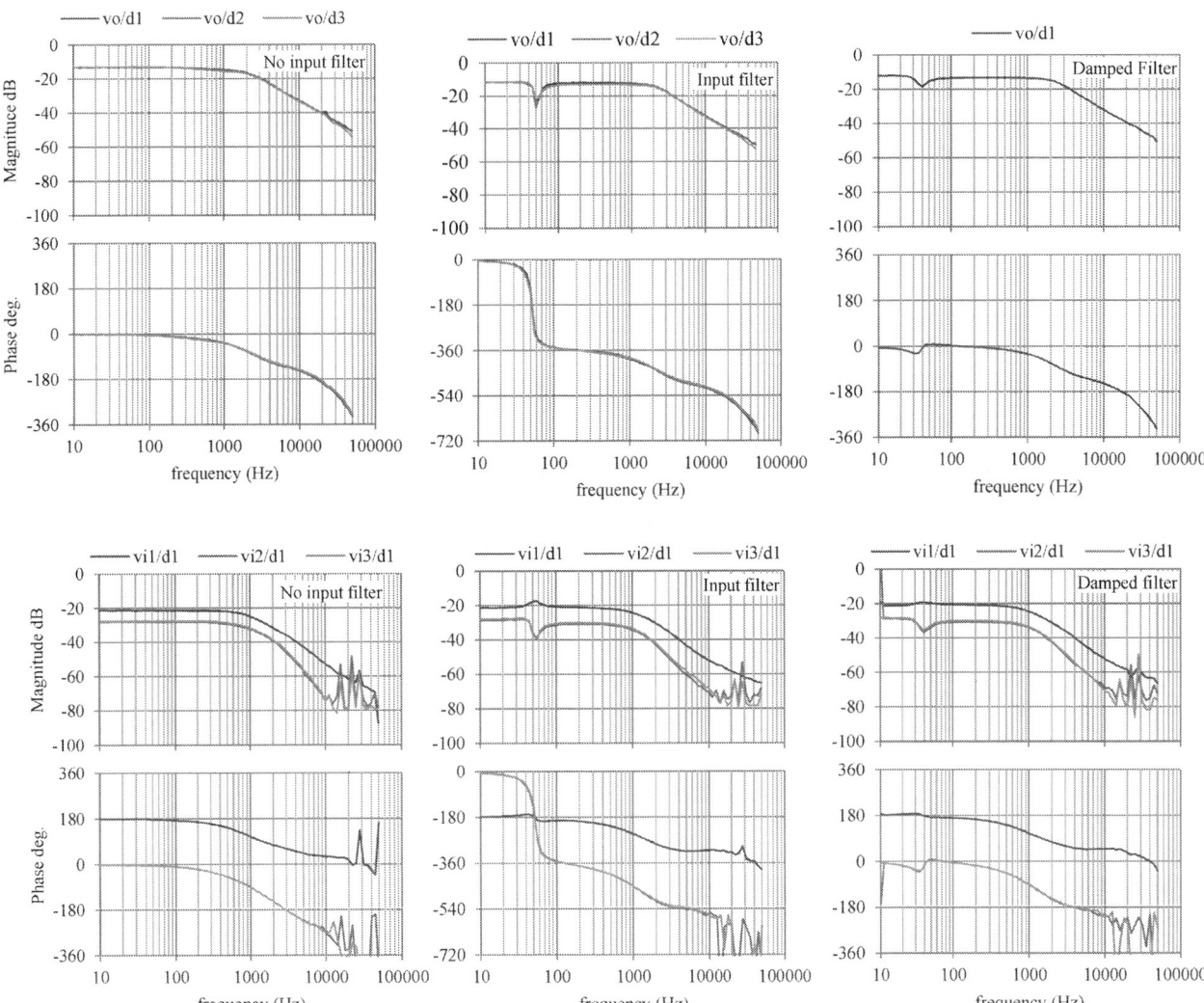

Fig. 7.   Transfer functions without input filter

Fig. 8.   Transfer functions with input filter

Fig. 9.   Trasnfer function with damped input filter

First, the relationships among the transfer functions can be considered. From the point of view of output voltage $\hat{v}_o$, it is important to check if it depends on the same way on the three duty cycles $\hat{d}_j$. It is shown in 0 and Fig. 8, agreeing with the described characteristic. In the case of input voltages it is important to check that the transfer function of the input voltage of the modules with the non perturbed duty cycle ($\hat{v}_{I2}/\hat{d}_1$ and $\hat{v}_{I3}/\hat{d}_1$) are similar, as it can be seen in Fig. 7, Fig. 8, and Fig. 9.

Second, the influence of the input filter can be analyzed. In the case of output voltage, when the input filter is used, the phase of the transfer function $\hat{v}_o/\hat{d}_1$ decreases 360° due to the input filter, as in the classical buck converter with input filter [14], degrading the dynamic performance of the system. It should be noted that the effective resonance frequency of the input filter is not only defined by the filter components L_f and C_f, but also by the series equivalent of the input capacitors C. If only L_f and C_f are considered, the resonance frequency is 85 Hz. However, if the equivalent of the input capacitors (1720 μF/3) is considered in parallel with C_f, the resulting resonance frequency is 56 Hz, agreeing with experimental measurements. From the point of view of input voltages, it should be noted that due to the presence of input filter, $\hat{v}_{I2}/\hat{d}_1$ and $\hat{v}_{I1}/\hat{d}_1$ are not proportional (constant difference in magnitude, and 180° or 0° difference in phase). Consequently, decoupling technique based on linear transformation (6) would not obtain good results.

In Fig. 9 the frequency response of the system is measured using an additional damping R-C branch in parallel with C_f. From the point of view of output voltage it is now possible to design a compensator above the input filter resonance frequency. From the point of view of input voltages, even if it is not the ideal situation, transfer functions are compatible with the decoupling technique to design control loops.

Finally in 0 measurements have been compared with the theoretical models. In this case, parasitic components have been estimated from prototype frequency response. Modulator delay has not been considered in the theoretical model, while the gains of the probe, modulator and analog to digital converter have been taken into account in the comparison. It should be noted that the agreement between theory and measurements is extended to a significant frequency range.

IV. CONCLUSIONS

In this paper, the effect of the input filter in a modular ISOP architecture based on Full Bridge Phase Shifted converters has been analyzed.

A small signal model has been obtained showing the importance of the input filter in the control strategy, from the point of view of the regulation output voltage and the distribution of the input voltage in each module. Classical decoupling techniques for the control of modular converter

have to be carefully considered when using input filter, since an additional coupling among the modules appears. Simple strategies to overcome the effect of the input filter can be applied, as filter damping or slow controller design.

Frequency response measurements have been performed over a prototype of three modules. The influence on the output voltage and input voltages of the input filter (undamped and damped) has been analyzed, showing a good agreement between the theoretical model and experimental measurements.

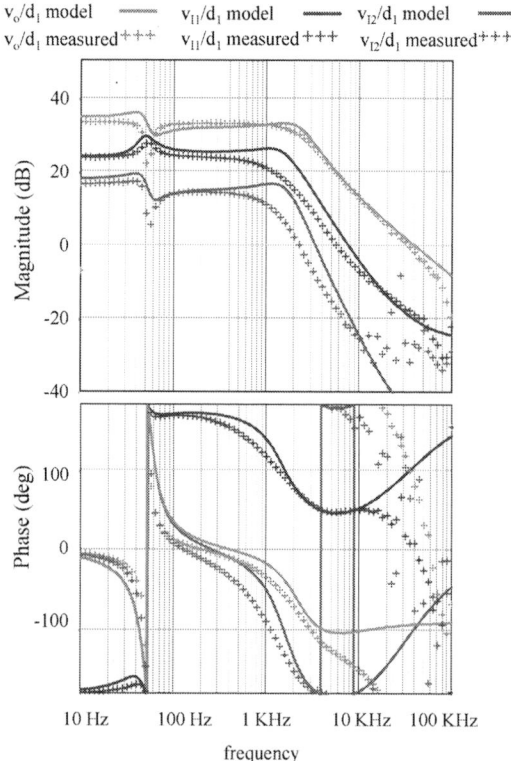

Fig. 10. Theoretical and measured trasnfer function comparison.

V. APPENDIX: SMALL SIGNAL MODEL DERIVATION

A. Input current and output voltage expression derivation

From the average circuit model of Fig. 1 and linearizing the response of the circuit, the output voltage $\hat{v}_o$ can be expressed as:

$$\hat{v}_o = \frac{K\left(\frac{V_G}{N}\left(\hat{d}_1 + \cdots + \hat{d}_N\right) + D\left(\hat{v}_{I1} + \cdots + \hat{v}_{IN}\right)\right)\frac{1}{z_L}}{\frac{N}{z_L} + \frac{1}{z_O}} \quad (12)$$

$$\hat{v}_o = \frac{K\left(\frac{V_G}{N}\left(\hat{d}_1 + \cdots + \hat{d}_N\right) + D\hat{v}_{IF}\right)}{N + \frac{z_L}{z_O}} \quad (13)$$

Where K is the turns ratio, V_G is the total input voltage, D is the nominal duty cycle, Z_L is the impedance of the output inductor (including parasitics), Z_O is the output impedance (output capacitor in parallel with load), $\hat{d}_i$ is the duty cycle of each module, $\hat{v}_{Ii}$ is the input voltage of a given module and $\hat{v}_{IF}$ is the input voltage of the modular converter after the input filter. In this expression, $\hat{v}_{IF}$ is not known, since it depends on the input current $\hat{i}_I$ and the filter impedance.

The input current $\hat{i}_I$ can be derived considering the input node of each converter. For the converter j, expression (14) corresponds to the input node.

$$\hat{i}_I = \hat{i}_{Cj} + K\left(D \cdot \hat{i}_{Lj} + I_L \cdot \hat{d}_j\right) \qquad (14)$$

Where $\hat{i}_{Ci}$ is the current though the input capacitor of module j, $\hat{i}_{Lj}$ is the variation of the inductor current, and I_L is the DC current though the inductor.

The addition of expression (14) applied to each module results in expressions (15) and (16)

$$N\hat{i}_I = \frac{1}{Z_C}(\hat{v}_{I1} + \cdots + \hat{v}_{IN}) +$$
$$+KD\frac{\hat{v}_o}{Z_O} + \frac{K\,V_o}{R\,N}(\hat{d}_1 + \cdots + \hat{d}_N) \qquad (15)$$

$$\hat{i}_I = \frac{1}{N\,Z_C}\hat{v}_{IF} + \frac{KD}{N}\frac{\hat{v}_o}{Z_O} + \frac{K\,V_o}{RN^2}(\hat{d}_1 + \cdots + \hat{d}_N) \qquad (16)$$

Where Z_C is the impedance of the input capacitor (including parasitic resistance) and R is the load resistance.

The input voltage after the filter $\hat{v}_{IF}$ can be related with the input current by means of expression (17).

$$\hat{v}_{IF} = -\hat{i}_I\frac{Z_{Lf} \cdot Z_{Cf}}{Z_{Lf} + Z_{Cf}} = -\hat{i}_I\,Z_F \qquad (17)$$

where Z_{Lf} and Z_{Cf} are the impedances of filter inductor and filter capacitor respectively. Z_F is the equivalent impedance of the input filter.

Substituting (16) in (15) and operating, the expression of the input current of the modular converter is obtained (19).

$$\hat{i}_I = \frac{\frac{V_O K}{R} + \frac{V_G D K^2}{(NZ_O+Z_L)}}{N + \frac{Z_F}{Z_C} + \frac{D^2 K^2 Z_F}{NZ_O + Z_L}}\frac{(\hat{d}_1 + \cdots + \hat{d}_N)}{N} \qquad (18)$$

$$\hat{i}_I = G_{IdF}\frac{(\hat{d}_1 + \cdots + \hat{d}_N)}{N} \qquad (19)$$

Output voltage can be now expressed as a transfer function depending only on the control signal $\hat{d}_i$

$$\hat{v}_o = \frac{K\left(V_G - D\frac{\frac{V_O K}{R} + \frac{V_G D K^2}{(NZ_O+Z_L)}}{N + \frac{Z_F}{Z_C} + \frac{D^2 K^2 Z_F}{NZ_O+Z_L}}Z_F\right)}{N + \frac{Z_L}{Z_O}}\frac{(\hat{d}_1 + \cdots + \hat{d}_N)}{N} \qquad (20)$$

$$\hat{v}_o = G_{VdF}\frac{(\hat{d}_1 + \cdots + \hat{d}_N)}{N} \qquad (21)$$

Analysis of expression (19) and (21) establish that from the point of view of total input current and total output voltage, the modular converter can be seen as a single converter where the control signal is the average of control signals of all modules.

B. Input voltage expression derivation

Expression (14) can be modified to obtain the input voltage of a module

$$\hat{v}_{Ij} = \left(\hat{i}_I - K\left(D \cdot \hat{i}_{Lj} + I_L \cdot \hat{d}_j\right)\right)Z_C \qquad (22)$$

The current through the output inductor in module I is

$$\hat{i}_{Lj} = \frac{1}{Z_L}\left(KD\hat{v}_{Ij} + \frac{KV_G}{N}d_j - \hat{v}_o\right) \qquad (23)$$

Substituting (23) in (22), expression (25) is obtained.

$$\hat{v}_{Ij} = \left(\hat{i}_I - K\left(\frac{V_o}{NR} + \frac{DKV_G}{NZ_L}\right)\hat{d}_j - \frac{D^2 K^2}{Z_L}\hat{v}_{Ij}\right. \\ \left. + \frac{DK}{Z_L}\hat{v}_o\right)Z_C \qquad (24)$$

$$\hat{v}_{Ij} = \frac{\left(\hat{i}_I - K\left(\frac{V_o}{NR} + \frac{DKV_G}{NZ_L}\right)\hat{d}_j + \frac{DK}{Z_L}\hat{v}_o\right)}{\frac{1}{Z_C} + \frac{D^2 K^2}{Z_L}} \qquad (25)$$

Substituting (19) and (21) in (25) and reordering terms, expression

$$\hat{v}_{Ij} = \frac{G_{IdF} + \frac{DK}{Z_L}G_{VdF}}{\frac{1}{Z_C} + \frac{D^2 K^2}{Z_L}}\frac{(\hat{d}_1 + \cdots + \hat{d}_N)}{N} - \\ - \frac{K\left(\frac{V_o}{NR} + \frac{DKV_G}{NZ_L}\right)}{\frac{1}{Z_C} + \frac{D^2 K^2}{Z_L}}\hat{d}_j \qquad (26)$$

$$\hat{v}_{Ij} = A_1\frac{(\hat{d}_1 + \cdots + \hat{d}_N)}{N} - A_2\hat{d}_j \qquad (27)$$

$$A_1 = \frac{G_{IdF} + \frac{DK}{z_L} G_{VdF}}{\frac{1}{z_C} + \frac{D^2 K^2}{z_L}} \qquad (28)$$

$$A_2 = \frac{K \left(\frac{V_O}{NR} + \frac{DKV_G}{Nz_L} \right)}{\frac{1}{z_C} + \frac{D^2 K^2}{z_L}} \qquad (29)$$

The analysis of expression (29) shows that the input voltage of every module depends on the average value of control signals, and on the control signal of this particular module.

VI. REFERENCES

[1] Guanghai Gong, Dominik Hassler, and Johann W. Kolar "A Comparative Study of Multicell Amplifiers for AC-Power-Source Applications" IEEE TRANSACTIONS ON POWER ELECTRONICS, VOL. 26, NO. 11, JANUARY 2011

[2] Raja Ayyanar, Ramesh Giri, and Ned Mohan "Active Input–Voltage and Load–Current Sharing in Input-Series and Output-Parallel Connected Modular DC–DC Converters Using Dynamic Input-Voltage Reference Scheme" IEEE Transactions on Power Electronics, vol. 19, no. 6, pp 1462,1473 Nov 2004

[3] Jung-Won Kim; Jung-Sik Yon; Cho, B. -H, "Modeling, control, and design of input-series-output-parallel-connected converter for high-speed-train power system," Industrial Electronics, IEEE Transactions on , vol.48, no.3, pp.536,544, Jun 2001

[4] Ting Qian; Lehman, B., "Coupled Input-Series and Output-Parallel Dual Interleaved Flyback Converter for High Input Voltage Application," Power Electronics, IEEE Transactions on , vol.23, no.1, pp.88,95, Jan. 2008

[5] Taotao Jin; Zhang, K.; Kan Zhang; Smedley, K., "A New Interleaved Series Input Parallel Output (ISIPO) Forward Converter With Inherent Demagnetizing Features," Power Electronics, IEEE Transactions on , vol.23, no.2, pp.888,895, March 2008

[6] Choudhary, V.; Ledezma, E.; Ayyanar, R.; Button, R.M., "Fault Tolerant Circuit Topology and Control Method for Input-Series and Output-Parallel Modular DC-DC Converters," Power Electronics, IEEE Transactions on , vol.23, no.1, pp.402,411, Jan. 2008

[7] X. Ruan, W. Chen, L. Cheng, C. Tse, Hong Yan y Tao Zhang , "Control Strategy for Input-Series–Output-Parallel Converters",IEEE Transactions on Industrial Electronics, vol. 56, no. 4, April 2009.

[8] Y. Huang. "General Control Considerations for Input-Series Output-Parallel Converters". *IEEE Trans. Ind. Electron.*, April 2009, VOL. 56, Nº4

[9] W. Cheng, X. Ruan, H. Yan, and C.K. Tse, "DC/DC Conversion system consisting of multiple converters modules: stability, control and experimental verification," *IEEE Trans. Power. Electron.*, vol. 24, no. 6, pp. 1463-1474, Jun. 2009.

[10] D. Sha, Z. Guo, and X. Liao, "Cross feedback output current sharing control for input-series–output-parallel modular dc–dc converters," *IEEE Trans. Power. Electron.*, vol. 25, no. 11, pp. 2762–2771, Nov. 2010.

[11] R. Giri, R. Ayyanar, N. Mohan, "Common duty ratio control of input series connected modular DC-DC converters with active input voltage and load current sharing," In Proc. IEEE APEC, 2003, pp.322-326.

[12] Zumel, P.; Ortega, L.; Lazaro, A.; Fernandez, C.; Barrado, A., "Control strategy for modular Dual Active Bridge input series output parallel," in Control and Modeling for Power Electronics (COMPEL), 2013 IEEE 14th Workshop on , vol., no., pp.1-7, 23-26 June 2013

[13] Zumel, P.; Ortega, L.; Lazaro, A.; Fernandez, C.; Barrado, A.; Rodriguez, A.; Hernando, M.M., "Modular dual active bridge converter architecture," in Applied Power Electronics Conference and Exposition (APEC), 2014 Twenty-Ninth Annual IEEE , vol., no., pp.1081-1087, 16-20 March 2014

[14] R.W. Erickson, D. Maksimovic, "Fundamental of Power Electronics", Springer Science & Business Media, 2007

Wide-input High Power Density Flexible Converter Topology for Dc-dc Applications

Parth Jain, Aleksandar Prodić

Laboratory for Power Management and Integrated SMPS
ECE Department, University of Toronto
10 King's College Rd., Toronto, ON, M5S 35G, CANADA
Email: parth.jain@mail.utoronto.ca, prodic@ele.utoronto.ca

Alexander Gerfer

Würth Elektronik eiSos GmbH & Co. KG
Max-Eyth-Str. 1, 74638 Waldenburg, GERMANY

Abstract— In this paper a highly flexible 2-phase step-down dc-dc converter topology designed for a wide range of point-of-load (PoL) applications is introduced. The new wide-input flying-capacitor based multi-level converter has smaller overall volume and better power processing efficiency than the conventionally used 2-phase interleaved buck solution, achieved through the reduction of voltage swings across its components. Compared to other multi-level solutions the new topology has much higher flexibility, allowing operating-condition dependent changes of the topology for efficiency optimization. Also, the silicon area required for the implementation of the switches is smaller than that of the other solutions, since all transistors are rated at ½ of the input voltage, further contributing to size reduction and power processing efficiency improvements. The advantages of the new topology have been experimentally verified with a wide input range, $3\ V \leqslant V_{in} \leqslant 48\ V$, wide output range, $1\ V \leqslant V_{out} \leqslant 12\ V$, output power $P_{out} \leqslant 72\ W$, 800 kilohertz switching frequency discrete experimental prototype. The results show that the introduced topology requires about 33% smaller inductor and 33% smaller output capacitor than the conventional 2-phase interleaved buck, while having higher power processing efficiency over the entire range of operation.

Keywords—wide-input dc-dc converter, 3-level converter

I. INTRODUCTION

In virtually all electronics devices today a large number of point-of-load (PoL) converters are used to provide well-regulated dc voltages for various functional blocks. Non-isolated PoL converters are usually supplied by a fixed or variable bus voltage, which, depending on the application, ranges from less than 5 V up to 48 V. To cover a large number of applications, numerous wide-input PoL converters have been developed [1]-[5]. These solutions, which are often based on the single or 2-phase buck converter, drastically simplify system level design but usually have significantly larger reactive components, in particular inductors, compared to the converters designed for a relatively narrow range of conversion ratios. Also, the wide-input converters suffer from significant efficiency drops for the cases when high step-down conversion ratios, for example 12 V to 1 V, are required. As a consequence, the system level solution based on multiple identical wide-input, i.e. universal, PoL modules, rather than custom ones, usually suffer from increased overall volume and,

This work of the Laboratory for Power Management and Integrated SMPS is supported by Würth Elektronik eiSos GmbH & Co, Germany

in some cases, reduced power processing efficiency.

The goal of this paper is to introduce the new highly flexible wide-input universal converter topology of Fig. 1 that, compared to existing solutions, has a much lower overall volume and increased power processing efficiency. The introduced 2-phase 7-switch flying capacitor (7SFC) based multi-level buck topology minimizes penalties existing in conventional solutions through the reduction of voltage stress/swing across the components to half of the input voltage. Additionally, depending on the operating conditions, the 7SFC converter can operate in various multi-level modes or as a conventional 2-phase 2-level buck converter, maximizing efficiency throughout the entire operating range.

II. PRINCIPLE OF OPERATION AND COMPARISON WITH OTHER TOPOLOGIES

The operation of the 7SFC multi-level buck and its advantages over conventional [1] and previously proposed alternatives [6,7], shown in Fig. 2, are explained in this section. Depending on the required conversion ratio and the load current, the converter of Fig. 1 can operate as the conventional 2-phase buck of Fig. 2(a), as the 2-phase double step-down

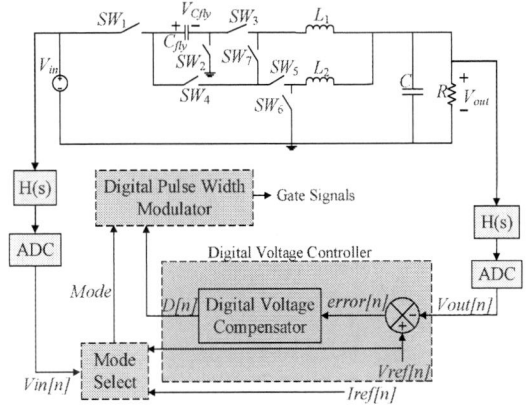

Fig. 1. Wide-input 7-switch flying capacitor (7SFC) multi-level buck and a complementary digital controller.

buck of Fig. 2(b), which was proposed by Nishijima [6]. The 7SFC can also operate as the single phase 3-level buck, introduced by Meynard [7], which 2-phase implementation is shown in Fig. 2(c). By operating in these different modes, and utilizing a digital controller, the 7SFC takes advantages of each of the topologies shown in Fig. 2.

A. 2-Phase Interleaved High Step-Down (HSD) Buck Mode

This mode of operation is introduced for improving efficiency for high step-down ratios under medium and heavy loads. The converter operates similar to the double step-down converter [6] of Fig. 2(b). In steady-state, the voltage across the flying capacitor, C_{fly}, is equal to $V_{in}/2$, where V_{in} is the input voltage of the converter. The switching sequence and key waveforms for this operating mode are shown in Fig. 3. The sequence consists of 4 states, where in *state* 1 the flying capacitor and inductor L_1 are charged with energy, *state* 2 is equivalent to the synchronous rectification of conventional converters, during *state* 3 the flying capacitor is discharged and inductor L_2 is charged, and *state* 4 is a repetition of *state* 2. Simple steady-state analysis [8] shows that, like in the double step-down buck case [6], the flying capacitor voltage is maintained at $V_{in}/2$, by the two inductors and that the conversion ratio is $M(D) = V_{out}/V_{in} = D/2$, where V_{out} is the output voltage and D is the duty ratio as depicted in the waveforms of Fig. 3(b). It can be seen that in this mode the inductors' voltage swings, i.e. the variations of the switching node voltages, v_{x1} and v_{x2}, are reduced by a half, compared to the conventional 2-phase buck, allowing for a significant

reduction of the inductance value [9]. It is clear that this reduction in the inductor value comes at the price of an extra capacitor. However, this is a very favorable tradeoff, since as shown in [10], for the applications of interest, the capacitors have about three orders of magnitude smaller volume for the same amount of stored energy than the inductors. Thus the overall volume of reactive components of the 7SFC buck is reduced compared to the conventional 2-phase buck. This has been demonstrated in [11], where a comparison of a conventional and a 3-level buck is performed.

By observing the 7SFC converter in this mode, it can be seen that all switches are blocking only a half of the input voltage. This means that, if the same silicon area is used for the two implementations, both switching and conduction losses for semiconductor components of this topology could potentially be smaller than that of the conventional 2-phase buck. This can

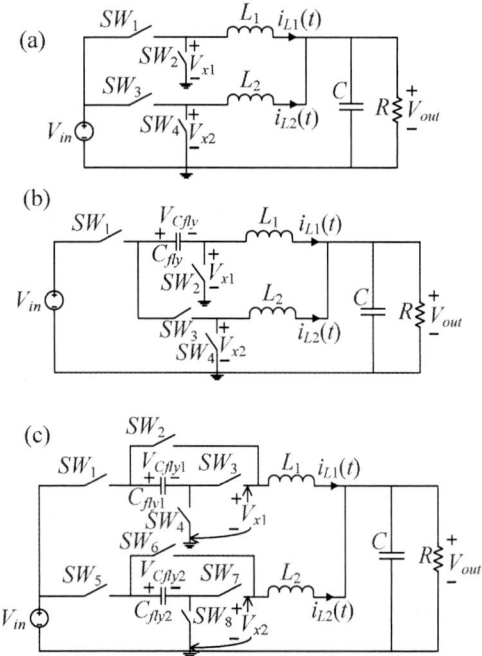

Fig. 2. (a) Conventional 2-phase interleaved buck converter, (b) Double step-down buck converter, (c) 2-phase 3-level buck converter

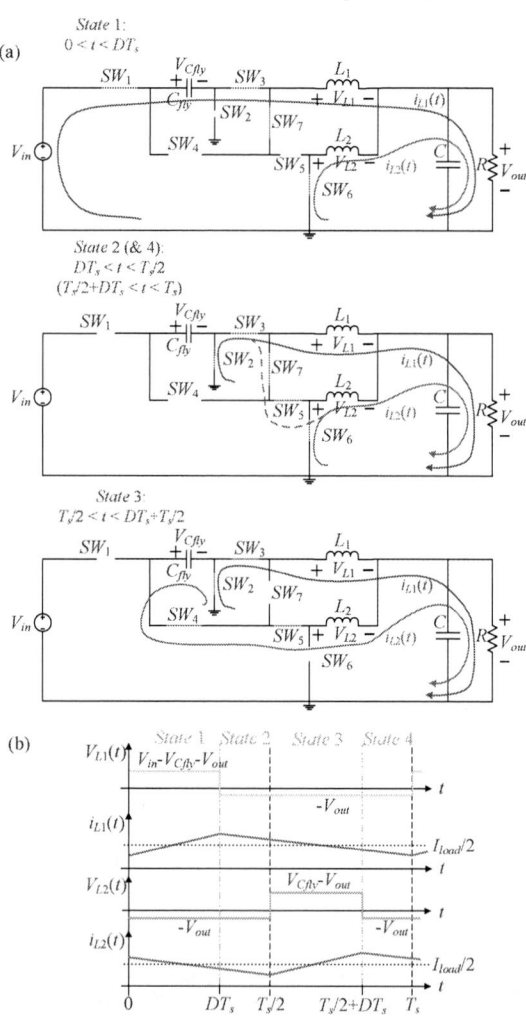

Fig. 3. (a) Switching sequence for the HSD buck mode, (b) Inductor voltage and current waveforms

978-1-4673-9551-9/16 $31.00 © 2016 IEEE

be confirmed using the analysis presented in [11].

The switch voltage stress reduction is also a major advantage of this topology over the previously presented double step down buck from the possible on-chip integration point of view, which for the targeted applications has been widely-adopted. In the converter of Fig. 2(b) switch SW_3 must be rated for the full input voltage which means that, even though the number of switches in that topology is smaller, the overall cost and silicon area required for the implementation of all switches can practically be equivalent to that for the 7SFC buck [12]. Thus, the 7SFC buck allows for potential full on-chip integration with a lower-voltage rating technology that has a better figure of merit (FOM), resulting in further reductions in conduction and switching losses.

While this mode is also advantageous for high step-down ratios, since it operates with a larger duty ratio than the conventional 2-phase buck [13], it has limitations. Like the standard double step down buck, in this mode the 7SFC buck cannot operate with duty ratios larger than 0.5 and cannot perform phase shedding [6], which is important for on-line efficiency optimization. To overcome these disadvantages, two additional modes of the converter are introduced.

B. Single-Phase 3-Level Buck Mode

The single-phase 3-level buck mode of operation is used under light load conditions, for the cases when a high step-down conversion, i.e. $V_{out}/V_{in} < 0.5$ is required. In this mode the 7SFC buck operates as a conventional 3-level flying capacitor buck converter, introduced by Meynard [7] (where the 2-phase version of the converter is shown in Fig. 2(c)). In this mode SW_5 and SW_7 are turned on at all times and the inductors are in parallel, practically creating a single-phase configuration of the topology shown in Fig. 2(c). The key waveforms and equivalent switching circuits of the 7SFC buck operating in the 3-level buck mode are shown in Fig. 4.

Again, each switching sequence has 4 states, where in *state* 1 the flying capacitor, C_{fly}, is charged through the inductors, *states* 2 and 4 correspond to synchronous rectification mode of operation, and in *state* 3, the flying capacitor is discharged to maintain approximately a constant $V_{in}/2$ voltage level. The main advantage of this mode over the conventional 2-phase buck is the reduced switching losses, since all of the switches operate with a half of the input voltage, $V_{in}/2$. This allows for significant efficiency improvements at lighter loads where the switching losses are dominant. Since in this mode of operation the effective inductance of the converter is reduced by a half, due to the output inductors appearing in parallel, it is not intended for operation under heavier load conditions where conduction losses dominate.

It should be mentioned that, by changing the gate driving sequence, this converter can also operate as a single phase 3-level buck for V_{out}/V_{in} ratios larger than 0.5. However, in the targeted applications, such modes are relatively rarely needed, compared to cases where a high step down is required. It should also be noted that the 2-phase implementation of the conventional 3-level buck of Fig. 2(c) has even larger functionality than the introduced 7SFC topology, while sharing the benefits of reduced voltage stress and lower voltage swing. It allows 3-level operation under heavy loads, for conversion ratios larger than 0.5. However, to achieve this relatively rarely used mode of operation, an extra flying capacitor and a switch , increasing the volume and cost of the system, would be needed.

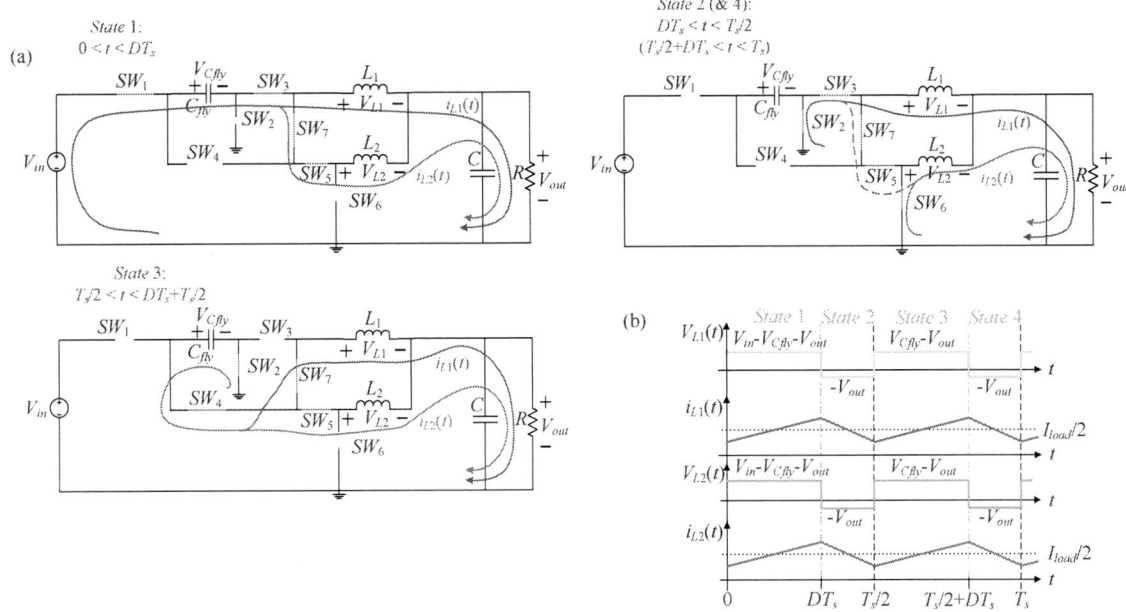

Fig. 4. (a) Switching sequence for the 3-level buck mode, (b) Inductor voltage and current waveforms.

C. 2-Phase Interleaved Buck Mode

The 2-phase interleaved buck mode of the 7SFC buck is required for providing conversion ratios where $V_{out}/V_{in} > 0.5$, due to the fact that neither of the other modes operates within that range. This mode is also utilized for $V_{out}/V_{in} < 0.5$ for medium step-down heavy load conditions, where the 3-level buck mode suffers due to conduction losses from the high inductor ripple. To keep all of the switch ratings at $V_{in,max}/2$, this mode is used for $V_{in} < V_{in,max}/2$, where $V_{in,max}$ is the highest allowable input voltage for the converter. With SW_1, SW_2 and SW_4 always kept 'ON' the 7SFC buck operates very similar to the conventional 2-phase interleaved buck with the flying capacitor C_{fly} held at the input voltage V_{in}. The 4 states of the switching sequence are shown in Fig. 5 for $D > 0.5$, where in *states* 1 and 3 both inductors are charged with V_{in} at the switching (v_x) nodes, in *state* 2 inductor L_1 continues charging

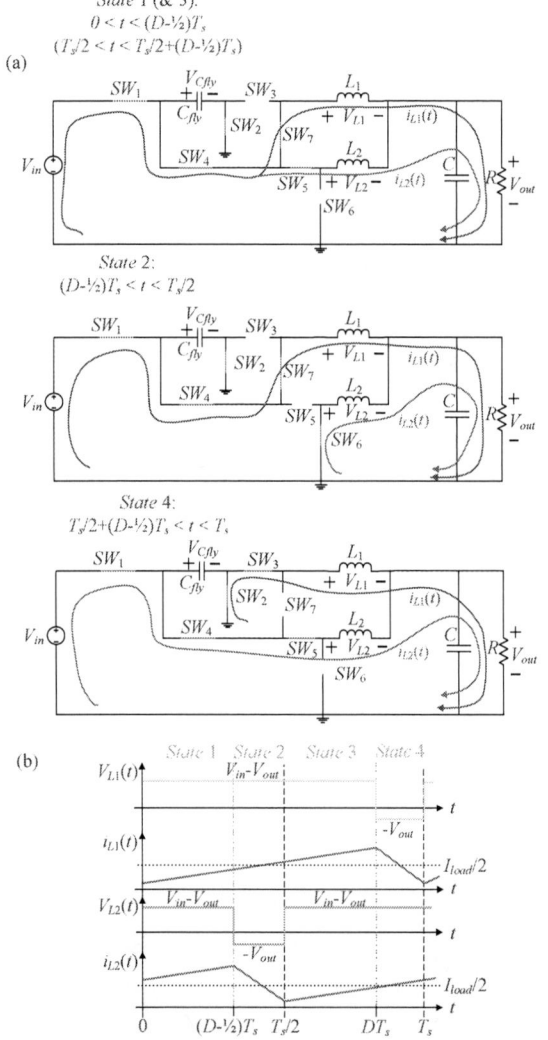

Fig. 5. (a) Switching sequence for the 2-phase interleaved Buck Mode, (b) Inductor voltage and current waveforms.

while L_2 is in synchronous rectification, and in *state* 4 the roles are reversed. Compared to the conventional 2-phase buck under the same operating conditions, the switches in the 7SFC buck are rated for a lower voltage, and the advantages of higher FOM semiconductors could potentially be utilized [12].

D. Digital Controller

The operation of the new topology is regulated by the digital controller of Fig. 1. The controller is a modification of a conventional digital voltage mode controller [8]. Here a digital pulse width modulator (DPWM) is modified to receive information about the mode of operation before creating the gating signals. Based on the input voltage and output voltage measurements, performed by sensors and corresponding analog-to-digital converters (ADCs), as well as the extracted inductor current value, the mode select logic determines the mode of operation of the 7SFC buck converter using a look-up table approach. For example, if the mode select logic determines the 7SFC should operate in the 2-phase interleaved buck mode, the DPWM will respond by keeping SW_1, SW_2 and SW_4 on while modulating the remaining switches according to the duty ratio sent from the compensator. In addition, the compensator can be designed with different parameters, depending on the mode of operation, to achieve the optimized dynamic performance for each mode.

III. EXPERIMENTAL VERIFICATION

To verify the performance of the introduced 7SFC multi-level buck converter, a wide input, $12\ \text{V} \leqslant V_{in} \leqslant 48\ \text{V}$, wide output, $1\ \text{V} \leqslant V_{out} \leqslant 12\ \text{V}$, 72 W, 800 kHz discrete experimental prototype was built. The performance of the 7SFC buck is compared with a conventional 2-phase buck

Fig. 6. Top view of the experimental prototype. Top PCB consists of the power stage components and bottom PCB consists of control circuitry.

under the same operating conditions and the same switching frequency. The control for the converters is performed using an FPGA based development board. Fig. 6 displays the previously mentioned discrete prototype, with the power stage PCB on top, containing the switches, gate drivers and reactive components, and the control circuitry on the bottom PCB, containing sensing circuits and ADCs. The components used for experimental testing are provided in Table I for the conventional 2-phase buck and the 7SFC buck.

TABLE I.   COMPONENTS USED FOR EXPERIMENTAL PROTOTYPE

Part	Role
CSD88539 MOSFET	7SFC: SW_1, SW_4, SW_5, SW_7
	Interleaved Buck: SW_1, SW_3
AO4264 MOSFET	7SFC: SW_2, SW_3, SW_6
	Interleaved Buck: SW_2, SW_4
MCP1407 low-side gate driver	7SFC: SW_2, SW_6
	Interleaved Buck: SW_2, SW_4
LTC4440 high-side gate driver	7SFC: SW_1, SW_3, SW_4, SW_5, SW_7
	Interleaved Buck: SW_1, SW_2
Wurth 1uH inductor: 744 373 490 10	7SFC inductors
Wurth 1.5uH inductor: 744 373 490 15	Interleaved buck inductors
2x10uF ceramic capacitor	C_{fly} for 7SFC
2x100uF ceramic capacitor	C for both converters (output capacitor)
Current sense wire for L_1 phase: 15mΩ	
Current sense wire for L_2 phase: 14mΩ	

Fig. 7 demonstrates two main advantages of the 7SFC converter over the conventional 2-phase interleaved buck, where both converters are operated under their worst case inductor current ripple operating condition, V_{in} = 48 V and V_{out} = 12 V, with 1uH inductors (Wurth: 744 373 490 10). From Fig. 7(a) it can be seen that in the conventional 2-phase buck the v_{x1} node varies between 0 V and 48 V, and the inductor ripple is approximately 10 A. On the other hand, Fig. 6(b) shows that the v_{x1} node of the 7SFC switches between 0 V and 24 V with an inductor ripple of about 6.6 A. This confirms that the 7SFC buck requires significantly smaller inductors (in this case about 33% smaller) to achieve the same current ripple,

and confirms the lower voltage stress of the components in the 7SFC buck. Thus, in further tests, the conventional 2-phase interleaved buck was tested with 1.5 uH inductors (Wurth: 744 373 490 15), to achieve the same worst case current ripple of 6.6A. Fig. 7(c) shows the 7SFC operating in the 3-level buck mode and confirms the reduced voltage swing by a half compared to that of the conventional 2-phase buck, as well as a further reduction in the inductor current ripple.

A. Volume Comparison

To compare the volume of the reactive components of the 7SFC buck to the conventional 2-phase buck the energy storage requirements are compared [10]. To achieve the same current ripple in the inductors, as shown previously, the inductance of the 7SFC buck can be reduced by 33%, resulting in 33% smaller inductors. This also results in a reduction of the output capacitor. This is because, for the targeted PoL applications, the dominant specification for selecting the output capacitor value is based on the transient performance in response to a load step, which depends on the current slew rate. Using the analysis presented in [14], which assumes time-optimal control during a transient, the output capacitor values were compared based on worst case operating conditions. Table II summarizes the required output capacitor values for both for a 50 mV output voltage deviation in response to a 3 A load transient. It is clearly shown that the required output capacitor for the 7SFC buck can be 33% smaller than the conventional 2-phase buck to achieve the same transient performance. The output capacitor reduction in the 7SFC buck is maintained when dealing with larger load transients as well.

In terms of silicon area, the 7SFC buck is expected to have comparable or smaller switch area, due to the fact that there are 7 switches, all rated at a half of the full input voltage, whereas the conventional 2-phase buck consists of 4 switches at the full input voltage. This estimation is based on the fact that, ideally, silicon area increases approximately with a squared relation to the blocking voltage of a switch, when considering a fixed switch on-resistance value [12].

B. Efficiency Comparison

The efficiency comparison between the two topologies was done by considering various loss components in each converter, as shown in Fig. 8 and Fig. 9.

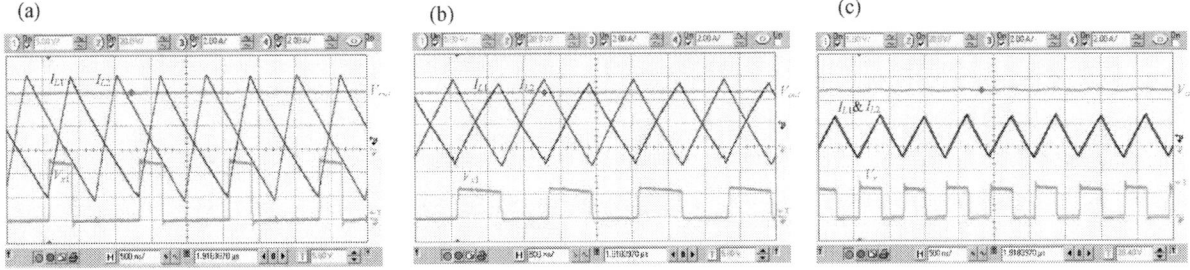

Fig. 7.   Conventional 2-phase interleaved buck and 7SFC buck waveforms for Vin = 48 V and Vout = 12 V with 1uH output inductors: (a) conventional 2-phase interleaved buck, (b) 7SFC high step-down buck mode (c) 7SFC 3-level buck mode.

TABLE II. OUTPUT CAPACITOR SELECTION BASED ON TRANSIENT PERFORMANCE

Topology	Transient Type	Operating Conditions	Capacitor Value
Conventional Interleaved Buck Converter	Step-Down	$V_{in} = 48V$, $V_{out} = 1V$	71.81uF
	Step-Up	$V_{in} = 3V$, $V_{out} = 2.875V$	544uF
2-Phase 3-Level Buck Converter	Step-Down	$V_{in} = 48V$, $V_{out} = 1V$	24.57uF
	Step-Up	$V_{in} = 3V$, $V_{out} = 2.875V$	182uF
7SFC based Multi-level Buck Converter	Step-Down	High Step-Down Mode $V_{in} = 48V$, $V_{out} = 1V$	49.15uF
	Step-Up	Interleaved Buck Mode $V_{in} = 3V$, $V_{out} = 2.875V$	364uF

The loss analysis demonstrated several important results. First, for cases where the high step-down mode of the 7SFC buck is used, most of the loss reduction is in the switching losses, as shown in Fig. 8. Compared to the conventional 2-phase buck, the switching losses are approximately halved, which is significant for large input voltages. For conversion ratios where the high step-down mode cannot be used, i.e. $0.25 < V_{out}/V_{in} < 0.5$, the loss breakdown for the 3-level buck mode and the interleaved buck mode are considered as shown in Fig. 9. It is clear that for light loads the loss savings are once again attributed to the switching loss reductions for the 3-level buck mode. Up to a certain load current level, however, the conduction losses dominate for the 3-level buck mode, and the interleaved buck mode is the most efficient. Efficiency results taken from the discrete prototype are shown for three operating points in Fig. 10. Fig. 10(a) clearly shows the efficiency improvements of using the high step-down mode of the 7SFC buck over the conventional 2-phase buck for the full range of load currents, due to the reduction in switching losses. In Fig. 10(b), for conversion close to 0.5, the 3-level buck mode of the 7SFC significantly improves light

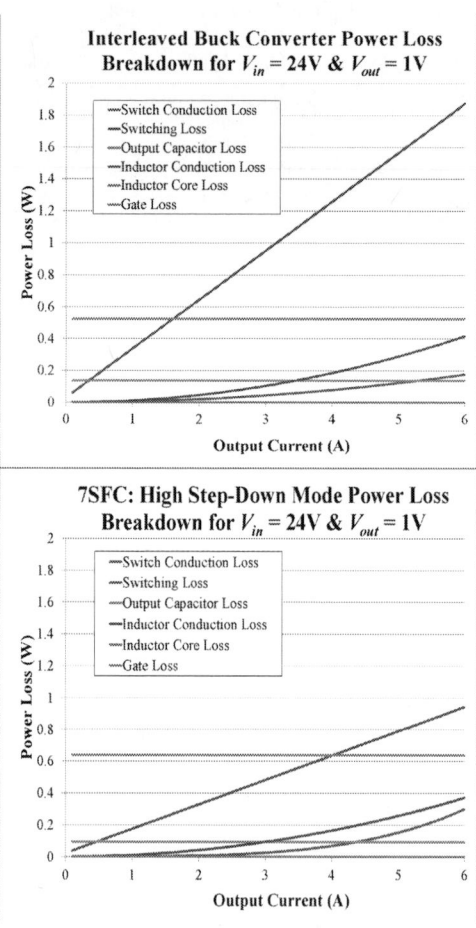

Fig. 8. Power loss breakdown for the conventional 2-phase buck and the 7SFC buck operating in the HSD mode for V_{in}=24V and V_{out}=1V

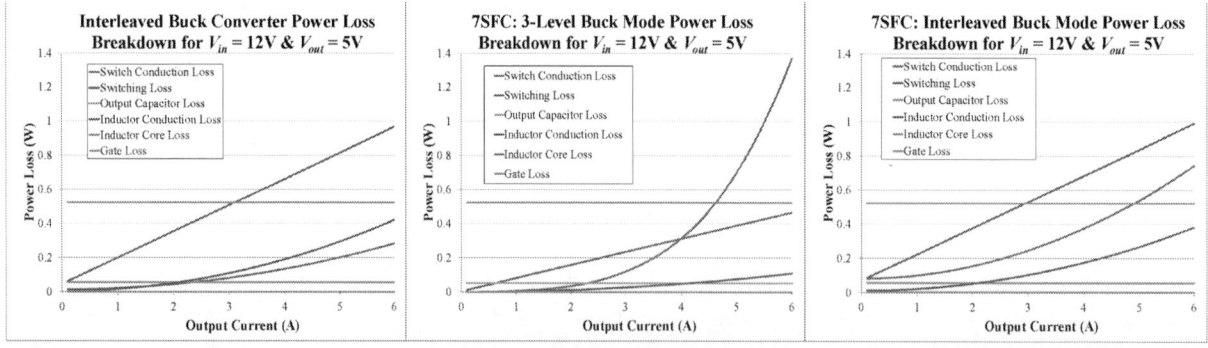

Fig. 9. Power loss breakdown for the conventional 2-phase buck and the 7SFC buck operating in the 3-level buck mode and the interleaved buck modes for V_{in}=12V and V_{out}=5V

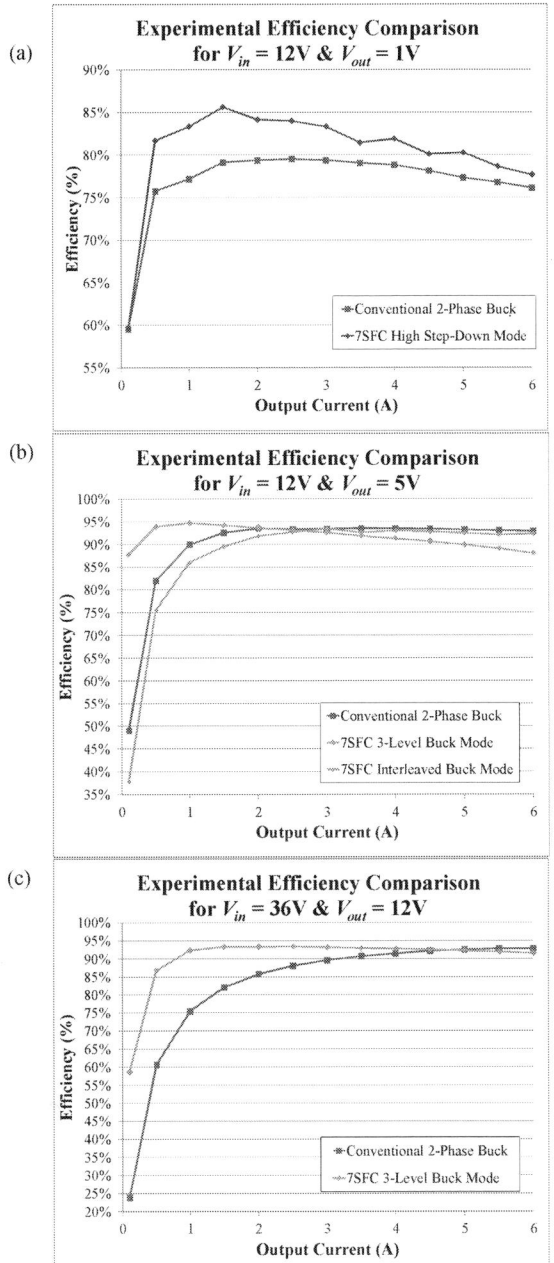

(a)

Experimental Efficiency Comparison for V_{in} = 12V & V_{out} = 1V

- Conventional 2-Phase Buck
- 7SFC High Step-Down Mode

(b)

Experimental Efficiency Comparison for V_{in} = 12V & V_{out} = 5V

- Conventional 2-Phase Buck
- 7SFC 3-Level Buck Mode
- 7SFC Interleaved Buck Mode

(c)

Experimental Efficiency Comparison for V_{in} = 36V & V_{out} = 12V

- Conventional 2-Phase Buck
- 7SFC 3-Level Buck Mode

Fig. 10. Experimental efficiency comparisons between conventional 2-phase buck and 7SFC buck for (a) V_{in}=12V, V_{out}=1V, (b) V_{in}=12V, V_{out}=5V and (c) V_{in}=36V, V_{out}=12V

load efficiency, and the interleaved buck mode should be used for load currents greater than 3A. The interleaved buck mode of the 7SFC has slightly reduced efficiency compared to the conventional 2-phase buck due to the increased conduction losses brought by the additional switches. Lastly, Fig. 10(c) demonstrates the advantage of using the 3-level buck mode of the 7SFC buck at light and medium loads for higher input

voltages (in this case V_{in} = 36 V). This once again confirms the advantage of having reduced switch blocking voltage and voltage swing. It can be seen that by properly selecting modes of the 7SFC, efficiency improvements are achievable over the full operating range.

IV. CONCLUSIONS

A novel high power density wide-input dc-dc highly flexible converter topology for a wide range of point of load applications is introduced. The 7-switch flying capacitor (7SFC) multi-level buck requires much smaller inductors compared to the conventional 2-phase buck and, at the same time, improves power processing efficiency. These advantages are achieved by reducing the voltage swing across the inductors and the voltage stress of all transistors to a half of full input voltage. Depending on the operating conditions, the 7SFC buck can operate in multiple 3-level buck converter modes or as a conventional 2-phase buck. For high step-down ratios it operates as a double step down buck, with the advantage of having all transistors rated for a low voltage, resulting in improved power processing efficiency and potentially allowing less costly on-chip integration. For light and medium loads the converter operates as a single phase 3-level buck and for conversion ratios larger than 0.5 as a conventional 2-phase buck. The advantages of the new topology are confirmed experimentally.

ACKNOWLEDGMENT

The authors of this paper would like to thank Tom Moiannou for his help.

REFERENCES

[1] D. Reusch, "High frequency, high power density integrated point of load and bus converters," Ph.D. Dissertation, Virginia Tech, Blacksburg, VA, USA, 2012

[2] A. Costabeber, P. Mattavelli, and S. Saggini, "Digital time-optimal phase shedding in multiphase buck converters," *IEEE Trans. Power Electron.*, vol.25, no.9, pp. 2242-2247, Sept. 2010

[3] J. Wang, K. Ng, T. Kawashima, M. Sasaki, H. Nishio, A. Prodić, and W.T. Ng,"A digitally controlled integrated dc-dc converter with transient suppression," *Proc. 22nd Int. Symp. Power Semicond. Devices IC's*, pp. 277-280, June 2010

[4] D. Reusch, F.C. Lee, D. Gilham, and S. Yipeng, "Optimization of a high density gallium nitride based non-isolated point of load module," *Proc. IEEE Energy Convers. Congr. Expo. (ECCE)*, pp.2914 -2920, 2012

[5] M. Shirazi, R. Zane, D. Maksimovic, L. Corradini, and P. Mattavelli, "Autotuning techniques for digitally-controlled point-of-load converters with wide range of capacitive loads," *Proc. IEEE Appl. Power Electron. Conf.*, pp.14 -20, Feb. 2007

[6] K. Nishijima, K. Harada, T. Nakano, T. Nabeshima, and T. Sato, "Analysis of double step-down two-phase buck converter for VRM," *Telecommunications Conference, 2005. INTELEC '05. Twenty-Seventh International* , pp.497,502, Sept. 2005

[7] T.A. Meynard, and H. Foch, "Multi-level conversion: high voltage choppers and voltage-source inverters," *Proc. Power Electron. Specialists Conf.*, vol. 1, pp.397 -403, 1992

[8] R. W. Erickson, and D. Maksimovic, "Fundamentals of power electronics." Springer Science & Business Media, 2007.

[9] P.S. Shenoy, M. Amaro, D. Freeman, and J. Morroni, "Comparison of a 12V, 10A, 3MHz buck converter and a series capacitor buck converter," *Proc. IEEE Applied Power Electron. Conf.*, pp. 461-468, Mar. 2015

[10] M. D. Seeman, "A design methodology for switched-capacitor dc-dc converters," Ph.D. Dissertation, University of California at Berkeley, 2009

[11] Y. Lei, W-C. Liu, and R. Pilawa-Podgurski, "An analytical method to evaluate flying capacitor multilevel converters and hybrid switched-capacitor convertors for large voltage conversion ratios," *Proc. 16th Workshop Comput. Power Electron. (COMPEL)*, pp. 1-7, 2015

[12] B. J. Baliga, "Fundamentals of power semiconductor devices." Springer Science & Business Media, 2010.

[13] J. Tucker, "Understanding output voltage limitations of DC/DC buck converters", Analog Applications Journal, Texas Instruments, pp. 11-14, 2008

[14] A. Bjeletić, L. Corradini, D. Maksimovic, and R. Zane, "Specifications-driven design space boundaries for point-of-load converters," *Proc. IEEE Applied Power Electron. Conf.*, pp. 1166-1173, Mar. 2011

High Efficiency *LLC* Converter Design for Universal Battery Chargers

Navid Shafiei, Seyed Ali Arefifar, Mohammad Ali Saket, and Martin Ordonez

Electrical and Computer Engineering, The University of British Columbia, Vancouver, BC V6T1Z4 Canada

Email: navid@ece.ubc.ca, arefifar@ieee.org, alisaket@ece.ubc.ca, mordonez@ieee.org

Abstract— In order to support different types of rechargeable batteries (e.g. Li-Ion, Lead-Acid, NiMh), the design of universal battery chargers must focus on wide conversion efficiency instead of traditional peak efficiency design. Wide efficiency is the ability to maintain high performance within the nominal output power while supporting the charging cycle voltage of different battery technologies. The objective of this paper is to tackle this new wide efficiency technical challenge and provide a design methodology that focuses on multiple operating points rather than obtaining peak efficiency at one operating point. The universal battery charger is expected to provide a demanding output voltage range between nominal and 1.5 times nominal and sustaining maximum power delivery with high efficiency. The proposed *LLC* converter design procedure successfully selects the resonant tank elements and operating frequencies to maximize efficiency for the maximum power region. The design procedure employs analytical equations and a Tabu Search algorithm (TS) for a $96VDC$, $960W$ universal battery charger implementation. The experimental results exhibit the excellent performance of the designed converter, which has an average efficiency of 96.1% within the nominal output power delivery range (between $96VDC$ and $144VDC$ output voltage range) with extreme regulation capability.

I. INTRODUCTION

Different types of rechargeable battery packs (Li-Ion, Lead-Acid, etc.) can be selected as storage systems for stationary power, Electric Vehicles (EVs), and other mobile applications. Depending on the battery type, the battery charger should deliver the energy to the battery pack according to the battery state of charge [1]. According to Fig. 1(a), the Li-Ion battery has two different modes in its charging profile which are constant current and constant voltage mode [2, 3]. During the constant current mode, the voltage of each cell is near its maximum (around 4V/cell) and therefore the battery charger is working around its maximum output power. According to Fig. 1(b), the Lead-Acid battery charge profile includes three main phases: Bulk, Absorption, and Finish [4–6]. During the Bulk phase, the maximum charging current is provided while monitoring the battery voltage. In the absorption mode, the charger increases the battery voltage to certain point, which is typically specified by the battery manufacturer (e.g. 2.35V/Cell). Due to the either over-discharging or extended storage time, the battery pack can be discharged completely and the voltage of the battery pack drops into dead-zone,

This work was supported in part by the Institute for Computing, Information and Cognitive Systems (ICICS) at UBC.

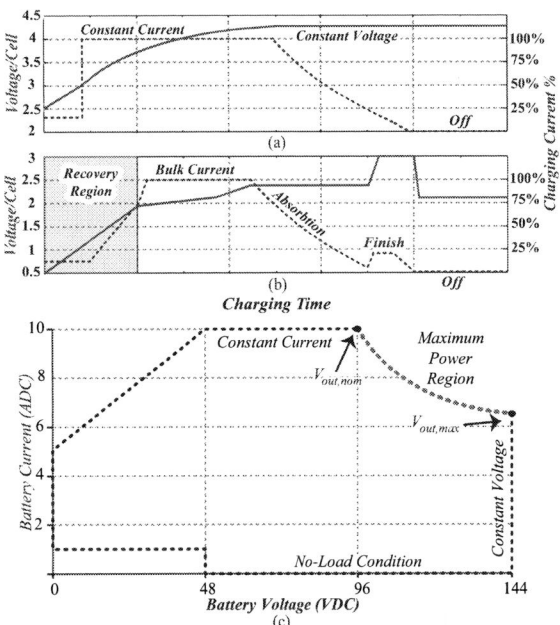

Fig. 1. (a) Charging cycle profile for a single Lithium-Ion battery cell, (b) Charging cycle profile for a single Lead-Acid battery including Recovery Region, (c) Desired V-I plane for a universal battery charger (e.g. $960W$ battery charger).

close to zero volts per cell (Recovery Region) [5]. Combining the charging profiles presented in Fig. 1(a) and (b) leads to a general V-I plane, presented in Fig. 1(c) that is essential to consider in universal battery chargers. According to Fig. 1(c), the battery charger should not only work in different load conditions, from absolutely zero to maximum output current, but also regulate the output voltage from near zero up to 1.5 times. Furthermore, wide conversion efficiency instead of traditional peak efficiency design during the maximum power delivery should be taken into account (e.g. between $96VDC$ and $144VDC$). Recently, various studies have been dedicated to developing reliable and efficient battery chargers for EVs. Among power converter topologies, the softly switched *LLC* resonant converter is one of the best battery charger designs, due to its ability to handle high power and produce variable voltage gains in different operating frequencies while providing soft switching for all semiconductor devices [5, 7, 8]. It is well understood that the optimization of resonant converters is a quite difficult task due to many degrees of freedom, and recently, the design procedure of resonant converters have been studied in the literature [5, 9–11]. An LLC resonant converter with a significant series resonant inductance is proposed for designing an adjustable wide-range regulated voltage source, but high amount of series resonant inductance in comparison with magnetizing inductance can lead to a voltage gain with a sharp slop which is not acceptable in the control circuit point of view [10]. An optimal design methodology has been studied in order to design battery charger for Li-Ion battery packs while considered maximum efficiency, but having wide regulation has not been considered during design procedure [11]. Musavi et al. designed and implemented an LLC resonant converter as a Lead-Acid battery charger, but wide conversion efficiency has not been considered [5]. In this paper, an advanced LLC design procedure is investigated in order to design an LLC resonant converter as an universal battery charger with wide conversion efficiency, which is able to provide advantageous extreme regulation and maximize the efficiency in the maximum power delivery region. Due to the many degrees of freedom (such as the selection of resonant circuit elements, operating switching frequency, transformer specification, maximum achievable efficiency), a Tabu Search algorithm (TS) is employed in order to find the optimum values of the design variables. The experimental results extracted from a $960W$ platform prove that the proposed design methodology can regulate the output voltage in a wide range while provide an average efficiency equal to 96.1% within the rated power delivery region.

II. OBJECTIVE FUNCTIONS, AND CONSTRAINTS

In order to design a high efficiency LLC resonant converter with wide output regulation, different objective functions and constraints should be defined. Fig. 2(a) presents the full bridge LLC resonant converter with a non-ideal transformer. In this circuit, the secondary leakage inductance and the magnetizing inductance of the transformer are considered L_{s2} and L_p, respectively, and the primary leakage inductance along with external inductor is considered L_{s1}. Fig. 2(b) presents the typical voltage gain of the LLC resonant converter. In this paper, two objective functions and three constraints are considered for the universal battery charger design. The two objective functions are as follows:

• Maximum output voltage regulation versus frequency variation: To respond to the charging cycle of different batteries, the charger should regulate the output voltage from near zero up to 1.5 times the nominal voltage under different loading conditions. In the LLC resonant converter, in order to Achieve a wide range of output voltage regulation, the switching frequency should be selected on the steep portion of the voltage gain curves around short circuit resonant

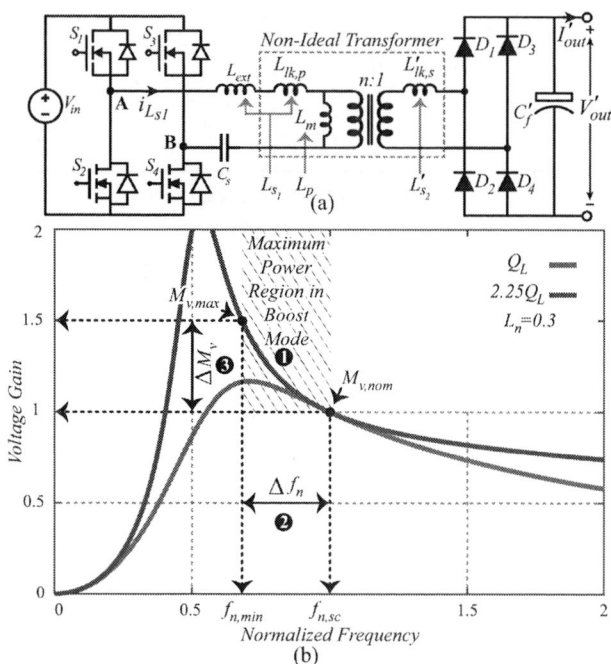

Fig. 2. (a) Full bridge LLC resonant converter schematic along with non-ideal transformer, (b) LLC resonant converter voltage gains. ① indicate the maximum power region in the LLC resonant voltage gains, ② presents the frequency variation in maximum power area, and ③ shows the additional voltage gain in the boost mode.

frequency, $f_{n,sc}$. Therefore, the first objective function should be presented by (1).

$$\left| \frac{\partial M_v}{\partial f_n} \right|_{f_n = f_{n,sc}} \rightarrow Maximize \qquad (1)$$

• Maximum obtainable efficiency at light loads: In order to have high part-load efficiency, the amplitude of the resonant circuit current should decrease with increasing load resistance. In other words, the less part load circulating current there is, the flatter the efficiency curve is in relation to the load. The second objective function is the relation of the resonant circuit current in full and half load conditions and is given by (2).

$$\left. \frac{I_{Ls1,FL}}{I_{Ls1,HL}} \right|_{f_n = f_{n,sc}} \rightarrow 2 \qquad (2)$$

Beside two defined objective functions, three most important constraints during design procedure are as follows:

• Input impedance phase of the LLC resonant converter: To guarantee the Zero Voltage Switching (ZVS) for the inverter Mosfets, the phase angle of the input impedance in all load conditions should be positive regarding to inverter output voltage ($v_{AB}(t)$). Ideally, the desire is to set this angle equal to zero, but due to practical limitation (e.g. the Drain-Source Mosfet capacitance, variation in the resonant elements), (3) is introduced for the input impedance angle.

$$\begin{cases} \varphi > 15° & \text{for practical margin} \\ \varphi < 25° & \text{for high efficiency achievement} \end{cases} \quad (3)$$

• Maximum obtainable gain of the LLC converter: Different selections of the resonant tank elements in the LLC resonant converter leads to a set of voltage gain curves versus loaded quality factor. During the design procedure, the resonant circuit elements should be selected some how to have around 50% increasing in the voltage gain for the loaded quality factor related to the maximum output voltage condition at maximum output power. The second constraint is related to the minimum voltage gain of the LLC resonant converter which is desirable in the boost mode and given by (4).

$$\frac{M_{v,max}|_{f_{n,min}}}{M_{v,nom}|_{f_{n,sc}}} \to 1.5 \quad (4)$$

• Maximum average efficiency at full power load in the boost mode: In the LLC resonant converter, the voltage gains are almost independent from the load when the switching frequency is around $f_{n,sc}$, resulting in minimum switching frequency variation versus load at nominal output voltage. This point is selected for nominal output voltage operation. Fig. 2(b) shows the voltage gain curves for two load conditions, which represent full load resistors at nominal output voltage and maximum output voltage conditions. According to this figure, in order to operate in the boost mode and increase the voltage gain by $\Delta M_v = 0.5$, the switching frequency should decrease by Δf_n, which is essential for covering the maximum power range, indicated in Fig. 1(c). Even though delivering the maximum output power at higher output voltage in the boost mode leads to lower output rectifier power losses, the lower frequency operation in the boost mode will have negative impact on the magnetics core losses (transformer and inductor core) [12]. Therefore, the last objective function which should be kept in mind is the efficiency of the charger in nominal and maximum output voltage at full load condition and presented by (5).

$$\frac{\eta_{Boost}}{\eta_{nom}}\bigg|_{F.L.} \to 1 \quad (5)$$

III. Tabu Search Optimization Algorithm

Due to the many degrees of freedom (such as the selection of resonant circuit elements, operating switching frequency, transformer specification, maximum achievable efficiency) employing an automatic optimization procedure is inevitable. In this paper, the Tabu Search algorithm (TS) is selected to find the optimum value of the charger parameters. The extracted normalized steady state equations of the LLC resonant converter along with transformer and output rectifier power losses are employed in the optimization process in order to find different parameters of the charger. The flowchart of the algorithm along with analytical equations base on First

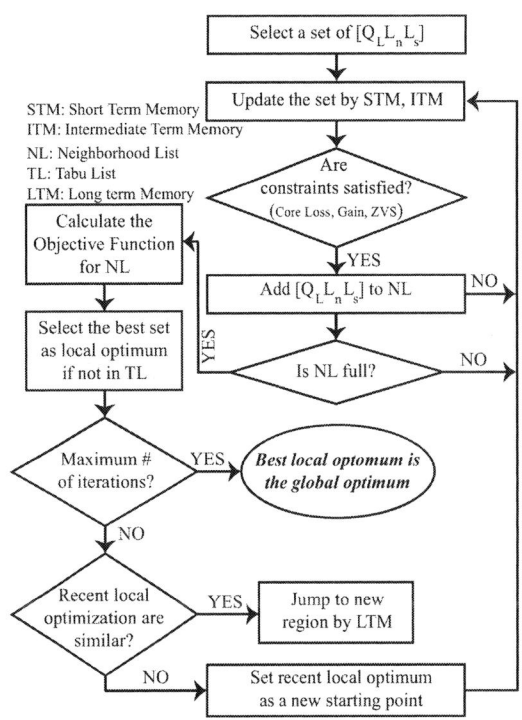

Fig. 3. Tabu Search Algorithm Flowchart.

Harmonic approximation (FHA) are presented in Fig. 3 and (6)-(12), respectively [8, 15].

$$Q_L = \frac{R_L}{Z_0}, Z_0 = \sqrt{\frac{L_{s_1}}{C_s}}, \omega_0 = \frac{1}{\sqrt{L_{s_1} C_s}},$$

$$\omega_n = \frac{\omega_s}{\omega_0}, L_s = \frac{L_{s_1}}{L_{s_2}}, L_n = \frac{L_{s_1}}{L_p} \quad (6)$$

$$\frac{\mathbf{Z_{in}}(j\omega_n)}{Z_0} = (j\omega_n + \frac{1}{j\omega_n}) + \frac{\frac{j\omega_n}{L_n}(\frac{j\omega_n}{L_s} + Q_L)}{\frac{j\omega_n}{L_n} + \frac{j\omega_n}{L_s} + Q_L} \quad (7)$$

$$|I_{L_{s_1}}| = \frac{4V_{in}}{\pi|Z_{in}(j\omega_n)|}, \; v_{AB_1}(t) = \frac{4V_{in}}{\pi} sin\omega_s t \quad (8)$$

$$M_v = \left| \frac{j\omega_n Q_L L_s}{Q_L L_n L_s + j\omega_n(L_n + L_s)} \right| \frac{1}{|\mathbf{Z_{in}}(j\omega_n)|} \quad (9)$$

$$P_{core} = k_i |\Delta B|^{\beta-\alpha} f_s | 2\Delta B|^{\alpha}(\frac{1}{f_s})^{1-\alpha}$$

$$k_i = \frac{k_c}{2^{\beta-1}\pi^{\alpha-1}(1.1044 + \frac{6.8244}{\alpha+1.354})}; \; \Delta B = \frac{V_{Tr,p}}{2N_p A_e f_s} \quad (10)$$

$$P_{Tr} = P_{core} + P_{Cu}, \; P_{Cu} = R_p I_{p,rms}^2 + R_s I_{s,rms}^2 \quad (11)$$

$$P_{rec} = 2V_F I_{s,rms} \quad (12)$$

The TS algorithm is a heuristic search method that employs different memory structures to guide the search to a good

TABLE I
PARAMETERS RANGES AND SELECTED VALUES

Parameters	Value/Range	Selected
Input Voltage	$200VDC$	***
Nominal Output Voltage	$96VDC$	***
Maximum Output Power	$960W$	***
Transformer Core	***	N87 E58/25/21
Rectifier Diode	***	MBR40250
L_{s1} to L_m Ratio (L_n)	$[0.1-1]$	0.99
L_{s1} to L_{s2} Ratio (L_s)	$[4-20]$	5
Loaded quality Factor (Q_L)	$[0.1-5]$	0.48
Switching Frequency	$[100\text{-}250]$	$150kHz$
First Series Inductance	***	$91\mu H$
Second Series Inductance	***	$18\mu H$
Parallel Inductance	***	$91\mu H$
Series Resonant Capacitor	***	$10nF$
Transformer Turn's Ratio	***	$16:6$

TABLE II
EXPERIMENTAL RESULTS

Parameters	Value	Selected
Output Current	$10ADC$	$6.7ADC$
Efficiency	96.2%	96%
Secondary Temperature	$99\,°C$	$90\,°C$
Diode Temperature	$74\,°C$	$63\,°C$
Core Temperature	$51\,°C$	$69\,°C$

solution, effectively [13, 14]. The TS starts with finding a feasible solution and continues iteratively until certain criterion, which is usually the maximum iteration numbers, is reached. In the present problem, each feasible solution is a vector of Q_L, L_n and L_s, where each component can be changed in a pre-specified range. The next step is to make sets of neighbors for the starting points. A neighbor can be defined in many ways; here each neighbor is selected by changing at least one component of the vector and checking the feasibility of the resulted design in terms of the three constraints mentioned in Section II. The next step is to calculate the objective functions for all neighborhoods, set the best neighbor as the new starting point and continue the process. Table I presents the range for the normalized parameters and the switching frequency along with the transformer and output rectifier specification. Also, the input-output specifications of the LLC resonant converter are presented in Table I. All the data presented in Table I are used in order to design the LLC resonant converter as a universal battery charger. Based on the defined objective functions and constraints, the selected Q_L, L_n and L_s along with the circuit elements specifications are presented in Table I. The final values of parameters have been selected after conducting simulations and taking into account practical considerations.

IV. EXPERIMENTAL RESULTS

In this section, the validation of the proposed design method for the LLC resonant converter as an universal battery charger is examined through a $960W$ prototype platform. As mentioned in the introduction, a universal battery charger must be able to respond to different modes of the charging cycle voltage and cover all of the V-I plane region (Fig. 1(c)). To extract the maximum output voltage regulation of the designed converter, the platform has been tested for different limitations on the boundaries of V-I plane, including constant current, constant voltage, constant power, and no-load conditions. The obtained experimental results shows that

the charger can cover almost all areas presented in Fig. 1(c). As shown in Fig. 1(c), the universal battery charger should be able to deliver the maximum output power to the battery in the range of nominal and 1.5 times the nominal output voltage and in order to run LLC resonant converter in the boost mode, the switching frequency should be decreased. See Fig. 2(b). As mentioned earlier, delivering the maximum output power at lower current in the boost mode leads to lower output rectifier power losses, but the lower operation frequency results in higher flux density which have negative impact on the magnetics core losses. Due to the importance of having high efficiency at maximum power delivery conditions, the experimental results related to constant power region are presented. Fig. 4(a) and 5(a) present the experimental results of the designed battery charger for the nominal and maximum output voltage. According to Fig. 4(a), the nominal switching frequency at $96VDC$ is $150kHz$ and the efficiency is measured equal to 96.2%. Fig. 5(a) presents the experimental results for the maximum output voltage. According to this figure, in order to provide the maximum output voltage ($144VDC$), the switching frequency decreases to $135kHz$ with the efficiency equal to 96%. It is worth while to mention during the design procedure, TS algorithm has put the minimum switching frequency near short circuit resonant frequency in order to maximize the efficiency at the maximum load condition. The temperature profile in steady state condition of the transformer and output rectifier is presented in Fig. 4(b) and 5(b) and summarized in Table II. Comparing the temperatures of different points shows that in higher output voltage, the temperature of the output rectifier and secondary winding are lower than nominal output voltage condition, but thanks to to limited frequency variation, the core temperature just increased by $18°C$. To sum up, the proposed methodology along with defined objective functions and constraints can be used for designing a universal battery charger with wide regulation and an average efficiency equal to 96.1% in the boost mode.

V. CONCLUSIONS

This paper proposed a new approach for designing an LLC resonant converter as a universal battery charger with wide conversion efficiency in full load condition. The design procedure was based on TS algorithm with two objective functions and three constraints which was able to find the optimum values of the normalized parameters of the

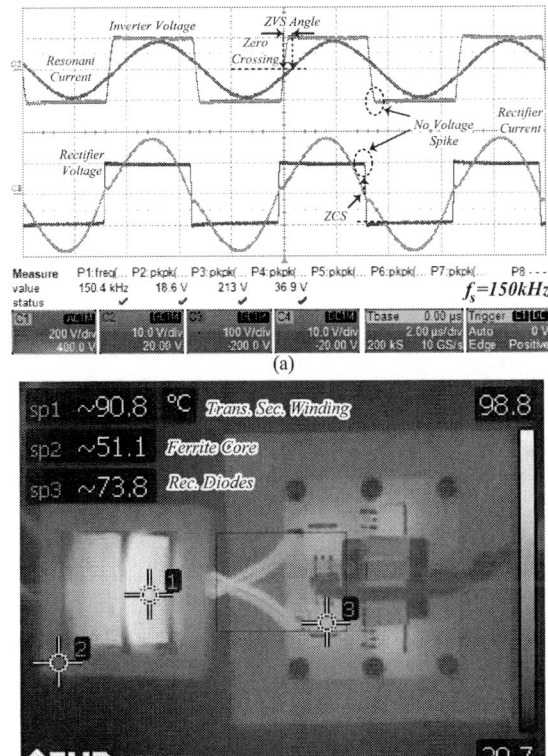
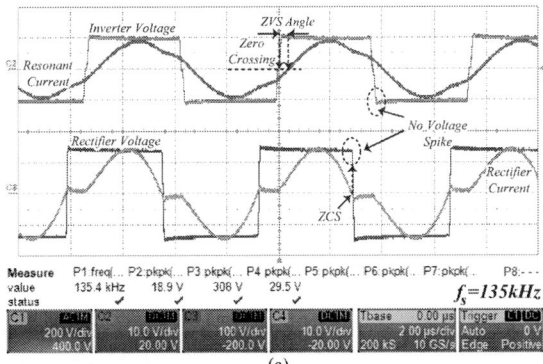
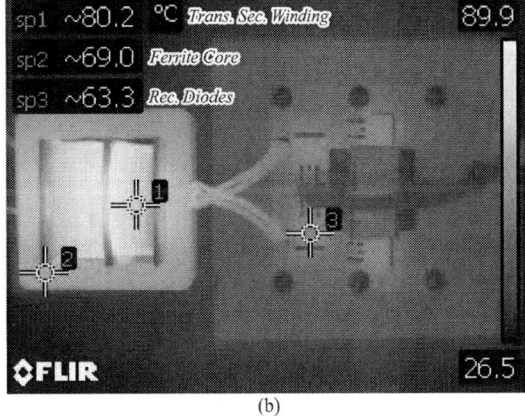

Fig. 4. Experimental Result for $P_{out} = 960W$ and $V_{out} = 96VDC$. (a) The voltage and current waveforms, and (b) The temperature gradient (after one hour test).

Fig. 5. Experimental Result for $P_{out} = 960W$ and $V_{out} = 144VDC$. (a) The voltage and current waveforms, and (b) The temperature gradient (after one hour test).

LLC resonant converter and the switching frequency with consideration of wide conversion efficiency at maximum output power delivery. The experimental results extracted from a $96VDC$, $960W$ power platform proved that the LLC resonant converter with this design procedure strategy covered almost all regions in the V-I plane, and had an average efficiency of 96.1% at the maximum power delivery range and can be employed as a universal battery charger.

REFERENCES

[1] M. Yilmaz, and P. Krein, "Review of battery charger topologies, charging power levels, and infrastructure for plug-in electric and hybrid vehicles," *IEEE Transaction on Power Electronics*, vol. 28, no. 5, pp. 2151-2169, May. 2013.

[2] J. Deng, S. Li, S. Hu, C. Mi, and R. Ma, "Design methodology of LLC resonant converters for electric vehicle battery chargers," *IEEE Transaction on Vehicular Technology*, vol. 63, no. , pp. 1581-1592, May 2014.

[3] Z. Fang, T. Cai, S. Duan, and C. Chen, "Optimal design methodology for LLC resonant converter in battery charging applications based on time-weighted average efficiency," *IEEE Transaction on Power Electronics*, vol. 30, no. 10, pp. 5469-5483, Oct. 2015.

[4] Torjan Battery Company, Deep-Cycle AGM type Batteries, *www.trojanbattery.com*

[5] F. Musavi, M. Craciun, D. Gautam, W. Eberle, and W. Dunford, "An LLC resonant DC-DC converter for wide output voltage range battery charging applications," *IEEE Transaction on Power Electronics*, vol. 28, no. 12, pp. 5437-5445, Dec. 2013.

[6] N. Shafiei, and M. Ordonez, "Improving the regulation range of EV battery chargers with L3C2 resonant converters," *IEEE Transaction on Power Electronics*, vol. 30, no. 6, pp. 3166-3184, May 2015.

[7] C. Chang, E. Chang, and H. Cheng, "A high-efficiency solar array simulator implemented by an LLC resonant DC-DC converter," *IEEE Transaction on Power Electronics*, vol. 28, no. 6, pp. 3039-3046, Jun. 2013.

[8] N. Shafiei, M. Ordonez, M. Craciun, C. Botting, and M. Edington, "Burst mode elimination in high power LLC resonant battery charger for electric vehicles," *IEEE Transaction on Power Electronics*, IEEE Early Access.

[9] J. Biela, U. Badstuebner, and J. Kolar, "Design of a 5-kW, 1-U, 10-kW/dm3 resonant DCDC converter for telecom applications," *IEEE Transaction on Power Electronics*, vol. 24, no. 7, pp. 1701-1710, Jul 2009.

[10] R. Beirvand, B. Rashidian, M. Zolghadri, and S. Alavi, "Using LLC resonant converter for designing wide-range voltage source," *IEEE Transaction on Industrial Electronics*, vol. 58, no. 5, pp. 1746-1756, May 2011.

[11] R. Yu, G. Ho, B. Pong, B. Ling, and J. Lam, "Computer-aided design and optimization of high-efficiency LLC series resonant Converter," *IEEE Transaction on Power Electronics*, vol. 27, no. 7, pp. 3243-3256, May 2012.

[12] W. G. Hurley and W. H. Wolfle, *Transformers and inductors for power electronics*, John Wiley and Sons, NY, 2013.

[13] Glover, F. "Tabu Search Part I", ORSA Journal on Computing 1989 1: 3, 190-206.

[14] Glover, F. "Tabu Search Part II", ORSA Journal on Computing 1990 2: 1, 4-32.

[15] M. K. Kazimierczuk and D. Czarkowski, *Resonant power converters*, John Wiley and Sons, NY, 1995.

A New High Power Density Modular Multilevel DC-DC Converter with Localized Voltage Balancing Control for Arbitrary Number of Levels

Ahmed Morsy, Yong Zhou, Prasad Enjeti
Department of Electrical and Computer Engineering
Texas A&M University
College Station, Texas, USA
enjeti@tamu.edu

Abstract—A new modular multilevel DC-DC converter with high power density and simplified localized voltage balancing control is proposed. In the proposed configuration, converter building blocks with the same power handling capability are connected in parallel in each row. Converter building block consists of an H-bridge and mutually coupled inductors whose total current is nearly ripple free. These features are shown to reduce the voltage ripple of DC-link capacitors significantly, leading to a smaller capacitance and size. An optimized control algorithm with voltage feedback PI loop is proposed resulting in the elimination of current sensors. This, reduces overall system complexity and increases cost-effectiveness. Significant ripple reduction of the inductor current and capacitor voltages is observed based on the simulation and prototype of multi-level systems. With a fully modular power stage module and localized control module, a system of arbitrary number of level can be built by stacking modules thereby contributing to enhanced system redundancy.

Keywords—Modular Multilevel Converter (MMC); DC-DC; High Power Density; Localized Control; Voltage Balancing; Converter Building Block; Controller Module; Integrated H-Bridge; Mutual Coupled Inductor; Current Cancellation; Voltage Ripple Reduction.

I. INTRODUCTION

Over the past few years, a lot of research has been done to address the technical challenges associated with the operation and control of modular multilevel converters (MMCs) [1], especially in high voltage (HV) and medium voltage (MV) applications. Fully modular converters are suitable for industrial applications whose power rating varies in a wide range. Generally, advantages of modular multilevel converters are: 1) "one time for all design" which is convenient for propagation; 2) predictive voltage and current stresses for components by incorporation of identical modules; 3) higher power density and better thermal management due to optimized design of each module and optimal combination of different modules; 4) reduced design, manufacturing, installation and maintenance costs due to the design standardization, etc. This section will review the following concerns: topology, modelling, control, operation characteristic and application.

The schematic diagram of a typical three-phase MMC is given in [1]. It consists of two arms per phase leg where each arm comprises N series-connected, nominally identical submodules (SMs), and a series inductor. Additionally, six popular structures for each SM are introduced, like half-bridge, full-bridge, flying capacitor and neutral point clamp (NPC), etc. Their possible voltage levels, DC-handling ability and power loss characteristics are analyzed and compared. The half-bridge SM has been widely accepted and applied because of the less complexity and higher efficiency. Similarly, a half-bridge submodule is used in a proposed bidirectional triangular modular multilevel DC-DC converter (TMMC) in [2]. The submodule is derived from a buck-boost converter. This topology is bidirectional such that high side voltage (V_{HV}) and low side voltage (V_{LV}) could be exchanged either as input or output. In [2], a two-level prototype system with three submodules is analyzed regarding current flow with different switching states.

Additionally, several other effective modular multilevel DC-DC converter have been proposed. In [3], a novel completely modular multilevel capacitor-clamped DC-DC converter is proposed with the comparison to the conventional flying capacitor topology. In [4], [5], the inductor-free design has high efficiency and flexible conversion ratio which is suitable for power management of fuel cell or automotive applications. Some merits in [6] are multilevel loads/sources support and fault tolerant capability. Furthermore, soft switching technology is researched focusing on efficiency increase and current/voltage spike reduction during switching transition. In [7] and [8], zero current switching and zero voltage switching are employed, respectively. Each one is able to achieve efficiency higher than 95%. In some applications requiring isolation, isolated DC-DC topologies will be used.

Some of the design and control issues regarding MMC are: SM capacitor voltage balancing, capacitor voltage/inductor current ripple reduction, centralized control, etc.

Similar to any other multilevel converter topology, the MMC needs an active voltage balancing strategy to balance and maintain the SM capacitor voltages at V_{DC}/N. In [2], an effective dual close-loop is proposed which can achieve not only capacitor voltage balancing, but equal inductor current sharing. The current reference of the internal current controller is provided by the outer voltage controller. This complicates the overall control algorithm. Some PWM techniques are discussed in [9], [10]. In [11], a voltage balancing strategy is achieved by assigning appropriate PWM pulses to the SMs of

978-1-4673-9551-9/16 $31.00 © 2016 IEEE

each arm. This does not require the measurement of arm currents and it simplifies the control loop and reduces the number of sensors. In [12], a modular controller concept is proposed which uses a closed-loop controller for each SM. Some other novel capacitor voltage balancing control strategies are also proposed, like predictive control and sorting method based control.

Regarding reducing SM capacitor voltage ripple, [2] proposed an interleaved operation technique which achieved significant reduction in the input current an output voltage ripples of a two-level prototype system. This is done by phase-shifting the corresponding gating signals 180 degree. Comparing with non-interleaved case, the experimental result shows a reduction of 38% and 45% for output voltage and input current ripples, respectively. Additionally, the efficiency is increased from 95.9% to 96.2%.

For the localized control, [8] proposed an effective PI control principle based on outer voltage loop and internal current loop by using TI DSP28335 for a two-level system. However, for systems with larger number of level, this principle needs to be re-evaluated and a higher controller cost may occur.

From the operation perspective, the capacitors in [3] are exposed to different steady-state voltage stresses, this makes the topology non-modular. In [2], the two-level MMC prototype system is verified by centralized DSP 28335 microcontroller. However, with the increased number of level, a centralized controller for all the modules is a potential problem.

II. MODELLING OF PROPOSED CONVERTER TOPOLOGY

The finalized converter is derived from the submodule configuration consisting of half-bridge and single inductor which is similar to the conventional synchronous buck converter. Then this module is redesigned and a new configuration with H-bridge and mutually coupled inductors is proposed. They are shown in Figure 1 (a) and (b), respectively.

In this paper, a modular five-level step-down DC-DC converter is set as example for discussion. However, this configuration can be generalized to any number of levels because of the modular design concept. The input can be constant voltage or current power sources. The load can be tapped to any voltage level, and multiple loads can be interfaced at different points. Also, the source side and load side are interchangeable which enables step-up operation as well. Due to the bi-directional conduction characteristic of power switches, like MOSFETs, the power flow in this configuration is also bi-directional. The detailed configuration of a modular five-level step-down (5:1) DC-DC converter by using building block of H-bridge with mutually coupled inductors is shown in Figure 2.

To simplify the analysis, the following assumptions are applied to the converter building block:

1) For the mutually coupled inductors, the self-inductance is L, mutual inductance is M and the leakage inductance is L_l, $L = M + L_l$. Coupling coefficient is k_M; r is the ESR for each

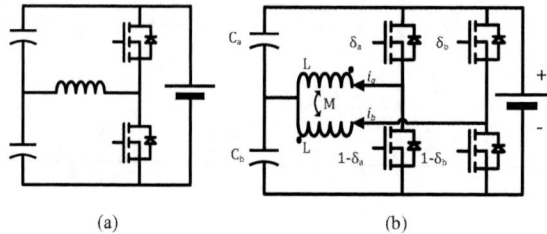

(a)   (b)

Fig. 1. Two submodules for MMC: (a) Half-bridge with single inductor. (b) H-bridge with mutually coupled inductors (proposed).

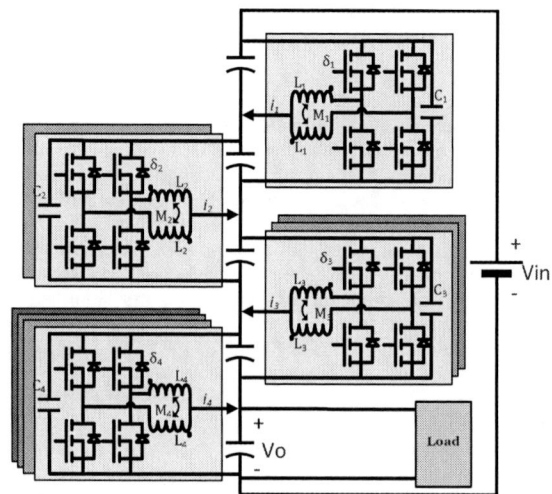

Fig. 2. Five level step down (5:1) DC-DC converter based on the final proposed building block

winding.

2) All the DC-link capacitor values are C;

3) Input source current is considered as constant, and the average value is express as I_s;

4) No phase shift between paralleled building blocks of the same row;

5) From top to bottom, the row number are 1st, 2nd, 3rd, etc.

Writing current equations for each building block as below:

$$L\,i_a - M\,i_b = \delta_a v_{C\,a} - (1-\delta_a)v_{C\,b} - r\,i_a \qquad (1)$$

$$L\,i_b - M\,i_a = \delta_b v_{C\,a} - (1-\delta_b)v_{C\,b} - r\,i_b \qquad (2)$$

Define total current i_t and current difference Δi, as shown in Equation (3) and (4), respectively.

$$i_t = i_a + i_b \qquad (3)$$

$$\Delta i = i_a - i_b \qquad (4)$$

During optimal operation, δ_a & δ_b are 50% duty cycle and 180° phase shifted, thus,

$$\delta_a + \delta_b = 1 \qquad (5)$$

By adding and subtracting (1) and (2), the following state space representation can be obtained.

$$\begin{bmatrix} i_t \\ \Delta i \end{bmatrix} = \begin{bmatrix} \frac{1}{L_l} & \frac{-1}{L_l} \\ \frac{2\delta_a-1}{2L-L_l} & \frac{2\delta_a-1}{2L-L_l} \end{bmatrix} \begin{bmatrix} v_{C\,a} \\ v_{C\,b} \end{bmatrix} + \begin{bmatrix} \frac{-r}{L_l} & 0 \\ 0 & \frac{-r}{2L-L_l} \end{bmatrix} \begin{bmatrix} i_t \\ \Delta i \end{bmatrix} \quad (6)$$

Rewriting, we get,

$$i_t^\circ = \frac{v_{C\,a}-v_{C\,b}}{L_l} - \frac{r}{L_l} i_t \quad (7)$$

$$\Delta i^\circ = \frac{(2\delta_a-1)(v_{C\,a}+v_{C\,b})}{2L-L_l} - \frac{r}{2L-L_l} \Delta i \quad (8)$$

The self-inductance L is calculated by (9) at 50% duty ratio.

$$L = \frac{1}{(1+k_M)} \frac{T_s/2}{\Delta i} \frac{2V_{DC}}{n} \approx \frac{V_{DC}}{n} \frac{T_s}{2} \frac{1}{k_i \, 2 \, I_s} \quad (9)$$

Where,

V_{DC} Input voltage;

n Number of levels;

T_s Switching period ($T_s = \frac{1}{f_s}$).

The self-inductance is proportional to SM capacitor voltage, and inversely proportional to switching frequency and current ripple percentage.

Observing Equations (7) and (8), the total current is driven by the voltage difference of the upper and lower capacitors (v_{Ca} and v_{Cb}) and the leakage inductance determines its dynamics. The current difference is driven by the duty cycle error compared to 50% for each leg, and the self-inductance determines its dynamics. In most cases, leakage inductance is very small, this requires the upper and lower capacitor voltages to be well balanced in order to keep the total current variation within an acceptable range. On the contrary, even if there is some duty cycle error, the current difference variation could be maintained within an acceptable level because of the relatively high self-inductance (leakage inductance is small compared to self-inductance). This reveals one critical point for control algorithm: achieving as accurate voltage balancing control as possible.

III.   PROPOSED LOCALIZED VOLTAGE BALACING CONTROL

The control algorithm is aiming at simplifying sensor circuit and also providing localized modular control of all converter building blocks. It's also focusing on hardware installation and future extension flexibility. Each controller module is only responsible for the converter building blocks of the same role. In this case, both the power stage building block and the controller module are really modular which is just as the name "modular multilevel DC-DC converter" shows. A brief block diagram of the whole system with controller modules is shown in Figure 3.

The modelling analysis illustrates that dynamic capacitor voltage balancing and duty cycle matching are the basis of proposed control algorithm. With this prerequisite, the equal current sharing between converter building blocks will be achieved automatically, equal power sharing as well.

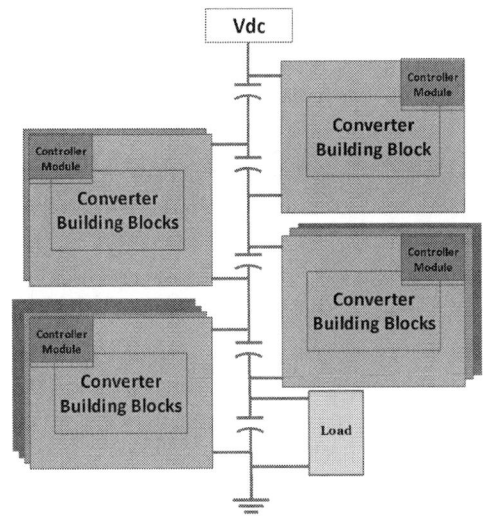

Fig. 3. Block diagram of the whole system with converter building blocks and localized controller modules

For two capacitors that are connected in series, if each capacitor voltage is half of the total voltage, the capacitor voltage balancing is achieved. The negative side of the lower capacitor could be regarded as the voltage reference, which is the floating ground of this row. From top to bottom, the floating ground of each row will be created in this manner. The detailed block diagram of proposed modular control algorithm of each row is expressed in Figure 4.

In this proposed control algorithm, current sensors and voltage references are eliminated. This greatly simplifies the control loop and improves the system robustness. The dynamic capacitor voltage balancing will always be achieved no matter how source voltage and load change. Additionally, by creating a floating ground for each row, this control algorithm can be applied to a system with any number of levels.

As clarified before, in each row, power stage modules are connected in parallel. There is no phase shift between each pair of gating signals for paralleled modules thanks to the current ripple cancellation strategy for the mutually coupled inductors. This leads to simpler control algorithm design and makes modular controller into reality.

For the implementation of control algorithm, digital feedback control with microcontroller is a great fit. The ADC inputs are DC which means the ADC sampling frequency can be relatively low. Maximum switching frequency is 20 KHz in this design. The resolution of PWM module should be relatively high in order to achieve good capacitor voltage balancing which is the basis of current balancing of coupled inductors. The microcontroller clock frequency should be fast enough for the sake of reducing the delay of digital feedback control loop. A very popular option is TI DSPs which are capable of achieving all the requirements discussed above. However, to demonstrate low cost solution for localized controllers dsPIC33FJ06GS102A from Microchip is chosen.

978-1-4673-9551-9/16 $31.00 © 2016 IEEE   2569

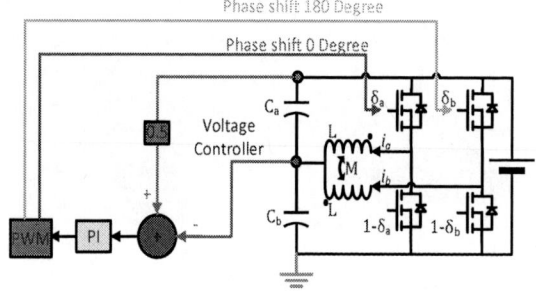

Fig. 4. The detailed block diagram of proposed modular control algorithm of each row

IV. SIMULATION ANALYSIS

A 5-level simulation example is implemented using PSIM. Primary parameter settings are shown in Table I. The simulation results of steady state are shown in Figure 5-8.

TABLE I. PRIMARY PARAMETER SETTINGS FOR A 5-LEVEL SIMULATION

System Power Rating	$P_o = 1.125kW$
Source Voltage (Constant)	$V_{in} = V_{DC} = 375V$
Source Current (Average)	$I_s = 3A$
Load	$R = 5\Omega, V_o = 75V, I_o = 15A$
Switch Frequency (Max)	$f_s = 20kHz$
Number of level	$n = 5$
Average current of each coil	$i_a \approx i_b \approx I_s = 3A$
Current ripple ratio for each coil	$k_i \leq 0.05$
Peak-peak current for each coil	0.15A
Current difference between two coils	$\Delta i = 0.3A$
Self-inductance of each coil	$L = 6.375mH$
Capacitor value	25uF

Figure 5 shows capacitor voltages waveforms. Figure 6 shows total inductor current for all building blocks of each row. Figure 7 shows the separate coil current waveform of the mutually coupled inductors. Figure 8 shows the power supply current waveform.

Based on the simulation results, the five DC-link capacitor voltages are balanced at exactly one fifth of the power supply voltage. The total current for mutually coupled inductors is 6A with negligible ripple. From top to bottom, the total current injected to the DC-link capacitor is proportionally to the number of paralleled converter building blocks of each row. The currents that are flowing through the two coupled coils are interleaved which results in current ripple cancellation for the total current. As for the power supply current, it is mainly of DC component value of 3A with some periodic ripples because of the switching operation of H-bridge.

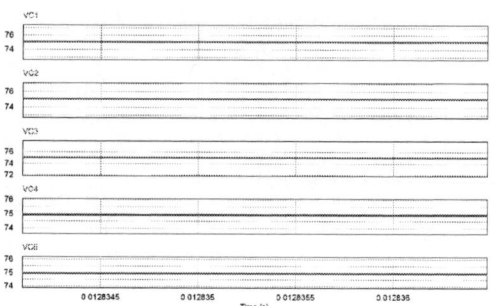

Fig. 5. Capacitor voltage waveform

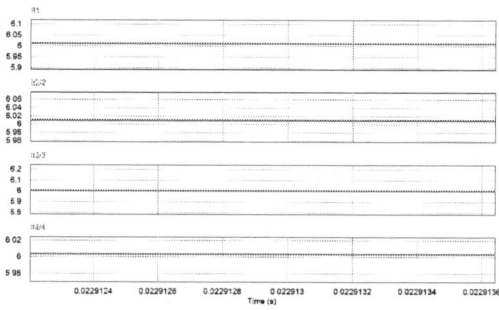

Fig. 6. Total inductor current for all building blocks of each row

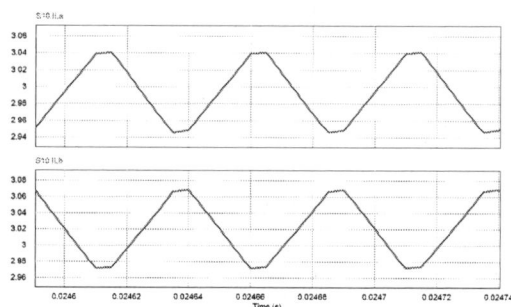

Fig. 7. Separate coil current of the mutually coupled inductors

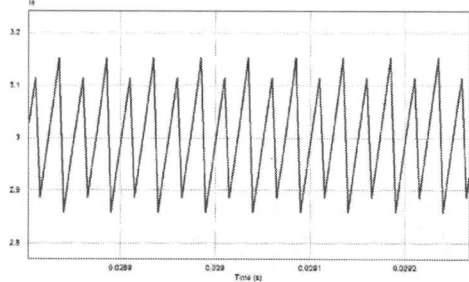

Fig. 8. Power supply current waveform

V. Prototype Experiment Result

Major component selection for the power stage building block is shown in Table II.

TABLE II. MAJOR COMPONENT SELECTION FOR THE POWER STAGE BUILDING BLOCK

Component	Value	Dimension	Description
Mutually Coupled Inductors	8 mH	1.45" x 1.35" x 0.8"	Triad Magnetics CMT-8112
Capacitor	25uF	1.65" x 1.1"	EPCOS (TDK) B32676E3256K
H-bridge	250V 4.6A	1.14" x 0.67"	International Rectifier IRSM505-084

Fig. 9. The 4-level hardware prototype system

The 4-level hardware prototype is shown in Figure 9. Figure 10 shows capacitor voltage waveforms V_{c1} and V_{c2}. Figure 11 shows capacitor voltage waveforms V_{c3} and V_{c4}. Figure 12 shows the total inductor current of all building blocks in each row. Figure 13 shows separate coil current of the mutually coupled inductors. Figure 14 shows the DC power supply current waveform.

Under the steady state operation of a 200V DC power supply, the capacitor voltages are evenly distributed across the four levels. The inductor current sharing ability is verified from waveforms in Figure 13. The ripple current cancellation property of mutually coupled inductors could be clarified from Figure 12. The transient peaks are caused by the resonance of parasitic circuit components. All the current and voltage waveforms are similar to the simulation results.

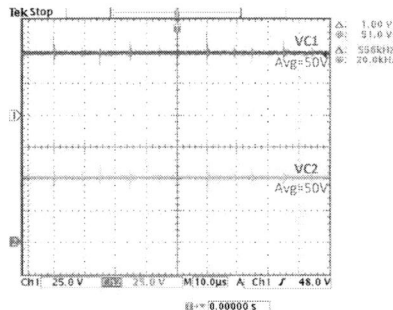

Fig. 10. Capacitor voltage waveforms V_{c1} and V_{c2}

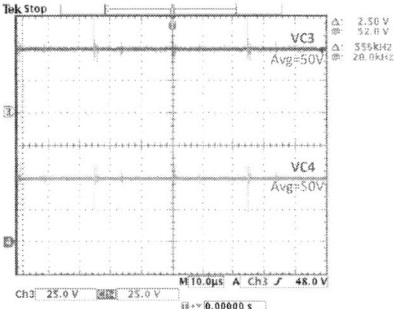

Fig. 11. Capacitor voltage waveform V_{c3} and V_{c4}

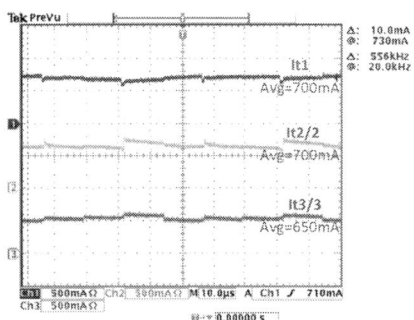

Fig. 12. Total inductor current for all building blocks of each row

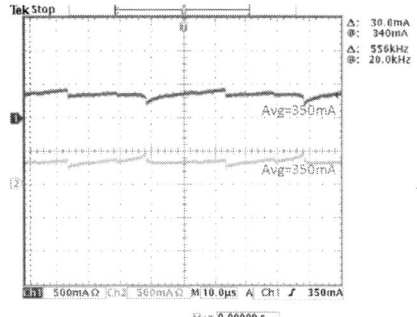

Fig. 13. Separate coil current of the mutually coupled inductors

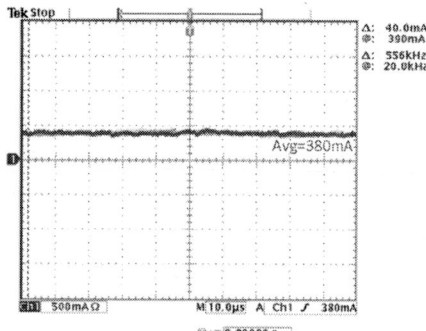

Fig. 14. DC power supply current waveform

VI. CONCLUSION

In this paper, a new modular multilevel DC-DC converter with simplified localized voltage balancing control is proposed. The size of the passive components is reduced and an integrated H-bridge is used which leads to a higher power density. The control algorithm is simplified and is capable of achieving both the voltage balancing of DC-link capacitors and current balancing of mutually coupled inductors. Due to the fully modular design, this topology can be applied to arbitrary number of levels. A four level prototype system is experimented whose results show similarity with the simulation waveforms. The effectiveness of this proposed topology is verified. Future experiments on increased power rating and higher number of levels could be conducted in order to extend this topology into MV and HV applications.

REFERENCES

[1] Debnath S, Qin J, Bahrani B, et al. Operation, control, and applications of the modular multilevel converter: A review [J]. Power Electronics, IEEE Transactions on, 2015, 30(1): 37-53.

[2] Filsoof K, Lehn P. A Bidirectional Modular Multilevel DC–DC Converter of Triangular Structure[J]. Power Electronics, IEEE Transactions on, 2015, 30(1): 54-64.

[3] Khan F H, Tolbert L M. A multilevel modular capacitor-clamped DC–DC converter[J]. Industry Applications, IEEE Transactions on, 2007, 43(6): 1628-1638.

[4] F. Khan and L. Tolbert, "Bi-directional power management and fault tolerant feature in a 5-kW multilevel dc-dc converter with modular architecture," IET Power Electron., vol. 2, no. 5, pp. 595–604, 2009.

[5] F. Khan and L. Tolbert, "A 5 kW bi-directional multilevel modular dc-dc converter (MMCCC) featuring built in power management for fuel cell and hybrid electric automobiles," in Proc. IEEE Veh. Power Propuls. Conf., 2007, pp. 208–214.

[6] F. Khan and L. Tolbert, "Multiple load-source integration in a multilevel modular capacitor clamped dc-dc converter featuring fault tolerant capability," in Proc. IEEE 22nd Annu. Appl. Power Electron. Conf., 2007, pp. 361–367.

[7] D. Cao and F. Z. Peng, "Zero-current-switching multilevel modular switched-capacitor dc-dc converter," in Proc. IEEE Energy Convers. Congr. Expo., 2009, pp. 3516–3522.

[8] D. Cao, X. Lu, X. Yu, and F. Z. Peng, "Zero voltage switching doublewing multilevel modular switched-capacitor dc-dc converter with voltage regulation," in Proc. IEEE 28th Annu. Appl. Power Electron. Conf. Expo., 2013, pp. 2029–2036.

[9] G. Konstantinou and V. Agelidis, "Performance evaluation of half-bridge cascaded multilevel converters operated with multicarrier sinusoidal PWM techniques," in Proc. IEEE Conf. Ind. Electron. Appl., 2009, pp. 3399–3404.

[10] M. Saeedifard and R. Iravani, "Dynamic performance of a modular multilevel back-to-back HVDC system," IEEE Trans. Power Del., vol. 25, no. 4, pp. 2903–2912, Oct. 2010.

[11] F. Deng and Z. Chen, "A control method for voltage balancing in modular multilevel converters," IEEE Trans. Power Electron., vol. 29, no. 1, pp. 66–76, Jan. 2014.

[12] M. Hagiwara and H. Akagi, "Control and experiment of pulsewidthmodulated modular multilevel converters," IEEE Trans. Power Electron., vol. 24, no. 7, pp. 1737–1746, Jul. 2009.

Design and Control of a Fault Tolerant Soft Switching DC-DC Converter for High Power High Voltage Applications

Tao Li, *Student Member IEEE*, Leila Parsa, *Senior Member IEEE*
Department of Electrical, Computer and Systems Engineering
Rensselaer Polytechnic Institute
Troy, New York 12180, USA
Email: lit4@rpi.edu, parsa@ecse.rpi.edu

Abstract— DC-DC converters are essential in high power high voltage applications such as HVDC transmission systems. Fault tolerant capability in such applications is highly desired since the cost of shutdown can be high. This paper presents a novel isolated DC-DC converter that can achieve soft switching for all active switches. The converter has an input-parallel-output-series (IPOS) topology, which increases the voltage transfer ratio and provides benefits of modularity, including online redundancy, interleaving capability, etc. The proposed converter offers two levels of fault tolerance: leg-level and module-level. It is also possible to retain soft switching after fault occurrence. The operation principle, design considerations, control strategy, and fault detection and reconfiguration are investigated in this paper. Experimental result verifies the expected normal operation and operation under fault of the proposed converter.

Index Terms—DC-DC power converters, fault tolerance, zero current switching, zero voltage switching.

I. INTRODUCTION

DC-DC converters are widely used in applications across all power levels, from sensor power supplies to solid state transformers and transportation electrification. With the increasing penetration of renewable energy and the development of high power IGBT, MW sized DC-DC converters have been proposed [1], [2]. Fault tolerant capability of DC-DC converter is highly desired, especially for high power high voltage applications where the cost of shutdown can be high. Of all faults in converters, a substantial portion is caused by failure of power devices. Short circuit fault (SCF) is usually the result of wrong gate voltage or an intrinsic failure. In many commercially available DC-DC converters and IGBT drivers, the detection and protection of SCF is a standard built-in feature [3]. An open circuit fault (OCF) may happen due to lifting of bonding wires, a driver fault or a SCF-induced IGBT rupture. Compared to SCF, OCF generally will not lead to a drastic system failure. However, the consequent DC voltage offset will result in reduced performance, transformer saturation and higher voltage stress on healthy switches.

Fault diagnosis and protection methods for IGBT are reviewed in [4]. More recently, [5] surveys on fault tolerant control methods, with emphasize on converter level fault tolerance. Most methods are developed for inverters. They utilize the information from Park's transformation and space vectors, and cannot be applied to DC-DC converters. In [6], a fault tolerant method for multilevel modular capacitor-clamped DC-DC converter (MMCCC) is proposed. However, this method requires redundant modules in offline mode. Many additional switches as well as one auxiliary leg are added to a multilevel H-bridge converter to achieve both SCF and OCF tolerance in [7]. Moreover, the converter is not suitable for high power applications. Fault detection and protection method based on monitoring flying capacitor voltage of three-level parallel resonant converter (PRC) is presented in [8]. Nevertheless, this method cannot identify the fault location and post-fault operation is not possible. Pei et al. [3, 9] proposed a fault tolerant control based on primary phase-shifted full bridge (PPS-FB) converters, dealing with OCF and SCF respectively.

Comparatively, this paper develops a fault tolerant control for secondary phase-shifted full bridge (SPS-FB) converter. The converter has larger soft-switching range and less circulating current than PPS-FB, as reported in [10]. The fact that diagonal switches operate in pairs makes the method suggested by Pei et al. not applicable. Also, input-parallel-output-series topology provides online redundancy and interleaving possibility. The output voltage/current dip caused by fault can be compensated via healthy modules, and no additional boosting stage is required.

The rest of the paper is organized as follows. Section II introduces the proposed IPOS converter and its normal operation. The fault scenarios analysis, detection, location and fault tolerant control are covered in section III. Simulation and experiment results and discussion are presented in section IV. Section V concludes this paper.

II. CONVERTER TOPOLOGY

The proposed converter topology is shown in Fig. 1. Each module is a zero-voltage zero-current-switching (ZVZCS) converter proposed in [11]. By connection modules in IPOS topology, higher voltage transfer ratio, lower input current stress and reduce ripple can be achieved without efficiency compromise.

978-1-4673-9551-9/16 $31.00 © 2016 IEEE

First, the operation of a single module is briefly explained. A simple control method for healthy operation is given afterward.

A. Opeartion principle of single module

The module has twelve half-cycle symmetric modes in steady state and its operation principle is buck-like. Diagonal switches on the primary side are operated at the same time, whilst secondary switches are complementary and phase-shifted by ϕ from corresponding primary switch pairs. The key waveforms during normal operation are shown in Fig. 2. By selecting sufficiently large filters, input voltage V_{in} and output current I_o can be assumed constant during a switching period. Soft switching is achieved via short resonance modes between transformer leakage inductance L_{lk} and a resonant capacitor C_r. Zero current switching (ZCS) for all secondary switches can be achieved for entire load range. In contrast, the following relationship should be met to ensure zero voltage switching (ZVS) for all primary switches:

$$\frac{L_{lk}}{C_r} > \left(\frac{V_{in}}{I_o}\right)^2. \tag{1}$$

Therefore, the size of C_r is reverse proportional to ZVS range. Instead of across the transformer primary winding, C_r can also be placed across each switch, so DC capacitors can be used. On the other hand, C_b is the DC blocking capacitor that protects transformer from biased voltage caused by asymmetric gate signals, etc. Therefore, C_b should be selected large enough that it would not affect the normal operation.

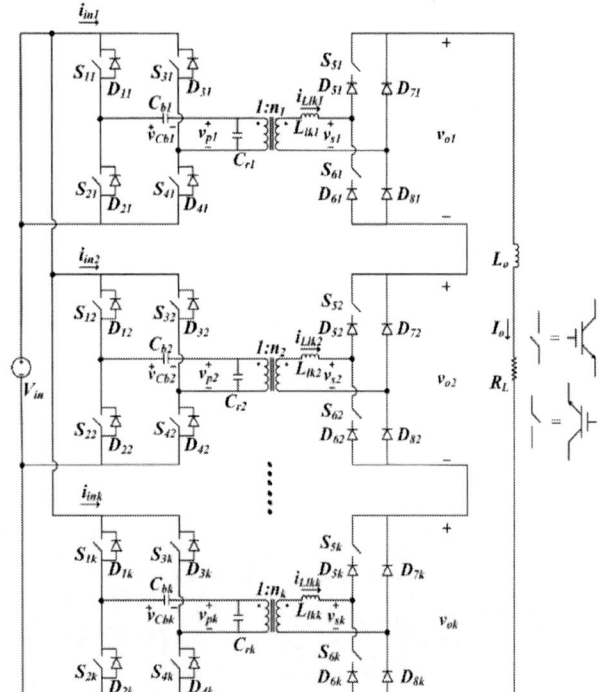

Figure1. Proposed IPOS ZVZCS converter

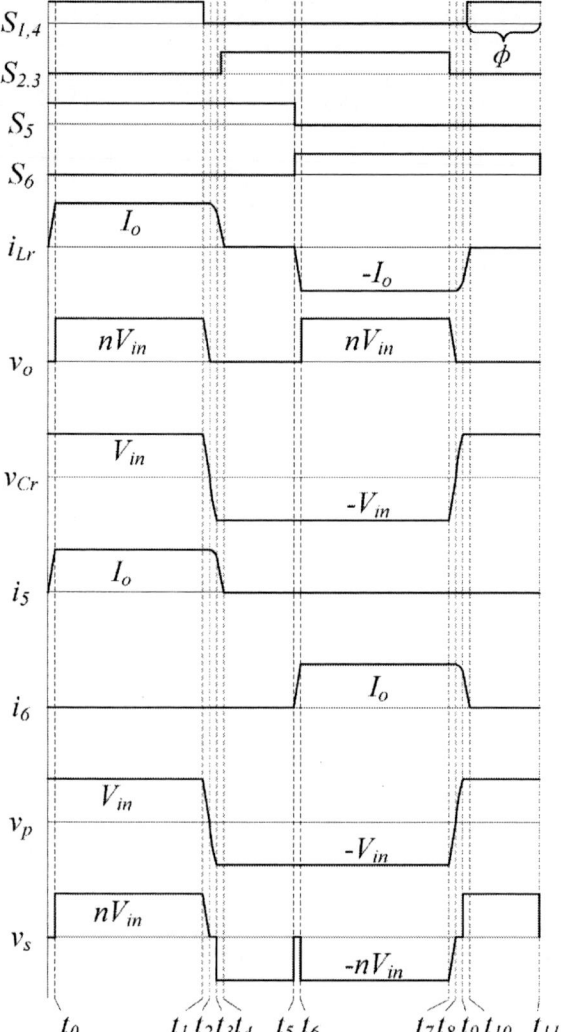

Figure 2. Normal operation waveforms for one module

Let us define duty ratio as the overlap between primary and secondary switch gating signals, i.e.:

$$D = \frac{t_7 - t_5}{T_s} \tag{2}$$

where T_s is the switching period. If there are no resonant modes, the converter becomes a full bridge isolated buck converter with duty ratio D. In reality, the effective duty ratio D_{eff} will be different. Based on energy balance, D_{eff} can be obtained as:

$$D_{eff} = D - D_{loss} < D \tag{3}$$

$$D_{loss} = \frac{\dfrac{3L_{lk}I_o}{V_{in}} - \dfrac{2C_rV_{in}}{I_o}}{2nT_s} \tag{4}$$

The inequality is found based on (1). As expected, D_{eff} is smaller than D, leading to a slightly decreased output voltage compared to hard switching case. For a specific design, V_{in}, L_{lk} and C_r are constant, whilst I_o may vary with the load level. Since D_{loss} increases with I_o, the maximum duty ratio loss $D_{loss,max}$ can be obtained at rated output current. To facilitate the following discussion, two sets of design parameters are listed in Table I. Using the values from Table I, $D_{loss,max}$ is 0.007 for full scale design and 0.012 for prototype design.

B. Controller design

In this converter, output current rather than output voltage is regulated. In this way, it is simple to connect several converter outputs in series to further increase the power rating. With proper design parameters, the transition modes that take place between $[t_1, t_4]$, $[t_5, t_6]$, $[t_7, t_{10}]$ and $[t_{11}, T_s+t_0]$ should be small enough to be ignored in control design process. For example, with the parameter sets in Table I, the transition modes only take up 2.1% and 6.6% of the switching period respectively.

Given the symmetry of the IPOS structure, the average model of the converter would be the similar to that of a single module. If constant resistive load is considered, based on output inductor volt-second balance, we have:

$$L_o \frac{di_o}{dt} + i_o R = \bar{V}_o = k(1-2\phi)nV_{in}. \tag{5}$$

Therefore, the transfer function from the control input ϕ to i_o is:

$$G_{\phi i(s)} = \frac{I_O(s)}{\Phi(s)} = -\frac{2knV_{in}}{sL_o + R}. \tag{6}$$

PI controller is designed to regulate output current according to reference value. The actual phase shift ϕ would be the difference between steady state phase shift and the output of controller. The PWM modulator then generates the carrier waves accordingly. The linear control-to-output relationship is plotted for full scale design parameters in Fig. 3. Output is less sensitive to slight variation of ϕ compared to dual active bridge derived topologies, reducing the resolution requirement for phase shift timer.

TABLE I. DESIGN PARAMETERS

	Full scale design	Prototype design
Power P	1.25 MW	1 kW
Input voltage V_{in}	1 kV	150 V
Output current I_o	208 A	10 A
Leakage inductance L_{lk}	30 μH	9 μH
Resonant capacitance C_r	100 nF	20 nF
Switching frequency f_s	2 kHz	20 kHz
Transformer turns ratio $n=N_s/N_p$	2.5	1
DC blocking capacitance C_b	1 mF	10 μF
Output filtering inductance L_o	50 mH	1.5 mH
Steady state phase shift φ_0	0.09	0.13
Number of modules k	3	3

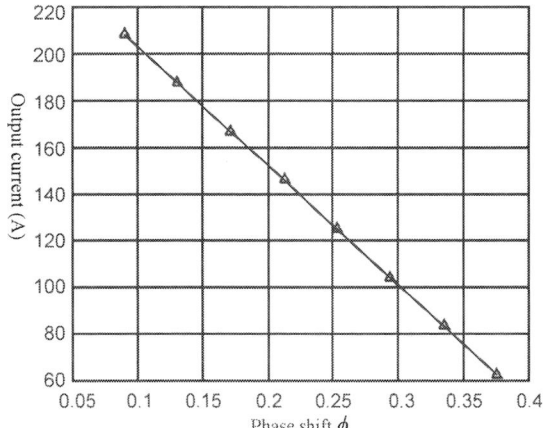

Figure 3.   Linear control-to-output relationship

This simple control method can provide stable operation for both steady state and transient. Additional balancing control is not necessary since the converter is immune to instabilities caused by parameter mismatch, unlike in input-series-output-series (ISOS) or input-parallel-output-parallel (IPOP) topologies. The average output voltages of different modules will be affected by transformer turns ratio mismatch. However with modern manufacturing technology, the mismatch is negligible [12].

III. FAULT TOLERANT CONTROL

Two levels of fault tolerances are offered by the proposed converter. Several types of fault can be tolerated intrinsically or via the leg-level tolerance. Other faults, including SCF of any primary switch or multiple component faults can be dealt with by module-level tolerance. For a converter with k modules, only $3k$ voltage sensors are required for fault detection and location. In order to design fault tolerant control, the consequences of different fault scenarios are first summarized.

A. Fault scenarios analysis

1)   OCF at secondary switch

When OCF occurs at one of the secondary switches, the leakage inductance current may be disrupted, while output current can still flow through other paths. The average current through leakage inductance will deviate from zero. However, the impact on the transformer voltage and the voltage across other switches are negligible. In steady state, the fault can be regarded as setting $\phi=0.5$ for half cycle. The power transferred can still be regulated by the phase shift of the other half cycle and the healthy modules. It is possible for output current to maintain the pre-fault value. However, the maximum power that can be transferred will decrease. Also, primary switches will no longer be zero-voltage-switched. Thermal design needs to take this scenario into consideration.

2)   SCF at secondary switch

Similarly, if SCF occurs at one of the secondary switches, the impact on the transformer voltage and the voltage across other switches are negligible. In steady state, the fault can be regarded as setting $\phi=0$ for half cycle. The power transferred can still be regulated by the phase shift of the other half cycle and other healthy modules. The minimum power that can be transferred will increase but the converter can operate safely at full load.

3) OCF at primary switch

When OCF occurs at one of the primary switches, voltage applied to primary side will have a DC component. The converter can still function, but DC blocking capacitor will be biased. Also, the maximum power that can be transferred will decrease. More seriously, if the anti-parallel diode of the faulty switch is also open, the other switch on the same leg will have to withstand much higher voltage spikes. This is because C_r voltage is no longer clamped by V_{in}. More discussion will be provided in the following section.

4) SCF at primary switch

When SCF takes place at one of the primary switches, the converter will experience a shoot-through fault, which needs immediate attention.

B. Fault detection and location

SCF protection circuit is usually included in gate drivers for industrial applications [3]. It will turn off the switch once potential SCF happens, making it an open circuit. Therefore, only OCF is considered in this paper. Moreover, the converter does not require any reconfiguration when OCF occurs at secondary. On the other hand, the converter with OCF at primary needs be reconfigured to avoid damage to healthy components or modules.

Fault detection is based on monitoring the DC component of the voltage across C_b. During normal operation, only minor DC bias due to imperfect gate signals, etc. will exist in v_{cb}. When OCF at primary happens at module #i, v_{cbi} will deviate from zero. The detectable faults and resultant v_{cbi} behavior are summarized in Table II.

Apparently, v_{cbi} alone does not provide enough insight for reconfiguration. Thus, voltages across the lower primary switches are also sensed to obtain further information about the fault. For example, assume it is S_{41} that suffers from OCF, as shown by red mark in Fig. 4. Then v_{cb1} will drop and average voltage across S_{41} will increase beyond normal. Now the fault can be identified as S_{41} OCF, and not S_{11} OCF. Once the fault

is detected and located, the converter can be reconfigured as will be described in the next section.

C. Reconfiguration method

If D_{41} is also open in the fault, C_{r1} voltage will no longer be clamped by input voltage during $[t_9, t_{10}]$. Peak voltage across the healthy switch on the same leg, i.e. S_{31} will increase if the fault is not dealt with.

Therefore, once OCF at S_{41} is identified, S_{41} will be turned off permanently, whilst S_{31}, S_{51} and S_{61} will be turned on permanently. Gate signals for S_{11} and S_{21} will remain unchanged. Then the faulty module will essentially be reconfigured into an asymmetric half bridge (AHB) converter operating in open loop, as shown in Fig. 4. The duty ratios of S_{11} and S_{21} remain at around 50% (with appropriate deadtime), transferring as much power as possible [13]. After the reconfiguration, the switches on the healthy leg, namely S_{11} and S_{21} will still remain zero voltage switched. Faults at other switches can be detected, located and tolerated in a similar manner. The leg-level fault tolerant control method is summarized in Fig 5.

If non-tolerable faults or multiple faults happen, the IPOS topology provides an additional level of fault tolerance. The entire faulty module can be easily bypassed by blocking all gate signals to the faulty module. Output current can freewheel through the diode leg formed by D_{71} and D_{81}, assuming module #1 is bypassed.

IV. RESULTS AND DISCUSSION

A. Simulation verification

Simulations are carried out in PLECS for both sets of parameters listed in Table I. The results for full scale design parameters are shown in Fig. 6.

The converter is started with zero initial condition and current reference $i_{ref} = 208A$ at t= 0. At t= 10ms, i_{ref} steps to 104A and at t=20ms it steps back to 208A. At t= 30ms, S_{41} OCF is applied by a timed relay switch. The fault is detected and located in approximately 3 ms, and reconfiguration method is applied. After the fault, the output current can no longer reach the reference value as ϕ has reached its boundary. At t=40 ms, module #1 is bypassed by blocking all the gate signals. Now i_o drops even further since module #1 is no longer transferring any power. The output current settles at around 163 A, approximately 80% of reference value. This is expected since now the entire converter is transferring around 2/3 of the rated power.

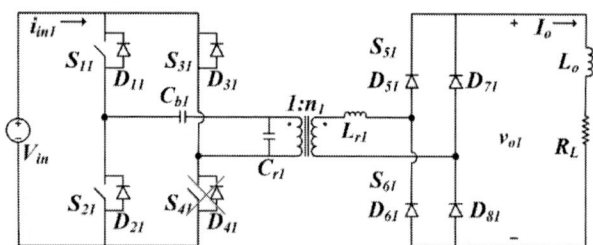

Figure 4.   Post-fault equivalent topology for module #1

TABLE II.   DETECTABLE FAULTS AND CORRESPONDING V_{Cbi} BEHAVIOR

Fault	v_{cbi}
S_{1i} OCF or S_{4i} OCF	Drop
S_{2i} OCF or S_{3i} OCF	Rise

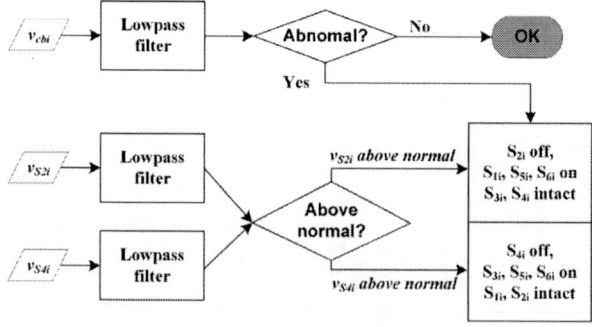

Figure 5.   Leg-level fault tolerant control

The gate signals and switch voltages are plotted in Fig 7. As we can see, after the OCF fault at S_{41}, the voltage stress on S_{31} increases. The switch may eventually breakdown as well if appropriate action is not taken. With the proposed reconfiguration method, v_{S31} is brought to zero.

B. Hardware experiment

A prototype with parameters in second column of Table I has been constructed, as shown in Fig. 8. Hardware-in-the-loop (HIL) fast prototyping method is used in development process. cRIO-9082 from National Instrument is the selected real-time (RT) controller. SPS-PWM and ADC are implemented on the FPGA chip inside the cRIO. Fault tolerant control and user interface are realized on RT system.

In Fig. 9, the gate signal of S_{13} and its drain to source voltage v_{S13} are shown, as well as the gate signal of S_{53} and its current i_{S53}. ZVS on primary switches and ZCS on secondary switches during normal operation are thus verified.

The leg-level fault tolerant control is tested and presented in Fig. 10. S_{43} OCF is triggered by blocking the gate signal. In Channel 1, the rising edge indicates the fault instant, whilst the falling edge shows the detecting instant. The fault is detected in 0.56 ms and is located in 0.96 ms. As expected, v_{ch3} drops following the fault. Output current i_o is supported by healthy modules and the dip is very small. This demonstrates the feasibility of the fault tolerant control method.

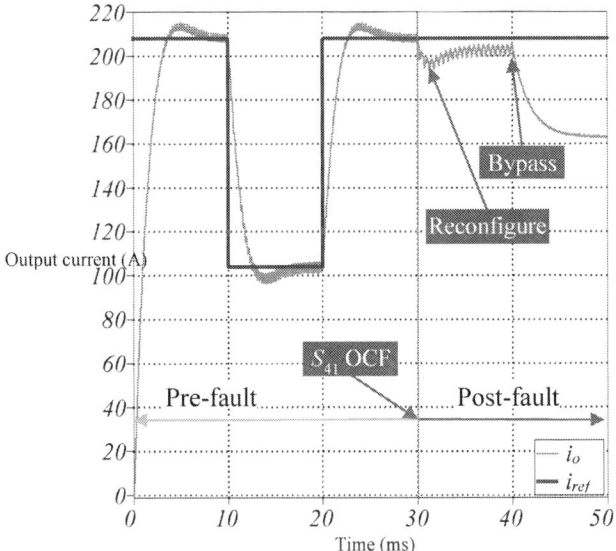

Figure 6. Pre- and post-fault output current step response

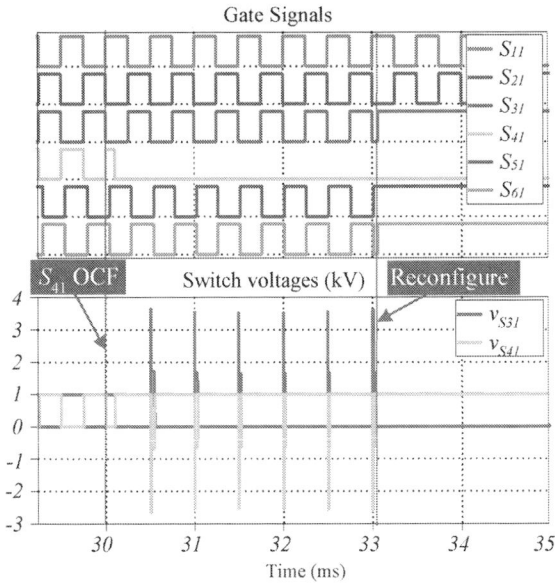

Figure 7. Gate signals and switch voltages of module #1 before and after OCF at S41

Figure 8. Converter prototype

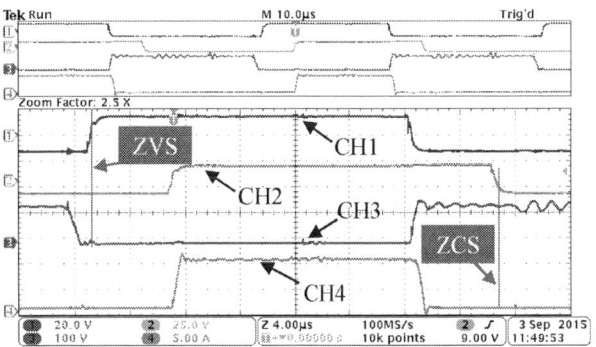

Figure 9. Soft switching during healthy operation (CH1: S13 gate signal, CH2: S53 gate signal, CH3: v_{S13}, CH4: i_{S53})

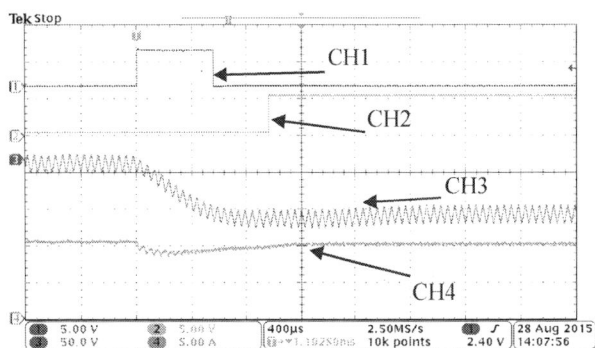

Figure 10. Leg-level fault tolerance (CH1: Fault triggered and detected, CH2: Fault located, CH3: v_{ch3}, CH4: i_o)

V. CONCLUSION

A novel fault tolerant isolated DC/DC converter topology is proposed. The converter can achieve soft switching for all active switches. IPOS structure provides two levels of tolerance. Leg-level tolerance not only prevents the fault from spreading but also allows full utilization of healthy components. The functioning switches within the faulty module still remain soft switched. Additionally, module-level tolerance can easily isolate a non-tolerable faulty module from the system. Converter closed loop operation, fault detection and location, as well as both levels of fault tolerant control are simulated and verified via HIL experiment.

ACKNOWLEDGMENT

This work was supported primarily by the Engineering Research Center Program of the National Science Foundation and the Department of Energy under NSF Award Number EEC-1041877 and the CURENT Industry Partnership Program.

REFERENCES

[1] G. Ortiz, J. Biela, D. Bortis, and J. W. Kolar, "1 Megawatt, 20 kHz, isolated, bidirectional 12kV to 1.2 kV DC-DC converter for renewable energy applications," in *Power Electronics Conference IPEC 2010 International*, 2010, pp. 3212–3219.

[2] W. Chen, A. Q. Huang, C. Li, G. Wang, and W. Gu, "Analysis and comparison of medium voltage high power DC/DC converters for offshore wind energy systems," *IEEE Trans. Power Electron.*, vol. 28, no. 4, pp. 2014–2023, Apr. 2013.

[3] X. Pei, S. Nie, Y. Chen, and Y. Kang, "Open-circuit fault diagnosis and fault-tolerant strategies for full-bridge DC–DC converters," *IEEE Trans. Power Electron.*, vol. 27, no. 5, pp. 2550–2565, May 2012.

[4] B. Lu and S. K. Sharma, "A literature review of IGBT fault diagnostic and protection methods for power inverters," *IEEE Trans. Ind. Appl.*, vol. 45, no. 5, pp. 1770–1777, Sept./Oct. 2009.

[5] W. Zhang, D. Xu, P. N. Enjeti, H. Li, J. T. Hawke, and H. S. Krishnamoorthy, "Survey on fault-tolerant techniques for power electronic converters," *IEEE Trans. Power Electron.*, vol. 29, no. 12, pp. 6319–6331, Dec. 2014.

[6] F. H. Khan and L. M. Tolbert, "Multiple load-source integration in a multilevel modular capacitor clamped DC-DC converter featuring fault tolerant capability," in *Conference Proceedings - IEEE Applied Power Electronics Conference and Exposition - APEC*, 2007, pp. 361–367.

[7] K. Ambusaidi, V. Pickert, and B. Zahawi, "New circuit topology for fault tolerant H-bridge DC-DC converter," *IEEE Trans. Power Electron.*, vol. 25, no. 6, pp. 1509–1516, Jun. 2010.

[8] H. Sheng, F. Wang, and C. W. Tipton IV, "A fault detection and protection scheme for three-level DC-DC converters based on monitoring flying capacitor voltage," *IEEE Trans. Power Electron.*, vol. 27, no. 2, pp. 685–697, Feb. 2012.

[9] X. Pei, S. Nie, and Y. Kang, "Switch short-circuit fault diagnosis and remedial strategy for full-bridge DC–DC converters," *IEEE Trans. Power Electron.*, vol. 30, no. 2, pp. 996–1004, Feb. 2015.

[10] T. Mishima and M. Nakaoka, "Practical evaluations of a ZVS-PWM DC-DC converter with secondary-side phase-shifting active rectifier," *IEEE Trans. Power Electron.*, vol. 26, no. 12, pp. 3896–3907, Dec. 2011.

[11] T. Li and L. Parsa, "Medium frequency soft switching DC/DC converter for HVDC transmission system," in *Industrial Electronics Society 40th Annu. Conf. of the IEEE*, 2014, pp. 1599–1605.

[12] R. Giri, V. Choudhary, R. Ayyanar, and N. Mohan, "Common-duty-ratio control of input-series connected modular DC-DC converters with active input voltage and load-current sharing," *IEEE Trans. Ind. Appl.*, vol. 42, no. 4, pp. 1101–1111, Jul. 2006.

[13] R. Oruganti, P. C. Heng, J. Tan, K. Guan, and L. A. Choy, "Soft-switched DC/DC converter with PWM control," *IEEE Trans. Power Deliv.*, vol. 13, no. 1, pp. 102–114, Jan. 1998.

Accurate Parametric Steady State Analysis and Design Tool for DC-DC Power Converters

Mohammad Daryaei*, Mohammad Ebrahimi[†] and S. Ali Khajehoddin[‡]

Department of Electrical & Computer Engineering
University of Alberta, Edmonton, AB, Canada
*daryayi@ualberta.ca,[†] m.ebrahimi@ualberta.ca, [‡]khajeddin@ualberta.ca

Abstract—**Accurate large signal analysis and modeling of Power Electronics converters are essential for achieving high performance and reliable designs. Converter topologies with large signal variations are conventionally analyzed using numerical methods, averaged or inaccurate analyses. In this paper, a mathematical theorem based on Laplace transform is developed to derive the steady state response of periodic signals with a switching input signal. It is shown that the proposed methodology provides accurate and parametric analysis tool for dc-dc power converters specially for resonant converters and has many applications in design and analysis of the converters and their control systems. The proposed method is used to analyze and model a few power circuit including full bridge Series Resonant Converter (SRC) topology where both CCM and DCM operating modes are analyzed. It is observed that the proposed analysis approach gives great insight and simplifies converter design. The proposed analysis and modeling approach is also validated by simulations and experimental results.**

Index Terms—**DC/DC converter, Analysis of Resonant Converters, large signal modeling, CCM and DCM analysis.**

I. INTRODUCTION

Due to the switching actions, large signal variations exist in many power converter topologies specifically in converters with high frequency isolation transformers where high frequency large switching type AC signal variations exist. In the analysis of such converters, several simplifying assumptions are normally made that lead to approximate results and inaccurate modeling of such converters. For example, different methods such as Fourier Series (FS) [1]–[3] and Fundamental Component Approximation (FCA) [4] have been utilized to analyze resonant converter topologies. Such methods are either too simplistic/inaccurate, or too complicated to achieve acceptable accuracy. The main drawback of such methods is that to achieve accurate analyses, infinite series have to be used to analyze the circuit. However, this is not a suitable approach for the classical optimum methods and analysis to be applied to converters.

Accurate modeling and analysis of converters have many applications in the design and control of converters and is widely adopted in the literature to predict and tackle practical issues. Authors in [5], [6] have used an accurate but tedious analysis approach for Dual Active Bridge (DAB) converter to achieve wider Zero Voltage Switching (ZVS) range and better efficiency. For a more complicated topology such as Dual Bridge Series Resonant Converter (DBSRC), conventional analysis approaches such as FCA or FS lead to inaccurate

or complicated analysis especially when asymmetrical gating schemes are used [7].

Modeling and analysis tools have also been used to achieve superior designs. Authors in [8] have considered the snubber capacitance of the switches to achieve a higher power density design. In spite of this achievement, since the converter is controlled with Asymmetrical Pulse Width Modulation (APWM) at low duty ratio, the conventional approach (FCA) has not provided accurate results. In another example in [9] an analysis of resonant converters is carried out to provide a sensor-less gating scheme for Synchronous Rectifiers (SR). It is shown that accurate current waveform estimation are essential for a successful design of the SR logic circuit.

In this paper an alternative method is proposed that will accurately predict and analyze the steady state large signal behavior of the switching waveforms of power converters. The proposed method provides a parametric solution that is computationally efficient compared with other existing methods. Although Laplace transform is normally used to analyze the transient behavior of circuits, in this paper a theorem is proposed that is based on Laplace transform to determine the steady state waveform of a periodic signal. This theorem is applied to different converter topologies to show its effectiveness in the design and large signal parametric analysis of power converters. It is shown that this method is specially useful for analysis of resonant converters. For each case study, simulation and experimental results are provided to validate the mathematical derivations. It is also shown that using the insight provided by these solutions, a more optimized converter with higher efficiency can be designed.

II. THE PROPOSED METHOD

Physical systems can be mathematically modeled by their governing Ordinary Differential Equations (ODEs). Since in most power converters a periodic and discontinuous switched waveform is applied to a passive circuit, in this paper only ODEs with periodic discontinuous inputs are discussed. As the input function is discontinuous, tedious stepwise methods are conventionally used to obtain the complete response (both transient and steady state response). If only the steady state response of the system is desired, Fourier series of the input function is traditionally used for the analysis of such converters. In these cases, the infinite Fourier series should be

978-1-4673-9551-9/16 $31.00 © 2016 IEEE

approximated by a finite number of summations, which in turn makes this analysis inaccurate.

In this paper, the proposed method uses Laplace transform based theorem to predict the accurate periodic solution signal, where the transient variations caused by the input function changes are accurately predicted in steady state conditions. In this method the initial conditions at the start of every cycle are determined such that the transient response caused by the input function is compensated and system response becomes purely periodic. Therefore, an accurate steady state response is achieved. The following theorem is proposed and used to find such initial conditions.

Theorem 1. *Assume $P(D)x(t) = f(t)$ is an ODE of a system where, $P(D) = \Sigma_0^n a_k D^k$ is the characteristic polynomial with n roots of s_j , $D = \frac{d}{dt}$ is the differentiation operator, and $f(t)$ is a periodic input with period T that can have any finite number of discontinuities.*

(i) Laplace transform of the desired response $x(t)$ is $X(s) = (F(s) + G(s))/P(s)$, where $F(s) = \left(\int_0^T f(t) e^{-sT} dt \right) / \left(1 - e^{-sT} \right)$ is the Laplace transform of the input function and $G(s)$ is a function of initial conditions of $x(t)$:

$$G(s) = x_0 a_n s^{n-1} + (x_0 a_{n-1} + x_1 a_n) s^{n-2} + \dots = \sum_0^{n-1} b_k s^k \quad (1)$$

(ii) By solving n equations of $G(s_j) = -F(s_j)$ for all $j = 1,\dots,n$, the n unknown initial conditions $x(0) = x_0$, $x^{(1)}(0) = x_1$, ..., $x^{(n-1)}(0) = x_{n-1}$ can be calculated so that $x(t)$ becomes periodic in steady state.

Proof. see Appendix. □

In summary, to apply the theorem to a converter, the following steps must be taken:

1) Obtain the ODE of the converter.
2) Find F(s), the Laplace transform of the input switching waveform f(t).
3) Find $G(s)$ function using (1) and assuming $x_0, \dots, x_{n-1}$ are the unknown initial conditions.
4) Form the following set of equations $G(s_j) = -F(s_j)$ for all roots of $P(s)$ and solve them to find $x_0, \dots, x_{n-1}$.
5) Now that the initial conditions at the start of each period at steady state are known, solve a new ODE $P(D)x_{tr}(t) = f_{tr}(t)$ using the calculated initial conditions to find $x_{tr}(t)$. Note that $x_{tr}(t)$ is one period of the solution and similarly $f_{tr}(t)$ is one period of the input and thus the new ODE can be easily solved.

Solution of the new ODE, $x_{tr}(t)$, is equal to desired signal $x(t)$ for $0 \le t < T$ and equal to zero anywhere else and thus $X(s) = X_{tr}(s)/\left(1 - e^{-sT}\right)$. Similarly $f_{tr}(t)$ is equal to $f(t)$ for $0 \le t < T$ and zero anywhere else. As it has been proved in the appendix $x_{tr}(t)$ will be equal to zero for $t > T$ because all n equations of $G(s_j) = -F(s_j)$ hold.

In order to illustrate the procedure of applying the proposed method to power circuits two simple examples are discussed

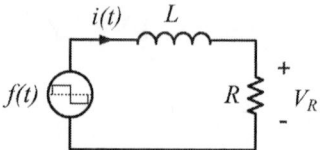

Fig. 1. Simple RL circuit with square wave input.

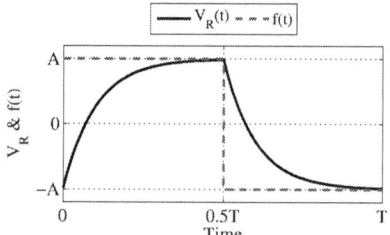

Fig. 2. Obtained V_R using the proposed method.

in here. Figure 1 shows an RL circuit with the square wave input. System ODE is $L\frac{di}{dt} + Ri(t) = f(t)$. If the input's amplitude and period are denoted by A and T, then $F(s) = \frac{A(1 - e^{-0.5Ts})}{s(1 + e^{-0.5Ts})}$. With $i(0) = I_0$ Laplace transform of the ODE results in $(Ls + R)I(s) - LI_0 = F(s)$. $s_j = -R/L$ is the only root of characteristic equation, then I_0 is found by solving the $F(s_j = -R/L) = LI_0$ as:

$$I_0 = \left(\frac{A}{R} \right) \frac{1 - e^{\frac{TR}{2L}}}{1 + e^{\frac{TR}{2L}}}. \quad (2)$$

Now if the ODE is solved for only one cycle, periodic solution is obtained. Since the input is half-wave symmetrical then the output would be half-wave symmetrical as well and only half cycle of the solution is enough to describe the system response. Thus $L\frac{di}{dt} + Ri(t) = A$ is solved using 2:

$$i(t) = \frac{A}{R} \left(1 - \frac{2e^{\frac{TR}{2L}}}{1 + e^{\frac{TR}{2L}}} e^{-\frac{R}{L}t} \right) 0 \le t < \frac{T}{2} \quad (3)$$

Figure 2 shows steady state voltage across R obtained by (3).

Buck converter is another example that is analyzed using the proposed method. The converter is shown in Fig. 3 where switches are considered ideal. Based on the values shown in Fig.3 system ODE is $10^{-10}\ddot{v}_o(t) + 10^{-5}\dot{v}_o(t) + v_o(t) = f(t)$. The switching input function has the frequency, amplitude and duty cycle of $100kHz$, $10V$ and $D = 0.5$ respectively. If the initial conditions are $v_o(0) = v$, $\dot{v}_o(0) = v'$ then $G(s)$ is found as $G(s) = \left(10^{-10}(sv + v') + 10^{-5}v \right)$. Characteristic equation has a pair of complex conjugate roots ($s_{1,2} = -50000 \pm 86602j$). Therefore, $G(s_1) = -F(s_1)$ provides two equations from equating imaginary and real parts that can be used to find $v = 4.97$, $v' = -1.25 \times 10^5$ unknown initial conditions. Now ODE with these initial conditions is solved. In this example due to the symmetry only half cycle of the solution is enough to describe the system response:

$$v_o(t) = -e^{-50000t} (4.35 \sin(86602t) + 5.03 \cos(86602t)) + 10 \quad (4)$$

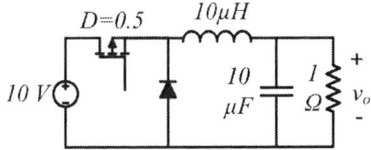

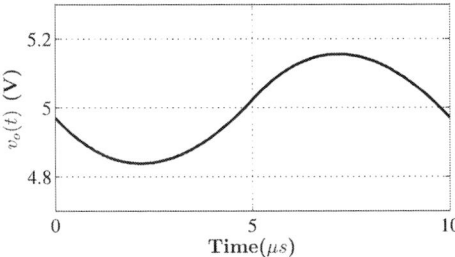

Fig. 3. Schematic of the Buck converter.

Fig. 4. Output voltage in a Buck converter obtained using the proposed method.

Figure 4 is a plot of the (4) to better show the ripple part of $v_o(t)$.

III. SERIES RESONANT CONVERTER (SRC) ANALYSIS USING THE PROPOSED METHOD

In this section the proposed analytical approach is applied to a full bridge Series Resonant Converter (SRC) and is validated by simulations and experimental results. Figure 5 shows the SRC converter and its simplified equivalent circuit. This converter is controlled using phase-shift PWM scheme as shown in Fig. 6. Depending on the circuit parameters, at small duty ratios, SRC may enter Discontinuous Conduction Mode (DCM) or it may remain in Continuous Conduction Mode (CCM). First, it is assumed that SRC remains in CCM mode and all the governing equations are obtained and then based on these equations DCM criteria is derived.

A. CCM Operation

CCM operation analysis is presented in this section while DCM analysis is covered in the next section. The SRC Operating under CCM is modeled by the following differential equation:

$$L\frac{di(t)}{dt} = V_{in}(t) - V_c(t) - V_o sign(i(t)). \quad (5)$$

Due to the $sign()$ function, (5) is nonlinear and discontinuous ODE and $v_{in}(t)$ is a discontinuous periodic switching input function. Conventionally, this equation was solved either by Fourier series or by tedious stepwise period by period solution to achieve the steady state response. In this paper the proposed method based on Laplace theorem is used to analyze this circuit.

Consider the equivalent circuit shown in Fig. 5(b). Superposition can be used to analyze this circuit. Figure 6 shows a predicted and general current waveform that has the unknown phase lag with respect to the input voltage denoted by θ. Using the proposed theorem, steady state initial conditions are

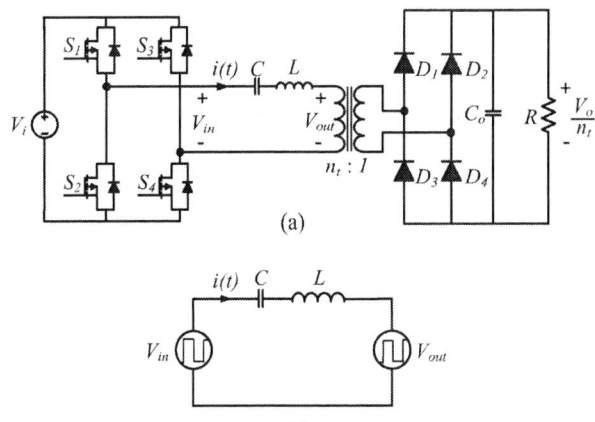

(a)

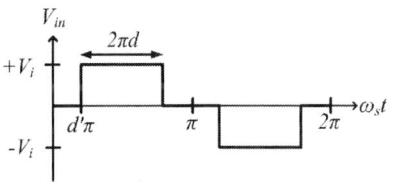

(b)

Fig. 5. (a) Schematic of the SRC, and (b) equivalent circuit.

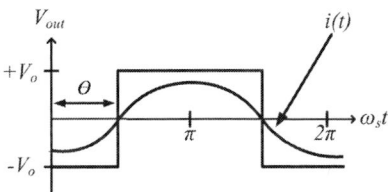

Fig. 6. Typical waveforms of SRC.

derived in terms of θ. Then, current equations are found based on the initial conditions. As the current at $\omega_s t = \theta$ is zero, $i(\theta) = 0$ is solved to find θ as follows:

$$\theta = \begin{cases} \frac{\pi}{2} - F\sin^{-1}\left(M\frac{\sin\frac{\pi}{2F}}{\cos\frac{\pi d'}{F}}\right) & \text{if } d'\pi < \theta, \\ F\cos^{-1}\left(M\frac{\sin\frac{\pi}{2F}}{\sin\frac{\pi d}{F}}\right) & \text{if } 0 \le \theta \le d'\pi \end{cases} \quad (6)$$

where ω_r is the resonant frequency, $M = V_o/V_i$ is the voltage gain, d is the duty ratio that varies from 0 to 50%, $d' = 0.5 - d$, $F = \omega_s/\omega_r$ (ω_s is the switching frequency), and $Q = (\omega_r L)/(n_t^2 R)$.

Equation (6) shows that θ is a function of the voltage gain M. In order to find the voltage gain, V_o should be found in terms of V_i. V_o is the product of load resistance multiplied by output DC current, which is the full-wave rectified of the resonant tank current waveform. DC component of the output current is calculated as a function of V_i, and M can be found in terms of d, F and Q as follows:

$$M = \begin{cases} \left(\frac{2F}{\pi Q}\right)\frac{AB\sqrt{A^2+B^2-1}-B^2}{A^2+B^2} & \text{if } d'\pi < \theta, \\ \frac{CD}{\sqrt{C^2+D^2}} & \text{if } 0 \le \theta \le d'\pi, \end{cases} \quad (7)$$

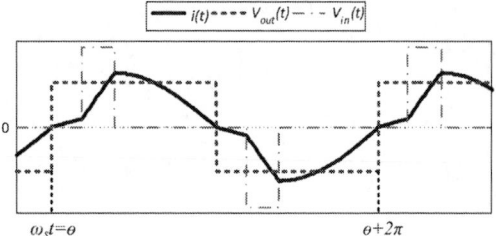

Fig. 7. Current waveform if converter remains in CCM for $d < d_{ZVS}$ (No ZVS).

$$A = \sin\left(\frac{\pi d}{F}\right)\tan\left(\frac{\pi}{2F}\right) + \cos\left(\frac{\pi d}{F}\right)$$

$$B = \frac{\pi Q \cos\left(\frac{\pi d'}{F}\right)}{2F \sin\left(\frac{\pi}{2F}\right)}$$

$$C = \frac{\sin\frac{\pi d}{F}}{\sin\left(\frac{\pi}{2F}\right)}$$

$$D = \frac{2F \sin\left(\frac{\pi d}{F}\right)}{\pi Q \cos\left(\frac{\pi}{2F}\right)}$$

If (7) is substituted in (6), θ will be calculated in terms of d, F, and Q.

$$\theta = \begin{cases} \frac{\pi}{2} - F\sin^{-1}\left(\frac{A\sqrt{A^2+B^2-1}-B}{A^2+B^2}\right) & \text{if } d'\pi < \theta, \\ F\cos^{-1}\left(\left(1+\left(\frac{\pi Q}{2F}\cot\frac{\pi}{2F}\right)^2\right)^{-\frac{1}{2}}\right) & \text{if } \theta \le d'\pi \end{cases} \quad (8)$$

It can be observed from (8) that for lower duty ratios where $d'\pi$ is larger than or equal to θ, the value of the θ is not a function of d anymore and it is a constant value if the tank parameters F and Q are constant. In other words, for any F and Q there exists a critical duty ratio d_{ZVS} where for all duty cycles smaller than that, θ is equal to this constant value.

$$d_{ZVS} = 0.5 - \frac{F}{\pi}\cos^{-1}\left(\left(1+\left(\frac{\pi Q}{2F}\cot\frac{\pi}{2F}\right)^2\right)^{-\frac{1}{2}}\right) \quad (9)$$

In fact if $d > d_{ZVS}$, then $d'\pi < \theta$ and converter is in ZVS mode and if $d < d_{ZVS}$, then $\theta \le d'\pi$ and ZVS is lost. Therefore, the $sign()$ function in (5) can be eliminated by dividing it into two linear ODEs according to d_{ZVS}. This set of ODE now can be solved for only one cycle to obtain steady state solution using the proposed theorem. Current waveform could be described using its half cycle, because it is half-wave symmetrical. For any F and Q, d_{ZVS} can be obtained. Then if $d > d_{ZVS}$ current waveform is obtained from (10), and if $d \le d_{ZVS}$ it is obtained from (11).

$$i(t) = \begin{cases} x_1(t) + x_3(t) & \text{if } 0 \le \omega_s t < d'\pi \\ x_2(t) + x_3(t) & \text{if } d'\pi \le \omega_s t < \theta \\ x_2(t) + x_4(t) & \text{if } d'\pi \le \omega_s t < (1-d')\pi \\ -x_1(t-\frac{\pi}{F\omega_r}) + x_4(t) & \text{if } (1-d')\pi \le \omega_s t < \pi \end{cases} \quad (10)$$

$$i(t) = \begin{cases} x_1(t) + x_3(t) & \text{if } 0 \le \omega_s t < \theta \\ x_1(t) + x_4(t) & \text{if } \theta \le \omega_s t < d'\pi \\ x_2(t) + x_4(t) & \text{if } \theta \le \omega_s t < (1-d')\pi \\ -x_1(t-\frac{\pi}{F\omega_r}) + x_4(t) & \text{if } (1-d')\pi \le \omega_s t < \pi \end{cases} \quad (11)$$

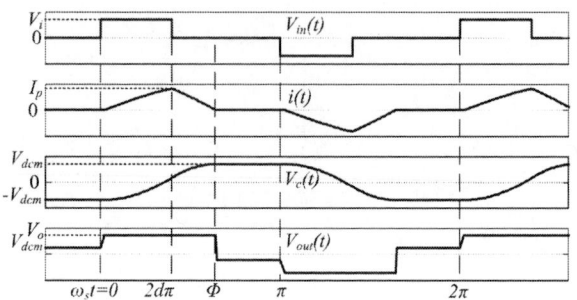

Fig. 8. DCM operation waveform.

$$x_1(t) = -v_i c \omega_r \frac{\sin\frac{\pi d}{F}}{\cos\frac{\pi}{2F}}\cos(\omega_r t)$$

$$x_2(t) = v_i c \omega_r \left(-\frac{\sin\frac{\pi d}{F}}{\cos\frac{\pi}{2F}}\cos(\omega_r t) + \sin(\omega_r t - \frac{d'\pi}{F})\right)$$

$$x_3(t) = M v_i c \omega_r \left(\tan\frac{\pi}{2F}\cos(\omega_r t - \frac{\theta}{F}) + \sin(\omega_r t - \frac{\theta}{F})\right)$$

$$x_4(t) = M v_i c \omega_r \left(\tan\frac{\pi}{2F}\cos(\omega_r t - \frac{\theta}{F}) - \sin(\omega_r t - \frac{\theta}{F})\right)$$

When $d > d_{ZVS}$, converter always operates in CCM and (10) is the only possible current waveform. But as discussed earlier for low duty ratios where $d \le d_{ZVS}$ converter may stay in CCM where current waveform is derived as (11) or it may enter DCM. In DCM, (5) does not describe its behavior anymore and another ODE that describes the system behavior under DCM should be derived. However, CCM current waveform can be used to find the boundary condition for the DCM operation. Figure 7 shows the CCM operation for $d \le d_{ZVS}$. If the slope of current is positive right after $\omega_s t = \theta$ then converter operates at CCM otherwise it has entered DCM and this current waveform is invalid. According to (11) slope of the current at $\omega_s t = \theta$ is $\dot{x}_1(\theta/\omega_s) + \dot{x}_4(\theta/\omega_s)$ which is non-positive if and only if $\pi Q \le 2F$. Then the necessary condition for converter to enter DCM is:

$$Q \le \frac{2}{\pi}F \quad (12)$$

B. DCM Analysis

Figure 8 shows the DCM operation waveform. Right before $\omega_s t = 0$ resonant tank current is zero but the tank capacitor has an unknown initial voltage $-V_{dcm}$. After applying the positive input voltage, the tank current starts rising and reaches its peak value at $\omega_s t = 2d\pi$. The capacitor voltage increases as long as the resonant tank current is positive. At $\omega_s t = 2d\pi$ input voltage is switched to zero when the energy stored in the inductor starts discharging into load and resonant capacitor. At an unknown instant $\omega_s t = \phi$, inductor current reaches zero and remains zero until the next half-cycle. When the current is zero none of the output diodes conducts and the load is isolated from the resonant circuit. During this period the primary voltage of the transformer will be equal to the capacitor voltage at $\omega_s t = \phi$ to keep the inductor current equal to zero. In other words from $\omega_s t = \phi$ to $\omega_s t = \pi$ both capacitor and transformer primary voltages are constant and equal to the capacitor voltage at $\omega_s t = \phi$.

Due to the symmetry, capacitor voltage at $\omega_s t = \pi$ must be equal to V_{dcm}. To analyze circuit, equivalent circuit of Fig.

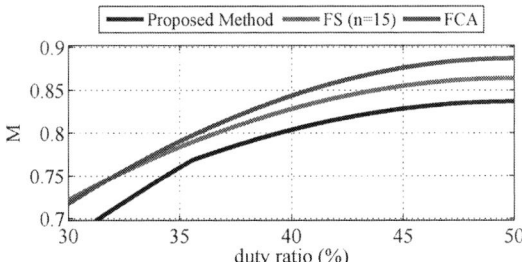

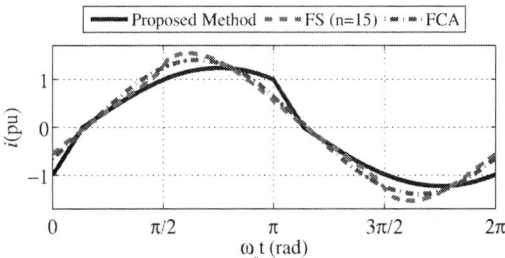

Fig. 9. Comparison of voltage gains calculated using the proposed method and other methods. $Q = 1 \& F = 1.23$

Fig. 11. Calculated current waveforms based on proposed method and other methods. $Q = 1 \& F = 1.23$

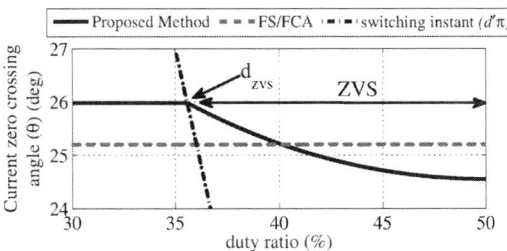

Fig. 10. Current zero crossing point (θ) and ZVS range based on proposed method and other methods. $Q = 1 \& F = 1.23$

Figure 10 shows the plot of the current zero crossing angle (θ), in terms of duty ratio. Due to the symmetry, FS and FCA methods provide similar results. Although both FS/FCA analysis predict that θ is independent of d, the proposed method shows that θ increases as d decreases. This is especially important in the prediction of zero voltage switching (ZVS) of the converter as ZVS directly depends on θ. Therefore, ZVS range calculated by FS/FCA analysis is smaller than the proposed method for this condition and it can be seen that the proposed method results in a better design. Figure 11 shows the current waveforms based on different methods. It is shown in the next section that the plot obtained from the proposed method matches the experimental and simulation results far better than other methods.

5(b) is used with $V_{in}(t)$ and $V_{out}(t)$ as shown in Fig. 8. In order to find unknown variables V_{dcm} and ϕ the proposed method is used and the steady state initial conditions are obtained in terms of V_{dcm} and ϕ. Since in steady state $i(0) = 0$ and $V_c(0) = -V_{dcm}$ two equations are formed from calculated initial conditions and are solved to find V_{dcm} and ϕ which are found as:

$$\phi = 2F \tan^{-1} \left(\frac{\sin(\frac{2d\pi}{F})}{\cos(\frac{2d\pi}{F}) + 2M - 1} \right) \qquad (13)$$

$$V_{dcm} = \frac{(M-1)\left(\cos(\frac{2d\pi}{F}) - 1\right)}{\cos(\frac{2d\pi}{F}) + 2M - 1} V_i \qquad (14)$$

Two variables V_{dcm} and ϕ are sufficient to analyze the converter. Due to the symmetry only half cycle of the current waveform is enough for the analysis:

$$i(t) = \begin{cases} -c\omega_r v_i \frac{(M-1)(2M)}{\cos(\frac{2\pi d}{F}) + 2M - 1} \sin(\omega_r t) & 0 \le \omega_s t \le 2d\pi \\ -c\omega_r v_i \left[\frac{(M-1)(2M)}{\cos(\frac{2\pi d}{F}) + 2M - 1} \sin(\omega_r t) + \right. \\ \qquad \left. \sin(\omega_r t - \frac{2\pi d}{F}) \right] & 2d\pi \le \omega_s t \le \phi \\ 0 & \phi \le \omega_s t \le \pi \end{cases} \qquad (15)$$

C. Comparison of the Proposed Method with FCA and FS

In this section the results of the analyses are plotted and compared with the FS/FCA. The experimental setup parameters ($F = 1.23$ & $Q = 1$) are used to compare the methods. For all methods, voltage gain is plotted in terms of duty ratio as shown in Fig. 9. It can be seen that for $d = 50\%$ the gain predicted by the proposed method is 0.83. While, the gain predicted by FS and FCA methods (considering 15 harmonics) are 0.86 and 0.88 respectively. Simulation and experimental results both confirm the prediction of the proposed method.

IV. SIMULATION AND EXPERIMENTAL RESULTS

To validate the proposed method, SRC circuit is experimentally tested. In this setup the following parameters are chosen: $L = 5.1 \mu H$, $C = 0.8 \mu F$ and $1 : 24$ transformer ratio. This SRC converter basically boosts up the input dc voltage of Fuel Cell or PV resources. In this design switching frequency and load resistance are chosen as $f = 97kHz$, $R = 1450\Omega$ to provide $F = 1.23$, and $Q = 1$. According to Fig. 9 it is predicted that this converter can boost a 20V input up to 400V output voltage at duty ratio equal to 50% which leads to 110W power transfer. Figure 12 shows the simulation waveforms using the PSIM software. It is worth mentioning that the calculated voltage gain $M = 0.83$ leads to $V_o = 16.6V$, which is validated by this simulation result. This scenario is also experimentally validated and results are shown in Fig. 13. It is worth noting that the proposed method calculates the current phase lag as 24.6 degrees (0.43 rad) or 720ns which exactly matches with experimental results in Fig. 13. Moreover, Fig. 14 compares the experimental and calculated current waveforms. It can be seen that the proposed method accurately predicts the simulations and experimental waveforms while providing closed form parametric equations for all the system variables.

In order to validate the derived equations for DCM operation, converter is tested in lower output power. According to (12) for $Q \le 0.78$ converter operates in DCM, therefore Q is selected as 0.5, or $R = 2.9k\Omega$ in this test. For $F = 1.23$ and $Q = 0.5$, critical duty ratio is $d_{ZVS} = 0.42$ and for duty

978-1-4673-9551-9/16 $31.00 © 2016 IEEE

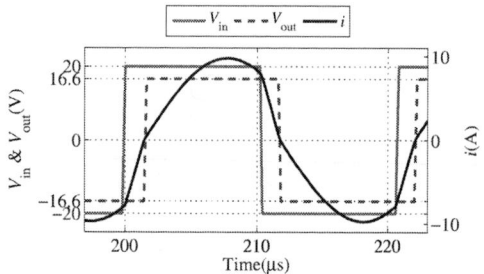

Fig. 12. PSIM simulation results for SRC with $V_{in} = 20V$, $d = 50\%$, $f = 97kHz$, $R = 1450\Omega$, and $n_t = 24$.

ratios less than 0.42 converter enters DCM, and we choose $d = 0.08$.

According to (13) and (14) current interruption instant ($\frac{\phi}{\omega_s}$) and capacitor voltage at the interruption instant (V_{dcm}) are calculated as $3.2\mu s$ and $4V$ respectively. Figure 15 shows the simulation results for the above scenario, where it can be seen that the current interruption time is $3.2\mu s$ and V_{dcm} equals $4V$, which are consistent with calculated results from the proposed method.

Figure 16 shows the experimental waveforms for this test. As it can be observed in this figure, at the zero crossing, current matches with the proposed method calculations. However, current continues to conduct through junction capacitors of the output diodes, since these capacitors have relatively large value in the experimental setup. Therefore, the current does not remain exactly zero after zero crossing point and the transformer primary voltage does not remain equal to the resonant capacitor voltage. In fact, because of large junction capacitors of the rectifier diodes, converter operation is similar to the LCC resonant converter in DCM mode, as also reported in [10]. Using the proposed method one can use a different equivalent circuit to further investigate the junction capacitor effects on the analysis or analyze the LCC converter in CCM mode.

It is worth mentioning that the inverter leading leg switches turn on and off with ZCS and lagging leg switches turn on with ZVS.

In order to cease the current conduction after zero crossing point of current, the gating signals of the inverter are changed in a way that the undesired current of the diodes' junction capacitance can not find a path to flow thorough the inverter. Figure 17 shows the current waveform in this test where the current remains zero after zero crossing point. Using this gating scheme it is possible to reduce the losses of the undesired leaking current while keeping soft switching of the inverter MOSFETs.

V. OTHER APPLICATIONS

A. Design Optimization

The exact solutions obtained by the proposed method for SRC could be used to achieve an optimum converter design. Depending on the application, requirements on the converter characteristics such as voltage gain, input range, ZVS range,

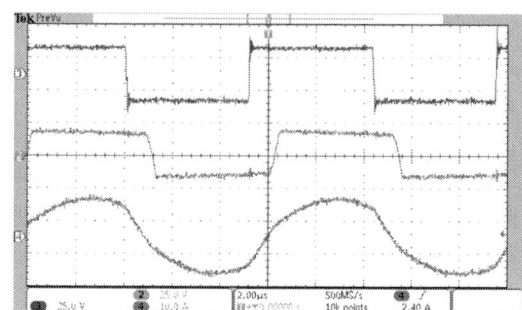

Fig. 13. Experimental results ($V_{in} = 20$, $d = 50\%$) Ch2: Transformer primary voltage (V_{out}) with 25V/div. Ch3: Inverter voltage (V_{in}) with 25V/div. Ch4: Series current (i) with 10A/div.

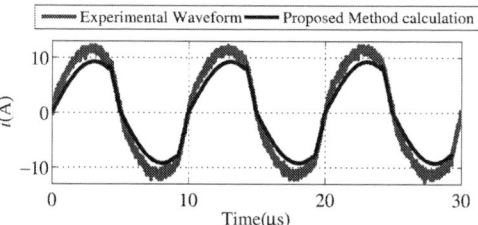

Fig. 14. Experimental Current waveform compared to the current waveform obtained based on proposed method.

etc. will be different. One of the main applications of the DC/DC SRC is PV systems, where soft switching operation of the converter is crucial for high power density design. In this application both the input voltage and power vary, but the output voltage is needed to be fixed. Therefore, large variation in the duty cycle is necessary, which may cause the loss of soft switching operation. For an optimized design of SRC tank parameters can be selected in a way that for any loading condition converter operates in either ZVS mode ($d > d_{ZVS}$) or in ZCS mode in DCM.

Consider the following SRC design problem. The output voltage must be fixed at $400V$, while input voltage range is $25V$ to $40V$, and power could vary from zero to $250W$. For such a large conversion ratio transformer turns is selected to be $1 : 24$ and for practical considerations $Z_0 = \sqrt{\frac{L}{C}}$ is selected to be 2.5Ω. Now that transformer turns ratio and output voltage are determined, every power level presents its corresponding Q. Therefore, d and F are two control variables that could be used to fix the output at $400V$ despite the input voltage and power variations.

As $d = 50\%$ is the maximum duty ratio that provides the largest M, then for every F the smallest input voltage that could be converted to $400V$ could be derived using (7). Also the ZVS violation duty cycle, d_{ZVS}, and its corresponding input voltage could be obtained for every F using (9) and (7) respectively. Figure 18 shows these boundaries for two values of F. Converter operates with ZVS between these two boundaries, and its DCM operation boundary is the Q value obtained using (12) and its corresponding power level. For $F = 1.2$, the striped regions in right and left show the

978-1-4673-9551-9/16 $31.00 © 2016 IEEE   2584

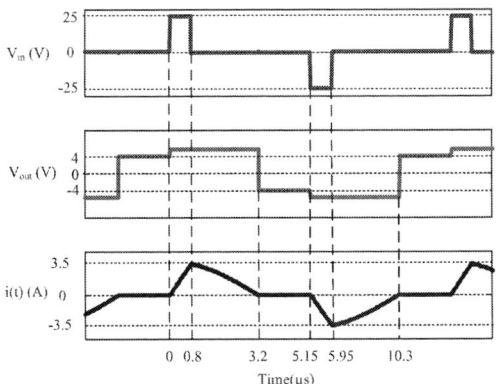

Fig. 15. Simulation results for DCM operation. $d = 8\%, Q = 0.5 \& F = 1.23$

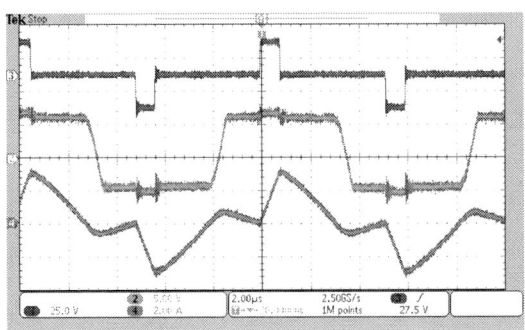

Fig. 16. Experimental results for DCM operation ($V_{in} = 25$, $d = 8\%$) Ch2: Transformer primary voltage (V_{out}) with 5V/div. Ch3: Inverter voltage (V_{in}) with 25V/div. Ch4: Series current (i) with 2A/div.

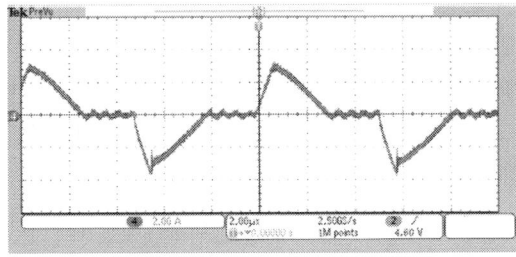

Fig. 17. Experimental current waveform for DCM operation with modified gating scheme ($V_{in} = 25$, $d = 8\%$) Ch4: Series current (i) with 5A/div.

ZVS and DCM operation regions respectively. SRC has soft switching operation for these regions but as it can be seen this single frequency does not cover all of the desired area. Thus a larger F value such as $F = 1.4$ as shown in the figure, should be added to extend the area for both ZVS and DCM operation regions. Using this procedure the SRC which is able to operate under soft switching that covers all the desired range is designed with four F values as $F_1 = 1.2$, $F_2 = 1.28$, $F_3 = 1.4$ and $F_4 = 1.54$.

B. Application in Control System Design

The proposed method can be used to obtain the exact relationship between the control input and output variables.

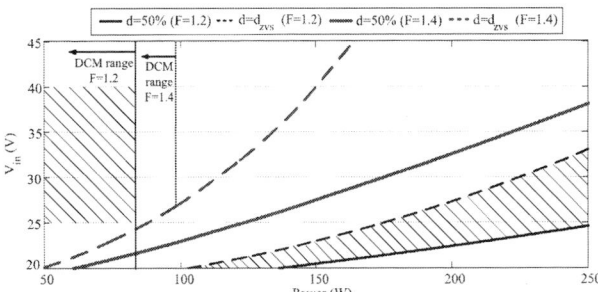

Fig. 18. SRC operation boundaries for $F = 1.2$ and $F = 1.4$

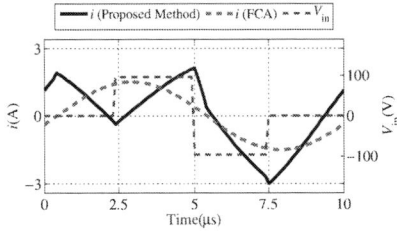

Fig. 19. Comparison of the current waveform of DBSRC obtained from proposed method and FCA.

This can be used to propose novel control systems. In a DBSRC circuit, the amount of transferred power, components stress, and ZVS operation are functions of d_1, d_2 and phase shift between the two bridges (φ). In the literature, φ is mostly used as the control variable to adjust the desired amount of transferred power. Therefore, the control scheme does not guarantee optimized component stress or maximum ZVS range for each power level. If all d_1, d_2, and φ are used as control variables, ZVS range and efficiency can be improved. If this control technique is used, waveforms are not close to sinusoidal and they may even have several steps in one switching cycle. Thus conventional FCA/FS analysis are not effective and/or accurate for this purpose.

In [7], DBSRC is controlled using both d_1 and φ, while d_2 is kept constant. Figure 19 shows the waveforms for FCA and the proposed method. Experimental current waveform of [7] and the calculated one from the proposed method are very close. The proposed method could be effectively used to control d_1, d_2, and φ such that the desired power is transferred under ZVS and optimized component stress.

VI. CONCLUSION

A mathematical theorem is proposed to find the closed form steady state periodic solution of a dynamic system with discontinuous and periodic inputs. The theorem is utilized to develop an effective analysis and design tool for converters with large signal variations. The proposed analysis tool is applied to few power circuits including full bridge Series Resonant Converter (SRC). Using this method, different operating modes of the resonant converter is discussed and analyzed and its effectiveness is shown by obtaining accurate circuit waveforms. It is shown that unlike the conventional approaches, the proposed analysis tool provides accurate closed form parametric solutions and equations. The proposed design approach is also validated by simulation and experimental results. In the end, other applications of the proposed analysis tool including converter design optimization and novel control system designs are also discussed.

APPENDIX

In order to prove the theorem it is assumed that $x(t)$ is the steady state periodic solution, then the necessary condition for this to happen is obtained. $x_{tr}(t)$ is defined to be the truncated version of $x(t)$ which is equal to $x(t)$ for $0 \le t < T$ and zero anywhere else. Since $x(t)$ is the steady state periodic solution its Laplace transform is:

$$X(s) = \frac{X_{tr}(s)}{1 - e^{-sT}} \quad \text{(A-1)}$$

$$X_{tr}(s) = X(s) \cdot \left(1 - e^{-sT} \right) \quad \text{(A-2)}$$

$$X(s) = \frac{Q(s)}{P(s)} + \frac{F(s)}{P(s)} \quad \text{(A-3)}$$

If (A-3) is substituted in (A-2):

$$X_{tr}(s) = \frac{Q(s)}{P(s)} \left(1 - e^{-sT} \right) + \left[\frac{1}{P(s)} \right] \cdot \left[F(s) \left(1 - e^{-sT} \right) \right] \quad \text{(A-4)}$$

Since $F(s) \left(1 - e^{-sT} \right) = F_{tr}(s)$:

$$X_{tr}(s) = \frac{Q(s)}{P(s)} \left(1 - e^{-sT} \right) + \left[\frac{1}{P(s)} \right] \cdot [F_{tr}(s)] \quad \text{(A-5)}$$

$$X_{tr}(s) = \frac{Q(s)}{P(s)} - \frac{Q(s)}{P(s)} e^{-sT} + \left[\frac{1}{P(s)} \right] \cdot [F_{tr}(s)] \quad \text{(A-6)}$$

Taking inverse Laplace from both sides results in:

$$x_{tr}(t) = L^{-1} \left\{ \frac{Q(s)}{P(s)} \right\} - L^{-1} \left\{ \frac{Q(s)}{P(s)} e^{-sT} \right\} + L^{-1} \left\{ \left[\frac{1}{P(s)} \right] \cdot [F_{tr}(s)] \right\} \quad \text{(A-7)}$$

Consider s_j are roots of the characteristic equation $P(s) = 0$, using the Heaviside expansion theorem followings are obtained.

$$L^{-1} \left\{ \frac{Q(s)}{P(s)} \right\} = \sum_{j=1}^{n} \frac{Q(s_j)}{P'(s_j)} e^{s_j t} u(t) \quad \text{(A-8)}$$

$$L^{-1} \left\{ \frac{1}{P(s)} \right\} = \sum_{j=1}^{n} \frac{1}{P'(s_j)} e^{s_j t} = h(t) \quad \text{(A-9)}$$

$$L^{-1} \left\{ \frac{1}{P(s)} F_{tr}(s) \right\} = h(t) * f_{tr}(t) = \int_0^t f_{tr}(\tau) h(t-\tau) d\tau = \int_0^t f_{tr}(\tau) \sum_{j=1}^{n} \frac{e^{s_j(t-\tau)}}{P'(s_j)} d\tau \quad \text{(A-10)}$$

If (A-8) and (A-10) are substituted in (A-7):

$$x_{tr}(t) = \sum_{j=1}^{n} \frac{Q(s_j)}{P'(s_j)} e^{s_j t} u(t) - \sum_{j=1}^{n} \frac{Q(s_j)}{P'(s_j)} e^{s_j(t-T)} u(t-T) + \int_0^t f_{tr}(\tau) \sum_{j=1}^{n} \frac{e^{s_j(t-\tau)}}{P'(s_j)} d\tau \quad \text{(A-11)}$$

For $t > T$, $f_{tr}(t) = 0$ and:

$$\int_0^t f_{tr}(\tau) \sum_{j=1}^{n} \frac{e^{s_j(t-\tau)}}{P'(s_j)} d\tau = \int_0^T f_{tr}(\tau) \sum_{j=1}^{n} \frac{e^{s_j(t-\tau)}}{P'(s_j)} d\tau =$$

$$\sum_{j=1}^{n} \frac{e^{s_j t}}{P'(s_j)} \int_0^T f_{tr}(\tau) e^{(-s_j \tau)} d\tau \quad \text{(A-12)}$$

Therefore $x_{tr}(t)$ for $t > T$ is:

$$x_{tr}(t) = \sum_{j=1}^{n} \frac{Q(s_j)}{P'(s_j)} e^{s_j t} - \sum_{j=1}^{n} \frac{Q(s_j)}{P'(s_j)} e^{s_j(t-T)} + \sum_{j=1}^{n} \frac{e^{s_j t}}{P'(s_j)} \int_0^T f_{tr}(\tau) e^{(-s_j \tau)} d\tau \quad \text{(A-13)}$$

$$x_{tr}(t) \sum_{j=1}^{n} \frac{e^{s_j t}}{P'(s_j)} \left[Q(s_j) \left(1 - e^{-s_j T} \right) + \int_0^T f_{tr}(\tau) e^{(-s_j \tau)} d\tau \right] \quad \text{(A-14)}$$

As it can be seen for $t > T$, $x_{tr}(t)$ is the summation of some exponential terms. Since $e^{s_j t}$ are linearly independent functions then $x_{tr}(t)$ is zero for $t > T$ if and only if all the coefficients of the exponential terms are equal to zero:

$$\left[Q(s_j) \left(1 - e^{-s_j T} \right) + \int_0^T f_{tr}(\tau) e^{(-s_j \tau)} d\tau \right] = 0 \quad \text{(A-15)}$$

$$Q(s_j) = -\frac{\int_0^T f_{tr}(\tau) e^{(-s_j \tau)} d\tau}{\left(1 - e^{-s_j T} \right)} \quad \text{(A-16)}$$

$$Q(s_j) = -F(s_j) \quad \text{(A-17)}$$

Therefore, $x(t)$ is the periodic solution of the ODE if and only if $Q(s_j) = -F(s_j)$ for $j = 1, \ldots, n$.

REFERENCES

[1] J. Hou, Q. Chen, X. Ren, X. Ruan, S.-C. Wong, and C. Tse, "Precise characteristics analysis of series/series-parallel compensated contactless resonant converter," *Emerging and Selected Topics in Power Electronics, IEEE Journal of*, vol. 3, no. 1, pp. 101–110, March 2015.

[2] A. Bhat, "A generalized steady-state analysis of resonant converters using two-port model and fourier-series approach," *Power Electronics, IEEE Transactions on*, vol. 13, no. 1, pp. 142–151, Jan 1998.

[3] P. Jain, H. Soin, and M. Cardella, "Constant frequency resonant dc/dc converters with zero switching losses," *Aerospace and Electronic Systems, IEEE Transactions on*, vol. 30, no. 2, pp. 534–544, Apr 1994.

[4] J. Martin-Ramos, J. Diaz, A. Pernia, J. Lopera, and F. Nuno, "Dynamic and steady-state models for the prc-lcc resonant topology with a capacitor as output filter," *Industrial Electronics, IEEE Transactions on*, vol. 54, no. 4, pp. 2262–2275, Aug 2007.

[5] H. Daneshpajooh, A. Bakhshai, and P. Jain, "Optimizing dual half bridge converter for full range soft switching and high efficiency," in *Energy Conversion Congress and Exposition (ECCE), 2011 IEEE*, Sept 2011, pp. 1296–1301.

[6] ——, "Modified dual active bridge bidirectional dc-dc converter with optimal efficiency," in *Applied Power Electronics Conference and Exposition (APEC), 2012 Twenty-Seventh Annual IEEE*, Feb 2012, pp. 1348–1354.

[7] H. Chen and A. Bhat, "Analysis and design of a dual-bridge series resonant dc-to-dc converter for capacitor semi-active battery-ultracapacitor hybrid storage system," in *Industrial Electronics (ISIE), 2014 IEEE 23rd International Symposium on*, June 2014, pp. 1788–1793.

[8] D. Tschirhart and P. Jain, "Design procedure for high-frequency operation of the modified series-resonant apwm converter to reduce size and circulating current," *Power Electronics, IEEE Transactions on*, vol. 27, no. 10, pp. 4181–4191, Oct 2012.

[9] ——, "Performance of adc for use in mixed-signal control of synchronous rectifiers in current-type resonant converters," in *Telecommunications Energy Conference, 2007. INTELEC 2007. 29th International*, Sept 2007, pp. 514–521.

[10] J. F. Lazar and R. Martinelli, "Steady-state analysis of the llc series resonant converter," in *Applied Power Electronics Conference and Exposition, 2001. APEC 2001. Sixteenth Annual IEEE*, vol. 2. IEEE, 2001, pp. 728–735.

Analysis of Multi-Output Half-Wave Semi-Synchronous Rectifier with A Uniform Magnetic Field Transmitter

Erdem Asa[1,3], Kerim Colak[2], Dariusz Czarkowski[3]

[1]Hevo Power Inc., New York, USA
[2]Istanbul Ulasim A.S., Istanbul, Turkey
[3]New York University, Polytechnic School of Engineering, New York, USA

Abstract— **This paper analyses a multi-output half-wave semi-synchronous rectifier (SSR) by using a uniform field transmitter. The presented multi-output half-wave rectifier topology is achieved by replacing half-wave rectifier lower diode with a synchronous switch and regulation by a phase-shifted PWM signal. The front end transmitter is intended to generate uniform magnetic fields over the coil area. To confirm the performance of the multi-output system, experimental results are provided by using funnel shaped air gap wireless transmitter (the radius of 20 cm) for which input 120 V with two receivers (the radius of 5 cm).**

Keywords— semi-synchronous, active rectifier, multi-output, phase shift, wireless, uniform field

I. INTRODUCTION

Wireless power transfer (WPT) is a proper technology that supplies the power to numerous potential applications without any physical contacts [1]-[6]. It enables portable electronic device charging from a single transmitter to multiple mutually coupled receivers over short distances. The multi output contactless energy transfer system must be well designed and organized through one transmitter to multi output receiver sides. However, power flow control of wireless multi pickup systems is an unresolved issue where the control settings (frequency, phase shift, or dc link control, etc.) from the primary side are not possible. A conventional multi pickup wireless energy transfer system is demonstrated in Fig. 1. As seen in the figure, the system consisted of two stages; a transmitter and receiver platforms [7]. The first stage is basically main part to deliver energy into the second stages. The various dc output voltage connected the load is provided by the second stage, high frequency rectifier and a non-isolated dc-dc conversion.

In a multiple receiver WPT system, magnetic coupling, circuit parameters, and load variations presents unique challenges in WPT system design. These factors result the

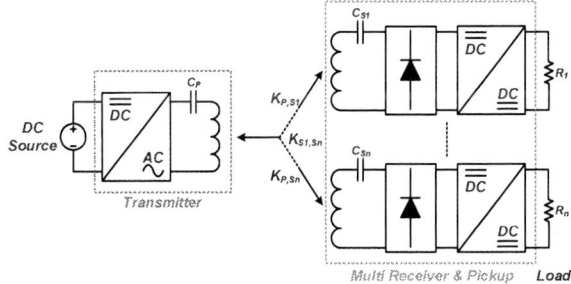

Fig. 1. The general multi pickup wireless energy transfer system.

problems for applications where a stable and constant output voltage is required. Current multiple receiver WPT techniques focus on new pickup circuit topologies, control algorithm, and compensation strategies as explored in the literature [8]-[16]. A common mode multi-phase half-wave semi-synchronous rectifier is investigated for applications in wireless energy transfer systems in [8]. A selective wireless power transfer technique using magnetic resonance coupling (MRC) is introduced for smart power delivery in a multiple-receiver system in [9]. Using cascaded boost and buck or cascaded buck and boost converter, researchers have studied reflected power to the transmitter side for low-middle power inductive power transfer applications in [10]-[11]. Authors in [12] demonstrate the power transfer via magnetic resonance coupling mechanism from a single coil to multiple resonant receivers. A novel phase shift control of semi-bridgeless active rectifier topology is proposed in [13]-[14]. Through the use of a multi frequency magnetic link, power mutually coupled multi receivers is proposed in [15]. A cascaded connected multi-output converter technology is presented for the dynamic wireless EV charger applications in [16].

In this project, multi-output half-wave semi-synchronous rectifier with a uniform field transmitter is investigated for the

978-1-4673-9551-9/16 $31.00 © 2016 IEEE

wireless power transfer applications. In this topology, it is possible to control the system receiver side without using any additional secondary side dc/dc converter or controller synthesis. With this property of the circuit, especially in the multi pickup implementation, independent control of each receiver ports can be achieved. In order to prove the idea and show the working principle, two receiver ports are considered in the paper. In the proposed secondary side multi-output rectifier topology, the half-bridge rectifier lower legs are modified with synchronous rectifier switches. The synchronous switches are driven by a phase-shifted signal to obtain a higher voltage gain without changing the operating frequency. The simple topology has brought a novel aspect and less complicated pickup technology. The converter model controllability is analyzed and the power function of the converter is derived. The system performance is confirmed with experimental results at 600 W output power in the laboratory conditions.

II. THE WIRELESS POWER LINK ANALYSIS

The phase control of multi-output half-wave semi-synchronous rectifier circuit topology for the inductive energy transfer system is shown in Fig. 2. It comprises a primary side dc-ac resonant inverter, a unified magnetic field transformer, and two single phase semi-synchronous half-wave rectifier in the secondary side. The active rectifier is consisted of one transistor in the lower leg with anti-parallel diode and one diode in the upper leg. The symmetrical phase-shift angle of the secondary side transistors regulates the output power in terms of variable load conditions.

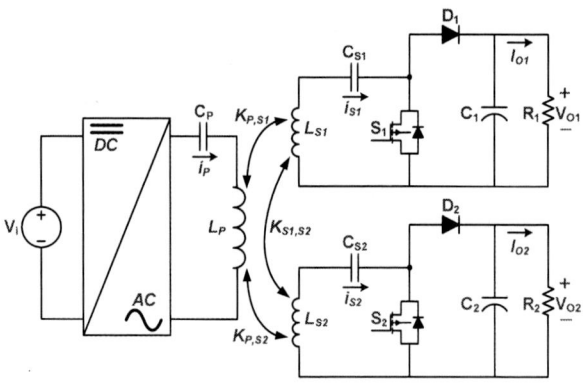

Fig. 2. The phase control multi-output semi-synchronous half-wave rectifier.

The proposed system theoretical analysis is simplified using the fundamental component approximation (FHA) method. Fig. 3 shows the equivalent circuit model of the proposed system between the primary and secondary coils, where $V_{i,1}$ is a fundamental component of voltage source V_i, R_P, R_{S1}, and R_{S2}

are the internal resistance of the primary and secondary coils, Z_{L1} and Z_{L2} are the load impedances of each output, C_P, C_{S1}, and C_{S2} are the capacitances of the primary, secondary first and second units, respectively. L_P, L_{S1}, and L_{S2} are the inductances of the primary, secondary first and second coils, respectively.

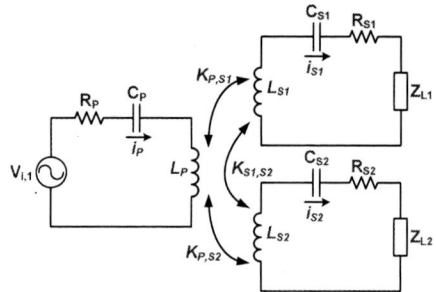

Fig. 3. The proposed system equivalent circuit diagram.

The coupling coefficients $K_{P,S1}$, $K_{P,S2}$, and $K_{S1,S2}$ between primary and secondary coils can be calculated as

$$K_{P,S1} = \frac{L_{M,P,S1}}{\sqrt{L_P L_{S1}}} \tag{1}$$

$$K_{P,S2} = \frac{L_{M,P,S2}}{\sqrt{L_P L_{S2}}} \tag{2}$$

$$K_{S1,S2} = \frac{L_{M,S1,S2}}{\sqrt{L_{S1} L_{S2}}} \tag{3}$$

where $L_{M,S1,S2}$ represents the mutual inductance between the secondary first and second coils. Similarly, $L_{M,P,S1}$ and $L_{M,P,S2}$ are the mutual inductances between the primary coil and the secondary first and second coils, respectively.

The primary equivalent impedance of Z_P and the equivalent impedances of the secondary first and second coils, Z_{S1} and Z_{S2} can be expressed as

$$Z_P = R_P + \frac{1}{j\omega C_P} + j\omega L_P \tag{4}$$

$$Z_{S1} = R_{S1} + \frac{1}{j\omega C_{S1}} + j\omega L_{S1} \tag{5}$$

$$Z_{S2} = R_{S2} + \frac{1}{j\omega C_{S2}} + j\omega L_{S2} \tag{6}$$

If we assume that the secondary both coils are tuned at the same resonant frequency ω_r as

$$\omega_r = \frac{1}{\sqrt{L_{S1} C_{S1}}} = \frac{1}{\sqrt{L_{S2} C_{S2}}} \tag{7}$$

The primary side single frequency current I_P flows through the coil and the leakage inductance and capacitor of the wireless link can be eliminated.

In order to describe the output power parameters, the following general equations can be written by using the primary and secondary side currents as

$$V_{i,1} = -j\omega L_{M,P,S1} I_{S1} - j\omega L_{M,P,S2} I_{S2} \qquad (8)$$

$$0 = -j\omega L_{M,P,S1} I_P - j\omega L_{M,S1,S2} I_{S2} + Z_{L1} I_{S1} \qquad (9)$$

$$0 = -j\omega L_{M,P,S2} I_P - j\omega L_{M,S1,S2} I_{S1} + Z_{L2} I_{S2} \qquad (10)$$

The described function can be stated in a matrix form as

$$\begin{bmatrix} V_{i,1} \\ 0 \\ 0 \end{bmatrix} = \begin{bmatrix} 0 & -j\omega L_{M,P,S1} & -j\omega L_{M,P,S2} \\ -j\omega L_{M,P,S1} & Z_{L1} & -j\omega L_{M,S1,S2} \\ -j\omega L_{M,P,S2} & -j\omega L_{M,S1,S2} & Z_{L2} \end{bmatrix} \begin{bmatrix} I_P \\ I_{S1} \\ I_{S2} \end{bmatrix} \qquad (11)$$

The primary coil current I_P and the secondary side currents I_{S1}, I_{S2} can be revealed as

$$I_P = \frac{V_O(e^2 - ab)}{ad^2 + bc^2 - 2cde} \qquad (12)$$

$$I_{S1} = \frac{I_P(bc - de)}{e^2 - ab} \qquad (13)$$

$$I_{S2} = \frac{I_P(ad - ce)}{e^2 - ab} \qquad (14)$$

where a, b, c, d, and e functions are described with the following equations as

$$a = Z_{L1} \qquad (15)$$

$$b = Z_{L2} \qquad (16)$$

$$c = -j\omega L_{M,P,S1} \qquad (17)$$

$$d = -j\omega L_{M,P,S2} \qquad (18)$$

$$e = -j\omega L_{M,S1,S2} \qquad (19)$$

The reflected equivalent impedances can be found by using first harmonic approximation (FHA) as

$$Z_{L1} = \frac{2R_1\left(\sin^3\frac{\beta_1}{2}\left[\sin\frac{\beta_1}{2} - j\cos\frac{\beta_1}{2}\right]\right)}{\pi^2} \qquad (20)$$

and

$$Z_{L2} = \frac{2R_2\left(\sin^3\frac{\beta_2}{2}\left[\sin\frac{\beta_2}{2} - j\cos\frac{\beta_2}{2}\right]\right)}{\pi^2} \qquad (21)$$

where β_1 and β_2 are the conduction angle of the half wave rectifiers in the receiver side, respectively. The equivalent resistances value can obtained by taking real portion of (20) and (21), as

$$R_{L1} = \frac{2R_1\left(\sin^4\frac{\beta_1}{2}\right)}{\pi^2} \qquad (22)$$

$$R_{L2} = \frac{2R_2\left(\sin^4\frac{\beta_2}{2}\right)}{\pi^2} \qquad (23)$$

Taking imaginary portion of the equivalent impedances, the corresponding reactance values are defined as

$$X_{L1} = -\frac{2R_1 \sin^3\frac{\beta_1}{2}\cos\frac{\beta_1}{2}}{\pi^2} \qquad (24)$$

$$X_{L2} = -\frac{2R_2 \sin^3\frac{\beta_2}{2}\cos\frac{\beta_2}{2}}{\pi^2} \qquad (25)$$

The out power $P_{O,L1}$ and $P_{O,L2}$ delivered to the loads is found to be as a function of circuit parameters as

$$P_{O,L1} = I_{S1}^2 R_1 \qquad (26)$$

$$P_{O,L1} = \frac{I_P^2 R_1 \left(\omega^2 L_{M,S1,S2} L_{M,P,S2} - Z_{L2} L_{M,P,S1}\right)^2}{\left(\omega^2 L_{M,S1,S2}^2 - Z_{L1} Z_{L2}\right)^2} \qquad (27)$$

and

$$P_{O,L2} = I_{S2}^2 R_2 \qquad (28)$$

$$P_{O,L2} = \frac{I_P^2 R_2 \left(\omega^2 L_{M,S1,S2} L_{M,P,S1} - Z_{L1} L_{M,P,S2}\right)^2}{\left(\omega^2 L_{M,S1,S2}^2 - Z_{L1} Z_{L2}\right)^2} \qquad (29)$$

The total out power $P_{O,T}$ can be exposed as

$$P_{O,T} = P_{O,L1} + P_{O,L2} = I_{S1}^2 R_1 + I_{S2}^2 R_2 \qquad (30)$$

III. EXPERIMENTAL RESULTS

In order to prove the idea and show the operating principle, two outputs semi-synchronous half-wave rectifier is considered to experimentally verify the proposed converter analysis as shown in Fig. 4. The proposed converter topology parameters are given in Table I.

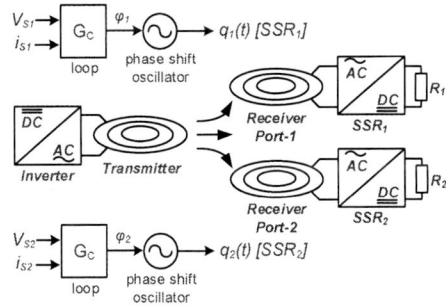

Fig. 4. The experimented system model with phase-shift control.

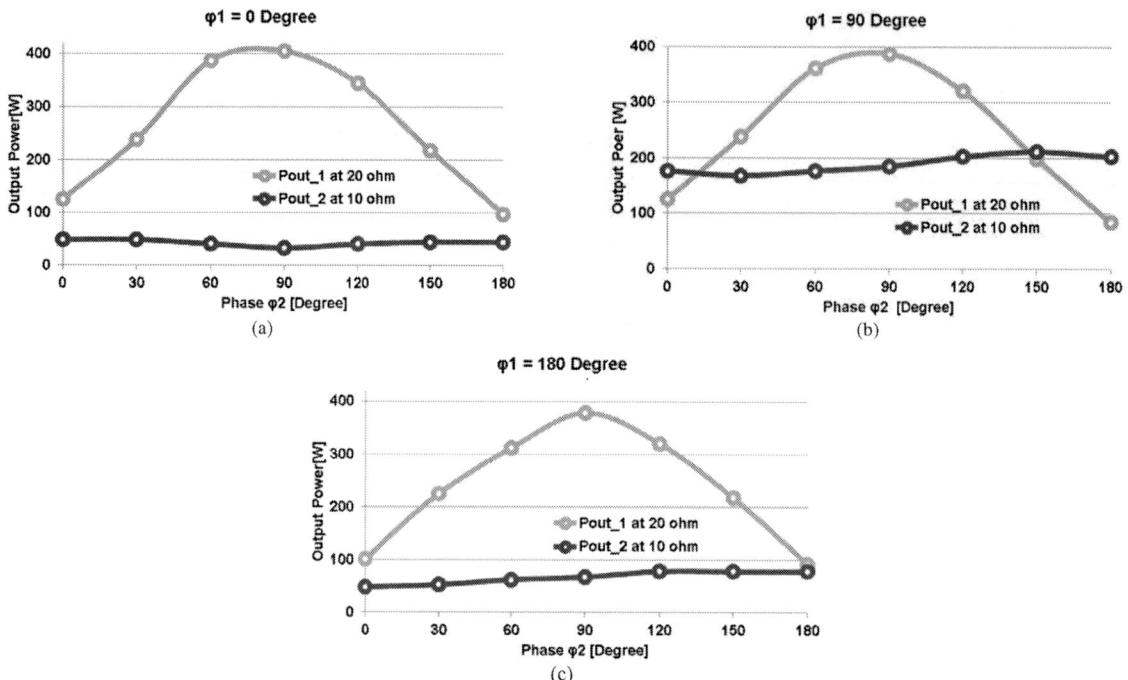

Fig. 5.   The output power characteristics with phase shift angle for a) $\varphi 1 = 0^\circ$, b) $\varphi 1 = 90^\circ$, c) $\varphi 1 = 180^\circ$.

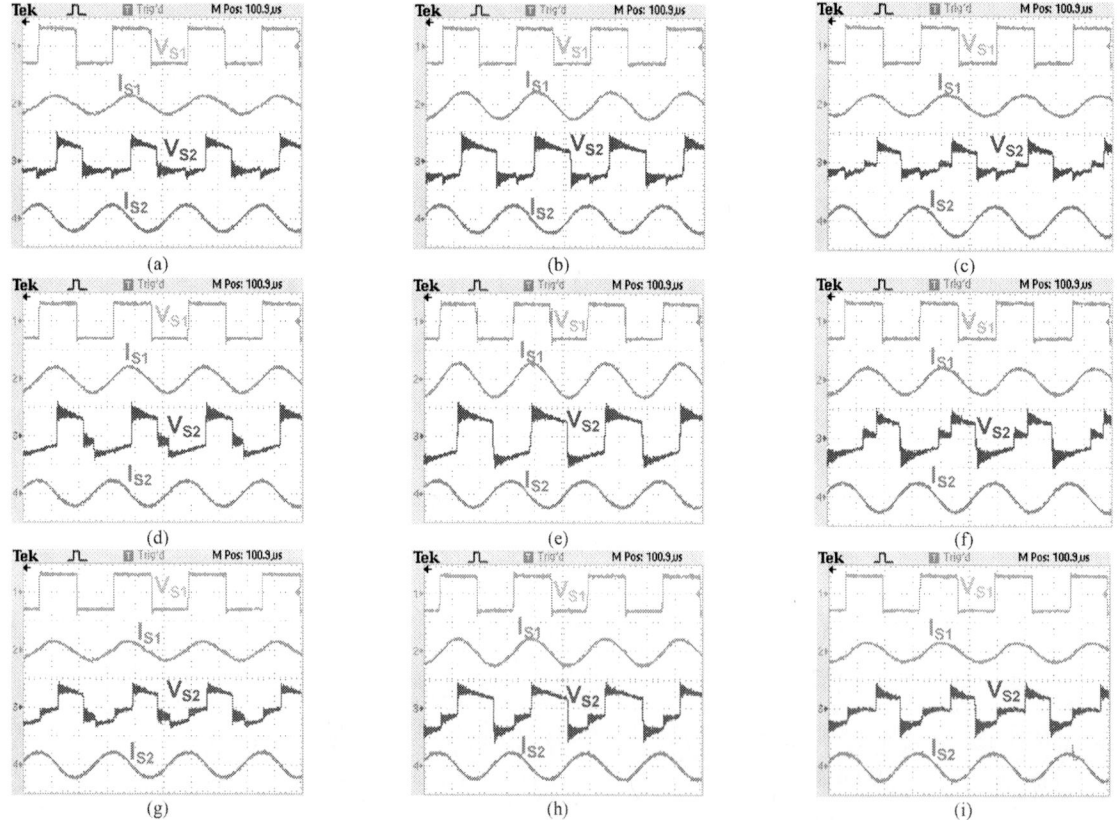

Fig. 6.   Receiver side first port voltage (V_{S1}), current (I_{S1}) at 20 Ω and second port voltage (V_{S2}), current (I_{S2}) at 10 Ω: V_{S1} (100V/div), I_{S1} (2A/div), V_{S2} (250V/div), I_{S2} (5A/div).

TABLE I

Symbol	Parameter	Values
V_i	DC Input Voltage	120 V
V_{O1}	Port-1 Output Voltage	0-100 V
V_{O2}	Port-2 Output Voltage	0-50 V
P_{OT}	Rated Power	600 W
C_P,	Primary Resonant Capacitor	40 nF
C_{S1}, C_{S2}	Secondary Resonant Capacitor	80 nF
L_P	Primary Self Inductance	25 µH
L_{S1}, L_{S2}	Secondary Self Inductance	12 µH
K	Coupling Factor	0.2
f_{sw}	Operating Frequency	150 kHz
t_{dead}	Dead Time	300 ns

The funnel shaped coreless transformer is tested for 120 V input voltage considering 10 Ω and 20 Ω output load conditions. The characteristic waveforms of the proposed converter described above are given to verify circuit operation with the waveforms. The output power conditions between output ports are presented with the different phase angle φ_1, φ_2, and different load conditions R_1 and R_2 in Fig. 5(a-c). As seen from the figures, a wide output and different power range can be achievable by controlling the phase angle value at the light and high load conditions in each port. Selected receiver side voltage and current waveforms at different load conditions are given in Fig. 6(a-i). The phase-shift angle controls the output waveforms in each port by changing the receiver side voltages as shown in the figures.

IV. CONCLUSIONS

In this work, a phase control of multi-output semi-synchronous half-bridge active rectifier is analyzed by using a uniform magnetic field transmitter for wireless power transfer applications. The power function of the converter is derived analytically. The system control adjusts the output power of the system by phase shift tuning of the synchronous switches in each output of the receiver side. The system performance is confirmed with theoretical and experimental results at different load conditions. To verify the proposed converter, a 600 W full power prototype is designed at 120 V input.

REFERENCES

[1] E. Asa, K. Colak, D. Czarkowski, B. Tamyurek, "An Efficiency Analysis of Bi-directional DC/DC Converter for Wireless Energy Transfer Applications," IEEE Energy Conversion Congress and Exposition (ECCE), pp.594-598, Sep. 2015.

[2] K. Colak, E. Asa, D. Czarkowski, H. Komurcugil, "A Novel Multi-level Bi-directional DC/DC Converter for Inductive Power Transfer Applications," 41th Annual Conference of IEEE Industrial Electronics Society (IECON), pp.1-5, Nov. 2015.

[3] M. Bojarski, E. Asa, D. Czarkowski, "A 25 kW Industrial Prototype Wireless Electric Vehicle Charger," IEEE Applied Power Electronics Conference and Exposition (APEC), pp.1-6, Mar. 2016.

[4] M. Nalbant, "Wireless Power Transmitter Having Low Noise and High Efficiency, and Related Methods," U.S. Patent, 2014/0132077 A1 May 15, 2014.

[5] E. Asa, K. Colak, M. Bojarski, D. Czarkowski, "A Novel Multi-Level Phase-Controlled Resonant Inverter with Common Mode Capacitor for Wireless EV Chargers," IEEE Transportation Electrification Conference and Expo (ITEC), pp.1-6, Jun. 2015.

[6] M. Bojarski, E. Asa, D. Czarkowski, "Effect of Wireless Power Link Load Resistance on the Efficiency of the Energy Transfer," IEEE International Electric Vehicle Conference (IEVC), Dec. 2014.

[7] K. Colak, E. Asa, D. Czarkowski, "A Novel Phase Control of Single Switch Active Rectifier for Inductive Power Transfer Applications," IEEE Applied Power Electronics Conference and Exposition (APEC), pp.1-6, Mar. 2016.

[8] K. Colak, E. Asa, M. Bojarski, D. Czarkowski, "A Novel Common Mode Multi-phase Half-wave Semi-synchronous Rectifier for Inductive Power Transfer Applications," IEEE Transportation Electrification Conference and Expo (ITEC), pp.1-6, Jun. 2015.

[9] Y. J. Kim, D. Ha, W. J. Chappell, P. P. Irazoqui, "Selective Wireless Power Transfer for Smart Power Distribution in a Miniature-Sized Multiple-Receiver System," IEEE Transactions on Industrial Electronics, Early Access, DOI: 10.1109/TIE.2015.2493142.

[10] K. Colak, M. Bojarski, E. Asa, D. Czarkowski, "A Constant Resistance Analysis and Control of Cascaded Buck and Boost Converter for Wireless EV Chargers," IEEE Applied Power Electronics Conference and Exposition (APEC), pp.3157-3161, Mar. 2015.

[11] M. Fu, C. Ma, X. Zhu, "A Cascaded Boost–Buck Converter for High-Efficiency Wireless Power Transfer Systems," IEEE Transactions on Industrial Informatics, vol.10, no.3, pp.1972-1980, Aug. 2014.

[12] R. Johari, J. V. Krogmeier, D. J. Love, "Analysis and Practical Considerations in Implementing Multiple Transmitters for Wireless Power Transfer via Coupled Magnetic Resonance," IEEE Transactions on Industrial Electronics, vol.61, no.4, pp.1774-1783, Apr. 2014.

[13] K. Colak, E. Asa, M. Bojarski, D. Czarkowski, O. C. Onar, "A Novel Phase Shift Control of Semi-Bridgeless Active Rectifier for Wireless Power Transfer" IEEE Transaction on Power Electronics, vol.30, no.11, pp.6288-6297, Nov. 2015.

[14] E. Asa, K. Colak, M. Bojarski, D. Czarkowski, "A Novel Phase Control of Semi Bridgeless Active Rectifier for Wireless Power Transfer Applications," IEEE Applied Power Electronics Conference and Exposition (APEC), pp.3225-3231, Mar. 2015.

[15] Z. Pantic, K. Lee, S. M. Lukic, "Receivers for Multifrequency Wireless Power Transfer: Design for Minimum Interference," IEEE Journal of Emerging and Selected Topics in Power Electronics, vol.3, no.1, pp.234-241, Mar. 2015.

[16] E. Asa, K. Colak, D. Czarkowski, "Analysis of Cascaded Multi-Output-Port Converter for Wireless Plug-in Hybrid On-Board EV Chargers," IEEE Applied Power Electronics Conference and Exposition (APEC), pp.1-5, Mar. 2016.

HIGH GAIN QZS DC/DC CONVERTER WITH COUPLED INDUCTOR

Rafael V. Silva[1], Antônio A. A. Freitas[2] , Marcus R. Castro[1], Fernando L. M. Antunes[3], Edilson M. Sá Jr[4].

e-mail: {vitor.crato, edilson.mineiro, alisson.alencar.freitas}@gmail.com, marcusdecastro@yahoo.com.br, fantunes@dee.ufc.br

[1]Department of Electrical Engineering, Federal University of Ceará, Sobral-CE, Brazil

[2] Department of Electrical Engineering, Federal University of Ceará, Fortaleza-CE, Brazil

[3]Department of Electrical Engineering, Federal Rural University of Semi-Árido, Caraúbas-RN Brazil

[4] Department of Industrial Process and Control, Federal Institute of Ceará, Sobral-CE, Brazil

Abstract— This paper proposes a non-isolated qZS converter to feed a frequency inverter applied on a standalone photovoltaic tricycle. The converter uses the clipping voltage of the power switch to be added to the output voltage, which allows reducing the duty cycle values and increase the gain of the converter. Some converter topologies, based on high gain coupled inductors and voltage multiplier cells cause large current ripple in the input and voltage spike at the power switch. As a consequence the lifespan of the components and the converter efficiency are reduced. The proposed converter showed a low current ripple in the input and the power switch was submitted to lower voltage. High gain qZS dc-dc converters operation principle and analysis are shown, and verified by simulation and experimental results as well as a losses analysis.

Keywords—Frequency Inverter, MIT, Photovoltaic Tricycle, quasi-Z-Source Converter.

I. INTRODUCTION (*Heading 1*)

The wheelchairs have been standing out as an important equipment of the assisted technology to help people with motor deficiencies [1]-[3]. Wheelchairs are divided in two groups: manual and powered. In most cases, powered wheelchairs are designed for indoor environments and limited spaces [4], [5].

Therefore, they are suitable for normal terrains and for small displacements, which results in a low range project. There are a number of projects being carried out to increase wheelchairs autonomy. Some studies are focusing on developing fully electric traction with using DC brushless motors or induction motors.

Nowadays, wheelchairs using three phase induction motor (MIT) manufactured are more due to the fact they are widely used in industry. However, applications using the MIT require a high-gain dc-dc converter that has high efficiency and reliability.

According to the literature, the classical boost converter cannot provide high voltage gains, several high gain converters are being studied [6]-[7]. Due to some limitations wheelchairs can be replaced by tricycle, because these are more resistant and support the weight of a standalone system [8] - [9].

The DC-DC converters qZS have been the focus of increasing attention of researchers [10]-[11]. Its static gain of 1 / (1-2D) is greater than the one of the boost converters 1 / (1-D). If coupled inductors are used, smaller values of the duty cycle and gain are required for the qZS converters, the converter which enables high efficiency values [12].

This paper proposes a non-isolated converter qZS to feed a frequency converter applied in an autonomous photovoltaic tricycle, which uses the clipping voltage of the power switch to be added to the converter output voltage, which allows reducing the cyclic ratio values and heighten the gain converter.

II. PROPOSED TOPOLOGY

The proposed converter is shown in Fig. 1. For the converter analysis, it will be considered that the values of the capacitances C1, C2, C3, C4 and C5 are high enough to keep its constant voltage as well as coupling coefficient. Losses in the diodes D1, D2, D3 and D4, conduction and switching losses on S1 and losses in the windings will all be disregarded.

The proposed converter basically has five operating stages. For the operation analysis of the converter stage will be working in continuous conduction mode.

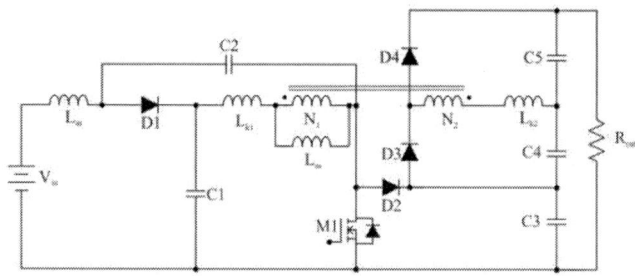

Fig. 1.   Proposed converter

First stage (t₀ - t₁). Fig. 2 shows the circuit for the first stage of operation. At this stage the S1 power switch and the diode D4 are still conducting. This is due to secondary leakage inductance have a residual energy polarizing the diode D4 for a very short time.

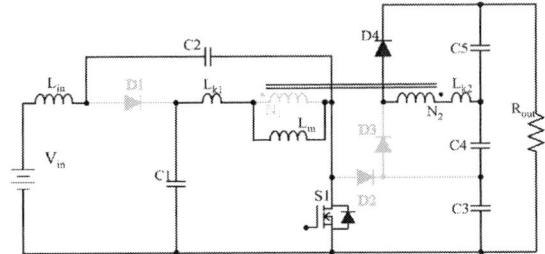

Fig. 2. First stage (t0 – t1).

Second stage ($t_1 - t_2$). Fig. 3 shows the circuit for the second stage of operation. At this stage, the diodes D1, D2 and D4 are blocked. The current in the inductor is linear. The diode D3 is directly polarized and charging C4. The voltage in C4 is equal to the reflected secondary voltage of the couple inductor VL2, which is due to the winding N2.

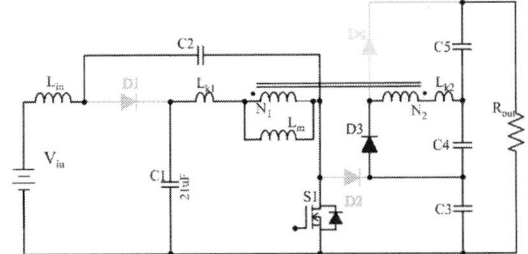

Fig. 3. Second stage (t1 – t2).

Third stage ($t_2 - t_3$). Fig. 4 shows the circuit for the third stage of operation. At this stage the power switch S1 is blocked and the diode D3 remains conducting due to the residual energy accumulated in the secondary leakage inductance polarizing D3 for a short period of time.

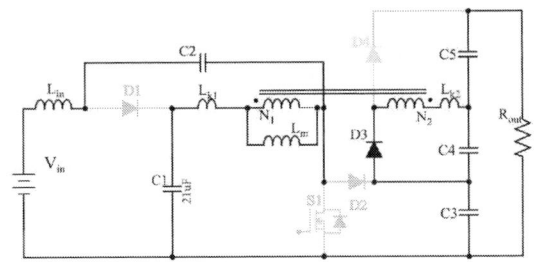

Fig. 4. Third stage ($t_2 - t_3$).

Fourth stage ($t_3 - t_4$). Fig. 5 shows the circuit for the fourth stage of operation. At this stage the power switch S1 is blocked and diode D1 enters in conduction. In this phase, the capacitor C2 is charged by a part of stored energy in inductance of magnetization. The load is supplied by capacitors C3, C4 and C5.

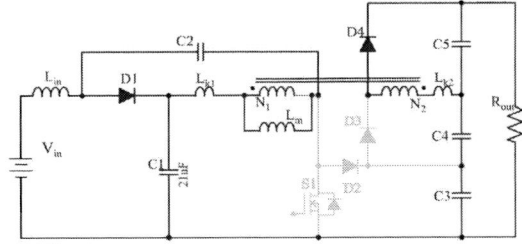

Fig. 5. Fourth stage ($t_3 - t_4$).

Fifth stage ($t_4 - t_5$). Fig. 6 shows the circuit for the fifth stage of operation. At this stage the power switch S1 remains blocked and the diode D2 comes into conduction. This occurs because the voltage in S1 is greater than the voltage at the capacitor C3. At the end of this step, C3 and C5 step will be charged.

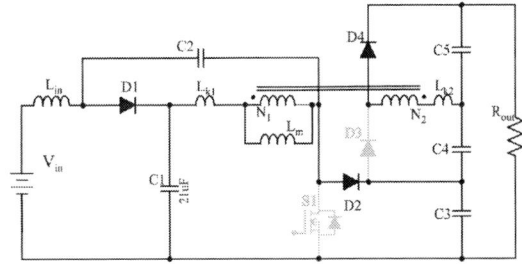

Fig. 6. Fifth stage ($t_4 - t_5$).

The Fig. 5 shows the ideal waveforms for the proposed converter.

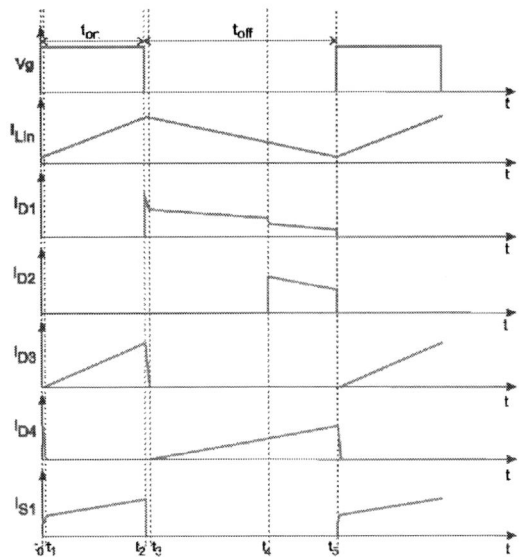

Fig. 7. Ideal waveforms.

III. QUANTITATIVE ANALYSIS

For quantitative analysis will be considered leakages inductances of L_1 and L_2. In addition, it is assumed that the coupling factor between the coupled inductors is not unitary.

$$n = \frac{N_2}{N_1} \qquad (1)$$

where:

n - transformation ratio;

N_1 - number of turns of the primary;

N_2- number of turns of the secondary.

The mutual inductance between the windings can be determined by (2) or by (3):

$$M = k\sqrt{L_1 . L_2} \qquad (2)$$

$$M = \frac{N_2}{N_1} . L_m = n.L_m \qquad (3)$$

where:

M - mutual inductance;

k - coupling factor;

L_1 - inductance of the primary;

L_2 - inductance of the secondary;

L_m - magnetizing inductance.

The inductance of the windings L_1 and L_2 are determined by:

$$L_1 = L_m + L_{k1} = \frac{M}{n} + L_{k1} \qquad (4)$$

$$L_2 = n^2 . L_m + L_{k2} = n.M + L_{k2} \qquad (5)$$

where:

L_{k1} - leakage inductance of the primary;

L_{k2} - leakage inductance of the secondary.

The coupling factor can be determined by.

$$k = \frac{L_m}{L_m + L_{k1}} \qquad (6)$$

A. First Stage (t0 – t1).

In the first stage, as already mentioned, the leakage inductance of the secondary has a residual energy that polarizes the diode D4.

B. Second Stage (t1 – t2).

During the second stage the voltage at L_{in} and L1 can be defined by:

$$V_{Lin} = V_{in} + V_{C2} \qquad (7)$$

$$V_{L1} = V_{C1} \qquad (8)$$

where:

V_{Lin} - voltage in the inductor L_{in};

V_{in} - input voltage;

V_{c2} - voltage in the capacitor V_{c2};

V_{L1} - voltage in the inductor L1;

V_{c1} - voltage in the capacitor V_{c1}.

The voltage in magnetizing inductance, V_{Lm}, can be defined by:

$$V_{Lm} = \frac{L_m}{L_m + L_{k1}} . V_{C1} \qquad (9)$$

Whereas the value of the current flowing through the inductor L2 is small, the voltage across the capacitor C4, at the end of the first stage, can be defined by:

$$V_{C4} = n \cdot k \cdot V_{C1} \qquad (10)$$

The value of C4 capacitance is high enough to consider the voltage ripple is practically zero at capacitor C4. Thus, V_{C4} be considered constant.

C. Third Stage (t2 – t3).

Similar to the first stage, the leakage inductance of the secondary has a residual energy that polarizes the diode D3.

D. Fourth Stage (t3 – t4).

During the fourth stage magnetizing inductance charges C2 and voltage in the inductor L1 can be defined by:

$$V_{L1} = -V_{C2} \qquad (11)$$

In this step, the voltage at the input inductor can be defined by:

$$V_{Lin} = V_{in} - V_{C1} \tag{12}$$

E. Fifth Stage (t2 – t3).

During the fifth stage the voltage at capacitor C3 and C5 can be defined by:

$$V_{C3} = V_{C1} + V_{C2} \tag{13}$$

$$V_{C5} = n.V_{L2} \tag{14}$$

F. Static Gain.

Analyzing the equations of the three stages of topology, it is possible to determine the static gain of the proposed converter. Knowing that the voltage average value in the coupled inductor is zero, the equation can be defined

$$\frac{V_{C2}}{V_{C1}} = \frac{D}{(1-D)} \tag{15}$$

where:

D - duty cycle.

Whereas the average value of voltage at the input inductor L_{in} is zero, the equation can be defined by:

$$\int_{0}^{DTs} V_{Lin}(t)dt + \int_{DTs}^{Ts} V_{Lin}(t)dt = 0 \tag{16}$$

Substituting (2) and (7) in (12) and simplifying it is possible to obtain the equation defined by.

$$D.V_{C2} + V_{in} - V_{C1} + D.V_{C1} = 0 \tag{17}$$

Substituting (11) and (13) and isolating V_{c1} and V_{c2} is possible to obtain the voltage at capacitor C1 and C2 respectively.

$$V_{C1} = \frac{(1-D)}{(1-2D)}.V_{in} \tag{18}$$

$$V_{C2} = \frac{D}{1-2D}.V_{in} \tag{19}$$

Substituting (14) and (15) in (8) and simplifying it is obtained voltage across the capacitor C3.

$$V_{C3} = \frac{1}{1-2D}.V_{in} \tag{20}$$

Substituting (14) in (5) and simplifying it is obtained voltage across the capacitor C4.

$$V_{C4} = n.\frac{(1-D)}{(1-2D)}.V_{in} \tag{21}$$

In the second stage the voltage at L2 when the switch is turned off can be approximated by the voltage on inductor reflected N2 magnetization. Thus, the voltage across the capacitor C5 may be defined by:

$$V_{C5} = \frac{n.D.V_{C1}}{1-D} \tag{22}$$

Substituting (14) in (18) and simplifying it is obtained voltage across the capacitor C5

$$V_{C5} = \frac{n.K.D}{1-2D}.V_{in} \tag{23}$$

The output voltage can be defined by:

$$V_{out} = V_{C3} + V_{C4} + V_{C5} \tag{24}$$

Substituting (16), (17) and (19) in (20) and simplifying it is obtained gain static of proposed converter.

$$\frac{V_{out}}{V_{in}} = \frac{1+n}{1-2D} \tag{25}$$

It can be seen that the static gain of the converter depends on the transformation ratio (n) and duty cycle. Observing that the transformation ratio and the coupling coefficient are constants, the static gain only depends on the variation of the duty cycle. In Figure 6 can be seen that proposed converter can achieve the high voltage gain value with lower duty cycle. This feature is essential for applications in wheelchairs by electric traction.

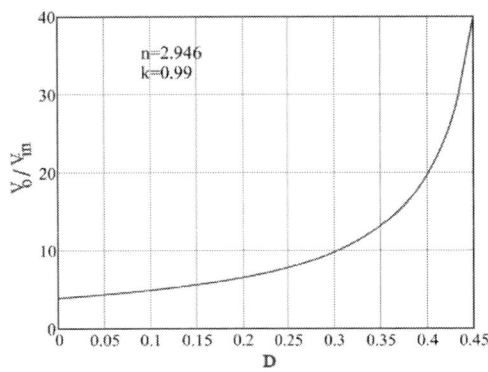

Fig. 8. Static gain of proposed converter.

IV. LOSSES ANALYSIS

The equations and procedures performed in this article for losses calculation can be found in [13] and [14]. In the study the main losses in a power converter were considered including the coupling coefficient. They were calculated in all the diodes, the power switch, the input inductor (Lin) and coupled inductor (Lm).

A. Losses in switches uncontrolled

The losses in the diodes can be divided into conduction losses and switching. The losses in diode D1, D2, D3 e D4 can be defined by.

$$P_{cond\_D} = V_{TO} \cdot I_{D\_avg} + R_T \cdot I_{D\_rms}^2 \qquad (26)$$

$$P_{sw} = C_{j\_D} \cdot f_s \cdot V_{R\_D}^2 \qquad (27)$$

$$P_{TOT\_D} = P_{sw} + P_{cond\_D} \qquad (28)$$

where:

P_{cond_D} – conduction losses;

V_{TO} – conduction threshold voltage;

I_{D_avg} – current average of diode;

R_T – intrinsic resistance of diode;

I_{D_rms} - current rms of diode;

P_{sw} – switching losses;

C_{j_D} – junction capacitance of diode;

f_s – frequency switching;

V_{R_D} – reverse voltage of diode;

P_{TOT_D} – total losses.

B. Losses in switches controlled

For power switch S1 were considered losses by conduction, turn-on and turn-off. The losses in S1 can be defined by.

$$P_{cond\_S1} = D \cdot R_{ds} \cdot I_{S1\_rms}^2 \qquad (29)$$

$$P_{turn\_on} = \frac{2}{3} \cdot f_s \cdot V_{pk}^2 \cdot C_{ds} \qquad (30)$$

$$P_{turn\_off} = \frac{4}{3} \cdot f_s \cdot V_{pk}^2 \cdot C_{ds} \qquad (31)$$

$$P_{TOT\_S1} = P_{cond\_S1} + P_{turn\_on} + P_{turn\_off} \qquad (32)$$

where:

P_{cond_S1} – conduction losses in S1;

D - duty cycle

R_{ds} – intrinsic resistance of S1;

I_{S1_rms} - current rms of S1;

P_{turn_on} – losses in conduction entry of S1;

V_{S1_pk} – peak voltage of S1;

C_{ds} – capacitance drain-source of S1;

P_{TOT_S1} – total losses in S1.

C. Losses in magnetic elements

The losses in the magnetic elements can be divided into core loss, and losses in the windings [15].

1) Losses in the core: The method used for the determination of core loss can be seen in [14]. Other theoretical and empirical methods are described in [15] and [16]. The methodology is based on design specifications, such as volume core. Thus, the core loss can be determined by.

$$P_{core} = (\rho \cdot V_e) \cdot P_p \qquad (33)$$

where:

P_{core} – core losses;

ρ - density of the material;

V_e – core volume;

P_p – core loss per unit mass.

The loss in the selected core must take into account the frequency operation, the change in magnetic flux as well as operating temperature.

2) Losses in the windings: The method used to determine the losses in the windings took into account only the losses by Joule effect. Therefore, the power dissipated in the windings can be determined by.

$$P_{winding} = n \cdot \frac{\rho \cdot l_e}{S} \cdot I_{rms}^2 \qquad (34)$$

where:

$P_{winding}$ – power dissipated in the winding;

l_e – average length of wire;

S – wire section area;

I_{rms} – rms current in the winding.

Table 01 show the data obtained in the calculations as well as commercial references chosen for each component.

TABLE I.   LOSSES COMPONENT IN THE PROPOSED CONVERTER

Component	Conduction Losses (W)	Switching Losses (W)	Reference
Mosfet S1	2.085	0.268	IRFPB4227
D1	6.235	0.192	V80100P
D2	0.619	0.04	SB5200
D3	1.042	0.117	C3D04060
D4	1.042	0.117	C3D04060
Component	Magnetic Losses (W)	Conductor Losses (W)	Reference
Lin	0.312	0.673	IP12R
Lm	1.106	1.776	IP12R

The total value of losses calculated in the converter was equal to 15.626 W assuming a duty cycle equal to 0.35 and a switching frequency equal to 100 kHz.

V.   EXPERIMENTAL RESULTS

The following will show the main waveforms of the high gain qZS dc-dc converter. The converter was implemented in laboratory dimensioned to 250W of power. The voltage of source of converter will be a battery, the minimum of operating voltage is 24V for no load.

The output of the converter must have a voltage level of 311V to allow the use of the frequency inverter source in the electric traction system. The Figure 9 shows the completed circuit the implemented converter. The Figure 10 shows the prototype implemented in laboratory.

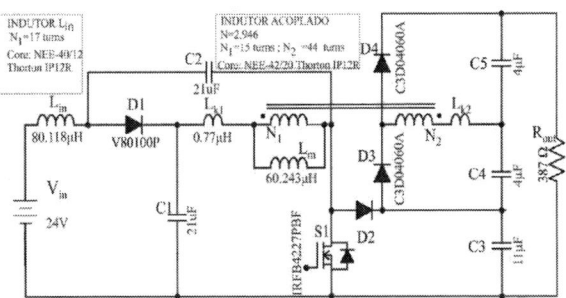

Fig. 9.   Complete circuit of implemented converter

Fig. 10. Protype implemented.

The Figure 11 shows the voltages at the terminals of capacitors C3, C4 and C5 and the input current ILin. The input voltage considered in the experiment was 24V. The converter output voltage is the sum of tensions in capacitors VC3, VC4 and VC5.

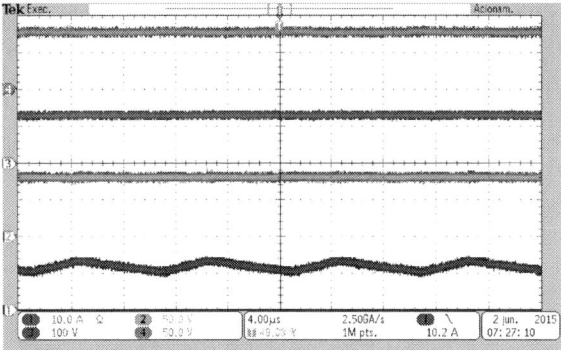

Fig. 11. ILin (dark blue) (10A/div); VC3 (pink) (50V/div); VC4 (green) (100V/div); VC5 (blue) (50V/div)..

The Fig 9 shows that the value of the output voltage was about 311V, thus validating the static gain shown in Equation (25). The input current flowing through the inductor Lin is shown to demonstrate that output voltage has almost no ripples

The Figure 12 shows the current and voltage at the terminals of switch S1 and the current in the diode D1.

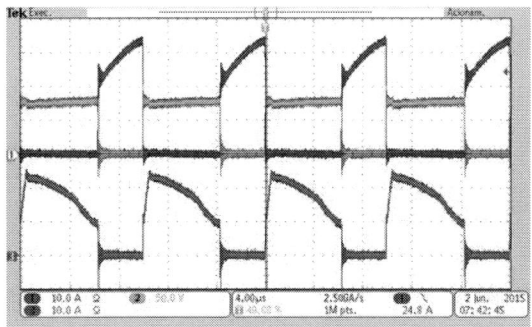

Fig. 12. IS1 (dark blue) (10A/div); VS1 (blue) (50V/div); ID1 (pink) (10A/div).

Analyzing the Figure 12, can be seen that there is no overvoltage in the MOSFET, which usually occurs in circuits with coupled inductors. The overvoltage are caused by dispersion. At the moment of S1 conduction the leakage inductance stores energy.

When the switch is turned off, the currents in the dispersion inductances would be blocked quickly, causing the abrupt discharge energy accumulated in the form of overvoltage on the switch, damaging it.

There is no overvoltage in S1 because when the switch is blocked dispersion is discharged into the capacitor C2. The Figure 13 shows the converter efficiency.

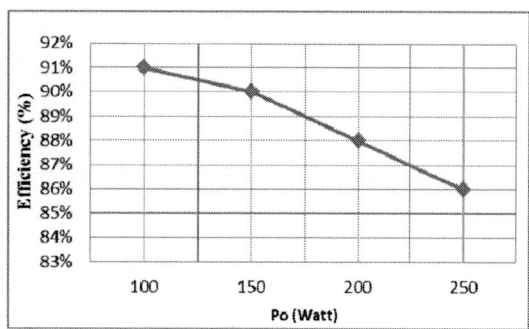

Fig. 13. Efficiency of proposed converter.

Figure 13 shows the converter efficiency considering the power of between 100 W and 250 W. For the nominal power, the efficiency obtained was 86%. With increased power the efficiency reduces slightly.

This is because there is an increase in the converter current and consequently an increase in conduction losses, especially, the diode D1 and Mosfet. The losses in the diode D1 can be reduced by using lower duty cycles values

A study of losses in qZSource converters was conducted and concluded that the 50% reduction in the voltage on the diode D1, you can increase the efficiency by at least 3%. This can be achieved by decreasing the duty cycle [17].The losses in Mosfet can also be reduced by reducing the duty cycle.

Considering these adjustments, the efficiency of the proposed converter can rise. It is important to consider that the voltage stresses which semiconductors are subject were low, in high power applications it may adapt well.

VI. CONCLUSIONS

This work has presented a new topology of high gain qZS dc-dc converter applied on a standalone photovoltaic tricycle. Qualitative, quantitative and losses analysis were presented. The experimental results validated the topology. The current and voltage efforts of MOSFET have been reduced. The reduction of the duty cycle allows the reduction of tension in the devices and therefore increasing the converter efficiency.

ACKNOLEDGEMENTS

The authors thank the Federal Institute of Ceará - Campus Sobral to have available the necessary infrastructure for the development of the experiment.

REFERENCE

[1] OMS 2008. "Guidelines on the Manual Wheelchairs supply in places with few resources". Available in:http://apps.who.int/iris/bitstream/10665/43960/38/9789241547482_por.pdf?ua=1.

[2] MEDOLA, F. O. "Development of a manual propulsion rim wheelchair access based on ergonomic concepts". 2010. Master. EESC/USP. São Carlos-SP

[3] PRESTES, R. C. "Assistive Technology: Producer of Design Attributes For Postural Custom Fit in position Sitting", 2011. Master. 2011. UFRS. Porto Alegre-RS.

[4] LOMBARDI JUNIOR, A. B. "Development and Collaborative Control Strategies Analysis for wheelchairs", 2005. Doctoral. UNICAMP. Campinas-SP

[5] Alvarenga, F. B. "A Methodological Approach for Inclusive Design Products". 2006. Doctoral. UNICAMP. Campinas-SP.

[6] KIM, K. D. et al. "Improved Non-isolated High Voltage Gain Boost Converter Using Coupled Inductors", Electrical Machines And Systems (ICEMS), 2011 International Conference On.

[7] ZHAO, Y. LI, W. HE, X. "Single-Phase Improved Active Clamp Coupled-Inductor-Based Converter With Extended Voltage Doubler Cell", IEEE Transactions on Power Electronics, vol. 27, no. 6, june 2012.

[8] TAKAHASHI, Y. e MATSUO, S. "Running Experiments of Electric Wheelchair Powered by Natural Energies", Industrial Electronics (ISIE), 2011 IEEE International Symposium on.

[9] CARUSO, M. et al. "A photovoltaic charging system of an electrically assisted tricycle for touristic purposes", AEI Annual Conference, 2013.

[10] TAKIGUCHI, T.; KOIZUMI, H. "Quasi-Z-source DC-DC Converter with Voltage-Lift Technique", Industrial Electronics Society, IECON 2013 - 39th Annual Conference Of The IEEE. Pp:1191 – 1196.

[11] LIIVIK, L.; VINNIKOV, D.; JALAKAS, T. "Synchronous Rectification in Quasi-Z-Source Converters: Possibilities and Challenges", Intelligent Energy and Power Systems (IEPS), 2014 IEEE International Conference on. Pp: 32-35.

[12] EVRAN, F.; AYDEMIR, M. T. "Z-source-based isolated high step-up converter", IEEE Transactions On Power Electronics, vol. 6, no.1, pp. 117-124,february 2012.

[13] KAZIMIERCZUK, MARIAN K. Pulse-width modulated DC-DC power converters. Chichester : Wiley, 2008. 782 p.. ISBN 978-0-470-77301-7

[14] THORNTON. Ferrites Catalog. São Paulo, 2008. 14p.

[15] BARBI, I.; FONT, C. H. I.; ALVES, R. L.,"Projeto Físico de Indutores e Transformadores". p. 0–10, 2002.

[16] BARBOSA, G. C. "Projeto de Transformador utilizado em uma Planta de Plasma Projeto de Transformador utilizado em uma Planta de Plasma". Universidade Federal do Rio Grande do Norte, 2012.

[17] LIIVIK, L.; VINNIKOV, D.; JALAKAS, "T. Synchronous Rectification in Quasi-Z-Source Converters: Possibilities and Challenges. Intelligent Energy and Power Systems" (IEPS), 2014 IEEE International Conference on. Pp: 32-35.

A Power Decoupling Method with Small Capacitance Requirement based on Single-Phase Quasi-Z-Source Inverter for DC Microgrid Applications

Dingyi He, *Student Member, IEEE*, Wen Cai, *Student Member, IEEE* and Fan Yi, *Student Member, IEEE*

Department of Electrical Engineering
The University of Texas at Dallas
Richardson, US
Email: {dingyi.he, wen.cai, fan.yi}@utdallas.edu

Abstract—This paper proposes a power decoupling method based on single-phase quasi-Z-source inverter (SPQZSI) for DC microgrid applications. Single-phase inverter would cause harmonic issues especially voltage ripple in DC microgrid because of low-frequency (100/120Hz) power ripple. Non-linear load or source would generate other harmonic on the common DC link as well. In order to achieve power decoupling, large electrolytic capacitor is required and connected to DC link in H-bridge inverters. However, electrolytic capacitor would affect the power density, reliability and lifetime of DC microgrid. Quasi-Z-source inverter is utilized instead for capacitance reduction. In terms of low-frequency ripple control with SPQZSI, an effective control method is necessary to regulate the current/voltage of both AC port and DC port. The theoretical analysis about capacitance requirement with power decoupling in SPQZSI is done first. Next, an advanced control method based on generalised predictive control is proposed to decouple low-frequency power ripple. The corresponding control law and predictor are designed as well. Experimental results with 500W prototype verify the stability, feasibility and superior performance of the proposed power decoupling method based on SPQZSI.

Keywords—DC microgrid, generalised predictive control, impedance-source, low-frequency ripple control, model predictive control, power decoupling, quasi-Z-source, single-phase inverter, Z-source

I. INTRODUCTION

DC microgrid has a promising future in the next generation of power grids. In DC microgrid, all kinds of sources and loads can be inserted, such as photovoltaic system, wind generation, battery, and AC drives [1-3]. For AC sources or AC loads, single-phase/three-phase inverters or rectifiers are necessary as shown in Fig. 1. In addition, inverters are also utilized in DC microgrid as active power filter (APF) [4] or state synchronous compensator (STATCOM) [5]. Despite their necessity, single-phase inverter will generate low-frequency power ripple on the DC link and causes harmonic issue (100/120Hz ripple). Moreover, other harmonics will appear at the DC port of three-phase inverter if the load is non-linear, such as uncontrolled rectifier or adjustable speed motor drives. Consequently, all devices in DC microgrid will bear power quality issue. Besides,

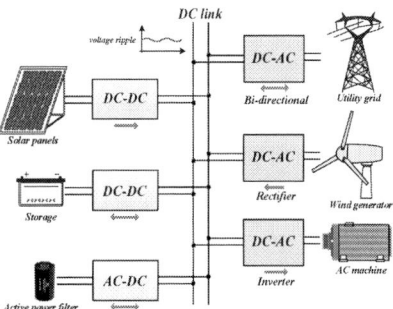

Fig. 1. Basic DC-bus-based microgrid with distributed converters.

DC link voltage control will be affected by such harmonics which might further cause system fault. Hence, it is desirable to keep the power ripple generated by inverters to a minimum, which is also referred to as power decoupling [6]. A simple way to restrain such power ripple is to apply a relatively large capacitance on the DC link of the inverter. Unfortunately, at high operating temperatures, the theoretical lifetime of electrolytic capacitors is much shorter than that of semiconductors and other components. Since electrolytic capacitor is an obstacle to the overall long-term reliability of DC microgrid [7, 8], it is worth to further study on how to cut off the low-frequency power ripple propagation with small capacitor in inverters.

In recent years, many researches have been launched in using either passive compensation approaches or active filters to cancel the undesirable power ripple instead of electrolytic capacitors. Considering the fact that the increase in size and cost brought by the passive methods limits its practical commercialization, active filters which are believed to be better solutions for power decoupling are used. They can be classified into two categories in general. One group is using power converters with advanced self-controlling method (SCM) to supply the pulsating power and to reduce the ripple [9-11]. Three different DC link configurations are compared in [12], and the module integrated converter with a line frequency DC-AC converter is proved to be the best choice. In these methods, no extra devices are needed and the capacitance requirement is

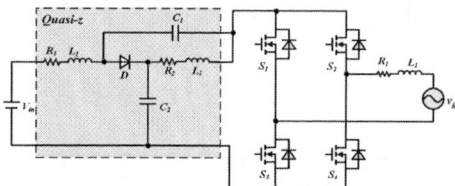

Fig. 2. Single-phase quasi-Z-source inverter.

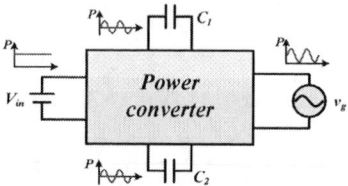

Fig. 3. Desired power flow in quasi-Z-source inverter.

reduced dramatically. Thus, the system efficiency can be guaranteed at an attractive low cost. Nevertheless, these methods are only applicable in two-stage power systems. [13] proposes a new four-switch three-port DC-DC-AC converter to decouple low-frequency power ripple. Even though only four semiconductor switches are necessary, their voltage stress is higher than that in H-bridge inverters. The other category is current ripple injecting method (CRIM) by using an extra device to afford the harmonic power and suppress the ripple [14-18]. Even though CRIMs need several additional devices, which increase power loss, cost and size, they are praised for their mobility and reliability in various applications since ripple reduction is realized by an auxiliary equipment.

In contrast to the two methods which have been presented and verified in the literature, this paper proposes to use the single-phase quasi-Z-source inverter (SPQZSI) to achieve power decoupling directly. On one hand, with this method, no extra circuits are necessary and the voltage/current stress stays the same, which guarantees its efficiency and power density. On the other hand, this method is not limited to two-stage converters because power ripple is compensated by the inverter itself. Thus, it can be easily modularized and used in any single-phase applications, which is desirable in DC microgrid. It is worth pointing out that this method can also be used in three-phase applications with non-linear loads. The Z-source inverter was introduced first in 2002 [19]. Following that, a series of semi-Z-source/quasi-Z-source/trans-Z-source inverters have been introduced and studied [20-23]. In 2013, the Γ-Z-source inverter series are proposed with higher operation gain [24]. Dead time is not necessary in Z-source inverters, which conduces to circuit protection and harmonic reduction [25]. Compared with conventional H-bridge inverter, the capacitors in quasi-Z-source unit are not connected in parallel with the DC source [26]. It affords the possibility to use these capacitors to compensate the low-frequency power ripple independently. In this paper, the voltage fed quasi-Z-source inverter (Fig. 2) with continuous input current is selected. The capacitance requirement for low-frequency power decoupling is analyzed and the required capacitance with SPQZSI is uncovered for hardware design. In order to access power decoupling, more complicated control method is needful since both the current through the DC port and the AC port should be controlled at the same time. In this multi-objective application, model predictive control method has several merits such as expendability and constraint handling. With the model of SPQZSI, an advanced control method based on generalised predictor is proposed to replace traditional multiple linear controllers. Afterwards, the control law is designed and calculated in detail. The stability of SPQZSI with discrete control system is analyzed using MATLAB. At last, experimental results indicate the steady-state and dynamic performances of SPQZSI with proposed control method, as well as power decoupling function with small capacitors.

II. CAPACITOR REQUIREMENT ANALYSIS WITH POWER DECOUPLING IN SPQZSI

As analyzed in the Introduction, the power from DC source is desired to be constant and the power provided to the AC port is fluctuating. The instantaneous power difference between the two ports has to be compensated by another one or two storage components such as capacitors. From Fig. 2, it can be seen that extra two capacitors (C_1 and C_2) have to be used to compensate such low-frequency power ripple. Fig. 3 indicates the desired power flow in SPQZSI with power decoupling function. Both capacitors C_1 and C_2 provide 100/120Hz power ripple which the AC port needs.

A. Low-frequency ripple analysis

[20] has addressed the steady-state equations for quasi-Z-source inverter. However, this is not enough for ripple analysis. The low-frequency power ripple would generate low-frequency voltage ripple on the capacitors and low-frequency current ripple on the inductors. If the duty cycle which the shoot through state occupies is denated by D_1, the voltages of capacitors C_1 and C_2 would be:

$$\begin{cases} u_1 = \dfrac{D_1}{1-2D_1} U_{in} \\ u_2 = \dfrac{1-D_1}{1-2D_1} U_{in} \end{cases} \qquad (1)$$

It can be seen from the above equations that the voltage of u_2 equals the sum of the voltage of u_1 and the input voltage. If the input voltage is constant, the voltage ripple of these two capacitors should be the same. Considering that the low-frequency power ripple is compensated by the two capacitors, the energy stored in the two capacitors in half of one cycle at 100/120Hz is calculated as:

$$\begin{aligned} E_c &= \frac{1}{2}C_1 u_1^2 + \frac{1}{2}C_2 u_2^2 = \frac{1}{2}C_1\left(\bar{u}_1 + \Delta u\right)^2 + \frac{1}{2}C_2\left(\bar{u}_2 + \Delta u\right)^2 \\ &= \left(\frac{1}{2}C_1\bar{u}_1^2 + \frac{1}{2}C_2\bar{u}_2^2\right) + \left(C_1\bar{u}_1 + C_2\bar{u}_2\right)\Delta u + \frac{1}{2}\left(C_1 + C_2\right)\Delta u^2 \end{aligned} \qquad (2)$$

In which Δu is the voltage ripple of the two capacitors. $\bar{u}_1, \bar{u}_2$ are the average voltage of C_1 and C_2, respectively. The energy ripple can be obtained as (3).

$$E_{r1} = \left(C_1\bar{u}_1 + C_2\bar{u}_2\right)\Delta u + \frac{1}{2}\left(C_1 + C_2\right)\Delta u^2 \qquad (3)$$

Assuming that the average voltage ($\bar{u}_1, \bar{u}_2$) is much higher than the voltage ripple (Δu), the energy ripple can be estimated by $E_{r1} \approx \left(C_1\bar{u}_1 + C_2\bar{u}_2\right)\Delta u$. Correspondingly, the energy ripple needed by the inverter can also be obtained as follows:

978-1-4673-9551-9/16 $31.00 © 2016 IEEE

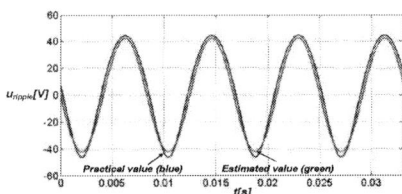

Fig. 4. Comparison of estimated value and practical value of capacitor voltage ripple.

$$E_{r2} = \int \left(u_{out} i_{out} - P_{ave} \right) dt$$

$$= \int \left[U_{ga} \sin(\omega t) I_{ga} \sin(\omega t + \theta) - \frac{U_{ga} I_{ga}}{2} \cos(\theta) \right] dt \quad (4)$$

$$= \int \left[-\frac{U_{ga} I_{ga}}{2} \cos(2\omega t + \theta) \right] dt = \frac{U_{ga} I_{ga}}{4\omega} \sin(2\omega t + \theta)$$

where, U_{ga}, I_{ga} are the amplitude of AC port voltage and current, and ω represents the angular frequency of AC port voltage. Comparing (3) and (4), the voltage ripple of the two capacitors can be calculated as:

$$E_{r1} = \left(C_1 \bar{u}_1 + C_2 \bar{u}_2 \right) \Delta u \approx E_{r2} = \frac{U_{ga} I_{ga}}{4\omega} \sin(2\omega t + \theta)$$
$$\Rightarrow \Delta u \approx \frac{U_{ga} I_{ga}}{4\omega \left(C_1 \bar{u}_1 + C_2 \bar{u}_2 \right)} \sin(2\omega t + \theta) \quad (5)$$

Based on the above equation, the magnitude of the capacitor voltage ripple is estimated as $u_{amp} = \dfrac{U_{ga} I_{ga}}{4\omega \left(C_1 \bar{u}_1 + C_2 \bar{u}_2 \right)}$.

The estimated ripple and the measured value have been shown in Fig. 4. The estimated value can track the measured voltage ripple on the capacitor.

B. Required capacitance calculation

From theoretical analysis above, it can be found that the capacitance requirement can be decreased by boosting the average voltage and the voltage ripple. However, the average voltage and the voltage ripple cannot increase infinitely. On one hand, the maximum voltage is limited by the capacitor material and fabrication. On the other hand, the minimum voltage should be high enough to output sinusoidal waveform. Otherwise, it couldn't approach voltage boost and further lead to harmonic at the AC port. The minimum voltage and maximum voltage of the two capacitors can be calculated as follows:

$$\begin{cases} M_{max1} = \bar{u}_1 + \dfrac{U_{ga} I_{ga}}{4\omega \left(C_1 \bar{u}_1 + C_2 \bar{u}_2 \right)} \\[2mm] M_{min1} = \bar{u}_1 - \dfrac{U_{ga} I_{ga}}{4\omega \left(C_1 \bar{u}_1 + C_2 \bar{u}_2 \right)} \\[2mm] M_{max2} = \bar{u}_2 + \dfrac{U_{ga} I_{ga}}{4\omega \left(C_1 \bar{u}_1 + C_2 \bar{u}_2 \right)} \\[2mm] M_{min1} = \bar{u}_2 - \dfrac{U_{ga} I_{ga}}{4\omega \left(C_1 \bar{u}_1 + C_2 \bar{u}_2 \right)} \end{cases} \quad (6)$$

According to the principle of operation in quasi-Z-source inverter, the sum of voltages across C_1 and C_2 needs to be larger

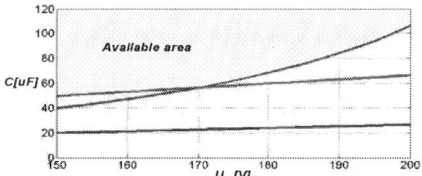

Fig. 5. Capacitance requirement in single-phase quasi-Z-source inverter.

than the magnitude of the AC output voltage (if the voltage drop of the inductor is ignored). Hence, the conditions listed in (7) have to be satisfied.

$$\begin{cases} C_1 \bar{u}_1 + C_2 \bar{u}_2 > \dfrac{U_{ga} I_{ga}}{2\omega \left(\bar{u}_1 + \bar{u}_2 - \dfrac{U_{ga}}{1 - D_1} \right)} \\[4mm] C_1 \bar{u}_1 + C_2 \bar{u}_2 > \dfrac{U_{ga} I_{ga}}{4\omega \left(U_{\lim 1} - \bar{u}_1 \right)} \\[4mm] C_1 \bar{u}_1 + C_2 \bar{u}_2 > \dfrac{U_{ga} I_{ga}}{4\omega \left(U_{\lim 2} - \bar{u}_2 \right)} \end{cases} \quad (7)$$

where, $U_{\lim 1}, U_{\lim 2}$ are the voltage limits dictated by the capacitor material and fabrication. (7) reveals the minimum values of the two capacitors which can help to guide hardware design. Assuming that the two capacitors are selected with the same capacitance and the parameters are $D_1 = 0.2$, $U_{in} = 150V$, $I_{ga} = 5A$, the required capacitance of C_1 and C_2 can be calculated according to (7). The relationship between the capacitance (C_1 and C_2) and the AC port voltage (U_{ga}) has been shown in Fig. 5. (7) also hints that by enlarging the capacitance of C_1 whose voltage rating is much lower than DC source voltage and the voltage rating of the capacitor C_2, the voltage ripple of C_2 could be decreased as well. The size of the low-voltage capacitor is much less than that of the high-voltage capacitor. Therefore, combination of low-voltage (large-capacitance) capacitor and high-voltage (low-capacitance) capacitor can improve performance and reliability of SPQZSI.

III. GENERALISED PREDICTIVE CONTROL

In conventional control method of SPQZSI, the voltage of the two capacitors only need be high enough to keep output voltage controlled. This control scheme is simple because only output voltage or current is regulated. However, in order to achieve power decoupling using SPQZSI, the input current has to be controlled at a constant value as well. This means that two control objectives (input current & output current/voltage) exist in this system. In addition, the voltage of the capacitors is desired to be controlled within a range in accordance to the requirement from output voltage. Under this condition, model predictive control is suitable because of its good expandability and constraint handling capability. It's known that there already exist several methods to approach model predictive control (MPC), generalised predictive algorithm (GPA) is chosen in this paper since it is easy to implement and its computational complexity is not overly high.

A. Principle

MPC is mainly consisted of two parts, namely predictor and optimizer. Predictor estimates the values of state variables based

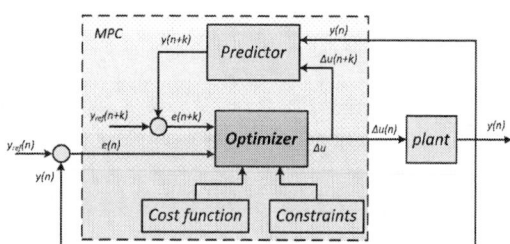

Fig. 6. Block diagram of model predictive control method.

on system model. Optimizer adjusts the input variables according to the difference of the estimated value and the detected value of the output variables. Fig. 6 shows the basic structure of the MPC. Cost function and constraints are supposed to derive the control law in optimizer.

It is worth pointing out that cost function and constraints for algorithm design of optimizer are among the important issues to secure good performance. Therefore, GPA is employed here to define cost function and design optimizer. While, constraints are not included in the basic GPA. In the next sub-section, another steady-state control loop is added in series with MPC which could handle constraints for better reliability.

B. MPC method based on GPA

Because of the complexity of MPC, the algorithm is mostly implemented with digital microprocessor. If the continuous state space equation is $\begin{cases} dx/dt = Ax + Bu \\ y = Cx \end{cases}$, the discretized state space equations can be calculated and written as the following form if the control period is T:

$$\begin{cases} x(k+1) = A_1 x(k) + B_1 u(k) \\ y(k) = C_1 x(k) \end{cases} \tag{8}$$

where, $A_1 = L^{-1}(sI-A)^{-1}, B_1 = \int_0^T e^{AT} B dt, C_1 = C$. Then, the

predictor can be obtained based on the discrete-time state space equations. Assuming that n_y steps are to be predicted, the predictive model can be written as:

$$\begin{bmatrix} x_{k+1} \\ x_{k+2} \\ x_{k+3} \\ \dots \\ x_{k+n_y} \end{bmatrix} = \begin{bmatrix} A_1 \\ A_1^2 \\ A_1^3 \\ \dots \\ A_1^{n_y} \end{bmatrix} x_k + \begin{bmatrix} B_1 & 0 & 0 & \dots \\ A_1 B_1 & B_1 & 0 & \dots \\ A_1^2 B_1 & A_1 B_1 & B_1 & \dots \\ \dots \\ A_1^{n_y-1}B_1 & A_1^{n_y-2}B_1 & A_1^{n_y-3}B_1 & \dots \end{bmatrix} \begin{bmatrix} u_k \\ u_{k+1} \\ u_{k+2} \\ \dots \\ u_{k+n_y-1} \end{bmatrix}$$

$$\begin{bmatrix} y_{k+1} \\ y_{k+2} \\ y_{k+3} \\ \dots \\ y_{k+n_y} \end{bmatrix} = \begin{bmatrix} C_1 A_1 \\ C_1 A_1^2 \\ C_1 A_1^3 \\ \dots \\ C_1 A_1^{n_y} \end{bmatrix} x_k + \begin{bmatrix} C_1 B_1 & 0 & 0 & \dots \\ C_1 A_1 B_1 & C_1 B_1 & 0 & \dots \\ C_1 A_1^2 B_1 & C_1 A_1 B_1 & C_1 B_1 & \dots \\ \dots \\ C_1 A_1^{n_y-1}B_1 & C_1 A_1^{n_y-2}B_1 & C_1 A_1^{n_y-3}B_1 & \dots \end{bmatrix} \begin{bmatrix} u_k \\ u_{k+1} \\ u_{k+2} \\ \dots \\ u_{k+n_y-1} \end{bmatrix} \tag{9}$$

The above model (9) can be simplified to become:

$$\underset{\rightarrow k}{x} = P_{xx} x_k + H_x \underset{\rightarrow k-1}{u}$$
$$\underset{\rightarrow k}{y} = P x_k + H \underset{\rightarrow k-1}{u} \tag{10}$$

In which, $\underset{\rightarrow k}{u}$, $\underset{\rightarrow k}{x}$ and $\underset{\rightarrow k}{y}$ are the *k*-step predicted values. The control law in optimizer can be derived according to GPA. In order to achieve zero tracking errors, one can use the cost function and optimization as:

$$J = \frac{1}{2}\left[\underset{\rightarrow}{x} - x_{ss}\right]^2 + \frac{1}{2}\lambda\left[\underset{\rightarrow}{u} - u_{ss}\right]^2 \Rightarrow$$
$$\underset{\underset{\rightarrow}{u}}{\min} J = \left[\underset{\rightarrow}{x} - x_{ss}\right]^T Q\left[\underset{\rightarrow}{x} - x_{ss}\right] + \lambda\left[\underset{\rightarrow}{u} - u_{ss}\right]^T R\left[\underset{\rightarrow}{u} - u_{ss}\right] \tag{11}$$

where, $Q = C_1^T C, R = D_1^T D_1$ and λ is a control weighting factor for input variable difference. D_1 is normally considered to be 0. Then, with derivation of (11), the minimization of J is known to give a state feedback control law in the form $u = -Kx$. The solution of this form can be rewritten as:

$$u = -K(x - x_{ss}) + u_{ss} \tag{12}$$

In which, x_{ss}, u_{ss} are the steady-state values of the state variables and the input variables. Consequently the control law which represents the optimizer in Fig. 6 can be obtained as:

$$u_k - u_{ss} = -e_1^T \left[H_x^T Q H_x + R \right]^{-1} H_x^T Q P_{xx} (x - x_{ss}) \tag{13}$$

Assuming the reference of the output variables is known as r, the steady-state values can be computed by solving the following equations.

$$\begin{cases} r = Cu_{ss} \\ x_{ss} = Ax_{ss} + Bu_{ss} \end{cases} \tag{14}$$

Even though GPA can help to derive control law easily and achieve good performance theoretically, there exist two limitations. First, the steady-state accuracy would depend on both the references of state variables and input variables which can be seen from (14). Unfortunately, it is difficult to obtain the required accurate commands for input variables, like duty cycles in power converters. This is because of the inherent uncertainty in ESRs of inductors, the collector-emitter saturation voltage of the semiconductors and the effect of dead time. Secondly, there is usually a preferred range for input variables like duty cycles. For instance, the duty cycle of buck converter should be in [0, 1]. Whereas, for boost converters, [0.3, 0.7] is desirable. However, when using GPA, these constraints are not included in optimizer [27].

In accordance to the particularity of SPQZSI, another linear control loop is added in front of the MPC based on GPA, as shown in Fig. 7. Because the voltage ripple on the capacitors is inevitable, its average value is detected and controlled to be constant. The amplitude of output voltage is idle and regulated as constant in an idle inverter. If the inverter works in grid-tied mode, the amplitude of output current can be inserted instead of the output voltage. Also, a phase-lock loop is necessary to observe the grid voltage phase. In MPC part, the input current (the inductor current, i_{L1}) and the output voltage (u_g) are the controlled variables. Meanwhile, the shoot through duty cycle and the inverter duty cycle are selected as the input variables.

C. Average model of quasi-Z-source inverter

In order to design the control parameters correctly, mode analysis of SPQZSI has been done first. There are three modes

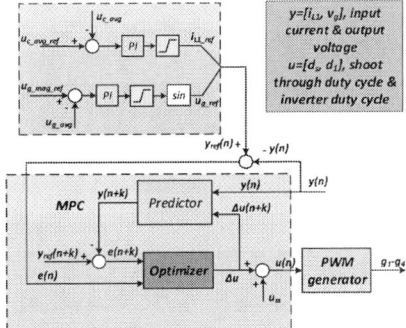

Fig. 7. Overall control scheme for the quasi-Z-source single-phase inverter.

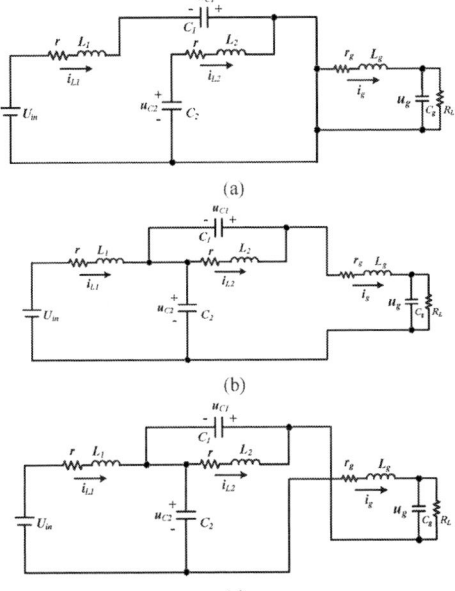

Fig. 8. Equivalent circuits: (a) Mode 1: shoot-through state (S1~S4 are on); (b) Mode 2: (S2, S3 are on, S1, S4 are off); (c) Mode 3: (S1, S4 are on, S2, S3 are off).

which have been listed below. Here, the two inductors are considered with the same inductance (L in Fig. 2) and ESR (r in Fig. 2) to simplify the theoretical analysis. The ESR of capacitors is ignored and the capacitance is assumed as C. The ESR and inductance of the inductor at the AC port are r_g and L_g. The input DC source is U_{in} and the AC voltage is assumed as u_g.

Mode 1 in Fig. 8 declares the shoot-through state when all of the four switches are turned on. Mode 2 and mode 3 represent the non-shoot-through states. In Mode 2, the switches S_1 and S_4 are turned off, and S_2 and S_3 are turned on. In mode 3, S_2 and S_3 are turned off, and S_1 and S_4 are turned on. Assuming that the duty cycle of Mode 1 is d_s and the duty cycle of Mode 2 is d_1, the duty cycle in Mode 3 would be (1-d_1-d_s) if the dead time is zero. Applying state-space average model to the three modes, and approaching small-signal analysis to state space equation, the state-space equations can be written as:

$$\begin{cases} \dfrac{dx}{dt} = Ax + Bu \\ y = Cx \end{cases} \quad (15)$$

In which,

$$x = \begin{bmatrix} i_{L1} & i_{L2} & v_{c1} & v_{c2} & i_g & u_g \end{bmatrix}^T, u = \begin{bmatrix} d_s \\ d_1 \end{bmatrix}, y = \begin{bmatrix} i_{L1} \\ u_g \end{bmatrix}$$

$$A = \begin{bmatrix} -\dfrac{r}{L} & 0 & \dfrac{1-D_S}{L} & \dfrac{D_S}{L} & 0 & 0 \\ 0 & -\dfrac{r}{L} & \dfrac{D_S}{L} & \dfrac{1-D_S}{L} & 0 & 0 \\ \dfrac{1-D_S}{C} & \dfrac{D_S}{C} & 0 & 0 & \dfrac{2D_1+D_S-1}{C} & 0 \\ \dfrac{D_S}{C} & \dfrac{1-D_S}{C} & 0 & 0 & \dfrac{2D_1+D_S-1}{C} & 0 \\ 0 & 0 & \dfrac{2D_1+D_S-1}{L_g} & \dfrac{2D_1+D_S-1}{L_g} & \dfrac{r_g}{L_g} & \dfrac{1}{L_g} \\ 0 & 0 & 0 & 0 & \dfrac{1}{C_g} & -\dfrac{1}{C_g R_L} \end{bmatrix}$$

$$B = \begin{bmatrix} \dfrac{U_{c1}-U_{c2}}{L} & 0 \\ \dfrac{U_{c1}-U_{c2}}{L} & 0 \\ \dfrac{I_{L1}+I_{L2}+I_g}{C} & \dfrac{2I_g}{C} \\ \dfrac{I_{L1}+I_{L2}+I_g}{C} & \dfrac{2I_g}{C} \\ \dfrac{2(V_{c1}+V_{c2})}{L_g} & \dfrac{U_{c1}+U_{c2}}{L_g} \\ 0 & 0 \end{bmatrix}, C = \begin{bmatrix} 1 & 0 & 0 & 0 & 0 & 0 \\ 0 & 0 & 0 & 0 & 0 & 1 \end{bmatrix}$$

Based on the above state-space equations, the control law and the optimizer for SPQZSI can be calculated.

D. Stability and performance analysis

As declared in Section III.B, the optimizer acts as the controller for the plant. Since the optimizer is a linear state feedback, the closed-loop transfer function between the output and the reference can be derived. If one transfers the discrete transfer function in (8) to matrix fraction description (MFD) which is defined as $D(z)y = N(z)u$, it can be rewritten as $y = D^{-1}Nu$ ((z) is omitted to simplify the expression). Meanwhile, the controller is declared in (12). All the necessary equations are listed as follows:

$$\begin{cases} r - y = C(x_{ss} - x) \\ y = D^{-1}Nu \\ u = -K(x - x_{ss}) + u_{ss} \end{cases} \quad (16)$$

By solving the equations in (19) and removing the state variables x, the relationship between outputs and references is represented as:

$$\left[N + DK\left(C^T C\right)^{-1} C^T \right] y = Du_{ss} + DK\left(C^T C\right)^{-1} C^T r \quad (17)$$

Here, u_{ss} is the reference of input variables which is considered as a constant value. Therefore, (20) can be rewritten as (21) which reveals the discrete transfer function between outputs and references. Based on (21), the dynamic performance can be addressed with Bode diagrams or Nyquist curve.

$$y(z) = \frac{DK\left(C^T C\right)^{-1} C^T}{N + DK\left(C^T C\right)^{-1} C^T} r(z) \quad (18)$$

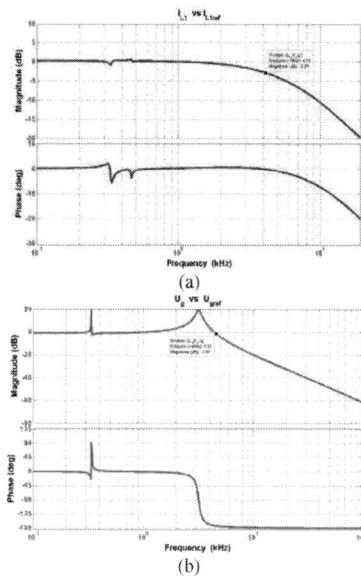

(a)

(b)

Fig. 9. Bode diagram of the closed-loop control system with GPA: (a) i_{L1} vs i_{L_ref}; (b) u_g vs u_{g_ref}.

With such control law, the closed-loop system can be computed later to estimate the system stability. Fig. 9 shows the corresponding Bode diagrams with designed control law (i_{L1} vs i_{L_ref} and u_g vs u_{g_ref}). From these bode diagrams, it can be seen that the stability can be guaranteed with the designed control law and the predictor. The closed-loop bandwidth for input current control is 4.15-kHz and open-loop phase margin is 71.1°. For output voltage control, the bandwidth is 4.61-kHz and open-loop phase margin is 90.8°.

IV. EXPERIMENTAL RESULTS

In order to verify the power decoupling function with single SPQZSI, a digitally controlled 500-W experimental prototype was built. Both steady-state and dynamic performance are tested based on this test-bed. In addition, non-linear load is also employed with SPQZSI to certify the feasibility of the proposed method. Performance comparison between conventional control method and proposed control method with input current regulation is developed in this section. Here, the two capacitors in quasi-Z-source unit are selected as $100\mu F$. The voltage rating of C_1 is $150V$ and the voltage rating of C_2 is $450V$. According to the theoretical analysis in Section II, the voltage ripple of the two capacitors would be $53.1V$ which is much higher than 5% of input voltage. Despite all that, the input current should be still controlled as constant. If conventional H-bridge inverter is used and 5% voltage ripple for input source is required, the capacitance has to be $2.3mF$ which is ten-times larger than the capacitors in SPQZSI.

The picture of the experimental test-bed is shown in Fig. 10. The overall system consists of three parts, including quasi-Z-source unit, single-phase H-bridge inverter and auxiliary power supply. The detailed hardware parameters have been listed in Table I. The switching frequency is set at 50-kHz. In consideration of the required output voltage, the input voltage is selected at $150V$. Here, CoolMos IPP60R099P6 from Infineon Technology is used with consideration of voltage and current

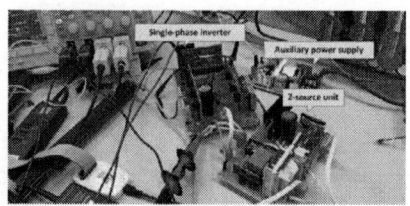

Fig. 10. 500W prototype based on SPQZSI.

rating. The proposed multi-objective model predictive control method as well as conventional control method is implemented via a 32-bit ARM Cortex MCU STM32F405 from STMicroelectronics. In the conventional control method, the shoot-through duty cycle is fixed.

A. Steady-state performance

The steady-state waveforms have been illustrated in Fig. 11. The PWM signals for all the four switches are in Fig. 11(a). It can be seen that the shoot through mode is achieved by turning on all the switches at the same time. In addition, there is no dead time for the two legs. Experimental results with fixed shoot through period are shown in Fig. 11(b). The voltage ripple of two capacitors C_1 and C_2 is distorted (Ch1 & Ch2). At the same time, the input current is displayed in Ch4. Large current ripple is contained in the input current (5.9A peak-peak value). In comparison with that, the experimental results with advanced control method are shown in Fig. 11(c). The voltage of the two capacitors is sinusoidal with low harmonic which matches the theoretical estimation in Section II. They are mainly used to compensate the low-frequency power ripple. The input current is shown in Ch4 as well. The current ripple is controlled within a small range (0.9A peak-peak value). The frequency distribution of the input current with different control methods is shown in Fig. 11(d). The DC components are almost the same. However, 90% of 120Hz current ripple has been restrained with the proposed control method. Meanwhile, the other parts (240Hz, 360Hz and 480Hz) are also removed.

TABLE I. HARDWARE PARAMETERS OF SPQZSI

Parameters	Value
V_{in}	150V DC
V_{out}	120V AC
I_{out}	5A AC
Diode (D)	RHRG3060
Inductor (L_1)	220μH, 0.1Ω
Inductor (L_2)	220μH, 0.1Ω
Capacitors (C_1)	100μF, 150V
Capacitors (C_2)	100μF, 450V
Switches	IPP60R099P6
Switching frequency	50-kHz
Output inductor (L_g)	470μH, 0.05Ω
Output capacitors (C_g)	20μF, 250V AC

978-1-4673-9551-9/16 $31.00 © 2016 IEEE

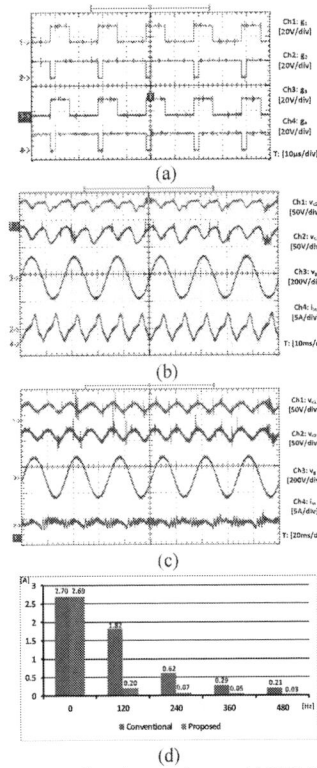

Fig. 11. Steady-state waveform in experiments: (a) PWM signals; (b) case 1: fixed duty cycle; (c) case 2: proposed method; (d) frequency distribution of input current.

B. Dynamic performance

The stability of the designed model predictive control scheme with power decoupling function is verified with dynamic response waveforms on the SPQZSI prototype. The experimental results are shown in Fig. 12. In the beginning, the load resistor is 80Ω. The RMS current is 1.5A. After that, another 80Ω resistor is connected in parallel on the output port, the output current steps from 1.5A to 3.0A. As shown in Fig. 12(a), the output voltage is still controlled as a sinusoidal wave. The capacitor voltage decreases because of the power unbalance. The waveform of a larger period is shown in Fig. 12(b). The average voltage of capacitor C_2 is regulated back to 200V. While, the voltage ripple is enlarged because of the fact that the required power ripple is increased. In terms of the input port, the input current is constant with small ripple. After load increase and 340ms transient process, the input current is controlled at a new constant value with small ripple again.

C. Non-linear AC load

The application under a non-linear load is also tested to verify the feasibility of the proposed solution and its performance. Here, 80Ω is still connected into the output port directly. At the same time, a diode rectifier with capacitor and resistor is used to act as non-linear load. The capacitor is $100\mu F$ and the load resistor is 80Ω as well. Fig. 13(a) shows the output current and the output voltage from the SPQZSI. They are not in phase and there is a large harmonic in current. Fig. 13(b) shows the waveforms with fixed shoot through period (d_l=0.2).

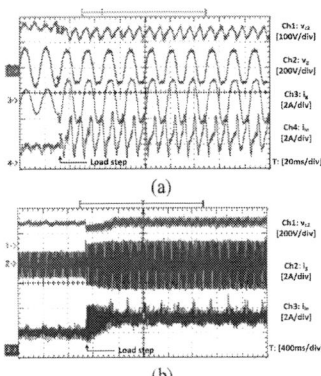

Fig. 12. Dynamic response in experiments: (a) small periodic waveform; (b) large periodic waveform.

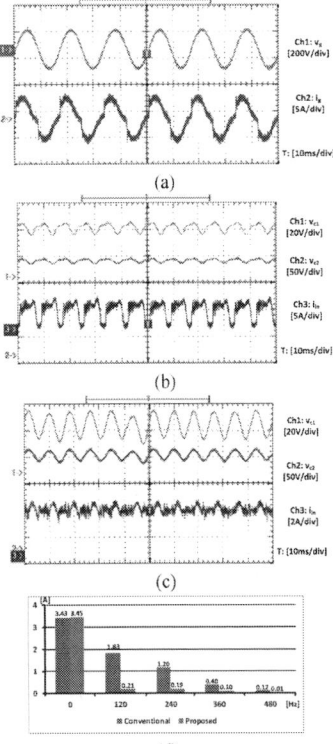

Fig. 13. Experimental results with non-linear load: (a) Output voltage & current; (b) case 1: fixed duty cycle; (c) case 2: proposed method; (d) frequency distribution of input current

The peak-to-peak value of the input current is 5.3A. Compared with that, the waveform with proposed control method is indicated in Fig. 13(c). Shoot through period is still regulated for input current control. The peak-to-peak value is 0.8A. The frequency distributions of two cases are listed in Fig. 13(d). The main component, 120Hz current ripple, has been compensated as well from 1.83A to 0.21A. In the meantime, because of the non-linear load, the 240Hz ripple current increases from 0.62A to 1.20A with fixed shoot through period. However, if the input current is controlled based on the proposed generalised predictive method and it can be suppressed to 0.19A.

The above experimental results demonstrate that with advanced control method based on single SPQZSI, more than 85% of input current ripple has been reduced which is good for DC sources and DC microgrid. It is worth pointing out that only two $100\mu F$ capacitors are used in these experiments, which means that 90% capacitance reduction is achieved.

V. CONCLUSIONS

A power decoupling method based on SPQZSI and generalised predictive algorithm is proposed for DC microgrid in this paper. Low-frequency power ripple is unavoidable in single-phase inverter which would affect the power quality of DC link and further cause severe harmonic issue. In order to achieve power decoupling with small capacitance, it is considered to compensate the low-frequency power ripple by using the capacitors in quasi-Z-source unit of SPQZSI. Theoretical analysis reveals that capacitance can be reduced by 90% compared with conventional H-bridge converter. Considering that multi-objective control is required, an advanced control method with combination of generalised predictive algorithm and linear controller is proposed and designed based on the circuit model of SPQZSI. Experimental results with linear and non-linear load demonstrate that more than 85% current ripple can be compensated on the DC link with 90% capacitance reduction. The feasibility of the proposed power decoupling method is also verified with load step.

ACKNOWLEDGMENT

This work is done in the Renewable Energy and Vehicular Technology (REVT) Laboratory, the University of Texas at Dallas. The authors would like to thank the founding director Dr. Babak Fahimi for his support.

REFERENCES

[1] R. H. Lasseter, "MicroGrids," in *IEEE Power Engineering Society Winter Meeting*, 2002, pp. 305-308 vol.1.

[2] B. S. Hartono, Y. Budiyanto, and R. Setiabudy, "Review of microgrid technology," in *QiR (Quality in Research), 2013 International Conference on*, 2013, pp. 127-132.

[3] X. Zhao, L. Zhang, X. Cui, C. Zheng, C.-Y. Lin, Y.-C. Liu, *et al.*, "A high-efficiency hybrid series resonant DC-DC converter with boost converter as secondary for photovoltaic applications," in *Energy Conversion Congress and Exposition (ECCE), 2015 IEEE*, 2015, pp. 5462-5467.

[4] P. Acuna, L. Moran, M. Rivera, J. Dixon, and J. Rodriguez, "Improved Active Power Filter Performance for Renewable Power Generation Systems," *Power Electronics, IEEE Transactions on*, vol. 29, pp. 687-694, 2014.

[5] L. Zhao, L. Bangyin, D. Shanxu, and K. Yong, "A Novel DC Capacitor Voltage Balance Control Method for Cascade Multilevel STATCOM," *Power Electronics, IEEE Transactions on*, vol. 27, pp. 14-27, 2012.

[6] T. Hirao, T. Shimizu, M. Ishikawa, and K. Yasui, "A modified modulation control of a single-phase inverter with enhanced power decoupling for a photovoltaic AC module," in *Power Electronics and Applications, 2005 European Conference on*, 2005, pp. 10 pp.-P.10.

[7] P. T. Krein, R. S. Balog, and M. Mirjafari, "Minimum Energy and Capacitance Requirements for Single-Phase Inverters and Rectifiers Using a Ripple Port," *IEEE Transactions on Power Electronics*, vol. 27, pp. 4690-4698, 2012.

[8] B. Gu, J. Dominic, J. Zhang, L. Zhang, B. Chen, and J.-S. Lai, "Control of electrolyte-free microinverter with improved MPPT performance and grid current quality," in *Applied Power Electronics Conference and Exposition (APEC), 2014 Twenty-Ninth Annual IEEE*, 2014, pp. 1788-1792.

[9] J. I. Itoh and F. Hayashi, "Ripple Current Reduction of a Fuel Cell for a Single-Phase Isolated Converter Using a DC Active Filter With a Center Tap," *IEEE Transactions on Power Electronics*, vol. 25, pp. 550-556, 2010.

[10] W. Ruxi, F. Wang, D. Boroyevich, R. Burgos, L. Rixin, N. Puqi, *et al.*, "A High Power Density Single-Phase PWM Rectifier With Active Ripple Energy Storage," *Power Electronics, IEEE Transactions on*, vol. 26, pp. 1430-1443, 2011.

[11] F. Yi and W. Cai, "Repetitive control-based current ripple reduction method with a multi-port power converter for SRM drive," in *Transportation Electrification Conference and Expo (ITEC), 2015 IEEE*, 2015, pp. 1-6.

[12] L. Quan and P. Wolfs, "A Review of the Single Phase Photovoltaic Module Integrated Converter Topologies With Three Different DC Link Configurations," *Power Electronics, IEEE Transactions on*, vol. 23, pp. 1320-1333, 2008.

[13] W. Cai, L. Jiang, B. Liu, S. Duan, and C. zou, "A Power Decoupling Method based on Four-Switch Three-Port DC/DC/AC Converter in DC Microgrid," *IEEE Transactions on Industry Applications*, vol. PP, pp. 1-1, 2014.

[14] W. Cai, B. Liu, S. Duan, and L. Jiang, "An Active Low-Frequency Ripple Control Method Based on the Virtual Capacitor Concept for BIPV Systems," *Power Electronics, IEEE Transactions on*, vol. 29, pp. 1733-1745, 2014.

[15] S. K. Mazumder, R. K. Burra, and K. Acharya, "A Ripple-Mitigating and Energy-Efficient Fuel Cell Power-Conditioning System," *Power Electronics, IEEE Transactions on*, vol. 22, pp. 1437-1452, 2007.

[16] L. Hongbo, Z. Kai, Z. Hui, F. Shengfang, and X. Jian, "Active Power Decoupling for High-Power Single-Phase PWM Rectifiers," *Power Electronics, IEEE Transactions on*, vol. 28, pp. 1308-1319, 2013.

[17] S. Mei, P. Pan, L. Xi, S. Yao, and Y. Jian, "An Active Power-Decoupling Method for Single-Phase AC/DC Converters," *IEEE Transactions on Industrial Informatics*, vol. 10, pp. 461-468, 2014.

[18] W. Rong-Jong and L. Chun-Yu, "Active Low-Frequency Ripple Control for Clean-Energy Power-Conditioning Mechanism," *Industrial Electronics, IEEE Transactions on*, vol. 57, pp. 3780-3792, 2010.

[19] F. Z. Peng, "Z-source inverter," in *Industry Applications Conference, 2002. 37th IAS Annual Meeting. Conference Record of the*, 2002, pp. 775-781 vol.2.

[20] J. Anderson and F. Peng, "Four quasi-Z-Source inverters," in *Power Electronics Specialists Conference, 2008. PESC 2008. IEEE*, 2008, pp. 2743-2749.

[21] L. Yuan, J. Shuai, J. G. Cintron-Rivera, and P. Fang Zheng, "Modeling and Control of Quasi-Z-Source Inverter for Distributed Generation Applications," *Industrial Electronics, IEEE Transactions on*, vol. 60, pp. 1532-1541, 2013.

[22] C. Dong, J. Shuai, Y. Xianhao, and P. Fang Zheng, "Low-Cost Semi-Z-source Inverter for Single-Phase Photovoltaic Systems," *Power Electronics, IEEE Transactions on*, vol. 26, pp. 3514-3523, 2011.

[23] W. Qian, F. Z. Peng, and H. Cha, "Trans-Z-source inverters," in *Power Electronics Conference (IPEC), 2010 International*, 2010, pp. 1874-1881.

[24] L. Poh Chiang, L. Ding, and F. Blaabjerg, "Γ-Z-Source Inverters," *Power Electronics, IEEE Transactions on*, vol. 28, pp. 4880-4884, 2013.

[25] Y. Liu, B. Ge, H. Abu-Rub, and D. Sun, "Comprehensive Modeling of Single-Phase Quasi-Z-Source Photovoltaic Inverter to Investigate Low-Frequency Voltage and Current Ripples," *Industrial Electronics, IEEE Transactions on*, vol. PP, pp. 1-1, 2014.

[26] W. Cai and F. Yi, "An Integrated Multi-Port Power Converter with Small Capacitance Requirement for Switched Reluctance Motor Drive," *Power Electronics, IEEE Transactions on*, vol. PP, pp. 1-1, 2015.

[27] T. T. C. Tsang and D. W. Clarke, "Generalised predictive control with input constraints," *Control Theory and Applications, IEE Proceedings D*, vol. 135, pp. 451-460, 1988.

Operation Analysis of High Efficiency Grid Connected Bi-Directional Power Conversion System for Various Storage Battery Systems with Bi-Directional Switch Circuit Topology

Go Yamada*, Takaaki Norisada*, Fumito Kusama*, Keiji Akamatsu* and Masakazu Michihira**

Email: yamada.goh@jp.panasonic.com, akamatsu.keiji@jp.panasonic.com
*Cooperate Engineering Division Automotive & Industrial Systems Company Panasonic Corporation
Moriguchi City, Osaka 570-8501, Japan
**Kobe city college of technology, Kobe City, Hyogo 651-2102, Japan

Abstract— In recent years, there is an increasing demand for various types of storage battery systems, such as home battery systems, V2H(Vehicle-to-Home)/V2G(Vehicle-to-Grid) systems, and UPS systems. Performance requirements for the power conversion system within these storage battery systems are small size, light weight and high efficiency. We propose a new power conversion system so called the GAP-D³ converter that is a bi-directional power control system, comprising of a bi-directional switch circuit topology with natural commutation switching mode. This paper describes the features of GAP-D³ converter including this new current compensation control with simulation and experimental results.

I. INTRODUCTION

In a power conversion system with storage battery systems, a wide variety of technical functions are required when connected to the electric power grid, such as the grid interconnection operation, autonomous operation, bi-directional power flow operation, and electrical insulation. In addition to these, the power conversion system must have small size, light weight, high efficiency, and high reliability [1].

We have been developing GAP-D³ converter (Grid-interconnection system by Asymmetrical Phase-shifted Double bi-Directional control isolated DC-AC converter) for the power conversion system within storage battery systems [2][3].

This paper describes fundamental operation analysis of GAP-D³ converter with grid interconnection. A good output tracking control and high efficiency characteristics of this system were demonstrated from both circuit simulation and prototype experimental results.

II. SYSTEM CONFIGURATION

Fig.1 and Fig.2 show the conventional bi-directional system and the proposed system, respectively. The proposed system of Fig.2 has a small switching loss due to the natural commutation switching mode, and the bi-directional switch circuit topology contributes to the reduction of power conversion stages and elimination of aluminum electrolytic capacitors. Therefore, downsizing the system is made possible.

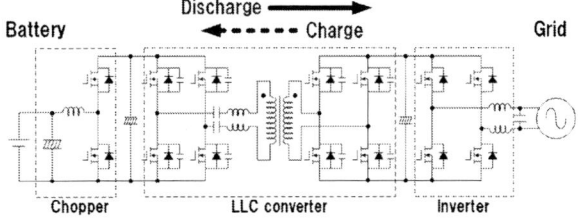

Fig.1. Conventional system

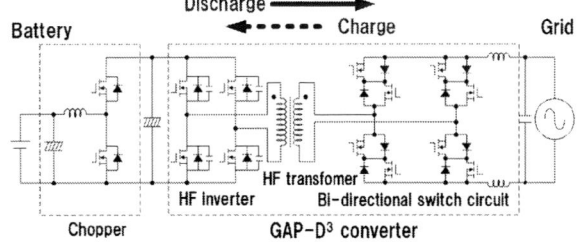

Fig.2. Proposed system

III. CONTROL SEQUENCE OF BI-DIRECTIONAL POWER FLOW

Fig.3 shows proposed bi-directional power control system connected to the electrical power grid.

Bi-directional switch circuit topology of secondary stage is operated with the natural communication switching mode by phase-shifted operation synchronized with the HF Inverter operation [4][5][6].

The switching sequence of bi-directional switch circuit topology is described at subsection-A.

The phase difference α is determined by two factors. One is feed-back control based on the conventional system. Another is the lagging current compensation for indigenous disturbance of the proposed system.

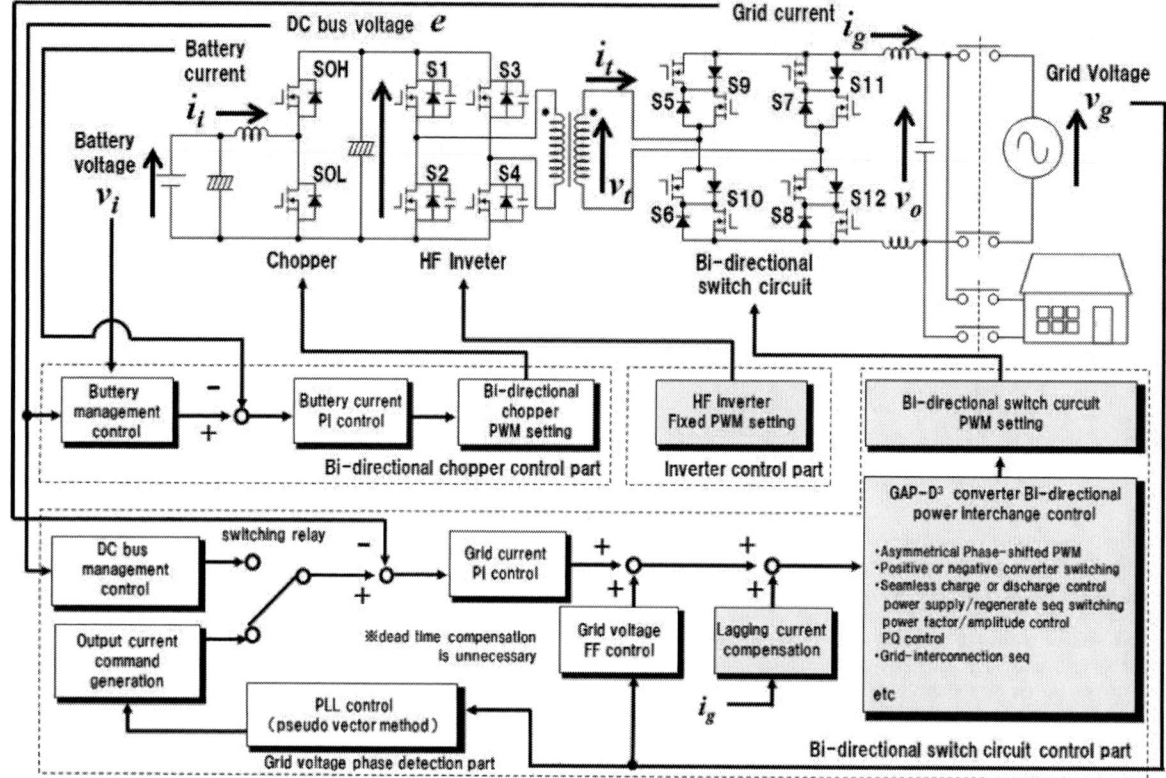

Fig.3. Control system of the proposed system

Control method of the lagging current compensation is described at subsection-B. The state-flow of power factor control for bi-directional power flow operation at the proposed system is described at subsection-C.

A. Swiching Sequence

The proposed system performs bi-directional power flow by exchanging three switching sequences depending on the state of power flow.

1) Discharge Sequence: Fig.4 shows the sequence at discharge from the battery system. Grid current i_g is controlled by operating bi-directional switch circuit as phase-shifted/synchronous rectification synchronized with HF-Inverter that switches at 50% duty. Switches of bi-directional circuit are exchanged depending on the polarities of output voltage of HF-transformer v_t, output AC voltage v_o and Grid current i_g. HF-Inverter drives in ZVS-operation by the partial resonance between exciting inductance of HF-transformer and loss-less snubber capacitor. Switches that operates phase-shifted/synchronous rectification are selected depending on the polarity of output voltage of HF-transformer v_t. When the polarity is positive, S5 and S9 are selected. On the other hand, S6 and S10 are selected when the polarity is negative. Therefore, GAP-D^3 converter is operated with the natural communication switching mode.

S7 and S8 are always on-state.

When the polarity of i_g is negative, it is controlled by each switch described within "()".

2) Charge Sequence: Fig.5 shows the sequence at charge to the battery system. Same as discharge sequence, switches that operates phase-shifted/synchronous rectification are selected depending on the polarity of output voltage of HF-transformer v_t. When the polarity is positive, S6 is selected. On the other hand, S5 is selected when the polarity is negative.

The difference between charge sequence and discharge sequence is the polarity of the HF-transformer. The v_t polarity is reverse polarity of i_t polarity. The difference of the its polarity is made by controlling α that is off-time of S7 and S8.

When the polarity of i_g is negative, it is controled by each switch described within "()".

3) Positive Negative Sequence: For good characteristic of i_g, it is necessary to remove the electrical charge of AC reactor. Therefore, in near the switching area of i_g polarity, another sequence is required between Discharge sequence and Charge sequence. Positive Negative Sequence move the charge of AC reactor to the DC side of the HF-Inverter.

This sequence improves distortion of polarity inversion of the i_g, and is original sequence for GAP-D^3 converter.

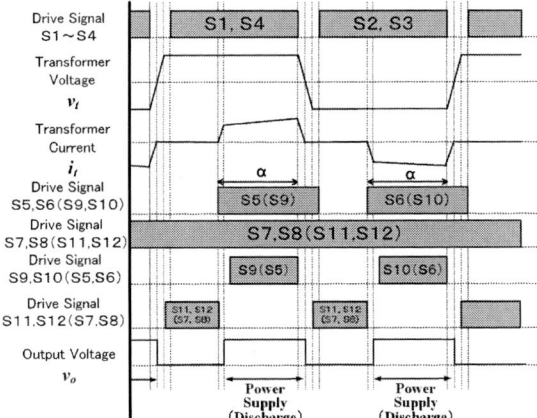

Fig.4. Switching sequence for discharge operation

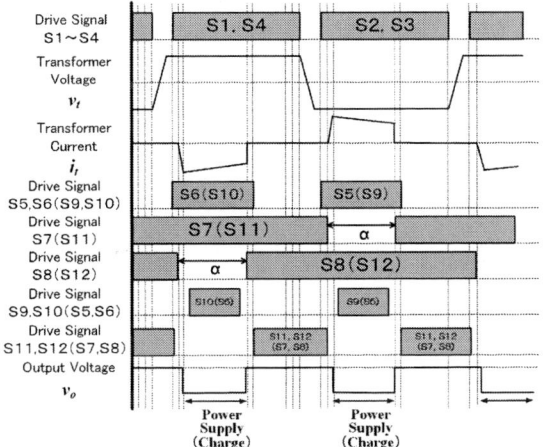

Fig.5. Switching sequence for charge sequence

B. Feedback Loop

We developed the feedback loop based on switching sequence of GAP-D^3 converter. (1) to (3) shows the electrical characteristic expression of the proposed system.

$$i_g = \frac{v_o - v_g}{L_{AC} - R_{AC}} \tag{1}$$

$$v_o = \alpha v_t - v_{Le} \tag{2}$$

$$v_t = e \frac{N_P}{N_S} \tag{3}$$

v_t is voltage of HF-transformer(secondary stage). v_o is output AC voltage. L_{AC} is inductance of AC reactor. R_{AC} is resistance of AC reactor. v_g is grid voltage.

N_P is turn of HF-transformer (primary stage). N_S is turn of HF-transformer (secondary stage). e is the middle link voltage. v_{Le} is disturbance voltage occurred by leakage inductance. v_{Le} is specific disturbance of the proposed system.

1) Influence of Leakage Inductance: Fig.6 shows Output (grid) current i_g characteristics by conventional feedback loop. There is the difference of current characteristic between the command and output. The cause of its difference is lag of power flow occurred by leakage inductance of HF-transformer.

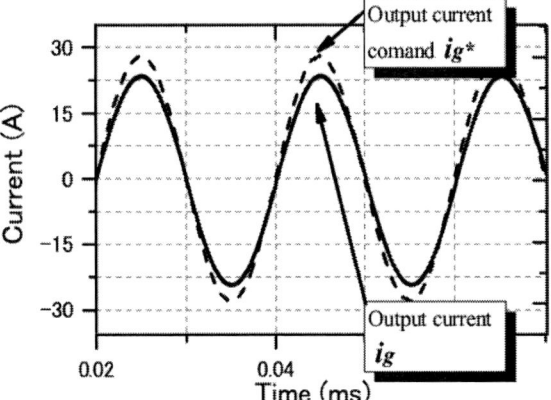

Fig.6. Current waveform without lagging current compensation

Lagging Current Compensation: Fig.7 shows new feedback loop for the proposed system. (4) shows that the influence of leakage inductance is added to (2).

$$v_o = \alpha v_t - i_g L_e \tag{4}$$

To compensate the lagging current by leakage inductance, it is need to add (5) to α. (5) is that disturbance voltage expression converted to time ratio.

$$\frac{i_g L_e}{v_t} = \frac{N_S i_g L_e}{N_P e} \tag{5}$$

Due to this compensation, feedback loop is able to be composed very simple. Controller *C(s)* is PI controller indicated as (6).

$$C(s) = K_p \left(1 + \frac{1}{T_i s} \right) \tag{6}$$

K_p is proportional gain of *C(s)*. Ti is integral time. K_p and T_i is indicated as (7) and (8).

$$K_p = \frac{N_S}{e K_p} (2\zeta \omega L_{AC} + R_{AC}) \tag{7}$$

$$T_i = \frac{2\zeta}{\omega} \tag{8}$$

ω is natural angular frequency of the proposed system. ζ is attenuation coefficient of system. (9) shows *G(s)* that is transfer function of closed loop for whole system.

$$G(s) = \frac{(T_i s + 1)\omega^2}{s^2 + 2\zeta \omega s + \omega^2} \tag{9}$$

Fig.8 shows the effect of the lagging current compensation. The actual output matches with the command value at all power range.

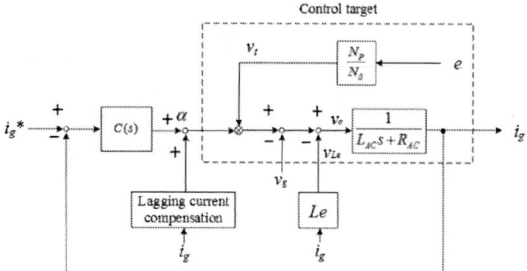

Fig.7. Re-modeling feedback loop

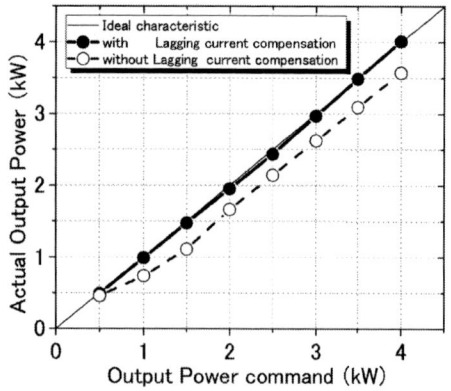

Fig.8. Effect of the lagging current compensation

C. State-flow

i_g is controlled to switch discharge/charge operation of battery system. Power factor control is essential for conforming to various standards such as grid interconnection examination. We developed state-flow control algorithm because switching sequence have to be switched depending on the state of power factor of i_g. State-flow discriminates five sequence numbers and four power factor states classified by v_g and i_g. Table.1 shows five sequence numbers (No.1 to No.5). Fig.9 shows four power factor states ((A) to (D)).

Sequence number is assigned according to the state of v_g and i_g. Four power factor states are (A)PF1mode, (B)leading PF mode, (C)lagging PF mode and (D)PF-1mode.

Fig.10 shows the state-flow in power factor control. At "Mode confirmation", the target of power factor state is determined from current state of voltage and current. At "Sequence confirmation", sequence number (No.1 to No.4) is selected from the state of v_g and i_g. At "Current reversal", i_g is judged whether $i_g \doteqdot 0$. If not $i_g \doteqdot 0$, State-flow moves to "Set PWM" to set PWM duty. If $i_g \doteqdot 0$, State-flow moves to "Choose sequence" to remove the electrical charge of AC reactor.

TABLE I. SEQUENCE NUMBER AND ITS VG, IG CONDITION

Sequence	v_g, i_g state	Sequence number
Discharge seq.	$v_g > 0$, $i_g > 0$	No.1
	$v_g < 0$, $i_g < 0$	No.2
Charge seq.	$v_g > 0$, $i_g < 0$	No.3
	$v_g < 0$, $i_g > 0$	No.4
Posi & Nega seq.	$i_g \doteqdot 0$	No.5

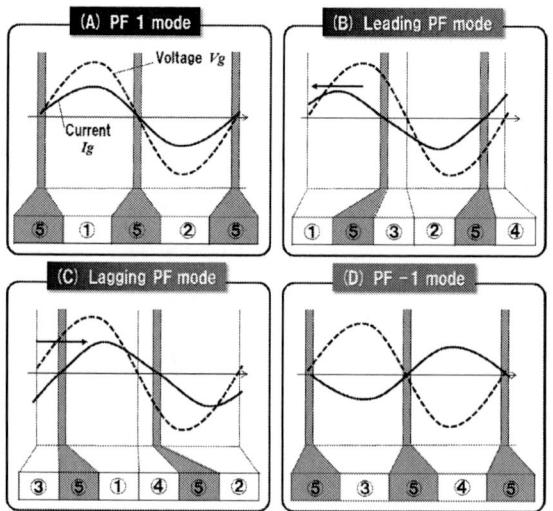

Fig.9. Voltage and current waveforms at each power factor condition

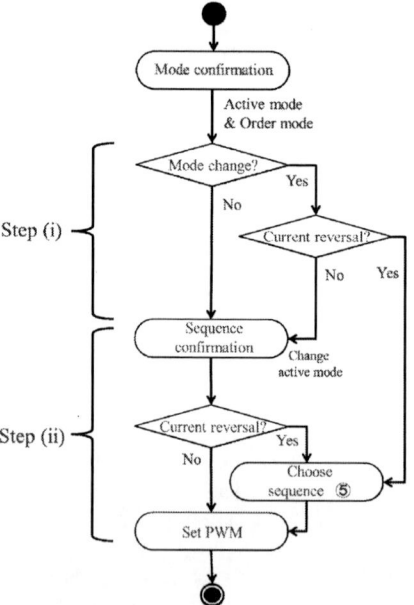

Fig.10. The State-flow in power factor control

IV. SIMULATION AND PROTOTYPE

A. Simulation and Prototype parameter

We tested bi-directional power flow operation in simulation and prototype. Table.2 shows simulation and prototype parameters. Table.3 shows components of prototype. Fig.11 shows the prototype appearance. An expected application of this prototype is the home battery systems.

TABLE II.   SIMULATION AND PROTOTYPE PARAMETERS

Item, Symbol		Value (unit)
Output power		2 (kW)
Grid Voltage, v_g		200 (Vrms)
Link DC Voltage, e		340 (V)
Battery Voltage, v_i		300 (V)
AC Filter, L_{ACf}, C_{ACf}		430 (uH), 5 (uF)
DC Filter, L_{DCf}, C_{DCf}		1.2 (mH), 900 (uF)
DC Bus Capacitor, C_{DC}		1680 (uF)
HF Inverter Operation Freq.		17.5kHz
HF Trans-former	Turn ratio	1 : 1
	Self Inductance, L_1, L_2	1.62 (mH), 1.62 (mH)
	Coupling Coefficient	0.999

TABLE III.   PROTOTYPE PARAMETERS AND COMPONENTS

Item, Symbol			Value (unit)
Power Device	Primary	SOH, SOL S1 ~ S4	Cree SiC-MOSFET C2M0025120D(1200V/90A)
	Secondary	S5 ~ S12	Cree SiC-MOSFET C2M0025120D(1200V/90A)
			Cree SiC-SBD C4D10120A(1200V/14A)
MPU			Renesas, RX63T

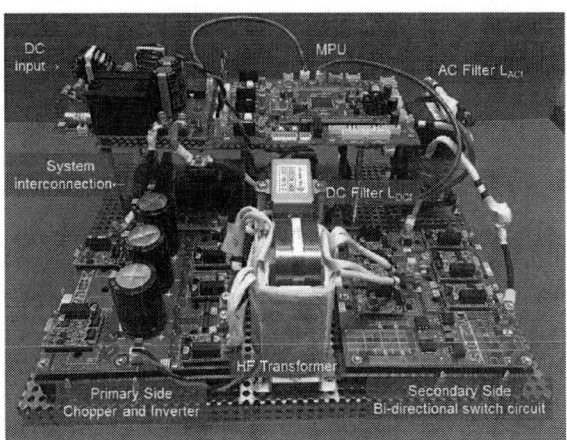

Fig.11. The prototype appearance

B. Operation analysis of GAP-D³ converter

Fig.12 shows efficiency characteristics of the proposed system. At battery voltage $v_i = 300V$, maximum efficiency is 94.5% at both charge and discharge. Fig.13 and Fig.14 show the waveform characteristics of i_g. Prototype waveforms are almost the same with simulation waveforms. Both at charge mode and discharge mode, THD is under 2.5% at 2kW output.

From this result, the validity of simulation results as well as the practicality of the proposed system is suggested.

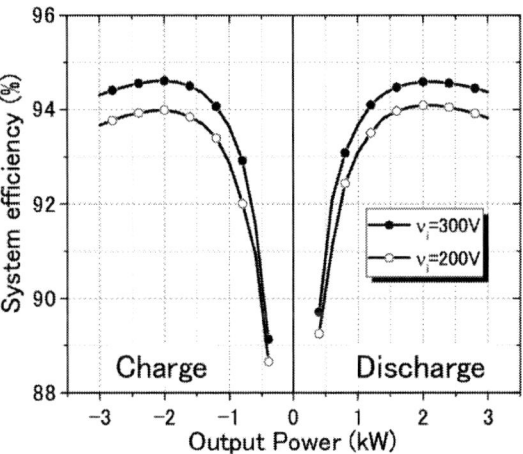

Fig.12. Efficiency characteristics of the proposed system

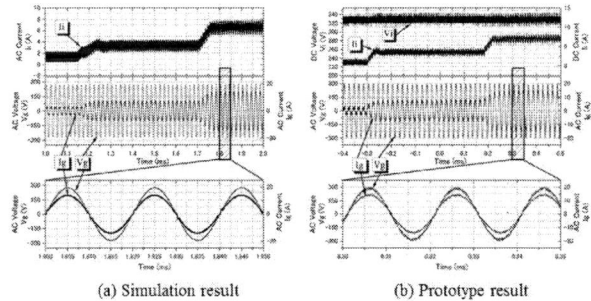

(a) Simulation result   (b) Prototype result

Fig.13. Discharge operation waveforms

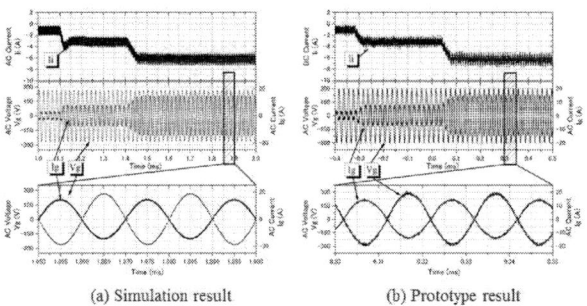

(a) Simulation result   (b) Prototype result

Fig.14. Charge operation waveforms

V. CONCLUSION

This paper proposed the GAP-D^3 converter and its new control scheme. Its output power characteristics and modes of operation were demonstrated from both the simulation and experimental analysis. Both at charge mode and discharge mode, maximum efficiency was 94.5%, and THD was under 2.5%. The results indicate that the GAP-D^3 converter possesses sufficient basic performance in electric power grid system for storage battery systems and that has a practicable performance.

REFERENCES

[1] Fuji-Keizai, ``The future outlook of energy, large storage battery and materials 2014"

[2] Investigating R&D Committee of Power Electronics Technologies Applied to Advanced Distribution Networks, "Power Electronics Technologies Applied to Advanced Distribution Networks (Smart Grid)", Aug 2014

[3] IEEJ, ``Power conversion system technology of grid interface", IEEJ Transactions on Industry Applications, Vol.1292, 2013

[4] T.Norisada, ``Operation Analysis of High Efficiency Grid Connected Bi-directional Power Conversion System for Various Storage Battery Systems with Bi-directional Switch Circuit Topology", IEEJ Transactions on Industry Applications, No.1-6, pp.I-57 I-60, 2015

[5] T.Norisada, ``Operation Analysis for Seamless Charge and Discharge Control of High Efficiency Grid Connected Bi-directional Power Conversion System for Various Storage Battery Systems with Bi-directional Switch Circuit Topology", JIPE-41-17, 2015

[6] M.Michihira, ``Operation analysis of quasi-resonant high-frequency transformer link DC-AC converter applied for phase-shift PWM control in secondary side", IEEJ Transactions on Industry Applications, Vol.117 , No.12, pp.1503-1510, 1997

Fault Tolerant Control of MMC with Redundant Sub-modules Based on Carrier Phase Shift Modulation

Kai Li, Zhengming Zhao, Liqiang Yuan, Sizhao Lu
State Key Laboratory of Power System, Department of
Electrical Engineering, Tsinghua University
Beijing 100084, China

Bing Pan, Zhengang Lu
State Grid Smart Grid Research Institute
Beijing 102209, China

Abstract—**Modular multilevel converter (MMC) is regarded as one of the most promising topologies for the high-voltage high-power applications. However, reliability is one of the most important challenges for MMC which is composed of a large number of power electronics sub-modules (SMs). In order to increase the system reliability, redundant SMs are often utilized. A strategy is proposed in this paper to implement the control of MMC with redundant SMs based on the carrier phase shift modulation. A rotating sliding choice box is adopted to select the working SMs and phase displacements in operation, which enables the equal burden of each SM. The failed SMs will be deleted from the selecting options, which guarantees the fault tolerant control. Simulations and experiments based on a 5-kW downscaled MMC hardware have verified the strategy.**

Keywords—MMC; redundant sub-modules; carrier phase shift modulation; fault tolerant control

I. INTRODUCTION

A lot of research work on MMC has been conducted in the past decade [1-4]. Compared with the conventional multilevel converters such as neutral point clamped (NPC) and flying capacitor converters, MMC has the advantages of modularity and scalability, and this enables energy conversion without filters or transformers [5]. Thus MMC has been considered as a very good choice in the high-power applications such as HVDC transmission, STATCOM, medium-voltage motor drive.

Generally a large number of components are used in MMC, including switching devices and capacitors. Each component can be seen as a potential failure point. Therefore, reliability is one of the most important challenges for MMCs [6]. In order to increase the system reliability, redundant or reserved SMs [7, 8] are often used in practice. Once one of the SMs fails, it will be replaced by a redundant SM and the converter will keep working normally without being interrupted until next regular maintenance. Therefore, the control of MMC with redundant SMs needs to be focused on. Ref. [9, 10] presented the design and control methods for fault-tolerant operations with redundant SMs based on nearest level modulation. Ref. [9] mentioned two schemes of managing redundant SMs: 1) cold reserve, which means that the redundant SMs are bypassed in normal operation; 2) hot reserve or spinning reserve, which means the redundant SMs operate in the same way as the other SMs. Scheme 1 has the drawbacks of long charging time for the cold reserved SMs and complex control system. Thus, scheme 2 is adopted in this

Supported by the project titled Research on the Failure Mechanism and Reliability Theory of Power Electronic Equipment Using Full-controlled Devices (SGRI-DL-71-15-009) and the Major Program of the National Natural Science Foundation of China (51490683).

paper, and a fault tolerant control method of MMC based on carrier phase shift (CPS) modulation is proposed. CPS modulation is commonly used in cascaded H-bridge converters, but it is also suitable for MMC with small number of SMs because it makes the loss evenly distributed among different switching devices, and it's consistent with the modularity and scalability of MMC [11].

II. CARRIER PHASE SHIFT MODULATION OF MMC

The basic working principle and control strategy of MMC have been introduced a lot in [12, 13]. In CPS modulation, the reference voltage for each arm can be given from the control system. After that, the reference voltage is distributed to each SM, adjusted a little for the capacitor voltage balancing propose, and then compared with a series of triangle carriers to generate the control signals for the switching devices. Single-phase MMC (shown as Fig. 1) is taken as an example to illustrate the CPS modulation. In Fig. 1(a), each arm has 4 SMs which are all half-bridge modules, and they are always in the working state, which means there is no redundant SMs. For comparison, Fig. 1(c) presents the case with 2 redundant SMs per arm, which will be used in the later section.

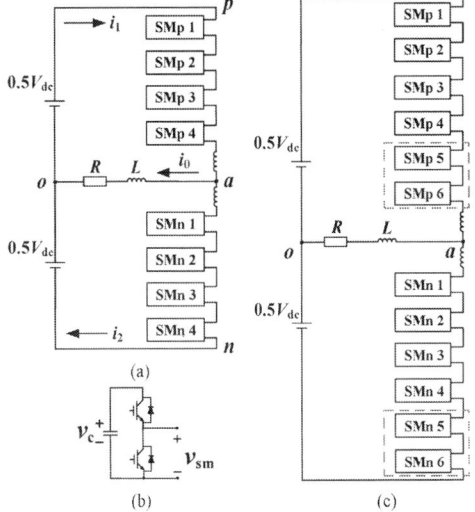

Fig. 1. Single-phase MMC: (a) 4 SMs per arm; (b) half-bridge SM; (c) 6 SMs per arm, with 2 redundant SMs shown in the dashed box.

For the single-phase MMC in Fig. 1(a), two schemes of CPS modulation are shown in Fig. 2, including the reference arm voltage waveforms, respective carriers, phase shift, the ideal output voltage and arm inductors' voltage waveforms. The difference between the two schemes lies in the different phase displacements between the upper and lower carriers, which will affect the interactions between the upper and lower arms, and thus determines the harmonic features of the output voltage and circulating current [11]. Furthermore, the phase displacements of the two schemes vary according to the odd or even number of SMs in each arm. From Fig. 2, it can be concluded that the phase shift between SMs in the same arm is T_s/N, and the displacement between upper and lower carriers can be 0 or $0.5T_s/N$, which distinguishes the two modulation schemes. When the SM number per arm is even (such as $N=4$ in Fig. 2), $0.5T_s/N$ phase displacement between upper and lower carriers will bring '$2N+1$' level output voltage, which minimizes the output voltage harmonics; otherwise, 0 phase displacement between upper and lower carriers will bring '$N+1$' level output voltage and higher harmonics, but it can cancel the circulating current harmonics. On the other hand, when the SM number per arm is odd, 0 and $0.5T_s/N$ phase displacement will bring '$2N+1$' level and '$N+1$' level output voltage respectively, and the circulating current harmonics feature can be derived correspondingly. Both the schemes are effective PWM solutions for MMCs, which one is adopted depends on the requirement to minimize the output voltage harmonics or to eliminate the switching harmonics of circulating current.

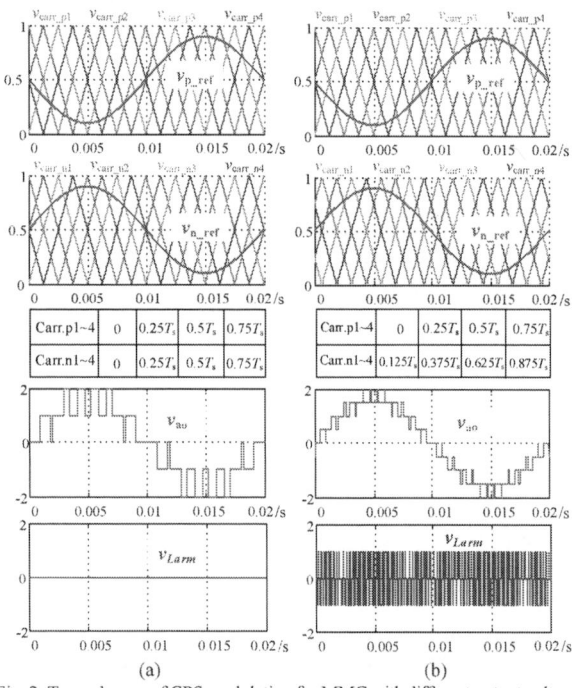

Fig. 2. Two schemes of CPS modulation for MMC with different output voltage levels: (a) '$N+1$' levels; (b) '$2N+1$' levels.

III. REDUNDANCY CONTROL BASED ON CPS MODULATION

For the single phase MMC in Fig. 1(c), only 4 SMs per arm can keep the converter operate properly. Therefore, there are 2 SMs which will be bypassed all the time. This section introduces

the proposed method to select the working SMs and the phase shift of the carriers, and then the whole control structure is provided for MMC with redundant SMs.

A. Selecting Working SMs and Phase Shift

There should be 4 SMs per arm which are in the working state all the time. In order to make the burden equally distributed in each SM, the 6 SMs in one arm need to be selected to work alternately. A rotating sliding choice box is used to select the working SMs and the corresponding phase shift, shown as Fig. 3. And the relative position of the phase shift is fixed to the choice box. The principle of Fig. 3 is introduced as follows. Firstly, a sector is defined as a rotating cycle, and it can be set as one or more switching cycles. When a shorter rotating cycle is selected, the voltage difference between different capacitors are smaller. Then the sector number is determined according to the current time. In each sector, 4 certain SMs are chosen to be the working SMs, and the chosen SMs with the phase shift will not change until the next sector. Secondly, as time goes by, the converter arm changes from one sector to another and the sliding choice box shifts. There should be $N+r=6$ kinds of sectors in total. $N=4$ is the number of necessary SMs to keep the converter operate normally, and $r=2$ is the number of the redundant SMs. In Fig. 3, after sector 6, the converter arm will return to sector 1. Therefore, the sector period is defined as the total time of 6 sectors. Thirdly, the chosen working SMs will join the control, which means they are working in the PWM mode, and their capacitor voltages need to be sampled and fed back. In contrast, the unselected SMs will receive bypassed signals from the control system. The control will be introduced in the next.

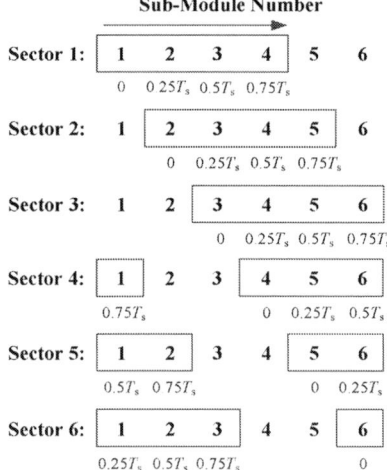

Fig. 3. Rotating sliding choice box to select the working SMs and phase shift

B. Redundancy Control Structures

Fig. 4 shows the redundancy control structure of MMC, in which capacitor voltages' feedback are adjusted because of the redundant SMs, and the adjustment is noted by the blue dashed box. The control system is briefly introduced in the following. The output current control in Fig. 4(a) adopts the deadbeat control, which is simple and accurate. The average capacitor voltage control in Fig. 4(b) guarantees the energy needed in the converter. The control in Fig. 4(c) with a repetitive controller can suppress the AC circulating current a lot [14, 15]. The individual and arm capacitor voltage balance control [12] in Fig. 4(d) and (e) make the voltage error between different capacitors very small. Fig. 4(f) shows the final voltage command for each SM respectively. Once the voltage command and the carrier phase shift (as Fig. 3) for each selected SM are obtained, the gate signals for each switching device can be generated.

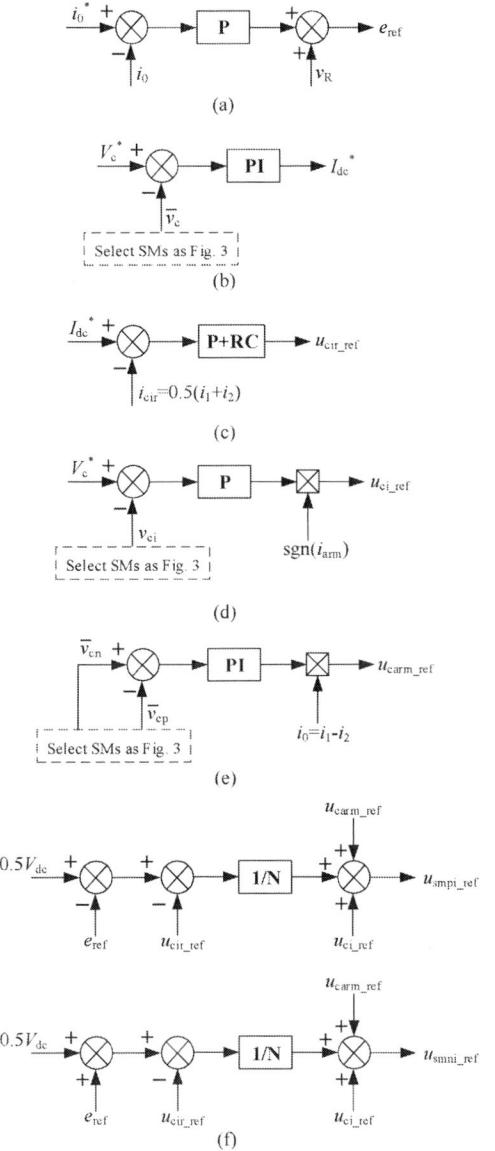

Fig. 4. Redundancy control structure of MMC: (a) output current control; (b) average capacitor voltage control; (c) circulating current control; (d) individual capacitor voltage balance control; (e) arm capacitor voltage balance control; (f) voltage command for each SM in the upper and lower arm.

IV. FAULT TOLERANT CONTROL

Once one SM in MMC fails because of devices' short-circuit or open-circuit, the fault can be detected and localized [6, 16], then the failed SM should be bypassed by the circuit breaker in

the SM, and the redundant SMs need to be activated. This section introduces the procedure of the fault tolerant control.

A. Delete the Failed SMs from the Selecting Options

Fig. 5 shows the rotating sliding choice box when the 3rd SM (in the red dashed box) in one arm fails. It can be seen that the 3rd SM is deleted from the selecting options, and the number of total sectors changes from 6 to 5, but the phase shift in different sectors are similar.

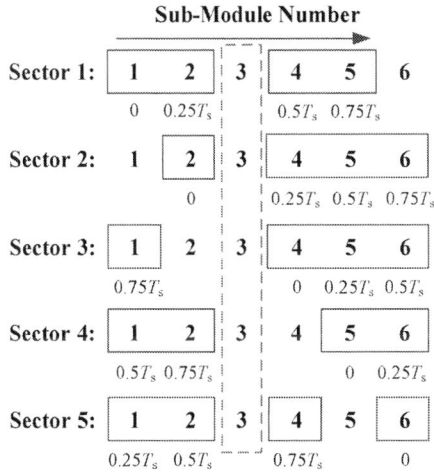

Fig. 5. Rotating sliding choice box when the 3rd SM of the arm fails.

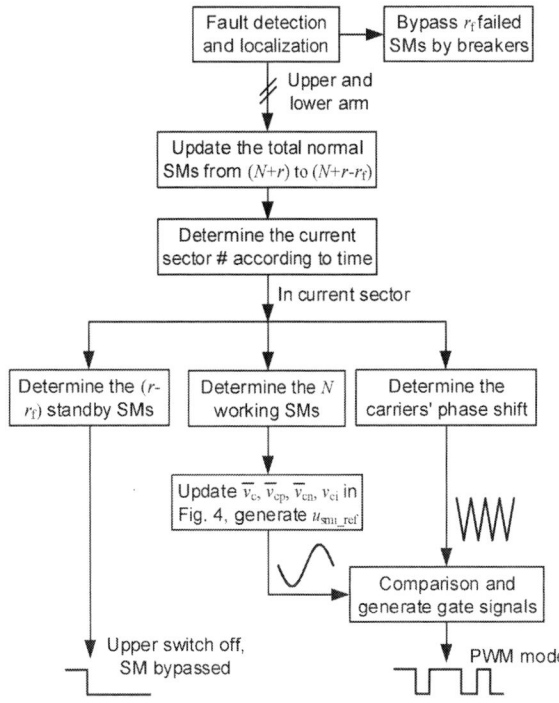

Fig. 6. Flowchart of the fault tolerant control of MMC.

B. Implementation of the Fault Tolerant Control

The flowchart of the fault tolerant control for MMC with redundant SMs is shown in Fig. 6. Once a failed SM is detected and localized, it will be bypassed by the circuit breaker in the SM immediately. At the same time, the failed SM will be deleted from the selecting options, thus the total number of normal SMs will decrease from $N+r$ to $N+r-r_f$, and the number of sectors decreases accordingly. After the fault, a new sector number can be decided, which determines the standby SMs, the working SMs and the carriers' phase shift.

The process to update the sector number, to choose the working SMs and to generate the carriers are described in detail in Fig. 7, in which 'm, n, k, i, j' are used as counters and 't' is the current time. 'WorkState' and 'Phase' are the final working state and phase shift for each SM, and 'State0' and 'Phase0' are used as temporary variables. When a SM is in standby mode or bypass mode (failed), its state and phase shift are all set as 0. The array 'Order' is used for storing the original SM number after the failed SMs are deleted from the selecting options.

In fact, Fig. 7 shows the case that all the SMs use 0 phase displacement carriers relative to the other arm. When the $0.5T_s/N$ phase displacement carriers are adopted for this arm, the phase shift for each SM should be set as follows instead.

$$\text{Phase}[\text{Order}[k]] = \text{Phase0}[k] \times T_s/N + 0.5T_s/N \qquad (1)$$

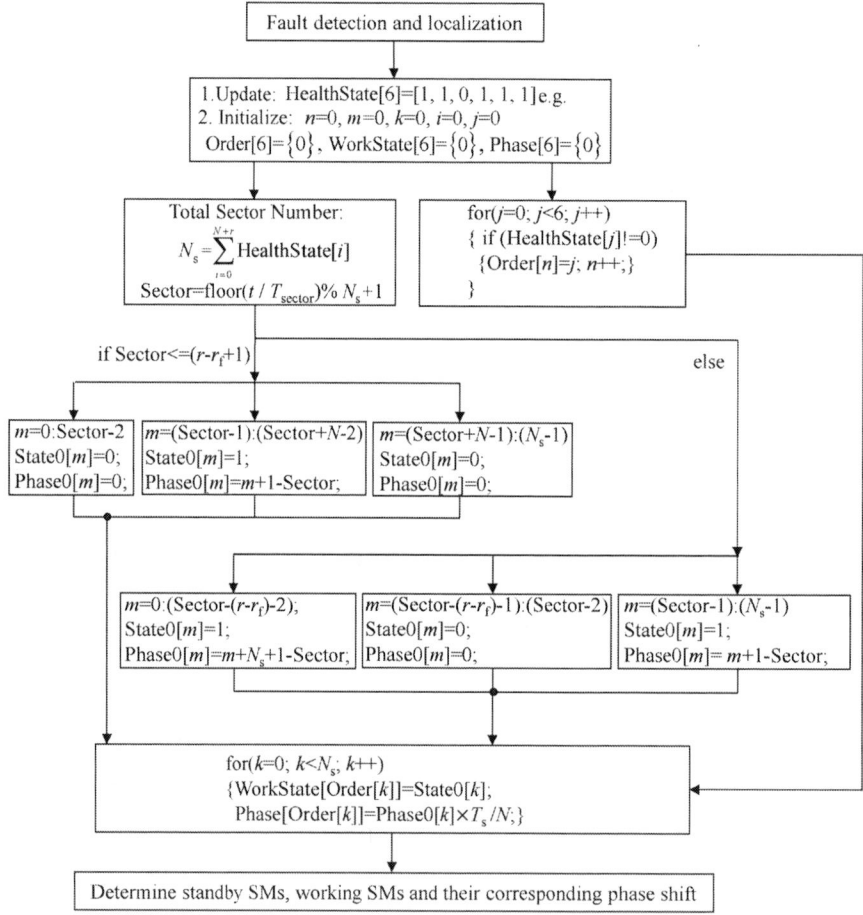

Fig. 7. Flowchart of determining the working state of each SM and its carrier phase shift.

V. SIMULATION VERIFICATION

Simulations are conducted in Matlab/ Simulink to verify the proposed redundancy and fault tolerant control method based on the CPS modulation. Sing-phase MMC with 6 SMs per arm (Fig. 1(c)) is chosen for simulation, and the parameters are listed in Table I, in which two discrete arm inductors are used, and the rotating cycle is set as one switching cycle.

The waveforms of MMC without redundant SMs utilizing the two CPS modulation schemes are shown in Fig. 8, which illustrates the different features. '$N+1$' and '$2N+1$' level output voltages are demonstrated, and the circulating currents in the two cases have little difference because AC components are both suppressed a lot. The waveforms of MMC with redundant SMs are shown in Fig. 9(b), which verifies the feasibility of the proposed redundancy control method, and the capacitor voltage ripple is also reduced, but the voltage difference between different capacitors increased, compared with waveforms without redundant SMs in Fig. 9(a).

TABLE I PARAMETERS OF SINGLE PHASE MMC FOR SIMULATION

Parameters	Values
DC-link voltage	V_{dc}=2400 V
No. of working SMs per arm	N=4
No. of redundant SMs per arm	r=2
SM rated capacitor voltage	U_c=600 V
SM capacitance	C=2 mF
Arm inductance	L=5 mH
Carrier frequency	f_c=5 kHz
Fundamental frequency	f_0=50 Hz
Load inductance	L_{Load}=3 mH
Load resistance	R_{Load}=12 Ohms
Output current mag.	I_{am}=80 A
Rotating cycle	T_{RC}=2e-4 s

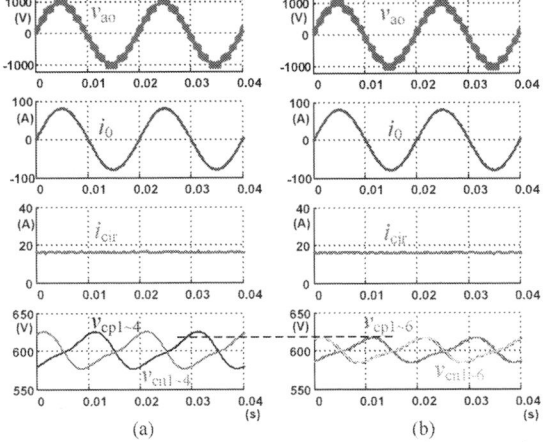

Fig. 8. Output voltage, output current, circulating current and capacitor voltages' waveforms for comparison of the two CPS modulation schemes: (a) 'N+1' level; (b) '$2N$+1' level.

Fig. 9. Output voltage, output current, circulating current and capacitor voltages' waveforms: (a) without redundant SMs; (b) with 2 redundant SMs per arm.

The circulating current and capacitor voltages' waveforms when some SMs are broken are presented in Fig. 10. At 0.1s, the 3rd SM in the upper arm fails; at 0.3s, the 5th and 6th SMs in

the lower arm fail at the same time; at 0.5s, the 5th SM in the upper arm fails. Capacitor voltage waveforms of the 1st SM in the upper arm and the 1st SM in the lower arm are given as well, which represent the waveforms of the healthy SMs. The capacitor voltages of the failed SMs change little after fault because once the failed SMs are bypassed, the capacitors will be discharged slowly through the paralleled bleeder resistors. The output voltage and current waveforms during the fault are just like those in Fig. 9, not repeated here. It can be pointed out that MMC can work normally even when all the redundant SMs fail, and the transient process is smooth, only with a very small and short fluctuation. This confirms the effectiveness of the fault tolerant control.

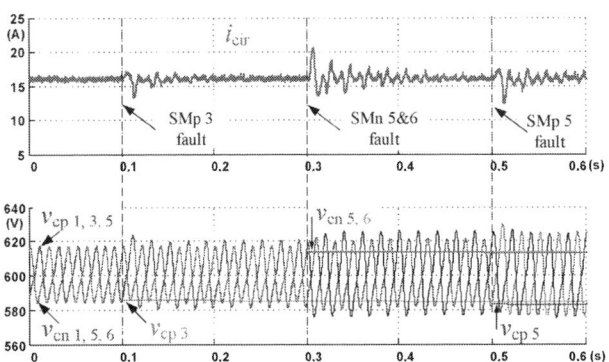

Fig. 10. Circulating current and capacitor voltages' waveforms when some SMs fail: (1) 0.1s, SMp 3 fails; (2) 0.3s, SMn 5 & 6 fail at the same time; (3) 0.5s, SMp 5 fails.

VI. EXPERIMENTAL VERIFICATION

Fig. 11 shows the picture of a 5-kW single-phase downscaled MMC hardware, with 6 SMs per arm, through which the proposed control strategies are verified by experiments. The parameters of the hardware prototype are listed in Table II, in which one center-tapped or coupled arm inductor (just as ref. [12, 17]) are used, and the rotating cycle is also set as one switching cycle here.

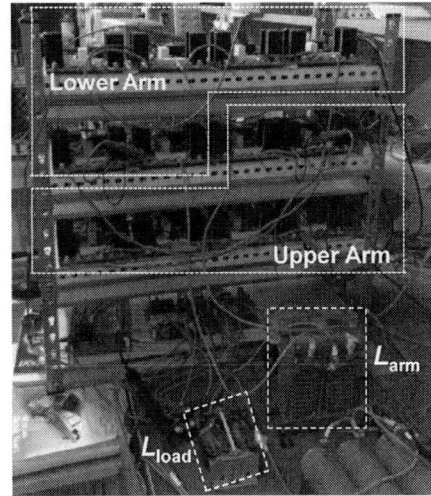

Fig. 11. Picture of the single-phase downscaled MMC hardware.

TABLE II PARAMETERS OF SINGLE PHASE MMC FOR EXPERIMENTS

Parameters	Values
DC-link voltage	V_{dc}=300 V
No. of working SMs per arm	N=4
No. of redundant SMs per arm	r=2
SM rated capacitor voltage	U_c=75 V
SM capacitance	C=3280 uF
Inductance of center-tapped inductor	L_{arm} =2.5 mH
Carrier frequency	f_c =6 kHz
Fundamental frequency	f_0 =50 Hz
Load inductance	L_{Load}=1 mH
Load resistance	R_{Load}=12 Ohms
Output current mag.	I_{am}= 10.2 A
Rotating cycle	T_{RC}=1/6000 s

The 'N+1' level modulation scheme is adopted as an example in the experiment, and the output voltage in the steady state is shown as Fig. 12, in which there are some mismatch pulses. It is proved that these mismatch pulses are caused by dead time setup and have only a little effect. Gate signals, capacitor voltages, arm currents, output current and output voltage waveforms when certain SMs fail are presented in Fig. 13, similar scenario as the previous simulation setup. At t_1, SMp3 fails; at t_2, SMn5 and SMn6 fail at the same time; at t_3, SMp5

fails. It can be seen that once a SM is in fault, its gate signal will change to zero immediately. Capacitor voltage waveforms of SMp1 and SMn1 represent the waveforms of the healthy SMs, and fluctuation always exists in these waveforms. In contrast, the capacitor voltages of the failed SMs change little after fault. The arm current, the output current and voltage waveforms during the fault are almost same as those in the normal condition. Therefore, it can be concluded that MMC can work continuously when any one of the redundant SMs fails. This result verify the effectiveness and stability of the proposed redundancy and fault tolerant control method.

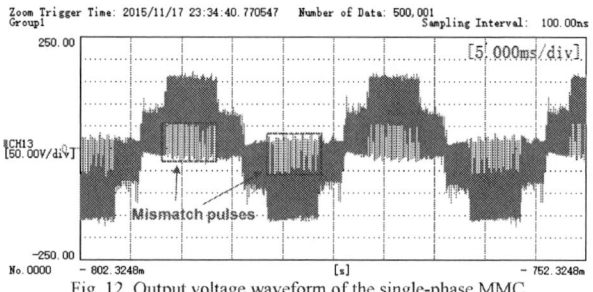

Fig. 12. Output voltage waveform of the single-phase MMC.

Fig. 13. Gate signals of the failed SMs, capacitor voltages, arm currents, output current and output voltage waveforms of MMC.

VII. CONCLUSION

With the comparison of two CPS modulation schemes which utilize different phase displacements between the upper and lower arms, a control method for MMC with spinning reserved SMs based on the CPS modulation has been proposed. Different working SMs are selected through the rotating sliding choice box according to different sectors. Phase shift angles for the carriers are determined at the same time. Adding the selecting mechanism, the whole control structure for MMC with redundant SMs is provided. Once one SM in the arm fails and is detected, it will be bypassed and deleted from the selecting options. Thus, this fault can be ridden through. The flowchart and the detailed process to realize the fault tolerant control are demonstrated. Simulations and experimental verification have confirmed the effectiveness and stability of the redundancy and fault tolerant control.

REFERENCES

[1] S. Debnath, Q. Jiangchao, B. Bahrani, M. Saeedifard, and P. Barbosa, "Operation, Control, and Applications of the Modular Multilevel Converter: A Review," Power Electronics, IEEE Transactions on, vol. 30, pp. 37-53, 2015-01-01 2015.

[2] M. A. Perez, S. Bernet, J. Rodriguez, S. Kouro, and R. Lizana, "Circuit Topologies, Modeling, Control Schemes, and Applications of Modular Multilevel Converters," IEEE Transactions on Power Electronics, vol. 30, pp. 4-17, 2015.

[3] J. Wang, R. Burgos and D. Boroyevich, "A survey on the modular multilevel converters-Modeling, modulation and controls", in Proc. IEEE ECCE, Denver, 2013, pp. 3894-3991.

[4] A. Lesnicar and R. Marquardt, "An innovative modular multilevel converter topology suitable for a wide power range," in Power Tech Conference Proceedings, 2003 IEEE Bologna, 2003, pp. 6 pp. Vol.3.

[5] N. Thitichaiworakorn, M. Hagiwara and H. Akagi, "Experimental Verification of a Modular Multilevel Cascade Inverter Based on Double-Star Bridge Cells," Industry Applications, IEEE Transactions on, vol. 50, pp. 509-519, 2014-01-01 2014.

[6] D. Fujin, C. Zhe, M. R. Khan, and Z. Rongwu, "Fault Detection and Localization Method for Modular Multilevel Converters," Power Electronics, IEEE Transactions on, vol. 30, pp. 2721-2732, 2015-01-01 2015.

[7] G. Liu, Z. Xu, Y. Xue, and G. Tang, "Optimized Control Strategy Based on Dynamic Redundancy for the Modular Multilevel Converter," IEEE Transactions on Power Electronics, vol. 30, pp. 339-348, 2015.

[8] B. Li, Y. Zhang, R. Yang, R. Xu, D. Xu, and W. Wang, "Seamless Transition Control for Modular Multilevel Converters When Inserting a Cold-Reserve Redundant Submodule," IEEE Transactions on Power Electronics, vol. 30, pp. 4052-4057, 2015.

[9] P. Hu, D. Jiang, Y. Zhou, Y. Liang, J. Guo, and Z. Lin, "Energy-balancing Control Strategy for Modular Multilevel Converters Under Submodule Fault Conditions," IEEE Transactions on Power Electronics, vol. 29, pp. 5021-5030, 2014.

[10] T. S. Gum, L. Hee-Jin, S. N. Tae, C. Yong-Ho, L. Uk-Hwa, B. Seung-Taek, H. Kyeon, and P. Jung-Wook, "Design and Control of a Modular Multilevel HVDC Converter With Redundant Power Modules for Noninterruptible Energy Transfer," Power Delivery, IEEE Transactions on, vol. 27, pp. 1611-1619, 2012-01-01 2012.

[11] B. Li, R. Yang, D. Xu, G. Wang, W. Wang, and D. Xu, "Analysis of the Phase-Shifted Carrier Modulation for Modular Multilevel Converters," IEEE Transactions on Power Electronics, vol. 30, pp. 297-310, 2015.

[12] M. Hagiwara, R. Maeda and H. Akagi, "Control and Analysis of the Modular Multilevel Cascade Converter Based on Double-Star Chopper-Cells (MMCC-DSCC)," Power Electronics, IEEE Transactions on, vol. 26, pp. 1649-1658, 2011-01-01 2011.

[13] Q. Song, W. Liu, X. Li, H. Rao, S. Xu, and L. Li, "A Steady-State Analysis Method for a Modular Multilevel Converter," IEEE Transactions on Power Electronics, vol. 28, pp. 3702-3713, 2013.

[14] B. Li, D. Xu and D. Xu, "Circulating Current Harmonics Suppression for Modular Multilevel Converters Based on Repetitive Control," Journal of Power Electronics, vol. 14, pp. 1100-1108, 2014-11-20 2014.

[15] M. Zhang, L. Huang, W. Yao, and Z. Lu, "Circulating Harmonic Current Elimination of a CPS-PWM-Based Modular Multilevel Converter With a Plug-In Repetitive Controller," IEEE Transactions on Power Electronics, vol. 29, pp. 2083-2097, 2014.

[16] S. Shuai, P. W. Wheeler, J. C. Clare, and A. J. Watson, "Fault Detection for Modular Multilevel Converters Based on Sliding Mode Observer," Power Electronics, IEEE Transactions on, vol. 28, pp. 4867-4872, 2013-01-01 2013.

[17] M. Hagiwara, R. Maeda and H. Akagi, "Theoretical analysis and control of the modular multilevel cascade converter based on double-star chopper-cells (MMCC-DSCC)," in Power Electronics Conference (IPEC), 2010 International, Sapporo, 2010, pp. 2029-2036.

978-1-4673-9551-9/16 $31.00 © 2016 IEEE

A New Topology of Multilevel VSC Converter for Hybrid HVDC Transmission System

Jae-Jung Jung[*], Shenghui Cui[†], and Seung-Ki Sul[*]

[*]Department of Electrical and Computer Engineering, Seoul National University, Seoul, Korea

[†]E.ON Energy Research Center, RWTH Aachen University, Germany

Abstract— **In this paper, the existing Modular Multilevel Converter (MMC) topologies for Line Commutated Converter (LCC)-Voltage Source Converter(VSC) connected as a hybrid High Voltage DC (HVDC) transmission system are reviewed and a new topology of multilevel converter for a hybrid HVDC is introduced. Among the existing MMC topologies for the hybrid HVDC, an MMC structure consisted of Half-Bridge SubModule(HBSM)s and Full-Bridge SubModule(FBSM)s has characteristics such as reduced system cost, low operation loss, but still keeping capability to cope with DC short circuit fault. However, it is very difficult for the conventional hybrid MMC structure, where each arm of MMC is consisted of mixed HBSM and FBSM, to balance the submodule capacitor voltages under sliding of DC bus voltage. For solving the defect of the conventional structure of MMC, an asymmetric MMC, where in a leg an arm is consisted of HBSM and other arm of FBSM, is devised. The proposed asymmetric MMC can regulate the DC bus voltage freely without uncontrollable submodule capacitor voltages. The problems of the conventional structure of MMC and the validity of asymmetric MMC are verified by both computer simulation and experiment results.**

Keywords—asymmetric mixed MMC; hybrid hvdc; high voltage dc transmission system; hybrid mmc; modular multilevel converter; submodule balancing

I. INTRODUCTION

For decades, a Line Commutated Converter (LCC) based HVDC has been developed and applied to most of HVDC transmission system. And nowadays, most HVDC systems in commercial operation employ the LCCs. It has several advantages such as higher reliability, higher power capability, excellent overload capability, and higher efficiency. However, it presents some drawbacks such as strong AC grid requirements, larger system size for harmonic and reactive filters, and lack of black starting capability. On the other hand, IGBT based Voltage Sourced Converter (VSC) has been recently developed, and it can cover the above demerits of LCC schemes. Among the VSC technologies, Modular Multilevel Converter (MMC) is a promising and competitive technology over two- or three-level VSC topologies. MMC presents many advantages such as very low harmonics, low dv/dt, modularity and simple scaling, high reliability and low switching loss, no necessity of series connection of power semiconductors, and the DC bus capacitor elimination, etc. Two types of submodule could be used in the MMC, which are a half-bridge chopper module, and a full-bridge inverter module. The Half-Bridge SubModule (HBSM) has been used widely to reduce the number of switching devices in a module and conduction loss. While, the output voltage of HBSM is confined to two levels, namely, null and DC link voltage of each module. While, that of Full-Bridge SubModule (FBSM) includes negative of DC link voltage, and the FBSM can synthesize AC output voltage even in the case that DC transmission voltage drops down to null due to DC short circuit faults and etc.

In some application, massive electricity may be transmitted from a strong AC grid, which consists of several large power plants, to several distributed power loads. In this case, the compactness and black starting capability in sending side may be not a crucial concern because of inherent large power plants at the site of HVDC converter. And, LCC type HVDC converter would be the best option at the sending side because of its technical maturity and higher operating efficiency. But in the receiving side, compact structure and black starting capability cannot be traded if the distributed loads are at the city centers or off-shore platforms. In such an application, a hybrid HVDC structure which contains a high power LCC-HVDC converter in source side and several medium power VSC-HVDC converters in distributed load sides would be a promising solution. For this reason, there have been several researches to accommodate LCC and VSC simultaneously in a HVDC transmission system, which is so called as a hybrid HVDC transmission system [1-3]. In recent years, a hybrid HVDC transmission system with LCC and MMC has become the most compatible candidate for future flexible DC transmission system. Since this configuration would combine the merits of the LCC as a single end in a large site and MMCs as distributed ends which have a compact structure, MMC could be installed in the distributed wind farms or offshore oil platforms, forming a multi-terminal system [4-5].

In this paper, the existing MMC topologies proposed in several literatures [4-9] are reviewed and the validity of them for hybrid HVDC transmission system is addressed. And, based on the results of review, a new MMC topology is proposed. It has the advantages such as cost saving and low loss, and it can also regulate the DC bus voltage quite freely and quickly for adjusting the DC transmission power in hybrid HVDC system. Also, it can deal with the black starting and DC fault ride through. Finally, the full scaled computer simulation studies and experiment results with scaled version of the proposed MMC topology in laboratory are provided to support the validity of the proposed topology.

978-1-4673-9551-9/16 $31.00 © 2016 IEEE

II. HYBRID HVDC CONFIGURATION AND FUNDAMENTAL PRINCIPLES

The basic structure of hybrid HVDC is shown in Fig. 1. The sending end is the conventional LCC based HVDC system. The receiving end is a VSC based HVDC system. In this paper, it is assumed that the LCC-HVDC system controls the DC current as constant and the VSC-HVDC system controls the DC bus voltage for regulating DC transmission power. The VSC on the receiving end has the turn-off capability. The VSC can maintain voltage and frequency stability in independence on the AC grid. So, active and reactive power may be controlled independently by VSC. The VSC not only requires no reactive power from AC grid but also can operate as STATCOM to compensate reactive power dynamically. This is to say, if VSC has enough capacity the hybrid HVDC can supply active power and reactive power to improve voltage and power angle stability when a fault occurs. No need for enhanced short circuit capacity takes place in the AC grid since the AC current of VSC is controllable. This is said that the relay protection does not need to revise after VSC-HVDC lines are added. Many VSCs may connect with a fixed polarity DC bus. It is easy to comprise multi-terminal HVDC

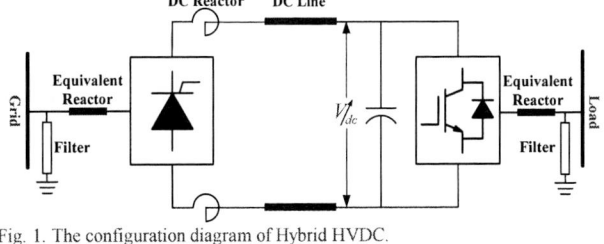

Fig. 1. The configuration diagram of Hybrid HVDC.

system that has the same topology with AC system and has flexible operating patterns. The sending converter adopts the conventional current source converter based HVDC rectifier system with perfect and mature technology and relatively at low cost. Therefore, the hybrid HVDC transmission system features both the well-developed technology and lower cost of LCC-HVDC and desirable regulating characteristics of VSC-HVDC. The candidates of the VSC-HVDC system can be two- or three-level converter, modular multilevel converter, and several modified modular structure based VSC high power converters.

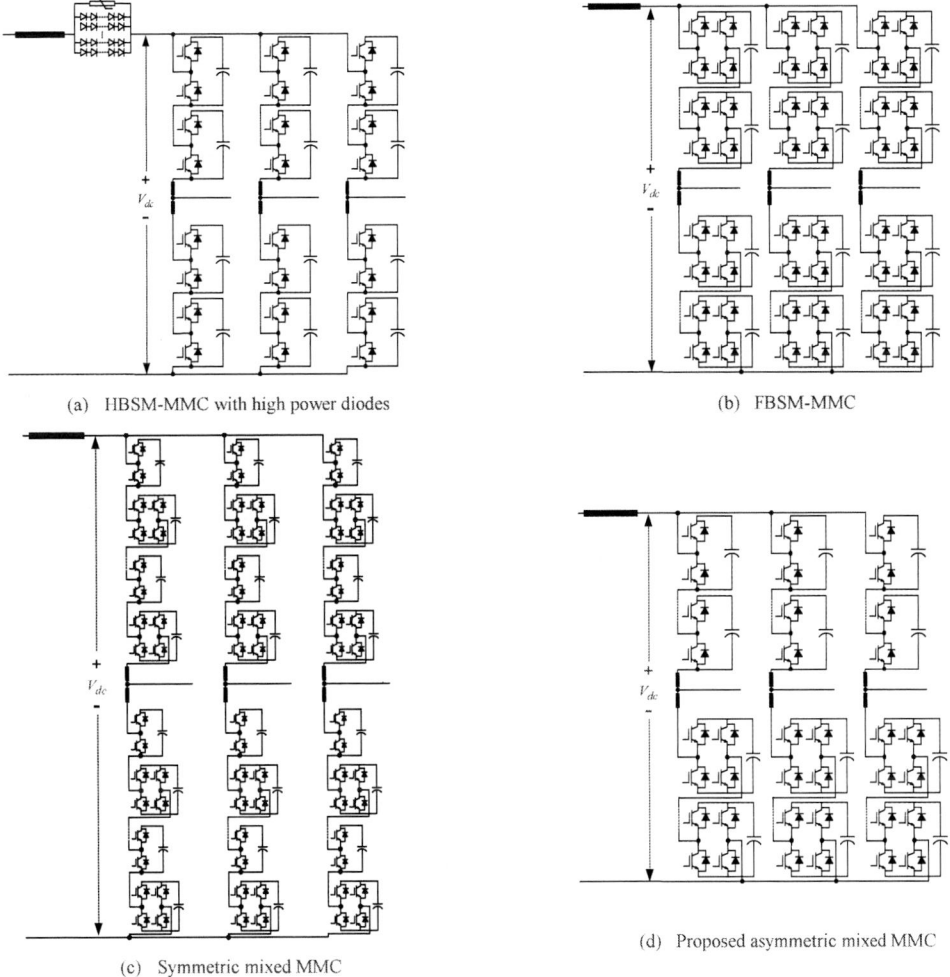

(a) HBSM-MMC with high power diodes

(b) FBSM-MMC

(c) Symmetric mixed MMC

(d) Proposed asymmetric mixed MMC

Fig. 2. The circuit diagrams of multilevel VSC converter topologies for hybrid HVDC transmission system.

III. Voltage Source Converter Topologies for Hybrid HVDC Transmission System

The HBSM-MMC could not ride through the DC short circuit fault without introducing AC or DC breakers. The circuit diagram of HBSM-MMC with high power diodes is shown like Fig. 2(a) [5]. A simple but effective method to block the DC fault current paths is to install the high power diodes in the overhead lines close to the MMC-based converters. The high power diodes consist of many rectifier diodes in series to withstand reverse blocking voltage during DC line faults. Given that the power flow of the proposed hybrid HVDC system is unidirectional, the high power diodes keep in conducting state in normal operation. However, the high power diode always causes not only the conduction loss in normal operation as well as fault operation, and but also additional installation and cost issue. Furthermore, the system has a disadvantage that DC bus voltage cannot be changed widely to vary the DC transmission power quantity because the HBSM-MMC has the narrow DC voltage variation abilities that is within output voltage margin range, near +Vdc (1p.u.).

The FBSM-MMC shown like Fig. 2(b) can make the DC side voltage from -1 p.u. to +1 p.u. by modulating output voltage of the FBSM without any restriction. So, it is very easy for FBSM-MMC to deal the black starting, power flow variation, and DC short circuit fault ride through. When DC short fault occurs, each submodule operates like a normal full-bridge and the DC bus voltage is synthesized as zero to clear the short circuit current of DC transmission line. Since a full-bridge inverter can output bipolar voltage, the converter can generate back-EMF to regulate AC side current during fault. Even though a full-bridge inverter can be modulated in bipolar or unipolar mode, to minimize switching loss, one of the lower switches should be normally ON and the other corresponding complementary switch should be normally OFF during normal operation. Then during normal operation, the full-bridge inverter based MMC operates like a conventional half-bridge chopper based MMC. As similar with aforementioned HBSM-MMC with high power diodes, the FBSM-MMC also causes extra loss while normal operation mode for most of its life.

Fig. 2(c) describes the circuit configuration of mixed HBSM and FBSM based MMC [6-8]. When the MMC operates at normal mode, the DC bus voltage is commonly

synthesized as +Vdc, and when a DC short circuit fault occurs the DC bus voltage should be synthesized as 0. Then, in fact, a half of the DC bus voltage output capability is redundant. However, it should be noted that the FBSM based MMC can output back-EMF to regulate AC side current while the DC bus is synthesized from –Vdc (-1p.u.) to +Vdc (+1p.u.). Meanwhile, in case of hybrid HVDC transmission system, the FBSM-MMC has the ability of power flow reversal. In conventional and practical operation of hybrid HVDC transmission system, it is more general that the power flow is unidirectional from LCC-HVDC to VSC-HVDC system. Therefore, a structure of mixed HBSM and FBSM can take full advantage of converter output voltage capability in the hybrid HVDC transmission system. During normal operating condition, the symmetric mixed converter operates like a conventional HBSM-MMC, and during DC short circuit fault the half-bridge choppers are bypassed and the converter operates like a full-bridge based MMC to ride through the fault. Especially, in case of hybrid HVDC transmission system, the amount of DC transmission power can be regulated and varied by controlling the DC bus voltage from 0 to +Vdc (+1p.u.).

Contrary to existing structure, where equal number of HBSMs and FBSMs are used in an arm of each leg of MMC, a circuit shown in Fig. 2(d) can be considered as another option, where in a leg one arm is consisted of fully HBSMs and the other of fully FBSMs. Based on the configuration of a leg of MMC, the former one can be called as symmetric mixed MMC and the later one as asymmetric mixed MMC. In Fig. 2(d), upper arm is made up of series connection of HBSMs and lower arm consists of series connection of FBSMs in one case of two compositions of asymmetric mixed MMC. The number of HBSMs and FBSMs are the same in both mixed MMCs as described in Fig. 2(c) and (d). In the proposed asymmetric mixed MMC shown in Fig. 2(d), the upper arm which is composed of HBSMs operates like as normal mode of HBSM-MMC in both normal operation and fault operation. The lower arm which is composed of FBSMs operates like HBSM-MMC in normal mode and operates like FBSM-MMC in fault mode and also in the low power transmission mode which have lower amount of DC power than 1p.u. Like the symmetric mixed MMC in Fig. 2(c), the DC transmission power can be regulated and varied by controlling the DC bus voltage from 0 to +Vdc

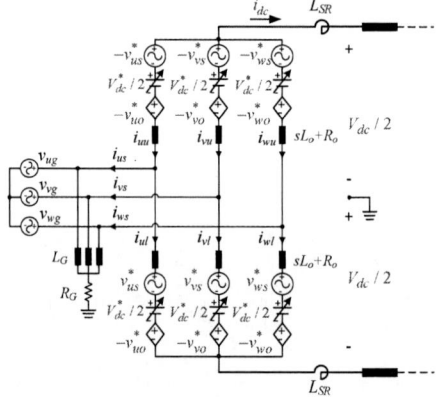

Fig. 3. Modeling of the symmetric mixed MMC

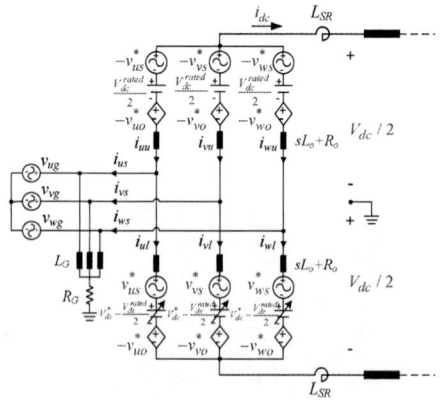

Fig. 4. Modeling of the asymmetric mixed MMC.

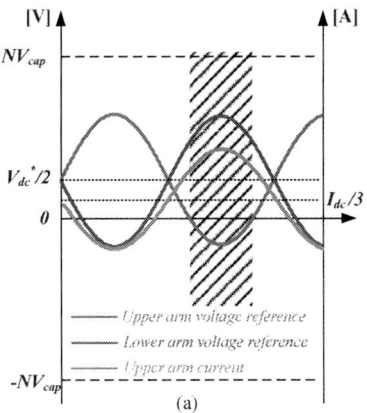

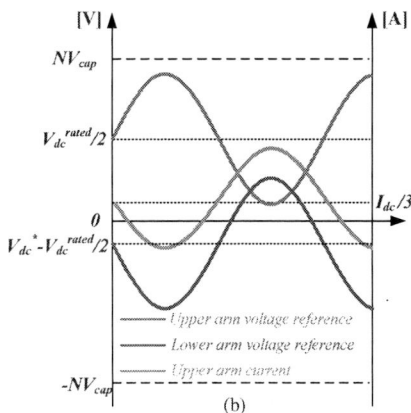

Fig. 5. The arm voltage references and arm current of (a) symmetric mixed MMC and (b) asymmetric mixed MMC when the DC bus voltage is lower than V_{dc}^{rated} (+1p.u.) for example in this figure, $V_{dc}^* = V_{dc}^{rated} / 2$.

(+1p.u.). The advantages and characteristics of asymmetric mixed MMC will be described in the subsequent sections compared to symmetric mixed MMC.

IV. COMPARISONS OF SYMMETRIC AND ASYMMETRIC MIXED MMC IN HYBRID HVDC TRANSMISSION SYSTEM

A. Operation Principles of the Symmetric mixed MMC

Modeling of the symmetric mixed MMC with the closed-loop indirect modulation [10] is shown in Fig. 3(a). The reference voltages (v_{xu}^* , v_{xl}^*) of upper and lower arm are derived by (1), where V_{dc}^* is DC bus voltage reference, v_{xs}^* is the phase voltage reference and x denotes a phase among u, v, and w. The leg internal voltage (v_{xo}^*) associated with circulating current for system balancing [11]. The leg internal voltage is omitted for simplification, by reason of its negligible magnitude as compared with that of DC bus voltage and phase voltage reference.

$$ v_{xu}^* = \frac{V_{dc}^*}{2} - v_{xs}^*, \qquad v_{xl}^* = \frac{V_{dc}^*}{2} + v_{xs}^*. \qquad (1) $$

Under the assumption that the number of submodules in one arm is N, each arm is composed of $N/2$ HBSMs and $N/2$ FBSMs. So, the capable range of an arm output voltage is from $-NV_{cap} / 2$ to NV_{cap}, assuming the capacitor voltages of all submodule are the same as V_{cap}.

B. Operating Principles of the Asymmetric mixed MMC

In modeling of the asymmetric mixed MMC of Fig. 4, the arm voltage references are given by (2) where V_{dc}^{rated} means the rated DC bus voltage in condition of the rated power transmission. And, the leg internal voltage is also excluded in (2).

$$ v_{xP}^* = \frac{V_{dc}^{rated}}{2} - v_{xs}^*, \qquad v_{xN}^* = (V_{dc}^* - \frac{V_{dc}^{rated}}{2}) + v_{xs}^*. \qquad (2) $$

In Fig. 2(d) and Fig. 4, the upper arm consists of N HBSMs and the lower arm is composed of N FBSMs. The capable range of an upper arm output voltage is from 0 to NV_{cap}, and that of an lower arm output voltage is from $-NV_{cap}$ to NV_{cap}. And, the upper arm voltage is always positive, and the lower arm voltage can be negative as well as positive.

C. Characteristics of the Symmetric and Asymmetric mixed MMC in DC Transmission Power Variation Mode in Hybrid HVDC System

In the case of the conventional hybrid HVDC system, LCC based HVDC system is sending end which operates as rectifier mode and VSC based HVDC system is receiving end which operates inverter mode [1-2]. The VSC based HVDC system regulates the DC bus voltage in case of point-to-point hybrid HVDC transmission system. The system operated as the DC bus voltage regulation mode can be named as voltage regulator. In conventional point-to-point hybrid HVDC system, LCC-HVDC controls the DC bus current to be constant and VSC-HVDC determines the quantity of the DC transmission power by regulating the DC bus voltage. Therefore, for controlling the DC bus voltage, the MMC-HVDC system should have the ability to regulate the DC bus voltage.

The arm voltage references and arm current of the symmetric mixed MMC while the DC voltage is lower than the rated DC bus voltage are depicted in Fig. 5(a). In this figure, as an example, $V_{dc}^* = V_{dc}^{rated} / 2$. Because the DC bus voltage is lower than the rated voltage, the arm voltage references have the negative value within the region which is marked in Fig. 5(a). If the upper arm voltage is positive, all submodules operate like the half-bridge chopper cells by unipolar mode of FBSMs. While the upper arm voltage is negative like the marked section of Fig. 5(a), all HBSMs of the upper arm are bypassed, and FBSMs in upper arm operate

TABLE I. COMPARISONS BETWEEN FOUR DIFFERENT MMC TOPOLOGIES FOR HYBRID HVDC TRANSMISSION SYSTEM.

	HBSM-MMC with high power diodes	FBSM-MMC	Symmetric mixed MMC	Asymmetric mixed MMC
Cells per arm	N	N	N	N
IGBTs per arm	2N	4N	3N	3N
DC short circuit fault blocking	Yes	Yes	Yes	Yes
DC fault ride through as a STATCOM	Yes	Yes	Yes	Yes
Conducting IGBTs per leg	2N	4N	3N	3N
Conduction loss	Low (when using low loss model of high power diodes)	High	Medium	Medium
DC power flow variation capability	No (Limited range around 1p.u.)	Yes (-1p.u.~1p.u.)	Yes (when 0 and 1p.u.) Difficult or imperfect (when 0<P_{dc}<1p.u.)	Yes (0~1p.u.)
Submodule balancing in power variation mode	Not available	Good	Not good	Good

in bipolar mode and make the negative arm voltage. Assuming that the power factor is unity and the arm voltage is negative and the arm current is positive as shown like Fig. 5(a), the

lower than the rated DC bus voltage can be depicted as Fig. 5(b). Again, in this figure, as an example, $V_{dc}^* = V_{dc}^{rated}/2$. As

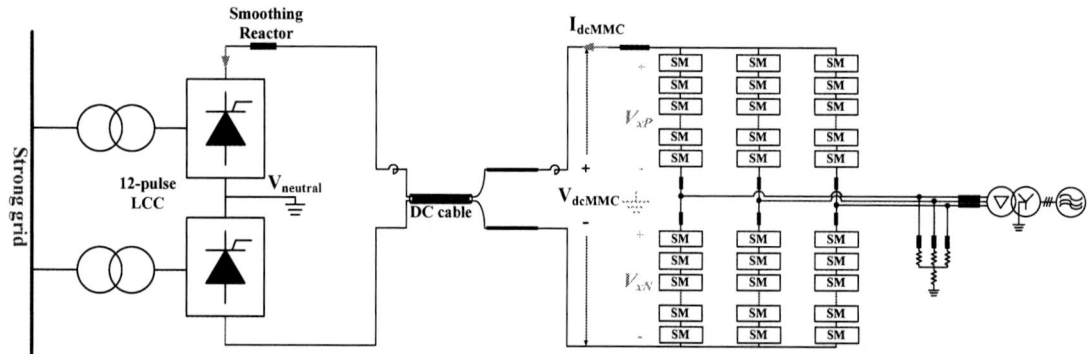

Fig. 6. Schematic diagram of the simulated hybrid HVDC transmission system with symmetric and asymmetric mixed MMCs.

capacitor voltages of FBSMs of upper arm decrease, whereas those of HBSMs of upper arm are unchanged, in accordance with the aforementioned modulation principle. Therefore, the voltage difference between FBSMs and HBSMs in the upper arm is getting larger in the interval during the arm voltage is negative. And the same difference may also happen in the lower arm when the arm voltage is negative and the arm current is positive. In this case, the capacitor voltages of FBSMs of lower arm decrease, whereas those of HBSMs of lower arm are unchanged. On the other hand, the voltage difference between FBSMs and HBSMs is getting smaller in the region during the arm voltage is positive, because FBSMs operate like half-bridge chopper module and thus the cell balancing algorithm applies equally to all submodules. However, as the interval of negative arm voltage is getting larger and the arm current is getting larger, the unbalance of the symmetric mixed MMC in each arm becomes worse and the system would be stalled finally.

The arm voltage references and arm current in case of the asymmetric mixed MMC system while the DC voltage is

shown like (2), the DC component of the upper arm voltage reference is fixed as a half of the rated DC bus voltage. While, the lower arm voltage reference determines the DC bus voltage and thus the DC component of the lower arm voltage reference can vary between $-V_{dc}^{rated}/2$ and $V_{dc}^{rated}/2$. The cell balancing algorithm applies equally to HBSMs in upper arm and FBSMs in lower arm respectively. Hence, the voltage unbalance between submodules cannot occur obviously.

Taking consideration of other MMC topologies, the characteristics of those can be summarized as described in Table I. And the intensive comparison and verification by simulation and experimental results between the symmetric and asymmetric mixed MMC are described in the next Section V and Section VI.

V. FULL SCALE SIMULATION RESULTS

A 400MVA MMC model has been established by using full scale computer simulations based on PSIM software. The schematic of the simulated system is illustrated in Fig. 6. The number of submodules per arm is 216, the rated DC bus voltage is 400kV, and the rated submodule capacitor voltage is 2200V. The detailed parameters for the simulation can be

TABLE II. PARAMETERS OF THE SIMULATED SYSTEM.

Quantity	Values
Grid line-to-line voltage (LCC side)	144 kV
Leakage Inductance of transformers (LCC)	13.8 mH (0.15p.u)
Inductance of Smoothing Reactor (LCC)	150 mH
Transmission line inductance	20 mH
Number of submodules per Arm	216
Rated DC bus voltage	400kV
Rated DC bus current	1500 A
Rated module capacitor voltage	2.2 kV
Capacitance of module capacitor	4.5 mF
Inductance of arm inductor	15.0 mH
Resistance of arm inductor	367.0 mΩ
Sampling frequency (MMC)	10.0 kHz
Rated MMC output voltage	180.5 kV

referred to Table II.

Fig. 7 shows the simulation results of the symmetric mixed MMC when the DC bus voltage is changed from 400kV (1 p.u.) to 140kV (0.35 p.u.) at 1.5s. And, the current is regulated to be -1500A as constant. As shown like the 3rd trace of Fig. 7, the DC components of the upper and lower arm voltage references are changed from 200kV to 70kV as aforementioned the arm voltage references like (1). The 5th and 6th traces of Fig. 7 show the averages of upper and lower arm HBSMs and FBSMs capacitor voltages of u-phase, respectively. When the DC bus voltage is the rated 400kV, all cell voltages are well balanced. However, when the DC bus voltage is reduced down to 140kV, the cell voltages between HBSMs and FBSMs are unbalanced each other. Fig. 8 shows the magnified waveforms of the section between 1.6s and 1.7s. From the moment that arm voltage reference becomes negative, the difference between HBSM and FBSM capacitor voltages starts to increase. Because the arm current is positive when the arm voltage is negative, capacitor voltages in FBSMs are discharged. When the arm voltage is positive, the cell balancing algorithm applies equally to all submodules. So, the difference between HBSM and FBSM voltages decreases, and the unbalance is resolved finally within one period.

Fig. 9 shows the simulation results when the DC transmission power decreases from 1 p.u. to 0.25 p.u. Because the DC bus voltage is stepped down to 100kV, the region where the arm voltage is negative is enlarged. The unbalance becomes more severe and the unbalance is not resolved even during the region of positive arm voltage. In other words, the discharged energy in FBSMs is larger than charged energy during one fundamental period. Hence, energy in FBSMs is depleted and thus the system is diverged finally. As shown like results of Fig. 9, it may have a severe effect on the stability and result in control and capacitor sizing issues of converter systems.

On the other hand, the simulation result of asymmetric MMC system is shown as Fig. 10. Because the upper and lower

arm voltage references follow (2), the upper arm voltage reference is always positive and the lower one regulates the DC component for producing exact DC bus voltage reference as shown by 3rd trace of the figure. From 5th trace of Fig. 10, the

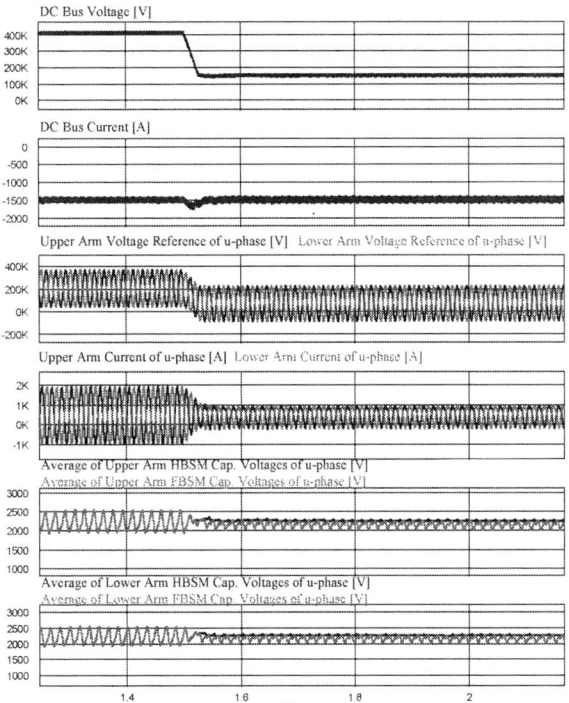

Fig. 7. Simulation results of the symmetric mixed MMC when the DC transmission power decreases from 1 p.u. to 0.35 p.u. by regulating the DC bus voltage.

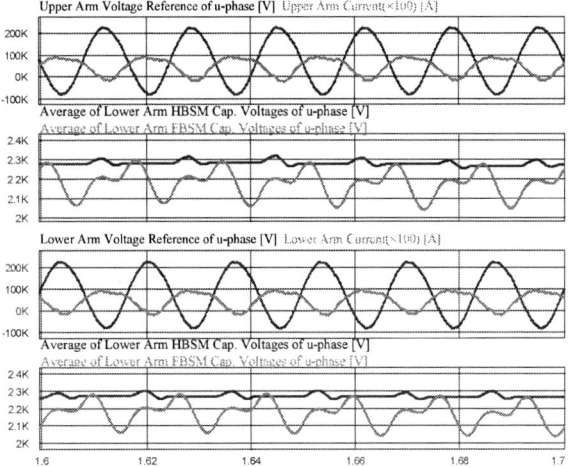

Fig. 8. Magnified waveforms of the section between 1.6s and 1.7s in Fig. 6.

submodule capacitor voltages in both upper and lower arm are well regulated below the allowable bound. Therefore, for regulating variable DC transmission power, the proposed asymmetric mixed MMC topology would be a promising option for LCC-VSC hybrid HVDC system.

VI. EXPERIMENT RESULTS

The comparison of the symmetric and asymmetric mixed MMC and the validity of the asymmetric mixed MMC are verified by the 10kVA reduced scale prototype hybrid HVDC system shown in Fig. 11(a). The number of cells in each arm, N, is 6, so there are a total of 36 cells used for the three-phase

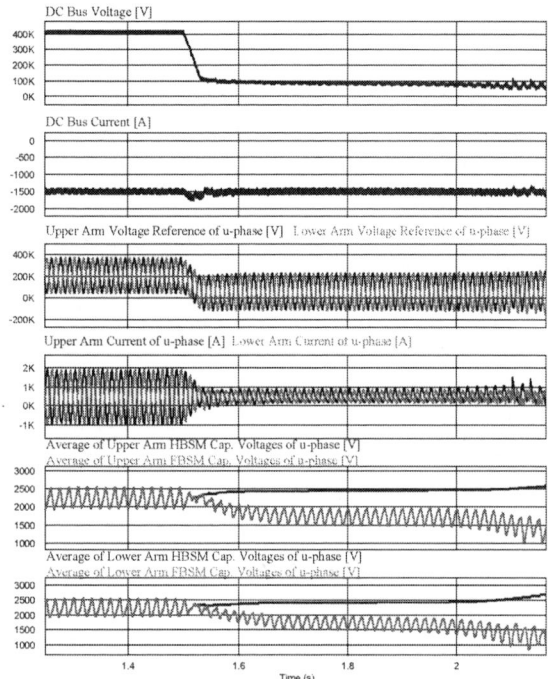

Fig. 9. Simulation results of the symmetric mixed MMC when the DC transmission power decreases from 1 p.u. to 0.25 p.u. by regulating the DC bus voltage.

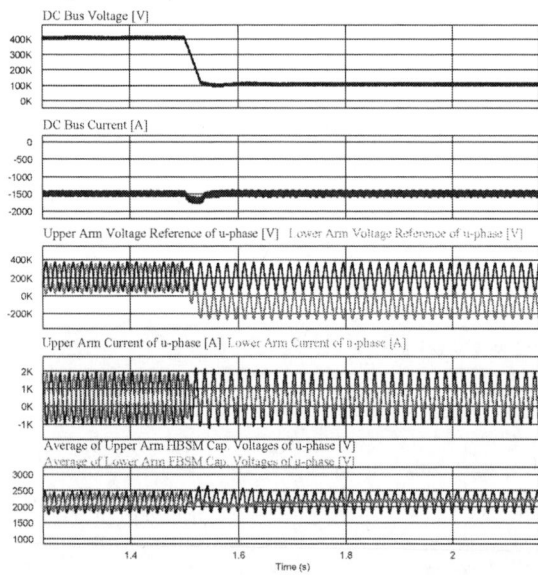

Fig. 10. Simulation results of the asymmetric mixed MMC when the DC transmission power decreases from 1 p.u. to 0.25 p.u. by regulating the DC bus voltage.

system. The parameters for the test setup are given in Table III. LCC is emulated by the full-bridge chopper in DC link of 2-level converter as shown like Fig. 11(b).

Fig. 12 shows the experiment results of the symmetric mixed MMC when the DC bus voltage decreases from 300V (1 p.u.) to 150V (0.5 p.u.) and 75V (0.25 p.u.) and the DC bus current is regulated as -5A constantly. As shown like the upper and lower arm voltage reference in Fig. 12, the DC components of the arm voltage references are changed from 150V to 75V and 37.5V as the aforementioned arm voltage references like (1). When the DC bus voltage is the rated 300V, all cell voltages are well balanced. However, when the DC bus voltage is 150V and 75V, the cell voltages between HBSMs and FBSMs are unbalanced each other.

From the moment when arm voltage reference is negative, the difference of HBSM and FBSM capacitor voltages begins to increase. When the DC voltage is 75V, the unbalance is more conspicuous.

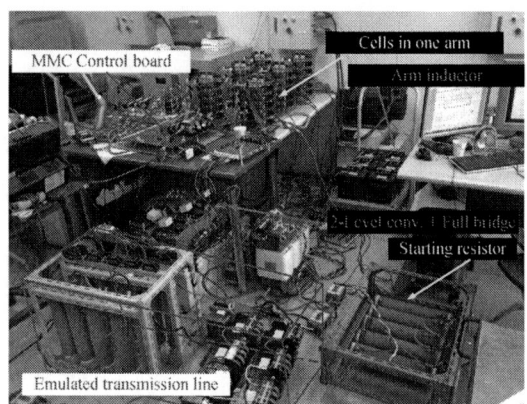

(a)

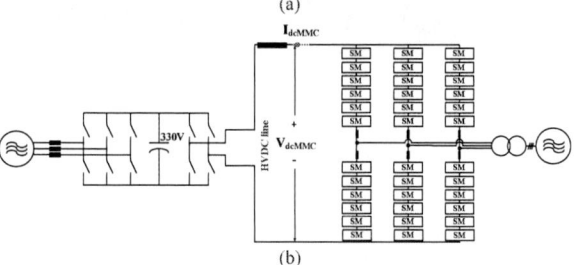

(b)

Fig. 11. Experimental setup: (a) a down scaled prototype hybrid HVDC system and (b) schematic diagram of experimental setup.

TABLE III. PARAMETERS OF PROTOTYPE HYBRID HVDC SYSTEM.

Quantity	Values
Transmission line inductance	8 mH
Transmission line resistance	0.5 Ω
Number of full-bridge modules per Arm	6
Rated DC bus voltage	300 V
Rated DC bus current	30 A
Rated module capacitor voltage	50 V
Capacitance of module capacitor	5.4 mF
Inductance of arm inductor	4 mH
Resistance of arm inductor	5 mΩ
Sampling frequency (MMC)	10.0 kHz
Rated MMC output voltage (line-to-line rms)	140 V

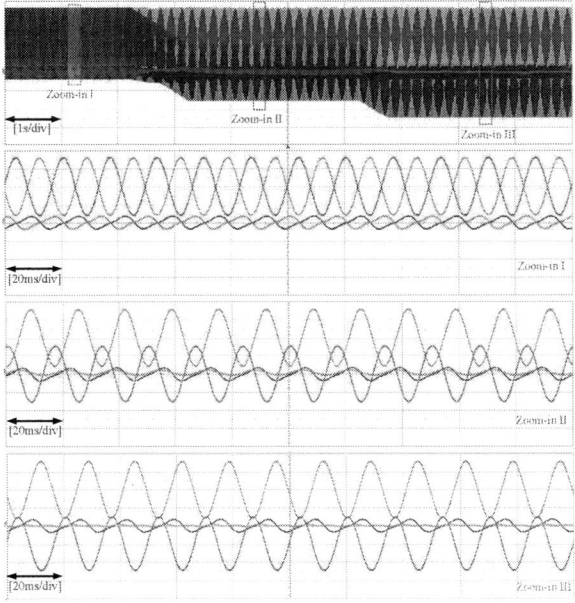

Fig. 12. Experiment results of the symmetric mixed MMC when the DC bus voltage decreases from 300V (1 p.u.) to 150V (0.5 p.u.) and to 75V (0.25 p.u.) and DC bus current is -5A as constant.

Fig. 13. Experiment results of thea symmetric mixed MMC when the DC bus voltage transfers from 300V (1 p.u.) to 150V (0.5 p.u.) and 75V (0.25 p.u.) and DC bus current is -5A as constant.

On the other hand, Fig. 13 shows the experiment results of the asymmetric mixed MMC when the DC bus voltage decreases from 300V to 150V and 75V and the DC bus current is controlled as -5A constantly. As the upper and lower arm voltage references follow (2), the upper arm voltage is modulated as positive and the lower arm voltage is modulated

for producing exact DC bus voltage. As shown in Fig. 13, the submodule capacitor voltages in both upper and lower arm are well controlled within allowable bound.

Fig. 14 and Fig. 15 show the experimental results in condition that the DC current reference of emulated LCC system is changed from -10A to -15A rapidly. In Fig. 14, after the moment of DC current change, the symmetric mixed MMC system becomes unstable and the system is stalled eventually. The capacitor voltages in FBSMs could not follow the reference value, 50V and energy in FBSMs was exhausted. However, the asymmetric mixed MMC system is stable even after DC current change from -10A to -15A as shown in Fig. 15. The reduced scale experimental results also support the validity of asymmetric mixed MMC topology in LCC-VSC hybrid HVDC transmission system.

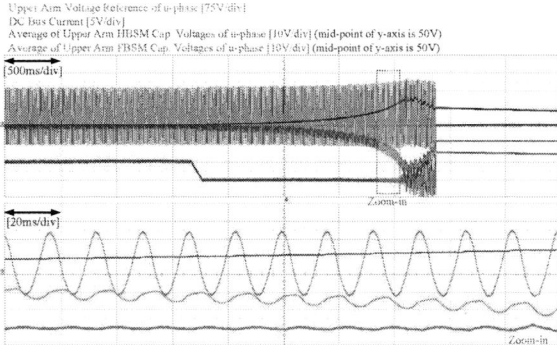

Fig. 14. Experiment results of the symmetric mixed MMC when the DC current is changed from -10A to -15A and the DC bus voltage is 75V (0.25 p.u.) as constant.

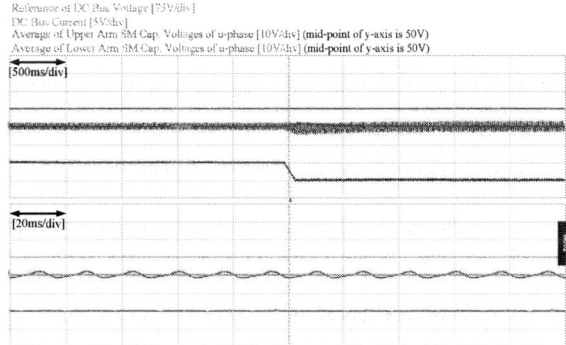

Fig. 15. Experiment results of the asymmetric mixed MMC when the DC current is changed from -10A to -15A and the DC bus voltage is 75V (0.25 p.u.) as constant.

VII. CONCLUSIONS

In this paper, the existing multilevel MMC topologies for LCC-VSC connected hybrid HVDC transmission system are reviewed and an asymmetric mixed MMC topology has been introduced. It has characteristics such as reduced system cost, less operational loss, and ability to cope with DC short circuit fault, as same with conventional symmetric mixed MMC. As contrast with the conventional mixed MMC, the asymmetric mixed MMC for hybrid HVDC transmission system can regulate the DC bus voltage freely without uncontrollable submodule capacitor voltages. The drawbacks of the conventional symmetric mixed MMC for hybrid HVDC have been identified and the validity of the asymmetric MMC system has been supported by the simulation studies and scaled version experimental results.

REFERENCES

[1] Z. Zhao and M. R. Iravani, "Application of GTO voltage source inverter in a hybrid HVDC link," Power Delivery, IEEE Transactions on, vol. 9, pp. 369-377, 1994.

[2] R. E. Torres-Olguin, M. Molinas, and T. M. Undeland, "A controller in d-q synchronous reference frame for hybrid HVDC transmission system," in Power Electronics Conference (IPEC), 2010 International, 2010, pp. 376-383.

[3] J.-H. Ying, H. Duchen, M. Karlsson, L. Ronstrom, and B. Abrahamsson, "HVDC with voltage source converters - a powerful standby black start facility," in Transmission and Distribution Conference and Exposition, 2008. T&D. IEEE/PES, 2008, pp. 1-9.

[4] Y. Lee, S. Cui, S. Kim, and S.-K. Sul, "Control of hybrid HVDC transmission system with LCC and FB-MMC," Energy Conversion Congress and Exposition (ECCE), 2014 IEEE , vol., no., pp.475,482, 14-18 Sept. 2014.

[5] G. Tang and Z. Xu, "A LCC and MMC hybrid HVDC topology with DC line fault clearance capability," International Journal of Electrical Power & Energy Systems, vol. 62, pp. 419-428, 11// 2014.

[6] S. Inoue, and S. Katosh, "Modular multilevel converter with DC fault protection," European patent application, Jun. 12, 2013.

[7] S. Cui, S. Kim, J.-J. Jung, and S.-K. Sul, "Principle, control and comparison of modular multilevel converters (MMCs) with DC short circuit fault ride-through capability," in Applied Power Electronics Conference and Exposition (APEC), 2014 Twenty-Ninth Annual IEEE , vol., no., pp.610-616, 16-20 March 2014.

[8] R. Zeng, L. Xu, L. Yao, "An improved modular multilevel converter with DC fault blocking capability," in PES General Meeting | Conference & Exposition, 2014 IEEE , vol., no., pp.1-5, 27-31 July 2014.

[9] R. Marquardt, "Modular Multilevel Converter topologies with DC-Short circuit current limitation," Power Electronics and ECCE Asia (ICPE & ECCE), 2011 IEEE 8th International Conference on , vol., no., pp.1425,1431, May 30 2011-June 3 2011.

[10] S. Debnath, J. Qin, B. Bahrani, M. Saeedifard, and P. Barosa, "Operation, control, and applications of the modular multilevel converter: A review," in IEEE Trans. on Power Electronics, vol. 30, pp. 37-53, 2015.

[11] A. Antonopoulos, L. Anguist, and H. P. Nee, "On dynamics and voltage control of the modular multilevel converter," in EPE European Conference on Power Electronics and Applications, pp. 1-10, 2009.

Performance of Solid State Transformers under Imbalanced Loads in Distribution Systems

Tao Yang, Ronan Meere, Cathal O'Loughlin and Terence O'Donnell

Electricity Research Centre, School of Electrical, Electronic and Communications Engineering
University College Dublin
Belfield, Dublin 4, Ireland
Email: tao.yang@ucdconnect.ie

Abstract— **This paper will compare the performance of two solid state transformer (SST) topologies in terms of their ability to deal with load imbalance in a four wire power distribution system. The topologies studied consist of an SST with a basic three phase output stage with three separate decoupled phases, and a topology with a four leg inverter output stage. Experimental validation of the control approach for the four leg inverter has been performed. It is shown that both topologies can maintain output voltage balance in the face of significant load imbalance, although the SST with three decoupled phases is best at maintaining output voltage balance. However because of the larger number of switches used it has lower efficiency than the other topologies. The four leg inverter topology is also shown to maintain good output voltage balance while having a lower number of switches and lower loss. Hence this topology presents a better a compromise between performance and efficiency.**

Keywords—Comparison, Dual Active Bridge, Faults, MF Transformer, Imbalanced Loads, Loss Calculation, SST.

I. INTRODUCTION

With the rapid development of power electronic technology in recent decades, the SST has received considerable interest as a potential replacement for the conventional transformer, especially for applications in traction systems, renewable energy distribution systems and smart grids. The SST has the potential to offer significant advantages over traditional low-frequency line transformers, with advanced power flow controllability, isolation, input-output decoupling and potential size reductions. Increases in power level and higher switching frequency in power electronic devices in recent decades has also motivated interest in the development of the SST for applications in the distribution system [1] [2] [3].

Imbalanced loads are quite common in the modern distribution grid with the growth of dynamically varying domestic loads and this imbalance is likely to increase in the future with the introduction of new loads such as electric vehicle charging points [4]. Imbalanced loads give rise to voltage imbalance in the network which results in problems such as the extra thermal loss due to the increased winding and core losses, and impacts negatively on system stability as discussed previously by Jouanne et al. [5]. In an SST-fed distribution system, it would be the responsibility of the SST to ensure that tolerances on phase voltage imbalance at its low voltage output terminals are met under imbalanced loading

conditions and that the efficiency of the SST is not adversely affected by such loading conditions. In contrast to the traditional low frequency transformer, the SST can in principle provide active voltage regulation to ensure output voltage balance and can decouple the output from the input so that imbalance does not propagate through the system. However to date this aspect of how an SST can deal with load imbalance has not been studied extensively. Very few publications have investigated the performance of the SST under imbalanced load conditions. Although Lai et al. [3] have proposed a three-phase SST with three single-phase modules that looked at system performance under power quality disturbances, including load transients and input voltage sag, the paper did not investigate performance under three-phase imbalanced loads with different degrees of imbalance. Alepuz et al. [6] have used a three-phase four-wire converter at the LV side to enable the SSTs to deal with imbalanced loading, however the paper did not investigate the efficiency of this approach with a quantitative comparison to other approaches. The question which this work addresses is how an SST can deal with load imbalance and what is the most suitable topology.

One commonly proposed approach for the low voltage side of the SST is the use of three single phase inverters, each fed from a separate DC link input, with the output connected in a Y-N connection to form a four wire system [3]. The four leg inverter is another suitable approach for dealing with unbalanced loads and has been proposed for application in motor drives and UPS [7]. In this work we compare the performance of these two different SST low voltage stage topologies under conditions of load imbalance. A suitable control approach for the four leg inverter is presented. Performance is compared in terms in the ability of the SST to maintain output voltage balance and the effect of imbalance on efficiency. The investigation is performed using Matlab/SimPowerSystems simulations with validation of the results for the four leg inverter performed on a 2 kVA three phase converter in hardware. The object of the study is to help determine the most suitable topology for handling the imbalance conditions considering the trade-off between system complexity and losses.

This work was supported in part by the UCD-Chinese Scholarship Council Scheme, in part by Research in Third Level Institutions (PRTLI), Cycle 5., and this work was conducted in the Electricity Research Centre, University College Dublin, Ireland, which is supported by the Electricity Research Centre's Industry Affiliates Programme (http://erc.ucd.ie/industry/). This material is based upon works supported by the Science Foundation Ireland, under Grant No. SFI/09/SRC/E1780.

II. METHODOLOGY

A. Overall SST Configuration

Similar to the traditional transformer, the SST provides a step-up or step-down voltage function, but with advanced functionality. The SST configuration used in this work is the three stage SST with medium voltage (MV) and low voltage (LV) DC links, which has been found to be one of the most suitable by Falcones et al. [8]. Fig.1. shows the basic configuration of the 3-stage SST consisting of an AC/DC rectifier, Dual Active Bridge (DAB) with a high frequency transformer and a DC/AC inverter. The rectifier converts a medium-voltage low-frequency three-phase AC input voltage into a medium DC voltage. The next step consists of a Dual Active Bridge (DAB) that transforms the medium DC voltage using a high frequency transformer, to a low voltage DC output. Finally, an inverter at the output stage converts the low DC voltage to a power frequency three-phase AC voltage connected to the load.

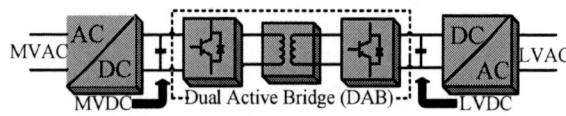

Fig.1. Basic configuration of three-stage SST

The comparison in this work is based on the example of a 330 kVA three-phase SST interfaced to an MV of 20 kV (line RMS) with a three-phase output of 380 V (line RMS). The AC-DC and DC-DC stages both have high input voltages which are beyond the rating capability of current commercial semiconductor switches. In order to deal with the high input voltage each phase rectifier and DAB consists of four series connected rectifier modules, connected to four input series output parallel connected DAB modules as shown in Fig.2.

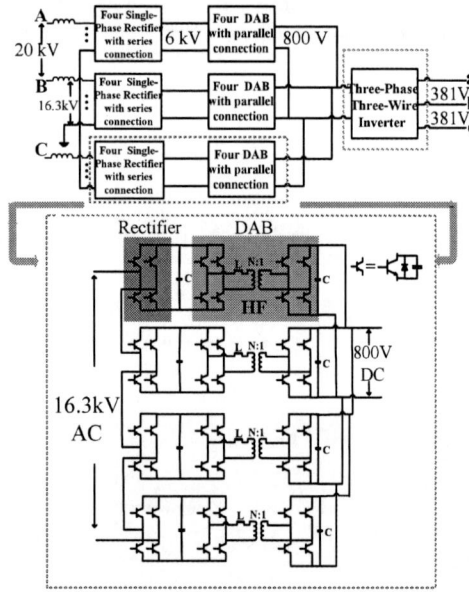

Fig. 2. Overall three-phase three-wire SST topology

The rectifier and DAB stages for each phase of SST configuration therefore consists of four series connected 16.3 kV input AC-DC rectifiers which each produce a 6 kV DC output, which is then connected to a DAB which converts the 6 kV to an 800 V DC output. The outputs from all four DAB modules are connected in parallel. Finally the inverter stage converts the 800 V DC to 50 Hz 311 V AC.

B. Two Topologies of Three Phase Four Wire SST

For investigating the ability of the SST to deal with imbalance in the four wire, low voltage network the topology and control of the output DC/AC stage is of most importance. Two approaches are investigated.

1) SST1: Three Phase Four Wire SST

The first two stages of the SST1 configuration are similar in design to the general SST. The output stage of SST1 consists of three single phase inverters, each fed from a separate 800 V DC input, with the output connected in a Y-N connection to form a four wire system, see Fig.3. The output voltage of each phase inverter is controlled independently as a single phase inverter. In order to avoid the steady state error associated with PI control in the natural reference frame, a single phase pseudo d-q control concept with an imaginary phase is used. Fig. 3 shows the control layout for the each inverter which is based on the work in [9]. The basic idea is to use the synchronous reference frame PI controller in an outer loop to regulate the output voltage, together with a simple inner loop capacitor current control with voltage-feed forward. In this dual-loop control strategy with voltage and current loop, the inner loop can use inductor current feedback or capacitor current feedback. The use of the capacitor current however helps improve the dynamic performance of the system in the presence of load disturbances and nonlinear loads [10]. As the capacitor is in parallel with the load, the load disturbance will directly lead to a change in capacitive current, and capacitor current feedback therefore provides a more direct response to the instantaneous load current changes. Sinusoidal pulse width modulation (SPWM) is used for the modulation in each phase.

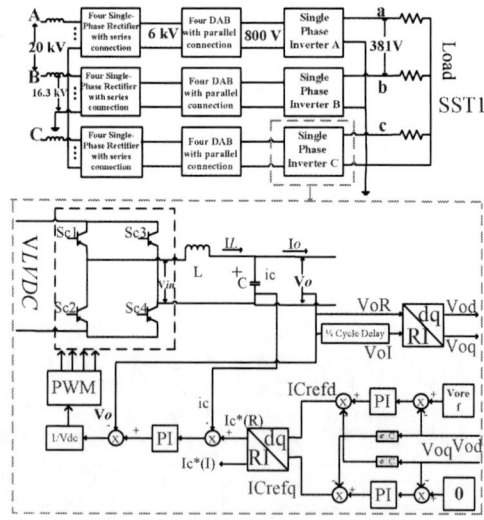

Fig. 3. Overall three-Phase four-Wire SST (SST1) topology

2) Three Phase Four Leg SST (SST2)

As seen in Fig.4 the SST2 design replaces the inverter stage with a three-phase four-leg inverter which uses a four-leg converter topology and ties the neutral point to the mid-point of the fourth neutral leg – see Zhang et al. [11] for further details.

For the three-phase four-leg SST2, the fourth leg is added to the conventional three-phase three-leg inverter to handle the neutral current, thus the conventional 2-D SPWM cannot be used in the three-phase four-leg inverter. An advanced three-dimensional space vector modulation (3-D SVPWM) is applied for the three-phase four-leg inverter in SST2, which has been described in detail in [11].

In general the imbalanced loads give rise to imbalanced load currents which result in voltage imbalance at the output of the SST. Conventionally, imbalanced 3-phase voltages can be represented by positive sequence, negative-sequence and zero-sequence components [12]. In the balanced case, only the positive-sequence components exist. The objective of the control loop is to maintain balanced output voltages, by regulating the positive sequence to the correct value, while eliminating the negative and zero-sequence. In the control approach for the four-leg inverter we attempt to control all three components separately. For traditional control of three-phase inverters, the rotating d-q-0 reference frame with anti-clockwise direction is used, which converts AC quantities into DC quantities. In order to deal with both positive and negative sequence components, Hsu et al. [13] have proposed anti-clockwise and clockwise rotating d-q reference frame control strategies to track the reference voltage for positive sequence voltages and to remove the negative sequence distortion due to the imbalanced loads. Note that Hsu's work dealt with a Δ-connection of the load so that zero-sequence components did not exist at the inverter output.

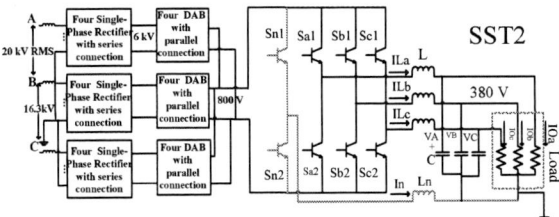

Fig. 4. Overall three-Phase four-leg SST (SST3) topology

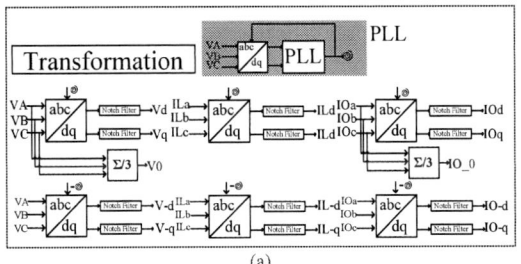

(a)

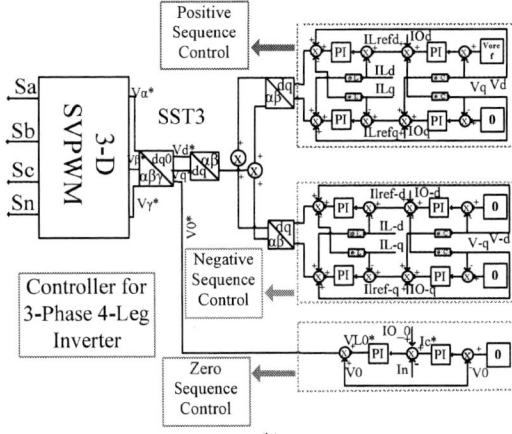

(b)

Fig. 5. Imbalance control for 3-Phase 4-Leg inverter

With this approach, both the positive and negative sequence components of the output will become DC quantities in an anti-clockwise and clockwise rotating d-q-0 reference frame respectively. With an appropriate controller in the negative sequence -d-q-0 reference frames, the negative sequence component can be regulated to zero and the imbalance eliminated.

Hsu's proposed control strategy has been found to be quite effective for three-phase inverters with a Δ-connection output by the experiments [13], however in this work we have a Y-connection of the load, which is also quite common in distribution systems. In this case Hsu's control strategy is unable to maintain balance among the output voltages, as both negative-sequence and zero-sequence components exists in Y-connected systems [14]. Thus for general three phase three wire SST with a Y-connection at the output side, we can only eliminate the negative-sequence distortion with Hsu's control strategy; the zero-sequence still exists in this system to affect its balance. Therefore, for the three-phase four-leg inverter in SST2 the addition of a control loop to handle the zero-sequence component is necessary to maintain balance.

As we can see from Fig.5, in SST2, this work adopts similar positive and negative sequence control which has been implemented in [13] by Hsu, but with the addition of zero-sequence control. All three sequences control have been implemented in the d-q-0 frame with P-I controllers. Equation (1) shows the general transformation from the a-b-c frame to the d-q-0 frame using the anticlockwise (+ω) rotating frame and (2) gives the resulting transformed positive, negative and zero-sequence components. Note under this transformation the positive-sequence components are dc quantities, the negative-sequence components rotate at 2ω and the zero-sequence rotate at ω. Note also that the positive and negative sequence components have both d and q-axis components, but the zero sequence has only got a component in the 0-axis. The coordinate transformation from the a-b-c coordinate to the anticlockwise (+ω) rotating d-q-0 coordinate is expressed in (3).

$$\begin{pmatrix} X_d \\ X_q \\ X_0 \end{pmatrix} = \frac{2}{3} \begin{bmatrix} \cos(\omega t) & \cos\left(\omega t - \frac{2}{3}\pi\right) & \cos\left(\omega t + \frac{2}{3}\pi\right) \\ -\sin(\omega t) & -\sin\left(\omega t - \frac{2}{3}\pi\right) & -\sin\left(\omega t + \frac{2}{3}\pi\right) \\ 1/2 & 1/2 & 1/2 \end{bmatrix} \cdot \begin{pmatrix} X_a \\ X_b \\ X_c \end{pmatrix} \quad (1)$$

$$\begin{pmatrix} V_{pd} \\ V_{pq} \\ V_{p0} \end{pmatrix} = \begin{pmatrix} V_p \cos\alpha_p \\ V_p \sin\alpha_p \\ 0 \end{pmatrix}, \begin{pmatrix} V_{nd} \\ V_{nq} \\ V_{n0} \end{pmatrix} = \begin{pmatrix} V_n \cos(2\omega t + \alpha_n) \\ V_n \sin(2\omega t + \alpha_n) \\ 0 \end{pmatrix},$$
$$\begin{pmatrix} V_{zd} \\ V_{zq} \\ V_{z0} \end{pmatrix} = \begin{pmatrix} 0 \\ 0 \\ V_z \cos(\omega t + \alpha_z) \end{pmatrix} \quad (2)$$

$$\begin{pmatrix} X_{-d} \\ X_{-q} \\ X_0 \end{pmatrix} = \frac{2}{3} \begin{bmatrix} \cos(\omega t) & \cos\left(\omega t + \frac{2}{3}\pi\right) & \cos\left(\omega t - \frac{2}{3}\pi\right) \\ -\sin(\omega t) & -\sin\left(\omega t + \frac{2}{3}\pi\right) & -\sin\left(\omega t - \frac{2}{3}\pi\right) \\ 1/2 & 1/2 & 1/2 \end{bmatrix} \cdot \begin{pmatrix} X_a \\ X_b \\ X_c \end{pmatrix} \quad (3)$$

$$\begin{pmatrix} V_{p-d} \\ V_{p-q} \\ V_{p-0} \end{pmatrix} = \begin{pmatrix} V_p \cos(2\omega t + \alpha_p) \\ V_p \sin(2\omega t + \alpha_p) \\ 0 \end{pmatrix}, \begin{pmatrix} V_{n-d} \\ V_{n-q} \\ V_{n-0} \end{pmatrix} = \begin{pmatrix} V_n \cos\alpha_n \\ V_n \sin\alpha_n \\ 0 \end{pmatrix},$$
$$\begin{pmatrix} V_{z-d} \\ V_{z-q} \\ V_{z-0} \end{pmatrix} = \begin{pmatrix} 0 \\ 0 \\ V_z \cos(\omega t + \alpha_z) \end{pmatrix} \quad (4)$$

Here, X_d, X_q, X_0 and X_a, X_b, X_c are the three-phase variable in d-q-0 and a-b-c respectively, ω is the line frequency for synchronized rotation. The coordinate transformation from the a-b-c coordinate to the clockwise ($-\omega$) rotating -d-(-q)-(-0) coordinate is expressed in (4). X_{-d}, X_{-q}, X_o are the three-phase variable in -d-(-q)-(-0) reference frame, V_{p-d}, V_{p-q}, V_{P-0}, V_{n-d}, V_{n-q}, V_{n-0}, V_{z-d}, V_{z-q}, V_{z-0} are the positive , negative and zero sequence components voltage in the clockwise d-q-0 frame respectively.

The positive and negative-sequence control, as seen in Fig.5 first converts the positive and negative-sequence quantities to the stationary DC d-q (Fig. 5(a) according to (2) and (4) respectively. A band-stop filter is used to remove the 2ω component, leaving only the relevant positive or negative sequence component. The second step (Fig. 5 (b)) uses an inner inductor current loop with feed-forward decoupling of inductor voltage and outer voltage loop with feed-forward decoupling of capacitor current with two closed loops to control the positive and negative sequence DC d-q quantities in the stationary frame.

The zero-sequence component exists independently in the 0-axis; therefore the control can be implemented independently in the 0-axis. As shown in the zero-sequence control blocks in Fig.5. We can eliminate and control the zero-sequence component distortions using two closed loops–an inductor current control loop and output voltage loop, with both strategies utilising direct PI control in the 0-axis.

III.   RESULTS

A. Validation of Four Leg Inverter Control

In order to verify the control approach of SST2 for unbalanced-load correction, a 220 V 2 kVA downscaled prototype of three-phase four-leg inverter has been implemented with the hardware in the loop real time simulation platform from Opal-RT. In this case the control algorithms and PWM are implemented in the OP5600 Series OPAL-RT simulators which are generated from the Matlab/Simulink models. The OP5600 Series OPAL-RT simulator also generates the firing pulses for the 4 leg inverter bridge which is based on the 8857-1 IGBT Chopper/Inverter from Lab-volt. This approach allows rapid testing of the control approach without building dedicated controllers. As the key stage for dealing with the unbalanced load for SST1 is the output stage, for the sake of simplicity, only the inverter stage of the SST is implemented in the experiment. Table I summarizes the main prototype parameters. Fig.6 shows the downscaled prototype of three-phase four-leg inverter.

TABLE I. PROTOTYPE PARAMETERS

Rated Power	6 kVA	*DC Bus voltage*	440 V
Load Resistors	25/39Ω	*Output Voltage*	220 V
Switching Frequency	2 kHz	*LC Filter*	60μF,150mH

Fig.6. The downscaled prototype

The imbalanced-load correction capability of SST2 is validated according to the experimental results shown in Fig.7. The waveforms of three phase output voltages and neutral current under the balanced-load state (25Ω, 25 Ω, 25 Ω) are shown in Fig. 7(a). Both of the three-phase voltages are balanced and there is no neutral current through the fourth leg. For the imbalanced load condition (25 Ω, 25 Ω, 39 Ω), as shown in Fig. 7(b), the balance of the three phase output voltages is successfully maintained. In this case the neutral current has increased from 0 to 3.5A as the zero-sequence current is carried by the fourth leg.

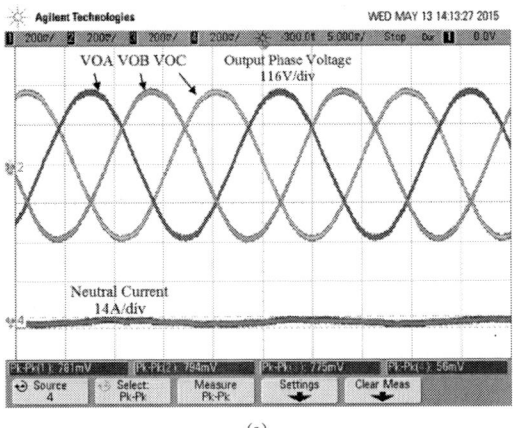

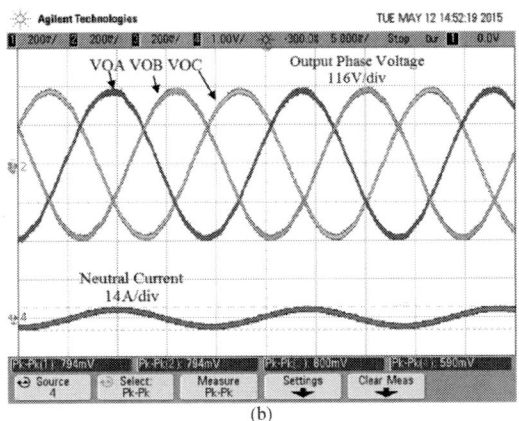

Fig. 7. Experimental results of SST2 under balanced and imbalanced load. (a) Under normal balanced loading conditions, (b) Under the 49% imbalanced loading conditions.

B. Comparison of SST Performance under Imbalanced Loads.

Having validated the control approach for the four leg inverter on the scaled hardware, the performance of the full scale SST configurations are now evaluated using the simulations to study their ability to handle imbalance and fault scenarios at the distribution side of the system. The three core aspects of the comparison concentrate on, ability to handle imbalance loads, fault handling, and efficiency. Note as regards efficiency this study only considers losses in the switches, it does not address transformer loss, losses in controller, gate drivers or other additional circuitry. The main SST parameters for the simulation study are given in Table II.

TABLE II. SIMULATION PARAMETERS

Input AC voltage	20 kV (RMS line-line)
MV filter inductor	25 mH
AC/DC switches	85 A/6500 V IGBTs
AC/DC switching frequency	1500 Hz
MV DC capacitors	750 uF
MV DC voltage	4*6000 V

DAB MV switches	85 A/6500 V IGBTs
DAB switching frequency	2500 Hz
Turn ratio of MF transformer	7.5:1
DAB LV switches	800 A/1700 V IGBTs
DAB LV DC voltage	800 V
LV DC capacitors	250 uF
DC/AC switches	800 A/1700 V IGBTs
DC/AC switching frequency	1500 Hz
Output AC voltage	220 V (RMS, phase-ground)
Rated output power	110*3 kW
Output DC/AC filter	L: 1.5 mH; C: 30 uF

1) Comparison for Handling Imbalanced Loads.

Imbalance conditions can occur on the distribution grid in a number of cases – one of the most common of these is for an open or short fault of single phase. Other scenarios may involve different impedances or varying power factors (PF) in each phase. For the comparison presented here, the performance of the two SST configurations is simulated under 0-10% Imbalance Degree Loads (IDL), the constant power and constant impedance loads are considered here. The IDL, given by Equation (5) is calculated using the MIL-STD-704E [15] standard based on the differences between the highest and lowest phase loads expressed as a fraction of rated load [15]. The study then determines (through simulation) the imbalance in the SST output voltage for any given IDL. The Imbalance Factor of the Output (IFO) given by Equation (6) is quantified using the IEC [16] definition of degrees of imbalance for a 3-phase system which is based on the difference between the maximum RMS phase voltage and the minimum RMS phase voltage expressed as a fraction of the rated RMS phase voltage. Fig.7 shows the IFO performance for the three SST configurations for imbalanced loads with imbalance degrees from 0-10%.

$$IDL = \frac{|Z_{Max(\text{phase amplitude})}| - |Z_{Min(\text{phase amplitude})}|}{|Z_{Rated(\text{phase amplitude})}|} \quad (5)$$

$$IFO = \frac{V_{a,b,c\ Max(RMS)} - V_{a,b,c\ Min(RMS)}}{V_{Rated(Phase\ RMS)}} \quad (6)$$

Both SST1 and SST2 have a fourth neutral wire with corresponding control to handle the imbalanced components which are caused by the imbalanced loads. The output voltage can be controlled independently by each phase in SST1 and the imbalanced components can be eliminated by the three-sequence component control strategy with the four-leg topology in SST2. As a result, the IFO for both SST1 and SST2 can be controlled to less than 1 %, which is within the standard guidelines established for imbalanced and distorted voltages in the power systems according to The American National Standards Institute (ANSI) and Institute of Electrical and Electronics Engineers (IEEE) [17][18].

Although under the high IDL, both SST1 and SST2's IFO can be controlled to below 1%, with the increase of the system's IDL, SST1 has an advantage over SST2 for handling the imbalanced load condition, as can be seen from Fig.8 below.

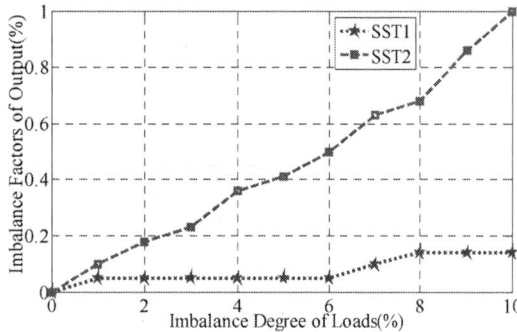

Fig.8. IFO of two SST under 0-10% imbalance loads

SST1 can maintain the IFO flat below 1%, while the IFO of SST2 is increased with the increase of the system's IDL. This is due to the fact that a steady-state error exists in the zero-sequence control loop of SST2 and increases with the increase of the system's IDL, as shown in equations (1)-(4) and Fig.9. Although, implementing the d-q-o transform reduces the three AC quantities of positive sequence and negative sequence components to two DC quantities respectively, the zero-sequence component only exists in 0-axis as an AC quantity. Here a PI controller in the natural reference frame has been used to eliminate the zero sequence, however with this the steady state error cannot be eliminated. The zero-sequence component will increase with the increase the system's IDL, resulting in an increase in the steady-state error. Although the results are acceptable for this application an improvement could be obtained by using a proportional resonant (PR) or pseudo-DQ controller for the zero-sequence component to reduce and eliminate the steady-state error.

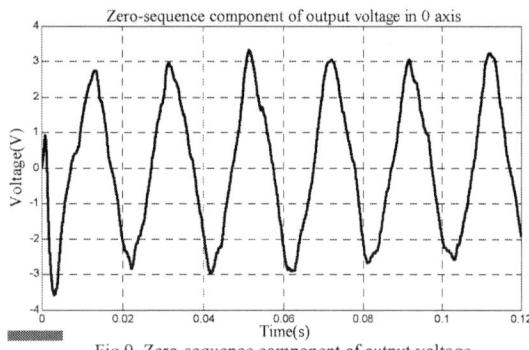

Fig.9. Zero-sequence component of output voltage

It is worth noting that although there is a difference between the two SST's outputs, there is no effect on the balance of the SST input voltages or currents because of the complete decoupling of the 3-phase input from the 3-phase output through the two intermediate DC busses.

2) Efficiency Comparison.

It is known that imbalance in the load causes increased losses in a traditional transformer due to the presence of imbalanced flux components in the core giving rise to extra core loss and the imbalanced currents giving rise to increased winding loss. For example in [19] an imbalance of 10% was shown to cause an increase in losses of 22%, mainly due to increased copper losses. It is therefore of interest to investigate how the imbalance affects the loss and efficiency of the SST.

For this comparison the switching loss for two SST configurations is evaluated using typical switch power loss characteristics which are provided in Table II. The switch loss calculation is for a 6500 V 85 A IGBT module for the high voltage side and 1700 V 800 A for the low voltage side switches. The switching frequency used for each stage of all three SST configurations is fixed at 2.5 kHz.

This analysis only includes the losses for the switches (switching and conduction) but not for the transformer and filters. The loss calculations of the switches are based on commercial specifications found on the manufacturer's datasheets [20][21] and implemented using the "half-bridge IGBT with thermal model & losses calculation" block provided in an SPS local library developed by Giroux [22]. These models are based on the thermal characteristics of the selected IGBT module, under both switching and conduction losses. The models developed by Giroux require as input the pre-switching value of the voltage across the device and post-switching value of the current flowing into the device. From these parameters the junction temperature dependent, energy loss is determined with the aid of a 3-D lookup table of the obtained from data in the manufacturer's data sheets. The energy losses are converted into a power pulse which is injected into the thermal network to obtain the junction temperature of the switch, as well as the turn-off conduction and diode losses.

TABLE III. LOSSES AND EFFICIENCY COMPARISON

Topology	Total Number of Switches	Balanced Loads		10% Imbalanced Loads	
		Total Loss (kW)	Power Efficiency (%)	Total Loss (kW)	Power Efficiency (%)
SST1	156	12.815	96.262	12.821	96.260
SST2	152	8.167	97.585	8.257	97.559

Table III displays the switch count, switch losses and efficiency for each topology obtained from running the switching models at unity input power factor and total rated power of 330 kVA with balanced and imbalanced loads (10% IDF: Phase A:115.5 kW, Phase B:110 kW, Phase C:104.5 kW). Fig.10 shows the breakdown of losses between the three stages of the SST at balanced load. We can clearly see that the power loss of the DAB and rectifier stage is almost the same among the two SST. As shown in Fig.10, the main differences between the two SST occurs in the inverter stage. SST1 has the lowest efficiency due to the fact that has the largest number of power switches of the two SST topologies (the number of switches which handle the current in the inverter for SST1 is two times that of SST2 under the balance loads, because under balanced loads the fourth leg of SST2 is not used.). However, as shown in Fig. 8 SST1 has the greatest capability to handle imbalanced loads conditions. Fig. 12 compares the losses of just the inverter stages of the two SST configurations under both balanced and 10% imbalanced load conditions. For SST2 under a balanced load situation, no switching loss occurs on the fourth leg in SST2. (With an additional fourth leg, it has no neutral current - see Fig. 11); Under the imbalanced loads situation, as expected, the losses of SST2 have increased when compared with balanced loads because the fourth leg has activated, to carry the zero-sequence current.

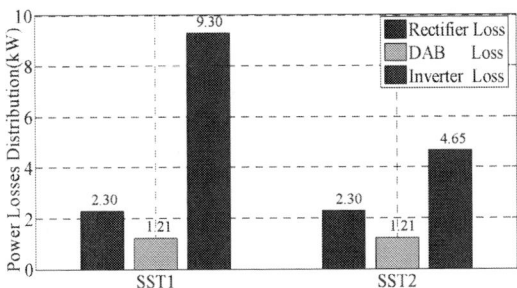

Fig.10. Losses distribution of two SST at balanced loads.

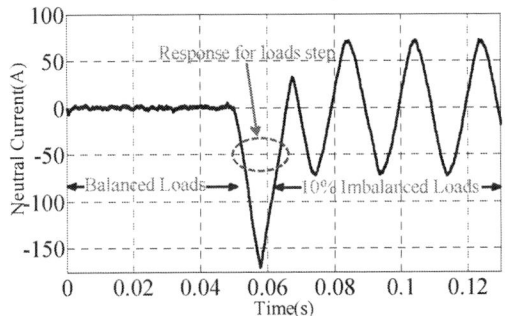

Fig.11. Neutral current of 3-phase 4-leg inverter in SST3

However as was shown earlier in Fig.8 SST1 has a greater capability to handle imbalance. Compared to SST2, which has similar capability to handle imbalance, SST2 has an overall 1.3% improvement in efficiency under both balanced and imbalance loads as shown in Table III. This is a relatively significant improvement in efficiency, however it should be noted that this doesn't account for the transformer loss and the filters loss. It is expected that those losses will account for up to 30% of the total loss, in which case the 1.3% efficiency improvement will be reduced when the transformer loss and the filters loss are accounted for.

Fig.13 shows the response of the two SST configurations at start-up and to a step change in load imbalance at 0.08 seconds. This shows that SST2 has the slowest start-up and transient response to a step change in load imbalance due to the complex control strategy of the four-leg topology at the inverter stage. In particular the 3-D SVPWM algorithm requires a coordinate transformation for the formulation in the $\alpha\beta\gamma$ frame, hence it has higher computational cost [23]. In addition both of the abc to dq0 transforms make use of a phase angle supplied by a PLL which tracks the grid phase as shown

in Fig.5 (a) and this PLL introduces a further delay [24] in the response for SST2. In comparison SST1 displays a fast start-up and transient response with lower overshoot.

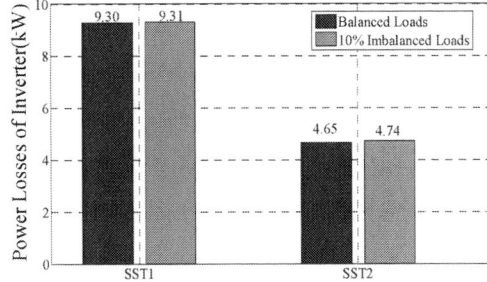

Fig.12. Inverter losses comparison among two SST

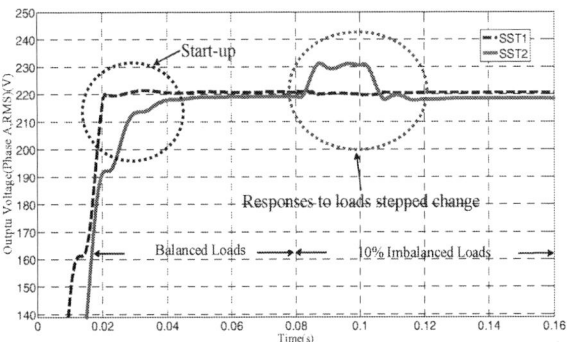

Fig.13. Three SST's transient respond to the step loads

IV. CONCLUSIONS

Two topologies of three-stage three-phase four-wire SSTs for distribution systems have been analyzed and are compared based on their system performance, switch count, switching losses and efficiency under the two most common situations at the output side, namely balanced loads and imbalanced loads conditions. A downscaled three-phase four-leg inverter prototype of SST2 has been implemented in order to verify the unbalanced-load correction performance of the SST2. It can be concluded that both SSTs have the ability to isolate the imbalance and faults from input to output, SST1 and SST2 have the capability to handle the imbalance loads so as to meet the requirements of the standards. With regard to efficiency, SST1 shows the poorest performance under both balanced and imbalance situations even though it has the best system performance. This is largely related to the fact that although system performance is enhanced by having independent phase control, losses are increased due to the requirement for a larger number of switches. Under imbalance, the losses of SST2 have increased compared with the balanced situation because the fourth leg has activated in order to take the zero-sequence current, resulting in the efficiency of SST1 with imbalanced loads being lower than balanced loads.

Reviewing the comparison presented here, SST2 is the best performing SST topology and has the best trade-off between cost, performance and efficiency for imbalance and fault scenarios in a distribution system. However, the relatively complex SST2 control strategy may cause a slower transient response for the system perhaps, indicating that a more advanced and dynamic control strategy/hardware may be needed in the future to compensate for this. Moreover, future research is required to focus on methods to improve the efficiency of SST–critical if the SST approach is to be considered as a viable alternative to conventional transformers.

REFERENCES

[1] E. Ronan, S. Sudhoff, S. Glover, and D. Galloway, A power electronic based distribution transformer, IEEE Trans. Power Delivery, vol. 17, no. 2, pp. 537–543, Apr. 2002.

[2] R. Hassan, G. Radman, "Survey on Smart Grid," IEEE Southeast Con 2010, Proceedings of the, vol., no., pp.210–213, 18–21 March 2010.

[3] Jib-Sheng Lai, A Maitra, A Mansoor and F. Goodman, "Multilevel Intelligent Universal Transformer for Medium Voltage Applications", Conference Record of Industry Applications Conference, Fourtieth IAS Annual Meeting, Oct, 2005, vol. 3, pp: 1893 – 1899.

[4] P. Richardson, D. Flynn, A. Keane, "Impact assessment of varying penetrations of electric vehicles on low voltage distribution systems", Proc. IEEE Power and Energy Soc. General Meeting, 2010, pp. 1 – 6.

[5] A. von Jouanne, B. Banerjee, "Assessment of Voltage Unbalance", IEEE Transactions on Power Delivery, vol. 16, no. 4, Oct. 2001.

[6] S. Alepuz, F. González, J. Martín-Arnedo, J. A. Martínez, "Solid state transformer with low-voltage ride-through and current unbalance management capabilities", Industrial Electronics Society, IECON 2013 - 39th Annual Conference of the IEEE, Vienna, 10-13 Nov. 2013.

[7] K. Matsuse, N. Kezuka, and K. Oka, "Characteristics of independent two induction motor drives fed by a four-leg inverter," IEEE Trans. Ind. Electron., vol. 47, no. 5, pp. 2125–2134, Oct. 2011.

[8] S.Falcones, X.L.Mao, and R.Ayyanar, "Topology comparison for solid state transformer implementation," in Proc.IEEE PES general meeting 2010, pp 1-8.

[9] S. Golestan, M. Monfared, J. M. Guerrero, and M. Joorabian,"A D-Q Synchronous Frame Controller for Single-Phase Inverters" 2nd Power Electrnics, Drive systems and Technologies Conference (PEDST 2011), Feb. 2011

[10] X. H. Wu and B.Y.Cheng. "Contrast Study on Two Instantaneous Feedback Control Techniques of Inverter System," Power and Energy Engineering Conference (APPEEC), 2010 Asia-Pacific.

[11] R. Zhang, V. H. Prasad, D. Boroyevich, and F. C. Lee, "Three-dimensional space vector modulation for four-leg voltage-source converters," IEEE Trans. Power Electron., vol. 17, no. 3, pp. 314–326, May 2002.

[12] Fortescue C L. Method of symmetrical co-ordinates applied to the solution of polyphase networks, AIEE Transactions, 1918, 37(2); 1027-1140.

[13] P. Hsu and M. Behnke, "A Three-Phase Synchronous Frame Controller forUnbalanced Load," Proceedings of the 1998 IEEE Power Electronics Specialists Conference, vol. 2, pp. 1369-1374.

[14] R. Zhang, "High performances power converter systems for nonlinear and unbalanced load/source," Ph.D. dissertation, Dept. Elect. Eng., Virginia Polytechnic Institute and State Univ., Virginia, 1998.

[15] MIL-STD -704E, "Military Standard 1991.

[16] "International Electrotechnical Vocabulary, Group 05, Fundamental Definitions," International Electrotechnical Commission (IEC) Publication 50 (05), 1954.

[17] NEMA MG1, "Motors and Generators," 1978.

[18] IEEE 446, "IEEE Recommended Practice for Emergency and Standby Power Systems for Industrial and Commercial Applications," 1987.

[19] TE Seiphetlho and APJ Rens. "Practical Evaluation of Voltage Unbalance at a Distribution Transformer Based on 50 Hz Negative Sequence Active Power" in AFRICON, 2011.

[20] 5SNA0400J650100 6500V/400A IGBT Module Data Sheet, ABB Switzerland Ltd, Semiconductors, Fabrikstrasse 3, CH-5600 Lenzburg, Switzerland.

[21] *5SNE0800M170100 1700V/800A IGBT Module Data Sheet*, ABB Switzerland Ltd, Semiconductors, Fabrikstrasse 3, CH-5600 Lenzburg, Switzerland.

[22] Pierre Giroux, Gilbert Sybille, Olivier Tremblay (12 Apr 2012). Loss Calculation in a 3-Phase 3-Level Inverter Using SimPowerSystems and Simscape. Hydro-Quebec Research Institute (IREQ). [Online]. Available: http://www.mathworks.co.uk/matlabcentral/fileexchange/36158-loss-calculation-in-a-3-phase-3-level-inverter-usingsimpowersystems-and-simscape.

[23] Lopez, O.; Alvarez, J.; Doval-Gandoy, J.; Freijedo, F.; Lago, A.; Penalver, C.M., "Four-dimensional space vector PWM algorithm for multilevel four-leg converters," Industrial Electronics, 2008. IECON 2008. 34th Annual Conference of IEEE, vol., no., pp.3252-3259, 10- 13 Nov. 2008.

[24] R.Ekström and M.Leijon, "FPGA Control Implementation of a Grid-Connected Current-Controlled Voltage-Source Inverter", Journal of Control Science and Engineering Volume 2013, Article ID 713293, 10 pages.

Steady-State Analysis of Modular Multilevel Converter (MMC) under Unbalanced Grid Conditions

Xiaojie Shi, Yalong Li, Zhiqiang Wang, Bo Liu, Leon M. Tolbert, and Fred Wang
Center for Ultra-wide-area Resilient Electric Energy Transmission Networks (CURENT)
Department of Electrical Engineering and Computer Science
The University of Tennessee
Knoxville, TN 37996-2250, USA
xshi5@vols.utk.edu

Abstract — This paper presents a steady-state analysis of the modular multilevel converter (MMC) for the second order voltage and current ripple prediction under unbalanced conditions, taking the impact of negative sequence current control into account. Using the circular relationship among current and voltage quantities, the magnitudes and initial phase angles of different circulating current components can be evaluated theoretically. With negative sequence phase current control, the positive, negative and zero sequence circulating currents are generated by more voltage sources and are no longer decoupled. Based on the generic inner relationship among current and voltage quantities, this steady state analysis is applicable to the MMC under both rectifier and inverter operating modes. Experimental results from a scaled down three-phase MMC system are provided to support the theoretical analysis and derived model.

I. INTRODUCTION

Compared to two level voltage source converters (VSCs), modular multilevel converters (MMCs) have lower switching frequency, smaller converter station footprint, modular design, high output voltage quality, etc., thus being regarded as a promising technology for long-distance high-voltage direct-current (HVDC) transmission systems to transmit power from offshore wind farms to main AC power grids [1] [2].

Under balanced conditions, only negative sequence circulating current exists, and constant dc voltage (current) can be achieved through dc voltage (active power) control [3] [4]. However, with unbalanced grids, other sequence circulating current will be generated, and thus second order harmonic current can also be observed in the dc side [5].

These additional current components may cause thermal issues in devices and passive components (e.g., arm inductors and submodule capacitance), while the second order dc side voltage could lead to ac voltage distortion through modulation process [6]. Therefore, an analytical analysis of the circulating current and voltage ripple would be helpful for semiconductor devices as well as passive components selection, which meanwhile can provide a straightforward insight into the operation characteristics of the MMC under unbalanced grid conditions.

However, most previous research efforts focused on proving the existence of zero sequence double line frequency active power, which will flow into the dc side and cause dc voltage and/or dc current ripple. References [7] and [8] give the relationship among instantaneous active/reactive power and fundamental frequency positive- as well as negative- sequence voltages and currents in ac terminals. According to this relationship, the current references for active power ripple elimination can be calculated. The relationship, however, does not take the 2nd order components in circulating current and capacitor voltages into account, and cannot fully describe the characteristics inside the MMC. As an improvement, instantaneous active power in each phase unit is derived in [9], considering negative sequence circulating current. Under unbalanced grid conditions, however, positive and zero sequence components also exist in circulating current, making the derivation inaccurate. Additionally, the magnitude of circulating current components and the dc voltage/current ripple remain unknown.

According to AC/DC decomposed circuit of a single phase MMC and active power balance between ac and dc terminals, a simple method is used for arm and circulating current calculation in [10]. Nevertheless, the dc voltage under unbalanced conditions contains double line frequency ripple (assumed to be constant in this paper), which is related with the unknown variables, i.e., circulating current. Besides, the impact arm inductors and submodule capacitors on the current and voltage magnitudes cannot be revealed.

This paper presents explicit analytical expressions for steady-state current and voltage quantities in a MMC under unbalanced grid conditions. Specifically, six equations for positive, negative, and zero sequence double line frequency circulating current can be derived based on the circular interaction of electrical quantities. By solving these equations, the steady state circulating current components can be obtained. Then the fundamental and low-order harmonic components of the arm currents, capacitor charging currents, capacitor voltages, and ac side phase voltages of the MMC can also be calculated. With the derived steady-state model, the impact of negative sequence current control on second order current and voltage ripple can be easily investigated.

II. STEADY-STATE MODEL FOR SECOND ORDER PHASE VOLTAGE PREDICTION

Under unbalanced grid voltage conditions, the classic dual current control, which consists of positive and negative sequence current control, is widely used [11]. To figure out the impact of this control on the 2nd order voltage and current components, two cases, with and without negative sequence current control, are studied.

978-1-4673-9551-9/16 $31.00 © 2016 IEEE

A. Without Negative Sequence Current Control

Assume a SLG fault occurs in the primary side of the Y/Δ transformer in MMC_1 station, which will create an unbalanced voltage in PCC_1. Due to this transformer, no zero sequence phase current will be observed in the converter side, as indicated by the upper and lower arm current in (1)

$$i_{au}(t) = \frac{\sqrt{2}}{2}I^+ \sin(\omega_1 t + \varphi_+) + \frac{\sqrt{2}}{2}I^- \sin(\omega_1 t + \varphi_-) + I_{cda} +$$
$$I_c^+ \sin(2\omega_1 t + \theta_+) + I_c^- \sin(2\omega_1 t + \theta_-) + I_c^0 \sin(2\omega_1 t + \theta_0)$$
$$i_{al}(t) = -\frac{\sqrt{2}}{2}I^+ \sin(\omega_1 t + \varphi_+) - \frac{\sqrt{2}}{2}I^- \sin(\omega_1 t + \varphi_-) + I_{cda} +$$
$$I_c^+ \sin(2\omega_1 t + \theta_+) + I_c^- \sin(2\omega_1 t + \theta_-) + I_c^0 \sin(2\omega_1 t + \theta_0) \quad (1)$$

where I^+, I^- are the amplitudes of the positive and negative sequence phase current, respectively; φ_+, φ_- are the corresponding initial phase angles; I_{cda} represents the dc component in arm currents of phase A; ω_1 represents the fundamental frequency angular frequency; I_c^+, I_c^- and I_c^0 denote the positive, negative and zero sequence circulating current, respectively, and θ_+, θ_-, and θ_0 are the corresponding initial phase angles. For simplicity, only the dominant second order harmonics are considered in this paper. Under certain operation conditions, I^+, I^-, φ_+, φ_- in (1) can be regarded as known parameters based on power system fault calculation [12], while I_c^+, I_c^-, I_c^0, θ_+, θ_-, and θ_0 are unknown variables and will be calculated in the following derivation.

In practice, the equivalent switching frequency of a MMC (Nf_c, f_c is the carrier frequency) is high enough such that high-frequency components in the switching functions can be neglected [13]. Therefore, an average switching function shown in (2) can be adopted in this analysis

$$S_{au\_av} = \frac{1}{2}[1 - M\sin(\omega_1 t)] \quad S_{al\_av}(t) = \frac{1}{2}[1 + M\sin(\omega_1 t)] \quad (2)$$

where M represents positive sequence modulation index; $S_{au\_av}(t)$ and $S_{al\_av}(t)$ are average switching functions of the SMs in the upper and lower arms, respectively.

Based on (1) and (2), the associated current and voltage quantities can be expressed as follows:

1) Capacitor Charging Current: according to [14], the average capacitor charging current in the upper (i_{c_u}) and lower arm (i_{c_l}) can be expressed as

$$i_{c_u}(t) = S_{au_{av}}(t) \cdot i_{au}(t)$$
$$= \frac{1}{2}I_{cda} - \frac{\sqrt{2}}{8}MI^+\cos\varphi_+ - \frac{\sqrt{2}}{8}MI^-\cos\varphi_- - \frac{1}{2}MI_{cda}\sin(\omega_1 t)$$
$$+ \frac{\sqrt{2}}{4}I^+ \sin(\omega_1 t + \varphi_+) + \frac{\sqrt{2}}{4}I^- \sin(\omega_1 t + \varphi_-) - \frac{1}{4}MI_C^+ \cos(\omega_1 t + \theta_+) - \frac{1}{4}MI_C^- \cos(\omega_1 t + \theta_-) - \frac{1}{4}MI_C^0 \cos(\omega_1 t + \theta_0)$$
$$+ \frac{\sqrt{2}}{8}MI^+ \cos(2\omega_1 t + \varphi_+) + \frac{\sqrt{2}}{8}MI^- \cos(2\omega_1 t + \varphi_-)$$
$$+ \frac{1}{2}I_c^+ \sin(2\omega_1 t + \theta_+) + \frac{1}{2}I_c^- \sin(2\omega_1 t + \theta_-) + \frac{1}{2}I_c^0 \sin(2\omega_1 t + \theta_0)$$
$$+ \frac{1}{4}MI_c^+ \sin(3\omega_1 t + \theta_+) + \frac{1}{4}MI_c^- \sin(3\omega_1 t + \theta_-)$$
$$+ \frac{1}{4}MI_c^0 \sin(3\omega_1 t + \theta_0)$$

$$i_{c_l}(t) = S_{al_{av}}(t) \cdot i_{al}(t)$$
$$= \frac{1}{2}I_{cda} - \frac{\sqrt{2}}{8}MI^+\cos\varphi_+ - \frac{\sqrt{2}}{8}MI^-\cos\varphi_- + \frac{1}{2}MI_{cda}\sin(\omega_1 t)$$
$$- \frac{\sqrt{2}}{4}I^+ \sin(\omega_1 t + \varphi_+) - \frac{\sqrt{2}}{4}I^- \sin(\omega_1 t + \varphi_-) + \frac{1}{4}MI_C^+ \cos(\omega_1 t + \theta_+) + \frac{1}{4}MI_C^- \cos(\omega_1 t + \theta_-) + \frac{1}{4}MI_C^0 \cos(\omega_1 t + \theta_0)$$

$$+ \frac{\sqrt{2}}{8}MI^+ \cos(2\omega_1 t + \varphi_+) + \frac{\sqrt{2}}{8}MI^-\cos(2\omega_1 t + \varphi_-)$$
$$+ \frac{1}{2}I_c^+ \sin(2\omega_1 t + \theta_+) + \frac{1}{2}I_c^- \sin(2\omega_1 t + \theta_-) + \frac{1}{2}I_c^0 \sin(2\omega_1 t + \theta_0)$$
$$- \frac{1}{4}MI_c^+ \sin(3\omega_1 t + \theta_+) - \frac{1}{4}MI_c^- \sin(3\omega_1 t + \theta_-)$$
$$- \frac{1}{4}MI_c^0 \sin(3\omega_1 t + \theta_0). \quad (3)$$

The dc component highlighted by pink color in (3) is zero during steady-state, which stands for a balanced active power between the ac and dc side. Solving the steady-state equation in (4) yields the dc component in circulating current

$$\frac{1}{2}I_{cda} - \frac{\sqrt{2}}{8}MI^+ \cos\varphi_+ - \frac{\sqrt{2}}{8}MI^- \cos\varphi_- = 0$$
$$\rightarrow I_{cda} = \frac{\sqrt{2}}{4}MI^+ \cos\varphi_+ + \frac{\sqrt{2}}{4}MI^- \cos\varphi_- \quad (4a)$$
$$\frac{1}{2}I_{cdb} - \frac{\sqrt{2}}{8}MI^+ \cos\varphi_+ - \frac{\sqrt{2}}{8}MI^- \cos(\varphi_- -120°) = 0$$
$$\rightarrow I_{cdb} = \frac{\sqrt{2}}{4}MI^+ \cos\varphi_+ + \frac{\sqrt{2}}{4}MI^- \cos(\varphi_- - 120°) \quad (4b)$$
$$\frac{1}{2}I_{cdc} - \frac{\sqrt{2}}{8}MI^+ \cos\varphi_+ - \frac{\sqrt{2}}{8}MI^- \cos(\varphi_- +120°) = 0$$
$$\rightarrow I_{cdc} = \frac{\sqrt{2}}{4}MI^+ \cos\varphi_+ + \frac{\sqrt{2}}{4}MI^- \cos(\varphi_- + 120°). \quad (4c)$$

Comparing (4a), (4b) and (4c), the dc components in the three phase arm current are no longer equal to $I_{dc}/3$ (which is $\sqrt{2}MI^+\cos\varphi_+/4$ in normal operation) under unbalanced conditions, indicating uneven active power contribution among the three phases. Moreover, such asymmetrical distribution will be more severe with higher negative sequence current I^-. Therefore, I_{cdk} (k = a, b, c) in (1) should be replaced by (4) during the following derivation.

2) Capacitor Ripple Voltage: Multiplying the capacitor reactance $1/(j\omega_1 C_{sub})$ and the fundamental frequency capacitor charging current in (3), the fundamental frequency capacitor ripple voltages in the upper and lower arms in phase A are obtained

$$u_{cu}^{(1)}(t) = \frac{1}{j\omega_1 C_{sub}}i_{cu}^{(1)} = \frac{MI_{cda}}{2\omega_1 C_{sub}}\cos(\omega_1 t) - \frac{\sqrt{2}}{4\omega_1 C_{sub}}I^+ \cos(\omega_1 t +$$
$$\varphi_+) - \frac{\sqrt{2}}{4\omega_1 C_{sub}}I^- \cos(\omega_1 t + \varphi_-) - \frac{1}{4\omega_1 C_{sub}}MI_c^+ \sin(\omega_1 t + \theta_+)$$
$$- \frac{1}{4\omega_1 C_{sub}}MI_c^- \sin(\omega_1 t + \theta_-) - \frac{1}{4\omega_1 C_{sub}}MI_c^0 \sin(\omega_1 t + \theta_0)$$

$$u_{cl}^{(1)}(t) = \frac{1}{j\omega_1 C_{sub}}i_{cl}^{(1)} = -\frac{MI_{cda}}{2\omega_1 C_{sub}}\cos(\omega_1 t) + \frac{\sqrt{2}}{4\omega_1 C_{sub}}I^+ \cos(\omega_1 t +$$
$$\varphi_+) + \frac{\sqrt{2}}{4\omega_1 C_{sub}}I^- \cos(\omega_1 t + \varphi_-) + \frac{1}{4\omega_1 C_{sub}}MI_c^+ \sin(\omega_1 t + \theta_+)$$
$$+ \frac{1}{4\omega_1 C_{sub}}MI_c^- \sin(\omega_1 t + \theta_-) + \frac{1}{4\omega_1 C_{sub}}MI_c^0 \sin(\omega_1 t + \theta_0). \quad (5)$$

Similarly, the second and third order capacitor ripple voltages are given in (6) and (7)

$$u_{cu}^{(2)}(t) = \frac{1}{j2\omega_1 C_{sub}}i_{cu}^{(2)} = \frac{\sqrt{2}MI^+}{16\omega_1 C_{sub}}\sin(2\omega_1 t + \varphi_+)$$
$$+ \frac{\sqrt{2}MI^-}{16\omega_1 C_{sub}}\sin(2\omega_1 t + \varphi_-) - \frac{I_c^+}{4\omega_1 C_{sub}}\cos(2\omega_1 t + \theta_+)$$
$$- \frac{I_c^-}{4\omega_1 C_{sub}}\cos(2\omega_1 t + \theta_-) - \frac{I_c^0}{4\omega_1 C_{sub}}\cos(2\omega_1 t + \theta_0)$$

$$u_{cl}^{(2)}(t) = \frac{1}{j2\omega_1 C_{sub}}i_{cl}^{(2)} = \frac{\sqrt{2}MI^+}{16\omega_1 C_{sub}}\sin(2\omega_1 t + \varphi_+)$$
$$+ \frac{\sqrt{2}MI^-}{16\omega_1 C_{sub}}\sin(2\omega_1 t + \varphi_-) - \frac{I_c^+}{4\omega_1 C_{sub}}\cos(2\omega_1 t + \theta_+)$$
$$- \frac{I_c^-}{4\omega_1 C_{sub}}\cos(2\omega_1 t + \theta_-) - \frac{I_c^0}{4\omega_1 C_{sub}}\cos(2\omega_1 t + \theta_0). \quad (6)$$

$$u_{cu}^{(3)}(t) = \frac{1}{j3\omega_1 C_{sub}} i_{c_u}^{(3)} = \frac{MI_c^+}{12\omega_1 C_{sub}} \cos(3\omega_1 t + \theta_+)$$
$$+ \frac{MI_c^-}{12\omega_1 C_{sub}} \cos(3\omega_1 t + \theta_-) + \frac{MI_c^0}{12\omega_1 C_{sub}} \cos(3\omega_1 t + \theta_0)$$

$$u_{cl}^{(3)}(t) = \frac{1}{j3\omega_1 C_{sub}} i_{c_l}^{(3)} = -\frac{MI_c^+}{12\omega_1 C_{sub}} \cos(3\omega_1 t + \theta_+)$$
$$- \frac{MI_c^-}{12\omega_1 C_{sub}} \cos(3\omega_1 t + \theta_-) - \frac{MI_c^0}{12\omega_1 C_{sub}} \cos(3\omega_1 t + \theta_0). \tag{7}$$

3) Phase voltage ripple: The ac side phase voltage ripple contributed by $u_{cu}^{(1)}$ and $u_{cl}^{(1)}$ can be expressed as

$$u_{ph_1}(t) = \frac{N}{2}[1 - M sin(\omega_1 t)] \cdot u_{cu}^{(1)}(t) + \frac{N}{2}[1 + M sin(\omega_1 t)] \cdot$$
$$u_{cl}^{(1)}(t) = -\frac{\sqrt{2}NMI^+}{8\omega_1 C_{sub}} sin\varphi_+ - \frac{\sqrt{2}NMI^-}{8\omega_1 C_{sub}} sin\varphi_- + \frac{NM^2 I_c^+}{8\omega_1 C_{sub}} \cos\theta_+$$
$$+ \frac{NM^2 I_c^-}{8\omega_1 C_{sub}} \cos\theta_- + \frac{NM^2 I_c^0}{8\omega_1 C_{sub}} \cos\theta_0 - \frac{NM^2 I_{cda}}{4\omega_1 C_{sub}} sin(2\omega_1 t)$$
$$+ \frac{\sqrt{2}NMI^+}{8\omega_1 C_{sub}} sin(2\omega_1 t + \varphi_+) + \frac{\sqrt{2}NMI^-}{8\omega_1 C_{sub}} sin(2\omega_1 t + \varphi_-)$$
$$- \frac{NM^2 I_c^+}{8\omega_1 C_{sub}} \cos(2\omega_1 t + \theta_+) - \frac{NM^2 I_c^-}{8\omega_1 C_{sub}} \cos(2\omega_1 t + \theta_-)$$
$$- \frac{NM^2 I_c^0}{8\omega_1 C_{sub}} \cos(2\omega_1 t + \theta_0). \tag{8a}$$

Phase voltage ripple contributed by $u_{cu,l}^{(2)}$ and $u_{cu,l}^{(3)}$ can be calculated as

$$u_{ph_2}(t) = \frac{\sqrt{2}NMI^+}{16\omega_1 C_{sub}} sin(2\omega_1 t + \varphi_+) + \frac{\sqrt{2}NMI^-}{16\omega_1 C_{sub}} sin(2\omega_1 t + \varphi_-)$$
$$- \frac{NI_c^+}{4\omega_1 C_{sub}} \cos(2\omega_1 t + \theta_+) - \frac{NI_c^-}{4\omega_1 C_{sub}} \cos(2\omega_1 t + \theta_-)$$
$$- \frac{NI_c^0}{4\omega_1 C_{sub}} \cos(2\omega_1 t + \theta_0) \tag{8b}$$

$$u_{ph_3}(t) = -\frac{NM^2 I_c^+}{24\omega_1 C_{sub}} \cos(2\omega_1 t + \theta_+) - \frac{NM^2 I_c^-}{24\omega_1 C_{sub}} \cos(2\omega_1 t + \theta_-)$$
$$- \frac{NM^2 I_c^0}{24\omega_1 C_{sub}} \cos(2\omega_1 t + \theta_0) + \frac{NM^2 I_c^+}{24\omega_1 C_{sub}} \cos(4\omega_1 t + \theta_+)$$
$$+ \frac{NM^2 I_c^-}{24\omega_1 C_{sub}} \cos(4\omega_1 t + \theta_-) + \frac{NM^2 I_c^0}{24\omega_1 C_{sub}} \cos(4\omega_1 t + \theta_0). \tag{8c}$$

As shown in (8), with arm current assumed in (1), the second and fourth order phase voltage components are generated, which in turn will create circulating current with the corresponding frequencies. Focusing on the dominant second order ripples and ignoring the fourth order items, the double line frequency phase voltage ripple in Phase A can be obtained by summing all the second order components in (8).

$$u_{ph_A}^{(2)}(t) = -\frac{\sqrt{2}NM^3 I^+}{16\omega_1 C_{sub}} \cos\varphi_+ sin(2\omega_1 t) + \frac{3\sqrt{2}NMI^+}{16\omega_1 C_{sub}} sin(2\omega_1 t + \varphi_+)$$
$$- \frac{NM^2 I_c^-}{6\omega_1 C_{sub}} \cos(2\omega_1 t + \theta_-) - \frac{NI_c^-}{4\omega_1 C_{sub}} \cos(2\omega_1 t + \theta_-)$$
$$- \frac{\sqrt{2}NM^3 I^-}{32\omega_1 C_{sub}} sin(2\omega_1 t + \varphi_-) + \frac{3\sqrt{2}NMI^-}{16\omega_1 C_{sub}} sin(2\omega_1 t + \varphi_-)$$
$$- \frac{NM^2 I_c^0}{6\omega_1 C_{sub}} \cos(2\omega_1 t + \theta_0) - \frac{NI_c^0}{4\omega_1 C_{sub}} \cos(2\omega_1 t + \theta_0)$$
$$- \frac{\sqrt{2}NM^3 I^-}{32\omega_1 C_{sub}} sin(2\omega_1 t - \varphi_-) - \frac{NM^2 I_c^+}{6\omega_1 C_{sub}} \cos(2\omega_1 t + \theta_+)$$
$$- \frac{NI_c^+}{4\omega_1 C_{sub}} \cos(2\omega_1 t + \theta_+). \tag{9}$$

Similarly, the second order voltage ripple in Phases B and C can be given in (10) and (11). The negative, zero, and positive sequence voltage ripples are distinguished by blue, red and black colors, respectively.

$$u_{ph_B}^{(2)}(t) = -\frac{\sqrt{2}NM^3 I^+}{16\omega_1 C_{sub}} \cos\varphi_+ sin(2\omega_1 t + 120°)$$
$$+ \frac{3\sqrt{2}NMI^+}{16\omega_1 C_{sub}} sin(2\omega_1 t + \varphi_+ + 120°) - \frac{NM^2 I_c^-}{6\omega_1 C_{sub}} \cos(2\omega_1 t + \theta_- + 120°)$$
$$- \frac{NI_c^-}{4\omega_1 C_{sub}} \cos(2\omega_1 t + \theta_- + 120°) - \frac{\sqrt{2}NM^3 I^-}{32\omega_1 C_{sub}} sin(2\omega_1 t + \varphi_-)$$

$$+ \frac{3\sqrt{2}NMI^-}{16\omega_1 C_{sub}} sin(2\omega_1 t + \varphi_-) - \frac{NM^2 I_c^0}{6\omega_1 C_{sub}} \cos(2\omega_1 t + \theta_0)$$
$$- \frac{NI_c^0}{4\omega_1 C_{sub}} \cos(2\omega_1 t + \theta_0) - \frac{\sqrt{2}NM^3 I^-}{32\omega_1 C_{sub}} sin(2\omega_1 t - \varphi_- - 120°)$$
$$- \frac{NM^2 I_c^+}{6\omega_1 C_{sub}} \cos(2\omega_1 t + \theta_+ - 120°) - \frac{NI_c^+}{4\omega_1 C_{sub}} \cos(2\omega_1 t + \theta_+ - 120°) \tag{10}$$

$$u_{ph_C}^{(2)}(t) = -\frac{\sqrt{2}NM^3 I^+}{16\omega_1 C_{sub}} \cos\varphi_+ sin(2\omega_1 t - 120°)$$
$$+ \frac{3\sqrt{2}NMI^+}{16\omega_1 C_{sub}} sin(2\omega_1 t + \varphi_+ - 120°) - \frac{NM^2 I_c^-}{6\omega_1 C_{sub}} \cos(2\omega_1 t + \theta_- - 120°)$$
$$- \frac{NI_c^-}{4\omega_1 C_{sub}} \cos(2\omega_1 t + \theta_- - 120°) - \frac{\sqrt{2}NM^3 I^-}{32\omega_1 C_{sub}} sin(2\omega_1 t + \varphi_-)$$
$$+ \frac{3\sqrt{2}NMI^-}{16\omega_1 C_{sub}} sin(2\omega_1 t + \varphi_-) - \frac{NM^2 I_c^0}{6\omega_1 C_{sub}} \cos(2\omega_1 t + \theta_0)$$
$$- \frac{NI_c^0}{4\omega_1 C_{sub}} \cos(2\omega_1 t + \theta_0) - \frac{\sqrt{2}NM^3 I^-}{32\omega_1 C_{sub}} sin(2\omega_1 t - \varphi_- + 120°)$$
$$- \frac{NM^2 I_c^+}{6\omega_1 C_{sub}} \cos(2\omega_1 t + \theta_+ + 120°) - \frac{NI_c^+}{4\omega_1 C_{sub}} \cos(2\omega_1 t + \theta_+ + 120°). \tag{11}$$

As can be observed in (9)-(11), the positive and zero sequence voltage ripples only exist under unbalanced conditions, while the negative sequence component exists as long as the positive sequence modulation index M and phase current I^+ are nonzero. Sharing the same expression as that under normal conditions, the negative sequence circulating current will not change much if if $M I^+$ has the same value as that in normal conditions.

B. With Negative Sequence Current Control

If dual current control is used, the negative sequence phase current can be eliminated, and thus the arm current in phase A becomes

$$i_{au}(t) = \frac{\sqrt{2}}{2} I^+ sin(\omega_1 t + \varphi_+) + I_{cda} + I_c^+ sin(2\omega_1 t + \theta_+)$$
$$+ I_c^- sin(2\omega_1 t + \theta_-) + I_c^0 sin(2\omega_1 t + \theta_0)$$

$$i_{al}(t) = -\frac{\sqrt{2}}{2} I^+ sin(\omega_1 t + \varphi_+) + I_{cda} + I_c^+ sin(2\omega_1 t + \theta_+)$$
$$+ I_c^- sin(2\omega_1 t + \theta_-) + I_c^0 sin(2\omega_1 t + \theta_0) \tag{12}$$

On the other hand, the negative sequence component will exist in the average switching function due to the introduction of the dual current control, as described in (13).

$$S_{au\_av}(t) = \frac{1}{2}[1 - M^+ sin(\omega_1 t) - M^- sin(\omega_1 t + \gamma_-)]$$
$$S_{al\_av}(t) = \frac{1}{2}[1 + M^+ sin(\omega_1 t) + M^- sin(\omega_1 t + \gamma_-)] \tag{13}$$

where M^+ denotes the positive sequence modulation index; M^- and γ_- represent the negative sequence modulation index and its initial phase angle; $S_{au\_av}(t)$ and $S_{al\_av}(t)$ are average switching functions of the SMs in the upper and lower arms, respectively.

Following the same derivation steps above, the steady-state dc component in three phase arm currents can be given as

$$\frac{1}{2} I_{cda} - \frac{\sqrt{2}}{8} M^+ I^+ \cos\varphi_+ - \frac{\sqrt{2}}{8} M^- I^+ \cos(\gamma_- - \varphi_+) = 0$$
$$\rightarrow I_{cda} = \frac{\sqrt{2}}{4} M^+ I^+ \cos\varphi_+ + \frac{\sqrt{2}}{4} M^- I^+ \cos(\gamma_- - \varphi_+)$$
$$I_{cdb} = \frac{\sqrt{2}}{4} M^+ I^+ \cos\varphi_+ + \frac{\sqrt{2}}{4} M^- I^+ \cos(\gamma_- - \varphi_+ - 120°)$$
$$I_{cdc} = \frac{\sqrt{2}}{4} M^+ I^+ \cos\varphi_+ + \frac{\sqrt{2}}{4} M^- I^+ \cos(\gamma_- - \varphi_+ + 120°). \tag{14}$$

According to (14), the negative sequence current control cannot equalize three phase dc arm currents. The double line frequency ac voltage ripple in phase A is derived in (15), where the negative, zero and positive sequence circulating three phase voltage ripples are still distinguished by blue, red and black colors. The voltage ripple for Phase B and C can be obtained in the same manner, with a phase shift of $\pm 2\pi/3$ in negative and positive sequence components, and no phase shift for zero sequence component.

As indicated in (15), with negative sequence phase current control, more combinations of current and modulation index serve as the sources of double line frequency phase voltage ripple. Moreover, compared to the result without negative sequence current control in (9), the circulating current components in (15) are no longer decoupled. For example, I_c^+ and I_c^0 contribute to negative sequence phase voltage ripple, and consequently are associated with I_c^-.

$$
\begin{aligned}
\boldsymbol{u}_{ph_A}^{(2)}(t) = & -\frac{\sqrt{2}N(M^+)^3 I^+}{16\omega_1 C_{sub}}\cos\varphi_+ \sin(2\omega_1 t) - \frac{\sqrt{2}NM^+(M^-)^2 I^+}{16\omega_1 C_{sub}}\sin(2\omega_1 t + \\
& \varphi_+) - \frac{\sqrt{2}N(M^-)^3 I^+}{32\omega_1 C_{sub}}\sin(2\omega_1 t + 3\gamma_- - \varphi_+) + \frac{3\sqrt{2}NM^+ I^+}{16\omega_1 C_{sub}}\sin(2\omega_1 t + \\
& \varphi_+) - \frac{N(M^+)^2 I_c^-}{6\omega_1 C_{sub}}\cos(2\omega_1 t + \theta_-) - \frac{N(M^-)^2 I_c^-}{6\omega_1 C_{sub}}\cos(2\omega_1 t + \theta_-) \\
& -\frac{NM^+M^- I_c^0}{6\omega_1 C_{sub}}\cos(2\omega_1 t + \theta_0 - \gamma_-) - \frac{NM^+M^- I_c^0}{6\omega_1 C_{sub}}\cos(2\omega_1 t + \theta_+ + \gamma_-) \\
& -\frac{NI_c^-}{4\omega_1 C_{sub}}\cos(2\omega_1 t + \theta_-) - \frac{\sqrt{2}NM^-(M^+)^2 I^+}{32\omega_1 C_{sub}}\sin(2\omega_1 t + \gamma_- - \varphi_+) \\
& -\frac{\sqrt{2}NM^-(M^+)^2 I^+}{8\omega_1 C_{sub}}\cos\varphi_+ \sin(2\omega_1 t + \gamma_-) - \frac{\sqrt{2}N(M^-)^3 I^+}{32\omega_1 C_{sub}}\sin(2\omega_1 t + \\
& \gamma_- + \varphi_+) + \frac{3\sqrt{2}NM^- I^+}{16\omega_1 C_{sub}}\sin(2\omega_1 t + \varphi_+ + \gamma_-) - \frac{N(M^+)^2 I_c^0}{6\omega_1 C_{sub}}\cos(2\omega_1 t + \\
& \theta_0) - \frac{NI_c^0}{4\omega_1 C_{sub}}\cos(2\omega_1 t + \theta_0) - \frac{NM^+M^- I_c^-}{6\omega_1 C_{sub}}\cos(2\omega_1 t + \theta_+ - \gamma_-) \\
& -\frac{NM^+M^- I_c^-}{6\omega_1 C_{sub}}\cos(2\omega_1 t + \theta_- + \gamma_-) - \frac{N(M^-)^2 I_c^0}{6\omega_1 C_{sub}}\cos(2\omega_1 t + \theta_0) \\
& -\frac{\sqrt{2}NM^-(M^+)^2 I^+}{32\omega_1 C_{sub}}\sin(2\omega_1 t + \varphi_+ - \gamma_-) - \frac{\sqrt{2}NM^-(M^-)^2 I^+}{16\omega_1 C_{sub}}\sin(2\omega_1 t - \\
& \varphi_+ + 2\gamma_-) - \frac{\sqrt{2}NM^-(M^-)^2 I^+}{16\omega_1 C_{sub}}\cos\varphi_+ \sin(2\omega_1 t + 2\gamma_-) \\
& -\frac{N(M^+)^2 I_c^+}{6\omega_1 C_{sub}}\cos(2\omega_1 t + \theta_+) - \frac{NM^+M^- I_c^-}{6\omega_1 C_{sub}}\cos(2\omega_1 t + \theta_- - \gamma_-) \\
& -\frac{NM^+M^- I_c^0}{6\omega_1 C_{sub}}\cos(2\omega_1 t + \theta_0 + \gamma_-) - \frac{N(M^-)^2 I_c^+}{6\omega_1 C_{sub}}\cos(2\omega_1 t + \theta_+) \\
& -\frac{NI_c^+}{4\omega_1 C_{sub}}\cos(2\omega_1 t + \theta_+).
\end{aligned} \tag{15}
$$

It is worth mentioning that the above analytical expressions are derived based on the generic inner relationship among current and voltage quantities, which are applicable to MMC under both rectifier and inverter modes.

III. STEADY-STATE MODEL FOR CIRCULATING CURRENT PREDICTION

With the derived steady-state model for second order phase voltage ripple, the circulating current can be predicted by solving the equivalent circuit shown in Fig. 1(a) and (b), where $u_{ph\_j}^+$ and $u_{ph\_j}^-$ (j = A, B, C) represent the corresponding three phase double line frequency voltage ripple in (3) and (7). The three-phase circulating current generated by the two voltage sources will add to zero and only circulate among phases.

For the zero sequence circulating currents generated by $u_{ph\_j}^0$, they cannot cancel with each other and have to flow into the dc bus. Consequently, in addition to arm inductance L_{arm} and its ESR R_{arm}, the equivalent dc load impedance Z_{dc_eq} will be incorporated into its equivalent circuit, as shown in Fig. 1(c).

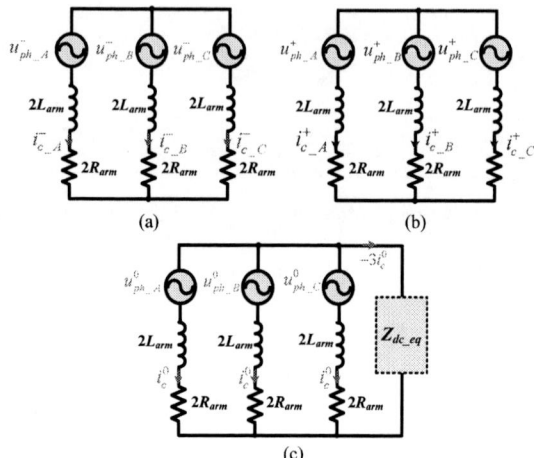

Fig. 1. Equivalent circuit of (a) negative sequence, (b) positive sequence, and (c) zero sequence circulating current.

Based on these equivalent circuits, the following equations hold

$$
i_{c\_k}^-(t) = I_c^- \sin(2\omega_1 t + \theta_-) = -\frac{u_{ph\_k}^-(t)}{j2\omega_1 \cdot 2L_{arm} + 2R_{arm}} \quad (k = A, B, C)
$$

$$
i_{c\_k}^+(t) = I_c^+ \sin(2\omega_1 t + \theta_+) = -\frac{u_{ph\_k}^+(t)}{j2\omega_1 \cdot 2L_{arm} + 2R_{arm}}
$$

$$
i_{c\_k}^0(t) = I_c^0 \sin(2\omega_1 t + \theta_0) = -\frac{u_{ph_k}^0(t)}{j2\omega_1 \cdot 2L_{arm} + 2R_{arm} + 3Z_{dc_{eq}}(t)}. \tag{16}
$$

From (16), circular interactions among double line frequency voltages and currents in the MMC can be found. Moreover, by substituting (9) and (15) into (16) and solving the obtained equations, magnitudes (I_c^+, I_c^-, I_c^0) and initial phase angles (θ_+, θ_-, θ_0) of circulating current with and without negative sequence current control can be calculated theoretically. With the calculated circulating current, all the previous equations will become explicit.

For simplicity, the solving process of the circulating current in a MMC rectifier with resistive load is analyzed.

A. Without Negative Sequence Current Control

The corresponding equations with a dc side resistive load R_L is given in (17), by replacing Z_{dc_eq} with R_L in (16). As mentioned above, the phase current amplitude I^+, I^-, phase angle φ_+, φ_-, and modulation index M are known variables. The magnitude and initial phase angle of different circulating current components (I_c^+, I_c^-, I_c^0, θ_+, θ_- and θ_0) can be easily calculated using (17). In addition, capacitor charging current, capacitor voltage ripple, and phase voltage ripple can be expressed by substituting the calculated circulating current components into (3) – (8).

$$
\begin{cases}
(B - 2A)I^+\cos\varphi_+ = -EI_c^-\sin\theta_- - 2R_{arm}I_c^-\cos\theta_- \\
BI^+\sin\varphi_+ = EI_c^-\cos\theta_- - 2R_{arm}I_c^-\sin\theta_-
\end{cases}
$$

$$
\begin{cases}
(B - A)I^-\cos\varphi_- = -EI_c^0\sin\theta_0 - (2R_{arm}+3R_L)I_c^0\cos\theta_0 \\
(B - A)I^-\sin\varphi_- = EI_c^0\cos\theta_0 - (2R_{arm}+3R_L)I_c^0\sin\theta_0
\end{cases}
$$

$$
\begin{cases}
-AI^-\cos\varphi_- = -EI_c^+\sin\theta_+ - 2R_{arm}I_c^+\cos\theta_+ \\
AI^-\sin\varphi_- = EI_c^+\cos\theta_+ - 2R_{arm}I_c^+\sin\theta_+
\end{cases} \tag{17}
$$

where,

$$
A = \frac{\sqrt{2}NM^3}{32\omega_1 C_{sub}} \quad B = \frac{3\sqrt{2}NM}{16\omega_1 C_{sub}} \quad C = \frac{NM^2}{6\omega_1 C_{sub}} \quad D = \frac{N}{4\omega_1 C_{sub}}
$$

$$
E = C + D - 4\omega_1 L_{arm}
$$

Under slightly unbalanced conditions, the magnitude of negative sequence circulating current changes very little since I^+ and M will not change much. Nevertheless, if a single-line-to-ground (SLG) fault occurs, M will decrease because of the lost phase voltage, while I^+ tends to increase to maintain dc voltage and/or output power. Therefore, the negative sequence circulating current will vary accordingly with different control schemes and current limit settings.

As for the positive and zero sequence circulating current, their magnitudes are proportional to the negative sequence phase current, and will reduce with lower M. Zero sequence circulating current, limited by dc load impedance, will be much lower than the other two sequence components in a MMC rectifier with a light load.

B. With Negative Sequence Current Control

By substituting (15) into (16) and replacing Z_{dc_eq} with R_L, the six equations used for circulating current calculation with a resistive load R_L are given in (18) when negative sequence current control is activated.

$$
\begin{cases}
\begin{aligned}
&-2AI^+cos\varphi_+ - BI^+cos\varphi_+ - CI^+\cos(3\gamma_- - \varphi_+) + DI^+M^+cos\varphi_+ + \\
&GI_c^0 sin(\theta_0 - \gamma_-) + GI_c^+sin(\theta_+ + \gamma_-) = -KI_c^-sin\theta_- - 2R_{arm}I_c^-cos\theta_- \\
&-BI^+sin\varphi_+ - CI^+sin(3\gamma_- - \varphi_+) + DI^+M^+sin\varphi_+ - GI_c^0 cos(\theta_0 - \gamma_-) \\
&-GI_c^+cos(\theta_+ + \gamma_-) = KI_c^-cos\theta_- - 2R_{arm}I_c^-sin\theta_-
\end{aligned}
\end{cases}
$$

$$
\begin{cases}
\begin{aligned}
&-PI^+\cos(\gamma_- - \varphi_+) - 4PI^+cos\varphi_+cos\gamma_- - CI^+\cos(\gamma_- + \varphi_+) \\
&+DI^+M^-\cos(\gamma_- + \varphi_+) + GI_c^+sin(\theta_+ - \gamma_-) + GI_c^-sin(\theta_- + \gamma_-) \\
&= -KI_c^0 sin\theta_0 - (2R_{arm} + 3R_L)I_c^0 cos\theta_0 \\
&-PI^+sin(\gamma_- - \varphi_+) - 4PI^+cos\varphi_+sin\gamma_- - CI^+sin(\gamma_- + \varphi_+) \\
&+DI^+M^-sin(\gamma_- + \varphi_+) - GI_c^+cos(\theta_+ - \gamma_-) - GI_c^-cos(\theta_- + \gamma_-) \\
&= KI_c^0 cos\theta_0 - (2R_{arm} + 3R_L)I_c^0 sin\theta_0
\end{aligned}
\end{cases}
$$

$$
\begin{cases}
\begin{aligned}
&-PI^+\cos(\varphi_+ - \gamma_-) - BI^+\cos(2\gamma_- - \varphi_+) - BI^+cos\varphi_+cos2\gamma_- \\
&+GI_c^-sin(\theta_- - \gamma_-) + GI_c^0 sin(\theta_0 + \gamma_-) \\
&= -KI_c^-sin\theta_+ - 2R_{arm}I_c^-cos\theta_+ \\
&-PI^+sin(\varphi_+ - \gamma_-) - BI^+sin(2\gamma_- - \varphi_+) - BI^+cos\varphi_+sin2\gamma_- \\
&-GI_c^-cos(\theta_- - \gamma_-) - GI_c^0 cos(\theta_0 + \gamma_-) \\
&= KI_c^+cos\theta_+ - 2R_{arm}I_c^+sin\theta_+
\end{aligned}
\end{cases}
\tag{18}
$$

where,

$$
A = \frac{\sqrt{2}N(M^+)^3}{32\omega_1 C_{sub}} \quad B = \frac{\sqrt{2}NM^+(M^-)^2}{16\omega_1 C_{sub}} \quad C = \frac{\sqrt{2}N(M^-)^3}{32\omega_1 C_{sub}}
$$

$$
D = \frac{3\sqrt{2}N}{16\omega_1 C_{sub}} \quad E = \frac{N(M^+)^2}{6\omega_1 C_{sub}} \quad F = \frac{N(M^-)^2}{6\omega_1 C_{sub}} \quad G = \frac{NM^+M^-}{6\omega_1 C_{sub}}
$$

$$
H = \frac{N}{4\omega_1 C_{sub}} \quad P = \frac{\sqrt{2}N(M^+)^2 M^-}{32\omega_1 C_{sub}} \quad K = E + F + H - 4\omega_1 L_{arm}.
$$

Compared to (17), the elimination of negative sequence current I^- does not remove the positive and zero sequence circulating current. In addition, they are generated not only by different combinations of positive sequence current I^+ and modulation index M^+, M^-, but also other circulating current components by a ratio of G. Due to such couplings, (18) becomes more difficult to solve, and the variation of circulating current magnitude and phase angles is not straightforward.

IV. SIMULATION AND EXPERIMENTAL VERIFICATION

A. Simulation Verification

To verify the derived steady state model, a three-phase modular multilevel rectifier with passive load is established in Matlab/Simulink. The main system parameters are listed in Table I. For high equivalent switching frequency and relatively simple implementation, the $N + 1$ level phase shift PWM (PS-PWM) modulation scheme is adopted. Additionally, the widely used capacitor voltage sorting and balancing control is applied to guarantee operation performance of MMC under both normal as well as unbalanced conditions.

Fig. 2 gives the comparison between simulation and calculation results of circulating current under SLG fault. When the negative sequence current control is not activated, a deviation appears between the calculated and tested magnitudes of circulating current. During the derivation of second order voltage ripple in (9), the negative sequence modulation index M^- is assumed to be zero. In simulation, nevertheless, a small M^- is introduced by the dc voltage feedback control, which mainly induces the mismatch in Fig. 2(a).

On the other hand, with negative sequence current control, since the assumption $I^- = 0$ in (13) always holds when dual current control operates, the mismatch between calculation and simulation results of positive and negative sequence circulating current in Fig. 2(b) is greatly reduced within the tested range.

Table I. Main circuit parameters of MMC inverter.

Parameter	Value	Parameter	Value
AC voltage v_{s1}	10 kV	Carrier frequency	2 kHz
Y/Δ transformer ratio	10 kV / 2.5 kV	Reactive power	0 MVar
DC Link voltage V_{dc}	5 kV	Active power	0.5 MW
No. of sub-modules in each arm N	4	Submodule capacitance	2000 μF
AC side inductance L_{AC1}, L_{AC2}	4.775 mH (0. 144 pu)	Arm inductance L_{arm}	5.635 mH (0. 17 pu)
ESR of AC inductors R_{AC}	0.09 Ω	ESR of arm inductors R_{arm}	0.106 Ω

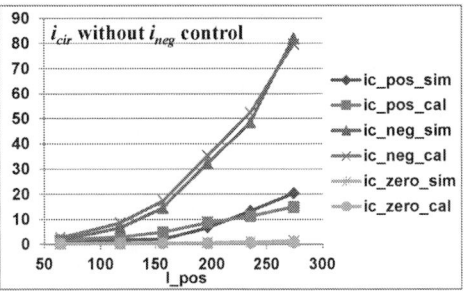

(a) Without negative sequence current control

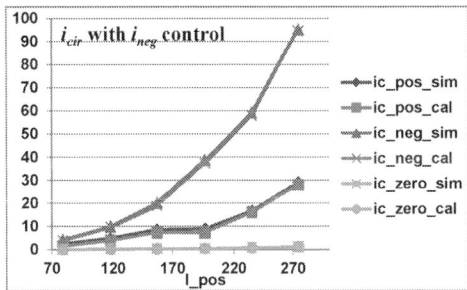

(b) With negative sequence current control

Fig. 2. Comparison between simulation and calculation results of circulating current in a modular multilevel rectifier under SLG fault.

Due to the large dc resistance, the magnitude of zero sequence circulating current is quite low (below 1.5 A). The negative sequence current control also leads to an apparent magnitude increase of negative and positive sequence circulating currents, as shown in Fig. 2(b).

B. Experimental Verification

As a further verification, a scaled-down three-phase MMC is built, as shown in Fig. 3. Given that the number of SMs has little impact on the model verification, four cells per phase is used in this paper to achieve low hardware requirement. The control schemes are implemented by dSPACE DS1103 and FPGA (Cyclone IV DE0-Nano).

To create serious unbalanced conditions, a SLG fault is emulated by setting phase C voltage to zero by an AC power supply (FCS Series II by California Instruments). The configuration of the modular multilevel rectifier is illustrated in Fig. 4, with its main circuit parameters given in Table II. To prevent potential damage or degradation of power semiconductor devices under fault conditions, the current reference is limited to 1.2 times of the rated phase current. Current flowing from dc to ac side is defined as the positive direction.

Fig. 5 shows selected experimental results with only positive sequence current control. When the SLG fault occurs, converter side phase current becomes unbalanced because of the negative sequence component.

Meanwhile, the zero sequence circulating current flows into the dc side, forming a second order dc current ripple. Its magnitude, however, is only 0.09 A (2.05 %) due to the large dc load resistance. The percentage values refer to comparison of current and voltage components to base dc voltage of 150 V and dc current of 4.98 A. The dc components in the three phase arm currents change from 1.66 A under normal condition to 2.33 A (140.4 %), 2.05 A (123.5 %) and 0.4 A (24.1 %) under SLG fault, which represents uneven active power contributions. Moreover, the three phase arm currents present different shapes and RMS values due to unbalanced phase and circulating current, resulting in unequal power losses across semiconductor devices. The second order ripple can also be observed in the dc voltage, which has a magnitude of 2.58 V (1.72 %). Additionally, due to the lack of dc central capacitors, the switching ripple cannot be filtered as in two level VSCs, which appears in dc voltage and current (Fig. 5(a) and (b)).

After the negative sequence current control is activated, converter side phase current becomes balanced, as shown in Fig. 6(a). This balanced phase current, however, results in a further drop of dc voltage, which reduces from 144.4 V (96.3 %) in Fig. 5(b) to 129.8 V (86.5 %) in Fig. 6(b).

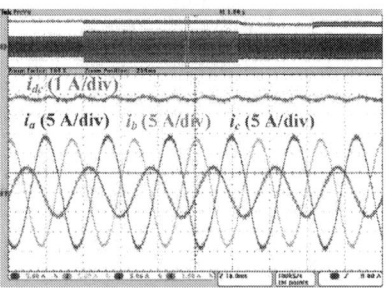

(a) Converter side phase current and dc current

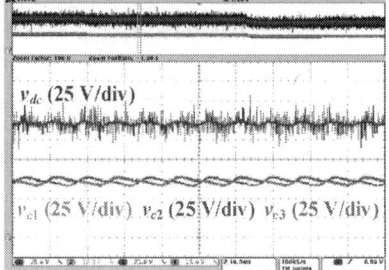

(b) Capacitor voltage in phase A and dc voltage

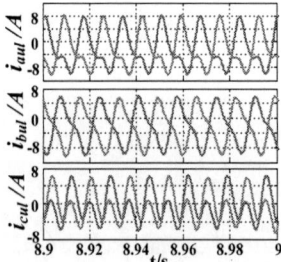

(c) Three phase arm currents

Fig. 5. Experimental results of a MMC rectifier under SLG fault, with only positive sequence current control.

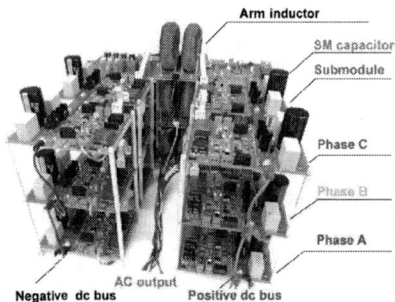

Fig. 3. Scaled down three phase MMC inverter.

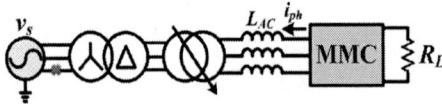

Fig. 4. Configuration of the experimental system.

Table II. Main circuit parameters of modular multilevel rectifier prototype.

Parameter	Value	Parameter	Value
DC Link voltage V_{dc}	150 V	Carrier frequency	1 kHz
AC voltage v_s	40 V	No. of sub-modules in each arm N	2
Load resistance R_L	30 Ω	Sub-module capacitance	2200 μF
AC inductance L_{AC}	1.8 mH	ESR of AC inductors	0.09 Ω
Arm inductance L_{arm}	0.74 mH	ESR of arm inductors	0.06 Ω

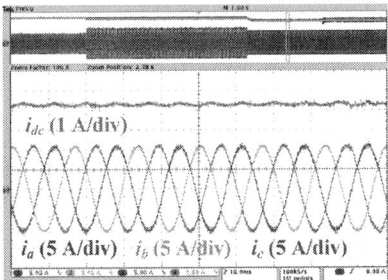

(a) Converter side phase current and dc current

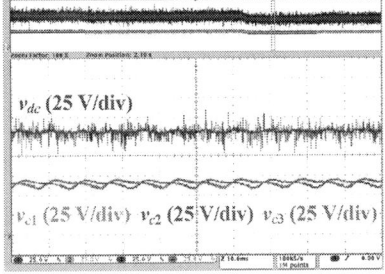

(b) Capacitor voltage in phase A and dc voltage

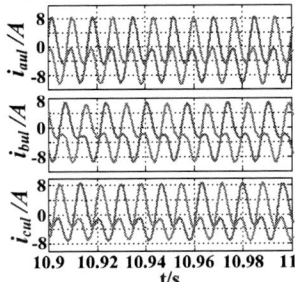

(c) Three phase arm currents

Fig. 6. Experimental results of a MMC rectifier under SLG fault,
with negative sequence current control.

Compared to Fig. 5, the second order ripple in both dc current and dc voltage are also reduced to 0.06 A (1.4 %) and 1.69 V (1.13 %) in this case. The dc components in the three phase arm currents drop to 1.89 A (113.9 %), 1.91 A (115.1 %), and 0.5 A (30.1 %).

When circulating current control is also activated in abc coordinates, the double line frequency positive, negative and zero sequence circulating current are greatly suppressed, enabling sinusoidal arm currents and a second order dc current ripple of 0.01 A (0.2 %) in Fig. 7. The dc components in three phase arm currents are almost the same as those in Fig. 6.

In spite of the elimination of circulating current, a second order dc voltage ripple with magnitude of 0.49 V (0.33 %) (Fig. 7(b)) still exists because of the first four items in the zero sequence components in (15), which may demand for special designed dc voltage ripple suppression control in real applications [5]. The additional switching actions induced by circulating current control add more switching ripple in the dc voltage and current [15].

Fig. 8 shows the comparison between experimental and calculation results of circulating current under SLG fault. Same as the simulation, when the negative sequence current control is not activated, a mismatch exists in the calculated and tested circulating current magnitudes. Otherwise, the mismatch is greatly reduced within the tested range in Fig. 8(b). In addition, the negative sequence current control also leads to higher negative and positive sequence circulating currents.

V. CONCLUSION

This paper presents a steady-state model of MMC to predict the circulating current under unbalanced conditions, considering the impact of the widely used negative sequence current control. The derived model is verified by simulation results obtained from a high voltage high power modular multilevel rectifier and experimental results from a scaled-down MMC prototype.

ACKNOWLEDGEMENT

This work was supported primarily by the Engineering Research Center Program of the National Science Foundation and Department of Energy under NSF Award Number EEC-1041877 and the CURENT Industry Partnership Program.

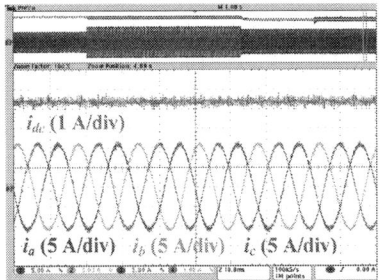

(a) Converter side phase current and dc current

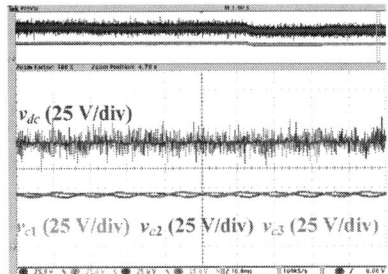

(b) Capacitor voltage in phase A and dc voltage

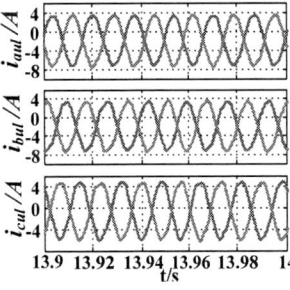

(c) Three phase arm currents

Fig. 7. Experimental results of a MMC rectifier under SLG fault, with negative sequence current and circulating current control.

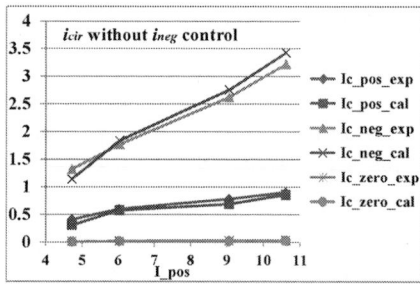

(a) Without negative sequence current control

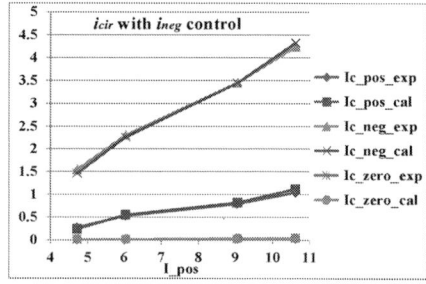

(b) With negative sequence current control

Fig. 8. Comparison between experimental and calculation results of circulating current in a MMC rectifier under SLG fault.

References

[1] J. Mei, Y Ji, X. Du, T. Ma, C. Huang, and Q. Hu, "Quasi-fixed-frequency hysteresis current tracking control strategy for modular multilevel converters," *IEEE Trans. Power Electron.*, vol. 14, no.6, pp. 1147-1156, Nov. 2014.

[2] Z. Zhao, J. Zhao and C. Huang, "An improved capacitor voltage balancing method for five-level diode-clamped converters with high modulation index and high power factor," *IEEE Trans. Power Electron.*, vol. PP, no. 99, pp. 1-14, 2015.

[3] N. Flourentzou, V. G. Agelidis, and G. D. Demetriades, "VSC-Based HVDC power transmission systems: an overview," *IEEE Trans. Power Electron.*, vol. 24, no. 3, pp. 592-602, Mar. 2009.

[4] M. Callavik, "ABB-HVDC grids for integration of renewable powersources," presentation at *EPRI* HVDC & *FACTS Users Meeting*, Oct., 2010, Palo Alto, CA. Available online: http://www05.abb.com.

[5] X. Shi, Z. Wang, B. Liu, Y. Liu, L. M. Tolbert, and F. Wang, "Characteristic investigation and control of modular multilevel converter based HVDC system under single-line-to-ground fault conditions," *IEEE Trans. Power Electron.*, vol. 30, no. 1, pp. 408–421, Jan. 2015.

[6] Y. Suh and T. A. Lipo, "Control scheme in hybrid synchronous stationary frame for PWM AC/DC converter under generalized unbalanced operating conditions," *IEEE Trans. Ind. Appl.*, vol. 42, no. 3, pp. 825–835, May/Jun. 2006.

[7] M. Guan and Z. Xu, "Modeling and control of a modular multilevel converter-based HVDC system under unbalanced grid conditions," *IEEE Trans. Power Electron.*, vol. 27, no. 12, pp. 4858–4867, Dec. 2012.

[8] Q. Tu, Z. Xu, Y. Chang, and L. Guan, "Suppressing DC voltage ripples of MMC-HVDC under unbalanced grid conditions," *IEEE Trans. Power Del.*, vol. 27, no. 3, pp. 1332–1338, July 2012.

[9] J. Moon, C. Kim, J. Park, D. Kang, and J. Kim, "Circulating current control in MMC under the unbalanced voltage," *IEEE Trans. on Power Del.*, vol. 28, no. 3, July 2013, pp. 1952–1959.

[10] Y. Zhou, D. Jiang, J. Guo, P. Hu, and Y. Liang, "Analysis and control of modular multilevel converters under unbalanced conditions," *IEEE Trans. Power Electron.*, vol. 28, no. 4, pp. 1986–1995, Oct. 2013.

[11] H. Song and K. Nam, "Dual current control scheme for PWM converter under unbalanced input voltage conditions," *IEEE Trans. Ind. Electron.*, vol. 46, no. 5, pp. 953–959, Oct. 1999.

[12] L. Xiao, S. Huang, and K. Lu, "DC-bus voltage control of grid-connected voltage source converter by using space vector modulated direct power control under unbalanced network conditions," *IET on Power Electronics*, vol. 6, no. 5, pp. 925–934, 2013.

[13] Q. Song, W. Liu, X. Li, H. Rao, S. Xu, and L. Li, "A steady-state analysis method for a modular multilevel converter," *IEEE Trans. Power Electron.*, vol. 28, no. 8, pp. 3702–3713, Aug. 2013.

[14] L. Harnefors, S. Norrga, A. Antonopoulos, and H.-P. Nee, "Dynamic modeling of modular multilevel converters," in *Proc. European Conference on Power Electronics and Applications*, 2011, pp. 1-10.

[15] Y. Li, E. A. Jones, and F. Wang, "Switching-frequency ripple on DC link voltage in a modular multilevel converter with circulating current suppressing control," in *IEEE Applied Power Electronics Conference and Exposition*, 2014, pp. 191–195.

978-1-4673-9551-9/16 $31.00 © 2016 IEEE

Design and Control of a Compensated Submodule Testing Scheme for Modular Multilevel Converter

Yuan Tang, Li Ran, Olayiwola Alatise, Philip Mawby

School of Engineering
The University of Warwick
Coventry, United Kingdom, CV4 7AL
yuan.tang@warwick.ac.uk

Abstract— **Modular multilevel converters (MMCs) for high power applications may contain over a thousand submodules. In situations such as submodule design or innovation, to build a full system with as many levels to test the submodule is time consuming and costly. To solve the problem, testing schemes have been developed to allow individual submodules to be tested without a complete MMC system. This paper presents a new submodule testing scheme that builds on a recent development to increase testing capability. The architectural approach is introduced along with design and control techniques. A prototype test platform, designed to test submodules with 400 V rated voltage and 14 A current, has been built and tested. It is shown that the output characteristics of the tested submodule (submodule capacitor voltage, electro-thermal characteristics of power switches etc.) are very close to what it would be in a complete MMC system.**

Keywords—Design validation, electromagnetic and electro-thermal characteristics, MMC, submodule, testing

I. Introduction

The modular multilevel converter (MMC) is a promising technology for the next generation of high-power voltage source converters (VSCs) [1,2]. Typically, as shown in Fig.1, a dc to 3-phase ac MMC would consist of six arms, each containing a string of submodules (SMs). In the case of SM design or topology innovation, it would be important to verify the design performance before final assembly, in terms of electro-magnetic, electro-thermal and electro-mechanical characteristics under practical operation conditions. As building a complete MMC for SM testing would be both time consuming and costly. Power supplies as well as galvanic isolations for such testing with voltage level up to a few hundred kilovolts are hard to achieve. As a solution, various methods were proposed to test the SM individually without a complete MMC [3-8]. Such single SM testing method can be applied to any well-balanced system with identical SMs. The test of one SM can well predict the performance of the rest. There are two major challenges in the single SM testing. First, the arm current that passes through the SM usually contains not only a sinusoidal component at the fundamental line frequency, but also a dc offset as well as low order harmonic circulating current components. Second, the switching frequency of the SM can be as low as the line frequency that is independent of the current passing through. Hence, to accomplish the test, a current source is required that is able to generate the reference

'arm current' without the cooperation (switching) of the SM under test. Most of the test methods presented in the past [3-7] failed to address the two requirements simultaneously.

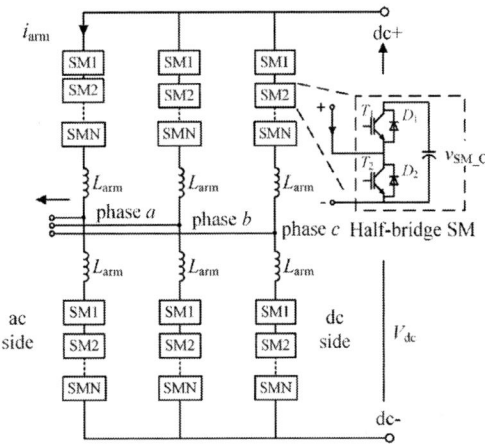

Fig. 1. Typical diagram of a modular multilevel converter (half-bridge SM)

To solve the problem, as shown in Fig.2, Tang *et al.* [8] proposed to use a full-bridge converter with a coupling inductor to emulate the reference 'arm current' through hysteresis control. Both the generated 'arm current' and the SM switching sequence are very close to the situation as if the SM were installed in a complete MMC. However, in order to generate the current with the hysteresis control, the dc supply voltage of the full-bridge must be higher than the peak SM capacitor voltage. In some cases, the required dc supply voltage would be more than double the rated voltage of the tested SM, which calls for much higher voltage withstanding capability of the power switch in the full-bridge converter. For tested SMs with rated voltage lower than 2 kV, the required dc supply voltage may be as high as 4 kV. In such a case, IGBT/diode power modules with higher voltage level (such as 6.5 kV) are still commercially available for the full-bridge converter. However, if the SM rated voltage is higher than 2 kV, leading to more than 4 kV dc supply voltage, none of the existing IGBT/diode power modules can be directly used. Series connection of power switches in the full-bridge may have to be adopted to withstand the much higher dc supply voltage, leading to complex circuitry and voltage balancing scheme.

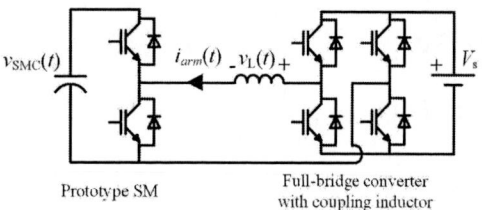

Fig. 2. Circuit diagram of the reported SM testing circuit extracted from [8]

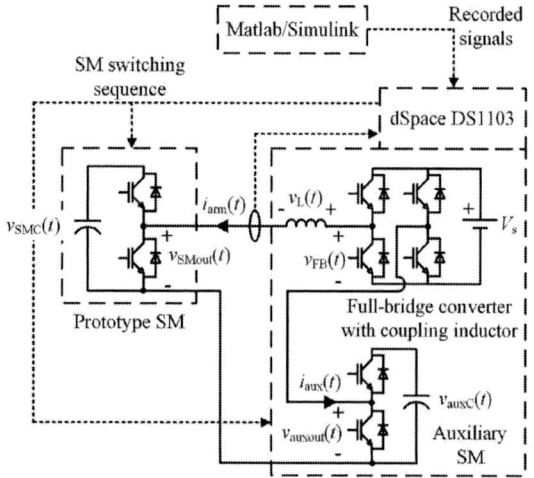

Fig. 3. Compensated SM testing scheme with half-bridge prototype SM

In order to improve the test capability of the existing test platform as in Fig.2, this paper proposes the use of an auxiliary SM to compensate the large dc voltage in the capacitor of the SM under test. By doing that, the dc supply only needs to be higher than the SM capacitor voltage ripple (ac variation) that is much lower than the SM rated voltage, leading to largely reduced demand of the dc supply as well as the voltage with-standing capability of the power switches in the full-bridge.

The rest of the paper is organized as follows. Section II explains the proposed compensated SM testing scheme and the control strategy. Components selection will be provided in Section III. Section IV presents the experimental results to show the effectiveness and accuracy of the testing scheme and finally Section V concludes the paper.

II. COMPENSATED MODEL ASSISTED SM TESTING SCHEME: METHOD AND CONTROL STRATEGY

A. Basic Concept and Circuit Diagram

Fig.3 shows the proposed compensated SM test platform with offline simulation as proposed in this paper. The Matlab/Simulink is used as the reference signal generator and a dSpace DS1103 as the interface between the master computer and the platform. Similar to the uncompensated test platform as in Fig.2, the current source consists of a full-bridge converter and a coupling inductor. As the difference, an auxiliary SM is connected in series with the prototype SM in a reverse way. Both SMs have the same current and voltage ratings. Note the

ratings of the auxiliary SM can be higher than the prototype. With the same switching sequence, the two SM capacitors will be switched into the circuit simultaneously and the dc component in both capacitor voltages will be cancelled out leaving only small ac variations (voltage ripples). The voltage ripple of the two SMs would have the same shape but opposite directions if identical capacitors are used. They are added together and the dc supply voltage V_s only needs to be higher than the peak total voltage ripple to achieve the hysteresis current control, which is usually much lower than the rated voltage of the prototype SM.

B. Control Strategy – Hysteresis Control

Fig.4 shows the control diagram of the proposed test platform. As discussed in [8], since the PI and the proportional-resonant (PR) controllers for single phase inverters can only operate for a single frequency, in order to generate the desired 'arm current' with not only the fundamental but also dc offset and other low order harmonics, multiple PI or PR controllers have to operate concurrently [9-11]. As a result, the controller design is complex. In contrast, the hysteresis current control offers a simple way to track the current reference with multiple components. This method is unconditionally stable with good transient response and accuracy [12,13]. It is therefore adopted. Simulation results showed that the hysteresis current controller is adequate, and the high frequency current tracking errors introduced by the hysteresis switching have negligible impacts on the SM under test.

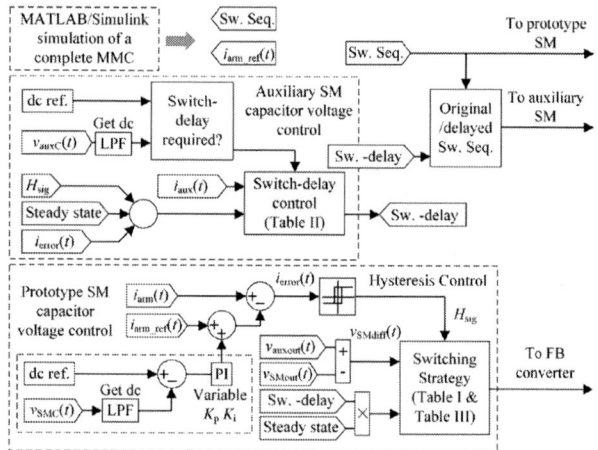

Fig. 4. Control diagram of the compensated SM test bench

The hysteresis current control is realized by changing the voltage on the coupling inductor through switching the full-bridge converter. In order to limit the di/dt of the generated 'arm current', the switching of the full-bridge shall not only consider the current tracking error $i_{error}(t)$, but also the outputs of the two SMs. In order to include such effect, $v_{SMdiff}(t)$ is defined that equals the output voltage difference of the two SMs due to the reversed-series-connection configuration. The other control variable that affects the switching of the full-bridge is related with capacitor voltage control of the auxiliary SM. In such a case, the two SMs will not be simultaneously switched in or out as to be explained later. Here, the case of simultaneous switching is analyzed first.

$i_{error}(t)$ is the difference between the actual 'arm current' $i_{arm}(t)$ achieved by the hysteresis switching and its reference $i_{arm\_ref}(t)$. The hysteresis band is defined as $\pm H_{band}$. The selection of H_{band} will be introduced later. When $i_{error}(t)$ is above the hysteresis band $+H_{band}$, the output signal of the hysteresis block will be $H_{sig}=1$, indicating that the actual 'arm current' is too large and needs to be reduced. When $i_{error}(t)$ is below $-H_{band}$, H_{sig} will be 0, indicating that the actual 'arm current' is too small and needs to be increased.

As shown in Fig.3, the voltage on the coupling inductor $v_L(t)$ equals

$$v_L(t) = v_{FB}(t) + v_{SMdiff}(t). \tag{1}$$

where $v_{FB}(t)$ is the output voltage of the full-bridge. $v_{SMdiff}(t)$ can be derived by

$$v_{SMdiff}(t) = v_{auxout}(t) - v_{SMout}(t) \tag{2}$$

where $v_{auxout}(t)$ is the output voltage of the auxiliary SM and $v_{SMout}(t)$ the output voltage of the prototype (see Fig.3 for reference directions). When the capacitor of the auxiliary or the prototype SM is switched into the circuit, the output voltage $v_{auxout}(t)$ or $v_{SMout}(t)$ will equal to its capacitor voltage $v_{auxC}(t)$ or $v_{SMC}(t)$. When the SM is bypassed, the output voltage is approximately zero.

The switching strategy of the full-bridge converter is simple when both SMs are bypassed. In such a case, the full-bridge will output either $+V_s$ or $-V_s$ when the 'arm current' is too small or too large. When the two SMs are switched into the circuit, as (1) shows, the inductor voltage is a combined effect of both the full-bridge output and the SM outputs. When the actual 'arm current' is too small and needs to be increased ($H_{sig}=0$), $v_L(t)$ must be positive. One way to generate the positive voltage is through switching the full-bridge to have $v_{FB}(t)=V_s$. This applies when $v_{SMdiff}(t)$ is negative or positive but with small amplitude. As $v_{SMdiff}(t)$ further increases, the total voltage ripple of the two SM capacitors is enough to increase the 'arm current' at a comparable rate. In such a case, the full-bridge converter will only output zero voltage to limit the value of $v_L(t)$, which helps to limit the di/dt of the current and also the switching frequency. As a result, the coupling inductance can be very small and the switching frequency of the full-bridge would only vary in a small range. The threshold voltage to decide whether the full-bridge converter shall output V_s or 0 is defined as V_{thres}. The situation is similar when the actual 'arm current' is too large and needs to be reduced ($H_{sig}=1$). Due to symmetry, the threshold voltage in this case will be $-V_{thres}$. Detailed output strategy for the full-bridge converter is concluded in Table I.

One way to select the threshold voltage V_{thres} is to ensure that the changing rate of the current when $v_{FB}(t)=0$ is always faster than the case when the full-bridge outputs $\pm V_s$. For instance, when $i_{arm}(t)$ is too small ($H_{sig}=0$), the slowest rising rate of $i_{arm}(t)$ in the case when $v_{FB}(t)=V_s$ corresponds to the minimum coupling inductor voltage as

$$V_{Lmin} = V_s + V_{SMdiffmin} \tag{3a}$$

$$V_{SMdiffmin} = -\left(\hat{V}_{SMCac} + \hat{V}_{auxCac}\right) \tag{3b}$$

where $\hat{V}_{SMCac}$ and $\hat{V}_{auxCac}$ are the amplitudes of the prototype and the auxiliary SM capacitor voltage ripples respectively. For simplicity, the capacitor voltage ripple is assumed symmetrical around its time average (dc component). The small difference between the positive and negative peaks is neglected. Two parameters, k_{SMCac} and k_{auxCac} are defined as the ratio between the peak-to-peak voltage ripple and the rated dc voltage. That is

$$k_{SMCac} = \frac{2\hat{V}_{SMCac}}{V_{SMCdc}} \tag{4a}$$

$$k_{auxCac} = \frac{2\hat{V}_{auxCac}}{V_{auxCdc}} \tag{4b}$$

where V_{SMCdc} and V_{auxCdc} are the rated dc voltage of the prototype and the auxiliary SMs respectively. In order to achieve better voltage cancellation effect, the rated dc voltage of the auxiliary SM is set equal to that of the prototype.

As stated above, when the full-bridge converter outputs zero voltage, the minimum coupling inductor voltage would equal to V_{thres}. The selection of V_{thres} must be large enough so that the rising rate of $i_{arm}(t)$ is faster than the slowest case given by (3a). On the other hand, since the maximum inductor voltage will be $(V_{thres}+V_{dc})$ when $v_{FB}(t)=V_s$, V_{thres} needs to be chosen as small as possible to limit the maximum $i_{arm}(t)$ rising rate. Hence, a suitable value for V_{thres} can be derived by substituting (3b), (4a) and (4b) into (3a)

$$V_{thres} = V_s - \frac{1}{2}(k_{SMCac} + k_{auxCac})V_{SMCdc} \tag{5}$$

where V_{auxCdc} equals V_{SMCdc} when needed. Due to symmetry, the negative threshold voltage is set at $-V_{thres}$. In some cases, when both capacitor voltage ripples are so small that the threshold voltage is never met, the full-bridge converter will always output $\pm V_s$.

TABLE I.   OUTPUT STRATEGIES FOR THE FULL-BRIDGE CONVERTER

Prototype and auxiliary SMs are simultaneously switched in/out			
i_{arm} is too small ($H_{sig}=0$)		$v_{SMdiff}(t) \leq V_{thres}$	$v_{SMdiff}(t) > V_{thres}$
	$v_{FB}(t)$	$+V_s$	0
	$v_L(t)$	$V_s+v_{SMdiff}(t)$	$v_{SMdiff}(t)$
i_{arm} is too large ($H_{sig}=1$)		$v_{SMdiff}(t) < -V_{thres}$	$v_{SMdiff}(t) \geq -V_{thres}$
	$v_{FB}(t)$	0	$-V_s$
	$v_L(t)$	$v_{SMdiff}(t)$	$-V_s+v_{SMdiff}(t)$

C. Control Strategy – Capacitor Voltage Control

In steady state, the time average (dc component) of a SM capacitor voltage will be constant for any given 'arm current' and switching sequence from computer simulation. In practice, due to the switch-ON dead-band, non-ideal characteristics of power switches, and current tracking error of the current control, the capacitor dc voltage may not be balanced anymore [8]. As shown in Fig.4, the prototype SM adopts the voltage regulator proposed in [8] through injecting small compensation current into the 'arm current'. The value of the injected current is derived by a PI controller with variable gains. The input of the PI controller is the error between the capacitor dc voltage reference and the actual measurement. During energizing and discharging of the SM capacitor, the proportional gain (K_p) is large and the integral gain (K_i) is very small for fast transient response. When the system reaches steady state, K_p is largely reduced and K_i is increased so that the PI controller will then be dominated by the integral part with very small bandwidth to limit noise effect. The advantage of this voltage regulator is that no time delay will be introduced between the 'arm current' and the switching sequence of the SM to ensure high accuracy of the test. In steady state, the injected current is usually less than 1% of the peak 'arm current' and it has been shown to have negligible impacts on the test results [8].

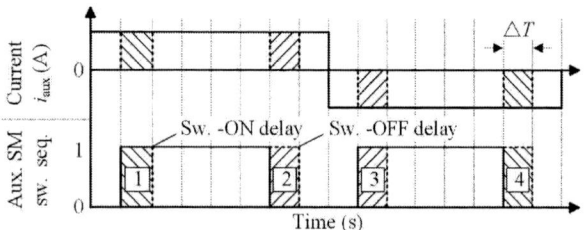

Fig. 5. Auxiliary SM (aux. SM) capacitor voltage regulator through switch delay

For the auxiliary SM, however, the aforementioned voltage regulator will have opposite effects on its capacitor dc voltage $v_{auxCdc}(t)$ due to the reversed-series-connection configuration. Positive injected current will actually reduce $v_{auxCdc}(t)$. To solve this problem, Fig.5 shows a switch-delay control for the auxiliary SM, which is independent from the prototype SM voltage regulator. The top graph is the current flowing into the auxiliary SM $i_{aux}(t)$ (see Fig.3) and the bottom shows its switching sequence, where '1' stands for SM switched-in and '0' means SM bypassed. ΔT is the sampling period of the test platform. When $i_{aux}(t)$ is positive, if a switch-ON delay is applied, the capacitor will not get charged during the delay period, and if a switch-OFF delay is applied, the capacitor will get further charged. In the other case, when $i_{aux}(t)$ is negative, the capacitor will be discharged less or get further discharged if switch-ON or –OFF delay is applied.

Note that when the switching of the auxiliary SM is delayed, the two SMs will be no longer switched simultaneously. $v_{SMdiff}(t)$ will equal to one full SM capacitor voltage, either $v_{SMC}(t)$ or $v_{auxC}(t)$, which is usually much higher than V_s. In such a case, the full-bridge converter will not be able to control the 'arm current' $i_{arm}(t)$ and the dynamics of $i_{arm}(t)$ will be dominated by $v_{SMdiff}(t)$. For instance, when $v_{auxCdc}(t)$ is too high, switch-OFF delay can be applied when $i_{aux}(t)$ is negative to get further discharged. At this moment, only the auxiliary SM capacitor is connected in the circuit and $v_{SMdiff}(t)$ equals $v_{auxC}(t)$ according to (2). As V_s is always smaller than $v_{auxC}(t)$, $i_{arm}(t)$ tends to increase in any event. In order to ensure that the current error is always within the allowable band, the switch-OFF delay can only be applied when $i_{arm}(t)$ actually needs to be increased ($H_{sig}=0$), and the current tracking error $i_{error}(t)$ is less than a threshold value I_{thres-}. I_{thres-} is usually negative and is set

to leave enough margin for $i_{error}(t)$ to increase. If the permitted maximum current tracking error is defined as ΔI_{max} and the maximum change of $i_{error}(t)$ during a switch-delay period is dI_{error_delay}. The current threshold can be derived as

$$I_{thres-} = \Delta I_{max} - dI_{error\_delay}. \tag{6a}$$

In the same case ($v_{auxCdc}(t)$ is too high), when $i_{aux}(t)$ is positive, if the switch-ON delay is applied and the auxiliary SM capacitor will not be charged during the delay period. At this moment, $v_{SMdiff}(t)$ would equal to -$v_{SMC}(t)$ and $i_{arm}(t)$ tends to decrease in any event. Hence, the switch-ON delay can only be applied when $i_{arm}(t)$ actually needs to be reduced ($H_{sig}=1$) and $i_{error}(t)$ is higher than a positive threshold current I_{thres+}. Similar to (6a), the value of I_{thres+} can be derived by

$$I_{thres+} = -\Delta I_{max} + dI_{error\_delay} \tag{6b}$$

Detailed switch-delay strategies for the auxiliary SM are concluded in Table II. As the conditions in the brackets would slow down the dynamics of the voltage control, they are only applied in steady state. In transient, the switch-delay is used whenever $v_{auxCdc}(t)$ is outside the permitted range. The period of the switch-delay is usually chosen as short as possible to limit the resulting large current error. In this paper, it equals the sampling period ΔT (50 μs) of the test platform dSpace 1103.

TABLE II.   SWITCH-DELAY STRATEGIES FOR THE AUXILIARY SM

$v_{auxCdc}(t)$	$i_{aux}(t) \leq 0$ (flow out)	$i_{aux}(t) > 0$ (flow in)
Too high	Switch-OFF delay ($H_{sig} = 0$ & $i_{error}(t) < I_{thres-}$)[a]	Switch-ON delay ($H_{sig} = 1$ & $i_{error}(t) > I_{thres+}$)[a]
Too low	Switch-ON delay ($H_{sig} = 1$ & $i_{error}(t) > I_{thres+}$)[a]	Switch-OFF delay ($H_{sig} = 0$ & $i_{error}(t) < I_{thres-}$)[a]

[a]. Only applied in steady state.

Table III offers another measure to limit the current error during the switch-delay period through compensating the large $v_{SMdiff}(t)$. The full-bridge converter will output $-V_s$ when the switch-OFF delay is used and will output $+V_s$ when the switch-ON delay is used.

TABLE III.   OUTPUT STRATEGIES FOR THE FULL-BRIDGE CONVERTER

Auxiliary SM switch-delay active (steady state)		
Sw.-OFF delay	$v_{FB}(t)$	-V_s
	$v_L(t)$	$-V_s + v_{auxC}(t)$
Sw.-ON delay	$v_{FB}(t)$	$+V_s$
	$v_L(t)$	$V_s - v_{SMC}(t)$

III.   COMPONENTS SELECTION FOR THE CURRENT SOURCE

The primary objective of the test platform is to provide the desired 'arm current' and inject it into the prototype SM. Due to the constant switching of the SM capacitor, the load conditions of the full-bridge converter vary frequently. Hence, in order to keep the current tracking error within an allowable range at all times, parameters of the test platform must be carefully chosen. In this section, dynamic characteristics of the current tracking error are firstly analyzed. Based on the understanding, a method is provided to select the coupling inductance L and the full-bridge converter dc supply voltage V_s.

A. Current Tracking Error

Fig.6 shows a typical waveform of the current tracking error $i_{error}(t)$ in the proposed test platform. The outermost dashed lines define the permitted maximum current error range $[-\Delta I_{max}, +\Delta I_{max}]$ that $i_{error}(t)$ must never exceed. The innermost double dashed lines are the thresholds for the hysteresis control. When $i_{error}(t) > +H_{band}$, $H_{sig}=1$ and when $i_{error}(t) < -H_{band}$, $H_{sig} = 0$. Due to the limited sampling frequency f_s of the test platform, the actual current error may be much larger than the defined hysteresis band. That is the reason a margin is left between the outer $\pm\Delta I_{max}$ and inner $\pm H_{band}$. The minimum width of this band equals the maximum change of the current error in a sampling period ΔT, i.e. dI_{error_max}. The remaining two double-dots dashed lines are the current thresholds I_{thres+} and I_{thres-} for the auxiliary SM switch-delay control.

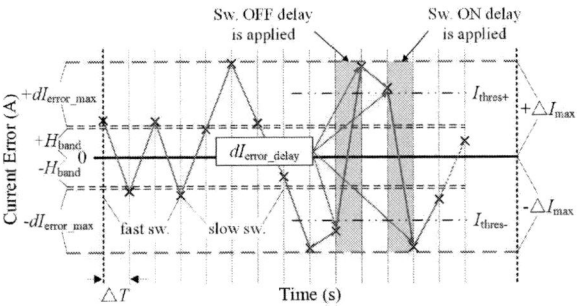

Fig. 6. A typical waveform of the current error $i_{error}(t)$

B. Selection of Control Parameters

The maximum permitted current error $\pm\Delta I_{max}$ defines the current tracking performance of the test platform. One way to determine $\pm\Delta I_{max}$ is according to the resulting error in the experiment outputs, such as losses in the power switches. Detailed selection process can be found in [8]. In order to quantify and compare the current tracking performance of different test platforms, ΔI_{max} can be written as

$$\Delta I_{max} = k_{\Delta I_{max}} A \qquad (7)$$

where A is the peak amplitude of the fundamental component in the reference 'arm current' and $k_{\Delta I_{max}}$ is defined as error constant. Usually, $k_{\Delta I_{max}}$ is around 0.1–0.15 depending on the current tracking accuracy requirement of certain tests.

Once the permitted current error $\pm\Delta I_{max}$ is decided, the hysteresis band $\pm H_{band}$ can be selected. As the hysteresis band directly decides the switching of the full-bridge converter, one design criterion of $\pm H_{band}$ is to ensure that the upper switching frequency limit of the full-bridge (f_{sw_max}) will never be exceeded for heat dissipation and EMC reasons. As shown by the first few current error segments in Fig.6, the full-bridge would switch at very sampling instant as the current error keeps falling outside the region $[-H_{band}, +H_{band}]$. In such a case, the actual switching frequency will be as high as half of the sampling frequency f_s. If the permitted switching frequency is lower than that value, the width of the hysteresis band must be larger than the maximum current error change dI_{error_max} in a sampling period that can be derived as

$$dI_{error\_max} = \frac{1}{L} V_{Lmax} \Delta T + \omega A \Delta T \qquad (8)$$

where V_{Lmax} is the maximum inductor voltage, ω the MMC ac side line frequency. Here, in estimation the harmonic circulating currents are neglected because of their much lower amplitudes. A fixed relation can be found between H_{band} and dI_{error_max} to ensure that f_{sw_max} is never exceeded as [8]

$$H_{band} = \frac{1}{2}\left\lfloor \frac{f_s}{2f_{sw\_max}} \right\rfloor dI_{error\_max} . \qquad (9)$$

C. Selection of Component Parameters

This section will focus on the parameter selection for the coupling inductance L and the dc supply voltage V_s. Firstly, in order to ensure the current error is always within the permitted range $[-\Delta I_{max}, +\Delta I_{max}]$, the minimum gap between $\pm\Delta I_{max}$ and $\pm H_{band}$ equals the maximum current error change dI_{error_max}, that is

$$\Delta I_{max} - H_{band} \geq dI_{error\_max} . \qquad (10)$$

Substituting (8) and (9) into (10) gives (11) as

$$\frac{V_{Lmax}}{L} \leq \frac{2\Delta I_{max} f_s}{2 + \left\lfloor \dfrac{f_s}{2f_{sw\_max}} \right\rfloor} - \omega A . \qquad (11)$$

Note that $\Delta T = 1/f_s$. (11) gives the upper limit of the ratio between V_{Lmax} and L. Another design criterion is that the rising speed of the 'arm current' must be faster than its reference at any time. With the harmonic components neglected, the fastest rising speed of the reference occurs when the current crosses its dc offset from underneath. At that instant, the slope equals to ωA and (12) must be satisfied

$$\frac{V_{Lmin}}{L} > \omega A \qquad (12)$$

where V_{Lmin} is the minimum inductor voltage.

Due to symmetry, only the case when $i_{arm}(t)$ is rising is considered. V_{Lmax} and V_{Lmin} are the maximum and minimum inductor voltage when it is positive. The above equations also apply when $i_{arm}(t)$ is decreasing.

According to (3)-(5) and Table I, V_{Lmax} and V_{Lmin} can be derived as

$$V_{Lmax} = 2V_s - \frac{1}{2}\left(k_{SMCac} + k_{auxCac}\right)V_{SMCdc} \qquad (13a)$$

$$V_{Lmin} = V_s - \frac{1}{2}\left(k_{SMCac} + k_{auxCac}\right)V_{SMCdc} . \qquad (13b)$$

Note that the auxiliary SM switch-delay control is assumed inactive here.

Substituting (13a) and (13b) into (11) and (12) leads to

$$\omega LA + \frac{1}{2}\left(k_{\text{SMCac}} + k_{\text{auxCac}}\right)V_{\text{SMCdc}} < V_{\text{s}} \leq$$

$$\frac{\Delta I_{\max} f_{\text{s}} L}{2 + \left\lfloor \dfrac{f_{\text{s}}}{2f_{\text{sw\_max}}} \right\rfloor} - \frac{\omega LA}{2} + \frac{1}{4}\left(k_{\text{SMCac}} + k_{\text{auxCac}}\right)V_{\text{SMCdc}} . \quad (14)$$

As the maximum value of (14) must be larger than the minimum, leading to the lower limit of the inductance L as

$$L > \frac{\dfrac{1}{2}\left(k_{\text{SMCac}} + k_{\text{auxCac}}\right)V_{\text{SMCdc}}}{\dfrac{2k_{\Delta I_{\max}} f_{\text{s}}}{2 + \left\lfloor \dfrac{f_{\text{s}}}{2f_{\text{sw\_max}}} \right\rfloor} - 3\omega} \cdot \frac{V_{\text{SMCdc}}}{A} . \quad (15)$$

where $\Delta I_{\max}$ equals $k_{\Delta I \max} A$ when needed.

(14) and (15) give the criteria to choose V_{s} and L for the current source. If the platform is required to test SMs with different ratings, the inductance can be chosen as the highest value for all cases. Then V_{s} can be decided by (14).

Note that, (13a) does not consider the auxiliary SM switch-delay. When all parameters are decided, the current thresholds $I_{\text{thres+}}$ and $I_{\text{thres-}}$ can be derived by (6a) and (6b). If the two thresholds are outside the permitted region $[-\Delta I_{\max}, +\Delta I_{\max}]$, the delay period shall be reduced or the supply voltage V_{s} shall be increased to further compensate the large inductor voltage due to the switch-delay control.

IV. PROTOTYPE DESIGN AND EXPERIMENT RESULTS

A hypothetic, 21-level, 106 kVA grid-connected MMC with half-bridge SMs is modelled in Matlab/Simulink. The converter is connected to a ±4 kV dc-link and a 4.7 kV (line-to-line rms), 50 Hz ac grid. There are 20 SMs in each arm with a 400 V rated SM dc voltage. In steady state when the system operates as a rectifier absorbing 106 kW active power from the grid with unity power factor, signals with a duration of one second of both arm current and switching sequence of one SM in the upper arm of phase a are recorded and then used to run the experiment.

In the experimental setup, the SM rated dc voltage is 400 V and the 'arm current' contains a -4.4 A dc offset and a fundamental frequency component with a peak amplitude of $A = 9.8$ A and other harmonics with lower amplitudes. The peak arm current is around -14 A and the permitted maximum current error $\pm\Delta I_{\max}$ is set at ±10% of it or ±1.4 A. The maximum switching frequency $f_{\text{sw\_max}}$ of the IGBT/diode power module in the full-bridge is set at 6 kHz. The sampling frequency f_{s} of the dSpace DS1103 platform is 20 kHz. The capacitor voltage ripple of the prototype SM is 15% peak-to-peak of the SM rated dc, or $k_{\text{SMCac}}=0.15$ and that of the auxiliary SM is 7.5% due to larger capacitance value. k_{auxCac} equals 0.075.

According to (15), the minimum L is 4.8 mH. If L is chosen to be 5 mH, V_{s} can be selected between 60.4 V and 61.5 V according to (14). If V_{s} is set to 61 V, when the switch-delay control is applied, V_{Lmax} changes to the difference between one

peak SM capacitor voltage and V_{s} that is 369 V (=430 V–61 V). $I_{\text{thres-}}$ is derived to be -2.44 A using (6a), when needed $dI_{\text{error\_delay}}$ is derived using (8), V_{Lmax} equals 369 V and ΔT=50 μs. Since $I_{\text{thres-}}$ is smaller than $-\Delta I_{\max}$ (-1.4 A), V_{s} needs to be increased. If L is chosen to be 10 mH, V_{s} can be chosen between 75.8 V to 100.4 V. V_{s} is set close to the upper limit 100 V to further compensate the SM output voltage mismatch during the switch-delay control. V_{Lmax} is derived to be 330 V and $I_{\text{thres-}}$ equals -0.4 A ($I_{\text{thres+}}$=0.4 A). The new $I_{\text{thres-}}$ is larger than $-\Delta I_{\max}$, leaving enough margin for the large current change due to the switch-delay control. Hence, V_{s}=100 V and L=10 mH are adopted. With the selected parameters, $dI_{\text{error\_max}}$ equals 0.93 A according to (8) when the switch-delay is inactive. Finally, the hysteresis band is derived to be ±0.47 A using (9). All parameters are listed in Table IV.

TABLE IV. PARAMETERS FOR THE EXPERIMENT

Item	Value
Current source	Full-bridge converter
V_{s}	100 V
L	10 mH @ 20 A$_{\text{dc}}$
V_{SMdc} (V_{auxCdc})	400 V
C_{SM}	373 μF
k_{SMCac}	0.15
C_{aux}	759 μF
k_{auxCac}	0.075
SM IGBT/diode module	Infineon IKW30N60T
Interface	dSpace DS1103

Fig.7 shows an oscilloscope snapshot of the experimental results. From top to bottom are the waveforms for the SM capacitor voltages (larger ripple – prototype and smaller ripple – auxiliary), the generated 'arm current' and the prototype SM switching sequence. The time average voltages of both SM capacitors are measured to be around the reference 400 V. The dc offset can be clearly seen in the 'arm current' generated by hysteresis switching.

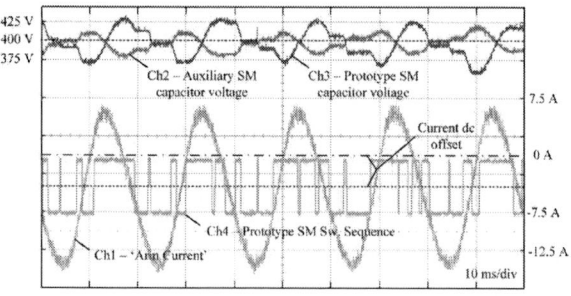

Fig. 7. Experiment results recorded by the oscilloscope. Horizontal Axis: 10 ms/div. Ch1-'arm current': 5 A/div, -0.5div offset (yellow). Ch2-auxiliary SM capacitor voltage: 50 V/div, -5.5div offset (red). Ch3-prototype SM capacitor voltage: 50 V/div, -5.5div offset (blue). Ch4-prototype SM switching seq-uence (green).

Fig.8 compares the experiment results to the complete model simulation. Fig.8(a) shows the prototype SM capacitor voltage in the experiment perfectly in line with the complete model simulation. In Fig.8 (b), the current flowing into the SM in the experiment tightly tracks the given reference. The two zoomed views give the details of the actual 'arm current' generated by the hysteresis switching. Fig.8(c) shows the actual injected current by the capacitor voltage regulator for the prototype SM. The current is nearly dc with a magnitude of around -130 mA, which is less than 1% of the peak 'arm current'. Current errors are shown in Fig.8(d) and (e). In the zoomed view in Fig.8(e), the largest current change in a sampling period is found to be 1.7 A due to the auxiliary SM switch-delay control. All current errors are within the designed ±1.4 A region at all times.

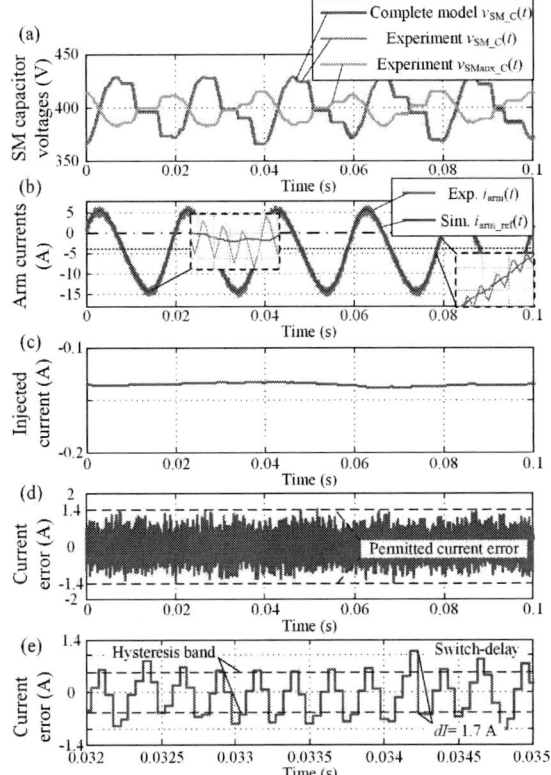

Fig. 8. Comparison between the experiment results and the complete model simulation: (a) SM capacitor voltages; (b) 'arm current'; (c) injected current; (d) current error; and (e) zoomed view of the current error

V. CONCLUSION

A compensated model assisted SM testing scheme for the MMC is proposed in this paper. During the test, when the SM is switched into the circuit, an additional auxiliary SM is used to compensate the dc component in the prototype SM capacitor voltage. As a result, the dc power supply of the full-bridge only needs to be higher than the total capacitor voltage ripple of the two SMs, which is much lower than the SM rated dc voltage. When compared with the original test method, to test SMs with

the same voltage and current ratings, the proposed method requires much lower dc supply voltage as well as much smaller coupling inductance. In addition, with the test equipment, the proposed method is able to test SMs with much higher rated voltage. In addition to the prototype SM capacitor voltage regulator, a new switch-delay controller is proposed to independently control the auxiliary SM capacitor (time-average) voltage. Experiment showed the effectiveness and performance of the proposed test method. During the test, the 'arm current' can be faithfully achieved, which contains not only the fundamental component but also the dc offset. The SM capacitor voltage is perfectly in line with the waveform given by simulation of a complete MMC system.

REFERENCES

[1] Y. Liu, A. Escobar-Mejia, C. Farnell, Y. Zhang, J. C. Balda, and H. A. Mantooth, "Modular multilevel converter with high-frequency transformers for interfacing hybrid DC and AC microgrid systems," in *Proc. 5th IEEE Int. Symp. Power Electron. Distrib. Generation Syst.*, 2014, pp. 1-6.

[2] Y. Tang, L. Ran, O. Alatise, and P. Mawby, "Capacitor Selection for Modular Multilevel Converter," in *Proc. IEEE ECCE*, Pittsburgh, PA, USA, Sep. 2014, pp. 2080-2087.

[3] C. Gao, X. Luo, X. Wei, and Z. Lv, "Steady-state operation test device of flexible direct current transmission MMC high-pressure submodule," China Patent CN201993425U, Sep. 28, 2011.

[4] G. Tang, K. Zha, C. Gao, X. Luo, and Y. Yang, "Steady-state operation test method for flexible direct-current power transmission modular multilevel converter (MMC) high-voltage sub-module," China Patent CN102175942(B), Jul. 2, 2014.

[5] P. Wang and Z. Chu, "Test device and test method for modularized multi-level current transformer sub-module," China Patent CN103018586A, Apr. 3, 2013.

[6] J. Feng, J. Ke, W. Deng, Z. Lv, and D. Liu, "Flexible direct current transmission sub-module test device and test method," China Patent CN103063945A, Apr. 24, 2013.

[7] T. Modeer, S. Norrga, and H. P. Nee, "Resonant test circuit for high-power cascaded converter submodules," in *Proc. 15th Eur. Conf. Power Electron. Appl.*, 2013, pp. 1–5.

[8] Y. Tang, L. Ran, O. Alatise, and P. Mawby, "A model assisted testing scheme for modular multilevel converter," *IEEE Trans. Power Electron.*, vol. 31, no. 1, Jan. 2016.

[9] U. A. Miranda, L. G. B. Rolim, and M. Aredes, "A DQ synchronous reference frame current control for single-phase converters," in *Proc. 36th IEEE PESC*, 2005, pp. 1377-1381.

[10] B. Bahrani, A. Rufer, S. Kenzelmann, and L. Lopes, "Vector control of single-phase voltage-source converters based on fictive-axis emulation," *IEEE Trans. Ind. Appl.*, vol. 47, no. 2, pp. 831-840, Mar./Apr. 2011.

[11] D. N. Zmood and D. G. Holmes, "Stationary frame current regulation of PWM inverters with zero steady-state error," *IEEE Trans. Power Electron.*, vol. 18, no. 3, pp. 814-822, May 2003.

[12] H. Mao, X. Yang, Z. Chen, and Z. Wang, "A hysteresis current controller for single-phase three-level voltage source inverters," *IEEE Trans. Power Electron.*, vol. 27, no. 7, pp. 3330-3339, Jul. 2012.

[13] K. M. Rahman, M. R. Khan, M. A. Choudhury, and M. A. Rahman, "Variable-band hysteresis current controllers for PWM voltage-source inverters," *IEEE Trans. Power Electron.*, vol. 12, no. 6, pp. 964-970, Nov. 1997.

A Voltage Independent Islanding Detection Method and Low Voltage Ride Through of a Two-Stage PV Inverter

Partha Pratim Das, Souvik Chattopadhyay, *Member, IEEE* and Shiladri Chakraborty
Department of Electrical Engineering
Indian Institute of Technology Kharagpur
Kharagpur - 721302, India
Email: partha08das@gmail.com, souvik@ee.iitkgp.ernet.in, shiladri@ee.iitkgp.ernet.in

Abstract—This paper presents an islanding detection method for a two stage PV inverter. Islanding condition is detected based on saturation of the PI controller of the outer voltage control loop. This makes the proposed detection algorithm more reliable since it is not immune to spurious misoperations due to sudden load changes. In addition, implementation strategy for low voltage ride through (LVRT) operation is also discussed. Numerical simulation results illustrate effectiveness of the proposed approach. Finally hardware results on a 2 KW laboratory prototype are presented, for experimental verification.

I. INTRODUCTION

Distributed generation (DG) and microgrid are getting increased attention day by day for their flexible power controlling capability and renewable energy applications. Photovoltaic (PV) cell is one of the most popular renewable sources of energy. One of the popular approaches for integrating PV panels to the grid is to use a two-stage inverter with a front-end boost (DC/DC) converter followed by an inverter [1]. The overall block diagram of such a two-stage grid-tied inverter system is shown in Fig. 1.

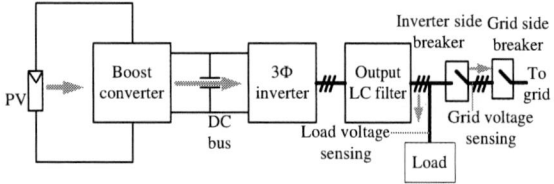

Fig. 1: Basic block diagram of the system.

As can be seen, there are two breakers, one each on the (local) load side and the utility supply side. In the event of grid non-availabilty due to upstream fault in the power system, it is likely that the utility-side breaker trips under the action of grid-side control. Under such a scenario, the inverter continues to pump power into the local load, resulting in islanding operation [2], [3]. It is sometimes desired for safety reasons that islanding operation should not be continued

This work is supported by Solar Energy Research Initiative (SERI) under Department of Science and Technology (DST), India at the Indian Institute of Technology Kharagpur.

and the inverter should be tripped [4], [5], [14]. This operating mode of grid connected inverters is known as anti-islanding.

It is clear that in the event of opening of the grid-side breaker following a grid outage, the grid-side voltage sensor starts sensing the inverter output voltage. Thus inherent detection of islanding condition is not possible by merely monitoring the grid-side sensor and specific detection mechanisms are necessary. Generally two approaches are adopted for this purpose. They are classified as - passive methods [6]-[8] and active methods [4]-[9]. Passive methods mostly sense the variation in voltage, frequency or harmonics for islanding detection. Those methods detect islanding very fast. However, passive methods are perceived to be less reliable as misoperation can occur in event of sudden load change [3]. In active methods, a small amount of reactive power or harmonics is continuously injected into the grid. Islanding is detected by sensing the change in voltage after injection of reactive power or harmonics. A major drawback of such active methods, is their detrimental effect on power quality. Moreover, sometimes a small perturbation of the injected variable into the grid is not enough to produce significant effect that can be monitored and acted upon.

In some weak grids, the grid voltage may decrease to a substantially lower voltage compared to the rated grid voltage, albeit for a small duration. In this situation, tripping the inverter based on instantaneous value is not a good solution as it would generate unnecessary interruptions to the local load. It is better to have a low voltage ride through (LVRT) functionality implemented in the inverter to handle such power system disturbances [11].

In this paper, an islanding detection method is proposed, where islanding is detected from saturation of the outer voltage control loop. An implementation of LVRT method is also discussed. The LVRT method rides through the low voltage that occurs for a small time. The LVRT method does anti-islanding when low voltage occurs for a long time.

The paper is organised as follows. After the introduction, proposed voltage independent islanding detection method is discussed. Implementation of LVRT method is also discussed in this section. The subsequent section discusses how islanding detection and LVRT work together to take care of all abnormal grid conditions. Simulations results are provided in the next section. Finally, experimental results of all the algorithms, tested in the laboratory prototype are provided.

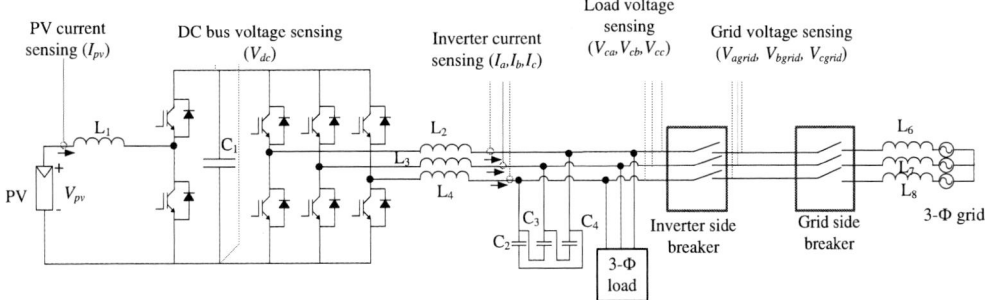

Fig. 2: Power circuit diagram of the system.

II. PROPOSED CONTROL TECHNIQUE

The power circuit diagram of a two stage grid connected PV inverter is shown in Fig.2. In normal operating condition, both the breakers remain connected. The grid in this condition works as a power balancing source to the PV inverter. The load voltage remains fixed by the grid.

the outer voltage control loop to its upper limit. The saturation block is also shown in the block diagram. The upper limit of the saturation block is the current reference coming from maximum power point tracking (MPPT) controller whereas the lower limit is zero. So, after the saturation, the boost converter operates with a current reference of I_{mpp}. The control loop and PLL are same as [13].

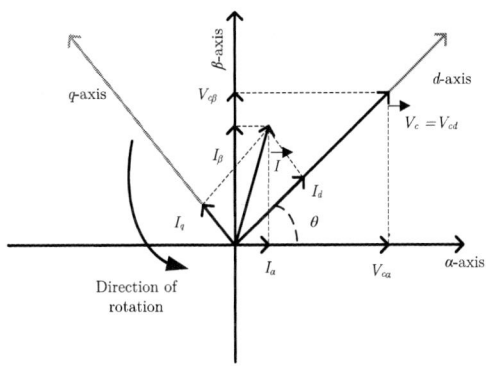

Fig. 3: d and q axis consideration.

The complete control block diagram of the system is shown in Fig. 5 [13]. In the grid connected (GC) mode of operation, the breaker remains connected. In both standalone (SA) and GC modes of operation, the DC voltage is controlled by the inverter. The DC bus voltage controller generates the d axis current reference.

The mutually orthogonal components of the unit vector ($\sin\theta$ and $\cos\theta$) are generated by SRF-PLL following an approach discussed in [13]. The d and q axis consideration is shown in Fig. 3. As shown in the figure, the d axis is taken along the filter capacitor voltage vector. So, V_{cq} is always zero. I_q^* is given zero for unity power factor operation. Now in GC mode of operation V_{cd} is same as $V_{d_{grid}}$, where $V_{d_{grid}}$ is the d axis grid voltage. $V_{q_{grid}}$ is zero as it is same with V_{cq} in GC mode of operation. As shown the block diagram, $V_{cd}^* = V_{d_{grid}} + \Delta V_d$ is given in GC mode, where ΔV_d is a small voltage. Now in GC mode, V_{cd} can never reach $V_{d_{grid}} + \Delta V_d$ as V_{cd} is same with $V_{d_{grid}}$. So, the reference sends a ΔV_d error to the PI controller of the outer voltage control loop. Eventually, this error saturates the PI controller of

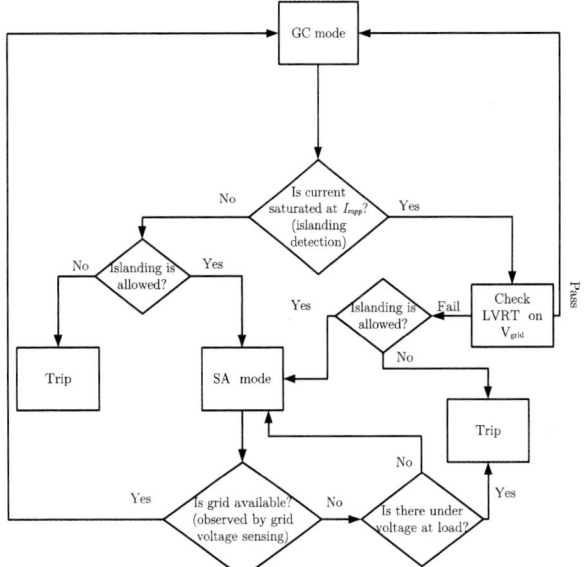

Fig. 4: Flowchart explaining operation of the system with islanding detection and LVRT.

Operation of the system with islanding detection and LVRT is shown in Fig. 4. As shown in Fig. 4, the system initially operates in GC mode. Now if islanding condition occurs, the system first detects it and goes into SA mode if operation in islanding mode is permitted. The system trips if the operation in islanding mode is not allowed. If islanding does not occur, the system checks for low voltage condition of the grid. If low voltage does not occur, the system continues to operate in GC mode. When some low voltage occurs, the system first does an LVRT test. If the system passes the LVRT test, it continues

978-1-4673-9551-9/16 $31.00 © 2016 IEEE   2653

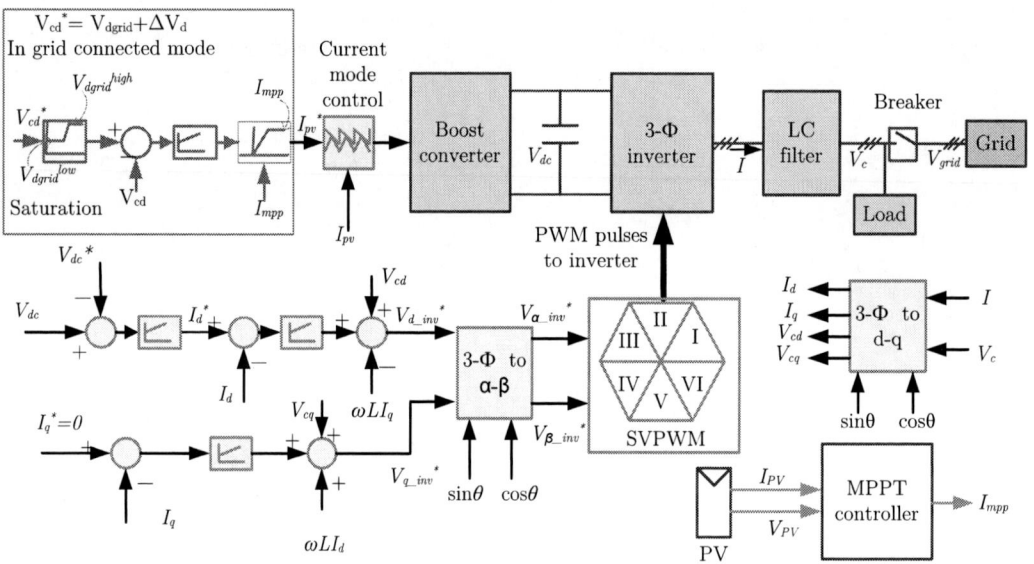

Fig. 5: Control block diagram of the system.

to operate in GC mode. If the system fails the LVRT test, the system goes into SA mode. If islanded operation is not allowed, system trips after LVRT test failure. After islanding detection or LVRT test failure, the system operates in SA mode, if the operation in islanding mode is allowed.

Sometimes the PV source does not have enough power to supply local load in SA mode. In such situations, low voltage occurs across the load. It is not safe to operate with low voltage as it can damage the local load. So, in SA mode if some low voltage occurs the system trips. The islanding detection method and LVRT are discussed later in this paper. Operation in islanding condition of a grid connected inverter is not always safe, as it can create safety problems for the workers or some equipment [14]. So, here we are considering that operation in islanding mode is not allowed. The system trips after islanding detection or LVRT test failure.

A. Islanding Detection

As already discussed, the PI controller of the outer voltage control loop eventually saturates in GC mode. After saturation, the boost converter operates with a current reference of I_{mpp} that comes from the MPPT controller of PV. So, in grid connected mode, system always operates on current control with I_{mpp} as the current reference. Now islanding condition occurs when the grid side breaker gets open. In this situation, the load voltage is not fixed by the grid anymore and can change. As the grid side breaker still remains connected, V_{cd} is same as $V_{d_{grid}}$. The command $V_{cd}^* = V_{d_{grid}} + \Delta V_d$ tries to increase the load voltage reference until the it reaches $V_{d_{grid}}^{high}$. Once the load voltage reference becomes same with $V_{d_{grid}}^{high}$, the load voltage reference will not increase anymore for the outer saturation block. After the load voltage becomes same with $V_{d_{grid}}^{high}$, PI controller of the outer voltage control loop comes out of saturation. Observing the PI controller coming out of saturation, islanding condition is detected. The control

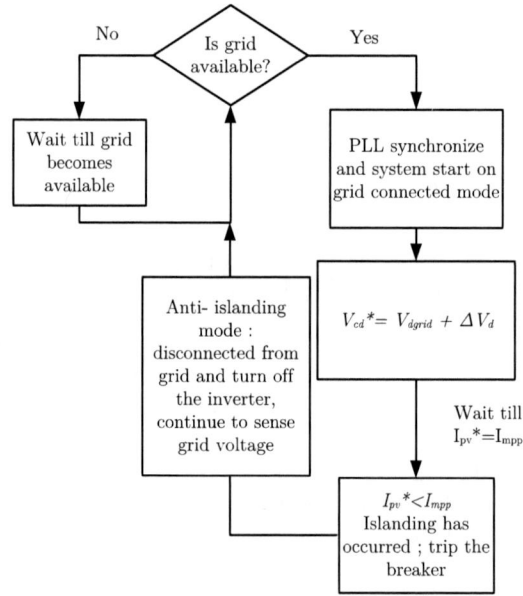

Fig. 6: Islanding detection algorithm.

algorithm is shown in Fig. 6. After grid becomes available, PLL synchronization is done according to [13]. Post phase synchronization, the inverter starts with a voltage reference of $V_{d_{grid}} + \Delta V_d$ and waits until the PI controller gets saturated. Thereafter, islanding detection starts. If the PI controller output is found to be outside saturation, it implies islanded condition has occurred and anti-islanding (i.e. disconnecting the inverter side breaker and withdrawing the gate pulses) is done.

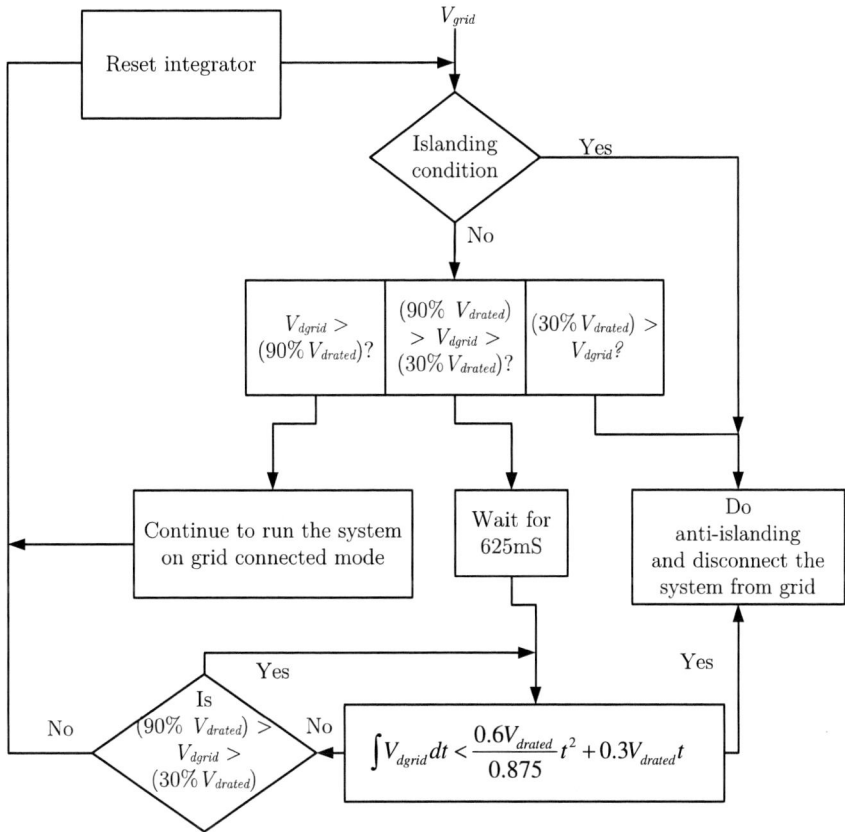

Fig. 7: Implementation logic for LVRT.

B. Low voltage ride through

Often a weak utility grid occurs where grid voltage becomes low and again comes back to its rated value. To help the grid in this situation, anti-islanding should not be done as soon as grid voltage falls below V_{grid}^{low}. Fig 8 shows voltage vs. time curve to be maintained during low grid voltage [11], [15].

If the grid voltage is below 30% of rated grid voltage, anti-islanding should be done instantly. If the grid voltage is above 90% of the rated grid voltage, the system should continue to run in grid connected mode. If the grid voltage is in between 30% and 90% of rated grid voltage, the system follows the curve to find allowable time to operate in this region. For finding the allowable time let us shift the origin to point 'a' (0,0.625). So from this point the equation of the curve can be written as

$$V_{d_{grid}} = \frac{0.6V_{d_{rated}}}{0.875}t + 0.3V_{d_{rated}} \; . \qquad (1)$$

If we take the moving average of grid voltage, it becomes

$$\frac{1}{t}\int V_{d_{grid}}dt = \frac{0.6V_{d_{rated}}}{0.875}t + 0.3V_{d_{rated}} \; . \qquad (2)$$

So, anti-islanding is done if the following condition is satisfied

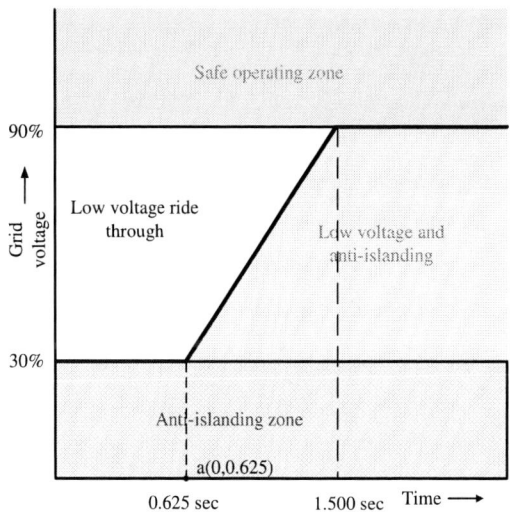

Fig. 8: Voltage vs. time curve for LVRT operation.

$$\int V_{d_{grid}} dt < \frac{0.6 V_{d_{rated}}}{0.875} t^2 + 0.3 V_{d_{rated}} t . \qquad (3)$$

The shift of the origin is done by giving a delay of 625 msec after low voltage occurs. The control algorithm of the system is shown in Fig. 7, where $V_{d_{rated}}$ is the rated grid voltage and $V_{d_{grid}}$ is the present grid voltage. As shown in the figure, the system first checks for islanding condition. If islanding occurs, the system does anti-islanding. If islanding does not occur, the system checks for low voltage. If the grid voltage is more than 90% of rated grid voltage, the system continues to operate in GC mode. If the grid voltage is less than 30% rated grid voltage, the system does anti-islanding. If the grid voltage is in-betweens 30% and 90% of rated grid voltage, system first waits for 0.625 sec. After 0.625 sec, the system checks for the condition given in (3).

If the condition is not satisfied, the system continues to work in GC mode, else the system does anti-islanding. After anti-islanding the system waits till the rated grid voltage becomes available again.

Under a faulty grid condition, two outcomes are possible - islanded operation and voltage sag. The islanded condition is detected using the first algorithm. On the other hand, if the voltage falls below grid voltage for the stipulated duration, LVRT algorithm comes into operation for doing anti-islanding. So, by performing islanding detection and LVRT method in conjunction, all faulty grid conditions can be detected.

III. SIMULATION RESULTS AND DISCUSSIONS

Proposed algorithms are verified using the PLECS simulation platform. The converter switching frequency is 10 KHz. For testing the islanding detection algorithm, the system is initially allowed to operate in the in GC mode for a small time so that the PI controller of the outer voltage control loop gets saturated. After the PI controller saturation, islanding detection can be started. Fig. 9 shows the simulation result of islanding detection. For generating the islanding condition, the grid side breaker is opened intentionally at 1.005 sec. So, at 1.005 sec the grid current becomes zero. After the grid side breaker gets opened, the reference increases till it reaches its upper limit. The load voltage also increases till it reaches its upper limit. Once it reaches its upper limit, the outer voltage control loop comes out of saturation. So I_{pv}^* starts decreasing. Detecting the current is out of saturation, islanding is detected. After islanding is detected, anti-islanding is done. So, the current becomes zero.

Fig. 10 - 12 shows the simulation results of LVRT. Fig. 10 shows the simulation results of LVRT operation where low voltage occurs for a small time. As shown in Fig. 10, the grid voltage decrease till 1.3 sec and then it goes back to its rated value. Here low voltage occurs for a small time. Here low voltage occurs for nearly 1 sec, which is less than allowable time in that condition. Hence, the system rides through the low voltage. The system continues to operate in grid connected mode in the same way. The signal of the inverter side breaker remains high. So, the system remains connected to the grid. Here,

$$x = \int V_{d_{grid}} dt , \qquad (4)$$

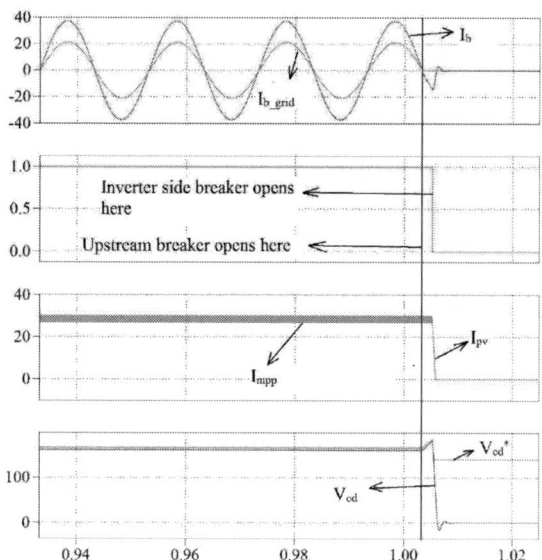

Fig. 9: Simulation results of islanding detection.

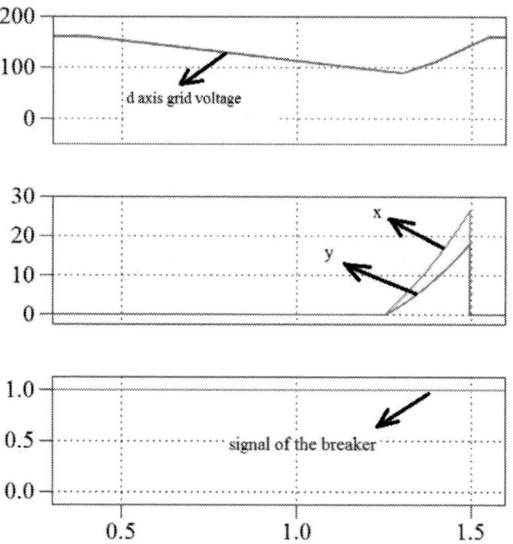

Fig. 10: LVRT operation when low voltage occurs for less time than allowable time.

$$y = \frac{0.6 V_{d_{rated}}}{0.875} t^2 + 0.3 V_{d_{rated}} t . \qquad (5)$$

Fig. 11 shows a situation where low voltage occurs for more time than the allowable time, as a result, the system disconnects itself from grid and anti-islanding is done. Here as shown in Fig. 11, after low voltage occurs, system first waits for 625 msec.

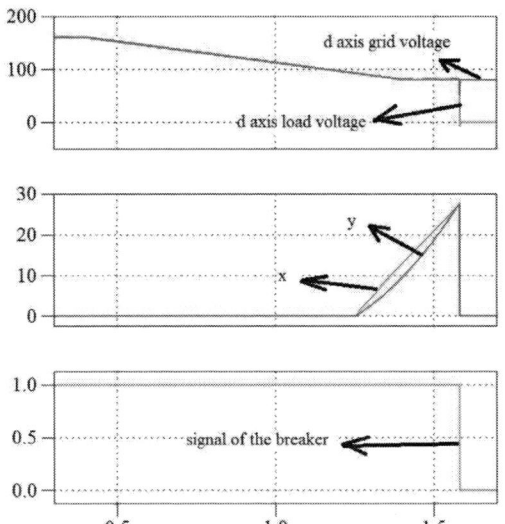

Fig. 11: LVRT operation when low voltage occurs for more time than allowable time.

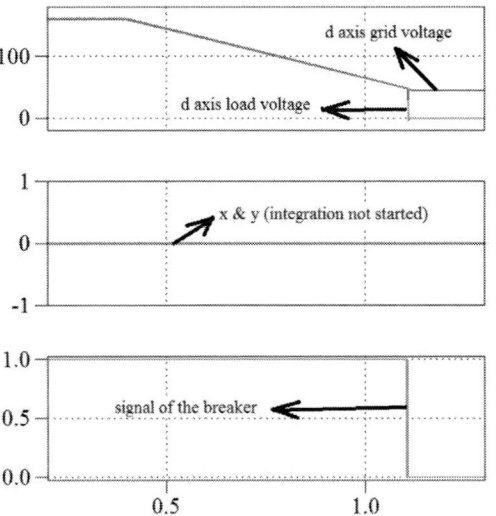

Fig. 12: LVRT operation when grid voltage becomes less than 0.3 p.u.

After that, the algorithm starts calculating the allowable time (here x and y calculation starts). When low voltage time becomes greater than the allowable time, anti-islanding is done. Here when x becomes less than y, anti-islanding is done. So, load voltage goes to zero.

Fig. 12 shows a situation where grid voltage becomes less than 30% rated grid voltage. As soon as the grid voltage falls below 30% rated grid voltage anti-islanding is done. Here the grid voltage becomes less than 30% rated grid voltage within 600 msec. So, the calculation of x and y does not start.

IV. EXPERIMENTAL RESULTS

The proposed control algorithm for islanding detection and LVRT are tested in the laboratory prototype. The experimental setup consists of one boost converter and one 3 phase inverter. Variable 3 phase resistive load is used as the local load. The control algorithm is implemented using TMS320F2812 digital signal processor. Switching frequency is chosen to be 10 kHz. The sampling frequency is chosen to be same as switching frequency. Fig. 13 shows the experimental result of islanding detection. Islanding is generated by opening the grid side breaker, so the grid current becomes zero. After sometimes islanding is detected and anti-islanding is done. So, load voltage and load current become zero. After anti-islanding, DC bus voltage falls with its dynamics as a small resistance is connected across the DC bus. The DC bus voltage saturates at the input voltage.

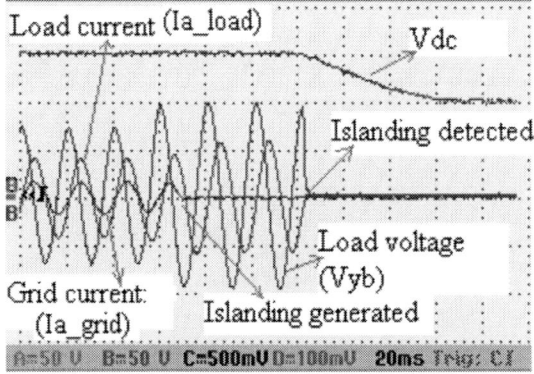

Fig. 13: Islanding detection and anti-islanding (current scale : 100 mV/A)

Fig. 14-16 shows the experimental results of LVRT operation. Fig. 14 shows a situation where low voltage occurs for less duration than allowable time. Here low voltage occurs for nearly 800 msec. The system rides through the low voltage and continues to operate in GC mode. The power of the PV source at MPP is constant. Hence as the grid voltage decreases, the grid current increases so that the power supplied by the inverter remains constant.

Fig. 15 shows a situation where low voltage occurs for more duration than allowable time. (In GC mode load voltage is same as grid voltage) After low voltage occurs, the system waits for 625 msec, after which calculation of x and y starts (x and y are as already defined in (4) and (5)). When y becomes more than x, the low voltage time exceeds the allowable low voltage time. Anti-islanding should be done at that moment. Here when y exceeds x, anti-islanding is done and the load voltage becomes zero. Fig. 16 shows a situation where grid voltage becomes less than 0.3 p.u. After low voltage occurs, system first wait for 625 msec before start calculating x and y. Here grid voltage becomes less than 0.3 p.u. before 625 msec. As soon as grid voltage becomes less than 0.3 p.u, anti-islanding is done. So, the load voltage becomes zero.

A picture of the complete experimental setup is given in Fig. 17.

978-1-4673-9551-9/16 $31.00 © 2016 IEEE

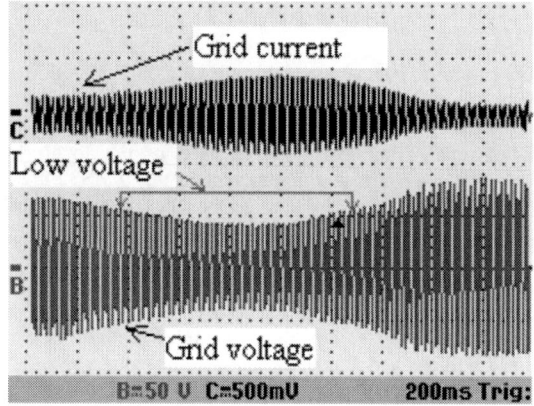

Fig. 14: LVRT operation where low voltage occurs for 800 msec (current scale : 100 mV/A)

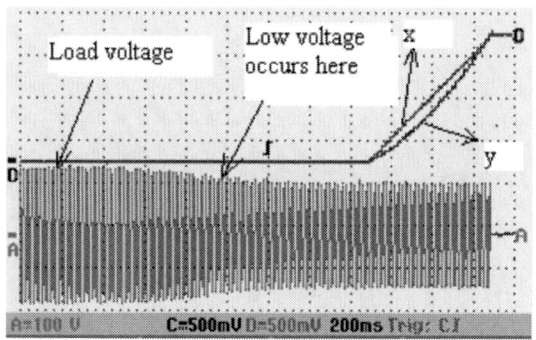

Fig. 15: LVRT operation where low voltage occurs for more time than allowable time.

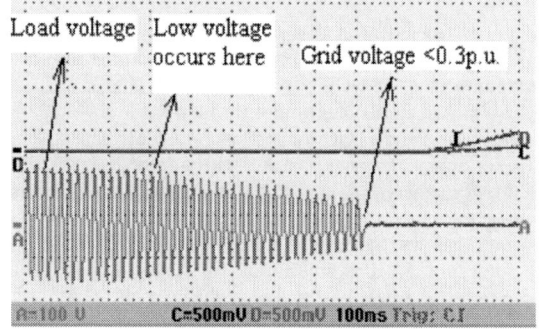

Fig. 16: LVRT when grid voltage is less than 0.3 p.u.

V. CONCLUSION

A control algorithm is developed for islanding detection and LVRT operation. Islanding is detected from the current saturation of the outer voltage control loop. The islanding detection algorithm does not directly depend upon the grid voltage sensing. An algorithm is also developed for implementation of LVRT operation of a two stage grid connected PV

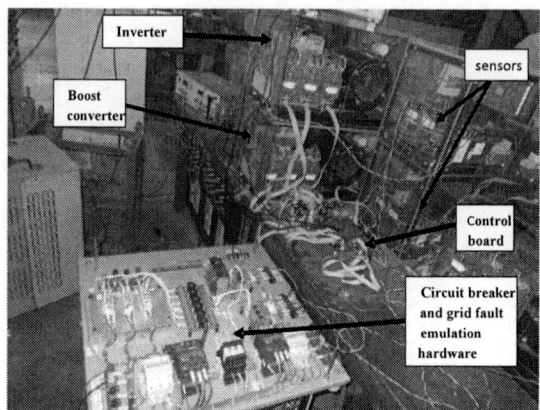

Fig. 17: Experimental test setup.

inverter. The system rides through a low voltage that occurs for less time than allowable time. Anti-islanding is done if low voltage occurs for a longer time than allowable time. The algorithm is tested in simulation. The total algorithm is also tested in laboratory prototype using a digital platform consisting of DSP TMS320F2812. The total algorithm ensures fast islanding detection and low voltage ride through with a reliable response.

REFERENCES

[1] F. Blaabjerg, R.Teodorescu,M. Liserre and A.V.Timbus, Overview of Control and Grid Synchronization for Distributed Power Generation Systems, *IEEE Transactions on Industrial Electronics* ,vol: 53, pp: 1398 - 1409.

[2] C.Jeraputra, and P.N. Enjeti, "Development of a robust anti-islanding algorithm for utility interconnection of distributed fuel cell powered generation," *IEEE Transactions on Power Electronics*, Sept. 2004 vol.19, no.5, pp.1163-1170.

[3] T.Funabashi, K. Koyanagi and R. Yokoyama, "A review of islanding detection methods for distributed resources," *IEEE Power Tech Conference Proceedings*, 2003 Bologna , June 2003 vol.2, Vol.2, pp:23-26.

[4] R.S.Kunte, and Wenzhong Gao, "Comparison and review of islanding detection techniques for distributed energy resources," *Power Symposium*, 2008. NAPS '08. 40th North American , Sept. 2008 vol.18, pp:28-30.

[5] Byunggyu Yu, Youngseok Jung, Junghun So, Hyemi Hwang and Gwonjong Yu, "A Robust Anti-islanding Method for Grid-Connected Photovoltaic Inverter," *IEEE World Conference on Photovoltaic Energy Conversion*, Conference Record of the 2006 4th, May 2006, vol.2., pp.2242-2245.

[6] H.Haeberlin and J. Graf, Islanding of Grid Connected PV Inverter: Test Circuits And Some Test Results *2nd world conference on photovoltaic solar energy conversion*, Vienna, Austria, 1998.

[7] H. Kobayashi, K. Takigawa, and E. Hashimoto, Method for preventing phenomenon of utility grid with a number of small scale PV systems, *IEEE Photovoltaic Specialists Conference*, 1991, pp.695-700 vol.1, 7-11 Oct 1991.

[8] A. Kitamura, M. Okamoto, F. Yamamoto, K. Nakaji, H. Matsuda, and K. Hotta, Islanding phenomenon elimination study at Rokko test center, *IEEE Photovoltaic Energy Conversion Conference*, 1994., pp.759-762 vol.1, 5-9 Dec 1994.

[9] M. E. Ropp, M. Begovic, and A. Rohatgi, Analysis and performance assessment of the active frequency drift method of islanding prevention, *IEEE Trans. On Energy Conv.*, vol. 14, pp. 810816, Sept. 1999.

[10] S. Ochs David , Sotoodeh Pedram and Mirafzal, Behrooz, A Technique for Voltage-Source Inverter Seamless Transitions Between Grid-

Connected and Standalone Modes *IEEE Twenty-Eighth Annual Applied Power Electronics Conference and Exposition* , 2013, pp 952 959.

[11] A.Marinopoulos, F.Papandrea, M.Reza, S. Norrga, F. Spertino, and R.Napoli Grid Integration Aspects of Large Solar PV Installations: LVRT Capability and Reactive power/Voltage support Requirements *IEEE Powertech conference*, 2011 pp:1-8.

[12] Li Jianlin, Zhuying, He Xiangtao and Xu Honghua, Study on Low Voltage Ride Through Characteristic of Full Power Converter Direct-Drive Wind Power System *IEEE Power Electronics and Motion Control Conference (IPEMC)* 2009., pp: 2213 2216.

[13] P. P. Das and S. Chattopadhyay "Smooth mode transition of A DC bus voltage controlled PV inverter using a novel phase locked loop method," *Applied Power Electronics Conference and Exposition (APEC), IEEE 2015*, pp.2133-2140, March 15-19 ,2015.

[14] J.Mulhausen, J.Schaefer, M.Mynam, A.Guzman, and M. Donolo, "Anti-islanding today, successful islanding in the future," *63rd Annual Conference for Protective Relay Engineers*, 2010, pp.1-8, March 29 2010-April 1 2010.

[15] Technical Guideline, Generating Plants Connected to the Medium-Voltage Network, Bundesverband der Energie- und Wasserwirtschaft e.V, BDEW, June 2008.

Low Cost and High Efficiency Topology for Flexible Integration of Multi-PV and Batteries in Resonant-based Converters

Ali Elrayyah

Qatar Environment and Energy Research Institute
Doha, Qatar
aelrayyah@qf.org.qa

Abstract—In this paper, a topology is proposed to integrate multi-PV modules and batteries with single resonant converter used to integrate PV sources with utility grids. The topology is proposed to reduce the cost of the converters while achieving maximum power point tracking for the individual modules. The structure provides high degree of flexibility as batteries can easily be interchanged with PV modules in the system without any need for hardware modifications. Three switches are needed per PV module and the highest stress over the switches does not exceed twice the voltage of a PV module. When batteries are integrated, they could operate in charging, discharging and bypassing modes through proper setting of the switches. The switches are turned ON under zero-voltage switching throughout the system operation to minimize the switching losses. The effectiveness of the proposed structure and the validity of its performance are verified through simulation studies.

Keywords—component; PV modules; resonant converters; PV-battery integration

I. INTRODUCTION

Nowadays, there is increasing interest in integrating more renewable energy sources (RES) with utility grids due to various technical and environmental purposes [1-4]. Among RESs, solar photovoltaic (PV) sources have witnessed a significant growth over the last few years [5]. The power electronics inverters that are used to interface PV modules with utility grids have great importance in the performance and cost of PV installations [6]. In central and string inverters, several PV modules are connected in series/parallel and single inverter interfaces them with the grid. However, the system efficiency could be deteriorated as it is affected by the production of underperforming modules. On the other hand, the system efficiency could be maximized by interfacing every PV module through a separate microinverter (MIC) [7-8]. Recently, resonant-based MICs have been capturing the interest since they reduce the cost, improve the efficiency and reduce the electrometric interference produced by MICs [9, 10].

Central and string inverters have relatively low cost since several PV modules share the same inverter. Utilizing this fact, in this paper, a topology is proposed to interface several PV modules with single resonant converter such that the system cost could be reduced. The proposed topology can be used to interface any number of modules which can be as low as two

and it can increase to reach any number. The topology is presented to serve as a universal converter that integrates PV modules and batteries. The switching between battery charging and discharging modes could easily be achieved through simple operation while maintaining the efficient operation. To keep the system cost at minimum level, the proposed topology requires low number of switches that experience low stresses.

II. CONVENTIONAL TOPOLOGY FOR MUTI-PV BASED RESONANT CONVERTERS

The converter in Fig. 1 is an effective candidate among the available topologies to integrate Multi-PV with single resonant converter. Every PV module has two switches: one switch is used to allow the current of the resonant circuit to flow over the module while the other is used to bypass that current by detaching the module from the resonant circuit. The operation of this circuit is simple. During every switching cycle, each of the PV modules is kept attached to the resonant circuit by turning ON the switch Q_{ax} for a duration that makes it track its maximum power point (MPP). Afterwards, the module gets detached by turning OFF Q_{ax} and turning ON Q_{dx}, $x=1, 2, .. N$. In this way, the current supplied by every PV module could be controlled separately allowing the MPP of the individual modules to be tracked precisely.

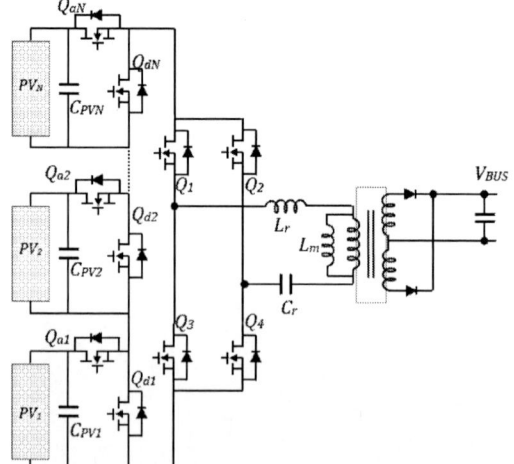

Figure 1. Conventional topology for multi-PV modules connected to the same resonant converter

---

This research is sponsored by Qatar Environment and Energy Research Institute for the project # GC5004.

In this topology, the required number of switches is *2N+4* where *N* is the total number of modules. The stress over the switches $Q1$-$Q4$ is $N \times V_{pv}$ where V_{pv} the voltage of a single module and therefore the rating these switches increases linearly with the value of *N*. Moreover, this fact limits the installation flexibly and scalability since as the number of modules increases the switched $Q1$-$Q4$ should be replaced to respond to the need of higher ratings devices. Note that in this topology, the power is allowed to flows in one direction only with respect to the connected modules. This feature prevents the flexibility of integrating batteries since the batteries require bidirectional current flow for charging and discharging modes of operation.

III. PROPOSED TOPOLOGY

Figure 2 shows a simple version of the proposed topology with only two modules to explain its operation principle. Two switches are connected across every module. When the resonant current i_r flows in its positive direction, one of the two switches across each modules is used to attach the module to the resonant circuit and the other switch is used to detach it. Unlike the topology in Fig. 1, the roles of the two switched is reversed when i_r becomes negative. This feature is very important to allow bidirectional current flow to charge and discharge batteries when they are integrated with the system.

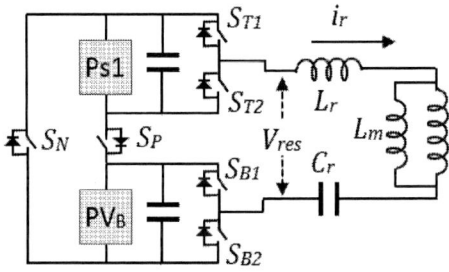

Figure 2. Simplest structure for the proposed topology

The operation of the proposed topology is shown in Fig. 3. Only the input side of the resonant converter is shown in these figures. Traditional circuits used at the outputs of resonant converters could be used at the output of the circuit in Fig. 3 since it does not affect the performance of the proposed topology.

As shown in Fig. 3(a), initially when i_r is in its positive half cycle, the switches S_{T1} and S_{B2} are turned ON while S_{T2} and S_{B1} are turned OFF. This makes the two modules get attached to the resonate converter and in this case the current flows through the switch S_P. Assume that the MPP operation of PV_T requires this module to get detached from the resonant circuit before PV_B. Then, by turning OFF S_{T1} the module PV_T will be bypassed by the diode of S_{T2} as shown in Fig. 3(b) and this switch can then be turned ON under zero-voltage switching (ZVS) to eliminate the diode conduction losses. In case that PV_B needs to be detached before PV_T, the switch S_{B2} can be turned OFF and the circuit will look like Fig. 3(c). In either case, when the module that remains attached to the resonant circuit get detached later, zero voltage will be applies across the resonant circuit as shown in Fig. 3(d). Based on the

selected switching frequency, the voltage across this circuit is required to be reversed at the end of the switching cycle. To apply the negative voltage the switch S_P needs to be turned OFF. In this case, the current flows over S_{T2} and S_{B1} and the diode of S_N, as shown in Fig. 3(e) which causes negative voltage polarity across the resonant circuit to be applied. S_N can then be turned ON under ZVS as indicated by Fig. 3(f). After the current changes polarity, the negative half cycle of the current starts. The same procedure is then repeated in the negative half cycle except that PV_T and PV_B are detached by turning OFF S_{T2} and S_{B1}, respectively.

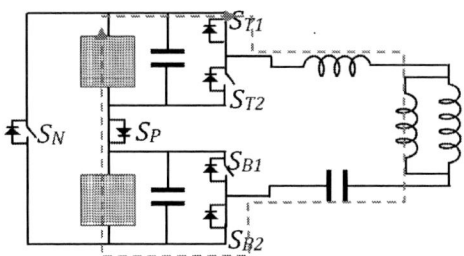

Fig. 3 (a)

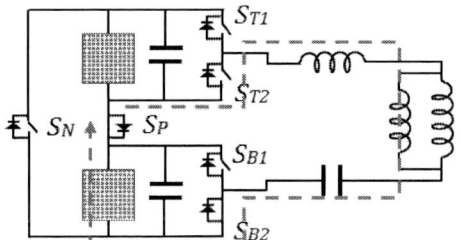

Fig. 3 (b)

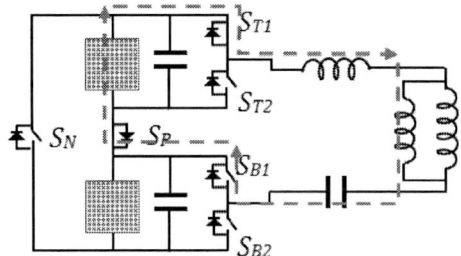

Fig. 3 (c)

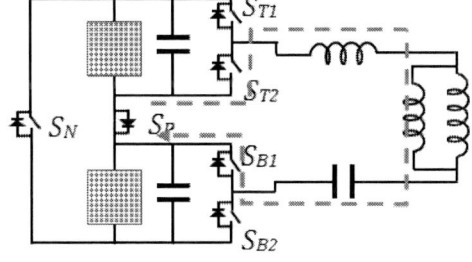

Fig. 3 (d)

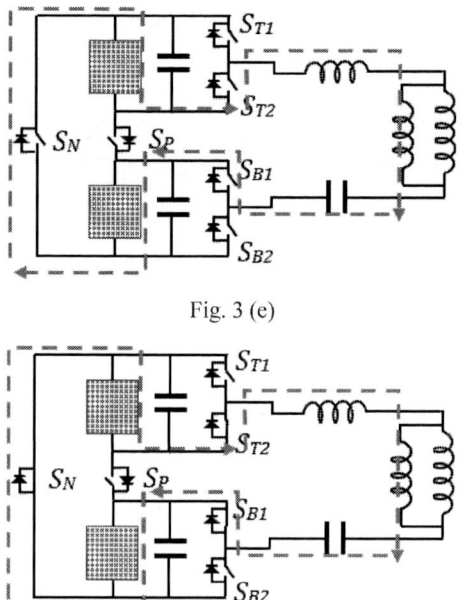

Fig. 3 (e)

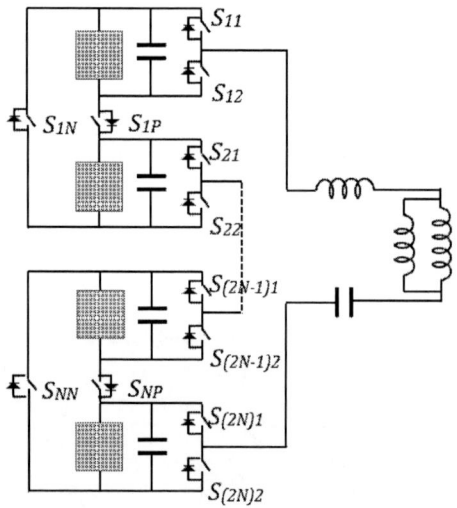

Fig. 3(f)

Figure 3. Operation of the proposed converter topology

IV. BATTERIES INTEGRATION IN THE PROPOSED TOPOLOGY

The structure shown in Fig. 3 shows the case of only two PV modules. To attached more than two modules, the generalized topology shown in Fig. 4 can be used. The circuit that represents the input to the resonant circuit in Fig. 3 is duplicated several times to attach more modules. Generally, six switches are needed to interface each pair of modules making the required number of switches to be $3N$ where N is the total number of PV modules.

Figure 4. Generalized structure of the proposed topology

By making a comparison study for the number of switches needed in the conventional topology shown in Fig. 1 and the one in Fig. 4 the following observation could be noticed. When two modules are integrated, only 6 switches are needed in the proposed structure which is two switches less than the number required by the topology in Fig. 1. For $N=4$, the two topologies requires the same number of switches and for $N>4$ the proposed topology requires more switches than the one in Fig. 1. However, unlike the topology in Fig. 1, the stresses over the switches S_{xP} and S_{xN} is fixed around $2V_{PV}$ unaffected by the value of N and this provides a design flexibility feature. To make the comparison on a common base, consider the case of using switches rated for the voltage of a single PV module and a number of them need to be connected in series to reach higher voltages. In the topology shown in Fig. 1, there is a need to use six switches per module; two switches for attaching and detaching the module and four for the H Bridge of the inverter. On the other hand, in the proposed topology shown in Fig. 3, two switches are needed for each of S_P and S_N as they experience a voltage stress of two modules. When these four switches are added to S_{T1}, S_{T2}, S_{B1} and S_{B2} the total number becomes eight switches for every pair of modules indicating four switches per module. Accordingly, the proposed topology offers reduction in the component count which leads to cost reduction.

Consider the case shown in Fig. 5 where two PV modules and two batteries are attached to the same circuit and no special arrangement is needed to integrate the batteries as they are treated just like any other PV module. The operation of the circuit in Fig. 5 is explained below.

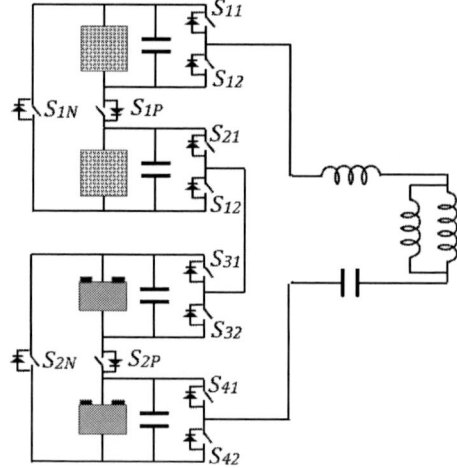

Figure 5. Batteries integration in the proposed topology

During the batteries discharging mode, the switches connected across the batteries are driven similar to the procedure described in the last section as shown in Fig. 6(a). In this case, the batteries are detached from the circuit on time durations that correspond to the required current to be supplied. Similar to the sequence shown in Fig. 3, the switches across the pair of batteries operate to supply current in positive half cycle and the zero voltage could be supplied and finally the opposite voltage is applied to start the negative half cycle.

978-1-4673-9551-9/16 $31.00 © 2016 IEEE

To operate the batteries in the charging mode, the operation sequence shown in Fig. 6(b)-(e) can be followed. In this case the operation of two switches across every battery is reversed. During the positive half cycle of i_r, the switches S_{32}, S_{41} and S_{2N} will be turned ON to allow the batteries to be charged by i_r as shown in Fig. 6(b) and when i_r is negative the switched S_{31}, S_{42} and S_{2P} are turned ON to allow the battery charging using this negative current as indicated in Fig. 6(e). As shown in Fig. 6(c) and (d), the transitions between the positive and negative half cycles is made by turning the switches ON under ZVS.

The batteries can also be operated under bypassing mode where no current is supplied by or to the batteries. This can be done very easily by turning S_{2P} ON during the positive half cycle while tuning S_{2N} ON during the negative half cycle. Moreover, it is also possible to operate one battery on charging or discharging while the other one is under bypassing mode. However, it is not possible to have one of the batteries pair operating under charging mode while the other under discharging mode at the same instant. Fortunately, this application could be achieved by making one battery in the bypassing mode and the other under charging/discharging mode and in the next cycles the roles could be interchanged.

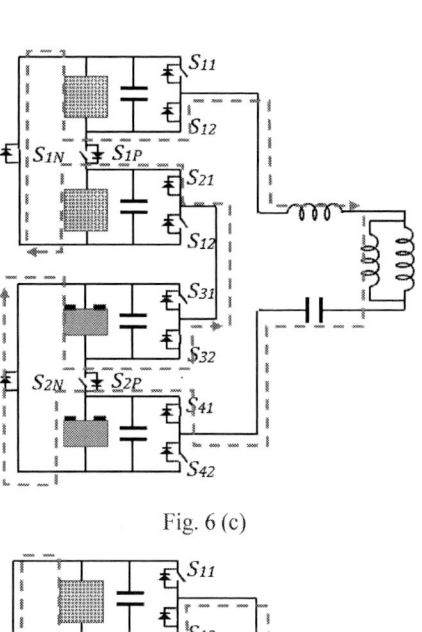

Fig. 6 (c)

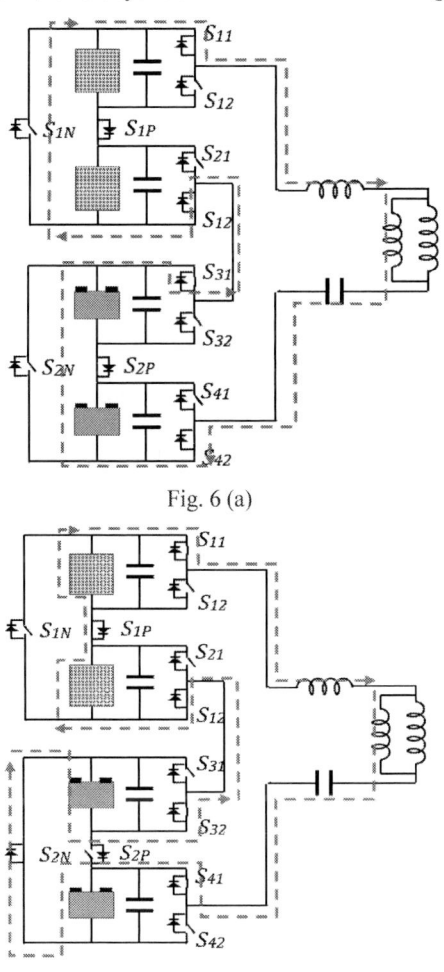

Fig. 6 (a)

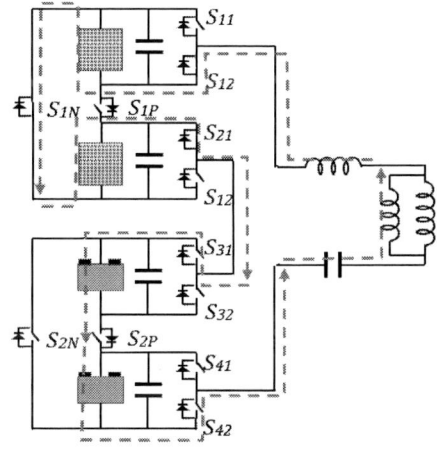

Fig. 6 (d)

Fig. 6 (b)

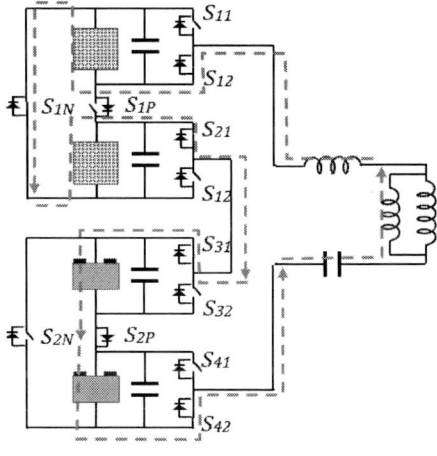

Fig. 6 (e)

Figure 6. Charging and discharging integrated batteries in the proposed system while maintaining high operation efficiency

The topology shown in Fig. 5 provides a simple example for explanation purpose. However, the connected number of PV modules and batteries can be adjusted freely accordingly to the requirement of the specific installation. Clearly, the number of connected PV modules and batteries could easily be modified in the field the thing that enhances the flexibility in any installation. Therefore, this topology could provide an effective, efficient and low cost method to integrate PV/battery hybrid sources with power systems.

V. SIMULATION RESULTS

The structure in Fig. 2 is simulated to verify the effectiveness of the proposed topology in achieving the MPP operation of the two modules while maintaining the efficient operation. The required value of the DC bus voltage in the secondary side of the transformer is 600 V. An LLC type resonant converter is considered where the turn ratio of the transformer is 1:25. The selected values for of the resonant inductor (L_r), the resonant capacitor (C_r) and the magnetization inductor (L_m) are taken as 950 nH, 1.36 μF and 80 μH, respectively. As known, LLC resonant converters have two resonant frequencies and the operation is preferred around the highest one which is formed by the resonant tank L_r and C_r. In this circuit, the highest resonant frequency is given by 140 KHz. To ensure ZVS in the converter the switching frequency is taken as of 116 KHz.

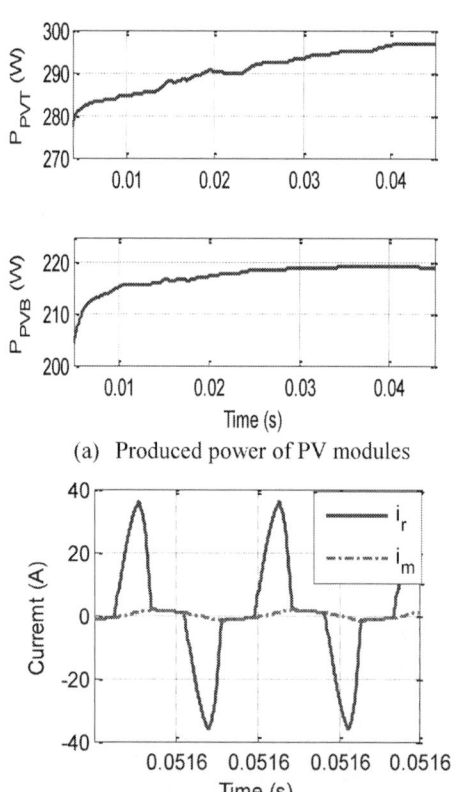

(a) Produced power of PV modules

(b) Current waveforms for the resonant circuit and the magnetization inductance

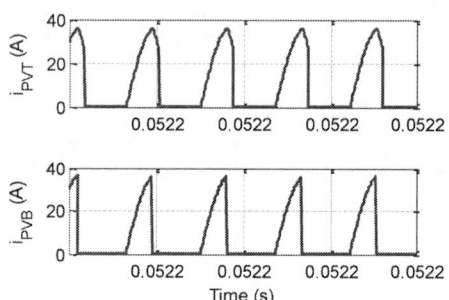

(c) Output Current waveforms of the PV modules

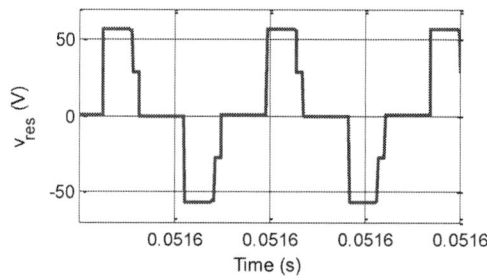

(d) Input voltage of the resonant circuit

Figure 7. Operation of the converter in Fig. 2 with two PV module having different power-voltage characteristics

The MPP of PV_T and PV_B are (297W, 30V) and (219W, 28V), respectively. The power produced by the modules is shown in Fig. 7(a) where a perfect MPP tracking for both modules could be achieved. The resonant circuit current is shown in Fig. 7(b). From the shape of the current, ZVS is always maintained. This can be noticed easily from the fact that at beginning of each switching cycle the current polarity is in the direction that makes ZVS.

The output currents of the two modules branches are shown in Fig. 7(c). The impact of detaching the module is shown by instances at which the current drops to zero. Since PV_T has higher power than PV_B, it remains attached to resonant circuit for a longer duration. Clearly, capacitors are connected across every module to absorb all the ripples in current shown in Fig.7(c) such that the PV modules produce constant DC currents. Figure 7(d) shows the input voltage applied to the resonant circuit. The staircase like waveform is produced as the PV modules get detached from the circuit at different time instances. As observed from the current waveforms, the voltage waveform shows that one module, PV_T, remains attached to the circuit for longer duration than the other one as it can produce more power at its MPP.

VI. CONCLUSIONS

The proposed topology in this paper allows several PV modules to share the same resonant converters such that individual module MPP tracking, high efficiency converter operation and low cost could all be achieved at the same time. Moreover, the proposed structure is extended to allow PV modules integration with batteries in the same topology without any hardware modification and with high degree of flexibility. The batteries could operate in the charging, discharging and bypassing modes and all transitions are achieved with minimum switching and conduction losses. Simulation studies were performed and they could validate the operation of the proposed topology. This topology helps in integrating hybrid PV/battery sources in any installation flexibly, efficiently and effectively.

REFERENCES

[1] G. Pepermans, J. Driesen, D. Haeseldonckx, R. Belmaans and W. D'haeseleer, "Distributed generation: definition, benefits and issues," Int. J. Energy Policy, vol. 33, no. 6, pp. 787-798, 2005.

[2] International Energy Agency, "Distributed generation in liberalised electricity markets," Paris, 2002.

[3] R. H. Lasseter and P. Piagi, "Microgrid: A conceptual solution," Proc. 35th PESC, vol. 6, pp. 4285-4290, 2004.

[4] M. McGranaghan, T. Ortmeyer, D. Crudele, J. Smith and P. Barker, "Renewable systems interconnection study: Advanced grid planning and operations," Sandia Natl. Lab. Albuq. Livermore, 2008.

[5] J. Traube, F. Lu, D. Maksimovic, J. Mossoba, M. Kromer, P. Faill, S. Katz, B. Borowy, S. Nichols, and L. Casey, "Mitigation of solar irradiance intermittency in photovoltaic power systems with integrated electric-vehicle charging functionality," IEEE Trans. Power Electron., vol. 28, no. 6, pp. 3058–3067, 2013.

[6] J. M. Carrasco, L. G. Franquelo, J. T. Bialasiewicz, E. Galvan, R. C. P. Guisado, M. A. M. Prats, J. I. Leon, and N. Moreno-Alfonso, "Power-electronic systems for the grid integration of renewable energy sources: A survey," IEEE Trans. Ind. Electron., vol. 53, no. 4, pp. 1002–1016, 2008.

[7] H. Hu , S. Harb , J. Shen and I. Batarseh "A review of power decoupling techniques for microinverters with three different decoupling capacitor locations in PV systems", IEEE Trans. Power Electron., vol. 28, no. 6, pp. 2711-2726, 2013.

[8] R. W. Erickson; A. P. Rogers "A Microinverter for Building Integrated Photovoltaic", Applied Power Electronics Conf. 2009, pp. 911-917.

[9] Bing, J.M., "Advances in commercial applications of photovoltaic technology in North America: 2009 update," IEEE Power and Energy Soc. Gen. Meet., pp. 1-5, 2010.

[10] C. Lin, A. Amirahmadi, Z. Qian, N. Kutkut and I. Batarseh, "Design and Implementation of Three-Phase Two-Stage Grid-Connected Module Integrated Converter," IEEE Trans . Power Electron., vol. 29, no. 8, pp. 3881-3892, 2014.

Real-time Integrated Model of a Micro-grid with Distributed Clean Energy Generators and their Power Electronics

Weiqiang Chen*, *Student member, IEEE*, Ali M. Bazzi, *Member, IEEE*, James Hare, *Student member, IEEE*,
Shalabh Gupta, *Member, IEEE*
Department of Electrical Computer Engineering, University of Connecticut, Storrs, CT, USA
*weiqiang.chen@uconn.edu

Abstract—This paper presents a real-time integrated model of a micro-grid to simulate its electrical energy infrastructure. This infrastructure includes two PV arrays, a fuel cell, and a diesel generator that support building loads when islanded from the utility grid. The paper reviews existing models, which are usually available either as 1) low-level dynamic models, along with power electronics for specific components, e.g. PV system or fuel cell, or 2) high-level such as with conventional power systems where power electronics are ignored due to their faster dynamics. The proposed modeling strategy for sustainable power generation is emphasized for both grid-connected and islanding modes and combines slow and fast dynamics where both a micro-grid and related power electronics dynamics are simulated to show how a high-fidelity model can be used in a dynamic micro-grid environment. A synchronized regulator for islanded mode is presented. The PV arrays and fuel cell are assumed to be always available and variable irradiance conditions (e.g. nighttime) and the change of hydrogen and oxygen densities of fuel cell are shown in the paper. The diesel generator is used for black start or when the utility grid voltage or frequency drop below a threshold during which potential grid collapse could occur and the micro-grid goes into the island mode. The micro-grid model is simulated using a real-time simulator so that longer case studies and scenarios can be studied without ignoring fast dynamics. The main contribution of this paper is that both these time scales (slow and fast) are integrated in this real-time simulation platform for more realistic performance analysis. The results show the ability of the integrated micro-grid simulation to function in both grid-tied and islanded modes.

I. INTRODUCTION

With the rise of concerns regarding traditional power systems' vulnerability to physical attacks, cyber-attacks, or failures due to natural disasters or aging, trends in power systems research have turned toward the development of distributed micro-grids. Micro-grids are small-scale power grids that integrate clean and/or conventional generation systems into a unified power system for robust, reliable, resilient, and sustainable load support. They provide an attractive ability to sense critical changes in the utility grid to facilitate an autonomous control action to disconnect from the utility

grid into the island mode. The ability of these micro-grids to disconnect and reconnect to the utility grid creates transient properties that must be controlled in order to maintain stable and sustainable power. When the micro-grid is grid-tied, it is common to have synchronous connection where all micro-grid voltages must be synchronized with the utility grid, and later operate at desired voltage and frequency in the island mode. Harmonics, synchronization, voltage sags, failure of rotating machinery, and many other anomalies could arise when voltage and frequency conditions are not met. Asynchronous interconnection and DC micro-grids are expected to be viable alternatives to mitigate the frequency and synchronization requirements. For existing and near-future micro-grids, the detailed simulation models, including high levels of clean energy penetration, must be established.

Throughout literature, modeling approaches to micro-grids have been very limited. One approach was through the development of inverter-based models [1, 2]. These assume that distributed generators output a DC voltage that gets synchronized to the utility grid through inverters even though some conventional sources with diesel and natural gas are AC. Others have modeled micro-grids with off-the-shelf libraries, e.g. Simulink [3], in islanded mode but did not analyze various system integration and time scale requirements, e.g. several hours of simulation time that can take several days or weeks to process. Other literature has focused on micro-grid control methods based on small signal models in state-space format [4, 5] where eigenvalues are utilized for control. Other approaches to modeling have tried studying the behavior of a micro-grid with electrical energy storage devices [6] under utility grid disturbances and for lifetime analysis of the micro-grid in island mode.

Most literature focuses on modeling one or two aspects of the micro-grid, but no model exists that integrates the electrical energy infrastructure as loads, distributed generation, energy conversion (power electronics), control at component levels, system integration, and real-time simulation. The power electronics' transients are usually ignored by some power

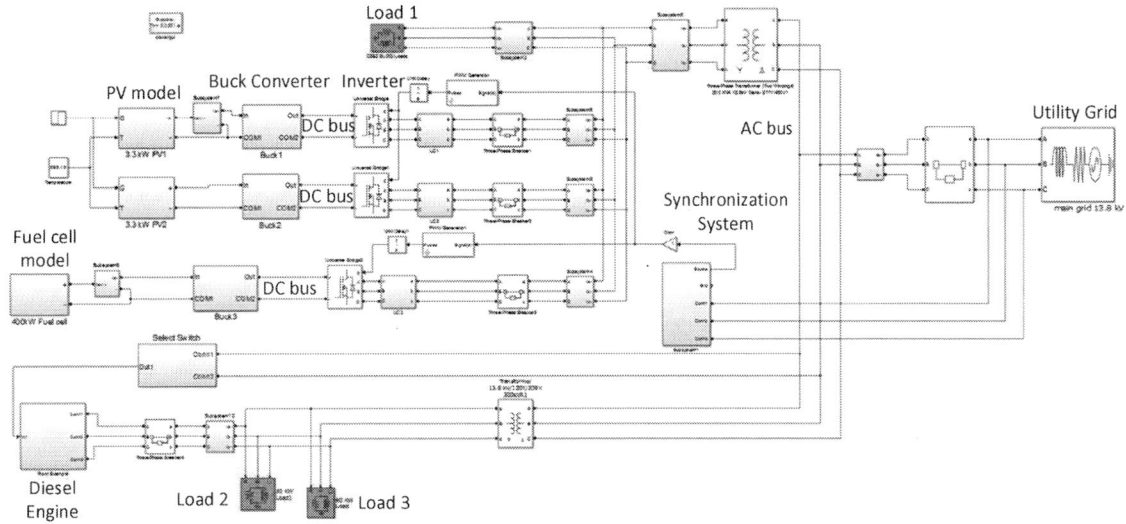

Fig 1. Simulated micro-grid system with diesel, PV, fuel cell, three building loads, and the utility grid

system simulators, meanwhile, the investigation of grid-tied power electronics usually ignores generator models and their transient impact, but both levels of systems and their transients are considered and analyzed in this paper to provide a more realistic model of a micro-grid. Note that the model presented here and which is formed by integrated several other models can be used as an example to apply a similar approach to this presented here for other micro-grids.

This paper aims to develop detailed models of a real micro-grid's electrical energy infrastructure. This includes clean and conventional generation systems: two PV arrays, a fuel cell, a diesel generator, inverters and their interconnections. Each generation system has voltage control with inverter voltages synchronized using phase lock loops (PLLs). Once all models are implemented into a larger micro-grid model, case studies are simulated on a Simulink Real Time platform where the simulated system is shown in Figure 1. Figure 2 shows a higher-level block diagram of the same system. Note that R-L loads shown in orange in Figure 1 were extracted as average values from real dynamic building loads.

II. MODELS OF DISTRIBUTED GENERATION

Each distributed generator is grid-tied through power electronics where a DC/DC stage regulates individual DC busses and a DC/AC stage inverts to synchronize with the grid or other sources, except for diesel generators. The PV arrays' voltage levels allow for the use of a buck converter to regulate higher voltage from the PV model to lower voltage at the DC bus. Local sources support building loads in grid-tied and island modes. Figures 3-5 show generation model block diagrams. Note that PCC is the point of common coupling.

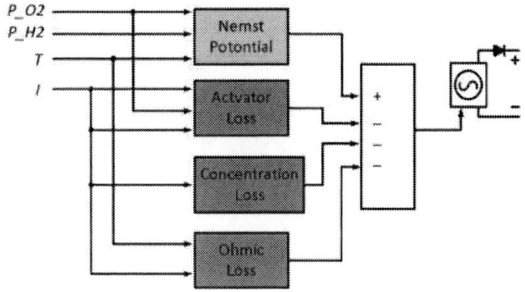

Fig.3. Non-ideal fuel cell model as a controlled voltage source taking into account various losses

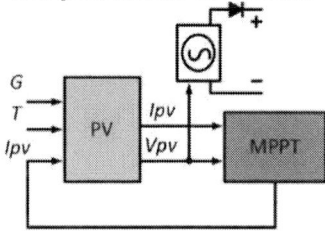

Fig.4. Non-ideal PV array model as a controlled voltage source

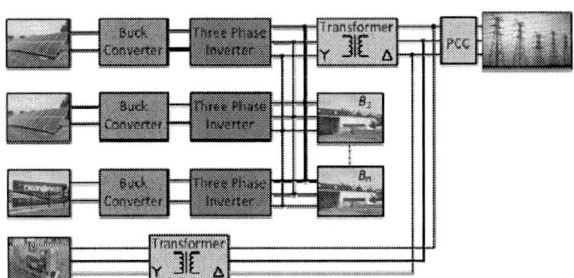

Fig.2. High-level block-diagram of the micro-grid

978-1-4673-9551-9/16 $31.00 © 2016 IEEE   2667

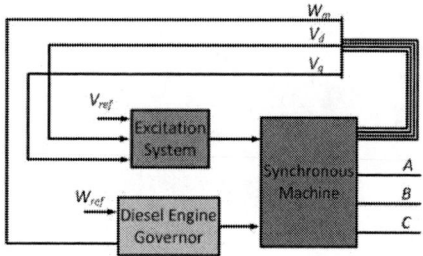

Fig.5. Diesel generator model

A. PV Array Model

Two parallel PV arrays are used, each array has 14 panels and each panel includes 60 cells. The PV array output power is not given much consideration since the PV arrays are modeled as controlled voltage sources to facilitate the PV converter/inverter control. Partial shading may occur during the operation of the panels which is not included in this paper but can be integrated as needed. Figure 4 provides a model of one array. Perturbation and observation (P&O) method is used to perform maximum power point tracking (MPPT) and a current sink is utilized as constant by which the buck converter can regulate the duty ratio to track the maximum power point. The output voltage of the PV array is modelled as

$$V_{PV} = \frac{A \times K \times T}{q} \log(\frac{I_{lg} - I_{PV}}{I_0} - 1) - I_{PV} \times R_s \quad (1)$$

where A is diode-ideality factor, K is Boltzmann constant, T is absolute temperature, q is the basic charge, I_{lg} is light generated current, I_0 saturation current, and R_S is series resistance. The output power is,

$$P_{PV} = I_{PV} \times V_{PV} \quad (2).$$

The P&O method regulates the duty cycle of the buck converter to track the maximum power point by comparing the old and new voltage and power being sampled as shown in Figure 6.

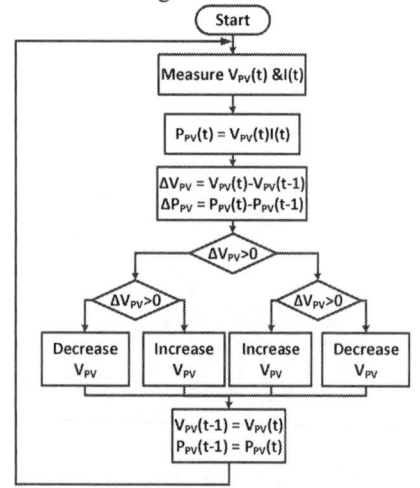

Fig. 6. Perturbation and observation method

The output voltage of the PV system is 430V DC and is controlled by a buck converter through a basic closed-loop voltage control to maintain a 350V DC into a three phase inverter. On the grid side, a Y-Δ three phase transformer connects to the grid.

B. Fuel Cell Model

For the fuel cell in Figure 3, a polymer electrolyte membrane (PEM) fuel cell is used [7]. PEM fuel cells have the advantages of operating at a low temperature, high power density, fast response and low emissions. They convert chemical energy of hydrogen and oxygen reactions into electrical energy. This paper models the fuel cell by assuming the hydrogen, oxygen, initial current and temperature are constant values [8], and losses are modeled as steady/dynamic state activation loss, ohmic loss and concentration loss [9] as shown in Figure 3. The model of a fuel cell can be represented by open circuit voltage E. The output voltage of the fuel cell is thus,

$$V_{fc} = E - V_{act} - V_{ohm} - V_{con} \quad (3)$$

The open circuit voltage is obtained as

$$E = 1.229 - 8.5 \times 10^{-4}(T - 298.15) + 4.308 \times 10^{-5}T(\ln P_{H_2} + 0.5\ln P_{O_2}) \quad (4)$$

The activation loss is

$$V_{act} = -0.9514 + 3.12 \times 10^{-3}T - 1.87 \times 10^{-4}T \ln I + 7.4 \times 10^{-5}T(\ln P_{H_2} + 0.5\ln P_{O_2}) \quad (5)$$

The ohmic loss is

$$V_{ohm} = -IR_{int} \quad (6)$$

The concentration loss is

$$V_{con} = B \ln(1 - I / I_{lim}) \quad (7)$$

where T is cell temperature, I is current, I_{lim} is limit current and B is 0.016. In equations (4) and (5), the partial pressures of hydrogen and oxygen are

$$P_{H_2} = 0.5 P_{H_2O}^{sat}[\exp(-\frac{1.635I / A}{T^{1.334}})\frac{P_a}{P_{H_2O}^{sat}} - 1] \quad (8)$$

$$P_{O_2} = P_{H_2O}^{sat}[\exp(-\frac{4.192I / A}{T^{1.334}})\frac{P_c}{P_{H_2O}^{sat}} - 1] \quad (9)$$

The concentration of oxygen can be obtained by

$$C_{O_2} = \frac{P_{O_2}}{5.08 \times 10^{-6}}\exp(\frac{498}{T}) \quad (10)$$

while the saturation pressure of water is calculated by

$$\log_{10} P_{H_2O}^{sat} = -2.18 + 2.95 \times 10^{-2}T_C - 9.18 \times 10^{-5}T_C^2 + 1.44 \times 10^{-7}T_C^3 \quad (11)$$

where $T_C = T - 273.15$.

The internal resistance is

$$R_{int} = 1.605 \times 10^{-2} - 3.5 \times 10^{-5}T + 8 \times 10^{-5}I \quad (12)$$

For the dynamic model, the equation of the transient activation loss is

$$\frac{dV_{act}^t}{dt} = \frac{I}{C} - \frac{V_{act}^t}{R_a C} \quad (13)$$

where C is the charge double layer capacitance, R_a is the equivalent resistance:

$$R_a = \frac{V_{con} - V_{act}}{I} \quad (14)$$

C. Diesel Generator Model

The diesel generator shown in Figure 2 and elaborated upon in Figure 5 consists of three parts: diesel engine governor, excitation system and synchronous machine [10, 11]. The excitation system provides the initial magnetic field to startup the synchronous machine, while the diesel engine governor utilizes a feedback mechanism to regulate and maintain the speed as needed to maintain electrical frequency by setting the reference speed w_f. The on and off state of the diesel engine can be controlled. The diesel generator model contains two main parts: Diesel engine governor and synchronous generator. The structure of the diesel engine governor is shown in Figure 7. The PI controller and actuator are modelled by transfer functions with time constant τ_1, τ_2, τ_3, τ_4 and PI parameters K_i and K_p.

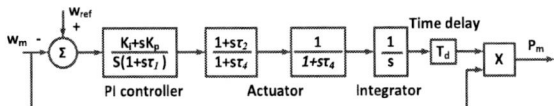

Fig. 7. Diesel engine governor model

The synchronous machine model considers the dynamics of the stator and field circuits:

$$L_s \frac{di_s^d(t)}{dt} = -r_s i_s^d(t) + w_e(t) L_s i_s^q(t) - v_s^d(t) - L_m \frac{di_f(t)}{dt} \quad (15)$$

$$L_s \frac{di_s^q(t)}{dt} = -r_s i_s^q(t) - w_e(t) L_s i_s^d(t) - v_s^q(t) - w_e(t) L_m i_f(t) \quad (16)$$

$$L_f \frac{di_f(t)}{dt} = -r_f i_f(t) + v_f(t) - L_m \frac{di_d(t)}{dt} \quad (17)$$

where i_s^d, i_s^q and V_s^d, V_s^q are stator currents and voltages in dq reference frame; r_s is stator resistance, L_s is stator inductance; i_f, V_f are field excitation current and voltage; The inductance of the field excitation circuit is L_f, the mutual inductance between the field circuit and the d axis of the stator is L_m. w_e is the electrical angular frequency and can be represented as:

$$w_e = N_{pp} w_r \quad (18)$$

where w_r is the speed of the rotor, N_{pp} is the number of pole pairs.

The diesel generator is a standby power source to supply the power system in the event when clean energy generation systems cannot produce sustainable power in island mode, or when the grid should be disconnected due to irregular voltage or frequency.

III. INVERTER CONTROL AND SYNCHRONIZATION

Closed-loop control is also applied to three phase inverter, different from classic PI control where the inverter output and balanced three-phase reference signals are synchronized as shown in Figure 8. To maintain this reference when the grid is lost, a synchronized source shown in Figure 9 is generated from the grid and synchronizes all necessary parameters, especially phase information, for all inverters on the micro-grid. After one cycle from the simulation start time, selection switches shown in Figure 8 switch to replace the utility grid generated ωt to create stand-alone sinusoidal control waveforms.

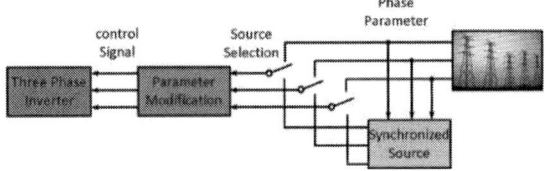

Fig. 8. Closed-loop control for three phase inverter

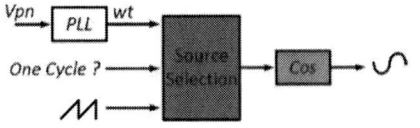

Fig. 9. Synchronized Source

Note that the models developed in the paper address a specific implemented micro-grid but other models can be integrated to the micro-grid simulation for more generation and load options. The main purpose of the models presented here is to achieve a robust and flexible real-time simulation platform for further research and development while capturing fast and slow dynamics in addition to control effects.

IV. REAL-TIME SIMULATION RESULTS

A Simulink Real-Time target that runs the model shown in Figure 1 is utilized for simulations. The simulation results for the micro-grid system have been successfully demonstrated where fast and slow dynamics and their effects could all be captured. This real-time implementation is able to reduce the run time of simulation significantly.

In order to monitor the system under different conditions and show that the PV and fuel cell output voltages are tightly regulated at 350V, the irradiance value in the PV model changes from $100W/m^2$ to $200W/m^2$ at 600s, and the hydrogen and oxygen pressures change between 1 standard atmosphere (atm) and 2 atm at 300s and 900s. Figure 10 shows different condition areas from A1 to A4 for 20 minutes or 1200 seconds, where each area has a unique fuel cell and PV parameter combination. The four areas include four conditions under islanding mode or grid-connected mode.

In Figure 10, P_{H2} is the effective partial pressure of hydrogen, P_{O2} is the effective partial pressure of oxygen whose change is reflected by the green dotted line. The different position of green dotted line means different combination of P_{O2} and P_{H2}. The irradiance change is reflected by the blue solid line. Three-phase PV and fuel cell output voltages and dc-bus voltages before and after the buck converters are shown in Figures 11 and 12. The voltage drop at 300s in Figure 11 indicates the change of maximum power point in PV curve changes, and the voltage increase between 300s and 900s in Figure 12 indicates the change of pressure settings in the fuel cell model. The simulation time is 1200s where 420s to 780s are in islanding mode where the micro-grid loses the grid. Transition from grid-connected to islanded mode is at 420s, and a transition back to grid-connected mode is at 780s. Even though the dc voltage of PV and fuel cell sources have some variations with the change of parameters, they are still well-regulated between grid-connected and island modes.

Diesel generator output is only connected once it senses that the utility grid is lost. In this simulation, to show more operation conditions, three different operation conditions which classified by the on and off states of diesel generator and grid: At 0s to 420s, grid is connected and diesel generator is off; at 420s to 780s, grid is disconnected and diesel generator is on; at 780s to 1200s, grid is connected again and diesel generator is off. AC bus, which ties the transformers and PCC, and diesel generator voltages are shown in Figures 13 and 14. In both figures, the three phase voltages in island mode almost keep the same as grid-connected mode even though in island mode, slight unbalance occurs.

Instantaneous power flow from/to the grid side is shown in Figure 15 (a), where the power becomes zero at 420s. After the system becomes in grid-connected mode again at 780s. The micro-grid side power flow is shown in Figure 15 (b) which is the averaged power flow to load1, and Figure 16 shows the power flow from the PV arrays and fuel cell where the power flow contributes more when the system goes into islanding mode. As shown in Figure 15 (b), even in islanding mode the power into the load is dropped, it still maintained at a relatively high level to continue supporting the building load. The power provided by the diesel generator is shown in Figure 17 where the generator is engaged in island mode. However, Figure 17 shows that some transient occurs at the generator's terminals when the micro-grid returns to grid-connected mode and this needs further refinement. It can be seen form the voltage and power figures that the DC bus, AC bus and power flow can be well-regulated and maintained.

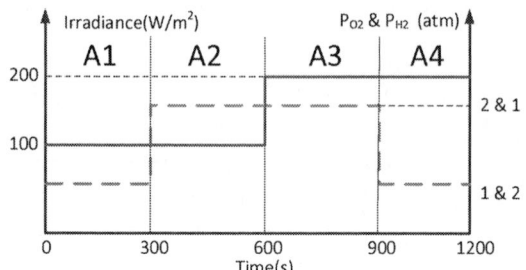

Fig.10. Operation conditions

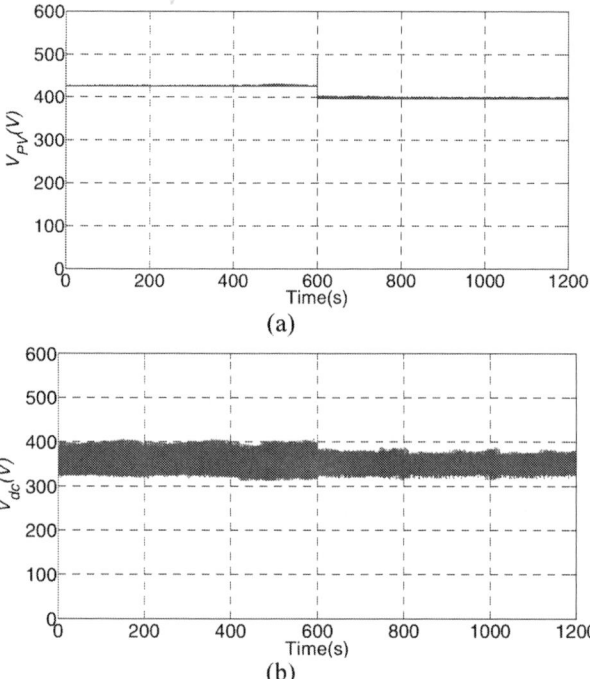

Fig. 11. PV (a) Output voltage, (b) dc-bus voltage

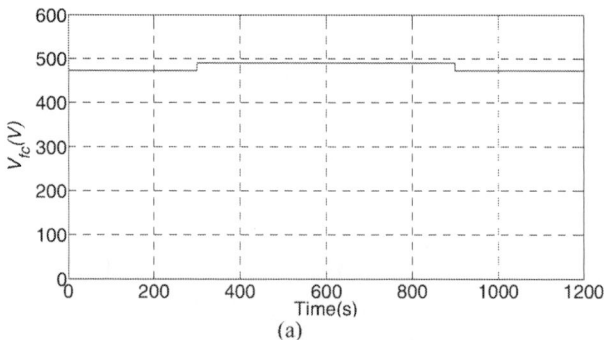

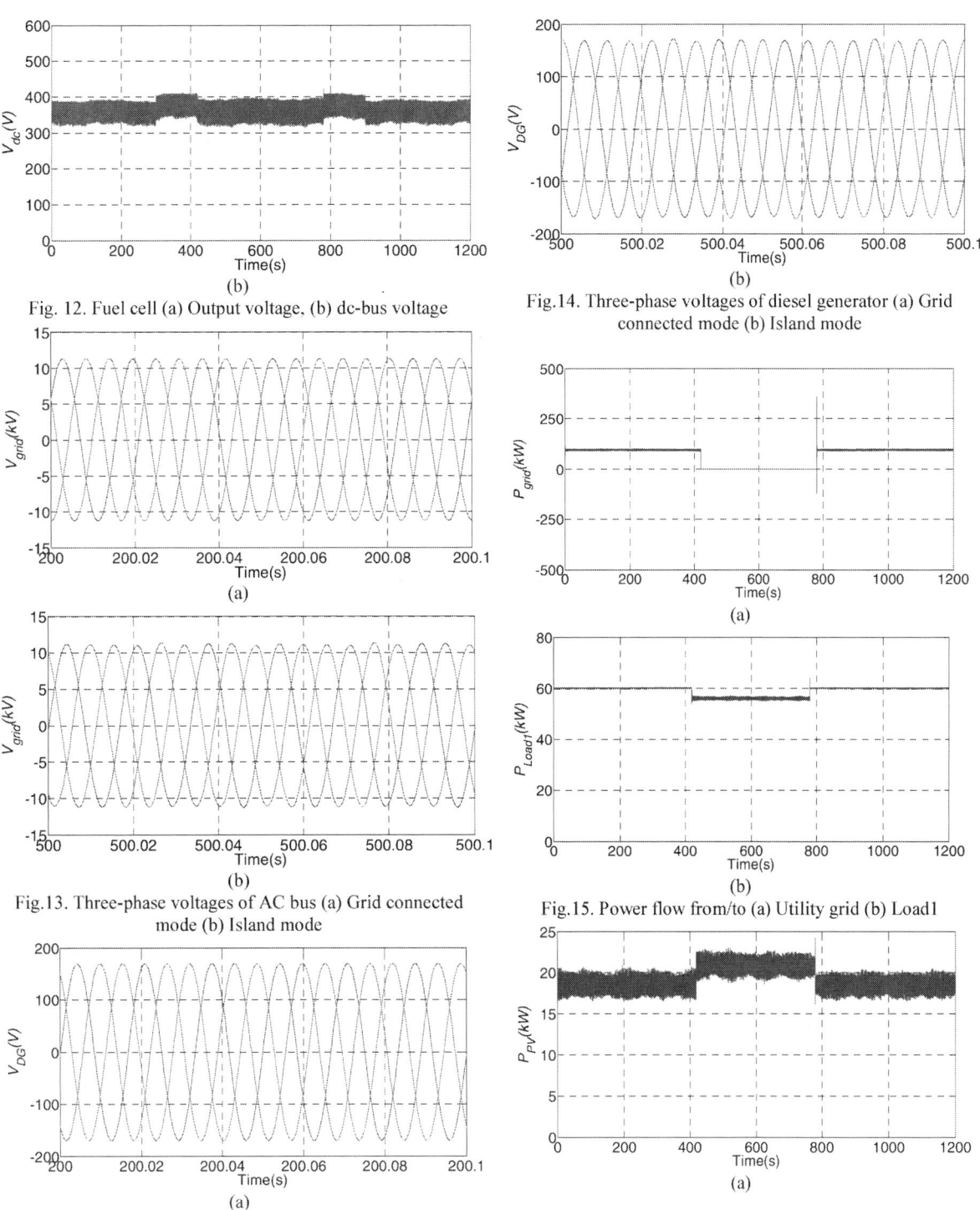

Fig. 12. Fuel cell (a) Output voltage, (b) dc-bus voltage

Fig.14. Three-phase voltages of diesel generator (a) Grid connected mode (b) Island mode

Fig.13. Three-phase voltages of AC bus (a) Grid connected mode (b) Island mode

Fig.15. Power flow from/to (a) Utility grid (b) Load1

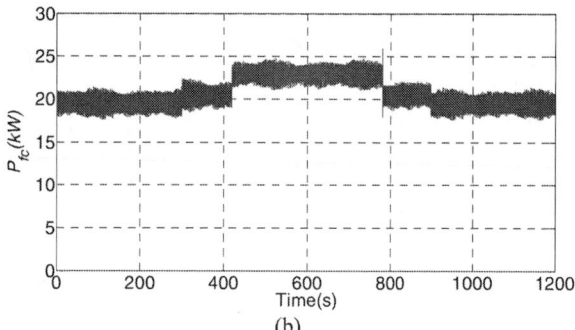

(b)

Fig.16. Power flow from (a) PV and (b) Fuel cell

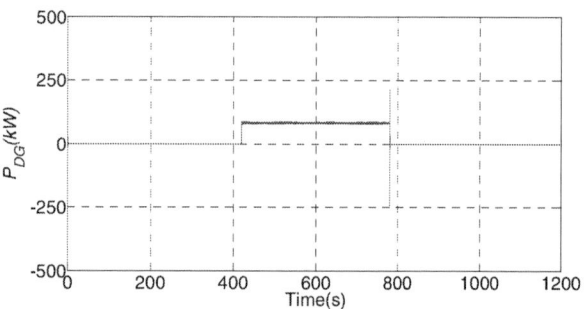

Fig.17. Power flow from diesel generator

V. CONCLUSION AND FUTURE WORK

In this paper, a micro-grid with both conventional generation and significant clean energy penetration is modeled and simulated in both grid-tied and island modes. All clean energy sources are used to support building loads and compensate for grid power when the main utility grid is disconnected. Models used utilize dynamic analytical equations, power electronics dynamics, converter-level control, and system-level integration. Both slow and fast dynamics of the micro-grid are captured in grid-connected and island modes. Results presented here show stiff DC bus regulation for clean energy sources and well-regulated power flow of clean energy sources and conventional diesel generator in both operating modes on a real-time simulation platform.

REFERENCES

[1] Pogaku, Nagaraju, Milan Prodanovic, and Timothy C. Green. "Modeling, analysis and testing of autonomous operation of an inverter-based micro-grid." In *IEEE Transactions on Power Electronics 22.2 (2007): 613-625.*

[2] Xiong, Xiaofu, and Jinxin Ouyang. "Modeling and transient behavior analysis of an inverter-based micro-grid." In *Proc. Electric Power Components and Systems 40.1 (2011): 112-130.*

[3] Brissette, Alexander, et al. "A micro-grid modeling and simulation platform for system evaluation on a range of time scales." In *Proc. Energy Conversion Congress and Exposition (ECCE), 2011 IEEE. IEEE, 2011.*

[4] Katiraei, F., and M. R. Iravani. "Power management strategies for a micro-grid with multiple distributed generation units.", In *IEEE Transactions on Power Systems, 21.4 (2006): 1821-1831.*

[5] Katiraei, F., M. R. Iravani, and P. W. Lehn. "Small-signal dynamic model of a micro-grid including conventional and electronically interfaced distributed resources." In *Proc. IET generation, transmission & distribution 1.3 (2007): 369-378.*

[6] Pawelek, R., et al. "Study on operation of energy storage in electrical power micro-grid-Modeling and simulation." In *Proc. 2010 14th International Conference on Harmonics and Quality of Power (ICHQP). IEEE, 2010.*

[7] Jia, J.; Li, Q.; Wang, Y.; Cham, Y. T.; Han, M., "Modeling and Dynamic Characteristic Simulation of a Proton Exchange Membrane Fuel Cell," *IEEE Transactions on Energy Conversion, vol.24, no.1, pp.283,291, March 2009*

[8] Younis, M. A A; Rahim, N.A; Mekhilef, S., "Fuel Cell Model for Three-Phase Inverter," In *Proc. IEEE International Power and Energy Conference, 2006. PEC on '06.,pp.399,404, 28-29 Nov. 2006*

[9] Yancheng Xiao; Agbossou, K., "Interface Design and Software Development for PEM Fuel Cell Modeling Based on Matlab/Simulink Environment," In *Proc. WRI World Congress on Software Engineering, 2009. WCSE '09.,pp.318,322, 19-21 May 2009*

[10] Theubou, T.; Wamkeue, R.; Kamwa, I, "Dynamic model of diesel generator set for hybrid wind-diesel small grids applications,", In *Proc. 2012 25th IEEE Canadian Conference on Electrical & Computer Engineering (CCECE), pp.1,4, April 29 2012-May 2 2012*

[11] Torres, M.; Lopes, L. A C, "Inverter-based virtual diesel generator for laboratory-scale applications," In *Proc. IECON 2010 - 36th Annual Conference on IEEE Industrial Electronics Society, pp.532,537, 7-10 Nov. 2010*

Minimization of Inter-Module Leakage Current in Cascaded H-Bridge Multilevel Inverters for Grid Connected Solar PV Applications

V. V. S. Pradeep Kumar and B. G. Fernandes
Department of Electrical Engineering
Indian Institute of Technology Bombay, India
email: vvspkumar@ee.iitb.ac.in, bgf@ee.iitb.ac.in

Abstract—Cascaded H-bridge multilevel inverters (CHB-MLIs) are well accepted for medium and high power grid connected photovoltaic (PV) applications. Independent MPPT and modularity are the advantages of these converters over the diode clamped and capacitor clamped MLIs. However, CHB-MLIs are associated with inter-module leakage current through the parasitic capacitor of PV panels even though an isolation transformer is used on the AC side. This necessitates the use of an additional isolated DC-DC converter in each power cell, which increases the cost and reduces efficiency. In order to address these limitations, a simple passive solution is proposed to minimize the intermodule leakage current. It comprises of several L-C filters placed among the power cells to mitigate the switching frequency components in voltage across the parasitic capacitors. Moreover, the proposed solution can be easily upgraded to higher number of power cells. In this paper, the expressions for inter-module leakage currents are derived for five level CHB-MLI. The required filter values are designed for various pulse width modulation (PWM) techniques and it is found that the modified carried disposition PWM and hybrid PWM techniques are superior in minimizing the leakage currents. Simulation studies validate the claims of proposed solution.

Index Terms—carrier phase shift PWM, cascaded H-bridge multilevel inverter, filter design, hybrid PWM, inter module leakage current, modified CD PWM, modified hybrid PWM, parasitic capacitor.

I. INTRODUCTION

Cascaded H-Bridge multilevel inverter is one of the popular topologies for the grid integration of medium and high power PV applications [1]–[3]. There are several single phase H-Bridge inverters connected in cascade, and each cell is supplied by an isolated DC supply. In case of the motor drive applications, the supply is ensured by using a multi winding transformer along with rectifiers. However, in case of PV applications, it is ensured by connecting separate PV panels at the input of each cell. Moreover, independent MPPT for each power cell can be easily achieved. These benefits make the CHB-MLI suitable for medium and high power grid connected PV systems. It has other advantages like: modularity, redundancy, and lower filter requirement compared to other multilevel inverters [4]. However, for PV applications, it has a limitation of intermodule leakage current due to the parasitic capacitance that exists between the power terminals of PV panel and the ground [5]. The main cause for this leakage current is the PWM nature of the voltage across these

parasitic capacitors. In case of a single phase grid connected inverter for roof-top PV applications, the undesired earth leakage current completes its path through the grid neutral as shown in Fig. 1 [6]. It can be minimized either by using an isolation transformer on ac side or by using any one of the transformerless inverter topologies proposed in literature [7]–[10].

However, the path for leakage current in case of a CHB-MLI is different from that of the conventional single phase grid connected inverter. The DC terminals of each power cell of the CHB-MLI are connected to a common ground through the parasitic capacitors of respective PV modules as shown in Fig. 2. This introduces several additional paths for the leakage current through all the power cells. Therefore, these leakage currents are termed as inter-module leakage currents. They do not interact with the utility grid and hence can not be avoided even if there is an isolation transformer at the output of inverter. Therefore, in order to minimize these leakage currents, an additional isolated DC/DC conversion stage is required in each power cell [11]. However, it increases the cost, size and losses of the converter.

Recently, few papers have reported the minimization of intermodule leakage currents without using any such isolation stage [5], [12]. In [12] a low frequency stepped voltage is achieved across the parasitic capacitors by modifying the level shift PWM technique. However, this solution is not generic and is applicable only for single phase five level CHB-MLI.

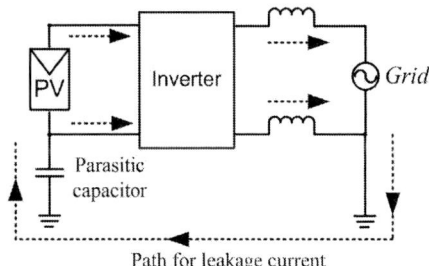

Fig. 1: Earth leakage current in case of a conventional grid connected inverter.

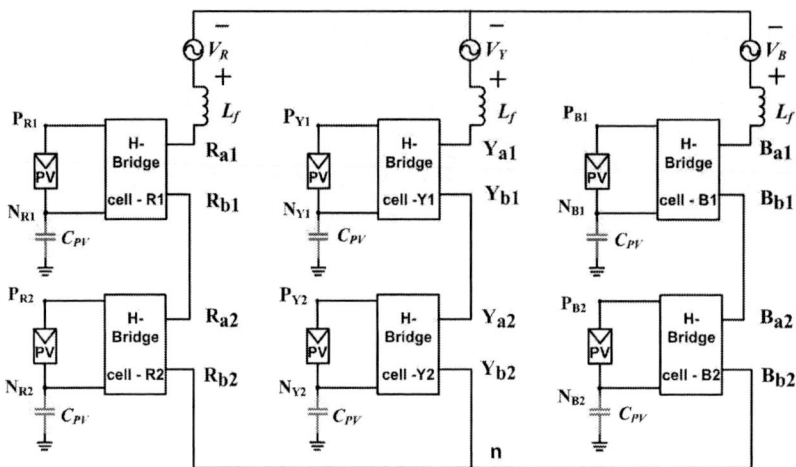

Fig. 2: Cascaded H-Bridge multilevel inverter with parasitic capacitor in each power cell.

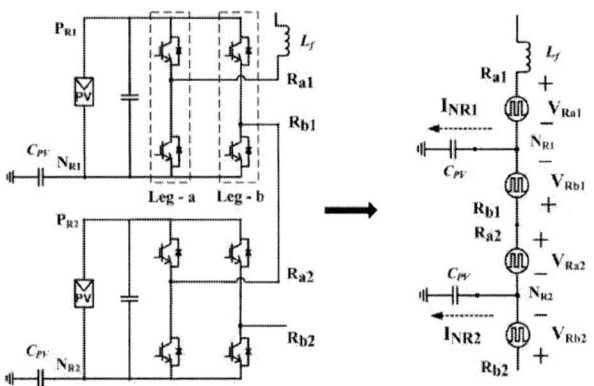

Fig. 3: Per phase equivalent circuit.

In [5], a generalized solution is proposed by introducing two common mode chokes and two capacitors in each power cell. However, the leakage current analysis was reported only for single phase CHB-MLI at low power level.

In this paper, the inter-module leakage currents (i_{IM}) are explained with analytical expressions for the three-phase five level CHB-MLI. A generalized filter configuration is proposed to filter out the high frequency PWM voltage across the parasitic capacitors. These filter components can also limit the output current THD with in the permissible limits without using any additional filter components.

The paper is organised as follows: The problem of inter module leakage current for five level CHB-MLI is explained in section-II. Section-III explains the proposed solution. In section IV, the filter components are designed and compared for various switching strategies. Section V presents the simulation results of the proposed solution and the conclusions are made in section VI.

Table I: Circuit Parameters

Parameter	Value
Output power	60 kVA
Cell output voltage	115 V
Cell output power	10 kVA
Switching frequency(f_s)	10 kHz
Parasitic capacitance (C_{PV}), for PV module with Aluminium frame on assembly stand [13]	0.1 μF
Filter inductance (L_{f1}, L_{f2}, L_{f3})	0.85 mH

II. INTER-MODULE LEAKAGE CURRENTS

In order to analyse the source for the inter module leakage current, an equivalent circuit for the three phase five level CHB-MLI is developed. The per-phase equivalent circuit of the inverter is shown in Fig. 3. Each leg of a H-Bridge is represented by an equivalent voltage source which is PWM in nature. The resultant three-phase circuit is analysed using the Superposition theorem and the voltage across the parasitic capacitors of each power cell (of R-phase) V_{NR1} and V_{NR2} are obtained as

$$V_{NR1} = \frac{5V_{Ra2b1} - 4V_{Rb2} - V_{Ya2b1}}{6} + \frac{2V_{Yb2} - V_{Ba2b1} + 2V_{Bb2}}{6}$$

$$V_{NR2} = \frac{-V_{Ra2b1} - 4V_{Rb2} - V_{Ya2b1}}{6} + \frac{2V_{Yb2} - V_{Ba2b1} + 2V_{Bb2}}{6}$$

(1)

where, V_{Xyi} is the equivalent voltage of the terminal Xyi to the respective DC bus negative terminal. Here, X represents the phase (R, Y, B) of the converter, and y represents the

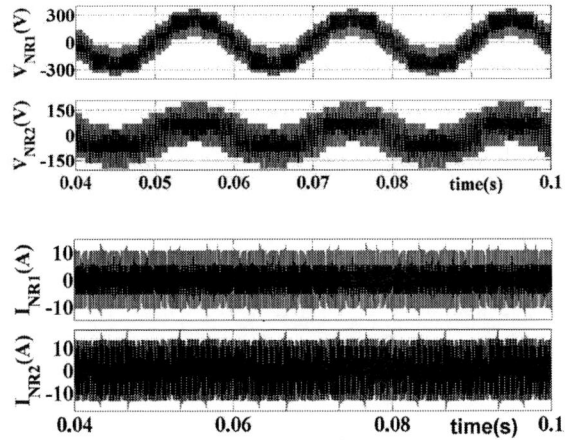

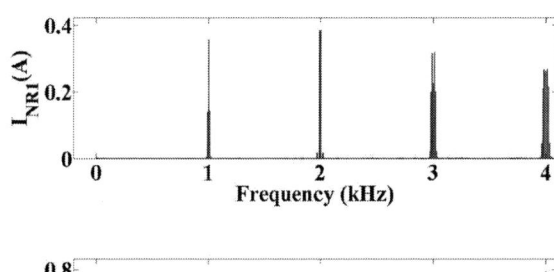

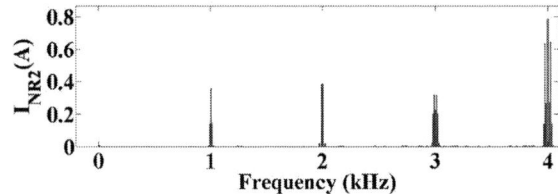

Fig. 4: (a) voltage across the parasitic capacitor (b) leakage current through the parasitic capacitor.

Fig. 5: FFT analysis of the leakage currents.

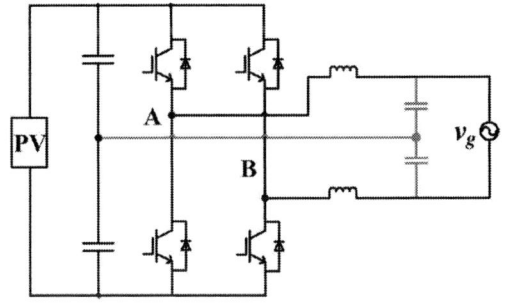

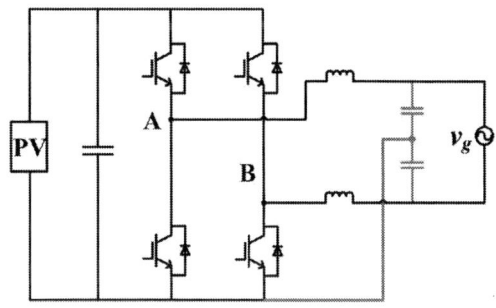

Fig. 6: Basic filter configurations to minimize the leakage current [9], [10].

leg (leg a or leg b) of the H-bridge, i represents the power cell (cell-X1 or cell-X2), and $V_{Ra2b1} = V_{Ra2} - V_{Rb1}$. All the components in the equation (1) are PWM in nature. This results in high frequency leakage current through each parasitic capacitor. To analyse the severity, a five level CHB-MLI is simulated with the parameters shown in Table-I. The PWM voltage and the high frequency leakage currents associated with these capacitors (I_{NR1} and I_{NR1}) are shown in Fig. 4. The FFT analysis of leakage current (as shown in Fig. 5) shows that their magnitude is significantly high and have predominant harmonic components at switching frequency and its multiple frequencies. According to the DIN standards, the total RMS value of leakage current should be limited below 30 mA for human safety and 300 mA for the safety from fire hazards [14]. Moreover, these leakage currents introduce additional EMI/EMC problems and interfere with the earth leakage current protective devices.

III. PROPOSED SOLUTION FOR MINIMIZING THE INTER-MODULE LEAKAGE CURRENTS

In case of a single phase inverter, the leakage current is minimized either by maintaining the common mode voltage constant [7], [8] or by modifying the existing LC filter configuration [9], [10]. As the solutions in [9], [10] do not require any additional switches, they can be adapted for CHB-MLI. In these methods, filter components are configured in such a way that the high frequency PWM voltages across the parasitic capacitors are filtered out. Two such configurations are shown in Fig. 6.

Based on this idea, a simple filter configuration is derived by placing several low pass LC filters among the power cells as shown in Fig. 7. From equation (1), it is clear that most of the leg voltages together contribute to the leakage current. Therefore, a filter inductor (L_{f2}) is connected between the two consecutive power modules of each phase and a filter capacitor (C_{f2}) is connected between the DC terminals of two consecutive power cells. In addition, there are two LC filters (L_{f1} - C_{f1} and L_{f3} - C_{f3}) connected at the two extreme terminals of each phase, respectively. This ensures the voltage across each parasitic capacitor to vary only at fundamental frequency. Similarly, for a n-cell CHB-MLI, there will be a total of $(n+1)$ LC filters per phase in which $(n-1)$ filters are placed between the two consecutive modules and the remaining two filters are connected with the extreme cells (cell-R1 and

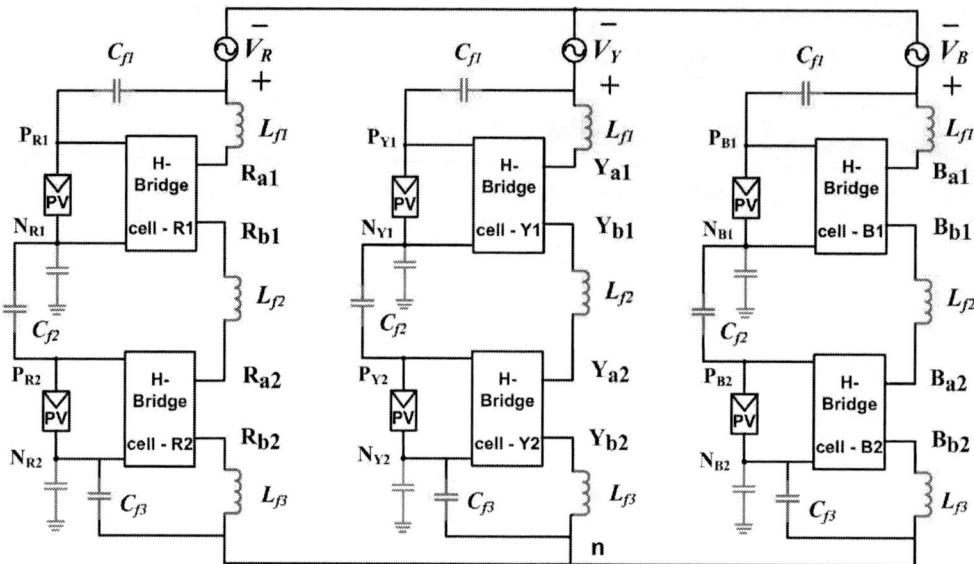

Fig. 7: Proposed solution to minimize the inter-module leakage current.

cell-Rn) as shown in Fig. 8.

IV. DESIGN OF THE FILTER COMPONENTS

From equation (1), V_{NR1} and V_{NR2} have the predominant harmonic component at switching frequency (f_s). Therefore, the filters are designed with the cut-off frequency (f_c) lower than the switching frequency of each cell. The proposed configuration is simulated with various values of f_c, and the critical value of f_c is determined to limit the leakage currents below 30 mA. Simulation studies are performed for various Sinusoidal pulse width modulation (SPWM) strategies which are commonly used for CHB-MLI [4]: carrier phase-shift PWM (CP PWM), modified level shift or modified carrier disposition PWM (CD PWM), hybrid PWM and modified hybrid PWM techniques. The carrier waves for these PWM strategies are shown in Fig. 9. The carrier wave frequency for modified CD PWM and hybrid PWM strategies are chosen in such a way that the average device switching remains same for all the PWM strategies. Further, the total value of filter inductance ($L_{f1} + L_{f2} + L_{f3}$) is chosen same as the filter inductance (L_f) that is used in case of a conventional CHB-MLI, and the filter capacitance is determined based on the cut-off frequency that is chosen.

The RMS values of leakage currents for these PWM techniques are plotted in Fig. 10 and the critical values of f_c are tabulated in Table II. It is observed that the modified CD PWM and hybrid PWM techniques have almost equal performance in minimizing the leakage currents. They are superior compared to CP-PWM and modified hybrid PWM techniques in mitigating the leakage currents. With modified CD PWM and hybrid PWM techniques, the filters can be tuned with the cut-off frequency lower than 3.8 kHz, where as

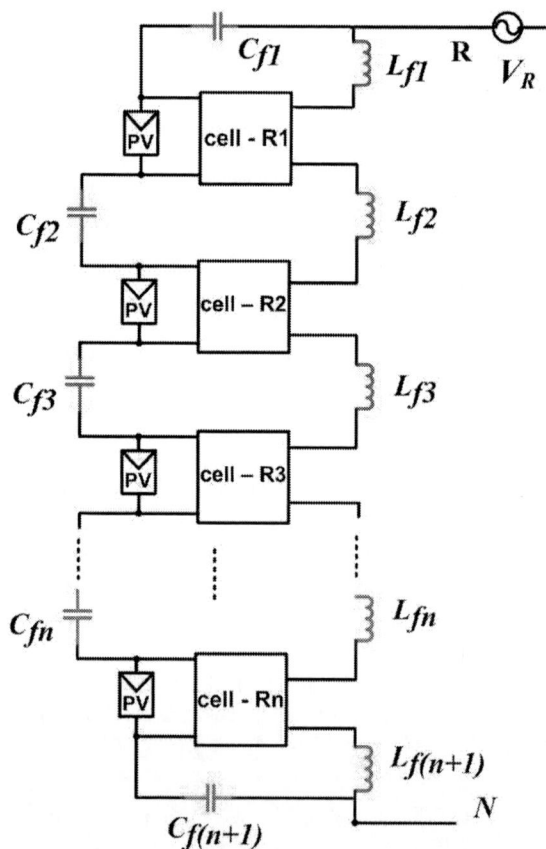

Fig. 8: Per phase filter arrangement for a n-cell CHB-MLI.

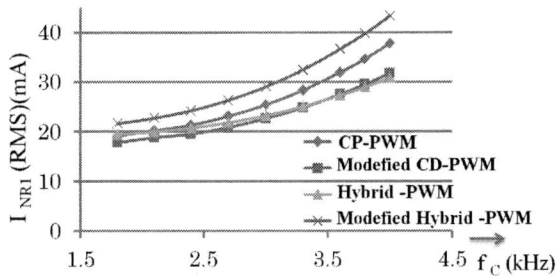
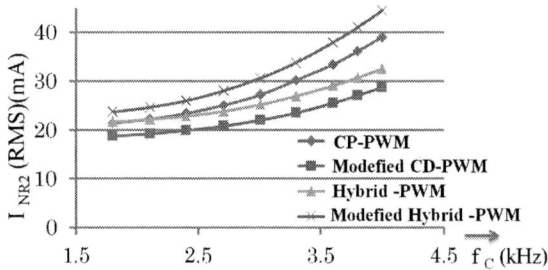

Fig. 10: RMS values for I_{NR1} and I_{NR2} for various values of cut-off frequency.

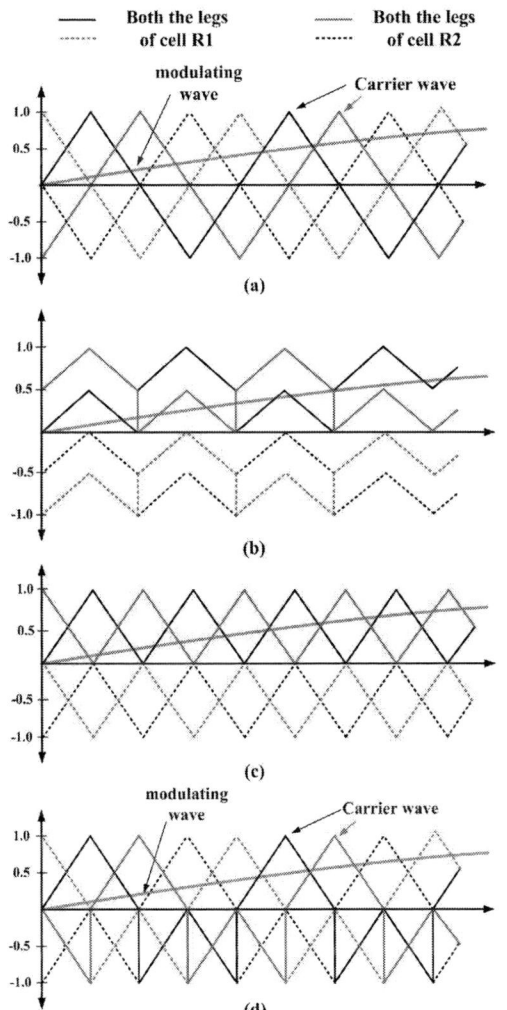

Fig. 9: Carrier waves for (a) CP PWM (b) modified CD PWM (c) hybrid PWM and (d) modified hybrid PWM techniques.

Table II: Critical values of cut-off frequency

PWM technique	Critical Cut-off frequency
PS-PWM	3.3 kHz
Modified CD-PWM	3.9 kHz
Hybrid PWM	3.84 kHz
Modified Hybrid PWM	3.05 kHz

with other PWM techniques (CP-PWM and modified hybrid PWM techniques) the filters have to be tuned with the cut-off frequency less than 3.33 kHz and 3.05 kHz respectively. Therefore, the additional cost due to filter components is minimum with modified CD PWM and hybrid PWM techniques. However, these filter components can also limit the line current THD without any additional filter requirement.

V. SIMULATION RESULTS

The CHB-MLI with the proposed filter configuration shown in Fig. 7 is simulated in MATLAB/Simulink with the parameters shown in Table-I, and the cut-off frequency is chosen as 3.3 kHz. The waveforms for V_{NR1}, V_{NR2} and the leakage current through the parasitic capacitors (I_{NR1} and I_{NR2}) are shown in Fig. 11 for the modified CD-PWM technique. As explained earlier, the voltage across each parasitic capacitor

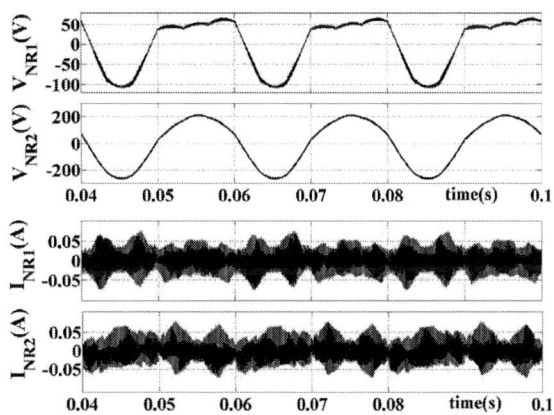

Fig. 11: Simulation results for (a) voltage across the parasitic capacitors (b) leakage currents.

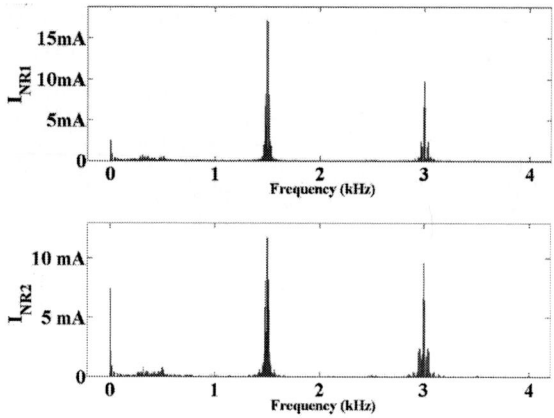

Fig. 12: FFT analysis for the leakage currents.

does not contain any high frequency ripple component. From the FFT analysis of the leakage currents as shown in Fig. 12, the high frequency ripple current is significantly reduced and the total RMS value is limited below 30 mA as per the requirement [14]. In addition to that, the line current THD of 0.93% is obtained without any additional filter components.

A. Limitations

The filter components should be designed with the cut-off frequency lower than switching frequency irrespective of the number of levels of the converter. This results in a large filter requirement at higher power level. Therefore, the proposed solution can be conveniently applicable at medium power level than high power level.

VI. CONCLUSIONS

Despite the advantages like independent MPPT and modularity, the CHB-MLI suffers from the limitation of inter module leakage currents which can not be avoided with an isolation transformer at the output. The main cause for these leakage currents are the PWM voltages appearing across the parasitic capacitors of PV panels. The analytical expressions for these PWM voltages are derived from the equivalent circuit of CHB-MLI. Further, a new filter configuration is proposed by placing several low pass filters among the power cells to mitigate the high frequency PWM voltage across the capacitors. The proposed solution is simulated with several PWM techniques, and it is found that the modified CD-PWM and hybrid PWM techniques are superior in minimizing the leakage currents with minimum filter requirement. The filter configuration is generic and it can be easily upgraded for higher number of power cells. Moreover, these filter components are sufficient to limit the THD of output current within the permissible limits.

REFERENCES

[1] Khajehoddin S. A., Bakhshai A., and Jain P., "The application of the cascaded multilevel converters in grid connected photovoltaic systems," in *Electrical Power Conference, EPC 2007*, pp. 296-301, 25-26 Oct. 2007.

[2] Alonso O., Sanchis P., Gubia E., and Marroyo L., "Cascaded H-bridge multilevel converter for grid connected photovoltaic generators with independent maximum power point tracking of each solar array," in *Power Electronics Specialist Conference, PESC 2003*, pp. 731-735, 15-19 June 2003.

[3] Yifan Yu, Konstantinou G., Hredzak B., and Agelidis V. G., "Optimal zero sequence injection in multilevel cascaded H-bridge converter under unbalanced photovoltaic power generation," in *ECCE-ASIA Power Electronics Conference IPEC-2014*, pp. 1458-1465, 18-21 May 2014.

[4] Sarkar I., and Fernandes B. G., "Modified hybrid multi-carrier PWM technique for cascaded H-Bridge multilevel inverter," in *40th Annual Conference of the IEEE Industrial Electronics Society, IECON 2014*, pp. 4318-4324, Oct. 29 2014 - Nov. 1 2014.

[5] Yan Zhou, and Hui Li, "Analysis and suppression of leakage current in cascaded-multilevel-inverter-based PV systems," *IEEE Trans. on Power Electron.*, vol. 29, no. 10, pp. 5265-5277, Oct. 2014.

[6] Lin Ma, Fen Tang, Fei Zhou, Xinmin Jin, and Yibin Tong, "Leakage current analysis of a single-phase transformer-less PV inverter connected to the grid," *IEEE Proceedings of ICSET 2008*, pp. 285-289, 24-27 Nov. 2008.

[7] S. B. Kjaer, J. K. Pedersen, and F. Blaabjerg, "A review of single-phase grid connected inverters for photovoltaic modules", *IEEE Trans. Ind. Appl.*, vol. 41, no. 5, pp. 1292-1306, Sep/Oct. 2005.

[8] Li Zhang, Kai Sun, Lanlan Feng, Hongfei Wu, and Yan Xing, "A family of neutral point clamped full-bridge topologies for transformerless photovoltaic grid-tied inverters," in *IEEE Trans. on Power Electron.*, vol. 28, no. 2, pp. 730-739, Feb. 2013.

[9] Dong Dong, Fang Luo, Boroyevich D., and Mattavelli P., "Leakage current reduction in a single-phase bidirectional AC-DC full-bridge inverter," *IEEE Trans. Power Electron.*, vol. 27, no. 10, pp. 4281-4291, Oct. 2012.

[10] Dong Dong, "Ac-dc bus-interface bi-directional converters in renewable energy system," Ph.D. dissertation, Virginia Polytech. Inst., Virginia State Univ, 2012.

[11] Wei Zhao, Hyuntae Choi, Konstantinou G., Ciobotaru M., and Agelidis V. G., "Cascaded H-bridge multilevel converter for large-scale PV grid-integration with isolated DC-DC stage", *in the Proc. of IEEE PEDe 2012*, pp. 849-856, 25-28 June 2012.

[12] Vazquez G., Martinez Rodriguez P. R., Sosa J.M., Escobar G., and Arau J., "A modulation strategy for single-phase HB-CMI to reduce leakage ground current in transformer-less PV applications", *IECON 2013 - 39th Annual Conference on IEEE Industrial Electronics Society*, pp. 210-215, 10-13 Nov. 2013.

[13] *Information on the design of transformerless inverters*, Application notes by SMA.

[14] *Automatic disconnection device between a generator and the public low-voltage grid*, Standard DIN V VDE V 0126-1-1, 2006.

Effect of Grid Inductance on Grid Current Quality of Parallel Grid-Connected Inverter System with Output LCL Filter and Closed-Loop Control

Wooyoung Choi, *Student Member, IEEE*, Woongkul Lee, *Student Member, IEEE*,
and Bulent Sarlioglu, *Senior Member, IEEE*
Electrical and Computer Engineering
Wisconsin Electric Machines and Power Electronics Consortium (WEMPEC)
University of Wisconsin-Madison
Madison, WI 53706 USA
sarlioglu@wisc.edu

Abstract—**Grid-connected inverters (GCIs) are power electronics interfaces to connect distributed energy resources into the grid. In order to improve the performance of distributed energy resources as well as maintain the high power quality of the grid, it is important to understand the characteristic of GCIs with appropriate controllers depending on its applications. This paper presents a study on the effect of grid inductance on quality of the grid current when there are multiple parallel-connected inverters to the grid. The model includes three-phase two-level voltage source inverters, output LCL filter, the grid, and voltage-oriented PI controller with a power controller. Parallel GCI systems are mathematically modeled. Frequency responses of grid current are investigated according to different grid inductances. A parallel grid-connected inverter system is simulated using MATLAB Simulink in order to present the effect of grid inductance on grid current and power quality.**

Index Terms—**Distributed energy resource (DER); grid-connected inverter (GCI); grid-connected inverter control; grid current quality; grid inductance; parallel grid-connected inverter; power quality; output LCL filter; weak grid**

I. INTRODUCTION

As the number of inverter-based distributed energy resources (DERs) increases in modern electric power system, the role of grid-connected inverters (GCIs) has become significant to improve the performance of DERs. Also, appropriate control of GCIs is important to satisfy standards of the grid, such as power quality, when DERs are interconnected to grid. According to the IEEE standard 1547 [1], there are requirements of DERs such as regulation, response to area electric power system abnormal condition, power quality, and islanding to be connected to the grid. Total demand distortion of grid current at point of common coupling (PCC) is limited to be less than 5 % including up to the 35th harmonics of fundamental power frequency.

In addition, to improve power quality of grid and to avoid instability, it is necessary to understand the overall system analysis with many parallel-connected inverters. It is noted that the performance of parallel GCI system with a consideration of grid inductance could be different from ideal case. Ideally, the grid is assumed to be stiff so that grid inductance is zero or very small value. However, in weak grid condition where distributed generation is located in remote area with long distribution wire, maximum grid inductance could be 10 % of base impedance. To assure that grid current injected into the grid satisfies the requirements, it is important to investigate the relationship between the number of parallel GCIs and the grid inductance. This paper presents the effect of grid inductance on the quality of grid current in parallel GCI system.

There are various contributions in literature studying parallel configuration of GCIs [2-8]. Stability and bandwidth of hard-coupled and soft-coupled cases are examined for digitally controlled parallel GCIs in [2]. Small-signal analysis is done for parallel GCIs in stand-alone ac supply system in [3], and parallel voltage-source converters are modeled and controlled for centralized vehicle-to-grid application in [4]. The study of parallel inverters in distribution system are presented in [5, 6]. Also, a modular control of parallel GCIs in standalone ac supply system is introduced in [7]. In reference [8], a digital controller with active damping is designed and the N-paralleled inverters are modeled including the effect of grid impedance. Different control strategies to improve power quality of GCIs [9-11] have also been examined. Regarding grid inductances, reference [8] presents the coupling effect of grid inductance, and dynamic performance in weak grid is introduced in [12, 13]. Robust control is implemented to mitigate uncertainties of grid inductance in [14, 15]. The effect of grid inductance on stability of GCI system are presented in [16]. The parallel GCIs are studied to investigate the stability and power quality of grid.

The objective of this paper is to fill the knowledge gap by contributing a study of the effect of grid inductance on quality of grid current when multiple GCIs are parallel connected to the grid. This paper presents mathematical model of parallel GCI system with a consideration of grid inductance. A single GCI system includes three-phase two-level voltage source inverter, output LCL filter, closed loop current controller, and power controller. Frequency responses of grid current with respect to inverter output voltage are presented. Moreover, parallel GCIs are simulated using MATLAB Simulink according to different grid inductances. There are two cases examined in this paper: the first case is when there are two GCIs parallel connected to the grid, the second case is when five GCIs parallel

978-1-4673-9551-9/16 $31.00 © 2016 IEEE

connected to the grid. Both cases are simulated for the different values of grid inductance from 0 pu to 0.05 pu.

II. ANALYTICAL RESULTS OF PARALLEL GRID-CONNECTED INVERTER SYSTEM

A. Parallel Grid-Connected Inverter System

A single three-phase grid-connected inverter system consists of a DER source, DC/DC converter, voltage source inverter (VSI), output LCL filter, power grid, and closed-loop controller. A topology of VSI selected for the study is three-phase two-level VSI, which is widely used in the vast majority of applications. An output LCL filter is used to attenuate the undesired harmonics of PWM switching of inverter. LCL filter is used rather than L filter due to its benefits of better attenuation property and smaller size. Closed-loop controller includes functions of grid synchronization, current control, and power control.

Multiple GCIs can be connected parallel to the grid such that the grid current is the sum of each output current of GCI with the same voltage level at PCC as shown in Fig 1. This parallel system happens in distributed generation system where many distributed energy resources are connected to the ac grid.

In Fig. 1, v_i is inverter output voltage, C_{dc} is DC link capacitor, L_1 is inverter-side inductance of filter, L_2 is grid-side inductance of filter, C_f is shunt capacitance of filter, and R_d is damping resistance of filter, v_g is phase grid voltage, i_g is phase grid current, R_g is grid resistance, and L_g is grid inductance. v_{fg} is phase voltage at PCC and N is number of parallel GCIs connected to grid.

In stiff grid condition, grid inductance is zero or close to zero and there is no coupling between parallel inverters and grid such that grid voltage, v_g, is equal to grid voltage at PCC, v_{fg}. Coupling effect exists when grid inductance is nonzero and the power quality and controller performance will be affected. The maximum grid impedance is assumed to be 10 % of the base impedance in weak grid condition. Table I shows specifications of grid, inverter, designed LCL output filter, and controller gains.

B. Effects of Grid Inductance on Frequency Response of Grid Current in Open-Loop System

The open-loop equation of frequency response of grid current with respect to grid voltage and inverter output voltage is shown in (1) when there is N-number of GCIs connected in parallel to grid with consideration of grid inductance and grid resistance.

$$I_g(s) = A_1(s)V_g(s) + A_2(s)V_i(s) \tag{1}$$

where $V_i(s)$ is output voltage of inverter, $V_g(s)$ is grid voltage, and $I_g(s)$ is grid current in frequency domain.

It is assumed that voltages sensed at each GCI are identical to grid voltage at PCC such that

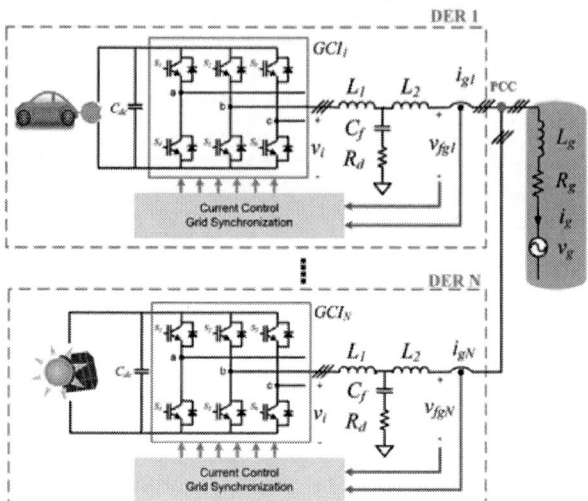

Fig. 1. Parallel-connected GCIs to the grid.

TABLE I. SPECIFICATIONS OF GRID , INVERTER, AND LCL OUTPUT FILTER, AND CONTROLLER

	Parameter	Value
Grid	Line-to-line grid voltage, V_g	240 Vrms
	Grid frequency, f_g	60 Hz
	Grid inductance, L_g	0 pu (stiff) ~ 0.1 pu (weak)
Inverter	Rated apparent power, S_R	14.14 kVA
	Switching frequency, f_{sw}	6 kHz
LCL Output Filter	Inverter-side inductance, L_1	990 μH
	Grid-side inductance, L_2	430 μH
	Shunt filter capacitor, C_f	20 μF
	Damping resistance, R_d	3.5 Ω
	Resonance frequency, f_{res}	2.05 kHz
Control	Proportional gain for d-axis, $K_p{}^d$	5
	Integral gain for d-axis, $K_i{}^d$	60
	Proportional gain for q-axis, $K_p{}^q$	25
	Integral gain for q-axis, $K_i{}^q$	100

$$v_{fg,1} = v_{fg,2} = \cdots = v_{fg,N} = v_{fg} \tag{2}$$

For simplicity, another assumption is made that the inverter voltage of each GCI is the same as shown in (3).

$$v_{i,1} = v_{i,2} = \cdots = v_{i,N} = v_i \tag{3}$$

Transfer functions of $A_1(s)$ and $A_2(s)$ in (1) are represented in (4) and (5), respectively. It is shown that there are additional terms involving N, which is different from single GCI system.

In equation (4) and (5), LC impedance, Z_{LC}, involving L_2 and C_f is defined as (6).

$$A_1(s) = \frac{-N\, Z_{LC}^2\left(L_1 C_f s^2 + R_d C_f s + 1\right)}{L_1\left(1 + N Z_{LC}^2 L_g C_f\right)s^3 + C_f\left\{L_1 R_d \omega_{res}^2 + N Z_{LC}^2\left(L_g R_d + R_g L_1\right)\right\}s^2 + \left\{L_1 \omega_{res}^2 + N Z_{LC}^2\left(L_g + R_g R_d C_f\right)\right\}s + N Z_{LC}^2 R_g} \tag{4}$$

$$A_2(s) = \frac{N\, Z_{LC}^2\left(R_d C_f s + 1\right)}{L_1\left(1 + N Z_{LC}^2 L_g C_f\right)s^3 + C_f\left\{L_1 R_d \omega_{res}^2 + N Z_{LC}^2\left(L_g R_d + R_g L_1\right)\right\}s^2 + \left\{L_1 \omega_{res}^2 + N Z_{LC}^2\left(L_g + R_g R_d C_f\right)\right\}s + N Z_{LC}^2 R_g} \tag{5}$$

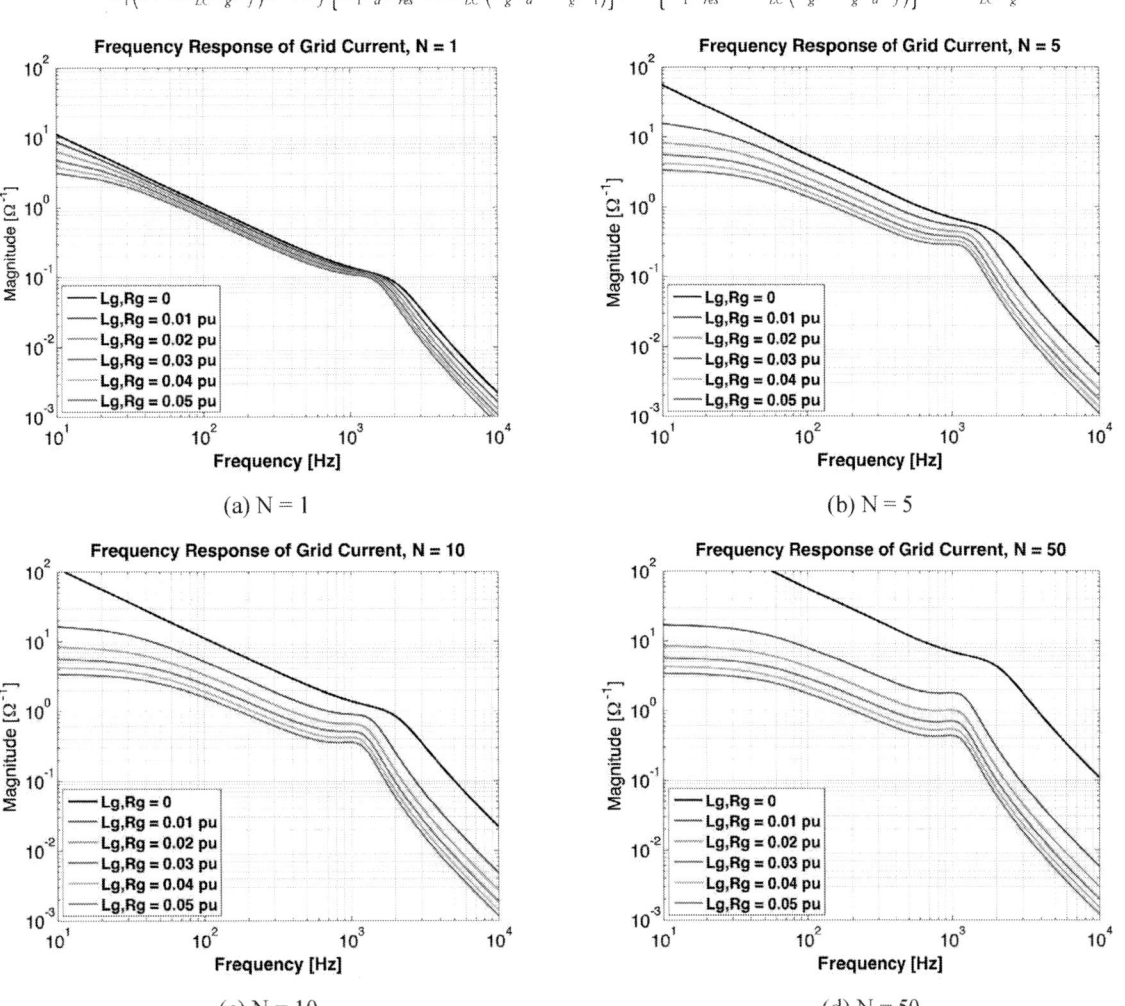

(a) N = 1

(b) N = 5

(c) N = 10

(d) N = 50

Fig. 2. Frequency responses of grid current with respect to inverter output voltage in open-loop system for different values of grid inductance when N-number of GCIs are connected to grid (a) N = 1, (b) N = 5, (c) N = 10, and (d) N = 50.

$$Z_{LC} = \sqrt{\frac{1}{L_2 C_f}} \tag{6}$$

Resonance frequency, f_{res}, is expressed as

$$f_{res} = \frac{1}{2\pi}\sqrt{\frac{L_1 + L_2}{L_1 L_2 C_f}} \tag{7}$$

The frequency responses of grid current with respect to output·voltage of inverter are shown in Fig. 2 when N-number of parallel GCIs are connected to grid including effect of grid inductance and resistance. When grid is ideal

(stiff grid condition), $L_g = 0$, $R_g = 0$, the magnitude of grid current in frequency domain increases over entire frequency range as number of parallel GCI increases. This result corresponds to (5), which has no coupling effect between grid and GCIs when grid inductance and resistance are zero.

However, frequency responses with nonzero grid inductance (weak grid condition) are different from one in stiff grid condition. For the frequency range greater than 100 Hz, the magnitude of frequency responses increases when L_g is 0.01 pu as number of GCIs increases but not as much as the ideal case. When L_g is larger than 0.01 pu, magnitude of frequency response of grid current with respect to inverter

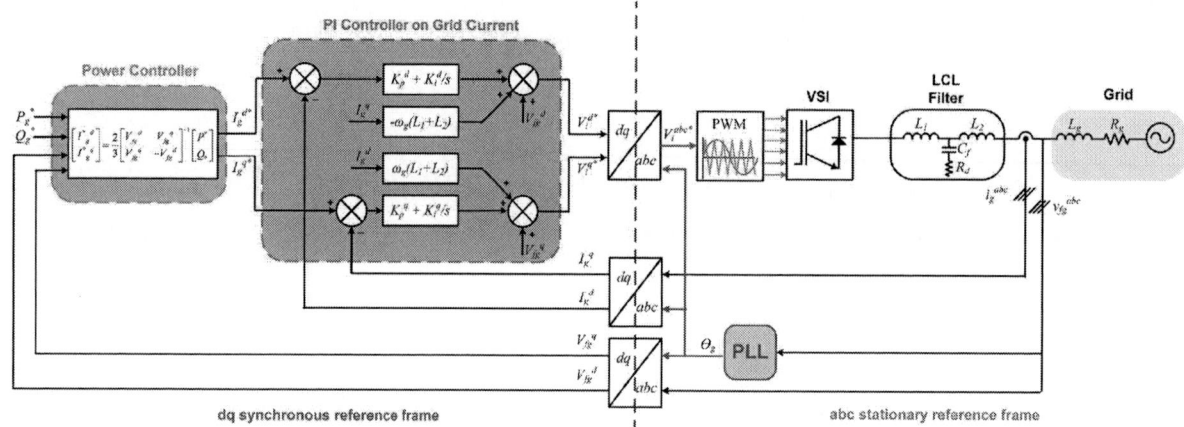

Fig. 3. Voltage oriented PI controller and power controller of single GCI system with LCL filter and grid.

output voltage does not change significantly as N increases. This is a clear indication of the impact of the grid impedance on the multiple GCIs system. Whether the grid is stiff or weak, the impact of the grid impedance is marginal when there is only one GCI connected to the grid as shown in Fig. 2 (a). However, the difference of the frequency responses for different grid inductances becomes dramatic when the number of GCIs connected to the grid increases, especially when the grid is in weak condition as shown in Fig. 2 (d). Therefore, frequency response of grid current depends on the number of parallel GCIs in stiff grid condition. In weak grid condition, frequency response is more dependent on grid inductance rather than the number of parallel GCIs.

C. Closed-Loop Control: Voltage oriented PI controller and Power Controller in Synchronous dq Reference Frame

Power controller and voltage oriented PI controller are used for current-controlled voltage-source inverters as shown in Fig. 3. Grid voltage at PCC is aligned with d-axis in synchronous *dq* reference frame. The grid current and grid voltage at PCC are sensed and fed back to the PI controller. Given command of active and reactive power, power controller provides command of *d*- and *q*-axis grid current. Coupling terms between *d*- and *q*-axis components are decoupled and feedforward of grid voltage at PCC is included.

III. SIMULATION RESULTS OF PARALLEL GCIS

The effects of coupling grid inductance on grid current are simulated using MATLAB Simulink and Simpower tool with specifications in Table I. It is assumed that DER and DC/DC converter are ideal and grid voltage is not distorted. Also, parallel GCIs are assumed to be identical such that inverter output voltages and power ratings of GCIs are the same. Two cases are examined: when there are two parallel GCIs (N = 2), and when there are five parallel GCIs (N = 5). The grid inductances examined are 0, 0.01, 0.03, and 0.05 pu, respectively. Controller's gains are set to be consistent with values in Table I for different cases and grid inductances.

A step change in command active power of each GCI from -5 kW to -10 kW at 0.15 s is simulated with command reactive power of inverter set to be zero. The sign of current flowing from the grid to the inverter is set to be positive. The negative sign of power means that the inverter provides power to the grid. When there are two GCIs connected (N = 2), the total command active power of the grid would change from -10 kW to -20 kW at 0.15 s and total command reactive power would be zero for the entire time range. The total command active power of the grid would step up from -25 kW to -50 kW for the case of N = 5. The simulated results for the case of N = 2 are shown in Fig. 4 – 8. Figure 9 – 13 are the results when there are five GCIs connected to the grid.

For the first case when N = 2, Fig. 4 shows waveforms of phase A grid current for different values of grid resistances and inductances from 0.15 s to 0.2 s. As the command active power of each GCI is changing from -5 kW to -10 kW, the current command of each GCI is also changing from 12 Arms to 24 Arms. Therefore, the grid current increases from 24 Arms to 48 Arms at 0.15 s. The waveforms of the grid currents are sinusoidal including a little of high frequency harmonics for 0, 0.01, and 0.03 pu of grid inductances. The waveform of the grid current is distorted for 0.05 pu of the grid inductance when two GCIs are connected. The waveforms of phase A current during one cycle are shown in Fig. 5 for different values of grid inductances. It is shown that the grid current for 0.05 pu of the grid inductance is different from the grid currents for 0, 0.01, and 0.03 pu of the grid inductance. When L_g is 0.05 pu, the controller does not track the command current well that the system becomes unstable ultimately leading to a poor grid current quality with a distorted waveform.

The instantaneous active power and reactive power of the grid are shown in Fig. 6 and 7, respectively. High frequency noise and 2nd order harmonic are included. The active power increases from -10 kW to -20 kW when N = 2. Regarding the reactive power of the grid, reactive power becomes zero after a few cycles when the grid inductance is zero. When L_g is 0.03 pu, the GCI flows more reactive power than

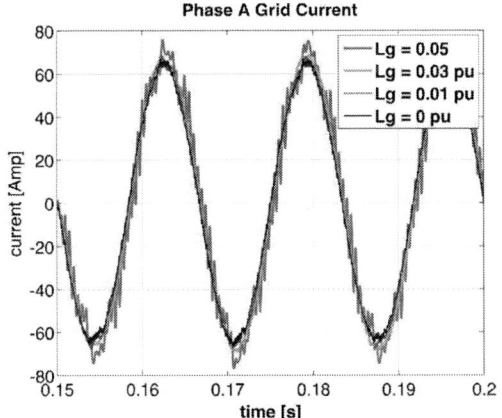

Fig. 4. Phase A grid current for different values of grid inductance when two GCIs connected parallel to grid during 0.15 s to 0.2 s (N = 2).

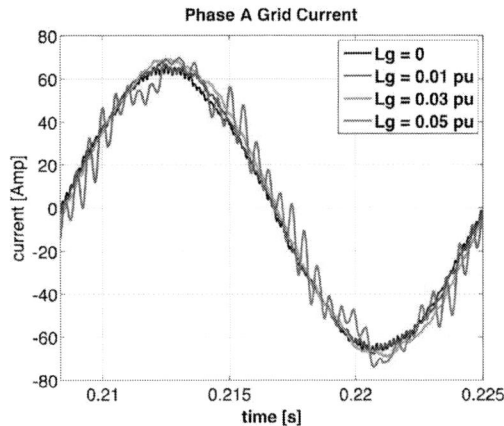

Fig. 5. Phase A grid current for different values of grid inductance when two GCIs connected parallel to grid during one cycle (N = 2).

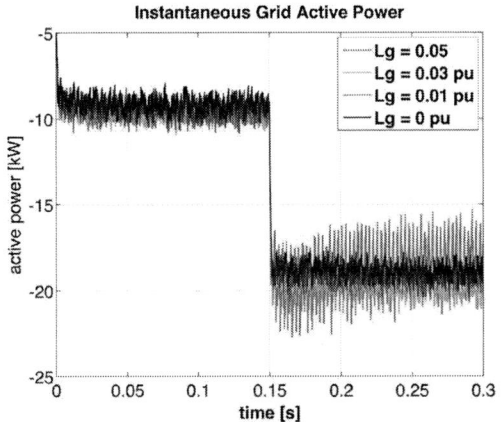

Fig. 6. Instantaneous active power of grid when two GCIs connected parallel to grid (N = 2).

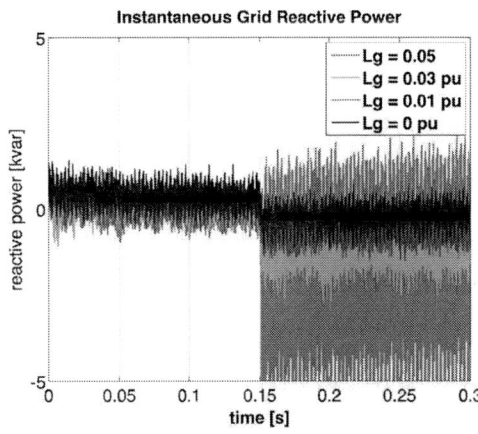

Fig. 7. Instantaneous reactive power of grid when two GCIs connected parallel to grid (N = 2).

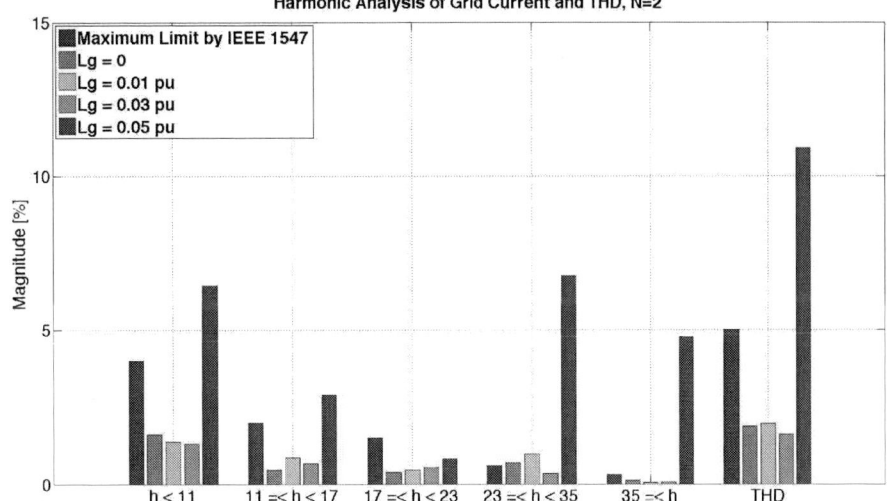

Fig. 8. Harmonic distortion and THD of phase A grid curernt for different values of grid inductance when two GCIs connected parallel to grid (N = 2).

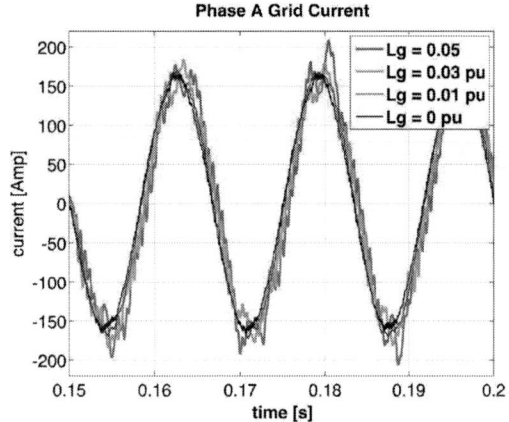

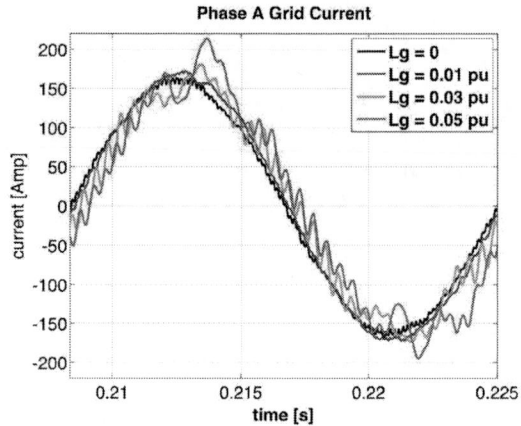

Fig. 9. Phase A grid current for different values of grid inductance when five GCIs connected parallel to grid during 0.15 s to 0.2 s (N = 5).

Fig. 10. Phase A grid current for different values of grid inductance when two GCIs connected parallel to grid during one cycle (N = 5).

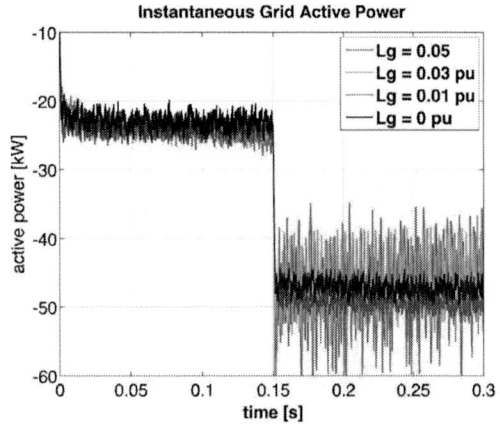

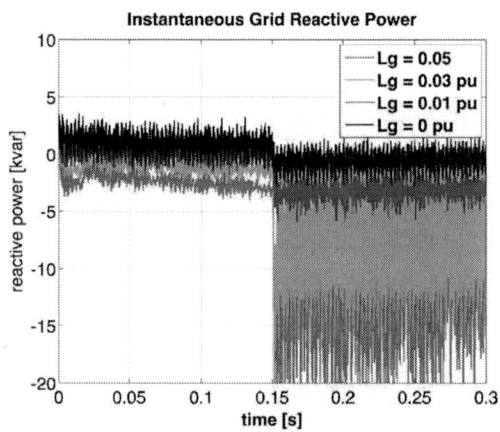

Fig. 11. Instantaneous active power of grid when five GCIs connected parallel to grid (N = 5).

Fig. 12. Instantaneous reactive power of grid when five GCIs connected parallel to grid (N = 5).

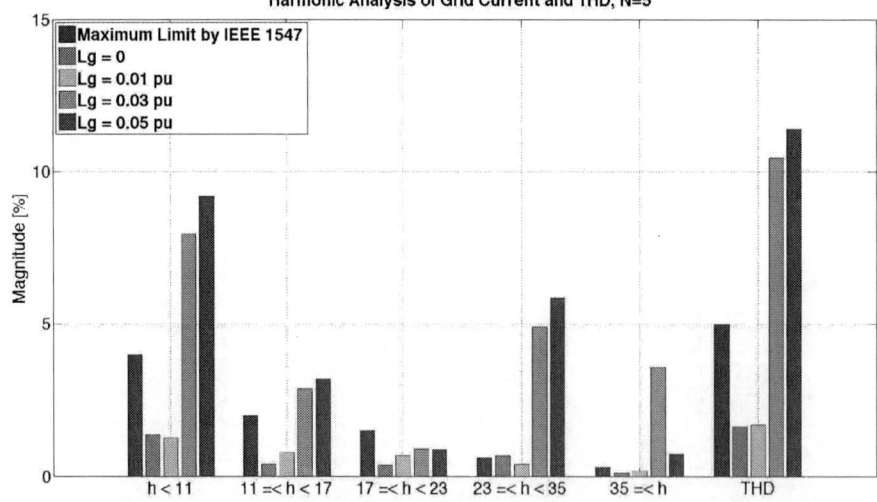

Fig. 13. Harmonic distortion and THD of phase A grid curernt for different values of grid inductance when five GCIs connected parallel to grid (N = 5).

command to the grid after 0.15 s. This is due to the lagging effect between grid voltage and grid current caused by the high grid inductance. As system becomes unstable for 0.05 pu of L_g, more active and reactive power flow into the grid leading to a poor power factor. In particular, lots of undesired reactive power flows into the grid, which is the condition to disconnect DER from the grid.

Figure 8 shows harmonics distortions according to different values of grid inductance when there are two GCIs connected parallel to the grid. The waveform during 8 cycles after 0.1584 s of phase A grid current is examined including up to 35th harmonic.

When calculating total harmonic distortion (THD), THD depends on RMS current even if harmonic components are not changing. Total demand distortion (TDD) is another terminology to describe harmonic distortions with a consideration of operating load condition. At full load condition, THD is identical to TDD. Since GCIs operate at 100 % of rated condition after 0.15 s, THD is identical to TDD. The x-axis is a range of harmonics only including odd harmonics: harmonics up to 10th, 11th to 16th harmonics, 17th to 22nd harmonics, 23rd to 34th harmonics, 35th harmonic, and THD including from the 2nd to the 35th harmonics. The dark blue colored bar is the maximum limit of harmonic distortion according to IEEE 1547 standard. The maximum limit of even harmonic distortion is 25 % of the maximum limit of odd harmonic distortion. The maximum THD up to the 35th harmonic component should be limited to 5 % by standard.

THDs of grid current are 1.87 %, 1.97 %, 1.62 %, and 10.9 % when L_g is 0 pu, 0.01 pu, 0.03 pu, and 0.05 pu, respectively. THDs are all less than 5 % for 0, 0.01, and 0.03 pu of grid inductances when N = 2. As shown in Fig. 4 and 5, THD of grid current for 0.05 pu of L_g is 10.9 %, which is higher than 5 % by standard. The harmonics of the 5th and 7th, and 33rd and 35th mainly cause high THD when L_g is 0.05 pu. The other observation is that THD of grid current with zero grid inductance is larger than one with 0.03 pu of grid inductance. This indicates that grid inductance acts as a low pass filter. However, as the grid inductance becomes larger and more dominant, it causes controller of GCI instable. This eventually leads to much higher THD as shown in the case where the grid inductance is 0.05 pu.

Figures 9 – 13 are simulated results when there are five GCIs connected parallel to the grid (N = 5). Figure 9 is phase A grid current from 0.15 s to 0.2 s. The waveforms of grid current become far from sinusoidal waveforms and are highly distorted for 0.05 pu and 0.03 pu of L_g. This is also shown in Fig. 10 showing phase A grid current during one cycle. For 0.03 pu and 0.05 pu of grid inductance, qualities of grid current are poor with 10.5 % and 11.4 % of THDs, respectively, which are above the limitation. THDs of grid current are 5% less for L_g = 0 pu and 0.01 pu.

The instantaneous grid active power and reactive power in time domain are presented in Fig. 11 and 12, respectively. The power quality of the grid is poor in weak grid condition. For 0.03 pu and 0.05 pu of L_g, undesired active and reactive power flow into the grid. In addition, more high frequency components are included that hurts the power factor. In this

case, DER should be disconnected to the grid to avoid the poor power quality unless there is no compensation device, which improves power quality.

Harmonic distortion analysis is done for N = 5 case and results are shown in Fig. 13. THDs of grid current are 1.64 %, 1.7 %, 10.5 %, and 11.4 % when L_g is 0 pu, 0.01pu, 0.03 pu, and 0.05 pu, respectively. In weak grid condition with 0.05 pu of L_g, the THD of grid current is 11.4 %, which is larger than 5 %. THD is above maximum limit of standard for all range of harmonics also. This amount of THD is against the standard so that DER cannot be connected to the grid. Especially, the 2nd, 9th, and 32nd harmonics are the major causes of the high THD when L_g is 0.05 pu. Only 0.01 pu of L_g satisfies the standard and functions as filtering. For 0.03 pu of grid inductance, harmonic distortions including harmonics of the 5th and 7th, and 33rd and 35th mainly cause of high THD.

The number of parallel-connected GCIs and the value of grid inductance bring about the coupling effect on the grid current quality. Even though the grid inductance is kept at the same value, the number of parallel GCIs and power rating of GCIs impact the grid current quality. When L_g is 0.03 pu, THD of grid current is 1.62 % for N = 2 case, whereas THD of grid current is 10.5 % for N = 5 case.

The relatively small grid inductance (L_g = 0.01 pu when N = 2) is good to act as a low pass filter. On the other hand, the relatively large grid inductance (L_g = 0.05 pu when N = 2) gives results to the detriment of grid current quality in terms of harmonic distortions especially when there are many GCIs parallel connected to the grid. Also, more reactive power flowing to the grid leads to the poor power factor as well as negatively impacting the DER.

IV. CONCLUSION

The effect of grid inductance and resistance on grid current quality is studied for parallel GCI system including three-phase two-level voltage source inverters, LCL output filter, grid, and voltage-oriented PI controller and power controller.

The N-number of parallel grid-connected inverter system is mathematically modeled and frequency responses of grid current are shown for different grid inductances. It is shown that frequency response of system is highly dependent on the number of parallel GCIs in weak grid condition.

The simulated results show that relatively small inductance improves grid current quality compared to case without grid inductance. However, grid currents are highly distorted with high THD more than 10 % for 0.05 pu of L_g when two and five GCIs connected to the grid. Harmonic distortion is examined including up to 35th harmonics to analyze frequency spectrum of grid current for different values of grid inductance. Especially, the 5th and 7th harmonics and the 33^{h} and 35th harmonics cause high THD for 0.05 pu of grid inductance. The results also validate that the larger grid inductance results in poorer grid current quality when there are many GCIs connected.

ACKNOWLEDGMENT

This research has been funded in part by the DOE Sunshot Initiative under award DE-0006341.

REFERENCE

[1] *IEEE Standard for Interconnecting Distributed Resources with Electric Power Systems,* IEEE Std 1547-2003, Jul., 2003.

[2] R. Turner, S. Walton, and R. Duke, "Stability and bandwidth implications of digitally controlled grid-connected parallel inverters," *IEEE Trans. Ind. Electron.,* vol. 57, no. 11, pp. 3685-3694, Nov. 2010.

[3] E. A. A. Coelho, P. C. Cortizo, and P. F. D. Garcia, "Small-signal stability for parallel-connected inverters in stand-alone AC supply systems," *IEEE Trans. Ind. Appl.,* vol. 38, no. 2, pp. 533-542, Mar./Apr. 2002.

[4] B. Zhang, X. Yan, X. Xiao, H. Liu, and Y. Li, "The VSC parallel structure and control technology for the centralized V2G system," in *Proc. ISIE'13,* 2013, pp. 1-6.

[5] J. H. Enslin and P. J. Heskes, "Harmonic interaction between a large number of distributed power inverters and the distribution network," *IEEE Trans. Power Electron.,* vol. 19, no. 6, pp. 1586-1593, Nov. 2004.

[6] M. Younis, N. Rahim, and S. Mekhilef, "Distributed generation with parallel connected inverter," in *Proc. ICIEA'09,* 2009, pp. 2935-2940.

[7] M. C. Chandorkar, D. M. Divan, and R. Adapa, "Control of parallel connected inverters in standalone AC supply systems," *IEEE Trans. Ind. Appl.,* vol. 29, no. 1, pp. 136-143, Jan./Feb. 1993.

[8] J. L. Agorreta, M. Borrega, J. López, and L. Marroyo, "Modeling and control of-paralleled grid-connected inverters with lcl filter coupled due to grid impedance in pv plants," *IEEE Trans. Power Electron.,* vol. 26, no. 3, pp. 770-785, Mar. 2011.

[9] F. Blaabjerg, R. Teodorescu, M. Liserre, and A. V. Timbus, "Overview of control and grid synchronization for distributed power generation systems," *IEEE Trans. Ind. Electron.,* vol. 53, no. 5, pp. 1398-1409, Oct. 2006.

[10] E. Twining and D. G. Holmes, "Grid current regulation of a three-phase voltage source inverter with an LCL input filter," *IEEE Trans. Power Electron.,* vol. 18, no. 3, pp. 888-895, May. 2003.

[11] M. Prodanovic and T. C. Green, "Control and filter design of three-phase inverters for high power quality grid connection," *IEEE Trans. Power Electron.,* vol. 18, no. 1, pp. 373-380, Jan. 2003.

[12] G. Ledwich and H. Sharma, "Connection of inverters to a weak grid," in *Proc. PESC'00,* 2000, pp. 1018-1022.

[13] J. O. G. Tande, "Exploitation of wind-energy resources in proximity to weak electric grids," *Applied Energy,* vol. 65, no. 1–4, pp. 395-401, 2000.

[14] I. J. Gabe, J. R. Massing, V. F. Montagner, and H. Pinheiro, "Stability analysis of grid-connected voltage source inverters with LCL-filters using partial state feedback," in *Proc. IEEE Eur. Conf. Power Electron. Appl.,* Sep. 2007, pp. 1-10.

[15] I. J. Gabe, V. F. Montagner, and H. Pinheiro, "Design and implementation of a robust current controller for VSI connected to the grid through an LCL filter," *IEEE Trans. Power Electron.,* vol. 24, no. 6, pp. 1444-1452, Jun. 2009.

[16] J. Sun, "Impedance-based stability criterion for grid-connected inverters," *IEEE Trans. Power Electron.,* vol. 26, no. 11, pp. 3075-3078, Nov. 2011.

Small Signal Modeling and Control of a Grid Tied Converter without a Syncronization Unit

Subhajyoti Mukherjee, Pourya Shamsi, Mehdi Ferdowsi

Department of Electrical and Computer Engineering
Missouri University of Science and Technology
Rolla, MO, USA
smbgb@mst.edu

Abstract— **In this paper, control of a grid-tied converter as a voltage controlled converter is investigated for distribution level grids. Small signal model of the converter is presented and a control structure is designed to deliver desired active and reactive power. This controller is using power flow equations to control the converter as a synchronous machine and without the need for a dedicated synchronization unit (PLL) during its normal operation. It is demonstrated that the controller ensures decoupled control and is immune to grid frequency fluctuations.**

Keywords— Grid connected converter, synchronization, phase-locked loop, bidirectional power flow.

I. INTRODUCTION

Control of grid connected converters has always been an attractive topic of research in the field of power electronics. Such converters are generally supplied from voltage sources and are generally operated as current-controlled converters which are referred to as voltage-source current-controlled converters [1]. The conventional control approach requires a phase locked loop (PLL). A PLL with fast dynamics and high noise rejection is difficult to design. Furthermore, a PLL can lead to unstable operations especially in weak grids [2]. Another approach in controlling grid tied converters is direct power control (DPC) [3]. DPC suffers from variable switching frequency and is not recommended for industrial applications. Very few studies have investigated controlling a grid connected converters as voltage-controlled converter [4]-[6]. Control of the grid connected converter as voltage-controlled converter relies on the power flow equations and does not require a PLL under normal operation conditions. Such inverters are mainly studied for high voltage grids where the grid is mainly inductive [4].

In this paper, applications of voltage-controlled grid-tied converters in distribution level grids are investigated. Section II analyses the power flow equations in a distribution level grid and presents the small signal modelling of the system. Based on this small signal model the controller design is presented in Section III. The results are demonstrated in Section IV, while the conclusion is presented in Section V.

II. POWER FLOW EQUATIONS AND SYSTEM MODELLING

A 3- phase converter connected to a distribution level grid is shown in Fig. 1. In the figure, L is the per phase filter inductance, r is the per phase winding resistance of the filter inductor, C is the per phase filter capacitance, Z_g is the per phase impedance of the cable connecting the converter to the grid, r_g is the resistive part of the impedance of the cable connecting the inverter to the grid, L_g is the inductive part of the impedance of the cable connecting the inverter to the grid, i_L is the per phase filter inductor current, i_o is the per phase load current. For a distribution level grid, according to [7], $r_g >> \omega L_g$. Active power P and reactive power Q delivered by the converter in such a grid are given by [8]

$$P \approx \frac{V_g}{r_g}(V_c \cos \delta - V_g) \tag{1}$$

$$Q \approx -\frac{V_c V_g}{r_g}\sin \delta \tag{2}$$

where, V_g is the per phase grid voltage, V_c is the per phase output voltage of an inverter (voltage of the filtering capacitor, C), δ is the phase angle between grid voltage and inverter output voltage.

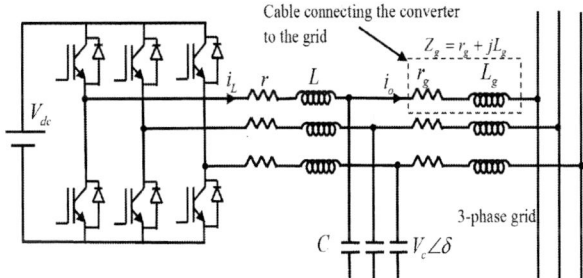

Fig. 1. 3-ph grid connected converter.

Generally power angle δ is small, so (1) and (2) can be modified as

$$P \approx \frac{V_g}{r_g}(V_c - V_g) \tag{3}$$

$$Q \approx -\frac{V_c V_g}{r_g}\delta \tag{4}$$

(3) and (4) suggest that the active power delivered, P, is primarily dependent on $(V_c - V_g)$, while the reactive power delivered, Q, is dependent on power angle δ. Therefore, active power injected to the grid can be controlled by controlling the magnitude of the converter voltage V_c, while the reactive power can be controlled by changing the power angle δ. To achieve this, a controller has to be implemented to control the required active and reactive power demand. An active power controller will generate the reference for the required voltage magnitude while a reactive power controller will generate the reference for the load angle and frequency. To design such controllers, small signal transfer function of $\Delta P/\Delta V_c$, $\Delta Q/\Delta\delta$ and $\Delta Q/\Delta f$ needs to be derived. This is done in a similar manner as in [4]. The equations are shown below.

In the synchronous frame of reference the active and reactive power is given by

$$P = V_{cd}i_{od} + V_{cq}i_{oq} \tag{5}$$

$$Q = V_{iq}i_{Ld} - V_{id}i_{Lq} \tag{6}$$

Linearizing (5) and (6) about an operating point

$$\Delta P = V_{cd}\Delta i_{od} + V_{cq}\Delta i_{oq} + \Delta v_{cd}I_{od} + \Delta v_{cq}I_{oq} \tag{7}$$

$$\Delta Q = V_{cq}\Delta i_{od} - V_{cd}\Delta i_{oq} + \Delta v_{cq}I_{od} - \Delta v_{cd}I_{oq} \tag{8}$$

In the synchronously rotating reference frame, the dynamics of the output current is given by

$$L_g \frac{di_{od}}{dt} + r_g i_{od} - \omega L_g i_{oq} = V_c \cos\delta - V_g \tag{9}$$

$$L_g \frac{di_{oq}}{dt} + r_g i_{oq} + \omega L_g i_{od} = V_c \sin\delta \tag{10}$$

Taking the Laplace transform of the linearized versions of (9) and (10) about an operating point

$$(sL_g + r_g)\Delta i_{od} - \omega L_g \Delta i_{oq} = \Delta V_c \cos\delta \tag{11}$$

$$(sL_g + r_g)\Delta i_{oq} - \omega L_g \Delta i_{od} = \Delta V_c \sin\delta \tag{12}$$

Solving (11) and (12) for Δi_{od} and Δi_{oq}

$$\Delta i_{od} = \frac{(sL_g + r_g)\cos\delta + \omega L_g \sin\delta}{(sL_g + r_g)^2 + (\omega L_g)^2}\Delta V_c \tag{13}$$

$$\Delta i_{oq} = \frac{(sL_g + r_g)\sin\delta - \omega L_g \cos\delta}{(sL_g + r_g)^2 + (\omega L_g)^2}\Delta V_c \tag{14}$$

Solving (9) and (10) for the steady state value I_{od} and I_{oq}

$$I_{od} = \frac{r_g(V_c \cos\delta - V_g) + \omega L_g V_c \sin\delta}{r_g^2 + (\omega L_g)^2} \tag{15}$$

$$I_{oq} = \frac{r_g V_c \sin\delta - (V_c \cos\delta - V_g)\omega L_g}{r_g^2 + (\omega L_g)^2} \tag{16}$$

Substituting (13) - (16) in (7) results

$$\frac{\Delta P}{\Delta V_c} = V_c \left[\frac{(s + \frac{r_g}{L_g})(s + \frac{1 + xr_g}{xL_g})}{(s + \frac{r_g}{L_g})^2 + \omega^2} \right] \tag{17}$$

Where,

$$x = \frac{\left(r_g V_c - V_g(r_g \cos\delta - \omega L_g \sin\delta)\right)}{\left(V_c r_g^2 + V_c(\omega L_g)^2\right)} \tag{18}$$

ΔP and ΔQ in above equations can suffer from ac components which appear due to unbalance or harmonic distortions. Hence the calculated P and Q are passed through low pass filters to generate P_{fil} and Q_{fil} to eliminate such effects. Considering a low pass filter with a cut-off frequency at ω_f, (17) is modified as

$$\frac{\Delta P_{fil}}{\Delta V_c} = V_c \left[\frac{(s + \frac{r_g}{L_g})(s + \frac{1 + xr_g}{xL_g})}{(s + \frac{r_g}{L_g})^2 + \omega^2} \right] \left(\frac{1}{s/\omega_f + 1} \right) \tag{19}$$

Similar mathematics can be done to derive the small signal transfer functions $\Delta Q_{fil}/\Delta\delta$, and $\Delta Q_{fil}/\Delta f$

$$\frac{\Delta Q_{fil}}{\Delta\delta} = V_c^2 \left[\frac{(s + \frac{r_g}{L_g})(s + \frac{1 + xr_g}{xL_g})}{(s + \frac{r_g}{L_g})^2 + \omega^2} \right] \left(\frac{1}{s/\omega_f + 1} \right) \tag{20}$$

$$\frac{\Delta Q_{fil}}{\Delta f} = \frac{V_c^2 * 2\pi}{s} \left[\frac{(s + \frac{r_g}{L_g})(s + \frac{1 + xr_g}{xL_g})}{(s + \frac{r_g}{L_g})^2 + \omega^2} \right] \left(\frac{1}{s/\omega_f + 1} \right) \tag{21}$$

Each of the transfer functions has two zeros as well as one real and a pair of complex conjugate poles. Also it can be noted that all of the three poles and one of the zeros are independent of the operating point while one zero is dependent on the operating point. In fact depending on the operating point, the system can have a right half plane zero

(RHPZ) and become a non-minimum phase system. Therefore, controller design should be done for the operating point where the RHPZ is at the lowest frequency. If the controller is able to ensure sufficient phase margin at this operating point, it will ensure stable operation of the system at all operating points. From (19) it can be concluded that the RHPZ arises when $x < 0$. Considering δ to be small

$$x = \left(r_g (V_c - V_g) \right) / \left(V_c r_g{}^2 + V_c (\omega L_g)^2 \right) \qquad (22)$$

Hence for x to be negative, $V_c < V_g$. Again from (3) if $V_c < V_g$, then $P < 0$. Or in other words the converter is consuming active power. It can be concluded that the right half plane zero arises only when the converter is consuming active power and its frequency decreases with the consumption of active power. The worst case operating point, from control point of view, is when the converter is consuming the rated active power and the controller design is done for this operating point. The frequency response of the transfer function of $\Delta P_{fil}(s)/\Delta V_c(s)$ for each of the case when the converter is delivering and consuming active power is shown in Fig. 2 and Fig. 3 respectively. It can be seen from Fig. 3 that even if the system represented by (19) has three poles and two zeros, the phase of the system goes to -270°. This is because one of the zeros has shifted to the right half plane, thereby nullifying the phase boost of the other zero. The locus of the moving zero is showing in Fig. 4.

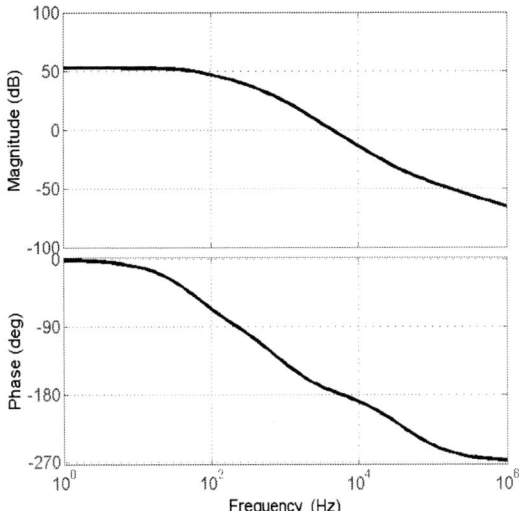

Fig. 3. Frequency response of $\Delta P_{fil}(s)/\Delta V_c(s)$ when consuming 5kW active power.

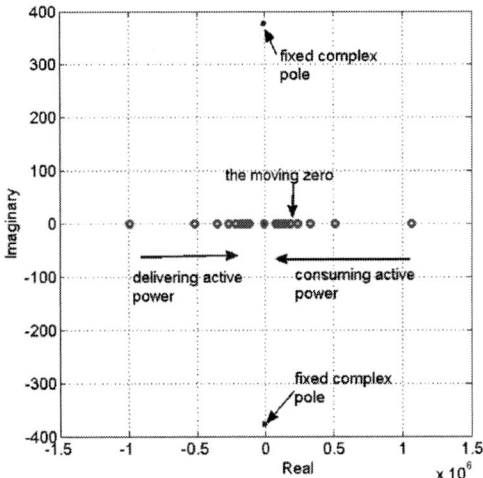

Fig. 4. The locus of the moving zero with change in active power.

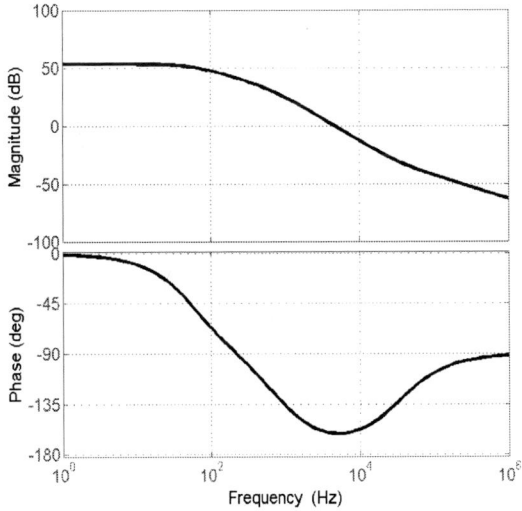

Fig. 2. Frequency response of $\Delta P_{fil}(s)/\Delta V_c(s)$ when delivering 5kW active power.

III. CONTROLLER DESIGN

The active and reactive power references being dc quantities in the synchronous frame of reference, a PI controller of the form $k_p(1+T_{ip}/s)$, is used to control each of the active and reactive power.. Similarly a PI controller is sufficient to control load angle δ. However, the overall loop gain of the frequency controller is very low. So the frequency controller needs a lead-lag compensator in addition to the PI controller to achieve desired phase margin. The proposed control structure is shown in Fig. 5. The frequency

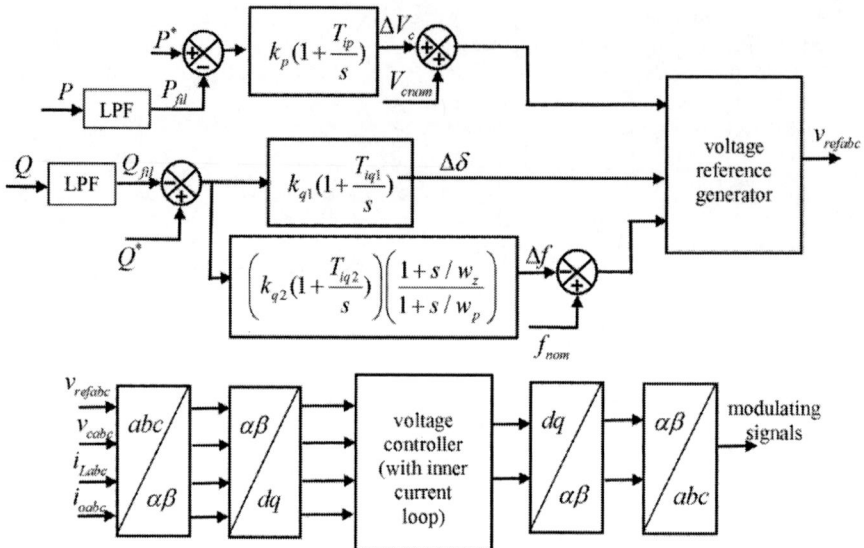

Fig. 5. The overall controller structure.

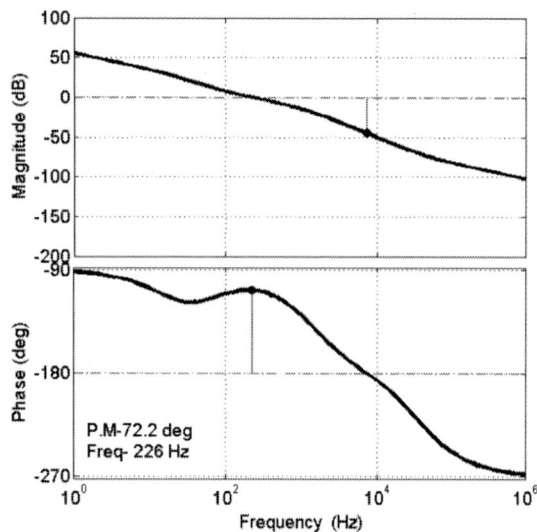

Fig. 6. Frequency response of the loop gain of $\Delta P_{fil}(s)/\Delta V_c$ (s) when consuming 5kW active power.

response of the plant and the controller for the active power controller is shown in Fig. 6. The voltage controller is a conventional PI controller with an inner current loop. The inner current controller is taken as a proportional controller. These two controllers (voltage and inner current) are studied in great detailed in existing literature and hence are not discussed in details here. Since a resistive grid inherently has sufficient damping, hence additional active damping is not required as was required for inductive grids as in [4]. This simplifies the controller structure.

IV. RESULTS

A. Simulation Study

The system shown in Fig. 1 is simulated for a 2 kVA converter. All the possible combinations of the active and reactive power are covered. Results presented in Fig. 7 confirm the stable operation for all such combinations. Note that the figures are for the actual power and not for the filtered power. The decoupled control of the active and reactive power is also evident from Fig. 7. The voltage and current during a transition in active power from 1.35 kW to −1.35 kW is shown in Fig. 8, while Fig. 9 shows the same for a transition in reactive power from 1.35 kVAR to −1.35 kVAR. Stable operation was noted in both cases without any significant distortion in voltage waveforms.

The grid frequency was changed from 60 Hz to 60.2 Hz with the converter delivering 1 kW of active power and 0.5 kVAR of reactive power. As seen from Fig. 10, the frequency controller was successful in tracking grid frequency variations. During this period, active power remains constant and is unaffected by frequency variations. However, slight variation in the reactive power is noted.

B. Experimental Study

Experimental results of the line voltage and phase current when the converter is delivering 1.25 kW of active power is shown in Fig. 11. The corresponding waveforms for the converter consuming −1.25 kW of active power are reported in Fig. 12. The waveforms while delivering reactive power are shown in Fig. 13.

978-1-4673-9551-9/16 $31.00 © 2016 IEEE

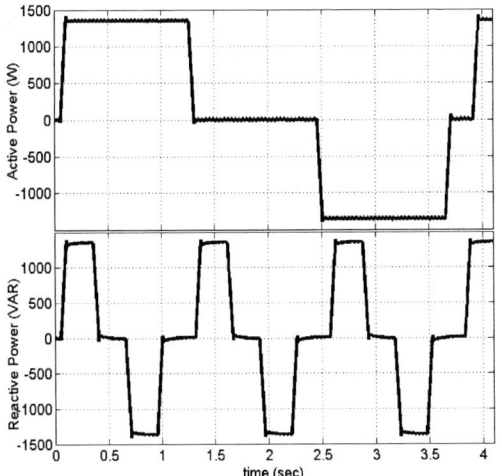

Fig. 7. Active and reactive power output from the converter.

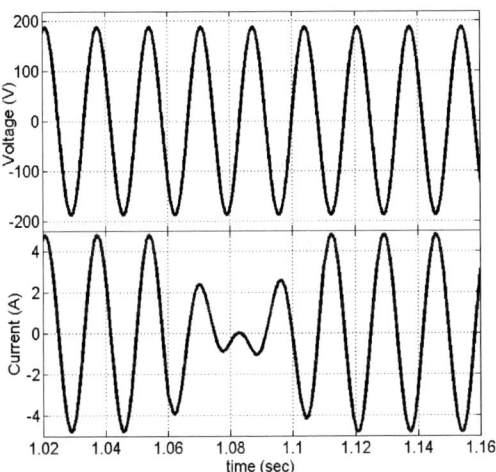

Fig. 8. Phase a voltage and current waveform during transition from P =1.35 kW to P = −1.35 kW.

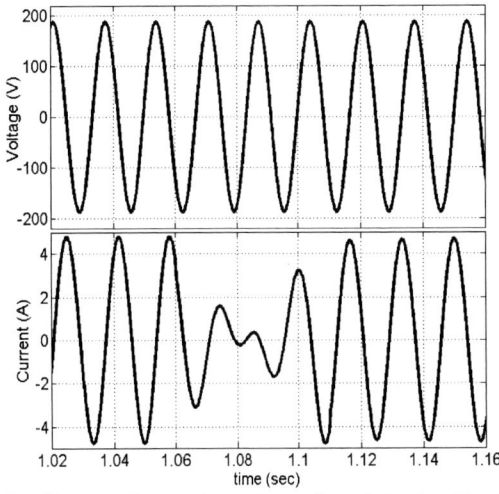

Fig. 9. Phase a voltage and current waveform during transition from Q=1.35 kVAR to Q= −1.35 kVAR .

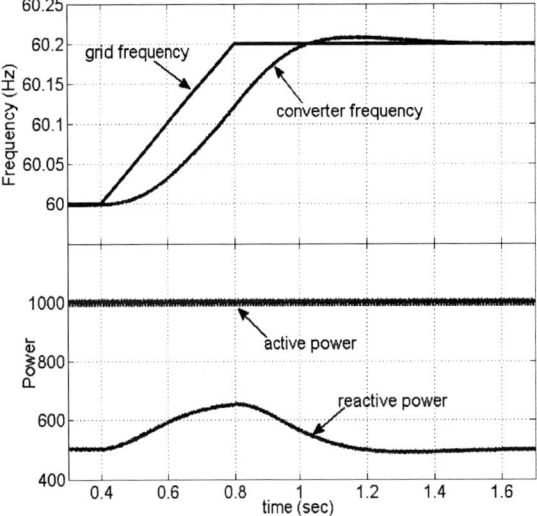

Fig. 10. Effect on active and reactive power due to change in grid frequency.

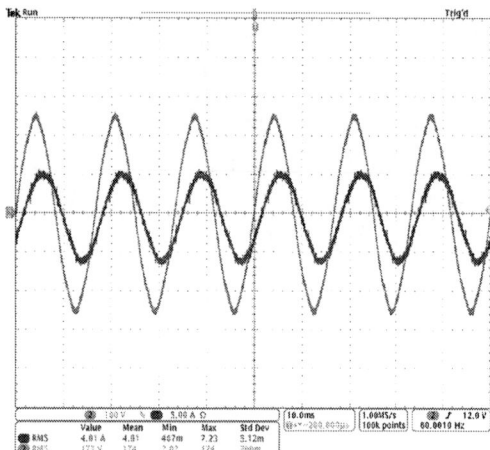

Fig. 11. Steady state line voltage and current for P = 1.25 kW.

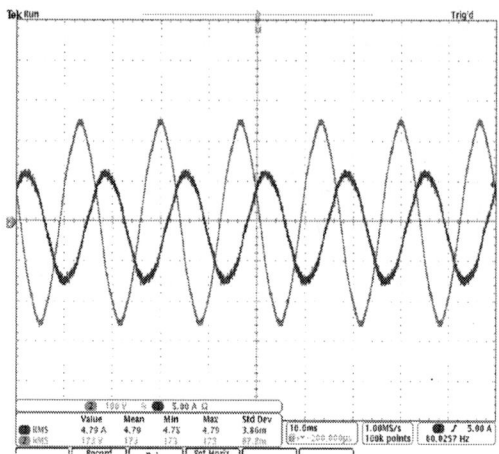

Fig. 13. Steady state line voltage and current for Q = 1.35 kVAR.

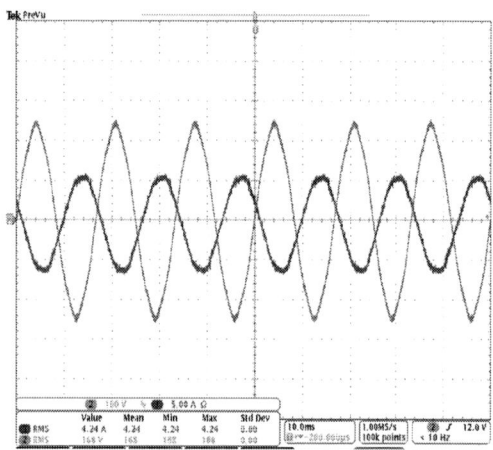

Fig. 12. Steady state line voltage and current for P = −1.25 kW.

V. CONCLUSION

The control of a grid connected converter as a voltage-controlled voltage-source converter for distribution level grids was presented in this paper. The small signal modelling of the system based on the power flow equations was presented and the controller was designed based on this model. Furthermore, normal operation of the system does not require a PLL which will reduce sensitivity of traditional designs to grid voltage harmonics and frequency variations. Simulation and experimental results confirm the decoupled nature of the control and its ability to track variations in the grid frequency.

REFERENCES

[1] M.P. Kazmierkowski, L. Malesani, "Current control techniques for three-phase voltage-source PWM converters: a survey," IEEE Trans. Ind. Electron., vol. 45, no. 5, pp. 691–703, Oct 1998.

[2] D. Dong, B. Wen, D. Boroyevich, P. Mattavelli, and Y. Xue, "Analysis of phase-locked loop low frequency stability in three-phase grid-connected power converters considering impedance interactions," IEEE Trans. Ind. Electron., vol. 62, no. 1, pp. 310–321, Jan. 2015.

[3] J. Noguchi, H. Tomiki, S. Kondo, and I. Takahashi, "Direct power control of PWM converter without power-source voltage sensors," IEEE Trans. Ind. Appl., vol. 34, no. 3, pp. 473–479, May/Jun. 1998.

[4] L. Zhang, L. Harnefors, and H.-P. Nee, "Power-synchronization control of grid-connected voltage-source converters," IEEE Trans. Power Syst.,vol. 25, no. 2, pp. 809–820, May 2010.

[5] Q.-C. Zhong and G. Weiss, "Synchronverters: Inverters that mimic synchronous generators," IEEE Trans. Ind. Electron., vol. 58, no. 4, pp. 1259–1267, Apr. 2011.

[6] Q. -C. Zhong, P. -L. Nguyen, Z. Ma, and W. Sheng, "Self-synchronized synchronverters: inverters without a dedicated synchronization unit," IEEE Trans. Power Electron., vol. 29, no. 2, pp. 617–630, 2014.

[7] A. Engler, "Applicability of droops in low voltage grids," DER J., vol. 1, pp. 1–5, Jan. 2005.

[8] X. Yu, A. M. Khambadkone, H. Wang, and S. T. S. Terence, "Control of parallel-connected power converters for low-voltage microgrid—Part I: A hybrid control architecture," IEEE Trans. Power Electron., vol. 25, no.12, pp. 2962-2970, Dec. 2010.

Bridgeless SEPIC PFC Converter for Low Total Harmonic Distortion and High Power Factor

Yasemin Onal
Department of Electrical and Electronic Engineering
Bilecik Seyh Edebali University, BILECIK, TURKEY
Email: yasemin.onal@bilecik.edu.tr

Yılmaz Sozer
Department of Electrical and Computer Engineering
University of Akron, OHIO
Email: ys@akron.edu

Abstract— **There is a need to improve the power quality of the grid as well as the power factor implied on the grid due to the nonlinear loads connected to it. A new single phase bridgeless AC/DC power factor correction (PFC) topology to improve the power factor as well as the total harmonic distortion (THD) of the utility grid is proposed in this research. By eliminating the input bridge in conventional PFC converters, the control circuit is simplified; the total harmonics distortion THD and power factor PF are improved. The controller operates in multi loop fashion as the outer control loop calculates the reference current through innovative filtering and signal processing. Inner current loop generates PWM switching signals through the PI controller. Analytical derivation of the proposed converter is presented in detail. Performance of the proposed PFC topology is verified for prototype using PSIM circuit simulations. The experimental system is developed, and the experimental results agree with simulation results.**

Keywords— *DSP; Bridgeless converter; power factor correction (PFC); Sepic converter; Total Harmonic Distortion THD*

I. INTRODUCTION

The request for developing power quality of the AC system has drawn excessive interest during the recent years. The increased usage of power electronic devices, such as variable speed drives, uncontrolled rectifiers and other switching devices, affects the power quality of the utility grid significantly. Standards similar to International Electro technical Commission (IEC) 61000-3-2 restrict the harmonics generated by these equipments [1]. To reduce harmonics in energy transmission lines, the research on active power factor correction (PFC) techniques has taken on an accelerated path [2-5].

Typical PFC converter topologies are boost [6,7], buck-boost [8], buck [9-11] and SEPIC [12-17]. The boost PFC converter is often used in practical applications, as the input current can be conveniently formed into a sinusoidal waveform to obtain unity power factor. However, the boost PFC converter has a restricted capability since the DC output voltage must be higher than the peak value of the AC input voltage [8]. On the other hand, the DC output voltage of the buck PFC is lower than the peak of the AC input voltage, which allows reducing components ratings and the cost. [11]. A buck PFC converter procures an alternative for low-voltage applications such as a 48V DC bus. Moreover, the buck PFC

can obtain high efficiency over the entire input voltage range with distorted input current that comfortably passes the limits imposed by IEC 61000-3-2 requirements [1].

The input current of the buck PFC converter has dead zones along the cycle, which requires extensive passive filtering to improve the power factor. There is a tradeoff between output voltage choice and power factor. To solve this problem, SEPIC or Cuk converters were proposed. A conventional SEPIC converter can supply a high power factor in wide range of voltage conditions [12-13]. The output voltage could be reduced or increased without the need of inversion with the SEPIC converters [14].

This paper presents a new topology for single phase bridgeless AC/DC PFC converters that reduces the THD and improves the PF of the operation. Proposed SEPIC converter combines the bridge and DC-DC stages into one stage. Section 2 analyses operation of the proposed SEPIC PFC converter. In Section 3, component selection and control circuit design are presented. The simulation results of the conventional and proposed SEPIC converter are presented in Section 4. Summary and future work is provided in Section 5.

II. PROPOSED BRIDGELESS SEPIC PFC CONVERTER

The conventional SEPIC PFC converter is shown in Figure 1 [15]. The operation of the circuit can be separated in to two modes concerning the position of the switches. When the switch Q_1 switched on, output diode D is reverse biased. The input inductor L_1 starts to charge, output inductor L_2 and AC input capacitor C_1 creates a resonant circuit. Here, load draws current from the output capacitor C_0. During this situation, the voltage of the input inductor will be same as the rectified AC voltage V_{ac}. Besides input capacitor's voltage and output inductor's voltage are equal to V_{ac} during this mode of operation. In the second mode, the switch is turned off, diode is forward biased and L_1, C_1, L_2 creates a loop. The load is directly connected to the inductors during this mode, which discharges them during the mode of operation.

The proposed bridgeless SEPIC PFC converter with three active switches is shown in Figure 2. When Q_1, Q_3 and Q_4 turn on, input inductor currents starts to increase linearly. The

978-1-4673-9551-9/16 $31.00 © 2016 IEEE

output inductor voltage is equal to the voltage of C_1 which was equal input voltage before the switches are turned on. Thereby, i_{L2} reduces linearly. This mode finishes by turning off Q_1, Q_3 and Q_4. By turning Q_1, Q_3 and Q_4 off, D starts to conduct. Input inductor current reduces linearly and i_{L2} increases linearly until the diode current extinguishes. When D turns off, output side is disengaged from the input side, the current through the inductors freewheel at the input side. Working modes for proposed SEPIC PFC converter is provided in Figure 3.

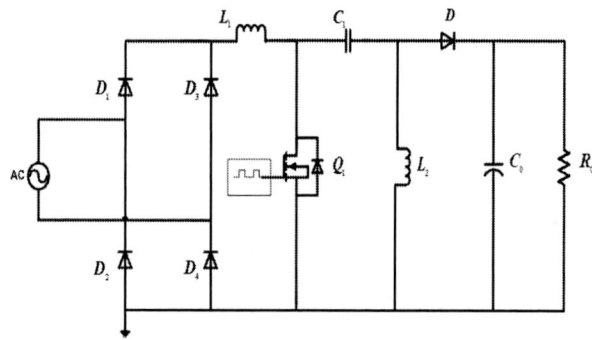

Fig. 1. Conventional SEPIC PFC converter

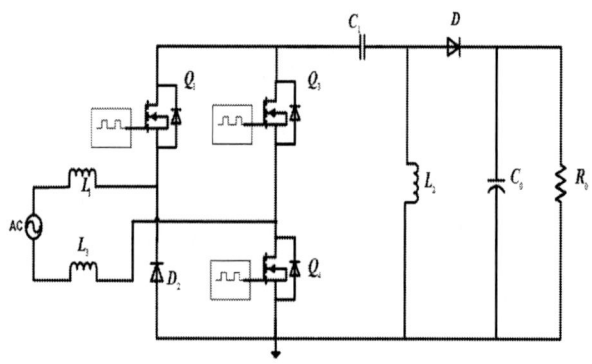

Fig. 2. Proposed SEPIC PFC converter circuits

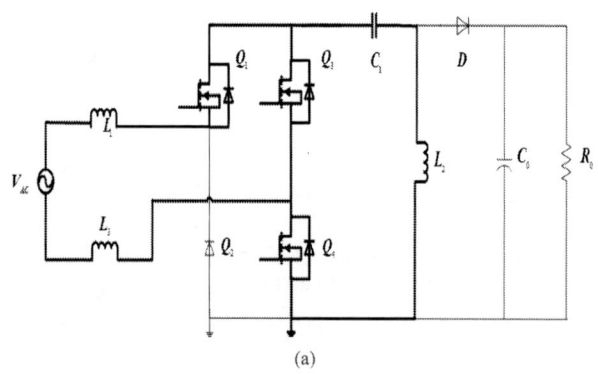

(a)

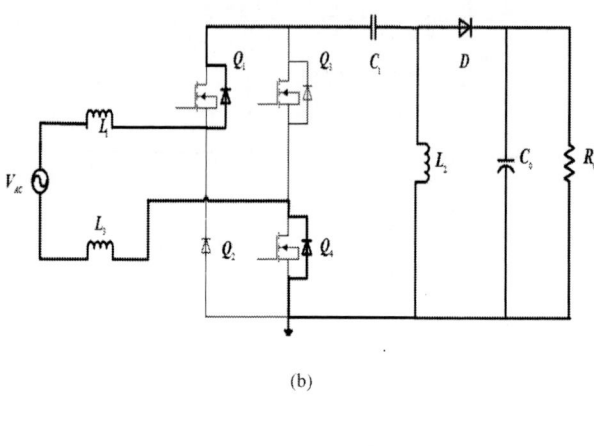

(b)

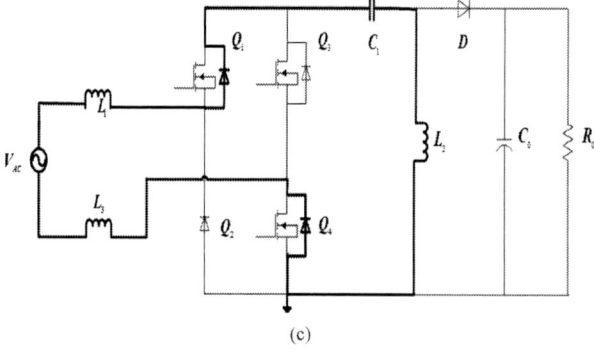

(c)

Fig. 3. Each operating mode for proposed SEPIC PFC converter circuits

III. PRINCIPLE OF OPERATION

Since the proposed SEPIC converter circuit comprises of two symmetrical structures as shown in Fig. 3, the circuit is investigated for the positive half cycle structure. Suggesting that the circuit working in a positive half cycle of a switching period T_s can be divided into three working modes, as shown in Fig. 3(a)–(c) and it can be defined as follows.

Mode 1: In this mode, Q_1, Q_3 and Q_4 switches are turned on, as shown in Fig. 3(a). In this mode, the input inductor currents increase and output inductor current decreases linearly at a rate proportional to the input voltage V_{ac}. The rate of increase of the input inductor currents and the rate of decrease of output inductor current are given by

$$\frac{di_{ac}}{dt} = \frac{di_{L1}}{dt} = \frac{di_{L3}}{dt} = \frac{V_{ac}}{L_{1,3}} \tag{1}$$

$$\frac{di_{L2}}{dt} = -\frac{V_{ac}}{L_2} \tag{2}$$

where $L_{1,3}$ value is given by

$$\frac{1}{L_{1,3}} = \frac{1}{L_1} + \frac{1}{L_3} \tag{3}$$

Thus, the switch current is given by

$$i_{Q3} = i_{ac} - i_{L2} = \frac{V_{ac}}{L_{1,3}} + \frac{V_{ac}}{L_2} \tag{4}$$

This mode ends when Q_1, Q_3 and Q_4 switches are turned off, starting the next mode.

Mode 2: In this mode, Q_1, Q_3 and Q_4 switches are turned off, but Q_1 and Q_4 are conducting through anti parallel body diodes. Fast diode D is turned on, providing a route for the input and output inductor currents. In this mode, the input inductor currents decrease linearly at a rate that is proportional to the output voltage V_{dc} and output inductor current increase linearly. The three inductor currents are given by

$$\frac{di_{L1}}{dt} = \frac{di_{L3}}{dt} = -\frac{V_{dc}}{L_{1,3}} \tag{5}$$

$$\frac{di_{L2}}{dt} = \frac{V_{dc}}{L_2} \tag{6}$$

This mode ends when the diode current distinguishes.

Mode 3: In this mode, all the active switches are turned off, as shown in Fig. 3(c). Q_1 and Q_4 are conducting through anti parallel body diodes. This mode ends by starting the next switching cycle. In this mode, the inductors L_1, L_2 and L_3 currents are equal. The switch voltage and diode voltage are equal input voltage V_{ac} and output voltage V_{dc} respectively. The duration of this mode is

$$\Delta_1 = \frac{V_{ac}}{V_{dc}} \times d \tag{7}$$

where, d is the duty cycle.

IV. DESIGN OF THE PROPOSED CONVERTER

The standard design equations for the main components of the AC/DC SEPIC PFC converter are provided in [15-17]. The proposed converter is designed for 25 V_{rms}, 60 Hz AC input voltage to generate at 10 V DC. The input current ripple is limited to 20% of the peak current I_{ac_peak} with the switching frequency f_s of 30 kH_z.

The following calculations are used to select the appropriate inductors for L_1 and L_2. For an approximate efficiency (η) of 95%, following equation can be derived

$$I_{ac} = I_{ac\_peak} \sin(\omega t) = \frac{2 \times P_0}{\eta \times V_{ac\_peak}} \sin(\omega t) \tag{8}$$

$$I_{ac\_peak} = 140mA$$

Input current ripple is

$$\Delta I_L = 20\% I_{ac\_peak} = 28mA$$
$$\Delta I_L = \frac{V_s \times d}{L_1 \times f_s} \tag{9}$$

The output current in a switching period is equal to the average of the fast diode current. The output average current switching period is obtained by

$$i_{dc\_avg} = 0.5 i_{dc\_peak} \Delta_1 \tag{10}$$

where, i_{dc_avg} is the peak current of fast diode and Δ_1 is the duty ratio of D, $\Delta_1 \langle 1-d$, and i_{dc_avg} can be calculated as:

$$i_{dc\_peak} = i_{L1} + i_{L2} = \left(\frac{1}{L_{1,3}} + \frac{1}{L_2}\right) V_{ac} d T_s \tag{11}$$

$$i_{dc\_avg} = 0.5 \left(\frac{V_{ac}^2}{\left(\frac{1}{L_{1,3}} + \frac{1}{L_2}\right) V_{dc}} d^2 T_s \right) \tag{12}$$

$$I_{dc\_avg} = (1/\pi) \int_0^\pi i_{dc\_avg} d\omega t = \frac{V_{ac\_peak}^2}{4 L_e V_{dc}} d^2 T_s \tag{13}$$

where, $L_e = \frac{L_{1,3} \times L_2}{L_{1,3} + L_2}$.

From (7), the duty cycle d is calculated as:

$$d \langle \frac{V_{dc}}{V_{ac} + V_{dc}} = 0.22 \tag{14}$$

selecting $d = 0.2$, we would get

$$L_e = \frac{V_{ac\_peak}^2 \times d^2}{4 \times V_{dc} \times f_s \times I_{dc\_avg}} = 180 \mu H \tag{15}$$

L_1 and L_3 can be obtained as

$$L_{1,3} = \frac{V_{ac\_peak} \times d}{f_s \times \Delta I_L} = 8.8mH \qquad (16)$$

$$L_1 = L_3 = L_{1,3} / 2 = 4.4mH$$

Therefore, L_2 can be obtained from the following equation

$$\frac{1}{L_2} = \frac{1}{L_e} - \frac{1}{L_{1,3}} \Rightarrow L_2 = 100\mu H \qquad (17)$$

The output capacitance needed to achieve desired current ripple can be calculated as

$$C_0 = \frac{P_{load}}{V_{dc} \times \Delta V_{dc}(\%) \times 4 \times f_{ac}} = 2.2mF \qquad (18)$$

The multi loop control is proposed for the converter, outer voltage controller generating the reference current to regulate the DC voltage and the inner PI controller generating the gating signals as shown in Figure 4. The high frequency switching of the converter produces switching ripples on the DC voltage. Thus the measured DC voltage is processed through a band stop filter to eliminate the noise on the measurements.

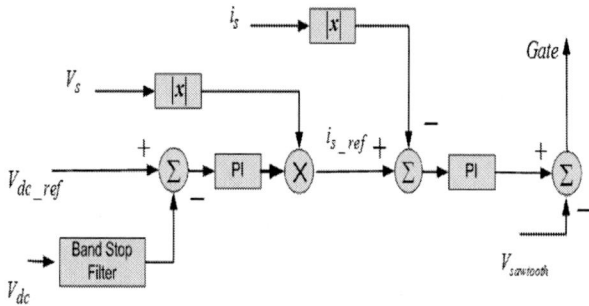

Fig. 4. Block diagram of controller

V. SIMULATION RESULTS

The proposed single phase bridgeless SEPIC topology is simulated by PSIM with the parameters based on the design provided in Section 4. Figure 5 presents the transient input voltage, input current, output voltage, output current and output power for the conventional bridgeless SEPIC PFC converter. Figure 6 presents the input voltage and input current for the conventional bridgeless SEPIC PFC converter.

Figure 7 presents the transient input voltage, input current, output voltage, output current and output power for the proposed bridgeless SEPIC PFC converter. Figure 8 presents the input voltage and input current for the proposed bridgeless SEPIC PFC converter. It can be seen from the Figure 8 that input current is in phase with input voltage and is sinusoidal with low THD and high PF values. Output voltage is obtained at about 10V, with a 120 Hz low frequency ripple.

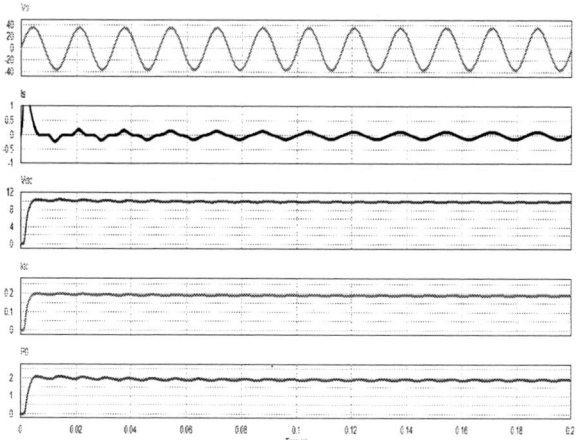

Fig. 5. The transient signals for conventional SEPIC PFC

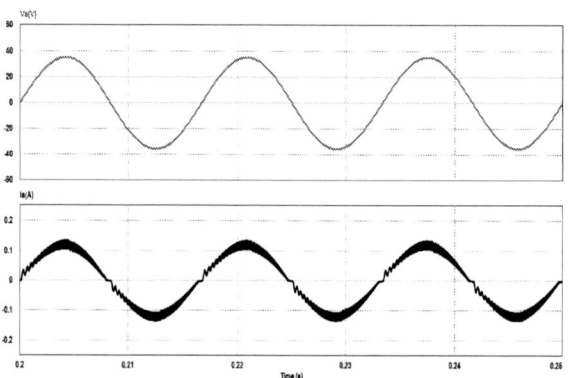

Fig. 6. The input voltage and current for conventional SEPIC PFC

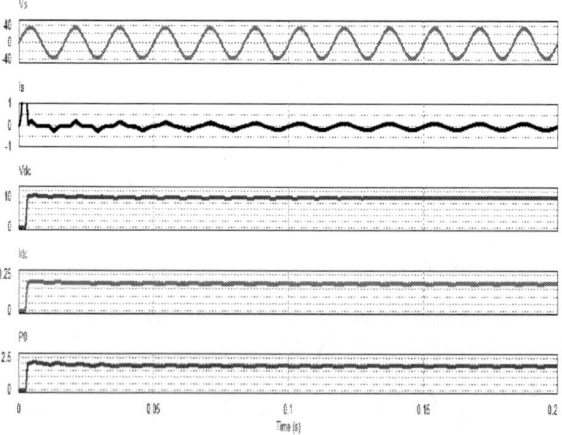

Fig. 7. The transient signals for proposed SEPIC PFC

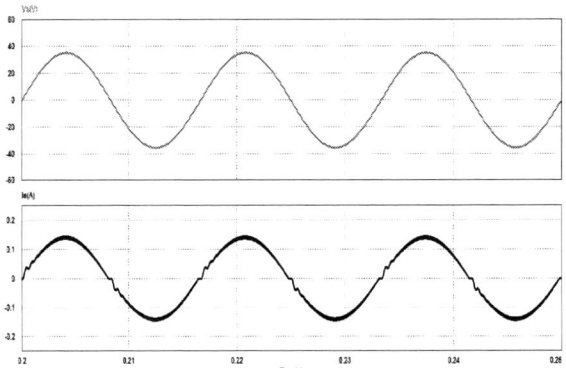

Fig. 8. The input voltage and current for proposed SEPIC PFC

The simulation results of the PF and THD values for a conventional SEPIC PFC converter, proposed bridgeless SEPIC PFC converter are provided in Table 1. The proposed converter is able to reduce the THD 3.23% from 8.93% and improve the power factor to 0.998. The proposed topology provides much better THD and PF compared to conventional one.

TABLE I.   COMPARISON OF BRIDGELESS SEPIC PFCs

	THD (%)	PF
Conventional SEPIC PFC	8.93 %	99.3%
Proposed bridgeless SEPIC PFC	3.23%	99.8%

VI. EXPERIMENTAL RESULTS

The experimental circuit of the proposed converter is developed for the design provided in Section 4. The experimental setup is provided in Figure 9.

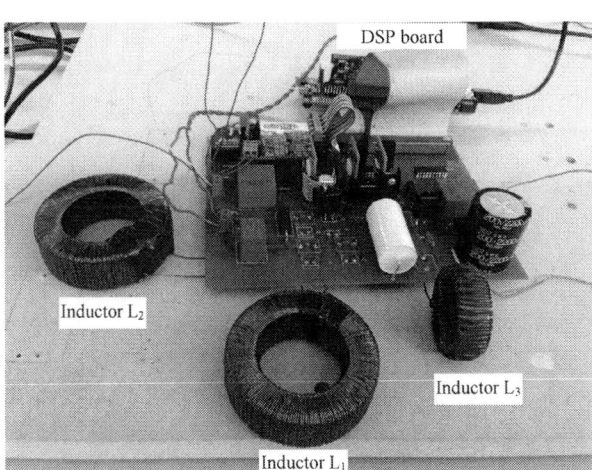

Fig. 9. The experimental prototype for proposed SEPIC converter

Input inductors L_1, L_3 and output inductor L_2 are made with toroidal core from Micro metals T300-60D (turn=205, wire= 23AWG) and T184-2 (turn=65, wire = 18AWG) respectively. IRF840PBF (500V, 8A) for switches and LXA06T600 (fast recovery, 600V, 6A, V_f=2.94V) for diodes are selected. The control circuit is implemented using a TMS320F28335 digital signal processing DSP.

The experimental results of input voltage, input current and output voltage for conventional SEPIC PFC are shown in Fig. 10 and 11. For the input voltage of 25 V_{rms}, output voltage of 10 V_{rms}, and the input current of 140 mA, the THD is measured to be 5.722%, with a power factor of 0.995

The experimental results of input voltage, input current and output voltage for the proposed SEPIC PFC are shown in Fig. 12 and 13. For the input voltage of 25 V_{rms}, output voltage of 10 V_{rms}, and the input current of 136 mA, the THD is measured to be 2.837%, with a power factor of 0.998. The output voltage ripple is obtained 0.15 V at 10 Vdc as it is shown in Fig. 13. The phase of the input current is similar to the input voltage and the obtained PF is near unity.

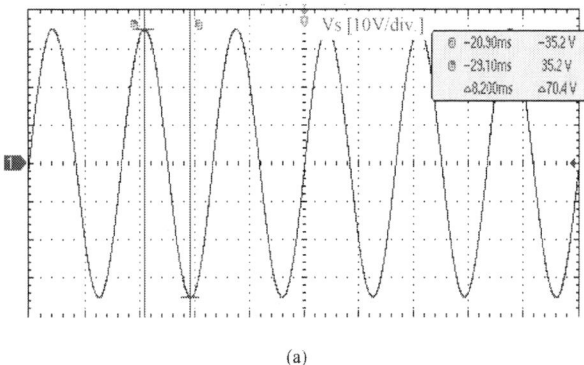

(a)

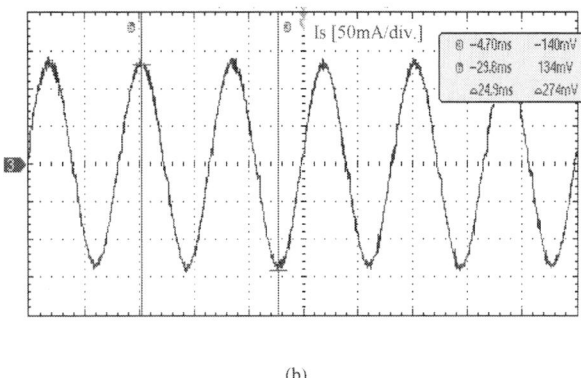

(b)

Fig. 10. Experimental result for conventional SEPIC PFC. a)Input voltage and b) input current.

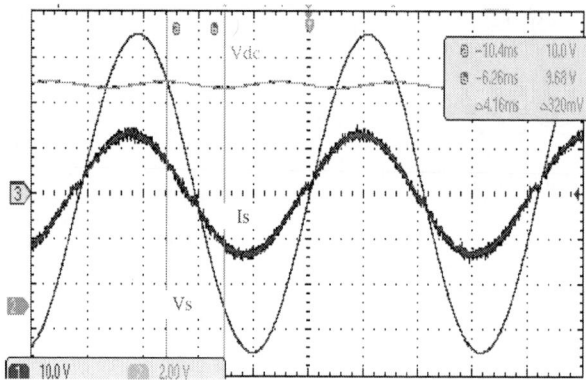

Fig. 11. Experimental result for conventional SEPIC PFC. *Vs*= 25 Vrms, Is = 140 mA with THD=5.722%, Vdc= 10 Vrms and PF=0.995

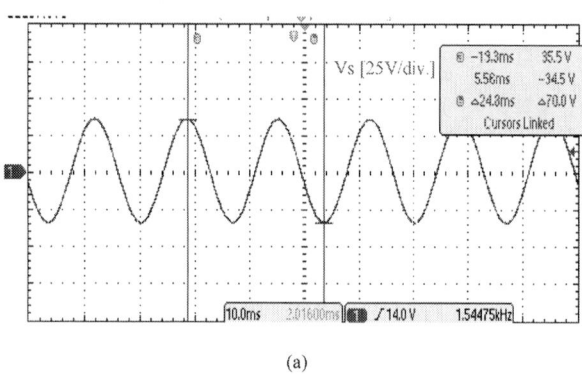

(a)

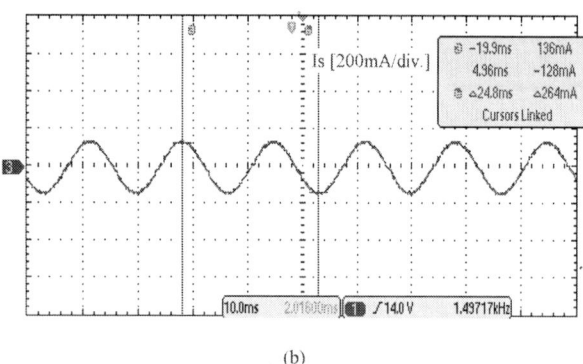

(b)

Fig. 12. Experimental result for proposed SEPIC PFC. a)Input voltage and b) input current.

VII. CONCLUSION

In this paper, a new single phase bridgeless SEPIC PFC converter topology is proposed, analyzed and verified with the simulations. In order to improve the power factor as well as the THD of the utility grid, the full bridge diode in input is removed. Through simulation and experimental studies the performance of the proposed SEPIC converter topology are

compared with the conventional SEPIC converter topology. The proposed converter is able to reduce the THD 2.83% from 5.72% and improve the power factor to 0.998. It is found that the proposed bridgeless SEPIC PFC converter topology provides much better performance than conventional SEPIC PFC converter. The topology is implemented on a converter operating from 25 V AC input to generate 10 V DC. The proposed converter topology is proved to be very good option for single phase bridgeless SEPIC PFC solution for lower power equipments especially those requiring high quality input power.

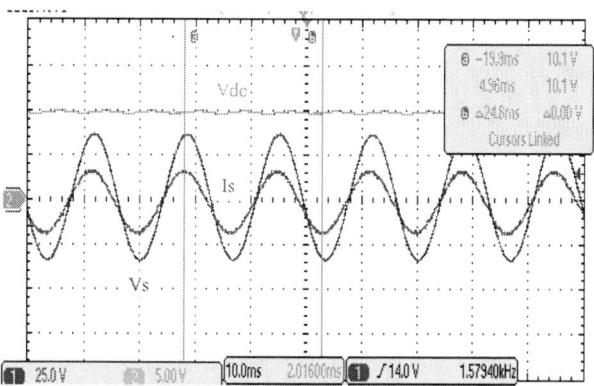

Fig. 13.   Experimental result for proposed SEPIC PFC. *Vs*= 25 Vrms, Is = 136 mA with THD=2.837, *Vdc*= 10 Vrms and PF=0.998.

ACKNOWLEDGMENT

This study is supported by the TUBITAK (The Scientific and Technological research Council of Turkey).

REFERENCES

[1]   IEC 61000-3-2, International Electro technical Commission, Geneva, Switzerland, 1998.

[2]   C. Qiao, K. M. Smedley, "A topology survey of single-stage power factor corrector with a boost type input-current-shaper", IEEE Trans. Power Electron., vol. 16, no. 3, pp. 360-368, May, 2001.

[3]   O. Gracia, J. A. Cobos, R. Prieto, J. Uceda, "Single phase power factor correction: A survey", IEEE Trans. Power Electron., vol. 18, no. 3, pp. 749-755, May, 2003.

[4]   M. M. Jovanovic, Y. Jang, "State-of-the-art, single-phase, active power-factor-correction techniques for high-power applications-An-overview", IEEE Trans. Ind. Electron., vol. 52, no. 3, pp. 701-708, Jun, 2005.

[5]   A. Villarejo, J. Sebastian, F. Soto, E. de Jódar, "Optimizing the design of single-stage power-factor correctors", IEEE Trans. Ind. Electron., vol. 54, no. 3, pp. 1472-1482, Jun. 2007.

[6]   L. Huber , Y. Jang and M. M. Jovanovic, "Performance evaluation of bridgeless PFC boost rectifiers", IEEE Trans. Power Electron., vol. 23, no. 3, pp.1381-1390, 2008.

[7]   M. Gopinanth, Prabakaran, S. Ramareddy, "A brief analysis on bridgeless boost PFC converter", Sustainable Energy and Intelligent Systems Conference, Chennai India, July 2011, pp. 242-246.

[8]   W. Wei , L. Hongpeng , J. Shigong, X. Dianguo "A novel bridgeless buck-boost PFC converter", Proc. IEEE Power Electron. Spec. Conf., Rhodes, 2008, pp.1304-1308.

[9] Y. Ohnuma, and J. Itoh, "A novel single-phase buck PFC AC-DC converter with power decoupling capability using an active buffer", IEEE IEEE Trans. on Ind. Appl., vol. 50, no. 3, pp. 1905-1914, June 2014.

[10] A. A. Fardoun, E. H. Ismail, N. Khraim, A. J. Sabzali, M. A. Al-Saffar, "Bridgeless High Power Factor Buck-Converter Operating in Discontinuous Capacitor Voltage Mode", IEEE Transactions on Industry Applications , vol. 50, no.5, pp.3457-3467, October 2014.

[11] L. Huber, L. Gang, M. M. Jovanovi´c, "Design-Oriented analysis and performance evaluation of buck PFC front-end", IEEE Trans. Power Electron., vol. 25, no. 1, pp. 85-94, Jan. 2010.

[12] E. H. Ismail, "Bridgeless SEPIC rectifier with unity power factor and reduced conduction losses", IEEE Trans. Ind. Electron., vol. 56, no. 4, pp. 1147-1157, Apr. 2009.

[13] M. R. Shaid, A. H. M. Yatim, T. Taufik, "A new ac–dc converter using bridgeless SEPIC", in Proc. Annu. Conf. IEEE Ind. Electro. Society, Glendale, AZ, 2010, pp. 286-290.

[14] M. Mahdavi, H. Farzanehfard, "Bridgeless SEPIC PFC rectifier with reduced components and conduction losses", IEEE Trans. Ind. Electron., vol. 58, no. 9, pp. 4153-4160, Sep. 2011.

[15] Danly E. M., Jyothi G. K., "Simulation of Bridgeless SEPIC Converter with Modified Switching Pulse ", Journal of Modern Engineering Research, vol. 4 , no. 3, pp.15-23, Mar. 2014.

[16] Athira K R., Rajan P T., Neena M., "Analysis of Bridgeless SEPIC Converter with Minimum Component Stress and Conduction Losses for the Speed Control of Dc Motor", International Journal of Advanced Research in Electrical, Electronics and Instrumentation Engineering, vol. 3, no 5, pp.639-649, December, 2014.

[17] J.-W. Yang "Bridgeless SEPIC converter with a ripple-free input current", IEEE Trans. Power Electron., vol. 28, no. 7, pp.3388 -3394, July, 2013.

Effectiveness of Pareto-Front Analysis Applied to the Design of a Single-Phase PFC Rectifier

Mahmoud Ibrahim[1], Luc Gonnet[1], Pierre Lefranc[2,3], David Frey[2,3], Jean-Paul Ferrieux[2,3], Sokchea Am[2,3]

[1]Eaton Power Quality France, EATON, 38330 Montbonnot St. Martin, France
[2]Univ. Grenoble Alpes, G2Elab, F-38000 Grenoble, France
[3]CNRS, G2Elab, F-38000 Grenoble, France

Abstract— The design of power converters becomes a challenge especially when several objectives and critical factors have to be considered, like high efficiency, high power density, cost and reliability. This paper addresses this issue for the design of PFC rectifier. The design is constructed as a bi-objectives optimization. The focus is set on efficiency and power density as principal targets. Then using Pareto-front analysis, optimization results are evaluated regarding additional design considerations. Models of power components are built to ensure high reliability of final solutions. By including temperature effect within components model, thermal performance is studied. Pareto front results of 3 levels 3kW PFC rectifier are presented, and an optimized design is selected and compared to conventionally designed converter. With an experimental prototype of the optimized design, a comparable power density and higher efficiency are achieved and they are close to the values predicted by the design procedure, that demonstrates the effectiveness of the design methodology and its associated models.

Keywords— *Single-Phase PFC Rectifiers; Design optimization; Pareto analysis; Electro-thermal modeling; PFC Boost Inductor Design.*

I. INTRODUCTION:

Nowadays, power electronics applications connected to the grid are in a constant growth. Therefore, power factor correction (PFC) rectifiers become very popular to ensure sinusoidal input currents and to fulfill harmonic standards for low and high frequencies. With the constant evolution of emerging technologies, and the development of robust optimization methodologies, these structures have been improved successively in order to achieve high efficiency and high power density.

The literature on the design of power converters with an optimization procedures arises from the capability of these methods to increase converter performances by optimizing the selection of design variables [1]-[7]. Meanwhile, the focus is often set on one critical design factor, like high efficacy, high power density or cost. Consequently, several methods have been proposed to address multi-objectives optimization [5][6]. Particularly, Pareto front analysis shows, in addition to find optimal solutions regarding predefined objectives, a high potential to evaluate optimization results in function of design variables [6][7].

The aim of this paper is to apply Pareto front analysis on the design of 3 levels PFC rectifier. Therefore, a methodology to design a PFC converter with an optimization process is proposed. The paper starts by introducing a first design, then an optimization procedure is developed based on bi-objectives problem. Therefore, optimization constraints and objective functions are defined. The models used to design magnetic components and to calculate both power losses and component temperatures are developed. Later, an analysis of the Pareto fronts is hold to find an improved optimal design. Finally, the validation of this methodology is done by the mean of an experimental prototype.

II. PFC RECTIFIER DESIGN

Before the optimization, a first design of PFC rectifier for UPS systems had been built. This design is based on a 3 levels boost topology operating in continuous conduction mode (CCM). This system has been designed using conventional design method to get out the best trade-off between efficiency and size. The system specifications were set to the values given in Table I. This first design has very good performances regarding all constraints applied on UPS applications with high efficiency and high power density. An experimental prototype of this design achieve a high power density of 3kW/L with an efficiency of 97.1%.

Based on this design, an optimization procedure will be carried out to enhance its performances.

TABLE I. SPECIFICATIONS OF THE FIRST DESIGN

Power (out)	3kW	Vdc (out)	400 V
Vac (in)	230 V	F (in)	50Hz
F_{sw}	40 (kHz)	ΔI	30 (%)
MOSFET	FCH76N60N	Boost Diode	RHRG75120
Input Diode	VS-60EPS12	Heat-Sink	ABL, 146AB
Inductor Core Wire	410(uH) T184-34 1.88(mm)	L(DM) Core	3.7(uH) T92-2
DC link capacitor	3*B437-850(uF)	C(DM)	0.480(uF)
Efficiency	97.1 (%)	Size	1.04 (L)

III. OPTIMIZATION PROCEDURE:

From optimization methodology design presented in [7], which build a multi-objectives optimization regarding two functions (efficiency and size) then another design factors, like cost, could be taken into account by a Pareto Front analysis. Comparing to others multi-objectives optimization methods [3], Pareto analysis sums up the best compromises between several objectives by giving several possible optimized solutions. Furthermore, the effect of design variable is integrated to Pareto front results to study the impact of these variable on final solutions [6][7].

A. Optimisation functions

The two objectives used to construct the Pareto Front are the converter volume and efficiency. The efficiency is calculated for a converter rated power by taking all component losses and the boxed volume for each component. The converter volume is then calculated by summing all component volumes and increasing the result by 30% to take into account the component placement on the PCB.

B. Optimisation variables and constraints

To demonstrate the effectiveness of the studied methodology regardless preselected components, some variables are fixed such as the diode (Fairchild, RHRG75120, 1200V, 75A), MOSFET (Fairchild, FCH76N60N, 600V, 76A), DC link capacitor (EPCOS, B437*0A5857M, 850μF, 450V) the heat sink profile (ABL, 146AB [16]) and the ambient temperature (40°C). On the other hand, other design variables are set to change within specific ranges listed in Table II.TABLE III. Therefore, optimization results are limited by preselected components performances.

To ensures technical feasibility of optimization results, some constraints are set to sort solutions regarding physical, operating conditions and geometric limits. The optimization constraints are summarized in Table III.

IV. PFC RECTIFIER DESIGN MODELS

A. Inductor Design and Modeling:

Characteristics of both magnetic core and winding wire have been taken into account, iron powder toroid cores have been chosen for their high power density ratio, and also solid wires and litz ones have been introduced. The model calculates require inductor by taking into account inductor

TABLE II. OPTIMISATION VARIABLES VARIATION RANGES

Optimization variable	Min value	Max value	Variable Type
F_{sw}: Switching frequency, kHz	10	75	Continuous
ΔI: Input current ripple,%	10	80	Continuous
L_{HS}: HeatSink length, mm	60	200	Continuous
Nb_{Cbus}: Number of parallel DC link capacitors	1	10	Discrete
$Core_{Boost}$: Boost inductor Core	Micrometal Core Catalog [8]		Discrete
W_{Boost}: Boost inductor wire	AWG Wires Catalog		Discrete
Nb_{DM_Stage}: Number of DM Filter Stages	1	2	Discrete
$Core_{DM}$: DM filter inductor Core	Micrometal Core Catalog (14 & 2) [8]		Discrete
W_{DM}: DM filter inductor wire	AWG Wires Catalog		Discrete
C_{DM}: DM filter capacitor, μF	0.33	4.7	Discrete

variations during main time period due to permeability variation of magnetic material as shown on Fig. 1 [7]. Then, the model computes inductor current waveforms to be used

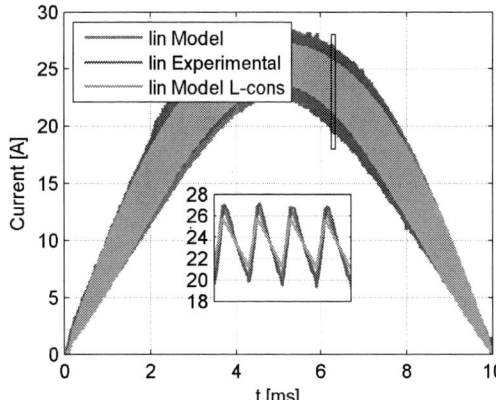

Fig. 1. Evolution of Inductor current waveform during half main period experimental versus models curves

to estimate power losses.

Using Dowell 1D model [9][10][11], conductor resistance as a function of harmonic frequency (Rac) are estimated taking into account both skin and proximity effect. Therefore, copper losses are driven considering RMS current for each high frequency harmonic of the inductor. Based on inductor current waveforms and on core datasheet, time variation of the flux density has been determined. Therefore, core losses are estimated based on the method derived in [12]. The inductor temperature is estimated by the thermal model developed in [18].

B. Input EMI-Filter:

The required filter attenuation is calculated by considering the total noise current as a single peak only at the switching frequency [17][18].. Consequently, differential mode (DM) inductor is determined with a given capacitor value, a given number of filter stages, and the required attenuation. Then, filter inductor is designed based on inductor model of the previous section. Inductor core permeability must be constant over frequency and magnetic operation range, which gives additional constraint to design DM inductor.

Therefore, based on inductor and capacitor models filter losses and volume are estimated.

TABLE III. CONSTRAINTS OF THE OPTIMIZATION PROBLEM

Physical Quantity	Constraint
Inductors Winding Factor	WF < 45%
Current Density	J < 6 A.mm^{-2}
Magntic Material Saturation	Sat < 80%
DM Filter Inductor Permeability Variations	$\mu_\% = 100\%$
Core Temerature Rise	ΔT_{core} < 70°C
Junction Temperatures	Tj < 150°C
Capacitors Current Ripple	Capacitors Datasheet
Output Voltage Ripple	ΔU_{DC} < 15%
Total Harmonic Distortion of Input Current	THDi < 5%

C. DC Link Capacitor:

The capacitor model evaluates each capacitor and determines the number of capacitor needed to respect the voltage variation limit, and to achieve the required hold up time capability. Several constraints are set on voltage and current variation tolerance within the capacitor. Capacitor losses are then calculated based on the ESR value provided in the datasheet.

D. Semiconductor and Heat-Sink:

Using MOSFET datasheet, the evolution of R_{ds}(on) as a function of junction temperature is extracted. Therefore, conduction losses are calculated. For switching losses, losses estimation based on datasheet is less reliable. Turn-on and turn-off times are given for specific conditions which is not adapted for optimization or even iterative processes. Moreover, switching energy curves as function of switched current losses are not always given. To demonstrate the error could be done by using models based only on datasheet values two losses models [13] and [14] have been compared to both physical simulation models and experimental data, Fig. 2(a). On this figure, one can see that using these models could under estimate switching losses. In other hand, physical models have good agreement with experimental results. Subsequently, switching losses models are built using simulation data extrapolated for converter voltage and current specifications. In addition, losses due to the reverse recovery of the diode [15] occupy a high percent of switching on losses as shown on Fig. 2 (b), where reverse recovery current of diode (RHRG75120) is plotted in function of switched current in (FCH76N60N) MOSFET derived by 10Ω gate resistance. At this point also, physical models are worthy enough to get switching losses results, Fig. 2(c).

From model developed in [7], diode conduction losses are calculated based on a fitting curve function based on diode datasheet values, where diode drop voltage is expressed as a function of forward current and junction temperature.

Heat sink is modeled by its thermal resistance given in datasheet as a function of its length. Therefore, using a simple electrical equivalent circuit, Fig. 3(b), junction temperatures of MOSFET and Diodes placed on the heat-sink; Fig. 3(a); are estimated. A worst case error of 5°C is observed on FEM simulations, using FloTHERM, which results from ignoring heat spreading for the selected profile.

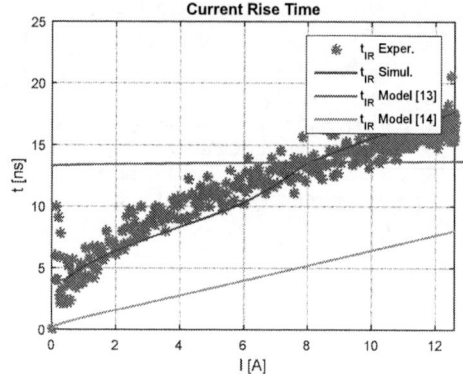

(a) Current rise time for MOSFET FCH76N60N with Rg=10Ω

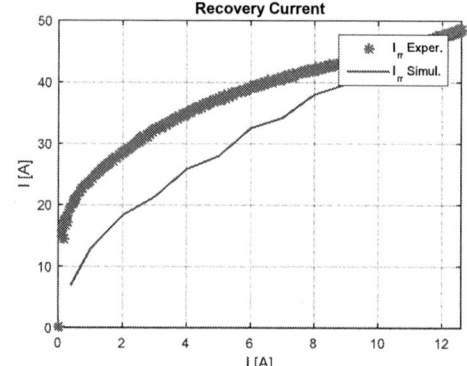

(b) Reverse recovery current of diode (RHRG75120)

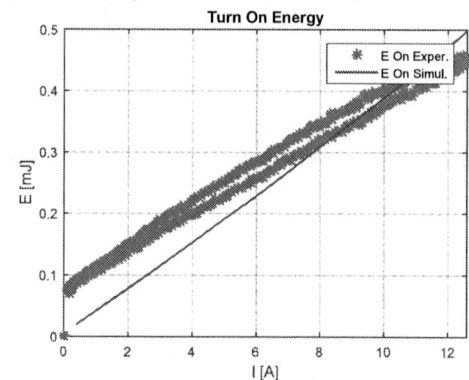

(c) Turn on energy of (FCH76N60N, Rg=10Ω) & (RHRG75120)

Fig. 2. Switching carcterstique of (FCH76N60N) & (RHRG75120)

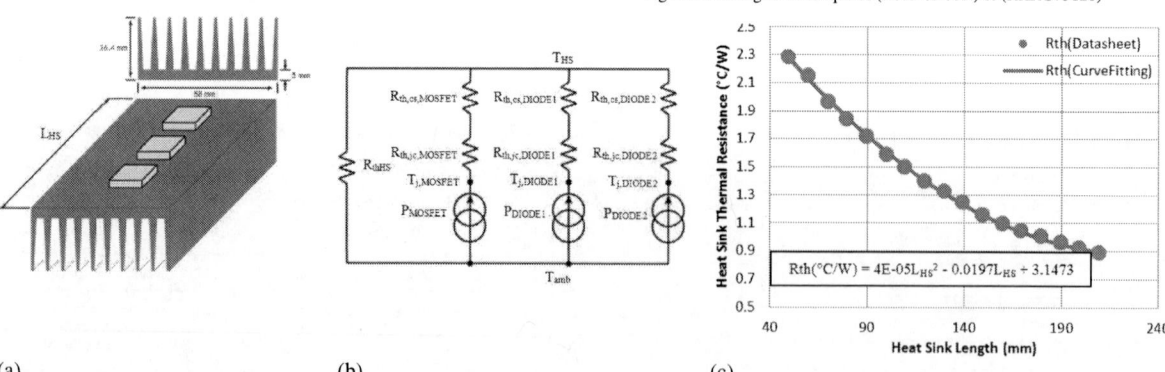

(a)   (b)   (c)

Fig. 3. (a)Heatsink dimensions and components placement on it, (b) Electrical equivalent circuit for thermal modelling of MOSFET, Diodes and Heatsink, (c)Heatsink thermal resistance as a function of heatsink's length, comparison between analytical equation and datasheet's values

V. OPTIMISATION RESULTS

A. Pareto Front:

Pareto Front results are a set of optimized solutions plotted in functions of defined optimization objectives, efficiency and volume in the studied design. So, each solution is the best solution regarding its efficiency and volume. Fig. 5(a) shows that solutions efficiency varies from 97.4 to 98.6% and theirs volume from 0.9 liter to 2.3 liter.

B. Pareto Front Analysis:

In order to choose the most adapted design among Pareto front results, further critical factors related to application target have to be considered; cost, weight, reliability, etc. Therefore, the following useful considerations will be presented to bring the designer to get out the best solution with respect to system performance.

1) Cost:

System final cost is difficult to be modeled or to be estimated in the pre-design phase. Nevertheless, it arises from design components selection. Consequently, known design performance as a function of components will determine expected cost. Fig. 4.(d) shows that inductor and heat-sink designs play an important role on target performances. If high efficiency is needed, cost will be high

as well, because the size of inductor and heat sink has to be increased.

2) Reliability:

Components working temperature is a good indicator of system reliability, if it increase it will reduce components life time [19]. On other hand, working temperature is directly related to system size as shown on Fig. 4.(c). Therefore to better select final design a compromise between system reliability and volume has to be done. Fig. 5Fig. 4.(c) shows that semiconductor temperatures do not exceed 80°C with the pre-selected heat-sink. As a result, if the reliability factor margin is respected, cost can be reduced by involving another heat sink with greater thermal resistance, or volume can be reduced by selecting a smaller heat sink.

VI. EXPERIMENTAL RESULTS

In order to verify optimization results, an optimized solution is selected among Pareto-front results based on the methodology explained in the previous section. In order to get comparable system size to first design presented in section II but with better efficiency, the optimized system shown on Fig. 4(a) is chosen.

The specifications differences between these two systems are shown in Table IV. Other specifications are discussed in section III.

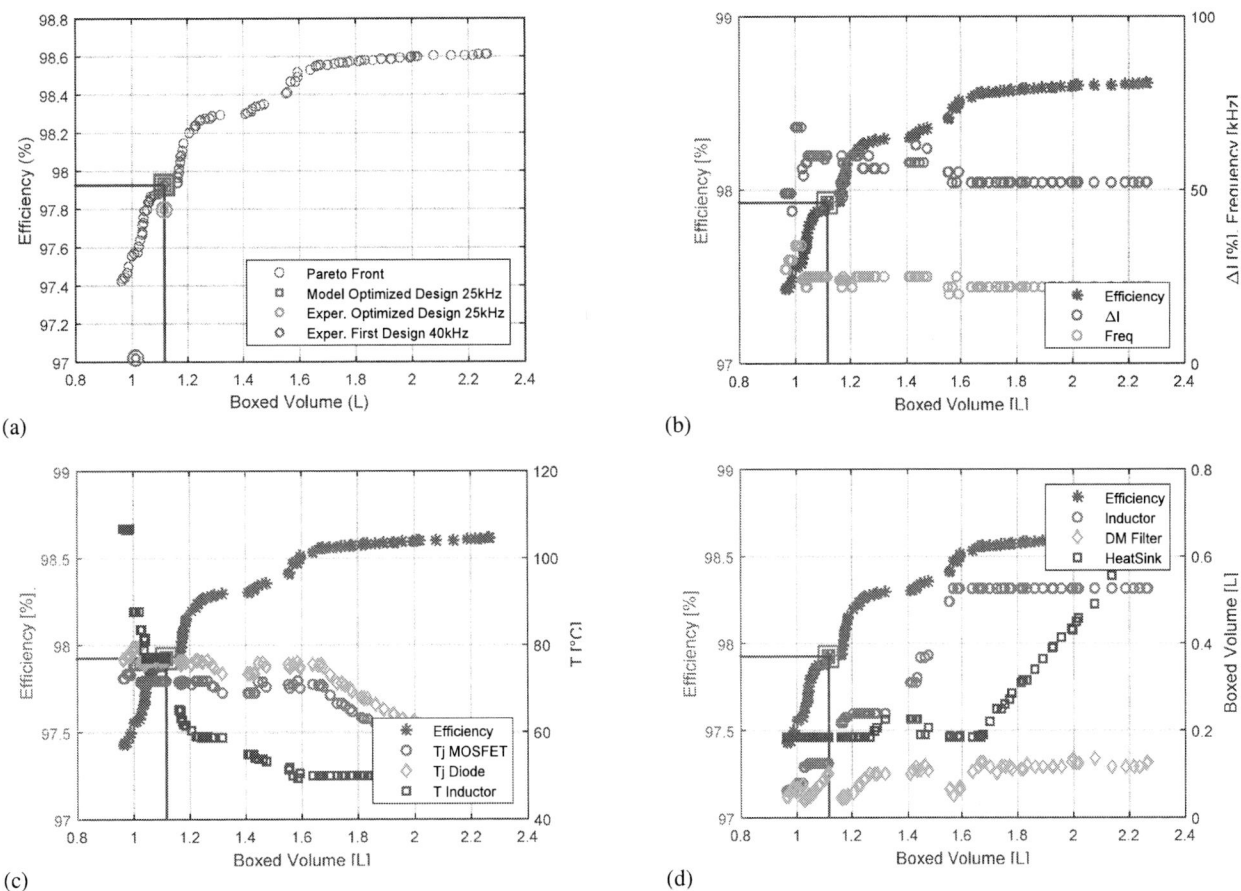

Fig. 4. Solutions characteristics on the Pareto front, a)Pareto front between efficiency and volume, b) Switching frequency & Current ripple, c) Diode and MOSFET junction temperatures & Boost inductor temperature rise, e) Boost inductor, HeatSink and DM filter volumes

TABLE IVIV. SPECIFICATIONS DIFFERENCES: FIREST DESIGN VS OPTIMZED DESIGN

	First Design	Optimized Design
F_{sw} (kHz)	40	25
ΔI(%)	30	60
Inductor (uH) Core Wire (mm)	410 T184-34 1.88	260 T200-2 2.12
LDM (uH) Core	3.7 T92-2	0.77 T68-2
CDM (uF)	0.480	3.3
Efficiency (%)	97.1	97.8
Size (L)	1.04	1.15

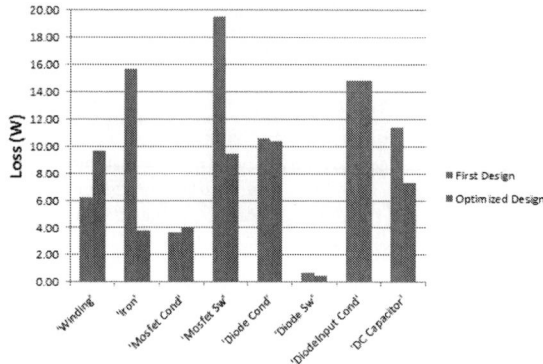

Fig. 6. Losses distribution: Optimized Design vs First Design

choose final optimal design by taking into account critical factors like cost, weigh and reliability.

Experimental optimal design is built to achieve 97.8% efficiency. Then, it was compared to conventional design with 97.1%, which prove the effectiveness of the proposed methodology. Finally losses distribution in power components has been compared between the two designs, demonstrating the importance of models precision.

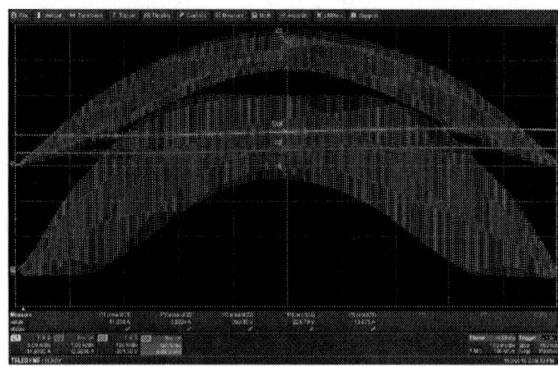

Fig. 5. Voltage and current measurements done with Optimized Design setup at full load P = 3kW, Fsw = 25kHz)

A prototype is built with specifications of the selected optimized design on Table IV. Fig. 5 shows waveforms of current in the inductor and input voltage without DM filter.

With the experimental prototype a system efficiency of 97.8% is achieved, which is close to the value predicted by the design procedure with an error of 0.1%.

On Fig. 6, the first design and optimized one are compared regarding losses distribution in power components. One can notice that MOSFET switching losses are almost divided by 2. In fact, thanks to high current ripple, so reduced turn on current, diode reverse recovery losses on the MOSFET are reduced. On the other hand, winding losses are increased due to high frequency current harmonics, which indicates the influence of high frequency resistance of winding on total losses. Whereby, the reliability of this model is significant.

Finally, iron losses decrease could be easily explained by the use of better iron powder material.

VII. CONCLUSIONS

In this paper, a methodology for optimizing PFC rectifier is proposed. Design and losses calculation models are presented. Considerations related to the reliability of design models are addressed. In particular, switching losses, reverse recovery and ripple current variations in the inductor, all contribute to better estimate power losses. Therefore, impacts of design parameters including temperature influence on the optimization procedure were discussed via a Pareto front analysis. This analysis help designers to better

REFERENCES

[1] J. W. Kolar, F. Krismer, Y. Lobsiger, J. Muhlethaler, T. Nussbaumer, and J. Minibock, "Extreme efficiency power electronics," in 7th International Conference on Integrated Power Electronics Systems (CIPS), 2012, pp. 1-22

[2] J. Kolar, J. Biela, and J. Miniboeck, "Exploring the pareto front of multi-objective single-phase pfc rectifier design optimization - 99.2% efficiency vs. 7kW/dm3 power density," in Proceedings of the 6th IEEE International Power Electronics and Motion Control Conference (IPEMC), 2009.

[3] K. Raggl , T. Nussbaumer , G. Doerig , J. Biela and J. Kolar "Comprehensive design and optimization of a high-power-density single-phase boost PFC", IEEE Trans. Ind. Electron., vol. 56, no. 7, pp.2574 -2587 2009

[4] H. Confortin Sartori, F. Beltrame, M.L. Martins, J.E. Baggio and J. Renes Pinheiro, "Evaluation of an optimal design for a single-phase boost PFC converter (CCM) considering different magnetic materials core", Power Electronics Conference (COBEP), 2013 Brazilian, pp. 1304 – 1310

[5] K. Deb, "Multi-Objective Optimization Using Evolutionary Algorithms", New York: Wiley, 2001.

[6] P. Lefranc, X. Jannot, and P. Dessante, "Virtual prototyping and pre-sizing methodology for buck DC-DC converters using genetic algorithms," IET Power Electronics, vol. 5, no. 1, pp. 41-52, 2012.

[7] M. Ibrahim, P. Lefranc, L. Gonnet D. Frey, J.P. Ferrieux, S. Am "Design Optimization of Single-Phase PFC Rectifier Using Pareto-Front Analysis and Including Electro-Thermal Modelling", 41st Annual Conference of the IEEE Industrial Electronics Society (IECON), 2015 Japon.

[8] [Online]. Micrometal website. Available at www.micrometals.com

[9] P. L. Dowell "Effects of eddy currents in transformer windings", Proc. IEE, vol. 113, pp.1387 -1394 1966.

[10] G. Lefevre , H. Chazal , J. Perrieux and J. Roudet "Application of Dowell method for nanocrystalline toroid high frequency transformers", Proc. PESC\'04, pp.899 -904 2004.

[11] Xi Nan and C. R. Sullivan, "An Improved Calculation of Proximity-Effect Loss in High-Frequency Windings of Round Conductors", IEEE Power Electronics Specialists Conference, June 2003, pp. 853–860.

[12] K. Venkatachalam, C. Sullivan, T. Abdallah, and H. Tacca, "Accurate prediction of ferrite core loss with nonsinusoidal waveforms using only Steinmetz parameters," in Proc. IEEE Workshop Comput. Power Electron., Jun. 3–4, 2002, pp. 36–41.

[13] F. Bjoerk, J. Hancock and G. Deboy, "How to make most beneficial use of the latest generation of super junction technology devices", Infineon application note: AN-CoolMOS-CP- 01, Version 1.1, February 2007.

[14] Jess Brown, "Power MOSFET Basics:Understanding Gate Charge and Using It To Assess Switching Performance" Vishay Siliconix: application note: AN608, 02-Dec-04.

[15] P. Haaf and J. Harper, "Understanding Diode Reverse Recovery and its Effect on Switching Losses", Fairchild Power Seminar 2007.

[16] [Online]. Abl web site. Available at http://www.abl-heatsinks.co.uk/index.php?page=extrudedproduct&product=9

[17] M. L. Heldwein and J. W. Kolar, "Design of minimum volume EMC input filters for an ultra-compact three-phase PWM rectifier," in Proc. 9th COBEP, Sep. 30–Oct. 4, 2007.

[18] K. Raggl, T. Nussbaumer, and J. W. Kolar, "Guideline for a simplified differential mode EMI filter design," IEEE Transactions on Industrial Electronics,no.99,2009

[19] Amerasekera, E.A., Najm, F.N.: 'Failure mechanisms in semiconductor devices' (Wiley, 1997, 2nd edn.), p. 18

State Space Analysis and Duty Cycle Control of a Switched Reactance based Center-Point-Clamped Reactive Power Compensator

Pankaj Kumar Bhowmik, Somasundaram Essakiappan, Madhav Manjrekar
University of North Carolina – Charlotte
9201, University City Blvd, Charlotte, NC 28223, USA
madhav.manjrekar@uncc.edu

Abstract— A new approach for static reactive power compensation has been presented in this paper. Conventional thyristor controlled Static VAr Compensators (SVC) have inherent disadvantages like slow response times and poor harmonic performance as these FACTS devices are based on slow switching power electronics devices. Alternately, Pulse Width Modulated (PWM) dc-ac inverters and direct ac-ac converter structures offer higher bandwidth and push the spectral content to higher switching frequencies that are easier to filter. However, the application space of this approach is limited by the low voltage blocking capability of power devices employed in these converters. A center-point-clamped ac-ac direct power converter has been reported recently in literature which operates on the principle of neutral-point-clamped dc-ac inverter. By clamping the grid voltage to its mid-point, the center-point-clamped converter structure reduces voltage stress on the bi-directional switches by 50%. Compared to the conventional two-level and multilevel dc-ac inverters, the proposed compensator based on direct ac-ac conversion has a simpler structure and control. The operating principle as well as dynamic analysis for the proposed VAr compensation approach has been presented in the paper. A feedback controller has been designed for closed loop control. Simulation results presented in the paper verify that proposed converter offers better control of reactive power, retrofit capability, and reduced voltage stress on the bi-directional switches. Furthermore, it has been shown that leading and lagging reactive compensation can be accomplished with a smooth control of the reactance through duty cycle modulation.

Keywords—reactive power compensator, switched reactance, state space analysis, closed loop control, center-point-clamping

I. INTRODUCTION

Increasing use of electronic power converter based loads, adjustable speed drives and the introduction of automated manufacturing processes have led to rise in issues related to reliability and quality of the power supplied by the utilities. Also, the penetration of renewable energy resources to compensate the energy demands of the customer has led to grid congestion and adversely affected system operation. In the early 20th century, in order to improve the power quality of the grid, reactive power compensation would be provided to the grid using synchronous condensers [1]. In the later years, advances in power semiconductors have facilitated the use of power semiconductor device based reactive power compensators to maintain the power quality as well as voltage profile of the grid within acceptable tolerances. But the employment of the VAr compensator topologies that have been reported in the literature remains challenging due to non-availability of power semiconductor devices rated at utility scale voltage and power levels. The topology of Center-Point-Clamped AC-AC Reactive Power Compensator [11] that has been introduced in this paper uses power semiconductor devices rated at half the utility scale voltages.

With the proposed Center-Point-Clamped Reactive Power Converter topology, it has been shown that one can reliably construct a circuit for 4.16kV applications that uses 3.3kV IGBTs. The following section of this paper presents a review of the shunt type switched reactance based reactive power compensators that have been reported in literature. A brief derivation and description of the center-point-clamped reactive power compensator has been highlighted in section III. Circuit topology and operating principles of this VAr compensation approach has also been discussed in this section. Section IV presents the dynamic analysis using state space averaging technique for the VAr compensator. The state space equations for power and freewheeling stroke have been presented and small signal transfer functions have been derived in this section. Based on the derived system transfer functions, a feedback compensator has also been designed and presented in this section. A simulation model of the Center-Point-Clamped Reactive Power Compensator has been created using Matlab-Simulink. Simulation results verifying the operating principle of the proposed VAr compensator and the command following performance of the designed feedback controller has been shown in section V. Further, an experimental prototype for the Center-Point-Clamped VAr compensator has been built and the experimental results examined in Section VI further validate the effectiveness of the proposed Center-Point-Clamped Reactive Power Compensator. The concluding section discusses the distinctive features of the proposed VAr compensator that makes it stand out from the other reactive VAr compensators that have been reported in literature.

978-1-4673-9551-9/16 $31.00 © 2016 IEEE

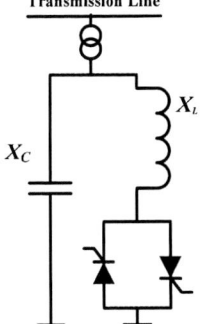

Fig. 1. Simplified circuit schematic of Thyristor based reactive power compensator

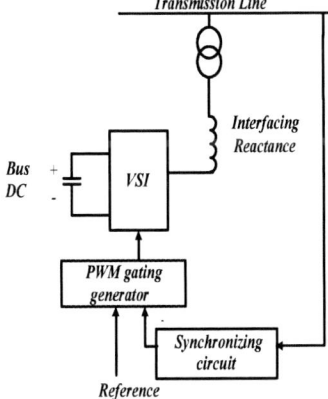

Fig. 2. Simplified circuit schematic of Voltage Source Inverter based reactive power compensator

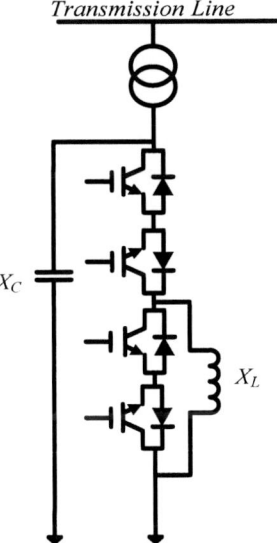

Fig. 3. Simplified circuit schematic of Switched Reactance based reactive power compensator

II. REVIEW OF SHUNT TYPE REACTIVE POWER CONVERTERS

Shunt type Static VAr Compensators (SVC) have been proposed by Gyugyi and Otto [2]-[4] in 1978 and a simplified architecture may be seen in Fig.1. Such compensators operate by means of Thyristor Controlled Reactors (TCR) which have inherent disadvantages like natural commutation and lower switching frequency. These phase controlled thyristors introduce low frequency harmonics that are difficult to filter and the overall performance suffers because they operate with a half-period delay. An alternative way to control reactive power by means of a shunt compensator is the STATCOM solution [4] where a Voltage Source Inverter (VSI) is connected to the ac grid via an interfacing reactance as shown in Fig. 2. In this approach, the reactive current is directly controlled by Pulse Width Modulation (PWM) of the inverter switches leading to higher bandwidth. However practical applicability of this method is limited by the voltage blocking capability of the switches in conventional two-level inverters, which leads to employment of multilevel structures [4]. A third alternative has been employment of switched reactor based static VAr compensator as shown in Fig.3 [5]. In this approach, the effective inductance and hence the injected reactive power is altered by means of duty ratio D. Such variable reactance solution is simpler to implement when compared to STATCOM. However, one of the principal drawbacks of this approach is that this circuit cannot be scaled to higher voltages because of lack of multilevel structures for direct ac/ac power converters [6]-[10],[12]-[13]. This paper introduces a Center-Point-Clamped Reactive Power Compensator topology that offers a means to clamp the grid voltage to half its value, thereby reducing the voltage stress on power electronic devices. The salient features of the proposed VAr compensator topology are:

- Semiconductor devices rated at half the utility scale voltages can used to build VAr compensators rated at utility voltages.

- The reduced voltage stress on semiconductor devices results in lower switching losses than higher-voltage devices.

- The architecture allows the use of higher switching frequencies, thereby leading to reduced filtering requirements and smaller passive components.

III. SWITCHED REACTANCE BASED REACTIVE POWER COMPENSATOR WITH CENTER-POINT-CLAMPING

A simplified circuit schematic of the proposed static reactive power compensator based on center-point-clamped ac-ac conversion [11] approach is shown in fig. 4. As shown in fig. 4, the voltage at O is clamped to the midpoint level of line voltage, v_{in}, which is one of the unique features of this VAr compensator topology. This midpoint level voltage clamping methodology reduces the voltage blocking capacity requirements of the bi-directional switches, S_1, S_2, S_3, and S_4. These switches are operated in such a manner that the voltage stress across any open semiconductor switch is limited to half of the input voltage v_{in}. The switching pattern is shown in Table I.

978-1-4673-9551-9/16 $31.00 © 2016 IEEE   2707

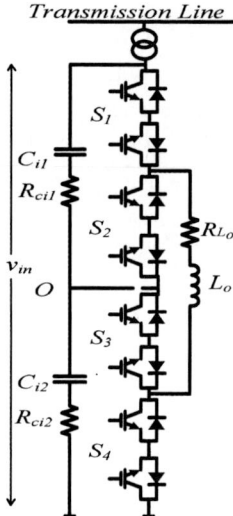

Fig. 4. Simplified circuit schematic of proposed Center-Point-Clamped Reactive Power Compensator

TABLE I: SWITCHING PATTERN OF CENTER-POINT-CLAMPED VAR COMPENSATOR

Switches→ States↓	S_1	S_2	S_3	S_4
State 1	closed	open	open	closed
State 2	open	closed	closed	open

The switching sequence involves two operating states. In the first operating state, switches S_1 and S_4 are closed, while switches S_2 and S_3 are open. Thus, the voltage stress across each of the bidirectional switches has been reduced to $v_{in}/2$ and the voltage across the inductor v_{Lo} is v_{in}. During operating state 2, the voltage stress across the switches still remains the same, $v_{in}/2$, but the voltage across the inductor is zero. Therefore, the inductor is either connected or disconnected from the grid, according to the switching duty ratio, to provide required amount of reactive power to the grid. Through this switched control of the inductor, an equivalent impedance of the compensator may be controlled according to the duty ratio of the switches. The equivalent input impedance can be derived as shown in (1).

$$|X_{in}| = \frac{|v_{in}|}{|i_{in}|} = \left(\frac{|X_{Lo}|}{D^2}\right) ||(|X_{ci1} + X_{ci2}|) \qquad (1)$$

where, D is the duty cycle, z_{Lo} is the equivalent impedance of inductor, L_o, z_{ci1} and z_{ci2} are the equivalent impedances of the capacitors, C_{i1} and C_{i2}, v_{in} is the transmission line voltage at the secondary side of the transformer, and i_{in} is the input current flowing into or from the switched reactor. Thus, the input impedance can be expressed as a function of the inductor impedance, capacitor impedance and switching duty cycle, D. The desired impedance for any operating condition of the grid may be generated by duty cycle control. Dynamic analysis of the VAr compensator is discussed in the following section.

IV. STATE SPACE ANALYSIS AND DUTY CYCLE CONTROLLER DESIGN

The proposed reactive power compensator uses four bidirectional switches S_1, S_2, S_3, and S_4. The dynamic analysis of the compensator is performed using the state space averaging technique. It is assumed that the compensator is operated in continuous conduction mode and the bi-directional switches are lossless. The parasitic elements associated with the key passive components are also taken into consideration for the power circuit design. The state variables are reactor current, i_{Lo}, capacitor voltages, v_{Ci1}, and v_{Ci2}. Input variable is input voltage, v_{in} and output variable is input current, i_{in}. The state space equations for the center-point-clamped reactive power compensator are given in (2) and (3).

$$\begin{bmatrix} \frac{di_{Lo}}{dt} \\ \frac{dv_{Ci1}}{dt} \\ \frac{dv_{Ci2}}{dt} \end{bmatrix} = \begin{bmatrix} \frac{-R_{Lo}}{Lo} & 0 & 0 \\ 0 & \frac{-1}{C_{i1}R_{Ci1}} & 0 \\ 0 & 0 & \frac{-1}{C_{i2}R_{Ci2}} \end{bmatrix} \begin{bmatrix} i_{Lo} \\ v_{Ci1} \\ v_{Ci2} \end{bmatrix} + \begin{bmatrix} \frac{D}{Lo} \\ \frac{1}{2C_{i1}R_{Ci1}} \\ \frac{1}{2C_{i2}R_{Ci2}} \end{bmatrix} v_{in} \qquad (2)$$

$$i_{in} = \begin{bmatrix} D & \frac{-1}{2R_{Ci1}} & \frac{-1}{2R_{Ci2}} \end{bmatrix} \begin{bmatrix} i_{Lo} \\ v_{Ci1} \\ v_{Ci2} \end{bmatrix} + \left(\frac{1}{4R_{Ci1}} + \frac{1}{4R_{Ci2}}\right) v_{in} \qquad (3)$$

The above state space equations are perturbed and then converted to frequency domain equations using Laplace transformation. The transfer function between the input current to duty ratio is given in (4).

$$G_{id} = \frac{3400}{0.012ss + 6.007} \qquad (4)$$

The input current to input voltage transfer function is derived as in (5).

$$G_{in} = \frac{3.757*10^{-7}s^3 + 1.765*10^{-3}s^2 + 2.291s + 819.7}{5.015*10^{-6}s^3 + 15.61*10^{-3}s^2 + 15.11s + 4281} \qquad (5)$$

Bode plots for the transfer functions, G_{id} and G_{in} are shown in fig.7 and fig.8 respectively. As seen in fig.7 and fig.8, the designed reactive power compensator is stable in operation, with satisfactory gain and phase margins. Fig.9 shows the step response for the input current to duty ratio transfer function. It may be seen from fig.9 that the dynamic performance of the designed VAr compensator with step change in input is fast and robust. The design of the feedback controller for the VAr compensator is discussed in the following section.

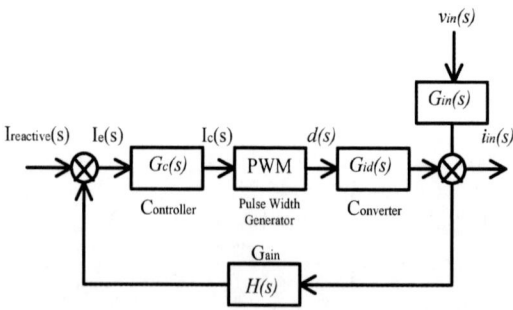

Fig. 5. Block diagram of of controller for current control in the center-point clamped compensator

978-1-4673-9551-9/16 $31.00 © 2016 IEEE   2708

A simplified block schematic of the feedback controller designed for the Center-Point-Clamped VAr compensator is shown in Fig.5. In the controller schematic, $I_{reactive}(s)$ is the reference reactive current input, $I_e(s)$ is current error signal, $I_c(s)$ is compensated signal, $d(s)$ is duty ratio, $G_{id}(s)$ is converter input current to duty ratio transfer function, $G_{in}(s)$ is converter input current to input voltage transfer function, $i_{in}(s)$ is the input current, $v_{in}(s)$ is the input voltage and $H(s)$ is the feedback gain in sensors, which is considered as unity in our design. Using the transfer function, $G_{id}(s)$, a feedback compensator has been designed using Matlab-Simulink SISO tool. The designed compensator, $G_c(s)$ offers fast transient response as well as improves line regulation. The compensator transfer function is shown in (6).

$$G_c = \frac{5.696*10^{-06}s + 2.589*10^{-07}}{5.8 \, s^2 + s} \tag{6}$$

Bode plots for the transfer function, $G_{id} * G_c$ have been shown in fig.10. It may be seen from fig. 10, that the implementation of the feedback controller in the center-point-clamped reactive power compensator has not affected the stability of the converter as phase margin after compensation is 90 degrees. In order to verify the operation of the compensator, a single phase of the compensator is modelled in Matlab-Simulink for leading as well as lagging 21MVA application. The designed simulation parameters are displayed in Table II. As may be seen from the simulation component design parameters, the parasitic elements associated with the key components are considered in our simulation.

TABLE II: SIMULATION COMPONENT DESIGN PARAMETERS

V_{source}	3400 V
C_{i1}	0.1473 mF
R_{Ci1}	6.673 Ω
C_{i2}	0.1473 mF
R_{Ci2}	6.673 Ω
Lo	12 mH
R_{Lo}	6.0069 Ω
L_transformer	12.2654 mH
f_{sw}	11.1 kHz
R_load	266.974 Ω
L_load	12 mH
C_load	57.365536 µH

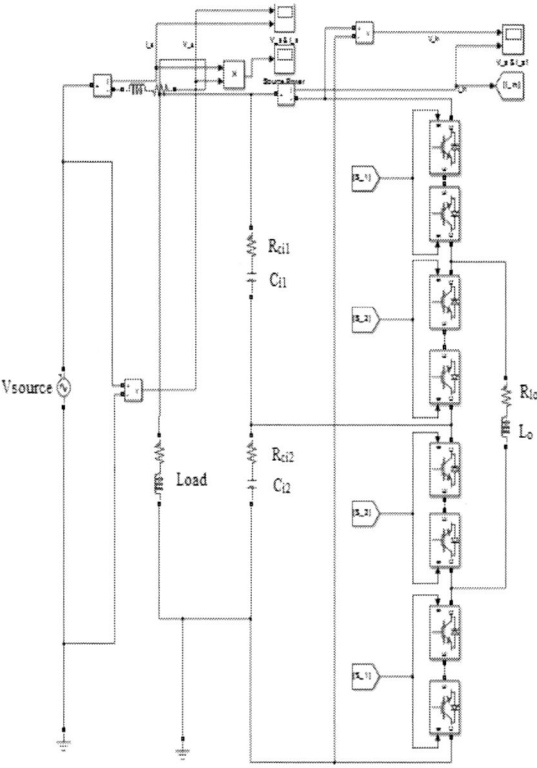

Fig. 6. Circuit Schematic of the Matlab-Simulink based simulation model of Center-Point-Clamped Reactive Power Compensator

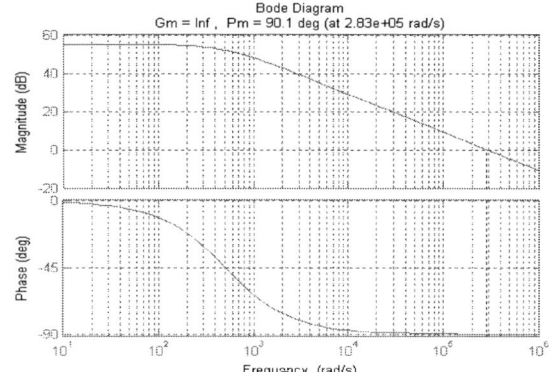

Fig. 7. Bode plot of uncompensated duty ratio to input current transfer function $G_{id}(s)$ (Gm = Inf dB, Pm = 90.1 deg)

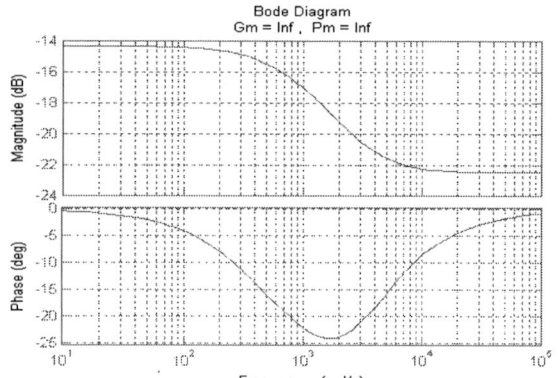

Fig. 8. Bode plot of uncompensated input current to inout voltage transfer function $G_{in}(s)$ (Gm = Inf dB, Pm=Inf deg)

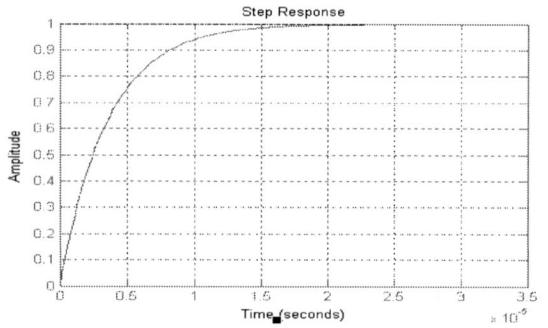

Fig. 9. Unit Step response for uncompensated duty ratio to input current transfer function $G_{id}(s)$ (Rise time = 1.5s)

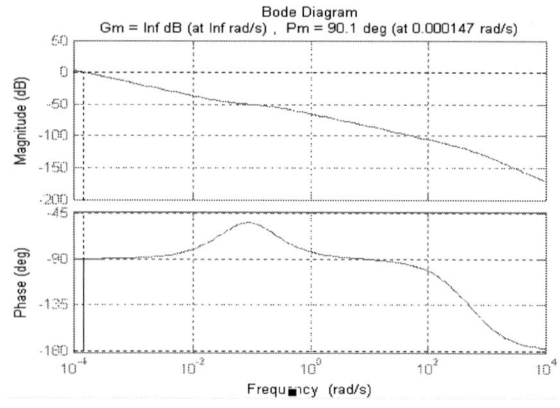

Fig. 10. Bode plot of compensated duty ratio to input current transfer function $G_{id}(s)*G_c(s)$ (Gm= Inf dB, Pm=90.1 deg)

V. SIMULATION RESULTS

Simulations are performed to verify the operation of proposed reactive power compensator for an R-L load as well as an R-C load. A simulation model rated at 21 MVA was designed using Matlab-Simulink. The compensator has been connected to the transmission line via a transformer, which has been modelled as an inductor, L_transformer. A case study of the various operating modes of the Center-Point-Clamped Reactive Power Compensator for an R-L load and an R-C load has been presented in Table III. Simulation results for an R-L load have been shown in figs. 11, 12 and 13.

TABLE III: GRID VOLTAGE AND CURRENT WAVEFORMS BASED ON DIFFERENT OPERATING CONDITIONS

Grid Operating Conditions	Grid voltage and current waveforms
Reactive power Compensator is disconnected from grid	V_s − − I_s(RL load) − − I_s(RC load)
Reactive Power Compensator is connected to grid with Duty Cycle D=1	V_s − − I_s(RL load) − − I_s(RC load)
Reactive Power Compensator is connected to grid and operated at unity power factor	V_s − − I_s(RL load) − − I_s(RC load)
Reactive Power Compensator is connected to grid with Duty Cycle D=0	V_s − − I_s(RL load) − − I_s(RC load)

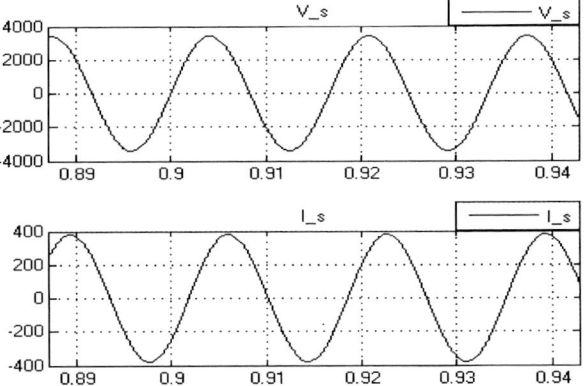

Fig. 11. Voltage and Current waveforms of the power drawn from the grid by an RL load, with maximum lagging reactive power injected into the grid from the compensator.

It may be seen from fig.11 that the grid current lags the voltage as the compensator is injecting the maximum possible lagging reactive power into the grid as the duty ratio of the switches is equal to 1.

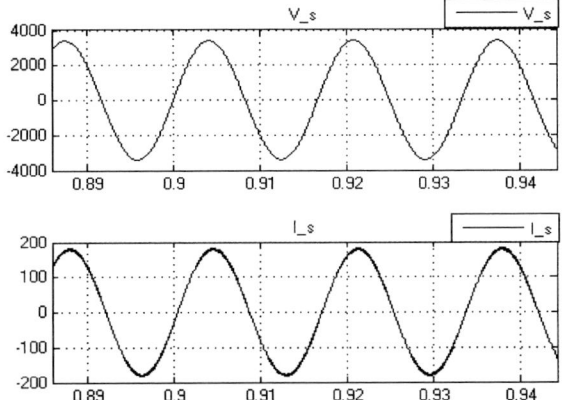

Fig. 12. Voltage and Current waveforms of the power drawn from the grid by an RL load, with required reactive power injected into the grid from the VAr compensator such that unity power factor is maintained at the grid.

It may be observed from fig. 12 that the grid current is in phase with the source voltage as the VAr compensator is injecting only the required leading reactive power.

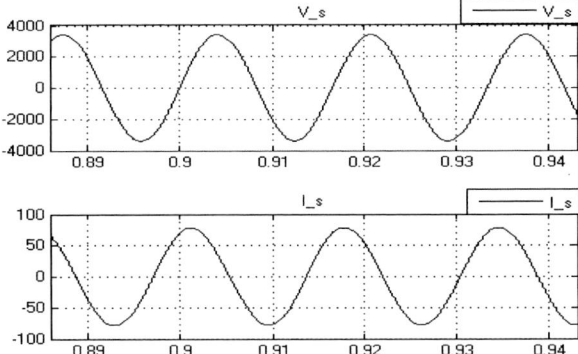

Fig. 13. Voltage and Current waveforms of the power drawn from the grid by an RL load, with maximum possible leading reactive power injected into the grid from the compensator.

As may be seen from fig.13, the grid current leads the grid voltage as maximum possible leading reactive power is being injected to the grid. It is so because the inductor has been disconnected from the grid as the duty ratio of the switches is equal to zero. Simulation results for an R-C load have been displayed in figs. 14, 15 and 16. It may be noted from fig.14 that maximum possible leading reactive power is being injected to the grid and as such the grid current leads the grid voltage. Fig. 15 demonstrates the unity power factor operation of the reactive power compensator as it injects only the required amount of reactive power such that the grid current is in phase with grid voltage. Fig. 16 shows the maximum possible lagging reactive power from the designed VAr compensator being injected into the grid. Hence, the grid current lags the grid voltage.

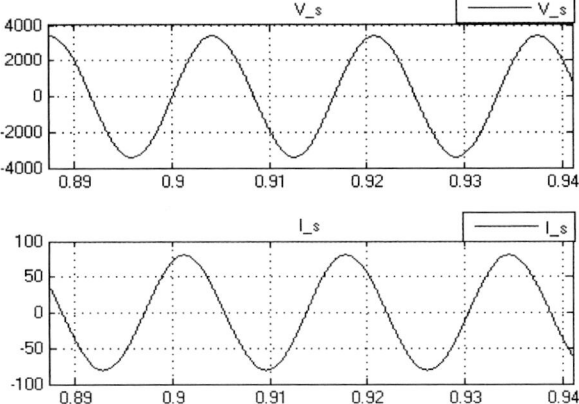

Fig. 14. Voltage and Current waveforms of the power drawn from the grid by an RC load, with maximum possible leading reactive power injected into the grid from the compensator.

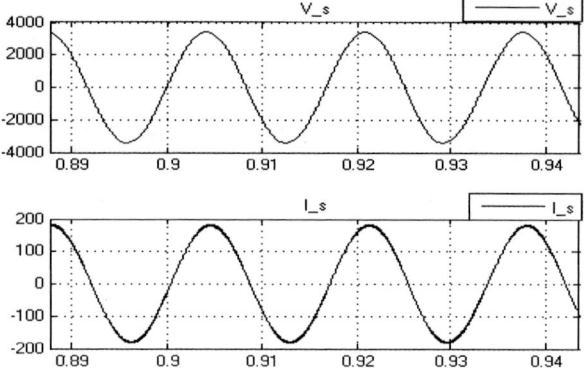

Fig. 15. Voltage and Current waveforms of the power drawn from the grid by an RC load, with required reactive power injected into the grid from the compensator such that unity power factor is maintained at the grid.

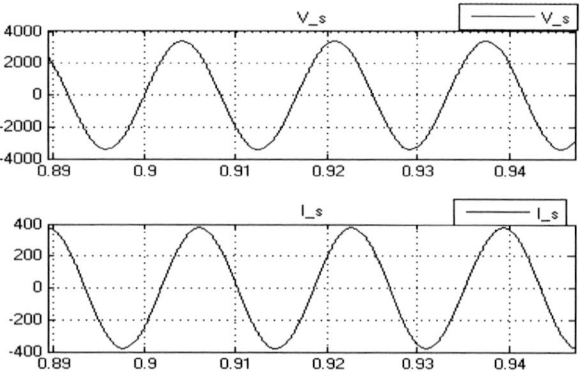

Fig. 16. Voltage and Current waveforms of the power drawn from the grid by an RC load, with maximum possible reactive power injected into the grid from the compensator.

Simulation results presented in figs. 14, 15 and 16 verify the operation of the designed feedback controller for the proposed VAr compensator for varying load power factor conditions.

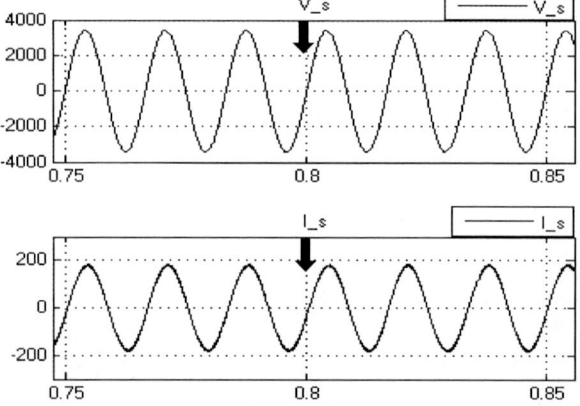

Fig. 17. Step Change of load from R-L load to R-C load at 0.8sec.

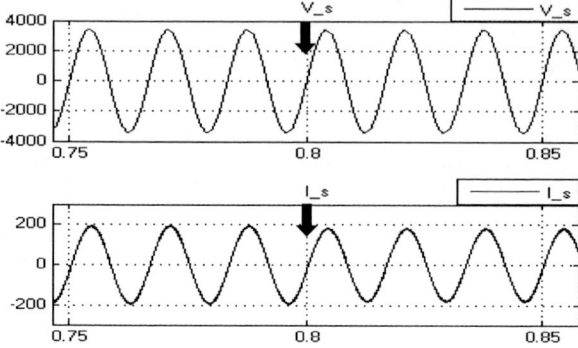

Fig. 18. Fig. 18: Step increase in load impedance level by 50 percent at 0.8sec.

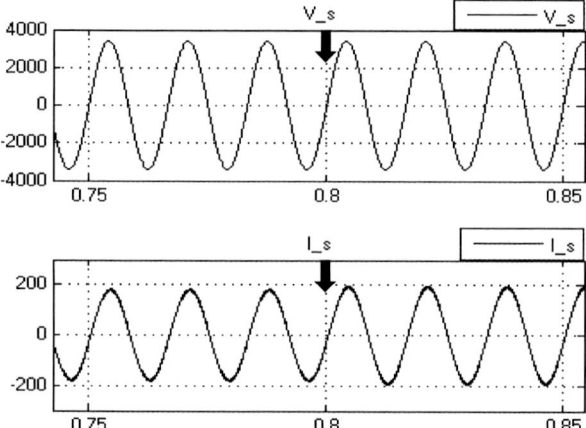

Fig. 19. Fig. 19: Step decrease in load impedance level by 50 percent at 0.8sec.

It may be seen from fig. 17 that with step change of load at 0.8sec the grid voltage and current may have a slight deviation from unity power factor but then settles to unity power factor within two fundamental cycles. Figs. 18 and 19 show the performance of the controller for step change in load level by 50 percent. It may be observed that grid voltage and current have small deviation from unity power factor operation. This verifies the tracking performance of the designed feedback controller. Thus, the controller can smoothly inject and absorb reactive power to and from the transmission line respectively through duty cycle control.

VI. EXPERIMENTAL RESULTS

A small scale single phase 120Vrms, 6A peak bench prototype has been built and preliminary experimental results have been presented in figs. 20 and 21. As seen from fig. 20, the displacement power factor of the input source power is poor as it is heavily inductive, but it may be seen from fig. 21 that with the introduction of the Center-Point-Clamped Reactive Power Compensator, the displacement power factor has improved to 0.98.

978-1-4673-9551-9/16 $31.00 © 2016 IEEE

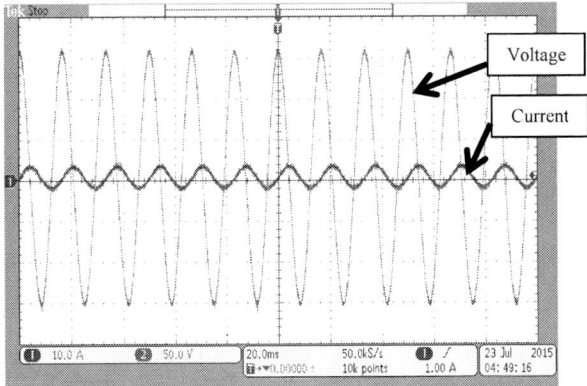

Fig. 20. Voltage and Current waveforms of the power drawn from the input source with negligible reactive power injected into the input source from the VAr compensator. (without VAr compensation)

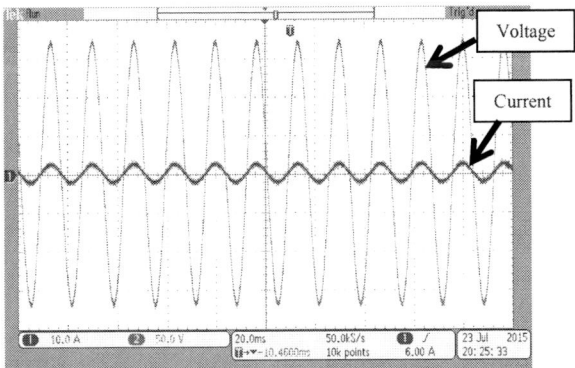

Fig. 21. Voltage and Current waveforms of the power drawn from the input source with required reactive power injected into the input source from the VAr compensator such that unity power factor is maintained at the input source. (with VAr compensation)

VII. CONCLUSIONS

A novel Center-Point-Clamped Reactive Power Compensator topology has been presented in this paper. The proposed VAr compensator offers a unique methodology to clamp the input voltage at half the magnitude of source voltage, thereby enabling the use of low voltage power electronic switches to operate with higher system voltages and improving the power transfer efficiency of the VAr compensator. The switching pattern is easier to implement than multilevel converter based VAr compensator topologies. Furthermore, dynamic analysis of this converter has enabled us to model a feedback controller for better controllability and superior output power quality. Simulation results presented in the paper verify the operation of the reactive power compensator in injecting or absorbing reactive power to or from the transmission line depending on the duty cycle control. Also, it is observed that the line current THD is

maintained at less than 5%. Preliminary experimental results from a small scale laboratory prototype have been presented.in the paper. This verifies the operation methodology of the Center-Point-Clamped Reactive Power Compensator for injection of reactive power to the grid. Also, it may be noted that the demonstrated experimental results follow closely the simulation results described in previous section.

ACKNOWLEDGMENTS

The authors gratefully acknowledge support from the Energy Production & Infrastructure Center (EPIC) at University of North Carolina – Charlotte.

REFERENCES

[1] G. Venkataramanan, B. Johnson, A. Sundaram, "An AC-AC Power Converter for Custom Power Applications," IEEE Trans. on Power Delivery, vol. 11, No. 3, July 1996, pp. 1666-1671.

[2] L. Gyugyi, R.A. Otto, T.H. Putman, "Principles and Applications of Static, Thyristor-Controlled Shunt Compensators", IEEE Trans. PAS, Vol. PAS-97, No. 5, Sept/Oct, 1978.

[3] L. Gyugyi, "Power electronics in electric utilities: Static VAr compensators," Proceedings of the IEEE, vol. 16, no. 4, pp. 483493, Apr. 1988.

[4] Dixon, J.; Moran, L.; Rodriguez, J.; Domke, R., "Reactive Power Compensation Technologies: State-of-the-Art Review," Proceedings of the IEEE, vol.93, no.12, pp.2144, 2164, Dec. 2005.

[5] H. Jin, G. Goós, and L. Lopes, "An efficient switched-reactor based static VAr compensator," IEEE Trans. Ind. Appl., vol. 30, no.4, pp. 997–1005, Jul./Aug. 1994.

[6] Z. Fedyczak, R. Strzelecki, R. Kasperek, and K. Skorski, "Three-phase self-commutated static VAr Compensator based on Cuk converter topology," in Proc. IEEE PESC '00, 2000, pp. 494-499.

[7] Liangzong He; Shanxu Duan; Fangzheng Peng "Novel family of quasi-z-source AC/AC converter with safe- (ECCE), 2011 IEEE, on page(s): 1417 – 1422.

[8] S. Arunprasanth, A. Arulampalam, P.J. Binduhewa, M.A.R.M. Fernando, and S.G. Abeyrathne, "Dynamic Reactive Power Compensator (DRPC) for Unbalance Load Reactive Power Compensation", 2013 IEEE 8th International Conference on Industrial and Information Systems, ICIIS 2013, Aug. 18-20, 2013, Sri Lanka.

[9] H.L. Jou, J. C Wu, J.J. Yang, W. P. Hsu, "Novel Circuit Configuration for Hybrid Reactive Power Compensator," CESIIEEE 5th International Power Electronics and Motion commutation", Energy Conversion Congress and Exposition Control Conference, 2006, IPEMC 2006, voL2, no., pp.I-6, 14-16 Aug. 2006.

[10] C.H. da Silva, R.R. Pereira, L. E. B. da Silva, G. L. Torres, R. B. Gonzatti, "A Hybrid Active VAr Compensator (HAVArC)," 14th International Conference on Harmonics and Quality of Power (ICHQP), 2010, vol., no., pp.I-5, 26-29 Sept 2010.

[11] Bhowmik, P.K.; Yellapragada, S.; Manjrekar, M., "Dynamic analysis and controller design for a center-point-clamped ac-ac converter," Applied Power Electronics Conference and Exposition (APEC), 2015 IEEE, pp. 535 – 542, 15-19 March 2015

[12] Gyugyi, Laszlo, "Reactive compensation. Control of shunt compensation with reference to new design concepts," in Generation, Transmission and Distribution, IEE Proceedings C , vol.128, no.6, pp.374-381, November 1981

[13] Manjrekar, M.; Venkataramanan, G., "Control strategies for a hybrid static reactive compensator," in Electrical and Computer Engineering, 1996. Canadian Conference on , vol.2, no., pp.834-837 vol.2, 26-29 May 1996

A SiC-based Power Converter Module for Medium-Voltage Fast Charger for Plug-in Electric Vehicles

Srdjan Srdic, Chi Zhang, Xinyu Liang, Wensong Yu, Srdjan Lukic
Department of Electrical and Computer Engineering
North Carolina State University
Raleigh, NC, USA
ssrdic@ncsu.edu, czhang17@ncsu.edu, xliang5@ncsu.edu, wyu2@ncsu.edu, smlukic@ncsu.edu

Abstract—This paper presents an isolated power converter module for medium-voltage (2.4 kV), high-power-quality (PF ≥ 0.98, current THD ≤ 2%), 50 kW fast charger for plug-in electric vehicles. The proposed high-efficiency (above 96%), and reduced-footprint converter module utilizes off-the-shelf Silicon Carbide (SiC) devices to step down the rectified single-phase medium-voltage input. The developed module can also serve as a building block for other medium voltage rectifier applications, including power supplies for data centers and other dc power distribution systems. Based on the system requirements, the appropriate unidirectional converter topology was selected, its operation was simulated and validated by experiments on the developed converter prototype.

Keywords—medium voltage rectifier, SiC MOSFET, electric vehicles

I. INTRODUCTION

The battery capacity of most of today's electric vehicles (EVs) meets the typical driver's daily needs. However, the relatively long charging times are still one of the main obstacles for the even broader adoption of PEVs. In order to enable charging of EV's batteries under 30 minutes, most of the modern off-board fast chargers have a power rating of around 50 kW. Those chargers are categorized by the Society of Automotive Engineers (SAE) as DC chargers (SAE J1772 Standard). According to publicly-available data, the highest-power EV chargers are Tesla's new 135 kW Superchargers installed in Europe. However, no other details about the Tesla's Supercharger (such as efficiency, power quality, etc.) are publicly available. Currently, commercially available DC fast chargers have efficiencies up to 95% and require a bulky low-frequency dedicated service transformer to provide 208/480 V ac three-phase input [1]-[4]. This significantly increases size, and weight of those chargers as well as their installation cost [5]. The other approach to fast chargers relies on Solid State Transformer (SST) technology, which does not include the service transformer. An example of the SST fast charger is a 50 kW Utility Direct Fast Charger (UDFC), developed by Virginia Tech and EPRI, which is connected directly to the MV (2.4 kV) utility network. This enables the UDFC to achieve higher power density and lower installation cost compared to the Low Voltage (LV) commercial chargers [5]. Many SST designs that could be used in PEV charging applications have been proposed in the literature [6]-[12], and some of them have been designed and fully tested. An extensive analysis of three unidirectional SST concepts, suitable for directly interfacing 400 V dc to medium-voltage ac distribution systems is given in [13]. Due to the relatively high input voltage (2.4 kV ac), a modular approach to design needs to be adopted if commercially available Silicon Carbide (SiC) devices are to be used. Based on the published data, four prospective modular MV converter topologies are selected and some of their basic features are summarized in Table I.

Due to its high power density, high efficiency, low cost, low number of switches and expected high reliability, the Multi-Cell Boost (MCB) topology with three input-series-output-parallel (ISOP)-connected dc/dc converter modules, shown in Fig 1, is finally adopted for implementing the fast charger. By using this topology, the losses can be reduced by 40%, compared to the state-of-the-art solution with low-frequency transformer and three-phase rectifiers [13].

TABLE I. BASIC FEATURES OF SELECTED MV FAST CHARGER MODULAR TOPOLOGY CANDIDATES

Metric	Matrix MMC [6]	IUT [7]	MCB [8]	FREEDM SST [11]
MOSFETs	96	24	18	48
MOSFETs for PFC	N/A	12	6	24
Switch utilization	Very Poor	Good	Best	Poor
HF Transformer Modules	1	1	3	6

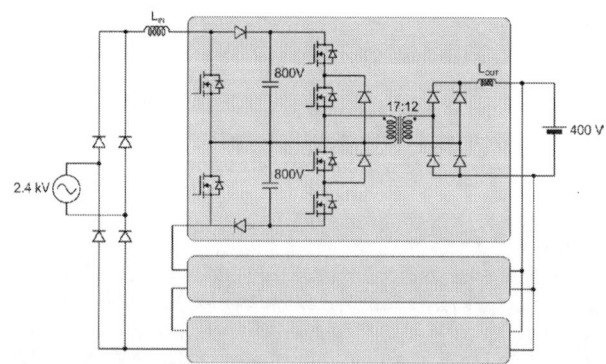

Fig. 1. Multi-Cell Boost (MCB) topology with three input-series-output-parallel (ISOP)-connected dc/dc converter modules.

From the control standpoint, the MCB topology has an advantage over the other topology candidates, since it simultaneously enables independent control of input and output stages, relatively simple control strategy for dc bus capacitor voltage balancing and the reduction of the input and output inductors by interleaving the input 3-level boost converters and the NPC converters at the output. Due to the interleaving on the input side, the effective switching frequency seen by the input inductor increases 6 times. This enables the reduction of switching frequency (for the same input current ripple as in the non-interleaved case) and therefore the reduction of switching losses in the boost converters. However, the reduction of the boost converters switching frequency will be limited by the input current THD requirement. The ISOP configuration of the MCB topology modules provides significant scalability in the input voltage and the output power ratings.

II. MCB MODULE COMPONENT SELECTION AND DESIGN

Each dc/dc module of the adopted MCB topology consists of a 3-level boost converter, dc-link capacitors, NPC inverter, high-frequency transformer and output diode rectifier.

A. Semiconductor devices selection

In order to reduce potential risks involved with using non-qualified devices, and to enable faster module deployment, one of the design decisions was to use 1.2 kV off-the-shelf SiC devices. The comparative summary of some basic features of the considered SiC MOSFETs is given in Table II and the summary for the SiC diodes is given in Table III. Both tables are color coded, where green tones mean better properties. The conditions for which the values are valid are shown in parentheses.

TABLE II. BASIC FEATURES OF PROSPECTIVE OFF-THE-SHELF SiC MOSFETs

Transistor	Package	I_D [A] $T_C = 100°C$	R_{DSon} [mΩ] $T_J = 150°C$	Q_G [nC] $V_{DS} = 800$ V	E_{ON} [mJ] $V_{DS} = 600$ V	E_{OFF} [mJ] $V_{DS} = 600$ V	R_{thJC} [K/W]
CREE C2M0025120D	TO-247 Single	60	43 ($I_D = 50$ A)	161 ($I_D = 50$ A)	1.10 ($I_D = 50$ A)	0.25 ($I_D = 50$ A)	0.24
ST Microelectronics SCT30N120 800V	TO-247 Single	34	90 ($I_D = 20$ A)	105 ($I_D = 20$ A)	0.50 ($I_D = 20$ A)	0.35 ($I_D = 20$ A)	0.65
ROHM SCH2080KE	TO-247 Single	28	125 ($I_D = 10$ A, $T_J = 125°C$)	106 ($V_{DS} = 400$ V, $I_D = 10$ A)	1.20 ($I_D = 30$ A)	0.36 ($I_D = 30$ A)	0.44
CREE CAS120M12BM2	Half-bridge (isolated 5 kV)	126	23 ($I_D = 120$ A)	378 ($I_D = 120$ A)	1.20 ($I_D = 50$ A)	0.4 ($I_D = 50$ A)	0.125
Microsemi APTMC120AM20CT1AG	Half-bridge (isolated 4 kV)	108	22 ($I_D = 100$ A)	360 ($I_D = 100$ A)	1.38 ($I_D = 50$ A)	0.65 ($I_D = 50$ A)	0.21
Microsemi APTMC120AM55CT1AG	Half-bridge (isolated 4 kV)	35.5	75 ($I_D = 40$ A)	98 ($I_D = 40$ A)	0.90 ($I_D = 50$ A)	0.5 ($I_D = 50$ A)	0.5

TABLE III. BASIC FEATURES OF PROSPECTIVE OFF-THE-SHELF SiC DIODES

Diode	Package	I_F [A] $T_C = 100°C$	V_F [V] $T_J = 125°C$, $I_F = 50$ A	Q_C [nC]	I_R [☐A] $T_J = 175°C$, $V_R = 1200$ V	R_{thJC} [K/W]
CREE C4D40120D	TO-247 Dual	80 (2 x 40)	2.15	198 ($I_F = 40$ A, $V_R = 800$ V)	400	0.29
GeneSiC GB50SLT12-247	TO-247 Single	75	2.3	247 (IF = 50 A, VR = 960 V)	100 typ. / 3000 max.	0.242
Global Power GDP50P120B	TO-247 Single	90	2.1	173 ($V_R = 1200$ V)	5598	0.2
Global Power GDP60Z120E	TO-247 Single	110	1.75	207 ($V_R = 1200$ V)	6717	0.16
Rohm SCS240KE2C	TO-247 Dual	56 (2 x 28)	1.9	132 ($V_R = 800$ V)	260	0.28
Microsemi APT2X50DC120J	SOT-227 Dual, isolated 2.5 kV	100 (2 x 50)	2.2	200 ($I_F = 50$ A, $V_R = 600$ V)	280 typ. / 5000 max.	0.32
Microsemi APT2X60DC120J	SOT-227 Dual, isolated 2.5 kV	120 (2 x 60)	1.9	240 ($I_F = 60$ A, $V_R = 600$ V)	336 typ. / 6000 max.	0.26
Microsemi APT40DC120HJ	SOT-227 (Graetz, isolated 2.5 kV)	4 x 40	2.5	160 ($I_F = 40$ A, $V_R = 600$ V)	224 typ. / 4000 max.	0.39

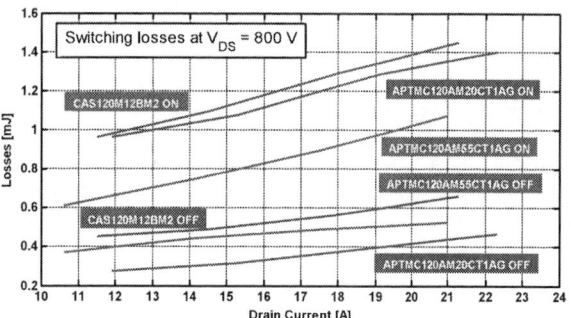

Fig. 2. The switching losses of isolated modules from Table II, at $V_{DS} = 800$ V.

Although both non-isolated and isolated devices are considered, a slight advantage was given to the isolated devices due to their better thermal properties. The switching losses in the isolated MOSFET module from Table II were experimentally determined from double pulse tests, since the loss data was not available for the entire current range from the datasheet. The results obtained at V_{DS} voltage of 800 V are shown in Fig. 2. The similar trend can be observed at V_{DS} voltages of 600, 700 and 900 V. As can be seen from Fig. 2, the APTMC120AM55CT1AG has the lowest total switching losses, among the tested devices and it is therefore selected to be used in the MCB module.

In total, 3 of these SiC power modules are needed to construct one MCB module. Similarly, SiC Power diodes APT2X60DC120J are selected since they exhibit the lowest conduction losses, combined with the good thermal properties, high voltage isolation of 2.5 kV RMS and relatively low leakage current. Since each diode module contains 2 diodes, 2 diode modules are required to construct the MCB module. For the output rectifier stage, 1.2 kV SiC diode bridge module APT40DC120HJ is selected.

B. Desription of used passive components

The input and output inductors were custom designed for the application. The input inductor was made by using 2 stacked sendust cores MS-520026-2, with 100 turns of 8 AWG litz wire. The measured open-circuit inductance was 1.1 mH, and the inductance at 20 A current was around 960 μH.

The output inductor was made by using single Kool Mμ core 0077908A7, with 61 turn of 10 AWG litz wire. The measured open-circuit inductance was 140 μH, and the inductance at 40 A current was approximately 76 μH. The output inductor was undersized as it heated up to 100°C during tests. In future iterations of system design, both inductors will be further optimized for better thermal performance and to further reduce losses and volume.

The electrolytic capacitors LNC2H222MSEG of 2200 mF/500 V were selected as the dc bus capacitors. Two capacitors are connected in series to support the full voltage of 800 V. An additional film capacitor of 1 μF was connected in parallel with the series connection of the electrolytic capacitors in order to decouple the parasitic inductance of the capacitors and connecting wires.

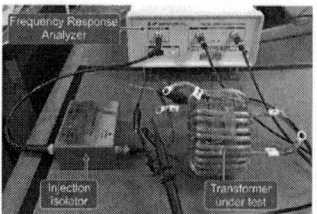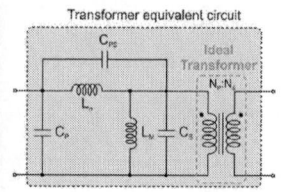

Fig. 3. Transformer test setup and equivalent circuit.

TABLE IV. HF TRANSFORMER PARAMETERS

Turns ratio	L_σ [μH]	L_M [mH]	C_P [pF]	C_S [pF]	C_{PS} [pF]
17:12	6.1	3.0	24.1	47.7	311.5

C. High-frequency transformer

The high-frequency transformer was made by using 4 N87 SIFERRIT material toroid cores, with the transfer ratio of 17:12 turns of 8 AWG litz wire. Both windings are wound in a single layer in order to reduce the parasitic capacitance within and between the windings. The primary and the secondary were completely interleaved. In order to determine the transformer parameters, the transformer was tested with the AP Instruments' Model 300 Frequency Response Analyzer, as shown in Fig. 3. The transformer equivalent circuit was also shown in Fig. 3 and its parameters were presented in Table IV.

The transformer leakage inductance needs to be carefully adjusted in order to ensure the soft switching of the outer NPC switches for the desired load current range. The inner NPC switches would be turned on at the reduced voltage, as explained in [14]. The transformer leakage inductance required for the soft switching was determined by simulations in PSIM.

In order to verify the simulation results, experiments were carried out on the realized module prototype, shown in Fig. 4. For this purpose, an additional variable inductor (not shown in Fig. 4) was added in series with the transformer primary. The simulation and experimental results are shown in Fig. 5.

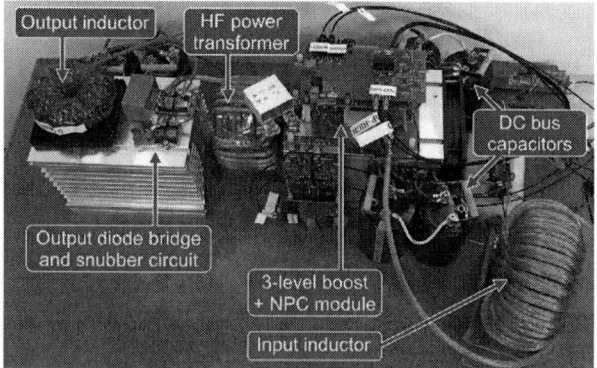

Fig. 4. A MCB module prototype.

III. EVALUATION OF THE MCB MODULE AND THE OVERALL SYSTEM EFFICIENCY

The MCB module operation was simulated in PLECS and the module efficiency was obtained by using thermal descriptions of the semiconductor components, obtained experimentally. The control structure, adopted for testing purposes, is presented in Fig. 6. Three control loops with PI controllers were used. The total voltage of the split dc bus is controlled by the dc bus main loop, the capacitor voltages were balanced by the Voltage balancing loop and the output voltage is controlled by the NPC loop. The obtained experimental waveforms are shown in Fig. 7.

The module was simulated in PLECS, and its efficiency was found by using thermal descriptions of the semiconductor components obtained experimentally. The simulation results of module efficiency and breakdown of losses are shown in Fig. 8. The corresponding experimental result is shown in Fig. 9. Since the input voltage of the used power analyzer (Yokogawa WT300) is limited to 1000 V, experimental validation of the PLECS thermal model is performed at 896 V dc input voltage (which is lower than the rated 1200 V, i.e. duty cycle of the 3-level boost is higher). The transformer leakage inductance was set to 20 µH, which enables soft switching of the outer NPC switches, as shown in Fig. 5. In order not to exceed the analyzer current limit of 30 A, the output power is measured on channels 1 and 2 and the input power is measured on channels 3 and 4. The efficiency is then calculated as:

$$\eta_1 = \frac{P_1 + P_2}{P_3 + P_4} \qquad (1)$$

According to Figs. 8 and 9, the simulation results from PLECS match well with the efficiency measured experimentally.

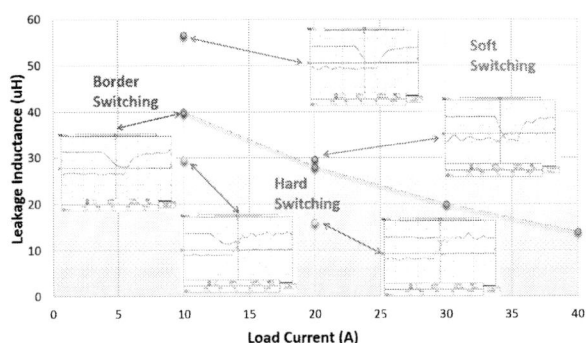

Fig. 5. The influence of the transformer leakage induction on the swithing behavior of the outer NPC switches. Blue dots represent the data from PSIM.

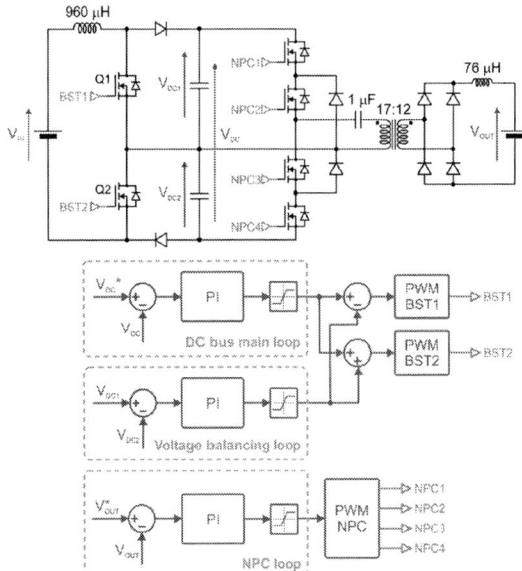

Fig. 6. The control structure adopted for the testing purposes.

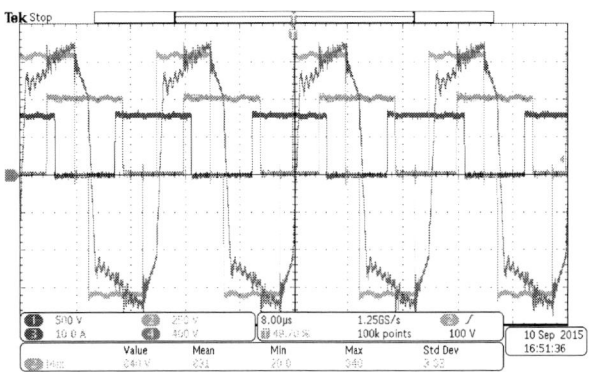

Fig. 7. Experimental waveforms recorded at $V_{IN} = 900$ V: (blue) Q1 Drain-Source voltage, (green) Q2 Drain-Source voltage, (cyan) transformer input voltage, (magenta) transformer input current.

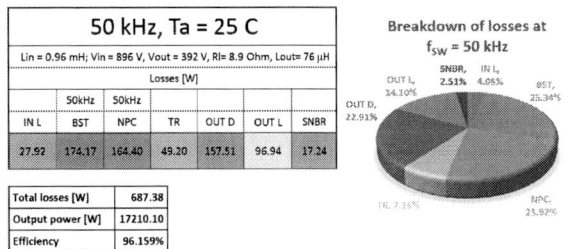

50 kHz, Ta = 25 C						
Lin = 0.96 mH; Vin = 896 V, Vout = 392 V, Rl= 8.9 Ohm, Lout= 76 µH						
Losses [W]						
	50kHz	50kHz				
IN L	BST	NPC	TR	OUT D	OUT L	SNBR
27.92	174.17	164.40	49.20	157.51	96.94	17.24

Total losses [W]	687.38
Output power [W]	17210.10
Efficiency	96.159%

Fig. 8. The simulated module efficiency and breakdown of power losses.

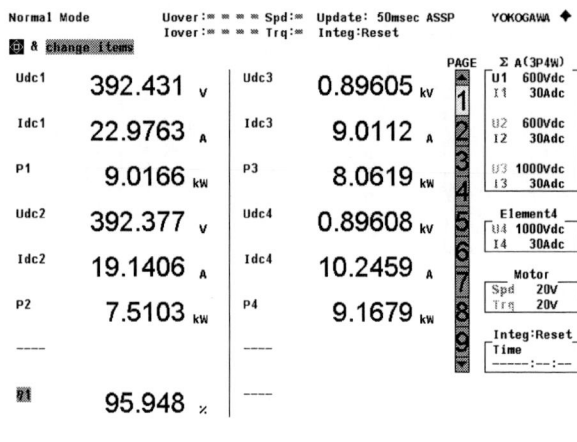

Fig. 9. The experimentally obtained module efficiency. Channels 1 and 2 measure output power, while channels 3 and 4 measure input power.

TABLE V. SIMULATION RESULTS FOR THE FAST CHARGER SYSTEM, AT 50 KW OUTPUT POWER AND THE REDUCED SWITCHING FREQUENCY OF THE INTERLEAVED BOOST (BST) CONVERTERS

f_{SW} [kHz]	THDi [%]	Max. inductor current ripple [A]	Inductor peak current [A]	Total BST losses [W]
50	1.653	0.586	30.71	496.20
45	1.697	0.837	30.76	456.10
40	1.727	0.916	30.80	420.60
35	1.775	1.078	30.85	384.90
30	1.851	1.255	30.92	349.50
25	1.979	1.495	31.02	313.80

In order to increase the overall system efficiency, the switching frequency of the interleaved boost converters can be reduced, while keeping the input current THD below 2%. Table V shows the simulation results for the system from Fig. 1, at 50 kW output power. The switching frequency was reduced down to 25 kHz, at which the THD reached the imposed limit of 2%.

Changing the switching frequency of the interleaved boost converters to 25 kHz and using optimized input and output inductors, the simulated efficiency of the entire 50 kW system would increase to 96.6%, as shown in Fig. 10.

A 3D rendering of the optimized system, with film capacitors instead of the electrolytic ones, shows a significant reduction in size, compared to the state-of-the-art low voltage chargers with low-frequency transformer [1]-[4].

IV. CONCLUSION

An isolated power converter module for medium-voltage (2.4 kV), high-power-quality (PF ≥ 0.98, current THD ≤ 2%), 50 kW fast charger for plug-in electric vehicles has been presented in this paper. The proposed high-efficiency (above 96%), and reduced-footprint converter module utilizes off-the-shelf Silicon Carbide (SiC) devices. Based on the initial design requirements, the multi-cell boost topology was selected.

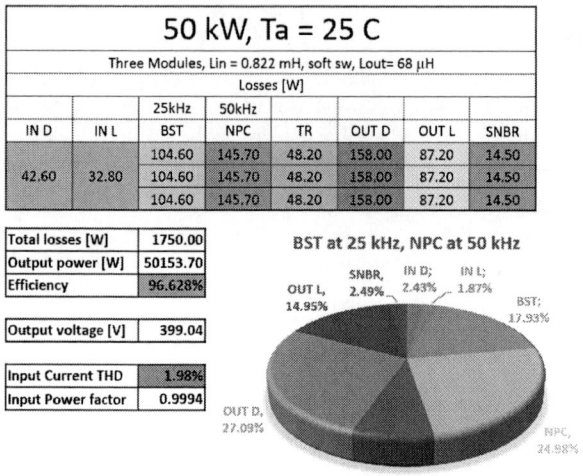

Fig. 10. Simulated efficiency of the 50 kW system with boost switching frequency reduced to 25 kHz and slightly improved input and output inductors.

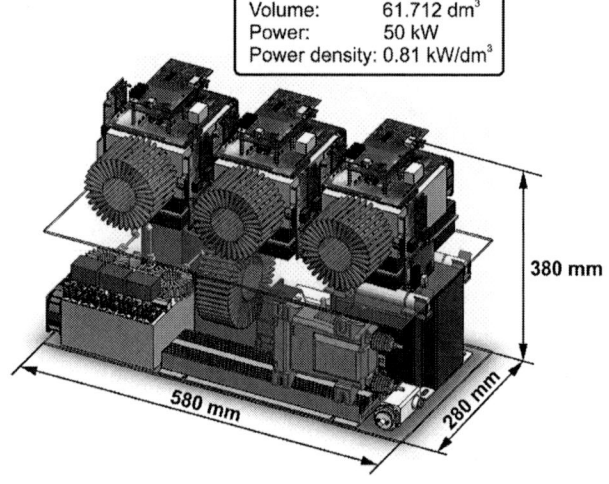

Fig. 11. A 3D rendering of the optimized EV fast charger.

Initial prototype was designed, the system efficiency was found to be around 96%. With further system optimization the efficiency can be increased by at least 0.5%. Significant reduction in volume, compared to the state-of-the-art, is expected for the final charger design.

ACKNOWLEDGMENT

The information, data, or work presented herein was funded in part by the Office of Energy Efficiency and Renewable Energy (EERE), U.S. Department of Energy, under Award Number DE-EE0006521 with North Carolina State University, PowerAmerica Institute.

DISCLAIMER

The information, data, or work presented herein was funded in part by an agency of the United States Government. Neither the United States Government nor any agency thereof, nor any of their employees, makes any warranty, express or implied, or assumes any legal liability or responsibility for the accuracy, completeness, or usefulness of any information, apparatus, product, or process disclosed, or represents that its use would not infringe privately owned rights. Reference herein to any specific commercial product, process, or service by trade name, trademark, manufacturer, or otherwise does not necessarily constitute or imply its endorsement, recommendation, or favoring by the United States Government or any agency thereof. The views and opinions of authors expressed herein do not necessarily state or reflect those of the United States Government or any agency thereof.

REFERENCES

[1] Delta EV DC Quick Charger, DELTA, Jan. 2013. [online]. Available: www.deltaww.com/evcs.

[2] Electric Vehicle Charging Infrastructure, Terra 53 multi-standard DC charging station, ABB, Jun. 2014. [online]. Available: www.abb.com/evcharging.

[3] DC Quick Charger for electric vehicles, Technical Data TD0EV00004E, EATON, Jan. 2014. [online]. Available: www.eaton.com.

[4] DC Fast Charger, Blink, Feb. 2014. [online]. Available: http://prod.blinknetwork.com/chargers-commercial-dc-fast.html.

[5] Utility Direct Medium Voltage DC Fast Charger Update: DC Fast Charger Characterization. EPRI, Palo Alto, CA: 2012. 1024106.

[6] M. Glinka, R. Marquardt, "A new AC/AC-multilevel converter family applied to a single-phase converter," in *The Fifth International Conference on Power Electronics and Drive Systems, PEDS 2003.*, vol. The switching losses

[7] Jih-Sheng Lai; A. Maitra, A. Mansoor, F. Goodman, "Multilevel intelligent universal transformer for medium voltage applications," in *Industry Applications Conference, 2005. Fourtieth IAS Annual Meeting. Conference Record of the 2005*, vol. 3, pp. 1893–1899, 2–6 Oct. 2005.

[8] W. van der Merwe, T. Mouton, "Solid-state transformer topology selection," in *IEEE International Conference on Industrial Technology, ICIT 2009.*, pp. 1–6, 10–13 Feb. 2009.

[9] R. Marquardt, "Modular Multilevel Converter: An universal concept for HVDC-Networks and extended DC-Bus-applications," in 2010 International Power Electronics Conference (IPEC), pp. 502–507, 21–24 June 2010.

[10] S. Falcones, Xiaolin Mao; R. Ayyanar, "Topology comparison for Solid State Transformer implementation," in *2010 IEEE Power and Energy Society General Meeting*, pp. 1–8, 25–29 July 2010.

[11] S. Bhattacharya, Tiefu Zhao, Gangyao Wang, S. Dutta, Seunghun Baek, Yu Du, B. Parkhideh, Xiaohu Zhou, A. Q. Huang, "Design and development of Generation-I silicon based Solid State Transformer," in *2010 Twenty-Fifth Annual IEEE Applied Power Electronics Conference and Exposition (APEC)*, pp. 1666–1673, 21–25 Feb. 2010.

[12] Xu She, A. Q. Huang, R. Burgos, "Review of Solid-State Transformer Technologies and Their Application in Power Distribution Systems," *IEEE Journal of Emerging and Selected Topics in Power Electronics*, vol. 1, no. 3, pp. 186–198, Sept. 2013.

[13] D. Rothmund, G. Ortiz, J. W. Kolar, "SiC-based unidirectional solid-state transformer concepts for directly interfacing 400V DC to Medium-Voltage AC distribution systems," in *2014 IEEE 36th International Telecommunications Energy Conference (INTELEC)*, pp. 1–9, Sept. 28–Oct. 2, 2014.

[14] R. A. Friedemann, F. Krismer, J. W. Kolar, "Design of a minimum weight dual active bridge converter for an Airborne Wind Turbine system," in *Applied Power Electronics Conference and Exposition (APEC), 2012 Twenty-Seventh Annual IEEE*, pp. 509–516, 5–9 Feb. 2012.

Shunt Active Power Filter Based on Cascaded Transformers Coupled with Three-Phase Bridge Converters

Gregory A. de Almeida Carlos[1,2,3], Cursino B. Jacobina[1,2], João Paulo R. A. Méllo[1,2] and Euzeli C. dos Santos Jr.[4]

[1]Post-Graduate Program in Electrical Engineering - PPgEE - Copele
Campina Grande, PB 58429-900 - Brazil
[2]Department of Electrical Engineering - Federal University of Campina Grande (UFCG)
Campina Grande, PB 58429-900 - Brazil
[3]Department of Control and Industrial Process - Federal Institute of Alagoas (IFAL)
Palmeira dos Índios, AL 57601-220 - Brazil
[4]Department of Electrical and Computer Engineering - Indiana University - Purdue University (IUPUI)
Indianapolis, IN 46202 - USA

Abstract—This paper proposes a multilevel shunt active power filter (SAPF) to deal with either harmonic currents compensation or reactive power compensation. Such a device can reduce the harmonic distortion at the grid currents provided by non-linear loads located in stiff systems. The proposed SAPF is based on three-phase bridge (TPB) converters connected to cascaded single-phase transformers. The transformers arrangement permits the compensator to use a single dc-link unit which simplifies the control strategy and number of sensors. The multilevel waveforms are generated by using a suitable PWM strategy associated with the transformers turns ratio. Modularity and simple maintenance make the proposed SAPF an attractive solution compared to some conventional configurations. The model, PWM technique and control strategy, are presented as well as studies considering harmonic distortion and semiconductor losses estimation. Simulation and experimental results are presented in order to validate theoretical approaches.

I. INTRODUCTION

Distribution power systems are suffering hard impact on their power quality. This is due to the intensive use of non-linear loads added with the growth of renewable energy sources. Such aspects have been leading electrical systems to poor power quality levels. Most common disturbances include: i) harmonic voltages/currents, ii) voltages imbalances, iii) voltage sags/swells, iv) flickers, v) transients and vi) interruptions. Non-linear loads and unbalanced loads are the main reasons of the injected current harmonics and reactive power components at the grid, leading to a low system efficiency with a poor power factor. Extensive surveys have been carried out to quantify these problems, as discussed in [1]. In this way, some custom power devices have been introduced and investigated in the literature. Initially, the harmonic current and reactive power compensations have been treated with passive L-C filters and capacitors. However, this solution has disadvantages associated with fixed compensation and resonance effect on the power system. Then, power companies/industries (e.g., ABB, Siemens, Fuji, Westinghouse, etc) started the development of active power filters (APFs) [1], [2]. There are three types of APFs most applied in the customized power quality market: i) series APF [3]; ii) shunt APF [4]–[6] and iii) universal APF or unified power quality conditioner (UPQC) [7], [8]. This paper considers the shunt APF (SAPF) technology for applications on three-phase three-wire systems.

The shunt active power filter (SAPF) is commonly composed by: i) voltage source converter (VSC) or current source converter (CSC), i) dc-link capacitors, iii) optional passive filters, iv) optional isolation transformers and v) protection circuits (e.g., bypass thyristors). In this paper, the SAPF considered is composed by VSC type with isolation transformers. There are several types of topologies to be used in VSC configuration [2], [9]–[12]. Generally, the two-level (2L) based VSC is commonly used for a 480V voltage rating (i.e., low power applications). Nevertheless, for high power applications, such as distribution voltages levels, the multilevel-based VSC becomes a more attractive solution [13]. In this way, some multilevel configurations have been studied and documented in technical literature [2], [13], [14]. However, those multilevel configurations have disadvantages associated to the high number of dc-link capacitors that, depending on the configuration, can increase the control complexity. For instance, cascaded neutral point-clamped (NPC) and cascaded flying capacitors have issues with imbalance dc-link voltages and power sharing in each cell [14]. Additionally, the lifetime of electrolytic dc-link capacitors is usually shorter than that of other components in power converters [15]. It is usually addressed as one of the most important issues in terms of failure rate in the field operation of power electronic systems [15]–[17]. Degradation failures occur more often than catastrophic failures (i.e., short or open circuits). In this case, critical failure mechanisms cover electrolytic vaporization and electromechanical reaction (e.g., degradation of oxide layer) [15], [16]. Then, a high number of dc-link capacitors tends to increase the failure probability

978-1-4673-9551-9/16 $31.00 © 2016 IEEE

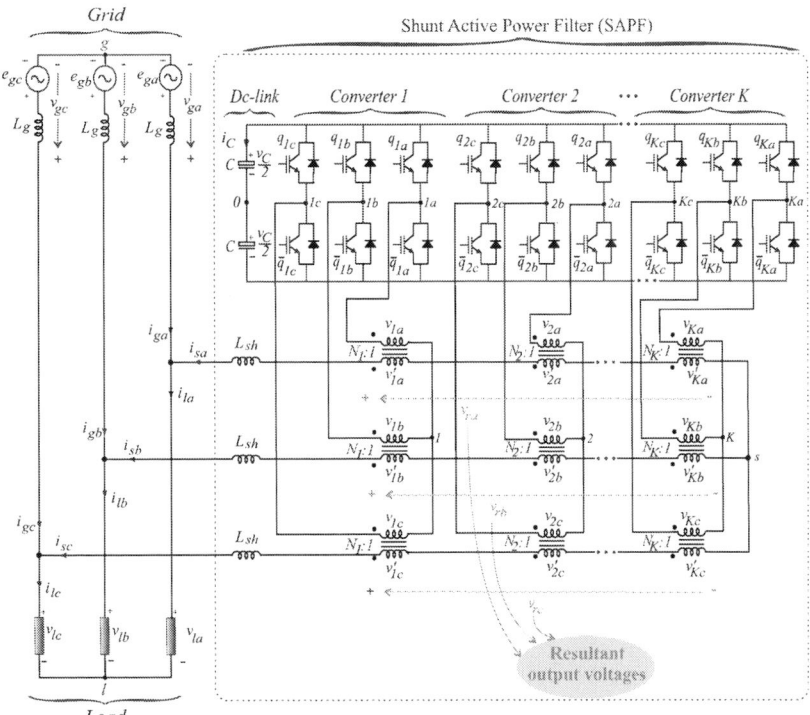

Fig. 1.   Proposed SAPF generalized with K-cascaded transformers per phase wye-connected with K-three-phase-bridge converters.

of the SAPF. Hence, cascaded transformers can be considered to deal with this issue, maintaining the multilevel feature with a single dc-link capacitor in its configuration. A conventional SAPF used with cascaded transformer coupled with H-bridge (HB) converters was presented in [10].

In this paper, a SAPF based on cascaded transformer coupled with three-phase-bridge (TPB) converters is proposed, see Fig. 1. Such a structure is generalized for K-stages in which K-transformers are coupled with K-TPB converters. Equivalent multilevel operation is achieved with reduced number of semiconductors devices if compared to conventional HB one [10]. The multilevel waveforms are generated by TPB converters by using suitable PWM strategy associated with the transformers turns ratio. The modularity and simple maintenance makes proposed SAPF an attractive solution in comparison with some conventional configurations. Model and PWM control are presented. Simulation and experimental results are presented for validation purposes.

II. PROPOSED SAPF MODEL

The configuration depicted in Fig. 1 is generalized for K-stages (i.e., K-transformers and K-three-phase-bridge converters). The converters legs are represented by K-power switches (i.e., $q_{1j}, \overline{q}_{1j}, q_{2j}, \overline{q}_{2j}, ..., q_{Kj}$, and $\overline{q}_{Kj}$) in which the subscript j is related to each phase (e.g., $j = a, b, c$). In addition, power switches q and $\overline{q}$ are complementary to each other. The conduction state of all power switches is represented by an homonymous binary variable, where $q = 1$ indicates a closed switch and $q = 0$ an open one.

The converter pole voltages (v_{1j0}, v_{2j0},..., v_{Kj0}), can be expressed as

$$v_{kj0} = (2q_{kj} - 1)\frac{v_C}{2} \tag{1}$$

where k corresponds for each stage (i.e., $k = 1, 2, 3, ..., K$), j is related for each phase ($j = a, b, c$) and v_C is the dc-link voltage.

A previous work in which the converter was considered as a series compensator has been presented in [18]. Since the converter equations are similar to that presented in [18], some equations are not detailed in this paper. Taking into consideration the leakage inductance of a transformer and external interfacing shunt inductance represented by l_{sh} and the load dependent loss of a transformer denoted by r_{sh}, a differential equation for shunt active power filter can be written as

$$v_{rj} - v_{gs} = l_{sh}\frac{di_{sj}}{dt} + r_{sh}i_{sj} - l_g\frac{di_{gj}}{dt} - r_g i_{gj} + e_{gj} \tag{2}$$

where v_{rj} are the resultant voltages of the converter related to the secondary voltages of the scaled transformers and v_{gs} is voltage between the neutral points g and s.

From the current node it can be written

$$i_{sj} = i_{lj} - i_{gj} \tag{3}$$

where the load currents i_{lj} are given by the load model.

Replacing (3) in (2) will give

$$\overbrace{v_{rj} - v_{gs} = -(l_g + l_{sh})\frac{di_{gj}}{dt} - (r_g + r_{sh})i_{gj}}^{\text{1st term}}$$

$$\underbrace{+ l_{sh}\frac{di_{lj}}{dt} + r_{sh}i_{lj} + e_{gj}}_{\text{2nd term}} \quad (4)$$

From the standpoint of control, the *2nd term* in (4) are perturbations which must be compensated by the controller.

The transformer voltages at the secondary side of each transformer (v'_{1j}, v'_{2j}, ..., v'_{Kj}) are associated with v_{rj} such that

$$v_{rj} = v'_{1j} + v'_{2j} + ... + v'_{Kj} \quad (5)$$

where $v'_{1j} = N_1(v_{1j0} - v_{10})$, $v'_{2j} = N_2(v_{2j0} - v_{20})$, ..., $v'_{Kj} = N_K(v_{Kj0} - v_{K0})$ in which N_1, N_2, ..., N_K are the transformer turns ratios associated with converters 1, 2,..., K, respectively.

Considering a perfect isolation from primary to secondary side of the transformers (i.e., ideal transformers), the output voltages (v_{rj}) of the resultant converter can be expressed as

$$v_{rj} = v_{rjo} - v_{ro} \quad (6)$$
$$v_{rjo} = N_1 v_{1j0} + N_2 v_{2j0} + ... + N_K v_{Kj0} \quad (7)$$
$$v_{ro} = N_1 v_{10} + N_2 v_{20} + ... + N_K v_{K0}. \quad (8)$$

Since the system is assumed to be a three-phase three-wire system (i.e., $v_{ka} + v_{kb} + v_{kc} = 0$ and $i_{ka} + i_{kb} + i_{kc} = 0$) the voltage v_{ro} is given by

$$v_{ro} = \frac{N_1}{3}\sum_{j=a}^{c} v_{1j0} + \frac{N_2}{3}\sum_{j=a}^{c} v_{2j0} + ... + \frac{N_K}{3}\sum_{j=a}^{c} v_{Kj0}. \quad (9)$$

Substituting (9) in (6) will give

$$v_{rj} = v_{rjo} - \frac{N_1}{3}\sum_{j=a}^{c} v_{1j0} - ... - \frac{N_K}{3}\sum_{j=a}^{c} v_{Kj0}. \quad (10)$$

It should be noted that the voltages v_{rj} can have a maximized number of levels if the voltages (v_{rjo}) assume a suitable sequence of the switching states. This is achieved by considering the transformer turns ratios (N_1, N_2, ..., N_K). A particular case, with 3 transformers per phase and 3 three-phase-bridge (TPB) converters has been described in [18] for series compensation application. It has shown that voltages v_{rj} can reach 8 different levels per phase according to the switching states. In this case, the converter must operate with different transformer turns ratios (i.e., $N_k = 2^{(k-1)}$). The one-dimension region of output voltage v_{rj} for each phase (i.e., $j = a, b, c$) associated with switching states [q_{1j}, q_{2j} and q_{3j}] was detailed described in [18]. Such a representation permits to easily synthesize the reference output voltage by using always the nearest switching states to the reference output voltage.

This approach has advantages to reduce the harmonic distortion of the power converter topology. No redundant levels are provided if the transformer turns ratios are considered to have the maximized number of different levels (i.e., 2^K). The redundancy levels (with more than one switching states

giving the same voltage level) can be obtained by choosing some equal transformer turns ratios. For instance, for 2 stages operation the redundancy is achieved by using $N_1 = N_2 = 1$. The redundancy property can be used in order to improve other features associated to the power converters operation such as the number of commutations or dc-link voltage balancing as discussed in work which has considered cascaded H-Bridge converters applied with one-dimension modulation approach [19].

III. PWM STRATEGY

The pulse width modulation (PWM) technique used in this work is based on level-shifted-carrier-based PWM (LSPWM). However, a simpler algorithm calculation can be obtained. It takes into consideration references for the resultant output voltage (v_{rjo}^*). The PWM strategy is similar to that one presented in [18] and is not detailed in this paper.

The references for the PWM strategy are given by

$$v_{rjo}^* = v_{rj}^{*''} + v_{rogs}^* \quad (11)$$

where $v_{rj}^{*''}$ are references provided by current controllers, v_{rjo}^* is similar as that one presented in (7), v_{rogs}^* is a degree of freedom from the system characteristics.

The reference voltage for v_{rogs}^* is calculated as

$$v_{rogs}^* = \mu_{rogs}^* v_{rogs\,\text{max}}^* + (1 - \mu_{rogs}^*)v_{rogs\,\text{min}}^* \quad (12)$$

where $0 \leq \mu_{rogs}^* \leq 1$ and

$$v_{rogs\,\text{min}}^* = -0.5v_C^*(N_1 + ... + N_K) - \min\{v_{rj}^{*''}\} \quad (13)$$
$$v_{rogs\,\text{max}}^* = 0.5v_C^*(N_1 + ... + N_K) - \max\{v_{rj}^{*''}\}. \quad (14)$$

where v_C^* is the dc-link voltage reference.

The reference voltages v_{rjo}^* are compared with $2^n - 1$ triangular waveforms, which are level-shifted carriers (v_{t1} - v_{t2^n-1}) placed according to the levels shown in details in [18]. The result of this comparison gives the switching states (q_{1j}, q_{2j}, ..., q_{Kj}) that are imposed for each TPB converter.

IV. HARMONIC DISTORTION ESTIMATION

The resultant output voltages (v_{rj}) have been evaluated by considering the weighted total harmonic distortion (WTHD). Such a parameter permits to quantify the waveform quality that will be processed by the converter of the SAPF. The following expression was considered for this evaluation

$$WTHD(p) = \frac{100}{a_1}\sqrt{\sum_{i=2}^{p}\left(\frac{a_i}{i}\right)^2} \quad (15)$$

where a_1 is the amplitude of the fundamental voltage, a_i is the amplitude of ith harmonic and p is the number of harmonics taken into consideration.

Then, considering a maximum injection of v_{rj} and the filter compensating only reactive power, the WTHD of the voltages v_{rj} for the proposed configuration can be observed in Table I. For this case, the switching frequency (f_s) was fixed in 10 kHz. From values observed in Table I, as expected, as far as the number of stages increases, the WTHD value decreases. The equivalent waveforms for that result, with 1, 2 and 3 stages, are shown in Figs. 3-5, respectively. Such an evaluation were also presented in [18] but without redundancy case $N_1 = N_2 = 1$.

TABLE I
WHTD OF v_{rj} FOR PROPOSED SAPF WITH TRANSFORMERS CONNECTED TO TPB IN A WYE-CONNECTION TYPE.

Number of stages	Tran. turns ratios	$WTHD$ (%) of v_{rj}
1 stage	1:1 ($N_1 = 1$)	0.210
2 stages	1:1 ($N_a = 1$) 1:1 ($N_b = 1$)	0.093
2 stages	1:1 ($N_1 = 1$) 2:1 ($N_2 = 2$)	0.060
3 stages	1:1 ($N_1 = 1$) 2:1 ($N_2 = 2$) 4:1 ($N_3 = 4$)	0.025

V. SEMICONDUCTOR LOSSES ESTIMATION

Power switch losses were estimated considering the proposed and conventional SAPF [10] having the same number of power switches (i.e., 6 legs). This leads to the proposed topology operating with 2 stages, whereas the conventional with HB operates with 1 stage. The estimation was implemented by using the thermal module, an existing tool in PSIM $v9.0$. Such a tool was used with calibration parameters that gives an equivalent loss estimation to that one presented in [20], achieved by experimental tests. The power switch losses model includes:

a) IGBT and diode conduction losses;

b) IGBT turn-on losses;

c) IGBT turn-off losses;

d) Diode turn-off energy.

This comparison is quite similar to that presented in [18]. However, in the comparative study presented in this paper, the converters were operated for a load power rated at 300 kW. Hence, a reduction up to 82% was observed for proposed configuration compared to the conventional one. Notice that the semiconductor losses comparison observed [18] was done for a lower power than 300 kW. Then, as expected, this permits to conclude that the proposed converter presents much more lower semiconductor losses estimation with the increasing of the power ratings.

TABLE II
LOSSES COMPARISON ESTIMATION BETWEEN PROPOSED OPERATING WITH 2 STAGES ($N_1 = 1$ AND $N_2 = 2$) AND CONVENTIONAL HB WITH 1 STAGE.

Configuration		Semiconductor Losses Estimation		
Topology	Stages	Switching	Conduction	Total
Conv. HB [10]	1	3.74 kW	0.46 kW	4.20 kW
Proposed	2	0.46 kW	0.3 kW	0.76 kW

VI. CONTROL STRATEGY

Fig. 2 presents the block diagram of the control system. The dc-link capacitor voltage v_C is controlled by means of the controller R_C, whose output is the reference amplitude current I_g^* of the three-phase system. The instantaneous grid reference

currents i_{ga} and i_{gb} are obtained by synchronizing their phase with e_{ga} and e_{gb}, provided by the block S_{in} in which has its input signal provided by the phase-locked loop PLL block. The PLL used in this work is based on fictitious electrical power (i.e., power-based PLL) and is detailed in [21]. The current controllers Ri_{ab} define the references voltages $v_{ra}^{*''}$ and $v_{rb}^{*''}$. The dc-link controller R_C is a conventional proportional-integral (PI) controller whereas the current controllers Ri_{ab} are double-sequence controllers (i.e., resonant PI controllers) [22]. From the reference voltages, the PWM strategy defines the state of the switches (q_{1j}, q_{2j}, ..., q_{Kj}).

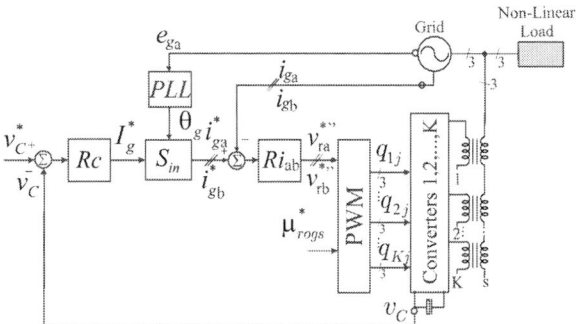

Fig. 2. Block diagram of overall control strategy of proposed SAPF.

VII. SIMULATION RESULTS

Simulation results trough PSIM $v9.0$ are presented in this section. Such outcomes show the resultant output voltages of the converter (v_{rj}) with PWM implementation accordingly to the PWM strategy presented earlier. The first set is shown by considering 1 stage as observed in Fig. 3.

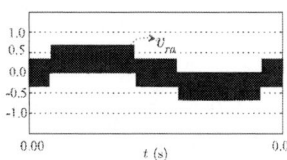

Fig. 3. Simulation result. Resultant phase-voltage (v_{ra}) of converter in phase a for converter operation with 1 stage.

The cases in which the converter operates with 2 stages are shown in Fig. 4. In these cases, only converters 1 and 2 with 2 single-phase transformers and 2 TPB connected in each phase are considered. It can be seen that the result presented in Fig. 4(a) is equivalent to that obtained with 3L-NPC converter or cascaded HB with equal dc-link voltages. It was considered $N_1 = N_2 = 1$, which gives more redundant switching states. The cases without redundancy improves the quality of the waveform with more level steps, which leads the voltage v_{rj} to lower harmonic distortion.

A result in which the converter operates with 3 stages is shown in Fig. 5. In this case, the implementation was done with converters 1, 2 and 3 as well as transformers turns ratio being $N_1 = 1$, $N_2 = 2$ and $N_3 = 4$.

978-1-4673-9551-9/16 $31.00 © 2016 IEEE

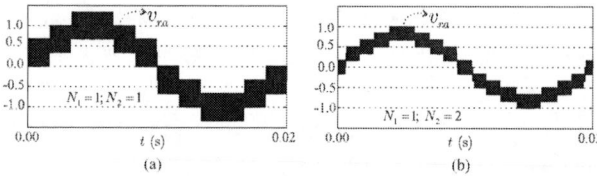

Fig. 4. Simulation result. Resultant phase-voltage (v_{ra}) of converter in phase a for converter operation with 2 stages. (a) Operation with redundancy, having $N_1 = N_2 = 1$. (b) Operation without redundancy, having $N_1 = 1$ and $N_2 = 2$.

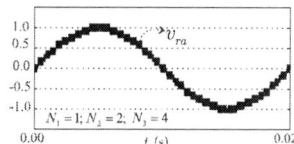

Fig. 5. Simulation result. Resultant phase-voltage (v_{ra}) of converter in phase a for converter operation with 3 stages.

VIII. EXPERIMENTAL RESULTS

The theoretical approaches were validated experimentally with a downscaled experimental platform, as observed on the left side of Fig. 6. The main components used in this platform include inverters considered as IGBTs from Semikron that are linked with the control strategy by means of a digital signal processor (DSP) TMS320F28335 with microcomputer equipped with appropriate plug-in boards and sensors. A zoomed view of the main devices can be observed on the right side of Fig. 6. The dc-link capacitance used was $C = 2200\mu F$ with voltage reference fixed as $v_C^* = 50V$. The switching frequency considered was $f_{sw} = 10kHz$

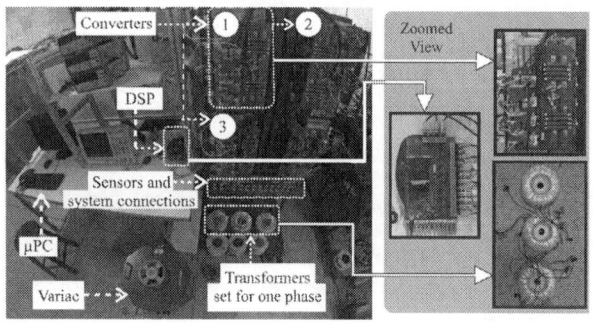

Fig. 6. Downscaled experimental platform. (On the left) Photograph of experimental setup. (On the rigth) Zoomed view of the main devices.

Fig. 7 shows the dynamic operation of the SAPF. In this case a load transient was applied. It can be seen that the dc-link voltage is regulated in its reference value (i.e., $v_C^* = 50V$) before and after the transient application. The steady state waveforms before the transient can be observed in a zoomed view. The results for the other phases are similar. The power factor at the grid can be observed in Fig. 8. It can be seen that the grid current and grid voltage are in phase, which means a power factor close to the unity.

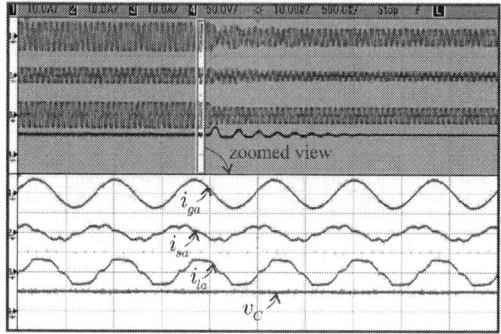

Fig. 7. Experimental results. Proposed SAPF operating with a load transient application. System currents in phase a. Grid current i_{ga}. SAPF current i_{sa}. Load current i_{la}. Dc-link voltage v_C.

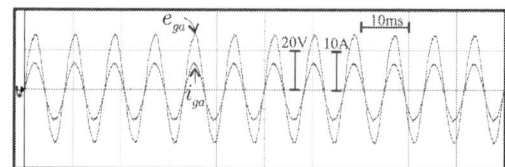

Fig. 8. Experimental results displayed in high resolution. Power factor correction at the grid. Grid current (i_{ga}) and grid voltage (e_{ga}) in phase a.

The harmonic spectrum and the total harmonic distortion (THD) of the currents at the grid and at the load can be observed in Fig. 9. It can be seen that SAPF compensation has provided the grid current with THD fixed in 2.58%. In this case, the load current THD was 8.10%.

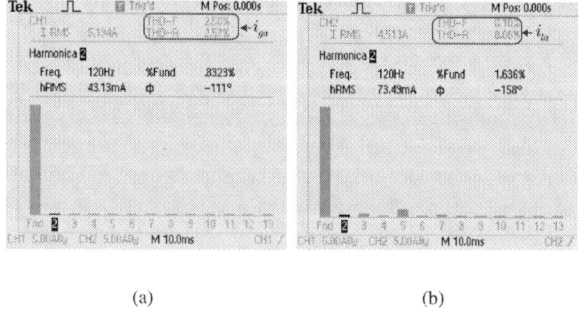

(a)   (b)

Fig. 9. Experimental results. (a) Grid current (i_{ga}) harmonic spectrum with THD fixed in 2.58%. (b) Load current (i_{la}) harmonic spectrum with THD fixed in 8.10%.

Fig. 10 shows the PWM implementation with proposed SAPF having 1 stage (i.e., which means using only converter 1). Fig. 11 shows the PWM implementation in which SAPF was considered with 2 stages with $N_1 = N_2 = 1$, Fig. 11(a) and with $N_1 = 1$, $N_2 = 2$, Fig. 11(b). Fig. 12 shows similar result, but in this case the SAPF is operating with 3 stages (i.e., with converters 1, 2 and 3). The results for the other phases are similar.

978-1-4673-9551-9/16 $31.00 © 2016 IEEE   2724

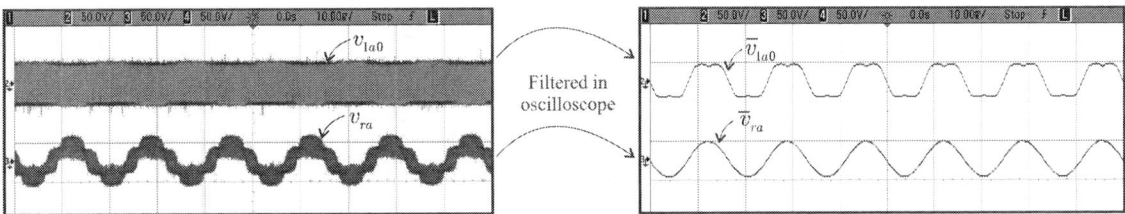

Fig. 10. Experimental result with 1 stage in which $N_1 = 1$. PWM converter voltages in phase a. The PWM implementation of proposed SAPF (on the left) and its fundamental signal filtered in oscilloscope (on the right).

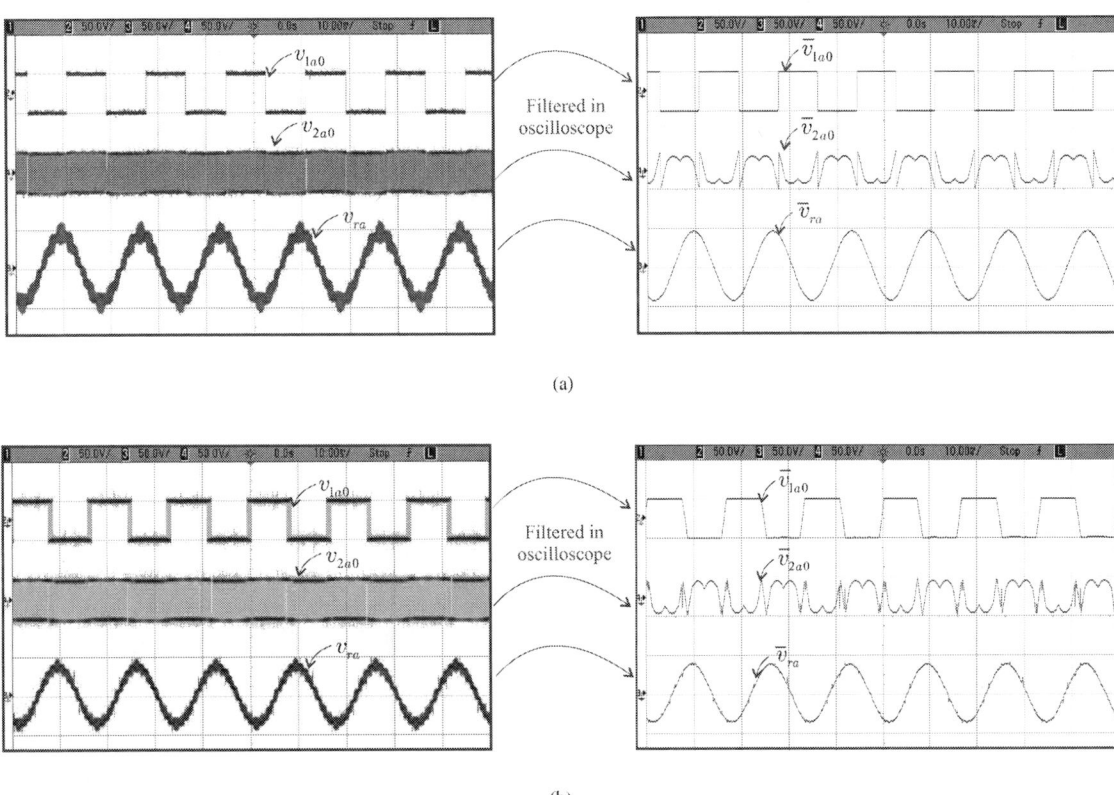

(a)

(b)

Fig. 11. Experimental results with 2 stages. (a) $N_1 = N_2 = 1$. (b) $N_1 = 1$ and $N_2 = 2$. PWM converter voltages in phase a with 1 stage in which $N_1 = 1$. The PWM implementation of proposed SAPF (on the left) and its fundamental signal filtered in oscilloscope (on the right).

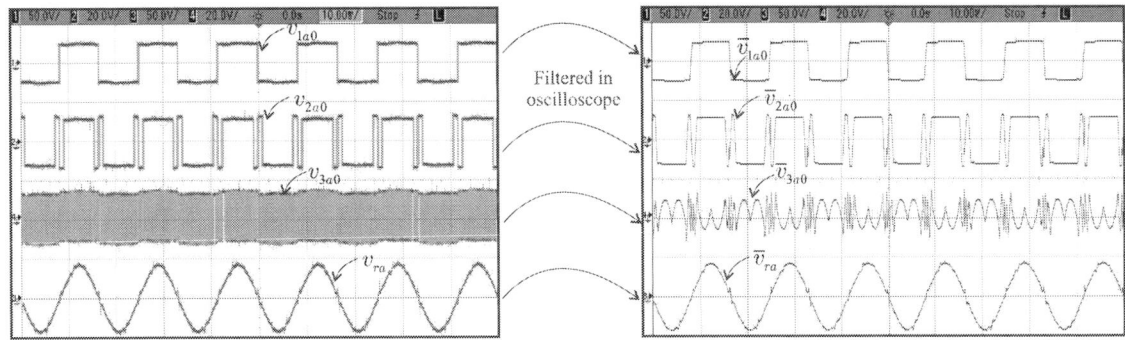

Fig. 12. Experimental results with 3 stages in which $N_1 = 1$, $N_2 = 2$ and $N_3 = 1$. PWM converter voltages in phase a. The PWM implementation of proposed SAPF (on the left) and its fundamental signal filtered in oscilloscope (on the right).

978-1-4673-9551-9/16 $31.00 © 2016 IEEE

IX. CONCLUSION

A shunt active power filter (SAPF) was studied in this paper. The configuration is based on cascaded transformers coupled with three-phase bridge (TPB) converters. The proposed SAPF has presented lower harmonic distortion content in comparison with the conventional one [10]. Such a reduced harmonic distortion has lead to compare the semiconductor losses by fixing the same WTHD value for proposed and conventional configurations. In this way, the proposed configuration could have its switching frequency decreased to match the same WTHD value obtained with the conventional one. Once the conventional topology considered in [10] has also isolation transformers, a semiconductor losses reduction close to 82% was observed in comparison with the conventional topology.

A generalization for K-cascaded TPB converters and K-transformers was presented. Such a generalization was validated experimentally for 1, 2 and 3 stages. The proposed SAPF has presented advantages because it provides the improvement of the quality at the signals generated by the PWM converter, by maintaining its modularity features and using simple dc-link control strategy since it need just a single dc-link unity.

Another remark is that the number of levels generated at the voltage v_{rj} for the proposed configuration is greater when compared to the conventional one [10], considering the same number of power switches. However, to match the same number of semiconductors losses, the topologies have to operate with different number of stages (i.e., different numbers of transformers). Hence, the proposed configuration will need one additional scaled-transformer for each phase. Nevertheless, such an additional transformer must have lower ratings in comparison with the transformers used for the previous stage. The outcomes have shown its accordance with the theoretical approaches. Detailed analyses comparing the transformer ratings will be investigated for an overall cost estimation between the proposed and conventional solutions.

ACKNOWLEDGMENT

The authors would like to thank CAPES, PPgEE/COPELE and DEE/UFCG for the financial support and research grants.

REFERENCES

[1] B. Singh, K. Al-Haddad, and A. Chandra, "A review of active filters for power quality improvement," *Industrial Electronics, IEEE Transactions on*, vol. 46, pp. 960–971, Oct 1999.

[2] H. Rudnick, J. Dixon, and L. Moran, "Delivering clean and pure power," *Power and Energy Magazine, IEEE*, vol. 1, pp. 32–40, Sep 2003.

[3] M. Hamad, M. Masoud, and B. Williams, "Medium-voltage 12-pulse converter: Output voltage harmonic compensation using a series apf," *Industrial Electronics, IEEE Transactions on*, vol. 61, pp. 43–52, Jan 2014.

[4] P. Jintakosonwit, H. Fujita, H. Akagi, and S. Ogasawara, "Implementation and performance of cooperative control of shunt active filters for harmonic damping throughout a power distribution system," *Industry Applications, IEEE Transactions on*, vol. 39, pp. 556–564, Mar 2003.

[5] L. Asiminoaei, F. Blaabjerg, S. Hansen, and P. Thogersen, "Adaptive compensation of reactive power with shunt active power filters," *Industry Applications, IEEE Transactions on*, vol. 44, pp. 867–877, May 2008.

[6] H. Akagi, H. Fujita, and K. Wada, "A shunt active filter based on voltage detection for harmonic termination of a radial power distribution line," *Industry Applications, IEEE Transactions on*, vol. 35, pp. 638–645, May 1999.

[7] R. Millnitz dos Santos, J. da Cunha, and M. Mezaroba, "A simplified control technique for a dual unified power quality conditioner," *Industrial Electronics, IEEE Transactions on*, vol. 61, pp. 5851–5860, Nov 2014.

[8] V. Khadkikar and A. Chandra, "Upqc-s: A novel concept of simultaneous voltage sag/swell and load reactive power compensations utilizing series inverter of upqc," *Power Electronics, IEEE Transactions on*, vol. 26, pp. 2414–2425, Sept 2011.

[9] J. Wen, L. Zhou, and K. Smedley, "Power quality improvement at medium-voltage grids using hexagram active power filter," in *Applied Power Electronics Conference and Exposition (APEC), 2010 Twenty-Fifth Annual IEEE*, pp. 47–57, Feb 2010.

[10] M. Ortuzar, R. Carmi, J. Dixon, and L. Moran, "Voltage-source active power filter based on multilevel converter and ultracapacitor dc link," *Industrial Electronics, IEEE Transactions on*, vol. 53, pp. 477–485, April 2006.

[11] G. de Almeida Carlos, C. Jacobina, E. dos Santos, E. Fabricio, and N. Rocha, "Shunt active power filter with open-end winding transformer and series-connected converters," *Industry Applications, IEEE Transactions on*, vol. 51, pp. 3273–3283, July 2015.

[12] M. Odavic, V. Biagini, M. Sumner, P. Zanchetta, and M. Degano, "Low carrier - fundamental frequency ratio pwm for multilevel active shunt power filters for aerospace applications," *Industry Applications, IEEE Transactions on*, vol. 49, pp. 159–167, Jan 2013.

[13] S. Kouro, M. Malinowski, K. Gopakumar, J. Pou, L. Franquelo, B. Wu, J. Rodriguez, M. Perez, and J. Leon, "Recent advances and industrial applications of multilevel converters," *Industrial Electronics, IEEE Transactions on*, vol. 57, pp. 2553–2580, Aug 2010.

[14] J.-S. Lai and F. Z. Peng, "Multilevel converters-a new breed of power converters," *Industry Applications, IEEE Transactions on*, vol. 32, pp. 509–517, May 1996.

[15] X.-S. Pu, T. H. Nguyen, D.-C. Lee, K.-B. Lee, and J.-M. Kim, "Fault diagnosis of dc-link capacitors in three-phase ac/dc pwm converters by online estimation of equivalent series resistance," *Industrial Electronics, IEEE Transactions on*, vol. 60, pp. 4118–4127, Sept 2013.

[16] H. Wang and F. Blaabjerg, "Reliability of capacitors for dc-link applications in power electronic converters - an overview," *Industry Applications, IEEE Transactions on*, vol. 50, pp. 3569–3578, Sept 2014.

[17] H. Wang, M. Liserre, and F. Blaabjerg, "Toward reliable power electronics: Challenges, design tools, and opportunities," *Industrial Electronics Magazine, IEEE*, vol. 7, pp. 17–26, June 2013.

[18] G. A. de Almeida Carlos and C. B. Jacobina, "Series compensator based on cascaded transformers coupled with three-phase bridge converters," in *Energy Conversion Congress and Exposition (ECCE), 2015 IEEE*, pp. 3414–3421, Sept 2015.

[19] J. Leon, R. Portillo, S. Vazquez, J. Padilla, L. Franquelo, and J. Carrasco, "Simple unified approach to develop a time-domain modulation strategy for single-phase multilevel converters," *Industrial Electronics, IEEE Transactions on*, vol. 55, pp. 3239–3248, Sept 2008.

[20] J. Dias, E. dos Santos, C. Jacobina, and E. da Silva, "Application of single-phase to three-phase converter motor drive systems with igbt dual module losses reduction," in *Power Electronics Conference, 2009. COBEP '09. Brazilian*, pp. 1155–1162, Sept 2009.

[21] R. Santos Filho, P. Seixas, P. Cortizo, L. Torres, and A. Souza, "Comparison of three single-phase pll algorithms for ups applications," *Industrial Electronics, IEEE Transactions on*, vol. 55, pp. 2923–2932, Aug 2008.

[22] C. Jacobina, M. Correa, T. Oliveiro, A. Lima, and E. Cabral da Silva, "Current control of unbalanced electrical systems," *Industrial Electronics, IEEE Transactions on*, vol. 48, pp. 517–525, Jun 2001.

Independent DC Link Voltage Control of Cascaded Multilevel PV Inverter

Qingyun Huang, Wensong Yu and Alex Q. Huang

FREEDM Systems Center, Department of Electrical and Computer Engineering

North Carolina State University

Raleigh, USA

Abstract—**For the independent DC link voltage control of the single phase cascaded multilevel PV inverter, this paper proposes an improved control strategy that consists only one total voltage loop and *n* feed-forward-based weighting factors. Actually, the multiple modulation signals have predictable ratios with each other because the series connection of the outputs of the H-bridges. Utilizing this feature, this proposed control strategy directly generates the weighting factors by the DC link input powers and voltage references which are independent with the output of the voltage feedback loop. The voltage feedback loop controls the sum of the DC link voltages while the feed-forward signals force the ratio between DC link voltages to be equal to the ratio between the references at steady state. Compared with the previous control strategies that contain at least *n* voltage controllers, the proposed control structure is much simplified. Besides, the small signal modeling and controller design of the inner current and outer voltage loop are also included in this paper. In addition, the adaptive gain is proposed to keep the loop gain unchanged even at the situation of weak irradiation. This control strategy is verified by the simulated and experimental results with accurate DC link control and more than 99.5% MPPT efficiency for each H-bridge even at severe irradiation mismatch conditions.**

Keywords—Multilevel inverter; cascaded H-bridge; PV system; individual MPPT; control strategy; DC link control.

I. INTRODUCTION

Among renewable energy power generation applications, grid-connected photovoltaic (PV) inverter systems have been given a high recognition during the last decades. In particular, the single phase grid-connected PV inverter systems including the string inverters (single string or multistring configuration) and the Micro-inverters have been the most popular products for the residential application [1, 2]. The state-of-the-art string inverter has one DC/DC converter for each string and the DC/DC parts are parallel to the central DC/AC inverter [1]. The DC/DC converter can achieve individual maximum power point tracking (MPPT) for each string [1]. The state-of-the-art micro-inverter is the interleaved Flyback with unfolding bridge which has peak 96.5% efficiency [3].

In the past few years, the multilevel technologies for grid-connected PV inverter systems have drawn more interests. Among the conventional multilevel topologies, the cascaded H-bridge multilevel grid-connected inverter as shown in Fig. 1 has become a major option [4]. If the PV cell in Fig. 1 is a PV

string, this topology should be an alternative for the multistring PV inverter. If the PV cell is a single PV panel, this converter can utilize most of the benefits of the string inverter and Micro-inverter. For the output side, this cascaded multilevel inverter topology can reach the output voltage level without transformer as featuring the series connection of the H-bridges' outputs. Besides, the output voltage can be synthesized as multilevel waveform to decrease the harmonic distortion. For the DC link sides, the independent MPPT can be realized to improve the system efficiency under mismatching conditions. In addition, due to the cascaded connection, the lower voltage rating semiconductors which feature a lower cost and better switching performance can be utilized [4].

The two control tasks of the cascaded multilevel inverter are the sinusoidal grid current injection and the independent DC link voltage control which is required by the individual MPPT for each PV cell. The DC link control strategies for the cascaded multilevel converter are classified into two categories according to the types of the DC links. The first one is the voltage balancing control for the identical DC link capacitors. The other one is the independent DC link control for different DC links such as this cascaded multilevel PV inverter.

For the voltage balancing control, as referred in [5, 6], the clustered balancing control and the individual balancing control for the cascaded STACOM system are discussed. And both of them consist of *n* individual voltage feedback loop controllers.

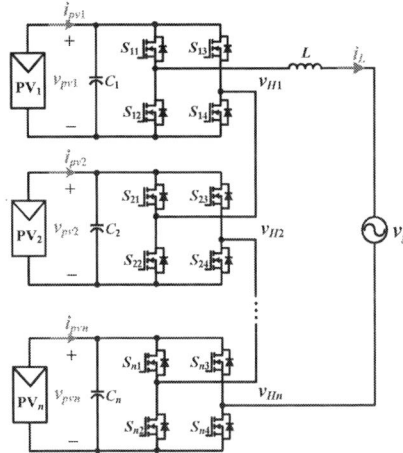

Fig. 1. Topology of single Phase cascaded H-bridge PV Inverter

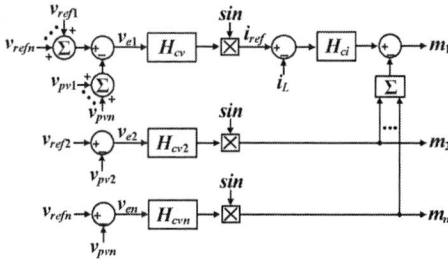

Fig. 2. Master-and-slave-type control strategy

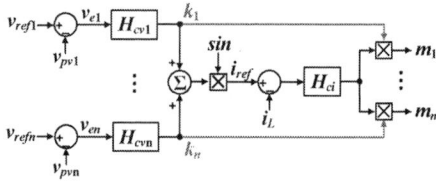

Fig. 3. The conventional non-master-and-slave-type control strategy with n voltage feedback loops and n feedback-based weighting factors

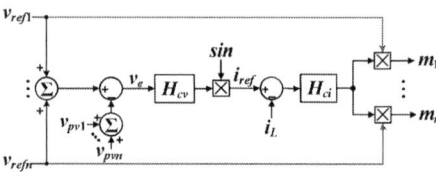

Fig. 4. The proposed non-master-and-slave-type control strategy with 1 voltage feedback loop and n feed-forward-based weighting factors

The control strategy proposed in [7] includes one total voltage feedback loop and n-1 individual voltage loops. Besides, the individual voltage balance control method presented in [8] has n individual voltage feedback controllers and n individual power feedback controllers. These existed voltage balancing control strategies all comprise at least n voltage feedback loops and their structures are very complex.

For the independent control of the cascaded multilevel PV inverter with different DC links, the state-of-the art control strategies are the master-and-slave type and the non-master-and-slave type. The master-and-slave type control strategy [9] consists of a total DC link voltage feedback loop controller and n-1 individual voltage loop controllers shown in Fig. (2). In particular, this strategy's n voltage loops are not completely equal due to the master-and-slave structure. The conventional non-master-and-slave type control strategy [10] shown in Fig. 3 comprises n DC voltage loop controllers and n feedback-loop-based weighting factors. Thus it realizes equal control for the DC link voltages. Actually, the concept of this weighting-factor-based control strategy is to get n modulation signals from the shared control signal (output of the current controller) multiplying with n weighting factors. However, the weighting factors are got by the outputs of the voltage loop compensators. This means that weighting factor generation mechanism will be fully coordinated with the n DC voltage feedback controllers.

In this paper, an improved weighting-factor-based control strategy for the independent and equal control of the DC link voltages has been proposed. This strategy consists of n feed-forward-based weighting factors and only one total DC link voltage feedback loop. It should be noted that these weighting factors are independent of the output of the voltage controller. These features significantly simplify the complexity of the control structure compared with the conventional methods. This control strategy could be extended to most of the other applications of cascaded multilevel converter system including identical and different DC links. For the current loop, a conventional PR compensator is adopted to inject the perfect sinusoidal current to the grid.

This paper is organized as follows. First of all, the system modeling and the description of control strategy are presented in Section II. Section III addresses the design criteria of the controller. Section IV shows the simulation and experimental results to demonstrate the independent MPPT and gird current shaping capabilities with the proposed DC link voltage control strategy. Finally, Section V gives the conclusions of this work.

II. SYSTEM MODELING AND PROPOSED CONTROL STARTEGY DESCRIPTION

A. Averaged Modeling of Cascaded H-bridge Inverter

In terms of the averaged switch modeling which removes the switching frequency ripple, the cascaded H-bridge inverter can be characterized in Fig. 5 as the system averaged model. The modulation signal for the kth H-bridge denoted as $m_k(t)$ is from -1 to 1. Because of the series connection of the outputs of the H-bridges, the controlled current sources $m_k(t)<i_L(t)>_T$ (k=1...n) share the same part $<i_L(t)>_T$. Besides, the controlled voltage sources $m_k(t)<v_{pvk}(t)>_T$ (k=1...n) are in series with the inductor and the grid. Accordingly, the system model can be described with the following two equations.

$$\left\langle i_{pvk}(t) \right\rangle_T + C_k \frac{d\left\langle v_{pvk}(t) \right\rangle_T}{dt} = m_k(t)\left\langle i_L(t) \right\rangle_T, \quad k=1...n \quad (1)$$

$$\left\langle v_g(t) \right\rangle_T + L \frac{d\left\langle i_L(t) \right\rangle_T}{dt} = \sum_{k=1}^{n} m_k(t)\left\langle v_{pvk}(t) \right\rangle_T \quad (2)$$

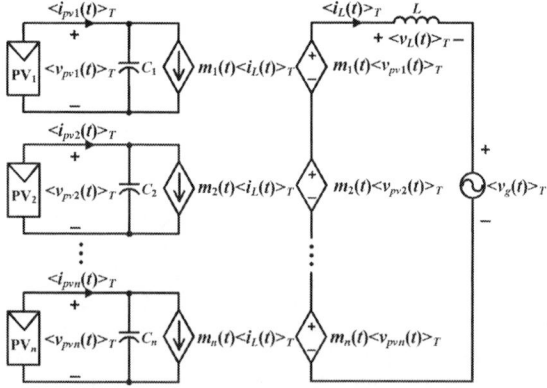

Fig. 5. Average model of the cascaded multilevel PV inverter system

978-1-4673-9551-9/16 $31.00 © 2016 IEEE

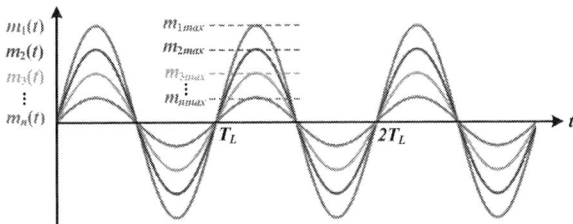

Fig. 6. Averaged modulation signals of multiple H-bridges

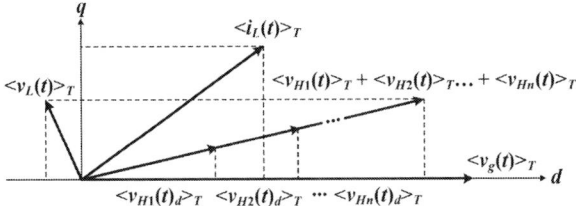

Fig. 7. Phasor diagram based on the system averaged model

Because the individual DC voltages is controlled equally and independently, if the system is stable, $m_k(t)$ ($k=1...n$) should be the same shape while the difference is the amplitude m_{kmax}. Fig. 6 and Equation (3) have illustrated this feature.

$$m_k(t) = m_{k\max}\sin(\omega_L t + \theta_d), \qquad k = 1...n \qquad (3)$$

The ω_L and T_L is the angular frequency and the period of the AC grid voltage. Besides, θ_d is the phase angle of $m_k(t)$.

B. Derivation of The Ratio between Multiple Modulation Signals $m_k(t)$

The averaged output voltage of each H-bridge is written as:

$$\left\langle v_{Hk}(t)\right\rangle_T = m_k(t)\left\langle v_{pvk}(t)\right\rangle_T, \qquad k = 1...n \qquad (4)$$

Because the outputs of the H-bridges are connected in series, assuming the power factor is 1, the ratio between $<v_{Hk}(t)>_T$ is equal to the ratio between $<P_{Hk}(t)>_T$ the averaged output power of the H-bridge ($k=1...n$) at steady state.

$$\left\langle v_{H1}(t)\right\rangle_T :....: \left\langle v_{Hn}(t)\right\rangle_T = \left\langle p_{H1}(t)\right\rangle_T :....: \left\langle p_{Hn}(t)\right\rangle_T \qquad (5)$$

Removing the periodical energy buffered in the DC capacitor, the ratio is equal to the ratio between $<P_{pvk}(t)>_T$ the averaged PV power.

$$\left\langle v_{H1}(t)\right\rangle_T :....: \left\langle v_{Hn}(t)\right\rangle_T = \left\langle p_{pv1}(t)\right\rangle_T :....: \left\langle p_{pvn}(t)\right\rangle_T \qquad (6)$$

Then assuming the power factor is less than 1, as shown in Fig. 7 the phasor diagram of the system, the ratio is determined as

$$\left\langle v_{H1}(t)\right\rangle_T :....: \left\langle v_{Hn}(t)\right\rangle_T = \left\langle v_{H1}(t)_d\right\rangle_T :....: \left\langle v_{Hn}(t)_d\right\rangle_T \qquad (7)$$

However $<v_{Hk}(t)_x>_T$ is still determined by the averaged output active power of the H-bridge. As a result, at steady state, the ratio between $<v_{Hk}(t)>_T$ is always equal to the ratio between $<P_{pvk}(t)>_T$ the averaged PV power when PF<=1 as (6).

Due to (4), the ratio between modulation signals $m_k(t)$

$$m_1(t):...:m_n(t) = \frac{\left\langle v_{H1}(t)\right\rangle_T}{\left\langle v_{pv1}(t)\right\rangle_T}:....:\frac{\left\langle v_{Hn}(t)\right\rangle_T}{\left\langle v_{pvn}(t)\right\rangle_T} \qquad (8)$$

Substituting by (6), equation (8) is rewritten as (9) or (10).

$$m_1(t):...:m_n(t) = \frac{\left\langle p_{pv1}(t)\right\rangle_T}{\left\langle v_{pv1}(t)\right\rangle_T}:....:\frac{\left\langle p_{pvn}(t)\right\rangle_T}{\left\langle v_{pvn}(t)\right\rangle_T} \qquad (9)$$

$$m_1(t):...:m_n(t) = \left\langle i_{pv1}(t)\right\rangle_T :....: \left\langle i_{pvn}(t)\right\rangle_T \qquad (10)$$

Therefore, the ratio between multiple modulation signals at steady state derived as (9), (10) is proportional to the PV power and in-proportional to PV voltage.

C. Description of The Proposed Control Strategy

The conventional weighting-factor-based control strategy utilizes the outputs of the voltage controllers to create the weighting factors which are dependent on the dynamics of the voltage loops. However, according to (9), this paper proposes to use the individual PV power and reference $v_{ref}(t)$ to directly generate the feed-forward signals $I_k(t)$ as the weighting factors shown in (11).

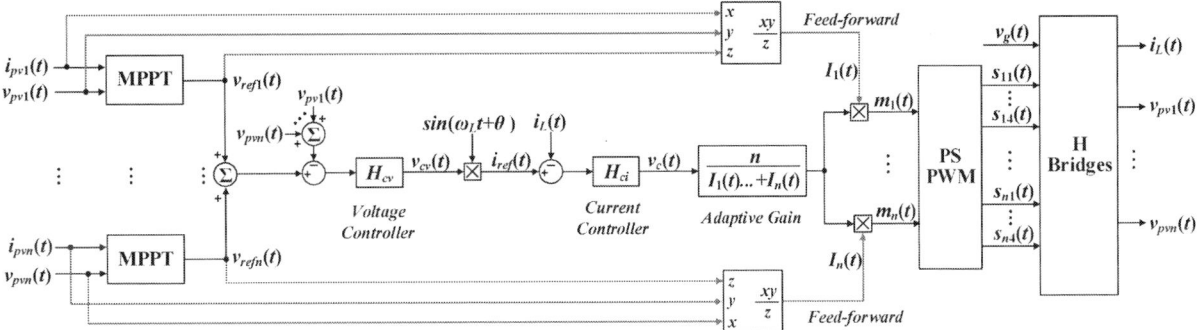

Fig. 8. Detailed description of the improved weighting-factor-based control strategy for the cascaded H-bridge PV inverter

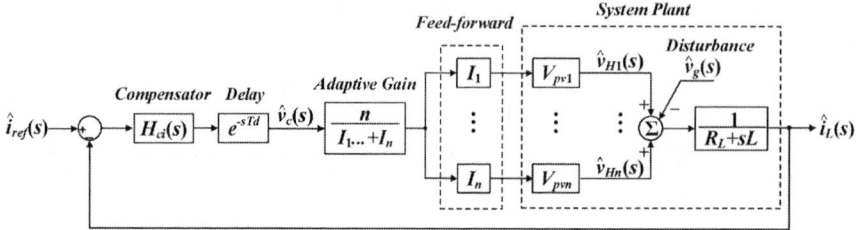

Fig. 9. Current loop diagram of the proposed control strategy for the cascaded H-bridge PV inverter

$$I_k(t) = \frac{p_{pvk}(t)}{v_{refk}(t)} = \frac{v_{pvk}(t)i_{pvk}(t)}{v_{refk}(t)} \qquad (11)$$

These feed-forward signals will directly force the ratio between modulation signals as (12), (13).

$$m_1(t):...:m_n(t) = I_1(t):...:I_n(t) = \frac{p_{pv1}(t)}{v_{ref1}(t)}:...:\frac{p_{pvn}(t)}{v_{refn}(t)} \qquad (12)$$

$$m_1(t):...:m_n(t) = \frac{v_{pv1}(t)i_{pv1}(t)}{v_{ref1}(t)}:...:\frac{v_{pvn}(t)i_{pvn}(t)}{v_{refn}(t)} \qquad (13)$$

With combing (9) with (12), at steady state, the ratio between $<v_{pvk}(t)>_T$ is forced to be the same with the ratio between voltage references $v_{refk}(t)$ as shown in (15).

$$\frac{\left\langle p_{pv1}(t)\right\rangle_T}{\left\langle v_{pv1}(t)\right\rangle_T}:...:\frac{\left\langle p_{pvn}(t)\right\rangle_T}{\left\langle v_{pvn}(t)\right\rangle_T} = \frac{p_{pv1}(t)}{v_{ref1}}:...:\frac{p_{pvn}(t)}{v_{refn}} \qquad (14)$$

$$\left\langle v_{pv1}(t)\right\rangle_T:...:\left\langle v_{pvn}(t)\right\rangle_T = v_{ref1}(t):...:v_{refn}(t) \qquad (15)$$

Therefore, the essential features of the proposed control strategy are: (a). the only one feedback voltage loop control the sum of the DC link voltages for stable operation and zero tracking error; (b) the n feed-forward-based weighting factors force the ratio between DC link voltages to be the same as the ratio between the references at steady state. With the following n equations (1 for sum, n-1 for the ratios) shown in (16), the n DC link voltages should be controlled independently and stably with zero steady state error.

$$\begin{cases} \sum_{k=1}^{k=n}\left\langle v_{pvk}(t)\right\rangle_T = \sum_{k=1}^{k=n}v_{refk}(t) \\ \left\langle v_{pv1}(t)\right\rangle_T:...:\left\langle v_{pvn}(t)\right\rangle_T = v_{ref1}(t):...:v_{refn}(t) \end{cases} \qquad (16)$$

Besides the feed-forward parts, there is an adaptive gain $1/I(t)$ in the control loop described in (17). It is the reciprocal of the average of $I_k(t)$. According to (11), the physical meaning of the weighting factor $I_k(t)$ is the PV current at steady state. Therefore, the variation of the irradiation and temperature will cause the variation of $I_k(t)$. With that adaptive gain, the system loop gain will be adaptive to the variation of $I_k(t)$. This feature will be included in the section of current controller's design.

$$\frac{1}{I(t)} = \frac{n}{I_1(t)...+I_n(t)} \qquad (17)$$

In addition, for the current control loop, the output of the voltage controller provides the maximum value of the current reference. The output of the current loop compensator is the shared control signal to be multiplied with n weighting factors to produce n modulation signals. The PWM modulator here is phase-shift PWM modulator to generate multilevel waveform, and its amplitude is V_m. The individual MPPT block which gives the voltage references adopts the algorithm of adaptive P&O [11]. This improved MPPT algorithm utilizes the variable step size to accelerate the MPPT speed and optimize the MPPT efficiency. Moreover, through changing the phase angle θ of the current reference, this control strategy also can support the inverter to accurately generate the reactive power. And this mechanism is similar to the conventional control strategies for single phase grid-connected inverter.

III. CONTROLLER DESIGN

The prosed control strategy for the cascaded multilevel H-bridge inverter still has a cascaded dual loop structure similar to the single H-bridge inverter. There is a basic assumption for this cascaded control structure that the bandwidth of the current loop is much larger (usually more than ten times) than the bandwidth of the voltage loop.

A. Current Loop Controller Design

The current loop small signal diagram of the system is shown in Fig. 9. Because the current loop is much faster than the voltage loop, the DC link voltages are taken as constants when constructing the current loop. Thus, DC link voltages are put as DC gains V_{pvk} in the loop. The feed-forward parts are all substitutes as DC values as DC gains I_k.

$$I_k = \frac{P_{pvk}}{V_{refk}} = \frac{V_{pvk}I_{pvk}}{V_{refk}} \qquad (18)$$

Assuming $V_{pvk} = V_{refk}$ at steady state,

$$I_k = I_{pvk} \qquad (19)$$

It means that the feed-forward gains are determined by the PV currents. To evaluate the adaptive gain $1/I$ denoted in (20) (I is the average value of I_k), the gain from the control signal to the outputs of the H-bridges is derived as (21).

978-1-4673-9551-9/16 $31.00 © 2016 IEEE

$$\frac{1}{I} = \frac{n}{I_{1k}...+I_{nk}} \quad (20)$$

$$\frac{\hat{v}_{H1}(s)...+\hat{v}_{Hn}(s)}{\hat{v}_c(s)} = \frac{n}{I_1(s)...+I_n(s)}(I_1(s)V_{pv1}...+I_n(s)V_{pvn}) \quad (21)$$

The variations of the PV voltages are small even when the irradiations and temperatures change a lot. Therefore, there is an assumption shown in (22). Equation (21) can be rewritten as (23), then that gain equals to a constant value nV_{pv} with small variance. As a result, with the adaptive gain $1/I$, the loop gain is adaptive to the variations of the PV currents.

$$V_{pv1} \approx V_{pv2}... \approx V_{pvn} \approx V_{pv} \quad (22)$$

$$\frac{\hat{v}_{H1}(s)...+\hat{v}_{Hn}(s)}{\hat{v}_c(s)} \approx nV_{pv} \quad (23)$$

For the design of the current loop compensator, compared with PI compensator, the PR compensator shown in (24) is more suitable because it provides near infinite gain at grid AC frequency. This feature significantly improves the current loop dynamic response and achieves near zero reference tracking error and disturbance error at the AC frequency [12].

$$H_{PR}(s) = K_{p1}(1+\frac{1}{\tau_{i1}}\frac{s}{s^2+\omega_o^2}) \quad (24)$$

The current loop gain is constructed as (25). T_d is the time delay that is assumed to be equal to the switching period due to the AD conversion, calculation and PWM modulation.

$$\frac{\hat{i}_L(s)}{\hat{i}_{ref}(s)} = H_{PR}(s)e^{-sT_d}I(\sum_{k=1}^{k=n}I_kV_{pvk})(\frac{1}{R_L+sL}) \quad (25)$$

The parameters are listed as follow: L=300µH, R_L=20mΩ, n=3, f_s=10kHz. In this paper, each PV cell is a single PV panel, and let $V_{pv1}=V_{pv2}=V_{pv3}$=30V. Let K_{p1}=0.019, τ_{i1}=0.00115. Bode plot of the current loop gain is shown in Fig. 10. The band-width is 1.4kHz, and the phase-margin is 45°. Moreover, with the adaptive gain, there is no change for the current loop gain when the irradiation is changed from 1000W/m² to 150W/m². However, if without the adaptive gain, the band-width will be reduced to 0.18kHz. For current loop, weak irradiation for all the panels at the same time is the worst case, because the loop gain is much lower. The other cases, like being under the mismatch irradiations, are better than that one.

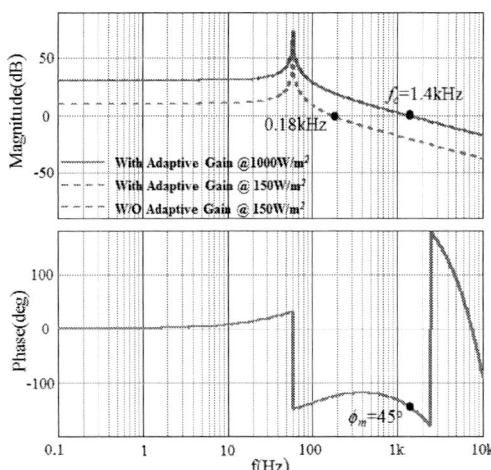

Fig. 10. Bode plot of current loop with PR compensator

B. Voltage Loop Controller Design

As described in II, the DC link voltages are independently controlled by the coordination of one total voltage feedback loop and n feed-forward signals. Therefore, for this voltage feedback loop, there is only one PI compensator needed here as shown in Fig. 11.

To model the voltage loop, there are two basic assumptions. Because of the wide-bandwidth inner current loop, the first assumption is that the AC current is controlled ideally with fast tracking speed and zero tracking error. With this assumption, the inductor current should be regarded as equal to the current reference as (26).

$$\hat{i}_L(s) = \hat{i}_{ref}(s) \quad (26)$$

Another assumption for the modeling of the voltage loop is that not only the switching frequency harmonics but also the two times AC line frequency variations of the DC voltage and current should be removed. This low frequency small signal model is effective under AC line frequency. The small signal equivalent circuit of the H-bridge is shown in Fig. 11 [13]. It should be noted that the linearized model of the PV cell (the string or a panel) is a voltage source in series with an internal resistance [14]. The dynamic model of the PV cell is illustrated in (27).

$$\hat{v}_{pvk}(s) = \hat{i}_{ref}(s)R_{pvk} \quad (27)$$

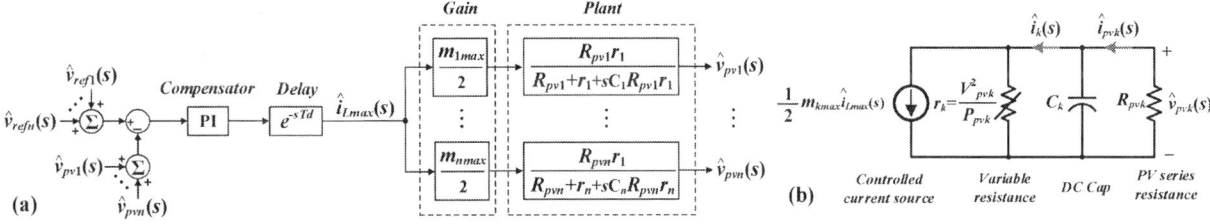

Fig. 11. (a). Voltage loop diagram of the proposed control strategy; (b). Low frequency small signal equivalent circuit of the kth H-bridge

In this model, for each H-bridge, there is a controlled current source that shares the common part $i_{Lmax}(s)$ the variance of the maximum inductor current with other H-bridges due to the series connection of the outputs of the H-bridges. The gain of the controlled current source is half of the maximum value $m_{kmax}/2$ of the individual modulation signal. Thus the circuit's transfer function is derived as (28).

$$\frac{\hat{v}_{pvk}(s)}{\hat{i}_{L\max}(s)} = \frac{m_{k\max}}{2}\frac{R_{pvk}r_k}{R_{pvk}+r_k+sC_kR_{pvk}r_k} \quad (28)$$

And r_k is the variable resistance in this model,

$$r_k = \frac{V^2_{pvk}}{P^2_{pvk}} \quad (29)$$

Both of V_{pvk} and P_{pvk} are the DC value of $v_{pvk}(t)$ and $p_{pvk}(t)$.

The voltage loop is constructed as Fig. 10 based on the low frequency model of the inverter. With the PI controller in (30), the loop gain of the system is built as (31) with $C_1=...C_n=C$.

$$H_{PI}(s) = K_{p2}\frac{1+s\tau_{i2}}{s\tau_{i2}} \quad (30)$$

$$\frac{\sum\limits_{k=1}^{k=n}\hat{v}_{pvk}(s)}{\sum\limits_{k=1}^{k=n}\hat{v}_{refk}(s)} = H_{PI}(s)e^{-sT_d}\sum\limits_{k=1}^{k=n}[\frac{m_{k\max}}{2}(\frac{R_{pvk}r_k}{R_{pvk}+r_k+sCR_{pvk}r_k})] \quad (31)$$

As referred in [14], the PV resistance R_{pvk} is varied from 1Ω to 300Ω with the various operating points on the P-V curve of PV panel. The resistance R_{pvk} around maximum power point is 3Ω to 15Ω. Besides, the value of r_k and m_{kmax} will also be changed by the variation of PV powers and DC link voltages. Let $V_g=50sin(\omega t)$, $C_k=C=13$mF, $K_{p2}=0.5$, $\tau_{i2}=0.01$. In Fig. 12, the operating points are changed simultaneously, R_{pvk} could be 1Ω, 2Ω, 5Ω, 15Ω and 300Ω. Another typical case is shown in Fig. 13. The PV$_1$ and PV$_2$ are operating around the maximum power point ($R_{pv1}=R_{pv2}=10$Ω). Then R_{pv3} is from 1Ω to 300Ω.

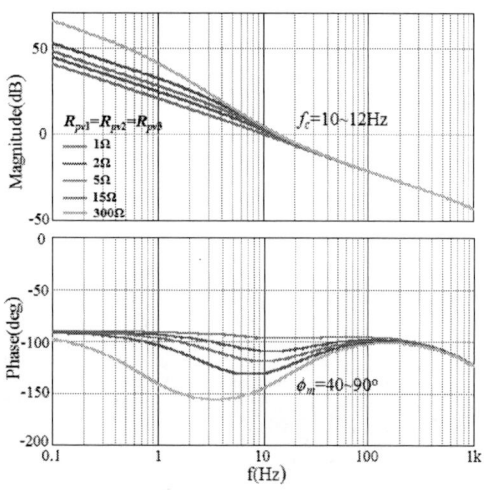

Fig. 12. Bode plot of voltage loop with various R_{pvk} (k=1,2,3)

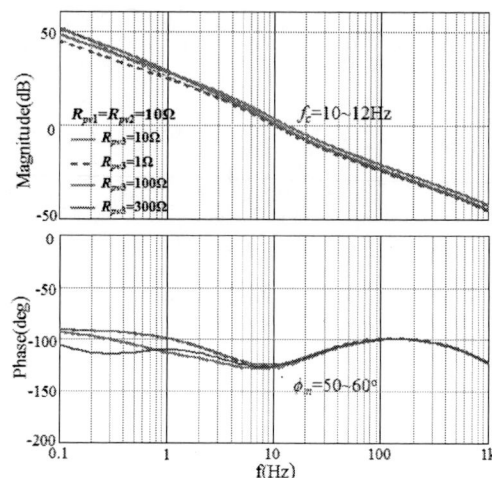

Fig. 13. Bode plot of voltage loop with various R_{pv3}

In the above worst cases, the bandwidth is kept from 10~12Hz and the phase margin is from 40~90°.

IV. SIMULATION AND EXPERIMENTAL RESULTS

A. Simulation Results

The simulation is taken in PSIM 10.0, and the parameters of the system are the same with that in last section: $L=300\mu$H, $C_k=13$mF, $f_s=10$kHz, $K_{p1}=0.019$, $\tau_{i1}=0.00115$, $K_{p1}=0.5$, $\tau_{i1}=0.01$. For Fig. 14~16, n=3, $V_g=50\sin(\omega t)$.

As shown in Fig. 14, V_{ref1}= 34V, V_{ref2}=30V, V_{ref3}=26V, the irradiation of PV$_2$ is 800W/m². The Fig. 14 shows that the DC link voltages and the filtered grid current track the references with zero errors. Besides, the individual modulation signals are varied with different power and DC link voltages shown in Fig.14. As shown in Fig. 15, the irradiation for the three PV panels remains constant while the V_{ref1} steps up and V_{ref3} steps down with 2V simultaneously. DC link voltages independently track the references fast in several grid cycles. As shown in Fig. 16, the voltage references keep constant, while the irradiation for PV$_2$ and PV$_3$ is changed to 700W/m² and the temperature for PV$_3$ is changed to 50°. Then all the DC link

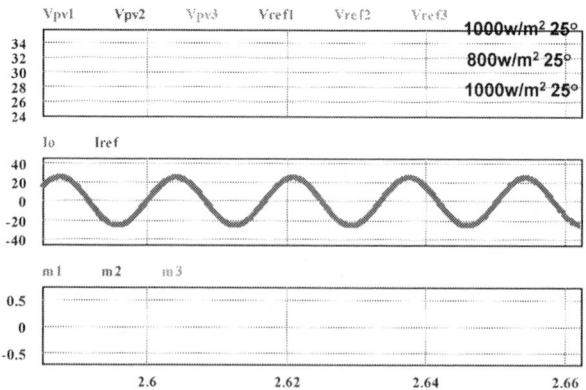

Fig. 14. Steady state simulation results with different DC link voltages

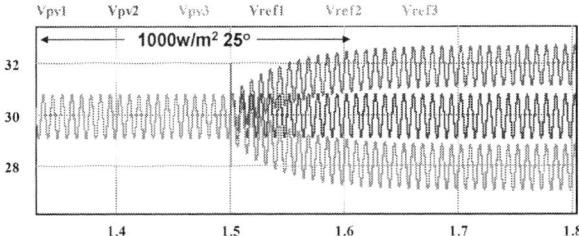

Fig. 15. Simulation results of DC link control with step change references

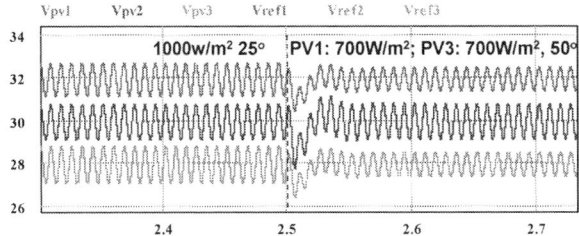

Fig. 16. Simulation results of DC link control with changed conditions

voltages can recover from the change of the conditions during several grid cycles.

For Fig. (17), $n=2$, $V_g=36sin(\omega t)$. The irradiation of PV_2 is changed from 1000 W/m^2 and 25° to 600W/m^2 and 50°C while PV_1's conditions remain the same. As shown in Fig. 17, the two PV panels can keep tracking the maximum power points within several step sizes at that conditions with this proposed independent DC link control method. With this adaptive P&O, during the dynamic, the large adaptive voltage step sizes accelerate the MPPT speed. Besides, at the steady state, there is a high MPPT efficiency because the voltage step size is close to zero.

B. Experimental Results

The experiments are taken with 2 PV simulators, and the other parameters are the same with the simulation. The steady state waveform of the system is shown in Fig. 18. It illustrates the multilevel waveform and sinusoidal current tracking of the cascaded multilevel H-bridge inverter.

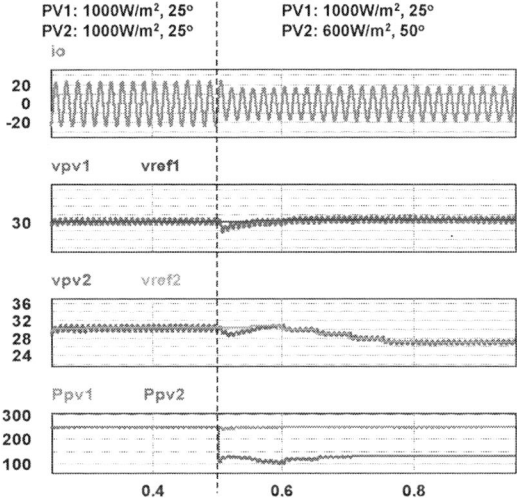

Fig. 17. Simulation results of individual MPPT based on the independent DC link control with changed irradition and temperature conditions

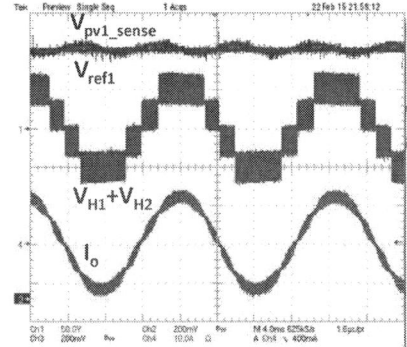

Fig. 18. Experimental Steady state waveforms of the system.

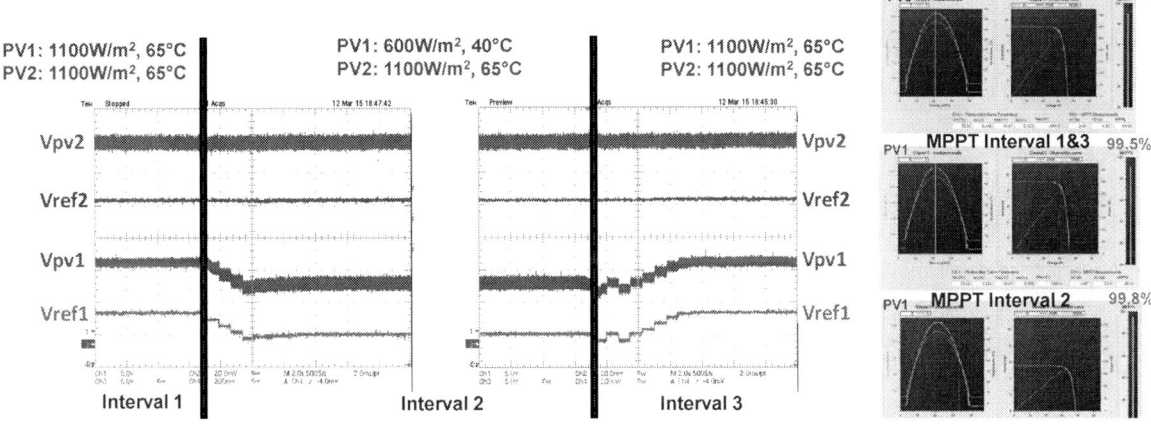

Fig. 19. Experimental results of the dynamic individual MPPT waveforms and steady state MPPT efficiency based on the independend DC link control.

The individual MPPT performance of the system is tested in Fig. 19. The conditions of PV panel 1 are changed from $1100W/m^2$ 65°C to $600W/m^2$ 40°C and then back to $1100W/m^2$ 65°C while the PV_2's conditions remain $1100W/m^2$ 65°C. V_{pv2} tracks the changing V_{ref2} while V_{pv1} is kept the same as shown in Fig. 19. As a result, within several steps, DC link voltage reaches the maximum power point. These dynamic DC link voltages waveforms show the stability and good performance of this proposed independent DC link voltage control strategy. The adaptive step sizes provided by the adaptive P&O MPPT algorithm are also shown in Fig. 19. At steady state, the step size is near to zero to keep the high MPPT efficiency.

In addition, the steady state MPPT efficiencies of PV_1 and PV_2 are higher than 99.5% during any interval. The highest MPPT efficiency is up to 99.8%. These results demonstrate the accuracy of the DC link control and the effects of the adaptive P&O MPPT.

V. CONCLUSIONS

This paper proposes an improved independent DC link control strategy for the single phase cascaded multilevel H-bridge PV inverter. The proposed control strategy consists only one total DC voltage feedback loop and n feed-forward-based weighting factors. The feedback loop controls the sum of the DC link voltages with zero tracking error, and the feed-forward signals force the ratio between DC link voltages to be the same with the ratio between the references at the steady state. In this cascaded multilevel topology, because the outputs of the H-bridges share the same output current, the multiple modulation signals have inherent ratios with each other. The proposed control strategy utilizes this feature, and directly generates the weighting factors by the DC link input powers and voltage references which are independent with the output of the voltage feedback loop. Compared with the conventional DC link voltage control strategies, the structure of the proposed one is much simplified. Besides, the bandwidth of the inner current loop remains constant due to the adaptive gain put in the loop even when there are large variations of conditions for PV cells. Moreover, this independent DC link control strategy can be extended to most of the applications with cascaded multilevel converters including identical and different DC links.

REFERENCES

[1] Kjaer, Soeren Baekhoej, John K. Pedersen, and Frede Blaabjerg. "A review of single-phase grid-connected inverters for photovoltaic modules." Industry Applications, IEEE Transactions on 41, no. 5 (2005): 1292-1306.

[2] Huang, Qingyun, Xu Yang, Frank Ren, Qian Ouyang, and Xiang Hao. "An improved constant on/off time control scheme for photovoltaic DC/DC MIC." InApplied Power Electronics Conference and Exposition (APEC), 2013 Twenty-Eighth Annual IEEE, pp. 738-743. IEEE, 2013.

[3] Enphase datasheet, "Enphase S280 Microinverter".

[4] Kouro, Samir, et al. "Recent advances and industrial applications of multilevel converters." Industrial Electronics, IEEE Transactions on 57.8 (2010): 2553-2580.

[5] Akagi, Hirofumi, Shigenori Inoue, and Tsurugi Yoshii. "Control and performance of a transformerless cascade PWM STATCOM with star configuration."Industry Applications, IEEE Transactions on 43, no. 4 (2007): 1041-1049.

[6] Fujii, Kenichi, and Rik W. De Doncker. "A novel DC-link voltage control of PWM-switched cascade cell multi-level inverter applied to STATCOM."Industry Applications Conference, 2005. Fourtieth IAS Annual Meeting. Conference Record of the 2005. Vol. 2. IEEE, 2005.

[7] Zhao, Tianjie, Guibin Wang, Surya Bhattacharya, and Alex Q. Huang. "Voltage and power balance control for a cascaded H-bridge converter-based solid-state transformer." Power Electronics, IEEE Transactions on 28, no. 4 (2013): 1523-1532.

[8] Barrena, Jon Andoni, Luis Marroyo, Miguel Á. Vidal, and José R. Torrealday Apraiz. "Individual voltage balancing strategy for PWM cascaded H-bridge converter-based STATCOM." Industrial Electronics, IEEE Transactions on 55, no. 1 (2008): 21-29.

[9] Villanueva, Elena, Pablo Correa, José Rodríguez, and Mario Pacas. "Control of a single-phase cascaded H-bridge multilevel inverter for grid-connected photovoltaic systems." Industrial Electronics, IEEE Transactions on 56, no. 11 (2009): 4399-4406.

[10] Chavarria, Javier, Domingo Biel, Francesc Guinjoan, Carlos Meza, and Juan J. Negroni. "Energy-balance control of PV cascaded multilevel grid-connected inverters under level-shifted and phase-shifted PWMs." Industrial Electronics, IEEE Transactions on 60, no. 1 (2013): 98-111.

[11] Lee, Kui-Jun, and Rae-Young Kim. "An Adaptive Maximum Power Point Tracking Scheme Based on a Variable Scaling Factor for Photovoltaic Systems." Energy Conversion, IEEE Transactions on 27, no. 4 (2012): 1002-1008.

[12] Shen, Guoqiao, et al. "A new feedback method for PR current control of LCL-filter-based grid-connected inverter." Industrial Electronics, IEEE Transactions on 57.6, pp 2033-2041, 2010.

[13] Erickson, Robert W., and Dragan Maksimovic. Fundamentals of power electronics. Springer Science & Business Media, 2007.

[14] Xiao, Weidong, William G. Dunford, Patrick R. Palmer, and Antoine Capel. "Regulation of photovoltaic voltage." Industrial Electronics, IEEE Transactions on 54, no. 3 (2007): 1365-1374.

New Active Damping Method for *LCL* Filter Resonance Based on Two Feedback System

Mahmoud A. Gaafar[1,2], Gamal M. Dousoky[1,3], Masahito Shoyama[1]

[1] Graduate School of Information Science & Electrical Engineering, Kyushu University, Fukuoka, 819-0395, Japan
[2] Electrical Engineering Dept., Aswan University, Aswan, 81542, Egypt
[3] Electrical Engineering Dept., Minia University, Alminia, 61517, Egypt
gaafar@ckt.ees.kyushu-u.ac.jp, dousoky@mu.edu.eg, shoyama@ees.kyushu-u.ac.jp

Abstract— Two feedback active damping technique for *LCL* filter resonance is proposed using the capacitor voltage and the grid current. The proposed method is derived in continuous time domain with discussion for its discrete implementation. Based on the proposed method, excitation of un-stable open loop pools, which implies non-minimum phase behavior, can be avoided over a wide range of resonant frequencies; Co-design procedures for both the active damping loops along with the fundamental current regulator are presented. Numerical example and simulation work are presented to confirm the performance of the proposed method at different resonant frequencies.

Keywords—Active damping, *LCL* filter resonance, high pass filter, grid.

I. INTRODUCTION

PWM converters are often used with L or *LCL* filters to connect the distributed energy sources to the utility grid; *LCL* filters are more interesting over *L* filters due to its higher attenuation for switching harmonics with lower weight and volume [1]; however, due to the resonance hazard, damping techniques are required. Many passive damping techniques have been introduced in the literature [1]; however it creates additional power losses and increases the filter size. Thus active damping by modifying the control algorithm is preferred where there is no additional power losses and more flexibility is obtained [2].

For resonant frequencies more than one-sixth of the sampling frequency, it was shown in [3] that system stability can be maintained with single grid current control loop without additional active damping. However, this technique is not suitable especially in weak grids where the resonant frequency varies significantly due to grid inductance variation; this, in turns, give rise instability if there is no external active damping is employed [4].

Number of active damping methods has been presented in the literature [4]-[15]. Using a cascaded filter in the current control loop was discussed in [5] and [6]; however, the bandwidth was largely reduced. Using state variables feedback is more preferred. It was proved that a proportional feedback from the filter capacitor current is equivalent to a shunt resistor with the filter capacitor [7]; Comprehensive study for this method was introduced in [3], [4], [8]. It was shown that excitation of un-stable open loop poles is mandatory for resonant frequencies more than one-sixth of the sampling

frequency to stabilize the closed loop system [4], [8]; This non-minimum phase can decline the overall system performance. This limit is extended to one-third of the sampling frequency by using a high pass filter (HPF) from the capacitor current [8]. However, a high precision current sensor or a complicated observer loop is required [9], [10]. A differentiation of the capacitor voltage can be used to produce the damping effect; however, this method causes noise amplification; to overcome this issue, a lead-lag compensator is used to behave as a differentiator around the resonant frequency [11], [12], [13]. However, it was indicated in [11] that the lead-lag compensator can behave effectively as a differentiator over limited range of resonant frequencies between 1/3.2 and 1/3.4 of the sampling frequency. An inner feedback loop from the grid current can be used for the damping purpose without the need to additional sensors or complicated control algorithms. Ideally this needs an s^2 term in the grid current feedback loop; however it cannot be implemented practically due to the noise amplification. A high pass filter (HPF) from the grid current has been used in [14], [15] to replace the s^2 term, however the design procedures is complicated especially if the co-design of this HPF along with the fundamental current regulator is considered.

To overcome the above-mentioned limitations in the previous methods, this work proposes a new active damping method for *LCL* filter resonance. The proposed method employs two feedback loops from the capacitor voltage and the grid current. The proposed method can behave effectively over a wide range of resonant frequencies. Moreover, straight-forward co-design steps for the active damping loops along with the fundamental current regulator can be used. Also, the unstable open loop behavior can be avoided over a wide range of resonant frequencies; this improves the system robustness and allows reducing the resonant peak below unity without losing the system stability. Virtual flux technique can be employed to estimate the capacitor voltage [13] for reduced number of sensors, however this is not considered here.

The paper is organized as follow: the proposed active damping method is derived in section II in continuous time domain; section III discuss the discrete implementation for the proposed method with presenting co-design procedures for the active damping loops along with the fundamental current regulator. Section IV presents a numerical example to verify the system performance at different resonant frequencies, also a simulation work is carried out; Finally, section V presents the conclusion.

978-1-4673-9551-9/16 $31.00 © 2016 IEEE

II. PROPOSED ACTIVE DAMPING METHOD

A. System Description

A single phase inverter connected to the grid through LCL filter is shown in Fig. 1. Using a proportional feedback (H_d) of the capacitor current as an active damping for filter resonance, Fig. 2 shows the control system block diagram. Proportional resonant (PR) regulator with a transfer function $G_c(s)$ – expressed in (1) – is used as a fundamental current regulator.

$$G_c(s) = K_p + \frac{2\omega_i K_r s}{s^2 + 2\omega_i s + \omega_o^2} \tag{1}$$

where ω_o and ω_i, in rad/sec, are the fundamental grid frequency and the resonant part bandwidth respectively.

Using capacitor current feedback, the actively damped filter transfer function can be expressed in (2) where $\omega_{res} = \sqrt{(L_i + L_g)/(CL_iL_g)}$ is the filter resonant frequency.

$$F_{ad}(s) = \frac{1}{CL_gL_iS\left(s^2 + \frac{H_d}{L_i}S + \omega_{res}^2\right)} \tag{2}$$

Using the signal flow graph manipulation, the system shown in Fig. 2 is manipulated in Figs. 3 (a) to 3 (d). From Fig. 3 (c), it is shown that the capacitor current feedback is equivalent to

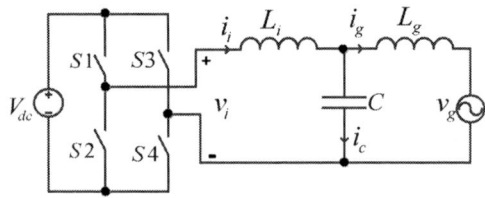

Fig. 1. Single phase inverter connected to the grid through LCL filter.

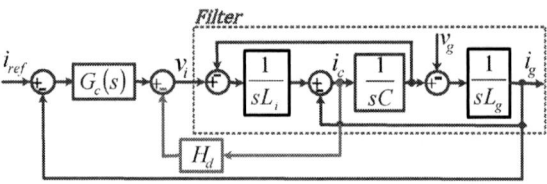

Fig. 2. System block diagram with capacitor current feedback.

using a three feedback loops from the input modulated voltage (v_i), the capacitor voltage and the grid current. This system is further manipulated in Fig. 3(d) where the feedback loop of the modulated voltage is augmented as a cascade transfer function of $G_h(s)$ expressed in (3) with $\omega_h = H_d/L_i$.

$$G_h(s) = \frac{1}{\omega_h} \cdot \frac{s}{1 + s/\omega_h} \tag{3}$$

This is a high pass filter (HPF) with cut off frequency of ω_h. Typical rang for ω_h can be determined by expressing the actively damped filter transfer function in (2) in terms of ω_h and writing it in a standard form as in (4)

$$F_{ad}(s) = \frac{1}{CL_gL_iS(s^2 + \omega_h S + \omega_{res}^2)}$$
$$= \frac{1}{CL_gL_iS(s^2 + 2\zeta\omega_{res}S + \omega_{res}^2)} \tag{4}$$

The typical value for the damping ratio ζ is around 0.7 [16], so the typical range for ω_h can be expressed as in (5)

$$\omega_{res} < \omega_h < 2\omega_{res} \tag{5}$$

Note that the feedback from the capacitor voltage is an integrator with a time constant of L_i, this integrator is denoted as $G_i(s)$ in Fig. 3 (d).

Due to the presence of the HPF (G_h) in the main control loop, it is expected for the system disturbance rejection capability to deteriorate. This can be verified easily by comparing the output transfer function to the disturbance input (grid voltage) for both the basic system (shown in Fig. 2) and the final manipulated one (shown in Fig. 3 (d)) taking into account the typical range of ω_h in (5).

B. Proposed Active Damping system

The proposed system is shown in Figure 4, and it is derived in two steps as follow:

1. The HPF, $G_h(s)$, is eliminated from the main control loop and inserted only in the active damping feedback loop. By multiplying it with the coefficient H_d, the resulted transfer function, G_{ad} (s), is still a HPF and expressed in (6). From Fig. 4, the actively damped filter ($F_{new}(s)$) is expressed as in (7) with $G_{ig}(s)$ expressed in (8). Substituting (8) into (7), $F_{new}(s)$ is expressed in (9).

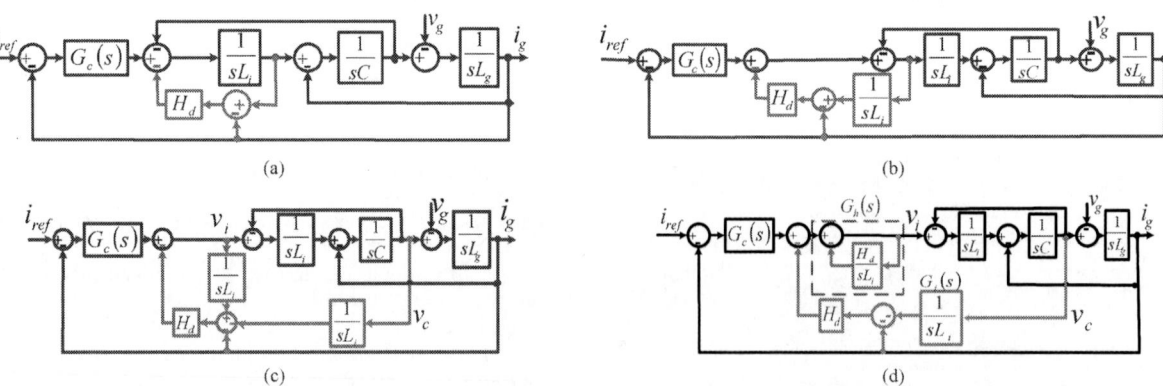

Fig. 3. Manipulation of the well-known active damping method which uses a proportional feedback of capacitor current.

978-1-4673-9551-9/16 $31.00 © 2016 IEEE   2736

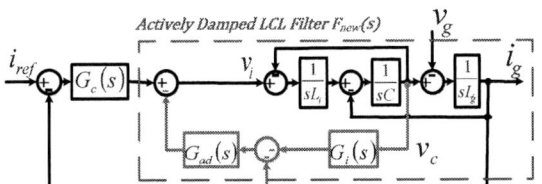

Fig. 4. System block diagram of the proposed system.

$$G_{ad}(s) = \frac{sL_i}{1+s/\omega_h} \qquad (6)$$

$$F_{new} = \frac{G_{ig}(s)}{1-G_{ad}(s)G_{ig}(s)\left(1+sL_gG_i(s)\right)} \qquad (7)$$

$$G_{ig}(s) = \frac{I_g(s)}{V_i(s)} = \frac{1}{L_iL_gCs(s^2+\omega_{res}^2)} \qquad (8)$$

$$F_{new}(s) = \frac{(1+s/\omega_h)}{s^2\left(\frac{CL_iL_g}{\omega_h}s^2+CL_iL_gs+\frac{(L_i+L_g)}{\omega_h}\right)} \qquad (9)$$

Introducing an s^2 term in $F_{new}(s)$ will cause a constant phase of -180° in the open loop bode plot; this, in turns, deteriorates the phase margin dramatically. So, more modifications are necessary.

2. Both the $G_{ad}(s)$ gain and the time constant of $G_i(s)$ are expressed in terms of a new variable (K_d) as in (10) and (11) respectively. Substituting by (10) and (11) into (7), $F_{new}(s)$ is re-written as in (12).

$$G_{ad}(s) = \frac{(K_d-L_g)s}{1+s/\omega_h} \qquad (10)$$

$$G_i(s) = \frac{1}{(K_d-L_g)s} \qquad (11)$$

$$F_{new} = \frac{(1+s/\omega_h)}{CL_iL_gs(s^2+\omega_r^2)(1+s/\omega_h)-sK_d}$$

$$= \frac{(1+s/\omega_h)}{\frac{CL_iL_g}{\omega_h}s^4+CL_iL_gs^3+\frac{L_i+L_g}{\omega_h}s^2+(L_i+L_g-K_d)s} \qquad (12)$$

Using Routh's criteria, to guarantee a stable open loop operation, and hence minimum phase behavior, K_d has to follow the constraint in (13).

$$0 < K_d < (L_i + L_g) \qquad (13)$$

To generalize the analysis, K_d is expressed in terms of its maximum limit $(L_i + L_g)$ as in (14).

$$K_d = \beta_d \left(L_i + L_g\right) \qquad (14)$$

with $0 < \beta_d < 1$ for stable open loop system.

Substituting (14) into (12), the final proposed form for the actively damped filter is expressed in (15).

$$F_{new}(s) = \frac{(1+s/\omega_h)}{CL_iL_gs(s^2+\omega_{res}^2)(1+s/\omega_h)-s\beta_d\left(L_i+L_g\right)} \qquad (15)$$

III. DISCRETE IMPLEMENTATION

A. System Discretization

The system discrete implementation is shown in Fig. 5 where the DSP delay is modeled by one sample delay and the discrete PR controller $G_c(z)$ is modeled using the Tustin approximation with pre-warping at the fundamental frequency as expressed in (16).

$$G_c(z) = K_p + K_r \frac{sin(\omega_oT_s)}{2\omega_o}\frac{z^2-1}{(z^2-2z\,cos(\omega_oT_s)+1)} \qquad (16)$$

In addition to $G_{ig}(s)$, expressed in (8), the following transfer functions are defined for discretization purpose:

- $G_{iv}(s)$ is the transfer function related the modulated input voltage to the capacitor voltage; it is expressed in (17).
- $G_{vg}(s)$ is the transfer function related the capacitor voltage to the injected grid current; it is expressed in (18).

$$G_{iv}(s) = \frac{1}{CL_i(s^2+\omega_{res}^2)} \qquad (17)$$

$$G_{vg}(s) = \frac{1}{sL_g} \qquad (18)$$

ZOH discretization is used to determine $G_{ig}(z)$ and $G_{iv}(z)$ as expressed in (19) and (20) respectively. $G_{vg}(z)$ is determined as $G_{ig}(z)/G_{iv}(z)$.

$$G_{ig}(z) = \frac{T_s}{(L_i+L_g)}\left(\frac{(1-\alpha)z^2-2(cos(\delta)-\beta)z+(1-\alpha)}{(z-1)(z^2-2z\,cos(\delta)+1)}\right) \qquad (19)$$

$$G_{iv}(z) = \frac{1}{L_iC\omega_{res}^2}\frac{(1-cos(\delta))(1+z)}{(z^2-2z\,cos(\delta)+1)} \qquad (20)$$

where $\delta = \omega_{res}T_s$ and $\alpha = \frac{sin(\omega_{res}T_s)}{\omega_{res}T_s}$

Tustin approximation is used to determine $G_i(z)$ and $G_{ad}(z)$ as expressed in (21) and (22) respectively.

$$G_i(z) = \frac{T_s}{2(K_d-L_g)}\frac{z+1}{z-1} \qquad (21)$$

$$G_{ad}(z) = K_{ad}\frac{z-1}{z+\omega_{ad}} \qquad (22)$$

where $K_{ad} = \frac{2\omega_h\left(\beta_d\left(L_i+L_g\right)-L_g\right)}{\omega_hT_s+2}$ and $\omega_{ad} = \frac{\omega_hT_s-2}{\omega_hT_s+2}$

Using the above expressions, the actively damped filter discrete model and the loop transfer function are expressed in (23) and (24) respectively.

$$F_{new}(z) = \frac{z^{-1}G_{ig}(z)}{1-z^{-1}G_{ig}(z)G_{ad}(z)\left(1+G_i(z)/G_{vg}(z)\right)} \qquad (23)$$

$$T_{loop}(z) = G_c(z)F_{new}(z) \qquad (24)$$

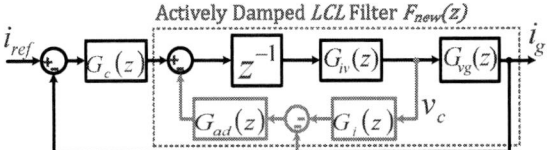

Fig. 5. Proposed System discrete implementation

978-1-4673-9551-9/16 $31.00 © 2016 IEEE

B. Control Parameters Design

For the tuning purpose, the equivalent s-domain representation, shown in Fig. 6, will be used; the DSP delay is represented by a transfer function of $G_d(s) = e^{-1.5sT_s}$ [17]. Based on this representation, both the actively damped transfer function (F_{new-d}) and the loop transfer function (T_{loop-d}) are represented in (25) and (26) respectively.

$$F_{new-d}(s) = \frac{(1+s/\omega_h)G_d(s)}{CL_iL_gs(s^2+\omega_{res}^2)(1+s/\omega_h)-s\beta_d\left(L_i+L_g\right)G_d(s)} \quad (25)$$

$$T_{loop-d}(s) = G_c(s)F_{new-d}(s) \quad (26)$$

It was indicated in [8] that the discrete implementation causes a variation in the resonant frequency; the new resonant frequency is denoted as ω_{res-ad} at which, the gain of F_{new-d} can be approximated in (27).

$$\left|F_{new-d}\left(j\omega_{res-ad}\right)\right| \cong \left|\frac{(1+j\omega_{res-ad}/\omega_h)}{-j\omega_{res-ad}\beta_d\left(L_i+L_g\right)}\right| \quad (27)$$

This implies that high values of ω_h should be adopted to acquire good damping performance. Theoretically, with discrete implementation, ω_h can be extended to $0.5\omega_s$ (Nyquist sampling theory, where ω_s is the sampling frequency in rad/sec); however, such value can deteriorate the discretization process. A value of $\omega_h = 0.4\omega_s$ is adopted; Based on this, only β_d has to be co-designed with the main regulator parameters to achieve certain performance.

Since the resonant gain of PR controller has negligible effect above the fundamental frequency, the PR controller can be reduced to its proportional gain (K_p). At the crossover frequency (ω_c), which should be adequately below both the resonant frequency and the adopted value of ω_h, the loop gain can be approximated as in (28).

$$T_{loop-d}\left(j\omega_c\right) = \frac{K_p}{\omega_c\left(L_i+L_g\right)}\frac{e^{-j1.5T_s\omega_c}}{\left(1-\beta_d e^{-j1.5T_s\omega_c}\right)}$$

$$\left|T_{loop-d}\left(j\omega_c\right)\right| = \frac{K_p}{\omega_c(L_i+L_g)}\left|\frac{1}{\left(1-\beta_d e^{-j1.5T_s\omega_c}\right)}\right| = 1 \quad (28)$$

Using Trigonometry, this gain can be reduced to (29).

$$\left|T_{loop-d}\left(j\omega_c\right)\right| = \frac{K_p}{\omega_c\left(L_i+L_g\right)}\left|\frac{1}{A_c e^{j\theta_c}}\right|$$

$$= \frac{K_p}{\omega_c(L_i+L_g)A} = 1 \quad (29)$$

where

$$A_c = \sqrt{1+\beta_d^2-2\beta_d^2\cos(1.5T_s\omega_c)}$$
$$\theta_c = \sin^{-1}\frac{\beta_d\sin(1.5T_s\omega_c)}{A_c} \quad (30)$$

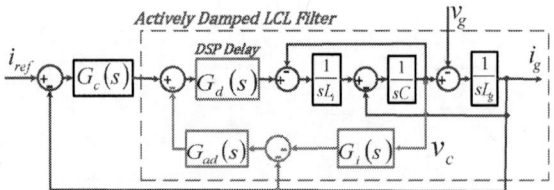

Fig. 6 System equivalent discrete implementation

Hence, to achieve a certain crossover frequency at certain value of β_d, K_p should be determined as in (31).

$$K_p = \omega_c\left(L_i+L_g\right)A_c \quad (31)$$

Substituting by (31) into (24), the loop transfer function can be expressed as follow

$$T_{loop}(z) = A_c\omega_c\left(L_i+L_g\right)F_{new}(z) \quad (32)$$

In a similar way, since the resonant gain of PR controller has the main effect at the fundamental frequency, the PR controller can be reduced to its resonant gain (K_r) at the fundamental frequency. The fundamental loop gain can be approximated as in (33).

$$\left|T_{loop-d}\left(j\omega_o\right)\right| = \frac{K_r}{\omega_o(L_i+L_g)A_o} \quad (33)$$

where $A_o = \sqrt{1+\beta_d^2-2\beta_d^2\cos(1.5T_s\omega_o)}$

It can be expressed in dB as in (34).

$$T_{fo} = 20\log_{10}\frac{K_r}{\omega_o(L_i+L_g)A_o} \quad (34)$$

From (34), to achieve a certain fundamental loop gain, K_r should be determined as in (35).

$$K_r = \omega_o\left(L_i+L_g\right)A_o \cdot 10^{\frac{T_{fo}}{20}} \quad (35)$$

Using above-addressed expressions, the following steps can be used to co-design the PR regulator along with the active damping loop.

1. For certain value of ω_c, plot the pole map for $F_{new}(z)$, expressed in (23) with variation of β_d. Identify the range of β_d for stable actively damped filter. Select a value of β_d to place the poles of $F_{new}(z)$ as far as possible inside the unit circle to achieve the best possible damping.
2. Determine the corresponding value of K_p from (31).
3. For certain value of fundamental loop gain (T_{fo}) along with the selected value of β_d, use (35) to determine K_r.
4. Plot the bode diagram for the loop transfer function expressed in (32). Check the resonant peak. If the resonant peak is more than 0 dB, then decrease the pre-specified crossover frequency (ω_c) and repeat the tuning process. Note that, low values of ω_c can deteriorate the system dynamic performance, so it may be advisable to allow some resonant peak more than 0 dB.

IV. VERIFICATION

A. Numerical Example

Table I presents the parameter values for the single phase inverter shown in Fig. 1. Three values of capacitances, corresponding to resonant frequencies of $0.15\omega_s$, $0.18\omega_s$ and $0.22\omega_s$, are used to verify the proposed system performance at different resonant frequencies with respect to the sampling frequency; these frequencies are denoted as ω_{res1}, ω_{res2} and ω_{res3} respectively. A cut off frequency (ω_h) of $0.4\omega_s$ is adopted for the HPF to reduce the resonant peak as possible, also, a value of 70 dB is adopted for the fundamental loop gain (T_{fo}). Finally, an initial value for crossover frequency of 0.3 of the corresponding resonant frequency is adopted.

Using the tuning steps in the last section, the pole-map of the actively damped filter is plotted with variation of β_d; these pole maps-plot are shown in Figs. 7 (a) and 8 (a) for the resonant frequencies ω_{res1} and ω_{res2} respectively. To achieve the best damping effect, values of β_d corresponding to the most inner resonant poles inside the unit circle are selected, these values are determined as 0.5 and 0.45 for ω_{res1} and ω_{res2} respectively; the corresponding values of K_p and K_r are determined form (31) and (35) respectively. Using the determined values of β_d, K_p and K_r, Figs. 7 (b) and 8 (b) show the bode plot for the loop transfer function, expressed in (24) at the ω_{res1} and ω_{res2} respectively. It is shown that the resonance peak (T_{res}) is less than 0 dB.

For ω_{res3}, with $\omega_c=0.3\omega_{res3}$, it is found that the frequency response will exhibit resonance peak more than of 0 dB. To overcome this issue, a reduction in the crossover frequency has to be adopted. It is found that a reduction of the resonant frequency of $0.05\omega_{res3}$ can reduce the resonant gain less than 0 dB. Figs.9 (a) and (b) show the pole-map at crossover frequencies of $0.3\omega_{res3}$ and $0.05\omega_{res3}$ respectively, also Fig. 10 shows the frequency response for the two cases. Table II summaries the designed control parameters.

B. Simulation Work

For the system shown in Fig. 1, Simulation work is carried out in PSIM environment using the system parameters listed in

Table I. A discrete model for the active damping loop and the PR regulator is constructed using the PSIM digital control modules. Fig. 11 and Fig. 12 show the grid current for ω_{res1} and ω_{res} respectively. On the other hand, Figs. 13 (a) and (b) show the injected grid current for ω_{res3} at the two considered crossover frequencies ($0.3\omega_{res3}$ & $0.05\omega_{res3}$) respectively.

TABLE I
SYSTEM PARAMETERS

Symbol	Quantity	Value
P	Rated power	500 W
V_g	Grid voltage	100 V
F_o	Grid Frequency	50 Hz
V_{dc}	DC Voltage	170 V
L_i	Inverter side inductance	1.5 mH
L_g	Grid side inductance	0.96 mH
C	Capacitance	4.7 µF, 3.3 µF, and 2.2 µF
F_{sw}	Switching Frequency	10 KHz
F_s	Sampling Frequency	20 KHz

TABLE II
DESIGNED CONTROL PARAMETERS

C (µF)	ω_c	β_d	K_p	K_r*10^3	T_{res} (dB)
4.7	$0.3\omega_{res1}$	0.5	8.2	1.22	-8.8
3.3	$0.3\omega_{res2}$	0.45	10.8	1.34	-2.84
2.2	$0.3\omega_{res3}$	0.25	16.7	1.83	14.2
	$0.05\omega_{res3}$	0.3	2.4	1.71	-2.7

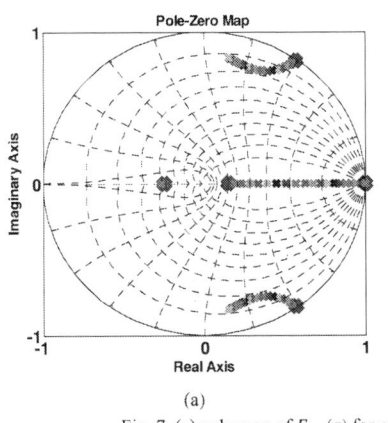
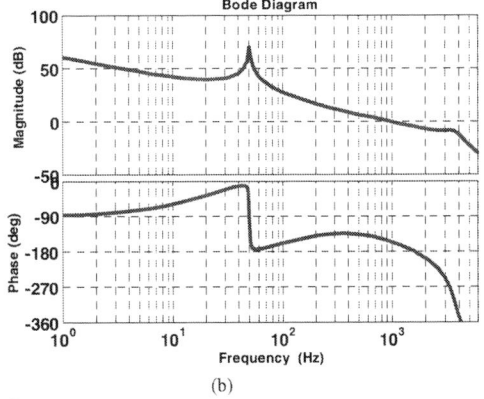

Fig. 7. (a) pole-map of $F_{new}(z)$ for ω_{res1} with variation β_d (b) corresponding bode plot for T_{loop} at β_d =0.5.

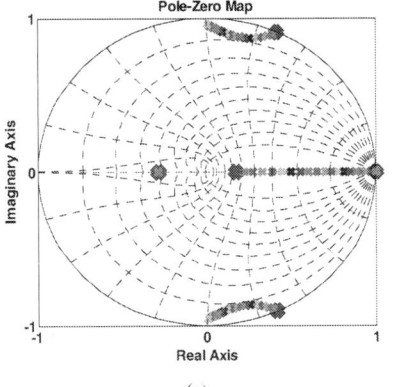
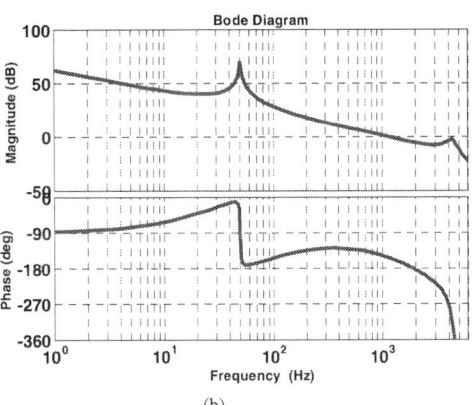

Fig. 8. (a) pole-map of $F_{new}(z)$ for ω_{rv2} with variation β_d, (b) corresponding bode plot for T_{loop} at β_d =0.45.

978-1-4673-9551-9/16 $31.00 © 2016 IEEE

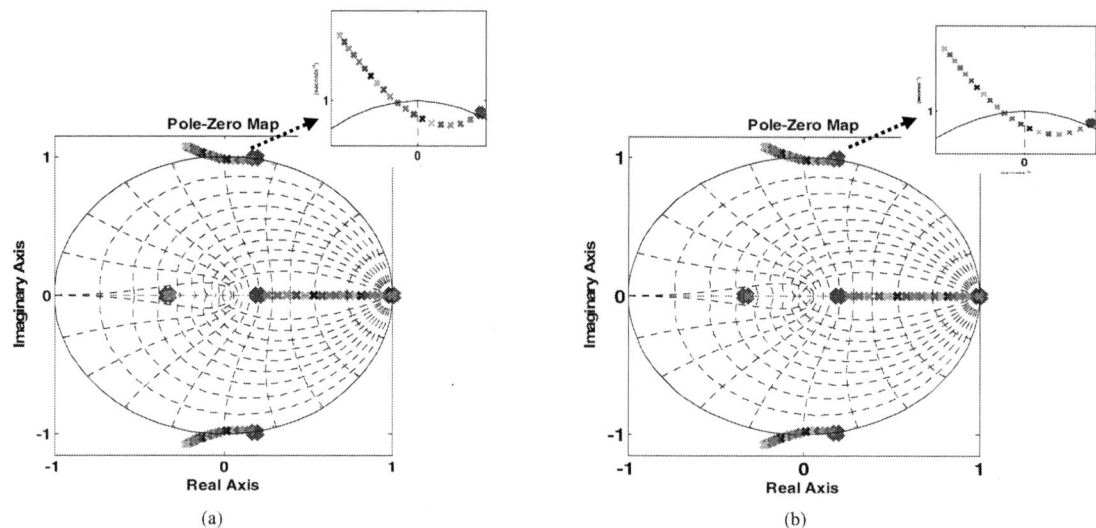

Fig. 9. pole-map of $F_{new}(z)$ for ω_{res3} with variation β_d, at (a) $\omega_c = 0.3\omega_{res3}$ and (b) $\omega_c = 0.05\omega_{res3}$.

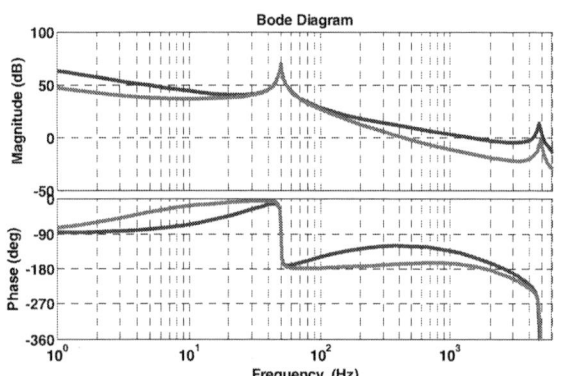

Fig. 10. Bode plot of $T_{loop}(z)$ for ω_{res3} at (▬▬) $\omega_c = 0.3\omega_{res3}$ & $\beta_d = 0.25$ and (▬▬) $\omega_c = 0.05\omega_{res3}$ & $\beta_d = 0.3$.

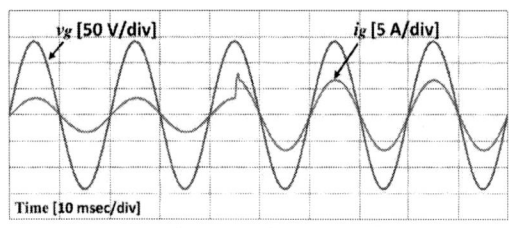

Fig. 11. Grid voltage (v_g) and grid current (i_g) for ω_{res1}.

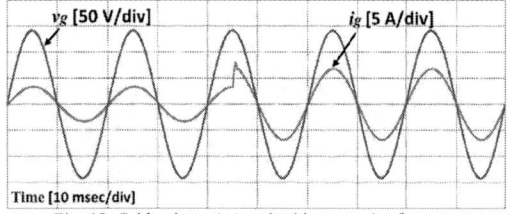

Fig. 12. Grid voltage (v_g) and grid current (i_g) for ω_{res2}.

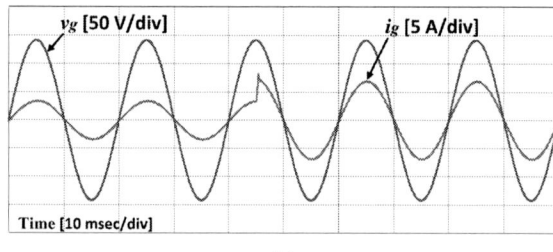

(a)

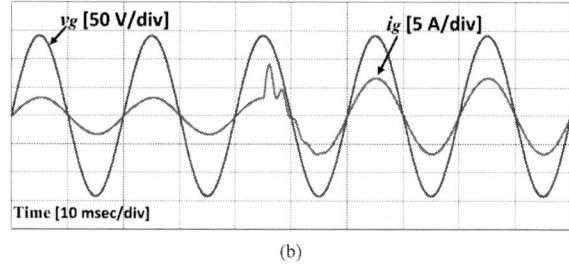

(b)

Fig. 13. Grid voltage (v_g) and grid current (i_g) for ω_{res2} (a) $\omega_c = 0.3\omega_{res3}$ & $\beta_d = 0.25$ and (b) $\omega_c = 0.05\omega_{res3}$ & $\beta_d = 0.3$.

V. CONCLUSION

This work proposes a new active damping method using two feedback loops from the capacitor voltage and the grid current. The proposed method can overcome the limitations in the previous methods; it can be behaved effectively over a wide range of resonant frequencies without non-minimum phase behavior; the cost is reduced by omitting the high cost current sensor; Moreover, straight-forward co-design steps for the active damping loops along with the fundamental current regulator is proposed.

VI. REFERENCES

[1] R. Beres, X. Wang, F. Blaabjerg, Cl. L. Bak, and M. Liserre, "A Review of Passive Filters for Grid-Connected Voltage Source Converters," in *Proc. IEEE APEC*, pp. 2208-2215, 2014.

[2] C. Wessels, J. Dannehl, and F. W. Fuchs, "Active Damping of LCL-Filter Resonance based on Virtual Resistor for PWM Rectifiers - Stability Analysis with Different Filter Parameters," in *IEEE ECCE*, pp. 3532-3538, 2008.

[3] S. G. Parker, B. P. McGrath, and D. G. Holmes, "Regions of active damping control for LCL filters," *IEEE Trans. Ind. Appl.*, vol. 50, no. 1, pp. 424-432, Jan./Feb. 2014.

[4] X.Wang, F. Blaabjerg, P. C. Loh, "Virtual RC damping of LCL-filtered voltage source converters with extended selective harmonic compensation," *IEEE Trans. on Power Electron.*, vol. 30, no. 9, pp. 4726-4737, Sept. 2014.

[5] J. Dannehl, M. Liserre, and F. W. Fuchs, "Filter-Based Active Damping of Voltage Source Converters With LCL Filter," *IEEE Trans. Ind. Electron.*, vol. 58, no. 8, pp. 3623-3633, 2011.

[6] Y. Y. X. Z. a. F. B. W. Yao, "Digital notch filter based active damping for LCL filters," in *Proc. of IEEE APEC*, pp. 2399 - 2406, 2015.

[7] P. A. Dahono, Y. R. Bahar, Y. Sato, and T. Kataoka, "Damping of transient oscillations on the output LC filter of PWM inverters by using a virtual resistor," in *IEEE Int. Conf. Power Electron. Drive Syst.*, pp. 403–407., 2001.

[8] D. Pan, X. Ruan, C. Bao, W. Li, and X. Wang, "Capacitor-Current-Feedback active damping with reduced computation delay for improving robustness of LCL-type grid-connected inverter," *IEEE Trans. on Power Electron.*, vol. Vol. 29, no. No. 7, pp. 3414-3427, 2014.

[9] M. A. Gaafar, M. Shoyam, "Active damping for grid-connected LCL filter based on optimum controller design using injected grid current feedback only," in *IEEE ECCE*, pp. 3628-3633, Sept. 2014.

[10] V. Miskovic, V. Blasko, T. M. Jahns, A. H. C. Smith, and C. Romenesko, "Observer-Based Active Damping of LCL Resonance in Grid-Connected Voltage Source Converters,"

IEEE Trans on Ind. Applications, vol. 50, no. 6, pp. 3977-3985, Nov./Dec 2014.

[11] R. P-Alzola, M. Liserre, F. Blaabjerg, R. Sebastián, J. Dannehl, and F. W. Fuchs, "Systematic Design of the Lead-Lag Network Method for Active Damping in LCL-Filter Based Three Phase Converters," *IEEE Trans. on Ind. Informatics,* vol. 10, no. 1, pp. 43-52, Feb. 2014.

[12] M. Malinowski and S. Bernet, "A Simple Voltage Sensorless Active Damping Scheme for Three-Phase PWM Converters With an LCL Filter," *IEEE Tans. on Ind. Electron.,* vol. 55, no. 4, pp. 1876-1880, April 2008.

[13] W. Gullvik, L. Norum, and R. Nilsen, "Active damping of resonance oscillationsin LCL-filters based on virtual flux and virtual resistor," in *Eur. Conf. Power Electron. Appl.,* pp 1-10, Sep.2–5,2007.

[14] J. Xu, S. Xie, and T. Tang, "Active damping-based control for grid-connected LCL-filtered inverter with injected grid current feedback only," *IEEE Trans. Ind Electron.,* vol. 61, no. 9, pp. 4746-4758, Sept. 2014.

[15] X. Wang, F. Blaabjerg, and P. Chiang Loh, "Grid-Current-Feedback active damping for LCL resonance in grid-connected voltage source converters," *IEEE Trans. on Power Electron.,* pp. 1-1, Mar. 2015.

[16] C. Bao, X. Ruan, X. Wang, W. Li, D. Pan, and K. Weng, "Step-by-step controller design for LCL-type grid-connected inverter with capacitor–current-feedback active-damping," *IEEE Trans. on Power Electron.,* vol. 29, no. 3, pp. 1239-1253, Mar. 2014.

[17] S. Buso and P. Mattavelli, Digital Control in Power Electronics, San Francisco: CA: Morgan & Claypool Publ., 2006.

Static Synchronous Generator Model for Investigating Dynamic behaviors and Stability Issues of Grid-tied Inverters

Liansong Xiong, Xiaokang Liu, Feng Wang, Fang Zhuo

School of Electrical Engineering, Xi'an Jiaotong University, Xi'an, Shaanxi, P.R.China 710049

Email: xiongliansong@163.com

Abstract—**By modeling the typical electromechanical transient of grid-tied PWM inverters, this paper firstly proves that PWM inverters and Rotational Synchronous Generators (RSGs) are similar in physical mechanism and equivalent in mathematical model, and the concept of Static Synchronous Generator (SSG) is thereby developed. Furthermore, the rationality and feasibility of migrating the concepts, tools and methods of RSG stability analysis to investigate the dynamic behaviors and stability issues of SSG is therefore confirmed. The criteria of small signal and transient stability of a grid-tied PWM inverter is put forward, providing clear physical interpretation about the dynamic characteristics and stability issues.**

I. INTRODUCTION

The PWM inverters in intermittent energy power generation are analogous to Rotational Synchronous Generators (RSGs) in traditional power generation from the functional point of view, yet their energy transfer media are different [1]-[3]. The grid-tied PWM inverters and RSGs, which are distinguished mainly through the mechanical behaviors, i.e. static or rotational, play an equal role in their respective system, and the behaviors of their energy transfer media can be described by the same fist-order kinetic equation [6]. Furthermore, both of their normal operation must be synchronized with grid. Accordingly, the grid-tied PWM inverter is referred to as Static Synchronous Generator (SSG) in this paper.

Existing methods for SSG stability analysis mainly comprise of the eigenvalue analysis [7]-[10] and impedance analysis [11]-[13]. Eigenvalue analysis establishes differential equations for the whole system and uses the eigenvalue theory to analyze the system stability [7]-[10]; however, the order of differential equations increases dramatically as the system complexity increases [11]-[13]. Accordingly, when this method is applied to the dynamic behavior analysis of a multi-inverter grid, the dynamic stability and the physical mechanism cannot be recognized easily because the high-order model can hardly show the clear physical meaning of the system [12], resulting in great inconvenience to achieve the best design and optimal control [14]. Impedance analysis method has

This work was supported by the National Basic Research Program of China(973 Program), P.R. China, No.2015CB251001 and the High Technology Research and Development Program of China (863 program), P.R. China, No.2015AA050606.

achieve a great success in DC power system stability analysis. Considering AC systems, however, many problems are still unsolved [13] and it is difficult to obtain a breakthrough in short term. In addition, the physical mechanism of instability of SSG can hardly be explained as well.

With the increasing penetration level of renewable power, the 'irresponsible' operation of SSG cannot meet the requirement of safe and stable operation of future grid. Consequently, technical improvements need to be made for grid-tied SSGs to cooperate in the aspects of grid power balance, voltage/frequency regulation, fault ride-through control and so on. For this purpose, the VSG technology is proposed, aiming to mimic the RSG operating characteristics for SSGs through control methods.

In this paper, several physical concepts related to a SSG are developed and analyzed, demonstrating the inherent unity of SSG and RSG in physical meaning and internal mechanism. In this sense, it is feasible to migrate the concepts, tools and methods of RSG stability analysis to SSG applications. Based on this conclusion, the virtual swing equation, which describes the electromechanical process of a SSG, is developed in this paper and harnessed to analyze the stability of SSG, providing a new methodology to analyze the dynamic behavior and stability issues of SSG.

II. GENERAL CONFIGURATION AND CONTROL OF SSG

Generally, a renewable energy generation system comprise of the following parts (see Fig.1): the primary energy source, the prime mover, the DC capacitor, the damper, the grid-tied PWM inverter, the passive filter and grid [15]-[16]. By analyzing the structure of inverter based generation system, it is apparent that the prime mover may differ for different primary energy and application. However, for any kind of renewable energy generation system, the DC capacitor, the damper, the grid-tied inverter and the passive filter are necessary to form the basic structure of SSG. In addition, the diverse inverters are identical in major functions and can be represented by an ideal PWM bridge inverter on the electromechanical time scale. Hence Fig. 2 is utilized to describe the SSG model.

The output power of L-type SSG can be obtained as:

$$P_e = P_{\max} \sin \delta = \frac{3}{2} \frac{U_s U_g}{X} \sin \delta \qquad (1)$$

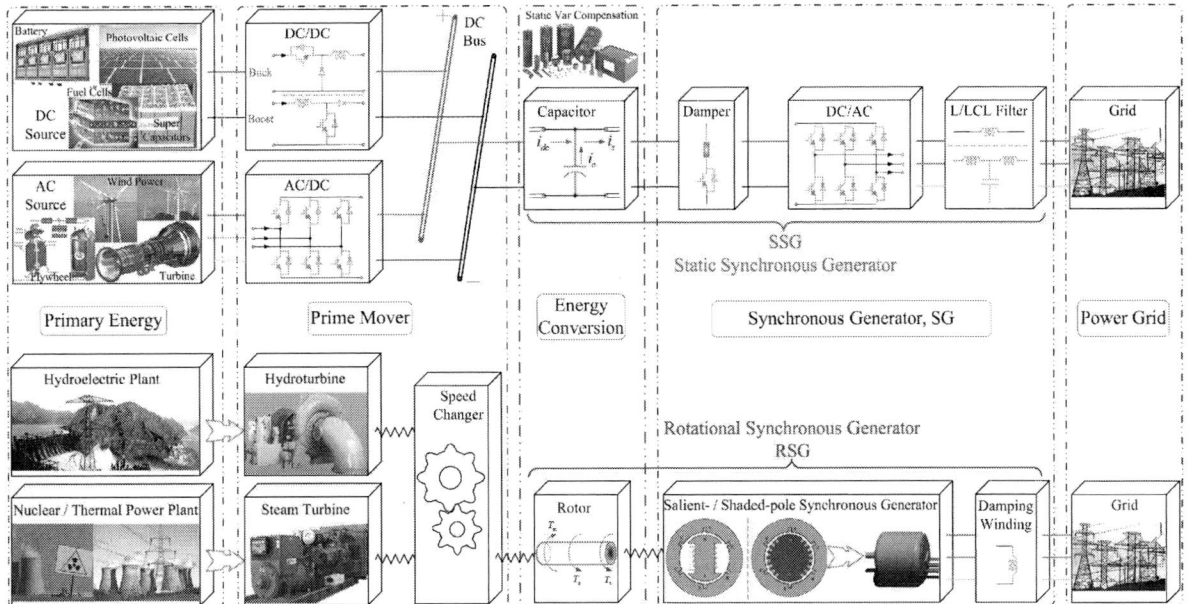

Fig. 1. General configuration and inherent relationship between SSG and RSG.

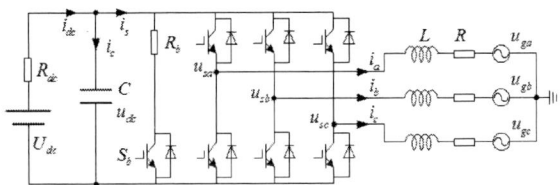

Fig. 2. Topology of SSG with an L-type filter.

RSG and SSG share strong similarities in dimensions of physical structures, characteristic parameters, energy transfer process, dynamic behaviors and mathematical models, and goes as follows. It is apparent from Fig. 1 that physical structures of conventional and renewable energy generation systems, which consist of the primary energy, the prime mover, and the energy transfer medium, the SG, damper and grid, are coherent (see Fig. 1). The physical structures of RSG and SSG are corresponding (see TABLE I). The energy transfer media of RSG and SSG are rotational rotor and static capacitor respectively, determining the electromechanical transient stability. The characteristic parameters of SSG and RSG are also coherent, as shown in TABLE II [6-7], [23], [25].

The similarity between RSG and SSG can also be viewed by energy transfer process (i.e. electromechanical dynamic process), which can be written in the generic form as

$$T_J \frac{dx}{dt} = P_{in} - P_e \qquad (2)$$

Obviously, the energy transfer process keeps its equilibrium when P_{in} balances P_e. The control strategies of SSG (including the fundamental and additional control) aim to achieve power balance, as well as the stable operation of the overall energy transfer process by controlling the prime mover power and/or the output electromagnetic power. Hence, the electromechanical transient process is the foundation of SSG dynamic behavior as well as stability analysis. The typical control (see Fig. 3) forms the foundation control framework, under which SSG is able to be tied to the grid and generate electricity stably [1], [3].

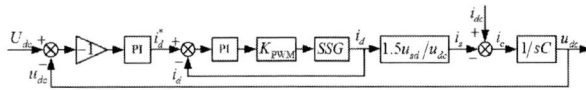

Fig. 3. Diagram of inner current and outer voltage control.

The stability of SSG electromechanical process is mainly dependent on the dynamic behavior of DC capacitor. Considering external disturbance, additional control strategies need to be implemented such that high performance control of DC capacitor voltage can be achieved. Whichever control schemes must help to realize power balance and capacitor voltage stability. Hence, the electromechanical dynamic is the key point to investigate the SSG stability.

III. SWING EQUATION OF SSG

The electromechanical dynamic process of RSG can be described by the standard dynamic equation, i.e.

$$\begin{cases} \dfrac{d\Delta\delta}{dt} = \Delta\omega \\ T_J \dfrac{d\Delta\omega}{dt} = -T_S\Delta\delta - T_D\Delta\omega \end{cases} \qquad (3)$$

978-1-4673-9551-9/16 $31.00 © 2016 IEEE

TABLE I
CORRESPONDENCE IN SYSTEM COMPONENTS

	SSG	RSG
Primary Energy	Renewable Energy, Battery, Wind, Solar,etc.	Conventional Energy, Water, Coal, Nucleus,etc.
Prime Mover, PM	Power Electronics Converter: DC/DC, AC/DC	Fluid Machine: Water Turbine, Steam Turbine
Energy Transfer Meium	Static Capacitor	Rotationa Rotor
Damper	Chopper Circuit(IGBT + Power Resistor)	Damping Winding
Synchronous Generator	L-type SGG	Shaded-pole RSG
SG	LCL-type SSG	Salient-pole RSG

TABLE II
CORRESPONDENCE IN CHARACTERISTIC PARAMETERS.

SSG	RSG
Passive Filter	Stator Circuit
Capacitance of Capacitor C	Moment of Inertia J Passive Filter Stator Circuit
DC Capacitor Voltage	Rotor Angle(P Controller)/Rotational Speed(I Controller)
Output Voltage of Inverter	Internal Voltage of RSG
Virtual Internal Voltage of SSG with LCL-type Filter	Virtual Internal VOltage of Salient-pole RSG
Output Voltage of Passive Filter	Terminal Voltage of RSG
Stored Electric Energy of Capacitor E_K	Rotational Kinetic Energy of Rotor W_K
Rated DC Capacitor Voltage	Synchromous Speed of Rotor
Input/Output/Charging Current of DC-side Capacitor	Mechanical/Electromagnetic/Accelerating Torque of Rotor

where, T_J, T_S and T_D denote the equivalent inertia, synchronizing and damping coefficients of RSG, respectively, and represent the inertia, synchronizing ability and damping effect of the generator. Following analysis demonstrates that RSG and SSG share the identical swing equation which mainly describes their dynamic behaviors at electromechanical time scale.

By choosing the rated capacitor voltage U_{dc} as the SSG DC-side base voltage U_{dcB}, the base current of DC-side is $I_B = S_B/U_{dcB} = P_B/U_{dc}$. Considering the damping effect of SSG, the electromechanical dynamic process can be written as

$$
\begin{aligned}
i_c &= C\frac{du_{dc}}{dt} = \frac{0.5 \times CU_{dc}^2}{0.5 \times CU_{dc}^2} \cdot C \cdot \frac{du_{dc}}{dt} \\
&= \frac{2E_K}{U_{dc}^2}\frac{du_{dc}}{dt} = i_{dc} - i_s - i_d
\end{aligned}
\tag{4}
$$

By dividing both sides of the equation by I_B, we get

$$
\begin{aligned}
\frac{2E_K}{I_B U_{dc}}\frac{d\left(u_{dc}/U_{dc}\right)}{dt} &= \frac{2E_K}{P_B}\frac{du_{dc}^*}{dt} = T_J\frac{du_{dc}^*}{dt} \\
&= \frac{i_{dc}}{I_B} - \frac{i_s}{I_B} - \frac{i_d}{I_B} \\
&= i_{dc}^* - i_s^* - i_d^*
\end{aligned}
\tag{5}
$$

Hence,

$$
T_J\frac{du_{dc}^*}{dt} = i_{dc}^* - i_s^* - i_d^*
\tag{6}
$$

which is the normalized **virtual swing equation of SSG**.

T_J is the time duration required for DC capacitor voltage u_{dc} to be raised from 0 to its nominal value U_{dc}, with rated charging current I_B. To simplify the symbols, * is omitted. As a simplified assumption, u_{dc} is kept constant 1 in that

the variations of u_{dc} are small with capacitor voltage control. Therefore, i_{dc} equals P_{dc} in per-unit value, so do is and P_s, i_d and P_d. Consequently, the standard model of swing equation in SSG can be obtained, by considering the relationship between the power angle and grid angular frequency ω, as (in $p.u.$)

$$
\begin{cases}
\dfrac{d\delta}{dt} = (\omega - 1)\,\omega_s \\
T_J\dfrac{du_{dc}}{dt} = P_{dc} - P_s - P_d
\end{cases}
\tag{7}
$$

By substituting the linearized SSG electromagnetic power in (1) into (7) and assuming that the input power of prime mover keeps constant ($\Delta P_{dc} = 0$), standard dynamic equation of SSG can be obtained as

$$
\begin{cases}
\dfrac{d\Delta\delta}{dt} = \Delta\omega \\
T_J\dfrac{d\Delta u_{dc}}{dt} = -T_S^{'}\Delta\delta - T_D^{'}\Delta u_{dc}
\end{cases}
\tag{8}
$$

In (8), it holds

$$
\begin{cases}
T_J = \dfrac{2E_K}{P_B} \\
T_S^{'} = 1.5KU_g \\
T_D^{'} = \dfrac{1}{R_d}
\end{cases}
\tag{9}
$$

By comparison of (8) and (3), it is evident that the swing equations of RSG and SSG are coherent in form with different state variables (ω and u_{dc}). Further study is needed about the relationship between ω, δ and u_{dc} in order to determine whether ω in RSG is the counterpart of u_{dc} in SSG.

For the basic SSG control scheme (see Fig. 3), the bandwidth of inner loop is designed overwhelmingly large such that the inner-loop transient is negligible compared with the outer-loop. Consequently, the inner-loop current yields

$$i_d = i_d^* = -\left(K_p + \frac{K_i}{s}\right)(U_{dc} - u_{dc}) \qquad (10)$$

Besides, from the phasor diagram of SSG, it is

$$i_d = \frac{U_s}{X}\sin\delta \Rightarrow \frac{U_s}{X}\sin\delta = -\left(K_p + \frac{K_i}{s}\right)(U_{dc} - u_{dc})$$
$$(11)$$

Accordingly, by linearizing (11) and substituting $s\Delta\delta$ by $\Delta\omega$, we have

$$\Delta\omega = \frac{1}{K}(sK_p + K_i)\Delta u_{dc} \qquad (12)$$

Rewriting (12) in Laplace domain and combining with (8), we have

$$\begin{cases} \dfrac{d\Delta\delta}{dt} = \Delta\omega \\ T_J\dfrac{d\Delta\omega}{dt} = -T_S\Delta\omega - T_D\Delta\omega \end{cases} \qquad (13)$$

In (13), it holds

$$\begin{cases} T_J = 2\dfrac{E_K}{P_B} \\ T_D = \dfrac{K_p}{K}T_S' + T_D' = 1.5K_pU_g + \dfrac{1}{R_d} \\ T_S = \dfrac{K_i}{K}T_S' = 1.5K_iU_g \end{cases} \qquad (14)$$

Obviously, (13) suggests that the damping effect, which provides SSG with the immunity to external disturbances by damping oscillations, is generated by chopper circuit and P regulator. I regulator provides SSG with the synchronization ability which helps SSG to work at the desired operating point.

IV. ANALYSIS AND CRITERION OF SSG STABILITY

The electromechanical dynamic process and the stability behavior of SSG are discussed as follows. Obviously, P_S equates P_{dc} in steady-state (see Fig. 4(a)), hence operating points a and b are present, satisfying power balance.

A. Small-signal stability criterion

In this case, only operating point a is able to operate stably, while operating point b is unstable (see Fig. 4(a)). Similarly, the region on the left part of critical point is small-signal stable. Therefore, the system operating at point b will either lose its stability or regain stability by moving to point a, i.e. point b is small-signal unstable. Similarly, the operating region on the right part of critical point is unstable. Therefore, the small-signal stability criterion of SSG is given by

$$\frac{dP_S}{d\delta} > 0 \qquad (15)$$

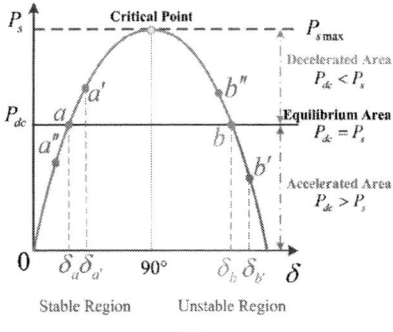

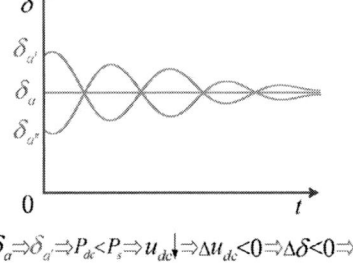

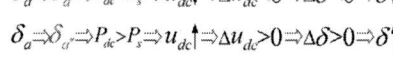

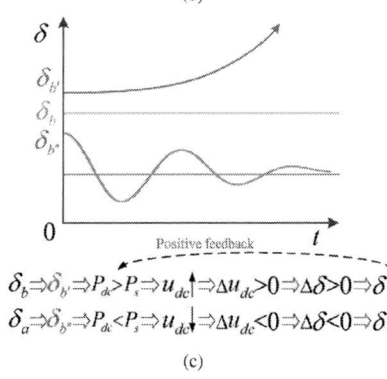

Fig. 4. Illustration of SSG's small-signal stability.

B. Transient stability criterion

Transient stability depends on both the initial operating state and the severity of disturbance. Usually, the system is altered so that the post-disturbance steady-state operation differs from that prior to the disturbance, as shown in Fig. 5. Power balance is achieved when SSG operates at point a, hence SSG is able to work stably under small disturbances. When a fault occurs, the curve drops from P_S to P_{sI}. Accordingly, the instantaneous output power will fall (or increase with another fault type) to point b. In this case, $P_{dc} > P_S$ and the capacitor will charge, leading to continuous increase of u_{dc} and δ. More power is generated to balance the charging power of capacitor. Assuming that fault is detected and instantaneously cleared when δ increases to δ_c, after that the curve elevated from P_{sI} to P_{sII} or P_s if normal system operation can be retrieved

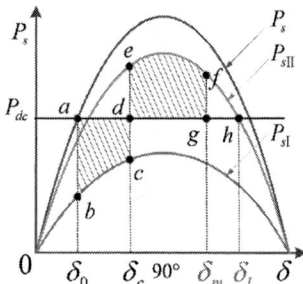

Fig. 5. Illustration of SSG power angle during transient.

after the fault is cleared. At the time instant when fault is cleared, the instantaneous output power rises to point e with δ_c at that time. At the same time, $P_{dc} < P_S$ and hence the capacitor will discharge, leading to continuous decrease of u_{dc}. However, u_{dc} is still greater than its nominal value U_{dc}, leading to continuous increase of δ such that more power can be generated to accelerate capacitor discharging. When δ increases to δ_m, u_{dc} drops to its nominal value U_{dc}. After that δ will continuously decrease to generate less power, thus lower the speed of capacitor discharging. Hence, δ_m is the extreme position of power angle during the transient. It is evident that δ_m must be strictly smaller than the limit power angle δ_l, otherwise a positive feedback is formed and SSG will lose its stability. Consequently, the transient stability criterion is given by $\delta_m < \delta_l$.

According to the analysis above, DC capacitor will absorb extra energy and charge during the fault; furthermore the increased stored energy equates the value of accelerating area S_+ (encircled by abcd). After the fault is cleared, the capacitor will release the extra energy during charging in order to return to its normal state. The released energy equates the decelerating area S_- (enclosed by defg). Obviously, the absorbed energy must be completely released to bring the capacitor voltage back to normal, otherwise the energy stored in capacitor will keep increasing and SSG will eventually lose its stability. Therefore, the transient stability criterion of SSG can also be described as $S_+ < S_-\text{max}$.

It is apparent that clearing the fault timely is of paramount importance in keeping the transient stability of SSG, elsewise the DC capacitor will be charged with too much energy that cannot be released thoroughly before δ_l is achieved and the system is unstable. The critical power angle $\delta_c m$ for fault clearing can be expressed as

$$\cos \delta_{cm} = \frac{P_{dc}(\delta_l - \delta_0) + P_{smax}\cos \delta_l - P_{smax}\cos \delta_0}{P_{smax} - P_{smax}} \tag{16}$$

V. SIMULATION VERIFICATION

Simulations are performed to verify the conclusions of stability studies. Fig. 6(a) illustrates the influence of P controller on the damping characteristic of SSG. Increasing K_p, the oscillation becomes smaller, indicating stronger damping

effect. When $K_p = 5$, SSG operates in a critical damping state and system is characterized by both fast dynamic and small overshooting. It is apparent that larger K_p indicates a better damping ability. Fig. 6(b) shows the influence of I controller on system synchronization ability. Increasing K_i, u_{dc} returns faster to the steady state, i.e. the greater K_i, the stronger regulation ability SSG has on the DC voltage, and the stronger synchronization ability SSG has.

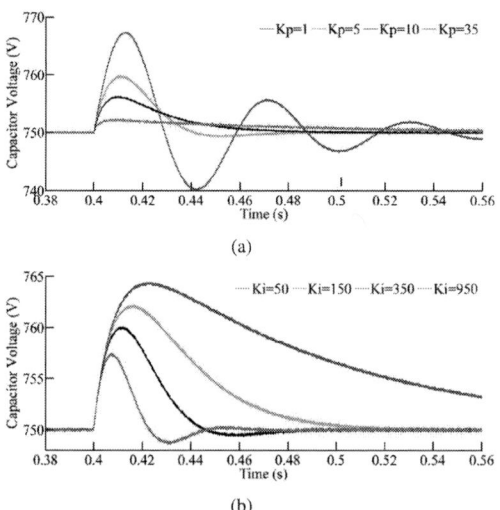

Fig. 6. Influence of K_p and K_i on dynamic behaviors and stability.

Besides, in the above cases when controller parameters are different, u_{dc} remained close to its rated value though the dynamic characteristics are not the same. This can be explained by the positive damping effect and synchronization capability provided by PI regulator, and hence SSG is small-signal stable.

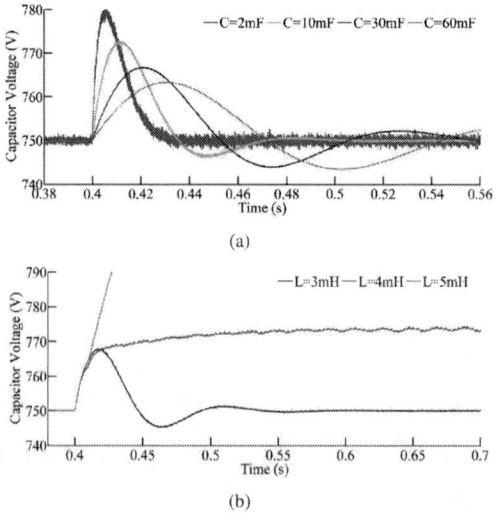

Fig. 7. Influence of C and L on dynamic behaviors and stability.

Fig. 7 demonstrates the dynamic behavior of u_{dc} with respect to C and L changes. From Fig. 7(a), the oscillation of u_{dc} decreases with the increase of C, indicating better resistance against external disturbances, i.e. higher inertia ability. Bigger inertia is beneficial for capacitor voltage stabilization, avoiding overvoltage or even worse fault in the case of large external disturbances. In normal operation, system is able to maintain small-signal stable.

VI. CONCLUSION

The electromechanical transient model and physical interpretation of SSG are presented to illustrate the inherent relationship and intrinsic differences between SSG and RSG, and results suggest that it is feasible to migrate the concepts, tools and methods of RSG stability analysis to investigate the dynamic behaviors and stability issues of SSG. Based on the developed SSG model, following main conclusions are obtained:

1) the inertia of SSG is proportional to the DC capacitance and capacitor voltage; the damping ability is dependent on the proportional gain of DC voltage loop; the synchronizing ability is determined mainly by the integral gain of DC voltage loop. The larger values of control parameters, the better regulation ability of capacitor voltage, thus better stability.

2) Larger linking inductance indicates weaker connection and thus the weaker stability. Increasing the virtual internal voltage and the DC-side capacitor voltage can enhance not only the real-power transmission ability but the system stability.

3) The larger power angle, the smaller stability margin system will have and thus the weaker system stability.

REFERENCES

[1] Y. S. Park, S. K. Sul, C. H. Lim, et al, "Asymmetric control of DC-link voltages for separate MPPTs in three-level inverters," *IEEE Trans. Power Electron.*, vol. 28, no. 6, pp. 2760-2769, Jun. 2013.

[2] J. Yao, H. Li, Z. Chen, et al, "Enhanced control of a DFIG-based wind-power generation system with series grid-side converter under unbalanced grid voltage conditions," *IEEE Trans. Power Electron.*, vol. 28, no. 7, pp. 3167-3181, Jul. 2013.

[3] Y. Zhou, and H. Li, "Analysis and suppression of leakage current in cascaded-multilevel-inverter-based PV systems," *IEEE Trans. Power Electron.*, vol. 29, no. 10, pp. 5265-5277, Oct. 2014.

[4] Q. C. Zhong, "Synchronverters: inverters that mimic synchronous generators," *IEEE Trans. Ind. Electron.*, vol. 58, no. 4, pp. 1259-1267, Apr. 2011.

[5] Pogaku N., Prodanovic M., Green T. C., "Modeling, analysis and testing of autonomous operation of an inverter based micro-grid," *IEEE Trans. Power Electron.*, vol. 22, no. 2, pp. 613-625, Feb. 2007.

[6] Barklund E., Nagaraju P., Green T. C., et al, "Energy management in autonomous micro-grid using stability contrained droop control of inverters," *IEEE Trans. Power Electron.*, vol. 23, no. 5, pp. 2346-2352, May. 2008.

[7] X. Q. Guo, Z. G. Lu, B. C. Wang, et al, "Dynamic phasors-based modeling and stability analysis of droop-controlled inverters for microgrid applications," *IEEE Trans. Smart Grid*, vol. 5, no. 6, pp. 2980 - 2987, Jun. 2014.

[8] Wildrick C. M., Lee F. C., "A method of defining the load impedance specification for a stable distributed power system," *IEEE Trans. Power Electron.*, vol. 10, no. 3, pp. 280-285, Mar. 1995.

[9] Liu J. J., Feng X., Lee F. C., "Stability margin monitoring for DC distributed power systems via perturbation approaches," *IEEE Trans. Power Electron.*, vol. 18, no. 6, pp. 1254-1261, Jun. 2003.

[10] Kim J., Guerrero J. M., Rodriguez P., "Mode adaptive droop control with virtual output impedances for an inverter based flexible AC microgrid," *IEEE Trans. Power Electron.*, vol. 26, no. 3, pp. 689-701, Mar. 2011.

[11] Z. Liu, J. J. Liu, W. H. Bao, and Y. L. Zhao, "Infinity-norm of impedance-based stability criterion for three-phase AC distributed power systems with constant power loads," *IEEE Trans. Power Electron.*, vol. 30, no. 6, pp. 3030-3043, Jun. 2015.

[12] F. C. Liu; J. J. Liu; H. D. Zhang; D. H. Xue, "Stability issues of Z+Z type cascade system in hybrid energy storage system (HESS)," *IEEE Trans. Power Electron.*, vol. 29, no. 11, pp. 5846-5859, Nov. 2014.

[13] B. Wu, S. Li, S. Singer, and K. Smedley, "Analysis of High Power Switched Capacitor Converter Regulation based on Charge-balance Transient-calculation Method," *IEEE Trans. Power Electron.* , vol. 30, no. 2, pp. 1-1, Jun. 2015.

[14] Q. C. Zhong, P. L. Nguyen, Z. Y. Ma, "Self-synchronized synchronverters: inverters without a dedicated synchronization unit," *IEEE Trans. Power Electron.*, vol. 29, no. 2, pp. 617-630, Feb. 2014.

[15] B. Wu, S. Li, Y. Liu, and K. Ma Smedley, "A New Hybrid Boosting Converter for Renewable Energy Applications," *IEEE Trans. Power Electron.*, vol. 31, no. 2, pp. 1203C1215, Feb. 2016.

[16] B. Wu, S. Li, K. Ma Smedley, and S. Singer, "A Family of Two-Switch Boosting Switched-Capacitor Converters," *IEEE Trans. Power Electron.*, vol. 30, no. 10, pp. 5413C5424, Oct. 2015.

Initial Orientation and Sensorless Starting Strategy of Wound-Rotor Synchronous Starter/Generator

Jichang Peng, Weiguo Liu, Jinhao Meng, Tao Meng, Guangzhao Luo

Institute of REPM Electrical Machines and Control Technology
School of Automation, Northwestern Polytechnical University
Xi'an, Shaanxi, P. R. China
linkpjc@gmail.com

Abstract—This paper proposes a sensorless control method for Wound-Rotor Synchronous Starter/Generator (WSSG) in starting process. It is necessary to identify the initial rotor position when uses sensorless control method to start-up AC machines. However, during the process of initial orientation the excitation current of the main generator contains severe pulsating components, which leads to the ineffectiveness of traditional position estimating methods. In order to improve the accuracy of estimation results, the specified transient pulse signal voltage injection and the associated signal processing method are propose in this paper. In starting process, the square-wave-type voltage signal injection method, which capably of eliminates the interference of pulsating excitation current and enhancing the dynamic performance of the system, is used in the starting process to acquire the rotor position by demodulating the induced high-frequency stator current. Experimental results proved the effectiveness and accuracy of the propose method.

Keywords—Starter/Generator; sensorless; AC machines

I. INTRODUCTION

With the fast development of more electric aircrafts (MEA) and all electric aircrafts (AEA), the WSSG is becoming increasingly important. Starter/Generator(S/G) performs two functions: starting the aero engine and generating the power in aircraft power system [1]-[2].The rotor position information which is only needed in the starting process. Thus, Sensorless control method for S/G makes this type of integrated system even more competitive .In the starting procedure, the excitation current of the main generator contains severe pulsating components. As the influence of the excitation current ripple, traditional position estimation algorithms failed to obtain the accurate initial rotor position [3],[4]. In order to improve the accuracy of estimation results, the paper uses Fourier series to fit the reflecting current curve. In addition, due to the ripples of excitation current, traditional sensorless control technologies cannot be readily applied to S/G[5]-[9].There are few literatures on the sensorless control of S/G. Reference [10] presents a sensorless method based on back electromotive force voltage estimation, which has good performance in the high speed region. But the scheme in low speed region is not given. Reference [11] injected a pulsating high-frequency voltages in estimated d-axis and acquired the rotor position from zero to engine ignition speed, however, the inherent time delay of LPF (Low pass filter) which used in the position estimated process will limits the performance of sensorless control. The paper proposes a square-wave-type voltage injecting method to estimate the rotor position of the S/G[5]. The proposes method eliminating the pulsation noise of excitation current substitute the finite difference for the LPF in the estimate process .The dynamic performance of the proposes method is largely increased compared with those method by the conventional sinusoidal-type voltage injection.

II. MACHINE MODELING

The structure of WSSG is shown in Fig. 1. The S/G consists of three stages: the pre-exciter (PE), the main exciter (ME) and the main generator (MG). The PE is a three-phase permanent magnet generator. The ME is a synchronous machine with a field winding and rotating three armature winding. The MG is an electrically excited synchronous machine. In starting process, the ME is fed with an external AC voltage and the MG works as a transformer when as the machine in is standstill. The electromotive force induces in the armature winding finally forms a DC field current for MG through the rotating diode rectifier. By theory deduction and algorithm validation, the armature voltage of the ME can be

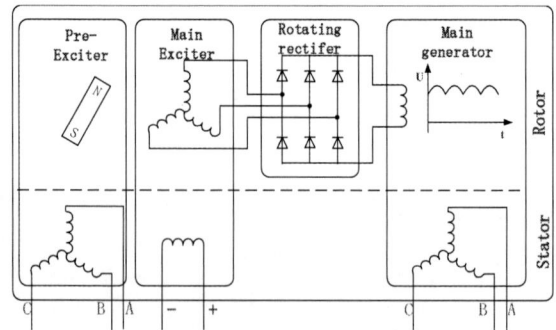

Fig.1 Schematic of WSSG

This work was supported by National Natural Science Foundation of China (51277152)

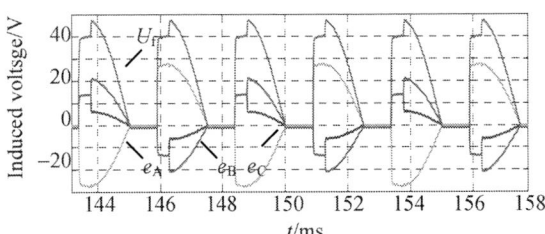

Fig. 2 Waveforms of induced voltage

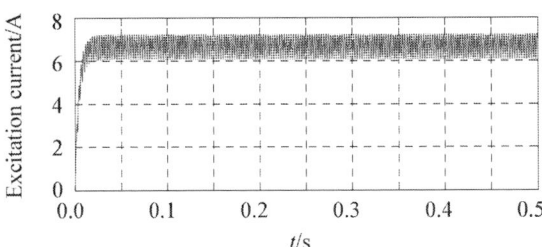

Fig.3 excitation current of the MG i_f

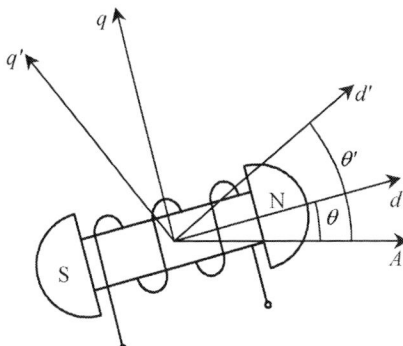

Fig. 4 Schematic of MG

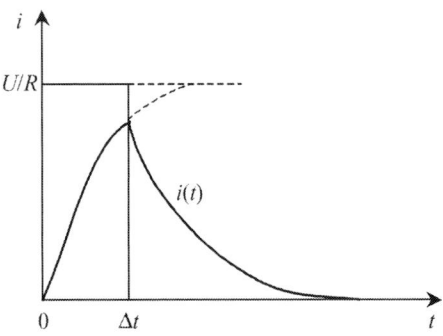

Fig. 5 Curve of induced current

described as:

$$\begin{cases} e_A = 2\sqrt{2}E_+ \sin(2\pi f_1 t)\cos\delta \\ e_B = 2\sqrt{2}E_+ \sin(2\pi f_1 t)\cos(\delta + 2\pi/3) \\ e_C = 2\sqrt{2}E_+ \sin(2\pi f_1 t)\cos(\delta + 4\pi/3) \end{cases} \quad (1)$$

Where $E_+ = 4.44 N_2 k_{N_2} f_+ \phi_m$, N_2 is the number of turns in each phase series winding, k_{N_2} is the winding coefficient, ϕ_m is the amplitude of flux per pole, $f_+ = f_1 - f$ where f_1 is the frequency of single-phase AC source, f is the machine frequency.

Fig. 2 shows the armature voltage of the ME e_A, e_B, e_C and the electromotive force U_f in field winding of MG. U_f influenced by the rotating rectifier contains severe pulsating components. Fig. 3 shows the excitation current of the MG i_f. In Fig. 3 a 210Vrms/200Hz single-phase AC fed ME in the stationary state and the excitation current of the MG still contains severe pulsating components influenced by diode commutation processes even in stable state. The pulsating components will lead to the failure of traditional initial position estimation and sensorless control technologies.

III. SENSORLESS VECTOR CONTROL OF WSSG

As illustrated in Fig. 4 the specified transient pulse signal voltage was injected into d'-axis(assuming the angle of u'_d is the estimated d-axis). The equation of the stator terminal voltage in d'-q' axis is as following:

$$\begin{bmatrix} u'_d \\ u'_q \end{bmatrix} = R\begin{bmatrix} i'_d \\ i'_q \end{bmatrix} + \begin{bmatrix} L+\Delta L\cos 2\sigma & \Delta L\sin 2\sigma \\ \Delta L\sin 2\sigma & L-\Delta L\cos 2\sigma \end{bmatrix} \cdot$$
$$D\begin{bmatrix} i'_d \\ i'_q \end{bmatrix} + \omega\phi_f\begin{bmatrix} \cos\sigma \\ \sin\sigma \end{bmatrix} \quad (2)$$

Where $L = (L_d + L_q)/2$, $\Delta L = (L_d - L_q)/2$, ϕ_f is the rotor fluxes, ω is the electrical rotor speed, θ' is the angle of the specified transient pulse signal voltage injected to the machine, θ is real rotor position, $\sigma = \theta - \theta'$.

When injected a specified transient pulse signal voltage in the stator terminal, the MG respond just like the zero state response of the RL circuit. As in Fig. 5, a transient pulse signal voltage is injected to the MG and the amplitude of response current is decided by the inductive impedance. In the stationary state, (2) can be rewritten as (3):

$$D\begin{bmatrix} i'_d \\ i'_q \end{bmatrix} = \frac{u'_d}{L^2 - (\Delta L)^2}\begin{bmatrix} L-\Delta L\cos 2\sigma \\ -\Delta L\sin\sigma \end{bmatrix} \quad (3)$$

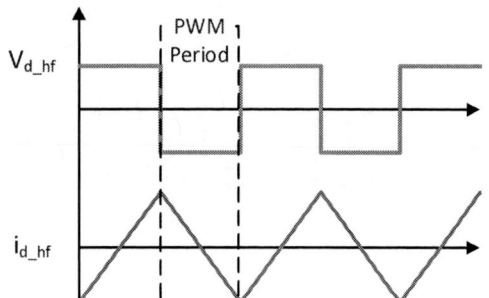

Fig. 6 d-axis square-wave voltage injection

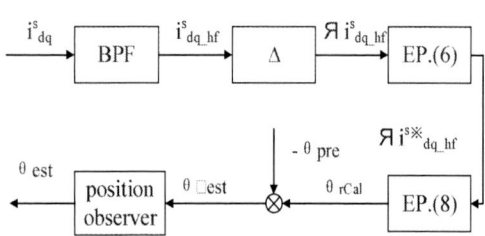

Fig. 7 Block diagram of the demodulation method

$$\begin{bmatrix} \dot{i_d} \\ \dot{i_q} \end{bmatrix} = \int_0^{\Delta t} D \begin{bmatrix} \dot{i_d} \\ \dot{i_q} \end{bmatrix} dt = \frac{u_d^{'} \Delta t}{L^2 - (\Delta L)^2} \begin{bmatrix} L - \Delta L \cos 2\sigma \\ -\Delta L \sin \sigma \end{bmatrix} \quad (4)$$

According to the above principle, 360 degrees transient pulse signal voltage is injected to the stator terminal, setting the step

size of $u_d^{'}$ to 5 angle degree. Then Clark, Park transformation are applied to get the current $i_d^{'}$. As the machine is salient, the angle of voltage $u_d^{'}$ corresponding to the maximum amplitude of $i_d^{'}$ is the real rotor position. Afterwards, by identifying the magnetic polarity, the positive direction of d-axis (N-pole) will be found. Decreasing step size will improve the estimation accuracy. The above method yields larger estimate errors when applied to the Starter/Generator. To improve the performance of position estimate, Fourier series are used to fit the reflecting current curve. The Fourier series expansion is as following:

$$\begin{cases} f(x) = \dfrac{a_0}{2} + \sum_{n-1}^{\infty} \left[a_n \cos(nx) + b_n \sin(nx) \right] \\ a_n = \dfrac{1}{\pi} \int_{-\pi}^{\pi} f(x) \cos(nx) dx \\ b_n = \dfrac{1}{\pi} \int_{-\pi}^{\pi} f(x) \sin(nx) dx \end{cases} \quad (5)$$

The method based on higher frequency square-wave-type voltage injection is proposed in this paper which is capable of eliminating the pulsation noise of i_f by the finite difference process. By removing LPF in the estimating process, this method can be implemented in full speed range. The injected voltage is described as :

$$V_{dq\_hf}^{\hat{r}} = V_{d\_hf}^{\hat{r}} + j V_{q\_hf}^{\hat{r}} = \pm V_{inj} + j \times 0 \quad (5)$$

Fig. 6 illustrates the square-wave injection in the estimate d-axis, the polarity of the square-wave changes each PWM duty cycle. The slope of reflecting current contains the rotor position information caused by the spatial saliency. The injection-induced current ripples in stationary frame can be deduced as follows:

$$\begin{bmatrix} \Delta i_{d\_hf}^s \\ \Delta i_{q\_hf}^s \end{bmatrix} = sign(V_h) V_h \Delta T \begin{bmatrix} \dfrac{\cos(\theta_r)\cos(\tilde{\theta}_r)}{L_{d\_hf}^r} + \dfrac{\sin(\theta_r)\sin(\tilde{\theta}_r)}{L_{q\_hf}^r} \\ \dfrac{\sin(\theta_r)\cos(\tilde{\theta}_r)}{L_{d\_hf}^r} - \dfrac{\cos(\theta_r)\sin(\tilde{\theta}_r)}{L_{q\_hf}^r} \end{bmatrix} (6)$$

Where $\Delta i_{d\_hf}^s$ and $\Delta i_{q\_hf}^s$ are the difference between the present value and the previous value of HF current. θ_r is the real rotor position, $\tilde{\theta}_r$ is the error between the real and the estimated rotor position. $L_{d\_hf}^r$ and $L_{q\_hf}^r$ are the high frequency inductance of d-q axis. If $\tilde{\theta}_r \approx 0$, the injection-induced current ripples in stationary frame is simplified as :

$$\begin{bmatrix} \Delta i_{d\_hf}^s \\ \Delta i_{q\_hf}^s \end{bmatrix} \approx \frac{V_h \Delta T}{L_{d\_hf}^r} \begin{bmatrix} \cos(\theta_r) \\ \sin(\theta_r) \end{bmatrix} \quad (7)$$

$$\theta_{rCal} = atan2\left(\Delta i_{q\_hf}^s, \Delta i_{d\_hf}^s \right) \quad (8)$$

Fig. 7 shows the signal processing process of the position estimation. A PI controller serves as the position observer. The estimate rotor position of the last cycle is θ_{pre} , the present estimate rotor position is θ_{est}. The position observer is shown in Fig. 8. The proposed position observer using a PI controller to get the signal of estimated position θ_{est} without using any low pass filter (LPF).

IV. EXPERIMENTAL RESULTS

The test bench consists a WSSG, a inverter. The control algorithms implemented on RTLAB real-time system is shown in Fig. 9. The sample frequency is 10 kHz, though the PWM

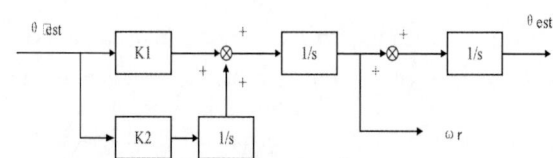

Fig. 8 Block diagram of position observer

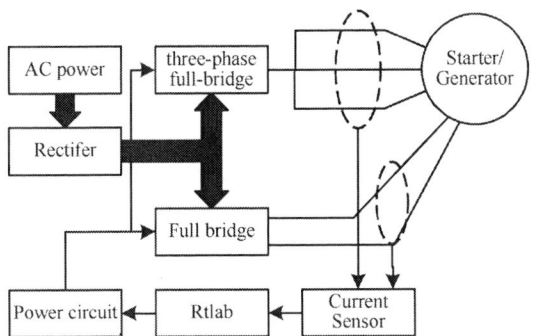

Fig. 9   Hardware structure of the experiment

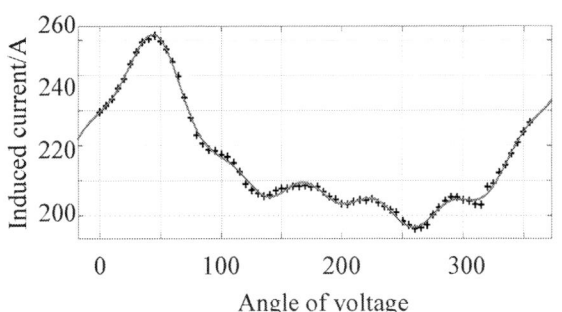

Fig. 11 Fitting by sixth-order Fourier series

frequency is 5 kHz.

In order to prevent machine from vibrating, the transient pulse signal voltage is injected to the stator terminal as show in Fig 10. The transient pulse signal voltage injects to the stator terminal in pairs. There is a 180° lag between the phase angle of the present and the previous transient pulse signal voltage. Before injecting transient pulse signal voltage to the stator terminal, the response current should be return to zero.

Fig. 11 shows the curve of current i'_d .The current in Fig. 11 which influenced by the magnetic saturation phenomenon is not change in accordance with the law in (Eq.4).When the angle of u'_d coincides with the real d-axis, the magnetic circuit in d-axis gets the most saturated status in the test which lead to the decrease of the inductance of d-axis and the current i'_d also reaches its maximum value. With the fading of saturation status

the response current i'_d decreases.

As magnetic circuit in q-axis is in the unsaturated, the response current in positive and negative direction of q-axis are unchanged. As a result, the maximum amplitude in the current curve of i'_d is real rotor position. Fourier series is adopted to fit the current curve in Fig. 11.

Then fit the currents with sixth-order Fourier series, the fitting curve matches the original data. The real d-axis is found without a second magnetic polarity identification process. The estimate rotor position is 47.6°in this experiment, the error between real rotor position and estimate position δ=49.38°-47.6°=1.78° .

It is difficult to implement the sixth-order Fourier series in real time embedded systems. For the purpose of decreasing high computational complexity of the fitting algorithm, a more effective algorithm is proposed.

The transient pulse signal voltage is injected to the stator terminal in pairs. The phase angle between the present and the previous transient pulse signal voltage is 180°. The pair of voltage is injected to the stator terminal successively with a short interval. First, the voltage in the direction $d'_{previous}$ is

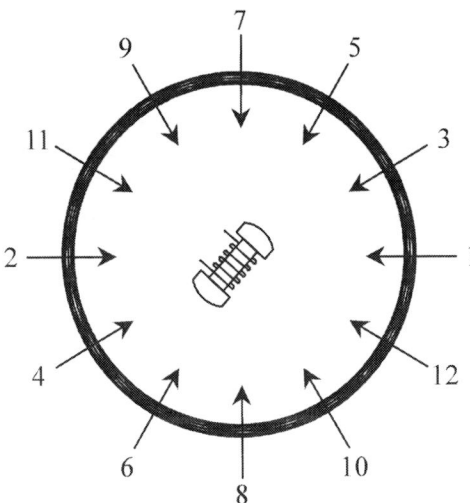

Fig. 10   The sequence of voltage vector

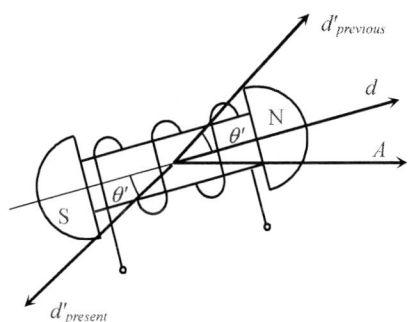

Fig. 12   The pair of voltage vector

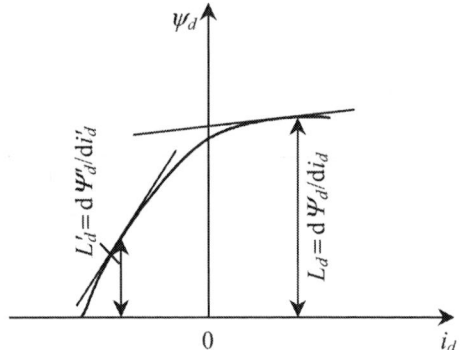

Fig. 13   Position estimation performance

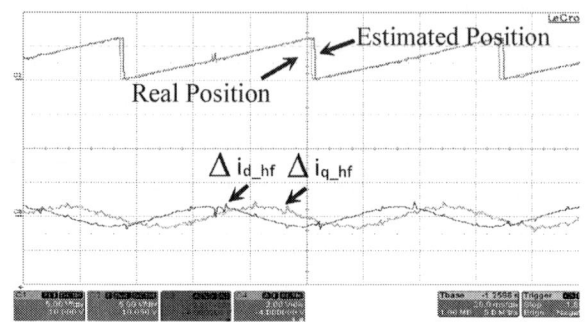

Fig. 15   Position estimation performance

injected to the stator terminal, then after 5 PWM duty cycles the second voltage which has same angle with $d'_{present}$ is injected to the stator terminal. The pair of voltage shows in Fig. 12.

Ignoring the magnetic polarity, the pair of voltage has the same angle as real d-axis. Those two voltages excite and weaken the rotor field respectively, as show in Fig. 13. A new current i''_d is obtained by the finite difference process which is capable of eliminating the pulsation noise of i_f.

$$i''_d = i'_{d\_present} - i'_{d\_previous}$$

Where i''_d is the new response current, $i'_{d\_present}$ is the present response current value, $i'_{d\_previous}$ is the previous current value of calculating instants.

The third-order Fourier series are used to fitting the new response current i''_d, as in Fig. 14. The estimate rotor position is 49.38° in this experiment, the error between real rotor position and estimate position is δ=49.38°-49.1°=0.28° . The error of estimated position is 0.18° less than the method which fitting the current curve by sixth-order Fourier series.

After the finite difference process a precise rotor position is found by third-order Fourier series. The proposed method use third-order Fourier series, which has characters of simple calculation and high accuracy, and easy to implement. The

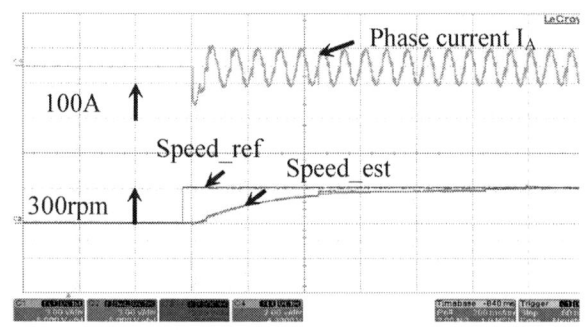

Fig. 16 Position estimation performance

Fig. 17   Experimental platform

finite difference process improves calculation accuracy by eliminate common mode noise.

Fig. 15 shows the position estimation performance on a constant speed operation at 300 r/min, the estimate position has virtually no time delay. The real position is obtained by a resolver in this experiment. $\Delta i^s_{q\_hf}, \Delta i^s_{d\_hf}$ which is influenced by noise and inverter nonlinearities contains severe pulsating components, However the position observer can eliminate those interferences and get a precise rotor position.

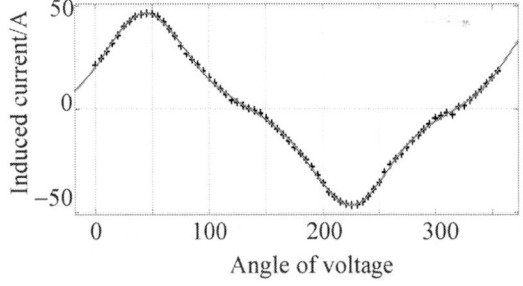

Fig. 14   Fitting by third-order Fourier series

978-1-4673-9551-9/16 $31.00 © 2016 IEEE

Fig. 16 shows the starting process from zero to 300 r/min, speed_ref is the given speed, and it can be seen the proposed method is stability in the low speed region. Fig. 17 is the experimental platform.

V. CONCLUSIONS AND FUTURE WORK

This paper has proposed an initial rotor position estimation method and sensorless control strategy for an AWSSG. For the reason of the excitation current of the main generator contains pulsating components, traditional initial rotor position estimate methods were failed. In this paper, Fourier analysis method is adopted to analysis the response currents which provides a successful initial position estimate solution. During the start-up, a control strategy based on the injection of square-wave-type voltage was proposed. The propose method enhanced the bandwidth and the dynamic characteristics of the controller in the sensorless drive.

The future research will focus on the potential sources of estimation inaccuracy include saturation and cross-saturation phenomena. Then, the sensorless control in full speed range will implement.

ACKNOWLEDGMENT

This work was supported by the National Natural Science Foundation of China (51277152).

REFERENCES

[1] Ningfei Jiao; Weiguo Liu; Jichang Peng; Shuai Mao; Hua Zhang, "Design and control strategy of a two-phase brushless exciter for three-stage starter/generator," Energy Conversion Congress and Exposition (ECCE), 2014 IEEE , vol., no., pp.5864,5869, 14-18 Sept. 2014

[2] Sarlioglu, B.; Morris, C.T., "More Electric Aircraft: Review, Challenges, and Opportunities for Commercial Transport Aircraft," Transportation Electrification, IEEE Transactions on , vol.1, no.1, pp.54,64, June 2015

[3] Gong, L.M.; Zhu, Z.Q., "Robust Initial Rotor Position Estimation of Permanent-Magnet Brushless AC Machines With Carrier-Signal-Injection-Based Sensorless Control," Industry Applications, IEEE Transactions on , vol.49, no.6, pp.2602,2609, Nov.-Dec. 2013.

[4] Hyunbae Kim; Kum-Kang Huh; Lorenz, R.D.; Jahns, T.M., "A novel method for initial rotor position estimation for IPM synchronous machine drives," Industry Applications, IEEE Transactions on , vol.40, no.5, pp.1369,1378, Sept.-Oct. 2004.

[5] Young-Doo Yoon; Seung-Ki Sul; Morimoto, S.; Ide, K., "High-Bandwidth Sensorless Algorithm for AC Machines Based on Square-Wave-Type Voltage Injection," Industry Applications, IEEE Transactions on , vol.47, no.3, pp.1361,1370, May-June 2011.

[6] Shih-Chin Yang, "Saliency-Based Position Estimation of Permanent-Magnet Synchronous Machines Using Square-Wave Voltage Injection With a Single Current Sensor," Industry Applications, IEEE Transactions on , vol.51, no.2, pp.1561,1571, March-April 2015.

[7] Zhe Chen; Jianbo Gao; Fengxiang Wang; Zhixun Ma; Zhenbin Zhang; Kennel, R., "Sensorless Control for SPMSM With Concentrated Windings Using Multisignal Injection Method," Industrial Electronics, IEEE Transactions on , vol.61, no.12, pp.6624,6634, Dec. 2014.

[8] Dongouk Kim; Yong-Cheol Kwon; Seung-Ki Sul; Jang-Hwan Kim; Rae-Sung Yu, "Suppression of injection voltage disturbance for High Frequency square-wave injection sensorless drive with regulation of induced High Frequency current ripple," Power Electronics Conference (IPEC-Hiroshima 2014 - ECCE-ASIA), 2014 International , vol., no., pp.925,932, 18-21 May 2014.

[9] Raca, D.; Garcia, P.; Reigosa, D.D.; Briz, F.; Lorenz, R.D., "Carrier-Signal Selection for Sensorless Control of PM Synchronous Machines at Zero and Very Low Speeds," Industry Applications, IEEE Transactions on , vol.46, no.1, pp.167,178, Jan.-feb. 2010.

[10] Maalouf, A.; Idkhajine, L.; Le Ballois, S.; Monmasson, E., "Field programmable gate array-based sensorless control of a brushless synchronous starter generator for aircraft application," Electric Power Applications, IET , vol.5, no.1, pp.181,192, January 2011.

[11] Griffo, A.; Drury, D.; Sawata, T.; Mellor, P.H., "Sensorless starting of a wound-field synchronous starter/generator for aerospace applications," Industrial Electronics, IEEE Transactions on , vol.59, no.9, pp.3579,3587, Sept. 2012.

A Novel Method for Polarity Detection of Non-salient PMSMs in Initial Position Estimation

Bing Liu*, Bo Zhou, Jiadan Wei, Long Wang, Tianheng Ni
Jiangsu Key Laboratory of New Energy Generation and Power Conversion
Nanjing University of Aeronautics and Astronautics
Nanjing, China
Emil: liubingnuaa@163.com

Abstract—This paper proposes a novel method for magnetic polarity detection in pulsating current injection based self-sensing control of surface-mounted permanent-magnet synchronous motors (SPMSM). When a proper high frequency current is superimposed in *d*-axis, the flux linkage saturates which can lead to the distortion of the resulting voltage. With the Fourier analysis, the 2nd harmonic of the resulting voltage in estimated *d*-axis includes the magnetic polarity information. A demodulation mechanism is designed to detect the magnetic polarity. This method can achieve fast and accurate initial position estimation and is verified by the simulations and experiments.

Keywords—*self-sensing; magnetic polarity detection; current injection; PMSM.*

I. INTRODUCTION

This paper focuses on the self-sensing control of permanent-magnet synchronous motors (PMSM). As is well-known, position estimation at zero and very low speeds is one of the difficulties. In recent years, high frequency signal injection-based methods have attracted a lot of interests [1-9]. To surface-mounted permanent-magnet synchronous motors (SPMSM), pulsating voltage injection works well in self-sensing control at low speed range [10, 11]. However, it requires 2 low-pass-filters (LPF) to filter out the high frequency component in the feedback of *d*- and *q*-axis currents which complicates the system, and the reliability of system suffers from the temperature-drift of stator resistance [12]. Pulsating current injection simplifies the system by removing 2 LPFs, and the reliability of system has nothing to do with the temperature-drift. Nevertheless, it injects high frequency current in *d*-axis. If a PI regulator is applied as the current regulator, the feedback contains amplitude attenuation and phase delay which enlarges the estimating error [13]. Applying a resonant control in current regulators has been proved an efficient way to track the high frequency given current. As initial position estimation is an essential process in self-sensing control, how to estimate initial position and detect magnetic polarity based on pulsating current injection need further study. Such problems restrict the application of pulsating current injection.

This paper proposes a novel method for magnetic polarity detection in initial position estimation based on high frequency

pulsating current injection. In part II, the principle of pulsating current injection is introduced. In part III the Fourier analysis of resulting voltage in estimated *d*-axis is carried out and the novel method for magnetic polarity detection is proposed. Simulations and experiments are carried out to verify the proposed method, and the results are presented in part IV. The conclusion is given in Part V.

II. PULSATING CURRENT INJECTION

A. Mathematical Model of SPMSM

Ignoring the eddy current and hysteresis losses, and assuming the magnetic field is spatially sinusoidal, the continuous-time model representing the electrical subsystem of a SPMSM can be written as follows:

$$
\begin{bmatrix} u_d \\ u_q \end{bmatrix} = \begin{bmatrix} r_s & -\omega L_q \\ \omega L_d & r_s \end{bmatrix} \begin{bmatrix} i_d \\ i_q \end{bmatrix} + \begin{bmatrix} L_d & 0 \\ 0 & L_q \end{bmatrix} p \begin{bmatrix} i_d \\ i_q \end{bmatrix} + \begin{bmatrix} 0 \\ \omega \psi_f \end{bmatrix} \quad (1)
$$

Where:

u_d, u_q — *d*- and *q*-axis voltages;

i_d, i_q — *d*- and *q*-axis currents;

L_d, L_q — *d*- and *q*-axis inductances;

r_s — Stator resistance;

ψ_f — Flux linkage of rotor magnet without stator current;

ω — Electrical frequency of rotor.

At zero and low speeds, the cross coupling term and the back EMF can be negligible, (1) is simplified as follows:

$$
\begin{bmatrix} u_d \\ u_q \end{bmatrix} = \begin{bmatrix} r_s & 0 \\ 0 & r_s \end{bmatrix} \begin{bmatrix} i_d \\ i_q \end{bmatrix} + \begin{bmatrix} L_d & 0 \\ 0 & L_q \end{bmatrix} p \begin{bmatrix} i_d \\ i_q \end{bmatrix} = \begin{bmatrix} Z_d & 0 \\ 0 & Z_q \end{bmatrix} \begin{bmatrix} i_d \\ i_q \end{bmatrix} \quad (2)
$$

Where $Z_d = r_s + \omega L_d$, $Z_q = r_s + \omega L_q$.

B. Principle of plusating current injection

SPMSMs have no spatial saliency due to their structure. However, the d-axis flux linkage saturates easily if injecting a proper current in d-axis. Then it differs from the q-axis flux linkage, and this makes it possible to estimate the rotor position by tracking the rotor saliency.

Define: the stationary reference frame is α-β axis; the actual rotor synchronous reference frame is d-q axis; the estimated rotor synchronous reference frame is $\hat{d}$-$\hat{q}$ axis. As shown in Fig.1.

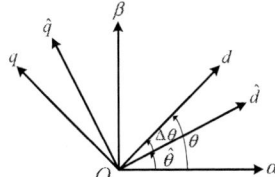

Fig. 1 Relationship among the reference frames.

The rotor position estimating error $\Delta\theta$ can be written as:

$$\Delta\theta = \theta - \hat{\theta} \tag{3}$$

According to (2) and (3), the voltage expression in the estimated d-q axis can be resolved as follows:

$$\begin{bmatrix} \hat{u}_d \\ \hat{u}_q \end{bmatrix} = \begin{bmatrix} Z + \Delta Z\cos 2\Delta\theta & \Delta Z\sin 2\Delta\theta \\ \Delta Z\sin 2\Delta\theta & Z - \Delta Z\cos 2\Delta\theta \end{bmatrix} \begin{bmatrix} \hat{i}_d \\ \hat{i}_q \end{bmatrix} \tag{4}$$

Where $Z = (Z_d + Z_q)/2$; $\Delta Z = (Z_d - Z_q)/2$.

A pulsating current with the amplitude I_m and frequency ω_h is superimposed in the estimated d-axis as follows:

$$\begin{bmatrix} \hat{i}_{dh} \\ \hat{i}_{qh} \end{bmatrix} = \begin{bmatrix} I_m\sin\omega_h t \\ 0 \end{bmatrix} \tag{5}$$

If the current regulator is well designed, the feedbacks equal to the given currents in the estimated d- and q-axis. Applying (5) to (4), the resulting voltages are as follows:

$$\begin{bmatrix} \hat{u}_{dh} \\ \hat{u}_{qh} \end{bmatrix} = \begin{bmatrix} (Z + \Delta Z\cos 2\Delta\theta)I_m\sin\omega_h t \\ \Delta Z\sin 2\Delta\theta I_m\sin\omega_h t \end{bmatrix} \tag{6}$$

The resulting voltage in the estimated q-axis is as follows:

$$\hat{u}_{qh} = \omega_h \Delta L I_m \sin 2\Delta\theta \cos \omega_h t \tag{7}$$

By demodulating $\hat{u}_{qh}$, the rotor position estimating error can be expressed as follows:

$$f_c(\Delta\theta) = LPF(\hat{u}_{qh} * \cos\omega_h t) = \omega_h \Delta L I_m \sin 2\Delta\theta / 2 \tag{8}$$

A closed-loop control system is constructed to keep $f_c(\Delta\theta)=0$ as shown in Fig. 2. Then $\sin 2\Delta\theta=0$, which means $\Delta\theta$ equals to either 0 or π. The magnetic polarity should be detected to ensure $\Delta\theta=0$. This will be discussed in the following part.

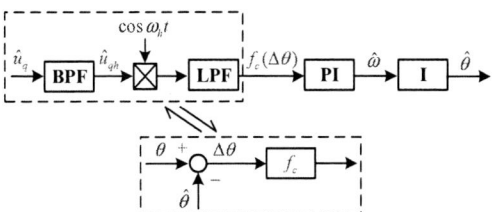

Fig. 2 Closed-loop control system of position.

III. Initial Position Estimation

A commonly used way to detect the magnetic polarity is injecting a positive and a negative pulse voltage into the estimated d-axis. The resulting currents are different in amplitude or some other characters as the d-axis flux linkage saturates. This is an effective method. However, it is somehow complicated because the superimposed signal in the estimated d-axis should be changed twice, from high frequency sine wave to pulses, and then back to high frequency sine wave. And moreover, the width and amplitude of the pulses should be selected carefully according to a particular motor as these parameters differ in different motors.

To simplify the magnetic polarity detection process, this paper presents a novel method. The superimposed signal will only be the high frequency current in initial position estimation and the time cost is significantly decreased.

A. Harmonics in the resulting voltage

Applying the closed-loop control system of position in Fig. 2, the rotor position estimating error is close to either 0 or π. According to (6), the resulting voltage in the estimated d-axis is as follows:

$$\hat{u}_{dh} = \sqrt{r_s^2 + (\omega_h L_d)^2}\, I_m \sin(\omega_h t + \arctan\frac{\omega_h L_d}{r_s}) \tag{9}$$

Due to the rotor structure of SPMSM, the d-axis flux linkage saturates when a proper current is superimposed. As a result, the d-axis inductance is smaller than the q-axis inductance. To simplify the analysis, define L_d^+ as the average inductance in d-axis when the current is positive; define L_d^- as the average inductance when the current is negative. Obviously, $L_d^- > L_d^+$.

As the rotor position estimating error can be close to 0 or π, first we assume it is around 0. According to (5) and (6), the resulting voltage in estimated d-axis can be expressed as follows:

$$\hat{u}_{dh} = \begin{cases} \left| Z_d^+ \right| I_m \sin(\omega_h t + \varphi_d^+), \ t \in (\dfrac{0+2k\pi}{\omega_h}, \dfrac{\pi+2k\pi}{\omega_h}] \\[3mm] \left| Z_d^- \right| I_m \sin(\omega_h t + \varphi_d^-), \ t \in (\dfrac{\pi+2k\pi}{\omega_h}, \dfrac{2\pi+2k\pi}{\omega_h}] \end{cases} \quad (10)$$

Where $\begin{cases} \left| Z_d^+ \right| = \sqrt{r_s^2 + (\omega_h L_d^+)^2} \\ \left| Z_d^- \right| = \sqrt{r_s^2 + (\omega_h L_d^-)^2} \end{cases}$, $\begin{cases} \varphi_d^+ = \arctan\left(\omega_h L_d^+ / r_s\right) \\ \varphi_d^- = \arctan\left(\omega_h L_d^- / r_s\right) \end{cases}$.

Applying Fourier analysis and ignoring the 3rd and more than 3rd harmonics, (10) can be further expressed as follows:

$$\hat{u}_{dh} = \frac{a_0}{2} + a_1 \cos(\omega_h t) + b_1 \sin(\omega_h t) \\ + a_2 \cos(2\omega_h t) + b_2 \sin(2\omega_h t) \quad (11)$$

Where $a_0 = 0$, $a_1 = \omega_h (L_d^+ + L_d^-) I_m / 2$, $b_1 = r_s I_{mh}$, $a_2 = 0$, $b_2 = \dfrac{4}{3\pi} \omega_h (L_d^+ - L_d^-) I_m$. Substituting these values into (11):

$$\hat{u}_{dh} = \frac{\omega_h (L_d^+ + L_d^-) I_m}{2} \cos(\omega_h t) + r_s I_{mh} \sin(\omega_h t) \\ - \frac{4}{3\pi} \omega_h (L_d^- - L_d^+) I_m \sin(2\omega_h t) \quad (12)$$

Obviously, the resulting voltage in the estimated d-axis includes not only the fundamental component but the 2nd harmonic.

If the rotor position estimating error is around π, (10) should be expressed as follows:

$$\hat{u}_{dh}^* = \begin{cases} \left| Z_d^- \right| I_m \sin(\omega_h t + \varphi_d^-), \ t \in (\dfrac{0+2k\pi}{\omega_h}, \dfrac{\pi+2k\pi}{\omega_h}] \\[3mm] \left| Z_d^+ \right| I_m \sin(\omega_h t + \varphi_d^+), \ t \in (\dfrac{\pi+2k\pi}{\omega_h}, \dfrac{2\pi+2k\pi}{\omega_h}] \end{cases} \quad (13)$$

Similarly, applying Fourier analysis, (13) can be further expressed as follows:

$$\hat{u}_{dh}^* = \frac{\omega_h (L_d^+ + L_d^-) I_m}{2} \cos(\omega_h t) + r_s I_{mh} \sin(\omega_h t) \\ + \frac{4}{3\pi} \omega_h (L_d^- - L_d^+) I_m \sin(2\omega_h t) \quad (14)$$

Define $s(\Delta\theta)$ as follows:

$$s(\Delta\theta) = \begin{cases} 1, \ \Delta\theta \approx 0 \\ -1, \ \Delta\theta \approx \pi \end{cases} \quad (15)$$

And then (11) and (14) can be further expressed as follows:

$$\hat{u}_{dh} = \frac{\omega_h (L_d^+ + L_d^-) I_m}{2} \cos(\omega_h t) + r_s I_{mh} \sin(\omega_h t) \\ - \frac{4}{3\pi} \omega_h (L_d^- - L_d^+) I_m s(\Delta\theta) \sin(2\omega_h t) \quad (16)$$

The 2nd harmonic of the resulting voltage in the estimated d-axis is as follows:

$$\hat{u}_{dh2} = -\frac{4}{3\pi} \omega_h (L_d^- - L_d^+) I_m s(\Delta\theta) \sin(2\omega_h t) \quad (17)$$

It is not hard to see, the magnetic polarity can be detected by demodulating $\hat{u}_{dh2}$. This will be discussed in the following part.

B. Magnetic ploarity detection

Apply a band-pass filter (BPF) and a low-pass filter (LPF) in the demodulating process of magnetic polarity detection as shown in Fig. 3. The magnetic polarity information can be expressed as follows:

$$g_{NS} = LPF[BPF(\hat{u}_d) * \sin(2\omega_h t)] \\ = -\frac{2}{3\pi} \omega_h (L_d^- - L_d^+) I_m \, s(\Delta\theta) \quad (18)$$

According to (15) and (18), if $\Delta\theta$ is around 0, then $g_{NS} < 0$. If $\Delta\theta$ is around π, then $g_{NS} > 0$. As a result, the sign of g_{NS} can be used to detect the magnetic polarity and set the proper value of compensation angle θ_c (0 or π).

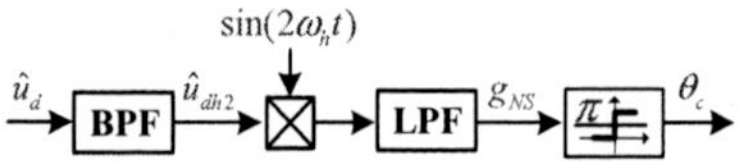

Fig. 3 Process of the magnetic polarity detection

The self-sensing vector control scheme is present in Fig. 4. The process of initial position estimation includes 2 steps. First the primary initial position is obtained from the resulting voltage in the estimated q-axis; second the magnetic polarity is detected by demodulating the 2nd harmonic of the resulting voltage in the estimated d-axis, and the compensation angle is obtained by detecting the sign of g_{NS}. After the 2 steps, the accurate initial position is obtained. In the step of magnetic polarity detection, no extra testing signal but high frequency current is needed. This method can achieve fast and accurate initial position estimation.

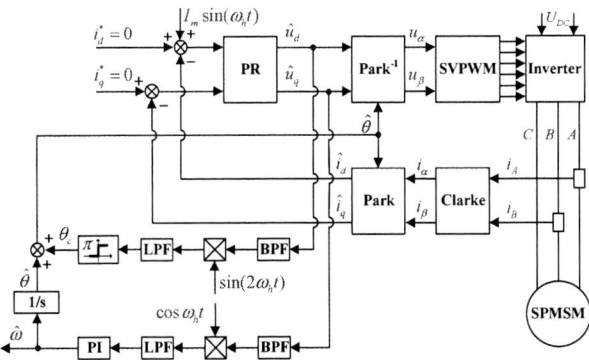

Fig. 4 The self-sensing vector control scheme

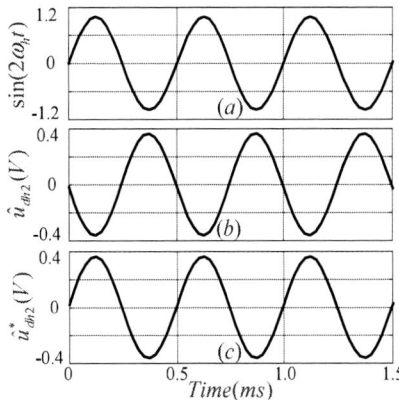

Fig. 5 The 2nd harmonic of resulting voltage in estimated d-axis (simulations)

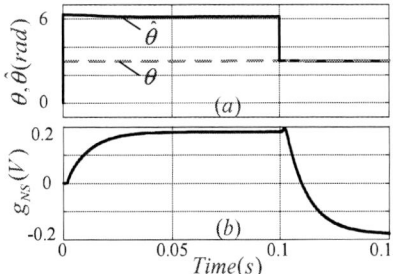

Fig. 6 Process of the initial position estimation (simulations)

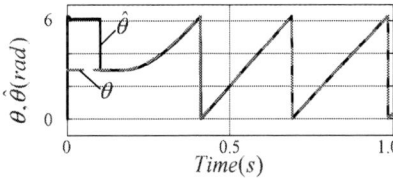

Fig. 7 The starting process (simulations)

IV. SIMULATIONS AND EXPERIMENTS

Simulations and experiments are carried out on a SPMSM to evaluate the proposed initial position estimating method. A RT-LAB is applied as the controller. The switching frequency is 20kHz, the frequency of the superimposed current is 1kHz, the amplitude of the superimposed current is 0.5A，the dead time is 2μs. Table I shown the parameters of the SPMSM.

TABLE I. PARAMETERS OF THE SPMSM

Character	Value	Character	Value
Nominal Output (P_N)	400W	Resistance (r_s)	3.2Ω
Nominal Speed (n_N)	6000r/min	d-axis Inductance (L_d)	8mH
Nominal Current (I_N)	2.1A	q-axis Inductance (L_q)	8mH

A. Simulations

Fig. 5 shows the 2nd harmonic of resulting voltage in estimated d-axis. Fig. 5(a) presents a reference sine wave with the frequency 2 kHz. Fig. 5(b) shows $\hat{u}_{dh2}$ which corresponds to the situation that $\Delta\theta$ is around 0. Fig. 5(c) shows $\hat{u}_{dh2}^*$ which corresponds to the situation $\Delta\theta$ is around π. According to Fig. 5, the resulting voltage in the estimated d-axis contains 2nd harmonic due to the saturation of d-axis flux linkage. The phase difference between $\hat{u}_{dh2}$ and $\hat{u}_{dh2}^*$ is π.

Fig. 6 shows the process of initial position estimation. Fig. 6(a) presents the real rotor position and the estimated rotor position. Fig. 6(b) shows the magnetic polarity information g_{NS}. At first the estimating error is close to π, and $g_{NS} > 0$. According to Fig. 3, the compensatory angle is π, At $t = 0.1$ s, the estimated position is compensated, after compensation, the estimated position is very close to the real position, and $g_{NS} < 0$.

Fig. 7 shows the starting process. With the proposed initial position estimating method, the motor starts quickly and smoothly.

B. Experiments

Fig.8 shows the 2nd harmonic of resulting voltage in estimated d-axis. Fig. 8(a) is the wave of $\hat{u}_{dh2}$ when $\Delta\theta$ is around 0. Fig. 8(b) is the wave of $\hat{u}_{dh2}^*$ when $\Delta\theta$ is around π. The phase difference between $\hat{u}_{dh2}$ and $\hat{u}_{dh2}^*$ is π rad obviously. Fig.9 shows the process of initial position estimation. θ, $\hat{\theta}$, $\Delta\theta$ and g_{NS} are all given in this figure. At first, the estimating error is around π, and $g_{NS} > 0$, according to Fig. 3, the compensatory angle is π. After compensation, the estimating error is around 0, and $g_{NS} < 0$. Fig. 10 shows the starting process. The given speed is 100r/min. Thanks to the accurate initial position, the motor starts quickly and smoothly with a very small estimating error.

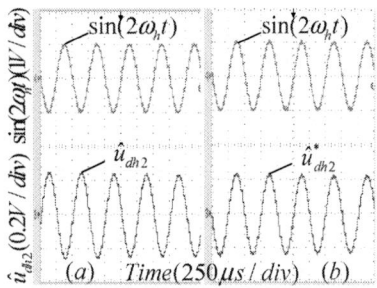

Fig. 8 The 2nd harmonic of resulting voltage in estimated d-axis (experiments)

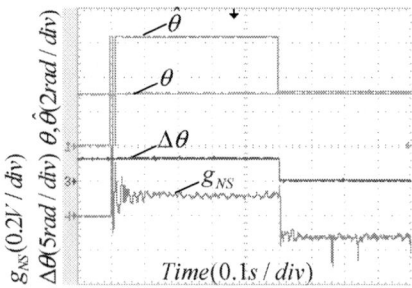

Fig. 9 Process of initial position estimation (experiments)

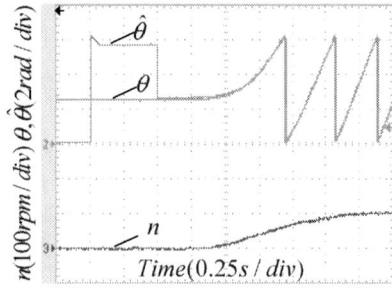

Fig. 10 The starting process (experiments)

V. CONCLUSION

This paper makes a contribution to pulsating current injection based self-sensing control of SPMSM at zero speed. A novel method to detect the magnetic polarity is proposed in the initial position estimation. The saturation of the flux linkage in estimated d-axis leads to the distortion of the resulting voltage. And the 2nd harmonic of the resulting voltage in estimated d-axis includes the magnetic polarity information. A demodulation mechanism of this information is designed. In the process of magnetic polarity detection, no extra testing signals but high frequency pulsating current is needed which achieves fast and accurate initial position estimation. Simulations and experiments both verify the proposed method.

ACKNOWLEDGMENT

This work was supported by the Specialized Research Fund for the Doctoral Program of Higher Education, MOE (20123218130002) and by Jiangsu province university outstanding science and technology innovation team project.

REFERENCES

[1] J. H. Jang, S. K. Sul, J. I. Ha, et al, "Sensorless drive of surface-mounted permanent-magnet motor by high-frequency signal injection based on magnetic saliency," *IEEE Transactions on Industry Applications*, vol. 39, no. 4, pp. 1031-1039, 2003.

[2] Y. Jeong, R. D. Lorenz, T. M. Jahns, et al, "Initial rotor position estimation of an interior permanent magnet synchronous machine using carrier-frequency injection methods," in *IEEE 2003 International Conference on Electric Machines and Drives*, 2003, pp. 1218-1223.

[3] D. Race, P. Garcia, D. D. Reigosa, et al, "Carrier-Signal Selection for Sensorless Controlof PM Synchronous Machines at Zero and Very Low Speeds," *IEEE Transactions on Industry Applications*, vol. 46, no. 1, pp. 167-178, 2010.

[4] C. C. Choi, J. K. Seok, "Compensation of Zero-Current Clamping Effects in High-Frequency-Signal-Injection-Based Sensorless PM Motor Drives," *IEEE Transactions on Industry Applications*, vol. 43, no. 5, pp. 1258-1265, 2007.

[5] O. A. Mohammed, A. A. Khan, "A Wavelet Filtering Scheme for Noise and Vibration Reduction in High-frequency Signal Injection-Based Sensorless Control of PMSM at Low Speed," *IEEE Transactions on Energy Conversion*, vol. 27, no. 2, pp. 250-260, 2012.

[6] J. M. Liu, Z. Q. Zhu, "Novel Sensorless Control Strategy With Injection of High-Frequency Pulsating Carrier Signal Into Stationary Reference Frame," *IEEE Transactions on Industry Applications*, vol. 50, no. 4, pp. 2574-2583, 2014.

[7] S. Medjmadj, D. Diallo, M. Mostefai, et al, "PMSM Drive Position Estimation: Contribution to the High-Frequency Injection Voltage Selection Issue," *IEEE Transactions on Energy Conversion*, vol. 30, no. 1, pp. 349-358, 2015.

[8] T. Szalai, G. Berger, J. Petzoldt, "Stabilizing Sensorless Control Down to Zero Speed by Using the High-Frequency Current Amplitude," *IEEE Transactions on Power Electronics*, vol. 29, no. 7, pp. 3646-3656, 2014.

[9] B. Liu, B. Zhou, J. D. Wei, et al, "A rotor initial position estimation method for sensorless control of SPMSM," in *40th Annual Conference of the IEEE Industrial Electronics Society*, 2014, pp. 354-359.

[10] S. C. Yang, R. D. Lorenz, "Comparison of resistance-based and inductance-based self-sensing controls for surface permanent-magnet machines using high-frequency signal injection," *IEEE Transactions on Industry Applications*, vol. 48, no. 3, pp. 977-986, 2012.

[11] O. A. Mohammed, A. A. Khan, A. M. Eltallawy, et al, "A Wavelet Filtering Scheme for Noise and Vibration Reduction in High-frequency Signal Injection-Based Sensorless Control of PMSM at Low Speed," *IEEE Transactions on Energy Conversion*, vol. 27, no. 2, pp. 250-260, 2012.

[12] Y. Liu, B. Zhou, Y. Feng, et al, "Sensorless control with two types of pulsating high frequency signal injection methods for SPMSM at low speed," in *IEEE 2011 International Conference on Power Electronics Systems and Applications*, 2011, pp. 1-5.

[13] J. M. Wang, X. M. Zhao, Z. Guo, "Analysis and Improvement of Pulsating Current Injection Based Sensorless Control of Permanent Magnet Synchronous Motor," in *IEEE 2014 International Conference on Electrical Machines and Systems*, 2014, pp. 716-721.

A Speed Adaptive Sensorless Flux Observer for the Induction Motor Drive using Sylvester Criterion Design

Mihai Comanescu
Department of Engineering
Penn State Altoona
Altoona, PA, USA

Abstract – **The paper presents the design of a speed and flux observer for the sensorless induction motor (IM) drive. The design is based on the full-order model of the IM in the stationary reference frame. The proposed method consists of an observer with linear feedback that yields the motor fluxes and also estimates the speed using an adaptive law. Convergence conditions are formulated based on Lyapunov's nonlinear stability theory - the design of the feedback gains is done using Sylvester criterion. While this observer structure was discussed in the literature previously, the novelty of the method is given by the Sylvester criterion formulation - this uncovers interesting inequalities and offers insight into the design choices and stability margins of the observer. The method can be used to obtain the field orientation angle and speed needed in a sensorless IM drive. The theory is supported by simulations and experiments.**

Keywords: **induction motor control, flux estimation, speed estimation, rotor position estimation, Lyapunov stability theory, adaptive control, sensorless control.**

I. INTRODUCTION

The induction motor (IM) drive is widely used in industrial processes, especially to drive fans, pumps and compressors but also in windmills, electric vehicles or in rail transportation. Field-oriented control is the preferred control method for the IM drive in high-performance applications – while stator-field or airgap-field orientation are possible [1][2], rotor-field orientation is often preferred. The rotor field-oriented scheme is simpler since it does not require steady-state decoupling. Based on that, the synchronous currents i_d, i_q can be easily controlled to obtain the desired flux and torque [3]-[6].

To control i_d and i_q, the angle of the rotor flux is needed. In addition to that, the flux magnitude and the motor speed are also required for control implementation. In a sensorless scheme, these physical quantities need to be obtained using only the electrical measurements (measured voltages and currents).

The typical approach for flux-speed estimation is to construct an observer using one of the models of the induction motor. There is a wide variety of approaches: estimation can be done with or without signal injection - a general review of the techniques was presented in [7]. The approaches include open-loop estimators [8][9], closed-loop observers [10][11] or Model Reference Adaptive Control [12][13].

The field orientation angle can also be obtained using signal-injection [14][15]; implementation is difficult, however, this is often the only approach that works at low speed.

If the stationary frame fluxes $\lambda_\alpha, \lambda_\beta$ are estimated using an observer (and are available), the rotor flux angle and the flux magnitude can be calculated using the Direct Field Orientation (DFO) method [10][11][16]. DFO is quite popular and is often used in speed sensorless IM drives.

Several DFO methods have been presented: some treat the IM as a time-invariant plant and use linear design methods [17]; other consider the IM a time-varying plant (since the speed varies) and use nonlinear design methods to design state or state-speed estimators [18]. The complexity of the estimation mathematics increases even more when, along with the IM state and speed, parameter estimation is also attempted [19][20].

A special family of sensorless methods that obtain the motor fluxes and speed corresponds to the adaptive observers described in [21]-[28]. These methods use an observer with linear feedback to obtain the states and employ an adaptive law to estimate the speed. The gain design initially proposed in [21][22] yields a closed-loop system whose poles are proportional with the open-loop poles. This type of gain design was also used in [23][26]; sometimes, the proportionality constant is zero and there are issues related to the instability of the speed estimate during low speed regeneration.

The design of the feedback gains of these adaptive observers follows from working with the characteristic equation of the closed-loop system. Since this is a polynomial of the 4th order, the calculations are difficult and the gains end up being complicated algebraic expressions.

This design approach has a few disadvantages: first, having closed-loop poles that are a multiple of the motor's natural poles is not necessarily a guarantee of good performance. Second, the approach does not provide any indication regarding the stability margins.

978-1-4673-9551-9/16 $31.00 © 2016 IEEE

A separate issue is that the gains depend on ω_r: since the speed is not measured, only a speed estimate $\hat{\omega}_r$ is available (this is used instead by assuming that $\hat{\omega}_r \approx \omega_r$). While $\hat{\omega}_r$ should match ω_r most of the time (for a converging design), the dynamics of the speed estimate depends on the tuning of the adaptation law; therefore, $\hat{\omega}_r$ can differ significantly from ω_r during fast transients or at speed reversal. Therefore, it would be necessary to understand how to make the observer less sensitive to the potential speed mismatch $\bar{\omega}_r$.

This paper presents a new design approach for the adaptive observer presented in [21][22] and its versions shown in [23]-[29]. In the same manner, the speed of the induction motor is obtained using an adaptive law that involves a PI controller.

The gain design process is based on Lyapunov's nonlinear stability method and uses the Sylvester criterion. This approach results in a series of inequalities that allow to formulate the stability conditions, to design the gains and also to obtain insight regarding the stability margins. This is the main novelty.

II. NOMENCLATURE

ω_r	Rotor electrical speed
ω_e	Synchronous speed
R_s, R_r	Stator, rotor resistances
L_m, L_s, L_r	Magnetizing, stator and rotor inductances
η	Inverse of rotor time constant
v_α, v_β	Voltages in stationary reference frame
i_α, i_β	Currents in stationary reference frame
$\lambda_\alpha, \lambda_\beta$	Rotor fluxes in stationary reference frame
$x, \hat{x}$	Real and estimated value of x
$\bar{x}$	Mismatch of x

III. STATE-SPACE MODEL OF THE INDUCTION MOTOR

The state-space model of the induction motor consists of the equations of the fluxes and currents in the stationary reference frame. The general form of the model is:

$$\begin{cases} \dot{x} = Ax + Bu \\ y = Cx \end{cases} \quad (1)$$

where $x = [i_\alpha \quad i_\beta \quad \lambda_\alpha \quad \lambda_\beta]^T$ is the corresponding state vector and $u = [v_\alpha \quad v_\beta]$ is the control. The matrices are:

$$A = \begin{bmatrix} -\gamma & 0 & \eta\beta & \omega_r\beta \\ 0 & -\gamma & -\omega_r\beta & \eta\beta \\ \eta L_m & 0 & -\eta & -\omega_r \\ 0 & \eta L_m & \omega_r & -\eta \end{bmatrix} \quad (2)$$

$$B = \frac{1}{\sigma L_s}\begin{bmatrix} 1 & 0 \\ 0 & 1 \\ 0 & 0 \\ 0 & 0 \end{bmatrix} \quad C = \begin{bmatrix} 1 & 0 & 0 & 0 \\ 0 & 1 & 0 & 0 \end{bmatrix} \quad (3)$$

Model (1) does not include the equation of the torque and the mechanical equation of the speed – this is because in the development shown, the equations of the torque and speed are not used. Matrix A is time-varying since it depends on ω_r. The parameters $\sigma, \beta, \gamma, \eta$ are:

$$\sigma = 1 - \frac{L_m^2}{L_s L_r} \quad \beta = \frac{L_m}{\sigma L_s L_r} \quad \gamma = \frac{1}{\sigma L_s}\left(\frac{L_m^2}{L_r^2}R_r + R_s\right) \quad \eta = \frac{R_r}{L_r} \quad (4)$$

and they are positive: $\sigma > 0, \beta > 0, \gamma > 0, \eta > 0$.

IV. SPEED ADAPTIVE OBSERVER DESIGN

Working with model (1), the voltages v_α, v_β and currents i_α, i_β are measured. The problem is to design an observer that estimates the state vector and the speed.

The form the observer is:

$$\dot{\hat{x}} = \hat{A}\hat{x} + Bu + GC(\hat{x} - x) \quad (5)$$

where $\hat{A} = \hat{A}(\hat{\omega}_r)$ is the estimate of the state transition matrix and G is a matrix of feedback gains whose generic form corresponds to the one in [22]:

$$G = \begin{bmatrix} g_1 & -g_2 \\ g_2 & g_1 \\ g_3 & -g_4 \\ g_4 & g_3 \end{bmatrix} \quad (6)$$

The estimation mismatches are defined as $\bar{x} = \hat{x} - x$. After subtraction, the dynamics of the mismatches is:

$$\dot{\bar{x}} = (A + GC)\bar{x} + \Delta A\hat{x} \quad (7)$$

In (7), the difference matrix ΔA depends on the speed mismatch $\bar{\omega}_r = \hat{\omega}_r - \omega_r$. The expression of ΔA is:

$$\Delta A = \begin{bmatrix} 0 & 0 & 0 & \bar{\omega}_r\beta \\ 0 & 0 & -\bar{\omega}_r\beta & 0 \\ 0 & 0 & 0 & -\bar{\omega}_r \\ 0 & 0 & \bar{\omega}_r & 0 \end{bmatrix} \quad (8)$$

The design of the observer requires to design the four feedback gains in (6) and also to obtain the speed estimate $\hat{\omega}_r$.
For that, select the Lyapunov function:

$$V = \bar{i}_\alpha^2 + \bar{i}_\beta^2 + \bar{\lambda}_\alpha^2 + \bar{\lambda}_\beta^2 + \frac{\bar{\omega}_r^2}{c} \quad (9)$$

where c is a constant $(c > 0)$ that needs to be selected.

It is clear that V is positive definite. Using the 4th order identity matrix I, this can be rewritten as:

$$V = \bar{x}^T I \bar{x} + \frac{\bar{\omega}_r^2}{c} \quad (10)$$

Function V is differentiated; the result is:

$$\dot{V} = \dot{\bar{x}}^T I \bar{x} + \bar{x}^T I \dot{\bar{x}} + 2\frac{\bar{\omega}_r \dot{\bar{\omega}}_r}{c} \quad (11)$$

After replacing the derivatives, the expression of $\dot{V}$ becomes:

$$\dot{V} = \bar{x}^T[(A + GC)^T + (A + GC)]\bar{x} + \hat{x}^T\Delta A^T\bar{x} + \bar{x}^T\Delta A\hat{x} + 2\frac{\bar{\omega}_r \dot{\bar{\omega}}_r}{c} \quad (12)$$

By rewriting the last three terms in (12), $\dot{V}$ becomes:

$$\dot{V} = \bar{x}^T[(A + GC)^T + (A + GC)]\bar{x} + 2\bar{\omega}_r(\hat{\lambda}_\alpha\bar{\lambda}_\beta - \hat{\lambda}_\beta\bar{\lambda}_\alpha) + 2\bar{\omega}_r(\hat{\lambda}_\beta\bar{i}_\alpha - \hat{\lambda}_\alpha\bar{i}_\beta) + 2\frac{\bar{\omega}_r \dot{\bar{\omega}}_r}{c} \quad (13)$$

Working with the expression in (13), the objective is to select the design variables such that $\dot{V}$ is negative definite ($\dot{V} < 0$).

First, note that the term $2\bar{\omega}_r(\hat{\lambda}_\alpha\bar{\lambda}_\beta - \hat{\lambda}_\beta\bar{\lambda}_\alpha)$ cannot be calculated. Since the real fluxes λ_α, λ_β are unknown quantities, $\bar{\lambda}_\alpha, \bar{\lambda}_\beta$ are also unknown. This term will be denoted as:

$$\epsilon = 2\bar{\omega}_r(\hat{\lambda}_\alpha\bar{\lambda}_\beta - \hat{\lambda}_\beta\bar{\lambda}_\alpha) \quad (14)$$

The term ϵ cannot be eliminated; the hope is that this term is small and plays a minor role in the performance of the proposed flux-speed estimator.

An adaptive law for the speed is chosen such that the last two terms in (13) cancel each other. The condition is:

$$\dot{\bar{\omega}}_r = -c(\hat{\lambda}_\beta\bar{i}_\alpha - \hat{\lambda}_\alpha\bar{i}_\beta) \quad (15)$$

After integrating (15), the speed mismatch $\bar{\omega}_r$ is obtained as:

$$\bar{\omega}_r = c\int(\hat{\lambda}_\alpha\bar{i}_\beta - \hat{\lambda}_\beta\bar{i}_\alpha) \quad (16)$$

In order to improve the dynamics of the speed estimate, it is typical to feed the term $(\hat{\lambda}_\alpha \bar{\imath}_\beta - \hat{\lambda}_\beta \bar{\imath}_\alpha)$ through a PI controller [21][22][23] rather than through a simple integrator. Then, assuming zero initial conditions, the speed estimate is:

$$\hat{\omega}_r = \left(K_p + \frac{K_i}{s}\right) c\left(\hat{\lambda}_\alpha \bar{\imath}_\beta - \hat{\lambda}_\beta \bar{\imath}_\alpha\right) \qquad (17)$$

In (17), the PI gains K_p, K_i need to be designed.

V. Design of the Feedback Gains

Working with expression (13), it is necessary to design the gains of matrix G such that the quadratic form $\bar{x}^T[(A + GC)^T + (A + GC)]\bar{x}$ will be negative. This is equivalent to saying that the matrix $-[(A + GC)^T + (A + GC)] = -A^*$ should be positive definite.

Using the expressions of C and G in (3),(6), matrix $-A^*$ is:

$$-A^* = \begin{bmatrix} 2(\gamma - g_1) & 0 & -l_3 & -l_4 \\ 0 & 2(\gamma - g_1) & l_4 & -l_3 \\ -l_3 & l_4 & 2\eta & 0 \\ -l_4 & -l_3 & 0 & 2\eta \end{bmatrix} \qquad (18)$$

where the elements l_3 and l_4 are defined as:

$$\begin{cases} l_3 = g_3 + \eta(L_m + \beta) \\ l_4 = g_4 + \omega_r \beta \end{cases} \qquad (19)$$

The feedback gains g_1, g_2, g_3, g_4 must be chosen such that matrix $-A^*$ in (18) is positive definite. The design is done using Sylvester Criterion. According to this, an nxn matrix is positive definite if all its $1x1$, $2x2$...nxn determinants (formed from the upper left corner and denoted as Δ_1, Δ_2...Δ_n) are positive. For matrix $-A^*$, the Sylvester determinants are:

$$\Delta_1 = 2(\gamma - g_1) \qquad (20)$$

$$\Delta_2 = \begin{vmatrix} 2(\gamma - g_1) & 0 \\ 0 & 2(\gamma - g_1) \end{vmatrix} = 4(\gamma - g_1)^2 \quad (21)$$

$$\Delta_3 = \begin{vmatrix} 2(\gamma - g_1) & 0 & -l_3 \\ 0 & 2(\gamma - g_1) & l_4 \\ -l_4 & l_4 & 2\eta \end{vmatrix} =$$
$$= 2(\gamma - g_1)[4\eta(\gamma - g_1) - (l_3^2 + l_4^2)] \quad (22)$$

$$\Delta_4 = \begin{vmatrix} 2(\gamma - g_1) & 0 & -l_3 & -l_4 \\ 0 & 2(\gamma - g_1) & l_4 & -l_3 \\ -l_3 & l_4 & 2\eta & 0 \\ -l_4 & -l_3 & 0 & 2\eta \end{vmatrix} =$$
$$= [4\eta(\gamma - g_1) - (l_3^2 + l_4^2)]^2 \quad (23)$$

Working with these determinants, (20) and (22) bring the following design conditions:

$$\gamma - g_1 > 0 \iff g_1 \in (-\infty, \gamma) \qquad (24)$$
$$4\eta(\gamma - g_1) - (l_3^2 + l_4^2) > 0 \qquad (25)$$

From (24), g_1 can be either positive (but no bigger than γ) or negative (in this case, it can be made as high as desired). Condition (25) can be rewritten as:

$$l_3^2 + l_4^2 < 4\eta(\gamma - g_1) \qquad (26)$$

In (26), the right hand side can be increased by choosing g_1 large and negative. This allows to increase the values of l_3 and l_4 if needed.

Inequality (26) can be satisfied in a number of ways: with a small l_3 and a large l_4, vice-versa, or with moderate values for both of them. To control the magnitude of $l_3^2 + l_4^2$ on the left side of (26), g_3 and g_4 are chosen using (19) as:

$$\begin{cases} g_3 = -\eta(L_m + \beta) \pm K_1 \\ g_4 = -\omega_r \beta \pm K_2 \end{cases} \qquad (27)$$

Then, the free constants K_1 and K_2 can be found such that:

$$(\pm K_1)^2 + (\pm K_2)^2 < 4\eta(\gamma - g_1) \qquad (28)$$

A few observations: according to (27), since η, β and L_m are known parameters, it is easy to select g_3; however, it is not clear whether K_1 should be positive or negative.

Also, note that g_4 depends on the speed ω_r. Since this is unknown (only a speed estimate $\hat{\omega}_r$ may be available), g_4 may be chosen as:

$$g_4 = -\hat{\omega}_r \beta \pm K_2 \qquad (29)$$

Again, from the conditions available so far, it is not clear whether K_2 in (29) should be positive or negative.

Another important observation is that there is no design condition for g_2 (this is because the half-symmetry of G in (6) makes the gain g_2 disappear from matrix $-A^*$). The effect of g_2 will be studied by simulations.

In conclusion, the gains g_1, g_3, g_4 correspond to (20),(27) and (29); the free constants K_1 and K_2 must satisfy (28). There is no design condition for g_2.

VI. Simulation Results

The proposed observer is simulated using the Matlab/Simulink software. The parameters of the induction motor used are given in Table I. The motor is operated in speed control mode; the simulation uses synchronous frame dq current control. Reference current i_q^* is obtained through speed feedback using a PI controller. The simulation uses a sampling time of 50 μs.

The motor is started, reaches a constant speed and then it accelerates. After that, the speed is reversed twice. The initial load torque is 0.4 Nm; this is increased to 0.6 Nm at t=1.3s and is reduced to 0.3 Nm at t=1.7s. The drive starts with $i_d^* = 0.85A$ and this increases to 1.35A at t=2s.

The gains of the speed adaptation law (17) are chosen by trial and error; for the simulations that follow, they are: $K_p = 0.01$, $K_i = 0.008$. This paper does not offer any indication on how to choose these gains; they are found by trying different values.

The simulations show that if K_p, K_i are improper, the speed estimate will be very inaccurate even if the gains g_1 through g_4 are properly designed according to (24),(25).

The simulation is repeated several times in order to assess the gain design. The simulations show the following:
- The observer works well when K_2 in (27) is small (almost zero) and when K_1 is large. Gain K_1 should be chosen close to the maximum limit given by inequality (28). Both K_1 and K_2 should be positive.
- If the observer is supposed to run in wide speed range, g_1 needs to be increased (it should be high and negative).
- The value of g_2 should be equal to zero. Any $g_2 \neq 0$ (positive or negative) worsens the behavior of the observer.

Fig.3 and Fig.4 show the estimates of the currents and fluxes when the motor runs at relatively low speed (a few hundred rpm), the gains are: $g_1 = \gamma - 5800$, $g_2 = 0$, $K_1 = 0.95K_{max}$, $K_2 = 0.02K_{max}$ ($K_{max} = 2\sqrt{\eta(\gamma - g_1)}$ as per (28)).

Note that the current and fluxes converge, the speed estimate matches the real speed and function V tends to zero.

Fig.5 and Fig.6 show the same waveforms when the drive runs at a top speed of 1500 rpm. The same gains as before are used; however, g_1 needed to be increased to $g_1 = \gamma - 10800$.

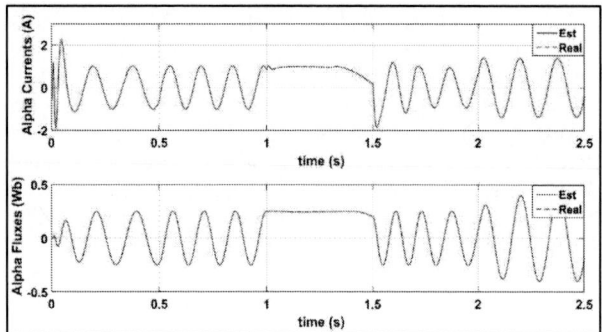

Fig.3. Estimated current and flux (α axis) at low speed

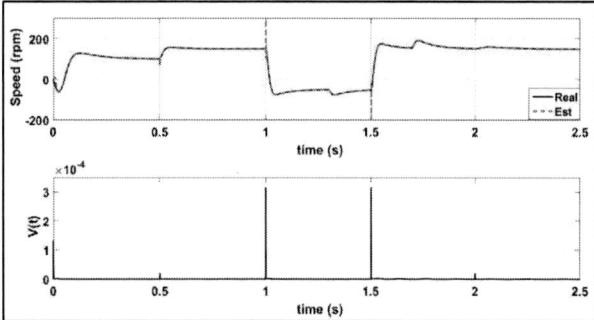

Fig.4. Estimated speed and Lyapunov function $V(t)$ at low speed

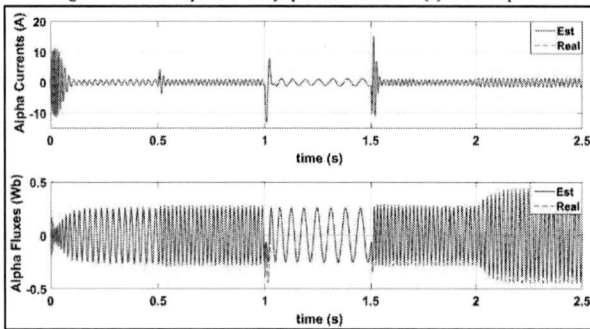

Fig.5. Estimated current and flux (α axis) at high speed

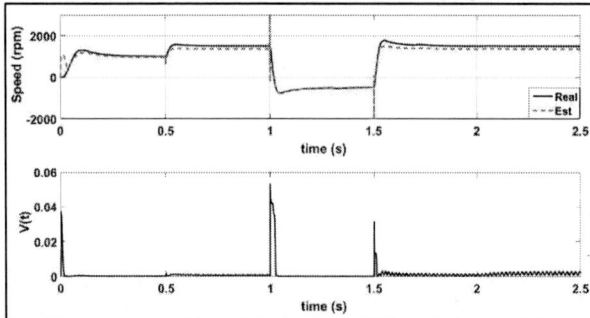

Fig.6. Estimated speed and Lyapunov function $V(t)$ at high speed

VII. EXPERIMENTAL RESULTS

The experimental setup (Fig.7) uses a three-phase squirrel cage induction motor (Dayton 2N863M), a Texas Instruments (TI) TMS320F2812 DSP, and a Spectrum Digital MOSFET inverter. The sampling time is $66\ \mu s$.

The experimental waveforms are obtained using the second PWM state machine of the TI chip (the variables to be displayed

are used as PWM duty cycles; the resulting waveforms are filtered with RC circuits and are captured with a scope).

The motor currents are measured using resistors mounted in the lower legs of the inverter. The voltages are computed from the measured dc bus voltage and the PWM duty cycles.

The setup shown does not contain an encoder; instead, an estimate of the speed is obtained from the slip equation:

$$\omega_r = \omega_e - \frac{L_m}{T_r}\frac{1}{\lambda_r^2}\left(\hat{\lambda}_\alpha i_\beta - \hat{\lambda}_\beta i_\alpha\right) \qquad (30)$$

The fluxes obtained from the adaptive observer are compared with the ones obtained from the Voltage Model Observer described in [30]. The base quantities are: $V_B = 179.7\ V$; $I_B = 3.75A$; $\omega_B = 377\ rad/s$; $\lambda_B = L_m I_B = 1.1137\ Wb$.

Fig.8 shows the estimated fluxes and currents of the adaptive observer at steady-state. Fig.9 shows the estimated rotor flux angle and the estimated speed for a 0.1 to 0.5 pu speed transient. Fig.10 shows the fluxes and speed for the 0.5 to 0.1 to 0.5 pu transient. The gains are $g_1 = 1$, $g_2 = 0$, $K_1 = 0$, $K_2 = -1$.

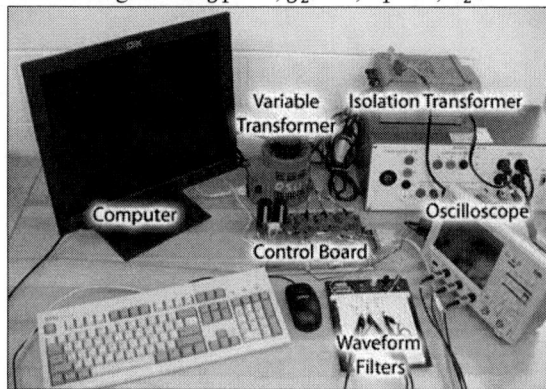

Fig.7. Experimental setup for the induction motor

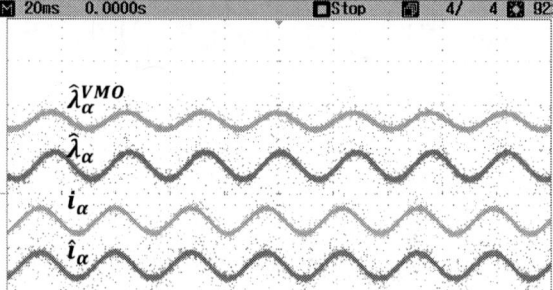

Fig.8. Estimated fluxes and currents of the adaptive observer

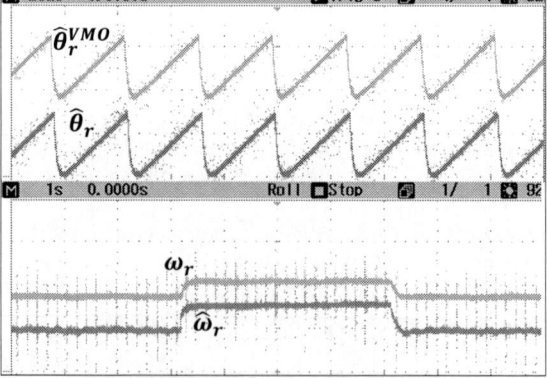

Fig.9. Estimated rotor flux angle and speed

978-1-4673-9551-9/16 $31.00 © 2016 IEEE

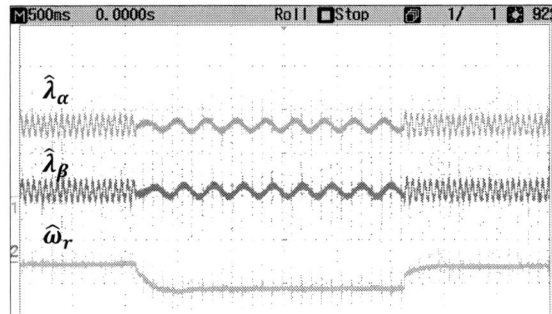

Fig. 10. Estimated fluxes and speed of the observer, 0.1-0.5 pu transient

VIII. Conclusions

The paper presents an adaptive flux and speed observer for the induction motor drive. The observer uses linear feedback to estimate the fluxes while the speed is obtained from an adaptive law. The design is done based on Lyapunov's nonlinear stability theory and using Sylvester criterion. The novelty is that this approach provides interesting design inequalities and insight into the stability margins. Simulations and experiments show that the proposed observer performs as expected: the estimates of the speed and fluxes are relatively accurate and they match closely the estimates obtained with other methods.

References

[1] B.K. Bose *"Modern Power Electronics and AC Drives,"* Upper Saddle River, NJ, Prentice Hall, 2002.

[2] R. De Doncker and D. Novotny "The universal field oriented controller," IEEE Transactions on Industry Applications, Vol. 30, 1994, pp. 92-100.

[3] C.F. Hu, R.B. Hong and C.H. Liu "Stability Analysis and PI controller tuning for a Speed Sensorless Induction Motor Drive," Industrial Electronics Society Conference IECON 2004, pp. 877-882.

[4] K. Ohyama and K. Shinohara "Small Signal Stability Analysis of Vector Control System of Induction Motor Without Speed Sensor Using Synchronous Current Regulator," IEEE Transactions on Industry Applications, Vol. 36, No.6, Nov-Dec. 2000.

[5] F. Briz, M.W. Degner and R.D. Lorenz "Analysis and design of current regulators using complex vectors," IEEE Transactions on Industry Applications, Vol. 36, No.3, May-June 2000, pp. 817-825.

[6] J. Jung, S. Lim and K. Nam "PI type decoupling control scheme for high speed operation of induction motors," Power Electronics Specialists Conference, PESC '97, June 1997, Vol. 2, pp.1082-1085.

[7] J. Holtz "Sensorless Control of Induction Machines – With or Without Signal Injection?" IEEE Transactions on Industrial Electronics, Vol. 53, No. 1, Feb. 2006, pp. 7-30.

[8] J. Holtz "Sensorless control of induction motors," Proceedings of IEEE, Vol. 90, No.8, Aug. 2002, pp. 1358-1394.

[9] J. Holtz "The Representation of AC machine dynamics by complex signal flow graphs," IEEE Transactions on Industrial Electronics, Vol. 42, No. 3, June 1995, pp. 263-271.

[10] P.L. Jansen, R.D. Lorenz and D.W. Novotny "Observer based direct field orientation – analysis and comparison of alternative methods," IEEE Trans. on Industry Applications, Vol. 30, 1994, pp. 945-953.

[11] E.K.K. Sng, L. Ah-Choy and T.A. Lipo "New observer based DFO scheme for speed sensorless field oriented drives for low-zero-speed operation," IEEE Transactions on Power Electronics, Vol. 13, 1998, pp. 959-968.

[12] C. Schauder "Adaptive speed identification for vector control of induction motors without rotational transducers," IEEE Transactions on Industry Applications, Vol. 28, No.5, pp. 1054-1061, 1992.

[13] H. Tajima and Y. Hori "Speed sensorless orientation control of induction machine," IEEE Transactions on Industry Applications, Vol. 29, No.1, 1993, pp. 175-180.

[14] J.I. Ha and S.K. Sul "Sensorless field-oriented control of an induction machine by high-frequency signal injection," IEEE Transactions on Industry Applications, Vol. 35, No.1, Feb. 1999, pp. 45-51.

[15] B.H. Bae, G.B. Kim and S.K. Sul "Improvement of low speed characteristics of railway vehicle by sensorless control using high-frequency injection," IEEE Industry Applications Society Annual Meeting, Rome, Oct. 2000, pp.1874-1880.

[16] L. Kreindler, J.C. Moreira, A. Testa and T.A. Lipo "Direct field orientation controller using the stator phase voltage third harmonic," IEEE Trans. on Industry Applications, Vol. 30, 1994, pp. 441-447.

[17] M. Hinkkanen "Analysis and Design of Full Order Flux Observers for Sensorless Induction Motors," IEEE Transactions on Industrial Electronics, vol. 51, no. 5, Oct. 2004, pp. 1033-1040.

[18] P. Vaclavek and P. Blaha "Lyapunov function based flux and speed observer for AC induction motor sensorless control and parameter estimation," IEEE Transactions on Industrial Electronics, vol. 53, no. 1, Feb. 2006, pp. 138-145.

[19] A.B. Proca and A. Keyhani "Sliding-mode flux observer with online rotor parameter estimation for induction motors," IEEE Transactions on Industrial Electronics, vol. 54, no.2, 2007, pp. 716-723.

[20] S. Rao, M. Buss and V.I. Utkin "Simultaneous State and Parameter Estimation in Induction Motors using First and Second Order Sliding Modes," IEEE Transactions on Industrial Electronics, vol. 56, no. 9, 2009, pp. 3369-3376.

[21] H. Kubota, K. Matsuse and T. Nakano "New Adaptive Flux Observer of Induction Motor for Wide Speed Range Motor Drives," IEEE Industrial Electronics Conference, IECON, 1990, pp. 921-926.

[22] H. Kubota, K. Matsuse and T. Nakano "DSP-Based Speed Adaptive Flux Observer of Induction Motor," IEEE Transactions on Industry Applications, Vol. 29, No.2, March/April 1993, pp. 344-348.

[23] G. Yang and T.H. Chin "Adaptive-Speed Identification Scheme for a Vector-Controlled Speed Sensorless Inverter-Induction Motor Drive," IEEE Transactions on Industry Applications, Vol. 29, No. 4, July/August 1993, pp. 820-825.

[24] H. Hashimoto, Y. Ohno, S. Kondo and F. Harashima "Torque Control of Induction Motor Using Predictive Observer," Power Electronics Specialist Conference, PESC, June 1989, Vol. 1, pp. 271-278.

[25] H. Hoffman and S.R. Sanders "Speed-Sensorless Vector Torque Control of Induction Machines using a Two-Time-Scale Approach," IEEE Trans. on Industry Applications, Vol. 34, No. 1, Jan/Feb. 1998, pp. 169-177.

[26] H. Kubota, I. Sato, Y. Tamura, K. Matsuse, H. Ohta and Y. Hori "Regenerating-Mode Low-Speed Operation of Sensorless Induction Motor Drive with Adaptive Observer," IEEE Transactions on Industry Applications, Vol. 38, No.4, July/Aug. 2002, pp. 1081-1086.

[27] S. Suwankawin and S. Sangwonwanich "A Speed-Sensorless IM Drive with Decoupling Control and Stability Analysis of Speed Estimation," IEEE Trans. on Ind. Electronics, Vol. 49, No. 2, Apr. 2002, pp. 444-455.

[28] J. Maes and J.A. Melkebeek "Speed-Sensorless Direct Torque Control of Induction Motors using an Adaptive Flux Observer," IEEE Transactions on Industry Applications, Vol. 36, No. 3, May/June 2000, pp. 778-785.

[29] M. Depenbrock and A. Steimel "Discussion of Regenerating Mode Low Speed Operation of Sensorless Induction Motor Drive with Adaptive Observer," IEEE Transactions on Industry Applications, Vol. 39, No. 1, Jan/Feb. 2003, pp. 19.

[30] C. Lascu, I. Boldea and F. Blaabjerg "A Modified Direct Torque Control for Induction Motor Sensorless Drive," IEEE Transactions on Industry Applications, Vol. 36, No. 1, 2000, pp. 120-130.

TABLE I. Induction Motor Specifications and Parameters

Rating	¼ hp	Pole #	4
Speed	1732 rpm	Voltage	220 V
R_s	10.9 Ω		
L_{ls}, L_{lr}	0.015 H		
L_m	0.30 H		
R_r	5.57 Ω		

Discontinuous PWM for Low Switching Losses in Indirect Matrix Converter Drives

Yeongsu Bak and Kyo-Beum Lee
Department of Electrical and Computer Engineering
Ajou University
Suwon, South Korea
wov2@ajou.ac.kr, kyl@ajou.ac.kr

Abstract—This paper proposes a discontinuous pulse width modulation (DPWM) method for an indirect matrix converter (IMC) driving induction motor (IM) in order to reduce switching losses. The IMC has advantages that long lifetime and reduced volume depending on the lack of capacitors. However, the IMC has many electric switching devices because it is composed of two stages such as a rectifier stage and an inverter stage for AC/AC power conversion. Therefore, the IMC has higher switching losses according to a number of switching devices consequently. The other topologies are researched for reduction in the number of switching devices in the rectifier stage such as sparse matrix converter. On the contrary, in this paper, the DPWM method is used to reduce the switching losses of the inverter stage. The effectiveness of the proposed method for reduce the switching losses of the inverter stage is verified by simulation results.

Keywords—Discontinuous PWM; Indirect matrix converter; Switching loss; Induction motor

I. INTRODUCTION

AC/AC power conversion systems are used in numerous applications such as power transmission, adjustable-speed motor drives, and renewable energy conversion system [1], [2]. Generally, the AC/AC power conversion systems have energy storage elements such as dc-link capacitors. The back-to-back (B2B) converter is common AC/AC power conversion system with energy storage elements. It is composed of a rectifier stage, an inverter stage, and the dc-link capacitors between the two stages. However, the B2B converter has crucial disadvantages including as high volume and short lifetime owing to the dc-link capacitors [3], [4]. In order to solve the disadvantages, researchers have been tried to remove the energy storage elements in the dc-link.

One of the AC/AC power conversion systems, an indirect matrix converter (IMC) is similar to the B2B converter structurally. The IMC does not have a capacitor in the dc-link contrary to the B2B converter. It has advantages such as small size, low weight, and durability owing to the lack of the dc-link capacitors. Moreover, the IMC has some characteristics. It is bidirectional power transmission system and guarantees sinusoidal input-output waveforms. Additionally, the IMC always operates in buck mode because the input-output maximum voltage transfer ratio of the IMC is restricted to 0.866 [5]. Although the IMC has many advantages, the IMC is still not enough to apply for the industrial applications because the IMC also has disadvantages such as complex control method and restricted application field depending on the low input-output maximum voltage transfer ratio [6]-[8]. Main

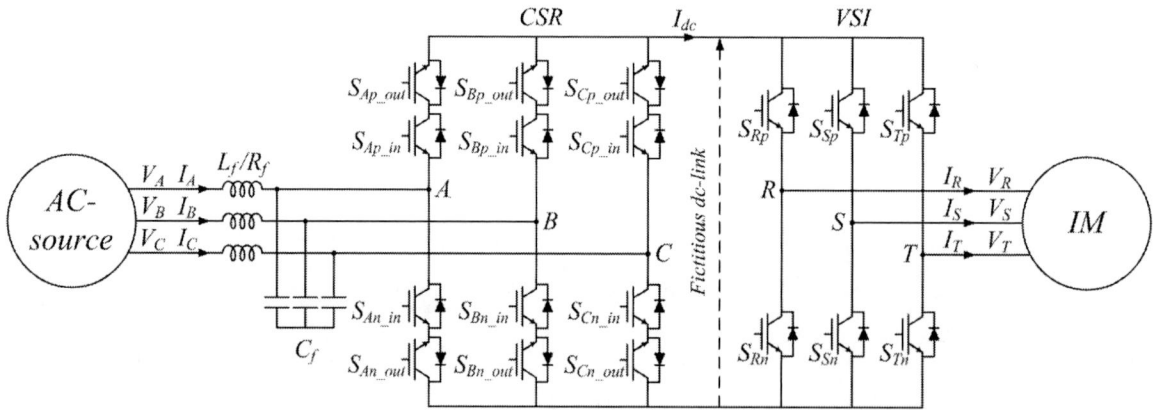

Fig. 1. Topology of the indirect matrix converter.

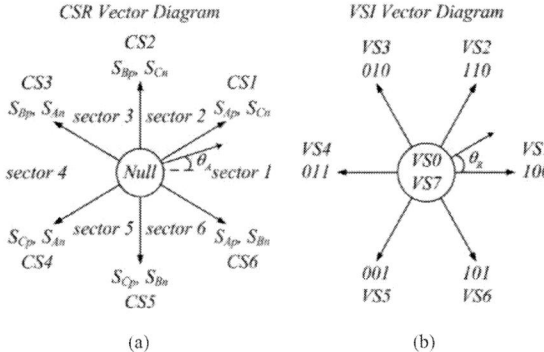

Fig. 2. Space vector diagrams of (a) CSR and (b) VSI.

disadvantages of the IMC topology is that the IMC has a larger number of switching devices than the other AC/AC power conversion system topologies.

The IMC is composed of two stages such as a rectifier stage, an inverter stage, and a fictitious dc-link without capacitors. The rectifier stage is usually composed of 12 IGBTs and 12 diodes, which are components of the bidirectional switches. In addition, the inverter stage is composed of 6 IGBTs and 6 diodes. The IMC has a number of switches compared to other AC/AC power conversion systems such as the diode rectifier-voltage source inverter composed 6 IGBTs and 12 diodes, the B2B converter composed 12 IGBTs and 12 diodes. Therefore, the IMC topology increases the cost and the switching losses of the converter according to a number of switches. In order to overcome the disadvantage of the higher switching losses, the sparse matrix converter (SMC) has been researched. The SMC is form eliminated the 3 IGBTs in the rectifier stage and it is composed of 9 IGBTs and 12 diodes in the rectifier stage. In addition, the very sparse matrix converter (VSMC) and the ultra sparse matrix converter (USMC) have been researched. The USMC is composed of 3 IGBTs and 12 diodes. It is cannot be operated to bidirectional power transmission [9]-[11].

This paper proposes a method for reduction of the switching losses in terms of the inverter stage. The discontinuous pulse width modulation (DPWM) method can reduce the number of the switching in the inverter stage by operating the only two-phase switches of the three-phase switches during one switching period. The crucial purpose using the DPWM method is reduction of the switching losses through the reduction of the number of the switching. The DPWM method has variable methods according to setting of the discontinuous sections. A 60 degree DPWM method of the variable DPWM methods can reduce the switching frequency by retaining switch state as turn ON or turn OFF during the 60 degree period that the magnitude of the phase voltage is largest. It can reduce the switching frequency of 67 % compared to the continuous pulse width modulation (CPWM) method because the switches in the inverter stage do not switching during the 1/3 period. Fundamentally, in order to minimize the switching losses, the switches must not be operated during the period when the magnitude of the phase current is highest. It is more effective than using the DPWM method during the period when the magnitude of the phase voltage is highest. Therefore,

in case that the power factor (PF) is changed by the difference of phase between the phase current and the phase voltage, the discontinuous period should be considered to minimize the switching losses [12]-[14]. This paper proposes a DPWM method for the low switching losses of the IMC driving the induction motor (IM). The validity of the proposed control schemes concerning the low switching losses is demonstrated by PSIM simulation results.

This paper is composed as follows: the topology of the IMC-fed IM and modulation methods of the VSR and the CSI are presented in Section II. The DPWM method for low switching losses of the VSI and its analysis are presented in Section III. Some simulation results are given in Section IV. Finally, Section V presents the conclusions to this paper.

II. TOPOLOGY AND MODULATION OF INDIRECT MATRIX CONVERTER DRIVING INDUCTION MOTOR

A. Topology of the Indirect Matrix Converter

Fig. 1 shows the circuit configuration of the IMC driving the IM. It is composed of 4 parts simply that a three-phase AC voltage source of the input stage, input filter capacitors, the IMC as power conversion system, and the IM of the output stage. The IMC is a two-stage AC/AC power conversion system and can be classified to two stages such as a current source rectifier (CSR) on the input side and a voltage source inverter (VSI) on the output side. The CSR is composed of 12-bidirectional switches and the VSI is composed of 6-bidirectional switches. The rectifier stage and the inverter stage are directly connected by the fictitious dc-link. The fictitious dc-link is existed between the CSR and the VSI because the IMC does not have the dc-link energy storage elements such as capacitor. Additionally, the filter capacitors are required to reduce a voltage ripple in the input stage and the IM as output stage is controlled by control strategy of the IMC.

B. Modulation Method of the CSR

The conventional modulation method of the rectifier stage in the fundamental IMC can be applied to the modulation of the CSR in this topology. Through the modulation method of the CSR, the maximum voltage generated by the AC-source is transmitted to the dc-link. In addition, the sinusoidal input current and the unity input power factor are maintained by the modulation of the CSR. The positive voltage of the dc-link is synthesized by the rectifier stage depending on variable switching states. The CSR is usually operated with one upper (S_{xp_in} and S_{xp_out} | $x = A$, B or C) and one lower (S_{xn_in} and S_{xn_out} | $x = A$, B or C) switches as ON state at all instant. The adjacent two switches such as S_{Ap_in} and S_{Ap_out} have equivalent switching states. However, the switches of the rectifier stage should be operated with a four-step commutation. The four-step commutation is important strategy because the rectifier circuit can be short or open circuit while the ON state switches of the upper or lower are converted to OFF state, simultaneously, OFF state switches of the upper or lower in the difference phase-leg are converted to ON state.

Fig. 2(a) shows the space vector diagrams of the CSR states. The vectors of the CSR are classified as 6-active states and 3-

null states depending on the ON state of the switches. In case the one upper and one lower ON state switches exist in the same phase-leg or different phase-leg, the CSR generates the null states or the active states respectively. The fictitious dc-link voltage is shorted to zero in the null state and the power is transferred to the load in the active states. A reference current phasor lying on the sector 1 is reproduced the reference using the nearest vectors the *CS1* and the *CS6* as in (1).

$$
\begin{aligned}
i_A^* &= I_m \cos\theta_A, \ \theta_A = \omega_o t \\
i_B^* &= I_m \cos\theta_B, \ \theta_B = \theta_A - 2\pi/3 \\
i_C^* &= I_m \cos\theta_C, \ \theta_C = \theta_A - 4\pi/3
\end{aligned}
\tag{1}
$$

where I_m is common amplitude and θ_A, θ_B, and θ_C are respective phase angle. The S_{Ap} is turned ON during the full switching period because the current phasor lies on the sector 1 between the *CS1* and the *CS6* including the S_{Ap}. The upper dc-link is connected to clamping phase A depending on the switching state of the S_{Ap}. In addition, in order to synthesize the current phasor, the S_{Bn} and the S_{Cn} are modulated using the duty ratios (d_x and d_y) as in (2). The lower dc-link is connected to clamping phase B and C alternately depending on the switching state of the S_{Bn} and the S_{Cn}. The reference current phasor in the other five sectors apply the same interpretation.

$$
\cos\theta_A + \cos\theta_B + \cos\theta_C = 0, \quad -\frac{\cos\theta_B}{\cos\theta_A} - \frac{\cos\theta_C}{\cos\theta_A} = 1 \tag{2}
$$
$$
d_x = -\cos\theta_B / \cos\theta_A, \quad d_y = -\cos\theta_C / \cos\theta_A.
$$

Additionally, the average dc-link voltage ($V_{dc(av)}$) of the fictitious dc-link is expressed by multiplying the duty ratios and the input line-to-line voltages (V_{AB} and V_{CA}) respectively as in (3). The $V_{dc(av)}$ is represented with the phase voltage amplitude V_m and the input power factor angle ϕ_o.

$$
\begin{aligned}
V_{dc(av)} &= d_x V_{AB} - d_y V_{CA} \\
&= \frac{3V_m}{2\cos\theta_A} \cdot \cos\phi_o, \quad -\frac{\pi}{6} \le \theta_A \le \frac{\pi}{6}.
\end{aligned}
\tag{3}
$$

C. Modulation Method of the VSI

The configuration of the VSI is similar to the common inverter having 6-switch bridges. Fig. 2(b) shows the space vector diagrams of the VSI states. The vectors of the VSI are classified as 6-active states and 2-null states depending on the ON state of the switches. The 2-null states are zero vectors, which are generated by turning ON the all upper switches (S_{Rp}, S_{Sp}, S_{Tp}) or all lower switches (S_{Rn}, S_{Sn}, S_{Tn}) of the VSI such as the *VS0* or the *VS7*. The VSI can be modulated by the space vector modulation after compensating for the floating $V_{dc(av)}$ and the carrier based PWM method. Therefore, the PWM signals are generated through the two-modulation signals with a triangular carrier signals. In the phase R, the two-modulation signals are written as in (4). The other sectors apply the same interpretation.

$$
\begin{aligned}
v_{R(upper)} &= -2d_y \cdot \frac{v_R + v_{offset}}{V_{dc(av)}} + d_x \\
v_{R(lower)} &= 2d_x \cdot \frac{v_R + v_{offset}}{V_{dc(av)}} - d_y
\end{aligned}
\tag{4}
$$

where v_R is voltage amplitude of the phase R and v_{offset} is offset voltage of the 3-phase voltages, which is written as in (5).

$$
v_{offset} = -\frac{1}{2} \cdot \left\{ max\left(v_R, \ v_S, \ v_T\right) + min\left(v_R, \ v_S, \ v_T\right)\right\}. \tag{5}
$$

III. Application of Discontinuous PWM Method for Low Switching Losses

A. Discontinuous PWM Method for the VSI

The VSI used to the IMC is similar to the general three-phase inverter having 6-switch bridges although it does not have the dc-link capacitors. Fundamentally, the CPWM method that all three-phase switching devices are always switched in the switching period is used to the voltage modulation method for the VSI. However, the CPWM method has a disadvantage regarding the losses of the switching devices. Therefore, to solve this problem in this paper, the DPWM method is used for reduction of the switching losses as the voltage modulation method. The switching losses of the VSI can be reduced by decreasing the number of switching of the VSI. The only two-phase switching devices are switched and the other-phase switching device is not switched, which is clamped upper or lower fictitious dc-link in the switching period.

B. Proposed Modulation Method of the VSI

The 60 degree DPWM method is used most often although there are various DPWM methods. Fig. 3 shows the three-phase reference phase voltages (v^*_{Rs}, v^*_{Ss}, v^*_{Ts}), the offset voltage ($v_{offset(60DPWM)}$) for the 60 degree DPWM method, and the reference pole voltages (v^*_{Rn}, v^*_{Sn}, v^*_{Tn}). The v^*_{Rs}, v^*_{Ss}, and the v^*_{Ts} of the VSI are generated by the current control of the output stage. Additionally, the reference currents of the output stage for the current control are generated by the speed control of the IM through the reference speed. The $v_{offset(60DPWM)}$ is calculated by the v^*_{Rs}, v^*_{Ss}, v^*_{Ts}, and the $V_{dc(av)}$ as in (6).

$$
\begin{aligned}
v_{offset(60DPWM)} &= \begin{cases} \dfrac{V_{dc(av)}}{2} - v_{max} & |v_{max}| \ge |v_{min}| \\[2mm] -\dfrac{V_{dc(av)}}{2} - v_{min} & |v_{max}| < |v_{min}| \end{cases} \\
v_{max} &= max\left(v^*_{Rs}, \ v^*_{Ss}, \ v^*_{Ts}\right) \\
v_{min} &= min\left(v^*_{Rs}, \ v^*_{Ss}, \ v^*_{Ts}\right)
\end{aligned}
\tag{6}
$$

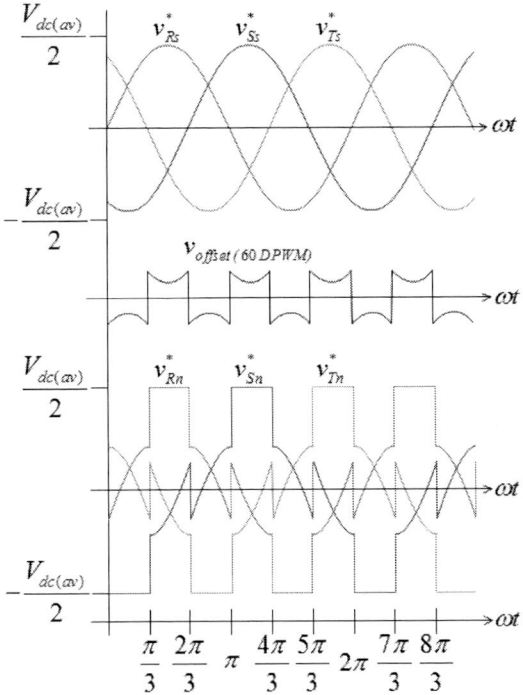

Fig. 3. Modulation signals of the 60 degree DPWM method.

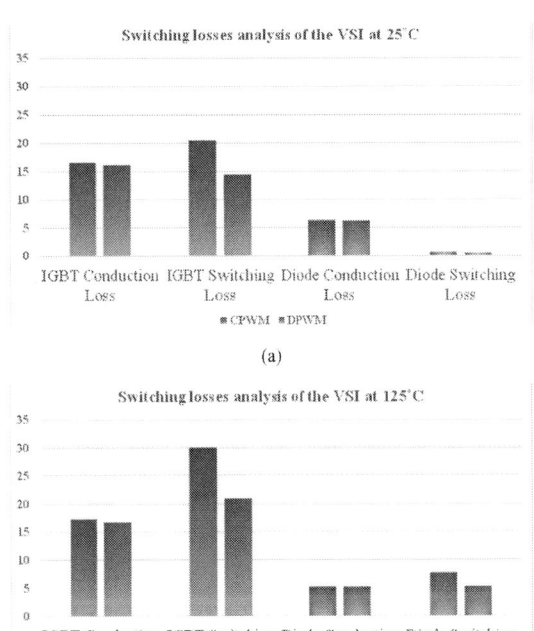

(a)

(b)

Fig. 4. Losses anaysis of the switches belong to the VSI at (a) 25 °C and (b) 125 °C.

where the $V_{dc(av)}$ is the average dc-link voltage of the fictitious dc-link as in (3). The v_{max} and v_{min} are the maximum and minimum voltages, respectively, among the v^*_{Rs}, v^*_{Ss}, and the v^*_{Ts}. The v^*_{Rn}, v^*_{Sn}, and v^*_{Tn} are calculated through the sum of the v^*_{Rs}, v^*_{Ss}, v^*_{Ts}, and the $v_{offset(60DPWM)}$ respectively.

In the general three-phase voltage source inverter having 6-switch bridges, the v^*_{Rn}, v^*_{Sn}, and v^*_{Tn} are used for the modulation signals of the 60 degree DPWM method. In addition, the PWM signals for the switching devices are generated by comparing with the triangular carrier signals. However, regarding the VSI of the IMC, the modulation signals compared to the triangular carrier signals in phase R are calculated as in (7). The other phases can be applied with the same interpretation.

$$
\begin{aligned}
v_{R(upper)(60DPWM)} &= -2d_y \cdot \frac{v^*_{Rs} + v_{offset(60DPWM)}}{V_{dc(av)}} + d_x \\
&= -2d_y \cdot \frac{v^*_{Rn}}{V_{dc(av)}} + d_x
\end{aligned} \tag{7}
$$

$$
\begin{aligned}
v_{R(lower)(60DPWM)} &= 2d_x \cdot \frac{v^*_{Rs} + v_{offset(60DPWM)}}{V_{dc(av)}} - d_y \\
&= 2d_x \cdot \frac{v^*_{Rn}}{V_{dc(av)}} - d_y
\end{aligned}
$$

where $v_{R(upper)(60DPWM)}$ and $v_{R(lower)(60DPWM)}$ are the modulation signals applied the 60 degree DPWM method for the low switching losses of the VSI. The PWM signals for the VSI are generated by comparing the modulation signals with the triangular carrier signals. In this proposed modulation method, the switching devices of the VSI are not switched during 60 degree range where the v^*_{Rs}, v^*_{Ss}, and the v^*_{Ts} have the largest amplitude in the switching period.

C. Switching Losses Analysis of the VSI

Fig. 4 shows the losses analysis of the switches belong to the VSI. In order to identify the low switching losses of the DPWM method compared to the CPWM method, the IGBT losses and the diode losses are considered. These losses are also divided into the conduction loss and the switching loss respectively. Additionally, two junction temperatures such as 25 °C and 125 °C are considered. The performing conditions are that the AC-source supplies the 60 Hz/220 V_{rms} line-to-line voltage, the motor speed of the IM is controlled to 1400 r/min, and the load torque is 8 N·m. The VSI using the DPWM method has low switching losses, which can be confirmed to the analysis shown in Fig. 4. The IGBT switching losses of the VSI using the DPWM method are decreased by 1/3 compared with the CPWM method because the 60 degree DPWM method is used for the VSI. In addition, the other losses such as IGBT conduction loss and diode losses of the VSI using the DPWM method are similar to the losses of the VSI using the CPWM method in case both having the 25 °C and 125 °C junction temperatures.

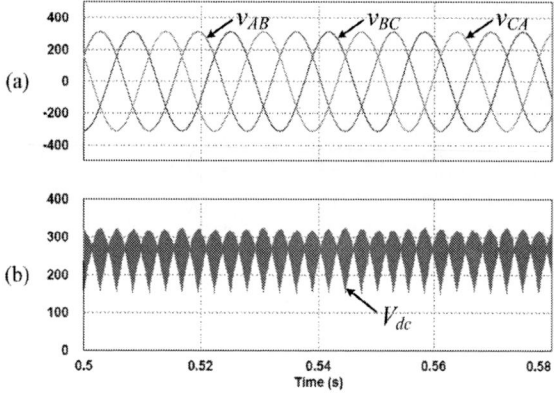

Fig. 5. Simulation results of the IMC input line-to-line voltages and dc-link voltage.

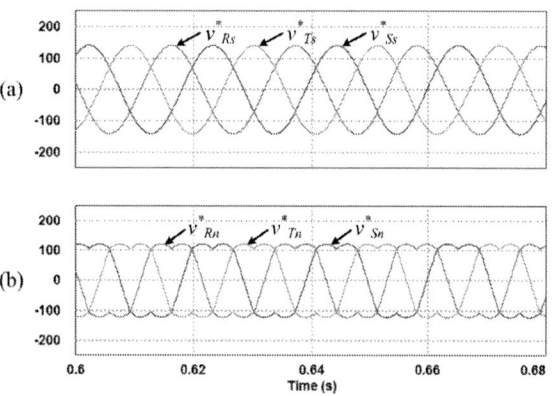

Fig. 6. Simulation results of the modulation signals using the fundamental CPWM method.

IV. SIMULATION

The IMC system driving the IM was simulated using PSIM software for verifying a performance of the proposed modulation method for the low switching losses. The AC-source supplies three-phase 60 Hz/220 V_{rms} line-to-line voltage. In the Fig. 1, the parameters of the elements were realized as L_f = 2 mH with the resistor R_f = 10 Ω added for damping purpose and C_f = 15 μF for filtering, which are not optimized. In addition, the parameters of the IM are appeared to table 1.

Fig. 5 shows the simulation results of the IMC input line-to-line voltages and dc-link voltage. The Fig. 5(a) shows the input line-to-line voltages, which is generated by the three-phase AC-voltage source. It is transmitted to the input stage of the IMC. Through the modulation method of the CSR, the maximum voltage generated by the AC-source is transmitted to the dc-link. This fictitious dc-link voltage is shown in Fig 5(b).

Fig. 6 shows the simulation results of the modulation signals using the fundamental CPWM method. The Fig. 6(a) shows the v^*_{Rs}, v^*_{Ss}, and v^*_{Ts} as three-phase reference phase voltages. In this case, the v_{offset} as in (5) is used to offset voltage.

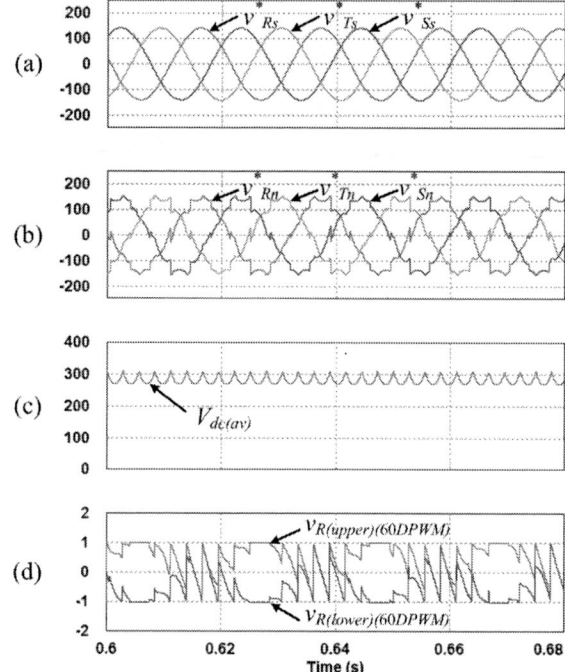

Fig. 7. Simulation results of the modulation signals using the 60 degree DPWM method.

Therefore, the v^*_{Rn}, v^*_{Sn}, and v^*_{Tn} as the reference pole voltages are shown in Fig. 6(b).

Contrary to the CPWM method, Fig. 7 shows the simulation results of the modulation signals using the 60 degree DPWM method. The Fig. 7(a) shows the three-phase reference phase voltages equivalent to the Fig. 6(a). However, in the Fig. 7(b), the v^*_{Rn}, v^*_{Sn}, and the v^*_{Tn} have a difference from the Fig. 6(b) in parts of the largest amplitude. The calculated $V_{dc(av)}$ as in (3) is shown in Fig. 7(c) because the IMC does not have dc-link capacitors. In this regard, the $V_{dc(av)}$ has an effect on the reference pole voltages because the $V_{dc(av)}$ is used to calculate the $v_{offset(60DPWM)}$ as in (6). Additionally, the Fig. 7(d) shows the modulation signals of the VSI compared with the triangular carrier signals.

TABLE I.   INDUCTION MOTOR PARAMETERS

Parameters	Value
Stator resistance	0.295 Ω
Rotor resistance	0.379 Ω
Stator inductance	0.001794 H
Rotor inductance	0.001794 H
Magnetizing inductance	0.059 H
Number of pole	4
Moment of inertia	0.00179 kg·m²

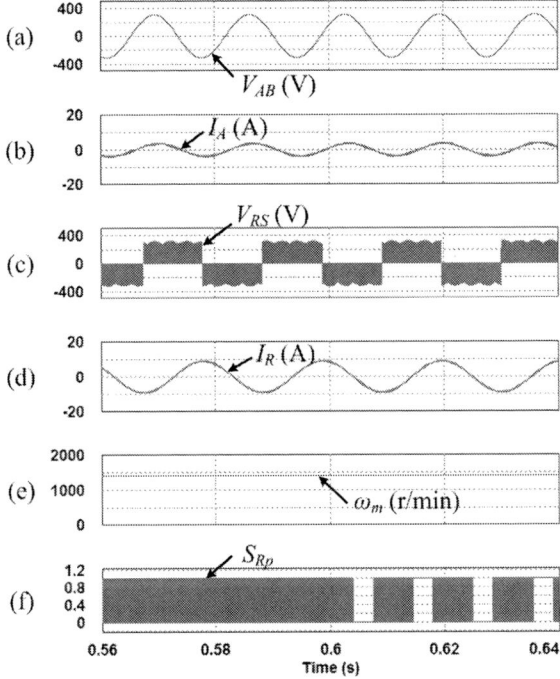

Fig. 8. Simulation results of the IMC system driving the IM with the 60 degree DPWM method at 0.6 s.

Fig. 8 shows the simulation results of the IMC system driving the IM with the 60 degree DPWM method. The Fig. 8(a) and (b) show the line-to-line voltage and phase current of the input stage. The Fig. 8(c) and (d) show the line-to-line voltage and phase current of the output stage. The input-output current of the IMC have sinusoidal waveforms, which is characteristic of the IMC. The Fig. 8(e) shows the motor speed of the IM and it is controlled to reference speed as 1400 r/min. Finally, the Fig. 8(f) shows the PWM signal used for the upper switch in the phase R of the VSI. The proposed 60 degree DPWM method with the IMC is used for low switching losses at 0.6 s. Therefore, the PWM signal is clamped during 1/3 period of the switching period depending on the 60 degree DPWM method. Although the 60 degree DPWM method is applied at 0.6 s, it does not have an effect on the control of the IM.

V. CONCLUSION

This paper has proposed the 60 degree DPWM method for low switching losses of the IMC driving the IM. The IMC has many electric switching devices in the two stages such as a rectifier stage and an inverter stage. Therefore, the IMC has higher switching losses according to a number of switching devices. In this paper, the 60 degree DPWM method is used to reduce the switching losses of the VSI. The modulation signals of the VSI is changed by the offset voltage depending on the 60 degree DPWM method. The PWM signal for the switches of the VSI are clamped during 1/3 period of the switching period. Through this method, the switching losses of the VSI are

decreased to 67 % compared to the CPWM method. The application of the DPWM method does not have effect on the control of the IMC driving the IM.

REFERENCES

[1] Y. Bak, E. Lee, and K.-B. Lee, "Indirect Matrix Converter for Hybrid Electric Vehicle Application with Three-Phase and Single-Phase Outputs," *Energies*, vol. 8, no. 5, pp. 3849-3866, Apr. 2015.

[2] K.-B. Lee and F. Blaabjerg, "Simple Power Control for Sensorless Induction Motor Drives Fed by a Matrix Converter," *IEEE Trans. Energy Convers.*, vol. 23, no. 3, pp. 781-788, Sep. 2008.

[3] J.-S. Lee, K.-B. Lee, and F. Blaabjerg, "Open-Switch Fault Detection Method of a Back-to-Back Converter Using NPC Topology for Wind Turbine Systems," *IEEE Trans. Ind. Appl.*, vol. 51, no. 1, pp. 325-335, Jan./Feb. 2015.

[4] T. Friedli, J. W. Kolar, J. Rodriguez, and P. W. Wheeler, "Comparative Evaluation of Three-Phase AC-AC Matrix Converter and Voltage DC-link Back-to-Back Converter System," *IEEE Trans. Ind. Electron.*, vol. 59, no. 12, pp. 4487-4510, Dec. 2012.

[5] E. Lee and K.-B. Lee, "Fault Diagnosis for a Sparse Matrix Converter using Current Patters," in *Proc. APEC Conf.*, 2012, pp. 1549-1554.

[6] M. Jussila and H. Tuusa, "Comparison of Simple Control Strategies of Space-Vector Modulated Indirect Matrix Converter Under Distorted Supply Voltage," *IEEE Trans. Power Electron.*, vol. 22, no. 1, pp. 139-148, Jan. 2007.

[7] J. Riedemann, R. Pena, and R. Blasco-Gimenez, "A resonant current control of an open-end winding induction motor fed by an indirect matrix converter," in *Proc. APEC Conf.*, 2015, pp. 2346-2350.

[8] C. Qi, X. Chen, and Y. Qiu, "Carrier-Based Randomized Pulse Position Modulation of an Indirect Matrix Converter for Attenuating the Harmonic Peaks," *IEEE Trans. Power Electron.*, vol. 28, no. 7, pp. 3539-3548, Jul. 2013.

[9] J. W. Kolar, F. Schafmeister, S. D. Round, and H. Ertl, "Novel Three-Phase AC-AC Sparse Matrix Converters," *IEEE Trans. Power Electron.*, vol. 22, no. 5, pp. 1649-1661, Sep. 2007.

[10] E. Lee, K.-B. Lee, J.-S. Lim, Y. Lee, and J.-H. Song, "Predictive Current Control for a Sparse Matrix Converter," in *Proc. IPEMC Conf.*, 2012, pp. 36-40.

[11] K. Park, K.-B. Lee, and F. Blaabjerg, "Improving Output Performance of a Z-Source Sparse Matrix Converter Under Unbalanced Input-Voltage Conditions," *IEEE Trans. Power Electron.*, vol. 27, no. 4, pp. 2043-2054, Apr. 2012.

[12] J.-H. Park, H.-G. Jeong, and K.-B. Lee, "An Improved DPWM Method for Reduction of Resonant Problem in the Inverter," in *Proc. ECCE Conf.*, 2014, pp. 1528-1533.

[13] J.-S. Lee and K.-B. Lee, "Carrier-Based Discontinuous PWM Method for Vienna Rectifiers," *IEEE Trans. Power Electron.*, vol. 30, no. 6, pp. 2896-2900, Jun. 2015.

[14] O. Ojo, "The Generalized Discontinuous PWM Scheme for Three-Phase Voltage Source Inverters," *IEEE Trans. Ind. Electron.*, vol. 51, no. 6, pp. 1280-1289, Dec. 2004.

Model Predictive Control for Extended Kalman Filter Based Speed Sensorless Induction Motor Drives

Jie Li
Xi'an University of Technology
No. 5, JinhuaNanlu
Xi'an, 710048, China
Email: lijie@xaut.edu.cn

Li-Heng Zhang
Xi'an University of Technology
No. 5, JinhuaNanlu
Xi'an, 710048, China

Ying Niu
Xi'an University of Technology
No. 5, JinhuaNanlu
Xi'an, 710048, China

Hai-Peng Ren
Xi'an University of Technology
No. 5, JinhuaNanlu
Xi'an, 710048, China
Email: renhaipeng@xaut.edu.cn

Abstract—**Model predictive control (MPC) strategy is a typical optimization predictive control method. MPC can achieve satisfying control performance with lower model parameter dependency, and it has been widely applied in process control systems. To improve the system reliability and reduce the hardware cost, extended Kalman filter (EKF) was introduced to achieve sensorless operation. A novel speed sensorless control system of induction motors which coordinates the MPC current controllers and the EKF state estimator is presented in this paper. The proposed scheme has been verified by simulations, and the results show that excellent speed control performance and state estimation performance over wide speed range and load torque range have been achieved either in dynamics or in steady states.**

Keywords—*model predictive control; extended Kalman filter; sensorless control; induction motor*

I. INTRODUCTION

Field oriented control (FOC) and direct torque control (DTC) are two mature control schemes of induction motor. These two schemes are simple, easy to be implemented and have fast dynamic response. But they depend on the model and are sensitive to the change of parameters. Model predictive control comes into being under this kind of demand [1]. MPC has advantages of its simple principle, easy modeling and handling nonlinear constraints, strong robustness [2-8]. Therefore, MPC has become the third choice of high performance drive system in addition to the FOC and DTC speed control schemes [9], attracting great research interest.

Compared with the traditional PI controller, the MPC has a very good performance and is very robust to the model parameters [10]-[11]. Compared to DTC, the predictive torque control reduces the flux and the torque ripples. At the same time, the very fast torque response of DTC is maintained [12]. In [13], the predictive current control of induction motors with low switching frequency shows fast dynamic response and reduces the current harmonic distortion and the switching losses.

The speed sensorless technology has advantages of reducing the cost of system hardware, increasing the adaptability and the reliability of the system. It has been widely used in modern AC speed control systems. However, there is not much research on the combination of MPC and the speed sensorless technology [14]. With the present DSP computing power and computing speed, the speed estimation algorithms such as EKF has been available in real time. In addition, some time margin in the PWM interrupt service routine can be used to realize the advanced control algorithms which are more complex than the PI algorithms. Based on this, we believe that the combination of MPC and EKF is the optimal combination for realization currently on the actual DSP platform. We have the experience of implementing the full order EKF, the EKF speed estimator will consume about 25μs. Considering the implementability, we chose the MPC algorithm presented in [10] and the classical full order EKF algorithm to construct our control system for the sensorless induction motor drive system.

In this paper, a detailed comparison between the performance of the proposed sensorless induction motor drive system and that of the conventional FOC sensorless induction motor drive system employing PI controllers and full order EKF speed estimator will be given. According to the simulation results, it is shown that the MPC+EKF system shows better speed control performance and state estimation performance over wide speed range and load torque range either in dynamics or in steady states, especially in low speed range.

II. MODEL PREDICTIVE CONTROLLERS FOR EXTENDED KALMAN FILTER BASED SPEED SENSORLESS INUDCTION MOTOR DRIVES

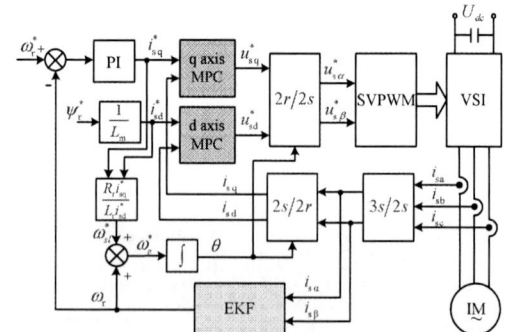

Fig. 1. The proposed control scheme based on MPC current controllers and EKF speed estimation

---

This work is supported in part by NSFC (50907054); SRFDP (20126118110008); IRT of Shaanxi Province (2013KCT-04);SRICP of Shaanxi Province (2013KW05-02); Research and Development Project of Beilin District of Xi'an City (GX1503); Collaborative Innovation Program of Xi'an City(CXY1509-19); SFKD of Shaanxi Province (00X901).

Fig.1 gives the proposed control scheme based on MPC current controllers and EKF speed estimation. Where the current inner loop employs d-axis MPC and q-axis MPC, and the speed feedback is provided by EKF estimator.

A. Model Predictive Controllers for Current Inner Loops

The stator voltage equation of induction motors in synchronous rotating coordinate system is as follows:

$$
\begin{cases}
u_{sd} = R_s i_{sd} + L_\sigma p i_{sd} - \omega_e L_\sigma i_{sq} + \dfrac{L_m}{L_r} p \psi_r \\
u_{sq} = R_s i_{sq} + L_\sigma p i_{sq} + \omega_e L_\sigma i_{sd} + \dfrac{L_m}{L_r} \omega_e \psi_r
\end{cases}
\tag{1}
$$

where u_{sd}, u_{sq} are the synchronous rotating frame d-axis stator voltage and q-axis stator voltage, respectively. i_{sd}, i_{sq} is the synchronous rotating frame d-axis stator current and q-axis stator current, respectively. R_s is the stator resistance. L_σ is the leakage inductance. ω_s is the synchronous angular speed. L_m is the mutual inductance. L_r is the rotor inductance. ψ_r is the magnitude of the rotor flux linkage. p is the differential operator.

Model predictive control should be used for an SISO system. In order to reduce the complexity of the model predictive control algorithm, the simplified two-dimensional current equations was employed during the procedure of developing this MPC algorithm. The cross coupling between the d and the q axis items can be regarded as a feedforward compensator for current control. Then, the d axis and the q axis currents are discretized as follows:

$$
i_{sd}(k+1) = (1 - \frac{R_s}{L_\sigma} T_s) i_{sd}(k) + \frac{T_s}{L_\sigma} u_{sd}
\tag{2}
$$
$$
= A_{MPC} i_{sd}(k) + B_{MPC} u_{sd}
$$

$$
i_{sq}(k+1) = (1 - \frac{R_s}{L_\sigma} T_s) i_{sq}(k) + \frac{T_s}{L_\sigma} u_{sq}
$$
$$
= A_{MPC} i_{sq}(k) + B_{MPC} u_{sq}
\tag{3}
$$

Rewrite (2) and (3) in vector form:

$$
\boldsymbol{i}_s(k+1) = A_{MPC} \boldsymbol{i}_s(k) + B_{MPC} \boldsymbol{u}_s
\tag{4}
$$

where $\boldsymbol{i}_s$ is the synchronous rotating frame stator current vector, $\boldsymbol{u}_s$ is the synchronous rotating frame stator voltage vector.

The MPC algorithm for the induction motors is as follows:

Firstly, set the initial value for the current predictive value vector $\boldsymbol{i}_{sm}(0) = 0$;

Then, for $k = 0,1,2,3,\ldots\ldots \infty$, the following steps will be executed in each loop:

(1) Calculate the predictive error vector:

$$
\boldsymbol{e}_s(k+1) = \boldsymbol{i}_s(k+1) - \boldsymbol{i}_{sm}(k+1|k)
\tag{5}
$$

where $\boldsymbol{e}_s$ is the synchronous rotating frame stator current predictive error vector. $\boldsymbol{i}_{sm}(k+i|k)$ is the synchronous rotating frame current predictive value vector at time $k+i$ according to the current predictive value vector at time k.

(2) Correct the predictive value:

$$
\begin{bmatrix}
\boldsymbol{i}_{sp}(k+1|k) \\
\boldsymbol{i}_{sp}(k+2|k) \\
\boldsymbol{i}_{sp}(k+3|k)
\end{bmatrix}
=
\begin{bmatrix}
\boldsymbol{i}_{sm}(k+1|k) \\
\boldsymbol{i}_{sm}(k+2|k) \\
\boldsymbol{i}_{sm}(k+3|k)
\end{bmatrix}
+ \boldsymbol{h} \cdot \boldsymbol{e}_s(k+1)
\tag{6}
$$

where $\boldsymbol{i}_{sp}$ is the corrected current predictive value vector, $\boldsymbol{h} = [h_1 \ h_2 \ h_3]^T$ is the correction vector.

(3) Time shift:

$$
\begin{bmatrix}
\boldsymbol{i}_{s0}(k+1|k) \\
\boldsymbol{i}_{s0}(k+2|k) \\
\boldsymbol{i}_{s0}(k+3|k)
\end{bmatrix}
=
\begin{bmatrix}
0 & 1 & 0 \\
0 & 0 & 1 \\
0 & 0 & 1
\end{bmatrix}
\begin{bmatrix}
\boldsymbol{i}_{sp}(k+1|k) \\
\boldsymbol{i}_{sp}(k+2|k) \\
\boldsymbol{i}_{sp}(k+3|k)
\end{bmatrix}
\tag{7}
$$

where subscript 0 represents time shifted vectors of the corrected current predictive value vector.

(4) Calculate the control:

$$
\boldsymbol{u}_s(k+1) = \boldsymbol{u}_s(k) + \Delta \boldsymbol{u}_s(k)
\tag{8}
$$

where $\Delta u_s(k)$ is the incremental value of the stator voltage vector at time k.

$$
\Delta \boldsymbol{u}_s(k) = (\boldsymbol{G}_{MPC}^T \boldsymbol{Q}_{MPC} \boldsymbol{G}_{MPC} + \boldsymbol{R}_{MPC})^{-1}
$$
$$
\boldsymbol{G}_{MPC}^T \boldsymbol{Q}_{MPC}
\begin{bmatrix}
\boldsymbol{i}_s^*(k+1|k) - \boldsymbol{i}_{sp}(k+1|k) \\
\boldsymbol{i}_s^*(k+2|k) - \boldsymbol{i}_{sp}(k+2|k) \\
\boldsymbol{i}_s^*(k+3|k) - \boldsymbol{i}_{sp}(k+3|k)
\end{bmatrix}
\tag{9}
$$

where

$$
\boldsymbol{G}_{MPC} =
\begin{bmatrix}
B_{MPC} & 0 & 0 \\
A_{MPC} \cdot B_{MPC} & B_{MPC} & 0 \\
A_{MPC}^2 \cdot B_{MPC} & A_{MPC} \cdot B_{MPC} & B_{MPC}
\end{bmatrix},
$$

$$
\boldsymbol{Q}_{MPC} = diag(q_1, q_2, q_3),
$$

$$
\boldsymbol{R}_{MPC} = diag(r_1, r_2, r_3),
$$

$\boldsymbol{Q}_{MPC}$ is the error weighting coefficient matrix, $\boldsymbol{R}_{MPC}$ is the control weighting coefficient matrix, $\boldsymbol{i}_s^*$ is the reference of the stator current vector.

(5) Update the current predictive value vector $\boldsymbol{i}_{sm}$:

$$
\begin{bmatrix}
\boldsymbol{i}_{sm}(k+1|k) \\
\boldsymbol{i}_{sm}(k+2|k) \\
\boldsymbol{i}_{sm}(k+3|k)
\end{bmatrix}
=
\begin{bmatrix}
\boldsymbol{i}_{s0}(k+1|k) \\
\boldsymbol{i}_{s0}(k+2|k) \\
\boldsymbol{i}_{s0}(k+3|k)
\end{bmatrix}
+ \boldsymbol{G}_{MPC} \cdot
\begin{bmatrix}
\Delta \boldsymbol{u}_s(k) \\
\Delta \boldsymbol{u}_s(k) \\
\Delta \boldsymbol{u}_s(k)
\end{bmatrix}
\tag{10}
$$

B. Extended Kalman Filter Based Speed Estimator

In the stationary frame, the synchronous speed ω_e disappears in the model of induction motors and the computational burden can be reduced. Therefore, according to the state space equation of induction machines on the stationary α-β axis, the on-line EKF speed estimator is developed. The state space equation of induction motors on the stationary frame is as follows:

$$
\dot{\boldsymbol{x}} = A_{EKF} \boldsymbol{x} + B_{EKF} \boldsymbol{u}
$$
$$
\boldsymbol{y} = C_{EKF} \boldsymbol{x}
\tag{11}
$$

where $\boldsymbol{x} = [i_{s\alpha} \ i_{s\beta} \ \psi_{r\alpha} \ \psi_{r\beta} \ \omega_r]^T$, $\boldsymbol{y} = [i_{s\alpha} \ \ i_{s\beta}]^T$, $\boldsymbol{u} = [u_{s\alpha} \ \ u_{s\beta}]^T$,

$$A_{EKF} = \begin{bmatrix} -(\frac{R_s}{\sigma L_s} + \frac{1-\sigma}{\sigma \tau_r}) & 0 & \frac{L_m}{\sigma L_s L_r \tau_r} & \omega_r \frac{L_m}{\sigma L_s L_r} & 0 \\ 0 & -(\frac{R_s}{\sigma L_s} + \frac{1-\sigma}{\sigma \tau_r}) & -\omega_r \frac{L_m}{\sigma L_s L_r} & \frac{L_m}{\sigma L_s L_r \tau_r} & 0 \\ \frac{L_m}{\tau_r} & 0 & -\frac{1}{\tau_r} & -\omega_r & 0 \\ 0 & \frac{L_m}{\tau_r} & \omega_r & -\frac{1}{\tau_r} & 0 \\ 0 & 0 & 0 & 0 & 0 \end{bmatrix},$$

$$B_{EKF} = \begin{bmatrix} \frac{1}{\sigma L_s} & 0 \\ 0 & \frac{1}{\sigma L_s} \\ 0 & 0 \\ 0 & 0 \\ 0 & 0 \end{bmatrix},$$

$$C_{EKF} = \begin{bmatrix} 1 & 0 & 0 & 0 & 0 \\ 0 & 1 & 0 & 0 & 0 \end{bmatrix},$$

where $\sigma = 1 - L_m^2/L_s L_r$ is the total leakage coefficient, $\tau_r = L_r/R_r$ is the rotor time constant. $u_{s\alpha}$, $u_{s\beta}$ are the stationary frame α-axis stator voltage and β-axis stator voltage, respectively. $i_{s\alpha}$, $i_{s\beta}$ is the stationary frame α-axis stator current and β-axis stator current, respectively. $\psi_{r\alpha}$, $\psi_{r\beta}$ are the stationary frame α-axis rotor flux and β-axis rotor flux, respectively. ω_r is the rotor electrical angular speed.

Equation (11) can be discretized as follows

$$x(k) = f(x(k-1), u(k-1))$$
$$= A_{EKFd} x(k-1) + B_{EKFd} u(k-1) \quad (12)$$
$$y(k) = C_{EKFd} x(k)$$

where $A_{EKFd} \approx I + A_{EKF} T_s$, $B_{EKFd} \approx B_{EKF} T_s$, $C_{EKFd} = C_{EKF}$, and T_s is the sampling time.

Substitute the above induction machine model into the EKF algorithm.

Initialize with

$$\hat{x}(0) = E[x(0)], \qquad P_x(0) = E[(x(0) - \hat{x}(0))(x(0) - \hat{x}(0))^T]$$

For $k = 1, 2, 3, \ldots \infty$, the time update equation for the state filter are

$$\hat{x}^-(k) = f(\hat{x}(k-1), u(k-1))$$
$$= A_{EKFd} \hat{x}(k-1) + B_{EKFd} u(k-1) \quad (13)$$

$$P_x^-(k) = G_{EKF}(k) P_x(k-1) G_{EKF}(k)^T + Q_{EKFx} \quad (14)$$

where

$$G_{EKF}(k) = \frac{\partial f}{\partial x}\bigg|_{x=\hat{x}(k-1)}$$

The measurement update equations for the state filter are

$$K_x(k) = P_x^-(k) C_{EKFd}^T (C_{EKFd} P_x^-(k) C_{EKFd}^T + R_{EKFx})^{-1} \quad (15)$$

$$\hat{x}(k) = \hat{x}^-(k) + K_x(k)(y(k) - C_{EKFd} \hat{x}^-(k)) \quad (16)$$

$$P_x(k) = (I - K_x(k) C_{EKFd}) P_x^-(k) \quad (17)$$

III. SIMULATION RESULTS

In order to verify the effectiveness of the proposed scheme of Fig. 1, simulations are carried out in MATLAB/Simulink environment. The parameters of the induction motor used for simulation are listed: P_N=1.1kW, U_N=380V, f_N=50Hz, R_s=5.27Ω, R_r=5.07Ω, L_{ls}=0.002H, L_{lr}=0.058H, L_m=0.421H, T_N=7.45Nm.

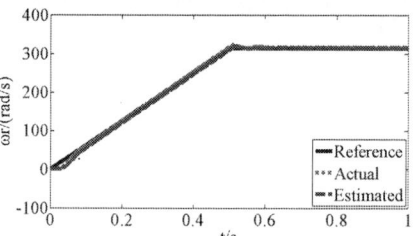

Fig. 2. Speed response of MPC+EKF under ramp speed reference

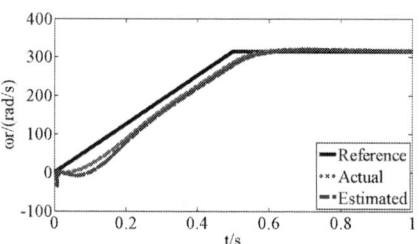

Fig. 3. Speed response of FOC+EKF under ramp speed reference

Figs. 2-3 show the performance of speed tracking behavior under ramp speed reference. The reference speed has been changed from 0 to rated speed 314rad/s at the time instant 0.5s. Fig. 3 presents that the FOC+EKF has a certain delay. Fig. 2 presents that the MPC+EKF can overcome this shortcoming. The estimation errors of MPC+EKF and FOC+EKF are shown in the Fig. 4. The response errors of MPC+EKF and FOC+EKF are shown in the Fig. 5. Figs. 4-5 show that MPC+EKF system has good control and tracking performance.

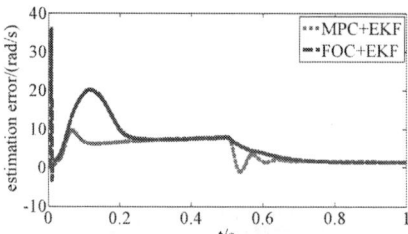

Fig. 4. Speed estimation errors of MPC+EKF and FOC+EKF

The low speed waveforms are presented in Figs. 6-7. The reference speed has been changed from 0 to rated speed 3.14rad/s at the time instant 0.5s. The estimation errors of MPC+EKF and FOC+EKF are shown in the Fig. 8. The response errors of MPC+EKF and FOC+EKF are shown in the

Fig. 9. In fig. 6, it is obvious that the MPC+EKF shows superior performance at low speed with faster dynamic response. In fig.7, it is also can be seen that the FOC+EKF system has the poor performance at low speed. Figs. 8-9 show that MPC+EKF system has excellent control and tracking performance even at low speed.

Figs. 10-11 present the wide speed range tracking performance. The speed reference alters from 314rad/s to 3.14rad/s, then to 188rad/s. The load torque is 0. The dynamic response time of MPC+EKF and FOC+EKF are 0.2s and 0.5s, respectively.

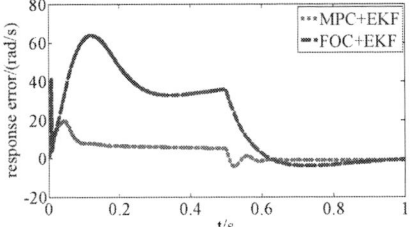

Fig. 5. Speed response errors of MPC+EKF and FOC+EKF

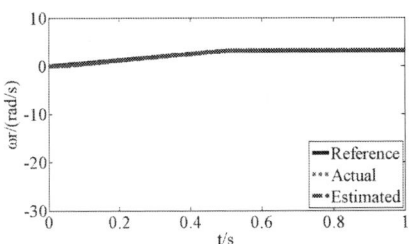

Fig. 6. Low speed performance of MPC+EKF

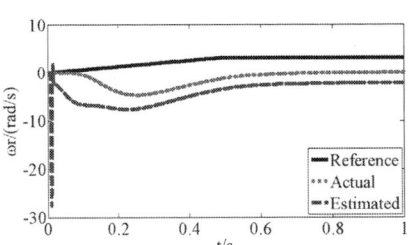

Fig. 7. Low speed performance of FOC+EKF

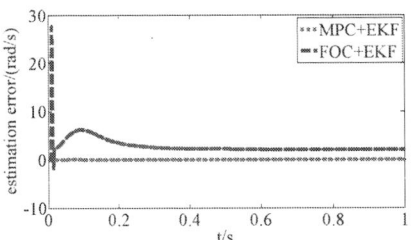

Fig. 8. Low speed estimation error of MPC+EKF and FOC+EKF

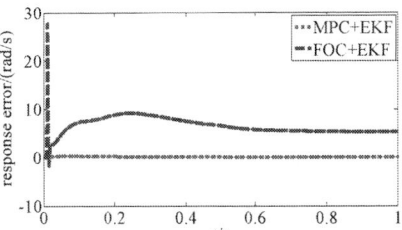

Fig. 9. Low speed response error of MPC+EKF and FOC+EKF

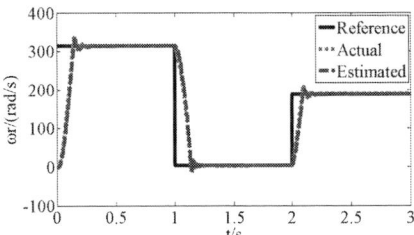

Fig. 10. Wide speed operation of MPC+EKF

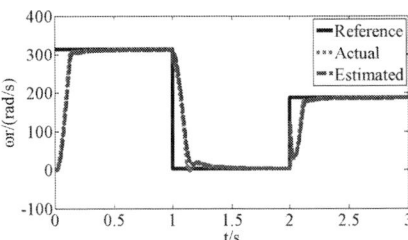

Fig. 11. Wide speed operation of FOC+EKF

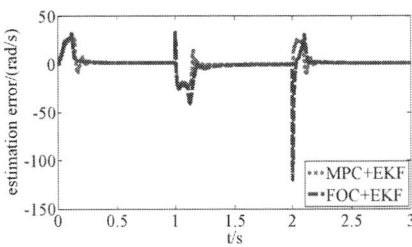

Fig. 12. Speed estimation error of MPC+EKF and FOC+EKF

Fig. 12 shows the estimation error of MPC+EKF and FOC+EKF. It is seen that the estimation performance of MPC+EKF is superior to the FOC+EKF. Fig. 13 shows the response error of MPC+EKF and FOC+EKF. The response error is almost similar. Figs. 10-13 show the good tracking performances in a wide speed range.

Figs. 14-15 show the load disturbance performance. The load alters from 0 to 1pu, then to 0. Fig. 16 shows the estimation error of MPC+EKF and FOC+EKF. Fig. 17 shows the response error of MPC+EKF and FOC+EKF. The estimation and response performance of MPC+EKF are superior to the FOC+EKF. Figs. 18-19 show the enlarged image of load increase of MPC+EKF and FOC+EKF. Figs. 20-21 show the enlarged image of load decrease of MPC+EKF and FOC+EKF. Figs. 14-21 show very good control and track performance during load disturbance.

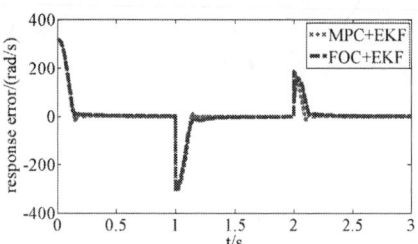

Fig. 13. Speed response error of MPC+EKF and FOC+EKF

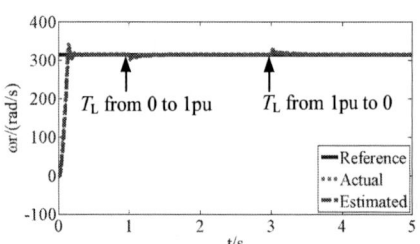

Fig. 14. Speed performance of MPC+EKF under load step up and down

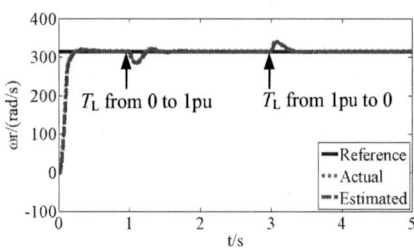

Fig. 15. Speed performance of FOC+EKF under load step up and down

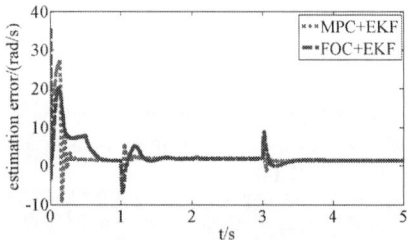

Fig. 16. Speed estimation error of MPC+EKF and FOC+EKF

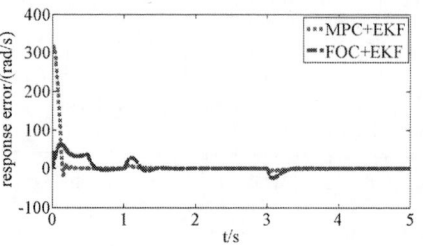

Fig. 17. Speed response error of MPC+EKF and FOC+EKF

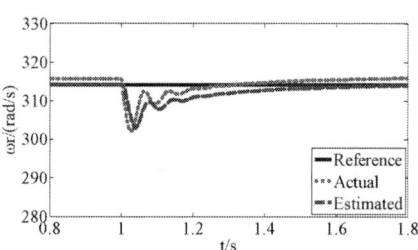

Fig. 18. Enlarged image of load step up of MPC+EKF

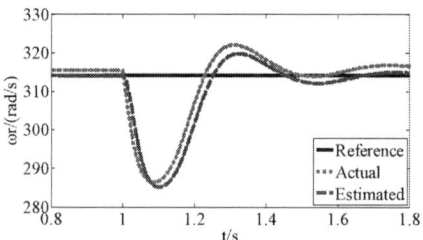

Fig. 19. Enlarged image of load step up of of FOC+EKF

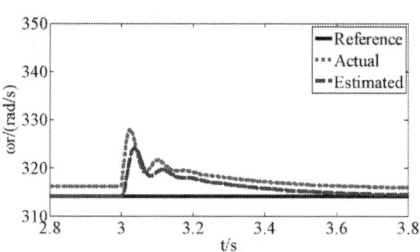

Fig. 20. Enlarged image of load step down of MPC+EKF

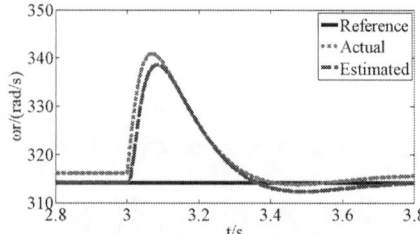

Fig. 21. Enlarged image of load step down of FOC+EKF

IV. CONCLUSION

This paper investigates the sensorless MPC controlled induction motor drives. It is feasible to combine the model predictive control with the extended Kalman filter algorithm for the high performance control of induction motors. The obtained simulation results confirm the good properties of the proposed scheme for speed sensorless induction motor drives. The proposed novel speed sensorless control system of induction motors which coordinates the MPC current controllers and the EKF state estimator has been verified by simulations, and the results show that excellent speed control performance and state estimation performance over wide speed range and load torque range have been achieved either in dynamics or in steady states.

REFERENCES

[1] Jose Rodriguez, Patricio Cortes. Predictive Control of Power Converters and Electrical Drives. *Wiley-IEEE Press*, Chichester, United Kingdom, 2012.

[2] Rodriguez J, Kennel R, Espinoza J, et al. "High performance control strategies for electrical drives: an experimental assessment," *IEEE Trans. on Industrial Electronics*, vol. 59, no. 2, pp. 812-820, 2012.

[3] Rodriguez J, Kazmierkowski M, Espinoza J, et al. "State of the art of finite control set model predictive contol in power electronics," *IEEE Trans. on Industrial Informatics*, vol. 9, no. 2, pp. 1003-1016, 2013.

[4] Mariethoz S, Domahidi A, Morari, M. "High-bandwidth explicit model predictive control of electrical drives," *IEEE Trans. on Industrial Electronics*, vol. 48, no. 6, pp. 1980-1992, 2012.

[5] Geyer T, Papafotiou G, Morari, M. "Model predictive direct torque control; part I: concept, algorithm, and analysis," *IEEE Trans. on Industrial Electronics*, vol. 56, no. 6, pp. 1894-1905, 2009.

[6] Miranda H, Cortes P, Yuz J, et al. "Predictive torque control of induction machines based on state-space models," *IEEE Trans. on Industrial Electronics*, vol. 56, no. 6, pp. 1916-1924, 2009.

[7] Geyer T, Mastellone S. "Model predictive direct torque control of a five-level anpc converter drive system," *IEEE Trans. on Industry Application*, vol. 48, no. 5, pp. 1565-1575, 2012.

[8] Cortes P, Kazmierkoeski M, Kennel R, et al. "Predictive control in power electrionics and drives," *IEEE Trans. on Industrial Electronics*, vol. 55, no. 12, pp. 4312-4324, 2008.

[9] Sergio Vazquez, Josei Leon, Leopoldo G, et al. "Model predictive control," *IEEE Trans. on Industrial Electronics*, 2014, pp. 16-31.

[10] Zheng Zedong, Wang Kui, Li Yongdong, et al. "Current controller for AC motors using model predictive control," *Transactions of China Electrotechnical Society*, vol. 28, no. 11, pp. 118-123, 2013.

[11] He Dongwei, Peng Xiafu, Jiang Xuecheng. "A current control strategy for the permanent magnet synchronous motor based on model predictive control," *Journal of Harbin Engineering University*, vol. 34, no. 12, pp. 1556-1565, 2013.

[12] Ruan Zhi-yong, Song Wen-xiang, Zhu Hong-zhi. "Model predictive direct torque control of induction motors," *Advanced Technology of Electrical Engineering and Energy*, vol. 33, no. 4, pp. 16-20, 2014.

[13] Ruan Zhi-yong, Song Wen-xiang, Zhu Hong-zhi, et al. "Model predictive direct current control of induction motor at low switching frequency," *Journal of Shanghai University (Natural Science)*, vol. 19, no. 6, pp. 647-653, 2013.

[14] Zhang Yongchang, Yang Haitao, "Model Predictive Control for Speed Sensorless Induction Motor Drive," *Proceedings of the CSEE*, vol. 34, no. 15, pp. 2422-2429, May 25, 2014. (in Chinese)

Research on Excitation Control Methods for the Two-Phase Brushless Exciter of Wound-Rotor Synchronous Starter/Generators in the Starting Mode

Ningfei Jiao, Weiguo Liu, Tao Meng, Jichang Peng and Shuai Mao
Institute of REPM Electrical Machines and Control Technology
Department of Electrical Engineering, Northwestern Polytechnical University
Xi'an, Shaanxi 710072, P. R. China
Emails: jiaoningfei@gmail.com, lwglll@nwpu.edu.cn, mengtao0504@163.com, linkpjc@gmail.com, maoshuai1989@126.com

Abstract—**Integrated starter/generator based on the wound-rotor synchronous machine is becoming increasingly popular in modern aircraft due to the advantages of high safety and low cost in maintenance. In this paper, three detailed excitation control methods for the two-phase brushless exciter of a wound-rotor synchronous starter/generator in the starting mode were proposed and compared. In order to get constant field current and/or minimum armature currents for the main generator during the start-up process, feedback control for the field currents of the two-phase brushless exciter and speed reference control for the excitation frequency and phase sequence were proposed. Experimental results verified the feasibility and effectiveness of these three excitation control methods.**

Keywords—two-phase brushless exciter; excitation control; wound-rotor synchronous machine (WRSM); integrated starter/ generators (ISG)

I. INTRODUCTION

In traditional aircraft propulsion and power system, the aero engine is started by a dedicated starter during the start-up process. And after the start-up process, the aero engine drags the brushless synchronous generator to generate power for the aircraft. Therefore, two electric machines, the dedicated starter and generator, are needed in the traditional aircraft power system. With the development of more electric aircraft (MEA), there is an increasing demand for the integrated starter/generator (ISG) system [1-10]. ISG system can start the aero engine during the start-up process and operate as the generator after the start-up. So only one electric machine is used in the ISG system, which can decrease the weight and volume of the aircraft power system. Various types of electric machines can be considered to operate as an ISG [1-5]. Because of advantages such as high safety and low cost in maintenance, brushless wound-rotor synchronous machine (WRSM) becomes an attractive candidate for ISG [5-10]. Early research on the ISG system based on a WRSM mainly focused on the construction design [8-10], basic operation principle [8-11] and start control algorithm [12-14]. The research on the detailed excitation control for the brushless exciter was rarely carried out and will be studied in this paper.

This work was supported by National Natural Science Foundation of China (51277152).

A two-phase brushless exciter was newly proposed to solve the weak excitation problem of the WRSM when it operates as a motor [15]. The structure of the wound-rotor synchronous starter/generator with a two-phase brushless exciter is shown in Fig. 1. And its operation principle is as follow: In the starting mode, the two-phase Main Exciter (ME) controller supplies two-phase AC excitation or DC excitation (The DC excitation is only for high-speed starting mode) for the ME, and then the rotor armature winding of the two-phase ME provides DC field current for the Main Generator (MG) through the rotating rectifier. Using particular start control method with the MG start controller, the MG will generate magnetic torque to start the aero engine as a motor in the starting mode. In the generation mode, the two-phase field winding of the ME is connected in series into a single winding and connected with the Generator Control Unit (GCU). The Pre-Exciter (PE), which is a permanent magnet generator, then provide DC excitation for the ME through the GCU. And the operation principle in the generation mode is the same as traditional brushless wound-rotor synchronous generator. Compared with single-phase brushless exciter and three-phase brushless exciter, the two-phase brushless exciter has the

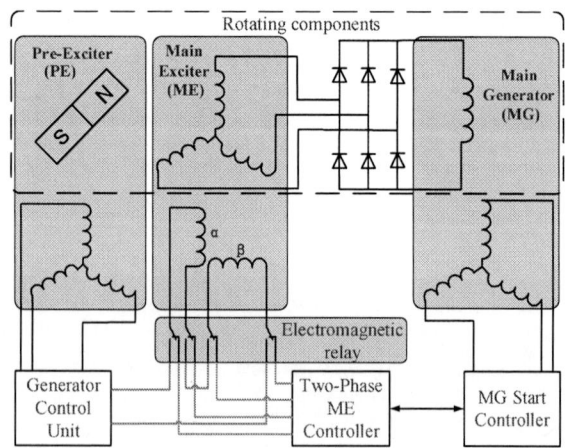

Fig. 1. Structure of the wound-rotor synchronous starter/generator with a two-phase brushless exciter.

advantages of high excitation efficiency in the starting mode and unchanged structure and control methods in the generation mode [15].

For the purpose of constant field current and/or minimum armature currents of the MG in the starting mode, three detailed excitation control methods for the two-phase ME were proposed in this paper, and comparison of these three methods were carried out. The test platform was built and experimental results verified the feasibility and effectiveness of these three excitation control methods.

II. EXCITATION CONTROL METHODS FOR THE TWO-PHASE BRUSHLESS EXCITER

In the starting mode, the field current of the MG is supplied by the ME through the rotating rectifier. On the one hand, the ME should provide big enough field current for the MG to start with large load. On the other hand, the field current of the MG should keep basically constant during the start-up process, which can result in simpler start control scheme for the MG. Unfortunately, as the field winding of the MG is mounted in the rotor part, the field current of the MG cannot be measured directly when the starter/generator rotates. As a result, direct regulation for the field current of the MG, to make it constant, cannot be achieved.

The output of the ME mainly relies on the magnetic field intensity and relative speed of the rotor (armature winding) and magnetic field. The magnetic field intensity is related to the field currents of the two-phase ME when the armature currents are supposed to be constant. And the rotating speed and direction of the magnetic field is decided by the excitation frequency and phase sequence of the two-phase field currents for the ME. So the field current of the MG can be indirectly regulated by the control of the magnitude (corresponding to magnetic field intensity), excitation frequency (corresponding to rotating speed of the magnetic field) and phase sequence (corresponding to rotating direction of the magnetic field) of the two-phase field currents for the ME .

The overall block diagram of the excitation control for the two-phase ME is illustrated in Fig. 2. The two-phase excitation voltages for the ME, denoted by u_α and u_β, are decided by the rotor speed and DC-filtered magnitude of the field currents through the excitation control scheme. The measured magnitude of the two-phase field currents, denoted by i_s, is calculated using

$$i_s = \sqrt{i_\alpha{}^2 + i_\beta{}^2} \tag{1}$$

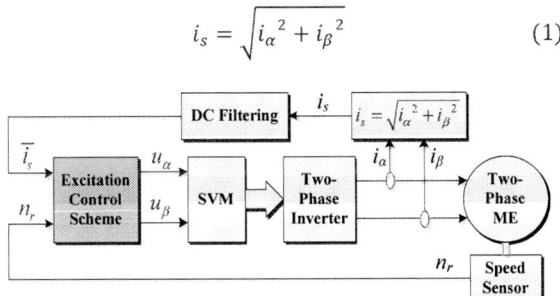

Fig. 2. Overall block diagram of the excitation control method for the two-phase ME in the starting mode.

and then filtered through a DC filter. Where i_α and i_β are instantaneous values of the two-phase field currents, $\overline{i_s}$ is the DC-filtered magnitude of the two-phase field currents.

Early study on the two-phase ME in the stationary status showed that the field current of MG rise linearly with the increase of the magnitude of the excitation voltage for ME [15]. So in order to provide the biggest field current for the MG in the beginning of the start-up process, the excitation voltage for the ME in the stationary status was set as the highest value that the two-phase inverter can provide under a limited DC bus voltage.

A. Method I: Two-Phase AC Excitation during the Entire Start-Up Process

Constant magnetic field intensity and relative speed of the rotor and magnetic field can result in basically constant output of the ME, hence basically constant field current for the MG. ('Constant' means the magnitude and frequency or speed is constant for AC or rotating variables.) Feedback control for the magnitude of the two-phase field currents of the ME can make sure the constant filed currents, hence constant magnetic field. And adjustment of the excitation frequency and phase sequence according to the rotor speed can keep the relative speed between the rotor and magnetic field constant.

The basic idea of the excitation control Method I for the two-phase ME is shown as follow: During the entire start-up process, two-phase AC excitation is used for the ME, and the magnitude of the two-phase field currents is regulated to be constant through a PI regulator. In the beginning of the start-up process, the rotating direction of the magnetic field is the opposite as that of the rotor. In order to keep the relative speed between the rotor and magnetic field constant, the rotating speed of the magnetic field, corresponding to the excitation frequency, decreases as the rotor speeds up. After the excitation frequency approaches zero as the rotor speed reaches a certain value (denoted as the switch speed by n_s), the rotating direction of the magnetic field is then changed to be the same as the rotor, and the excitation frequency increases as the rotor speed gets higher.

To keep the relative speed constant during the entire start-up process, the excitation frequency for the two-phase ME is calculated from the rotor speed by (2).

$$f_e = f_0 - \frac{p_n}{60}n_r \tag{2}$$

Where f_e represents the excitation frequency, f_0 is the excitation frequency in the stationary status, n_r is the rotor speed, and p_n represents the number of pole pairs of the two phase ME.

As the rotor speed increases, according to (2), f_e will be a negative value after the switch speed. The negative value of f_e means the phase sequence of the two-phase field currents is opposite as defined. In this research, a positive value of f_e means the rotating direction of the magnetic field is the opposite as the rotor, and the negative one means the same direction.

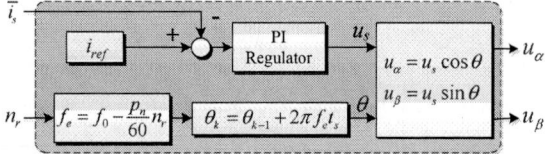

Fig. 3. Block diagram of the excitation control scheme using Method I.

By setting $f_e = 0$ in (2), the switch speed can be calculated as

$$n_s = 60 f_0 / p_n. \tag{3}$$

In the excitation control scheme of Method I, the magnitude of the excitation voltage vector is decided by the PI regulation for the field currents of the ME. And the phase angle of the excitation voltage vector can be calculated by the integral of the angular frequency. In digital implementation, the integral of the angular frequency can be converted into discrete expression. Then the phase angle of the excitation voltage vector can be obtained by

$$\theta_k = \theta_{k-1} + 2\pi f_e t_s \tag{4}$$

Where θ_k is the phase angle of the current voltage vector, θ_{k-1} is the previous phase angle, and t_s is the period time.

The block diagram of the excitation control scheme using Method I is shown in Fig. 3. The PI regulator for the field currents is used to keep the magnetic field intensity constant and it determines the magnitude of the excitation voltage vector for the ME (denoted by u_s). The speed reference control for the excitation frequency is aimed to make the relative speed constant and it determines the phase angle of the excitation voltage vector (denoted by θ).

B. Method II: Counter-Rotating Two-Phase AC Excitation and DC Excitation

Using Method I for the excitation control for the ME can make the field current of the MG basically constant during the entire start-up process. However, after the switch speed, the excitation frequency increases as the rotor speed get higher and higher. And as a result of the PI regulation for the field current, the excitation voltage will also increase. If the excitation voltage has reached its maximum value before the engine accelerates to its self-sustain speed (the end of the start-up process), the excitation voltage cannot go higher as the rotor speed keeps increasing. In this situation, the PI regulation for the field current cannot keep the field current constant anymore, and this excitation control method will be invalid. Therefore, the Method I can only be used in situations that the self-sustain speeds of the engines are not very high.

In order to solve the disadvantages of the speed limitation of Method I, excitation control Method II was proposed, and the details are shown below.

Before the switch speed, the excitation control Method II is the same as Method I. When the excitation frequency decreases to 0, two-phase AC excitation for the ME switches to DC excitation, and DC excitation will be used in the remaining start-up process.

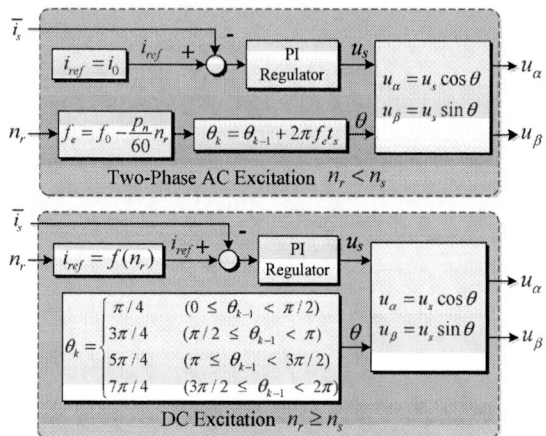

Fig. 4. Block diagram of the excitation control scheme using Method II.

In the DC excitation stage, the DC field currents in the two-phase filed windings should be equal so that the heating condition of the ME can be balanced. Because of the symmetry of the two-phase field windings, equal DC voltages for the two-phase filed windings can result in equal DC field currents. So in the DC excitation stage, the phase angle of the excitation voltage vector is fixed at 45°, 135°, 225° or 315° to make the DC field currents in the two-phase filed windings equal. Besides, the switch from the two-phase AC excitation to DC excitation should be as smooth as possible in order to reduce abrupt change of the field current. Because the magnitude of the excitation voltage for the ME is small just before and after the switch from two-phase AC to DC excitation, there is not an abrupt and large change for the excitation voltage. As for the phase angle of the excitation voltage vector, it will change directly from a random value to a fixed value (45°, 135°, 225° or 315°). To minimize the phase angle change, an optimal switching method, based on the quadrant position of the voltage vector before the switch, is proposed to make the switch as smooth as possible: The phase angle of the excitation voltage vector jump to the nearest one among those four values (45°, 135°, 225° or 315°) in the DC excitation stage. This optimal switching method for the phase angle of the excitation voltage vector during the switch can be illustrated as follow

$$\theta_k = \begin{cases} \pi/4 & when\ 0 \le \theta_{k-1} < \pi/2 \\ 3\pi/4 & when\ \pi/2 \le \theta_{k-1} < \pi \\ 5\pi/4 & when\ \pi \le \theta_{k-1} < 2\pi/2 \\ 7\pi/4 & when\ \pi3/2 \le \theta_{k-1} < 2\pi. \end{cases} \tag{5}$$

In order to make the field current of the MG still constant in the DC excitation stage, the reference value of the field current in the feedback control loop of the field current of the ME should be adjusted in different speeds. The relationship between the rotor speed and the reference value of the field current can be decided off-line by point-by-point test and then proper function fitting can be made, which will be used in the excitation control of the ME.

The block diagram of the excitation control scheme using Method II was shown in Fig. 4. Where $i_{ref} = f(n_r)$ is the function fitting result of the relationship between the rotor speed and the reference value of the field current.

C. Method III: Co-Rotating Two-Phase AC Excitation and DC Excitation

The Method I and Method II are aimed to make the field current of the MG constant during the start-up process. However, as the rotating directions of the magnetic field and the rotor are opposite in the beginning, the two-phase ME will generate a negative magnetic torque for the MG, which will result in extra armature currents for the MG. In order to minimize armature currents of the MG in the beginning of the start-up process, in Method III, the rotating direction of the magnetic field is the same as the rotor in the beginning, so that the ME can generate a positive magnetic torque.

However, if the rotating direction of the magnetic field is the same as the rotor in the beginning of the start-up process, in order to keep the relative speed constant, the frequency of the field currents should become higher and higher as the rotor speed increases. In this situation, the magnitude of the excitation voltage for the two-phase ME should also increase to make the magnitude of the field currents constant as a PI-regulation result. However, the excitation voltage supplied to the ME in the stationary status have already been chosen to be the highest value that the two-phase inverter can provide, and there is no possibility to increase the excitation voltage any more as the rotor speed increases. Therefore, it is very difficult to make the field current of the MG keep constant when choosing the rotating direction of the magnetic field same as that of the rotor in the beginning.

A compromising method for this situation is that the excitation type for the ME in the stationary status is persistently used as the rotor speeds up, and no regulation is made for the field currents and frequency. In this situation, the field current of the MG will slightly decrease as the rotor speed get higher because the relative speed of the armature winding and magnetic field get smaller. When the rotor speed reaches a particular value where DC excitation for the ME can provide bigger field current for the MG than co-rotating two-phase AC excitation, AC excitation for the ME will switch to DC excitation, and after that, the DC excitation will be adopted until the end of the start-up process as Method II. So in Method III, before the switch, co-rotating two-phase AC excitation is used for the ME, and after the switch, DC excitation is used.

In Method II, before the switch, the excitation frequency gradually decreases to 0, and the magnitude of the excitation voltage is already very small when the switch begins. Whereas in Method III, the excitation frequency and voltage magnitude keeps constant, which are pretty high, before the switch. And after the switch, the excitation frequency is 0 and the DC excitation voltage is very small in order to limit the DC field currents of the ME. So in Method III, if the switch from AC excitation to DC excitation is processed directly, there will be an abrupt change of excitation voltage and frequency during the switch, which may result in serious problems.

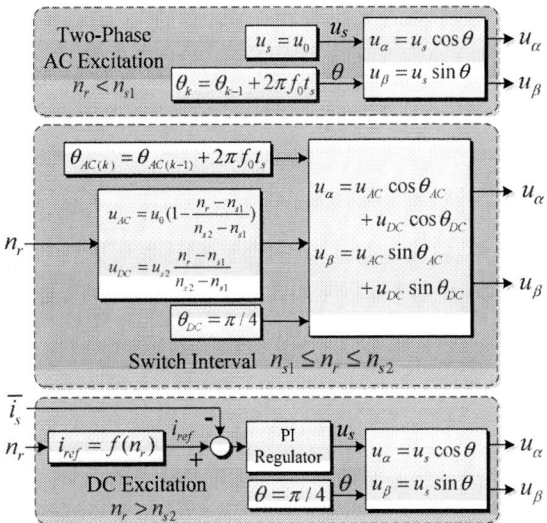

Fig. 5. Block diagram of the excitation control scheme using Method III.

To solve this problem, the switch from two-phase AC excitation to DC excitation should be gradually carried out. Details are shown below.

A proper switch interval, not a switch point, should be decided first. Before the switch interval, two-phase AC excitation is used, and after the interval, DC excitation is used. During the interval, the combination of two-phase AC excitation and DC excitation is used, and the two-phase AC component decreases from 1 in the beginning of the switch interval to 0 in the end of the switch interval gradually, while the DC component increases gradually from 0 to 1 at the same time. So during the switch interval, the two-phase excitation voltage for the ME can be calculated as:

$$u_\alpha = u_{AC}cos\theta_{AC} + u_{DC}cos\theta_{DC}$$
$$u_\beta = u_{AC}sin\theta_{AC} + u_{DC}sin\theta_{DC} \tag{6}$$
$$u_{AC} = u_0 \left(1 - \frac{n_r - n_{s1}}{n_{s2} - n_{s1}}\right)$$
$$u_{DC} = u_{s2} \frac{n_r - n_{s1}}{n_{s2} - n_{s1}} \tag{7}$$

Where u_{AC} and u_{DC} are the magnitudes of two-phase AC component and DC component, respectively. θ_{AC} and θ_{DC} (constant value) are the phase angles of two-phase AC component and DC components, respectively. n_{s1} and n_{s2} are the speeds in the beginning and end of the switch interval, respectively. u_0 and u_{s2} are the magnitudes of voltage vectors in the stationary status and speed of n_{s2}, respectively. In order to keep the two-phase field currents balanced during and after the switch interval, the value of θ_{DC} is set as 45°, 135°, 225° or 315° (45° was chosen in this research).

After the switch interval when the ME is excited by DC, the same control method can be used to make the field current of MG keep constant as the Method II.

Fig. 6 Experiment platform.

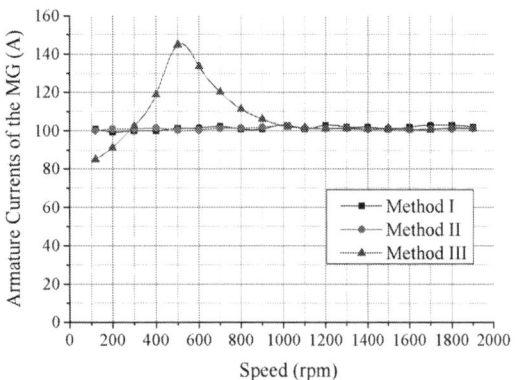

Fig. 7 Armature currents of the MG at different speeds.

The block diagram of the excitation control scheme using Method III was shown in Fig. 5.

III. EXPERIMENTAL RESULTS

Excitation control Method I and II were aimed to keep the field currents of the MG constant during the start-up process. However, as the field winding of the MG is in the rotating part, the field current of the MG cannot be measured directly. As a result, direct experimental results (field currents of the MG) cannot be obtained to verify the feasibility and effectiveness of these proposed excitation control methods.

An alternative method was used to verify these excitation control methods: The armature currents of the MG were measured and used to refer the information of the field current when the same load and same control method were used in the start control for the MG. It can be referred that the field currents of the MG are constant if the armature currents of the MG are constant during the start-up process.

The experiment platform was built as shown in Fig. 6. A load platform was set at a constant torque value and connected to the three-stage wound-rotor synchronous starter/generator. A MG controller was used to drive the starter/generator to start with load, while the two-phase ME was controlled by a ME controller during the start-up process with proposed excitation control Method I, II and III, respectively. The experiments were carried out from standstill to 1900 rpm. For

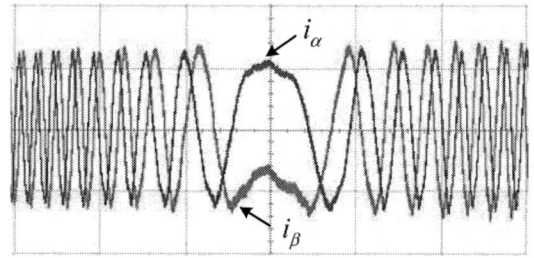

Fig. 8 Field currents of the ME using Method I.

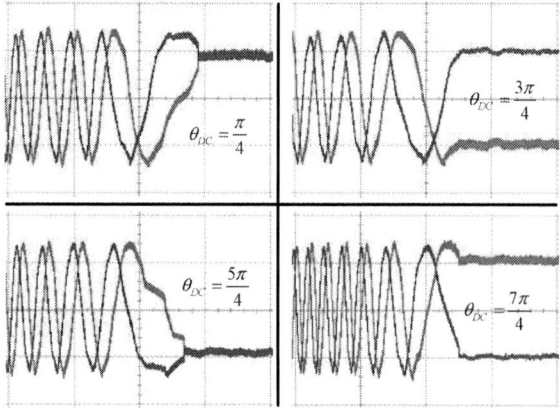

Fig. 9 Field currents of the ME using Method II.

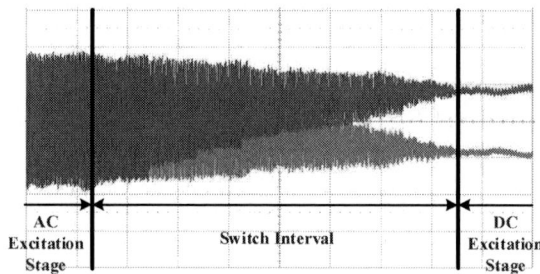

Fig. 10 Field currents of the ME using Method III.

excitation control Method I and II, the switch speeds were 1000 rpm. And for Method III, the switch interval was from 500 rpm to 600 rpm. The armature currents of the MG and field currents of the ME were measured and used to verify these excitation control methods.

The experimental results of the armature currents of the MG at different speeds using these proposed excitation control methods were shown in Fig. 7. The constant armature currents of the MG using Method I and Method II can refer the constant field currents of the MG during the start-up process. And the armature current of the MG using Method III was smaller than others at the beginning of the start-up process, but not constant in two-phase AC excitation stage.

TABLE I.   COMPARISON OF THESE PROPOSED EXCITATION CONTROL METHODS

	Description	Advantages	Disadvantages
Method I	Two-Phase AC Excitation during the Entire Start-Up Process	Constant field current for the MG; Do NOT need off-line test.	Invalid in high speed range; Two-phase ME generate negative torque for the MG.
Method II	Counter-Rotating Two-Phase AC Excitation and DC Excitation	Constant field current of the MG; High-speed operation range.	Need off-line test first; Two-phase ME generate negative torque for the MG.
Method III	Co-Rotating Two-Phase AC Excitation and DC Excitation	Minimum armature currents of the MG in the beginning.	Need off-line test first; Non-constant field current for the MG in AC excitation stage.

The field currents of the ME around the switch speed/ interval using Method I, Method II and Method III were shown in Fig. 8, Fig. 9 and Fig. 10, respectively. In Method II, four switch situations were all illustrated, and all of those had smooth switch process.

IV.   CONCLUSIONS

Three detailed excitation control methods for the two-phase brushless exciter of a wound-rotor synchronous starter/generator in the starting mode have been presented in this paper. Method I and Method II were aimed to obtain constant field current for the MG, while Method III was designed to get minimum armature currents for the MG, during the start-up process. The experiment platform was built and these proposed excitation control methods were verified.

According to the analysis and test results of these three excitation control methods, the comparison of these excitation control methods about the advantages and disadvantages was shown in Table I.

References

[1] B. S. Bhangu and K. Rajashekara, "Electric Starter Generators: Their Integration into Gas Turbine Engines," Industry Applications Magazine, IEEE, vol. 20, pp. 14-22, 2014.

[2] C. A. Ferreira, S. R. Jones, W. S. Heglund, and W. D. Jones, "Detailed design of a 30-kW switched reluctance starter/generator system for a gas turbine engine application," Industry Applications, IEEE Transactions on, vol. 31, pp. 553-561, 1995.

[3] A. K. Jain, S. Mathapati, V. T. Ranganathan, and V. Narayanan, "Integrated starter generator for 42-V powernet using induction machine and direct torque control technique," Power Electronics, IEEE Transactions on, vol. 21, pp. 701-710, 2006.

[4] C. Zhihui, W. Huizhen, and Y. Yangguang, "A Doubly Salient Starter/Generator With Two-Section Twisted-Rotor Structure for Potential Future Aerospace Application," Industrial Electronics, IEEE Transactions on, vol. 59, pp. 3588-3595, 2012.

[5] G. Friedrich and A. Girardin, "Integrated starter generator," Industry Applications Magazine, IEEE, vol. 15, pp. 26-34, 2009.

[6] R. Wrobel, A. Griffo, A. Mlot, J. Yon, P. Mellor, J. Turner, et al., "Design study of a three-phase brushless exciter for aircraft starter/generator," in Energy Conversion Congress and Exposition (ECCE), 2011 IEEE, 2011, pp. 3998-4004.

[7] W. Jiadan, Z. Qingqing, S. Mingming, Z. Bo, and L. Jie, "The excitation control strategy of the three-stage synchronous machine in the start mode," in Applied Power Electronics Conference and Exposition (APEC), 2014 Twenty-Ninth Annual IEEE, 2014, pp. 2469-2474.

[8] A. Griffo, R. Wrobel, P. H. Mellor, and J. M. Yon, "Design and Characterization of a Three-Phase Brushless Exciter for Aircraft Starter/Generator," Industry Applications, IEEE Transactions on, vol. 49, pp. 2106-2115, 2013.

[9] W. Jiadan, Z. Qingqing, and Y. Yiwei, "Integrated AC and DC excitation method for brushless synchronous machine," in Energy Conversion Congress and Exposition (ECCE), 2012 IEEE, 2012, pp. 2322-2325.

[10] J. Ningfei, L. Weiguo, P. Jichang, M. Shuai, and Z. Hua, "Design and control strategy of a two-phase brushless exciter for three-stage starter/generator," in Energy Conversion Congress and Exposition (ECCE), 2014 IEEE, 2014, pp. 5864-5869.

[11] Y.-n. Li, B. Zhou, J.-d. Wei, and C. Han, "Modeling of Starter/Generator Based on Three-stage Brushless Synchronous Machines," in Electrical and Control Engineering (ICECE), 2010 International Conference on, 2010, pp. 5345-5348.

[12] A. Griffo, D. Drury, T. Sawata, and P. H. Mellor, "Sensorless starting of a wound-field synchronous starter/generator for aerospace applications," Industrial Electronics, IEEE Transactions on, vol. 59, pp. 3579-3587, 2012.

[13] A. Maalouf, L. Idkhajine, S. Le Ballois, and E. Monmasson, "Field programmable gate array-based sensorless control of a brushless synchronous starter generator for aircraft application," Electric Power Applications, IET, vol. 5, pp. 181-192, 2011.

[14] A. Maalouf, S. Le Ballois, L. Idkhajine, E. Monmasson, J. Midy, and F. Biais, "Sensorless control of brushless exciter synchronous starter generator using Extended Kalman Filter," in Industrial Electronics, 2009. IECON '09. 35th Annual Conference of IEEE, 2009, pp. 2581-2586.

[15] N. Jiao, W. Liu, T. Meng, J. Peng, and S. Mao, "Design and Control of a Two-Phase Brushless Exciter for Aircraft Wound-Rotor Synchronous Starter/Generator in the Starting Mode," Power Electronics, IEEE Transactions on, to be published, 2015.

A High Performance Speed Regulator Design for AC Machines

Adil Khurram
American University of Sharjah, UAE
Email:b00056654@aus.edu

Habibur Rehman
American Univerity of Sharjah, UAE
Email: rhabib@aus.edu

Shayok Mukhopadhyay
American Univerity of Sharjah, UAE
Email: smukhopadhyay@aus.edu

Abstract—**Proportional Integral controllers tuned using classical methods may exhibit poor performance under external disturbances or sudden load changes. Fractional Order Proportional Integral (FO-PI) controllers are expected to perform better in terms of load disturbance rejection due to their Iso-damping property. This paper investigates the potential of an FO-PI controller for the speed control of an indirect field oriented (IFO) induction motor drive system with cascaded current control. The motor is first approximated with a first order plus dead time (FPDT) model and consequently an optimal FO-PI controller is designed. The FO-PI controller is implemented in both simulation and experimentation and shows better performance than its integer order counterparts.**

I. INTRODUCTION

The widespread use of induction motors in applications such as hybrid electric vehicles' traction motor drive system, wind power and automation and process control etc. are all well recognized. Decoupled control of flux and torque using field orientation is the most popular method for speed control. The latest research work on closed loop speed control of induction motors has been focused on intelligent control schemes such as model predictive control [1]–[6], auto disturbance rejection controllers (ADRC) [7] and fuzzy logic based controllers [8]. Model predictive control schemes adjust the control input based on the time evolution of the system model. Auto disturbance rejection controllers (ADRC) can compensate for the external disturbances and parameter variations. Although several parameter estimation techniques exist, the incentive to adopt ADRC is their ability to reject disturbance without knowledge of the actual plant [7]. Similarly, adaptive fuzzy sliding mode controllers have been used to avoid the chattering problem [8]. Measurement based methods, as opposed to model based approaches, have been used in [9], and a nonlinear controller is developed in [10].

The above control schemes perform better but increase the computational complexity of the drive system. For model predictive control, an optimal control problem must be solved at each sampling time instant [11]. Fuzzy logic controlled drives increase the computational power demand by the controller, while the non-linear controller introduces additional adaptation dynamics to the system [12]. The potential of FO-PI has been investigated mostly for DC motors control while controlling the induction motors requires a different set of challenges for which no relevant experimental validation showing the performance of FO-PI exists. The performance of an induction machine using an FO-PI controller for speed control and direct torque control has been investigated in [13]

and [14] respectively. However, this work has been done in simulation environment only and does not specify any tuning procedure. An effort has been made [15] to find the gains of an FO-PI controller using gain and phase margin yet no easily usable tuning rules are provided. This paper presents a simpler procedure for designing an FO-PI controller for the speed control of an induction machine using the tuning rules developed in [16]. The decoupling capability of indirect field oriented control is used to approximate the model of induction motor as a first order plus dead time (FPDT) model. The performance of both FO-PI and IO-PI controllers are compared both in the simulations and experimentation. Also the performance of the designed controller is investigated for disturbance rejection.

II. INDIRECT FIELD ORIENTED DRIVE SYSTEM

Indirect field oriented control of induction motor is used for the speed control of induction motor. The torque and speed of the induction motor is controlled by regulating the quadrature axis current (i_{qs}^*) while the flux in the motor is controlled independently by regulating the direct axis i_{ds}^* current. The drive system shown in Fig. 1 is constructed in both simulation and experimental environment to test the performance of the proposed and conventional proportional integral controllers. Two cascaded current control loops are used to regulate the speed (ω_r^*) and the quadrature current (i_{qs}^*) in the induction motor. The direct axis current (i_{ds}^*) is regulated by a separate PI controller. The relation described in Eq. 1 is used to calculate the slip, ω_{sl} and is added to the rotor speed, ω_r to generate ω_e. The integral of ω_e provides the necessary θ_e which is used for the conversion between synchronous and stationary reference frames.

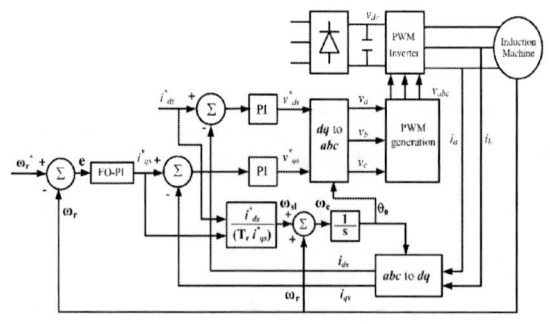

Fig. 1. Indirect field oriented drive dystem

$$\omega_{sl} = \frac{i_{ds}^*}{T_r i_{qs}^*} \qquad (1)$$

$$\omega_e = \omega_{sl} + \omega_r \qquad (2)$$

III. FO-PI CONTROLLER DESIGN

A. First order plus dead time (FPDT) of an IFO Drive

The rotor flux (λ_{dr}) in an induction machine working under field orientation is completely characterized by Eq. 3, and it depends on mutual inductance (L_m), rotor time constant (τ_r) and the flux producing current (i_{ds}). The electromagnetic torque (T_e) is the product of i_{ds} and torque producing current (i_{qs}), as shown as in Eq. 4 where L_r is the rotor inductance. The open loop transfer function of the motor with i_{qs} and output speed (ω_r) is given in Eq. 5. This transfer function can be approximated with a first order plus dead time (FPDT) model given by G_{approx} in Eq. 6 [17].

$$\lambda_{dr} = \frac{i_{ds}L_m}{1 + \tau_r s} \qquad (3)$$

$$T_e = \frac{pL_m^2}{2L_r}i_{ds}^* i_{qs}^* \qquad (4)$$

$$G_p(s) = \frac{\omega_r}{i_{qs}} = \frac{\frac{p}{2}\frac{L_m^2}{L_r}i_{ds}^*}{Js + B} \qquad (5)$$

$$G_{approx}(s) = \frac{K_{prcs}}{Ts + 1}e^{-Ls} \qquad (6)$$

The open loop step response of the induction motor is obtained using the experimental setup because an actual test on the motor accomodates the nonlinear nature of the drive system such as the friction and actual inetia of the motor. Therefore, this model is used to design the FO-PI controller. The same controller is then tested in both simulation and experimentation. In order to obtain the step response, the flux in the machine is first established with a command i_{ds}^*, followed by an i_{qs}^* step command. The Control System toolbox in MATLAB is used to determine the relation between the rotor speed ω_r and the input current, i_{qs}^*. The dead time 'L', time constant 'T' and the process gain 'K_{prcs}' are calculated to be 0.03062, 9.43 s and 609.43 rad/sec respectively. The actual step response matches the FPDT model response as shown in Fig. 2 which confirms that induction motor operating under field orientation can infact be modeled using a FPDT model.

B. Tuning Rules

An FO-PI controller is obtained by replacing the integer order integral in a PI controller by an integral of arbitrary order [18] resulting in Equation 7, where $u(t)$ is the control output, $e(t)$ is the error between the actual speed and the reference speed, and α is the order of the integral operator. The order α is an additional parameter which can be tuned along with the gains of the FO-PI controller K_p and K_i. The differintegral operator, ${}_a^C D_t^\alpha$, is used for the evaluation

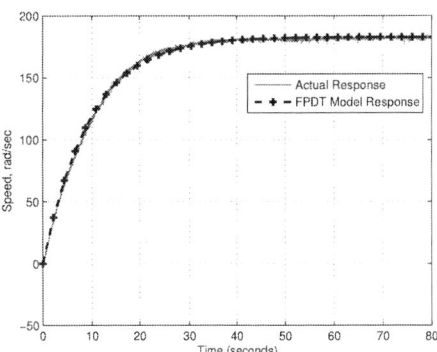

Fig. 2. Step response for obtaining first order plus dead time model

of the integral of arbitrary order and we use the fractional order Caputo derivative based implementation with a, t as the terminals (please consult [19], [20] for more details). The tuning rules used for the design of an optimum FO-PI controller were developed in [16] and are a generalized form of maximum sensitivity (M_s) constrained integral gain optimization (MIGO) based tuning method. They have been modified for FO-PI controllers and are called fractional MIGO (F-MIGO) tuning rules [16]. Also in the tuning rules in Eq. 8, 'τ' is the dead time.

$$u(t) = K_p e(t) + K_i \left({}_a^C D_t^\alpha e(t) \right)$$

$$
{}_a^C D_t^\alpha f(t) = \frac{1}{\Gamma(n-\alpha)} \int_a^t \frac{f^{(n)}(\tau)}{(t-\tau)^{\alpha+1-n}} d\tau \qquad (7)
$$

$$(n-1 < \alpha < n), (n \in \Re)$$

$$
\tau = \frac{L}{T+L}; \alpha = \begin{cases} 1.1 & \tau \geq 0.6 \\ 1.0 & 0.4 \leq \tau \geq 0.6 \\ 0.9 & 0.1 \leq \tau \geq 0.4 \\ 0.7 & \tau < 0.1 \end{cases}, \qquad (8)
$$

$$K_p = \frac{1}{K}\left(\frac{0.2978}{\tau + 0.000307} \right)$$

$$\frac{K_p}{K_i} = T_i = T\left(\frac{0.8578}{\tau^2 - 3.402\tau + 2.405} \right)$$

TABLE I.   INDUCTION MOTOR PARAMETERS

415 V		0.4 A	175 W
Lls = 145.5 mH, Llr = 122.5 mH		Lm = 750.9 mH	1475 RPM
Rs = 50		Rr = 16	4 poles

IV. SIMULATION RESULTS

An induction motor IFO drive system with the underlying sine triangle pulse width modulation (SPWM) methodology is constructed in SIMULINK. The parameters of the induction motor used in the simulations are given in table I. The FPDT model obtained in the previous section is used to design the fractional order controller. The tuning rules defined in Eq. 8 give the optimal proportional and integral gains. The fractional order and integer order controllers parameters are tabulated in Table II. The integral order obtained for FO-PI controller is 0.7 which indicates that the system is nonlinear and integer

978-1-4673-9551-9/16 $31.00 © 2016 IEEE

order controllers are not able to incorporate this nonlinearity. In order to test the performance of the proposed FO-PI controller and compare it to the integer order (IO-PI) controller, the step response and trapezoidal wave tracking are used.

TABLE II.   PI CONTROLLER GAINS

Parameters	FO-PI	Trial Error
α	0.7	1
K_p	1.3038	0.4251
K_i	0.8	0.1

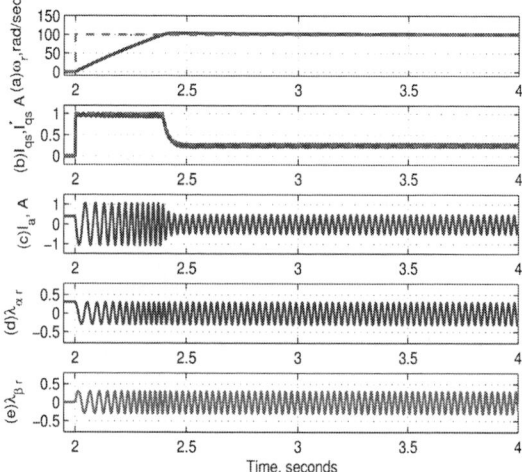

Fig. 3.   Step response of FO-PI controller in simulation

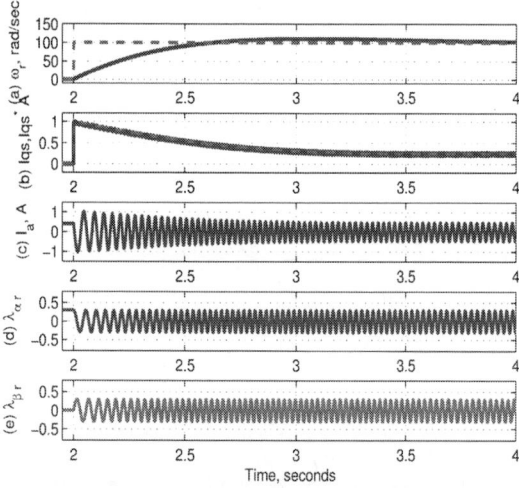

Fig. 4.   Step response of IO-PI controller in simulation

The step response of the drive system for a reference input of 100 rad/sec is plotted in Fig. 3 and 4 for FO-PI controller and IO-PI controller respectively. In these plots, the step is applied at 2 seconds. It is observed that the step response of FO-PI controller is faster than the IO-PI controller. The

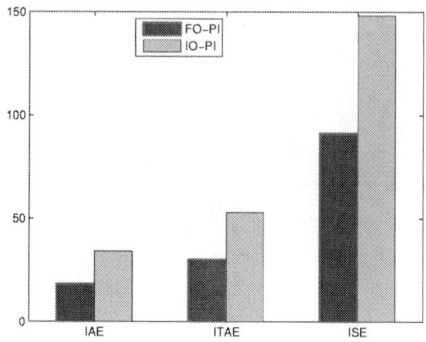

Fig. 5.   Comparison of IAE, ITAE and ISE at 100 rad/sec in simulation

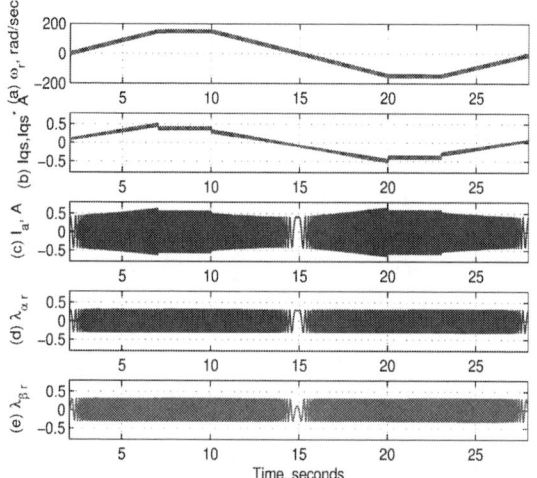

Fig. 6.   Trapezoidal speed tracking of FO-PI controller

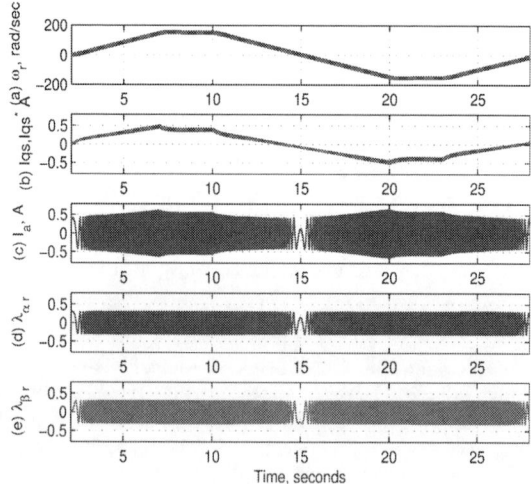

Fig. 7.   Trapezoidal speed tracking of IO-PI controller

settling time and percentage overshoot recorded for the FO-PI controller is less than the that of the IO-PI controller. The percentage overshoot observed for IO-PI controller is 10.72% while the settling time is approximately 2.1s. The FO-PI controller performs better with the percentage overshoot of 3.74% and the settling time of approximately 0.96s. The Integral of Absolute Error (*IAE*), Integral Squared Error (*ISE*), and Integral Time Absolute Error (*ITAE*), which are defined in Eq. 7, are used to comapre the performance of the two controllers. The comparison of IAE, ITAE and ISE is shown in Fig. 5 for both FO-PI and IO-PI controllers. The IAE, ITAE and ISE observed in the case of FO-PI controller are 18.3, 912 and 30 respectively. Whereas in the case of integer order controller, these parameters are found to be 34.2, 1482.45 and 53 respectively. It can be seen that the FO-PI controller shows superior performance than the IO-PI controller.

$$IAE = \int \|e(t)\| \, dt$$
$$ISE = \int (e(t))^2 \, dt \qquad (9)$$
$$ITAE = \int t \cdot \|e(t)\| \, dt$$

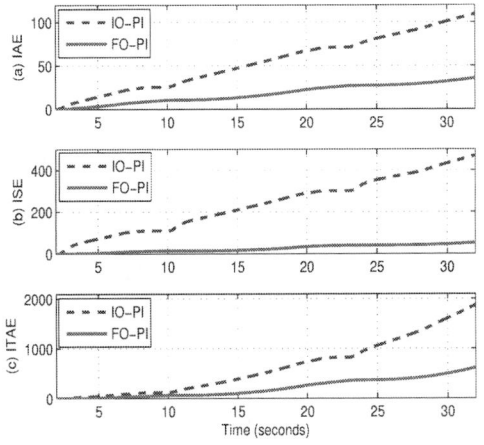

Fig. 8. IAE, ISE and ITAE comparison of FO-PI and IO-PI controllers for trapezoidal wave reference

The system is then tested for trapezoidal wave tracking with both controllers. The reference varies linearly from 0 to 150 rad/sec in 5 seconds. The reference then remains at 150 rad/sec for 3 seconds followed by a linear decrease from +150 rad/sec to -150 rad/sec in 10 seconds. The FO-PI controller and IO-PI controller response is plotted in Fig. 3 and 4. Both controllers succesfully followed the reference. The absolute mean of the control effort required by FO-PI controller is 0.2505 which is approximately equal to the control effort required by IO-PI controller i.e. 0.2482. However, a comparison of IAE, ISE and ITAE highlightes that enhanced performance of FO-PI controller as shown in Fig. 8. The IAE, ISE and ITAE are comparatively lower in the case of FO-PI controller than the IO-PI controller which shows that FO-PI controller performs

better than IO-PI controller while tracking the same trapezoidal reference.

V. EXPERIMENTAL RESULTS

The proposed controller is tested on a prototype 175 W test induction motor using dSPACE DSP system. The parameters of the induction motor are tabulated in table II. The FPDT model for this motor is obtained from the open loop step response of the system and consequently the controller is designed as described in the previous section. These parameters are used to analyze the behavior of the two controllers.

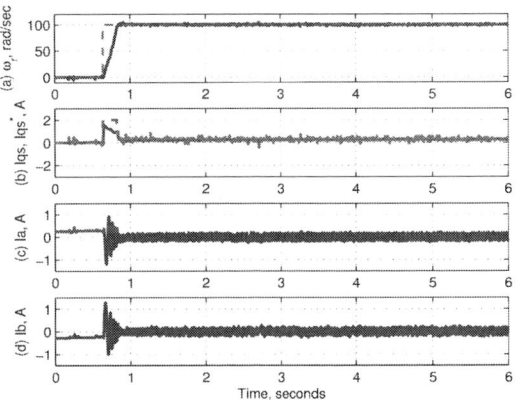

Fig. 9. Step response of FO-PI controller at 100 rad/sec

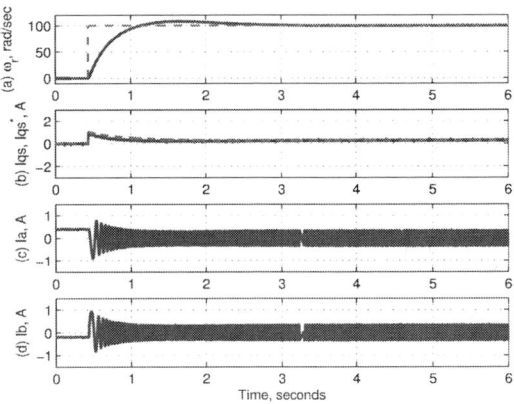

Fig. 10. Step response of IO-PI controller at rad/sec

The step response of the FO-PI controller is plotted in Fig. 9 for the step input of 100 rad/sec. The maximum overshoot observed for this step command is 4.72 rad/sec which is 4.72% while the settling time is determined to be 0.17s. As compared to FO-PI controller, the IO-PI controller which has been tuned using the trial and error tuning procedure shows a higher overshoot and settling time. The step response of the IO-PI controller is plotted in Fig. 10. According to this figure, the maximum overshoot observed is 8.34% while the settling time is calculated to be 2.28s. The overshoot is approximately twice the overshoot observed in the case of FO-PI controller. Similarly, the settling time is approximately 13 times more than the settling time in the case of FO-PI controller.

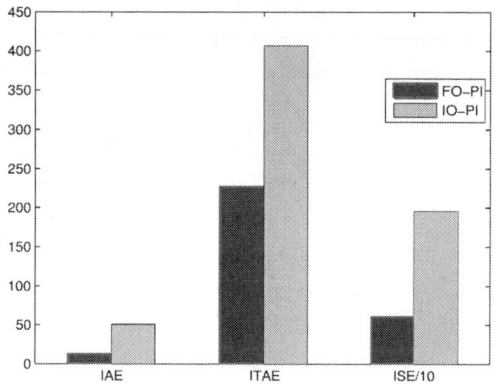

Fig. 11.   Comparison of IAE, ITAE and ISE at 100 rad/sec

The parameters, IAE, ITAE, ISE defined in Eq. 9 are used to compare the performance of both controllers for a step input of 100 rad/sec. The IAE. ITAE and ISE for IO-PI controller are found to be 50.74, 406.72 and 1952.875 respectively. Whereas, in the case of FO-PI controller, the IAE, ITAE and ISE are found to be 12.4226, 603.8819 and 226.0181 respectively which are less than the ones observed for IO-PI controller. These lower values for FO-PI controller are due to the faster dynamic response of the controller as compared to IO-PI controller. This rapid response of the FO-PI controller drives the error to zero faster than the IO-PI controller resulting in lower accumulated error.

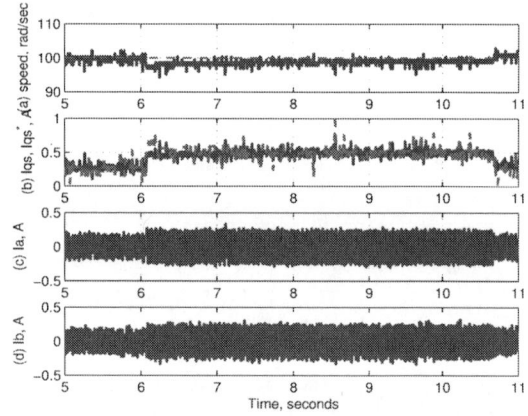

Fig. 12.   Disturbance rejection of FO-PI controller at 100 rad/sec

A. Disturbance Rejection

A step load of approximately 0.5 N.m is added when the motor is running at 100 rad/sec followed by removal of the load after some time. The speed response of the FO-PI controller, plotted in Fig. 12, shows that the controller completely rejected

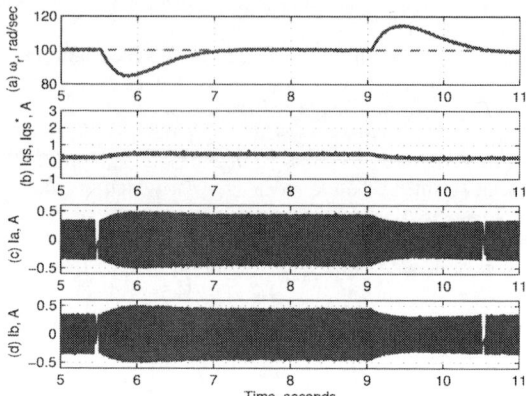

Fig. 13.   Disturbance rejection of IO-PI controller at 100 rad/sec

the disturbance. However, the motor speed decreases in the case of IO-PI controller when the disturbance is added at approximately 5.5 seconds as plotted in Fig. 13. It is worth mentioning that IO-PI controller was specifically tuned for 100 rad/sec while FO-PI gains were kept the same at various speeds. When the disturbance is removed at approximately 9.1 seconds, the motor speed overshoots the set reference. These results provide evidence of the disturbance rejection capabilities of the fractional order controllers.

VI. CONCLUSION

An optimal FO-PI controller is designed for a field oriented induction motor drive system. The tuning procedure for induction machine controller is implemented. The proposed controller performs better than an IO-PI controller in terms of the dynamic response for tracking a step reference in simulations. Moreover, the IAE, ISE and ITAE are lower in the case of FO-PI controller than the IO-PI controller while tracking at trapezoidal reference. Experimental results analyze the step response of the two controllers. The comparison of the step response in terms of percentage overshoot and settling time has shown that the FO-PI controller performs better than the IO-PI controller. Detailed experimental results are under progress and the results will be shared in the future publications.

ACKNOWLEDGMENT

The authors would like to thank the Office of Research and Graduate Studies (ORGS) at the American University of Sharjah, Sharjah, UAE for funding this work through the research grant FRG14-2-25.

REFERENCES

[1] J. Rodriguez, J. Pontt, C. Silva, P. Correa, P. Lezana, P. Cortes, and U. Ammann, "Predictive current control of a voltage source inverter," *IEEE Trans. Industrial Electron.*, vol. 54, no. 1, pp. 495–503, Feb 2007.

[2] R. Vargas, J. Rodriguez, U. Ammann, and P. Wheeler, "Predictive current control of an induction machine fed by a matrix converter with reactive power control," *IEEE Trans. Industrial Electron.*, vol. 55, no. 12, pp. 4362–4371, Dec 2008.

978-1-4673-9551-9/16 $31.00 © 2016 IEEE

[3] S. Chai, L. Wang, and E. Rogers, "A cascade MPC control structure for a PMSM with speed ripple minimization," *IEEE Trans. Industrial Electron.*, vol. 60, no. 8, pp. 2978–2987, Aug 2013.

[4] H. Liu and S. Li, "Speed control for PMSM servo system using predictive functional control and extended state observer," *IEEE Trans. Industrial Electron.*, vol. 59, no. 2, pp. 1171–1183, Feb 2012.

[5] E. Fuentes, D. Kalise, J. Rodriguez, and R. Kennel, "Cascade-free predictive speed control for electrical drives," *IEEE Trans. Industrial Electron.*, vol. 61, no. 5, pp. 2176–2184, May 2014.

[6] P. Alkorta, O. Barambones, J. Cortajarena, and A. Zubizarrreta, "Efficient multivariable generalized predictive control for sensorless induction motor drives," *IEEE Trans. Industrial Electron.*, vol. 61, no. 9, pp. 5126–5134, Sept 2014.

[7] J. Li, H.-P. Ren, and Y. ru Zhong, "Robust speed control of induction motor drives using first-order auto-disturbance rejection controllers," *IEEE Trans. Ind. Appl.*, vol. 51, no. 1, pp. 712–720, Jan 2015.

[8] A. Saghafinia, H. W. Ping, M. Uddin, and K. Gaeid, "Adaptive fuzzy sliding-mode control into chattering-free IM drive," *IEEE Trans. Ind. Appl.*, vol. 51, no. 1, pp. 692–701, Jan 2015.

[9] M. Fnaiech, S. Khadraoui, H. Nounou, M. Nounou, J. Guzinski, H. Abu-Rub, A. Datta, and S. Bhattacharyya, "A measurement-based approach for speed control of induction machines," *IEEE J. Emerging Sel. Topics Power Electron.*, vol. 2, no. 2, pp. 308–318, June 2014.

[10] G. Konstantopoulos, A. Alexandridis, and E. Mitronikas, "Bounded nonlinear stabilizing speed regulators for VSI-fed induction motors in field-oriented operation," *IEEE Trans. Control Syst. Technol.*, vol. 22, no. 3, pp. 1112–1121, May 2014.

[11] H. Michalska and D. Mayne, "Receding horizon control of nonlinear systems," in *Proc. 28th IEEE Conf. Decision Control*, Dec 1989, pp. 107–108 vol.1.

[12] L. Amezquita-Brooks, J. Liceaga-Castro, and E. Liceaga-Castro, "Speed and position controllers using indirect field-oriented control: A classical control approach," *IEEE Trans. Industrial Electron.*, vol. 61, no. 4, pp. 1928–1943, April 2014.

[13] M. A. Duarte-Mermoud, F. J. Mira, I. S. Pelissier, and J. C. Travieso-Torres, "Evaluation of a fractional order PI controller applied to induction motor speed control," in *Proc. 8th IEEE Int. Conf. Control Autom.*, 2010, pp. 573–577.

[14] L. Bouras, Y. Zennir, and F. Bourourou, "Direct torque control with svm based a fractional controller: Applied to the induction machine," in *Proc. of 3rd IEEE International Conference on Systems and Control (ICSC)*, 2013, pp. 702–707.

[15] C. Li, M. Chen, and S. Gao, "Fractional order PI speed control for permanent magnet synchronous motor drives," in *Proc. of 11th World Congress on Intelligent Control and Automation (WCICA)*, June 2014, pp. 4681–4685.

[16] Y. Chen, T. Bhaskaran, and D. Xue, "Practical tuning rule development for fractional order proportional and integral controllers," *Journal of Computational and Nonlinear Dynamics*, vol. 3, no. 2, p. 021403, 2008.

[17] T. Hägglund and K. J. Åström, "Revisiting the Ziegler-Nichols tuning rules for PI control," *Asian Journal of Control*, vol. 4, no. 4, pp. 364–380, 2002.

[18] S. Manabe, "The non-integer integral and its application to control systems," *JIEE (Japanese Institute of Electrical Engineers) Journal*, vol. 80, no. 860, pp. 589–597, 1960.

[19] H. Koivo and J. Tanttu, "Tuning of PID controllers: Survey of SISO and MIMO techniques," in *Proceedings of the IFAC Intelligent Tuning and Adaptive Control Symposium*, 1991, pp. 75–80.

[20] S. Yamamoto and I. Hashimoto, "Present status and future needs: The view from Japanese industry," in *Proceedings of the 4th International Conference on Chemical Process Control, Texas*, 1991.

Zero-sequence Current Suppression for Open-end Winding Induction Motor Drive with Resonant Controller

Hajime Kubo, Yasuhiro Yamamoto and Takeshi Kondo
MEIDENSHA CORPORATION
Numazu, Japan
Email: kubo-ha@mb.meidensha.co.jp

Kaushik Rajashekara and Bohang Zhu
University of Texas at Dallas
Richardson, USA

Abstract—**Open-end winding topology with a common DC source has path for zero-sequence current. Zero-sequence current needs to be suppressed because it does not contribute to the drive torque but has harmful effects such as loss and torque ripple. Both inverter and motor have sources of the zero-sequence component. The zero-sequence source on the inverter side is the voltage error due to dead time in switching. The zero-sequence source on the motor side is the zero-sequence component of the back EMF which consists of the third harmonic.**

In this paper, the two zero-sequence sources are investigated both theoretically and experimentally for an open-end winding induction motor drive system, and a method to suppress the zero-sequence current suppression is proposed. The voltage error is compensated by feedforward to the reference voltage. The zero-sequence component of the back EMF is compensated by a proportional and resonant controller because its frequency is known while its amplitude and phase offset are unknown.

I. INTRODUCTION

Dual inverter drive with open-end winding topology has several advantages such as increased rating power, three-level voltage and doubled effective switching frequency, compared with single inverter drive [1]. The open-end winding topology has two major DC source configurations, which are two isolated DC sources and a common DC source. Common DC source configuration has the advantage of lower cost and smaller size but it has a drawback of having path for zero-sequence current. Zero-sequence current needs to be suppressed because it does not contribute to the drive torque but has harmful effects such as conducting loss and torque ripple. To suppress the zero-sequence current, PWM methods using space vectors which contain no zero-sequence component [2], [3] or keeping the average zero-sequence voltage zero [4] are used. Zero-sequence current cannot be eliminated only by these PWM methods because of the voltage error due to device nonlinearity and dead time. To reduce this voltage error, automatic regulation by PI controller and dead time compensation by feed forward to the reference voltage are used [5], [6]. These researches of zero-sequence current suppression mainly focus on the inverter side. However, in motor drive application, the third harmonic of back EMF also affects zero-sequence component and results in zero-sequence current. In [7], zero-sequence current due to the third harmonic of

the back EMF by the inherent flux harmonic component is analyzed for the application in permanent magnet synchronous generator (PMSG) systems and the necessity of active zero-sequence voltage for open-end winding PMSG to compensate the zero-sequence component of the back EMF is shown.

In this paper, the causes of the zero-sequence current are analyzed for an open-end winding induction motor (IM) drive application. The influences of the two zero-sequence sources are examined in an experimental study. The experimental results show that the open-end winding IM drive also needs active zero-sequence voltage to compensate the third harmonic of the back EMF. The open-end winding IM control strategy proposed in this paper can effectively suppress the zero-sequence current.

II. METHODOLOGY

A. Open-end winding topology and its sources of zero-sequence current

The open-end winding topology with a common DC link is shown in Fig. 1. In this topology, zero-sequence current can flow through the DC link connection between the two inverters. The variables in *a-b-c* frame are transformed to the variables in the rotating *d-q-0* frame. The mathematical model of the zero-sequence current is expressed as follows [8].

$$L_l \frac{\mathrm{d}i_{s,0}}{\mathrm{d}t} = v_{s,0} - R_s i_{s,0} - v_{emf,0} \qquad (1)$$

Where L_l is leakage inductance, $i_{s,0}(= i_{s,a} + i_{s,b} + i_{s,c})$ is zero-sequence current, $v_{s,0}(= v_{s,a} + v_{s,b} + v_{s,c})$ is zero-sequence component of output voltage, R_s is stator winding resistance and $v_{emf,0}$ is zero-sequence component of back EMF. Zero-sequence components are decoupled from d-q components and no d-q components appear in (1). The output voltage $v_{s,0}$ is controlled and generated by the inverters. However the output voltage may contain error from the reference voltage provided by the controller. Considering the error, the equation (1) is transformed as follows.

$$L_l \frac{\mathrm{d}i_{s,0}}{\mathrm{d}t} = v_{s,0}^* - v_{s,0err} - R_s i_{s,0} - v_{emf,0} \qquad (2)$$

978-1-4673-9551-9/16 $31.00 © 2016 IEEE

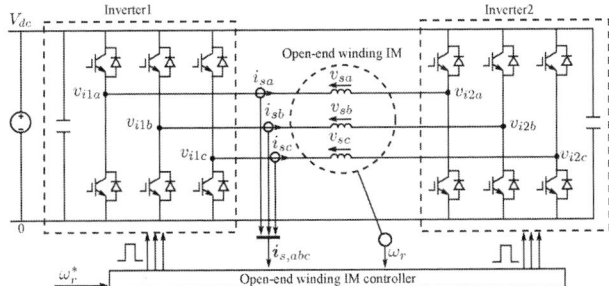

Fig. 1. Open-end winding IM drive system with a common DC link shared by two inverters

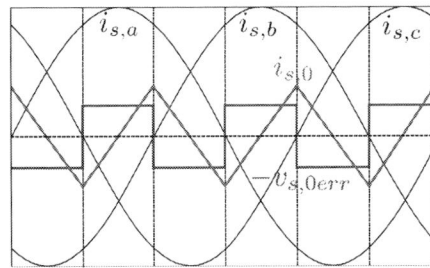

Fig. 2. Zero-sequence voltage and current due to the dead time.

Where $v_{s,0}^*$ is the reference zero-sequence voltage and $v_{s,0err}$ is the error voltage which is defined as $v_{s,0err} = v_{s,0}^* - v_{s,0}$. As seen from (2), the error voltage $v_{s,err}$ and the zero-sequence component of back EMF $v_{emf,0}$ are the two causes of the zero-sequence current. Therefore, the reference voltage needs to compensate the two to keep the zero-sequence current to be zero.

B. The effect of the dead time on zero-sequence current

The dead time provided in switching causes error in the pole voltages of the inverters. The polarity of the error depends on the direction of the current and the magnitude of the error depends on the ratio between the dead time and the time duration of one cycle of the triangular carrier wave. The error voltage in one phase due to the dead time is as follows.

$$
\begin{aligned}
v_{serr} &= v_{i1err} - v_{i2err} \\
&= \mathrm{sgn}(i_s)\frac{V_{dc}T_{dead}}{T_c} - \mathrm{sgn}(-i_s)\frac{2V_{dc}T_{dead}}{T_c} \\
&= \mathrm{sgn}(i_s)\frac{2V_{dc}T_{dead}}{T_c}
\end{aligned}
\tag{3}
$$

Where T_{dead} is the duration of the dead time and T_c is the duration of one cycle of the triangular carrier wave. Therefore, the zero-sequence error voltage is described as follows.

$$
v_{s,0err} = (\mathrm{sgn}(i_{sa}) + \mathrm{sgn}(i_{sb}) + \mathrm{sgn}(i_{sc}))\frac{2V_{dc}T_{dead}}{T_c}
\tag{4}
$$

The zero-sequence error voltage becomes a square wave whose polarity changes at every zero-crossing points of the three phase currents as shown in Fig. 2. From (2), the zero-sequence current caused by the square wave voltage error becomes a triangular wave which has peaks at the zero-crossing points of the three phase currents. Since this error is known from the polarities of phase currents, it is easily compensated by the feedforward of (4) to the reference voltage.

C. The effect of the third harmonic of back EMF on zero-sequence current

The back EMF from the motor has harmonics due to the machine spatial harmonics. In many cases, sinusoidal back EMF is assumed and the harmonics are ignored. However, the third harmonic cannot be neglected in open-end winding with a common DC source configuration because it composes zero-sequence. In this study, electrical angle θ_e is defined as the angle between the rotor flux ϕ_r and the α-axis on stationary frame as shown in Fig. 3. The fundamental components of the three phase back EMF are described as follows.

$$
\begin{bmatrix} v_{emf1,a} \\ v_{emf1,b} \\ v_{emf1,c} \end{bmatrix} = \begin{bmatrix} V_{emf1}\sin((\theta_e + \pi)) \\ V_{emf1}\sin((\theta_e + \pi) - \frac{2\pi}{3}) \\ V_{emf1}\sin((\theta_e + \pi) - \frac{4\pi}{3}) \end{bmatrix}
\tag{5}
$$

Where V_{emf1} is the amplitude of the fundamental component of the back EMF. Then, the third harmonic components of the three phase back EMF are described as follows.

$$
\begin{bmatrix} v_{emf3,a} \\ v_{emf3,b} \\ v_{emf3,c} \end{bmatrix} = \begin{bmatrix} V_{emf3}\sin(3(\theta_e + \pi) + \theta_3) \\ V_{emf3}\sin(3(\theta_e + \pi) + \theta_3) \\ V_{emf3}\sin(3(\theta_e + \pi) + \theta_3) \end{bmatrix}
\tag{6}
$$

Where V_{emf3} is the amplitude of the third harmonic component of the back EMF and θ_3 is the phase offset. The zero-sequence component of back EMF is the summation of them.

$$
\begin{aligned}
v_{emf,0} &= \sum_{i=a,b,c}(v_{emf1,i} + v_{emf3,i}) \\
&= 3V_{emf3}\sin(3(\theta_e + \pi) + \theta_3) \\
&= -3V_{emf3}\sin(3\theta_e + \theta_3)
\end{aligned}
\tag{7}
$$

The fundamental components are cancelled in (7) and the zero-sequence component of the back EMF consists of the third harmonic components. The waveforms of the zero-sequence component of the back EMF and the zero-sequence current caused by it are shown in Fig. 4. In this figure, the amplitude and phase offset are assumed as $3V_{emf3} = 1$ and $\theta_3 = 0$. It is difficult to compensate $v_{emf,0}$ by feedforward to the reference voltage because its amplitude and phase offset are usually unknown.

D. Open-end winding IM controller with zero-sequence compensation

The block diagram of the proposed open-end winding IM controller is shown in Fig. 5. The motor is controlled by conventional indirect field oriented control (FOC). The speed and the currents in synchronous d-q frame are controlled by proportional and integral (PI) controllers. The reference stator

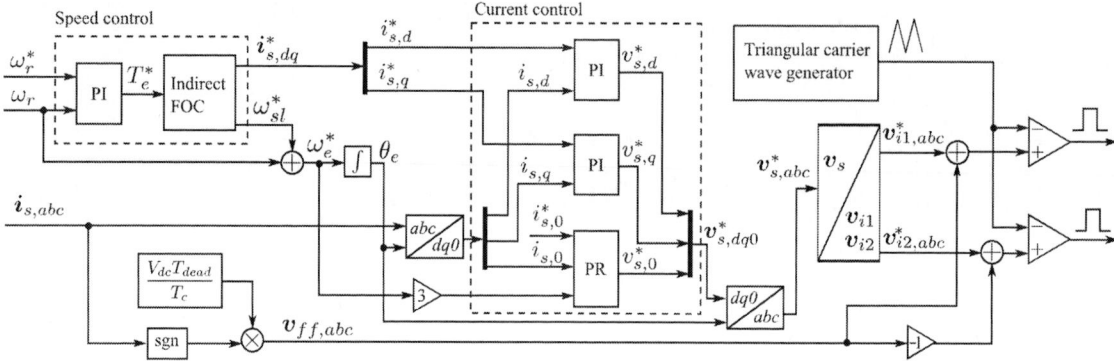

Fig. 5. Open-end winding IM controller with zero-sequence current suppression

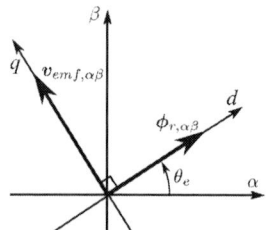

Fig. 3. Definition of the electrical angle

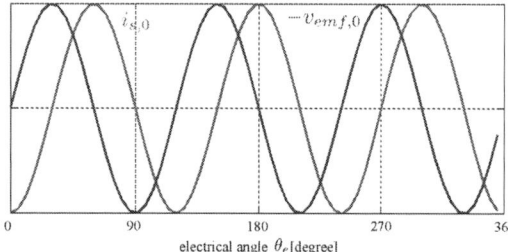

Fig. 4. Zero-sequence voltage and current due to the third harmonic of back EMF

TABLE I
MODEL AND SPECIFICATION OF THE EXPERIMENTAL EQUIPMENT

Inverters:	VT240S 7.5kW (MEIDENSHA)
Controllers:	VT240S (MEIDENSHA)
Open-end winding IM:	VTF0-FKK (Hitachi)
	5.5kW, 200V, 50Hz, 1470RPM
PMSM:	HTIB05-ZFPS (MEIDENSHA)
	5.5kW, 200V, 50Hz, 1500RPM

is used to compensate this high frequency zero-sequence component because a PR controller has large gain for its target frequency [9]. The target frequency of the PR controller is set to the three times of the fundamental. The reference zero-sequence voltage $v_{s,0}^*$ calculated in PR block becomes equal to the zero-sequence component of the back EMF $v_{emf,0}$, if the zero-sequence current is controlled to be zero.

III. EXPERIMENT

A. Experimental system

The connection diagram of the experimental system is shown in Fig. 6. An open-end winding IM is coupled with a permanent magnet synchronous motor (PMSM) which is used as a load. Three inverters are used, two for the open-end winding IM drive and one for the PMSM drive. All the three inverters are sharing a common DC link. The waveforms of the stator phase winding voltage $v_{s,abc}$ and the current $i_{s,abc}$ are recorded using oscilloscopes. The variables in the controller, such as the reference zero-sequence voltage $v_{s,0}^*$ and the electrical angle θ_e, are monitored through DA converters on the controller. The experimental setup is shown in Fig. 7 and specifications are listed in Table I.

The speed of the motor is controlled to be at 1000 RPM. The carrier frequency is set to 4 kHz. The dead time is set to 4 μs. The data are recorded at 1 MS/s. The average value in one carrier cycle is obtained by smoothing the recorded data with a window of length 250 points.

B. Result

The waveforms of three phase current, zero-sequence current, reference zero-sequence voltage and zero-sequence volt-

winding voltage $v_{s,abc}^*$ is converted into the reference inverter pole voltages $v_{i1,abc}^*$ and $v_{i2,abc}^*$ as

$$
\begin{aligned}
v_{i1,abc}^* &= 0.5v_{s,abc}^*, \\
v_{i2,abc}^* &= -0.5v_{s,abc}^*.
\end{aligned} \tag{8}
$$

The reference inverter pole voltages are compared with a triangular carrier wave and synthesized by PWM. The voltage error caused by the dead time is compensated by the feed-forward voltage $v_{ff,abc}$. To keep the zero-sequence current to be zero ($i_{s,0}^* = 0$), the zero-sequence current controller needs to compensate the zero-sequence component of the back EMF whose frequency is known but amplitude and phase offset are unknown. As seen from (7), the zero-sequence component of the back EMF has frequency three times as high as the fundamental, which is too high for a PI controller to compensate. A proportional and resonant (PR) controller

978-1-4673-9551-9/16 $31.00 © 2016 IEEE

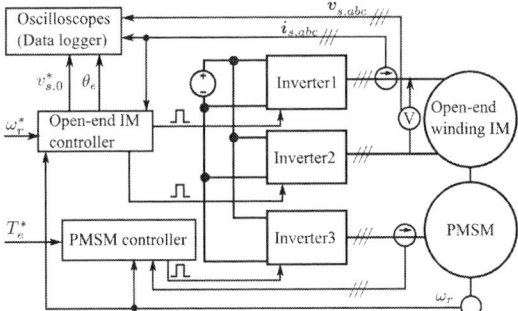

Fig. 6. Connection diagram of the experimental setup.

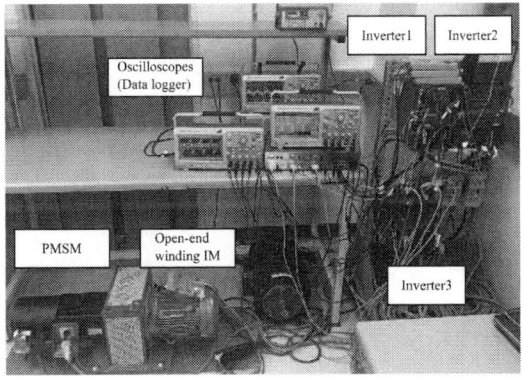

Fig. 7. Photograph of the experimental setup.

age are shown in Fig. 8-11. These four figures have different conditions for zero-sequence component compensation: (i)"No compensation", (ii)"Deadtime compensation", (iii)"Deadtime compensation + PI" and (iv)"Deadtime compensation + PR". The horizontal axis of the figures stands for the electrical angle. The vertical axis on the left is for current and the vertical axis on the right is for voltage. The waveforms are measured for three different load conditions; 0%, 30% and 60% for each compensation condition. Larger current scales are used for heavier load condition while the voltage scales are fixed. The zero-sequence current waveforms are enlarged six times.

Fig. 8 shows the waveforms when neither dead time compensation nor zero-sequence current control is applied. Although the reference zero-sequence voltage $v_{s,0}^*$ is fixed to zero, the inverter output zero-sequence voltage $v_{s,0}$ becomes a square wave which changes its polarity at every zero-crossing of the three phase current. This waveform matches well with the waveform derived from (4). This voltage waveform does not change with load because it is affected only by the polarity of the current and not by the magnitude of the current. In no load condition, the waveform of the zero-sequence current is a triangular wave, which is same as derived in section II-B. In loaded condition, larger zero-sequence current flows and it has different phase from the triangular wave of no load condition. As the current increases, the effect of the zero-

sequence component of the back EMF on the zero-sequence current becomes dominant.

Fig. 9 shows the waveforms when the dead time compensation is applied. The output zero-sequence voltage follows the reference zero-sequence voltage which is fixed to zero, except at the zero-crossing of the three phase currents. The zero-sequence current is suppressed in no load condition. In loaded condition, zero-sequence current is same as that of no compensation.

Fig. 10 shows the waveforms when the dead time compensation is applied and a PI controller is used for zero-sequence current control. Since the input to the PI controller $(i_{s,0}^* - i_{s,0})$ has frequency three times that of the fundamental, the proportional control mainly works. As a result, the zero-sequence voltage is 180 degree phase shifted from the measured zero-sequence current. However, the zero-sequence component of the back EMF is 90 degree phase shifted from the zero-sequence current as shown in Fig. 4. Therefore, PI controller cannot compensate the zero-sequence component of the back EMF effectively and zero-sequence current still flows in the windings.

Fig. 11 shows the waveforms when the dead time compensation is applied and a PR controller is used for zero-sequence current control. The zero-sequence voltage becomes larger as the load is increased. The zero-sequence current is suppressed in all the three load conditions. The waveform of the zero-sequence component of the motor back EMF $v_{emf,0}$ is estimated from the inverter zero-sequence voltage $v_{s,0}$ when $v_{emf,0}$ is compensated by $v_{s,0}$ and the zero-sequence current is maintained at zero value. In Fig. 12, the zero-sequence component of the back EMF is estimated from the reference zero-sequence voltage in "Deadtime compensation + PR" condition and the zero-sequence current is measured in "Deadtime compensation" condition. In section II-C, the phase offset of the third harmonic θ_3 is assumed to be zero and the waveforms in Fig. 4 is derived. In this experimental system the phase offset is $\pi/2$ which is measured from Fig. 12. The relationship between the zero-sequence component of the back EMF and the magnitude of the current was investigated further by varying the load from 10% to 80% at 10% interval. Fig. 13 shows the reference zero-sequence voltage obtained from the PR controller for eight different load conditions. The zero-sequence current was found to be suppressed for all the conditions, and the zero-sequence component of the back EMF is estimated to be equal to the zero-sequence voltage. The amplitude of the zero-sequence voltage increases monotonically with the increase in load, while the phase of the zero-sequence voltage does not change. This trend is also seen in Fig. 14 which is a plot of amplitude of the zero-sequence voltage versus amplitude of the phase current. The amplitude change and the phase offset of the third harmonic of the back EMF are unknown before these elaborative investigation. However, zero-sequence current feedback control with PR controller can effectively compensate the third harmonic of the back EMF and suppress zero-sequence currents without knowing the amplitude and the phase offset.

IV. CONCLUSION

In this paper, the two main sources of zero-sequence current flow in open-end winding IM with common DC link are investigated theoretically and experimentally. The one source is the voltage error due to dead time in switching. The effect of this voltage error is constant in any load condition because it only depends on the polarity of the phase current. The amount of this error is theoretically known so it is easily compensated by feedforward to the reference voltage. The other source is the zero-sequence component from the motor, which consists of the third harmonic of the back EMF. The effect of this zero-sequence component increases with load. Feedforward compensation of this zero-sequence component is difficult because its amplitude and phase offset are usually unknown. However, feedback control with PR controller can effectively compensate this zero-sequence component and suppress zero-sequence current.

The PWM method that only uses space vectors without zero-sequence component is popular for open-end windings with a common DC link. However, this method does not work for motor drive application because inverters need to actively output zero-sequence voltage to cancel the third harmonic of the back EMF from the motor. Therefore, a simple carrier based PWM method that can synthesize the average active zero-sequence voltage is used in this research.

REFERENCES

[1] H. Stemmler and P. Guggenbach, "Configurations of high-power voltage source inverter drives," 1993, pp. 7–14 vol.5.

[2] V. Somasekhar, K. Gopakumar, E. Shivakumar, and S. Sinha, "A space vector modulation scheme for a dual two level inverter fed open-end winding induction motor drive for the elimination of zero sequence currents," *EPE Journal*, vol. 12, no. 2, pp. 26–36, 2002.

[3] M. Baiju, K. Mohapatra, K. Gopakumar, and R. Kanchan, "A dual two-level inverter scheme with common mode voltage elimination for an induction motor drive," *Power Electronics, IEEE Transactions on*, vol. 19, pp. 794– 805, 2004.

[4] V. T. Somasekhar, K. K. Kumar, and S. Srinivas, "Effect of zero-vector placement in a dual-inverter fed open-end winding induction-motor drive with a decoupled space-vector pwm strategy," *Industrial Electronics, IEEE Transactions on*, vol. 55, pp. 2497–2505, 2008.

[5] F. Senicar, C. Junge, S. Gruber, and S. Soter, "Zero sequence current elimination for dual-inverter fed machines with open-end windings." IEEE, 2010, pp. 853–856. [Online]. Available: http://ieeexplore.ieee.org/xpl/articleDetails.jsp?arnumber=5675175

[6] A. Kolli, O. Bethoux, A. De Bernardinis, E. Laboure, and G. Coquery, "Sensitivity analysis of the control of a three-phase open-end winding h-bridge drive." IEEE, 2014, pp. 1–6. [Online]. Available: http://ieeexplore.ieee.org/xpl/articleDetails.jsp?arnumber=6861843

[7] Y. Zhou and H. Nian, "Zero-sequence current suppression strategy of open-winding pmsg system with common dc bus based on zero vector redistribution," *Industrial Electronics, IEEE Transactions on*, vol. 62, pp. 3399–3408, 2015. [Online]. Available: http://ieeexplore.ieee.org/xpl/articleDetails.jsp?arnumber=6945324

[8] P. Krause, O. Wasynczuk, S. Sudhoff, and S. Pekarek, *Symmetrical Induction Machines*. Wiley-IEEE Press, 2013, p. 608. [Online]. Available: http://ieeexplore.ieee.org/xpl/articleDetails.jsp?arnumber=6739383

[9] D. N. Zmood and D. G. Holmes, "Stationary frame current regulation of pwm inverters with zero steady-state error," *Power Electronics, IEEE Transactions on*, vol. 18, pp. 814– 822, 2003.

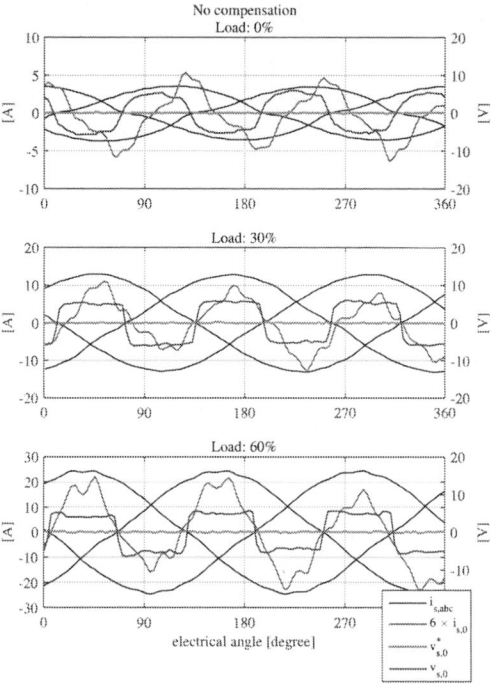

Fig. 8. Zero-sequence voltage and current waveform without compensation.

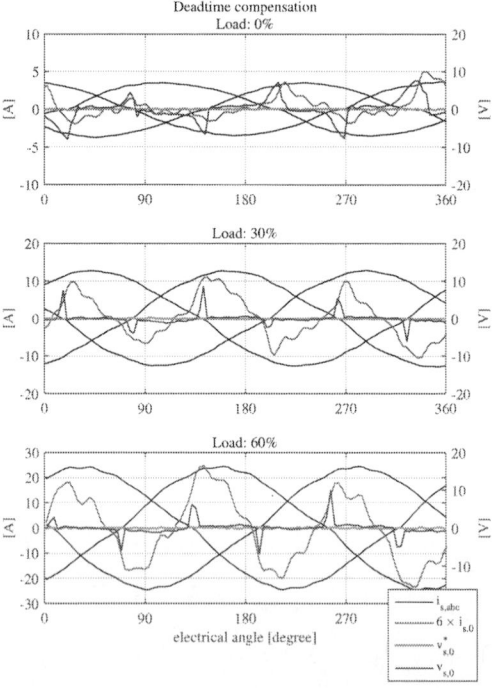

Fig. 9. Zero-sequence voltage and current waveform with dead time compensation.

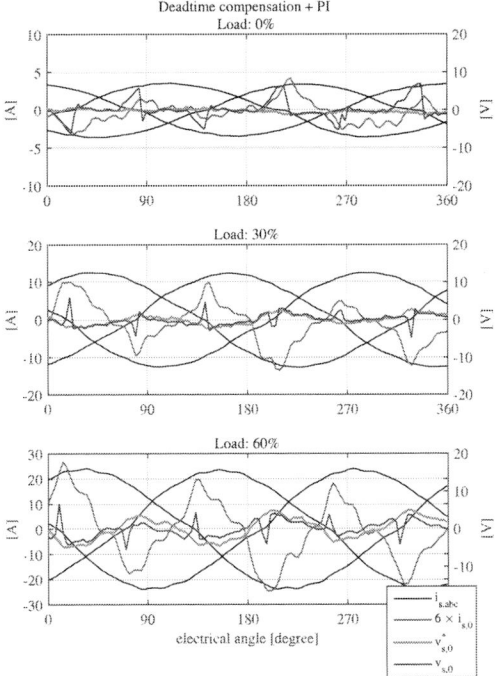

Fig. 10. Zero-sequence voltage and current waveform with dead time compensation and PI control.

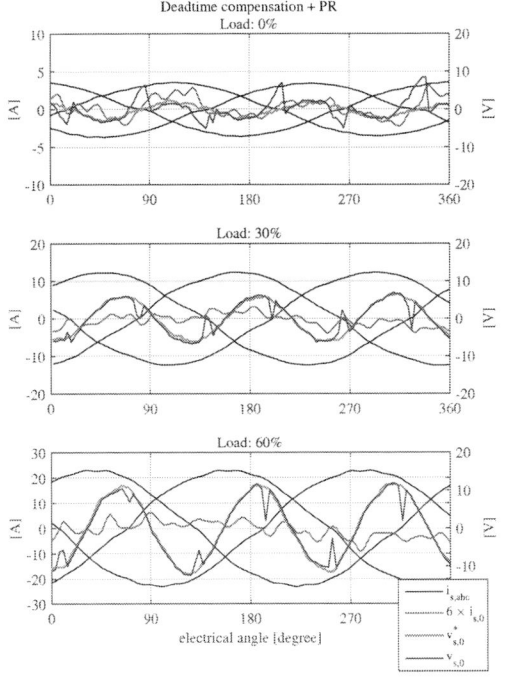

Fig. 11. Zero-sequence voltage and current waveform with dead time compensation and PR control.

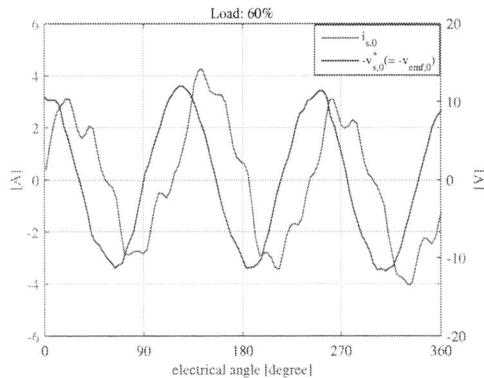

Fig. 12. Zero-sequence current and the zero-sequence component of back EMF estimated from the output of PR controller

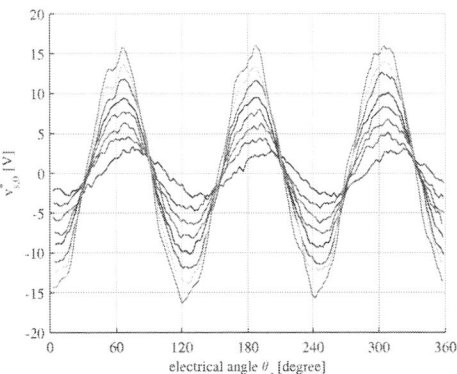

Fig. 13. Zero-Sequence component of back EMF estimated from reference zero-sequence voltage at different load condition from 10% to 80%.

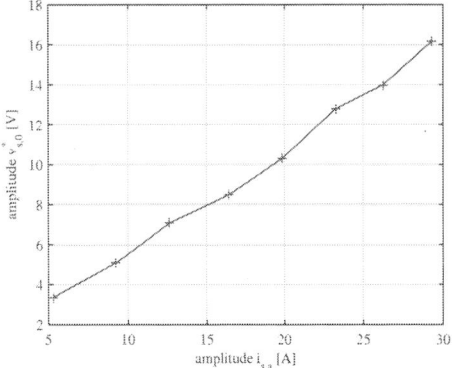

Fig. 14. Phase current amplitude versus zero-sequence voltage amplitude

Optimized Control of High-Performance Servo-Motor Drives in the Field-Weakening Region

Jack Bermingham, Gerard O'Donovan and Ray Walsh
Moog Ireland Ltd, Cork, Ireland
j.bermingham@umail.ucc.ie

Michael Egan, Gordon Lightbody and John G. Hayes
Power Electronics Research Laboratory
School of Engineering
University College Cork, Ireland

Abstract--Magnetic saturation may cause the inductance of a servo motor to deviate from values defined during the design of a drive's control system. Modern servo-drives are equipped with field-weakening strategies that control the trajectory of the motor's current vectors to produce the optimum levels of torque within the defined limit of current and voltage. The objective of this paper is to develop a 'plug and play' control scheme to integrate industrial inverters and machines without extensive characterization of the matched set, while operating within the operating specifications of the drive components. In this paper, an approach to torque optimization is presented in which the motor-terminal voltage-vector magnitude is regulated at high motor speeds while producing torque-optimizing current-vector commands. This method differs from other field-weakening solutions due to the active control of the voltage vector trajectory on the dq-voltage plane across an extended motor speed range. In a decoupled cascaded control strategy, the voltage-control loops produce current-vector trajectory commands in order to realize the voltage set points. The resulting current vector continuously conforms to the defined voltage and current limits of the servo drive without characterization of magnetic saturation. Results of the successful hardware implementation of the method in an industrial drive are presented.

Keywords— Field weakening, magnetic saturation, permanent-magnet synchronous machine, torque optimization.

I. INTRODUCTION

The effects of saturation have become a limiting factor in the current-loop performance of modern drives. This is due to the complexity of the issue and the difficulty in characterizing the cross-coupled and non-linear behavior of inductance across current operating points [1,2]. Ignoring magnetic saturation can result in errors that affect the performance of the drive system in terms of the tracking accuracy of performance-optimization strategies, such as field weakening (FW) [3-7]. As discussed in Section II, FW algorithms map the operating limits of the current and voltage vectors across an extended speed range of the motor. These operating limits are translated into current-trajectory commands to generate

optimum levels of torque as a function of motor speed. A change in the voltage vs. current relationships due to parameter variation can result in a breach of system limits, and a consequent loss of current control and performance due to unrealizable current trajectory commands. In the absence of a full characterization of magnetic saturation within the servo motor, the problem is usually overcome by adding a voltage-headroom margin to the control calculations which reduces the available power and torque/speed performance of the drive. A review of existing and proposed FW solutions has revealed a need for the development of a control strategy that (i) can operate at the limit of the available voltage supply and (ii) is free of a dependency on motor characterization [8-13]. The proposed voltage-controlled FW technique and the associated control strategy are presented in this paper. This strategy has been successfully implemented and tested within Moog Inc. industrial hardware, the results of which are documented and discussed in Section V.

II. PRINCIPLES OF FIELD WEAKENING

The electromagnetic torque of a servo motor is controlled at a high level to achieve and maintain the speed set-points commanded of the servo-drive speed controller [14]. Expressed as a dc model on a direct-, d, and quadrature-, q, axis rotating-reference frame, the equation for electromagnetic torque in a non-salient synchronous machine, such as the surface-mounted permanent magnet machine (SPM), is purely a function of q-axis current, i_q, as follows,

$$T_{em} = \frac{3}{2}\frac{p}{2}\lambda_m i_q \qquad (1)$$

where p is the number of poles in the machine and λ_m is the flux linkage due to the permanent magnet [15]. The potential amount of q-axis current produced in a SPM is limited by the available d- and q-axis voltages, v_d and v_q, at the terminals of the motor and the rotational frequency of the machine, ω_e. The voltage equations are as follows,

$$v_d(t) = R_s i_d(t) + L_d \frac{di_d(t)}{dt} - L_q \omega_e i_q(t) \qquad (2)$$

$$v_q(t) = R_s i_q(t) + L_q \frac{di_q(t)}{dt} + L_d \omega_e i_d(t) + \omega_e \lambda_m \qquad (3)$$

where R_s is the stator resistance per phase and L_d and, L_q are the dq inductances [16]. Although d-axis current, i_d, has no contribution to the electromagnetic torque produced in a SPM, it can be used to reduce the total magnitude of the speed-dependent voltage terms. This allows for an increase in i_q magnitude across an extended speed range of the motor as it operates at the limit of the available voltage supply, V_{max}, and within the rated current limit I_{max}, which are as follows:

$$I_{max} = \sqrt{i_d^2 + i_q^2} \tag{4}$$

and
$$V_{max} = \sqrt{v_d^2 + v_q^2} \tag{5}$$

The current-vector trajectory is controlled by an algorithm to produce an optimum torque-producing current within the limits of the available voltage supply. This is the principle of field weakening, but in reality, this is made difficult to accurately achieve as the motor inductances of (2) and (3) are non-linear and cross-coupled functions of motor current. A particular model for the dq inductances is as follows,

$$L_{dd}(i_d, i_q) = L_d(1 - 6\alpha_{3,0} L_d^2 i_d - 12\alpha_{4,0} L_d^3 i_d^2 - 2\alpha_{2,2} L_d L_q^2 i_q^2) \tag{6}$$

$$L_{dq}(i_d, i_q) = L_{qd}(i_d, i_q) = 2L_d L_q^2 i_q (\alpha_{1,2} + 2\alpha_{2,2} L_d i_d) \tag{7}$$

$$L_{qq}(i_d, i_q) = L_q(1 - 2\alpha_{1,2} L_q L_d i_d - 2\alpha_{2,2} L_q L_d^2 i_d^2 - 12\alpha_{0,4} L_q^3 i_q^2) \tag{8}$$

where $\alpha_{i,j}$ are the saturation parameters, unique to each motor model[1,2].

This method of modeling magnetic saturation, as shown in Fig. 1 and Fig. 2, are used to analyze the variation of L_{dd}, L_{qq} and the cross-coupled inductance, L_{dq}, with normalized d- and q-axis currents for the example SPM machine discussed in [2], defined as SPM1. The saturation model of inductance enables the illustration of the relationship between motor current and voltage on the dq- current plane, as shown in Fig. 3, by substituting (6), (7) and (8) for the inductance terms in (2) and (3), solving for i_q and i_d in (5) and plotting the voltage limit on the i_d vs. i_q plane. This reveals an accurate map of realizable current commands for a given motor speed and illustrates how FW algorithms that ignore the effects of saturation and estimate a linear model of the current vs. voltage may result in tracking errors, a loss of available power and, possibly, a loss of current control. A solution would be to characterize the effects of parameter variation for an individual machine and application, but this is both labor and computationally intensive. The alternative approach, presented in this paper, is to monitor and control the dq voltage vector at the voltage limit and to generate current commands to achieve optimum torque levels, based on the variable relationship between current and voltage under the effects of saturation.

III. Voltage Vector Control

The proposed control strategy monitors and regulates the motor-voltage vector and utilizes decoupled PI compensation to compute realizable torque-optimizing current commands to the drive's existing current controller. The proposed FW

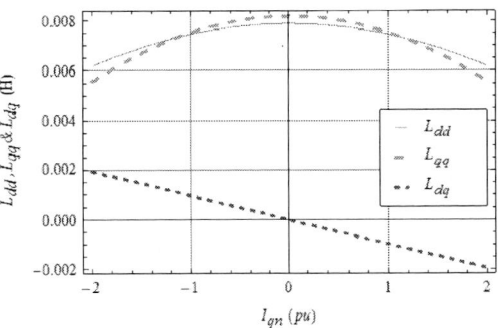

Fig. 1. L_{dd}, L_{qq} and L_{dq} varying with q-axis current, normalized to 5 A.

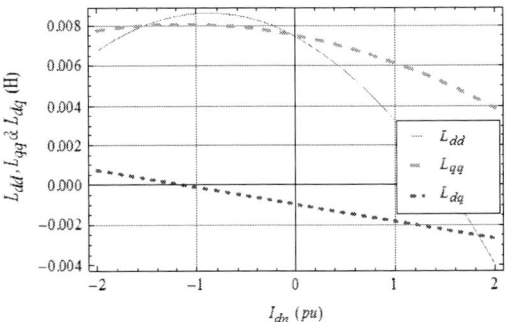

Fig. 2. L_{dd}, L_{qq} and L_{dq} varying with d-axis current, normalized to 5 A.

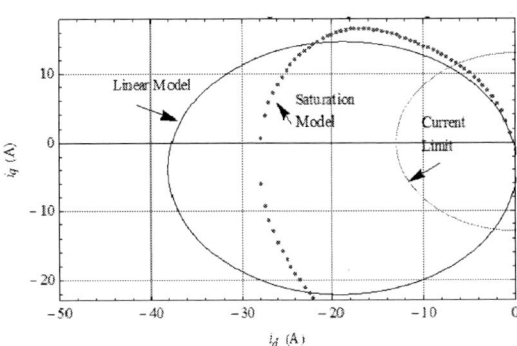

Fig. 3. Models of the maximum voltage limit of SPM1 with and without the effects of magnetic saturation mapped to the i_d vs i_q plane at 5500 rpm.

algorithm generates the voltage-vector trajectory commands necessary to achieve optimum torque levels under varying speed and motor parameters. This continuously maintains a constant motor-voltage magnitude at the limit of the supply and eliminates the risk of control loss or performance degradation in the field-weakening region. The following work features the introduction of the proposed control strategy to a Moog servo-drive system that features a SPM with parameters and ratings featured in Table 1. This machine is defined as SPM2.

The proposed algorithm generates voltage commands under four modes of operation that exist within the speed and current

TABLE 1

PARAMETERS AND RATING OF SPM2

Motor	SPM
Rated Power	784 W
Peak Rated Current I_n	50 A_{pk}
Bus Voltage	21 V
Rated Speed	1650 rpm
Rated Torque	4.54 Nm
Motor Poles p	12
Resistance R_s	0.0235 Ω
Magnet Flux Linkage λ_m	0.0111 Wb
d-axis Inductance L_d	0.082 mH
q-axis Inductance L_q	0.086 mH

operating ranges of the machine. The calculations are solved as line equations on the d- and q-axis voltage plane, where maximum torque lines are a tangent to the current or voltage limit in the FW regions.

Mode 1 is defined as operation within a motor speed range that results in a voltage-vector magnitude of less than the maximum limit of the system. This is defined as the constant-torque region, referring to the torque/speed profile. This mode is not in the FW region and external torque commands are realized directly by applying scaled i_q^* commands to the existing current controller with no i_d contribution.

Mode 2 is defined as operation at the voltage limit, but below the current limit, where external torque commands are realizable with the introduction of negative i_d components to the motor. In this mode of operation, the touch-point of the command torque line and the voltage limit are tracked with a change in speed, as illustrated in Fig. 4. The external torque command can be realized until the current limit is reached or until the command-torque line exits the voltage limit.

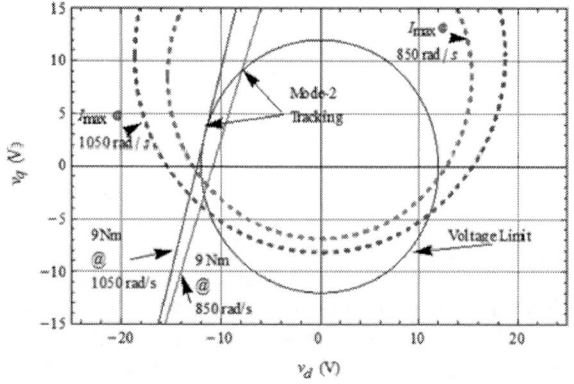

Fig. 4. Mode-2 operation on the voltage plane. The voltage trajectory is controlled to follow the touch-point of the voltage limit and a torque-command line.

In order to express the current limit in terms of voltage, it is necessary to express the d- and q-axis currents in terms of voltage. This is achieved by treating the d- and q-axis voltage expressions, (2) and (3), as simultaneous equation,

$$\frac{v_d R_s}{L_q \omega_e} = \frac{R_s^2 i_d}{L_q \omega_e} - i_q R, \quad v_q = R_s i_q + \omega_e L_d i_d + \omega_e \lambda_m \quad (9)$$

Adding the equations presents an expression without i_q,

$$v_q + \frac{v_d R_s}{X_q} - E_f = i_d \left(L_d \omega_e + \frac{R_s^2}{X_q} \right) \quad (10)$$

where $X_q = \omega_e L_q$ and $E_f = \omega_e \lambda_m$.

With some equation manipulation and assuming $L_d \approx L_q$, an expression for i_d in terms of v_d and v_q is presented,

$$i_d = \frac{v_q + \frac{v_d R_s}{X_q} - E_f}{\left(X_q + \frac{R_s^2}{X_q} \right)} = \frac{v_d R_s + X_q \left(v_q - E_f \right)}{Z^2} \quad (11)$$

where $Z^2 = X_q^2 + R_s^2$.

The same method is used to present an expression for i_q in terms of v_d and v_q.

$$i_q = \frac{R_s \left(v_q - E_f \right) - v_d X_q}{Z^2} \quad (12)$$

The expression for i_q can then be used to develop a torque-line equation in terms of voltage by substituting into (1) and expressing the torque value T_{em} in terms of q-axis current command i_q^*. A line equation for v_q as a function of v_d can then be expressed,

$$v_q(v_d) = i_q \frac{Z^2}{R_s} + v_d \frac{X_q}{R_s} + E_f \quad (13)$$

Mode 2 operates at the voltage limit and therefore (5) can be used to express v_d in terms of v_q and V_{max}.

$$v_d = \sqrt{V_{max}^2 - v_q^2} \quad (14)$$

This expression can then be introduced to (13), leaving a quadratic function of v_q. When solved, this provides an expression for the q-axis element of the voltage trajectory that tracks the touch-point of the torque-command line and the voltage limit on the dq-voltage plane:

$$v_q^*(\omega_e, i_q^*) = \frac{E_f R^2 + i_q^* R Z^2}{Z^2}$$
$$+ \frac{\sqrt{X_q^2 \left(-E_f^2 R^2 + Z^2 \left(V_{max}^2 - 2 E_f i_q^* R - i_q^{*2} Z^2 \right) \right)}}{Z^2} \quad (15)$$

Mode 3 operation occurs at the limits of both the current and voltage. The touch-point of the two limit boundaries is tracked on the voltage plane to optimize torque production within the system limits, as illustrated in Fig. 6. The limit touch-point changes with motor speed and can be tracked on the voltage plane by expressing the maximum-current

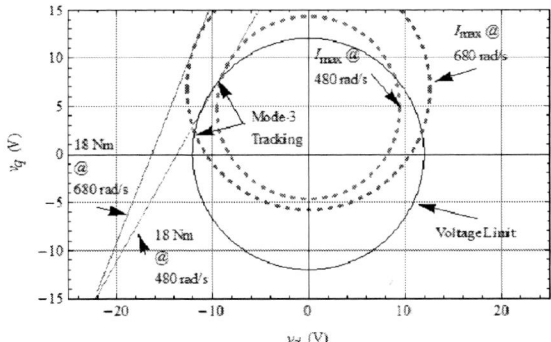

Fig. 5. Mode-3 operation. The voltage vector trajectory is controlled to follow the touch-point of the current and voltage limits.

equation (4) and by using (11) and (12) to express i_d and i_q in terms of voltage,

$$I_{\max}^{2} = \left(\frac{v_d R_s + X_l \left(v_q - E_f \right)}{Z^2} \right)^2 + \left(\frac{R_s \left(v_q - E_f \right) - v_d X_l}{Z^2} \right)^2 \quad (16)$$

Expanding this equation reveals a simplified expression for I_{max},

$$I_{\max}^{2} = \frac{v_d^2 + \left(v_q - E_f \right)^2}{Z^2} \quad (17)$$

Mode 3 operates at the voltage limit and therefore (14) can be substituted for the v_d terms. By solving for v_q, an expression for the q-axis element of the voltage trajectory to track the touch-point of the system limits on the voltage plane can be presented,

$$v_q(\omega_e) = \frac{V_{\max}^2 + E_f^2 - I_{\max}^2 Z^2}{2 E_f} \quad (18)$$

Mode 4 operates at the point of maximum torque on the voltage limit. This trajectory exists where a torque-line equation is a tangent to the circular voltage limit, as illustrated in Fig. 7. The equation for the tangent to the voltage limit can be defined by differentiating the maximum-voltage equation (5) with respect to v_d,

$$2 v_d + 2 v_q \frac{dv_q}{dv_d} = 0. \quad (19)$$

This reveals a useful relationship between the axis-voltage elements,

$$\frac{dv_q}{dv_d} = -\frac{v_d}{v_q} \quad (20)$$

Using (13) and (1) to obtain the $v_q(v_d)$ relationship for a specified electrical torque, T_{em}, yields,

$$\frac{dv_q(v_d)}{dv_d} = \frac{d\left(\frac{4 T_{cm} Z^2}{\lambda_m 3 p R_s} + v_d \frac{X_q}{R_s} + E_f \right)}{dv_d} = -\frac{v_d}{v_q} \quad (21)$$

Differentiating (14) with respect to v_d reveals the torque line

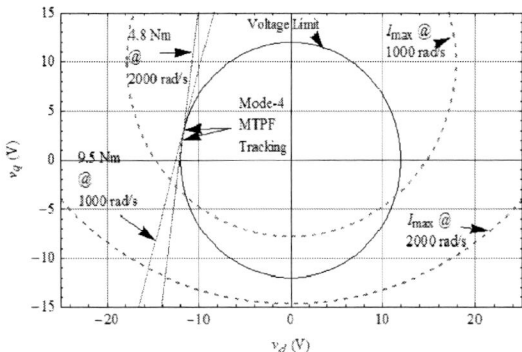

Fig. 6. Mode-4 Operation. The voltage trajectory is controlled to track the point of maximum torque on the voltage limit.

equation as a tangent to the circle in terms of dq voltage,

$$\frac{X_q}{R_s} = -\frac{v_d}{v_q} \quad (22)$$

Operating on the voltage limit, equation (5) can again be substituted for the v_d term. This reveals an expression for the q-axis element of the voltage trajectory that tracks the point of maximum torque production on the circular voltage limit of the voltage plane,

$$v_q^{*}(\omega_e) = \frac{V_{\max} R_s}{Z} \quad (23)$$

By defining the boundaries between the modes of operation, an algorithm is created that generates the developed voltage commands as functions of motor speed and current command.

IV. VOLTAGE CONTROL STRATEGY

Fig. 11, located in the Appendix, presents the proposed digital control strategy to implement the developed voltage-vector commands. While the strategy initially appears complex, it is a simple one-loop addition to a standard decoupled dq current controller. The bandwidth of the voltage loops is orders of magnitude lower than that of the high performance current loops and the introduced decoupling of the voltage loop commands prevents sensitivity or interaction between the voltage and current control loops. The voltage loops control a plant that consists of the servo motor and the digital current controller of the drive, as illustrated in Fig. 7. The current loops are not engaged in the constant torque region, leaving the high performance current loops to realize current. In the field-weakening region, where the drive current commands cannot otherwise be realized, the voltage loops are engaged to regulate the motor voltage at the defined limit of the drive, while optimizing torque production across an extended motor speed range.

The voltage-loop controllers decouple the voltage commands by feeding back i_d and i_q, and the filtered output signal of the current controllers is fed back to close the control loop. The output signals of the current controllers are defined as the decoupled-voltage terms v_{dq},

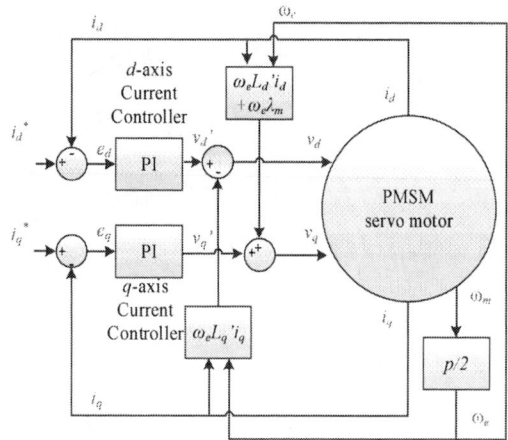

Fig 7. Control plant, consisting of the existing current controller and SPM.

$$v_d' = v_d + \omega_e L_q i_q \tag{24}$$

$$v_q' = v_q - \omega_e L_q i_q - \omega_e \lambda_m \tag{25}$$

This decoupling process ensures that the q-axis voltage-control loop controls the q-axis current-loop command exclusively and that the d-axis voltage loop controls the d-axis current-loop command exclusively. Also, characterization errors resulting from any parameter variation in the system are isolated and compensated for by the PI controllers in the voltage loops. The control plant, consisting of the SPM and the current-control loops, can be obtained by linearization as the linearized state-space model [15],

$$\frac{d}{dt}\begin{pmatrix} i_d \\ i_q \end{pmatrix} = \begin{pmatrix} \dfrac{-R}{L_d} & 0 \\ 0 & \dfrac{-R}{L_q} \end{pmatrix}\begin{pmatrix} i_d \\ i_q \end{pmatrix} + \begin{pmatrix} \dfrac{1}{L_d} & 0 \\ 0 & \dfrac{1}{L_q} \end{pmatrix}\begin{pmatrix} v_d' \\ v_q' \end{pmatrix}$$

$$\underline{y} = \begin{pmatrix} 1 & 0 \\ 0 & 1 \end{pmatrix}\begin{pmatrix} i_d \\ i_q \end{pmatrix} \tag{26}$$

Transforming to the digital domain with a sampling time T_s,

$$A_d = e^{AT_s} = \Phi(T_s) \tag{27}$$

$$B_d = \int_0^{T_s} \Phi(\eta) B d\eta \tag{28}$$

reveals the digital state-space representation of the plant,

$$\begin{pmatrix} i_d(k+1) \\ i_q(k+1) \end{pmatrix} = \begin{pmatrix} a_{d11} & a_{d12} \\ a_{d21} & a_{d22} \end{pmatrix}\begin{pmatrix} i_d(k) \\ i_q(k) \end{pmatrix} + \begin{pmatrix} b_{d11} & 0 \\ 0 & b_{d22} \end{pmatrix}\begin{pmatrix} v_d'(k) \\ v_q'(k) \end{pmatrix}$$

$$\underline{y}(k) = \begin{pmatrix} 1 & 0 \\ 0 & 1 \end{pmatrix}\begin{pmatrix} i_d(k) \\ i_q(k) \end{pmatrix} \tag{29}$$

The decoupled voltage terms are then expressed as the outputs of the current controllers, where K_p is the proportional gain of the current controller and ξ_{dq} is the integration of the loop-feedback error,

$$v_d'(k) = K_p\left(i_d^*(k) - i_d(k)\right) + \zeta_d(k) \tag{30}$$

$$v_q'(k) = K_p\left(i_q^*(k) - i_q(k)\right) + \zeta_q(k) \tag{31}$$

This results in a system with 4 states, where the inputs are the current-controller commands i_{dq}^* and the outputs are the decoupled voltage terms v_{dq}',

$$\begin{pmatrix} i_d(k+1) \\ i_q(k+1) \\ \zeta_d(k+1) \\ \zeta_q(k+1) \end{pmatrix}$$
$$= \begin{pmatrix} a_{d11} - b_{d11}K_p & 0 & b_{d11} & 0 \\ 0 & a_{d22} - b_{d22}K_p & 0 & b_{d22} \\ -K_i & 0 & 1 & 0 \\ 0 & -K_i & 0 & 1 \end{pmatrix}\begin{pmatrix} i_d(k) \\ i_q(k) \\ \zeta_d(k) \\ \zeta_q(k) \end{pmatrix}$$
$$+ \begin{pmatrix} b_{d11}K_p & 0 \\ 0 & b_{d22}K_p \\ K_i & 0 \\ 0 & K_i \end{pmatrix}\begin{pmatrix} i_d*(k) \\ i_q*(k) \end{pmatrix} \tag{32}$$

$$\begin{pmatrix} v_d'(k) \\ v_q'(k) \end{pmatrix} = \begin{pmatrix} -K_p & 0 & 1 & 0 \\ 0 & -K_p & 0 & 1 \end{pmatrix}\begin{pmatrix} i_d(k) \\ i_q(k) \\ \zeta_d(k) \\ \zeta_q(k) \end{pmatrix}$$
$$+ \begin{pmatrix} K_p & 0 \\ 0 & K_p \end{pmatrix}\begin{pmatrix} i_d*(k) \\ i_q*(k) \end{pmatrix}$$

The root-locus method is used to design the d- and q-axis voltage controllers to deliver a desired damping and bandwidth response [8]. The root-locus method provides useful information about the interaction of the voltage-controller design parameters with the performance of the inner current loop by plotting the closed loop gains of the inner and outer loops simultaneously. The discrete transfer functions used in the controller design are obtained by the discrete transfer function matrix [15],

$$G(z) = C(zI - A_d)^{-1}B_d + D \tag{33}$$

This reveals the decoupled plant models for the d and q axes,

$$\frac{V_d'(z)}{I_d^*(z)} = \frac{K_p z^2 - K_p(a_{d11}+1)z - a_{d11}K_p + K_i}{z^2 + z\left(b_{d11}K_p - a_{d11} - 1\right) + b_{d11}\left(K_i - K_p\right) + a_{d11}} \tag{34}$$

$$\frac{V_q'(z)}{I_q^*(z)} = \frac{K_p z^2 - K_p(a_{d22}+1)z - a_{d22}K_p + K_i}{z^2 + z\left(b_{d22}K_p - a_{d22} - 1\right) + b_{d22}\left(K_i - K_p\right) + a_{d22}} \tag{35}$$

V. EMBEDDED IMPLEMENTATION WITHIN AN INDUSTRIAL DRIVE

After successful simulated testing of the control strategy, the control algorithm and voltage controllers were implemented within the digital-control software of a high-performance Moog drive system featuring the SPM2 servo motor. Torque/speed tests were carried out on the servo-drive system across a wide range of speed and current operating points to examine the performance of the strategy in servo-drive hardware in a laboratory environment.

Initially, the strategy was compared to the drives existing FW solution which relied on a large voltage margin to maintain stability under the effects of magnetic saturation. Fig. 8 displays how the proposed strategy produces superior torque/speed performance by controlling the drive at the very limit of the supply. This improvement results in an increased speed range and increased torque per speed set point, increasing the versatility of the motor product. Fig. 9 (a)-(b) display the control performance of the drives traditional FW solution with the voltage margin removed. The error produced by commanding an inaccurate current trajectory for the available voltage within the system is considerable in this servo-drive and loss of current control occurs at a modest motor speed 1300 rpm. Fig. 10 (a)-(b) display the control performance of the proposed strategy, operating at the very limit of the drive supply voltage. The results show how current control is maintained across the operating speed range of the test and how the voltage loops maintain the vector magnitude below the defined maximum-voltage limit. For a variation in motor parameters, the voltage PI controllers integrate the control error and adjust the current command accordingly. In this way the strategy offers a robust and flexible solution to counteract behavior that is extremely difficult to characterize.

VI. CONCLUSIONS AND FURTHER WORK

A proposed alternative FW strategy introduces a cascaded control scheme, where the voltage vector of the motor is monitored and controlled to produce torque optimizing current trajectories within the voltage and current limits of the drive system. An algorithm to produce the torque-optimizing voltage-trajectory commands and the control strategy to implement these commands has been developed. The results of a laboratory test of the performance and effectiveness of the strategy were presented, revealing that the principle goals of the proposed system have been successfully achieved. The proposed control strategy has been shown to successfully optimize torque production within the constraints of voltage

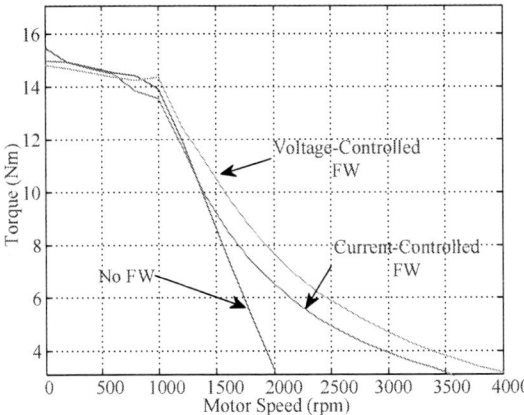

Fig 8. Torque/Speed Performance Improvement of Proposed Strategy.

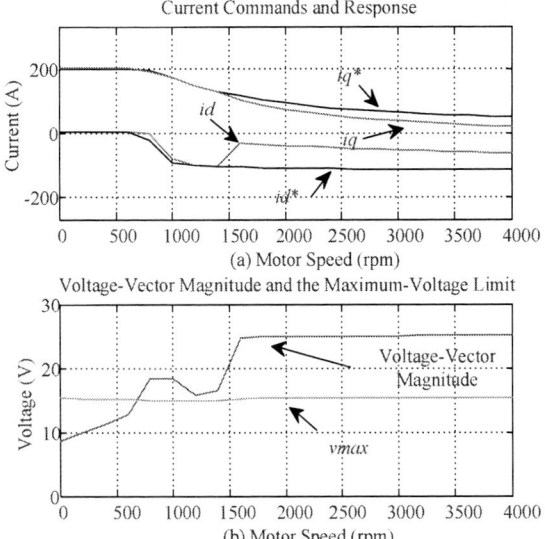

Fig. 9. Laboratory torque/speed performance results of traditional field–weakening algorithm without characterization of magnetic saturation.

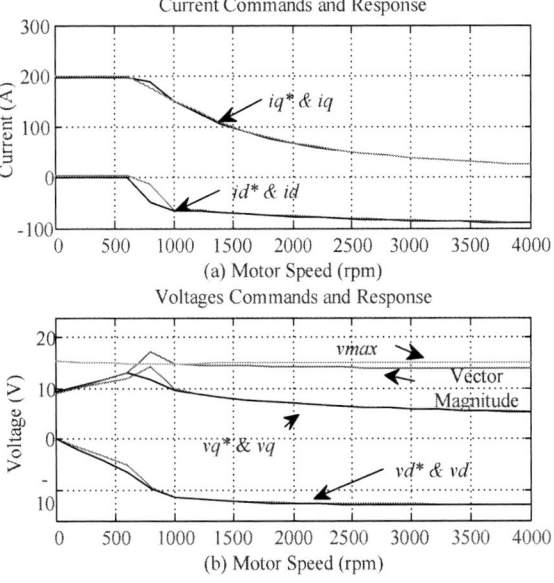

Fig. 10. Laboratory torque/speed performance results of the proposed Voltage-Controlled field–weakening strategy, also without characterization of magnetic saturation.

and current limits, remaining robust to the effects of large parameter variation. The system offers an alternative to the characterization of saturation effects and other parameter variations in each motor model controlled by a drive system product. This flexible system continuously evaluates the relationship between the voltage and current trajectories as motor parameters vary in magnitude and then produces the optimal level of torque for the given voltage/current relationship. The objective of this study has been to develop a 'plug and play' control scheme to integrate industrial inverters and machines without extensive characterization of the matched set, while operating within the operating specifications of the drive components.

ACKNOWLEDGMENT

This project and paper submission was sponsored by Moog Ireland Ltd. and Moog Inc. Information about Moog services and products can be found at [18].

REFERENCES

[1] A. Jebai, F. Malrait, P. Martain, and P. Rouchon, "Sensorless position estimation of permanent-magnet synchronous motors using a saturation model," *IEEE International Conference on Electrical Machines*, 2012, pp. 2245-2251.

[2] A. Jebai, F. Malrait, P. Martain, and P. Rouchon, "Estimation of saturation of permanent-magnet synchronous motors through an energy-based model," *IEEE International Electrical Machines Drives Conference*, 2011, pp. 1316-1321.

[3] P. Vaclavek, P. Blaha, "Interior permanent magnet synchronous machine high speed operation using field weakening control strategy," *12th WSEAS International Conference on SYSTEMS*, 2008, pp. 581-585.

[4] D. S. Maric, S Hiti, C.C. Stancu, J.M. Nagashima, D.B. Rutledge, "Two flux weakening schemes for surface-mounted permanent-magnet synchronous drives and transition response considerations," *IEEE International Symposium on Industrial Electronics*, vol. 2, 1999, pp. 673-678.

[5] N. Mohan, "Analysis of current-regulated voltage-source inverters for permanent magnet synchronous motor drives in normal and extended speed ranges," *IEEE Transactions on Energy Conversion*, vol. 5, No. 1, March 1990.

[6] R. Krishnan, "Control and operation of PM synchronous motor drives in the field-weakening region," *IEEE Conference on Industrial Electronics, Control and Instrumentation*, vol. 2, 1993, pp. 745-750.

[7] T.M. Jahns, "Flux-weakening regime operation of an interior permanent-magnet synchronous motor drive," *IEEE Transactions on Industry Applications*, vol. 1A-23, no.4, July/August 1987.

[8] N.V Olarescu, M. Weinmann, S. Zeh, S. Musuroi, C. Sorandaru, "Optimum current reference generation algorithm for four quadrant

operation for PMSMS drive system without regenerative unit," *IEEE International Symposium on Industrial Electronics*, 2010.

[9] R. Nalepa, T. Orlowska-Kowalska, "Optimum trajectory control of the current vector of a non-salient-pole PMSM in the field-weakening region," *IEEE Transactions on Industry Applications*, vol. 59, no.7, July 2012.

[10] T. Miyajima, H. Fujimoto, M. Fujitsuna, "Model-based design of voltage phase controller for SPMSM in field-weakening region" *IEEE Applied Power Electronics Conf.* July 2013, pp. 2266-2272.

[11] S-H. Kim, S-K. Sul, "Voltage control strategy for maximum torque operation of an induction machine in the field weakening region" *IEEE Transactions on Industrial Electronics*, vol. 44, no. 4, 1997, pp. 512-518.

[12] B-H. Bae, N. Patel, S. Schulz, S-K. Sul, "New field weakening technique for high saliency interior permanent magnet motor" *IEEE Industry Applications Conference*, vol. 2, 2003, pp. 898-905.

[13] P. Fajkus, B. Klima, P. Hutak, "High speed range field oriented control for permanent magnet synchronous motor," *IEEE International Symposium on Power Electronics, Electrical Drives, Automation and Motion*, 2012, pp. 225-230.

[14] F. Briz, M.W. Degner, R.D. Lorenz, "Analysis and design of current regulators using complex vectors," *IEEE Transactions on Industry Applications*, vol. 36, no.3, May/June 2000.

[15] J.F. Moynihan, "Motor Modelling for Current Control Purposes," in *Aspects of Digital Current Control for AC Drives*. PhD thesis, UCC, 1996.

[16] C-M. ONG, "Basics of electric machines and transformations," in *Dynamic Simulation of Electric Machinery using Matlab/Simulink*, 1st ed. Prentice Hall.

[17] G. F. Franklin, J. D. Powell, *Digital Control of Dynamic Systems*, 2nd ed. London, UK: Addison Wesley, ISBN, 1980.

[18] *Moog Products.* (2015, November,1). Retrieved from http://www.moog.com/products

APPENDIX

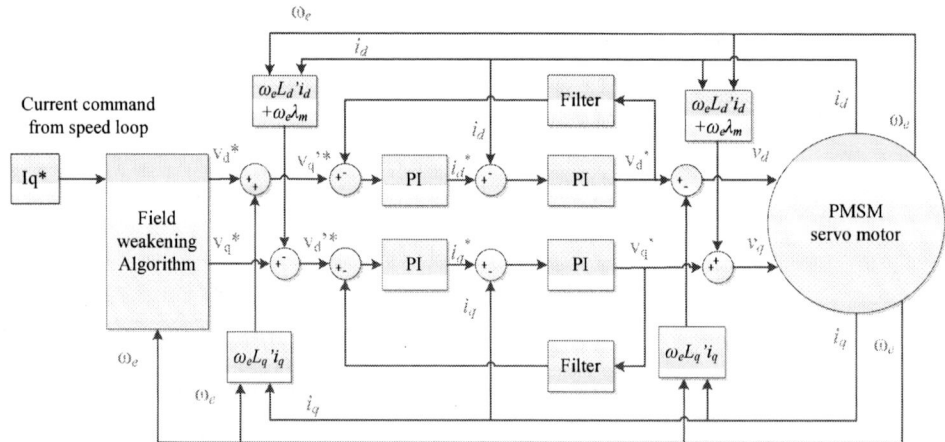

Fig.11. Decoupled voltage control strategy to implement the proposed FW algorithm

978-1-4673-9551-9/16 $31.00 © 2016 IEEE

Motor Current Reference Generation for Reducing Motor Currents in Drive Systems with Single-Phase Diode Rectifier and Small DC-link Capacitor

Young-Ho Chae and Jung-Ik Ha
Department of Electrical & Computer Engineering
Seoul National University
Seoul, Korea
E-mail: chaeho0531@gmail.com, jungikha@snu.ac.kr

Abstract—**In this paper, new method for generating motor current references is proposed in motor drive systems with single-phase diode rectifier and small dc-link capacitor in order to reduce the amount of the motor currents. Unlike the conventional large dc-link capacitor, small dc-link capacitor cannot work as an energy buffer. Therefore the motor currents, d and q-axis currents in synchronous reference frame, as well as output power are pulsating. The previous studies for generating motor current references in small dc-link capacitor systems did not consider the derivation terms of current in voltage equation. This paper presents new motor current reference generation algorithm considering the derivation terms so that the current can be saved where the dc-link voltage is enough and on-line calculation of current references can be possible. With the proposed method, the level of the motor current is reduced compared to the previous one and the efficiency of the system increases consequently. The effectiveness of the proposed method was verified with the simulation results.**

Keywords—*Small DC-link capacitor system, electrolytic capacitor-less system, motor current reference generation,*

I. INTRODUCTION

There have been many studies about motor drive system with a small dc-link capacitor [1]-[9]. Studies for small dc-link capacitor systems are divided into two classifications, three-phase and single-phase grid connected system. Single-phase grid connected system is widely used in home appliance application. Thus there are various studies about small dc-link capacitor system with the single-phase diode rectifier [3]-[9]. Fig. 1 shows the configuration of the small dc-link capacitor motor drive system with single-phase diode rectifier. However, this type of small capacitor systems has a drawback that the flux weakening current is required excessively when the dc-link voltage falls to zero, which is represented by blue-line shown in Fig. 2(a). To improve the immoderate d-axis current problem, methods that set a lower limit of the dc-link voltage as shown in Fig. 2(a) were proposed so that the requisite d-axis current for driving motor can be reduced [7], [8].

There are two systems that have minimum value of dc-link voltage. A typical small capacitor system and ADLC(Active DC Link Capacitor) system which has a compensator circuit connected in parallel with dc-link capacitor as shown with dotted lines in Fig 1. The output power references in these

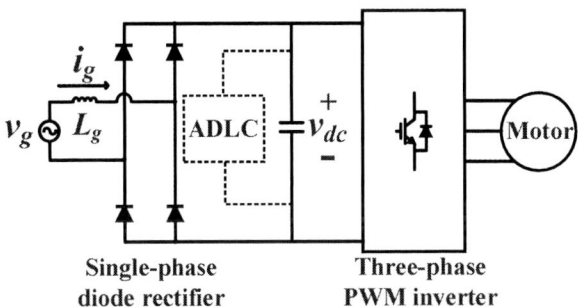

Fig. 1. Configuration of small dc-link capacitor system with single-phase diode rectifier.

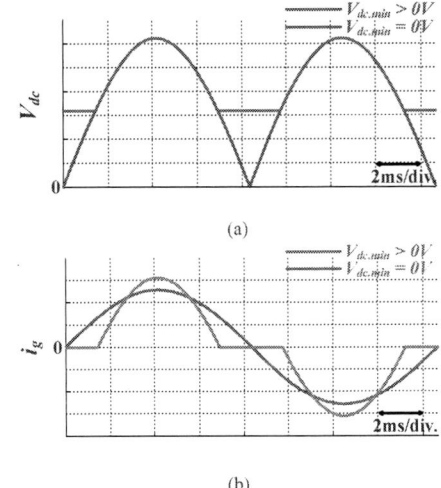

(a)

(b)

Fig. 2. (a) dc-link voltage (b) grid current in the small dc-link capacitor systems which have minimum dc-link voltage

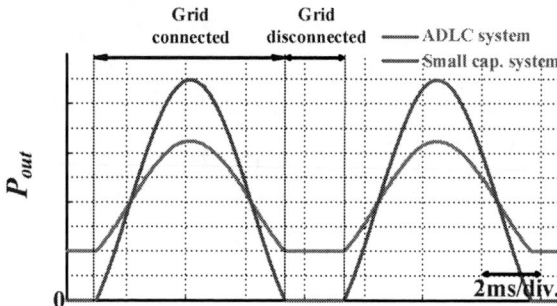

Fig. 3. Output power references in the small dc-link systems which have the minimum dc-link voltage.

systems are pulsating as shown in Fig. 3. Therefore the motor current references in synchronous reference frame would be no longer constant in contrast with those of the conventional large dc-link capacitor system. The erstwhile controls for motor current references in the small capacitor system are mostly constant d-axis current control [4], [5]. However, the current references generated by those methods are not considering the magnitude of the motor current. Therefore, those current references are not necessarily proper solutions in terms of reducing the magnitude of the inverter current. Another method is offline calculation of current references by numerical analysis [6]. But this method has a huge calculation burden despite employing the approximated current and current references should be calculated in each operation condition.

In this paper, new current control algorithm considering the derivation terms is proposed in the small dc-link capacitor system. Generation methods of the motor current references are analyzed on both grid connected section and disconnected section depicted in Fig. 3. With the proposed method, the motor currents would be decreased compared to the previous one because the voltage constraint condition is first considered to generate the current references. Simulation results demonstrate the feasibility of the proposed method.

II. MOTOR CURRENT REFERENCE GENERATION METHOD

The voltage equation and the load torque equation of the interior permanent magnet synchronous motor (IPMSM) in rotor reference frame and voltage constraint condition are as follows:

$$v_{ds}{}^r = R_s i_{ds}{}^r + L_d \frac{d}{dt} i_{ds}{}^r - \omega_r L_q i_{qs}{}^r$$
$$v_{qs}{}^r = R_s i_{qs}{}^r + L_q \frac{d}{dt} i_{qs}{}^r + \omega_r (L_d i_{ds}{}^r + \lambda_{pm}) \tag{1}$$

$$T_e = 1.5 PP((L_d - L_q)i_{ds}{}^r + \lambda_{pm})i_{qs}{}^r \tag{2}$$

$$v_{ds}{}^{r\,2} + v_{qs}{}^{r\,2} \le \frac{v_{dc}}{\sqrt{3}} \tag{3}$$

where $v_{ds}{}^r$, $v_{qs}{}^r$, $i_{ds}{}^r$, $i_{qs}{}^r$, R_s, L_d, L_q, λ_{pm}, PP, ω_r, and T_e are voltages and currents of the motor in rotor reference frame, the stator resistor, the d and q-axis inductances, the permanent magnet magnetic flux, the pole pair, the angular speed of the motor, and the load torque.

A. Conventional current reference generation method

In the small dc-link capacitor system, it is difficult to analyze the voltage equation including derivation terms of current because the current references are pulsating. Hence, most of the small dc-link capacitor systems fed by single-phase ac grid adopted the constant d-axis current control. Another method is offline calculation of current references based on assumptive waveforms. In the both methods, the current references are chosen first and then checked whether the voltage constraint and output power conditions are fulfilled. The selected current references are one of the many solutions which satisfy operating conditions of the inverter. This paper proposes the current reference generation method with priority consideration of the voltage constraint containing the derivation terms of current and the real-time calculation method.

B. Proposed current reference generation method

Depending on the dc-link minimum voltage, the operating modes of the motor drive system can be divided into two parts, the grid connected and disconnected sections. In addition, the output power references also can be different according to the dc-link minimum voltage and the type of the small dc-link capacitor systems as shown in Fig 3. However, the proposed method deals with the general approaches to generating the motor current references regardless of the operating modes and the output power references.

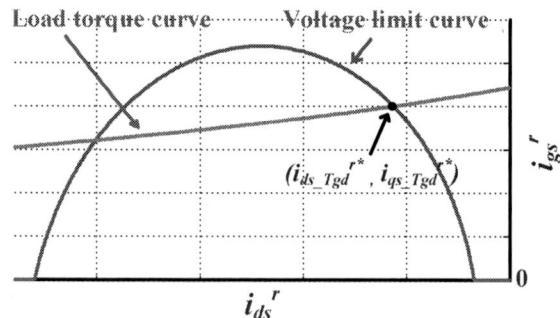

Fig. 4. Voltage limit curve and load torque curve for the grid disconnected section, where $v_{dc} = V_{dc.min}$

$$i_{qs}{}^r = \frac{ -\omega_r R_s ((L_d - L_q)i_{ds}{}^r + \lambda_{pm}) + \sqrt{ (\omega_r R_s ((L_d - L_q)i_{ds}{}^r + \lambda_{pm}))^2 - (R_s{}^2 + (\omega_r L_q)^2)((R_s i_{ds}{}^r)^2 + (\omega_r (L_d i_{ds}{}^r + \lambda_{pm}))^2 - \frac{V_{dc\_min}{}^2}{3}) } }{ R_s{}^2 + (\omega_r L_q)^2 } \tag{4}$$

978-1-4673-9551-9/16 $31.00 © 2016 IEEE

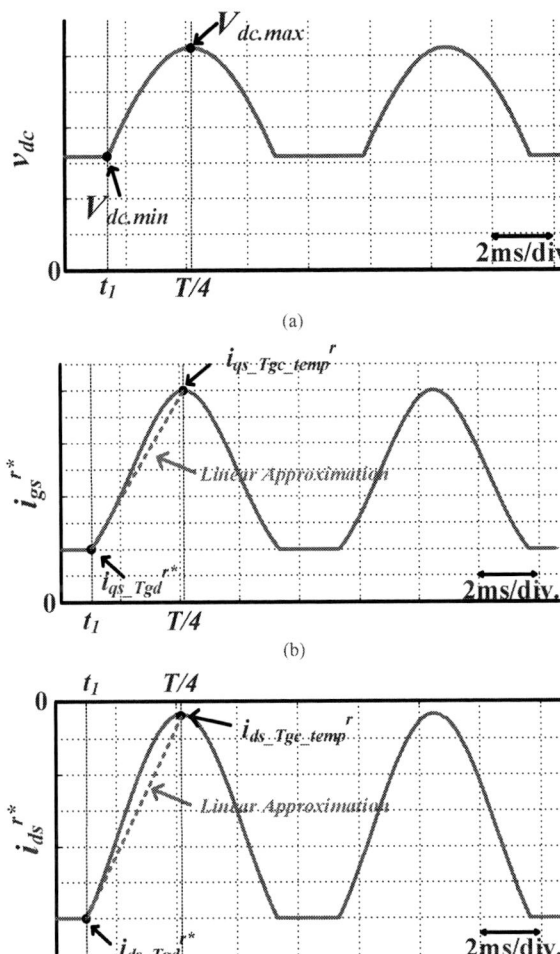

Fig. 5. Conceptual diagrams of the proposed method: (a) dc-link voltage (b) q-axis current (c) d-axis current.

(i) *Grid disconnected section*

When the grid is disconnected, the dc-link voltage and the output power reference are both constant as shown in Fig. 2(a) and Fig. 3. Therefore, the motor current can be constant and the derivation terms of current can be neglected. By using (1) and (3), the voltage limit curve equation can be obtained as (4) and the load torque curve equation is derived by (2). The voltage limit curve and the load torque curve are shown in Fig. 4. In Fig. 4, one of the two intersection points two curves make, a set of $i_{ds\_Tgd}^{r*}$ and $i_{qs\_Tgd}^{r*}$, is selected as the current references of the grid disconnected section in order to obtain less current.

(ii) *Grid connected section*

On the other hand, while the grid is connected, the output power and the dc-link voltage are fluctuated. Under these operating conditions, there could be many combinations of d and q-axis currents for the grid connected time. However each of the combinations has different magnitude. Thus, it is necessary to find a solution which has relatively small amount of the current by considering the voltage limit condition incorporating the derivation terms of current first.

In the equation (1) and (3), there are two difficulties for dealing with the equations. The first is the consideration of the derivation terms of unknown currents and the second is the problem of solving the biquadratic if i_{ds}^r is calculated because i_{qs}^r is the inverse function of i_{ds}^r from (2). To simplify the equations, it is assumed that the q-axis current has a certain values, $i_{qs\_Tgc\_temp}^r$ shown in Fig. 5(b), at the time when the dc-link voltage reference and the output power reference are equal to the maximum value of the individual, i.e. $v_{dc}=V_{dc.max}$ and $p_{out}=P_{max}$. The temporary value $i_{qs\_Tgc\_temp}^r$ is defined as follows:

$$i_{qs\_Tgc\_temp}^r = \frac{P_{max}}{1.5\,\omega_r\,\lambda_{pm}\,\alpha} \tag{5}$$

where α is defined as $(1+(L_d-L_q)i_{ds\_Tgc\_temp}^r/\lambda_{pm})$ and is a kind of control factor. Since the q-axis current term i_{qs}^r becomes a constant value, the voltage constraint equation is a quadratic equation. If there is a solution of the voltage constraint equation, $i_{ds\_Tgc\_temp}^r$, the derivation terms can be approximated linearly between $i_{ds\_Tgc\_temp}^r$ and $i_{ds\_Tgd}^{r*}$ on an average. The final equations with the approximated derivation term and the equation (6) are written as follows:

$$v_{ds}^r = R_s i_{ds\_Tgc\_temp}^r + L_d \frac{i_{ds\_Tgc\_temp}^r - i_{ds\_Tgd}^{r*}}{0.25T-t_1} - \omega_r L_q i_{qs\_Tgc\_temp}^r$$

$$v_{qs}^r = R_s i_{qs\_Tgc\_temp}^r + L_q \frac{i_{qs\_Tgc\_temp}^r - i_{qs\_Tgd}^{r*}}{0.25T-t_1} + \omega_r (L_d i_{ds\_Tgc\_temp}^r + \lambda_{pm}) \tag{6}$$

where t_1 is defined as $sin(V_{dc.min}/V_{dc.max})^{-1}/\omega$. By substituting (6) into (3) and calculating the voltage limit equation, $i_{ds\_Tgc\_temp}^r$ depicted in Fig. 5(c) can be calculated by (7)

The d-axis reference can be generated by connecting $i_{ds\_Tgc\_temp}^r$ and $i_{ds\_Tgd}^{r*}$. In the small capacitor systems with the dc-link minimum voltage, the output power is shaped based on the square of sine wave and the dc-link voltage based on the sine wave. Because the q-axis current is proportional to the output power, the derivation term of the q-axis current gives rise to insufficient dc-link voltage in the front part of the grid connected section. Thus, the d-axis reference should be a shape that has the small slope in the front and the large slope in the rear side for the half period of the grid connected time.

$$i_{ds\_Tgc\_temp}^r = \frac{-(C_{d_1}C_{d_2}+C_{q_1}C_{q_2})+\sqrt{(C_{d_1}C_{d_2}+C_{q_1}C_{q_2})^2-(C_{d_1}^2+C_{q_1}^2)-(C_{d_2}^2+C_{q_2}^2-V_{dc.max}^2/3)}}{(C_{d_1}^2+C_{q_1}^2)} \tag{7}$$

$$where\ C_{d_1}=R_s+\frac{L_d}{0.25T-t_1},\quad C_{d_2}=-(\frac{L_d i_{ds\_Tgd}^{r*}}{0.25T-t_1}+\omega_r L_q i_{qs\_Tgc\_temp}^r),\quad C_{q_1}=\omega_r L_d,\quad and\ C_{q_2}=R_s i_{qs\_Tgc\_temp}^r+L_q \frac{i_{qs\_Tgc\_temp}^r-i_{qs\_Tgd}^{r*}}{0.25T-t_1}+\omega_r \lambda_{pm}$$

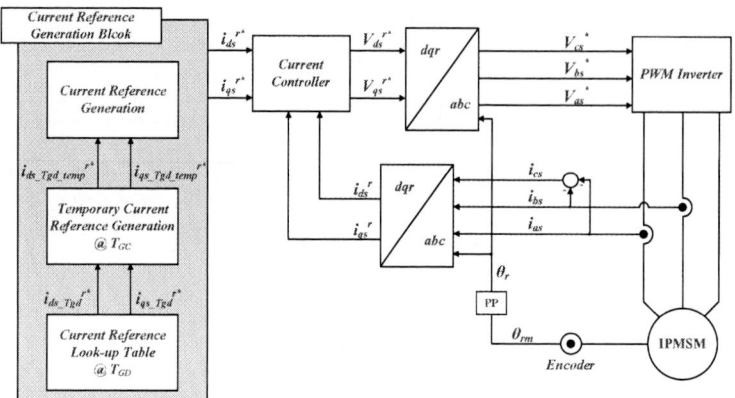

Fig. 6. Iterative algorithm to find the current references that satisfy the voltage constraint and voltage limit curve and load torque curve where $v_{dc} = V_{dc.max}$.

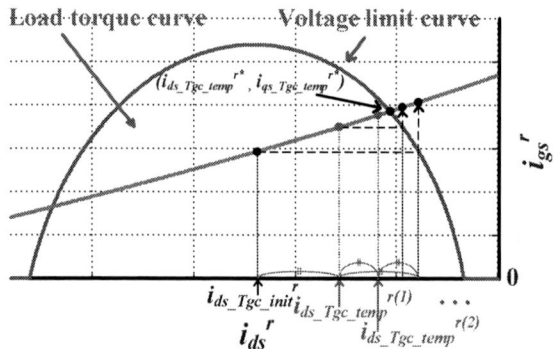

Fig. 7. Iterative algorithm to find the current references that satisfy the voltage constraint and voltage limit curve and load torque curve where $v_{dc} = V_{dc.max}$.

TABLE I. PARAMETERS OF THE IPMSM

Parameter	Values
Rated power, P_{out}	2 [kW]
Rated speed, w_r	6000 [r/min]
Stator resistance, R_s	0.68 [Ω]
d-axis inductance, L_d	4.27 [mH]
q-axis inductance, L_q	6.55 [mH]
Back-EMF constant, λ_{pm}	0.097 [V·s/rad]

In this paper, the d-axis current reference, called as $i_{ds\_Tgc}^{r*}$, is shaped based on the square of sine wave.

The q-axis reference can be re-calculated by (2) with the calculated d-axis current reference $i_{ds\_Tgc}^{r*}$ and is defined as $i_{qs\_Tgc}^{r*}$. But $i_{ds\_Tgc}^{r*}$ and $i_{qs\_Tgc}^{r*}$ can be changed depending on α and it is possible to exceed the given voltage limit. Therefore, this paper employs the iterative algorithm to find the optimal point during a half period of the grid connected time. The iterative algorithm is illustrated in Fig. 7. When $i_{qs\_Tgc\_temp}^{r}$ is decided at first, $i_{ds\_Tgc\_init}^{r}$ is automatically calculated according to (2). By using (6) and (7), $i_{ds\_Tgc\_temp}^{r}$ is obtained in the first calculation. Since $i_{ds\_Tgc\_temp}^{r}$ can be located on the outside of the voltage limit curve, $i_{ds\_Tgc\_temp}^{r(1)}$ is defined as the point

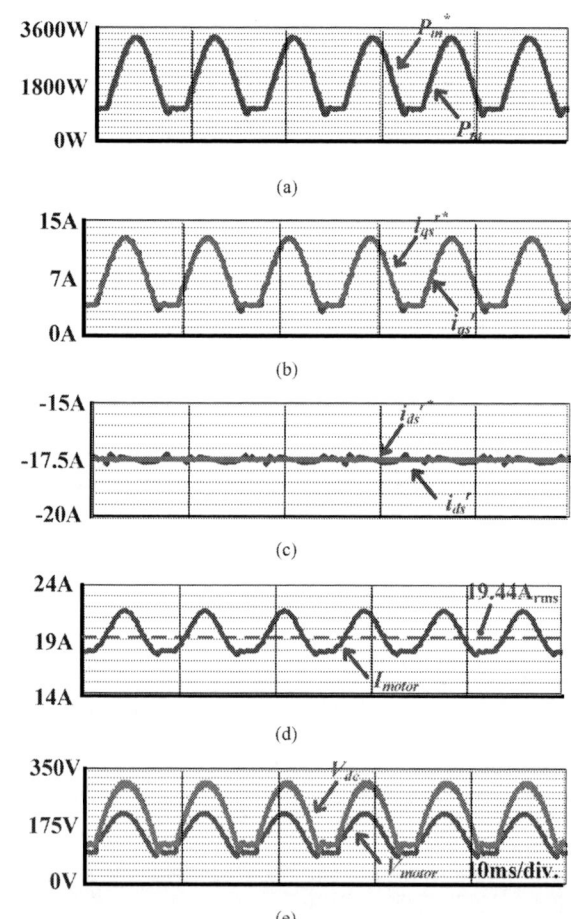

Fig. 8. Simulation results of the constant d-axis current control: (a) output power (b) q-axis current (c) d-axis current (d) motor current (e) dc-link voltage and motor voltage.

which divides the distance between $i_{ds\_Tgc\_init}^{r}$ and $i_{ds\_Tgc\_temp}^{r}$ in half and $i_{ds\_Tgc\_temp}^{r(i)}$ is corresponded to a new q-axis temporary

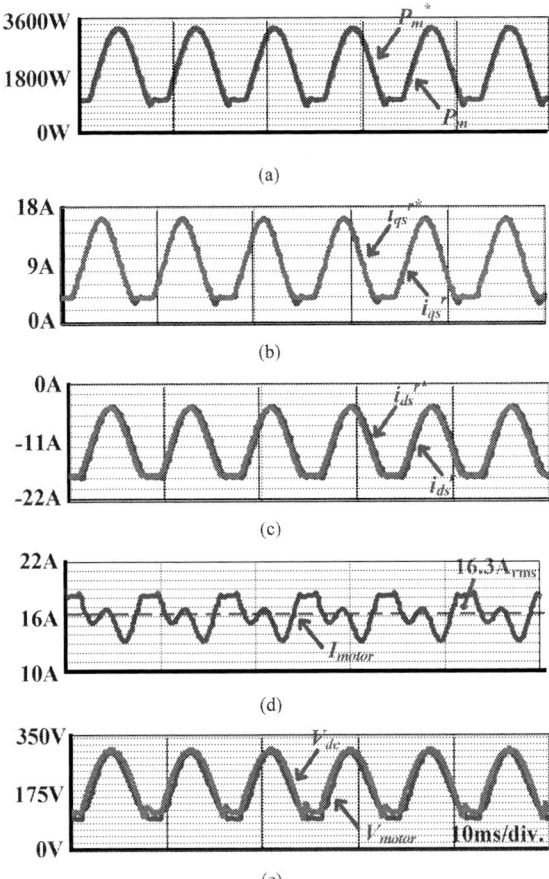

(a)

(b)

(c)

(d)

(e)

Fig. 9. Simulation results of the proposed current control: (a) output power (b) q-axis current (c) d-axis current (d) motor current (e) dc-link voltage and motor voltage.

value in the subsequent calculations. After three or four repeated processes, the iterative algorithm can achieve the current reference $i_{ds\_Tgc\_temp}^{r*}$ and $i_{qs\_Tgc\_temp}^{r*}$, which are located on the cross point of the voltage limit curve and load torque curve. However, if the dc-link voltage at its maximum point is enough to drive motor, $i_{ds\_Tgc\_temp}^{r*}$ can be a field boosting current. In other words, it is possible to use motor current more than necessary. In this case, $i_{ds\_Tgc\_temp}^{r*}$ is selected as one of the d-axis currents on the MTPA satisfying the load torque condition curve. Finally, $i_{ds\_Tgc}^{r*}$ is shaped based on the square of the sine wave by using $i_{ds\_Tgd}^{r*}$ and $i_{ds\_Tgc\_temp}^{r*}$ and $i_{qs\_Tgc}^{r*}$ is obtained by (2). Combining the current references of two sections, the motor current reference, i_{ds}^{r*} and i_{qs}^{r*}, are generated and controller block diagram is represented in Fig. 6.

III. SIMULATION RESULTS

The simulations are conducted with 2-kW IPMSM and ADLC motor drive system. The parameters of the motor used in the simulations are listed in Table I. Fig. 8 shows the simulation results of the conventional constant d-axis current control where the output power and the motor speed are its

rated values, 2-kW and 6000 r/min. In Fig. 8(d), the RMS value of the motor current is 19.44 A. However, in Fig. 8(e), the dc-link voltage is enough to drive motor in the grid connected section, which alludes to the possibility of reduction in motor current. Fig. 9 shows the simulation results of the proposed current control where the output power and the motor speed are same with those of the conventional one. When the d-axis current is shaped based on the square of sine wave as shown in Fig. 9(c), the RMS value of the motor current is 16.3A as depicted in Fig 9(d), which is about 16.2% reduction versus the conventional one. In the proposed method, the dc-link voltage is better utilized as shown in Fig 9(e) and the overall motor current can be reduced. Consequently, the conduction loss of the system is decreased and efficiency of the system can be increased.

IV. CONCLUSION

This paper proposes novel motor current reference generation algorithm in the small dc-link capacitor system with single-phase diode rectifier. Conventional methods have not considered the magnitude of the current in motor side when the current references are generated because it is difficult to calculate the derivation terms of current in the voltage equation. The proposed method generates the motor current references that have less amount of current with primary consideration of the voltage constraint condition which contains the derivation terms of current. The simulation results demonstrate about 16.2% motor current reduction compared with conventional method. As a result, the efficiency of the system can be increased.

REFERENCES

[1] W.-J. Lee and S.-K. Sul, "DC-Link voltage stabilization for reduced DC-link capacitor inverter," IEEE Trans. Ind. Appl., vol. 50, no. 1, pp. 404-414, Jan.-Feb. 2014.

[2] H. Yoo and S.-K. Sul, "A novel approach to reduce line harmonic current for a three-phase diode rectifier-fed electrolytic capacitor-less inverter," in Proc. IEEE APEC, 2009, pp. 1897-1903.

[3] H. Lamsahel and P. Mutschler, "Permanent magnet drives with reduced DC-link capacitor for home appliances," in Proc. 35th IEEE IECON, 2009, pp. 725-730.

[4] H.-S. Jung, S.-J. Chee, S.-K. Sul, Y.-J. Park, H.-S. Park and W.-K. Kim, "Control of three-phase inverter for AC motor drive with small DC-link capacitor fed by single-phase AC source," IEEE Trans. Ind. Appl., vol. 50, no. 2, pp. 1074-1081, Mar.-Apr. 2014.

[5] K. Inazuma, K. Ohishi, and H. Haga, "High-power-factor control for inverter output power of IPM motor driven by inverter system without electrolytic capacitor," in Proc. IEEE ISIE, 2011, pp. 619-624.

[6] Y. Son and J.-I. Ha, "Direct power control of a three-phase inverter for grid input current shaping of a single-phase diode rectifier with a small DC-link capacitor," IEEE Trans. Power Electron., vol. 30, no. 7, pp. 3794-3803, July 2015.

[7] H. Shin and J.-I. Ha, "Active dc-link circuit for single-phase diode rectifier system with small capacitance," in Proc. IEEE PEAC, 2014, pp. 875-880.

[8] Y. Son and J.-I. Ha, "Efficiency improvement in motor drive system with single phase diode rectifier and small DC-link capacitor," in Proc. IEEE ECCE, 2014, pp. 3171-3178.

[9] H. Utsugi, K. Ohishi, and H. Haga, "Reduction in current harmonics of electrolytic capacitor-less diode rectifier using inverter-controlled IPM motor," in Proc. 38th IEEE IECON, 2012, pp. 6206-6211.

A Simple Double Mapping Based SVPWM Method for Balancing DC-Link Capacitor Voltages of Five-Level Diode-Clamped Converters

Aparna Saha[1]
as157@zips.uakron.edu

Ali Elrayyah[2]
aelrayyah@qf.org.qa

Yilmaz Sozer[1]
ys@zips.uakron.edu

[1] Department of Electrical and Computer Engineering
The University of Akron, Akron, Ohio, USA

[2] Qatar Environment and Energy Research Institute
Doha, Qatar

Abstract—DC capacitor voltage imbalances are one of the major technical concerns for higher level diode-clamped converters (DCC). In this paper, a simple double mapping based space vector pulse width modulation (SVPWM) algorithm is proposed for balancing dc-link capacitor voltages of five-level DCC utilizing the switching redundancies. By choosing the appropriate switching pattern and its duration simultaneously, the voltage across each capacitor can be controlled adaptively. The developed mapping approach considers minimum loss SVPWM to address capacitor voltage equalization at high modulation indices. After having all the obtainable switching combinations, the five-levelDCC can be mapped into any lower level DCC structure depending on capacitor voltages to increase available switching redundancy. The duty cycles are calculated by taking into account only two-level SVPWM and can be updated for any higher level converter structure by adding required level number. In the proposed scheme dc-capacitor voltage balancing is attained with minimal switching operation, without any requirement of additional auxiliary power circuits. The simulation results validate the capability of mapping based SVPWM strategy to regulate the dc-link capacitor voltages for five-level DCC at different operating conditions.

Keywords—diode-clamped converters(DCC); space vector pulse width modulation (SVPWM), Five-level (5-level)

I. INTRODUCTION

Multi-level converters have gained significant attention in high power high voltage applications in recent times. Compared withtwo-level converters, these converters provide numerous advantages including better power quality, reduced switching losses and lower common-mode voltages [1]-[4].

Among diverse multilevel topologies, the diode-clamped converter (DCC) topology has been analyzed extensively in literature because of its control flexibility. DCC allows operating with higher dc-link voltages with lower voltage-rated devices. However, inherent capacitor voltage imbalancesdegrade the stable performance of the DCC. Voltage unbalances produce lower order harmonics on ac-

side currents and introduce unnecessary voltage stresses on switching devices. For the three-level DCC, a number of techniques have already been adopted[5]-[8] for capacitor voltage balancing. But complexity rises with the increased number of the converter levels.Regardless of the various technical merits, the dc-capacitor voltage drift phenomenon is one of the major technical concerns for DCC having higher than three levels.

Several approaches have been suggested to balance the dc capacitor voltages of DCC [7]-[19]. In general these schemes can be categorized into three major types: (i) addition of auxiliary balancing circuits [9] [10], (ii) advanced back-to-back converter topology [11]-[13] and (iii) utilization of redundant switching vectors by software realization [14]-[19]. The first scheme involves additional power circuitry to transfer energy from the higher capacitor voltage to an adjacent lower capacitor voltage which incurs further hardware costs. In second approach where voltage balancing is achieved with the help of double converterseem to be impractical for the stand-alone DCC applications. The third method involving the switching redundancy has attained more interest recently for the stand-alone DCCs.Because this technique does not add further hardware cost and also implementation becomes more practical with the advancement of high performance DSPs.

For the third software balancing method, variety of control strategies has been investigated in literature [14]-[19]. These can be classified as object function optimization, zero sequence voltage injection, virtual space-vector PWM and reference vector decomposition strategy. The voltage balancing using object function optimization scheme [14] assumes the sum of the instantaneous dc-link capacitor currents as zero which could change due to the non-idealities in the DCC architecture. In [16] the duty cycles calculation for the redundant states are adjusted between two zero vectors to control the capacitor voltages. A comprehensive study in [15] reveals that voltage balancing become more challenging athigh modulation

978-1-4673-9551-9/16 $31.00 © 2016 IEEE

index for 5-level DCC. The virtual SVM scheme ensures voltage balancing in each cycle [5]; however the control complexity rises extensively with the increase in converter levels and introduces more switching losses. The redundancy balancing at high modulation index for 5-level DCC is accomplished in [17], but the SVM structure needs to be generated twice by moving the origin which makes the algorithm more complicated and also switching mechanism may be violated at certain sampling instances, which will undermine the reliable operation of the converter.

In this paper, the voltage balancing is achieved for five-level DCC at high modulation index, m_a (>0.5) by applying simple dual stage mapping based minimum loss SVPWM (MLSVPWM) strategy. The technique extensively utilizes the switching redundancies in the existing SVPWM structure. Depending on capacitor voltages, the five-level DCC structure can be mapped into lower level configurations in two steps without violating switching mechanism and at the same time adding to a number of sufficient switching states to address voltage imbalance. By selecting the suitable switching patterns and their corresponding duty cycles voltage balancing can be obtained effectively at any modulation indices. As the stepping down to a lower level DCC is done through systematic mapping approach; it facilitates to reduce total harmonic distortion (THD) on ac-side line voltages even at m_a>0.5. In the proposed method since there is no switching for one phase in each switching cycle; total switching losses can be reduced by one third.

The rest of the paper is organized as follows: Section II demonstrates the basic operating principle of the 5-level DCC structure and minimum loss SVPWM in brief. Section III explains the proposed voltage balancing scheme with detail example of mapping. In Section IV the simulation results are given to validate the feasibility of the developed method. Finally, Section V outlines the conclusions.

II. OPERATING PRINCIPLE OF FIVE-LEVEL DCC

A. Basics of Operation

A general three phase 5-level DCC structure primarily consists of four series connected capacitors on the dc side of the converter as shown in Fig. 1. A net dc voltage source of V_{dc} is connected in parallel with series connected capacitors C_1, C_2, C_3 and C_4 For ideal case the voltage across each capacitor will be $E=V_{dc}/4$. Each phase leg has four complementary pair of switches respectively. For example, at phase leg A, the lower four switches i.e. S_{A1}', S_{A2}', S_{A3}' and S_{A4}' are complementary of the upper four switches, S_{A1}, S_{A2}, S_{A3} and S_{A4}. There are five distinct switching states for each phase of a five-level DCC. The switching states and corresponding voltage levels and line voltages for phase A is shown in Table I.

B. SVPWM Scheme

The voltage space vector arrangement of a three phase five-level DCC including all 125 switching states is shown in

Fig. 2. Space Vector Modulation (SVM) algorithm for multi-level DCC is an extension of SVM for general two-level DCC. In Minimum Loss SVPWM (MLSVPWM) method, for the realization of a particular reference voltage vector (V_{ref}), three nearest switching vectors are selected such that one of the phases is kept on or off for the whole switching period and make only one switching event for the other two phases [21] and eventually reduces switching losses in

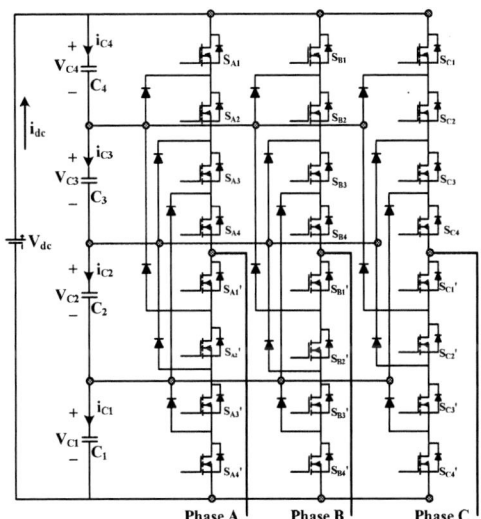

Fig. 1. A Five-Level DCC Structure.

TABLE I. Switching states for 5 level DCC		
Switching State $S_{A1}\ S_{A2}\ S_{A3}\ S_{A4}$ $S_{A1}'\ S_{A2}'\ S_{A3}'\ S_{A4}'$	**Voltage Level**	**Line Voltage**
11110000	4	V_{dc}
01111000	3	$3V_{dc}/4$
00111100	2	$V_{dc}/2$
00011110	1	$V_{dc}/4$
00001111	0	0

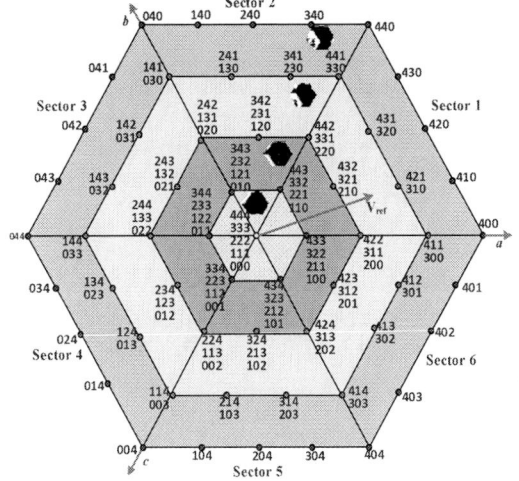

Fig. 2. Space Voltage Vectors for Five-Level DCC.

eachelectrical cycle.Execution of the MLSVPWM for 3-level and 5- level DCC have been presented in [19] and [21]. The five-level DCC has four concentric hexagons as shown in Fig. 2. The inner-most hexagon (H1) represents two-level DCC and third and fourth hexagons (H3) and (H4) represent four-level and five-level DCC respectively. The duty cycle calculation for one switching sequence is conductedconsidering a 60° coordinate system i.e. gh plane [20]. The detailed duty cycle computation for 5-level DCC is explained in [22].

III. PROPOSED VOLTAGE BALANCING TECHNIQUE

In space vector configuration of 5-level DCC from Fig. 2, at the center point of the hexagonal structure there are four possible switching states (000), (111), (222), (333) and (444) and the redundancy decreases by one as the selection move towards the outer layer. Now if the switching selection starts with any vector such as (444), (000), (111), (222) or (333) at the center, the MLSVPWM algorithm will generate different switching patterns and will charge capacitors with different rates. This freedom is the key point in the proposed balancing scheme, as the switching patterns can be changed periodically based on the requirement.

Thereference voltage vector of 5-level DCC can lie inside among any of the four concentric hexagons for different modulation indices. The available switching instances on each hexagon affect capacitor voltages in a different way. The switching vectors on this first hexagon (H1) can charge/discharge one of the capacitor voltages at each cycle. Similarly, the second hexagon (H2) corresponds to three-level DCC and each vector on this layer can affect two capacitor voltages in the same fashion. Likewise, third and fourth hexagons (H3) and (H4) can control three capacitor voltages and four capacitor voltages respectively. For modulation index less than 0.5, capacitor voltage balancing can be attained easily using the existing redundant switching vectors for any level of DCC structure. But when the reference space vector lies outside the H2 at high m_a (0.5 to 1), capacitor voltage balancing becomes an issue as there are no redundant vectors on the outer H4 and just one insufficient redundant vector exist on H3. To attend this difficulty, in the proposed balancing technique analyzing four dc-capacitor voltages 5-level DCC is mapped to any lower level DCC structure in two steps to reduce the switching losses and have less harmonics on line voltages. After that necessary switching sequences and duty cycle correspond to that level is evaluated and finally switching states are transformed into for 5-level DCC structure. This

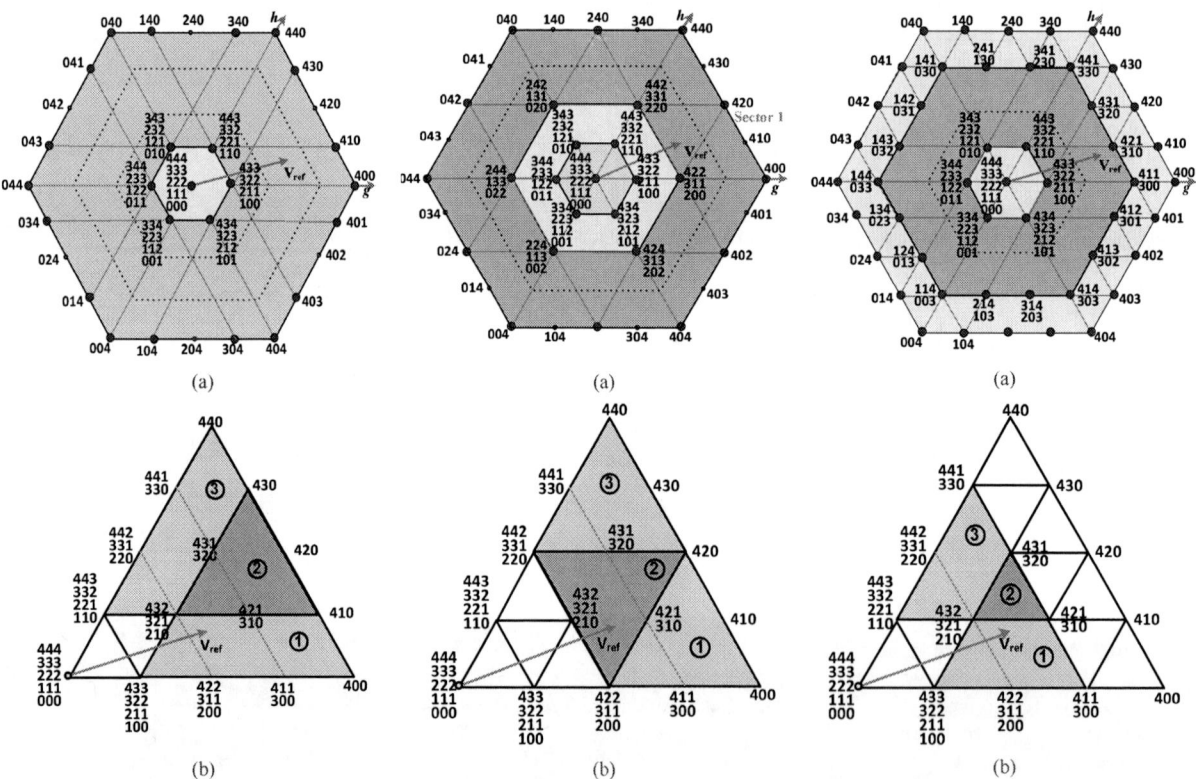

(a)   (a)   (a)

(b)   (b)   (b)

Fig. 3. 5-Level DCC mapped into 3-level DCC (a) By grouping H2, H3 and H4. (b) Detailed switching vectors of sector 1.

Fig. 4. 5-Level DCC mapped into 4-level DCC (a) By grouping H3 and H4. (b) Detailed switching vectors of sector 1.

Fig. 5. 5-Level DCC mapped into 4-level DCC (a) By grouping H2 and H3. (b) Detailed switching vectors of sector 1.

simplifies the balancing algorithm for outer hexagons H3 and H4 by adding sufficient switching redundancies with existing MLSVPWM methods.

A. The Mapping Concept

The voltages across the dc-link capacitors for the five-level DCC in Fig. 1 are stated as V_{c1}, V_{c2}, V_{c3} and V_{c4} and theexpected voltage after one switching cycle of T across the k_{th}capacitor C_k, where k=1......4 is

$$ v'_{ck} = v_{ck} + \frac{i_{c(k)}}{c_k} T \dots\dots\dots (1) $$

where,$i_{c(k)}$ is the average value of the currents in capacitor $C_k$after one switching cycle T. The proposed balancing method requires the measurement of four dc-capacitor voltages. By investigating the capacitor voltage distribution at any particular sample time, the five-level DCC can be mapped into any lower level DCC as per capacitor loading requirement. For low modulation index (m_a<0.5) the general 5-level MLSVPWM method is followed because it will allow to control over each capacitor (C_1, C_2, C_3, C_4) to balance V_{c1}, V_{c2}, V_{c3} and V_{c4} individually. But when V_{ref} stays in the outer most two layers (H3 and H4) i.e. for the $m_a$higher than 0.5, in approach [17] the six vectors on hexagon 1 is considered as new centers and the vectors on the hexagons H3 and H4 are rearranged with six two-level vector spaces. But there is a drawback with the system that it violates the regular switching method considerations. For instance, at the end of certain switching cycle for V_{ref} at triangle 1 of Fig. 3(b), the selected switching state is (433) that means V_{c4} is stressed more and assuming that V_{c1}> E. In the next switching cycle it is forced to activate the switching state (100) without any transitional states which implies harsh impact on switching generation scheme.

To address these voltage drifting at high modulation index (m_a>0.5) in the proposed method, a double mapped SVPWM scheme is considered. The main idea is to restructure the general SVM structure of 5-level DCC in two steps without jumping abruptly into lower levels. First, the 5-level DCC structure configures as 4-level by reorganizing the switching activation and providing the generated duty cycles from 4-level conventional MLSVPWM algorithm. Then another level of mapping is established by reshaping the SVM structure i.e. ignoring inner hexagon layers depending on capacitor voltage requirement.

For the first phase of mapping from 5-level to 4-level there are two possible switching arrangements: 1) consider the top switches S_{x1}(where $x = a$, b, c) of each phase leg turned off that means loading (C_2, C_3, C_4); and 2) make the bottom switches (S_{x4}) above the neutral point of each phase leg turned on for the entire switching pattern for loading (C_1, C_2, C_3). While topmost switches (S_{x1}) of each phase are turned off the switching pattern follows the path of selecting lower most switching states on each hexagon. For the other option of turning on the bottom switch (S_{x4}) of each phase leg the switching pattern tracks the top most switching vector at each point of selecting the nearest vectors. The duty cycle generation for both of these switching patterns follows the

previously developed MLSVPWM for 4-level which makes the control simple. Alternating between these two switching patterns for the two above mentioned switching selections and analyzing the capacitor voltages, it is clear that this mapping stage can make control over only the middle two capacitor voltages i.e. V_{c2} and V_{c3}. Because for switching case 1) it discharges only voltage V_{c4} and for case 2) it stresses the voltage V_{c1} across C_1.

To ensure overall balancing the second stage of mapping is introduced in the paper. It eliminates the inner layers of hexagon to provide adequate switching states for stabilizing capacitor voltages. Fig. 3(a) and (b) demonstrate the new SVM configuration for 3-level mapping and detailed view of sector 1 respectively. It clusters H2, H3 and H4 to continue from 1st stage 4-level mapping to equalize (V_{c1} and V_{c4}). It will assist only if the voltages (V_{c2} and V_{c3}) are already balanced by 1st stage 4-level mapping. Now if it is required to control the summation of (V_{c1} and V_{c2}) with (V_{c3} and V_{c4}) it is advantageous to choose another 4-level mapping as a second stage mapping as shown in Fig. 4(a) and (b) or 5(a) and (b). Fig. 4 eliminates H3 while Fig. 5 removes H2 hexagons to facilitate 2nd stage 4-level mapping.

B. Algorithm Steps

Utilizing the redundant vectors embedded in the 5-level SVM structure, balancing of the four capacitor voltages is attained in a systematic manner by simple dual stage

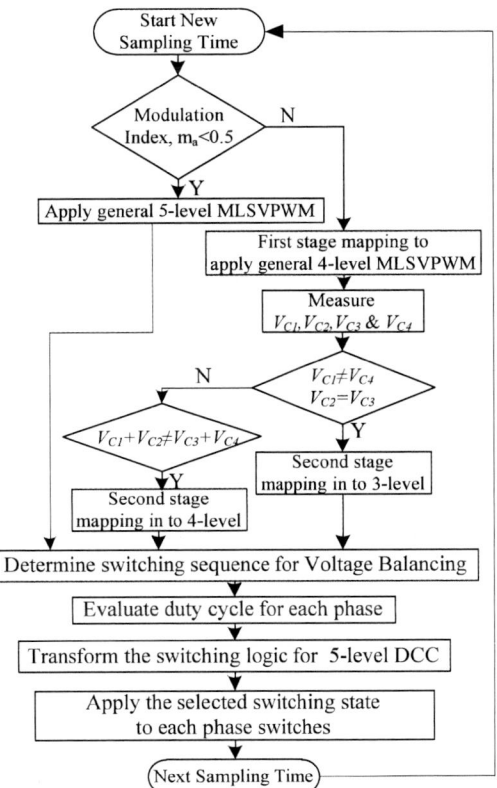

Fig. 6. Flow diagram of the proposed algorithm.

mapping. A flow diagram of the proposed technique is shown in Fig. 6. Briefly the proposed algorithm can be summarized as follows:

i. Determine the duty cycle for each phase for 2-level DCC allowing MLSVPWM algorithm and develop for any higher level DCC by increasing the level number [21].
ii. For modulation index, $m_a > 0.5$ choose 1st stage mapping by applying 4-level MLSVPWM method.
iii. Measure and analyze four dc-capacitor voltages (V_{c1}, V_{c2}, V_{c3} and V_{c4}) to decide which capacitors can be grouped together.
iv. Based on decision evaluate the mapped-level for 2nd stage mapping.
v. Transform the switching states of the mapped-level to the four switches of the three-phase 5-level DCC.
vi. Apply the switching states and alternate between the necessary switching sequences for certain period to gain the balancing for the capacitor voltages.

IV. VALIDATION RESULTS

The performance of the proposed controller with double mapping based SVPWM algorithm is evaluated for the 5-level DCC in this section. Circuit simulations are implemented in MATLAB/Simulink environment for a 500V DC/AC inverter operating at a switching frequency of 20 kHz for different modulation indices. Each of the capacitor voltages connected in series are to be regulated at 125 V through the proposed voltage balancing algorithm. Modulation indices of 0.55 and 0.82 are taken to demonstrate the effectiveness of the algorithm at high modulation indices.To validate the proposed method simulation results are shown for 5-level mapped into both 4-level and 3-levelas second stage mapping for a particular sample time below.

Fig. 7 shows the four capacitor voltage balancing at modulation index, m_a=0.55 for 4-level mapping and Fig. 8 reflects corresponding resultant V_{ref}. Similarly, Fig. 9 demonstrates voltage balancing at high modulation index, m_a=0.82 for 3-level mapping of the 5-level DCC. Fig.10 illustrates respective V_{ref}. Switching actions for each phase A, B and C are illustrated in Fig. 11, 12 and 13 and there are no switching for one third of a cycle for each phase which ensures minimum switching loss. Fig. 14and 15 present three phase load currents and output line to line voltage respectively at m_a=0.82. The results show that proposed mapping method performed effectively for voltage balancing at high modulation indices with existing switching redundancies.

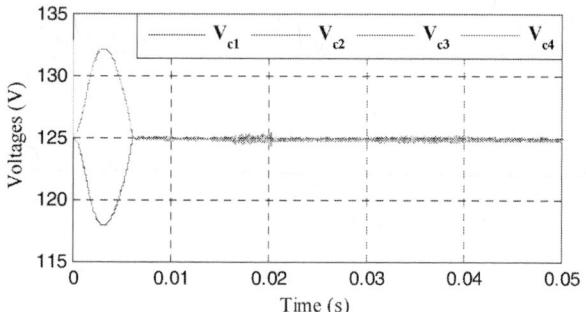

Fig. 7.DC capacitor voltages at m_a=0.25.

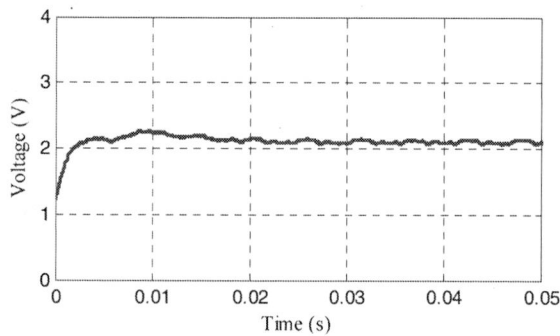

Fig. 8.Resultant reference voltage vector, V_{ref}.atm_a=0.55.

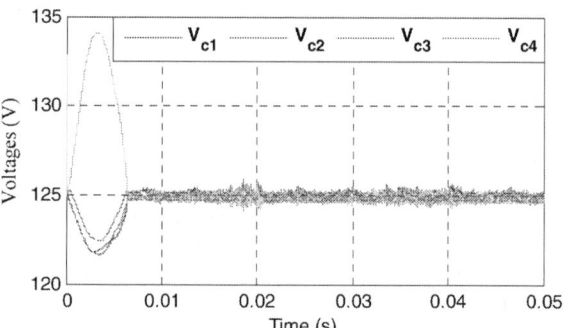

Fig. 9.DC capacitor voltages at m_a=0.82.

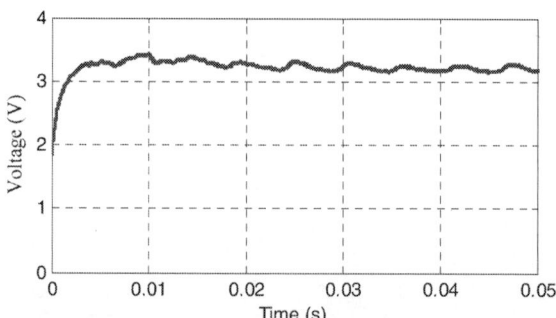

Fig. 10.Resultant reference voltage vector, V_{ref}.atm_a=0.82.

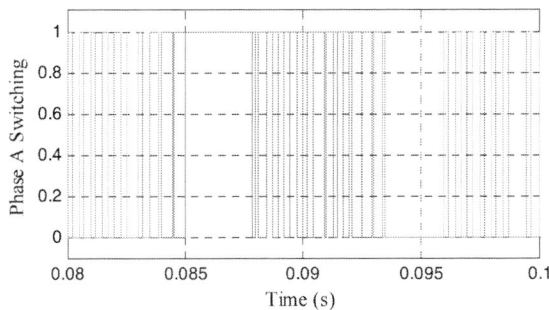

Fig. 11. Phase A switching actions.

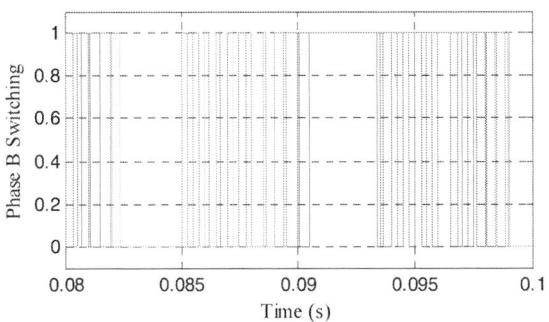

Fig. 12. Phase B switching actions.

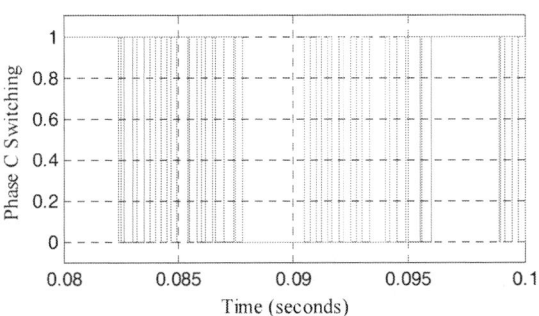

Fig. 13. Phase C switching actions.

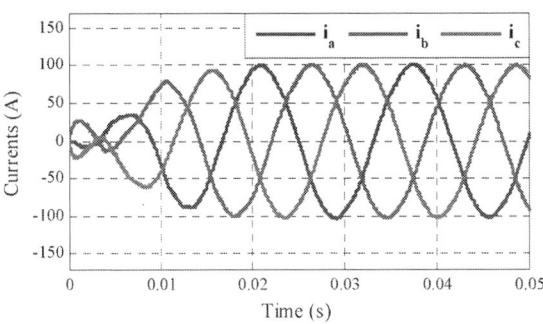

Fig. 14. Three phase ac-sideload currents.

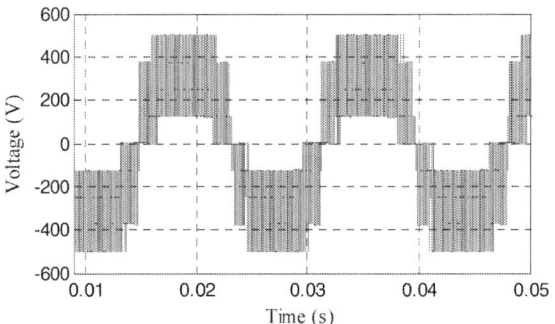

Fig. 15. Output line to line voltage.

V. CONCLUSION

To address the capacitor voltage imbalance issue for higher level DCC, this paper proposes a simple double mapping based SVPWM method for dc-capacitor voltage balancing at high modulation index for the five-level DCC. This dual stage mapping scheme helps to reduce total harmonic distortion (THD) at the line voltages on the AC sides of DCC and also lessen switching losses by lowering the mapping level in a systematic manner. Simulation results confirm that this presented scheme is capableof capacitor voltage balancingathigh modulation index for5-level DCC with utilizing only the existing switching redundancy in the system. The same control technique can be extended easily for higher level DCC.Overall, capacitor voltage balancing can be achieved without any supplementary power circuits and complex object function estimations while maintaining minimum switching losses for each electrical cycle.Therefore the developed mapping algorithmwill decrease the hardware costs and energy losses significantly, as the number of converter level raises.

REFERENCES

[1] Jih-Sheng Lai, F. Z.Peng, "Multilevel Converters-A New Breed of Power Converters," *IEEE Trans. Ind. Appl.*, vol. 32, no. 3, pp. 509-517, May/Jun. 1996.

[2] A. Nabae, I. Takahashi, H. Akagi, "A New Neutral-Point-Clamped PWM Inverter," *IEEE Trans. Ind. Appl.*, vol.IA-17, no.5, pp.518-523, Sept. 1981.

[3] S. Busquets-Monge, R. Maheshwari, J. Nicolas-Apruzzese, E. Lupon, S. Munk-Nielsen, J.Bordonau,"Enhanced DC-Link Capacitor Voltage Balancing Control of DC–AC Multilevel Multileg Converters," *IEEE Transactions Ind. Electron.*, vol.62, no.5, pp.2663-2672, May 2015.

[4] X. M. Yuan,I. Barbi,"Fundamentals of a new diode clamping multilevel inverter," *IEEE Trans. Power Electron.*, vol.15, no.4, pp.711-718, Jul 2000.

[5] S. Busquets-Monge, J. Bordonau, D. Boroyevich, S. Somavilla, "The Nearest Three Virtual Space Vector PWM - A Modulation for the Comprehensive Neutral-Point Balancing in the Three-Level NPC Inverter," *IEEE Power Electron.Letters*, vol. 2, no. 1, pp. 11-15, Mar. 2004.

[6] N. Celanovic, D. Boroyevich, "A comprehensive study of neutral-point voltage balancing problem in three-level neutral-point-clamped

voltage source PWM inverters," *IEEE Trans. Power Electron.*, vol.15, no.2, pp.242-249, Mar 2000.

[7] S. A. Khajehoddin, A. Bakhshai, P. K. Jain, "A Simple Voltage Balancing Scheme for m-Level Diode-Clamped Multilevel Converters Based on a Generalized Current Flow Model,"*IEEE Trans. Power Electron.*, vol. 23, no. 5, pp. 2248-2259, Sept. 2008.

[8] M. Saeedifard, R. Iravani, J. Pou, "Analysis and Control of DC-Capacitor-Voltage-Drift Phenomenon of a Passive Front-End Five-Level Converter," *IEEE Trans. Ind. Electron.*, vol. 54, no. 6, pp. 3255-3266, Dec. 2007.

[9] K. Hasegawa, H. Akagi, "A New DC-Voltage-Balancing Circuit Including a Single Coupled Inductor for a Five-Level Diode Clamped PWM Inverter," *IEEE Trans. Ind. Appl.*, vol. 47, no. 2, pp. 841-852, Mar.-Apr. 2011.

[10] Z.L. Shu, X. Q. He, Z. Y. Wang, D. Q. Qiu, Y. Z. Jing, "Voltage Balancing Approaches for Diode-Clamped Multilevel Converters Using Auxiliary Capacitor-Based Circuits,"*IEEE Trans. Power Electron.*, vol.28, no.5, pp.2111-2124, May 2013.

[11] Zhiguo Pan, Fang Zheng Pen, "Voltage Balancing Control of Diode-Clamped Multilevel Rectifier/Inverter Systems," *IEEE Trans. Ind. Appl.*, vol. 41, no. 6, pp. 1698-1706, Nov.-Dec. 2005.

[12] T. Ishida, T. Miyamoto, T. Oota, K. Matsuse, K. Sasagawa, Lipei H., "A Control Strategy for a Five-Level Double Converter with Adjustable DC Link Voltage," in *Proc.Ind. Appl. Conf.*, pp. 530-536, 2002.

[13] M. Saeedifard, R. Iravani, J. Pou, "A Space Vector Modulation Strategy for a Back-to-Back Five-Level HVDC Converter System," *IEEE Trans. Ind. Electron.*, vol. 56, no.2, pp.452-466, Feb. 2009.

[14] S. Li, N.Li and Y. Wang, "A novel DC voltage balancing scheme of five-level converters based on reference-decomposition SVPWM," in *Proc.IEEEAppl. Power Electron. Conf. and Expo. (APEC)*, pp.1597-1603, 2012.

[15] Zhiguo Pan, F. Z. Peng, "A Sinusoidal PWM Method With Voltage Balancing Capability for Diode-Clamped Five-Level Converters," *IEEE Trans. Ind.Appl.*, vol.45, no.3, pp.1028-1034, May-Jun 2009.

[16] Y. Deng, Y. Wang, K.H. Teo, R.G. Harley, "A Simplified Space Vector Modulation Scheme for Multilevel Converters," *IEEE Trans. Power Electron.*, vol.31, no.3, pp.1873-1886, Mar. 2016.

[17] H. A. Hotait, A. M. Massoud, S. J. Finney, B. W. Williams, "Capacitor Voltage Balancing Using Redundant States of Space Vector Modulation for Five-Level Diode Clamped Inverters," *IETPower Electron.*, vol. 3, no. 2, pp. 292-313, March 2010.

[18] Q. Jiangchao, M. Saeedifard, "Capacitor voltage balancing of a five-level Diode-Clamped Converter based on a predictive current control strategy," in *Proc.IEEEAppl. Power Electron. Conf. and Expo. (APEC)*, pp.1656-1660, 2011.

[19] A. Saha, Y. Sozer, A. Elrayyah, "Capacitor voltage balancing of a five-level diode-clamped converter using minimum loss SVPWM algorithm for wide range modulation indices," in *Proc.IEEEEnergy Convers. Cong. and Expo. (ECCE)*,pp.227-233, 2014.

[20] N. Celanovic, D. Boroyevich, "A Fast Space-Vector Modulation Algorithm for Multilevel Three-Phase Converters," *IEEE Trans. Ind. Appl.*, vol. 37, no. 2, pp. 637-641, Mar.-Apr.2001.

[21] Y. Sozer, D. A. Torrey, A. Saha, H. Nguyen, N. Hawes, "Fast minimum loss space vector pulse-width modulation algorithm for multilevel inverters," *IET Power Electron.* , vol. 7, no. 6, pp. 1590-1602,June2014.

[22] A. Saha, Y. Sozer, "Capacitor voltage balancing using minimum loss SVPWM for a five-level diode-clamped converter," in *Proc.IEEEAppl. Power Electron. Conf. and Expo. (APEC)*, pp.225-230, Mar. 2014.

Capacitor-Clamped Inverter Based Transient Suppression Method for Azimuth Thruster Drives

Shantha Gamini Jayasinghe,
Viknash Shagar,
and Hossein Enshaei
Australian Maritime College
University of Tasmania
Tasmania, Australia

Danyal Mohammadi
Department of Electrical and Computer
Engineering
Boise State University
Boise, USA

Mahinda Vilathgamuwa
School of Electrical Engineering and
Computer Science
Queensland University of Technology
Brisbane, Australia

Abstract—**The more-electric trend is there in almost all the corners of the automotive industry. The same trend is followed by the maritime transportation industry as well and as a result, conventional mechanical transmission based propulsion systems are gradually being overtaken by electric power transmission based propulsion systems. An azimuth thruster driven by an electric motor is a common configuration found in modern electric propulsion systems. Due to the tight speed control and stiff drivetrain in these propulsion systems load transients easily get propagated into the dc-link of the motor drive and subsequently into the upstream power bus as well. These transients can cause disturbances to the other loads connected to the power system. In the worst case, stability of the shipboard power system gets affected by the transients. This paper proposes to use the capacitor-clamped inverter based motor drive itself to absorb such transients and thereby prevent the propagation into the power bus. The efficacy of the proposed concept is verified through computer simulations. Simulation results show that the capacitor-clamped inverter is capable of absorbing load transients without passing them to the upstream power bus.**

Keywords—azimuth thrusters; capacitor-clamped inverter; electric propulsion; transient suppression

I. INTRODUCTION

Shipping is considered as the linchpin of the global economy as it accounts for more than 90% of the goods transported locally as well as internationally [1]. This figure is on the rise as the world population and economies continue to grow. With this foreseeable growth, the demand for fuel oil increases and as a result the price is expected to show a steady rise in the long run. On the other hand, the global share of greenhouse gas emissions from ships is on the rise. In this context, more-electric technologies such as electric propulsion are widely adopted to improve the fuel efficiency and thereby reduce the fuel cost and emissions [2]. With this global trend, more and more ships are being fitted with electric propulsion systems such as azimuth thrusters and podded propulsion systems [3].

Azimuth thrusters get the name from the fact that the housing where the propeller is being fitted can be rotated in any horizontal direction. The drivetrain of azimuth thrusters come in two forms as the *L*-drive configuration and the *Z*-drive configuration. The *L*-drive is generally chosen for electric propulsion where the motor is connected to the vertical shaft and the propeller connects to the horizontal shaft. There is a bevel gear to link the two shafts. The Z-drive is preferred in direct mechanical power transmission where the engine is mounted in the horizontal direction. In this configuration both the engine shaft and the propeller shaft are horizontal. There is a vertical shaft and two bevel gears to link the two horizontal shafts.

The propeller experiences extreme hydrodynamic forces and transient conditions mainly caused by waves, ice-interaction, ventilation and propeller racing [4]. The fatigue caused by the repetition of these conditions lead to premature failures in the mechanical drivetrain [5], [6]. In order to prevent such failures, shafts and gears of the drivetrain are designed to be stiff and thus transients acting on the propeller straightaway propagate into the motor drive. If the motor drive is not designed to absorb such transients they appear as distortions in the dc-link voltage of the power converter. This can cause instabilities, tripping and failures in the power converter. Moreover, the transients can propagate further into the upstream power bus. Given the fact that loads and generators in ships are comparable in size, this propagation can even lead to blackout in the shipboard power system. Therefore, it is important to contain transient energy within the converter with the use of a proper transient energy absorbing mechanism.

The simplest ways of absorbing transients are fitting passive dampers to the drivetrain or adding energy storage elements to the dc-link of the motor drive through interfacing converters. However, these add additional hardware and power electronics to the system increasing the losses, cost and complexity. Therefore, in this paper, authors have explored the possibility of using the energy storage capability of the capacitor-clamped inverter to absorb transients. This approach eliminates the need for additional energy storage elements, interfacing power converters and other associated hardware.

The advantages of the proposed system come with challenges and limitations as well, mainly due to the change in clamping-capacitor voltage with the absorption of transient energy. Pulse width modulation (PWM) of the converter is the major challenge which has been overcome in this study through modifications introduced to the sinusoidal PWM (SPWM) method. The modifications ensure the delivery of the required current without significant distortions even under

978-1-4673-9551-9/16 $31.00 © 2016 IEEE

varying capacitor voltage conditions. The proposed concept and modified SPWM method are verified through computer simulations. The results show that the capacitor-clamped inverter is capable of absorbing transients without passing them to the shipboard power system. Moreover, the modified SPWM method found to be capable of delivering required current even under variable capacitor voltage conditions without introducing significant distortions to the output current.

Section II of the paper describes the proposed transient suppression method followed by the system modelling in Section III. The modifications introduced to SPWM are presented in Section IV. Challenges and implementation issues of the proposed method are discussed in Section V followed by the simulation results in Section VI.

II. THE PROPOSED TRASIENT SUPPRESSION METHOD

Over the last decade, there has been a steady growth in the interest among shipping companies for larger ships as compared to deploying more number of small ships [7]. With the increase of the ship size electrical power demand also increases and hence shipboard power systems move from low voltage systems to high voltage systems [8]. The traditional two-level converter based electrical drives are not able to meet high voltage and high power levels of the large vessels [9]. Advanced converters technologies such as cascaded H-bridge converters, neutral-point-clamped converters and capacitor-clamped (flying capacitor) converters have emerged as promising alternatives. Out of these alternatives, capacitor-clamped converter topology has good dynamic performance and a relatively simple modulation process [10-20]. Moreover, due to the presence of capacitors, the capacitor-clamped converter has the capability to absorb transient energy itself [21]. The proposed concept is based on this special feature which has not yet been explored in relation to propulsion drive systems.

Fig. 1 illustrates the schematic of the proposed azimuth thruster drive system where the clamping-capacitors of the inverter are supposed to absorb load transients. The challenge in letting the clamping-capacitors to absorb transients is the change of their voltages which in turn create unbalanced conditions. The obvious effect of unbalanced capacitor voltages is the increase in the total harmonic distortion (THD) in the output current. The solution to this issues is the incorporation of an appropriate compensation into the modulation scheme which is discussed in detail in section IV.

III. SYSTEM MODELING

A. PMSM Model

A permanent magnet synchronous motor (PMSM) is used as the driving motor of the azimuth thruster considered in this study. The synchronous reference frame based PMSM model, shown in Fig. 2, is used in the simulation. Based on this model, two expressions can be derived for d-q axis voltages as in (1) and (2) respectively.

$$v_d = i_d R_s - \omega_e \varphi_q + L_d \frac{di_d}{dt} \tag{1}$$

$$v_q = i_q R_s + \omega_e \varphi_d + L_q \frac{di_q}{dt} \tag{2}$$

where v_d and v_q are d-q axis voltages, i_d and i_q are d-q axis currents, R_s is stator resistance, L_d and L_q are d-q axis inductances, ω_e is electrical rotational speed and φ_d and φ_q are magnetic flux components in d-q axes respectively. Magnitudes of φ_d and φ_q are given in (3) and (4). φ_m in (3) is the flux produced by permanent magnets of the motor. The electric torque, T_e, produced by the motor is given in (5) where p is the number of pole pairs in the motor.

$$\varphi_d = L_d i_d + \varphi_m \tag{3}$$

$$\varphi_q = L_d i_q \tag{4}$$

$$T_e = \frac{3}{2} p \left(\phi_d i_q - \varphi_q i_d \right) \tag{5}$$

Angular acceleration of the rotor shaft and the relationship between the electrical rotational speed and the mechanical rotational speed are given in (6) and (7) respectively.

$$\dot{\omega}_m = \frac{1}{J} \left(T_e - T_L \right) \tag{6}$$

$$\omega_e = p \omega_m \tag{7}$$

where J is the inertia of the load, ω_m is the mechanical rotational speed of the rotor and T_L is the load torque.

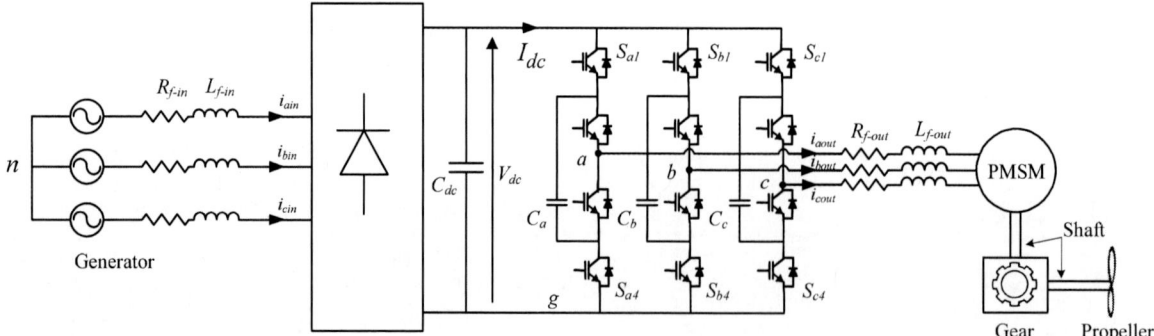

Fig. 1. Capacitor-clamped inverter based azimuth thruster drive system

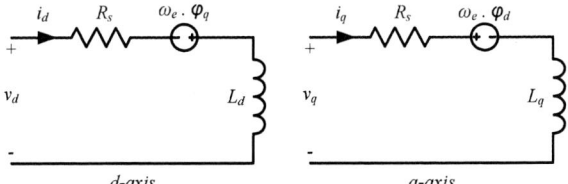

Fig. 2. PMSM model in the synchronous reference frame

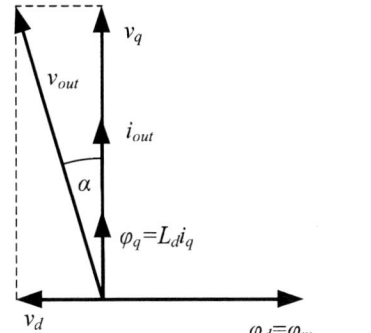

Fig. 3. Vector diagram showing zero direct axis current control of the PMSM

B. Drivetrain Model

The drive train of L-drive azimuth thrusters contain two shafts and a bevel gear. In order to simplify the analysis the shafts are assumed to be identical and the bevel gear is assumed to be an ideal gear with 1:1 gear ratio. The equivalent model of the two shafts can be expressed as in (8) where K_s and B are the stiffness and damping coefficient of the shaft respectively. The torque transmitted through the shaft varies with the angle of twist between the two ends, $(\theta_m - \theta_p)$, where θ_m is the angle at the drive end and θ_p is the angle at the load end. The corresponding block diagram is shown in Fig. 4 which includes the inertia of the load, J_p, as well.

$$T_p = K_s \int (\omega_m - \omega_p) dt + B(\omega_m - \omega_p) \qquad (8)$$

IV. MODULATION AND CONTROL

A. Modulation Method

Modulation and control of capacitor-clamped three level inverters are comprehensively discussed in literature [10-20]. Nevertheless, in all these publications, balanced or equal conditions are assumed, or some techniques are used to balance capacitor voltages. Conventional carrier based PWM methods or space vector PWM methods can directly be used as the modulation method in those systems. However, as explained in the section II, voltage unbalance is an unavoidable phenomenon in the proposed system. If the conventional modulation methods are directly used in such situations, output current get distorted.

In [21] authors have proposed a modified space vector PWM (SVPWM) scheme to reduce the effects of unbalanced voltages and deliver required current even at unbalanced

conditions. Similarly, the standard SPWM method can also be modified to suite unbalanced conditions [22]. The block diagram shown in Fig. 5 summarizes the modifications introduced to the standard SPWM method. The fundamental concept of comparing the three-phase reference voltages generated by the controller against triangular carriers is common for the modified SPWM method as well. The difference comes in the way of generating carrier waveforms. Generally, only two symmetrical carriers are sufficient to implement SPWM for a balanced capacitor-clamped inverter. But, in the proposed system, each inverter leg require an individual set of carriers due to the dynamic changes in clamping-capacitor voltages. In addition to that, each set should consist of four carriers as shown in Fig. 5. Therefore, altogether there should be 12 different carriers to implement the modified SPWM method. Modern processors that are tailored for motor drive applications generally come with large number of PWM units and thus the need for 12 PWM units does not create practical implementation issues.

The need for four carriers for each inverter leg can be justified as follows with reference to the leg 'a' of the inverter. In the proposed system a given reference voltage for the leg 'a' can be synthesized in two alternate ways. The first method uses the three voltage levels of 0, V_{Ca}, and V_{dc}. Similarly, in the second method the voltage levels 0, V_{dc} -V_{Ca} and V_{dc} are used (the corresponding switching states and gate signals are given in Table I). Since there are three voltage levels associated with each method two carriers are required for each method. As a result, each leg of the converter requires four different carrier waveforms. In addition to that, amplitudes of these carriers should be varied according to the changes in capacitor voltages. The equations for calculating the corresponding carrier amplitudes are given in Table II.

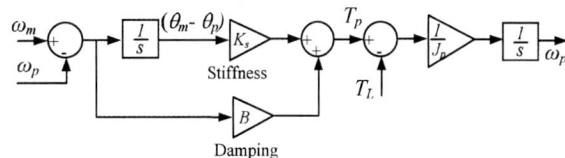

Fig. 4. Drivetrain and load model

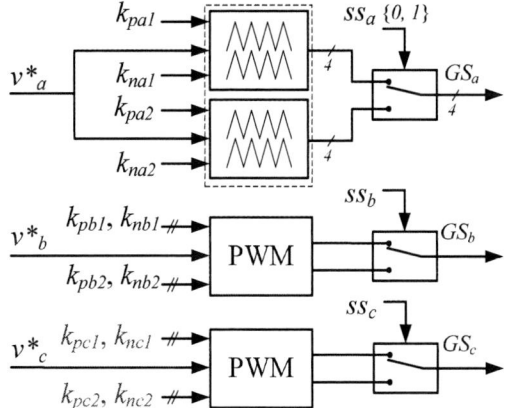

Fig. 5. Block diagram of the modified SPWM method

TABLE I. Switching States and Line to Ground Voltages for Leg A

Switching state (S_a)	Gate signals ($GS_{a1}..GS_{a4}$)	v_{ag}
0	0011	0
1	0101	V_{Cc}
2	1010	$V_{dc} - V_{Ca}$
3	1100	V_{dc}

TABLE II. Amplitudes of Modified Carriers

Carrier amplitudes for the leg 'a'		Carrier amplitudes for the leg 'b'		Carrier amplitudes for the leg 'c'	
k_{pa1}	$2(1-V_{Ca}/V_{dc})$	k_{pb1}	$2(1-V_{Cb}/V_{dc})$	k_{pc1}	$2(1-V_{Cc}/V_{dc})$
k_{na1}	$2V_{Ca}/V_{dc}$	k_{nb1}	$2V_{Cb}/V_{dc}$	k_{nc1}	$2V_{Cc}/V_{dc}$
k_{pa2}	$2V_{Ca}/V_{dc}$	k_{pb2}	$2V_{Cb}/V_{dc}$	k_{pc2}	$2V_{Cc}/V_{dc}$
k_{na2}	$2(1-V_{Ca}/V_{dc})$	k_{nb2}	$2(1-V_{Cb}/V_{dc})$	k_{nc2}	$2(1-V_{Cc}/V_{dc})$

B. Capacitor Charge/Discharge control

Charge/discharge controller for the capacitor C_a attached to the leg 'a' of the inverter, is shown in Fig. 6. The same controller and the following analysis can equally be used for the other two phases as well. The controller output SS_a selects the suitable output voltage synthesizing method out of the two possibilities shown in Fig. 5. If SS_a is permanently held at '0', the capacitor C_a get discharged during the positive half cycle of the a-phase current and get charged in the negative half cycle. Similarly, the opposite happens when SS_a is held permanently at '1'. Therefore, if SS_a is tied to 0 or 1, the average current flow through the capacitor is zero and hence the average voltage of the capacitor will not get affected. This indicates that the controller output SS_a should be changed at each and every half cycle to obtain an average charging or discharging current. In order to obtain a net discharging current, SS_a should be held at '0' during the positive half cycle and '1' at the negative half cycle. Similarly, if SS_a is held at '1' during the positive half cycle and '0' at the negative half cycle, a net charging current can be obtained. These two settings would produce maximum rates of discharging and charging for the capacitor C_a respectively. An intermediate rate can be obtained by switching between the above two settings. This switching is achieved through PWM as shown in Fig. 6.

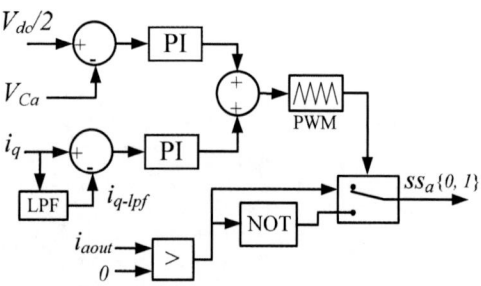

Fig. 6. Charge/discharge controller for the capacitor Ca attached to the leg 'a' of the inverter

As shown in Fig. 6 reference signal for the PWM unit is generated by adding the output of two PI controllers. The first PI controller sets the reference to bring the capacitor voltage to the half of the dc-link voltage. The second PI controller adjust the reference signal to absorb transients. As discussed in the following section, the q-axis current, i_q, varies with the load and thus changes in i_q reflect load transients. Taking this into account, the low pass filtered value of i_q is subtracted from the instantaneous value to obtain the error signal for the second controller. The gains of the two controller are selected in a way that the second controller dominates during transients and the first controller dominates during the steady state. As a result, even if the clamping-capacitor voltages vary during transients they gradually return to the balanced condition during steady state operation.

C. Motor Controller

The synchronous reference frame based controller is used for the speed control of the motor. The corresponding controller block diagram is shown in Fig. 7. The zero d-axis current control technique is used to control the speed of the PMSM. In this method the d-axis component of the stator current is maintained at zero. As a result, the d-axis component of the magnetic flux becomes equal to the flux produced by the permanent magnet of the motor as shown in Fig. 3. As expressed in (9), this flux, together with the q-axis current, produces the electrical torque which is analogues to the operation of dc motors. Therefore, in the speed controller shown in Fig. 7 the q-axis current is controlled to control the speed of the motor. In the steady state operation the speed controller compares the actual speed of the motor with the reference speed and the error is passed through a PI controller. The output of the PI controller is the torque reference which is used in (10) to derive the required q-axis current. The d-axis current reference is kept at zero. Based on these current references, required voltage components can be calculated using (11) and (12). Equations (13) and (14) are used to determine the amplitude and the angle of the inverter output voltage vector [23]. Equations (15-17) are used to generate three-phase reference voltages for the PWM unit. This is a standard speed controller for PMSMs. Apart from this standard controller there is an additional controller added in the speed controller shown in Fig. 7 which alters the q-axis current reference (torque reference) to minimize the speed difference between the two ends of the shaft. This in turn reduces stresses in the shaft during load transients [24]. However, due to the influence of this additional controller the transients straightaway get passed into the motor drive.

$$T_e = \frac{3}{2} p \phi_m i_q \tag{9}$$

$$i_q^* = \frac{2T_e^*}{3 p \phi_m} \tag{10}$$

$$v_d^* = -\omega_e \varphi_q = -\omega_e L_d i_q \tag{11}$$

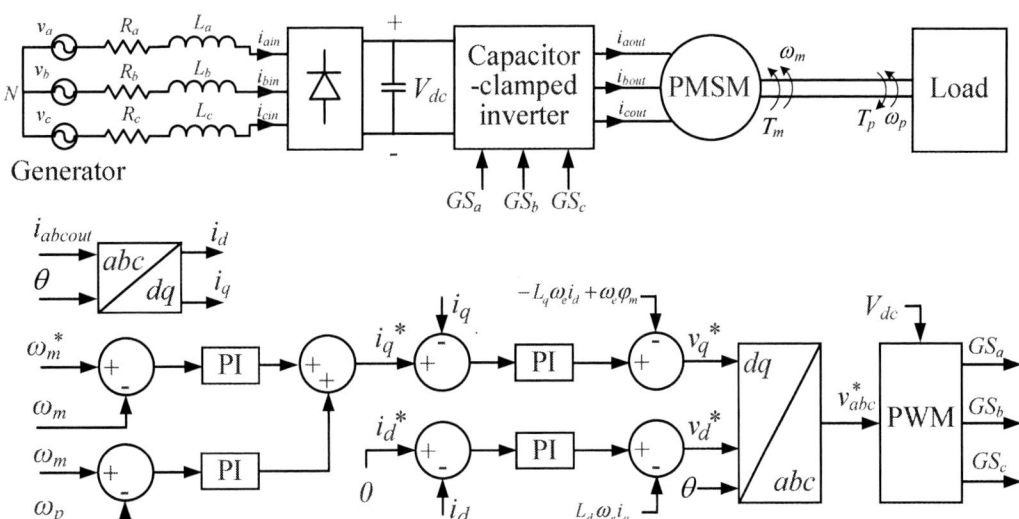

Fig. 7. Block diagram of the speed controller

$$v_q^* = i_q R_s + \omega_e \varphi_d = i_q R_s + \omega_e \varphi_m \qquad (12)$$

$$v_s^* = \sqrt{v_d^{*2} + v_q^{*2}} \qquad (13)$$

$$\alpha = \tan^{-1}\left(\frac{v_d^*}{v_q^*}\right) \qquad (14)$$

$$v_a^* = v_s \sin(\omega_e t + \alpha) \qquad (15)$$

$$v_b^* = v_s \sin\left(\omega_e t + \alpha - \frac{2\pi}{3}\right) \qquad (16)$$

$$v_c^* = v_s \sin\left(\omega_e t + \alpha + \frac{2\pi}{3}\right) \qquad (17)$$

V.   CHALLENGES AND LIMITATIONS

A.   Challenges in Capacitor Implementation

The proposed transient suppression method requires the capacity of the floating capacitors to be increased so as to absorb transients in the system. This would imply that the capacitors would then be expected to be heavier and of a higher volume. The extent of this added weight and volume has to be carefully considered because in the case of marine applications, the space onboard a ship is extremely limited and costly.

A survey on the capacitors rated for 600V indicate that the weight and volume of the capacitors experience a sudden increase above the 100KVar [25]. The volume of the capacitor bank enclosure stays the same for KVar ratings from 50 to 100. However, this volume increases 8-fold for the capacitors rated from 100KVar to 150KVar [25].

The weight also doubles when capacity is more than 100KVar [25]. These large changes in weight and volume can be attributed to the electrolyte material and arrangement of plates used for capacitors with higher capacitance. The weight of the capacitor bank increases by about 0.2kg for every 10mF increase in rating of the capacitor that supply 100KVar or less. The corresponding value for capacitors supplying more than 100KVar is 0.19kg [25]. In this study, the reactive power supplied by the capacitors is between 50KVar to 150KVar therefore aforementioned weight and volume increases are significant. A table summarizing how the physical characteristics of capacitor banks vary with the electrical parameters can be found in Table III [25].

B.   Trends in Improving the reliability of Capacitors

The reliability of capacitors is a major factor in the operation of the capacitor-clamped inverter. Therefore, any failure of capacitors can potentially cause a short circuit in the

TABLE III.   CAPACITOR PROPERTIES

KVar Rating	Capacitive Reactance (ohms)	Capacitance (mF)	Weight (kg)	Volume (m³)
50	7.20	368	10.0	0.0289
60	6.00	442	14.5	0.0289
70	5.14	516	15.0	0.0289
80	4.50	589	17.3	0.0289
90	4.00	663	17.3	0.0289
100	3.60	737	17.3	0.0289
110	3.27	810	39.5	0.249
120	3.00	884	39.5	0.249
130	2.77	957	39.5	0.249
140	2.57	1032	40.8	0.249
150	2.40	1105	45.0	0.249

system which can lead to catastrophic results especially at high voltage levels. In addition to that, marine application involve strict safety regulations as well. Therefore, increased reliability of the capacitors is mandatory to promote capacitor-clamped converters as a competitive power converter for the motor drive in azimuth thrusters. Moreover, in order to bring down the maintenance costs of the system, it is desired that system components such as capacitors have longer lifetime.

In view of this, there are a number of technologies designed to improve the lifetime of capacitors. Dry film technology for capacitors has recently entered the market and it has been touted to be self-healing. It is gradually replacing aluminium electrolytic capacitors at medium to high voltage levels. The voltage range of this technology is from 600V to 1350V and a dry film capacitor has a voltage gradient of up to 500V per micrometre during discharge. High capacitance values of up to 48mF have been reported using this technology [26]. The self-healing is achieved by coating the di-electric film of a capacitor with a thin metallic layer that evaporates due to the heat produced when a defect in the film occurs, thus, isolating the defect from the rest of the capacitor and preventing its spread. This leads to reduced maintenance for the capacitors. It has been found that capacitors using dry film technology have only a 2% drop in capacitance after 100000 hours of operation. Also, a large voltage gradient means that the capacitor can handle high ripple currents and voltage surges up to twice the rated voltage [26]. In comparison, aluminium electrolytic capacitors suffer chemical breakdowns when the rated voltage has been exceeded by 50%.

Heat is often the most crucial factor for capacitor failure. Therefore, efforts have been underway to reduce the losses caused by current flow into a capacitor. One way is to reduce the equivalent series resistance of capacitors by using multiple laser welded electrode tabs. This is effective because a higher number of metal tabs connecting the outer electrodes to the capacitor winding using advanced laser welding reduces the resistance of a capacitor [27]. This results in the capacitor experiencing lesser internal heating and a higher capability to withstand ripples.

Electrolyte evaporation has also been cited as a common cause for decreased lifetime of electrolytic capacitors. This problem has been tackled by reducing the evaporative capability of the electrolyte and improving the sealing of the electrolyte to reduce electrolyte loss. Solvents containing ethylene glycol have been found to reduce electrolyte evaporation at higher temperatures. Better compositions for the rubber sealant as well as double sealing have been explored. A combination of the above measures have been found to double the lifetime of capacitors from 10000hrs to 20000hrs [28].

The above improvements in capacitor performance show that the reliability and longevity of capacitor can be enhanced in light of further advances in electrical and chemical technologies in the near future. This indicates that the use of capacitor-clamped inverter as the motor drive and integrating the proposed method into it is feasible from the practical implementation point of view.

C. *Techniques to avoid capacitor short circuit in the event of IGBT short circuit failure*

Maritime applications demand for enhanced reliability in power electronic converter systems which are used for powering up the essential loads such as propulsion systems. An investigation on power electronic converter failures has revealed that 60% of the failures are due to capacitor failures [29]. Moreover, capacitor failure incidents and resultant damages to ship power systems are reported in marine accident investigation branch (MAIB) repots [30]. This highlights the need for detection, identification and isolation of failures as soon as possible. In multilevel inverters such as capacitor-clamped inverter, there are more switches which increases the availability and make the system to be more tolerable in case of a fault [31]. There are several methods proposed in literature to isolate faults such as IGBT short-circuit faults and thereby avoid catastrophic consequences. Fast acting fuses and thyristors are key elements that are used in these methods to isolate the fault from the whole system [32], [33]. The advancement of fault detection, identification and isolation techniques together with the development of fault accommodation topologies help improve the safety and availability of capacitor-clamped inverter based motor drives.

VI. SIMULATION RESULTS

A computer simulation on the MATLAB/Simulink platform was carried out to show the efficacy of the proposed transient suppression method. Two step changes are introduced to the load as shown in Fig. 8(a) to create a load gain scenario and a load drop scenario. The load gain is introduced at 100ms by increasing the propeller load from 100Nm to 400Nm. The step increase of the load results in a speed drop at the load end of the shaft as shown in Fig. 8(b) by the trace marked as 'Load end speed'. This speed drop propagates to the other end of the shaft as well with a slight drop in the magnitude as shown in the same graph by the trace marked as 'Drive end speed'. The speed controller reacts to this variation and changes the output current to restore the speed to the set value. The two speed variations shown in Fig. 8(b) is an indication of tight speed regulation which in turn passes load changes in the drive train directly into the motor drive. The speed difference between the two ends of the shaft accounts for the transient energy absorb within the shaft. If the shaft is flexible the speed difference between the two ends becomes large resulting in more transient energy absorb within the shaft. Nevertheless, flexible shafts introduce torsional oscillations which are detrimental to the shaft life. Therefore, shafts are made to be stiff which, together with the tight speed control, passes the transients straightaway to the motor drive. Moreover, the additional controller mentioned in Section IV reduces the speed difference between the two ends of the shaft by modifying the q-axis current reference. Therefore, even if the

978-1-4673-9551-9/16 $31.00 © 2016 IEEE   2818

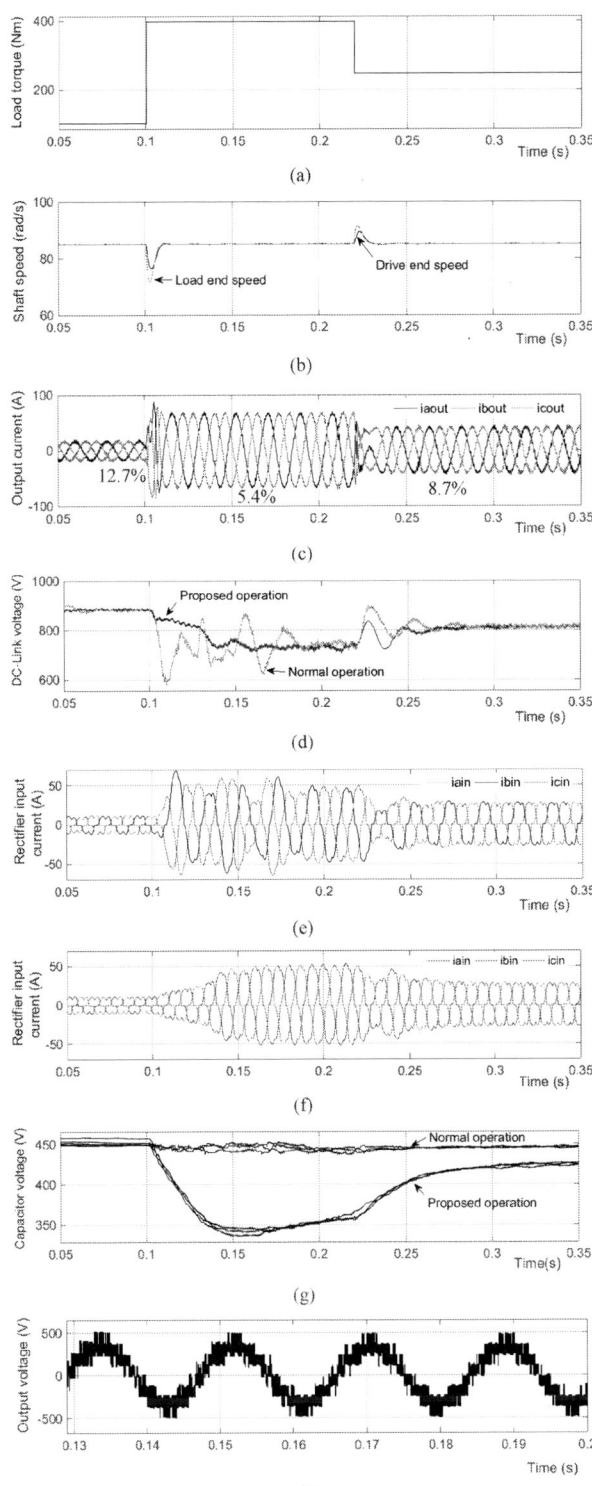

Fig. 8. Simulation results (a) load torque, (b) shaft speed at the drive end and load end, (c) output current, (d) dc-link voltage, (e) rectifier input current in the normal operation, (f) rectifier input current in the proposed operation, (g) clamping capacitor votlage, (h) inverter output voltage (a-phase)

shaft is flexible, all the transients get passed into the motor drive. The transients in the output current, shown in Fig. 8(c), is an indication of this transient propagation into the power converter. As a result, the dc-link voltage of the frequency converter gets affected as shown in Fig. 8(d) by the trace marked as 'Normal operation'. The transients can propagate further into the upstream power bus as shown by the rectifier input current shown in Fig. 8(e). If the generators are not capable of reacting against this kind of fast changes the shipboard power system becomes unstable.

With the proposed method these transients get absorbed by the clamping-capacitors and thus the dc-link voltage changes slowly as shown in Fig. 8(d) by the trace marked as 'Proposed operation'. As a result, the input currents to the rectifier also show smooth changes as shown in Fig. 8(f) which gives enough time for generators to react to the changes and thus maintain the stability of the power system. As a result of absorbing the transients, clamping-capacitor voltages get affected. The corresponding clamping-capacitor voltage variations are shown in Fig. 8(g). As shown in Fig. 8(c) the modified SPWM method is capable of producing required output currents even under variable capacitor voltage conditions. The THD values of the output current at different places are given in Fig. 8(c) to show that even if the clamping capacitor voltages vary THD in output currents still remain at an acceptable level for marine applications [34]. The a-phase voltage waveform of the inverter output is shown in Fig. 8(h) to illustrate the capability of the modified modulation method to produce 3-level output voltage waveforms even under unbalanced conditions. System parameters of the simulation setup are given in Table IV which are arbitrarily chosen to demonstrate the operation of the proposed concept.

TABLE IV. SYSTEM PARAMETERS

Nominal voltage of the generator ($Vll\ rms$)	690V
Frequency of the generator output voltage	50Hz
Resistance of the generator (R_{abc})	0.01Ω
Inductance of the generator (L_{abc})	15mH
DC-link capacitance (C_{dc})	0.5mF
Capacitance of the Clamping capacitors (C_{abc})	4.7mF
Filter resistance (R_{fout})	0.01Ω
Filter inductance (L_{fout})	1mH
Resistance of the motor (R_m)	0.01Ω
Inductance of the generator (L_m)	1mH
Number of pole pairs in the motor (p)	4
Rotor inertia of the motor (J_m)	0.017kgm2
Stiffness of the shaft (K)	20000Nm/rad
Damping coefficient of the shaft (B)	10Nm/rads-1
Inertia of the load (J_p)	0.025kgm2

978-1-4673-9551-9/16 $31.00 © 2016 IEEE

VII. CONCLUSIONS

This paper proposes to use the clamping-capacitors of the capacitor-clamped three-level inverter to absorb load transients in azimuth thrusters and thereby prevent the transient propagation into the shipboard power system. This makes the clamping-capacitors to deviate from the balanced condition. The challenge of delivering required output current under unbalanced conditions is achieved through a modified SPWM method. Simulation results verify the ability of the capacitor-clamped inverter to absorb load transients and thereby prevent the propagation into the upstream power bus. Moreover, the results verify the capability of the modified SPWM method in delivering desired current even in unbalanced conditions while keeping the THD level low to suite maritime applications.

REFERENCES

[1] Report of the UN Framework Convention on Climate Change, "Control of greenhouse gas emissions from ships engaged in international trade," Nov. 2011, available online at http://unfccc.int/resource/docs/2011/smsn/igo/142.pdf.

[2] S.D.G. Jayasinghe, G. Lokuketagoda, V. Shagar, D. Ranmuthugala, and H. Enshaei, "Electro-technologies for Energy Efficiency Improvement and Low Carbon Emission in Maritime Transport," in *Proc.* IAMU AGA, pp.119-123, Sept. 2015.

[3] C.A. Reusser, and H. Young, "Full electric ship propulsion based on a flying capacitor converter and an induction motor drive," in *Proc.* International Conference on Electrical Systems for Aircraft, Railway, Ship Propulsion and Road Vehicles (ESARS 2015), 3-5 March 2015.

[4] Ø.N. Smogeli, "Control of Marine Propellers: From Normal to Extreme Conditions," Ph.D. dissertation, Department of Marine Technology, Norwegian University of Science and Technology (NTNU), Trondheim, Norway, Sept. 2006.

[5] M. Fonte, L. Reis, and M. Freitas, "Failure analysis of a gear wheel of a marine azimuth thruster," *Engineering Failure Analysis*, Volume 18, Issue 7, pp. 1884-1888, Oct. 2011.

[6] T. Rauti, "Gear failures: Lessons learned." in *Proc.* Dynamic Positioning Conferenc, Oct. 2013.

[7] "Review of Maritime Transport 2014," United nations conference on trade and development (UNCTAD 2014), pp.27-48, 2014.

[8] D.T. Hall, "Practical Marine Electrical Knowledge," 3rd Edition, Witherby Publishers, London, 2014.

[9] K. Thantirige, A.K. Rathore, S.K. Panda, G. Jayasinghe, M.A. Zagrodnik, and A.K. Gupta, "Medium voltage multilevel converters for ship electric propulsion drives," in *Proc. Intl. Conf.* on *Elec. Sys. for Aircraft, Railway, Ship Prop. and Road Vehicles*, pp.1-7, Mar. 2015.

[10] W. Li, Q. Luo, Y. Mei, S. Zong, X. He, and C. Xia, "Flying-Capacitor Based Hybrid LLC Converters with Input Voltage Auto-Balance Ability for High Voltage Applications," *IEEE Trans. Power Electron.*, vol.31, no.3, pp.1908-1920, Mar. 2016.

[11] A. Ashraf Gandomi, K. Varesi, and S.H. Hosseini, "Control strategy applied on double flying capacitor multi-cell inverter for increasing number of generated voltage levels," *IET Power Electronics*, vol.8, no.6, pp.887,897, June 2015.

[12] L. Ziyou, A.I. Maswood, and G.H.P. Ooi, "Modular-Cell Inverter Employing Reduced Flying Capacitors With Hybrid Phase-Shifted Carrier Phase-Disposition PWM," *IEEE Trans. Ind. Electron.*, vol.62, no.7, pp.4086,4095, July 2015.

[13] V. Dargahi, A.K. Sadigh, M. Abarzadeh, S. Eskandari, and K.A. Corzine, "A New Family of Modular Multilevel Converter Based on Modified Flying-Capacitor Multicell Converters," *IEEE Transactions on Power Electronics*, vol.30, no.1, pp.138,147, Jan. 2015.

[14] P.R. Kumar, R.S. Kaarthik, K. Gopakumar, J.I. Leon, and L.G. Franquelo, "Seventeen-Level Inverter Formed by Cascading Flying

Capacitor and Floating Capacitor H-Bridges," *IEEE Transactions on Power Electronics*, vol.30, no.7, pp.3471,3478, July 2015.

[15] L. Zhang, and S. J. Watkins, "Capacitor voltage balancing in multilevel flying capacitor inverters by rule-based switching pattern selection," *IET Trans. Electric Power Appl.*, vol.1, no.3, pp.339-347, May 2007.

[16] B. R. Lin, and Chun-Hao Huang, "Implementation of a Three-Phase Capacitor-Clamped Active Power Filter Under Unbalanced Condition," *IEEE Trans. Ind. Electron.*, vol.53, no.5, pp.1621-1630, Oct. 2006.

[17] Jing Huang, and K. A. Corzine, "Extended operation of flying capacitor multilevel inverters," *IEEE Trans. Power Electron.*, vol.21, no.1, pp. 140- 147, Jan. 2006.

[18] K. Dae-Wook, L. Byoung-Kuk, J.H. Jeon, T.J. Kim, and D.S. Hyun, "A symmetric carrier technique of CRPWM for voltage balance method of flying-capacitor multilevel inverter," *IEEE Trans. Ind. Electron.*, vol.52, no.3, pp. 879- 888, June 2005.

[19] K. Xiaomin, K.A. Corzine, and Y.L. Familiant, "A unique fault-tolerant design for flying capacitor multilevel inverter," *IEEE Trans. Power Electron.* vol.19, no.4, pp. 979- 987, July 2004.

[20] D.W. Kang, W.K. Lee, and D.S. Hyun, "Carrier-rotation strategy for voltage balancing in flying capacitor multilevel inverter," *IEE Trans. Electric Power Appl.*, vol.151, no.2, pp. 239- 248, Mar 2004.

[21] S.D.G Jayasinghe, and D.M. Vilathgamuwa, "Flying Supercapacitors as Power Smoothing Elements in Wind Generation," *IEEE Trans. Ind. Electron.*, vol.60, no.7, pp.2909-2918, July 2013.

[22] D.M. Vilathgamuwa, S.D.G. Jayasinghe, and U.K. Madawala, "Battery clamped three-level inverter for renewable energy systems," in *Proc. IEEE Ind. Electron. Society Conf.*, pp.3105-3110, 7-10 Nov. 2011.

[23] D.M. Vilathgamuwa, S.D.G. Jayasinghe, F.C. Lee, and U.K. Madawala,"A unique battery/supercapacitor direct integration scheme for hybrid electric vehicles," in *Proc. IEEE Ind. Electron. Society Conf.*, pp.3020-3025, 7-10 Nov. 2011.

[24] H. Geng, D. Xu, Bin Wu, and Geng Yang, "Active Damping for PMSG-Based WECS With DC-Link Current Estimation," *IEEE Trans. Industrial Electronics*, vol.58, no.4, pp.1110,1119, April 2011.

[25] (Product Catalogue Style) *Capacitors, Arrestors and Harmonic Filters*, General Electric, Bloomington, IL, 2014. Available online at: http://www.geindustrial.com/catalog/buylog/24_BuyLog2013_Capacitrs ArrestrsHarmonFiltrs.pdf?omni_key=PG-BL24

[26] Gilles Terzulli, "Film Technology to replace Electrolytic Technology in Wind Power Applications," *AVX, A Kyocera Group Company*. Available online at: http://web.arrownac.com/sites/default/files/pdfs/FilminWindPower.pdf

[27] "Life-Limiting factors in Electrolytic Capacitors," *EVOX RIFA PASSIVE COMPONENTS*, 2001. Available online at: http://www.efo-power.ru/pub/power/Capacitors/articles/kak_otsenit_srok_cond.pdf

[28] "Long Life Technology for Aluminium non-solid Electrolytic Capacitors," *Dempa Shimbun. High Technology*. Available online at: http://www.rubycon.co.jp/en/products/topics/img/t003_04.pdf

[29] Mirafzal, B., "Survey of Fault-Tolerance Techniques for Three-Phase Voltage Source Inverters," in *Industrial Electronics, IEEE Transactions on* , vol.61, no.10, pp.5192-5202, Oct. 2014

[30] Report on the investigation of the catastrophic failure of a capacitor in the aft harmonic filter room on board RMS Queen Mary 2. Dec. 2011. Available online at: https://assets.digital.cabinet-office.gov.uk/media/547c6fa6ed915d4c10000031/QM2Report.pdf

[31] Fuchs, F.W., "Some diagnosis methods for voltage source inverters in variable speed drives with induction machines - a survey," in *Industrial Electronics Society, 2003. IECON '03. The 29th Annual Conference of the IEEE* , vol.2, no., pp.1378-1385 Vol.2, 2-6 Nov. 2003.

[32] Bolognani, S.; Zordan, M.; Zigliotto, M., "Experimental fault-tolerant control of a PMSM drive," in *Industrial Electronics, IEEE Transactions on* , vol.47, no.5, pp.1134-1141, Oct. 2000.

[33] B.A. Welchko, T.A. Lipo, T.M. Jahns, and S.E. Schulz, "Fault tolerant three-phase AC motor drive topologies: a comparison of features, cost, and limitations," *IEEE Trans. Power Electron.*, vol.19, no.4, pp.1108-1116, July 2004.

[34] T. Hoevenaars, I. Evans, and A. Lawson, "New marine harmonic standards," *IEEE Ind. Appl. Mag.*, vol.16,no.1, pp.16-25, Jan.-Feb. 2011.

Active Common-Mode Voltage Reduction in a Fault-Tolerant Three-Phase Inverter

Danyal Mohammadi, Said Ahmed-Zaid
danyalmohammadi@u.boisestate.edu
Boise State University
Boise, ID, USA

Abstract—**A fault-tolerant topology in a three-phase four-leg inverter which is capable of reducing the common-mode voltage (CMV) during the post-fault condition is presented. The CMV during both post-fault and pre-fault is investigated. This paper proposes a topology to reduce the common-mode voltage during pre- and post-fault operation of the inverter by using the healthy switches. The accompanying simulation results verify the common-mode current reduction during the fault period.**

I. INTRODUCTION

The reliability and quality of power electronic systems are of great interest to the electric drive community. Having a healthy and continuous operation of a power electronic system such as an adjustable speed AC motor drive, unified power quality correction, or a hybrid electric vehicle is becoming more important. The operation of three-phase inverters in adjustable speed AC motor drives has a direct impact on the driven electric machines. The dynamic performance of electrical machines has been improved with the emergence of fast switching semiconductor devices. However, the switching patterns introduce a common-mode voltage (CMV) which causes conducted and radiated electromagnetic interference (EMI) emissions [1] and bearing breakdowns [2]. Moreover, the switching semiconductor devices in power electronics are also prone to failure. The main causes for faults in power electronics systems are semiconductor switch and capacitor failures. A fault-tolerant system has remedial topologies for different type of faults such as a switch short circuiting, a switch unable to open, a permanently open switch, or a failure of the gate driver circuit of the switch. In a fault-tolerant topology, any remedial solution, for either a short-circuit or an open-circuit fault adds more components to the power electronic system. In the case of both faults, the number of extra components required for the remedial topology increases substantially [3]. This paper reviews the current state-of-the-art of fault-tolerant strategies for reliability concerns as well as common-mode voltage reduction methods for quality issues.

II. COMMON-MODE VOLTAGE REDUCTION METHODS

One of the main quality concerns of three-phase inverters is reducing the CMV to avoid EMI pollution and bearing breakdowns. Common-mode voltage (CMV) is defined as the instantaneous sum of the voltages from the stator winding phases to the midpoint of the DC link. Many studies have

been proposed to reduce CMV [4]–[14]. They can be categorized as utilizing common-mode choke/transformers, cascaded inverters [4], filters [5], [6], active cancellation using an extra leg [7], [8], and software approaches [9]–[14].

Non-software methods [4]–[8] reduce the CMV, but they increase the cost and complexity of the system. Software methods such as modulation techniques requires less complexity and change in the system [9]–[14]. A proper modulation technique reduces CMV to a limited voltage. Three modulation techniques were proposed by [9]. Only Near State Pulse Width Modulation (NSPWM3) is suitable for practical implementation. Active zero-states are used to create a zero voltage as shown in Figure 1(a). This modulation uses only non-zero states in order to limit the CMV to $\pm \frac{V_{dc}}{6}$ as shown in Fig. 1(b). The same modulation technique is called Active Zero-State Pulse Width Modulation (AZSPWM1) in [12] where a comparison among the three methods, especially in regard to modulation index range, has been reported. Some modified versions of these techniques have been reported in [10] which focus on the mitigation of the dead-time effect. The reduced

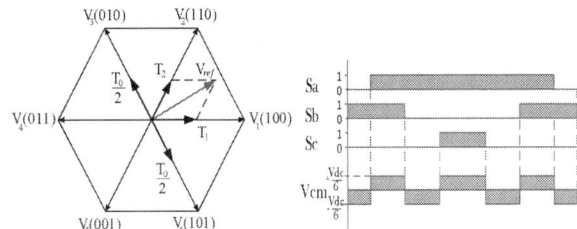

(a) Active Zero-State Pulse Width Modulation

(b) Common-Mode Voltage and Switches Status

Fig. 1. AZSPWM Modulation in Sector I [8]

CMV can be eliminated completely in combination with filters as it has been proposed in [8]. In this paper, a modified version of the proposed method in [8] is implemented for CMV reduction as shown in Fig 2(a). This method requires the satisfaction of the following equation (1) during each sampling period.

$$V_1 + V_2 + V_3 + V_4 = 0 \qquad (1)$$

where V_i and $i = 1, ..4$ are the pole voltages of the each leg. This equation cannot be satisfied using the conventional space modulation where the zero vector voltage is used. NSPWM3

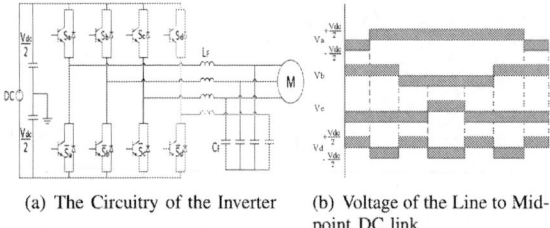

(a) The Circuitry of the Inverter (b) Voltage of the Line to Mid-point DC link

Fig. 2. Common-Mode Voltage Reduction Proposed in [8]

(or AZSPWM) is used to satisfy the equation (1). The statuses of the gates S_d and $\bar{S}_d$ as shown in Fig. 2(b) are used to satisfy this equation. They can be calculated using the following equation,

$$S_d = S_a \oplus S_b \oplus S_c \qquad (2)$$

where $\oplus$ represents the xor operation and also S_d and $\bar{S}_d$ is complementary. The maximum CMVR is possible if both switches are working. The top switch, S_d applies $+\frac{V_{dc}}{2}$ and $-\frac{V_{dc}}{2}$ by the bottom switch to help balance the phase voltage at the fundamental frequency. Filters are used to ensure that the sum of each phase voltage will tend to zero and filter any high switching frequency. Figure 3 shows the magnitude and the harmonic analysis of the CMV for three-leg and four-leg inverters.

III. FAULT-TOLERANT SYSTEMS

As fault-tolerant systems are increasingly being investigated and implemented, many studies [15], [16] reported detection methods for diagnosing a fault in the system. These methods identify and localize one or multiple faulted switches in a three-phase voltage source inverter. Many remedial strategies have been presented that improve the reliability of the system after the detection of a fault [17]–[19]. Isolating a leg and connecting the neutral of the three-phase winding to midpoint of the DC link has been proposed in [17]. One-phase and two-phase operation of induction machine can be done in these topologies. This method can produce the same electromagnetic torque which requires increasing both the DC link voltage and the switch ratings by $\sqrt{3}$. In this topology and its variants studied in [18], [19], access to the neutral of the motor stator winding and midpoint DC link is required. Another remedial topology is to use an extra leg which can be used during the post-fault period. Employment of an extra leg in three-phase inverters can provide full-load operation during the post-fault period without increasing the DC link voltage or requiring access to the midpoint of the DC link. An extra leg can also help reduce the CMV as mentioned in the previous section. A thorough output capacity for inverter faults for different fault-tolerant topologies is studied in [19]. The only topology that is able to isolate a single switch from the circuit is the topology proposed in [20]. In this method, a faulted switch can be isolated by firing its corresponding thyristor.

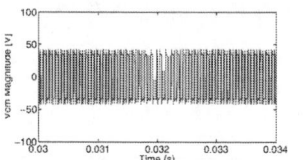 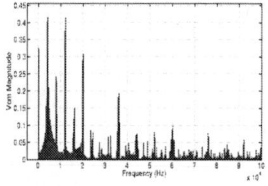

(a) Three-Leg Topology Magnitude (b) Three-Leg Normalized Harmonics

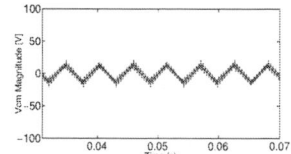 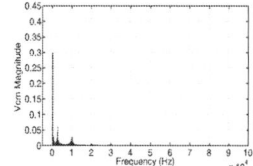

(c) Four-Leg Topology Magnitude (d) Four-Leg Normalized Harmonics

Fig. 3. Common-Mode Voltage Using AZSPWM in Different Topologies

IV. PROPOSED REMEDIAL STRATEGY

In this paper, a fault-tolerant topology [20] is combined with a common-mode voltage reduction topology [8] in such a way that CMVR can be done even if there is a faulted switch in the four-leg inverter or a fault in the motor phases. The priority of choosing a fault-tolerant system is to have a system cover all four fault types which include a short-circuited switch (short-switch), an open-circuited switch (open-switch), a short-circuited phase (short-phase), and an open-circuited phase (open-phase). The proposed topology is shown in Fig 4. This scheme may be best understood by discussing the modes of operation in the following subsections.

A. Pre-Fault Operation Mode

Figure 4 shows components of this topology consist of switches, filters, fault thyristors, and load. When the system is healthy, the fault thyristors, T_a+, T_b+, and T_c+ are on, and thyristors T_a-, T_b-, and T_c- are off. This allows the three-phase inverter to operate normally. During normal operation, the three legs, a, b and c, are supplying the load and the fourth leg, d, is reducing the common-mode current substantially as shown in Fig. 3(c). In this mode, CMV and the harmonic content of CMV are greatly reduced as shown in Fig. 3(c) and 3(d).

B. Reconfiguration Mode

In this mode, the circuit is reconfigured so that the inverter can continue supplying three-phase voltages to the load with minimum interruption. To avoid any damage and for safety reasons, all switches are turned off when a fault is detected. As an example, switch S_a is isolated if there is a failure in that device by firing the SCRa-. This will cause the fuse Fa+ to burn out. After isolation of the faulted switch in phase a, the fault thyristors shown in Fig. 4, associated with the faulted phase, need to be reconfigured. In this example, fault

978-1-4673-9551-9/16 $31.00 © 2016 IEEE 2822

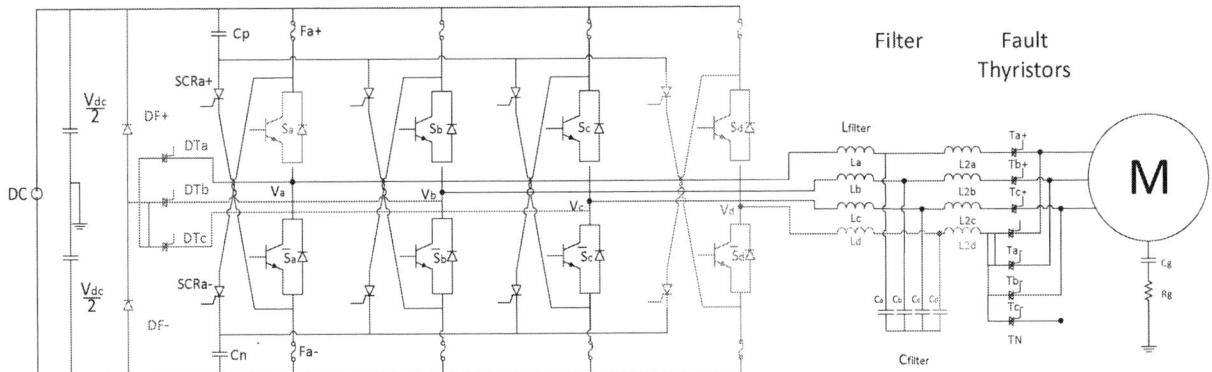

Fig. 4. Proposed Topology

thyristor T_a+ will be turned off and T_a- will be turned on. Figure 5(b) shows a post-fault circuit of short-switch fault after reconfiguration mode.

C. Post-Fault Operation Mode

After reconfiguring the fault thyristors, the fourth leg has been replaced with the faulty leg and the gating signals of the phase a will now control the switches in leg d. This allows full operation of the inverter. Using this scheme, C_a, L_a, and $\bar{S}_a$ can now be utilized in order to reduce the CMV. By switching $\bar{S}_a$, there will be some insignificant currents in this leg (at point Va). Both positive and negative currents can flow through diodes DF+ and DF-. These diodes are involved in the circuitry by firing one of the thyristors DTa, DTb, or DTc in the reconfiguration mode. The power ratings of the diodes are not high because of the low current magnitudes as shown in the next section. The increase in cost by adding these diodes will be insignificant.

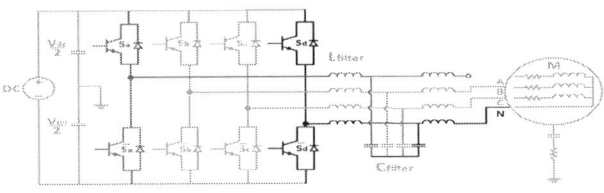

(a) Post-Fault circuit for phase-open (two-phase operation)

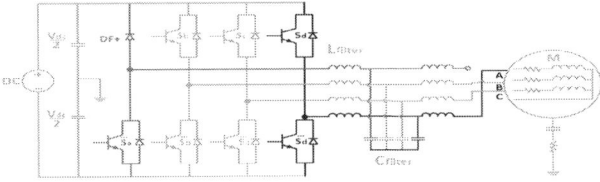

(b) Post-Fault circuit for switch-short

Fig. 5. Post-Fault Circuits

D. Filter Design

Filters are used to reduce the current distortion injected to a load or to the grid utility in case of grid-connected systems.

An LCL filter is used in this design to attenuate higher order harmonics and to reduce EMI noise. Thorough analysis of LCL filter designs are studied in [21]. In this design topology, considering allowable current ripple of 14%, L_{1a-d} can be found as given

$$L_{a-d} = \frac{V_{dc}}{6f_s(14\%)I_{max}} \quad (3)$$

where I_{max} is the rated current and f_s is the switching frequency. The resonance frequency calculated is within the constraint of $10f < f_{res} < 0.5f_s$ and the resonant frequency can be calculated using the following equation

$$2\pi f_{res} = \sqrt{\frac{L_{a-d} + L_{2a-d}}{L_{a-d}L_{2a-d}C_{a-d}}} \quad (4)$$

where $L_{2a-d} = 0.4L_{a-d}$. The obtained values using $I_{max} = 100A$ are given in Table I.

V. SIMULATION RESULTS

A three-phase load has been simulated using three parallel resistors and inductances. Figure 6 shows simulation of DC link voltage, CMV, common-mode current (CMC), and current responses to short-switch, open-switch, and open-phase failures. Table I provides the parameters used for these simulations. Typical detection time for an open-circuit fault is in order of milliseconds and that of a short-circuit fault is in order of microseconds. Since the purpose of this paper is not fault diagnosis, the detection time of 3 ms for an open-circuit failure and 3 μs for a short-circuit failure are chosen. In all of the case studies, the faults are cleared at $t = 0.04\ s$. In the following subsections, faults in phase and faults in switch are presented, respectively.

A. Phase Faults

In case of an open-phase or short-phase fault in phase a of motor, thyristor T_a+ will be off so that the faulted phase is isolated as shown in Fig 5(a). Two-phase and single-phase operations are the two possible remedial topologies when a

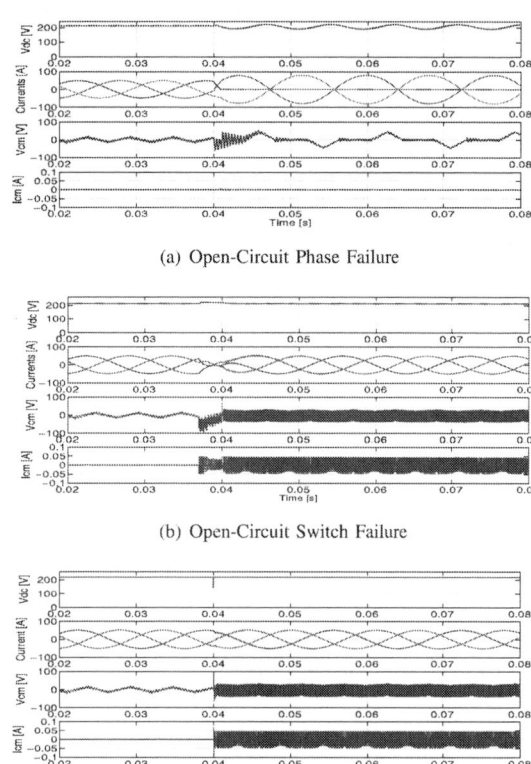

(a) Open-Circuit Phase Failure

(b) Open-Circuit Switch Failure

(c) Short-Circuit Switch Failure

Fig. 6. Simulation of the Fault-Tolerant System Under Open-Phase, Open-Switch, and Short-Switch Failure

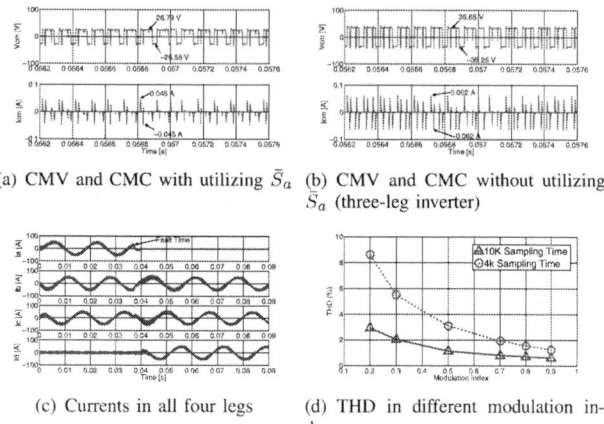

(a) CMV and CMC with utilizing $\bar{S}_a$ (b) CMV and CMC without utilizing $\bar{S}_a$ (three-leg inverter)

(c) Currents in all four legs

(d) THD in different modulation indexes

Fig. 7. Open-Switch post-fault simulation of CMV, Current, and THD in different topologies

TABLE I
SIMULATION PARAMETERS AND VALUES FOR FAULT AND CMV

Parameters	Value	Parameters	Value
C_{dc}	2×1 mF	Switching Frequency (f_s)	10 kHz
L_{a-d}	0.25 mH	C_g	3 nF
$L2_{a-d}$	0.1 mH	Load Inductance	1.04 mH
C_{a-d}	15 μF	Load Resistance	0.81 Ω
V_{dc}	210 V	R_g	900 Ω

be the same as $\bar{S}_d$ during pre-fault. In Fig 7(c), the magnitude of the current in each leg is shown. Figure 7(d) shows THD of the load current at different modulation indexes.

VI. CONCLUSION

The proposed fault-tolerant topology is able to clear open-switch, short-switch, short-phase, and open-phase faults. The common-mode voltage and current are minimum during pre-fault. Using this topology, common-mode voltage and current are eliminated during post-fault operation of the inverter for open-phase and short-phase faults and reduced for open-siwtch and short-switch faults.

REFERENCES

[1] H. Akagi and T. Shimizu, "Attenuation of conducted EMI emissions from an inverter-driven motor," *Power Electronics, IEEE Transactions on*, vol. 23, no. 1, pp. 282–290, Jan 2008.

[2] D. Busse, J. Erdman, R. Kerkman, D. Schlegel, and G. Skibinski, "Bearing currents and their relationship to PWM drives," *Power Electronics, IEEE Transactions on*, vol. 12, no. 2, pp. 243–252, Mar 1997.

[3] Y. Song and B. Wang, "Survey on reliability of power electronic systems," *Power Electronics, IEEE Transactions on*, vol. 28, no. 1, pp. 591–604, Jan 2013.

[4] Z. Yang, X. Xu, S. Xu, and J. Zhao, "Research on eliminating common-mode voltage of cascaded high-voltage inverter based on svm," in *Power and Energy Engineering Conference (APPEEC), 2010 Asia-Pacific*, March 2010, pp. 1–4.

[5] M. Swamy, K. Yamada, and T. Kume, "Common mode current attenuation techniques for use with PWM drives," *Power Electronics, IEEE Transactions on*, vol. 16, no. 2, pp. 248–255, Mar 2001.

[6] X. Dianguo, G. Qiang, and W. Wei, "Design of a passive filter to reduce common-mode and differential-mode voltage generated by voltage-source PWM inverter," in *IEEE Industrial Electronics, IECON 2006 - 32nd Annual Conference on*, Nov 2006, pp. 2483–2487.

fault occurs in one the phases in a three-phase motor. Two-Phase operation requires connecting neutral of the motor to an extra leg which can be done by thyristor T_N. In this type of operation, CMV does not exist. However, in one-phase operation, when the neutral of the motor is not available, the CMV has one of the values of 0, $+\frac{V_{dc}}{2}$, or $-\frac{V_{dc}}{2}$ which can be eliminated by a proper switching of leg a. The gating signals of switches in leg a can remain the same as in pre-fault period. However, during this fault, changing the signals S_a to $\bar{S}_b$ and S_d to $\bar{S}_c$ yields to a lower switching frequency of the switches. The CMV during post-fault remains zero as shown in Fig 6(a).

B. Switch Faults

Once the fault has been diagnosed, the faulty switch is isolated from the faulted leg and the healthy switch in each faulted leg can still be used to reduce the common-mode voltage as shown in Fig. 5(b). The magnitude and number of spikes in CMC has been reduced due to the reduced magnitude and the number of dv/dt in CMV. The CMV and CMC are reduced by about 25% in magnitude and around 50% reduction in the number of dv/dt caused by CMV compared to the three-leg topology. The effect of using the healthy switch, in this example $\bar{S}_a$, on CMV and CMC can be seen in the detailed Fig. 7(a) and 7(b). The gating signal of switch $\bar{S}_a$ now will

[7] Z. Liu, J. Liu, and J. Li, "Modeling, analysis, and mitigation of load neutral point voltage for three-phase four-leg inverter," *Industrial Electronics, IEEE Transactions on*, vol. 60, no. 5, pp. 2010–2021, May 2013.

[8] A. Julian, G. Oriti, and T. Lipo, "Elimination of common-mode voltage in three-phase sinusoidal power converters," *Power Electronics, IEEE Transactions on*, vol. 14, no. 5, pp. 982–989, Sep 1999.

[9] Y.-S. Lai, "Investigations into the effects of pwm techniques on common mode voltage for inverter-controlled induction motor drives," in *Power Engineering Society 1999 Winter Meeting, IEEE*, vol. 1, Jan 1999, pp. 35–40 vol.1.

[10] R. Tallam, R. Kerkman, D. Leggate, and R. Lukaszewski, "Common-mode voltage reduction PWM algorithm for AC drives," *Industry Applications, IEEE Transactions on*, vol. 46, no. 5, pp. 1959–1969, Sept 2010.

[11] Y.-S. Lai and F.-S. Shyu, "Optimal common-mode voltage reduction PWM technique for inverter control with consideration of the dead-time effects-part I: basic development," *Industry Applications, IEEE Transactions on*, vol. 40, no. 6, pp. 1605–1612, Nov 2004.

[12] A. Hava and E. Un, "A high-performance PWM algorithm for common-mode voltage reduction in three-phase voltage source inverters," *Power Electronics, IEEE Transactions on*, vol. 26, no. 7, pp. 1998–2008, July 2011.

[13] K. Li, T. Lu, Z. Zhao, L. Yin, F. Liu, and L. Yuan, "Carrier based implementation of reduced common mode voltage PWM strategies," in *ECCE Asia Downunder (ECCE Asia), 2013 IEEE*, June 2013, pp. 578–584.

[14] Y. Zhang, G. Kuang, and L. Long, "Research on reduced common-mode voltage nonzero vector pulse width modulation technique for three-phase inverters," in *Power Elect. and Motion Control Conference (IPEMC), 2012 7th Int.*, vol. 4, June 2012.

[15] M. Trabelsi, M. Boussak, and M. Gossa, "Multiple IGBTs open circuit faults diagnosis in voltage source inverter fed induction motor using modified slope method," in *Electrical Machines (ICEM), 2010 XIX International Conference on*, Sept 2010, pp. 1–6.

[16] T. Orlowska-Kowalska and P. Sobanski, "Simple sensorless diagnosis method for open-switch faults in SVM-VSI-fed induction motor drive," in *Industrial Electronics Society, IECON 2013 - 39th Annual Conference of the IEEE*, Nov 2013, pp. 8210–8215.

[17] T. Elch-Heb and J. Hautier, "Remedial strategy for inverter-induction machine system faults using two-phase operation," in *Power Electronics and Applications, 1993., Fifth European Conference on*, Sep 1993.

[18] B. MIRAFZAL, "Survey of fault-tolerance techniques for three-phase voltage source inverters," *Industrial Electronics, IEEE Transactions on*, vol. 61, no. 10, pp. 5192–5202, Oct 2014.

[19] B. Welchko, T. Lipo, T. Jahns, and S. Schulz, "Fault tolerant three-phase AC motor drive topologies: a comparison of features, cost, and limitations," *Power Electronics, IEEE Transactions on*, vol. 19, no. 4, pp. 1108–1116, July 2004.

[20] S. Bolognani, M. Zordan, and M. Zigliotto, "Experimental fault-tolerant control of a PMSM drive," *Industrial Electronics, IEEE Transactions on*, vol. 47, no. 5, pp. 1134–1141, Oct 2000.

[21] W. Cao, K. Liu, Y. Ji, Y. Wang, and J. Zhao, "Design of a four-branch lcl-type grid-connecting interface for a three-phase, four-leg active power filter," *Energies*, vol. 8, no. 3, pp. 1606–1627, 2015. [Online]. Available: http://www.mdpi.com/1996-1073/8/3/1606

Power Cycling Lifetime Improvement of Three-Level NPC Inverters with an Improved DPWM Method

Jiangbiao He[*], Lixiang Wei[**], Nabeel A.O. Demerdash[*]

[*]Department of Electrical and Computer Engineering
Marquette University
Milwaukee, Wisconsin, 53233
jiangbiao@ieee.org

[**]Standard Drives Business
Rockwell Automation
Mequon, Wisconsin, 53092
lwei@ra.rockwell.com

Abstract—This paper investigates the power cycling lifetime of a three-phase three-level neutral-point-clamped (NPC) inverter used in a 50-kVA adjustable speed drive (ASD). Considering the short lifetime of NPC inverters under low output frequency conditions, an improved discontinuous pulse width modulation (DPWM) method is introduced to extend the inverter lifetime. This improved DPWM method is achieved by injecting a zero-sequence signal with relative high frequency into the voltage reference signals, which is in order to obtain lower swing value of the junction temperatures in the Insulated Gate Bipolar Transistors (IGBTs). Both simulation and experimental results are presented to verify the effectiveness of this solution.

Keywords—Neutral-point clamped inverters; power cycling lifetime; discontinuous PWM method; zero-sequence signal injection, swing of junction temperature.

I. INTRODUCTION

Multilevel inverters have been widely applied in low-voltage and medium-voltage energy conversion equipment, such as adjustable speed drives (ASDs) [1], renewable energy generation [2][3], uninterruptable power supplies (UPSs) [4], and the like. Compared with conventional two-level power inverters, multilevel inverters have numerous advantages including lower input and output harmonic distortion, lower change rate of output voltage (dv/dt), lower common mode voltage, and higher dc-bus voltage withstanding capabilities.

However, like most solid-state power inverters, multilevel inverters also suffer from a sharp decline of power device lifetime when operating under the condition of low output frequencies and high output current. Such phenomenon is due to the mismatch of the Coefficient of Thermal Expansion (CTE) between different materials in the power devices, such as aluminum bond wires, silicon chips, and alloying solder inside the device packages [5-7]. The mismatch of the CTEs in combination of the large swing of the device junction temperatures will impose significant thermal-mechanical stress on the bond wires and soldering joints, which will eventually lead to device wear-out failures. For the utilization of multilevel inverters in adjustable speed drives (ASD), such operating conditions refer to low-speed high-torque regions, which is a typical operating mode for ASDs used in EVs/HEVs, elevators, machine tools, variable-speed wind turbines, and the like. Therefore, it is of paramount importance to introduce effective

and practical solutions to improve the lifetime of power devices used in multilevel power converters.

The power cycling lifetime of an IGBT device can be described by a well-known empirical model proposed in [8], namely, the so-called "Coffin-Manson-Arrhenius" model, and its mathematical expression is given as follows [8]:

$$N_f = A \cdot \Delta T_j^{-\alpha} \cdot \exp[Q/R \cdot T_m] \qquad (1)$$

where, N_f is the number of cycles to failure. A, α, Q and R are all module-dependent positive constants; ΔT_j and T_m are the swing value and mean value of device junction temperatures in one power cycle, as illustrated in Fig. 1. Most semiconductor device manufactures follow the Coffin-Manson-Arrhenius model for predicting the remaining lifetime of power converters and hence the corresponding constants are generally readily available. Also, it can be seen from Equation (1) that the lifetime of IGBT devices is mainly determined by ΔT_j and T_m. This indicates that the lifetime of IGBT inverter can be improved by reducing the value of ΔT_j or T_m through control or modulation techniques.

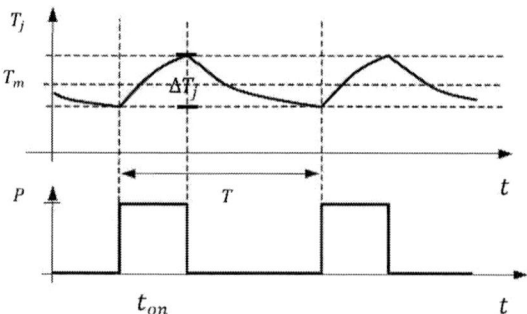

Fig. 1 IGBT Junction Temperature profile for periodic load.

In the literature, a number of approaches have been proposed to extend the lifetime of IGBT power converters [9-17]. All these approaches can be classified into two categories, namely, hardware-based device packaging approaches and firmware-based control/modulation methods. The methods fall into the first category are generally achieved by selecting materials with better thermal and mechanical matching/compatibility as well as

978-1-4673-9551-9/16 $31.00 © 2016 IEEE

comparative CTEs during the packaging process of the IGBTs. For instance, replacing the silicon (Si) chips with silicon carbide (SiC) material [9], replacing aluminum (Al) bond wire with copper (Cu) material [10], or replacing the "Al+AlN" substrate material with "Cu+Si₃N₄" [11], and so forth. However, the adoption of these proposed material substitutes such as SiC, Cu and "Cu+ Si₃N₄" in the IGBT packaging generally involves several factors to be concurrently considered, such as thermal conductivity, convenience of processing, and material cost, which is quite challenging to make a compromise in practice. Therefore, lifetime extension methods that are implemented through control or modulation firmware are more preferred.

Among these firmware-based existing approaches, either the switching frequencies, load currents or output voltages of the power inverters are regulated to reduce the dissipated device losses and subsequently obtain longer lifetime of such IGBT power converters. In [12], a PWM frequency hysteresis control approach was proposed, in which the PWM switching frequency was reduced to its lowest level when the swing magnitudes of the IGBT junction temperatures are higher than their preset upper limit. In other words, under normal operation, if the swing magnitudes of the IGBT junction temperatures are lower than the preset upper limit, all the control and PWM strategies will remain in the normal mode. As a result, the overall lifetime of the IGBT inverter was significantly improved. However, one drawback of this method is the dramatic increase of the harmonic distortion in the output when the PWM switching frequency is reduced. Such deterioration in harmonic distortions may not be acceptable in certain applications requiring high-precision motion control and low acoustic noise, unless the output filter of the inverter is overdesigned.

Similarly, another method based on the manipulation of switching frequencies and load currents to adjust the IGBT losses was introduced in [13]. Two proportional-integral (PI) regulators were utilized to reduce the switching frequencies and load currents, respectively, when the IGBT junction temperature was approaching its upper limit of 110°C [13]. The inputs for these PI regulators are the actual switching frequencies or load currents, and the output is the estimated IGBT junction temperatures, which serve as the feedback variables for the PI regulators.

In addition, novel PWM methods were investigated in the literature to reduce the conduction or switching losses in the IGBT modules. Once the device losses are reduced, this results in lower IGBT junction temperatures, and subsequently the IGBT lifetime would be extended. In [14], a discontinuous PWM (DPWM) method and a space vector PWM (SVPWM) method were selectively utilized during the converter modulation, namely, the so-called "hybrid modulation method". Such a modulation method was developed for improving the lifetime of the generator-side converter in a wind turbine generation system [14]. Specifically, when the wind speed is below certain threshold value, the conventional SVPWM method will be employed. Once the wind speed is higher than the threshold reference value, the DPWM method will be utilized to mitigate the thermal stress on the associated IGBTs. This DPWM method has been well-known for the reduction of IGBT switching losses, although the harmonic distortions generated under certain modulation indices are generally higher

than those yielded under the SVPWM method. By selectively using the DPWM and SVPWM methods, a longer IGBT lifetime can be achieved, although the harmonic distortion in the converter outputs will be inevitably increased.

Another DPWM method was proposed to improve the IGBT lifetime in a two-level voltage source inverter under low output frequency conditions [15]. In this new DPWM method, a zero-sequence signal with frequency higher than the time constants of IGBT junction-to-case thermal impedance was injected into the voltage reference signals. As a result, lower IGBT junction temperatures were achieved and correspondingly the lifetime of the IGBT inverter was extended. According to the simulation and experimental results given in [15], a reduction of 15% of the IGBT junction temperatures were achieved for a 480V/65A ASD system, while the harmonic distortions in the output voltages and currents almost stay the same, compared to the results obtained under the conventional SVPWM modulation.

As for multilevel neutral-point-clamped (NPC) inverters which are being investigated in this paper, existing solutions in the literature to improve their IGBT lifetime mainly focused on the utilization of redundant voltage space vectors to actively redistribute the losses from the overloaded devices to other cooler devices [16][17]. In [16][17], redistribution of the semiconductor losses in a three-level NPC inverter by taking advantage of the redundant zero voltage vectors and small voltage vectors were carried out for both low and high modulation indices, respectively. A trade-off between the loss redistribution and the neutral-point voltage control freedom was considered in these investigations for purposes of ensuring the proper operation of the control system for such NPC inverters.

In this paper, the DPWM method proposed in [15] will be further extended for improving the power cycling lifetime of a three-phase three-level NPC inverter, which is one of the most widely used multilevel converter topology in industrial applications. The circuit topology of the three-level NPC inverter is shown in Fig. 2. The characteristics and working principle of the NPC inverter was detailed in [18], and therefore will not be repeated here. The remainder content of this paper is organized as follows. In Section II, the improved DPWM method for three-level NPC inverter will be introduced. In section III, thermal modeling of the IGBT-based NPC inverter and the related simulation will be carried out in PLECS software environment, and the simulation results will be given to confirm the efficacy of this improved DPWM method. In section IV, experimental results acquired from a 50-kVA experimental ASD prototype based on the NPC inverter topology will be presented to verify the performance of this improved DPWM method. Finally, conclusions are drawn in Section V.

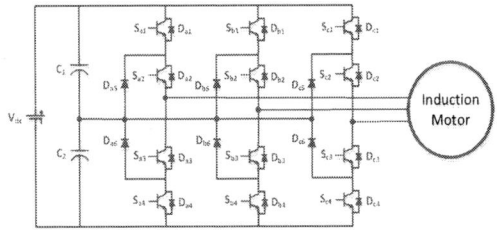

Fig. 2 Circuit topology of a three-phase three-level NPC inverter.

II. THE PROPOSED DPWM METHOD

One important factor contributing the short lifetime of power inverters at low output frequency lies in the fact that the thermal time constant of semiconductor devices is much lower than the inverter output fundamental period. Thus, the lifetime of IGBT devices in NPC inverters could be improved upon injecting a zero sequence signal with higher frequency than the thermal time constant of semiconductor devices. Here, an improved DPWM method introduced in [15] is extended to multilevel NPC inverter to reduce the variation of the IGBT junction temperatures at low-frequency operation. Details of this novel PWM scheme and its relationship to the conventional SVPWM are given next.

Assuming the duty ratio for each phase of an NPC inverter under carrier-based SPWM can be written as follows:

$$\begin{cases} d_u = m_a \cos(\theta) \\ d_v = m_a \cos(\theta - 2\pi/3) \\ d_w = m_a \cos(\theta - 4\pi/3) \end{cases} \quad (2)$$

where, m_a, is the amplitude modulation index. θ is the initial phase angle. d_u, d_v, d_w are the duty ratios for the three phases. It follows that the instantaneous maximum and minimum duty ratio will be:

$$d_{max} = \mathrm{MAX}(d_u, d_v, d_w) \quad (3)$$

$$d_{min} = \mathrm{MIN}(d_u, d_v, d_w) \quad (4)$$

In conventional SVPWM, the injected zero-sequence signal is defined as:

$$d_0 = -(d_{max} + d_{min})/2 \quad (5)$$

With the injection of such zero-sequence signal, the duty ratio for each phase of an NPC inverter under the SVPWM method can be written as:

$$\begin{cases} d_{u,SV} = m_a \cos(\theta) + d_0 \\ d_{v,SV} = m_a \cos(\theta - 2\pi/3) + d_0 \\ d_{w,SV} = m_a \cos(\theta - 4\pi/3) + d_0 \end{cases} \quad (6)$$

However, to mitigate the swing of the IGBT junction temperature, a zero-sequence signal with a fundamental frequency higher than that of the IGBT junction-to-case time constants are defined as follows:

$$d_{zss} = \begin{cases} 1 - d_{max}, & \mathrm{SIN}(2\pi f_{cm} t) \geq 0 \\ -1 - d_{min}, & \mathrm{SIN}(2\pi f_{cm} t) < 0 \end{cases} \quad (7)$$

where, f_{cm} is the fundamental frequency of the injected zero-sequence signal.

Therefore, the duty ratio for each phase under this novel discontinuous PWM method, namely, NDPWM, will be:

$$\begin{cases} d_{u,NDPWM} = m_a \cos(\theta) + d_{zss} \\ d_{v,NDPWM} = m_a \cos(\theta - 2\pi/3) + d_{zss} \\ d_{w,NDPWM} = m_a \cos(\theta - 4\pi/3) + d_{zss} \end{cases} \quad (8)$$

A graphic illustration of the SVPWM and NDPWM methods is shown in Fig. 3 (a) and (b), respectively, in which the red and green triangular signals are the upper and lower carrier signals,

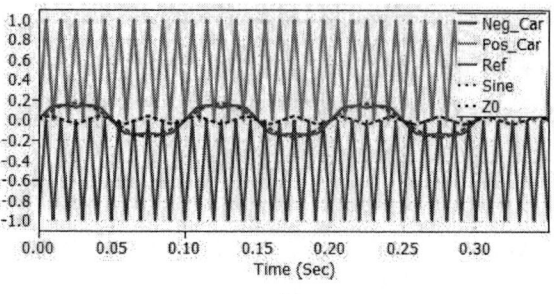

(a)

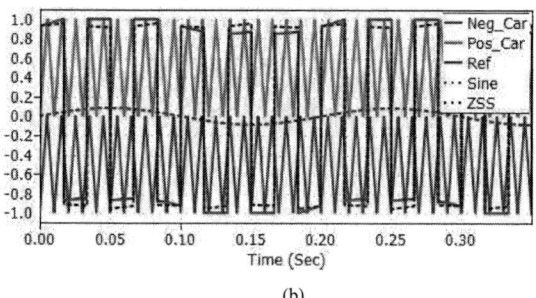

(b)

Fig. 3 PWM methods for the three-level NPC inverter at low frequency (a) SVPWM (b) The improved DPWM.

respectively. The purple sinusoidal signal is the voltage reference signal, and the blue signals in both figures are the duty ratios for one of the three phase legs in the NPC inverter. It can be seen from Fig. 3(b) that the voltage reference signal in this NDPWM method exhibits a sinusoidal envelop in the waveform profile but with higher frequency. In the following section, the effectiveness of this NDPWM method in reducing the IGBT junction temperatures will be elaborated through simulation results.

III. SIMULATION RESULTS

In the PLECS simulation software, both the SVPWM and NDPWM methods for a 50 kVA NPC-inverter-based ASD are implemented. In all the simulations, the input power supply is a three-phase 480V/60Hz source. The switching frequency in both the SVPWM and NDPWM methods is set at 4 kHz. The initial ambient temperature for all device thermal models is assumed to be 40°C. The output frequency of the ASD is regulated ranging from 2 Hz to 60 Hz under open-loop constant volts/Hertz (V/Hz) scalar control. All the parameters for the subject ASD investigated in the simulation are summarized in Table 1.

TABLE 1 SPECIFICATIONS OF THE 50 KVA ASD BASED ON A THREE-PHASE THREE-LEVEL NPC INVERTER.

Control mode	Open-loop scalar control (V/Hz)		
Input	Line-to-line voltage	480	V$_{RMS}$, three phase
	Frequency	60	Hz
	Switching frequency	4	kHz
	DC bus voltage	650	V$_{dc}$, max
	Initial ambient temp.	40	°C
Output	Rated power	50	kVA
	Overload Current	180	A$_{RMS}$, max
	Line-to-line Voltage	480	V$_{RMS}$, three phase

For each specific output frequency, the device losses and the related swing magnitudes of the junction temperatures of each device are accordingly obtained from the simulation. The junction temperature profiles of each device in the NPC inverter under the NDPWM method are compared to these under the SVPWM method at low frequencies. These profiles are given in Fig. 4 through Fig. 6. It should be noted that, considering the symmetries of the NPC inverter topology, only half of the semiconductor devices in one phase leg are investigated (S_{x1}, S_{x2}, D_{x1}, D_{x2}, and D_{x5}, $x\equiv$a, b, or c, as shown in Fig. 2), and the other half of the complimentary devices have the identical thermal performance.

Since the lifetime of the NPC inverter is mainly determined by the most thermally-stressed component, which in this case happens to be the clamping diodes at low output frequency operation. Accordingly, the number of cycles to failure, N_f, can be computed based on substituting the ΔT_j and T_m of the clamping diodes into the Coffin-Manson-Arrhenius model given in Equation (1). According to the power cycling data provided by the manufacturer [19], the IGBT module constants in Equation (1) are obtained through curve fitting techniques. In this work, the values of these constants used in this model are listed as follows: A=$3.3125*10^6$, $\alpha = 5.039$, Q=9.89×10^{-20}, and R=1.38066×10^{-23} [19]. Thus, the values of N_f for the NPC inverter can be calculated at various frequencies, which are depicted in Fig. 7. As can be seen in Fig. 7, the number of cycles to failure, N_f, decreases dramatically at low output frequencies under the SVPWM method. For instance, the value of N_f is reduced to 200,000 cycles at 2 Hz, which may result in a remaining useful lifetime (RUL) of just a few days if one is to keep operating the NPC inverter at such a lower frequency or near dc condition. However, with the implementation of the NDPWM method, the number of cycles to failure, N_f, is significantly improved at both the low and the high output frequency operations, as again demonstrated in Fig. 7. As can be seen, the value of N_f is improved up to 1,130,200 cycles by using NDPWM at the 2 Hz of low-frequency. This is almost 6 times higher than the value of N_f of the NPC inverter modulated by the SVPWM method.

In addition, at low frequencies, the thermal stress is most severe in the clamping diodes of the NPC inverter modulated by the SVPWM method. Such thermal stress is significantly mitigated through the use of this proposed NDPWM method, as can be seen from Fig. 4 to Fig. 6. Taking the comparison between the two modulation methods at 2 Hz as an example, the average junction temperatures of IGBTs S_{a1} and S_{a2} are 40.4°C and 64.3°C, respectively, under the use of the SVPWM method, which become 55.4°C and 55.5°C, respectively, under the use of the proposed DPWM method. This indicates a better thermal balance among the IGBT devices. More importantly, the swing value of the device junction temperature is significantly reduced, which contributes the extension of the NPC inverter. For instance, comparing Fig. 4(a) to (b), the ΔT_j of the inner IGBT device in Phase-A leg, S_{a2}, is approximately 70°C under the condition of using the SVPWM method, which is attenuated to 56°C by using the NDPWM method. The similar attenuation of the ΔT_j can be observed in Fig. 5 and Fig. 6.

Based on all the analysis and simulation results above, a conclusion can be drawn that the NDPWM method can effectively increase the number of cycles to failure for the most stressed power devices in the NPC inverter, and the thermal distribution balancing among their power devices is further improved, which will in turn extend the reliability of the whole ASD systems. However, it should be noted that the NDPWM method is mainly recommended for NPC inverter operating at low output frequency due to the effectiveness in lifetime improvement. At higher output frequencies, for instance, above 10 Hz, IGBT lifetime is not a severe concern anymore, and correspondingly SVPWM method will be recommended, which is due to the lower harmonic distortion in the output voltages.

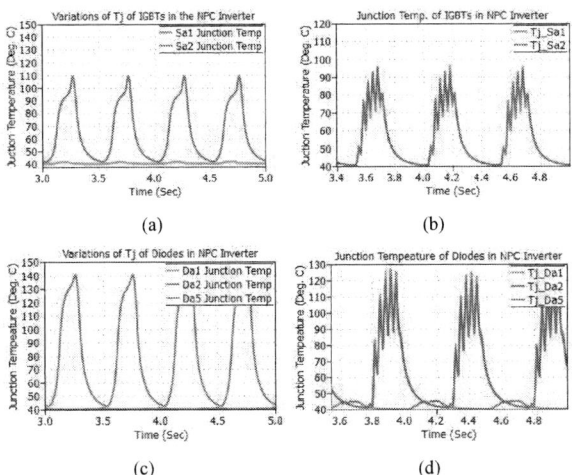

Fig. 4 Comparison of the device junction temperature profiles between the SVPWM method and the proposed NDPWM method at the output frequency of 2 Hz (a) T_j of IGBTs under SVPWM (b) T_j of IGBTs under the NDPWM (c) T_j of the diodes under the SVPWM (d) T_j of the diodes under the NDPWM.

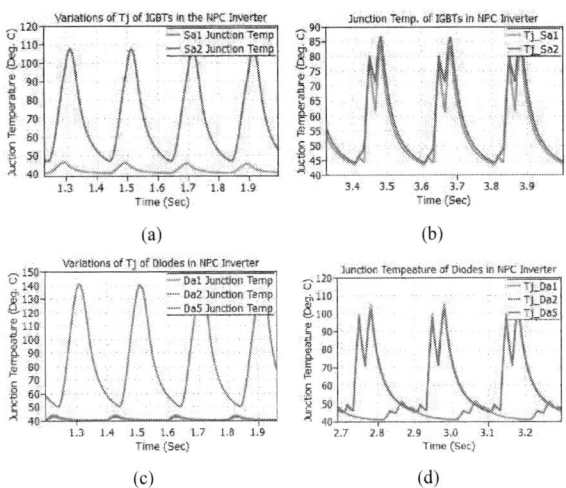

Fig. 5 Comparison of the device junction temperature profiles between the SVPWM method and the proposed NDPWM method at the output frequency of 5 Hz (a) T_j of IGBTs under SVPWM (b) T_j of IGBTs under the NDPWM (c) T_j of the diodes under the SVPWM (d) T_j of the diodes under the NDPWM.

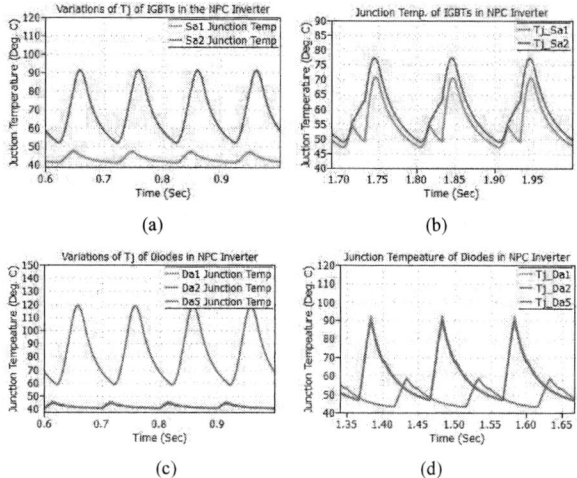

(a)   (b)

(c)   (d)

Fig. 6 Comparison of the device junction temperature profiles between the SVPWM method and the proposed NDPWM method at the output frequency of 10 Hz (a) T_j of IGBTs under the SVPWM method (b) T_j of IGBTs under the NDPWM method (c) T_j of the diodes under the SVPWM method (d) T_j of the diodes under the NDPWM method.

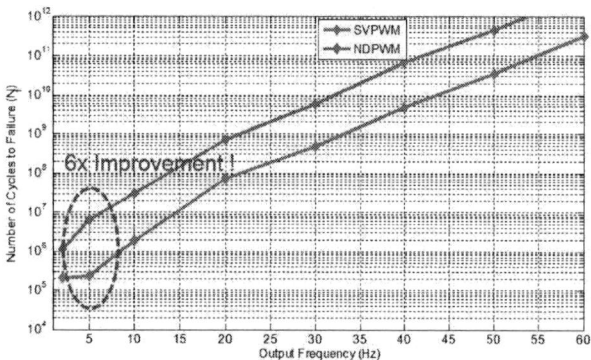

Fig. 7 Comparison of the number of cycles to failure (N_f) of the three-level NPC inverter between operating under the condition of the SVPWM and the proposed DPWM methods.

IV. EXPERIMENTAL RESULTS

Experimental verifications have been carried out to evaluate the performance of this NDPWM method, mainly including the harmonic distortion in the output voltages and currents, as well as the influence on the balancing of the dc-bus capacitor voltages of the NPC inverter. The necessity to investigate the balancing of dc-bus capacitor voltages is due to the fact that any unbalance between the upper dc-bus capacitor voltage and the lower dc-bus capacitor voltage in the NPC inverter will directly degrade the output current/voltage waveform quality and the lifetime of dc-bus capacitors. The mitigation of the IGBT junction temperatures cannot be directly measured, thus cannot be demonstrated here through experimental results. A 50-kVA NPC-inverter-based ASD prototype is shown in Fig. 8. In this drive prototype, three Infineon IGBT three-level NPC module, F3L200R07PE4 (200A/650V) [19], are used to constitute the three-phase three-level NPC inverter. A 32-bit floating-point DSP TMS320F28335 evaluation board from Texas Instrument

is utilized as the control board, in which all the control and PWM programming is embedded. Regarding the load side, two three-phase 5-hp induction machines are utilized as the motor and generator, as shown in Fig. 9, and an ABB adjustable speed drive, namely, ACS800 [20], is employed to apply torque to the inductor machine.

Based on the NDPWM method embedded into the DSP microcontroller, three-phase currents and dc-bus capacitor voltages were captured at very low output frequencies of 2 Hz and 5 Hz and are shown in Fig. 10 and Fig. 11, respectively. To make a comparison with the conventional SVPWM method, three-phase load currents and dc-bus capacitor voltages were also captured under the SVPWM method at the same load conditions, which are shown in Fig. 12 and Fig. 13. It should be mentioned that all these experimental waveforms were captured without any LC filter connected at the output of the ASD system. Also, the frequency of the injected zero-sequence signal is 30Hz, which is much higher than the fundamental output frequencies of 2 Hz and 5 Hz. The modulation index is set as 0.8 p.u. It can be seen that the harmonic distortions in the output currents between using the NDPWM method and the SVPWM method are close to each other. At 2 Hz of output frequencies, the dc-bus balancing under the NDPWM method is slightly better than that under the SVPWM method. The maximum unbalance voltage between the upper and the lower dc-bus capacitors is around 50 V, which accounts for 7.7% of the rated dc-bus voltage of 650V. Such unbalance extent is negligible in general industrial drive systems. It is worth noting that the injected zero-sequence signals in the NDPWM method is closely related with the dc-bus natural-point voltage potential. The balancing of the dc-bus capacitor voltages through the control of these injected zero-sequence signals will be presented in future work.

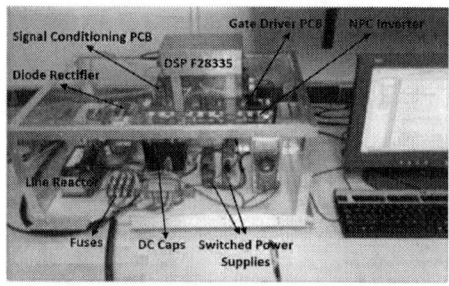

Fig. 8 The customized 50-kVA ASD lab-scale prototype based on the three-phase three-level NPC inverter topology.

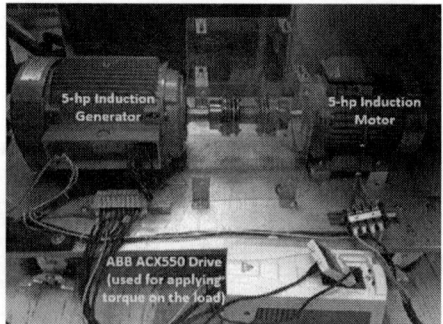

Fig. 9 Dynamometer setup used in the experiments.

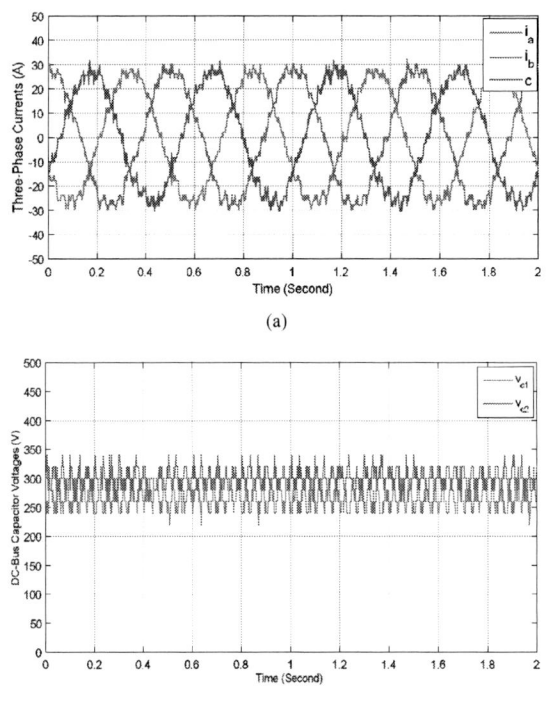

(a)

(b)

Fig. 10 Measured three-phase currents and dc-bus capacitor voltages from the 50kVA ASD under the NDPWM method at 2 Hz of output frequencies (a) measured three-phase currents (b) measured dc-bus capacitor voltages.

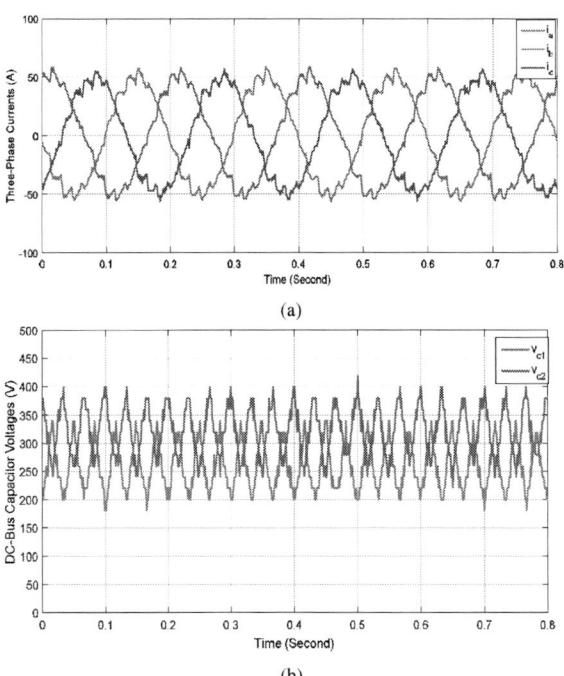

(a)

(b)

Fig. 11 Measured three-phase currents and dc-bus capacitor voltages from the 50kVA ASD under the NDPWM method at 5 Hz of output frequencies (a) measured three-phase currents (b) measured dc-bus capacitor voltages.

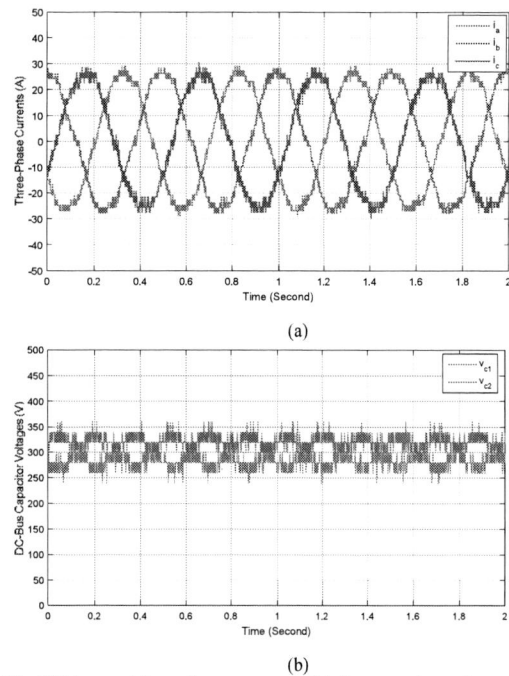

(a)

(b)

Fig. 12 Measured three-phase currents and dc-bus capacitor voltages from the 50kVA ASD under the SVPWM method at 2 Hz of output frequencies (a) measured three-phase currents (b) measured dc-bus capacitor voltages.

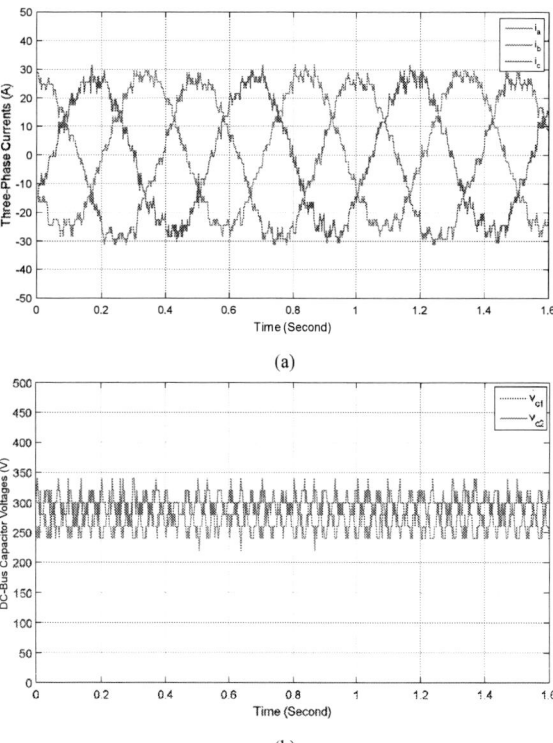

(a)

(b)

Fig. 13 Measured three-phase currents and dc-bus capacitor voltages from the 50kVA ASD under the SVPWM method at 5 Hz of output frequencies (a) measured three-phase currents (b) measured dc-bus capacitor voltages.

V. CONCLUSIONS

In this paper, an improved DPWM modulation method, namely, the NDPWM method, was introduced to extend the power cycling lifetime of a three-level NPC inverter during low-frequency operations. Through the injection of a zero-sequence signal with relatively high frequency, the swing value of the device junction temperature is effectively reduced. Correspondingly, the lifetime of the NPC inverter is extended. In addition, the thermal distribution (i.e., junction temperatures) among all the power devices in the NPC inverter is better balanced. Simulation results obtained from thermal modeling in the PLECS simulation software have verified the effectiveness of this DPWM method. Experimental results based on a 50-kVA ASD prototype demonstrate that the harmonic distortions in the output voltages and currents as well as the balance of the dc-bus capacitor voltages are acceptable under the using of this improved DPWM method. The further balancing of the dc-bus capacitor voltages through the manipulation of the injected zero-sequence signal will be presented in future work.

ACKNOWLEDGMENT

The authors would like to express thanks to the U.S. National Science Foundation (NSF) for financial support of the work in this paper through NSF-GOALI Grant No. 1028348. Also, the licenses of the PLECS simulation software provided by Plexim, Inc. is appreciated by the authors of this paper.

REFERENCES

[1] S. Kouro, J. Rodriguez, B. Wu, S. Bernet, and M. Perez, "Powering the future of industry: high-power adjustable speed drive topologies," IEEE Applications Magzine, vol. 18, no. 4, pp. 26-39, 2012.

[2] X. Jing, J. He and N. A. O. Demerdash, "Application and Losses Analysis of ANPC Converters in Doubly-Fed Induction Generator Wind Energy Conversion System," in Proc. IEEE International Machines and Drives Conference, Chicago, May 2013, pp.131-138.

[3] N. Huang, J. He and N. A. O. Demerdash, "Sliding mode observer based position self-sensing control of a direct-drive PMSG wind turbine systemfed by NPC converters," in Proc. IEEE International Machines and Drives Conference, Chicago, May 2013, pp. 919-925.

[4] A. Lega, S. Munk-Nielsen, F. Blaabjerg, and D. Casadei, "Multilevel converters for UPS applications: comparison and implementation," in Proc. 2007 European Conference on Power Electronics and Applications, Aalborg, Sep. 2007, pp.1-9.

[5] M. H. Poech, K. J. Dittmer, and D. Gabisch, "Investigations on the damage mechanisms of aluminum wire bonds used for high power applications," in proceedings of EUPAC, 1996, pp. 128-131.

[6] M. Ciappa, and W. Fichtner, "Life-time prediction of IGBT modules for traction applications," in proceedings of International Reliability Physics Symposium (IRPS), San Jose, California, 2000.

[7] M. Ciappa, et al, "Lifetime prediction and design of reliability tests for high-power devices in automotive applications," in proceedings of International Reliability Physics Symposium (IRPS), Dallas, Texas, 2003.

[8] M. Held, et al, "Fast power cycling test for IGBT modules in traction applications," in proceedings of 1997 International conference on power electronics and drive systems, May 26-29, 1997, pp. 425-430.

[9] R. Wang, Z. Chen, D. Boroyevich, L. Jiang, Y. Yao, and K. Rajashekara, "A novel hybrid packaging structure for high-temperature SiC power module," IEEE Trans. on Industry Applications, vol. 49, no. 4, pp. 1609-1618, 2013.

[10] T. Dagdelen, Failure analysis of thick wire bonds, M.S. thesis, University of Waterloo, Ontario, Canada, 2013.

[11] K. Sasaki, M. Hiyoshi, and K. Horiuchi, "Small size, low thermal resistance and high reliability packaging technologies of IGBT module for wind power applications," [Online] http://www.hitachi-power-semiconductor-device.co.jp/en/product/igbt/, 2014.

[12] L. Wei, J. McGuire and R. A. Lukaszewski, "Analysis of PWM frequency control to improve the lifetime of PWM inverter," IEEE Trans. on Industry Applications, vol. 47, no. 2, pp. 922-929, 2011.

[13] D. A. Murdock, J. E. R. Torres, J. J. Connors and R. D. Lorenz, "Active thermal control of power electronic modules," IEEE Trans. on Industry Applications, vol. 42, no. 2, pp. 552-558, 2006.

[14] X. Du, G. Li, P. Sun, L. Zhou and H. M. Tai, "A hybrid modulation method for lifetime extension of power semiconductors in wind power converters," in Applied Power Electronics Conference and Exposition (APEC), Conference Record of the 2014 IEEE, Charlotte, NC, 2015, pp. 2565-2570.

[15] L. Wei, J. McGuire and J. Hu, "Novel discontinuous PWM control method to improve IGBT reliability at low speed," in Energy Conversion Congress and Exposition (ECCE), Conference Record of the 2014 IEEE, Pittsburgh, PA, 2013, pp. 3819-3825.

[16] T. M. Phan, G. J. Riedel, N. Oikonmou and M. Pacas, "Active thermal protection and lifetime extension in 3L-NPC-inverter in the low modulation range," in Proc. Applied Power Electronics Conference and Exposition (APEC), 2015 IEEE, Charlotte, NC, 2015, pp. 2269-2276.

[17] T. M. Phan, G. J. Riedel, N. Oikonmou and M. Pacas, "PWM for active thermal protection in three level neutral point clamped inverters," in Energy Conversion Congress and Exposition (ECCE), Conference Record of the 2013 IEEE, Melbourne, VIC, 2013, pp. 906-911.

[18] A. Nabae, I. Takahashi, and H. Akagi, "A new neutral-point-clamped PWM inverter," IEEE Tran. on Industry Applications, vol. IA-17, no. 5, pp. 518-523, 1981.

[19] [Online] Infienon Technologies AG, https://www.infineon.com/, 2015.

[20] [Online] ABB, Manual of ACS 800, http://www.abb.com/, 2015.

Synchronous Optimal Pulsewidth Modulation Digital Implementation Concept for Multilevel Converters

Jackson Lago and Marcelo Lobo Heldwein

Federal University of Santa Catarina (UFSC) – Electrical Engineering Department (EEL)
Power Electronics Institute (INEP) – www.inep.ufsc.br - Phone:+55(48)3721-7464
88040-970 — PO box: 5119 — Florianópolis, SC, BRAZIL
E-mail: jacksonl@inep.ufsc.br ; heldwein@inep.ufsc.br

Abstract—This work proposes a digital implementation concept for a synchronous optimal pulse width modulator (SOP) exemplarily used in a three–phase 5–level NPC H–bridge inverter driving a permanent magnet synchronous motor at variable speed. The numeric optimization is performed offline and the resulting information that defines the inverter output waveforms are stored in a look-up table. The real time portion of the modulator recomposes the pre-computed optimal waveforms and chooses the switching states that, both, synthesize the waveforms and perform voltage balance for the three bipolar dc-buses of the power converter. The proposed modulation algorithm main feature is that changes in the number of commutations per cycle are transparently handled without the need for the dynamic reconfiguration of the digital modulator, which makes it very suitable for synchronous modulation techniques in variable speed applications.

Index Terms—Modulation, Multilevel Converters, High power ac drives.

I. INTRODUCTION

Several industrial processes rely on high–power electric motors that are typically fed by medium voltage (MV) distribution networks. The multilevel voltage source converters proved to be suitable for these applications and today this technology is widely adopted by the industry [1]–[10]. However, the growing demand for increasing power and voltage levels and the current limitations of high-voltage power semiconductor devices leads to increased interest on the area of low switching frequency modulation techniques [11]–[15]. High voltage electronic devices generate high switching losses and, thus, pose limits to practical switching frequencies. This limitation and the use of conventional modulation techniques typically lead to motor currents with relevant harmonic distortion. This leads to additional motor losses, torque ripple and mechanical fatigue. Even though, synchronous optimal pulse width modulation (SOP) and its basis emerged in the late '70s, at that time to deal with the frequency limitation in low voltage two–level inverters, only recently it has been used in the modulation of multilevel inverters [16]–[27]. In this technique, the commutation instant of each of the power devices is computed so as to optimize in some sense the motor currents so to make the best use possible of the few commutations allowed per cycle.

The computation of the switching angles in modulation techniques such as SHE (Selective Harmonic Elimination) and SOP are very time consuming to be performed in real time. The common practice in these applications is to solve the optimization problems that define the switching angles offline, for all possible operational conditions, and store the optimal angles in a table [28]–[34]. The microprocessor that controls the converter accesses this (offline computed) table and generates the gate signals for all controlled power devices based on its optimal switching angles for the desired condition in real time. This is straightforward for two–level inverters. However, it can be a complex task for multilevel ones with high number of switching devices. Apart from the high number of gate signals that must be generated, multilevel converters often present many redundant states that are commonly used to ensure internal balances. These must also be controlled in real time.

Other practical difficulties arise from the the use of synchronous modulators in variable speed applications since the number of switching transitions per cycle must be increased or decreased as a function of the inverter fundamental output frequency in order to keep the switching frequency of each power switch within a predefined range [26], [34].

This work addresses the main implementation issues for SOP, exemplarily for a synchronous optimal pulse width digital modulator that commands a 5–level NPC H–bridge inverter driving an electric motor at variable speed (see Fig. 1). The work proposes a novel digital implementation for a SOP modulator that does not need to be reconfigured for different system operating conditions, including the motor speed variation, i.e., modulation index and angular frequency real time adjustments. The proposed modulator is also able to perform the dc-link voltages balancing functions required by the NPC H-bridge converter.

II. SYNCHRONOUS OPTIMAL PULSEWIDTH MODULATION

The basic idea of the synchronous optimal pulsewidth modulation technique is to obtain the switching angles that command all controlled power semiconductor devices and fully define the output voltage waveforms of a power electronics

978-1-4673-9551-9/16 $31.00 © 2016 IEEE

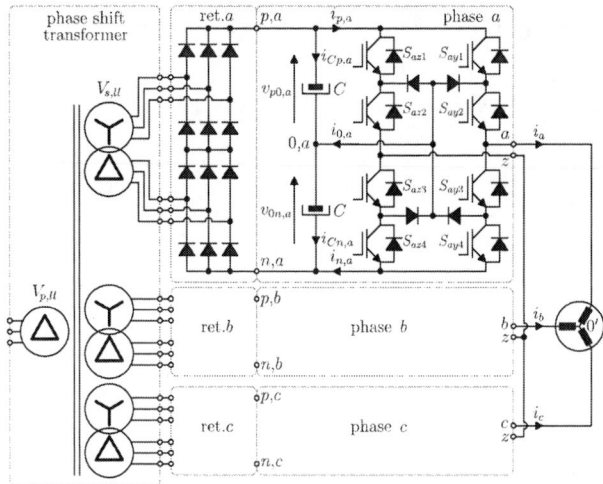

Fig. 1. Variable speed drive based on the NPC H–bridge converter.

converter from an optimization problem that simultaneously optimizes some of the aspects of the driven system. The typical aim for motor drives is to minimize the harmonic distortion of the motor currents and, thus, torque ripple.

The mathematical formulation for the optimization of two–level and three–level waveforms is well establish and extensively studied [35]–[41]. In these cases the voltage output waveforms can assume only two levels in a half–cycle, and consequently, consecutive switching just toggles the level of the output waveform such that the direction of each step transition is imposed and well defined. For multilevel inverters with a number of levels L greater than three, each voltage step transition from intermediary levels can assume two directions. That results in additional degrees of freedom in the modulation algorithm [16], [27], [42]. Therefore, for the modulation of a multilevel inverter it is not enough to define the switching angles of the output waveform. The definition of the direction of each transition step is also required. The complexity due to the need to determine the direction of each step transition in addition to its phase angle is analyzed in [27], which also presents a solution to this problem achieved by an adaptation in the formulation of the optimization problem through an expansion of the search space of the optimization variables. In this formulation the harmonic spectrum of a generic waveform with quarter–wave symmetry is defined by

$$\hat{l}_h\left(\boldsymbol{\gamma}\right) = \frac{2}{\pi h}\sum_{k=1}^{N}\sin(h\gamma_k), \qquad (1)$$

where $\hat{l}_h$ is the amplitude of the harmonic component of order h, N is the number of commutation of the waveform within $0 \leq \theta < \pi/2$, and $\boldsymbol{\gamma} = \{\gamma_1, \gamma_2, \ldots, \gamma_N\}$ is the set of variable to be obtained by solving the optimization problem, composed of the angles that fully defines the commutations in the waveform. Unlike the approach adopted from other works, which use the set of the switching angles $\boldsymbol{\theta} = \{\theta_1, \theta_2, \ldots, \theta_N\}$

that only contains information related to the instant of each switching as optimization variable, the set $\boldsymbol{\gamma}$ contains information related, both, to the instant of each waveform step transition as well as its direction. This information can be extracted from $\boldsymbol{\gamma}$ through

$$\begin{aligned} \gamma_k &= \delta_k \theta_k & , -\pi/2 < \gamma_k < \pi/2 \\ \theta_k &= \mathrm{abs}(\gamma_k) & , 0 < \theta_k < \pi/2 \\ \delta_k &= \mathrm{sign}(\gamma_k) & , \delta_k \in \{-1, +1\} \end{aligned} \qquad (2)$$

where θ_k is the angle of the k–th step transition and δ_k is its direction, i.e., the waveform level is increased or decreased at the instant $\theta(t) = \theta_k$.

The optimal values for $\boldsymbol{\gamma}$ that minimize the Weighted Total Harmonic Distortion (WTHD) of the inverter output voltage is obtained by solving the optimization problem defined by

$$\min_{\boldsymbol{\gamma}} \sum_{\substack{h=6n\mp1 \\ n\in\mathbb{N}^*}}^{H} \frac{4}{\pi^2 h^4}\left[\sum_{k=1}^{N}\sin(h\gamma_k)\right]^2$$

$$\text{subject to} \begin{cases} -\pi/2 \leq \gamma_k \leq \pi/2 & , \\ \frac{4}{\pi}\sum_{k=1}^{N}\sin(\gamma_k) = M \\ 0 \leq l_k(\boldsymbol{\gamma}) \leq (L-1)/2 \end{cases} \qquad (3)$$

where H is the highest considered harmonic order, h is the index representing the harmonic order, M is the modulation index, L is the maximum number of levels that can be synthesized by the converter power topology structure, N is the number of commutations performed in a quarter–wave and l_k is the level of the waveform at the angle θ_k. The first constraint of the problem (3) ensures the validity of the range of the found angles, the second one sets the amplitude of the fundamental component to achieve the desired modulation index and the third one ensures the solution do not use more levels that the inverter can physically synthesize.

For a two–level VSI, one can derive the gate signals for all controlled switches directly from the optimal switching angles once the optimal output voltage waveform is defined. This is because there is no intra–phase redundancies and the only state redundancy is related to the zero vector that does not affect any major operational aspect of the converter. Multilevel inverters, on the other hand, tend to present a higher number of redundancies and these are very useful in order to ensure internal balances, often essential for the proper operation of the power structure. For instance, the 5–level NPC H–bridge shown in Fig. 1 presents two intra–phase redundant states for the output voltage levels $l = \pm1$, and each of those affects the dc-bus voltage balance in the opposite direction. Thus, these redundancies can be used to balance the dc-buses. The valid phase states for this converter and how each state affects the voltage balance are given in Table I.

Therefore, for multilevel inverters it is appropriate to adopt a different approach and split the modulation problem into two steps in a similar way to what is done in techniques such as SVM (Space Vector Modulation). The first step determines

978-1-4673-9551-9/16 $31.00 © 2016 IEEE

TABLE I
ESTADOS DE COMUTAO PARA UMA FASE DO CONVERSOR HNPC.

#	s_{1z}	s_{4z}	s_{1y}	s_{4y}	l	i_p	i_n	i_0
E_4	1	0	1	0		0	0	0
E_6	0	1	0	1	0	0	0	0
E_5	0	0	0	0		0	0	0
E_2	0	0	0	1	+1	0	$+i_y$	$+i_y$
E_3	1	0	0	0		$+i_y$	0	$-i_y$
E_7	0	0	1	0	-1	$-i_y$	0	$+i_y$
E_8	0	1	0	0		0	$-i_y$	$-i_y$
E_1	1	0	0	1	+2	$+i_y$	$+i_y$	0
E_9	0	1	1	0	-2	$-i_y$	$-i_y$	0

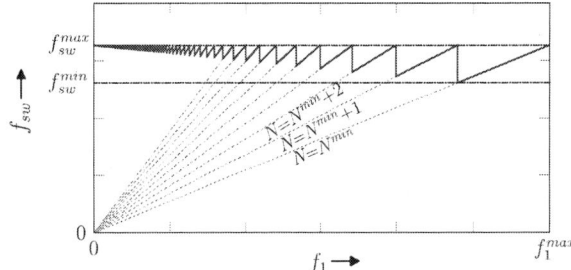

Fig. 2. Maximum switching frequency limitation as a function of the load fundamental frequency f_1.

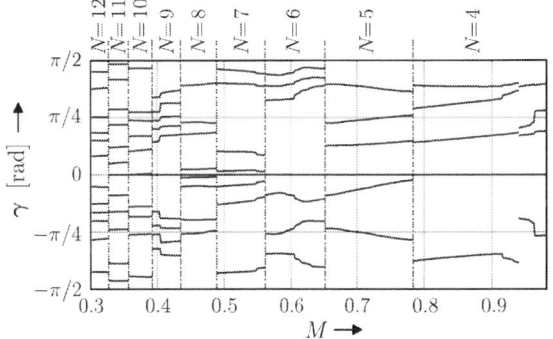

Fig. 3. Optimal switching angles with variable number of swithings per cycle.

the optimal output voltage waveform for a given operation condition without worrying about how this waveforms can be synthesized by the power structure and neither with the required internal balances. The second step is responsible for, in real time, generating the gate signals for the controlled switches that synthesize this predefined optimal waveforms and, at the same time, providing the necessary balances by choosing the redundant states accordingly.

A. SOP to Variable Speed Drive

The majority of multilevel inverters are used within electric motor drives that require a wide range output fundamental voltage and frequency in order to drive the motor with variable speed, from zero up to its rated speed. The wide range for the output voltage is no problem for the SOP modulation technique since the switching angles required for that can be obtained by solving the optimization problem (3) multiple times until the full range of modulation indexes M is covered.

The wide range of the output frequency however imposes a problem since the SOP modulation is synchronous with respect to fundamental component of the output voltage. This implies that the variation of the fundamental frequency leads to varying the switching frequency. Thus, it is common to implement this sort of modulator with a variable number of commutations per cycle [16], [26], [27], [34], [42], [43] in order to limit the switching frequency for, both, maximum and minimum values. Thus, N varies according to the desired fundamental frequency. Thus, for low fundamental output frequency the number of commutation per cycle can be increased in order to ensure low distortion without create extra switching losses for the semiconductor devices. Even though N can vary, it should remain an integer due to the synchronous nature of this modulation technique, more specifically $N \in \mathbb{N}^*$. A possible (but not the only [26]) way to limit the switching device frequency is to define N as [34]

$$N(f_1) = \text{floor}\left[\frac{(L-1)}{2}\frac{f_{sw}^{max}}{f_1}\right] \qquad (4)$$

where f_{sw}^{max} is the imposed limitation of the maximum switching frequency for an individual device and f_1 is the fundamental frequency. This leads to a fixed and known relationship between the switchings per quarter-cycle and the

fundamental frequencies as shown in Fig. 2. It is clear from Fig. 2 that (4) leads to a very large N for low fundamental frequency and that N is changed very often for small changes of the modulation index M and fundamental output frequency. This can result in several N changes within the period of the fundamental depending on the increasing ratio of the output frequency during the acceleration of the motor.

A constant switching frequency modulation technique is typically used to solve this problem while operating under very low fundamental frequency. The SOP modulator is used only for higher values of f_1. In this work an In Phase Disposition (IPD) modulation modified for the NPC H–brigde [44] (mIPD) was used for low values of f_1.

The use of variable N with the SOP modulation requires that the optimization problem (3) is solved several times, for all values of N and for the entire range of M. However, when a constant flux (V/f) control scheme is used, the output fundamental amplitude and frequency are no longer independent of each other. In that case the range of modulation indexes M for which the optimization problem must be solved for each value of N is reduced. The values of the optimal angles table to be used with a constant V/f control scheme with $N = 12, 11, 10, \ldots, 4$ is shown in Fig. 3.

III. MULTILEVEL SOP DIGITAL MODULATOR IMPLEMENTATION

The modulation algorithm is split into two steps so that the optimal angles table stores only the information related with the optimal output voltage waveforms and this information alone is not sufficient to generate the gate signals for all controlled switches. The additional information is acquired in real time from measurements of voltages and currents in order to balance the dc-buses voltages. The output voltages are not affected by this process and the voltages synthesized at the inverter output have the optimal waveform computed offline as desired since this balance can be achieved by selecting redundant states.

The control algorithm passes two variables as references to the modulator during real time operation: the angle θ and the modulation index M that respectively define the instantaneous phase and the amplitude of the fundamental voltage that must be synthesized at the inverter output. The operation to generate the gate signals for the controlled switches is to compute, in real time, the instantaneous reference for the optimal output voltage waveforms $l_a(\theta)$, $l_b(\theta)$ and $l_c(\theta)$, where $l_x(\theta) \in \mathbb{Z}$ is the level of the output voltage that must be synthesized by the phases $x \in \{a, b, c\}$ at the instant $\theta(t)$. The algorithm that generates these signals from the references θ and M and from the optimal angles table can be directly derived from the approach used in [27] to model the output waveform.

Reference [27] decomposes a generic L–level waveform as a series of N elementary 3-level signals $\xi(\theta, \gamma_k)$ in order to generalize the optimization problem with a single commutation angle γ_k within $0 \leq \theta < \pi/2$ and quarter-wave symmetry. Thus,

$$l_a(\theta, \boldsymbol{\gamma}) = \sum_{k=1}^{N} \xi(\theta, \gamma_k) \qquad (5)$$

where these elementary 3–level signals are described as

$$\xi(\theta, \gamma_k) = -\text{sign}\,(\theta - \gamma_k) - \text{sign}\,(\theta - \pi + \gamma_k)\,, \qquad (6)$$

where $\boldsymbol{\gamma}(M) = \{\gamma_1, \gamma_2, \dots \gamma_N\}$ are the set of optimal switching angles stored in the modulation index look-up table.

From (5), (6) and from the imposed quarter–wave symmetry it is possible to establish an algorithm to instantly reconstitute the optimal waveform l_a that must be synthesized at the inverter output at any given time. The signals l_b and l_c are derived in a similar way, differing only by a $\pm 2\pi/3$ rad phase shift in θ.

The gate signals can be obtained from Table I once the instantaneous values of the signals l_a, l_b and l_c are defined. This is done by choosing the redundant state that tends to balance the buses voltage based on the instantaneous measurements of the values of these voltages and phase current direction. The complete modulator algorithm is shown in Fig. 4.

Implementing the real time portion of the modulator as described in Fig. 4 has the advantage of not requiring any dynamic reconfiguration of the modulator when the number of commutations per cycle must be changed in order to change the fundamental frequency. It can be noted from (5) that any angle $\gamma_k = 0$ does not contribute to the composition of the output voltage waveform. Thus, the implementation of the modulator algorithm just needs to ensure that the vector $\boldsymbol{\gamma}$ is large enough to bear the maximum number of angles from the optimal angles table (N_{max}), corresponding to the lower frequency of operation. For higher output frequencies, with lower number of switching angles per cycle, the additional angles of the vector $\boldsymbol{\gamma}$ need just to be filled with zeros since this leads to not add any additional commutation to the reconstructed waveform. Thus, the algorithm presented in Fig. 4 is able to operate with any value of $N < N_{max}$, i.e., with the entire optimal angles table shown in Fig. 3 without any need for modulator reconfiguration when N is increased or decreased. This is a great advantage over implementations that employ either real time reconfiguration or that run several modulators simultaneously.

Fig. 5 shows the experimental setup used to test the SOP modulator. The power electronics hardware is composed of three single-phase NPC H–bridge converters in Y-configuration. Each of these modules has its own local microcontroller. This is responsible for the command of the eight controlled switches of the module, the local protections, as well as to acquire the voltages and current measurements required to balance the module dc-buses. Current measurements could also be communicated by the central controller.

The control and modulation algorithms are centralized and run in a central control board based on a DSC (Digital Signal Controller) and an FPGA (Field Programmable Gate Array). The local microcontrollers are connected to this central control board through serial communication channels via optic fiber links. A similar control/modulation platform was used in [45]. The gate signals of all controlled switches are generated by the centralized modulator, in which the algorithm described in Fig. 4 is implemented and signals are transmitted to the local controllers. Any change in the switching states triggers a new message from the central controller to the local ones with the information of the according new state. The local controllers are responsible for verifying if the received state is a valid one, then applying the new state to the gate signals, adding dead time as required, and report back to the central control board confirming the change of the switching state. In addition, the local controllers regularly acquire the local measured variables and send them to the central control platform to be used by the modulator algorithm.

The practical implementation further includes a variety of protection functions. The most basic ones are implemented locally by the local module controller, namely: IGBT desaturation, over current, over voltage, voltage unbalance, over temperature and a variety of communication errors detected by this end. Besides ensuring local protection, each local controller reports the detection of any fault to the central controller in order for this to disable the other phase modules since there is no direct communication between the phase modules. These error messages exchanged between the central controller and the local ones are sent through the same communication channels used to send the the gate signals and measurements, but the ones related to failures have higher priority in both directions.

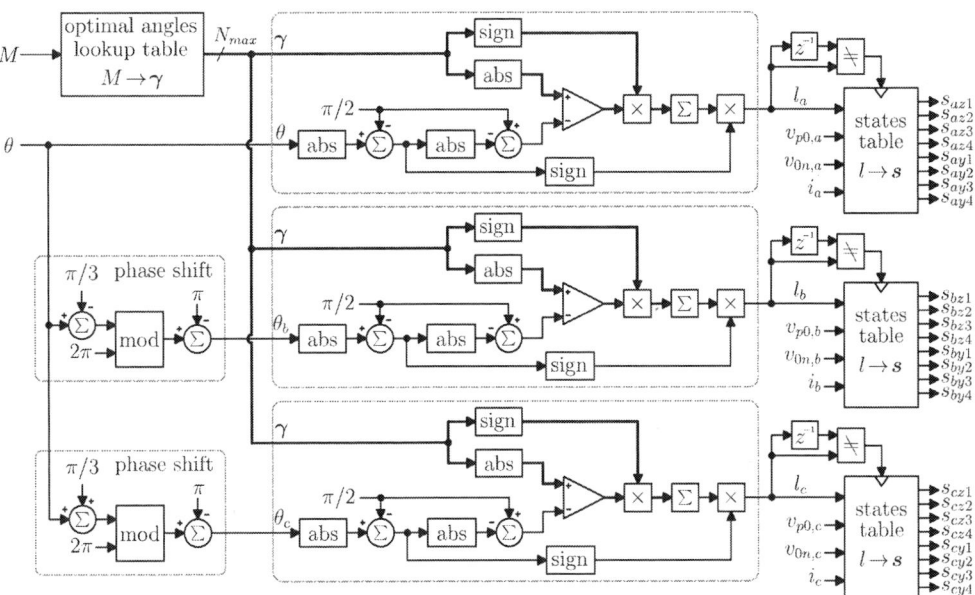

Fig. 4. SOP modulator algorithm.

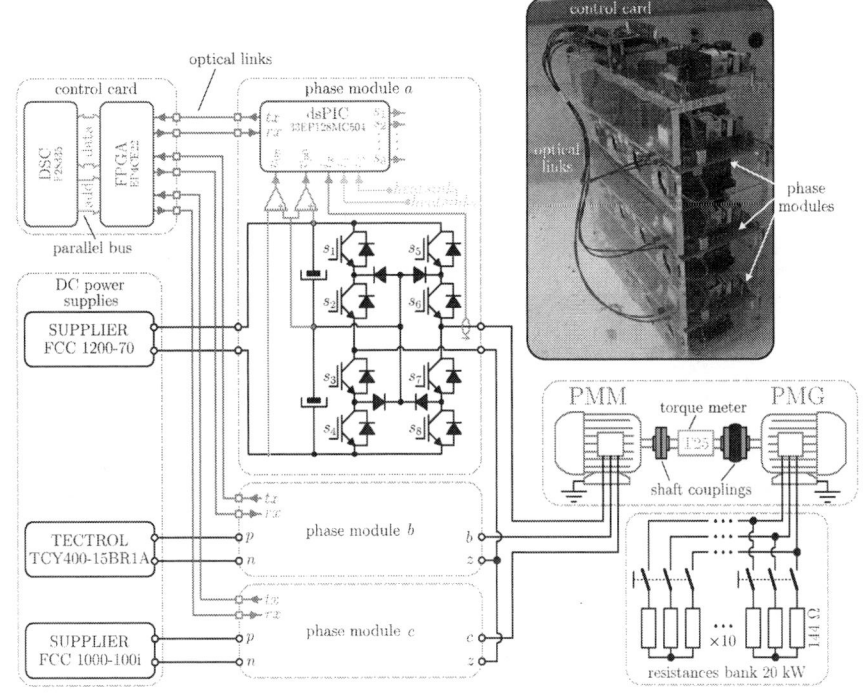

Fig. 5. Experimental setup.

IV. EXPERIMENTAL RESULTS

The experimental setup is composed of a three–phase NPC H–bridge converter supplied by three insulated dc power supplies as in Fig. 5. The load of the inverter is a 15 hp, 380 V, 90 Hz and 1800 rpm permanent magnet synchronous motor whose shaft is connected to a permanent magnet generator of similar power. In this setup the mechanical load to the motor shaft is controlled by switching resistances at the generator terminals.

At very low speeds the inverter is modulated by a carrier

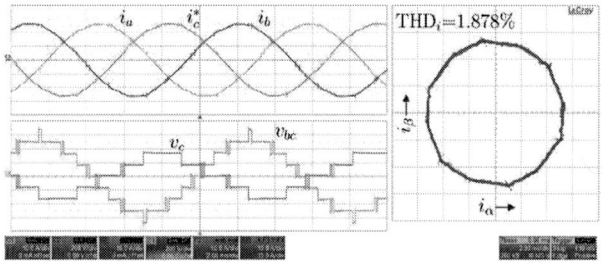

Fig. 6. Experimental results for the system at rated conditions.

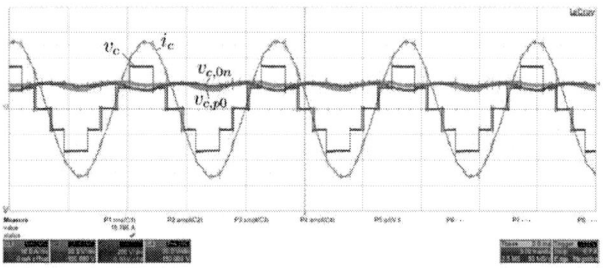

Fig. 7. Experimental results showing the dc–bus voltage balance.

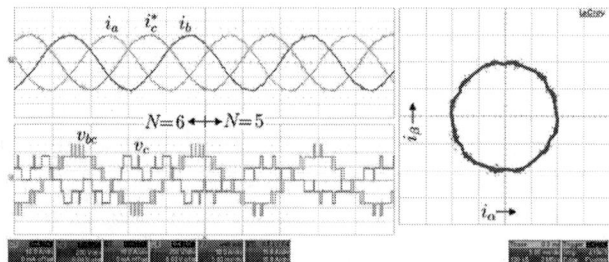

Fig. 8. Experimental results showing the system behavior at the instant of change of the number of switchings per cycle from $N = 6$ to $N = 5$.

based technique. For output frequencies higher then a predefined minimum the inverter operates with the proposed SOP modulator, with the optimal angles table shown in Fig. 3.

Fig. 6 shows the experimental results for the system at rated conditions, with 380 V and 90 Hz fundamental output voltage and mechanical shaft power of 11.5 kW at 1800 rpm. At this frequency the SOP modulator woks with $N = 4$, which for this power structure corresponds to operating each switch only twice per cycle, i.e. with $f_{sw} = 180$ Hz for $f_1 = 90$ Hz. Despite the low switching frequency the measured harmonic distortion (power analyzer YOKOGAWA WT1800) showed values of approximately 1.878%, which demonstrates the effectiveness of the implemented modulation technique.

Fig. 7 shows the DC bus voltage balance with the system under similar conditions to those shown in Fig. 6.

Fig. 8 shown the system behavior at the instant of change of the number of switchings per cycle. In this particular case from $N = 6$ to $N = 5$. This acquisition shows that these transitions occur smoothly, without disturbing the motor currents. This behavior is observed throughout the whole operation range.

V. CONCLUSIONS

This work proposed a digital implementation concept for a synchronous optimal pulse width modulator for a three–phase 5–level NPC H–bridge inverter driving a permanent magnet synchronous motor at variable speed. The optimization part of the SOP technique was performed offline and the information that defines the optimal output waveforms is stored in the form of a look-up table. The real time portion of the modulator recomposes the pre-computed optimal waveforms and chooses the switching states that synthesize the optimal waveforms and

lead to the voltage balance of the three bipolar dc-buses of the power structure.

Even though a centralized modulation scheme was used, each phase module has its own local microprocessor and a optic communication link with the central control unit that enables module-wise local measurements, supervision and protection that further increases the system modularity and reliability.

The proposed digital SOP modulation algorithm is such that changes in the number of commutations per cycle are transparently handled without the need for dynamic reconfiguration of the modulator. This makes it very suitable for synchronous modulation techniques in variable speed applications and reduces the risks for the converter operation when compared with a modulator that uses dynamic reconfiguration.

REFERENCES

[1] S. Kouro, M. Malinowski, K. Gopakumar, J. Pou, L. Franquelo, B. Wu, J. Rodriguez, M. Perez, and J. Leon, "Recent advances and industrial applications of multilevel converters," *IEEE Transactions on Industrial Electronics*, vol. 57, no. 8, pp. 2553–2580, 2010.

[2] J.-S. Lai and F. Z. Peng, "Multilevel converters-a new breed of power converters," *Industry Applications, IEEE Transactions on*, vol. 32, no. 3, pp. 509–517, May 1996.

[3] J. Rodriguez, L. Franquelo, S. Kouro, J. Leon, R. Portillo, M. Prats, and M. Perez, "Multilevel converters: An enabling technology for high-power applications," *Proceedings of the IEEE*, vol. 97, no. 11, pp. 1786–1817, Nov 2009.

[4] K. Kamiyama, T. Ohmae, and T. Sukegawa, "Application trends in ac motor drives," in *Proceedings of the 1992 International Conference on Industrial Electronics, Control, Instrumentation, and Automation, 1992. Power Electronics and Motion Control*, 1992, pp. 31–36 vol.1.

[5] B. Wu, *High-Power Converters and AC Drives*. USA: Wiley Interscience, 2006.

[6] J. Rodriguez, S. Bernet, B. Wu, J. Pontt, and S. Kouro, "Multilevel voltage-source-converter topologies for industrial medium-voltage drives," *IEEE Transactions on Industrial Electronics*, vol. 54, no. 6, pp. 2930–2945, 2007.

[7] R.-D. Klug and N. Klaassen, "High power medium voltage drives - innovations, portfolio, trends," in *2005 European Conference on Power Electronics and Applications*, 2005, pp. 10 pp.–P.10.

[8] C. Dietrich, S. Gediga, M. Hiller, R. Sommer, and H. Tischmacher, "A new 7.2kv medium voltage 3-level-npc inverter using 6.5kv-igbts," in *2007 European Conference on Power Electronics and Applications*, 2007, pp. 1–9.

[9] H. Abu-Rub, A. Lewicki, A. Iqbal, and J. Guzinski, "Medium voltage drives - challenges and requirements," in *2010 IEEE International Symposium on Industrial Electronics (ISIE)*, 2010, pp. 1372–1377.

[10] M. Malinowski, K. Gopakumar, J. Rodriguez, and M. Perez, "A survey on cascaded multilevel inverters," *Industrial Electronics, IEEE Transactions on*, vol. 57, no. 7, pp. 2197–2206, July 2010.

[11] A. Soualmi, F. Dubas, D. Depernet, A. Randria, and C. Espanet, "Study of copper losses in the stator windings and pm eddy-current losses for pm synchronous machines taking into account influence of pwm harmonics," in *Electrical Machines and Systems (ICEMS), 2012 15th International Conference on*, Oct 2012, pp. 1–5.

[12] R. Liu, C. Mi, and D. Gao, "Modeling of iron losses of electrical machines and transformers fed by pwm inverters," in *Power Engineering Society General Meeting, 2007. IEEE*, June 2007, pp. 1–7.

[13] E. Nicol Hildebrand and H. Roehrdanz, "Losses in three-phase induction machines fed by pwm converter," *Energy Conversion, IEEE Transactions on*, vol. 16, no. 3, pp. 228–233, Sep 2001.

[14] H. Penrose and Z. Zych, "Impact of rotor bar bridges in torque development in integral horsepower machines in pwm inverter environments," in *Electrical Insulation Conference (EIC), 2011*, June 2011, pp. 32–35.

[15] S. D. Robertson and K. Hebbar, "Torque pulsations in induction motors with inverter drives," *Industry and General Applications, IEEE Transactions on*, vol. IGA-7, no. 2, pp. 318–323, March 1971.

[16] R. Rathore, H. Holtz, and T. Boller, "Generalized optimal pulsewidth modulation of multilevel inverters for low-switching-frequency control of medium-voltage high-power industrial ac drives," *IEEE Transactions on Industrial Electronics*, vol. 60, no. 10, pp. 4215–4224, 2013.

[17] N. Oikonomou and J. Holtz, "Closed-loop control of medium-voltage drives operated with synchronous optimal pulsewidth modulation," *IEEE Trans. Ind. Appl.*, vol. 44, no. 1, pp. 115–123, Jan 2008.

[18] Y. Liu, Z. Du, A. Huang, and S. Bhattacharya, "An optimal combination modulation strategy for a seven-level cascade multilevel converter based statcom," in *IEEE Ind. Appl. Conf. 2006. 41st IAS Annual Meeting.*, vol. 4, Oct 2006, pp. 1732–1737.

[19] J. Holtz and X. Qi, "Optimal control of medium-voltage drives — an overview," *IEEE Trans. Ind. Electron.*, vol. 60, no. 12, pp. 5472–5481, Dec 2013.

[20] J. Holtz, G. da Cunha, N. Petry, and P. Torri, "Control of large salient-pole synchronous machines using synchronous optimal pulsewidth modulation," *IEEE Trans. Ind. Electron.*, vol. 62, no. 6, pp. 3372–3379, June 2015.

[21] A. Ghias, J. Pou, V. Agelidis, and M. Ciobotaru, "Optimal switching transition-based voltage balancing method for flying capacitor multilevel converters," *IEEE Trans. Power Electron.*, vol. 30, no. 4, pp. 1804–1817, April 2015.

[22] A. Edpuganti and A. Rathore, "New optimal pulsewidth modulation for single dc-link dual-inverter fed open-end stator winding induction motor drive," *Power Electronics, IEEE Transactions on*, vol. 30, no. 8, pp. 4386–4393, Aug 2015.

[23] ——, "Fundamental switching frequency optimal pulsewidth modulation of medium-voltage nine-level inverter," *IEEE Trans. Ind. Electron.*, vol. 62, no. 7, pp. 4096–4104, July 2015.

[24] ——, "Optimal low-switching frequency pulsewidth modulation of medium voltage seven-level cascade-5/3h inverter," *IEEE Trans. Power Electron.*, vol. 30, no. 1, pp. 496–503, Jan 2015.

[25] T. Boller, J. Holtz, and A. Rathore, "Neutral-point potential balancing using synchronous optimal pulsewidth modulation of multilevel inverters in medium-voltage high-power ac drives," *IEEE Trans. Ind. Appl.*, vol. 50, no. 1, pp. 549–557, Jan 2014.

[26] A. Edpuganti and A. Rathore, "Fundamental switching frequency optimal pulsewidth modulation of medium-voltage cascaded seven-level inverter," *Industry Applications, IEEE Transactions on*, vol. 51, no. 4, pp. 3485–3492, July 2015.

[27] J. Lago and M. Lobo Heldwein, "Multilevel synchronous optimal pulsewidth modulation generalized formulation," in *Control and Modeling for Power Electronics (COMPEL), 2014 IEEE 15th Workshop on*, June 2014, pp. 1–7.

[28] G. S. Buja and G. B. Indri, "Optimal pulsewidth modulation for feeding ac motors," *IEEE Transactions on Industry Applications*, vol. IA-13, no. 1, pp. 38–44, 1977.

[29] A. Edpuganti and A. Rathore, "A survey of low switching frequency modulation techniques for medium-voltage multilevel converters," *Industry Applications, IEEE Transactions on*, vol. PP, no. 99, pp. 1–1, 2015.

[30] N. Mittal, B. Singh, S. Singh, R. Dixit, and D. Kumar, "Multilevel inverters: A literature survey on topologies and control strategies," in *Power, Control and Embedded Systems (ICPCES), 2012 2nd International Conference on*, Dec 2012, pp. 1–11.

[31] J. Huber and A. Korn, "Optimized pulse pattern modulation for modular multilevel converter high-speed drive," in *2012 15th International Power Electronics and Motion Control Conference (EPE/PEMC)*, sept. 2012, pp. LS1a–1.4–1 –LS1a–1.4–7.

[32] J. Holtz and B. Beyer, "Optimal synchronous pulsewidth modulation with a trajectory tracking scheme for high dynamic performance [inverter control]," in *Applied Power Electronics Conference and Exposition. 1992. APEC '92. Conference Proceedings 1992., Seventh Annual*, Feb 1992, pp. 147–154.

[33] J. Pontt, J. Rodriguez, P. Newman, A. Liendo, and J. Holtz, "Network friendly low-switching frequency high-power three-level pwm rectifier," in *Power Electronics and Applications, 2005 European Conference on*, Sept 2005, pp. 9 pp.–P.9.

[34] A. Rathore, J. Holtz, and T. Boller, "Optimal pulsewidth modulation of multilevel inverters for low switching frequency control of medium voltage high power industrial ac drives," in *Energy Conversion Congress and Exposition (ECCE), 2010 IEEE*, Sept 2010, pp. 4569–4574.

[35] H. S. Patel, "Thyristor inverter harmonic elimination using optimization techniques," PhD, Dep. Elec. Eng., University of Missouri, Columbia, 1971.

[36] H. S. Patel and R. Hoft, "Generalized techniques of harmonic elimination and voltage control in thyristor inverters: Part i–harmonic elimination," *IEEE Transactions on Industry Applications*, vol. IA-9, no. 3, pp. 310–317, 1973.

[37] ——, "Generalized techniques of harmonic elimination and voltage control in thyristor inverters: Part ii — voltage control techniques," *IEEE Transactions on Industry Applications*, vol. IA-10, no. 5, pp. 666–673, 1974.

[38] F. C. Zach, R. Martinez, S. Keplinger, and A. Seiser, "Dynamically optimal switching patterns for pwm inverter drives (for minimization of the torque and speed ripples)," *IEEE Transactions on Industry Applications*, vol. IA-21, no. 4, pp. 975–986, 1985.

[39] J. Wells, B. Nee, P. Chapman, and P. Krein, "Optimal harmonic elimination control," in *2004. PESC 04. 2004 IEEE 35th Annual Power Electronics Specialists Conference*, vol. 6, 2004, pp. 4214–4219 Vol.6.

[40] J. Hobraiche, J. Vilain, and C. Plasse, "Offline optimized pulse pattern with a view to reducing dc-link capacitor application to a starter generator," in *2004. PESC 04. 2004 IEEE 35th Annual Power Electronics Specialists Conference*, vol. 5, 2004, pp. 3336–3341 Vol.5.

[41] C. Namuduri and P. Sen, "Optimal pulsewidth modulation for current source inverters," *IEEE Transactions on Industry Applications*, vol. IA-22, no. 6, pp. 1052–1072, 1986.

[42] A. Rathore, J. Holtz, and T. Boller, "Synchronous optimal pulsewidth modulation for low-switching-frequency control of medium-voltage multilevel inverters," *IEEE Transactions on Industrial Electronics*, vol. 57, no. 7, pp. 2374–2381, 2010.

[43] P. J. Torri, G. Cunha, T. Boller, K. A. Rathore, J. Holtz, and N. Oikonomou, "Optimal pulse width modulation for multi-level systems," Patent EP Patent 2 312 739, April, 20, 2011.

[44] Y. Pei, L. Chongjian, G. gang, Z. Chunyi, L. Zhiming, W. Chengsheng, D. Wei, and Y. Qiongtao, "Investigation on the control strategy of high power npc/h-bridge inverter," in *ICEMS Int. Conf. Electrical Machines and Systems*, Oct 2013, pp. 1670–1673.

[45] J. Lago, G. Sousa, and M. Heldwein, "Digital control/modulation platform for a modular multilevel converter system," in *Power Electronics Conference (COBEP), 2013 Brazilian*, Oct 2013, pp. 271–277.

Analytical Determination of Conduction Losses for Modified Flying Capacitor Multicell Converters

Vahid Dargahi*, *Student Member, IEEE*, Arash Khoshkbar Sadigh[†], *Member, IEEE*, and Keith Corzine*, *Senior Member, IEEE*

Microgrid and Power Electronics Laboratory, Holcombe Department of Electrical and Computer Engineering, Clemson University, Clemson, SC 29634, USA
[†]*Extron Electronics, Anaheim, CA 92805, USA*

Abstract— Multilevel converters are mostly applied for medium-voltage high-power applications. Flying-capacitor-based multilevel converters such as flying capacitor multicell (FCM) and modified FCM (MFCM) are promising breeds of multilevel converters. Considering the advantages of MFCM converter over conventional FCM converter, and noting that conduction power loss investigation can be very advantageous in the design phase of multilevel converters, this paper presents an analytical approach to calculate and analyze conduction power losses in MFCM converters. First, the rms and average currents of the insulated-gate bipolar transistors (IGBTs) and anti-parallel diodes are analytically calculated by considering the associated duty cycle of each IGBT/diode in terms of the converter modulation index, load current, and load power factor. Numerical results of the derived closed-form equations to calculate the rms and average currents of IGBTs/diodes are compared with simulation results. All the simulation and analytic results agree well with each other which validate the derived closed-form equations. Afterwards, the obtained equations for rms and average current computations are utilized to calculate the conduction power losses in a 12.4MVA 3.3kV 9-level (line-to-line) MFCM converter. A 2.5kV 1.5kA IGBT module from ABB is considered as the power switch in the performed study for MFCM converter.

Index Terms- Multilevel Converter, Modified Flying Capacitor Multicell (MFCM) Converter, IGBT Average and RMS Current Ratings, Conduction Power Loss.

I. INTRODUCTION

Power conversion through power electronic converters employing semiconductor devices exhibiting internal inherent characteristics such as on-state resistance and forward voltage drop account for conduction power losses and consequently slightly dropped efficiency and some energy loss inside these converters. The conduction power losses are calculated with obtaining the required characteristics of the on-state resistance and forward voltage drop from datasheet of the insulated-gate bipolar transistor (IGBT) and diode through linearization method and corresponding average and rms currents of the phase current waveform [1]–[4].

For appropriate design of multilevel converters, it is a concern of utmost importance and practical interest to calculate rms and average current ratings of IGBTs/diodes in order to: 1) select IGBTs/diodes (power switches) properly 2) calculate conduction power loss of converter. It is noteworthy that analytical calculation of rms and average currents of IGBTs/diodes also results in analytical investigation of the conduction power losses which is advantageous in design procedure. Available related articles in this topic have investigated switching and conduction power loss of multilevel converters through simulation/experimental results. Using simulation/experimental results for investigation of the conduction power losses is time consuming, cumbersome, and arduous task because converter circuit needs to be run for numerous cases considering different values of modulation index and load power factor [5]–[13].

Thus, it is required to run the circuit for each case individually even if we do the tasks automated. Lets say for 10 values of modulation indices and 10 different values of power factors which means $10 \times 10 = 100$ case studies. This significantly takes time to run the circuit for all 100 cases (keep in mind that it is needed to let the converter reach to the steady state for each case). While the analytical closed-form equations derived in this paper can provide all conduction power losses data in less than couple of seconds while running each case of circuit through simulation can take easily couple of seconds. Besides this, also consider that it might be required to test different switches with different characteristics (different curves of v_{CE} vs i_C) for all operation range.

Hence, this paper derives closed-form equations to calculate average and rms current of each IGBT and diode in terms of converter modulation index, load current, and load power factor. Finally, the derived closed-form equations are utilized to analytically investigate the conduction power loss in MFCM multilevel converters.

II. CONDUCTION POWER LOSSES IN IGBTs AND DIODES

The IGBT collector-emitter voltage drop v_{CE} when it is conducting can be approximated very well as follows [14]–[18]:

$$v_{CE} = V_{CE0} + R_C \cdot i_C \tag{1}$$

where V_{CE0} represents IGBT on-state zero-current collector-emitter forward voltage drop and R_C is collector-emitter on-state resistance. The same approximation can be used for the anti-parallel diode, giving:

$$v_F = V_{F0} + R_F \cdot i_F \tag{2}$$

where V_{F0} represents the anti-parallel diode on-state zero-current forward voltage drop and R_F is anti-parallel diode on-state resistance. These important parameters can be obtained directly from the IGBT datasheet. The instantaneous values of the IGBT conduction losses ($p_{CT}(t)$) and the average losses (P_{CT}) are:

$$
\begin{aligned}
p_{CT}(t) &= v_{CE}(t) \cdot i_C(t) \\
&= V_{CE0} \cdot i_C(t) + R_C \cdot i_C^2(t)
\end{aligned} \tag{3}
$$

$$
\begin{aligned}
P_{CT} &= \frac{1}{2\pi} \int_0^{2\pi} [p_{CT}(t)]d(\omega t) \\
&= \frac{1}{2\pi} \int_0^{2\pi} [V_{CE0} \cdot i_C(t) + R_C \cdot i_C^2(t)]d(\omega t)
\end{aligned} \tag{4}
$$

$$P_{CT} = V_{CE0} \cdot I_{C,avg} + R_C \cdot I_{C,rms}^2 \tag{5}$$

where $I_{C,avg}$ and $I_{C,rms}$ are the average and rms currents of IGBT, respectively. Similar to the IGBT, the average value of the diode conduction losses P_{CD} is:

$$P_{CD} = V_{F0} \cdot I_{D,avg} + R_F \cdot I_{D,rms}^2 \tag{6}$$

where $I_{D,avg}$ and $I_{D,rms}$ are the average and rms currents of anti-parallel diode, respectively. As it can be seen in Eq. 5 and Eq. 6, it is required to obtain the average and rms current flowing through IGBT and diode in order to calculate their conduction power losses. The main contribution of this paper is to derive the analytical equations expressing average and rms currents of semiconductors, i.e., IGBT and diode, in terms of load power factor, load peak current and modulation index which are explained in following sections.

III. ANALYTICAL CALCULATION OF AVERAGE AND RMS CURRENTS OF IGBTS AND DIODES IN MFCM CONVERTER

In this section, analytical approach is presented to calculate average and rms currents flowing through IGBTs/diodes in terms of load power factor, load peak current, and modulation index in MFCM converter which is shown in Fig. 1 [19], [20]. Duty cycle of power switches of (S_x) can be expressed

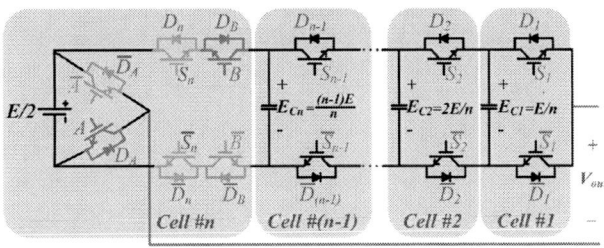

Fig. 1. Generalized topology of an n-cell $n + 1$-level MFCM converter.

according to the following equations.

$$D(t) = \frac{1}{2}(1 + M\sin(\omega t + \varphi)) \tag{7}$$

where M is modulation index, $\omega = 2\pi f$ is angular frequency and f is output voltage frequency. The modulation index represents the normalized voltage, and is between zero and one. For the sake of simplicity in deriving analytical equations for average and rms currents, the phase current can be assumed to be sinusoidal. However, the actual current waveform is slightly distorted by PWM high-frequency ripple current and motor non-linearity. The phase current for an induction motor normally lags the phase voltage by the phase angle φ. Because the current is a simple sine function, the math works out to be much easier if the voltage is assumed to lead the current by φ, and integrating over the current waveform. The resulting relationships are the same using either method. Thus, phase current ($i_\phi(t)$) is defined according to the following equations where I_P is peak current:

$$i_\phi(t) = I_P\sin(\omega t) \tag{8}$$

Average current of IGBT in power switch of (S_x) can be calculated as follows:

$$
\begin{aligned}
I_{C,avg} &= \frac{1}{2\pi} \int_0^{\pi} [i_\phi(t) \cdot D(t)]d(\omega t) \\
&= I_P(\frac{1}{2\pi} + \frac{M}{8}\cos(\varphi))
\end{aligned} \tag{9}
$$

By following the same procedure, average current of diode in power switch of (S_x) can be calculated as follows:

$$
\begin{aligned}
I_{D,avg} &= \frac{1}{2\pi} \int_\pi^{2\pi} [-i_\phi(t) \cdot D(t)]d(\omega t) \\
&= I_P(\frac{1}{2\pi} - \frac{M}{8}\cos(\varphi))
\end{aligned} \tag{10}
$$

As the next step, rms current of IGBT in power switch of (S_x) can be calculated as follows:

$$
\begin{aligned}
I_{C,rms} &= \sqrt{\frac{1}{2\pi} \int_0^{\pi} [i_\phi^2(t) \cdot D(t)]d(\omega t)} \\
&= I_P\sqrt{\frac{1}{8} + \frac{M}{3\pi}\cos(\varphi)}
\end{aligned} \tag{11}
$$

By following the same procedure, rms current of diode in power switch of (S_x) can be calculated as follows:

$$
\begin{aligned}
I_{D,rms} &= \sqrt{\frac{1}{2\pi} \int_\pi^{2\pi} [(-i_\phi(t))^2 \cdot D(t)]d(\omega t)} \\
&= I_P\sqrt{\frac{1}{8} - \frac{M}{3\pi}\cos(\varphi)}
\end{aligned} \tag{12}
$$

Based on MFCM converter switching technique, following equations can be written for IGBTs/diodes of power switches (A and B) in n^{th} switching-power-cell.

$$I_{C,avg,A} = I_{C,avg} = I_P(\frac{1}{2\pi} + \frac{M}{8}\cos(\varphi)) \tag{13}$$

$$I_{D,avg,A} = I_{D,avg} = I_P(\frac{1}{2\pi} - \frac{M}{8}\cos(\varphi)) \tag{14}$$

$$I_{C,rms,A} = I_{C,rms} = I_P \sqrt{\frac{1}{8} + \frac{M}{3\pi} \cos(\varphi)} \qquad (15)$$

$$I_{D,rms,A} = I_{D,rms} = I_P \sqrt{\frac{1}{8} - \frac{M}{3\pi} \cos(\varphi)} \qquad (16)$$

$$I_{C,avg,B} = I_{D,avg} = I_P \left(\frac{1}{2\pi} - \frac{M}{8} \cos(\varphi) \right) \qquad (17)$$

$$I_{D,avg,B} = I_{C,avg} = I_P \left(\frac{1}{2\pi} + \frac{M}{8} \cos(\varphi) \right) \qquad (18)$$

$$I_{C,rms,B} = I_{D,rms} = I_P \sqrt{\frac{1}{8} - \frac{M}{3\pi} \cos(\varphi)} \qquad (19)$$

$$I_{D,rms,B} = I_{C,rms} = I_P \sqrt{\frac{1}{8} + \frac{M}{3\pi} \cos(\varphi)} \qquad (20)$$

Finally, obtained analytical equations for average and rms current flowing through IGBTs and diodes can be substituted in Eq. 5 and Eq. 6 to calculate conduction power losses in MFCM converter.

IV. VALIDATION OF DERIVED CLOSED-FORM EQUATIONS

In order to verify the derived equations for calculation of average and rms currents flowing through IGBTs and diodes in MFCM converters, numerical computation results of the derived equations are compared against simulation results. Due to Eq. 13-20, verification of the derived closed-form solutions is done for power switch of (S_x), and they are not repeated for power switches of (A and B).

The numerical computation and simulation studies are done for a 12.4MVA 3.3kV three-phase 9-level (line-to-line) MFCM converter. In this study, a dc link of 2.7kV (instead of 5.4kV for an FCM) is used for MFCM converter. All IGBTs/diodes in power switches of (S_x) should withstand 1.35kV whereas IGBTs/diodes in power switch of (A and $\overline{A}$) should withstand 2.7kV. Since the utilization factor regarding the voltage of the high-power medium-voltage switches is practically around 50% to 60%, 2.5kV switches are needed to withstand and block 1.35kV. For this purpose, ABB 5SNA 1500E250300 HiPak 2.5kV 1.5kA IGBT module with parameters of $V_{CE0} = 1.2$V, $R_C = 1m\Omega$, $V_{F0} = 1.1$V, $R_F = 0.4m\Omega$ is considered for all power switches of (S_x). it is worthy of mentioning that two of the aforementioned IGBT modules being connected in series are utilized for power switch of (A and $\overline{A}$). The maximum of line peak current I_P is 3060A in all studies to avoid overcurrent situation in all IGBTs and diodes. In order to verify the derived equations, numerical computation results are compared against simulation results. This comparison is shown in Fig. 2 illustrating almost zero error between numerical and simulation results which validates derived analytical equations. It is worth mentioning that in this case study, load peak current is assumed 3060A for all modulation indices and various power factors. So, the load impedance is not considered constant. According to the derived closed-form analytical equations and also simulation results in Fig. 2, it is worth mentioning that by decreasing the modulation index, average and rms current decreases in IGBTs/diodes of MFCM converter.

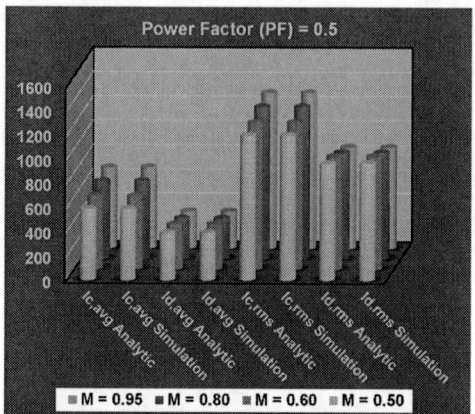

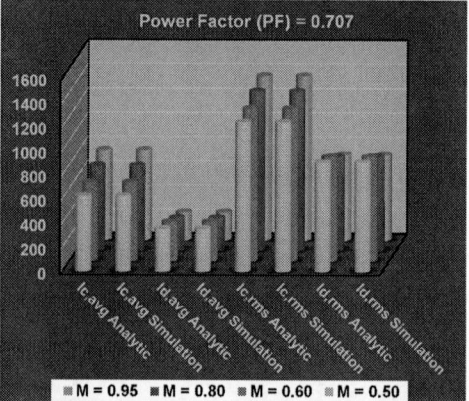

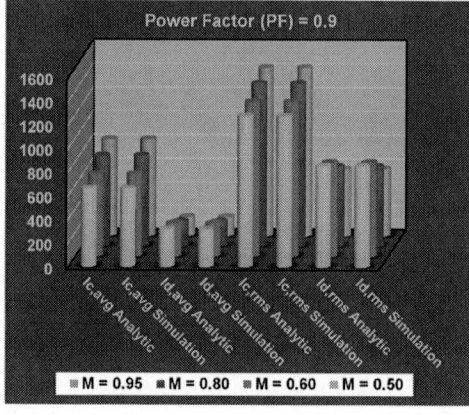

Fig. 2. Comparison between numerical computation and simulation results of the average and rms currents of IGBTs and diodes in 12.4MVA 3.3kV 9-level (line-to-line) MFCM converter considering constant load current (Ip=3060A).

V. CONDUCTION POWER LOSS INVESTIGATION

After verifying derived equations to calculate average and rms current of IGBTs/diodes in MFCM converter through simulation and experimental results, closed-form equations are utilized to calculate conduction losses of IGBTs, diodes, and switching-power-cells as well as whole three-phase MFCM

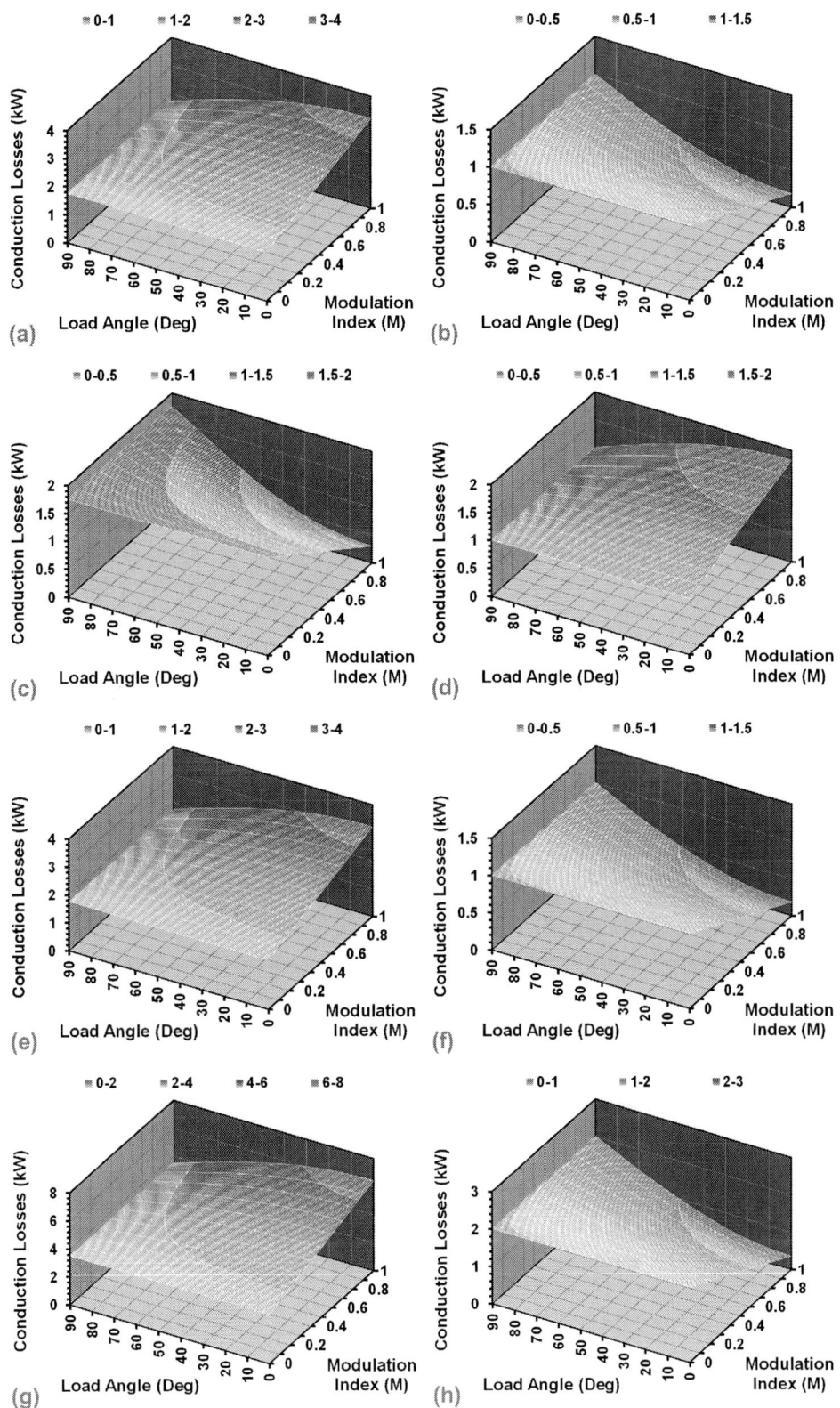

Fig. 5. Cell components conduction power losses versus modulation index and power factor in a three-phase 9-level (line-to-line) MFCM converter considering constant load current (Ip=3060A): (a) IGBT losses of switch S_1; (b) anti-parallel diode losses of D_1; (c) IGBT losses of switch B; (d) anti-parallel diode losses of D_B; (e) IGBT losses of switch S_4; (f) anti-parallel diode losses of D_4; (g) IGBT losses of switch A; (h) anti-parallel diode losses of D_A.

978-1-4673-9551-9/16 $31.00 © 2016 IEEE

converter. The obtained results as a function of converter modulation index and load power factor for 12.4MVA 3.3kV 9-level (line-to-line) MFCM converter are shown in various cases in Figs. 3-6.

The load impedance is considered constant in Figs. 3-4; hence, its current varies linearly with converter modulation index while the load peak current (Ip=3060A) in Figs. 5-6 is considered constant. It can be concluded from Fig. 3 that by increasing the load power factor, conduction power losses of IGBTs increase whereas anti-parallel diodes conduction power losses drop in MFCM converter. Moreover, conduction losses of IGBT rather than anti-parallel diode in MFCM increases significantly by increasing the modulation index, as shown in Fig. 3. An interesting phenomenon is that the conduction power losses of switching-power-cell as well as whole converter in MFCM converter is almost the same for different values of load power factor as shown in Fig. 4. Moreover, it can be inferred from Fig. 4 that conduction power losses of the switching-power-cell which encompasses dc voltage source in MFCM converter is higher than that value for the other switching-power-cells. The reason lays in the fact that switching-power-cell encompassing dc voltage source requires two IGBTs/diodes in series (for each switch of A and $\overline{A}$) to withstand dc-link voltage, and two IGBTs/diodes (switches of B and $\overline{B}$) for proper operation of natural balancing during MFCM start-up. This fact results in more conduction power in the switching-power-cell encompassing dc voltage source in comparison with other cells, illustrated in Fig. 6.

VI. CONCLUSION

This paper presented analytic equations to calculate rms and average current of switches in terms of converter modulation index, load current, and load power factor for MFCM multi-level converters. Numerical results of the obtained analytical equations for calculation of rms and average currents of IG-BTs/diodes were compared with simulation results. The excellent match between analytic and simulation results validates the derived closed-from equations. Afterwards, the obtained closed-form solutions for calculation of the rms and average currents flowing through IGBTs/diodes were utilized to calculate the conduction power losses in a 12.4MVA 3.3kV 9-level (line-to-line) MFCM converter. The rms and average current ratings, and associated conduction power losses for switches and converter were illustrated in detail and the obtained numerical results were provided.

REFERENCES

[1] D. Andler, R. Alvarez, S. Bernet, and J. Rodriguez, "Switching loss analysis of 4.5-kv-5.5-ka igcts within a 3l-anpc phase leg prototype," *Industry Applications, IEEE Transactions on*, vol. 50, no. 1, pp. 584–592, Jan 2014.

[2] ——, "Experimental investigation of the commutations of a 3l-anpc phase leg using 4.5-kv-5.5-ka igcts," *Industrial Electronics, IEEE Transactions on*, vol. 60, no. 11, pp. 4820–4830, Nov 2013.

[3] P. Alemi and D.-C. Lee, "Power loss comparison in two- and three-level pwm converters," in *Power Electronics and ECCE Asia (ICPE ECCE), 2011 IEEE 8th International Conference on*, May 2011, pp. 1452–1457.

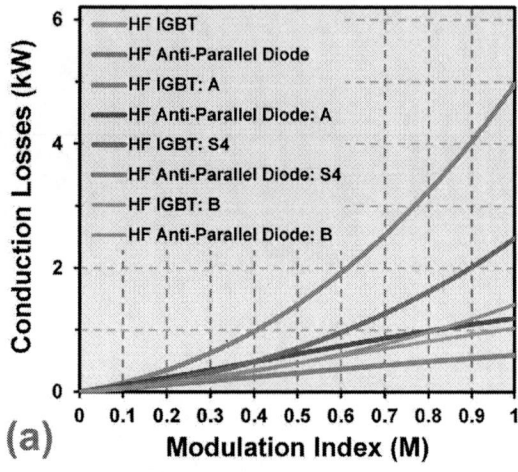

(a)

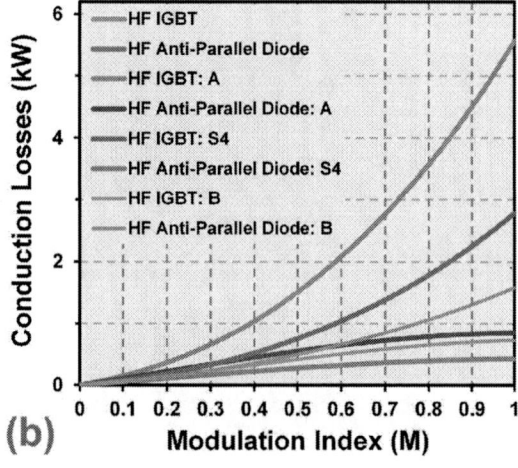

(b)

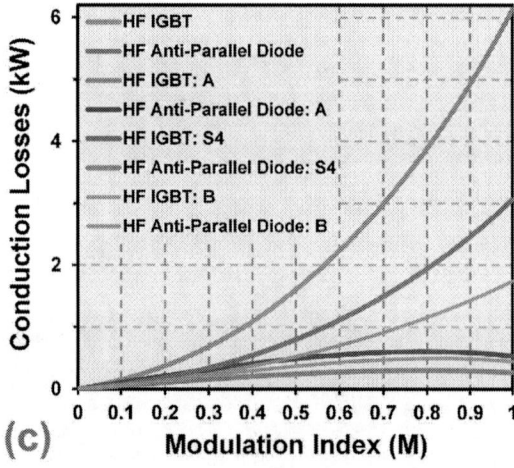

(c)

Fig. 3. Conduction power losses of IGBTs and diodes in 12.4MVA 3.3kV 9-level (line-to-line) MFCM converter considering constant load impedance for load power factors of: (a) 0.5; (b) 0.707; (c) 0.9.

978-1-4673-9551-9/16 $31.00 © 2016 IEEE   2844

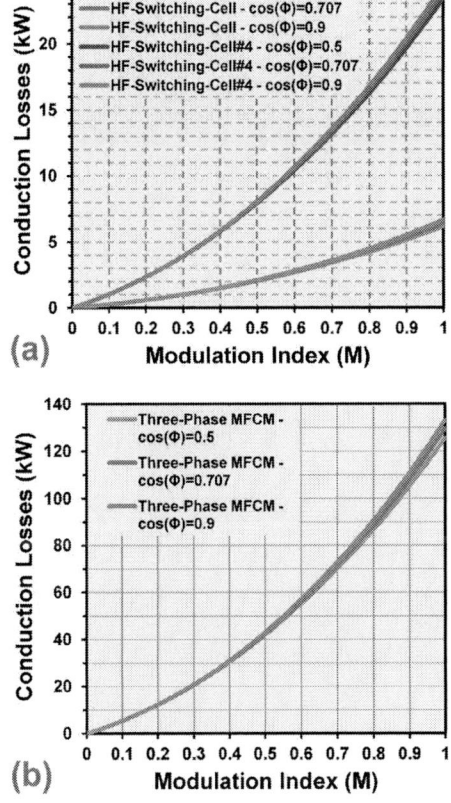

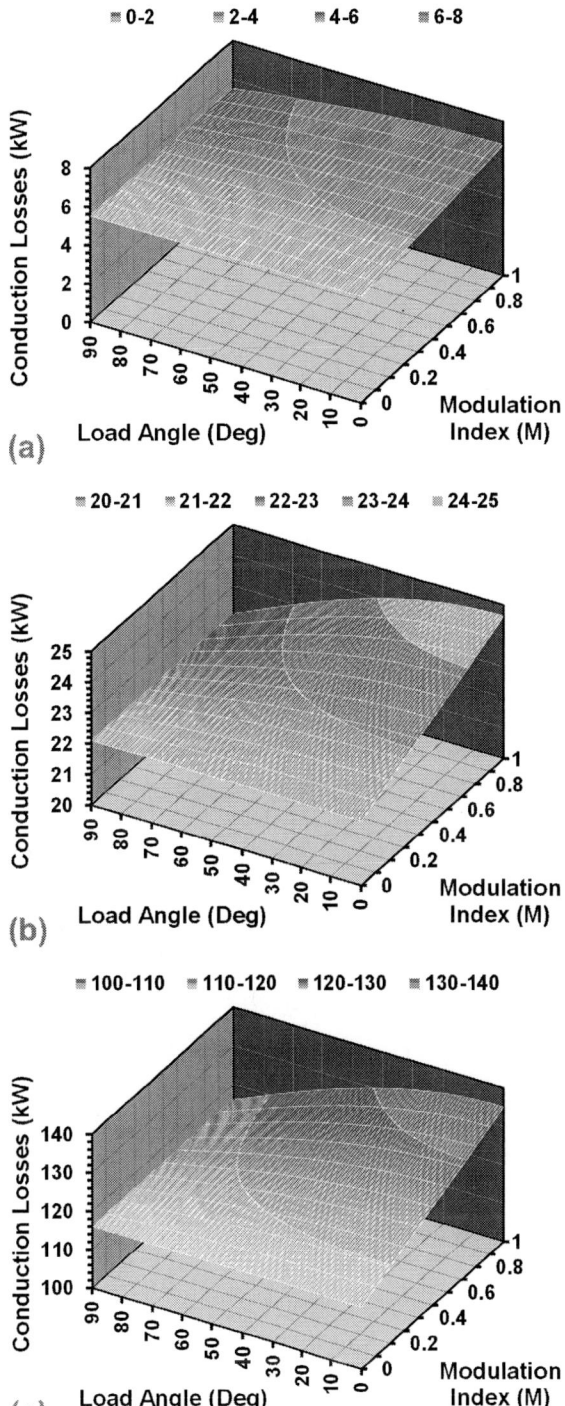

Fig. 4. Conduction power losses of switching-power-cell and whole three-phase converter in 12.4MVA 3.3kV 9-level (line-to-line) MFCM converter considering constant load impedance (a) switching-power-cell; (b) three-phase converter.

[4] Q. Tu and Z. Xu, "Power losses evaluation for modular multilevel converter with junction temperature feedback," in *Power and Energy Society General Meeting, 2011 IEEE*, July 2011, pp. 1–7.

[5] O. Senturk, L. Helle, S. Munk-Nielsen, P. Rodriguez, and R. Teodorescu, "Power capability investigation based on electrothermal models of press-pack igbt three-level npc and anpc vscs for multimegawatt wind turbines," *Power Electronics, IEEE Transactions on*, vol. 27, no. 7, pp. 3195–3206, July 2012.

[6] S. Rohner, S. Bernet, M. Hiller, and R. Sommer, "Modulation, losses, and semiconductor requirements of modular multilevel converters," *Industrial Electronics, IEEE Transactions on*, vol. 57, no. 8, pp. 2633–2642, Aug 2010.

[7] M. Schweizer and J. Kolar, "Design and implementation of a highly efficient three-level t-type converter for low-voltage applications," *Power Electronics, IEEE Transactions on*, vol. 28, no. 2, pp. 899–907, Feb 2013.

[8] S. Rohner, S. Bernet, M. Hiller, and R. Sommer, "Modulation, losses, and semiconductor requirements of modular multilevel converters," *Industrial Electronics, IEEE Transactions on*, vol. 57, no. 8, pp. 2633–2642, Aug 2010.

[9] A. Bazzi, P. Krein, J. Kimball, and K. Kepley, "Igbt and diode loss estimation under hysteresis switching," *Power Electronics, IEEE Transactions on*, vol. 27, no. 3, pp. 1044–1048, March 2012.

[10] F. Cazakevicius, R. Krug, H. Figueira, R. Beltrame, and H. Hey, "Loss and thermal analysis of semiconductor devices applied to an electric circuit simulator," in *Power Electronics Conference (COBEP), 2011 Brazilian*, Sept 2011, pp. 1050–1055.

Fig. 6. Cell conduction power losses versus modulation index and power factor in a three-phase 9-level (line-to-line) MFCM converter considering constant load current (Ip=3060A): (a) switching-power-cell#1 conduction losses; (b) switching-power-cell#4 conduction losses; (c) converter total conduction losses.

[11] M. Aleenejad, H. Iman-Eini, and S. Farhangi, "Modified space vector modulation for fault-tolerant operation of multilevel cascaded h-bridge inverters," *Power Electronics, IET*, vol. 6, no. 4, pp. 742–751, April 2013.

[12] M. Aleenejad, P. Moamaei, H. Mahmoudi, and R. Ahmadi, "Unbalanced selective harmonic elimination for fault-tolerant operation of three phase multilevel cascaded h-bridge inverters," in *Applied Power Electronics Conference and Exposition (APEC), 2015 IEEE*, March 2015, pp. 1589–1594.

[13] M. Aleenejad, H. Iman-Eini, and S. Farhangi, "A minimum loss switching method using space vector modulation for cascaded h-bridge multilevel inverter," in *Electrical Engineering (ICEE), 2012 20th Iranian Conference on*, May 2012, pp. 546–551.

[14] V. Dargahi, A. Khoshkbar-Sadigh, and K. Corzine, "Analytic determination of conduction power losses in flying capacitor multicell power converter," in *Applied Power Electronics Conference and Exposition (APEC), 2015 IEEE*, March 2015, pp. 2358–2364.

[15] A. Sadigh, V. Dargahi, and K. Corzine, "Calculation of conduction power losses in double flying capacitor multicell converter," in *Applied Power Electronics Conference and Exposition (APEC), 2015 IEEE*, March 2015, pp. 2351–2357.

[16] D. Graovac, M. Pursche, and A. Kniep, "Mosfet power losses calculation using the data-sheet parameters," *Infineon Technologies*, 2008.

[17] A. Khoshkbar-Sadigh, V. Dargahi, and K. Corzine, "Analytical determination of conduction power losses in flying-capacitor-based active neutral-point-clamped multilevel converter," *Power Electronics, IEEE Transactions on*, vol. 31, pp. 1–22, 2016.

[18] A. Sadigh, V. Dargahi, and K. Corzine, "New multilevel converter based on cascade connection of double flying capacitor multicell converters and its improved modulation technique," *Power Electronics, IEEE Transactions on*, vol. 30, no. 12, pp. 6568–6580, Dec 2015.

[19] V. Dargahi, A. Sadigh, M. Abarzadeh, S. Eskandari, and K. Corzine, "A new family of modular multilevel converter based on modified flying-capacitor multicell converters," *Power Electronics, IEEE Transactions on*, vol. 30, no. 1, pp. 138–147, Jan 2015.

[20] A. Sadigh, V. Dargahi, M. Abarzadeh, and S. Dargahi, "Reduced dc voltage source flying capacitor multicell multilevel inverter: analysis and implementation," *Power Electronics, IET*, vol. 7, no. 2, pp. 439–450, February 2014.

Comparison of Electrical Losses in an Inverter-Fed Five-Phase and Three-Phase Permanent Magnet Assisted Synchronous Reluctance Motor

AKM Arafat and Seungdeog Choi

Electrical and Computer Engineering
University of Akron
Akron, USA, schoi@uakron.edu

Abstract— This paper is to present a comparison of the electrical losses that occur in the five-phase and three-phase permanent magnet assisted synchronous reluctance motors (PMa-SynRM). Minimizing the losses has been the best practice to get maximum efficiency from high performance motor drive systems. This paper investigated four types of electrical losses which are common in a motor drive. Comparative loss analysis are done for both machines that includes core losses, winding losses, switching losses and conduction losses. The efficiency have been calculated and compared for five-phase and three-phase machines. Extensive theoretical analysis has been done through the MATLAB Simulink and finite element analysis (FEA) to make the comparison. The experimental results are done for the five-phase system under maximum torque per ampere (MTPA) condition by utilizing 5hp dynamo system controlled with TI DSP (F28335) and five-phase inverter system.

Keywords— Five-phase, PMa-SynRM, Effciency.

Nomenclature

n	Number of phases (3 or 5)
P_{cn}	Core loss
P_{wn}	Winding loss
P_{hn}	Hysteresis Loss
P_{edn}	Eddy current loss
P_{sw}	Switching loss
P_{Tn}	Conduction loss
Idn	d-axis current
Iqn	q-axis current
R_{cn}	Core resistance
C_{Ln}	Resistance model for conduction loss
SW_n	Resistance model for switching loss
I_{cdn}, I_{cqn}	Currents in core components
RPM	Revolution per minute

I. Introduction

The growing applications of electric motors warrants the need to improve their operating performance and efficiency. Although the efficiency of the induction machine is lower [1], it has been used in the industries due to its lower cost and higher availability. Three-phase permanent magnet motors considerably earned enormous popularity due to their higher efficiency compared to the conventional electric motors [2]. In literatures [3-5], several design and control methods have been discussed to improve the efficiency of inverter-fed three-phase interior permanent synchronous motors (IPMSMs). However, due to the greater impact of the harmonics at higher speed, the inverter side losses and core losses become prominent which eventually decreases its overall performance. Three-phase permanent magnet assisted synchronous reluctance motors (PMa-SynRMs) are special type of motors in which the magnet size has been reduced compared to other IPMSMs [6-7]. These motors have been suggested in low cost integrated motor drive systems with higher constant power speed range (CPSR) [6]. In addition, the magnet loss and core loss is reduced due to its smaller magnet size. The application of such high performance motors are equally evident in other industries, especially in electric and hybrid vehicular applications [7]. In these cases, the efficiency is the key performance indicator (KPI) to select type of motor in long term service operation. Hence, to operate with maximum efficiency the losses of the total system needs to be identified.

Multi-phase machines are gaining attention in industries due to having higher fault tolerant capability [8-9]. Among the many configuration of the multi-phase systems, five-phase PMa-SynRMs are suggested as a better candidate in terms of optimal design and control [10-13].These motors are an ideal option with improved torque-speed characteristics, higher torque density, reduced torque ripple and higher reliability [10-13]. Even though five-phase machines are higher fault tolerant the efficiency of the overall five-phase motor drive needs to be justified in comparison with a same rated three-phase configuration.

In this paper, a comparison study of the electrical losses found in inverter-fed three-phase and five-phase PMa-SynRM has been attempted. Different losses such as switching losses,

$$T(\theta)=J\begin{bmatrix} \mathrm{A}\cos(\theta) & \mathrm{B}\cos\left(\theta-\dfrac{2\pi}{n}\right) & \mathrm{C}\cos\left(\theta-\dfrac{4\pi}{n}\right) & \mathrm{D}\cos\left(\theta-\dfrac{6\pi}{n}\right) & \mathrm{E}\cos\left(\theta+\dfrac{2\pi}{n}\right) \\[2mm] \mathrm{A}\sin(\theta) & \mathrm{B}\sin\left(\theta-\dfrac{2\pi}{n}\right) & \mathrm{C}\sin\left(\theta-\dfrac{4\pi}{n}\right) & \mathrm{D}\sin\left(\theta-\dfrac{6\pi}{n}\right) & \mathrm{E}\sin\left(\theta+\dfrac{2\pi}{n}\right) \\[2mm] \dfrac{\mathrm{A}}{\sqrt{2}} & \dfrac{\mathrm{B}}{\sqrt{2}} & \dfrac{\mathrm{C}}{\sqrt{2}} & \dfrac{\mathrm{D}}{\sqrt{2}} & \dfrac{\mathrm{E}}{\sqrt{2}} \end{bmatrix} \quad (2)$$

conduction losses, winding and core losses are investigated and compared in three-phase and five-phase PMa-SynRMs. As the stray losses are independent of the control and negligible compared to other losses they are not included in this study. Detail theoretical analysis has been carried to support the comparison study. MATLAB Simulink has been used to defend the mathematical explanation. The comparison has been further evaluated using the Finite element analysis. The experimental results are given utilizing the 5 hp dynamo system that are controlled at the MTPA condition by utilizing the TI DSP (F28335) and multiphase inverter system.

II. MODELING OF THE THREE-PHASE AND FIVE-PHASE PMA-SYNRM

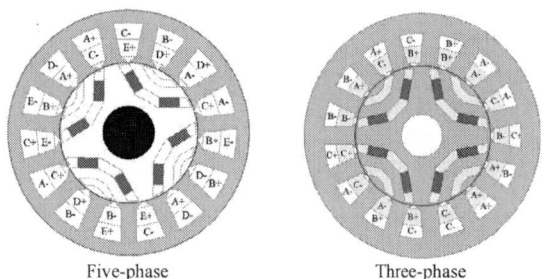

Five-phase   Three-phase

Fig.1. FEA models.

Fig.1 shows the FEA models of the five-phase and three-phase machine that are utilized in the finite element analysis. Both of the three-phase & five-phase models have been developed considering same rated output (3 kW), rated speed (1800 rpm) ,outer dimension (diameter 190 mm) and winding topology (4 poles, winding factor 0.94). Both of the machines are optimized with same objective function (optimize the motor for reduced torque ripple, current, and cost) which is discussed in [14-15]. The same materials are used (core- S18 and, AWG-13) for both machines. These models have been utilized in the finite element analysis to compare the core losses. The mathematical model for three-phase and five-phase motor are derived in reference to the rotating frame (d-q). Irrespective of the number of phases, the voltage, current and flux of three-phase and five-phase PMa-SynRM are derived as follows:

$$
\begin{aligned}
V_d &= R_a I_d + p L_d I_d - \omega_r (L_q \cdot I_q - \lambda_{PM}) \\
V_q &= R_a I_q + p L_q I_q + \omega_r (L_d \cdot I_d) \\
\lambda_q &= L_q \cdot I_q - \lambda_{PM} \\
\lambda_d &= L_d \cdot I_d
\end{aligned}
\qquad (1)
$$

Where, R_a is the winding resistance, p is the time derivative operator, V_d and I_d is the d-axis voltage and current, V_q and I_q is the q-axis voltage and current, L_d is the d-axis inductance, L_q is the q-axis inductance, λ_{PM} is the permanent magnet flux linkage, λ_d, λ_q are the d and q axis

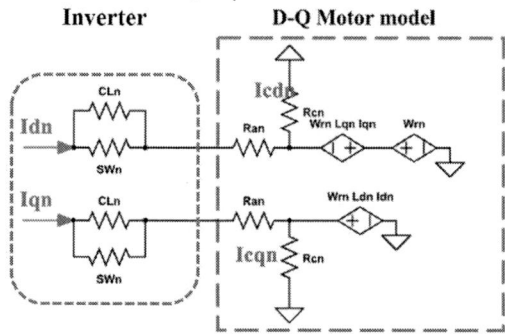

Fig.2. Resistive model of the drive system.

flux linkage. The d-q axis current and voltage can be derived using the equation (2).

Equation (2) shows the transformation matrix which is utilized to transform the five-phase and three-phase currents to the d-q axis components simultaneously. In this equation (2), J is $2/(A+B+C+D+E)$ and A, B, C, D, E represent the phases, n=3 or 5. For five-phase n=5 and A=B=C=D=E=1. For three-phase system n=3 and A=B=C=1, D=E=0.

Fig. 2 shows the d-q equivalent circuits for the motors. In the equivalent circuit, the common notation "n" is used to differentiate the number of phases. The d-q axis currents are presented as Idn and Iqn respectively in Fig. 2. The winding losses are presented as Ran. The core loss is a combination of the hysteresis loss and eddy current loss which are frequency dependent, has been depicted as Rcn. The conduction loss and switching losses are presented as CLn and SWn respectively in Fig. 2.

III. ELECTRICAL LOSSES IN THREE-PHASE AND FIVE-PHASE MOTOR

A. Winding loss

The winding loss varies with the resistance of the copper wire inserted in the machines and the magnitude of the current flows through it. In a five-phase motor there are two more phases compared to the three-phase motor. The loss analysis of the three-phase and five-phase machines can be done by starting from the inverter side and then using the d-q equivalent circuits of the motor. In the inverter side, considering both the machines provides same amount of load power, the phase current is I_5 and I_3 for five-phase and three-phase machines. The modulation index can be kept same for both systems as

$$m_{a5} = m_{a3} = \hat{V}_{control} / \hat{V}_{tri}$$

The phase voltage is,

$$V_{ao} = \tfrac{1}{2} E \, , (\hat{V}_{ao})_{rms} = (m_a . \tfrac{1}{2} E) / \sqrt{2} \qquad (3)$$

where E is the DC link voltage. The total phase power generated in the five-phase (P_5) and three-phase (P_3) motor are follows:

$$P_5 = 5I_5 .(m_a . \tfrac{1}{2} E) / \sqrt{2}$$
$$P_3 = 3I_3 .(m_a . \tfrac{1}{2} E) / \sqrt{2} \qquad (4)$$

By assuming the power supplied by the five-phase and three-phase are same ($P_5 = P_3$), from the equation (4), the current ratio comes as,

$$\frac{I_5}{I_3} = \frac{3}{5} \qquad (5)$$

From equation (5), mathematically it can be seen that, the phase current in five-phase is smaller than the three-phase systems by the factor of 0.6. Though in a practical system, the phase currents are not exactly related as in equation (5), the phase currents in five-phase tends to be smaller with same loading condition (shown in the simulation section V). Using the equation (5), the winding losses in the machines can be estimated as follows:

$$\frac{P_{w5}}{P_{w3}} = \frac{\sqrt{I_{d5}{}^2 + I_{q5}{}^2}.R_{a3}}{\sqrt{I_{d3}{}^2 + I_{q3}{}^2}.R_{a3}} = \frac{3}{5} \qquad (6)$$

where, P_{w5} and P_{w3} are the winding loss in the five-phase and three-phase system, R_{a5} and R_{a3} are the winding resistance for five-phase and three-phase motor respectively.

B. Core Loss

There has been two types of core losses are – hysteresis loss and eddy current loss. In the Fig.2, the core losses are modeled as the resistances R_{cn} where n stands for number of phases. Considering the same core materials in three-phase and five-phase machines, the losses mostly depend on the current that goes through the parallel branches of the d-q model. The d and q axis currents are I_{cdn} and I_{cqn} which can be derived as follows:

$$I_{cdn} = (-\omega_{rn} L_{qn} I_{qn} + \omega_{rn} \lambda_{PMn}) / R_{cn} \qquad (7)$$

$$I_{cqn} = \omega_{rn} L_{dn} I_{dn} / R_{cn} \qquad (8)$$

The d and q axis inductances are almost similar for three-phase and five-phase machines. Considering the same speed operation under similar load power, the core branch currents (I_{cdn}, I_{cqn}) depend on the d and q axis currents (I_{dn}, I_{qn}). The core losses in five-phase (P_{c5}) and three-phase (P_{c3}) can be estimated as in equation (9).

$$P_{c5} = I_{cd5}{}^2 .R_{cd5} + I_{cq5}{}^2 .R_{cq5}$$
$$P_{c3} = I_{cd3}{}^2 .R_{cd3} + I_{cq3}{}^2 .R_{cq3} \qquad (9)$$

Where, R_{cd5} and R_{cq5} are the core resistance components in the d-axis and q axis of the five-phase system, R_{cd3} and R_{cq3} are the core resistance components in the d-axis and q axis of the three-phase system. Form the equation (5) and (9), it is found that the core losses in five-phase is smaller than the three-phase system.

Being a function of frequency, core loss can be analyzed in a different way using the fundamental equations of hysteresis loss and eddy current loss. It has been observed that, these losses change with the fundamental frequency variation and deteriorate with the higher order harmonics. Considering the machines operate below the saturation level, the hysteresis loss for five-phase (P_{h5}) and three-phase (P_{h3}) machines can be derived as in equation (10) and (11).

$$P_{h5} = K_h (f_0 .(B_{m0})^{1.6} + f_9 .(B_{m9})^{1.6} + f_{11} .(B_{m11})^{1.6} + ...)$$
$$P_{h3} = K_h (f_0 .(B_{m0})^{1.6} + f_5 .(B_{m5})^{1.6} + f_7 .(B_{m7})^{1.6} + ...) \qquad (10)$$

Similarly, the eddy current loss for five-phase (P_{ed5}) and three-phase (P_{ed3}) can be estimated as in equation (12).

$$P_{ed5} = K_e (f_0{}^2 .(B_{m0})^2 + f_9{}^2 .(B_{m9})^2 + f_{11}{}^2 .(B_{m11})^2 + ...)$$
$$P_{ed3} = K_e (f_0{}^2 .(B_{m0})^2 + f_5{}^2 .(B_{m5})^2 + f_7{}^2 .(B_{m7})^2 + ...) \qquad (11)$$

Where K_h is the hysteresis loss coefficient, K_e is the eddy current losses coefficient, p is the harmonic number, f_p is the frequency and B_{mp} is the peak magnetic flux density. The order of the harmonics and the amplitude are different for five-phase and three-phase motor which eventually make the core losses significantly different from each other. The magnetic flux density ($B_{mp} = \mu H = \mu N I$) depends on the current flows through the machines. Due to the presence of the current harmonics the core loss characteristics inside the machine are different for different motors.

IV. INVERTER-SIDE LOSSES IN THREE-PHASE AND FIVE-PHASE MOTOR DRIVE

A. Switching loss

Switching loss is different in a five-phase system (more number of switches are used) from a three-phase system. Considering the fundamental components in currents switching loss is estimated in [16] for inverter application. Considering higher order harmonics, switching losses can be estimated as in equation (12).

$$P_{sw} = \frac{1}{6} \sum_{i=0}^{U} \sum_{j=1}^{V} v_{ij} i_{ij} (T_{on} + T_{off}) f_{swi} \qquad (12)$$

where V is total number of switching cycles in one period, j is the jth switching, i is the harmonic number, f_{swi} is the switching frequency of ith harmonic, T_{on} is the turn-on time, T_{off} is the turn-off time, and v_{ij}, i_{ij} is the instantaneous voltage and current of the ith harmonic at j^{th} switching. In the equation (12), the switching loss can be reduced by reducing the switching frequency and by reducing the instantaneous value of the current or voltage. However, reducing the switching frequency increases the harmonics in the system.

B. Conduction loss

Using the piecewise linear approximation of the currents and voltage, the conduction loss is shown in equation (13),

$$P_T = m.(\overline{I_T}.V_T + \left|I_T\right|^2_{rms}.r_T) \qquad (13)$$

where I_T is the current through the transistor, V_T is the constant voltage drop, r_T is the transistor resistance, and m is the no of IGBTs. The conduction loss in five-phase (P_{T5}) and three-phase (P_{T3}) can be presented as in equation (14).

$$\begin{aligned} P_{T5} &= 10.(\overline{I_{T5}}.V_T + \left|I_{T5}\right|^2_{rms}.r_T) \\ P_{T3} &= 10.(\overline{I_{T3}}.V_T + \left|I_{T3}\right|^2_{rms}.r_T) \end{aligned} \qquad (14)$$

In an inverter system, ten switches are used for five-phase and six switches are used in three-phase motor. Using the equation (5) and (14), the ratio of the conduction losses can be effectively presented as in equation (15).

$$\frac{P_{T5}}{P_{T3}} = \frac{P_{T5}}{P_{T5} + \left|I_{T5}.\right|^2_{rms}.\frac{2}{3}r_T)} < 1 \qquad (15)$$

Equation (15) clearly shows that the conduction losses in five-phase system is smaller than the three-phase system.

V. SIMULATION AND EXPERIMENTAL RESULTS

Table I Specification of PMa-SynRM.

Parameter	Specifications
Rated current (rms)(A)	15.17
Rated voltage 3-Ø (rms) (V)	85
Rated voltage 5-Ø (rms) (V)	67
Power (hp)	5
Rated speed (rpm)	1800
Rated Torque (Nm)	15

In this section, the simulation and experimental results are presented. The mathematical models and the FEA models (PMa-SynRM) of the both machines take the following specification as in Table I.

A. Simulation result:Simulink

In the MATLAB Simulink, the three-phase and five-phase machines are mathematically modeled using the equation (1). Both the machines are operated at same operating condition (1500 rpm and 5 Nm). Fig. 3 shows the Simulink models of the five-phase and three-phase system. The switching frequency is used 10 kHz. Sine PWM method has been adopted for both systems. The phase currents of the five-phase and three-phase systems have been taken to the fast Fourier transformation (FFT) to see the higher order harmonics effects. Fig. 4 and 5 show the per unit FFT results of the five-phase and three-phase system respectively.

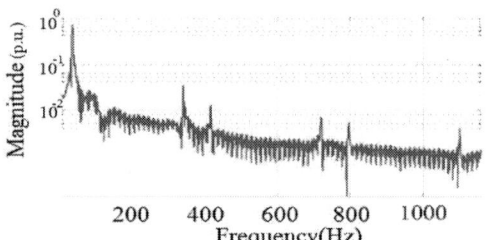

Fig.4. Current harmonics in five-phase.

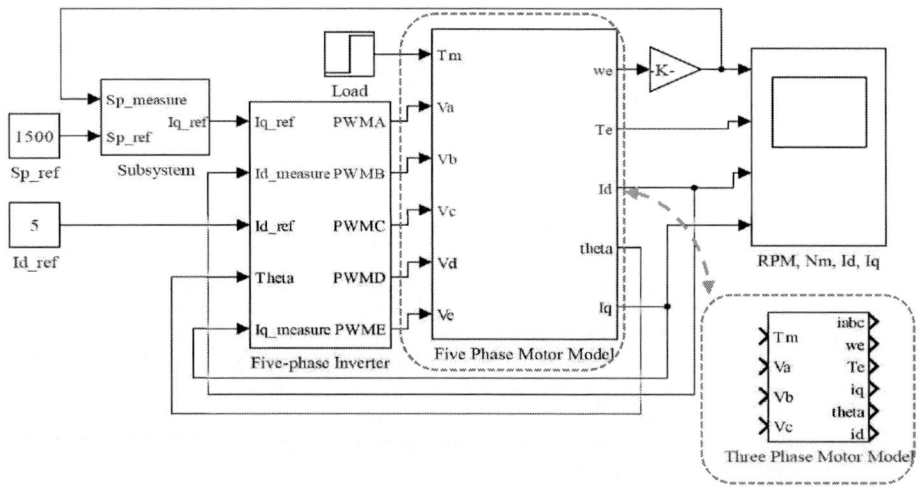

Fig.3. Simulink model for five-phase and three-phase system.

978-1-4673-9551-9/16 $31.00 © 2016 IEEE

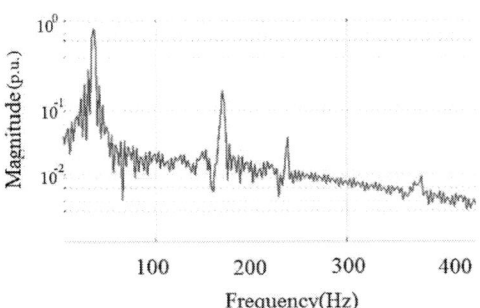

Fig.5. Current harmonics in three-phase.

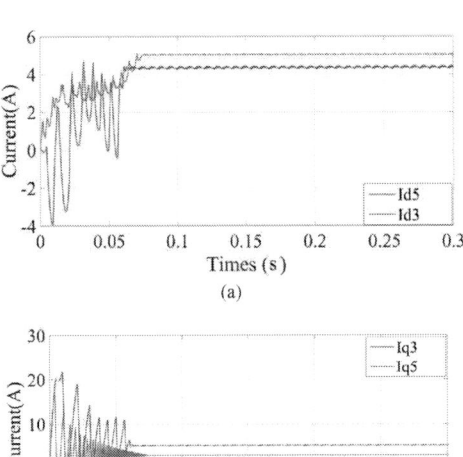

(a)

(b)

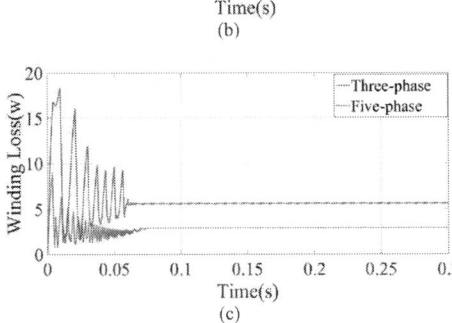

(c)

Fig.6. Comparison of: a) d-axis current, b) q-axis current, c) Winding losses.

Fig. 4 shows that first, ninth and eleventh harmonics are present in five-phase system. The magnitudes of the harmonics are 0.7, 0.05 and .02 in p.u. respectively. For three-phase system, Fig. 5 shows that first, fifth and seventh harmonics are present in three-phase system. The magnitude of the harmonics are of 0.7, 0.2 and .05 in p.u. respectively. It is observed from the Fig. 4 and 5 that the lower order harmonic effects are more noticeable in three-phase system than the five-phase system which are eventually responsible for higher core losses as shown in equation (10) and equation (11).

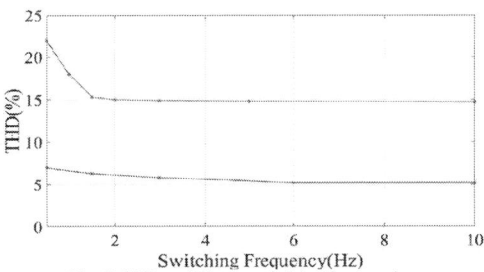

Fig. 7. THD at different switching frequencies.

The comparison of the d-q axis currents (Id3 and Iq3 for three-phase and Id5 and Iq5 for five-phase) and corresponding winding losses are shown in Fig.6. In Fig. 6 (a), the d- axis currents in three-phase and five-phase system are found as 4.1 and 5 amperes respectively. In Fig. 6(b), the q-axis currents in three-phase and five-phase system are found as 3.3 and 5.2 amperes respectively. Using the equation (7), the winding losses in three-phase and five-phase system are shown in Fig. 6(c). The losses are calculated as ~6W and ~3W for three-phase and five-phase system respectively.

The five-phase and three-phase systems operated with different switching frequency. Fig. 7 shows the total harmonic distortion (THD %) versus switching frequencies. It is observed that the, the THD for five-phase system remain closely 5%~7% even in the lower switching frequency (<2 kHz). For three-phase system, the THD is observed as 15% .For this three-phase system THD is even deteriorated at lower switching frequencies.

B. Simulation result:FEA

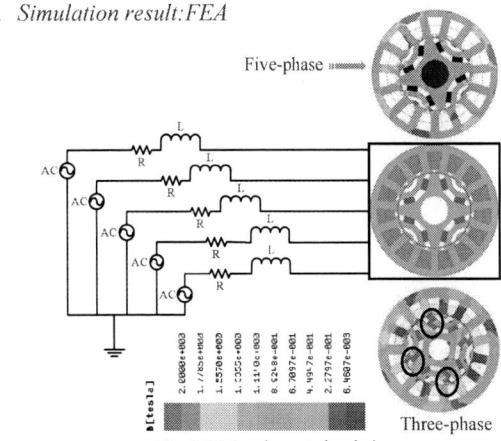

Fig. 8. Finite element simulation.

To compare the core losses in five-phase and three-phase machine finite element methods are obtained. Ansys Maxwell is used to do the finite element analysis. The model in the Fig. 1 is simulated under rated voltage excitation as shown in Fig. 8. The Flux plots of the five-phase and three phase is shown in the top and bottom in Fig 8. It is observed that the three-phase flux plot gets more saturated (circled in black) at the rated condition than the five-phase system. The core losses and the efficiencies of the five-phase and three-phase machine are given in the Fig.9.

978-1-4673-9551-9/16 $31.00 © 2016 IEEE

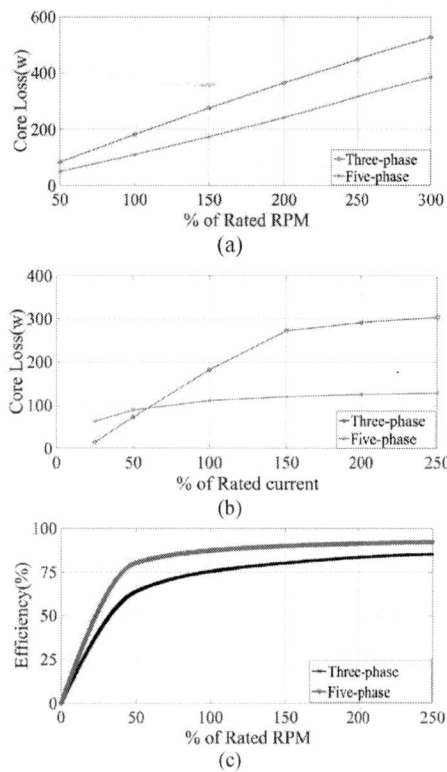

Fig.9. Comparison of: a) core losses on changing RPM , b) core losses on changing phase-current, c) Efficiency

Fig.9 (a) shows the core losses in the five-phase and three-phase motors under different rated speed. It is observed that under 100% rated the core losses are 160W and 190W in five-phase and three-phase motors respectively. Even the core losses increase with the speed, three-phase system always show higher values than the five-phase system. Fig.9 (b) shows the core loss variation under different phase currents. It is observed that, under 100% rated current the core losses are 110W and 190W in five-phase and three-phase motors respectively.

Fig.9 (c) shows the efficiency for five-phase and three-phase motors under same operating conditions. It is observed that, under 100% rated RPM the five-phase motor and three-phase motor show 88% and 75% efficiency respectively. Even though at higher speed (> rated rpm) the efficiency increases for three-phase motor, it reamins below than the efficiency of the five-phase motor.

C. Experimental result

The experimental testing has been performed to calculate the different losses in the five-phase systems including the five-phase inverter and PMa-SynRM. The machine with the parameter given in Table I has been fabricated for testing as shown in Fig. 10 (b) and (c) which are the five-phase stator and the rotor of PMa-SynRM. The five-phase inverter system

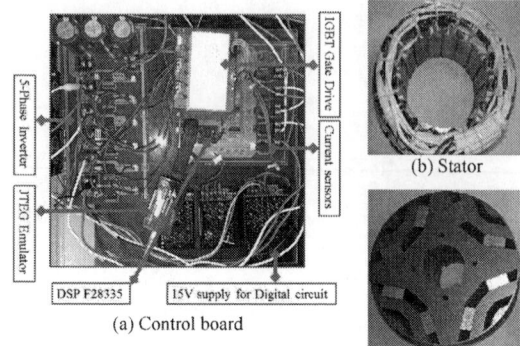

(a) Control board

(b) Stator

(c) Rotor

Fig. 10. Five-phase motor controller with fabricated motor.

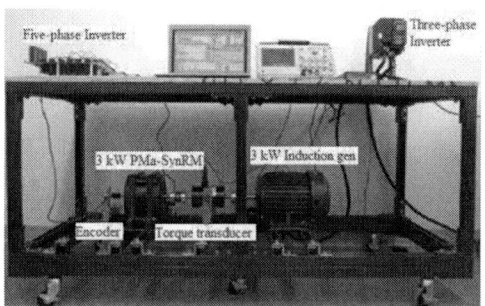

Fig. 11. Five hp dynamo system.

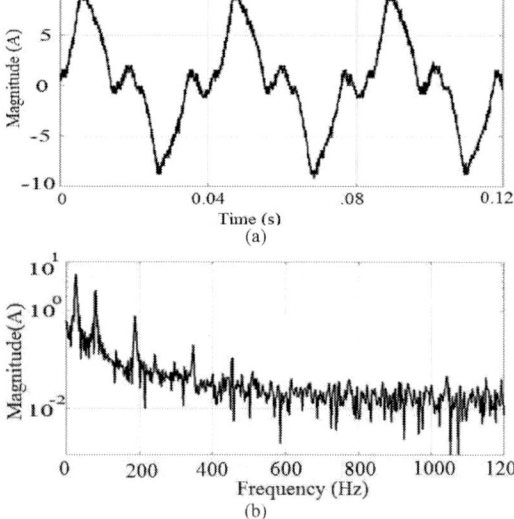

Fig.12. Five-phase results: (a) Phase currents, (b) Harmonics of the current,

is shown in Fig. 10 (a). A 5 hp dynamo system with the motor (5-phase PMa-SynRM) has been utilized as shown in Fig. 11. In the experiment, the five-phase system has been operated at vector control under MTPA condition by utilizing that 5 hp dynamo system integrated with TI DSP (F28335). The five-phase motor has been operated with 30% of the rated load. The current in phase A and the corresponding harmonics are shown in Fig. 12(a) and Fig.12 (b). The rms of the phase current is measured as 5.8A. As the machine was operated at

978-1-4673-9551-9/16 $31.00 © 2016 IEEE

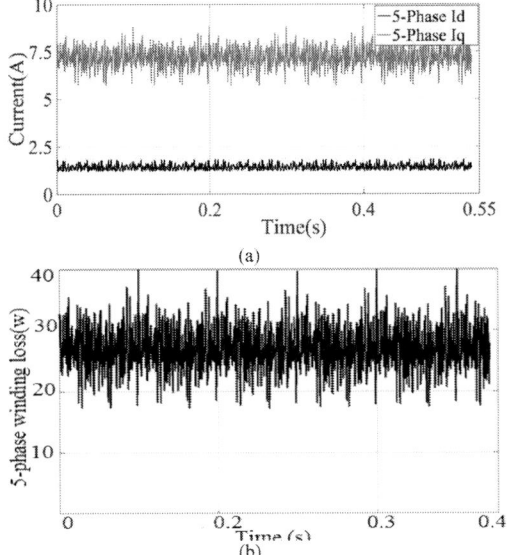

Fig.12. Five-phase results: (a) d-q axis current, (b) winding loss.

low load and low speed (720 rpm), the third harmonic is observed in the phase currents in Fig. 12(b) which is expected to be lower at full rated condition.

Fig. 13(a) shows the d-q axis currents of the five-phase system. Using the d-q axis currents, the winding losses are presented in Fig. 13 (b). The winding loss is observed ~28W.

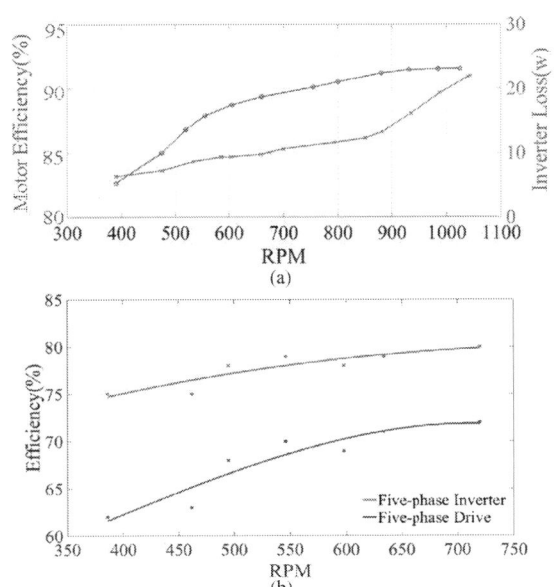

Fig.14. (a) inverter loss and motor efficiency and (b) Inverter efficiency and total drive efficiency.

Fig. 14(a) shows the inverter losses (switching loss and conduction losses) and the efficiency of the motor and drive system. The inverter loss is found as 22W at 1030 RPM. Fig.14 (a) also shows the efficiency of the motor system is measured as ~91% at 1030 RPM. Fig.14 (b) shows the

efficiency of the five-phase inverter and the total drive. The efficiency of the five-phase inverter and the total drive system are measured as 77% and 70% at 720 rpm.

VI. CONCLUSION

In this paper, a comparison study has been done on the electrical losses that occur in the inverter-fed three-phase and five-phase motor. Four types of electrical losses, such as winding loss and core loss in the machine-side, switching loss and conduction losses in the inverter-side are investigated and compared for both three-phase and five-phase machines. It has been observed that, for similar rated machines, the electrical losses occurred in the three-phase system are higher than five-phase system. Hence, the efficiency in the five-phase system has been found higher than the three-phase. The theoretical analysis has been proved by simulating the systems in MATLAB Simulink and finite element analysis. The experimental results for the five-phase system have been provided in this paper.

ACKNOWLEDGMENT

This work was partially supported by the Ohio Third Frontier Technology Validation and Startup Fund. The statements made herein are solely the responsibility of the authors.

REFERENCES

[1] E. B. Agamloh, A. Boglietti and A. Cavagnino, "The Incremental Design Efficiency Improvement of Commercially Manufactured Induction Motors," *IEEE Transactions on Industry Applications*, vol. 49, no. 6, 2013.

[2] K. Yamazaki, "Torque and Efficiency Calculation of an Interior Permanent Magnet Motor Considering Harmonic Iron Losses of Both the Stator and Rotor," *IEEE Transactions On Magnetics*, vol. 39, no. 3, 2003.

[3] M. Barcaro, N. Bianchi and M. Freddy Magnussen, "Permanent-Magnet Optimization in Permanent-Magnet-Assisted Synchronous Reluctance Motor for a Wide Constant-Power Speed Range," *IEEE Transactions On Industrial Electronics*, vol. 59, no. 6, 2012.

[4] L. Fang, J.-w. Jung, J.-P. Hong and a. J.-H. Lee, "Study on High-Efficiency Performance in Interior Permanent-Magnet Synchronous Motor With Double-Layer PM Design," *IEEE Transactions On Magnetics*, vol. 44, no. 11, 2008.

[5] C. Mademlis, I. Kioskeridis and N. Margaris, "Optimal Efficiency Control Strategy for Interior Permanent-Magnet Synchronous Motor Drives," *IEEE Transactions On Energy Conversion*, vol. 19, no. 4, 2004.

[6] P. Guglielmi, B. Boazzo, E. Armando, G. Pellegrino and A. Vagati, "Permanent-Magnet Minimization in PM-Assisted Synchronous Reluctance Motors for Wide Speed Range," *IEEE Transactions on Industry Applications*, vol. 49, no. 1, 2013.

[7] S. S. R. Bonthu, J. Baek and S. Choi, "Comparison of Optimized Permanent Magnet Assisted Synchronous Reluctance Motors with Three-phase and Five-phase Systems," *IEEE Energy Conversion Congress and Exposition (ECCE)*, September 2014.

[8] Suman Dwari and Leila Parsa, "Optimum Fault-Tolerant Control of Multi-phase Permanent Magnet Machines for Open-Circuit and Short-Circuit," *IEEE Applied Power Electronics Conference (APEC)*, February-March 2007.

[9] A. Mohammadpour, S. Mishra, and Leila Parsa, "Fault-Tolerant Operation of Multiphase Permanent-Magnet Machines Using Iterative

Learning Control," *IEEE Journal of Emerging and Selected Topics in Power Electronics*, vol. 2, no. 2, December 2013.

[10] A. Arafat and S. Choi, "Optimal Sustainable Fault Tolerant Control of Five-Phase Permanent Magnet Assisted Synchronous Reluctance Motor," *IEEE International Electric Machines and Drives Conference (IEMDC)*, May 2015.

[11] Leila Parsa and A. Toliyat, "Fault-Tolerant Five-Phase Permanent Magnet Motor Drives," *IEEE industry application Conference*, vol.2, October 2004.

[12] Nicola Bianchi, Silverio Bolognani, and Michele Dai Pré, "Strategies for the Fault-Tolerant Current Control of a Five-Phase Permanent-Magnet Motor," *IEEE Transactions on industry applications*, vol. 43, no. 4, July-August 2007.

[13] A. Mohammadpour and Leila Parsa, "A Unified Fault-Tolerant Current Control Approach for Five-Phase PM Motors With Trapezoidal Back EMF Under Different Stator Winding Connections," *IEEE transaction on Power Electronics*, vol. 28, no.7, October 2012.

[14] M. Islam, S. S. R. Bonthu, and S. Choi, "Obtaining Optimized Designs of Multi-Phase PMa-SynRM Using Lumped Parameter Model Based Optimizer", *IEEE International Conference on Machines and Drives (IEMDC)*, May 2015.

[15] M. Islam, and S. Choi, "Design of Five-phase Ferrite Magnet Assisted Synchronous Reluctance Motor using Lumped Parameter Model Based Optimizer and FEA," *IEEE International Conference on Machines and Drives (IEMDC)*, May 2015.

[16] B.kaku, I. Miyashita, and S.Sone, "Switching Loss Minimised Space Vector PWM Method for IGBT Three-Level Inverter," *IEE Electric Power Applications*, vol. 144, no.3, 1997.

978-1-4673-9551-9/16 $31.00 © 2016 IEEE

A Hybrid Adaptive Observer for the Speed and Flux Estimation of Induction Motors

Mihai Comanescu
Department of Engineering
Penn State Altoona
Altoona, PA, USA

Abstract – The paper presents a speed adaptive observer for the induction motor (IM) drive that can be used in a sensorless implementation. The observer is based on the model of the IM in the stationary reference frame: it uses linear feedback terms to estimate the fluxes and an adaptive law to estimate the speed. The feedback gains are designed using Lyapunov's nonlinear stability method. The investigation finds that the initial design only performs in limited speed range – as the speed increases, the speed estimate of the observer becomes inaccurate. Based on the initial design, the paper develops a hybrid observer: this uses a speed signal as an additional input (this input speed signal is relatively inaccurate - it can be an initial estimate of the motor speed or an estimate of the synchronous speed). It is shown that the hybrid design can be tuned to perform well in wide speed range: it yields relatively accurate estimates of the fluxes and speed. The theoretical developments are supported with simulations and experiments.

Keywords: induction motor control, state estimation, rotor flux angle estimation, speed estimation, flux estimation, adaptive observers, sensorless control.

I. INTRODUCTION

Field-orientation is the preferred control method for the induction motor (IM) drive in high-performance applications. The induction motor is used in many industries, mainly to drive fans, pumps and compressors. The IM is also used in traction applications, in electric or hybrid-electric vehicles and in windmills.

To implement the field-oriented control, the angle of the rotor flux vector is required [1]-[4] – this needs to be either measured or estimated. Sensorless control is attractive because the speed sensor (usually, an optical encoder) ruins the ruggedness of the IM drive – it also increases the cost, is sensitive to vibrations and performs poorly at high speed.

Sensorless control has been studied widely in the last two decades and is currently a relatively mature technology. The main method of obtaining the field-orientation angle and speed is to use an observer that is designed along one of the motor models. The estimation can be done with or without signal injection – a review of IM sensorless control was

presented in [1][6]; injection schemes were shown in [7][8].

There are various methods that can be used to estimate the fluxes and speed: the typical state observer [9], third-harmonic based estimation method [10], full-order observer [11] or sliding mode observers [12][13].

A special class of methods corresponds to the adaptive observers shown in [12]-[25]. These methods use the model of the IM in the stationary reference frame and design observers with linear feedback that estimate the state vector and speed. The speed estimate is obtained from an adaptive law that follows from the stability analysis of the observer. In practice, the adaptive law uses a PI controller.

A big deal related to the adaptive observers mentioned is the design of the feedback gains. In particular, [12][15] propose a set of gains such that the closed-loop poles of the system are a multiple of the open-loop poles. The gains are obtained from working with the characteristic equations of the two systems – since these are polynomials of the 4th order, the coefficients are rather complicated algebraic expressions.

This paper presents an alternative design of the adaptive observer: the feedback gains are obtained using a simple approach based on Lyapunov's nonlinear stability method. The speed estimate is obtained from an adaptive law that is similar to the one in [12][15].

It is shown through simulations that the basic design only works well in limited speed range. At low speed, the fluxes converge and the speed is accurate. However, when the speed range is extended, the speed estimate becomes inaccurate.

To solve this problem, the paper develops a hybrid observer: this takes a speed signal as an additional input - this initial speed signal (which may not be 100% accurate) is fed into the sensitive terms of the adaptive observer.

It is shown that the hybrid observer works reasonably well in wide speed range: it estimates the fluxes and yields a speed estimate that matches the real speed. Experimentally, it is found that the flux estimates have magnitudes that are larger than expected.

II. Nomenclature

ω_r, ω_e	Rotor electrical speed, synchronous speed
R_s, R_r	Stator, rotor resistances
L_m, L_s, L_r	Magnetizing, stator and rotor inductances
λ_r	Rotor flux magnitude
η	Inverse of rotor time constant
v_α, v_β	Voltages in stationary reference frame
i_α, i_β	Currents in stationary reference frame
$\lambda_\alpha, \lambda_\beta$	Rotor fluxes in stationary reference frame
$x, \hat{x}$	Real and estimated value of x
$\bar{x}$	Mismatch of x

III. Induction Motor Model

The model of the induction motor in the stationary reference frame consists of the equations of the fluxes and currents:

$$\begin{cases} \dfrac{d\lambda_\alpha}{dt} = -\eta\lambda_\alpha - \omega_r\lambda_\beta + \eta L_m i_\alpha \\[6pt] \dfrac{d\lambda_\beta}{dt} = \omega_r\lambda_\alpha - \eta\lambda_\beta + \eta L_m i_\beta \\[6pt] \dfrac{di_\alpha}{dt} = \eta\beta\lambda_\alpha + \omega_r\beta\lambda_\beta - \gamma i_\alpha + \dfrac{1}{\sigma L_s}v_\alpha \\[6pt] \dfrac{di_\beta}{dt} = -\omega_r\beta\lambda_\alpha + \eta\beta\lambda_\beta - \gamma i_\beta + \dfrac{1}{\sigma L_s}v_\beta \end{cases} \quad (1)$$

The equations of the torque and speed are omitted. The parameters $\sigma, \beta, \gamma, \eta$ are defined as:

$$\sigma = 1 - \frac{L_m^2}{L_s L_r} \quad \beta = \frac{L_m}{\sigma L_s L_r} \quad \gamma = \frac{1}{\sigma L_s}\left(\frac{L_m^2}{L_r^2}R_r + R_s\right) \quad \eta = \frac{R_r}{L_r} \quad (2)$$

IV. Adaptive Observer for the Speed and Flux of the IM

Working with model (1), the problem is to estimate the state vector $[\lambda_\alpha \quad \lambda_\beta \quad i_\alpha \quad i_\beta]^T$ and the speed ω_r. For this task, the voltages v_α, v_β and currents i_α, i_β are measured. The proposed observer uses a speed estimate $\hat{\omega}_r$; the equations are:

$$\begin{cases} \dfrac{d\hat{\lambda}_\alpha}{dt} = -\eta\hat{\lambda}_\alpha - \hat{\omega}_r\hat{\lambda}_\beta + \eta L_m i_\alpha + l_{11}\bar{\imath}_\alpha + l_{12}\bar{\imath}_\beta \\[6pt] \dfrac{d\hat{\lambda}_\beta}{dt} = \hat{\omega}_r\hat{\lambda}_\alpha - \eta\hat{\lambda}_\beta + \eta L_m i_\beta + l_{21}\bar{\imath}_\alpha + l_{22}\bar{\imath}_\beta \\[6pt] \dfrac{d\hat{\imath}_\alpha}{dt} = \eta\beta\hat{\lambda}_\alpha + \hat{\omega}_r\beta\hat{\lambda}_\beta - \gamma i_\alpha + \dfrac{1}{\sigma L_s}v_\alpha + l_{31}\bar{\imath}_\alpha + l_{32}\bar{\imath}_\beta \\[6pt] \dfrac{d\hat{\imath}_\beta}{dt} = -\hat{\omega}_r\beta\hat{\lambda}_\alpha + \eta\beta\hat{\lambda}_\beta - \gamma i_\beta + \dfrac{1}{\sigma L_s}v_\beta + l_{41}\bar{\imath}_\alpha + l_{42}\bar{\imath}_\beta \end{cases} \quad (3)$$

The mismatches are defined as: $\bar{\lambda}_\alpha = \hat{\lambda}_\alpha - \lambda_\alpha$; $\bar{\lambda}_\beta = \hat{\lambda}_\beta - \lambda_\beta$, $\bar{\imath}_\alpha = \hat{\imath}_\alpha - i_\alpha$, $\bar{\imath}_\beta = \hat{\imath}_\beta - i_\beta$. After subtraction:

$$\begin{cases} \dfrac{d\bar{\lambda}_\alpha}{dt} = -\eta\bar{\lambda}_\alpha - \omega_r\bar{\lambda}_\beta - \bar{\omega}_r\hat{\lambda}_\beta + l_{11}\bar{\imath}_\alpha + l_{12}\bar{\imath}_\beta \\[6pt] \dfrac{d\bar{\lambda}_\beta}{dt} = \bar{\omega}_r\hat{\lambda}_\alpha + \omega_r\bar{\lambda}_\alpha - \eta\bar{\lambda}_\beta + l_{21}\bar{\imath}_\alpha + l_{22}\bar{\imath}_\beta \\[6pt] \dfrac{d\bar{\imath}_\alpha}{dt} = \eta\beta\bar{\lambda}_\alpha + \omega_r\beta\bar{\lambda}_\beta + \bar{\omega}_r\beta\hat{\lambda}_\beta + l_{31}\bar{\imath}_\alpha + l_{32}\bar{\imath}_\beta \\[6pt] \dfrac{d\bar{\imath}_\beta}{dt} = -\hat{\omega}_r\beta\hat{\lambda}_\alpha - \omega_r\beta\bar{\lambda}_\alpha + \eta\beta\bar{\lambda}_\beta + l_{41}\bar{\imath}_\alpha + l_{42}\bar{\imath}_\beta \end{cases} \quad (4)$$

The speed mismatch is $\bar{\omega}_r = \hat{\omega}_r - \omega_r$. The gains l_{11} through l_{42} in (3) need to be designed. For that, select the positive definite Lyapunov function:

$$V = \frac{1}{2}\left(\bar{\imath}_\alpha^2 + \bar{\imath}_\beta^2 + \bar{\lambda}_\alpha^2 + \bar{\lambda}_\beta^2\right) + \frac{\bar{\omega}_r^2}{2\lambda} \quad (5)$$

where λ is a positive constant. After differentiating V and replacing the derivatives from (4), the result is:

$$\dot{V} = -\eta\bar{\lambda}_\alpha^2 - \omega_r\bar{\lambda}_\alpha\bar{\lambda}_\beta - \bar{\omega}_r\bar{\lambda}_\alpha\hat{\lambda}_\beta + l_{11}\bar{\lambda}_\alpha\bar{\imath}_\alpha + l_{12}\bar{\lambda}_\alpha\bar{\imath}_\beta + \\ \omega_r\bar{\lambda}_\alpha\bar{\lambda}_\beta - \eta\bar{\lambda}_\beta^2 + \bar{\omega}_r\hat{\lambda}_\alpha\bar{\lambda}_\beta + l_{21}\bar{\lambda}_\beta\bar{\imath}_\alpha + l_{22}\bar{\lambda}_\beta\bar{\imath}_\beta + \eta\beta\bar{\lambda}_\alpha\bar{\imath}_\alpha + \\ \omega_r\beta\bar{\lambda}_\beta\bar{\imath}_\alpha + \bar{\omega}_r\beta\hat{\lambda}_\beta\bar{\imath}_\alpha + l_{31}\bar{\imath}_\alpha^2 + l_{32}\bar{\imath}_\alpha\bar{\imath}_\beta - \omega_r\beta\bar{\lambda}_\alpha\bar{\imath}_\beta + \\ \eta\beta\bar{\lambda}_\beta\bar{\imath}_\beta - \bar{\omega}_r\beta\hat{\lambda}_\alpha\bar{\imath}_\beta + l_{41}\bar{\imath}_\alpha\bar{\imath}_\beta + l_{42}\bar{\imath}_\beta^2 \quad (6)$$

In (6), the terms alike are combined or simplified, some of them are desirable and will be retained; the others should be eliminated if possible.

Working with (6), the design gains should be chosen such that $\dot{V}$ is negative. If $\dot{V} < 0$ always, V decays, the mismatches tend to zero and the observer is stable.

The terms $-\eta\bar{\lambda}_\alpha^2 - \eta\bar{\lambda}_\beta^2 + l_{31}\bar{\imath}_\alpha^2 + l_{42}\bar{\imath}_\beta^2$ help make $\dot{V}$ negative; the gains are chosen as:

$$l_{31} = l_{42} = -k \qquad k > 0 \quad (7)$$

The terms in $\bar{\lambda}_\alpha\bar{\imath}_\alpha$ and the terms in $\bar{\lambda}_\beta\bar{\imath}_\beta$ can be eliminated; for that, we need:

$$l_{11} = -\eta\beta \qquad l_{22} = -\eta\beta \quad (8)$$

The terms in $\bar{\imath}_\alpha\bar{\imath}_\beta$ are not useful, they are eliminated with:

$$l_{32} = l_{41} = 0 \quad (9)$$

The terms in $\bar{\lambda}_\alpha\bar{\imath}_\beta$ and $\bar{\lambda}_\beta\bar{\imath}_\alpha$ are grouped together and they yield $(l_{12} - \omega_r\beta)\bar{\lambda}_\alpha\bar{\imath}_\beta$ and $(l_{21} + \omega_r\beta)\bar{\lambda}_\beta\bar{\imath}_\alpha$. Since these terms have unknown signs, they should be eliminated (however, ω_r is not measured). To eliminate or reduce them, the hope is that the speed estimate (that this method obtains) matches the real speed closely. Then, the gains l_{12} and l_{21} are chosen as:

$$l_{12} = \hat{\omega}_r\beta \qquad l_{21} = -\hat{\omega}_r\beta \quad (10)$$

With this set of gains, the expression of $\dot{V}$ becomes:

$$\dot{V} = -\eta\left(\bar{\lambda}_\alpha^2 + \bar{\lambda}_\beta^2\right) - k\left(\bar{\imath}_\alpha^2 + \bar{\imath}_\beta^2\right) + \bar{\omega}_r\beta\left(\bar{\lambda}_\alpha\bar{\imath}_\beta - \bar{\lambda}_\beta\bar{\imath}_\alpha\right) + \\ \bar{\omega}_r\left(\hat{\lambda}_\alpha\bar{\lambda}_\beta - \bar{\lambda}_\alpha\hat{\lambda}_\beta\right) - \bar{\omega}_r\beta\left(\hat{\lambda}_\alpha\bar{\imath}_\beta - \bar{\lambda}_\beta\hat{\imath}_\alpha\right) + \frac{1}{\lambda}\bar{\omega}_r\dot{\bar{\omega}}_r \quad (11)$$

Working with (11), $\dot{\bar{\omega}}_r$ is chosen such that the last two terms cancel each other. Since $\dot{\bar{\omega}}_r = \dot{\hat{\omega}}_r - \dot{\omega}_r$, assuming that the speed varies slowly ($\dot{\omega}_r \approx 0$), the adaptive law for the speed estimate is:

$$\dot{\hat{\omega}}_r = \lambda\beta\left(\bar{\lambda}_\alpha\bar{\imath}_\beta - \hat{\lambda}_\beta\bar{\imath}_\alpha\right) \quad (12)$$

Note that the estimates of the fluxes and the current mismatches in (12) are known – therefore, the expression on the right hand side can be calculated and then integrated.

The resulting expression of $\dot{V}$ is:

$$\dot{V} = -\eta\left(\bar{\lambda}_\alpha^2 + \bar{\lambda}_\beta^2\right) - k\left(\bar{\imath}_\alpha^2 + \bar{\imath}_\beta^2\right) + \bar{\omega}_r\beta\left(\bar{\lambda}_\alpha\bar{\imath}_\beta - \bar{\lambda}_\beta\bar{\imath}_\alpha\right) + \\ \bar{\omega}_r\left(\hat{\lambda}_\alpha\bar{\lambda}_\beta - \bar{\lambda}_\alpha\hat{\lambda}_\beta\right) \quad (13)$$

In (13), the terms $\bar{\omega}_r\beta\left(\bar{\lambda}_\alpha\bar{\imath}_\beta - \bar{\lambda}_\beta\bar{\imath}_\alpha\right)$ and $\bar{\omega}_r\left(\hat{\lambda}_\alpha\bar{\lambda}_\beta - \bar{\lambda}_\alpha\hat{\lambda}_\beta\right)$ cannot be calculated (because $\bar{\lambda}_\alpha$ and $\bar{\lambda}_\beta$ are unknown). These terms cannot be eliminated – the only hope is that the design of the observer will make them small. The observer settles in an equilibrium point given by $\dot{V} = 0$. This corresponds to:

$$\eta\left(\bar{\lambda}_\alpha^2 + \bar{\lambda}_\beta^2\right) + k\left(\bar{\imath}_\alpha^2 + \bar{\imath}_\beta^2\right) = \bar{\omega}_r\beta\left(\bar{\lambda}_\alpha\bar{\imath}_\beta - \bar{\lambda}_\beta\bar{\imath}_\alpha\right) + \\ \bar{\omega}_r\left(\hat{\lambda}_\alpha\bar{\lambda}_\beta - \bar{\lambda}_\alpha\hat{\lambda}_\beta\right) \quad (14)$$

If the right hand side of (14) is small, then, at the equilibrium point, $\bar{\lambda}_\alpha, \bar{\lambda}_\beta, \bar{\imath}_\alpha, \bar{\imath}_\beta, \bar{\omega}_r$ will be small. From a formal point of view, the adaptive observer shown does not converge – instead, it settles into equilibrium in a point given by (14) that, hopefully, is close to the origin.

978-1-4673-9551-9/16 $31.00 © 2016 IEEE

Note that the use of the adaptive law in (11) (that eliminates a corresponding term) does not guarantee that $\hat{\omega}_r \to \omega_r$. The only way $\bar{\omega}_r \to 0$ is if the equilibrium point given by (14) can be stretched towards the origin. In conclusion, the equations of the observer with the proposed gains are:

$$
\begin{cases}
\frac{d\hat{\lambda}_\alpha}{dt} = -\eta\hat{\lambda}_\alpha - \hat{\omega}_r\hat{\lambda}_\beta + \eta L_m i_\alpha - \eta\beta\bar{\imath}_\alpha + \hat{\omega}_r\beta\bar{\imath}_\beta \\
\frac{d\hat{\lambda}_\beta}{dt} = \hat{\omega}_r\hat{\lambda}_\alpha - \eta\hat{\lambda}_\beta + \eta L_m i_\beta - \hat{\omega}_r\beta\bar{\imath}_\alpha - \eta\beta\bar{\imath}_\beta \\
\frac{d\hat{\imath}_\alpha}{dt} = \eta\beta\hat{\lambda}_\alpha + \hat{\omega}_r\beta\hat{\lambda}_\beta - \gamma i_\alpha + \frac{1}{\sigma L_s}v_\alpha - k\bar{\imath}_\alpha \\
\frac{d\hat{\imath}_\beta}{dt} = -\hat{\omega}_r\beta\hat{\lambda}_\alpha + \eta\beta\hat{\lambda}_\beta - \gamma i_\beta + \frac{1}{\sigma L_s}v_\beta - k\bar{\imath}_\beta
\end{cases} \quad (15)
$$

V. SIMULATION RESULTS

The proposed observer is simulated using Simulink. The nameplate data and parameters of the induction motor used correspond to Table II. The motor is operated in speed control mode using synchronous reference frame PI current control. Current i_q^* is obtained through speed feedback. The observer is simulated with a 50 μs sampling time.

The motor is started, it then goes through acceleration and speed reversal. The load torque is 0.4 Nm; then 0.6 Nm at t=1.3s and then 0.3 Nm at t=1.7s. The drive starts with $i_d^* = 0.85A$ and this increases to 1.35A at t=2s.

The observer (15) is discretized using Euler's method and is implemented in discrete-time. The adaptive law (12) does not provide a fast enough dynamics of the speed estimate; instead the term $\lambda\beta(\hat{\lambda}_\alpha\bar{\imath}_\beta - \hat{\lambda}_\beta\bar{\imath}_\alpha)$ is fed through a PI controller. The speed estimate is of the form:

$$
\hat{\omega}_r = \left(K_p + \frac{K_i}{s}\right)\lambda\beta\left(\hat{\lambda}_\alpha\bar{\imath}_\beta - \hat{\lambda}_\beta\bar{\imath}_\alpha\right) \quad (16)
$$

The gains K_p, K_i, and λ are chosen through trial and error: $K_p = 0.00001$, $K_i = 0.0001$, $\lambda = 10^6$. The performance of the observer is tested for several values of k – this is the only feedback gain that can be chosen freely. The observer is simulated using a speed pattern that involves acceleration and speed reversal – the speed envelope is increased gradually.

Fig.1 and Fig.2 show the estimates (currents, fluxes, speed) at relatively low speed (the drive goes up to 300 rpm and reverses to -200 rpm) with $k = 30000$. The Lyapunov function V tends to zero and all the estimates are accurate – this result confirms the gain design.

The simulation show that, as the speed range increases, k must also be increased for the observer to perform well. However, at some point, this gain increase leads to instability or inaccuracy.

Fig.3 shows the behavior at higher speed (around 1200 rpm). It can be seen that the flux estimates and the speed are inaccurate and unusable. The currents still converge.

Disappointingly, this high speed behavior cannot be changed using the design gains. A further increase of k leads to instability; $k = 30000$ is already pretty high compared with $\eta = 17.8$ in (13). The attempt to change the adaptation parameters K_p and K_i does not improve the high speed estimation behavior. In conclusion, the adaptive observer with this gain design does not perform in wide speed range.

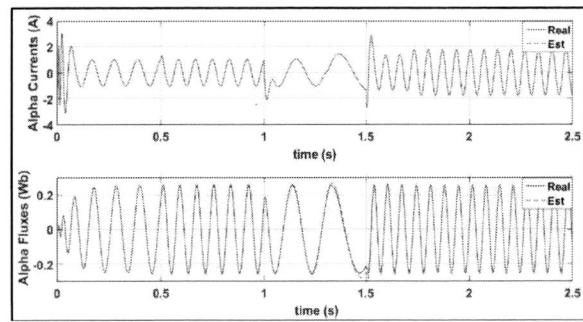

Fig.1. Real versus estimated currents and fluxes at low speed

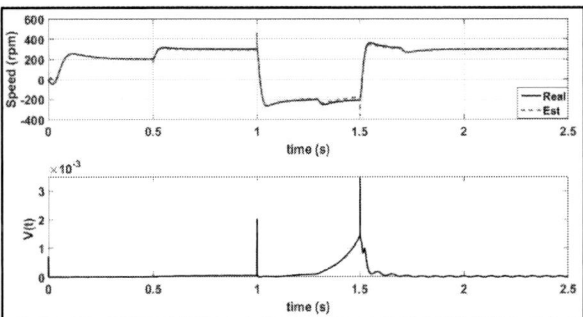

Fig.2. Estimated speed and Lyapunov function $V(t)$

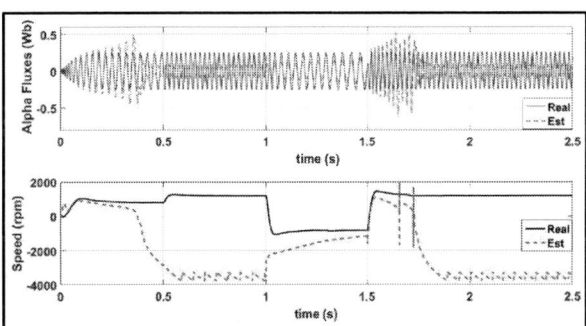

Fig.3. Estimated flux and speed at higher speed (1200 rpm)

VI. HYBRID ADAPTIVE OBSERVER WITH ADDITIONAL SPEED INPUT

This section examines the proposed adaptive observer and offers an alternative design – it will be shown that this performs better in wide speed range.

Examining the observer equations (15), note that the speed estimate $\hat{\omega}_r$ appears in three different roles: first, as a coefficient in the equations of fluxes [A], second, as a feedback gain in the flux dynamics [B] (terms $\hat{\omega}_r\beta\bar{\imath}_\beta$ and $-\hat{\omega}_r\beta\bar{\imath}_\alpha$); also, in the equations of the currents [C].

When the observer runs at high speed, the estimate $\hat{\omega}_r$ drifts away significantly from ω_r (Fig.3). When this highly inaccurate $\hat{\omega}_r$ is fed back into [A], [B] and [C], the observer does not perform. The investigation is looking to establish which of the three destinations is critical. For that, an accurate speed signal (the measured speed) is fed in each of these destinations and the estimation behavior is evaluated.

The main idea is that, if we identify which destination needs a precise value of $\hat{\omega}_r$, it is then possible to construct a hybrid observer: the sensitive destinations are fed with an input signal that is equal or almost equal to the real speed while the others still function with the speed estimate $\hat{\omega}_r$.

In the simulation, the speed is measured and is fed as an extra input in the observer – it replaces $\hat{\omega}_r$ in $[A],[B],[C]$. The combinations tested are shown in Table I.

It is found that the observer performs best when the input speed is fed into $[A]$ and $[B]$. In this case, the estimated fluxes are in phase with the real ones (the magnitudes are slightly higher but relatively close), the speed estimate is slightly smaller than the real speed. The waveforms are omitted.

TABLE I. HYBRID OBSERVER COMBINATIONS

Observer Performance	Destination for ω_r		
	$[A]$	$[B]$	$[C]$
Fluxes and speed are accurate	x		
Does not perform		x	
Speed estimate diverges			x
Fluxes and speed are accurate	x	x	
Does not perform		x	x
Does not perform	x		x

VII. HYBRID ADAPTIVE OBSERVER WITH SPEED ESTIMATE

Based on Table I, the proposal is to construct an observer that is fed with a speed estimate at the input (in addition to v_α, v_β and i_α, i_β); this observer uses the previous adaptation law and gain design.

The observer can be fed with a speed estimate (obtained from another source), or, for example, it can be fed with an estimate of the synchronous speed. It is reasonable to assume that the speed estimate available is within $\pm 20\%$ of the real speed. For this hybrid design, the incoming speed signal $\hat{\omega}_r^i$ is fed into destinations $[A]$ and $[B]$.

A block diagram of this observer is shown in Fig.4. To increase the flexibility, a tuning gain K_ω is included and this is used to slightly modify the value of the input speed.

Fig.5 shows the estimated currents and fluxes when the hybrid adaptive observer is fed with the synchronous speed; $K_\omega = 1.1$, $k = 30000$, $K_p = 0.00001$, $K_i = 0.0001$. The drive runs at 1200 rpm. Note that the estimates are accurate.

Fig.6 shows the waveforms of the speed estimate for various values of K_ω. It is clear that the output $\hat{\omega}_r$ depends on the input $K_\omega \hat{\omega}_r^i$. Depending on the input speed signal and on the intended speed range, K_ω can be tuned such that $\hat{\omega}_r$ is accurate. For the situation shown in Fig.6, the observer is fed with an estimate of the synchronous speed and performs quite well for $K_\omega = 1.2$. The waveforms of $\hat{\omega}_r$ are also shown for $K_\omega = 0.9$ and $K_\omega = 1.1$.

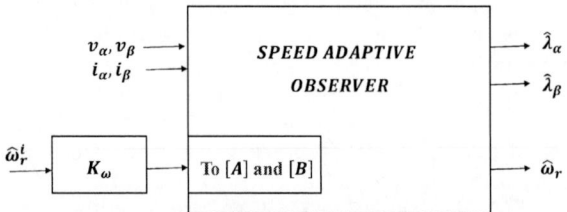

Fig.4. Block diagram of the adaptive observer with input speed signal

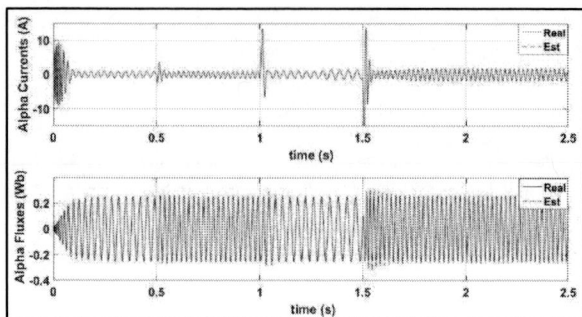

Fig.5. Estimated currents and fluxes of the hybrid observer

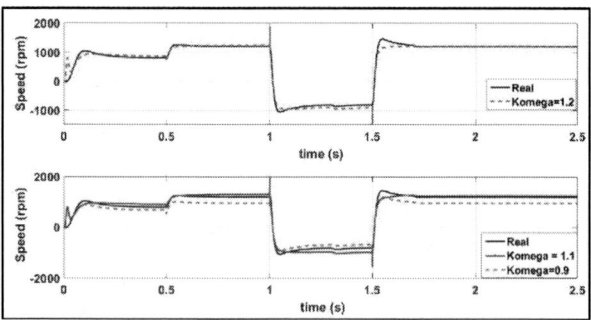

Fig.6. Estimated speed of the hybrid observer for several values of K_ω

VIII. EXPERIMENTAL RESULTS

The experimental setup uses a Dayton 2N863M motor, a TI TMS320F2812 digital signal processor (DSP), and a MOSFET inverter (Fig.7).

The waveforms are obtained using the second PWM state machine of the DSP chip (the variables to be displayed are used as PWM duty cycles; the output waveforms are RC filtered and are captured on a scope).

The currents are measured using resistors mounted in the lower legs of the inverter. The voltages are computed from the measured dc bus voltage and the PWM duty cycles.

The control algorithm is implemented using a single interrupt. The switching frequency of the inverter is $15\ KHz$ and the sampling time of the control system is $66\ \mu s$.

The experimental setup does not have an encoder; instead, the speed is calculated from the slip equation:

$$\omega_r = \omega_e - \frac{L_m}{T_r}\frac{1}{\lambda_r^2}\left(\hat{\lambda}_\alpha i_\beta - \hat{\lambda}_\beta i_\alpha\right) \qquad (17)$$

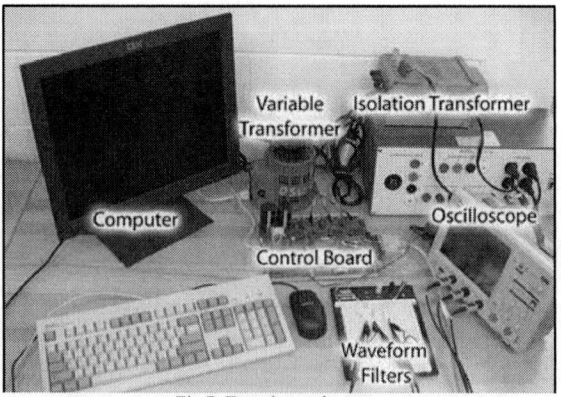

Fig.7. Experimental setup

978-1-4673-9551-9/16 $31.00 © 2016 IEEE

The fluxes obtained from the proposed observer are compared with the estimated fluxes obtained from the Voltage Model Observer described in [26] (this serves as reference).

Three adaptive observers have been implemented. The basic design given by (15) and the hybrid observer where $\hat{\omega}_r^i$ is fed in [A] did not perform – the waveforms are omitted.

The only design that performed corresponds to row 4 in Table I ($\hat{\omega}_r^i$ is fed into [A] and [B]). Fig.8 and Fig.9 show the waveforms of this observer, $\hat{\omega}_r^i$ was obtained with (17) and $K_\omega = 1$. It can be seen that the currents converge, the fluxes are in phase with the VMO fluxes (but the magnitude is higher –as seen in the simulations). The speed estimate corresponds qualitatively with the reference speed (the transient in Fig.8a corresponds to a 0-1000 rpm start-up and then a shut-down). In Fig.8b, the rotor flux angle obtained by this observer matches very well the one of the VMO.

Fig.10 is obtained with $K_\omega = 0.9$ – note that the magnitude of the estimated flux is way larger than the VMO flux and is almost approaching the per unit saturation limit of the DSP. The currents converge. For this value of K_ω, the speed estimate is inaccurate in transient (waveforms are not shown).

In the experiment, the feedback gain of the adaptive observer was $k = 0.05$ and $K_p = 5$, $K_i = 0.2$.

In principle, the proposed observer seems to yield only two of the three quantities required for IM sensorless control: the flux angle is accurate, the speed is relatively accurate; however, the flux magnitude is not. Changing the value of K_ω does not appear to solve this problem; also, working with $K_\omega < 1$ yields fluxes that have higher magnitude.

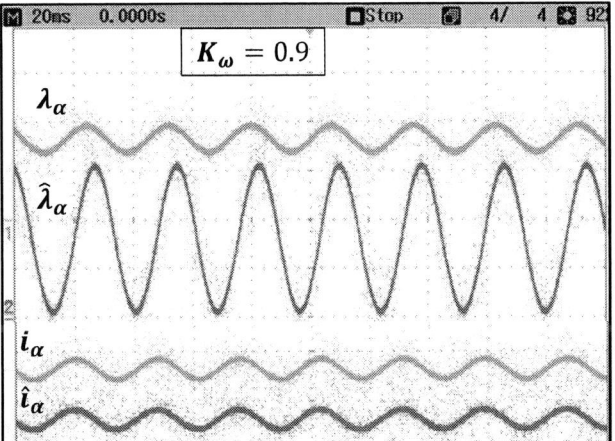

Fig.10. Estimated flux and currents of the hybrid observer (α axis), $\hat{\omega}_r^i$ is fed into [A] and [B], $K_\omega = 0.9$

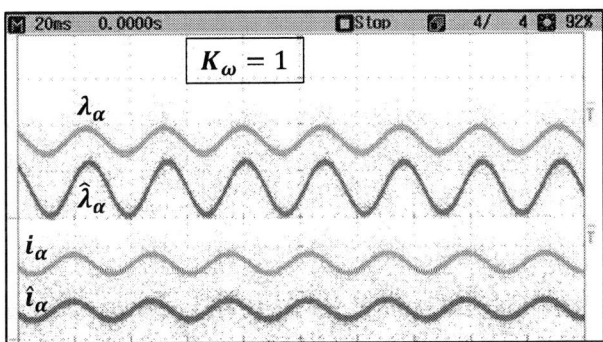

Fig.8. Estimated fluxes and currents of the hybrid observer (α axis), $\hat{\omega}_r^i$ is fed into [A] and [B], $K_\omega = 1$

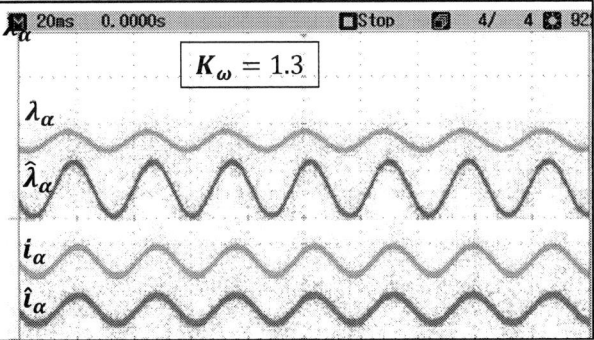

Fig.11. Estimated flux and currents of the hybrid observer (α axis), $\hat{\omega}_r^i$ is fed into [A] and [B], $K_\omega = 1.3$

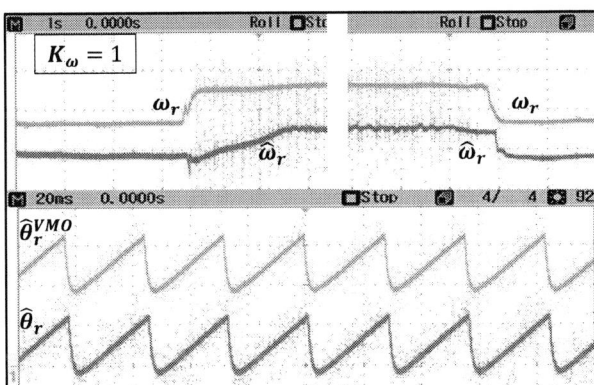

Fig.9. Estimated speed and rotor flux angle of the hybrid observer, $\hat{\omega}_r^i$ is fed into [A] and [B], $K_\omega = 1$

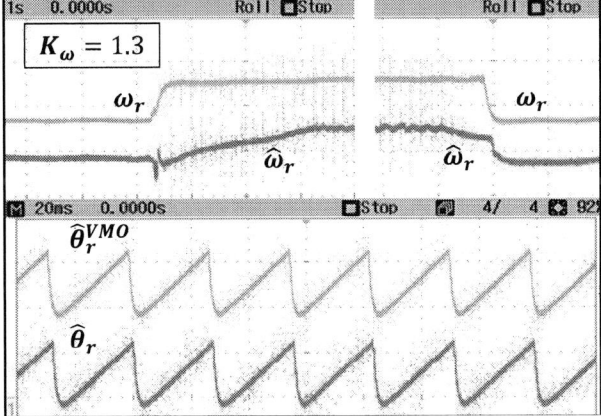

Fig.12. Estimated speed and rotor flux angle of the hybrid observer, $\hat{\omega}_r^i$ is fed into [A] and [B], $K_\omega = 1.3$

Fig.11 and Fig.12 are obtained for $K_\omega = 1.3$. Again, the currents converge and the magnitude of the flux estimate is higher than real but is in phase with the VMO flux. The speed estimate matches the real speed although the acceleration is slow (Fig.12); the estimated rotor flux angle corresponds with the one of the VMO.

IX. CONCLUSIONS

The paper presents a simple design of a speed adaptive observer for the induction motor drive. The observer is based on the model of the IM in the stationary reference frame - it uses linear feedback terms to estimate the fluxes and an adaptive law to estimate the speed. The gains are designed using Lyapunov's nonlinear stability method. The investigation finds that, in the initial form, the observer only seems to work well at low speed – as the speed range is extended, the estimates becomes inaccurate. To solve this problem, a hybrid observer is developed: this uses an incoming speed signal as an additional input and estimates the motor speed and fluxes. It is shown that the hybrid observer can be tuned to work well in wide speed range. Experimentally, this observer yields fluxes that are in phase with the expected ones; however, they have inaccurate magnitudes. The method is applicable to a sensorless IM drive where the speed and the field-orientation angle are needed for implementation of the control algorithm.

REFERENCES

[1] B.K. Bose *"Modern Power Electronics and AC Drives,"* Upper Saddle River, NJ, Prentice Hall, 2002.
[2] R. De Doncker and D. Novotny "The universal field oriented controller," IEEE Transactions on Industry Applications, Vol. 30, 1994, pp. 92-100.
[3] C.F. Hu, R.B. Hong and C.H. Liu "Stability Analysis and PI controller tuning for a Speed Sensorless Induction Motor Drive," Industrial Electronics Society Conference IECON 2004, pp. 877-882.
[4] K. Ohyama and K. Shinohara "Small Signal Stability Analysis of Vector Control System of Induction Motor Without Speed Sensor Using Synchronous Current Regulator," IEEE Transactions on Industry Applications, Vol. 36, No.6, Nov-Dec. 2000.
[5] J. Holtz "Sensorless Control of Induction Machines – With or Without Signal Injection?" IEEE Transactions on Industrial Electronics, Vol. 53, No. 1, Feb. 2006, pp. 7-30.
[6] J. Holtz "Sensorless control of induction motors," Proc. IEEE, vol. 90. No.8, Aug. 2002, pp. 1358-1394.
[7] J.I. Ha and S.K. Sul "Sensorless field-oriented control of an induction machine by high-frequency signal injection," IEEE Transactions on Industry Applications, Vol. 35, No.1, Feb. 1999, pp. 45-51.
[8] B.H. Bae, G.B. Kim and S.K. Sul "Improvement of low speed characteristics of railway vehicle by sensorless control using high-frequency injection," IEEE Industry Applications Society Annual Meeting, Rome, Oct. 2000, pp.1874-1880.
[9] P.L. Jansen, R.D. Lorenz and D.W. Novotny "Observer based direct field orientation – analysis and comparison of alternative methods," IEEE Trans. on Industry Applications, Vol. 30, 1994, pp. 945-953.
[10] L. Kreindler, J.C. Moreira, A. Testa and T.A. Lipo "Direct field orientation controller using the stator phase voltage third harmonic," IEEE Trans. on Industry Applications, Vol. 30, 1994, pp. 441-447.
[11] M. Hinkkanen "Analysis and Design of Full Order Flux Observers for Sensorless Induction Motors," IEEE Transactions on Industrial Electronics, vol. 51, no. 5, Oct. 2004, pp. 1033-1040.
[12] A.B. Proca and A. Keyhani "Sliding-mode flux observer with online rotor parameter estimation for induction motors," IEEE Transactions on Industrial Electronics, vol. 54, no.2, 2007, pp. 716-723.
[13] S. Rao, M. Buss and V.I. Utkin "Simultaneous State and Parameter Estimation in Induction Motors using First and Second Order Sliding Modes," IEEE Transactions on Industrial Electronics, vol. 56, no. 9, 2009, pp. 3369-3376.
[14] H. Kubota, K. Matsuse and T. Nakano "New Adaptive Flux Observer of Induction Motor for Wide Speed Range Motor Drives," IEEE Industrial Electronics Conference, IECON, 1990, pp. 921-926.
[15] H. Kubota, K. Matsuse and T. Nakano "DSP-Based Speed Adaptive Flux Observer of Induction Motor," IEEE Transactions on Industry Applications, Vol. 29, No.2, March/April 1993, pp. 344-348.

[16] G. Yang and T.H. Chin "Adaptive-Speed Identification Scheme for a Vector-Controlled Speed Sensorless Inverter-Induction Motor Drive," IEEE Transactions on Industry Applications, Vol. 29, No. 4, July/August 1993, pp. 820-825.
[17] H. Hashimoto, Y. Ohno, S. Kondo and F. Harashima "Torque Control of Induction Motor Using Predictive Observer," Power Electronics Specialist Conference, PESC, June 1989, Vol. 1, pp. 271-278.
[18] H. Hoffman and S.R. Sanders " Speed-Sensorless Vector Torque Control of Induction Machines using a Two-Time-Scale Approach," IEEE Transactions on Industry Applications, Vol. 34, No. 1, Jan/Feb. 1998, pp. 169-177.
[19] H. Kubota, I. Sato, Y. Tamura, K. Matsuse, H. Ohta and Y. Hori "Regenerating-Mode Low-Speed Operation of Sensorless Induction Motor Drive with Adaptive Observer," IEEE Transactions on Industry Applications, Vol. 38, No.4, July/Aug. 2002, pp. 1081-1086.
[20] S. Suwankawin and S. Sangwonwanich "A Speed-Sensorless IM Drive with Decoupling Control and Stability Analysis of Speed Estimation," IEEE Transactions on Industrial Electronics, Vol. 49, No. 2, Apr. 2002, pp. 444-455.
[21] J. Maes and J.A. Melkebeek "Speed-Sensorless Direct Torque Control of Induction Motors using and Adaptive Flux Observer," IEEE Transactions on Industry Applications, Vol. 36, No. 3, May/June 2000, pp. 778-785.
[22] M. Depenbrock and A. Steimel "Discussion of Regernerating Mode Low Speed Operation of Sensorless induction Motor Drive with Adaptive Observer," IEEE Transactions on Industry Applications, Vol. 39, No. 1, Jan/Feb. 2003, pp. 19.
[23] H. Kubota " Closure to Discussion of Regenerating _Mode Low-Speed operation of Sensorless Induction Motor Drive with Adaptive Observer," IEEE Transactions on Industry Applications, Vol. 39, No. 1, Jan/Feb. 2003, pp. 20.
[24] F. Hoffman and S. Koch "Steady State Analysis of Speed Sensorless Control of Induction Machines," Proceedings of the 24th Annual Conference of IEEE Industrial Electronics Society, Aug/Sept. 1998, Vol. 3, pp. 1626-1631.
[25] C. Heising, V. Staudt and A. Steimel "Speed-sensorless stator-flux-oriented control of induction motor drives in traction," 2010 Symposium on Sliding Mode Control for Electrical Drives, July 2010, pp. 100-106.
[26] C. Lascu, I. Boldea and F. Blaabjerg "A modified direct torque control for induction motor sensorless drive," IEEE Transactions on Industry Applications, Vol. 36, No. 1, 2000, pp. 120-130.

TABLE II. INDUCTION MOTOR SPECIFICATIONS AND PARAMETERS

Rating	¼ hp	Pole #	4
Speed	1732 rpm	Voltage	220 V
R_s	10.9 Ω		
L_{ls}, L_{lr}	0.015 H		
L_m	0.30 H		
R_r	5.57 Ω		

Determination of CM Choke Parameters for SiC MOSFET Motor Drive Based on Simple Measurements and Frequency Domain Modeling

Di Han, Casey Morris, Woongkul Lee, Bulent Sarlioglu, *Senior Member, IEEE*
Wisconsin Electric Machines and Power Electronic Consortium (WEMPEC)
University of Wisconsin-Madison
Madison, WI 53706 USA
sarlioglu@wisc.edu

Abstract—The adoption of silicon carbide (SiC) MOSFETs in variable speed motor drives makes it possible to increase the inverter switching frequency up to several hundred kilohertz without incurring excessive inverter loss. As a result, the harmonic currents and related losses in the machine can be significantly reduced, and the dynamic performance of motor will also be improved. However, the increased switching frequency of SiC drives will increase the ground leakage current in the common mode (CM) path, presenting new challenges on CM choke design. This paper aims at understanding the CM choke design under this new circumstance. First, a simple and accurate frequency domain CM circuit modeling approach suitable for SiC motor drives is proposed and subsequently verified through experimental tests. Based on the model, required choke parameters are then determined through analytical calculation. Through comparative analysis, the impact of increased switching frequency on CM choke design is studied.

Keywords—*common mode choke (CMC); electromagnetic interference (EMI); motor drive; silicon carbide*

I. INTRODUCTION

Silicon carbide (SiC) based power switching devices offer superior properties such as lower semiconductor loss, higher switching speed, and higher temperature capabilities, compared to conventional silicon based devices [1]-[3]. By replacing conventional Si IGBTs with SiC MOSFETs in motor drives, the semiconductor loss in power converters will decrease considerably, resulting in heatsink reduction and efficiency improvement [4]-[5]. On the other hand, it is also feasible to increase the switching frequency of SiC inverters from 20 kHz up to 200 kHz [4], in order to reduce the harmonic power loss and improve the dynamic performance of motors [6]. The ability to increase switching frequency has been proven to be extremely beneficial for high speed or high fundamental frequency motor drives [7]. However, the increased switching frequency also leads to an increase in the electromagnetic interference (EMI) noise emission from the motor drives, with common mode (CM) ground leakage current (Fig. 1) as one of the primary concern. Typically, CM choke inductors are employed as the simplest and most cost-effective means to suppress the ground leakage current. However, the CM chokes utilized in conventional low switching frequency drives will no longer be suitable for SiC drives operating at high switching frequency, and need to be redesigned. Hence, this paper investigates the CM choke design for high switching frequency SiC MOSFET based motor drives.

The CM noise path and filters for variable speed drives have been extensively studied in the literature over the past twenty years [8]-[14]. Various aspects of CM inductor design and performance have been analyzed [15]-[18]. The required characteristics of CM chokes for induction motor drive have been specified in [15]. Reference [16] investigated the influence of CM voltage-second on CM inductor saturation and highlighted the resonance in CM circuit. Reference [17] studied the modulation impact on CM choke performance. In reference [18], a simple CM choke design method has been proposed. However, previous studies only focus on the Si IGBT based drives, and the switching frequencies mentioned in the literature hardly exceed 20 kHz. The influence of increased switching frequency on CM inductor design has not been well

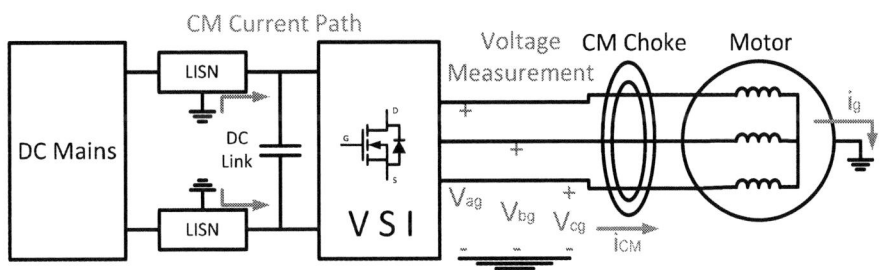

Fig. 1. Circuit diagram of a DC fed motor drive with CM choke.

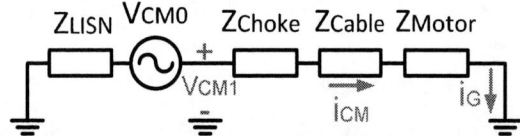

Fig. 2. CM model of motor drive with CM choke.

Fig. 3. Conducted EMI test bench for SiC motor drive.

investigated and is not well understood.

In this paper, the CM choke inductor design for SiC based inverter motor drive operating at 200 kHz switching frequencies is studied. To facilitate the analysis, a simple and accurate frequency domain modeling approach is introduced. With the proposed model, the inductance value and volt-second requirements of the CM choke are determined. As a comparison, the same analysis is done for the conventional 20 kHz switching frequency case, so that influence of increased switching frequency is clearly understood.

II. FREQUENCY DOMAIN MODELING BASED ON SIMPLE MEASUREMENTS

Modeling of EMI noise propagation is generally difficult because of the ubiquitous parasitic components in the practical systems. A widely applied strategy to cope with the difficulty of modeling EMI is to extract the "dominant" parasitics and form a simplified lumped element circuit model for analysis [14], [16], [17]. However, the extraction and simplification process usually requires priori knowledge of the problem to be analyzed; hence, the accuracy is not always guaranteed. Instead, the modeling approach proposed in this paper treats each part of the system as a frequency-dependent impedance, which completely characterizes the circuit behavior with all parasitics included.

Fig. 1 shows the diagram of a SiC inverter motor drive system. The proposed CM model of the system is illustrated in Fig. 2, where all quantities (voltage, current, and impedance) are expressed in frequency domain. CM impedance values can be determined through measurements using impedance analyzer or theoretical calculation over the frequency range of interest. CM voltage can be obtained through no load

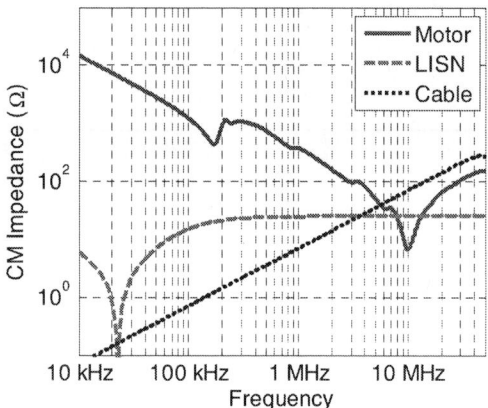

Fig. 4. CM impedances of motor drive system.

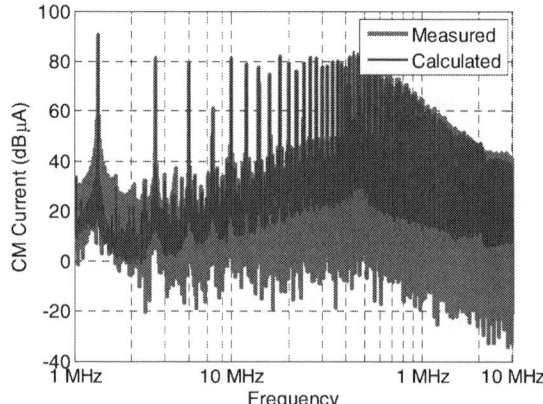

Fig. 5. Comparison of calculated and measured CM current.

measurements and transformed into frequency domain. Therefore, the system behavior can be fully characterized.

To demonstrate and verify the effectiveness of the proposed modeling approach, a SiC MOSFET (CMF20120) based induction motor drive is setup for conducted EMI testing as shown in Fig. 3. The system is configured the same way as shown in Fig. 1 except that the CM choke is not present. For easy current measurement, the motor is grounded through a grounding wire. The motor is set to run at no load condition and the switching frequency of SiC inverter is set to be 200 kHz.

The CM impedance of each part is measured from 10 kHz to 50 MHz with an impedance analyzer (Wayne Kerr 6500B) and the magnitude values are plotted in Fig. 4 (phase angle not shown). CM voltage can be acquired by measuring three line-to-ground voltages at inverter terminals,

$$V_{CM1} = (V_{ag} + V_{bg} + V_{cg}) / 3. \qquad (1)$$

However, most papers fail to notice that the CM voltage measured at inverter terminals V_{CM1} is not the source CM voltage V_{CM0} when machine is grounded as shown in Fig. 2,

$$V_{CM0} = (Z_{LISN} / Z_{Load} + 1) V_{CM1}, \qquad (2)$$

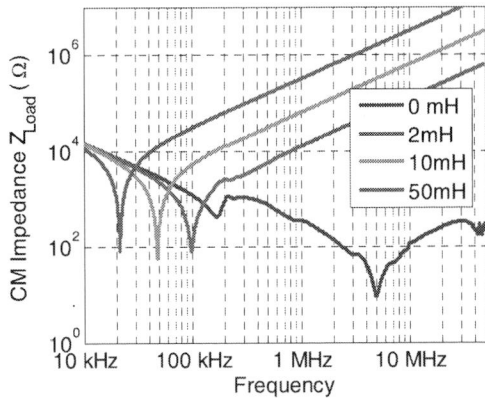

Fig. 6. Load CM impedance after the insertion of ideal CM chokes.

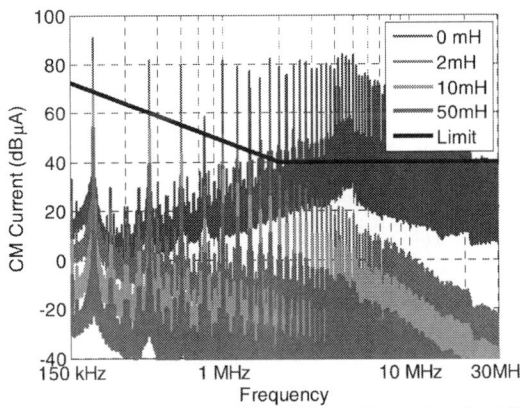

Fig. 7. Calculated CM current with/without CM chokes for 200 kHz switching frequency case.

where
$$Z_{Load} = Z_{Choke} + Z_{Cable} + Z_{Motor}. \tag{3}$$

Then the CM current can be calculated as

$$I_{CM} = I_G = V_{CM0} / (Z_{LISN} + Z_{Load}). \tag{4}$$

The calculated and measured CM currents are compared in Fig. 5. As can been seen, the two spectra match very well with each other over the entire conducted EMI frequency range (150 kHz – 30 MHz). Thus, the accuracy of the model is verified.

III. DETERMINATION OF CM CHOKE PARAMETERS WITH PROPOSED MODEL

A. Case 1: 200 kHz Switching Frequency

With the model built in previous section, performances of different CM chokes can be simply examined by including the Z_{choke} term in the equations. In this section, it is assumed that the CM impedance of a given choke is purely inductive in the frequency range of interest, and its inductance does not vary with frequency, thus is characterized by $j2\pi f L$. It is also assumed that the CM current limit is $72 - 40$ dBμA for 150 kHz – 2 MHz, respectively, and 40 dBμA for 2 MHz – 30 MHz, as suggested in [19].

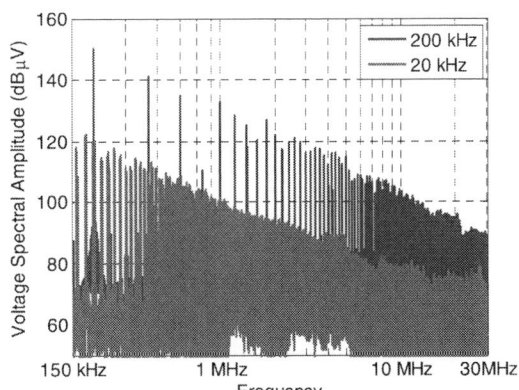

Fig. 8. Comparison of CM voltage V_{CM0} for 200 kHz and 20 kHz.

To determine the optimal CM choke parameters, a sweep of inductance can be carried out. In this study, three different chokes, with CM inductance values of 2, 10, and 50 mH, are analyzed. The load CM impedance Z_{Load} after the insertion of chokes is shown in Fig. 6. It is seen that the CM chokes not only increase the circuit impedance, but also shift the resonant point leftwards to avoid the EMI frequency range and the switching frequency. The resulted CM current spectra are calculated and shown in Fig. 7. As can be seen, the highest amplitude component is always located at 200 kHz, i.e. the switching frequency. Only the 50 mH choke can successfully attenuate the 200 kHz current component to below the given limit.

The CM volt-seconds impressed on the choke can be estimated using the highest amplitude current component according to,

$$\lambda = \int V_L dt = 2LI_{pk}. \tag{5}$$

The resulted values are 48, 50, and 50 μVs for the 2, 10, 50 mH chokes, respectively. Hence, the volt-second stresses for all three chokes are roughly the same. This is due to the fact that resonant frequencies are shifted to far below the switching frequency and all source CM volt-seconds are dropped on the choke inductors. Hence, the 50 mH choke will have the largest number of turns according to (6), and yield the smallest flux density in magnetic core according to (7).

$$L = n^2 \mu A / l, \tag{6}$$

$$\lambda = 2nB_{pk}A. \tag{7}$$

In other words, among the three chokes, the 50 mH choke will need the largest window area to fit the windings, but the smallest core cross sectional area to prevent saturation.

B. Case 2: 20 kHz Switching Frequency

In this part, 20 kHz is used as the switching frequency to represent the typical value used for a regular motor drive nowadays. The CM voltage V_{CM1} is measured the same way as introduced previously. V_{CM0} is calculated and compared with the 200 kHz case as shown in Fig. 8. It is shown that the CM

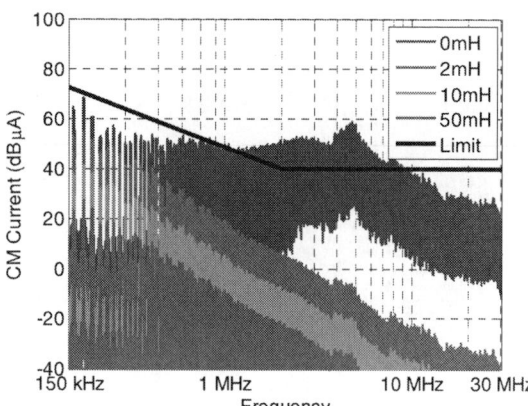

Fig. 9. Calculated CM current with/without CM chokes for 20 kHz switching frequency case.

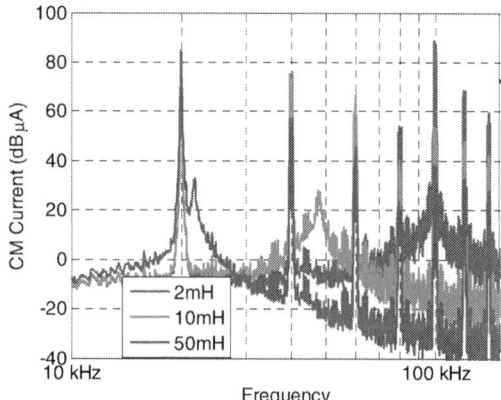

Fig. 10. Calculated CM current spectra around switching frequency (20 kHz) and its low order harmonics.

voltage is reduced by 20 to 30 dB in the frequency range of interest, as switching frequency is decreased by 10 times.

The CM current with and without CM chokes are also calculated and plotted in Fig. 9. Due to the reduced CM voltage, the CM current of 20 kHz case is also 20 dB lower than the 200 kHz case. It can be observed that by simply adding a 2 mH CM inductor, the CM current will be suppressed to below the limit.

Another important distinction from the 200 kHz case is that 20 kHz is below or around the resonant frequency of the system as shown in Fig. 6. Hence, the highest-amplitude CM current component would be located at 20 kHz or its low order harmonics as plotted in Fig. 10. It can be observed that resonant current peak occurs at 100, 40, and 20 kHz for the 2, 10, and 50 mH chokes, respectively. Even though these low frequency current components do not fall into the conducted EMI frequency range, they do determine the CM volt-second stress on CM chokes. Volt-second on each choke can be evaluated according to (5). The resulted values are 2205, 560, and 1650 µVs for the 2, 10, and 50 mH chokes, respectively. Hence, the 10 mH choke would have the lowest flux density among the three chokes, and requires the smallest core area.

TABLE I
PARAMETERS OF THE DESIGNED CM CHOKES

No.	1	2	3	4	5	6
Switching frequency f (kHz)	20	20	20	200	200	200
CM inductance L_{cm} (mH)	2	10	50	2	10	50
Phase current I_{DM} (Apk)	3	3	3	3	3	3
AWG	21	21	21	21	21	21
CM Volt-sec λ (µVs)	2205	560	1650	48	50	50
Magnetic core part No.	T60006-L2040-W452	T60006-L2025-W523	T60006-L2063-W517	T60006-L2025-W380	T60006-L2050-W516	T60006-L2090-W518
Outer diameter (mm)	40	25	63	25	50	90
Inner diameter (mm)	32	20	50	16	40	60
Height (mm)	15	10	25	10	20	20
Volume (mL)	26.99	7.66	101.74	7.64	52.60	176.56
Total weight (g)	38	10.4	161	17	79	395
No. of turns N	14	20	32	16	38	63
Leakage inductance per phase L_{dm} (µH)	14.9	13.2	112	57.1	96.8	414
Flux density B_{pk} (T)	0.660	0.482	0.746	0.079	0.124	0.097

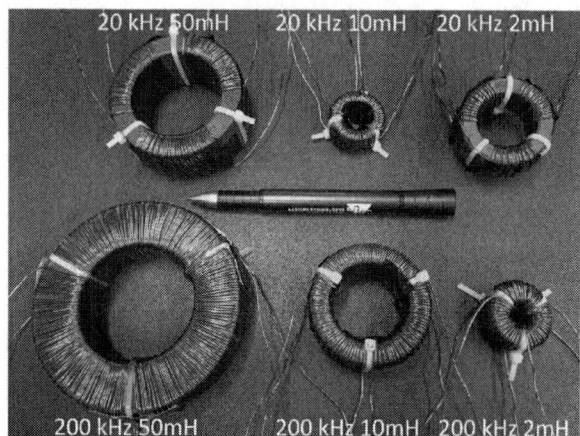

Fig. 11. CM chokes for different switching frequency and inductance.

IV. CM CHOKE DESIGN AND TESTING

A. Magnetic Design of CM Choke

With the given inductances, current rating, and the volt-second stress obtained from the previous section, the six CM chokes are designed according to the procedure and analytical equations detailed in [20] and [21].

Nanocrystalline material VITROPERM 500F [22] from VAC are chosen as the core material due to its superior properties, such as high permeability, high saturation flux density, low loss, and low temperature sensitivity. Off-the-shelf toroid cores are used for the design. Single layer of windings are also required to minimize the parasitic capacitance and the self-resonance frequency of chokes [23]. Critical parameters of the designed CM chokes are listed in Table I. According to the design, six chokes are manufactured in the lab, as shown in Fig. 11.

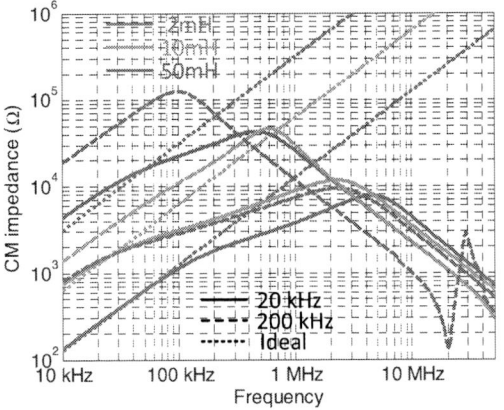

Fig. 12. Measured impedance of CM chokes. Red 2 mH, green 10 mH, magenta 50 mH; solid line 20 kHz, dashed line 200 kHz, dotted line ideal.

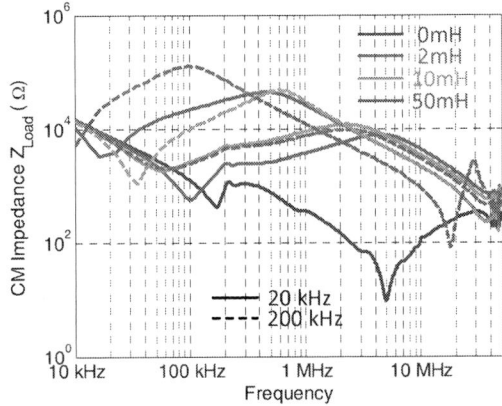

Fig. 13. Measured load CM impedance after the insertion of CM chokes. Blue 0 mH, red 2 mH, green 10 mH, magenta 50 mH; solid line 20 kHz, dashed line 200 kHz.

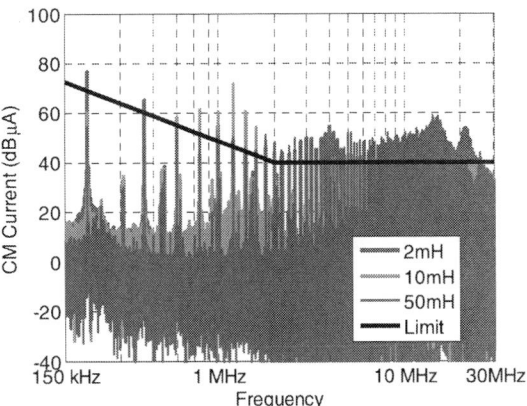

Fig. 14. Measured CM current with CM chokes for 200 kHz switching frequency case.

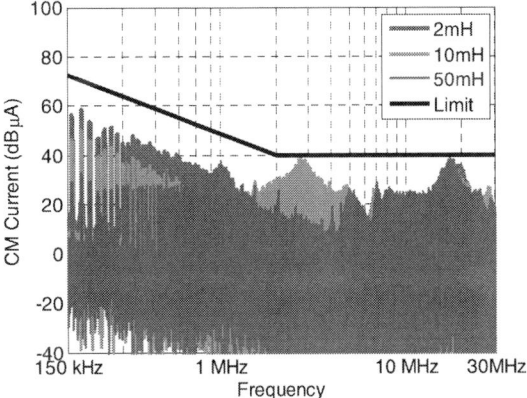

Fig. 15. Measured CM current with CM chokes for 20 kHz switching frequency case.

From Table I and Fig.11, the following observations can be made. The sizes of 200 kHz chokes are predominantly determined by core window size to fit the number of winding turns, while the flux densities in the cores are around only 0.1 T, which is far from saturation (0.8 T). Hence, the sizes of 200 kHz chokes increase with increasing CM inductance, and the 50 mH choke is the biggest among three.

Conversely, 20 kHz chokes are sized such that core cross-sectional areas are large enough to prevent core saturation, while the windings cannot fully cover the inner circumferences of the cores. This is due to the large volt-second stress on 20 kHz chokes as has been analyzed in the previous section. As a result, the 10 mH choke is the smallest among the three 20 kHz CM chokes.

Comparing the same inductance CM chokes of different frequencies, say, 20 kHz 50 mH choke and 200 kHz 50 mH choke, it can be noticed that the latter has more turns. This is due to the fact that the permeability of core material decreases at high frequency.

B. Measurement with CM Chokes

In the previous section, it is assumed that the CM chokes are purely inductive and their inductance does not vary with frequency. However, this is far from the truth. The CM impedances of the six chokes are measured and plotted in Fig. 12 along with their ideal values. The measured load CM impedances after insertion of the CM chokes are also plotted in Fig. 13 as compared to the ideal curves in Fig. 6.

It is shown that CM inductances of chokes decrease quickly with increasing frequency after 30-100 kHz due to the decreasing material permeability. The CM impedances peak at the self-resonance frequency of the chokes around a few hundred of kilohertz then start to decrease linearly with frequency due to the winding capacitances. In addition, the self-resonance frequency decreases with designed CM inductance, due to increasing number of winding turns. In the case of 200 kHz, 50 mH CM choke, the self-resonance frequency is as low as 100 kHz.

Due to the discrepancy between measured and ideal choke impedances in mid-to-high frequency range, the resulted CM current spectra are expected to be different from the calculated

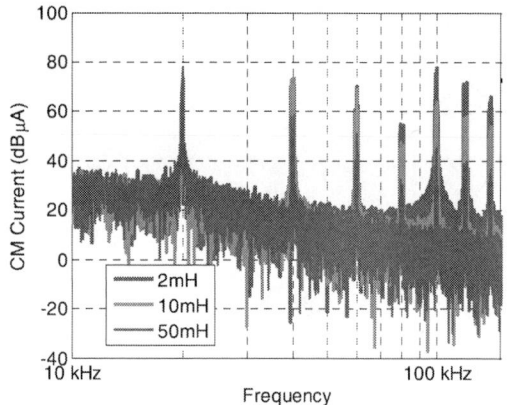

Fig. 16. Measured CM current spectra around switching frequency (20 kHz) and its low order harmonics.

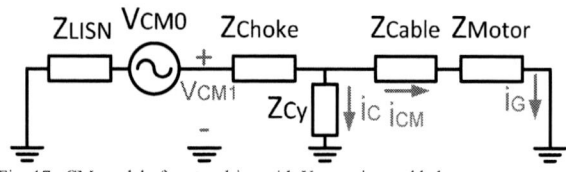

Fig. 17. CM model of motor drive with Y capacitors added.

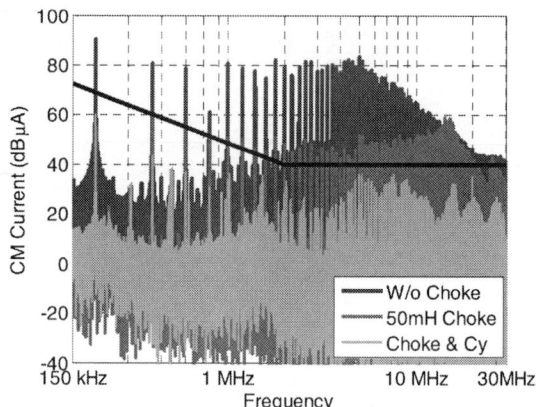

Fig. 18. Comparison of CM current spectra when (a) no choke added, (b) 50 mH choke added, and (c) both 50 mH choke and Y capacitors added.

spectra shown earlier. Hence, the CM currents with the chokes are measured and shown in Fig. 14 and Fig. 15, for 200 kHz and 20 kHz switching frequency, respectively. As can be observed from the figures, the measured spectra are 20-60 dB higher than the calculated values above 2 MHz.

In the case of 200 kHz switching frequency, none of the three CM chokes can reduce the CM current to below the given limit. Even though 50 mH choke provides the highest attenuation below 300 kHz, it has the lowest impedance above 2 MHz, among the three chokes. Hence, its corresponding current spectra cannot pass the limit in high frequency range. However, in the case of 20 kHz switching frequency, all three measured spectra can pass the given limit despite the discrepancy in high frequency range, due to the large margin left in Fig. 9. The measured CM current spectra in the low frequency range (10 kHz – 150 kHz) for 20 kHz chokes are also plotted in Fig. 16, which matches well with calculated spectra in Fig. 10. Hence, the excessive volt-second stresses on the 20 kHz chokes are verified, and the minimal size of 10 mH choke is justified.

C. Further Attenuation of CM Current in 200 kHz Switching Frequency Drive with Y Capacitors

As observed earlier, none of the three CM chokes can attenuate the CM current in 200 kHz switching frequency drive to below the given limit. Hence, in this part, Y-capacitors are added to the motor side of the 50 mH CM choke to provide additional attenuation. Three 22nF film capacitors are connected to the three phase cable in Y connection, with their neutral point grounded through an 8 Ω damping resistor. The new CM circuit model after adding the Y-capacitors are shown in Fig. 17. The Y-capacitors provide a lower impedance path to

ground compared to the load, thus diverting some of the CM current from the cable and motor.

The measured load CM current with Y-capacitors added are compared with the spectra without Y-capacitors and without chokes in Fig. 18. As shown in the figure, the original current spectrum without chokes surpasses the limit by 20 - 40 dB in all frequency range. Adding 50 mH choke manages to suppress the CM current spectrum below 1 MHz to within the limit, but spectrum above 1 MHz still goes over the limit by up to 20 dB. Finally adding additional Y-capacitors further attenuates the CM current over the whole frequency range of interest, which eventually passes the given limit.

V. CONCLUSION

This paper investigates the output CM choke parameters for SiC MOSFET based inverter motor drive, with a special emphasis on the high switching frequency operation.

A frequency domain based modeling approach is proposed and verified experimentally to facilitate further analysis. Based on the proposed model, the performances of three CM chokes (2 mH, 10 mH, and 50 mH) are evaluated for the SiC based motor drive for 200 kHz and 20 kHz switching frequencies separately. It is found that at 200 kHz switching frequency, only 50 mH choke can meet the chosen conducted EMI limit. On the contrary, at 20 kHz, all three chokes can meet the chosen EMI limit, and the 10 mH choke has the least volt-second stress among three chokes. Comparing two switching frequencies, it is further observed that volt-second stress on 20 kHz chokes is at least 10 times higher than the 200 kHz chokes due to CM circuit resonance.

Based on the volt-second stress values obtained from the modeling analysis, six CM chokes (three for each switching frequency) are designed and fabricated with off-the-shelf magnetic toroid cores. It is shown that the sizes of 200 kHz chokes are predominantly determined by core window size to fit the number of winding turns. As a result, the sizes of 200 kHz chokes increase with increasing CM inductance, and the 50 mH choke is the largest among three. On the contrary, 20 kHz chokes are sized such that core cross-sectional areas are large enough to prevent core saturation. Hence, the choke with smallest volt-second stress, i.e. the 10 mH choke, yields the smallest physical size among the three.

Finally, the six-chokes are tested with the motor drive. It is noted that non-idealities, such as the decreasing permeability and the winding capacitance, makes the CM current spectra 20 - 60 dB higher than predicted by the model. In this case, the motor drive with 20 kHz switching frequency can pass the limit with any of the three chokes, with the 10 mH choke as the best solution. However, none of the three 200 kHz chokes is able to help the 200 kHz motor drive to meet the requirement. Hence, the 50 mH choke with three 22 nF Y-capacitors are proposed as a viable solution.

ACKNOWLEDGEMENT

Support for this research has been funded in part by Mid-West Energy Research Consortium. The authors would also like to acknowledge Vacuumschmelze (VAC) for their generous donation of magnetic cores.

REFERENCES

[1] A. Elasser and T. P. Chow, "Silicon carbide benefits and advantages for power electronics circuits and systems," *Proc. IEEE*, vol. 90, no. 6, pp. 969- 986, June 2002.

[2] J. Millan, P. Godignon, X. Perpina, A. Perez-Tomas, and J. Rebollo, "A survey of wide bandgap power semiconductor devices," *IEEE Trans. Power Electron.*, vol. 29, no. 5, p. 2155-2163, May 2014.

[3] D. Han, J. Noppakunkajorn, and B. Sarlioglu, "Comprehensive efficiency, weight, and volume comparison of SiC and Si-based bidirectional DC-DC converters for hybrid electric vehicles," *IEEE Trans. Veh. Technol.*, vol. 63, no. 7, pp. 3001-3010, Sept. 2014.

[4] D. Han, J. Noppakunkajorn, and B. Sarlioglu, "Analysis of a SiC three-phase voltage source inverter under various current and power factor operations," in *Proc. Ind. Electron. Soc. Ann. Conf.*, pp.447,452, 10-13 Nov. 2013.

[5] H. Zhang, L. M. Tolbert, and B. Ozpineci, "Impact of SiC devices on hybrid electric and plug-in hybrid electric vehicles," *IEEE Trans. Ind. Appl.*, vol. 47, no. 2, pp. 912-921, March-April 2011.

[6] J. W. Kolar, H. Ertl, and F. C. Zach, "Influence of the modulation method on the conduction and switching losses of a PWM converter system," *IEEE Trans. Ind. Appl.*, vol. 27, no. 6, pp. 1063-1075, Nov/Dec 1991.

[7] D. Han, Y. Li, and B. Sarlioglu, "Analysis of SiC based power electronic inverters for high speed machines," in *Proc. IEEE Appl. Power Electron. Conf.*, March 2015, pp. 2344-2350.

[8] E. Zhong, and T. A. Lipo, "Improvements in EMC performance of inverter-fed motor drives," *IEEE Trans. Ind. Appl.*, vol. 31, no. 6, pp. 1247-1256, Nov/Dec 1995.

[9] T. Guo, D. Chen, and F. C. Lee, "Separation of common-mode and differential-mode conducted EMI noise," *IEEE Trans. Power Electron.*, vol. 11, no. 3, pp. 480-488, May 1996.

[10] D. Rendusara, and P. Enjeti, "An improved inverter output filter configuration reduces common and differential modes dv/dt at the motor terminals in PWM drive systems," *IEEE Trans. Power Electron.*, vol. 13, no. 6, pp. 1135-1143, Nov 1998.

[11] L. Ran, S. Gokani, J. Clare, K. J. Bradley, and C. Christopoulos, "Conducted electromagnetic emissions in induction motor drive systems. II. Frequency domain models," *IEEE Trans. Power Electron.*, vol. 13, no. 4, pp. 768-776, Jul 1998.

[12] M. Cacciato, A. Consoli, G. Scarcella, A. Testa, "Reduction of common-mode currents in PWM inverter motor drives," *IEEE Trans. Ind. Appl.*, vol.35, no.2, pp.469,476, Mar/Apr 1999.

[13] P. Chen, and Y. Lai, "Effective EMI filter design method for three-phase inverter based upon software noise separation," *IEEE Trans. Power Electron.*, vol. 25, no. 11, pp. 2797-2806, Nov. 2010.

[14] H. Akagi and T. Shimizu, "Attenuation of conducted EMI emissions from an inverter-driven motor," *IEEE Trans. Power Electron.*, vol. 23, no. 1, pp. 282–290, Jan. 2008.

[15] C. Mei, J. C. Balda, W. P. Waite, and K. Carr, "Analyzing common-mode chokes for induction motor drives," in *Proc. Power Electron. Spec. Conf.*, 2002, vol. 3, pp.1557-1562.

[16] F. Luo, S. Wang, F. Wang, D. Boroyevich, N. Gazel, Y. Kang, and A. C. Baisden, "Analysis of CM volt-second influence on CM inductor saturation and design for input EMI filters in three-phase DC-fed motor drive systems," *IEEE Trans. Power Electron.*, vol. 25, no.7, pp. 1905-1914, July 2010.

[17] D. Jiang, F. Wang, and J. Xue, "PWM impact on CM noise and AC CM choke for variable-speed motor drives," *IEEE Trans. Ind. Appl.*, vol. 49, no. 2, pp. 963-972, March-April 2013.

[18] A. Muetze, and C. R. Sullivan, "Simplified design of common-mode chokes for reduction of motor ground currents in inverter drives," *IEEE Trans. Ind. Appl.*, vol. 47, no. 6, pp. 2570-2577, Nov.-Dec. 2011.

[19] X. Zhang, D. Boroyevich, P. Mattavelli, J. Xue, and F. Wang, "EMI filter design and optimization for both AC and DC side in a DC-fed motor drive system," in *Proc. Appl. Power Electron. Conf. Expo*, March 2013, pp. 597-603.

[20] F.Luo, D. Boroyevich, P. Mattevelli, and N. Gazel, "A comprehensive design for high power density common mode EMI inductor," in *Porc. Energy Conver. Congress Expo.* Sept. 2011, pp. 1861-1867.

[21] M. L. Heldwein, L. Dalessandro, and J. W. Kolar, "The Three-Phase Common-Mode Inductor: Modeling and Design Issues," *IEEE Trans. Ind. Electron.*, vol. 58, no. 8, pp. 3264-3274, Aug. 2011.

[22] *Nanocrystalline Vitroperm EMC Components*, Vacuumschmelze (VAC) GmbH Co., Hanau, Germany, 2004.

[23] A. Massarini, and M. K. Kazimierczuk, "Self-capacitance of inductors," *IEEE Trans. Power Electron.*, vol.12, no.4, pp.671-676, Jul 1997.

An Improved Model Predictive Current Control of Permanent Magnet Synchronous Motor Drives

Yongchang Zhang, Suyu Gao
Power Electronics and Motor Drives
Engineering Research Center of Beijing
North China University of Technology
Beijing, China
Email: yozhang@ieee.org

Wei Xu
School of Electrical & Electronic Engineering
Huazhong University of Science and Technology
Wuhan, China
Email: weixuforhappy@foxmail.com

Abstract—Model predictive current control (MPCC) is widely recognized as a high performance control strategy of permanent magnet synchronous machine (PMSM) drives due to its quick response and simple principle. However, conventional MPCC relies heavily on the model of system to predict the future behavior of stator currents for each voltage vector. The one minimizing the current error is selected as the best voltage vector. As only one voltage vector is applied during one control period, it fails to give satisfactory performance due to the limited voltage vectors, especially in the case of two-level converters. This paper proposes an improved MPCC strategy for PMSM drives, which firstly estimates the back electromotive force (EMF) based on the past value of stator voltage and currents, and then applies the estimated EMF in the stator current prediction. As a result, the robustness against machine parameter variations is enhanced. Furthermore, to achieve steady state performance improvement, a null vector along with the active vector obtained from conventional MPCC is applied during one control period. Meanwhile, the dynamic response is not affected. Both simulation and experimental results are presented to confirm the effectiveness of the proposed method.

I. INTRODUCTION

Variable speed drives have been widely used in industrial applications. In particular, PMSM drives are receiving more and more attention due to its high efficiency, high torque density and low volume [1]. Various methods have been proposed to achieve high performance control of PMSM drives and the most notable control methods are vector control (VC) and direct torque control (DTC) [2], [3]. VC can achieve good steady state performance and quick dynamic response over a wide speed range, but it requires fine tuning work on internal current loop and rotary transformation [3]. Furthermore, as the bandwidth of internal current loop is limited, VC may be not suitable for applications requiring very high dynamic response. DTC eliminates the current loop hence its tuning work and pulsewidth modulation (PWM) block by directly obtaining the final voltage vector from a predefined switching table, based on the torque and flux error signs and the position of stator flux [4]. Very quick response is achieved in DTC with simple structure. However, it suffers from the high torque and flux ripple and variable switching frequency, and at the same time it exists undesired acoustic noises. Hence, various improved methods have been proposed to overcome the drawbacks of

conventional VC and DTC [1], [5]–[7].

Recently model predictive control (MPC) is emerging as a powerful scheme for high performance control of PMSM drives [8]–[11]. Based on the internal model of system, MPC predicts the future behavior of controlled variables, such as current, torque and stator flux. By minimizing the error between the reference value and the predicted value, the best voltage vector can be obtained. Compared to DTC, the selected voltage vector from MPC is more accurate and effective. Compared to VC, MPC achieves much quicker dynamic response. To achieve high performance torque control, usually torque and stator flux are selected as the control variables and the corresponding MPC is called mode predictive torque control (MPTC) [12]–[14]. Better performance than DTC in terms of torque ripple and current harmonics can be obtained in MPTC. As torque and stator flux have different units, when combining both torque error and flux error into one single cost function, special care has to be taken on the weighting factor of stator flux, which requires some tuning work and makes MPTC not very universal [15]–[18].

A more natural selection of control variable is stator current and the corresponding MPC is called model predictive current control (MPCC) [19], [20]. Currently MPCC for PMSM drives has been widely recognized as a high performance control strategy with quick response and simple principle [21]. Compared to MPTC, which requires an estimator or observer to obtain the value of torque and stator flux and some tuning work in the weighting factor, MPCC is much simpler because the control variable (current) in MPCC can be directly measured. Furthermore, the tuning work for weighting factor in MPTC is eliminated in MPCC. Thus, for PMSM drives, MPCC is a simpler solution than MPTC. However, the performance of MPCC relies heavily on the system model and machine parameters, especially in the stage of prediction. The machine parameters may vary with the operating point and temperature and failing to consider this parameter variations will affect the accuracy of prediction [22]. Furthermore, conventional MPCC usually applies only one voltage vector during one control period among all basic voltage vectors [19], which may not reduce the current error to a minimal value. In DTC it has been widely known that the steady state performance can be

978-1-4673-9551-9/16 $31.00 © 2016 IEEE

improved by applying one active vector and one null vector during one control period [1], because generally the null vector produces small current variations. Enlightened by this fact, it is also possible to employ this principle in MPCC to achieve further performance improvement.

This paper proposes an improved MPCC with enhanced robustness against machine parameter variations and better steady state performance than conventional single-vector-based MPCC. The electromotive force (EMF) is firstly estimated from the historical value of stator voltage and current and then applied in the stage of stator current prediction [23]. As a result, the proposed MPCC improves the robustness of system significantly. By using one active vector from conventional MPCC and an additional null vector during one control period [24], less current harmonics and better current control accuracy are obtained. The duration of the active vector is obtained based on the principle of deadbeat control [25], which is simple to implement. The proposed MPCC is comparatively studied with conventional MPCC and its effectiveness is confirmed by both simulation and experimental results.

II. MACHINE EQUATIONS OF PMSM

To simplify the analysis, the model of PMSM is established under the following assumptions. The magnetic saturation is neglected, the back electromagnetic force is supposed sinusoidal and the eddy current, magnetic hysteresis losses and the clogging torque are very small and then also neglected. According to the theory in [26], by using the concept of "active flux", any salient-pole rotor machines can be equivalently described by a non salient-pole rotor machines. For PMSM, the model in rotor synchronous coordinate is most popular because all the machine parameters become constant.

In this paper, to reduce the control complexity, the model of PMSM in stationary $\alpha\beta$ frame is preferred to avoid complicated coordinate transformation. To be specific, the model of PMSM can be expressed in stationary $\alpha\beta$ frame using complex vectors as [26]:

$$\begin{aligned} \boldsymbol{u}_s &= R_s \boldsymbol{i}_s + L_q \frac{d\boldsymbol{i}_s}{dt} + \frac{d\boldsymbol{\psi}_x}{dt} \\ &= R_s \boldsymbol{i}_s + L_q \frac{d\boldsymbol{i}_s}{dt} + \boldsymbol{e}_x \end{aligned} \quad (1)$$

$$\boldsymbol{\psi}_x = [\psi_f + (L_d - L_q)i_d]e^{j\theta_e} \quad (2)$$

$$T_e = \frac{3}{2}p\left(\boldsymbol{\psi}_x \otimes \boldsymbol{i}_s\right) \quad (3)$$

where $\boldsymbol{u}_s$, $\boldsymbol{i}_s$, $\boldsymbol{\psi}_x$ and $\boldsymbol{e}_x$ are stator voltage vector, stator current vector, equivalent active flux and equivalent back EMF, respectively; i_d and i_q are the d-axis and q-axis current in synchronous frame, respectively; R_s, L_d, L_q, ψ_f, ω and θ_e represent stator resistance, d-axis inductance, q-axis inductance, permanent magnet flux, electrical rotor speed and electrical rotor position ($\frac{d\theta_e}{dt} = \omega$), respectively; T_e is electromagnetic torque and p is the number of pole pairs.

Finally, the mechanical equations of PMSM is expressed as

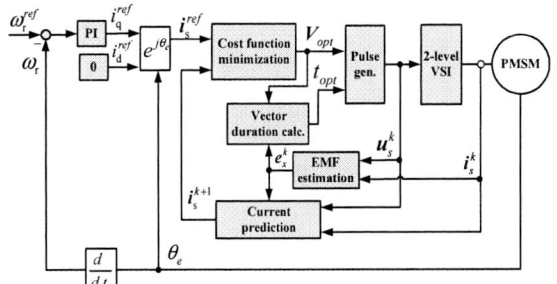

Fig. 1. Control diagram of the proposed MPCC

$$T_e - T_L = J\frac{d\omega}{dt} + B\omega \quad (4)$$

where J is the machine inertia, B is viscous friction coefficient and T_L is the load torque.

III. PRINCIPLE OF PROPOSED MPCC

The control diagram of the proposed MPCC is shown in Fig. 1, which is mainly composed of five parts: estimation of EMF, current prediction, cost function minimization, vector duration calculation and pulse generation. An outer speed loop is employed to generate the q-axis current current and the d-axis current reference is set to zero to achieve maximum torque per ampere (MTPA) operation. The stator current reference is transformed from synchronous frame to stationary frame and the rest part of controller is implemented in stationary frame for the sake of simplicity. The detailed introduction of the control diagram will be elaborated in the following text.

A. EMF Estimation and Current Prediction

From (1), the current equation of PMSM can be rearranged as follows:

$$\frac{d\boldsymbol{i}_s}{dt} = \frac{1}{L_q}\left(\boldsymbol{u}_s - R_s\boldsymbol{i}_s - \boldsymbol{e}_x\right) \quad (5)$$

According to (5), the stator current at the next control period can be predicted using forward Euler discretization as:

$$\boldsymbol{i}_s^{k+1} = \boldsymbol{i}_s^k + \frac{T_s}{L_q}\left(\boldsymbol{u}_s^k - R_s\boldsymbol{i}_s^k - \boldsymbol{e}_x^k\right) \quad (6)$$

where T_s is the control period.

The main difficulty in (6) is that the back EMF $\boldsymbol{e}_x^k$ cannot be directly measured. In theory, $\boldsymbol{e}_x^k$ can be obtained from (1) and (2), but it requires the accurate knowledge of L_d, L_q, ψ_f, ω and θ_e. As the machine parameters may vary due to temperature, saturation and skin effect, it is desirable to estimate the EMF directly and reduce the use of machine parameters. As the mechanical time constant of the motor is much larger compared to the electromagnetic time constant, generally motor speed can be considered as constant during several control period, which means $\omega(k+1) \approx \omega(k)$. The EMF $\boldsymbol{e}_x^k$ is roughly proportional to rotor speed ω, hence we can assume that $\boldsymbol{e}_x^k = \boldsymbol{e}_x^{k-1} = \boldsymbol{e}_x^{k-2}$. As a result, we can estimate the EMF at $(k-1)$th instant using the past value of stator voltage and current as:

$$e_x^{k-1} = u_s^{k-1} - R_s \frac{i_s^k + i_s^{k-1}}{2} - \frac{L_q}{T_s}\left(i_s^k - i_s^{k-1}\right) \quad (7)$$

Similarly, the EMF at $(k-2)$th instant and $(k-3)$th instant can be estimated as:

$$e_x^{k-2} = u_s^{k-2} - R_s \frac{i_s^{k-1} + i_s^{k-2}}{2} - \frac{L_q}{T_s}\left(i_s^{k-1} - i_s^{k-2}\right) \quad (8)$$

$$e_x^{k-3} = u_s^{k-3} - R_s \frac{i_s^{k-2} + i_s^{k-3}}{2} - \frac{L_q}{T_s}\left(i_s^{k-2} - i_s^{k-3}\right) \quad (9)$$

The final estimated value of e_x^k is obtained as a mean value of the past EMF, which is expressed as

$$e_x^k = \frac{1}{3}\left(e_x^{k-1} + e_x^{k-2} + e_x^{k-3}\right) \quad (10)$$

It can be seen that the use of rotor speed, rotor position and permanent flux is eliminated when deriving the EMF. The stator resistance is small and its influence can be neglected. The remaining parameter is L_q. However, following the stability analysis method in [27], it is found that the proposed EMF estimation is robust to the variation of L_q.

B. Current prediction

The controller output cannot be applied instantaneously, because there are various delays, including the inherent sampling delay, filtering delay and other factors. In digital implementation, the commanding voltage vector is obtained at kth instant, but not applied until the $(k+1)$th instant due to the updating mechanism of modern digital signal processors (DSPs) and micro controllers. As a result, the selected voltage vector at kth instant may be not the best choice. The impact of the one step delay is especially serious when the sampling frequency is low. Therefore, in order to improve the control accuracy, it is necessary to compensate the one step delay [1]. In this paper, the second-order Euler discretization is used. After ignoring some tedious deduction process, the formula is expressed as

$$i_{sp}^{k+1} = i_s^k + \frac{T_s}{L_q}\left(u_s^k - R_s i_s^k - e_x^k\right) \quad (11)$$

$$i_s^{k+1} = i_{sp}^{k+1} + \frac{-R_s(i_{sp}^{k+1} - i_s^k)T_s}{2 \cdot L_q} \quad (12)$$

C. Cost Function Minimization

After obtaining the back EMF from (7) to (10), the current at the next control period can be predicted using (12). For conventional MPCC, the control aim is to select the best voltage vector to minimize the following cost function:

$$J = \left|i_s^{ref} - i_s^{k+1}\right| \quad (13)$$
$$s.t. u_s \in \{u_0, u_1......u_6, u_7\}$$

where $i_s^{ref} = (0 + j \cdot i_q)e^{j\theta_e}$ is the current reference in stationary frame. For two-level inverter, there are only eight

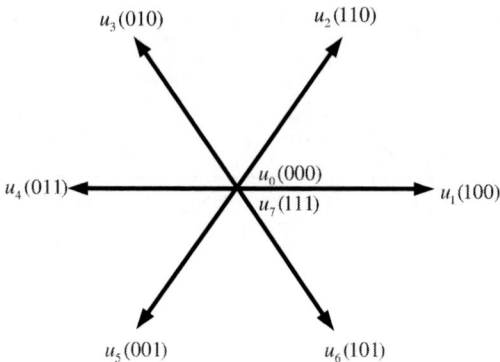

Fig. 2. Voltage vectors of a two-level inverter.

switching states (including two null vectors), as shown in Fig. 2. It is feasible to predict the value of i_s^{k+1} for each voltage vector and find the best voltage vector minimizing (13).

The conventional MPCC has better performance in terms of torque ripple and current harmonics compared to DTC, but the steady state performance still has to be improved. In this paper, a null vector will also be applied during one control period to achieve performance improvement, so it is only necessary to evaluate the non-zero voltage vectors for the cost function shown in (13).

D. Vector Duration

After obtaining the EMF in Section III-A, the current slope can be easily obtained. For simplicity, the current slope caused by a zero vector u_0 and the best non-zero voltage vector u_{opt} minimizing (13) are represented by s_0 and s_1 respectively, which are expressed as:

$$s_0 = \frac{di_s}{dt}\Big|_{u_s = u_0} = \frac{1}{L_q}\left(-R_s i_s - e_x\right) \quad (14)$$

$$s_1 = \frac{di_s}{dt}\Big|_{u_s = u_{opt}} = s_0 + \frac{1}{L_q}u_{opt} \quad (15)$$

From (14), it is found that the zero vector will always decrease the current. Hence, by introducing zero vector during one control period, the effects of voltage on current can be regulated more moderately, diminishing the ripples in current. As the vectors have been selected, the remaining work is to determine the optimal duration t_{opt} of u_{opt}.

In this paper, the principle of deadbeat control is applied to calculate the optimal duration t_{opt} of u_{opt}, as illustrated in Fig.3. From Fig.3, it is not difficult to find that if only one basic voltage vector is used in the whole control period, the current at the end of control period will be well above or below the reference value, producing large ripples. By introducing the zero voltage vector, it is possible to minimize the error between i_s^{ref} and predicted stator current i_s^{k+1}.

The current at the end of control period can be expressed as:

$$i_s^{k+1} = i_s^k + s_1 \cdot t_{opt} + s_0 \cdot (t_s - t_{opt})$$

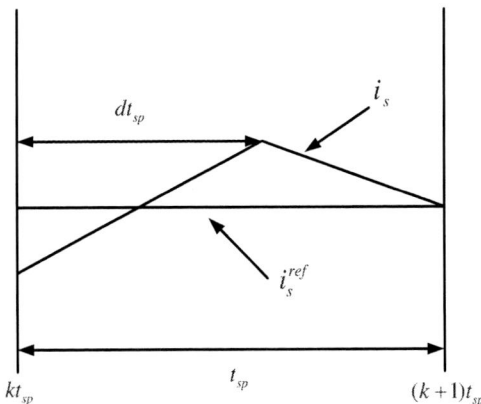

Fig. 3. duty of deadbeat MPCC

TABLE I
PARAMETERS OF PMSM

Power [kW]	P_n	1.5
Number of pole pairs	p	4
Permanent magnet flux [wb]	ψ_f	0.2125
Stator resistance [Ω]	R_s	0.4
d-axis and q-axis inductance [mH]	L_d, L_q	4.06
rated speed [rpm]	n_N	1500
rated torque [NM]	T_n	9.55
DC bus voltage [V]	U_{dc}	380
Interia [$kg \cdot m^2$]	J	0.002

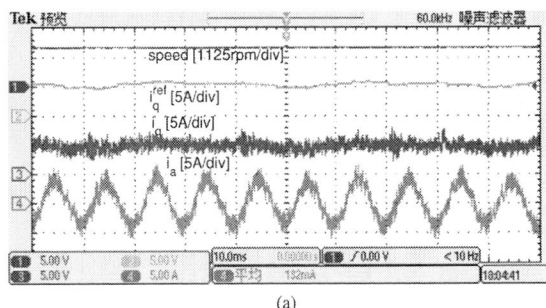

(a)

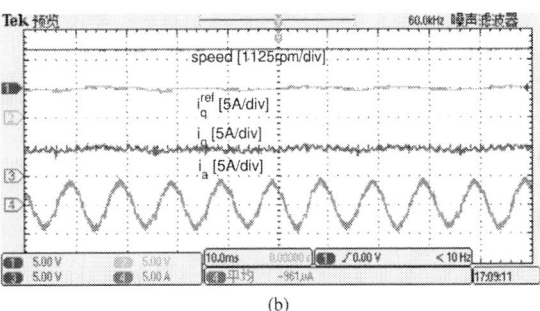

(b)

Fig. 4. Steady state responses at 1500 rpm with rated load for (a) conventional MPCC and (b) the proposed MPCC

By minimizing the cost function $\left| i_s^{ref} - i_s^{k+1} \right|$, the optimal duration t_{opt} can be obtained as

$$t_{opt} = \frac{\left(i_s^{ref} - i_s^k - s_0 T_s \right) \odot (s_1 - s_0)}{\left| s_1 - s_0 \right|^2} \quad (16)$$

where $\odot$ means dot product of two complex vectors.

E. Pulse Generation

After selecting the two vectors and determining their durations, the switching pulses can be generated for the gating of power switches. For two-level inverters there are two null vectors and the one produces less switching jumps is selected following the active vector. For example, if the selected active vector is $u_2(110)$, then $u_7(111)$ rather than $u_0(000)$ should be followed.

IV. EXPERIMENTAL RESULTS

To confirm the effectiveness of the proposed MPCC, some experimental tests are carried out on a two-level inverter-fed PMSM drive platform. The machine and control parameters are listed in Table I.. The conventional single-vector-based MPCC is also implemented for the aim of comparison. To achieve similar average switching frequency, the sampling frequencies of the proposed MPCC and conventional MPCC are 20 kHz and 30 kHz, respectively. During the tests, all variables are displayed on oscilloscope via on-board DA converter except the stator current, which is measured directly by a current probe.

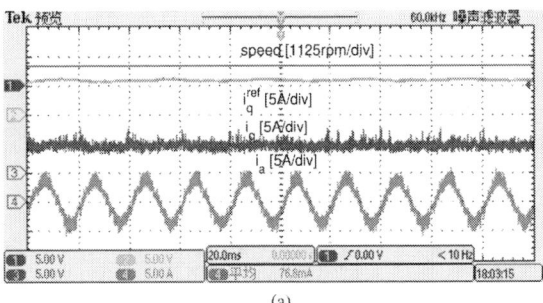

(a)

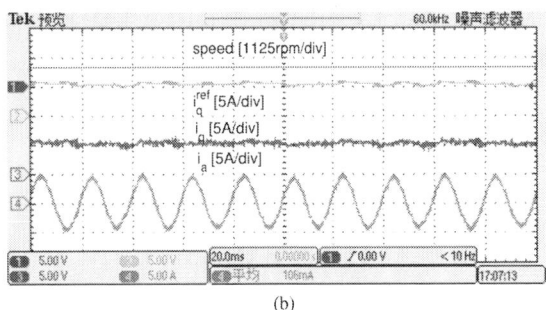

(b)

Fig. 5. Steady state responses at 750 rpm with rated load for (a) conventional MPCC and (b) the proposed MPCC

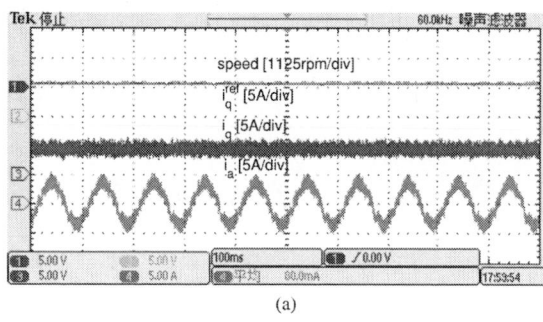

(a)

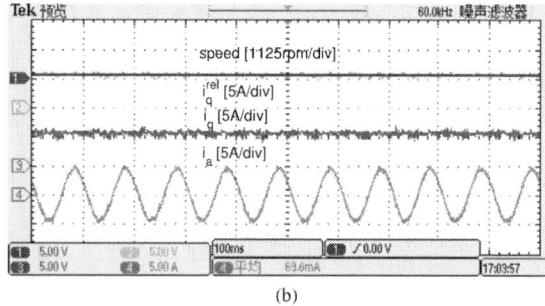

(b)

Fig. 6. Steady state responses at 150 rpm with rated load for (a) conventional MPCC and (b) the proposed MPCC

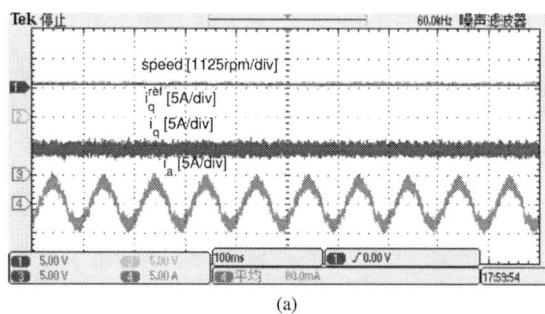

(a)

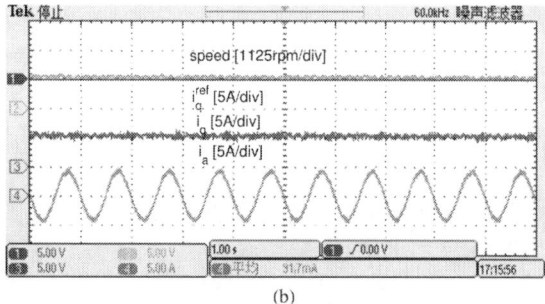

(b)

Fig. 7. Steady state responses at 15 rpm with rated load for (a) conventional MPCC and (b) the proposed MPCC

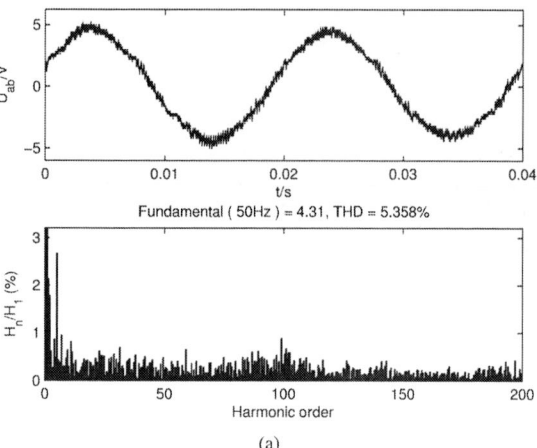

(a)

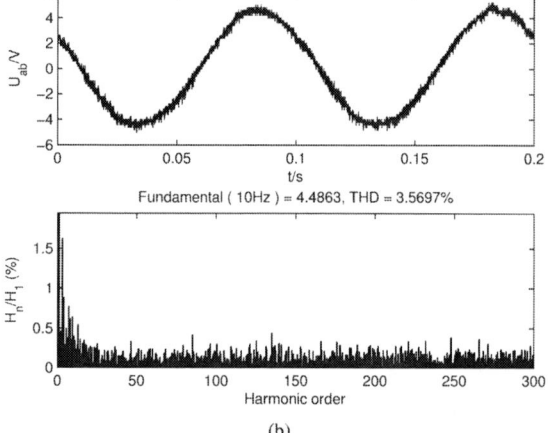

(b)

Fig. 8. Harmonics spectrum of stator current for the proposed MPCC, (a) 750 rpm, (b) 150 rpm.

Firstly the steady state responses at 100% rated speed with rated load are presented in Fig. 4, with the conventional MPCC shown in the left half and the proposed MPCC in the right half. From top to bottom, the curves shown in Fig. 4 are rotor speed, q-axis current reference, q-axis current and one-phase stator current. It is clearly seen that in the proposed MPCC, the q-axis current tracks the reference value more accurately and there are much less harmonics in the stator current than that in conventional MPCC, validating its effectiveness in improving steady state performance. Similar results can be also observed at the medium speed of 750rpm, low speed of 150rpm and very low speed of 15 rpm (1% rated speed), as shown in Fig. 4 to 7. These results confirm that the the proposed MPCC is effective in obtaining better steady state performance than conventional MPCC in the whole speed range.

For further quantitative analysis of the steady state performance, the harmonics spectrum of stator current for the proposed MPCC at 750 rpm and 150 rpm are shown in Fig.8. It can be seen that the current THD at 750 rpm and 150 rpm are

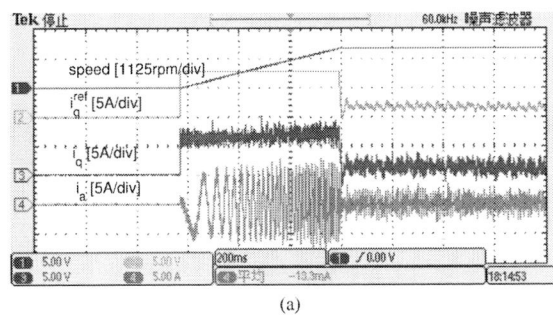

(a)

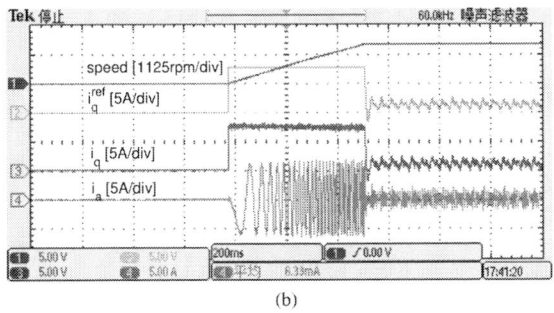

(b)

Fig. 9. Starting responses from standstill to rated speed for (a) conventional MPCC and (b) the proposed MPCC.

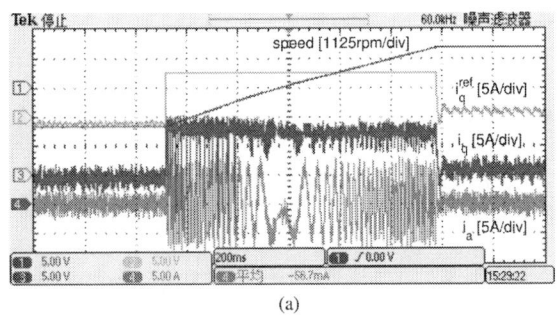

(a)

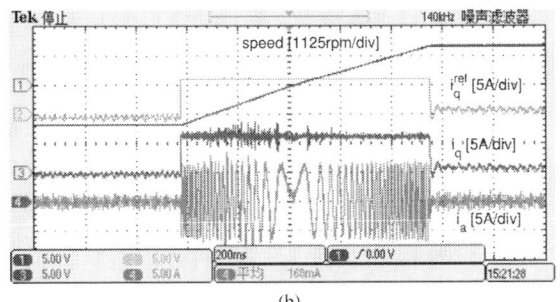

(b)

Fig. 10. The dynamic responses of speed reversion of ± 1500 rpm for (a) conventional MPTC and (b) the proposed MPTC

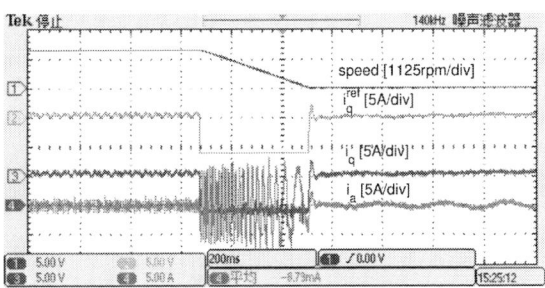

Fig. 11. Dynamic responses of the proposed MPCC when decelerating the speed from 1500 rpm to 75 rpm.

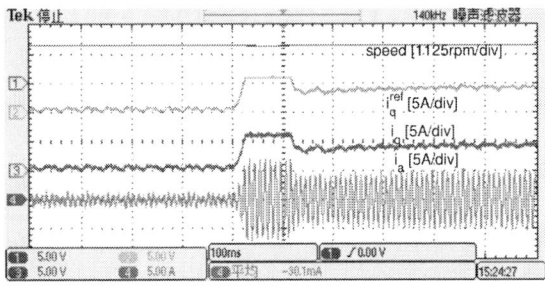

Fig. 12. Dynamic responses of the proposed MPCC for stepped change in load torque at 1500 rpm

5.358% and 3.5697%, respectively. Both of them are within the acceptable range.

Apart from the steady-state performance comparison, Fig. 9 further presents the dynamic response of starting from standstill to rated speed for conventional MPCC and the proposed MPCC. It is seen that the dynamic response of the proposed MPCC is very similar to that of conventional MPCC. However, the proposed MPTC presents lower torque ripple and less current harmonics. This confirms that the proposed MPCC can achieve better steady-state performance without degrading the dynamic performance.

The dynamic responses of speed, q-axis current reference, q-axis current and one-phase stator current during speed reversal of ± 1500 rpm are presented in Fig. 10. It is seen that during the dynamic process, both of the two methods have very similar response and there are much lower ripple in the proposed MPCC than the conventional MPCC.

Finally, the dynamic response of decreasing the speed from 1500 rpm to 75 rpm and the responses to 100% rated external load disturbance are presented in Fig. 11 and 12. There is a slight speed fall when the external load is applied and then the motor speed returns to its normal value quickly, exhibiting strong robustness against external load disturbance. These results validate the good dynamic performance of the proposed method.

V. CONCLUSION

This paper proposes an improved MPCC with duty ratio optimization to achieve ripple reduction without degrading

978-1-4673-9551-9/16 $31.00 © 2016 IEEE

the dynamic performance for the high performance control of PMSM drives. By selecting the stator current as control variable, the weighting factor tuning work is eliminated, which is usually required in MPTC. The robustness against machine parameter variations is enhanced by estimating the EMF from previous value of stator voltages and currents. After selecting the best non-zero voltage vector from conventional MPCC, a null vector is subsequently applied during one control period to achieve steady state performance improvement. The duration of the non-zero vector is determined based on the principle of current error minimization. A comparative study is presented for conventional MPCC and the proposed MPCC and both simulation and experimental results validate the effectiveness of the proposed method. It is concluded that the proposed MPCC can improve steady state performance with excellent dynamic performance.

ACKNOWLEDGMENT

This work was supported in part by the National Natural Science Foundation of China under Grants 51505003, 51207003 and 51347004, in part by Beijing Nova Program under Grant xx2013001, in part by the Scientific Research Foundation for the Returned Overseas Chinese Scholars, State Education Ministry, in part by the Chongqing Natural Science Foundation under Grant cstc2012jjB107, and in part by the Open Research Fund from Key Laboratory of Special Power Supply under Grant MSPS2012-03 and MSPS2013-02.

REFERENCES

[1] Y. Zhang and J. Zhu, "Direct torque control of permanent magnet synchronous motor with reduced torque ripple and commutation frequency," *IEEE Trans. Power Electron.*, vol. 26, no. 1, pp. 235 –248, jan. 2011.

[2] F. Korkmaz, I. Topaloglu, M. Cakir, and R. Gurbuz, "Comparative performance evaluation of foc and dtc controlled pmsm drives," in *Power Engineering, Energy and Electrical Drives (POWERENG), 2013 Fourth International Conference on*, May 2013, pp. 705–708.

[3] D. Casadei, F. Profumo, G. Serra, and A. Tani, "Foc and dtc: two viable schemes for induction motors torque control," *IEEE Trans. Power Electron.*, vol. 17, no. 5, pp. 779 – 787, sep 2002.

[4] Y. Inoue, S. Morimoto, and M. Sanada, "Comparative study of pmsm drive systems based on current control and direct torque control in flux-weakening control region," in *Proc. IEEE Int. Electric Machines & Drives Conf. (IEMDC)*, 2011, pp. 1094–1099.

[5] J. Beerten, J. Verveckken, and J. Driesen, "Predictive direct torque control for flux and torque ripple reduction," *IEEE Trans. Ind. Electron.*, vol. 57, no. 1, pp. 404 –412, jan. 2010.

[6] Y. Zhang and J. Zhu, "A novel duty cycle control strategy to reduce both torque and flux ripples for dtc of permanent magnet synchronous motor drives with switching frequency reduction," *IEEE Trans. Power Electron.*, vol. 26, no. 10, pp. 3055 –3067, oct. 2011.

[7] M. Sharifian, T. Herizchi, and K. Firouzjah, "Field oriented control of permanent magnet synchronous motor using predictive space vector modulation," in *Industrial Electronics Applications, 2009. ISIEA 2009. IEEE Symposium on*, vol. 2, oct. 2009, pp. 574 –579.

[8] H. Miranda, P. Cortes, J. Yuz, and J. Rodriguez, "Predictive torque control of induction machines based on state-space models," *IEEE Trans. Ind. Electron.*, vol. 56, no. 6, pp. 1916 –1924, june 2009.

[9] Y. Zhang and H. Yang, "Model predictive torque control of induction motor drives with optimal duty cycle control," *IEEE Trans. Power Electron.*, vol. 29, no. 12, pp. 6593–6603, 2014.

[10] Y. Zhang, J. Zhu, W. Xu, and Y. Guo, "A simple method to reduce torque ripple in direct torque-controlled permanent-magnet synchronous motor by using vectors with variable amplitude and angle," *IEEE Trans. Ind. Electron.*, vol. 58, no. 7, pp. 2848–2859, july 2011.

[11] C.-K. Lin, T.-H. Liu, J.-t. Yu, L.-C. Fu, and C.-F. Hsiao, "Model-free predictive current control for interior permanent-magnet synchronous motor drives based on current difference detection technique," *IEEE Trans. Ind. Electron.*, vol. 61, no. 2, pp. 667–681, 2014.

[12] T. Geyer, G. Beccuti, G. Papafotiou, and M. Morari, "Model predictive direct torque control of permanent magnet synchronous motors," in *Energy Conversion Congress and Exposition (ECCE), 2010 IEEE*, sept. 2010, pp. 199 –206.

[13] P. Landsmann and R. Kennel, "Saliency-based sensorless predictive torque control with reduced torque ripple," *Power Electronics, IEEE Transactions on*, vol. 27, no. 10, pp. 4311–4320, Oct 2012.

[14] Y. Zhang, J. Zhu, and W. Xu, "Predictive torque control of permanent magnet synchronous motor drive with reduced switching frequency," in *Electrical Machines and Systems (ICEMS), 2010 International Conference on*, oct. 2010, pp. 798 –803.

[15] C. Rojas, J. Rodriguez, F. Villarroel, J. Espinoza, C. Silva, and M. Trincado, "Predictive torque and flux control without weighting factors," *Industrial Electronics, IEEE Transactions on*, vol. 60, no. 2, pp. 681–690, Feb 2013.

[16] Y. Zhang and H. Yang, "Two-vector-based model predictive torque control without weighting factors for induction motor drives," *IEEE Trans. Power Electron.*, vol. 31, no. 2, pp. 1381–1390, 2016.

[17] ——, "Generalized two-vector-based model-predictive torque control of induction motor drives," *IEEE Trans. Power Electron.*, vol. 30, no. 7, pp. 3818–3829, 2015.

[18] ——, "Model-predictive flux control of induction motor drives with switching instant optimization," *IEEE Trans. Energy Convers.*, vol. 30, no. 3, pp. 1113–1122, 2015.

[19] Y. Zhang and H. Lin, "Simplified model predictive current control method of voltage-source inverter," in *2011 IEEE 8th International Conference on Power Electronics and ECCE Asia (ICPE ECCE)*, 2011, pp. 1726–1733.

[20] D.-H. Yim, B.-G. Park, R.-Y. Kim, and D.-S. Hyun, "A predictive current control associated to ekf for high performance ipmsm drives," in *Applied Power Electronics Conference and Exposition (APEC), 2011 Twenty-Sixth Annual IEEE*. IEEE, 2011, pp. 1010–1016.

[21] F. Morel, X. Lin-Shi, J.-M. Retif, B. Allard, and C. Buttay, "A comparative study of predictive current control schemes for a permanent-magnet synchronous machine drive," *IEEE Trans. Ind. Electron.*, vol. 56, no. 7, pp. 2715 –2728, july 2009.

[22] C. Lin, T. Liu, J. Yu, L. Fu, and C. Hsiao, "Model-free predictive current control for interior permanent magnet synchronous motor drives based on current difference detection technique," 2013.

[23] Y. Zhang, J. Zhu, and W. Xu, "Analysis of one step delay in direct torque control of permanent magnet synchronous motor and its remedies," in *Electrical Machines and Systems (ICEMS), 2010 International Conference on*, oct. 2010, pp. 792 –797.

[24] Y. Zhang and X. Wei, "Torque ripple rms minimization in model predictive torque control of pmsm drives," in *Electrical Machines and Systems (ICEMS), 2013 International Conference on*, Oct 2013, pp. 2183–2188.

[25] S.-M. Yang and C.-H. Lee, "A deadbeat current controller for field oriented induction motor drives," *IEEE Trans. Power Electron.*, vol. 17, no. 5, pp. 772 – 778, sep 2002.

[26] I. Boldea, M. Paicu, and G. Andreescu, "Active flux concept for motion-sensorless unified ac drives," *IEEE Trans. Power Electron.*, vol. 23, no. 5, pp. 2612–2618, Sept 2008.

[27] W. Wang and X. Xi, "Current control method for pmsm with high dynamic performance," in *Electric Machines & Drives Conference (IEMDC), 2013 IEEE International*. IEEE, 2013, pp. 1249–1254.

Analysis of Magnet Defect Faults in Permanent Magnet Synchronous Motors through Fluxgate Sensors

*Taner Goktas, Kun Wang Lee, Mohsen Zafarani, and Bilal Akin
The University of Texas at Dallas, Electrical Engineering Department, USA, TX 75252
*Corresponding author: taner.goktas@utdallas.edu

Abstract— **This paper presents magnet defect faults detection through fluxgate sensors simply by monitoring the leakage flux content around permanent magnet synchronous motors (PMSMs). As known, flux spectrums of electric machines contain most critical and direct fault related information to monitor and characterize various faults and their progressions. In this paper, a remote/on case fault monitoring prototype is prepared which includes fluxgate sensor, signal sensing/conditioning circuit, and a microcontroller for leakage flux data streaming. In order to identify magnet defect faults in PMSMs, faults patterns in the leakage flux spectrum are exhaustively analyzed at different torque-speed profiles. Simulation and experimental results show that deployment of fluxgate sensor in magnet defect fault detection yields better results than the classical stator current analysis in PMSMs both in time and frequency domains.**

Keywords—Condition monitoring; fault diagnosis; magnet defect, fluxgate sensors, permanent magnet synchronous motor

I. INTRODUCTION

In order to avoid costly shutdowns, minimize safety concerns and obtain high performance in long period of time, electric motors should be monitored carefully. The existing commercial monitoring solutions for electrical machines are not meant for real time monitoring, bulky, costly and may not practical to use in many applications. In industry, there is an increasing tendency to use permanent magnet synchronous motors (PMSMs) due to their high efficiency, torque-current ratio and power density. Although the phase current information is mostly preferred motor variable to detect the faults in PMSMs; our studies have shown that the fault signatures in the current spectrum depend on stator winding configuration and motor structure. In order to eliminate these shortcomings, leakage flux spectrum can be used to detect the magnet related faults such as broken magnets, demagnetization as it facilitate reliable diagnostic and enables remote or on-case monitoring options through small foot-print fluxgate sensors.

There are some studies in the literature addressing the leakage flux measurement for condition monitoring in electric machines [1-11]. The leakage flux sensors-coils are used to detect stator winding fault [1], rotor faults [2] rotor slot harmonics and speed [3] in induction machines. Besides, the search coils are mounted around each tooth to detect faults in PMSMs [4]. However, using external search coils has some major drawbacks such as size, noise and installation. In [5], stray flux analysis considering both stator end-winding and

segments is presented for stator and rotor faults to define the low frequency electromagnetic emission free regions for humans [6]. In [7], the stray flux around the motor is used to detect the inter-turn short circuits faults in low voltage induction motors. It is shown that stray flux is more efficient than stator current to detect fault particularly at no-load condition. The leakage flux losses are also analyzed to improve the performance of the permanent magnet generator [8]. The transverse component in external magnetic field [9] is used to detect short-circuit fault without the knowledge of healthy state. In [10], the stray flux spectrum of a three-phase induction machine is analyzed to investigate the gearbox behavior in induction machine. It is shown that mesh and its related frequencies which cannot be easily identified in stator current show up in the EMF stray flux spectrum. Moreover, the bearing faults [11] especially the deformation of the seal are detected using external flux probe at different sensor locations in induction machines.

Intensive research has been conducted to detect demagnetization faults in PMSMs [12-16]. In [12], local demagnetization and the effect of the stator winding configuration for demagnetization faults are investigated through the FEM simulations. It is shown that, detecting the demagnetization of the motors with fractional windings is easier than symmetrical windings configuration. In [13], the amplitudes of 1st and 5th harmonics are taken into account to detect the demagnetization at high speeds and Hilbert-Huang transforms (HHT) is used to extract the fault components at low speeds. Choi-Williams distribution (CWD) [14], a fault severity index [15], and inverse transform [16] are also suggested to determine magnet defect faults in PMSMs.

In this study, it is aimed to detect the magnet defect faults through leakage flux analysis using very small footprint fluxgate sensors providing resolution in the order of nT and has very high noise / offset immunity. For this purpose, a remote fault monitoring prototype including fluxgate sensor and microcontroller is constructed to stream leakage flux data. In order to analyze the characteristic harmonics 2D-Time Stepping Finite Element Method (TSFEM) is employed. The obtained findings and simulation results are validated through the experiments. Results have shown that there are many advantages using fluxgate sensor to detect magnet defect faults in PMSMs such as easy mount on the motor frame, providing absolute flux value, and both frequency and time domain analysis remote sensing, speed and torque independent fault detection.

978-1-4673-9551-9/16 $31.00 © 2016 IEEE

II. Fluxgate Sensors and Modeling Permanent Magnet Synchronous Motor

A. Fluxgate Sensors

There are many applications using fluxgate sensors such as navigation systems, parking systems in automotive and testing defects on surface of the material. Fluxgate sensors include excitation coil, sensing coil and ferromagnetic core [17]. Sensing and excitation coil are wounded around the ferromagnetic core as shown in Fig. 1. The sensor is triggered through periodic modulation of the permeability of ferromagnetic core.

Extremely sensitive fluxgate sensors with small form factor have strong potential to monitor the condition of the motor by reducing the size, cost and false alarm rate significantly. Built-in closed-loop magnetic core degaussing procedure and offset calibration can increase dynamic range up to six decades while supporting higher system-level accuracy. A handheld wand prototype is prepared including fluxgate sensor and data streaming units to justify the 2D-FEM findings experimentally.

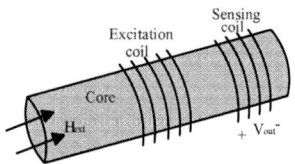

Fig. 1. Conceptual fluxgate sensor

B. Modeling Motor

In order to model PMSM, ANSYS@Maxwell software providing finite element solutions is used. In the proposed 2-D TSFEM, all mechanical, electrical and magnetic parts, such as slots, materials, air-gap and precise dimensions of these parts are taken into account. The model variables in FEM are updated a time dependent motion equation component in the z-axis (1):

$$\nabla \times \upsilon \nabla \times A = J_s - \sigma \frac{dA}{dt} - \sigma \nabla V + \nabla \times H_c \qquad (1)$$

where A is magnetic vector potential, υ is the reluctivity, J_s is the source of current density, V is the electric potential, H_c is the coercivity of the permanent magnet.

The motor model shown in Fig. 2 has distributed double layer winding with 9 stator slots, 8 poles and is fed by three phase balanced phase voltages. In order to minimize the eddy current effects, each magnet in the rotor consist two pieces along with the shaft axis. 7% of one of these magnets is removed along the z-axis to generate magnet defect fault. This introduces the fault pattern defined in (2) which can clearly be seen on leakage flux spectrums yet with significantly different dynamics.

$$f_{pattern} = \left(1 \pm \frac{k}{p}\right) f_s \qquad k = 1, 2, 3, ... \qquad (2)$$

where f_s is supply frequency, k is an integer constant, p is the number of pole pairs.

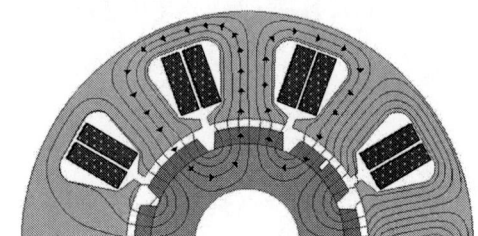

Fig. 2. The cross section of Finite Element Model, motor with magnet defect fault

III. Detection of MAgnet Defect Fault Through Leakage Flux Analysis

The magnetic field distribution plays a fundamental role in motor performance. In particular, knowing the air-gap magnetic field helps to determine the back-emfs, torque and other type of flux such as leakage flux, stator flux etc. In PMSMs, permanent magnets (PMs) placed in the rotor produce strong magnetic field (up to 1 Tesla) and higher flux density in the air-gap due to its material characteristic. Therefore, huge quantity of flux density is provided from PMs. When rotor spins at rotational frequency, an alternating magnetic field produced by PMs is generated depending on rotor rotational frequency. Thus, the interaction of stator and rotor magnetic fields drives the demanded torque on the shaft in PMSMs.

In order to analyze the flux, the magnetic equivalent circuit including all possible reluctances is introduced in PMSMs. In this model, the rotor and stator cores are modeled simply as reluctances R_r and R_s respectively. The flux source of PMs (ϕ_r), magnet reluctance (R_m) and the air-gap reluctance (R_g) are taken into account to mimic the magnetic behavior in PMSMs. In addition, the leakage flux crossover from one magnet to the next one (ϕ_{ml}), and the flux links rotor and stator, but passes through the vicinity of the motor called leakage flux (ϕ_l) are depicted as shown in Fig. 3a. Among the flux, the leakage flux (ϕ_l) do not have any contribution to generate torque and they have small amplitudes compare to air-gap flux (ϕ_g). However, they have valuable information about the motor condition and can be used to detect magnet defect fault in PMSMs. In order to simplify the analysis of magnetic circuit, the magnet leakage flux (ϕ_{ml}) can be eliminated since the magnet flux (main flux) following the leakage path are very small than that of air-gap. Thus, the air-gap flux can be written as

$$\phi_g = k_{ml}.\phi \qquad (3)$$

where k_{ml} is a magnet leakage factor ($k_{ml}<1$). By eliminating magnet leakage flux, rotor and air-gap reluctances are in series. By using basic circuit and magnetic theorems, ϕ_r remains unchanged but R_m doubles (see Fig. 3b).

The stator and rotor reluctances (R_s and R_r) have nonlinear characteristics due to saturation of the materials. Therefore, these reluctances should be eliminated from the magnetic equivalent circuit model to provide the analytic solution. As well known, R_s and R_r have relatively high permeability than air which means that they can be eliminated from the corresponding circuit.

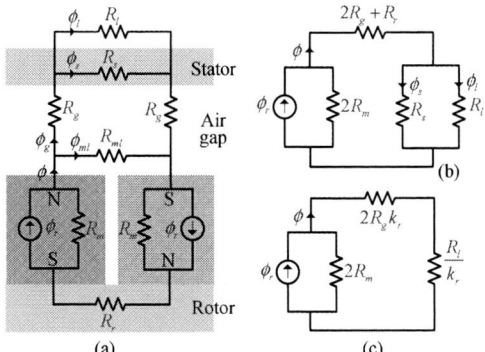

Fig. 3. Magnetic equivalent circuit for PMSM

In order to eliminate the reluctances, a reluctance factor, $k_r>1$, is defined as shown in Fig. 3c. Thus, the leakage flux, ϕ_l, can be written as;

$$\phi_l = k_l.\phi_g = k_l.k_{ml}.\phi \qquad (4)$$

where k_l is a leakage factor ($k_l<<1$). Therefore, the flux ϕ can be calculated from the basic circuit laws as follows:

$$\phi = \frac{2R_m}{2R_m + 2R_g k_r + \dfrac{R_l}{k_r}}\phi_r = \frac{1}{1 + k_r \dfrac{R_g}{R_m} + \dfrac{1}{k_r}\dfrac{R_l}{2R_m}}\phi_r \quad (5)$$

The general formulas for R_m, R_g and R_l are defined as

$$R_m = \frac{l_m}{\mu_r \mu_0 A_m}; \; R_g = \frac{g}{\mu_0 A_g}; \; R_l = \frac{l_l}{\mu_0 A_l} \qquad (6)$$

where l_m, g and l_l are the length of magnet, air-gap and leakage flux, respectively. A_m, A_g and A_l are the area of corresponding reluctances. Depending on flux concentration factors($C_{\phi x}$) and permanent coefficients (P_{cx}) where index x represent the corresponding flux, the leakage flux ϕ_l can be written as

$$C_{\phi x} = A_m / A_x$$
$$P_{cx} = l_m / g.C_{\phi x} \qquad (7)$$

$$\phi_l = \frac{k_l.k_{ml}}{1 + k_r \dfrac{\mu_r}{P_{cg}} + \dfrac{1}{2k_r}\dfrac{\mu_r}{2P_{cl}}}\phi_r \qquad (8)$$

As seen from Eq. (8), the leakage flux (ϕ_l) is a function of permanent coefficients and magnet material. One should note that C_{cl} is lower than C_{cg} due to cross section areas. The possible failures on PMs cause appearing some characteristic harmonics (2) in the air-gap flux. The reflection of these harmonics can also be observed in the leakage flux. The leakage flux spectrums modify with physical effects both on stator and rotor current frequency components such as magnet defects, off-centered rotor, asymmetry of supply voltages and various types of other abnormal situations or failures.

If there is a current in the stator windings, the magnetic field produced by stator windings can be seen on the leakage flux. Thus, the amplitude of leakage flux increases with demanded torque. But, the effects of magnetic flux produced by stator windings are much lower than that of PMs since permanent magnets have very strong magnetic field compare

to stator windings which will be discussed in the next section. Thanks to direction sensitive fluxgate sensors with nT level resolution, the exact amplitude of leakage flux and relative changes in the flux spectrum can be monitored very accurately at any point around the motor such as behind tooth and behind slot.

IV. EXPERIMENTAL SETUP

The main purpose in the proposed experimental setup is to measure actual leakage values specifications with low-cost, portability, online remote monitoring. In order to measure leakage flux, a footprint fluxgate sensor (DRV421) is used. It has high linearity, high dynamic range and wireless option to get and analyze the measured data remotely as well introduced the industry's first magnetic sensing integrated circuit (IC) with a fully integrated fluxgate sensor and compensation coil driver, along with the entire required signal conditioning circuitry [18]. The dimension of DRV421 is only in a 4-mm-by-4-mm quad flat no-lead (QFN) package. In experiments, the PM motor is coupled with hysteresis brake as shown in Fig.5 to adjust and measure the load level. The measured data is transferred Labview to analyze data. System interface is shown as Fig. 5. The leakage flux around the motor is monitored at several points and the results corresponding to two different sensor locations are presented in this paper; behind slot and behind tooth.

V. RESULTS

Simulations and experiments are carried out using 0.4 kW star connected PMSM controlled through closed loop vector control. First, in order to examine the rotor effects on the signature content, rotor is run at fixed speed while stator is open circuit. Fig. 4 shows the sensor locations used in simulations and experiments.Thus, the leakage flux is investigated at two different directions; radial (ψ_r) and tangential (ψ_t) as shown in Fig. 4.

A. Results for Open-Circuit Stator

First, simulation and experimental results are analyzed in time domain when the stator is open circuit to decouple the effects of current and magnets on the flux content. Around the motor, both the radial (ψ_r) and tangential (ψ_t) leakage flux components have been investigated. Due to direction of flux paths (see Fig. 4) in PMSM, radial flux in behind tooth, tangetial flux in behind slot are choosen around the motor to analyze the leakage flux Fig. 6. and Fig. 7. show the radial flux direction (ψ_r) in behind tooth, and tangential flux direction in behind slot at 200Hz supply frequency, respectively. As shown in Fig. 6 and Fig. 7.,

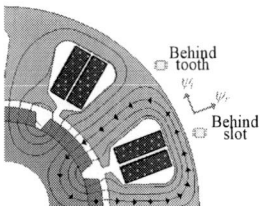

Fig. 4 . Sensor locations around the motor

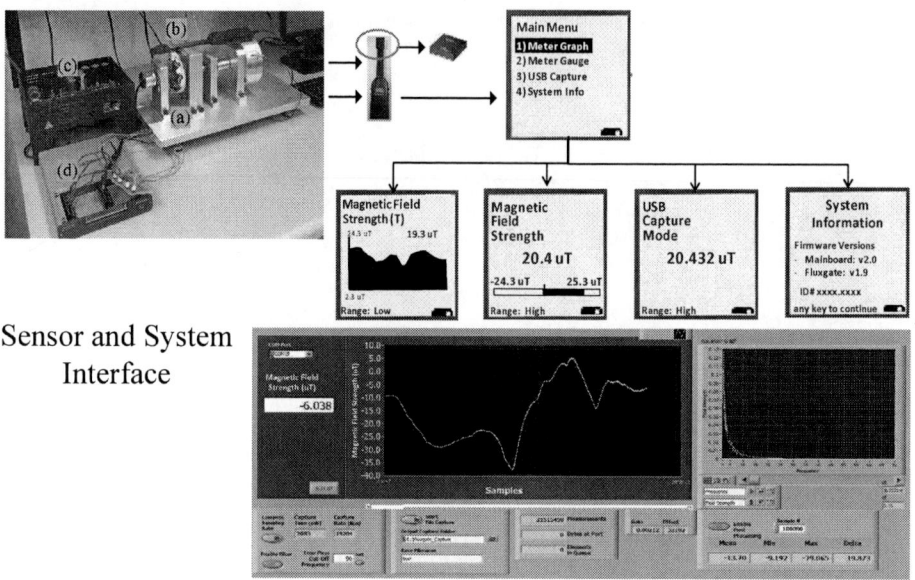

Sensor and System Interface

Fig. 5 . Test rig setup a) PMSM b) Fluxgate sensor c) Motor Drive d) DAQ card

when there is a magnet defect fault on the rotor, the leakage flux decreases across the defected magnet in both simulation and experimental results. The decreased flux due to magnet defect can be observed at every mechanical cycle ($p/2$ where p is number of pole). Besides, the leakage flux becomes unsymmetrical particularly adjacent fluxes of the defected magnets in the case of magnet defect fault. After several tests, it is observed that the magnet defect at one or multiple locations can be monitored clearly with minimum signal processing support. It means that the fault detection and the location of defected magnet can be achieved without any knowledge of signal proccesing. Only one or more rotor cycle may be enough to monitor the magnet defect fault.

In order to investigate the speed effects on the fault signatures, the back-emf spectrums are investigated while stator is open-circuit. FFT (Fast Fourier Transform) is used to obtain the leakage flux spectrums. Fig. 8 shows the spectrum of radial leakage flux components behind the tooth and Fig. 9 shows the tangential leakage flux components behind the slot at different speeds when stator is open-circuit. As shown in Fig. 8 and Fig. 9., the fault patterns given in (2) is observed in the spectrums and the fault signatures do not depend on speed which provides speed independent fault detection as a fundamental advantage for fault monitoring in adjustable speed drives. After several tests, it is also observed that some of the fault signatures are most dominant from the others such as 0.25th and 0.5th harmonics.

B. Results for Excited Stator

In order to analyze the dynamic behavior of magnet fault signatures when the stator coils are excited, the tests are repeated at different torque/speed profiles. In this case, behind tooth and behind slot location are chosen with different leakage flux directions to show the diagnostic capability of the proposed approach. Firstly, PMSM are loaded 20% and 100% of full-load to investigate the torque effects on leakage

spectrums. In these tests, rotor speed is set to 3000rpm. Fig. 10 and Fig. 11. show the spectrum of radial (for behind tooth) and tangential (for behind slot) leakage flux under different load conditions [20% and 100% of full-load]. As shown in Fig 10 and Fig. 11, the signatures show up clearly particularly 0.25th and 0.5th harmonics in the case of magnet defect fault and increase with torque but negligibly. This is because that the magnetic flux on PMs have the higher percentage than that of the stator winding as mentioned section III. If motor is loaded, then the current will get increase on the stator windings. So, this current creates more flux but the amplitude of these fluxes are always much lower than the magnetic flux of PMs. Therefore, to change the torque negligible affects the amplitude of each harmonic on the leakage flux spectrums.

In order to show the speed effects on the fault signatures when stator is excited, the motor is run at two different speeds set point; 600 rpm and 3000 rpm both in simulation and experimental tests. In this case motor is loaded at full-load. As shown in Fig. 12 and Fig. 13, the fault related harmonics defined in (2) are almost speed independent here,- as the case of stator open circuit. Actually the reason of this contrubution comes from using fluxgate sensor. Because the fluxgate sensor is directly measure the leakage flux, not back-emf voltage due to leakage flux. Based on the faraday laws, if the flux deviation ($d\phi/dt$) is changed, the back-emf voltage will change. But in this study, the leakage flux is directly measured using fluxgate sensors. Therefore, the deviations of the speed do not affect the amplitude of flux on the leakage flux spectrums, but affect the leakage flux cycling time in time domain.

VI. Conclusions

In this paper, the magnet defect fault monitoring is studied through the leakage flux information using a very small footprint fluxgate sensor. For this purpose, handheld wand prototypes is designed which includes fluxgate sensor, data streaming units and a simple screen for user interface.

978-1-4673-9551-9/16 $31.00 © 2016 IEEE

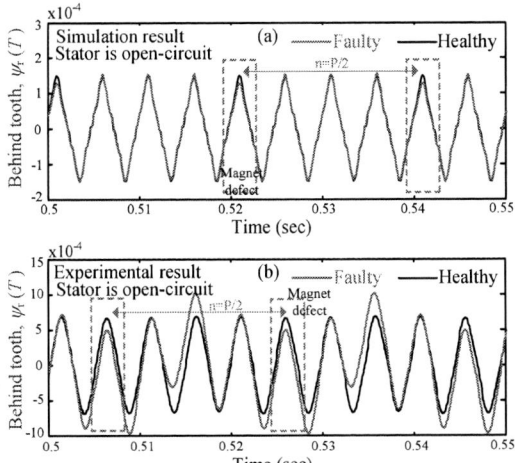

Fig. 6. Radial flux in time domain, behind tooth (stator is open circuit), healthy and faulty motor a) Simulation result b) Experimental result

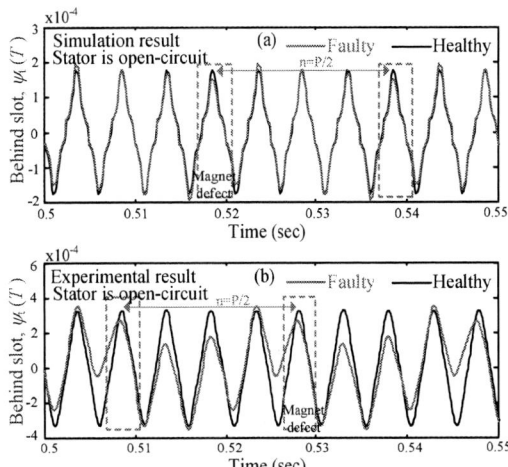

Fig. 7. Tangential flux in time domain, behind slot (stator is open circuit), healthy and faulty motor a) Simulation result b) Experimental result

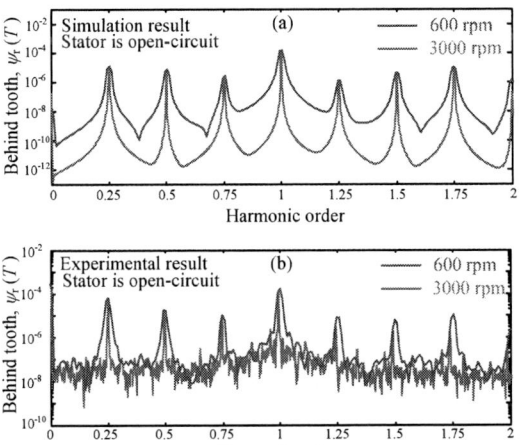

Fig. 8. Radial flux spectrums, behind tooth (stator is open circuit), faulty motor with different speeds a) Simulation result b) Experimental result

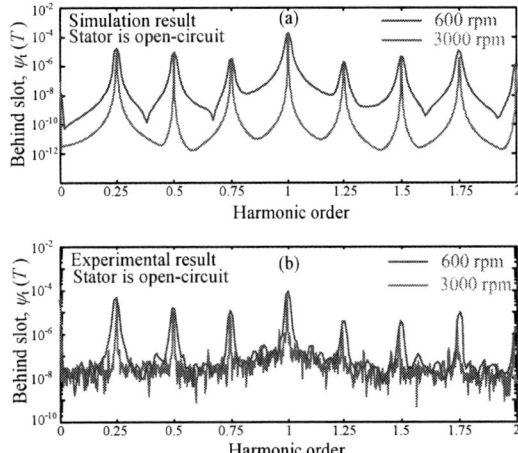

Fig. 9. Tangential flux spectrums, behind slot (stator is open circuit), faulty motor with different speeds a) Simulation result b) Experimental result

The dynamic behaviors of the flux spectrum components and effects of sensor location are discussed in detail. It is shown that PM motor leakage flux content can be used for monitoring magnet defects both in frequency and time domains, and provides speed and torque independent result which is an essential advantage for adjustable speed drives.

ACKNOWLEDGMENT

The authors wish to acknowledge the support of Texas Instruments Inc. and SRC/TxACE Center under task 1836.149.

REFERENCES

[1] Henao, H.; Demian, C.; Capolino, G.-A., "A frequency-domain detection of stator winding faults in induction machines using an external flux sensor," *Industry Applications, IEEE Transactions on*, vol.39, no.5, pp.1272,1279, Sept.-Oct. 2003.

[2] Yazidi, A.; Henao, H.; Capolino, G.-A., "Broken rotor bars fault detection in squirrel cage induction machines," *Electric Machines and Drives, 2005 IEEE International Conference on* , vol., no., pp.741,747, 15-15 May 2005

[3] Ertan, H.B.; Keysan, O., "External search coil as a means of measuring rotor speed of an induction motor," *Advanced Electromechanical Motion Systems & Electric Drives Joint Symposium, 2009.*

[4] Yao Da; Krishnamurthy, M., "Novel fault diagnostic technique for permanent Magnet Synchronous Machines using electromagnetic signature analysis," *Vehicle Power and Propulsion Conference (VPPC), 2010 IEEE* , vol., no., pp.1,6, 1-3 Sept. 2010.

[5] Henao, H.; Capolino, G.A.; Martis, C., "On the stray flux analysis for the detection of the three-phase induction machine faults," *Industry Applications Conference, 2003. 38th IAS Annual Meeting. Conference Record of the* , vol.2, no., pp.1368,1373, 12-16 Oct. 2003

[6] Adam, A.A.; Korolu, S.; Gulez, K., "Stray electromagnetic field distribution around permanent magnet synchronous motor drive," *Industrial Electronics, 2009. IECON '09. 35th Annual Conference of IEEE* , vol., no., pp.1782,1787, 3-5 Nov. 2009

[7] Frosini, L.; Borin, A.; Girometta, L.; Venchi, G., "A novel approach to detect short circuits in low voltage induction motor by stray flux measurement," *Electrical Machines (ICEM), 2012 XXth International Conference on* , vol., no., pp.1538,1544, 2-5 Sept. 2012

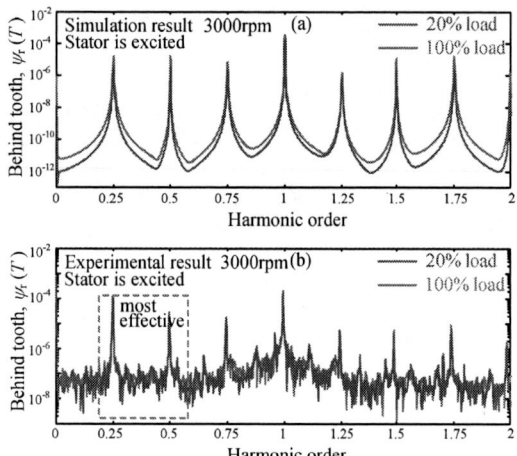

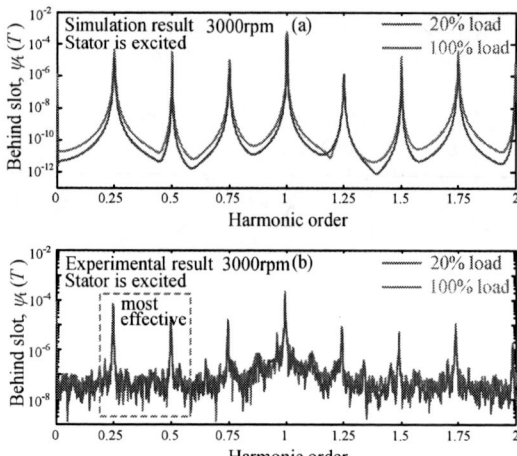

Fig. 10. Radial flux spectrums, behind tooth (stator is excited), faulty motor with different torques a) Simulation result b) Experimental result results

Fig. 11. Tangential flux spectrums, behind slot (stator is excited), faulty motor with different torques a) Simulation result b) Experimental result

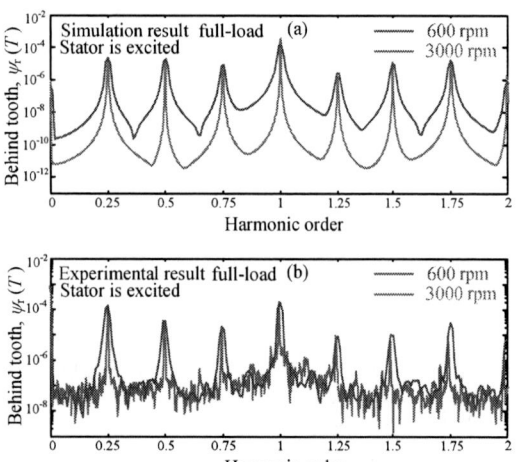

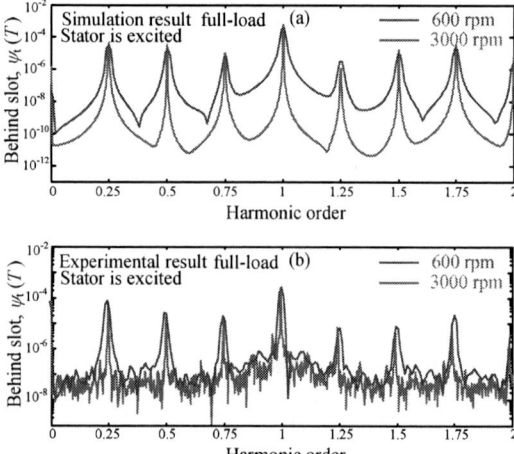

Fig. 12. Radial flux spectrums, behind tooth (stator is excited), faulty motor with different speeds a) Simulation result b) Experimental result

Fig. 13. Tangential flux spectrums, behind slot (stator is excited), faulty motor with different speeds a) Simulation result b) Experimental result

[8] Dobzhanskyi, O.; Mendrela, E.E.; Trzynadlowski, A.M., "Analysis of Leakage Flux Losses in the Transverse Flux Permanent Magnet Generator," *Green Technologies Conference (IEEE-Green), 2011 IEEE*, vol., no., pp.1,6, 14-15 April 2011

[9] Pusca, R.; Romary, R.; Ceban, A., "Detection of inter-turn short circuits in induction machines without knowledge of the healthy state," *Electrical Machines (ICEM), 2012 XXth International Conference on*, vol., no., pp.1637,1642, 2-5 Sept. 2012

[10] Rastegar Fatemi, S.M.J.; Henao, H.; Capolino, G.A., "Gearbox monitoring by using the stray flux in an induction machine based electromechanical system," *Electrotechnical Conference, 2008. MELECON 2008. The 14th IEEE Mediterranean*, vol., no., pp.484,489, 5-7 May 2008

[11] Frosini, L.; Harliska, C.; Szabó, L., "Induction Machine Bearing Fault Detection by Means of Statistical Processing of the Stray Flux Measurement," *Industrial Electronics, IEEE Transactions on*, vol.62, no.3, pp.1846,1854, March 2015

[12] D. Casadei, F. Filippetti, C. Rossi, A. Stefani, "Magnets faults characterization for permanent magnet synchronous motors," *IEEE Diagnostics for Electric Machines, Power Electronics and Drives (SDEMPED'09), Cargese*, pp. 1-6, Aug. 2009.

[13] A. G. Espinosa, A. Rosero, J. Cusido, L. romeral, and J. A. Ortega, "Fault detection by means of Hilbert-huang transform of the stator current in a PMSM with demagnetization," ult diagnosis in inverter fed induction motors," *IEEE Transactions on Energy Conversion,* vol. 25, no. 2, pp. 312-318, June. 2010.

[14] M. D. Prieto, A. G. Espinosa, J. R. Ruiz, J. C. Urresty, and J. A. Ortega, "Feature extraction of demagnetization faults in permanent-magnet synchronous motors based on box-counting fractal dimension," *IEEE Transactions on Industrial Electronics*, vol. 58, no. 5, pp. 1594-1605, May. 2011.

[15] J. Urresty, J. Riba, and L. Romeral, "A back-emf based method to detect magnet failures in PMSMs," *IEEE Trans.on Magnetics*, vol. 49, no. 1, pp. 591-598, Jan 2013.

[16] K. Abbaszadeh, S. Saied, S. Hemmati, A. Tenconi, "Inverse transform method for magnet defect diagnosis in permanent magnet machines," *IET Electr. Power Appl.,*vol. 8, Iss. 3, pp 98-107, Oct. 2013.

[17] O. Zorlu, "Orthogonal fluxgate type magnetic microsensors with wide linear operation range," *Ph.D. dissertation,* École Polytechnique Fédérale De Lausanne, 2008.

[18] DRV421 integrated magnetic fluxgate sensor for closed-loop current sensing datasheet, Texas Instruments, SBOS704A, July, 2015.

Performance Comparison of Transfer Switch Topologies in Switched-Doubly-Fed Machine Drives

Arijit Banerjee, Steven B. Leeb, and James L. Kirtley

Department of Electrical Engineering and Computer Science, Massachusetts Institute of Technology
Cambridge, Massachusetts, USA
arijit@mit.edu, sbleeb@mit.edu, kirtley@mit.edu

Abstract—**Switched doubly-fed machines (DFM) creates variable speed drives with reduced power electronics requirements compared to shaft power, and with the additional benefit of controllable stator power factor. A solid-state transfer switch is a critical component of a switched-DFM drive that not only allows seamless shaft control across the full speed range but also permits effective grid interaction. This paper presents and compares two transfer switch topologies—twelve-thyristor-based and eight-thyristor-based that reconfigures the DFM connections on-the-fly with different shaft torque-speed demands. Reducing the number of thyristors allows the operation of the DFM with lesser commutation constraints. Appropriate control input is derived that maximizes the damping of the stator flux during the mode transition.**

I. INTRODUCTION

Variable speed drives (VSD) have wide range of applications in industrial processes, electric propulsion systems, and power generation plants [1]. In these VSDs, power converters that control the electromechanical energy conversion are mostly rated to handle the full shaft power. At higher shaft power, the power converter design becomes overly challenging due to limited available component ratings and allowable device switching frequency. For example, currently the voltage rating of IGBTs that are typically used for medium voltage drives ranges between 3.3 kV and 6.5 kV with an allowable switching frequency of a few hundred Hz [2]. Stacking multiple devices in series/parallel—as in multilevel converters—becomes a standard approach for processing power at levels beyond the available ratings of the individual devices [3], [4], [5].

Alternatively, for limited-speed range applications, high-power VSDs can be created using a reduced-sized power converter with a wound-rotor induction/doubly-fed machine (DFM). The configuration has been widely used in the wind power generation [6] where for the limited-speed range operation the power processed by the power converter is typically one-third of the full shaft power. For full-speed range applications, including electric propulsion drives, the flexibility of the DFM can be exploited with on-the-fly reconfiguration of the power connections to the stator - a configuration known as a switched-DFM drive [7]. A typical switched-DFM drive operates over a speed range of ± 1.5 p.u (normalized relative to the ac source synchronous speed) with a rotor converter power rating of a third of the maximum shaft power [8]. The reduction in the converter rating improves drive efficiency,

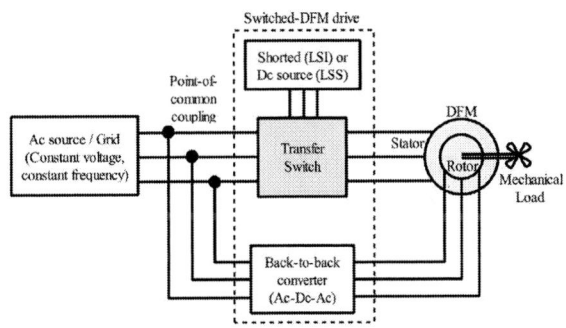

Figure 1. Typical configuration of a switched-doubly-fed machine drive.

lowers the fundamental drive frequency, reduces filter size, improves machine-side and source-side harmonics, and reduces converter cooling requirements [9]. Additionally, the switched DFM drive offers seamless interface, controllable power factor, and reactive power support to the ac grid [10]. Figure 1 shows a switched-DFM drive with a transfer switch that connects the stator windings of the DFM to multiple sources or shorts the stator windings together based on the operating speed while a back-to-back power converter controls the electromechanical energy conversion from the rotor.

The switched-DFM drive can operate either in the low-speed induction topology (LSI) or in the low-speed synchronous topology (LSS) [11], [12], [13] based on the possible stator winding configurations. At lower drive speed, the DFM stator is either shorted in the LSI topology or is connected to a low-voltage dc source in the LSS topology. This mode of operation of the switched-DFM drive is referred to as "low-speed" mode. At higher drive speed, in both topologies, the DFM stator is connected to the ac source/grid, which is denoted as "high-speed" mode. The mode transition speed depends on the topology as well as on the required drive torque-speed characteristic.

The transfer switch is the key enabler for a switched-DFM drive that reconfigures the stator connection on-the-fly. Although a mechanical transfer switch can potentially be used for the stator reconfiguration [11], [14], [15], this choice results in poorer reliability and transfer performance for applications that need frequent reconfigurations and seamless

978-1-4673-9551-9/16 $31.00 © 2016 IEEE

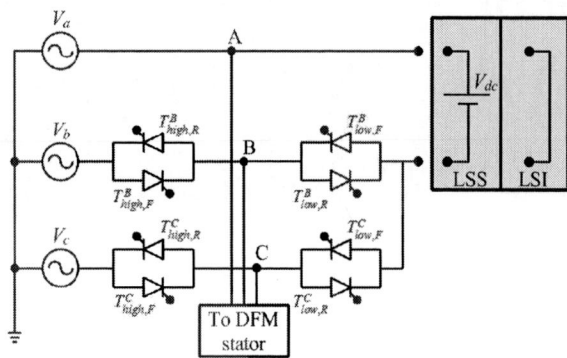

Figure 2. Eight thyristor-based transfer switch (ETB) - A-phase of the ac source is the reference for the transfer switch operation [19].

performance. Of course, for the configuration shown in this paper, a mechanical transfer switch could be used in parallel with a semiconductor transfer switch to reduce steady-state losses to a minimum [16]. Reference [17] proposed a twelve-thyristor-based (TTB) transfer switch that offers seamless transition and drive performance with an appropriate rotor-side control [18]. Alternative configurations of the thyristor-based transfer switch using eight thyristors are proposed in [19]. This paper compares the performance of the eight-thyristor-based transfer switch (ETB) with the twelve-thyristor-based transfer switch specifically under lightly-loaded conditions in the low-speed synchronous topology. Analysis shows reduction in number of thyristors for creating a transfer switch opens up the range of feasible operating conditions of the DFM over which the mode transition is accomplished seamlessly in addition to simplifying the transfer switch and decreasing the cost. An optimal non-linear controller is presented that minimizes the stator flux perturbation during the mode transition through the maximum allowable damping. Finally, an alternative transfer switch topology is proposed that is applicable for the LSS topology with even lesser number of thyristors eliminating the limitations on operating conditions for a seamless transition in the low-speed-to-high-speed mode transition. Experimental results are presented to verify the operation of the eight-thyristor-based transfer switch using an off-the-shelf 1 HP, 220V/150V, 4 pole, 1800 rpm doubly-fed machine.

II. EIGHT-THYRISTOR-BASED TRANSFER SWITCH OPERATION AND CONTROL IN THE LSS TOPOLOGY

Typically, fully controllable switches such as IGBTs add degrees of freedom in a power electronic converter design. On the contrary, semi-controllable switches such as thyristors (SCR) act as a constraint when used to design systems by using the turn-off functionality. Eliminating SCRs from the TTB transfer switch improves the operational flexibility of the switched-DFM drive. The eight-thyristor-based (ETB) transfer switch is shown in Fig. 2. The A-phase of the DFM is chosen as the reference for the transfer switch operation based on the

advantages presented in [19]. SCRs $T_{high,F}^B$, $T_{high,R}^B$, $T_{high,F}^C$, and $T_{high,R}^C$ form the high-speed SCR bank for the transfer switch while SCRs $T_{low,F}^B$, $T_{low,R}^B$, $T_{low,F}^C$, and $T_{low,R}^C$ form the low-speed SCR bank. Turing ON the high-speed SCR bank connects the DFM stator to the ac source. The SCRs conduct for half of the ac source fundamental time period. The losses in these SCRs in steady state are predominantly the conduction losses that can be minimized using parallel mechanical switches. Alternatively, turning the low-speed SCR bank ON either connects the DFM stator to a dc source in the LSS topology or shorts the stator together in the LSI topology. In the LSS topology, for a balanced three phase stator winding of the DFM, the current in the A-phase is twice that of the currents in B and C-phases. Removing the anti-parallel SCRs from the A-phase of the TTB transfer switch reduces the conduction loss in the transfer switch in the LSS topology by two-thirds. In the LSI topology, the conduction loss in the ETB transfer switch is two-thirds in comparison to the TTB transfer switch. However, the SCRs in the ETB transfer switch must be rated to withstand the ac source line-to-line voltages instead of the ac source phase voltages as in the TTB transfer switch. In Sec. II-A an analytical framework is developed to compare the performance of the two transfer switches during a low-speed-to-high-speed mode transition in the LSS topology. Sections II-B and II-C shows the operational flexibilities of the ETB transfer switch over the TTB transfer switch under different drive torque conditions. Sec. II-D presents an optimal control policy for the rotor d-axis current that maximizes the stator flux damping during the mode transition.

A. Analytical framework for evaluation of the low-speed-to-high-speed mode transition in the LSS topology

The feasibility of a seamless low-speed-to-high-speed mode transition in the LSS topology is dependent on the drive torque demand. This section introduces an analytical framework that is used in the later sections to evaluate and compare the performances of the ETB and the TTB transfer switches. With the DFM stator initially connected to the dc source, the low-speed SCR-bank is ON. The stator current vector $\bar{i}_s$ and the stator voltage are stationary relative to the ABC stator winding axis as shown in Fig. 3. Connecting the A-phase of the stator to the positive polarity and B and C-phases to the negative polarity of the dc source enforces the stator voltage vector orientation along the A-phase axis. Under steady-state condition, the stator current vector is also directed towards the A-phase axis. The incoming ac source voltage vector $\overline{v_{ac}}$ rotates in ABC-phase sequence at the grid frequency ω_{ac} relative to the stationary reference frame. The blue sector represents the feasible region for the incoming ac source voltage vector that ensures natural commutation for all outgoing SCRs of the low-speed SCR bank in the TTB transfer switch. Reducing the number of thyristors to eight extends the feasible region to the green sector as described in [19]. Prior to the mode transition, the orientation of the stator flux vector $\bar{\psi}_s$ is dependent on the operating drive torque τ, the normalized dc source voltage v_{dc} and the designed stator flux magnitude

978-1-4673-9551-9/16 $31.00 © 2016 IEEE

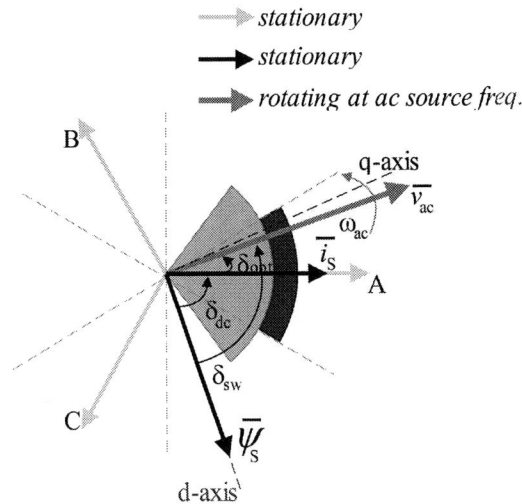

Figure 3. Commutation diagram during low-speed-to-high-speed mode transition under rated drive torque condition. Turning ON the high-speed SCR bank of the ETB transfer switch when the ac source voltage vector is within the green sector naturally turns OFF the low-speed SCR bank. The corresponding region for the TTB transfer switch is shown in blue.

in the low-speed mode ψ_s and is given by

$$\delta_{dc} = \arcsin \frac{\tau r_s}{v_{dc} \psi_s}, \tag{1}$$

where δ_{dc} represents the angle between the stator flux vector and the stator voltage vector in the low-speed mode.

Minimum perturbation in the stator flux during mode transition is achieved by connecting the ac source to the stator at the instant when the incoming ac voltage vector is at an angle δ_{sw} relative to the stator flux vector $\bar{\psi}_s$ [17], where

$$\delta_{sw} = \arccos\left(v_{dc}\cos\delta_{dc}\right). \tag{2}$$

Therefore, the desired angle for the incoming ac voltage vector relative to the A-phase axis at the instant of mode transition is given by

$$\delta_{opt} = \delta_{sw} - \delta_{dc}. \tag{3}$$

Based on the chosen transfer switch topology, the feasible region for δ_{opt} is bounded by the requirement of the natural commutation of the outgoing SCRs in the low-speed SCR bank, which is represented as

$$\delta_{opt} \in [-\epsilon, \epsilon]. \tag{4}$$

For the TTB transfer switch, $\epsilon = 30°$ [17] as represented by the blue sector while for the ETB transfer switch [19]

$$\epsilon = 60° - \arcsin\left(\frac{v_{dc}}{\sqrt{3}}\right), \tag{5}$$

as represented by the green sector where the peak phase voltage of the ac source is chosen as the base voltage for normalization. Assuming a high current controller bandwidth for the rotor currents and a high mechanical time constant, the trajectory of the stator flux after the dc-to-ac source transition is governed by [20]

$$\frac{1}{\omega_{ac}}\frac{d}{dt}\begin{bmatrix}\psi_s\\\delta\end{bmatrix} = \begin{bmatrix}-\frac{r_s}{x_s}\psi_s + \cos\delta\\1 - \frac{1}{\psi_s}\sin\delta + \frac{1}{\psi_s^2}r_s\tau\end{bmatrix} + \begin{bmatrix}\frac{r_s x_m}{x_s}i_{rd}\\0\end{bmatrix} \tag{6}$$

where the state variables are the stator flux magnitude ψ_s and the angle between the stator flux vector and the ac source voltage δ. The parameters of the state-space representation are the normalized stator resistance r_s, stator inductance x_s, and the mutual inductance x_m of the DFM. To ensure a bumpless transition at the shaft, the drive torque τ needs to be maintained constant during the mode transition. The rotor d-axis current i_{rd} is the only available control input that affects the state trajectory during the mode transition. The steady-state operating point of the DFM in the high-speed mode can be calculated equating (6) to zero.

During the mode transition, the DFM must be operated within the allowable stator and rotor currents ratings that implies

$$i_{sd}^2 + i_{sq}^2 \le 1 \tag{7}$$

and

$$i_{rd}^2 + i_{rq}^2 \le I_{rpu}^2, \tag{8}$$

where I_{rpu} represents the normalized rated rotor current relative to the stator current rating. The rotor q-axis current component is represented by i_{rq} while i_{sd} and i_{sq} represent the d and q-axis current components for the stator. Additionally, the maximum voltage that can be impressed on the rotor is limited by the designed voltage rating of the rotor power electronic converter based on the steady-state drive design [7]. Considering v_{rd} and v_{rq} as the d and q-axis components of the rotor voltage, the limit on the allowable rotor voltage V_r enforces

$$v_{rd}^2 + v_{rq}^2 \le V_r^2. \tag{9}$$

Constraint (7) is mapped as a function f in the $\delta - \psi_s$ plane using the DFM machine model in the stator flux orientation as given in [7] by

$$f\left(\psi_s, \tau, i_{rd}\right) = \left(\frac{\psi_s - x_m i_{rd}}{x_s}\right)^2 + \left(\frac{\tau}{\psi_s}\right)^2 - 1 \le 0 \tag{10}$$

The constraint is dependent on the stator flux magnitude, drive torque, and the rotor d-axis current input. Similarly, constraint (8) is mapped as a function g in the $\delta - \psi_s$ plane by

$$g\left(\psi_s, \tau, i_{rd}\right) = i_{rd}^2 + \left(-\frac{x_s \tau}{x_m \psi_s}\right)^2 - I_{rpu}^2 \le 0. \tag{11}$$

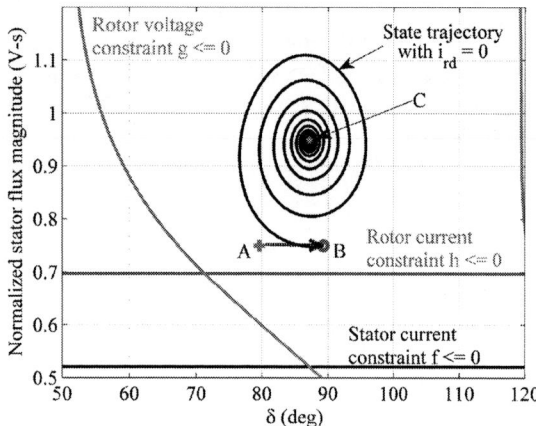

Figure 4. Stator flux trajectory during dc-to-ac source transition under rated drive torque condition lies within the constraint boundaries set by the stator current, rotor current, and the rotor power electronics voltage rating. A: Initial operating point in the $\delta - \psi_s$ plane while the DFM is operated in the low-speed mode. B: Operating point right after the mode transition. C. Steady-state operating point in the high-speed mode.

Finally, the individual components of the rotor voltage is calculated using

$$v_{rd} = r_e i_{rd} + \frac{x_m}{x_s} \cos\delta - \frac{r_s x_m}{x_s^2}\psi_s - x_e\left(\omega_s - \omega_T\right)\left(-\frac{x_s\tau}{x_m\psi_s}\right) \quad (12)$$

and

$$v_{rq} = r_r\left(-\frac{x_s\tau}{x_m\psi_s}\right) + \left(\omega_s - \omega_T\right)\left[x_e i_{rd} + \frac{x_m\psi_s}{x_s}\right] \quad (13)$$

where ω_T represents the low-speed-to-high-speed mode transition speed. The stator flux frequency is represented by ω_s and is calculated by

$$\omega_s\left(\psi_s, \delta, \tau\right) = \frac{\sin\delta}{\psi_s} - \frac{r_s\tau}{\psi_s^2}. \quad (14)$$

r_e and x_e represents the equivalent resistance and inductance and is given by

$$r_e = r_r + r_s\frac{x_m^2}{x_s^2}; \quad x_e = x_r - \frac{x_m^2}{x_s}. \quad (15)$$

Using (12) and (13), the rotor voltage constraint (9) is mapped as a function h in the $\delta - \psi_s$ plane and is represented by

$$h\left(\psi_s, \delta, \tau, i_{rd}\right) = v_{rd}^2 + v_{rq}^2 - V_r^2 \leq 0. \quad (16)$$

In the following sections, the developed model along with the constraints projected on the $\delta - \psi_s$ plane are used to compare the performance of the ETB and the TTB transfer switch during the mode transition under different operating drive torque conditions.

B. Low-speed-to-high-speed mode transition at rated drive torque condition

At the rated drive torque condition during the desired mode transition, τ equals 0.498 p.u for the example DFM as derived in [7]. The parameters of the example DFM are also given in [7]. With a dc source voltage of 0.068 p.u and a designed stator flux magnitude of 0.75 p.u, full drive torque demand sets the operating point of the DFM in the low-speed mode at $(80°, 0.75)$ denoted by A in the $\delta - \psi_s$ plane as shown in Fig. 4. The optimum transition instant δ_{opt} for minimizing the stator flux perturbation is calculated as $10°$ using (3). The optimum transition instant lies well within the constraint ϵ of the TTB transfer switch to enable a natural commutation of the outgoing SCRs in the low-speed SCR bank. For the ETB transfer switch, ϵ is calculated as $\sim 55°$ using (5) implying that the dc to ac source transition is equally feasible at the optimum transition instant. After the source transition, the operating point is placed instantly at $(90°, 0.75)$ denoted by B in the $\delta - \psi_s$ plane. The stator flux transition dynamics follows (6) to reach the steady state in the ac source connected mode denoted by C. The stator flux transition trajectory shown in Fig. 4 is with the d-axis rotor current input set to zero.

The constraints from the stator current rating, rotor current rating, and the voltage rating of the rotor power electronic drive, using (10), (11), and (16) respectively, are mapped to the $\delta - \psi_s$ plane. The entire state trajectory remains within the constraint boundaries during the mode transition ensuring that the low-speed-to-high-speed mode transition is achievable seamlessly for the example DFM under rated drive torque condition using either of the ETB or the TTB transfer switches. Additionally, the rotor d-axis current can be commanded to damp the oscillations in the stator flux magnitude using a full-state feedback controller [18] or an optimally damped control law as will be discussed in Sec. II-D.

C. Low-speed-to-high-speed mode transition in the LSS topology at low drive torque condition

Under lightly loaded condition, the demanded drive torque is lower than the rated value for the DFM. For example, assuming a drive torque demand τ to be 0.174 p.u (35% of the rated torque in the low-speed mode) for the example DFM, the operating point of the DFM in the low-speed mode is at $(20°, 0.75)$ denoted by D in the $\delta - \psi_s$ plane as shown in Fig. 5. The optimum transition instant δ_{opt} for minimizing the stator flux perturbation is calculated as $66°$ using (3). The optimum transition instant is not feasible to be achieved using either of the TTB or the ETB transfer switches as this would violate the condition required to satisfy the natural commutation for the outgoing SCRs. The condition is depicted in the vector diagram shown in Fig. 6 where the optimum transition instant from the perspective of minimizing stator flux transients is achieved when the ac source voltage vector is at $\overline{v_{ac}}$ while lying outside the feasible sectors for natural commutation of the outgoing SCRs. To ensure natural commutation of the SCRs, the transition must be initiated at $\overline{v_{ac}'}$ for the ETB transfer switch and at $\overline{v_{ac}''}$ for the TTB transfer switch.

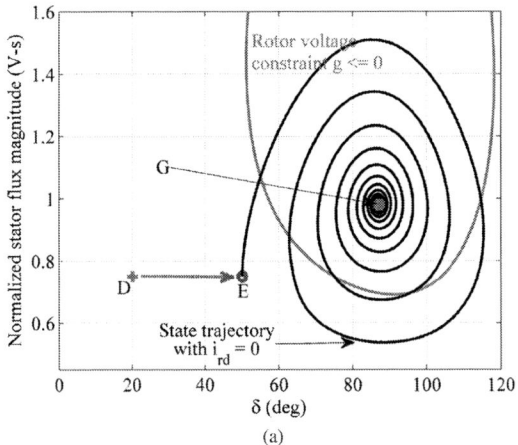

(a)

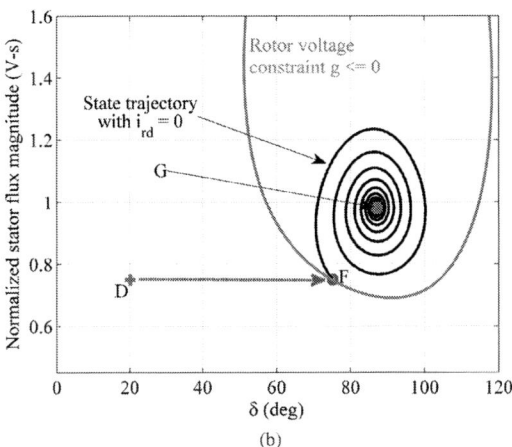

(b)

Figure 5. Stator flux trajectory during dc-to-ac source transition under low drive torque condition (35% of the rated drive torque). D: Initial operating point in the $\delta - \psi_s$ plane while the DFM is operated in the low-speed mode. G: Steady-state operating point in the high-speed mode. (a) TTB transfer switch: E is the operating point right after the mode transition. $DE = 30°$ (b) ETB transfer switch: F is the operating point right after the mode transition. $DF = 55°$

The natural commutation of the SCR constraint enforces the operating point of the DFM after the source transition to be at $(50°, 0.75)$ denoted by E for the TTB transfer switch and at $(75°, 0.75)$ denoted by F for the ETB transfer switch as shown in Fig. 5(a) and (b) respectively.

Using (6), the stator flux trajectory with zero d-axis rotor current input for these two initial conditions are also shown in Fig. 5. The stator flux magnitude swing is 20% higher for the TTB transfer switch in comparison to the ETB transfer switch. A higher flux swing results in saturation of the magnetic circuit of the DFM affecting the overall performance of the DFM. Moreover, the TTB transfer switch will not be able to provide a seamless torque at the shaft during the mode transition. This can be observed by superimposing the rotor converter voltage constraint of $V_r = 0.52$ p.u in the $\delta - \psi_s$ plane of

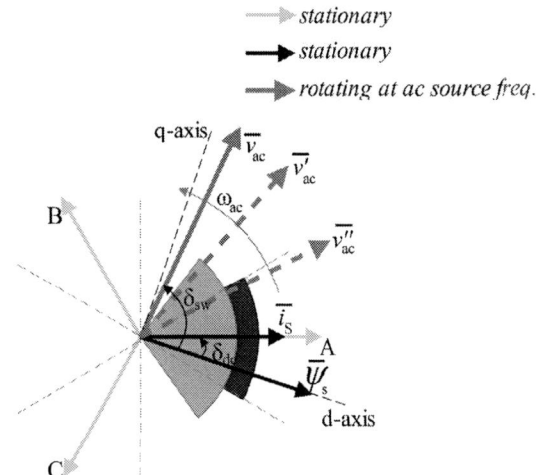

Figure 6. Commutation diagram during low-speed-to-high-speed mode transition under low drive torque condition. From the DFM perspective, a dc-to-ac source transition instant when the ac source voltage vector is at $\overline{v_{ac}}$ incurs least perturbation in the stator flux. However, constraints for the natural commutation of the outgoing SCRs in the low-speed SCR bank enforces the transition instant to be at $\overline{v'_{ac}}$ for the ETB transfer switch and at $\overline{v''_{ac}}$ for the TTB transfer switch respectively.

Fig. 5(a). Multiple points in the state trajectory including the initial condition of E lies outside the rotor voltage constraint boundary implying that the rotor voltage will be limited during the mode transition. The limited rotor voltage leads to limited q-axis rotor current and consequently a bump in the drive torque. The constraints for the stator current rating and the rotor current rating are much relaxed in the $\delta - \psi_s$ plane and do not directly affect the drive performance under low drive torque condition. For the ETB transfer switch, all the points in the state trajectory including the initial condition of F lies within the rotor voltage constraint boundary implying that a seamless transition can be achieved with zero d-axis rotor current command.

D. Optimum rotor d-axis current input to minimize stator flux transients

The analysis shown in Sec. II-B and Sec. II-C explicitly sets the rotor d-axis current to zero to emphasize on the state trajectories during the mode transition with zero control input. However, appropriate rotor d-axis current damps the oscillations in the stator flux within the allowable bounds of the stator current, rotor current and the rotor converter voltage rating. Multiple approaches can be used to optimize the d-axis rotor current input based on the system-specific requirement. For example, one approach is to maximize the damping of the stator flux during a low-speed-to-high-speed mode transition.

Two steps are involved to compute the optimum rotor d-axis current that maximizes the stator flux damping. First, identification of the correct polarity of i_{rd} that can maintain stability of the DFM during the mode transition and second,

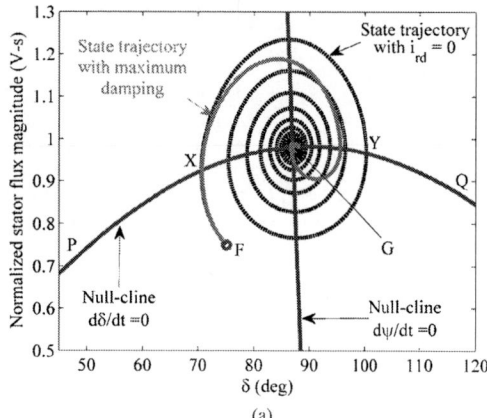

(a)

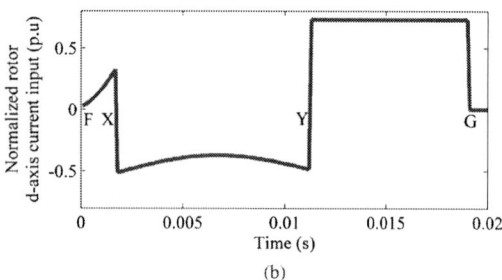

(b)

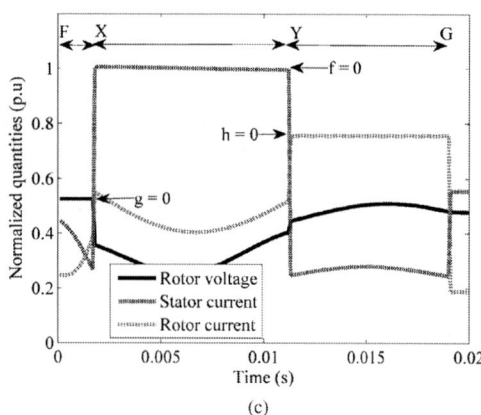

(c)

Figure 7. Maximizing rotor d-axis current input to damp stator flux oscillation during dc-to-ac source transition. (a) Comparison of the stator flux trajectories with and without d-axis rotor current damping under light drive torque condition. (b) Optimal rotor d-axis current input ensuring that the constraints of the stator current, rotor current, and rotor converter voltage rating are satisfied at all operating points of the damped trajectory. (c) The constraint functions f, g, and h during the dc-to-ac source transition that remain within the respective limits for the example DFM.

maximizing i_{rd} that can be impressed along the state trajectory to damp the stator flux within the allowable limits. As the rotor d-axis current affects the state trajectory only along the ψ_s axis, as evident from (6), the stabilizing polarity of i_{rd} at each instant is such that the state trajectory is driven inwards

towards the steady-state operating point G in Fig. 5. The null-cline obtained by equating $d\delta/dt$ in (6) to zero is shown as by the line segment PQ in Fig. 7(a). For the operating points below the null-cline PQ, a non-negative rotor d-axis current results in forcing the state flux to inner trajectories in the phase plane. Conversely for the operating points above the null-cline PQ, a non-positive rotor d-axis current ensures that the stator flux is drawn towards the inner trajectories in the phase plane.

The maximum rotor d-axis current that can be impressed to damp the stator flux transients is governed by the constraints described by (10), (11), and (16). For the example DFM with a 35% drive torque demand during the mode transition, the optimum d-axis rotor current is shown in Fig. 7(b). During the sections FX and YG in the state trajectory, a positive d-axis rotor current is commanded while in the section XY, a negative rotor d-axis current is used to damp the stator flux oscillation. The magnitude of the rotor d-axis current is governed by the rotor voltage constraint in the section FX as shown in Fig.7(c). In the section XY, the magnitude of the rotor d-axis current is set by the stator current constraint while in the section YG, the limit on the rotor current plays the dominant role in determining the magnitude of the rotor d-axis current. The above outlined procedure can be repeated to find the optimal d-axis rotor current that provides maximum stator flux damping for other drive torque conditions.

III. ALTERNATIVE THYRISTOR-BASED TRANSFER SWITCH FOR LSS TOPOLOGY: OPERATION AND CONTROL

The minimum drive torque requirement for a torque-bumpless transition from low-speed mode to high-speed mode can be eliminated for the LSS topology using an alternative transfer switch topology as shown in Fig. 8. In this transfer switch configuration, two of the stator windings (for example, A and B-phases) conduct equal and opposite currents as set by dc source while the current in the C-phase is zero in the low-speed mode. In steady-state condition, the stator current vector $\bar{i}_s^l$ in the low speed mode is given by

$$\bar{i}_s^l = \left(\frac{3}{2} - j\frac{\sqrt{3}}{2}\right)\frac{v_{dc}'}{2r_s} = \frac{\sqrt{3}}{2}\frac{v_{dc}'}{r_s}\angle -30° \quad (17)$$

where v_{dc}' is the dc source voltage. On the contrary, the stator current vector in the low-speed mode for the TTB or the ETB transfer switch is given by

$$\bar{i}_s = \frac{v_{dc}}{r_s}. \quad (18)$$

This requires that the dc source voltage v_{dc}' is marginally increased such that the magnitude of the stator current vector remains identical for all the transfer switch topologies achieving identical stator flux magnitude. Therefore,

$$v_{dc}' = \frac{2}{\sqrt{3}}v_{dc} \quad (19)$$

The operation of the DFM in the low-speed mode remains same as the TTB or the ETB transfer switch except that the

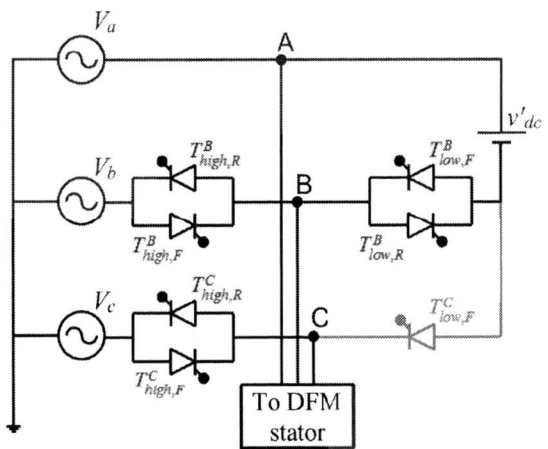

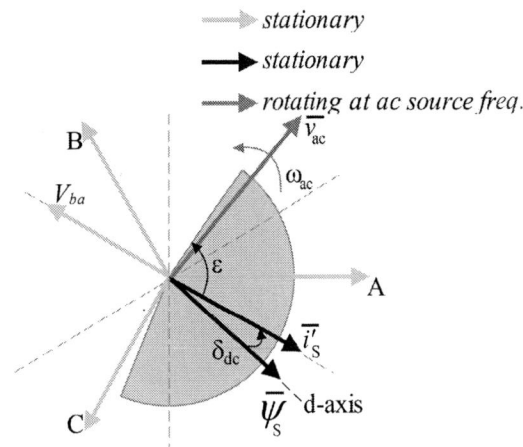

Figure 8. An alternative transfer switch topology for a switched-DFM drive in the LSS topology that achieves seamless dc-to-ac-source transition over a wider range of drive torque demand compared to the ETB transfer switch with $\epsilon \approx 90°$. SCR $T_{low,F}^C$ does not conduct in the low-speed mode but aids in current commutation during ac-to-dc source transition.

Figure 9. Commutation diagram during low-speed-to-high-speed mode transition under low drive torque condition for the alternative proposed transfer switch. A wide range of allowable ac source voltage vector location is feasible during a dc-to-ac source transition shown by the yellow sector to meet the natural commutation requirement of the SCR $T_{low,R}^B$. The dc source voltage vector and the steady state stator current vector is at an angle $-30°$ relative to the stator A-phase winding axis.

stator current vector $\bar{i}_s'$ is oriented along $-30°$ relative to the stator A-phase winding axis as shown in Fig. 9. With the current in the B-phase stator winding being negative, for a successful commutation of the $T_{low,R}^B$ thyristor during low-speed-to-high-speed mode transition, the required constraint on the ac source voltage is given by

$$v_{ba} < -v_{dc}'. \tag{20}$$

The possible commutation region for the incoming ac source voltage vector is shown by the yellow sector in Fig. 9 with the ϵ being calculated as

$$\epsilon = 90° - \arcsin\left(\frac{v_{dc}'}{\sqrt{3}}\right). \tag{21}$$

Equation (21) is similar to (5) but with a wider range of allowable stator voltage vector location during low-speed-to-high-speed mode transition. For the example DFM, ϵ is calculated as $87°$ implying that the dc-to-ac source transition can be initiated for any positive drive torque.

In the absence of the SCR $T_{low,F}^C$ in Fig. 8, the ac-to-dc source transition can only be performed when the stator current in the C-phase is zero. Presence of the SCR $T_{low,F}^C$ assists during the ac-to-dc source transition without affecting the dc-to-ac source transition.

IV. EXPERIMENTAL RESULT

The experimental setup used to verify the operation of the proposed transfer switch has been described in [19]. The 146 V (line-to-line, rms), 40 Hz ac grid created using two parallel-operated synchronous generators provides the overall power to the switched-DFM drive. The stator of the DFM is connected to a prototype ETB transfer switch that can toggle the stator

connection to the ac source during high-speed mode operation. The drive is configured in the LSS topology where a separate 20 V dc source supplies the stator during low-speed mode. In the transfer switch, diodes are placed in series with each SCR to enable faster turn off during mode transition. An RC snubber of 330 Ω and 6.8 nF is connected in parallel with the SCRs to limit the dv/dt stress. The rotor of the DFM is connected to the ac grid using a back-to-back converter configuration using two Texas Instruments High Voltage Motor Control & PFC Developer's Kits. The DFM is connected to a load generator (PMSG), which can operate with different load torque profiles. The mode transition is initiated by a hysteresis comparator that compares the rotor speed to high and low threshold set points.

Figure 10 shows the stator flux transition in the $\delta - \psi_s$ plane during dc-to-ac source change over in the LSS topology. A_1, A_2, and A_3 represent three initial operating points in the low-speed mode under three different drive torque demands, which are 56%, 74%, and 94% of the low-speed drive torque capability respectively. The ETB transfer switch enforces the operating points to move to B_1, B_2, and B_3 after the stator of the DFM is connected to the ac grid such that the outgoing low-speed SCR bank undergoes natural commutations. Finally, the stator flux settles down to steady-state operating points C_1, C_2, and C_3 respectively. As expected from the analytical framework, experimental results show similar behavior in the stator flux transition during low-speed-to-high-speed mode transition. Under lower drive torque demand, A_1 in this case, the transition instance is set by the SCR commutation requirement rather than optimum instance for least perturbation of the

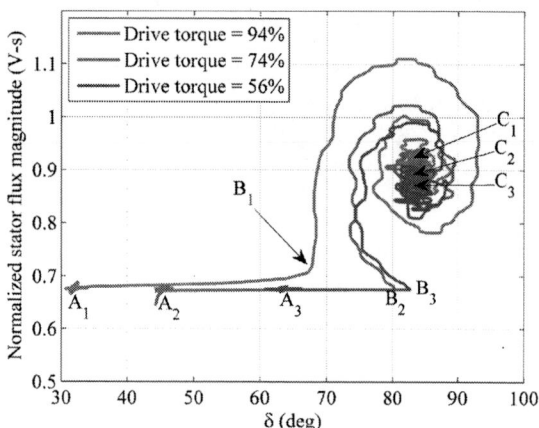

Figure 10. Experimental result: Stator flux transition during low-speed-to-high-speed mode transition under different drive torque condition. A_1, A_2, and A_3: Initial operating point with the DFM stator connected to the dc source. B_1, B_2, and B_3: Operating point post transition to the ac source. C_1, C_2, and C_3: Steady-state operating point in the high speed mode.

stator flux after transition.

V. CONCLUSIONS

This paper presented performance comparison of three different topologies of solid-state transfer switches for a switched-DFM drive under low drive torque condition in the LSS topology. The seamless transition of the DFM along with the natural commutation of the SCR-based transfer switch must be simultaneously satisfied for a smooth or "bumpless" mechanical operation of the switched-DFM drive over a wide range of drive torque-speed requirements. Reducing the number of SCRs in the transfer switch result in removing turn-off constraints that enables seamless mode transition over a wide range of drive torque demands. Optimal d-axis rotor current command damps stator flux transition within the allowable limits as set by the stator current, rotor current, and the rotor converter voltage rating. As the associated power converter for a switched-DFM drive required to handle only one-third of the shaft mechanical power over the complete speed range, the overall drive scheme becomes attractive for many high power applications including propulsion. Many variations on the transfer switch are possible, including parallel mechanical switches or relays to reduce steady state power consumption. The analytical framework in this paper can be extended to other arrangements to provide convenient transfer switch properties for any application.

ACKNOWLEDGMENT

This research was performed with support from the Skoltech-MIT SDP Program, and The Grainger Foundation.

REFERENCES

[1] S. Kouro, J. Rodriguez, B. Wu, S. Bernet, and M. Perez, "Powering the future of industry: High-power adjustable speed drive topologies," *Industry Applications Magazine, IEEE*, vol. 18, no. 4, pp. 26–39, July 2012.

[2] J. Sayago, T. Bruckner, and S. Bernet, "How to select the system voltage of mv drives - a comparison of semiconductor expenses," *Industrial Electronics, IEEE Transactions on*, vol. 55, no. 9, pp. 3381–3390, Sept 2008.

[3] J. Rodriguez, S. Bernet, B. Wu, J. Pontt, and S. Kouro, "Multi-level voltage-source-converter topologies for industrial medium-voltage drives," *Industrial Electronics, IEEE Transactions on*, vol. 54, no. 6, pp. 2930–2945, Dec 2007.

[4] J. Rodriguez, B. Wu, S. Bernet, N. Zargari, J. Rebolledo, J. Pontt, and P. Steimer, "Design and evaluation criteria for high power drives," in *Industry Applications Society Annual Meeting, 2008. IAS '08. IEEE*, Oct 2008, pp. 1–9.

[5] S. Bernet, "Recent developments of high power converters for industry and traction applications," *Power Electronics, IEEE Transactions on*, vol. 15, no. 6, pp. 1102–1117, Nov 2000.

[6] Z. Chen, J. Guerrero, and F. Blaabjerg, "A review of the state of the art of power electronics for wind turbines," *Power Electronics, IEEE Transactions on*, vol. 24, no. 8, pp. 1859–1875, Aug 2009.

[7] A. Banerjee, M. Tomovich, S. B. Leeb, and J. L. Kirtley, "Power converter sizing for a switched doubly fed machine propulsion drive," *Industry Applications, IEEE Transactions on*, vol. 51, no. 1, pp. 248–258, Jan 2015.

[8] A. Banerjee, S. B. Leeb, and J. L. Kirtley, "Switched doubly-fed machine for ship propulsion," *Electric Machines Technology Symposium, ASNE*, May 2014.

[9] B. Wu, *High-Power Converters and AC Drives*. IEEE Press, March 2006.

[10] A. Banerjee, S. B. Leeb, and J. L. Kirtley, "Seamless grid interaction for a switched doubly-fed machine propulsion drive," in *Electric Machines Drives Conference (IEMDC), 2015 IEEE International*, May 2015.

[11] L. Morel, H. Godfroid, A. Mirzaian, and J.-M. Kauffmann, "Double-fed induction machine: converter optimisation and field oriented control without position sensor," *Electric Power Applications, IEE Proceedings -*, vol. 145, no. 4, pp. 360–368, Jul 1998.

[12] S. B. Leeb, J. L. Kirtley, W. Wichakool, Z. Remscrim, C. N. Tidd, J. A. Goshorn, K. Thomas, R. W. Cox, and R. Chaney, "How much dc power is necessary?" *Naval Engineers Journal*, vol. 122, no. 2, pp. 79–92, 2010.

[13] A. Banerjee, S. B. Leeb, and J. L. Kirtley, "A comparison of switched doubly-fed machine drive topologies for high power applications," in *Electric Machines Drives Conference (IEMDC), 2015 IEEE International*, May 2015.

[14] F. Bonnet, L. Lowinsky, M. Pietrzak-David, and P.-E. Vidal, "Doubly fed induction machine speed drive for hydro-electric power station," in *Power Electronics and Applications, 2007 European Conference on*, Sept 2007, pp. 1–9.

[15] X. Yuan, J. Chai, and Y. Li, "A converter-based starting method and speed control of doubly fed induction machine with centrifugal loads," *Industry Applications, IEEE Transactions on*, vol. 47, no. 3, pp. 1409–1418, May 2011.

[16] B. Tian, C. Mao, J. Lu, D. Wang, Y. He, Y. Duan, and J. Qiu, "400 v/1000 kva hybrid automatic transfer switch," *Industrial Electronics, IEEE Transactions on*, vol. 60, no. 12, pp. 5422–5435, Dec 2013.

[17] A. Banerjee, A. Chang, K. Surakitbovorn, S. B. Leeb, and J. L. Kirtley, "Bumpless automatic transfer for a switched-doubly-fed-machine propulsion drive," *Industry Applications, IEEE Transactions on*, vol. 51, no. 4, pp. 3147–3158, July 2015.

[18] A. Banerjee, M. Tomovich, S. B. Leeb, and J. L. Kirtley, "Control architecture for a switched doubly fed machine propulsion drive," *Industry Applications, IEEE Transactions on*, vol. 51, no. 2, pp. 1538–1550, March 2015.

[19] A. Banerjee, S. B. Leeb, and J. L. Kirtley, "Solid state transfer switch topologies for a switched doubly-fed machine drive," *Power Electronics, IEEE Transactions on*, vol. PP, no. 99, pp. 1–1, 2015.

[20] ——, "Transient performance comparison of switched doubly-fed machine propulsion drives," in *Transportation Electrification Conference and Expo (ITEC), 2015 IEEE*, June 2015.

Multilevel Converter Topologies for High-Power High-Speed Switched Reluctance Motor: Performance Comparison

Devendra Patil, Shiliang Wang and Lei Gu
Department of Electrical Engineering
The University of Texas at Dallas TX 75080, USA
drp140230@utdallas.edu

Abstract— In this paper, two topologies for SRM drive is proposed, namely five level asymmetric neutral point clamped and asymmetric modular multilevel converter for high speed and high power SRM drive. For high speed operation, back EMF of the motor increases to high levels. Therefore, input voltage should be more than back EMF in order to force current into the winding. Multilevel converter can apply variable dc input voltage over a wide high speed range. Additionally, it can shape the phase current easily with multiple dc voltage levels to reduce high frequency torque ripple, thereby minimizing vibration. Operation of the proposed converters is verified using Matlab Simulink and performance comparison in terms of torque ripple and efficiency is evaluated. Also, performance of the drive is studied for different dc link voltage levels and the corresponding effect on efficiency is evaluated.

Keywords— *Multilevel; Switched reluctance motor; modular multilevel; neutral point clamped converter; torque ripple; vibration.*

I. INTRODUCTION

Large motors in range of 1000HP are generally used in industrial drives such as compressor with operating speed in range of 15000rpm to 20,000rpm. It is estimated that motor running at fixed speed without any inverter to control speed in response to load variation can waste significant amount of energy in mechanical transmission [1]. Therefore to improve the efficiency of the complete system, adjustable speed motor running at speeds higher than 15000rpm are being considered which does not require gearbox. Fig.1 shows the typical application of high speed motor drive. Notably, a variable speed drive is connected to a high speed motor which is in tandem with a compressor. Switched Reluctance Machine (SRM) is best suited for high speed application as it portrays highly efficient, robust structure, inherent mechanical strength, low cost and free from rotor winding and permanent magnet [2, 3]. However, challenges with SRM are high torque ripple and the fact that it requires high dynamic response controller (hysteresis controller) to control the phase current. At high speed, each phase turn on for short period of time. Therefore, to control the current maximum switching frequency can be very high. However, if hysteresis controller is used with large hysteresis band, the current controller will switch only few time in a cycle, commonly referred to as single pulse control. In addition, back EMF of the motor at high speed increases to high value. Therefore, to force current in to the machine the applied dc voltage has to be more than back EMF of the machine in constant torque region. Consequently, at high speed, controlling current is difficult without reducing the hysteresis band of the controller, provided sufficient dc link voltage is applied. Therefore, at higher speed there is a tradeoff

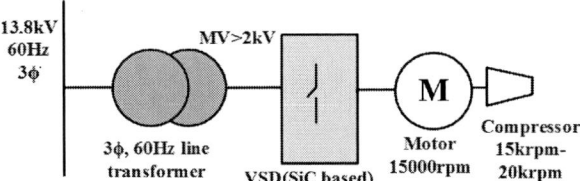

Fig. 1. Variable speed drive without gear box to drive compressor [1]

between hysteresis band and efficiency of the drive. Conversely, if back EMF of motor is more than dc link voltage the motor will operate in single pulse control and the motor can operate in constant power region.

With conventional Silicon devices for the drive with lower hysteresis band, efficiency of power electronic system will be low. To enhance system efficiency, power converter efficiency has to be very high for wide variation of the speed. With evolution of silicon carbide device for high voltage and high power application it is now possible to reduce the switching losses and thereby increasing the switching frequency of the SRM drive [1, 4, 5].

In the literature, various topologies are reported for switched reluctance motor for low voltage application (<1kV) [6]. However, very limited work is reported for operation of SRM at high voltage (>1kV) and high power (>100kW). Multilevel converters are suitable for this application, to reduces the voltage stress across devices and to allow commercially available devices to be used [7, 8]. Multilevel converters can also apply different dc voltage levels to SRM, which can give better flexibility to control current at high speeds. In [7], authors have discussed neutral point clamped, flying capacitor and cascaded multilevel converter for SRM. However, only three level multilevel converters for low speed SRM was reported and performance comparison of topology was not discussed. Dong *et. al* [8] have proposed an advanced multi-level converter for 4-phase SRM drive, which can boost the voltage applied to phase winding to improve performance at high speed. In [9], a four level converter was investigated to reduce the torque ripple using torque distribution control. Tomczewski *et. al.* [10] proposed a quasi-three level converter for SRM to reduce rising and falling time of the phase current, in order to reduce time of commutation and to increase the motor average torque at high speed. In [11], an asymmetric three level neutral point diode clamped converter for switched reluctance motor is investigated for low voltage application and performance comparison in terms of cost and efficiency is reported. Overall, multilevel converter for SRM in literature

978-1-4673-9551-9/16 $31.00 © 2016 IEEE

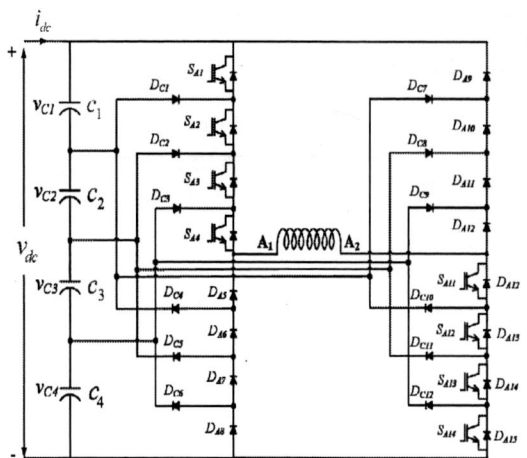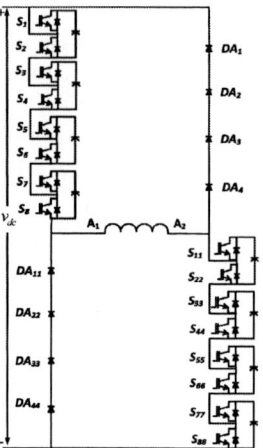

(a)   (b)

Fig. 2 Proposed (a) Diode clamped five level asymmetric SRM drive shown for single phase and (b) Asymmetric Modular Multilevel Converter (AMMC) SRM drive shown for single phase

are reported to improve the performance of the system at high speed by providing boosting voltage greater than back EMF of the motor and for low voltage SRM, multilevel converter provide benefits in terms of reduction in switching frequency and improvement in efficiency[7,8-12].

In this paper, a five level asymmetric diode clamped multilevel converter and asymmetric modular multilevel converter are proposed for SRM drive. Detailed operation and simulation of the proposed converters is provided. Further, performance comparison in terms of current ripple, torque ripple and efficiency of both topologies is provided for 1MW. In addition, advantage of applying different voltage levels on SRM is discussed.

II.   REQUIREMENT OF HIGH SPEED SRM CONVERTER

Switched reluctance motor drive requires specific requirement in order to be used at high speed operation [14]:

1. Forcing voltage have to be greater than back EMF of the motor, in order to inject current rapidly into the motor phase.
2. Converter should be able to apply negative demagnetizing voltage, to diminish tail current quickly.
3. Converter have to be capable of handling efficiently energy recycling during demagnetizing interval.
4. Torque ripple have to be less, to reduce capacitor requirement of the converter.
5. Independent control over each phase. To demagnetize one phase and magnetize other phase in motor having overlap between phase currents.

III.   PROPOSED TOPOLOGIES

Fig.2 (a) shows the diode clamped five level Asymmetric Neutral Point Clamped (ANPC) Multilevel converter for switched reluctance motor and Fig. 2(b) shows five level

Asymmetric Modular Multilevel Converter (AMMC) implementation for switched reluctance motor.

Table I. Switching logic for ANPC

V_{out}	S_{A1}	S_{A2}	S_{A3}	S_{A4}	S_{A11}	S_{A12}	S_{A13}	S_{A14}	S_{A15}
V_{dc}	1	1	1	1	1	1	1	1	1
$3V_{dc}/4$	0	1	1	1	1	1	1	1	1
$V_{dc}/2$	0	0	1	1	1	1	1	1	1
$V_{dc}/4$	0	0	0	1	1	1	1	1	1
$-V_{dc}/2$	0	0	1	1	0	0	0	0	0
$-V_{dc}/4$	0	0	0	1	0	0	0	0	0
$-3V_{dc}/4$	0	1	1	1	0	0	0	0	0
$-V_{dc}$	0	0	0	0	0	0	0	0	0

A. Asymmetric Neutral Point Clamped Multilevel converter

The proposed ANPC is similar to conventional neutral point clamped multilevel converter with the exception that the bottom part of the half bridge and top part of second half bridge have diodes instead of switches. Switching logic for ANPC is shown in Table I. Both positive and negative output voltage can be applied to SRM by switching different switch combinations. Fig. 3 shows the different switching combinations to generate different output voltage level. Fig. 3(a) shows that for generating output voltage of $+V_{dc}$ across winding, switches, S_{A1}, S_{A2}, S_{A3}, S_{A4} have to turned on for the upper half and S_{A11}, S_{A12}, S_{A13} and S_{A14} have to be turned on for lower half leg. In this mode, all switches are turned on and current flows through all switches; and all clamping diodes are reverse biased. Similarly, Fig. 3(b) indicates that the upper half switches S_{A2}, S_{A3}, S_{A4} and lower leg switches S_{A11}, S_{A12}, S_{A13}, S_{A14} have to be turned on to develop and to develop output voltage of $+0.75V_{dc}$. Load current in this scenario flows through the diode D_{C1}-S_{A2}-S_{A3}-S_{A4}-A_1-A_2-S_{A11}-S_{A12}-S_{A13}-S_{A14} and then back to the capacitor C_4, C_3 and C_2. Negative voltage of $-0.5 V_{dc}$ across output terminal can be generated by turning on switch S_{A3} and S_{A4} for upper leg and S_{A11}, S_{A12}, S_{A13} and S_{A14} have to be turned on for lower leg as shown in Fig. 3(c).

978-1-4673-9551-9/16 $31.00 © 2016 IEEE   2890

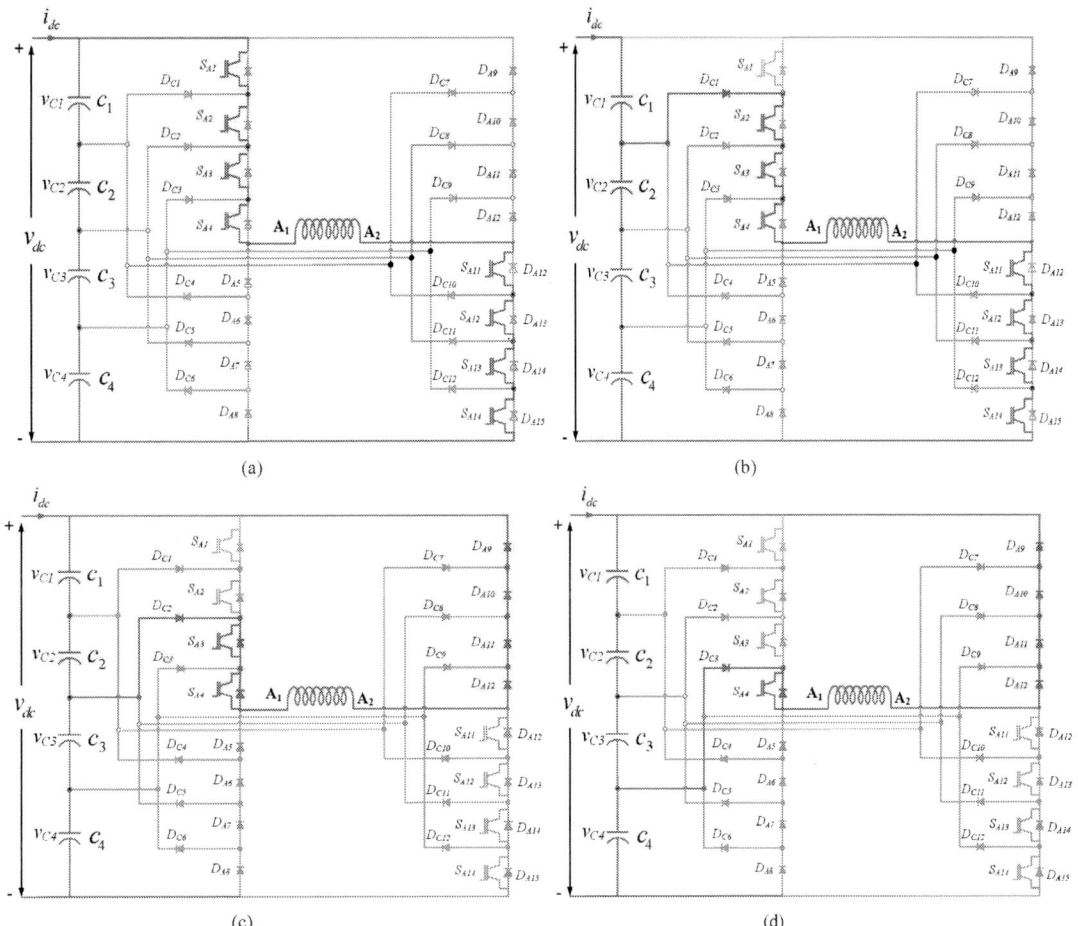

(a)   (b)

(c)   (d)

Fig. 3 ANPC switching operation for (a) output voltage of V_{dc} ;(b) output voltage of $0.75V_{dc}$;(c) output voltage of $-0.5V_{dc}$ and (d) output voltage of - $0.75V_{dc}$

To apply negative voltage across winding, all bottom switches are turned off and all top diodes are forward biased. Similarly, Fig. 3(d) shows the switching combination to generate V_{out}=- $0.75V_{dc}$. Similarly, to apply different negative voltage different upper switches have to be turned on or off and all bottom switches have to be turned off.

B. Asymmetric Modular Multilevel Converter

Fig.2 (b) shows AMMC for SRM. As compared to conventional Modular Multilevel Converter (MMC) bottom switches in first half bridge and top switches in second half bridges are not required. Also, in conventional MMC converter circulating current in the bridge is the main problem and to minimize the circulating current an arm inductor is connected in series with sub-modules in each arm. To compensate voltage difference between the phase leg voltage and dc link voltage. In AMMC, circulating current problem is not existent as bottom switches are not present. AMMC control logic for switches is different as compared to conventional MMC converter. Switch S_1 and S_2 operate complimentary to each other and each sub module is given the same duty cycle. Fig. 4 shows different operating states of

each module and its equivalent voltage and current direction. Table II shows the switching logic to generate different voltage levels across the winding. Fig. 5(a) shows the switch combination to be turned on to generate output voltage of $+V_{dc}$, switch S_2, S_4 ,S_6 and S_8 are turned on from upper legs and S_{22}, S_{44} , S_{66} and S_{88} are turned on from lower legs. Similarly, to develop output voltage of $+0.75V_{dc}$, switches S_6 and S_8 have to be turned on for upper half and S_{22}, S_{44}, S_{66} and S_{88} are turned on from lower leg as depicted in Fig. 5(b). Negative voltage of $-0.5V_{dc}$ across output can be generated by turning on, switch S_{11}, S_{33} and S_{55} and S_{77} have to be turned on

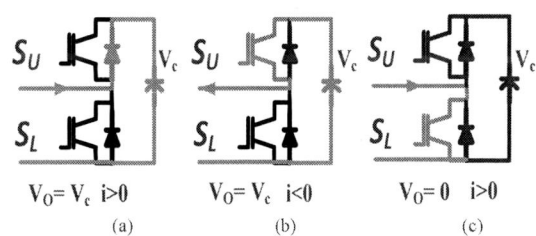

(a)   (b)   (c)

Fig. 4 AMMC module operating modes (a) D_U ON; (b) S_U ON and (c) S_L ON

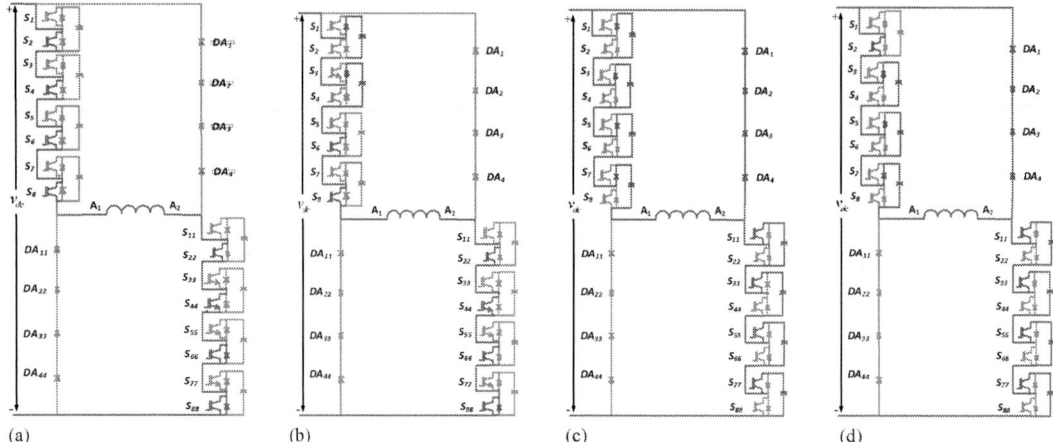

	(a)	(b)	(c)	(d)

Fig. 5 AMMC switching operation for (a) output voltage of V_{dc} ;(b) output voltage of $0.75V_{dc}$;(c) output voltage of $-0.5V_{dc}$ and (d) output voltage of $-0.75V_{dc}$

Table II. Switching logic for AMMC

V_{out}	S_1	S_2	S_3	S_4	S_5	S_6	S_7	S_8	S_{11}	S_{22}	S_{33}	S_{44}	S_{55}	S_{66}	S_{77}	S_{88}
V_{dc}	0	1	0	1	0	1	0	1	0	1	0	1	0	1	0	1
$3V_{dc}/4$	0	0	0	1	0	1	0	1	0	1	0	1	0	1	0	1
$V_{dc}/2$	0	0	0	0	0	0	0	0	0	1	0	1	0	1	0	1
$V_{dc}/4$	0	0	0	0	0	0	0	0	0	0	0	0	0	1	0	1
$-V_{dc}/2$	0	0	0	0	0	0	0	0	1	0	1	0	1	0	1	0
$-V_{dc}/4$	0	1	0	1	0	0	0	0	1	0	1	0	1	0	1	0
$-3V_{dc}/4$	0	1	0	0	0	0	0	0	1	0	1	0	1	0	1	0
$-V_{dc}$	0	0	0	0	0	0	0	0	0	0	0	0	0	0	0	0

for lower leg as shown in Fig. 5(c). Bottom side switches are turned on to provide forward bias voltage to upper leg diodes. Phase current in this mode of operation flows through all upper leg capacitor and upper leg diodes. Similarly, to generate $-0.75V_{dc}$ switch, S_2 from upper leg and switch S_{11}, S_{33} and S_{55} and S_{77} have to be turned on for lower leg as shown in Fig. 5(d). To apply different negative dc voltage across phase, upper module have to be either bypassed or turned off.

C. Control implementation

The control strategy for constant speed control is shown in Fig. 6, which is based on current control technique and speed controller is used to control the speed to desired value. However, at higher speeds just by changing the current reference speed cannot be tracked. Due to fact that at higher speed time for which each phase is turned on is very short Therefore, current cannot reach desired reference value. Also, back EMF of the motor is large which limits the current rise. Therefore, to control speed at higher reference value, phase has to be turned on and off at optimized angle and as the output load torque changes this angles have to be dynamically controlled. At high speed, hysteresis current controller can be used, although it has issue of variable switching frequency. But it offers good dynamic response.

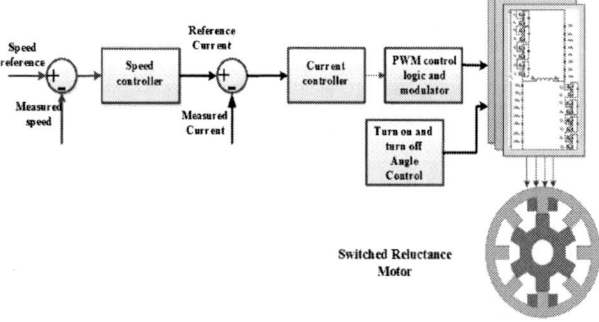

Fig. 6 Control block diagram implementation of SRM drive

TABLE III. DETAILS OF THE CONVERTER

Parameters	Values
SRM	6/4
Power rating	1MW
DC link Voltage	6000V
Load Torque	640N-m
Speed	15000rpm
Unaligned inductance	0.7mH
Aligned inductance	13mH
IGBT	ABB 5SNE 0800M170100

IV. SIMULATION RESULTS

In order to analyze performance of the proposed converter topologies extensive simulation is performed in Matlab Simulink. The parameters of SRM for high speed is obtained by performing simulation in ANSYS Maxwell. The parameters obtained from ANSYS Maxwell are used to build a model in Matlab Simulink. Parameters of the motor and converter are provided in Table III.

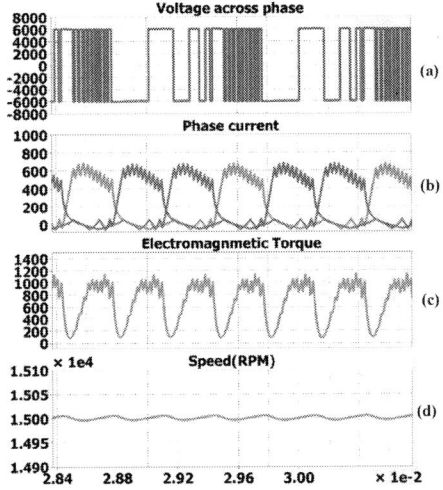

Fig. 7 AMMC (a) phase voltage; (b) phase current; (c) electromagnetic torque developed and (d) speed of motor

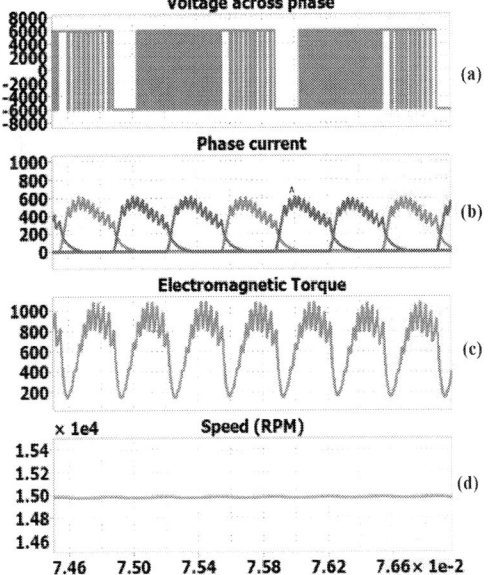

Fig. 8 ANPC (a) phase voltage; (b) phase current; (c) electromagnetic torque developed and (d) speed of motor

A. AMMC simulation results

Fig. 7(a) shows the voltage across one of the phases of the SRM. It can be clearly observed that both positive and negative voltages are applied to the phase. Negative voltage expedites the demagnetizing of the phase and reduce the tail

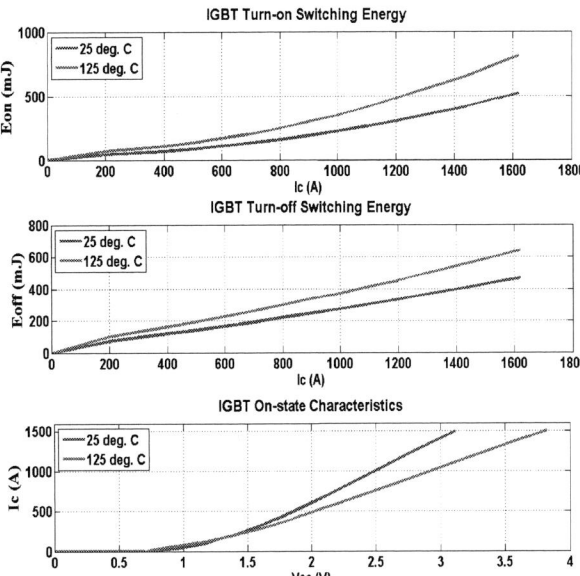

Fig. 9 Shows the IGBT (a)turn on switching loss vs collector current; (b) turn off switching loss vs collector current and (c) collector current vs collector emitter voltage

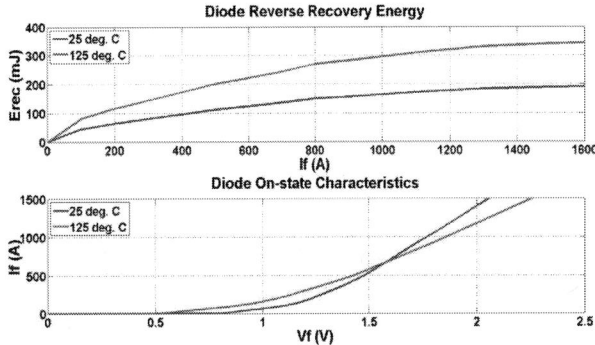

Fig. 10 Shows the IGBT anti parallel diode (a) reverse recovery loss vs collector current and (b) voltage drop vs collector current.

current. Fig. 7(b) depicts the phase current through each phase. It is seen that phase current is almost kept constant, which is in contrast with conventional high speed motor control i.e single pulse control. It was also observed that a low value of current is flowing through phase, when phase is deactivated. Low value of current observed is the capacitor charging current. Fig. 7(c), shows the developed electromagnetic torque of the motor. A large torque ripple is observed in the developed torque (due to phase commutation). Fig. 7(d) shows the speed is almost kept constant at 15krpm.

B. ANPC simulation results

Fig. 8(a) shows the voltage across one of the phase of SRM. It can be observed that ANPC is also capable of applying both positive and negative voltage to each phase. Fig. 8(b) shows the phase current for SRM, unlike AMMC the phase current does not have any low value of the current flowing through

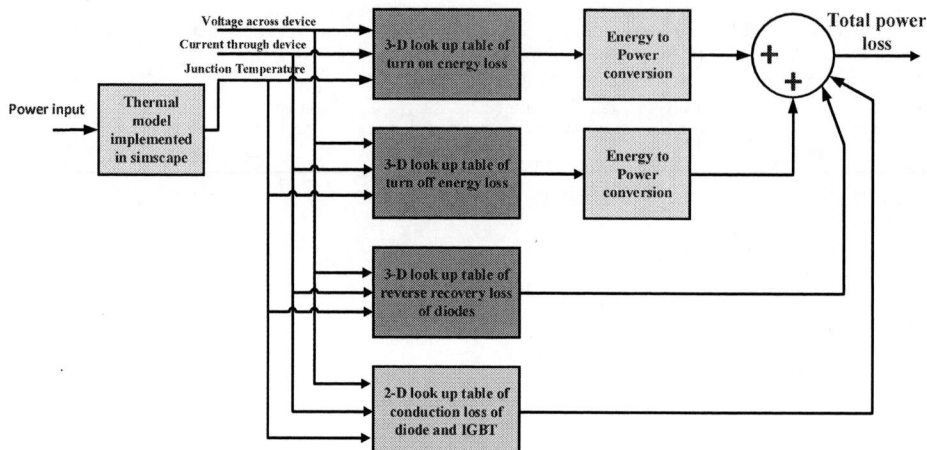

Fig. 11 Shows the loss model implemented in Matlab

phase during turn-off condition. Fig. 8(c) shows the developed electromagnetic torque and Fig. 8(d) depicts the speed is being constant at 15krpm.

V. DISCUSSION AND PERFORMANCE COMPARISION

This section will compare ANPC and AMMC converter for various parameters. To compare the converter in terms of efficiency, a Matlab based loss model of IGBT is developed, input to the model is simulated voltage and current waveform across devices.

A. Silicon IGBT loss model

Fig.11 shows block diagram of developed loss model based on thermal characteristics of IGBT and plots provided in datasheet(Fig.8 and Fig.9). Input to loss model is voltage across device and current through the device. A 3-D lookup table is developed in Matlab to calculate turn on and turn off switching loss and 2-D lookup table is developed to calculate conduction loss. Similarly, a 3-D lookup table for calculating reverse recovery loss is developed. The total loss is addition of IGBT switching loss, conduction loss and diode reverse recovery loss and diode conduction loss.

B. Torque ripple and vibration

Major problems in SRM are high an torque ripple and high vibration. Therefore, torque ripple is important parameter for deciding the topology. A high torque ripple will lead to increase in value of the dc link capacitor for power electronics and increases the vibration and acoustic noise created by harmonics of electromagnetic torque. In literature many methods are proposed to reduce this problem by changing the machine design or by current profiling [15, 16]. To shape phase current at high speed motor requires high switching frequency, which leads to high switching losses. However, with multilevel converter various levels of voltages can be easily applied to each phase which can shape the current easily with low switching frequency. Fig. 12(a) shows the torque waveform for AMMC and ANPC. It can be observed that torque ripple is 980N-m and 820N-m in AMMC and ANPC respectively at 15000rpm. Fig.13(a) shows the current

waveform for AMMC which has less current ripple as compared to ANPC in Fig.13 (b). However, each sub module capacitor charges when phase is turned off. Therefore we can observe current in phase even though it is not energized.

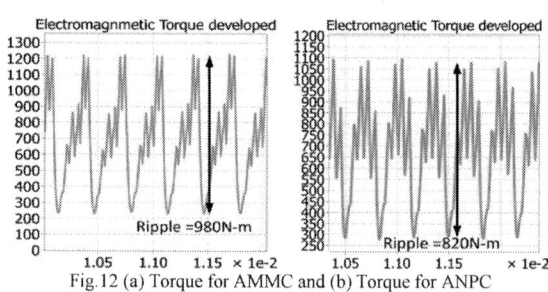

Fig.12 (a) Torque for AMMC and (b) Torque for ANPC

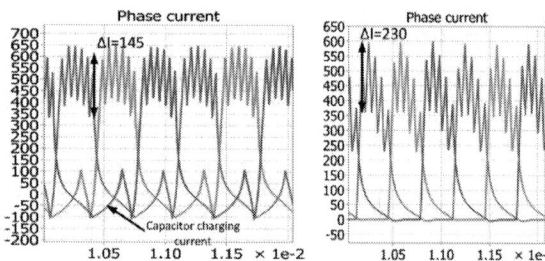

Fig.13 (a) Current in each phase for AMMC and (b) Current in each phase for ANPC

C. Current ripple

Fig. 14(a) shows current waveform for ANPC at 15000rpm and load torque of 640N-m. It is observed that switching pulses to control phase current are less at high speed. Fig.14 (b) shows waveforms for 20000rpm speed and 640N-m load torque using a hysteresis controller band of 50A and Fig.14(c) shows the same results for hysteresis band of 25A. It can be observed that at higher speed the hysteresis band has to be lower in order to control the current effectively. The maximum switching frequency in 25A hysteresis band case is 59khz, whereas for 50A band is only 38kHz. Therefore, wide

978-1-4673-9551-9/16 $31.00 © 2016 IEEE   2894

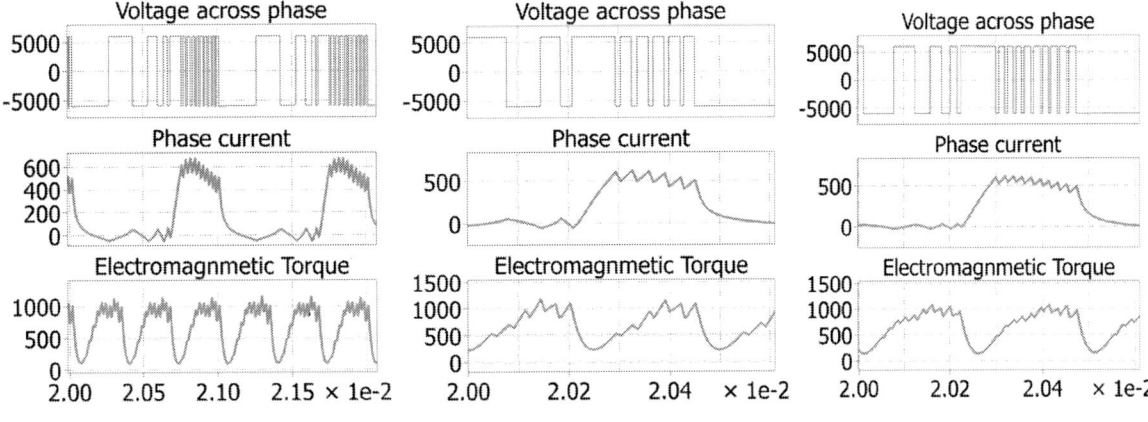

(a)   (b)   (c)

Fig.14 Simulated AMMC (a) waveforms for speed of 15000rpm; (b) waveforms for speed of 20000rpm and hysteresis band of 50A;(c)) waveforms for speed of 20000rpm and hysteresis band of 25A

band gap devices are the only option in order to achieve high efficiency at this speed.

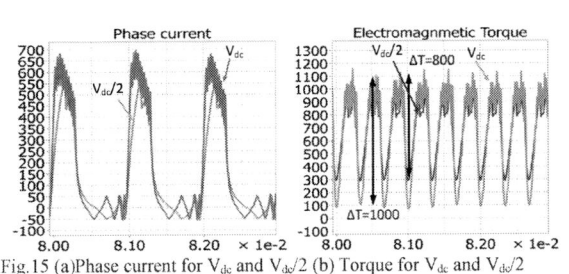

Fig.15 (a)Phase current for V_{dc} and $V_{dc}/2$ (b) Torque for V_{dc} and $V_{dc}/2$

Table IV. Performance comparison for different voltage level

	V_{dc}	$V_{dc}/2$
Speed	15000rpm	15000rpm
Torque ripple	1000N-m	800N-m
Efficiency	97.765%	99.1%

D. Multiple voltage level

One of the advantages of multilevel converter for SRM is different voltage levels that can be used to energies phase winding. Fig. 15(a) shows the phase current for dc voltage of V_{dc} and $V_{dc}/2$, it can be observed that with full dc voltage across winding, at high speed the forcing voltage is more than enough to control the current. Therefore, switches are turned on and off many times in a cycle, hence the switching losses are high. When half dc voltage is applied to phase, the forcing voltage is not that high as compared to back EMF of the phase. Therefore, switches are operated only two times in each complete cycle, which leads to low switching losses. Table IV shows the comparison of half voltage and full voltage for different parameters. From simulation it was observed that for constant load torque (640N-m) and speed (15000rpm), torque ripple in full dc voltage mode is high as compared to half dc

voltage mode. Also, efficiency of half voltage mode is 99.1% as compared to full dc voltage mode of 97.76%, which is almost an efficiency improvement of 1.34%. Fig.16 shows the efficiency plot for AMMC and ANPC for different voltage levels applied to phase winding. It is observed that AMMC has higher efficiency as compared to ANPC. Also, for $V_{dc}/2$ level efficiency of both converter is high which is due to low switching loss as compared to higher voltage.

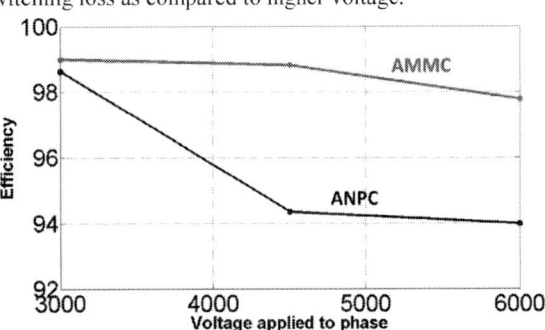

Fig.16 Efficiency plot for variation in dc level voltage applied to phase

Table V. Performance comparison of ANPC and AMCC for different parameters

Parameters	ANPC	AMMC
Voltage stress across device	1500V	1500V
RMS Current rating of devices	265.11	250.532
Number of capacitor	4	8
Number of switches	8	16
Number of diodes	12	8
Efficiency	94%	97.9%
Cost	Low	High

VI. CONCLUSIONS

This paper proposed two different multilevel converter, asymmetrical neutral point clamped converter and asymmetrical modular multilevel converter converters for high power, high voltage and high speed SRM drive. Key features are ability to work over wide speed range, low device

978-1-4673-9551-9/16 $31.00 © 2016 IEEE   2895

voltage stress, high efficiency and multiple level of dc voltage can be applied to phase winding, which can shape the current easily with multiple levels of dc voltage to reduce torque ripple and to decrease the rise time and fall time. Detailed operation of proposed multilevel converters is presented. Additionally, performance comparison of ANPC and AMMC is discussed for torque ripple, current stress and number of devices. It was observed that AMMC has higher efficiency compared to ANPC. However, number of semiconductor devices is less in ANPC. Further, number of capacitor required are more in AMMC. Therefore, high reliability as compared to AMMC is achieved. Operation of the system is also verified through simulation.

Acknowledgment

This work is done in the Renewable Energy and Vehicular Technology (REVT) Laboratory, the University of Texas at Dallas under guidance of Dr. Babak Fahimi. I would like to thank him for all support and providing funding for this work.

References

[1] "Next generation electric machines: megawatt class motors", Department Of Energy (DOE) Office Of Energy Efficiency and Renewable Energy (EERE), Available online: https://eere-exchange.energy.gov/

[2] Li, Haoding, et al. "Performance evaluation of a high-speed high-power switched reluctance motor drive." 2015 IEEE Applied Power Electronics Conference and Exposition (APEC), 2015.

[3] Hofmann, Andreas G et al.. "High-speed switched reluctance drives: A promising alternative to power electric vehicles." 2011 IEEE 8th International Conference on Power Electronics and ECCE Asia (ICPE & ECCE), 2011.

[4] Waqas Arshad, et. al., "High MW Converters Role of SiC",ABB, Corporate Research, Available online :http:// www.nist.gov/pml/ high_megawatt/upload/Approved-Arshad-DoE_HMW-VSD.pdf

[5] Biela, J. et. al., "SiC versus Si—Evaluation of Potentials for Performance Improvement of Inverter and DC–DC Converter Systems

by SiC Power Semiconductors," IEEE Trans. on Ind. Electron., vol.58, no.7, pp.2872-2882, July 2011.

[6] Barnes, M.; Pollock, C., "Power electronic converters for switched reluctance drives", IEEE Trans. on Power Electron., vol.13, no.6, pp.1100- 1111, Nov 1998.

[7] Watkins, S.J. et. al, "Multilevel asymmetric power converters for switched reluctance machines," IEEE International Conference on Power Electronics, Machines and Drives, pp.195-200, June 2002.

[8] Dong-Hee Lee, Huijun Wang and Jin-Woo Ahn, "An advanced multi-level converter for four-phase SRM drive," IEEE Power Electronics Specialists Conference, pp.2050-2056, June 2008.

[9] Dong-Hee Lee and Jin-Woo Ahn, "A Novel Four-Level Converter and Instantaneous Switching Angle Detector for High Speed SRM Drive," IEEE Trans. on Power Electron., vol.22, no.5, pp.2034-2041, Sept. 2007.

[10] K. Tomczewski and K. Wrobel,"Quasi-three-level converter for switched reluctance motor drives reducing current rising and falling times", IET Power Electron., vol. 5, no. 7, pp.1049 -1057 2012.

[11] Peng, Fei, et. al. "An asymmetric three-level neutral point diode clamped converter for switched reluctance motor drives." IEEE Transportation Electrification Conference and Expo (ITEC), 2015.

[12] Dahmane, M., Meibody-Tabar, F. and Sargos, F.-M., "An adapted converter for switched reluctance motor/generator for high speed applications," IEEE Ind. Appl. Conf., vol.3, no., pp.1547,1554 vol.3, 2000.

[13] So-Yeon Ahn, Jin-Woo Ahn and Dong-Hee Lee, "A novel torque controller design for high speed SRM using negative torque compensator," IEEE 8th International Conference on Power Electronics and ECCE Asia , pp.937-944, 2011.

[14] MacMinn, Stephen R. and Jones, W.D., "A very high speed switched-reluctance starter-generator for aircraft engine applications," Proceedings of the IEEE in National Aerospace and Electronics Conference,pp.1758-1764 vol.4, May 1989.

[15] So-Yeon Ahn; Jin-Woo Ahn; Dong-Hee Lee, "A novel torque controller design for high speed SRM using negative torque compensator," 2011 IEEE 8th International Conference on Power Electronics and ECCE Asia (ICPE & ECCE), pp.937,944, May 30 2011-June 3 2011.

[16] Fahimi, B., Suresh, G., Rahman, K.M.and Ehsani, M., "Mitigation of acoustic noise and vibration in switched reluctance motor drive using neural network based current profiling," in IAS Annual Meeting Industry Applications Conference, pp.715-722, Oct. 1998.

Bidirectional Magnetically Coupled T-Source Inverter for Extra Low Voltage Application

Thomas Baier and Bernhard Piepenbreier, *Senior Member, IEEE*
Chair of Electrical Drives and Machines, University of Erlangen-Nuremberg, Germany
Email: thomas.baier@fau.de

Abstract—**This paper investigates the performance of a magnetically coupled T-Source-Inverter (TSI) for extra low voltage and high current applications. As the TSI is part of the well-known group of inverters with impedance source networks, it is capable to boost the output voltage beyond the level of the input voltage within a single stage. This boost operation is achieved by inserting a new operational state called 'shoot through'. By using a proper turns ratio for the transformer design, the TSI can theoretically reach the same boost factor compared to inverters with impedance source networks without coupled inductors (e.g. Z-Source and Quasi-Z-Source) while operating with a smaller shoot through duty cycle. Considering the fact that it is not possible to build an ideal transformer, the coupling coefficient of the transformer respectively the leakage inductance determines its performance. For that reason, this paper focuses on a practical extra low voltage and high current application of the TSI and its limitations which are exposed in the section of the experimental results.**

Keywords—Z-Source, T-Source, Impedance Source Network, Magnetically Coupled Inverter

I. INTRODUCTION

Traditional Voltage Source Inverters (VSI) are composed of a voltage source and the main inverter circuit consisting of a half bridge for every single output phase. The supplying voltage source is usually supported by a relatively large capacitor and every half bridge itself consists of two active switches with two antiparallel diodes for bidirectional power flow [1], [2]. This kind of inverters face certain limitations or problems. One limitation is concerning the output voltage which cannot exceed the level of the dc link voltage without using an extra power converting stage to boost the input voltage. However, such an extra converting stage would increase the system costs and the complexity and in addition would lower the efficiency. Another aspect is that the upper and lower transistor of each phase leg cannot be switched on simultaneously either by purpose or by electromagnetic interference (EMI). On the contrary, the inverter would be destroyed by huge short circuit currents induced by the shoot through. Especially the vulnerability to EMI noise is a major problem in inverter reliability. In order to overcome these restrictions, the Z-Source inverter was proposed in [3], [4]. With its special impedance source network build with two capacitors and two inductors arranged with a diode and an antiparallel switch this inverter type is able to provide ac-ac, ac-dc, dc-ac and dc-dc power conversion with buck-boost ability within a single stage. Based on the idea and functionality of the Z-Source inverter topology, rearrangements were made and new

impedance source network inverters like the Quasi-Z-Source came up [5], [6]. In 2009, a new inverter with a magnetically coupled impedance source network, called T-Source Inverter, was proposed [7], [8], [9]. The turns ratio of magnetically coupled inductors provides an additional degree of freedom for the design of the inverter. In the last few years, more impedance source network inverters with magnetically coupled inductors have been proposed like the Trans-Z-Source[10], Y-Source [11], [12], TZ-Source [13], [14], LCCT-Z-Source [15], [16], Γ-Z-Source Inverter [17], [18].

This paper focuses on the magnetically coupled T-Source Inverter that is shown in Fig. 1 with a DC/AC power conversion. In contrast to [7], [8], [9], a switch S_0 is inserted antiparallel to the input diode to permit a bidirectional power flow. The equivalent circuit of the transformer consists of magnetizing $L_m = k^2 L_1$ and a leakage inductance $L_l = \left(1 - k^2\right)L_1$ and the conversion factor $t = k\sqrt{L_1/L_2}$ where k is the coupling coefficient. The winding ration is called $n = N_1/N_2$.

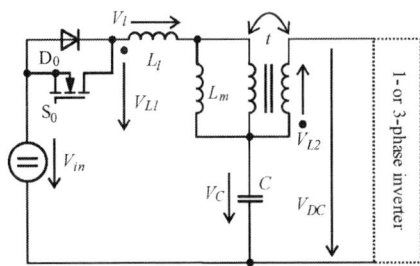

Fig. 1. Magnetically coupled T-Source Inverter

II. ANALYTICAL ANALYSIS

The analytical circuit analysis and the derivation of the mathematical equations are performed under ideal conditions. Voltage drops at the semiconductors are neglected and the transformer is assumed to be perfectly coupled with $k = 1$. Moreover, the inverter is operating in steady state. Due to this fact, the average inductor voltages $\bar{V}_{L1}$, $\bar{V}_{L2}$ and the average capacitor current $\bar{I}_C$ respectively the capacitor charge $\bar{Q}_C$ in one switching time period T_S are equal to zero:

$$\bar{V}_{L1} = \bar{V}_{L2} = 0 \tag{1}$$

$$\bar{I}_C = 0 \tag{2}$$

978-1-4673-9551-9/16 $31.00 © 2016 IEEE

TABLE I.   SWITCHING STATES AND OUTPUT VOLTAGES

States	State of the Switches			Output Voltage V_{out}[V]
	S_1	S_2	S_0	
Active State	On	Off	On	$(V_c(n+1) - V_{in})/n$
Zero State	Off	On	On	0
Shoot-Through	On	On	Off	0

Fig. 2 and Fig. 3 show the DC model of the TSI. The inverter side consists of one phase leg and the load is represented by a constant current source. For operations in buck mode the capacitor voltage equals the input voltage $V_C = V_{in}$, in boost mode the capacitor voltage exceeds the input voltage $V_C > V_{in}$.

According to Table I in active state the switches S_0 and S_1 are switched on, the switch S_2 remains off. The load is connected to V_{DC} and power is being transmitted to the load. Fig. 2 shows the TSI in the active state in buck or boost operation, the current conducting paths are marked in red.

While executing a shoot through, the input diode is reverse biased and the output is short circuited by turning the switches S_1 and S_2 on simultaneously. In this state, V_C equals V_{L2} which causes the output current to rise very quickly, increasing the energy of the magnetic field of the transformer at the same time. The load current remains freewheeling, using the lower output diode. V_{out} is equal to zero. By terminating the shoot through state and entering a new state (active state or zero state), the capacitor voltage rises due to the previously stored energy. The shoot through state is pictured in Fig. 3. The current conducting parts are marked in red.

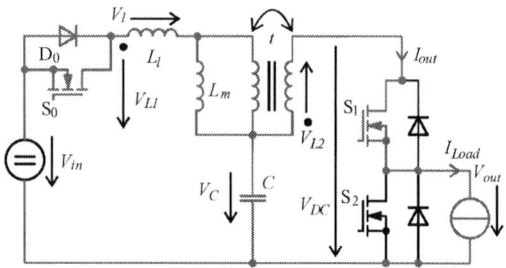

Fig. 2.   DC model of the TSI operating in active state (buck or boost mode), current conducting parts are marked in red

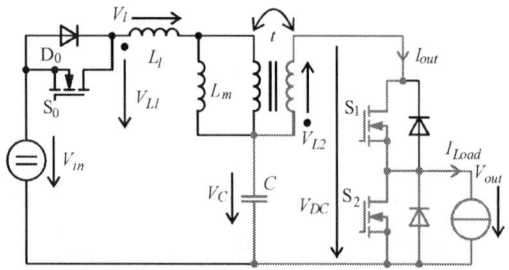

Fig. 3.   DC model of the TSI operating in shoot through state (boost mode), current conducting parts marked in red

Table I shows the possible switching states and the output voltages for the circuit analysis. Zero state and shoot through state produce the same output voltage. Steady state analysis

and state space averaging of the inductor voltage V_{L2} for one switching time period $T_S = T_{ST} + T_{AS} + T_{ZS}$, T_{ST} being the shoot through time, T_{AS} being the active state time and T_{ZS} being the zero state time according (3), leads to equation (4). t_{ST}, t_{AS} and t_{ZS} are the relative time periods in relation to the switching time period T_S.

$$\bar{V}_{L2} = [T_{ST}V_C + (T_{AS} + T_{ZS})\,(V_{in} - V_C)/n]/T = 0 \quad (3)$$

$$V_C/V_{in} = (1 - t_{ST})/(1 - (n+1)t_{ST}) \quad (4)$$

Considering the denominator in (4), the maximum shoot through duty cycle can be calculated:

$$t_{ST} < 1/(n+1) \quad (5)$$

Using (4), the output voltage in active state that can be found in Table I respectively the DC link voltage in active state can be rewritten and the average output voltage can be calculated:

$$V_{DC,AS} = V_{in}/(1 - (n+1)t_{ST}) \quad (6)$$

$$\bar{V}_{out} = (1 - t_{ST} - t_0)V_{in}/(1 - (n+1)t_{ST}) \quad (7)$$

In order to achieve the maximum average output voltage, the zero state is skipped and one switching period solely consists of active state and shoot through state $t_{ST} + t_{AS} = 1$. Under this condition, comparing (4) and (7), the average output voltage equals the capacitor voltage:

$$\bar{V}_{out} = V_C \quad (8)$$

Assuming the before mentioned ideal conditions and neglecting losses, the average input current in active state and the average output current in shoot through state can be found:

$$\bar{I}_{in,AS} = I_{load}/(1 - (n+1)t_{ST}) \quad (9)$$

$$\bar{I}_{out,ST} = (n+1)(1 - t_{ST})I_{load}/(1 - (n+1)t_{ST}) \quad (10)$$

Fig. 4 and Fig. 5 are illustrating the output voltage and the dc link voltage in relation to the input voltage. According to Fig. 4, different shoot through factors, depending on the winding ratio of the TSI, are necessary to achieve the same output voltage. The higher the winding ratio, the smaller the shoot through factor to reach the same output voltage. For a winding ratio $n > 1$, the TSI exceeds the boost ability of the ZSI and QZI [3], [4], [5], [6]. The different winding ratios respectively shoot through factors denote different dc link voltages as well, which is represented in Fig. 5. The smaller the winding ratio, the higher is the dc link voltage to reach the same output voltage. As one can see, the winding ratio of the TSI is an important parameter that needs to be considered with regard to the inverter design to avoid unnecessary voltage stress for the semiconductors.

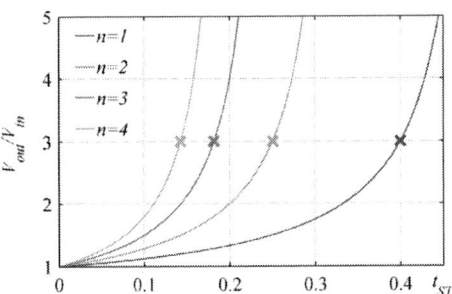

Fig. 4. $\bar{V}_{out}/V_{in}$ of the TSI for different winding ratios $n = N_1/N_2$. $\bar{V}_{out}/V_{in} = 3$ is marked with an x.

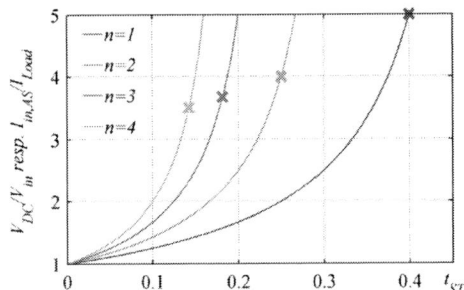

Fig. 5. $V_{DC,AS}/V_{in}$ respectively $\bar{I}_{in,AS}/I_{Load}$ of the TSI for different winding ratios $n = N_1/N_2$. $\bar{V}_{out}/V_{in} = 3$ is marked with an x.

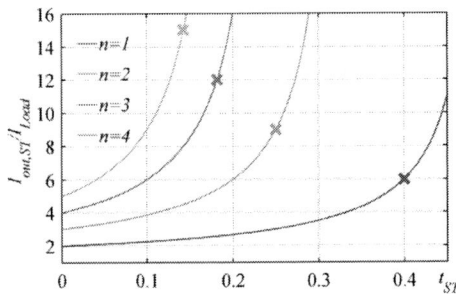

Fig. 6. $\bar{I}_{out,ST}/I_{Load}$ of the TSI for different winding ratios $n = N_1/N_2$. The marks are showing the shoot through factor that produces $\bar{V}_{out}/V_{in} = 3$.

Another requirement for the semiconductors of the inverter operating in boost mode is the ability to conduct the high shoot through currents. Under ideal conditions, the inverter output current jumps on a high current level by switching into shoot through state. The level of the shoot through current is determined by the desired output voltage respectively the shoot through factor, the load current, and the winding ratio (10). Like mentioned before, different winding ratios need different shoot through factors to reach the same output voltage. Fig. 6 illustrates the average output current in shoot through state for different shoot through factors, $V_{out}/V_{in} = 3$ is marked again. High winding ratios with small shoot through factors lead to which are high in amplitude but short in time. These shoot through currents need to be conducted by the semiconductors. If one weights the amplitude of the shoot through current with the duration there are advantages for a TSI with a high turns ratio.

III. LEAKAGE INDUCTANCE

In the chapter of the analytical analysis it was pointed out that the turns ratio of the transformer has a significant influence on the voltage and current levels of the TSI which is expressed in (6), (7), (9), (10). Hence the transformer is the part that determines the performance of the T-Source Inverter [7], [10]. In boost mode and shoot through state, energy is stored in the magnetic field of the transformer. In order to avoid saturation and achieve enough capacitance to store energy, a big air gap between the transformer cores is necessary. Taking extra low input voltage and high current into account, the cross section of transformer windings needs to be well designed to prevent overheating. However these two requirements influence the leakage inductance of the transformer in a bad way. Assuming ideal conditions, the input current steps down directly to zero when the inverter state is switched from active state to shoot through state. No time is needed to switch from one state into another. But taking a certain leakage inductance respectively a not perfectly coupled transformer into account, the inverter cannot switch from active state to shoot through state immediately. Before the shoot through state is entered, the input current has to be zero and the input diode needs to be reverse biased. The leakage inductance determines the transition between these two states. According to (12), the duration of shoot through state is linearly decreased by the leakage inductance of the transformer. The winding ratio of a transformer $n = N_1/N_2$ increases linearly with the number of windings of the primary side. In contrast to this, the primary inductance itself increases quadratically. In the same way the leakage inductance increases assuming a constant coupling coefficient. Comparing different non-ideal TSI with different turns ratios, but same coupling coefficients, different shoot through factors are necessary to achieve the same output voltage according to Fig. 4. The higher the turns ratio, the higher the leakage inductance and the transition time between active state and shoot through state, but the smaller the shoot through factor that is needed. Expressed in a different way, the higher the turns ratio is, the more time is lost from an even smaller shoot through time. Consequently, the required output voltage cannot be reached anymore with the calculated shoot through factor. Depending on the load current and the shoot through time, it is possible that the actual shoot through state cannot even be reached. Taking the leakage inductance into account, the good characteristics of a TSI with a high turns ratio according to the analytical analysis cannot be confirmed. Quite the contrary, a high turns ratio is likely to deteriorate the performance of the TSI.

Another problem referring to the existence of transformer leakage inductance are voltage spikes due to hard switching operation transiting from shoot through state to active state. Since the load current in active state is smaller than the shoot through current, an alternative path must be provided to avoid the destruction of the semiconductors because of the voltage spikes. This path can be realized by using an additional snubber network [19], [23].

$$V_l = L_l di/dt \qquad (11)$$

$$\Delta t_{ST}T_S = L_l I_{Load}/((1 - (n + 1)t_{ST})V_l) \qquad (12)$$

IV. EXPERIMENTAL RESULTS

TABLE II. CONFIGURATION OF THE EXPERIMENTAL SETUP

Induction Motor
$P_N = 7,5$ kW
$U_N = 110$ V Y
$I_N = 51$ A
$\cos \varphi_N = 0,87$
$n_N = 2950$ Hz
T-Source Network
$V_{in} = 60$ V
$N_1/N_2 = 2$
$L_1 = 61,22$ µH
$k = 0,984$
Semiconductors
Mosfet SKM 180A020

The configuration of the experimental setup can be found in Table II The induction motor was used in delta configuration and the 3-phase inverter was controlled by a modified space vector modulation [20], [21], [22]. The modified space vector modulation can be realized very easily by shifting the switching actions from zero state to active state 1 and from active state 1 to zero state of the traditional space vector modulation about the shoot through time. In that way, no additional switching actions are necessary. Fig. 7 illustrates the sequence of the modified space vector modulation. In order to prevent the destruction of the semiconductors due to the leakage inductance of the transformer, a RCD snubber circuit was used [23]. Fig. 8, Fig. 9, and Fig. 10 are showing the TSI in buck operation providing an output voltage $\hat{V}_{out} \approx 50$ V. The variation of the load conditions can be found in the horizontal axis with the increasing inverter output current. The efficiency in an output range from 2 to 5 kW is about 92% (2 kW) and 90% (5 kW). The decline of the efficiency is caused by the increasing output current.

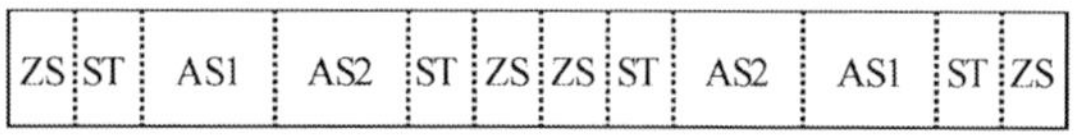

ZS	ST	AS1	AS2	ST	ZS	ZS	ST	AS2	AS1	ST	ZS

Fig. 7. Switching sequence of the modified space vector modulation.

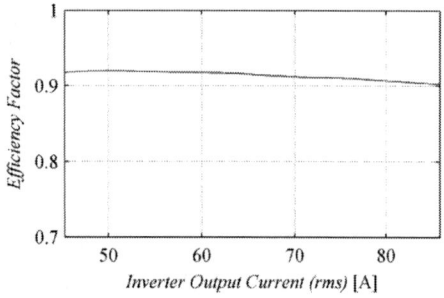

Fig. 8. Efficiency factor of the TSI in buck operation at an output voltage of $\hat{V}_{out} \approx 50$ V. The enhancement of the load is represented by the inverter output current in the horizontal axis.

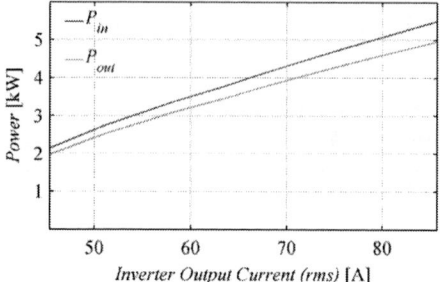

Fig. 9. Input and output power as basis for the calculaton of the efficency factor in Fig. 8.

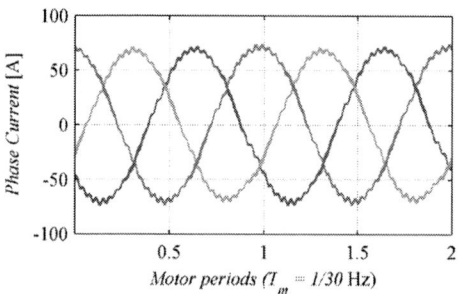

Fig. 10. Phase currents of the motor for an output voltage $\hat{V}_{out} \approx 50$ V and maximum load conditions referring to Fig. 8 and Fig. 9.

Fig. 11, Fig. 12, Fig. 13, and Fig. 14 are illustrating the TSI in boost operation mode under constant light load conditions and an increasing requested voltage V_{out}^* in the horizontal axis. The shoot through was performed as a one phase shoot through, the upper and the lower switch of only one half bridge are turned on simultaneously, and a three phase shoot through, the switches of all the half brides are switched on simultaneously.

Fig. 11 is executed with the shoot through factor $t_{ST,control}$ calculated according to the ideal control equations and depicted in Fig. 12. Due to the leakage inductance of the transformer, the actual shoot through time is shortened and the requested voltage V_{out}^* cannot be reached. Implementing a three phase shoot through enhances the performance of the inverter because less system losses are produced. The output voltage still rises clearly above the level of the input voltage, which shows that boost operation is still possible. Up to an output voltage of $V_{out}^* \approx 66$ V, it is not possible to use the three phase shoot through because of the timewise short shoot through.

The lack of output voltage in Fig. 11 can be compensated by adding an extra shoot through factor that can be found in Fig. 12. $t_{ST,1ph}$ is used to increase the voltage using one phase shoot through, $t_{ST,3ph}$ belongs to the three phase shoot through operation. In that way, it is possible to compensate the influence of the transformer leakage and reach the desired voltage. Due to the higher losses of the one phase shoot through, the added factor $t_{ST,1ph}$ exceeds $t_{ST,3ph}$.

Fig. 13 and Fig. 14 are showing the efficiency factors and the input and output powers as basis for its calculation for one phase and three phase shoot through. The efficiency in an output range between 2.3 kW and 3 kW under one phase shoot through conditions is about 82.5% and 61%, respectively. By using the three phase shoot through, the efficiency can be clearly enhanced.

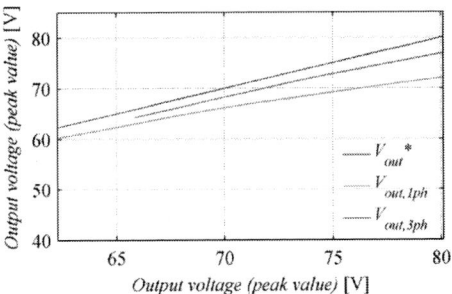

Fig. 11. Requested output voltage V_{out}^* and achieved output voltages with one phase shoot through ($V_{out,1ph}$) and three phase shoot through ($V_{out,3ph}$) under light load conditions. The Shoot through factor is calculated according to the control equation.

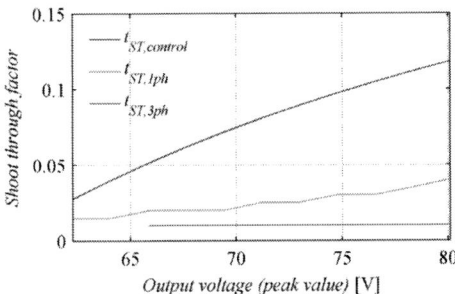

Fig. 12. Calculated shoot through factor according to the control equation and shoot through factors to compensate the lack of voltage for one phase and three phase shoot through referring to Fig. 11.

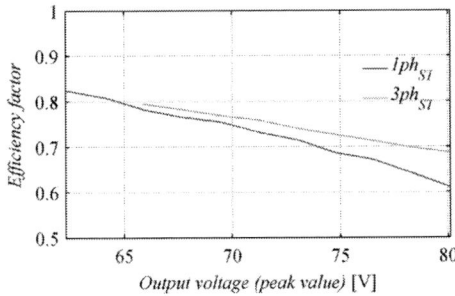

Fig. 13. Efficiency factor for the TSI operating with one phase and 3 phase shoot through. Output voltages are compensated with the additional shoot through factors according to Fig. 12 under light load conditions.

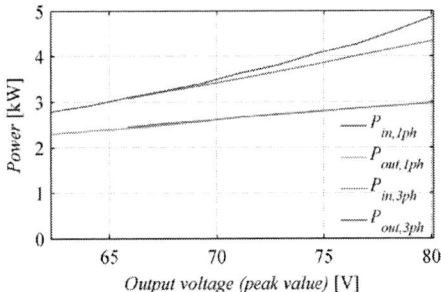

Fig. 14. Input and output power referring to the efficiency factors in Fig. 13.

Fig. 15, Fig. 16, and Fig. 17 are picturing an operation point in boost mode with a requested output voltage of $V_{out}^* \approx 71$ V and constant high load conditions. Fig. 15 illustrates the effect of adding an extra shoot through factor to the calculated one referring to the control equation. For the three phase shoot through, adding extra shoot through results in an increasing output voltage and the requested voltage can be reached. For the case of operating with a one phase shoot through the output voltage processes different with an increasing shoot through factor. First the output voltage rises with an increasing shoot through factor until it reaches a maximum, afterwards more shoot through results in a decreasing output voltage. The fact can be explained by the low efficiency respectively the high losses. The higher the shoot through factor, the higher the losses that are produced by the shoot through and the higher the input power at a constant output power. Higher input power at a constant input voltage results in a higher input current. Like described before in the analytical analysis, the current through the diode needs to be zero before the shoot through state can be reached. The lost shoot through time because of the higher input currents overweighs the effect of additional shoot through.

Fig. 16 and Fig. 17 are illustrating the efficiency factor, the input and output power according to Fig. 15. At an efficiency about 60%, the decrease of the output voltage with an increasing shoot through factor begins.

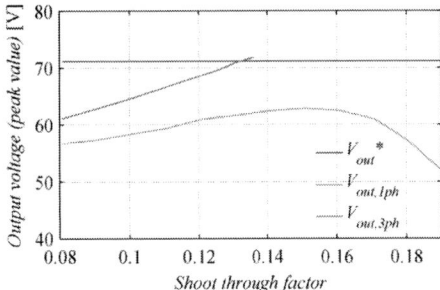

Fig. 15. Effect of additional shoot through on the actual achieved output voltage for one phase and three phase shoot through. Starting with the shoot through factor according to the control equation. Requested output voltage $V_{out}^* \approx 71$ V at constant high load conditions.

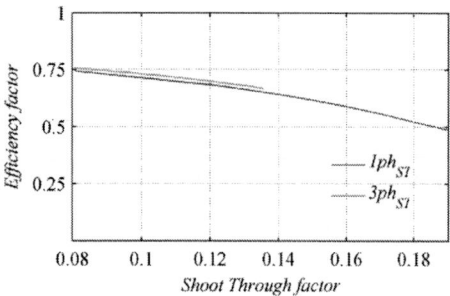

Fig. 16. Efficiency factor referring to Fig. 15.

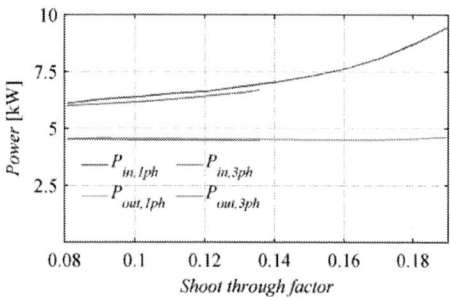

Fig. 17. Input and output power referring to Fig. 15 and Fig. 16.

At this point, the performance of the TSI is investigated both in different output voltages and different load conditions to get an idea of its complete properties. As before stated in the chapter of the leakage inductance and already seen in the previous experimental results, there is a great dependence of the efficiency on the shoot through factor. It can be stated the longer the shoot through duration, the higher the losses. Moreover, especially for non-ideal TSI, there is a great influence of the load current. Therefore, low efficiency can be expected the higher the requested output voltage respectively the higher the shoot through factor and the higher the load current.

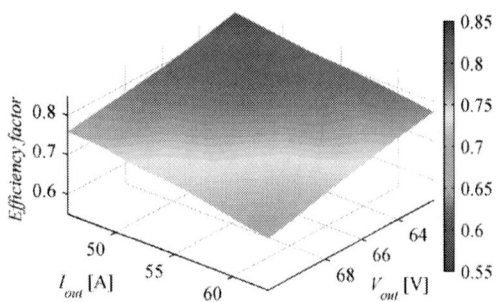

Fig. 18. Efficiency factor of the TSI for three phase shoot through in an operation range of $62\,V < V_{out} < 71\,V$ (peak value) and $45\,A < I_{out} < 63\,A$ (rms value).

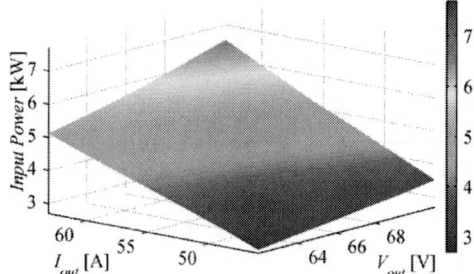

Fig. 19. Input power of the TSI for three phase shoot through referring to Fig. 18.

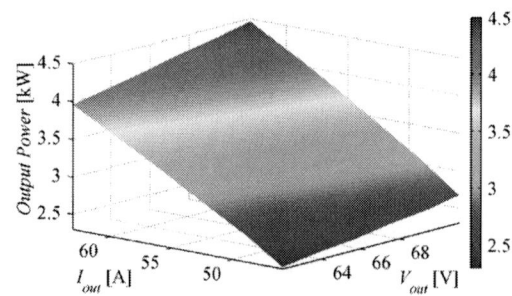

Fig. 20. Output power of the TSI for three phase shoot through referring to Fig. 18

Fig. 18 and Fig. 21 show the efficiency of the TSI for one phase (Fig. 21) and three phase (Fig. 18) shoot through in an operation range $62\,V < V_{out} < 71\,V$ (peak value) and $45\,A < I_{out} < 63\,A$ (rms value). As expected, the efficiency is high for small output voltages and load currents. In this area, there is almost no difference between the one phase and the three phase shoot through and the efficiency is around 83%. In contrast, there is a great difference when it comes to high shoot though factors and high load conditions. By using a 3 phase shoot through and splitting up the shoot through current equally on the three phase legs, the conducting losses of the switches can be kept small in comparison to a one phase shoot through. In that way, the desired output voltage for high load conditions can be reached with an efficiency around 68%. By operating with a one phase shoot through it is not possible to achieve the same output voltage. Like already state the reason for that are the high losses. The efficiency for one phase shoot through is on a significant lower level of 58%. Fig. 19 and Fig. 22 are picturing the input power referring to the efficiencies of the one and three phase shoot through operation. In case of the three phase shoot through, the input power is in a range between 2.8 kW and 6.5 kW, for one phase shoot through between 2.8 kW and 7.7 kW. The maximum input power for one phase shoot through exceeds the input power for three phase shoot through about 1.2 kW. As the output powers are pretty much equal referring to Fig. 20 and Fig. 23, the 1.2 kW are pure losses.

978-1-4673-9551-9/16 $31.00 © 2016 IEEE

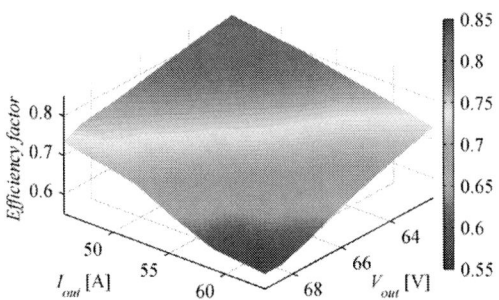

Fig. 21. Efficiency factor of the TSI for one phase shoot through in an operation range of $62\,V < V_{out} < 71\,V$ (peak value) and $45\,A < I_{out} < 63\,A$ (rms value).

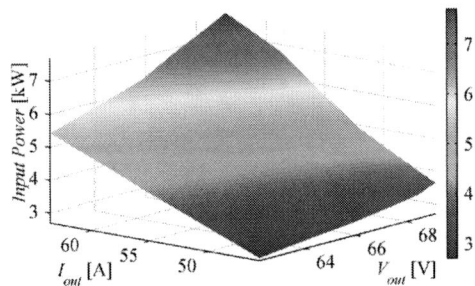

Fig. 22. Input power of the TSI for one phase shoot through referring to Fig. 21

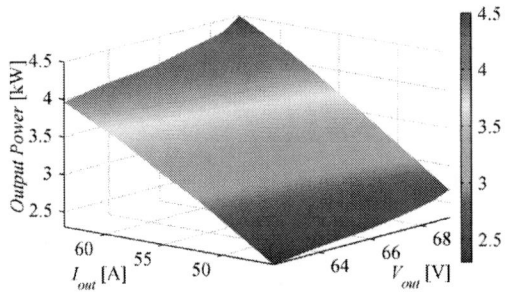

Fig. 23. Output power of the TSI for one phase shoot through referring Fig. 21

V. CONCLUSION

This paper presents an investigation on bidirectional T- Source inverter for extra low voltage and high current application. The T- Source inverter has been theoretically investigated and established in an experimental setup. Theoretically, the performance of the TSI overcomes previous proposed inverters with impedance source networks (e.g. Z-Source or Quasi-Z-Source) but the experimental setup shows some drawbacks of the inverter as it comes to real application. Nevertheless, the functionality of the inverter is proven and the experimental results are in agreement with the theoretical considerations if attention is payed to transformer leakage as well. The experimental results show that three phase shoot through enhances the performance of the TSI for extra low

voltage and high current application in comparison with a one phase shoot through. In order to improve the performance of the TSI, it is necessary to use an active snubber. However, this method would increase the number of active switches of the setup. The most important criteria for the functionality of the T-Source inverter is the output current but not the boost factor since the first determines the design of the transformer.

Overall, it can be stated that T-Source inverter is unsuitable for an extra low voltage and high current application.

REFERENCES

[1] D. G. Holmes and T. A. Lipo, Pulse Width Modulation for Power Converters, John Wiley & Son, 2003.

[2] D. Schröder, Leistungselektronische Schaltungen, Berlin Heidelberg: Springer-Verlag, 2008.

[3] F. Z. Peng, "Z-Source Inverter," Industry Applications Conference, vol. 2, pp. 775-781, 2002.

[4] F. Z. Peng, "Z-Source Inverter," IEEE Transactions on Industry Applications, vol. 39, no. 2, pp. 504-510, 2003.

[5] J. Anderson und F. Z. Peng, „A Class of Quasi-Z-Source Inverters," Industry Applications Society Annual Meeting, pp. 1-7, 2008.

[6] J. Anderson und F. Z. Peng, „Four Quasi-Z-Source Inverters," Power Electronics Specialists Conference, pp. 2743-2749, 2008.

[7] R. Strzelecki, M. Adamowicz, N. Strzelecka und W. Bury, „New Type T-Source Inverter," Conference on Compatibility and Power Electronics, pp. 191-195, 2009.

[8] R. Strzelecki, W. Bury, M. Adamowicz und N. Strzelecka, „New Alternative Passive Networks to Improve the Range Output Voltage Regulation of the PWM Inverters," Applied Power Electronics Conference and Exposition, pp. 857-863, 2009.

[9] S. P. Kumar und P. Shailaja, „T-Shaped Z-Source Inverter," International Journal of Engineering Research Technology (IJERT), Bd. 1, Nr. 9, pp. 1-6, 2012.

[10] W. Qian, F. Z. Peng und H. Cha, „Trans-Z-Source Inverters," IEEE Transactions on Power Electronics, Bd. 26, Nr. 12, pp. 3453-3463, 2011.

[11] Y. Siwakoti, P. C. Loh, F. Blaabjerg und G. Town, „Y-Source Impedance Network," Applied Power Electronics Conference and Exposition (APEC), pp. 3362-3366, 2014.

[12] Y. P. Siwakoti, P. C. Loh, F. Blaabjerg, S. J. Andreasen und G. E. Town, „Y-Source Boost DC/DC Converter for Distributed Generation," IEEE Transactions on Industrial Electronics, Bd. 62, Nr. 2, pp. 1059-1069, 2015.

[13] M.-K. Nguyen, Y.-C. Lim und Y.-G. Kim, „TZ-Source Inverters," IEEE Transactions on Industrial Electronics, Bd. 60, Nr. 12, pp. 5686-5695, 2013.

[14] A. Mostaan, S. Sharifi Malfejani, M. Soltani und A. Baghramian, „Novel T-Z source inverter with high voltage gain and reduced transformer turn ratio," Power Electronics, Drives Systems & Technologies Conference (PEDSTC), pp. 178-182, 2015.

[15] M. Adamowicz, J. Guzinski, R. Strzelecki, F. Z. Peng und H. A. Rub, „High step-up continuous input current LCCT-Z-source inverters for fuel cells," IEEE Energy Conversion Congress and Exposition (ECCE), pp. 2276-2282, 2011.

[16] M. Adamowicz, R. Strzelecki, F. Z. Peng, J. Guzinski und H. A. Rub, „New type LCCT-Z-source Inverters," European Conference on Power Electronics and Applications (EPE 2011), pp. 1-10, 2011.

[17] W. Mo, P. Loh und F. Blaabjerg, „Voltage type Γ-source inverters with continuous input current and enhanced voltage boost capability," International Power Electronics and Motion Control Conference (EPE/PEMC), pp. LS5d.2-1-LS5d.2-8, 2012.

[18] W. Mo, P. C. Loh und F. Blaabjerg, „Asymmetrical Γ-Source Inverters," IEEE Transactions on Industrial Electronics, Bd. 61, Nr. 2, pp. 637-647, 2014.

[19] Y. Siwakoti, P. Loh, F. Blaabjerg und G. Town, „Effects of leakage inductances on magnetically-coupled impedance-source networks,"

European Conference on Power Electronics and Applications (EPE'14-ECCE Europe), pp. 1-7, 2014.

[20] M. von Zimmermann, M. Lechler und B. Piepenbreier, „Z-source drive inverter using modified SVPWM for low Output Voltage and regenerating Operation," *European Conference on Power Electronics and Applications*, pp. 1-10, 2009.

[21] M. von Zimmermann, S. Labusch und B. Piepenbreier, „Bi-directional AC-AC Z-source inverter with active rectifier and feedforward control," *IEEE Energy Conversion Congress and Exposition (ECCE)*, pp. 3180-3186, 2010.

[22] Y. Liu, B. Ge und H. Abu-Rub, „Theoretical and experimental evaluation of four spacevector modulations applied to quasi-Z-source inverters," *IET Power Electronics*, Bd. 6, Nr. 7, pp. 1257-1269, 2013.

[23] S. Dong, Q. Zhang, C. Zhou und S. Cheng, „Analysis and design of snubber circuit for Z-source inverter," *European Conference on Power Electronics and Applications (EPE'14-ECCE Europe)*, pp. 1-10, 2014.

[24] M. Ismeil, M. Orabi, R. Kennel, O. Ellabban und H. Abu-Rub, „Experimental studies on a three phase improved switched Z-source inverter," *Applied Power Electronics Conference and Exposition (APEC)*, pp. 1248-1254, 2014.

[25] A. Battiston, E.-H. Miliani, J.-P. Martin, B. Nahid-Mobarakeh, S. Pierfederici und F. Meibody-Tabar, „A Control Strategy for Electric Traction Systems Using a PM-Motor Fed by a Bidirectional Z -Source Inverter," *IEEE Transactions on Vehicular Technology*, Bd. 63, Nr. 9, pp. 4178-4191, 2014.

Active Virtual Ground: Single Phase Grid-Connected Voltage Source Inverter Topology

River TinHo Li
Department of Power and Control
ABB (China) Ltd., Corporate Research Center
Beijing, China
river-tinho.li@cn.abb.com

Carl Ngai-Man Ho
Department of Electrical Engineering
University of Manitoba
Winnipeg, Canada
carl.ho@umanitoba.ca

Abstract — An efficient single-phase grid-connected voltage source inverter (VSI) topology by using the proposed Active Virtual Ground (AVG) technique is presented. With AVG, the conventional output L filter can be reconfigured to LCL structure without adding additional inductor. High frequency differential mode (DM) current ripple can be significantly suppressed comparing to the prior single-phase grid-connected inverter topologies. Additionally, strong attenuation to the high frequency common mode (CM) voltage is achieved. It is particularly important for some applications such as photovoltaic (PV). High efficiency can be achieved due to fewer component involved in the conduction loss. Cost of the inverter can be minimized since the required inductance of the filter is small. Performance of the proposed inverter has been evaluated analytically. Experimental verification is performing on a 50W, 100V input, and 60Vrms output prototype.

Keywords—inverter; common mode current; differential mode current

I. INTRODUCTION

Direct Current (DC) energy sources and storage elements, such as Photovoltaic (PV) and Battery cells, are critical components in the modern power systems [1]-[2]. The way to transform DC energy to alternative current (AC) and further inject to a public AC grid is typically using a grid-connected Voltage Source Inverter (VSI) [3]. However, it is well known that a simple full bridge VSI using unipolar switching scheme (3-Level switching scheme) will induce a high frequency Common-Mode (CM) voltage for grid-connected applications [4]. PV and Battery cells are suffered from the CM voltage to shorten the life time of the cells. Besides, the CM voltage causes EMI problems to affect other the performance apparatuses which are connected in the same grid. Furthermore, the CM voltage generates a high-frequency leakage current at the ground terminal in the grid network which would trigger circuit breakers to shut off the whole network. Using a line-frequency transformer to isolate the grid and the VSI is a conventional method to avoid the CM voltage problems since the transformer breaks the path of the leakage current loop. But commercial products are usually used tranformerless VSI topologies due to lower cost, higher efficiency and power density. Therefore, there is an industrial standard (DIN VDE 0126-1-1) to limit the leakage current of commercial power electronics transformerless inverter to maximum 300mA to avoid the CM voltage issues.

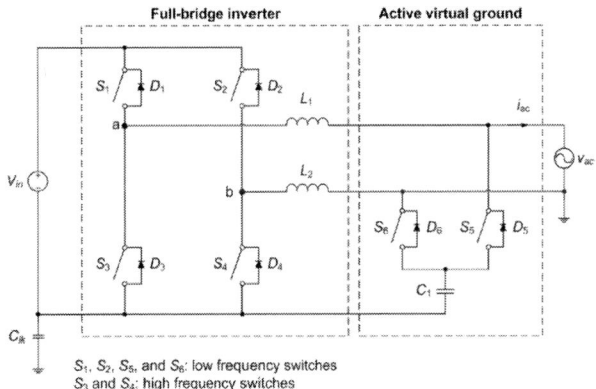

Fig. 1. The proposed inverter topology.

$S_1, S_2, S_5,$ and S_6: low frequency switches
S_3 and S_4: high frequency switches

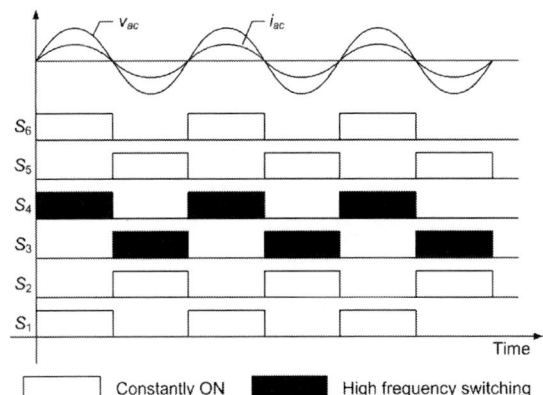

Fig. 2. Switching pattern of S1 – S6.

By only considering PV inverter systems, manufacturers and researchers have proposed various methods to get rid of the CM voltage problems with a simple full bridge inverter due to cost reason. The conventional method is to use bipolar switching scheme (2-level switching scheme) [3], however, this scheme causes high *di/dt* at the output current then a high inductance is required for a output inductor; besides, high switching losses are generated due to the use of IGBTs as main semiconductor switches. H5 inverter topology was proposed to solve the CM voltage issues with using unipolar switching scheme [5]. An additional power semiconductor switch is

978-1-4673-9551-9/16 $31.00 © 2016 IEEE

connected in series of the energy source and the full bridge inverter. Switching actions of the additional power switch is synchronized with the main high frequency switches in the inverter, as a result, the additional power switch breaks the leakage current path when the current is circulating in the main semiconductor switches during the zero states. The unipolar switching scheme can be used with a MOSFET and a fast reverse recovery diode can form a very fast switching cell to reduce switching losses. However, the main drawback of the topology is a full rated current and voltage semiconductor switch is used as the additional switch and it conducts the rated current of the inverter. A high conduction loss can be expected of the switch. Another approach is to put an additional path in parallel with the inverter for the zero state current circulation, HERIC is one of the topologies [6]. The approach successfully isolated the input energy source and the output grid network during the zero states. The leakage current cannot be generated due to no closed circuit. But it requires two more semiconductor switches to form a bidirectional blocking switch, it increases component cost and conduction losses. Other topologies have been proposed to solve the same problems [7] - [15], but they are all based on the above concepts, adding semiconductors to break or to bypass the leakage current path, with semiconductors in the circuit. Virtual Ground (VG) technology has been proposed for PV systems to solve the common mode voltage issues by using a three-phase three-wire inverter with controlling a potential difference of its mid-point of dc input capacitors and an unconnected neutral terminal to be almost zero voltage [16]. This method can effectively reduce the leakage current for three phase inverters, but it cannot be applied to single phase systems since the neutral wire is always connected to the inverter.

A technology namely Active Virtual Ground (AVG) is proposed in this paper to mitigate the high frequency common mode voltage. Two bidirectional blocking semiconductors and a capacitor form the AVG circuit which is connecting to the line terminal, the neutral terminal and a terminal of the input dc link. Fig. 1 shows the configuration and connections of the system. The circuit always keeps one of the grid terminals connecting to the input dc link with the AVG capacitor. High frequency current components are forced to return to the input of VSI from the output of VSI with a low impedance path, as a result, they do not pass through the grid network to return to the input. This complete inverter topology has the following advantages.

1) The topology gives low high-frequency leakage current to the AC grid network.

2) Unipolar switching scheme can be used to reduce the inductances of output filter.

3) The combination of a MOSFET and an IGBT with an anti-parallel diode can be used in a phase leg. It gives low switching losses of semiconductors.

4) The output filter is a LCL fitter but without additional grid side inductors. The output inductor of switching legs takes different roles, inverter side inductor or grid side inductor, in different operating modes. It reduces the component cost and increases the overall system power density.

5) As only high-frequency current ripple pass through the AVG circuit, additional conduction losses are minimized.

The operating principles of the proposed topology and mathematical steady-state characteristics of the converter system will be given. The proposed topology is successfully applied to a 1000W, 400V input, 50Hz grid-connected VSI prototype with a digital controller. The steady-state common mode voltage characteristics of the VSI are studied. Experimental results show that the inverter can deliver low high-frequency components output current to the grid. The theoretical prediction and experimental results are in good agreement.

II. OPERATING PRINCIPLE

A. The proposed inverter topology

Fig. 1 shows the circuit schematic of the proposed inverter with the AVG circuit. $S_1 - S_4$, L_1, and L_2 form an ordinary single-phase full bridge inverter; S_5, S_6, and C_1 form the AVG circuit. Fig. 2 shows the gate signals of all switches and output waveforms of the proposed inverter at unity power factor operating condition. The inverter is employing unipolar switching scheme (USS). Based on the switching scheme, S_1 and S_2 are switching at grid frequency and alternatively; S_3 and S_4 are switching at high frequency to shape the output current waveform as sinusoidal. Switching actions of S_5 and S_6 are synchronized with S_2 and S_1, respectively, for the case of unity power factor operation. Fig. 3(a) shows the operating mode for the positive half line cycle. S_1 is continuously turned-on in the complete half line cycle; S_4 is modulated by an average current mode controller and which is switched at high frequency, i.e. switching frequency. It can be observed that an equivalent high frequency switching voltage source, namely v_b, with magnitude changing between V_{in} and zero is applied across node b and the negative terminal of V_{in}. In order to minimize the differential mode (DM) voltage which could worsen the EMI and the filter inductor design, S_6 has to be conducted and S_5 has to be opened. With the aforementioned switching pattern, the structure of the output filter towards v_b becomes an LCL structure. The high frequency equivalent circuit of the proposed inverter for this operating mode is presented in Fig. 3(b). L_2, C_1, and L_1 form an LCL filter towards v_b and L_1 is series connected to the AC grid. In other words, L_1 and L_2 represent a grid side inductor and an inverter side inductor, respectively. As a result, the attenuation to the DM voltage, as well as the DM current, along the ac grid by the filter is 60dB/decade. Most of the prior single-phase grid-connected inverter topologies cannot provide output LCL filter unless adding an additional grid side inductor per phase leg. Therefore, the DM voltage attenuation of the proposed inverter is 40dB higher comparatively. A significant reduction on the required inductance for a given current ripple magnitude is achieved. The operating mode for the negative half line cycle is similar. An LCL output filter is formed by conducting S_5 and opening S_6. L_1 becomes the inverter side inductor and L_2 acts as the grid side inductor; S_2 is continuously turned-on in the complete half line cycle; S_3 is modulated by an average current mode controller and which is switched at high frequency. The corresponding high frequency equivalent circuit with an

978-1-4673-9551-9/16 $31.00 © 2016 IEEE   2906

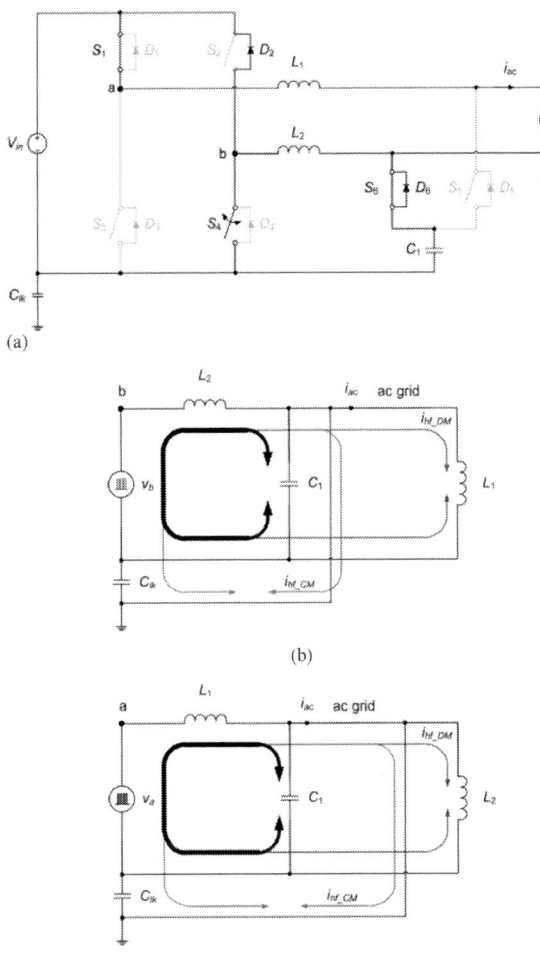

(a)

(b)

(c)

Fig. 3. (a) Operating mode for the positive half line cycle, (b) positive half line cycle high frequency equivalent circuit, and (c) negative half line cycle high frequency equivalent circuit.

equivalent high frequency switching voltage source, v_a, is shown in fig. 3(c).

The required filter inductance can be reduced due to the high order filter structure. Fig. 4 shows an example in comparing the frequency response of L and LCL filter. With $L_1 = L_2 = 800\mu H$, $C_1 = 2.2\mu F$, the LCL filter gives more than 24dB in attenuation at the switching frequency, 20 kHz in this example. Fig. 5 shows a comparison on the minimum required inductance for 2 kW single-phase inverters with the AVG circuit and inverter with capacitor-center-tapped filter (CCT) [17] with 10% current ripple. It can be seen that inverter with the AVG circuit provides high flexibility on the optimization of cost and efficiency. By keeping the switching frequency unchanged, the required inductance is only 800μH instead of 4mH for CCT. This design results in smaller inductor and lower cost. On the other hand, by keeping the inductance at 4mH, the switching frequency of inverter with the AVG circuit can be reduced to 7.5 kHz. This design results in higher efficiency. Hence, the proposed inverter

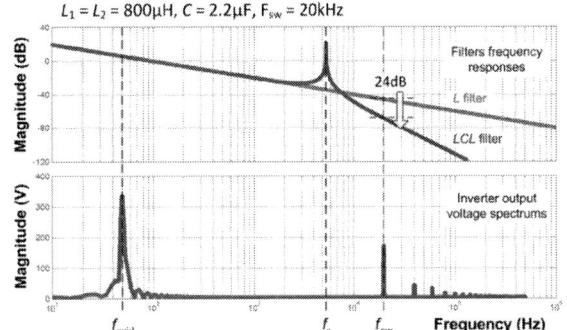

Fig. 4. Frequency response of L and LCL filters.

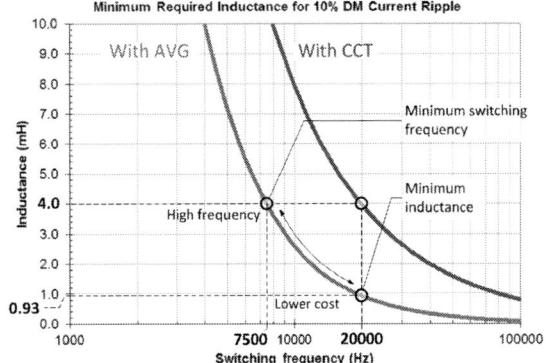

Fig. 5. Required inductance of inverter with AVG and CCT.

gives high flexibility on achieving an efficient and low cost design.

B. High frequency (HF) CM voltage attenuation

CM voltage is an important issue for some applications such as motor drives and PV. The proposed inverter is also providing high attenuation ability to the CM voltage. As shown in fig. 3(b) and (c), C_1 is connected in parallel with the leakage capacitor, C_{lk}, such as the parasitic capacitor across the ball bearing in a motor and across the frame and humidity of a PV panel [17]. Therefore, the high frequency CM voltage is stabilized by C_1 and the high frequency CM current is reduced accordingly. The relationship between the CM voltage and C_1 is given by eq. (1),

$$V_{hf\_CM}(t) = \sum_{n=1,3,\dots} \left| \frac{V_{a\_n} \sin n\omega_{sw} t}{2 - (n\omega_{sw})^2 L(C_{lk}+C_1)} \right| \qquad (1)$$

where L is the inductance of and L_1 and L_2 and setting $L_1 = L_2 = L$, and n is the harmonic order of the high frequency switching voltage source v_a or v_b dependence of the polarity of the line cycle. Fig. 6 shows the graphical expression of eq. (1), it can be seen that the attenuation to the HF CM voltage is higher with larger C_1. As a results, the HF CM current of the proposed inverter can be adjusted by the value of C_1.

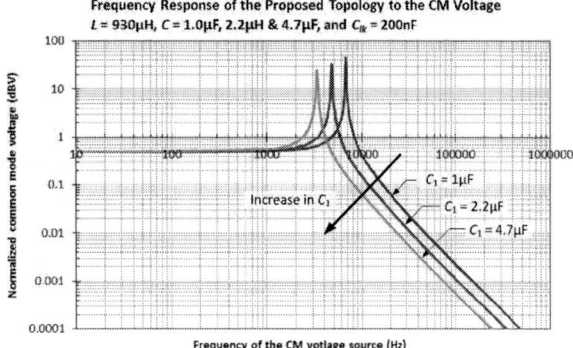

Fig. 6. Frequency response of different topology to the CM voltage.

C. Grid frequency ground leakage current of full bridge inverter using USS with AVG

AVG is not affecting the Grid Frequency (GF) CM voltage of a typical full bridge (FB) with USS. Fig. 7 shows the equivalent circuits for investigating the GF CM voltage by shorting the inductors and opening the high frequency leg. C_{lk} is connected in parallel with the circuit that is highlighted in blue. Therefore, the total voltage across the blue path is the GF CM voltage across the C_{lk} in either positive or negative half line cycle. Therefore, the CM voltage can be described as follow.

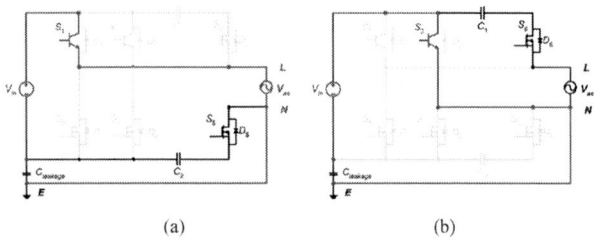

Fig. 7. Equivalent circuits for investigating the GF CM voltage (a) positive half line cycle and (b) negative half line cycle.

For positive half line cycle,

$$v_{CM\_GF\_AVG\_+ve} = V_{in} - \widehat{V_{ac}} \sin \omega t \quad (2)$$

For negative half line cycle,

$$v_{CM_{GF\_AVG}\_-ve} = V_{in} \quad (3)$$

Fig. 8 shows the plot of GF CM voltage of FB inverter using USS with AVG by using Equations (2) and (3).

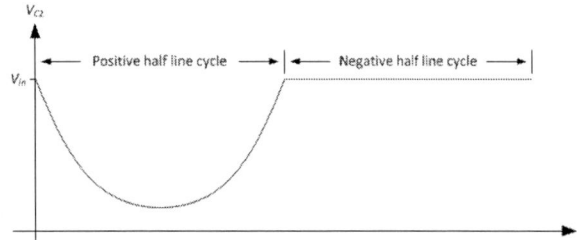

Fig. 8. GF CM voltage of FB USS AVG

The GF CM voltage can be expressed as follow by the observation of Fig. 8.

$$v_{CM\_GF\_AVG} = V_{in} - \frac{\widehat{V_{ac}}}{2}(\sin \omega t + |\sin \omega t|) \quad (4)$$

$$v_{CM\_GF\_AVG} = V_{in} - \frac{\widehat{V_{ac}}}{2}\left(\frac{2}{\pi} + \sin \omega t - \frac{4}{\pi}\sum_{n=2,4,...}^{\infty}\frac{\cos n\omega t}{n^2-1}\right) \quad (5)$$

The first, second and third terms inside the bracket of Equation (5) are the DC, fundamental and the even order harmonics terms of the GF CM voltage. Therefore, expression of the corresponding GF ground leakage current is shown as follow,

$$i_{CM\_GF\_AVG} = -\frac{\widehat{V_{ac}}C_{leakage}}{2}\left(\omega \cos \omega t + \frac{4n\omega}{\pi}\sum_{n=2,4,...}^{\infty}\frac{\sin n\omega t}{n^2-1}\right)$$

From Equation (6), it can be seen that the ground leakage current is not inference by the AVG circuit.

III. SIMPLIFIED DESIGN GUIDELINE AND BENCHMARKING

In this section, a simplified design procedure of the AVG circuit is presented. It is basically governed by the required HF DM and HF CM mode current magnitude.

A. Design of output filter inductors with AVG

The output filter inductor is designed by considering the maximum HF DM current ripple. By setting the grid side HF DM current of FB with CCT equals to the one with AVG, $L_1 = L_2 = L$,

$$\Delta i_{CCT}\frac{1}{2} = \Delta i_{AVG}\frac{1}{1+\omega_{sw}^2 LC_1} \quad (8)$$

Δi_{CCT} is the inverter side DM current ripple of FB with CCT and USS. The grid side DM ripple is about half of the inverter side. Δi_{AVG} and $\Delta i_{AVG}\frac{1}{1+\omega_{sw}^2 LC_1}$ are the inverter side and grid side DM current ripple respectively of FB with AVG and USS.

$$\frac{v_{L_{CCT}}}{L_{CCT}}\Delta t\frac{1}{2} = \frac{v_{L_{AVG}}}{L_{AVG}}\Delta t\frac{1}{1+\omega_{sw}^2 L_{AVG}C_1} \quad (9)$$

As $v_{L_{CCT}} \cong v_{L_{AVG}}$, refer to CCT, the required inductance of FB with AVG can be determined as follow.

$$L_{AVG}^2 + \frac{L_{AVG}}{\omega_{sw}^2 C_1} - \frac{2L_{CCT}}{\omega_{sw}^2 C_1} = 0 \quad (10)$$

The change in permeability against magnetic field strength has to be considered after determining required inductance with refer to the magnetic material. But it is out of the scope of this paper.

B. Design of filter capacitor with AVG

The design of AVG capacitor is governed by the maximum value of the HF CM current. By using Fig. 3(b), the HF CM current can be determined as follow,

$$i_{HF\_CM}(t) = i_{L_1}(t) - i_{L_2}(t) - i_{C_1}(t) \quad (11)$$

Because of the impedance of L_2 to the HF CM current is much higher than C_{lk} and C_1, $i_{L_2\_HF\_CM}(t)$ is assumed close to zero. Therefore, C_1 and C_{lk} are connected in parallel and worked as a current divider to i_{L1}. The capacitance C_1, as well as C_2, can be determined with the information of the maximum value of the HF CM as follow,

$$C_1 = C_2 = C_{lk}\left(\frac{V}{8L_1 f_{sw} i_{HF\_CM}} - 1\right) \quad (12)$$

Fig. 9 shows a design example of the inductance and capacitance by using (10) and (12) with respect to the switching frequency. The red line shows the minimal required total inductance, i.e. $L_1 + L_2$, and the blue line shows the minimum required capacitance of C_1. For the switching frequency equals to 20 kHz, the inductance $L_1 + L_2 \geq 930\mu H$ and $C_1 \geq 1.94\mu F$. Finally, $L_1 = L_2 = 500\mu H$ and $C_1 = 2.2\mu F$ are selected for the hardware prototype.

C. Benchmarking to The Prior-Art Single-phase Grid-connected Inverter Topologies

Table I shows a comparison of the proposed AVG topology to the prior-art solutions which are used to solve CM voltage issues for PV inverter applications. It can be seen that the proposed topology has the advantages of 1) low conduction loss since only two semiconductor devices are connected in series, 2) fully utilization of output filter inductors and high order output filter structure, it results smaller inductance requirement for a specified DM current ripple thus smaller inductor size, moreover, 3) no complicated control since the AVG's switches is switching at low frequency and synchronizing with the low frequency main switches. As a result, the proposed inverter is a cost-efficient topology with low CM voltage.

IV. SIMULATION AND EXPERIMENTAL RESULTS

In order to verify the CM voltage suspension capability by using the proposed AVG circuit, a simulation has been performed between the proposed AVG and CCT topologies. The value of the output filters, including the filter inductor and capacitor, are identical for the 2 topologies in this comparison. Fig. 10(a) shows the simulation results of both CM and DM currents. The top window shows the CM current, the current magnitude of the full-bridge inverter with AVG circuit is about two times larger than CCT's. To reduce the CM current for the inverter with AVG, it can be realized by using a larger filter capacitor according to eq. (1). The bottom window shows the

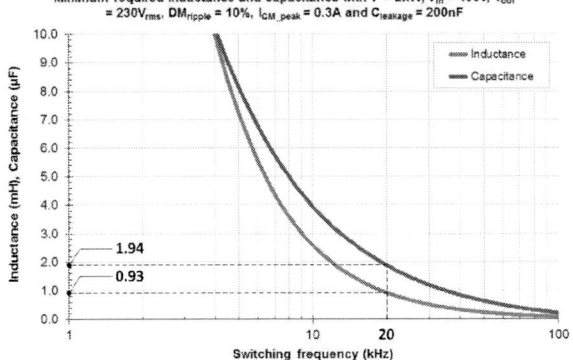

Fig. 9. Minimum requirements of L and C of FB with AVG and USS

TABLE I. Benchmarking of the prior single-phase grid-connected inverters

Inverter topology	No. of switches	No. of devices conduct in energy transfer or freewheeling states	O/P filter
HERIC [6]	6	2	L
Hybrid frequnecy phase legs [7]	4	2	L
H5 [5]	5	3	L
H6 [14]	6	4	L
H6 type [8]	6	transfer: 3, freewheeling: 2	L
Neutral wire – PV panel direct connection [10]	6	2	L
Neutral wire – PV panel direct connection [12]	6	transfer: 3, freewheeling: 5 (equivalent)	L
Neutral wire – PV panel direct connection [13]	2	2	L
Neutral wire – PV panel direct connection [15]	3	Positive half-line cycle: 1, negative half-line cycle: 3	L
NPC type [9], [11]	6	transfer: 3, freewheeling: 2	L
The proposed inverter with AVG	6	2	LCL

Table II. Use of component in the hardware prototype

Item	Value / Model	Item	Value / Model
L_1 and L_2	500μH @ 0A (NPS300060)	S_1 & S_2	IKWT75N60T
C_1	2.3μH (MLCC)	S_3 & S_4	IPW60R041C6
DSP	TMS28335	S_5 & S_6	STGB7NC60HD
Fsw	20kHz	Gate driver (S_3, S_4)	IR2113S
		Gate driver (S_1, S_2, S_5, S_6)	TLP350

DM current ripple of both topologies. The maximum current ripple for the one with AVG is only about 0.96A which is about 4 times smaller than CCT's. It can be proved that AVG topology is more effective to filter out HF DM current at the output and provide effective suppression to the HF CM current.

A 1kW, 20kHz switching, 400V input and 220Vrms output full bridge inverter with the proposed AVG circuit hardware prototype has been built for experimental verifications. The key components in prototype is tabled in Table II. The prototype was tested with resistive load and with and without the AVG circuit as shown in Fig. 1. Fig. 12(a) shows the experimental waveforms of the configuration without the AVG circuit. The output voltage can be seen that carrying a large HF DM ripple component. Fig. 12(b) shows the one with the proposed AVG circuit. It can be seen that the HF DM ripple is highly suppressed by simply adding the proposed AVG circuit without changing the modulation. Fig. 13 shows the voltage waveform across the switches S_5 and S_6. It can be seen that the voltage across S_5 and S_6 is the same as the output voltage and no high frequency switching action. Therefore, no higher voltage rating switch is needed and only conduction loss has to be considered in selecting the semiconductor switch.

Fig. 14 shows the design of the converter prototype. All semiconductor switches in the prototype are controlled by a DSP. It can be seen that the AVG switches, S_5 and S_6, adopt small surface mounted package MOSFETs, since the power dissipation is low. Two output filter inductors are designed as identical, and each inductor connects to one individual current sensor. This is because the operating modes of the inductors are change in every half line cycle, it requires different inductor current feedback signals to control the semiconductor switches in the inverter to shape the output current in the positive line cycle and the negative line cycle.

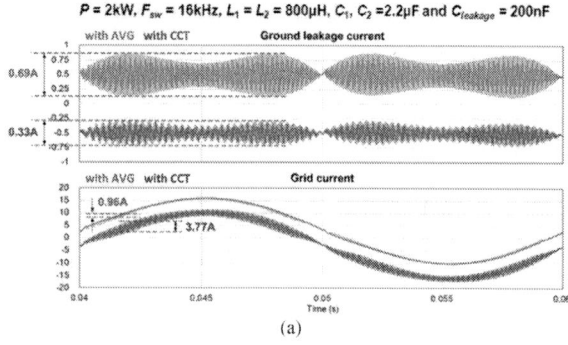

(a)

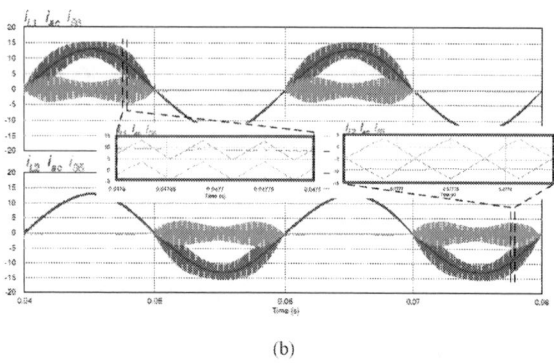

(b)

Fig. 10. Simulation results (a) FB inverter with CCT and the AVG, (b) current via the AVG switches, S5 and S6.

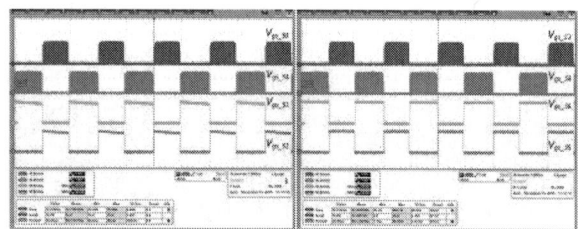

Fig. 11. Gate signals of the proposed FB inverter with USS and AVG.

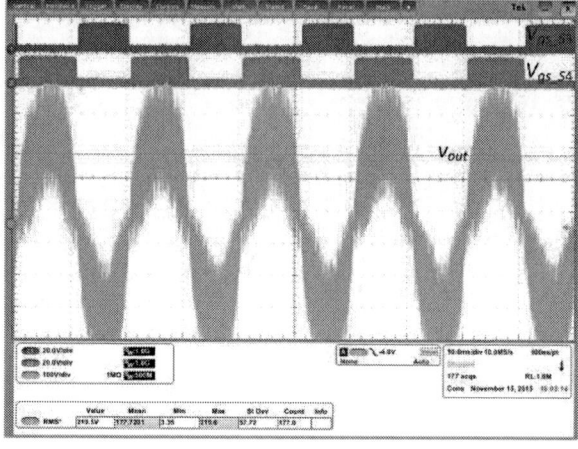

(a)

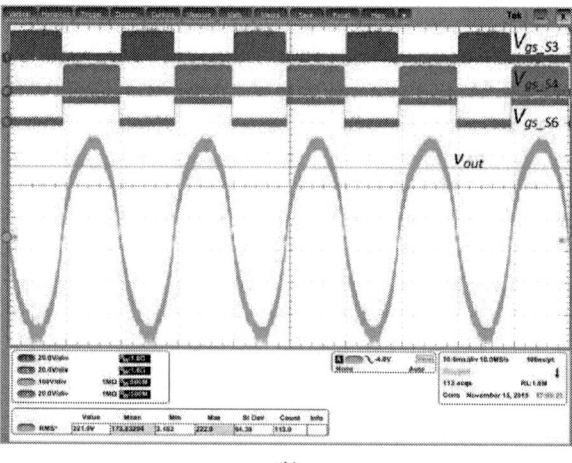

(b)

Fig. 12. Gate signals and output voltage experimental waveforms of (a) FB inverter with USS, and (b) FB inverter with USS and the proposed AVG.

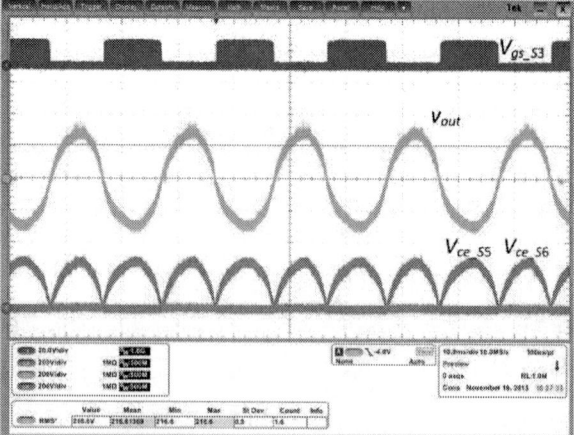

Fig. 13. AVG switches voltage experimental waveforms.

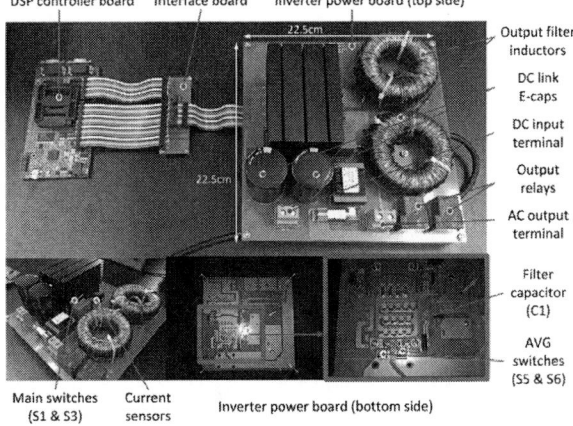

Fig. 14. Photo of the laboratory prototype of the proposed inverter with AVG.

V. CONCLUSIONS

A single-phase grid-connected inverter with Active Virtual Ground (AVG) that provides high attenuation ability on the HF CM and DM voltage is presented in this paper. The proposed inverter is a cost-efficient topology in comparing with the prior-art solutions for specified applications. The key merit is changing the output filter from L to LCL without adding additional inductor. Thus, the filter attenuation ability is 40dB/decade higher and the required inductance can be reduced. The operating principle of AVG towards HF CM and DM voltage are studied; the theoretical prediction is verified by computer simulations and experimental results. The theoretical prediction and experimental results are in good agreement. Besides, circuit steady-state characteristics and simplified design guidelines have been presented in the paper.

REFERENCES

[1] J. M. Carrasco, L.G. Franquelo, J.T. Bialasiewicz, E. Galvan, R.C.P. Guisado, A.M. Prats, J.I. Leon, and N. Moreno-Alfonso, "Power-Electronic Systems for the Grid Integration of Renewable Energy Sources: A Survey," *IEEE Trans. on Industrial Electronics,* vol.53, no.4, pp.1002-1016, June 2006.

[2] W. Jiang and B. Fahimi, "Active Current Sharing and Source Management in Fuel Cell–Battery Hybrid Power System," *IEEE Trans. on Industrial Electronics,* vol.57, no.2, pp.752-761, Feb. 2010.

[3] C. Ho, V. Cheung and H. Chung, "Constant-frequency hysteresis current control of grid-connected VSI without bandwidth control", *IEEE Trans. on Power Electronics,* Vol. 24, No. 11, pp. 2484 – 2495, Nov. 2009.

[4] S.B. Kjaer, J.K. Pedersen, and F. Blaabjerg, "A review of single-phase grid-connected inverters for photovoltaic modules," *IEEE Trans. on Industry Applications,* vol.41, no.5, pp.1292-1306, Sept.-Oct. 2005

[5] M. Victor, F. Greizer, S. Bremicker, and U. Hübler, "Method of converting a direct current voltage from a source of direct current voltage, more specifically from a photovoltaic source of direct current voltage, into a alternating current voltage", US Patent, No. US7411802 B2, 12/08/2008.

[6] H. Schmidt, C. Siedle, and J. Ketterer, "DC/AC converter to convert direct electric voltage into alternating voltage or into alternating current," US Patent, No. US7046534 B2, 16/05/2006.

[7] LI et al., "Transformer-free unilateral inductor grid-connected inverter circuit," Chinese Patent, No. CN102255331A, 23/11/2011.

[8] W. Yu, J. S. Lai, H. Qian, and C. Hutchens, "High-Efficiency MOSFET Inverter with H6-Type Configuration for Photovoltaic Nonisolated AC-Module Applications, *IEEE Trans. Power Electronics,* vol.26, no. 4, pp. 1253-1260, April 2011.

[9] X. Li, "Transformer-less grid-connected inverting circuit," Chinese Patent, No. CN102195507A, 21/11/2011.

[10] H. Jochen, "Direct current voltage converting method for use in inverter, involves clocking switch units such that high potential and input direct current voltage lie at inputs of storage reactor in magnetized and free-wheel phases, respectively," German Patent, No. DE102006010694B4, 8/03/2006.

[11] B. Sven et al., "Inverter, more specifically for photovoltaic plants," US Patent, US7843714B2, 30/11/2010.

[12] Z. Peter, "Inverter for Feeding Electric Energy into a Power System," US Patent Application, No. US20100135054A1, 3/06/2010.

[13] Z. Peter et al., "Inverter," US Patent, No. US7813153B2, 12/10/2010.

[14] G. Senosiain et al., "Single-Phase Circuit to Condition and Transform Direct Current Electric Power into Alternating Current Electric Power, European Patent, No. EP2053730B1, 17/06/2015.

[15] W. Richard, "Monopolar DC to Bipolar to AC Converter," US Patent, No. US7064969B2, 26/08/2004.

[16] G. Escobar, N. Ho, and S. Pettersson, "Method and apparatus for zero sequence damping and voltage balancing" " US Patent, No. US9030854 B2, 12/05/2015.

[17] D. Dong, F. Luo, D. Boroyevich, and P. Mattavelli, "Leakage Current Reduction in a Single-Phase Bidirectional AC–DC Full-Bridge Inverter", *IEEE Trans. on Power Electronics,* vol. 27, no. 10, Oct. 2012, pp. 4281 – 4291.

Design and Evaluation of 30kVA Inverter Using SiC MOSFET for 180°C Ambient Temperature Operation

Feng Qi, Miao Wang, and Longya Xu
The Ohio State University
Columbus, Ohio, U.S.
E-mail: feng.qi4academic@gmail.com

Bo Zhao, Zhe Zhou, and Xizhou Ren
Smart Grid Research Institution of SGCC
Beijing, China

Abstract—A 30kVA SiC MOSFET inverter is designed and evaluated for 180°C ambient temperature operation. The entire inverter system is designed for high temperature (HT) except the DSP control circuit in room temperature environment. The power structure is designed using SiC MOSFETs and HT capacitors. The gate driver circuits with protections are also designed for the HT environment. The prototype has been built and tested in 180°C environment at 30kVA for 5 hours accumulatively without any problems. The inverter is also tested in -10°C environment at its full capacity of 30kVA for competed evaluation.

Keywords—high temperature; low temperature; SiC MOSFET; three phase inverter

I. INTRODUCTION

SiC power devices have great potentials to work in HT environment[1]. After many years of efforts, both SiC power MOSFETs and JFETs are now commercially available. As normally-off device is preferred in power circuits, this paper focuses on the SiC MOSFET's application for future electrified transportations. Compared to Si MOSFET, SiC MOSFET shows superior advantages in conductivity, breakdown voltage and high temperature capabilities [2]. In today's market, SiC MOSFET module has been developed to the rating of 1200V/300A with a maximum junction temperature of 150°C [3]. With proper thermal management, it is possible to develop a three-phase SiC inverter operating in a harsh environment with an ambient temperature of 150°C or higher.

Many efforts have been made to build HT power electronic and associated control circuits [4 – 6]. In the work reported in [4], with commercially available HT silicon-on-insulator (SOI) integrated circuits (IC), the mother board of the power circuits has been successfully tested in a thermal chamber of 150°C for one hour. With an in-house built HT power module, the mother board and the power circuit were tested at 1.4kW outside the thermal chamber and the junction temperature of the in-house built power module reaches 250°C. In [5], an alternative approach was taken to design HT power electronic circuits. Thermoelectric cooling technology is integrated into the signal electronics controller to provide proper operating environment for the control electronics. HT fan is developed to cool the power circuits built with commercially available SiC JFETs. Both the control and the power circuits were tested in a 120°C thermal chamber up to 10kW. To push the HT boundary of power electronic circuits, researchers in [6] have

developed a 200°C gate driver IC based on SOI and a SiC power module with junction temperature greater than 225°C. For this circuit, thermal chamber test at 200°C was performed on the gate driver board. For the continuous test, the gate driver board and the power circuit are tested outside thermal chamber. The junction temperature of the power module is elevated up to 232°C through self-heating. The power level achieved by this work is 1.25kW.

The published literature indicates that there are two major challenges for HT power electronics technology development. The first is the capability of the microelectronics for controls, such as the gate driver circuits. The second is the thermal management of the HT power devices. Though many solutions to the challenges have been proposed, efforts are still needed to investigate the tradeoffs and identify the optimal solutions among the possibilities. For example, in HT microelectronics, though HT SOI technology can significantly increase the temperature capability of the gate driver circuits, the costs of HT SOI technology are prohibitively high. Alternative solutions have been studied to design the cost-effective HT gate driver circuits with HT Si discrete components [7– 9] and potentially the component costs can be reduced by a factor of ten but with an increased size. Meanwhile, it is also pointed out that with proper design, the HT gate driver circuit and protections can be made in a very compact size[9].

In this paper, we report our continued efforts to design and evaluate a HT 30kVA three-phase inverter. The cost-effective gate driver circuits and protections are designed to overcome the challenge for HT microelectronics in controls. In addition, commercially available SiC power module and HT capacitors are chosen to build the power circuit. The HT thermal management problem is solved by modifying the cooling system instead of forcing the SiC power module to endure a junction temperature beyond its specification (150°C). The HT 30kV inverter system has been designed, including the control circuit, power circuit and cooling system. Selection of HT components is discussed. HT capacitor and power module are evaluated focusing on performance at an elevated temperature. The entire inverter including the gate drivers and protection has been placed into a thermal chamber of 180°C for laboratory testing for accumulatively 5 hours. The inverter has also been tested in -10°C environment for 1 hour. The results of testing and analysis are presented in the paper to verify the successful design and operation of the SiC based HT inverter.

978-1-4673-9551-9/16 $31.00 © 2016 IEEE

II. SYSTEM DESIGN

A. System Configuration

According to our state-of-the-art survey, when junction temperature greater than 150°C is required, most microelectronic components, such as microcontrollers and sensors, become either very expensive or unavailable. On the other hand, in many cases, it is not necessary to expose all microelectronics to HT environment since control and sensor signals can be transmitted over a relatively long length of wire to avoid hostile conditions. However, in any power converter design, it is not realistic to transmit control signals over a long wire from the gate drivers to the power devices because of the EMI and EMC issues. In effect, placing the gate drivers as close as possible to the power devices is mandatory to minimize the gate loop stray inductance. Hence, we have divided the microelectronic circuits into two stages: one operating in room temperature and another HT such as the gate drivers. Since the microelectronics for measurement instruments cannot survive in HT environment, the test points in HT environment are connected to the measurement instruments in room temperature through HT wires.

The overall configuration of the designed HT inverter system and its test bed are illustrated in Fig. 1. In the figure, the yellow blocks represent lab equipment. The green blocks are the microelectronic circuits and the red blocks the power circuit. The liquid cooling system is shown in blue. Fig. 2 is the 3D drawing of the HT circuits, including the HT gate driver boards, HT DC-link (HT capacitor bank and DC bus-bar), SiC power modules, liquid-cooled cold plate and system frame.

For power structure, though SiC MOSFET dies can be operated at a temperature higher than 150°C [2, 6], the junction temperature is limited to 150°C due to its package limitations. However, it is possible to operate the commercially available SiC power module at higher environment temperature when proper thermal management is applied. It is also to be noted that most DC-link capacitors will fail at temperature higher than 125°C [5]. If the thermal management needs to cover DC-link capacitors, the whole power structure will become very bulky. To solve this problem, HT DC-link in the design is constructed with HT capacitors. To address the challenges that the HT capacitors usually have very limited capacitance, we have divided the DC-link capacitors into two groups – one group for room temperature and another for HT to achieve sufficient capacitance. The room temperature group is composed by commonly used film and electrolytic capacitors with a temperature limit of 125°C. The HT group uses HT ceramic capacitors with a much higher maximum operating temperature.

B. HT Gate Driver

The gate driver circuits for controlling SiC MOSFET power module in the design are organized as shown in Fig. 3. The orange part represents for the HT Board. In the design, the gate driver and protection circuits are designed for HT conditions. A HT pulse transformer circuit is developed to achieve galvanic isolation for the PWM signal transmission. The pulse transformer and its receiver circuit are implemented on the HT board with the gate driver and protection circuits.

The transmitter circuit is placed close to the main control board in room temperature. Commercially available isolated power supply is chosen to power the HT board. Design details are discussed and can be found in [9].

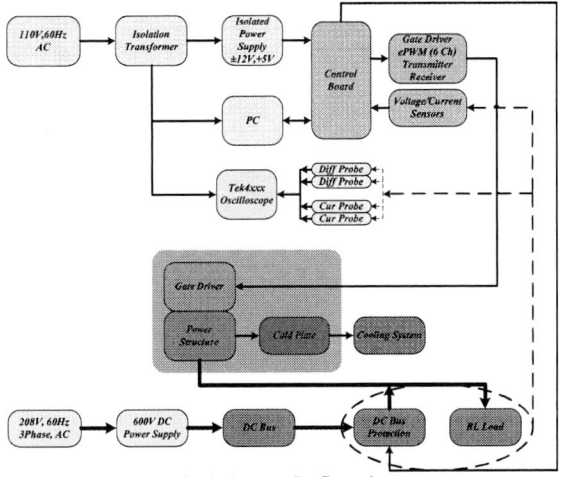

Fig. 1. System Configuration

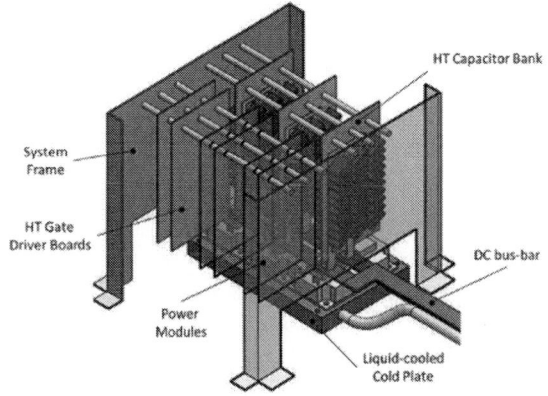

Fig. 2. 3D drawing of HT inverter

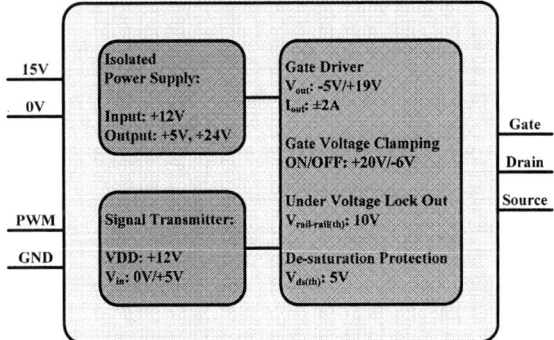

Fig. 3. Organization of HT gate driver and protection circuit

C. HT Power Structure

For the overall power structure, 150°C SiC MOSFET power modules CREE CAS300M12BM2 and 200°C HT ceramic capacitors AVX SXP47C105KAA are chosen. The schematic of the HT inverter power circuit is shown in Fig. 4. For testing convenience, the DC bus-bar connecting the room temperature capacitors and HT capacitors needs to be designed longer than 20 inch, which causes large stray inductance in the DC-link. To analyze the negative effect caused by the large stray inductance, a simplified model for the whole DC-link is established as shown in Fig. 5.

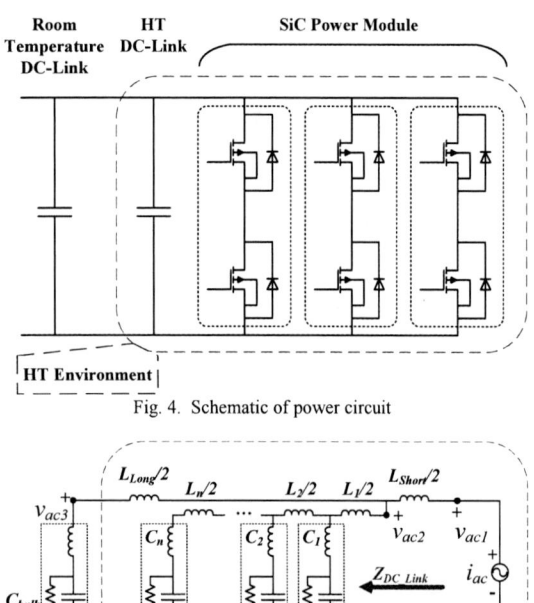

Fig. 4. Schematic of power circuit

Fig. 5. Simplified model for DC-link Analysis

As well known, the large stray inductance can cause sharp voltage spike when high di/dt is encountered. To study stray inductance effect, three phase legs of the inverter are simplified to a single AC current source. L_{short} represents the stray inductance from power module terminal connecting the power chips to the capacitor terminal. L_{long} is the stray inductance of the long bus bar connecting the power module to the room temperature capacitors. HT capacitor bank is connected to power module directly and L_i ($1 \leq i \leq n$) stands for the inductance from connection between HT capacitors. For capacitors, both of ESL and insulation resistance are considered in the model. The DC voltage source for the inverter, which is a short circuit for AC current source, is considered to be far away from the DC-link and the AC impedance between the DC voltage source and DC-link is infinite in the frequency range of this analysis. Since the analysis is focusing on the effect from stray inductance instead of the model itself, for convenience of analysis, model

parameters are not directly using measurement results but made based on measurement results. Table I shows all model parameters using in the analysis.

TABEL I DC-LINK MODEL PARAMETERS

Component		Value	Unit
L_{long}	L	0.01/0.1/1/10	μH
L_i	L	0.01	μH
L_{short}	L	0.01	μH
C_{bulk}	C	1000	μF
	ESL	0.01	μH
	$R_{insulation}$	1	MΩ
C_i	C	1	μF
	ESL	0.01	μH
	$R_{insulation}$	1	MΩ

In this design, based on measurement, the range of L_{long} is from 0.1μH to 1μH. To analyze this design, an ideal design is adopted as a reference to compare and its L_{long} is set to 0.01 μH. Only one HT capacitor is included in the comparison. Fig. 6 (a) shows significant increment of DC-link impedance when L_{long} increases. The difference of DC-link impedance can be as high as a factor of 100 when studying the case $L_{long} = 1$μH and the case $L_{long} = 0.01$μH. To neutralize the effect from large stray inductance, the number of HT capacitors is added to 100 in further comparison shown in Fig. 6 (b). As a result, compared to the first case ($L_{long} = 1$ μH, $n=1$), the DC-link impedance is reduced by a factor around 10. It can be found that 100 times of HT capacitance is not able to fully neutralize the effect from 100 times of L_{long}. In the third comparison, L_{long} is further increased to 10μH, which emulates that the room temperature capacitors is placed very far away. In other words,

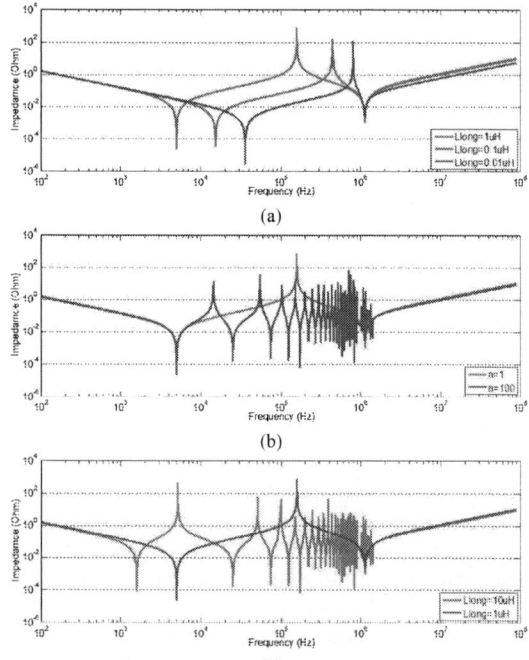

(a)

(b)

(c)

Fig. 6. DC-link impedance Analysis

978-1-4673-9551-9/16 $31.00 © 2016 IEEE   2914

the DC-link is only supported by 100 HT capacitors. From Fig. 6 (c), though DC-link impedance in high frequency region is still low, the impedance in low frequency region turns to be around 10 times higher than the case (L_{long} = 1μH, n=1). If DC-link is only supported by 100 HT capacitors, large voltage ripple around 10kHz should be expected. On the other hand, more HT capacitors is needed in the DC-link to make the room temperature capacitor obsolete. However, limited by space and cost, it is unrealistic to build a DC-link with only HT capacitors.

In this design, performance is the top priority and cost the subordinate. Limited by space, HT DC-link is built by a total of 138 HT capacitors which are divided into 3 groups. Each group is directly mounted on top of a power module to minimize stray inductance.

D. Thermal Management

In HT environment, effective thermal management is critical for the safe operation of power modules. Since the target ambient temperature (180°C) will be higher than the allowed junction temperature (150°C) of the SiC power modules, the thermal management system should be able to remove the heat generated by power modules itself and the heat absorbed by the power module from the HT environment. On the other hand, to maximize the thermal management system efficiency, the thermal management system must minimize direct heat exchange between the cold plate and HT environment.

Though air cooling is the most widely used cooling method used in inverter systems [5]. However, this method does not fit into the situation where the ambient temperature is equal to or greater than junction temperature. Fig. 7 shows a typical image of heat flow in an air-cooled inverter in the 150°C ambient temperature condition. The heat generated by the power modules is transferred to the heat sink and then carried away by the air blowing through the heat sink. Due to the thermal resistance existing between power device junction and the ambient air, a large temperature rise would be found in the junction area. Given that most commercially available power modules are rated at a maximum junction temperature of 150°C, the air-cooled system cannot provide an adequate cooling capability to ensure the safe operation at or higher than 150°C.

To solve this problem, a liquid-cooling system is adopted. A commercial aluminum cold plate manufactured by MaxQ is chosen to build the liquid-cooling system. DI water is selected as the coolant and the inlet temperature is set at 55°C. Given that aluminum is a good thermal conductor, to minimize direct heat exchange with HT environment, an insulation layer made of silicone sponge and ceramic fiber is applied on the bottom and around the cold plate. Thermal simulation is performed to predict the thermal performance of liquid-cooled inverter in HT environment. In the simulation, the ambient temperature is set to 180°C and the HT inverter is simplified with main thermal conductive parts. A fixed temperature condition is added to the center area of the cold plate to emulate the stable temperature of flowing liquid. A 200W thermal source is assigned on the top surface of each base plate to represent the power loss at steady state. Vertical copper frame is added to

emulate the connection between the SiC power module and HT capacitors. The plastic case of power module and the thermal

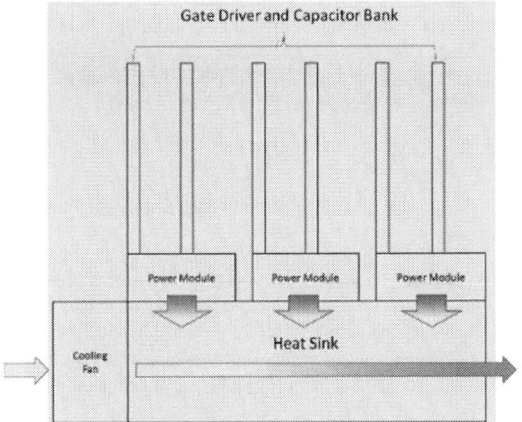

Fig. 7 Heat flow in air-cooled inverter

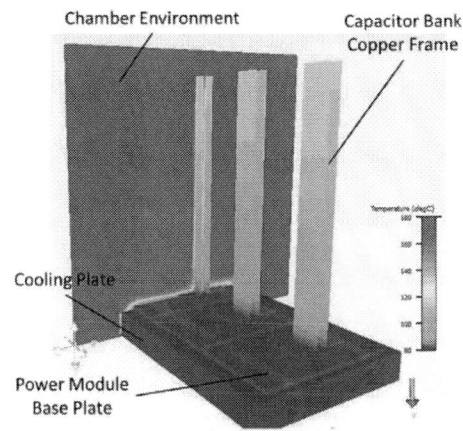

(a) 3D model overview

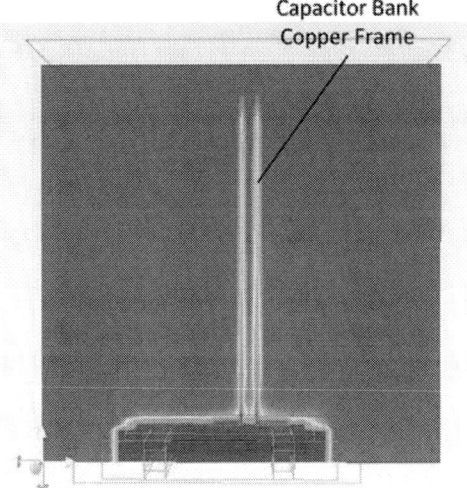

(b) Temperature distribution in model cross section
Fig. 8. Thermal simulation result

insulation material are removed to simplify the simulation. Simulation result is shown in Fig. 8 and as the thermal image indicates, the cold plate keeps the base plate of the power module around 80°C. The simulation results also indicate that the junction temperature also maintained around this temperature. At the top along the copper frame, the temperature rises from 80°C to 120°C. The simulation results verify that the liquid-cooling system can keep power module junction temperature lower than the hostile ambient temperature of 180°C.

III. EXPERIMENTAL EVALUATION

A. Key Component Evaluation

This section focuses on evaluation of the key HT components at high temperatures. In the survey, it is found out that for the SiC power device CAS300M12BM2, its switching loss stays almost unchanged at high temperature as compare to that at room temperature but its on-resistance at 150°C is 1.8 times of that at 25°C [3]. As a result, more power loss is expected at high temperature. Though the device conduction loss increase at high temperature results in a decreased system energy efficiency, HT operation, on the other hand, could avoid using a liquid cooling system to improve the overall system efficiency. If junction temperature can be elevated to a higher temperature, such as 250°C, simple convection cooling could be sufficient to keep power electronics safely running in 180°C environment. In this way, complexity of inverter system can be significantly reduced and system efficiency improved. In this design, the commercial power module package limits the junction temperature to 150°C and liquid cooling is still required to make the power module survive in 180°C environment.

For capacitor component SXP47C105KAA, its capacitance is measured at different temperatures by impedance analyzer, Agilent 4294A. In the measurement, the capacitor is placed on the top of a hot plate and covered by a one inch layer of ceramic fiber sheet to keep the high temperature stable and uniformly distributed. Connection is made by HT wire between the capacitor to be measured and the analyzer. To minimize measurement error due to the stray inductance, the capacitance is measured at 10 kHz. This is because at 10 kHz, the impedance of 1μF capacitor is 1000 times larger than that of 1μH inductor. The error caused by stray inductance is negligibly small since the stray inductance is smaller than 1μH. Fig. 9 shows the temperature dependency of the HT capacitor SXP47C105KAA.

As indicated by the testing results, with temperature increasing, capacitance of the HT capacitor starts to drop rapidly, especially at the temperature higher than 150°C. At 200°C, only 55% of capacitance at 25°C remains. As a result, degradation of DC-link performance is expected in 180°C environment. Such a large temperature drift of capacitance is caused by the dielectric material, X7R/X9U. If a smaller capacitance drift is required at high temperatures, capacitors made by C0G material will be the right choice. However, the disadvantage of C0G capacitors is that for the same rated voltage, the capacitance of X7R/X9U capacitor is 20 times higher than that of C0G capacitor [10]. In this design, to achieve high capacitance in a limited space, X7R/X9U capacitor is chosen.

B. System Evaluation

The HT 30kVA three-phase inverter prototype is designed and built as shown in Fig. 10 where (a) is the HT portion and (b) the room temperature portion of the inverter. To control the inverter, space vector pulse width modulation (SVPWM) is

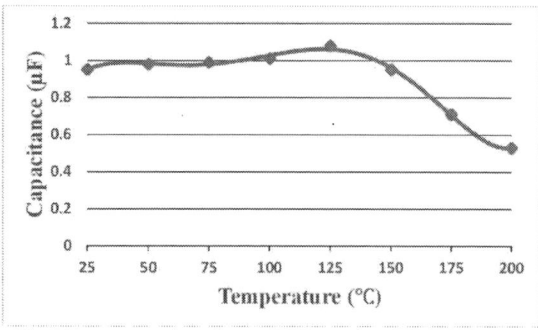

Fig. 9 Temperature dependency of HT capacitor, SXP47C105KAA

(a) HT section

(b) Room temperature section
Fig. 10. Prototype of HT 30kVA three-phase inverter

applied to the HT inverter switching. For load testing, a three phase Y-connected inductor bank is connected to the inverter. During load testing, it is found that the inductor core becomes saturated when the load current is high and the load impedance becomes smaller than the nominal values. Due to the inductor core saturation, HT inverter actually reaches 31kVA with a DC-link voltage of 300V. On the load side the frequency is maintained at 90Hz, phase voltage 97V_{rms}, and phase current 108A_{rms}. For the completed testing, both HT and low temperature tests are conducted to fully test the operating range of the inverter. The HT test temperature profile is measured and recorded by OMEGA data logger RDXL4SD as shown in Fig. 11.

As can be noted in the figures, at the beginning of the testing, the inverter is powered up to 300V at DC-link after the thermal chamber temperature is stable at 25°C. Then, the HT inverter is adjusted to 31kVA. After a half hour operation at 25°C, the thermal chamber is set to 180°C in four steps: 75°C, 125°C, 150°C and 180°C. Figs. 12 (a) and (b) present the three-phase current waveforms of the HT inverter operated at 31kVA

in 25°C and 180°C environment respectively. The current aliasing is caused by the inductor saturation and observed in both waveforms. The power module and DC-link voltage waveforms are shown in Fig. 13. As can be noted, voltage spikes caused by the high di/dt have been observed for voltages V_{DS} (device drain-source) and V_{DC} (DC-link) in Figs. 13 (a) through (c). The voltage spikes are quite dense especially in Fig. 13 (c). Although the high speed switching of SiC power

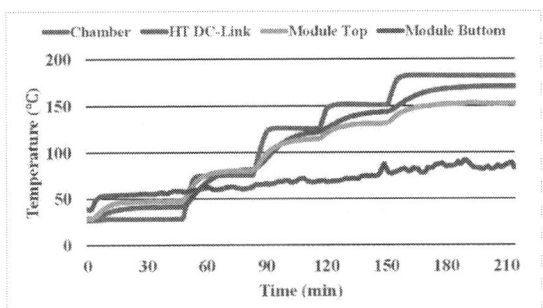

Fig. 11 Recorded temperature profile in HT evaluation

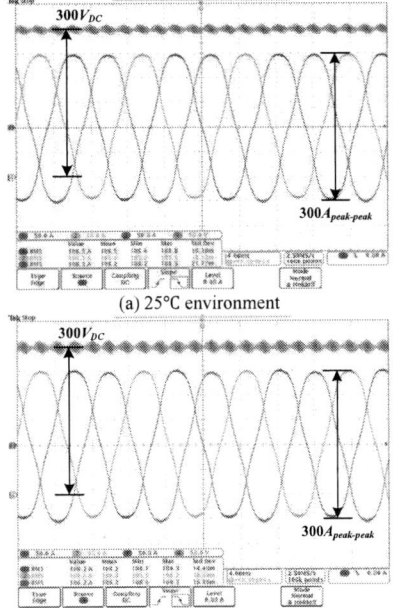

(a) 25°C environment

(b) 180°C environment

Fig. 12 Three-phase current waveform at 31kVA

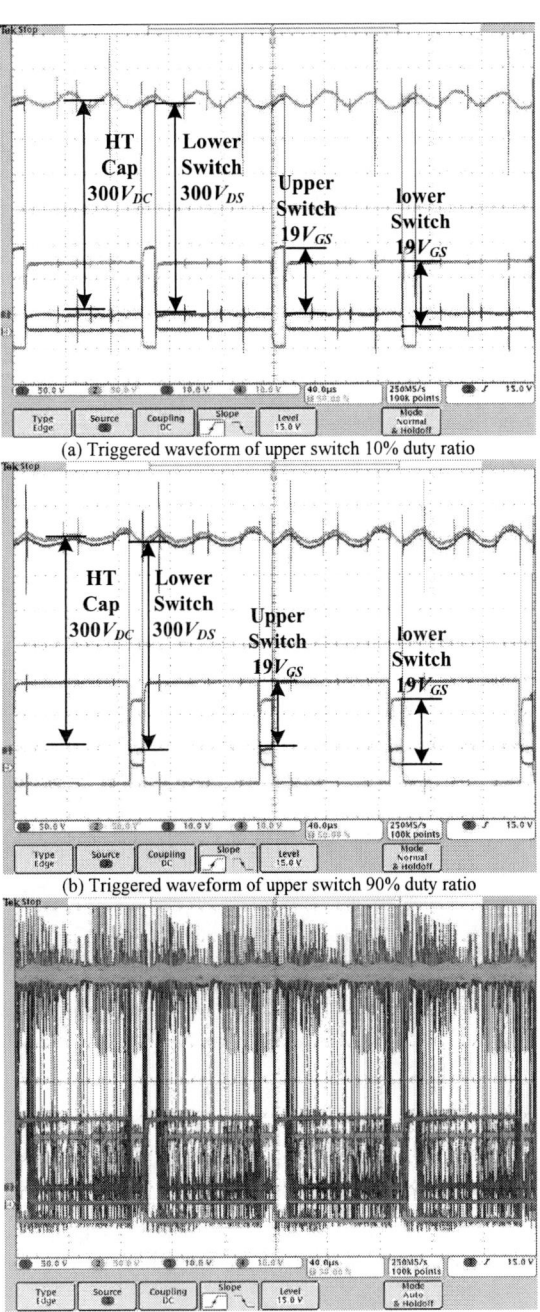

(a) Triggered waveform of upper switch 10% duty ratio

(b) Triggered waveform of upper switch 90% duty ratio

(c) Continuous refreshing waveform

Fig. 13. Switching waveform at 180°C 31kVA

devices produces such dense and strong noises, the HT inverter is still running well without any faults. The robustness of the HT inverter is evidently testified in such a harsh environment. After one hour operation at 180°C and 31kVA, the HT inverter is powered off and the thermal chamber is set back to the room temperature.

In low temperature testing, the recorded temperature profile is shown in Fig. 14. At the beginning, the HT inverter is powered up to 300V at DC-link after the thermal chamber temperature is stable at -10°C. Then, the HT inverter is adjusted to 31kVA. Under such a condition, the current waveform aliasing is also observed as shown in Fig. 15. After one hour operation at -10°C and 31kVA, the thermal chamber is set back to 25°C in a single step. At the end of testing, the inverter is powered off after one hour operation at 25°C and 31kVA.

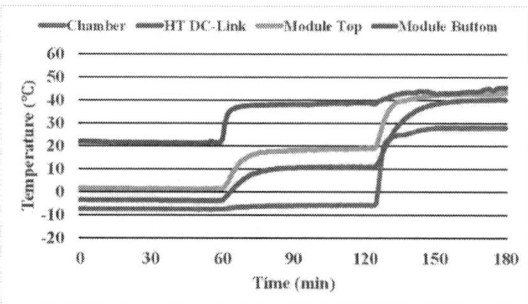

Fig. 14 Recorded temperature profile in low temperature evaluation

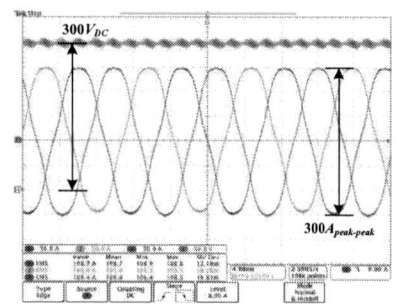

Fig. 15 Three-phase current waveform at -10°C 31kVA

IV. CONCLUSIONS

In this paper, the work is reported for a 30kVA three-phase inverter using SiC MOSFETs designed and evaluated in HT environment up to 180°C. The design involves two major portions: the power structure with peripheral circuits to survive at HT and the thermal management system to address the major challenges for HT operation of SiC modules with a junction temperature limit of 150°C.

The HT gate driver is also designed and evaluated based on discrete component instead of SOI IC to achieve a cost reduction of 90%. HT ceramic capacitors are combined with room temperature film-electrolytic capacitors to provide stable DC-link voltage. Thermal management system is designed to allow power module to stay within its junction temperature limit of 150°C in the HT environment chamber of 180°C.

Experimental results and analysis are conducted to validate the design. The lab testing results verify that the designed inverter is capable of operating at the designed power rating, 30kVA, and in the HT environment of 180°C. Additionally, the inverter is also successfully tested in -10°C environment. The cost-effective gate driver circuits and HT DC-link can work with SiC power modules CAS300M12BM2, in 30kVA 180°C conditions. The thermal management system stabilizes the power device junction temperature within its limit. Overall, the HT 30kVA three-phase inverter is successfully designed, built and experimentally evaluated at 30kVA in a wide temperature range from -10°C to 180°C.

ACKNOWLEDGMENT

The work is supported by SGCC Grant 5355DD130003.

REFERENCES

[1] Dreike, P.L.; Fleetwood, D.M.; King, D.B.; Sprauer, D.C.; Zipperian, T.E., "An overview of high-temperature electronic device technologies and potential applications," Components, Packaging, and Manufacturing Technology, Part A, IEEE Transactions on , vol.17, no.4, pp.594,609, Dec 1994

[2] Feng Qi; Lixing Fu; Longya Xu; Ping Jing; Guoliang Zhao; Jiangbo Wang, "Si and SiC power MOSFET characterization and comparison," Transportation Electrification Asia-Pacific (ITEC Asia-Pacific), 2014 IEEE Conference and Expo , vol., no., pp.1,6, Aug. 31 2014-Sept. 3 2014

[3] Cree, CAS300M12BM2, 1200 V, 300 A Silicon Carbide Half-Bridge Module. (2014). [Online]. Available: http://www.cree.com/

[4] Ruxi Wang; Boroyevich, D.; Puqi Ning; Zhiqiang Wang; Fei Wang; Mattavelli, P.; Ngo, K.D.T.; Rajashekara, K., "A High-Temperature SiC Three-Phase AC - DC Converter Design for > 100 °C Ambient Temperature," Power Electronics, IEEE Transactions on , vol.28, no.1, pp.555,572, Jan. 2013

[5] Wrzecionko, B.; Bortis, D.; Kolar, J.W., "A 120 °C Ambient Temperature Forced Air-Cooled Normally-off SiC JFET Automotive Inverter System," Power Electronics, IEEE Transactions on , vol.29, no.5, pp.2345,2358, May 2014

[6] Zhiqiang Wang; Xiaojie Shi; Tolbert, L.M.; Wang, F.F.; Zhenxian Liang; Costinett, D.; Blalock, B.J., "A High Temperature Silicon Carbide mosfet Power Module With Integrated Silicon-On-Insulator-Based Gate Drive," Power Electronics, IEEE Transactions on , vol.30, no.3, pp.1432,1445, March 2015

[7] Feng Qi; Longya Xu; Guoliang Zhao; Jiangbo Wang, "Transformer isolated gate drive with protection for SiC MOSFET in high temperature application," Energy Conversion Congress and Exposition (ECCE), 2014 IEEE , vol., no., pp.5723,5728, 14-18 Sept. 2014

[8] Qi, Feng; Xu, Longya; Zhao, Bo; Zhou, Zhe, "High current gate drive circuit with high temperature potential for SiC MOSFET module," in Energy Conversion Congress and Exposition (ECCE), 2015 IEEE , vol., no., pp.7031-7037, 20-24 Sept. 2015

[9] Qi, Feng; Xu, Longya; Zhao, Bo; Zhou, Zhe, "A high temperature de-saturation protection and under voltage lock out circuit for SiC MOSFET," in Energy Conversion Congress and Exposition (ECCE), 2015 IEEE , vol., no., pp.6169-6174, 20-24 Sept. 2015

[10] AVX, SMPS Molded Radial MLC Capacitors - SXP Style for High Temperature Applications up to 200°C, (2014). [Online]. Available: http://www.avx.com/

A DC to Three-Phase Boost-Buck Inverter with Stored Energy Modulation and a Tiny DC Link Capacitor

Mahima Gupta, *Member, IEEE*, Giri Venkataramanan, *Senior Member, IEEE*

Abstract—Three phase voltage source inverters typically feature a significant amount of dc energy storage to maintain a stiff dc bus. In these cases, the dc link capacitor is sized to store enough energy to maintain several tens of cycles of ac output at the rated power. In this paper, we propose to reduce the size of the dc link capacitor dramatically, to store enough energy to provide just one high frequency switching cycle of ac output power. The dc bus is no longer stiff and hence classical sinusoidal pulse width modulation cannot be used. But, the *stored energy modulation* (SEM) concept that we propose herein synthesizes high quality sinusoidal output voltage waveforms even with such tiny dc link capacitors. In SEM, the switching intervals of the interconnecting switches are carefully determined non-linear functions of various operating parameters such as reactive component sizing, switching frequency, load levels, etc. This paper presents the analytical and detailed design development of the *stored energy modulation* approach along with circuit simulation, experimental results and comparative evaluation of a dc to three phase system example.

Index Terms—DC-AC power converters, Voltage source inverter, Buck-boost, Three phase, Stored Energy Modulation, Pulse Width Modulation

I. Introduction

Voltage source type of three phase inverters form the workhorse for various motor drive applications, wind/solar generation systems and uninterruptible power supply systems [1–3]. While this inverter topology is derived from the buck converter, many emerging dc-ac power conversion applications demand a voltage boost function as well [4, 5]. For such applications, it is common to derive a cascaded topology of a boost converter followed by the classical voltage source inverter, although there has been considerable academic interest in exploring the Z-source converter as well [6]. The intermediate capacitive dc link of these topologies, commonly referred to as the 'bulk' capacitor tends to dominate the weight and volume of the system. In order to improve the power density of the conversion system, there has been considerable interest in reducing the size of the reactive components in the system by increasing the switching frequency [7, 8]. While this approach leads to reduction of output filter components, it does not necessarily lead to reduction of dc link energy storage requirements [9, 10]. In most of the cascaded power converters, the amount of dc energy storage can typically provide several (say, 10-100) cycles of ac output at the rated

M. Gupta and G. Venkataramanan are with the Department of Electrical and Computer Engineering, University of Wisconsin-Madison, Madison, WI, 53706 USA e-mail: mgupta29@wisc.edu

power. Such large amounts of energy storage leads to a 'stiff' dc bus, which allows ac waveform synthesis using linear pulse width modulation approaches [11]. In linear pulse width modulation, the duty ratio of the interconnecting switches are a linear function of the ratio of time varying ac output and the stiff dc voltage [12].

This paper presents a radically different approach wherein the energy storage in the dc bus is just sufficient to provide one high frequency switching cycle of ac output at the rated power. The energy stored in the dc bus is returned to zero at the end of every *high frequency switching period*. Three phase output waveform synthesis is to be performed using *stored energy modulation* (SEM) concept that is proposed in this paper. In this modulation approach, the duty ratio of the throw switches that form the three phase inverter are carefully regulated, but the voltage waveform across the output are *not* formed of pulses of sinusoidally varying width, with constant amplitude. Rather, the voltage waveforms are of varying shapes (triangular or trapezoidal), that provide the appropriate amount of average voltage during the high frequency switching cycle. The major and significant consequence of this approach is the *tiny* size of the dc link capacitor, in comparison to conventional approaches. The operational approach and other features of this topology have been described in the following sections.

II. Topology and Operation

The circuit topology of a dc-dc boost converter cascaded with a three phase voltage source inverter is illustrated in Figure 1. The throws of the switches are realized using IGBTs and diodes just as in classical approaches. In this topology, the size of the input inductor and output inductive filter components follow the typical design norms chosen to feature negligible amount of ripple current. On the other hand, use of the SEM allows the size of the dc link capacitor C to be orders of magnitude less than traditional designs.

The operation of the converter within each switching period follows a predetermined energy charge-discharge pattern at the dc link. During the charge interval, the boost switch S_U is connected to the positive throw t_{U+} and the current in the input inductor L_i transfers energy to the dc link capacitor C. The output inverter is held in zero state. Therefore, the three phase output does not receive any power from the dc link during the charge interval. The voltage (v_C) across the capacitor C increases linearly from zero. Soon after the

978-1-4673-9551-9/16 $31.00 © 2016 IEEE

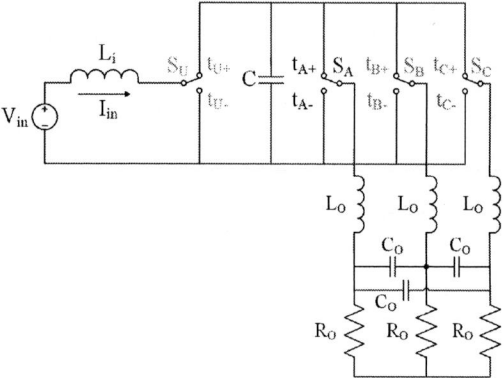

Fig. 1: Schematic of the proposed topology

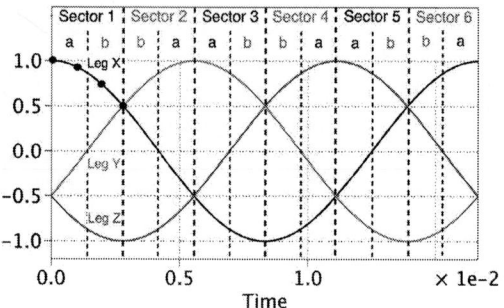

Fig. 2: A typical three-phase sinusoidal waveform

energy stored in dc link capacitor C reaches a sufficient and predetermined value, the input boost switch S_U is connected to the negative throw t_{U-}. The discharge interval begins after this instant.

During the discharge interval, the input stage does not exchange any power with the dc link and the three phase inverter switches $S_A - S_C$ are modulated appropriately to provide a discharge path for the energy storage in the dc link capacitor C. The excess period when the converter is not in either discharge or charge state, the dc link is in the idle state, with dc link voltage maintained either at zero, at its peak or at an intermediate value.

The operation of the modulator is defined by drawing a predefined packet of energy from the source during part of the switching period, and subsequently transferring to the load during the remaining switching period. It is this distinguishing feature of the modulator that leads to the nomenclature, *stored energy modulation*, or *SEM*.

III. STORED ENERGY MODULATION

Consider the three phase balanced set of sinusoidal voltage waveforms (A, B and C) desired at the output, shown in Figure 2. The time interval of one period of the three phase output waveforms can be divided into 6 sectors, depending on which of the phase voltages is the most positive and which of the phase voltages is the most negative. Each of the 6 sectors may be further classified into 2 sub-sectors each, depending on either having two of the phase voltages negative or having two of the phase voltages positive, which are labeled *Subsector a* and *Subsector b* respectively in the figure. It may be observed that the three phase waveforms in these sectors are either symmetrical or mirror images of each other. Due to the waveform symmetry, the SEM developed for any one sector can be extended to the remaining sectors by mapping the solution through appropriate translations and sign reversals.

The detailed operation of SEM during a particular sector is described in the following subsections. The following notation is throughout the paper. The switching functions of the various throws are defined as

$$h_k(t) = \begin{cases} 1, & \text{throw } t_k \text{ conducts} \\ 0 & \text{otherwise} \end{cases} \quad (1)$$

and d_k, is the time-averaged value of $h_k(t)$ over the switching period T_S, where k can take the values of various throws labeled in Figure 1 such as $A-$, $A+$, $U+$, $U-$, etc.

A. Charge interval

The charge interval defines the time interval $(D_{U+}T_S)$ during which the capacitor is charged by the current through L_i. It may be observed that during the time $D_{U+}T_S$, S_{U+} is always connected to its positive throw. This has been illustrated by the green trace of h_{U+} in Figure 3. The output inverter is maintained in zero state, alternating between Z_+ (throws t_{X+}, t_{Y+} and t_{Z+} are enabled together), and Z_- (throws t_{X-}, t_{Y-}, and t_{Z-} are enabled together) for every other switching period as illustrated in the same figure. The quantitative analysis of the charge interval can be obtained as illustrated further.

The power P, supplied by the dc voltage source V_{in}, to the capacitor C can be derived as given by (2).

$$P = V_{in}I_{in} = \frac{1}{2}V_{C-peak}I_{in}D_{U+} \quad (2)$$

where, V_{C-peak} is the peak capacitor voltage at the end of the time interval $D_{U+}T_S$, and I_{in} is the dc source current with negligible ripple.

The peak capacitor voltage can also be written in terms of dc input current (3). The charge interval duty ratio, D_{U+}, defines the average power input (P) to the dc link capacitor during the entire switching period T_S, from V_{in} as shown in (4). P and thus D_{U+} can simply be obtained by eliminating V_{C-peak} from (2) and (3).

$$I_{in} = C\frac{dv_C}{dt} = C\frac{V_{C-peak}}{D_{U+}T_S} \quad (3)$$

$$P = \frac{2V_{in}^2 C}{T_S D_{U+}^2} \quad (4)$$

$$D_{U+} = V_{in}\sqrt{\frac{2C}{T_S P}} \quad (5)$$

It can be observed that D_{U+}, given by (5), is independent of time as long as input voltage, power throughput and switching frequency are maintained constant. The expression for D_{U-} can simply be obtained as given by (6).

$$D_{U-} = 1 - D_{U+} \quad (6)$$

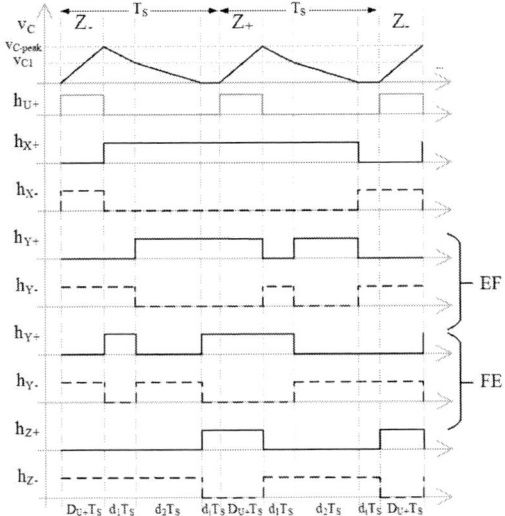

Fig. 3: Zoomed-in waveforms of switching functions of various throws of the converter under SEM

Once the capacitor has received a predetermined packet of energy from the dc source, it has to be discharged appropriately in order to obtain the desired voltages at the output load. The following subsection details the idle and discharge interval of *SEM*.

B. Idle interval

An idle interval ($d_i T_S$) may be distributed as a dead time between the charge and discharge intervals as desired without affecting the waveform synthesis. During the idle interval, the inverter is neither in a discharge or a charge state and the capacitor current is zero, and the voltage is maintained constant at zero, at its peak V_{C-peak}, or at an intermediate value.

C. Discharge interval

For the sake of notational book keeping, let the sum of the duty ratios of the charge interval and the idle interval be labeled as or d_{C+i} as shown by (7).

$$d_{C+i} = D_{U+} + d_i \qquad (7)$$

The power stored in the dc capacitor is transferred to the load during the remaining interval which is denoted as the discharge interval and can be expressed as $1 - d_{C+i}$.

The division of a three phase sinusoidal voltage waveform into sectors, defined in Figure 2, has been done depending on which of the phase voltages is most positive and which of the phase voltages is most negative. The notation of phases X, Y and Z has been adopted in this section to represent the three phases, where they may take the value of one of the three phases A, B and C depending on the sector of operation as explained further.

The phase with the most positive voltage may be denoted as phase X while the phase with the most negative voltage may

be denoted as phase Z. The phase which lies between phase X and phase Z may be denoted as phase Y. The mathematical description of this mapping is shown by (8)-(11).

$$X = \phi[\max\{V_A, V_B, V_C\}] \qquad (8)$$

$$Z = \phi[\min\{V_A, V_B, V_C\}] \qquad (9)$$

where the phase function is defined as,

$$\phi[V_k] = k \qquad (10)$$

$$Y = \in \{A, B, C\} \neq [X \ or \ Z] \qquad (11)$$

As an illustration, phases X, Y and Z can be mapped to the phases B, A and C of Sector 2 of Figure 2 respectively. Similarly, in Sector 3, phases X, Y and Z can be mapped to the phases B, C and A respectively.

While distributing the energy storage from the dc link among the three phases, the throws t_{X+} and t_{Z-} of phases X and Z are always enabled during the entire discharge interval. Hence, in this interval, phase X is always connected to the positive bus while phase Z is always connected to the negative bus. Furthermore, during the charge interval and the idle interval, Z_+ and Z_- are alternatively enabled, in order to minimize the common mode neutral voltages at the three phase output. This modulation of phase X and Z has been illustrated by h_{X+}, h_{X-}, h_{Z+} and h_{Z-} traces of Figure 3. The mathematical description of the duty ratios of the various throws of phases X and Z is given by (12)-(15).

$$d_{X+} = 1 - \frac{1}{2}d_{C+i} \qquad (12)$$

$$d_{X-} = 1 - d_{X+} = \frac{1}{2}d_{C+i} \qquad (13)$$

$$d_{Z+} = \frac{1}{2}d_{C+i} \qquad (14)$$

$$d_{Z-} = 1 - d_{Z+} = 1 - \frac{1}{2}d_{C+i} \qquad (15)$$

The turn-on intervals of the throws of Phase Y may be divided into two subsequent intervals $d_1 T_S$ and $d_2 T_S$ respectively, as shown in Figure 3. During $d_1 T_S$ interval, let the capacitor voltage be discharged from V_{C-peak} to V_{C1}, by a current equal to I_1, corresponding to a particular position of the switch S_Y. Subsequently, during $d_2 T_S$ interval, let the capacitor voltage be discharged from V_{C1} to 0, by a current equal to I_2, corresponding to the complementary position of the switch S_Y. The power transferred to and from the capacitor during various intervals can be derived in terms of the duty intervals as illustrated further.

The power transferred from the capacitor by current I_1 can be derived as given by (16) and (17).

$$I_1 = C \frac{V_{C-peak} - V_{C1}}{d_1 T_S} \qquad (16)$$

and,

$$P_1 = \frac{C}{2T_S}[V_{C-peak}^2 - V_{C1}^2] = \frac{I_{in} I_1 d_{U+} d_1 T_S}{C} - \frac{I_1^2 d_1^2 T_S}{2C} \qquad (17)$$

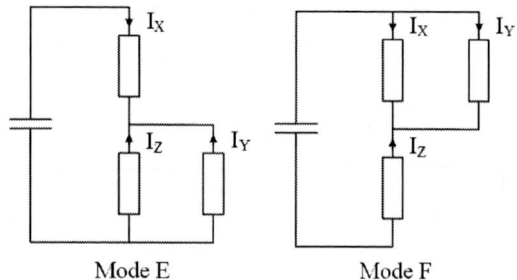

Fig. 4: Mode of operation of Phase Y

where P_1 is the net power transferred from the capacitor during first discharge interval $d_1 T_S$. By substituting the value of V_{C-peak} from Equation 3, a relationship for d_1 may be determined as

$$d_1 = \sqrt{\frac{2CP}{I_1^2 T_S}}[1 - \sqrt{1 - \frac{P_1}{P}}] \quad (18)$$

Similarly, the power transferred from the capacitor by current I_2 to completely discharge it to zero can be derived as given by (19) and (20).

$$I_2 = C\frac{V_{C1}}{d_2 T_S} \quad (19)$$

and,

$$P_2 = \frac{1}{T_S}\frac{1}{2}C[V_{C1}^2] = \frac{I_2^2 d_2^2 T_S}{2C} \quad (20)$$

where P_2 is the net power transferred from the capacitor during second discharge interval $d_2 T_S$. Equation (20) can be simplified further to obtain an expression of d_2.

$$d_2 = \sqrt{\frac{2CP_2}{I_2^2 T_S}} \quad (21)$$

The particular position of S_Y during $d_1 T_S$ and $d_2 T_S$ determine the values of I_1, I_2, P_1, and P_2, and lead to the actual value of duty ratios of its throws. The connection of Phase Y to the negative bus is denoted as Mode E, and the connection of the positive bus is denoted as Mode F, as illustrated in Figure 4.

Let us assume that the power transferred to Phase X during modes E and F are P_{X_E} and P_{X_F} respectively. Similarly, the power transferred to the Phases Y and Z during modes E and F can be denoted by P_{Y_E} and P_{Y_F}, and P_{Z_E} and P_{Z_F} respectively. Hence,

$$P_{X_E} + P_{X_F} = I_X^2 R \quad (22)$$

$$P_{Y_E} + P_{Y_F} = I_Y^2 R \quad (23)$$

$$P_{Z_E} + P_{Z_E} = I_Z^2 R \quad (24)$$

During mode E and mode F, illustrated in Figure 4, the phase powers can be written in terms of the phase currents as shown by (25) and (26) respectively

$$\frac{P_{Y_E}}{P_{Z_E}} = \frac{I_Y}{I_Z} \quad (25)$$

$$\frac{P_{X_F}}{P_{Y_F}} = \frac{I_X}{I_Y} \quad (26)$$

The power distribution between the phases during mode E and mode F will be as shown by (27)-(28) and (29)-(30) respectively.

$$\frac{2}{3}P_E = P_{X_E} \quad (27)$$

$$\frac{1}{3}P_E = P_{Y_E} + P_{Z_E} \quad (28)$$

$$\frac{2}{3}P_F = P_{Z_F} \quad (29)$$

$$\frac{1}{3}P_F = P_{X_F} + P_{Y_F} \quad (30)$$

where, P_E and P_F are the total powers transferred by the capacitor to the three phases during modes E and F respectively. Equations (22)-(30) can be solved simultaneously to determine powers P_E and P_F in terms of the phase currents which are shown by (31) and (32) respectively.

$$P_E = I_X^2 R - I_X I_Y R \quad (31)$$

$$P_F = I_Z^2 R - I_Z I_Y R \quad (32)$$

The order of sequencing the E and F modes during the $d_1 T_S$ and $d_2 T_S$ may be arbitrary. In the EF scheme, mode E precedes mode F. In other words, throw t_{Y-} is enabled for initial discharge time interval $d_1 T_S$ and subsequently throw t_{Y+} is enabled for time interval $d_2 T_S$. This modulation technique has been shown by the first pair of h_{Y+} and h_{Y-} traces in Figure 3. In the FE scheme, mode F precedes mode E. In other words, throw t_{Y+} is enabled for initial discharge time interval $d_1 T_S$ and subsequently throw t_{Y-} is enabled for time interval $d_2 T_S$. This modulation technique has been shown by the second pair of h_{Y+} and h_{Y-} traces in Figure 3. The mapping of the currents I_1, I_2, powers P_1, P_2, and the duty ratios d_1 and d_2 for the two schemes are illustrated in Table I. The mapping may be discerned by examining the equivalent circuits of the two modes illustrated in Figure 4. It may be observed from the table that each of the duty ratio terms for the particular throws d_{Y+}, and d_{Y-} as distinct from the d_1 and d_2 duty ratios include the sum of idle intervals and zero intervals expressed by $d_{C+i}/2$.

In order to arrive at sinusoidal voltage output, the currents and powers in (18) and (21) can be substituted with the corresponding sinusoidal functions with appropriate phases. As an illustration, in Sector 1, for Scheme EF, these expressions can be simplified to (33) and (34).

$$d_{Y+}^{EF} = 3V_{out}\sqrt{\frac{C}{T_S P}}[\sqrt{\frac{4sin\theta}{\sqrt{3}cos\theta + 3sin\theta}}] + \frac{1}{2}d_{C+i} \quad (33)$$

$$d_{Y-}^{EF} = 3V_{out}\sqrt{\frac{C}{T_S P}}[\frac{1 - \sqrt{sin^2\theta + \frac{sin\theta cos\theta}{\sqrt{3}}}}{cos\theta}] + \frac{1}{2}d_{C+i} \quad (34)$$

where, V_{out} is the RMS voltage at the load and θ is the electrical angle of the output voltage waveform.

TABLE I: Mapping of the duty ratios, currents and powers of the two discharge intervals d_1 and d_2 to the Schemes EF and FE

Scheme	I_1	I_2	P_1	P_2	d_1	d_2
EF	$|I_X|$	$|I_Z|$	P_E	P_F	$d_{Y-} - \frac{1}{2}d_{C+i}$	$d_{Y+} - \frac{1}{2}d_{C+i}$
FE	$|I_Z|$	$|I_X|$	P_F	P_E	$d_{Y+} - \frac{1}{2}d_{C+i}$	$d_{Y-} - \frac{1}{2}d_{C+i}$

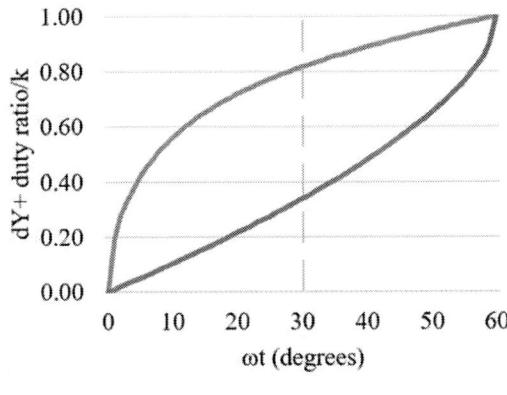

Fig. 5: Plot of Phase Y positive throw duty ratio divided by a factor of k where k is $3V_{out}\sqrt{\frac{C}{T_S P}}$. ($\frac{1}{2}d_{C+i}$ has not been accounted)

Similarly, for Scheme FE, the corresponding expressions (18) and (21) become (35) and (36).

$$d_{Y+}^{FE} = 3V_{out}\sqrt{\frac{C}{T_S P}}\left[\frac{1 - \sqrt{cos^2\theta - \frac{sin\theta cos\theta}{\sqrt{3}}}}{\frac{1}{2}cos\theta + \frac{\sqrt{3}}{2}sin\theta}\right] + \frac{1}{2}d_{C+i} \quad (35)$$

$$d_{Y-}^{FE} = 3V_{out}\sqrt{\frac{C}{T_S P}}\sqrt{1 - \frac{tan\theta}{\sqrt{3}}} + \frac{1}{2}d_{C+i} \quad (36)$$

The duty ratio of positive throw of the switch S_Y given by (33) and (35) for Scheme EF and FE respectively, can be plotted for one complete sector, as illustrated in Figure 5. It can be observed that d_{Y+} is zero at the entry instant of Sector 1 which will be the case as phase B and phase C voltages are the same while the phase A voltage is twice the magnitude of phase B or phase C voltage. For simplicity, the term $\frac{1}{2}d_{C+i}$, is not shown in the plot. Another interesting observation from Figure 5 is that the scheme FE is segment of a straight line during Sector 1a when two out of the three phases are negative. On the other hand, Scheme EF is a straight line segment during Sector 1b when two out of the three phases are positive. This provides for an opportunity to adopt scheme FE for Sector 1a and scheme EF for Sector 1b for a simple implementation of duty ratio functions.

D. Summary

Table II shows the mapping of the phases X, Y and Z to the inverter phases, A, B and C for the complete sinusoidal output. The variation of the duty ratios of the different throws

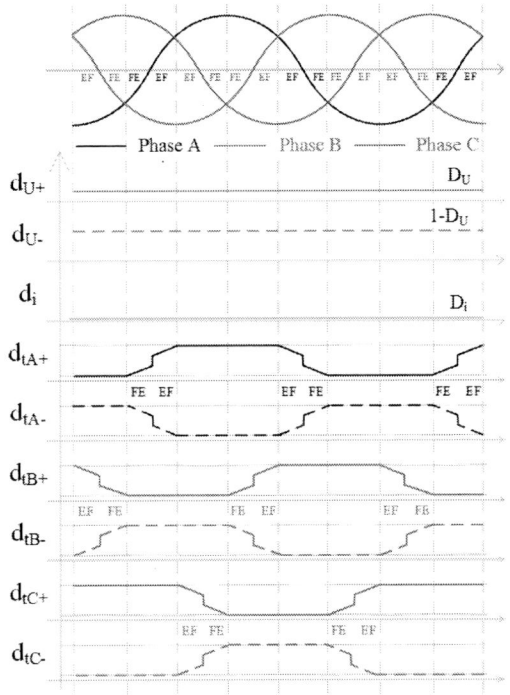

Fig. 6: Duty ratios of various throws of the converter under SEM

during all the sectors and sub-sectors are plotted in Figure 6. It can be noticed that the duty ratios of the various throws are maintained at a steady maximum value or a steady minimum value during certain sectors, depending on the particular phase voltages being most positive or most negative. During the time interval when the phase voltage lies between the most positive and the most negative phase voltages, the duty ratios follow the expressions (33) and (35), which approximate linearly increasing or decreasing functions of time, with a discontinuous jump that occurs at the zero crossings. In the table and figure, scheme FE is illustrated for a sub-sectors and scheme EF is illustrated for b sub-sectors. This choice is rather convenient although, either of the schemes FE or EF may be adopted without compromising waveform fidelity.

IV. DESIGN EXAMPLE AND SIMULATION RESULTS

In order to verify the proposed approach with tiny dc link capacitor using SEM, an example design of a photo-voltaic dc to 3.3 kW, 208V 3 phase ac converter application has been developed and compared with conventional PWM approach. The output LC filter components have been designed in order to limit the output voltage ripple to within 3%. Table IV shows the design parameters for the two cases for comparison.

TABLE II: Mapping of S_A-S_C switches to S_X, S_Y and S_Z

Sector No.	S_X	S_Z	S_Y	
			d_1	d_2
1a	Phase A: t_{A+} conducts	Phase C: t_{C-} conducts	Phase B: t_{B+} conducts	Phase B: t_{B-} conducts
1b			Phase B: t_{B-} conducts	Phase B: t_{B+} conducts
2a	Phase B: t_{B+} conducts	Phase C: t_{C-} conducts	Phase A: t_{A-} conducts	Phase A: t_{A+} conducts
2b			Phase A: t_{A+} conducts	Phase A: t_{A-} conducts
3a	Phase B: t_{B+} conducts	Phase A: t_{A-} conducts	Phase C: t_{C+} conducts	Phase C: t_{C-} conducts
3b			Phase C: t_{C-} conducts	Phase C: t_{C+} conducts
4a	Phase C: t_{C+} conducts	Phase A: t_{A-} conducts	Phase B: t_{B-} conducts	Phase B: t_{B+} conducts
4b			Phase B: t_{B+} conducts	Phase B: t_{B-} conducts
5a	Phase C: t_{C+} conducts	Phase B: t_{B-} conducts	Phase A: t_{A+} conducts	Phase A: t_{A-} conducts
5b			Phase A: t_{A-} conducts	Phase A: t_{A+} conducts
6a	Phase A: t_{A+} conducts	Phase B: t_{B-} conducts	Phase C: t_{C-} conducts	Phase C: t_{C+} conducts
6b			Phase C: t_{C+} conducts	Phase C: t_{C-} conducts

TABLE III: Design parameters of the reactive components for a 200V DC to 208V 3 phase AC boost-buck converter, switching at 10kHz

Parameters	Conventional PWM	Proposed SEM
Input Inductor L_i	$5mH$	$5mH$
Link Capacitor C	$1000\ \mu F$	$0.68\ \mu F$!
Output Inductor L_o	$7.5mH$	$7.5mH$
Output Capacitor C_o	$0.55\mu F$	$1.5\mu F$

The ripple current through the capacitors in the two design examples in order to meet the required load demand is about 11 A. This results in the voluminous size of the electrolytic capacitor in the case of conventional PWM example [13]. On the other hand, SEM can give the desired output waveforms with a film capacitor of only 0.68μF.

The operation of the converter was simulated in PLECS environment with the proposed SEM and conventional PWM, and selected results from the simulation of the converter are presented in Figures 7a and 7b respectively. The top traces of Figure 7 illustrate the variation of the three phase output voltages. The high quality of the output voltage in spite of the small size of dc link capacitor is readily apparent. The lower traces of Figure 7 illustrates the high frequency behavior of the input current and capacitor voltage. All these simulation results confirm the operation of the proposed modulation approach, and the design solutions.

V. EXPERIMENTAL RESULTS

In order to further validate the proposed approach, a laboratory scale proof of concept hardware prototype of a 100W, 9.5V DC to 24V AC inverter with design parameters L_i = $2mH$, C = $1\mu F$, L_o = $4mH$, C_o = $1\mu F$ and F_s = $7.8kHz$, operating under SEM, was developed. Figure 7c shows the corresponding waveforms of three phase output voltages, input current and capacitor voltage, that can be compared with the simulation waveforms presented in Figure 7a. The experimental results clearly validate the operation of the proposed SEM.

VI. LOSS AND VOLUME COMPARISON

A preliminary comparative analysis of the converter operation under SEM and PWM has been performed by extending the simulation in order to incorporate the thermal models of the reactive and active devices. Table IV lists the selected manufacturer part numbers of the device components for efficiency comparison. Figure 8 illustrates the loss percentages in the selected device components.

High speed single IGBT modules with continuous current rating of 15A, manufactured by Infineon Technologies, have been selected for the two designs. While in the case of conventional PWM inverter, the collector to emitter voltage rating can be 600V, SEM operated inverter requires a minimum of 1200V rated modules [14]. This leads to an expected increase in the conduction and switching losses in the case of 1200V rated modules. However, it can be observed that SEM facilitates zero-voltage switching. When an output phase leg is Phase X or Phase Z, it witnesses ZVS during half of the switching instances. When an output phase leg is Phase Y, it witnesses ZVS during 25% of the switching instances (Figure 3). Schottky Diodes from Cree have been selected for S_{U+} throw since the power transfer in the paper illustrated is unidirectional [15]. The conduction losses in the diodes in case SEM is higher due to higher voltage stress across the device.

The difference in the losses between the output LC filter components of the two designs is negligible as the RMS current and the voltage stress is similar in the two cases. On the other hand, the significant difference in terms of the link capacitor losses is clearly visible. SEM allows the use of a single film capacitor as opposed to using an electrolytic capacitor which has lower ESR. In conclusion, it can be observed that in the discussed example, the efficiency of SEM modulated inverter drops by only 0.25% despite of a higher voltage stress with a significantly reduced dc link capacitor size.

Table V illustrates the volume occupied by the device components in the two example inverter designs. Due to the higher voltage stress across the active components, the $IGBT$ modules occupy higher volume in the case of SEM. However, the link capacitor volume can be reduced by as much as

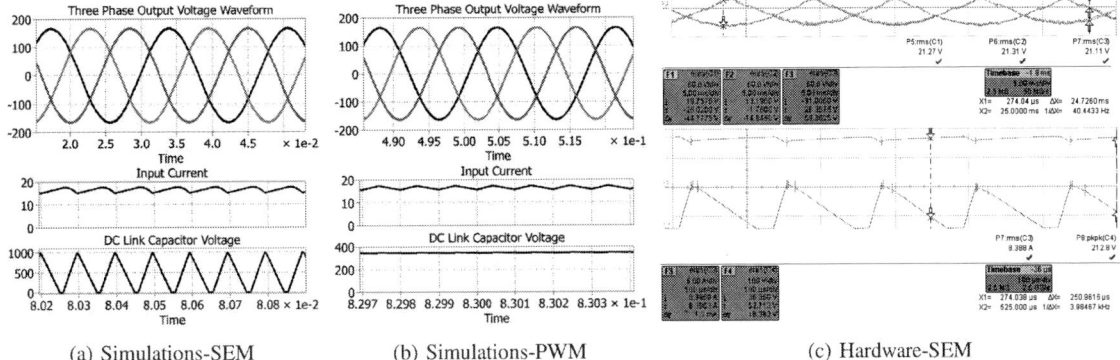

| (a) Simulations-SEM | (b) Simulations-PWM | (c) Hardware-SEM |

Fig. 7: Example dc to three phase inverter waveforms of three phase output, input current and dc link voltage from simulation model using (a) proposed SEM and (b) conventional PWM and (c) experimental prototype using SEM

TABLE IV: Selected device components for efficiency comparison

Device	Conventional PWM	Proposed SEM
Input Inductor L_i	DKIL-0231-1505	DKIL-0231-1505
Input Switch S_{U-}	IKP15N60T	IKW15N120H3
Input Diode S_{U+}	C3D10065E	C4D10120E
Link Capacitor C	ALS312102ND500	C4BSPBX3680Z
Output Switches S_o	IKP15N60T	IKW15N120H3
Output Inductors L_o	DKLP-0229-0430	DKLP-0229-0430
Output Capacitors C_o	C1825C274KBRACTU	CKG57NX7T2W155M500

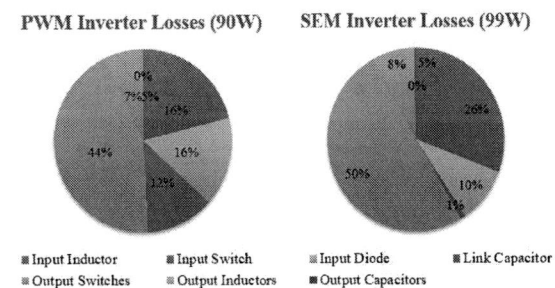

Fig. 8: Comparison of loss components between conventional PWM and proposed SEM

90%. A SEM modulated photo-voltaic inverter will hence occupy 20% less space as compared to PWM modulated inverters. More importantly, as power semiconductor technology advances, allowing for higher switching frequency and higher voltage devices simultaneously, the input and output filters can be reduced further, without having be dominated by the dc link capacitor sizing.

The operational life of film capacitors are typically much higher than that of electrolytic capacitors. In the case of the capacitor components selected for the discussed example, an electrolytic capacitor is expected to last for 5,000 operational hours at an ambient temperature of 85°C at rated ripple current. On the other hand, the operational hours for film capacitors at rated ac RMS voltage is 30,000 hours. A more detailed analysis of lifetime of the entire converter is beyond the scope of this paper, and will be developed in a future publication.

VII. CONCLUSIONS

This paper presents SEM as a radically different modulation strategy for three phase inverters that allows the energy storage in the dc link capacitor to be reduced by *several* orders of magnitude ($.1\mu F$ against $100 - 2000\mu F$). The proposed approach is based on careful determination of duty ratio of the switches that are nonlinear functions of load conditions, switching frequency, and dc link capacitor value, that is based on modulating the energy stored in the dc link. The

TABLE V: Volume comparison of the device compoents selected for comparison

Device	Conventional PWM (in cm^3)	Proposed SEM (in cm^3)
Input Inductor L_i	325.12	325.12
Input Switch S_{U-}	0.76	1.77
Input Diode S_{U+}	0.18	0.18
Link Capacitor C	349.25	33.20 !
Output Switches S_o	4.53	10.64
Output Inductors L_o	851.56	851.56
Output Capacitors C_o	0.35	0.50
Total volume	1531.74	1222.95

dc link voltage and energy storage returns to zero during each switching cycle. The waveforms corresponding to the modulator have been used to develop design equations for converter components and verified using experimental results presented in the paper.

Detailed analytical development of the modulation strategy is presented in the paper along with a preliminary comparative evaluation, simulation results and experimental results. Not surprisingly, the orders of magnitude reduction of capacitor size is accompanied by an increase in voltage stress across the dc link by a factor of about two to three. On balance, the proposed *SEM* leads to a significant reduction in size of the inverter and the dc link components. The proposed technique can replace the use of bulk electrolytic capacitors with tiny film capacitors which can increase the inverter lifespan.

The concept of SEM has been introduced for the case of a dc to three phase converter operating under a unity power factor load. However, the SEM approach can be extended to operation in rectifier mode, ac to ac converter and general power factor conditions with suitably adaptation of the algorithms. In general, SEM allows for a radical reexamination of the concept of 'dc' link, by dramatically relaxing the stiffness requirements of reactive elements that form the energy transfer process in power converters. Successful analytical modeling and application of the SEM principle along with suitable refinements are expected to lead to power converters with radically increased power density in the future.

ACKNOWLEDGMENT

The authors would like to thank the Wisconsin Electric Machines and Power Electronics Consortium (WEMPEC) for supporting the work at the University of Wisconsin-Madison.

REFERENCES

[1] T. Jahns and V. Blasko, "Recent advances in power electronics technology for industrial and traction machine drives," *Proceedings of the IEEE*, vol. 89, no. 6, pp. 963–975, Jun 2001.

[2] H. Deng, R. Oruganti, and D. Srinivasan, "Modeling anxd control of single-phase UPS inverters: A survey," in *International Conference on Power Electronics and Drives Systems, PEDS 2005.*, vol. 2, Nov 2005, pp. 848–853.

[3] F. Blaabjerg, K. Ma, and Y. Yang, "Power electronics - the key technology for renewable energy systems," in *2014 Ninth International Conference on Ecological Vehicles and Renewable Energies (EVER)*, March 2014, pp. 1–11.

[4] F. Bradaschia, M. Recife, Brazil Cavalcanti, P. Ferraz, F. Neves, and C. Diniz Neto, "Comparative study of topologies for three-phase transformerless photovoltaic systems," *Power Electronics Conference (COBEP), 2013 Brazilian*, pp. 493 – 500, Oct 2013.

[5] M. Olszewski, "Evaluation of 2004 Toyota Prius Hybrid Electric Drive System," Oak Ridge National Laboratory, U.S. Department of Energy Freedom, CAR and Vehicle Technologies, EE-2G 1000 Independence Avenue, S.W. Washington, D.C. 20585-0121, Tech. Rep. ORNL/TM-2006/423, may 2005.

[6] M. C. Cavalcanti, M. T. M. Neto, F. Bradaschia, L. R. Limongi, and E. Bueno, "Three-state three-phase Z-source inverter for transformerless photovoltaic systems," *Power Electronics Conference (COBEP), 2013 Brazilian*, pp. 509 – 516, oct 2013.

[7] J. Kolar, U. Drofenik, J. Biela, M. Heldwein, H. Ertl, T. Friedli, and S. Round, "PWM Converter Power Density Barriers," in *Power Conversion Conference - Nagoya, 2007. PCC '07*, April 2007, pp. P–9–P–29.

[8] R. Cuzner, D. Drews, and G. Venkataramanan, "Power Density and Efficiency Comparisons of System Compatible Drive Topologies," *IEEE Transactions on Industrial Electronics*, vol. PP, no. 99, pp. 1–1, 2014.

[9] T. Friedli, S. Round, D. Hassler, and J. Kolar, "Design and performance of a 200-kHz all-SiC JFET Current DC-Link Back-to-Back Converter," *IEEE Transactions on Industrial Electronics*, vol. 45, no. 5, pp. 1868–1878, Sept 2009.

[10] J. Xu and Y. Sato, "An investigation of minimum DC-link capacitance in PWM rectifier-inverter systems considering control methods," in *2012 IEEE Energy Conversion Congress and Exposition (ECCE)*, Sept 2012, pp. 1071–1077.

[11] J. Holtz, "Pulsewidth modulation-A survey," *IEEE Transactions on Industrial Electronics*, vol. 39, no. 5, pp. 410–420, Oct 1992.

[12] J. Holtz, "Pulsewidth modulation for electronic power conversion," *Proceedings of the IEEE*, vol. 82, no. 8, pp. 1194–1214, Aug 1994.

[13] "KEMET Electronics Corporation, Screw Terminal Aluminum Electrolytic Capacitors," http://www.kemet.com/Lists/ProductCatalog/Attachments /389/KEM-A4031-ALS30-31.pdf, sep 2015, pg. 10.

[14] "Infineon Technologies, Short Form Catalog - High Power Semiconductors for Industrial Applications," http://www.infineon.com/dgdl?folderId=db3a304412b407 950112b4095af601e2&fileId=db3a30431a47d73d011a52 9661267822, 2015, pg. 12-13.

[15] "Cree, SiC Schottky Diodes," http://www.wolfspeed. com/Power/ProductsSiCSchottkyDiodes, 2015.

Drive Circuits for Ultra-fast and Reliable Actuation of Thomson Coil Actuators used in Hybrid AC and DC Circuit Breakers

Chang Peng,
Alex Huang and Iqbal Husain
FREEDM Systems Center
North Carolina State University
Raleigh, NC 27606, USA
Email: cpeng3@ncsu.edu

Bruno Lequesne
E-Motors Consulting, LLC
Menomonee Falls, WI 53051, USA
Email: bruno.lequesne@ieee.org

Roger Briggs
Energy Efficiency Research, LLC
Colgate, WI 53017, USA
Email: roger.briggs@eer-llc.com

Abstract—Thomson coil actuators (also known as repulsion coil actuators) are well suited for vacuum circuit breakers when fast operation is desired such as for hybrid AC and DC circuit breaker applications. This paper presents investigations on how the actuator drive circuit configurations as well as their discharging pulse patterns affect the magnetic force and therefore the acceleration, as well as the mechanical robustness of these actuators.

Comprehensive multi-physics finite-element simulations of the Thomson coil actuated fast mechanical switch are carried out to study the operation transients and how to maximize the actuation speed. Different drive circuits are compared: three single switch circuits are evaluated; the pulse pattern of a typical pulse forming network circuit is studied, concerning both actuation speed and maximum stress; a two stage drive circuit is also investigated. A 630 A, 15 kV / 1 ms prototype employing a vacuum interrupter with 6 mm maximum open gap was developed and tested. The total moving mass accelerated by the actuator is about 1.2 kg. The measured results match well with simulated results in the FEA study.

Index Terms—Fast mechanical switch, Hybrid circuit breaker, DC circuit breaker, Thompson coil actuator, repulsion coil actuator

I. INTRODUCTION

The past two decades have witnessed a resurgence of research in high voltage DC circuit breakers [1]. In a manner that departs from high voltage DC (HVDC) transfer switches that are used in classical HVDC systems [2, 3], voltage source converter (VSC) based HVDC systems requires much faster protections in order to provide high reliability and availability of power transfer in the HVDC networks [4, 5]. Among the proposed schemes, the so-called "hybrid configuration" which combines mechanical and electronic switches is the most promising [4, 6–8]. Although driven by the needs in DC applications, such concepts could also be of advantage for AC systems. One of the most desirable features of the

fast mechanical switches (FMS) for hybrid circuit breakers is a fast opening speed, which led to selecting the Thomson coil actuator design over other electro-mechanical actuator configurations.

A. Operation principle of the Thomson coil actuator

The FMS employing a Thomson coil actuator has four main parts, as shown in Fig. 1: the interrupter, the operating mechanism, the energy storage and control circuit, and the damping and holding mechanism. Referring to the control circuit in Fig. 1, when the FMS is to open, the control switch (the SCR in Fig. 1) turns on and allows the capacitor bank to discharge through the opening coil (the upper one in the figure). The fast rising discharge current in the coil induces current in the conductive (typically copper) disk which results in a strong repulsive force between the coil and copper disk. The force drives a copper disk downwards and opens the FMS. The role of the diode in this particular circuit is to limit the rate of the current decay after it has peaked. This opening movement is stopped by a disc spring. The closing operation is accomplished in a similar way by turning on a second switch, not shown in this figure, that controls the discharge through the closing coil (the lower one), then the copper disk moves upward to close the contacts.

B. Problem description

One of the key challenges of the FMS is to actuate the moving mass as fast as possible. Fig. 2 - 6 show a number of drive circuits described in the literature for the repulsion coil application (Fig. 1 and the above description used the simplest of these drive circuits, Circuit 1, shown in Fig. 2). Both simulations and experiments indicate that a capacitor bank charged to a higher voltage drives the moving mass to a higher speed [9–17]. However, this is only a reflection of systems starting with more stored energy. What still needs to be elucidated is the effectiveness of the energy transfer from storage to motion. For instance, the peak force may be as important a parameter as force rise, and duration of that

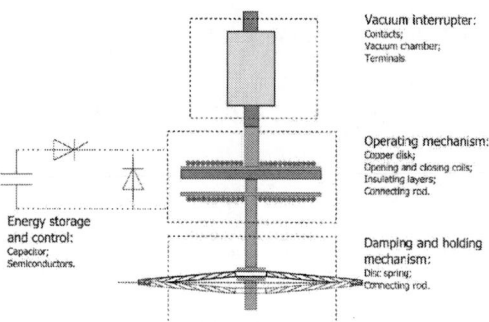

Fig. 1: Structure of the FMS.

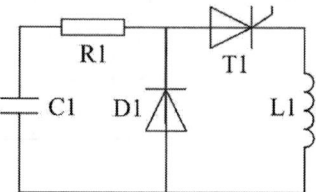

Fig. 3: Circuit 2, the second single pulse drive circuit.

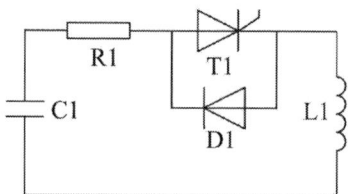

Fig. 4: Circuit 3, the bidirectional drive circuit.

peak, in terms of inducing fast motion. Furthermore, a fair design comparison must be based on similar ratings of both the mechanical and electrical components, some of which are rated based on peak values. This is the case with the peak force which generally corresponds to the peak stress level in the moving parts. It is also critical to use the stored energy wisely, to minimize the overall size of the energy storage, the rating of the control components and the dimensions of the actuator structure. The focus here is on design efficacy, that is, how to get the fastest motion from a given design and set of parameter and ratings. It is not a question of efficiency, understood as energy used per motion, which is not of much concern since breakers operate very infrequently. Therefore this paper aims to investigate different drive circuits regarding the aforementioned considerations.

Most implementations of Thomson coil based FMS [9–16, 18–20], employ a single pulse circuit, perhaps because of simplicity. Its variations include Circuit 1, 2 and 3 as shown in Fig. 2 to 4. But reference [21] has proposed a pulse forming network as shown in Fig. 5 (Circuit 4), to generate multiple pulses into the coil. Reference [22] has developed a fast acting circuit breaker using Circuit 5 shown in Fig. 6, which is called "two-stage fast actuator power supply" using only one control switch but two capacitors pre-charged to different voltage levels. Circuit 4 and 5 have more components and thus require more detailed optimization. Because of the problem complexity, this paper investigates the problems by finite element analysis (FEA); and the FEA model has been validated with experimental measurement.

II. FEA MODELING

The finite element modeling of the actuator consists of a coil as the stationary part, and a shaft and copper disk as the moving part. The actuator is then co-simulated with an external drive circuit with lumped parameters and sequence control. The drive circuit can be controlled by switches to discharge the coil in desired patterns. Electromagnetic, thermal, and structural physics are coupled into the model, and the multiphysics equations are solved simultaneously in time domain. The software used is COMSOL Multiphysics. For additional details regarding the method of the FEA modeling, the readers are referred to [17].

The system baseline consists of a copper disk with outer diameter of 80 mm and a thickness of 4 mm. The coil has an inner diameter of 12 mm with 11 turns of 2.31 mm by 4.62 mm wires. The gap between the disk and coil is 3 mm. The payload of the actuator is 0.7 kg; and the total moving mass is about 1.2 kg. The geometry is built in axis-symmetrical two-dimensions because of the symmetry of the actuator structure. This reduces the required simulation time for this complex multi-physics problem. To obtain accurate results, the mesh is refined and more than 40,000 elements are created; the time domain step size is 2 us. Fig. 7 shows a typical three-dimensional plot of current density for the actuator at a given position and moment. Fig. 8 corresponds to a stress distribution in the moving structure.

III. SINGLE CAPACITOR DRIVE CIRCUITS

Circuits that deliver a single pulse into the coil, shown in Fig. 2 - 4 are most common in the prior art. There are some subtle differences between Circuit 1 and 2; Circuit 3 is quite different as the current oscillates bidirectionally. Nevertheless, these differences have not been discussed by earlier publications and will be investigated in this section.

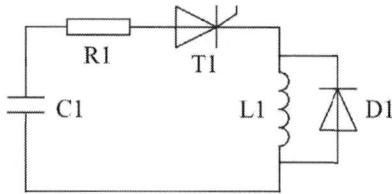

Fig. 2: Circuit 1, the first single pulse drive circuit.

978-1-4673-9551-9/16 $31.00 © 2016 IEEE

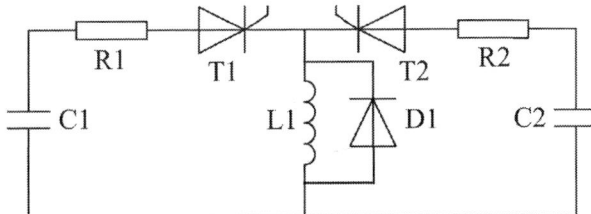

Fig. 5: Circuit 4, the pulse forming network drive circuit.

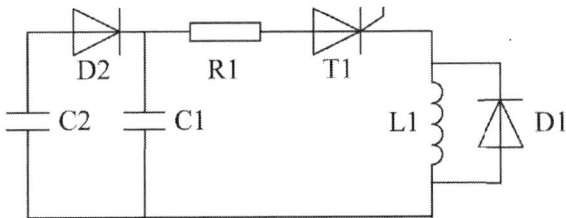

Fig. 6: Circuit 5, the two-stage drive circuit.

A. Simulation

Circuit 1, 2 and 3 were all modeled with the same components. C1, 2 mF and pre-charged to 300 V, represents the capacitor bank, charged with the same amount of energy in all 3 cases; R1, 0.002 Ω, represents the equivalent series resistance (ESR) of the capacitor; T1 and D1 are the thyristor switch and diode, respectively. Both T1 and D1 are modeled as a voltage drop in series with a certain resistance. By FEA simulation, the operation transients for all three single pulse drive circuits are compared in Fig. 9 and 10.

B. Circuit 1 and 2 - unidirectional circuits

Based on Fig. 9 and 10, it is apparent that the performance of the actuator driven by Circuit 1 is slightly better than Circuit 2. In both Circuit 1 and 2, the current and force rises and peaks are the same. After the current peak, its decay is limited by the voltage drop in the diode as in Circuit 1 and in a similar manner in the diode and thyristor in series as in Circuit 2. Therefore the current decays slightly faster in Circuit 2, and the control switch in Circuit 2 needs to withstand more current surge than in Circuit 1. This is because, after the current peaks in Circuit 1, the voltage across the coil is equal to the diode voltage drop, thus forcing the current to zero in the capacitor/thyristor branch of the circuit and making the coil current flow exclusively through the diode. While a similar voltage limitation across the coil is in operation in Circuit 2, it does not lead to an interruption of the current in the thyristor switch. Considering the high current levels in such applications, it is therefore desirable that the current stress in the control switch should be minimized as is achieved in Circuit 1.

C. Circuit 3 - a bidirectional circuit

In contrast to Circuit 1 and 2, Circuit 3 generates a damped sinusoidal current in the coil, see solid green curve in Fig. 9.

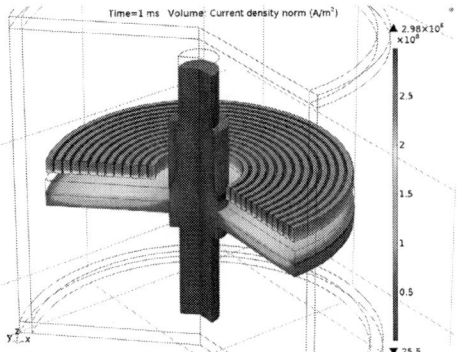

Fig. 7: FEA simulation model, 3-D view.

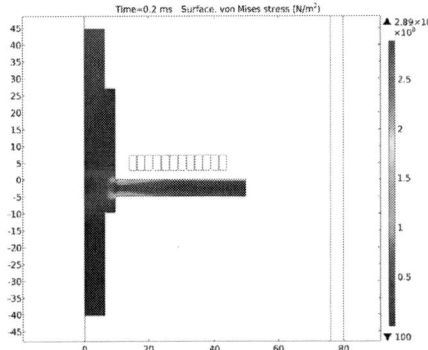

Fig. 8: Von Mises stress distribution, 2-D view.

It should be noted that the polarity of the current in the solid disk follows that of the coil current (with some delay), so there is a repulsion force most or all of the time, regardless of coil current polarity. Accordingly, as the current goes from positive to negative and back, multiple force pulses, with decreasing magnitudes, are observed, see broken green line in Fig. 9. The force in this case is therefore periodically higher and periodically lower than those observed with Circuits 1 and 2. However, overall, the effect of the multiple force pulses on the acceleration is weaker than with Circuits 1 and 2, as seen in Fig. 10. The forces are the same until the peak in all 3 cases, but with Circuit 3, the peak is followed by a temporary, but sharp decrease in the force during which the disk loses momentum, a loss it never recovers from completely despite the higher forces later on. Concerning the circuit design, the bidirectional current requires that the capacitor be unpolarized, an added cost for Circuit 3. Further, the force pattern may induce vibrations in the disk that are largely avoided with Circuits 1 and 2.

IV. PULSE FORMING NETWORK (PFN) DRIVE CIRCUIT

The pulse forming network (PFN) shown in Fig. 5 generates multiple pulses into the coil. In essence, this circuit decouples the peak and frequency of the current waveform, whereas these two parameters are linked in Circuits 1 to 3. In general, there can be as many pulses as possible in a PFN to drive the

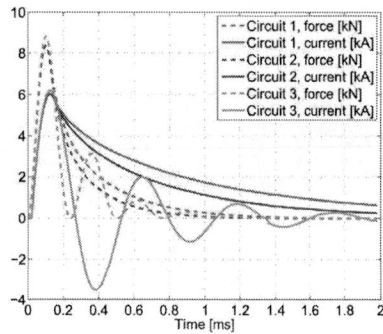

Fig. 9: Currents and forces of Circuit 1, 2, 3.

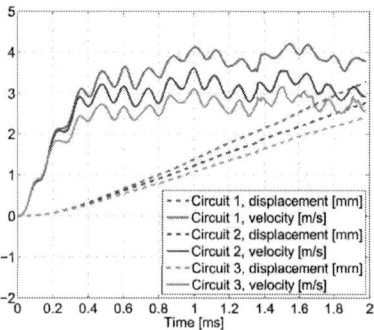

Fig. 10: Displacements and velocities of Circuit 1, 2, 3.

actuator. However, a two-pulse circuit is sufficient to show their differences to other circuit, and there fore is evaluated in this section.

A. Simulation

A set of simulations was conducted to find out how the delay between two discharge pulses in a PFN with two capacitors (C1 and C2) affects the transients behavior of the actuation. Both C1 and C2 are 1 mF and pre-charged to 300 V, therefore the total stored energy is the same as in Circuits 1 to 3. In stead of dumping the energy into the coil in the same time, C1 and C2 are separately controlled by two thyristor switches, T1 and T2 as shown in Fig. 5. The delay between firings of T1 and T2 are swept from 0 to 250 us.

B. Discussion

Fig. 13 plots motion versus time for various values of delay in the firing of Thyristor T2, with a delay of 0 making Circuit 4 equivalent in principle to Circuit 1. One can see in the figure that, from the displacement point of view, the delay does not help; on the contrary, the delay slows down the displacement compared with the circuit with no delay. At the same time, the forces profiles shown in Fig. 12 and the currents profiles shown in Fig. 11 also indicate that peak force and peak current is higher with no delay, which is consistent with the faster acceleration.

While faster accelerations are obviously desirable, they can come at a price if the higher force is much larger, in proportion, than the motion is faster. In other words, this raises the question of the effectiveness of the system to provide a given acceleration and a given motion pattern for a set level of peak rating, both on the electrical and mechanical components. In this example, when driven with no delay, the peak force is 9.4 kN and the device travels to 4 mm in 2 ms; with a 150 us delay, the peak force is 7.5 kN and the device travels to 3.7 mm at 2 ms. Comparing these two cases, if driven with no delay, the peak force is more than 25% higher, but the displacement is less than 10% faster. Accordingly, from the mechanical point of view the maximum stress levels also increase significantly, Fig. 14, approximately in proportion of the peak force. From a system design point of view, a fair comparison must be made with comparable rating levels, including similar mechanical stress levels. Consequently, a design with a single pulse may require a reinforced, thus heavier, moving disk as compared to a system with a PFN circuit, which leads to a larger moving mass. This will result in a slower acceleration, especially if the mass of the moving disk is important compared to that of the payload. Similarly the electrical component rating must be increased thus leading to higher cost.

In conclusion, circuits with multiple control switches and discharging pulses would not necessarily result in faster actuation than a single discharge circuit when the total energy stored in the capacitors are the same. Allowing the energy to be delivered into the actuator in a few steps helps reduce the maximum stress and therefore improve the mechanical reliability of the switch unit. In the same manner, even if the mechanical design is strong enough, releasing higher energy in a single shot would generate tens of kA current that could exceed the rating of the most powerful thyristor switches available and therefore require more complicated designs.

V. TWO-STAGE DRIVE CIRCUIT

Kimblin [22] has proposed a fast acting circuit breaker using Circuit 5, which is called "two-stage fast actuator power supply", shown in Fig. 6. In that paper, C1 is 110 uF pre-charged to 5 kV while C2 is 2 mF pre-charged to 2 kV. This two stage drive circuit can be considered as a special form of the PFN: the delay of the second discharge is dictated by the discharge time of the first capacitor. The second capacitor C2 is blocked by Diode D2 until the voltage of the first capacitor C1 falls below the voltage across C2. Since Circuit 5 uses 1 thyristor and 2 diodes, as opposed to 2 thyristors and 1 diode in Circuit 4, Circuit 5 is smaller and cheaper than Circuit 4.

A. Simulation

In this paper, FEA simulations were conducted to explore how this two capacitors should be selected to optimize the operation of the actuator and how the performance differs with different combinations of those two capacitors. In the simulation, Capacitor C2 of the two stage drive circuit is assumed 1 mF and pre-charged to 300 V, while Capacitor C1 has a variable capacitance but pre-charged with a voltage such

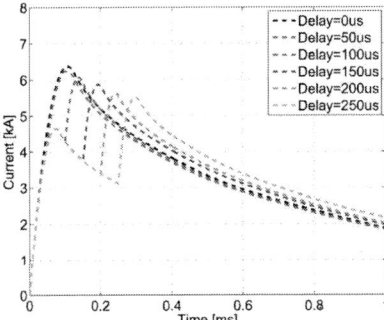

Fig. 11: Currents when driven by PFN with different delays.

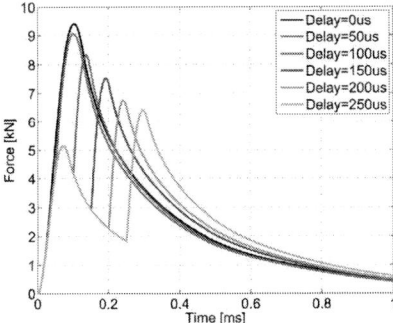

Fig. 12: Forces when driven by PFN with different delays.

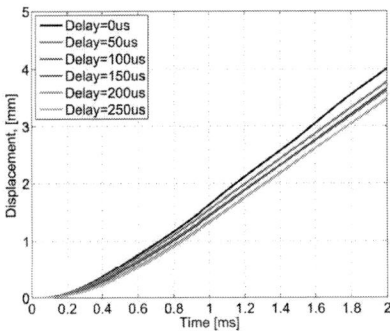

Fig. 13: Displacements when driven by PFN with different delays.

that the stored energy in C1 is twice as much as that in C2. C1 varies from 2 mF to 0.125 mF, and the voltage U1 across C1 from 300 V to 1200 V. When C1 is 2 mF and U1 is 300 V (green curves in Fig. 15 and 16), this is equivalent to Circuit 1 with a 3 mF capacitor pre-charged to 300 V.

B. Discussion

The simulation results shown in Fig. 15 and 16 indicate that for the setup above, the currents and forces profiles shift to the right (later times) with lower peak values when C1 increases and U1 decreases. It is also observed that a higher U1 results in

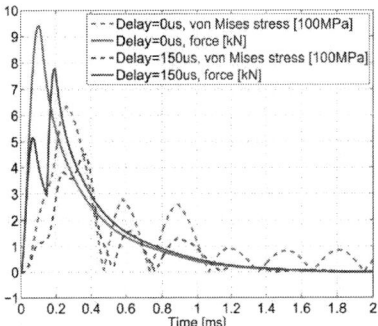

Fig. 14: Mechanical stresses when driven by PFN with different delays.

slightly higher current, but significantly larger force. However, after reaching the peaks, both the current and force drop sooner in the case with a lower C1 and a higher U1 than in the case with a larger C1 and a lower U1. This results in different accelerations in the periods before and after their peak values in each of the cases with different C1 and U1 combinations. In other words, the combinations that lead to higher current and force peaks also lead to earlier, and narrower, peaks.

In order to determine the impact of these force patterns on motion, two plots are presented: Fig. 17 with the entire displacement for all the cases studied, and Fig. 18 with a zoom-in view for two of these cases: Case 1, C1=0.125 mF, U1=1200 V; Case 2, C1=2 mF, U1=300 V (Case 2 is also equivalent to Circuit 1). Referring first to Fig. 17, motion is affected but not very much across these various cases. Since one of them (Case 2, green curves) also corresponds to Circuit 1, a first conclusion is that the complexity of Circuit 5 will result in only minor improvement in performance, if improved at all.

Comparing the various cases studied for Circuit 5, a first impression could be that higher capacitance and lower voltage values lead to both lower stresses (Fig. 15 and 16) and faster motion (at 2 ms on Fig. 17). However, the motion pattern is more complex, as revealed by Fig. 18 which shows that these motion curves actually cross one another. For Case 1 and 2, the cross over occurs at round time 0.9 ms. This indicates that combinations with a higher U1 and lower C1 do give a faster initial acceleration, but later on the acceleration becomes slower.

For a given actuator mechanical design, the values of C1, U1, C2 and U2 will give different displacement curves. Some values will produce faster initial movement with slower ending speeds, while other values will result in slow initial displacement followed by faster acceleration later. This can be seen in Fig. 18 where the plots of two sets of values cross. Since each design will have a time target, the values for capacitances and voltages should be selected based on which gives the greater displacement at that time target while still staying within the stress limits of all the components.

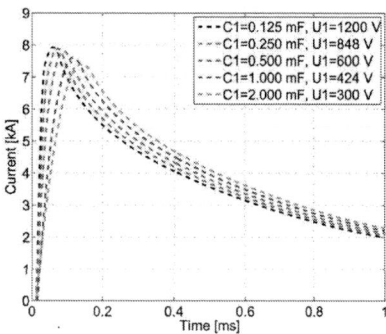

Fig. 15: Currents of Circuit 5 with different C1 and U1.

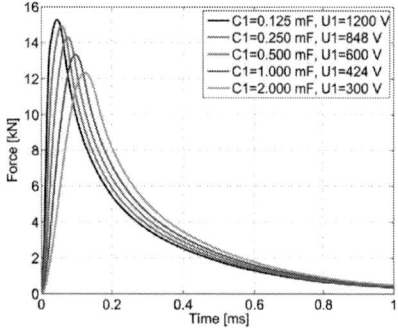

Fig. 16: Forces of Circuit 5 with different C1 and U1.

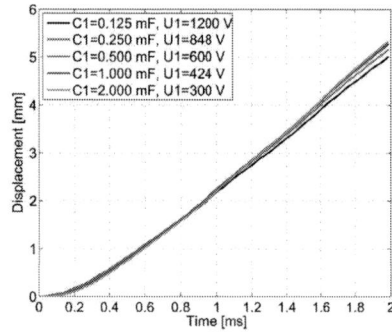

Fig. 17: Displacements of Circuit 5 with different C1 and U1.

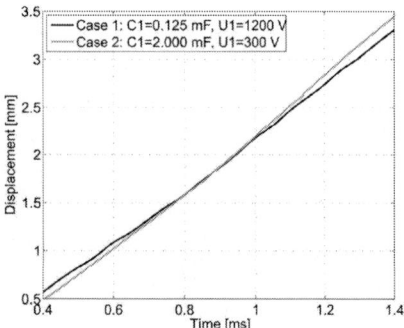

Fig. 18: Zoomed-in displacements of Case 1 and Case 2.

Circuit 5 is more likely to be favorable when the time target is really short. It will give slightly faster acceleration within that time target at the cost of an additional diode and a increased stress level. In Case 1 and 2, for example, if the time target is 0.6 ms, Case 1 with Circuit 5 gives a faster acceleration; if the time target is 1.2 ms, however, Case 2 with Circuit 1 shows better performance.

VI. EXPERIMENTAL VERIFICATION

A prototype was fabricated to validate the accuracy of the model and simulation. It is shown in Fig. 19. Details regarding the electrical and structural design can be found in [23]. The prototype has the same dimensions and materials used in the FEA model.

Fig. 19: FMS prototype assembly.

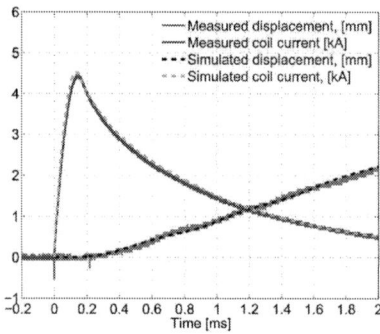

Fig. 20: Simulation and measurement results.

This model in the finite element domain has been verified by measurements. Fig. 20 shows the comparison of the test results of the prototype driven by a 2 mF capacitor bank pre-charged to 300 V, and the simulation results obtained from FEA simulation of the modeled prototype driven based on Circuit 1. Both the discharging current in the coil and the displacement of the movable parts match well with each other in the simulation and experiment. The current peaks at 4.5 kA at about 160 us and then drops slowly down to 1.5 kA at 1 ms, and 0.5 kA at 2 ms. The movable parts start moving at around 200 us; they reach 0.9 mm at 1 ms and 2.2 mm at 2 ms. The experimental data has verified the FEA model presented in the previous sections and therefore the study of drive circuit analysis based on such model is supported.

VII. CONCLUSIONS

This paper has presented an investigation of different electric drive circuits used in Thomson coil actuators. Five different drive circuits are evaluated. Comprehensive FEA simulation, confirmed by experimental studies, have revealed that:

1) The circuits come in two broad categories, single and multiple pulse circuits. With two or more pulses, the PFN decouples the peak value and the frequency of the current pulse, providing an additional degree of freedom for the designer. With slower peaks, the PFN circuits result in slightly lower actuation speed compared to single pulse circuits, with the same amount of stored energy. However, they significantly reduce the mechanical stress on the structure, as well as the peak current the electrical components are exposed to. Whether single of multiple pulses is better depends therefore on rating margins. For instance if the load being moved is large compared to the moving disk, the disk can be made heavier to sustain higher forces, resulting in single-pulse circuits being better, and vice versa.

2) The single switch circuits such as Circuit 1 and Circuit 2 with a single pulse can drive the actuator to a comparable speed; however, the control switch in Circuit 1 is exposed to less current surge stress.

3) Circuit 3 generates a bidirectional current. It does not drive the actuator any faster than Circuits 1 and 2. The current oscillation can cause vibration of the moving structure and the circuit requires an unpolarized capacitor.

4) The PFN drive circuit is an attractive solution when the discharge current expected from a single pulse exceeds the current rating of the available semiconductor switches. The reduction in peak force can also be advantageous in reducing the moving disk mass, for a given stress level.

5) The two-stage drive circuit is more complex than the single-pulse circuit options. It will provide slightly better performance if only in specific situations where travel is particularly short and the moving disk can be reinforced to accommodate higher stress levels, without impacting acceleration in a significant way.

6) The FEA simulation has been verified by a prototype. The measured movement and current transients match very well with the simulated results based on the same FEA model used in the drive circuit study.

REFERENCES

[1] C. M. Franck, "HVDC circuit breakers: A review identifying future research needs," *Power Delivery, IEEE Transactions on*, vol. 26, no. 2, pp. 998–1007, 2011.

[2] C. Peng, J. Wen, G. Ma, X. Wang, Z. Liu, and K. Yu, "Analysis and simulation on current commutation of the DC transfer switches in UHVDC transmission systems," *Proceedings of the Chinese Society for Electrical Engineering*, vol. 31, no. 36, pp. 1–7, 2011.

[3] C. Peng, J. Wen, W. Xiuhuan, Z. Liu, Z. Shen, and K. Yu, "Development of DC Transfer Switch in Ultra High Voltage DC Transmission Systems," *Proceedings of the Chinese Society for Electrical Engineering*, vol. 16, p. 020, 2012.

[4] J. Häfner and B. Jacobson, "Proactive Hybrid HVDC Breakers-A key innovation for reliable HVDC grids," *CIGRE paper*, vol. 264, 2011.

[5] C. Peng and A. Q. Huang, "A protection scheme against DC faults VSC based DC systems with bus capacitors," in *Applied Power Electronics Conference and Exposition (APEC), 2014 Twenty-Ninth Annual IEEE*, March 2014, pp. 3423–3428.

[6] A. Burnett, C. Oates, and C. Davidson, "High voltage dc circuit breaker apparatus," Sep. 6 2013, WO Patent App. PCT/EP2012/053,574.

[7] Y. Wang and R. Marquardt, "Future HVDC-grids employing modular multilevel converters and hybrid DC-breakers," in *Power Electronics and Applications (EPE), 2013 15th European Conference on*, 2013, pp. 1–8.

[8] C. Peng, A. Q. Huang, M. A. Rezaei, X. Huang, and M. Steurer, "Development of Medium Voltage Solid-State Fault Isolation Device for Ultra-Fast Protection of Distribution System," in *The 40th Annual Conference of the IEEE Industrial Electronics Society*. IEEE, Oct. 2014.

[9] S. Basu and K. Srivastava, "Analysis of a fast acting circuit breaker mechanism part i: Electrical aspects," *Power Apparatus and Systems, IEEE Transactions on*, no. 3, pp. 1197–1203, 1972.

[10] ——, "Analysis of a fast acting circuit breaker mechanism, part ii : Thermal and mechanical aspects," *Power Apparatus and Systems, IEEE Transactions on*, vol. PAS-91, no. 3, pp. 1203–1211, May 1972.

[11] R. Rajotte and M. G. Drouet, "Experimental analysis of a fast acting circuit breaker mechanism: Electrical aspects," *Power Apparatus and Systems, IEEE Transactions on*, vol. 94, no. 1, pp. 89–96, 1975.

[12] A. M. S. Atmadji, "Direct current hybrid breakers: A design and its realization," Ph.D. dissertation, Technische Universiteit Eindhoven, Eindhoven, The Netherlands, May 2000.

[13] B. Roodenburg, A. Taffone, E. Gilardi, S. Tenconi, B. Evenblij, and M. Kaanders, "Combined zvs–zcs topology for high-current direct current hybrid switches: design aspects and first measurements," *Electric Power Applications, IET*, vol. 1, no. 2, pp. 183–192, 2007.

[14] M. Tsukima, T. Takeuchi, K. Koyama, and H. Yoshiyasu, "Development of a high-speed electromagnetic repulsion mechanism for high-voltage vacuum circuit breakers," *Electrical Engineering in Japan*, vol. 163, no. 1, pp. 34–40, 2008.

[15] A. Bissal, J. Magnusson, E. Salinas, G. Engdahl, and A. Eriksson, "On the design of ultra-fast electromechani-

cal actuators: A comprehensive multi-physical simulation model," *Electromagnetic Field Problems and Applications (ICEF), 2012 Sixth International Conference on*, pp. 1–4, June 2012.

[16] V. Puumala and L. Kettunen, "Electromagnetic design of ultrafast electromechanical switches," *Power Delivery, IEEE Transactions on*, vol. 30, no. 3, pp. 1104–1109, June 2015.

[17] C. Peng, I. Husain, and A. Huang, "Evaluation of Design Variables in Thompson Coil based Operating Mechanisms for Ultra-Fast Opening in Hybrid AC and DC Circuit Breakers," in *Applied Power Electronics Conference and Exposition, Thirtieth Annual IEEE*, 2015.

[18] Y. Kishida, K. Koyama, H. Sasao, N. Maruyama, and H. Yamamoto, "Development of the high speed switch and its application," in *Industry Applications Conference, 1998. Thirty-Third IAS Annual Meeting. The 1998 IEEE*, vol. 3, Oct 1998, pp. 2321–2328 vol.3.

[19] M. Steurer, K. Frohlich, W. Holaus, and K. Kaltenegger, "A novel hybrid current-limiting circuit breaker for medium voltage: principle and test results," *Power Delivery, IEEE Transactions on*, vol. 18, no. 2, pp. 460–467, 2003.

[20] J.-M. Meyer and A. Rufer, "A DC hybrid circuit breaker with ultra-fast contact opening and integrated gate-commutated thyristors (IGCTs)," *Power Delivery, IEEE Transactions on*, vol. 21, no. 2, pp. 646–651, 2006.

[21] Z. Wang, J. He, X. Yin, J. Lu, D. Hui, and H. Zhang, "10kV High Speed Vacuum Switch With Electromagnetic Repulsion Mechanism," *Transactions of China Electrotechnical Society*, vol. 24, no. 11, pp. 68–75, 2009.

[22] C. W. Kimblin, "Development of current limiter using vacuum arc current commutation. Phase 2: Maximizing the current rating of a single 72kV device using a minimum amount of parallel capacitance." *NASA STI/Recon Technical Report N*, vol. 77, p. 33427, Oct. 1979.

[23] C. Peng, I. Husain, A. Huang, B. Lequesne, and R. Briggs, "A Fast Mechanical Switch for Medium Voltage Hybrid DC and AC Circuit Breakers," in *IEEE Energy Conversion Conference and Expo, 2015*, 2015.

Improved Transformerless Dual Buck Inverters with Buffer Inductors

Liwei Zhou, Feng Gao
School of Electrical Engineering
Shandong University
Jinan, China
18769785783@163.com

Abstract—**In an inversion system, high reliability is one of the main targets pursuing. Some problems will threaten the reliability of the system, such as the shoot through issue and the failure of reverse recovery. The dual buck inverters can solve the above problems without adding dead time but the low magnetic utilization increases the volume and weight of the system. This paper firstly reviews the traditional dual buck topologies and a single inductor dual buck inverter which can make full use of the inductance. Then a kind of H3 dual buck phase leg with buffer inductor is proposed to improve the reliability of the MOSFET inverter. The novel method maintains the dual buck topologies' advantage of high reliability and can make full use of the inductance. Also, compared to the single inductor dual buck topology, the novel method can achieve lower conducting loss and simpler controlling strategy. Finally, the simulation and experimental results verified the theoretical analysis.**

Keywords—dual buck topology; reverse recovery; buffer inductor; common mode voltage

I. INTRODUCTION

The rapid development of the clean energy power generation requires the inverters to be more and more reliable. Yet shoot through problem is a major threaten to the reliability. As is known, a traditional method to solve the shoot through issue is by setting dead time. However, the dead time will cause a distortion of the output current. Also, during the dead time, the current may flow through the body diode of the switch which can cause the failure of the reverse recovery [1].

In order to solve the above problems, the dual buck topologies are proposed in a lot of research. By combining two unidirectional buck circuits, the dual buck inverters can avoid shoot through problem and the freewheeling current will flow through the independent diodes which can solve the reverse recovery problem of the MOSFET's body diodes. However, the major drawback of the dual buck topologies is the magnetic utilization. Only half of the inductance is used in the working modes which increase the weight and volume of the system [2]-[4].

To improve the magnetic utilization of the dual buck inverter, a kind of single inductor dual buck topology was proposed [5]. Two extra switches are applied in the inverter. The single inductor topology can make full use of the

inductance, but the conducting loss is largely increased because four switches are flown through during the power delivering modes. Also, a kind of HERIC-based inverter is proposed in [6] to improve the reliability and magnetic utilization of the inversion system.

This paper proposed a novel H3 dual buck phase leg with buffer inductor to highly improve the reliability of the inverter, especially for the MOSFET inverter [6]. Applying the proposed phase leg to the full bridge inverters, the novel topologies have the following advantages: firstly, maintain the reliability of the traditional dual buck inverters, secondly, make full use of the inductance which is close to the single inductor dual buck topology, thirdly, the proposed inverters cost less switches than the single inductor topology, thus the novel topologies will achieve a lower conducting loss and a simpler controlling strategy. The simulation and experimental results

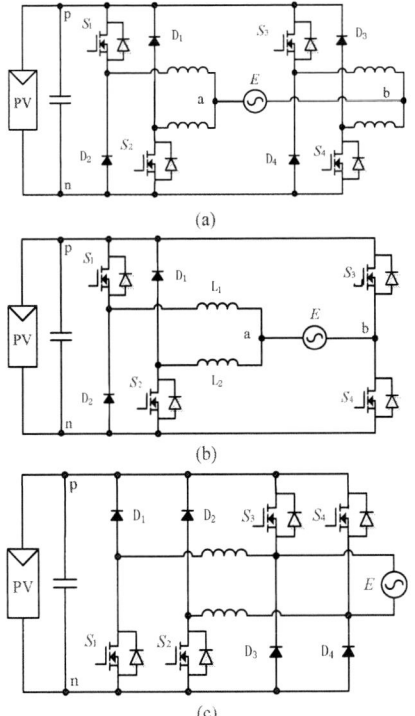

(a)

(b)

(c)

Fig. 1. Traditional dual buck and dual boost full bridge inverters.

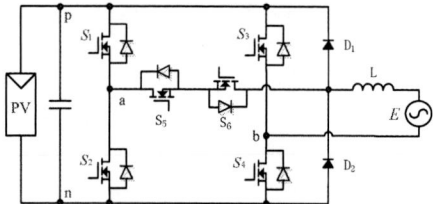

Fig. 2. Dual buck full bridge inverter with single inductor.

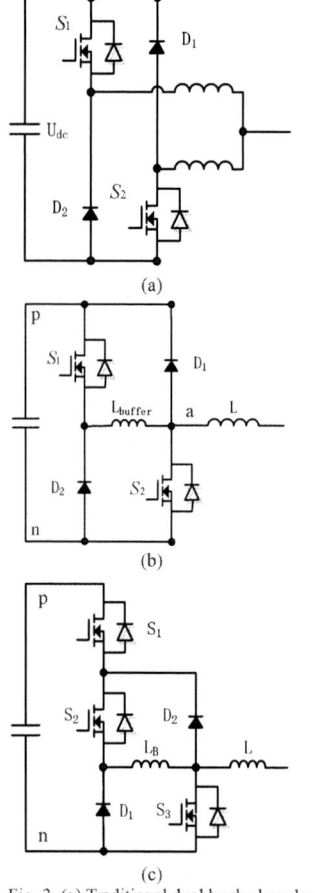

(a)

(b)

(c)

Fig. 3. (a) Traditional dual buck phase leg (b) phase leg in [6] (c) Novel dual buck H3 phase leg with buffer inductor.

have verified the analysis.

II. TRADITIONAL DUAL BUCK TOPOLOGIES

Fig. 1 shows the traditional dual buck and dual boost inverters [7]-[8]. The most attractive advantage of the dual buck topologies is the high reliability. Firstly, without adding the extra dead time, the dual buck topologies can solve the shoot through problem. Secondly, compared to the traditional H-bridge inverter, the current will not flow through the body diodes of the switches in the dual buck topologies which means no reverse recovery problem exists in the MOSFET phase legs. Considering the above two aspects, the dual buck topologies

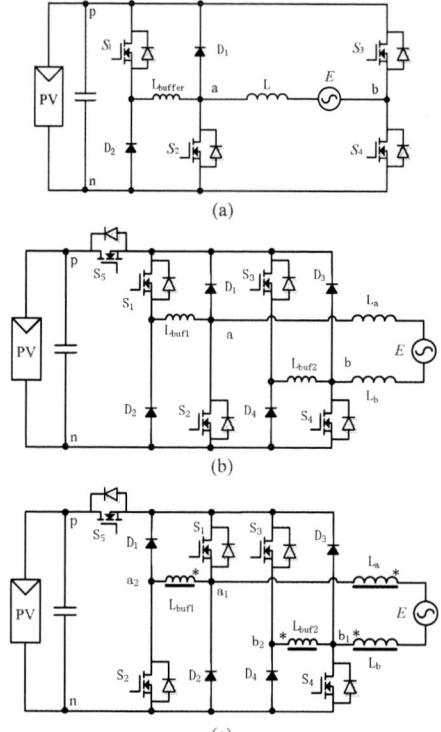

(a)

(b)

(c)

Fig. 4. Proposed Dual buck full bridge inverter (a) with one buffer inductor (b) with H3 phase legs (c) with coupled H3 phase legs.

can achieve high reliability without the shoot through and reverse recovery issues.

However, the main drawback of the dual buck topologies is the low magnetic utilization. In each power delivering and freewheeling modes, the current only flows through half of the inductance, which means the other half of the inductance is wasted in each working condition. The low utilization of the inductance makes the increasing of the weight and volume for the whole system. To solve this problem, a concept of single inductor dual buck full bridge inverter is proposed. Fig. 2 shows the single inductor topology. The novel topology includes six switches and two diodes. Comparing to the traditional dual buck full bridge inverter, the single inductor topology can save half of the inductance. And the novel topology retains the original advantages of high reliability. Also, there is no need to add the dead time in the high frequency unipolar switching strategy. The inductance can be fully utilized in the single inductor inverter. However, a high level of conduction loss is the main drawback of the novel topology. During the power delivering mode, the current flows through four switches which is a lot more than the traditional full bridge inverters. Besides, compared to the traditional H-bridge inverters, the extra two switches make controlling strategy more complex. And in the dual buck single inductor inverter, the current will flow through the body diodes of the series MOSFET switches which can cause the problem of reverse recovery.

978-1-4673-9551-9/16 $31.00 © 2016 IEEE   2936

To solve the problem of traditional H-bridge inverter, including the shoot through issue and the reverse recovery of the MOSFET, a kind of highly reliable and efficient dual buck inverter with buffer inductor is proposed in this paper. The newly proposed topologies maintain the advantages of dual buck inverter and solve the problem of low magnetic utilization. Also, the proposed topologies will not invite extra switches which means a simpler controlling strategy and lower conducting loss compared to the dual buck single inductor full bridge inverter.

III. HIGHLY RELIABLE DUAL BUCK INVERTERS WITH BUFFER INDUCTORS

In Fig. 2, the single inductor dual buck full bridge inverter improves the magnetic utilization of the traditional dual buck topologies, thus it saves the weight and volume of the system. However, this method will cause the extra conduction loss . In this section, a novel topology to solve the reverse recovery issue of the MOSFET phase leg is proposed. A buffer inductor can be applied to the traditional MOSFET phase leg. Fig. 3 shows the traditional and the proposed H3 dual buck phase leg with buffer inductor inserted in the middle of the dual buck phase leg. Applying the novel phase leg in Fig. 3(c) to the full bridge inverters, the novel dual buck topologies with the buffer inductors are proposed in Fig. 4. It can be indicated that by inserting the buffer inductor in the middle of the dual buck phase leg [6], the novel topology in Fig. 4(a) is a combination of the high-frequency phase leg with the buffer inductor and the traditional half bridge switching in power frequency. Also, applying the H3 dual buck bridge in Fig. 3(c), the novel H5 dual buck inverter is shown in Fig. 4(b), where the extra two buffer inductors have the function of preventing the freewheeling current flowing through the body diodes of the upper MOSFET. Thus the problem of reverse recovery in H5 inverter is solved and the magnetic utilization is largely improved compared to the traditional dual buck topologies. Also, when the inverter is used for the PV application, the common-mode behavior is good because the common-mode voltage is constant in the whole grid period. And in order to reduce the core loss of the inductors, the buffer inductors and the filters can be designed into the coupled inductors which is shown in Fig. 4(c). The dotted terminals of the coupled inductors are labeled with asterisks.

The reason why buffer inductor can prevent the freewheeling current from flowing through the body diode of the MOSFET is that the current through the buffer inductor will not change in a sudden. And also the buffer inductor can solve the shoot through problem among the switching commutation of the power frequency. Thus, neither the switching-frequency dead time nor the power-frequency dead time is needed in the proposed topologies.

To briefly illustrate the operational principle of proposed single-phase inverters with buffer inductor, the topology of Fig. 4(c) is assumed as an example, which has four operation modes. Fig. 5 shows the specific current flowing paths during the energy transferring modes and the freewheeling modes. Also, the common-mode characteristic of the proposed inverter is briefly analyzed.

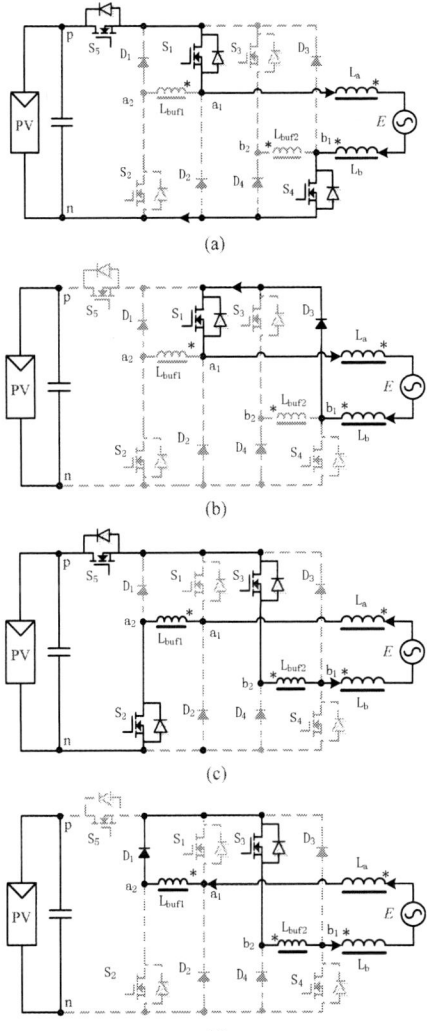

(a)

(b)

(c)

(d)

Fig. 5. Four working modes of the proposed dual buck full bridge inverter with coupled H3 phase legs in Fig. 4(c).

A. Operational Principle of the Proposed Inverter

Mode 1: Mode 1: During positive half period, S_4 and S_5 are modulated in high frequency, while S_1 is always ON. When S_4 and S_5 are on, the current flows through S_5, S_1, grid and S_4 successively. The common-mode voltage [9] is

$$u_{cm} = \frac{u_{a_1n} + u_{b_1n}}{2} = \frac{u_{dc} + 0}{2} = \frac{u_{dc}}{2} \quad (1)$$

Mode 2: When S_4 and S_5 are off, the current flows through D_3, S_1 and grid successively. As is shown in Fig. 5(b), the current through buffer inductor, L_{buf2}, cannot change in a sudden. So that it has the function of preventing the current flowing through the body diode of S_3, which solves the problem of reverse recovery. The common-mode voltage is

$$u_{cm} = \frac{u_{a_1n} + u_{b_1n}}{2} = \frac{u_{dc}/2 + u_{dc}/2}{2} = \frac{u_{dc}}{2} \quad (2)$$

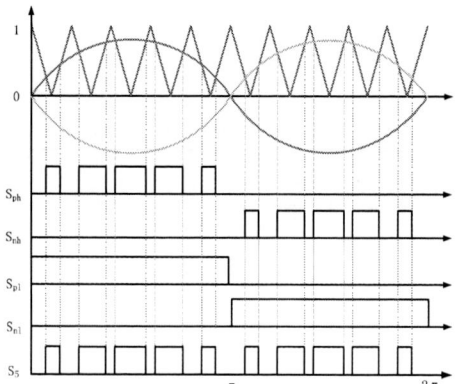

Fig. 6. The switching signals of the proposed inverters.

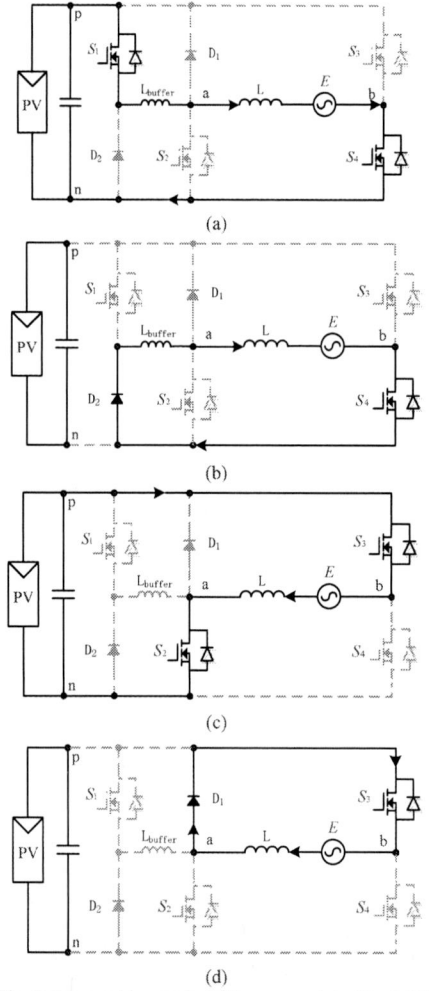

(a)

(b)

(c)

(d)

Fig. 7. Four working modes of the proposed dual buck full bridge inverter in Fig. 4(a).

Mode 3: During negative half period, S_2 and S_5 are modulated in high frequency, while S_3 is always ON. When S_2

and S_5 are on, the current flows through S_5, S_3, grid and S_2 successively. The common-mode voltage is

$$u_{cm} = \frac{u_{a_2 n} + u_{b_2 n}}{2} = \frac{0 + u_{dc}}{2} = \frac{u_{dc}}{2} \qquad (3)$$

Mode 4: When S_2 and S_5 are off, the current flows through D_1, S_3 and grid successively. As is shown in Fig. 5(d), the current through buffer inductor, L_{buf1}, cannot change in a sudden neither. And the current flowing through L_{buf1} is decreasing during the freewheeling period which makes the potential of S_1's drain terminal higher than the source terminal. So that it also has the function of preventing the current flowing through the body diode of S_1, which solves the problem of reverse recovery. The common-mode voltage is

$$u_{cm} = \frac{u_{a_2 n} + u_{b_2 n}}{2} = \frac{u_{dc}/2 + u_{dc}/2}{2} = \frac{u_{dc}}{2} \qquad (4)$$

Also, the operational principle of the proposed topology in Fig. 4(a) is shown in Fig. 7 which includes the specific current flowing paths during the energy transferring modes and the freewheeling modes. The corresponding modulation strategies of the inverters in Fig. 4(a) and (c) are both shown in Fig. 6, where S_{ph}, S_{nh} represent the high frequency switches in positive and negative grid period and S_{nh}, S_{nl} represent the low frequency switches in positive and negative period respectively. It is noted that the proposed topology in Fig. 4(a) has less device cost and lower conducting loss while the topology in Fig. 4(c) has a better common-mode behavior which is suitable for the PV applications.

B. Comparison with Other Inverters

The proposed dual buck topologies with buffer inductors firstly maintain the traditional dual buck inverter's advantages, such as high reliability and high efficiency. The reverse recovery problem is solved by the buffer inductor which is inserted in the middle of the MOSFET phase leg. Secondly, the buffer inductor can be very small compared to the inductance of the filter (the buffer inductor is one-tenth of the filtering inductance). So the magnetic utilization is closed to the single inductor dual buck topology, which is a lot higher than the traditional dual buck inverter. Specifically, the magnetic utilization of the proposed dual buck inverter with buffer inductors is 95.5%, which is almost twice as the traditional dual buck topologies. Thirdly, because only one extra device is applied compared to the single inductor topology in Fig. 2, the conducting loss and the device cost of the proposed topologies are lower. Finally, the common-mode voltage of the proposed inverter in Fig. 4(c) is constant during the whole grid period which means the common-mode leakage current is of low level. So the proposed topology can be a good application in the PV system. (The specific analysis of the common-mode behavior is shown in the next section.)

IV. COMMON MODE ANALYSIS OF THE PROPOSED TOPOLOGY

The transformerless photovoltaic (PV) grid-connected system is an important application for the single phase inverter. However, in a transformerless PV system, the fluctuation of the common mode voltage will excite leakage current in the common mode path which may cause the safety problems and distort the output current. The value of the leakage current

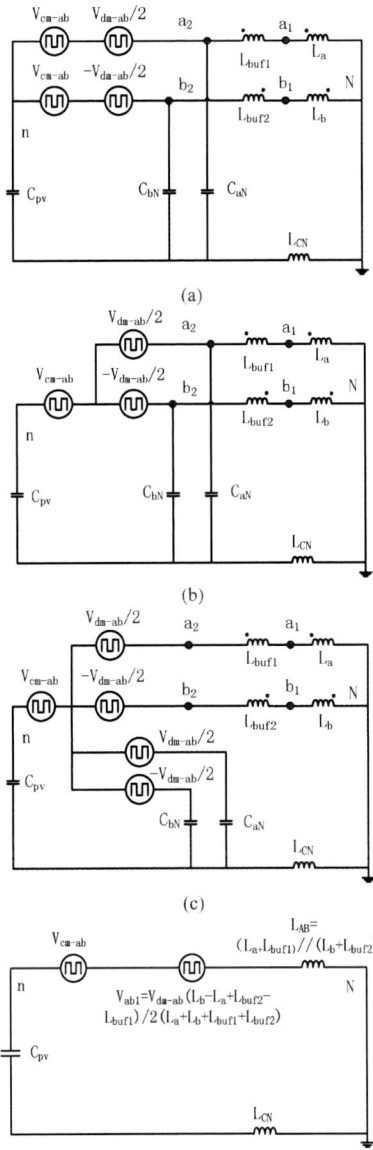

(a)

(b)

(c)

(d)

Fig. 8. The equivalent common-mode circuits of the Novel dual buck inverter with coupled H3 phase legs.

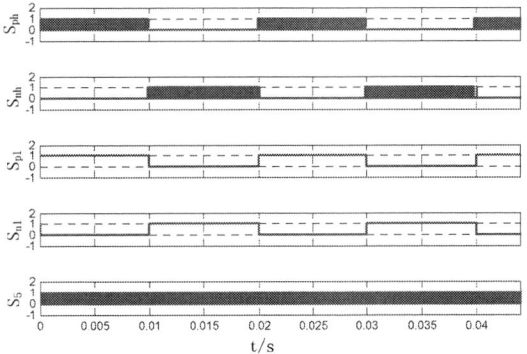

Fig. 9. The simulated switching waveforms of the proposed inverter in Fig. 4(c).

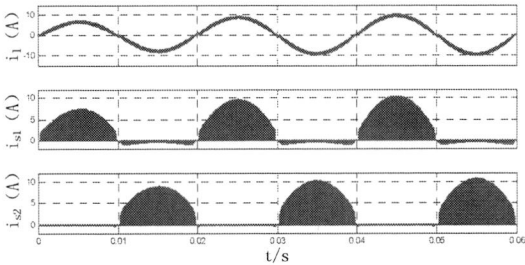

Fig. 10. The filtering current and the switching current of the proposed inverter in Fig. 4(a).

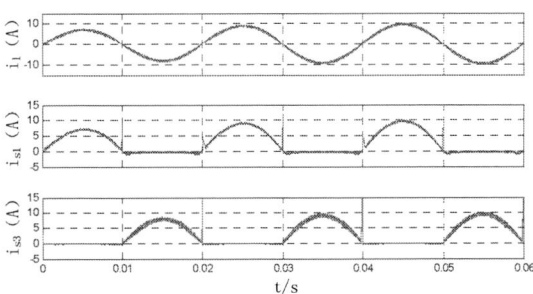

Fig. 11. The filtering current and the switching current of the proposed inverter in Fig. 4(c).

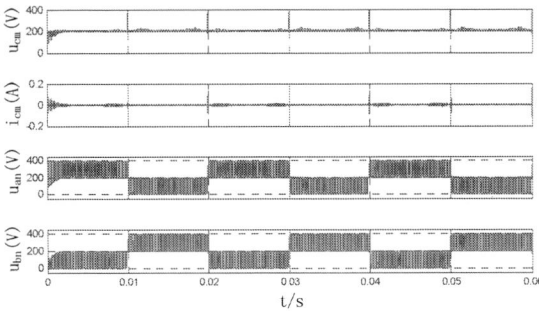

Fig. 12. The common-mode characteristics of the proposed inverter in Fig. 4(c).

depends on the fluctuating frequency of the common-mode voltage, u_{Cpv} where the C_{pv} represents the equivalent stray capacitance of the PV panel.

The equivalent common-mode circuit of the proposed inverter in Fig. 4(c) is specifically analyzed. The common-mode voltage [10] is defined as:

$$u_{cm} = \frac{u_{an} + u_{bn}}{2} \qquad (5)$$

And the differential mode voltage is defined as:

$$u_{dm} = u_{an} - u_{bn} \qquad (6)$$

So, the equivalent common-mode circuit of the inverter is illustrated in Fig. 8, where a1(b1) and a2(b2) represent the two terminals of the buffer inductors, point n represents the

978-1-4673-9551-9/16 $31.00 © 2016 IEEE   2939

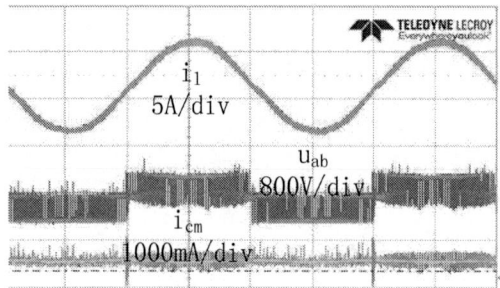

Fig. 13. The experimental waveforms of the proposed inverter in Fig. 4(a).

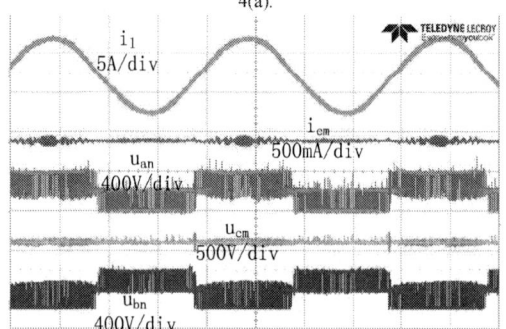

Fig. 14. The experimental waveforms of the proposed inverter in Fig. 4(c).

negative side of the DC rail, point N represents the ground. From Fig.8(a) to Fig. 8(d), the simplified circuit only consists of the common-mode voltage source, u_{cm-ab}, the differential mode component of the voltage source, u_{ab1}, and the equivalent capacitance and inductors as indicated in Fig. 8(d).

The differential mode component of the voltage source can be expressed as:

$$u_{ab1} = u_{dm} \times \frac{(L_b - L_a + L_{buf2} - L_{buf1})}{2(L_a + L_b + L_{buf1} + L_{buf2})} \quad (7)$$

And the equivalent output filtering inductor can be expressed as:

$$L_{AB} = \frac{(L_a + L_{buf1})(L_b + L_{buf2})}{(L_a + L_b + L_{buf1} + L_{buf2})} \quad (8)$$

Thus, according to the finally simplified common-mode circuit in Fig. 8(d) and the calculation of the u_{cm} during the four operating modes, the leakage current can be eliminated only when the differential mode component of the voltage source, u_{ab1}, is zero. So from equation (7), the two buffer inductors, L_{buf1} and L_{buf2}, should be equal and the filtering inductors, L_a and L_b, should be symmetric. Then, the common mode voltage u_{cm} is constant and the common mode current i_{cm} indicated in (9) will be attenuated in a low level.

$$i_{cm} = C_{pv} \frac{du_{cm}}{dt} \quad (9)$$

So, according to the common-mode analysis of the proposed inverter in Fig. 4(c), the differential mode component of the voltage source, u_{ab1}, can be eliminated in the common-mode circuit only when the buffer inductors and the filters are symmetric. Also, from the equations (1) to (4), the u_{cm} is constant, which means the extra buffer inductors will not

influence the common-mode behavior of the inverter. Thus, the common-mode behavior of the proposed inverter is satisfying and the application of PV system is suitable for this inverter because of the high reliability without the threaten of common-mode leakage current.

V. SIMULATION AND EXPERIMENTAL RESULTS

The simulation and experimental results are shown in this section. The proposed inverters in Fig. 4(a) and Fig. 4(c) were simulated in Matlab/Simulink. The DC voltage is 400V, and the grid voltage is 220V/50Hz. The switching frequency is 10kHz. The output inductors are both 2mH. And the buffer inductors in Fig 4(c) are both 0.2mH. The grid current is controlled by a conventional PR controller. Fig. 9 shows the simulated switching waveforms of the proposed inverters in Fig. 4(c). The S_{ph} and S_{nh} represent the high-frequency switch in positive grid period and negative grid period respectively. And the S_{pl} and S_{nl} represent the low-frequency switch in positive grid period and negative grid period respectively. Fig. 10 and 11 show the filtering current and switching current of the proposed inverters in Fig. 4(a) and Fig. 4(c) respectively. The current waveforms of the switches are all unidirectional which indicate that no freewheeling current is flowing through the body diodes of the MOSFET. So the proposed inverters will not be threatened by the reverse recovery issue, thus the reliability of the inversion system is largely improved. As is shown in Fig. 12, the common-mode voltage, u_{cm}, and leakage current, i_{cm}, of the proposed inverter in Fig. 4(c) is of low level, which means a good common mode behavior.

In order to further validate the proposed topologies and the modulation strategies, the 1kW experimental prototypes of the proposed inverters were built with the same parameters as the simulation model. The experimental results are shown in Fig. 13 and Fig. 14, where i_l represents the grid current, u_{ab} represents the output voltage, u_{an} and u_{bn} represent the bridge voltage, u_{cm} and i_{cm} represent the common mode voltage and common mode current respectively.

VI. CONCLUSION

This paper reviews the already published dual buck topologies. The advantages and disadvantages of the dual buck inverters are specifically analyzed. In order to solve the main drawback of low magnetic utilization, a kind of dual buck phase leg with buffer inductor is proposed. By applying the novel phase legs to the full bridge inverters, the new topologies maintain the high reliability of the traditional dual buck inverter and the magnetic utilization is largely improved. Also, compared to the traditional single inductor dual buck inverter, the novel topologies has the advantages in conducting loss and controlling complexity. The simulation and experimental results verified the performance of proposed inverters.

REFERENCES

[1] T. Kerekes, R. Teodorescu, P. Rodriguez, G. Vazquez, E. Aldabas, "A new high-efficiency single-phase transformerless PV inverter topology," IEEE Trans. Ind. Electron., vol. 58, no. 1, pp. 184-191, Jan. 2011.

[2] Zhu, Chenghua, Fanghua Zhang, and Yangguang Yan, "A novel split phase dual buck half bridge inverter", in Proc. 20th IEEE Applied Power Electronics Conference and Exposition, 2005, vol.2, pp.845-849.

[3] Hong Feng, Ying Pei-pei, Wang Cheng-hua, "Decoupling Control of Input Voltage Balance for Diode-Clamped Dual Buck Three-Level Inverter", in Proc. 28th Annual IEEE Applied Power Electronics Conference and Exposition, Long Beach, California, USA, March 17-21, 2013, pp.482-488.

[4] Liu Miao, Hong Feng, Wang Cheng-hua. A Novel Flying-Capacitor Dual Buck Three-Level Inverter [C], in Proc. 28th Annual IEEE Applied Power Electronics Conference and Exposition, Long Beach, California, USA, March 17-21, 2013, pp.502-506.

[5] Hong Feng, Liu Jun, Ji Baojian, Zhou Yufei, and Wang Jianhua, "Single Inductor Dual Buck Full-Bridge Inverter", IEEE Trans. Ind. Electron., vol. 62, no. 8, pp. 4869–4877, Aug 2015.

[6] B. F. Chen, B. Gu, L. H. Zhang, Z. U. Zahid, Z. L. Liao, J.-S. Lai, Z. L. Liao and R. X. Hao, "A High Efficiency MOSFET Transformerless Inverter for No-isolated Micro-inverter Application," IEEE Trans. Power Electron., vol. 30, no. 7, pp. 3610–3622, July. 2015.

[7] B. F. Chen, P. W. Sun, C. Liu, C-L. Chen, J.-S. Lai, and W. Yu, "High efficiency transformerless photovoltaic inverter with wide-range power factor capability", in Proc. of IEEE 27th Applied Power Electronics Conference and Exposition, Orlando, FL, Feb. 2012.

[8] B. Gu, J. Dominic, J.-S. Lai, C-L. Chen, T, LaBella, and B. F. Chen, "High Reliability and Efficiency Single-Phase Transformerless Inverter for Grid-Connected Photovoltaic Systems," IEEE Trans. Power Electron, vol.28, no. 5, pp.2235-2245. May, 2013.

[9] T. Kerekes, R. Teodorescu, P. Rodriguez, G. Vazquez, and E. Aldabas, "A new high-efficiency single-phase transformerless PV inverter topology," IEEE Trans. Ind. Electron., vol. 58, no. 1, pp. 184–191, Jan. 2011.

[10] T. Kerekes, R. Teodorescu, and M. Liserre. "Common mode voltage in case of transformerless PV inverters connected to the grid," IEEE International Symposium on Industrial Electronics, 2008, pp. 2390-2395.

A 99% Efficiency SiC Three-phase Inverter Using Synchronous Rectification

Shan Yin, K. J. Tseng, C. F. Tong
Rolls-Royce@NTU Corporate Lab, School of Electrical and
Electronic Engineering
Nanyang Technological University
Singapore

Rejeki Simanjorang, C. J. Gajanayake, Amit K. Gupta
Advanced Technology Centre
Rolls-Royce Singapore Pte. Ltd.
Singapore

Abstract—The reactive power in power converter with inductive load (motor drive e.g.) requires a current commutation path for the freewheeling current. Due to the high voltage drop of body diode of SiC MOSFET, a SiC Schottky diode is normally recommended as the anti-parallel freewheeling diode for SiC MOSFET to suppress the conduction of body diode. However, since the MOSFET can work as synchronous rectifier, the freewheeling diode only conducts during the dead time, leading to a low utilization rate of device. In this work, the three-phase SiC inverter using synchronous rectification is investigated. The analytical model for inverter power loss with and without freewheeling diode is built. Based on the switching characterization, the inverter with synchronous rectification permits a surprising higher efficiency than that with freewheeling diode due to the reduced current overshoot at turn-on. And a 5 kW prototype of three-phase inverter is developed, which shows a 99% high efficiency at the switching frequency of 40 kHz. This work confirms the possibility to remove the freewheeling diode in SiC inverter without degrading the efficiency.

Keywords—efficiency, inverter, SiC, synchronous rectification

I. INTRODUCTION

The synchronous rectification (SR) has been extensively implemented in the low voltage (< 100 V) DC/DC converters to replace the diode rectifier for lower conduction loss. It is because the MOSFET channel can work in the reverse conduction when applying a positive gate voltage. Compared with the forward voltage drop across the PN junction of diode, it permits a smaller voltage drop at a certain range of current. During the dead time, the freewheeling current will flow through the intrinsic body diode of MOSFET working as SR, which is a bipolar PiN diode. Once the MOSFET working as the main switch is triggered on, the minority carrier stored in the body diode has to be swiped out of the drift region, leading to a current overshoot across the main switch due to the reverse recovery process of PiN diode.

When extending the concept of SR to higher voltage applications, like the motor drive, grid connected inverter and telecommunication power supply, it suffers from the high reverse recovery current and switching loss. The conventional strategies to suppress the reverse recovery current include anti-parallel Schottky diode [1] and dead time optimization [2, 3]. However, the additional Schottky diode only conducts during

the dead time, leading to a low utilization rate of device. And the minimized dead time tends to increase the risk of shoot-through.

The presence of wide bandgap material silicon carbide (SiC) allows the MOSFET to compete with Si CoolMOS and IGBT in the in the range of 600 V and above. Due to the 3x bandgap of SiC compared with the Si counterpart, the body diode of SiC MOSFET has a forward voltage drop up to 2.5 V, which is significantly higher than that of Si MOSFET (typically 0.7 V). Hence, the body diode of SiC MOSFET is normally regarded as a poor rectifier. Therefore, an external SiC Schottky diode is recommended by the vendor [4], it is also included in the existing commercial SiC power module [5]. However, the body diode of SiC MOSFET has been proven with excellent reverse recovery process, which shows apparently improved switching performances compared with that of Si counterpart over a wide range of temperature [6, 7]. Since the intrinsic body diode already shows the reduced switching loss as well as EMI issue, the use of anti-parallel SiC Schottky diode becomes not so indispensable to suppress the reverse recovery current. In addition, considering the high speed switching that SiC MOSFET can achieve, the dead time can be minimized to reduce the conduction loss dissipated by the body diode. Hence, it is of great interest to investigate the benefits to remove the SiC Schottky diode from the existing SiC power converter.

Hence, some works were conducted to investigate the SiC power converter with SR recently. The feasibility of SR for SiC MOSFET in the DC/DC converter was investigated in [8]. And it was concluded that the body diode of SiC MOSFET has the comparable reverse recovery characteristics to the SiC Schottky diode. A 5 kW, 100 kHz resonant full bridge inverter was demonstrated in [7], which shows a 99% efficiency. In [9], a three-phase inverter was investigated, which shows a high efficiency up to 98% at a full load of 10 kW. And the power loss dissipated by the body diode only accounts for less than 5% of the total loss. These early works still lack the complete system-level investigation on the benefits of applying SiC power converter with SR, including the cost, power loss and efficiency.

In this work, the application of SR in the three-phase SiC voltage source inverter (VSI) is investigated. In section II, the analytical models for the power loss of VSI with SR and

978-1-4673-9551-9/16 $31.00 © 2016 IEEE

external freewheeling diode (FWD) are developed. In section III, the details for the experimental setup are demonstrated. In section IV, the switching characterizations of SiC MOSFET half bridge configuration with and without FWD are conducted. Then the inverter efficiencies with SR as well as FWD are derived and compared. Finally a 5 kW inverter prototype with SR is developed and tested.

II. POWER LOSS CALCULATION FOR SiC THREE-PHASE INVERTER

The ideal three-phase two-level VSI consists of 6 switches, as Fig. 1 shows. The sinusoidal reference wave is modulated by a triangle carrier wave to control the high side (HS) and low side (LS) switches to conduct alternatively. Considering the turn-on and turn-off times of power semiconductor devices, a dead time is required to avoid the arm shoot-through. Although the dead time tends to deteriorate the output harmonics, it ensures the safe operation of power converter.

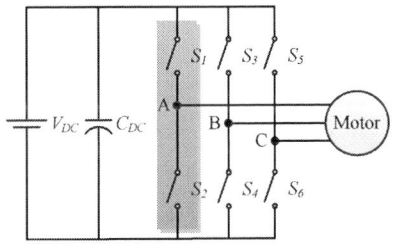

Fig. 1. Schematic of three-phase inverter for motor drive application.

To evaluate the efficiency of SiC power converter, the power loss needs to be accurately calculated based on the operating conditions, which is also beneficial for heat sink design. There are two strategies for power loss calculation: simulation and analytical correlation. The simulation-based strategy requires an accurate circuit simulation model, and is normally implemented in the circuit simulator like PSpice [10]. However, an extremely high CPU resource is required since the power loss is calculated during each switching cycle. Hence, the analytical correlation based on averaging method is adopted in this work [11-13].

A. Ideal Three-phase VSI with Zero Dead Time

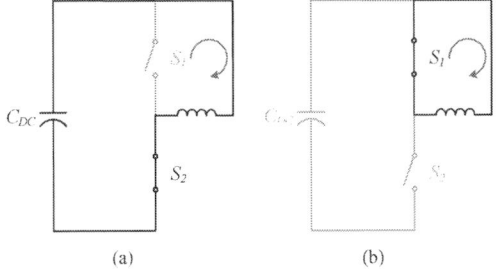

Fig. 2. Operation principle of chopper: (a) S_1 off and S_2 on, (b) S_1 on and S_2 off.

Firstly, the power loss of ideal three-phase VSI with zero dead time can be analyzed by the equivalent chopper shown by Fig. 2. During the positive half-wave cycle, S_1 and S_2 conduct alternatively. When S_1 is off and S_2 is on, the DC-link capacitor starts to charge the inductor via S_2. When S_1 is on and S_2 is off,

the inductor current freewheels via S_1. Considering the circuit symmetry, only the positive half-wave cycle is discussed in the following work. Since the switching frequency f_s is apparently higher than the modulating frequency f_0, the reference voltage and current flowing via the devices during one switching cycle can be approximated by constant, as Fig. 3(a) shows. Then the continuous reference voltage and sinusoidal output current can be discretized into

$$v_r(n) = V_r \sin\theta_n \qquad (1)$$

$$i_o(n) = I_P \sin(\theta_n - \phi) \qquad (2)$$

$$\theta_n = \frac{2\pi}{N}n, \; n = 1, 2, ..., N \qquad (3)$$

where V_r is the magnitude of reference signal, I_P is the peak value of phase current, N is the frequency modulation ratio and given by $N = f_s/f_0$, and ϕ is the phase angle. From Fig. 3(b), the duty cycles are derived without consideration of the dead time

$$D_n' = \frac{V_c - v_r(n)}{2V_c} = \frac{1}{2}(1 - m_a \sin\theta_n) \qquad (4)$$

$$D_n = 1 - D_n' = \frac{1}{2}(1 + m_a \sin\theta_n) \qquad (5)$$

where m_a is the amplitude modulation ratio and given by $m_a = V_r/V_c$.

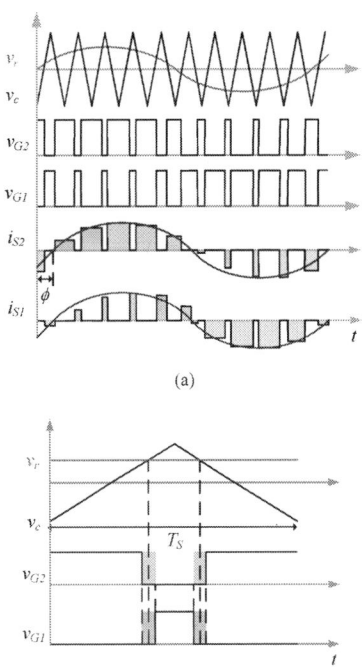

Fig. 3. Switching waveforms: (a) entire fundamental cycle, (b) one switching cycle during the positive half-wave cycle.

During the positive half-wave cycle, the conduction energy of S_2 is a sum of loss for the $N/2$ switching cycles, as given by

$$E_{cond,S2} = \sum_{1}^{N/2} i_o\left(n\right)^2 R_{DS,on} D_n T_s = I_P^2 R_{DS,on} T_0 \left(\frac{1}{8} + \frac{m_a \cos\phi}{3\pi} \right)$$

(6)

where $R_{DS,on}$ is the on-resistance of SiC MOSFET operating in the 1st quadrant. Similarly, the conduction energy of S_1 is derived

$$E_{cond,S1} = \sum_{1}^{N/2} i_o\left(n\right)^2 R_{SD,on} D_n{}' T_s = I_P^2 R_{SD,on} T_0 \left(\frac{1}{8} - \frac{m_a \cos\phi}{3\pi} \right)$$

(7)

where $R_{SD,on}$ is the on-resistance of SiC MOSFET operating in the 3rd quadrant. The total conduction loss of the SiC-based three-phase inverter is six times of the sum of Eq.(6) and (7), then times the modulating frequency

$$P_{cond} = 6\left(E_{cond,S2} + E_{cond,S1} \right) f_0$$
$$= 6 I_P^2 \left[\frac{1}{8}\left(R_{DS,on} + R_{SD,on} \right) + \frac{m_a \cos\phi}{3\pi}\left(R_{DS,on} - R_{SD,on} \right) \right]$$

(8)

For the switching loss calculation, the physical-based switching loss model proposed in [12-14] lacks the accuracy due to the neglect of stray inductance. In addition, it requires several physical parameters of power devices, which are rarely provided by the device manufacturer. Hence, the switching loss calculation mostly relies on the switching energy provided by the datasheet [15]. Assuming the switching energy is a linear function of both voltage and current, it is a scaled value of that from the datasheet. It provides a rough and fast estimation for switching loss calculation. To further improve the accuracy, the switching loss should be obtained from the switching characterization.

The total switching energy of a switch (e.g. S_2) during one switching cycle can be approximated by a polynomial function of output current

$$E_{sw}\left(n\right) = ai^2\left(n\right) + bi\left(n\right) + c$$

(9)

where a, b and c are the fitting parameters. The switching energy is a sum of loss for the $N/2$ switching cycles

$$E_{sw,S_2} = \sum_{1}^{N/2} E_{sw}\left(n\right) = N\left(\frac{aI_P^2}{4} + \frac{bI_P}{\pi} + \frac{c}{2} \right)$$

(10)

The total switching loss of switch is therefore six times of Eq.(10), then times the modulating frequency

$$P_{sw} = 6 E_{sw,S_2} f_0 = 6\left(\frac{aI_P^2}{4} + \frac{bI_P}{\pi} + \frac{c}{2} \right) f_s$$

(11)

B. Three-phase VSI with Synchronous Rectification

The operation principle of the three-phase VSI with SR (VSI-SR) can be approached by the chopper shown by Fig. 4. During the dead time when both M_1 and M_2 are off, the inductor current freewheels via body diode of M_1. Due to the large forward voltage of body diode of SiC MOSFET, the conduction

loss dissipated by the body diode becomes no longer negligible, as given by

$$E_{cond,BD1} = 2\sum_{1}^{N/2} i_o\left(n\right)\left[V_{BD} + i_o\left(n\right) R_{BD} \right] T_d$$
$$= N\left(\frac{2I_P V_{BD}}{\pi} + \frac{I_P^2 R_{BD}}{2} \right) T_d$$

(12)

where V_{BD} is the knee voltage, R_{BD} is the on-resistance of body diode, and the factor 2 is because there are two times of current commutation in one switching cycle. In real situation, the conduction time of body diode should be smaller than the dead time T_d after considering the switching time of MOSFET. To simplify analysis, the worst assumption is made in this work. And the total conduction loss of body diode is given by

$$P_{cond,BD} = 6 E_{cond,BD1} f_0$$
$$= 6\left(\frac{2I_P V_{BD}}{\pi} + \frac{I_P^2 R_{BD}}{2} \right) T_d f_s$$

(13)

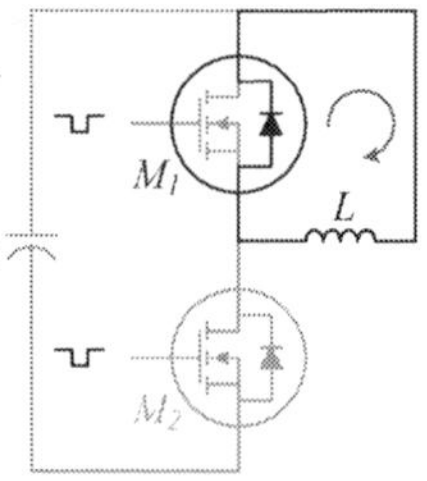

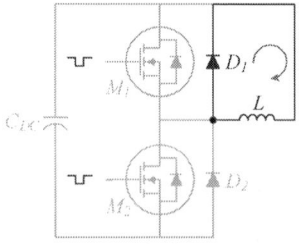

Fig. 4. Operation principle with synchronous rectification.

C. Three-phase VSI with Freewheeling Diode

The operation principle of the three-phase VSI with FWD (VSI-FWD) also can be approached by the chopper shown by Fig. 5. During the dead time when both M_1 and M_2 are off, the inductor current freewheels via D_1. Similar to previous discussion, the total conduction loss dissipated by FWD is given by

$$P_{cond,D} = 6\left(\frac{2I_P V_D}{\pi} + \frac{I_P^2 R_D}{2} \right) T_d f_s$$

(14)

where V_D is the knee voltage, R_D is the on-resistance of FWD.

Fig. 5. Operation principle with freewheeling diode.

III. EXPERIMENTAL SETUP

In this work, the 1.2 kV, 20 A SiC MOSFET C2M0080120D from Wolfspeed is used [16]. The design details for gate driver, converter and DC-bus are presented as follow.

A. Gate Driver Design

For high voltage application above 600 V, individual gate driver with galvanically isolated power supply instead of the cost-saving bootstrap power supply is preferred for consideration of reliability and switching frequency [17]. In this work, the power supply of gate driver is provided by isolated DC/DC converter. The 4 A output current gate drive optocoupler from Avago is adopted as the driver. Since the driver IC only has single output terminal, the turn-off diode is adopted to separate turn-on and turn-off paths, as Fig. 6 shows. The high dv/dt ratio resulted from the high-speed switching of SiC MOSFET will induce perturbation to logic circuits via Miller capacitor and coupling capacitor. Hence, an isolation barrier is required to minimize the coupling capacitance [18].

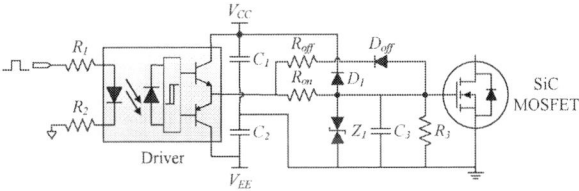

Fig. 6. Schematic of gate driver.

B. Converter Layout Design

The layout design plays a key role in successfully operating SiC power converter in high frequency. As Fig. 7 shows, a bypass capacitor with low equivalent series inductance (ESL) needs to be put as close as possible to the converter. In addition, the current loop should be minimized to reduce the loop stray inductance as well as EMI effect. In this work, the 900 V, 4 μF through-hole film capacitor is used with 21 nH ESL measured from impedance analyzer. Based on the parasitic extraction tool, the stray inductances contributed by the layout and package are estimated to be 17 and 18 nH, respectively.

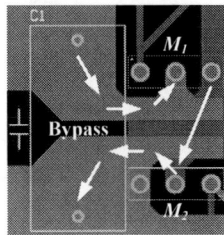

Fig. 7. Converter layout with current flow direction (white arrow).

C. DC-bus Design

The laminated structure is widely used in bus bar to reduce the stray inductance contributed by bus layout [19, 20]. The positive- and negative-planes are overlapped with each other, with an isolation layer between the two planes. Hence, the two planes are strongly coupled to give a large mutual inductance. In this work, the laminated DC-bus is implemented by double-layer PCB with 1.6 mm FR4 and 2 oz copper. And the DC-link capacitors consist of 10 parallel 900 V, 4 μF film capacitors.

IV. RESULTS AND DISCUSSION

A. Switching Characterization of SiC MOSFET

Firstly, a double pulse testing (DPT) experiment is set up to test the switching characteristics of SiC MOSFET, as Fig. 8 shows. The control signal is generated by the TI DSP TMS320F28335. A single-layer winding, air core inductor with low equivalent parallel capacitance is used as the current load. To measure the current, a scaled-down current transformer (CT) is made, which comprises 10 turns of isolated wire wound on a ferrite toroid. And it is installed on the source pin of M_2 directly.

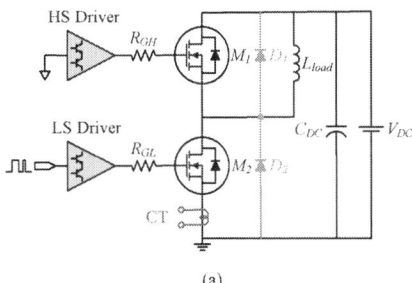

(a)

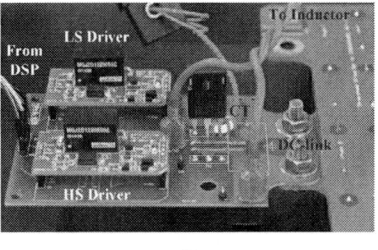

(b)

Fig. 8. Schematic (a) and experimental setup (b) of double pulse testing.

To compare the switching loss of VSI-SR and VSI-FWD, the switching characterization for half bridge configuration with and without anti-parallel FWD is conducted. The 1.2 kV, 20 A SiC Schottky diode C4D20120A from Wolfspeed is used as the FWD [21], and it is directly soldered on the pin of MOSFET to minimize the loop inductance, as Fig. 9 shows.

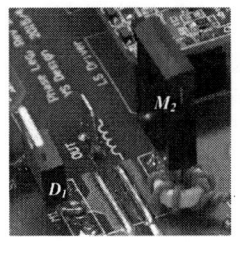 

(a)   (b)

Fig. 9. Top side (a) and back side (b) of double pulse testing experiment with anti-parallel freewheeling diode.

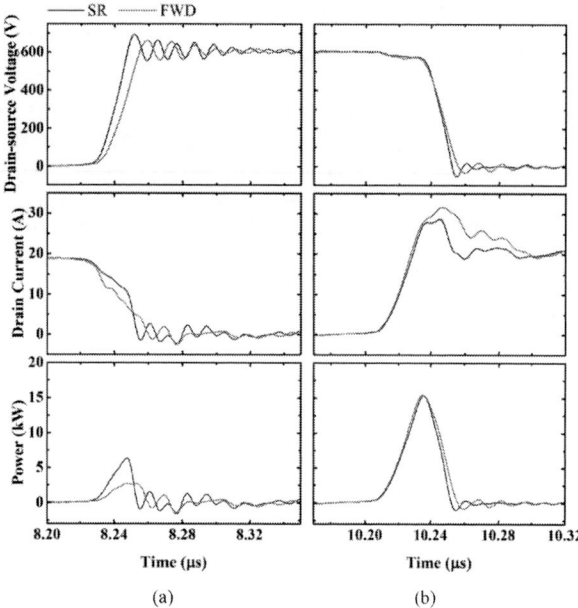

Fig. 10. Switching waveforms of half bridge configuration with synchronous rectification and freewheeling diode: (a) turn-off, (b) turn-on.

The presence of anti-parallel FWD will increase the effective output capacitance, leading to a smaller voltage overshoot at turn-off while a larger current overshoot at turn-on, as Fig. 10 shows. It also tends to speed up di/dt at turn-on transition, while slow down the rest of ratios, as Table 1 shows. In addition, the reverse recovery process of body diode of SiC MOSFET shows comparable performance with the SiC Schottky diode. Hence, the power loss of reverse recovery process is neglected in the following work.

TABLE 1. DV/DT AND DI/DT RATIOS.

	Unit	SR		FWD	
		Turn-off	Turn-on	Turn-off	Turn-on
dv/dt	V/ns	37	36	30	31
di/dt	A/ns	2.1	1.1	0.92	1.3

Applying FFT to the switching waveforms, the resonant frequencies with SR is 80 MHz. Considering the output capacitance of C2M0080120 is 81 pF at 600 V, the loop stray inductance is calculated to be 49 nH according to Eq.(15). Hence, the stray inductance contributed by the DC-link is derived to be 14 nH, which is smaller than the ESL of bypass capacitor due to the effect of laminated DC-bus design.

$$f = \frac{1}{2\pi\sqrt{L_s C_{oss}}} \qquad (15)$$

The relationship between the switching loss and load current is given by Fig. 11. The turn-off loss of VSI-FWD is always smaller than that of VSI-SR, but the turn-on loss is opposite. For low load current, the total loss of VSI-FWD is higher, and the two curves tend to overlap at high load current. The polynomial correlations between the total switching loss and load current is given by

$$E_{tot,SR} = 0.467i^2 + 8.71i + 81.9 \ (\mu J) \qquad (16)$$

$$E_{tot,FWD} = 0.487i^2 + 5.59i + 144 \ (\mu J) \qquad (17)$$

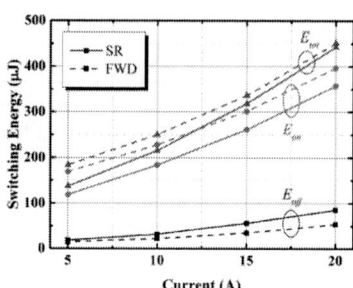

Fig. 11. Switching loss versus load current, solid lines are with synchronous rectification, dashed lines are with freewheeling diode.

B. Efficiency Evaluation

The output power of three-phase VSI is given by

$$P_{out} = 3V_{P,rms}I_{P,rms}\cos\phi \qquad (18)$$

where $V_{P,rms}$ and $I_{P,rms}$ are the rms values of phase voltage an current, respectively. Converting the rms value to the peak value, the output power can be rewritten as

$$P_{out} = \frac{3}{2}V_P I_P \cos\phi \qquad (19)$$

where V_P is the peak value of phase voltage and is given by [22]

$$V_P = m_a \frac{V_{DC}}{2} \qquad (20)$$

The total power loss of VSI-SR is a sum of Eq.(8), (11), (13)

$$P_{loss} = P_{cond} + P_{sw} + P_{cond.BD} \qquad (21)$$

Similarly, the total power loss of VSI-FWD is given by

$$P_{loss} = P_{cond} + P_{sw} + P_{cond,D} \qquad (22)$$

And the inverter efficiency is given by

$$\eta = \frac{P_{out}}{P_{in}} = \frac{P_{out}}{P_{out} + P_{loss}} \qquad (23)$$

TABLE 2. OPERATION CONDITIONS.

Parameter	Value
V_{DC}	600 V
$\cos\phi$	1
M_a	0.8
f_0	400 Hz
T_d	200 ns

TABLE 3. DEVICE PARAMETERS OF SiC MOSFET C2M0080120D AND SiC SCHOTTKY DIODE C4D20120A.

	Parameter	Value
C2M0080120D	$R_{DS,on}$	89 mΩ
	$R_{SD,on}$	78 mΩ
	V_{BD}	2.2 V
	R_{BD}	38 mΩ
C4D20120A	V_D	0.9 V
	R_D	31 mΩ

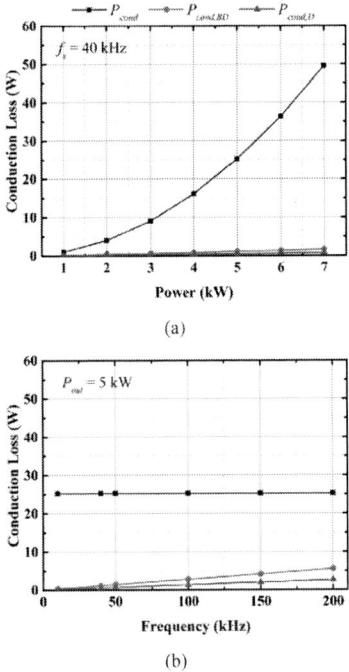

(a)

(b)

Fig. 12. Conduction loss at different output power (a) and switching frequency (b).

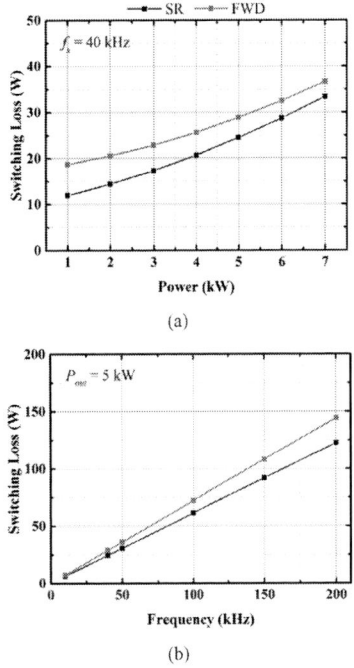

(a)

(b)

Fig. 13. Switching loss at different output power (a) and switching frequency (b).

The operation conditions of inverter are given by Table 2. A maximum power factor of 1 is assumed. And a narrow dead time of 200 ns is set according to the switching characterization so as to avoid arm shoot-through during current commutation. The device parameters for efficiency calculation are list in Table 3.

According to the previous discussion, the fractions of conduction loss and switching loss for both VSI-SR and VSI-FWD are shown by Fig. 12 and Fig. 13, respectively. Due to the high forward voltage drop of body diode, the conduction loss of body diode is always higher than that of FWD. And this difference continues to increase with the increasing of output power as well as switching frequency. However, both are negligible compared with the conduction loss of MOSFET. By removing FWD, the VSI-SR shows the advantage in switching loss, especially at light load and high frequency.

The VSI-SR always shows a higher efficiency than the VSI-FWD no matter at different output power or switching frequency, as shown by Fig. 14. Even though the conduction loss of body diode is larger than that of FWD during the dead time, it is still much smaller than the switching loss. Hence, it can be easily offset by the lower switching loss of VSI-SR compared with that of VSI-FWD. The two inverters show a similar efficiency at full load, however, the VSI-SR gradually differs from VSI-FWD with frequency increasing.

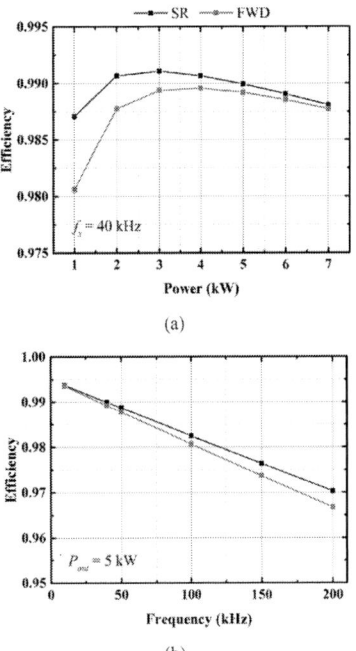

(a)

(b)

Fig. 14. Inverter efficiency at different output power (a) and switching frequency (b).

C. Converter Testing Results

Fig. 15. Prototype of three-phase inverter.

The prototype of the three-phase VSI-SR is developed, as Fig. 15 shows. It consists of one DC-bus, three phase legs and six gate drivers, all of which are connected with each other by PCB connectors or screws. A clear separation between the controller side and converter side is made so as to avoid the EMI issue. The open-loop test is conducted with SPWM control and implemented by TI DSP TMS320F28335. The prototype is tested up to 5 kW, with 14.5 Ω/2 mH RL load and the operation conditions given by Table 2. The experimental waveforms of phase current and line-to-line voltage are shown by Fig. 16.

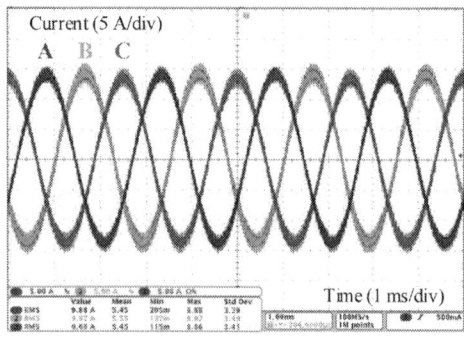

(a)

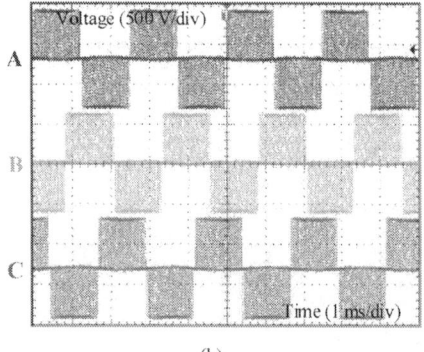

(b)

Fig. 16. Experimental current (a) and voltage (b) waveforms.

V. CONCLUSION

The excellent reverse recovery characteristics of body diode of SiC MOSFET makes it possible to remove the anti-parallel FWD in the SiC three-phase VSI. Compared with the conventional inverter topology with FWD, it shows reduced switching loss, especially at light load and high switching frequency. Although the high forward voltage drop of body diode will increase the conduction loss during dead time, the high-speed switching advantage of SiC MOSFET makes it practical to minimize the dead time and body diode conduction loss, which occupies less than 3% of total loss. Hence, the use of anti-parallel FWD in off-the-shelf SiC power module becomes unnecessary. By removing the anti-parallel SiC Schottky diode, the power density in terms of weight as well as volume will increase. In addition, it brings additional benefits including reduced cost, improved system reliability and simplified converter layout design.

ACKNOWLEDGEMENT

This work was conducted within the Rolls-Royce@NTU Corporate Lab with support from the National Research Foundation (NRF) Singapore under the Corp Lab@University Scheme.

REFERENCES

[1] H. Kim, T. M. Jahns, and G. Venkataramanan, "Minimization of reverse recovery effects in hard-switched inverters using CoolMOS power switches," in *Proc. IEEE Ind. Appl. Soc. Annu. Meeting*, 2001, pp. 641-647.

[2] V. Yousefzadeh and D. Maksimovic, "Sensorless optimization of dead times in DC-DC converters with synchronous rectifiers," *IEEE Trans. Power Electron.*, vol. 21, pp. 994-1002, 2006.

[3] T. Reiter, D. Polenov, H. Probstle, and H. Herzog, "Optimization of PWM dead times in DC/DC-converters considering varying operating conditions and component dependencies," in *Proc. IEEE Euro. Conf. Power Electron. Appl.*, 2009, pp. 1-10.

[4] B. Callanan, "Application considerations for silicon carbide MOSFETs," *Cree Inc. Application Note*, 2011.

[5] SiC half bridge module CAS120M12BM2, Wolfspeed Power [Online]. Available: http://www.wolfspeed.com/Power/Products/SiC-Power-Modules/SiC-Modules/CAS120M12BM2

[6] Z. Wang, J. Ouyang, J. Zhang, X. Wu, and K. Sheng, "Analysis on reverse recovery characteristic of SiC MOSFET intrinsic diode," in *Proc. IEEE Energy Convers. Congr. Expo.*, 2014, pp. 2832-2837.

[7] J. Jordan, V. Esteve, E. Sanchis-Kilders, E. J. Dede, E. Maset, J. B. Ejea, et al., "A comparative performance study of a 1200 V Si and SiC MOSFET intrinsic diode on an induction heating inverter," *IEEE Trans. Power Electron.*, vol. 29, pp. 2550-2562, 2014.

[8] T. Funaki, M. Matsushita, M. Sasagawa, T. Kimoto, and T. Hikihara, "A study on SiC devices in synchronous rectification of dc-dc converter," in *Proc. IEEE Appl. Power Electron. Conf. Expo.*, 2007, pp. 339-344.

[9] H. Liu, H. Wu, Y. Lu, Y. Xing, and K. Sun, "A high efficiency inverter based on SiC MOSFET without externally antiparalleled diodes," in *Proc. IEEE Appl. Power Electron. Conf. Expo.*, 2014, pp. 163-167.

[10] S. Yin, T. Wang, K. J. Tseng, J. Zhao, and X. Hu, "Electro-thermal modeling of SiC power devices for circuit simulation," in *Proc. IEEE Ind. Electron. Soc. Conf.*, 2013, pp. 718-723.

[11] K. Berringer, J. Marvin, and P. Perruchoud, "Semiconductor power losses in AC inverters," in *Proc. IEEE Ind. Appl. Soc. Annu. Meeting*, 1995, pp. 882-888.

[12] B. Ozpineci, L. M. Tolbert, S. K. Islam, and M. Hasanuzzaman, "Effects of silicon carbide (SiC) power devices on HEV PWM inverter losses," in *Proc. IEEE Ind. Electron. Soc. Conf.*, 2001, pp. 1061-1066.

[13] H. Zhang, "Electro-thermal modeling of SiC power electronic systems," Ph.D. Dissertation, University of Tennessee, Knoxville, 2007.

[14] A. Q. Huang and B. Zhang, "Comparing SiC switching power devices: MOSFET, NPN transistor and GTO thyristor," *Solid-State Electron.*, vol. 44, pp. 325-340, 2000.

[15] Y. Kashihara and J.-i. Itoh, "The performance of the multilevel converter topologies for PV inverter," in *Proc. IEEE Int. Conf. Integrated Power Electron. Syst.*, 2012, pp. 1-6.

[16] SiC MOSFET C2M0080120D, Wolfspeed Power [Online]. Available: http://www.wolfspeed.com/Power/Products/MOSFETs/TO247/C2M008 0120D

[17] J. Strydom, M. De Rooij, and J. Van Wyk, "A comparison of fundamental gate-driver topologies for high frequency applications," in *Proc. IEEE Appl. Power Electron. Conf. Expo.*, 2004, pp. 1045-1052.

[18] M. Z. Choo, "Keep hybrid powertrain drives noise free by rejecting dvdt noise with isolated-gate drivers," *Avago Technologies Application Note*, 2011.

[19] M. C. Caponet, F. Profumo, R. W. De Doncker, and A. Tenconi, "Low stray inductance bus bar design and construction for good EMC performance in power electronic circuits," *IEEE Trans. Power Electron.*, vol. 17, pp. 225-231, 2002.

[20] K. Wada, M. Ando, and A. Hino, "Design of DC-side wiring structure for high-speed switching operation using SiC power devices," in *Proc. IEEE Appl. Power Electron. Conf. Expo.*, 2013, pp. 584-590.

[21] SiC Schottky diode C4D20120A, Wolfspeed Power [Online]. Available: http://www.wolfspeed.com/Power/Products/Diodes/TO220/C4D20120A

[22] N. Mohan and T. M. Undeland, *Power electronics: converters, applications, and design*: John Wiley & Sons, 2007.

Comparison and Evaluation of Common Mode EMI Filter Topologies for GaN-based Motor Drive Systems

Casey T. Morris, Di Han, *Student Member, IEEE*, Bulent Sarlioglu, *Senior Member, IEEE*
Wisconsin Electric Machines and Power Electronic Consortium (WEMPEC)
University of Wisconsin-Madison
Madison, WI 53706 USA
sarlioglu@wisc.edu

Abstract--As many industries look to wide bandgap (WBG) devices to replace Si devices in power applications, their effects on the power converter need to be first understood in full. While the benefits of WBG devices are widely known, including high switching frequency operating capability, the special characteristics of these new devices have the potential to cause other issues, in particular electromagnetic interference/compatibility (EMI/EMC). In this paper, multiple EMI filter topologies are investigated for a motor drive system utilizing WBG devices, GaN HEMTs. The different topologies are simulated using LTSpice circuit simulation software and the results are analyzed and complied with MATLAB. The tradeoffs of different filter topologies, namely attenuation versus efficiency and ground current, are demonstrated, providing crucial information for system level designers implementing WBG power switching devices.

Keyword—common mode, electromagnetic interference, filter topologies, GaN, motor drives, wide bandgap

I. INTRODUCTION

In recent years, special interest has been paid to wide bandgap (WBG) semiconductor based power switching devices, as they have many inherent benefits compared to their silicon counterparts. Examples of these benefits include reduced on state resistance, higher switching speed capabilities, and higher temperature characteristics [1], [2]. WBG devices, such as silicon carbide (SiC) MOSFETs and gallium nitride (GaN) HEMTs, have been researched and studied in a variety of applications and have been shown to reduce weight and increase efficiency [3], [4]. Despite these benefits, it is necessary to understand the whole spectrum of impact that the implementation of these new devices will have on switching power converters and drives. For instance, the impact of dead-time on GaN switches was investigated in [5]. Another area of particular interest is conducted electromagnetic interference (EMI).

Although the enhanced switching speed capability of WBG devices has been shown to reduce output filter sizes [6],

TABLE I. PARAMETER VALUES USED IN STUDY

Filter Specs.				Motor Specs.		
Name	Value	Unit		Name	Value	Unit
V_{bus}	100	V		C_m	1.2	nF
f_s	200	kHz		R_m	1	Ω
C_{dc}	30	μF		R_s	10.6	Ω
L_{dc_cm}	1.2	mH		L_s	6.7	mH
C_{cm}	100	nF		L_r	10	mH
L_{ac_cm}	2.7	mH		L_m	76.4	mH
R_{dm}	7	Ω		R_r/s	5.2	Ω

the EMI induced in the converter also increases [7]. This is a result of the increased dv/dt and di/dt transients, which lead to more noise production. To safely and reliably use these new converters at the higher switching speeds, some form of EMI mitigation is required, usually in the form of passive filters [8], [9]. Research has focused on the development of various EMI filters for power converters, both with Si components and WBG devices [10]–[12]. Different design methods have been proposed [7], [13], [14], as well as research into other related practical issues, such as inductor saturation [15] and grounded heat sink induced parasitics [16].

The contribution of this paper is to achieve a comprehensive study of the EMI of GaN-based power converters and potential filter topologies needed. Due to the coupling effects between the differential and common mode filters shown in [17], [18], only the common mode (CM) noise and filter are considered. Different CM filter topologies are studied and characterized in terms of their effectiveness and complexity in a GaN motor drive application, providing EMI filter designers a basis to develop the best filter for their needs.

978-1-4673-9551-9/16 $31.00 © 2016 IEEE

The layout of this paper is as follows: Section II describes the modeling and measurement of common mode noise; Section III presents the filter topologies that are studied; Section IV compares the filters in terms of attenuation, ground current and their efficiency; and finally, Section V draws conclusions.

II. CM EMI EMISSION, MODELING, AND MEASUREMENT

CM noise is the noise signal between the neutral phase and the ground node, i.e. it is the high frequency current traveling from the drive and motor to the ground plane. It is caused by many of the parasitics in the system, including the motor winding to case capacitance, device to heatsink capacitance, etc. [16], [19], [20]. These elements provide an alternative path to ground and can negatively impact the power supplies, inverter, and motors. The additional current flowing through the inverter increases losses and results in the need of additional cooling. In the motor, the result of the CM noise can be seen in the bearing currents, which are potentially threatening to the health of the machine. Thus, eliminating this noise is beneficial; however, a large challenge is modelling and predicting the EMI of a physical system.

To facilitate the design of the EMI filter, modelling of the inverter, motor and its parasitics is crucial, as measuring EMI noise before filter design is time consuming and laborsome [21], [18]. Some modeling methods have been suggested, [16], [20], and in this paper, the method discussed in [16] to model the parasitics is used. This model is sufficient for the purpose of this study as it lends insights into the dominating factors to be considered when designing an EMI filter.

The machine and its parasitics studied in this paper are modelled as in Figure 1. The values of the components are displayed in Table I, many of which were derived from experimentally testing an induction motor in the lab. The study is performed at steady state operating conditions. The standardized method to measure EMI is through a line impedance stabilization network (LISN), which is connected between the source and inverter as shown in Figure 2. The common mode voltage is then calculated as:

$$V_{cm} = \frac{V_{pp} + V_{nn}}{2} \tag{1}$$

where V_{pp} and V_{nn} are the voltages between the DC side lines and ground, labeled in Figure 2. The measurements are then transformed to the frequency domain and their limits are defined depending on application.

III. FILTER DERIVATIONS

The study was performed comparing 3 different EMI filter topologies to a baseline case with no filters. The parameters of the study are shown in Table I, and the various iterations of the filters are shown in Figure 3. The study was performed using the LTSpice simulation software, and each filter is

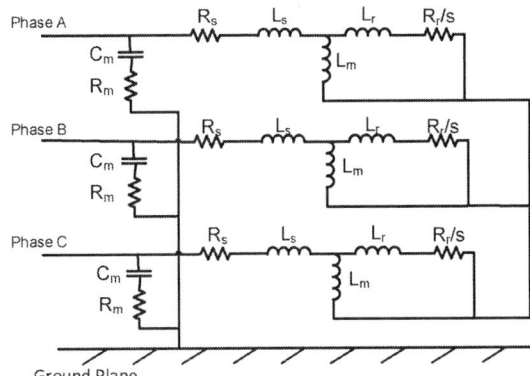

Figure 1. Steady state model of induction machine with parasitics included at input.

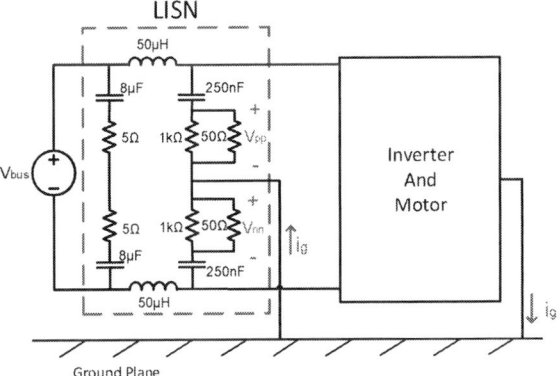

Figure 2. Configuration of LISN to measure CM and DM

simulated. In all of the simulations, an DM filter comprised of a floating LC connection is present and is the same for all CM filter iterations. The values are designed based on the work done in [18].

A. GaN Motor Drive Characteristics

The inverter used is a 3 phase, two level VSI, utilizing sine-PWM. The devices used in the study are the EPC2010 GaN HEMT, of which EPC publishes a model for simulation purposes. As previously mentioned, GaN devices have many inherent electrical benefits, such as increased switching speed capability due to their fast turn-on and turn-off functionality. However, these increased turn-on and turn-off times increase the di/dt and dv/dt transients, which contribute to the CM EMI noise generation. Because of this, special consideration needs to be paid to the CM EMI noise of motor drives utilizing GaN devices operating at 200 kHz, as presented in this study.

In order to most accurately derive valid values of the CM EMI filter topologies to compare, it is important to understand how much attenuation is needed. With the measurements of the induction machine presented earlier, as well as the addition of common parasitics modeled in the previous section, a baseline simulation was performed to see how much CM noise was inherent to the system. In order to better understand the severity of the CM noise with the GaN motor drive, the measured noise is compared to an EMI common mode emission standard, CISPR Class B, which is commonly used for industrial and commercial drives. This result is shown in Figure 4.

As shown in Fig. 4, the CM noise without a filter exceeds the acceptable limit by greater than 60 dBμV towards the lower end of the spectrum, namely near the first few

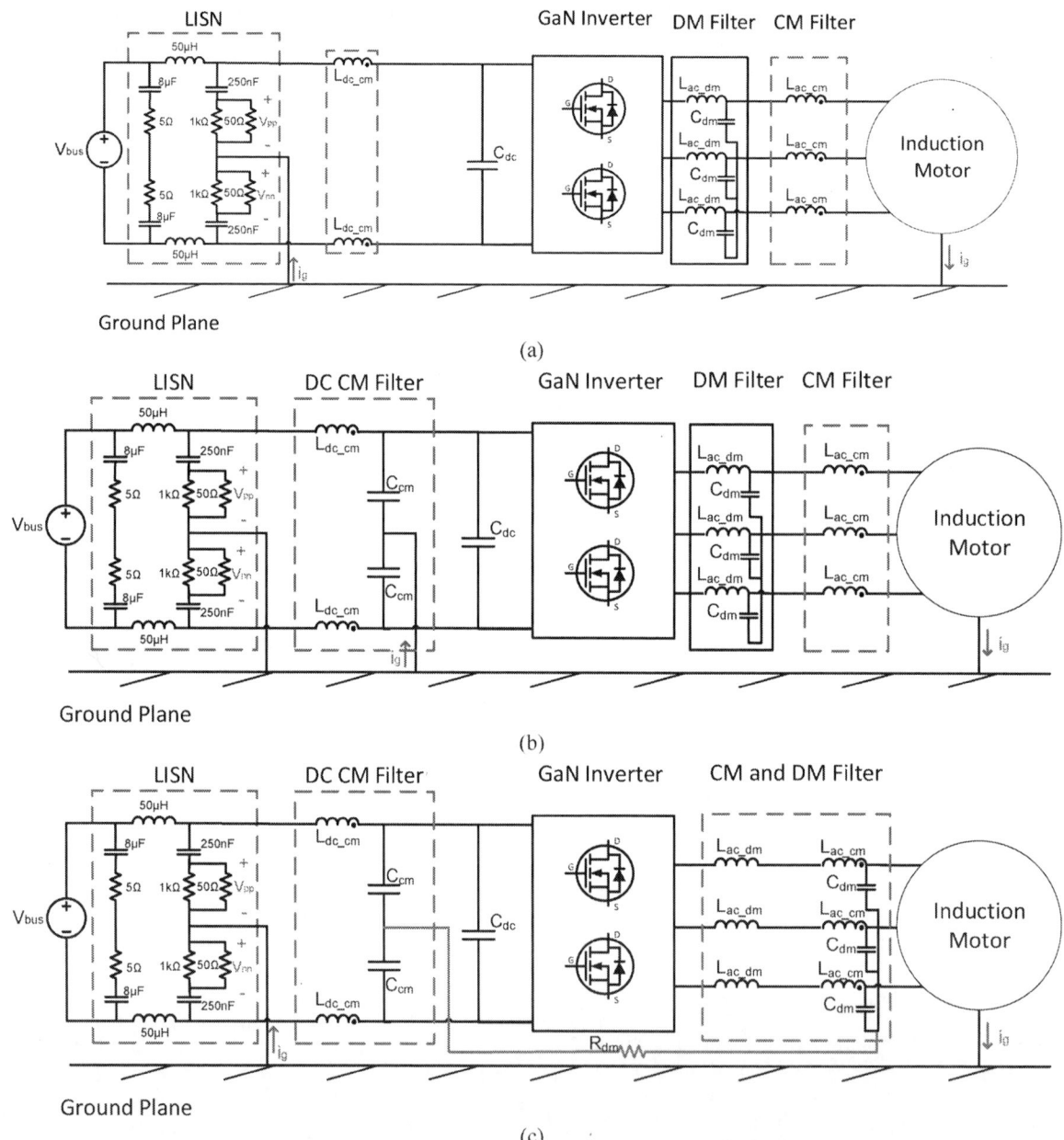

Figure 3. Configuration of various EMI filter topologies. (a) Filter 1 (b) Filter 2 and (c) Filter 3.

harmonics of the switching frequency. In this frequency range (<2 MHz), the noise must be greatly attenuated.

B. Filter Topology 1

The first filter topology used to do this is shown in the red dashed lines of Fig. 3(a). These common mode inductors (chokes) act like a low pass filter at the inputs (DC side) and outputs (AC side), attempting to suppress the high frequency noise. The values chosen (1.2 mH for the DC side and 2.7 mH for the AC side) are in the range commonly used in industry. They serve as very low impedance (~2 Ω) in the range of operating frequency (in this study a few hundred of Hz), but are high impedance (>kΩ) in the frequency range of needed attenuation.

C. Filter Topology 2

While Filter 1 can be effective, if further attenuation is needed, increasing the inductance values would be necessary. This is not ideal, not only because it might begin to have a noticeable effect at the power frequency, but also because the increased inductor size becomes too large. To counter this, Filter 2 is applied (Fig. 3(b)). The main difference between this filter and the previous filter is the addition of the small capacitors on the input side between the dc power lines and the ground plane.

The purpose of this capacitor is to enhance the attenuation of the filter at a more specific frequency, generally in the region of the switching frequency where reduction of noise is most needed. Thus, the value of this capacitor (100 nF) is selected to resonate with the previous input side inductor, providing higher impedances around the switching frequency.

D. Filter Topology 3

As mentioned in the introduction, a main concern for CM EMI noise in machine drives is the induced bearing currents, which are closely related to the ground currents. While the

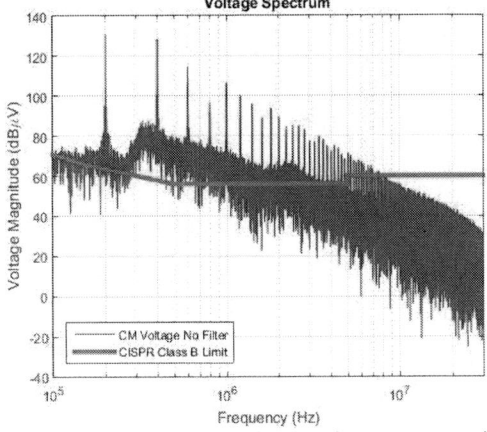

Figure 4. Unfiltered CM voltage noise spectrum with CISPR Class B limits.

previous two filters attenuate the CM voltage noise, it is also often important to reduce the circulating ground current.

There are two main differences between Filter 2 and Filter 3 shown in Figure 3(c): (1) The capacitors previously associated with the DM Filter are moved and placed after the CM inductor, and (2) the midpoint of the two common mode capacitors on the DC side is no longer connected to the ground plane, but instead, is connected to the neutral point of the AC output CM and DM filter capacitor through a small damping resistor. This connection provides a lower impedance path of the CM current to return to the input side rather than going through the machine and ground. While this may have a small adverse effect on the attenuation of the voltage noise, the main benefit is the potential bearing current being shunted away from the machine. The DM filter capabilities will also slightly change in this configuration; however, if designed well and with plenty of margin, it will not compromise the whole converter system.

IV. COMPARISON OF FILTERS

As mentioned, there are multiple perspectives with which to view the effectiveness of a filter. This section considers three different aspects: attenuation, ground current, and efficiency.

A. Attenuation

As shown in Fig. 4 and 5(a), the measurement of the CM EMI noise without a filter shows that the noise exceeds the acceptable limit by greater than 60 dBμV. Filter 1 attempts to attenuate the noise, but it is not completely effective, attenuating the CM noise by only 30 dBμV at the peak of the spectrum. Fig. 5(b) shows the CM noise exceeds the limits for the first 7 harmonics of the switching speed, an improvement, but not adequate.

Filter 2, Fig 5(c), is shown to have the most attenuation, providing at least 10 dBμV of margin in the entire range of interest. The addition of the small capacitor has had a great impact as expected and prevented the need to further increase the CM choke size.

Filter 3, Fig. 5(d), attenuates the CM noise below the limit in the entire range, with a few dBμV of margin, although less than Filter 2 at the lower end of the spectrum. While having a higher margin is generally preferred, there are a few other factors to consider before determining what type of CM EMI filter is the best solution for a given application.

B. Ground Current

Because the focus of this study is with respect to an electric motor drive application, reduction of the ground current is greatly emphasized. A reduction in ground current has the potential to lead to increased lifetime and reliability of the motor drive system.

Figure 6 and Table II both show the comparison of the filters with respect to the measured ground current. As shown,

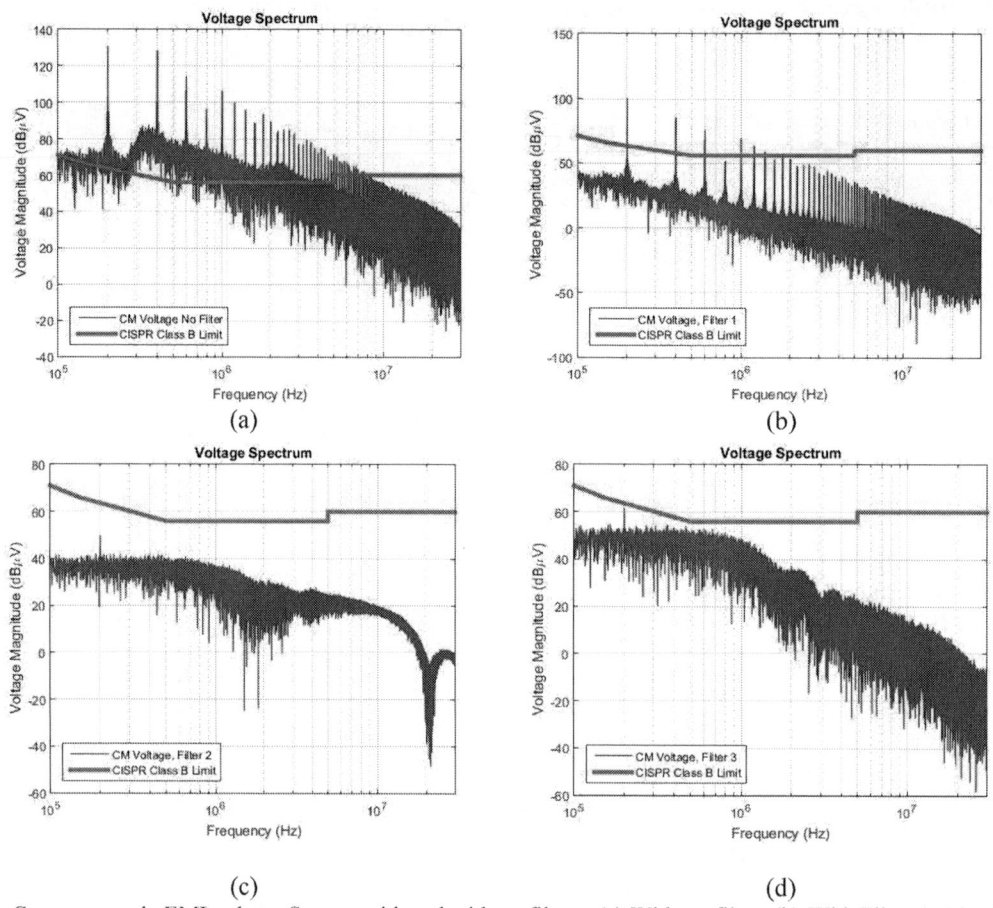

Figure 5. Common mode EMI voltage Spectra with and without filters. (a) With no filter. (b) With Filter 1, (c) with Filter 2, and (d) with Filter 3. Filter 2 has best attenuation, but benefits of Filter 3 are shown in Figure 6.

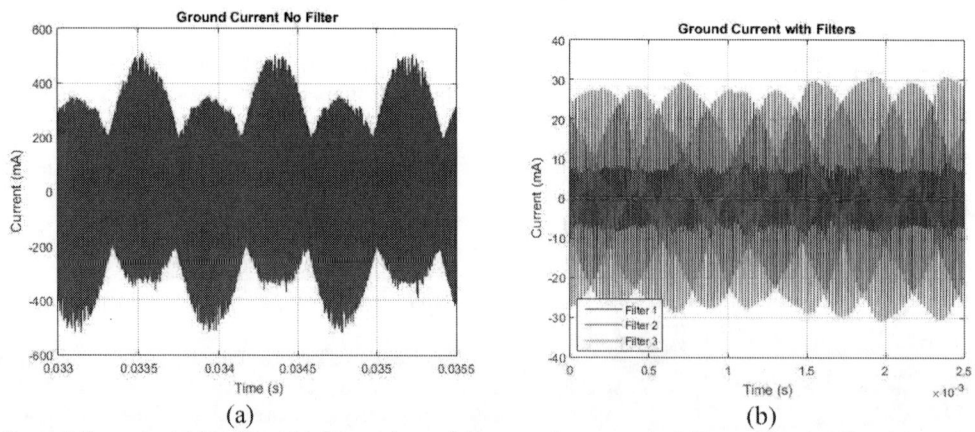

Figure 6. Ground Currents. (a) No filter (b) Comparison of the ground currents and filters. Clearly Filter 3 has a much more desirable ground current than Filter 2.

all filters greatly reduce the ground current relative to the pre-filtered scenario. Because Filter 1 was shown to not meet the attenuation requirements, it can be eliminated from further discussion when deciding which filter is most appropriate.

The tradeoff between Filter 2 and Filter 3, however, is shown in Fig. 6(b) and Table II, as the ground current is even worse than that of Filter 1, increasing from 4.6 mA 14.4 mA. However, Filter 3 has the added benefit of greatly reducing the ground current to 0.26 mA. This result is as expected, as some of the CM noise current is shunted away from the machine. Depending on the overall system, this filter topology may be more beneficial; however, a comparison must be made to ensure the power capability of the machine and the efficiency of the machine are not compromised by the addition of the filters.

TABLE II. GROUND CURRENT RMS VALUES.

Filter Type	Ground Current (mA$_{rms}$)
No Filter	231.55000
Filter 1	4.62510
Filter 2	14.40300
Filter 3	0.25678

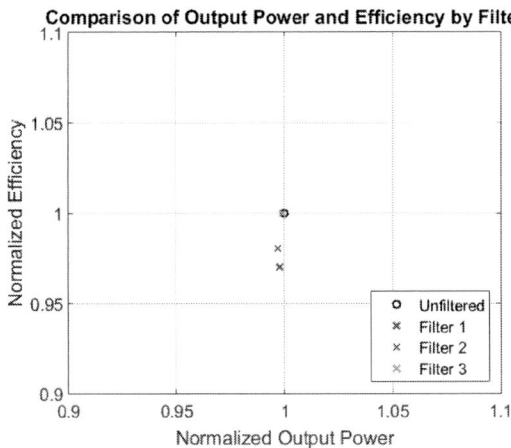

Figure 7. Output power and efficiency by filter.

C. Output Power and Efficiency

This analysis is crucial to ensure that the filters work as expected and do not attenuate any of the energy at the power frequencies and do not degrade the efficiency of the system. One drawback of the LTSpice simulation tool is its ability to accurately predict the core loss of the filters. Thus, the efficiency number is not the overall expected system efficiency. However, it is still useful and accurate to compare the loss of the switching devices as well as the output power

of the machine.

Figure 7 plots the output power of each filter topology and the loss of the GaN switching devices, both normalized to the un-filtered performance. As can be seen, none of the filters have a noticeable impact on either output power or efficiency. It should be noted that these are also dependent on the operating and load conditions, but similar trends will emerge. As a result, the attenuation and ground current circulation are the critical factors in the selection the filters for this study.

V. CONCLUSION

It has been shown that different filters have a variety of tradeoffs. The type of tradeoff favoring Filter 3 can be beneficial as reduced ground current circulating in the motor may result in extended lifetime and reliability. If efficiency is not a critical deciding point, i.e. if the efficiency is the same for all filters as in this case, Filter 3 may be the filter selected for this application. However, it is still important to sweep this analysis over the range of operating conditions to ensure that the margin presented for each filter holds. If it does not, further tuning of the filter values would be required.

Thus, this paper shows the impact of the various EMI filters needed for a GaN HEMT converter on the CM EMI emissions in a motor drive system. The various advantages and disadvantages of different filter topologies are compared, describing the techniques and considerations crucial to the design of an EMI filter early in the product development life cycle.

REFERENCES

[1] J. Millan, P. Godignon, X. Perpina, a Perez-Tomas, and J. Rebollo, "A Survey of Wide Bandgap Power Semiconductor Devices," *IEEE Trans. Power Electron.*, vol. 29, no. 5, pp. 2155–2163, 2014.

[2] M. Ostling, R. Ghandi, and C.-M. Zetterling, "SiC power devices - Present status, applications and future perspective," *in Proc. Power Semiconductor Devices and ICs (ISPSD) 23rd International Symposium on*, 2011, pp. 10–15.

[3] A. Elasser and T. P. Chow, "Silicon carbide benefits and advantages for power electronics circuits and systems," *Proc. IEEE*, vol. 90, no. 6, pp. 969–986, 2002.

[4] D. Han, J. Noppakunkajorn, and B. Sarlioglu, "Comprehensive Efficiency , Weight , and Volume Comparison of SiC- and Si-Based Bidirectional DC – DC Converters for Hybrid Electric Vehicles," *IEEE Trans. Veh. Technol.*, vol. 63, no. 7, pp. 3001–3010, 2014.

[5] D. Han and B. Sarlioglu, "Dead-Time Effect on GaN-Based Synchronous Boost Converter and Analytical Model for Optimal Dead-Time Selection," *IEEE Trans. Power Electron.*, vol. 8993, no. c, pp. 1–1, 2015.

[6] H. Kosai, J. Scofield, S. McNeal, B. Jordan, and B. Ray, "Design and performance evaluation of a 200C interleaved boost converter," *IEEE Trans. Power Electron.*, vol. 28, no. 4, pp. 1691–1699, 2013.

[7] H. F. Huang, L. Y. Deng, B. J. Hu, and G. Wei, "Techniques for improving the high-frequency performance of the planar CM EMI filter," *IEEE Trans. Electromagn. Compat.*, vol. 55, no. 5, pp. 901–908, 2013.

[8] P. Bogónez-Franco and J. B. Sendra, "EMI comparison between Si and SiC technology in a boost converter," *IEEE Int. Symp. Electromagn. Compat.*, pp. 1–4, 2012.

[9] Q. Ji, X. Ruan, and Z. Ye, "The Worst Conducted EMI Spectrum of Critical Conduction Mode Boost PFC Converter," *IEEE Trans. Power Electron.*, vol. 8993, no. 3, pp. 1–1, 2014.

[10] M. Jin and M. Weiming, "Power Converter EMI Analysis Including IGBT Nonlinear Switching Transient Model," *IEEE Trans. Ind. Electron.*, vol. 53, no. 5, pp. 1577–1583, 2006.

[11] B. Toure, J. Schanen, L. Gerbaud, Y. Avenas, R. Ruelland, R. Demaglie, and T. Meynard, "EMI study of a 70kW interleaved three-phase inverter for aircraft applications," in *Proc. IEEE Energy Conversion Congress and Exposition*, 2012, no. 1, pp. 623–628.

[12] K. Kostov, J. Rabkowski, and H. P. Nee, "Conducted EMI from SiC BJT boost converter and its dependence on the output voltage, current, and heatsink connection," in *Proc. 5th IEEE Annual International Energy Conversion Congress and Exhibition*, 2013, pp. 1125–1130.

[13] F. Luo, D. Dong, D. Boroyevich, P. Mattavelli, and S. Wang, "Improving High-Frequency Performance of an Input Common Mode EMI Filter Using an Impedance-Mismatching Filter," *IEEE Trans. Power Electron.*, vol. 29, no. 10, pp. 5111–5115, 2014.

[14] P. Pairodamonchai, S. Suwankawin, and S. Sangwongwanich, "Design and implementation of a hybrid output EMI filter for high-frequency common-mode voltage compensation in PWM inverters," *IEEE Trans. Ind. Appl.*, vol. 45, no. 5, pp. 1647–1659, 2009.

[15] F. Luo, S. Wang, F. Wang, D. Boroyevich, N. Gazel, and Y. Kang, "Common mode voltage in DC-Fed motor drive system and its impact on the EMI filter," in *Proc. IEEE Applied Power Electronics Conference and Exposition*, 2010, pp. 1272–1278.

[16] H. Akagi and T. Doumoto, "A passive EMI filter for preventing high-frequency leakage current from flowing through the grounded inverter heat sink of an adjustable-speed motor drive system," *IEEE Trans. Ind. Appl.*, vol. 41, no. 5, pp. 1215–1223, 2005.

[17] R. M. Tallam, G. L. Skibinski, T. a. Shudarek, and R. a. Lukaszewski, "Integrated differential-mode and common-mode filter to mitigate the effects of long motor leads on AC drives," *IEEE Trans. Ind. Appl.*, vol. 47, pp. 2075–2083, 2011.

[18] X. Zhang, D. Boroyevich, P. Mattavelli, J. Xue, and F. Wang, "EMI filter design and optimization for both AC and DC side in a DC-fed motor drive system," in *Proc. IEEE Applied Power Electronics Conference and Exposition*, 2013, pp. 597–603.

[19] J. Sun and L. Xing, "Parameterization of three-phase electric machine models for EMI simulation," *IEEE Trans. Power Electron.*, vol. 29, no. 1, pp. 36–41, 2014.

[20] O. a. Mohammed and S. Ganu, "FE-circuit coupled model of electric machines for simulation and evaluation of EMI issues in motor drives," *IEEE Trans. Magn.*, vol. 46, no. 8, pp. 3389–3392, 2010.

[21] F. Y. Shih, D. Y. Chen, Y. P. Wu, and Y. T. Chen, "A procedure for designing EMI filters for AC line applications," *IEEE Trans. Power Electron.*, vol. 11, no. 1, pp. 170–181, 1996.

Analysis of Thermal Cycling Stress on Semiconductor Devices of the Modular Multilevel Converter for Drive Applications

Xiangyu Han, Qichen Yang, Liyao Wu, and Maryam Saeedifard

School of Electrical and Computer Engineering
Georgia Institute of Technology
Atlanta, GA, 30332
xhan70@gatech.edu

Abstract—**Thermal cycling stress is one of the most potential factors challenging the reliability/lifetime of semiconductor devices and cause of their failures. This paper analyzes and investigates the impacts of various control strategies on thermal cycling stress of the semiconductor devices of the Modular Multilevel Converter (MMC) for drive applications. Various control strategies based on injecting a common-mode voltage and a circulating current are investigated and their impacts on power loss distribution and thermal cycling of the semiconductor devices of the MMC-based drive systems are evaluated. Simulation results in the MATLAB/SIMULINK software environment are presented to evaluate the accuracy of the analysis and impacts of three control strategies on power losses and thermal distribution of semiconductor devices of the MMC-based drive system.**

Keywords—Modular Multilevel Converter; Drive System; Thermal Cycling

I. INTRODUCTION

THE DC-AC Modular Multilevel Converter (MMC) has become the most promising converter topology for high-voltage applications because of its salient features, i.e., high efficiency, scalability/modularity, and superior harmonic performance [1]-[3]. Over the past few years, extensive research efforts have been made to address the technical challenges associated with the operation and control of the MMC and to improve its performance for various applications. Those applications mainly include High-Voltage DC (HVDC) transmission systems [2]-[6], variable speed drives [7]-[11], flexible AC transmission systems [12], [13], grid integration of energy storage units [14], [15], and railway systems [16]. One of the most important issues for medium/high-voltage application domains is the reliability of the power electronic systems as the cost of failure of those systems is significant. As the MMC circuit is built based upon stacking a large number of Submodules (SMs), its operation is vulnerable to any SM failure. The MMC SM circuit, under normal operation, suffers from unequal power loss distribution among its semiconductor devices. Consequently some of the devices undergo more thermal cycling stress, which among all stressors, has been identified as the most prominent killers of power devices [17]. Temperature variations create thermo-mechanical stress, e.g., creep and fatigue of die attach materials and bond-wire lift off of the IGBT modules, and in the long term, result in device

degradation and eventually failure [17], [18]. Failure of the device is related to the magnitude and frequency of temperature cycles, which vary based on application. Recent work on this subject [19]-[20] mainly focuses on the application of MMC for grid integration of wind and influence of wind speed and circuit topology on thermal distribution of its semiconductor devices. Nevertheless, thermal cycling and uneven power loss distribution among the semiconductor devices, which can compromise the reliability of the MMCs, have neither been studied nor investigated for drive applications.

This paper is focused on analysis of thermal cycling of the MMC when using in drive systems and evaluation of the impacts of different control strategies on the thermal cycling. Based on injecting a common-mode voltage and circulating current, the impacts of various common-mode voltage and circulating current control strategies on the power losses and thermal cycling of the semiconductor devices are investigated. Simulation results in the MATLAB/SIMULINK software environment are presented to evaluate the accuracy of the analysis and impacts of three control strategies on power losses and thermal distribution of semiconductor devices of the MMC-based drive system.

II. BASICS OF MMC

Fig. 1 shows a schematic diagram of a three-phase DC-AC MMC. As shown in Fig. 1, the MMC consists of two arms per phase-leg, where each arm comprises N series-connected, nominally identical SMs, and a series inductor l. While the SMs in each arm are controlled to generate the required AC phase voltage, the arm inductor suppresses the high-frequency components in the arm current. The upper (lower) arm of three phase-legs is represented by subscript "p" ("n"). Depending on the state of its two complementary switches, each SM of Fig. 1 can provide two voltage levels at its terminal, i.e., $v_{SM}^{xi,j} = 0$ or $v_{SM}^{xi,j} = v_C^{xi,j}$. $v_C^{xi,j}$ is the capacitor voltage of the ith SM in the phase-leg j, where $i = 1,2,\dots,2N; j = a,b,c$. The two switching states of SM-i in arm-x of phase-j are: 1) $S_1 = 1$ and $S_2 = 0$, corresponding to the ON-state or inserted, and 2) $S_1 = 0$ and $S_2 = 0$, corresponding to the OFF-state or bypassed. The required number of inserted SMs with each arm of the converter is determined by a pulse width modulation (PWM) strategy. Proper operation of the MMC of Fig. 1

978-1-4673-9551-9/16 $31.00 © 2016 IEEE

necessitates an active voltage balancing strategy, e.g., the sorting algorithm to maintain the SM capacitor voltages balanced at their nominal values, i.e., V_{dc}/N. In the MMC of Fig. 1, the upper and lower arm currents of phase-j, $j = a, b, c$, i.e., $i_{arm}^{p,j}$ and $i_{arm}^{n,j}$, are expressed by:

$$i_{arm}^{p,j} = i_{\text{circ},j} + \frac{i_j}{2}, \tag{1a}$$

$$i_{arm}^{n,j} = i_{\text{circ},j} - \frac{i_j}{2}, \tag{1b}$$

where $i_{\text{circ},j}$ and i_j represent the circulating and AC-side currents in phase-j, respectively. Therefore, the circulating current in phase-j can be expressed by

$$i_{\text{circ},j} = \frac{i_{arm}^{p,j} + i_{arm}^{n,j}}{2}. \tag{2}$$

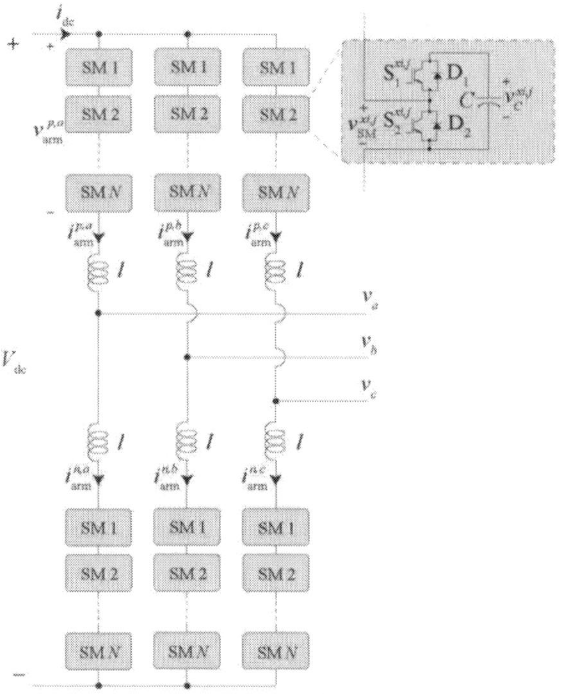

Fig. 1. Circuit diagram of a DC-AC MMC.

Circulating currents are dominantly negative-sequence second-order harmonic currents and do not impact the ac-side currents of the MMC. However, if not suppressed, they increase the peak and RMS values of the arm currents, leading to additional power losses, an increased amplitude of SM capacitor voltage ripple, and increased ratings of components. To address the aforementioned issues, in the technical literature, various circulating current suppression techniques have been proposed/investigated to suppress the circulating currents for MMC-HVDC applications in which the MMC is connected to a grid/load with fixed frequency [2]-[6]. Nevertheless, for MMC-based drive systems in which the MMC is connected to a variable-frequency load, the magnitude and frequency of the circulating currents are

controlled to a certain value such that the large magnitude of the SM capacitor voltage ripple at low frequencies are mitigated. The existing capacitor voltage ripple reduction techniques are based on the mitigation of low-frequency components in the capacitor voltage ripple by injecting a common-mode voltage at the AC-side voltages and a circulating current within the phase-legs of the MMC. The most viable technique is based on injecting a square-wave common-mode voltage and a sinusoidal circulating current (with or without a third harmonic component) [8]-[11]. The advantage of this technique over the other techniques is the reduction in peak/RMS value of circulating currents and the SM capacitor voltage ripple [8]. The common-mode reference voltage waveform and the circulating current are given by:

$$m_{\text{cm}} = \begin{cases} M_{\text{cm}} & \text{if } 0 < t \le \dfrac{1}{2f_{\text{cm}}} \\[2mm] -M_{\text{cm}} & \text{if } \dfrac{1}{2f_{\text{cm}}} < t \le \dfrac{1}{f_{\text{cm}}} \end{cases}, \tag{3}$$

$$i_{\text{circ},j} = i_j \left(m_1 \sin\omega_{\text{cm}}t + m_3 \sin3\omega_{\text{cm}}t \right) \left(\frac{1 - m_j^2}{\frac{4}{\pi}M_{\text{cm}}} \right) + \frac{m_j i_j}{2} - \frac{i_{dc}}{3}$$

Subject to $1 - m_1 - \dfrac{m_3}{3} = 0$, $\tag{4}$

where m_1 and m_3 in (4) are determined to minimize the RMS value of the circulating current and to mitigate the low-frequency components of the SM capacitor voltage ripple. Two of the solutions for (4) include: a) $m_1 = 1$, $m_3 = 0$, and b) $m_1 = 0.9$, $m_3 = 0.3$ [8]. The two solutions provide a trade-off in terms of the peak/RMS value of circulating current and SM capacitor voltage ripple. Therefore, the solution and the common-mode frequency chosen for a particular application are based on a Pareto optimal front based on the performance parameters.

III. ELECTRO-THERMAL MODELING

To analyze the thermal cycling stresses of the semiconductor devices, an electro-thermal model, consisting of an electrical model coupled with a thermal model is required so that their power losses along with their junction temperatures can be mutually and accurately estimated.

Semiconductor Device Model and Power Loss: The semiconductor power losses include the conduction and switching losses. The adapted power loss calculation method is based on curve fitting of the characteristic curves of semiconductor devices from their datasheets and simulated currents and voltages of the MMC.

Power Losses: To calculate the conduction losses of the IGBTs, the relationship between the conduction loss and device current is obtained by multiplying the collector current i_c by the corresponding forward voltage drop [9]. To obtain an approximate formula for the transistor conduction loss $P_{T,cond}$, a polynomial fitting curve is developed in the form of

$$P_{T,cond} = a(T_j)\, i_c^2 + b(T_j)\, i_c + c(T_j), \tag{3}$$

where a, b and c are fitting coefficients, which vary with junction temperature T_j to consider thermal impacts on device conduction losses. A similar method is applied to calculate the conduction loss $P_{D,cond}$ of the anti-parallel diode of the IGBT module.

Switching Losses: To calculate the switching losses, the switching loss energies are estimated by [21]

$$E_{T,on} = \left[a_{on}\left(T_j\right)i_c^3 + b_{on}\left(T_j\right)i_c^2 + c_{on}\left(T_j\right)i_c \right]\frac{V_{ds,B}}{V_{ds,ref}}, \quad (4)$$

$$E_{T,off} = \left[a_{off}\left(T_j\right)i_c^3 + b_{off}\left(T_j\right)i_c^2 + c_{off}\left(T_j\right)i_c \right]\frac{V_{ds,B}}{V_{ds,ref}}, \quad (5)$$

where $E_{T,on}$ and $E_{T,off}$ are the turn-on and turn-off switching loss energy, respectively, and $v_{ds,B}$ represents the real-time blocking voltage and $V_{ds,ref}$ is the blocking voltage under which the energy-current relationship is obtained. To account for the impacts of junction temperature, the fitting coefficients here are also functions of T_j. A similar method is applied to calculate the reverse-recovery energy loss $E_{D,rev}$ of the anti-parallel diode of the IGBT module. The coefficients of the above power loss model are listed in Table I. The switching power losses of the devices are calculated as the summation of all switching on/off energy losses within a switching interval. The total losses of an IGBT module are the summation of conduction and switching losses of the IGBT as well as its anti-parallel diode, given by

$$P_{TOTAL} = P_{T,cond} + P_{T,on} + P_{T,off} + P_{D,cond} + P_{D,rev}. \quad (6)$$

To evaluate the power losses of the MMC system, the aforementioned algorithm is embedded into the MMC system simulation model. The voltages and currents are measured and the switching transitions are recorded to calculate the power losses of each SM and, consequently, the semiconductor losses of the MMC system.

Table I
Coefficients of the power loss model

	Conduction Loss (W)		
	a	b	c
	0.001989485	2.790983821	-50.51645571
IGBT temp=25	Turn-on energy (mJ)		
	a	b	c
	8.58575E-06	-0.014284388	13.1841207
	Turn-off energy (mJ)		
	a	b	c
	3.51828E-07	-0.000790901	7.206399113
	Cond. Loss (W)		
	a	b	c
Diode temp=25	0.00187823	2.850026031	-84.10932374
	Reverse recovery loss energy (mJ)		
	a	b	c
	4.58079E-06	-0.010356233	7.527727024

Thermal Model: The thermal model uses a 4th order Cauer model shown in Fig. 2. The input to this model is the power dissipation calculated in the previous part, which is represented as a current source in the RC network; the output is junction temperature, which is represented as voltage. The parameters R and C of the model are extracted from the device datasheet [22]. The junction temperature is given by

$$T_h = Z_{h-a} \cdot (P_{loss-IGBT} + P_{loss-Diode}), \quad (7a)$$

$$T_c = Z_{c-h} \cdot P_{loss} + T_h, \quad (7b)$$

$$T_j = Z_{h-c} \cdot P_{loss} + T_c. \quad (7c)$$

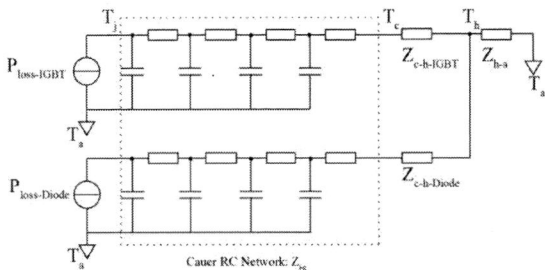

Fig. 2. Thermal model based on a 4th order Cauer network.

IV. SIMULATION RESULTS

Simulation results are carried out in the MATLAB/SIMULINK software environment for a MMC-based drive system with 6 SMs per arm, supplying a Permanent-Magnet Synchronous Machine (PMSM), with parameters listed in Table II. The MMC-based drive system is operating in steady state with a 1.5 MNm torque and a rotating speed of 0.07 rpm. The magnitude and frequency of the common mode voltage, i.e. M_{cm} and f_{cm} in (3), are 2.5 kV and 50 Hz, respectively. The device model coefficients are listed in Table I and Cauer model parameters are extracted based on the device datasheet [22].

Table II
Parameters of the MMC-Based PMSM Drive

Parameter	Value
Rated capacity S	9 MVA
Rated voltage (line-to-line)	8 kV
Rated frequency	35.77 Hz
Stator winding resistance r_s	72.2 mΩ
Stator leakage inductance L_{ls}	1.55 mH
d-axis inductance L_d	4.33 mH
q-axis inductance L_q	4.33 mH
Number of pole pairs	145
Number of SM	6
Arm resistor R	0.8 Ω
Arm inductor L	9 mH
SM Capacitance	8 mF
Device	Infineon FZ750R65KE3

In this paper, the impacts of three control strategies, i.e., Strategy I with a fundamental-frequency sine-wave circulating current and a square-wave common-mode voltage, Strategy II with a fundamental-frequency sine-wave circulating current and a sine-wave common-mode voltage, and Strategy III with a fundamental and 3rd-order circulating current and a square-wave common-mode voltage, as summarized in Table III, are investigated.

In the MMC-based drive system operating based on Strategy I, as shown in Figs. 3(a) and (b), the currents flowing through the semiconductor devices of each SM are different, leading to unequal power loss distribution among the devices. Unlike grid-connected MMC [19], in the first half cycle of waveforms shown in Fig. 3(b), D_1 and S_2 have the highest power losses while in the second half cycle, D_2 has the highest value followed by D_1. As shown in Fig. 3(c), the junction temperature profile of each SM in the MMC-based PMSM drive system consists of a 50 Hz and a 1 Hz cycle. The lower frequency cycle, which is due to injection of circulating current, dominates the magnitude of junction temperature variations.

Table III
Summary of three investigated control strategies

Strategy	Common Mode Voltage	Circulating Current
I	Square wave	1st order component
II	Sine wave	1st order component
III	Square wave	1st order plus 3rd order component

The current, power loss and junction temperature of each semiconductor device in one SM when Strategies II and III are adopted, are shown in Figs. 4(a)-(c) and Figs. 5 (a)-(c), respectively. In addition, the rain flow counting algorithm is applied to count the junction temperature cycles within 1 second. Since S_2, D_1, and D_2 are the most stressed switches, the counting results of temperature cycles of S_2, D_1 and D_2 for all 3 control strategies are shown in Figs. 6 (a)-(c), Figs. 7 (a)-(c) and Figs.8 (a)-(c).

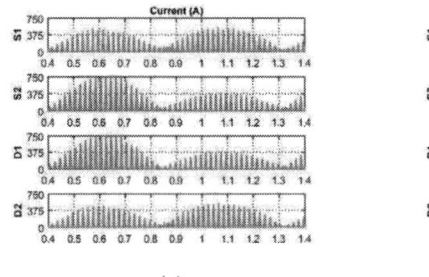 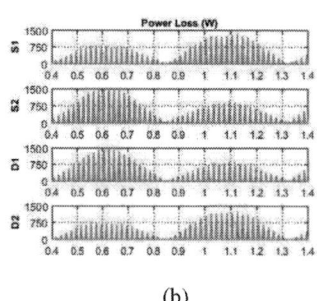 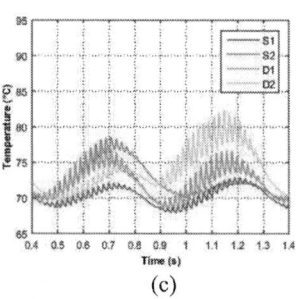

| (a) | (b) | (c) |

Fig. 3. Simulated waveforms of a single SM of the MMC-based PMSM drive system using Strategy I: (a) current of each device, (b) power loss of each device, and (c) temperature of each device.

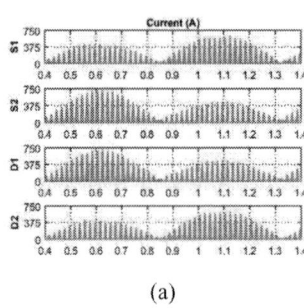 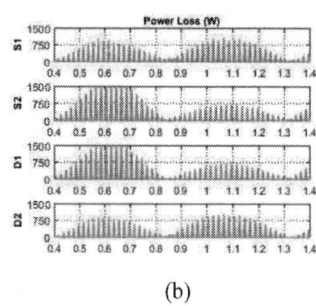 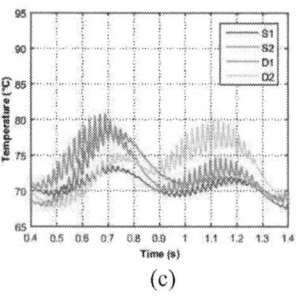

| (a) | (b) | (c) |

Fig. 4. Simulated waveforms of a single SM of the MMC-based PMSM drive system using Strategy II: (a) current of each device, (b) power loss of each device, and (c) temperature of each device.

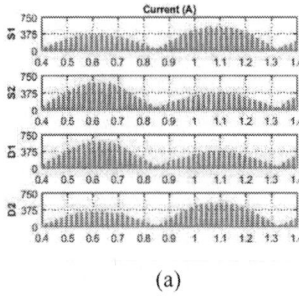 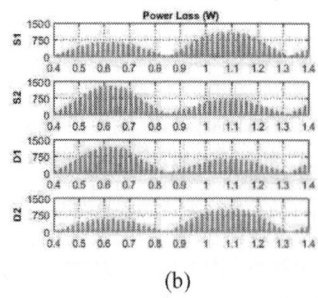 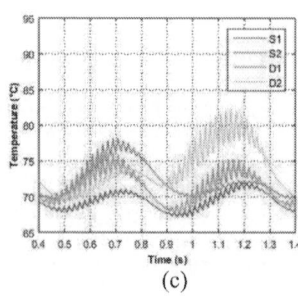

| (a) | (b) | (c) |

Fig. 5. Simulated waveforms of a single SM of the MMC-based PMSM drive system using Strategy III: (a) current of each device, (b) power loss of each device, and (c) temperature of each device.

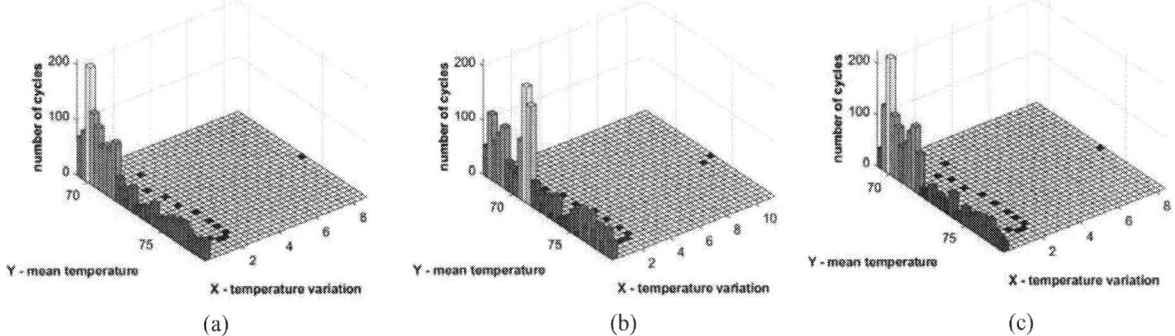

Fig. 6. Rain flow counting result of temperature cycles of S_2 for (a), (b), and (c) Strategies I, II, and III, respectively.

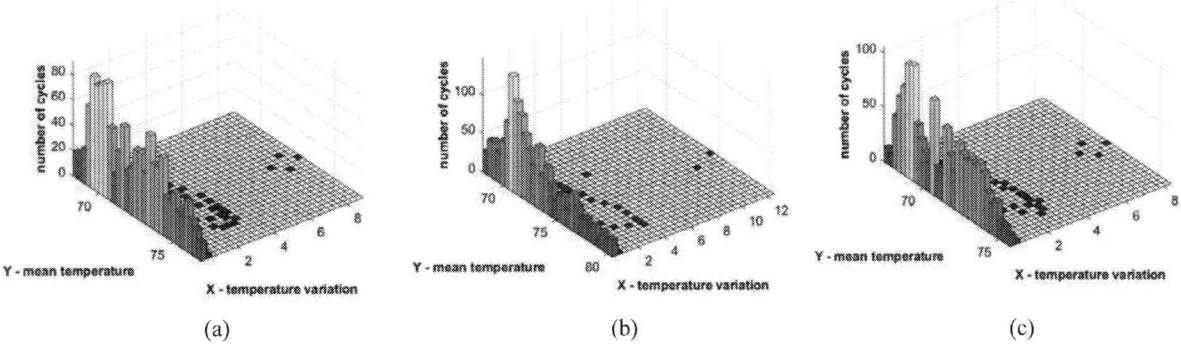

Fig. 7. Rain flow counting result of temperature cycles of D_1 for (a), (b), and (c) Strategies I, II, and III, respectively.

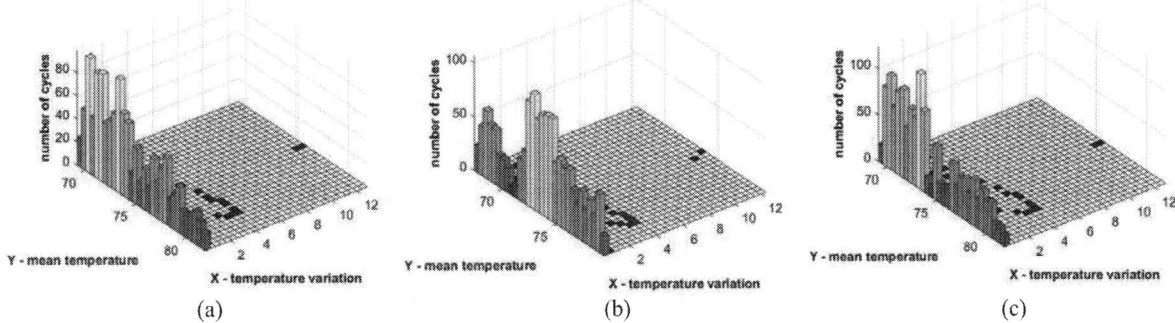

Fig. 8. Rain flow counting result of temperature cycles of D_2 for (a), (b), and (c) Strategies I, II, and III, respectively.

By comparison of Fig. 3(c) with Fig. 4(c) and Figs. 6(a), 7(a), 8(a) with Figs. 6(b), 7(b), 8(b), the impact of common-mode voltage can be analyzed. When using Strategy II, the thermal cycling stresses on S_2, D_1 and D_2 are balanced while using Strategy I, S_2 and D_1 are less stressed than D_2. However, the stress levels of all of the switches, when Strategy II is adopted, are higher than of those from Strategy I. Nevertheless, the average temperatures of the largest temperature cycles are almost the same for all the 3 strategies. Therefore, a sine common-mode voltage can stress the devices in a more balanced way, although with increases magnitude of cycles.

By comparison of Fig. 3(c) with Fig. 5(c) and of Figs. 6(a), 7(a), 8(a) with Figs. 6(c), 7(c), 8(c), the impact of circulating current can be analyzed. As shown in the figures, there is no sensible difference between Strategy I and Strategy III, which means the added 3rd-order circulating current has no impacts on the temperature cycles.

V. CONCLUSION

This paper analyzes the impacts of various control strategies on the thermal cycling of semiconductor devices of the MMC in drive applications. Unlike the grid-connected MMC, the MMC-based drive system has a low frequency

978-1-4673-9551-9/16 $31.00 © 2016 IEEE

thermal cycle, which is dominant. For each control strategy, the unequal power loss distributions among the semiconductor devices lead to different temperature cycles. As analyzed in this paper, the injected common-mode voltage has a significant impact on power loss and temperature distribution, while injection of 3rd-order circulating current has negligible influence.

REFERENCES

[1] S. Debnath, J. Qin, B. Bahrani, M. Saeedifard, and P. Barbosa, "Operation, control, and applications of the modular multilevel converter: A review," *IEEE Trans. Power Electron.*, vol. 30, no. 1, pp. 37–53, Jan. 2015.

[2] J. Liang, F. Dijkhuizen, and G.D. Demetriades, "Modular Multilevel Converters for HVDC Applications: Review on Converter Cells and Functionalities," *IEEE Trans. Power Del.*, vol. 30, no. 1, pp. 18-36, 2015.

[3] H. Liu, K. Ma, Z. A. Qin, P. C. Loh, F. Blaabjerg, "Lifetime Estimation of MMC for Offshore Wind Power HVDC Application," *Journal of Emerging and Selected Topics in Power Electronics*, 2015.

[4] Q. Tu, Z. Xu, Y. Chang, and L. Guan, "Suppressing dc voltage ripples of MMC-HVDC under unbalanced grid conditions," *IEEE Trans. Power Del.*, vol. 27, no. 3, pp. 1332–1338, July 2012.

[5] S. Li, X. Wang, Z. Yao, T. Li, and Z. Peng, "Circulating current suppressing strategy for MMC-HVDC based on non-ideal proportional resonant controllers under unbalanced grid conditions," *IEEE Trans. Power Electron.*, vol. 30, no. 1, pp. 387–397, Jan 2015.

[6] X. Shi, Z. Wang, B. Liu, Y. Liu, L. Tolbert, and F. Wang, "Characteristic investigation and control of a modular multilevel converter-based HVDC system under single-line-to-ground fault conditions," *IEEE Trans. Power Electron.*, vol. 30, no. 1, pp. 408–421, Jan 2015.

[7] A. Korn, M. Winkelnkemper, and P. Steimer, "Low output frequency operation of the modular multi-level converter," in *Proc. IEEE Energy Convers. Congr. Expo.*, Sep. 2010, pp. 3993–3997.

[8] S. Debnath, J. Qin, M. Saeedifard, "Control and Stability Analysis of Modular Multilevel Converter Under Low-Frequency Operation," *IEEE Trans. Ind. Electron.*, vol.62, no.9, pp.5329-5339, Sept. 2015

[9] M. Hagiwara, K. Nishimura, and H. Akagi, "A medium-voltage motor drive with a modular multilevel PWM inverter," *IEEE Trans. Power Electron.*, vol. 25, no. 7, pp. 1786–1799, Jul. 2010.

[10] M. Hagiwara, I. Hasegawa and H. Akagi, "Start-up and low-speed operation of an electric motor driven by a modular multilevel cascade inverter," *IEEE Trans. Ind. Appl.*, vol. 49, no. 4, pp.1556–1565, 2013.

[11] A. Antonopoulos, L. Angquist, S. Norrga, K. Ilves, L. Harnofes, and H.P. NEE, "Modular multilevel converter ac motor drives with constant torque from zero to nominal speed," *IEEE Trans. Ind. Appl.*, vol. 50, no. 3, pp.1982–1993, 2014.

[12] H. Mohammadi and M. Bina, "A transformerless medium-voltage STATCOM topology based on extended modular multilevel converters," *IEEE Trans. Power Electron.*, vol. 26, no. 5, pp. 1534–1545, May 2011.

[13] S. Du and J. Liu, "A study on DC voltage control for chopper-cell based modular multilevel converters in D-STATCOM application," *IEEE Trans. Power Del.*, vol. 28, no. 4, pp. 2030–2038, Oct. 2013.

[14] M. Vasiladiotis and A. Rufer,"Analysis and Control of Modular Multilevel Converters With Integrated Battery Energy Storage," *IEEE Trans. On Power Electron., Vol. 30, no. 1*, pp. 163–175, 2015.

[15] A. Hillers and J. Biela, "Optimal design of the modular multilevel converter for an energy storage system based on split batteries," in *Proc. Eur. Conf. Power Electron. Appl.*, 2013, pp. 1–11.

[16] M.Winkelnkemper, A. Korn, and P. Steimer, "A modular direct converter for transformerless rail interties," in *Proc. IEEE Int. Symp. Ind. Electron.*, 2010, pp. 562–567.

[17] U.-M. Choi, F. Blaabjerg, and K.-B. Lee, "Study and handling methods of power IGBT module failures in power electronic converter systems," *IEEE Trans. Power Electron.*, vol. 30, no. 5, pp. 2517-2533, 2015.

[18] H. Wang, K. Ma, F. Blaabjerg, "Design for reliability of power electronic systems," *Proc. of IECON*, pp. 33-44, 2012.

[19] K. Ma, F. Blaabjerg, and M. Liserre, "Thermal Analysis of Multilevel Grid-Side Converters for 10-MW Wind Turbines Under Low-Voltage Ride Through," *IEEE Trans. Ind. Appl.*, vol. 49, No.2, pp. 909-921, 2013

[20] Q. Tu, Z. Xu, and L. Xu, "Reduced switching-frequency modulation and circulating current suppression for modular multilevel converters," *IEEE Trans. Power Del.*, vol. 26, no. 3, pp. 2009–2017, Jul. 2011.

[21] L. Wu, J. Qin, M. Saeedifard, O. Wasynczuk, and K. Shenai, "Efficiency evaluation of the modular multilevel converter based on Si and SiC switching devices for medium/high-voltage applications," *IEEE Trans. Electron Devices*, vol. 62, no. 2, pp. 286-293,2015.

[22] FZ750R65KE3 datasheet, available from: https://www.infineon.com/

Fault Tolerant Topologies of Five-Level Active Neutral-Point-Clamped Converters

Jun Li

ABB US Corporate Research Center
Raleigh, NC, USA
jun.li@us.abb.com

Abstract—This paper proposed the fault tolerant topologies (FTTs) of five-level active neutral-point-clamped (5L-ANPC) converters. In the proposed approaches, the traditional 5L-ANPC converters are modified by adding a low count of low-cost components and are operated with the proposed control and modulation schemes. Simulation results show that the proposed 5L-ANPC FTTs can achieve fault tolerant operation under device open or short failure conditions, while generate the same maximum modulation index and the rated output power as normal operation. The proposed concept improves the reliability and availability of the 5L-ANPC converter systems.

Keywords—five-level; active neutral-point-clamped, 5L-ANPC; fault tolerant; topologies; device open failure; device short failure

I. INTRODUCTION

The five-level active neutral-point-clamped (5L-ANPC) converter, shown in Figure 1, is an emerging technology for MV high power applications, such as MV drives and FACTs. During normal operation, by properly switching the devices according to Table 1, the 5L-ANPC phase-leg can generate five-level output voltage waveforms. Meanwhile, the phase capacitor (Cf) voltages can be regulated by using the redundant switching states (RSSs) of **V5/V6** and **V1/V2** [1] [2].

Many researchers have investigated fault tolerant operation of various multilevel converter topologies under device failure conditions, such as 3L-NPC topology [3] [4], 3L-ANPC topology [5], flying capacitor topology [6], cascaded H-bridge topology [7], modular multilevel converter (MMC) topology [8] and etc [9]. Fault tolerant control schemes were previously

TABLE I. SWITCHING STETES OF 5L-ANPC TOPOLOGY

T8	T7	T6	T5	T4	T3	T2	T1	Output voltage	Effect on V_{Cf}		Switching state
									$i_p>0$	$i_p<0$	
0	1	0	1	0	1	0	1	V	n/a	n/a	V7 (4)
0	1	0	1	0	1	1	0	V/2	+	-	V6 (3)
0	1	0	1	1	0	0	1	V/2	-	+	V5 (3)
0	1	0	1	1	0	1	0	0	n/a	n/a	V4 (2)
1	0	1	0	0	1	0	1	0	n/a	n/a	V3 (2)
1	0	1	0	0	1	1	0	-V/2	+	-	V2 (1)
1	0	1	0	1	0	0	1	- V/2	-	+	V1 (1)
1	0	1	0	1	0	1	0	-V	n/a	n/a	V0 (0)

'+' charge, '-' discharge, 'n/a' no impact

introduced for 5L-ANPC converters under device short and open failure conditions [10] [11], showing that the 5L-ANPC converters can achieve fault tolerant operation with the proposed control schemes. However, for most device failure conditions in [10] and [11], the 5L-ANPC converters have to operate in derating mode because the maximum modulation index is limited to 0.575 during the fault tolerant operation. This drawback also limits their applications, for example, grid connection where the maximum modulation in fault tolerant mode is still required to be kept at the same level as normal operation, which is 1.15 without over modulation.

In this paper, the fault tolerant topologies (FTTs) are proposed for 5L-ANPC converters by adding a low count of low-cost components. Moreover, by applying the proposed control and modulation schemes to the introduced 5L-ANPC FTTs, the converters can generate the rated output voltage amplitude under device short or open failure conditions, thus it does not require derating operation. Therefore, the proposed approaches greatly improve the reliability of 5L-ANPC converters and expand their applications.

II. OPERATION OF 5L-ANPC CONVERTER UNDER DEVICE FAILURE MODE

In [10] and [11], detailed operation of traditional 5L-ANPC converters under device open or short failure conditions have been illustrated thoroughly. Due to the space limit of the paper,

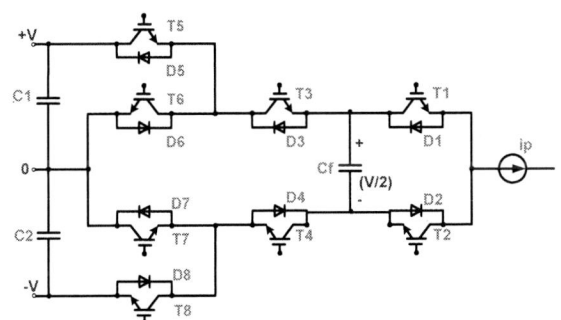

Fig.1. Phase-leg circuit of 5L-ANPC converters

978-1-4673-9551-9/16 $31.00 © 2016 IEEE

the operation of traditional 5L-ANPC converters under device fault scenarios are not reiterated here.

The impact of device failures on 5L-ANPC converter operation is summarized as follows. Some of the phase output voltage levels become unavailable after the device fails. In case of device open failure, this is caused by the missing of correct current conduction path. In case of device short failure, it becomes even worse since the shorted device may cause short circuit of DC-link capacitors or phase capacitors, resulting in more device damages due to overvoltage or overcurrent. In such conditions, the 5L-ANPC converter cannot operate properly with the conventional control and modulation schemes. By applying the proposed control schemes in [10] and [11], the 5L-ANPC converter can generate symmetrical output phase currents and achieve fault tolerant operation. However, one disadvantage is that, for most device failure conditions, the maximum modulation index is limited to 0.575, which is only half of that in normal operation. Therefore, the converters have to operate in derating mode. Moreover, they cannot be applied for the applications where full output voltage amplitude is required during fault tolerant operation, such as grid connection applications.

III. PROPOSED FAULT TOLERANT TOPOLOGIES (FTTs) OF 5L-ANPC CONVERTERS

In this paper, the fault tolerant topologies (FTTs) and the control and modulation schemes of 5L-ANPC converters are proposed. The objective is to add a low count of low-cost components to the standard 5L-ANPC converter to achieve fault tolerant operation under device open or short failure conditions, while the same maximum output voltage amplitude (or modulation index) and rated output power can be generated as those in normal operation of 5L-ANPC converters. This section firstly presents the proposed FTTs of 5L-ANPC converters for device short failure, device open failure, and a unified FTT for both short and open device failure. Then, the control and modulation schemes of the proposed FTTs and the voltage vector diagrams are analyzed and discussed.

A. FTT of 5L-ANPC Converter For Device Short Failure

Figure 2 shows the proposed FTT of 5L-ANPC converters for device short failure. Compared with the standard 5L-ANPC topology in Figure 1, two modifications are made. First, one bi-

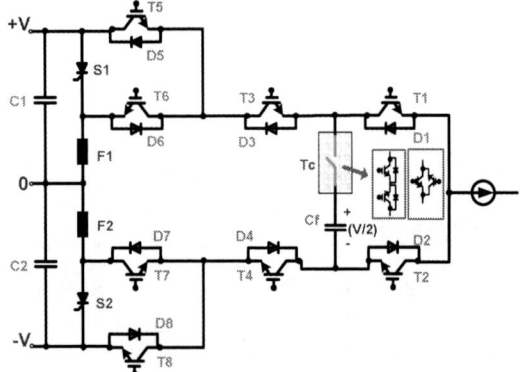

Fig. 2. FTT of 5L-ANPC converter for device short failure

directional switch **Tc** is connected in series with the phase capacitor Cf. In normal operation, switch **Tc** is always closed (in ON state). Therefore, a device with low conduction loss is preferred. For example, the bi-direction device can consist of two IGBTs (with anti-parallel diodes) connected in opposite direction or a reverse blocking IGBT (RG-IGBT). During fault tolerant operation, the switch **Tc** is operated in a controlled way to generate the proper output voltage levels. Since the switch **Tc** is operated much less frequently than the other devices of the converter, it can be considered as fault-free device in the proposed FTT topology. Second, a SCR and a fuse are inserted between the DC links and the NP of the converter, as shown in Figure 2. The SCR can be fired to blow the fuse to disconnect the NP from T6/D6 or T7/D7 for fault tolerant operation under short failure of T6/D6 or T7/D7.

B. FTT of 5L-ANPC Converter For Device Open Failure

Figure 3 shows the proposed FTT of 5L-ANPC converters for device open failure. Compared with the standard 5L-ANPC topology, a pair of SCRs is connected with each device in the phase leg in parallel except T6/D6 and T7/D7. When a device fails open, its paralleled SCRs are fired. Thus, the failed device is bypassed and behaves as short failure. As shown in Figure 3, no additional SCRs or fuses are required for T6/D6 or T7/D7 open failure. This is because open failure of T6/D6 or T7/D7 naturally disconnects them from the NP clamping path of the converter. Furthermore, a bi-directional switch **Tc** is added and operates in a similar way as that in Figure 2.

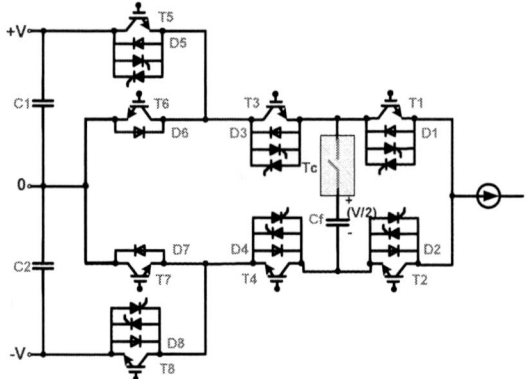

Fig. 3. FTT of 5L-ANPC converter for device open failure

C. Unified FTT of 5L-ANPC Converter For Device Failure

To combine the aforementioned two FTTs of 5L-ANPC converters together, a unified FTT is proposed and shown in Figure 4, which allows comprehensive fault tolerant operation under both device short and open failure.

In Figure 4, when one of the devices fails, except T6/D6 and T7/D7, the pair of the SCRs in parallel with that device are fired to make the faulty device behave as short failure. When T6/D6 or T7/D7 fails, the corresponding SCR will be fired to blow the fuse to disconnect that failed device from the NP of the converter. For example, when T6/D6 fails, the SCR **S1** will be fired to blow the fuse **F1** to disconnect T6/D6 from the NP. The switch **Tc** operates in a similar way as that in Figure 2 and Figure.3.

978-1-4673-9551-9/16 $31.00 © 2016 IEEE

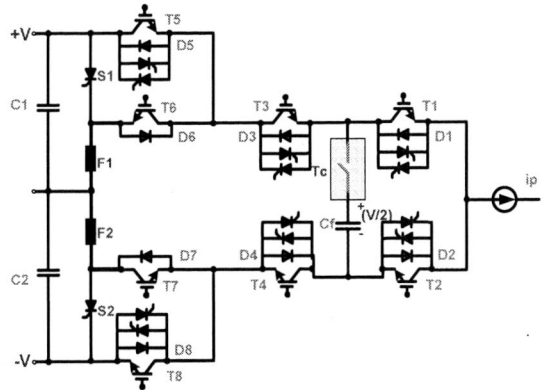

Fig. 4. Unified FTT of 5L-ANPC converter for device open or short failure

After the location of the failed device is identified, the unified FTT can continue operating to generate balanced output phase currents and output phase voltages by applying the proposed control and modulation schemes, as presented in the next section.

D. Control and Modulation Schemes of The Proposed FTTs of 5L-ANPC Converters

To achieve fault tolerant operation of the proposed FTTs, the applied switching states need to meet two basic requirements: (a) the switching state does not short the DC-link capacitors or phase capacitors; (b) the phase capacitor voltage can still be regulated by the redundant switching state (RSS) of the faulty phase leg. For example, if switching state **V1** in Table I cannot be applied to the FTT during fault tolerant operation, the use of its RSS **V2** should also be avoided.

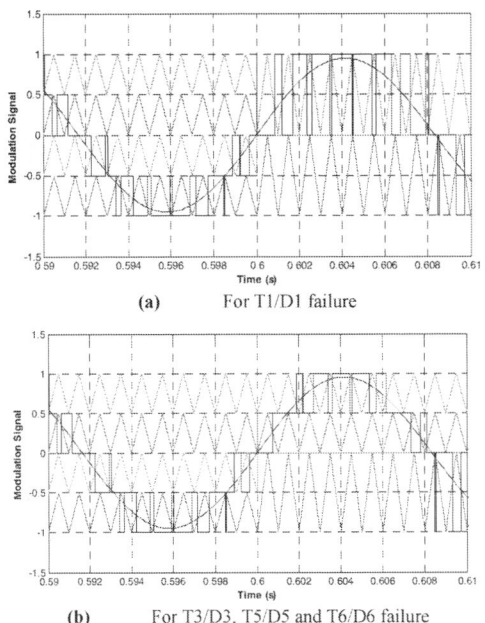

(a)   For T1/D1 failure

(b)   For T3/D3, T5/D5 and T6/D6 failure

Fig. 5. Modulation signals of the proposed FFTs of 5L-ANPC converter

According to the principles above, the modified switching states for the proposed FTTs of 5L-ANPC converters are summarized in Table II, which is valid for all the three FTTs as presented above. '0' and '1' mean ON and OFF of the corresponding device respectively. Due to the symmetrical characteristics of the proposed FTTs, Table II only shows the failure scenarios of T1/D1, T3/D3, T5/D5 and T6/D6. For the other device failure scenarios, the switching state table can be easily derived, thus it is omitted in this paper.

According to Table II, the faulty phase can always generate the output voltage level of ±V and 0 under any single device failure condition, where V is half DC-link voltage. This indicates that the proposed FTTs can generate the same maximum modulation index as that in normal operation. Moreover, for the failure case of T3/D3, T5/D5 and T6/D6, the output voltage level V/2 can also be generated because the RSSs **V5** and **V6** are both available to regulate the balance of phase capacitor voltage in the faulty phase.

From Table II, it is observed that, for T3/D3 failure condition, all the five output voltage levels are available, which indicates that the operation of the FTTs behaves close to that in normal operation. However, it should be mentioned that, in order to generate switching state **V1**, the device T5/D5 has to block the voltage 3V/2 instead of V as in a standard 5L-ANPC converter. Thus, T5/D5 in the FTTs needs higher voltage rating than that in a stander 5L-ANPC converter. In this paper, this voltage level and thus the switching state **V1** are not considered. Furthermore, because switching state **V1** is unavailable, its RSS **V2** is not considered neither for phase capacitor voltage balance purpose.

Regarding to the voltage rating of the switch **Tc**, it is observed that when T5/D5 fails, in order to generate output voltage level 0 by using switching state **V4**, the device T3/D3 and **Tc** should block a total voltage of 3V/2. Since the blocking voltage of T3/D3 in a standard 5L-ANPC converter is V/2, the proposed FTTs require the switch Tc to block voltage V.

Figure 5 shows the carrier-based PWM modulation signals of the proposed FTTs. After the device failure, e.g. occurring at 0.6 sec, the carriers are modified as shown in Figure 5. During fault tolerant operation, two carriers are required for T1/D1 failure and three carriers are required for other device failure in order to generate the output voltage levels in Table II.

E. Voltage Vector Diagram of The Proposed FTTs

According to Table II, the voltage vector diagrams of the proposed 5L-ANPC FTTs under device failure modes are shown in Figure 6, where the faulty phase is assumed to be phase-A. As observed in Figure 6, most of the 61 voltage vectors of 5L-ANPC converters are still available while only 4 vectors are lost for T1/D1 failure, and 2 vectors are unavailable for other device failure. This indicates the FTTs can still generate 9-level line-to-line voltage waveforms and achieve the same maximum modulation index (1.15) as normal operation. Moreover, some of the redundant voltage vectors still remain, which can be used to balance the NP voltage of DC-link and optimize the operation performance of the FTTs (e.g. harmonic mitigation, power loss reduction, common mode voltage mitigation, etc.) [12] [13].

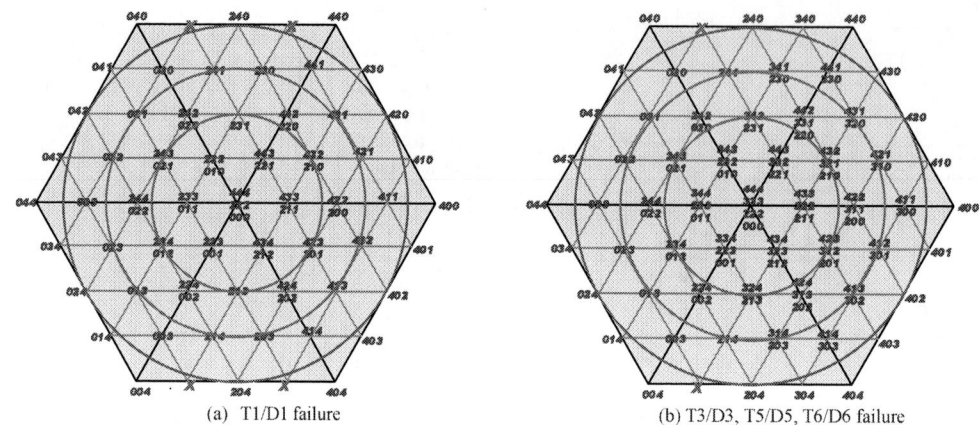

(a) T1/D1 failure   (b) T3/D3, T5/D5, T6/D6 failure

Fig. 6. Voltage vector diagram of the proposed FFTs of 5L-ANPC converter under device failure conditions

TABLE II.   PROPOSED SWITCHING STATE FOR FTTS OF 5L-ANPC CONVERTERS

	T8	T7	T6	T5	T4	T3	T2	T1	Tc	Output voltage	Effect on V_{Cf}		Switching State
											$i_p>0$	$i_p<0$	
T1/D1 short /open failure	0	1	0	1	0	1	0	X	0	V	n/a	n/a	V7 (4)
	0	1	0	1	1	0	1	X	0	0	n/a	n/a	V4 (2)
	1	0	1	0	0	1	0	X	0	0	n/a	n/a	V3 (2)
	1	0	1	0	1	0	1	X	0	-V	n/a	n/a	V0 (0)
T3/D3 short /open failure	0	1	0	1	0	X	0	1	1	V	n/a	n/a	V7 (4)
	0	1	0	1	0	X	1	0	1	V/2	+	-	V6 (3)
	0	1	0	0	1	X	0	1	1	V/2	-	+	V5 (3)
	0	1	0	0	1	X	1	0	1	0	n/a	n/a	V4 (2)
	1	0	1	0	0	X	0	1	1	0	n/a	n/a	V3 (2)
	1	0	1	0	0	X	1	0	1	-V/2	+	-	V2 (1)
	1	0	0	0	1	X	0	1	1	-V/2	-	+	V1 (1)
	1	0	1	0	1	X	1	0	0	-V	n/a	n/a	V0 (0)
T5/D5 short /open failure	0	1	0	X	0	1	0	1	1	V	n/a	n/a	V7 (4)
	0	1	0	X	0	1	1	0	1	V/2	+	-	V6 (3)
	0	1	0	X	1	0	0	1	1	V/2	-	+	V5 (3)
	0	1	0	X	1	0	1	0	1	0	n/a	n/a	V4 (2)
	1	0	0	X	1	0	1	0	0	-V	n/a	n/a	V0 (0)
T6/D6 short /open failure	0	1	X	1	0	1	0	1	1	V	n/a	n/a	V7 (4)
	0	1	X	1	0	1	1	0	1	V/2	+	-	V6 (3)
	0	1	X	1	1	0	0	1	1	V/2	-	+	V5 (3)
	0	1	X	1	1	0	1	0	1	0	n/a	n/a	V4 (2)
	1	0	X	0	1	0	1	0	1	-V	n/a	n/a	V0 (0)

IV. SIMULATION RESULTS

In this paper, the proposed FTTs of 5L-ANPC converters and their control and modulation schemes are verified through simulation. The simulation parameters are listed in Table III.

In Figure 7 and Figure 8, the simulation waveforms of the proposed FTTs are presented for T1 short failure in the FTT of Figure 2, and T6 open failure in the FTT of Figure 3 respectively. In each figure, from top to bottom, it shows phase capacitor voltage, DC-link voltage, phase currents, output line-line voltages and output phase voltages. In the simulation, the device failure occurs at 0.6 sec.

TABLE III.   SIMULATION PARAMETERS

DC-bus voltage	4 kVdc
Fundamental frequency	60 Hz
Modulation index	0.95
Carrier frequency	1 kHz
Loads	R=10 Ω, L=10 mH
DC-link capacitance	C1=C2=1 mF
Phase capacitance	Cf=0.5 mF

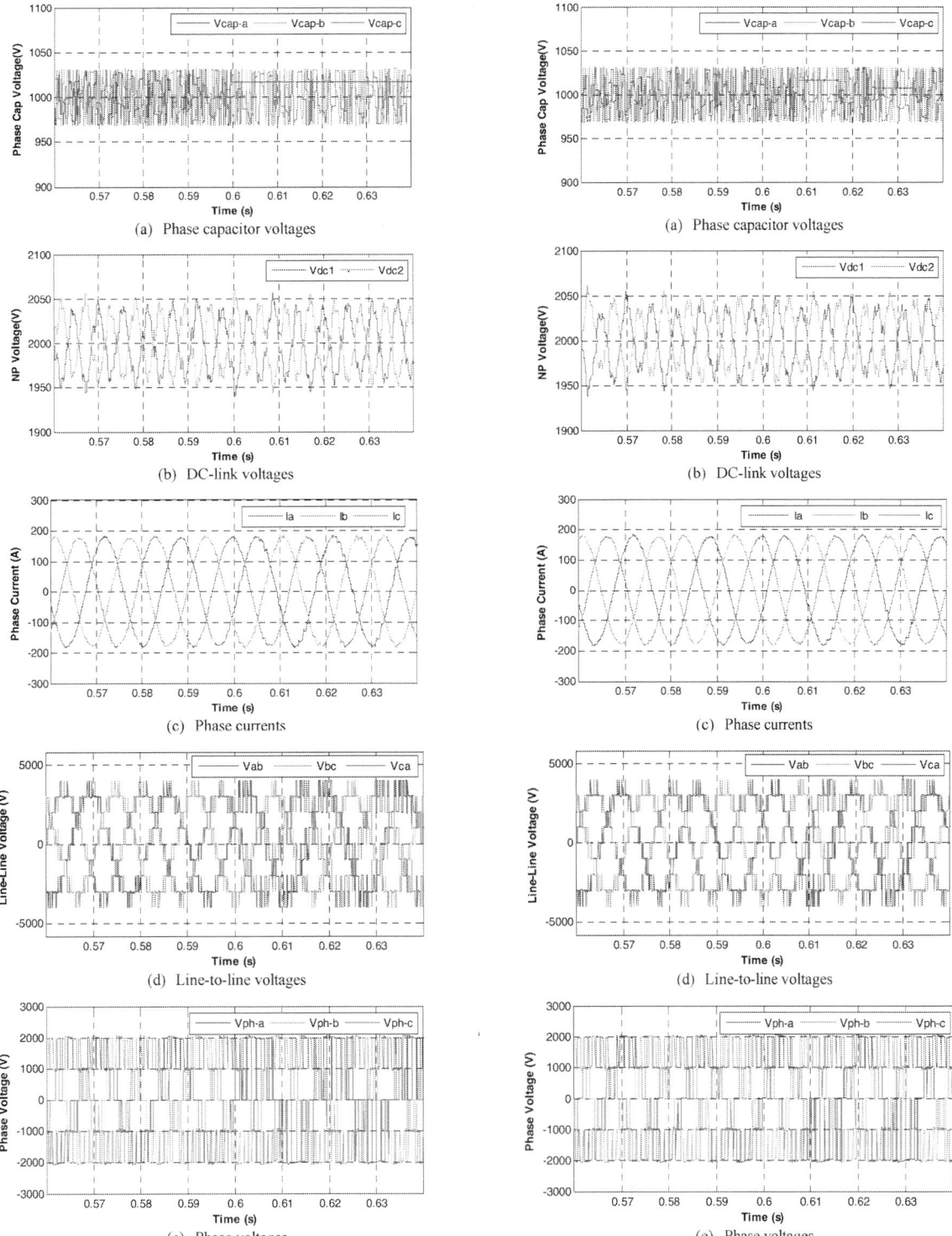

(a) Phase capacitor voltages

(b) DC-link voltages

(c) Phase currents

(d) Line-to-line voltages

(e) Phase voltages

Fig. 7. Waveforms of proposed FTT converters under T1 short failure

(a) Phase capacitor voltages

(b) DC-link voltages

(c) Phase currents

(d) Line-to-line voltages

(e) Phase voltages

Fig. 8. Waveforms of proposed FTT converters under T6 open failure

After the device failure, the proposed control and modulation schemes, which were presented in Section III (*D*) and (*E*), are applied to the proposed FTTs respectively. In this paper, it is assumed that the device failure is detected immediately after the failure occurs, and the FTT operation transits from normal operation mode to fault tolerant operation mode accordingly.

As observed, successful fault tolerant operation is achieved by the proposed FTTs after the device fails. The balance of the phase capacitor voltages and the NP voltage is maintained. The converters can continue generating symmetrical output currents and output line-line voltages at their full amplitude. Therefore, no derating operation is required during the fault tolerant operation of the proposed FTTs.

V. CONCLUSIONS

This paper proposed the fault tolerant topologies (FTTs) of 5L-ANPC converters. A low count of low-cost components are added to the conventional 5L-ANPC topologies. The proposed FTTs, together with the proposed control and modulation schemes, are able to achieve continuous fault tolerant operation under device open or short failure. The maximum output voltage amplitude is the same as that in normal operation. Thus, no derating operation is required during fault tolerant operation of the FTTs. Simulation results proved the correctness of the proposed concepts.

REFERENCES

[1] P. Barbosa, P. Steimer, J. Steinke, M. Winkelnkemper and N. Celanovic, "Active-neutral-point-clamped (ANPC) multilevel converter topology," *EPE conference 2015*, pp. 1-10, 2005.

[2] F. Kieferndorf, M. Basler, L. A. Serpa, J.-H. Fabian, A. Coccia and G. A. Scheuer, "ANPC-5L technology applied to medium-voltage variable-speed drives applications," *IEEE SPEEDAM Conference 2010*, pp. 513-518, Jun. 2010.

[3] S. Li and L. Xu, "Strategies of fault tolerant operation for three level PWM inverters," *IEEE Trans. Power Electron.*, vol. 21, no. 4, pp. 933-940, Jul. 2006.

[4] S. Ceballos, J. Pou, E. Robles, J. Zaragoza, and J. L.Martin, "Performance evaluation of fault tolerant neutral-point-clamped converters," *IEEE Trans. Ind. Electron.*, vol. 55, no. 3, pp. 982-995, Mar. 2008.

[5] J. Li, A. Q. Huang, S. Bhattacharya, and G. Tan, "Three-level active neutral-point-clamped (ANPC) converter with fault tolerant ability," *IEEE APEC 2009*, pp. 840-845. Feb. 2009.

[6] F. Richardeau, P. Baudesson, and T. A. Meynard, "Failure-tolerance and remedial strategies of a PWM multicell inverter," *IEEE Trans. Power Electron.*, vol. 17, no. 6, pp. 905-912, Nov. 2002.

[7] W. Song and A. Q. Huang, "Control strategy for fault-tolerant cascaded multilevel converter based STATCOM," *IEEE APEC* 2007, pp. 1073-1076, Feb. 2007.

[8] K.Shen, B.Xiao, J.Mei and L.M.Tolbert, "A modulation reconfiguration based fault-tolerant control scheme for modular multilevel converters," *IEEE APEC 2013*, pp. 3251-3255, Mar. 2013.

[9] P. Lezana, J. Pou, T. A. Meynard, J. Rodriguez, S. Ceballos, and F. Richardeau, "Survey on fault operation on multilevel inverters," *IEEE Trans. on Ind. Electron.*, vol. 57, no. 7, pp. 2207-2218, Jul. 2010.

[10] J.Li, J.Xu, L.Qi and R.Burgos, "Fault tolerant operation of 5L-ANPC converter," *IEEE IECON 2014*, pp. 4464-4470, Nov. 2014

[11] J.Li, J.Xu, L.Qi, Z.Pan and R.Burgos, "Analysis and control of fault tolernat operation of five-level ANPC inverters," *IEEE COMPEL 2014*, pp. 1-6, Jun. 2014.

[12] J.Li, Z.Pan and R.Burgos, "A new control scheme of five-level active NPC converters for common mode voltage mitigation in medium voltage drives," *IEEE ECCE 2014*, pp. 234-241, Sept. 2014.

[13] G.Tan, Q.Deng and Z.Liu, "An optimized SVPWM strategy for five-level active NPC (5L-ANPC) converters," *IEEE Trans. Power Electron.*, vol. 29, no. 1, pp. 386-395, Jan. 2014.

Dynamic Characterization of the Input and Reverse Transfer Capacitances in Power MOSFETs under High Current Conduction

Cristino Salcines
Ingmar Kallfass
University of Stuttgart
Institute of Robust Power Semiconductor Systems (ILH)
Stuttgart, Germany
cristino.salcines@ilh.uni-stuttgart.de
ingmar.kallfass@ilh.uni-stuttgart.de

Hisao Kakitani
Atsushi Mikata
Keysight Technologies International Japan
Semiconductor Test Division R/D
Hachioji Tokyo, Japan
hisao_kakitani@keysight.com
atsushi_mikata@keysight.com

Abstract—This paper introduces a low-complexity, novel dynamic measurement technique to extract C_{iss} and C_{rss} of a power MOSFET under high current conduction, providing an accurate description of the transistor dynamic characteristics while hard-switching an inductive load. The results of our dynamic measurements revealed larger capacitance values when compared to those extracted from conventional static measurements.

Index Terms—Terminal Capacitances, Inter-Electrode Capacitances, Ciss, Crss, CV, Dynamic Measurement, Inductive Load, Switching Losses, Power MOSFET

I. INTRODUCTION

Power transistors are mainly used as switches in power electronics, but their behavior actually differs from that of an ideal lossless switch. Two different sources of losses are commonly distinguished in a power MOSFET, conduction or static losses and switching or dynamic losses. The former arises from the resistive nature of a power MOSFET and occurs during the time interval when the transistor is conducting current. A parameter which describes this effect is Rdson. The latter is originated during the switching transients and is mainly due to the energy stored in the inter-electrode capacitances of the power MOSFET. These capacitances actually define the transistor switching capabilities.

Basically, the size of the active area of a power device defines its value of on resistance R_{dson}, blocking voltage and terminal capacitances. Large devices have lower on resistances but with the drawback of larger terminal capacitances whereas small devices possess smaller terminal capacitances but larger on resistances. This means that there is always a trade-off between static and dynamic losses when it comes to selecting the optimal size for a transistor.

New wide-bandgap (WBG) materials such as Gallium Nitride (GaN) or Silicon Carbide (SiC) hold superior electrical properties than Silicon (Si), allowing to reduce the size of transistors while maintaining the same value of on-resistance. This improvement permits a drastic reduction of the switching

power losses in the total losses budget. Alternatively, higher switching frequencies can be used while maintaining the total power losses, reducing the size of passive components and therefore improving the compactness of the final circuitry.

Since the terminal capacitances define the switching capabilities of a transistor, their characterization is of utmost importance to predict switching power losses. Datasheets commonly offer Capacitance-Voltage (CV) curves which are used by circuit designers to roughly estimate the transistor switching capabilities. CV graphs typically specify the value of C_{iss}, C_{oss} and C_{rss} for a wide range of drain source voltages V_{ds}. These capacitance values are always extracted when the transistor is not conducting current (i.e. depletion mode), whereas in the application, the transistor operates under high current conduction (i.e. enhancement mode). What means that the bias point used to extract the device capacitances is different from that used in the application.

Figure 1 depicts the operation points of a transistor hard-switching an inductive load and the bias points where conventional measurement techniques extract the value of the terminal capacitances. In an inductive load switching scenario, V_{ds} drops from high blocking voltages V_{ds_OFF} to low conduction voltages V_{ds_ON} while the transistor channel is conducting the maximum current I_{dmax}. In contrast, conventional CV measurements extract the capacitances values when no current flows through the transistor.

Inter-electrode capacitances are depending not only on V_{ds} but also on V_{gs} [1], [2]; indeed, they increase with V_{gs}. Therefore, measurements extracted for a depleted channel are not fairly representing, but underestimating, the real switching capabilities of a power MOSFET under high current conduction.

Conventionally, an LCR-meter can be used to extract the terminal capacitances of a transistor [3]. The measurements are performed with a fixed gate-source voltage V_{gs} below the threshold voltage V_{th} for different V_{ds} voltages. This technique

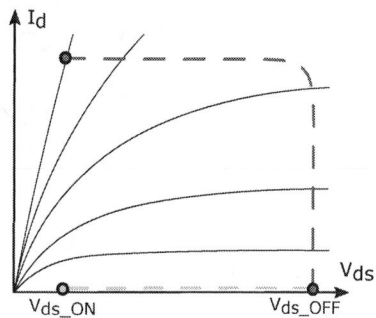

Fig. 1. IV contours of a power MOSFET hard-switching an inductive load (red trace) and operation points of conventional CV measurements (yellow trace)

is simple and accurate, but doesnt allow to perform any measurement under high current conduction without damaging the DUT.

A dynamic measurement of C_{oss} and C_{rss} has been recently presented [4]. However, an active switch in parallel to the DUT is taking the inductor's current during the turn-on and the gate-source voltage V_{gs} is fixed to zero during the measurement. Consequently, the dynamic measurement is still measuring capacitances when no current flows through the transistor channel.

C. Delm et al. [5], [6] have already presented a measurement technique to measure MOSFETs inter-electrode capacitances under high current conduction. However, this approach requires a complex circuitry to carry out the measurements.

A low-complexity, novel dynamic measurement technique to extract C_{iss} and C_{rss} under high current conduction while hard-switching an inductive load, is discussed and analyzed in the following sections. We believe this measurement will allow a more accurate switching losses calculation as well as an improvement in capacitance models of power transistor models used by circuit simulators.

II. MEASUREMENT SETUP AND THEORETICAL ANALYSIS

In order to analyze the dynamic characteristics of a power MOSFET, a conventional test fixture is a Double Pulse Tester (DPT) [7]. A DPT reproduces the current-voltage waveforms typical of an inductive load hard-switching. The measurements introduced in this work were carried out with a B1506A unit, which implements a novel measurement fixture, alternative to a conventional DPT [8], [9]. The idea behind is to replace the load inductor by a wide SOA, low on-resistance transistor acting as a current source. This concept permits a safer measurement for the DUT and provides the measurement setup of dynamism, repeatability and a wide range of current-voltages configuration on a single test fixture.

Figure 2 shows a basic schematic of the setup used in our measurements and the turn-on waveforms of a power MOSFET.

Information about C_{iss} and C_{rss} can be extracted from the gate current I_g injected in the transistor during a switching cycle. For the sake of simplicity the turn-on transient of the switching cycle is divided into five different regions.

In the region 1, between t0 and t1, V_{gs} starts rising from 0 V to the threshold voltage V_{th}. A constant current injected in the gate I_g charges the input capacitance C_{iss}. As long as the drain-source voltage V_{ds} remains constant, and V_{gs} rises linearly, the value of C_{iss} for the blocking voltage V_{ds_OFF} when no current flows through the transistor channel can be simply extracted as

$$C_{iss\_OFF} = \frac{I_g}{\frac{\partial V_{gs}}{\partial t}} \Big|_{t2}^{t1} = \frac{\partial Q_g}{\partial V_{gs}} \Big|_{t2}^{t1} \qquad (1)$$

At the beginning of region 2, t1, the transistor gate reaches V_{th} and the drain current I_d starts rising until t2, the time when it reaches its maximum value I_{dmax}. Between t2 and t1, we can extract a dynamic measurement of C_{iss} while I_d rises by applying (1).

The region 3 or plateau starts at t2. The transistor is already conducting the maximum current I_{dmax} and V_{ds} falls from the blocking voltage V_{ds_OFF} to the conduction voltage V_{ds_ON}. The gate-source voltage V_{gs} is clamped to the plateau voltage $V_{plateau}$ due to the Miller effect. The gate source capacitance C_{gs} of a vertical power MOSFET working in enhancement mode is hardly affected by V_{ds} [10]. Therefore, the small amount of current injected in C_{gs}, due to its small V_{ds} dependence, can be neglected. Consequently, we consider all of the current injected in the transistors gate during the plateau is exclusively discharging (region 3a) and charging (region 3b) C_{rss}. Thus, we extract C_{rss} when the transistor is conducting I_{dmax} by applying

$$C_{rss\_ON} = \frac{I_g}{\frac{\partial V_{dg}}{\partial t}} \Big|_{t4}^{t2} = \frac{\partial Q_g}{\partial V_{dg}} \Big|_{t4}^{t2} \qquad (2)$$

Region 4 starts at t4, when V_{ds} has finally fallen to V_{ds_ON} and the gate-source voltage V_{gs} can continue increasing lin-

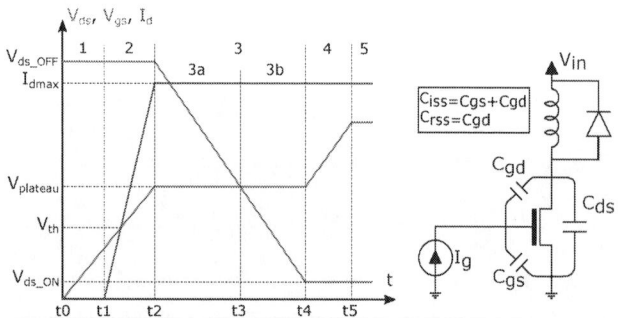

Fig. 2. Transient waveforms of an inductive load hard switching (left) and equivalent schematic of our setup (right)

978-1-4673-9551-9/16 $31.00 © 2016 IEEE

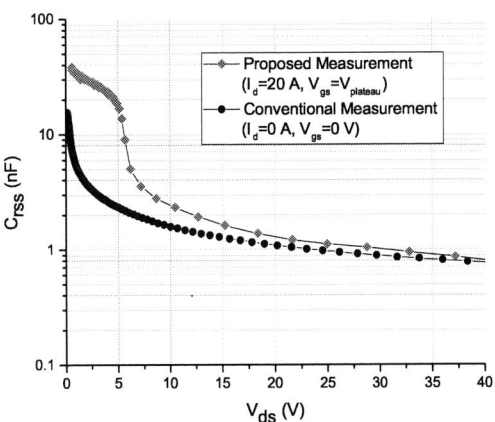

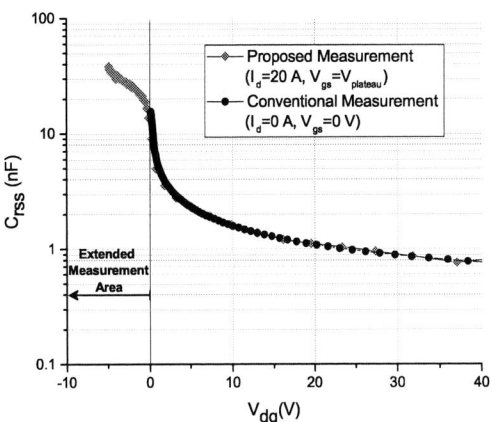

Fig. 3. Reverse transfer capacitance C_{rss} versus drain-source voltage V_{ds} when no current flows through the transistor (black) and when the transistor is conducting current (red)

Fig. 4. Reverse transfer capacitance C_{rss} versus drain-gate voltage V_{dg}. Values offered by the datasheet (black) and proposed extended measurement area (red)

early. Applying (3) in this region, we extract C_{iss} for V_{ds_ON} while conducting I_{dmax}

$$C_{iss\_ON} = \frac{I_g}{\frac{\partial V_{gs}}{\partial t}}\bigg|_{t5}^{t4} = \frac{\partial Q_g}{\partial V_{gs}}\bigg|_{t5}^{t4} \qquad (3)$$

III. MEASUREMENT ANALYSIS

Conventional CV measurements performed for $V_{gs} = 0$ V (i.e. same values as shown in the transistor's datasheet) were also carried out and compared to the capacitance values extracted from the proposed dynamic measurement technique. A Si power MOSFET 100V/20A/11mΩ in a TO-247 package is selected as DUT for this experiment.

A. Reverse Transfer Capacitance C_{rss}

The C_{rss} value extracted using (2), while the transistor is conducting a current of $I_{dmax} = 20$ A (i.e. $V_{gs} = V_{Plactau}$), is plotted in Figure 3 versus V_{ds} together with the conventional CV measurement ($I_d = 0$ A, $V_{gs} = 0$ V). $C_{rss} = 35$ nF is measured when the transistor is conducting current whereas a $C_{rss} = 15$ nF is extracted from conventional measurements. From the comparison in Figure 3, we see the value of C_{rss} when I_{dmax} flows through the transistor is higher by a factor of 2 compared to the value provided in the datasheet, extracted when no current flows through the transistor.

Figure 4 depicts the values of C_{rss} vs. the gate-drain voltage V_{dg}. The dynamic measurement technique extends the standard CV graphs to bias points where the transistor is operated in the application.

B. Input Capacitance C_{iss}

The measured C_{iss} values are plotted together with I_d versus V_{gs} in Figure 5. C_{iss} drastically changes when the transistor channel is enhanced and is conducting current.

The value of C_{iss} extracted with the proposed measurement technique is plotted in Figure 6 together with conventional CV graphs extracted for $V_{gs} = 0$ V. The measurement of C_{iss} when the transistor is not conducting current has the same value as in the datasheet, $C_{iss} = 18$ nF. However, at V_{ds_ON}, when the transistor is conducting $I_{dmax} = 20$ A ($V_{gs} = 15$ V), $C_{iss} = 50$ nF, i.e. almost 2 times larger than the value $C_{iss} = 32$ nF extracted when the transistor is off ($V_{gs} = 0$ V).

C. Comparison of Dynamic and Static Measurements

Table I summarizes and compares the different values of C_{iss} and C_{rss} extracted from the proposed dynamic measurement,

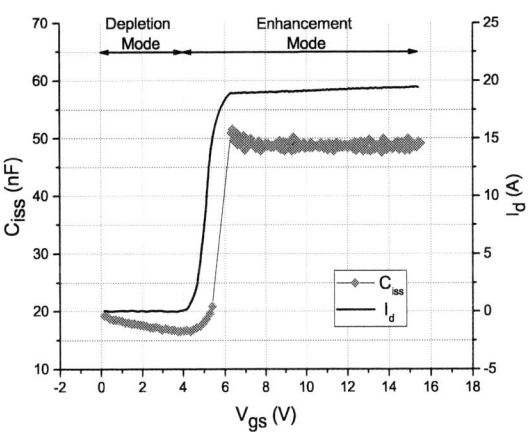

Fig. 5. Dynamic measurement of the input capacitance C_{iss} and the drain current I_d versus the gate-source voltage V_{gs}

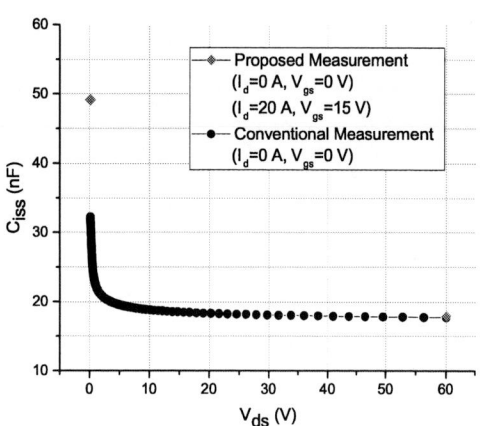

Fig. 6. Dynamic (red) and conventional (black) measurement of the input capacitance C_{iss} versus the drain-source voltage V_{ds}

TABLE I
COMPARISON OF CAPACITANCE VALUES EXTRACTED FROM TRANSISTOR DATASHEET, CONVENTIONAL AND DYNAMIC MEASUREMENTS

		Dynamic Measurement ($I_d = 20$ A, $V_{gs} > V_{th}$)	Conventional Measurement ($I_d = 0$ A, $V_{gs} = 0$ V)	Datasheet ($I_d = 0$ A, $V_{gs} = 0$ V)
V_{ds_ON}	C_{rss}	**35** nF	15 nF	15 nF
	C_{iss}	**50** nF	32 nF	32 nF

conventional measurements and the values provided in the datasheet.

IV. CONCLUSION

An accurate characterization of the dynamic behavior of a power MOSFET's terminal capacitances is important to predict its behavior in the application. The switching capabilities of a transistor, and so its power switching losses, are defined by its terminal capacitances. These are dependent not only on V_{ds} but also on V_{gs}.

Conventional characterization techniques only consider a single value of V_{gs} below the threshold voltage to extract the terminal capacitances vs. V_{ds}. However, in the application, the transistor might be operating for $V_{gs} > V_{th}$ (i.e. conducting current) by the time V_{ds} falls from the blocking voltage to the on voltage. A novel low complex dynamic measurement technique was presented to measure the value of C_{rss} and C_{iss} in a power MOSFET under high current conduction. This technique doesnt require a complex measurement setup and can be applied by simply looking into the injected gate current and terminal voltages of the DUT during the switching cycle while switching an inductive load. The dynamic measurements revealed a larger value of C_{iss} and C_{rss} than that expected from the datasheet or extracted with conventional measurement

techniques. Therefore, we believe that considering conventional datasheet CV graphs the transistor terminal capacitances are underestimated and so the calculated switching power losses.

In the presented measurements, the effect of V_{ds} on C_{gs} is neglected during the C_{rss} extraction due to the characteristics of our DUT (i.e. a vertical power MOSFET). This effect might have to be considered with a different DUT, especially with devices with high non-linear dependency of C_{gs} with V_{ds}.

The proposed method is expected to improve the switching losses prediction of power transistors as well as the accuracy of power MOSFET models for a more accurate reproduction of their switching capabilities in electrical simulators.

ACKNOWLEDGMENT

The authors would like to thank Keysight Technologies for funding, supporting and contributing to this research work and for providing the measurement equipment to carry out all the experiments.

REFERENCES

[1] Y. Lembeye, J. L. Schanen, J. P. Keradec, *"Experimental Characterization of Insulated Gate Power Components: Capacitive Aspects,"* IEEE Industry Applications Society, October 1997
[2] Y. Lembeye, J. P. Keradec, J. L. Schanen, *"Experimental Characterization of a Non Linear Electrostatic Quadripole: Application to Insulated Gate Power Components,* IEEE Instrumentation and Measurement Technology Conference (IMTC), vol. 1, pp. 525-529, 1998
[3] T. Funaki, N. Phankong, T. Kimoto, T. Hikihara, *"Measuring Terminal Capacitance and its Voltage Dependency for High-Voltage Power Devices,"* IEEE Transactions on power electronics, vol 24, no. 6, June 2009
[4] X. Song, A.Q. Huang, M. Lee, G. Wang, *"A Dynamic Measurement Method for Parasitic Capacitances of High Voltage SiC MOSFETs,"* IEEE Energy Conversion Congress and Exposition (ECCE), pp. 935-941, 2015
[5] C. Delm, *Input and Reverse Transfer Capacitance Measurement of MOS-Gated Power Transistors under High Current Flow,* IEEE Transactions on industry applications, vol. 37, no. 4, July/August 2001
[6] C. Delm, K. Hoffmann, *"Gate-Drain Capacitance Behavior of the DMOS Power Transistor under High Current Flow,"* Power Electronics Specialists Conference(PESC), vol. 2, pp. 1716-1719, 1998
[7] Cree, Application Note CPWR-AN09, *"SiC MOSFET Double Pulse Fixture,"* http://www.cree.com/~/media/Files/Cree/Power/Application%20Notes/CPWRAN09.pdf
[8] A. Mikata, *"A Novel Gate Charge Measurement Method for High-Power Devices,"* IEEE Applied Power Electronics Conference and Exposition (APEC), 2015
[9] H. Kakitani, R. Takeda, Keysight Technologies White Paper, *"Selecting Best Device for Power Circuit Design Through Gate Charge Characterization,"* http://literature.cdn.keysight.com/litweb/pdf/5991-4405EN.pdf
[10] P. Ralston, T. H. Duong, Y. Nanyin, D.W. Berning, *"High-Voltage Capacitance Measurement System for SiC Power MOSFETs,"* IEEE Energy Conversion Congress and Exposition (ECCE), Sept. 2009

AUTHOR INDEX

A

Abdelmoaty, Ahmed	2437
Abdul Azeez, Najath	3140
Abe, Seiya	1640, 2422
Abedinpour, Siamak	3669
Abramov, Eli	111, 692
Abramson, Rose A.	1138
Abu Qahouq, Jaber A.	1868, 2114, 3611, 3684
Abu-Rub, Haitham	1214, 3663
Acero, J.	3020, 3026, 3566
Achanta, Prasanta K.	3273
Adhikari, Jeevan	9
Adragna, Claudio	564
Afridi, Khurram K.	1138, 1392, 1947, 2395
Afsharian, Jahangir	33, 899, 2312, 2320
Agamy, Mohammed	3403
Agelidis, Vassilios G.	236, 1702
Agostinelli, M.	339, 350
Agostini, Francesco	472
Agrawal, Neeraj	951
Aguilar, A.	2540
Ahmed, Emad M.	1505
Ahmed, Ibrahim	1882
Ahmed, Mahrous	1505
Ahmed, Shamim	1646
Ahmed, Shehab	936
Ahmed-Zaid, Said	2821
Ahn, Jung-Hoon	163, 1273, 2161
Ahsanuzzaman, S.M.	2497
Akagi, Hirofumi	1163
Akamatsu, Keiji	2607
Akin, Bilal	505, 1096, 1176, 2108, 2875
Alatise, Olayiwola	253, 2645
Al-Durra, Ahmed	1941
Alexandrov, Peter	2973
Al-Hallaj, Said	3128
Alharbi, Mahmood	3333
Ali, Kawsar	2491
Allard, Bruno	524
Allen, Scott	979
Allmeling, Jost	1108
Alonso, J. Marcos	1115
Alonso, R.	3020

Alonso, Rafael 3026
Alou, P. 2409
Al-Shyoukh, Mohammad 2437
Am, Sokchea 2401, 2700
Amaro, Mike 66
Ambacher, Oliver 2083
Amirabadi, Mahshid 3704
Amirahmadi, Ahmadreza 3333
Amon, Cristina 1350
Amouzandeh, Maryam S. 329
Andersen, Michael A.E. 1090, 1430, 1541, 1842, 2252, 2473
Andersen, Thomas 1430, 1842
Ando, Masato 2986
Aniruddhan, Sankaran 1878
Anthon, Alexander 1235, 2252
Antunes, Fernando L.M. 2592
Anwar, Saeed 424
Anwar, Usama 1947
Arafat, A.K.M. 1123
Arafat, Akm 2847
Arefifar, Ali 2561
Arias, Andrea 79
Arias, M. 1823
Arias, Manuel 822
Arnold, Cory 1597, 3273
Asa, Erdem 1323, 1756, 1767, 2587
Asensi, R. 1624
Ayers, Curtis 3529
Ayyanar, Raja 432, 3364
Azcondo, F.J. 2389
Azuma, Katsunori 283

B

Badawey, Mohammed 392
Baek, Jeihoon 3004
Bagawade, Snehal 544
Bahman, Amir Sajjad 261, 3012
Bahmani, M.A. 3043
Bai, Hua 529
Bai, Yongjiang 766, 3623
Baier, Thomas 2897
Bak, Claus Leth 3051
Bak, Yeongsu 2764, 3416
Baker, Michael W. 1597
Bakhshai, Alireza 460
Bakker, Cas 2457
Balasubramanian, Bharat 1868
Balda, Juan Carlos 143, 362, 651, 1387, 3712
Ball, Roy 2122, 3038

Bandyopadhyay, Santanu	3286
Banerjee, Arijit	2881
Baranwal, Rohit	2043
Barbosa, A.U.	3231
Bari, Syed	3259
Barlow, Matthew	1646
Barner, Alexander	106
Barrado, A.	2545, 3090
Barth, Christopher	1512
Barthelmebs, Clement	453
Batarseh, Issa	1381, 3333
Bawohl, Melanie	3069
Bayhan, Sertac	3663
Bazzi, Ali M.	2666
Bęczkowski, Szymon	704, 974, 3101
Bede, Lorand	1702, 2264
Beres, Remus	3051
Bergman, Joshua	79
Bermejo, M.	3090
Bermingham, Jack	2794
Berzoy, Alberto	928, 3200
Betz, Vaughn	1882
Bezdenezhnykh, Yevgeny	308
Bhalla, Anup	2973
Bhangu, Bicky	715
Bhardwaj, Manish	505
Bhattachaarjee, Parijat	1344
Bhattacharya, Subhashish	295, 601, 778, 886, 1497, 1632, 2076
Bhattacharya, Tanmoy	199
Bhowmik, Pankaj Kumar	2706
Biglarbegian, Mehrdad	2998
Biswas, Suvankar	1934
Bizjak, Luca	1663
Blaabjerg, Frede	221, 229, 261, 288, 370, 1253, 1872, 1941, 1995, 2011, 2154, 2207, 2215, 2264, 3012, 3051, 3416, 3431, 3500
Blalock, Benjamin J.	684, 893, 1569, 3255
Blasko, Vladimir	2167
Böcker, Joachim	1547
Bodano, Emanuele	1663
Bojarski, Mariusz	1756
Bonthu, Sai Sudheer Reddy	1131
Bonyadi, Roozbeh	253
Bonyadi, Yeganeh	253
Borges, Beatriz	3637
Born, Rachael	1148, 3243
Boroyevich, Dushan	177, 516, 524, 739, 1024, 1315, 1561
Botting, Chris	854
Boynuegri, Ali R.	207
Braga, A.P.S.	3231

Brandt, Tobias 3172
Brar, Berinder 79
Breaz, Elena 3476
Briggs, Roger 2927
Brohlin, Paul 838
Brothers, John A. 990, 1967
Brown, Alan 529
Burdío, J.M. 1040, 1762, 3020, 3026, 3566
Burgos, Rolando 177, 516, 524, 1024, 1561
Buticchi, Giampaolo 2449, 3493, 3629
Buttay, Cyril 524

C

Cai, Wen 1057, 1861, 2599
Campbell, S.L. 1307
Canacsinh, Hiren 3637
Canales, Francisco 472
Cao, Dong 3553
Cao, Jiankun 1371
Cao, Wenchao 2229
Cao, Yuan 1868, 3684
Carlos, Gregory A.A. 3641
Carretero, C. 663, 3020, 3026, 3566
Casady, Jeffrey 979
Castro, Ignacio 25, 822
Castro, Marcus R. 2592
Ceballos, Salvador 236
Cervera, Alon 111, 692, 2298
Cha, Hanju 511, 1708, 3598
Chae, Young-Ho 2801
Chakraborty, Shiladri 1954, 2652, 3389
Challingsworth, Mark 3069
Chang, C.-H. 951
Chatterjee, Urmimala 1183
Chattopadhyay, Ritwik 778
Chattopadhyay, Souvik 1954, 2652, 3389
Chawda, Pradeep 3266
Chee, Seung-Jun 1206, 2370
Chen, Alian 3453, 3465
Chen, Changdong 2981
Chen, Cheng-Po 3255
Chen, Chingchi 1554
Chen, Di 529
Chen, Fang 177
Chen, Guipeng 1450
Chen, Guodong 499
Chen, Guoliang 1227
Chen, Hao 1437
Chen, Hua 1947

Chen, Jie	3453
Chen, Min	138
Chen, Minjie	1138, 1443
Chen, Qianhong	2518
Chen, Runruo	1045
Chen, Weiqiang	2666
Chen, Wenjie	493, 766, 3115, 3623
Chen, Woei-Luen	3471
Chen, Xinwen	1358
Chen, Xuling	1788
Chen, Yang	899, 2304, 2312, 2320
Chen, Yang-Lin	558
Chen, Yaow-Ming	558
Chen, Ying	2071
Chen, Yuxiang	499, 1462
Chen, Zhe	3431
Cheng, Chun Sing	1795
Cheng, Kuang-Yao	118
Cheung, Chun	1616
Chi, Yongning	1462
Chinthavali, M.	1307
Cho, Bo-Hyung	487, 1416
Cho, Shin Young	3690
Choe, Songbaek	2051
Choi, Beomseok	1947
Choi, Byeung G.	1773
Choi, Hee-Su	3153
Choi, Seungdeog	631, 1123, 1131, 1748, 2847, 3004
Choi, Sewan	859
Choi, Sung-Jin	3153
Choi, Wooin	1416
Choi, Wooyoung	2679
Chou, Derek	1512
Chow, Jeff Po Wa	1795
Chowdhury, Md Asif Mahmood	207
Chub, Andrii	2533
Chun, Chang Yoon	3322
Chung, Henry Shu-Hung	1795, 1807, 2154
Chung, Steven	1350
Church, Ron	786
Ci, Song	3189
Ciobotaru, Mihai	1702
Cobos, J.A.	2409
Coelho, Ernane A.A.	3585
Colak, Kerim	1323, 1756, 1767, 2587
Colmenares, Juan	746, 1018
Comanescu, Mihai	2759, 2855
Connaughton, A.M.	355
Conway, Thomas	1670

Cook, M. 3537
Correa, Maurício B.R. 3641, 1032
Corzine, Keith 720, 1191, 1481, 2187, 2840
Cosetin, Marcelo 1115
Costa, Levy F. 2449
Costa, Louelson A. 1032
Costa, Paulo Junior Silva 2376
Costinett, Daniel 424, 872, 893, 1010, 1569, 2441, 3255, 3577
Craciun, Marian 854
Cui, Shenghui 2620
Cui, Yutian 893
Curuvija, Boris 3553
Cuzner, Robert 1577
Czarkowski, Dariusz 1323, 1756, 1767, 2587
Czwickla, Christoph 3069

D

Dahan, Nadav 802
Dai, Ke 1358
Dai, Zhiyong 3134
Dai, Ziwei 1358
Dally, William J. 86
Dang, Zhigang 2114, 3684
Daniel, Michael T. 1695
Dargahi, Vahid 720, 1191, 1481, 2187, 2840
Daryaei, Mohammad 2579
Das, Partha Pratim 2652
Das, Pritam 552, 2491
Dashmiz, Shadi 3297
Davari, Pooya 221, 229
Davletzhanova, Zarina 253
de Almeida Carlos, Gregory A. 2720
de Almeida, Bruno Ricardo 60
De Carne, Giovanni 3493
De Doncker, Rik W. 643
de Oliveira Pacheco, Juliano 3346
de Rooij, Michael 2292
de Souza Oliveira Jr., Demercil 60, 3231, 3346
De, Ankan 295, 1632
Debnath, Suman 1528
Deboy, Gerald 3570
Degner, Michael W. 241
Delhotal, J. 1926
Demerdash, Nabeel A.O. 1065, 2826
Deng, Cheng 143, 362, 651
Deng, Hao 816
Deng, Lu 3521
Deng, Yan 1450
Dias Jr., A.J.S. 3231

Diaz Reigosa, Paula 288
Diaz, Nelson 1227
Dimarino, Christina 516
Ding, Pengling 1371
Ding, Weisheng 440
Dinulovic, Dragan 3097
Ditze, Stefan 864
Divan, Deepak 1437
Dix, Jeffery 684
do Prado, Ricardo N. 1115
Dobmeier, Christian 1741
Domoto, Kazuhide 2422, 2465
Dong, Dong 3403
Dong, Zhou 73, 2518
Doolla, Suryanarayana 2245, 3376
Dorn-Gomba, Lea 453
dos Santos Jr., Euzeli C. 2720
Dos Santos, Gutemberg G. 1032
Dou, Manfeng 3134
Dou, Qingyun 2272
Dousoky, Gamal M. 2735
Driesen, Johan 1183
Driessen, Anton 2457
Drofenik, Uwe 472
Du, Weijing 1002
Du, Weijing 2334
Du, Xiong 2992
Duarte, J.L. 3158
Dujic, Drazen 156, 1108
Dumais, Alex 3219
Dusmez, Serkan 505, 1176, 2108
Dutta, Atanu 3012

E

Eberle, Wilson 1286
Ebrahimi, Mohammad 2579, 3207
Edpuganti, Amarendra 402, 943
Egan, Michael 2794
Ehrlich, Stefan 1741
Einspieler, Sascha 759
Ekhtiari, Marzieh 1430
Elrayyah, Ali 392, 2660, 2806
Elsayed, Ahmed T. 1267
El-Taweel, Nader A. 830
Emadi, Ali 453, 1300
Engelmann, Georges 643
Eni, Emanuel-Petre 974, 3101
Enjeti, Prasad 936, 1695, 2567, 3545
Enshaei, Hossein 2813

Enslin, Johan 2998
Erickson, Robert 1947
Ertl, H. 1
Escobar-Mejía, Andrés 362
Eskandari, Soheila 2127
Essakiappan, Somasundaram 2706
Eum, Hyunchul 2355
Evzelman, Michael 1603
Ezra, Ofer 308

F

Fabricio, Edgard L.L. 3641
Fan, Bo 334
Faraci, Eric 838
Fard, Miad 1403
Farhang, Peyman 733
Farley, Kathleen Blair 1737, 3526
Farnell, Chris 143
Faulkner, Bryan 54
Fayed, Ayman 2437
Fedison, J.B. 247
Fei, Chao 322
Feng, Junjie 1534, 2334
Ferdowsi, Mehdi 1962, 2687
Fernandes, B.G. 2245, 2673, 3376
Fernandes, Darlan A. 1032
Fernandez, A. 1823
Fernandez, C. 2545, 3090
Ferrieux, Jean-Paul 2700
Figge, Heiko 1547
Flankl, Michael 623
Foulkes, Thomas 1512
Francés, A. 1624
Francis, A. Matt 1646
Freitas, Antônio A.A. 2592
Freitas, Luiz C.G. 3585
Frey, David 2401, 2700
Friedrichs, Daniel 3577
Fröhleke, Norbert 1547
Fu, Lixing 1554, 1967
Fu, Shihang 1475
Furukawa, Keita 1336

G

Gaafar, Mahmoud A. 2735
Gafford, James 1577
Gajanayake, C.J. 2942, 3058
Gakhar, Vikram 1878
Galiano Zurbriggen, Ignacio 386

Gan, Yiliang	3560
Gandikota, Srikant	1051
Gao, Fei	3476
Gao, Feng	410, 536, 921, 2259, 2935
Gao, Mingzhi	138
Gao, Rui	3383
Gao, Sugu	2868
Gao, Xieping	3185
Gao, Yabiao	1737, 3526
Gao, Yikai	1861
Garces, Luis	3403
Garcia Rodriguez, Luciano Andres	651
Garcia, Jorge	3508
Garcia, O.	2409, 1624
Garcia, Pablo	3508
Garcia, Virginia	3069
Garcia-Rodriguez, Luciano A.	362, 3712
Gavagsaz-Ghoachani, Roghayeh	446, 3397
Ge, Baoming	1214
Ge, Hongjuan	1424
Ge, Ting	668
Ge, Xiongxuan	3080
Geng, Shengbao	3560
Georgious, Ramy	3508
Gerfer, Alexander	2553, 3097
Gerling, Dieter	215
Ghaffarzadeh, Hooman	3353
Ghandi, Reza	3255
Ghat, Mahendra B.	2342
Ghias, Amer M.Y.M.	236
Giezendanner, Florian	1018
Ginart, Antonio	1737, 3526
Glavanovics, Michael	759
Glover, S.	3537
Goetz, Stefan M.	2349
Gohil, Ghanshyamsinh	1702, 2264
Goktas, Taner	1096, 2875
Gong, Bing	33
Gonnet, Luc	2700
Gonzalez, S.	1926
Gonzalez-Llorente, Jesus	3712
Gorla, Naga Brahmendra Yadav	2491
Gotovac, Ante	1663
Gou, Ruifeng	2071
Gray, C. Thomas	86
Greer III, Thomas H.	86
Gritti, Giovanni	564
Grosse, Thorben	643
Gu, Bin	838

Gu, Dong-Jie 1243
Gu, Lei 2889
Guerrero, Josep M. 398, 1227, 1376, 3459, 3697
Gui, Han-Dong 1243
Guirguis, David 1350
Gulbudak, Ozan 3248
Gunasekaran, Deepak 1045, 2525
Gundel, Paul 3069
Gunter, Samantha J. 1138
Guo, Ben 1010
Guo, Feng 1682
Guo, Suxuan 2063
Guo, Xiaoqiang 398
Gupta, Amit K. 715, 2942, 3058
Gupta, Ankit 1344
Gupta, Mahima 2919
Gupta, Ranjan K. 1520
Gupta, Shalabh 2666
Gurusinghe, Nicoloy 2479

H

Ha, Jung-Ik 193, 487, 1398, 2801, 3717
Hadjidemetriou, Lenos 3500
Hafez, Bahaa 936
Halivni, Bar 111
Hameyer, Kay 643
Han, Di 2861, 2950
Han, Jung Kyu 3690
Han, Xiangyu 2957
Han, Yang 816
Hang, Lijun 3560
Hanrahan, Robert 3266
Hanson, Alex J. 98
Haque, Moinul Shahidul 3004
Hare, James 2666
Harfman-Todorovic, Maja 3403
Hariharan, K. 315
Hariya, Akinori 2430
Harris, Richard Kyle 3255
Harrison, M.J. 247
Hartmann, M. 1
Haryani, Nidhi 1024, 1561
Hasan, Iftekhar 638
Hasegawa, Kazunori 3032
Hata, Katsuhiro 1731
Hata, Yuki 468
Hatae, Shinji 468
Hattori, Yoshiyuki 3146
Haug, Martin 3097

Hayakawa, Seiichi	283
Hayes, John G.	2794
He, Dingyi	2599
He, Haibing	3185
He, Jiangbiao	1065, 1084, 2826
He, Jinwei	1249
He, Ruirui	766
He, Xiangning	499, 1450, 1462, 1475
He, Xiaobin	2182, 2194
Heckel, Thomas	864
Heldwein, Marcelo Lobo	2833
Henke, M.	700
Henkenius, Carsten	1547
Herbert, Joseph	1123, 3004
Hernando, Marta M.	822
Higaki, Yusuke	1713
Hilber, Patrik	746
Hinken, Reiner	303
Hitoshi, Ishii	676
Ho, Carl Ngai-Man	2905
Ho, Kwun Yuan Godwin	2328
Hofmann, Heath	1721, 1726
Hofmann, Wilfried	2175
Holmes, Grahame	2252
Hopkins, Douglas C.	295, 2141
Hori, Yoichi	1731
Hosseini, Rasoul	1577
Hou, Dongbin	657
Hou, Ruoyu	1300
Hsiehu, H.-C.	951
Hu, Haibing	2182, 2194
Hu, Ji	253
Hu, Sheng	3310
Hu, Xiaolei	1071, 3409, 3591
Hu, Zhiyuan	899, 2320
Huang, Alex Q.	132, 269, 983, 2063, 2365, 2727, 2927, 3383, 3648, 3677
Huang, J.-W.	1364
Huang, Kuohsien	2355
Huang, Qingjun	3521
Huang, Qingyun	2727, 3648
Huang, Xiucheng	1002, 1534, 1853, 2334
Huang, Yi	1616
Huang, Ying	1888, 1894
Huang, Yung-Ting	1900
Huang, Zhengrong	1847
Huber, Laszlo	38, 46
Huh, Sungjae	2370
Hui, S.Y. Ron	169, 913, 1169, 1888, 1894, 2328, 3302, 3481
Hui, Zhao	2019

Hull, Brett 979
Husain, Iqbal 2141, 2927, 3383
Hwu, K.I. 2415

I

Iannuzzo, Francesco 288
Ibrahim, Mahmoud 2401, 2700
Iijima, Ryuji 3722
Illa Font, Carlos Henrique 2376
Ilves, Kalle 276
Imura, Takehiro 1731
Inaba, Masamitsu 283
Inokuchi, Seiichiro 468
Inoue, Shuntaro 3146
Ishikuro, Hiroki 1802
Ishizuka, Yoichi 2422, 2430, 2465
Islam, Md. Zakirul 631
Islam, Rakib 3279
Isobe, Takanori 3722
Isurin, Alexander 880
Itagaki, Atsushi 2465
Itakura, Tetsuro 1907
Itoh, Jun-Ichi 1336, 1911
IV, Prasanna 9
Iyer, Vishnu Mahadeva 295
Izuka, Arata 468

J

Jacobina, Cursino B. 2720, 3641
Jahns, Thomas 2167
Jain, Parth 2553
Jain, Praveen 378, 460, 544
Jang, Yujin 3690
Jang, Yungtaek 595, 1292
Jayasinghe, Shantha Gamini 2813
Jedtberg, Holger 3629
Jensen, Scott 1947
Jerinic, Vladan 303
Ji, Junpeng 493, 3115
Ji, Lin 3446
Ji, Shiqi 1456
Jia, Xiaoyu 398
Jiang, Dan 1788
Jiang, Dong 3616
Jiang, Ling 872
Jiang, Qirong 1468
Jiang, W.Z. 2415
Jiang, Xinjian 907
Jiao, Ningfei 2776

Jin, Qian	2409
Jo, Hyunsik	511
Jo, Jongmin	1708
Joffe, Christopher	1741
John, Vinod	951, 2200, 3439
Johnson, J.	1926
Jones, David C.	3273
Jones, Edward A.	1010, 2441
Jones, Vinson	1387
Jourdan, Charlie	3333
Jovanović, Milan M.	38, 46, 1292
Jung, Jae-Jung	2620
Jung, Jee-Hoon	3213
Jung, Kyungsub	17

K

Kakitani, Hisao	2969
Kallfass, Ingmar	2969
Kang, Taeyong	3545
Kang, Yong	3521
Kapat, Santanu	315, 2504, 3224, 3237
Karbalaye Zadeh, Mehdi	446, 3397
Karimi-Ghartemani, Masoud	3165
Karki, Ujjwal	2525
Kashyap, Avinash	3255
Katzir, Liran	3655
Kawai, Yasufumi	2051
Kawajiri, Toru	1802
Kawase, Daisuke	283
Kazama, Taisuke	1585
Ke, Haotao	295
Ke, Xugang	94
Ke, Ziwei	241
Kelly, Anthony	1591, 1670
Kerekes, Tamas	974, 1702, 2264
Khajehoddin, S. Ali	2579, 3207
Khaligh, Alireza	54, 440
Kharezy, M.	3043
Khayat, Joseph	66
Khoshkbar Sadigh, Arash	720, 1191, 1481, 2187, 2840
Khurram, Adil	2782
Kieferndorf, Frederick	472
Kikuchi, Naoto	3146
Kim, Byeong-Heon	1220
Kim, Dong-Hee	1273
Kim, Hyeokjin	1947
Kim, Hyeon-Sik	1206
Kim, Jae-Gu	2161
Kim, Ji-Min	3690

Kim, Jin-Woong ... 1398
Kim, Jonghoon ... 1690, 3322
Kim, Minjae ... 859
Kim, Nari ... 1273
Kim, Youngjong ... 2355
Kim, Yun-Sung ... 163
Kirshenboim, Or ... 111, 802
Kirtley, James L. ... 2881
Knott, Arnold ... 1541, 1842
Ko, Youngjong ... 3629
Koga, Tomoya ... 2430
Kolar, Johann W. ... 615, 623, 1198
Kolluri, Sandeep ... 552, 2491
Koltsov, H. ... 339
Kondo, Ryota ... 1713
Kondo, Takeshi ... 2788
Konishi, Kyohei ... 1780
Konrad, Werner ... 3570
Kostov, Konstantin ... 1018
Kou, Lei ... 1489, 2278
Kouchaki, Alireza ... 2382
Krischan, K. ... 355, 759
Krishna Moorthy, Radha Sree ... 794
Krishnamurthy, Mahesh ... 880, 3128
Kshirsagar, Parag ... 2027, 3616
Kubendran, S. ... 663
Kubo, Hajime ... 2788
Kudva, Sudhir S. ... 86
Kularatna, Nihal ... 2479
Kulkarni, Abhijit ... 2200, 3439
Kulkarni, Onkar Vitthal ... 2245, 3376
Kulkarni, S. ... 663
Kulothungan, Gnana Sambandam ... 402, 943
Kumar, Ashish ... 1392, 1947, 2395
Kumar, Misha ... 38, 46, 1292
Kumar, Nikhil ... 1344
Kumar, V. Inder ... 3224, 3237
Kumar, V.V.S. Pradeep ... 2673
Kurokawa, Fujio ... 754
Kusaka, Keisuke ... 1336
Kusama, Fumito ... 2607
Kwak, Sangshin ... 1748
Kwon, Yong-Cheol ... 1206
Kyriakides, Elias ... 3500

L

Lago, Jackson ... 2833
Lai, Jih-Sheng ... 1148, 1974
Lai, Jih-Sheng Jason ... 3243

Lai, Wei-Han	1974
Lam, John	786, 830
Lamar, Diego G.	25, 822, 1823, 2545
Lamo, Paula	2389
Lamoureux, Carl	1882
Langham, Jeff	334
Langmaack, N.	700
Lashway, Christopher R.	1267
Lave, M.	1926
Lazaro, A.	2545, 3090
Lazaro, Orlando	66
Lazzarin, Telles Brunelli	2376
Le, Hoai Nam	1911
Lee, Albert T.L.	169
Lee, Byoung-Kuk	163, 1273, 2161
Lee, C.K.	913
Lee, Eun S.	1773
Lee, Fred C.	322, 343, 657, 1002, 1534, 1608, 1847, 1853, 2334, 3259
Lee, Hyun-jun	1690
Lee, Jaedo	3598
Lee, Jae-Hyun	3086
Lee, Jian-Hsing	1900
Lee, June-Seok	3416
Lee, Junwon	511
Lee, Kevin	2003
Lee, Kun Wang	2875
Lee, Kyo-Beum	2764, 3416
Lee, Kyu-Chan	487
Lee, Moonhyun	1416
Lee, Woongkul	2679, 2861
Lee, Yong-Duk	125
Lee, Yongjae	193
Lee, Younggi	2370
Leeb, Steven B.	2881
Lefranc, Pierre	2401, 2700
Lehman, Brad	417, 2122, 2286, 3038
Lei, Yang	2063
Lei, Yutian	1512
Lemmon, Andrew	1577
Leng, Mingzhi	1554
Leng, Siyu	1941
Lenz, Kevin	303
Leong, K.K.	355
Lequesne, Bruno	2927
Leubner, Martin	2175
Levy, Aron	138
Li, Chendan	3459
Li, Dan	595
Li, David K.W.	1350

Li, Guojie	3560
Li, He	990, 1554, 1967
Li, Helong	704, 3101
Li, Hongxu	1657
Li, Hui	1675, 2237
Li, Jie	2770
Li, Jun	2963
Li, Kai	2613
Li, Kaiyuan	3422
Li, Peide	3697
Li, Qiang	322, 343, 657, 1002, 1534, 1608, 1847, 1853, 2334, 3259
Li, River Tin-Ho	2905
Li, Rui	1675
Li, Sinan	169
Li, Tao	2573
Li, Tengfei	2992
Li, Virginia	343
Li, Wenyu	1450
Li, Wuhua	499, 1462
Li, Xing	728
Li, Xinlei	3623
Li, Xuan	2063
Li, Xueqing	2973
Li, Yalong	2637
Li, Yan	1462
Li, Yan-Cun	1853
Li, Yaohua	3080
Li, Yongdong	3317
Li, Yuan	417, 2525
Li, Yun Wei	1249
Li, Yungui	3560
Li, Yunwei	185
Li, Zhiqing	1329
Li, Zhongxi	2349
Li, Zhongyu	3697
Liang, Beihua	1249
Liang, Lin	2981
Liang, Tsorng-Juu	1900
Liang, Xinyu	2349, 2714
Liao, Yi-Hung	1831
Liao, Zitao	1512
Lidow, Alex	587
Lightbody, Gordon	2794
Liivik, Liisa	2533
Lim, Changjin	2370
Lim, Seungbum	98
Lima, Gustavo B.	3585
Lin, Hua	728, 1078
Lin, L.-C.	951, 1364

Lin, Ni	3189
Lin, P.-H.	1364
Liserre, Marco	2449, 3493, 3629
Lisi, G.	2540
Liu, Baojin	3328, 3370
Liu, Bing	843, 2754
Liu, Bo	966, 2441, 2637
Liu, Fuxin	1788
Liu, Gang	38, 46, 595, 1292
Liu, Haichun	1371
Liu, Haoyan	3180
Liu, Hongpeng	1253
Liu, Jingbo	2147
Liu, Jinjun	739, 2272, 3193, 3328, 3370
Liu, Liming	990
Liu, Pei-Hsin	343
Liu, Pengkun	132, 983
Liu, Sucheng	1489, 2278
Liu, Teng	2272
Liu, Tianshu	899
Liu, Tingting	1410
Liu, Weiguo	1726, 2748, 2776
Liu, Wenbo	899, 2095, 2320
Liu, Wen-Chuen	1512
Liu, Xianzhuo	3317
Liu, Xiaohu	3403
Liu, Xiaokang	2742
Liu, Yan-Fei	899, 1243, 1489, 2087, 2095, 2278, 2304, 2312, 2320
Liu, Yang	959
Liu, Yunting	1045
Liu, Yushan	1214
Liu, Yushi	1392
Liu, Yusi	143
Liu, Zeng	739, 2272, 3328, 3370
Liu, Zhengyang	1847, 1853
Liu, Zhichao	1155
Loh, Poh Chiang	229, 1253, 1872, 1995, 2011, 2207
Lomonova, E.A.	3158
Lope, I.	3020
López del Moral, D.	3090
López, Felipe	2389
Lopez, Ozzie	3065
Lorenz, Robert D.	215, 2055, 2167
Lotfi, Ashraf	1882
Lu, Daorong	2194
Lu, Fei	1721, 1726
Lu, Jie	1392, 1947
Lu, Juncheng	529
Lu, Minghui	1941

Lu, Sizhao 2613
Lu, Ting 1456
Lu, Yong 2215
Lu, Zhengang 2613
Lu, Zhengyu 2003
Lu, Zhigang 398
Lu, Zhouyu 1243
Lucia, Oscar 1040, 1762, 3566
Lukic, Srdjan M. 2349, 2714
Luna, Adriana 1227
Luo, Fang 709, 2981
Luo, Guangzhao 2748
Luo, Haoze 499
Luo, Min 1108
Luo, Tianyi 3065
Lynch, Brian 66

M

Ma, Cong 3279
Ma, Dongsheng 94
Ma, Hongbo 3243
Ma, Jun 499
Ma, Ke 261
Ma, Weizhong 816
Ma, Yingxian 2497
Ma, Yiwei 966, 1261, 2229, 3121
Madhusoodhanan, Sachin 886, 1497, 1632, 2076
Madsen, Mickey P. 1842
Magne, Pierre 453
Mahajan, Anirudh 601
Mahdavikhah, Behzad 329, 3297
Mahmoodzadeh, Zahra 3353
Mainali, Krishna 886, 1497, 1632, 2076
Maitra, Arindam 1974
Makhdoomi Kaviri, Sajjad 378
Makoschitz, M. 1
Maksimović, Dragan 580, 1392, 1947, 2292, 3273
Malcolm, Doug 2087, 2095
Mallik, Ayan 54
Manabe, Shinya 2465
Mandal, Arindam 3237
Mandi, Bipin Chandra 2504
Manjrekar, Madhav 2706
Mansour, Makram 3266
Mantooth, H. Alan 143, 1646, 3012, 3180
Mao, Shuai 2776
Marsili, S. 339
Martín, Kevin 25
Martineau, Donatien 524

Martínez, Gilberto	1115
März, Martin	864, 1741
Mátéfi-Tempfli, Stefan	733
Mathew, Dinto	3286
Matsumori, Hiroaki	676, 3051
Matsuura, Ken	2430
Mattavelli, Paolo	1315
Mauerer, M.	1198
Mawby, Philip	253, 2645
Mazhari, Iman	2998
Mazumder, Paromita	1344
Mazumder, Sudip K.	1344, 1989
Mazzola, Michael	1577
McAmmond, Matt	529
McCann, Roy A.	143
McCue, Benjamin M.	3255
McDonald, Brent	329, 334, 3297
McGrath, Brendan	2252
McHugh, Colin	1947
McIntrye, W.	2540
McKenzie, Craig	3038
McRae, T.	2540
Meder, Dirk	2083
Meere, Ronan	2629
Megyei, George	3255
Mehrizi-Sani, Ali	3353
Mehrotra, Vivek	79
Mekhilef, S.	1163
Méllo, João Paulo R.	2720
Meng, Jinhao	2748
Meng, Peipei	1102
Meng, Tao	2748, 2776
Meola, Marco	1591
Mertens, Axel	3172
Meyer, Jeffrey	1947
Mi, Chris	1721, 1726
Michihira, Masakazu	2607
Mikata, Atsushi	2969
Mikulla, Michael	2083
Miraoui, Abdellatif	3476
Mishima, Tomokazu	1780
Mishra, Richa	2342
Miskovic, Vlatko	2167
Mitra, Rakesh	3279
Miwa, Brett	1597, 3273
Miyazaki, Koutarou	1640
Miyazaki, Takayuki	1907
Modes, Christina	3069
Moeini, Amirhossein	2019

Mohamed, A.A.S. 928
Mohammad, Mostak 1748
Mohammadi, Danyal 2813, 2821
Mohammadi, Mehdi 848
Mohammadpour, Bahador 378
Mohammed, Osama 928, 1267, 3200
Mohan, Ned 1051, 1520, 1934, 1982, 2043
Mojab, Alireza 1989
Molinas, Marta 446, 3397
Mønster, Jakob D. 1842
Moon, Gun-Woo 3690
Moon, Intae 1512
Moon, Seung-Ryul 1974
Morgan, Adam 295, 2141
Moroto, Takahiro 1802
Morris, Casey 2861, 2950
Morsy, Ahmed 2567
Mosesian, Jerry 2122
Moss, Jim 668
Motto, Eric R. 468
Moury, Sanjida 786
Mu, Xianmin 1381
Muetze, A. 355, 759, 3570
Mukherjee, Subhajyoti 2687
Mukhopadhyay, Shayok 2782
Mukhopadhyay, Siddhartha 315
Munk-Nielsen, Stig 288, 704, 974, 1376, 3101
Murmann, Boris 1650
Musavi, Fariborz 772
Musumeci, Salvatore 3669
Muyeen, S.M. 1941

N

Na, Woonki 3322
Nadarajan, Sivakumar 715
Nademi, Hamed 3291
Nagai, Shuichi 2051
Nahid-Mobarakeh, Babak 446, 3397
Nakano, Toshiya 468
Nakao, Hiroshi 754
Nakaoka, Mutsuo 1780
Nakashima, Yoshiyasu 754
Nan, Chenhao 432
Narasimhan, Sneha 2043
Nasr, Miad 1350
Navarro, Angel 3508
Nawaz, Muhammad 276
Nee, Hans-Peter 746, 1018
Neely, J. 1926, 3537

Neft, Charles 79
Negoro, Noboru 2051
Ngo, Khai 668
Nguyen, Duy T. 1773
Ni, Tianheng 2754
Ni, Xijun 983
Niapour, S.A.Kh. Mozaffari 3704
Nikolaidis, Ilias 3069
Ning, Puqi 3080
Ninomiya, Tamotsu 2422, 2430
Nishizawa, Shin-Ichi 3032
Niu, He 2055
Niu, Ying 2770
Noh, Shinyoung 859
Nomura, Katsuya 3146
Nondahl, Thomas A. 2147
Noquil, Jonathan 3065
Norisada, Takaaki 2607
Norum, Lars Einar 3291
Nowak, Torsten 3069
Nymand, Morten 609, 2382

O

O'Donnell, Terence 2629
O'Donovan, Gerard 2794
Ogawa, Taichi 1907
Oh, Chang-Yeol 163
Oh, Jaeyoon 2370
Ojo, Olorunfemi 2035
Okubo, Hiizu 2465
Oliver, J.A. 2409
O'Loughlin, Cathal 2629
O'Mathuna, C. 663
Omura, Ichiro 1640, 3032
Oña, E. 2545
Onal, Yasemin 2693
Onar, O.C. 1307
Orabi, Mohamed 1505
Ordonez, Martin 386, 848, 854, 2561
Orikawa, Koji 1336, 1911
Orr, Ray 1350
Ortiz-Gonzalez, Jose 253
Ortiz-Rivera, Eduardo I. 3712
Otsuka, Masafumi 1350
Ouyang, Ziwei 2473
Ozimek, Patrick E. 1084
Ozpineci, Burak 3529

P

Padhee, Varsha 1982
Pagano, Rosario 3669
Pahlevani, Majid 378
Pala, Vipindas 979
Palaniappan, Vishal 1350
Palmour, John 979
Pam, Srikanth 3266
Pan, Bing 2613
Panda, S.K. 9, 552, 715, 2491
Park, Hwa-Pyeong 3213
Park, Joung-hu 1690
Park, Sung-Yeul 125
Parkhideh, Babak 2998
Parsa, Leila 2573
Parvez, M. 1163
Patel, Ankur 150
Pathan, Abrar Ahmed 2497
Patil, Devendra 2889
Patra, Amit 2504
Pavlick, Stephanie A. 1138
Pavlovic, Z. 663
Paz, Francisco 386
Pedersen, Jeppe A. 1541, 1842
Peixoto, Paulo P. 3346
Peng, Chang 132, 269, 983, 2927
Peng, Fang Z. 959, 1045, 2525
Peng, Hao 1450
Peng, Jichang 2748, 2776
Peng, Kang 2127
Peng, Li 1358
Perales, Mico 1967
Perdigão, Marina 1115
Peretz, Mor Mordechai 111, 308, 692, 802, 2298
Perez, Aday 215
Pérez-Tarragona, Mario 1762
Perreault, David J. 98, 1138
Perrin, Remi 524
Perry, Jeff 3266
Persons, Ryan 3069
Pervaiz, Saad 1947, 2395
Peterchev, Angel V. 2349
Pevere, Alessandro 1183
Phillips, Evan 3684
Piepenbreier, Bernhard 2897
Pierfederici, Serge 446, 3397
Pigazo, Alberto 2389
Pilawa-Podgurski, Robert C.N. 1512
Ping, Dinggang 38, 46

Piya, Prasanna ... 3165
Pong, M.H. Bryan ... 2328
Poshtkouhi, Shahab ... 1350, 1403
Pou, Josep ... 236
Praça, Paulo P. ... 60, 3231
Pramod, Prerit ... 3279
Prasai, Anish ... 1437
Preciat, Philippe ... 524
Prieto, R. ... 1624
Prodić, Aleksandar ... 329, 1597, 2497, 2540, 2553, 3297
Puukko, Joonas ... 990

Q

Qi, Feng ... 2912
Qian, Qiang ... 1919, 3446
Qiao, Wei ... 3514
Qin, Jiangchao ... 1528
Qin, Liang ... 2525
Qin, Shibin ... 1512
Qin, Xianhui ... 843
Qiu, Maohang ... 138
Qiu, Yajie ... 2320
Qu, Liyan ... 3279, 3514
Qu, Xiaohui ... 2154
Quan, Zhongyi ... 185
Quay, Rüdiger ... 2083
Quentin, Nicolas ... 524

R

Raciti, Angelo ... 3669
Rahnamaee, Arash ... 1989
Raizada, Shirish ... 1344
Rajashekara, Kaushik ... 2788
Ramachandran, Rakesh ... 609
Ramadass, Yogesh ... 838
Ramani, Ramanathan ... 66
Rambal-Vecino, Andres ... 3712
Ramezani, Medhi ... 2035
Ramos, Francisco ... 215
Ramu, Krishnan ... 2027
Ran, Li ... 253, 1443, 2645
Ranstad, Per ... 1018
Rao, Yuan ... 1585
Rashkin, L. ... 3537
Rathore, Akshay K. ... 402, 794, 943
Ravey, Alexandre ... 3476
Redondo, Luís M. ... 3637
Rehman, Habibur ... 2782
Reiner, Richard ... 2083

Reitz, Jessica 3069
Remus, Nico 2175
Ren, Hai-Peng 2770
Ren, Ren 2441
Ren, Xiaoyong 73, 2518, 3488
Ren, Xizhou 2912
Ren, Yu 2071, 2102
Rengifo, Johnny 3200
Renjit, Ajit A. 1682
Reusch, David 587
Riazmontazer, Hossein 1989
Rim, Chun T. 1773
Roasto, Indrek 2533
Robbins, William 1934
Roberts II, Charles 3255
Rodrigues, Danillo B. 3585
Rodriguez, A. 1823
Roehrs, Benjamin D. 3255
Rogers, Daniel J. 1650
Romero, David 1350
Rosahl, Thoralf 106
Roßkopf, Andreas 1741
Round, W. Howell 2479
Ruan, Xinbo 1788, 3488
Ruiz, Juan M. 1292

S

Sá Jr., Edilson M. 2592
Saasaa, Raed 1286
Sadik, Diane-Perle 746, 1018
Saeed, Sarah 3508
Saeedifard, Maryam 1528, 2136, 2957
Safaee, Alireza 460
Saha, Aparna 2806
Sahoo, Ashish Kumar 1982
Sahoo, Saroj Kumar 199
Saito, Katsuaki 283
Saito, Shoji 468
Saket, Mohammad Ali 854, 2561
Sakurai, Takayasu 1640
Salameh, Mohamad 3128
Salcines, Cristino 2969
Salem, Ahmed 1505
Sandoval, José Juan 3545
Sangwongwanich, Ariya 370
Sankman, Joseph 94
Santi, Enrico 2127, 3248
Santiago-González, Juan A. 98
Santos de Moura, Diogo Cesar 3553

Santos Guimarães, Jéssica	3346
Sanz, M.	2545, 3090
Sariri, Kouros	3255
Sarlioglu, Bulent	2679, 2861, 2950
Sarnago, Hector	1040, 1762, 3566
Satija, Yudhister	3266
Sato, Masaki	2465
Saublet, Louis-Marie	3397
Saur, Michael	215
Savaghebi, Mehdi	1227, 3697
Scandrett, Brad	417
Schmidt, Peter B.	2147
Schubert, Michael	643
Schweitzer, Ben	3128
Sebastián, Javier	25, 822, 1823
Seeman, Michael	838
Seltzer, Daniel	1947
Sen, Paresh C.	1489, 2278
Senanayake, Thilak	3722
Senol, Murat	643
Seo, Gab-Su	487
Sepahvand, Alihossein	580, 1947
Serrano, J.	3020
Setyawan, Leonardy	3409
Severson, Eric	2043
Shafiei, Navid	848, 854, 2561
Shagar, Viknash	2813
Shah, Neel	2998
Shahbazi, Caitlin	3069
Shamsi, Pourya	1962, 2687
Shang, Fei	880
Shao, Jianwen	3659
Sharkh, Suleiman M.	2223
Sharma, Ratnesh	1682
Shen, Ang	1962
Shen, Guangtong	2003
Shen, Zhiyu	516, 1561
Sheng, Su	417, 2286
Shenoy, Pradeep S.	66
Shi, Baoping	843
Shi, Jianjiang	1475
Shi, Xiaojie	2637
Shi, Yuxiang	1675
Shimizu, Toshihisa	676, 3051
Shiu, T.-H.	1364
Shmilovitz, Doron	3655
Shousha, Mahmoud	3097
Shoyama, Masahito	2735
Shrivastav, Ashish	601

Shu, Zhan 2223
Shukla, Anshuman 2342, 3286
Silva, J. Fernando 3637
Silva, Paulo R. 3585
Silva, Rafael V. 2592
Simanjorang, Rejeki 2942, 3058
Singh, Amandeep 3140
Singh, Shikhar 601
Singh, Surinder P. 1585
Sinha, Sreyam 1947
Siwakoti, Yam P. 1872
Sleik, Roland 759
Sohn, Hoon 3690
Soltani, Hamid 229
Somani, Apurva 1520
Somani, Utsav 3333
Son, Yeongrack 3717
Son, Young-Kwang 2370
Song, Xiaoqing 132, 269, 983
Soni, Harshit 1344
Sozer, Yilmaz 207, 392, 638, 2693, 2806
Srdic, Srdjan 2714
Srinivasan, Dipti 402, 943
Srivastava, Vineet 786
Stack, David 1670
Stamm, Thomas 3659
Steenis, Joel 3219
Steyn-Ross, D. Alistair 2479
Stillwell, Andrew 1512
Strydom, Johan 587, 2292
Stübig, Marc 2175
Styles, Julian 529
Su, Yipeng 118
Subotic, Ivan 623
Suh, Yongsug 17
Sul, Seung-Ki 1206, 1220, 2370, 2620
Sun, Bo 1376
Sun, Kai 1227, 1410, 2182, 2194
Sun, Lei 505
Sun, Lejia 810
Sun, Libing 1227
Sun, Pengju 2992
Sun, Wei 1462
Sung, Won-Yong 163
Sveum, Peter 3128

T

T T, Anandha Ruban 1878
Tabata, Osamu 2051

Tadano, Hiroshi	3722
Tadeparthy, Preetam	1878
Tai, Heng-Ming	2992
Takamiya, Makoto	1640
Takano, Koushi	676
Talebi Khanmiri, Dawood	2122, 3038
Tan, Kai	132, 983
Tan, Linlin	2102
Tan, Nadia M.L.	1163
Tan, Pingan	3185
Tan, Siew-Chong	169, 913, 1169, 1888, 1894, 3302, 3481
Tang, Yi	1071, 3591
Tang, Yichao	440
Tang, Yuan	1443, 2645
Tareilus, G.	700
Tayebi, S. Milad	1381
Teixeira, Carlos	2252
Tekgun, Burak	207
Teodorescu, Remus	974, 1702, 2264
Tewari, Saurabh	1520, 2043
Thiringer, T.	3043
Thone, Jef	3097
Tian, Shuilin	1608
Tian, Ye	499
Ting, Lo Pang-Yen	1900
Tkachov, Sergii	350, 1663
Tolbert, Leon M.	893, 966, 1261, 1307, 1569, 2637, 3121
Tomas-Manez, Kevin	1235
Tomioka, Satoshi	2430
Tong, C.F.	2942, 3058
Trabelsi, Mohamed	3663
Tran, Yan-Kim	156
Trento, Brad	3577
Trescases, Olivier	1350, 1403, 1882
Trintis, Ionut	1376
Tripathi, Awneesh	886, 1497, 1632, 2076
Tse, Zion Tsz Ho	1737, 3526
Tseng, King Jet	1071
Tseng, K.J.	2942, 3058, 3107, 3422, 3591
Tsukuda, Masanori	1640
Tung, Chung-Pui	1807
Tüysüz, Arda	615, 623, 1198

U

Uceda, J.	1624
Uddin, Md Wasi	638
Ueda, Tesuzo	2051
Ueno, Takeshi	1907
Ugur, Enes	1176

Urteaga, Miguel 79

V

Vasić, M. 2409
Vásquez, Juan C. 1227, 3459
Vázquez, A. 25, 2545
Vechalapu, Kasunaidu 295, 886, 1497, 1632, 2076
Vekslender, Timur 308
Venkataramanan, Giri 2919
Venkateswaran, Muthusubramanian 1878
Vermulst, B.J.D. 3158
Vermulst, Bas 2457
Vesti, S. 339
Vilathgamuwa, Mahinda 2813
Villarejo, J.A. 1823
Vinnikov, Dmitri 2533
Vitorino, Montiê A. 1032
Vrankovic, Zoran 1084
Vu, Trong Tue 1835, 2485
Vukadinović, Nenad 1597

W

Wada, Keiji 1640, 2986
Walsh, Ray 2794
Waltereit, Patrick 2083
Wang, Chengshan 1249
Wang, Fan 334
Wang, Feng 810, 2742
Wang, Fred 424, 893, 966, 1010, 1261, 1456, 1569, 2229, 2441, 2637, 3121
Wang, Gangyao 979
Wang, Guo-Xiang 1123
Wang, Haoyu 480, 1280, 1329
Wang, Hongliang 899, 1489, 2278, 2304, 2312, 2320
Wang, Huai 370, 2154
Wang, Jiangfeng 2182
Wang, Jin 990, 1554, 1967
Wang, Jun 516
Wang, Kui 3317
Wang, Kun 1450
Wang, Laili 899, 2087, 2095, 2320
Wang, Li 2365
Wang, Long 2754
Wang, Meilin 1078
Wang, Meng-Jie 3471
Wang, Mengqi 3648
Wang, Miao 709, 2912
Wang, Ming-Hao 3302
Wang, N. 663
Wang, Peng 1071, 3409, 3591

Wang, Qin 1657
Wang, Shike 3328, 3370
Wang, Shiliang 2889
Wang, Shuo 2019, 3603
Wang, Wei 1253
Wang, Xiaoping 572
Wang, Xingwei 1078
Wang, Xiongfei 704, 1253, 1941, 2011, 2207, 2215, 3051, 3101, 3431
Wang, Yanbo 3431
Wang, Yi 1815
Wang, Zhenxiong 3358
Wang, Zhiqiang 2637
Watanabe, Hiroki 1336
Watanabe, Yoshitoshi 3146
Wattes, J.L. 3231
Weber, Bastian 3172
Wei, Chun 3514
Wei, Jiadan 843, 2754
Wei, Lixiang 1065, 2826
Wei, Yingdong 1468
Weise, Nathan 1065
Weiss, Beatrix 2083
Wen, Changyun 3409
Wen, Lucheng 990
Wen, Xuhui 3080
Wens, Mike 3097
Wespel, Matthias 2083
Wicht, Bernhard 106
Wijnands, C.G.E. 3158
Wiktor, Wlodek 66
Williamson, Sheldon S. 3140
Wilson, D. 3537
Winterhalter, Craig 1084
Wittmann, Jürgen 106
Wu, Dalei 3189
Wu, Hongfei 1410, 1424
Wu, John 1967
Wu, Liyao 2136, 2957
Wu, Qunfang 1657
Wu, T.-F. 951, 1364
Wu, Teng 3328, 3370
Wu, Tong 3529

X

Xia, Yinglai 3364
Xiao, Guochun 2215
Xiao, Jianfang 3409
Xiao, Lan 1657
Xiao, Xi 1424, 3338

Xie, Shaojun	1371, 1919, 3446
Xie, Xiaogao	816
Xie, Yicong	2360
Xin, Zhen	1995, 2207, 3697
Xing, Xiangyang	3453, 3465
Xing, Yan	1410, 1424, 2182, 2194
Xiong, Liansong	2742
Xiong, Song	1888, 1894
Xu, Chen	1358
Xu, Dewei David	33
Xu, Dianguo	1253
Xu, Jialin	1657
Xu, Jing	990
Xu, Jinming	1919, 3446
Xu, Longya	709, 2912
Xu, Qianwen	3409
Xu, Tao	921
Xu, Wei	2868
Xu, Yang	2141
Xue, Fei	3677
Xue, Lingxiao	1315

Y

Yadav Gorla, Naga Brahmendra	552
Yadav, Akshat	2076
Yamada, Go	2607
Yamada, Masaki	1713
Yamaguchi, Koji	3075
Yamaguchi, Masahiro	2465
Yamamoto, Keiichi	283
Yamamoto, Yasuhiro	2788
Yan, Shuo	913
Yan, Xingda	2223
Yanagi, Hiroshige	2430
Yang, Ching-Chieh	558
Yang, Enxing	499
Yang, Heya	1462
Yang, Hongbin	1078
Yang, Jianwei	3134
Yang, Liu	1261, 3121
Yang, Pengzhi	1554
Yang, Qichen	2957
Yang, Shuitao	959
Yang, Shunfeng	1071, 3591
Yang, Tao	2629
Yang, Tianbo	913
Yang, Xu	493, 766, 2071, 2102, 3115, 3623
Yang, Yang	1243
Yang, Yong	1788

Yang, Yongheng 221, 370, 1253, 3500
Yang, Yuanyu 843
Yang, Yuchen 1534
Yang, Yun 1169, 3481
Yang, Zhihua 33, 899, 2312, 2320
Yao, Chengcheng 990, 1554
Yao, Jianhui 2182, 2194
Yao, Kai 572, 1815
Yao, Wenxi 2003
Yau, Y.T. 2415
Yazdanian, Mehrdad 3353
Ye, Qing 2237
Ye, Zichao 1512
Yeo, H.L. 3107
Yeong, Lee Meng 3409
Yi, Fan 1057, 1861, 2599
Yi, Hao 3358
Yin, Shan 2942, 3058
Yonezawa, Yu 754
Young, George 1835, 2485
Yu, Hualong 1456
Yu, Ruiyang 3677
Yu, Sheng-Yang 2511
Yu, Wensong 2063, 2141, 2365, 2714, 2727, 3648
Yu, Xinyu 1468
Yu, Yanqi 1286
Yuan, Huawei 907
Yuan, Liqiang 2613

Z

Zafarani, Mohsen 1096, 2875
Zagrodnik, Michael 1071, 3591
Zane, Regan 1603
Zare, Firuz 221, 229
Zefran, Milos 1989
Zeltser, Ilya 802
Zeng, Hulong 1045
Zeng, Xiangjun 2102
Zhak, Serhii M. 3273
Zhan, Xiaohai 1424
Zhan, Xiaoqing 2154
Zhang, Bin 1155
Zhang, Binfeng 1919
Zhang, Canhui 138
Zhang, Chengduo 2349
Zhang, Chenghui 3453, 3465
Zhang, Chi 2714
Zhang, Dan 766, 3623
Zhang, Fan 1947, 2071, 2102

Zhang, Hao 2973
Zhang, Haojiong 2167
Zhang, Hua 1721, 1726
Zhang, Hui 3338
Zhang, Jianqiu 595
Zhang, Julia 241
Zhang, Jun 2992
Zhang, Junfang 572
Zhang, Ke 3476
Zhang, Lanhua 1148, 1974, 3243
Zhang, Li 3488
Zhang, Li-Heng 2770
Zhang, Lin 1868
Zhang, Liqi 269, 2063, 2365
Zhang, Shuoting 966, 3121
Zhang, Weimin 424, 893
Zhang, Wenli 524, 1002
Zhang, Xiangming 1102
Zhang, Xuan 990, 1967, 2229
Zhang, Xuning 177, 1024, 1561
Zhang, Yongchang 2868
Zhang, Yuanzhe 580, 2292
Zhang, Yuzhi 3180
Zhang, Zhe 1090, 1235, 1430, 2252, 3514
Zhang, Zhen 3134
Zhang, Zheyu 684, 1010, 1569, 2441
Zhang, Zhigang 3358
Zhang, Zhiliang 73, 1243, 2518
Zhang, Zhiyu 1475
Zhang, Zicheng 3453, 3465
Zhao, Bo 2912
Zhao, Chongwen 3577
Zhao, Dongdong 3134
Zhao, Hengyang 1475
Zhao, Hui 3603
Zhao, Rende 1995, 2207, 3697
Zhao, Shuze 1882
Zhao, Tao 33
Zhao, Xiaonan 1148, 3243
Zhao, Xin 295
Zhao, Zhengming 1456, 2613
Zheng, Sheng 966
Zheng, Yue 417
Zheng, Zedong 3317
Zhi, Na 3338
Zhou, Bo 843, 2754
Zhou, Daming 3476
Zhou, Jinping 2360
Zhou, Keliang 1155

Zhou, Liwei	410, 2259, 2935
Zhou, Luowei	2992
Zhou, Min	2360
Zhou, Qi	536
Zhou, Sizhan	3193
Zhou, Yong	2567
Zhou, Yuan	73, 2518
Zhou, Zhe	2912
Zhu, Bohang	2788
Zhu, Donghai	3521
Zhu, Guorong	3310
Zhu, Ke	990
Zhu, Qianlai	2365
Zhu, Tianhua	810
Zhuo, Fang	810, 2742, 3358
Zojer, Bernhard	996
Zou, Juan	651
Zou, Ke	1554
Zou, Xudong	3521
Zou, Xuewen	73, 2518
Zou, Zhi-Xiang	3493
Zumel, P.	2545, 3090

IEEE
445 Hoes Lane
Piscataway, NJ 08854-4141

ISBN 978-1-4673-9551-9